Flora of North America

Contributors to Volume 7

Ihsan A. Al-Shehbaz
Patrick J. Alexander
George W. Argus
C. Donovan Bailey
James B. Beck
Mark A. Beilstein
Liv Borgen
Steve Boyd
Thomas F. Daniel
James E. Eckenwalder
Reidar Elven

Sara Fuentes-Soriano
John F. Gaskin
Barbara E. Goodson
James G. Harris
Walter C. Holmes
Hugh H. Iltis
Robert W. Kiger
Marcus Koch
Judita Lihová
Karol Marhold
Santiago Martín-Bravo

Juan B. Martínez-Laborde
Nancy R. Morin
Steve L. O'Kane Jr.
Mark E. Olson
James E. Rodman
Robert F. Thorne
Gordon C. Tucker
Staria S. Vanderpool
Suzanne I. Warwick
Michael D. Windham

Editors for Volume 7

David E. Boufford
Taxon Editor for Brassicaceae, Limnanthaceae, Moringaceae, Resedaceae, Salicaceae, and Tropaeolaceae

Craig C. Freeman
Taxon Editor for Bataceae

Kanchi Gandhi
Nomenclatural Editor

Martha J. Hill
Technical Editor

Robert W. Kiger
Bibliographic Editor

Jackie M. Poole
Taxon Editor for Caricaceae

Heidi H. Schmidt
Managing Editor

Leila M. Shultz
Taxon Editor for Capparaceae, Cleomaceae, and Koeberliniaceae

John L. Strother
Reviewing Editor

James L. Zarucchi
Editorial Director

Volume Composition

Pat Harris
Compositor and Editorial Assistant

Cassandra L. Howard
Compositor and Editorial Assistant

Kristin Pierce
Compositor and Editorial Assistant

Heidi H. Schmidt
Production Coordinator and Managing Editor

Salix ovalifolia

Flora of North America

North of Mexico

Edited by FLORA OF NORTH AMERICA EDITORIAL COMMITTEE

VOLUME 7

Magnoliophyta: Salicaceae to Brassicaceae

NEW YORK OXFORD · OXFORD UNIVERSITY PRESS · 2010

Oxford University Press, Inc., publishes works that further
Oxford University's objective of excellence
in research, scholarship, and education.

Oxford New York
Auckland Cape Town Dar es Salaam Hong Kong Karachi
Kuala Lumpur Madrid Melbourne Mexico City Nairobi
New Delhi Shanghai Taipei Toronto

With offices in
Argentina Austria Brazil Chile Czech Republic France Greece
Guatemala Hungary Italy Japan Poland Portugal Singapore
South Korea Switzerland Thailand Turkey Ukraine Vietnam

Copyright © 2010 by Flora of North America Association

Published by Oxford University Press, Inc.
198 Madison Avenue, New York, New York 10016

www.oup.com

Oxford is a registered trademark of Oxford University Press

All rights reserved. No part of this publication may be reproduced,
stored in a retrieval system, or transmitted, in any form or by any means,
electronic, mechanical, photocopying, recording, or otherwise,
without the prior written permission of the Flora of North America Association.

Library of Congress Cataloging-in-Publication Data
(Revised for Volume 7)
Flora of North America North of Mexico
edited by Flora of North America Editorial Committee.
Includes bibliographical references and indexes.
Contents: v. 1. Introduction—v. 2. Pteridophytes and gymnosperms—
v. 3. Magnoliophyta: Magnoliidae and Hamamelidae—
v. 22. Magnoliophyta: Alismatidae, Arecidae, Commelinidae (in part), and Zingiberidae—
v. 26. Magnoliophyta: Liliidae: Liliales and Orchidales—
v. 23. Magnoliophyta: Commelinidae (in part): Cyperaceae—
v. 25. Magnoliophyta: Commelinidae (in part): Poaceae, part 2—
v. 4. Magnoliophyta: Caryophyllidae (in part): part 1—
v. 5. Magnoliophyta: Caryophyllidae (in part): part 2—
v. 19, 20, 21. Magnoliophyta: Asteridae (in part): Asteraceae, parts 1–3—
v. 24. Magnoliophyta: Commelinidae (in part): Poaceae, part 1—
v. 27. Bryophyta, part 1—
v. 8. Magnoliophyta: Paeoniaceae to Ericaceae—
v. 7. Magnoliophyta: Salicaceae to Brassicaceae

ISBN: 978-0-19-531822-7 (v. 7)
1. Botany—North America.
2. Botany—United States.
3. Botany—Canada.
I. Flora of North America Editorial Committee.
QK110.F55 2002 581.97 92-30459

1 2 3 4 5 6 7 8 9
Printed in the United States of America
on acid-free paper

Contents

Dedication vii
Founding Member Institutions viii
Donors ix
Editorial Committee x
Project Staff xi
Contributors xii
Taxonomic Reviewers xiii
Regional Reviewers xiii
Preface xvi

Introduction xvii

Salicaceae 3
Tropaeolaceae 165
Moringaceae 167
Caricaceae 170
Limnanthaceae 172
Koeberliniaceae 184
Bataceae 186
Resedaceae 189
Capparaceae 194
Cleomaceae 199
Brassicaceae 224

Literature Cited 747
Index 775

This volume of the

Flora of North America North of Mexico

is dedicated to the memory of

Grady L. Webster

1927 – 2005

FOUNDING MEMBER INSTITUTIONS
Flora of North America Association

Agriculture and Agri-Food Canada
Ottawa, Ontario

Arnold Arboretum
Jamaica Plain, Massachusetts

Canadian Museum of Nature
Ottawa, Ontario

Carnegie Museum of
 Natural History
Pittsburgh, Pennsylvania

Field Museum of Natural History
Chicago, Illinois

Fish and Wildlife Service
United States Department of
 the Interior
Washington, D.C.

Harvard University Herbaria
Cambridge, Massachusetts

Hunt Institute for Botanical
 Documentation
Carnegie Mellon University
Pittsburgh, Pennsylvania

Jacksonville State University
Jacksonville, Alabama

Jardin Botanique de Montréal
Montréal, Québec

Kansas State University
Manhattan, Kansas

Missouri Botanical Garden
St. Louis, Missouri

New Mexico State University
Las Cruces, New Mexico

The New York Botanical Garden
Bronx, New York

New York State Museum
Albany, New York

Northern Kentucky University
Highland Heights, Kentucky

Université de Montréal
Montréal, Québec

University of Alaska
Fairbanks, Alaska

University of Alberta
Edmonton, Alberta

The University of British Columbia
Vancouver, British Columbia

University of California
Berkeley, California

University of California
Davis, California

University of Idaho
Moscow, Idaho

University of Illinois
Urbana-Champaign, Illinois

University of Iowa
Iowa City, Iowa

The University of Kansas
Lawrence, Kansas

University of Michigan
Ann Arbor, Michigan

University of Oklahoma
Norman, Oklahoma

University of Ottawa
Ottawa, Ontario

University of Southwestern
 Louisiana
Lafayette, Louisiana

The University of Texas
Austin, Texas

University of Western Ontario
London, Ontario

University of Wyoming
Laramie, Wyoming

Utah State University
Logan, Utah

For their support of the preparation of this volume,
we gratefully acknowledge and thank:

Chanticleer Foundation

The Andrew W. Mellon Foundation

an anonymous foundation

ESRI

For sponsorship of illustrations included in this volume,
we express sincere appreciation to:

John Edmondson - Liverpool, England
Stanleya viridiflora

Jennifer H. Richards - Miami, Florida
Cynophylla flexuosa

The Arboretum at Flagstaff - Flagstaff, Arizona
Salix arizonica

Anonymous - in honor of the 200th anniversary of Wislizenus's birth (1810)
Wislizenia californica

Nancy R. Morin - Point Arena, California - in honor of James Shevock
Boechera shevockii

FLORA OF NORTH AMERICA EDITORIAL COMMITTEE (as of 16 October 2009)

VASCULAR PLANTS

Guy Baillargeon
Liaison to Governmental Agencies
Agriculture and Agri-Food Canada
Le Bic, Québec

David E. Boufford
Taxon Editor and Regional Coordinator,
Northeastern United States
Harvard University Herbaria
Cambridge, Massachusetts

Luc Brouillet
Taxon Editor and Regional Coordinator,
Eastern Canada and Greenland
Institut de recherche en biologie végétale
Université de Montréal
Montréal, Québec

Wayne J. Elisens
Taxon Editor
University of Oklahoma
Norman, Oklahoma

Bruce A. Ford
Taxon Editor and Regional Coordinator,
Western Canada
University of Manitoba
Winnipeg, Manitoba

Craig C. Freeman
Taxon Editor and Regional Coordinator,
North Central United States
R. L. McGregor Herbarium
The University of Kansas
Lawrence, Kansas

Kanchi Gandhi
Nomenclatural Editor
Harvard University Herbaria
Cambridge, Massachusetts

David E. Giblin
University of Washington
Seattle, Washington

Lynn J. Gillespie
Taxon Editor
Canadian Museum of Nature
Ottawa, Ontario

Ronald L. Hartman
Taxon Editor and Regional Coordinator, Rocky Mountains
University of Wyoming
Laramie, Wyoming

Steffi M. Ickert-Bond
Regional Coordinator, Alaska and Yukon
University of Alaska, Museum of the North
Fairbanks, Alaska

Robert W. Kiger
Bibliographic and Taxon Editor
Hunt Institute for Botanical Documentation
Carnegie Mellon University
Pittsburgh, Pennsylvania

Geoffrey A. Levin
Taxon Editor
Illinois Natural History Survey
Champaign, Illinois

Barney L. Lipscomb
Botanical Research Institute of Texas
Fort Worth, Texas

Timothy K. Lowrey
University of New Mexico
Albuquerque, New Mexico

James A. Macklin
Harvard University Herbaria
Cambridge, Massachusetts

James S. Miller
The New York Botanical Garden
Bronx, New York

Nancy R. Morin
Taxon Editor and Regional Coordinator,
Southwestern United States
Point Arena, California

Robert F. C. Naczi
The New York Botanical Garden
Bronx, New York

Jackie M. Poole
Taxon Editor and Regional Coordinator,
South Central United States
Texas Parks and Wildlife Department
Austin, Texas

Richard K. Rabeler
Taxon Editor
University of Michigan
Ann Arbor, Michigan

Jay A. Raveill
Taxon Editor
University of Central Missouri
Warrensburg, Missouri

Heidi H. Schmidt
Managing Editor
Missouri Botanical Garden
St. Louis, Missouri

Leila M. Shultz
Taxon Editor
Utah State University
Logan, Utah

John L. Strother
Reviewing and Taxon Editor
University of California
Berkeley, California

Janet R. Sullivan
Taxon Editor
University of New Hampshire
Durham, New Hampshire

Debra K. Trock
Taxon Editor
California Academy of Sciences
San Francisco, California

Gordon C. Tucker
Taxon Editor
Eastern Illinois University
Charleston, Illinois

Michael A. Vincent
Taxon Editor
Miami University
Oxford, Ohio

Alan S. Weakley
Taxon Editor and Regional Coordinator,
Southeastern United States
University of North Carolina
Chapel Hill, North Carolina

James L. Zarucchi
Editorial Director
Missouri Botanical Garden
St. Louis, Missouri

BRYOPHYTES

Terry T. McIntosh
Taxon Editor
University of British Columbia
Vancouver, British Columbia

Dale H. Vitt
Taxon Editor
Southern Illinois University
Carbondale, Illinois

David H. Wagner
Taxon Editor
Northwest Botanical Institute
Eugene, Oregon

Alan T. Whittemore
Taxon Editor
United States National Arboretum
Washington, D.C.

Richard H. Zander
Bryophyte Coordinator and Taxon Editor
Missouri Botanical Garden
St. Louis, Missouri

Emeritus/a Members of the Editorial Committee

George W. Argus
Merrickville, Ontario

Mary E. Barkworth
Logan, Utah

Marshall R. Crosby
Sugar Grove, North Carolinia

Claudio Delgadillo M.
Mexico City, Mexico

Marie L. Hicks
Moab, Utah

Marshall C. Johnston
Austin, Texas

Aaron I. Liston
Corvallis, Oregon

John McNeill
Edinburgh, Scotland

Barbara M. Murray
Fairbanks, Alaska

David F. Murray
Fairbanks, Alaska

J. Scott Peterson
Baton Rouge, Louisiana

Alan R. Smith
Berkeley, California

Richard W. Spellenberg
Las Cruces, New Mexico

Lloyd R. Stark
Las Vegas, Nevada

Barbara M. Thiers
Bronx, New York

Rahmona A. Thompson
Ada, Oklahoma

Frederick H. Utech
Pittsburgh, Pennsylvania

*Project Staff — past and present
involved with the preparation of this volume*

Barbara Alongi, *Illustrator*
Michael Blomberg, *Scanning Specialist*
Ariel Roads Buback, *Manuscript Specialist*
Trisha K. Distler, *GIS Analyst*
A. Michele Funston, *Editorial Assistant*
Pat Harris, *Editorial Assistant and Compositor*
Linny Heagy, *Illustrator*
Martha J. Hill, *Technical Editor*
Cassandra L. Howard, *Editorial Assistant and Compositor*
Fred Keusenkothen, *Scanning Supervisor*
Ruth T. King, *Editorial Assistant*
John Myers, *Illustrator and Illustration Compositor*
Kristin Pierce, *Editorial Assistant and Compositor*
Hong Song, *Programmer*
Yevonn Wilson-Ramsey, *Illustrator*

Contributors to Volume 7

Ihsan A. Al-Shehbaz
Missouri Botanical Garden
St. Louis, Missouri

Patrick J. Alexander
New Mexico State University
Las Cruces, New Mexico

George W. Argus
Merrickville, Ontario

C. Donovan Bailey
New Mexico State University
Las Cruces, New Mexico

James B. Beck
Duke University
Durham, North Carolina

Mark A. Beilstein
Harvard University Herbaria
Cambridge, Massachusetts

Liv Borgen
Natural History Museum
University of Oslo
Oslo, Norway

Steve Boyd
Rancho Santa Ana Botanic Garden
Claremont, California

Thomas F. Daniel
California Academy of Sciences
San Francisco, California

James E. Eckenwalder
University of Toronto
Toronto, Ontario

Reidar Elven
Natural History Museum
University of Oslo
Oslo, Norway

Sara Fuentes-Soriano
University of Missouri
St. Louis, Missouri

John F. Gaskin
United States Department of Agriculture
Northern Plains Agricultural Research Laboratory
Sidney, Montana

Barbara E. Goodson
The University of Texas
Austin, Texas

James G. Harris
Utah Valley State College
Orem, Utah

Walter C. Holmes
Baylor University
Waco, Texas

Hugh H. Iltis
University of Wisconsin
Madison, Wisconsin

Robert W. Kiger
Hunt Institute for Botanical Documentation
Carnegie Mellon University
Pittsburgh, Pennsylvania

Marcus Koch
Max-Planck-Institute for Chemical Ecology
Jena, Germany

Judita Lihová
Slovak Academy of Sciences
Bratislava, Slovakia

Karol Marhold
Slovak Academy of Sciences
Bratislava, Slovakia

Santiago Martín-Bravo
Universidad Pablo de Olavide
Seville, Spain

Juan B. Martínez-Laborde
Universidad Politécnica de Madrid
Madrid, Spain

Nancy R. Morin
Point Arena, California

Steve L. O'Kane Jr.
University of Northern Iowa
Cedar Falls, Iowa

Mark E. Olson
Universidad Nacional Autónoma de México
Mexico City, Mexico

James E. Rodman
Silverdale, Washington

Robert F. Thorne
Rancho Santa Ana Botanic Garden
Claremont, California

Gordon C. Tucker
Eastern Illinois University
Charleston, Illinois

Staria S. Vanderpool
Arkansas State University
State University, Arkansas

Suzanne I. Warwick
Agriculture and Agri-Food Canada
Ottawa, Ontario

Michael D. Windham
Duke University
Durham, North Carolina

Taxonomic Reviewers

Hugh H. Iltis
University of Wisconsin
Madison, Wisconsin

Richard C. Keating
Southern Illinois University
Edwardsville, Illinois

Staria S. Vanderpool
Arkansas State University
State University, Arkansas

Dieter H. Wilken
Santa Barbara Botanic Garden
Santa Barbara, California

Regional Reviewers

ALASKA / YUKON

Alan Batten
University of Alaska,
 Museum of the North
Fairbanks, Alaska

Bruce Bennett
Yukon Department of
 Environment
Whitehorse, Yukon

Robert Lipkin
Alaska Natural Heritage Program
University of Alaska
Anchorage, Alaska

David F. Murray
University of Alaska,
 Museum of the North
Fairbanks, Alaska

Mary Stensvold
U.S.D.A. Forest Service
Sitka, Alaska

PACIFIC NORTHWEST

Edward R. Alverson
The Nature Conservancy
Eugene, Oregon

Adolf Ceska
British Columbia Conservation
 Data Centre
Victoria, British Columbia

Richard R. Halse
Oregon State University
Corvallis, Oregon

Eugene N. Kozloff
Shannon Point Marine Center
Anacortes, Washington

Jim Pojar
British Columbia Forest Service
Smithers, British Columbia

Cindy Roché
Talent, Oregon

Peter F. Zika
University of Washington
Seattle, Washington

SOUTHWESTERN UNITED STATES

H. David Hammond
Northern Arizona University
Flagstaff, Arizona

G. Frederic Hrusa
Courtland, California

James D. Morefield
Nevada Natural Heritage Program
Carson City, Nevada

REGIONAL REVIEWERS

Donald J. Pinkava
Arizona State University
Tempe, Arizona

Jon P. Rebman
San Diego Natural History Museum
San Diego, California

Margriet Wetherwax
University of California
Berkeley, California

WESTERN CANADA

William J. Cody†
Agriculture and Agri-Food Canada
Ottawa, Ontario

Lynn J. Gillespie
Canadian Museum of Nature
Ottawa, Ontario

A. Joyce Gould
Alberta Tourism, Parks and Recreation
Edmonton, Alberta

Vernon L. Harms
University of Saskatchewan
Saskatoon, Saskatchewan

Elizabeth Punter
University of Manitoba
Winnipeg, Manitoba

ROCKY MOUNTAINS

Curtis R. Björk
University of Idaho
Moscow, Idaho

Bonnie Heidel
University of Wyoming
Laramie, Wyoming

B. E. Nelson
University of Wyoming
Laramie, Wyoming

NORTH CENTRAL UNITED STATES

William T. Barker
North Dakota State University
Fargo, North Dakota

Anita F. Cholewa
University of Minnesota
St. Paul, Minnesota

Neil A. Harriman
University of Wisconsin
Oshkosh, Wisconsin

Bruce W. Hoagland
University of Oklahoma
Norman, Oklahoma

Robert B. Kaul
University of Nebraska
Lincoln, Nebraska

Deborah Q. Lewis
Iowa State University
Ames, Iowa

Ronald L. McGregor
The University of Kansas
Lawrence, Kansas

Lawrence R. Stritch
U.S.D.A. Forest Service
Shepherdstown, West Virginia

George Yatskievych
Missouri Botanical Garden
St. Louis, Missouri

SOUTH CENTRAL UNITED STATES

David E. Lemke
Southwest Texas State University
San Marcos, Texas

EASTERN CANADA

Sean Blaney
Atlantic Canada Conservation Data Centre
Sackville, New Brunswick

Jacques Cayouette
Agriculture and Agri-Food Canada
Ottawa, Ontario

Frédéric Coursol
Mirabel, Québec

William J. Crins
Ontario Ministry of Natural Resources
Peterborough, Ontario

John K. Morton
University of Waterloo
Waterloo, Ontario

Marian Munro
Nova Scotia Museum of Natural History
Halifax, Nova Scotia

Michael J. Oldham
Natural Heritage Information Centre
Peterborough, Ontario

Claude Roy
Université Laval
Québec, Québec

GREENLAND

Geoffrey Halliday
University of Lancaster
Lancaster, England

NORTHEASTERN UNITED STATES

Ray Angelo
New England Botanical Club
Cambridge, Massachusetts

Tom S. Cooperrider
Kent State University
Kent, Ohio

Arthur Haines
Bowdoin, Maine

Edward G. Voss
University of Michigan
Ann Arbor, Michigan

Kay Yatskievych
Missouri Botanical Garden
St. Louis, Missouri

SOUTHEASTERN UNITED STATES

Mac H. Alford
University of Southern Mississippi
Hattiesburg, Mississippi

J. Richard Carter Jr.
Valdosta State University
Valdosta, Georgia

L. Dwayne Estes
Austin Peay State University
Clarksville, Tennessee

Paul J. Harmon
West Virginia Natural Heritage
 Program
Charleston, West Virginia

W. John Hayden
University of Richmond
Richmond, Virginia

John B. Nelson
University of South Carolina
Columbia, South Carolina

Bruce A. Sorrie
University of North Carolina
Chapel Hill, North Carolina

R. Dale Thomas
Seymour, Tennessee

Lowell E. Urbatsch
Louisiana State University
Baton Rouge, Louisiana

Thomas F. Wieboldt
Virginia Polytechnic Institute and
 State University
Blacksburg, Virginia

B. Eugene Wofford
University of Tennessee
Knoxville, Tennessee

FLORIDA

Loran C. Anderson
Florida State University
Tallahassee, Florida

Bruce F. Hansen
University of South Florida
Tampa, Florida

Richard P. Wunderlin
University of South Florida
Tampa, Florida

Preface for Volume 7

Since the publication of *Flora of North America* volume 8 (the fifteenth volume to have been published in the *Flora* series) during the latter half of 2009, the membership of the Flora of North America Association Board of Directors has undergone changes. Marshall R. Crosby, Claudio Delgadillo M., Aaron I. Liston, J. Scott Peterson, Lloyd R. Stark, and Frederick H. Utech have retired from the board. New board members include: David E. Giblin, Timothy K. Lowrey, Terry T. McIntosh (Taxon Editor), Robert F. C. Naczi, Janet R. Sullivan (Taxon Editor), David H. Wagner (Taxon Editor), and Alan T. Whittemore (Taxon Editor). The Board succeeded the former Editorial Committee as the result of a reorganization finalized in 2003, but for the sake of continuity of citation, authorship of *Flora* volumes is still to be cited as "Flora of North America Editorial Committee, eds."

The vast majority of editorial processing for this volume of the *Flora* was undertaken at the Missouri Botanical Garden in St. Louis where occurrence maps were also prepared based on taxon distribution statements found in the treatments. Pre-press production for the volume, including typesetting and layout, plus coordination for all aspects of planning, executing, and scanning the illustrations, was done at the St. Louis center. Art panel composition and labeling was completed in Gaston, Oregon.

Line drawings published in this volume were executed by four talented artists: Barbara Alongi prepared illustrations for various Brassicaceae (*Anelsonia*, *Armoracia*, *Aubrieta*, *Draba*, *Iodanthus*, *Isatis*, *Lepidium*, *Microthlaspi*, *Myagrum*, *Nevada*, *Rorippa*, *Synthlipsis*, and most taxa of *Boechera* – except those listed below that were prepared by Wilson-Ramsey), Limnanthaceae, Moringaceae, Resedaceae, and Tropaeolaceae; Linny Heagy prepared illustrations for Caricaceae; John Myers illustrated Salicaceae plus he composed and labeled all of the line drawings in this volume and prepared the color frontispiece depicting *Salix ovalifolia*; and, Yevonn Wilson-Ramsey prepared illustrations for Bataceae, most Brassicaceae (including *Boechera evadens*, *B. lemmonii*, *B. nevadensis*, *B. rollei*, *B. rollinsiorum*, *B. schistacea*, *B. serpenticola*, *B. shevockii*, *B. shockleyi*, *B. sparsiflora*, and *B. villosa*), Capparaceae, Cleomaceae, and Koeberliniaceae.

Starting with volume 8 published in 2009, users will find that the circumscription and ordering of some families within the *Flora* have been modified so they mostly reflect that of the Angiosperm Phylogeny Group [APG] rather than the previously followed Cronquist organizational structure. The groups of families found in this and other yet-to-be published volumes in the series are mostly ordered following E. Haston et al. (2007); since APG views of relationships and circumscriptions have evolved, and will certainly change further through time, some discrepancies in organization will occur. Volume 30 of the *Flora of North America* will contain an overall index to the published volumes.

The Flora of North America Association remains deeply grateful to the many people who continue to help create, encourage, and sustain the *Flora*.

Introduction

Scope of the Work

Flora of North America North of Mexico is a synoptic account of the plants of North America north of Mexico: the continental United States of America (including the Florida Keys and Aleutian Islands), Canada, Greenland (Kalâtdlit-Nunât), and St. Pierre and Miquelon. The *Flora* is intended to serve both as a means of identifying plants within the region and as a systematic conspectus of the North American flora.

The *Flora* will be published in 30 volumes. Volume 1 contains background information that is useful for understanding patterns in the flora. Volume 2 contains treatments of ferns and gymnosperms. Families in volumes 3–26, the angiosperms, were first arranged according to the classification system of A. Cronquist (1981) with some modifications, and starting with the publication of Volume 8, the circumscriptions and ordering of families generally follow those of the Angiosperm Phylogeny Group [APG] (see E. Haston et al. 2007). Bryophytes are being covered in volumes 27–29. Volume 30 will contain the cumulative bibliography and index.

The first two volumes were published in 1993, Volume 3 in 1997, and Volumes 22, 23, and 26, the first three of five volumes covering the monocotyledons, appeared in 2000, 2002, and 2002, respectively. Volume 4, the first part of the Caryophyllales, was published in late 2003. Volume 25, the second part of the Poaceae, was published in mid 2003, and Volume 24, the first part, was published in January 2007. Volume 5, completing the Caryophyllales plus Polygonales and Plumbaginales, was published in early 2005. Volumes 19–21, treating Asteraceae, were published in early 2006. Volume 27, the first of two volumes treating mosses in North America, was published in late 2007. Volume 8, Paeoniaceae to Ericaceae, was published in September 2009. The correct bibliographic citation for the *Flora* is: Flora of North America Editorial Committee, eds. 1993+. Flora of North America North of Mexico. 16+ vols. New York and Oxford.

Volume 7 treats 923 species in 125 genera contained in 11 families. For additional statistics please refer to Table 1 on p. xviii.

Contents · General

The *Flora* includes accepted names, selected synonyms, literature citations, identification keys, descriptions, phenological information, summaries of habitats and geographic ranges, and other biological observations. Each volume contains a bibliography and an index to the taxa included in that volume. The treatments, written and reviewed by experts from throughout the systematic botanical community, are based on original observations of herbarium specimens and, whenever possible, on living plants. These observations are supplemented by critical reviews of the literature.

Table 1. *Statistics for Volume 7 of Flora of North America.*

Family	Total Genera	Endemic Genera	Introduced Genera	Total Species	Endemic Species	Introduced Species	Conservation Taxa
Salicaceae	4	0	1	123	72	15	8
Tropaeolaceae	1	0	1	1	0	1	0
Moringaceae	1	0	1	1	0	1	0
Caricaceae	1	0	1	1	0	1	0
Limnanthaceae*	2	2	0	8	8	0	10
Koeberliniaceae	1	0	0	1	0	0	0
Bataceae	1	0	0	1	0	0	0
Resedaceae	2	0	1	5	0	4	0
Capparaceae	3	0	0	4	0	0	0
Cleomaceae	12	2	6	34	15	10	2
Brassicaceae	97	13	37	744	498	106	176
Totals	**125**	**17**	**48**	**923**	**593**	**138**	**196**

* = endemic to flora area
Italic = introduced

Basic Concepts

Our goal is to make the *Flora* as clear, concise, and informative as practicable so that it can be an important resource for both botanists and nonbotanists. To this end, we are attempting to be consistent in style and content from the first volume to the last. Readers may assume that a term has the same meaning each time it appears and that, within groups, descriptions may be compared directly with one another. Any departures from consistent usage will be explicitly noted in the treatments (see References).

Treatments are intended to reflect current knowledge of taxa throughout their ranges worldwide, and classifications are therefore based on all available evidence. Where notable differences of opinion about the classification of a group occur, appropriate references are mentioned in the discussion of the group.

Documentation and arguments supporting significantly revised classifications are published separately in botanical journals before publication of the pertinent volume of the *Flora*. Similarly, all new names and new combinations are published elsewhere prior to their use in the *Flora*. No nomenclatural innovations will be published intentionally in the *Flora*.

Taxa treated in full include extant and recently extinct native species, hybrids that are well established (or frequent), and waifs or cultivated plants that are found frequently outside cultivation and give the appearance of being naturalized. Taxa mentioned only in discussions include waifs or naturalized plants now known only from isolated old records and some non-native, economically important or extensively cultivated plants, particularly when they are relatives of native species. Excluded names and taxa are listed at the ends of appropriate sections, e.g., species at the end of genus, genera at the end of family.

Treatments are intended to be succinct and diagnostic but adequately descriptive. Characters and character states used in the keys are repeated in the descriptions. Descriptions of related taxa at the same rank are directly comparable.

With few exceptions, taxa are presented in taxonomic sequence. If an author is unable to produce a classification, the taxa are arranged alphabetically and the reasons are given in the discussion.

Treatments of hybrids follow that of one of the putative parents. Hybrid complexes are treated at the ends of their genera, after the descriptions of species.

We have attempted to keep terminology as simple as accuracy permits. Common English equivalents usually have been used in place of Latin or Latinized terms or other specialized terminology, whenever the correct meaning could be conveyed in approximately the same space, e.g., "pitted" rather than "foveolate," but "striate" rather than "with fine longitudinal lines." See *Categorical Glossary for the Flora of North America Project* (R. W. Kiger and D. M. Porter 2001; also available online at http://huntbot.andrew.cmu.edu) for standard definitions of generally used terms. Very specialized terms are defined, and sometimes illustrated, in the relevant family or generic treatments.

References

Authoritative general reference works used for style are *The Chicago Manual of Style*, ed. 14 (University of Chicago Press 1993); *Webster's New Geographical Dictionary* (Merriam-Webster 1988); and *The Random House Dictionary of the English Language*, ed. 2, unabridged (S. B. Flexner and L. C. Hauck 1987). *B-P-H/S. Botanico-Periodicum-Huntianum/Supplementum* (G. D. R. Bridson and E. R. Smith 1991) has been used for abbreviations of serial titles, and *Taxonomic Literature*, ed. 2 (F. A. Stafleu and R. S. Cowan 1976–1988) and its supplements by F. A. Stafleu et al. (1992–2009) have been used for abbreviations of book titles.

Graphic Elements

All genera and approximately 31 percent of the species in this volume are illustrated. The illustrations may show diagnostic traits or complex structures. Most illustrations have been drawn from herbarium specimens selected by the authors. Data on specimens that were used and parts that were illustrated have been recorded. This information, together with the archivally preserved original drawings, is deposited in the Missouri Botanical Garden Library and is available for scholarly study.

Specific Information in Treatments

Keys

Dichotomous keys are included for all ranks below family if two or more taxa are treated. More than one key may be given to facilitate identification of sterile material.

Nomenclatural Information

Basionyms of accepted names, with author and bibliographic citations, are listed first in synonymy, followed by any other synonyms in common recent use, listed in alphabetical order, without bibliographic citations.

The last names of authors of taxonomic names have been spelled out. The conventions of *Authors of Plant Names* (R. K. Brummitt and C. E. Powell 1992) have been used as a guide for including first initials to discriminate individuals who share surnames.

If only one infraspecific taxon within a species occurs in the flora area, nomenclatural information (literature citation, basionym with literature citation, relevant other synonyms) is given for the species, as is information on the number of infraspecific taxa in the species and their distribution worldwide, if known. A description and detailed distributional information are given only for the infraspecific taxon.

Descriptions

Character states common to all taxa are noted in the description of the taxon at the next higher rank. For example, if sexual condition is dioecious for all species treated within a genus, that character state is given in the generic description. Characters used in keys are repeated in the descriptions. Characteristics are given as they occur in plants from the flora area. Characteristics that occur only in plants from outside the flora area may be given within square brackets, or instead may be noted in the discussion following the description. In families with one genus and one or more species, the family description is given as usual, the genus description is condensed, and the species are described as usual. Any special terms that may be used when describing members of a genus are presented and explained in the genus description or discussion.

In reading descriptions, the reader may assume, unless otherwise noted, that: the plants are green, photosynthetic, and reproductively mature; woody plants are perennial; stems are erect; roots are fibrous; leaves are simple and petiolate; flowers are bisexual, radially symmetric, and pediceled; perianth parts are hypogynous, distinct, and free; and ovaries are superior. Because measurements and elevations are almost always approximate, modifiers such as "about," "circa," or "±" are usually omitted.

Unless otherwise noted, dimensions are length × width. If only one dimension is given, it is length or height. All measurements are given in metric units. Measurements usually are based on dried specimens but these should not differ significantly from the measurements actually found in fresh or living material.

Chromosome numbers generally are given only if published, documented counts are available from North American material or from an adjacent region. No new counts are published intentionally in the *Flora*. Chromosome counts from nonsporophyte tissue have been converted to the $2n$ form. The base number ($x = $) is given for each genus. This represents the lowest known haploid count for the genus unless evidence is available that the base number differs.

Flowering time and often fruiting time are given by season, sometimes qualified by early, mid, or late, or by months. Elevations over 50 m generally are rounded to the nearest 100 m; those 50 m and under are rounded to the nearest 10 m. Mean sea level is shown as 0 m, with the understanding that this is approximate. Elevation often is omitted from herbarium specimen labels, particularly for collections made where the topography is not remarkable, and therefore precise elevation is sometimes not known for a given taxon.

The term "introduced" is defined broadly to refer to plants that were released deliberately or accidentally into the flora and that now exist as wild plants in areas in which they were not recorded as native in the past. The distribution of non-native plants is often poorly documented and presence of the plants in the flora may be ephemeral.

If a taxon is globally rare or if its continued existence is threatened in some way, the words "of conservation concern" appear before the statements of elevation and geographic range.

Criteria for taxa of conservation concern are based on NatureServe's (formerly The Nature Conservancy)—see http://www.natureserve.org—designations of global rank (G-rank) G1 and G2:

G1 Critically imperiled globally because of extreme rarity (5 or fewer occurrences or fewer than 1000 individuals or acres) or because of some factor(s) making it especially vulnerable to extinction.

G2 Imperiled globally because of rarity (5–20 occurrences or fewer than 3000 individuals or acres) or because of some factor(s) making it very vulnerable to extinction throughout its range.

The occurrence of species and infraspecific taxa within political subunits of the *Flora* area is depicted by dots placed on the outline map to indicate occurrence in a state or province. The Nunavut boundary on the maps has been provided by the GeoAccess Division, Canada Centre for Remote Sensing, Earth Science. Authors are expected to have seen at least one specimen documenting each geographic unit record (except in rare cases when undoubted literature reports may be used) and have been urged to examine as many specimens as possible from throughout the range of each taxon. Additional information about taxon distribution may be presented in the discussion.

Distributions are stated in the following order: Greenland; St. Pierre and Miquelon; Canada (provinces and territories in alphabetic order); United States (states in alphabetic order); Mexico (11 northern states may be listed specifically, in alphabetic order); West Indies; Bermuda; Central America (Belize, Costa Rica, El Salvador, Guatemala, Honduras, Nicaragua, Panama); South America; Europe, or Eurasia; Asia (including Indonesia); Africa; Atlantic Islands; Indian Ocean Islands; Pacific Islands; Australia; Antarctica.

Discussion

The discussion section may include information on taxonomic problems, distributional and ecological details, interesting biological phenomena, and economic uses.

Selected References

Major references used in preparation of a treatment or containing critical information about a taxon are cited following the discussion. These, and other works that are referred to in discussion or elsewhere, are included in Literature Cited at the end of the volume.

CAUTION

The Flora of North America Editorial Committee **does not encourage, recommend, promote, or endorse** any of the folk remedies, culinary practices, or various utilizations of any plant described within this volume. Information about medicinal practices and/or ingestion of plants, or of any part or preparation thereof, has been included only for historical background and as a matter of interest. Under no circumstances should the information contained in these volumes be used in connection with medical treatment. Readers are strongly cautioned to remember that many plants in the flora are toxic or can cause unpleasant or adverse reactions if used or encountered carelessly.

Key to boxed codes following accepted names:

- C of conservation concern
- E endemic to the flora area
- F illustrated
- I introduced to the flora area
- W weedy, based mostly on R. H. Callihan et al. (1995) and/or D. T. Patterson et al. (1989)

Flora of North America

SALICACEAE Mirbel
• Willow Family

George W. Argus
James E. Eckenwalder
Robert W. Kiger

Shrubs or trees, heterophyllous or not, sometimes clonal, forming clones by root shoots, rhizomes, layering, or stem fragmentation; glabrous or glabrescent to pubescent; branching monopodial or sympodial. **Stems** erect to pendent; branched. **Leaves** persistent, deciduous or marcescent, alternate (opposite or subopposite in *Salix purpurea*), spirally arranged, simple; stipules present or not; petiole present; blade margins toothed or entire, sometimes glandular. **Peduncles** present or absent. **Inflorescences** racemose or spicate, usually catkins, unbranched, sometimes fasciculate or racemelike cymes, flowering before or as leaves emerge or year-round; floral bract (1) subtending each flower, displaced onto pedicel or distinct, scalelike, apex entire, toothed, or laciniate; bract subtending pistillate flower deciduous or persistent. **Pedicels** present or absent. **Flowers** usually unisexual, sometimes bisexual, usually staminate and pistillate on different plants; sepals present or absent, or perianth modified into 1 or 2 nectaries, or a non-nectariferous disc; stamens 1–60(–70); filaments distinct or connate basally, slender; anthers longitudinally dehiscent; ovary 1, 2–7[–10]-carpellate, 1–7[–10]-locular; placentation usually parietal, sometimes axile on intruded, fused placentae; ovules 1–25 per ovary; style 1 per carpel, distinct or connate; stigmas 2–4, truncate, notched-capitate, or 2- or 3-lobed. **Fruits** capsular, baccate, or drupaceous. **Seeds** sometimes surrounded by arillate coma of relatively long, silky hairs; endosperm scant or absent.

Genera 50+, species ca. 1000 (4 genera, 123 species in the flora): nearly worldwide.

Taxonomic placement of the Salicaceae and the genera included in it have varied greatly. Some botanists (H. G. A. Engler and K. Prantl 1887–1915) treated it as a primitive member of the Dicotyledoneae and grouped it with other families having simple, apetalous, unisexual flowers arranged in catkins, the "Amentiferae." At about the same time, others (C. E. Bessey 1915) took a different view, regarding the simple flowers as the result of reduction, and placed the taxa in Caryophyllales. As early as 1905, H. Hallier could see that there were similarities between Salicaceae and Flacourtiaceae; at the time, he was vigorously challenged by E. Gilg (1915). A. D. J. Meeuse (1975) summarized evidence for a close relationship between these families, including wood anatomy, phytochemistry, host-parasite relationships (including rust fungi), and morphology. He concluded that the Salicaceae could be combined with the Flacourtiaceae, "perhaps as a tribe." A. Cronquist (1988) and R. F. Thorne (1992b) placed the Salicaceae, in a narrow sense, in Violales near Flacourtiaceae.

Molecular studies support a close relationship between Salicaceae and Flacourtiaceae in Malpighiales and show that Flacourtiaceae, in a broad sense, is paraphyletic. Based on a study of plastid rbcL DNA sequences, *Salix* and *Populus* were nested within a subset of 52 genera of Flacourtiaceae (M. W. Chase et al. 2002). Chase et al. proposed moving some genera from broadly circumscribed Flacourtiaceae to Salicaceae. Other studies, based on different gene sequences, came to the same conclusion (O. I. Nandi et al. 1998; V. Savolainen et al. 2000; K. W. Hilu et al. 2003; Angiosperm Phylogeny Group 2003). The discovery of the extinct fossil genus *Pseudosalix* (L. D. Boucher et al. 2003), from the Eocene Green River Formation of Utah, provided further support for placing some members of Flacourtiaceae in Salicaceae. The well-preserved *Pseudosalix* fossils, in which reproductive structures are directly associated with the leaves, occur intermixed with *Populus* fossils. The leaves are slender and have salicoid teeth, inflorescences are cymose, flowers are unisexual, pedicellate, tetrasepalous, and 3- or 4-carpellate, and seeds are comose, i.e., having characteristics intermediate between Salicaceae and Flacourtiaceae.

The presence, in both families, of salicoid teeth is often cited in support of their close relationship (W. S. Judd 1997b; O. Nandi et al. 1998; M. W. Chase et al. 2002; H. P. Wilkinson 2007). Salicoid teeth were first recognized and defined as having the tip of the medial vein (seta) of the tooth retained as a dark, but not opaque, non-deciduous spherical callosity fused to the tooth apex and were reported to occur in Salicaceae and *Idesia* of the Flacourtiaceae (L. J. Hickey and J. A. Wolfe 1975). Nandi et al. reported that a broad survey of angiosperm leaves showed that salicoid teeth occur outside of Flacourtiaceae and Salicaceae only in Tetracentraceae.

Isozyme and cytological evidence show that *Populus* and *Salix* are ancient polyploids (D. E. Soltis and P. S. Soltis 1990; Wang R. and Wang J. 1991). All *Salix* and *Populus* species contain salicin (R. T. Palo 1984).

The genera often included in Salicaceae, in the narrow sense, are *Chosenia*, *Populus*, *Salix* (A. K. Skvortsov 1999), and, sometimes, *Toisusu*. Molecular studies (E. Leskinen and C. Alström-Rapaport 1999; T. Azuma et al. 2000) show that *Chosenia* is nested within *Salix*. H. Ohashi (2001) treated *Toisusu* as *Salix* subg. *Pleuradinea* Kimura and *Chosenia* as *Salix* subg. *Chosenia* (Nakai) H. Ohashi.

SELECTED REFERENCES Fisher, M. J. 1928. The morphology and anatomy of the flowers of the Salicaceae. Amer. J. Bot. 15: 307–326, 372–395. Floderus, B. G. O. 1923. Om Grönlands salices. (On the Salicaceae of Greenland.) Meddel. Grønland 63: 61–204. Judd, W. S. 1997b. The Flacourtiaceae in the southeastern United States. Harvard Pap. Bot. 10: 65–79. Leskinen, E. and C. Alström-Rapaport. 1999. Molecular phylogeny of Salicaceae and closely related Flacourtiaceae: Evidence from 5.8 S, ITS 1 and ITS 2 of the rDNA. Pl. Syst. Evol. 215: 209–227.

1. Flowers in catkins; sepals absent; fruits capsules.
    2. Buds 3–10-scaled (usually resinous); leaf blades usually less than 2 times as long as wide, venation ± palmate (basal secondary veins strong, paired, except in *Populus angustifolia*); stipules caducous; catkins pendulous, sessile; floral bracts: apex deeply or shallowly cut, pistillate floral bracts deciduous after flowering; flowers without nectaries (with a non-glandular, cup- or saucer-like disc); stamens 6–60(–70); stigmas 2–4; capsules 2–4-valved, narrowly ovoid to spherical . . . . 1. *Populus*, p. 5
    2. Buds 1-scaled (oily in *Salix barrattiana*); leaf blades often more than 2 times as long as wide, venation usually pinnate; stipules persistent or absent; catkins erect, spreading, or ± pendulous, sessile or terminating flowering branchlets; floral bracts: apex entire, erose, 2-fid, or irregularly toothed, pistillate floral bracts persistent or deciduous after flowering; flowers: perianth reduced to adaxial nectary (rarely also with abaxial nectary, then distinct or connate into shallow cup); stamens 1, 2, or 3–10; stigmas 2; capsules 2-valved, obclavate to ovoid or ellipsoid . . . . 2. *Salix*, p. 23
1. Flowers not in catkins; sepals present; fruits drupes or berries.
    3. Flowers in racemelike cymes or solitary; fruits drupes, 18–25 mm . . . . 3. *Flacourtia*, p. 163
    3. Flowers in fascicles; fruits berries, 4–7 mm . . . . 4. *Xylosma*, p. 163

1. POPULUS Linnaeus, Sp. Pl. 2: 1034. 1753; Gen. Pl. ed. 5, 456. 1754 • [Latin *populus*, the people, many fanciful allusions supposed but none certain]

James E. Eckenwalder

Trees, usually heterophyllous, usually clonal, clones formed by root shoots; branching usually monopodial [or sympodial]. **Stems** not spinose. **Buds** 3–10-scaled (resinous or not, terminal buds present [or absent]). **Leaves** deciduous; stipules present (caducous, usually minute, sometimes prominent on sucker shoots); petiole not glandular; (blade usually less than twice as long as wide, venation ± palmate, basal secondary veins strong, paired, except in *Populus angustifolia*, margins subentire or crenate, basilaminar glands 0–6). **Inflorescences** axillary or terminal, catkins, pendulous, sessile, unbranched, (leafless, flowering before leaves emerge); floral bract caducous, apex deeply or shallowly cut, (sometimes ciliate, usually glabrous, except pubescent abaxially in *P. heterophylla*); pistillate bract deciduous after flowering. **Pedicels** present. **Flowers:** perianth modified into non-nectariferous disc, (persistent, caducous in *P. heterophylla*), cup- or saucer-shaped; stamens 6–60(–70); filaments distinct; ovary 2–4-carpellate; ovules (1 or) 2–25 per ovary; styles distinct; stigmas 2–4, cylindrical to platelike, often rolled or convoluted, entire or 2-lobed. **Fruits** capsular, (2–4-valved, ovoid or spherical). **Seeds:** aril present. $x = 19$.

Species ca. 30 (8 in the flora): worldwide, mostly in northern hemisphere in moist to wet habitats, Arctic Circle to s Mexico, Asia (s China, n India, s Arabian Peninsula), n Africa, outlier in Kenya.

Populus has six well-marked sections, of which four occur in the flora area: swamp poplars [sect. *Leucoides* Spach (*P. heterophylla*)]; balsam poplars [sect. *Tacamahaca* Spach (*P. angustifolia*, *P. balsamifera*, and *P. trichocarpa*)]; cottonwoods [sect. *Aigeiros* Duby (*P. deltoides* and *P. fremontii*)]; and aspens [sect. *Populus* (*P. grandidentata* and *P. tremuloides*)]. Species within a section usually have separate distributions and hybridize freely where they come in contact. Species of different sections often have overlapping ranges and do not hybridize, except that members of sect. *Aigeiros* hybridize with all species of both sect. *Leucoides* and sect. *Tacamahaca* with which they are sympatric (J. E. Eckenwalder 1984). All known natural hybrids in the flora area are discussed under their parent species. Although some were originally described as species, they are not self-perpetuating. Because they can persist for decades by clonal growth, they can often be found in the absence of one or both parents.

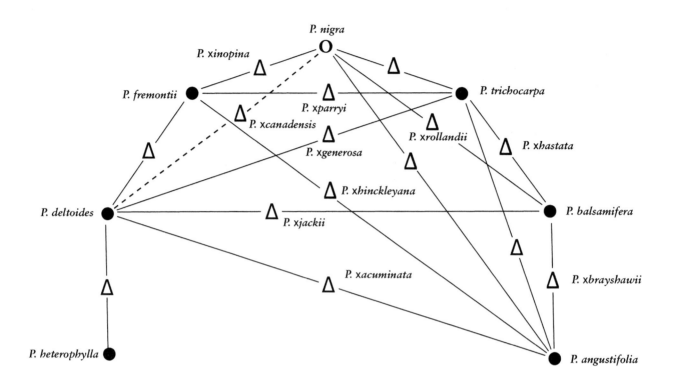
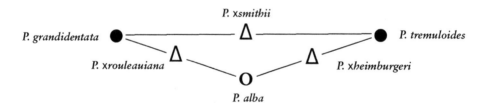

- ● native parent species
- ○ introduced parent species
- △ hybrids
- - - hybrids known in Europe and introduced in North America

Natural hybridization among species of *Populus* in North America

All poplars are capable of clonal expansion, producing new trees from sprouts of root systems (soboliferous habit). Different species vary greatly in their propensity for cloning. The cottonwoods, *Populus deltoides* and *P. fremontii*, rarely produce root-borne shoots under natural conditions, with most clonal suckers arising from buried branches (S. B. Rood et al. 1994); the aspens, *P. grandidentata* and *P. tremuloides*, regularly form characteristic, dome-shaped clonal groves, some hectares in extent, with thousands of individual stems (B. V. Barnes 1966). Other species fall between these extremes. As a consequence of clonal growth, whole patches of trees may be solely staminate or pistillate and have uniform leaf morphology and phenology, sometimes making it difficult to find individuals of both sexes (Barnes 1969).

Collecting representative materials and identifying poplars is also complicated by seasonal heterophylly (variation in leaves along a shoot over the course of a growing season; W. B. Critchfield 1960; J. E. Eckenwalder 1980c, 1996). Two principal categories of leaves may differ in some characteristics, including size, shape, and marginal teeth. Preformed leaves (also known as early leaves) overwinter in buds as usually 3–10 tiny rolled up leaves before expanding with spring flush following flowering. Neoformed leaves (also known as late leaves) are initiated and expand during the growing season as shoots continue to lengthen after spring flush. Some neoformed leaves of rapidly growing suckers and young trees are the largest produced by each species, often more than twice as long as the largest leaves of mature trees. Preformed leaves are clustered at shoot bases, separated by internodes less than 1 cm. On long shoots (and all leaves on clonal sucker shoots in their first year), neoformed leaves are produced with internodes of 2 cm or more throughout the growing season until formation of winter buds, which contain next year's preformed leaves. By the time winter buds form, most preformed leaves may already have fallen, leaving only neoformed leaves. As a result of this seasonality and unisexuality, individual staminate and pistillate trees should be marked and collected on three separate occasions: 1) at flowering; 2) when preformed leaves are mature (and when capsules are just opening on pistillate trees); and 3) with formation of mature winter buds, but before leaves have turned color and developed their abscission layer.

Poplar trees vary greatly in proportion of preformed and neoformed leaves during their life cycle and in how distinct these two leaf types are. Ironically, *Populus heterophylla* is one of the least heterophyllous species; its name alluding to variation among preformed leaves. In contrast, species with large teeth on preformed leaves, such as *P. fremontii* and *P. grandidentata*, are often strongly heterophyllous, with smaller and more numerous teeth on neoformed leaves (N. L. Britton 1886; J. E. Eckenwalder 1996). The most dramatic example of heterophylly in poplars occurs in the introduced Eurasian *P. alba*, which has angular-ovate preformed leaves and maplelike, palmately 5-lobed neoformed leaves that are unique in the genus. In all species, teeth of neoformed leaves have larger glands at the tips than those of preformed leaves. These glands are connected to leaf venation and, in combination with basilaminar glands on the adaxial side of a leaf blade at its junction with the petiole, secrete resins onto young, expanding leaves (J. D. Curtis and N. R. Lersten 1978).

Poplars are often grown for ornament, shelterbelts, timber, pulp, and specialty wood products. Breeding of poplars, especially using interspecific crosses, has become an important source of stock for plantation forestry. Selected clones of some crosses, such as *Populus* ×*generosa* (discussed under *P. trichocarpa*) and *P.* ×*canadensis* (discussed under *P. deltoides*), are among the fastest growing and most productive temperate trees. All native species (except *P. heterophylla*) and some natural hybrids are cultivated to some extent within their native ranges. In addition to native poplars, some Eurasian species and hybrids are commonly cultivated and may persist and spread clonally to form prominent landscape components. Because most of these taxa are represented in North America almost entirely by single clones and are solely pistillate or staminate, they are not truly naturalized here, despite their occasional prominence in the landscape. Eurasian taxa represented exclusively by staminate individuals produce no seed, and any seeds dispersed by the taxa that are solely pistillate of necessity have arisen from pollination by related (usually native) species. The most frequently encountered cultivated Eurasian species are included in the keys and are briefly described and discussed, along with any known natural hybrids, in the discussion for the most closely related native species.

Publication of the complete draft sequence of the *Populus trichocarpa* genome (G. A. Tuskan et al. 2006) has helped consolidate the position of *Populus* among the elite ranks of "model" organisms used for genetic, evolutionary, developmental, physiological, plant pathological, and herbivory related studies, among others. The published genome sequence may also lead to improved understanding of relationships within the genus but has not yet influenced molecular phylogenetic studies, which have suffered from limited sampling, although some interesting insights on wider past hybridization, than any seen today, have emerged (R. L. Smith and K. J. Sytsma 1990).

When A. Cronquist (1968, 1981) formulated the classification used in much of this flora, treating Salicaceae in a monotypic Salicales associated with Violales, it was already strongly suspected, based on floral development, leaf architecture, and shared secondary metabolites, that *Populus* and *Salix* were derived from within the predominantly tropical family Flacourtiaceae, as traditionally circumscribed (A. D. J. Meeuse 1975; W. S. Judd et al. 1994; J. E. Eckenwalder 1996). There are strong similarities between *Populus* and the monotypic eastern Asian genera *Idesia*, *Itoa*, and *Poliothyrsis*, and also with the more widespread *Xylosma*, found in all tropical regions except Africa, all are members of tribe Flacourtieae in traditional circumscriptions (D. E. Lemke 1988; D. J. Mabberly 1997). Molecular studies have amply confirmed these relationships and greatly extended them in showing that these and other flacourt genera are more closely related to *Populus* and *Salix* than they are to many other genera within traditional Flacourtiaceae (M. W. Chase et al. 2002). In fact, Flacourtiaceae are so heterogeneous that the family has been split into two, about half of the tribes being aligned with *Populus* and *Salix* in an enlarged Salicaceae (the earlier name, even though *Flacourtia* itself is included in the family). Chase et al. restricted tribe Saliceae to *Populus* and *Salix*, but their tribe Flacourtieae is paraphyletic and some of its genera (including *Itoa* and *Poliothyrsis*) will have to be transferred to Saliceae to establish consistent monophyly of groups within the family. *Populus* and *Salix* are sister taxa within this tribe but, of the two genera, some poplar species have retained more plesiomorphic features than have any of the willows, including more stamens and carpels, a less modified perianth, and broader leaves with more complex venation. Most of these species, belonging to two generally more southerly sections not found in the flora (sect. *Abaso*, with *P. mexicana*, and sect. *Turanga*, with *P. euphratica*, *P. ilicifolia*, and *P. pruinosa*), have strongly heteroblastic developmental leaf change, in which broad adult leaves expected through familiarity with the common temperate representatives of the genus, including all species in the flora except *P. angustifolia*, are preceded by willowlike, narrow, seedling and juvenile leaves (Eckenwalder 1980c, 1996b).

The earliest known fossil poplars, found in latest Paleocene and Eocene sediments within the flora area, are assigned to sect. *Abaso* and run the full gamut of leaf widths, some seemingly heteroblastic and others with narrow, fully adult leaves (J. E. Eckenwalder 1980c; S. R. Manchester et al. 2006). The common ancestor of *Populus* and *Salix* was probably somewhat more similar to *Populus* than to *Salix*, and this is partially corroborated by the recently described Eocene genus *Pseudosalix* (L. D. Boucher et al. 2003). Based on its morphology, this plant is the closest known relative of *Populus* and *Salix* and co-occurs with both genera in the Green River Formation of Utah and Colorado, so it is too late to represent their common ancestor. Nonetheless, its features are reasonably intermediate between traditional Salicaceae and their newly accepted relatives among extant traditional Flacourtiaceae. *Populus* strongly supports the concept of an enlarged Salicaceae and equally so an enlarged tribe Saliceae that would include some genera retained by M. W. Chase et al. (2002) in tribe Flacourtieae against their own evidence, which, admittedly, did not contain all relevant genera. The botanical community already recognizes a different set of family relationships than presented in the *Flora of North America North of Mexico*.

Keys. Three different keys are provided here to permit more effective identification throughout the annual phenological cycle in *Populus*: a key to flowering specimens that may also be used (but not optimally) for leafless, winter dormant specimens; a key to fruiting specimens; and a key to specimens with mature leaves.

Flowering in early spring is short-lived in poplars, with abscission of staminate catkins and pistillate floral bracts, and shrivelling of stigmas all occurring before emergence of the leaves, much as in precocious *Salix* species (in which bracts are not caducous). During anthesis, some winter characters involving twig colors and textures, leaf scars, and features of buds generally remain intact, so such features are used as supplementary characters in the key to flowering specimens. Characters of flower buds can also be helpful in winter identification but they are absent or obscured with emergence of the inflorescences at anthesis and are omitted from this key. The key includes characters of both staminate and pistillate individuals, which generally do not occur together on a single specimen. The characters of those rare individuals with mixed catkins or hermaphroditic flowers may be anomalous in other ways as well, and such individuals may not be readily identifiable.

At the time of fruit maturation, in late spring to early summer, staminate individuals are not distinguishable from non-fruiting pistillate individuals. Furthermore, while preformed leaves may then be expanded enough to reveal their mature characteristics, there may well not yet be any neoformed leaves present. Because the time of fruit maturation is not optimal for identifying non-fruiting individuals of *Populus* by their vegetative characters, the key to leafy specimens is designed primarily for late summer use, after the seeds and fruiting catkins have been shed. Because trees of the commonly cultivated, introduced *P. nigra*, *P. simonii*, and *P.* ×*canadensis* are rarely anything but staminate in North America (except in arboreta, botanical gardens, or poplar plantations), these taxa are excluded from the fruiting key.

Leafy specimens are best identified in late summer or early autumn when winter buds are forming and neoformed leaves are present and preformed leaves have not yet been shed. The key to leafy specimens is most effective at this time but should prove useful whenever mature, or nearly mature, leaves are present.

Three commonly encountered introduced Eurasian species (*Populus alba*, *P. nigra*, and *P. simonii*) are included in the keys but are not given full treatments in the text, where they are discussed following the most closely related native species. Other, less common, introduced species that are only found clearly under cultivation are omitted from the keys as are the relatively numerous different hybrid combinations (except for three introduced hybrids that have much the same status as the prominent introduced species: *P.* ×*canadensis* 'Eugenei,' *P.* ×*canescens,* and *P.* ×*jackii* 'Balm of Gilead'). There are more different natural hybrid poplar combinations in the flora than there are poplar species. Although these hybrids can be found across the continent, they are usually much less common than their parents and typically grow with both, which gives a clue to their identity. The natural hybrids are not included in the keys because their numbers, additional variability, and general intermediacy between the parent species would greatly reduce ease of use of the keys and confidence in identification for the sake of relatively few specimens. Instead, hybrids are discussed following the descriptions of their parent species. A specimen that does not quite key out in the appropriate seasonal key here may well be a hybrid and the discussion of the species it comes closest to should be consulted for distinguishing characteristics. Overlapping parental distributions are also helpful, although, on rare occasions, hybrids may be found far beyond the natural range of one (or both) of their parent species.

SELECTED REFERENCES Eckenwalder, J. E. 1977. North American cottonwoods (*Populus*, Salicaceae) of sections *Abaso* and *Aigeiros*. J. Arnold Arbor. 58: 193–208. Eckenwalder, J. E. 1984. Natural intersectional hybridization between North American species of *Populus* (Salicaceae) in sections *Aigeiros* and *Tacamahaca*. II. Taxonomy. Canad. J. Bot. 62: 325–335. Eckenwalder, J. E. 1996. Systematics and evolution of *Populus*. In: R. F. Stettler et al., eds. 1996. Biology of *Populus* and Its Implications for Management and Conservation. Ottawa. Pp. 7–32. Sudworth, G. B. 1934. Poplars, Principal Tree Willows and Walnuts of the Rocky Mountain Region. Washington. [U.S.D.A., Techn. Bull. 420.]

Key to flowering specimens of *Populus*

1. Floral bracts ciliate; stamens 6–12; flower discs narrowly cup-shaped, oblique; stigmas 2, filiform; capsules narrowly ovoid to lanceoloid; winter buds not or slightly resinous.
    2. Branchlets and terminal buds densely to sparsely tomentose; floral bracts: apex shallowly cut.
        3. Hairs bright white . . . . *Populus alba* [p. 21]
        3. Hairs grayish, brownish, or dirty white . . . . *Populus ×canescens* [p. 21]
    2. Branchlets and terminal buds glabrous or pubescent; floral bracts: apex deeply cut.
        4. Winter buds pubescent proximally (dull); catkins (4–)6–10 cm; flowers: discs shallowly toothed . . . . 7. *Populus grandidentata*
        4. Winter buds glabrous proximally (shiny); catkins (1.7–)4–7 cm; flowers: discs entire . . . . .8. *Populus tremuloides*
1. Floral bracts not ciliate; stamens 10–60(–70); flower discs broadly cup- or saucer-shaped, not oblique; stigmas 2–4, not filiform; capsules ovoid to spherical; winter buds resinous or not.
    5. Floral bracts pubescent abaxially; anthers apiculate; catkins 10–15(–45)-flowered; flowers: discs toothed (caducous); winter buds slightly resinous . . . . 1. *Populus heterophylla*
    5. Floral bracts glabrous abaxially; anthers usually truncate; catkins (3–)15–150(–175)-flowered; flowers: discs entire; winter buds resinous throughout or partly.
        6. Winter buds with yellow resin.
            7. Stamens 30–40(–55); pedicels 1–10(–17 in fruit) mm; ovaries ovoid, discs saucer-shaped, 1–3(–4) mm diam.; branchlets usually glabrous or thinly long-hairy. . . . . 5. *Populus deltoides*
            7. Stamens (30–)40–60(–70); pedicels 1–4(–5.5 in fruit) mm; ovaries spherical, discs cup-shaped, (2.5–)4–7(–9) mm diam.; branchlets glabrous, glabrate, or hairy. . . . . .6. *Populus fremontii*
        6. Winter buds with red resin.
            8. Ovaries usually 3- or 4-carpelled; stamens 30+.
                9. Ovaries spherical, hairy to glabrate; discs 4–6 mm diam.; plants pistillate or staminate (dioecious). . . . . .2. *Populus trichocarpa*
                9. Ovaries broadly ovoid, glabrous; discs 1–3.5 mm diam.; plants pistillate . . . .*Populus ×jackii* [p. 15]
            8. Ovaries usually 2-carpelled; stamens 10–30.
                10. Terminal buds (3–)6–9(–13) mm; branchlets round or 5-angled, whitish tan by third year; catkins 3–8 cm.
                    11. Stamens 10–20; plants staminate or pistillate (dioecious) . . . . . . .4. *Populus angustifolia*
                    11. Stamens usually fewer than 12; plants staminate. . . . . *Populus simonii* [p. 15]
                10. Terminal buds usually (8–)12–16(–20) mm; branchlets usually round, tan or gray by third year; catkins (4–)7–15 cm.
                    12. Branchlets grayish brown by third year, first year reddish brown; resin of winter buds red, fragrant (balsamic); plants staminate or pistillate (dioecious) . . . . .3. *Populus balsamifera*
                    12. Branchlets tan by third year, first year bright orange-brown to reddish brown; resin of winter buds orange-red, fragrant (not balsamic); plants staminate only.

13. Stamens 12–20(–30); first year branchlets reddish brown; habit fastigiate, branchlets parallel or nearly so . . . . *Populus nigra* [p. 17]
13. Stamens (15–)20–30; first year branchlets orange-brown; habit somewhat spreading, branchlets divergent . . . . *Populus* ×*canadensis* [p. 17]

Key to fruiting specimens of *Populus*

1. Capsules usually 2-valved.
    2. Capsules ovoid to spherical, 3–8 mm.
        3. Capsules (3–)5–8 mm; seeds 15–22 per placenta; petioles (0.2–)1.5–5 cm . . . . 3. *Populus balsamifera*
        3. Capsules 3–5 mm; seeds (2–)4–7(–9) per placenta; petioles 0.2–0.8(–1.7) cm . . .
        . . . . 4. *Populus angustifolia*
    2. Capsules lanceoloid or narrowly ovoid, 2–5(–7) mm.
        4. Seeds (1 or) 2 (or 3) per placenta; floral bracts densely tomentose . . . . *Populus alba* [p. 21]
        4. Seeds (3–)5–8 (or 9) per placenta; floral bracts glabrous or tomentose.
            5 Leaf blades: abaxial surface tomentose at emergence . . . . *Populus* ×*canescens* [p. 21]
            5. Leaf blades: abaxial surface glabrous or densely silky at emergence.
                6. Discs shallowly toothed; catkins (4–)6–10(–14 in fruit) cm; preformed blades densely silky at emergence, with (1–)5–12(–16) teeth on each side
                . . . . 7. *Populus grandidentata*
                6. Discs entire; catkins 4–7(–12.5 in fruit) cm; preformed blades glabrous or sparsely sericeous at emergence, with (12–)18–30(–42) teeth on each side
                . . . . 8. *Populus tremuloides*
1. Capsules usually 3- or 4-valved.
    7. Seeds 6–9 per placenta; branchlets tomentose to glabrate; discs caducous . . . . 1. *Populus heterophylla*
    7. Seeds (3–)7–15(–25) per placenta; branchlets glabrous, glabrate, or long-hairy; discs persistent.
        8. Discs cup-shaped, 4–8 mm diam.; capsules spherical, (6–)7–9 mm.
            9. Capsules densely hairy to glabrate; blades: abaxial surface white to grayish white or greenish white (with red resin stain); margins not ciliate; preformed blade margins with (20–)35–40(–50) teeth on each side, sinuses 0.1–0.4 mm deep . . . . 2. *Populus trichocarpa*
            9. Capsules glabrous; blades: abaxial surface yellowish green (red resin stain not evident); margins ciliate; preformed blade margins with 3–10(–15) teeth on each side, sinuses (0.2–)0.5–4(–5.5) mm deep . . . . 6. *Populus fremontii*
        8. Discs saucer-shaped, 1–4 mm diam.; capsules ovoid, (4–)8–11(–16) mm.
            10. Pedicels to 3 mm; branchlets and petioles hairy; preformed blade margins with 20–45 teeth on each side, sinuses to 1.5 mm deep . . . . *Populus* ×*jackii* [p. 15]
            10. Pedicels 1–13 mm; branchlets and petioles glabrous or thinly long-hairy; preformed blade margins with (3–)5–15(–30) teeth on each side, sinuses (0.4–)0.7–5(–7) mm deep . . . . 5. *Populus deltoides*

Key to leafy specimens of *Populus*

1. Leaf blades usually 5–8(–10) cm, abaxial surface densely tomentose when young, retaining dense tomentum on at least some intervein regions.
    2. Leaf blades: abaxial surface tomentum bright white; neoformed blades (3 or) 5-lobed
    . . . . *Populus alba* [p. 21]
    2. Leaf blades: abaxial surface tomentum dull grayish, tannish, or dirty white; neoformed blades irregularly toothed . . . . *Populus* ×*canescens* [p. 21]
1. Leaf blades (1–)3–20(–27.5) cm, abaxial surface glabrous, glabrate, densely hairy, silky, or pubescent (not densely tomentose when young, not tomentose on intervein regions).
    3. Petioles round, cylindrical, or slightly flattened distally to plane of blades.

4. Leaf blades: base deeply cordate to subsagittate, apex obtuse to apiculate, abaxial surface pubescent to partly glabrate, retaining tomentum at least basally and on midvein .... 1. *Populus heterophylla*
4. Leaf blades: base acute, cuneate, rounded, truncate, or shallowly cordate, apex obtuse, acute, or acuminate, abaxial surface glabrous or hairy (hairs short and stiff).
    5. Petioles 0.2–0.8(–1.7) cm; leaf blades: abaxial surface whitish green (not obviously stained with reddish resin).
        6. Leaf blades usually lanceolate to narrowly ovate .... 4. *Populus angustifolia*
        6. Leaf blades elliptic-rhombate to obovate .... *Populus simonii* [p. 15]
    5. Petioles 0.2–9.5 cm; leaf blades: abaxial surface white to grayish or greenish white, stained with reddish resin.
        7. Petioles densely pubescent, at least distally .... *Populus* ×*jackii* [p. 15]
        7. Petioles glabrous or sparsely pubescent.
            8. Leaf blades usually triangular-ovate or narrowly ovate to cordate, base rounded to cordate; petioles often markedly swollen distally; w and s of Continental Divide .... 2. *Populus trichocarpa*
            8. Leaf blades usually narrowly ovate to ovate (rarely broadly ovate), base rounded to broadly cuneate or subcordate; petioles not conspicuously swollen distally; e and n of Continental Divide .... 3. *Populus balsamifera*

[3. Shifted to left margin.—Ed.]

3. Petioles flattened at right angle to plane of blades distally.
    9. Leaf blade: margins not translucent, not ciliate.
        10. Preformed blade: margins coarsely serrate, teeth (1–)5–12(–16) on each side, sinuses 0.3–4.5(–6) mm deep .... 7. *Populus grandidentata*
        10. Preformed blade: margins subentire to finely crenate-serrate, teeth (12–)18–30(–42) on each side, sinuses 0.1–1 mm deep .... 8. *Populus tremuloides*
    9. Leaf blade: margins translucent, ciliate.
        11. Leaves: basilaminar glands (0 or) 1–6.
            12. Basilaminar glands 0 or 1; blade bases broadly cuneate, apices gradually acuminate .... *Populus* ×*canadensis* [p. 17]
            12. Basilaminar glands 0–6; blade bases truncate to subcordate, apices abruptly acuminate .... 5. *Populus deltoides* (in part)
        11. Leaves: basilaminar glands 0.
            13. Largest preformed blade: margins with sinuses less than 1.2 mm deep ...... *Populus nigra* [p. 17]
            13. Largest preformed blade: margins with sinuses 2(–7) mm deep.
                14. Neoformed blade: margins with (10–)25–40(–55) teeth on each side; preformed blade: margins with (3–)5–15(–30) teeth on each side .... 5. *Populus deltoides* (in part)
                14. Neoformed blade: margins with (10–)20–30(–45) teeth on each side; preformed blade: margins with 3–10(–15) teeth on each side .... 6. *Populus fremontii*

1. Populus heterophylla Linnaeus, Sp. Pl. 2: 1034. 1753

• Swamp or downy poplar [E] [F]

Plants to 28 m, 12 dm diam.; not obviously heterophyllous. Bark reddish brown to brownish gray, deeply furrowed. Branchlets reddish brown, becoming grayish by third year, round to 5-angled, 3–6 mm diam., very coarse, thinly tomentose to glabrate. Winter buds reddish brown, pubescent, slightly resinous; terminal buds 4–7 mm; flowering buds separated on branchlets, 4–7 mm. Leaves: petiole round distally, (1–)4–8(–12) cm, ½ blade length, (tomentose to glabrate); blade ovate, (3.5–)9–20(–24) × (3.5–)7.5–12.5(–19) cm, w/l = 2/3–3/4, base deeply cordate to subsagittate, basilaminar glands 0 or 2, round, margins not translucent, not ciliate, apex obtuse to apiculate, abaxial surface pale green, pubescent to partly glabrate, retaining tomentum at least basally and on midvein, adaxial dark green, glabrous; preformed and neoformed blade margins finely and unevenly crenate-serrate throughout, teeth 30–60 on each side, sinuses 0.3–1 mm deep. Catkins sparsely 10–15(–45)-flowered, 4.5–8(–18 in fruit) cm; floral bract apex deeply cut, not ciliate (pubescent

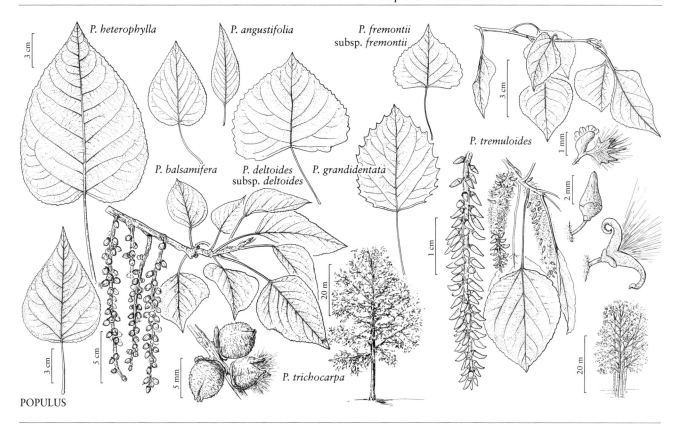

POPULUS

abaxially). **Pedicels** (1–)5–10(–18 in fruit) mm. **Flowers:** discs (caducous), cup-shaped, not obviously oblique, toothed, 1.5–2.5(–4) mm diam.; stamens 15–35; anthers apiculate; ovary 3-carpelled, ovoid to spherical; stigmas 2–4, convoluted, expanded, erect. **Capsules** ovoid, 8–14 mm, glabrous, 3-valved. **Seeds** 6–9 per placenta.

Flowering Apr; fruiting May–Jun. *Nyssa-Taxodium* swamps, drainage ditches, natural and artificial wet depressions, coastal plains, central lowlands, piedmont along major streams; 0–200(–400) m; Ont.; Ala., Ark., Conn., Del., D.C., Fla., Ga., Ill., Ind., Ky., La., Md., Mich., Miss., Mo., N.J., N.Y., N.C., Ohio, S.C., Tenn., Va.

Populus heterophylla is widespread and uncommon in most of its range, which is entirely included within that of *P. deltoides*. Reports of swamp poplars from upland habitats are based on misidentified sucker shoots of *P. grandidentata*. Hybrids of *P. heterophylla* and *P. deltoides* are found in disturbed habitats, such as drainage ditches. Hybrids have glabrous leaves more elongate than those of *P. deltoides*.

2. **Populus trichocarpa** Torrey & A. Gray, Icon. Pl. 9: plate 878. 1852 • Black cottonwood, baumier de l'ouest [F]

Populus balsamifera Linnaeus subsp. *trichocarpa* (Torrey & A. Gray) Brayshaw; *P. trichocarpa* var. *ingrata* (Jepson) Parish

Plants to 75 m, 26 dm diam.; moderately heterophyllous. **Bark** grayish brown, deeply furrowed. **Branchlets** reddish brown, becoming gray by third year, round, 1.5–3(–7) mm diam., coarse, usually densely hairy. **Winter buds** red, sparsely hairy or glabrous, resinous (resin red, abundant, very fragrant); terminal buds 8–15(–20) mm; flowering buds clustered distally on branchlets, 18–20 mm. **Leaves:** petiole cylindrical or distally slightly flattened in plane of blade (often markedly swollen distally), 1–5(–9.5) cm, ½ blade length, (glabrous or sparsely pubescent); blade usually triangular-ovate or narrowly ovate to cordate, (3–)5–9(–15) × (1–)2.5–6(–10) cm, w/l = 1/2–2/3, base rounded to cordate, basilaminar glands 0–2, round, margins not translucent, not ciliate, apex obtuse to acute, abaxial surface white to grayish white or greenish white with red resin stains, sparsely pubescent, adaxial dark green, glabrous; preformed blade margins very finely, evenly crenate-serrate throughout, teeth (20–)35–40

(–50) on each side, sinuses 0.1–0.4 mm deep; neoformed blade margins finely, evenly crenate-serrate throughout, teeth (25–)40–60 on each side, sinuses 0.2–0.6 mm deep. **Catkins** densely (10–)25–50(–90)-flowered, (4.5–)7–10(–17 in fruit) cm; floral bract apex deeply cut, not ciliate. **Pedicels** 0.5–2.5(–3 in fruit) mm. **Flowers:** discs broadly cup-shaped, not obviously oblique, entire, 4–6 mm diam.; stamens 30–50(–60); anthers truncate; ovary 3- or 4-carpelled, spherical, (hairy); stigmas 2–4, platelike, expanded, spreading. **Capsules** spherical, (6–)7–9 mm, densely hairy to glabrate, 3- or 4-valved. **Seeds** (6–)10–15(–19) per placenta. $2n = 38$.

Flowering early spring. Floodplains, lake margins, mesic areas, taluses and other slopes to subalpine tree line; 0–2600(–3000) m; Alta., B.C.; Alaska, Calif., Idaho, Mont., Nev., Oreg., Utah, Wash., Wyo.; Mexico (Baja California).

Populus trichocarpa has been mistakenly reported from North Dakota based on specimens of *P.* ×*jackii* (*P. balsamifera* × *P. deltoides*). It hybridizes with *P. balsamifera* to form *P.* ×*hastata* Dode along the northern Rocky Mountain axis (Alaska, Alberta, British Columbia, and Idaho). Hybrids have capsules with 2–4 glabrous or sparsely hairy valves. The extent of hybridization has led to treatment of *P. trichocarpa* as a subspecies of *P. balsamifera* (T. C. Brayshaw 1965; L. A. Viereck and J. M. Foote 1970); these two balsam poplars are more closely related to Asian members of sect. *Tacamahaca* than they are to each other (J. E. Eckenwalder 1996). Mountain ranges of the Intermountain Region (California, Idaho, Nevada, Oregon, and Utah) have trees intermediate between *P. trichocarpa* and *P. angustifolia*. These hybrids have narrower leaves with shorter petioles and sparsely hairy capsules with 2–3 valves.

In addition to hybridizing with other North American species of sect. *Tacamahaca*, *Populus trichocarpa* also hybridizes with both native species of sect. *Aigeiros*. *Populus* ×*generosa* A. Henry (synonym *P.* ×*interamericana* van Broekhuizen), a hybrid between *P. trichocarpa* and *P. deltoides*, is rare in the far western area of the range for *P. deltoides* subsp. *monilifera*, where it overlaps with the more drought-tolerant inland *P. trichocarpa* (Idaho, Montana, Washington, and Wyoming) (J. E. Eckenwalder 1984). This hybrid has also been grown artificially, and such hybrids between coastal *P. trichocarpa* and *P. deltoides* subsp. *deltoides* are becoming increasingly important plantation trees in the Pacific Northwest from northern Oregon to British Columbia, as well as in Europe. They are perhaps the fastest growing of all poplars in volume, with the rapid height growth of *P. trichocarpa* added to the steady diameter growth of *P. deltoides* (R. F. Stettler et al. 1988).

Populus ×*parryi* Sargent, a hybrid between *P. trichocarpa* and *P. fremontii*, is commonly found in a wide variety of mesic habitats throughout the region of sympatry between its parents in California and Nevada (and beyond the range of *P. trichocarpa* in Mohave County, Arizona; J. E. Eckenwalder 1992). It can be found particularly in canyons where its parents are elevationally separated but overlap as permanent streams spill out into lower elevations (Eckenwalder 1984, 1984b). A morphologically and ecologically distinctive race of *P. trichocarpa* in coastal southern California with heart-shaped leaves may have arisen through this kind of hybridization (Eckenwalder 1984c). This race includes the type of *P. trichocarpa* from Ventura County.

Populus maximowiczii A. Henry is an Asian balsam poplar that is sometimes cultivated as an ornamental, but usually as a plantation tree or parent of plantation hybrids. It is distinguished from *P. trichocarpa* and *P. balsamifera* by its elliptic leaves with rugose adaxial surfaces.

3. **Populus balsamifera** Linnaeus, Sp. Pl. 2: 1034. 1753 • Balsam poplar, hackmatack, bam tree, tacamahaca, baumier E F W

Populus balsamifera var. *subcordata* Hylander; *P. candicans* Aiton; *P. tacamahacca* Miller

Plants to 40 m, 21 dm diam.; weakly heterophyllous. **Bark** reddish gray, furrowed. **Branchlets** reddish brown, becoming grayish brown by third year, round, 1.5–3.5(–5) mm diam., coarse, glabrous or glabrate to densely hairy. **Winter buds** reddish, glabrous, resinous (resin red, abundant, very fragrant, balsamic); terminal buds (8–)12–16(–20) mm; flowering buds clustered distally on branchlets, 15–19 mm. **Leaves:** petiole cylindrical or distally slightly flattened in plane of blade, (0.2–)1.5–5 cm, 1/3–1/2 blade length (usually glabrous); blade usually narrowly ovate to ovate, rarely broadly ovate, (2.5–)5–9(–15) × (0.7–)3–5.5(–9) cm, w/l = 1/2–2/3, base rounded to broadly cuneate or subcordate, basilaminar glands 0 or 2(–5), round, margins not translucent, not ciliate, apex obtuse to acute, abaxial surface often with reddish orange resin stains, glaucous, (veins prominent), adaxial dark green, glabrous; preformed blade margins subentire to very finely, evenly crenate-serrate throughout, teeth (9–)20–35(–45) on each side, sinuses 0.1–0.4 mm deep; neoformed blade margins finely crenate-serrate throughout, teeth (20–)30–45(–60) on each side, sinuses 0.2–0.6 mm deep. **Catkins** moderately loosely (35–)50–70(–80)-flowered, 7.5–

15(–18 in fruit) cm; floral bract apex deeply cut, not ciliate. **Pedicels** 0.5–2(–3.5 in fruit) mm. **Flowers:** discs shallowly cup-shaped, not obviously oblique, entire, 2–3(–4) mm diam.; stamens 20–30; anthers truncate; ovary 2-carpelled, ovoid to spherical; stigmas 2–4, platelike, expanded, reflexed. **Capsules** ovoid, (3–) 5–8 mm, glabrous, 2-valved. **Seeds** 15–22 per placenta. $2n = 38$.

Flowering Mar–Jun; fruiting May–Jul. Open, rich, low woods, cool, seasonally wet soils, bog margins in boreal forests, aspen parklands, montane streamsides, rocky slopes, gallery forests within tundra; 0–2900 (–3700) m; St. Pierre and Miquelon; Alta., B.C., Man., N.B., Nfld. and Labr. (Nfld.), N.W.T., N.S., Nunavut, Ont., P.E.I., Que., Sask., Yukon; Alaska, Colo., Conn., Del., Ill., Ind., Iowa, Maine, Mass., Mich., Minn., Mont., N.H., N.Y., N.Dak., Ohio, Pa., S.Dak., Vt., W.Va., Wis., Wyo.

Populus balsamifera has been reported in error from Nebraska based on incorrectly localized specimens and from Idaho, Oregon, and Utah based on vegetative specimens of *P. trichocarpa* or intergrades. The two species can be difficult to separate vegetatively, particularly in their region of overlap and hybridization along the northern Cordilleran axis from southern Alaska to northwestern Wyoming. Their hybrids, named *P.* ×*hastata* Dode, have intermediate leaf shapes and also differ from *P. balsamifera* in having capsules with 2–4 sparsely hairy or glabrous valves (T. C. Brayshaw 1965). *Populus balsamifera* also hybridizes and intergrades with another native species of sect. *Tacamahaca*, *P. angustifolia*, to form *P.* ×*brayshawii* B. Boivin where the margins of their ranges overlap (Brayshaw 1965b). *Populus* ×*brayshawii* is intermediate in some respects between its parents. It is most similar to *P. angustifolia*; it differs in longer petioles and darker twigs, characteristics in which it approaches *P. balsamifera*. Some trees of *P. balsamifera* from North Dakota (Bottineau and Divide counties) also seem to show an influence of *P. angustifolia*, although they are far from the present range of the latter species.

Populus ×*jackii* Sargent (synonyms *P.* ×*andrewsii* Sargent, *P.* ×*bernardii* B. Boivin, *P.* ×*dutillyi* Lepage, and *P.* ×*gileadensis* Rouleau) is an intersectional hybrid with *P. deltoides* (sect. *Aigeiros*) and is moderately common in riparian and other wet habitats throughout the broad range of overlap between these two species (Alberta, Colorado, Illinois, Indiana, Iowa, Manitoba, Massachusetts, Michigan, Minnesota, Montana, Nebraska, New York, North Dakota, Ontario, Quebec, Pennsylvania, Saskatchewan, South Dakota, West Virginia, and Wisconsin) (W. G. Ronald et al. 1973; J. E. Eckenwalder 1984). A pistillate clone that probably arose from this hybrid by segregation or backcrossing ('Balm-of-Gilead') has been widely cultivated since at least the eighteenth century for its bud resin, used in treating coughs (E. Rouleau 1948). Capsules rarely, if ever, mature and trees do not appear to produce fertile seed, but persist and spread by root sprouts in waste places and at edges of woods. It is cultivated mostly in southeastern Canada and eastern United States to the Great Plains, chiefly in mountains in southeastern United States. This clone is more balsam poplarlike than first generation hybrids and has differences from wild hybrids in North America. It may have arisen in Europe through hybridization and backcrossing between *P. balsamifera* and southern *P. deltoides* subsp. *deltoides*. Most wild hybrids have *P. deltoides* subsp. *monilifera* as the cottonwood parent.

Similar to the other North American balsam poplars, *Populus balsamifera* hybridizes sporadically with the introduced Eurasian *P. nigra*. That hybrid, *P.* ×*rollandii* Rouleau, which was originally thought to have involved *P.* ×*canadensis*, itself a hybrid offspring of *P. nigra*, has been collected in the vicinity of Montreal, Quebec (E. Rouleau 1944). Its leaves are similar in shape to those of *P.* ×*canadensis* but have the reddish resin stains and cylindrical petioles of *P. balsamifera*.

Populus balsamifera does not hybridize naturally with *P. tremuloides*, as sometimes reported (E. Lepage 1961; F. G. Bernard 1968). Specimens that formed the basis for those reports are either *P.* ×*jackii* or slender sucker shoots of *P. tremuloides* bearing correspondingly narrow leaves.

Populus ×*jackii* has branchlets that are short-haired or pubescent, petioles densely pubescent, at least distally, preformed blade margins with 20(–45) teeth on each side, sinuses to 1.5 mm deep, pedicels to 3 mm, discs saucer-shaped, 1–4 mm diam., ovaries 3- or 4-carpelled and glabrous, capsules usually 3- or 4-valved, ovoid, (4–)8–11(–16) mm, and seeds (6 or) 7–15(–25) per placenta. It is similar to *P. simonii* in having winter buds with red resin, petioles to 2 cm, round, cylindrical, or distally slightly flattened in the plane of blade, leaf blades lighter green abaxially, elliptic-rhombate to ovate, (1–)3–20(–27.5) cm, bases acute, cuneate, rounded, truncate, or shallowly cordate, apices obtuse, acute, or acuminate, and surfaces not obviously resin-stained, with abaxial surface glabrous or with short, stiff hairs. The flowers are similar with discs entire, persistent, and not obviously oblique, catkins with floral bracts not ciliate and glabrous abaxially, 10–60(–70) stamens (*P. simonii* with fewer than 12), anthers usually truncate, stigmas 2–4 and expanded, and ovaries ovoid to spherical. In addition to these traits, *P. simonii* has terminal buds that are usually less than 12 mm, branchlets whitish tan by the third year, catkins 3–8 cm, and stamens usually fewer than 12.

4. Populus angustifolia E. James, Account Exped. Pittsburgh 1: 497. 1823 • Narrowleaf or mountain cottonwood, liard amer [F]

Populus ×*sennii* B. Boivin; *P. tweedyi* Britton

Plants to 20 m, 7 dm diam.; moderately heterophyllous. **Bark** light brown, shallowly furrowed. **Branchlets** orange-brown, becoming whitish tan by third year, round or 5-angled, 1.5–2.5(–3.5) mm diam., not coarse, glabrous. **Winter buds** reddish brown, glabrous, resinous (resin red, fragrant); terminal buds (3–)6–9(–13) mm; flowering buds clustered distally on branchlets, 8–12(–18) mm. **Leaves:** petiole round, adaxially slightly channeled distally, 0.2–0.8(–1.7) cm, 1/8–1/5 blade length, (glabrous); blade usually lanceolate to narrowly ovate, (1.5–)4–8(–13.5) × 0.8–2.5(–4) cm, w/l = 1/5–1/2, base acute to rounded, basilaminar glands 0, margins not translucent, not ciliate, apex acute, abaxial surface whitish green, weakly glaucous, adaxial dark green, glabrous; preformed blade margins subentire to minutely, evenly crenate-serrate throughout, teeth (14–)23–35(–65) on each side, sinuses 0.1–0.3 mm deep; neoformed blade margins finely crenate-serrate throughout, teeth 35–65(–80) on each side, sinuses 0.1–0.6(–1.5) mm deep. **Catkins** ± densely 35–50-flowered, 3–8(–9 in fruit) cm; floral bract apex deeply cut, not ciliate. **Pedicels** 0.5–1.5(–3 in fruit) mm. **Flowers:** discs shallowly cup-shaped, not obviously oblique, entire, 1–1.5(–3) mm diam.; stamens 10–20; anthers truncate; ovary 2-carpelled, ovoid to spherical; stigmas 2–4, broad, expanded. **Capsules** broadly ovoid to spherical, 3–5 mm, glabrous, 2-valved. **Seeds** (2–)4–7(–9) per placenta. $2n = 38$.

Flowering Apr–May; fruiting Jun–Jul. Streamsides in mountains and foothills; 1500–2400(–3300) m; Alta., Sask.; Ariz., Colo., Idaho, Mont., Nebr., Nev., N.Mex., S.Dak., Tex., Utah, Wyo.; Mexico (Chihuahua, Coahuila, Sonora).

Populus angustifolia has been mistakenly reported from eastern California and eastern Oregon based on intergrades with, and narrow-leaved specimens of, *P. trichocarpa* in those arid regions. It is a characteristic species of the Rocky Mountains, extending onto the plains and overlapping in canyon mouths as they exit the mountains with two North American species of sect. *Aigeiros*, *P. deltoides* and *P. fremontii*, and hybridizing with each of them. It also hybridizes with the other two native species of sect. *Tacamahaca*, *P. balsamifera* and *P. trichocarpa*. The hybrid with *P. balsamifera*, *P.* ×*brayshawii* B. Boivin, differs most obviously in longer petioles, at least 2.5 cm, and is increasingly common from Colorado northwards, largely replacing *P. angustifolia* in southern Alberta (T. C. Brayshaw 1965b; S. B. Rood et al. 1985). The hybrid with *P. trichocarpa* (unnamed) is uncommon and largely confined to the Great Basin region, including Montana (W. W. White 1951). *Populus angustifolia* does not hybridize naturally with *P. tremuloides*, as sometimes reported (B. Boivin 1966b). The specimens that formed the basis for that report are long shoots of *P. angustifolia* with relatively coarsely toothed neoformed leaves.

Populus ×*acuminata* Rydberg is the intersectional hybrid of *P. angustifolia* with *P. deltoides* (sect. *Aigeiros*) that occurs on floodplains of major streams, primarily along the foot of the Rocky Mountain Front Ranges where these species grow together, but also extends onto the plains and Colorado Plateau (Alberta, Arizona, Colorado, Montana, Nebraska, New Mexico, North Dakota, South Dakota, Texas, Utah, and Wyoming) (J. E. Eckenwalder 1984). As with other cloning hybrids, it can often occur without one or both parents. It differs from *P. angustifolia* in larger, ovate leaves with coarser teeth, less color differentiation between abaxial and adaxial surfaces, and longer petioles that are slightly flattened side to side near the junction with the blade. Because of its frequency and morphological consistency, *P.* ×*acuminata* was first described as a species and is often treated as such in local and regional floras. It was long suspected of being a hybrid, and its hybrid origin was amply confirmed by multiple lines of evidence in the 1970s and 1980s (D. J. Crawford 1974; A. G. Jones and D. S. Seigler 1975; S. B. Rood et al. 1985). The name has also been widely misapplied to intersectional hybrids involving other combinations of balsam poplar and cottonwood parents (Eckenwalder).

Populus ×*berolinensis* Dippel is a similar cultivated, introduced, intersectional hybrid between two Eurasian species, *P. laurifolia* Ledebour (sect. *Tacamahaca*) and *P. nigra* Linnaeus (sect. *Aigeiros*). It has an upright growth habit and leaves similar in shape to those of *P.* ×*acuminata*, but more balsam poplar-like with smaller teeth and greater color differentiation between abaxial and adaxial surfaces. Pistillate individuals may hybridize with native species of both parent sections (W. G. Ronald and J. W. Steele 1974).

Populus ×*hinckleyana* Correll (synonyms *P.* ×*acuminata* nothomorph *rehderi* Sargent and *P.* ×*intercurrens* S. Goodrich & S. L. Welsh) is the intersectional hybrid of *P. angustifolia* with *P. fremontii* (sect. *Aigeiros*) and is moderately common in canyons throughout their range of sympatry (Arizona, Idaho, Nevada, New Mexico, Texas, Utah, and Mexico [Chihuahua, Sonora]). At the type locality in the Davis Mountains, Texas, *P. angustifolia* is now absent, and the cottonwood parent is *P. fremontii* subsp. *mesetae*, but other occurrences have arisen through hybridization with *P. fremontii* subsp. *fremontii* (G. C. Bennion et al. 1961; J. E. Eckenwalder 1984; P. Keim et al. 1989). It differs from *P. angustifolia* in hairy young shoots and a broader floral disc.

A related Eurasian species, *Populus simonii* Carrière, is moderately commonly cultivated, especially in the form of a broadly pyramidal staminate clone ('Pyramidalis'). It shares the slender, often 5-angled twigs and relatively small leaves of *P. angustifolia*, but differs most noticeably in elliptic blades of preformed leaves and obovate blades of neoformed leaves.

5. **Populus deltoides** W. Bartram ex Marshall, Arbust. Amer., 106. 1785 (as deltoide) • Eastern cottonwood, cotonier [F] [W]

Plants to 55 m, 35 dm diam.; moderately to strongly heterophyllous, (often 2 or more trunks near base). **Bark** light brown, deeply furrowed. **Branchlets** yellow-brown, becoming tan by third year, round or 5-angled, coarse or not, (1–)2–3.5(–6) mm diam., glabrous or thinly long-hairy. **Winter buds** greenish yellow, glabrous or stiffly hairy, resinous (resin yellow, moderately fragrant); terminal buds (6–)8–15(–21) mm; flowering buds separated on branchlets, (8–)14–20(–28) mm. **Leaves:** petiole distally flattened at right angle to plane of blade, (1–)3–8(–13) cm, about equaling blade length, (glabrous); blade broadly triangular-ovate, (1–)3–9(–14) × (1.5–)3–9(–16.5) cm, w/l = 4/5–6/5, base truncate to cordate or broadly cuneate, basilaminar glands 0–6, round or tubular, margins translucent, ciliate, apex abruptly short- or long-acuminate, surfaces grayish green to bright green, glabrous (or visibly pilose only at emergence); preformed blade margins coarsely crenate-serrate midblade, teeth (3–)5–15(–30) on each side (graded, rounded), sinuses (0.4–)0.7–5(–7) mm deep; neoformed blade margins crenate-serrate, teeth (10–)25–40(–55) on each side (graded), sinuses (0.1–)0.5–1.5(–3) mm deep. **Catkins** loosely (3–)15–40(–55)-flowered, (0.7–)5–18(–24 in fruit) cm; floral bract apex deeply cut, not ciliate. **Pedicels** 1–13(–17 in fruit) mm. **Flowers:** discs saucer-shaped, not obviously oblique, entire, 1–3(–4) mm diam.; stamens 30–40(–55); anthers truncate; ovary (3- or)4-carpelled, ovoid; stigmas 2–4, platelike, spreading. **Capsules** ovoid, (4–)8–11(–16) mm, glabrous, (3- or)4-valved. **Seeds** (3–)7–10(–23) per placenta. $2n = 38$.

Subspecies 3 (3 in the flora): North America, n Mexico.

Populus deltoides hybridizes with *P. fremontii*, the other native species of sect. *Aigeiros*, in the Colorado Plateau region (Arizona, New Mexico, and Utah) and trans-Pecos Texas. These hybrids involve *P. deltoides* subsp. *wislizeni* with both subspecies of *P. fremontii* and are difficult to distinguish because the parent species are so similar. They have shallowly cup-shaped discs 3–5 mm diam., pedicels 4–6 mm, and, usually, sparsely pubescent branchlets. *Populus deltoides* hybridizes also with three native members of sect. *Tacamahaca*. All three hybrids, *P.* ×*generosa* A. Henry (*P. trichocarpa* × *P. deltoides*), *P.* ×*jackii* Sargent (*P. balsamifera* × *P. deltoides*), and *P.* ×*acuminata* Rydberg (*P. angustifolia* × *P. deltoides*), are distinguished from *P. deltoides* by their buds with reddish resin, fewer triangular leaves with finer teeth, less flattening of the petiole, and a distinctly paler, slightly whitened abaxial leaf surface (J. E. Eckenwalder 1984). Individual hybrids have distinct ranges corresponding to their parental regions of sympatry and may be distinguished from each other by relative leaf width (less than two-thirds as wide as long in *P.* ×*acuminata* and more than two-thirds as wide as long in the other two) and base shapes (cordate in *P.* ×*jackii* and cuneate in the other two). Hybrids with members of the other two sections are rare or unknown. Hybridization with *P. heterophylla* (sect. *Leucoides*) is apparently rare and very local in South Carolina, even though the region of sympatry of these two species occupies essentially the entire range of *P. heterophylla*. Reported hybrids with *P. tremuloides* (sect. *Populus*) named as *P.* ×*bernardii* B. Boivin (T. C. Brayshaw 1965b; B. Boivin 1966b) are actually individuals of *P.* ×*jackii* (Eckenwalder).

Two related members of sect. *Aigeiros*, *Populus nigra* Linnaeus and *P.* ×*canadensis* Moench, are often planted as staminate clones, often persist after cultivation, and spread by root suckers but never become naturalized. Most individuals of Eurasian *P. nigra* cultivated in North America are Lombardy poplars (cv. Italica), an unmistakable, narrowly columnar, staminate clone with heavily buttressed trunk, rhombic preformed leaves, and triangular-ovate neoformed leaves broader than wide. This tree has been known since the eighteenth century and is widely (over-)planted throughout the temperate portion of the flora area as an accent tree. It hybridizes sporadically here with the three native balsam poplars, *P. angustifolia*, *P. balsamifera*, and *P. trichocarpa*; hybrids are discussed under those species. A rare hybrid with *P. fremontii* (*P.* ×*inopina* Eckenwalder) apparently originated from an uncommon pistillate tree of *P. nigra* (J. E. Eckenwalder 1982). It resembles *P. fremontii* in leaf shape but has dark reddish brown winter buds.

Populus nigra and *P.* ×*canadensis* are both staminate and are similar in having winter buds usually 12+ mm with a balsamic fragrance and orange-red resin. The branchlets are round and bright orange-brown to reddish brown in the first year, turning tan by the third year. The petioles are distally flattened at a right angle to the plane of blade. The margins of the leaf blade are translucent and ciliate; the leaf surfaces are glabrous or glabrate to pubescent but not tomentose. The catkins usually have more than 15 flowers, (4–)7–15 cm. The floral disc is entire, persistent, broadly cup- or saucer-shaped, and not obviously oblique. The anthers are usually truncate. The 2-carpelled ovary is ovoid to spherical and the 2–4 stigmas are expanded. The floral bracts are not ciliate and are glabrous abaxially. The two taxa differ in that *P. nigra* has branchlets that are nearly parallel, leaf blades without basilaminar glands, preformed blade margins with sinuses no more than 1.2 mm deep, and 12–20(–30) stamens; *P.* ×*canadensis* has divergent branchlets, branching at 50° or more, leaf blades with no more than 1 basilaminar gland, preformed blades with the base broadly cuneate and apex gradually acuminate, and (15–)20–30 stamens.

Populus ×*canadensis* (*P.* ×*euramericana* Guinier [illegitimate name]; B. K. Boom 1957) is an intercontinental hybrid that first arose spontaneously between *P. deltoides* and *P. nigra* after the former was introduced into Europe in the late seventeenth century. Deliberate new hybrids of this parentage are one of the mainstays of Europe's growing commercial poplar plantations. They are also important in eastern North America but are often replaced by *P.* ×*generosa* (*P. trichocarpa* × *P. deltoides*) in commercial plantations in British Columbia, Oregon, and Washington. Only one clone is commonly, and very widely, grown horticulturally, the Carolina poplar ('Eugenei'), a staminate clone with a fairly narrow habit inherited from its staminate parent, the Lombardy poplar. It is often confused with *P. deltoides*, with narrower preformed leaves, often slightly longer than wide, with more numerous, smaller teeth, and with bases obtuse or rounded, rather than truncate or subcordate. It differs further from *P. deltoides* subspp. *deltoides* and *monilifera* in having 0–1 basilaminar glands rather than 2–6.

1. Leaf blades: apices long-acuminate, bases usually with 2 round basilaminar glands; pedicel lengths uniform, 1–6(–8 in fruit) mm . . . . 5b. *Populus deltoides* subsp. *monilifera*
1. Leaf blades: apices short-acuminate, bases usually with 0 or 3–6 tubular basilaminar glands; pedicel lengths progressively graded or uniform, 1–13 (–17 in fruit) mm.
    2. Winter buds usually glabrous; leaves: basilaminar glands 3–6, abaxial surface pilose at emergence; neoformed blades: lengths usually distinctly greater than widths; pedicel lengths graded (shorter from base to apex) . . . . .5a. *Populus deltoides* subsp. *deltoides*
    2. Winter buds pubescent; leaves: basilaminar glands 0, abaxial surface glabrous at emergence; neoformed blades: lengths usually less than widths; pedicel lengths uniform. . . . . 5c. *Populus deltoides* subsp. *wislizeni*

5a. Populus deltoides W. Bartram ex Marshall subsp. **deltoides** • Southern cottonwood [E][F]

Populus deltoides var. *missouriensis* (A. Henry) A. Henry

Plants to 55 m. **Winter buds** usually glabrous. **Leaves:** blade base usually with 3–6 tubular basilaminar glands, apex short-acuminate, abaxial surface pilose at emergence; preformed blade with (6–)12–20(–30) teeth on each side; neoformed blade lengths usually distinctly greater than widths. **Pedicels:** lengths graded, shorter from base to apex, 1–13(–17 in fruit) mm. **Capsules** with thin, flexible valves.

Flowering Mar–Apr (fruiting Apr–Jul). Floodplains, low wet areas, secondary woodlands; 0–400 m; Ont., Que.; Ala., Ark., Conn., Del., D.C., Fla., Ga., Ill., Ind., Iowa, Ky., La., Md., Mass., Miss., Mo., N.H., N.J., N.Y., N.C., Ohio, Okla., Pa., S.C., Tenn., Tex., Vt., Va., W.Va.

Southern cottonwoods are nearly ubiquitous, although varying greatly in abundance, throughout the southeastern United States. They are among the fastest growing and largest trees in the region and are the basis for local poplar plantation forestry rather than hybrids used in other regions. Subspecies *deltoides* intergrades rather freely with subsp. *monilifera* up the Mississippi River drainage system, and traces of its morphological influence may be found as far north as Minnesota and Wisconsin (E. Marcet 1962).

5b. Populus deltoides W. Bartram ex Marshall subsp. **monilifera** (Aiton) Eckenwalder, J. Arnold Arbor. 58: 204. 1977 • Plains cottonwood [E]

Populus monilifera Aiton, Hort. Kew. 3: 406. 1789; *P. besseyana* Dode; *P. deltoides* var. *occidentalis* Rydberg; *P. sargentii* Dode; *P. sargentii* var. *texana* (Sargent) Correll

Plants to 40 m. **Winter buds** pubescent, hairs relatively short, stiff. **Leaves:** blade base usually with 2 round basilaminar glands, apex long-acuminate, abaxial surface glabrous at emergence; preformed blade with (3–)7–15(–21) teeth on each side; neoformed blade lengths about equaling widths. **Pedicels:** lengths uniform, 1–6(–8 in fruit) mm. **Capsules** with thin, flexible valves.

Flowering Mar–Jun; fruiting May–Jul. Along streams and lakes, usually in mesic soils, on dunes, also cultivated; 50–2200 m; Alta., Man., Ont., Sask.; Colo., Idaho, Ill., Ind., Iowa, Kans., Mich., Minn., Mo., Mont., Nebr., N.Mex., N.Y., N.Dak., Ohio, Okla., Oreg., Pa., S.Dak., Tex., Wash., Wis., Wyo.

Subspecies *monilifera* is the abundant cottonwood of the Great Plains and extends sparsely through the Rocky Mountains to the Columbia Plateau. It intergrades with subsp. *deltoides* in a band along the southeastern edge of the Great Plains to the Great Lakes region. These intermediates have longer pedicels and may have 3–5 elongate basilaminar glands. It intergrades also with subsp. *wislizeni* along the southwestern margin of the Great Plains, where there are plants with longer pedicels and preformed leaves with fewer teeth and without basilaminar glands.

5c. Populus deltoides W. Bartram ex Marshall subsp. **wislizeni** (S. Watson) Eckenwalder, J. Arnold Arbor. 58: 205. 1977 • Rio Grande cottonwood, álamillo

Populus fremontii S. Watson var. *wislizeni* S. Watson, Amer. J. Sci. Arts, ser. 3, 15: 136. 1878 (as fremonti); *P. wislizeni* (S. Watson) Sargent

Plants to 35 m, usually less than 20 m. **Winter buds** pubescent, hairs relatively short, stiff. **Leaves:** blade base with 0 basilaminar glands, apex short-acuminate, abaxial surface glabrous at emergence; preformed blade with (3–)5–10 teeth on each side; neoformed blade lengths usually less than widths. **Pedicels:** lengths uniform, (5–)8–12(–15 in fruit) mm. **Capsules** with thick, stiff valves.

Flowering Mar–May; fruiting May–Jul. Floodplains, permanent streams, near springs, usually in moist soil, often planted near ranches, irrigation ditches, in towns; 1000–2300 m; Ariz., Colo., N.Mex., Tex., Utah, Wyo.; Mexico (Chihuahua).

Subspecies *wislizeni* is the common cottonwood of the Rio Grande and Colorado Plateau regions, where it forms relatively small, scattered groves. It is a fairly homogeneous and distinctive southwestern part of the species. It intergrades with subsp. *monilifera* on the eastern and northern margins of its range. *Populus sargentii* Dode var. *texana* Correll is based upon such intergrades.

6. Populus fremontii S. Watson, Proc. Amer. Acad. Arts 10: 350. 1875 • Fremont cottonwood [F]

Plants to 30 m, 37 dm diam.; strongly heterophyllous, (often 2 or more trunks near base). **Bark** pale tan, deeply furrowed. **Branchlets** tannish brown, becoming paler tan to bone-white by third year, round, 1–3(–5) mm diam., slender to coarse, glabrous, glabrate, or sparsely to densely hairy, (yellowish). **Winter buds** yellow-brown, usually densely stiffly hairy, resinous (resin yellow); terminal buds (4–)7–11(–14) mm; flowering buds separated on branchlets, (5–)11–18(–22) mm. **Leaves:** petiole distally flattened at right angle to plane of blade, 1–6(–9) cm, 3/5–3/4 blade length; blade rhombic-ovate to broadly triangular-ovate, (1.5–)4–8 (–14) × (1.5–)3–8(–11) cm, w/l = 3/5–1/1, base cuneate to truncate or cordate, basilaminar glands 0, margins translucent, ciliate, apex short- to long-acuminate, surfaces yellowish green, resin stains not evident, glabrous or densely hairy; preformed blade margins coarsely crenate-serrate midblade, teeth 3–10(–15) on each side (graded, rounded), sinuses (0.2–)0.5–4(–5.5) mm deep; neoformed blade margins finely crenate-serrate much of margin, teeth (10–)20–30(–45) on each side, sinuses 0.1–1 mm deep. **Catkins** loosely (10–)15–25(–35)-flowered, (3–)4.5–10(–14 in fruit) cm; floral bract apex deeply cut, not ciliate. **Pedicels** 1–4 (–5.5 in fruit) mm. **Flowers:** discs broadly cup-shaped, not obviously oblique, entire, (2.5–)4–7(–9) mm diam.; stamens (30–)40–60(–70); anthers truncate; ovary 2–4-carpelled, spherical; stigmas 2–4, flat, platelike, expanded. **Capsules** spherical, (5–)6–11 mm, glabrous, 2–4-valved. **Seeds** 9–15(–25) per placenta. $2n = 38$.

Subspecies 2 (2 in the flora): sw United States, n Mexico.

Populus fremontii hybridizes with *P. deltoides* subsp. *wislizeni* where they come in contact in the Colorado Plateau region and trans-Pecos Texas. Hybrids have longer pedicels and narrower discs than does *P. fremontii*, and less densely hairy twigs in regions where *P. fremontii* is densely pubescent. It also hybridizes with the two members of sect. *Tacamahaca* with which it overlaps, *P. trichocarpa* (*P.* ×*parryi* Sargent), and *P. angustifolia* (*P.* ×*hinckleyana* Correll) (synonyms *P.* ×*acuminata* Rydberg nothomorph *rehderi* Sargent and *P.* ×*intercurrens* S. Goodrich & S. L. Welsh). Both hybrids have smaller, more numerous teeth on preformed leaves than does *P. fremontii*, with blades clearly paler abaxially than adaxially, and buds with reddish resin (J. E. Eckenwalder 1984). Preformed leaves are broadly ovate to heart-shaped in *P.* ×*parryi* (Eckenwalder 1984b) and usually ovate in *P.* ×*hinckleyana*; trees from the type locality in the Davis Mountains, Texas, had round leaves (D. S. Correll 1960). They could be distinguished from *P. ×parryi* by their smaller leaf blades and hairy petioles. Trees of *P. ×parryi* have been found in Mohave County, Arizona, east of the present range of *P. trichocarpa* (Eckenwalder 1992).

Populus ×inopina Eckenwalder is a hybrid between *P. fremontii* and *P. nigra* known only from the type locality along Coyote Creek, San Jose, California, where *P. nigra* was the seed parent (J. E. Eckenwalder 1982). Trees of the same parentage were produced artificially by A. B. Stout and E. J. Schreiner (1933); none appears to have survived. They differ from *P. nigra* in broader habit, twigs thicker, buds thicker and less red, broader and more cordate leaves with larger teeth, and more numerous stamens. They differ from *P. fremontii* in narrower habit, twigs more orange, buds without a green cast, longer leaf apices, more numerous teeth, and fewer stamens. If pistillate trees occur, they would be expected to have 2 or 3 carpels and discs 3–5 mm diam., both intermediate between those of the parents.

1. Neoformed blade: usually about as wide as long, rarely wider, bases truncate or cordate, apices short-acuminate; branchlets glabrous or glabrate to densely hairy ....
.... 6a. *Populus fremontii* subsp. *fremontii*
1. Neoformed blade: longer than wide, bases cuneate or truncate, apices long-acuminate; branchlets usually densely hairy.....................
.... 6b. *Populus fremontii* subsp. *mesetae*

6a. Populus fremontii S. Watson subsp. fremontii [F]

Populus arizonica Sargent; *P. fremontii* var. *arizonica* (Sargent) Jepson; *P. fremontii* var. *macdougalii* (Rose) Jepson; *P. fremontii* var. *pubescens* Sargent; *P. fremontii* var. *thornberi* Sargent; *P. fremontii* var. *toumeyi* Sargent; *P. macdougalii* Rose

Plants to 30 m. **Branchlets, leaf blades, and petioles** glabrous or glabrate to densely hairy, hairs whitish. **Neoformed leaves:** blade broadly triangular-ovate, usually about as wide as long, rarely wider, base truncate or cordate, apex short-acuminate. **Capsules** (6–)7.5–11 mm.

Flowering Feb–May; fruiting Mar–Jun. Floodplains, canyons, springs, other moist places; (-60–)0–1500 (–2200) m; Ariz., Calif., Idaho, Nev., N.Mex., Utah; Mexico (Baja California, Baja California Sur, Sonora).

Subspecies *fremontii* is the common cottonwood of southwestern North America west of the Continental Divide. It intergrades with subsp. *mesetae* along a line from southeastern Arizona south along the border between Chihuahua and Sonora.

6b. Populus fremontii S. Watson subsp. mesetae Eckenwalder, J. Arnold Arbor. 58: 201, fig. 2B. 1977 • Plateau cottonwood, álamo cimarron

Populus deltoides Bartram ex Marshall var. *mesetae* (Eckenwalder) Cronquist

Plants to 20 m. **Branchlets, leaf blades, and petioles** usually densely hairy, hairs (soft) yellowish. **Neoformed leaves:** blade rhombic-ovate, longer than wide, base cuneate or truncate, apex long-acuminate. **Capsules** 5–8(–10) mm.

Flowering Feb–Apr; fruiting Apr–Jun. Stream banks, canyons, seeps, springs, pools, moist soils; 800–2400 m; Tex.; Mexico (Chihuahua, Coahuila, Mexico, Nuevo León, Puebla).

Subspecies *mesetae* forms groves and local patches along major streams, otherwise primarily in arroyos and canyons in foothills, and also is commonly planted near ranches and in towns. It intergrades with subsp. *fremontii* along a line from southeastern Arizona south along the boundary between Chihuahua and Sonora.

7. Populus grandidentata Michaux, Fl. Bor.-Amer. 2: 243. 1803 • Bigtooth aspen, grand tremble [E] [F]

Populus tremula Linnaeus subsp. *grandidentata* (Michaux) Á. Löve & D. Löve

Plants to 35 m, 14 dm diam.; strongly heterophyllous. **Bark** dark grayish brown, furrowed only basally on large trees, (light gray and smooth otherwise). **Branchlets** reddish brown, becoming reddish gray by third year, round, 1.3–2.5 (–5) mm diam., moderately coarse, thinly tomentose to glabrate. **Winter buds** reddish, proximally pubescent, (dull), not evidently resinous; terminal buds 2.5–7 (–10) mm, (glabrous or pubescent); flowering buds separated on branchlets or clustered distally, 6–9(–13) mm. **Leaves:** petiole distally flattened at right angle to plane of blade, 1.5–6(–11) cm, ½–¾ blade length; blade ovate, (2–)4–10(–27.5) × (2–)3–8(–28.5) cm, w/l = 3/4, base broadly cuneate to subcordate, basilaminar glands (1 or) 2(–4), cup-shaped, margins not translucent, not ciliate, apex acute, abaxial surface greenish-white, resin stains absent, (glaucous), densely silky, (hairs white, relatively long, appressed) at emergence, soon becoming glabrate, adaxial bright dark green, glabrous; preformed blade margins coarsely serrate midblade, teeth (1–)5–12 (–16) on each side (graded, sharp), sinuses 0.3–4.5(–6) mm deep; neoformed blade margins finely crenate-serrate throughout, teeth (5–)15–50(–138) on each side (rounded), sinuses 0.8–1.5(–2.5) mm deep. **Catkins** densely (30–)50–150(–175)-flowered, (4–)6–10(–14 in fruit) cm; floral bract apex deeply cut, ciliate. **Pedicels** 0.2–1.5(–2 in fruit) mm. **Flowers:** discs narrowly cup-shaped, obviously oblique, shallowly toothed, 1–2 mm diam.; stamens 6–12; anthers truncate; ovary 2-carpelled; stigmas 2, filiform, erect. **Capsules** narrowly ovoid, 2–5(–6) mm, glabrous, 2-valved. **Seeds** (3–)6–8(–9) per placenta. $2n = 38$.

Flowering Mar–May; fruiting May–Jun. Dry to moist, open to closed upland woodlands and forests; 0–1000 m; Man., N.B., N.S., Ont., P.E.I., Que.; Conn., Del., D.C., Ill., Ind., Iowa, Ky., Maine, Md., Mass., Mich., Minn., Mo., N.H., N.J., N.Y., N.C., Ohio, Pa., R.I., Tenn., Vt., Va., W.Va., Wis.

Populus grandidentata is a successional species that regenerates after fires by suckering from living rootstocks. The exclusively neoformed leaves on such suckers are much larger than those found on mature trees, are conspicuously pubescent abaxially, and are similar enough to preformed and neoformed leaves of *P. heterophylla* that they are responsible for incorrect published reports of the latter in upland sites. Once suckers reach their second or third year and begin to branch, they start to bear at least some preformed leaves that clearly identify them as Bigtooth aspen. As far as is known, *P. grandidentata* and *P. heterophylla* never grow together at a single site.

Bigtooth aspen hybridizes sporadically with the other native aspen, *Populus tremuloides*, to form *P.* ×*smithii* B. Boivin (synonym *P.* ×*barnesii* W. H. Wagner) throughout their large region of sympatry (Frère Marie-Victorin 1930; S. S. Pauley 1956; B. V. Barnes 1961; W. H. Wagner Jr. 1970). Leaves of the hybrids have more numerous, smaller, more rounded teeth than those of *P. grandidentata*. They may be found as far west as Niobrara River valley, Nebraska, 350 km west of the nearest present station of *P. grandidentata*.

The related Eurasian white poplar, *Populus alba* Linnaeus, is commonly and widely planted throughout temperate North America as a pistillate clone with a spreading crown or, less often, as a columnar staminate clone, the Bolleana poplar ('Pyramidalis'), both of which can persist after cultivation and even spread to a limited extent by root sprouts in old garden sites, roadsides, waste places, hedgerows, and edges of woods. This species differs from *P. grandidentata* (and all other species of the genus) in having neoformed leaf blades palmately 5-lobed and, along with the petioles, densely white-tomentose abaxially. Unlike all native species of *Populus*, white poplar has floral bract apices only shallowly cut; these are ciliate like those of native aspens. *Populus alba* hybridizes commonly with both *P. grandidentata*, forming *P.* ×*rouleauiana* B. Boivin, and *P. tremuloides*, forming *P.* ×*heimburgeri* B. Boivin, in southeastern Canada and the northeastern United States; the hybrids are progressively uncommon southward (E. L. Little Jr. et al. 1957; T. A. Spies and B. V. Barnes 1982). Although their leaves are tomentose abaxially, they differ from *P. alba* in not having the deeply 5-lobed neoformed leaves. Those of *P.* ×*heimburgeri* are shallowly 3-lobed, with apical lobes much larger than lateral ones, and those of *P.* ×*rouleauiana* are irregularly and compoundly toothed. With their prominently white-tomentose leaves abaxially, both hybrids are often misidentified as white poplar and are the basis for published reports of naturalized *P. alba* in sites away from present or former cultivation.

The gray poplar, *Populus* ×*canescens* (Aiton) Smith, a natural hybrid between *P. alba* and the Eurasian aspen, *P. tremula* Linnaeus, is common and variable in Europe but usually represented by a single pistillate clone in North America. It is widely cultivated, persisting and spreading by root sprouts at former homesites, in waste places, and edges of woods. It is usually less frequent than *P. alba*, but mostly replaces white poplar in southeastern United States, where it is more widely established under semi-natural conditions. Neoformed leaves have a thin, grayish tomentum abaxially and are irregularly and coarsely toothed rather than 5-lobed, like those of *P. alba*. *Populus* ×*tomentosa* Carrière, from China, is a similar hybrid white poplar with larger leaves and two prominent, cup-shaped basilaminar glands that is rarely planted in southeastern United States in the form of a pistillate clone. It is derived from hybridization between *P. alba* and a Chinese aspen, *P. adenopoda* Maximowicz.

Populus alba is similar to *P.* ×*canescens* in having resinous winter buds, branchlets and terminal buds that are densely to sparsely tomentose, leaf blade surfaces densely tomentose abaxially when young, retaining their dense tomentum on at least some intervein regions, and usually 5–8(–10) cm. Flowers are similar with discs narrowly cup-shaped, obviously oblique, catkins with floral bracts that are ciliate and apices shallowly cut, 6–12 stamens, the 2 stigmas filiform, and ovaries narrowly ovoid to lanceoloid. Capsules are similar in that they are usually 2-valved, lanceoloid or narrowly ovoid, and 2–7(–9) mm, with seeds (1 or) 2 (or 3) per placenta. The two taxa differ by *P. alba* having white hairs, neoformed blade margins deeply to shallowly (3 or) 5-lobed, catkin floral bracts densely tomentose, and (1 or) 2 (or 3) seeds per placenta; *P.* ×*canescens* has grayish, tannish, or dirty white (tomentose) hairs, neoformed blade margins irregularly toothed, and (3–)5–8 (or 9) seeds per placenta.

8. **Populus tremuloides** Michaux, Fl. Bor. Amer. 2: 243. 1803 • Trembling or quaking aspen, tremble, álamo temblón [F] [W]

Populus aurea Tidestrom; *P.* ×*polygonifolia* F. G. Bernard; *P. tremula* Linnaeus subsp. *tremuloides* (Michaux) Á. Löve & D. Löve; *P. tremuloides* var. *aurea* (Tidestrom) Daniels; *P. tremuloides* var. *magnifica* Victorin; *P. tremuloides* var. *vancouveriana* (Trelease) Sargent

Plants to 35 m, 10 dm diam.; moderately heterophyllous. **Bark** dark gray, shallowly furrowed only basally on large trees, (greenish or yellowish white to gray and smooth otherwise). **Branchlets** reddish brown, becoming grayish yellow by third year, round, 1.2–3.5(–5) mm diam., coarse or not, glabrous. **Winter buds** reddish brown, glabrous, (shiny), slightly resinous; terminal buds (2.5–)4–6(–9) mm, (glabrous); flowering buds separated on branchlets or clustered distally, (4.5–)6–10(–11) mm. **Leaves**: petiole distally flattened at right angle to plane of blade, (0.7–)1–6 cm, about equaling blade length; blade somewhat circular to ovate, (1–)3–7(–12) × (0.5–)3–7(–10.5) cm, w/l = ca. 1, base shallowly cuneate to subcordate, shouldered, basilaminar glands (0 or) 1 or 2, round, margins not translucent, not ciliate, apex acuminate to acute, abaxial surface whitish green, resin stains not obvious, (slightly glaucous), glabrous, adaxial dark green, glabrous; preformed blade margins subentire to finely crenate-serrate throughout, teeth (12–)18–30(–42) on each side, sinuses 0.1–1 mm deep, (surfaces glabrous or sparsely sericeous); neoformed blade margins finely crenate-serrate throughout, teeth (20–)25–40(–50) on each side, sinuses 0.1–1.3 mm deep. **Catkins** densely (20–)50–65(–130)-flowered, (1.7–)4–7(–12.5 in fruit) cm; floral bract apex deeply cut, ciliate. **Pedicels** 0.5–1.5(–2 in fruit) mm. **Flowers**: discs narrowly cup-shaped, obviously oblique, entire, 1.3–1.8(–3 in fruit) mm diam.; stamens 6–12; anthers truncate; ovary 2-carpelled; stigmas 2, filiform, basal lobes expanded, erect. **Capsules** narrowly ovoid, (2–)2.5–4.5(–7) mm, glabrous, 2-valved. **Seeds** (3–)5–7(–9) per placenta. $2n = 38, 57, 76$.

Flowering Mar–Jun; fruiting May–Jul. Dry to wet, open to closed woodlands and forests, edges of meadows and prairies, talus-slopes and canyon-heads, sites of human disturbance, timber cuts, mine tailings, gravel pits, quarries, roadsides; 0–3000(–4000) m; St. Pierre and Miquelon; Alta., B.C., Man., N.B., Nfld. and Labr. (Nfld.), N.W.T., N.S., Ont., P.E.I., Que., Sask., Yukon; Alaska, Ariz., Calif., Colo., Conn., Del., Idaho, Ill., Ind., Iowa, Maine, Md., Mass., Mich., Minn., Mo., Mont., Nebr., Nev., N.H., N.J., N.Mex., N.Y., N.C., N.Dak., Ohio, Oreg., Pa., R.I., S.Dak., Tex., Utah, Vt., Va., Wash., W.Va., Wis., Wyo.; Mexico (Baja California, Chihuahua, Coahuila, Nuevo León, Tamaulipas, south to Hidalgo and the state of Mexico).

Clonal aspen groves develop rapidly following fires and other disturbances and may quickly decay in their absence as infections are transmitted through the connecting root system. *Populus tremuloides* is the most widely distributed tree in North America, found throughout cold and cool-temperate regions from coast to coast and from within the Arctic Circle to the north rim of the Valley of Mexico. It ranges from sea level in the north and east to the north slopes of high mountains in the southernmost part of its range. The southerly locations, like that on Mt. Livermore in the Davis Mountains, Texas, the most southerly stand in the flora area, may be Pleistocene relicts. Groves are often occupied by single clones and show no sexual reproduction but persist and spread by root suckers. Clone formation commonly results also in striking differences in appearance and phenology of adjacent groves or blocks of trees (B. V. Barnes 1969). Some individuals display a particularly rich, yellow autumn coloration that makes them a standout among North American trees, particularly in the West, where this richness was the basis for segregation of *P. tremuloides* var. *aurea*. There do not appear to be consistent regional differences within the species that would justify recognition of subspecies or varieties (Barnes 1975). Instead, there is as much variation from clone to clone within a region as there is among regions.

Populus tremuloides hybridizes with both the native *P. grandidentata* (*P.* ×*smithii* B. Boivin) and the Eurasian *P. alba* (*P.* ×*heimburgeri* B. Boivin) in southeastern Canada and the northeastern United States (B. V. Barnes 1961; T. A. Spies and Barnes 1982). *Populus* ×*smithii* occurs as far west as the Niobrara River valley, Nebraska, ca. 350 km west of the nearest present populations of *P. grandidentata*. Preformed leaves are more ovate than those of *P. tremuloides* and have larger teeth. *Populus* ×*heimburgeri* has transiently tomentose twigs, buds, and abaxial leaf surfaces. Contrary to some published reports (E. Lepage 1961; T. C. Brayshaw 1965b; B. Boivin 1966b; F. G. Bernard 1968), *P. tremuloides* does not hybridize naturally with *P. angustifolia*, *P. balsamifera*, or *P. deltoides*. The correct identification of such specimens is discussed under each of the purported parents.

The closely related Eurasian aspen, *Populus tremula* Linnaeus, is sometimes cultivated in North America, particularly as a columnar staminate clone ('Erecta'). Its leaves are very similar in shape to those of *P. tremuloides* and usually have slightly larger teeth. Buds are often minutely hairy. Artificial hybrids between *P. tremula* and *P. tremuloides*, *P.* ×*wettsteinii* Hämet-Ahti, are sometimes grown for plantation forestry, particularly in the Great Lakes region.

2. SALIX Linnaeus, Sp. Pl. 2: 1015. 1753; Gen. Pl. ed. 5, 447. 1754, name conserved
• Willow, saule [Latin name for willow]

George W. Argus

Shrubs or trees, slightly heterophyllous, clonal or not, clones formed by root shoots, rhizomes, layering, or stem fragmentation; branching sympodial. **Stems** not spinose. **Buds** 1-scaled (oily in *S. barrattiana*), margins connate into calyptra or distinct and overlapping adaxially, scale inner membranaceous layer usually not separating from outer layer, sometimes free and separating). **Leaves** deciduous or marcescent; stipules persistent, caducous, or absent (varying in presence and size on early and late leaves); petiole glandular-dotted or lobed distally; (blade often more than twice as long as wide, venation usually pinnate, margins entire, crenulate, crenate, serrate, serrulate, or spinulose-serrulate, teeth gland-tipped). **Inflorescences** axillary or subterminal, catkins, erect, spreading, or ± pendulous, sessile or terminating flowering branchlets, unbranched (except in subg. *Longifoliae*); floral bract apex entire, erose, 2-fid, or irregularly toothed; pistillate bract persistent or deciduous after flowering. **Pedicels** present or absent. **Flowers:** (sessile), perianth reduced to adaxial nectary (rarely also abaxial nectary, then distinct or connate into shallow cup); stamens 1, 2, or 3–10; filaments distinct or connate; ovary (stipitate or sessile), 2-carpellate; ovules (2–)4–24(–42) per ovary; styles usually connate, sometimes distinct distally; stigmas 2, entire or 2-lobed (less than 2 mm). **Fruits** capsular, (2-valved, obclavate to ovoid or ellipsoid). **Seeds:** aril present. $x = 19$.

Species ca. 450 (113 in the flora): North America, Mexico, West Indies, Central America, South America, Europe, Asia (Malaysia), Africa, Atlantic Islands; mostly in arctic, boreal, and temperate regions; introduced in Australasia, Oceania.

Species of *Salix* have been studied by taxonomists, morphologists, anatomists, geneticists, cytologists, chemists, ecologists, arborists, entomologists, and others. Classification of the genus and identification of specimens remains difficult. C. K. Schneider (1919b) stated, "In determining willows one is only too often entirely misled at first, [but] even by a slow and careful examination it is not always possible to determine the proper identity of the plant."

Classification. Traditionally, the subgeneric classification of *Salix* was based on morphological characteristics; recent molecular studies have begun to provide useful insights. The first classification of New World *Salix* (C. K. Schneider 1921) recognized 23 sections and arranged them in linear order corresponding to usually recognized subgenera. The first classification to use subgenera (R. D. Dorn 1976) recognized two: subg. *Salix* (including tree willows and sect. *Longifoliae*) and subg. *Vetrix* (including shrubby and dwarf arctic-alpine willows). G. W. Argus (1997) recognized four subgenera: *Chamaetia*, *Longifoliae*, *Salix*, and *Vetrix*. In the present classification, five subgenera are recognized, with subg. *Salix* being divided into subg. *Protitea* (bud-scales with distinct, overlapping margins and flowers with multiple stamens) and subg. *Salix* (bud-scales with connate margins and flowers usually with two stamens; see discussion under 2a. subg. *Protitea*).

There are two published studies of *Salix* classification based on molecular data (see discussion under subgenera *Protitea* and *Longifoliae*). In both studies, the number of *Salix* species included is relatively low. E. Leskinen and C. Alström-Rapaport (1999) studied phylogeny of Salicaceae and Flacourtiaceae and sought to determine the relationship of *Chosenia* to *Salix*. Parsimony analysis showed little resolution within *Salix* and bootstrap support was strong for only three relatively small groups. *Salix* and *Chosenia* were placed in a single clade with two major branches: 1) the first included *S. exigua* (subg. *Longifoliae*); 2) the second included two subgroups: 2a) *S. amygdaloides* (subg. *Protitea*) and *S. alba*, *S. euxina*, and *S. pentandra* (subg.

Salix); 2b) all the other species (subg. *Chamaetia* and *Vetrix*). The species in this subgroup were unresolved. T. Azuma et al. (2000) sought to determine the taxonomic position of *Chosenia* and *Toisusu*, as well as the classification of *Salix*. Within *Salix* two major clades were recognized. Clade 1 consisted of three major branches: 1a) included *S. interior* (subg. *Longifoliae*) along with *S. amygdaloides* and *S. nigra* (placed here in subg. *Protitea*); 1b) included Asian *S. chaenomelioides* (subg. *Pleuradinea*); and 1c) included African and Asian *S. safsaf* and *S. tetrasperma* (subg. *Protitea*). Clade 2 also had three major branches: 2a) *S. triandra* (as *S. subfragilis*); 2b) unresolved members of subg. *Chamaetia* and *Vetrix*; and 2c) Asian genera *Toisusu* and *Chosenia* (now placed in subg. *Pleuradinea*). Molecular data, although not conclusive because of unresolved species, lend support for recognition of subg. *Longifoliae* and subg. *Protitea*. Further studies are needed to refine the subgeneric classification.

Biology. *Salix* are pioneer or early succession species well-adapted to disturbance. Each pistillate plant can produce hundreds to thousands of seeds annually. At maturity, the seeds can be lifted into the air surrounded by a parachute of fine hairs that can carry them tens to hundreds of meters from the parent plant. Seeds that land on water can float for several days because of the flotation capability of the hairy hilar aril surrounding each seed (E. M. A. Steyn et al. 2004). Willow seeds have no food reserves; most will perish within days unless they land and germinate in a suitable habitat. Some arctic and subarctic species (R. A. Densmore and J. C. Zasada 1983) and some members of sect. *Salicaster* (see 12. *S. serissima*) are able to survive through the winter and germinate in the spring. Some tropical species flower year-round (P. Parolin et al. 2002) by producing sylleptic catkins and, thus, are able to take advantage of newly disturbed habitats at any time.

Seedling success depends mainly on an adequate supply of moisture and the absence of shading (C. F. Sacchi and P. W. Price 1992). When such habitat is available, seeds will germinate almost immediately upon arrival. During the Pleistocene, such conditions were present on a large scale because repeated glacial and interglacial periods provided suitable environments for some northern *Salix* to acquire their present circumpolar or transcontinental distributions. Willow seedlings most commonly occur on riparian sand and gravel bars, old burns, landslides, drained lakes and wetlands, and in open, unstable arctic and alpine habitats. They also are common along the vast network of roads that crisscross even some of the most remote wilderness areas. Road margins, ditches, and gravel and sand borrow pits provide favored habitats. Even minor disturbances, such as upturned tree roots, ungulate tracks in wet meadows and mires, cryogenic frost boils and cracks in tundra, and animal diggings, can provide willow habitat. Because of this need for open habitats for reproduction by seed, large stands of mature willows growing in stable habitats such as marshes, fens, bogs, treed riverbanks, and even active sand dunes, have become established in these habitats before a closed cover had developed. While individual plants can sometimes invade closed vegetation, large stands of willows require large disturbances.

The zonation of willows on floodplains is a function of seed dispersal timing and water level fluctuations (L. R. Walker et al. 1986; I. Van Splunder et al. 1995). Seedling success depends on abiotic factors, such as erosion by flooding later in the season and siltation in subsequent years, and on biotic factors, mainly herbivory by moose and snowshoe hares. The colonization of glacial moraines of different ages (G. W. Argus 1973) showed that all species of *Salix* in the area colonized the earliest moraines. Over time, the dwarf and presumably less shade-tolerant species, *S. arctica*, *S. reticulata*, and *S. stolonifera*, were eliminated, followed by the low to mid shrubs until only tall shrubs and trees of *S. sitchensis* remained along streams and in openings in the spruce-fir forests.

Some willows are adapted for vegetative reproduction by stem fragmentation, layering, or root shoots, and all species can collar-sprout at or below ground level (P. Del Tredici 2001). Vegetative reproduction by stem fragmentation is characteristic of riparian species, some of which have brittle branches, which can be dispersed by wind and water to where they may become lodged and root. Species that spread by root shoots or layering are more limited in their dispersal potential but often form distinctive clones.

Most willows can be propagated by cuttings; some root more easily than others. Adventitious root primordia are initiated mainly at the base of cuttings, sometimes at or between nodes, under stimulus of auxin or other compounds that migrate to the basal end (B. E. Haissig 1974). Riparian species (e.g., *Salix alaxensis*, *S. lasiandra*, *S. pseudomyrsinites*) usually have preformed primordia and root along the entire cutting; non-riparian species (e.g., *S. bebbiana*, *S. glauca*, and *S. scouleriana*), usually described as rooting poorly, do not have preformed primordia (R. A. Densmore and J. C. Zasada 1978). This led Densmore and Zasada to suggest that preformed root primordia are an adaptation to the flooding and siltation that occur in riparian habitats. The formation of root primordia is more complex than that. Willows of non-riparian habitats (bogs, fens, prairies, sand dunes, tundra, and upland forests) display diverse rooting patterns. All of these habitats may have aggrading surfaces due to siltation, moss and peat accumulation, sand drifting, etc., that can encourage the formation of adventitious roots. Bog and fen species (*S. candida*, *S. fuscescens*, and *S. pedicellaris*) typically root along their stems, as do plants growing in sand dunes (*S. brachycarpa*, *S. cordata*, and *S. silicicola*). Even upland species (*S. bebbiana*, *S. humilis*, and *S. scouleriana*), usually regarded to root poorly, can root prolifically when growing in wet or aggrading habitats.

Morphology. Salix are woody plants varying from trees reaching 30 m, to dwarf arctic-alpine shrubs less than 5 mm. They often form clones by stem fragmentation, layering, rhizomes, or root shoots. Branchlets are current-year stems, and branches are stems more than one-year old. Buds have a single scale. The bud-scale margins usually are connate; in subg. *Protitea* they are distinct and overlapping. Because shoot growth in *Salix* is sympodial, buds at shoot apices are subterminal. Vegetative and reproductive buds vary in size, shape, and position. Three general types of bud size and shape gradation are recognized; namely, *alba*-type, *arctica*-type, and *caprea*-type; there are intermediates (A. K. Skvortsov 1999). Plants with *alba*-type bud gradation have buds that are very similar in size and shape along branchlets (monomorphic), but floral and vegetative buds cannot be distinguished from one another. Plants with *arctica*-type bud gradation usually have relatively few buds. The two or three (sometimes to five) distal buds are the largest, diminishing in size proximally. Usually only the larger buds open and those buds may be either floral or vegetative. Plants with *caprea*-type bud gradation have floral buds that are strikingly different in size and shape from vegetative buds (dimorphic). Usually, the distal two or three buds are vegetative, the next three to six (or more) are floral, and proximal to them are smaller vegetative buds.

Stipules borne on either side of the petiole may be foliaceous, minute rudiments, or absent. Those on early (preformed) leaves often are rudimentary; those on late (neoformed) leaves often are foliaceous. Although not all leaves are differentiated in the winter buds (E. Moore 1909), it is difficult to determine precisely which leaves are neoformed. While it is probable that morphological differences occur among leaves on an individual plant, as in some *Populus* species (W. B. Critchfield 1960), only stipule differences have been noted. Petioles sometimes have glandular-spherical dots or lobes at the distal end just proximal to the blade. Sometimes the petioles are ventricose or inflated around the subtended floral buds.

Three types of leaves are recognized in *Salix*: large medial blades are the "normal" leaves; proximal blades are the first two to four reduced true-leaves at the base of branchlets or on catkin-bearing shoots and differ from distal (late) ones in shape, indumentum, dentition, and prominence of stipules; and juvenile blades are young unfolding leaves at the distal end of branchlets. These leaves vary in shape from linear to subcircular, bases are cuneate to cordate, margins are entire or crenate to spinulose-serrulate, and leaf teeth are gland-tipped. Glands on teeth, or entire margins, may be marginal, submarginal (blade edge viewed from abaxial surface has a margin slightly revolute or thickened), or well up on the adaxial surfaces (epilaminal). Abaxial blade surfaces, and sometimes adaxial, are often glaucous with a dull, waxy coating; blade surfaces may be glabrous or hairy, leaf hairs (trichomes) are usually white, sometimes ferruginous (rust-colored). Syllepsis, the opening of buds without a rest period, is common in subg. *Longifoliae*, as well as some *Populus*, and has been recorded in 19 species of *Salix* representing all subgenera. Sylleptic leaf morphology sometimes differs from proleptic leaves (see 2c. subg. *Longifoliae*).

The inflorescences are catkins (aments), each of which consists of a flower-bearing rachis (essentially a spike of unisexual, apetalous, sessile flowers, each subtended by a floral bract), and a peduncle. A catkin may be sessile on a branch or borne on a relatively short, vegetative flowering branchlet (a shoot bearing three or more green leaves). Catkins arise from lateral or subterminal buds. They flower before leaves emerge (precocious), as leaves emerge (coetaneous), or throughout the season. For use of the term serotinous, see 12. *Salix serissima*. The flowering rachis is usually unbranched; in subg. *Longifoliae* secondary or tertiary branching can occur. After anthesis, the rachis of a pistillate catkin continues to elongate but rachises of staminate catkins do not. A floral bract (scale) subtends each flower. Pistillate floral bracts are usually persistent in fruit; in some subgenera they are deciduous. Each flower consists of an adaxial nectary (a reduced perianth, according to M. J. Fisher 1928) located between the stamens or pistil and rachis axis; in some taxa there is also an abaxial nectary located between the floral bract and fertile structures; the two nectaries may be distinct or connate into a cup-like structure. Staminate flowers usually have two stamens, but the number can be one or three to ten. Pistillate flowers have a single pistil, which is sessile or borne on a stipe (Fisher; S. Sugaya 1960). There are two styles, usually connate, each terminated by a two-branched stigma. Stigma lobes are: 1) flat abaxially, papillate adaxially and with a rounded or pointed tip; 2) slenderly cylindrical (length greater than four times the width) or broadly cylindrical (length less than four times the width); or 3) subspherical (plump). The number of ovules per ovary can be determined by counting funiculi remaining in mature capsules after seeds have been shed. For definitions of *Salix* terminology, see G. W. Argus (2007).

Variability. Some species of *Salix* are highly variable and closely related species may be only subtly distinct. Underlying most identification problems is morphological variability, some of which is related to biology of the genus. Some phenotypic variability can result from habitat modification. For example, shade conditions can reduce the density of leaf and branch glaucescence (R. D. Dorn 2003), as well as leaf thickness. The most important sources of morphological variation are hybridization, introgression, and allopolyploidy.

In describing the 'gloss' of stems and leaves in *Salix*, three classes are used: dull, slightly glossy, and highly glossy. These three classes intergrade, but, usually, they can be separated. The modifier 'highly' is used with the word glossy to emphasize that this condition is a distinct extreme of glossy, even though in many species there is gradation from one to the other. It is best to understand the distinctions by example; the adaxial leaf surface of *S. petiolaris* is glossy or dull, whereas in *S. serissima* it is highly glossy (analogous to a varnished surface or freshly waxed floor). Only a few species of *Salix* (*S. caroliniana*, *S. floridana*, *S. maccalliana*, *S. nummularia*, *S. ovalifolia*, *S. pentandra*, *S. phlebophylla*, *S. planifolia*, *S. rotundifolia*, *S. serissima*, *S. stolonifera*, *S. tyrrellii*) have leaves that are described as only highly glossy. In these species, the 'gloss' of the adaxial leaf surface is a diagnostic character.

Hybridization. Approximately 120 *Salix* hybrids have been recognized in the North American flora, and about half of these are relatively common. Others are either putative hybrids in which one parent may be uncertain or unconfirmed, and/or they are doubtful hybrids. North American botanists, in general, have been conservative in their recognition of hybrids, probably in reaction to some European botanists who readily recognized not just simple hybrids but multiple-species hybrids. In Greenland, B. G. O. Floderus (1923) recognized five pure *Salix* species and seven interspecific hybrids, some of which were three-species hybrids. Working in a similar flora, H. M. Raup (1943, 1959) argued against an uncritical recognition of hybrids and suggested that intermediate specimens should be given the name of the species they resemble most. These views were shared by A. K. Skvortsov (1999), who agreed that willows are inherently variable and that a better understanding of species variability would reduce the number of presumed hybrids.

There are barriers to hybridization, including differences in flowering time (A. Mosseler and C. S. Papadopol 1989), pollen-stigma incompatibility (Mosseler 1989), and F_1 hybrid inviability. Nevertheless, hybridization among *Salix* species can be an important source of variability. Hybridization, clonal reproduction, and the ability of hybrids to backcross may be accompanied by introgression (J. Salick and E. Pfeffer 1999). Hybrids can sometimes be recognized by discordant character variations, such as the occurrence of partially hairy ovaries within species characterized by glabrous ovaries, or by having leaf surfaces glaucous abaxially within species that characteristically lack leaf glaucescence. Sometimes such variation may occur along with teratological flowers, or other evidence of infertility and reproductive imbalance. Hybrids, which can be difficult to recognize in the herbarium, sometimes are recognizable in the field as being different from other individuals nearby.

A morphological and molecular study of hybridization and introgression between *Salix eriocephala* and *S. sericea* (T. M. Hardig et al. 2000) found that ca. one-third of plants originally identified as *S. eriocephala* were possible introgressants. Other plants showed unequivocal evidence of backcrossing with *S. sericea*; inter- and intra-specific chloroplast diversity found within a hybrid zone suggested both historic introgression, perhaps in a glacial refugium, and contemporary hybridization. Hardig et al. found that hybrids might not be readily recognized in either the field or herbarium and wrote that, "If major distinguishing characters are under the control of one or two dominant genes, hybridization may go unrecognized. Important taxonomic characters that are quantitative might result in recognizably intermediate hybrids but … hybrids may be imperfectly intermediate or highly variable, resulting in an interpretation that unrecognized hybrids are merely part of the morphological variation in one of the species." The practical taxonomic message is that the interpretation of species variation as either inherent or due to hybridization must be made carefully. While it may be unwise to mistake hybridization for species variability, it is equally unwise to mistake species variability for hybridization.

Polyploidy. The widespread occurrence of polyploidy in *Salix* is an important indication of the evolutionary importance of hybridization. Among the 99 native *Salix* species in the flora area, 70% have chromosome counts and, of those, 47% are polyploid. It can be assumed that 50% or more of the native *Salix* are polyploid. It is probable that most of these are allopolyploids, inasmuch as there is little evidence of autoploidy in *Salix* (W. Buechler, pers. comm.). Some of the most variable species may have evolved through hybridization and polyploidy. For example,

S. arctica and *S. glauca* each display several ploidy levels, as well as many hybrids; both species probably evolved through repeated hybridization and backcrossing with each other and other species. The possibility that recurrent polyploidy has contributed to variability in these and other polyploid species (R. J. Abbott and C. Brochmann 2003) needs study.

Collection and identification. Because *Salix* are dioecious, a single individual cannot provide the full range of reproductive and vegetative structures needed for identification. Some species flower well-before leaves emerge; reproductive structures, especially staminate catkins, and foliage may not be available simultaneously. Ideal specimens for identification include flowering, fruiting, vegetative, and winter twigs. It is possible to gain an in-depth understanding of seasonal morphological variability by tagging plants and making collections at different developmental stages. Insights into population variability, hybridization, and introgression can be gained by attempting to identify every plant in a stand (A. K. Skvortsov 1999). At a minimum, well-collected and pressed specimens are essential. All available plant parts should be collected: leaves (including juvenile leaves), catkins, and twigs. Sprouts, or compensatory shoots, are not commonly included in keys or descriptions. If collected at all, they should be to supplement normal shoots and labeled as such. To avoid loss of glaucescence (wax on stems or leaves), specimens should be dried as rapidly as possible but without using excessive heat. Plant habit and evidence of vegetative reproduction should be noted.

Keys. Variability makes writing dichotomous keys to *Salix* taxa difficult. At best, the keys often are cumbersome to use and may only account for a relatively small part of variability within a taxon. The preparation of separate keys to staminate, pistillate, and vegetative specimens is useful but such keys are not usually provided (G. W. Argus 1986; A. Cronquist and R. D. Dorn 2005). One of the best ways to identify specimens is to use interactive keys (M. J. Dallwitz et al., http://delta-intkey.com; R. J. Pankhurst 1991). An interactive key to New World *Salix* using Intkey is available (G. W. Argus, http://aknhp.uaa.alaska.edu/willow). It can be used not only to identify specimens, but to describe or compare species, or to list species by state, province, or taxonomic group.

Uses. Willows play major roles in ecosystems by rehabilitating disturbed sites through stabilization to prevent erosion, to improve soil, to remove pollutants and heavy metals, and to provide wildlife food and habitat. Willows are used widely as ornamentals. Most introduced species are ornamental cultivars. In many parts of the world, willows are used in basketry, as sources of tannins, and in apiaries as food for brood rearing and making honey. Traditionally, willows were used in medicines; salicin (a component of aspirin) was first derived from *Salix*. Their use as a source of energy biomass is being investigated worldwide. Indigenous peoples have used willows for fuel, construction, basketry, medicines, tools and weapons, and ceremonially.

Conservation. Inasmuch as willows are pioneer species, they present special conservation problems. Attempts to protect them by preventing habitat disturbance will be counterproductive. Although they do not spread without disturbance, once established they may require protection against biotic factors, such as browsing, so that they can produce propagules and disperse them to nearby disturbed sites (see 59. *Salix arizonica*). Given the opportunity, some non-native willows can be very aggressive (see 81. *S. cinerea*). In Australia and New Zealand, where *Salix* is not native, some introductions have been so successful that they are regarded as invasive weeds (C. J. West 1994), and control measures are being implemented. In the flora area, some introductions, such as *S.* ×*fragilis*, have spread so readily by stem fragmentation that, although they rarely produce seed, they appear to be part of the native flora.

Notes on Style. In the descriptions of taxa and in the keys, quantitative morphological data have been given in three ways, depending on degree of variation and sample size. For example: 10–100 [minimum–maximum]; (3–)10–75(–100) [(rare extreme–)usual minimum–usual maximum(–rare extreme)]; 10–20–100 [minimum–mean–maximum]. For *Salix* distribution maps with more detailed ranges, see G. W. Argus (2007).

SELECTED REFERENCES Argus, G. W. 1973. The Genus *Salix* in Alaska and the Yukon. Ottawa. [Natl. Mus. Nat. Sci. Publ. Bot. 2.] Argus, G. W. 1986. The genus *Salix* in the southeastern United States. Syst. Bot. Monogr. 9. Argus, G. W. 1986b. Studies in the *Salix lucida* Muhl. and *S. reticulata* L. complexes in North America. Canad. J. Bot. 64: 541–551. Argus, G. W. 1995. Arizona Salicaceae: Willow family. Part two: *Salix*. J. Arizona-Nevada Acad. Sci. 29: 39–62. Argus, G. W. 1997. Infrageneric classification of *Salix* L. in the New World. Syst. Bot. Monogr. 52. Argus, G. W. 2007. *Salix* L. (Salicaceae): Distribution maps and a synopsis of their classification in North America, north of Mexico. Harvard Pap. Bot. 12: 335–368. Azuma, T., T. Kajita, J. Yokoyama, and H. Ohashi. 2000. Phylogenetic relationships of *Salix* (Salicaceae) based on rbcL sequence data. Amer. J. Bot. 87: 67–75. Dorn, R. D. 1976. A synopsis of American *Salix*. Canad. J. Bot. 54: 2769–2789. Dorn, R. D. 1977b. Willows of the Rocky Mountain states. Rhodora 79: 390–429. Dorn, R. D. 1997. Rocky Mountain Region Willow Identification Field Guide. Denver. Raup, H. M. 1943. The willows of the Hudson Bay Region and the Labrador Peninsula. Sargentia 4: 81–135. Raup, H. M. 1959. The willows of boreal western America. Contr. Gray Herb. 185: 1–95.

Key to Subgenera of *Salix*

1. Bud-scale margins distinct, overlapping; stamens 3–7[–9]; pistillate bracts deciduous after flowering (except *S. floridana*, sometimes *S. bonplandiana*); usually trees . . . . . . 2a. *Salix* subg. *Protitea*, p. 30
1. Bud-scale margins connate; stamens usually 2, sometimes 1 (3–10 in sect. *Salicaster*); pistillate bracts persistent (except in sect. *Salicaster* and in subg. *Longifoliae*); shrubs or trees.
    2. Stamens 3–10; pistillate bracts deciduous after flowering . . . .
        . . . . 2b. *Salix* subg. *Salix* (sects. *Salicaster* and *Triandrae*), p. 38
    2. Stamens 2 or 1; pistillate bracts persistent after flowering (except in subg. *Longifoliae*, *S. alba*, and *S. euxina*).
        3. Pistillate bracts deciduous after flowering; leaf blades usually linear to narrowly elliptic; catkins sometimes branched; syllepsis common; plants clonal by root shoots; petioles not glandular distally . . . . 2c. *Salix* subg. *Longifoliae*, p. 50
        3. Pistillate bracts persistent after flowering (except in *S. alba* and *S. euxina*); leaf blades usually not linear, rarely narrowly elliptic; catkins not branched; syllepsis uncommon; plants not clonal by root shoots (except *S. setchelliana*); petioles glandular or not distally.
            4. Petioles usually glandular-lobed or -dotted distally; branches and branchlets brittle at base . . . . 2b. *Salix* subg. *Salix* (in part), p. 38
            4. Petioles not glandular distally or glands simple, spherical; branches flexible at base (sometimes brittle in subg. *Vetrix*).
                5. Buds *arctica*-type or transitional or *alba*-type; catkins arising from subterminal buds (sects. *Chamaetia*, *Herbella*, *Myrtosalix*) as well as lateral buds; shrubs 0.005–6 m; juvenile blade hairs usually white, rarely ferruginous (in *S. athabascensis*); pistillate catkins always on flowering branchlets; largest medial blades (relatively broad) 0.8–5.5 times as long as wide; ovaries sometimes glaucous (sects. *Chamaetia*, *Diplodictyae*, *Myrtilloides*, *Ovalifoliae*), hairs sometimes flat and ribbonlike (sect. *Myrtosalix*); most staminate and some pistillate flowers with abaxial and adaxial nectaries; inner membranaceous bud-scale layer not separating from outer layer . . . . 2d. *Salix* subg. *Chamaetia*, p. 60
                5. Buds *alba*-type or intermediate; catkins arising from lateral buds (subterminal buds vegetative); shrubs or trees, 0.1–20 m; juvenile blade hairs mostly white, sometimes ferruginous; pistillate catkins sessile or borne on flowering branchlets; largest medial blades (relatively narrower) 0.7–13.7 times as long as wide; ovaries not glaucous, hairs not flat and ribbonlike; staminate and pistillate flowers usually without abaxial nectaries; inner membranaceous bud-scale layer sometimes distinct and separating from outer layer . . . . 2e. *Salix* subg. *Vetrix*, p. 93

2a. **Salix** Linnaeus subg. **Protitea** Kimura, Bot. Mag. (Tokyo) 42: 290. 1928

Shrubs or trees, 1–30 m, not clonal, or clonal by stem fragmentation. **Stems** usually erect; branches flexible to highly brittle at base or throughout, glaucous or not. **Buds** *alba*-type, scale margins distinct, overlapping. **Leaves:** stipules on early ones usually rudimentary, sometimes absent or foliaceous, on late leaves foliaceous, rarely rudimentary (deciduous early or autumn); petiole usually deeply grooved adaxially (except *S. floridana*), not usually glandular distally (sometimes with basilaminar, spherical, or foliaceous glands); largest medial blade usually hypostomatous, sometimes amphistomatous, narrowly oblong, oblong, narrowly ovate, lorate, lanceolate, narrowly elliptic, or linear, 2.5–13 times as long as wide, angles of base and apex less than 90°, surface hairs usually white, sometimes also ferruginous; juvenile blade hairs white, sometimes also ferruginous. **Catkins** usually flowering as leaves emerge (sometimes throughout season, axillary, sessile in *S. bonplandiana*), from lateral buds; staminate from flowering branchlet, slender to stout; pistillate from flowering branchlet, usually loosely flowered, slender to stout; floral bract usually tawny, sometimes greenish, apex entire, erose, 2-fid, or irregularly toothed; pistillate bract deciduous (persistent in *S. floridana*, sometimes *S. bonplandiana*) after flowering. **Staminate flowers:** abaxial nectary present; stamens 3–7[–9]; filaments usually distinct, sometimes connate less than ½ their lengths, usually hairy basally, sometimes on proximal ½; anthers yellow, usually globose. **Pistillate flowers:** abaxial nectary absent, (adaxial nectary shorter than stipe); ovary not glaucous, usually glabrous, sometimes villous, beak slightly bulged below or tapering to styles; ovules 4–24 per ovary; styles usually connate, sometimes distinct distally; stigmas usually flat, abaxially non-papillate, tip usually rounded, or stigmas slenderly cylindrical, or 2 plump lobes.

Species 33 (7 species in the flora): North America, Mexico, West Indies, Central America, Asia.

Two of the three sections recognized in this subgenus are found in the flora area.

Subgenus *Protitea* includes the New World members of sects. *Floridanae* and *Humboldtianae*. These sections often are placed in subg. *Salix* along with sect. *Salicaster* (R. D. Dorn 1976; G. W. Argus 1997; A. K. Skvortsov 1999); including them in a separate subgenus better emphasizes their position within the genus. Taxa in subg. *Protitea* are characterized by bud-scales with distinct, overlapping adaxial margins, flowers with 3–7[–9] stamens, and, usually, a tropical distribution. When subg. *Protitea* was described, it did not receive strong support. The debate was whether the multistaminate willows should be grouped on stamen number (T. Nakai 1928) or bud-scale characteristics (A. Kimura 1928). In a cladistic study of multistaminate *Salix* (Zhang M. L. 1994), members of subg. *Protitea* were in a clade that included two members of sect. *Salicaster* (subg. *Salix*). In a phenetic study of New World *Salix* (Argus), subg. *Protitea* appeared as a separate branch but was closely associated with a branch including members of subg. *Salix*. Two molecular studies provided equivocal results. A study based on ribosomal DNA (E. Leskinen and C. Alström-Rapaport 1999) included only one species from subg. *Protitea* (*S. amygdaloides*); it grouped with *S. alba* (subg. *Salix*). A study based on cpDNA, *rbc*L gene sequence (T. Azuma et al. 2000) placed members of subg. *Protitea* in a clade along with taxa from subg. *Salix* and subg. *Longifoliae*. Within this clade, species included here in subg. *Protitea* appeared on different branches. These results argue against the recognition of subg. *Protitea* but they were not definitive because the molecular cladistic studies included too few species of the subgenus and the morphological phenetic study did not account for polyploidy. All members of subg. *Protitea* for which we have counts are diploid; species in sect. *Salicaster*, the other section with multistaminate species, are polyploid. It is likely that sect. *Salicaster*, which is placed here in subg. *Salix*, originated through hybridization between members of subg. *Protitea* and subg. *Salix* followed by polyploidy. This would account for the occurrence of both multiple stamens and calyptrate bud-scales in the polyploid members of subg. *Salix*. Placing the diploid, multistaminate *Salix* with distinct, overlapping bud-scales in a subgenus separate from the polyploid, multistaminate *Salix* (sect. *Salicaster*) with calyptrate bud-scales would make subg. *Protitea* a more natural group. A study including a large sample of the 33 species in this subgenus would help resolve this question.

1. Largest medial blades: abaxial surface not glaucous; pistillate bracts deciduous after flowering.
    2. Largest medial blades usually linear, 10–28.6 times as long as wide, adaxial surface dull; stipules on late leaves usually rudimentary; ovaries with distinct, often raised veins . . . . *Salix humboldtiana* [p. 34]
    2. Largest medial blades narrowly elliptic or narrowly lanceolate, sometimes linear, 4.7–13.2 times as long as wide, adaxial surface slightly glossy; stipules on late leaves usually foliaceous; ovaries without distinct, raised veins.
        3. Branches red-brown to yellow-brown; capsules 3–5 mm; stipules on early leaves rudimentary or foliaceous, on late leaves usually foliaceous, apex usually acuminate or acute, glands few or absent adaxially; ovaries usually glabrous, rarely pilose; anthers with strongly recurved axes; branches highly brittle at base; pistillate adaxial nectaries oblong (swollen) . . . . 5. *Salix nigra*
        3. Branches yellow-brown to gray-brown; capsules 6–7 mm; stipules on early leaves broad rudiments or foliaceous, on late leaves foliaceous, apex rounded to convex, glands numerous adaxially; ovaries usually glabrous, sometimes villous; anthers with straight axes; branches flexible to ± brittle at base; pistillate adaxial nectaries square (flattened) . . . . 6. *Salix gooddingii*
1. Largest medial blades: abaxial surface glaucous; pistillate bracts deciduous or persistent after flowering.
    4. Pistillate bracts persistent after flowering.
        5. Largest medial blades 38–55 mm wide, 2.5–4 times as long as wide, bases rounded to cordate, margins usually spinulose-serrulate, abaxial surface tomentose, adaxial highly glossy; petioles glandular-lobed distally; stipes 3.2–5.6 mm; styles 0.3–0.4 mm; Alabama, Florida, Georgia [2a1. *Salix* sect. *Floridanae*] . . . . 1. *Salix floridana*
        5. Largest medial blades 7–27 mm wide, 4.5–10.7 times as long as wide, bases cuneate to convex, margins serrulate to crenulate, or entire, abaxial surface glabrous or glabrescent, adaxial dull or slightly glossy; petioles usually not glandular, rarely with spherical glands distally; stipes 0.4–2.4 mm; styles 0.2–0.3 mm; Arizona . . . . 2. *Salix bonplandiana*
    4. Pistillate bracts deciduous after flowering.
        6. Branches, branchlets, petioles, and leaves glabrous; largest medial blades sometimes amphistomatous, adaxial surface dull; stigmas 0.24–0.4 mm . . . . 7. *Salix amygdaloides*
        6. Branches, branchlets, petioles, and leaves hairy, sometimes glabrate; largest medial blades hypostomatous, adaxial surface slightly to highly glossy; stigmas 0.16–0.28 mm.
            7. Largest medial blades: margins often entire, sometimes crenulate or serrulate; stipules rudimentary or foliaceous; floral bracts toothed; pistillate flowering branchlets 3–14 mm; stipes 1.4–2.8 mm; west of 102d meridian . . . . 3. *Salix laevigata*
            7. Largest medial blades: margins serrate or serrulate; stipules of late leaves foliaceous; floral bracts entire or erose; pistillate flowering branchlets 3–35 mm; stipes 1.3–5.3 mm; east of 102d meridian . . . . 4. *Salix caroliniana*

2a1. SALIX Linnaeus (subg. PROTITEA) sect. FLORIDANAE Dorn, Canad. J. Bot. 54: 2775. 1976 [E]

Shrubs or trees, 2–8 m. **Branches** brittle throughout. **Leaves**: petiole with stalked, spherical, or foliaceous glands distally; largest medial blade relatively broad (38–55 mm wide), margins slightly revolute, (glands submarginal), abaxial surface tomentose. **Pistillate catkins** stout (17–27 mm wide). **Pistillate flowers**: ovary glabrous.

Species 1: se United States.

1. Salix floridana Chapman, Fl. South. U.S., 430. 1860 • Florida willow [C][E][F]

Stems: branches red-brown, pubescent to glabrescent; branchlets yellow-brown or red-brown, sparsely velvety or pubescent. **Leaves**: stipules absent or rudimentary on early ones, foliaceous on late ones, apex rounded; petiole (shallowly grooved adaxially), 13–20 mm, puberulent or velvety adaxially; largest medial blade narrowly oblong, oblong, lanceolate, or narrowly ovate, 100–170 × 38–55 mm, 2.5–3(–4) times as long as wide, base rounded, convex, or subcordate (cordate on larger leaves), margins serrulate or spinulose-serrulate, apex acute, acuminate, or convex, abaxial surface sparsely tomentose, hairs straight, adaxial highly glossy, sparsely villous or pilose to glabrescent (midrib remaining villous), hairs white and ferruginous; proximal blade margins entire; juvenile blade sparsely pubescent to very densely villous or pilose abaxially, hairs white. **Catkins**: staminate 29–72 × 12–15 mm, flowering branchlet 1–10 mm; pistillate 50–81 × 17–27 mm, flowering branchlet 5–30 mm; floral bract (tawny, sometimes greenish), 2–3.6 mm, apex rounded, entire, abaxially sparsely hairy, hairs wavy; pistillate bract persistent after flowering. **Staminate flowers**: abaxial nectary 0.4–0.8 mm, adaxial nectary ovate, 0.5–1.1 mm, nectaries distinct or connate and cup-shaped; stamens 3–7; filaments hairy basally; anthers 0.4–0.5 mm. **Pistillate flowers**: adaxial nectary square, 0.5–0.9 mm; stipe 3.2–5.6 mm; ovary obclavate to ellipsoidal, (rarely puberulent), beak gradually tapering to styles; ovules 4 per ovary; styles 0.3–0.4 mm; stigmas 0.16–0.17–0.2 mm. **Capsules** 6–7 mm. $2n = 38$.

Flowering mid Feb–early Apr. Swamps, marshy shores of streams in woodlands, calcareous areas, shade tolerant; of conservation concern; 10–40 m; Ala., Fla., Ga.

The closest relatives of *Salix floridana* are in the Old World sect. *Tetraspermae*. A detailed discussion of the distribution, taxonomy, and relationships of this uncommon subtropical endemic was given by G. W. Argus (1986).

2a2. SALIX Linnaeus (subg. PROTITEA) sect. HUMBOLDTIANAE Andersson in A. P. de Candolle and A. L. P. P. de Candolle, Prodr. 16(2): 199. 1868

Shrubs or trees, 1–30 m. **Branches** somewhat brittle, highly brittle, or flexible at base. **Leaves**: petiole with pairs or clusters of spherical glands distally; largest medial blade relatively narrow (6–37 mm wide), margins flat, (glands marginal), abaxial surface glabrous, glabrescent, puberulent, pubescent, or pilose. **Pistillate catkins** slender, (5–16 mm wide). **Pistillate flowers**: ovary glabrous or villous.

Species 15 (6 in the flora): North America, Mexico, West Indies, Central America, Asia.

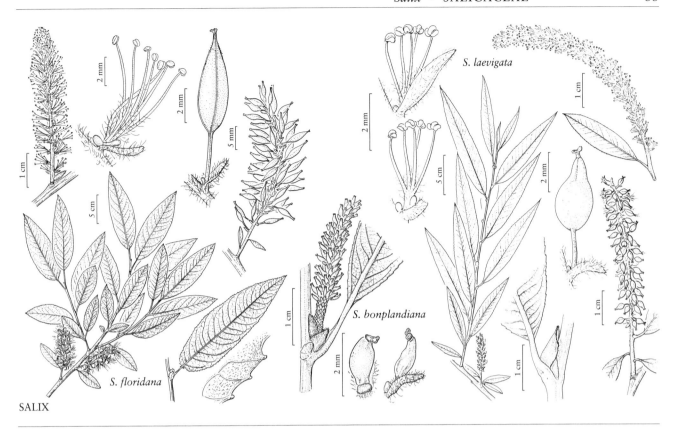

2. Salix bonplandiana Kunth in A. von Humboldt et al., Nov. Gen. Sp. 2(qto.): 24; 2(fol.): 20; plates 101, 102. 1817 • Bonpland's willow [F]

Salix bonplandiana var. *toumeyi* (Britton) C. K. Schneider

Trees, 1–13 m. **Stems:** branches yellow-brown to red-brown, glabrous; branchlets yellowish, streaked with red or red-brown, glabrous or puberulent, nodes hairy. **Leaves:** (marcescent), stipules absent or rudimentary on early ones, foliaceous on late ones, apex rounded, convex, or acute; petiole (rarely with spherical glands distally), 4–16 mm, puberulent or pubescent to glabrescent adaxially; largest medial blade lorate to narrowly lanceolate, 58–155 × 7–27 mm, 4.5–10.7 times as long as wide, base cuneate to convex, margins serrulate to crenulate or entire, apex acuminate to acute, abaxial surface glabrous or glabrescent, hairs appressed, adaxial dull or slightly glossy, glabrous or pilose; proximal blade margins entire; juvenile blade glabrous, puberulent, pilose, or sparsely long-silky abaxially, hairs white. **Catkins:** (usually flowering throughout season and axillary, sessile), staminate 24–131 × 3–10 mm, flowering branchlet 0–12 mm; pistillate (densely to loosely flowered), 24–47 × 6–12 mm, flowering branchlet 0–10 mm; floral bract 0.6–2.2 mm, apex rounded to convex, irregularly toothed or entire, abaxially sparsely to moderately densely hairy proximally, hairs irregularly curly; pistillate bract persistent after flowering. **Staminate flowers:** abaxial nectary 0.2–0.6 mm, adaxial nectary oblong, square, or ovate, 0.2–0.6 mm, nectaries distinct or connate and cup-shaped; stamens 3–7; filaments hairy basally; anthers 0.3–0.5 mm. **Pistillate flowers:** adaxial nectary square to oblong, 0.3–0.6 mm; stipe 0.4–2.4 mm; ovary pyriform to obturbinate, beak slightly bulged below or tapering to styles; ovules 8–18 per ovary; styles 0.2–0.3 mm; stigmas (sometimes slenderly cylindrical), 0.18–0.27–0.32 mm. **Capsules** 3–6 mm. $2n = 38$.

Flowering Feb–Apr and throughout year. Riparian forests, along streams, dry washes; 700–2000 m; Ariz.; Mexico (Baja California, Chiapas, Chihuahua, Distrito Federal, Durango, Guanajuato, Guerrero, Hidalgo, Jalisco, Michoacán, Morelos, Nuevo León, Oaxaca, Puebla, Querétaro, Sonora, Tlaxcala, Veracruz, Zacatecas); Central America (Guatemala).

Salix bonplandiana and *S. laevigata* are closely related and are sometimes treated as varieties (R. D. Dorn 1994). Their ranges overlap in Arizona and in northern Baja California, Mexico.

3. Salix laevigata Bebb, Amer. Naturalist 8: 202. 1874

• Red or polished willow F W

Salix bonplandiana Kunth var. laevigata (Bebb) Dorn; S. laevigata var. angustifolia Bebb; S. laevigata var. araquipa (Jepson) C. R. Ball; S. laevigata var. congesta Bebb

Trees, 2–15 m. Stems: branches flexible to highly brittle at base, gray-brown to yellow-brown, glabrous or villous; branchlets yellow-brown or red-brown, glabrous, densely villous, velvety, or pilose, nodes hairy. Leaves: stipules rudimentary grading into foliaceous or absent on early ones, usually foliaceous on late ones, apex acute, acuminate, rounded or convex; petiole (shallowly or deeply grooved adaxially, margins sometimes touching, sometimes with basilaminar glands, thickening), 3.5–18 mm, pubescent to glabrescent adaxially; largest medial blade lorate, narrowly oblong, narrowly elliptic, lanceolate, or obovate, 53–190 × 11–35 mm, 2.8–9 times as long as wide, base convex, subcordate, rounded, or cuneate, margins crenate, entire, or finely serrulate, apex acuminate, acute, or caudate, abaxial surface glabrous or pubescent, hairs spreading, white and/or ferruginous, adaxial highly or slightly glossy, glabrous or pubescent, midrib sometimes villous; proximal blade margins entire; juvenile blade glabrous or moderately densely long-silky to pilose abaxially, hairs white and/or ferruginous. Catkins: staminate 31–83 × 7–13 mm, flowering branchlet 2–26 mm; pistillate 28–79 × 6–11 mm, flowering branchlet 3–14 mm; floral bract 1.6–3.4 mm, apex rounded or acute, irregularly toothed or entire, abaxially sparsely to moderately densely hairy proximally or throughout, hairs wavy; pistillate bract deciduous after flowering. Staminate flowers: abaxial nectary 0.4–0.6 mm, adaxial nectary oblong, square, or ovate, 0.3–0.6 mm, nectaries distinct; stamens 3–7; filaments (sometimes basally connate), hairy on proximal ½ or basally; anthers 0.4–0.6 mm. Pistillate flowers: adaxial nectary square, 0.2–0.6 mm; stipe 1.4–2.8 mm; ovary pyriform, obturbinate, or ellipsoidal, beak slightly bulged below styles; ovules 12–24 per ovary; styles 0.1–0.2 mm; stigmas 0.2–0.23–0.28 mm. Capsules 3–5.5 mm.

Flowering (Dec–)Feb or mid Apr–early Jun. Riparian forests along streams, seepage areas, springs, subalkaline or brackish lakeshores, canyons, ditches; 0–2200 m; Ariz., Calif., Nev., Oreg., Utah; Mexico (Baja California, Baja California Sur).

Hybrids:

Salix laevigata forms natural hybrids with S. gooddingii.

Related Species:

Salix humboldtiana Willdenow: Humboldt willow is not known to occur in the flora area. It is characterized by: trees, 4–25 m; branches highly brittle at base, bud-scale margins distinct and overlapping adaxially; stipules on late leaves rudimentary or foliaceous; largest medial leaf blade usually linear, abaxial surface not glaucous, adaxial dull; pistillate bract deciduous after flowering; stamens 3–7; capsules with distinct, often raised, white veins. It occurs throughout much of Mexico to central Chile.

Salix humboldtiana is closely related to S. nigra in its generally narrow leaf blades, which are not glaucous abaxially. The two differ in the following characters: S. humboldtiana has leaf blades linear to sometimes narrowly oblong (10–28.6 times as long as wide), ovaries usually ovoid to ellipsoid, ovary walls often stomatiferous and with raised, white veins, and capsule valves relatively thick, slightly recurved. S. nigra has leaf blades usually narrowly lanceolate (6–13 times as long as wide), ovaries pyriform to obclavate, ovary walls neither stomatiferous nor notably veined, and capsule valves relatively thin and strongly recurved. Both species occur in Chihuahua, Mexico.

The report by R. I. Lonard et al. (1991) that specimens identified as Salix nigra from the lower Rio Grande, Texas, resemble S. humboldtiana in having strongly veined capsules suggests that S. humboldtiana, or intergrades with that species, may occur in Texas. Attempts to locate a voucher specimen were unsuccessful; because strongly veined capsules are diagnostic, further field study is indicated.

An earlier name, Salix chilensis Molina, has been applied to this species; it does not seem to pertain to this taxon (C. K. Schneider 1918).

4. Salix caroliniana Michaux, Fl. Bor.-Amer. 2: 226. 1803

• Carolina or coastal plain willow F W

Salix longipes Shuttleworth ex Andersson var. venulosa (Andersson) C. K. Schneider; S. longipes var. wardii (Bebb) C. K. Schneider

Trees, 5–10 m. Stems: branches ± brittle at base, gray-brown to red-brown, glabrous, villous, or tomentose; branchlets yellow-brown to red-brown, glabrous, sparsely or densely villous or tomentose. Leaves: stipules rudimentary or foliaceous on early ones, foliaceous on late ones, apex convex to acute; petiole (with spherical glands distally), (3–)4.5–14(–22) mm, tomentose or pilose adaxially; largest medial blade lorate or lanceolate to narrowly lanceolate, (50–)75–115(–220) × 10–22(–35) mm, 5–10

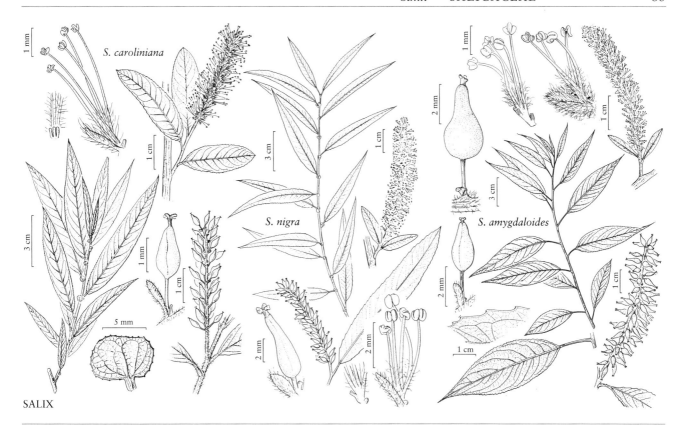

times as long as wide, base usually convex or cuneate, sometimes rounded to cordate, margins serrate or serrulate, apex acuminate, acute, or caudate, abaxial surface glabrous or sparsely tomentose on midribs, hairs white and/or ferruginous, wavy, adaxial highly glossy, glabrous or pilose, hairs white and/or ferruginous; proximal blade margins entire or serrulate; juvenile blade glabrous, or moderately densely tomentose or silky abaxially, hairs white and ferruginous. **Catkins:** staminate 28–97 × 5–11 mm, flowering branchlet 4–25 mm; pistillate 33–93 × 7–15 mm, flowering branchlet 3–35 mm; floral bract 1–3 mm, apex acute or rounded, entire or erose, abaxially sparsely hairy, hairs wavy; pistillate bract deciduous after flowering. **Staminate flowers:** abaxial nectary 0.3–0.5 mm, adaxial nectary oblong to narrowly oblong, 0.3–0.6 mm, nectaries distinct; stamens 4–7; filaments (sometimes connate less than ½ their lengths), hairy basally; anthers 0.4–0.6 mm. **Pistillate flowers:** adaxial nectary oblong, square, or ovate, 0.3–0.7 mm; stipe 1.3–5.3 mm; ovary pyriform to obclavate, beak slightly bulged below styles; ovules 12–16 per ovary; styles (sometimes distinct distally), 0.1–0.2 mm; stigmas 0.16–0.2–0.28 mm. **Capsules** 4–6 mm.

Flowering (south) Dec–early May, (north) mid Apr–early Jun. Alluvial woods on floodplains, swamps, hummocks, marshes, wet interdunal depressions, rocky or gravelly streambeds, ditches, canals, usually on calcareous substrates; 0–600 m; Ala., Ark., D.C., Fla., Ga., Ill., Ind., Kans., Ky., La., Md., Miss., Mo., N.C., Ohio, Okla., Pa., S.C., Tenn., Tex., Va., W.Va.; Mexico (Nuevo León); West Indies (Cuba); Central America (Guatemala).

Hybrids:

Salix caroliniana forms natural hybrids with *S. nigra*. Hybrids with *S. amygdaloides* have been reported (N. M. Glatfelter 1898); no convincing specimens have been seen.

Salix caroliniana × *S. nigra*: This hybrid is characterized by stipes to 1.3 mm and leaves glaucous; it probably occurs wherever the two parents come into contact. In the southeastern United States, it occurs from northern Florida to West Virginia and Maryland with intergradation mainly on the Atlantic coastal plain from northern Florida and southern Georgia (G. W. Argus 1986). Reports (N. M. Glatfelter 1898) of it from the vicinity of St. Louis, Missouri, are unconfirmed.

5. Salix nigra Marshall, Arbust. Amer., 139. 1785 • Black willow F W

Salix nigra var. *falcata* (Pursh) Torrey; *S. nigra* var. *lindheimeri* C. K. Schneider

Trees, 5–20+ m. Stems: branches highly brittle at base, red-brown to yellow-brown, glabrous; branchlets gray-brown to red-brown, glabrous or pilose to villous. Leaves: stipules rudimentary or foliaceous on early ones, usually foliaceous on late ones, (glands few or absent adaxially), apex acuminate, acute, or rounded; petiole (margins covering groove, with spherical glands distally), (2–)3–10(–15) mm, glabrous or pilose adaxially; largest medial blade (sometimes amphistomatous), very narrowly elliptic, lanceolate to narrowly lanceolate, or linear to lorate, (50–)70–103(–190) × (6–)7.5–17(–23) mm, 6–13 times as long as wide, base cuneate to convex, margins serrulate, apex acuminate, acute, or caudate, abaxial surface (not glaucous), glabrous or pilose, hairs white or ferruginous, wavy, adaxial slightly glossy, glabrous or pilose (especially on midribs); proximal blade margins serrulate; juvenile blade glabrous or pilose abaxially, hairs white and/or ferruginous. Catkins: staminate 35–83 × 7–13 mm, flowering branchlet 5–35 mm; pistillate 23–74(–80 in fruit) × 5–10 mm, flowering branchlet 6–35 mm; floral bract 1–3 mm, apex acute or rounded, entire, abaxially sparsely hairy, hairs wavy; pistillate bract deciduous after flowering. Staminate flowers: abaxial nectary 0.3–1 mm, adaxial nectary oblong to ovate, 0.2–0.6 mm, nectaries distinct or connate and shallowly cup-shaped; stamens 4–6; filaments hairy on proximal ½ or basally; anthers 0.4–0.6 mm, (axes strongly recurved). Pistillate flowers: adaxial nectary oblong, (swollen), 0.2–0.5 mm; stipe 0.5–1.5 mm; ovary pyriform to obclavate, (rarely pilose), beak slightly bulged below styles; ovules 12–16 per ovary; styles 0.1–0.3 mm; stigmas (sometimes 2 plump lobes), 0.2–0.28–0.36 mm. Capsules 3–5 mm. $2n = 38$.

Flowering (south) early Feb–early May, (north) late Mar–early Jul. Floodplains, edges of ponds and lakes, swamps, marshes, white cedar bogs, wet meadows, open fields, roadside ditches, mixed upland deciduous woods along streams; 10–1400 m; N.B., Ont., Que.; Ala., Ark., Conn., Del., D.C., Fla., Ga., Ill., Ind., Iowa, Kans., Ky., La., Maine, Md., Mass., Mich., Minn., Miss., Mo., Nebr., N.H., N.J., N.Y., N.C., Ohio, Okla., Pa., R.I., S.C., Tenn., Tex., Vt., Va., W.Va., Wis.; Mexico (Chihuahua).

Hybrids:

Salix nigra forms natural hybrids with *S. alba*, *S. amygdaloides*, *S. caroliniana*, *S. gooddingii*, and *S. lucida*.

6. Salix gooddingii C. R. Ball, Bot. Gaz. 40: 376, plate 12, figs. 1, 2. 1905 (as gooddingi) • Goodding's black willow W

Salix gooddingii var. *vallicola* (Dudley) C. R. Ball; *S. gooddingii* var. *variabilis* C. R. Ball; *S. nigra* Marshall var. *vallicola* Dudley

Trees, 3–30 m. Stems: branches flexible to ± brittle at base, yellow-brown to gray-brown, pubescent to glabrescent; branchlets usually yellowish or yellow-green, sometimes reddish brown, puberulent or pubescent to glabrescent. Leaves: stipules broad rudiments or foliaceous on early ones, foliaceous on late ones, (glands numerous adaxially), apex rounded to convex; petiole (sometimes with spherical glands distally), 4–10 mm, pilose adaxially; largest medial blade (sometimes amphistomatous), narrowly elliptic, very broadly oblong, lorate, or linear, 67–130 × 9.5–16 mm, 4.7–12.4 times as long as wide, base cuneate to convex, margins serrulate to serrate, apex acuminate, caudate, or acute, abaxial surface (usually not glaucous, rarely thinly so), glabrous or puberulent, hairs wavy, adaxial slightly glossy, pilose to glabrescent; proximal blade margins entire or shallowly serrulate; juvenile blade sparsely velvety to pilose abaxially, hairs white. Catkins: staminate 19–80 × 6–10 mm, flowering branchlet 2–23 mm; pistillate 23–82 × 6–15 mm, flowering branchlet 2–48 mm; floral bract 1.4–2.4 mm, apex acute or rounded, entire or toothed, abaxially sparsely to moderately densely hairy, hairs wavy; pistillate bract deciduous after flowering. Staminate flowers: abaxial nectary (0.2–)0.3–0.6 mm, adaxial nectary square to ovate, 0.2–0.6 mm, nectaries distinct; stamens 4–6(–8); filaments (sometimes basally connate), hairy on proximal ½; anthers 0.4–0.5 mm, (axes straight). Pistillate flowers: adaxial nectary square (flattened), 0.2–0.6 mm; stipe 1.2–3.2 mm; ovary pyriform, (sometimes villous), beak slightly bulged or abruptly tapering to styles; ovules 12–18 per ovary; styles 0.1–0.3 mm; stigmas 0.2–0.29–0.32 mm. Capsules 6–7 mm. $2n = 38$.

Flowering late Mar–Jun. Riparian forests, springs, seepage areas, washes, meadows; -40–500(–2500) m; Ariz., Calif., Colo., Nev., N.Mex., Okla., Tex., Utah; Mexico (Baja California, Chihuahua, Coahuila, Guerrero, Sinaloa, Sonora).

Salix gooddingii and *S. nigra* are closely related and are sometimes treated as conspecific (C. R. Ball 1950). *Salix gooddingii* has yellow-brown or pale gray-brown branches, capsules 6–7 mm, and ovaries usually glabrous but pilose in ca. 20% of specimens. *Salix nigra* has red-brown to dark gray-brown branches, capsules 3–5 mm, and ovaries almost always glabrous. A single plant with pilose ovaries was found in Ontario, Canada; reports (W. A. Archer 1965) of *S. nigra* with hairy ovaries in Alabama, Arkansas, Illinois, Iowa, Kansas, Massachusetts, Oklahoma, and Texas could not be confirmed. Ranges of these taxa overlap in west-central Texas, where there is evidence of intergradation; they rarely occur in the same population. The map by E. L. Little Jr. (1971), who treated them as conspecific, shows a significant range disjunction between the two.

Catkins of *Salix gooddingii* flowering in March and early April are sometimes borne in leaf axils. This suggests that the sylleptic condition, typical of *S. bonplandiana*, is sometimes ecotypic.

Hybrids:

Salix gooddingii forms natural hybrids with *S. amygdaloides* and *S. nigra*. Hybrids with *S. lasiandra* have been reported (C. K. Schneider 1921); no convincing specimens have been seen.

Salix gooddingii × *S. laevigata*: In Arizona, a population of young plants displays intermediate characteristics. They have leaf blades sparsely glaucous abaxially, as in *S. laevigata*, but narrow, often amphistomatous, and with petioles sometimes not glandular distally, as in *S. gooddingii*. Both parental species occur in the region. This hybrid was also reported by C. K. Schneider (1921) from California.

Salix gooddingii × *S. nigra*: This hybrid may occur in western Texas where the parental species overlap. Some specimens from that area seem to be "intermediate" in branch color, but the differences are subtle.

7. **Salix amygdaloides** Andersson, Öfvers. Kongl. Vetensk.-Akad. Förh. 15: 114. 1858 • Peach leaf willow E F

Trees, 4–20 m. Stems: branches flexible to ± brittle at base, yellow to gray-brown, glabrous; branchlets yellow-brown, gray-brown, or red-brown, glabrous. Leaves: stipules absent or rudimentary on early ones, foliaceous or rudimentary on late ones, apex rounded; petiole (margins covering groove, not glandular or with spherical glands distally), 7–21 mm, glabrous or puberulent adaxially; largest medial blade (sometimes amphistomatous), very narrowly elliptic, elliptic, lanceolate, or lorate, 55–130 × 24–37 mm, 2.8–6 times as long as wide, base convex, cuneate, or cordate, margins serrulate, apex acuminate to caudate, abaxial surface glaucous, glabrous, adaxial dull, glabrous or sparsely pubescent along midrib; proximal blade margins entire or shallowly serrulate; juvenile blade glabrous or pubescent abaxially, hairs white and/or ferruginous. **Catkins:** staminate 23–80 × 5–12 mm, flowering branchlet 3–28 mm; pistillate 41–110(–127 in fruit) × 8–16 mm, flowering branchlet 17–35 mm; floral bract 1.5–2.8 mm, apex acute to rounded, entire or toothed, abaxially sparsely to moderately densely hairy proximally, hairs wavy; pistillate bract deciduous after flowering. **Staminate flowers:** abaxial nectary 0.2–0.7 mm, adaxial nectary narrowly oblong to square, 0.3–0.8 mm, nectaries distinct; stamens 3–7; filaments hairy on proximal ½ or basally; anthers 0.5–0.6 mm. **Pistillate flowers:** adaxial nectary square, 0.1–0.6 mm; stipe 1.4–3.2 mm; ovary pyriform, beak slightly bulged below styles; ovules 16–18 per ovary; styles 0.2–0.4 mm; stigmas 0.24–0.31–0.4 mm. **Capsules** 3–7 mm. $2n = 38$.

Flowering early Apr–Jun. Moist to mesic floodplains, shores of lakes on sandy, silty, or gravelly substrates, marshes, wet sand dune slacks; 60–2400 m; Alta., B.C., Man., Ont., Que., Sask.; Ariz., Colo., Idaho, Ill., Ind., Iowa, Kans., Ky., Mich., Minn., Mo., Mont., Nebr., Nev., N.Mex., N.Y., N.Dak., Ohio, Okla., Oreg., Pa., S.Dak., Tex., Utah, Wash., Wis., Wyo.

Presence of *Salix amygdaloides* in Massachusetts, New Hampshire, and Vermont has not been verified; its occurrence in those New England states was reported by H. A. Gleason and A. Cronquist (1991), and by M. L. Fernald (1950).

Hybrids:

Salix amygdaloides forms natural hybrids with *S. gooddingii* and *S. nigra*. Hybrids with *S. caroliniana* (N. M. Glatfelter 1898) and *S. eriocephala* (M. L. Fernald 1950) have been reported; no convincing specimens have been seen. Controlled pollination between *S. amygdaloides* and *S. eriocephala*, *S. interior*, and *S. petiolaris* set no seed; controlled pollination with *S. lucida* produced a few seeds; some seedlings suffered necrosis in the cotyledon stage (A. Mosseler 1990).

Salix amygdaloides × *S. gooddingii* (*S.* ×*wrightii* Andersson): This hybrid occurs throughout the Rio Grande Valley, Texas, and New Mexico (C. K. Schneider 1919; C. R. Ball 1961), and at Happy and Rio Frio, Texas, and Virgil Run, Arizona. The leaves are somewhat glaucous abaxially, as in *S. amygdaloides*, but they are linear to narrowly elliptic and branchlets are sparsely pubescent as in *S. gooddingii*.

Salix amygdaloides × *S. nigra* (*S.* ×*glatfelteri* C. K. Schneider) resembles *S. amygdaloides* in leaves somewhat glaucous abaxially, but usually linear or narrowly elliptic, as in *S. nigra*. The stipules are not as prominent as in *S. nigra* but are foliaceous on late leaves; it should be expected wherever the ranges of the two species overlap. The hybrid is common in Missouri, where N. M. Glatfelter (1894) estimated that ca. 40% of the populations were hybrids, and in Illinois (R. H. Mohlenbrock 1980; G. Wilhelm, pers. comm.). Putative hybrids occur also in Ontario. Narrow leaves are typical of juvenile plants of *S. amygdaloides* but even at that stage they tend to be broadest at the midpoint or toward the base rather than in a midzone as in *S. nigra*.

2b. Salix Linnaeus subg. Salix

Shrubs or trees, 1–25 m, usually not clonal, or clonal by stem fragmentation. **Stems** usually erect (pendulous in *S. babylonica*); branches flexible to highly brittle at base, not glaucous, (usually slightly or highly glossy). **Buds** *alba*-type, scale margins connate. **Leaves**: stipules on early ones usually absent or rudimentary, sometimes foliaceous, on late ones rudimentary or foliaceous, (deciduous spring or autumn); petiole convex, flat, or shallowly to deeply grooved adaxially, not glandular, or with spherical or foliaceous glands distally or throughout; largest medial blade amphistomatous, hemiamphistomatous, or hypostomatous, lorate, narrowly oblong, narrowly elliptic, elliptic, lanceolate, narrowly ovate, or oblanceolate, 2–14.4 times as long as wide, angle of base and of apex less than 90°, surface hairs white and/or ferruginous; juvenile blade hairs white and/or ferruginous. **Catkins** flowering as leaves emerge (sometimes before in *S. babylonica*), from lateral buds; staminate on flowering branchlet or sessile, usually slender or stout, sometimes subglobose; pistillate on flowering branchlet, densely or loosely flowered, slender, stout, or subglobose; floral bract usually tawny, sometimes greenish, apex usually entire, toothed, or erose, rarely sinuate; pistillate bract deciduous or persistent after flowering. **Staminate flowers**: abaxial nectary usually present; stamens 2–10; filaments distinct or connate less than ½ their lengths, hairy on proximal ½ or basally, rarely glabrous; anthers usually yellow, sometimes reddish turning yellow. **Pistillate flowers**: abaxial nectary absent or present; ovary not glaucous, glabrous or, rarely, villous, beak usually gradually tapering to styles; ovules 2–36 per ovary; styles connate or distinct; stigmas usually flat, abaxially non-papillate, tip rounded, sometimes stigmas cylindrical, or 2 plump lobes.

Species 85 (9 species in the flora): North America, Mexico, South America, Eurasia.

Five of the eight sections recognized in this subgenus are found in the flora area.

See 2a. *Salix* subg. *Protitea* for the rationale for including the multistaminate species of sect. *Salicaster* in subg. *Salix*.

1. Trees; stems pendulous; pistillate bracts persistent after flowering; pistillate adaxial nectaries longer than stipes; styles 0.15–0.3 mm; stipes 0–0.2 mm; ovules 2–8 per ovary; capsules 1–3.5 mm; stamens 2 [2b1. *Salix* sect. *Subalbae*].
    2. Branches yellowish, yellow-green, or yellow-brown . . . . *Salix* ×*sepulcralis* (in part) [p. 40]
    2. Branches yellow-brown to red-brown or gray-brown.
        3. Pistillate catkins on flowering branchlets (0–)2–4 mm; ovaries: beak abruptly tapering to styles; anthers 0.4–0.5 mm . . . . 8. *Salix babylonica*
        3. Pistillate catkins on flowering branchlets 3–14 mm; ovaries: beak gradually tapering to styles; anthers 0.5–0.8 mm.
            4. Petioles short-silky adaxially; branches yellowish, yellow-green, or yellow-brown; staminate catkins moderately densely flowered, slender, nectaries distinct . . . . *Salix* ×*sepulcralis* (in part) [p. 40]
            4. Petioles glabrous, pilose, or velvety to glabrescent adaxially; branches yellow-brown, gray-brown, or red-brown; staminate catkins loosely flowered, stout, nectaries connate and shallowly cup-shaped . . . . *Salix* ×*pendulina* [p. 41]

1. Shrubs or trees; stems erect; pistillate bracts deciduous after flowering (persistent in
   S. maccalliana, sometimes persistent until fruiting in *S. triandra*); pistillate adaxial nectaries
   shorter than or equal to stipes; styles 0.1–1.2 mm; stipes 0.2–4 mm; ovules 8–30 per ovary;
   capsules 3.5–12 mm; stamens 2–10.
   5. Petioles without spherical glands or lobes distally; stamens 2; anthers purple turning
      yellow in age; ovaries densely villous; pistillate bracts persistent after flowering; leaf
      blades: abaxial surface not glaucous [2b4. *Salix* sect. *Maccallianae*] . . . . 15. *Salix maccalliana*
   5. Petioles with spherical glands or lobes distally; stamens 2 or 3–10; anthers yellow;
      ovaries glabrous; pistillate bracts deciduous after flowering (sometimes persistent in
      S. triandra); leaf blades: abaxial surface glaucous or not.
      6. Largest medial blades: abaxial surface usually not glaucous; stamens 3–10
         [2b3. *Salix* sect. *Salicaster* (in part), 2b5. *Salix* sect. *Triandrae*].
         7. Largest medial blades: adaxial surface dull or ± slightly glossy; styles
            0.2–0.3 mm . . . . 16. *Salix triandra* (in part)
         7. Largest medial blades: adaxial surface slightly or highly glossy; styles
            0.3–1 mm.
            8. Stipules on early leaves foliaceous; juvenile blade hairs white and
               ferruginous; branchlets usually hairy, sometimes glabrescent. . . . . 13. *Salix lucida*
            8. Stipules on early leaves absent or rudimentary; juvenile blades glabrous;
               branchlets glabrous.
               9. Largest medial blade margins crenate or crenulate; stamens 2; floral
                  bracts 0.8–1.3 mm; capsules 3–5 mm . . . . 10. *Salix euxina*
               9. Largest medial blade margins serrulate; stamens 3–10; floral bracts
                  1.2–4 mm; capsules 6–12 mm.
                  10. Branches highly glossy; pistillate catkins slender to stout; floral
                      bracts sparsely hairy proximally; ovaries: beak gradually tapering
                      to styles . . . . 11. *Salix pentandra*
                  10. Branches dull or slightly glossy; pistillate catkins stout to globose;
                      floral bracts moderately densely hairy; ovaries: beak slightly
                      bulged below or abruptly tapering to styles . . . . 12. *Salix serissima*
      6. Largest medial blades: abaxial surface glaucous (except usually not glaucous in
         S. triandra); stamens 2 or 3–6.
         11. Juvenile and largest medial blades with white hairs; stamens 2 [2b2. *Salix* sect.
             Salix].
             12. Largest medial blades persistently silky on both surfaces; petioles long-
                 silky adaxially; styles 0.2–0.4 mm; branches flexible to ± brittle at base . . . . 9. *Salix alba*
             12. Largest medial blades glabrescent or glabrous on both surfaces; petioles
                 glabrous or puberulent adaxially; styles 0.4–1 mm; branches highly brittle
                 at base . . . . *Salix ×fragilis* [p. 42]
         11. Juvenile and largest medial blades with white and ferruginous hairs; stamens
             usually more than 2 [2b3. *Salix* sect. *Salicaster* (in part)].
             13. Petioles convex to shallowly grooved adaxially; leaf blades: abaxial
                 surface long-silky to glabrescent; stipes 0.3–0.6 mm; pistillate flowering
                 branchlets 8–16 mm . . . .*Salix ×jesupii* [p. 43]
             13. Petioles deeply grooved adaxially; leaf blades: abaxial surface usually
                 glabrous or pubescent; stipes 0.8–2 mm; pistillate flowering branchlets
                 6–56 mm.
                 14. Stipules: apex convex or rounded; petioles: margins not touching;
                     stamens 3–5. . . . . 14. *Salix lasiandra*
                 14. Stipules: apex acuminate or acute; petioles: margins covering groove;
                     stamens 3 . . . . 16. *Salix triandra* (in part)

2b1. SALIX Linnaeus (subg. SALIX) sect. SUBALBAE Koidzumi, Bot. Mag. (Tokyo) 27: 88. 1913

Trees, 4–20 m. **Stems:** (pendulous), branches somewhat brittle at base. **Leaves:** stipules on late ones and vigorous shoots minute rudiments; petiole not glandular, or with spherical glands or lobes distally; largest medial blade hemiamphistomatous, abaxial surface glaucous; juvenile blade glabrous, puberulent, or long-silky, hairs white. **Pistillate floral bracts** persistent after flowering. **Staminate flowers:** stamens 2; anthers (dry), 0.4–0.5 mm. **Pistillate flowers:** adaxial nectary longer than stipe; stipe 0–0.2 mm; ovary glabrous.

Species 17 (1 in the flora): introduced; Eurasia; introduced also in Mexico, South America.

8. **Salix babylonica** Linnaeus, Sp. Pl. 2: 1017. 1753
   • Weeping willow

Stems: branches yellow-brown to red-brown; branchlets sparsely to moderately densely tomentose, especially at nodes. **Leaves:** stipules absent or rudimentary on early ones; petiole convex to flat or shallowly to deeply grooved adaxially, 7–9 mm, tomentose abaxially; largest medial blade lanceolate, narrowly oblong, or narrowly elliptic, 90–160 × 5–20 mm, 5.5–10.5 times as long as wide, base cuneate, margins flat, spinulose-serrulate or serrulate, apex acuminate, caudate, or acute, surfaces glabrous or sparsely short-silky, hairs straight, dull adaxially; proximal blade margins entire; juvenile blade reddish or yellowish green. **Catkins** (flowering just before leaves emerge); staminate 13–35 mm, flowering branchlet 1–6 mm; pistillate densely flowered, stout or subglobose, 9–27 × 2.5–7 mm, flowering branchlet (0–)2–4 mm; floral bract 1.1–1.8 mm, apex acute, rounded, or truncate, entire, abaxially sparsely hairy throughout or proximally, hairs wavy. **Staminate flowers:** abaxial nectary 0.2–0.6 mm, adaxial nectary oblong or ovate, 0.4–0.7 mm, nectaries distinct or connate and cup-shaped; filaments distinct, hairy on proximal ½ or basally; anthers (sometimes reddish turning yellow), ellipsoid or globose. **Pistillate flowers:** adaxial nectary oblong, square, ovate, or obovate, 0.4–0.8 mm; ovary ovoid or obturbinate, beak (sometimes pilose proximally), slightly bulged below or abruptly tapering to styles; ovules 2–4 per ovary; styles distinct or connate ½ their lengths, 0.2–0.3 mm; stigmas flat, abaxially non-papillate with rounded tip, or 2 plump lobes (almost capitate), 0.2–0.3 mm. **Capsules** 2–2.7 mm. $2n = 76$.

Flowering spring. Around settlements; ca. 50 m; introduced; Ala., Ark., Calif., Del., D.C., Fla., Ga., Ky., La., Md., N.C., S.C., Tenn., Va.; Asia; introduced also in Mexico (Mexico City), South America.

Little is known about the origin of the strongly weeping cultivar of *Salix babylonica*. It was described by Linnaeus (1737[1738]) based on young garden specimens (W. J. Bean 1970–1988, vol. 4). It is thought to have originated in China, although it no longer occurs in the wild and its origin is uncertain. Selections are thought to have been transported to Europe along the trade route from China. In Tajikistan, there are three cultivated clones, one of which is staminate (A. K. Skvortsov 1999). Taxonomic treatments of *S. babylonica* are variable. Some botanists recognize a single species, including both pendulous and non-pendulous forms (Skvortsov), while others recognize four species: *S. babylonica*, with a weeping habit, *S. capitata* Y. L. Chou & Skvortsov, *S. pseudolasiogyne* H. Léveillé, and the commonly cultivated *S. matsudana* Koidzumi (Fang Z. F. et al. 1999), with an erect or spreading habit. Here, *S. babylonica* is treated in a narrow sense, including only weeping forms.

Salix babylonica is not cold tolerant and is not commonly grown in Europe (R. D. Meikle 1984) or in northern North America. In the flora area, cultivated trees with strongly pendulous branches and branchlets have been identified as *S. babylonica* (G. W. Argus 1985, 1986, 1993), but many are hybrids with *S. alba* (*S.* ×*sepulcralis*) or *S. euxina* (*S.* ×*pendulina*). *Salix* ×*sepulcralis*, especially nothovar. *chrysocoma*, with bright yellow branchlets, is the most commonly grown of these hybrids. All reported occurrences of *S. babylonica* need verification.

Hybrids:

Salix ×*sepulcralis* Simonkai: Weeping willow, *S. alba* × *S. babylonica*, is introduced from Europe and widely naturalized throughout the world. Synonyms include *S.* ×*salamonii* Carrière ex Henry and *S.* ×*sepulcralis* nothovar. *chrysocoma* (Dode) Meikle. It is characterized by: trees, to 12 m, stems pendulous; branches somewhat to highly brittle at base, yellowish, yellow-green, or yellow-brown; branchlets yellowish, yellow-green, or golden; stipules rudimentary or foliaceous on late leaves; petiole not glandular or with pairs or clusters of spherical glands distally or scattered throughout, short-silky adaxially; largest medial blade amphistomatous or hemiamphistomatous, narrowly elliptic to very narrowly so, margins finely serrulate or spinulose-serrulate, abaxial surface glaucous, adaxial glaucous, sparsely long-silky to glabrescent, hairs white or white and ferruginous, adaxial surface slightly glossy; catkins on distinct flowering branchlet 3–14 mm; staminate moderately densely flowered, slender, 23–53 × 3–9 mm; pistillate moderately densely to loosely flowered, slender to stout, 18–30 × 3–8 mm, flowering branchlet 3–14 mm; pistillate bracts persistent after flowering; staminate abaxial and adaxial nectaries distinct; stamens 2; anthers 0.5–0.8 mm; pistillate nectary longer than stipe; stipe 0–0.2 mm; ovaries gradually tapering to styles; ovules 4 per ovary; styles 0.15–2 mm; capsules 1–2 mm. In the flora area, it occurs in: British Columbia, New Brunswick, Nova Scotia, Ontario, Quebec; Alaska, Arizona, Arkansas, California, Connecticut, District of Columbia, Illinois, Iowa, Kentucky, Louisiana, Maine, Maryland, Massachusetts, Michigan, Missouri, Nevada, New Hampshire, New Jersey, New Mexico, New York, North Carolina, Ohio, Oregon, Pennsylvania, Tennessee, Utah, Virginia, and West Virginia.

The most commonly cultivated, and sometimes escaped, weeping willow with golden or yellow-green branchlets is *Salix* ×*sepulcralis* nothovar. *chrysocoma* (Dode) Meikle. It probably originated as *S. alba* var. *vitellina* × *S. babylonica* (R. D. Meikle 1984). According to F. S. Santamour Jr. and A. J. McArdle (1988), *S.* ×*sepulcralis* cv. Salamonii has a broadly pyramidal crown and is only slightly pendulous. It is not clear just how this cultivar differs from *S.* ×*pendulina*. For a discussion of the taxonomy of these and other weeping willows see J. Chmelař (1983).

Salix ×*pendulina* Wenderoth: Weeping willow, *S. babylonica* × *S. euxina*, is introduced from Europe and grown throughout the world. It is characterized by: trees, 2.5–12 m, stems pendulous; branches highly brittle at base, yellow-brown, gray-brown, or red-brown; branchlets yellowish to brownish; stipules foliaceous on late leaves; petioles glabrous, pilose, or velvety to glabrescent adaxially; largest medial blade amphistomatous or hypostomatous, very narrowly elliptic to lanceolate, or linear, margins serrulate, irregularly so, or spinulose-serrulate, abaxial surface glaucous, adaxial slightly glossy or dull; catkins on distinct flowering branchlet, 3–14 mm; staminate loosely flowered, stout, 16–34 × 7–11 mm; pistillate densely or moderately densely flowered, slender or stout, 20–36 × 3.5–11 mm; pistillate bract persistent after flowering; staminate abaxial and adaxial nectaries connate and shallowly cup-shaped; stamens 2; anthers 0.5–0.6 mm; pistillate nectary longer than stipe; stipe 0 mm; styles 0.2–0.6 mm; ovules 4–8 per ovary; capsules 1.8–3.5 mm. In the flora area, it occurs in: Ontario; California, Connecticut, District of Columbia, Georgia, Illinois, Indiana, Maine, Maryland, Massachusetts, Michigan, Missouri, Nebraska, New Jersey, New Mexico, New York, North Carolina, Ohio, Oregon, Pennsylvania, Texas, Virginia, Washington, and West Virginia.

Reports of this hybrid in British Columbia and California are undocumented. Plants of *Salix* ×*pendulina* with prominent, caudate stipules are var. *blanda* (Andersson) Meikle; those with ovaries with patchy or streaky hairiness are var. *elegantissima* (K. Koch) Meikle.

2b2. SALIX Linnaeus (subg. SALIX) sect. SALIX

Trees, 3–25 m. **Stems:** (erect), branches somewhat to highly brittle at base or flexible. **Leaves:** stipules on late ones and vigorous shoots foliaceous or minute rudiments, apex acute or acuminate; petiole with spherical glands or lobes distally; largest medial blade amphistomatous or hypostomatous, abaxial surface glaucous or not; juvenile blade glabrous, pubescent, or short- to long-silky abaxially, hairs white. **Pistillate floral bracts** deciduous after flowering. **Staminate flowers:** stamens 2; anthers 0.4–0.7(–0.8) mm. **Pistillate flowers:** adaxial nectary shorter than or equal to stipe; stipe 0.2–1.5 mm; ovary glabrous.

Species 8 (2 in the flora): introduced; Eurasia.

9. **Salix alba** Linnaeus, Sp. Pl. 2: 1021. 1753 • White willow ⒤ Ⓦ

Trees, 10–25 m. **Stems:** branches flexible or ± brittle at base, yellow, gray-brown, or red-brown, glabrous or hairy; branchlets yellowish or gray to red-brown, pilose, densely villous, or long-silky. **Leaves:** stipules rudimentary or absent on early ones, rudimentary or foliaceous on late ones, apex acute; petiole shallowly grooved adaxially, 3–13 mm, with pairs or clusters of spherical glands or lobes distally, long-silky abaxially; largest medial blade amphistomatous, narrowly oblong, very narrowly elliptic, narrowly elliptic, or lanceolate, 63–115 × 10–20 mm, 4.2–7.3 times as long as wide, base cuneate or convex, margins flat, serrate or serrulate, apex acuminate, caudate, or acute, abaxial surface very densely long-silky to glabrescent, hairs straight, (glaucous or obscured by hairs), adaxial dull, sparsely long-silky; proximal blade margins entire; juvenile blade yellowish green or reddish, very densely long-silky abaxially. **Catkins:** staminate 27–60 × 6–10 mm, flowering branchlet 2–8 mm; pistillate loosely flowered, slender, 31–51 × 4–8 mm, flowering branchlet 3–14 mm; floral bract 1.6–2.8 mm, apex rounded, entire, abaxially sparsely hairy, hairs straight. **Staminate flowers:** adaxial nectary oblong to square, 0.3–0.7 mm, nectaries usually distinct (rarely connate); filaments distinct, hairy on proximal ½ or basally; anthers (purple, turning yellow), shortly cylindrical to globose, 0.5–0.7 mm. **Pistillate flowers:** adaxial nectary square, 0.3–0.7 mm, equal to or shorter than stipe; stipe 0.2–0.8 mm; ovary obclavate to pyriform, beak slightly bulged below styles; ovules 8 or 9 per ovary; styles connate, 0.2–0.4 mm; stigmas flat, abaxially non-papillate with rounded tip, or broadly cylindrical, 0.3–0.6 mm. **Capsules** 3.5–5 mm. $2n = 76$.

Flowering early May–late Jun. Riverbanks, sandy beaches, fens, old fields, roadsides, gravel pits; 70–2000 m; introduced; N.B., Ont., Que., Sask.; Ariz., Ark., Calif., Colo., Conn., Del., D.C., Ga., Idaho, Ill., Ind., Ky., Maine, Md., Mass., Mich., Minn., Mo., Mont., Nebr., Nev., N.H., N.Y., N.C., Ohio, Pa., R.I., Tenn., Vt., Va., W.Va., Wis.; Eurasia.

Reports that *Salix alba* is naturalized in Alberta, British Columbia, California, and Washington are undocumented.

The variants of *Salix alba*, commonly cultivated in the flora area, are often treated as subspecies (K. H. Rechinger 1993) or varieties (R. D. Meikle 1984) but they are all cultivars. The most common ones are: *S. alba* cv. Sericea (*S. alba* var. *sericea* Gaudin) with densely and persistently long-silky leaves and branchlets; *S. alba* cv. Vitellina (*S. alba* var. *vitellina* (Linnaeus) Stokes) with yellow to yellow-brown branchlets and branches; *S. alba* cv. Caerulea (*S. alba* var. *caerulea* (Smith) Smith) with dark brown branchlets and leaves coarsely toothed and sparsely silky abaxially; and *S. alba* cv. Chermesina (*S. alba* var. *chermesina* Hartig) with reddish twigs. Plants referred to in the literature as *S. alba* var. *vitellina* cv. Pendula are treated here as *S.* ×*sepulcralis*.

Hybrids:

Salix alba forms natural hybrids with *S. lucida* and *S. nigra*. Hybrids with *S. petiolaris* have been reported (M. L. Fernald 1950) but no convincing specimens have been seen.

Salix alba × *S. nigra* is an infrequent hybrid that is reported to have the catkins of *S. alba* and the foliage of *S. nigra* (M. S. Bebb 1895). Buds of this hybrid should be examined to see if they have the partially connate bud-scales characteristic of crosses between species with connate and distinct margins (J. Chmelař 1978).

Salix alba × *S. euxina*. See *S.* ×*fragilis* below.

Salix alba × *S. lucida*. See *S.* ×*jesupii* [p. 43].

Salix alba var. *vitellina* × *S. babylonica*. See *S.* ×*sepulcralis* Simonkai [p. 40].

Salix ×*fragilis* Linnaeus: The hybrid white willow, *S. alba* Linnaeus × *S. euxina* I. Belyaeva, a European introduction, is the most commonly cultivated and naturalized tree-willow in the flora area. It is characterized by: trees, 3–20 m, stems erect or drooping; branches highly brittle at base; petioles with spherical or foliaceous glands distally, pilose or villous adaxially; largest medial leaf blade amphistomatous, very narrowly elliptic or narrowly elliptic, margins uniformly serrate or serrulate, abaxial surface glaucous, both surfaces sparsely long-silky to glabrescent, adaxial surface slightly glossy or dull; juvenile leaves at first densely long-silky soon glabrous; pistillate bract deciduous after flowering; stamens 2; anthers yellow; pistillate adaxial nectary shorter than or equal to stipe; stipe 0.3–0.5 mm; ovary pyriform, glabrous; ovules 6–12 per ovary; styles 0.4–1 mm; capsules 4.5–6 mm; $2n = 57, 76$. Flowering is in late May–early June. Individual trees can persist for years by trunk suckering and spread vegetatively by shoot fragmentation along stream margins, shingle and sand beaches, sedge meadows, hardwood forests, and sand pits. It occurs from 0 to 2500 m in Alberta, British Columbia, Manitoba, New Brunswick, Newfoundland, Nova Scotia, Ontario, Prince Edward Island, Quebec, Saskatchewan; Alaska, Arizona, Arkansas, California, Colorado, Connecticut, Delaware, District of Columbia, Georgia, Idaho, Illinois, Indiana, Iowa, Kentucky, Maine, Maryland, Massachusetts, Michigan, Minnesota, Missouri, Montana, Nebraska, Nevada, New Hampshire, New Jersey, New Mexico, New York, Ohio, Oregon, Pennsylvania, Rhode Island, South Dakota, Tennessee, Utah, Vermont, Virginia, Washington, West Virginia, Wisconsin, and Wyoming.

A study of *Salix* ×*fragilis* in Colorado (as *S.* ×*rubens*) showed that 2172 of 2175 trees were pistillate. Occasionally seed was set, possibly fertilized by *S. alba* (P. B. Shafroth et al. 1994). There are at least five clones of *S.* ×*fragilis* (as *S.* ×*rubens*) in cultivation (T. Berg in B. Jonsell and T. Karlsson 2000+, vol. 1); the pistillate are sterile but the staminate produce viable pollen. The hybrid plants are often misidentified as *S.* "*fragilis*" or as *S. nigra*. In the flora area, reproduction of the hybrid seems to be mainly by stem fragmentation.

Prior to the lectotypification of *Salix fragilis* Linnaeus and the description of *S. euxina* (I. V. Belyaeva 2009), the name *S.* "*fragilis*" was often inadvertently used for both the pure species and for its hybrids with *S. alba*. Thus all herbarium specimens under the names "*fragilis*" and "×*rubens*" need to be revised.

Salix ×*fragilis* can be separated from *S. euxina* by having branches and branchlets hairy or glabrescent in age versus glabrous; leaf blades not glaucous abaxially versus glaucous; leaves amphistomatous versus hypostomatous or with stomata only along veins and at apex; and pistillate catkins slender and loosely flowered versus stout and moderately densely flowered.

Several molecular studies have been designed to understand the nature of this hybrid. H. Beissmann et al. (1997), using AFLP markers, were able to recognize three clusters: *Salix alba*, *S. euxina* (as *S. fragilis*), and *S.* ×*fragilis* (as *S.* ×*rubens*); but a study by K. De Cock et al. (2003), also using AFLP markers, was unable to resolve *S. alba* and *S.* ×*fragilis* (as *S.* ×*rubens*). They recommended the use of experimental hybridization to study the genesis of this hybrid. Molecular and genetic studies by L. L. Triest (2001) and coworkers concluded that in modern open agricultural situations in Belgium, hybridization was of low occurrence, and that morphologically intermediate plants were not necessarily genetically intermediate. These studies saw different facets of the question. Clearly there are three entities, *S. alba*, *S. euxina*, and their hybrid but, because *S. euxina* may be rare outside of cultivation, natural hybridization may not occur and the question of whether *S.* ×*fragilis* can be backcrossed with *S. alba* remains to be studied. The specimens used in these molecular studies require reidentification.

Salix ×*jesupii* Fernald: *S. alba* × *S. lucida*; *S.* ×*ehrhartiana* of authors, not G. Meyer. The origin of this hybrid, between a European and a native species, is unknown. It is characterized by: shrubs or trees, 7–10 m; branchlets red-brown; petioles convex to shallowly grooved adaxially, 3–11 mm, with pairs or clusters of spherical or stalked glands distally; largest medial leaf blade amphistomatous or hemiamphistomatous, abaxial surface glaucous, sparsely long-silky to glabrescent, hairs white and ferruginous; floral bract apex acute to rounded, toothed or entire; pistillate flowering branchlet 8–16 mm, bract deciduous after flowering; stamens 3–5; stipes 0.3–0.6 mm; ovary pyriform, glabrous; ovules 12–14 per ovary; styles 0.3–0.8 mm; capsules 4–5 mm. Flowering is late April–mid May. It occurs in the flora area on the edges of streams and lakes, in wet deciduous woods, sand dunes, and wet railroad rights-of-way, at 10–300 m, in Ontario, Quebec, Prince Edward Island, Saskatchewan; Illinois, Kentucky, Massachusetts, New Hampshire, New York, Ohio, Pennsylvania, Vermont, Virginia, Washington, West Virginia, and Wisconsin. It is possibly introduced in Washington.

Salix ×*jesupii* was named by M. S. Bebb (1895) as a formula hybrid, *S. alba* × *S. lucida*, and Fernald based his binomial on Bebb's exsiccatae. In North America, it was mistaken for the European *S.* ×*ehrhartiana* G. Meyer (*S. alba* Linnaeus × *S. pentandra* Linnaeus) (G. W. Argus 1986). Study of nectary morphology confirmed that the North American plants were not the same as the European hybrid. In herbaria, plants of *S.* ×*jesupii* often are misidentified as *S. alba*, *S. euxina*, *S.* ×*fragilis*, or *S. lucida*.

10. **Salix euxina** I. V. Belyaeva, Taxon 58: 1345. 2009 • Crack or brittle willow [I]

Salix fragilis Linnaeus var. *sphaerica* Hryniewiecki

Trees, 6–18 m. **Stems:** branches highly brittle at base, yellow-green, yellow-brown, or gray-brown, (highly glossy), glabrous; branchlets yellow-green, yellow-brown, or red-brown, (highly glossy), glabrous. **Leaves:** stipules rudimentary on early ones, rudimentary or foliaceous on late ones, (early deciduous to marcescent), apex acuminate or acute; petiole deeply grooved adaxially (margins usually touching), 6–8.6 mm, with pairs of spherical glands distally, glabrous or puberulent adaxially; largest medial blade hemiamphistomatous or hypostomatous, lanceolate, narrowly oblong, or narrowly elliptic, 60–120 × 14–30 mm, 3.6–6.5 times as long as wide, base convex or rounded, margins flat or slightly revolute, serrate, crenate, or crenulate, apex acute, acuminate, or caudate, abaxial surface (not glaucous), glabrous, adaxial slightly or highly glossy, glabrous; proximal blade margins entire; juvenile blade yellowish green or reddish, glabrous. **Catkins:** staminate 24–40 × 7–15 mm, flowering branchlet 12–15 mm; pistillate moderately densely flowered, stout, 24–35 mm, flowering branchlet 12–15 mm; floral bract (greenish or tawny), 0.8–1.3 mm, apex convex or rounded, entire or erose, abaxially hairy, hairs straight or wavy. **Staminate flowers:** abaxial nectary present, adaxial nectary oblong, 0.3–0.6 mm, nectaries distinct; filaments distinct or connate less than ½ their lengths, hairy at base; anthers ellipsoid, 0.4–0.5 mm. **Pistillate**

flowers: adaxial nectary oblong or square, 0.3–0.6 mm, shorter than stipe; stipe 0.6–1.2 mm; ovary pyriform, beak gradually tapering to styles; styles connate, 0.3–0.5 mm; stigmas flat, abaxially non-papillate with rounded tip, or 2 plump lobes, 0.3–0.5 mm. **Capsules** 3–5 mm. $2n = 76$.

Flowering late spring. Cultivated or riparian; introduced; Ont., Que.; n, nw Asia; introduced also in Europe.

Distribution of *Salix euxina* (formerly *S. fragilis*) in the flora area is uncertain. It may occur throughout southern Canada and the United States; often known as the cultivar "Bullata," it rarely escapes. *Salix euxina* is native to the northern Black Sea and the Transcaucausian regions.

2b3. SALIX Linnaeus (subg. SALIX) sect. SALICASTER Dumortier, Fl. Belg., 14. 1827

Salix sect. *Pentandrae* (Borrer) C. K. Schneider

Shrubs or trees, 1–15 m. **Stems:** branches somewhat to highly brittle at base or flexible. **Leaves:** stipules on late ones and vigorous shoots foliaceous or minute rudiments, apex convex to rounded; petiole with spherical glands or lobes distally or throughout; largest medial blade hypostomatous, hemiamphistomatous, or amphistomatous, abaxial surface glaucous or not; juvenile blade glabrous, villous, or long-silky abaxially, hairs white or white and ferruginous. **Pistillate floral bracts** deciduous after flowering. **Staminate flowers:** stamens 3–10; anthers (dry), 0.5–1 mm. **Pistillate flowers:** adaxial nectary shorter than or equal to stipe; stipe 0.5–4 mm; ovary glabrous.

Species 9 (4 in the flora): North America, Eurasia.

11. **Salix pentandra** Linnaeus, Sp. Pl. 2: 1016. 1753
• Bay-leaf or bay or laurel willow I W

Shrubs or trees, 5–15 m. **Stems:** branches flexible at base, brownish or yellow-green, highly glossy, glabrous; branchlets yellow-green, red-brown, or brownish, glossy, glabrous. **Leaves:** stipules absent or rudimentary on early ones, rudimentary or foliaceous on late ones, apex rounded; petiole deeply to shallowly grooved adaxially, 5–15 mm, with pairs or clusters of spherical glands distally or throughout, glabrous adaxially; largest medial blade hypostomatous, narrowly elliptic, elliptic, or lanceolate, 50–135 × 20–50 mm, 2–4 times as long as wide, base convex, margins slightly revolute or flat, serrulate, apex acuminate, abaxial surface pale not glaucous, glabrous, adaxial highly glossy, glabrous; proximal blade margins entire or serrulate; juvenile blade reddish, glabrous abaxially. **Catkins:** staminate 27–81 × 9–13 mm, flowering branchlet 9–21 mm; pistillate moderately to densely flowered, slender or stout, 29–68 × 7–15 mm, flowering branchlet 9–42 mm; floral bract 2–4 mm, apex acute or rounded to truncate, entire or toothed, abaxially sparsely hairy (mainly proximally), hairs wavy or straight. **Staminate flowers:** abaxial nectary 0.6–1.7 mm, adaxial nectary square, ovate, or oblong, 0.5–1.5 mm, nectaries distinct or ± connate and cup-shaped; stamens 4–10; filaments distinct, hairy on proximal ½; anthers ellipsoid or globose, 0.5–0.6 mm. **Pistillate flowers:** (abaxial nectary present or absent), adaxial nectary oblong, square, or ovate, 0.4–0.8 mm, (nectaries distinct or connate and shallowly cup-shaped), shorter than or equal to stipe; stipe 0.5–1.6 mm; ovary pyriform, beak bulged below or tapering to styles; ovules 18–22 per ovary; styles connate or distinct, 0.4–0.6 mm; stigmas flat, abaxially non-papillate with rounded tip, or slenderly cylindrical, 0.4–0.6 mm. **Capsules** 6–9 mm. $2n = 76$.

Flowering late May–mid Jun. Shores of streams and lakes, marshes, roadsides, waste places; 0–2300 m; introduced; Alta., B.C., Man., N.B., Nfld. and Labr. (Nfld.), N.S., Ont., Que., Sask.; Alaska, Colo., Conn., D.C., Ill., Iowa, Ky., Maine, Md., Mass., Minn., Mont., Nebr., N.H., N.J., N.Y., N.C., Ohio, Pa., R.I., S.Dak., Vt., Va., Wis., Wyo.; Eurasia.

The Ohio occurrence is based on information from T. Cooperrider (pers. comm.).

Only pistillate plants of *Salix pentandra* are known to occur in the flora area.

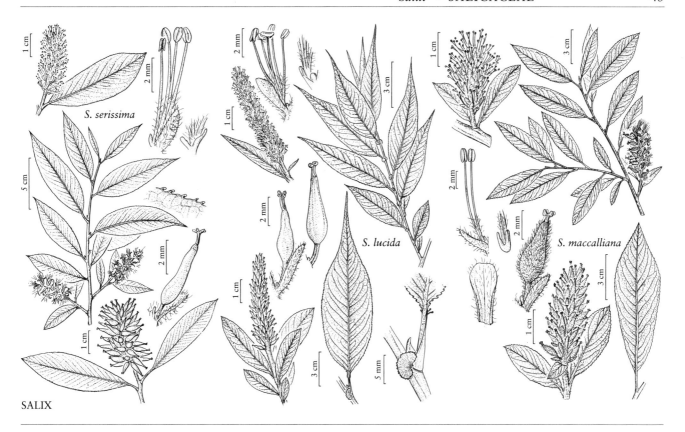

12. Salix serissima (L. H. Bailey) Fernald, Rhodora 6: 6. 1903 • Autumn willow E F

Salix lucida Muhlenberg var. serissima L. H. Bailey, Bull. Geol. Nat. Hist. Surv. Minnesota 3: 19. 1887

Shrubs, 1–5 m. **Stems**: branches usually flexible at base, sometimes brittle, yellow-brown, red-brown, or gray-brown, glabrous, slightly glossy or dull; branchlets yellow-brown or red-brown, glabrous, slightly or highly glossy. **Leaves**: stipules absent or rudimentary; petiole shallowly to deeply grooved adaxially, 3–13 mm, with pairs of spherical glands distally or throughout, glabrous adaxially; largest medial blade hypostomatous or hemiamphistomatous, narrowly oblong, very narrowly elliptic, elliptic, lanceolate, or narrowly ovate, 43–110 × 9–33 mm, 2.4–6 times as long as wide, base convex or cuneate, margins flat, serrulate, apex acuminate, caudate, or acute, abaxial surface usually not glaucous, sometimes thinly so (appearing pale green), slightly glossy, glabrous, adaxial highly glossy, glabrous; proximal blade margins serrulate or entire; juvenile blade reddish or yellowish green, glabrous abaxially. **Catkins**: staminate (stout), 25–53 × 12–16 mm, flowering branchlet 5–14 mm; pistillate (fruiting in autumn, often persistent) moderately densely to loosely flowered, stout to globose, 17–42(–65 in fruit) × 11–22 mm, flowering branchlet 5–32(–65 in fruit) mm; floral bract (sometimes greenish tawny), 1.2–4 mm, apex acute, rounded, or truncate, glandular-toothed, abaxially moderately densely hairy, hairs straight or wavy. **Staminate flowers**: abaxial nectary 0.5–1.1 mm, adaxial nectary oblong or ovate, 0.4–1.1 mm, nectaries distinct or connate and cup-shaped; stamens 3–9; filaments distinct or basally connate, hairy on proximal ½ or basally; anthers ellipsoid or shortly cylindrical, 0.5–0.7 mm. **Pistillate flowers**: adaxial nectary ovate, oblong, or flask-shaped, 0.3–1.1 mm, shorter than stipe; stipe 1.2–2.4 mm; ovary pyriform to obclavate, beak slightly bulged below or abruptly tapering to styles; ovules 12–16 per ovary; styles connate, 0.3–1 mm; stigmas flat, abaxially non-papillate with rounded tip, or slenderly cylindrical, 0.4–0.7 mm. **Capsules** 7–12 mm. $2n = 76$.

Flowering early Jun–early Jul. Wet thickets, fens, brackish marshy strands, marly lakeshores, treed bogs, gravelly stream banks, lakeshores; 10–3000 m; Alta., B.C., Man., N.B., Nfld. and Labr., N.W.T., Nunavut, Que., Sask.; Colo., Conn., Ill., Ind., Mass., Mich., Minn., Mont., N.J., N.Y., N.Dak., Ohio, Pa., S.Dak., Vt., Wis., Wyo.

Salix serissima is found in Nunavut only on Akimiski Island in James Bay.

Flowering of *Salix serissima* is often described as serotinous (i.e., long after leaves emerge), but actually, they flower just as leaves emerge. Although they flower only a little later in spring than related species, they set fruit in late summer, and fruiting catkins often persist throughout winter. Their seeds remain dormant throughout the winter and germinate in the spring, thus enabling them to invade fens by completing their first annual growth before the sedges and grasses are tall enough to shade them out. This strategy has been reported to occur also in the related *S. pentandra* (A. K. Skvortsov 1999).

North American *Salix serissima* is closely related to Eurasian *S. pseudopentandra* (Floderus) Floderus (A. K. Skvortsov 1999), which is known in China as *S. pentandra* var. *intermedia* Nakai and possibly also *S. humaensis* Y. L. Chou & R. C. Chou (Fang Z. F. et al. 1999). The relationship of *S. serissima* and *S. pseudopentandra* is similar to that of *S. arbusculoides* and *S. boganidensis* (G. W. Argus 1997). These two species, along with the amphiberingian *S. vestita*, are relictual members of former panboreal distributions.

Hybrids:

Hybrids between *Salix lucida* and *S. serissima* have been reported (M. L. Fernald 1950); no convincing specimens have been seen.

13. **Salix lucida** Muhlenberg, Ges. Naturf. Freunde Berlin Neue Schriften 4: 239, plate 6, fig. 7. 1803 • Shining willow E F

Pleiarina lucida (Muhlenberg) N. Chao & G. T. Gong; *Salix lucida* var. *angustifolia* (Andersson) Andersson; *S. lucida* var. *intonsa* Fernald

Shrubs or trees, 4–6 m. Stems: branches flexible to highly brittle at base, yellow-brown, gray-brown, or red-brown, slightly to highly glossy, glabrous or villous to glabrescent; branchlets yellow-brown, gray-brown, or red-brown, glabrous, pilose, densely villous, or velvety, hairs spreading, straight, wavy, or crinkled. **Leaves:** stipules foliaceous, apex convex to rounded; petiole shallowly to deeply grooved adaxially, 5–13 mm, with clusters of spherical or foliaceous glands distally, glabrous, pilose, or densely villous adaxially; largest medial blade usually hypostomatous or hemiamphistomatous, rarely amphistomatous, lorate, very narrowly elliptic, narrowly elliptic, or lanceolate, (24–)55–133 × 11–43 mm, 2.5–6.2 times as long as wide, base convex or cuneate, margins flat, serrulate, apex acuminate to caudate, abaxial surface usually not glaucous (rarely so), glabrous, pilose, or moderately densely villous or long-silky, hairs appressed or spreading, white and/or ferruginous, straight or wavy, (coarse, caducous), adaxial (secondary veins flat or protruding), slightly or highly glossy, glabrous, pilose, or long-silky, hairs white and/or ferruginous; proximal blade margins entire and glandular-dotted, or serrulate or crenulate; juvenile blade reddish or yellowish green, glabrous or densely villous or long-silky abaxially, hairs white and ferruginous. **Catkins:** staminate 19–69 × 4–14 mm, flowering branchlet 5–23 mm; pistillate (fruiting in summer), moderately densely to loosely flowered, slender to stout, 23–56(–70 in fruit) × 8–12 mm, flowering branchlet 8–25 mm; floral bract 1.5–3 mm, apex convex or rounded, entire or toothed, abaxially sparsely hairy throughout or proximally, hairs wavy. **Staminate flowers:** abaxial nectary 0.5–1.1 mm, adaxial nectary square or ovate, 0.3–0.9 mm, nectaries connate and cup-shaped; stamens 3–6; filaments distinct, hairy on proximal ½ or basally; anthers ellipsoid, shortly cylindrical, obovoid, or globose, 0.6–0.8 mm. **Pistillate flowers:** adaxial nectary square or ovate, (swollen), 0.2–0.5 mm, shorter than stipe; stipe 0.5–2 mm; ovary pyriform, beak slightly bulged below or gradually tapering to styles; ovules 18–24 per ovary; styles connate or distinct ½ their lengths, 0.5–0.8 mm; stigmas flat, abaxially non-papillate with rounded tip, broadly cylindrical, or 2 plump lobes, 0.24–0.31–0.42 mm. **Capsules** 5–7 mm. $2n = 76$.

Flowering early May–mid Jul. Sandy or gravelly floodplains, lake margins, sedge meadows, vernal pools, alvars, open fens, marl bogs, treed bogs; 0–600 m; St. Pierre and Miquelon; Man., N.B., Nfld. and Labr., N.S., Ont., P.E.I., Que., Sask.; Conn., Del., Ill., Ind., Iowa, Maine, Md., Mass., Mich., Minn., N.H., N.J., N.Y., N.Dak., Ohio, Pa., R.I., S.Dak., Vt., Va., W.Va., Wis.

The Virginia plants of *Salix lucida* are introduced (G. W. Argus 1986).

The *Salix lucida* complex is a group of three weakly delimited taxa, *S. lasiandra* var. *caudata*, *S. lasiandra* var. *lasiandra*, and *S. lucida*. The morphological characters used to separate them (leaves amphistomatous or hypostomatous and blades glaucous abaxially or not) are usually geographically correlated, but there are exceptions. G. W. Argus (1986b) proposed, based on principal components analysis of morphological data, to treat them as a single species consisting of three subspecies. The geographic overlap of the northeastern *S. lucida* and the western *S. lasiandra* is a relatively small area in central Saskatchewan. Evidence of intergradation was based on cultivation of a plant that, in the wild, had leaves that were not glaucous abaxially but were glaucous in cultivation. It seems best to treat them as two species, *S. lucida* and *S. lasiandra*, the latter with two varieties, var. *lasiandra* and var. *caudata*.

Hybrids:

Salix lucida forms natural hybrids with *S. alba* and *S. nigra*. Hybrids with *S. serissima* have been reported (M. L. Fernald 1950) but no convincing specimens have been seen. Attempts to hybridize *S. lucida* with members of subg. *Protitea* (*S. amygdaloides*), subg. *Longifoliae* (*S. interior*), and subg. *Vetrix* (*S. discolor, S. eriocephala*, and *S. petiolaris*) were unsuccessful (A. Mosseler 1990).

Salix lucida × *S. nigra* (*S.* ×*schneideri* B. Boivin) seems to be a rare intersubgeneric hybrid between tetraploid *S. lucida* and diploid *S. nigra*. It is known only from the type specimen, an infertile plant, growing with both parents. It resembles *S. lucida* in bud-scale margins connate, in petiolar glands stalked or foliaceous, and in leaf shape, and *S. nigra* in stipules rudimentary on proximal leaves and sometimes even on early leaves, stipule apex acute, pistillate catkins relatively long and slender, and styles relatively short.

14. Salix lasiandra Bentham, Pl. Hartw., 335. 1857 [E]

Salix lucida Muhlenberg subsp. *lasiandra* (Bentham) A. E. Murray

Shrubs or trees, 1–9(–11) m. **Stems:** branches flexible to highly brittle at base, yellow-brown, gray-brown, or red-brown, slightly to highly glossy, glabrous or pilose to glabrescent; branchlets yellow-brown, gray-brown, or red-brown, glabrous, pilose, villous, or velvety, hairs spreading, straight, wavy, or crinkled, (inner membranaceous bud-scale layer free and separating from outer layer). **Leaves:** stipules usually foliaceous, apex convex or rounded; petiole deeply grooved adaxially, (1–)4–30 mm, with pairs or clusters of spherical or foliaceous glands distally, glabrous or pilose adaxially; largest medial blade hypostomatous, hemiamphistomatous, or amphistomatous, narrowly oblong, very narrowly to narrowly elliptic, narrowly lanceolate to lanceolate, or oblanceolate, 53–170 × 9–31 mm, 3.1–9.8 times as long as wide, base convex or rounded, margins flat, serrulate, apex caudate to acuminate, abaxial surface glaucous or not, glabrescent or pilose, hairs white, sometimes also ferruginous, straight, wavy, or curved, adaxial slightly or highly glossy, glabrous, pilose, or long-silky, hairs white and ferruginous; proximal blade margins entire and glandular-dotted or shallowly serrulate; juvenile blade reddish or yellowish green, moderately to very densely villous, long-silky, or glabrous abaxially, hairs white and ferruginous. **Catkins** (fruiting in summer, persistent); staminate 21–78 × 8–15 mm, flowering branchlet 3–27 mm; pistillate moderately to very densely flowered, slender, stout, or subglobose, 18.5–103 × 6–17 mm, flowering branchlet 6–56 mm; floral bract 1.7–4 mm, apex rounded, entire, toothed, or erose, abaxially hairy throughout or proximally, hairs wavy. **Staminate flowers:** abaxial nectary (usually present), (0–)0.4–0.8 mm, adaxial nectary square or ovate, 0.2–0.6 mm, nectaries distinct or connate and shallowly cup-shaped; stamens 3–6; filaments distinct, hairy on proximal ½ or basally; anthers ellipsoid, shortly cylindrical, obovoid, or globose, 0.6–1 mm. **Pistillate flowers:** adaxial nectary square or ovate, 0.2–0.6 mm, shorter than stipe; stipe 0.8–4 mm; ovary pyriform, beak slightly bulged below or gradually tapering to styles; ovules 16–30 per ovary; styles connate, 0.2–0.8 mm; stigmas broadly cylindrical or 2 plump lobes, 0.2–0.4 mm. **Capsules** 4–11 mm.

Varieties 2 (2 in the flora): w North America.

Varieties of *Salix lasiandra* at the south end of Kootenay Lake, British Columbia, are sympatric. In that area, they differ not only in the usual characteristics, leaf glaucescence and presence of stomata in adaxial epidermis, but plants of var. *lasiandra* are heavily infested with sawfly galls, whereas var. *caudata* are not, and leaves of var. *lasiandra* are stiffer than those of var. *caudata*.

1. Largest medial blades hypostomatous or hemiamphistomatous (rarely amphistomatous), surfaces usually glaucous abaxially (rarely not glaucous), bases convex to rounded; staminate flowers: nectaries distinct; petioles with clusters of spherical or foliaceous glands distally . . . . 14a. *Salix lasiandra* var. *lasiandra*
1. Largest medial blades amphistomatous, surfaces not glaucous abaxially, bases convex; staminate flowers: nectaries usually distinct, sometimes connate and shallowly cup-shaped; petioles with pairs or clusters of spherical glands distally . . . . 14b. *Salix lasiandra* var. *caudata*

14a. Salix lasiandra Bentham var. **lasiandra** • Pacific willow [E]

Salix lasiandra var. *lyallii* Sargent; *S. lasiandra* var. *macrophylla* (Andersson) Little; *S. lasiandra* var. *recomponens* Raup

Leaves: stipules sometimes rudimentary on early ones, foliaceous on late ones, apex convex; petiole (2–)4–30 mm, with clusters of spherical or foliaceous glands distally; largest medial blade usually hypostomatous or hemiamphistomatous (rarely amphistomatous), narrowly oblong, very narrowly elliptic, narrowly elliptic, lanceolate, narrowly lanceolate, or oblanceolate, base convex to rounded, abaxial surface usually glaucous (rarely not glaucous), glabrous or glabrescent, hairs white, sometimes also ferruginous, straight or wavy (secondary veins protruding abaxially and adaxially). **Catkins:** staminate 21–78 × 9–14 mm, flowering branchlet 5–24 mm; pistillate moderately to very densely flowered, slender, stout, or subglobose, 18.5–103 × 6–17 mm, flowering branchlet 6–56 mm; floral bract 1.7–4 mm. **Staminate flowers:** abaxial nectary 0.4–0.6 mm, adaxial nectary 0.2–0.5 mm, nectaries distinct. **Pistillate flowers:** adaxial nectary 0.2–0.6 mm; stipe 0.8–2 mm; ovules 16–24 per ovary; styles 0.2–0.8 mm. $2n = 76$.

Flowering late Mar–early Jun. Thickets, silty, sandy to gravelly alluvium along streams; 0–2700 m; Alta., B.C., Man., N.W.T., Sask., Yukon; Alaska, Ariz., Calif., Colo., Idaho, Mont., Nev., N.Mex., Oreg., Utah, Wash.

Hybrids:

Variety *lasiandra* forms natural hybrids with *Salix eastwoodiae*. Hybrids with *S. gooddingii* have been reported (C. K. Schneider 1921), but no convincing specimens have been seen.

14b. Salix lasiandra Bentham var. **caudata** (Nuttall) Sudworth, Bull. Torrey Bot. Club 20: 43. 1893 • Tail-leaf willow E

Salix pentandra Linnaeus var. *caudata* Nuttall, N. Amer. Sylv. 1: 61, plate 18. 1842; *S. lucida* Muhlenberg subsp. *caudata* (Nuttall) A. E. Murray

Leaves: stipules foliaceous, apex rounded; petiole (1–)4–15 mm, with pairs or clusters of spherical glands distally; largest medial blade amphistomatous, lorate, very narrowly elliptic, narrowly elliptic, or narrowly lanceolate to lanceolate, base convex, abaxial surface not glaucous, pilose, hairs white and ferruginous, straight or curved, (secondary veins protruding abaxially, impressed adaxially). **Catkins:** staminate (stout), 24–50 × 8–15 mm, flowering branchlet 3–27 mm; pistillate moderately densely flowered, slender or stout, 30–63(–70 in fruit) × 9–15 mm, flowering branchlet 10–30 mm; floral bract 2.8–4 mm. **Staminate flowers:** abaxial nectary (0–)0.4–0.8 mm, adaxial nectary 0.3–0.6 mm, nectaries often distinct, sometimes connate and shallowly cup-shaped. **Pistillate flowers:** (abaxial nectary rarely present, then nectaries distinct or connate and shallowly cup-shaped), adaxial nectary 0.2–0.5 mm; stipe 0.8–4 mm; ovules 28–30 per ovary; styles 0.2–0.6 mm. $2n = 76$.

Flowering late May–late Jun. Riparian thickets, openings in woods, silty, sandy to gravelly alluvium, along streams, wet meadows, lakeshores; 30–3100 m; Alta., B.C., N.W.T., Sask., Yukon; Alaska, Calif., Colo., Idaho, Mont., Nev., Oreg., S.Dak., Utah, Wash., Wyo.

2b4. Salix Linnaeus (subg. **Salix**) sect. **Maccallianae** Argus, Salix Alaska Yukon, 38. 1973 E

Shrubs, 1–5 m. **Stems:** branches flexible at base. **Leaves:** stipules on late ones and vigorous shoots foliaceous or minute rudiments, apex acute; petiole not glandular distally; largest medial blade hypostomatous or amphistomatous, abaxial surface not glaucous; juvenile blade densely short-silky or tomentose abaxially, hairs white and ferruginous. **Pistillate floral bracts** persistent after flowering. **Staminate flowers:** stamens 2; anthers 0.8–1 mm. **Pistillate flowers:** adaxial nectary shorter than stipe; stipe 0.8–2 mm; ovary densely villous.

Species 1: n North America.

15. Salix maccalliana Rowlee, Bull. Torrey Bot. Club 34: 158. 1907 • MacCalla's willow [E] [F]

Stems: branches dark red-brown, slightly or highly glossy, glabrous; branchlets red-brown or yellow-brown, puberulent to glabrescent, hairs wavy, curved, straight, or geniculate. **Leaves:** stipules absent or rudimentary on early ones; petiole convex to flat, or shallowly grooved adaxially, 4–15 mm, pilose or pubescent adaxially; largest medial blade lorate or narrowly oblong, 40–85 × 8–25 mm, 2.9–4.8(–5.7) times as long as wide, base convex or cuneate, margins flat, entire, serrulate, or crenate, apex acute to sometimes acuminate, abaxial surface (pale), glabrous or sparsely pubescent, hairs white and ferruginous, straight, relatively short and stiff, adaxial highly glossy, glabrous, puberulent, or sparsely tomentose, hairs white and ferruginous; proximal blade margins serrulate; juvenile blade reddish. **Catkins:** staminate (stout or subglobose), 15.5–42 × 9–16 mm, flowering branchlet 1.5–11 mm; pistillate densely flowered, slender, stout, or subglobose, 25–50 × 10–20 mm, flowering branchlet 3–12 mm; floral bract 1.6–3.6 mm, apex rounded to truncate, entire, abaxially hairy throughout or proximally, hairs white and ferruginous, wavy. **Staminate flowers:** abaxial nectary (0–)0.6–0.9 mm, adaxial nectary oblong or narrowly oblong, 0.5–1 mm, nectaries distinct or connate and cup-shaped; filaments distinct, hairy on proximal ½; anthers (purple turning yellow), ellipsoid or shortly cylindrical. **Pistillate flowers:** (abaxial nectary present), adaxial nectary oblong, 0.4–1 mm, (nectaries distinct); ovary pyriform; ovules 12–16 per ovary; styles connate or distinct ½ their lengths, 0.8–1.2 mm; stigmas flat, abaxially non-papillate with rounded tip, or slenderly cylindrical, 0.3–0.6 mm. **Capsules** 7–11 mm. $2n$ = ca. 190, ca. 228.

Flowering early May–early Jul. Sedge meadows, shrubby fens, marly or bouldery lakeshores, string bogs, treed bogs, *Calamagrostis* grasslands; 0–1500 m; Alta., B.C., Man., N.W.T., Ont., Que., Sask., Yukon; Minn., N.Dak., Wash.

The decaploid to dodecaploid chromosome number for *Salix maccalliana*, highest in the genus, suggests a complex origin. Relationships with subg. *Chamaetia* and subg. *Salix* were suggested by Rowlee and by H. M. Raup (1959). Staminate flowers with abaxial nectaries, tawny and persistent bracts, and villous ovaries suggest a link with *S. glauca*; leaves with coarse, ferruginous hairs and serrate margins suggest *S. lucida* (Rowlee). Although *S. maccalliana* is phenetically closer to sect. *Salicaster* than to (subg. *Chamaetia*) sect. *Glaucae* (G. W. Argus 1997), it is probable that because it incorporates genomes from more than one subgenus, its subgeneric placement is arbitrary.

2b5. SALIX Linnaeus (subg. SALIX) sect. TRIANDRAE Dumortier, Bijdr. Natuurk. Wetensch. 1: 58. 1826 [I]

Salix sect. *Amygdalinae* W. D. J. Koch

Shrubs or trees, 2–7(–10) m. **Stems:** (bark flaking in plates), branches ± brittle at base. **Leaves:** stipules on late ones and vigorous shoots foliaceous, apex acute to acuminate; petiole with paired, clustered, or stalked spherical glands or foliaceous glands distally; largest medial blade hypostomatous, abaxial surface glaucous or not; juvenile blade puberulent or pubescent abaxially, hairs white and ferruginous. **Pistillate floral bracts** persistent or deciduous after flowering. **Staminate flowers:** stamens 3 or, rarely, 2; anthers (dry), 0.4–0.6 mm. **Pistillate flowers:** adaxial nectary shorter than stipe; stipe 1–2 mm; ovary glabrous.

Species 5 (1 in the flora): introduced; Eurasia.

A study of genetic relationships in *Salix* using AFLP (S. Trybush et al. 2008) showed that the genetic similarity of *S. triandra* to subg. *Salix* and subg. *Vetrix* was similar and greater than the genetic similarity between these subgenera. Further study may support treating sect. *Triandrae* as a distinct subgenus.

16. Salix triandra Linnaeus, Sp. Pl. 2: 1016. 1753 · Almond leaf willow [I]

Salix amygdalina Linnaeus; *S. amygdalina* var. *discolor* Wimmer & Grabowski; *S. triandra* subsp. *discolor* (Wimmer & Grabowski) Arcangeli

Stems: branches glabrous or glabrescent; branchlets yellow-brown, red-brown, or brownish, usually glabrous, rarely pilose. **Leaves:** stipules rudimentary to foliaceous on early ones (absent on proximal ones); petiole deeply grooved adaxially, margins covering groove, 4–26 mm, pubescent or puberulent to glabrescent adaxially; largest medial blade oblong, narrowly oblong, narrowly elliptic, elliptic, or lanceolate to obovate, 53–114 × 14–35 mm, 2.7–6.3 times as long as wide, base convex or cuneate, margins flat or slightly revolute, crenate or serrulate, apex acuminate, acute, or ± caudate, abaxial surface glabrous or glabrescent, adaxial dull or slightly glossy, glabrous or glabrescent; proximal blade margins crenate or crenulate; juvenile blade reddish. **Catkins:** staminate 20–60 × 5.5–10 mm, flowering branchlet 3–17 mm; pistillate moderately to very densely flowered, slender to stout, 20–60 × 5–8 mm, flowering branchlet 5–31 mm; floral bract 1–2.3 mm, apex rounded or acute, abaxially hairy (mainly proximally), hairs wavy. **Staminate flowers:** abaxial nectary 0.2–1.1 mm, adaxial nectary oblong, square, or ovate, 0.2–0.6 mm, distinct; filaments distinct, hairy on proximal ½; anthers ellipsoid. **Pistillate flowers:** adaxial nectary obovate to square, 0.3–0.5 mm; ovary pyriform, beak gradually tapering to or slightly bulged below styles; ovules 30–36 per ovary; styles distinct ½ their lengths, 0.2–0.3 mm; stigmas flat, abaxially non-papillate with rounded tip, 0.1–0.2 mm. **Capsules** 3–6 mm. $2n$ = 38 (44), 57, or 88.

Flowering late spring. Stream banks, waste places; 10–40 m; introduced; N.S., Ont.; D.C., Maine, Ohio, Va.; Eurasia.

Salix triandra usually has been overlooked in North American floras. At one time, it was a very important basket willow and probably was introduced into North America for that purpose. Some authors treat the glaucous and nonglaucous forms as subspecies (F. Martini and P. Paiero 1988; K. H. Rechinger 1993); A. K. Skvortsov (1999) noted that, although the two have somewhat distinct ranges, both kinds occur throughout the species and sometimes can be found in the same population. His suggestion that genetic inheritance of this character should be studied has not been taken up. The species is characterized by bark that is dark gray, smooth, and flaking in large irregular patches, as in *Platanus* ×*acerifolia*. The ovary-style transition is so indistinct that styles are often described as absent, but there are two, distinct styles, each terminating in a short stigma. A color change, later in the season, between the styles and ovary suggests that the tip of the ovary and the two distinct styles are both stylar tissues. In general, it appears that the styles are connate proximally and distinct distally.

Collections of *Salix triandra* made in 1934–35 by H. Hyland along the Penobscot River, Orono, Maine, were labeled by him as "introduced," but they could have spread from cultivation or have been naturalized (A. Haines, pers. comm.). Recent naturalized occurrences are known from Toronto, Ontario, and Wolfville, Nova Scotia. Specimens identified as *S. triandra* by C. R. Ball are from Virginia and the District of Columbia. *Salix triandra* is reported to occur in Ohio (T. D. Sydnor and W. F. Cowen 2000) but voucher specimens were not found.

2c. SALIX Linnaeus subg. LONGIFOLIAE (Andersson) Argus, Syst. Bot. Monogr. 52: 57. 1997

Salix [unranked] *Longifoliae* Andersson, Öfvers. Kongl. Vetensk.-Akad. Förh. 15: 116. 1858

Shrubs or trees, 0.5–17 m, clonal by root shoots. **Stems** erect; branches flexible at base, not or weakly glaucous. **Buds** *alba*-type, scale margins connate. **Leaves:** stipules on early ones absent, rudimentary, or foliaceous, on late ones foliaceous or rudimentary, rarely absent, (usually deciduous in autumn); petiole usually shallowly grooved, sometimes flat to convex adaxially, not glandular; largest medial blade usually amphistomatous, sometimes hypostomatous, linear, lorate, narrowly elliptic, or narrowly oblanceolate, 2.8–37.5 times as long as wide, angle of base and of apex less than 90°, surface hairs white; juvenile blade hairs white. **Catkins** flowering as leaves emerge or throughout growing season by syllepsis from lateral buds (branched or unbranched); staminate on flowering branchlet, slender to stout; pistillate on flowering branchlet, loosely to densely flowered, slender or stout; floral bract usually tawny (sometimes brown or greenish), apex entire, toothed, or erose; pistillate bract deciduous after flowering. **Staminate flowers:** abaxial nectary present or absent; stamens 2; filaments distinct, hairy; anthers usually yellow, sometimes reddish turning yellow. **Pistillate flowers:** abaxial nectary sometimes present; ovary not glaucous, hairy or glabrous, beak usually abruptly tapering to or bulged below styles; ovules 12–36 per ovary; styles usually connate, sometimes distinct; stigmas usually flat, abaxially non-papillate with rounded tip, or stigmas slenderly or broadly cylindrical, or 2 plump lobes.

Species 8 (7 in the flora): w North America, Mexico.

Subgeneric rank for *Longifoliae* is based on molecular, anatomical, developmental, chemical, genetic, and morphological evidence. In a molecular study using ribosomal DNA, E. Leskinen and C. Alström-Rapaport (1999) found that *Salix interior* (as *S. exigua*) fell well outside all other *Salix* species included in their study and suggested that it may have diverged early. A study based on chloroplast-encoded *rbc*L gene (T. Azuma et al. 2000) did not strongly support the uniqueness of *S. interior* other than to group it with subg. *Protitea* rather than with subg. *Vetrix* and to suggest that its two-stamened condition was independently derived. A cladistic and genetic distance study based on isozyme data (D. K. X. Chong et al. 1995) showed that *S. interior* (as *S. exigua*) was about equally divergent from both subg. *Protitea* (as subg. *Salix*) and subg. *Vetrix*. An anatomical study by A. K. Skvortsov and M. D. Golysheva (1966) showed that the completely isolateral leaves of *S. interior* and *S. microphylla*, with a bilateral, chlorophyll-deficient, one-layer hypodermis, resemble leaves in subg. *Chosenia* (as *S. chosenia*) and *Populus* subg. *Turanga*. Also, the leaf epidermis of the former, which consists of unequal-sized cells, resembles that of some *Populus* and *Salix* sects. *Humboldtianae* and *Triandrae*. W. Büchler (1996) reported that proximal leaves of *S. exigua* and *S. interior* both have an opposite decussate phyllotaxis, indicating that they are out of place in subg. *Salix* and confirming their morphologically isolated position within Salicaceae. Cyanogenesis, not positively documented for any other *Salix*, was found to occur in living and herbarium material of *S. interior* from central United States (A. M. Brinker et al. 1987). Pollen-stigma incongruity data (A. Mosseler 1989) showed that *S. interior* is more compatible with members of subg. *Vetrix* than with those of subgenera *Protitea* and *Salix*. Mosseler (1990) also found that interspecific hybrids between *S. interior* (as *S. exigua*) and species of subg. *Vetrix* were more viable.

Distinctive morphological characteristics of *Longifoliae* include the presence of root shoots, known in Salicaceae only in some species of *Populus* and in *Salix setchelliana* (subg. *Chamaetia*), branched catkins, in which the proximal two floral bracts, or the leaves on flowering branchlets, subtend secondary catkins (G. W. Argus 1997; A. K. Skvortsov 1999), the frequent production of sylleptic shoots, and the occasional occurrence of tricarpellate ovaries in *S. exigua*. Secondary buds flanking the axillary buds, which appear in some *S. exigua* and *S. thurberi*, may have possible taxonomic significance.

Subgenus *Longifoliae* originated in the New World, probably in riparian habitats in the semiarid regions of Mexico or Central America (S. J. Brunsfeld et al. 1992). Its xeromorphic leaf morphology is highly adaptive in that region (A. K. Skvortsov 1999). Species in Central Asia with similar xeromorphic leaf morphology, such as *Salix linearifolia* Wolf of sect. *Helix*, are cases of convergence.

The species of subg. *Longifoliae* are taxonomically difficult; they seem to form a syngameon of poorly resolved semispecies (V. Grant 1981; S. J. Brunsfeld et al. 1991). These taxa are not only highly variable, but they produce clones through root shoots (rhizoblasts), hybridize and introgress freely, and often produce sylleptic vegetative and reproductive shoots. Sylleptic shoots, arising from buds without a dormant period, can differ morphologically from the primary shoots, which arise from the previous year's buds. In *Salix*, leaves on sylleptic branchlets usually are more densely hairy and more prominently toothed than those of the proleptic primary branchlets. Catkins of subg. *Longifoliae*, described in the literature as borne on relatively long, flowering branchlets, often are sylleptic shoots terminated by a catkin, and the branchlet length given may include the primary branchlets on which these secondary branchlets are borne. Sometimes syllepsis occurs after defoliation by insects, but usually it occurs without the loss of primary leaves. The factors that stimulate syllepsis in subg. *Longifoliae* are unknown, but it is common in *Populus*, where it has been shown to increase light capture and carbon production (R. Ceulemans et al. 1990). Because syllepsis is common and may have an influence on leaf morphology, taxonomists must be careful not to confuse proleptic and sylleptic shoots.

A study of genetic variation by S. J. Brunsfeld et al. (1991) revealed four major elements within subg. *Longifoliae* in North America: 1) *Salix interior*, 2) *S. taxifolia*, 3) the *S. exigua* group, and 4) *S. melanopsis*. Within the *S. exigua* group, they recognized four major geographic entities: *S. hindsiana* (treated here as *S. exigua* var. *hindsiana*), *S. sessilifolia*, and a northern and a southern race in the Intermountain West. The two races were not named but they include *S. exigua* and possibly *S. thurberi*. Brunsfeld et al. also noted that *S. columbiana* (as *S. fluviatilis*) may be of hybrid origin. All of these taxa are recognized here, although the ranks do not always correspond to those proposed. A second study by Brunsfeld et al. (1992) revealed incongruencies between molecular genetic and morphological data, as well as evidence of long-distance gene transfer.

Taxonomic problems in subg. *Longifoliae*, including the practical problem of specimen identification, cannot be solved by field study and herbarium specimens alone. This group requires an interdisciplinary approach including molecular genetics, cytology, and common garden studies where synthetic hybridization and morphological observations of the same plants can be made in all stages of development.

SELECTED REFERENCES Argus, G. W. and C. L. McJannet. 1992. A taxonomic reconsideration of *Salix taxifolia* sensu lato (Salicaceae). Brittonia 44: 461–474. Brunsfeld, S. J., D. E. Soltis, and P. S. Soltis. 1992. Evolutionary patterns and processes in *Salix* section *Longifoliae*: Evidence from chloroplast DNA. Syst. Bot. 17: 239–256. Dorn, R. D. 1998. A taxonomic study of *Salix* section *Longifoliae* (Salicaceae). Brittonia 50: 193–210.

1. Ovaries glabrous or glabrescent.
    2. Largest medial blades lorate, narrowly oblong, narrowly elliptic, narrowly oblanceolate, or linear, 2.4–8(–15.1) times as long as wide; juvenile blades reddish or yellowish green, villous abaxially; proximal blade margins entire or serrulate; branchlets gray-brown to dark red-brown; floral bracts tawny or brown .... 21. *Salix melanopsis*
    2. Largest medial blades linear to lorate, 4.6–37.5 times as long as wide; juvenile blades yellowish green, silky abaxially; proximal blade margins entire; branchlets yellowish, yellow-brown, or red-brown; floral bracts tawny (or greenish).
        3. Largest medial blade margins entire, not glandular-dotted; juvenile blades short-silky abaxially; pistillate catkins densely flowered .... 18. *Salix exigua* (in part)
        3. Largest medial blade margins spinulose-serrulate or, if entire, glandular-dotted; juvenile blades long-silky abaxially; pistillate catkins loosely flowered.
            4. Largest medial blades: abaxial surface usually densely silky or pilose, adaxial sparsely silky to glabrescent, margins usually entire, slightly revolute; petioles pubescent adaxially; staminate adaxial nectaries 0.3–0.8 mm; capsules 3–10 mm. .... 18. *Salix exigua* (in part)
            4. Largest medial blades: abaxial surface usually glabrescent, sometimes densely villous or long-silky, adaxial usually glabrescent, sometimes densely villous to pilose, margins remotely spinulose-serrulate, flat; petioles glabrous or villous adaxially; staminate adaxial nectaries 0.6–1.4 mm; capsules (4–)5–8(–10) mm .... 19. *Salix interior* (in part)

1. Ovaries pilose, villous, or long-silky.
   5. Largest medial blades 5.6–24(–42 in *S. taxifolia*) mm; catkins 4–23 mm.
      6. Largest medial blades lorate, narrowly oblong or narrowly oblanceolate, 2.1–13.3 times as long as wide, margins entire or serrulate; stipules on late leaves foliaceous; stigmas 0.6–0.92 mm; staminate abaxial nectaries absent; stipes 0.2–1.2 mm (see 17. *S. taxifolia*) .... *Salix microphylla* [p. 60]
      6. Largest medial blades linear, lorate, or narrowly oblanceolate, 5.8–24.6 times as long as wide, margins usually entire, rarely spinulose-serrulate; stipules on late leaves absent or rudimentary; stigmas 0.4–0.6 mm; staminate abaxial nectaries 0.1–0.6 mm; stipes 0–0.3 mm ....17. *Salix taxifolia* (in part)
   5. Largest medial blades 30–160 mm; catkins 9–90 mm.
      7. Petioles glabrous, puberulent, pubescent, or villous adaxially.
         8. Branchlets and juvenile blades moderately to densely short-silky-villous; pistillate catkins densely or moderately densely flowered; stipes 0–0.2 mm .... 18. *Salix exigua* (in part)
         8. Branchlets and juvenile blades sparsely to densely long-silky; pistillate catkins loosely flowered; stipes 0.2–0.9.
            9. Largest medial blades: abaxial surface silky or pilose, adaxial sparsely silky to glabrescent, margins usually entire, slightly revolute; petioles pubescent adaxially; staminate adaxial nectaries 0.3–0.8 mm; capsules 4–8 mm . . . . 18. *Salix exigua* (in part)
            9. Largest medial blades: abaxial surface glabrous, glabrescent, densely villous, or long- or short-silky, adaxial usually glabrescent, sometimes moderately densely villous to pilose, margins usually spinulose-serrulate, flat; petioles glabrous or villous adaxially; staminate adaxial nectaries 0.6–1.4 mm; capsules 6–10 mm.
               10. Ovaries glabrous or sparsely long-silky; largest medial blades moderately densely hairy or glabrous; staminate abaxial nectaries present; floral bracts sparsely hairy; capsules (4–)5–8(–10) mm ....19. *Salix interior* (in part)
               10. Ovaries moderately to very densely short-silky or villous; largest medial blades moderately to very densely appressed hairy; staminate abaxial nectaries absent; floral bracts moderately to very densely hairy; capsules (2.5–)4–7 mm .... 20. *Salix thurberi* (in part)
      7. Petioles short- or long-silky adaxially.
         11. Largest medial blades usually less than 42 mm; beak of ovary gradually tapering to style or slightly bulged below style.........................17. *Salix taxifolia* (in part)
         11. Largest medial blades usually greater than 42 mm; beak of ovary abruptly tapering to style.
            12. Plants of Texas .... 20. *Salix thurberi* (in part)
            12. Plants of California, Oregon, Washington, British Columbia.
               13. Petioles long-silky adaxially (hairs 0.5 mm or longer); juvenile blades villous or long-silky; largest medial blades with submarginal glands .... 23. *Salix sessilifolia*
               13. Petioles short-silky adaxially (hairs shorter than 0.5 mm); juvenile blades short-silky; largest medial blades with marginal glands.
                  14. Branchlets pubescent or villous with soft spreading, sometimes appressed, hairs; stigmas slenderly cylindrical; largest medial blade margins entire or remotely spinulose-serrulate; stipes 0–0.2 mm; proximal blade margins entire; shrubs or trees 1–17 m.... .... 18. *Salix exigua* (in part)
                  14. Branchlets glabrous, pubescent or puberulent, hairs appressed; stigmas broadly cylindrical; largest medial blade margins remotely spinulose-serrulate or sinuate; stipes 0.2–0.7 mm; proximal blade margins entire or remotely spinulose-serrulate; shrubs 2–6.5 m .... 22. *Salix columbiana*

17. Salix taxifolia Kunth in A. von Humboldt et al., Nov. Gen. Sp. 2(qto.): 22; 2(fol.): 18. 1817

• Yew-leaf willow

Salix exilifolia Dorn; *S. taxifolia* var. *lejocarpa* Andersson; *S. taxifolia* var. *limitanea* I. M. Johnston; *S. taxifolia* var. *sericocarpa* Andersson

Shrubs or trees, 2–16 m. Stems: branches red-brown, yellow-brown, or gray-brown, hairy or glabrous; branchlets yellow-brown, very densely long-silky, villous, or short-silky to glabrescent. **Leaves:** stipules absent or rudimentary; petiole (sometimes deeply grooved adaxially) 0.2–1.5 mm, long- to short-silky adaxially; largest medial blade linear, lorate, or narrowly oblanceolate, 13–42 × 1.1–4.4 mm, 5.8–24.6 times as long as wide, base cuneate, margins flat, usually entire, rarely remotely spinulose-serrulate, apex acute to acuminate (apiculate), abaxial surface glaucous or not (sometimes obscured by hairs), densely long-silky, or villous to glabrescent, hairs appressed, straight, adaxial slightly glossy or dull, moderately densely long- or short-silky to glabrescent; proximal blade margins entire; juvenile blade yellowish green (color obscured by hairs), very densely long-silky abaxially. **Catkins:** staminate 6.5–18 × 4–7 mm, flowering branchlet 1–38(–75) mm; pistillate moderately densely or loosely flowered, slender or stout, 6–16 × 4–7 mm, flowering branchlet 8–13(–120) mm; floral bract 1.3–2.8 mm, apex acute or convex, toothed or entire (glandular-dotted), abaxially hairy throughout or proximally, hairs straight. **Staminate flowers:** abaxial nectary 0.1–0.6 mm, adaxial nectary oblong or narrowly oblong, 0.5–1.1 mm, nectaries distinct; filaments hairy on proximal ½ or basally; anthers 0.4–0.7 mm. **Pistillate flowers:** adaxial nectary square, narrowly oblong, ovate, or oblong, 0.4–1.4 mm, longer than stipe; stipe 0–0.3 mm; ovary pyriform, long-silky or pilose, beak gradually tapering to or bulged below styles; ovules 16–26 per ovary; styles 0–0.3 mm; stigmas slenderly cylindrical, 0.4–0.6 mm. **Capsules** 3–6 mm.

Flowering throughout year, mostly early Mar–late Jun. Silty to sandy floodplains, gravelly arroyos, dry washes; 400–2000 m; Ariz., N.Mex., Tex.; Mexico (Baja California, Baja California Sur, Chiapas, Chihuahua, Coahuila, Sonora).

The name *Salix taxifolia* is used here in the sense of G. W. Argus and C. L. McJannet (1992). R. D. Dorn (1998) used that name instead for what Argus and McJannet called *S. microphylla*, and for the present species used *S. exilifolia*. Dorn did not cite specimens of *S. exilifolia*, but Texas plants, possibly annotated by him, were mapped by B. L. Turner et al. (2003). The distribution of *S. taxifolia* parallels that of *S. thurberi* Rowlee, recognized here and by Dorn as a distinct species.

18. Salix exigua Nuttall, N. Amer. Sylv. 1: 75. 1842
E F W

Salix fluviatilis Nuttall var. *exigua* (Nuttall) Sargent; *S. longifolia* Muhlenberg var. *exigua* (Nuttall) Bebb

Shrubs or trees, 0.5–5(–17) m. Stems: branches gray-brown, red-brown, or yellow-brown, villous, or tomentose to glabrescent; branchlets yellowish, yellow-brown, or red-brown, pubescent or puberulent, tomentose, short- or long-silky villous. **Leaves:** stipules absent or rudimentary on early ones, foliaceous or rudimentary on late ones; petiole 1–5(–10) mm, pubescent or puberulent, villous, or short-silky adaxially; largest medial blade (sometimes hypostomatous), linear or lorate, (glands marginal), 30–136–143 × 2–14 mm, 6.5–28(–37.5) times as long as wide, base cuneate, margins slightly revolute, entire or remotely spinulose-serrulate, apex acuminate or acute, abaxial surface glaucous (sometimes obscured by hairs), densely long-silky, villous or pilose to glabrescent, hairs appressed or spreading, straight or wavy, adaxial slightly glossy, sparsely or densely long-silky to glabrescent; proximal blade margins entire; juvenile blade yellowish green, densely long-silky-villous abaxially. **Catkins:** staminate 7–54 × 2–10 mm, flowering branchlet 1.5–56 mm; pistillate loosely to densely flowered, slender or stout, 14.5–70 × 3–12 mm, flowering branchlet 2–55 mm; floral bract 1.2–2.6 mm, apex acute, convex, or rounded, entire or erose, hairs wavy, straight, or crinkled, abaxially hairy throughout or proximally, or glabrate. **Staminate flowers:** abaxial and adaxial nectaries present and distinct; filaments hairy on proximal ½; anthers (sometimes reddish turning yellow). **Pistillate flowers:** adaxial nectary oblong, ovate, or flask-shaped, relative adaxial nectary/stipe length variable even within same

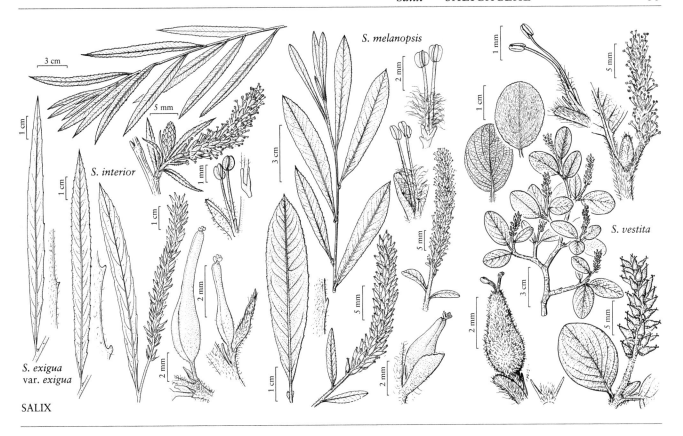

SALIX

catkin; stipe 0–0.9 mm; ovary obclavate or pyriform, glabrous, pilose, or villous, or sometimes beak hairy, beak bulged below or abruptly tapering to styles; ovules 12–30 per ovary; styles (sometimes distinct), 0–0.5 mm; stigmas flat, abaxially non-papillate with rounded tip, or 2 plump lobes, 0.1–0.5 mm. **Capsules** 4–8 mm. $2n = 38$.

Varieties 2 (2 in the flora): w North America.

1. Branchlets densely short-silky-tomentose or short-silky-villous; juvenile blades densely long-silky abaxially, hairs usually appressed, sometimes a few spreading; stipes 0.2–0.9 mm; styles connate, 0–0.2 mm; ovaries usually glabrous, sometimes beak hairy; stigmas flat, abaxially non-papillate with rounded tip, or 2 plump lobes. . . . . 18a. *Salix exigua* var. *exigua*
1. Branchlets pubescent or densely long-silky-villous; juvenile blades densely long-silky-villous abaxially, hairs spreading, appressed; stipes 0–0.2 mm; styles connate or ± distinct, 0.1–0.5 mm; ovaries pilose or villous to glabrescent; stigmas flat, abaxially non-papillate with pointed tip, or slenderly cylindrical . . . 18b. *Salix exigua* var. *hindsiana*

18a. Salix exigua Nuttall var. **exigua** • Narrow-leaf or coyote willow E F

Salix argophylla Nuttall; *S. exigua* var. *luteosericea* (Rydberg) C. K. Schneider; *S. exigua* var. *nevadensis* (S. Watson) C. K. Schneider; *S. exigua* var. *stenophylla* (Rydberg) C. K. Schneider; *S. exigua* var. *virens* Rowlee; *S. fluviatilis* Nuttall var. *argophylla* (Nuttall) Sargent; *S. hindsiana* Bentham var. *tenuifolia* (Andersson) C. K. Schneider; *S. longifolia* Muhlenberg var. *argophylla* (Nuttall) Andersson; *S. longifolia* var. *opaca* Andersson; *S. luteosericea* Rydberg; *S. nevadensis* S. Watson; *S. stenophylla* Rydberg

Shrubs or trees, 0.5–5(–10) m. **Stems:** branches yellow-brown to red-brown; branchlets yellowish, yellow-brown, or red-brown, puberulent, densely short-silky-tomentose or short-silky-villous, hairs appressed, sometimes few spreading. **Leaves:** stipules absent or rudimentary on early ones, foliaceous on late ones; petiole pubescent adaxially; largest medial blade lorate, 30–136–143 × 2–8 mm, 10–16–28 times as long as wide, abaxial surface densely silky-villous or pilose, hairs appressed or spreading, straight or wavy, adaxial sparsely long-silky to glabrescent; juvenile blade very densely long-silky abaxially (hairs usually appressed, sometimes a few spreading). **Catkins:** staminate 13–54 × 2–10 mm, flowering branchlet 1.5–16 mm; pistillate loosely flowered, slender or stout, 14.5–70 × 4–12 mm, flowering branchlet 2–16 mm; floral bract 1.2–1.6 mm, apex rounded, entire, abaxially hairy (or glabrate), hairs wavy or straight. **Staminate flowers:** abaxial nectary 0.2–0.8 mm, adaxial nectary oblong, 0.3–0.8 mm; anthers 0.5–1 mm. **Pistillate flowers:** adaxial nectary oblong or ovate, 0.3–0.9 mm shorter than stipe; stipe 0.2–0.9 mm; ovary obclavate, usually glabrous, sometimes beak pilose, beak bulged below or abruptly tapering to styles; ovules 13–30 per ovary; styles 0–0.2 mm; stigmas flat, abaxially non-papillate with rounded tip, or 2 plump lobes, 0.3–0.5 mm. $2n = 38$.

Flowering early Mar–late Jun or mid Jul. Riparian, bars and shores of streams and lakes, silty, sandy, or gravelly substrates; 600–2800 m; Alta., B.C.; Ariz., Calif., Colo., Idaho, Mont., Nebr., Nev., N.Mex., Okla., Oreg., Tex., Utah, Wash., Wyo.

Hybrids:

Variety *exigua* forms natural hybrids with var. *hindsiana*, *Salix columbiana*, *S. interior*, and *S. melanopsis*.

Variety *exigua* × var. *hindsiana* was reported by R. D. Dorn (1998). Inasmuch as the two varieties are subtly distinct, hybrids are difficult to recognize.

Variety *exigua* × *Salix interior* probably occurs throughout their area of overlap; it is known to me from Alberta and Nebraska, where there are plants with leaves indistinctly toothed and more silky than in *S. interior*. R. D. Dorn (1998) reported it from Alberta, British Columbia, Colorado, Montana, Nebraska, New Mexico, Oklahoma, South Dakota, Texas, and Wyoming.

Variety *exigua* × *Salix melanopsis* is intermediate, with juvenile leaves densely hairy proximally, truncate floral bracts, and prominent stipules (R. D. Dorn 1998).

18b. Salix exigua Nuttall var. **hindsiana** (Bentham) Dorn, Brittonia 50: 203. 1998 • Hind's willow [E]

Salix hindsiana Bentham, Pl. Hartw., 335. 1857; *S. exigua* var. *parishiana* (Rowlee) Jepson; *S. hindsiana* var. *leucodendroides* (Rowlee) C. R. Ball; *S. hindsiana* var. *parishiana* (Rowlee) C. R. Ball; *S. macrostachya* Nuttall var. *leucodendroides* Rowlee; *S. parishiana* Rowlee; *S. sessilifolia* Nuttall var. *hindsiana* (Bentham) Andersson ex Bebb; *S. sessilifolia* var. *leucodendroides* (Rowlee) C. K. Schneider

Shrubs or trees, 1–5(–17) m. **Stems:** branches gray-brown or red-brown, glabrous or hairy; branchlets yellow-brown or red-brown, pubescent or densely long-silky-villous, hairs spreading, appressed. **Leaves:** stipules rudimentary on early ones, foliaceous or rudimentary on late ones; petiole puberulent, villous, or short-silky adaxially; largest medial blade (sometimes hypostomatous) linear, 39–96 × 5.3–14 mm, 6.5–31 times as long as wide, abaxial surface glabrous or densely silky-villous to glabrescent, hairs appressed or spreading, straight, adaxial sparsely to very densely long-silky-villous; juvenile blade densely silky-villous abaxially (hairs relatively long, spreading, appressed). **Catkins:** staminate 7–42 × 4–9 mm, flowering branchlet 2–56 mm; pistillate densely or moderately densely flowered, slender to stout, 22–43 × 3–11 mm, flowering branchlet 2–55 mm; floral bract 1.4–2.6 mm, apex acute or convex, entire or erose, abaxially hairy throughout or proximally, hairs wavy, straight, or crinkled. **Staminate flowers:** abaxial nectary 0.5–0.9 mm, adaxial nectary narrowly oblong, square, or ovate, 0.4–1.1 mm; anthers 0.6–1.1 mm. **Pistillate flowers:** adaxial nectary oblong, ovate, or flask-shaped, 0.4–1.3 mm, longer than stipe; stipe 0–0.2 mm; ovary pyriform, pilose or villous to glabrescent, beak abruptly tapering to styles; ovules 12–24 per ovary; styles (± distinct) 0.2–0.5 mm; stigmas flat, abaxially non-papillate with pointed tip, or slenderly cylindrical, 0.3–1 mm.

Flowering mid Apr–mid May. Riparian, sandy-gravel floodplains; 0–600 m; Calif., Oreg.

Variety *hindsiana* often is not distinguished from var. *exigua*. The two are difficult to separate morphologically even though var. *hindsiana* has been shown to have highly diverged chloroplast DNA (S. J. Brunsfeld et al. 1992). It can be distinguished from var. *exigua* by its soft, spreading hairs on branchlets and juvenile leaves (S. Brunsfeld, pers. comm.).

Hybrids:

Variety *hindsiana* forms natural hybrids with var. *exigua*.

19. Salix interior Rowlee, Bull. Torrey Bot. Club 27: 253. 1900 • Sandbar willow [F][W]

Salix rubra Richardson in J. Franklin, Narr. Journey Polar Sea, 752. 1823, not Hudson 1762; *S. exigua* Nuttall var. *exterior* (Fernald) C. F. Reed; *S. exigua* subsp. *interior* (Rowlee) Cronquist; *S. exigua* var. *pedicellata* (Andersson) Cronquist; *S. exigua* var. *sericans* (Nees) Dorn; *S. fluviatilis* Nuttall var. *sericans* (Nees) B. Boivin; *S. interior* var. *exterior* Fernald; *S. interior* var. *pedicellata* (Andersson) C. R. Ball; *S. interior* var. *wheeleri* Rowlee; *S. linearifolia*

Rydberg; *S. longifolia* Muhlenberg var. *interior* (Rowlee) M. E. Jones; *S. longifolia* var. *pedicellata* Andersson; *S. longifolia* var. *sericans* Nees; *S. longifolia* var. *wheeleri* (Rowlee) C. K. Schneider; *S. wheeleri* (Rowlee) Rydberg

Shrubs or trees, 4–9 m. **Stems:** branches gray-brown to red-brown, glabrous or villous; branchlets yellow brown to red-brown, densely tomentose or villous to glabrescent. **Leaves:** stipules absent or rudimentary on early ones, rudimentary or foliaceous on late ones; petiole 1–5(–9) mm, glabrous or sparsely villous adaxially; largest medial blade linear to lorate, 60–160 × 4–11 mm, (6.5–)11–19(–31) times as long as wide, base cuneate, margins flat, remotely spinulose-serrulate (teeth 2–5 per cm), apex acute or subacuminate, abaxial surface thinly glaucous, densely villous or long-silky to glabrescent, adaxial slightly glossy, pilose or densely villous to glabrescent; proximal blade margins entire; juvenile blade reddish or yellowish green, moderately densely to sparsely long-silky abaxially. **Catkins** (flowering throughout season); staminate 20–61 × 4–10 mm, flowering branchlet 3–20 mm; pistillate loosely flowered, slender or stout, 20–67 × 5–9 mm, flowering branchlet 3–19 mm; floral bract (sometimes greenish), 1.5–3.5 mm, apex acute, acuminate, or rounded, entire, erose, or toothed, abaxially hairy either proximally or distally, hairs wavy. **Staminate flowers:** abaxial nectary 0.5–1.1 mm, adaxial nectary ovate, narrowly oblong, or flask-shaped, 0.6–1.4 mm, nectaries distinct; filaments hairy; anthers 0.4–0.9 mm. **Pistillate flowers:** adaxial nectary narrowly oblong, 0.4–1.1 mm, shorter to longer than stipe; stipe 0.4–0.8 mm; ovary obclavate to pyriform, glabrous, glabrescent, or long-silky, beak abruptly tapering to styles; ovules 16–36 per ovary; styles 0–0.2 mm; stigmas flat, abaxially non-papillate with pointed tip, or broadly cylindrical, 0.3–0.7 mm. **Capsules** (4–)5–8(–10) mm. $2n = 38$.

Flowering early Apr–early Jul. Sandy to silty flood plains, margins of lakes, ponds, and prairie sloughs, dry prairie sand hills, marshes, disturbed areas; 10–1800 m; Alta., B.C., Man., N.B., N.W.T., Ont., Que., Sask., Yukon; Alaska, Ark., Colo., Conn., Del., D.C., Ill., Ind., Iowa, Kans., Ky., La., Maine, Md., Mich., Minn., Miss., Mo., Mont., Nebr., N.J., N.Y., N.Dak., Ohio, Okla., Pa., S.Dak., Tenn., Tex., Va., W.Va., Wis., Wyo.; Mexico (Tamaulipas, Veracruz).

Sometimes *Salix interior* is treated as a subspecies of *S. exigua* (R. D. Dorn 1998). *Salix exigua* and *S. interior* hybridize and apparently intergrade in the western Great Plains; because the area of overlap is relatively small and distinctiveness of the two taxa is not compromised by hybridization and introgression, it is best to treat them as separate species.

Leaves on sylleptic shoots are usually very densely silky. *Salix interior* sometimes has shoots that arise from buds on either side of the normal axillary bud. They do not seem to be directly related to the stipules because they are enclosed by the petiole. Catkins with both staminate and pistillate flowers are rare in *S. interior*, but a Quebec specimen had some catkins predominantly pistillate and others staminate; most were a mixture. The flowers were not teratological, but a mature capsule contained aborted ovules.

Hybrids:

Salix interior forms natural hybrids with *S. exigua* var. *exigua*. Controlled pollinations using *S. interior* (as *S. exigua*) from southern Ontario (A. Mosseler 1990) successfully produced F_1 hybrids with *S. bebbiana*, *S. discolor*, *S. eriocephala*, and *S. petiolaris*. Seed production was usually relatively low, except in crosses with *S. discolor*. In general, F_1 viability was relatively low in crosses with these members of subg. *Vetrix*. No seeds were produced in crosses with members of subgenera *Protitea* or *Salix*. Morphology of the hybrids usually was intermediate between the two parents, but when *S. petiolaris* was used as the maternal parent, the F_1s more closely resembled that species. J. Salick and E. Pfeffer (1999) extended these findings to show that, although crosses between *S. interior* (as *S. exigua*) and *S. eriocephala* are partially sterile, their clonal growth parameters (sprouting, shoot length, and biomass production) are strong and thus permit these partially sterile hybrids to exist as successful individuals and perhaps to "... make a contribution to interspecific gene flow over time." Of particular taxonomic interest is that, in this cross, the staminate parent has a significant influence on leaf shape, whereas in the cross *S. eriocephala* × *S. petiolaris* it is the pistillate parent that is significant for leaf shape. Relatively few hybrids resembling those produced by Mosseler have been recognized in nature, but it is possible that the unusually broadly leaved plants named *S. interior* var. *exterior* and var. *wheeleri*, from northern Maine, Nebraska, New York, and West Virginia, and probably elsewhere, may be hybrids. Phenological isolation may be strong enough to prevent crosses in nature (A. Mosseler and C. S. Papadopol 1989) with the earlier flowering *S. eriocephala* and *S. petiolaris*, a barrier that even an occasional period of overlap cannot breach.

20. Salix thurberi Rowlee, Bull. Torrey Bot. Club 27: 252. 1900 • Thurber's willow

Salix exigua Nuttall var. angustissima (Andersson) Reveal & C. R. Broome; S. interior Rowlee var. angustissima (Andersson) Dayton; S. longifolia Muhlenberg var. angustissima Andersson

Shrubs or trees, 4–10 m. **Stems**: branches red-brown, glabrous or glabrescent; branchlets yellow-green, red-brown, or violet, tomentose or pubescent to glabrescent. **Leaves**: stipules absent or rudimentary; petiole 0–4(–8) mm, pubescent or short-silky adaxially; largest medial blade linear, 66–95(–140) × 2–16 mm, 11–35 times as long as wide, base cuneate, margins flat, remotely spinulose-serrulate, apex acute to subacuminate, abaxial surface very thinly glaucous, sparsely short-silky (especially along midrib), to glabrescent, adaxial slightly glossy, short-silky, pilose to glabrescent; proximal blade margins entire; juvenile blade reddish or yellowish green, moderately densely to sparsely long-silky to glabrescent abaxially. **Catkins**: staminate 8–35 × 6 mm, flowering branchlet 3–55 mm; pistillate loosely flowered, slender, 15–40(–50) × 3.2–8 mm, flowering branchlet 7–28 mm; floral bract 2–4 mm, apex acute or acuminate, entire, abaxially hairy, hairs wavy. **Staminate flowers**: abaxial nectary absent, adaxial nectary ovate, narrowly oblong, or flask-shaped, 0.6–1.4 mm; filament hairy; anthers 0.3–0.8 mm. **Pistillate flowers**: adaxial nectary narrowly oblong to ovate, 0.4–0.7 mm, shorter to longer than stipe; stipe 0–0.8 mm; ovary obclavate to pyriform, densely long-silky or villous, beak abruptly tapering to styles; ovules 16–36 per ovary; styles 0–0.2 mm; stigmas flat, abaxially non-papillate with rounded tip, or broadly cylindrical, 0.3–0.7 mm. **Capsules** (2.5–)4–7 mm.

Flowering Mar–Dec. Sandy to silty floodplains, disturbed areas; 0–1600 m; Tex.; Mexico (Chihuahua, Coahuila, Hidalgo, Nuevo León, Tamaulipas, Veracruz).

The description above is based on specimens supplemented by published descriptions (W. W. Rowlee 1900; C. R. Ball 1961; R. D. Dorn 1998). Ball noted that aside from some characters, Salix thurberi is similar to S. interior. Ball's and Dorn's concepts of the species were not the same because Ball gave the Texas distribution as Brewster, Cameron, El Paso, Guadalupe, Hidalgo, Jeff Davis, Matagorda, Starr, and Val Verde counties, but Dorn recognized it only in Pecos and Val Verde counties.

21. Salix melanopsis Nuttall, N. Amer. Sylv. 1: 78, plate 21. 1842 • Dusky willow [E] [F]

Salix bolanderiana Rowlee; S. exigua Nuttall var. gracilipes (C. R. Ball) Cronquist; S. exigua subsp. melanopsis (Nuttall) Cronquist; S. exigua var. tenerrima (L. F. Henderson) C. K. Schneider; S. fluviatilis Nuttall var. tenerrima (L. F. Henderson) Howell; S. longifolia Muhlenberg var. tenerrima L. F. Henderson; S. melanopsis var. bolanderiana (Rowlee) C. K. Schneider; S. melanopsis var. gracilipes C. R. Ball; S. melanopsis var. kronkheitii Kelso; S. melanopsis var. tenerrima (L. F. Henderson) C. R. Ball; S. parksiana C. R. Ball; S. sessilifolia Nuttall var. vancouverensis Brayshaw; S. tenerrima (L. F. Henderson) A. Heller

Shrubs, 0.8–4 m. **Stems**: branches gray-brown or red-brown, glabrous or hairy; branchlets gray-brown to dark red-brown, glabrous, puberulent, densely long-silky, or villous to glabrescent. **Leaves**: stipules absent, rudimentary, or foliaceous on early ones, foliaceous on late ones (apex acuminate); petiole 1.5–8 mm, glabrous adaxially; largest medial blade lorate, narrowly oblong, narrowly elliptic, narrowly oblanceolate, or linear, 30–133 × 5–20 mm, 3.4–8–15 times as long as wide, base cuneate or convex, margins flat, spinulose-serrulate or entire, apex acute, acuminate, or convex, abaxial surface glaucous or not, pilose, villous, or long-silky to glabrescent, hairs appressed or spreading, wavy, adaxial slightly glossy, villous to glabrescent; proximal blade margins entire or serrulate; juvenile blade reddish or yellowish green, densely villous abaxially. **Catkins**: staminate 18–48 × 5–13 mm, flowering branchlet 3–15 mm; pistillate moderately densely flowered, slender or stout, 22–58 × 4–9 mm, flowering branchlet 4–12 mm; floral bract (sometimes brown), 1.3–2.8 mm, apex rounded (sometimes truncate), entire or erose, abaxially hairy mainly proximally, hairs wavy. **Staminate flowers**: abaxial nectary 0.3–0.9 mm, adaxial nectary narrowly oblong, oblong, or flask-shaped, 0.4–1.2 mm, nectaries distinct; filaments densely hairy on proximal ½; anthers 0.55–0.7–0.9 mm. **Pistillate flowers**: adaxial nectary ovate, oblong, or flask-shaped, 0.4–1.1 mm, longer than stipe, nectaries distinct or connate and cup-shaped; stipe 0–0.7 mm; ovary obclavate or pyriform, glabrous, beak abruptly tapering to styles; ovules 13–22 per ovary; styles 0–0.14–0.5 mm; stigmas slenderly cylindrical or 2 plump lobes, 0.2–0.5 mm. **Capsules** 4–5 mm.

Flowering early May–mid Jul. Riparian, floodplains, stream banks, subalpine meadows, coarse-textured substrates, silt; 600–3100 m; Alta., B.C.; Calif., Colo., Idaho, Mont., Nev., Oreg., Wash., Wyo.

Salix fluviatilis Nuttall, long used for a Columbia River endemic (see 22. S. columbiana), is a rejected name.

Hybrids:

Salix melanopsis forms natural hybrids with *S. exigua* var. *exigua*, *S. sessilifolia*, and *S. sitchensis* (R. D. Dorn 1998).

22. **Salix columbiana** (Dorn) Argus, Harvard Pap. Bot. 12: 359. 2007 • Columbia River willow [E]

Salix exigua Nuttall var. *columbiana* Dorn, Brittonia 50: 204. 1998

Shrubs, 2–6.5 m. **Stems:** branches red-brown, glabrous or hairy at nodes; branchlets yellow-brown or red-brown, pubescent, puberulent, or glabrous, (hairs appressed, straight or wavy). **Leaves:** stipules usually absent or rudimentary, sometimes foliaceous on early ones (apex acute), rudimentary or foliaceous on late ones; petiole 2–5 mm, short-silky adaxially; largest medial blade (sometimes hypostomatous, glands marginal), linear or very narrowly elliptic, 58–115 × 5–17 mm, 5.7–12.8 times as long as wide, base acute or cuneate, margins flat to slightly revolute, remotely spinulose-serrulate or sinuate, apex acuminate, abaxial surface glaucous (sometimes obscured by hairs), sparsely to very densely short- or long-silky, hairs appressed to slightly spreading, straight, adaxial dull, pubescent or long-silky; proximal blade margins entire or remotely denticulate; juvenile blade color obscured by hairs, densely short-silky abaxially. **Catkins:** staminate 20–83 × 5–13 mm, flowering branchlet 6–125(–200) mm; pistillate densely to moderately densely flowered, slender or stout, 35–90 × 5–12 mm, flowering branchlet 11–160 mm; floral bract 1.4–4 mm, apex acute, rounded, retuse, or truncate, entire or toothed, abaxially hairy throughout or proximally, hairs straight or wavy. **Staminate flowers:** abaxial nectary 0.3–0.9 mm, adaxial nectary narrowly oblong, oblong, ovate, or flask-shaped, 0.5–1 mm, nectaries distinct or connate and shallowly cup-shaped; filaments hairy on proximal ½, throughout, or basally; anthers 0.8–1.3 mm. **Pistillate flowers:** (abaxial nectary 0.3–0.4 mm), adaxial nectary narrowly oblong, ovate, or flask-shaped, 0.4–1.3 mm, longer than stipe, nectaries distinct or connate and shallowly or partially cup-shaped; stipe 0.2–0.7 mm; ovary pyriform, long-silky, beak abruptly tapering to styles; ovules 18–30 per ovary; styles 0.1–0.2(–0.4) mm; stigmas flat, abaxially non-papillate with rounded tip, or broadly cylindrical, (0.3–)0.5–1.1 mm. **Capsules** 3.4–5.6 mm. $2n = 38$.

Flowering May–late Jul. Riparian, sandy-silty to sandy gravel floodplains, old beach dunes, rocky fill along streams; 5–40 m; Oreg., Wash.

Hybrids:

Salix columbiana forms natural hybrids with *S. exigua* var. *exigua* and *S. sessilifolia*. Both hybrids are reported from Oregon and Washington (R. D. Dorn 1998).

23. **Salix sessilifolia** Nuttall, N. Amer. Sylv. 1: 68. 1842 • Northwest sandbar willow [E]

Salix exigua Nuttall var. *sessilifolia* (Nuttall) Dorn; *S. fluviatilis* Nuttall var. *sessilifolia* (Nuttall) Scoggan; *S. longifolia* Muhlenberg var. *sessilifolia* (Nuttall) M. E. Jones; *S. macrostachya* Nuttall; *S. macrostachya* var. *cusickii* Rowlee; *S. sessilifolia* var. *villosa* Andersson

Shrubs, 3–5 m. **Stems:** branches gray-brown or red-brown, hairy or glabrous; branchlets red-brown, densely villous. **Leaves:** stipules absent or rudimentary on early ones, rudimentary or foliaceous on late ones (apex acute); petiole 1–6 mm, long-silky adaxially; largest medial blade (sometimes hypostomatous, glands submarginal), linear, lorate, narrowly elliptic, or narrowly oblanceolate, 40–120 × 7.5–16 mm, 3–8.5 times as long as wide, base cuneate or convex, margins flat, remotely spinulose-serrulate, apex acuminate, acute, or caudate, abaxial surface not glaucous, densely to sparsely long- or short-silky or villous, hairs appressed or spreading, straight or wavy, adaxial dull, sparsely long-silky or villous; proximal blade margins entire; juvenile blade yellowish green, very densely long-silky or villous abaxially. **Catkins:** staminate 38–45 × 6–7 mm, flowering branchlet 25–45 mm; pistillate densely flowered, slender, 40–73 × 6–7 mm, flowering branchlet 20–32 mm; floral bract (sometimes brown), 2.4–3.2 mm, apex acute or rounded, entire or irregularly toothed, abaxially hairy, hairs wavy. **Staminate flowers:** abaxial nectary 0.4–1 mm, adaxial nectary narrowly oblong, oblong, or flask-shaped, 0.5–1.1 mm, nectaries distinct; filaments hairy on proximal ½; anthers 0.8–1.2 mm. **Pistillate flowers:** adaxial nectary oblong, square, ovate, or flask-shaped, 0.6–1 mm, longer than stipe; stipe 0.2–0.7 mm; ovary pyriform, villous or long-silky, beak abruptly tapering to styles; ovules 24–36 per ovary; styles 0.1–0.7 mm; stigmas flat, abaxially non-papillate with pointed tip, or slenderly cylindrical, 0.4–1 mm. **Capsules** 4–5.5 mm.

Flowering late Apr–late May. Riparian, streamshores, floodplains, sandy or gravelly substrates; 0–200 m; B.C.; Oreg., Wash.

Hybrids:

Salix sessilifolia forms natural hybrids with *S. columbiana* and *S. melanopsis*.

Excluded species:

Salix microphylla Schlechtendal & Chamisso occurs only in Mexico and Guatemala. (See 17. *S. taxifolia* for comments on its nomenclature.) It is characterized by: shrubs or trees, 0.8–6 m; stipules on late leaves foliaceous; largest medial blade amphistomatous, lorate, narrowly oblong, or narrowly oblanceolate, 5.6–24 × 1.2–3.8 mm, 2.1–13.3 times as long as wide, margins entire or serrulate; staminate abaxial nectary absent; stipes 0.2–1.2 mm; ovary pyriform, long-silky; ovules 21–43 per ovary; style 0.2–0.3 mm; stigmas persistent, slenderly cylindrical lobes, 0.6–0.75–0.92 mm; capsules 4–7 mm (Mexico, Central America [Guatemala]).

2d. SALIX Linnaeus subg. CHAMAETIA (Dumortier) Nasarow in V. L. Komarov et al., Fl. URSS 5: 31. 1936

Salix sect. *Chamaetia* Dumortier, Bijdr. Natuurk. Wetensch. 1: 56. 1826

Shrubs, 0.005–6 m, clonal by layering or rhizomes, rarely root shoots, or not clonal. **Stems** erect, trailing, or decumbent; branches flexible at base, usually not glaucous (usually slightly or highly glossy). **Buds** usually *arctica*-type (*alba*-type in *S. athabascensis*), scale margins connate. **Leaves:** stipules usually absent or rudimentary, sometimes foliaceous; petiole shallowly or deeply grooved, convex, or flat adaxially, often not glandular, sometimes with 1 or 2 pairs of spherical glands distally; largest medial blade amphistomatous, hemiamphistomatous, or hypostomatous, (usually pinnately veined), narrowly to broadly elliptic, obovate, subcircular, circular, oblanceolate, narrowly oblong, oblong, or broadly obovate, 0.8–5.5 times as long as wide, angle of base and of apex usually less than or greater than 90°, (abaxial surface usually glaucous), surface hairs usually white, rarely ferruginous; juvenile blade (usually yellowish green), hairs usually white, rarely ferruginous. **Catkins** flowering as leaves emerge, usually from lateral buds (sometimes subterminal); staminate on flowering branchlet or sessile; pistillate on flowering branchlet, usually stout, globose, or subglobose, sometimes slender; floral bract brown, tawny, or bicolor, apex usually entire, sometimes toothed; pistillate bract usually persistent after flowering. **Staminate flowers:** abaxial nectary present or absent; stamens 2 (1 in *S. uva-ursi*); filaments distinct or connate, glabrous or hairy; anthers usually purple, or red turning yellow. **Pistillate flowers:** abaxial nectary absent or present; ovary usually not glaucous, hairy or glabrous, hairs flattened, ribbonlike, or cylindrical, beak abruptly or gradually tapering to styles or slightly bulged below styles; ovules 4–23 per ovary; styles usually connate; stigmas flat, abaxially non-papillate with rounded tip, or stigmas slenderly or broadly cylindrical, or 2 plump lobes.

Species ca. 133 (27 species in the flora): North America, Eurasia.

Eight of the 14 sections recognized in this subgenus are found in the flora area.

Subgenus *Chamaetia* is difficult to separate from subg. *Vetrix* by morphological characters. It has diverged in habit, which is often dwarfed and rhizomatous; in leaf venation, which often is almost palmate with several veins that diverge from near the base of the blade; in buds, which often contain all the vegetative and reproductive structures that will be produced during the growing season; and less-needed structures such as cataphylls, and in some axillary buds, which are highly reduced. It is possible, as A. K. Skvortsov (1999) pointed out, that this reduction has gone so far that relationships are obscured. The primary characters that distinguish subg. *Chamaetia* are adaptations to arctic-alpine environments that involve simplification through reduction (Skvortsov). Molecular studies have not resolved any species usually placed in subg. *Chamaetia* from subg. *Vetrix* (E. Leskinen and C. Alström-Rapaport 1999; T. Azuma et al. 2000); two unpublished DNA studies also have not revealed a "*Chametia*" clade. A phenetic study (G. W. Argus 1997) revealed a distinct group separated at a level similar to other subgenera. The exception to this was sect. *Glaucae*, which sometimes clustered with subg. *Chamaetia* and other times with subg. *Vetrix*. This inconsistent clustering may be due to high polyploidy in sect. *Glaucae*, which may have evolved through hybridization with members of subg. *Vetrix*, perhaps sect. *Hastatae*. Although it is clear that subg. *Chamaetia* is more closely related to subg. *Vetrix* than to subg. *Salix*, it is highly probable that it is polyphyletic and does not deserve subgeneric rank. Until more information is available, it is taxonomically useful to treat it as a subgenus.

1. Plants 0.08–6 m, (relatively low to tall) not dwarf.
    2. Ovaries glabrous.
        3. Plants forming clones by root shoots; branchlets woolly; largest medial blades amphistomatous [2d2. *Salix* sect. *Setchellianae*] . . . . 27. *Salix setchelliana*
        3. Plants not clonal or forming clones by layering; branchlets glabrous or velvety; largest medial blades hypostomatous [2d7. *Salix* sect. *Myrtilloides* (in part)].
            4. Branches red-brown, weakly glaucous; petioles 1–3.5 mm; stipes 0–0.4 mm; staminate abaxial nectaries present (Gaspe Peninsula, Quebec) . . . . 46. *Salix chlorolepis*
            4. Branches yellow-brown or gray-brown, not glaucous; petioles 2–9 mm; stipes 0.4–3.2 mm; staminate abaxial nectaries present or absent.
                5. Stipules rudimentary; largest medial blades dull, glaucous or sparsely short-silky adaxially; proximal blade margins entire; staminate abaxial nectaries absent; stipes 2.1–3.2 mm; styles 0.1–0.2 mm; pistillate flowering branchlets 7–25 mm; plants clonal by layering . . . . 44. *Salix pedicellaris*
                5. Stipules foliaceous; largest medial blades slightly glossy, not glaucous adaxially; proximal blade margins serrulate; staminate abaxial nectaries present; stipes 0.4–1.2 mm; styles 0.6–0.8 mm; pistillate flowering branchlets 4–7 mm; plants not clonal . . . . 47. *Salix raupii*
    2. Ovaries hairy (sometimes glabrescent in *S. fuscescens*).
        6. Juvenile blades glabrous; branchlets glabrous; largest medial blades glabrous abaxially, margins toothed proximally . . . . 33. *Salix fuscescens*
        6. Juvenile blades hairy; branchlets hairy or glabrescent; largest medial blades hairy abaxially, margins entire or toothed throughout.
            7. Juvenile and largest medial blades: hairs usually white, sometimes also ferruginous; branchlets pubescent . . . . Rapaport . . . . 45. *Salix athabascensis*
            7. Juvenile and largest medial blades: hairs white; branchlets pilose, villous, tomentose, woolly, or long-silky to glabrescent.
                8. Largest medial blades: veins strongly impressed-reticulate adaxially, margins strongly revolute; branches dull; petioles with 2 dark spherical glands distally; catkins from subterminal buds . . . . 24. *Salix vestita*
                8. Largest medial blades: veins not strongly impressed-reticulate adaxially, margins slightly revolute or flat; branches slightly to highly glossy; petioles without glands distally; catkins from lateral buds.
                    9. Petioles 1–27 mm, much longer than subtended buds; stipes 0.3–2.8 mm; largest medial blades narrowly elliptic, elliptic, oblanceolate, or obovate, apex acute, acuminate, convex, or rounded . . . . 50. *Salix glauca*
                    9. Petioles 1.3–5 mm, often shorter than or barely exceeding subtended buds; stipes 0–0.6 mm; largest medial blades narrowly oblong, oblong, narrowly to broadly elliptic, narrowly oblanceolate or oblanceolate, obovate or ovate, apex acuminate, acute, rounded, or convex.

10. Stipules absent or rudimentary on early leaves; ovules 2–10 per ovary; pistillate catkins stout to globose, densely flowered, 0.3–20 mm; floral bracts usually tawny; largest medial blades with flat margins .... 48. *Salix brachycarpa*
10. Stipules foliaceous on early leaves; ovules 8–20 per ovary; pistillate catkins usually slender, sometimes stout or subglobose, moderately densely flowered, 16–69 mm; floral bracts tawny, brown, or black; largest medial blades with slightly revolute margins .... 49. *Salix niphoclada*

1. Plants 0.005–0.15 m, dwarf.
    11. Catkins from subterminal buds.
        12. Largest medial blades glaucous abaxially; staminate abaxial nectaries present.
            13. Plants 3–15 cm; stipules often rudimentary; largest medial blades amphistomatous or hemiamphistomatous, oblong to obovate or circular, usually broader, 1–1.5 times as long as wide, margins crenulate or entire; anthers 0.3–0.4 mm; filaments hairy throughout or on proximal ½; ovaries obclavate to pyriform; capsules 4.5–5 mm .... 25. *Salix reticulata*
            13. Plants 1–4 cm; stipules usually absent; largest medial blades hypostomatous, elliptic to broadly elliptic, usually narrower, 1.1–2.8 times as long as wide, margins entire; anthers 0.4–0.6 mm; filaments glabrous or hairy basally; ovaries obturbinate; capsules 3–4 mm .... 26. *Salix nivalis*
        12. Largest medial blades not glaucous abaxially; staminate abaxial nectaries present or absent.
            14. Largest medial blades: margins crenulate, bases subcordate or cordate; styles 0.2–0.4 mm; pistillate abaxial nectaries present; floral bracts 0.5–1.5 mm .... 28. *Salix herbacea*
            14. Largest medial blades: margins entire, bases cuneate, rounded, or convex; styles 0.5–1.2 mm; pistillate abaxial nectaries absent; floral bracts 1.5–2.8 mm.
                15. Ovaries villous or pilose; leaves deciduous (in autumn); catkins more than 15-flowered; petioles 1.3–10 mm; branches red-brown ...... 30. *Salix polaris* (in part)
                15. Ovaries glabrous or puberulent (hairs in patches, especially on beaks); leaves usually marcescent; catkins 2–15-flowered; petioles 0.4–5.5 mm; branches yellow-green, yellow-brown, or gray-brown .... 31. *Salix rotundifolia*
    11. Catkins from lateral buds.
        16. Ovaries glabrous.
            17. Largest medial blades not glaucous abaxially .... 29. *Salix nummularia*
            17. Largest medial blades glaucous abaxially.
                18. Largest medial blades highly glossy adaxially.
                    19. Stamens usually 1; largest medial blades 4–23 mm; petioles 2–6.5 mm; ovules 4–9 per ovary; ovaries not glaucous; arctic-alpine northeastern North America and Greenland .... 35. *Salix uva-ursi* (in part)
                    19. Stamens 2; largest medial blades 8.5–46 mm; petioles 1.1–20 mm; ovules 10–15 per ovary; ovaries glaucous or not; arctic northwestern North America.
                        20. Branches trailing or erect, not or weakly glaucous; ovaries: beak gradually tapering to styles; styles 0.6–2 mm; largest medial blades amphistomatous or hemiamphistomatous; plants often rhizomatous .... 37. *Salix stolonifera*
                        20. Branches trailing, not glaucous; ovaries: beak abruptly tapering to styles; styles 0.2–0.8 mm; largest medial blades hypostomatous; plants not rhizomatous .... 38. *Salix ovalifolia* (in part)
                18. Largest medial blades slightly glossy adaxially.
                    21. Leaves and stipules marcescent; juvenile leaves yellow-green; stamens 1; staminate abaxial nectaries absent; largest medial blades: margins serrulate or crenulate. .... 35. *Salix uva-ursi* (in part)
                    21. Leaves and stipules deciduous; juvenile leaves reddish; stamens 2; staminate abaxial nectaries present or absent; largest medial blades: margins entire.

22. Staminate catkins 5.5–14 mm; ovaries obnapiform, beak abruptly tapering to styles; Newfoundland . . . . .39. *Salix jejuna* (in part)
22. Staminate catkins 21–53 mm; ovaries obclavate to pyriform, beak gradually tapering to styles or slightly bulged distally; Alaska, Northwest Territories . . . . 42. *Salix sphenophylla* (in part)

[16. Shifted to left margin.—Ed.]

16. Ovaries hairy.
    23. Juvenile leaves reddish.
        24. Staminate catkins 21–53 mm, flowering branchlets 8–20 mm; ovaries: beak gradually tapering to or slightly bulged below styles; largest medial blades 19–52 mm . . . . 42. *Salix sphenophylla* (in part)
        24. Staminate catkins 4.8–15 mm, flowering branchlets 1–7 mm; ovaries: beak abruptly tapering to styles; largest medial blades 8–25 mm.
            25. Capsules 5.2–9.6 mm; largest medial blades hypostomatous, highly glossy adaxially; pistillate abaxial nectaries sometimes present; petioles 1.1–7(–16) mm; Alaska, Northwest Territories, Yukon . . . . 38. *Salix ovalifolia* (in part)
            25. Capsules 3–5 mm; largest medial blades amphistomatous or hemiamphistomatous, slightly glossy adaxially; pistillate abaxial nectaries absent; petioles 1.5–14 mm; Newfoundland . . . . .39. *Salix jejuna* (in part)
    23. Juvenile leaves yellowish green.
        26. Largest medial blades not glaucous abaxially.
            27. Largest medial blades deciduous, slightly glossy adaxially; petioles glabrous adaxially; floral bracts 1.5–2.5 mm; pistillate nectaries longer than stipes . . . . 30. *Salix polaris* (in part)
            27. Largest medial blades marcescent (becoming skeletonized), highly glossy adaxially; petioles hairy adaxially; floral bracts 1–1.3 mm; pistillate nectaries shorter than or equal to stipes . . . . 36. *Salix phlebophylla*
        26. Largest medial blades glaucous abaxially.
            28. Ovary hairs ribbonlike, usually crinkled (refractive); staminate abaxial nectaries absent; largest medial blades: margins closely and prominently serrulate or crenulate (sometimes entire in *S. arctophila*).
                29. Largest medial blades: margins closely and prominently serrulate or spinulose-serrulate (teeth 7–14 per cm); stipes 0.2–0.4 mm; pistillate nectaries equal to or longer than stipes . . . . 32. *Salix chamissonis*
                29. Largest medial blades: margins inconspicuously crenulate or entire (teeth 1–8 per cm); stipes 0.8–1.4 mm; pistillate nectaries shorter than stipes . . . . 34. *Salix arctophila*
            28. Ovary hairs flattened, not crinkled (white, not refractive); staminate abaxial nectaries present or absent; largest medial blades: margins entire.
                30. Largest medial blades 9–26 × 3.8–7.5 mm, petioles 1.5–5 mm; plants forming rhizomatous mats; pistillate catkins 10–23 mm, 15–35(–43)-flowered; stipes 0–0.6 mm; staminate abaxial nectaries present . . . . 43. *Salix cascadensis*
                30. Largest medial blades 10–85 × 5.5–60 mm, petioles 2–35 mm; plants not forming rhizomatous mats (often trailing and rooting); pistillate catkins 20–145 mm, 18–80-flowered; stipes 0.2–1.6 mm; staminate abaxial nectaries present or absent.
                    31. Branchlets usually villous or pilose, sometimes glabrous; floral bracts brown or black; stipules usually foliaceous; stipes 0.2–1.6 mm; styles 0.6–2.2 mm; plants 3–25 cm; arctic and northern cordillera . . . . 40. *Salix arctica*
                    31. Branchlets usually glabrous, sometimes pilose to glabrescent; floral bracts tawny or light brown; stipules absent or rudimentary; stipes 0.2–0.8 mm; styles 0.4–1.6 mm; plants 2–10 cm; southern cordillera . . . . 41. *Salix petrophila*

2d1. SALIX Linnaeus (subg. CHAMAETIA) sect. CHAMAETIA Dumortier, Bijdr. Natuurk. Wetensch. 1: 56. 1826

Plants 0.01–1.5 m, sometimes clonal by layering or rhizomes. **Largest medial blades** hypostomatous, amphistomatous, or hemiamphistomatous, abaxial surface glaucous. **Catkins** from subterminal buds. **Staminate flowers:** filaments glabrous or hairy. **Pistillate flowers:** abaxial nectary sometimes present, then distinct, or connate to adaxial and forming a cup; ovary not glaucous, sparsely to very densely short-silky, hairs white or white and ferruginous, cylindrical or flattened.

Species 4 (3 in the flora): North America, Eurasia.

The branchlets in all members of sect. *Chamaetia* rarely elongate to produce late (neoformed) leaves. Flowering branchlets usually are not differentiated from vegetative branchlets and often have the same number of leaves.

24. Salix vestita Pursh, Fl. Amer. Sept. 2: 610. 1813

• Rock willow [F]

Salix leiolepis Fernald; *S. vestita* subsp. *leiolepis* (Fernald) Argus; *S. vestita* var. *psilophylla* Fernald & H. St. John

Plants 0.2–1.5 m. **Stems** erect; branches brownish or red-brown, (dull), glabrous, long-silky, or villous to glabrescent; branchlets yellow-brown or gray-brown, long-silky, pilose, or moderately densely villous. **Leaves:** stipules absent or rudimentary; petiole (shallowly to deeply grooved adaxially), 2–8 mm, (with 2 spherical glands distally, dark brown, sometimes basilaminar, sparsely pubescent or glabrous adaxially); largest medial blade hypostomatous, (veins strongly impressed-reticulate), broadly elliptic, subcircular, or obovate, 18–67 × 10–40 mm, 1.1–2.3 times as long as wide, base rounded, convex, or subcordate, margins strongly revolute, crenate or subentire, apex rounded, convex, retuse, or toothed, abaxial surface sparsely to densely villous or long-silky, veins often with long, straight hairs, adaxial slightly glossy, glabrous or sparsely long-silky; proximal blade margins entire or crenate; juvenile blade (yellowish green), abaxially very densely long-silky. **Catkins:** staminate 13–48 × 4–6.5(–8) mm, flowering branchlet 3–31(–50) mm; pistillate densely flowered, slender or stout, 18–56 × 4–10 mm, flowering branchlet 3–27(–40) mm; floral bract tawny, 0.8–1.6 mm, apex rounded, entire, abaxially densely hairy, hairs straight. **Staminate flowers:** abaxial nectary 0.6–0.8 mm, adaxial nectary narrowly oblong, 0.5–1.2 mm, nectaries connate and shallowly cup-shaped, or distinct; filaments distinct, hairy on proximal ½; anthers ellipsoid or globose, 0.3–0.5 mm. **Pistillate flowers:** abaxial nectary (0–)0.8–0.9 mm, adaxial nectary oblong, ovate, or narrowly oblong to almost filiform, 0.7–1.4 mm, shorter to longer than stipe, nectaries distinct or connate and cup-shaped; stipe 0.4–1.2 mm; ovary pyriform or obnapiform, densely short-silky, hairs cylindrical, beak abruptly tapering to styles; ovules 13–15 per ovary; styles connate ½ their lengths to almost distinct, 0.2–0.4 mm; stigmas flat, abaxially non-papillate with rounded tip, slenderly cylindrical, or 2 plump lobes, 0.2–0.28–0.36 mm. **Capsules** 3–5 mm. $2n = 38$.

Flowering mid Jun–late Jul. Moist to dry open forests and rocky streamsides, in upper montane and subalpine zones, rarely alpine; 0–2400 m; Alta., B.C., Man., Nfld. and Labr., N.S., Nunavut, Ont., Que.; Mont., Oreg., Wash.; Asia (China [Xinjiang], Mongolia, Russia, e, c Siberia).

Salix vestita is an ancient amphiberingian species. Its distribution includes a series of isolated, disjunct populations in Central Siberia, the northern Rocky Mountains, the west coast of Hudson Bay, and the northeastern arctic and subarctic. Occurrence in Nunavut is on Akpatok Island in Ungava Bay and on the Belcher Islands in Hudson Bay. It may be extirpated in Washington.

The flowering and vegetative branchlets sometimes have relatively short internodes. In subsequent years, branches have the appearance of short shoots similar to those in *Alnus*. Short shoots do not appear on all branches or in all years. The formation of short shoots may be related to adverse growing conditions.

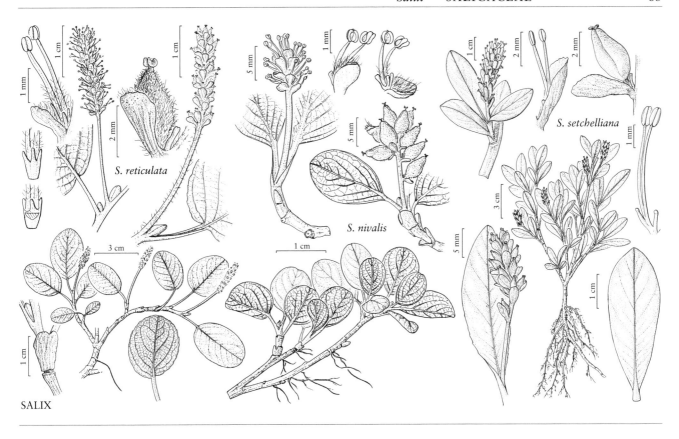

25. Salix reticulata Linnaeus, Sp. Pl. 2: 1018. 1753

• Net-leaf willow F

Salix reticulata var. *gigantifolia* C. R. Ball; *S. reticulata* subsp. *glabellicarpa* Argus; *S. reticulata* var. *semicalva* Fernald

Plants 0.03–0.15 m, (dwarf, forming clones by layering). **Stems** trailing; branches and branchlets yellow-brown or red-brown, glabrous. **Leaves:** stipules absent or rudimentary; petiole 3–46 mm, (sometimes glandular distally); largest medial blade amphistomatous or hemiamphistomatous, (2 pairs of secondary veins arising at or close to base, arcing toward apex), oblong, broadly oblong, broadly elliptic, subcircular, or circular, (8–)12–66 × 8–50 mm, 1–1.5 times as long as wide, base convex, rounded, subcordate, or cordate, margins slightly revolute, entire or crenulate (glandular-dotted), apex rounded, convex, or retuse, abaxial surface sparsely long-silky to glabrescent, adaxial (venation deeply impressed), slightly or highly glossy, glabrous or pilose; proximal blade margins entire; juvenile blade glabrous. **Catkins:** staminate 11–54 × 4–9 mm, flowering branchlet 2–28 mm; pistillate densely flowered (more than 6 flowers), slender or stout, 11–79 × 3–8 mm, flowering branchlet 2–37 mm; floral bract tawny, 0.8–1.8 mm, apex rounded to retuse, entire, abaxially glabrous. **Staminate flowers:** abaxial nectary 0.5–0.9 mm, adaxial nectary oblong or ovate, 0.5–1 mm, nectaries connate and cup-shaped; filaments distinct, hairy on proximal ½ or throughout; anthers ellipsoid or globose, 0.3–0.4 mm. **Pistillate flowers:** abaxial nectary (0–)0.3–0.5 mm, adaxial nectary narrowly oblong, 0.5–1 mm, equal to or longer than stipe, nectaries distinct or connate and cup-shaped; stipe 0–0.8 mm; ovary pyriform or ovoid, short-silky, hairs flattened, beak abruptly tapering to styles; ovules 8–18 per ovary; styles connate to distinct ½ their lengths, 0.2–0.3 mm; stigmas flat, abaxially non-papillate with rounded tip, broadly cylindrical, or 2 plump lobes, 0.2–0.26–0.36 mm. **Capsules** 4.5–5 mm. $2n = 38$.

Flowering early Jun–mid Aug. Arctic-alpine, polygonal tundra, dry tussock tundra, partially stabilized sand dunes, sedge meadows, *Dryas* tundra on alpine cliffs and ledges, snowbeds, stabilized talus slopes, white spruce woods, treed bogs; 0–3500 m; Alta., B.C., Man., Nfld. and Labr., N.W.T., Nunavut, Ont., Que., Sask., Yukon; Alaska, Colo.; Europe; Asia (Chukotka, Russian Far East, arctic, e Siberia, Spitzbergen).

The reported occurrence of *Salix reticulata* in Colorado (R. D. Dorn 1997) needs further study.

Salix reticulata occurs in Europe in northern Scotland, northern Scandinavia, the Alps and other European mountains, and arctic Eurasia. The species is circumpolar except for Greenland and Iceland.

A population of *Salix reticulata* on the Queen Charlotte Islands, with consistently glabrous ovaries, was named subsp. *glabellicarpa*. Some southeastern Alaska populations have plants with glabrous, partially hairy, and completely short-silky ovaries growing together. The possibility that subsp. *glabellicarpa* may be a hybrid or a simple mutation needs study.

26. Salix nivalis Hooker, Fl. Bor.-Amer. 2: 152. 1838 • Dwarf snow willow E F

Salix nivalis var. *saximontana* (Rydberg) C. K. Schneider; *S. reticulata* Linnaeus subsp. *nivalis* (Hooker) Á. Löve, D. Löve & B. M. Kapoor; *S. reticulata* var. *saximontana* (Rydberg) Kelso

Plants 0.01–0.04 m, (dwarf, forming clonal mats by rhizomes). **Stems** trailing or erect; branches yellow-brown or red-brown, glabrous or pubescent; branchlets yellow-brown or red-brown, glabrous or pilose. **Leaves:** stipules absent or rudimentary; petiole 1.5–7 mm (sometimes glandular distally or throughout); largest medial blade hypostomatous, (veins impressed-reticulate, 2 pairs of secondary veins arising at or close to base, arcing toward apex,), elliptic to broadly elliptic, 6–22 × 4–15 mm, 1.1–2.8 times as long as wide, base convex, rounded, subcordate, or cuneate, margins slightly revolute, entire (glandular-dotted), apex convex, rounded, or retuse, abaxial surface glabrous or with long-silky hairs, adaxial slightly glossy, glabrous; proximal blade margins entire; juvenile blade glabrous. **Catkins:** staminate 7–19 × 2.5–6 mm, flowering branchlet 0.5–17 mm; pistillate densely to loosely flowered (4–17 flowers), stout, subglobose or globose, 7–21 × 2–9 mm, flowering branchlet 1–10 mm; floral bract tawny or light rose, 0.8–1.8 mm, apex rounded, entire, abaxially glabrous. **Staminate flowers:** abaxial nectary 0.5–1.3 mm, adaxial nectary narrowly oblong, oblong, or square, 0.5–1.2 mm, nectaries connate and cup-shaped; filaments distinct, glabrous or hairy basally; anthers ellipsoid or shortly cylindrical, 0.4–0.6 mm. **Pistillate flowers:** abaxial nectary (0–)0.2–0.5 mm, adaxial nectary oblong, 0.2–1 mm, longer than stipe, nectaries distinct or connate and shallowly cup-shaped; stipe 0–0.8 mm; ovary obturbinate, short-silky, hairs flattened, beak abruptly tapering to styles; ovules 8–10 per ovary; styles distinct to connate ½ their lengths, 0.2–0.4 mm; stigmas flat, abaxially non-papillate with rounded tip, 0.2–0.26–0.36 mm. **Capsules** 3–4 mm. $2n = 38$.

Flowering late Jun–late Aug. Alpine tundra, cirques, lake basins, rocky slopes and ridges, fellfields; 1900–4000 m; Alta., B.C.; Calif., Colo., Idaho, Mont., Nev., N.Mex., Oreg., Utah, Wash., Wyo.

Because geographic overlap is small and evidence of intergradation is tenuous, *Salix nivalis* is best treated as a species separate from *S. reticulata*; *S. nivalis* was previously treated as a subspecies of *S. reticulata* (G. W. Argus 1986b, 1991).

2d2. SALIX Linnaeus (subg. CHAMAETIA) sect. SETCHELLIANAE Argus, Syst. Bot. Monogr. 52: 62. 1997 E

Plants 0.08–0.3 m, clonal by root shoots. **Largest medial blades** amphistomatous, abaxial surface glaucescent. **Catkins** from lateral buds. **Staminate flowers:** filaments glabrous or hairy. **Pistillate flowers:** abaxial nectary absent; ovary not glaucous, glabrous.

Species 1: nw North America.

27. Salix setchelliana C. R. Ball, Univ. Calif. Publ. Bot. 17: 410, plate 72. 1934 • Setchell's willow E F

Stems erect or semi-prostrate; branches gray-brown or red-brown, glabrous or woolly to glabrescent; branchlets reddish, densely woolly (hairs spreading). **Leaves:** stipules absent or rudimentary; petiole (shallowly grooved adaxially), 0–3 mm; largest medial blade narrowly oblong, narrowly elliptic, elliptic, oblanceolate, or obovate, 25–87 × 10–30 mm, 2–3.9 times as long as wide, base cuneate or convex, margins flat, entire or serrulate, apex rounded to convex, abaxial surface glabrous, adaxial dull, glabrous; proximal blade margins entire or serrulate; juvenile blade glabrous. **Catkins** (sometimes branched); staminate 12–27 × 6–10 mm, flowering branchlet 3–8 mm; pistillate densely flowered, stout or subglobose, 20–34 × 6–13 mm, flowering branchlet 5–19 mm; floral bract tawny or greenish, translucent, 2–3.6 mm, apex rounded or truncate, erose, sinuate, irregularly toothed, or entire, abaxially glabrous. **Staminate flowers:** abaxial nectary 0.4–0.8 mm, adaxial nectary narrowly oblong, oblong, or ovate, 0.6–1 mm, nectaries distinct, or connate and shallowly cup-shaped;

filaments distinct, glabrous or hairy basally or on proximal ½; anthers long-cylindrical, 0.6–0.8 mm. **Pistillate flowers:** adaxial nectary oblong, 0.8–1.3 mm, equal to or longer than stipe; stipe 0–0.6 mm; ovary obclavate or ovoid, beak abruptly tapering to styles; ovules 16–23 per ovary; styles distinct, 0.3–0.4 mm; stigmas flat, abaxially non-papillate with pointed tip, or slenderly cylindrical, 0.32–0.47–0.6 mm. **Capsules** 3.6–10 mm. $2n = 38$.

Flowering late May–late Jun. Pioneer on sandy to gravelly beaches, bars along glacial streams, glacial moraine; 10–1100 m; B.C., Yukon; Alaska.

Salix setchelliana shares some unique characters with members of subg. *Longifoliae* and some *Populus*. It produces shoots from roots, the catkins are sometimes branched, and the leaves are isolateral, with hypodermis present on both sides of the blade (W. Buechler, pers. comm.). While it is possible that these characters evolved independently, it is more likely that they were derived from a common ancestor.

Salix setchelliana is a highly successful colonizer of temporary gravel bar habitats. This is made possible by rapid expansion of its clones by root shoots and its ability to become established in new locations by both seedlings and clonal fragments (D. A. Douglas 1989).

Hybrids:

Salix setchelliana forms natural hybrids with *S. niphoclada*.

2d3. SALIX Linnaeus (subg. CHAMAETIA) sect. HERBELLA Seringe, Exempl. Rév. Salix, 14. 1824

Salix sect. *Retusae* A. Kerner

Plants 0.005–0.09 m, clonal by layering or rhizomes. **Largest medial blades** amphistomatous, abaxial surface not glaucous. **Catkins** from subterminal or lateral buds. **Staminate flowers:** filaments glabrous or hairy. **Pistillate flowers:** abaxial nectary sometimes present, then distinct or connate to adaxial and forming a cup; ovary not glaucous, glabrous, puberulent, or villous to pilose.

Species 7 (4 in the flora): North America, Eurasia, Atlantic Islands.

28. **Salix herbacea** Linnaeus, Sp. Pl. 2: 1018. 1753
 • Snowbed willow [F]

Plants 0.005–0.05 m, (dwarf), forming clonal mats by rhizomes. **Stems** erect; branches red-brown to violet, (sometimes weakly glaucous), glabrous; branchlets yellow-brown or red-brown, glabrous. **Leaves:** stipules absent; petiole (convex or flat to deeply grooved adaxially), 1.5–6(–7) mm; largest medial blade (2 pairs of secondary veins arising at or close to base, arcing) circular, subcircular or broadly elliptic, 6–21(–34) × 6–17(–31) mm, 0.9–1.4 times as long as wide, base usually subcordate or cordate, sometimes convex or rounded, margins flat, crenulate or crenate, apex rounded, convex, retuse, or toothed, abaxial surface (not glaucous), glabrous, adaxial slightly glossy to almost dull, glabrous; proximal blade margins crenulate; juvenile blade glabrous. **Catkins:** from subterminal buds; staminate 3–7.5 × 1.5–5 mm, flowering branchlet 0.3–2 mm; pistillate loosely flowered (2–11 flowers), stout to globose, 3.3–13 × 2–10 mm, flowering branchlet 0.8–3.5 mm; floral bract tawny, light rose, or brown, 0.5–1.5 mm, margins ciliate, apex rounded, retuse, or truncate, entire, abaxially glabrous. **Staminate flowers:** abaxial nectary 0.5–0.8 mm, adaxial nectary oblong or ovate, 0.6–1.1 mm, nectaries distinct, or connate and shallowly cup-shaped; filaments distinct, glabrous, or hairy on proximal ½; anthers shortly cylindrical or globose, 0.3–0.6 mm. **Pistillate flowers:** abaxial nectary (0–)0.2–0.3 mm, adaxial nectary narrowly oblong or oblong, 0.3–1.1 mm, longer or shorter than stipe, nectaries distinct or connate and shallowly cup-shaped; stipe 0.3–1.1 mm; ovary pyriform or ovoid, glabrous, beak abruptly tapering to styles; ovules 11–18 per ovary; styles connate to distinct, 0.2–0.4 mm; stigmas broadly cylindrical or 2 plump lobes, 0.08–0.24–0.32 mm. **Capsules** 2.2–7.5 mm. $2n = 38$.

Flowering late Jun–mid Aug. Snowbeds and places with good snow protection, well-drained riverbanks, sandy beaches, granite boulder ridges, steep bouldery slopes, or in marshes, usually on non-calcareous substrates, places exposed to sea-spray; 0–1700 m; Greenland; Man., Nfld. and Labr., N.W.T., Nunavut, Que.; Maine, N.H.; Europe (British Isles, Russia, Scandinavia, Spitzbergen); Atlantic Islands (Iceland).

Salix herbacea is the only willow with an amphi-Atlantic distribution. Disjunct populations occur as far

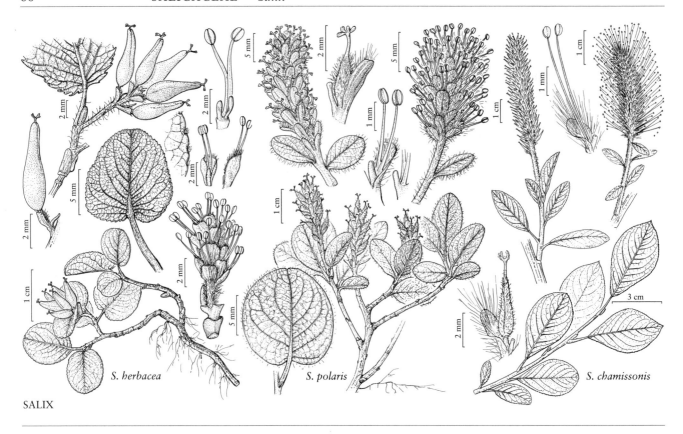

west as Great Bear and Great Slave lakes, Northwest Territories. Macrofossils show that, during the late-Wisconsinan period, it occurred in North America along the glacial margin between Minnesota (R. G. Baker et al. 1999) and Cambridge, Massachusetts (G. W. Argus and M. B. Davis 1962). D. J. Beerling (1998) provided a comprehensive review of its biology and ecology.

Hybrids:

Salix herbacea forms natural hybrids with *S. arctica*, *S. argyrocarpa*, *S. fuscescens*, and *S. uva-ursi*.

Salix herbacea × *S. uva-ursi* (*S.* ×*peasei* Fernald) was described from Mt. Washington, New Hampshire, but occurs also in northern Quebec (G. W. Argus, unpubl.). It is morphologically intermediate between the parents. Its distinctly crenulate, broadly obovate leaves are similar to those of *S. herbacea*, its catkins are smaller and have fewer flowers than those of *S. uva-ursi* but more flowers than those of *S. herbacea*, its leaves are sparsely glaucous abaxially, and it has stems stouter than those of *S. uva-ursi*.

29. **Salix nummularia** Andersson in A. P. de Candolle and A. L. P. P. de Candolle, Prodr. 16(2): 298. 1868 • Coin-leaf willow

Salix nummularia subsp. *tundricola* (Schljakov) Á. Löve & D. Löve

Plants 0.01–0.03 m, (dwarf), forming clones by layering. **Stems** trailing; branches yellow-brown or red-brown, glabrous; branchlets yellow-brown or red-brown, pubescent, pilose, or glabrescent. **Leaves:** stipules absent or rudimentary; petiole 1.5–2 mm (glabrous or pubescent adaxially); largest medial blade (2 pairs of secondary veins arising at or close to base, arcing toward apex) broadly elliptic, subcircular, broadly ovate, or elliptic, 9–22(–30) × 7.5–14(–19) mm, 1.2–2 times as long as wide, base rounded, subcordate, cordate, or convex, margins flat or slightly revolute, entire or serrulate, apex convex, rounded, or retuse, abaxial surface glabrous, adaxial highly glossy, glabrous; proximal blade margins entire or serrulate; juvenile blade pilose or puberulent abaxially. **Catkins:** from lateral buds; staminate (3–8 flowers), 3.2–6.6 × 2–5.2 mm, flowering branchlet

0.8–4.6 mm; pistillate loosely flowered (3–5 flowers), shape indeterminate, 7.5–13 × 3–10 mm, flowering branchlet 0.5–10 mm; floral bract tawny, 0.6–1.4 mm, apex rounded or truncate, entire, abaxially glabrous or sparsely hairy. **Staminate flowers:** abaxial nectary (0.3–)0.5–0.7 mm, adaxial nectary narrowly oblong or oblong, 0.7–1.1 mm, nectaries distinct, or connate and cup-shaped; filaments distinct or connate less than ½ their lengths, glabrous; anthers ellipsoid, 0.4–0.5 mm. **Pistillate flowers:** abaxial nectary (0–)0.6–0.9 mm, adaxial nectary narrowly oblong or oblong, 0.6–1.4 mm, longer than stipe, nectaries connate and shallowly cup-shaped; stipe 0–0.7 mm; ovary pyriform, glabrous, beak slightly bulged below styles; ovules 8–10 per ovary; styles 0.2–1 mm; stigmas flat, abaxially non-papillate with rounded tip, or broadly cylindrical, 0.2–0.27–0.32 mm. **Capsules** 3.5–7.5 mm. $2n$ = 38.

Flowering late Jun–early Aug (based on Russian collections). Exposed, relatively dry, stony, moss-lichen, and moss tundra, polygonal tundra, outcrops, marine sediments, sand dunes, restricted to snow-free areas, usually on acidic substrates; 0–1900 m; Alaska; Asia (China [Jilin], Chukotka, Japan [Hokkaido], North Korea, Russian Far East, arctic, e Siberia).

Salix nummularia occurs in Alaska on St. Paul Island.

30. Salix polaris Wahlenberg, Fl. Lapp., 261, plate 13, fig. 1. 1812 • Polar willow [F]

Salix polaris subsp. *pseudopolaris* (Floderus) Hultén; *S. polaris* var. *selwynensis* Raup

Plants 0.01–0.09 m, (dwarf), forming clones by rhizomes. **Stems** erect; branches red-brown, (often glaucous, dull), glabrous; branchlets brownish, glabrous. **Leaves:** stipules absent or rudimentary; petiole (deeply grooved), 1.3–10 mm, (glabrous adaxially); largest medial blade (deciduous in autumn, 2 pairs of secondary veins arising at or close to base, arcing toward apex), elliptic, broadly elliptic, obovate, or subcircular, 5–32 × 7–18 mm, 1.1–2.8 times as long as wide, base convex, rounded, or cuneate, margins slightly revolute or flat, entire, ciliate, apex usually rounded or convex, sometimes retuse, abaxial surface (rarely glaucous), glabrous or pilose, adaxial slightly glossy, glabrous; proximal blade margins entire; juvenile blade glabrous. **Catkins** sometimes from subterminal buds; staminate 9–34 × 6–12 mm, flowering branchlet 1.5–14 mm; pistillate densely or moderately flowered (more than 15 flowers), stout to globose, 10–50 × 7–13 mm, flowering branchlet 1–12 mm; floral bract brown, black, or bicolor, 1.5–2.5 mm, apex rounded or convex, entire, abaxially sparsely hairy, hairs straight or wavy, (exceeding bract by 0.6–1.12–1.8 mm). **Staminate flowers:** abaxial nectary 0.3–0.7 mm, adaxial nectary oblong, narrowly oblong, square, or ovate, 0.5–1.4 mm, nectaries distinct; filaments usually distinct, sometimes connate proximally, glabrous; anthers ellipsoid or ovoid, 0.4–0.6 mm. **Pistillate flowers:** abaxial nectaries absent, adaxial nectary narrowly oblong, oblong, or ovate, 0.8–1.8 mm, longer than stipe; stipe 0.2–0.7 mm; ovary obclavate or pyriform, densely villous to pilose, hairs flattened, beak gradually tapering to or slightly bulged below styles; ovules 12–17 per ovary; styles connate to distinct ½ their lengths, 0.7–1.2 mm; stigmas flat, abaxially non-papillate with pointed tip, or slenderly to broadly cylindrical, 0.3–0.6(–0.7) mm. **Capsules** 4.8–8.25 mm. $2n$ = 76, 114.

Flowering mid Jun–early Aug. Arctic-alpine, moist late snowbed and snowflush areas, talus and scree slopes, sides of depressed center frost polygons, sedge meadows, and mud boils, calcareous tills, sandy marine silts; 0–1800 m; B.C., N.W.T., Nunavut, Yukon; Alaska; Eurasia (Chukotka, Novaya Zemlya, Russian Far East, arctic Siberia, Spitzbergen, and Sweden).

The sectional placement of *Salix polaris* is uncertain. It was placed in sect. *Myrtosalix* (G. W. Argus 1997) but more recently Argus et al. (1999) placed it in sect. *Herbella*. This polyploid species may be an intersectional hybrid.

Hybrids:

Salix polaris forms natural hybrids with *S. arctica* and perhaps *S. rotundifolia*.

Salix polaris × *S. rotundifolia*: This putative hybrid occurs in Alaska and the Yukon. Many plants previously identified as *S. rotundifolia* but that have ovaries with hairs on the beaks or in patches, leaves not commonly marcescent, and catkins with fewer flowers than in *S. polaris* may be this hybrid.

31. Salix rotundifolia Trautvetter, Nouv. Mém. Soc. Imp. Naturalistes Moscou 2: 304, plate 11. 1832

Plants 0.005–0.05 m, (dwarf), forming clones by rhizomes. **Stems** erect; branches yellow-green, yellow-brown, or gray-brown, glabrous; branchlets yellow-brown or red-brown, glabrous; branches and branchlets sometimes weakly glaucous. **Leaves** (marcescent but not skeletonized), stipules usually absent or rudimentary, rarely present on late ones; petiole (convex, or shallowly to deeply grooved, flat), 0.4–4.6(–5.5) mm, (glabrous adaxially); largest medial blade (2 pairs of secondary veins arising at or close to base, arcing toward apex) broadly elliptic, subcircular, or circular, 1.9–16.3

× 3–10.5 mm, 0.84–1.17(–2.53) times as long as wide, base rounded or convex, margins flat, entire, ciliate, apex retuse, rounded, convex, or acute, abaxial surface glabrous, adaxial highly glossy, glabrous; proximal blade margins entire; juvenile blade glabrous or puberulent. **Catkins** from subterminal buds; staminate subglobose, stout, or indeterminate, 3.3–18.5 × 2.5–12 mm, flowering branchlet 0.5–9 mm; pistillate moderately densely to loosely flowered (2–15 flowers), stout, subglobose, globose, or indeterminate, 4.5–35 × 2–17 mm, flowering branchlet 0.5–22 mm; floral bract brown, 1.6–2.8 mm, apex rounded or retuse, entire, abaxially sparsely hairy or ciliate, hairs usually wavy, crinkled or curly, rarely straight. **Staminate flowers:** abaxial nectary 0.5–1 mm, adaxial nectary narrowly oblong or oblong, 0.8–1.4 mm, nectaries distinct; filaments distinct or connate less than ½ their lengths, glabrous; anthers ellipsoid or globose, 0.4–0.6 mm. **Pistillate flowers:** abaxial nectary present or absent, adaxial nectary usually narrowly oblong or oblong, sometimes flask-shaped, 0.8–2 mm, longer than stipe; stipe 0.4–0.8 mm; ovary pyriform, glabrous or puberulent, (hairs in patches, especially on beak), beak slightly bulged below styles; ovules 7–17 per ovary; styles connate or slightly distinct distally, 0.5–1 mm; stigmas flat, abaxially non-papillate with pointed tip, or slenderly or broadly cylindrical, 0.28–0.6 mm. **Capsules** 3.8–8.3 mm.

Varieties 2 (2 in the flora): North America, e Asia.

Salix rotundifolia is closely related to *S. polaris*, from which it can be separated by its glabrous ovaries and fewer-flowered catkins. They also differ somewhat in leaf venation: *S. rotundifolia* typically having three main veins arising from the leaf base, often only one or two pair of secondary veins, and no or indistinct tertiary veins; *S. polaris* typically having pinnate venation, multiple secondary veins, and distinct tertiary veins. *Salix rotundifolia* consists of two varieties, the diploid var. *dodgeana* and the hexaploid var. *rotundifolia*. In general, var. *dodgeana* is a high alpine species in the southern cordillera of Wyoming and Montana, the St. Elias Mountains in Alaska and Yukon, the Mackenzie Mountains, Northwest Territories, and the Richardson Mountains, Yukon Territory. A diploid specimen of *S. rotundifolia* in the Cherski Mountains, Yakutia, Russia (B. A. Jurtzev and P. G. Zhukova 1982), which fits var. *dodgeana* in its 2–3-flowered catkins, relatively small leaves (3.5 × 3.9 mm), and small stomata (490 μm²), may represent an ancestral population. Variety *rotundifolia* usually occurs at lower elevations in Alaska and in easternmost Chukotka and Wrangel Island, Russia, but elevation separation is not distinct. There is a general correlation between stomatal size and ploidal level (W. Buechler, pers. comm.), but relatively large stomata in some diploid specimens of *S. rotundifolia* indicates a need for further cytological study. For the present, it is best to recognize the two cytotypes as varieties.

Hybrids:

Salix rotundifolia forms natural hybrids with *S. arctica*, *S. phlebophylla*, and *S. polaris*.

1. Pistillate catkins: (3–)4–7–15 flowers; largest medial blades 4.5–8–16.3 mm, 0.92–1.23–2.27 times as long as wide; petioles 0.5–2–4.6(–5.5) mm; floral bracts: hairs usually wavy, some straight, curly, or crinkled, exceeding bract by 0.32–0.71–1.25(–2.4) mm; pistillate flowers: abaxial nectaries present or absent; $2n = 114$ . . . . . . 31a. *Salix rotundifolia* var. *rotundifolia*
1. Pistillate catkins: 2–4–9 flowers; largest medial blades 2.9–6.1–7.4 mm, 0.84–1.5–2.2 times as long as wide; petioles 0.4–1.1–2.8 mm; floral bracts: hairs usually wavy, crinkled, or curly, rarely straight, exceeding bract by 0.1–0.37–0.75 mm; pistillate flowers: abaxial nectaries absent; $2n = 38$ . . . . 31b. *Salix rotundifolia* var. *dodgeana*

31a. Salix rotundifolia Trautvetter var. rotundifolia

- Round-leaf or least willow

Salix leiocarpa (Chamisso) Coville; *S. polaris* Wahlenberg var. *leiocarpa* Chamisso

Plants 0.01–0.05 m. **Leaves:** petiole 0.5–2–4.6(–5.5) mm; largest medial blade 4.5–8–16.3 × 3–5.9–10.5 mm, 0.92–1.23–2.27(–2.53) times as long as wide. **Catkins:** staminate subglobose or stout, 9.5–18.5 × 5–12 mm, flowering branchlet 1–9 mm; pistillate: (3–)4–7–15 flowers, stout, subglobose, or globose, 13–35 × 6–17 mm, flowering branchlet 2–22 mm; floral bract hairs usually wavy, some straight, curly, or crinkled, exceeding bract by 0.32–0.71–1.25(–2.4) mm. **Staminate flowers:** abaxial nectary 0.5–0.6 mm. **Pistillate flowers:** abaxial nectary present or absent, oblong or narrowly oblong, 0.9–2 mm; stigmas slenderly or broadly cylindrical, 0.36–0.47–0.6 mm. **Capsules** 4–8.3 mm. $2n = 114$.

Flowering Jul–early Aug. Arctic-alpine, *Dryas* tundra and willow thickets along streams, sedge-grass tundra, sandy, saline coastal flats, beach cobbles, *Empetrum*-lichen heath on wet slopes, upland shrubby tundra, polygonal tundra and raised center polygons, snowbeds, scree and colluvial slopes, substrates from clay to coarse rubble, both calcareous limestone and acidic; 0–2200 m; N.W.T., Yukon; Alaska; e Asia (Chukotka, Russian Far East, Wrangel Island).

31b. Salix rotundifolia Trautvetter var. **dodgeana** (Rydberg) A. E. Murray, Kalmia 13: 30. 1983

• Dodge's willow

Salix dodgeana Rydberg, Bull. New York Bot. Gard. 1: 277. 1899; *S. rotundifolia* subsp. *dodgeana* (Rydberg) Argus

Plants 0.005–0.02 m. **Leaves:** petiole 0.4–1.1–2.8 mm; largest medial blade 2.9–6.1–7.4 × 3–3.4–5.6 mm, 0.84–1.5–2.2 times as long as wide. **Catkins:** staminate shape indeterminate, 3.3–8 × 2.5–6 mm, flowering branchlet 0.5–3 mm; pistillate: 2–4–9 flowers, shape indeterminate, 4.5–10.5 × 2–7 mm, flowering branchlet 0.5–3 mm; floral bract hairs usually wavy, crinkled, or curly, rarely straight, exceeding bract by 0.1–0.37–0.75 mm. **Staminate flowers:** abaxial nectary 0.6–1 mm. **Pistillate flowers:** abaxial nectary absent, adaxial nectary oblong, narrowly oblong, or flask-shaped, 0.8–1.9 mm; stigmas flat, abaxially non-papillate with pointed tip, or slenderly cylindrical, 0.28–0.34–0.44 mm. **Capsules** 3.8–5.5 mm. $2n = 38$.

Flowering Jul. Arctic-alpine, *Dryas* tundra, *Empetrum* tussock tundra, sedge-shrub tundra, grassy tundra, moss tundra on alluvial fans and floodplains, dry and windswept ridges, moist scree, talus slopes, limestone substrates; 150–3400 m; N.W.T., Yukon; Alaska, Mont., Wyo.; e Asia (Chukotka, Russian Far East, Wrangel Island).

2d4. Salix Linnaeus (subg. Chamaetia) sect. Myrtosalix A. Kerner, Verh. K. K. Zool.-Bot. Ges. Wien 10: 203. 1860

Salix subsect. *Myrtilloides* C. K. Schneider

Plants 0.01–0.55 m, clonal by layering or rhizomes. **Largest medial blades** hypostomatous, amphistomatous, or hemiamphistomatous, abaxial surface glaucous or not. **Catkins** from subterminal or lateral buds. **Staminate flowers:** filaments glabrous or hairy. **Pistillate flowers:** abaxial nectary sometimes present, not connate to adaxial nectary; ovary not glaucous, puberulent or pilose to very densely villous or silky, or glabrous, hairs white, gray, or ferruginous, flattened or ribbonlike.

Species 21 (5 in the flora): North America, Europe, Asia.

32. Salix chamissonis Andersson in A. P. de Candolle and A. L. P. P. de Candolle, Prodr. 16(2): 290. 1868

• Chamisso willow [F]

Plants 0.03–0.1 m, (dwarf), forming clones by layering. **Stems** long-trailing; branches red-brown, glabrous; branchlets yellow-green, glabrous. **Leaves:** stipules foliaceous; petiole 5–13 mm, (sometimes with 1–2 pairs of spherical glands distally); largest medial blade hypostomatous, broadly elliptic, subcircular, or obovate, 30–50 × 17–30 mm, (1.1–)1.6–1.9(–2.1) times as long as wide, base cuneate, margins flat, closely and prominently serrulate or spinulose-serrulate, (teeth 7–14 per cm), apex acuminate, convex, acute, or rounded, abaxial surface glaucous, glabrous, adaxial slightly glossy, glabrous; proximal blade margins entire, closely gland-dotted, or serrulate; juvenile blade glabrous or sparsely long-silky abaxially. **Catkins:** staminate 30–64 × 12–22 mm, flowering branchlet 4–28 mm; pistillate densely or moderately densely flowered, stout, 32–73 (–105 in fruit) × 10–17 mm, flowering branchlet 4–28 mm; floral bract brown or black, 1.2–2.8 mm, apex convex or rounded, entire, abaxially moderately densely hairy, hairs straight. **Staminate flowers:** abaxial nectary absent, adaxial nectary square, 0.5–0.9 mm; filaments distinct, glabrous; anthers ellipsoid or shortly cylindrical, 0.5–0.6 mm. **Pistillate flowers:** abaxial nectary absent, adaxial nectary square or oblong, 0.3–1 mm, equal to or longer than stipe; stipe 0.2–0.4 mm; ovary obclavate, pilose or villous, hairs ribbonlike, (sometimes in patches or streaks, refractive), beak gradually tapering to styles; ovules 12–18 per ovary; styles 0.8–1.2 mm; stigmas flat, abaxially non-papillate with rounded or pointed tip, or slenderly cylindrical, 0.4–0.7 mm. **Capsules** 5–7 mm. $2n = 114$.

Flowering mid–late Jun. Arctic-alpine, *Dryas* heath tundra, dwarf birch-lichen tundra, sandy lakeshores, snowbeds, rock stripes or gravel, wet seepage areas, sedge meadows, willow-dwarf birch-sphagnum bogs,

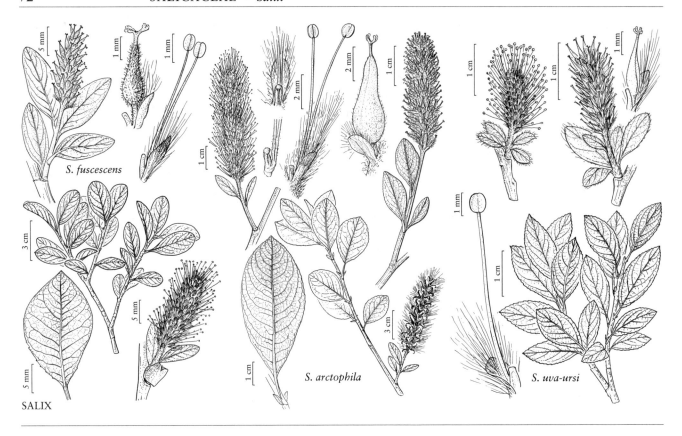

limestone and shale substrates; 0–1500 m; N.W.T., Yukon; Alaska; e Asia (Chukotka, Commander Islands, Russian Far East, disjunct in Sakhalin).

Salix chamissonis is disjunct on Attu Island in Alaska.

33. **Salix fuscescens** Andersson, Monogr. Salicum, 97. 1867 • Alaska bog willow [F]

Salix fuscescens var. *reducta* C. R. Ball

Plants 0.15–0.55 m, forming clones by layering. **Stems** decumbent or trailing; branches yellow-brown, glabrous; branchlets red-brown, gray-brown, or yellow-brown, glabrous. **Leaves:** stipules absent or rudimentary; petiole 2–5.6(–6.4) mm; largest medial blade hypostomatous, narrowly oblong, oblong, obovate, or broadly obovate, (14–)17–27(–45) × 7–21 mm, 1.4–2.5 times as long as wide, base cuneate or convex, margins slightly revolute or flat, entire, or serrulate to crenate proximally, (glands marginal), apex convex, acuminate, or rounded, abaxial surface glaucous, glabrous, adaxial slightly or highly glossy, glabrous; proximal blade margins entire, (sometimes hairs ferruginous abaxially); juvenile blade glabrous. **Catkins:** staminate 8–58 × 5–19 mm, flowering branchlet 0.5–15 mm; pistillate moderately densely to loosely flowered, slender, stout, or subglobose, 13.5–64(–70 in fruit) × 6.5–15 mm, flowering branchlet 4–18 mm; floral bract brown or bicolor, 0.8–1.6 mm, apex rounded, entire, abaxially sparsely hairy, hairs (usually white, sometimes also ferruginous), wavy or straight. **Staminate flowers:** abaxial nectary absent, adaxial nectary oblong, 0.5–0.8 mm; filaments distinct or connate less than ½ their lengths, glabrous; anthers ellipsoid or shortly cylindrical, 0.3–0.4 mm. **Pistillate flowers:** abaxial nectary absent, adaxial nectary oblong, 0.4–0.9 mm, shorter than stipe; stipe 0.8–2.5 mm; ovary obclavate, pubescent or short-silky to glabrescent, hairs (often ferruginous), flattened or ribbonlike, beak abruptly tapering to styles; ovules 8–12 per ovary; styles connate or distinct distally, 0.1–0.4(–0.65) mm; stigmas slenderly or broadly cylindrical, 0.24–0.3–0.68 mm. **Capsules** 5.5–8 mm. $2n$ = 38.

Flowering Jun–late Jul. Bogs, treed bogs, sedge fens, poorly drained lakeshores, wet tundra, silt or fine sandy-gravel substrates; 50–1000 m; Man., N.W.T., Nunavut, Yukon; Alaska; e Asia (Chukotka, Japan [Hokkaido], North Korea, Russian Far East, arctic, e Siberia).

Hybrids:

Salix fuscescens forms natural hybrids on the arctic coast of Alaska with *S. arctica*, *S. ovalifolia*, and *S. phlebophylla*, and in continental Nunavut with *S. herbacea*. These hybrids are not usually recognized but they appear sporadically.

Salix fuscescens × *S. herbacea* has crenate margins and is often confused with the latter species.

Some specimens identified as *Salix fuscescens* × *S. ovalifolia* are similar to hybrids with *S. phlebophylla*, but they lack marcescent leaves.

Salix fuscescens × *S. phlebophylla* has obovate leaves with one or two pairs of serrations proximally, and glaucous abaxially as in *S. fuscescens*, but it grows in relatively dry tundra and has the marcescent, sometimes skeletonized, leaves of *S. phlebophylla*.

34. **Salix arctophila** Cockerell ex A. Heller, Cat. N. Amer. Pl. ed. 3, 89. 1910 • Northern willow E F

Salix arctophila var. *lejocarpa* (Lange) C. K. Schneider; *S. groenlandica* Lundström var. *lejocarpa* Lange

Plants 0.03–0.15 m, (dwarf), forming clones by layering. **Stems** prostrate, long-trailing; branches yellow-brown, red-brown, or green-brown, glabrous; branchlets yellow-green or yellow-brown to red-brown, (sometimes weakly glaucous), glabrous, (inner membranaceous bud-scale layer free, not separating from outer layer). **Leaves:** stipules rudimentary, absent, or foliaceous on early ones, foliaceous or rudimentary on late ones; petiole 3–7.8–15 mm; largest medial blade hypostomatous or hemiamphistomatous, elliptic, obovate, broadly elliptic, broadly obovate, subcircular, or oblanceolate, 15–31–60 × 6.5–16–35 mm, 1.2–3–4.3 times as long as wide, base cuneate, convex, or rounded, margins slightly revolute, inconspicuously crenulate or entire, apex usually acute or convex, sometimes rounded, abaxial surface glaucous, glabrous, adaxial slightly or highly glossy, glabrous; proximal blade margins entire or serrulate; juvenile blade glabrous. **Catkins:** staminate 19–54 × 7–16 mm, flowering branchlet 4–20 mm; pistillate densely to moderately densely flowered, slender to subglobose, 30–79(–130 in fruit) × 10–20 mm, flowering branchlet 8–57 mm; floral bract brown, black, or bicolor, 0.8–2.4 mm, apex rounded or acute, entire, abaxially densely hairy, hairs straight. **Staminate flowers:** abaxial nectary absent, adaxial nectary oblong, square, narrowly oblong, or ovate, 0.4–1 mm; filaments distinct or connate less than ½ their lengths, glabrous, or hairy on proximal ½; anthers ellipsoid or long-cylindrical, 0.5–0.7 mm. **Pistillate flowers:** abaxial nectary absent, adaxial nectary oblong or narrowly oblong, 0.5–0.9 mm, shorter than stipe; stipe 0.8–1.4 mm; ovary pyriform or obclavate, pubescent or short-silky, (refractive), hairs (white, grayish, or ferruginous), crinkled, often refractive, ribbonlike, beak gradually tapering to styles; ovules 8–16 per ovary; styles connate or distinct ½ their lengths, 0.6–1.4 mm; stigmas slenderly or broadly cylindrical, 0.24–0.47–0.72 mm. **Capsules** 5–9 mm. $2n = 76$.

Flowering late May–late Jul. Arctic-alpine, subarctic, hummocks in wet, mossy, grass or sedge meadows, margins of streams or ponds, among granite boulders, on alluvial plains, sometimes in snowbeds; 40–600 m; Greenland; Man., Nfld. and Labr., N.W.T., Nunavut, Ont., Que., Sask., Yukon; Alaska, Maine.

Salix arctophila occurs in western Greenland.

Hybrids:

Salix arctophila forms natural hybrids with *S. arctica*, *S. glauca* var. *cordifolia*, and *S. uva-ursi*.

Salix arctophila × *S. glauca* var. *cordifolia*: Plants with villous leaves and moderately densely hairy branchlets and branches suggest this hybrid. Putative hybrids are rare but have been seen from Kuujjuaq and Ivujivik, Quebec (G. W. Argus, unpubl.), and are reported to be common in West Greenland (T. W. Böcher 1952).

Salix arctophila × *S. uva-ursi* is a rare hybrid. The plants often have ovaries with patches of hairs, some of which are ribbonlike, as in *S. arctophila*, but their habit is compact, as in *S. uva-ursi*, rather than long-trailing as in *S. arctophila*. Some specimens are infertile and are evidently hybrids, but there is little to confirm *S. uva-ursi* as the second parent. N. Polunin (1940b) also expressed some uncertainty about plants intermediate between *S. arctophila* and *S. uva-ursi*, and A. K. Skvortsov (1971) discounted this hybrid but noted that there were a few somewhat doubtful specimens.

35. **Salix uva-ursi** Pursh, Fl. Amer. Sept. 2: 610. 1813 • Bearberry willow E F

Salix ivigtutiana Lundström; *S. myrsinites* Linnaeus var. *parvifolia* Lange

Plants 0.01–0.05 m, (dwarf), forming clonal compact mats by layering. **Stems** prostrate, short-trailing; branches red-brown, gray-brown, or yellow-brown, glabrous; branchlets yellow-green or yellow-brown, glabrous or puberulent. **Leaves:** stipules (sometimes marcescent) absent, rudimentary, or foliaceous on early ones, foliaceous on late ones; petiole (shallowly to deeply grooved adaxially), 2–6.5 mm; largest medial blade (marcescent), amphistomatous or hypostomatous, ovate, broadly obovate, oblanceolate, or elliptic, 4–23 × 3.5–10 mm, 1.7–3.6 times as long as wide, base convex or cuneate, margins flat, serrulate or crenulate, apex convex, acuminate, acute, or retuse, abaxial surface glaucous, usually glabrous (rarely few hairs), adaxial slightly or highly glossy, usually glabrous (rarely a few hairs); proximal blade margins entire or serrate; juvenile blade glabrous, pilose, or puberulent abaxially. **Catkins:** staminate 9–19 × 5–8 mm, flowering branchlet 0.5–9 mm; pistillate densely flowered, slender to subglobose, 11–47(–55 in fruit) × 6–10 mm, flowering branchlet 2–10 mm; floral bract brown, black, tawny, light rose, or bicolor, 1.1–1.8 mm, apex rounded or acute, entire, abaxially sparsely hairy, hairs straight or wavy. **Staminate flowers:** abaxial nectary absent, adaxial nectary narrowly oblong or oblong, 0.4–0.9 mm; filaments distinct, glabrous; stamens usually 1, rarely 2; anthers ellipsoid or shortly cylindrical, 0.4–0.7 mm. **Pistillate flowers:** abaxial nectary absent, adaxial nectary narrowly oblong or oblong, 0.5–0.8 mm, shorter to longer than stipe; stipe 0.3–1.6 mm; ovary ovoid or pyriform, glabrous, beak gradually tapering to styles; ovules 4–9 per ovary; styles 0.4–1 mm; stigmas flat, abaxially non-papillate with rounded tip, or slenderly cylindrical, 0.1–0.23–0.4 mm. **Capsules** 3–5 mm. $2n = 38$.

Flowering mid Jun–early Aug. Exposed, often dry or moist, calcareous, serpentine, dioritic, and granitic rocks, boulders, gravel, sand on beaches, outcrops, in snowbeds; 10–1200 m; Greenland; St. Pierre and Miquelon; Nfld. and Labr., N.S., Nunavut, Que.; Maine, N.H., N.Y., Vt.

Hybrids:

Salix uva-ursi forms natural hybrids with *S. herbacea*.

36. **Salix phlebophylla** Andersson, Proc. Amer. Acad. Arts 4: 72. 1858 • Skeleton-leaf willow

Plants 0.01–0.07 m, (dwarf), forming clonal mats by rhizomes. **Stems** trailing; branches red-brown or yellow-brown, glabrous; branchlets red-brown, glabrous. **Leaves** (prominently marcescent, becoming skeletonized); stipules absent or rudimentary; petiole (deeply grooved or convex to flat adaxially), 1.2–3.2(–4.8) mm, (sparsely pubescent adaxially); largest medial blade amphistomatous, (2 pairs secondary veins arising at or close to base, arcing toward apex), elliptic, broadly elliptic, obovate, subcircular, or circular, 7–15 × 3–11 mm, 1.1–2.5(–3.5) times as long as wide, base convex or cuneate, margins flat, entire, sometimes ciliate, apex convex, rounded, or retuse, abaxial surface not glaucous, glabrous, midrib sometimes pilose, hairs long, straight, wavy, or crinkled, adaxial highly glossy, glabrous; proximal blade margins entire; juvenile blade (green), glabrous abaxially or ciliate. **Catkins:** staminate 10–35 × 6–10 mm, flowering branchlet 1–11 mm; pistillate moderately densely flowered, stout or subglobose, 12–38 × 5–11 mm, flowering branchlet 3–14 mm; floral bract brown, black, or bicolor, 1–1.3 mm, apex rounded, entire, abaxially sparsely hairy, hairs straight or wavy. **Staminate flowers:** abaxial nectary absent, adaxial nectary narrowly oblong or oblong, 0.4–1.1 mm; filaments distinct or connate less than ½ their lengths, glabrous; anthers ellipsoid or obovoid, 0.3–0.5 mm. **Pistillate flowers:** abaxial nectary absent, adaxial nectary oblong or narrowly oblong, 0.4–1.6 mm, shorter than or equal to stipe; stipe 0.4–1.4 mm; ovary pyriform, sparsely to moderately densely short-silky or villous, at least on beaks, hairs ribbonlike, beak slightly bulged below styles; ovules 12 per ovary; styles connate or slightly distinct distally, 0.3–1 mm; stigmas flat, abaxially non-papillate with rounded tip, or slenderly to broadly cylindrical, 0.16–0.34–0.52 mm. **Capsules** 2.9–4.8 mm. $2n = 38$.

Flowering Jun–Jul. Arctic-alpine, dry *Dryas*-lichen tundra, polygonal tundra with stone stripes and dry raised centers, scree and colluvial slopes, grass-sedge tussock tundra, sedge meadows in drainage ways, dwarf birch thickets, granitic and sandstone substrates; 0–2100 m; N.W.T., Yukon; Alaska; e Asia (Chukotka, Russian Far East, arctic, e Siberia).

Hybrids:

Salix phlebophylla forms natural hybrids with *S. arctica*, *S. fuscescens*, and *S. rotundifolia*.

Salix phlebophylla × *S. rotundifolia* has hairy ovaries, some skeletonized leaves, a compact growth form that may lack rhizomes, and catkins with more than 15 flowers. It occurs on the Alaska arctic slope, outside the range of *S. polaris*.

2d5. Salix Linnaeus (subg. Chamaetia) sect. Ovalifoliae (Rydberg) C. K. Schneider in C. S. Sargent, Pl. Wilson. 3: 140. 1916

Salix [unranked] *Ovalifoliae* Rydberg, Bull. New York Bot. Gard. 1: 274. 1899

Plants 0.01–0.1 m, (dwarf), not clonal or clonal by layering or rhizomes. **Largest medial blades** hypostomatous, amphistomatous, or hemiamphistomatous, abaxial surface glaucous. **Catkins** from lateral buds. **Staminate flowers:** filaments glabrous. **Pistillate flowers:** abaxial nectary sometimes present, then distinct or connate to adaxial and forming a cup; ovary sometimes glaucous, glabrous, tomentose, or pubescent to very densely villous.

Species 5 (3 in the flora): North America, Asia.

37. Salix stolonifera Coville, Proc. Wash. Acad. Sci. 3: 333, plate 41, fig. 1. 1901 • Creeping willow [E]

Plants 0.02–0.1 m, forming clones by layering or rhizomes. **Stems** trailing or erect; branches red-brown, (sometimes weakly glaucous), glabrous; branchlets yellow-brown or greenish brown, glabrous. **Leaves:** stipules (sometimes marcescent) absent or rudimentary on early ones, rudimentary or foliaceous on late ones; petiole (deeply to shallowly grooved adaxially), 3–9–20 mm, (ciliate, glabrous adaxially); largest medial blade amphistomatous or hemiamphistomatous, (sometimes with 2 pairs of secondary veins arising at or close to base, arcing toward apex), elliptic, broadly elliptic, or subcircular, 16–42 × 12–30(–38) mm, 1–2 times as long as wide, base convex, cuneate, or rounded, margins flat or slightly revolute, entire or serrulate, ciliate (hairs wavy), apex convex, acuminate, rounded, or retuse, abaxial surface glabrous, adaxial highly glossy, glabrous; proximal blade margins entire or irregularly serrulate; juvenile blade pilose to glabrescent abaxially. **Catkins:** staminate 13–31 × 7–11 mm, flowering branchlet 1–15 mm; pistillate moderately densely to loosely flowered, stout to globose, 15–54(–90 in fruit) × 6–15 mm, flowering branchlet 2–42 mm; floral bract brown, 1.6–2 mm, apex rounded, entire, abaxially sparsely hairy throughout or distally, hairs straight or wavy. **Staminate flowers:** abaxial nectary (0–)0.2–0.7 mm, adaxial nectary oblong, narrowly oblong, or ovate, 0.6–1.3 mm, nectaries distinct; filaments distinct (glabrous); anthers ellipsoid, 0.5–0.6 mm. **Pistillate flowers:** abaxial nectary absent, adaxial nectary oblong, 0.5–1.4 mm, longer than stipe; stipe 0.2–0.8 mm; ovary pyriform, sometimes glaucous, glabrous, beak gradually tapering to styles; ovules 12–13 per ovary; styles connate to distinct ½ their lengths, (0.6–)0.8–2 mm; stigmas flat, abaxially non-papillate with rounded or pointed tip, or slenderly cylindrical, 0.32–0.5–0.88 mm. **Capsules** 4–10 mm.

Flowering early Jun–early Jul. Arctic, subarctic, and alpine, wet sedge meadows, hummocky tundra, raised center polygons, *Dryas*-willow-sedge tundra, *Dryas* mats on dry ridge tops; 0–1000 m; Alta., B.C.; Alaska.

Hybrids:

Salix stolonifera forms natural hybrids with *S. arctica* and *S. barclayi*.

38. Salix ovalifolia Trautvetter, Nouv. Mém. Soc. Imp. Naturalistes Moscou 2: 306, plate 13. 1832 • Arctic seashore or oval-leaf willow [F]

Plants 0.02–0.05 m, not clonal or forming clones by layering. **Stems** trailing; branches yellow-brown, gray-brown, or red-brown, glabrous or hairy; branchlets yellow-green, yellow-brown, or red-brown, glabrous or pilose. **Leaves:** stipules usually absent or rudimentary, rarely foliaceous; petiole (deeply to shallowly grooved adaxially), 1.1–16 mm, (glabrous); largest medial blade hypostomatous, narrowly to broadly elliptic, circular, subcircular, or obovate, 13–46 × 7–20 mm, 1–3.4 times as long as wide, base subcordate, cordate, rounded, or convex, margins slightly revolute or flat, entire, sometimes ciliate, apex convex, rounded, acuminate, acute, or retuse, abaxial surface glabrous, villous, long-silky, pubescent, or pilose, hairs wavy or straight, adaxial highly glossy, usually glabrous; proximal blade margins entire; juvenile blade (reddish or yellowish green), pilose, villous, or long-silky abaxially. **Catkins:** staminate 4.8–46 × 5–11 mm, flowering branchlet 1.5–24 mm; pistillate moderately densely flowered, stout, subglobose, globose, or slender, 6.3–50 × 5–28 mm, flowering branchlet 2.5–22 mm; floral bract brown, greenish, or bicolor, 1.2–2 mm, apex rounded, entire or, sometimes, 2-fid, abaxially hairy, hairs straight or wavy. **Staminate flowers:** abaxial nectary 0.6–1 mm, adaxial nectary oblong or ovate, 0.6–1.6 mm, nectaries distinct or connate and

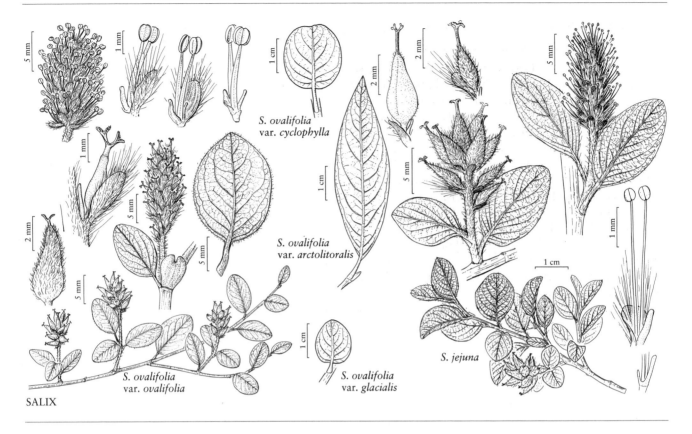

cup-shaped; filaments distinct or connate less than ½ their lengths (glabrous); anthers elliptic, short-cylindrical, or globose, 0.3–0.5(–0.6) mm. **Pistillate flowers:** abaxial nectary (0–)0.4–0.6 mm, adaxial nectary longer than stipe; stipe 0.2–1.4 mm; ovary obclavate or pyriform, glaucous or not, usually glabrous or tomentose, sometimes pubescent or villous, beak abruptly tapering to styles; ovules 10–15 per ovary; styles 0.2–0.8 mm; stigmas flat, abaxially non-papillate with pointed tip, or slenderly cylindrical, 0.32–0.41–0.64 mm. **Capsules** 5.2–9.6 mm. $2n = 38$.

Varieties 4 (4 in the flora): n, nw North America, Asia.

The varieties of *Salix ovalifolia* are relatively minor variants; their ranges overlap and their differences in leaf shape and ovary indumentum intergrade (G. W. Argus 1969, 1973). The only one with a more or less distinctive geographical distribution is var. *cyclophylla*; where its range overlaps with var. *ovalifolia* there is intergradation. Variety *arctolitoralis*, which is characterized by larger leaves and catkins, may be an ecotype. Variety *glacialis* is known only from near Point Barrow, Alaska. E. Hultén (1968) suggested, probably based on its often tomentose ovaries, that it is *S. arctica* × *S. ovalifolia*. All varieties of the species have some plants with hairy ovaries, but the suggestion that this character is an indication of hybridization deserves study.

Hybrids:

Salix ovalifolia forms natural hybrids with *S. arctica* and *S. fuscescens*.

1. Ovaries usually tomentose, sometimes glabrous; largest medial blades 8.5–14 mm; endemic to Point Barrow, Alaska . . . . 38d. *Salix ovalifolia* var. *glacialis*
1. Ovaries usually glabrous, sometimes pubescent or villous; largest medial blades 13–46 mm; not endemic to Point Barrow, Alaska.
    2. Largest medial blades subcircular to circular, 1–1.5 times as long as wide . . . . 38b. *Salix ovalifolia* var. *cyclophylla*
    2. Largest medial blades narrowly to broadly elliptic, obovate, or subcircular, 1.1–3.4 times as long as wide.
        3. Largest medial blades elliptic, broadly elliptic, or subcircular, 13–28 mm . . . . 38a. *Salix ovalifolia* var. *ovalifolia*
        3. Largest medial blades narrowly to broadly elliptic or obovate, 25–46 mm . . . . 38c. *Salix ovalifolia* var. *arctolitoralis*

38a. Salix ovalifolia Trautvetter var. ovalifolia [F]

Salix flagellaris Hultén

Leaves: petiole 2–7 mm; largest medial blade elliptic, broadly elliptic, or subcircular, 13–28 × 7–18 mm, 1.1–2.2 times as long as wide, base subcordate, cordate, rounded, or convex, apex convex, rounded, or acute, abaxial surface glabrous, pilose, sparsely villous, or long-silky. **Catkins:** staminate 9.5–21 × 5–10 mm, flowering branchlet 2–7 mm; pistillate stout, subglobose, globose, or slender, 11.5–50 × 6–18 mm, flowering branchlet 3–21 mm. **Staminate flowers:** abaxial nectary 0.6–1 mm, adaxial nectary 0.6–1.6 mm, nectaries distinct or connate and cup-shaped. **Pistillate flowers:** abaxial nectary present or absent, adaxial nectary ovate, oblong, or narrowly oblong, 0.5–2 mm, nectaries connate and cup-shaped; stipe 0.2–1.4 mm; ovary glaucous or not, usually glabrous, sometimes pubescent, or villous. **Capsules** 5.2–6.5 mm.

Flowering early Jun–early Aug. Arctic and subarctic, coastal sandy gravel spits; 0–150 m; N.W.T., Yukon; Alaska; Asia (Chukotka, Russia).

38b. Salix ovalifolia Trautvetter var. cyclophylla (Rydberg) C. R. Ball, Proc. Natl. Acad. Sci. U.S.A. 21: 184. 1935 [F]

Salix cyclophylla Rydberg, Bull. New York Bot. Gard. 1: 274. 1899; *S. ovalifolia* subsp. *cyclophylla* (Rydberg) Jurtsev & V. V. Petrovsky

Leaves: petiole 2–7 mm; largest medial blade subcircular to circular, 13–28 × 7–18 mm, 1–1.5 times as long as wide, base subcordate, cordate, or convex, apex convex, rounded, or retuse, abaxial surface glabrous, pilose or villous. **Catkins:** staminate 12.5–15.5 × 8–10 mm, flowering branchlet 3–12 mm; pistillate stout, subglobose, or globose, 10–31 × 6–12 mm, flowering branchlet 3.5–15 mm. **Staminate flowers:** abaxial nectary 0.8–1.1 mm, adaxial nectary 0.7–1.1 mm, nectaries connate and cup-shaped. **Pistillate flowers:** abaxial nectary absent, adaxial nectary ovate, 0.5–2.1 mm; stipe 0.2–1.4 mm; ovary glaucous or not, usually glabrous, sometimes pubescent, or villous. **Capsules** 5.2–6.5 mm.

Flowering mid May–late Jul. Arctic and possibly subarctic, shores of lakes and lagoons, *Empetrum* tundra, in moss on rocky pavements, meadows, beach ridges; 10–600 m; Alaska; Asia (Chukotka, Russia).

38c. Salix ovalifolia Trautvetter var. arctolitoralis (Hultén) Argus, Canad. J. Bot. 47: 795. 1969 [E][F]

Salix arctolitoralis Hultén, Svensk Bot. Tidskr. 34: 373, fig. 1. 1940

Leaves: petiole 4–16 mm; largest medial blade narrowly to broadly elliptic or obovate, 25–46 × 10–20 mm, 1.6–3.4 times as long as wide, base convex, apex convex, acute, acuminate, or rounded, abaxial surface glabrous, sparsely pubescent, or pilose. **Catkins:** staminate 23–46 × 8–11 mm, flowering branchlet 3–24 mm; pistillate stout to subglobose, 28–48 × 10–28 mm, flowering branchlet 5–22 mm. **Staminate flowers:** abaxial nectary 0.8–1.1 mm, adaxial nectary 0.7–1.1 mm, nectaries connate and cup-shaped. **Pistillate flowers:** abaxial nectary absent, adaxial nectary ovate, 0.5–1.9 mm; stipe 0.2–1.4 mm; ovary often glaucous, glabrous. **Capsules** 5.2–9.6 mm.

Flowering early Jul–late Aug. Arctic, coastal beach ridges, sand spits, tundra meadows; 10–30 m; N.W.T., Yukon; Alaska.

38d. Salix ovalifolia Trautvetter var. glacialis (Andersson) Argus, Canad. J. Bot. 47: 798. 1969 [E][F]

Salix glacialis Andersson, Öfvers. Kongl. Vetensk.-Akad. Förh. 15: 131. 1858; *S. ovalifolia* subsp. *glacialis* (Andersson) Jurtsev & V. V. Petrovsky

Leaves: petiole 1.1–3.2 mm; largest medial blade elliptic to subcircular, 8.5–14 × 4.5–9 mm, 1.1–2.2 times as long as wide, base subcordate, cordate, or convex, apex usually convex (rarely acute), abaxial surface pilose. **Catkins:** staminate 4.8–15 × 5–9 mm, flowering branchlet 1.5–7 mm; pistillate stout to globose, 6.3–15 × 5–12 mm, flowering branchlet 2.5–8 mm. **Staminate flowers:** abaxial nectary 0.5–0.9 mm, adaxial nectary 0.6–1.1 mm, nectaries distinct or connate and shallowly cup-shaped. **Pistillate flowers:** abaxial nectary present [(0–)0.4–0.6 mm], adaxial nectary ovate, 0.5–1.6 mm, nectaries distinct or connate and cup-shaped; stipe 0.2–0.8 mm; ovary glaucous or not, usually tomentose, sometimes glabrous. **Capsules** 5.2–6.5 mm.

Flowering Jul. Arctic, coastal sandy gravel spits; 0–10 m; Alaska.

39. Salix jejuna Fernald, Rhodora 28: 177. 1926
• Barrens willow [C] [E] [F]

Plants 0.01–0.04 m, usually forming clones by layering or rhizomes. Stems erect, decumbent, or trailing; branches red-brown, glabrous; branchlets yellow-brown or red-brown, glabrous. Leaves: stipules absent or rudimentary on early ones, foliaceous on late ones; petiole 1.5–4.9–14 mm; largest medial blade amphistomatous or hemiamphistomatous, elliptic, subcircular, or circular, 8–25 × 4–22 mm, 0.8–2.3 times as long as wide, base convex or rounded, margins slightly revolute, entire, sometimes ciliate, apex rounded or convex, abaxial surface glabrous, adaxial slightly glossy, glabrous; proximal blade margins entire; juvenile blade (reddish), glabrous, often ciliate. Catkins: staminate 5.5–14 × 5–10 mm, flowering branchlet 1–6 mm; pistillate moderately densely flowered, globose, subglobose, or stout, 8–29 × 6–14 mm, flowering branchlet 1.5–20 mm; floral bract brown, 1–1.8 mm, apex rounded, abaxially sparsely hairy or ciliate, hairs straight or wavy. Staminate flowers: abaxial nectary 0.6–1 mm, adaxial nectary narrowly oblong to ovate, 0.6–1.3 mm, nectaries distinct; filaments distinct; anthers ellipsoid to globose, 0.4–0.6 mm. Pistillate flowers: abaxial nectary absent, adaxial nectary narrowly oblong, ovate, or irregularly square, 0.6–1.4 mm, longer than stipe; stipe 0.2–1 mm; ovary obnapiform, not glaucous, usually glabrous, sometimes pubescent or short-silky, with hairs in patches or streaks, beak abruptly tapering to styles; ovules 9–18 per ovary; styles connate or distinct ½ their lengths, 0.5–1.4 mm; stigmas flat, abaxially non-papillate with rounded tip, or slenderly cylindrical, 0.24–0.32–0.4 mm. Capsules 3–5 mm.

Flowering late Jun–mid Jul. Wet, thin, sandy-gravel soil on limestone barrens; of conservation concern; 0–20 m; Nfld. and Labr. (Nfld.).

Salix jejuna, known from the Northern Peninsula of Newfoundland, is closely related to the western arctic *S. ovalifolia*. It is similar to *S. ovalifolia* in that leaves are hypostomatous, glabrous even when young, and with secondary veins raised on both surfaces. Its ovaries also are usually glabrous, but somewhat hairy plants sometimes occur in the same population. Plants with hairy ovaries may be confused with *S. arctica* but they are separable on the characters mentioned. Because of its small range adjacent to a highway, *S. jejuna* is at risk.

2d6. SALIX Linnaeus (subg. CHAMAETIA) sect. DIPLODICTYAE C. K. Schneider in C. S. Sargent, Pl. Wilson. 3: 136. 1916

Plants 0.02–0.25 m, (dwarf), not clonal or clonal by layering or rhizomes. Largest medial blades hypostomatous, hemiamphistomatous, or amphistomatous, abaxial surface glaucous. Catkins from lateral buds. Staminate flowers: filaments glabrous. Pistillate flowers: abaxial nectary absent; ovary glaucous or not, pilose to very densely villous or glabrous, hairs white, flattened.

Species 4 (4 in the flora): North America, Eurasia, Atlantic Islands.

40. Salix arctica Pallas, Fl. Ross. 1(2): 86. 1788
• Arctic willow [F]

Salix arctica var. *antiplasta* (C. K. Schneider) Fernald; *S. arctica* var. *araioclada* (C. K. Schneider) Raup; *S. arctica* var. *brownei* Andersson; *S. arctica* subsp. *crassijulis* (Treviranus ex Trautvetter) A. K. Skvortsov; *S. arctica* var. *kophophylla* (C. K. Schneider) Polunin; *S. arctica* var. *pallasii* (Andersson) Kurtz; *S. arctica* subsp. *torulosa* (Treviranus ex Trautvetter) Hultén; *S. crassijulis* Treviranus ex Trautvetter

Plants 0.03–0.25 m, not clonal or forming clones by layering. Stems erect, decumbent, or trailing; branches yellow-brown, gray-brown, or red-brown, (strongly glaucous or not, slightly glossy), glabrous; branchlets yellow-brown, red-brown, or violet, (strongly glaucous or not), usually densely villous or pilose (usually appearing unkempt), sometimes glabrous, (inner membranaceous bud-scale layer free, not separating from outer layer). Leaves: stipules absent, rudimentary, or foliaceous (0.2–1.5–10 mm); petiole 2–35 mm, (longer than subtended bud, puberulent or glabrous adaxially); largest medial blade hypostomatous or hemiamphistomatous, narrowly to broadly elliptic, subcircular, circular, oblanceolate, obovate, or broadly obovate, 10–85 × 5.5–60 mm, 1–3.6(–4.9) times as long as wide, base cuneate, convex, or rounded, margins slightly revolute or flat, entire, apex acuminate, acute, convex, or rounded, abaxial surface pilose or midrib sparsely short-silky, or apex long-silky bearded, hairs usually straight or wavy, adaxial slightly glossy or dull, glabrous, pilose or long-

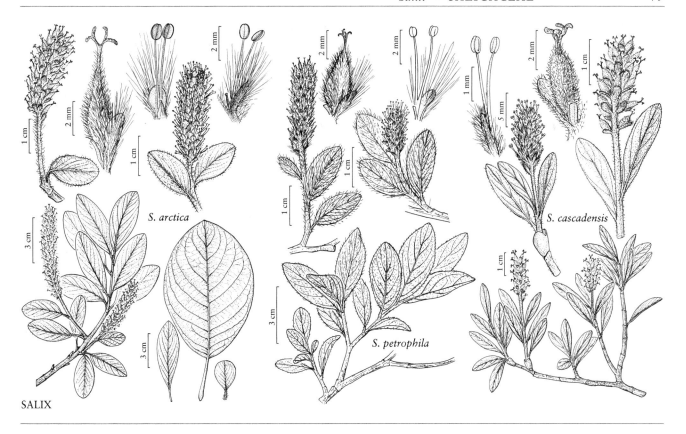

silky margin; proximal blade margins entire; juvenile blade glabrous or sparsely villous abaxially, hairs straight, oriented toward apex. **Catkins:** staminate 14–65 × 5–18 mm, flowering branchlet 2–36 mm; pistillate densely to moderately densely flowered (30+ flowers), slender, stout, or subglobose, 20–145 × 8–22 mm, flowering branchlet 2–40 mm; floral bract brown or black, 1.6–3.7 mm, margins sometimes sinuate, apex broadly rounded, convex, or retuse, entire, sinuate, or 2-fid, abaxially hairy, hairs straight. **Staminate flowers:** abaxial nectary (0–)0.3–0.8 mm, adaxial nectary narrowly oblong, oblong, or square, 0.5–1.2 mm, nectaries distinct; filaments distinct; anthers ellipsoid, 0.3–0.9 mm. **Pistillate flowers:** adaxial nectary oblong, ovate, or narrowly oblong, 0.4–1.8 mm, much longer than stipe; stipe 0.2–1.6 mm; ovary obclavate or pyriform, villous, beak abruptly to gradually tapering to or slightly bulged below styles; ovules 12–15 per ovary; styles (connate to distinct ½ their lengths), 0.6–2.2 mm; stigmas slenderly cylindrical, 0.35–0.56–0.88(–1.13) mm. **Capsules** 4–9 mm. $2n = 76, 114$.

Flowering early Jun–mid Aug. Arctic-alpine, wet to mesic or dry habitats, including hummocks in wet *Sphagnum* bogs and sedge meadows, polygonal tundra, solifluction slopes, snowbeds, margins of pools, beach ridges, shale and gypsum ridges, gneissic cliffs, colluvial slopes, talus slopes, glacial moraines, imperfectly drained calcareous silty till, muddy salt flats, frost-heaved clay polygons, dry calcareous gravel, coarse sandy soil; 0–2000 m; Greenland; Alta., B.C., Nfld. and Labr., N.W.T., Nunavut, Ont., Que., Yukon; Alaska, Wash.; Eurasia (China, Chukotka, Novaya Zemlya, Russian Far East, arctic, e Siberia); Atlantic Islands (Iceland).

Salix arctica is polymorphic and nomenclaturally confusing. E. Hultén (1967, 1971) recognized three subspecies: 1) subsp. *arctica* (circumpolar from Iceland and the Faeroe Islands across northern Russia, Alaska and Canada to Greenland, south to the Hudson Bay shores of Ontario and the Gaspe Peninsula); 2) subsp. *crassijulis* (a North Pacific race ranging from Kamchatka and the Russian Far East to the Aleutian Islands, south central and southeastern Alaska along the coast to northern Washington); and 3) subsp. *torulosa* (ranging from the mountains of central Asia to Kamchatka and the Bering Straits, the Brooks Range and the Rocky Mountains in Alaska, south in the cordillera to southern British Columbia and Alberta). While formal recognition of the three races is appealing, they are actually very difficult or impossible to separate morphologically and have strongly overlapping ranges. Some of the variability may be due to environmental modification (D. B. O. Savile 1964; G. W. Argus 1973; J. H. Soper and J. M. Powell 1985). On Attu Island, Alaska, there are plants to 2 m along with dwarf plants (C. Parker, pers. comm.). Their tall stature cannot be accounted for by habitat alone. The possibility that the complex morphological variability within *S. arctica* may be ecophenic or ecotypic deserves study.

T. E. Dawson (1987) showed that the pistillate-biased sex ratios in *Salix arctica* are environmentally controlled; pistillate plants are significantly over-represented in mesic-wet, more fertile sites with low soil temperature, whereas staminate plants are predominant in drier, less fertile sites.

The oldest reported age of *Salix arctica* is 236 years in East Greenland (H. M. Raup 1965) and 85 years on Ellesmere Island (D. B. O. Savile 1979b).

In northern Greenland, *Salix arctica* is one of the primary foods for musk ox, arctic hare, and collared lemming (D. R. Klein and C. Bay 1991).

Hybrids:

Salix arctica forms natural hybrids with *S. arctophila*, *S. barclayi*, *S. fuscescens*, *S. glauca* varieties, *S. herbacea*, *S. ovalifolia*, *S. pedicellaris*, *S. phlebophylla*, *S. polaris*, *S. rotundifolia*, and *S. stolonifera*.

Salix arctica × *S. arctophila* (*S.* ×*hudsonensis* C. K. Schneider) resembles *S. arctophila* in its trailing habit and toothed leaf margins, and *S. arctica* in its hairy leaves and petioles, particularly proximal leaves, and branchlets that are both hairy and glaucous. Ovaries have the flattened, refractive hairs of *S. arctica*, rather than the ribbonlike hairs of *S. arctophila*. Some plants have undeveloped ovaries, suggesting infertility. The plants are often sterile. Other putative hybrid specimens resemble *S. arctica* in having staminate flowers with abaxial and adaxial nectaries and anthers ca. 0.36 mm, and *S. arctophila* in having glabrous leaves and branchlets. N. Polunin (1940) described intermediates between these species from Burwell, Labrador, and Wakeham Bay, Quebec, but the available specimens were re-identified as either *S. arctica* or *S. arctophila*. Polunin (1943) wrote that in Greenland these two species do not "hybridize at all frequently" even where growing "side by side."

Salix arctica × *S. barclayi* grows with both parents in tundra above the Tokositna Glacier, Alaska. It resembles *S. barclayi* in having buds with a distinct and separating inner membrane and stigmas broad and short, and *S. arctica* in its trailing habit (tundra shrub to 0.23 m) and in stipules absent. It is intermediate in having ovaries hairy only on beaks, and leaves with scattered marginal teeth. It also occurs in southeastern Alaska (G. W. Argus 1973).

Salix arctica × *S. fuscescens* has the growth form, habitat, and proximal leaves often with ferruginous hairs of *S. fuscescens*, but ovaries are densely hairy and long-styled as in *S. arctica*.

Salix arctica × *S. glauca* varieties: Hybrids are characterized by combinations of characteristics of *S. arctica* and *S. glauca* (G. W. Argus 1965, 1973; C. Bay 1992). They resemble *S. arctica* in its prostrate habit, glaucous branches and buds, sparse branchlet hairiness, dark-colored bracts with long, straight hairs, leaves with long, straight hairs on the abaxial surfaces projecting in a 'beard' at the apex, capsules reddish with long stigmas, and dark-colored anthers. They resemble *S. glauca* in its erect habit, leaves less oblanceolate and without the cuneate base of *S. arctica*, shorter petioles, bracts light-colored with shorter, wavy hairs, and divided styles. The presence of glaucous branches, as an indication of *S. arctica*, should be used with caution because *S. pellita* and *S. planifolia* also have glaucous branches, but these early flowering species may be phenologically isolated from *S. glauca*. On Baffin Island, Nunavut, the hybrid may be particularly difficult to recognize because of environmentally induced morphological convergence of *S. arctica* and *S. glauca*. On the Belcher Islands, on the east side of Hudson Bay, many plants have brownish to pinkish floral bracts with wavy or straight hairs, glabrate branches, and pilose branchlets, suggesting a hybrid swarm. Within the geographic overlap of *S. arctica* and *S. glauca* var. *cordifolia*, the name *S.* ×*waghornei* Rydberg can be applied to this hybrid (Argus 2003). See also A. K. Skvortsov (1971), Argus (1973), and Bay.

Salix arctica × *S. herbacea* occurring in northern Quebec, the eastern Canadian Arctic Archipelago, and eastern continental Nunavut resembles *S. arctica* in stems stouter, plants usually not rhizomatous, stipules often present, leaves sparsely glaucous abaxially, sometimes with long straight hairs, catkins more than 20-flowered and arising from lateral buds, and styles to 1.4 mm; and *S. herbacea* in plants sometimes rhizomatous, leaf margins crenulate, apex sometimes retuse, catkins few-flowered and sometimes arising from subterminal buds, ovaries glabrous, styles sometimes 0.2 mm. These hybrids set some seeds although the parental species have different chromosome numbers (tetraploid or hexaploid versus diploid, respectively). A variant of this hybrid with entire leaf margins was collected in eastern Greenland where both parents grow together. This hybrid resembles *S. herbacea* in leaf blades 0.5–1.1 times as long as wide with ciliate margins, floral bracts 0.9–1.2 mm, and stigmas 0.3 mm; and *S. arctica* in leaf margins entire and densely villous ovaries. It is intermediate between the two parents in leaf blades 11.5–14 mm, adaxial surfaces sparsely glaucous, floral bracts sparsely short-hairy, and styles 0.4–0.5 mm.

Salix arctica × *S. ovalifolia*: An intermediate specimen from Nuvagapak Point, Alaska, it resembles *S. arctica* in its growth form, leaf shape, and catkin size, and *S. ovalifolia* in ovaries sparsely hairy and glaucous and stigmas with a flat, non-papillate abaxial surface and pointed tip. It is very different from *S. ovalifolia* var. *glacialis*, which E. Hultén (1967) suggested may be the same hybrid.

Salix arctica × *S. pedicellaris* (*S.* ×*hebecarpa* (Fernald) Fernald, based on *S. fuscescens* var. *hebecarpa* Fernald), occurs in a raised sphagnum bog on Mt. Albert, Quebec. The population includes typical *S. pedicellaris* and the hybrids. Hybrids exhibit varying degrees of ovary and stipe hairiness, some juvenile leaf hairiness, and petioles that are deeply grooved and with margins that touch and overlap the groove. All of the specimens clearly indicate that *S. pedicellaris* is one parent, but the identity of *S. arctica* as the second parent is not established with certainty.

Salix arctica × *S. phlebophylla* is an apparently sterile hybrid that resembles *S. phlebophylla* in its general growth form and small, skeletonized leaves, and *S. arctica* in leaves thinly glaucous abaxially and ovaries hairy throughout. No seeds were set.

Salix arctica × *S. polaris*: Dwarf plants, presumed to be this hybrid, are characterized by being rhizomatous, as in *S. polaris*, but with leaves distinctly glaucous abaxially as in *S. arctica*. In general, the leaves and catkins are smaller than in typical *S. arctica*, and sometimes the styles also are shorter, but there is considerable variability. Ovaries are often sparsely hairy throughout, or glabrous at base or throughout except for the beak. The hybrids rarely form seed and may be infertile. Staminate plants may have anthers 0.9–1 mm, as in *S. arctica*, and leaves not glaucous abaxially, as in *S. polaris*. There is no single character that defines this hybrid and some plants from mixed populations appear to be intermediate. A specimen from Cornwallis Island, Northwest Territories, which was not rhizomatous but had relatively small, glaucous leaves and catkins with ovaries with patches of hairs especially on beaks, was presumed to be this hybrid, but it could simply be a dwarfed or abnormal specimen of *S. arctica*. In the Canadian Arctic Archipelago, this hybrid appears on Banks, Cornwallis, and Victoria islands. It is also known from Alaska, British Columbia, Northwest Territories, and Yukon.

Salix arctica × *S. rotundifolia* resembles *S. rotundifolia* except for its partially hairy ovaries and leaves glaucous abaxially.

Salix arctica × *S. stolonifera* occurs in swarms wherever the two parents grow together. The hybrids, and possible intergrades, have ovaries with bare patches or with hairs only on beaks.

41. Salix petrophila Rydberg, Bull. New York Bot. Gard. 1: 268. 1899 • Rocky Mountain willow E F

Salix arctica Pallas var. *petraea* Andersson in A. P. de Candolle and A. L. P. P. de Candolle, Prodr. 16(2): 287. 1868, not *S. petraea* Andersson 1829; *S. arctica* subsp. *petraea* (Andersson) Á. Löve, D. Löve & B. M. Kapoor; *S. arctica* var. *petrophila* (Rydberg) Kelso; *S. petrophila* var. *caespitosa* (Kennedy) C. K. Schneider

Plants 0.02–0.1 m, forming clones by layering. **Stems** decumbent or trailing; branches yellow-brown, gray-brown, or red-brown, (often weakly glaucous, dull or slightly glossy), glabrous; branchlets yellow-green or yellow-brown, usually glabrous, sometimes pilose to glabrescent. **Leaves:** stipules absent or rudimentary; petiole 2–13 mm (longer than subtended bud); largest medial blade usually amphistomatous, very narrowly to broadly elliptic, oblanceolate, or obovate, 19–44 × 7–21 mm, 1.5–4.6 times as long as wide, base cuneate or convex, margins flat, entire, apex acute, acuminate, convex, or rounded, abaxial surface pilose to glabrescent, hairs wavy, adaxial dull or slightly glossy, pilose or sparsely villous to glabrescent; proximal blade margins entire; juvenile blade villous or pilose abaxially, mainly on margin. **Catkins:** staminate (ca. 50 flowers), 18–32 × 6–13 mm, flowering branchlet 3–13 mm; pistillate moderately densely to loosely flowered, (18–80 flowers), slender or stout, 18–59(–70 in fruit) × 6–15 mm, flowering branchlet 2–40 mm; floral bract tawny or brown, 0.5–3.6 mm, apex acute or rounded, entire or 2-fid, abaxially hairy or ciliate, hairs straight or wavy. **Staminate flowers:** abaxial nectary 0–0.2 mm, adaxial nectary oblong, narrowly oblong, or square, 0.4–1.2 mm, nectaries distinct; filaments distinct or connate less than ½ their lengths; anthers ellipsoid, short-cylindric, or globose, 0.4–0.8 mm. **Pistillate flowers:** adaxial nectary oblong, square, narrowly oblong, or ovate, 0.5–1.2 mm, equal to or longer than stipe; stipe 0.2–0.8 mm; ovary pyriform or obturbinate, sparsely to densely villous, beak gradually tapering to styles; ovules 6–12 per ovary; styles 0.4–1.6 mm; stigmas flat, abaxially non-papillate with rounded or pointed tip, or broadly to slenderly cylindrical, 0.28–0.36–0.6 mm. **Capsules** 3.6–5 mm. $2n = 76$.

Flowering Jul–Aug. Alpine snowbeds, meadows, talus slopes and open dry pool beds in spruce-fir forests; 1700–4000 m; Alta., B.C.; Calif., Colo., Idaho, Mont., Nev., N.Mex., Oreg., Utah, Wash., Wyo.

Salix petrophila is often included in *S. arctica* (G. W. Argus 1993), but southern cordilleran populations, extending as far north as southern British Columbia and Alberta, seem to be a distinct taxon (Argus 1997).

The exact northern limit of this species still needs to be established, but in Alberta it does not seem to extend north of Waterton Lakes National Park, except for a population on springy slopes above Agness Lake, Banff National Park. Suitable alpine habitats between Waterton Lakes and Banff national parks, e.g., Mt. Armstrong, Tornado Mountain, and Crowsnest Pass, should be explored for *S. arctica* and *S. petrophila*.

42. **Salix sphenophylla** A. K. Skvortsov, Spisok Rast. Gerb. Fl. S.S.S.R. Bot. Inst. Vsesoyuzn. Akad. Nauk 16: 62. 1966 • Wedge-leaf willow

Salix sphenophylla subsp. *pseudotorulosa* A. K. Skvortsov

Plants 0.03–0.12 m, not clonal or forming clones by layering. **Stems** trailing and rooting; branches yellow-brown or brownish, glabrous; branchlets yellow-brown, glabrous. **Leaves:** stipules absent or rudimentary; petiole 4–25 mm, (glabrous or pilose adaxially); largest medial blade hypostomatous, narrowly elliptic, broadly elliptic, obovate, or very broadly obovate, 19–52 × 10–28 mm, 1–3 times as long as wide, base cuneate or convex, margins flat or slightly revolute, entire, apex convex, retuse, or rounded, abaxial surface glabrous, pilose or sparsely long-silky to glabrescent, hairs straight or wavy, adaxial slightly glossy, glabrous or pilose; proximal blade margins entire; juvenile blade (reddish), very sparsely long-silky abaxially. **Catkins:** staminate 21–53 × 7–13 mm, flowering branchlet 8–20 mm; pistillate loosely to densely flowered, slender or stout, 32–79 × 7–18 mm, flowering branchlet 4–27 mm; floral bract brown or black, 1.1–2 mm, apex rounded, entire, abaxially hairy or ciliate, hairs straight. **Staminate flowers:** abaxial nectary absent, adaxial nectary oblong, 0.6–1 mm; filaments distinct; anthers ellipsoid or shortly cylindrical, 0.4–0.6 mm. **Pistillate flowers:** adaxial nectary oblong or ovate, 0.7–1.6 mm, equal to or longer than stipe; stipe 0.5–1.4 mm; ovary obclavate or pyriform, glabrous, or patchy or streaky pilose or villous, especially on beak, beak gradually tapering to or slightly bulged below styles; ovules 10–18 per ovary; styles 0.6–1.8 mm; stigmas flat, abaxially non-papillate with pointed tip, or slenderly cylindrical, 0.32–0.5–0.68 mm. **Capsules** 4–12 mm. $2n$ = 38, 57.

Flowering mid Jun–late Jul (early Aug). Stony or gravelly substrates on talus, rocky outcrops, dry, stony tundra, sandy and moss tundra; 10–900 m; N.W.T., Yukon; Alaska; e Asia (Chukotka, Russian Far East, e Siberia).

The patchy or streaky indumentum on the ovaries of some plants suggests that they may be hybrids.

43. **Salix cascadensis** Cockerell, Muhlenbergia 3: 9. 1907 • Cascade willow [E] [F]

Salix tenera Andersson in A. P. de Candolle and A. L. P. P. de Candolle, Prodr. 16(2): 288. 1868, not A. Braun ex Unger 1850; *S. cascadensis* var. *thompsonii* Brayshaw

Plants 0.03–0.1 m, forming clones by rhizomes. **Stems** erect or trailing; branches yellow-brown or gray-brown, (sometimes weakly glaucous), glabrous; branchlets yellow-green or yellow-brown, glabrous or puberulent. **Leaves** (marcescent); stipules absent; petiole 1.5–5 mm; largest medial blade usually amphistomatous, narrowly elliptic or elliptic, 9–26 × 3.8–7.5 mm, 2.4–4.3 times as long as wide, base cuneate, margins flat, entire, ciliate, apex acute, acuminate, or convex, abaxial surface glabrous or pilose, hairs wavy, adaxial slightly glossy, glabrous or pilose; proximal blade margins entire; juvenile blade glabrous or sparsely villous abaxially. **Catkins:** staminate (20–50 flowers), 12.5–26.5 × 5.5–9 mm, flowering branchlet 1–11 mm; pistillate moderately densely to loosely flowered, (15–35(–43) flowers), stout or subglobose, 10–23(–30 in fruit) × 5–8 mm, flowering branchlet 2–9 mm; floral bract brown, 1.6–2.6 mm, apex rounded, entire, abaxially sparsely hairy or ciliate, hairs wavy or straight. **Staminate flowers:** abaxial nectary (0–)0.2–0.6 mm, adaxial nectary oblong or square, 0.4–1.2 mm, nectaries distinct; filaments distinct or basally connate; anthers ellipsoid, 0.3–0.6 mm. **Pistillate flowers:** adaxial nectary oblong or square, 0.5–1.2 mm, longer than stipe; stipe 0–0.6 m; ovary pyriform, usually densely villous, beak abruptly tapering to or slightly bulged below styles; ovules 6–10 per ovary; styles 0.3–1 mm; stigmas slenderly cylindrical, 0.28–0.39–0.56 mm. **Capsules** 3.5–5 mm.

Flowering early Jul–early Aug. Mesic to dry rocky slopes, ridges, high subalpine and alpine tundra; 2200–3900 m; B.C.; Colo., Utah, Wash., Wyo.

The morphological variability of *Salix cascadensis* is not well understood. Typically, it has leaves that are narrow, sharply pointed, and glaucous abaxially, catkins 15–43-flowered, dark brown floral bracts, and ovaries very densely hairy. Specimens with leaves not glaucous abaxially, catkins relatively few-flowered, and ovaries either sparsely hairy throughout, hairy only on beaks, or hairy in streaks, may be hybrids but the glabrous ovaried *S. cascadensis* var. *thompsonii* shows no obvious signs of hybridization.

Hybrids:

Salix cascadensis is suspected to hybridize with *S. barclayi*.

Salix cascadensis × *S. rotundifolia* var. *dodgeana* resembles the former in its narrow, acute to acuminate leaves and the latter in leaves not glaucous abaxially, tawny floral bracts, relatively short catkins, and glabrous ovaries. The catkins are intermediate in being 6–7-flowered.

2d7. SALIX Linnaeus (subg. CHAMAETIA) sect. MYRTILLOIDES (Borrer) Andersson in A. P. de Candolle and A. L. P. P. de Candolle, Prodr. 16(2): 229. 1868

Salix [unranked] *Myrtilloides* Borrer in J. C. Loudon, Arbor. Frutic. Brit. 3: 1587. 1838

Plants 0.15–1.8 m, not clonal or clonal by layering. **Largest medial blades** hypostomatous, abaxial surface glaucous. **Catkins** from lateral buds. **Staminate flowers:** filaments glabrous or hairy. **Pistillate flowers:** abaxial nectary sometimes present, then distinct or connate to adaxial and forming a cup; ovary glaucous or not, glabrous or puberulent to very densely long-silky, hairs white or white and ferruginous, flattened.

Species 5 (4 in the flora): North America, Eurasia.

Salix myrtilloides Linnaeus is found in Eurasia.

44. Salix pedicellaris Pursh, Fl. Amer. Sept. 2: 611. 1813 • Bog willow [E] [F]

Salix myrtilloides Linnaeus var. *hypoglauca* (Fernald) C. R. Ball; *S. pedicellaris* var. *hypoglauca* Fernald; *S. pedicellaris* var. *tenuescens* Fernald

Plants 0.2–1.5 m, forming clones by layering. **Stems** erect, decumbent or trailing; branches gray-brown, glabrous; branchlets yellow-brown or red-yellow, glabrous or puberulent (hairs straight, minute, inner membranaceous bud-scale layer free, not separating from outer layer). **Leaves:** stipules absent or rudimentary; petiole (deeply to shallowly grooved adaxially), 3–8 mm, (glabrous or puberulent); largest medial blade narrowly oblong, oblong, narrowly to broadly elliptic, narrowly oblanceolate or oblanceolate, 19–53(–69) × 5–20 mm, 1.8–4.9 times as long as wide, base convex or rounded, margins flat or slightly revolute, entire, apex acute, convex, or rounded, abaxial surface glabrous, adaxial dull, glaucous, glabrous (rarely very sparsely short-silky, hairs usually white, sometimes also ferruginous); proximal blade margins entire; juvenile blade (reddish or yellowish green), glabrous, puberulent, or sparsely pubescent abaxially, (hairs usually white, sometimes also ferruginous). **Catkins:** staminate 11–21 × 4–8 mm, flowering branchlet 3–12 mm; pistillate loosely flowered, stout or subglobose, 14–37 × 5–14 mm, flowering branchlet 7–25 mm; floral bract tawny or light rose, 0.8–1.6 mm, apex rounded, entire, abaxially very sparsely hairy distally, hairs straight or wavy. **Staminate flowers:** abaxial nectary absent, adaxial nectary oblong or narrowly oblong, 0.5–1.1 mm; filaments distinct or connate less than ½ their lengths, glabrous, or hairy basally or on proximal ½; anthers (yellow), ellipsoid, 0.4–0.6 mm. **Pistillate flowers:** abaxial nectary absent, adaxial nectary oblong, 0.2–1.4 mm, shorter than stipe; stipe 2.1–3.2 mm; ovary obclavate, often glaucous, glabrous, beak abruptly tapering to styles; ovules 4–6 per ovary; styles connate or distinct ½ their lengths, 0.1–0.2 mm; stigmas flat, abaxially non-papillate with rounded tip, or 2 plump lobes, 0.2–0.25–0.36 mm. **Capsules** 4–8 mm. $2n$ = 38, 57, 76.

Flowering mid Apr–mid Jul. *Sphagnum* bogs, fens, black spruce treed bogs; 0–1400 m; St. Pierre and Miquelon; Alta., B.C., Man., N.B., Nfld. and Labr., N.S., Nunavut, Ont., Que., Sask., Yukon; Conn., Idaho, Ill., Ind., Iowa, Maine, Mass., Mich., Minn., N.H., N.J., N.Y., N.Dak., Ohio, Oreg., Pa., R.I., Vt., Wash., Wis.

Salix pedicellaris is found in Nunavut on Akimiski Island in James Bay and on the Belcher Islands in Hudson Bay.

Salix pedicellaris is very distinct with decumbent habit, leathery, glabrous leaves that are glaucous on both surfaces, loosely flowered catkins, ovaries reddish, glabrous and often glaucous, and stipes 2.1–3.2 mm. In the flora area, it hybridizes with six other species (see below). This compares with the closely related European *S. myrtilloides* Linnaeus, which is reported (B. Jonsell

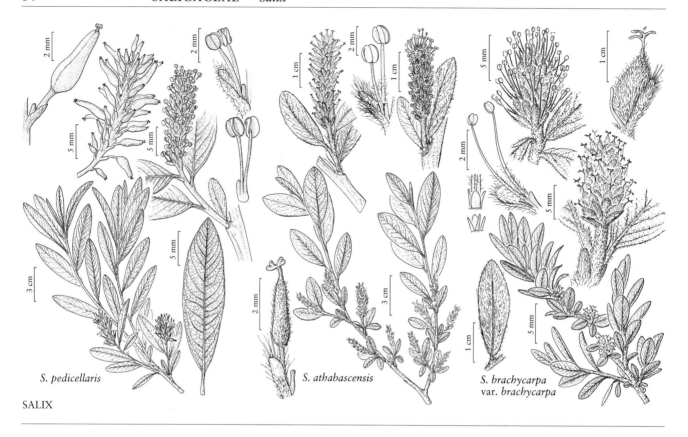

and T. Karlsson 2000+, vol. 1) to hybridize with seven species. The distinctive appearance may make hybrids easily recognizable but it is complex cytologically.

Hybrids:

Salix pedicellaris forms natural hybrids with *S. arctica*, *S. argyrocarpa*, *S. athabascensis*, *S. chlorolepis*, *S. glauca* var. *cordifolia*, *S. pedicellaris*, and *S. pellita*. Hybrids with *S. candida* and *S. eriocephala* have been reported (M. L. Fernald 1950) but no convincing specimens have been seen.

Salix pedicellaris × *S. pellita* (*S.* ×*jamesensis* Lepage) was described from the west coast of James Bay, Ontario. It is to be expected throughout northern Ontario and Quebec. According to E. Lepage (1964), its yellowish midveins and the hairiness of floral bracts superficially resemble those of *S. pellita* forma *psila* C. K. Schneider (syn. *S. pellita* forma *denudata*), but its reticulate leaf venation resembles that of *S. pedicellaris*.

45. Salix athabascensis Raup, Rhodora 32: 111, plate 202. 1930 • Athabasca willow [E] [F]

Salix fallax Raup; *S. pedicellaris* Pursh var. *athabascensis* (Raup) B. Boivin

Plants 0.6–1.3 m, not clonal. **Stems** erect; branches gray-brown, hairy; branchlets red-brown, sparsely or moderately densely pubescent, (buds *alba*-type). **Leaves:** stipules absent or rudimentary on early ones, usually rudimentary, rarely foliaceous, on late ones; petiole (shallowly grooved adaxially), 3–10 mm, (puberulent or villous); largest medial blade oblong, narrowly elliptic, elliptic, oblanceolate, or obovate, 17–50 × 8–18 mm, 1.9–3.2 times as long as wide, base cuneate or convex, margins flat or slightly revolute, entire, apex acuminate or convex, abaxial surface glabrescent or sparsely silky, hairs appressed or somewhat spreading, (usually white, sometimes also ferruginous), straight or wavy, adaxial dull or slightly glossy, glabrous, glabrescent, pilose, or sparsely long-silky along midribs and margin, (hairs usually white, sometimes also ferruginous, appressed); proximal blade margins entire; juvenile blade sparsely to moderately densely villous or long-silky abaxially (hairs usually white, sometimes also ferruginous). **Catkins:** staminate 14–31 × 8–18 mm, flowering branchlet 1.5–9 mm; pistillate loosely flowered, stout to globose, 10–58 × 7–25 mm, flowering branchlet 3.5–26 mm; floral bract tawny, 1–1.6 mm, apex rounded, entire, abaxially sparsely hairy, almost glabrous, hairs wavy. **Staminate flowers:** abaxial nectary (0–)0.3–0.6 mm, adaxial nectary oblong or ovate, 0.4–1.2 mm, nectaries distinct; filaments distinct, hairy basally or on proximal ½; anthers globose, 0.4–0.6 mm. **Pistillate flowers:** abaxial nectary absent, adaxial nectary oblong, 0.4–1.3 mm, shorter than stipe; stipe 0.8–1.3 mm; ovary pyriform, very densely long-silky, beak gradually tapering to or slightly bulged below styles; ovules 6–14 per ovary; styles 0.5–1 mm; stigmas flat, abaxially non-papillate with rounded tip, or broadly to slenderly cylindrical, 0.28–0.35–0.48 mm. **Capsules** 5.6–7.2 mm. $2n = 76, 95, 114$.

Flowering late May–late Jul. Fens, bogs, and treed bogs; 0–1800 m; Alta., B.C., Man., N.W.T., Sask., Yukon; Alaska.

The three polyploid chromosome numbers reported for *Salix athabascensis*, as well as the presence of leaves with ferruginous hairs, otherwise unknown in sect. *Myrtilloides*, are indicators of allopolyploidy.

Hybrids:

Salix athabascensis forms natural hybrids with *S. pedicellaris*. These hybrids combine the characteristics of the parents. The ovaries may be moderately densely villous or glabrous, but commonly have hairs in patches, or the stipes may be hairy and the ovaries glabrous; juvenile blades, and sometimes mature leaves, are hairy with white and ferruginous hairs. Some plants that resemble *S. athabascensis* have leaves glaucous adaxially, as in *S. pedicellaris*. The ovaries often appear to be infertile.

46. Salix chlorolepis Fernald, Rhodora 7: 186. 1905 • Green-bract willow [C][E]

Plants 0.15–0.2 m, not clonal. **Stems** erect; branches red-brown, (weakly glaucous), glabrous; branchlets yellow-brown, glabrous. **Leaves:** stipules absent or rudimentary; petiole (deeply to shallowly grooved adaxially), 1–3.5 mm; largest medial blade elliptic, oblanceolate, or obovate, 14–33 × 7–12 mm, 1.9–3.4 times as long as wide, base cuneate or convex, margins flat or slightly revolute, entire, ciliate, apex acute, convex, or rounded, abaxial surface glabrous, adaxial slightly glossy, glabrous; proximal blade margins entire; juvenile blade (sometimes reddish), glabrous, ciliate. **Catkins:** staminate 6.5 × 5–8 mm, flowering branchlet 1–3 mm; pistillate densely flowered, stout or subglobose, 7–12 × 3–6 mm, flowering branchlet 1.8–12 mm; floral bract tawny, brown, or greenish, 1–2.6 mm, apex broadly rounded to retuse, entire, abaxially glabrous. **Staminate flowers:** abaxial nectary 0.3–0.4 mm, adaxial nectary oblong, 0.5–0.8 mm, nectaries distinct; filaments distinct, glabrous; anthers ellipsoid, 0.4–0.6 mm. **Pistillate flowers:** adaxial nectary narrowly oblong, 0.9–1.6 mm, longer than stipe, nectaries distinct or connate and shallowly cup-shaped; stipe 0–0.4 mm; ovary pyriform, glabrous, beak abruptly tapering to or slightly bulged below styles; ovules 8–10 per ovary; styles connate to distinct ½ their lengths, 0.5–1.3 mm; stigmas flat, abaxially non-papillate with rounded tip, or slenderly cylindrical, 0.3–0.6 mm. **Capsules** 3.6–6 mm. $2n = 38$.

Flowering Jul–early Aug. Wet *Sphagnum* bog on alpine, serpentine barrens; of conservation concern; 1000–1200 m; Que.

Salix chlorolepis, known from Mt. Albert, is characterized by its general glabrousness. It seems to have a relationship with *S. brachycarpa* similar to that of *S. raupii* to *S. glauca*. Both may have originated through mutation or hybridization.

Hybrids:

Salix chlorolepis forms natural hybrids with *S. brachycarpa* var. *brachycarpa*.

Salix chlorolepis × *S. pedicellaris* is a putative hybrid that has relatively small, glabrous leaves. Both parents occur together on Mt. Albert, Quebec.

47. Salix raupii Argus, Canad. J. Bot. 52: 1303, plate 1. 1974 • Raup willow [C][E]

Plants 1.2–1.8 m, not clonal. **Stems** erect; branches gray-brown, glabrous; branchlets yellow-brown, glabrous. **Leaves:** stipules foliaceous; petiole 5–9 mm; largest medial blade narrowly elliptic, 32–58 × 12–19 mm, 2–3.3 times as long as wide, base cuneate or convex, margins slightly revolute, entire, apex acute to acuminate, abaxial surface glabrous, adaxial slightly glossy, glabrous; proximal blade margins shallowly serrulate; juvenile blade glabrous. **Catkins:** staminate 17.5–42 × 5–13 mm, flowering branchlet 6–7 mm; pistillate moderately densely flowered, stout, 20–40 × 6–12 mm, flowering branchlet 4–7 mm; floral bract tawny or bicolor, 1.3–2.5 mm, apex rounded, entire, abaxially glabrous. **Staminate flowers:** abaxial nectary 0.3–0.8 mm, adaxial nectary narrowly oblong, 0.6–1 mm, nectaries distinct; filaments distinct, glabrous; anthers ellipsoid, shortly cylindrical, or globose, 0.4–0.7 mm. **Pistillate flowers:** abaxial nectary absent, adaxial nectary narrowly oblong or oblong, 0.5–1.1 mm, equal to or longer than stipe; stipe 0.4–1.2 mm; ovary pyriform, usually glabrous, rarely puberulent, beak slightly bulged below styles; ovules 12 per ovary; styles 0.6–0.8 mm; stigmas flat, abaxially not papillate with rounded tip, or broadly cylindrical, 0.3–0.5 mm. **Capsules** 4.4–8 mm.

Flowering late Jun. Thickets in moist, open forests, gravel floodplains; of conservation concern; 800–1500 m; Alta., B.C., N.W.T., Yukon.

Salix raupii resembles glabrous *S. glauca* var. *villosa*. Thin-layer chromatography of leaf phenolics in *S. raupii* revealed a pattern similar to those of *S. glauca* vars. *villosa* and *acutifolia* and *S. athabascensis* (G. W. Argus, unpubl.). Based on overall similarity, its nearest neighbors are *S. glauca*, in a broad sense, and *S. athabascensis* (Argus 1997). The sectional placement of this species is uncertain. It is placed here in sect. *Myrtilloides* because it clusters with *S. athabascensis*, but it is evidently close to *S. glauca* and may be a species of intersectional hybrid origin.

2d8. SALIX Linnaeus (subg. CHAMAETIA) sect. GLAUCAE (Fries) Andersson in A. P. de Candolle and A. L. P. P. de Candolle, Prodr. 16(2): 273. 1868

Salix tribe *Glaucae* Fries, Physiogr. Sällsk. Årsberätt. 2: 35. 1825

Plants 0.2–6 m, not clonal or clonal by layering. **Largest medial blades** hypostomatous, hemiamphistomatous, or amphistomatous, abaxial surface glaucous (sometimes obscured by hairs). **Catkins** from lateral buds. **Staminate flowers:** filaments glabrous or hairy. **Pistillate flowers:** abaxial nectary sometimes present, then distinct or connate to adaxial one and forming a cup; ovary pubescent or moderately to very densely villous, tomentose, woolly, or silky, hairs white, cylindrical or flattened.

Species 8 (3 in the flora): North America, Eurasia.

Section *Glaucae* is placed here in subg. *Chamaetia* but it clusters with members of subg. *Vetrix* (G. W. Argus 1997) and could equally well be included there. It includes high polyploids, which probably incorporate genes from members of other sections and subgenera.

48. **Salix brachycarpa** Nuttall, N. Amer. Sylv. 1: 69. 1842 · Small-fruit willow [E] [F]

Plants 0.2–1.5 m, not clonal. **Stems** erect or decumbent; branches gray-brown or red-brown, villous or short-silky to glabrescent; branchlets red-brown, long-silky, villous, or woolly. **Leaves:** stipules absent or rudimentary on early ones, foliaceous or rudimentary on late ones; petiole (deeply to shallowly grooved adaxially), (0.5–)1–3(–4) mm, (usually shorter than or barely exceeding subtended bud); largest medial blade hypostomatous, narrowly oblong, oblong, narrowly elliptic, elliptic, narrowly oblanceolate, ovate, or obovate, base rounded, convex, cordate, or subcordate, margins flat, entire, apex rounded, acute, or convex, abaxial surface (sometimes obscured by hairs), densely villous, woolly, or long-silky, adaxial slightly glossy, pilose, villous, or long-silky to glabrescent, (hairs straight or wavy); proximal blade margins entire; juvenile blade very densely long-silky abaxially. **Catkins:** staminate 5.3–24 × 4–10 mm, flowering branchlet 0.3–24(–43) mm; pistillate (and staminate) densely flowered, globose, subglobose, or stout, 6–25 × 4–15 mm, flowering branchlet 0.3–11 mm; floral bract tawny or greenish, 1–3 mm, apex rounded or convex, entire, abaxially hairy, hairs straight or wavy. **Staminate flowers:** abaxial nectary (0–)0.5–1.5 mm, adaxial nectary 0.5–1.4 mm, nectaries distinct or connate and cup-shaped; filaments distinct or connate less than ½ their lengths, glabrous or hairy throughout or on proximal ½; anthers ellipsoid, shortly cylindrical, or globose, 0.3–0.8 mm. **Pistillate flowers:** abaxial nectary often present, adaxial nectary longer than stipe; stipe 0–0.6 mm; ovary pyriform, very densely villous, tomentose, or woolly, beak slightly bulged below styles; ovules 2–10 per ovary; styles connate to distinct ½ their lengths, 0.4–1.5 mm; stigmas broadly or slenderly cylindrical, 0.24–0.3–0.48 mm. **Capsules** 3–6.5 mm.

Varieties 2 (2 in the flora): North America.

1. Branchlets and largest medial blade surfaces moderately densely villous or long-silky; largest medial blades (1.5–)2.8–3(–4) times as long as wide; anthers 0.3–0.5 mm; catkins globose, subglobose, or stout . . . .
. . . . 48a. *Salix brachycarpa* var. *brachycarpa*
1. Branchlets and largest medial blade surfaces very densely woolly, villous, or long-silky; largest medial blades 1.4–2.6 times as long as wide; anthers 0.5–0.8 mm; catkins stout or subglobose (Lake Athabasca sand dunes, Saskatchewan) . . .
. . . . 48b. *Salix brachycarpa* var. *psammophila*

48a. Salix brachycarpa Nuttall var. brachycarpa E F

Salix brachycarpa var. *antimima* (C. K. Schneider) Raup; *S. brachycarpa* var. *glabellicarpa* C. K. Schneider; *S. brachycarpa* var. *sansonii* C. R. Ball

Stems: branches short-silky or villous to glabrescent; branchlets usually moderately densely long-silky, villous, or woolly. **Leaves:** petiole (0.5–)1–3(–4) mm; largest medial blade narrowly oblong, oblong, narrowly elliptic, elliptic, narrowly oblanceolate, or obovate, (10–)23–30(–40) × 5–16 mm, (1.5–)2.8–3(–4) times as long as wide, base rounded, convex, or subcordate, apex acute or convex, abaxial surface moderately densely villous or long-silky, adaxial pilose, villous, or long-silky to glabrescent, hairs straight or wavy. **Catkins:** staminate 5.3–21 × 4–10 mm, flowering branchlet 0.3–10 mm; pistillate globose, subglobose, or stout, 6–20 × 4–15 mm, flowering branchlet 0.3–11 mm; floral bract tawny, 1–3 mm. **Staminate flowers:** abaxial nectary 0.5–1.1 mm, adaxial nectary 0.5–1.4 mm; filaments distinct or connate less than ½ their lengths, glabrous, or hairy on proximal ½; anthers ellipsoid or globose, 0.3–0.5 mm. **Pistillate flowers:** abaxial nectary (0–)0.4–1.4 mm, adaxial nectary oblong, 0.8–2 mm; stipe 0–0.6 mm; ovules 2–5 per ovary; styles connate to distinct ½ their lengths, 0.5–1.5 mm. **Capsules** 3–6 mm. $2n = 38$.

Flowering mid Jun–late Aug. Moist to mesic open forests, sedge fens, seepage on limestone, scree slopes, gravel floodplains; 0–4000 m; Alta., B.C., Man., N.W.T., Nunavut, Ont., Que., Sask., Yukon; Calif., Colo., Idaho, Mont., N.Mex., Oreg., Utah, Wash., Wyo.

Variety *brachycarpa* occurs in Nunavut only on islands in James Bay.

Hybrids:

Variety *brachycarpa* forms natural hybrids with *Salix arizonica*, *S. barclayi*, *S. boothii*, *S. candida*, *S. chlorolepis*, *S. glauca* var. *villosa*, and *S. planifolia*.

Placement of specimens from Anticosti Island, Quebec, and North Point, James Bay, Ontario, with densely villous branchlets and relatively short petioles, thought to be hybrids with *Salix glauca* var. *cordifolia*, is dubious.

Variety *brachycarpa* × *Salix candida* (*S.* ×*argusii* B. Boivin) is infrequent in Manitoba, Quebec, and Saskatchewan.

Variety *brachycarpa* × *Salix chlorolepis* (*S.* ×*gaspeensis* C. K. Schneider) resembles var. *brachycarpa* but has leaves only slightly pilose and ovaries with hairs only on the beaks (G. W. Argus 1965).

Variety *brachycarpa* × *Salix glauca* var. *villosa* (*S.* ×*wyomingensis* Rydberg) is a frequent hybrid in southern Rocky Mountains. It is characterized by stipes 0.3 mm or longer, long-cylindrical catkins, ovaries with relatively long beaks, petioles more than three times the length of buds, and leaves sparsely hairy. The extent and nature of this hybridization needs to be studied (G. W. Argus 1965).

Variety *brachycarpa* × *Salix planifolia* "var. *monica*" occurs in Steens Mountains, Oregon.

48b. Salix brachycarpa Nuttall var. psammophila Raup, J. Arnold Arbor. 17: 230, plate 191. 1936

• Small-fruit sand dune willow C E

Stems: branches densely villous; branchlets very densely woolly, villous, or long-silky. **Leaves:** petiole 0.8–4 mm, (densely villous to long-silky adaxially); largest medial blade oblong, elliptic to ovate, 15–34 × 7.5–19 mm, 1.4–2.6 times as long as wide, base subcordate, cordate, or convex, apex rounded or convex, abaxial surface very densely woolly, villous, or long-silky, adaxial moderately densely villous or long-silky. **Catkins:** staminate 9.5–24 × 6–10 mm, flowering branchlet 2–24(–43 in later-flowering plants) mm; pistillate stout or subglobose, 18–28.5 × 7–14 mm, flowering branchlet 4–8 mm; floral bract tawny or greenish, 1.2–2.4 mm. **Staminate flowers:** abaxial nectary (0–)0.9–1.5 mm, adaxial nectary 0.5–2 mm; filaments distinct, hairy throughout or on proximal ½; anthers (yellow), ellipsoid or shortly cylindrical, 0.5–0.8 mm. **Pistillate flowers:** abaxial nectary 0.3–0.6 mm, adaxial nectary narrowly oblong or ovate (with slender tips), 0.8–1.8 mm; stipe 0–0.2 mm; ovules 8–10 per ovary; styles 0.4–1.2 mm. **Capsules** 4.5–6.5 mm.

Flowering late Jun–early Aug. Open sand dunes; of conservation concern; 200–300 m; Sask.

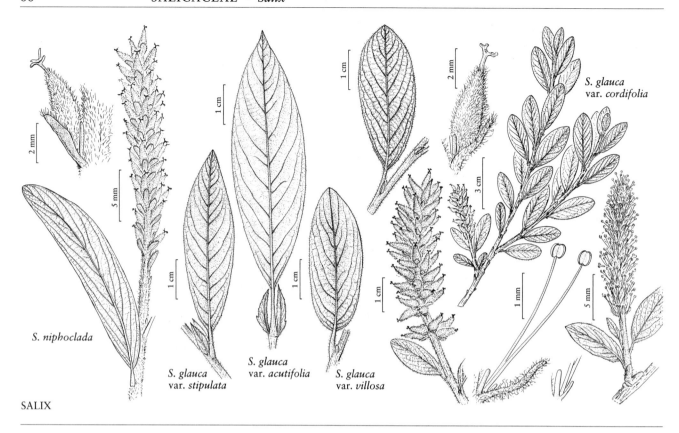

Previously, I (G. W. Argus 1965) did not recognize var. *psammophila*, but field experience showed that this sand dune endemic deserves formal taxonomic status. It is mainly characterized by very dense hairiness and slightly broader leaf blades. *Salix brachycarpa* var. *psammophila* is found only on Lake Athabasca sand dunes.

Hybrids:

Variety *psammophila* forms natural hybrids with *Salix pyrifolia*, *S. silicicola*, and *S. turnorii*. All of these taxa occur together in the sand dunes at Lake Athabasca in northwestern Saskatchewan.

Variety *psammophila* × *Salix pyrifolia* resembles *S. pyrifolia* in its reddish petioles and midribs, margins somewhat toothed, styles 0.4 mm, and stipes 0.4–1.1 mm; and *S. brachycarpa* var. *psammophila* in its sparsely hairy branches, branchlets, and ovaries.

Variety *psammophila* × *Salix turnorii* (*S.* ×*brachypurpurea* B. Boivin) resembles var. *psammophila* in floral bracts tawny, petioles relatively very short, and branchlets and juvenile leaves very densely long-silky; and *S. turnorii* in leaves amphistomatous, leaf margins serrulate, and branches yellow-brown. It is intermediate in ovaries varying from moderately densely hairy to glabrescent. The hybrids appear to be infertile (G. W. Argus 1965).

49. **Salix niphoclada** Rydberg, Bull. New York Bot. Gard. 1: 272. 1899 • Barren-ground or snow willow [F]

Salix brachycarpa Nuttall subsp. *fullertonensis* (C. K. Schneider) Á. Löve & D. Löve; *S. brachycarpa* var. *fullertonensis* (C. K. Schneider) Argus; *S. brachycarpa* var. *mexiae* C. R. Ball; *S. brachycarpa* subsp. *niphoclada* (Rydberg) Argus; *S. fullertonensis* C. K. Schneider; *S. glauca* Linnaeus var. *niphoclada* (Rydberg) Wiggins; *S. muriei* Hultén; *S. niphoclada* var. *fullertonensis* (C. K. Schneider) Raup; *S. niphoclada* var. *mexiae* (C. R. Ball) Hultén; *S. niphoclada* var. *muriei* (Hultén) Raup

Plants 0.3–1.5(–3) m, not clonal. **Stems** erect, decumbent, or trailing; branches gray-brown, yellow-brown, or red-brown, pilose, villous, or long-silky to glabrescent; branchlets violet, red-brown, or yellow-brown, pilose, densely villous, long-silky, or tomentose. **Leaves:** stipules (sometimes marcescent), foliaceous, (sometimes obscured by hairs); petiole (shallowly grooved adaxially), 2–5.5 mm, (usually shorter than or barely exceeding subtended bud, villous adaxially); largest medial blade hypostomatous or amphistomatous, narrowly oblong, narrowly to broadly elliptic, lanceolate, or obovate, 13–64 × 6–22 mm, 1.6–5.5 times as long as wide, base convex, margins slightly revolute, entire (obscured by hairs), apex acuminate or acute, abaxial surface densely villous or woolly to pilose, hairs straight or wavy, adaxial slightly glossy, moderately densely to sparsely villous; proximal blade margins entire; juvenile blade abaxially densely long-silky. **Catkins:** staminate 12–42 × 4–14 mm, flowering branchlet 0–20 mm; pistillate densely flowered, slender, stout, or subglobose, 16–69 × 4–13 mm, flowering branchlet 4–30 mm; floral bract tawny, brown, or black, 1.2–3.2 mm, apex rounded, entire, abaxially sparsely to moderately densely hairy, hairs wavy. **Staminate flowers:** abaxial nectary 0.5–1.2 mm, adaxial nectary narrowly oblong to ovate or flask-shaped, 0.5–1.5 mm, nectaries distinct or connate and cup-shaped; filaments distinct, glabrous, or sparsely hairy on proximal ½; anthers ellipsoid or globose, 0.3–0.5 mm. **Pistillate flowers:** abaxial nectary rarely present, 0.7 mm, adaxial nectary narrowly oblong, ovoid, or flask-shaped, 0.5–2 mm, longer than stipe; stipe 0–0.5 mm; ovary pyriform, very densely villous or long-silky, beak slightly bulged below or abruptly tapering to styles; ovules 8–20 per ovary; styles connate to almost distinct, 0.2–0.8 mm; stigmas slenderly to broadly cylindrical, 0.2–0.36–0.6 mm. **Capsules** 4–6 mm.

Flowering Jun–Jul. Wet to moderately well-drained calcareous, gravelly or sandy floodplains, terraces, eskers, drumlins, fine, silty loess deposits, dry to mesic stony alpine slopes and saline flats, limestone talus, sand blowouts, plains; 10–2300 m; B.C., N.W.T., Nunavut, Yukon; Alaska; Asia (Russia).

Salix niphoclada was treated as *S. brachycarpa* subsp. *niphoclada* (G. W. Argus 1965, 1973) because where the ranges of the two taxa overlap in northern British Columbia they appeared to intergrade. Species rank is used here, however, because there is no evidence to suggest that intergradation is common or extends beyond the small area of overlap.

Salix brachycarpa var. *fullertonensis*, which mainly occurs in southern Nunavut and the western side of Hudson Bay, is characterized by smaller leaves and catkins, broad, reddish floral bracts, and more sparsely hairy branchlets (G. W. Argus 1965). Its intergradation with *S. niphoclada* is so extensive that it is not recognized here as a separate taxon.

Hybrids:

Salix niphoclada forms natural hybrids with *S. glauca* var. *acutifolia* and *S. setchelliana*.

Salix niphoclada × *S. setchelliana* is a putative hybrid from Sheep Mountain, Kluane, Yukon, at 1180 m. It resembles *S. niphoclada* in leaf shape and indumentum, stigmas 0.6 mm, floral bracts oblong, 1.8 mm, and in the shape and length of pistillate catkins, but it has the glabrous, reddish ovaries of *S. setchelliana*. Both putative parents occur in the area.

50. Salix glauca Linnaeus, Sp. Pl. 2: 1019. 1753 • Gray or gray-leaf willow [F]

Plants 0.2–6 m, not clonal. **Stems** erect or decumbent; branches brownish, yellow-brown, gray-brown, or red-brown, villous or pilose to glabrescent; branchlets yellow-brown or red-brown, sparsely to densely villous or tomentose to glabrescent. **Leaves:** stipules (marcescent or not), foliaceous or rudimentary on early and late ones; petiole (usually deeply to shallowly grooved adaxially), 1–27 mm, (much longer than subtended bud); largest medial blade usually hypostomatous, sometimes hemiamphistomatous or amphistomatous, usually narrowly elliptic, elliptic, usually oblanceolate or obovate, sometimes narrowly oblong or obovate, 27–82 × 6–39 mm, 1.4–4.8 times as long as wide, base usually cuneate or convex, sometimes rounded, rarely subcordate, margins slightly revolute or flat, usually entire, apex acute, acuminate, convex, or rounded, abaxial surface densely villous or villous-silky, tomentose, short- or long-silky, or pilose, hairs usually wavy or straight, sometimes curved, adaxial usually slightly glossy, sometimes dull, moderately densely villous, pilose, or long-silky to glabrescent; proximal blade margins entire or serrulate; juvenile blade sparsely or densely villous, tomentose, or long-silky abaxially. **Catkins:** staminate 14–53 × 5–17 mm, flowering branchlet 1–25 mm; pistillate densely to sometimes loosely flowered, slender, stout, subglobose, or globose, 15–83 × 7–21 mm, flowering branchlet 2–37 mm; floral bract tawny, brown, bicolor, or greenish, 1–3.4 mm, apex convex or rounded, entire, abaxially hairy, hairs wavy, crinkled, or straight. **Staminate flowers:** abaxial nectary 0.1–1 mm, adaxial nectary narrowly oblong, oblong, square, ovate, or flask-shaped, 0.5–1.3 mm, nectaries distinct, or connate and cup-shaped; filaments distinct or slightly or partly connate, glabrous, or hairy on proximal ½; anthers 0.4–0.8 mm. **Pistillate flowers:** abaxial nectary absent, adaxial nectary usually narrowly oblong, oblong, or ovate, sometimes flask-shaped, 0.4–1.8 mm, shorter to longer than stipe; stipe 0.3–2.8 mm; ovary pyriform or obclavate, densely villous, tomentose, short-silky, or pubescent, beak usually gradually tapering to styles, sometimes gradually tapering to or slightly bulged below styles; ovules 6–22 per ovary; styles connate to distinct ½ their lengths or more, 0.3–1.6 mm; stigmas flat, abaxially non-papillate with rounded tip, or slenderly or broadly cylindrical, 0.2–0.8 mm. **Capsules** 4.5–9 mm. $2n = 76, 95, 114, 152$.

Varieties 5 (4 in the flora): North America, Eurasia (China [Altay Shan], Chukotka, Mongolia, Novaya Zemlya, Russian Far East, Sakhalin, Scandinavia, arctic, e, w Siberia).

Occurrence of *Salix glauca* is disjunct in western Siberia; var. *glauca* is known from Scandinavia.

Salix glauca is ubiquitous, highly polymorphic, and polyploid. It appears to have four major variations that are treated in different ways. European floras have treated it as an undivided polymorphic species (A. K. Skvortsov 1999), as several distinct species (K. H. Rechinger 1964b), or as comprising several subspecies (Rechinger 1993; G. W. Argus et al. 1999; B. Jonsell and T. Karlsson 2000+, vol. 1). In North America, Argus (1965) treated the four central tendencies as widely intergrading "phases" of a single species, later adopting varietal rank, whereas E. Hultén (1968) recognized them as subspecies. In view of the fact that the four elements are confluent over wide areas, there may be some merit in recognizing them as informal phases, but varietal rank is used here.

All specimens from Iceland named *Salix glauca* are *S. arctica* and those from Svalbard, Norway, are *S. lanata* Linnaeus.

The major reason for the high variability within *Salix glauca* seems to be high, and probably recurrent, polyploidy. Tetraploids, pentaploids, hexaploids, and octoploids are known in the species, and two of the subspecies include three ploidal levels. There also are intergrading geographical variations that are the basis for the recognition of infraspecific taxa.

Hybrids:

Salix glauca forms natural hybrids with *S. arctica*, *S. arctophila*, *S. ballii*, *S. barclayi*, *S. boothii*, *S. brachycarpa*, *S. eastwoodiae*, *S. myricoides*, *S. niphoclada*, *S. pedicellaris*, and *S. planifolia*.

The following key will help identify the varieties, but there are extensive areas of overlap among them and many intermediates.

1. Stipules often marcescent, foliaceous on late leaves, usually prominent, linear, narrowly elliptic, or lanceolate, 2–17 mm; branches red-brown or gray-brown or brownish; branchlets densely villous to glabrescent; proximal blades: margins entire; floral bracts 1.2–2.5 mm; staminate flowering branchlets 1–14 mm; ovaries obclavate or pyriform; Alaska, Canada.
   2. Shrubs, 0.3–1 m; branchlets densely villous; petioles 1–9 mm; largest medial blades: apex acute, sometimes acuminate, or convex, adaxial surface moderately densely villous, or long-silky to glabrescent; staminate catkins 14–26 mm; filaments distinct or partly connate, glabrous or hairy on proximal ½; pistillate catkins stout to subglobose, flowering branchlets 2–19 mm; stipes 0.4–1.8 mm; Alaska, Northwest Territories, w Nunavut, n Yukon . . . . 50a. *Salix glauca* var. *stipulata*
   2. Shrubs, 0.3–6 m; branchlets soon becoming pilose or glabrescent; petioles 4–27 mm; largest medial blades: apex acuminate to convex, adaxial surface often sparsely long-silky or pilose; staminate catkins 19–45 mm; filaments distinct, glabrous; pistillate catkins slender to stout, flowering branchlets 3–37 mm; stipes 0.5–2.8 mm; c Alaska, British Columbia, Northwest Territories, Yukon . . . . .50b. *Salix glauca* var. *acutifolia*
1. Stipules sometimes marcescent or rudimentary, if foliaceous, usually inconspicuous, oblong, elliptic, or ovate, 1–8(–10) mm; branches usually yellow-brown or red-brown; branchlets usually sparsely villous or tomentose; proximal blades: margins entire or serrulate; floral bracts 1–3.4 mm; staminate flowering branchlets 1.5–25 mm; ovaries pyriform; northern Canada east of Mackenzie River, Greenland, Rocky Mountains.
   3. Stipules 0.9–2.8–8(–10) mm, sometimes marcescent; floral bracts usually brownish; largest medial blades: petiole 3–14 mm; Alberta, British Columbia, Rocky Mountains, Saskatchewan to New Mexico . . . . 50c. *Salix glauca* var. *villosa*
   3. Stipules 1–2.1–4 mm, usually deciduous (in autumn); floral bracts usually tawny; largest medial blades: petiole 2–9 mm; c, e Canada, Greenland, St. Pierre and Miquelon . . . . 50d. *Salix glauca* var. *cordifolia*

50a. Salix glauca Linnaeus var. **stipulata** Floderus in C. A. M. Lindman, Sv. Fanerogamfl. ed. 2, 205. 1926 F

Salix stipulifera Floderus ex Hayren

Plants 0.3–1 m. **Stems:** branches gray-brown to red-brown, densely villous; branchlet color obscured by hairs, densely villous. **Leaves:** stipules sometimes marcescent, foliaceous (sometimes obscured by hairs) on early ones, foliaceous and prominent on late ones, linear to narrowly elliptic, 2–17 mm, apex acute to caudate; petiole 1–9 mm, villous adaxially; largest medial blade (hemiamphistomatous), sometimes narrowly oblong, 27–74 × 8–24 mm, 2.1–3.9 times as long as wide, base rounded, apex acuminate, convex, or acute, abaxial surface densely villous or villous-silky, hairs straight or wavy, adaxial moderately densely villous, or long-silky to glabrescent; proximal blade margins entire; juvenile blade densely villous or long-silky. **Catkins:** staminate 14–26 × 8–14 mm, flowering branchlet 1–14 mm; pistillate stout or subglobose, 15–54 × 7–15 mm, flowering branchlet 2–19 mm; floral bract tawny, light brown or bicolor, 1.2–2.2 mm, apex convex, hairs straight, wavy, or crinkled. **Staminate**

flowers: abaxial nectary 0.3–1 mm; filaments distinct or partly connate, glabrous, or hairy on proximal ½. **Pistillate flowers:** adaxial nectary sometimes flask-shaped, (0.4–)0.6–1.8 mm; stipe 0.4–1.8 mm; ovary obclavate or pyriform, very densely villous or tomentose, beak abruptly tapering to styles; ovules 10–20 per ovary; styles connate to distinct, 0.4–1 mm; stigmas slenderly cylindrical, 0.3–0.8 mm. **Capsules** 4.5–7.5 mm. $2n$ = 76, 114, 152.

Flowering mid Jun–mid Jul. Black spruce treed bogs, white spruce woods, floodplains, subarctic thickets, alpine tundra; 0–1000 m; N.W.T, Nunavut, Yukon; Alaska; Eurasia (China, n Mongolia, Russia, Scandinavia).

Hybrids:

Variety *stipulata* forms natural hybrids with *Salix arctica*, *S. arctophila*, and probably others.

50b. Salix glauca Linnaeus var. **acutifolia** (Hooker) C. K. Schneider, Bot. Gaz. 66: 327. 1918 [E] [F]

Salix villosa Barratt var. *acutifolia* Hooker, Fl. Bor.-Amer. 2: 145. 1838; *S. desertorum* Richardson; *S. glauca* subsp. *acutifolia* (Hooker) Hultén; *S. glauca* var. *alicea* C. R. Ball; *S. glauca* subsp. *desertorum* (Richardson) Andersson ex Hultén; *S. glauca* var. *perstipula* Raup; *S. glauca* var. *poliophylla* (C. K. Schneider) Raup; *S. glauca* var. *sericea* Hultén

Plants 0.3–6 m. **Stems:** branches red-brown or brownish, villous or pilose to glabrescent; branchlets densely villous to glabrescent. **Leaves:** stipules marcescent, rudimentary or foliaceous on early ones, foliaceous on late ones, usually prominent, linear to lanceolate, 2–17 mm, apex acuminate; petiole 4–27 mm, pilose adaxially; largest medial blade oblanceolate or obovate to narrowly elliptic, 43–82 × 12–39 mm, 2.2–4.8 times as long as wide, base cuneate or convex, apex acuminate or convex, abaxial surface very densely to sparsely villous-tomentose or short-silky, hairs straight, wavy, or curved, adaxial long-silky or pilose to glabrescent; proximal blade margins entire; juvenile blade densely villous or long-silky. **Catkins:** staminate 19–45 × 9–17 mm, flowering branchlet 2–9 mm; pistillate slender to stout, 24–83 × 8–17 mm, flowering branchlet 3–37 mm; floral bract tawny, light brown, or bicolor, 1.5–2.5 mm, apex convex to rounded, hairs wavy. **Staminate flowers:** abaxial nectary 0.1–0.6 mm, adaxial nectary narrowly oblong, oblong, or square, 0.6–0.9 mm; filaments distinct, glabrous. **Pistillate flowers:** adaxial nectary sometimes flask-shaped; stipe 0.5–2.8 mm; ovary obclavate or pyriform, densely tomentose, short-silky, or pubescent, beak gradually tapering to styles; ovules 12–22 per ovary; styles connate to distinct ½ their lengths, 0.3–1.4 mm; stigmas slenderly to broadly cylindrical, 0.4–0.59–0.8 mm. **Capsules** 5–9 mm. $2n$ = 76, 95, 114.

Flowering late May–early Jul. Wet to mesic thickets, black spruce treed bogs, white spruce woods, floodplains, fens, swamps, subarctic thickets, alpine tundra; 0–1200 m; B.C., N.W.T., Yukon; Alaska.

Hybrids:

Variety *acutifolia* forms natural hybrids with *Salix arctica* and *S. niphoclada*.

Variety *acutifolia* × *Salix niphoclada*, known from Alaska and Yukon, is intermediate between the parents. It combines the petioles, the narrowly elliptic-oblanceolate leaves with acute-attenuate apices, the stipules, and the narrow, loosely flowered catkins of *S. niphoclada*, and the relatively long petioles, the larger oblanceolate leaves, the larger stipules, and the densely flowered and broadly cylindrical catkins of *S. glauca* (G. W. Argus 1965).

50c. Salix glauca Linnaeus var. **villosa** Andersson, Proc. Amer. Acad. Arts 4: 68. 1858 [E] [F]

Salix villosa D. Don ex Hooker, Fl. Bor.-Amer. 2: 144. 1838, not Schleicher 1815; *S. glauca* var. *glabrescens* (Andersson) Hultén; *S. pseudolapponum* Seemen; *S. wolfii* Bebb var. *pseudolapponum* (Seemen) M. E. Jones

Plants 0.3–2 m. **Stems:** branches yellow-brown or red-brown, villous, soon glabrescent; branchlets sparsely to densely villous or tomentose. **Leaves:** stipules sometimes marcescent, rudimentary or foliaceous on early ones, foliaceous on late ones, usually inconspicuous, narrowly elliptic to ovate, 0.9–2.8–8 mm, apex acute; petiole 3–14 mm, villous or pilose adaxially; largest medial blade (often amphistomatous), narrowly elliptic, elliptic, oblanceolate, or obovate, 29–80 × 8–24 mm, 2.2–3.9 times as long as wide, base sometimes rounded, apex acuminate, acute, or convex, abaxial surface pilose or moderately densely villous to glabrescent, hairs wavy, adaxial pilose or moderately densely villous to glabrescent; proximal blade margins entire or serrulate; juvenile blade sparsely to densely villous. **Catkins:** staminate 19–53 × 9–14 mm, flowering branchlet 1.5–20 mm; pistillate slender to stout, 19–56(–60 in fruit) × 7–18 mm, flowering branchlet 2–27 mm; floral bract tawny, greenish, brown, or bicolor, 1–3.4 mm, apex convex to rounded, hairs wavy. **Staminate flowers:**

abaxial nectary 0.1–0.8 mm, adaxial nectary oblong or ovate, 0.6–1.3 mm; filaments distinct, glabrous, or hairy on proximal ½. **Pistillate flowers:** adaxial nectary sometimes flask-shaped, 0.6–1.4 mm; stipe 0.3–1.5 mm; ovary pyriform, densely villous or tomentose, beak gradually tapering to or slightly bulged below styles; ovules 6–15 per ovary; styles connate to distinct ½ their lengths, 0.4–1.4 mm; stigmas flat, abaxially non-papillate with rounded tip, or slenderly to broadly cylindrical, 0.2–0.4–0.64 mm. **Capsules** 5–8 mm. $2n = 114$.

Flowering mid May–late Jul. Riparian in subalpine and boreal forests, forest openings, sedge meadows, treed bogs, talus slopes, boulder fields, snowflush areas, alpine tundra, limestone shale, schist, granite, quartzite substrates; 150–3800 m; Alta., B.C., Man., N.W.T., Sask., Yukon; Colo., Mont., N.Mex., Oreg., Utah, Wash., Wyo.

Variety *villosa* is often confused with *Salix orestera* in the western United States. See discussion under 55. *S. orestera*.

Hybrids:

Variety *villosa* forms natural hybrids with *Salix brachycarpa* and *S. boothii*.

50d. Salix glauca Linnaeus var. **cordifolia** (Pursh) Dorn, Phytologia 90: 315. 2008 E F

Salix cordifolia Pursh, Fl. Amer. Sept. 2: 611. 1813; *S. callicarpaea* Trautvetter; *S. cordifolia* var. *callicarpaea* (Trautvetter) Fernald; *S. cordifolia* var. *eucycla* Fernald; *S. cordifolia* var. *intonsa* Fernald; *S. cordifolia* var. *macounii* (Rydberg) C. K. Schneider; *S. cordifolia* var. *tonsa* Fernald; *S. glauca* subsp. *callicarpaea* (Trautvetter) Böcher; *S. labradorica* Rydberg

Plants 0.2–2.5 m. **Stems:** branches yellow-brown or red-brown, villous to glabrescent; branchlets densely villous to glabrescent. **Leaves:** stipules not marcescent, rudimentary or foliaceous on early and late ones, usually inconspicuous, oblong to elliptic or ovate, 1–2.1–4 mm, apex acute to caudate; petiole 2–9 mm, pilose adaxially; largest medial blade (sometimes hemiamphistomatous), sometimes broadly obovate, 17–63 × 6–28 mm, 1.4–3.5 times as long as wide, base sometimes rounded, rarely subcordate, apex sometimes rounded, abaxial surface pilose, moderately densely villous or long-silky to glabrescent, hairs straight or wavy, adaxial (sometimes dull), pilose or villous to glabrescent; proximal blade margins entire or serrulate; juvenile blade sparsely to densely villous or tomentose. **Catkins:** staminate 10–48 × 5–14 mm, flowering branchlet 2–25 mm; pistillate subglobose or globose, 18–56(–60 in fruit) × 7–21 mm, flowering branchlet 2–26 mm; floral bract tawny, brown, or bicolor, 2–3 mm, apex rounded, hairs wavy or straight. **Staminate flowers:** abaxial nectary 0.3–1 mm, adaxial nectary narrowly oblong, oblong, ovate, or flask-shaped, 0.5–1.3 mm; filaments distinct or slightly connate, glabrous, or hairy on proximal ½. **Pistillate flowers:** adaxial nectary 0.4–1.4 mm; stipe 0.3–1.3 mm; ovary pyriform, densely villous or tomentose, beak gradually tapering to styles; ovules 10–18 per ovary; styles connate to distinct ½ their lengths or more, 0.7–1.6 mm; stigmas slenderly cylindrical, 0.36–0.48–0.72 mm. **Capsules** 6–7.5 mm.

Flowering late May–late Jul. Sand and cobbles among granitic boulders, sandy alluvium, on exposed eskers, scree slopes, *Sphagnum* bogs, *Empetrum* heaths, snowbeds; 0–3200 m; Greenland; St. Pierre and Miquelon; Man., Nfld. and Labr., N.W.T., N.S., Nunavut, Ont., Que.

The taxonomy of var. *cordifolia* in the Canadian arctic islands and Greenland is confusing. T. W. Böcher (1952) noted what he thought were two ecologically and morphologically different types of *Salix glauca* occurring in western Greenland: a glabrous type corresponding to *S. glauca* subsp. *callicarpaea* and a hairy one corresponding to *S. cordifolia* var. *intonsa*. G. W. Argus (1965) proposed that all Greenland material belonged to the eastern phase of *S. glauca* (corresponding to *S. glauca* var. *cordifolia*). Böcher later recognized two taxa on Greenland, *S. glauca* subsp. *callicarpaea* (including var. *intonsa*) and subsp. *glauca*, which he thought resembled the European *S. glauca* (Böcher et al. 1968). A. K. Skvortsov (1971) did not agree that European *S. glauca* occurs on Greenland. He pointed out that the Greenland plants differ by their usually shorter, broader, more obtuse leaves, adaxial leaf surfaces bare and even somewhat slightly glossy, flowering branchlets as long as the catkins themselves, bracts often brownish, stamens often with glabrous filaments, and a more spreading growth form.

Hybrids:

Variety *cordifolia* forms natural hybrids with *Salix arctophila*, *S. ballii*, *S. brachycarpa*, *S. myricoides*, and *S. pedicellaris*. Placement of specimens from Anticosti Island, Quebec, and North Point, James Bay, Ontario, with densely villous branchlets and relatively short petioles, sometimes thought to be hybrids with *S. glauca* var. *cordifolia*, is dubious. Hybrids with *S. planifolia* have been reported (C. K. Schneider 1921) but no convincing specimens have been seen.

Variety *cordifolia* × *Salix myricoides* (*S.* ×*amoena* Fernald, *S. glauca* var. *tonsa*) is a hybrid swarm characterized by buds with ferruginous hairs, leaves serrulate-crenulate and sparsely hairy, except on margins, and ovaries with flattened, refractive hairs. At the type locality on Ha-Ha Mountain, Newfoundland, it is growing in a population of var. *cordifolia*; *S. planifolia*, *S.* ×*pedunculata* Fernald, and *S. candida* were growing nearby. The second parent, *S. myricoides*, occurs some distance south along the coast. It is possible that the unusual vegetative characteristics were derived from *S. planifolia*, but they best fit *S. myricoides*.

Variety *cordifolia* × *Salix pedicellaris* occurs in northern Quebec. It has the general appearance of var. *cordifolia*, but the ovaries are mostly glabrous with patches or streaks of hairs, or, sometimes, mostly hairy, and with glabrous basal patches. These hybrids usually are fertile and evidently a cross between a species with hairy ovaries and one with glabrous ovaries. The identification of *S. pedicellaris* as the second parent is based on the hybrids having leaf surfaces glaucous adaxially, ovaries and capsules glaucous, and stipes as long as 1.6–1.8 mm. Plants without these diagnostic characteristics often have shorter styles (0.4–0.8 mm) and floral bracts (1.8–2.4 mm) than is usual for var. *cordifolia*.

2e. SALIX Linnaeus subg. VETRIX (Dumortier) Dumortier, Bull. Soc. Roy. Bot. Belgique 1: 141. 1862

Salix sect. *Vetrix* Dumortier, Bijdr. Natuurk. Wetensch. 1: 55. 1826

Shrubs or trees, 0.05–20 m, not clonal or, sometimes, clonal by layering or stem fragmentation. **Stems** usually erect or decumbent; branches usually flexible at base, usually not glaucous, (dull to highly glossy). **Buds** usually *alba*-type, sometimes *caprea*-type or intermediate, scale margins connate, (inner membranaceous layer sometimes free and separating from outer layer). **Leaves:** stipules on early ones absent, rudimentary, or foliaceous, on late ones usually foliaceous, sometimes rudimentary or absent; petioles usually convex to flat, or shallowly grooved adaxially, sometimes deeply grooved, usually not glandular distally; largest medial blade usually hypostomatous, sometimes amphistomatous or hemiamphistomatous, (glands, if present, usually marginal or submarginal), linear, lorate, narrowly oblong, oblong, narrowly to broadly elliptic, oblanceolate, obovate, or broadly obovate, 0.7–13.7 times as long as wide, angle of base and of apex less or greater than 90°, surface (usually not glaucous abaxially), hairs usually white, sometimes also ferruginous or gray; juvenile blade hairs white, sometimes also ferruginous. **Catkins** flowering as or just before leaves emerge, from lateral buds; staminate on flowering branchlet or sessile, usually stout, or slender to globose; pistillate on flowering branchlet or sessile, usually densely flowered, sometimes loosely, usually stout or subglobose to globose, sometimes slender; floral bract usually brown, black, tawny, bicolor, or sometimes light rose, apex usually entire; pistillate bract usually persistent after flowering. **Staminate flowers:** abaxial nectary usually absent (except *S. wolfii*, rarely *S. argyrocarpa* and *S. breweri*); stamens 2, (1 in sect. *Sitchenses*); filaments distinct or connate, glabrous or hairy; anthers usually purple or red turning yellow. **Pistillate flowers:** abaxial nectary usually absent; ovary not glaucous, glabrous or hairy, hairs usually flattened, beak usually gradually tapering to or slightly bulged below styles; ovules 2–37 per ovary; styles usually connate, sometimes distinct; stigmas flat, abaxially non-papillate with rounded or pointed tip, or stigmas slenderly or broadly cylindrical, or 2 plump lobes. $2n$ = 38 (29), 57 (1), 76 (16), 95 (1), 114 (4), unknown (21).

Species ca. 204 (63 species in the flora): North America, Eurasia.

Nineteen of the 29 sections recognized in this subgenus are found in the flora area.

SELECTED REFERENCES Dorn, R. D. 1975. A systematic study of *Salix* section *Cordatae* in North America. Canad. J. Bot. 53: 1491–1522. Dorn, R. D. 1995. A taxonomic study of *Salix* section *Cordatae* subsection *Luteae* (Salicaceae). Brittonia 47: 160–174. Dorn, R. D. 2000. A taxonomic study of *Salix* sections *Mexicanae* and *Viminella* subsection *Sitchenses* (Salicaceae) in North America. Brittonia 52: 1–19.

1. Plants presumed introduced and widely naturalized in the flora area.*
   2. Stipules on late leaves absent or rudimentary.
      3. Leaves and catkins alternate; branchlets densely pubescent or tomentose; largest medial blades: abaxial surface densely tomentose or woolly; floral bracts tawny or brown; ovaries glabrous; stipes 0.3–0.5 mm. . . . . 99. *Salix elaeagnos*
      3. Leaves and catkins often opposite or subopposite; branchlets glabrous; largest medial blades: abaxial surface glabrous; floral bracts bicolor or black; ovaries short-silky; stipes 0–0.1 mm. . . . . 113. *Salix purpurea*
   2. Stipules on late leaves foliaceous.
      4. Petioles shallowly grooved adaxially.
         5. Branches yellow-brown, gray-brown, or yellowish; largest medial blades: abaxial surface apparently glaucous but obscured by hairs, margins strongly revolute, glands epilaminal; stigmas 0.72–1.8 mm . . . . 98. *Salix viminalis*
         5. Branches red-brown or gray-brown; largest medial blades: abaxial surface distinctly glaucous or not, margins sometimes slightly revolute or flat, glands submarginal or marginal; stigmas 0.3–0.68 mm.
            6. Branches not glaucous; stipules reniform, not adnate to petiole; largest medial blades broader, 1.1–2.8 times as long as wide, elliptic, broadly elliptic, subcircular or broadly obovate, apex abruptly acuminate or acute, abaxial surface often not glaucous apically . . . . 75. *Salix myrsinifolia* (in part)
            6. Branches strongly glaucous; stipules lanceolate to ovate, often adnate to petiole; largest medial blades 2.5–6.4 times as long as wide, lorate, oblong, narrowly elliptic or elliptic, apex gradually acuminate, abaxial surface uniformly and distinctly glaucous . . . . 112. *Salix daphnoides*
      4. Petioles flat to convex adaxially.
         7. Largest medial blades with veins strongly impressed adaxially; branchlets sometimes weakly glaucous, tomentose . . . . 83. *Salix aurita*
         7. Largest medial blades flat or with veins slightly impressed adaxially; branchlets not glaucous, puberulent, pubescent, pilose, villous, or velvety.
            8. Petioles puberulent or villous adaxially; branches red-brown, glabrous; catkins flowering as leaves emerge; ovaries glabrous or sparsely hairy . . . . 75. *Salix myrsinifolia* (in part)
            8. Petioles glabrescent, pubescent, tomentose, or velvety adaxially; branches yellowish, yellow-brown, or gray-brown, pubescent, pilose, villous, or tomentose to glabrescent; catkins flowering before leaves emerge; ovaries short- or long-silky.
               9. Peeled wood smooth or with striae to 6 mm; ovaries: beak gradually tapering to styles.
                  10. Largest medial blades usually 2–3 times as long as wide; branches flexible at base; stipes 2–2.5 mm; styles 0.3–0.6 mm; stigma lobes 0.4–0.6 mm . . . . 80. *Salix caprea*
                  10. Largest medial blades usually 2.8–6.4 times as long as wide; branches ± brittle at base; stipes 0.9–2 mm; styles 0.5–1.2 mm; stigma lobes 0.6–1.2 mm [2e4. *Salix* sect. *Cinerella*]
                  . . . . *Salix* ×*smithiana* [p. 132]
               9. Peeled wood with striae to 62 mm; ovaries: beak slightly bulged below styles.
                  11. Shrubs, 3–7(–10) m; largest medial blades: abaxial surface with white hairs; branches brownish, to 62 mm . . . . 81. *Salix cinerea*
                  11. Shrubs or small trees, 3–12 m; largest medial blades: abaxial surface with white and ferruginous hairs; branches yellow-brown, gray-brown, or red-brown, to 45 mm . . . . 82. *Salix atrocinerea*

---

*Although it may be difficult to be certain if a plant is native or introduced, certain assumptions can be made. If it is growing in a wilderness area (alpine, arctic, boreal, desert) or far from human habitation, the probability that it is an introduction is very small; but if it is growing in a settled area, particularly on anthropogenically disturbed sites, or growing in rows suggesting planting along fencerows or as a windbreak, then it could be suspected to be an introduction. The possibility that it is a natural hybrid between native species cannot be excluded and both leads may need to be followed.

1. Plants presumed native in the flora area.
    12. Flowering before leaves emerge (sometimes just before in S. irrorata, S. myricoides, and S. tracyi); catkins usually sessile or on very short flowering branchlets (staminate 0–5 mm, to 9 mm in S. myricoides, pistillate 0–14 mm, to 35 mm in S. hookeriana).
        13. Ovaries glabrous.
            14. Stipules usually marcescent [2e9. Salix sect. Lanatae].
                15. Largest medial blades: apex acuminate, acute, or rounded; stipules 1–6–14 mm, 1.1–1.8–4.4 times as long as wide, usually ovate to oval or narrowly elliptic, pressing flat . . . . .92. *Salix calcicola*
                15. Largest medial blades: apex acute, acuminate, or convex; stipules 3–35 mm, 1.1–6 times as long as wide, shape variable, pressing with a pleat.
                    16. Largest medial blades slightly glossy adaxially; floral bracts moderately densely hairy; pistillate adaxial nectaries 2+-lobed; stipules 3–12–35 mm, 1.7–3.6–6 times as long as wide, lanceolate or narrowly elliptic, sometimes ovate or oval . . . . 93. *Salix richardsonii*
                    16. Largest medial blades dull adaxially; floral bracts sparsely hairy; pistillate adaxial nectaries unlobed; stipules 3.5–10–23 mm, 1–1.4–2.8 times as long as wide, lanceolate, ovate, or suborbiculate . . . . 94. *Salix tweedyi*
            14. Stipules deciduous (in autumn).
                17. Largest medial blades dull adaxially.
                    18. Ovaries: beak slightly bulged below styles; styles 0.3–0.6 mm; floral bracts moderately to very densely hairy; east of 102d meridian. . . . . 68. *Salix eriocephala* (in part)
                    18. Ovaries: beak gradually tapering to styles; styles 0.5–1.8 mm; floral bracts sparsely hairy; usually west of 102d meridian.
                        19. Petioles usually reddish; catkins usually sessile, flowering branchlets sometimes to 5 mm; juvenile blades glabrous or pubescent; stigmas 0.1–0.29 mm; stipes 0.5–3 mm; Canada to Montana, Black Hills of South Dakota, n Wyoming . . . . 65. *Salix pseudomonticola* (in part)
                        19. Petioles green; catkins on distinct flowering branchlets 0.5–7 mm; juvenile blades villous or long-silky; stigmas 0.24–0.56 mm; stipes 0.5–1.6 mm; ne Arizona, Colorado, nw New Mexico, Utah, s Wyoming . . . . 66. *Salix monticola* (in part)
                17. Largest medial blades slightly to highly glossy adaxially.
                    20. Largest medial blades: margins often entire, sometimes irregularly serrate or crenate.
                        21. Largest medial blades usually 1.5–5.2 times as long as wide, bases often rounded to cordate; pistillate flowering branchlets 0–20 mm; styles distinct to ½ their lengths or connate, 0.3–2.3 mm; capsules 5–10 mm.
                            22. Pistillate catkins loosely flowered; pistillate flowering branchlets 1.5–13 mm; filaments glabrous; styles 0.3–1.3 mm; e North America. . . . . 67. *Salix myricoides* (in part)
                            22. Pistillate catkins densely flowered; pistillate flowering branchlets 0–20 mm; filaments hairy or glabrous; styles 0.6–2.3 mm; w North America .77. *Salix hookeriana* (in part)
                        21. Largest medial blades usually 1.9–9.6 times as long as wide, bases usually cuneate or convex; pistillate flowering branchlets 0–6 mm; styles connate, 0.1–0.9 mm; capsules 2.5–5.5 mm [2e15. *Salix* sect. *Mexicanae*].

23. Branches strongly glaucous; juvenile blades glabrous or sparsely villous; largest medial blades glabrous or pilose adaxially, hairs white .... 104. *Salix irrorata*

23. Branches usually not glaucous, sometimes weakly so; juvenile blades sparsely to very densely tomentose, woolly or silky; largest medial blades glabrescent, tomentose, or short-silky adaxially, hairs usually white, sometimes also ferruginous .... 105. *Salix lasiolepis*

20. Largest medial blades: margins regularly toothed.
  24. Glands on largest medial blades submarginal.
    25. Pistillate catkins loosely flowered; pistillate flowering branchlets 1.5–13 mm; filaments glabrous; styles 0.3–1.3 mm; e North America .... 67. *Salix myricoides* (in part)
    25. Pistillate catkins densely flowered; pistillate flowering branchlets 0–20 mm; filaments hairy or glabrous; styles 0.6–2.3 mm; w North America .... 77. *Salix hookeriana* (in part)
  24. Glands on largest medial blades marginal.
    26. Petioles short-silky or velvety adaxially; staminate catkins not on flowering branchlets; floral bracts with straight hairs; ovules 18 per ovary .... 65. *Salix pseudomonticola* (in part)
    26. Petioles villous, tomentose, pilose, or pubescent adaxially; staminate catkins on flowering branchlets; floral bracts usually with wavy or curly hairs; ovules 12–16 per ovary.
      27. Largest medial blades with relatively very thick glaucescence abaxially, usually 2–5.2 times as long as wide; floral bracts brown to black; styles 0.3–1.3 mm; pistillate flowering branchlets 1.5–13 mm .... 67. *Salix myricoides* (in part)
      27. Largest medial blades with relatively thin glaucescence abaxially, usually 2.3–8 times as long as wide; floral bracts tawny or light brown; styles 0.3–0.6 mm; pistillate flowering branchlets 2–10 mm .... 68. *Salix eriocephala* (in part)

[13. Shifted to left margin.—Ed.]

13. Ovaries hairy.
  28. Stipules on early leaves absent or rudimentary.
    29. Ovaries obturbinate or squat, flask-shaped (serpentine endemics, California, Oregon) [2e17. *Salix* sect. *Sitchenses* (in part)].
      30. Floral bracts usually tawny, densely hairy; largest medial blades lorate, very narrowly elliptic or oblanceolate, apex acuminate or acute; ovaries squat, flask-shaped, beak abruptly tapering to styles; branches flexible at base .... 110. *Salix breweri* (in part)
      30. Floral bracts brown, sparsely to moderately densely hairy; largest medial blades elliptic or obovate, apex convex or rounded; ovaries obturbinate, beak gradually tapering to styles; branches highly brittle at base ..... 111. *Salix delnortensis* (in part)
    29. Ovaries pyriform or obclavate (not serpentine endemics).
      31. Largest medial blades highly glossy adaxially.
        32. Ovaries villous, tomentose, or woolly; stipes 0.5–2.2 mm; anthers yellow (Alaska to n California).......................... 77. *Salix hookeriana* (in part)
        32. Ovaries silky; stipes 0.2–1.1 mm; anthers purple turning yellow in age [2e6. *Salix* sect. *Phylicifoliae* (in part)].

33. Largest medial blades usually 4.2–11.3 times as long as wide, linear, lorate or narrowly elliptic, abaxial surface usually very densely villous, short-silky, tomentose, or woolly (sometimes sparsely so), glands marginal or epilaminal; petioles sometimes with 2 spherical glands distally; branches brittle at base . . . . .86. *Salix pellita* (in part)
33. Largest medial blades usually 1.5–5.1 times as long as wide, obovate, oblanceolate, narrowly oblong, elliptic, or narrowly elliptic, abaxial surface glabrous or sparsely silky, glands marginal or submarginal; petioles without glands at base of blade; branches flexible at base.
  34. Largest medial blades usually hypostomatous (sometimes hemiamphistomatous). . . . . 88. *Salix planifolia* (in part)
  34. Largest medial blades amphistomatous . . . . .89. *Salix tyrrellii* (in part)
31. Largest medial blades dull or slightly glossy adaxially.
  35. Stems decumbent, 0.3–3 m, layering, not glaucous or weakly so; largest medial blades: hairs white or gray; styles 0.2–0.4 mm; stigmas 0.2–0.56 mm. . . . . .78. *Salix humilis* (in part)
  35. Stems erect, 0.5–6 m, not layering, often strongly glaucous; largest medial blades: hairs white, sometimes also ferruginous; styles 0.3–1.5 mm; stigmas 0.32–0.76 mm.
    36. Largest medial blades: abaxial surface often obscured by dense hairs [2e6. *Salix* sect. *Phylicifoliae* (in part)].
      37. Petioles villous or velvety adaxially; largest medial blades usually 3–6.2 times as long as wide, glands submarginal; w North America . . . . 85. *Salix drummondiana*
      37. Petioles glabrous or pubescent adaxially; largest medial blades usually 4.2–11.3 times as long as wide, glands submarginal or epilaminal; e North America. . . . . .86. *Salix pellita* (in part)
    36. Largest medial blades: abaxial surface not obscured by hairs.
      38. Largest medial blades 4.2–11.3 times as long as wide, linear to narrowly elliptic; capsules 3.5–6.5 mm . . . . .86. *Salix pellita* (in part)
      38. Largest medial blades 1.5–4.5 times as long as wide, narrowly elliptic to obovate or broadly so; capsules 4.5–11 mm [2e4. *Salix* sect. *Cinerella* (in part)].
        39. Petioles and branchlets velvety; styles 0.2–0.6 mm; largest medial blades usually oblanceolate; ovaries densely long-silky . . . . .79. *Salix scouleriana* (in part)
        39. Petioles and branchlets not velvety; styles 0.3–2.3 mm; largest medial blades elliptic, oblanceolate, or obovate; ovaries moderately densely short-silky or densely villous, tomentose, or woolly.
          40. Stipes 1.6–2.7 mm; ovaries short-silky, hairs straight; styles 0.3–1 mm; pistillate flowering branchlets 0–10 mm; largest medial blades dull or slightly glossy adaxially, margins flat; primarily e North America to c British Columbia. . . . . 76. *Salix discolor* (in part)
          40. Stipes 0.5–2.2 mm; ovaries villous, tomentose, or woolly, hairs wavy; styles 0.6–2.3 mm; pistillate flowering branchlets 0–20 mm; largest medial blades slightly to highly glossy adaxially, margins slightly revolute; Alaska to n California . . . . .77. *Salix hookeriana* (in part)

[28. Shifted to left margin.—Ed.]
28. Stipules on early leaves foliaceous (rarely so in *S. drummondiana* and *S. pellita*).
  41. Pistillate nectaries usually shorter than or equal to stipes.
    42. Largest medial blades amphistomatous. . . . . .89. *Salix tyrrellii* (in part)
    42. Largest medial blades usually hypostomatous (or hemiamphistomatous).
      43. Stipes 0.3–0.8 mm . . . . 88. *Salix planifolia* (in part)
      43. Stipes 0.8–2.7 mm [2e4. *Salix* sect. *Cinerella*].

44. Largest medial blades with revolute margins.
    45. Shrubs, 0.3–3 m, forming clones by layering; juvenile blades tomentose to glabrescent; anthers 0.4–0.6 mm; ovaries sparsely to moderately densely short-silky................................78. *Salix humilis* (in part)
    45. Shrubs or trees, 3–10(–20) m, not clonal; juvenile blades villous or silky; anthers 0.7–1.2 mm; ovaries very densely long-silky ....79. *Salix scouleriana* (in part)
44. Largest medial blades with flat margins.
    46. Ovaries short-silky, hairs straight; stipes 1.6–2.7 mm; primarily e North America to c British Columbia................... 76. *Salix discolor* (in part)
    46. Ovaries villous, tomentose, or woolly, hairs wavy; stipes 0.5–2.2 mm; Alaska to n California..77. *Salix hookeriana* (in part)
41. Pistillate nectaries as long as or longer than stipes.
    47. Floral bracts tawny or light rose; styles 0.4–0.8 mm; stigmas 0.16–0.24 mm; ovaries squat, flask-shaped................................ 110. *Salix breweri* (in part)
    47. Floral bracts brown or black; styles 0.5–2.3 mm; stigmas 0.2–1.28 mm; ovaries obturbinate, obclavate, or pyriform.
        48. Ovaries long- or short-silky, hairs straight.
            49. Largest medial blades 29–54 mm wide; branchlets and petioles velvety; branches highly brittle at base................. 111. *Salix delnortensis* (in part)
            49. Largest medial blades 5–28 mm wide; branchlets and petioles glabrous, pilose, puberulent, villous, or short-silky; branches flexible at base [2e6. *Salix* sect. *Phylicifoliae* (in part)].
                50. Stipules usually marcescent, foliaceous, linear or lanceolate, 2–9.8–23 mm; juvenile blades glabrous or pilose; largest medial blades narrowly elliptic, elliptic, or obovate .... 87. *Salix pulchra*
                50. Stipules sometimes marcescent, rudimentary or foliaceous, ovate, oblong, or narrowly elliptic, 1–2.5(–4.5) mm; juvenile blades glabrous, puberulent, pubescent, or densely long-silky; largest medial blades narrowly oblong, narrowly elliptic, elliptic, or oblanceolate .... 88. *Salix planifolia* (in part)
        48. Ovaries tomentose, villous, or woolly, hairs wavy.
            51. Largest medial blades: abaxial surface glaucous or sparsely to moderately densely hairy, hairs white, sometimes also ferruginous, slightly or highly glossy adaxially; stipes 0.5–2.2 mm; branches highly brittle at base ....77. *Salix hookeriana* (in part)
            51. Largest medial blades: abaxial surface very densely hairy, hairs white, dull or slightly glossy adaxially; stipes 0–0.6 mm; branches flexible at base.
                52. Stipules resinous; floral bracts 2.8–5.2 mm; largest medial blades slightly glossy adaxially, margins flat; petioles weakly ventricose around floral buds...97. *Salix barrattiana*
                52. Stipules not resinous; floral bracts 1.5–3 mm; largest medial blades dull adaxially, margins revolute; petioles strongly ventricose around floral buds.
                    53. Largest medial blades: abaxial surface villous to tomentose, adaxially sparsely to moderately densely hairy; floral bracts: apex often acute to convex, sometimes crenate; stipules sometimes marcescent .... 95. *Salix alaxensis*
                    53. Largest medial blades: abaxial surface tomentose to woolly, adaxially moderately to very densely hairy; floral bracts: apex convex to rounded, entire; stipules deciduous (in autumn) ....96. *Salix silicicola*

[12. Shifted to left margin.—Ed.]
12. Flowering as leaves emerge (staminate catkins sometimes emerge just before in *S. irrorata*, *S. myricoides*, and *S. tracyi*), catkins usually on distinct flowering branchlets (staminate 0–30 mm, pistillate 0.5–38 mm).
    54. Ovaries hairy.

Salix · SALICACEAE

[55. Shifted to left margin.—Ed.]
55. Plants east of 102d meridian (longitudal line passing through the ne corner of Saskatchewan, along e border of Colorado).
  56. Stipules on late leaves usually foliaceous.
    57. Branchlets, petioles adaxially, abaxial surface of largest medial blade densely woolly or tomentose; ovaries tomentose or woolly .... 91. *Salix candida* (in part)
    57. Branchlets, petioles adaxially, abaxial surface of largest medial blade glabrous, puberulent, villous, or long-silky; ovaries short-silky.
      58. Largest medial blades: margins flat, entire, crenate, or irregularly serrate, adaxial surface dull or slightly glossy; ovaries obclavate; floral bracts 1.2–3.2 mm; stipes 2–6 mm .... 84. *Salix bebbiana* (in part)
      58. Largest medial blades: margins slightly revolute, serrulate, adaxial surface slightly or highly glossy; ovaries pyriform; floral bracts 0.8–1.2 mm; stipes 0.6–0.9 mm .... 90. *Salix arbusculoides* (in part)
  56. Stipules on late leaves usually absent or rudimentary.
    59. Largest medial blades: abaxial surface short-silky, hairs usually white (rarely ferruginous); branches highly brittle at base; floral bracts brown to black; ovaries ovoid; styles 0.2–0.4 mm; stipes 0.6–1.5 mm .... 107. *Salix sericea*
    59. Largest medial blades: abaxial surface long-silky, villous, pilose, or pubescent to glabrescent, hairs white, often with some ferruginous; branches flexible at base; floral bracts tawny to brown; ovaries pyriform or obclavate; styles 0–0.9 mm; stipes 1–6 mm.
      60. Shrubs delicate, 0.2–1 m; largest medial blades with strongly revolute margins; pistillate catkins densely flowered; capsules 2–4 mm; styles 0.4–0.9 mm; floral bracts 0.7–1.2 mm .... 100. *Salix argyrocarpa*
      60. Shrubs coarse, 0.5–10 m; largest medial blades with flat or slightly revolute margins; pistillate catkins loosely flowered; capsules 5–9 mm; styles 0–0.5 mm; floral bracts 1–3.2 mm.
        61. Largest medial blades 1.7–3.9 times as long as wide, narrowly oblong to obovate, hairs white or gray, margins crenate or entire, adaxial surface finely impressed-reticulate; floral bracts tawny; ovaries obclavate .... 84. *Salix bebbiana* (in part)
        61. Largest medial blades 5–9 times as long as wide, lorate to very narrowly elliptic, hairs white, sometimes also ferruginous, margins serrulate, serrate, or entire, adaxial surface not finely impressed-reticulate; floral bracts brown, tawny, light rose or bicolor; ovaries pyriform .... 101. *Salix petiolaris* (in part)
55. Plants west of 102d meridian (longitudal line passing through the ne corner of Saskatchewan, along e border of Colorado).
  62. Petioles glabrescent or puberulent adaxially.
    63. Branchlets glabrous or puberulent; juvenile blades very densely long-silky; stipes 0.6–0.9 mm; pistillate adaxial nectaries equal to or as long as stipes; ovaries very densely short-silky, beaks gradually tapering to styles .... 90. *Salix arbusculoides* (in part)
    63. Branchlets pubescent, villous, or velvety; juvenile blades pilose, tomentose, or sparsely or moderately densely long-silky; stipes 1.5–6 mm; pistillate adaxial nectaries shorter than stipes; ovaries densely to sparsely hairy, beaks abruptly tapering to or slightly bulged below styles.
      64. Largest medial blades narrowly oblong, narrowly elliptic, elliptic, oblanceolate, or obovate, 1.7–3.9 times as long as wide, hairs on abaxial surface divergent, wavy; branchlets villous to glabrescent; ovaries obclavate, beaks slightly bulged below styles .... 84. *Salix bebbiana* (in part)
      64. Largest medial blades lorate or very narrowly elliptic, 5–9 times as long as wide, hairs on abaxial surface appressed, straight; branchlets pubescent or velvety; ovaries pyriform, beak abruptly tapering to styles .... 101. *Salix petiolaris* (in part)
  62. Petioles pubescent, pilose, villous, tomentose, woolly, short-silky, long-silky, or velvety adaxially.

[65. Shifted to left margin.—Ed.]
65. Largest medial blades not glaucous abaxially.
    66. Branchlets yellow-green or red-brown, pilose or villous; anthers 0.5–0.9 mm; floral bracts 1.4–2.8 mm; pistillate catkins 11–51 mm . . . . .53. *Salix eastwoodiae*
    66. Branchlets yellowish or yellow-brown, pubescent or long-silky; anthers 0.3–0.5 mm; floral bracts 0.8–2 mm; pistillate catkins 8.5–19 mm . . . . 54. *Salix wolfii* (in part)
65. Largest medial blades glaucous abaxially or surfaces obscured by hairs.
    67. Largest medial blades: abaxial surface densely tomentose or woolly, with crinkled and interwoven hairs, adaxial surface often floccose, hairs white. . . . . 91. *Salix candida* (in part)
    67. Largest medial blades: abaxial surface without crinkled, interwoven hairs, adaxial surface not floccose, hairs white or, sometimes, also ferruginous.
        68. Largest medial blades: abaxial surface very densely hairy; peeled wood on branches inconspicuous (sometimes conspicuous in *S. geyeriana*).
            69. Largest medial blades 3.6–11.3 times as long as wide, linear, or lorate to very narrowly elliptic; pistillate catkins globose or subglobose; anthers 0.4–0.5 mm; stamens 2. . . . . 102. *Salix geyeriana* (in part)
            69. Largest medial blades 2.1–7.7 times as long as wide, lorate, elliptic, narrowly oblanceolate, oblanceolate, or obovate; pistillate catkins slender to stout; anthers 0.5–0.8 mm; stamens 1 or 2 [2e17. *Salix* sect. *Sitchenses* (in part)].
                70. Largest medial blades 2.1–3.1–4 times as long as wide, margins strongly or slightly revolute, entire or toothed, glands submarginal or epilaminal, abaxial surface woolly, short-silky, or silky-woolly; staminate catkins 22–54 mm, slender or stout; stamens 1; filaments glabrous; pistillate adaxial nectaries square, ovate, or flask-shaped . . . . 108. *Salix sitchensis*
                70. Largest medial blades (2.5–)3.3–5.3–7.3 times as long as wide, margins slightly revolute or flat, entire, glands submarginal, abaxial surface short-silky; staminate catkins 16–16.5 mm, stout or subglobose; stamens 1 or 2; filaments hairy on proximal ½; pistillate adaxial nectaries narrowly oblong, oblong, or flask-shaped. . . . . 109. *Salix jepsonii*
        68. Largest medial blades: abaxial surface sparsely or moderately densely hairy; peeled wood on branches conspicuous.
            71. Largest medial blades: margins entire, abaxial surface sparsely to moderately densely short- or long-silky, adaxial surface usually dull; filaments hairy basally; anthers purple turning yellow. . . . . 55. *Salix orestera*
            71. Largest medial blades: margins crenate, serrate, serrulate, or entire, abaxial surface glabrous, pilose, tomentose, villous, woolly, or short- or long-silky, adaxial surface slightly or highly glossy; filaments glabrous or hairy; anthers yellow or purple turning yellow.
                72. Largest medial blades 1.5–4.2 times as long as wide, 18–63 mm wide, margins entire, irregularly serrate, crenate, or sinuate, adaxial surface pilose, villous, or tomentose; staminate catkins 26–73 mm; pistillate catkins 36–117 mm; styles 0.6–2.3 mm. . . . . 77. *Salix hookeriana* (in part)
                72. Largest medial blades 3.4–12 times as long as wide, 5.4–22 mm wide, margins entire or serrulate, adaxial surface silky to glabrescent; staminate catkins 11–28 mm; pistillate catkins 8–44 mm; styles 0.1–1 mm.
                    73. Catkins globose to subglobose; floral bracts usually tawny, sometimes brown; anthers 0.4–0.5 mm, yellow or purple turning yellow; styles 0.1–0.6 mm; stipules absent or rudimentary; proximal blades with entire margins. . . . . 102. *Salix geyeriana* (in part)
                    73. Catkins stout; floral bracts dark brown or bicolor; anthers 0.5–0.9 mm, yellow; styles 0.3–1 mm; stipules foliaceous; proximal blades with entire or serrulate margins. . . . . 103. *Salix lemmonii*

[54. Shifted to left margin.—Ed.]
54. Ovaries glabrous [*Salix* sects. 2e1. *Hastatae* and 2e2. *Cordatae* (in part)].
    74. Plants east of 102d meridian (longitudal line passing through the ne corner of Saskatchewan, along e border of Colorado).
        75. Largest medial blades not glaucous abaxially.
            76. Juvenile blades densely villous on abaxial surface or midrib long-silky; largest medial blades usually villous on abaxial surface, hairs white; petioles tomentose adaxially; pistillate adaxial nectaries 0.4–1.3 mm; floral bracts 1–2.6 mm . . . . . 52. *Salix cordata*
            76. Juvenile blades glabrous on abaxial surface or midrib sparsely pubescent to short-silky; largest medial blades glabrous or pilose on abaxial surface, hairs white and, sometimes, ferruginous; petioles glabrous, pubescent, or villous adaxially; pistillate adaxial nectaries 0.2–0.4 mm; floral bracts 0.4–1.1 mm.
                77. Shrubs decumbent, 0.1–0.6(–1) m; stipules on late leaves rudimentary to foliaceous, 0.2–1.8(–5) mm; styles 0.3–0.7 mm; proximal blades with crenate margins . . . . 56. *Salix myrtillifolia* (in part)
                77. Shrubs erect, 1–7 m; stipules on late leaves foliaceous, 0.5–6 mm; styles 0.4–1.6 mm; proximal blades with entire or serrulate margins . . . . 57. *Salix pseudomyrsinites* (in part)
        75. Largest medial blades glaucous abaxially.
            78. Floral bracts tawny; ovaries obclavate . . . . 62. *Salix pyrifolia* (in part)
            78. Floral bracts usually brown or black; ovaries pyriform.
                79. Largest medial blades 1.4–2.8 times as long as wide, elliptic to obovate, apex usually convex or rounded, sometimes acute; stipules on early leaves absent or rudimentary, on late leaves foliaceous; proximal blades with serrulate or crenulate margins . . . . 58. *Salix ballii*
                79. Largest medial blades 2–8 times as long as wide, usually narrowly oblong, very narrowly elliptic or oblanceolate (sometimes elliptic or obovate), apex acuminate or acute; stipules on all leaves foliaceous; proximal blades with entire or serrulate margins.
                    80. Largest medial blades very thickly glaucous abaxially, surface hairs often ferruginous; stigmas 0.24–0.56 mm; pistillate catkins loosely flowered . . . . 67. *Salix myricoides* (in part)
                    80. Largest medial blades thinly or moderately glaucous abaxially, surface hairs usually white; stigmas 0.2–0.3 mm; pistillate catkins densely or moderately densely flowered . . . . 68. *Salix eriocephala* (in part)
    74. Plants west of 102d meridian (longitudal line passing through the ne corner of Saskatchewan, along e border of Colorado).
        81. Largest medial blades not glaucous abaxially [*Salix* sects. 2e1. *Hastatae* and 2e2. *Cordatae* (in part)].
            82. Largest medial blades to 60 mm, margins entire, not glandular-dotted; pistillate catkins 8.5–19 mm; stipules deciduous (in autumn); staminate flowers often with abaxial and adaxial nectaries . . . . 54. *Salix wolfii* (in part)
            82. Largest medial blades to 102 mm, margins usually toothed, if entire then also glandular-dotted; pistillate catkins 11–73 mm; stipules often marcescent; staminate flowers without abaxial nectary.
                83. Largest medial blades: surfaces dull adaxially.
                    84. Branchlets pilose, villous, or woolly.
                        85. Juvenile blades tomentose, woolly, or long-silky; largest medial blades 1.5–3.4 times as long as wide, hairs white; pistillate flowering branchlets 3–15 mm; floral bracts: apex rounded to acute . . . . 51. *Salix commutata* (in part)
                        85. Juvenile blades villous to pilose; largest medial blades 2–5.2 times as long as wide, hairs usually white and ferruginous; pistillate flowering branchlets 1–9 mm; floral bracts: apex rounded to retuse . . . . 60. *Salix boothii* (in part)

84. Branchlets glabrous or puberulent.
    86. Juvenile blades pilose to villous abaxially; stigmas slenderly cylindrical, flat, abaxially non-papillate with rounded tip, or 2 plump lobes; styles 0.3–1.4 mm; largest medial blades often amphistomatous . . . . 60. *Salix boothii* (in part)
    86. Juvenile blades glabrous or short- to long-silky abaxially; stigmas broadly cylindrical or flat, non-papillate abaxially, tip rounded; styles 0.2–0.5 mm; largest medial blades hypostomatous . . . . 73. *Salix monochroma*
83. Largest medial blades: surfaces slightly glossy adaxially.
    87. Floral bracts 0.4–1.1 mm; pistillate nectaries 0.2–0.4 mm; stipules on early leaves absent, rudimentary, or foliaceous; petioles deeply to shallowly grooved adaxially, usually glabrous.
        88. Shrubs decumbent, 0.1–0.6(–1) m; stipules on late leaves rudimentary to foliaceous, 0.2–1.8(–5) mm; styles 0.3–0.7 mm; proximal blades with crenate margins. . . . . 56. *Salix myrtillifolia* (in part)
        88. Shrubs erect, 1–7 m; stipules on late leaves foliaceous, 0.6–8 mm; styles 0.4–1.6 mm; proximal blades with entire or serrulate margins. . . . . 57. *Salix pseudomyrsinites* (in part)
    87. Floral bracts 0.7–3 mm; pistillate nectaries 0.3–1 mm; stipules foliaceous; petioles flat to shallowly grooved adaxially, pilose, tomentose, villous, or pubescent to glabrescent.
        89. Juvenile blades tomentose or long-silky; largest medial blades: abaxial surface usually moderately densely tomentose or villous (sometimes pilose to glabrescent), adaxial surface pilose or moderately densely villous to glabrescent; floral bracts tawny, brown, or bicolor; pistillate flowering branchlets 3–30 mm; stipules sometimes marcescent .51. *Salix commutata* (in part)
        89. Juvenile blades glabrous or pilose; largest medial blades: abaxial surface usually glabrous, pilose, or moderately densely short-silky, adaxial surface pilose or glabrous; floral bracts brown, black, or bicolor; pistillate flowering branchlets 1–10 mm; stipules deciduous (in autumn).
            90. Largest medial blades 1.6–3.6 times as long as wide, elliptic to broadly so, abaxial surface glabrous or pilose, hairs white; floral bracts: apex acute to convex; stipes 0.2–1 mm . . . . 59. *Salix arizonica*
            90. Largest medial blades 2–5.2 times as long as wide, lorate, narrowly oblong, narrowly elliptic to elliptic or broadly so, abaxial surface pilose to densely short-silky, hairs white, sometimes also ferruginous, sometimes glabrous; floral bracts: apex rounded or retuse; stipes 0.5–2.5 mm . . . . 60. *Salix boothii* (in part)
[81. Shifted to left margin—Ed.]
81. Largest medial blades glaucous abaxially.
    91. Largest medial blades: surfaces slightly or highly glossy adaxially.
        92. Floral bracts tawny.
            93. Branches red-brown; stipules early deciduous . . . . .62. *Salix pyrifolia* (in part)
            93. Branches yellowish to yellow-brown or gray-brown; stipules deciduous (in autumn).
                94. Largest medial blades with distinctly toothed margins; pistillate catkins slender to stout; juvenile blades pilose, villous or glabrous . . . . . . 69. *Salix famelica* (in part)
                94. Largest medial blades with entire or indistinctly toothed margins; pistillate catkins subglobose to stout; juvenile blades long-silky or glabrous . . .72. *Salix lutea* (in part)
        92. Floral bracts brown, black, or bicolor.
            95. Largest medial blades usually with white and ferruginous hairs on midrib of adaxial surface.
                96. Juvenile leaves reddish; nw California, adjacent Oregon . . . . 106. *Salix tracyi* (in part)
                96. Juvenile leaves yellowish green; Alaska, Alberta, British Columbia, Idaho, Montana, Northwest Territories, Oregon, Wyoming, Yukon.

97. Largest medial blades narrowly to broadly elliptic, or narrowly ovate to ovate; pistillate nectaries square or obovate; stipules on early leaves foliaceous; branches not glaucous; floral bracts: apex acute or rounded; Alaska, nw Northwest Territories, Yukon . . . . 63. *Salix hastata* (in part)
97. Largest medial blades narrowly elliptic to elliptic; pistillate nectaries oblong or ovate; stipules on early leaves absent or rudimentary (sometimes foliaceous); branches strongly to weakly glaucous or not; floral bracts: apex rounded; cordillera in Alberta, British Columbia, Idaho, Montana, Oregon, Northwest Territories, Wyoming, Yukon . . . . 64. *Salix farriae* (in part)
95. Largest medial blades with white hairs on midrib of adaxial surface (*S. hookeriana* may have some ferruginous hairs on surfaces but not restricted to midrib).
98. Floral bracts dark brown to black, 1.1–3.6 mm; styles 0.6–2.5 mm; largest medial blades 1.5–4.2 times as long as wide; stigmas usually slenderly or broadly cylindrical, 0.28–0.74 mm; anthers yellow.
99. Largest medial blades: abaxial surface glabrous or glabrescent, hairs white, margins serrulate, slightly revolute or flat; branches flexible at base . . . . 61. *Salix barclayi*
99. Largest medial blades: abaxial surface pilose, villous, tomentose, or woolly, hairs often white and ferruginous, margins entire, sinuate to serrulate, or crenate, slightly revolute; branches highly brittle at base . . . . 77. *Salix hookeriana* (in part)
98. Floral bracts tawny to light brown, 0.6–1.6 mm; styles 0.1–0.6 mm; largest medial blades 2.6–7 times as long as wide; stigmas flat, abaxially non-papillose, tip round, or 2 plump lobes, 0.12–0.32 mm; anthers purple turning yellow.
100. Branches usually red-brown, sometimes gray-brown (nw California, adjacent Oregon) . . . . 106. *Salix tracyi* (in part)
100. Branches yellowish or grayish.
101. Largest medial blades with distinctly toothed margins; pistillate catkins slender to stout; juvenile blades pilose, villous, or glabrous . . . . 69. *Salix famelica* (in part)
101. Largest medial blades with entire or indistinctly toothed margins; pistillate catkins subglobose to stout; juvenile blades long-silky or glabrous . . . . 72. *Salix lutea* (in part)

[91. Shifted to left margin—Ed.]
91. Largest medial blades dull adaxially.
102. Largest medial blades usually with some ferruginous hairs on midrib of adaxial surface.
103. Largest medial blades narrowly to broadly elliptic or narrowly ovate to ovate; pistillate adaxial nectaries square or obovate; stipules on early leaves foliaceous; branches not glaucous; floral bracts: apex acute or rounded; Alaska, nw Northwest Territories, Yukon. . . . . 63. *Salix hastata* (in part)
103. Largest medial blades narrowly elliptic to elliptic; pistillate adaxial nectaries oblong or ovate; stipules on early leaves absent or rudimentary (sometimes foliaceous); branches strongly to weakly glaucous or not; floral bracts: apex rounded to convex; cordillera in Alberta, British Columbia, Idaho, Montana, Oregon, Wyoming . . . . 64. *Salix farriae* (in part)
102. Largest medial blades with white hairs only on midrib of adaxial surface (*S. hookeriana* may have ferruginous hairs on adaxial surface but not restricted to midrib).
104. Largest medial blades amphistomatous.
105. Branches pilose or villous; filaments hairy basally; ovaries: beak slightly bulged below styles . . . . 70. *Salix turnorii*
105. Branches glabrous; filaments glabrous; ovaries: beak gradually tapering to styles.

106. Styles 0.6–1.1 mm; juvenile blades villous to long-silky, with white, and sometimes ferruginous, hairs; largest medial blades 2–3.9 times as long as wide, abaxial surface glabrous; stipes 0.5–1.6mm . . . . 66. *Salix monticola* (in part)

106. Styles 0.13–0.6 mm; juvenile blades glabrous or long-silky, hairs white; largest medial blades 2.8–5.6 times as long as wide, abaxial surface long-silky to pilose or glabrous; stipes 0.9–3.8 mm . . . . .72. *Salix lutea* (in part)

[104. Shifted to left margin.—Ed.]

104. Largest medial blades usually hypostomatous, sometimes hemiamphistomatous.

107. Largest medial blades slightly glossy adaxially.

108. Largest medial blades with distinctly toothed margins; pistillate catkins slender to stout; juvenile blades pilose, villous, or glabrous . . . . 69. *Salix famelica* (in part)

108. Largest medial blades with entire or indistinctly toothed margins; pistillate catkins subglobose or stout; juvenile blades long-silky or glabrous . . . . . .72. *Salix lutea* (in part)

107. Largest medial blades dull adaxially.

109. Branches yellow-gray, yellow-brown, or gray-brown; floral bracts tawny or brown; ovaries pyriform or ovoid . . . . .72. *Salix lutea* (in part)

109. Branches usually red-brown, rarely yellow-gray or yellow-brown; floral bracts brown or black; ovaries pyriform.

110. Styles 0.6–1.1 mm . . . . .66. *Salix monticola* (in part)

110. Styles 0.2–0.7 mm.

111. Largest medial blades lorate to narrowly elliptic, 2.9–6.4 times as long as wide, margins usually serrulate or serrate, rarely apparently entire; stipes 0.9–2.5 mm . . . . .71. *Salix ligulifolia*

111. Largest medial blades narrowly oblong, lanceolate or obovate, 2.4–4.5 times as long as wide, margins prominently serrate, serrulate, or spinulose-serrulate; stipes 1.3–4.2 mm . . . . .74. *Salix prolixa*

2e1. SALIX Linnaeus (subg. VETRIX) sect. HASTATAE (Fries) A. Kerner, Verh. K.-K. Zool.-Bot. Ges. Wien 10: 241. 1860

Salix tribe *Hastatae* Fries, Physiogr. Sällsk. Årsberätt 2: 34. 1825; *Salix* subsect. *Hastatae* (Fries) Dorn

Shrubs, 0.1–7 m. **Buds** alba-, arctica-, or caprea-type, or intermediate. **Leaves:** stipules on late ones minute rudiments or foliaceous; largest medial blade (sometimes amphistomatous), 1.2–7 times as long as wide, adaxial surface not glaucous. **Staminate flowers:** filaments glabrous or hairy. **Pistillate flowers:** adaxial nectary shorter than, equal to, or longer than stipe; stipe 0–3.5 mm; ovary glabrous or sparsely to very densely hairy; stigmas with flat, non-papillate abaxial surface, cylindrical, or plump, 0.1–0.8 mm.

Species 24 (17 in the flora): North America, Eurasia.

This section was treated by R. D. Dorn (1975) as sect. *Cordatae*. In 1995, he recognized two subsections: *Cordatae* and *Luteae*. These two subsections are treated here as sections *Hastatae* and *Cordatae*, respectively.

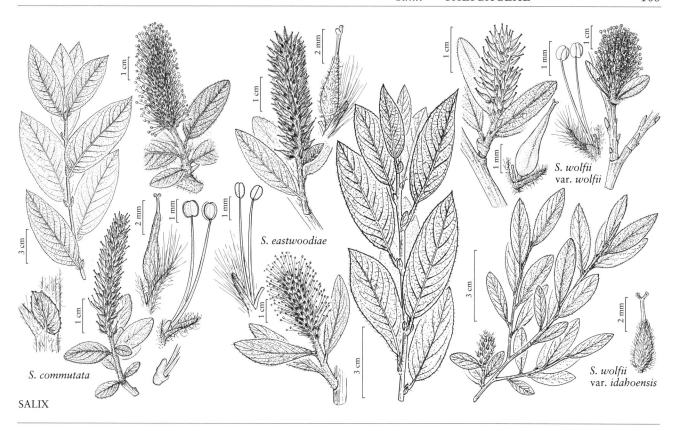

51. Salix commutata Bebb, Bot. Gaz. 13: 110. 1888

• Under-green willow [E] [F]

Salix commutata var. denudata Bebb; S. commutata subsp. mixta Piper; S. commutata var. puberula Bebb; S. commutata var. sericea Bebb

Plants 0.2–3 m. **Stems:** branches yellow-brown, gray-brown, or red-brown, not or weakly glaucous, pilose; branchlets yellow-green, yellow-brown, or red-brown, pilose to densely villous or woolly. **Leaves:** stipules (sometimes marcescent), foliaceous, apex rounded or acute; petiole shallowly grooved or convex to flat adaxially, 1.5–11 mm, pilose or tomentose adaxially; largest medial blade (sometimes amphistomatous), narrowly oblong, oblong, elliptic, or broadly elliptic, 10–100 × 5–44 mm, 1.5–3.4 times as long as wide, base convex, rounded, subcordate, or cordate, margins flat or slightly revolute, entire or serrulate, apex acuminate, acute, or convex, abaxial surface not glaucous, moderately densely tomentose, villous, or pilose to glabrescent, hairs wavy or straight, adaxial dull or slightly glossy, pilose or moderately densely villous to glabrescent; proximal blade margins entire or shallowly serrulate; juvenile blade yellowish green, sparsely to densely long-silky-tomentose abaxially, hairs white. **Catkins** flowering as leaves emerge; staminate stout or subglobose, 15–37 × 8–20 mm, flowering branchlet 2–33 mm; pistillate densely or moderately densely flowered, slender, stout, or subglobose, 17–60 × 7–15 mm, flowering branchlet 3–15(–30) mm; floral bract tawny, brown, or bicolor, 1–3 mm, apex acute or rounded, abaxially hairy, hairs straight or wavy. **Staminate flowers:** adaxial nectary oblong or square, 0.2–0.8 mm; filaments distinct, glabrous; anthers yellow or purple turning yellow, 0.4–1 mm. **Pistillate flowers:** adaxial nectary oblong, square, or ovate, 0.3–0.7 mm, shorter than stipe; stipe 0.3–2 mm; ovary pyriform or obclavate, glabrous, beak gradually to abruptly tapering to styles; ovules 10–28 per ovary; styles 0.5–1.5 mm; stigmas flat, abaxially non-papillate with rounded tip, broadly cylindrical, or 2 plump lobes, 0.16–0.34–0.4 mm. **Capsules** 3.5–8 mm. $2n = 38$.

Flowering late May–mid Aug. Rocky alpine and subalpine slopes, glacial moraine, open spruce woods, streamsides, gravel benches along streams, wet fens; 0–2400 m; Alta., B.C., N.W.T., Sask., Yukon; Alaska, Idaho, Mont., Oreg., Wash.

The rare occurrence in *Salix commutata* of plants with ovary indumentum composed of divergent, straight or wavy, flattened hairs may be hybrids with 53. *S. eastwoodiae* (see for discussion and comparison).

Hybrids:

Salix commutata forms natural hybrids with *S. barclayi*, *S. barrattiana*, and *S. eastwoodiae*.

52. Salix cordata Michaux, Fl. Bor.-Amer. 2: 225. 1803 · Heart-leaf or sand dune willow [E]

Salix adenophylla Hooker; *S. syrticola* Fernald

Plants 0.4–3 m, (often forming clones by layering or stem fragmentation). **Stems:** branches (sometimes ± brittle at base), red-brown, not glaucous (slightly glossy), tomentose to glabrescent; branchlets red-brown, moderately to very densely villous. **Leaves:** stipules foliaceous, apex acute or rounded; petiole shallowly grooved adaxially, 1–13 mm, tomentose adaxially; largest medial blade narrowly oblong, or narrowly to broadly elliptic, 33–88 × 13–45 mm, 1.6–3.2 times as long as wide, base cordate, rounded, or convex, margins flat or slightly revolute, serrulate or spinulose-serrulate, apex acuminate, abaxial surface not glaucous, moderately densely villous to glabrescent, midrib remaining hairy, hairs straight or wavy, adaxial dull or slightly glossy, very densely villous to glabrescent, midrib remaining hairy; proximal blade margins entire or serrulate; juvenile blade yellowish green, abaxially villous or midrib long-silky, hairs white. **Catkins** flowering as leaves emerge; staminate stout, 17–40 × 8–15 mm, flowering branchlet 1–8 mm; pistillate moderately densely flowered, stout or slender, 27–65 × 8–19 mm, flowering branchlet 3–16 mm; floral bract brown, 1–2.6 mm, apex acute or rounded, abaxially hairy, hairs straight or wavy. **Staminate flowers:** adaxial nectary oblong or narrowly oblong, 0.3–1.3 mm; filaments distinct, glabrous; anthers yellow, (ellipsoid or shortly cylindrical), 0.6–0.8 mm. **Pistillate flowers:** adaxial nectary oblong or narrowly oblong, 0.4–1.3 mm, about same length as stipe; stipe 0.5–1.4 mm; ovary pyriform or obclavate, glabrous, beak slightly bulged below styles; ovules 11–24 per ovary; styles 0.7–1.6 mm; stigmas flat, abaxially non-papillate with rounded tip, or 2 plump lobes, 0.2–0.3–0.36 mm. **Capsules** 3.6–7 mm. $2n = 38$.

Flowering mid Apr–early Jul. Sand dunes and beaches; 0–200 m; Nfld. and Labr., Ont., Que.; Ill., Ind., Mich., N.Y., Pa., Wis.

Salix cordata seems to be rare in Labrador, Newfoundland, and Quebec; the species needs further study in those areas.

53. Salix eastwoodiae Cockerell ex A. Heller, Cat. N. Amer. Pl. ed. 3, 89. 1910 (as fastwoodiae) · Sierra willow [E][F]

Salix californica Bebb, Willows Calif., 89. 1879, not Lesquereux 1878

Plants 0.6–4 m. **Stems:** branches yellow, red, or violet, not to strongly glaucous (slightly glossy), pilose; branchlets yellow-green or red-brown, pilose to villous (inner membranaceous bud-scale layer free, separating from outer layer). **Leaves:** stipules foliaceous, apex acute; petiole convex to flat, or shallowly grooved adaxially, 3–8–17 mm, pilose or villous adaxially; largest medial blade narrowly oblong, oblong, or elliptic, 21–57–99 × 6–20–37 mm, 1.9–2.9–5 times as long as wide, base rounded, convex, subcordate, or cordate, margins flat or slightly revolute, entire or serrulate (with relatively short, slender teeth), apex acuminate, acute, or convex, abaxial surface not glaucous, pilose, short-silky, or densely woolly-tomentose to glabrescent, hairs wavy, adaxial dull or slightly glossy, sparsely to densely silky-tomentose, midrib remaining hairy; proximal blade margins entire or serrulate; juvenile blade yellowish green, very densely long-silky or woolly abaxially, hairs white (sometimes yellowish). **Catkins** flowering as leaves emerge; staminate stout or subglobose, 9.5–36.5 × 7–15 mm, flowering branchlet 1.5–7 mm; pistillate densely or moderately densely flowered, stout or subglobose, 11–51 × 8–16 mm, flowering branchlet 2–12 mm; floral bract brown or black, 1.4–2.8 mm, apex rounded or acute, abaxially hairy, hairs straight or wavy. **Staminate flowers:** adaxial nectary narrowly oblong to oblong, 0.5–1.1 mm; filaments distinct, glabrous or hairy basally; anthers yellow or purple turning yellow, 0.5–0.9 mm. **Pistillate flowers:** adaxial nectary narrowly oblong to oblong, 0.5–1.1 mm, longer than or equal to stipe; stipe 0.2–1.6 mm; ovary pyriform, short- or long-silky to glabrescent, beak gradually tapering to or slightly bulged below styles; ovules 12–16 per ovary; styles 0.5–1.5 mm; stigmas flat, abaxially non-papillate with rounded tip, slenderly or broadly cylindrical, or 2 plump lobes, 0.18–0.39–0.76 mm (evidentially two size classes). **Capsules** 4–10 mm. $2n = 76$.

Flowering mid May–late Jul. Alpine and subalpine meadows, streams, lakeshores, talus slopes, granite substrate; 1600–3800 m; Calif., Idaho, Mont., Nev., Oreg., Wash., Wyo.

Salix eastwoodiae and *S. commutata* are distinct species with different ploidal levels, the former tetraploid and the latter diploid; where they come into contact in the Pacific Northwest, hybrids occur and vegetative plants are often difficult to separate. See comparison below. The most important difference is that ovaries of *S. eastwoodiae* usually are silky turning glabrescent in age and those of *S. commutata* are glabrous. Populations occur in Oregon with both glabrous and hairy ovaries without any other evident differences. There are also unusual specimens, which are often tentatively identified as *S. eastwoodiae*, that have glabrous ovaries and patches of hairs at the base and on the sutures. The possibility that they are hybrids between *S. eastwoodiae* and *S. boothii*, *S. commutata*, or *S. lemmonii* needs study.

Salix commutata is distinguished from *S. eastwoodiae* by having leaf blades sometimes amphistomatous, 1.5–3.4 times as long as wide, teeth 0–19 per cm, adaxial surfaces glabrous or pilose to villous, floral bracts tawny to brown, staminate and pistillate adaxial nectaries oblong to square, and ovaries glabrous; *S. eastwoodiae* has leaf blades hypostomatous, 1.8–5 times as long as wide, teeth 0–10 per cm, adaxial surfaces tomentose or long-silky, floral bracts brown to black, staminate and pistillate adaxial nectaries narrowly oblong to oblong, and ovaries silky to glabrescent.

Hybrids:

Salix eastwoodiae forms natural hybrids with *S. arizonica*, *S. boothii*, and *S. commutata*.

Salix eastwoodiae × *S. lasiandra* was found in Sierra County, California, growing with both parents in a wetland along a disturbed roadside. It had leaf indumentum and hair color of *S. eastwoodiae* and leaf shape and margins of *S. lasiandra*. Catkins of this intersubgeneric hybrid were teratological and presumably infertile.

54. **Salix wolfii** Bebb in J. T. Rothrock, Rep. U. S. Geogr. Surv., Wheeler, 241. 1879 • Idaho willow [E] [F]

Plants 0.1–2 m. Stems: branches red-brown, violet, yellow-gray, or yellow-brown, pubescent or pilose to glabrescent; branchlets yellowish, yellow-brown, red-brown, or yellow-green, sparsely or moderately densely pubescent, or densely long-silky, (inner membranaceous bud-scale layer free, separating from outer layer). Leaves: stipules rudimentary or foliaceous on early ones, foliaceous on late ones, apex rounded, acuminate, or acute; petiole convex to flat, or shallowly grooved adaxially, 3–12 mm, pubescent, long-silky, or villous adaxially; largest medial blade narrowly oblong, narrowly elliptic, elliptic, or oblanceolate, 26–56 × 8–16.5 mm, 2.5–3.7–5.6 times as long as wide, base cuneate, convex, or rounded, margins flat, entire, apex acute, acuminate, or convex, abaxial surface not glaucous, pubescent, short-silky, or villous, hairs appressed or spreading, straight or wavy, adaxial dull, sparsely to densely silky or villous; proximal blade margins entire; juvenile blade yellowish green, densely short- or long-silky or villous abaxially, hairs white. Catkins flowering as leaves emerge; staminate stout or subglobose, 9.5–16 × 6–12 mm, flowering branchlet 1–5.5 mm; pistillate moderately or very densely flowered, stout, subglobose or globose, 8.5–38 × 5–12 mm, flowering branchlet 1–11 mm; floral bract brown, black, or bicolor, 0.8–2 mm, apex rounded or acute, abaxially hairy, hairs wavy, straight, or curly. Staminate flowers: (abaxial nectary 0–0.2 mm), adaxial nectary oblong, 0.4–1.1 mm, (nectaries distinct); filaments distinct, glabrous; anthers yellow, 0.3–0.5 mm. Pistillate flowers: adaxial nectary oblong, ovate, or flask-shaped, 0.4–1.1 mm, shorter to longer than stipe; stipe 0.2–0.9 mm; ovary pyriform, glabrous or hairy, beak gradually tapering to or slightly bulged below styles; ovules 8–16 per ovary; styles 0.3–1 mm; stigmas flat, abaxially non-papillate with rounded or pointed tip, or slenderly or broadly cylindrical, 0.24–0.3–0.4 mm. Capsules 3–5 mm.

Varieties 2 (2 in the flora): w United States.

The two varieties of *Salix wolfii* are distinguished mainly by ovary hairiness; other characters in the key overlap. Ovaries of the typical var. *wolfii* are glabrous and those of var. *idahoensis* are hairy. In the latter variety, ovaries are sometimes hairy throughout, but most have hairs in streaks or in a patch at the base of the ovary and on the stipes. These plants usually do not set seed and may be infertile hybrids. Occasional occurrence of staminate flowers with abaxial nectaries suggests that this variety may be a hybrid with *S. glauca* or *S. brachycarpa*, although it could also be with *S. eastwoodiae*, as suggested by S. J. Brunsfeld and F. D. Johnson (1985). The presence of both abaxial and adaxial nectaries in staminate flowers of *S. wolfii* (staminate plants cannot be identified to variety) is an unusual character in subg. Vetrix; it rarely occurs in *S. argyrocarpa*, *S. breweri*, and *S. orestera*, but is common in *S. wolfii*. Both hairy ovaries and abaxial nectaries could have been acquired through hybridization and introgression, or polyploidy, with *S. glauca* or *S. brachycarpa*. Cytological study of *S. wolfii* may help answer this question.

1. Ovaries glabrous; pistillate adaxial nectaries 0.4–0.8 mm; stipes 0.2–0.9 mm . . . . 54a. *Salix wolfii* var. *wolfii*
1. Ovaries pubescent or tomentose (hairs in streaks or patches); pistillate adaxial nectaries 0.4–1.1 mm; stipes 0–0.4 mm . . . . . . 54b. *Salix wolfii* var. *idahoensis*

54a. Salix wolfii Bebb var. wolfii [E] [F]

Plants 0.5–2 m. Stems: branches red-brown or violet, pubescent to glabrescent; branchlets yellow-brown or yellowish, sparsely pubescent or densely long-silky, hairs straight, wavy, curved, or crinkled. Leaves: petiole convex to flat, or shallowly grooved adaxially, 3–5.6 mm, pubescent or long-silky adaxially; largest medial blade narrowly oblong, narrowly elliptic, elliptic, or oblanceolate, apex acute, acuminate, or convex, abaxial surface pubescent, short-silky, or villous, adaxial sparsely to densely short- to long-silky, or villous; juvenile blade short- to long-silky or villous abaxially. Catkins: pistillate moderately or very densely flowered, subglobose or globose, 8.5–19 × 7–12 mm, flowering branchlet 1–7 mm; floral bract 0.8–1.6 mm. Staminate flowers: abaxial nectary (0–)0.1–0.2 mm, adaxial nectary 0.4–0.5 mm. Pistillate flowers: adaxial nectary oblong, 0.4–0.8 mm, shorter to longer than stipe; stipe 0.2–0.9 mm; ovary glabrous; ovules 10–12 per ovary; stigmas flat, abaxially non-papillate with rounded or pointed tip. $2n = 38$.

Flowering early–mid Jun. Stream banks, springs, wet meadows, bogs; 2000–3800 m; Colo., Idaho, Mont., Nev., Oreg., Utah, Wyo.

Hybrids:

Variety *wolfii* forms natural hybrids with *Salix boothii*.

54b. Salix wolfii Bebb var. idahoensis C. R. Ball, Bot. Gaz. 40: 378. 1905 [E] [F]

Salix idahoensis (C. R. Ball) Rydberg

Plants 0.1–2 m. Stems: branches yellow-gray or yellow-brown, pubescent or pilose; branchlets yellow-green or red-brown (darker in age), sparsely to moderately densely pubescent, hairs wavy or geniculate. Leaves: petiole shallowly grooved adaxially, 3–10 mm, pubescent or villous adaxially; largest medial blade very narrowly elliptic to elliptic, or narrowly oblanceolate, apex acute to acuminate, abaxial surface villous, adaxial densely silky or villous; juvenile blade long-silky abaxially. Catkins: pistillate very densely flowered, stout or subglobose, 8.5–38 × 5–12 mm, flowering branchlet 1–11 mm; floral bract 1–2 mm. Staminate flowers: abaxial nectary 0–0.2 mm, adaxial nectary 0.6–1.1 mm. Pistillate flowers: adaxial nectary oblong, ovate, or flask-shaped, 0.4–1.1 mm, longer than or, rarely, equal to stipe; stipe 0–0.4 mm; ovary pubescent or tomentose, hairs in streaks or patches; ovules 8–16 per ovary; stigmas flat, abaxially non-papillate with rounded or pointed tip, or slenderly or broadly cylindrical. $2n$ = unknown.

Flowering early–mid Jun. Sedge meadows along stream and lake margins, drainageways, around springs; 2100–3100 m; Idaho, Mont., Utah, Wyo.

55. Salix orestera C. K. Schneider, J. Arnold Arbor. 1: 164. 1920 • Gray-leaf Sierra willow [E]

Salix commutata Bebb var. *rubicunda* Jepson; *S. glauca* Linnaeus subsp. *orestera* (C. K. Schneider) Youngberg; *S. glauca* var. *orestera* (C. K. Schneider) Jepson

Plants 0.5–2 m. Stems: branches dark red-brown or yellow-brown, not to strongly glaucous, glabrous; branchlets yellow-brown or red-brown, (not or weakly glaucous), pilose or pubescent, hairs straight, wavy, or geniculate, (inner membranaceous bud-scale layer free, separating from outer layer). Leaves: stipules foliaceous, rudimentary, or absent on early ones, foliaceous on late ones, apex acute; petiole convex to flat, or shallowly grooved adaxially, 4–9 mm, pilose adaxially; largest medial blade (sometimes amphistomatous), lorate, narrowly oblong, narrowly elliptic, or oblanceolate, 35–95 × 7.5–20 mm, 3.4–7.1 times as long as wide, base cuneate or convex, margins flat or slightly revolute, entire, apex acute, acuminate, or convex, abaxial surface glaucous (sometimes obscured by hairs), sparsely to moderately densely long- to short-silky or pubescent, hairs (white, sometimes also ferruginous), straight or wavy, adaxial dull or slightly glossy, sparsely or moderately densely pubescent or long- to short-silky, (hairs white, sometimes also ferruginous); proximal blade margins entire or serrulate; juvenile blade densely long-silky abaxially, hairs white, sometimes also ferruginous. Catkins flowering as leaves emerge; staminate stout, 15.5–34 × 7–14 mm, flowering branchlet 1–8 mm; pistillate moderately densely flowered, stout, 20–55(–65 in fruit) × 11–13 mm, flowering branchlet 2–15 mm; floral bract dark brown or bicolor, 1.2–2.5 mm, apex acute or rounded, abaxially hairy, hairs straight or wavy. Staminate flowers: adaxial nectary oblong or ovate, 0.6–1.1 mm; filaments distinct or connate less than ½ their lengths, hairy basally; anthers purple turning yellow, 0.6–1 mm. Pistillate flowers: adaxial nectary oblong or flask-shaped, 0.7–1.3 mm, shorter than or equal to stipe; stipe 0.8–2 mm; ovary obclavate or pyriform, short-silky-villous, beak gradually tapering to or slightly bulged below styles; ovules 15–16 per ovary; styles 0.6–1 mm; stigmas flat, abaxially non-papillate

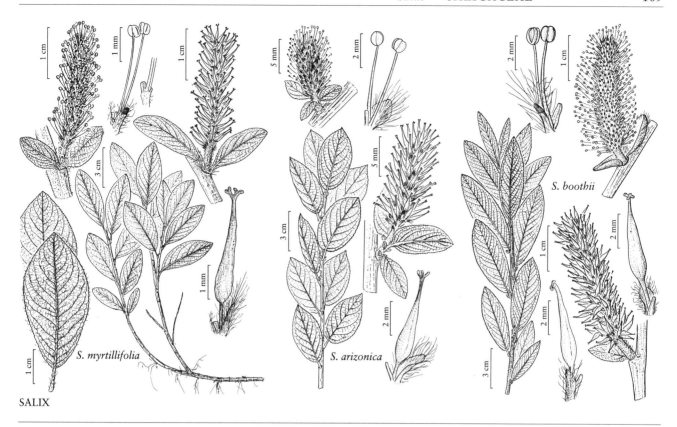

with rounded or pointed tip, or broadly cylindrical, 0.24–0.32–0.44 mm. **Capsules** 5–10 mm.

Flowering late May–late Aug. Subalpine meadows, slopes, lakes, streams, granite substrates; 2100–4000 m; Calif., Nev., Oreg.

Salix orestera is so similar to *S. glauca* var. *villosa* that it is sometimes included in *S. glauca*. It is phenetically most similar to *S. eastwoodiae* and *S. lemmonii* (G. W. Argus 1997), and the possibility that it is a hybrid or an alloploid involving those species needs study.

Salix orestera can be distinguished from *S. glauca* var. *villosa* in having largest medial blades usually narrower, 3.4–7.1 times as long as wide, abaxial surfaces distinctly silky with appressed hairs pointing toward apex, hairs sometimes ferruginous, secondary veins raised abaxially and adaxially, branches often strongly glaucous, and staminate flowers without abaxial nectaries; *S. glauca* var. *villosa* has largest medial blades usually broader, 2.2–3.9 times as long as wide, abaxial surfaces usually glabrescent or, if hairy, hairs unkempt, slightly spreading, hairs always white, secondary veins raised abaxially, flat or impressed adaxially, branches not glaucous, staminate flowers often with abaxial nectaries.

56. Salix myrtillifolia Andersson, Öfvers. Kongl. Vetensk.-Akad. Förh. 15: 132. 1858 • Low blueberry willow E F

Plants 0.1–0.6(–1) m, (forming clones by layering). **Stems** (decumbent); branches gray-brown, red-brown, or yellow-brown, not to strongly glaucous (dull or slightly glossy), pubescent; branchlets gray-brown, red-brown, or yellow-brown, sparsely pubescent. **Leaves:** stipules rudimentary, foliaceous, or absent on early ones, foliaceous on late ones (0.2–1.8(–5) mm); petiole deeply to shallowly grooved adaxially, 1.5–8 mm, glabrous or pubescent adaxially; largest medial blade (sometimes amphistomatous), elliptic, narrowly elliptic, obovate, or broadly obovate, 17–74 × 8–30 mm, 1.2–4.5 times as long as wide, base cuneate, convex, or subcordate, margins flat, serrulate, crenulate, or sinuate, apex acute, convex, or acuminate, abaxial surface not glaucous, glabrous, adaxial slightly glossy, glabrous; proximal blade margins crenate; juvenile blade reddish or yellowish green, glabrous. **Catkins** flowering as leaves emerge; staminate stout, 11.5–39 × 5–14 mm, flowering branchlet 0.5–6 mm; pistillate moderately densely flowered or densely flowered, slender or stout, 16–46(–50 in fruit) × 4–15 mm, flowering branchlet 1.5–12 mm; floral bract brown, black, tawny, or bicolor,

0.4–1.1 mm, apex retuse or acute, abaxially hairy throughout or proximally, hairs curly or wavy. **Staminate flowers:** adaxial nectary oblong, ovate, or square, 0.2–0.34–0.4 mm; filaments distinct, glabrous; anthers purple turning yellow, 0.3–0.6 mm. **Pistillate flowers:** adaxial nectary square, oblong, or ovate, 0.2–0.4 mm, shorter than stipe; stipe 0.6–1.7 mm; ovary pyriform, glabrous, beak gradually tapering to or slightly bulged below styles; ovules (6–)10–14 per ovary; styles connate or distinct ½ their lengths, 0.3–0.7 mm; stigmas flat, abaxially non-papillate with rounded tip, or 2 plump lobes, 0.16–0.23–0.32 mm. **Capsules** 4–6 mm. $2n = 38$.

Flowering early May–late Jul. Treed bogs, fens, stream banks, subalpine spruce thickets, *Pinus contorta* woods, sand dunes, coal spoils; 90–2800 m; Alta., B.C., Man., N.B., N.W.T., Nunavut, Ont., Sask., Yukon; Alaska, Colo., Wyo.

Salix myrtillifolia occurs in Nunavut on Akimiski Island in James Bay.

The complex of species related to *Salix myrtillifolia* includes *S. arizonica*, *S. ballii*, *S. boothii*, and *S. pseudomyrsinites*. Two are diploid (*S. arizonica* and *S. myrtillifolia*), and two are tetraploid (*S. boothii* and *S. pseudomyrsinites*); the chromosome number of *S. ballii* is unknown. They have been treated taxonomically in different ways, but are relatively distinct in their morphology, ecology, and geography. *Salix myrtillifolia* has outlying populations represented by single collections each from Colorado, Quebec, and Wyoming. Specimens attributed to this species from the Gaspe Peninsula, Quebec, and the Northern Peninsula, Newfoundland, all have evidence of leaf glaucescence and are *S. ballii*. See 57. *S. pseudomyrsinites* and 58. *S. ballii* for more description.

Hybrids:

Salix myrtillifolia forms natural hybrids with *S. candida*.

57. **Salix pseudomyrsinites** Andersson, Öfvers. Kongl. Vetensk.-Akad. Förh. 15: 129. 1858 (as pseudomyrsinites) • Tall blueberry willow [E]

Salix myrtillifolia Andersson var. *cordata* (Andersson) Dorn; *S. myrtillifolia* var *pseudomyrsinites* (Andersson) C. R. Ball ex Hultén; *S. novae-angliae* Andersson var. *cordata* Andersson

Plants 1–7 m. **Stems:** branches gray-brown, red-brown, or yellow-brown, glaucous to strongly so (slightly to highly glossy), villous to glabrescent; branchlets gray-brown, red-brown, yellow-brown, or yellow-green, (not or weakly glaucous), pilose, densely villous, or tomentose. **Leaves:** stipules rudimentary or foliaceous on early ones, foliaceous on late ones, (0.6–8 mm), apex acute or obtuse; petiole shallowly to deeply grooved adaxially, 2.5–8 mm, glabrous or villous adaxially; largest medial blade (sometimes amphistomatous), narrowly to broadly elliptic, oblong to oblanceolate or obovate, 32–109 × 10–47 mm, 1.8–4.8 times as long as wide, base convex, cuneate, or subcordate, sometimes cordate, margins flat, entire, crenate, or serrulate, apex acute, convex, or acuminate, abaxial surface not glaucous, glabrous or pilose, hairs (white, sometimes also ferruginous), wavy, adaxial slightly glossy, glabrous, pilose, pubescent, moderately densely short-silky, or velvety, midrib remaining pilose or short-hairy, (hairs sometimes also ferruginous, straight and geniculate); proximal blade margins serrulate or entire; juvenile blade sometimes reddish or yellowish green, abaxially glabrous, or midrib sparsely pubescent, or densely villous or short-silky, hairs white, sometimes also ferruginous. **Catkins** flowering as leaves emerge; staminate stout, 16.5–35.5 × 7–15 mm, flowering branchlet 0.5–12 mm; pistillate moderately or densely flowered, slender or stout, 10.5–68 × 5–20 mm, flowering branchlet 0.5–10 mm; floral bract brown, black, tawny, or bicolor, 0.6–1.1 mm, apex retuse, abaxially hairy, hairs long-wavy or curly. **Staminate flowers:** adaxial nectary oblong square, 0.2–0.4–0.6 mm; filaments distinct, glabrous; anthers purple turning yellow, 0.4–0.7 mm. **Pistillate flowers:** adaxial nectary square or oblong, 0.2–0.4 mm, shorter than stipe; stipe 0.8–1.4 mm; ovary pyriform, glabrous, beak slightly bulged below styles; ovules 11–18 per ovary; styles 0.4–1.6 mm; stigmas flat, abaxially non-papillate with rounded tip, or 2 plump lobes, 0.16–0.24–0.32 mm. **Capsules** 4.4–6.4 mm. $2n = 76$.

Flowering early May–early Jul. Shores of lakes and streams, dwarf-birch thickets, fens, marl bogs, rarely in treed bogs; 40–1000 m; Alta., B.C., Man., N.W.T., Nunavut, Ont., Sask., Yukon; Alaska.

Salix pseudomyrsinites occurs in Nunavut on Akimiski Island in James Bay.

Salix pseudomyrsinites and *S. myrtillifolia*, although sometimes treated as conspecific (R. D. Dorn 1975), deserve species rank. They differ in chromosome number and are distinct in habit, habitat, and general appearance, including glossiness of leaves, as well as a number of technical characteristics (L. A. Viereck and E. L. Little Jr. 1972; G. W. Argus 1973, 1997). There is no field evidence of hybridization, but some herbarium specimens appear to be intermediates, having the habit or habitat of one species and the leaf hairiness of the other.

Salix myrtillifolia is distinguished from *S. pseudomyrsinites* by having shrubs low, decumbent, 0.1–0.6 m, rarely to 1 m, of treed bogs and fens, juvenile and mature leaves typically glabrous, stipules usually rudimentary, 0.2–1.8(–5) mm, and styles often shorter, 0.3–0.7 mm; *S. pseudomyrsinites* has shrubs tall, erect, 1–7 m, of riparian habitats, juvenile leaves pubescent with hairs persisting on mature leaves, at least on adaxial midrib, stipules usually prominent and foliaceous, 0.6–8 mm, and styles often longer, 0.4–1.6 mm.

The nomenclature of these species is confusing. When treating them as varieties E. Hultén (1968) used the name *Salix myrtillifolia* var. *pseudomyrsinites* and R. D. Dorn (1975) used the name *S. myrtillifolia* var. *cordata*. At the species level, the name *S. novae-angliae* was used (L. A. Viereck and E. L. Little Jr. 1972; G. W. Argus 1973), but that name is illegitimate (Dorn) and is replaced by *S. pseudomyrsinites* (Argus 1997).

Hybrids:

Salix pseudomyrsinites forms natural hybrids with *S. barrattiana*.

Salix pseudomyrsinites × *S. barrattiana* is a rare hybrid that combines the characters of the two parents.

58. **Salix ballii** Dorn, Canad. J. Bot. 53: 1501. 1975
• Ball's willow E

Salix myrtillifolia Andersson var. *brachypoda* Fernald, Rhodora 16: 172. 1914, not *S. brachypoda* (Trautvetter & C. A. Meyer) Komarov 1923

Plants 0.2–1.2 m. **Stems:** branches red-brown or yellow-brown, not glaucous (dull or slightly glossy), pubescent; branchlets red-brown or yellow-brown, (not or strongly glaucous), pubescent, villous, or short-silky, (inner membranaceous bud-scale layer free, not separating from outer layer). **Leaves:** stipules absent or rudimentary on early ones, foliaceous on late ones, apex acute or convex; petiole shallowly to deeply grooved adaxially, 2.5–7.5 mm, pubescent; largest medial blade elliptic to obovate, 23–63 × 10–35 mm, 1.4–2.8 times as long as wide, base convex or rounded, sometimes cordate or subcordate, margins flat, serrulate or sinuate, apex convex, rounded, acute, or acuminate, abaxial surface glaucous, glabrous, adaxial slightly glossy, glabrous or sparsely pubescent on midrib, (hairs white, sometimes also ferruginous); proximal blade margins serrulate or crenulate; juvenile blade sometimes reddish, glabrous, or midrib sparsely pubescent abaxially, hairs white, sometimes also ferruginous. **Catkins** flowering as leaves emerge; staminate stout, 17–29.5 × 8–11 mm, flowering branchlet 3.5–12 mm; pistillate moderately densely flowered, slender, stout, or subglobose, 10–37.5 (–45 in fruit) × 5–12 mm, flowering branchlet 2.5–16 mm; floral bract brown or bicolor, 0.8–1.6 mm, apex rounded, convex or retuse, abaxially hairy throughout or proximally, hairs straight, curly, or wavy. **Staminate flowers:** adaxial nectary oblong, square, or ovate, 0.3–1 mm; filaments distinct, glabrous; anthers yellow, 0.4–0.8 mm. **Pistillate flowers:** adaxial nectary square or oblong, 0.2–0.6 mm, shorter than stipe; stipe 0.8–2 mm; ovary pyriform, glabrous, beak gradually tapering to or slightly bulged below styles; ovules 12–18 per ovary; styles 0.4–1 mm; stigmas flat, abaxially non-papillate with pointed tip, or broadly cylindrical, 0.2–0.3–0.36 mm. **Capsules** 3–6 mm.

Flowering late Jun–early Jul. Coastal barrens, terraces, ravines, talus slopes, coastal dunes, floodplains, *Carex* meadows, scrubby *Picea mariana* woods, dwarfed *Abies balsamea* thickets, *Picea mariana*-lichen-feathermoss woods, limestone and calcareous substrates; 0–400 m; Nfld. and Labr., Nunavut, Ont., Que.

Occurrence of *Salix ballii* in Nunavut is on Charlton Island in James Bay.

Salix ballii differs from *S. myrtillifolia* in having leaves that are distinctly glaucous abaxially. It was described as *S. myrtillifolia* var. *brachypoda* by Fernald, who noted that among the characters that distinguish it from *S. myrtillifolia* only the presence of leaf glaucescence does not occur elsewhere in *S. myrtillifolia*. This character may be lost when dried over excessive heat. For example, the only specimen supporting the occurrence of *S. myrtillifolia* on the Gaspe Peninsula, Quebec, is a badly damaged, poorly dried collection that may have lost its glaucescence in drying. A single character difference such as this usually would not recommend a taxon for species rank, but in this case it may be justified inasmuch as *S. ballii* and *S. myrtillifolia* are allopatric. A specimen from Île Couture, Lac Mistassini Region, Quebec, may be an exception but confirmatory collections are needed.

Hybrids:

Salix ballii forms natural hybrids with *S. glauca* var. *cordifolia* (*S.* ×*ungavensis* Lepage). This sterile hybrid with aborted ovaries is known only from the type locality in northern Quebec. It generally resembles *S. glauca* var. *cordifolia* but its ovaries are glabrous except for hairy patches at the base and on the stipe. It is evidently a hybrid involving *S. glauca* var. *cordifolia* and a species with glabrous ovaries. E. Lepage (1962) was correct in suggesting that the latter was *S. ballii* (as *S. myrtillifolia* var. *brachypoda*). Both taxa grow together in the area; style and floral bract lengths fall within the range of *S. ballii*, except for a slight overlap with *S. glauca* var. *cordifolia*, and its serrulate to crenulate leaf margins are characteristic of *S. ballii*.

59. Salix arizonica Dorn, Canad. J. Bot. 53: 1499. 1975 · Arizona willow [C][E][F]

Plants 0.3–2.6 m. Stems: branches red-brown or yellow-brown, not or weakly glaucous, glabrous or pilose at nodes; branchlets yellow-green, red-brown, or brownish, pilose. Leaves: stipules foliaceous, apex convex or rounded; petiole convex to flat, or shallowly grooved adaxially, 2–7.5 mm, villous or pubescent to glabrescent adaxially; largest medial blade (sometimes amphistomatous), elliptic or broadly elliptic, 20–50 × 10–31 mm, 1.6–2–2.8(–3.6) times as long as wide, base convex, rounded, or cordate, margins flat, serrulate or entire, apex acute, convex, or acuminate, abaxial surface not glaucous, pilose or glabrous, hairs wavy, adaxial slightly glossy, pilose or glabrous; proximal blade margins entire, serrulate, or crenulate; juvenile blade green, glabrous or pilose abaxially, hairs white. Catkins flowering as leaves emerge; staminate stout, subglobose, or globose, 7–17 × 6–10 mm, flowering branchlet 1–3 mm; pistillate densely or moderately densely flowered, stout or subglobose, 12–38 × 6–12 mm, flowering branchlet 1.5–10 mm; floral bract brown, black, or bicolor, 1–2 mm, apex acute or convex, abaxially hairy, hairs wavy. Staminate flowers: adaxial nectary narrowly oblong to oblong, 0.4–0.8 mm; filaments distinct, glabrous; anthers purple turning yellow, 0.4–0.6 mm. Pistillate flowers: adaxial nectary narrowly oblong to oblong, 0.4–1 mm, shorter to longer than stipe; stipe 0.2–1 mm; ovary pyriform, glabrous, beak gradually tapering to or slightly bulged below styles; ovules 8–12 per ovary; styles 0.5–1.2 mm; stigmas broadly cylindrical, 0.14–0.21–0.36 mm. Capsules 3.2–4.5 mm. $2n = 38$.

Flowering late May–late Jun. Subalpine sedge meadows, along streams, wet drainageways, cienegas; of conservation concern; 2600–3400 m; Ariz., Colo., N.Mex., Utah.

Salix arizonica is very similar to S. boothii. They were separated by Dorn on the presence of five flavonoid compounds identified in S. boothii not found in S. arizonica. Some morphological differences were noted but the characters used to separate them are quite variable. The most important feature seems to be the usually broader leaves of S. arizonica. In addition, the diploid chromosome number for S. arizonica separates it from the tetraploid S. boothii. Although the two are distinct species, the overlap in their morphological characters suggests that positive identification needs to be based on chromosome number or chromatographic analysis.

Salix arizonica is distinguished from S. boothii by having stipule apices convex to rounded, petioles 3–7.5 mm, juvenile blades glabrous or hairy, hairs white, largest medial blades elliptic or broadly elliptic, 20–50 mm, 1.6–3.6 times as long as wide, abaxial surfaces with white hairs, staminate catkins 1–1.7 times as long as broad, pistillate catkins 1.2–2.8 times as long broad, floral bract apices acute to convex, nectaries narrowly oblong to oblong, staminate nectaries 0.4–0.8 mm, anthers 0.4–0.6 mm, filaments distinct, glabrous, pistillate nectaries shorter to longer than stipes, stipes 0.2–1 mm, stigmas broadly cylindrical, and capsules 3.2–4.5 mm; S. boothii has stipule apices acuminate or acute to rounded, petioles 3–17 mm, juvenile blades hairy, hairs white, sometimes also ferruginous, largest medial blades lorate to narrowly or broadly elliptic, 26–102 mm, 2–5.2 times as long as wide, abaxial surfaces with white, sometimes also ferruginous, hairs, staminate catkins 1.2–3.1 times as long as broad, pistillate catkins 1.4–4.1 times as long as broad, floral bract apices rounded or retuse, nectaries narrowly oblong to ovate or flask-shaped, staminate nectaries 0.6–1.5 mm, anthers 0.48–0.8 mm, filaments distinct to connate about half their lengths, glabrous or hairy, pistillate nectaries shorter than stipes, stipes 0.5–2.5 mm, stigmas flat, abaxially non-papillate with rounded tip, slenderly cylindrical or plump, and capsules 2.5–6 mm.

Salix arizonica, originally known from Mt. Baldy in east-central Arizona, was proposed for listing under the United States Endangered and Threatened Wildlife and Plants Act, but the proposal was withdrawn when additional populations were discovered in southern Utah and in New Mexico; it is now listed as a "sensitive species" (K. Decker, www.fs.us/r2/projects/scp/assessments/salixarizonica.pdf). Major conservation concerns are from browsing by cattle and elk (J. Maschinski 2001) and Melampsora infection (M. L. Fairweather 1993; R. A. Obedzinski et al. 2001). The Arizona populations receive minimal protection from cattle and wildlife browsing by exclosures and by introduction into new localities. The Arizona and Utah populations have a genetic similarity of ca. 37%, which has been attributed to a period of panmixis followed by a long period of isolation in regions with different environments (J. T. Thompson et al. 2003).

Salix arizonica is in the Center for Plant Conservation's National Collection of Endangered Plants.

Hybrids:

Salix arizonica forms natural hybrids with S. brachycarpa, S. eastwoodiae, and S. lutea.

Salix arizonica × S. brachycarpa var. brachycarpa occurs in southern Utah. Its parentage is supported by chromatographic data (E. D. McArthur, pers. comm.).

60. Salix boothii Dorn, Canad. J. Bot. 53: 1505. 1975 • Booth's willow E F

Salix novae-angliae Andersson var. aequalis Andersson; S. pseudocordata (Andersson) Andersson ex Rydberg var. aequalis (Andersson) C. R. Ball ex C. K. Schneider

Plants 0.3–6 m. **Stems:** branches yellow-gray, yellow-brown, or red-brown, not glaucous, glabrous, pilose, or villous; branchlets yellow-brown, gray-brown, or red-brown, (not or weakly glaucous), glabrous, pilose, or moderately densely villous. **Leaves:** stipules foliaceous, apex acute, acuminate, or rounded; petiole convex to flat, or shallowly grooved adaxially, 3–17 mm, pilose, villous, or pubescent adaxially; largest medial blade (often amphistomatous), lorate, narrowly oblong, or narrowly to broadly elliptic, 26–102 × 8–30 mm, 2.5–3–5.2 times as long as wide, base convex, rounded, or subcordate, margins flat or slightly revolute, (thickened), entire or serrulate, apex acute or acuminate, abaxial surface not glaucous, glabrous, pilose, or moderately densely short-silky, hairs (white, sometimes also ferruginous), wavy, adaxial dull or slightly glossy, glabrous or moderately densely pilose; proximal blade margins entire, crenate, or serrulate; juvenile blade green, pilose to densely villous abaxially, hairs white, sometimes also ferruginous. **Catkins** flowering as or just before leaves emerge; staminate stout or subglobose, 7–37 × 5–12 mm, flowering branchlet 0.5–5.6 mm; pistillate densely or moderately densely flowered, stout, 12–62 × 7–17 mm, flowering branchlet 1–9 mm; floral bract brown, 0.7–2.1 mm, apex rounded or retuse, abaxially hairy throughout or proximally, hairs straight, wavy, or curly. **Staminate flowers:** adaxial nectary narrowly oblong, oblong, ovate, or flask-shaped, 0.6–1.5 mm; filaments distinct or connate ca. ½ their lengths, glabrous, hairy on proximal ½; anthers yellow or purple turning yellow, 0.5–0.8 mm. **Pistillate flowers:** adaxial nectary oblong, ovate, or flask-shaped, 0.3–0.8 mm, shorter than stipe; stipe 0.5–2.5 mm; ovary pyriform, glabrous, beak gradually to abruptly tapering to or slightly bulged below styles; ovules 10–18 per ovary; styles 0.3–1.4 mm; stigmas flat, abaxially non-papillate with rounded tip, or slenderly cylindrical, or 2 plump lobes, 0.2–0.26–0.48 mm. **Capsules** 2.5–6 mm. $2n = 76$.

Flowering early Apr–early Jul. Wet subalpine meadows, seepages, streams, lakeshores; 1500–3200 m; Alta., B.C., Sask.; Ariz., Calif., Colo., Idaho, Mont., Nev., Oreg., Utah, Wash., Wyo.

Hybrids:

Salix boothii forms natural hybrids with S. brachycarpa var. brachycarpa, S. eastwoodiae, S. glauca var. villosa, and S. wolfii. There are numerous intermediate specimens of S. boothii that suggest hybridization with S. arizonica, S. brachycarpa, S. eastwoodiae, S. lutea, or S. wolfii, but further study is needed. A DNA study of S. arizonica showed that a specimen from southwestern Utah previously identified as S. arizonica × S. wolfii probably was S. boothii × S. wolfii (J. T. Thompson et al. 2003).

In Mountain Park, Alberta, and the Steens Mountains, Oregon, the putative hybrid Salix boothii × S. glauca var. villosa grew in thickets with both parents. Ovaries were sparsely hairy on the distal half or on the beak, leaves were glaucous abaxially, and floral bracts were mostly glabrous abaxially, but some with hairs proximally. One plant in fruit produced copious seed hairs but no seed.

61. Salix barclayi Andersson, Öfvers. Kongl. Vetensk.-Akad. Förh. 15: 125. 1858 • Barclay's willow E F

Plants (0.3–)1–3(–5) m. **Stems:** branches usually red-brown, sometimes yellow-brown, not to strongly glaucous, glabrous or villous; branchlets yellow-green, yellow-brown, or red-brown, densely villous to pubescent, (buds caprea-type, inner membranaceous bud-scale layer free, separating from outer layer). **Leaves:** stipules foliaceous, (2–14 mm), apex rounded or acuminate; petiole shallowly grooved, or convex to flat adaxially, 3–14(–20) mm, villous or pilose adaxially; largest medial blade oblong, narrowly elliptic, elliptic, oblanceolate, or obovate, 33–70(–100) × 12–35(–48) mm, 1.6–2.8(–4) times as long as wide, base usually rounded or convex, sometimes subcordate, margins slightly revolute or flat, serrulate, apex acute, acuminate, or convex, abaxial surface glaucous, glabrous or glabrescent, hairs straight, adaxial slightly glossy or dull, glabrous or pilose, midrib pilose; proximal blade margins serrulate or entire; juvenile blade sometimes reddish, densely villous or glabrous abaxially, hairs white. **Catkins** flowering as leaves emerge; staminate stout, 13–60 × 10–25 mm, flowering branchlet 0–17 mm; pistillate moderately densely flowered, stout, subglobose, or slender, 26–80 × 9–18 mm, flowering branchlet 4–24 mm; floral bract brown or black, 1.6–2.8 mm, apex acute or rounded, abaxially hairy, hairs straight, wavy, or curly. **Staminate flowers:** adaxial nectary oblong, 0.5–1 mm; filaments distinct, glabrous; anthers yellow, 0.6–1 mm. **Pistillate flowers:** adaxial nectary oblong or ovate, 0.4–0.8 mm, shorter than or equal to stipe; stipe 0.4–1.5 mm; ovary

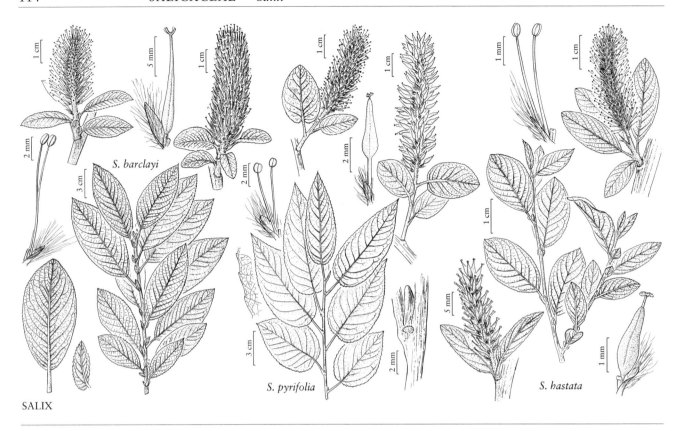

SALIX

obclavate or pyriform, glabrous, beak gradually tapering to styles; ovules 18–24 per ovary; styles 0.6–2.5 mm; stigmas slenderly cylindrical, 0.28–0.48–0.72 mm. **Capsules** 3–8 mm. $2n = 76$ (based on putative *Salix barclayi* × *S. barrattiana*, see below).

Flowering late May–early Aug. Lake and streamshores, fens, moist to mesic forest openings, subalpine and alpine slopes, glacial moraines; 0–2800 m; Alta., B.C., N.W.T., Yukon; Alaska, Idaho, Mont., Oreg., Wash., Wyo.

Vegetative specimens of *Salix hastata* often are misidentified as *S. barclayi*. The characters that indicate *S. hastata*, or sometimes possible hybrids, are the presence of ferruginous hairs on juvenile leaves or adaxial midribs of mature leaves, leaf margins often entire or with teeth present only distally, and buds with inner membranaceous layer not separating from outer layer.

Salix barclayi is distinguished from *S. farriae* and *S. hastata* by having bud-scales with inner membranaceous layer separating from outer layer, juvenile blades glabrous, pilose, or moderately densely villous, largest medial blades oblong, narrowly elliptic, elliptic, oblanceolate, or obovate, margins always toothed, pistillate flowering branchlets 4–24 mm, staminate flowering branchlets 0–17 mm, floral bracts brown to black, moderately densely hairy, anthers 0.6–1 mm, styles 0.6–2.5 mm, and stipes 0.4–1.5 mm; *S. farriae* has bud-scales with inner membranaceous layer separating from outer layer, juvenile blades glabrous or sparsely villous, largest medial blades narrowly elliptic to elliptic, margins usually entire, pistillate flowering branchlets 1.5–14 mm, staminate flowering branchlets 1–5 mm, floral bracts bicolor, brown, or black, sparsely hairy, anthers 0.3–0.6 mm, styles 0.3–1.2 mm, and stipes 0.5–1.2 mm; and *S. hastata* has bud-scales with inner membranaceous layer not separating from outer layer, juvenile blades sparsely pubescent, largest medial blades narrowly to broadly elliptic, narrowly ovate, or ovate, margins usually entire, pistillate flowering branchlets 1.5–9 mm, staminate flowering branchlets 1–7 mm, floral bracts brown or bicolor, sparsely hairy, anthers 0.4–0.6 mm, styles 0.2–0.48 mm, and stipes 0.4–1.2 mm.

Hybrids:

Salix barclayi is morphologically highly variable. While it is possible that much of this variability is inherent, hybridization and introgression have played an important role.

Salix barclayi forms natural hybrids with *S. arctica*, *S. barrattiana*, *S. brachycarpa*, *S. cascadensis*, *S. commutata*, *S. farriae*, *S. hastata*, *S. hookeriana*, *S. richardsonii*, and *S. stolonifera*.

Salix barclayi × *S. barrattiana* is characterized by oily buds that stain the pressing paper yellow, and leaves often entire, or nearly so, and closely gland-dotted, and not glaucous abaxially. Some hybrids resemble *S. barclayi* in having ovaries glabrous and leaves serrulate and slightly glaucous abaxially, while others resemble *S. barrattiana* in having stipules marcescent and leaves short-hairy on both surfaces and not glaucous abaxially, but it is the presence of the oily buds characteristic of *S. barrattiana* that indicates their hybridity. The hybrids usually were from populations that included both parents. An important ramification of the discovery of this hybrid is that the specimen on which the chromosome number for *S. barclayi* (R. D. Dorn 1976) is based is a plant with oily buds, which has leaves with a mixture of weakly developed teeth and closely gland-dotted, entire margins, and is putatively identified as this hybrid. Further chromosome counts for *S. barclayi* are needed. Hybrids are known in British Columbia from the mountains in the Peace and Liard river basins, and in Mt. Robson National Park. The hybrid should be suspected wherever the parents occur together.

Salix barclayi × *S. cascadensis*, collected in Washington in alpine mats growing with both parents, has ovaries with hairy beaks indicating hybridization.

Salix barclayi × *S. commutata* resembles *S. barclayi* in having leaves glaucous abaxially, and *S. commutata* in having leaves hairy on both surfaces (northern British Columbia, Unalaska Island, and southeastern Alaska).

Salix barclayi × *S. farriae* is represented by a series of specimens resembling *S. barclayi* but with ferruginous hairs on juvenile leaves or on adaxial midribs of mature leaves, found from the Rocky Mountains of Alberta to southern Yukon and adjacent Northwest Territories. Similar specimens from Alaska may be the hybrid *S. barclayi* × *S. hastata*. Both of these hybrids require further study. See 63. *S. hastata* and 64. *S. farriae*.

Salix barclayi × *S. hookeriana* hybrids or introgressants at Tanis Lake, Alaska, have subentire leaves and stipes glabrous or hairy.

Salix barclayi × *S. richardsonii* hybrids and intergrades are relatively common in Alaska, British Columbia, and Yukon. They resemble *S. barclayi* in having catkins borne on distinct flowering branchlets, small stipules, anthers yellow and longer, and *S. richardsonii* in having marcescent stipules, branchlets and branches coarsely villous with spreading hairs, pistillate nectaries longer than stipes, and catkins sometimes on relatively short flowering branchlets.

Salix barclayi × *S. scouleriana* has the general appearance of *S. barclayi* but the densely hairy ovaries and the relatively long stigmas (0.9–0.92 mm) of *S. scouleriana*. It was growing with *S. barclayi* in openings created in white spruce by bark beetles.

Salix barclayi × *S. stolonifera* is characterized by relatively small leaves with irregularly serrulate margins, some young leaves reddish green, and a prostrate habit (G. W. Argus 1973).

62. Salix pyrifolia Andersson, Monogr. Salicum, 162, plate 8, fig. 93. 1867 • Balsam willow [E] [F]

Salix balsamifera Barratt ex Andersson; *S. pyrifolia* var. *lanceolata* (Bebb) Fernald

Plants 0.4–4 m. **Stems:** branches red-brown, not glaucous, (highly glossy), glabrous; branchlets red-brown, yellow-brown, or yellowish, (not or weakly glaucous), glabrous, (inner membranaceous bud-scale layer free, not separating from outer layer). **Leaves:** stipules (early deciduous), foliaceous or rudimentary on early ones, foliaceous on late ones, apex acute, convex, or rounded; petiole convex to flat, or shallowly grooved adaxially, 7–20 mm, glabrous or sparsely velvety adaxially; largest medial blade narrowly oblong, oblong, elliptic, or broadly elliptic, 30–103 × 19–40 mm, 1.5–3.4 times as long as wide, base cordate, subcordate, rounded or convex, margins flat or slightly revolute, serrulate, irregularly serrate, crenate, or sinuate, apex acute or acuminate, abaxial surface glaucous, glabrous, adaxial slightly or highly glossy, glabrous; proximal blade margins serrulate or entire; juvenile blade translucent, glabrous or pilose abaxially, hairs white. **Catkins:** staminate flowering just before leaves emerge, pistillate flowering as leaves emerge; staminate stout or slender, 18.5–63 × 7–15 mm, flowering branchlet 1–5 mm; pistillate loosely flowered, stout or slender, 25–85 × 8–20 mm, flowering branchlet 2–22 mm; floral bract tawny, 1–2.4 mm, apex acute or convex, abaxially sparsely to moderately densely hairy, hairs straight or wavy. **Staminate flowers:** adaxial nectary square or ovate, 0.3–0.5 mm; filaments distinct, glabrous or sparsely hairy basally; anthers yellow, (ellipsoid or shortly cylindrical), 0.5–0.8 mm. **Pistillate flowers:** adaxial nectary narrowly ovate or square, 0.3–0.7 mm, shorter than stipe; stipe 1.8–3.5 mm; ovary obclavate, glabrous, beak slightly bulged below styles; ovules 10–19 per ovary; styles 0.4–0.5 mm; stigmas flat, abaxially non-papillate with rounded tip, or 2 plump lobes, 0.2–0.23–0.32 mm. **Capsules** 7–8 mm. $2n = 38$.

Flowering May–mid Jun(–mid Jul in subalpine). Fens, wet lake and slough margins, treed bogs; 0–300 m (–1600 m in subalpine); Alta., B.C., Man., N.B., Nfld. and Labr., N.W.T., N.S., Ont., P.E.I., Que., Sask., Yukon; Maine, Mich., Minn., N.H., N.Y., Vt., Wis.

Salix pyrifolia is characterized by juvenile leaves membranaceous and translucent, and largest medial leaves subcoriaceous, with abaxial surfaces reticulate, and with bases often cordate. Buds and foliage are reported to have a balsam-like fragrance.

Hybrids:

Salix pyrifolia forms natural hybrids with *S. brachycarpa* var. *psammophila*. Hybrids with *S. discolor* have been reported (C. K. Schneider 1921; M. L. Fernald 1950) but no convincing specimens have been seen.

63. **Salix hastata** Linnaeus, Sp. Pl. 2: 1017. 1753 • Halberd willow F

Salix farriae C. R. Ball var. *walpolei* Coville & C. R. Ball; *S. hastata* subsp. *subintegrifolia* (Floderus) Floderus; *S. walpolei* (Coville & C. R. Ball) C. R. Ball

Plants 0.2–4 m. **Stems:** branches reddish brown, not glaucous, (slightly glossy), pilose; branchlets yellow-brown or red-brown, villous or pilose. **Leaves:** stipules foliaceous, apex acute to acuminate; petiole convex to flat, or shallowly grooved adaxially, 2–6(–9) mm, pilose or villous adaxially; largest medial blade narrowly to broadly elliptic or narrowly ovate to ovate, 25–92 × 10–45 mm, 1.5–2.6(–3.4) times as long as wide, base convex, cuneate, or rounded, margins slightly revolute or flat, shallowly serrulate or entire, apex acuminate, acute, or convex, abaxial surface glaucous, sparsely pubescent, hairs wavy, adaxial dull to slightly glossy, pilose, sparsely pubescent or glabrous, midrib hairy, (hairs white and ferruginous); proximal blade margins entire or finely serrulate; juvenile blade sometimes reddish, sparsely pubescent abaxially, hairs white, sometimes also ferruginous. **Catkins** flowering as leaves emerge; staminate slender, stout, or subglobose, 14.5–34.5 × 8–12 mm, flowering branchlet 1–7 mm; pistillate moderately densely or loosely flowered, slender or stout, 21–59 × 6–16 mm, flowering branchlet 1.5–9 mm; floral bract brown or bicolor, 1.2–1.8 mm, apex acute to rounded, abaxially glabrate to hairy, hairs straight to wavy. **Staminate flowers:** adaxial nectary oblong or square, 0.3–0.7 mm; filaments distinct or basally connate, glabrous; anthers purple turning yellow, 0.4–0.6 mm. **Pistillate flowers:** adaxial nectary square or obovate, 0.3–0.6 mm, usually shorter than stipe; stipe 0.4–1.2 mm; ovary pyriform, glabrous, beak gradually tapering to styles; ovules 12–22 per ovary; styles connate (sometimes distinct ½ their lengths), 0.2–0.5 mm; stigmas flat, abaxially non-papillate with rounded tip, or broadly cylindrical, or 2 plump lobes, 0.2–0.32–0.44 mm. **Capsules** 3.2–8 mm. $2n = 38$.

Flowering early Jun–late Jul. Sandy and gravelly river bars and floodplains, lakeshores, sand dunes and blowouts, *Dryas* tundra, alpine sedge meadows, balsam poplar thickets, openings in upland spruce-willow forests; 0–1200 m; N.W.T., Yukon; Alaska; Eurasia (Norway, Russia, Siberia).

Salix hastata is characterized by branches pilose with short, curved hairs; juvenile and mature leaves with white or ferruginous hairs adaxially, glaucous or not abaxially; and ovaries often reddish. See 61. *S. barclayi* and 64. *S. farriae*.

Subspecies *subintegrifolia*, characterized by entire or subentire leaf margins, is recognized in *Flora Nordica* as the major northern Eurasian race of *Salix hastata*, and is described as occurring across Russia into northwestern North America (B. Jonsell and T. Karlsson 2000+, vol. 1). Because leaf marginal toothing is highly variable throughout the species range (E. Hultén 1967), the subspecies is not recognized here or in Russia (A. K. Skvortsov 1999).

Hybrids:

Salix hastata forms natural hybrids with *S. barclayi*.

64. **Salix farriae** C. R. Ball, Contr. U.S. Natl. Herb. 22: 321. 1921 (as farrae) • Farr's willow E

Salix farriae var. *microserrulata* C. R. Ball; *S. hastata* Linnaeus var. *farriae* (C. R. Ball) Hultén

Plants 0.2–1.5(–2) m. **Stems:** branches red-brown, not glaucous to strongly glaucous on buds, glabrous or puberulent at nodes; branchlets yellow-brown or red-brown, (sometimes weakly glaucous), glabrous or puberulent, (inner membranaceous bud-scale layer free, separating from outer layer). **Leaves:** stipules absent, rudimentary or foliaceous on early ones, foliaceous on late ones, apex acute; petiole shallowly grooved, or convex to flat adaxially, 5–8 mm, puberulent adaxially; largest medial blade narrowly elliptic or elliptic, (20–)30–65(–75) × (8–)10–30(–35) mm, 1.8–3.7 times as long as wide, base convex, rounded, or cuneate, margins slightly revolute or flat, entire or shallowly serrulate, apex acute, acuminate, or convex, abaxial surface glaucous, glabrous or glabrescent, adaxial slightly glossy or dull, glabrous or pilose, midrib sparsely pubescent, hairs short, white, and ferruginous; proximal blade margins entire or serrulate; juvenile blade green, glabrous, or midrib sparsely villous abaxially, hairs usually white and ferruginous. **Catkins** flowering as leaves emerge; staminate stout, 11–25 × 6–11 mm, flowering branchlet 1–5 mm; pistillate densely or loosely flowered, stout, 14–38.5 × 8–14 mm, flowering branchlet 1.5–14 mm; floral bract brown, black, or bicolor, 0.7–2 mm, apex rounded to convex, abaxially hairy, hairs wavy. **Staminate flowers:** adaxial nectary oblong, square, or ovate, 0.2–0.9 mm; filaments distinct, glabrous; anthers yellow, 0.3–0.6 mm. **Pistillate flowers:** adaxial nectary oblong or ovate, 0.4–0.8 mm, shorter than stipe; stipe 0.5–1.2 mm; ovary pyriform, glabrous, beak gradually tapering to styles; ovules 12–19 per ovary; styles 0.3–1.2 mm; stigmas flat, abaxially non-papillate with rounded tip, or 2 plump lobes, 0.2–0.3–0.56 mm. **Capsules** 3–7 mm.

Flowering late May–late Jul. Wet montane to subalpine meadows, stream banks; 600–2700 m; Alta., B.C., N.W.T., Yukon; Idaho, Mont., Oreg., Wyo.

Salix farriae is a cordilleran species ranging from Wyoming to central British Columbia with disjunct occurrences in northwestern British Columbia, western Northwest Territories, and southern Yukon. It is related to *S. hastata*, an amphiberingian species ranging from Scandinavia to southwestern Yukon and northwestern Northwest Territories. There may be reasons for treating these slightly different plants as *S. hastata* var. *farriae*, but R. D. Dorn (1975) maintained them as a species based on flavonoid differences. In a phenetic study (G. W. Argus 2007), the two taxa had dissimilarity values at the same level as other closely related species. They are treated here as species, primarily because their ranges are disjunct. They can be separated as follows: *Salix farriae* is distinguished from *S. hastata* by having largest medial blades narrowly elliptic to elliptic, pistillate nectaries oblong or ovate, stipules on early leaves absent or rudimentary (sometimes foliaceous), branches strongly to weakly glaucous or not, floral bract apices rounded, and plants of the cordillera in Alberta and British Columbia, in Idaho, Montana, Oregon, and Wyoming; *S. hastata* has largest medial blades narrowly elliptic to broadly elliptic or broadly obovate, pistillate nectaries square, stipules on early leaves foliaceous (sometimes rudimentary), branches not glaucous, floral bract apices acute or rounded, and plants of Alaska, Northwest Territories, and Yukon.

Salix farriae and *S. barclayi* are sympatric in western Canada and the Pacific Northwest, where they are difficult to separate. *Salix farriae* can often be recognized by its largest medial leaves with at least some minute, ferruginous hairs on the adaxial midrib or blade surfaces; ferruginous hairs do not occur in *S. barclayi*. Its leaf margins also tend to be more nearly entire, but relatively short teeth are not infrequent. Such plants are sometimes interpreted as intergrades between *S. farriae* and *S. barclayi* (R. D. Dorn 1975). The variable leaf toothing also occurs in *S. hastata* and may not be a reliable indicator of intergradation. *Salix farriae* also differs from *S. barclayi* in usually having shorter anthers, 0.3–0.6 mm versus 0.6–1 mm in *S. barclayi*. See 61. *S. barclayi*.

Hybrids:

Salix farriae forms natural hybrids with *S. barclayi*.

65. **Salix pseudomonticola** C. R. Ball, Contr. U.S. Natl. Herb. 22: 321. 1921 • False mountain willow [E] [F]

Salix barclayi Andersson var. *pseudomonticola* (C. R. Ball) Kelso

Plants 1–6 m. **Stems:** branches red-brown or yellow-brown, not or weakly glaucous, (slightly or highly glossy), glabrous or glabrescent; branchlets yellow-green, red-brown, or brownish, glabrous, pilose, or densely villous, (inner membranaceous bud-scale layer free, separating from outer layer). **Leaves:** stipules foliaceous, (prominent, 5–9–15 mm), apex rounded to acute, sometimes acuminate; petiole (usually reddish), shallowly grooved, or convex to flat adaxially, 6–20 mm, short-silky or velvety adaxially; largest medial blade broadly to narrowly elliptic, or ovate to broadly obovate, (25–)30–86(–118) × 12–51 mm, 1.4–3 times as long as wide, base convex, rounded, cuneate, subcordate, cordate, or sometimes irregularly lobed, margins flat, serrulate or crenate, apex acute, acuminate, or convex, abaxial surface glaucous, glabrous, pubescent, or pilose, hairs wavy, adaxial slightly glossy or dull, glabrous, puberulent, pubescent, or pilose, midrib hairy; proximal blade margins entire or serrulate; juvenile blade reddish, glabrous or pubescent abaxially, hairs white, sometimes also ferruginous. **Catkins** flowering before leaves emerge; staminate stout, 16–39 × 10–12 mm, flowering branchlet 0 mm; pistillate densely or moderately densely flowered, slender to globose, 17–73 × 8–20 mm, flowering branchlet 0–5 mm; floral bract brown or black, 1–2.4 mm, apex rounded or acute, abaxially hairy, hairs straight. **Staminate flowers:** adaxial nectary oblong, 0.3–1 mm; filaments distinct or connate less than ½ their lengths, glabrous; anthers purple turning yellow, 0.4–0.5 mm. **Pistillate flowers:** adaxial nectary oblong, flask-shaped, 0.3–0.8 mm, shorter than stipe; stipe 0.5–0.8–3 mm; ovary pyriform or obclavate, glabrous, beak gradually tapering to styles; ovules 18 per ovary; styles (0.5–)0.7–1.8 mm; stigmas flat, abaxially non-papillate with rounded tip, or 2 plump lobes, 0.1–0.21–0.29 mm. **Capsules** 4–7 mm. $2n = 38$.

Flowering late Apr–early Jun. Moist fens in drainageways in white spruce forests, treed bogs, balsam poplar forests, floodplains; 0–2500 m; Alta., B.C., Man., N.W.T., Ont., Que., Sask., Yukon; Alaska, Idaho, Minn., Mont., S.Dak., Wash., Wyo.

Salix pseudomonticola is characterized by precocious flowering; catkins sessile; juvenile leaf blades, petioles, and proximal midribs reddish; stipules prominent; and leaves and branchlets sparsely hairy. Branches older than two years have a distinctive pattern, which consists of a series of longitudinal splits in epidermis produced as

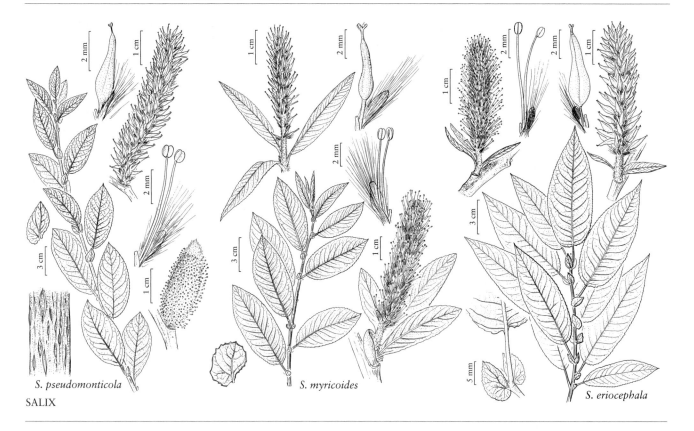

S. pseudomonticola · SALIX · S. myricoides · S. eriocephala

the branch expands. The edge of epidermis around the split, where it has separated from the branch, is yellow and contrasts with the red-brown branch to which the epidermis still adheres.

Vegetative specimens of *Salix pseudomonticola* with yellow-brown branches can be confused with *S. famelica*. They may be separated by their juvenile leaf margins prominently and closely gland-dotted; stipules usually prominent, sometimes early deciduous; leaves broader (1.4–3 times as long as wide versus 2.6–7 in *S. famelica*); and petioles slender and often longer in relation to blade length. The possibility of hybridization needs study.

Vegetative specimens of *Salix pseudomonticola* can be distinguished from *S. pyrifolia* by juvenile leaves reddish and almost always with some ferruginous hairs, versus yellowish-green and glabrous or with white hairs, and mature leaves usually dull adaxially versus glossy.

66. Salix monticola Bebb in J. M. Coulter, Man. Bot. Rocky Mt., 336. 1885 • Mountain or park or serviceberry willow [E]

Salix barclayi Andersson var. *padophylla* (Rydberg) Kelso; *S. cordata* Muhlenberg var. *monticola* (Bebb) Kelso; *S. padophylla* Rydberg; *S. pseudomonticola* C. R. Ball var. *padophylla* (Rydberg) C. R. Ball

Plants 1.5–6 m. **Stems:** branches yellow-brown or red-brown mottled with green, not or weakly glaucous, glabrous; branchlets red-brown to yellow-brown, glabrous or puberulent, pilose, or villous, (buds *caprea*-type). **Leaves:** stipules foliaceous or rudimentary on early ones, foliaceous on late ones, apex acute or acuminate; petiole shallowly grooved, or convex to flat adaxially, 5.5–14 mm, pilose, villous, or velvety to glabrescent adaxially; largest medial blade (sometimes amphistomatous), narrowly oblong to oblong, narrowly elliptic to elliptic, lanceolate, oblanceolate, or obovate, 35–95 × 11–33 mm, 2–3.9 times as long as wide, base convex, rounded, or subcordate, margins slightly revolute or flat, serrulate, serrate, or sinuate, apex acute to acuminate, abaxial surface glaucous, glabrous, adaxial dull, glabrous or pilose, midrib pilose to villous; proximal blade margins entire or serrulate; juvenile blade sometimes reddish, villous or long-silky abaxially, hairs white, sometimes also ferruginous. **Catkins:** staminate flowering before or just before leaves emerge, pistillate as leaves emerge; staminate stout, 14–39 × 9–17 mm, flowering branchlet 0.5–7 mm; pistillate densely flowered, stout, 21–60 × 8–16 mm, flowering branchlet 0.5–8 mm; floral bract brown, black, or bicolor, 1–2 mm, apex rounded to acute, abaxially hairy, hairs wavy, straight, or curly. **Staminate flowers:** adaxial nectary narrowly oblong, 0.6–1.1 mm; filaments distinct or connate less than ½ their lengths, glabrous; anthers purple turning yellow, 0.4–0.8 mm. **Pistillate flowers:** adaxial nectary narrowly oblong, oblong, or flask-shaped, 0.4–1 mm, shorter than or equal to stipe; stipe 0.5–1.6 mm; ovary pyriform, glabrous, beak gradually tapering to styles; ovules 11–15 per ovary; styles 0.6–1.1 mm; stigmas flat, abaxially non-papillate with rounded tip, or broadly cylindrical, 0.24–0.36–0.56 mm. **Capsules** 4–7 mm. $2n = 114$.

Flowering late Apr–early Jul. Streams, cienegas, meadows, springs; 1700–3500 m; Ariz., Colo., N.Mex., Utah, Wyo.

67. **Salix myricoides** Muhlenberg, Ges. Naturf. Freunde Berlin Neue Schriften 4: 235, plate 6, fig. 2. 1803 • Blue-leaf willow E F

Salix glaucophylla Bebb var. *albovestita* C. R. Ball;
S. *glaucophylloides* Fernald;
S. *glaucophylloides* var. *albovestita* (C. R. Ball) Fernald;
S. *glaucophylloides* var. *glaucophylla* C. K. Schneider;
S. *myricoides* Muhlenberg var. *albovestita* (C. R. Ball) Dorn

Plants 0.3–5 m, (sometimes forming clones by stem fragmentation or layering). **Stems:** branches (sometimes highly brittle at base), red-brown or yellow-brown, not or weakly glaucous, (slightly or highly glossy), glabrous or villous; branchlets red-brown or yellow-brown, glabrous or sparsely to very densely villous, (buds *caprea*-type, inner membranaceous bud-scale layer free and separating or not). **Leaves:** stipules rudimentary or foliaceous on early ones, foliaceous on late ones, apex acute or acuminate; petiole shallowly grooved, or convex to flat adaxially, 3.5–7.3–13 mm, (sometimes with 2 spherical glands distally), villous, tomentose, pilose, or pubescent adaxially; largest medial blade narrowly oblong, narrowly elliptic, elliptic, or oblanceolate, 35–61.3–110 × 11–16–46 mm, 2–2.7–5.2 times as long as wide, base convex, rounded, subcordate, or cuneate, margins flat or slightly revolute, (thickened and raised), crenulate or serrulate, apex acuminate, acute, or convex, abaxial surface usually very thickly glaucous, glabrous or pilose, midribs pubescent to tomentose, hairs (white, often also ferruginous), curved, wavy, or straight, adaxial slightly glossy, glabrous or pilose, midribs sparsely pubescent (hairs white, sometimes also ferruginous); proximal blade margins entire or serrulate; juvenile blade translucent, reddish or yellowish green, glabrous or sparsely pubescent abaxially, midribs often densely hairy, hairs white, sometimes also ferruginous. **Catkins:** staminate flowering before leaves emerge, pistillate as leaves emerge; staminate stout or slender, 23.5–35.6–51 × 9–12.7–22 mm, flowering branchlet 1–3.9–10 mm; pistillate loosely flowered, stout or slender, 19–42–62 (–85 in fruit) × 8–13–18 mm, flowering branchlet 1.5–5.9–13 mm; floral bract brown or bicolor, 1.2–1.8–3 mm, apex rounded or acute, sometimes toothed, abaxially hairy, hairs straight or wavy. **Staminate flowers:** adaxial nectary narrowly oblong, oblong, square, or ovate, 0.44–0.48–1.4 mm; filaments distinct, glabrous; anthers yellow, (ellipsoid or globose), 0.52–0.69–0.76 mm. **Pistillate flowers:** adaxial nectary oblong, narrowly oblong, square, or flask-shaped, 0.56–0.8–1.4 mm, shorter than stipe; stipe 0.96–1.7–3.4 mm; ovary pyriform, glabrous, beak slightly bulged below styles; ovules 12–14 per ovary; styles (sometimes distinct), 0.3–0.8–1.3 mm; stigmas flat, abaxially non-papillate with rounded or pointed tip, or slenderly cylindrical, 0.24–0.43–0.56 mm. **Capsules** 5–7–11 mm.

Flowering early Apr–early Jul. Stream and lake shores, gravel bars, subalpine conifer forests, alkaline fens, sea cliffs, dry limestone talus, swamps, tidal meadows, sand dunes; 0–1100 m; N.B., Nfld. and Labr. (Nfld.), Ont., Que.; Ill., Ind., Maine, Mich., Ohio, Pa., Wis.

Plants with densely villous branchlets and branches have been named var. *albovestita*. Branchlet indumentum varies widely in the species and seems to be continuous, with both villous and glabrous variants sometimes occurring in the same area. This characteristic is more common in populations on the shores of the Great Lakes and on the western coast of James Bay, but even these populations are variable.

Reports of *Salix myricoides* from Akimiski Island, Nunavut, were based on misidentified S. *planifolia*. See 68. S. *eriocephala* for differences.

Hybrids:

Salix myricoides forms natural hybrids with S. *bebbiana*, S. *discolor*, and S. *glauca* var. *cordifolia*. Hybrids with S. *eriocephala* have been reported (M. L. Fernald 1950) but no convincing specimens have been seen. Inasmuch as S. *eriocephala* and S. *myricoides* are very similar, hybrids between the two would be very difficult to identify.

2e2. Salix Linnaeus (subg. Vetrix) sect. Cordatae Barratt ex Hooker, Fl. Bor.-Amer. 2: 149. 1838 E

Salix subsect. *Luteae* Dorn

Shrubs or trees, 0.2–8 m. **Buds** *alba*-type and intermediate. **Leaves:** stipules on late ones foliaceous; largest medial blade (rarely amphistomatous), 2–8 times as long as wide, adaxial surface not glaucous. **Staminate flowers:** filaments glabrous or hairy basally. **Pistillate flowers:** adaxial nectary shorter than stipe; stipe 0.6–4.2 mm; ovary glabrous; stigmas with flat, non-papillate abaxial surface, or stigmas cylindrical, or plump, 0.1–0.4 mm.

Species 7 (7 in the flora): North America.

All of the species in sect. *Cordatae*, with the exception of *Salix turnorii*, were treated by R. D. Dorn (1995) as subspecies or varieties of *S. eriocephala*. These taxa have relatively distinct ranges and, although they hybridize where they come into contact, their integrity is not jeopardized. For those reasons, they are treated here as species.

68. **Salix eriocephala** Michaux, Fl. Bor.-Amer. 2: 225. 1803 • Missouri or diamond or heart-leaf willow E F

Salix angustata Pursh; *S. cordata* Muhlenberg; *S. cordata* Muhlenberg var. *abrasa* Fernald; *S. missouriensis* Bebb; *S. rigida* Muhlenberg; *S. rigida* var. *angustata* (Pursh) Fernald; *S. rigida* var. *vestita* (Andersson) C. R. Ball

Shrubs, 0.2–6 m, (sometimes forming clones by stem fragmentation). **Stems:** branches (sometimes highly brittle at base), red-brown, not glaucous, glabrous or glabrescent; branchlets yellow-brown to red-brown, pilose, moderately to densely velvety, pubescent, or villous, (inner membranaceous bud-scale layer free). **Leaves:** stipules foliaceous, (4.5–13 mm), apex rounded or acute; petiole shallowly grooved adaxially, 3–18 mm, tomentose adaxially; largest medial blade narrowly oblong, very narrowly elliptic or obovate, 58–96–136 × 9–21–36 mm, 2.3–4.6–8 times as long as wide, base cordate, convex, rounded, subcordate, or, sometimes, cuneate, margins flat, serrate or serrulate, apex acute to acuminate, abaxial surface thickly glaucous, glabrous, puberulent, sparsely pubescent or short-silky, adaxial highly glossy, glabrous or sparsely villous (hairs white, sometimes also ferruginous); proximal blade margins entire or shallowly serrulate; juvenile blade reddish or yellowish green, glabrous, pilose, or villous abaxially, hairs white. **Catkins:** staminate flowering just before leaves emerge, pistillate as leaves emerge; staminate slender or stout, 19–44 × 7–14 mm, flowering branchlet 0.5–5 mm; pistillate densely or moderately densely flowered, slender or stout, 22–65 × 7–14 mm, flowering branchlet 2–10 mm; floral bract dark brown or bicolor, 0.8–1.6 mm, apex rounded, abaxially hairy, hairs wavy. **Staminate flowers:** adaxial nectary narrowly oblong, oblong, or ovate, 0.2–1 mm; filaments distinct or connate less than ½ their lengths, glabrous; anthers yellow or purple turning yellow (ellipsoid or shortly cylindrical), 0.4–0.6 mm. **Pistillate flowers:** adaxial nectary oblong or flask-shaped, 0.3–0.8 mm, shorter than stipe; stipe 1.2–2.8 mm; ovary pyriform, glabrous, beak slightly bulged below styles; ovules 12–16 per ovary; styles 0.3–0.6 mm; stigmas flat, abaxially non-papillate with rounded tip, or broadly cylindrical, or 2 plump lobes, 0.16–0.28 mm. **Capsules** 3.5–7 mm. $2n = 38$.

Flowering early Apr–mid Jun. Gravelly or rocky stream banks, marshy fields, in mixed mesophytic woods on alluvium; 0–1200 m; N.B., Nfld. and Labr. (Nfld.), N.S., Ont., P.E.I., Que.; Ala., Ark., Conn., Del., D.C., Fla., Ga., Ill., Ind., Iowa, Kans., Ky., Maine, Md., Mass., Mich., Minn., Mo., Nebr., N.H., N.J., N.Y., N.Dak., Ohio, Pa., R.I., S.Dak., Tenn., Vt., Va., W.Va., Wis.

Salix eriocephala sometimes is very difficult to separate from *S. myricoides*. Some of the confusion may be due to hybridization but no definite hybrids have been seen.

Salix eriocephala can be distinguished from *S. myricoides* by having stipules on early leaves foliaceous, apices acute to rounded, largest medial blades 4.9–23.3 times as long as petiole, abaxial surface usually thickly glaucous (stomata can be seen through the grayish wax), margins serrulate or serrate, floral bracts 0.8–1.6 mm, moderately to very densely hairy, styles 0.3–0.6 mm, and stigmas 0.16–0.28 mm; *S. myricoides* has stipules on early leaves rudimentary or foliaceous, apices acuminate or acute, largest medial blades 4.7–13.4 times as long as petiole, abaxial surface usually with very thick wax (stomata cannot be seen through the wax), margins crenulate to serrulate, floral bracts 1.2–3 mm, sparsely to moderately densely hairy, styles 0.3–1.3 mm, and stigmas 0.28–0.56 mm.

Hybrids:

Salix eriocephala forms natural hybrids with *S. candida*, *S. famelica*, *S. humilis*, *S. interior*, *S. lasiandra*, *S. petiolaris*, and *S. sericea*. Hybrids with *S. amygdaloides*, *S. bebbiana*, *S. myricoides*, and *S. pedicellaris* have been reported (M. L. Fernald 1950) but no convincing specimens have been seen. Controlled pollinations made with *S. discolor* had low success and many seedlings were abnormal (A. Mosseler 1990). In controlled pollination using *S. eriocephala* as the maternal parent, seeds were rarely produced due to pollen-stigma incompatibility (Mosseler 1989).

Salix eriocephala × *S. famelica*: Hybrids and intergrades occur in the area of overlap (R. D. Dorn 1995). Specimens from a population in Douglas County, Nebraska, which included successive collections and cultivated specimens, have branches with yellow-mottled coloration of *S. famelica* and villous indumentum of *S. eriocephala*; they may be this hybrid.

Salix eriocephala × *S. petiolaris*: Controlled pollinations (A. Mosseler 1990) had low seed-set but a high percent of seed germination and seedling survival. Because reproductive barriers between these species are weak, it was suggested that their morphological variability may be due to interspecific gene flow (Mosseler). Natural hybrids are known from Illinois, Maine, Massachusetts, Michigan, Missouri, New York, Ontario, Quebec, and West Virginia.

Salix eriocephala × *S. sericea*: This hybrid is relatively common wherever the ranges of the parents overlap. It has been studied in the southeastern United States (G. W. Argus 1986) and in eastern Canada. The results of a molecular study (T. M. Hardig et al. 2000) have been discussed already under the genus. In general, the hybrids resemble *S. eriocephala* but have leaves that are sparsely to moderately densely short-silky on abaxial surfaces and ovaries hairy as in *S. sericea*. Foliaceous stipules are often present on late leaves and sometimes even on early leaves, as in *S. eriocephala*, but they are not as prominent. In *S. sericea* stipules usually are lacking or rudimentary, but on late leaves they may be foliaceous. Petioles and branchlets of hybrids are finely velvety as in *S. sericea*. This hybrid was described from Maine (O. W. Knight 1907), where it was noted that the catkins were usually abortive but sometimes produced one or two fertile seeds.

Salix eriocephala is distinguished from *S. sericea* in having stipules on early and late leaves foliaceous, 4–6.2–8.3 × 2.5–3.6–4.6 mm, 1.5–2 times as long as wide, ovaries glabrous, juvenile blades glabrous or sparsely hairy, hairs white, largest medial blade abaxial surfaces glabrous, puberulent, sparsely pubescent, or short-silky, stipes 1.2–2.8 mm, and capsules 3.5–7 mm; *S. sericea* has stipules on early leaves absent or rudimentary, on late leaves rudimentary to foliaceous, 1.1–1.6–2.1 × 0.4–0.6–0.8 mm, 2.3–3 times as long as wide, ovaries short-silky, juvenile blades very densely short-silky, hairs white, sometimes also ferruginous, largest medial blade abaxial surfaces densely short-silky, stipes 0.6–1.5 mm, and capsules 2.5–4 mm.

69. Salix famelica (C. R. Ball) Argus, Harvard Pap. Bot. 12: 361. 2007 • Hungry willow [E]

Salix lutea Nuttall var. *famelica* C. R. Ball, Bot. Gaz. 71: 426. 1921; *S. eriocephala* Michaux var. *famelica* (C. R. Ball) Dorn

Shrubs or trees, 1.5–7 m. **Stems:** branches yellow, yellow-gray, or yellow-brown, not glaucous, glabrous (tomentose at nodes) or pubescent; branchlets yellow-brown or red-brown, (sometimes weakly glaucous with sparkling wax crystals), usually glabrous, sometimes pilose, pubescent, or moderately densely villous. **Leaves:** stipules rudimentary or foliaceous on early ones, foliaceous on late ones, apex acute or rounded; petiole convex to flat, or shallowly grooved adaxially, 3–27 mm, puberulent, pubescent, tomentose, velvety, or glabrous adaxially; largest medial blade (sometimes hemiamphistomatous), narrowly oblong, narrowly elliptic, or lanceolate, 28–116 × 10–30 mm, 2.6–3.5–7 times as long as wide, base subcordate, convex, or rounded, margins flat or slightly revolute, serrate, shallowly serrulate or crenulate, apex acuminate to acute, abaxial surface glabrous, pilose, or pubescent, midribs hairy, adaxial dull or slightly glossy, glabrous, pilose, sparsely long-silky, or tomentose, midribs hairy; proximal blade margins entire or shallowly serrulate; juvenile blade reddish or yellowish green, glabrous, pilose, or villous abaxially, hairs white. **Catkins:** staminate flowering before or just before leaves emerge, pistillate as leaves emerge; staminate slender, stout, or subglobose, 15–44 × 8–14 mm, flowering branchlet 0.5–4 mm; pistillate loosely or moderately densely flowered, slender or stout, 16–74(–115 in fruit) × 7–15 mm, flowering branchlet 0.5–9 mm; floral bract dark brown or tawny, 0.8–1.6 mm, apex rounded or convex, abaxially hairy throughout or proximally, hairs straight or wavy. **Staminate flowers:** adaxial nectary narrowly oblong, oblong, or flask-shaped, 0.8–1 mm; filaments distinct or connate less than ½ their lengths, glabrous; anthers yellow or purple turning yellow, (ellipsoid), 0.5–0.7 mm. **Pistillate flowers:** adaxial nectary 0.6–0.8 mm, shorter than stipe; stipe 0.7–2.4(–2.75) mm; ovary pyriform or obclavate, glabrous, beak sometimes slightly bulged below styles; ovules 12–18 per ovary; styles 0.2–0.6 mm; stigmas flat, abaxially non-papillate with rounded tip, or 2 plump lobes, 0.12–0.22–0.32 mm. **Capsules** 5–6 mm.

Flowering mid Apr–mid Jun. Riparian willow thickets on silty, sandy-clay, gravelly, or bouldery banks and floodplains, sand dunes, alluvial fans, wet meadows, rich fens, prairie depressions, balsam poplar thickets; 0–2000 m; Alta., Man., N.W.T., Ont., Sask.; Colo., Iowa, Kans., Minn., Mo., Mont., Nebr., N.Dak., S.Dak., Wyo.

Salix famelica is a Great Plains taxon that was recognized by R. D. Dorn (1995) as *S. lutea* var. *famelica*. It is separable from the other members of sect. *Cordatae* mainly by its yellow-brown to gray-brown branches and contrasting red-brown branchlets. It is recognized here as a species because, although it intergrades with other taxa in the complex, it has a relatively large, allopatric distribution.

Hybrids:

Salix famelica forms natural hybrids with *S. candida*, *S. eriocephala*, *S. petiolaris*, and *S. pseudomonticola*.

Salix famelica × *S. petiolaris* resembles *S. famelica* in having foliaceous stipules on late leaves and yellow-brown branches, and *S. petiolaris* in having ferruginous hairs on juvenile leaves. It is intermediate in leaf shape, in having stipules rudimentary on early leaves, and in having ovaries with patches of hairs at the base.

Salix famelica × *S. pseudomonticola*: Saskatchewan specimens combine the characters of the parental species.

70. **Salix turnorii** Raup, J. Arnold Arbor. 17: 234, plate 193. 1936 • Turnor's willow [C] [E]

Salix lutea Nuttall var. *turnorii* (Raup) B. Boivin

Shrubs, 1–2.5 m, (forming clones by layering). **Stems:** branches yellow-brown or yellow-gray, not or weakly glaucous, (with sparkling wax crystals, dull or slightly glossy), pilose or villous; branchlets gray-brown or red-brown, pubescent, villous, or velvety. **Leaves:** stipules foliaceous, apex acute or convex; petiole shallowly grooved adaxially, 4–13 mm, villous or pubescent adaxially; largest medial blade (amphistomatous), narrowly oblong, narrowly elliptic, elliptic, oblanceolate, or lanceolate, 26–47 × 7.5–15 mm, 2.8–4.1 times as long as wide, base convex, rounded, or subcordate, margins slightly revolute or flat, serrate or serrulate, apex acuminate to acute, abaxial surface glaucous, glabrous, pilose, villous, or long-silky, hairs straight or wavy, adaxial dull, sparsely or moderately densely pilose or long-silky, especially on midrib; proximal blade margins entire or serrulate; juvenile blade reddish or yellowish green, sparsely to moderately densely long-silky or pubescent abaxially, hairs white. **Catkins** flowering just before or as leaves emerge; staminate stout, 16–30 × 8–10 mm, flowering branchlet 1.5–4 mm; pistillate loosely flowered, stout, 18–22 × 9–11 mm, flowering branchlet 3–4 mm; floral bract brown or tawny, 1.2–1.6 mm, apex acute or rounded, abaxially hairy throughout or proximally, hairs straight or wavy. **Staminate flowers:** adaxial nectary narrowly oblong or ovate, 0.7–1.1 mm; filaments distinct or connate, hairy basally; anthers yellow, 0.6–0.8 mm. **Pistillate flowers:** adaxial nectary oblong, flask-shaped, or narrowly ovate, 0.4–1 mm, shorter than stipe; stipe 2–4 mm; ovary pyriform, glabrous, beaks slightly bulged below styles; ovules 14–18 per ovary; styles 0.3–0.5 mm; stigmas flat, abaxially non-papillate with rounded or pointed tip, 0.16–0.23–0.28 mm. **Capsules** 2.5–5 mm.

No flowering time data are available (probably May or Jun). Active sand dunes; of conservation concern; 200–300 m; Sask.

Salix turnorii is known from the Lake Athabasca sand dunes in northwestern Saskatchewan. *Salix famelica* in the Great Sand Hills, southern Saskatchewan, is very similar morphologically and may have been the source of populations ancestral to *S. turnorii* that moved into northern Saskatchewan during the warm Holocene Hypsithermal Period (ca. 9000–6000 yrs. B.P.).

Hybrids:

Salix turnorii forms natural hybrids with *S. brachycarpa* var. *psammophila*.

71. **Salix ligulifolia** (C. R. Ball) C. R. Ball ex C. K. Schneider, J. Arnold Arbor. 2: 188. 1922 • Strap-leaf willow [E]

Salix lutea Nuttall var. *ligulifolia* C. R. Ball, Bot. Gaz. 71: 428. 1921; *S. eriocephala* Michaux var. *ligulifolia* (C. R. Ball) Dorn

Shrubs, 1–8 m. **Stems:** branches yellow-brown, gray-brown, or red-brown, not glaucous, glabrous or villous; branchlets yellow-green or yellow-brown, glabrous, sparsely to densely villous, or velvety, (inner membranaceous bud-scale layer free, not separating from outer layer). **Leaves:** stipules foliaceous, apex rounded, convex, acute or acuminate; petiole convex to flat, or shallowly grooved adaxially, 3–18 mm, glabrous, pilose, or velvety to glabrescent adaxially; largest medial blade lorate, narrowly oblong, or narrowly elliptic, 60–133 ×

12–30 mm, 2.9–6.4 times as long as wide, base rounded, convex, or subcordate, margins flat, usually serrulate or serrate, rarely (apparently) entire, apex acuminate to acute, abaxial surface glaucous, glabrous, sparsely short-silky or pubescent, hairs straight or wavy, adaxial dull, glabrous, sparsely short-silky, or midrib pubescent; proximal blade margins entire, serrulate, or crenulate; juvenile blade reddish or yellowish green, glabrous or sparsely to moderately densely pilose or puberulent abaxially, hairs white. **Catkins** flowering as or just before leaves emerge; staminate stout, 20.5–34 × 8–11 mm, flowering branchlet 0–3 mm; pistillate moderately densely flowered, slender to subglobose, 15.5–49 × 8–18 mm, flowering branchlet 1–6 mm; floral bract brown or bicolor, 0.8–1.6 mm, apex acute or rounded, abaxially hairy throughout or proximally (hairs usually arising from rachis), hairs wavy or curly. **Staminate flowers:** adaxial nectary narrowly oblong, oblong, flask-shaped, or triangular, 0.3–0.8 mm; filaments distinct or connate less than ½ their lengths (or appearing as a single stamen), glabrous or hairy basally; anthers purple or red turning yellow, (ellipsoid or globose), 0.5–0.8 mm. **Pistillate flowers:** adaxial nectary oblong, narrowly oblong, or flask-shaped, 0.3–0.9 mm, shorter than stipe; stipe 0.9–2.5 mm; ovary pyriform, glabrous, beak sometimes slightly bulged below styles; ovules 12–21 per ovary; styles 0.2–0.6 mm; stigmas flat, abaxially non-papillate with rounded tip, or slenderly cylindrical, 0.16–0.25–0.4 mm. **Capsules** 4–6 mm. $2n = 38$.

Flowering late Mar–mid Jun(–late Jul). Banks and floodplains, cienegas, sandy-clay or gravelly substrates; 0–3100 m; Ariz., Calif., Colo., N.Mex., Oreg., Utah, Wyo.

Salix ligulifolia sometimes has leaf teeth that are so short that it is referred to as "entire-leaved." That condition is uncommon and, even when some leaves appear to be entire, others with fine serrulations can be found on the same plant.

Hybrids:

Salix ligulifolia forms natural hybrids with *S. geyeriana*.

72. Salix lutea Nuttall, N. Amer. Sylv. 1: 63, plate 19. 1842 • Yellow willow [E] [F]

Salix cordata Muhlenberg var. *watsonii* Bebb; *S. eriocephala* Michaux var. *watsonii* (Bebb) Dorn; *S. lutea* var. *watsonii* (Bebb) Jepson; *S. rigida* Muhlenberg var. *watsonii* (Bebb) Cronquist

Shrubs, 3–7 m, (sometimes forming clones by stem fragmentation). **Stems:** branches (sometimes ± brittle at base) yellow-gray, yellow-brown, or gray-brown, (sometimes weakly glaucous, with sparkling wax crystals), glabrous; branchlets red-brown or brownish, glabrous or pilose, (inner membranaceous bud-scale layer free, separating from outer layer). **Leaves:** stipules rudimentary or foliaceous on early ones, foliaceous on late ones, apex acute or rounded; petiole convex to flat, or shallowly grooved adaxially, 4–19 mm, pilose, velvety, or pubescent to glabrescent adaxially; largest medial blade (sometimes amphistomatous), lorate, narrowly elliptic, elliptic, lanceolate, or narrowly oblanceolate, 42–90 × 8–32 mm, 2.8–3.9–5.6 times as long as wide, base rounded, convex, or subcordate, margins flat, entire, serrulate, crenulate, or sinuate, apex acuminate to acute, abaxial surface glaucous, glabrous, pilose, or sparsely long-silky, hairs straight, adaxial dull or slightly glossy, glabrous, pilose, sparsely long-silky, especially midrib; proximal blade margins entire, serrulate, or crenulate; juvenile blade reddish or yellowish green, glabrous or sparsely to moderately densely long-silky throughout, hairs white. **Catkins** flowering as leaves emerge; staminate stout, slender, or subglobose, 10–45 × 6–12 mm, flowering branchlet 0.5–2 mm; pistillate loosely to densely flowered, stout or subglobose, 13.5–38 × 7–15 mm, flowering branchlet 0.5–7 mm; floral bract brown, tawny, or bicolor, 0.6–1.2 mm, apex acute or rounded, abaxially glabrous or sparsely hairy, hairs curly. **Staminate flowers:** adaxial nectary narrowly oblong, oblong, square, or flask-shaped, 0.4–0.9 mm; filaments distinct or connate less than ½ their lengths, glabrous; anthers yellow or purple turning yellow, (ellipsoid or globose), 0.4–0.8 mm. **Pistillate flowers:** adaxial nectary oblong, square, or ovate, 0.3–0.9 mm, shorter than stipe; stipe 0.9–3.8 mm; ovary pyriform or ovoid, glabrous, beak gradually tapering to styles; ovules 12–24 per ovary; styles 0.1–0.6 mm; stigmas flat, abaxially non-papillate with rounded tip, or 2 plump lobes, 0.14–0.2–0.3 mm. **Capsules** 3–5 mm. $2n = 38$.

Flowering Mar–May. Banks of streams, meadows, hillsides, gullies, sandy-clay, sandy or rocky substrates; 600–3100 m; Ariz., Calif., Colo., Idaho, Mont., Nev., Oreg., Utah, Wyo.

The possible occurrence of *Salix lutea* in Ginkgo Petrified Forest Park, Washington, needs to be investigated.

Hybrids:

Salix lutea forms natural hybrids with *S. arizonica*.

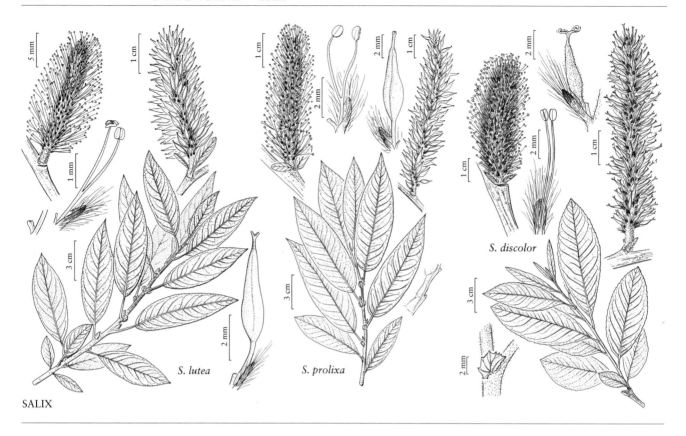

73. Salix monochroma C. R. Ball, Bot. Gaz. 71: 431, fig. 1. 1921 • One-color willow Ⓔ

Salix eriocephala Michaux var. *monochroma* (C. R. Ball) Dorn

Shrubs, 2–4 m. **Stems**: branches yellow-brown, gray-brown, or red-brown, not glaucous, glabrous; branchlets yellow-brown or red-brown, (sometimes weakly glaucous, with sparkling wax crystals), glabrous or puberulent. **Leaves**: stipules foliaceous, (prominent), apex rounded or acute; petiole convex to flat, or shallowly grooved adaxially, 6–16 mm, velvety or pilose adaxially; largest medial blade narrowly oblong, elliptic, or broadly elliptic, 54–100 × 18–35 mm, 1.8–3.6 times as long as wide, base convex, rounded, subcordate, or cordate, margins flat, serrate or serrulate (teeth sometimes elongate), apex acuminate to acute, abaxial surface not glaucous, glabrous, adaxial dull, glabrous or pilose, midrib hairy; proximal blade margins entire or serrulate; juvenile blade reddish or yellowish green, glabrous or sparsely short- or long-silky abaxially, hairs white. **Catkins** flowering as or just before leaves emerge; staminate stout or slender, 24–43(–52 in fruit) × 6–11 mm, flowering branchlet 1–4 mm; pistillate moderately densely to loosely flowered, stout, 33–73 × 8–16 mm, flowering branchlet 2–12 mm; floral bract brown, 0.7–1.6 mm, apex rounded, convex, or acute, abaxially hairy throughout or proximally, hairs wavy or curly. **Staminate flowers**: adaxial nectary narrowly oblong, 0.6–1.1 mm; filaments distinct or slightly connate, glabrous; anthers purple turning yellow, (ellipsoid or shortly cylindrical), 0.5–0.7 mm. **Pistillate flowers**: adaxial nectary narrowly oblong, oblong, or flask-shaped, 0.3–1.1 mm, shorter than stipe; stipe 1.4–3.6 mm; ovary pyriform, glabrous, beak sometimes slightly bulged below styles; ovules 18–22 per ovary; styles 0.2–0.5 mm; stigmas flat, abaxially non-papillate with rounded tip, or broadly cylindrical, 0.2–0.23–0.26 mm. **Capsules** 3.6–5.2 mm.

Flowering early–mid May. Silty or stony stream margins; 150–2200 m; Idaho, Oreg., Wash.

74. Salix prolixa Andersson, Monogr. Salicum, 94, plate 5, fig. 52. 1867 • Mackenzie's willow [E] [F]

Salix cordata Muhlenberg var. *mackenzieana* Hooker; *S. eriocephala* Michaux subsp. *mackenzieana* (Hooker) Dorn; *S. mackenzieana* (Hooker) Barratt ex Andersson; *S. mackenzieana* var. *macrogemma* C. R. Ball; *S. rigida* Muhlenberg var. *mackenzieana* (Hooker) Cronquist; *S. rigida* var. *macrogemma* (C. R. Ball) Cronquist

Shrubs, 1–5 m. **Stems:** branches gray-brown or red-brown, not or weakly glaucous, (with sparkling wax crystals), glabrous or moderately villous; branchlets yellow-green, or yellow-brown to red-brown, glabrous or sparsely to moderately densely velvety, (inner membranaceous bud-scale layer free, separating from outer layer). **Leaves:** stipules foliaceous, apex acuminate to rounded; petiole convex to flat adaxially, 6–12 mm, glabrous or pilose adaxially; largest medial blade narrowly oblong, narrowly elliptic, lanceolate, or obovate, 50–150 × 10–53 mm, 2.4–4.5 times as long as wide, base cordate, subcordate, rounded, or convex, margins flat, serrate, serrulate, or spinulose-serrulate, apex acuminate to acute, abaxial surface glaucous, glabrous, adaxial dull, glabrous, pilose, or sparsely pubescent, (sometimes adaxial stomata only along veins or apically); proximal blade margins entire; juvenile blade reddish or yellowish green, glabrous, pilose or sparsely long-silky abaxially, hairs white. **Catkins** flowering as leaves emerge; staminate slender or stout, 16–41 × 8–12 mm, flowering branchlet 0.5–3 mm; pistillate moderately densely to loosely flowered, slender or stout, 19–66 × 8–18 mm, flowering branchlet 0.5–6 mm; floral bract brown, 0.8–1.6 mm, apex acute or rounded, abaxially hairy mainly proximally, hairs straight or wavy. **Staminate flowers:** adaxial nectary oblong to narrowly oblong, 0.8–1 mm; filaments distinct or connate less than ½ their lengths, glabrous; anthers purple turning yellow, (ellipsoid), 0.5–0.6 mm. **Pistillate flowers:** adaxial nectary oblong, square, or flask-shaped, 0.3–0.8 mm, shorter than stipe; stipe 1.3–4.2 mm; ovary pyriform, glabrous, beak sometimes slightly bulged below styles; ovules 12–22 per ovary; styles 0.3–0.7 mm; stigmas flat, abaxially non-papillate with rounded tip, or 2 plump lobes, or slenderly cylindrical, 0.16–0.28–0.4 mm. **Capsules** 4–6 mm.

Flowering late Mar–late Jun. Along streams, lakes, springs, marsh margins, sandy-gravel, sandy or silty substrates; 100–2300 m; Alta., B.C., N.W.T., Yukon; Alaska, Calif., Idaho, Mont., Nev., Oreg., Wash., Wyo.

2e3. Salix Linnaeus (subg. **Vetrix**) sect. **Nigricantes** A. Kerner, Verh. K. K. Zool.-Bot. Ges. Wien 10: 235. 1860 [I]

Shrubs or trees, 1–5 m. **Buds** *caprea*-type. **Leaves:** stipules on late ones foliaceous; largest medial blade 1–3 times as long as wide, adaxial surface not glaucous. **Staminate flowers:** filaments glabrous or hairy basally. **Pistillate flowers:** adaxial nectary shorter than stipe; stipe 0.8–3.5 mm; ovary glabrous or pubescent; stigmas broadly cylindrical, 0.3–0.6 mm.

Species 3 (1 in the flora): introduced, Ontario; Eurasia.

75. Salix myrsinifolia Salisbury, Prodr. Stirp. Chap. Allerton, 394. 1796 • Dark-leaved willow [I]

Salix nigricans Smith

Stems: branches dark red-brown or gray-brown, not glaucous, glabrous or hairy, (peeled wood smooth or striate with relatively few, short striae); branchlets red-brown, moderately to very densely pubescent or velvety. **Leaves:** stipules usually foliaceous, sometimes minute rudiments on early ones, foliaceous on late ones, (ca. 4 mm), apex acute; petiole convex to flat, or shallowly grooved adaxially, 3.5–12(–15) mm, villous to puberulent adaxially; largest medial blade (sometimes hemiamphistomatous), broadly obovate, elliptic, broadly elliptic, or subcircular, 24–52(–100) × 12–45 mm, base concave, rounded, subcordate, cordate, or cuneate, margins sometimes slightly revolute, serrulate, or crenulate to subentire, apex abruptly acuminate or acute, abaxial surface glaucous (tip often not glaucous), sparsely to moderately densely puberulent, or silky to glabrescent, hairs appressed or spreading, straight or wavy, adaxial slightly glossy, glabrescent or sparsely to moderately densely puberulent, pubescent, or short-silky (especially midrib); proximal blade margins serrulate, crenulate, or entire; juvenile blade sometimes reddish, long-silky, villous, tomentose (at least on midrib), or glabrous abaxially, hairs usually white, rarely somewhat ferruginous. **Catkins** flowering as leaves emerge; staminate (densely flowered), stout or subglobose, 17–35 mm, flowering branchlet 1–3 mm; pistillate densely flowered, stout, 9–11(–30)(–80 in fruit) mm, flowering branchlet 2–5.5 mm; floral bract pale brown, 1–1.9(–2.8) mm, apex acute, convex, or rounded, abaxially sparsely hairy, hairs straight or wavy. **Staminate flowers:** adaxial nectary 0.5–0.7 mm; filaments distinct; anthers purple turning yellow, (ellipsoid or shortly cylindrical), 0.5–0.8 mm. **Pistillate flowers:** adaxial nectary ovate, square, or flask-shaped, 0.4–0.6(–1) mm; ovary pyriform, pubescent throughout or in patches or streaks, or glabrous (hairs refractive), beak gradually tapering to styles; ovules 12–14 per ovary; styles 0.6–1.5 mm. **Capsules** 6–10 mm. $2n = 114$.

Flowering mid Mar–mid Jun. Roadsides, waste places; ca. 100 m; introduced; Ont.; Eurasia.

Salix myrsinifolia may be naturalized in the vicinity of Ottawa, Ontario, but that needs confirmation.

2e4. SALIX Linnaeus (subg. VETRIX) sect. CINERELLA Seringe, Exempl. Rév. Salix, 2. 1824

Shrubs or trees, 0.3–20 m. **Buds** *caprea*-type. **Leaves:** stipules on late ones absent, minute rudiments, or foliaceous; largest medial blade 1.5–9 times as long as wide, adaxial surface not glaucous. **Staminate flowers:** filaments glabrous or hairy. **Pistillate flowers:** adaxial nectary shorter or longer than stipe; stipe 0.5–2.7 mm; ovary glabrous or sparsely to very densely hairy; stigmas slenderly to broadly cylindrical, 0.2–1 mm.

Species 37 (8 in the flora): North America, Eurasia, Africa.

76. **Salix discolor** Muhlenberg, Ges. Naturf. Freunde Berlin Neue Schriften 4: 236, plate 6, fig. 1. 1803
    • Pussy or large pussy willow ⓔ Ⓕ

Salix ancorifera Fernald; *S. discolor* var. *overi* C. R. Ball; *S. discolor* var. *prinoides* (Pursh) Andersson

Shrubs, 2–4(–8) m, (sometimes forming clones by stem fragmentation). **Stems:** branches dark red-brown or yellow-brown, not to strongly glaucous, villous to glabrescent, (peeled wood smooth or striate, striae sometimes very dense, to 10 mm); branchlets yellowish, red-brown, or yellow-brown, or dark brown, moderately densely velvety, velutinous, or tomentose to glabrescent. **Leaves:** stipules rudimentary on early ones, foliaceous on late ones, (0.8–12.5 mm), apex acute to acuminate; petiole convex to flat adaxially, 6–17 mm, tomentose adaxially; largest medial blade narrowly elliptic, elliptic, oblanceolate, or obovate, 30–80(–135) × 12–33 mm, (2.3–)3–3.5(–4.5) times as long as wide, base convex or cuneate, margins flat, crenate, irregularly toothed, sinuate, or entire, apex acute, convex, or acuminate, abaxial surface glaucous, glabrous, pilose, sparsely pubescent or long-silky, midrib glabrous or densely pubescent, hairs (white, sometimes also ferruginous), wavy, adaxial dull or slightly glossy, glabrous or pilose, (hairs rarely ferruginous); proximal blade margins entire or serrulate; juvenile blade reddish or yellowish green, pilose, tomentose or moderately densely short-silky abaxially, hairs white and ferruginous. **Catkins** flowering before leaves emerge; staminate stout or subglobose, 23–52 × 12–22 mm, flowering branchlet 0–3 mm; pistillate densely flowered (loose in fruit), slender or stout, 25–108(–135 in fruit) × 12–33 mm, flowering branchlet 0–10 mm; floral bract brown, black, or bicolor, 1.4–2.5 mm, apex acute or convex, abaxially hairy, hairs straight. **Staminate flowers:** adaxial nectary oblong, 0.6–1.1 mm; filaments distinct, glabrous or hairy basally; anthers yellow or purple turning yellow, ellipsoid or short- or long-cylindrical, 0.5–1 mm. **Pistillate flowers:** adaxial nectary oblong or ovate, 0.7–1.3 mm, shorter than stipe; stipe 1.6–2.7 mm; ovary obclavate or pyriform, short-silky (hairs straight), beak sometimes slightly bulged below styles; ovules 6–16 per ovary; styles 0.3–1 mm; stigmas slenderly or broadly cylindrical, 0.48–0.64–0.88 mm. **Capsules** 6–11 mm. $2n = 76, 95, 114$.

Flowering early Apr–late May. Marshy margins of ponds, streams, and open alluvial woods, fens, seepage areas, peaty substrates; 0–2400 m; Alta., B.C., Man., N.B., Nfld. and Labr., N.W.T., N.S., Ont., P.E.I., Que., Sask.; Colo., Conn., Del., Idaho, Ill., Ind., Iowa, Ky., Maine, Md., Mass., Mich., Minn., Mo., Mont., N.H., N.J., N.Y., N.C., N.Dak., Ohio, Pa., R.I., S.Dak., Vt., W.Va., Wis., Wyo.

Salix discolor is introduced in North Carolina.

Vegetative specimens of *Salix discolor* can be difficult to distinguish from *S. planifolia*, but there are two, somewhat variable, characters that can be used. *Salix discolor* usually has leaves dull adaxially, with arcuate secondary veins widely and irregularly spaced; *S. planifolia* has leaves slightly or highly glossy adaxially, with straight secondary veins closely and regularly spaced.

Salix discolor in northeastern United States can be difficult to distinguish from widely naturalized *S. atrocinerea* and *S. cinerea*. Useful diagnostic characters are: tertiary leaf veins, which are irregular in *S. discolor* but close and parallel in introduced species, and raised striae on peeled 3–5-year old branches, which are absent or indistinct and relatively short in *S. discolor*, but long and very prominent in the introductions.

Hybrids:

Salix discolor forms natural hybrids with *S. humilis*, *S. interior*, *S. myricoides*, *S. pellita*, and *S. planifolia*. Reports of hybrids with *S. candida* and *S. eriocephala* (M. L. Fernald 1950), and *S. bebbiana* and *S. pyrifolia* (C. K. Schneider 1921; Fernald), are not based on convincing specimens. Synthetic hybrids with *S. bebbiana* could not be made (G. W. Argus 1974; A. Mosseler 1990) and those made with *S. eriocephala* had low seed viability (Mosseler).

Salix discolor × *S. humilis* has tomentose leaves of *S. humilis* and longer catkins and styles of *S. discolor* (G. W. Argus 1986). These species readily hybridize and produce abundant seed (Argus 1974). The hybrids are fertile and backcross. Specimens of *S. discolor* with densely villous branchlets may be hybrids or introgressants with *S. humilis*. The two species usually are ecologically isolated; *S. discolor* occurs in wetland thickets and *S. humilis* in dry, sandy upland forests. Where the two habitats come into proximity, hybrids occur but large swarms have not been observed.

Salix discolor × *S. myricoides* (*S.* ×*laurentiana* Fernald, syn. *S. paraleuca* Fernald) usually resembles *S. myricoides* but has hairy ovaries (R. D. Dorn 1975, 1976). This hybrid was originally described as a species, from lower St. Lawrence River, Quebec. Its most distinctive feature is that hairs appear on ovaries in patches, at the base or, sometimes, only on the stipes. A similar ovary indumentum pattern appears in other hybrids or species of hybrid origin, e.g., *S. hookeriana*. Characteristics of *S. discolor* found in *S.* ×*laurentiana* include epidermis with gray-margined splits, leaf margins entire or sinuate, leaves with 2–4 teeth per cm, anthers yellow or purple, filaments hairy on proximal half or basally, ovaries hairy, greenish brown or green with red sutures, and adaxial pistillate nectaries ovate. Characteristics of *S. myricoides* include inner bud-scale membranes separating from the outer ones, stipules more prominent, catkins on distinct flowering branchlets, and longer styles sometimes distinct about half their lengths. This hybrid occurs throughout the area of overlap of the parents. All three taxa often are intermixed but few hybrids seem to produce well-developed seed.

Salix discolor × *S. pellita* (*S.* ×*pedunculata* Fernald) is characterized by juvenile leaves with infolded or sometimes revolute margins, ovaries with patches of hairs relatively short, flattened, crinkled, and refractive, and catkins borne on distinct flowering branchlets 2–10 mm. This sporadic hybrid does not seem to be fertile. It occurs in Newfoundland, Quebec, and Saskatchewan. Although it has been collected at few localities, it probably is more common and should be expected wherever the two species grow together. The type and other collections compare very well with synthetic hybrids (A. Mosseler 1990), which were reported to show a high hybridization success rate, high F_1 pollen viability, and high seedling viability. It was suggested that variability within these species may be due to interspecific gene flow. In interpreting the parentage of the wild hybrids it is not possible to rule out hybridization of *S. planifolia* or *S. myricoides* with *S. pellita*, or that these hybrids may be *S. myricoides* × *S. planifolia*, as suggested by B. G. O. Floderus (1939). *Salix* ×*pellicolor* Lepage is a later synonym of this hybrid.

77. Salix hookeriana Barratt ex Hooker, Fl. Bor.-Amer. 2: 145, plate 180. 1838 • Coastal or Hooker's willow [E]

Salix amplifolia Coville; *S. hookeriana* var. *laurifolia* J. K. Henry; *S. hookeriana* var. *tomentosa* J. K. Henry ex C. K. Schneider; *S. piperi* Bebb

Shrubs or trees, (0.6–)2–8 m, (sometimes forming clones by layering or stem fragmentation). **Stems:** branches (highly brittle at base), yellow-brown, gray-brown, red-brown, or violet, not or weakly glaucous, glabrous, tomentose, woolly, or sparsely villous to glabrescent (nodes hairy); branchlets gray-brown, red-brown, or yellow-brown (sometimes color obscured by hairs), glabrous, pilose, moderately densely villous, tomentose, or woolly, scale with inner membranaceous layer free, not separating from outer layer). **Leaves:** stipules rudimentary or absent on early ones, foliaceous (early deciduous) or rudimentary (sometimes obscured) on late ones, (2.5–7.8–18 mm), apex acuminate, acute, or rounded; petiole convex to flat, or shallowly grooved adaxially, 4–29 mm, villous, woolly, pilose, or tomentose adaxially; largest medial blade (sometimes hemiamphistomatous), narrowly to broadly elliptic, oblanceolate, or obovate to broadly obovate, 36–123 × 18–63 mm, 1.5–4.2 times as long as wide, base convex, rounded, subcordate, cordate, or cuneate, margins slightly revolute, crenate, serrate, shallowly serrulate, sinuate, or entire, apex acuminate, acute, or convex, abaxial surface glaucous, pilose, moderately densely tomentose, villous, or woolly, midrib hairy, hairs (white, sometimes also ferruginous), wavy or straight, adaxial highly or slightly glossy, glabrous, pilose, villous, or moderately densely tomentose, midrib and veins hairy (hairs white, sometimes also ferruginous); proximal blade margins entire or shallowly serrulate; juvenile blade yellowish, reddish green (sometimes obscured by hairs), pilose or sparsely to densely long-silky, tomentose, woolly, or villous abaxially, hairs white, sometimes also ferruginous, or yellowish. **Catkins** flowering before or as leaves emerge; staminate slender or stout, 26–73 × 10–27 mm, flowering branchlet 0–10 mm; pistillate densely flowered, slender or stout, 36–92(–140 in fruit) × 10–25 mm, flowering branchlet 0–20 mm; floral bract brown, black, or bicolor, 1.1–3.6 mm, apex convex, rounded, or acute, abaxially hairy, hairs straight or wavy. **Staminate flowers:** adaxial nectary oblong, ovate, or narrowly oblong, 0.5–1.4 mm; filaments distinct or slightly basally connate, glabrous or hairy on proximal ½ or basally; anthers yellow, cylindrical or ellipsoid, (0.5–)0.7–1 mm. **Pistillate flowers:** adaxial nectary narrowly oblong, oblong, or square, 0.5–1.4 mm, shorter than stipe; stipe 0.5–1.8(–2.8) mm; ovary obclavate or pyriform, glabrous, tomentose, villous, or woolly (hairs wavy), beak sometimes abruptly tapering to styles; ovules 12–20 per ovary; styles 0.6–2.3 mm; stigmas broadly to slenderly cylindrical, 0.3–0.8 mm. **Capsules** 5–10 mm. $2n = 114$.

Flowering mid Apr–mid Jun. Marine coastal beaches and sand dunes, interdunal depressions, coastal marshes, pine barrens, floodplains, ravines, wet sedge meadows, lakeshores, morainal flats, sandy or gravelly substrates; 0–1800 m; B.C.; Alaska, Calif., Oreg., Wash.

Salix hookeriana is primarily a coastal species occurring from northern California northward to Oregon, Washington, and southern Vancouver Island, with disjunct populations on Queen Charlotte Islands, British Columbia, and northward to Yakatut Bay, Turnagain Arm, and Kodiak, Alaska. It was treated by G. W. Argus (1973) and R. D. Dorn (2000) in a broad sense because of an absence of strong distinguishing characters and intergradation in characters that could be used to divide it. It is highly variable and three very similar taxa have been named: *S. amplifolia*, *S. hookeriana* (including vars. *tomentosa* and *laurifolia*), and *S. piperi*. Although extremes of these taxa sometimes are recognizable, the intergradation displayed is so great that even attempts to recognize them as varieties are thwarted. The *amplifolia* variant in Alaska is characterized by having only white leaf hairs, hairy ovaries, no stipules, and catkins often borne on distinct flowering branchlets, but variation can occur within the same population, and typical *S. hookeriana* on Vancouver Island sometimes displays the same characteristics. The *piperi* variant, an inland population in western Oregon and Washington, is usually recognized by local botanists as different from coastal populations. It is characterized by leaves and branchlets soon becoming glabrate and stipules prominent. These characteristics, however, sometimes appear in northern California coastal populations, and some inland populations in Oregon include very hairy individuals that are indistinguishable from coastal variants of *S. hookeriana*. In general, very hairy populations of *S. hookeriana* are probably an adaptation to marine coastal environments, but some variation may be due to hybridization and introgression with *S. scouleriana*. Inland populations suggest the influence of *S. lasiolepis*. Two hexaploid chromosome numbers reported for *S. hookeriana* from Vancouver Island (R. L. Taylor and S. Taylor 1977) and Queen Charlotte Islands (R. L. Taylor and G. A. Mulligan 1968), British Columbia, indicate that hybridization has played a role in the evolution of this complex. It is possible that each variant of *S. hookeriana* has had a different, possibly even recurrent, polyploid origin. Further cytological and genetic study is indicated.

The following comparisons may help to distinguish *Salix hookeriana*, *S. lasiolepis*, and *S. scouleriana*.

Vegetative specimens of *Salix hookeriana* can be distinguished from *S. lasiolepis* by having floral buds ellipsoid, beaks distinctly long-tapered, densely long-hairy (villous), red-brown, blades usually pilose, villous, or woolly on abaxial surfaces, usually 18–63 mm wide, and 1.5–4.2 times as long as wide; *S. lasiolepis* has floral buds ovoid, beaks inconspicuous and blunt, sparsely to moderately densely short-hairy (velvety), yellowish to red-brown, blades usually tomentose on abaxial surfaces, usually 6–32 mm wide, and 3.2–9.6 times as long as wide.

Salix hookeriana is distinguished from *S. scouleriana* by having branchlets with spreading hairs (woolly or tomentose to glabrate), petioles usually pilose to tomentose, blades typically narrowly elliptic but variable, stigmas 0.3–0.74, short in relation to styles (0.6–2.3 mm), and pistillate nectaries 0.5–1.4 mm, shorter or longer than stipes; *S. scouleriana* has branchlets usually with short, erect hairs (velutinous), sometimes spreading (villous or tomentose), petioles velvety or villous adaxially, blades typically oblanceolate but variable, stigmas 0.4–1.04 mm, long in relation to styles (0.2–0.6 mm), and pistillate nectaries 0.2–0.8 mm, shorter than stipes.

Hybrids:

Salix hookeriana forms natural hybrids with *S. barclayi* and *S. scouleriana*. Variation in some *S. hookeriana* populations suggests hybridization with *S. lasiolepis* but no positive identifications have been made. R. D. Dorn (2000) doubted that hybridization in California between these species with different chromosome numbers was possible, but species with different chromosome numbers do hybridize [for example, *S. athabascensis* ($4x$) × *S. pedicellaris* ($2x$)]; synthetic hybridization studies are indicated.

Salix hookeriana × *S. scouleriana*: Plants from southern British Columbia with leaves similar to *S. hookeriana* but with prominent stipules, catkins both erect and recurving, and relatively long stigmas were thought by J. K. Henry (1915) to be this hybrid.

78. Salix humilis Marshall, Arbust. Amer., 140. 1785 • Gray or small pussy or upland willow [E] [F] [W]

Shrubs, 0.3–3 m, (forming clones by layering). **Stems**: branches dark red-brown, not or weakly glaucous, tomentose to glabrescent; branchlets red-brown, yellow-brown, or greenish brown, moderately to very densely villous, tomentose, or velvety-tomentose to glabrescent. **Leaves**: stipules absent or rudimentary on early ones, absent, rudimentary or foliaceous on late ones, apex acute; petiole convex to flat, or shallowly grooved adaxially, 0.5–7(–12) mm, velvety, pilose, or villous adaxially; largest medial blade (sometimes hemiamphistomatous), narrowly oblong, narrowly elliptic, elliptic, oblanceolate, obovate, or broadly obovate, (13–)20–90(–135) × 3–23(–35) mm, 2.3–9 times as long as wide, base cuneate or convex, margins revolute or flat, entire, crenate, or sinuate, (glands submarginal), apex acuminate or convex, abaxial surface glaucous, sparsely to densely tomentose or woolly, hairs erect or spreading, wavy, adaxial slightly or highly glossy, glabrous, pubescent, tomentose, or pilose; proximal blade margins entire or serrulate; juvenile blade green, densely tomentose to glabrescent abaxially, hairs white, sometimes also ferruginous. **Catkins** flowering before leaves emerge; staminate 6.5–34 × 5–19 mm, flowering branchlet 0–1 mm; pistillate (and staminate) moderately to very densely flowered, stout, subglobose, or globose, 9–47(–55 in fruit) × 5.5–19 mm, flowering branchlet 0–4 mm; floral bract brown, black, or bicolor, 0.8–2 mm, apex rounded or acute, abaxially moderately densely hairy, hairs (white), straight or wavy. **Staminate flowers**: adaxial nectary oblong or square, 0.2–0.7 mm; filaments distinct, glabrous or hairy basally; anthers purple turning yellow, ellipsoid or cylindrical, 0.4–0.6 mm. **Pistillate flowers**: adaxial nectary square, 0.4–0.8 mm, shorter than stipe; stipe 1–2.5 mm; ovary obclavate or pyriform, moderately densely to sparsely short-silky-villous (hairs refractive), beak slightly bulged below styles, (valves recurving in fruit); ovules 6–12 per ovary; styles (sometimes slightly distinct distally), 0.2–0.4 mm; stigmas slenderly to broadly cylindrical, 0.2–0.56 mm. **Capsules** 5–12 mm.

Varieties 2 (2 in the flora): North America.

Varieties *humilis* and *tristis* are almost sympatric. In Maine they occur in the same populations without evident intermediate forms, while in other places, e.g., Florida, Georgia, and Nebraska, they apparently intergrade. Usually, they are distinct in habit, leaf and catkin size, and, most conspicuously, the presence of foliaceous stipules in var. *humilis* and their absence in var. *tristis*.

1. Stipules on late leaves foliaceous; largest medial blades (20–)50–90(–135) mm; petioles (1.5–)3–7(–12) mm; staminate catkins 14.5–34 mm; pistillate catkins longer than 20 mm; peeled wood smooth or striate, striae dense, to 20 mm .... 78a. *Salix humilis* var. *humilis*
1. Stipules on late leaves absent or rudimentary; largest medial blades (13–)20–50(–70) mm; petioles 0.5–3(–6) mm; staminate catkins 6.5–13.5 mm; pistillate catkins shorter than 20 mm; peeled wood smooth or striate, striae sparse, to 2 mm .... 78b. *Salix humilis* var. *tristis*

78a. Salix humilis Marshall var. humilis • Prairie willow [E] [F]

Salix humilis var. *hyporhysa* Fernald; *S. humilis* var. *keweenawensis* Farwell; *S. humilis* var. *rigidiuscula* (Andersson) B. L. Robinson & Fernald

Mid shrubs, 0.3–3 m. **Stems** usually erect, sometimes decumbent; branches tomentose to glabrescent, peeled wood smooth or striate, striae sometimes very dense, to 20 mm; branchlets red-brown or greenish brown. **Leaves**: stipules usually foliaceous (rarely rudimentary) on late ones; petiole (1.5–)3–7(–12) mm, velvety or pilose adaxially; largest medial blade narrowly oblong, narrowly elliptic, elliptic, oblanceolate, obovate, or broadly obovate, (20–)50–90(–135) × (7–)13–23(–35) mm, 2.3–4–7.5 times as long as wide, margins strongly revolute to flat, abaxial surface with hairs rarely also ferruginous, adaxial slightly or highly glossy, glabrous, pubescent, or pilose (hairs white, sometimes also ferruginous); proximal blade margins usually entire, sometimes serrulate. **Catkins**: staminate 14.5–34 × 7–19 mm, flowering branchlet 0 mm; pistillate 9–47 (–55 in fruit) × 5.5–19 mm, flowering branchlet 0–4 mm; floral bract 1.2–2 mm. **Staminate flowers**: filaments glabrous. **Pistillate flowers**: ovary obclavate; stigmas 0.24–0.33–0.56 mm; ovules 6–12 per ovary. **Capsules** 7–12 mm. $2n$ = 38 and 76.

Flowering (north) early Mar–early Jun, (south) late Jan–late Apr. Dry mixed woods and forests, *Picea mariana*-lichen woods, *Picea glauca-Abies balsamea* forests, wet to dry prairies, grassy balds, loess bluffs, sandy stream terraces, coastal barrens, *Carex-Typha* meadows, fine sand to rocky granitic, gneissic, limestone, and serpentine substrates; 20–1600 m; Man., N.B., Nfld. and Labr., N.S., Ont., P.E.I., Que.; Ala., Ark., Conn.,

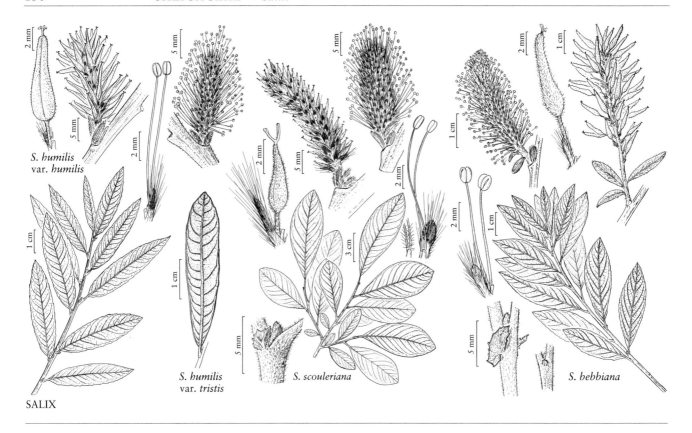

Del., D.C., Fla., Ga., Ill., Ind., Iowa, Kans., Ky., La., Maine, Md., Mass., Mich., Minn., Miss., Mo., Nebr., N.H., N.J., N.Y., N.C., N.Dak., Ohio, Okla., Pa., R.I., S.C., S.Dak., Tenn., Tex., Va., W.Va., Wis.

See 79. *Salix scouleriana* and 82. *S. atrocinerea* for comparative descriptions.

Hybrids:

Variety *humilis* forms natural hybrids with *Salix bebbiana*, *S. discolor*, *S. eriocephala*, and *S. planifolia*. Hybrids with *S. discolor* or *S. planifolia* often are suggested by glabrous or glabrate leaves.

78b. Salix humilis Marshall var. **tristis** (Aiton) Griggs, Proc. Ohio Acad. Sci. 4: 301. 1905 • Dwarf prairie willow E F

Salix tristis Aiton, Hort. Kew. 3: 393. 1789; *S. humilis* var. *microphylla* (Andersson) Fernald

Low to mid shrubs, 0.3–1 m. **Stems** decumbent; branches tomentose, peeled wood smooth or striate, striae sparse, to 2 mm; branchlets yellow-brown. **Leaves:** stipules absent or rudimentary on late ones; petiole 0.5–3(–6) mm, velvety or villous adaxially; largest medial blade narrowly oblong, narrowly elliptic, oblanceolate, or obovate, (13–)20–50(–70) × 3–13 mm, 2.3–9 times as long as wide, margins strongly revolute, abaxial surface hairs gray throughout, adaxial slightly glossy, moderately densely tomentose; proximal blade margins entire. **Catkins:** staminate 6.5–13.5 × 5–10 mm, flowering branchlet 0–1 mm; pistillate 11–17.5 × 5.5–12 mm, flowering branchlet 0–1.5 mm; floral bract 0.8–1.4 mm. **Staminate flowers:** filaments glabrous or hairy basally. **Pistillate flowers:** ovary pyriform; ovules 6 per ovary; stigmas 0.2–0.24–0.32 mm. **Capsules** 5–9 mm.

Flowering early Mar–late May. Moist limestone and serpentine barrens, open heath balds, open pine woods, moist prairies, swampy areas in open deciduous woods, stream banks; 60–1600 m; Ark., Conn., Del., D.C., Ga., Ill., Ind., Kans., Ky., La., Maine, Md., Mass., Mich., Minn., Miss., Mo., N.H., N.J., N.Y., N.C., N.Dak., Ohio, Okla., Pa., R.I., S.C., Tenn., Va., W.Va., Wis.

Hybrids:

Although variety *tristis* is reported to hybridize with *Salix eriocephala*, *S. humilis* var. *humilis*, and *S. petiolaris* (G. W. Argus 1986), hybrids have not been confirmed.

79. **Salix scouleriana** Barratt ex Hooker, Fl. Bor.-Amer. 2: 145. 1838 • Scouler's or mountain willow [F]

Salix scouleriana var. *poikila* C. K. Schneider

Shrubs or trees, 1–10(–20) m. **Stems:** branches gray-brown, yellow-brown, or red-brown, not glaucous, glabrous or tomentose; branchlets yellow-green or yellow-brown, sparsely to densely villous, tomentose, or velvety. **Leaves:** stipules absent, rudimentary, or foliaceous on early ones, foliaceous on late ones, (1–16 mm), apex acute or acuminate; petiole convex to flat adaxially, 2–13 mm, velvety or villous adaxially; largest medial blade usually oblanceolate, sometimes narrowly elliptic, elliptic or obovate, 29–100 × 9–37 mm, 1.7–3.9 times as long as wide, base cuneate or convex, margins strongly to slightly revolute or flat, entire, remotely serrate, crenate, or sinuate, (glands submarginal or epilaminal), apex acuminate, convex, or rounded, abaxial surface glaucous, sparsely to densely short- to long-silky or woolly, hairs (white, sometimes also ferruginous), wavy or straight, adaxial slightly glossy, pilose or moderately densely short-silky, midrib velutinous or villous, (hairs white, sometimes also ferruginous); proximal blade margins entire, serrulate, or crenulate; juvenile blade reddish or yellowish green, sparsely to densely villous, short- or long-silky abaxially, hairs white, sometimes also ferruginous. **Catkins** flowering before leaves emerge; staminate stout or subglobose, 18–40.5 × 8–22 mm, flowering branchlet 0–4 mm; pistillate very densely flowered, slender or stout, 18–60(–90 in fruit) × 10–22 mm, flowering branchlet 0–8 mm; floral bract brown, black, or bicolor, 1.5–4.5 mm, apex rounded or acute, abaxially hairy, hairs straight. **Staminate flowers:** adaxial nectary oblong or square, 0.4–0.9 mm; filaments distinct, glabrous or hairy on proximal ½; anthers purple turning yellow, ellipsoid to shortly cylindrical, 0.7–1.2 mm. **Pistillate flowers:** adaxial nectary oblong or square, 0.2–0.8 mm, shorter than stipe; stipe 0.8–2.3 mm; ovary pyriform or obclavate, densely long-silky, beak slightly bulged below styles; ovules 10–18 per ovary; styles 0.2–0.6 mm; stigmas slenderly cylindrical, 0.4–0.82–1.04 mm. **Capsules** 4.5–11 mm. $2n = 76$.

Flowering late Feb–mid Jun. Dry conifer forests, mature woods on edges of streams and lakes, treed bogs, meadows, subalpine slopes, springs, pine barrens, openings in old burns, arroyos and disturbed sites, sandy, silty-clay, or gravelly, igneous substrates; 0–3500 m; Alta., B.C., Man., N.W.T., Sask., Yukon; Alaska, Ariz., Calif., Colo., Idaho, Mont., Nev., N.Mex., Oreg., S.Dak., Utah, Wash., Wyo.; Mexico (Chihuahua, Sonora).

Western *Salix scouleriana* and eastern *S. humilis* are closely related and are sometimes difficult to separate. Although there is an apparent range disjunction between them in western Manitoba, it may be a collecting gap. In general, *S. scouleriana* differs from *S. humilis* in being a taller shrub, sometimes even tree-like, with broader leaves and longer catkins, floral bracts, stigmas, and styles, but these quantitative characteristics all overlap. The apparent difference in anther length (*S. scouleriana* 0.7–1.2 mm; *S. humilis* 0.4–0.6 mm) may be correlated with a difference in chromosome number. *Salix scouleriana* is tetraploid (Y. Suda and G. W. Argus 1968); *S. humilis* has been reported to be both diploid (Suda and Argus; L. Zsuffa and Y. Raj, unpubl.) and tetraploid (R. D. Dorn 1976). The latter count was from the same population as the one by Suda and Argus. Further chromosome counts are indicated.

See 77. *Salix hookeriana* for comparative descriptions.

Hybrids:

Salix scouleriana forms natural hybrids with *S. hookeriana*, *S. planifolia*, and *S. pulchra*.

80. **Salix caprea** Linnaeus, Sp. Pl. 2: 1020. 1753 • Goat or hoary willow [I] [W]

Shrubs or trees, 8–15 m. **Stems:** branches brownish, not glaucous, pubescent to glabrescent, (peeled wood smooth or striate, striae sparse, to 6 mm); branchlets yellow-brown or gray-brown, sparsely to densely villous, velvety, or pubescent. **Leaves:** stipules rudi-mentary on early ones, foliaceous on late ones, apex acute or convex; petiole convex to flat adaxially, 7–25 mm, tomentose to glabrescent adaxially; largest medial blade narrowly elliptic, broadly elliptic, oblanceolate, obovate, or broadly oblong, 50–130 × 25–80 mm, 2–3 times as long as wide, base cuneate or convex, margins slightly revolute, entire, crenate, or sinuate, (glands submarginal or epilaminal), apex acuminate or convex, abaxial surface glaucous, sparsely tomentose or pubescent, hairs erect, wavy, adaxial dull or slightly glossy, sparsely pubescent; proximal blade margins entire; juvenile blade green, densely tomentose abaxially, hairs white, sometimes also ferruginous. **Catkins** flowering before leaves emerge; staminate subglobose or globose, 16–39 × 12–30 mm, flowering branchlet 0–3 mm; pistillate densely flowered, stout or subglobose, 27–64 × 10–25 mm, flowering branchlet 0–7 mm; floral bract dark brown or black, 2–4 mm, apex acute or rounded, abaxially hairy, hairs (white), straight. **Staminate flowers:** adaxial nectary oblong or square, 0.4–0.7 mm; filaments distinct, glabrous; anthers yellow, ellipsoid or shortly cylindrical, 0.7–1.1 mm. **Pistillate flowers:** adaxial nectary oblong, narrowly oblong, or square, 0.4–0.9 mm, shorter than stipe; stipe 2–2.5 mm; ovary pyriform, densely short-silky, beak gradually tapering to styles; ovules 12–14 per ovary; styles 0.3–0.6 mm; stigmas slenderly cylindrical, 0.4–0.55–0.6 mm. **Capsules** 6–12 mm. $2n = 38$.

Flowering mid Mar–mid Jun. Thickets and roadsides; 20–4600 m; introduced; Ont.; Ala., Ark., Conn., Del., D.C., Ill., Md., Mass., Mich., Nebr., N.Y., N.C., Ohio, Pa., Wash.; Europe.

Hybrids:

Salix ×*smithiana* Willdenow: *S. caprea* × *S. viminalis* is an introduced European hybrid commonly naturalized in eastern Canada (New Brunswick, Newfoundland, Nova Scotia, and Prince Edward Island), where it was introduced for coarse basketry. Only pistillate specimens are known and seed-set has not been noted. Its spread is apparently by cultivation and stem fragmentation. It is characterized by: shrubs or trees, 2–10 m; branches yellow-brown, brownish, or gray-brown, pubescent to glabrescent, ± brittle at base, peeled 3–5 year-old branches smooth or striate, striae few, 0–6 mm; branchlets not glaucous, moderately densely pubescent; largest medial leaf blades very narrowly elliptic, narrowly elliptic, or narrowly oblong, 2.8–4.9(–6.4) times as long as wide, margins entire or sinuate, adaxial surfaces pubescent; catkins flowering before leaves emerge; stipes 0.9–2 mm; ovaries pyriform to obclavate, moderately to very densely short-silky; styles 0.5–1.2 mm; stigmas slenderly cylindrical, lobes 0.6–1.2 mm; and capsules 5–7 mm.

Hybrids *Salix caprea* × *S. viminalis* and *S. cinerea* × *S. viminalis* are commonly recognized in North American floras. They are difficult to separate and their nomenclature is confusing. I am following G. Larsson (1995), who typified *S.* ×*smithiana* and cited *S.* ×*sericans* (formerly *S. cinerea* × *S. viminalis*) and *S. caprea* × *S. viminalis* as synonyms. These two *S. viminalis* hybrids are very similar. *Salix caprea* × *S. viminalis* has wood with striae 2–6 mm and leaves usually broadest in the middle; *S. cinerea* × *S. viminalis* has decorticated wood with striae 10+ mm and leaves usually broader toward the tip (B. Jonsell and T. Karlsson 2000+, vol. 1). Authentic *S. cinerea* × *S. viminalis* has not been seen from the flora area.

See 86. *Salix pellita* and 98. *S. viminalis* for more discussion of morphologies.

81. **Salix cinerea** Linnaeus, Sp. Pl. 2: 1021. 1753
  • Large gray or gray willow

Shrubs, 3–7 m. **Stems:** branches brownish, not glaucous, pilose, villous, or tomentose to glabrescent, (peeled wood with striae to 62 mm); branchlets yellow-brown, pilose, velvety, or densely villous. **Leaves:** stipules rudi-mentary or foliaceous on early ones, foliaceous on late ones, apex acute or rounded; petiole convex to flat adaxially, 4–15 mm, tomentose adaxially; largest medial blade elliptic, broadly elliptic, oblanceolate, or obovate, 65–105 × 22–52 mm, 2–3 times as long as wide, base convex or cuneate, margins slightly revolute, entire, crenate, or sinuate, (glands submarginal), apex acuminate or convex, abaxial surface glaucous, tomentose, hairs erect or spreading, curly, adaxial dull or slightly glossy, pubescent or tomentose; proximal blade margins entire; juvenile blade yellowish green, sparsely to densely tomentose abaxially, hairs white. **Catkins** flowering before leaves emerge; staminate stout or subglobose, 26–39 × 12–26 mm, flowering branchlet 0–5 mm; pistillate densely flowered, stout or subglobose, 27–54(–75 in fruit) × 4–15 mm, flowering branchlet 1–5(–10) mm; floral bract dark brown, black, or bicolor, 2–3 mm, apex acute or convex, abaxially hairy, hairs straight. **Staminate flowers:** adaxial nectary oblong or ovate, 0.5–1 mm; filaments distinct, glabrous or hairy basally; anthers yellow or purple turning yellow, ellipsoid or shortly cylindrical, 0.7–1 mm. **Pistillate flowers:** adaxial nectary oblong or square, 0.4–1 mm, shorter than stipe; stipe 1.2–2.7 mm; ovary pyriform, long-silky, beak slightly bulged below styles; ovules 12 per ovary; styles 0.2–0.5 mm; stigmas slenderly or broadly cylindrical, 0.3–0.6 mm. **Capsules** 5–5.6 mm. $2n = 76$.

Flowering mid Mar–late May. Stream shores, mesic woodlands, gravelly or sandy beaches, waste ground; 0–700 m; introduced; Ont.; Ala., Conn., D.C., Ga., Iowa, Ky., La., Md., Mass., Mich., Mo., N.J., N.Y., N.C., Ohio, Pa., R.I., S.C., S.Dak., Tenn., Utah, Va., W.Va., Wis.; Eurasia.

The Ohio occurrence of *Salix cinerea* is based on information from T. Cooperrider (pers. comm.).

Salix cinerea and *S. atrocinerea* are very closely related. Their occurrence in the flora area, as naturalized introductions, is not well understood, probably because they usually are introduced under the name *S. caprea*, and that species often is not treated in North American floristic literature (e.g., C. K. Schneider 1921; M. L. Fernald 1950). They probably are introductions of long-standing brought to the New World for their value as ornamentals and bee-plants. *Salix atrocinerea* was first documented in the southeastern United States (G. W. Argus 1986) after plants with ferruginous hairs and prominently striate wood were found in North Carolina; since that time, it has been found in other states and provinces. In the northeastern states, *S. atrocinerea* and *S. cinerea* are thought to be invasive species. The species do reproduce by seed and hundreds of seedlings were observed in a drained reservoir (A. Zinovjev, pers. comm.) and on sandy pond shores (T. Rawinski, pers. comm.), where they are thought to compete with native species.

The presence of long, prominent, striae on the peeled wood of 4–5 year old branches is commonly used in European literature (K. H. Rechinger 1993; A. K. Skvortsov 1999) to separate *Salix cinerea* and *S. atrocinerea* from *S. caprea* etc., in which the wood is smooth or with fewer, shorter striae. In the flora area, long striae also occur in *S. bebbiana*, *S. discolor*, and *S. humilis*, but usually they are not as long as or as prominent in *S. cinerea* and *S. atrocinerea*. Some floras (e.g., F. Martini and P. Paiero 1988) use the relative prominence of striae to separate *S. cinerea* and *S. atrocinerea*, but their separation remains difficult. The presence of ferruginous hairs on the leaves of *S. atrocinerea* is the best diagnostic characteristic, but they are not always present or easily observed. For a comparison of these species, see the key to species under subg. *Vetrix*. For further discussion of morphologies, see *Salix* ×*smithiana* Willdenow [p. 132] and 76. *S. discolor*.

82. Salix atrocinerea Brotero, Fl. Lusit. 1: 31. 1804 (as atro-cinerea) • Rusty willow [I]

Salix cinerea Linnaeus var. *atrocinerea* (Brotero) O. Bolòs & Vigo; *S. cinerea* subsp. *oleifolia* Macreight

Shrubs, 3–12 m. **Stems:** branches yellow-brown, gray-brown, or red-brown, not glaucous, pilose or villous to glabrescent, (peeled wood often with many striae, to 45 mm); branchlets gray-brown or yellow-brown, puberulent, pilose, villous, or velvety. **Leaves:** stipules (sometimes marcescent) foliaceous, apex acute; petiole convex to flat adaxially, 3–15 mm, tomentose, or velvety to glabrescent adaxially; largest medial blade narrowly to broadly elliptic, oblanceolate, obovate, broadly obovate, 29–105 × 14–52 mm, 1.8–4.3 times as long as wide, base cuneate or convex, margins slightly revolute, entire, crenate, or sinuate, (glands submarginal or epilaminal), apex acute, convex, or acuminate, abaxial surface glaucous, tomentose or coarsely villous to glabrescent, hairs (white, sometimes also ferruginous), erect, spreading, or appressed, wavy or curved, adaxial dull or slightly glossy, pubescent or pilose, (hairs white, sometimes also ferruginous); proximal blade margins entire; juvenile blade yellowish green or reddish, glabrous, tomentose, or long-silky abaxially, hairs white, sometimes also ferruginous. **Catkins** flowering before leaves emerge; staminate stout, 11–16 mm, flowering branchlet 0–5 mm; pistillate densely to loosely flowered, stout, 11–18 mm, flowering branchlet 0–3 mm; floral bract brown, black, or bicolor, 1–3 mm, apex acute, convex, or rounded, abaxially hairy, hairs straight or wavy. **Staminate flowers:** adaxial nectary narrowly oblong, oblong, or ovate, 0.4–0.9 mm; filaments distinct, glabrous or hairy on proximal ½ or basally; anthers yellow, shortly cylindrical or ovoid, 0.6–1 mm. **Pistillate flowers:** adaxial nectary oblong, square, or obovate, 0.4–0.9 mm, shorter than stipe; stipe 1.2–2.7 mm; ovary pyriform or obclavate, tomentose or short-silky, beak slightly bulged below styles; ovules 12 per ovary; styles 0.2–0.5 mm; stigmas broadly cylindrical, 0.23–0.43–0.63 mm. **Capsules** 5–7 mm. $2n = 76$.

Flowering mid Mar–mid May. Wooded wetlands, marshes, sandy beaches, mesic prairies, edges of birch-maple or oak woodlands; 0–700 m; introduced; Ont.; Conn., Maine, Mass., Mo., Nebr., N.J., N.Y., N.C., Pa., R.I., Wis.; Europe.

Hairs on abaxial leaf surfaces of *Salix atrocinerea* often are spreading to erect, and curly as in *S. humilis*. Usually, *S. atrocinerea* can be recognized by its closely spaced parallel tertiary venation. See 76. *S. discolor* and 81. *S. cinerea* for further comparative descriptions.

83. Salix aurita Linnaeus, Sp. Pl. 2: 1019. 1753 • Eared willow [I]

Shrubs, 1–3 m. **Stems:** branches brownish, not glaucous, pubescent to glabrescent, (peeled wood often with very dense striae, to 21 mm); branchlets red-brown or yellow-brown, (weakly glaucous), sparsely tomentose. **Leaves:** stipules foliaceous, apex acute or convex; petiole convex to flat adaxially, 2–9 mm, velvety adaxially; largest medial blade obovate, broadly obovate, or elliptic, 27–85 × 14–35 mm, 1.5–2.8 times as long as wide, base convex or cuneate, margins slightly revolute, entire, remotely or irregularly serrate, or crenate, (glands submarginal), apex acuminate or convex, abaxial surface glaucous, pubescent or pilose, hairs (white, sometimes also ferruginous) spreading or erect, wavy or crinkled, adaxial dull or slightly glossy, pubescent or pilose to glabrescent, veins more hairy, (hairs white, sometimes also ferruginous); proximal blade margins entire; juvenile blade reddish or yellowish green, densely tomentose to glabrescent abaxially, hairs white. **Catkins** flowering before leaves emerge; staminate subglobose or globose, 15.5–21.5 × 10–15 mm, flowering branchlet 0.5–4 mm; pistillate loosely to moderately densely flowered, 15–37 × 9–20, flowering branchlet 2.5–7 mm; floral bract brown, tawny, or bicolor, 1–2.2 mm, apex acute or tapering and rounded, abaxially hairy, hairs straight. **Staminate flowers:** adaxial nectary oblong or square, 0.3–0.7 mm; filaments distinct, glabrous or hairy on proximal ½ or basally; anthers purple turning yellow, ellipsoid or shortly cylindrical, 0.5–0.8 mm. **Pistillate flowers:** adaxial nectary oblong or square, 0.3–0.7 mm, shorter than stipe; stipe 1.4–2.6 mm; ovary pyriform, densely short-silky, hairs wavy or crinkled, beak sometimes slightly bulged below styles (long-beaked); ovules 10–12 per ovary; styles 0–0.3 mm; stigmas broadly cylindrical, 0.25–0.37–0.5 mm. **Capsules** 4–13 mm. $2n = 76, 38$.

Flowering early Apr–early Jun. Wet thickets, swamps; 10–300 m; introduced; Mass., Pa.; Europe.

2e5. **Salix** Linnaeus (subg. **Vetrix**) sect. **Fulvae** Barratt, Salices Amer., sect. VII, unnumb. p. 1840

Salix subsect. *Substriatae* (Goertz) A. K. Skvortsov

Shrubs or trees, 0.5–10 m. **Buds** *alba*-type. **Leaves:** stipules on late ones rudimentary or foliaceous; largest medial blade 1.7–4 times as long as wide, adaxial surface not glaucous. **Staminate flowers:** filaments glabrous or hairy. **Pistillate flowers:** adaxial nectary shorter than stipe; stipe 2–6 mm; ovary short-silky; stigmas cylindrical, 0.32–0.44–0.64 mm.

Species 7 (1 in the flora): North America, Asia.

84. **Salix bebbiana** Sargent, Gard. & Forest 8: 463. 1895 • Gray or Bebb's or long-beaked willow [F]

Salix rostrata Richardson in J. Franklin et al., Narr. Journey Polar Sea, 753. 1823, not Thuillier 1799; *S. bebbiana* var. *capreifolia* (Fernald) Fernald; *S. bebbiana* var. *depilis* Raup; *S. bebbiana* var. *luxurians* (Fernald) Fernald; *S. bebbiana* var. *perrostrata* (Rydberg) C. K. Schneider; *S. bebbiana* var. *projecta* (Fernald) C. K. Schneider; *S. depressa* Linnaeus subsp. *rostrata* (Andersson) Hiitonen

Stems: branches divaricate, sometimes ± brittle at base, yellow-brown to dark red-brown, not or weakly glaucous, pilose to glabrescent, peeled wood often with very dense striae, to 25 mm; branchlets yellow-green or red-brown, moderately to very densely villous to glabrescent. **Leaves:** stipules rudimentary or absent on early ones, apex acute, acuminate, or convex; petiole convex to flat adaxially, 2–5.5–13 mm, pubescent adaxially; largest medial blade narrowly oblong, narrowly elliptic, elliptic, oblanceolate, or obovate, 20–44–87 × 10–16–45 mm, base cuneate, convex, or rounded, margins flat, entire, crenate, or irregularly serrate, glands submarginal, apex acute, acuminate, or convex, abaxial surface glaucous, moderately densely pubescent or long-silky to glabrescent, hairs white or gray, wavy, adaxial finely impressed-reticulate, dull or slightly glossy, moderately densely pubescent, sparsely short-silky, or glabrescent, hairs white or gray; proximal blade margins entire, gland-dotted; juvenile blade yellowish green or reddish, pilose or sparsely to moderately densely tomentose or long-silky abaxially, hairs white. **Catkins:** staminate flowering just before leaves emerge, pistillate flowering as leaves emerge; staminate stout to globose, 10–42 × 7–16 mm, flowering branchlet 0.5–11 mm; pistillate loosely flowered, stout, slender, or subglobose, 16.5–85 × 9–32 mm, flowering branchlet 1–26 mm; floral bract tawny, 1.2–3.2 mm, apex rounded, abaxially hairy to glabrescent, hairs straight or wavy. **Staminate flowers:** adaxial nectary oblong or ovate, 0.3–0.8 mm; filaments distinct or connate less than ½ their lengths, glabrous or hairy on proximal ½; anthers yellow or purple turning yellow, ellipsoid or shortly cylindrical, 0.5–0.8 mm. **Pistillate flowers:** adaxial nectary oblong or square, 0.3–0.8 mm; ovary obclavate, beak slightly bulged below styles (long-beaked); ovules 6–16 per ovary; styles 0.1–0.4 mm; stigmas slenderly to broadly cylindrical. **Capsules** 5–9 mm. $2n = 38$.

Flowering early Apr–late Jun. Riparian and upland conifer forests, wet lowland thickets, *Picea mariana* treed bogs, stream margins, lakeshores, prairie margins, dry south-facing slopes, cienegas, seeps, disturbed areas; 0–3300 m; Alta., B.C., Man., N.B., Nfld. and Labr., N.W.T., N.S., Nunavut, Ont., P.E.I., Que., Sask., Yukon; Alaska, Ariz., Calif., Colo., Conn., Idaho, Ill., Ind., Iowa, Maine, Md., Mass., Mich., Minn., Mont., Nebr., Nev., N.H., N.J., N.Mex., N.Y., N.Dak., Ohio, Oreg., Pa., R.I., S.Dak., Utah, Vt., Wash., Wis., Wyo.; Asia.

Salix bebbiana occurs in Nunavut on Akimiski Island in James Bay.

Hybrids:

Salix bebbiana forms natural hybrids with *S. candida*, *S. geyeriana*, *S. humilis,* and *S. petiolaris*. Reports of hybrids with *S. discolor* (C. K. Schneider 1921; M. L. Fernald 1950) are not based on convincing specimens, and synthetic hybrids could not be made (G. W. Argus 1974). Reports of hybrids with *S. eriocephala* and *S. myricoides* (Fernald) are unverified. Controlled pollinations with *S. eriocephala* and *S. interior* had low seed viability (A. Mosseler 1990).

Salix bebbiana × *S. candida* (*S.* ×*cryptodonta* Fernald, as species) is intermediate between parental species. It resembles *S. candida* in having juvenile leaves densely woolly, mature leaves sparsely to moderately woolly abaxially, margins strongly revolute to crenulate, and ovaries woolly; and *S. bebbiana* in having stipes 2.8–3 mm and capsules long-beaked, 8–9 mm. The hybrid commonly occurs in Newfoundland.

Salix bebbiana × *S. geyeriana*: A plant with the pistillate catkins and flowers of *S. bebbiana* and the narrow, entire or slightly serrulate leaves with white and ferruginous hairs of *S. geyeriana* was collected by R. D. Dorn in a mixed population in Montana (Beaverhead County).

Salix bebbiana × *S. humilis*: Reported by C. K. Schneider (1921) and M. L. Fernald (1950) and successfully synthesized by G. W. Argus (1974, 1986).

Salix bebbiana × *S. petiolaris* is known from Ontario, based on an infertile pistillate specimen, and from Alberta and Saskatchewan, where it is relatively uncommon. It was successfully synthesized (G. W. Argus 1974, 1986) and controlled pollinations showed high seed viability (A. Mosseler 1990).

2e6. SALIX Linnaeus (subg. VETRIX) sect. PHYLICIFOLIAE (Fries) Andersson in A. P. de Candolle and A. L. P. P. de Candolle, Prodr. 16(2): 240. 1868

Salix tribe *Phylicifoliae* Fries in C. F. Hornschuch, Syll. Pl. Nov. 2: 36. 1828; *Salix* subsect. *Bicolores* A. K. Skvortsov

Shrubs or trees, 0.1–9 m. **Buds** *arctica-* or *caprea-*type, or intermediate. **Leaves:** stipules on late ones absent, rudimentary, or foliaceous; largest medial blade (sometimes amphistomatous), 1.5–11 times as long as wide, adaxial surface not glaucous. **Staminate flowers:** filaments glabrous or hairy basally. **Pistillate flowers:** adaxial nectary shorter than, equal to, or longer than stipe; stipe 0.2–2 mm; ovary sparsely to very densely silky; stigmas with flat, non-papillate abaxial surface, or stigmas cylindrical, 0.3–1.1 mm.

Species 11 (5 in the flora): North America, Eurasia.

85. **Salix drummondiana** Barratt ex Hooker, Fl. Bor.-Amer. 2: 144. 1838 • Drummond's willow [E]

Salix drummondiana var. *bella* (Piper) C. R. Ball; *S. drummondiana* var. *subcoerulea* (Piper) C. R. Ball; *S. subcoerulea* Piper

Shrubs, 1–5 m, (sometimes forming clones by stem fragmentation). **Stems:** branches (highly to ± brittle at base), yellow-brown or red-brown, usually strongly glaucous, (slightly glossy), glabrous or glabrescent; branchlets red-brown or mottled yellow-brown, (strongly to not glaucous), glabrous, puberulent, pilose, or velvety, (buds *caprea*-type or intermediate). **Leaves:** stipules usually rudimentary or absent, or foliaceous, then small and ovate or slender, apex acute; petiole convex to flat, or shallowly grooved adaxially, 2–12 mm, villous or velvety adaxially; largest medial blade lorate, narrowly elliptic, elliptic, or oblanceolate, 40–85 × 9–26 mm, 3–6.2 times as long as wide, base cuneate or convex, margins slightly revolute, entire, or shallowly crenate to sinuate, apex acute, acuminate, or convex, abaxial surface glaucous (obscured by hairs), densely short- to long-silky, hairs (white, sometimes also ferruginous), straight or wavy, adaxial slightly glossy or dull, sparsely short-silky to glabrescent, (hairs white, sometimes also ferruginous); proximal blade margins entire; juvenile blade green, very densely short-silky abaxially (sparsely so adaxially), hairs white, sometimes also ferruginous. **Catkins** flowering before leaves emerge; staminate stout, 19–40 × 8–20 mm, flowering branchlet 0 mm; pistillate densely flowered, slender or stout, 22–87(–105 in fruit) × 8–18 mm, flowering branchlet 0–3(–6) mm; floral bract brown or black, 1.2–2.8 mm, apex acute or rounded, abaxially hairy, hairs straight. **Staminate flowers:** adaxial nectary oblong, 0.3–0.6 mm; filaments distinct, glabrous; anthers purple turning yellow, ellipsoid to shortly cylindrical, 0.4–0.6 mm. **Pistillate flowers:** adaxial nectary narrowly oblong, oblong, or ovate, 0.4–1 mm, shorter to longer than stipe; stipe 0.3–2 mm; ovary pyriform, short-silky, beak gradually tapering to styles; ovules 6–17 per ovary; styles 0.5–1.5 mm; stigmas flat, abaxially non-papillate with pointed tip, or slenderly to broadly cylindrical, 0.32–0.43–1 mm. **Capsules** 2.5–6 mm. $2n$ = 38, 57, 76.

Flowering late Apr–early Jul. Subalpine and montane forests and thickets, open spruce forests, streamsides, gravelly floodplains; 200–3400 m; Alta., B.C., N.W.T., Sask., Yukon; Calif., Colo., Idaho, Mont., Nev., N.Mex., Oreg., Utah, Wash., Wyo.

Western American *Salix drummondiana* and eastern *S. pellita* have the same close relationship as do *S. scouleriana* and *S. humilis*. In Saskatchewan, where their ranges overlap, separation is difficult. The characters shown in the comparison below often overlap but they will usually serve to separate the two. A useful diagnostic vegetative character is the frequent presence in *S. pellita* of epilaminal glands, which are borne on adaxial leaf surfaces well in from margins.

Salix drummondiana is distinguished from *S. pellita* by having stipules on early leaves absent or rudimentary, sometimes foliaceous, staminate nectaries 0.3–0.6 mm, largest medial blade margins with submarginal glands, surfaces glabrescent, short-silky, dull or slightly glossy adaxially, petioles villous or velvety adaxially, and juvenile blades short-silky; *S. pellita* has stipules on early leaves absent or rudimentary, staminate nectaries 0.6–1 mm, largest medial blade margins with epilaminal or submarginal glands, surfaces glabrous, glabrescent, pubescent, or villous, and slightly or highly glossy adaxially, petioles glabrous or pubescent adaxially, and juvenile blades glabrous, glabrescent, pubescent, or tomentose.

Salix drummondiana is distinguished from the similar, but unrelated, *S. sitchensis* by having branches often strongly glaucous, branchlets sparsely hairy, largest medial blades lorate, narrowly elliptic, elliptic, or oblanceolate, usually narrower, 3–6.2 times as long as wide, margins slightly revolute, and surfaces with white hairs, sometimes also ferruginous; *S. sitchensis* has branches not glaucous or weakly so, branchlets usually moderately to very densely hairy, largest medial blades elliptic, narrowly oblanceolate, oblanceolate, or obovate, usually slightly broader, 2.1–3.1–4 times as long as wide, margins strongly revolute, and surfaces with white hairs.

Vegetative specimens of *Salix drummondiana* are distinguished from *S. geyeriana* by having largest medial blade margins revolute, surfaces usually densely short-silky adaxially, and midribs glabrous; *S. geyeriana* has largest medial blade margins flat, surfaces usually moderately densely long-silky adaxially, and midribs silky or glabrous.

Salix drummondiana and *S. lemmonii* can be separated on the basis of variable characters including: branch glaucousness, leaf size, blade hair density and color, catkin size and shape, anther length, petiole length, and chromosome number. Hybridization is rare but may occur in Lassen and Sierra counties, California.

The diploid and tetraploid chromosome numbers for *Salix drummondiana* have been reported from Wyoming (R. D. Dorn 1975b) and the triploid count from a disjunct population in the Lake Athabasca sand dunes, northern Saskatchewan (Y. Suda and G. W. Argus 1968).

Hybrids:

Salix drummondiana forms natural hybrids with *S. alaxensis* var. *alaxensis*, *S. irrorata*, and *S. planifolia*.

86. **Salix pellita** (Andersson) Bebb, Bot. Gaz. 16: 106. 1891 • Satiny willow [E] [F]

Salix chlorophylla Andersson [unranked] *pellita* Andersson, Monogr. Salicum, 139, plate 7, fig. 72*. 1867

Shrubs, 0.5–6 m, (sometimes forming clones by stem fragmentation). **Stems:** branches (highly to ± brittle at base), red-brown, violet or yellow-brown, usually strongly glaucous, glabrescent; branchlets yellow-brown or red-brown, (usually strongly glaucous), glabrous or densely to sparsely pubescent or tomentose, (buds *caprea*-type). **Leaves:** stipules absent or rudimentary on early ones, rudimentary, absent, or foliaceous on late ones; petiole convex to flat, or shallowly grooved adaxially, 3–6.3–14 mm, (sometimes dark spherical glands distally), glabrous or pubescent adaxially; largest medial blade linear, lorate, or narrowly elliptic 40–79–123 × 6–12–20 mm, (2.3–)4.2–7.2–11.3 times as long as wide, base convex or cuneate, margins strongly or slightly revolute, entire, sinuate or sometimes crenulate, (glands submarginal or epilaminal), apex acuminate to acute, abaxial surface glaucous (sometimes obscured by hairs), densely villous, short-silky, woolly, or tomentose to glabrescent, hairs (white, sometimes also ferruginous), straight or wavy, adaxial slightly to highly glossy, glabrous, sparsely villous or pubescent (hairs white, sometimes also ferruginous); proximal blade

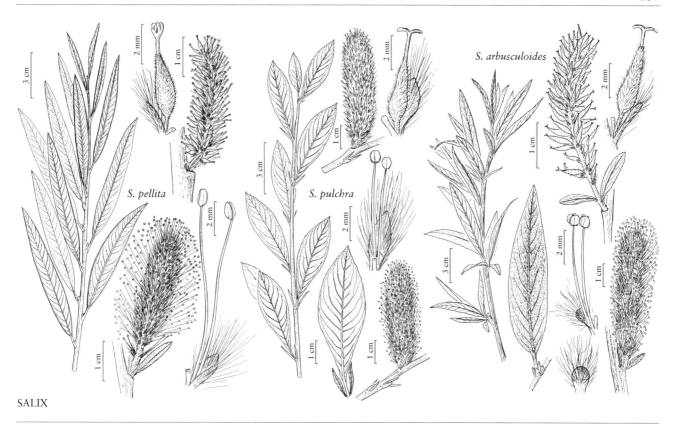

margins entire; juvenile blade reddish or yellowish green, densely tomentose, short-silky, pubescent, or glabrous abaxially, hairs white, sometimes also ferruginous. **Catkins** flowering before leaves emerge; staminate stout, 20–39 × 7–20 mm, flowering branchlet 0–2 mm; pistillate densely flowered, slender, stout, or subglobose, 19–65(–80 in fruit) × 7–17 mm, flowering branchlet 0–7 mm; floral bract tawny, brown, or black, 1–2.6 mm, apex acute, convex, or rounded, abaxially hairy, hairs straight. **Staminate flowers:** adaxial nectary oblong or narrowly oblong, 0.6–1 mm; filaments distinct, glabrous or hairy basally; anthers purple turning yellow, ellipsoid, 0.4–0.6 mm. **Pistillate flowers:** adaxial nectary oblong to depressed-ovate, 0.3–1 mm, shorter than or equal to stipe; stipe 0.5–1.1 mm; ovary pyriform, short-silky, beak sometimes slightly bulged below styles; ovules 10–18 per ovary; styles 0.6–1.5 mm; stigmas slenderly cylindrical, 0.4–0.55–0.76 mm. **Capsules** 3.5–6.5 mm. $2n = 38$.

Flowering late Apr–late Jun. Sandy or gravelly floodplains, stream and lake margins, marshes, fens, coastal dunes, metamorphic or calcareous substrates; 0–800 m; St. Pierre and Miquelon; Man., N.B., Nfld. and Labr., N.S., Ont., Que., Sask.; Maine, Mich., Minn., N.H., Vt., Wis.

Salix pellita sometimes has foliaceous stipules on late leaves. When present, they are correlated with hairy rather than glabrous branchlets. There is no clear evidence of hybridity in such specimens but further study is indicated.

Vegetative specimens of *Salix pellita* can be difficult to separate from *S. viminalis* and *S.* ×*smithiana* in eastern Canada, where the latter were introduced for coarse basketry and have become naturalized. The introduced species usually are tall shrubs to small trees, with branches usually flexible at base and not strongly glaucous, and their leaves tend to be broader. In contrast, *S. pellita* is a mid shrub rarely to 6 m but never tree-like, its branches usually are highly brittle at the base and often strongly glaucous, and its leaves tend to be narrower.

See 85. *Salix drummondiana* for further comparative descriptions.

Hybrids:

Salix pellita forms natural hybrids with *S. alaxensis* var. *alaxensis*, *S. discolor*, *S. pedicellaris*, *S. petiolaris*, and *S. planifolia*.

Salix pellita × *S. petiolaris*: Leaves of this hybrid are distinctly serrate and flat, as in *S. petiolaris*, but branchlets are glaucous and bud gradation is *caprea*-type as in *S. pellita*. It is uncommon in eastern Saskatchewan.

Salix pellita × *S. planifolia*: This cross is suspected to occur in Labrador, Newfoundland, Quebec, and Ontario. Both parents are tetraploids and flower at the same time. The name *S. pellita* forma *psila* may apply to this hybrid.

87. Salix pulchra Chamisso, Linnaea 6: 543. 1831

• Diamond-leaf willow F

Salix divaricata Pallas subsp. pulchra (Chamisso) Voroschilov; S. phylicifolia Linnaeus subsp. pulchra (Chamisso) Hultén; S. phylicifolia var. subglauca (Andersson) B. Boivin; S. planifolia Pursh subsp. pulchra (Chamisso) Argus; S. planifolia var. yukonensis (C. K. Schneider) Argus; S. pulchra var. yukonensis C. K. Schneider

Shrubs, 0.1–3(–4.5) m. **Stems:** branches yellow-brown or red-brown, not or weakly glaucous, (slightly or highly glossy), villous in patches to glabrescent; branchlets yellow-green or brownish, glabrous, puberulent, or densely villous, (buds *caprea*- or *arctica*-type). **Leaves:** stipules (usually marcescent), foliaceous, (linear or lanceolate, 2–9.8–23 mm), apex acuminate; petiole convex to flat adaxially, 2.8–10(–15) mm, glabrous or puberulent adaxially; largest medial blade narrowly elliptic, elliptic, or obovate, 22–75 × 8–26 mm, 1.7–4.7 times as long as wide, base cuneate, margins flat to slightly revolute, entire, crenate, or serrulate, apex acuminate, acute, or convex, abaxial surface glaucous, glabrous or pubescent, midrib pilose, hairs (white, sometimes also ferruginous), straight, adaxial highly to slightly glossy, midrib glabrous or sparsely pubescent; proximal blade margins entire; juvenile blade green, glabrous or pubescent abaxially, hairs white, sometimes also ferruginous. **Catkins** flowering before leaves emerge; staminate stout or subglobose, 21–53 × 12–19 mm, flowering branchlet 0–3 mm; pistillate densely flowered, slender to stout, 27–82 × 8–20 mm, flowering branchlet 0–3(–8) mm; floral bract brown or black, 1.6–2.8 mm, apex acute or rounded, abaxially hairy distally, hairs straight. **Staminate flowers:** adaxial nectary narrowly oblong to oblong, 0.6–1 mm; filaments distinct or connate less than ½ their lengths, glabrous; anthers purple turning yellow, ellipsoid to shortly cylindrical, 0.4–0.8 mm. **Pistillate flowers:** adaxial nectary narrowly oblong to oblong, 0.4–1.6 mm, longer than stipe; stipe 0.2–0.8 mm; ovary pyriform, long-silky, beak slightly bulged below styles; ovules 12–16 per ovary; styles 1–1.8 mm; stigmas slenderly cylindrical, 0.44–0.63–0.96 mm. **Capsules** 3.2–5.6 mm. $2n = 76$.

Flowering mid Apr–late Jul. Arctic boreal and subalpine thickets, stream and lake margins, tundra, black spruce-lichen woodlands, open white spruce-dwarf birch woodlands; 0–2000 m; B.C., N.W.T., Nunavut, Yukon; Alaska; Asia (Chukotka, Kamchatka, Lena-Kolyma, Okhotia, Russia [Anadyr], Siberia).

Plants with branchlets densely villous with white, gray, or, sometimes, ferruginous hairs have been named var. *yukonensis*. This hairy variant occurs scattered throughout populations in both North America and Russia and does not seem to warrant taxonomic recognition. The closely related *Salix planifolia* shows similar variation.

Hybrids:

Salix pulchra forms natural hybrids with *S. planifolia* and *S. scouleriana*.

Salix pulchra × *S. scouleriana* is known from Alaska. It resembles *S. pulchra* in its long, linear stipules, general leaf shape, and relatively long styles; and *S. scouleriana* in its juvenile leaves with abundant ferruginous hairs, petioles densely villous to velvety, and relatively short stigmas.

88. Salix planifolia Pursh, Fl. Amer. Sept. 2: 611. 1813

• Tea-leaf willow E

Salix monica Bebb; S. phylicifolia Linnaeus var. monica (Bebb) Jepson; S. phylicifolia subsp. planifolia (Pursh) Hiitonen; S. planifolia var. monica (Bebb) C. K. Schneider

Shrubs or trees, 0.1–9 m, (sometimes forming clones by layering). **Stems** (sometimes decumbent); branches yellow-brown, red-brown, or violet, not to strongly glaucous, glabrous or pubescent; branchlets yellow-brown, red-brown, or violet, glabrous, pilose, pubescent, moderately densely villous, or short-silky, (buds *caprea*-type). **Leaves:** stipules (sometimes marcescent), rudimentary or foliaceous (small and usually brownish) on early ones, rudimentary or foliaceous on late ones, (narrowly ovate to oblong, 1–2.5(–4.5) mm), apex acute; petiole shallowly grooved adaxially, 2–9(–13) mm, glabrous, pilose, or short-silky adaxially; largest medial blade (sometimes hemiamphistomatous), narrowly oblong, narrowly elliptic, elliptic, or oblanceolate, 20–36–65 × 5–13–23 mm, 1.7–2.8–4.7 times as long as wide, base cuneate or convex, margins sometimes slightly revolute basally, entire, or, sometimes, crenulate or serrulate, apex acute, acuminate, or convex, abaxial surface glaucous, glabrous or sparsely silky, hairs (white, sometimes also ferruginous) straight or wavy, adaxial highly glossy, glabrous or sparsely short-silky; proximal blade margins entire; juvenile blade reddish or yellowish green, glabrous, puberulent, pubescent, or densely long-silky abaxially, hairs white, sometimes also ferruginous. **Catkins** flowering before leaves emerge; staminate stout, subglobose, or globose, 12–41 × 10–20 mm, flowering branchlet 0–4 mm; pistillate densely flowered, slender, or stout to globose, 15–67 (–70 in fruit) × 8–18 mm, flowering branchlet 0–6 mm; floral bract dark brown or black, 1–3.2 mm, apex acute, convex, or rounded, sometimes 2-fid, abaxially hairy, hairs straight. **Staminate flowers:** adaxial nectary narrowly oblong or oblong, 0.4–1.1 mm; filaments distinct, glabrous or sparsely hairy basally; anthers purple turning yellow, shortly cylindrical, 0.5–0.7 mm. **Pistillate flowers:** adaxial nectary oblong, square, or ovate, 0.4–1.3 mm, shorter to longer than stipe; stipe 0.3–0.8 mm; ovary pyriform, short- to long-silky, sometimes slightly bulged below styles; ovules 11–16 per ovary; styles 0.5–2 mm; stigmas slenderly to broadly cylindrical, 0.36–0.52–1.1 mm. **Capsules** (2.5–)5.5–6 mm. $2n$ = 76, 57.

Flowering early May–late Jun. Arctic, alpine, subalpine, and boreal meadows and riverbanks, streams, seeps, snowflush areas, treed bogs, fens, sandy-loam, rocky igneous and limestone substrates; 100–4000 m; St. Pierre and Miquelon; Alta., B.C., Man., Nfld. and Labr., N.W.T., Nunavut, Ont., Que., Sask., Yukon; Alaska, Ariz., Calif., Colo., Idaho, Maine, Mich., Minn., Mont., Nev., N.H., N.Mex., Oreg., S.Dak., Utah, Vt., Wash., Wis., Wyo.

Variety *monica* applies to the diminutive alpine form that sometimes is recognized in the southern Rocky Mountains (S. J. Brunsfeld and F. D. Johnson 1985); it occurs at higher elevations (2200–4000 m) and is characterized by low growth form (0.14–1 m) and smaller, slightly broader leaves. Although it can be distinctive, it is morphologically confluent with the typical species. B. G. O. Floderus (1939) may be correct in characterizing it as an alpine ecotype.

Salix planifolia and *S. pulchra* are closely related. Their ranges overlap in northwestern Canada, from northern British Columbia across the southern quarter of the Yukon and northeastward into the Great Bear Lake area. Specimens identified as *S. pulchra* occur as far northeastward as Coppermine and northeast of Bathurst Inlet; *S. planifolia* has been recognized in the Mackenzie Delta and Eskimo Lake regions, Northwest Territories. Outlying records should be treated with caution because identification of individual specimens out of context may not be definitive. G. W. Argus (1969, 1973) treated these taxa as subspecies based on their intergradation in northwestern British Columbia, their tetraploid chromosome number, and their similar leaf flavonoid chromatographic patterns, but this taxonomy needs reconsideration.

The primary differences between the two species are stipule size, shape, and persistence and the pubescence on juvenile leaves. Stipules of *Salix planifolia* are oblong to narrowly elliptic or obovate, 0.8–3 mm (or –4.5 mm at Back River, Northwest Territories), distinctly shorter than petioles, and rarely marcescent for more than one year; stipules of *S. pulchra* are linear to narrowly oblong, 3–32 mm, usually longer than petioles, and usually marcescent for two or more years. Juvenile leaves of *S. planifolia* are usually more densely hairy, but vary from glabrescent to sparsely or very densely pubescent or long-silky, whereas juvenile leaves of *S. pulchra* are usually glabrous or, sometimes, sparsely hairy. The occurrence of rhombic mature leaf blades in *S. pulchra* sometimes is distinctive, but overlap in leaf shape between the two taxa is very great.

The area of geographic overlap in Yukon and western Northwest Territories is large, but evidence suggests that there the two species may be separated by elevation. In the vicinity of Whitehorse, Yukon, *Salix pulchra* occurs at higher elevations (1400–1900 m) than *S. planifolia* (ca. 1000 m); no mixed populations were seen. In Nahanni National Park, Northwest Territories, where *S. planifolia* is more common than *S. pulchra*, the latter occurs only in alpine and subalpine habitats (1200–1400 m). Evidence from both localities indicates an elevational separation of the two taxa. Within the area of overlap there is little evidence of intergradation except that *S. planifolia* has stipules that tend to be more marcescent (40% are marcescent) and sometimes longer (2–3.5 mm) than is usual outside the area of overlap. Nevertheless, specimens from the area of overlap can be easily assigned to one taxon or the other with only a few apparent intermediates. The problem in recognizing intermediacy is that there are only a few, variable characters that separate the two. In contrast, in 1973, G. W. Argus described evidence of hybridization and introgression along the Haines Road in northwestern British Columbia. This was based on variation in stipule size, presence, and persistence in what appeared to be a hybrid swarm. Further data are needed to answer questions about actual hybridization. Are the species separated by habitat or elevation, and are there reproductive barriers? Answers could be gained by population studies and controlled hybridization. Until that is done it is best to treat these taxa as species.

See 76. *Salix discolor* and 95a. *S. alaxensis* var. *alaxensis* for comparative descriptions.

Hybrids:

Salix planifolia forms natural hybrids with *S. alaxensis* var. *alaxensis*, *S. argyrocarpa*, *S. brachycarpa* var. *brachycarpa*, *S. candida*, *S. drummondiana*, *S. humilis*, *S. pellita*, *S. pulchra*, and *S. scouleriana*. Hybrids with *S. glauca* var. *cordifolia* have been reported (C. K. Schneider 1921) but no convincing specimens have been seen.

89. Salix tyrrellii Raup, J. Arnold Arbor. 17: 231, plate 192. 1936 • Tyrrell's willow [E]

Salix planifolia Pursh subsp. *tyrrellii* (Raup) Argus

Shrubs, 0.6–3.5 m, (sometimes forming clones by layering). Stems: branches and branchlets red-brown, not to strongly glaucous, glabrous, (buds *caprea*-type or intermediate). Leaves: stipules rudimentary or foliaceous (often brownish) on early ones, foliaceous or rudimentary on late ones, (ca. 4 mm), apex acute; petiole convex to flat, or shallowly grooved adaxially, 1–3.4–16 mm, glabrous, pilose, or short-silky adaxially; largest medial blade (amphistomatous), narrowly elliptic, elliptic, oblanceolate, or obovate, 15–29–65 × 3.5–8.8–18 mm, 2.3–3.3–4.4 times as long as wide, base cuneate or convex, margins strongly to slightly revolute, entire, or very shallowly serrulate or shallowly serrulate-crenulate, apex acute, acuminate, or convex, abaxial surface glaucous, sparsely long-silky to glabrescent, hairs (ferruginous), straight, adaxial highly glossy, glabrous or sparsely short-silky (hairs white, sometimes also ferruginous); proximal blade margins entire; juvenile blade yellowish green or sometimes reddish, sparsely long-silky abaxially, hairs white and ferruginous. Catkins flowering before leaves emerge; staminate stout, 14–35 × 12–16 mm, flowering branchlet 0 mm; pistillate densely flowered, stout, 17–51 × 10–13–22 mm, flowering branchlet 0–4 mm; floral bract brown, black, or bicolor, 1–3.7 mm, apex acute to acuminate or rounded, abaxially hairy, hairs long, straight. Staminate flowers: adaxial nectary oblong, 0.8–1.1 mm; filaments distinct, glabrous or hairy basally; anthers purple turning yellow, ellipsoid, 0.4–0.7 mm. Pistillate flowers: adaxial nectary oblong or flask-shaped, 0.6–1.1 mm, equal to or shorter than stipe; stipe 0.2–1 mm (–1.4 mm in cultivation); ovary pyriform, long-silky, beak gradually tapering to styles; ovules 12–16 per ovary; styles 0.6–1.2 mm; stigmas flat, abaxially non-papillate with rounded tip, or slenderly cylindrical, 0.44–0.55–0.75 mm. Capsules 3.6–5 mm.

Flowering mid Jun–mid Jul. Active sand dunes, shrubby tundra; 200–600 m; Alta., N.W.T., Nunavut, Sask.

Salix tyrrellii, first described from Lake Athabasca sand dunes in northwestern Saskatchewan and adjacent Alberta, is characterized by slender, amphistomatous leaves and relatively long, slender branchlets (G. W. Argus and J. W. Steele 1979). It probably evolved, in the past 10,000 years, from the widespread boreal *S. planifolia*. Originally, it was thought to be endemic to Lake Athabasca but recent collections from Nunavut and the Northwest Territories suggest that it may have a much wider range. Plants from Nunavut and the Northwest Territories do not have the long, slender branchlets found in sand dune populations, and identification is primarily based on the presence of amphistomatous leaves. This character, however, may be unreliable. For example, *S. planifolia* from 2610 m in the Glass Mountains, California, have amphistomatous leaves, suggesting that this character may be under environmental influence. The appropriate rank for this taxon remains uncertain.

2e7. SALIX Linnaeus (subg. VETRIX) sect. ARBUSCELLA Seringe, Exempl. Rév. Salix, 15. 1824

Salix subsect. *Arbusculae* (Hayek) Dorn

Shrubs or trees, 1–6 m. **Buds** *alba*-type. **Leaves**: stipules on late ones rudimentary or foliaceous; largest medial blade (hypostomatous to amphistomatous), 3–6.5 times as long as wide, adaxial surface not glaucous. **Staminate flowers**: filaments glabrous. **Pistillate flowers**: adaxial nectary equal to or longer than stipe; stipe 0.6–0.9 mm; ovary very densely short-silky; stigmas slenderly to broadly cylindrical, 0.16–0.29–0.44 mm.

Species 14 (1 in the flora): North America, Eurasia.

90. Salix arbusculoides Andersson, Monogr. Salicum, 147, plate 8, fig. 81. 1867 • Little-tree willow E F

Stems: branches gray-brown to red-brown, not glaucous, glabrous; branchlets red-brown, glabrous or puberulent. **Leaves:** stipules rudimentary on early ones, apex acute; petiole shallowly grooved adaxially, 3–11 mm, puberulent to glabrescent adaxially; largest medial blade very narrowly elliptic to elliptic, 38–78 × 7–18 mm, base cuneate or convex, margins slightly revolute, serrulate, apex acuminate, acute, or convex, abaxial surface glaucous (sometimes obscured by hairs), sparsely to densely long-silky, hairs (white, sometimes also ferruginous), straight, adaxial highly or slightly glossy, glabrous; proximal blade margins entire; juvenile blade yellowish green, very densely long-silky abaxially, hairs white, sometimes also ferruginous. **Catkins** flowering as or just before leaves emerge; staminate stout or slender, 17–43 × 5–10 mm, flowering branchlet 0–2.5 mm; pistillate densely to loosely flowered, stout to slender, 20–46 × 6–15 mm, flowering branchlet 0–6 mm; floral bract tawny or brown, 0.8–1.2 mm, apex convex to rounded, abaxially hairy, hairs straight or wavy. **Staminate flowers:** adaxial nectary oblong, 0.6–0.9 mm; filaments distinct; anthers purple turning yellow, ellipsoid to globose, 0.3–0.6 mm. **Pistillate flowers:** adaxial nectary oblong or ovate, 0.6–1 mm; ovary pyriform, beak gradually tapering to styles; ovules 16–18 per ovary; styles 0.3–0.5 mm. **Capsules** 4–6 mm. $2n = 38$.

Flowering mid May–early Jul. Stream margins, lakeshores, openings in white spruce forests, treed bogs, sedge fens, edges of alpine and arctic tundra; 0–2000 m; Alta., B.C., Man., N.W.T., Nunavut, Ont., Que., Sask., Yukon; Alaska.

Glands on leaf teeth of *Salix arbusculoides* are sometimes covered with fine crystals of sulphur, calcium, potassium, and silicon (R. Cooper, pers. comm.), indicating that they can function as hydathodes as well as resin glands.

2e8. Salix Linnaeus (subg. Vetrix) sect. Candidae C. K. Schneider, Ill. Handb. Laubholzk. 1: 46. 1904

Shrubs, 0.3–2.5 m. **Buds** *alba*-type. **Leaves:** stipules on late ones foliaceous; largest medial blade 3.3–12 times as long as wide, adaxial surface not glaucous. **Staminate flowers:** filaments glabrous. **Pistillate flowers:** adaxial nectary shorter to longer than stipe; stipe 0.1–1.2 mm; ovary very densely woolly or tomentose; stigmas with flat, non-papillate abaxial surface and rounded tip, 0.4–0.45–0.52 mm.

Species 2 (1 in the flora): North America, Asia.

Salix krylovii E. L. Wolf is known from Asia.

91. Salix candida Flüggé ex Willdenow, Sp. Pl. 4: 708. 1806 • Sage or sage-leaf willow E F

Salix candida var. *denudata* Andersson

Plants often forming clones by layering. **Stems:** branches dark gray-brown to yellow-brown, not glaucous, woolly in patches or floccose to glabrescent; branchlets yellow-brown to red-brown or gray-brown, densely (white) woolly or tomentose, sometimes floccose. **Leaves:** stipules rudimentary or foliaceous on early ones, late ones 2–3.6 mm, apex acute; petiole shallowly to deeply grooved adaxially, 3–10 mm, tomentose or densely (white) woolly adaxially (obscured by hairs); largest medial blade lorate, narrowly elliptic or oblanceolate, 47–103 × 5–20 mm, base convex or cuneate, margins strongly to slightly revolute, entire, or sinuate, apex acute or convex, abaxial surface glaucous (generally obscured by hairs), very densely to sparsely tomentose-woolly (cobwebby in age), hairs dull white, crinkled, adaxial dull or slightly glossy, moderately densely to sparsely tomentose, floccose, hairs dull white; proximal blade margins entire; juvenile blade yellowish green, very densely tomentose abaxially, hairs white. **Catkins** flowering as leaves emerge; staminate stout or subglobose, 17–39 × 8–16 mm, flowering branchlet 0.5–7 mm; pistillate densely to moderately densely flowered, stout or slender, 20–66 × 9–18 mm, flowering branchlet 1–24 mm; floral bract tawny or brown, 1.2–1.8 mm, apex rounded or acute, abaxially hairy, hairs straight to wavy. **Staminate flowers:** adaxial

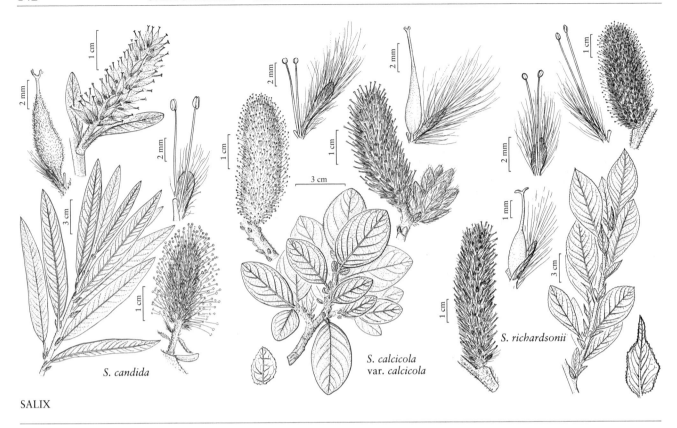

nectary narrowly oblong to oblong, 0.6–1 mm; filaments distinct or connate less than ½ their lengths; anthers purple turning yellow, ellipsoid, long-cylindrical, or globose, 0.4–0.6 mm. **Pistillate flowers:** adaxial nectary oblong, 0.4–1 mm; ovary pyriform, beak sometimes slightly bulged below styles; ovules 12–18 per ovary; styles 0.3–1.9 mm. **Capsules** 4–6 mm. $2n = 38$.

Flowering mid Apr–early Jul. Floodplains, marl bogs, fens, and meadows, calcareous substrates; 10–2800 m; St. Pierre and Miquelon; Alta., B.C., Man., N.B., Nfld. and Labr., N.W.T., N.S., Nunavut, Ont., P.E.I., Que., Sask., Yukon; Alaska, Colo., Conn., Idaho, Ill., Ind., Iowa, Maine, Mass., Mich., Minn., Mont., N.H., N.J., N.Y., N.Dak., Ohio, Pa., S.Dak., Vt., Wash., Wis., Wyo.

Occurrence of *Salix candida* in Nunavut is on Akimiski Island in James Bay.

Salix candida is geographically wide-ranging but limited to calcareous habitats and, for that reason, it is quite local or even rare in some parts of its range.

Hybrids:

Salix candida forms natural hybrids with *S. bebbiana*, *S. brachycarpa* var. *brachycarpa*, *S. calcicola*, *S. eriocephala*, *S. famelica*, *S. myrtillifolia*, *S. petiolaris*, and *S. planifolia*. Hybrids with *S. discolor*, *S. petiolaris*, and *S. sericea* have been reported (the latter also by C. K. Schneider 1921; M. L. Fernald 1950) but no convincing specimens have been seen. *Salix candida* hybrids are recognized from their woolly indumentum that often is conspicuous on leaves, stems, and ovaries. In hybrids, these characters, especially woolly patches on ovaries, stand out as discordant variation.

Salix candida × *S. eriocephala* (*S.* ×*rubella* Bebb ex C. K. Schneider) was described by W. W. Rowlee and K. M. Wiegand (1896) as *S. candida* × *S. cordata*. In addition to woolly patches on the ovaries, they noted that buds of the hybrids usually are shorter, more divergent, and blunter than those in *S. eriocephala*, and are glabrous or hairy. Known from New York and Newfoundland; it should be expected throughout the sympatric range of the parental species.

Salix candida × *S. famelica*: The Saskatchewan specimen resembles *S. famelica* but has the leaf indumentum of *S. candida*.

Salix candida × *S. myrtillifolia*: Saskatchewan specimens combine characters of the two parents.

Salix candida × *S. petiolaris*: Intermediates between these species are known from Michigan and New York (W. W. Rowlee and K. M. Wiegand 1896) as well as Ontario and Saskatchewan, but can be expected wherever the two grow together. The invalid name "*S.* ×*clarkei*" is sometimes used for this hybrid.

The glabrescent form of *Salix candida*, forma *denudata* (Andersson) Rouleau, may be of hybrid origin.

2e9. Salix Linnaeus (subg. Vetrix) sect. Lanatae (Andersson) Koehne, Deut. Dendrol., 87, 93. 1893

Salix [unranked] *Lanatae* Andersson in A. P. de Candolle and A. L. P. P. de Candolle, Prodr. 16(2): 273. 1868

Shrubs, 0.05–6.5 m. **Buds** *caprea-* or *arctica-*type. **Leaves**: stipules on late ones foliaceous; largest medial blade 0.7–4.2 times as long as wide, adaxial surface not glaucous. **Staminate flowers**: filaments glabrous or hairy basally. **Pistillate flowers**: adaxial nectary shorter than, equal to, or longer than stipe; stipe 0.2–1.5 mm; ovary glabrous; stigmas slenderly to broadly cylindrical, 0.2–1 mm.

Species 5 (3 in the flora): North America, Eurasia.

92. Salix calcicola Fernald & Wiegand, Rhodora 13: 251. 1911 • Limestone willow E F

Salix richardsonii Hooker var. *macouniana* Bebb, Bot. Gaz. 14: 50, plate 9. 1889; *S. lanata* Linnaeus subsp. *calcicola* (Fernald & Wiegand) Hultén

Plants 0.05–1.3 m, (gnarled, sometimes forming clones by layering). **Stems**: branches yellow-brown, gray-brown, or red-brown, not to weakly glaucous (dull or slightly glossy), villous to glabrescent; branchlets yellow-brown or red-brown, moderately densely to sparsely villous, (buds *caprea-* or *arctica-*type, scale inner membranaceous layer free, separating from outer layer in var. *calcicola*). **Leaves**: stipules (sometimes marcescent), foliaceous, (1–6–14 mm, 1.1–2–4.4 times as long as wide, usually ovate, oval, or narrowly elliptic, flat when pressed), apex acute, convex, or rounded; petiole shallowly grooved, or convex to flat adaxially, 1.5–3.8–9 mm, villous or pilose adaxially; largest medial blade (sometimes hemiamphistomatous), narrowly to broadly elliptic or subcircular, 16–31.5–61 × 10–21.8–44 mm, 0.7–1.5–2.6 times as long as wide, base rounded or convex, sometimes cordate or cuneate, margins (sometimes purplish), flat, entire or serrulate, apex acuminate, acute, or rounded, abaxial surface glaucous, sparsely villous or pilose to glabrescent, hairs straight, adaxial dull or slightly glossy, sparsely villous or pilose; proximal blade margins entire; juvenile blade reddish or yellowish green, moderately densely villous abaxially, hairs white. **Catkins** flowering before leaves emerge; staminate (no data on var. *glandulosior*) stout, 18–45 × 13–21 mm, flowering branchlet 0 mm; pistillate (no data on var. *glandulosior* except flowering branchlet length) densely flowered, stout, 32–75(–100 in fruit) × 12–25 mm, flowering branchlet 0–5 mm; floral bract brown or black, 1.2–3.2 mm, apex rounded or acute, abaxially hairy, hairs straight. **Staminate flowers**: adaxial nectary oblong or narrowly oblong, 0.5–1.2 mm; filaments distinct, glabrous; anthers purple turning yellow, ellipsoid or shortly cylindrical, 0.5–0.7 mm. **Pistillate flowers**: adaxial nectary oblong, 0.2–1.1 mm, longer than or equal to stipe (rarely shorter); stipe 0.2–1.2 mm; ovary pyriform or obclavate, beak gradually tapering to styles; ovules 13–20 per ovary; styles 0.9–3 mm; stigmas broadly cylindrical, 0.2–0.56 mm. **Capsules** 4–8 mm.

Varieties 2 (2 in the flora): n, c North America.

1. Styles 1.6–3 mm; largest medial blades: margins often purplish, teeth or glands 6–18 per cm; stipes 0.2–1.2 mm; e Canadian Arctic. . . . . 92a. *Salix calcicola* var. *calcicola*
1. Styles 0.9–1.8 mm; largest medial blades: margins green, teeth or glands 5–22 per cm; stipes 0.3–0.6 mm; Alberta, Colorado. . . . . 92b. *Salix calcicola* var. *glandulosior*

92a. Salix calcicola Fernald & Wiegand var. calcicola E F

Salix calcicola var. *nicholsiana* Polunin

Leaves: largest medial blade margins often purplish, teeth or glands 6–18 per cm, apex acuminate, acute, or rounded, adaxial surface dull. **Pistillate flowers**: adaxial nectary 0.6–1 mm; stipe 0.2–1.2 mm; ovary pyriform or obclavate; styles 1.6–3 mm; stigmas 0.2–0.38–0.56 mm. $2n = 38$.

Flowering early Jun–early Aug. Marine coastal shores, rubble above high tide, sand dunes, sandy and silty streamshores, clay frost boils, stony or gravelly calcareous substrates; 0–300 m; Man., Nfld. and Labr., Nunavut, Ont., Que.

Hybrids:

Variety *calcicola* forms natural hybrids with *Salix alaxensis* var. *alaxensis*, *S. candida*, and *S. richardsonii*.

Variety *calcicola* × *Salix candida* (*S.* ×*wiegandii* Fernald): This hybrid is usually intermediate between its parents. It resembles *S. candida* in having leaves not marcescent, with crinkled leaf hairs and slightly glossy adaxial surfaces, and in flowering as leaves emerge. It resembles var. *calcicola* primarily in general leaf size and shape, in being sparsely to moderately densely hairy, in having its staminate catkins without a flowering branchlet, and floral bracts moderately to densely hairy. It differs from both parents in having ovaries with hairs in streaks or patches, a characteristic of many hybrids between species with glabrous and hairy ovaries. It is known only from the Northern Peninsula, Newfoundland, where it is common on coastal limestone barrens on tundra cliffs, growing on edges of pools and streams, and in sedge fens. Treated here as a hybrid, it could equally well be treated as a species of hybrid origin. The latter view is supported by its absence in northern Quebec, where the two parents also grow together.

Variety *calcicola* × *Salix richardsonii*: Polunin described *S. richardsonii* var. *mckeandii* (see 93. *S. richardsonii* for discussion) and *S. calcicola* var. *nicholsiana*, which he thought were hybridizing and intergrading on Baffin and Southampton islands, Nunavut. There is some suggestion of hybridization between *S. calcicola* and *S. richardsonii* on Southampton Island, but no clear evidence has been shown of hybridization elsewhere in Nunavut or Northwest Territories. At Churchill, Manitoba, where *S. calcicola* is common and *S. richardsonii* rare, specimens intermediate in leaf length, petiole length, and general leaf shape, and having stipules of *S. richardsonii* were recognized as putative hybrids, but hybrids between these species are uncommon.

92b. Salix calcicola Fernald & Wiegand var. **glandulosior** B. Boivin, Naturaliste Canad. 75: 221. 1948 [E]

Leaves: largest medial blade margins green, teeth or glands 5–22 per cm, apex acuminate or acute, adaxial surface slightly glossy or dull. **Pistillate flowers:** adaxial nectary 0.2–1.1 mm; stipe 0.3–0.6 mm; ovary pyriform; styles 0.9–1.8 mm; stigmas 0.2–0.28–0.36 mm.

No data are available on flowering time (probably early). Boreal floodplains, alpine wet meadows, dwarf birch thickets, limestone substrates; 1200–3700 m; Alta.; Colo.

93. Salix richardsonii Hooker, Fl. Bor.-Amer. 2: 147, plate 182. 1838 (as *richardsoni*) • Richardson's willow [F]

Salix lanata Linnaeus subsp. *richardsonii* (Hooker) A. K. Skvortsov; *S. richardsonii* var. *mckeandii* Polunin

Plants 0.3–6.5 m. **Stems:** branches red-brown, yellow-brown, or violet, not to strongly glaucous (slightly or highly glossy), villous to glabrescent with persistent patches of hairs; branchlets red-brown or yellow-brown, pilose to densely villous, (buds *caprea*-type, scale inner membrane free, separating from outer layer). **Leaves:** stipules (sometimes marcescent), foliaceous, (3–12–35 mm, 1.7–3.6–6.6 times as long as wide, lanceolate or narrowly elliptic, sometimes ovate or oval, base broad, abruptly tapering), apex acuminate or acute; petiole convex to flat or shallowly grooved adaxially, 2–7.3–27 mm, villous to glabrescent adaxially; largest medial blade broadly to narrowly elliptic, or obovate, 23–45–100 × 10–22–55 mm, 1.2–2.2–4.2 times as long as wide, base cuneate, convex, or concave, margins slightly revolute or flat, entire or serrulate, apex acute, acuminate, or convex, abaxial surface glaucous, glabrous or pilose, hairs (white, sometimes also ferruginous), curved, adaxial slightly glossy, moderately densely villous or pilose to glabrescent; proximal blade margins usually entire, sometimes serrulate; juvenile blade yellowish green, abaxially densely villous or pilose, sometimes glabrous, hairs white, sometimes also ferruginous. **Catkins** flowering before leaves emerge; staminate stout or subglobose, 22–53 × 16–24 mm, flowering branchlet 0 mm; pistillate densely flowered, slender or stout, 25–69(–85 in fruit) × 13–20 mm, flowering branchlet 0–2 mm; floral bract brown or black, 2.4–3.2 mm, apex acute, rounded, or retuse, abaxially moderately densely hairy, hairs straight. **Staminate flowers:** adaxial nectary narrowly oblong to oblong, 0.7–1.8 mm; filaments distinct to connate ca. ½ their lengths, glabrous; anthers purple turning yellow, ellipsoid to shortly cylindrical, 0.5–0.8 mm. **Pistillate flowers:** adaxial nectary narrowly oblong, oblong, or flask-shaped, 0.6–1.4 mm (2+-lobed), longer than stipe; stipe 0.4–1 mm; ovary pyriform, beak gradually tapering to styles; ovules 22–37 per ovary; styles 1–2.8 mm; stigmas slenderly cylindrical, 0.32–0.63–0.86 mm. **Capsules** 4.5–6.8 mm. $2n = 38$.

Flowering late May–early Jul. Arctic, subarctic, subalpine and boreal, stream terraces and floodplains, open forests, wet sedge meadows, fens, drumlin fields, gravel ridges, bogs, cliff ledges, snowbeds, silt, sand and gravel, calcareous or igneous substrates; 10–1800 m; B.C., Man., N.W.T., Nunavut, Yukon; Alaska; e Asia (Chukotka, Kamchatka, Russia [Anadyr], e Siberia).

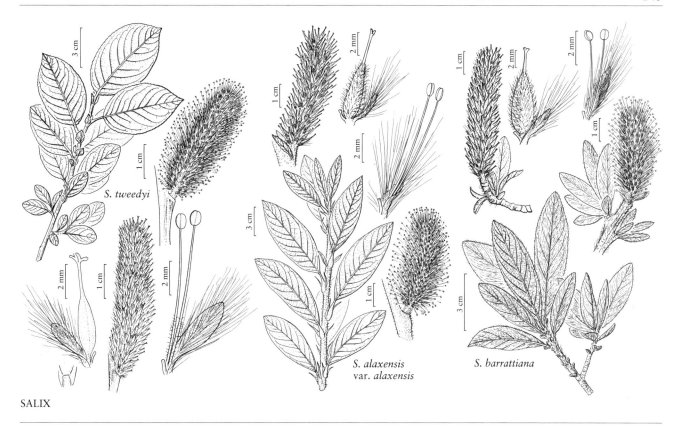

Hybrids:

Salix richardsonii forms natural hybrids with *S. barclayi* and *S. calcicola*.

94. Salix tweedyi (Bebb ex Rose) C. R. Ball, Bot. Gaz. 40: 377. 1905 • Tweedy's willow E F

Salix barrattiana Hooker var. *tweedyi* Bebb ex Rose, Contr. U.S. Natl. Herb. 3: 572. 1896

Plants 1–3 m. **Stems:** branches red-brown, not glaucous, villous to glabrescent; branchlets yellow-brown, (sometimes strongly glaucous), pilose or moderately densely villous, (buds *caprea*- or *arctica*-type, scale inner membranaceous layer free, separating from outer layer). **Leaves:** stipules (sometimes marcescent), foliaceous, (3.5–10–23 mm, 1–1.4–2.8 times as long as wide, usually lanceolate, ovate, or suborbiculate, (pleated when pressed), apex rounded, convex, or acute; petiole shallowly grooved adaxially, 3–11.3–26 mm, villous adaxially; largest medial blade elliptic to broadly elliptic, 36–65–100 × 17–35.5–85 mm, 1.1–1.9–2.9 times as long as wide, base convex, rounded, subcordate, or cordate, margins thickened, serrulate or spinulose-serrulate (sometimes teeth absent, margins with spherical glands), apex acuminate, acute, or convex, abaxial surface glaucous or not, pilose or villous, hairs curved, adaxial dull, pilose or villous to glabrescent; proximal blade margins entire (glands only) or serrulate; juvenile blade green, glabrous, pilose, or moderately densely villous abaxially, hairs white. **Catkins** flowering before leaves emerge; staminate stout, 39–56 × 14–22 mm, flowering branchlet 0–2 mm; pistillate densely flowered, stout, 39–70(–110 in fruit) × 13–22 mm, flowering branchlet 0–2 mm; floral bract dark brown, 1.8–4 mm, apex acute or rounded, abaxially sparsely hairy, hairs straight or wavy. **Staminate flowers:** adaxial nectary oblong, 0.6–1.2 mm; filaments distinct, glabrous or hairy basally; anthers yellow or purple turning yellow, 0.6–1 mm. **Pistillate flowers:** adaxial nectary oblong, narrowly oblong, or square, 0.5–1 mm, (unlobed), shorter to longer than stipe; stipe 0.4–1.5 mm; ovary pyriform, beak gradually tapering to styles; ovules 18–30 per ovary; styles 1.1–2.8 mm; stigmas slenderly cylindrical, 0.32–0.53–1 mm. **Capsules** 4–7 mm.

Flowering Jul. Subalpine and alpine, streamsides and bottoms, lakeshores, marshes, spruce bogs, talus slopes, tundra, quartzite, granite, or, sometimes, limestone substrates; 1400–4000 m; B.C.; Idaho, Mont., Wash., Wyo.

2e10. SALIX Linnaeus (subg. VETRIX) sect. VILLOSAE (Andersson) Rouy, Fl. France 12: 200. 1910

Salix [unranked] *Villosae* Andersson in A. P. de Candolle and A. L. P. P. de Candolle, Prodr. 16(2): 275. 1868

Shrubs or trees, 0.3–7 m. **Buds** *caprea*-type. **Leaves**: stipules on late ones foliaceous; largest medial blade 2–5 times as long as wide, adaxial surface not glaucous. **Staminate flowers**: filaments glabrous. **Pistillate flowers**: adaxial nectary longer than stipe; stipe 0–0.6 mm; ovary sparsely to very densely villous; stigmas slenderly cylindrical, 0.3–1.3 mm.

Species 6 (3 in the flora): North America, Asia.

95. Salix alaxensis (Andersson) Coville, Proc. Wash. Acad. Sci. 2: 280. 1900 • Felt-leaf willow F

Salix speciosa Hooker & Arnott var. *alaxensis* Andersson in A. P. de Candolle and A. L. P. P. de Candolle, Prodr. 16(2): 275. 1868

Shrubs or trees, 1–7 m. **Stems**: branches yellow-brown or red-brown, not glaucous, glabrous or villous; branchlets gray-brown or red-brown, glabrescent or villous. **Leaves**: stipules (sometimes marcescent), foliaceous, (3–23 mm), apex acuminate to acute; petiole convex to flat, or shallowly grooved adaxially, 3–20 mm, tomentose adaxially, (strongly ventricose around floral buds); largest medial blade broadly oblong, narrowly oblong, narrowly elliptic to elliptic, narrowly oblanceolate, oblanceolate, obovate, or broadly obovate, 50–110 × 13–35 mm, 2–4 times as long as wide, base cuneate or convex, margins strongly revolute, entire or crenate, apex acuminate, acute, or convex, abaxial surface glaucous or not (usually obscured by hairs, midrib yellowish), densely tomentose or villous-tomentose, hairs wavy, adaxial dull, sparsely or moderately densely villous (floccose) to glabrescent, (hairs white or gray); proximal blade margins entire; juvenile blade reddish or yellowish green (color often obscured by hairs), very densely woolly-tomentose abaxially, hairs white. **Catkins** flowering before leaves emerge; staminate stout, 23–55 × 13–27 mm, flowering branchlet 0–6 mm; pistillate densely flowered, slender to stout, 33–103 × 8–22 mm, flowering branchlet 0–13 mm; floral bract brown or black, 1.5–2–2.5 mm, apex acute to convex, abaxially sparsely hairy, hairs straight. **Staminate flowers**: adaxial nectary narrowly oblong to oblong, 0.5–1.4 mm; filaments distinct; anthers purple turning yellow, long-cylindrical, 0.6–0.9 mm. **Pistillate flowers**: adaxial nectary narrowly oblong, 0.6–1.6 mm; stipe 0–0.4 mm; ovary pyriform, (hairs refractive, wavy), beak gradually tapering to styles; ovules 14–18 per ovary; styles 1.3–2.8 mm; stigmas 0.4–0.99–1.28 mm. **Capsules** 4–5 mm.

Varieties 2 (2 in the flora): n North America, Asia (n, e Siberia).

Salix alaxensis is often used in northern regions for revegetation of disturbed sites and for wildlife habitat restoration (R. A. Densmore et al. 1987).

1. Branchlets not noticeably glaucous, very densely villous; largest medial blades: midrib evident, moderately densely tomentose to sparsely pubescent, abaxial surface not noticeably glaucous . . . . 95a. *Salix alaxensis* var. *alaxensis*
1. Branchlets strongly glaucous, glabrescent or sparsely villous; largest medial blades: midrib prominent, sparsely pubescent to glabrescent, abaxial surface noticeably glaucous or not . . . . 95b. *Salix alaxensis* var. *longistylis*

95a. Salix alaxensis (Andersson) Coville var. alaxensis F

Shrubs or trees, 1–7 m. **Stems**: branches red-brown, densely villous; branchlets not noticeably glaucous, very densely villous, hairs white or yellowish. **Largest medial leaves**: midrib evident, moderately densely tomentose to sparsely pubescent, abaxial surface not noticeably glaucous. **Catkins**: pistillate 33–85(–90 in fruit) × 10–22 mm, flowering branchlet 0–2 mm. $2n = 38$.

Flowering mid Apr–mid Jul. Stream and lake shores, terraces on coarse, calcareous gravel, well-watered scree slopes, well-drained to wet sand plains and dune remnants on deltas, wet alpine and subalpine meadows and thickets; 0–2000 m; Alta., B.C., Man., N.W.T., Nunavut, Que., Yukon; Alaska; Asia (n, e Siberia).

Variety *alaxensis* is one of the tallest "trees" in the Canadian Arctic. An extensive stand of willows, some of which reach tree-size, occurs in a valley on deep marine, lake, or river gravels and sands southeast of Deception Bay in northern Ungava, Quebec (P. F. Maycock and J. B. Matthews 1966). The largest of the dominant willows of tree stature, var. *alaxensis* and *Salix planifolia* were ca. 5 m tall and 20 cm in diameter. The maximum age for a stem 10 cm in diameter was 60 years, but at this age there was heartwood decay. An outlier stand of tree-sized willows on Victoria Island, Northwest Territories, reached 6–8 m, to 81 years of age (S. A. Edlund and P. A. Egginton 1984).

Hybrids:

Variety *alaxensis* forms natural hybrids with *Salix calcicola*, *S. drummondiana*, *S. pellita*, and *S. planifolia*.

Variety *alaxensis* × *Salix calcicola* was discovered by M. Blondeau at Kangigsujuak, Quebec. The plants resemble *S. calcicola* in having broad leaves and stipules, and reddish styles, and var. *alaxensis* in having densely villous leaves and branchlets, and relatively short pistillate flowering branchlets. The ovaries have hairy beaks or are sparsely hairy throughout.

Variety *alaxensis* × *Salix drummondiana*: In the northern Rocky Mountains, plants resembling *S. drummondiana* but with stipules prominent, linear or lanceolate, and foliaceous, and leaves abaxially densely woolly may be this hybrid.

Variety *alaxensis* × *Salix pellita* occurs in the Churchill, Manitoba, region where it grows with var. *alaxensis*, *S. pellita*, and *S. planifolia*. The plants resemble var. *alaxensis* in having very densely villous branchlets, and leaves with short wavy hairs on abaxial surfaces. The leaf indumentum is sparse, and ferruginous hairs often occur on juvenile and late leaves. Strongly revolute margins suggest *S. pellita* as the second parent.

Variety *alaxensis* × *Salix planifolia* occurs in the Churchill, Manitoba, region where it grows with the two parental species. It resembles var. *alaxensis* in its very densely villous branchlets and leaves with short wavy hairs on abaxial surfaces. Leaf indumentum is sparse, and ferruginous hairs often occur on juvenile and late leaves.

95b. Salix alaxensis (Andersson) Coville var. **longistylis** (Rydberg) C. K. Schneider, J. Arnold Arbor. 1: 225. 1920 E

Salix longistylis Rydberg, Bull. New York Bot. Gard. 2: 163. 1901; *S. alaxensis* subsp. *longistylis* (Rydberg) Hultén

Shrubs or trees, 1–4 m. **Stems:** branches yellow-brown or red-brown, glabrous or sparsely villous; branchlets strongly glaucous, glabrescent or sparsely villous. **Largest medial leaves:** midrib prominent, sparsely pubescent to glabrescent, abaxial surface noticeably glaucous or not. **Catkins:** pistillate 37–103 × 8–20 mm, flowering branchlet 0–13 mm.

Flowering late Mar–early Jul. Pioneer thickets on river alluvium and glacial moraines, subalpine thickets, alpine tundra, moist to mesic talus slopes, forest openings; 0–2000 m; B.C., Man., N.W.T., Yukon; Alaska.

Variety *longistylis* differs from var. *alaxensis* mainly in its glaucous and less densely hairy branchlets. Some specimens of var. *alaxensis*, in which the indumentum has been partially removed by wind erosion, show that they, too, have glaucous branchlets. The only fundamental difference between the two may be density of branchlet indumentum. Other differences, which appear in varietal descriptions, may represent inadequate sampling. These taxa sometimes intergrade and var. *longistylis* may not be worthy of taxonomic recognition. It is included here only because of very conspicuous differences between the extremes.

Hybrids:

Variety *longistylis* forms natural hybrids with *Salix sitchensis*.

Variety *longistylis* × *Salix sitchensis* grows with both parents at the mouth of the Twentyfour Mile River, south-central Alaska. The plants resemble var. *longistylis* in having relatively long styles and stigmas, oblong nectaries, and sparsely hairy branchlets, and *S. sitchensis* in having brittle branches, weakly glaucous branchlets, hairs sparse and crinkled, and ovaries with straight hairs. Fruits were set but no seed developed.

96. Salix silicicola Raup, J. Arnold Arbor. 17: 236, plate 194. 1936 • Blanket-leaf willow C E

Salix alaxensis (Andersson) Coville var. *silicicola* (Raup) B. Boivin

Shrubs, 1–3 m, (forming clones by layering). **Stems:** branches yellow-brown, gray-brown, or red-brown, not glaucous, villous in patches to glabrescent; branchlets gray-brown or red-brown, very densely villous. **Leaves:** stipules foliaceous, apex acute to acuminate; petiole convex to flat adaxially, 5–12 mm, villous or tomentose adaxially, (strongly ventricose around floral buds); largest medial blade (apparently hypostomatous but surfaces obscured by hairs), narrowly oblong, narrowly elliptic, elliptic, or obovate, 36–84 × 19–40 mm, 1.8–3.6 times as long as wide, base convex or cuneate, margins slightly revolute, entire, apex convex or acuminate, abaxial surface very densely woolly-tomentose, hairs wavy, adaxial dull, moderately to very densely, villous-tomentose; proximal blade margins entire; juvenile blade color obscured by hairs, very densely tomentose-woolly abaxially, hairs white. **Catkins** flowering before leaves emerge; staminate stout, 40–56 × 14–15 mm, flowering branchlet 0 mm; pistillate densely flowered, slender, 35–125(–130 in fruit) mm, flowering branchlet 0 mm; floral bract brown or black, 2–3 mm, apex convex to rounded, abaxially hairy, hairs straight. **Staminate flowers:** adaxial nectary oblong to narrowly oblong, 0.6–1.5 mm; filaments distinct; anthers yellow, ellipsoid, 0.6–0.7 mm. **Pistillate flowers:** adaxial nectary oblong or obtriangular, 0.5–1.1 mm; stipe 0–0.3 mm; ovary pyriform, beak gradually tapering to styles; ovules 12–14 per ovary; styles 1.2–2.2 mm; stigmas 0.52–0.75–1 mm. **Capsules** 4–7 mm. $2n = 38$.

No data are available on flowering time in the wild; in cultivation flowering is early May. Active sand dunes; of conservation concern; 20–500 m; Nunavut, Sask.

Comparisons of genetic variation in *Salix alaxensis* var. *alaxensis* from British Columbia and *S. silicicola* from Lake Athabasca sand dunes, Saskatchewan, based on isozyme loci, fit predicted relationships between progenitor and derived taxon (B. G. Purdy and R. J. Bayer 1995). Allelic diversity of *S. silicicola* was a subset of that of *S. alaxensis*, there was less genetic variation in *S. silicicola* than in *S. alaxensis*, and interspecific genetic variation within the two species was similar and relatively very high. This suggested a recent origin for the derived *S. silicicola*.

Salix silicicola is a uniform population that differs from *S. alaxensis* in its very densely villous or tomentose leaves and branchlets. These characters seem to be an adaptation to reduce sand abrasion and water loss in a sand dune environment. It is unlikely that it would have evolved *in situ* but probably derived from a pre-adapted source such as the one represented by specimens of putative *S. silicicola* from Pelly Lake, Nunavut. The isozyme study did not include specimens from that population or of *S. alaxensis* from Northwest Territories from which *S. silicicola* is likely to have been derived. Occurrence of *S. silicicola*-like plants in northern continental Nunavut suggests that during the late Pleistocene, it had a wider range, which now is represented by two disjunct populations. The question of appropriate taxonomic rank for the derived taxon is still unresolved. Although *S. silicicola* is different from *S. alaxensis* in its general appearance, they are very similar genetically, and argument could be made for treating them as varieties (B. Boivin 1966b).

Hybrids:

Salix silicicola forms natural hybrids with *S. brachycarpa* var. *psammophila*.

97. Salix barrattiana Hooker, Fl. Bor.-Amer. 2: 146, plate 181. 1838 • Barratt's willow [E] [F]

Salix barrattiana var. *marcescens* Raup

Shrubs, 0.3–1.5 m. **Stems:** branches red-brown, not or weakly glaucous, glabrous or villous in patches; branchlets red-brown or violet, moderately densely and coarsely villous, (bud-scale oily). **Leaves:** stipules (sometimes marcescent), foliaceous, (resinous, 1.5–7.5 mm), apex acute; petiole shallowly grooved, or convex to flat adaxially, 4–15 mm, villous or puberulent adaxially, (weakly ventricose around floral buds); largest medial blade narrowly to broadly elliptic, oblanceolate, or obovate, 35–95 × 10–29 mm, 2.2–4.2(–5) times as long as wide, base usually convex, rounded, or cuneate, sometimes cordate, margins flat, entire, apex acute, convex, or acuminate, abaxial surface not glaucous, moderately to very densely long-silky tomentose to glabrescent, hairs wavy, adaxial slightly glossy, sparsely villous or pubescent to glabrescent; proximal blade margins entire; juvenile blade color sometimes obscured by hairs, very densely long-silky abaxially, hairs white. **Catkins** flowering before leaves emerge; staminate stout or subglobose, 20–57 × 13–18 mm, flowering branchlet 0–2 mm; pistillate densely flowered, slender to stout, 28–92(–105 in fruit) × 12–19 mm, flowering branchlet 0–5 mm; floral bract brown or black, 2.8–5.2 mm, apex acute to convex, abaxially hairy, hairs straight. **Staminate flowers:** adaxial nectary narrowly oblong to oblong, 0.4–1.8 mm; filaments distinct; anthers yellow or purple turning yellow, ellipsoid or shortly cylindrical, 0.4–0.6 mm. **Pistillate flowers:** adaxial nectary narrowly oblong to oblong, 0.6–1.3 mm; stipe 0.2–0.6 mm; ovary pyriform, (hairs refractive), beak gradually tapering to styles; ovules 16–21 per ovary; styles 0.6–1.8 mm; stigmas 0.28–0.47–0.64 mm. **Capsules** 4.5–6 mm.

Flowering late May–late Jul. Moist to wet gravel bars, fans and terraces, stream banks, shrub fens, thickets and meadows, wet alpine tundra, often on limestone substrates; 150–3200 m; Alta., B.C., N.W.T., Yukon; Alaska, Mont., Wyo.

Salix barrattiana has stipules and buds that are typically strongly oily or resinous, staining pressing sheets yellow. The Montana and Wyoming records are the basis for a conservation assessment by J. A. R. Ladyman (www.fs.fed.us/r2/projects/scp/assessments/salixbarrattiana.pdf). The Wyoming population is represented by three staminate specimens. Their buds and stipules are not conspicuously oily and further verification is needed.

Salix barrattiana is placed here in sect. *Villosae*, but it is morphologically similar also to members of sect. *Lanatae*. The most conspicuous difference is its hairy ovaries. Inconsistent phenetic clustering of this species (G. W. Argus 1997) shows the difficulty in determining its sectional placement. The suggestion that it may link the two sections by hybridization and introgression, or by alloploidy, still remains to be assessed.

Hybrids:

Salix barrattiana forms natural hybrids with *S. barclayi*, *S. commutata*, and *S. pseudomyrsinites*.

Salix barrattiana × *S. commutata*: This hybrid, known from Alberta and the Yukon, usually resembles *S. barrattiana*, but ovaries are hairy in patches, stipes at ca. 1 mm are long for that species, and buds and stipules are not oily. A British Columbia specimen has glabrous ovaries and oily stipules.

Salix barrattiana × *S. pseudomyrsinites* is a rare putative hybrid that combines characteristics of the two parents.

2e11. SALIX Linnaeus (subg. VETRIX) sect. VIMINELLA Seringe, Exempl. Rév. Salix, 10. 1824

Shrubs, 3–7 m. **Buds** *caprea*-type. **Leaves:** stipules on late ones foliaceous; largest medial blade 4.7–13.7 times as long as wide, adaxial surface not glaucous. **Staminate flowers:** filaments glabrous. **Pistillate flowers:** adaxial nectary longer than stipe; stipe 0.1–0.5 mm; ovary very densely long-silky; stigmas slenderly cylindrical, 0.72–1.3–1.8 mm.

Species 10 (1 in the flora): introduced; Europe.

98. **Salix viminalis** Linnaeus, Sp. Pl. 2: 1021. 1753
   • Osier, basket willow, silky osier

Stems: branches yellow-brown, gray-brown, or yellowish, not glaucous, glabrous or puberulent; branchlets yellow-brown or yellowish (sometimes color obscured by hairs), glabrous, densely to sparsely villous, velvety, or puberulent. **Leaves:** stipules (not adnate to petioles), rudimentary or absent on early ones, (late ones sometimes brownish, linear, 5.4–10.4 mm), apex acuminate; petiole shallowly grooved adaxially, 4–13 mm, villous, puberulent, or velvety adaxially; largest medial blade linear, lorate, narrowly oblong, or narrowly elliptic, 53–130 × 5–33 mm, base cuneate, margins strongly revolute, sinuate or apparently entire, (glands epilaminal), apex acuminate, acute, or convex, abaxial surface apparently glaucous (obscured by hairs), densely short-silky, woolly, or tomentose, (midribs prominent, yellowish, and hairy), hairs appressed, spreading or erect, straight or wavy, adaxial dull or slightly glossy, sparsely or moderately densely pubescent, hairs gray; proximal blade margins entire; juvenile blade yellowish green, very densely tomentose or short-silky abaxially, hairs white. **Catkins** flowering just before or as leaves emerge; staminate stout, 24–48 mm, flowering branchlet 0–2 mm; pistillate densely flowered, 23–55 mm, flowering branchlet 0–6 mm; floral bract brown or tawny, 1.6–2.2 mm, apex convex or rounded, abaxially hairy, hairs straight. **Staminate flowers:** adaxial nectary narrowly oblong or oblong, 0.6–1.5 mm; filaments distinct; anthers purple turning yellow, ellipsoid to shortly cylindrical, 0.6–0.8 mm. **Pistillate flowers:** adaxial nectary narrowly oblong or oblong, 0.9–1.4 mm; ovary pyriform, beak gradually tapering to styles; ovules 12–18 per ovary; styles 0.6–1.8 mm. **Capsules** 4–6 mm. $2n = 38$.

Flowering Apr–early May. Sandy, open woods, cobble rivershores, lake margins, and roadsides; 0–300 m; introduced; N.B., Nfld. and Labr. (Nfld.), N.S., Ont., P.E.I., Que.; Conn., Ind., Iowa, Maine, Mass., N.J., N.Y., Ohio, R.I., Vt.; Europe.

Salix ×*smithiana* (*S. caprea* × *S. viminalis*) is distinguished from *S. viminalis* by having leaf blades usually broad, 2.8–4.9(–6.4) times as long as wide, stipes 0.9–2 mm, ovaries short-silky, branches ± brittle at base, and petioles flat to convex adaxially; *S. viminalis* has leaf blades usually very narrow, 4.7–13.7 times as long as wide, stipes 0.1–0.5 mm, ovaries long-silky, branches flexible at base, and petioles shallowly grooved adaxially.

See *Salix* ×*smithiana* [p. 132] and 86. *S. pellita* for further comparative descriptions.

2e12. SALIX Linnaeus (subg. VETRIX) sect. CANAE A. Kerner, Verh. K. K. Zool.-Bot. Ges. Wien 10: 222. 1860 [I]

Shrubs or trees, 1–6(–15) m. **Buds** *caprea*-type. **Leaves:** stipules on late ones rudimentary; largest medial blade 6.6–16.7–29 times as long as wide, adaxial surface not glaucous. **Staminate flowers:** filaments glabrous or hairy basally. **Pistillate flowers:** adaxial nectary shorter or longer than stipe; stipe 0.3–0.5 mm; ovary glabrous; stigmas slenderly cylindrical, 0.2–0.5 mm.

Species 1: introduced; Europe.

99. Salix elaeagnos Scopoli, Fl. Carniol. ed. 2, 2: 257. 1772 • Hoary willow [I]

Plants sometimes multistemmed. **Stems:** branches yellow-brown or red-brown, not glaucous, glabrous; branchlets yellow-brown or red-brown, densely pubescent or tomentose. **Leaves:** stipules absent or rudimentary on early ones; petiole shallowly grooved adaxially, 2–5 mm, tomentose or velvety adaxially; largest medial blade linear, narrowly oblong, narrowly oblanceolate, or narrowly elliptic, 5–160 × 3–10(–20) mm, base cuneate, margins strongly or slightly revolute, entire or serrulate, (glands submarginal or epilaminal), apex acuminate or acute, abaxial surface glaucous, densely tomentose or woolly, (midribs yellowish, prominent), hairs appressed, curved, adaxial dull or slightly glossy, sparsely pubescent to glabrescent; proximal blade margins entire; juvenile blade yellowish green, densely tomentose abaxially, hairs white or gray. **Catkins** flowering just before or as leaves emerge; staminate stout or slender, 26–34 × 6–10 mm, flowering branchlet 1–1.5 mm; pistillate moderately to very densely flowered, slender or stout, 19–40 × 3–10 mm, flowering branchlet 0.3–3.5 mm; floral bract light brown or tawny, 1.5–4 mm, apex rounded, truncate, or acute, entire or toothed, abaxially hairy, hairs straight. **Staminate flowers:** adaxial nectary oblong or square, 0.3–0.9 mm; filaments connate less or more than ½ their lengths; anthers yellow, 0.5–0.7 mm. **Pistillate flowers:** adaxial nectary oblong or square, 0.3–0.7 mm; ovary obclavate, beak gradually tapering to styles; ovules 2 per ovary; styles 0.7–0.9 mm. **Capsules** 3–5 mm. $2n = 38$.

Flowering Apr–mid May. Cultivated; 0–200 m; introduced; N.S., Ont., Que.; Conn., Maine, Mass., S.C., Wis.; Europe.

Occurrence of naturalized *Salix elaeagnos* in the flora area is based on late nineteenth and early twentieth century collections. There is no evidence that it is now either cultivated or naturalized.

2e13. SALIX Linnaeus (subg. VETRIX) sect. ARGYROCARPAE Fernald, Rhodora 48: 44. 1946 [E]

Shrubs, 0.2–1 m. **Buds** *caprea*-type. **Leaves:** stipules on late ones rudimentary or foliaceous; largest medial blade (2.6–)3.3–5.9 times as long as wide, adaxial surface not glaucous. **Staminate flowers:** filaments glabrous or, sometimes, hairy basally. **Pistillate flowers:** adaxial nectary shorter than stipe; stipe 1–2.2 mm; ovary moderately to very densely short-silky; stigmas with flat, non-papillate abaxial surface and rounded tip, 0.2–0.26–0.4 mm.

Species 1: e North America.

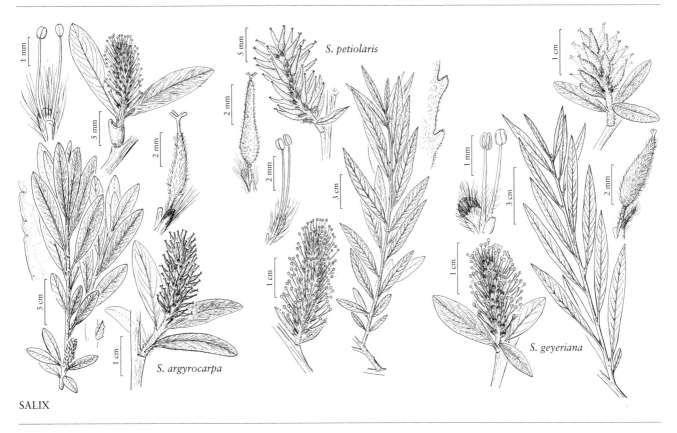

100. Salix argyrocarpa Andersson, Monogr. Salicum, 107, plate 6, fig. 60. 1867 • Labrador willow E F

Plants sometimes forming clones by layering. **Stems:** branches red-brown or brownish, not or weakly glaucous, (highly glossy), pubescent to glabrescent; branchlets yellow-brown or red-brown, sparsely pubescent. **Leaves:** stipules absent or rudimentary on early ones; petiole shallowly grooved adaxially, 3–8 mm, (sometimes glands present distally), pubescent adaxially; largest medial blade narrowly elliptic, narrowly oblong, or oblanceolate 25–65 × 7–15 mm, base cuneate or convex, margins strongly revolute, entire or crenulate, (glands submarginal or epilaminal), apex acute, convex, or acuminate, abaxial surface glaucous (sometimes obscured by hairs), pilose or densely long-silky or villous, (midribs yellow, prominent, glabrous or pubescent), hairs (white, sometimes also ferruginous), straight or wavy, adaxial slightly glossy, glabrous or sparsely pubescent, especially midrib, (hairs white, sometimes also ferruginous); proximal blade margins entire; juvenile blade yellowish green, sparsely to moderately densely long-silky abaxially, hairs white. **Catkins** flowering as leaves emerge; staminate stout or subglobose, 10–21.5 × 6–10 mm, flowering branchlet 1–8 mm; pistillate densely to moderately densely flowered, stout to subglobose, 11–20.5(–25 in fruit) × 5.5–17 mm, flowering branchlet 1–13 mm; floral bract tawny, brown, or bicolor, 0.7–1.2 mm, apex rounded, abaxially hairy, hairs straight. **Staminate flowers:** (abaxial nectary 0–0.6 mm), adaxial nectary oblong, narrowly oblong, or square, 0.4–1 mm, (nectaries usually distinct, sometimes cupulate); filaments distinct; anthers purple turning yellow, 0.4–0.5 mm. **Pistillate flowers:** adaxial nectary narrowly oblong, oblong, or square, 0.3–1.1 mm; ovary pyriform, beak gradually tapering to styles; ovules 12–13 per ovary; styles 0.4–0.9 mm. **Capsules** 2–4 mm. $2n = 76$.

Flowering early Jun–early Aug. Floodplains, lake and stream margins, wet snow flush areas, snowbeds, sedge meadows, treed bogs, shrubby tundra, subarctic and subalpine conifer forests, granitic, sandstone, or limestone substrates; 10–1800 m; Nfld. and Labr., Nunavut, Que.; Maine, N.H.

Salix argyrocarpa occurs in Nunavut on the Belcher Islands in Hudson Bay.

In the field, *Salix argyrocarpa* can be confused with *S. glauca*. They are not closely related but both have staminate flowers with both abaxial and adaxial nectaries, and tawny floral bracts. In subg. *Vetrix*, this characteristic occurs also in *S. wolfii* and sometimes in *S. orestera*, where it may be attributable to hybridization. See 54. *S. wolfii* for comment.

Vegetative specimens of *Salix argyrocarpa* and *S. pellita* are sometimes difficult to separate. There is no evidence that they hybridize.

Salix argyrocarpa is distinguished from *S. pellita* by having plants 0.2–1 m, stems delicate, largest medial blades 25–65 mm, 3.3–5.9 times as long as wide, branches highly glossy, not or weakly glaucous, flexible at base, and juvenile blades long-silky; *S. pellita* has plants 0.5–6 m, stems coarse, largest medial blades 40–123 mm, 4.2–11.3 times as long as wide, branches slightly glossy, often strongly glaucous, highly to ± brittle at base, and juvenile blades glabrous or pubescent, tomentose, or short-silky.

Hybrids:

Salix argyrocarpa forms natural hybrids with *S. herbacea*, *S. pedicellaris*, and *S. planifolia*.

Salix argyrocarpa × *S. herbacea* has leaf shape and margin dentition of *S. herbacea* but resembles *S. argyrocarpa* in having leaves glaucous abaxially, along with some white, silky hairs (especially on proximal leaves), juvenile leaves often revolute or infolded, and ovaries glabrous or with patches of hair, hairs appressed, short, straight or slightly curved, and flattened (having a saberlike appearance). These hybrids are sometimes misidentified as *S. arctophila* × *S. herbacea*, but ovary hair type and other characters suggest that *S. argyrocarpa* is the second parent. Occasional specimens with ferruginous hairs on the leaves suggest the influence of *S. pellita* or *S. planifolia*.

Salix argyrocarpa × *S. pedicellaris* (*S.* ×*dutillyi* Lepage) resembles *S. pedicellaris* in leaf shape and size and in having ovaries usually glabrous, although with patches or streaks of hair, and *S. argyrocarpa* in having leaves sparsely long-silky abaxially, in margins sparsely crenulate, and in proximal leaves with long-silky hairs abaxially. This hybrid is widespread in northern Quebec (G. W. Argus, unpubl.), where backcrosses seem to occur.

Salix argyrocarpa × *S. planifolia* (*S.* ×*grayi* C. K. Schneider): The collector, C. E. Faxon, noted that it could be distinguished at a distance from *S. planifolia* by its dull white color and upright branching, and from *S. argyrocarpa* by being taller. It resembles *S. argyrocarpa* in having juvenile leaves yellow-green, in catkins shorter and borne on longer flowering branchlets, and in stigmas purplish red. It resembles *S. planifolia* in having juvenile leaves with ferruginous hairs, and in the general appearance of the catkins (M. S. Bebb 1890).

2e14. SALIX Linnaeus (subg. VETRIX) sect. GEYERIANAE Argus, Syst. Bot. Monogr. 52: 85. 1997

Shrubs, 0.6–6 m. **Buds** *alba*- or *caprea*-type, or intermediate. **Leaves**: stipules on late ones absent, rudimentary, or foliaceous; largest medial blade (sometimes amphistomatous), 3–12 times as long as wide, adaxial surface not glaucous. **Staminate flowers**: filaments glabrous or hairy. **Pistillate flowers**: adaxial nectary shorter than stipe; stipe 0.4–4 mm; ovary sparsely to moderately densely short-silky; stigmas with flat, non-papillate abaxial surface, or stigmas cylindrical, 0.2–0.8 mm.

Species 4 (3 in the flora): North America, e Mexico.

Salix cana M. Martens & Galeotti is known from eastern Mexico.

101. Salix petiolaris Smith, Trans. Linn. Soc. 6: 122. 1802 • Meadow or skeleton-leaf willow [E] [F]

Salix gracilis Andersson; *S. gracilis* var. *textoris* Fernald; *S. petiolaris* var. *gracilis* (Andersson) Andersson; *S.* ×*subsericea* (Andersson) C. K. Schneider

Plants 1–6 m. **Stems**: branches red-brown or violet, not or weakly glaucous, (dull or slightly glossy), puberulent; branchlets yellow-green to red-brown, sparsely pubescent or moderately densely velvety, (buds *alba*-type or intermediate). **Leaves**: stipules rudimentary or absent; petiole shallowly grooved adaxially, 3–11 mm, pubescent, or velvety to glabrescent adaxially; largest medial blade lorate or very narrowly elliptic, 38–110 × 6–19 mm, 5–9 times as long as wide, base cuneate or convex, margins flat to slightly revolute, entire, serrate, serrulate, or spinulose-serrate, apex acute to acuminate, abaxial surface glaucous, densely long-silky to glabrescent, hairs (white, sometimes also ferruginous), adaxial dull or slightly glossy, glabrous or sparsely pubescent, (hairs white, sometimes also ferruginous); proximal blade margins sometimes serrulate; juvenile blade moderately densely long-silky abaxially, hairs white, sometimes also ferruginous. **Catkins** flowering as leaves emerge; staminate stout to globose, 12–29 × 6–17 mm, flowering branchlet 0.8–3 mm; pistillate loosely flowered, stout to globose, 12–39 × 6–18 mm, flowering branchlet 1–11 mm; floral bract brown, tawny, light rose, or bicolor, 1–2 mm, apex rounded, abaxially sparsely hairy, hairs straight. **Staminate flowers:** adaxial nectary square, ovate, or oblong, 0.3–0.7 mm; filaments distinct, hairy basally; anthers purple turning yellow, ellipsoid or globose, 0.4–0.6 mm. **Pistillate flowers:** adaxial nectary oblong to ovate, 0.3–0.9 mm; stipe 1.5–4 mm; ovary pyriform, beak abruptly tapering to styles; ovules 6–12 per ovary; styles 0–0.5 mm; stigmas slenderly to broadly cylindrical, 0.26–0.4–0.8 mm. **Capsules** 5–9 mm. $2n = 38$.

Flowering mid Apr–mid Jun. Sedge meadows, openings in moist, low, rich deciduous woods, sandy or peaty wet prairies, lakeshores; 10–2700 m; Alta., B.C., Man., N.B., N.W.T., N.S., Ont., P.E.I., Que., Sask.; Colo., Conn., Ill., Ind., Iowa, Maine, Mass., Mich., Minn., Nebr., N.H., N.J., N.Y., N.Dak., Ohio, Pa., S.Dak., Vt., Wis.

See 107. *Salix sericea* for a comparative description.

Because reproductive barriers between *Salix petiolaris* and *S. eriocephala* are weak, A. Mosseler (1990) suggested that their morphological variability may be due to interspecific gene flow.

Hybrids:

Salix petiolaris forms natural hybrids with *S. bebbiana*, *S. candida*, *S. eriocephala*, *S. famelica*, *S. pellita*, and *S. sericea*. Hybrids with *S. alba* have been reported (M. L. Fernald 1950) but no convincing specimens have been seen. Controlled pollinations with *S. discolor* produced no seed (A. Mosseler 1990).

Reports of *Salix petiolaris* × *S. sericea* from Massachusetts and Pennsylvania (C. K. Schneider 1921) probably refer to the densely sericeous variant of *S. petiolaris*. It is sometimes named *S.* ×*subsericea* (Andersson) C. K. Schneider but does not seem to be a hybrid (G. W. Argus 1965, 1986; E. G. Voss 1972–1996, vol. 2).

102. **Salix geyeriana** Andersson, Öfvers. Kongl. Vetensk.-Akad. Förh. 15: 122. 1858 • Geyer's willow E F

Salix geyeriana var. *argentea* (Bebb) C. K. Schneider; *S. geyeriana* var. *meleina* J. K. Henry

Plants 0.6–5 m, (sometimes forming clones by stem fragmentation). **Stems:** branches (sometimes ± brittle at base), yellow-green, gray-brown, red-brown, or violet, usually glaucous, glabrous or sparsely tomentose; branchlets yellowish, yellow-brown, red-brown, or violet, (strongly glaucous or not), glabrous or sparsely to moderately densely pubescent, (buds *caprea*-type). **Leaves:** stipules usually absent or rudimentary (rarely foliaceous); petiole convex to flat, or shallowly to deeply grooved adaxially, 2–9 mm, velvety, short-silky, or pubescent adaxially; largest medial blade lorate, narrowly elliptic, or linear, 32–89 × 5.5–14 mm, 3.6–8.4(–11.3) times as long as wide, base cuneate or convex, margins flat or slightly revolute, entire or distantly and shallowly serrulate, apex acute to acuminate, abaxial surface glaucous, glabrous or densely short- or long-silky, hairs (white, sometimes also ferruginous), straight, adaxial slightly glossy, densely short- or long-silky, especially midrib, to glabrescent, (hairs white, sometimes also ferruginous); proximal blade margins entire or distantly and shallowly serrulate; juvenile blade reddish or yellowish green, densely to sparsely long- or short-silky abaxially, hairs white, sometimes also ferruginous. **Catkins** flowering (before or) as leaves emerge; staminate globose, (1–1.1–)11–18 × 6–11 mm, flowering branchlet 1–5 mm; pistillate densely to loosely flowered, subglobose to globose, 8–21 × 7–17 mm, flowering branchlet 0.5–8 mm; floral bract tawny or brown (black), 1.2–2.8 mm, apex rounded or acute, abaxially hairy, hairs short, wavy or straight. **Staminate flowers:** adaxial nectary oblong, ovate, or square, 0.3–0.8(–0.9) mm; filaments distinct, glabrous or hairy on proximal ½ or basally; anthers yellow or purple turning yellow, ellipsoid or globose, 0.4–0.5(–0.6) mm. **Pistillate flowers:** adaxial nectary oblong, square, or ovate, 0.2–1 mm; stipe (0.4–)1–2.8 mm; ovary pyriform (gourd-shaped), beak gradually tapering to styles; ovules 6–12 per ovary; styles 0.1–0.2(–0.6) mm; stigmas flat, abaxially non-papillate with rounded or pointed tip, 0.2–0.3–0.44 mm. **Capsules** (3–)4–6 mm. $2n = 38$.

Flowering late Apr–late Jun. Lowland wet streamsides, lakeshores, sedge meadows, springs, seepages, swamps, cienegas, fine-textured substrates; 10–3300 m; B.C.; Ariz., Calif., Colo., Idaho, Mont., Nev., N.Mex., Oreg., Utah, Wash., Wyo.

Salix geyeriana is characterized by its dark gray appearance, slender, dark branches, narrow leaves long-silky on both surfaces, general absence of stipules, and small, subglobose catkins. Plants in the Pacific Northwest with foliaceous stipules may be hybrids or introgressants, but the other parent is unknown.

Hybrids:

Salix geyeriana forms natural hybrids with *S. bebbiana*, *S. irrorata*, *S. lemmonii*, *S. ligulifolia*, and *S. pedicellaris*. Alleged hybrids with *S. sitchensis*, based on plants from British Columbia with broader, more hairy leaves, and catkins longer than in *S. geyeriana*, but with the short stipes of *S. sitchensis* (J. K. Henry 1915), are unconvincing.

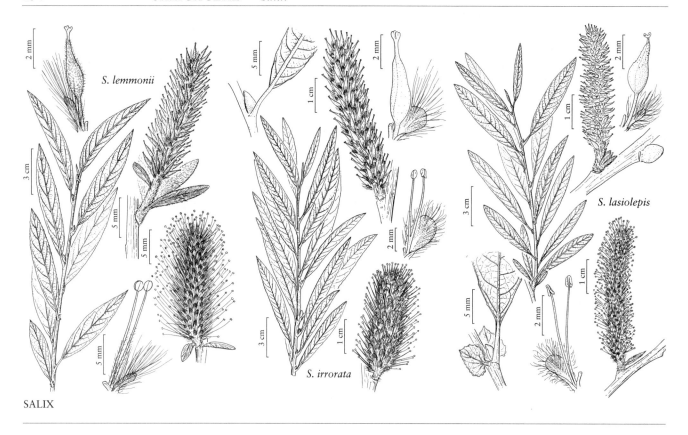

SALIX

Salix geyeriana × *S. irrorata*: A series of specimens from Arizona usually resemble *S. geyeriana* but have some characters of *S. irrorata*: foliaceous stipules, toothed leaf margins, catkins flowering before leaves emerge, shorter stipes (0.4–1.2 mm), and longer styles (0.2–0.6 mm). They also have some unique characters: proximal leaves sometimes serrulate, leaves sometimes amphistomatous, adaxial leaf surfaces mostly with ferruginous hairs, and ovaries sometimes gourd-shaped.

Salix geyeriana × *S. lemmonii* is uncommon but in mixed stands of the parental species some plants resemble *S. geyeriana* in having relatively short, subspherical catkins, small anthers, and petioles sometimes with petiolar glands; and *S. lemmonii* in having leaf blades amphistomatous, margins serrulate, and foliaceous stipules on early leaves. Because the species have different chromosome numbers, hybrids may be infertile, but occasional seeds have been seen. This hybrid is known from California (Lassen and Sierra counties), Oregon (Jefferson and Lane counties), and near Victoria, British Columbia.

Salix geyeriana × *S. ligulifolia*: Plants in Arizona with rudimentary stipules and leaves with ferruginous hairs may be this hybrid.

Salix geyeriana × *S. pedicellaris* occurs in Washington. It has the white and ferruginous hairs on leaves and ovaries of *S. geyeriana*, and leaves glaucous adaxially with prominent 2 and 3 veins of *S. pedicellaris*.

103. **Salix lemmonii** Bebb, Willows Calif., 88. 1879 (as lemmoni) • Lemmon's willow [E] [F]

Plants 1.2–4 m, (sometimes forming clones by stem fragmentation). **Stems:** branches (sometimes highly brittle at base), yellow-brown or red-brown, not to sometimes strongly glaucous, (dull to slightly glossy), glabrous or puberulent at nodes; branchlets yellow-brown to red-brown, (weakly or strongly glaucous or not), puberulent, (buds *alba*-type or intermediate, scale inner membranaceous layer free, not separating from outer layer). **Leaves:** stipules absent or rudimentary on early ones, foliaceous on late ones, apex acute; petiole convex to flat adaxially, 5–16 mm, tomentose or velvety adaxially; largest medial blade (sometimes amphistomatous), lorate or very narrowly to narrowly elliptic, 44–110 × 6–22 mm, 3.4–9.9(–12) times as long as wide, base convex or cuneate, margins flat to slightly revolute, entire or shallowly serrulate, apex acuminate to acute, abaxial surface glaucous, sparsely short- or long-silky to glabrescent, hairs (white, also ferruginous), straight or wavy, adaxial slightly glossy, sparsely short-silky, (hairs also ferruginous); proximal blade margins entire or shallowly serrulate; juvenile blade reddish or yellowish green, densely long-silky abaxially, hairs white and ferruginous. **Catkins**

flowering just before or, sometimes, as leaves emerge; staminate stout, (1.4–2.7–)16–28 × 9–17 mm, flowering branchlet 1–3 mm; pistillate loosely flowered, stout, (1.5–2.7–)19–44(–65 in fruit) × 10–18 mm, flowering branchlet 1–6 mm; floral bract brown or bicolor, 1.1–2.4 mm, apex rounded or convex, abaxially hairy, hairs straight. **Staminate flowers:** adaxial nectary narrowly oblong, oblong, or ovate, 0.3–0.9 mm; filaments distinct or connate less than ½ their lengths, glabrous or hairy on proximal ½ or basally; anthers yellow, ellipsoid, shortly cylindrical, or globose, 0.5–0.9 mm. **Pistillate flowers:** adaxial nectary oblong, narrowly oblong, or ovate, 0.4–1 mm; stipe 1.1–2.1 mm; ovary pyriform, beak sometimes slightly bulged below styles; ovules 12 per ovary; styles 0.3–1 mm; stigmas flat, abaxially non-papillate with rounded or pointed tip, 0.2–0.3–0.48 mm. **Capsules** 5–7 mm. $2n = 76$.

Flowering early Mar–late Jun. Streams, lakeshores, wet meadows, springs, old burns in subalpine conifer forests, sandy, granite substrates; 1400–3500 m; B.C.; Calif., Idaho, Mont., Nev., Oreg., Wyo.

See 85. *Salix drummondiana* for comments on differences.

Hybrids:

Salix lemmonii forms natural hybrids with *S. geyeriana*.

2e15. SALIX Linnaeus (subg. VETRIX) sect. MEXICANAE C. K. Schneider, J. Arnold Arbor. 3: 71. 1922

Shrubs or trees, 1–10 m. **Buds** *caprea*-type. **Leaves:** stipules on late ones rudimentary or foliaceous; largest medial blade 1.9–9.6 times as long as wide, adaxial surface not glaucous. **Staminate flowers:** filaments glabrous. **Pistillate flowers:** adaxial nectary shorter than stipe; stipe 0.4–2.4 mm; ovary glabrous; stigmas with flat, non-papillate abaxial surface, or stigmas plump, 0.1–0.6 mm.

Species 5 (3 in the flora): North America, Mexico.

104. **Salix irrorata** Andersson, Öfvers. Kongl. Vetensk.-Akad. Förh. 15: 117. 1858 • Blue-stem or dewy-stemmed willow [F]

Shrubs, 2–7 m, (multistemmed, sometimes forming clones by stem fragmentation). **Stems:** branches (highly brittle or not at base), red-brown to violet, usually glaucous, glabrous; branchlets yellow-brown, (strongly glaucous or not), glabrous, sparsely velvety, or tomentose. **Leaves:** stipules absent or rudimentary on early ones, foliaceous or rudimentary on late ones, apex acute; petiole convex to flat adaxially, 4–14 mm, velvety adaxially; largest medial blade lorate, narrowly oblong, narrowly elliptic, or narrowly oblanceolate, 47–115 × 8–22 mm, 3.5–7.7 times as long as wide, base cuneate or convex, margins flat to slightly revolute, entire and gland-dotted, serrulate or crenate, apex acuminate, acute, or convex, abaxial surface glaucous, glabrous, sparsely tomentose, or short-silky, hairs wavy, adaxial slightly to highly glossy, glabrous or pilose; proximal blade margins entire or serrulate; juvenile blade yellowish green or reddish, glabrous or sparsely villous abaxially, hairs white. **Catkins** flowering before or just before leaves emerge; staminate stout or subglobose, 15–34 × 8–22 mm, flowering branchlet 0(–2) mm; pistillate densely flowered, stout or slender, 14–43 × 7–12 mm, flowering branchlet 0–4 mm; floral bract brown or black, 1.3–2.5 mm, apex rounded, abaxially hairy, hairs straight or wavy. **Staminate flowers:** adaxial nectary narrowly oblong to oblong, 0.3–0.8 mm; filaments connate less than to more than ½ their lengths; anthers yellow or purple turning yellow, ellipsoid or shortly cylindrical, 0.4–0.7 mm. **Pistillate flowers:** adaxial nectary narrowly oblong to oblong, 0.3–0.7 mm; stipe 0.4–1 mm; ovary pyriform, beak slightly bulged below styles; ovules 9–12 per ovary; styles 0.3–0.9 mm; stigmas flat, abaxially non-papillate with rounded tip, 0.15–0.28–0.6 mm. **Capsules** 3.5–4 mm. $2n = 38$.

Flowering mid Mar–mid May. Streams, wet meadows; 1400–3000 m; Ariz., Colo., N.Mex., Wyo.; Mexico (Baja California, Chihuahua, Coahuila, Durango, Sonora).

Salix irrorata is very closely related to *S. lasiolepis*. The two are here maintained as separate species primarily because *S. irrorata* is a diploid and *S. lasiolepis* a tetraploid, and also because of their largely allopatric ranges (G. W. Argus 2007). Studies of sawflies (*Euura* Newman) by P. W. Price (pers. comm.) show that the same species of *Euura* can successfully reproduce on either willow. The only morphological character that separates the two is that in *S. irrorata* branchlets and branches are very strongly glaucous, whereas in S. lasiolepis they are not glaucous. Some plants have weakly glaucous stems (wax not visible except by polishing or only as sparkling crystals); this may be infraspecific variability or evidence of hybridization.

Hybrids:

Salix irrorata forms natural hybrids with S. drummondiana, S. geyeriana, and S. lasiolepis var. lasiolepis.

Salix irrorata × *S. lasiolepis* var. *lasiolepis*: This putative hybrid occurs in Arizona and New Mexico. It is characterized mainly by weakly glaucous branches.

105. Salix lasiolepis Bentham, Pl. Hartw., 335. 1857 • Arroyo willow

Salix bakeri Seemen; *S. bigelovii* Torrey; *S. franciscana* Seemen; *S. lasiolepis* var. *bracelinae* C. R. Ball; *S. lasiolepis* var. *sandbergii* (Rydberg) C. R. Ball; *S. lutea* Nuttall var. *nivaria* Jepson

Shrubs or trees, 1.5–10 m, (sometimes forming clones by stem fragmentation). **Stems:** branches (sometimes flexible to highly brittle at base), yellow-brown to red-brown, not or weakly glaucous, glabrous, tomentose, pubescent (appearing dusty); branchlets yellowish, yellow-brown, or red-brown, sparsely to densely villous, tomentose, or velvety to glabrescent, (inner membranaceous bud-scale layer free, separating from outer layer). **Leaves:** stipules usually foliaceous, or rudimentary or absent on early ones, foliaceous on late ones, apex acute; petiole convex to flat, or shallowly grooved adaxially, 3–16 mm, tomentose or velvety adaxially; largest medial blade lorate, narrowly oblong, narrowly elliptic, oblanceolate, or obovate to broadly obovate, 36–125 × 6–32 mm, 1.9–9.6 times as long as wide, base cuneate or convex, margins slightly to strongly revolute, entire or remotely or irregularly serrate, sinuate, (glands submarginal or epilaminal), apex acute, acuminate, convex or rounded, abaxial surface glaucous, sparsely pubescent, moderately densely tomentose or woolly-tomentose, short- or long-silky to glabrescent, hairs (white, sometimes also ferruginous), wavy, adaxial slightly or highly glossy, moderately densely tomentose or short-silky to glabrescent, (hairs white, sometimes also ferruginous); proximal blade margins entire or serrulate; juvenile blade color sometimes obscured by hairs, silky, tomentose or very densely woolly-tomentose abaxially, hairs white, sometimes ferruginous. **Catkins** flowering before or just before leaves emerge; staminate slender or stout, 18–88 × 5–15 mm, flowering branchlet 0–5 mm; pistillate densely flowered, slender or stout, 18–72 × 7–12 mm, flowering branchlet 0–6 mm; floral bract 1–2.4 mm, apex broadly rounded, abaxially hairy, hairs straight or wavy. **Staminate flowers:** adaxial nectary narrowly oblong, oblong, or ovate, 0.5–1.2 mm; filaments distinct or connate less than ½ their lengths; anthers purple turning yellow, ellipsoid or shortly cylindrical, 0.4–0.7 mm. **Pistillate flowers:** adaxial nectary oblong to flask-shaped, 0.2–1.1 mm; stipe 1–1.7 mm; ovary pyriform, beak slightly bulged below styles; ovules 10–18 per ovary; styles (sometimes slightly distinct distally), 0.1–0.6 mm; stigmas flat, abaxially non-papillate with rounded or pointed tip, or 2 plump lobes, 0.1–0.3 mm. **Capsules** 2.5–5.5 mm. $2n = 76$.

Flowering mid Jan–mid Jun. Streamshores, marshes, meadows, springs, coastal headlands, rocky bluffs, sand dunes, salt marshes, silty, sandy, gravelly, or rocky substrates, dolomite; 0–2800 m; Ariz., Calif., Idaho, Nev., N.Mex., Oreg., Tex., Utah, Wash.; Mexico (Baja California, Chiapas, Chihuahua, Coahuila, Durango, San Luis Potosí, Sonora).

Salix lasiolepis is polymorphic. Variety *bigelovii* has been recognized in coastal California and Oregon (G. W. Argus 1993). It differs mainly in density of leaf indumentum and in having leaves tending to be slightly broader; it may be a coastal ecotype and is not formally recognized here.

Hybrids:

Salix lasiolepis forms natural hybrids with S. irrorata and S. breweri. Hybrids with S. hookeriana are suspected but unconfirmed. The distinctly serrate leaves in some specimens from Arizona and New Mexico may be part of the species variability, but could also be due to hybridization.

106. Salix tracyi C. R. Ball, Univ. Calif. Publ. Bot. 17: 403, plates 69, 70. 1934 • Tracy's willow E

Shrubs, 1–6 m. **Stems:** branches gray-brown to red-brown, strongly to weakly glaucous, glabrous; branchlets yellow-brown to red-brown, (weakly glaucous), glabrous or sparsely to moderately densely velvety or tomentose. **Leaves:** stipules absent, rudimentary, or foliaceous on early ones, foliaceous on late ones, (2.5–9 mm), apex rounded or convex; petiole convex to flat, or shallowly grooved adaxially, 5–11 mm, tomentose or velvety adaxially; largest medial blade lorate, oblanceolate, or elliptic, 55–96 × 15–34 mm, 2–3.7 times as long as wide, base convex, margins slightly revolute, entire, sinuate or serrulate, apex acuminate to acute, abaxial surface glaucous, sparsely pubescent to moderately densely tomentose or glabrous, hairs (white, sometimes also ferruginous), straight or curved, adaxial slightly glossy, sparsely tomentose to glabrescent; proximal blade margins entire or serrulate; juvenile blade reddish, very densely tomentose or puberulent to glabrescent abaxially, hairs white, sometimes also ferruginous. **Catkins** flowering as or just before leaves emerge; staminate stout, 17–30 × 7–9 mm, flowering branchlet 1.5–5 mm; pistillate very densely flowered, stout to slender, 17–42 × 6–11 mm, flowering branchlet 1.5–3 mm; floral bract brown or bicolor, 0.8–1.6 mm, apex rounded to truncate, abaxially hairy throughout or proximally, hairs wavy or straight. **Staminate flowers:** adaxial nectary narrowly oblong, oblong, or ovate, 0.3–0.8 mm; filaments connate less to more than ½ their lengths; anthers purple turning yellow, ellipsoid, shortly cylindrical, or globose, 0.36–0.44 mm. **Pistillate flowers:** adaxial nectary oblong, or narrowly oblong to flask-shaped, 0.2–0.6 mm; stipe 1–1.7(–2.4) mm; ovary pyriform, beak sometimes slightly bulged below styles; ovules 12 per ovary; styles 0.1–0.6 mm; stigmas flat, abaxially non-papillate with rounded or pointed tip, 0.1–0.3 mm. **Capsules** 2.4–3.6 mm.

Flowering early Apr–early May. Shores and floodplains, sandy, gravelly, or rocky substrates, often serpentine; 90–500 m; Calif., Oreg.

Ball described *Salix tracyi* as part of a group of "more or less localized endemic willows of northern California and southern Oregon." A. K. Skvortsov (1971), followed by G. W. Argus (1993), thought that this species was a stunted form, perhaps shade-growth, of *S. lasiolepis*. R. D. Dorn (2000) studied the taxon in the field and considered it to be a distinct species. He noted that it could "usually" be separated from *S. lasiolepis* by its shorter styles, longer stipes, less hairy floral bracts, shorter catkins, thinner, less hairy leaves, and reddish juvenile leaves. He also noted differences in flowering times but they overlap completely. The differences are subtle and, although not diagnostic, should be studied further. *Salix tracyi* is treated here as a species because of its relatively large, localized range and interesting biological and morphological characteristics. Local botanists making collections from tagged individuals could provide useful information on habitat, elevation, flowering time, and morphology. Experimental study of this and the other serpentine endemic *Salix* in the region (see also 110. *S. breweri* and 111. *S. delnortensis*) could provide insights into their evolution and their relationship with *S. lasiolepis*.

A numerical taxonomic study by T. J. Crovello (1968) found that *Salix tracyi* clustered with members of sect. *Cordatae*, but both G. W. Argus (1997) and R. D. Dorn (2000) placed it in sect. *Mexicanae* with *S. lasiolepis*.

2e16. SALIX Linnaeus (subg. VETRIX) sect. GRISEAE (Borrer) Barratt ex Hooker, Fl. Bor.-Amer. 2: 148. 1838 E

Salix [unranked] *Griseae* Borrer in W. J. Hooker, Brit. Fl., 419. 1830

Shrubs, 0.5–4 m. **Buds** *caprea*-type. **Leaves:** stipules on late ones rudimentary or foliaceous; largest medial blade 3.6–5.3–7.7(–11) times as long as wide, adaxial surface not glaucous. **Staminate flowers:** filaments glabrous or hairy basally. **Pistillate flowers:** adaxial nectary shorter than stipe; stipe 0.6–1.5 mm; ovary moderately to very densely short-silky; stigmas with flat, non-papillate abaxial surface and rounded tip, 0.12–0.19–0.26 mm.

Species 1: e North America.

107. Salix sericea Marshall, Arbust Amer., 140. 1785 · Silky willow E F W

Salix coactilis Fernald

Plants sometimes forming clones by stem fragmentation. **Stems:** branches (highly brittle at base), gray-brown or violet, not glaucous, tomentose to glabrescent; branchlets red-brown, violet, or mottled yellow-brown, sparsely to densely velvety. **Leaves:** stipules absent or rudimentary on early ones, late ones 1.2–4 mm, apex acute or acuminate; petiole convex to flat, or shallowly grooved adaxially, 3.5–12(–21) mm, (sometimes 2 spherical glands distally), velvety adaxially; largest medial blade lorate, narrowly oblong, or narrowly elliptic, (48–)54–82–100(–125) × (7–)9–16.7–25 mm, base cuneate or convex, margins flat, serrulate or crenulate, apex acute, acuminate, or convex, abaxial surface glaucous (sometimes obscured by hairs), densely short-silky, hairs straight, adaxial dull, sparsely pubescent to glabrescent; proximal blade margins entire; juvenile blade reddish or yellowish green, very densely short-silky abaxially, hairs white, sometimes also ferruginous. **Catkins** flowering as or just before leaves emerge; staminate stout, 13.5–40 ×

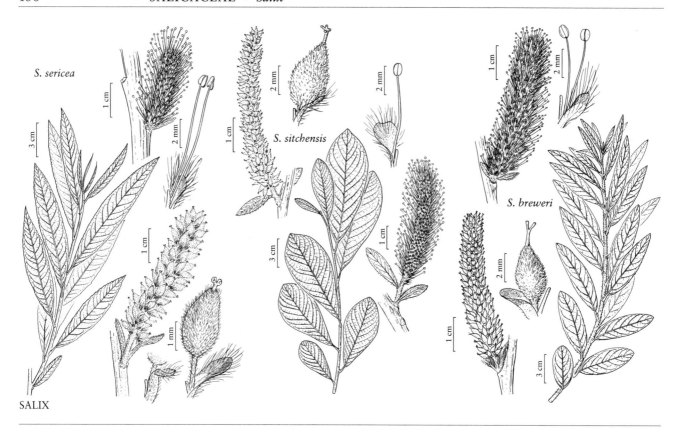

4–9 mm, flowering branchlet 0–2 mm; pistillate loosely to moderately densely flowered, slender to stout, 18–43 × 5–12 mm, flowering branchlet 1–3 mm; floral bract dark brown, black, or bicolor, 0.8–1.5 mm, apex rounded, abaxially hairy, hairs straight or wavy. **Staminate flowers:** adaxial nectary ovate to oblong, 0.3–0.8 mm; filaments distinct or connate less than ½ their lengths; anthers purple turning yellow, 0.4–0.6 mm. **Pistillate flowers:** adaxial nectary oblong, ovate, or flask-shaped, 0.4–0.7 mm; ovary ovoid, beak abruptly tapering to styles; ovules 6 per ovary; styles 0.2–0.4 mm. **Capsules** 2.5–4 mm.

Flowering early Mar–early Jun. Wet, boggy shores, sandy terraces, ledges along streams, low woods, sedge meadows, acid bogs, open seepages, rocky, silty, sandy, or peaty substrates, possibly also on serpentine soils; 5–1300 m; N.B., N.S., Que.; Ala., Ark., Conn., Del., D.C., Ga., Ill., Ind., Iowa, Ky., Maine, Md., Mass., Mich., Mo., N.H., N.J., N.Y., N.C., Ohio, Pa., R.I., S.C., Tenn., Vt., Va., W.Va., Wis.

Some specimens of *Salix sericea* are difficult to separate from *S. petiolaris* (G. W. Argus 1986). *Salix sericea* can be distinguished from *S. petiolaris* in having branches highly brittle at base, stipules on early leaves rudimentary or absent, on late leaves usually foliaceous, juvenile blades usually with white hairs, rarely some ferruginous, ovaries ovoid, beaks abruptly tapering to styles, and capsules 2.5–4 mm; *S. petiolaris* has branches flexible at base, stipules on all leaves absent or rudimentary, juvenile blades usually with conspicuous ferruginous hairs, ovaries pyriform, beaks gradually tapering to styles, and capsules 5–9 mm.

See 68. *Salix eriocephala* for comments on differences.

Hybrids:

Salix sericea forms natural hybrids with *S. eriocephala* and *S. petiolaris*. Reports of hybrids with *S. humilis* (C. K. Schneider 1921; M. L. Fernald 1950) are not based on convincing specimens.

2e17. SALIX Linnaeus (subg. VETRIX) sect. SITCHENSES (Bebb) C. K. Schneider, J. Arnold Arbor. 1: 91. 1919 [E]

Salix [unranked] *Sitchenses* Bebb, Bot. Gaz. 16: 105. 1891; *Salix* sect. *Breweriana* C. K. Schneider; *Salix* subsect. *Sitchenses* (Bebb) Dorn

Shrubs or trees, 1–8 m. **Buds** *alba*- or *caprea*-type, or intermediate. **Leaves:** stipules on late ones rudimentary or foliaceous; largest medial blade 1.5–8 times as long as wide, adaxial surface sometimes glaucous. **Staminate flowers:** (stamens 1 or 2), filaments glabrous or hairy. **Pistillate flowers:** adaxial nectary shorter than, equal to, or longer than stipe; stipe 0–1.4 mm; ovary moderately to very densely hairy; stigmas with flat, non-papillate abaxial surface, or stigmas broadly cylindrical, or plump, 0.16–0.4 mm.

Species 4 (4 in the flora): w North America.

108. **Salix sitchensis** Sanson ex Bongard, Mém. Acad. Imp. Sci. St.-Pétersbourg, Sér. 6, Sci. Math. 2: 162. 1832 • Sitka willow [E] [F]

Salix coulteri Andersson; *S. sitchensis* var. *parvifolia* (Jepson) Jepson

Shrubs or trees, 1–8 m, (sometimes forming clones by stem fragmentation). **Stems:** branches (sometimes highly brittle at base), yellow-brown or red-brown, not glaucous, glabrous or pilose; branchlets yellow-brown, gray-brown, or red-brown, densely short-silky, velvety, or villous, (buds *caprea*-type or intermediate). **Leaves:** stipules absent or rudimentary on early ones, rudimentary or foliaceous on late ones, apex acute; petiole convex to flat, or shallowly grooved adaxially, 3–13(–16) mm, tomentose or velvety adaxially; largest medial blade elliptic, narrowly oblanceolate, oblanceolate, or obovate, 31–70–120 × 17–48 mm, 2.1–3.1–4 times as long as wide, base cuneate or convex, margins slightly revolute or flat, strongly revolute proximally, entire, irregularly serrate, or sinuate, (glands submarginal or epilaminal), apex acuminate or convex, abaxial surface not evidently glaucous, (obscured by hairs), densely short-silky, woolly, or silky-woolly, hairs straight, wavy, or curved, adaxial slightly glossy (sometimes dull and glaucous), pilose or moderately densely short-silky; proximal blade margins entire or shallowly serrulate; juvenile blade green, densely long-silky or woolly abaxially, (sparsely silky-tomentose adaxially), hairs white. **Catkins** flowering just before or as leaves emerge; staminate slender or stout, (17–)22–54 × 8–15 mm, flowering branchlet 1–9 mm; pistillate moderately densely flowered, slender to stout, 25–73(–115 in fruit) × 5–15 mm, flowering branchlet 1–20 mm; floral bract tawny to dark brown, 1.4–2.4 mm, apex rounded or acute, abaxially hairy, hairs straight or wavy. **Staminate flowers:** adaxial nectary narrowly oblong, oblong, ovate, or flask-shaped, 0.4–1.3 mm; stamens 1; filaments distinct, glabrous; anthers purple turning yellow, shortly cylindrical, 0.5–0.7 mm. **Pistillate flowers:** adaxial nectary square, ovate, or flask-shaped, 0.5–0.9 mm, shorter to longer than stipe; stipe 0.4–1.4 mm; ovary ovoid to pyriform, long- or short-silky or villous, beak sometimes slightly bulged below styles; ovules 14–20 per ovary; styles 0.4–0.8 mm; stigmas flat, abaxially non-papillate with rounded tip, or broadly cylindrical, 0.16–0.28–0.4 mm. **Capsules** 3.5–5.6 mm. $2n = 38$.

Flowering early Apr–mid Jun (Mar in California). Tidal swamps and marshes, coastal fog belts and headlands, sand dunes, springs, gravelly streambeds and deltas, glacial moraines, avalanche tracks, dry canyons, clearings and edges of forests, shade tolerant; 0–1800 m; Alta., B.C.; Alaska, Calif., Idaho, Mont., Oreg., Wash.

Ovary hairiness in some *Salix sitchensis* populations varies from uniformly hairy to glabrescent, with intermediates with patchy or streaky hairiness. All three variations can occur together and do not seem to indicate hybridization.

Both *Salix sitchensis* and *S. scouleriana* have similar variants with leaves having very densely curly hairs on abaxial surfaces [*S. sitchensis* forma *coulteri* (Andersson) Jepson and *S. scouleriana* forma *poikila* (C. K. Schneider) C. K. Schneider]. Plants resembling *S. drummondiana* but with similar indumentum probably are hybrids with *S. alaxensis* (see 84. *S. drummondiana*). The *coulteri* taxon resembles *S. delnortensis* in having stipules with adaxial surfaces glabrous and very sparsely glandular toward the base, densely hairy abaxially, and with gland-dotted margins; its branchlets have wavy to crinkly hairs. The possible hybrid origin of *S. delnortensis* needs study (R. D. Dorn 2000).

Hybrids:

Salix sitchensis forms natural hybrids with *S. alaxensis* var. *longistylis* and *S. melanopsis*. Hybridization with *S. geyeriana* reported by J. K. Henry (1915) is not based on convincing specimens.

109. Salix jepsonii C. K. Schneider, J. Arnold Arbor. 1: 89. 1919 • Jepson's willow E

Salix sitchensis Sanson ex Bongard var. *angustifolia* Bebb, Willows Calif., 87. 1879, not *S. angustifolia* Willdenow 1806; *S. pellita* (Andersson) Bebb var. *angustifolia* (Bebb) B. Boivin; *S. sitchensis* var. *ralphiana* (Jepson) Jepson

Shrubs, 1–3 m, (sometimes forming clones by stem fragmentation). **Stems:** branches (highly brittle at base), yellow-brown or red-brown, weakly glaucous or not, (dull or slightly glossy), velvety or short-silky to glabrescent; branchlets gray-brown or red-brown, sparsely or moderately densely short-silky or velvety, (buds *caprea*-type or intermediate). **Leaves:** stipules usually rudimentary, sometimes foliaceous on early ones, foliaceous on late ones, apex acute; petiole shallowly grooved, or convex to flat adaxially, 3–12 mm, short-silky or velvety adaxially; largest medial blade lorate, narrowly oblanceolate, or oblanceolate, 43–74–103 × 8–25 mm, 2.9–4.8–7.7 times as long as wide, base convex or cuneate, margins slightly revolute to flat, entire, apex acuminate, convex, or acute, abaxial surface glaucous (sometimes obscured by hairs), densely short-silky, hairs straight, adaxial dull (sometimes thinly glaucous) to slightly glossy, sparsely short- or long-silky; proximal blade margins entire; juvenile blade yellowish green or reddish, densely long-silky abaxially, hairs white. **Catkins** flowering as leaves emerge; staminate stout or subglobose, 16–16.5 × 11–13 mm, flowering branchlet 2.5–3 mm; pistillate moderately densely flowered, slender to stout, 33–55 × 10–11 mm, flowering branchlet 1.5–7 mm; floral bract brown or tawny, 0.8–2 mm, apex rounded or acute, abaxially hairy, hairs straight. **Staminate flowers:** adaxial nectary narrowly oblong to oblong, 0.4–0.8 mm; stamens 1 or 2; filaments distinct or connate less than ½ their lengths, hairy on proximal ½; anthers purple turning yellow, ellipsoid or shortly cylindrical, 0.6–0.8 mm. **Pistillate flowers:** adaxial nectary oblong to narrowly oblong or flask-shaped, 0.3–0.6 mm, shorter than or equal to stipe; stipe 0.4–1.2 mm; ovary pyriform, densely long-silky, beak gradually tapering to styles; ovules 13–18 per ovary; styles 0.4–0.6 mm; stigmas flat, abaxially non-papillate with rounded to pointed tip, 0.2–0.3 mm. **Capsules** 3–5 mm.

Flowering mid–late Jun. Margins of lakes and streams, wet meadows, gravel, rocky, or bouldery substrates, granite; 1000–3400 m; Calif., Nev., Oreg.

R. D. Dorn (2000) justified treating *Salix jepsonii* as *S. sitchensis* var. *angustifolia* based on specimens of *S. jepsonii* from southern Oregon. *Salix jepsonii* does occur in southern Oregon but is distinct enough from *S. sitchensis* to be treated as a species.

110. Salix breweri Bebb, Willows Calif., 88. 1879 • Brewer's willow E F

Shrubs, 1–4 m. **Stems:** branches yellow-brown or red-brown, not or weakly glaucous, sparsely to densely tomentose; branchlets yellowish to yellow-brown, densely velvety, long-silky, or tomentose, (buds *caprea*-type or intermediate). **Leaves:** stipules rudimentary on early ones, foliaceous on late ones, apex acuminate, acute, or rounded; petiole shallowly grooved adaxially, 3–7 mm, tomentose adaxially; largest medial blade lorate, very narrowly elliptic, or oblanceolate, 58–144 × 11–30 mm, 2.8–7 times as long as wide, base convex or cuneate, margins slightly revolute, entire, irregularly serrate, serrulate, or sinuate, apex convex, acute, or acuminate, abaxial surface glaucous (sometimes obscured by hairs), sparsely to densely tomentose or woolly, hairs wavy, adaxial slightly glossy or dull, sparsely tomentose or villous; proximal blade margins entire; juvenile blade reddish or yellowish green, densely long-silky abaxially, hairs white or gray. **Catkins** flowering before leaves emerge; staminate slender, stout, or subglobose, 12–47 × 7–12 mm, flowering branchlet 0–1 mm; pistillate moderately densely flowered, slender, 19–59(–80 in fruit) × 5–10 mm, flowering branchlet 0–1 mm; floral bract tawny or light rose, 0.8–1.5 mm, apex rounded, abaxially hairy, hairs straight. **Staminate flowers:** adaxial nectary very narrowly oblong, 0.8–1.4 mm; filaments distinct or connate basally, glabrous; anthers yellow, ellipsoid or globose, 0.4–0.6 mm. **Pistillate flowers:** (abaxial nectary rarely present), adaxial nectary narrowly oblong to oblong, 0.6–1.5 mm, longer than stipe, (nectaries distinct); stipe 0–0.4 mm; ovary squat, flask-shaped, very densely tomentose, beak abruptly tapering to styles; ovules 4–12 per ovary; styles (sometimes distinct), 0.4–0.8 mm; stigmas flat, abaxially non-papillate with rounded tip, 0.16–0.24 mm. **Capsules** 4–6 mm.

Flowering early Mar–mid Apr. Streamshores, rocky or gravelly substrates, serpentine soils; 300–1300 m; Calif.

Hybrids:

Salix breweri apparently hybridizes with *S. lasiolepis*. The plants in Yolo County have less hairy leaves, long petioles, and capsules that are relatively long and slender, glabrescent, and distinctly stipitate.

See 111. *Salix delnortensis* for a discussion of origin.

111. Salix delnortensis C. K. Schneider, J. Arnold Arbor. 1: 96. 1919 • Del Norte willow [E]

Salix breweri Bebb var. *delnortensis* (C. K. Schneider) Jepson

Shrubs, 1–2 m, (sometimes forming clones by stem fragmentation). **Stems:** branches (highly brittle at base), red-brown, not glaucous, tomentose or velvety to glabrescent; branchlets red-brown or yellow-brown, densely velvety, (buds *caprea*-type). **Leaves:** stipules absent or rudimentary on early ones, rudimentary or foliaceous on late ones, apex acute; petiole convex to flat, or shallowly grooved adaxially, 6–16 mm, velvety or tomentose adaxially; largest medial blade elliptic or obovate, 53–102 × 29–54 mm, 1.3–2.8 times as long as wide, base cuneate or convex, margins slightly revolute, entire or sinuate, apex convex, rounded, or acute, abaxial surface glaucous (sometimes obscured by hairs), densely to sparsely velvety, tomentose, villous, or short-silky, hairs erect or spreading, wavy, adaxial dull, sparsely tomentose or short-silky; proximal blade margins entire; juvenile blade green, very densely velvety or long-silky abaxially, hairs white or gray. **Catkins** flowering before leaves emerge; staminate stout, 25–30 × 6–13 mm, flowering branchlet 0–5 mm; pistillate moderately densely flowered, slender to stout, 17–53 × 6–8 mm, flowering branchlet 1–3 mm; floral bract brown, 1.2–2.2 mm, apex rounded, abaxially hairy, hairs straight. **Staminate flowers:** adaxial nectary narrowly oblong to oblong, 0.5–1 mm; filaments distinct or slightly connate, glabrous; anthers purple turning yellow, ellipsoid, 0.6–0.7 mm. **Pistillate flowers:** adaxial nectary narrowly oblong to oblong, 0.4–0.9 mm, longer than stipe; stipe 0–0.3 mm; ovary obturbinate, short-silky or densely pubescent, beak gradually tapering to styles; ovules 14–18 per ovary; styles 0.6–1.2 mm; stigmas flat, abaxially non-papillate with rounded tip, or 2 plump lobes, 0.2–0.32–0.4 mm. **Capsules** 4 mm.

Flowering late Mar–early May. Streamshores, gravel to boulder substrates, serpentine soils; 90–500 m; Calif., Oreg.

The origin of the serpentine endemics *Salix breweri* and *S. delnortensis* is still unresolved (G. W. Argus 1997; R. D. Dorn 2000). The possibility that they are of hybrid origin, perhaps involving *S. lasiolepis* and *S. sitchensis* (Dorn 1976), cannot be resolved by field and herbarium studies alone; cytological and experimental methods need to be employed.

2e18. SALIX Linnaeus (subg. Vetrix) sect. DAPHNELLA Seringe, Exempl. Rév. Salix, 10. 1824 [I]

Shrubs or trees, 2–10 m. **Buds** *caprea*-type. **Leaves:** stipules on late ones foliaceous; largest medial blade (sometimes amphistomatous), 2.5–6.5 times as long as wide, adaxial surface not glaucous. **Staminate flowers:** filaments glabrous. **Pistillate flowers:** adaxial nectary shorter than, equal to, or longer than stipe; stipe 0.6–0.8 mm; ovary glabrous; stigmas with flat, non-papillate abaxial surface and rounded tip, 0.4–0.7 mm.

Species 6 (1 in the flora): introduced; Europe.

112. Salix daphnoides Villars, Prosp. Hist. Pl. Dauphiné, 51. 1779 • Violet or daphne willow [I]

Stems: branches red-brown, strongly glaucous (losing glaucescence in age but remaining so at nodes), glabrescent; branchlets yellow-brown, (not glaucous, except in age), usually glabrescent, sometimes sparsely or moderately densely tomentose. **Leaves:** stipules (often adnate to petioles), usually rudimentary on early ones, late ones lanceolate to ovate, apex acuminate or acute, often adnate to petiole; petiole shallowly grooved, or convex to flat adaxially, 5–18 mm, tomentose to glabrescent adaxially; largest medial blade oblong, lorate, narrowly elliptic, or elliptic, 50–96(–120) × 1–35(–40) mm, base cuneate to concave, margins slightly revolute, serrate to crenate, apex acuminate, abaxial surface glaucous, glabrescent or midrib sparsely tomentose, hairs (white, sometimes also ferruginous), spreading, straight, long or short, adaxial slightly glossy, (midrib sparsely tomentose or throughout); proximal blade margins entire, closely gland-dotted; juvenile leaves green, sparsely to moderately densely long-silky abaxially, hairs white, sometimes some ferruginous. **Catkins** flowering before or just before leaves emerge; staminate stout, 30–47 × 9–20 mm, flowering branchlet 0–1.2 mm; pistillate densely flowered, stout, 20–50 mm, flowering branchlet ca. 1.5 mm; floral bract dark brown or bicolor, 2.8–3 mm, apex acute to convex, moderately densely hairy throughout, hairs straight. **Staminate flowers:** adaxial nectary narrowly oblong to flask-shaped, 0.5–1 mm; filaments distinct or connate basally; anthers purple turning yellow, short- to long-cylindrical or ellipsoid, 0.5–0.7 mm. **Pistillate flowers:** adaxial nectary oblong to square, 0.4–0.9 mm; ovary pyriform, beak gradually tapering to styles; ovules 4–6 per ovary; styles 0.6–1.5 mm. **Capsules** 3.2–4.4 mm. $2n = 38$.

Flowering late Mar–early May. Disturbed habitats; 60–600 m; introduced; Alta., Ont., Que., Sask.; Mass., Minn.; Europe.

Salix daphnoides is cultivated on prairies for windbreaks and elsewhere for its ornamental catkins. Its stipules are unusual in being attached to petiole bases, even when rudimentary. As the petiole dilates around reproductive buds, the stipules become adnate to it.

2e19. SALIX Linnaeus (subg. VETRIX) sect. HELIX Dumortier, Bijdr. Natuurk. Wetensch. 1: 56. 1826 ☐

Shrubs or trees, 1.5–5 m, (branching opposite or subopposite). **Buds** *caprea*-type. **Leaves:** stipules on late ones absent; largest medial blade (amphistomatous), 2.8–9.2 times as long as wide, adaxial surface glaucous. **Staminate flowers:** filaments hairy basally. **Pistillate flowers:** adaxial nectary longer than stipe; stipe 0–0.1 mm; ovary very densely short-silky; stigmas with flat, non-papillate abaxial surface and rounded tip, 0.16–0.2–0.24 mm.

Species 30 (1 in the flora): introduced; Europe.

113. **Salix purpurea** Linnaeus, Sp. Pl. 2: 1017. 1753
• Purple or basket willow, purple osier ☐

Plants sometimes forming clones by stem fragmentation. **Stems:** branches (sometimes ± brittle at base), yellow-brown or olive-brown, not or weakly glaucous, glabrous; branchlets yellow-brown or olive-brown, violet tinged, glabrous. **Leaves** (sometimes opposite or subopposite); stipules absent; petiole shallowly grooved adaxially, 2–7 mm, glabrous adaxially; largest medial blade lorate, narrowly oblong, narrowly oblanceolate, oblanceolate, 35–77 × 5–20 mm, base convex or rounded, margins strongly revolute, entire or serrulate, apex acute, acuminate, or convex, abaxial surface glaucous, glabrous, adaxial dull to sublustrous, glabrous; proximal blade margins entire; juvenile blade yellowish green or reddish, glabrous or sparsely pubescent abaxially, hairs white, sometimes also ferruginous. **Catkins** flowering before leaves emerge, (subopposite, recurved); staminate stout or subglobose, 25–33 × 6–10 mm, flowering branchlet 0 mm; pistillate densely flowered, slender or stout, 13.5–34.5(–35 in fruit) × 3–7 mm, flowering branchlet 0.5–3 mm; floral bract black or bicolor, 0.8–1.6 mm, apex rounded, abaxially hairy, hairs straight or wavy. **Staminate flowers:** adaxial nectary oblong, square, or ovate, 0.4–0.8 mm; filaments connate; anthers (distinct), purple turning yellow, ellipsoid or globose, 0.4–0.5 mm. **Pistillate flowers:** adaxial nectary ovate, 0.3–0.7 mm; ovary obturbinate, beak gradually tapering to styles; ovules 6 per ovary; styles 0.2–0.3 mm. **Capsules** 2.5–5 mm. $2n = 38$.

Flowering mid Mar–mid May. Floodplains and shores, fens, swamps, alder thickets, sandy and limestone beaches, low dunes; 0–900 m; introduced; N.B., Nfld. and Labr. (Nfld.), N.S., Ont., P.E.I., Que.; Calif., Conn., Del., D.C., Ga., Ill., Iowa, Ky., Maine, Md., Mass., Mich., Minn., Mo., N.H., N.Y., N.C., Ohio, Oreg., Pa., R.I., Utah, Vt., Va., W.Va., Wis.; Europe.

Salix purpurea occurrence in Ohio is based on information from T. S. Cooperrider (pers. comm.).

3. FLACOURTIA Commerson ex L'Héretier, Stirp. Nov., 59, plates 30, 30 bis. 1786 • [For Étienne de Flacourt, 1607–1660, Governor of Madagascar]

Robert W. Kiger

Trees or shrubs, usually not heterophyllous, not clonal; branching sympodial. **Stems** usually spinose, sometimes unarmed, spines simple and/or compound. **Leaves** usually persistent, sometimes deciduous; stipules minute; petiole not glandular. **Inflorescences** axillary or terminal (on lateral twigs), few-flowered, racemelike cymes [2-flowered fascicles], pistillate flowers sometimes solitary. **Pedicels** articulate near base. **Flowers** (unisexual and/or bisexual, fragrant [not]); sepals 4–7; disc lobed; stamens ca. 15–30; filaments distinct; ovary [2–]5–7[–10]-carpellate (and -locular, placentation axile, placentae intruded and fused); styles (persistent) connate basally [entirely and columnar]; stigma notched-capitate or truncate. **Fruits** drupaceous. **Seeds**: aril absent. $x = 11$.

Species ca. 15 (1 in the flora): introduced, Florida; s Asia, c, s Africa, Indian Ocean islands, Pacific Islands.

1. Flacourtia indica (Burman f.) Merrill, Interpr. Herb. Amboin., 377. 1917 • Governor's or Indian or Madagascar plum [F] [I]

Gmelina indica Burman f., Fl. Indica, 132, plate 39, fig. 5. 1768; *Flacourtia ramontchi* L'Héretier

Trees or shrubs, 3–5(–10) m. **Leaves**: petiole 1–2 cm; blade red to pink when immature, ovate to orbiculate, 8–12 cm, becoming coriaceous, margins glandular-serrate or -crenate, surfaces glabrous or sparsely to densely pubescent. **Peduncles** 5–10 mm. **Pedicels** 5–10 mm. **Flowers**: bisexual ones sometimes on some branches of otherwise pistillate plants; sepals (persistent) slightly connate, greenish, ovate-orbiculate, 1.5–2.5 mm, apex acute to rounded, surfaces pubescent; filaments pubescent at base; ovary ovoid; styles spreading. **Drupes** reddish to purple or red-black at maturity, globose or ellipsoid, 1.8–2.5 cm. **Seeds** ca. 4–10, obovoid, 8–10 mm; testa crustaceous, rugose. $2n = 22$ (India, cult. Cuba), 44 (Africa).

Flowering and fruiting year-round. Roadsides, grassy areas, hammock edges; 0–10 m; introduced; Fla.; s Asia (India); Africa; introduced also in tropical and subtropical regions elsewhere.

Flacourtia indica has been cultivated in southern Florida for a century or more and has become naturalized there in Broward, Collier, Lee, Miami-Dade, and Monroe counties, the fruits being dispersed by birds (W. S. Judd 1997b). Throughout its wide range, it is highly variable in thorniness, pubescence, and leaf shape; various combinations of extremes have been described as separate species, although the morphological variation seems to be continuous and does not correlate with geography (Judd). Trying to recognize those segregate taxa among the plants introduced in North America seems futile.

4. XYLOSMA G. Forster, Fl. Ins. Austr., 72. 1786, name conserved • [Greek *xylon*, wood, and *osme*, odor, alluding to fragrant wood of some Pacific species]

Robert W. Kiger

Shrubs [trees], often ± heterophyllous, not clonal; branching sympodial. **Stems** usually spinose, sometimes unarmed, spines simple and/or compound. **Leaves** usually persistent, sometimes ± deciduous (sometimes congested at apices of relatively short lateral branches); stipules absent; petiole not glandular. **Inflorescences** axillary, fasciculate [racemose], 1 or 2 per axil. **Pedicels** articulate. **Flowers**: sepals 4–6 (± persistent, connate proximally, imbricate); disc lobed (lobes extrastaminal, ± confluent [distinct]); stamens [8–]16–24[–50+] (usually exserted); filaments distinct; ovary 2- or 3-carpellate; style indistinct [relatively short]; stigmas 2 or 3, expanded, obcompressed, ± lobed. **Fruits** baccate. **Seeds**: aril absent. $x = 10$.

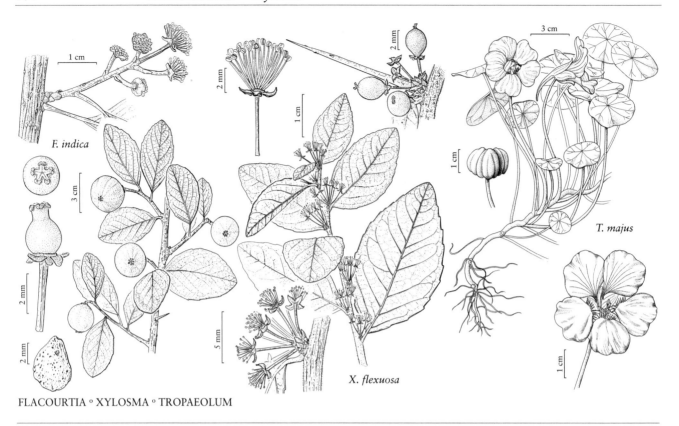

FLACOURTIA ∘ XYLOSMA ∘ TROPAEOLUM

Species 80–90 (1 in the flora): Texas, Mexico, West Indies, Central America, South America, se Asia, Pacific Islands.

The infrageneric taxonomy of New-World *Xylosma* is confused and perplexing. Relatively few characters vary significantly and they seem to do so independently, in a complex pattern of intergrading morphologies. Of species that have been recognized, relatively few are unequivocally distinct.

SELECTED REFERENCES Kiger, R. W. 2001. *Xylosma*. In: R. McVaugh and W. R. Anderson, eds. 1974+. Flora Novo-Galiciana: A Descriptive Account of the Vascular Plants of Western Mexico. 8+ vols. Ann Arbor. Vol. 3, pp. 328–334. Sleumer, H. 1980b. *Xylosma*. In: Organization for Flora Neotropica. 1968+. Flora Neotropica. 98+ nos. New York. No. 22, pp. 128–182.

1. **Xylosma flexuosa** (Kunth) Hemsley, Biol. Cent.-Amer., Bot. 1: 57. 1879 • Brush-holly, coronillo(a) [F]

Flacourtia flexuosa Kunth in A. von Humboldt et al., Nov. Gen. Sp. 7(fol.): 185; 7(qto.): 239. 1825; *F. celastrina* Kunth; *Xylosma blepharodes* Lundell; *X. celastrina* (Kunth) Gilg; *X. pringlei* B. L. Robinson

Shrubs [trees] to 2 [12] m. **Leaves:** petiole to 5 mm; blade usually ovate to obovate, sometimes rhombic, to 8 cm, chartaceous to ± coriaceous, base cuneate to rounded, margins usually crenate to serrate, sometimes subentire, apex shortly acuminate to rounded, sometimes retuse, surfaces usually glabrous, sometimes glabrescent on vasculature abaxially. **Inflorescences** sessile or obscurely pedunculate, 4–18-flowered. **Pedicels** to 5(–8) mm at anthesis, articulate proximal to middle, usually near base. **Flowers:** pistillate sometimes with sterile, ± reduced stamens; sepals white, yellowish, or greenish, sometimes reddish adaxially, ovate, 1–1.5 mm, margins short-ciliate, surfaces usually shortly pubescent, sometimes glabrate; stamens 16–24; filaments glabrous; ovary ovoid-oblong, glabrous. **Berries** reddish to red-black at maturity, subglobose, 4–7 mm, glabrous. **Seeds** 2–6, broadly ovoid, angular-compressed, sometimes reniform, 3–4 mm.

Flowering and fruiting year-round. Brushland and palm groves, often on slopes, often in moist soils; 0–100 m; Tex.; Mexico; West Indies (Curaçao); Central America; South America (Venezuela).

In the flora area, *Xylosma flexuosa* is found on the Gulf coastal plain from Cameron to Nueces counties.

TROPAEOLACEAE Jussieu ex de Candolle
• Nasturtium Family

Gordon C. Tucker

Herbs, annual [perennial]; usually glabrous; roots fibrous [tuberous], (producing glucosinolates). **Stems** often trailing [climbing]; not branched. **Leaves** alternate, simple; usually stipules present; petiole present (often twining); blade (peltate), margins entire or sinuate. **Inflorescences** axillary, flowers solitary; bracts absent. **Pedicels** present. **Flowers** zygomorphic; sepals 5, imbricate, adaxial 1 (or 3) prolonged into slender nectariferous spur; petals 5 (sometimes fewer by abortion), imbricate, clawed, distal 2 smaller than the others; stamens 8, (barely exserted) in 2 whorls, unequal; anthers 2-celled, dehiscence longitudinal; pistil 1; ovary 3-carpellate, 3-locular; placentation axile; ovule (1 per locule, pendulous from apex), anatropous, bitegmic; style 1 (apical); stigmas 3, linear. **Fruits** schizocarps (mericarps), 1-seeded, fleshy or dry. **Seeds** 1, (greenish), ovoid; embryo straight; endosperm absent.

Genus 1, species ca. 90 (1 in the flora): introduced; Mexico, Central America, South America; introduced also in Europe, Asia, Africa, Australia.

Tropaeolaceae have sometimes been included in Geraniaceae; they differ in having distinct stamens, fruits that split into 1-seeded mericarps, and persistent stylar beaks. Molecular phylogenies have indicated placement in Brassicales, with which they share the production of glucosinolates. The nearest relative of Tropaeolaceae is, putatively, the Akaniaceae, a family of woody plants in eastern Asia and eastern Australia (L. Watson and M. J. Dallwitz, http://delta-intkey.com, 29 July 2006). The family has been divided into three genera by some (e.g., A. Cronquist 1981): *Magallana* and *Trophaeastrum*, both monotypic and restricted to southern South America, and *Tropaeolum* with the remainder of the species. L. Andersson and S. Andersson (2000), based on molecular sequence data, concluded that the former two genera are imbedded within *Tropaeolum*, and made the needed transfers so that the family consists now of a single genus.

SELECTED REFERENCE Andersson, L. and S. Andersson. 2000. A molecular phylogeny of Tropaeolaceae and its systematic implications. Taxon 49: 721–736.

1. TROPAEOLUM Linnaeus, Sp. Pl. 1: 345. 1753; Gen. Pl. ed. 5, 162. 1754 • Nasturtium [Greek *tropaion*, trophy, alluding to a tree trunk on which were fixed the shields and helmets of defeated enemies; Linnaeus was reminded of this on seeing the type species growing on a post, the leaves representing the shields and the flowers the helmets] [I]

Leaves: blade orbiculate or reniform [palmately lobed or divided]. **Pedicels** axillary. **Flowers** relatively large, torus cupuliform. $x = 14$.

Species ca. 90 (1 in the flora): introduced; Mexico, Central America, South America; introduced also in Europe, Asia, Africa, Australia.

1. **Tropaeolum majus** Linnaeus, Sp. Pl. 1: 345. 1753 (as minus, corr. in errata) • Common or garden nasturtium, Indian cress [F] [I] [W]

Plants trailing, 15–100(–250) cm; glabrous or glabrate. **Leaves:** petiole 5–25 cm; blade with ca. 9 veins radiating from petiole, 3–10(–12) cm diam., abaxial surface usually papillose. **Pedicels** 6–13(–18) cm. **Flowers** 2.5–6 cm diam.; sepals light brown, oblong-lanceolate, 1.5–2 × 0.5–0.7 cm; spur straight or slightly curved, 2.5–3.5 cm; petals yellow, orange, purple, maroon, creamy white, or varicolored, mostly rounded, apex sometimes acuminate or emarginate, proximal 3 petals 2 × 2 cm, distal 2 petals usually entire, 2.5–5 × 1–1.8 cm, claw to 1.5 cm, claw margin deeply fringed; stamens 0.5–0.6 mm; anthers 0.5 mm; style 0.5–0.6 cm; stigmas 0.6–0.9 mm. **Fruits** oblate, 1.5–2 cm diam. **Seeds** 5–8 mm diam. $2n = 28$.

Flowering (May–)Jun–Oct, fruiting Jul–Oct. Disturbed roadsides, waste places, coastal bluffs, upper edges of beaches; 0–200 m; introduced; Calif., Conn., Mass., N.H., N.Y., Pa.; South America; introduced also in Europe, Asia, Africa, Australia.

Tropaeolum majus is frequently cultivated and often escapes; thoroughly naturalized and an invasive in coastal California, it occurs as a waif elsewhere.

Tropaeolum majus is often planted in gardens, and dozens of cultivars are available in the horticultural trade. Leaves of this plant have a peppery flavor and can be added to salads and sandwiches, and the flowers can be used to decorate salads and other dishes. The foliage is rich in vitamin C, and the plant also has diuretic and antibacterial properties. The plant is also used to treat wounds and infections of the urinary tract, and for problems associated with the respiratory tract (e.g., bronchitis and flu). Benzyl isothiocyanate, which can cause irritation, is responsible for the antibiotic action. The common name, Indian cress, refers to its cultivation in India.

Tropaeolum minus Linnaeus, *T. peltophorum* Bentham, *T. peregrinum* Linnaeus, and *T. tuberosum* Ruiz & Pavón also are cultivated in North America, but none is reported escaped. *Tropaeolum tuberosum* is the source of the root crop añu (ysaño), an important staple in the Andes.

MORINGACEAE Martinov
• Drumstick Tree Family

Mark E. Olson

Trees or shrubs; glabrous or puberulent; roots tuberous [massive, fibrous, or shrubs canelike with underground tubers], (producing glucosinolates and isothiocyanates, especially in roots and leaves; strongly odoriferous). **Stems** erect to pendent; unbranched [branched]. **Leaves** (drought-deciduous) alternate, (imparipinnate), compound; stipules present (glandular); petiole present; leaflets opposite, blade margins entire. **Inflorescences** axillary, paniculate, with (2–)3–4(–5) orders of branching; bracts (and bracteoles) usually present (glandular). **Pedicels** present. **Flowers** strongly zygomorphic [nearly actinomorphic]; perianth and androecium perigynous; sepals 5, distinct; petals 5 (bannerlike petal borne abaxially), distinct; stamens 5, opposite petals (inserted on rim of hypanthium); staminodes [3–]5, similar to filaments, alternate with stamens; anthers monothecal, bisporangiate; ovary 3-carpellate, 1-locular, (on gynophore); placentation probably marginal (sometimes interpreted as medial); ovules anatropous; style 1 (hollow, with a gaping stigmatic aperture). **Fruits** capsular, valvate, laxly dehiscent, fibrous. **Seeds** 10–35, brown, globular, winged or not; cotyledons oily.

Genus 1, species 13 (1 in the flora): introduced; sw Asia, sw, ne Africa, Indian Ocean Islands (Madagascar); introduced also pantropically.

Moringaceae are closely related to Caricaceae, with which they share glandular appendages at the apex of the petiole.

The bark of Moringaceae is conspicuous with bundles of phloem, schizogenous gum ducts are present in the pith and also on trauma in the bark, and exudes a straw- or pinkish-colored gum. The wood is white or yellowish and the plants are fibrous to mostly parenchymatous.

SELECTED REFERENCES Olson, M. E. 2002. Intergeneric relationships within the Caricaceae-Moringaceae clade (Brassicales), and potential morphological synapomorphies of the clade and its families. Int. J. Pl. Sci. 163: 51–65. Olson, M. E. 2002b. Combining data from DNA sequences and morphology for a phylogeny of Moringaceae. Syst. Bot. 27: 55–73. Olson, M. E. 2003. Developmental origins of floral bilateral symmetry in Moringaceae. Amer. J. Bot. 90: 49–71. Verdcourt, B. 1985. A synopsis of the Moringaceae. Kew Bull. 40: 1–23.

1. MORINGA Adanson, Fam. Pl. 2: 318, 579. 1763 • [Tamil *murungai*, twisted pod, alluding to young fruit]

Trees or shrubs, [massive pachycauls, baobab-like with water-storing trunk], slender-trunked. Leaves: stipules with nectaries at growing tip; rachis articulation with stalked glands; [1-pinnate](2–)3–4(–5)-pinnate; leaflet blade membranous [subcoriaceous], [lanceolate, oblanceolate, linear] round or oval, venation sometimes conspicuous abaxially, apex glandular, surfaces [pubescent] puberulent or glabrous. Flowers: parts usually with hairs forming a barrier distal to the nectariferous hypanthium; 1 sporangium initiated in anther ontogeny. Capsules 2-valved, often constricted between seeds. Seeds [1]2–3 cm, winged [not winged], sometimes with spongy seed coat, shed by gravity. $x = 11$.

Species 13 (1 in the flora): introduced; Asia (Bangladesh, India, Oman, Pakistan, Saudi Arabia, Yemen), sw, ne Africa, Indian Ocean Islands (Madagascar); introduced also pantropically.

Little is known about breeding systems in *Moringa*; *M. longituba* Engler appears incapable of self-pollination, and flowers with sterile anthers have been reported in *M. concanensis* Nimmo ex Dalzell & Gibson. All species are used medicinally locally; *M. stenopetala* (Baker f.) Cufodontis is used as a leaf vegetable in northwestern Kenya and southwestern Ethiopia.

1. **Moringa oleifera** Lamarck in J. Lamarck et al., Encycl. 1: 398. 1785 • Drumstick or horseradish or Ben tree

Guilandina moringa Linnaeus, Sp. Pl. 1: 381. 1753; *Hyperanthera moringa* (Linnaeus) Vahl

Plants 1–10 m, to 40 cm diam. Roots tuberous when young, woody with age. Bark pale gray or tan, smooth or finely rugose. Stems often canelike, becoming pendent with age, glabrous or finely puberulent. Leaves with pungent odor of horseradish; 30–60 cm, leaflets distributed on 4–8 pairs of pinnae; pinnae largest near base of leaf, 2 or 3 pinnate; leaflets 75–150, distalmost pairs represented by pairs of single leaflets along main rachis; blades bright to dark green, (0.5–)1–2(–3) × (0.3–) 0.5–1.5(–2) mm, base rounded to cuneate, apex rounded to emarginate, glands 3–5 mm (smaller at blade apex). Panicles (5–)10–25(–35) cm, each flower subtended by glandular bract. Pedicels 5–10(–20) mm; bracteoles 2. Flowers sweet-scented, 2–3 cm; sepals 10–20 × 3–4 mm, proximal ones usually reflexed, usually puberulent, distalmost pair usually largest, ± erect, enclosing banner petal, or ± reflexed; petals cream, 1–2 cm, distalmost banner petal ± erect, others usually ± reflexed; filaments and staminodes 7–10 mm, basally pubescent, adherent distally proximal to banner petal and anthers in a 3-tiered presentation; receptacle cup-shaped, 3–4 mm; gynophore 2–3 mm, appressed to banner petal; ovary 3–5 mm, with 3 ridges. Capsules tan, 10–30(–55) × 1.5–3 cm, apex beaked, 3 (or 4)-angled; valves silvery inside. Seeds pale to dark brown, globular, 3-winged; cotyledons exuding oil when compressed.

Flowering when leafless at end of dry season, fruiting as leaves emerge. Roadways, disturbed areas; 0–1000 m; introduced; Ariz., Calif., Fla.; Asia (India); introduced widely elsewhere.

Moringa oleifera is probably native to lowland dry tropical forests of northwestern India; recent collection information is lacking. It is cultivated in tropical countries as an ornamental and agricultural crop. It is occasionally reported as re-seeding along roadways and in other disturbed areas; there are no reports of *M. oleifera* invading intact habitats.

Moringa oleifera is often mistaken for a papilionoid legume or for a member of the Bignoniaceae. It is easily distinguished from both by stalked glands at the leaf base and at rachis articulations and by its pungent horseradish odor. The three-valved fruits with three-winged seeds also readily distinguish *M. oleifera* from both of those families. All parts of the plant are of economic importance: leaves are highly nutritious, flowers are edible, seeds contain large quantities of

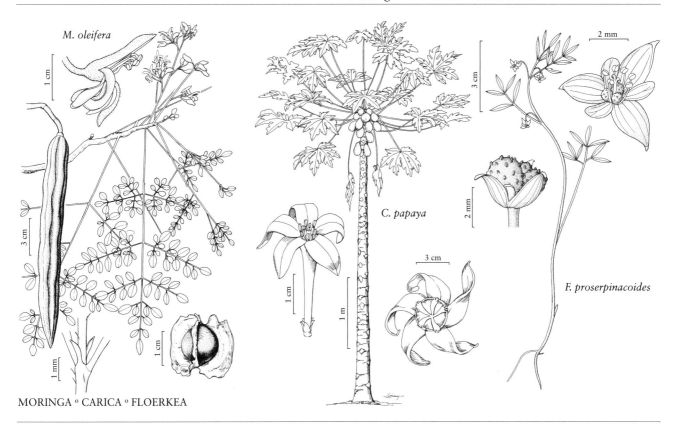

MORINGA ○ CARICA ○ FLOERKEA

high-quality oil, and presscake remaining after oil extraction contains one of the most powerful plant-derived flocculants known, used for clarifying turbid water. Roots are used as a horseradishlike condiment. *Moringa oleifera* is extremely fast growing (to 7 m in first year from seed), with fruit yields ca. 10 tons/ha/yr.

Moringa pterygosperma Gaertner is a commonly cited synonym of *M. oleifera*. However, the name is illegitimate as it is based on *Guilandia moringa*.

CARICACEAE Dumortier
• Papaya Family

Walter C. Holmes

Trees [**rarely herbs**], wood soft, sap milky. **Stems** erect; usually unbranched. **Leaves** alternate (borne at branch tips), palmately lobed [simple]; stipules absent; petiole present; blade margins entire or lobed. **Inflorescences** usually axillary, paniculate [cymose-paniculate, cymose, or racemose]; bracts present. **Pedicels** present or absent. **Flowers** usually unisexual, rarely bisexual, staminate and pistillate usually on different plants, 5-merous; calyces rotate, campanulate, or tubular, 5-toothed. **Staminate flowers:** corolla funnelform [tubular, salverform], tube elongate, 5-lobed, lobes oblong to linear [ovate]; stamens 10, in 2 series, borne at orifice of corolla tube, alternating longer and shorter; anthers dehiscing by longitudinal slits, introrse, distinct or connate basally, connective often projecting beyond anther sacs; ovary vestigial or absent. **Pistillate flowers:** petals distinct or connate basally, oblong to linear; ovary (1–)5-carpellate, 1-locular; placentation parietal; ovules 100+, anatropous, bitegmic; styles 0 or 1; stigmas 5, divided into 2 or more lobes. **Fruits:** berries. **Seeds** brown to black, ovoid to compressed, smooth or warty; aril gelatinous; embryo linear, cotyledons flat, broad.

Genera 6, species ca. 30 (1 in the flora): introduced; Central America, South America, w Africa; tropical regions; introduced pantropically.

Caricaceae consists mainly of soft-wood trees containing little secondary xylem. Any "wood" present is usually produced from phloem. Economically, the most important plant is *Carica papaya*, the source of papaya fruit. The fruits of *Jarilla* Rusby and *Jacaratia* A. de Candolle are locally grown and eaten in Mexico. Papaya has great variation in the inflorescence, especially in the pistillate flowers, probably as a result of being under extensive cultivation.

Caricaceae may be related to Passifloraceae, or to Cucurbitaceae.

1. CARICA Linnaeus, Sp. Pl. 2: 1036. 1753; Gen. Pl. ed. 5, 458. 1754 • Papaya [Alluding to imagined resemblance of leaves or fruits to those of a fig, *Ficus carica*, erroneously thought to be from Caria in southwestern Asia Minor]

Trees relatively short-lived. **Leaves** crowded distally on branches; glabrous. **Inflorescences:** staminate 100+-flowered, elongate; pistillate 1–several-flowered. **Flowers** each borne in axil of bract. **Berries** slightly 5-angled. $x = 9$.

Species, in the traditional sense, ca. 20 (1 in the flora): introduced, Florida; Central America, South America; introduced also pantropically.

V. M. Badillo (2000) considered *Carica* to consist of only one species (*C. papaya*), others being reassigned to the genus *Vasconcellea* A. Saint-Hilaire.

Some species of *Carica* in the traditional sense are grown for their edible fruits or sweet and juicy seed coverings (arils), the most important being *C. papaya*.

1. **Carica papaya** Linnaeus, Sp. Pl. 2: 1036. 1753
    • Papaya, papaw, pawpaw, lechosa

Plants ca. 6 m. **Stems** to ca. 20 cm diam., bark green to gray or brown, leaf scars prominent, smooth. **Leaves** drooping proximally, erect or spreading distally; petiole 35–70 cm, hollow; blade round in general contour, 20–60 cm diam., shallowly to deeply palmately 5–9-lobed, lobes lanceolate to ovate, margins entire or pinnately divided into lanceolate to ovate lobes. **Inflorescences** 15–60 cm; bracts ± rhombic. **Staminate flowers** sessile; calyx green, short-tubular, ca. 1.5 mm, teeth triangular to linear; corolla white, 2–3.5 cm, lobes narrowly oblong, ca. 1.3 cm; stamens yellow. **Pistillate flowers** subsessile; calyx yellow, short-tubular, 5–10 mm, teeth triangular to linear; petals lanceolate-oblong, white to pale yellow, 3–5 cm; ovary ellipsoidal to rounded, 2–3 cm. **Berries** hanging from trunk near summit, green to orange, ellipsoidal to oblong, rounded distally, 8–45 cm. **Seeds** blackish. $2n = 18$.

Flowering and fruiting year round. Disturbed areas, waste places, hummocks, roadsides; 0–100 m; introduced, Fla.; Central America; nw South America; introduced also pantropically.

Carica papaya is widely cultivated in tropical regions worldwide for its large melonlike fruit, one of the most popular tropical fruits, and occasionally as an ornamental. Papaya fruit is a good source of calcium, and vitamins A and C. Except for the ripe fruit, all parts of the plant possess a milky sap that contains the proteolytic enzymes papain and chymopapain, widely used to tenderize meat. Immature fruit, leaves, and flowers are also used as a cooked vegetable. Papaya also has industrial and pharmaceutical applications, including use in chewing gums, brewing, drugs for digestive ailments, treatment of gangrenous wounds, use in the textile industry, and use in production of soaps and shampoos.

The species is cultivated in extreme southern Texas, where occasional seedlings arise from seeds discarded in waste places or dumps but are never persistent.

LIMNANTHACEAE R. Brown
• Meadowfoam Family

Gordon C. Tucker

Herbs, annual; glabrous or pubescent (not glandular; producing glucosinolates). **Stems** erect, decumbent, or sprawling; unbranched. **Leaves** alternate, simple or compound; venation pinnate; stipules absent; petiole present; blade margins entire or pinnately lobed, bipinnate, or ternate. **Inflorescences** axillary, flowers solitary; bracts absent. **Pedicels** present. **Flowers** usually bisexual, usually actinomorphic, rotate; perianth and androecium hypogynous; sepals usually persistent, 3 or 5 (4 in *Limnanthes macounii*), distinct or slightly connate basally, equal or unequal; petals same number as sepals, convolute in bud, distinct, equal; nectary glands present; stamens 3, 6, 8, or 10 (same or twice the number of sepals); filaments distinct, glabrous; anthers dehiscing by longitudinal slits, introrse or extrorse, tetrasporangiate, pollen shed in single grains, binucleate, 2–4-aperturate, colpate or colporate; disc absent; gynophore absent; pistil 1; ovary 2–5-carpellate, syncarpous basally (united by gynobasic style); placentation basal; ovules 1 per locule, anatropous, unitegmic; style 1 (gynobasic); stigmas (2 or) 3–5 (dry, papillate). **Fruits** schizocarps (mericarps or nutlets), tuberculate, ridged, smooth, or rugulose. **Seeds** 1; not arillate; endosperm absent.

Genera 2, species 8 (8 in the flora): North America.

Placement of Limnanthaceae in Brassicales is supported by molecular and chemical data, especially the presence of mustard oils. Traditionally (A. Cronquist 1981), the family has been associated with the Geraniaceae in the Geraniales because of the similarity of habit and floral structure, especially the fruits that separate into mericarps, as do those in Geraniaceae.

Limnanthaceae are endemic to North America. The greatest diversity is in California, where most of the species of *Limnanthes* occur. *Floerkea*, with a single species, occurs widely across the continent; it is barely present in the southeastern United States.

SELECTED REFERENCES Link, D. A. 1992. The floral nectaries of Limnanthaceae. Pl. Syst. Evol. 179: 235–243. Ornduff, R. 1971. Systematic studies of Limnanthaceae. Madroño 21: 103–111. Ornduff, R. and T. J. Crovello. 1968. Numerical taxonomy of Limnanthaceae. Amer. J. Bot. 55: 173–182. Parker, W. H. and B. A. Bohm. 1979. Flavonoids and taxonomy of Limnanthaceae. Amer. J. Bot. 66: 191–197. Plotkin, M. S. 1998. Phylogeny and Biogeography of Limnanthaceae. M.S. thesis. University of California, Davis.

1. Petals and sepals 3; petals shorter than sepals; stamens 3 or 6 . . . . 1. *Floerkea*, p. 173
1. Petals and sepals (4 or) 5; petals usually longer than sepals; stamens 8 or 10 . . . . 2. *Limnanthes*, p. 173

1. FLOERKEA Willdenow, Ges. Naturf. Freunde Berlin Neue Schriften 3: 448. 1801 • [For Heinrich Gustav Floerke, 1764–1835, German lichenologist] E

Gordon C. Tucker

Stems decumbent to erect. **Leaves** ternate or pinnately lobed; leaflet blade oblanceolate to elliptic, sometimes 2–3-parted, margins entire. **Pedicels** spreading or nodding. **Flowers:** sepals 3, lanceolate to ovate-lanceolate; petals 3, shorter than sepals; nectary glands antisepalous; stamens 3 or 6; ovary 3-lobed. **Fruits** mericarps, ovoid-globose, tuberculate or rugulose (especially distally). $x = 5$.

Species 1: North America.

SELECTED REFERENCE Houle, G., M. F. McKenna, and L. Lapointe. 2001. Spatiotemporal dynamics of *Floerkea proserpinacoides* (Limnanthaceae), an annual plant of the deciduous forest of eastern North America. Amer. J. Bot. 88: 594–607.

1. Floerkea proserpinacoides Willdenow, Ges. Naturf. Freunde Berlin Neue Schriften 3: 449. 1801 • False mermaid, floerkée fausse E F

Cabomba pinnata (Pursh) Schultes & Schultes f.; *Nectris pinnata* Pursh

Plants (3–)5–30(–38) cm, herbage glabrous. **Leaves:** petiole 0.5–4 cm; rachis 2.5–5 mm; leaflets 3–7, 7–12 × 1–3.5 mm, apex obtuse or acute. **Pedicels** 0.5–2 cm (± equal to subtending petiole at anthesis, elongating 1.5 times by fruit maturity). **Flowers:** sepals green, 5-nerved, 2–6 mm; petals white, greenish white, or pale pink, oblanceolate to narrowly elliptic, 1.4–2 mm; filaments filiform, 1–1.5 mm; anthers yellowish, broadly ellipsoid, 0.2–0.3 mm. **Mericarps** green to brown, somewhat fleshy, 2–3.5 × 2–3 mm (often only 1 seed maturing per flower). $2n = 10$.

Flowering spring–early summer. Floodplain forests, swamps, wet-mesic coniferous or broadleaf woods, alpine meadows, pastures, moist areas in sagebrush or desert washes; 50–2600(–3200) m; B.C., N.S., Ont., Que.; Calif., Colo., Conn., Del., D.C., Idaho, Ill., Ind., Iowa, Ky., La., Md., Mass., Mich., Minn., Mo., Mont., Nev., N.J., N.Y., N.Dak., Ohio, Oreg., Pa., Tenn., Utah, Vt., Va., Wash., W.Va., Wis., Wyo.

The stems of *Floerkea proserpinacoides* have a spicy flavor and are eaten in salads. A flower of *F. proserpinacoides* is the logo for the Flora of North America project.

2. LIMNANTHES R. Brown, London Edinburgh Philos. Mag. & J. Sci. 3: 71. 1833, name conserved • Meadowfoam [Greek *limne*, marsh, and *anthe*, flower, alluding to habitat] E

Nancy R. Morin

Stems decumbent, erect, or ascending. **Leaves** usually pinnately lobed or compound; leaflet blade linear, lanceolate, elliptic, ovate, narrowly obovate, or oblong, margins entire or deeply lobed. **Pedicels** erect. **Flowers:** sepals (4 or) 5, ovate, obovate, lanceolate, linear-lanceolate, or lanceolate-ovate; petals (4 or) 5, usually longer than sepals; nectary glands at base of antisepalous stamens; stamens 8 or 10; ovary 4 or 5-lobed. **Fruits** nutlets, obovoid or subglobose, smooth, ridged, or tuberculate. $x = 5$.

Species 7 (7 in the flora): w North America.

Limnanthes was divided into two sections based on whether the petals flex outward (sect. *Reflexae* [= sect. *Limnanthes*]) or inward (sect. *Inflexae*) in fruit. The sections are supported by analysis of DNA (M. S. Plotkin 1998) and petal flavonoids (W. H. Parker and B. A. Bohm 1979), although not by analysis of whole-plant flavonoids, which Parker and Bohm suggested were linked to changes in the breeding system. R. Ornduff (1971) found that hybrids did not form in crosses attempted between the two sections.

In each section, there are highly allogamous, outbreeding taxa, and highly autogamous, inbreeding taxa. Flowers are produced acropetally (proximally to distally) along stems. In *Limnanthes alba*, an outbreeding species, the sepals and petals open during the day and, if not visited by a pollinator, flowers close at night, reopening on subsequent days until pollination occurs (T. R. Jahns et al., http://extension.oregonstate.edu/catalog/html/em/em8666-e). The stamens are in two whorls—antisepalous and antipetalous. In at least the outbreeding taxa, the anthers initially are pressed outward to the petals; the antisepalous ones move to the center of the flower, dehiscing first, followed by the antipetalous ones. The stigma is positioned well proximal to the anthers at anthesis; the style elongates over one to four days, with the stigmatic branches separating and becoming receptive and beadlike after the anthers have dehisced. In *L. floccosa*, a mostly autogamous species, anthers and stigmas are at the same level at anthesis and pollen is shed directly on the stigmas. Pollen of *Limnanthes* has an apertural band dividing the grain into two unequal portions (K. L. Huynh 1982). Floral nectaries are formed abaxially at the bases of the antisepalous stamens (D. A. Link 1992). Some populations of *L. douglasii* subsp. *nivea* have plants that are pollen-sterile (R. V. Kesseli and S. K. Jain 1984, 1987).

W. H. Parker (1981) reported that the corollas of *Limnanthes douglasii* subsp. *douglasii* and *L. macounii* have a distinctive bull's-eye pattern in visible light that is mirrored under ultraviolet light in *L. douglasii* subspp. *nivea* and *rosea*. Corollas of *Limnanthes douglasii* subspp. *rosea* and *striata* have longitudinal lines in visible light that serve as nectar guides; similar lines are visible under ultraviolet light in *L. alba* subsp. *parishii* and *L. floccosa* subsp. *bellingeriana*. Longitudinal rows of hairs on the base of each petal form a five-pointed star under ultraviolet light in most taxa. The ultraviolet lines are especially well-developed in *L. montana*, which also has unique ultraviolet reflective petal hairs.

J. M. Leong and R. W. Thorp (2005) found high levels of bee diversity in vernal pool populations of *Limnanthes douglasii* subsp. *rosea* in the Jepson Prairie near Sacramento, California. Of the 1970 bees trapped, 1598 were the solitary bee *Andrena limnanthis* Timberlake (= *Andrena pulverea* Viereck), which specializes on *L. douglasii* subsp. *rosea* and other taxa of the genus (e.g., *L. alba*, *L. douglasii* subsp. *striata*, *L. montana*, and *L. vinculans*). In total, 32 species of bees were foraging on the patches of *L. douglasii* subsp. *rosea* that were studied.

The carpels of *Limnanthes* become one-seeded nutlets. In most species, the nutlets are dispersed as individual units; in *L. floccosa* subspp. *bellingeriana*, *californica*, *floccosa*, and *grandiflora*, the nutlets are shed together with the enveloping calyx and corolla. H. H. Hauptli et al. (1978) identified eight distinct types and four subtypes of sculpturing on the nutlet surface, ranging from smooth to highly tuberculate; the tubercles platelike, conical, or filamentous. They found that taxa with the least tuberculate nutlets occurred in comparatively xeric sites, such as meadows and grasslands, whereas those with highly tuberculate nutlets occurred in wetter sites and had nutlets that floated longer than smoother nutlets.

A light, colorless, triglyceride oil composed of highly stable, long-chain fatty acids is derived from the seeds of *Limnanthes*. The oil is used commercially in personal care products and may have other applications. Interest in growing *Limnanthes* for its seeds on a commercial scale, and for use in genetic engineering, has stimulated studies of population genetics (e.g., V. K. Kishore et al. 2004), morphology and physiology (e.g., S. Krebs and S. K. Jain 1985), and cultivation requirements.

Limnanthes alba is the obligate host for the meadowfoam fly, *Scaptomyza apicalis* Hardy (S. Panasahatham et al. 1999). The larvae feed on the crowns and the stems, and later tunnel into the flower buds.

Limnanthes alba subspp. *parishii* and *gracilis*, *L. bakeri*, *L. douglasii* subsp. *sulphurea*, *L. floccosa* subspp. *bellingeriana*, *pumila*, *grandiflora*, and *californica*, *L. macounii*, and *L. vinculans* are all taxa of conservation concern. Most are restricted to vernal pools, an endangered habitat in California, Oregon, and British Columbia.

SELECTED REFERENCES Hauptli, H. H., B. D. Webster, and S. K. Jain. 1978. Variation in nutlet morphology in *Limnanthes*. Amer. J. Bot. 65: 615–624. Kesseli, R. V. and S. K. Jain. 1987. Origin of gynodioecy in *Limnanthes*. Theor. Appl. Genet. 74: 379–386. Mason, C. T. 1952. A systematic study of the genus *Limnanthes* R. Br. Univ. Calif. Publ. Bot. 25: 455–512.

Key to Sections:

1. Petals flexed outward (reflexed) as fruits mature . . . . 2a. *Limnanthes* sect. *Limnanthes*, p. 175
1. Petals flexed inward (curving over fruit) as fruits mature . . . . 2b. *Limnanthes* sect. *Inflexae*, p. 179

2a. LIMNANTHES R. Brown sect. LIMNANTHES E

Limnanthes sect. *Reflexae* C. T. Mason

Herbage glabrous. **Flowers** usually 5-merous (4 in *L. macounii*); sepals not accrescent; petals flexed outward (reflexed) as fruits mature, white, bright to pale yellow, or yellow with white tips, veins translucent, purplish, pink, rose, white, or cream, sometimes aging or drying lilac, pink, chartreuse, or pale yellow, longer than sepals, with marginal hairs basally.

Species 4 (4 in the flora): w North America.

SELECTED REFERENCE Kesseli, R. V. and S. K. Jain. 1984b. New variation and biosystematic patterns detected by allozyme and morphological comparison in *Limnanthes* sect. *Reflexae* (Limnanthaceae). Pl. Syst. Evol. 147: 133–164.

1. Sepals 4; petals 4; stems decumbent (sometimes upcurved apically) . . . . 3. *Limnanthes macounii*
1. Sepals 5; petals 5; stems erect or ascending.
    2. Leaflet blade margins usually toothed or lobed, sometimes entire; sepals 4–10 mm
        . . . . 4. *Limnanthes douglasii*
    2. Leaflet blade margins usually entire (rarely 2- or 3-lobed); sepals 5–7 mm.
        3. Petals 7–9 mm; styles 2–3 mm . . . . 1. *Limnanthes bakeri*
        3. Petals 12–18 mm; styles 4.5–6.5 mm . . . . 2. *Limnanthes vinculans*

1. Limnanthes bakeri J. T. Howell, Leafl. W. Bot. 3: 206. 1943 • Baker's meadowfoam C E

Plants 10–40 cm. Stems erect or ascending. Leaves 3–10 cm; leaflets 3–9, blade elliptic to ovate, margins usually entire (rarely 2- or 3-lobed). Flowers funnel- to bell-shaped; sepals lanceolate, 5–7 mm; petals pale yellow with white tips, cuneate, 7–9 mm, apex truncate, erose; filaments 2.5–4 mm; anthers cream, 0.5 mm; style 2–3 mm. Nutlets dark brown, 3–3.5 mm, tuberculate, tubercles light brown or pinkish, rounded. $2n = 10$.

Flowering Apr–May. Vernal pools, marshy margins of pools; of conservation concern; 100–900 m; Calif.

Limnanthes bakeri is known from the Inner Coast Ranges of Mendocino County. It is easily recognized by having relatively few leaves with broad, mostly entire leaflets. The stamens and styles are about equal in length and are pressed together by a funnel-shaped corolla, facilitating self-pollination (R. V. Kesseli and S. K. Jain 1984b). Combined ITS, *trn*L, and morphological analyses placed *L. bakeri* in a basal position with respect to the *Limnanthes* (as sect. *Reflexae*) clade, and most closely allied with *L. vinculans* (M. S. Plotkin 1998). W. H. Parker and B. A. Bohm (1979) suggested that *Floerkea proserpinacoides* and *L. bakeri* separated from the family line long before the separation of other species.

2. Limnanthes vinculans Ornduff, Brittonia 21: 11, fig. 1. 1969 • Sebastopol meadowfoam [C][E]

Plants 5–30 cm. Stems erect or ascending. Leaves 2–10 cm; leaflets 3–5, blade narrowly obovate, margins entire. Flowers bell-shaped to rotate; sepals widely lanceolate, 6–7 mm; petals white (aging or drying yellowish basally), cuneate, 12–18 mm, apex emarginate; filaments 5–7 mm; anthers white, 1.5–2 mm; style 4.5–6.5 mm. Nutlets dark brown, 3–4 mm, tuberculate, tubercles light brown or pinkish, rounded. $2n = 10$.

Flowering Apr–May. Wet meadows; of conservation concern; 0–300 m; Calif.

Limnanthes vinculans is the only member of the genus in which the first few leaves produced by the seedling are entire rather than lobed. It is known only from Sonoma and Napa counties, where it occurs in the centers of vernal pools. Some populations are sympatric with *L. douglasii* subsp. *nivea*, which tends to grow on the drier edges of vernal pools. *Limnanthes vinculans* is in the Center for Plant Conservation's National Collection of Endangered Plants.

3. Limnanthes macounii Trelease, Mem. Boston Soc. Nat. Hist. 4: 85. 1888 • Macoun's meadowfoam [C][E]

Plants 2–7(–15) cm. Stems decumbent (sometimes upcurved apically). Leaves 1–7 cm; leaflets 3–15, blade ovate, margins irregularly toothed to 5-lobed. Flowers bowl- to bell-shaped; sepals (4) ovate, 3–4 mm; petals (4) white, cuneate to obovate, 4–5 mm, apex emarginate (reflexed in fruit); filaments 2–2.5 mm; anthers cream, 0.3 mm; style 3 mm. Nutlets light brown, 3 mm, tuberculate, tubercles light brown, broadly conical. $2n = 10$.

Flowering Mar–early May. Depressions in shallow soil on rocks, seepage areas, rocky coastal areas, open forests; of conservation concern; 0–200 m; B.C.; Calif.

Limnanthes macounii is native on Vancouver Island and adjacent islands in British Columbia from East Sooke Park to Victoria, Inskip, Chatham, and Trial islands, to Yellow Point, Saltspring, Gabriola, and Hornby islands. It grows in seasonally moist depressions (including edges of vernal pools) in acidic soils that often have a high nutrient content. The population in California is in an agricultural field in San Mateo County; first discovered there in 1998 (E. G. Buxton 1998), it appears to be persisting. Phylogenetic analyses suggest this may be an aberrant population of *L. douglasii* (S. Meyers, pers. comm.)

The angular and comparatively massive tubercles of the nutlets of *Limnanthes macounii*, with the base equaling 1/4–1/3 the width of the nutlet, are unique in the genus (H. H. Hauptli et al. 1978). Allozyme studies by R. V. Kesseli and S. K. Jain (1984b) showed that *L. macounii* had alleles found in no other taxa at three loci. M. S. Plotkin (1998) concluded that *L. macounii* should be included in *L. douglasii* based on ITS analysis. Because of its unique characteristics and highly disjunct distribution, *L. macounii* is maintained as a species here pending further study.

SELECTED REFERENCE Committee on the Status of Endangered Wildlife in Canada. 2004. COSEWIC assessment and update status report on the Macoun's meadowfoam *Limnanthes macounii* in Canada. Ottawa.

4. Limnanthes douglasii R. Brown, London Edinburgh Philos. Mag. & J. Sci. 3: 71. 1833 (as douglassii) [E][F]

Plants 3–35(–100) cm. Stems erect or ascending. Leaves 3–7(–25 cm); leaflets 5–13, blade linear, ovate, or wide-ovate, margins usually toothed or lobed, sometimes entire. Flowers cup-, bell-, or funnel-shaped; sepals narrowly lanceolate-ovate, lanceolate, or linear-lanceolate, 4–10 mm (glabrous or ciliate); petals white, yellow, or yellow with white tips (sometimes greenish yellow basally), cuneate, obovate, spatulate, or obovate-spatulate, 7–18 mm, apex usually emarginate, sometimes truncate or obcordate; filaments 2–7 mm; anthers usually cream or yellow, sometimes dark pink, orange-red, or nearly black, 0.8–1.5 mm; style 3–8 mm. Nutlets dark brown, black, or gray, 2–5 mm, smooth, ridged, or tuberculate (at least apically), tubercles or ridges brown, pinkish, gray, or whitish, lamellar, planar, conical, rounded, or blunt.

Subspecies 5 (5 in the flora): w United States.

Limnanthes douglasii has five subspecies with core distributions as follows: subsp. *sulphurea* with two populations near the coast in Marin County and in southwestern San Mateo County; subsp. *douglasii* primarily coastal and Outer Coast Ranges with outlying populations in the northern Sierra Nevada foothills; subsp. *nivea* primarily Inner Coast Ranges and Bay Area with Central Valley and northern Sierra Nevada foothills populations; subsp. *rosea* in the Inner Coast Ranges, Central Valley, and Sierra Nevada foothills; and subsp. *striata* in the central Sierra Nevada foothills with outlying populations in Trinity County.

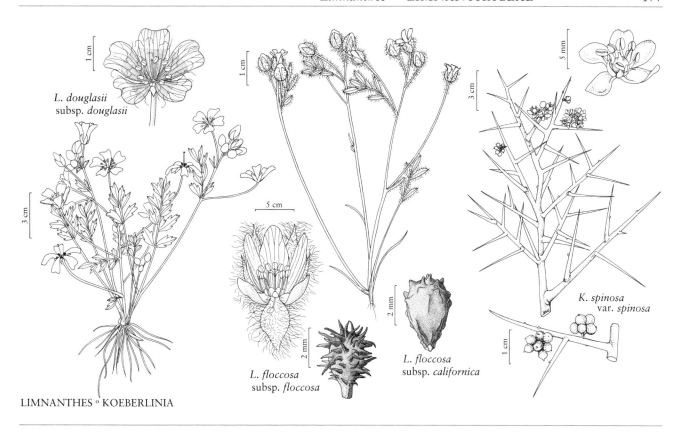

1. Flowers funnel-shaped; petals white with greenish yellow bases (veins usually dark); anthers usually cream, rarely dark......................
   .... 4d. *Limnanthes douglasii* subsp. *striata*
1. Flowers cup- or bell-shaped; petals yellow and/or white (veins rose, white, yellow, or purplish, not dark); anthers cream, yellow, dark pink, orange-red, reddish brown, or nearly black.
   2. Petals yellow or yellow with white tips.
      3. Petals yellow with white tips; anthers cream, yellow, or reddish brown........
         ...... 4a. *Limnanthes douglasii* subsp. *douglasii*
      3. Petals yellow; anthers yellow ....
         ......4b. *Limnanthes douglasii* subsp. *sulphurea*
   2. Petals white (sometimes aging or drying pale pink or yellowish).
      4. Petals 7–13 mm (veins white or purplish); anthers cream to yellow; sepals 4–7 mm
         .... 4c. *Limnanthes douglasii* subsp. *nivea*
      4. Petals 12–18 mm (veins rose); anthers cream, yellow, dark pink, orange-red, or nearly black; sepals 7–8 mm...........
         .... 4e. *Limnanthes douglasii* subsp. *rosea*

4a. Limnanthes douglasii R. Brown subsp. **douglasii**
E F

Leaflets: blade linear to ovate, margins entire, irregularly toothed, or lobed. **Flowers** cup- or bell-shaped; sepals narrowly lanceolate-ovate, 6–9 mm; petals yellow with white tips (veins yellow), obovate-cuneate, 12–18 mm, (apex deeply emarginate); filaments 6–7 mm; anthers cream, yellow, or reddish brown, oval, 1.1–1.5 mm. **Tubercles** absent or not (or 1 row of whitish globules on ridges and crests), brown, lamellar, planar, or conical. $2n = 10$.

Flowering Mar–Jun. Wet meadows; 0–700 m; Calif., Oreg.

Subspecies *douglasii* is common in vernal pools and moist areas in the Coast Ranges from San Benito County to Humboldt County in California, and in the Umpqua Valley. A disjunct population occurs in the Sierra Nevada foothills near Honcut in Butte County, California.

R. V. Kesseli and S. K. Jain (1984b) found clusters of populations in the Inner Coast Ranges, predominantly Mendocino and Lake counties ("inland"), that they considered to be intermediate between *Limnanthes douglasii* subspp. *douglasii* and *nivea*, and clusters of those two subspecies plus subsp. *sulphurea* in coastal areas ("coastal"). Inland plants had open, usually smaller flowers than the coastal group, with petals ca. 13 mm, white or pale, or dark yellow-centered, the yellow extending from the base over half of the petal length and fading to white, and apical notch rounded. These populations flowered later and had narrower leaflets than the coastal cluster. The coastal cluster had a high frequency of one allele that was absent in the inland cluster. The coastal group had larger flowers, petals ca. 14.7–16 mm, obcordate, completely dark yellow- to yellow-centered (the yellow extending from the base less than half the length of the petal, sharply demarcated from the white part) or white, with earlier flowering and broader leaflets. These studies indicate that the easily recognizable flower color patterns used to characterize subsp. *douglasii* may not reflect genetic affinities. Additional research is warranted. Kesseli and Jain found also that the Oregon populations of subsp. *douglasii* had alleles that were rare or nonexistent in all other subspecies of *L. douglasii*, and that these populations had distinctive nutlets (possibly more similar to *L. bakeri* and *L. vinculans*) and shorter stamens and styles than usual in subsp. *douglasii*.

4b. Limnanthes douglasii R. Brown subsp. **sulphurea** (C. T. Mason) C. T. Mason, Madroño 36: 50. 1989
 • Point Reyes meadowfoam [C] [E]

Limnanthes douglasii var. *sulphurea* C. T. Mason, Univ. Calif. Publ. Bot. 25: 477, fig. 7, plates 43, 44, 45b. 1952

Leaflets: blade ovate, margins irregularly toothed or lobed. **Flowers** cup- or bell-shaped; sepals lanceolate-ovate, 8–10 mm; petals yellow (veins yellow), obovate-cuneate, 12–18 mm (apex deeply emarginate); filaments 4 mm; anthers yellow, oval, 1.5 mm. **Tubercles** absent or not (without surface globules, appearing slick), brown, lamellar, planar, or conical. $2n = 10$.

Flowering Apr–Jun. Wet meadows of coastal prairies; of conservation concern; 0–100 m; Calif.

Subspecies *sulphurea* is easily recognized by its entirely yellow petals. It is known only from Point Reyes, Marin County, and the Butano Creek area, San Mateo County (R. Ornduff 1971).

4c. Limnanthes douglasii R. Brown subsp. **nivea** (C. T. Mason) C. T. Mason, Madroño 36: 50. 1989 [E]

Limnanthes douglasii var. *nivea* C. T. Mason, Univ. Calif. Publ. Bot. 25: 477, fig. 7, plates 43, 45c. 1952

Leaflets: blade linear to ovate, margins entire or shallowly to deeply lobed. **Flowers** cup- or bell-shaped; sepals narrowly lanceolate-ovate, 4–7 mm; petals white (sometimes aging or drying pale yellow basally, veins white or purplish), cuneate, 7–13 mm; filaments 4–6 mm; anthers cream to yellow, oblong-sagittate, 1 mm. **Tubercles** absent or not (or 1 row of whitish globules on ridges and crests), brown, lamellar, planar, or conical. $2n = 10$.

Flowering Mar–Jun. Wet meadows, edges of vernal pools, ephemeral streams; 0–1000 m; Calif.

Subspecies *nivea* occurs in the Coast Ranges from San Luis Obispo County to Humboldt County, with disjunct populations in Butte County.

4d. Limnanthes douglasii R. Brown subsp. **striata** (Jepson) Morin, J. Bot. Res. Inst. Texas 1: 1017. 2007 [E]

Limnanthes striata Jepson, Fl. Calif. 2: 411. 1936

Leaflets: blade linear to ovate, margins entire, irregularly toothed, or 2- or 3-lobed. **Flowers** funnel-shaped; sepals linear-lanceolate, 4–6(–7) mm; petals white with greenish yellow bases (veins usually dark), spatulate (with narrow claw and narrowly ovate blade), 8–17 mm (apex truncate to slightly emarginate); filaments 2–4 mm; anthers usually cream, rarely dark, almost round, 0.8–1 mm. **Tubercles** absent or not, light brown, pinkish, or gray, rounded, blunt, lamellar, or conic. $2n = 10$.

Flowering Apr. Vernal pools, stream edges; 0–800 m; Calif.

C. T. Mason (1952) treated *Limnanthes striata* as a distinct species and suggested that it is closely related to *L. douglasii* and possibly should be included there. R. V. Kesseli and S. K. Jain (1984b) found the "*striata*" cluster to be an integral component of the "*douglasii*" complex. They identified two populations that keyed as subsp. *striata* but did not share alleles with other populations of subsp. *striata*. Of the two populations, the one in the eastern foothills of the Coast Range near Junction City, Trinity County, had fleshy sepals and long pedicels. The other population is in Bear Valley, Mariposa County, at the highest and southernmost point of the distribution of subsp. *striata*. The Bear Valley population has relatively small flowers with relatively short petals, but relatively longer sepals and dense hairs on the sepals and leaves. Kesseli (pers. comm.) reported that he was unable to cross subsp. *striata* with other *L. douglasii* subspecies and had not seen hybrids in the field. Subspecies *striata* warrants further study.

4e. Limnanthes douglasii R. Brown subsp. **rosea** (Hartweg ex Bentham) C. T. Mason, Madroño 36: 50. 1989 [E]

Limnanthes rosea Hartweg ex Bentham, Pl. Hartw., 302. 1849; *L. douglasii* var. *rosea* (Hartweg ex Bentham) C. T. Mason

Leaflets: blade usually linear, sometimes wide-ovate, margins entire or deeply, irregularly lobed. **Flowers** cup- or bell-shaped; sepals lanceolate, 7–8 mm; petals white (sometimes aging or drying pale pink or yellow, veins rose), cuneate to obovate, 12–18 mm (apex sometimes obcordate); filaments 5 mm; anthers cream, yellow, dark pink, orange-red, or nearly black, oval, 1.5 mm. **Tubercles** present, whitish or pinkish, lamellar, planar, or blunt. $2n = 10$.

Flowering Mar–May. Wet meadows, edges of vernal pools; 0–800 m; Calif.

Subspecies *rosea* occurs in the Inner Coast Ranges, Central Valley, and Sierra Nevada Foothills. C. T. Mason (1952) noted that populations in the northern part of the range tend to have white petals with pink veins, stamens, and pistils; those in the southern part of the range have petals with cream bases and yellow stamens and pistils.

2b. LIMNANTHES R. Brown sect. **INFLEXAE** C. T. Mason, Univ. Calif. Publ. Bot. 25: 473. 1952 [E]

Herbage glabrous or glabrate to densely hairy. **Flowers** 5-merous; sepals accrescent or not; petals flexed inward (curving over fruit) as fruits mature, usually white or greenish white, sometimes yellowish proximally, veins sometimes purplish or dark yellow proximally, aging or drying white, lilac, or pink, usually longer than sepals (shorter to slightly longer in *L. floccosa*), usually with marginal hairs basally (except *L. floccosa* subspp. *bellingeriana* and *floccosa*).

Species 3 (3 in the flora): w United States.

The species of sect. *Inflexae* are distributed from southwestern Oregon to southern California (San Diego County). *Limnanthes montana* occurs in the central Sierra Nevada. *Limnanthes floccosa* has one widespread subspecies, *floccosa*, occurring at lower elevations of the northern Sierra Nevada and Klamath Mountains, with some outliers in the Central Valley and Inner Coast Range, and four other subspecies of very limited distribution in northern California and southern Oregon. *Limnanthes alba* has two widespread subspecies, *alba* in the foothills of the Sierra Nevada, and *versicolor* at higher elevations in the Sierra Nevada and Klamath Mountains, and two localized subspecies, *gracilis* in southwestern Oregon, and *parishii* in San Diego County.

SELECTED REFERENCE   McNeill, C. I. and S. K. Jain. 1983. Genetic differentiation studies and phylogenetic inference in the plant genus *Limnanthes* (section *Inflexae*). Theor. Appl. Genet. 66: 257–269.

1. Petals 0.5–1.1 times longer than sepals . . . . 7. *Limnanthes floccosa*
1. Petals 1.25–2.25 times longer than sepals.
   2. Petals ca. 2.75 times as long as wide, aging or drying white; anthers 0.5–0.8(–1) mm . . . . 5. *Limnanthes montana*
   2. Petals 1–1.5 times as long as wide, aging or drying pink or lilac; anthers 1–2 mm . . . . 6. *Limnanthes alba*

5. Limnanthes montana Jepson, Fl. Calif. 2: 412. 1936 E

Plants 10–40 cm; herbage glabrous or sparsely hairy (hairs long). Stems ascending to erect. Leaves 3–15 cm; leaflets 7–11, blade linear to ovate, margins entire, or shallowly 2-lobed to deeply 3-lobed. Flowers funnel-shaped; sepals not accrescent, lanceolate to ovate-lanceolate, 3–6 mm; petals white (sometimes yellowish basally, or veins purplish, aging or drying white), cuneate to obovate, 7–12 mm, ca. 2.75 times as long as wide, 2–2.5 times longer than sepals, apex truncate, emarginate; filaments 2.5–4 mm; anthers (cream), 0.5–0.8(–1) mm; style 2.5–4 mm. Nutlets gray, 2–3 mm, tuberculate, tubercles gray, conic. $2n = 10$.

Flowering (Feb–)Mar–Jun. Wet meadows, stream edges; 200–1800 m; Calif.

C. T. Mason (1952) recognized *Limnanthes montana* as a species and *L. gracilis* as having two subspecies: *gracilis* and *parishii*. He suggested that *L. montana* might represent a central remnant of an earlier, widespread species, with *L. alba* occupying the central part and *L. gracilis* subspp. *gracilis* and *parishii* being the northern and southern relictual populations, respectively. C. I. McNeill and S. K. Jain (1983), based on allozyme studies and morphology, concluded that subspp. *gracilis* and *parishii* are more closely related to *L. alba* and that *L. montana* is more distant. This conclusion was supported by combined morphological and molecular analyses (M. S. Plotkin 1998). R. Ornduff (1971) found that *L. montana* did not hybridize with subspp. *gracilis* or *parishii*.

Nutlet morphology (densely packed, small, conic tubercles), flower shape (funnel-form), and the larger number of leaflets of *Limnanthes montana* are distinctive. *Limnanthes montana* usually has extremely small anthers, averaging ca. 0.5 mm; populations in Mariposa and Madera counties have anthers ca. 0.8 mm, rarely to 1 mm. *Limnanthes montana* may have some hairs on the veins of the petals; *L. alba* subsp. *versicolor* has hairs scattered throughout the petal surface. Anther size and hairs on the petals are useful characters to separate *L. montana* from *L. alba*.

6. Limnanthes alba Hartweg ex Bentham, Pl. Hartw., 301. 1849 E

Plants 8–40 cm; herbage glabrous or sparsely to densely hairy. Stems erect. Leaves 2–10 cm; leaflets 5–9, blade oblong, ovate, lanceolate, or linear-lanceolate, margins entire or shallowly 2-lobed to deeply 3-lobed. Flowers bowl- to bell-shaped; sepals accrescent or not, lanceolate, ovate, or lanceolate-ovate, 4–8 mm; petals white or cream (sometimes cream basally, aging or drying pink or lilac), obovate, obovate-cuneate, or obcordate, 8–16 mm, 1–1.5 times as long as wide, 1.25–2.1 times longer than sepals, apex usually emarginate, sometimes truncate; filaments 3–6 mm; anthers 1–2 mm; style 2–6 mm. Nutlets gray or dark brown, 3–4 mm, tuberculate or not, sometimes ridged, tubercles gray or dark brown, relatively low and wide. $2n = 10$.

Subspecies 4 (4 in the flora): w United States.

Subspecies *gracilis* and *parishii*, previously in *Limnanthes gracilis*, are more closely related to subsp. *alba* than they are to each other. The genetic distance between them and other subspecies of *L. alba* is sufficient that an argument could be made for recognizing both *parishii* and *gracilis* as species (R. V. Kesseli, pers. comm.).

SELECTED REFERENCE Ritland, K. and S. K. Jain. 1984. A comparative study of floral and electrophoretic variation with life history variation in *Limnanthes alba* (Limnanthaceae). Oecologia 63: 243–251.

1. Herbage sparsely to densely hairy; sepals densely hairy; petals 10.5–16 mm . . . . 6a. *Limnanthes alba* subsp. *alba*
1. Herbage glabrous; sepals glabrous or sparsely hairy; petals 8–15 mm.
   2. Petals 8–10 mm, 1 times as long as wide, obovate; styles 2–3 mm; San Diego County . . . . 6c. *Limnanthes alba* subsp. *parishii*
   2. Petals (8–)12–15 mm, 1–1.3 times as long as wide, obcordate or obovate; styles 3–4 mm; central Sierra Nevada to southern Oregon.
      3. Leaflet blade narrowly oblong; petals cream, aging or drying lilac, obcordate, sparsely hairy (hairs long) . . . . 6b. *Limnanthes alba* subsp. *versicolor*
      3. Leaflet blade ovate to linear-lanceolate; petals usually white, sometimes cream basally, aging or drying pink, obovate, glabrous . . . . 6d. *Limnanthes alba* subsp. *gracilis*

6a. Limnanthes alba Hartweg ex Bentham subsp. alba [E]

Herbage sparsely to densely hairy. **Leaflets:** blade oblong to lanceolate. **Flowers:** sepals not accrescent, ovate, 6–8 mm, densely hairy; petals white, aging or drying pink, broadly obovate-cuneate, 10.5–16 mm, 1.2–1.5 times as long as wide (apex truncate or slightly emarginate, with very broad notch), sparsely hairy; filaments 4–6 mm; anthers cream, ± 2 mm; style 5–6 mm. **Nutlets** ca. 4 mm, smooth or ridged (ridges relatively wide).

Flowering Mar–May. Winter-wet grasslands, woodlands; 0–1400 m; Calif., Oreg.

Subspecies *alba* occurs on the east side of the Sacramento Valley and in the foothills of the Sierra Nevada. It has escaped from cultivation in the Willamette Valley of western Oregon on roadsides.

6b. Limnanthes alba Hartweg ex Bentham subsp. versicolor (Greene) C. T. Mason, Madroño 36: 51. 1989 [E]

Floerkea versicolor Greene, Erythea 3: 62. 1895; *Limnanthes alba* var. *versicolor* (Greene) C. T. Mason

Herbage glabrous. **Leaflets:** blade narrowly oblong. **Flowers:** sepals not accrescent, broadly lanceolate, 6–7 mm, glabrous or sparsely hairy; petals cream, aging or drying lilac, obcordate, 12–15 mm, 1.2–1.3 times as long as wide (apex recurved), sparsely hairy (hairs long); filaments ca. 4 mm; anthers cream or pinkish, ± 1 mm; style ± 4 mm. **Nutlets** 3.5 mm, smooth, tuberculate, or ridged.

Flowering Apr–May(–Jul). Winter-wet grasslands, woodlands, edges of vernal pools, edges of ephemeral streams; 150–2100 m; Calif.

6c. Limnanthes alba Hartweg ex Bentham subsp. parishii (Jepson) Morin, J. Bot. Res. Inst. Texas 1: 1017. 2007 [C] [E]

Limnanthes versicolor (Greene) Rydberg var. *parishii* Jepson, Fl. Calif. 2: 412. 1936; *L. gracilis* Howell subsp. *parishii* (Jepson) R. M. Beauchamp; *L. gracilis* var. *parishii* (Jepson) C. T. Mason

Herbage glabrous. **Leaflets:** blade lanceolate to ovate. **Flowers:** sepals accrescent, lanceolate-ovate, 4–8 mm, glabrous; petals usually white, sometimes cream basally, aging or drying pink, obovate, 8–10 mm, 1 times as long as wide (apex recurved), glabrous or with scattered hairs on veins; filaments 3–4 mm; anthers cream to yellow, ± 1 mm; style 2–3 mm. **Nutlets** 3–4 mm, tuberculate (tubercles relatively low and wide).

Flowering Mar–May(–Jun). Wet meadows, ephemeral stream edges; of conservation concern; 600–2000 m; Calif.

Subspecies *parishii* has been found only in Riverside and San Diego counties.

6d. Limnanthes alba Hartweg ex Bentham subsp. gracilis (Howell) Morin, J. Bot. Res. Inst. Texas 1: 1017. 2007 [C] [E]

Limnanthes gracilis Howell, Fl. N.W. Amer., 108. 1897

Herbage glabrous. **Leaflets:** blade ovate to linear-lanceolate. **Flowers:** sepals accrescent, lanceolate-ovate, 4–8 mm, glabrous; petals usually white, sometimes cream basally, aging or drying pink, obovate, 8–10 mm, as long as wide, glabrous; filaments 3–4 mm; anthers cream, 1 mm; style 3–4 mm. **Nutlets** ca. 3 mm, tuberculate (tubercles relatively low and wide).

Flowering Mar–May. Seasonally wet meadows, rocky slopes and basins, often on serpentine soils; of conservation concern; 150–1700 m; Oreg.

Subspecies *gracilis* occurs in the Rogue River Valley in Josephine and Jackson counties and in Douglas County.

7. Limnanthes floccosa Howell, Fl. N.W. Amer., 108. 1897 • Woolly meadowfoam E F

Plants 3–25 cm; herbage glabrous or sparsely to densely hairy. Stems erect to ascending or decumbent. Leaves 1–8 cm; leaflets 5–11, blade linear to ovate-elliptic, margins entire, irregularly toothed, or lobed. Flowers urn-, cup-, or bell-shaped; sepals accrescent or not, ovate, obovate, lanceolate, oblong-lanceolate, or lanceolate-ovate, 4–10 mm; petals white, obovate, oblong, or obovate-cuneate, 4.5–10 mm, 1.6–1.8 times as long as wide, 0.5–1.1 times longer than sepals, apex retuse, obtuse, erose, truncate, or emarginate; filaments 2–7 mm; anthers (yellow) 0.4–1.5 mm; style 1.5–4 mm. Nutlets dark brown or gray, 3–4.5 mm, tuberculate, tubercles straw-colored, platelike, conic, or awl-shaped.

Subspecies 5 (5 in the flora): w United States.

SELECTED REFERENCE Arroyo, M. T. K. 1973. A taxometric study of infraspecific variation in autogamous Limnanthes floccosa (Limnanthaceae). Brittonia 25: 177–191.

1. Flowers bell- or urn-shaped (petals without marginal hairs basally); anthers usually dehiscing introrsely.
   2. Sepals abaxially and adaxially densely villous; nutlet tubercles awl-shaped ....
      .... 7a. Limnanthes floccosa subsp. floccosa
   2. Sepals abaxially and adaxially glabrous or sparsely hairy; nutlet tubercles platelike.....
      ...... 7b. Limnanthes floccosa subsp. bellingeriana
1. Flowers cup-shaped (petals with marginal hairs basally); anthers usually dehiscing extrorsely (introrsely in subsp. pumila).
   3. Herbage densely hairy; sepals abaxially densely hairy ....
      .... 7e. Limnanthes floccosa subsp. californica
   3. Herbage glabrous or sparsely hairy; sepals abaxially glabrous or sparsely hairy.
      4. Sepals not accrescent, 7.5–8 mm, adaxially glabrous.......................
         ....7c. Limnanthes floccosa subsp. pumila
      4. Sepals accrescent, 8.5–9 mm, adaxially densely hairy......................
         ..... 7d. Limnanthes floccosa subsp. grandiflora

7a. Limnanthes floccosa Howell subsp. floccosa E F

Herbage densely hairy. Stems erect to ascending. Flowers bell- to urn-shaped; sepals accrescent, ovate, 4–9 mm, abaxially and adaxially densely villous; petals obovate, 4.5–8.5 mm, apex rounded (without marginal hairs basally); filaments 2–4.5 mm; anthers 0.4–1 mm, usually dehiscing introrsely; style 1.5–3 mm. Nutlets 1–3, tubercles (as long as wide) awl-shaped.

Flowering Mar–May. Moist meadows, vernal pools; 0–600 m; Calif., Oreg.

The distribution of subsp. floccosa overlaps those of all the other subspecies. The densely hairy leaves and sepals distinguish it from the glabrous or sparsely hairy subspp. bellingeriana, grandiflora, and pumila. Nutlets with relatively long and narrow tubercles rather than relatively short, conic ones separate subsp. floccosa from subsp. californica.

7b. Limnanthes floccosa Howell subsp. bellingeriana (M. Peck) C. T. Mason, Univ. Calif. Publ. Bot. 25: 501. 1952 C E

Limnanthes bellingeriana M. Peck, Proc. Biol. Soc. Wash. 50: 93. 1937

Herbage glabrous. Stems erect to ascending. Flowers urn-shaped; sepals accrescent, narrowly ovate, 5.5–8 mm, abaxially and adaxially glabrous or sparsely hairy; petals oblong, 5.5–7.5 mm, apex retuse, (without marginal hairs basally); filaments 2–3.5 mm; anthers ± 0.5 mm, usually dehiscing introrsely; style 2.3–2.5 mm. Nutlets 1 or 2 (3), tubercles platelike.

Flowering Apr–May. Vernal pool edges in shallow soil of rocky meadows; of conservation concern; 300–1200 m; Calif., Oreg.

D. Southworth and J. Seevers (1997) compared pollen of subspp. bellingeriana and floccosa. Pollen of Limnanthes floccosa has a striated zone perpendicular to the apertural band; in subsp. bellingeriana, the striations are criss-crossed and in subsp. floccosa the striations are parallel. Southworth and Seevers reported that the two subspecies could be easily distinguished in the field in mixed populations based on the density of hairs on the sepals. In subsp. floccosa, the hairs are so dense that they prevent the sepals from fully opening; in subsp. *bellingeriana*, the sepal hairs are relatively short (50 μm) and sparse. Based on the absence of intermediates in these characters, Southworth and Seevers recommended elevating subsp. *bellingeriana* to species level. Allozyme studies (C. I. McNeill and S. K. Jain 1983) group subspp. *bellingeriana* and *floccosa* together, and subspp. *californica*, *grandiflora*, and *pumila* together. Also, in ITS analysis (M. S. Plotkin 1998), subsp. *bellingeriana* is nested within the *floccosa* clade. Crosses between subspp. *bellingeriana* and *floccosa* resulted in viable hybrids (R. Ornduff 1971). Subspecies *bellingeriana* is found in shaded, damp edges of meadows; sympatric subsp. *floccosa* occurs on the upper, dry, exposed edges of meadows (M. T. K. Arroyo 1973). On balance, I prefer to retain this taxon as a subspecies within *L. floccosa*. Subspecies *bellingeriana* is in the Center for Plant Conservation's National Collection of Endangered Plants.

SELECTED REFERENCE Southworth, D. and J. Seevers. 1997. Taxonomic status of *Limnanthes floccosa* subsp. *bellingeriana* (Limnanthaceae). In: T. N. Kaye et al., eds. 1997. Conservation of Native Plants and Fungi. Corvallis. Pp. 147–152.

7c. Limnanthes floccosa Howell subsp. **pumila** (Howell) Arroyo, Brittonia 25: 187. 1973 • Dwarf meadowfoam C E

Limnanthes pumila Howell, Fl. N.W. Amer., 108. 1897; *L. floccosa* var. *pumila* (Howell) C. T. Mason

Herbage glabrous. **Stems** decumbent. **Flowers** cup-shaped; sepals not accrescent, lanceolate to lanceolate-ovate, 7.5–8 mm, abaxially and adaxially glabrous; petals oblong, 7.5–8.5 mm, apex obtuse (not emarginate); filaments 4.2–4.7 mm; anthers ± 1 mm, usually dehiscing introrsely; style 3–3.5 mm. **Nutlets** 3–5, tubercles broad-based, platelike cones.

Flowering Mar–May. Edges of deep vernal pools on lava flows; of conservation concern; 600 m; Oreg.

Subspecies *pumila* is found only on the summits of Upper Table Rock and Lower Table Rock, in Jackson County.

7d. Limnanthes floccosa Howell subsp. **grandiflora** Arroyo, Brittonia 25: 188. 1973 • Large-flowered dwarf meadowfoam C E

Herbage sparsely hairy. **Stems** erect to ascending. **Flowers** cup-shaped; sepals accrescent, oblong-lanceolate to lanceolate-ovate, 8.5–9 mm, abaxially glabrous or sparsely hairy, adaxially densely hairy; petals obovate, 4.5–8.5 mm, apex obtuse, not emarginate, sometimes erose; filaments 4.5–5 mm; anthers ca. 0.8 mm, usually dehiscing extrorsely; style 3.2–4 mm. **Nutlets** 3–5, tubercles broad-based, platelike cones.

Flowering Apr–May. Wet, inner edges of vernal pools; of conservation concern; 300–400 m; Oreg.

Subspecies *grandiflora* occurs primarily in the Agate Desert of Jackson County.

7e. Limnanthes floccosa Howell subsp. **californica** Arroyo, Brittonia 25: 187. 1973 • Shippee meadowfoam C E F

Herbage densely hairy. **Stems** erect to ascending. **Flowers** cup-shaped; sepals accrescent, obovate, 7.5–10 mm, abaxially and adaxially densely hairy; petals broadly obovate-cuneate, 8–10 mm, apex truncate or slightly emarginate; filaments 3–7 mm; anthers 1–1.5 mm, usually dehiscing extrorsely; style 3.5–4 mm. **Nutlets** 3–5, tubercles broad-based, platelike cones.

Flowering Mar–Apr. Edges of vernal pools; of conservation concern; 0–100 m; Calif.

Subspecies *californica* occurs in Butte County on the eastern edge of the Sacramento Valley and on Table Mountain. It is sympatric with *Limnanthes alba* subsp. *alba*; it flowers earlier. J. A. Dole and M. Sun (1992) found no evidence of hybridization between them.

SELECTED REFERENCE Dole, J. A. and M. Sun. 1992. Field and genetic survey of endangered Butte County meadowfoam—*Limnanthes floccosa* subsp. *californica* (Limnanthaceae). Conservation Biol. 6: 549–558.

KOEBERLINIACEAE Engler

• Allthorn Family

Gordon C. Tucker

Shrubs or trees (deciduous); (thorny) spines absent: glabrous or glabrate. **Stems** erect; profusely branched. **Leaves** alternate, spirally arranged, (minute, scalelike); venation pinnate; stipules present (caducous, scalelike); petiole absent. **Inflorescences** axillary, racemose, corymbose, or flowers solitary; pedunculate; bud scales usually persistent; bracts usually absent, sometimes present. **Pedicels** present. **Flowers** bisexual, actinomorphic, rotate to crateriform, campanulate, or urceolate; perianth and androecium hypogynous; sepals persistent, 4, distinct or connate basally; petals 4, attached directly to receptacle, imbricate, distinct, equal; intrastaminal nectary-discs or glands usually present; stamens 8; filaments distinct, glabrous or pubescent; anthers dehiscing by longitudinal slits; pistil 1; ovary 1-carpellate, 2-locular; placentation axile; ovules anatropous, bitegmic; style 1 (straight, short and thick); stigma 1, capitate, unlobed. **Fruits:** berries, indehiscent (spheroidal). **Seeds** (1 or) 2–4, blackish; not arillate; endosperm scanty or absent, sometimes a persistent perisperm present; cotyledons incumbent. $x = 11$.

Genus 1, species 2 (1 in the flora): sw North America, n Mexico, South America (Bolivia).

Koeberlinia has been included in the traditional Capparaceae by some authors, including A. Cronquist (1981). Recent molecular studies support its removal from Capparaceae and a relationship with Bataceae and Salvadoraceae (P. F. Stevens, www.mobot.org/MOBOT/research/APweb).

SELECTED REFERENCE Holmes, W. C., K. L. Yip, and A. E. Rushing. 2008. Taxonomy of *Koeberlinia* (Koeberliniaceae). Brittonia 60: 171–184.

1. KOEBERLINIA Zuccarini, Flora 15(2, Beibl.): 73. 1832 • Allthorn, junco, crown-of-thorns, corona-de-Cristo [For Christoph Ludwig Koeberlin, 1794–1862, German clergyman and botanist]

Plants unscented. **Leaves** evanescent. **Flowers:** sepals each often subtending a nectary; filaments thickened in middle; anthers ellipsoid; gynophore slender, elongating in fruit; style straight;

stigma minute. **Berries** globose, ± fleshy. **Seeds** cochleate-reniform or orbicular to reniform, slightly rugulose.

Species 2 (1 in the flora): sw United States, n Mexico, South America (Bolivia).

Koeberlinia holacantha W. C. Holmes, K. L. Yip & Rushing is known only from Departamento Santa Cruz, Bolivia. The Seri tribe of coastal Sonora, Mexico, traditionally fumigate their huts with the black, oily smoke from burning wood of *Koeberlinia* (R. S. Felger and M. B. Moser 1985).

1. **Koeberlinia spinosa** Zuccarini, Flora 15(2, Beibl.): 74. 1832 F

Plants 0.5–10 m, spreading, globose, compact to somewhat open. **Stems** ± terete, green twigs thorn-tipped, 25–70 cm. **Leaves** 0.2–1.5 × 0.5 mm. **Inflorescences** 3–15 mm. **Pedicels** 3–6 mm. **Flowers:** sepals greenish white, ovate-deltate, 1–2 mm, glabrous; petals greenish white or cream, obovate or oblanceolate, 3–4.8 × 0.8–1.4 mm; filaments brownish, 2.8–4 mm, flat; anthers 0.8–1 mm, slightly curved; gynophore (stipe) 0.3–0.8 mm; ovary ovoid, 1–1.2 mm; style 0.4–1.4 mm. **Fruits** ca. 5 mm, mucronate from persistent style. **Seeds** 3–3.5 mm. $2n = 44, 88$.

Varieties 3 (3 in the flora): sc, sw United States, n Mexico.

Infraspecific variation has long been noted in the genus. W. C. Holmes et al. (2008) recognized three largely allopatric varieties.

1. Stipes 0.3–0.5 mm; styles 0.4–0.7 mm; mature thorns less than 15 times longer than base width . . . . 1c. *Koeberlinia spinosa* var. *wivaggii*
1. Stipes 0.4–0.8 mm; styles 0.8–1.4 mm; mature thorns 20+ times longer than base width.
   2. Plants less than 0.5–2.5 m; young thorns glabrous or finely puberulent; petals 3–4 mm . . . . 1a. *Koeberlinia spinosa* var. *spinosa*
   2. Plants 1–10 m; young thorns densely puberulent; petals 4–4.8 mm . . . . 1b. *Koeberlinia spinosa* var. *tenuispina*

1a. Koeberlinia spinosa Zuccarini var. **spinosa** F

Koeberlinia spinosa var. *verniflora* Bogusch

Plants 0.5–2.5 m; young thorns puberulent; mature thorns 30–80 mm, 0.7–2 mm at base (20+ times longer than base width). **Flowers:** petals 3–4 mm; style 0.8–1.4 mm, distally bent, curved or hooked when dried; gynophore (stipe) 0.4–0.8 mm.

Flowering Feb–May. Brushlands, gullies, hillsides and flats; 0–2100 m; Tex.; Mexico (Coahuila, Guanajuato, Hidalgo, Nuevo León, Querétaro, San Luis Potosí, Tamaulipas).

1b. Koeberlinia spinosa Zuccarini var. **tenuispina** Kearney & Peebles, J. Wash. Acad. Sci. 29: 486. 1939

Koeberlinia spinosa subsp. *tenuispina* (Kearney & Peebles) A. E. Murray

Plants 1–10 m; young thorns densely puberulent; mature thorns 33–82 mm, 1.2–2 mm at base (20+ times longer than base width). **Flowers:** petals 4–4.8 mm; style 1–1.2 mm, straight (sometimes uncinate or S-shaped when dry); gynophore (stipe) 0.5–0.8 mm.

Flowering Mar–Jul. Desert areas; 0–900 m; Ariz., Calif.; Mexico (Baja California, Baja California Sur, Sonora).

Variety *tenuispina* is a component of the Sonoran Desert flora, occurring in La Paz and Yuma counties in southwest Arizona, Imperial County in southeastern California, and adjacent northwest Mexico.

1c. Koeberlinia spinosa Zuccarini var. **wivaggii** W. C. Holmes, K. L. Yip & Rushing, Brittonia 60: 180, figs. 2D, 4. 2008

Plants 1–3 m; young thorns densely puberulent; mature thorns 20–60 mm, 2–3.4 mm at base (15 or less times longer than base width). **Flowers:** petals ca. 3 mm; style 0.4–0.7 mm, straight (sometimes slightly curved when dry); gynophore (stipe) 0.3–0.5 mm.

Flowering Jun–Jul. Desert flats, hillsides, arroyos; 100–2100; Ariz., N.Mex., Tex.; Mexico (Chihuahua, Coahuila, Durango, Nuevo León, San Luis Potosí, Sonora, Zacatecas).

Variety *wivaggii* is primarily a taxon of the Chihuahuan Desert, occurring in a region between the more easterly var. *spinosa* and the more westerly var. *tenuispina*.

BATACEAE Martius ex Perleb
• Saltwort Family

Robert F. Thorne

Subshrubs; (halophytic); spines absent; largely glabrous. **Stems**: branches arching or prostrate; branchlets drooping or erect. **Leaves** (cauline), opposite, simple (each with a basal-abaxial appendage); stipules caducous, (paired, minute); petioles absent; blade (succulent), margins entire. **Inflorescences** axillary or terminal, conelike spikes (catkins) [lax and spikelike or solitary flowers]. **Flowers** unisexual, staminate and pistillate on different plants [same plant in different inflorescences], actinomorphic; bracts present. **Staminate flowers** each initially enclosed in sac splitting into 2 lobes (spathella); tepals 4 or 5, distinct, white, equal, base long-clawed [tapered]; androecium hypogynous; nectary glands absent; stamens 4 or 5, alternating with and longer than tepals; filaments distinct; anthers dehiscing by longitudinal slits. **Pistillate flowers** bracteate; tepals absent; pistil 1, 2-carpellate; ovary falsely 4-loculed; placentation basal-parietal; ovules anatropous, bitegmic; style absent; stigmas 2-lobed. **Fruits** drupaceous syncarps [drupes], ellipsoid. **Seeds** 1–4, narrow, flattened.

Genus 1, species 2 (1 in the flora): s North America, Mexico, West Indies, Central America, n South America, Pacific Islands (Galapagos Islands, Hawaii, s New Guinea), n Australia.

Bataceae are warm-temperate, subtropical, and tropical, coastal halophytes. Because of their proclivity to inhabit saline habitats and because of morphological similarities to halophytic members of Chenopodiaceae, early taxonomists tended to place Bataceae in the Chenopodiales (Centrospermae). The absence of betalains (T. J. Mabry and B. L. Turner 1964), dissimilarity of sieve-element plastid types (H.-D. Behnke and Turner 1971), and dissimilar chromosome numbers (P. Goldblatt 1976) rule out that relationship. Phylogeneticists now accept Bataceae as a member of Brassicales (Capparales) based on the presence of glucosinolates (Mabry 1976) and myrosinase, and molecular similarities (J. E. Rodman et al. 1998). In Brassicales, there are similarities to the Australian Gyrostemonaceae and Salvadoraceae (B. Prijanto 1970; S. Carlquist 1978; Rodman et al.).

The fruits and seeds are water-dispersed. The succulent syncarps can float for one or two weeks before releasing the one-seeded pyrenes, which also can float up to three months before germinating after reaching land (H. B. Guppy 1903–1906, vol. 2).

SELECTED REFERENCES Bayer, C. and O. Appel. 2003. Bataceae. In: K. Kubitzki et al., eds. 1990+. The Families and Genera of Vascular Plants. 9+ vols. Berlin, etc. Vol. 5, pp. 30–32. Rogers, G. K. 1982c. The Bataceae in the southeastern United States. J. Arnold Arbor. 63: 375–386.

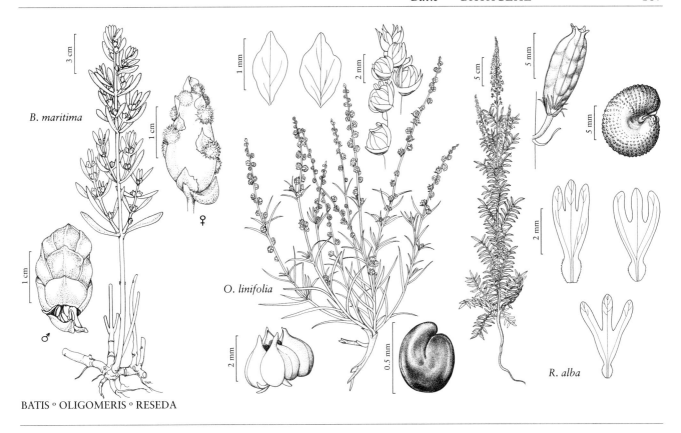

BATIS · OLIGOMERIS · RESEDA

1. BATIS P. Browne, Civ. Nat. Hist. Jamaica, 356. 1756 • Saltwort [probably from Greek via Latin for another coastal plant, or possibly Greek *batos*, bramble]

Plants relatively low, sprawling. **Leaf blades** obovoid to oblanceoloid. **Spikes** subsessile, ellipsoid, subglobose, or turbinate [lax, bracteate, flowers solitary]. **Flowers** anemophilous; filaments slender, or sometimes winged; anthers versatile, dorsifixed; stigmas sessile, papillate. **Syncarps** each with 1–4 seeds (pyrenes). **Seeds** narrow, flattened; coats thin. $x = 11$.

Species 2 (1 in the flora): s North America, Mexico, West Indies, Central America, South America, Pacific Islands (Galapagos Islands, Hawaii, s New Guinea), Australia.

Batis argillicola P. Royen occurs along the coasts of southern New Guinea and northern Australia. It differs from *B. maritima* in being monoecious and having solitary flowers, shorter leaves, tepals of staminate flowers each with one vascular bundle and a tapered base, and winged filaments. Wood anatomy and habit of the two species are similar. There was an earlier dubious chromosome count of $x = 9$.

1. Batis maritima Linnaeus, Syst. Nat. ed. 10, 2: 1289. 1759 • Vidrillos [F]

Batis californica Torrey

Plants usually less than 1 m. **Main stems** 2–4 cm diam. at base; branches rooting at nodes and tips. **Leaf blades** 5–20 × 2–3 mm, apex rounded to acuminate, adaxial surface flattened. **Spikes** 6 × 4 mm, flowers inconspicuous. **Staminate flowers:** tepals spatulate or tapered, 3 × 3 mm; filaments 2 mm; anthers 0.7 mm. **Pistillate flowers** connate at maturity, subsessile, 6 × 4 mm. **Syncarps** green, 10 × 6–7 mm, spongy. **Seeds** 3 × 1 mm. $2n = 22$.

Flowering (Jan–)Apr–Sep [year-round in the tropics]. Saline or brackish, often wet, open, maritime shores, salt marshes, dune swales, saline or brackish lagoons, sandy shell plateaus, shore hummocks, muddy or sandy tidal flats, often with mangroves, especially *Avicennia*; 0–10 m; Ala., Calif., Fla., Ga., La., N.C., S.C., Tex.;

Mexico; West Indies; Central America; n South America; Pacific Islands (Galapagos Islands, Hawaii).

Batis maritima probably is naturalized in Hawaii (W. Hillebrand 1888). Plants frequently cover large areas in dense, tangled stands and are tolerant of very high salinity and water-logging, including inundation for considerable periods (R. F. Thorne 1954).

The salty leaves and stems of *Batis maritima* have been eaten in salads, used as potherbs, or pickled (G. K. Rogers 1982c). According to K. D. Perkins and W. W. Payne (1978), when eaten in large quantities by livestock, *B. maritima* is thought to be poisonous. Like other halophytes, it has been burned for ash. Also, it has been used medicinally to prevent or treat skin problems, ulcers, kidney and bladder stones, and other medicinal problems (Rogers).

RESEDACEAE Martinov
• Mignonette Family

Santiago Martín-Bravo
Gordon C. Tucker
Thomas F. Daniel

Herbs [shrubs], annual, biennial, or perennial; usually glabrous, sometimes puberulent (producing glucosinolates). **Stems** erect or ascending; unbranched or branched. **Leaves** cauline and sometimes basal; rosulate or not; alternate, sometimes fasciculate, simple; venation pinnate; stipules present (modified into glands, interpetiolar, distinct); petiole present; blade (sometimes pinnately lobed, base somewhat decurrent on petioles) margins entire, pinnatisect, or with 1–2 hyaline teeth. **Inflorescences** usually terminal, sometimes axillary, racemes or spikes; bracts present. **Pedicels** present or absent. **Flowers** bisexual [unisexual or polygamous], ± zygomorphic (almost actinomorphic); perianth and androecium hypogynous [rarely perigynous]; sepals persistent or deciduous, 2–6[–8], distinct or basally connate, equal or unequal, margins usually hyaline; petals [0–]2–6(–8), imbricate, distinct or connate, rotate, attached to receptacle, heteromorphic, lateral and abaxial (anterior) usually progressively smaller, adaxial (superior) ones larger, clawed, margins entire or incised; intrastaminal nectary-discs present or absent, asymmetrical, simple [double]; stamens 3–40 (adaxial ones shorter); filaments persistent or deciduous, distinct or basally connate, glabrous or scaberulous; anthers dehiscing by longitudinal slits, introrse, tetrasporangiate (pollen shed as single grain), binucleate, commonly 3-aperturate, colpate or colporate; pistils [2–]3–4[–6(–8)], connate [distinct or basally connate] (± open apically), each with an apical tooth (style and stigma) containing stigmatic tissue; ovary [2 or] 3 or 4 [6(–8)]; (placentation parietal [basal-axial or marginal], placenta usually entire, sometimes apically forked; ovules 1–many per locule, campylotropous, bitegmic. **Fruits** capsules [sometimes free carpidia], valvate [fleshy], cylindric, ovoid-oblong, subglobose, or ovoid, ± open apically; gynophore present (sometimes minute, hidden by surrounding disc). **Seeds** 3–ca. 30, brownish or dark brown to black, reniform, papillose, rugose, or smooth; not arillate; endosperm scanty or none; cotyledons incumbent.

Genera 6, species ca. 85 (2 genera, 5 species in the flora): North America, n Mexico, Europe, sw, c Asia, n, s Africa, Atlantic Islands; temperate and subtropical regions; introduced in c Mexico, South America, Australia.

Placement of Resedaceae in the Capparales or Brassicales is well settled, based on morphological, anatomical, and molecular criteria, and the shared presence of glucosinolates (mustard oil glucosides). The family has been traditionally considered closest to Brassicaceae and Capparaceae. Molecular data (J. C. Hall et al. 2002, 2004) revealed an unexpected relationship to the Australian endemic Gyrostemonaceae and two genera of Capparaceae (*Forchhammeria* and *Tirania*). The foundations of the taxonomy of Resedaceae were established in the nineteenth century primarily through the work of J. Müller Argoviensis (1857, 1868), who recognized three tribes within the family (Astrocarpeae, Cayluseae, Resedeae) based on ovary and placentation type. Validity of his tribal classification was confirmed by later morphological and molecular studies (M. S. Abdallah and H. C. D. de Wit 1978; S. Martín-Bravo et al. 2007). Apparently, the only species of the family native to the New World is *Oligomeris linifolia* (see below).

SELECTED REFERENCES Abdallah, M. S. and H. C. D. de Wit. 1978. The Resedaceae: A taxonomical revision of the family (final installment). Meded. Landbouwhoogeschool 78. Bolle, F. 1936. Resedaceae. In: H. G. A. Engler et al., eds. 1924+. Die Natürlichen Pflanzenfamilien..., ed. 2. 26+ vols. Leipzig and Berlin. Vol. 17b, pp. 659–693. Carlquist, S. 1998b. Wood anatomy of Resedaceae. Aliso 16: 127–135. Edmonson, J. R. 1993. Resedaceae. In: T. G. Tutin et al., eds. 1993+. Flora Europaea, ed. 2. Cambridge and New York Vol. 1, pp. 417–420. El Naggar, S. M. 2002. Taxonomic significance of pollen morphology in some taxa of Resedaceae. Feddes Repert. 113: 518–527. González Aguilera, J. J. and A. M. Fernández Peralta. 1984. Phylogenetic relationships in the family Resedaceae. Genetica 64: 185–197. Hennig, L. 1929. Beiträge zur Kenntnis der Resedaceen—Blüte und Frucht. Planta 9: 507–563. Kubitzky, K. 2003. Resedaceae. In: K. Kubitzky et al., eds. 1990+. The Families and Genera of Vascular Plants. 9+ vols. Berlin etc. Vol. 5, pp. 334–338. Martín-Bravo, S. et al. 2007. Molecular systematics and biogeography of Resedaceae based on ITS and *trn*L-F sequences. Molec. Phylogen. Evol. 44: 1105–1120. Mitra, K. and S. N. Mitra. 1979. Pollen morphology in relation to taxonomy and geography of Resedaceae. Bull. Bot. Surv. India 18: 194–202. Sobick, U. 1983. Blutenentwicklungsgeschichtliche Untersuchungen an Resedaceen unter besonderer Berücksichtigung von Androeceum und Gynoeceum. Bot. Jahrb. Syst. 104: 203–248. Weberling, F. 1968. Über die Rudimentärstipeln der Resedaceae. Acta Bot. Neerl. 17: 360–372.

1. Leaf blades 0.8–5 × 0.05–0.2 cm, margins usually entire, sometimes toothed near base with 1–2 hyaline teeth; inflorescences usually dense spikes (or spikelike racemes); pedicels absent or nearly so; petals 2–3; stamens 3–4; intrastaminal nectary-discs absent . . . . 1. *Oligomeris*, p. 190
1. Leaf blades 3–15 × 0.5–5 cm, margins entire, pinnatisect, or lobed; inflorescences racemes; pedicels present; petals 4–6; stamens 10–40; intrastaminal nectary-discs present . . . . 2. *Reseda*, p. 191

1. OLIGOMERIS Cambessèdes in V. Jacquemont, Voy. Inde 4(Bot.): 23. 1839, name conserved • Whitepuff [Greek *oligos*, few, and *meros*, part, alluding to fewer stamens and petals than in other genera of family]

Herbs, annual [biennial], or perennial. Stems erect to ascending, branched basally. Leaves not rosulate; alternate (sometimes fasciculate); petiolate, petiole shorter than blade, winged; blade margins usually entire, sometimes toothed near base, teeth 1–2, hyaline. Inflorescences usually dense spikes (or spikelke racemes). Pedicels absent or nearly so. Flowers: sepals persistent, 2–4(–5), distinct, equal; petals 2–3[–5], distinct or basally connate, margins entire or shallowly incised; intrastaminal nectary-discs absent; stamens 3–4[–10]; filaments persistent, distinct or basally connate; ovaries 4(–5)-carpelled. Capsules erect, angled, subglobose to obovoid, walls membranous. Seeds smooth (glossy). $x = 15$.

Species 3 (1 in the flora): sw North America, n Mexico, sw, c Asia, n, s Africa, n Atlantic Islands.

Molecular research has shown that *Oligomeris* is a monophyletic group nested within *Reseda* in tribe Resedeae and sister to *Reseda* sect. *Glaucoreseda* (S. Martín-Bravo et al. 2007). Nonetheless, it seems preferable to maintain the genus; it has clear-cut diagnostic morphological characteristics. Hyaline teeth at the base of the leaves may be vestiges of a laminar segmentation of the leaves (F. Weberling 1968); some have interpreted the teeth as stipular.

1. **Oligomeris linifolia** (Vahl) J. F. Macbride, Contr. Gray Herb. 53: 13. 1918 • Flaxleaf-whitepuff, desert-spike [F]

Reseda linifolia Vahl in J. W. Hornemann, Hort. Bot. Hafn. 2: 501. 1815

Plants usually annual, rarely perennial, 8–40 cm. **Leaves:** blade linear to oblanceolate, 0.8–5 × 0.05–0.2 cm, sometimes with hyaline basal teeth 0.2–0.4 mm, apex obtuse or acute, surfaces glabrous or scabrid. **Inflorescences** 2–12(–25) cm; bracts persistent, linear to lanceolate, 1–1.5 mm. **Flowers:** sepals lanceolate to deltate, 0.7–1.5(–2) mm; petals whitish, ovate, slightly shorter than sepals, 0.7–1(–1.7) mm, usually distinct, sometimes barely connate basally; filaments 1.5(–2) mm, barely connate basally; anthers 0.3–0.5 mm. **Capsules** sessile or subsessile, compressed-subglobose, 1.7–3 × 2–3.4 mm, apically 4-toothed, teeth acute, 0.5–0.7 mm. **Seeds** black or brown, 0.5–0.7 mm. $2n = 30$.

Flowering (Jan–)Feb–Jul(–Dec). Pastures, desert washes, sand dunes, creosote scrub, mesquite thickets, volcanic soil, sometimes in brackish or alkaline soils; -60–1200(–1400) m; Ariz., Calif., Nev., N.Mex., Tex., Utah; n Mexico (Baja California, Baja California Sur, Chihuahua, Coahuila, Durango, Nuevo León, Sonora, Zacatecas); sw, c Asia; n Africa; n Atlantic Islands.

The native status of *Oligomeris linifolia* in North America has long been controversial; its presence represents the most remarkable disjunction in the family. Recent molecular research (S. Martín-Bravo et al. 2009) strongly suggests that it is native here.

Oligomeris linifolia has been reported to be toxic to cattle. Occurrence of this species in Utah is based on the following collection: "southern Utah," 1877, *E. Palmer 47* (US, WIS). "Cambess" has appeared in literature as a common name for *O. linifolia*, apparently resulting from confusion with an abbreviation of the author Cambessèdes. "Lineleaf whitepuff" has also appeared as a common name for this species; the epithet *linifolia* should be translated as "flax-leaved."

SELECTED REFERENCE Martín-Bravo, S. et al. 2009. Is *Oligomeris* (Resedaceae) indigenous to Noth America: Molecular evidence for a natural colonization from the Old World. Amer. J. Bot. 96: 507–518.

2. **RESEDA** Linnaeus, Sp. Pl. 1: 448. 1753; Gen. Pl. ed. 5, 207. 1754 • Mignonette, réséda [Latin *re-*, again, and *sedo*, calm, assuage, alluding to medicinal properties attributed by Pliny the Elder to plants growing close to Rimini (Italy)] [I]

Herbs [subshrubs], annual, biennial, or perennial. **Stems** erect to ascending (ribs longitudinally marked), simple or branched (usually distally). **Leaves:** usually rosulate; alternate; usually petiolate (sometimes cauline subsessile); petiole (slender) much shorter than blade, sometimes winged; blade margins entire, pinnatisect, or lobed, lobes 1–11(–15) on each side. **Inflorescences** racemes (sometimes lateral from distal axils). **Pedicels** present (sometimes only 1 mm). **Flowers:** (rarely almost actinomorphic) sepals persistent or deciduous, 4–6[–8], basally connate (alternating with petals [sometimes accrescent]), usually unequal (adaxial larger); petals 4–6[–8], distinct or basally connate, margins usually incised, sometimes entire; intrastaminal nectary-discs present; stamens [7–]10–40; filaments persistent or deciduous, basally connate; ovaries [2–]3–4[–5]-carpelled (stigmatic teeth as many as carpels, placenta sometimes forked apically). **Capsules** erect or pendulous, angled, cylindric, ovoid-oblong, subglobose, or ovoid, walls membranous or chartaceous. **Seeds** (10–ca. 30) papillose, rugose, or smooth (sometimes with caruncles). $x = 5, 6, [7, 8]$.

Species ca. 65 (4 in the flora): introduced; Europe, sw Asia, n Africa, n Atlantic Islands; introduced also in n, c Mexico, South America, c Asia, s Africa, Australia.

The four species of *Reseda* in the flora area represent four sections of the genus, which facilitates identification within this taxonomically complex group. *Reseda alba* and *R. lutea* are the most polymorphic species within their respective sections. The great morphological variability within these taxa helps to explain the specific and infraspecific taxa that have been described, most of which are of little taxonomic value. This polymorphism has been related to polyploidy.

Reseda is of limited economic importance. Some species, including three treated here, are cultivated as ornamental annuals. *Reseda odorata* has been cultivated as a fragrance plant since Roman times. *Reseda luteola* has been used as a dye plant and *R. alba* as a potherb in some regions of Europe.

Reseda phyteuma Linnaeus, a close relative of *R. odorata*, is known in the flora area only from nineteenth century collections around ballast from ports in New Jersey, New York, and Pennsylvania (e.g., *A. Brown, 1878*, NY). It has not been found in the flora area for more than 100 years and was never widely present.

SELECTED REFERENCES Arber, A. 1942. Studies in flower structure. VII. On the gynaecium of *Reseda*, with a consideration of paracarpy. Ann. Bot. (Oxford), n. s. 6(21): 43–48. Avetisian, E. M. and A. K. Mekhakian. 1980. Palynology of the genus *Reseda* L. Biol. Zhurn. Armenii 23: 472–479. González Aguilera, J. J. and A. M. Fernández Peralta. 1983. The nature of polyploidy in *Reseda* sect. *Leucoreseda* (Resedaceae). Pl. Syst. Evol. 142: 223–237. Gori, C. 1957. Sull'embriologia e citologia di alcune specie del genere *Reseda*. Caryologia 10: 391–401.

1. Capsules 4-carpelled, apically 4-toothed; leaf blades pinnatisect (lobes 4–15 pairs); stamens 10–14. . . . . 1. *Reseda alba*
1. Capsules 3-carpelled, apically 3-toothed; leaf blades entire, subentire, or ternately to biternately lobed (lobes 1–3 pairs); stamens 12–40.
    2. Sepals 4; leaf blade margins entire or subentire; placentas forked distally; filaments persistent. . . . . 3. *Reseda luteola*
    2. Sepals (5–)6; leaf blade margins entire or ternately to biternately lobed; placentas entire; filaments deciduous.
        3. Seeds glossy, smooth; capsules usually erect; sepals not reflexed in fruit; petals yellow, lateral lobes of adaxial petals entire or incised . . . . 2. *Reseda lutea*
        3. Seeds dull, undulate-rugose; capsules deflexed or pendent; sepals reflexed in fruit; petals white or light yellow, lateral lobes of adaxial petals deeply laciniate. . . . . . . 4. *Reseda odorata*

1. Reseda alba Linnaeus, Sp. Pl. 1: 449. 1753 • White mignonette, mignonette-blanche, réséda blanc [F][I][W]

Plants annual or biennial (perennial), 30–90 cm, glabrous. **Stems** erect, usually branched. **Leaves** (cauline shorter distally); blade ovate to ovate-oblong, pinnatisect (lobes 4–15 pairs, lanceolate-oblong), 3–15 × 3–5 cm, (base attenuate), margins entire or repand to raggedly toothed, surfaces glabrous. **Racemes** (dense) 20–40 cm; bracts persistent, lanceolate-linear, 3–3.5 mm. **Pedicels** 2–8 mm. **Flowers:** sepals persistent, 5(–6), not reflexed in fruit, lanceolate-linear, 2–2.5 mm; petals 5(–6), white, (3.5–)4–6 mm, subrounded-clawed, adaxial ones 3-lobed; stamens 10–14; filaments persistent, 2–3.5 mm, glabrous; intrastaminal nectary-discs papillose; anthers 1–2 mm; placenta entire. **Capsules** erect, 4-carpelled, cylindric to ovoid-oblong, 8–14 × 4–6 mm, apically 4-toothed, glabrous or ribs papillose. **Seeds** 1–1.3 mm, dull, finely papillose. $2n = 40$.

Flowering May–Aug(–Nov). Muddy shores, dunes, waste places, railroad ballasts, roadsides, basic soils; 0–800 m; introduced; B.C., Man., Ont., Que., Sask.; Ark., Calif., Conn., Del., Ill., Kans., Maine, Mass., Mich., N.H., N.J., N.Y., Ohio, Oreg., Pa., Vt., Wash.; s Europe; sw Asia; n Africa; introduced also in South America, s Africa, Australia.

2. Reseda lutea Linnaeus, Sp. Pl. 1: 449. 1753 • Yellow or wild or cutleaf mignonette [I][W]

Plants annual or perennial, 30–80 cm, glabrous, papillose-scabrid, or hirtellous. **Stems** erect or ascendent, branched. **Leaves:** blade obovate, 4–10 × 1–2.5 cm, margins entire or ternately to biternately lobed (lobes 1–3 pairs, narrow-oblong or lanceolate-linear, margins flat or repand), surfaces scabrous or glabrous. **Racemes** 10–20(–50) cm; bracts caducous, oblong-linear, 2–4 mm. **Pedicels** 3–7 mm. **Flowers** (rarely polygamous); sepals persistent or tardily deciduous, (5–)6, not reflexed in fruit, linear-oblong, 3–4 mm; petals (5–)6, yellow, 2–5 mm, rounded-clawed, adaxial ones trisect, lateral lobes semilunate or falcate, margins entire or incised; stamens (12–)14–18; filaments deciduous, 1.5–3 mm, scabrid-papillose; intrastaminal nectary-discs pilose; anthers 0.9–1.5 mm; placenta entire. **Capsules** usually erect, 3-carpelled, cylindric or ovoid to subglobose, 7–15 × 4–6 mm, apically 3-toothed, glabrous or papillose, ribs scabrid. **Seeds** 1.4–2 mm, glossy, smooth (carunculate). $2n = 48$.

Flowering (Apr–)May–Sep. Grasslands, pastures, gravelly or shaley slopes, railroads, roadsides, mortared crevices in stone walls, disturbed areas, ballast ground, agricultural fields, usually on basic soils; 0–1500 (–2500) m; introduced; B.C., Man., Ont., Sask.; Ala., Ark., Calif., Colo., Conn., D.C., Iowa, Kans., Ky., Maine, Md., Mass., Mich., Mo., Mont., Nebr., N.H., N.J., N.Y., Ohio, Oreg., Pa., R.I., Utah, Vt., W.Va., Wis., Wyo.; Europe; sw Asia; n Africa; introduced also in South America, c Asia, s Africa, Australia.

In Australia, *Reseda lutea* is considered an invasive introduction that causes damage to crops; research is being carried on for means to control its expansion. It has been used as a dye plant, to a lesser extent than *R. luteola*. It has been found to have antibacterial activity against some pathogens. There is also potential utility for phytoremediation in soils contaminated with copper.

SELECTED REFERENCE Harris, J. D., E. S. Davis, and D. M. Wichman. 1995. Yellow mignonette (*Reseda lutea*) in the United States. Weed Technol. 19: 196–198.

3. **Reseda luteola** Linnaeus, Sp. Pl. 1: 448. 1753
  • Dyer's-rocket, mignonette-jaunâtre [I]

Plants annual or biennial, (20–)40–100(–150) cm, glabrous. Stems erect, simple or branched. Leaves (cauline subsessile distally); blade oblong-spatulate, 5–10(–15) × 0.5–2 cm, margins entire or subentire (flat or crispate), surfaces glabrous or, sometimes, with 1–2 conical glands basally. Racemes 10–50 cm; bracts persistent, lanceolate-attenuate, 2–3.5 mm (4–5 mm in fruit). Pedicels 1–3 mm. Flowers: sepals persistent, 4, not reflexed in fruit, lanceolate-ovate, 1–2.5 mm; petals 4, yellowish, 2–4 mm, rounded-clawed, adaxial ones irregularly lobed; stamens 20–40; filaments persistent, 2–3 mm, glabrous; intrastaminal nectary-discs glabrous; anthers 0.5–0.6 mm; placenta forked distally. Capsules erect, 3-carpelled, ovoid to subglobose, 3–5 × 4–6 mm, apically 3-toothed, usually glabrous. Seeds 0.6–1 mm, glossy, smooth. $2n = 24, 26$.

Flowering (Jan–)Mar–Sep(–Dec). Waste ground, roadsides, fields, railway yards, ballast ground, basic and sandy soils; 0–2900 m; introduced; B.C., N.S.; Calif., Colo., Conn., Del., Ill., Ind., Iowa, Md., Mass., Mo., N.J., N.Y., Ohio, Oreg., Pa., R.I., Tex., Wash.; Europe; sw Asia; n Africa; n Atlantic Islands; introduced also in n, c Mexico.

Reseda luteola is a traditional Old World dye plant, used since Roman times. It contains a high amount of the flavonoid luteolin, which yields one of the most brilliant yellow dyes. When combined with woad (*Isatis tinctoria*, Brassicaceae), it yields "Saxon Green." In the nineteenth century *R. luteola* was widely growing, which favored its spreading through many parts of the world; today, it has fallen into disuse. Its potential as a crop for natural dyeing of textiles is being re-evaluated. It is also grown as an ornamental; the appealing rosettes of yellowish green leaves acquire a reddish blush in cool weather.

4. **Reseda odorata** Linnaeus, Amoen. Acad. 3: 51. 1756
  • Fragrant or garden or sweet mignonette, mignonette-odorante [I]

Plants annual 25–60(–80) cm, glabrous or puberulent. Stems erect or ascendent, branched. Leaves: blade spatulate to obovate, 3–5(–7) × 1–2 cm, margins entire or ternately lobed (distal cauline lobes 1–3 pairs). Racemes (3–)5–20 cm; bracts persistent, lanceolate-attenuate, 2–3 mm. Pedicels 4–6 mm (7–16 mm in fruit). Flowers (very fragrant); sepals persistent, 6, reflexed in fruit, narrowly elliptic to spatulate, 2.5–4.5(–6.5) mm; petals 6, white or light yellow, 2.5–4.5 mm, clawed, adaxial ones trisect, lateral lobes deeply laciniate; stamens 20–25; filaments deciduous, 2–3 mm, usually papillose; intrastaminal nectary-discs puberulous to velutinous; anthers 1.5–2 mm; placenta entire. Capsules deflexed or pendent, 3-carpelled, broadly cylindric to subglobose, 5–9 × 4–6 mm, apically 3-toothed, walls glabrous, ribs minutely scabrid. Seeds 1.5–1.8(–2.2) mm, dull, undulate-rugose. $2n = 12$.

Flowering (Feb–)Jun–Oct. Disturbed soils; 0–100 m; introduced; B.C.; Calif., Conn., Maine, Mass., N.J., N.Y., N.C., Pa., S.C., Vt., Wis.; n Africa.

Reseda odorata is known primarily as cultivated plants, probably of artificial hybrid origin (S. Martín-Bravo et al. 2007). Probably originating in the southeastern Mediterranean basin (M. S. Abdallah and H. C. D. de Wit 1978), it has been grown in gardens for centuries for its fragrant flowers. The perfume industry has used the essence of the flowers. Now subcosmopolitan, *R. odorata* escapes from cultivation occasionally but is rarely found naturalized. It is far less common in the flora area than the other species of the genus.

CAPPARACEAE Jussieu

• Caper Family

Gordon C. Tucker

Shrubs or trees (deciduous or evergreen); spines absent (usually without thorns except *Atamisquea*); glabrous or puberulent with stellate trichomes or lepidote scales (producing glucosinolates). **Stems** erect or spreading; sparsely to profusely branched. **Leaves** alternate and distichous or spirally arranged, simple [palmately compound]; venation pinnate; stipules caducous, scalelike or absent; petioles present (pulvinus absent, nectaries present or absent, [petiolar spines present]); blade margins entire. **Inflorescences** terminal or axillary, usually racemose, sometimes corymbose or flowers solitary; pedunculate; bud scales usually persistent; bracts absent. **Pedicels** present. **Flowers** bisexual (sometimes appearing unisexual), actinomorphic or slightly zygomorphic, rotate to crateriform, campanulate, or urceolate; perianth and androecium hypogynous; sepals usually persistent (deciduous in *Quadrella incana*), 4, distinct; petals 4, attached directly to receptacle, imbricate, distinct, equal; intrastaminal nectary-discs or glands present or absent; stamens 6–250; filaments distinct, glabrous or pubescent; anthers dehiscing by longitudinal slits, pollen shed in single grains, binucleate, commonly tricolporate; pistil 1; ovary 1-carpellate, 2-locular; placentation parietal; ovules anatropous, bitegmic; style 1, (straight, relatively short and thick); stigma 1, capitate, unlobed. **Fruits** capsules or berries, valvate, elongate, dehiscent or not by 2 lateral valves (stipitate from elongation of gynophore). **Seeds** 1–38 or many, usually tan to yellowish brown or brown, sometimes green; arillate or not; endosperm scanty or absent, sometimes a persistent perisperm present; cotyledons incumbent to accumbent, (radicle-hypocotyl relatively short and conical). $x = 8, 10$.

Genera ca. 28, species ca. 650 (3 genera, 4 species in the flora): worldwide; tropical and warm-temperate regions.

The broad circumscription of Capparaceae followed by A. Cronquist (1981) was similar to that of F. Pax and K. Hoffmann (1936). Traditionally, the approximately 45 genera and 800 species of Capparaceae in a broader sense have been classified into two major subfamilies, Capparoideae and Cleomoideae. Molecular and morphological analyses of the family and its relatives reveal that Capparaceae as traditionally circumscribed is paraphyletic, with the larger, mostly temperate family Brassicaceae embedded within it (J. E. Rodman et al. 1993, 1996; W. S. Judd et al. 1994; J. C. Hall et al. 2002, 2004). Chloroplast sequences strongly support the monophyly of each of the three lineages Brassicaceae, Capparaceae, and Cleomaceae, with strong support for a sister relationship of Cleomaceae to Brassicaceae (Hall et al. 2002, 2004). Rather than merging the three families into one, all-inclusive Brassicaceae (in the sense of Angiosperm Phylogeny Group 1998, 2003), it might be more acceptable to recognize the three clades as separate, amply distinct families (Zang M. L. and G. C. Tucker 2008). This bears out the proposal of the family by H. K. Airy Shaw (1964), who noted that recognition of Cleomaceae was "a logical necessity." Cleomaceae can be distinguished from Capparaceae as follows:

Trees or shrubs; inflorescence bracts usually absent; leaves simple; fruits capsules or berries; seeds 4–30 mm, globose to reniform, usually arillate; cotyledons incumbent to accumbent, radicle-hypocotyl relatively short and conical . . . . Capparaceae

Herbs or shrubs; inflorescence bracts usually present; leaves usually palmately compound; fruits capsules, nutlets, or schizocarps; seeds 0.5–4 mm (except *Peritoma* with seeds to 15 mm), subglobose, triangular, oblong, or horseshoe-shaped, usually not arillate (except *Hemiscola*); cotyledons incumbent, radicle-hypocotyl elongated . . . . Cleomaceae

SELECTED REFERENCES Ernst, W. R. 1963b. The genera of Capparidaceae and Moringaceae in the southeastern United States. J. Arnold Arbor. 44: 81–95. Hall, J. C., H. H. Iltis, and K. J. Sytsma. 2004. Molecular phylogenetics of core Brassicales, placement of orphan genera *Emblingia, Forchhammeria, Tirania*, and character evolution. Syst. Bot. 29: 654–669. Hall, J. C., K. J. Sytsma, and H. H. Iltis. 2002. Phylogeny of Capparaceae and Brassicaceae based on chloroplast sequence data. Amer. J. Bot. 89: 1826–1842. Kers, L. E. 2003. Capparaceae. In: K. Kubitzky et al., eds. 1990+. The Families and Genera of Vascular Plants. 9+ vols. Berlin etc. Vol. 5, pp. 36–56.

1. Plants usually glabrous or glabrescent, or rarely puberulent (hairs minute, stellate or simple) on branches, inflorescences, or abaxial surface of leaves . . . . 2. *Cynophalla*, p. 196
   Plants lepidote or hairy (hairs stellate) on leaves, branchlets, and abaxial surface of sepals.
   2. Leaf blades narrowly oblong to linear-oblong; inflorescences racemes or flowers solitary (in axils of distal leaves); stamens ca. 6 . . . . 1. *Atamisquea*, p. 195
   2. Leaf blades ovate to ovate-elliptic or narrowly to broadly elliptic or lanceolate; inflorescences racemes or corymbs; stamens 8–30 . . . . 3. *Quadrella*, p. 197

1. ATAMISQUEA Miers ex Hooker & Arnott, Bot. Misc. 3: 142. 1833 • [For Atamisco region of Chile]

Shrubs or trees, deciduous; pubescent with tufted, stellate, stellulate, or multicellular, unbranched hairs, or echinoid lepidote-peltate scales. **Stems** spreading (bluntly thorn-tipped). **Leaves:** alternate; petiole relatively short; nectaries absent; blade narrowly oblong to linear-oblong, margins entire. **Inflorescences** terminal or axillary from distal blades, racemes or solitary flowers (subtended by leaf blades). **Flowers:** sepals equal or 2 unequal pairs, each often subtending a nectary; stamens ca. 6; filaments inserted on discoid or conical receptacle (androgynophore); anthers ellipsoid; gynophore slender, elongating in fruit. **Capsules** or berries, usually dehiscent, ovoid, ± fleshy. **Seeds** 1(2 or 3), globose to reniform, usually not arillate. $x = 8$.

Species 1: sw United States, nw Mexico, s South America.

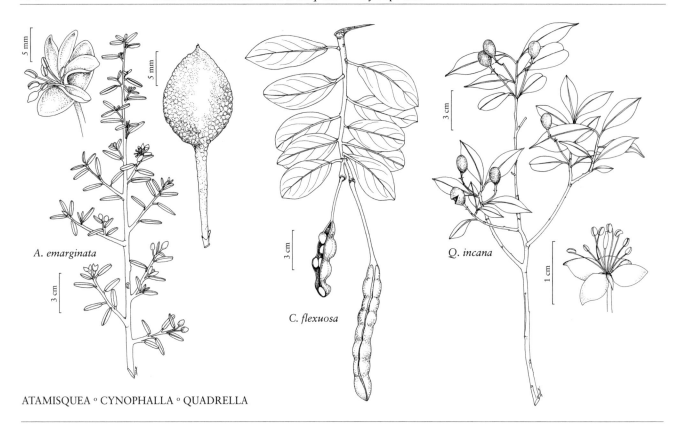

ATAMISQUEA ○ CYNOPHALLA ○ QUADRELLA

1. **Atamisquea emarginata** Miers ex Hooker & Arnott, Bot. Misc. 3: 143. 1833 • Palo-zorillo [F]

Capparis atamisquea Kuntze

Shrubs or multi-trunked trees, 1–8 m. **Stems** branched at right angles, ± terete; branchlets with lepidote scales. **Leaves:** petiole 1–1.2 mm; blade 1–3(–5) × 0.2–0.6 cm, base rounded, apex shallowly emarginate to rounded, abaxial surface densely lepidote, adaxial surface smooth. **Inflorescences** terminating spur shoots, 1–3 cm, lepidote. **Pedicels** 3–4 mm. **Flowers:** sepals (reflexed at anthesis), ovate, proximal pair 0.8–1.5 mm, distal pair 3–3.5 × 2.5 mm, abaxially lepidote, adaxially tomentose, hairs simple; petals white, oblong, 3–6.5 × 2–3 mm; staminodia 0–3; gynophore 3–8 mm. **Capsules** 8–11 × 5–6 mm, dehiscing into 2–4 segments, constricted between seeds, mesocarp red. **Seeds** 3.8–5 mm. $2n = 16$.

Flowering early summer. Desert scrub, arroyos or flats; 50–200 m; Ariz.; Mexico (Baja California, Baja California Sur, Sinaloa, Sonora); South America (Argentina, Bolivia, Chile).

2. CYNOPHALLA (de Candolle) J. Presl in F. Berchtold and J. S. Presl, Prir. Rostlin 2: 275. 1825 • Caper-tree [Greek *kynos*, dog, and *phallos*, penis, alluding to brilliant red color inside rupturing fruits, which reminded early botanists of a dog's penis]

Capparis Linnaeus sect. *Cynophalla* de Candolle in A. P. de Candolle and A. L. P. P. de Candolle, Prodr. 1: 249. 1824; *Capparis* subg. *Cynophalla* (de Candolle) Eichler

Shrubs [trees or rarely vines], evergreen; usually glabrous, rarely puberulent (trichomes multicellular, tufted, stellate, stellulate, or unbranched). **Stems** erect. **Leaves** ± distichous (stipules supra-axillary glands, serially arranged in leaf axils just distal to petioles on branchlets, increasing in size distally); petiole relatively long or short, nectaries present; blade obovate, oblong, or broadly to narrowly elliptic, margins entire. **Inflorescences** terminal or axillary, racemes. **Flowers:** sepals [equal] 2 unequal pairs (distal pair smaller), each often subtending a nectary; stamens [28–]120–150[–ca. 250]; filaments inserted on a discoid or conical receptacle (androgynophore); anthers ellipsoid; gynophore slender, elongating in fruit. **Capsules** [berries] dehiscent [indehiscent], linear-cylindric, ± fleshy. **Seeds** [1–]10–30[–many], usually reniform, arillate, (embryo green). $x = 8, 10$.

Species ca. 20 (1 in the flora): se United States, Mexico, West Indies, Central America, n South America.

Cynophalla is a clearcut, unique entity; its species lack the distinctive lepidote indumentum of the vegetative parts that characterizes the other woody caper genus, *Quadrella*, which also occurs in southern Florida and the Caribbean region. Plants of *Cynophalla* are usually glabrous; if pubescent, hairs are minute, stellate or simple on branches, inflorescences, or abaxially on new leaves. Some authors would place *Cynophalla* in the unigeneric subfamily *Cynophalloideae*. It is a taxonomically difficult group of species, but well-circumscribed by distinctive characteristics, such as a biseriate calyx with distal sepals smaller, and supra-axillary glands (1–3 per axil) on young shoots.

1. **Cynophalla flexuosa** (Linnaeus) J. Presl in F. Berchtold and J. S. Presl, Přir. Rostlin 2: 275. 1825 • Bay-leaved caper-tree, limber-caper, mimbre del monte F

Morisonia flexuosa Linnaeus, Pl. Jamaic. Pug., 14. 1759; *Capparis brevisiliqua* de Candolle; *C. flexuosa* (Linnaeus) Linnaeus

Plants 2–4 m. **Stems** ± terete, usually glabrous or glabrescent, rarely puberulent. **Leaves:** petiole 4–10 mm; blade 3–7(–9) × 2–4(–5) cm, coriaceous, base narrowly cuneate to rounded, apex emarginate, round, or acute, abaxial surface usually glabrescent, adaxial surface glabrous. **Inflorescences** (1–)2–6 cm. **Pedicels** 7–12 mm, (densely felted). **Flowers** nocturnal, fragrant; sepals orbiculate, decussate, proximal pair 5–7 × 10–12 mm, distal pair 3.5–5 × 1–1.5 mm, falling before anthesis, glabrous; petals white, becoming yellowish or reddish white, oblong-obovate, (10–)15–20 × 12 mm; stamens 40–60(–120) mm; gynophore 4–6(–8) cm. **Capsules** reddish brown to yellowish, irregularly constricted between seeds, 5–15(–28) × 9–13(–17) mm, seeds in 1 or 2 rows. **Seeds** green, 7–14 mm, covered with oily, white pulp. $2n = 16$.

Flowering spring–late summer. Coastal thickets, hummocks, marl flats; 0–10 m; Fla.; Mexico; West Indies; Central America; n South America.

Cynophalla flexuosa is wide-ranging and polymorphic; Floridian plants are sometimes recognized as subsp. *flexuosa*. Some South American plants are distinguished as subsp. *polyantha* (Triana & Planchon) H. H. Iltis.

3. **QUADRELLA** (de Candolle) J. Presl in F. Berchtold and J. S. Presl, Přir. Rostlin 2: 260. 1825 • [Latin *quadra*, a square, and *-ella*, diminutive, alluding to perianth]

Capparis Linnaeus sect. *Quadrella* de Candolle in A. P. de Candolle and A. L. P. P. de Candolle, Prodr. 1: 251. 1824; *Capparis* subg. *Quadrella* (de Candolle) Eichler

Shrubs or trees [rarely vines], evergreen [deciduous]; lepidote (scales peltate) and/or hairy (trichomes stellate). **Stems** erect. **Leaves:** alternate; petiole relatively long or short, nectaries present or not; blade ovate to ovate-elliptic or narrowly to broadly elliptic, or lanceolate, margins entire. **Inflorescences** terminal or axillary, racemes or corymbs. **Flowers:** sepals equal [2 unequal pairs], each often subtending a nectary; stamens 8 or 18–30[–60]; filaments inserted on ± flat receptacle; anthers ellipsoid; gynophore slender, elongating in fruit. **Capsules** dehiscent or not, linear-cylindric, ovoid, obovoid, or globose (thick-walled). **Seeds** 1–many, usually reniform, not arillate. *x* = 8 [10].

Species ca. 25 (2 in the flora): s, se United States, Mexico, West Indies, Central America, n South America.

Quadrella is found mostly in arid places. It is characterized by indument of peltate scales or stellate hairs, and closed, valvate calyx aestivation with sepals that tend to recurve and fall off at or after anthesis.

1. Leaf blades ovate to ovate-elliptic, apices usually emarginate, sometimes acuminate or acute; petals 12–18 mm; capsules linear-cylindric, (7–)20–38 cm, constricted between seeds; Florida. . . . . 1. *Quadrella jamaicensis*
1. Leaf blades narrowly to broadly elliptic, or lanceolate, apices acuminate, acute, or barely obtuse; petals 5–6(–8) mm; capsules ovoid to obovoid or globose, 1.3–1.7(–2) cm, not constricted between seeds; Texas. . . . . 2. *Quadrella incana*

1. **Quadrella jamaicensis** (Jacquin) J. Presl in F. Berchtold and J. S. Presl, Prir. Rostlin 2: 260. 1825 • Jamaican caper tree, shrubby dog caper, zebrawood, black willow

Capparis jamaicensis Jacquin, Enum. Syst. Pl., 23. 1760; *C. emarginata* A. Richard; *C. longifolia* Swartz

Shrubs or trees, 1–6 m. **Stems** ± terete, lepidote (scales peltate). **Leaves**: petiole 6–21 (–25) mm, nectaries present; blade ovate to ovate-elliptic, 5–15(–21) × 2.5–5(–8) cm, coriaceous, base narrowly cuneate to rounded, apex usually emarginate, sometimes acuminate or acute, abaxial surface lepidote, adaxial surface glabrous. **Inflorescences** (1–)2–6 cm. **Pedicels** 7–18(–35) mm, lepidote. **Flowers** nocturnal, fragrant; sepals (7–)12–18 × 6–7 mm, abaxially reddish lepidote; petals creamy white to purple, elliptic-obovate, 12–18 × 10 mm; stamens 18–30, 40–50 mm (woolly basally); gynophore (1–)6–7(–9) cm. **Capsules** indehiscent, linear-cylindric, (7–)20–38 × 0.8–1.2 cm, ± regularly constricted between seeds, lepidote, scales peltate. **Seeds** (3–)6–16 (–38), 7–8 mm. 2*n* = 16.

Flowering spring–early summer. Coastal hammocks, disturbed coastal scrub, on bare rock; 0–10 m; Fla.; Mexico (Campeche, Quintana Roo); West Indies (Cuba, Jamaica); Central America.

Quadrella jamaicensis is an ornamental with attractive flowers and mostly emarginate leaves. It is characteristic of coastal areas from central Florida (Cape Canaveral, St. Petersburg) to the Keys, Cuba, and Jamaica; it is rarely sympatric with its closest relative, *Q. cynophallophora* (Linnaeus) Linnaeus, in the Bahamas and Hispaniola.

2. **Quadrella incana** (Kunth) H. H. Iltis and X. Cornejo, Novon 17: 452. 2007 • Caper F

Capparis incana Kunth in A. von Humboldt et al., Nov. Gen. Sp. 5(fol.): 73; 5(qto.): 94. 1821; *C. karwinskiana* Schlechtendal; *C. pauciflora* C. Presl; *Octanema incana* (Kunth) Rafinesque

Shrubs, 0.5–2.5[–8] m. **Stems** ± terete, tomentose, trichomes stellate. **Leaves**: petiole 7–14 (–18) mm, nectaries absent; blade narrowly to broadly elliptic or lanceolate, 3–5(–9) × 1–3 cm, not coriaceous, base cuneate, apex acuminate, acute, or barely obtuse, abaxial surface stellate-pubescent, adaxial surface glabrous. **Inflorescences** 2–6 cm. **Pedicels** 4–6 mm, tomentose. **Flowers** nocturnal, fragrant; sepals deciduous, linear-oblong, 2–3 × 0.5–1 mm, abaxially with stellate trichomes; petals white, ovate, 5–6(–8) × 3–4 mm; stamens 8, 6–8 mm; gynophore 4–5 mm. **Capsules** dehiscing into 2–4 valves, ovoid to obovoid or globose, 13–17(–20) × 10–13 mm, not constricted between seeds. **Seeds** 1, (cochleate-reniform), 11–13 mm (embryo white). 2*n* = 16.

Flowering spring. Coastal scrub; 0–5 m; Tex.; Mexico (Campeche, Oaxaca, Tabasco, Tamaulipas, Veracruz, Yucatán); Central America (Belize, Guatemala, Honduras).

CLEOMACEAE Berchtold & J. Presl
• Spiderflower Family

Gordon C. Tucker

Staria S. Vanderpool

Herbs or shrubs, annual or perennial (usually deciduous, evergreen in *Peritoma arborea*); spines usually absent (present in *Hemiscola* and *Tarenaya*); glabrous or glandular-pubescent, hairs stalked or sessile (producing glucosinolates). **Stems** usually erect, sometimes spreading or procumbent; branched or unbranched. **Leaves** alternate, spirally arranged (usually palmately compound, sometimes simple); venation pinnate; stipules usually present (usually caducous, sometimes deciduous, 3–8-palmatifid, linear, threadlike, minute, scalelike, or absent, nodal (stipular) spines present in *Tarenaya* and *Hemiscola*); petiole present (pulvinus usually present, nectaries absent, petiolar spines sometimes present, petiolules present); blade margins entire, serrate, or serrulate. **Inflorescences** terminal or axillary, usually racemose, sometimes flat-topped, or flowers solitary (usually elongated in fruit); bud scales absent; bracts present or absent (unifoliate, often trifoliate proximally, bracteoles absent). **Pedicels** present. **Flowers** usually bisexual (developmentally unisexual within sections of racemes), actinomorphic or slightly zygomorphic, rotate to crateriform, campanulate, or urceolate; perianth and androecium hypogynous; sepals persistent or deciduous, 4, distinct or connate basally; petals 4, attached directly to receptacle, imbricate, distinct, equal or unequal; intrastaminal nectary-discs, scales, or glands present or absent; stamens [4–]6–27[–35]; filaments free or basally adnate to gynophore (or along proximal $^1/_3$–$^1/_2$ in *Gynandropsis*) or androgynophore, glabrous or pubescent; anthers dehiscing by longitudinal slits, pollen shed in single grains, binucleate, commonly tricolporate; gynophore present or absent; pistil 1; ovary 1-carpellate (except 2 in *Oxystylis*), 2-locular; placentation parietal; ovules 1–18(–26+) per locule, anatropous, bitegmic; style 1 (straight, relatively short, thick, not spinelike in fruit, except in *Oxystylis*, sometimes in *Wislizenia*); stigma 1, capitate, unlobed. **Fruits** capsular or nutlets (usually stipitate from elongation of gynophore, erect to divergent, usually not inflated), valvate, elongate (± dehiscent by 2 lateral valves, except in *Polanisia*), or schizocarps (inflated in *Peritoma arborea*), indehiscent or dehiscent. **Seeds** 1–65[–200], tan, yellowish brown, light brown, pale green, brown, reddish brown, silver-gray, or gray to black (papillose or tuberculate); arillate or not; endosperm scanty or absent, persistent perisperm sometimes present; cotyledons incumbent, (radicle-hypocotyl elongated).

Genera 17, species ca. 150 (12 genera, 34 species in the flora): nearly worldwide; tropical and temperate regions.

A discussion of the status of Cleomaceae and its segregation from Capparaceae (in the narrow sense) appears under the latter. Throughout this treatment, style length indicates the length in fruit; in some species, the style elongates after anthesis. The key to genera, in some cases, include characteristics of species, in addition to those of genera. This circumscription of Cleomaceae includes Oxystylidaceae Hutchinson.

SELECTED REFERENCES Ernst, W. R. 1963b. The genera of Capparidaceae and Moringaceae in the southeastern United States. J. Arnold Arbor. 44: 81–95. Hall, J. C., H. H. Iltis, and K. J. Sytsma. 2004. Molecular phylogenetics of core Brassicales, placement of orphan genera *Emblingia*, *Forchhammeria*, *Tirania*, and character evolution. Syst. Bot. 29: 654–669. Hall, J. C., K. J. Sytsma, and H. H. Iltis. 2002. Phylogeny of Capparaceae and Brassicaceae based on chloroplast sequence data. Amer. J. Bot. 89: 1826–1842. Holmgren, P. K. and A. Cronquist. 2005. Cleomaceae. In: A. Cronquist et al. 1972+. Intermountain Flora. Vascular Plants of the Intermountain West, U.S.A. 6+ vols. in 7+. New York and London. Vol. 2, part B, pp. 160–174. Iltis, H. H. 1952. A Revision of the Genus *Cleome* in the New World. Ph.D. dissertation. Washington University. Iltis, H. H. 1957. Studies in the Capparidaceae. III. Evolution and phylogeny of the western North American Cleomoideae. Ann. Missouri Bot. Gard. 44: 77–119. Iltis, H. H. 1960. Studies in the Capparidaceae. VII. Old World cleomes adventive in the New World. Brittonia 12: 279–294. Kers, L. E. 2003. Capparaceae. In: K. Kubitzki et al., eds. 1990+. The Families and Genera of Vascular Plants. 9+ vols. Berlin etc. Vol. 5, pp. 36–56. Sánchez-Acebo, L. 2005. A phylogenetic study of the New World *Cleome* (Brassicaceae, Cleomoideae). Ann. Missouri Bot. Gard. 92: 179–201. Vanderpool, S. S., W. J. Elisens, and J. R. Estes. 1991. Pattern, tempo, and mode of evolutionary and biogeographic divergence in *Oxystylis* and *Wislizenia* (Capparaceae). Amer. J. Bot. 78: 925–937. Woodson, R. E. Jr. 1948. *Gynandropsis*, *Cleome*, and *Podandrogyne*. Ann. Missouri Bot. Gard. 35: 139–148.

1. Shrubs (evergreen); leaflets 3; fruits inflated or not . . . . 2. *Peritoma* (in part), p. 205
1. Herbs; leaflets 1–9; fruits usually not inflated.
   2. Inflorescences axillary, racemes (flat-topped); style spinelike in fruit . . . . 6. *Oxystylis*, p. 215
   2. Inflorescences terminal or axillary (from distal leaves), racemes or corymbs, or flowers solitary; style not spinelike in fruit (except sometimes in *Wislizenia*).
      3. Stamens 8–32; gynophore usually 0–14 mm in fruit; fruits dehiscent in distal ½ . . . . 1. *Polanisia*, p. 201
      3. Stamens 6 (except *Arivela* with 14–25); gynophore usually 0.5–85 mm in fruit (in *Arivela viscosa* fruits sessile or absent); fruits dehiscent ± entire lengths or indehiscent.
         4. Fruits schizocarps; seeds 2(–4), 0.5 mm . . . . 5. *Wislizenia*, p. 213
         4. Fruits capsules; seeds 1–40, 1.2–3.5 mm.
            5. Fruits 2–8 mm, as long as or shorter than wide . . . . 4. *Cleomella*, p. 209
            5. Fruits (2–)12–150 mm, much longer than wide.
               6. Plants with stipular spines (sometimes petioles with spines).
                  7. Petals 11–30(–45) mm; gynophore 45–80 mm in fruit . . . . 9. *Tarenaya*, p. 218
                  7. Petals 5–10 mm; gynophore 1–4 mm in fruit . . . . 10. *Hemiscola*, p. 220
               6. Plants without stipular spines (petioles without spines).
                  8. Filaments adnate basally to gynophore (scars evident in fruiting specimens near midpoint of gynophore) . . . . 12. *Gynandropsis*, p. 222
                  8. Filaments free from gynophore (or gynophore obsolete) or inserted on androgynophore.
                     9. Bracts 1–18 mm, unifoliate. . . . . 8. *Cleoserrata*, p. 216
                     9. Bracts 1–25 mm, unifoliate or trifoliate (sometimes expanded).
                        10. Gynophore obsolete; stems, leaflet surfaces, and fruits glandular-hirsute . . . . 11. *Arivela*, p. 221
                        10. Gynophore 1–25 mm (in fruit); stems, leaflet surfaces, and fruits usually not glandular (sometimes pubescent).
                            11. Sepals distinct; gynophore 2–5 mm in fruit; anthers 3–5 mm . . . . 3. *Carsonia*, p. 208
                            11. Sepals partly connate or distinct; gynophore 1–25 mm; anthers 0.6–2.6 mm.

[12. Shifted to left margin.—Ed.]

12. Filaments inserted on cylindric androgynophore (usually expanded adaxially into gibbous or flattened appendage); leaflets conduplicate and flat . . . . 2. *Peritoma* (in part), p. 205
12. Filaments inserted on discoid or conical androgynophore; leaflets flat. . . . . 7. *Cleome*, p. 215

1. POLANISIA Rafinesque, Amer. J. Sci. 1: 378 1819 • Clammyweed [Greek *polys*, many, and *anisos*, unequal, alluding to stamens]

Gordon C. Tucker

Cristatella Nuttall

Herbs, annual or perennial (unscented). **Stems** sparsely or profusely branched; glandular-pubescent, viscid. **Leaves:** stipules absent or minute; petiole without pulvinus distally; leaflets 3. **Inflorescences** terminal or axillary (from distal leaves), racemes (flat-topped); bracts present (pedicels green to purple, glandular-pubescent). **Flowers** slightly zygomorphic; (receptacle prolonged into gland adaxially); sepals deciduous, distinct, equal; petals unequal (adaxial pair larger); stamens 8–32; filaments inserted on gynophore, (unequal), glabrous; anthers (ellipsoid), not coiling as pollen is released; gynophore ascending in fruit. **Fruits** capsules, dehiscent in distal ½, cylindric, oblong. **Seeds** 12–65, globose or subglobose, not arillate, (cleft fused between ends). $x = 10$.

Species 5 (5 in the flora): North America, Mexico.

SELECTED REFERENCE Iltis, H. H. 1958. Studies in the Capparidaceae. IV. *Polanisia* Raf. Brittonia 10: 33–58.

1. Leaflet blades obovate to oblanceolate or broadly elliptic, 0.5–2(–3) cm wide; styles deciduous in fruit, 5–40 mm.
   2. Plants usually annual, rarely perennial; seeds roughened or tuberculate-rugose, 2–2.3 mm; stamens 10–20; styles 5–17 mm. . . . . 1. *Polanisia dodecandra*
   2. Plants perennial; seeds smooth, 0.7–2 mm; stamens 20–27; styles 20–40 mm (New Mexico, Texas). . . . . 5. *Polanisia uniglandulosa*
1. Leaflet blades linear or narrowly elliptic, 0.05–0.2 cm wide; styles persistent in fruit, 2.5–4.5 mm.
   3. Adaxial petals 6–11 mm; nectary glands 1–5.5 mm; gynophore 3–14 mm in fruit . . . . . 3. *Polanisia erosa*
   3. Adaxial petals 3–5 mm; nectary glands 0.5–1 mm; gynophore 1–4 mm in fruit.
      4. Capsules 8–30 mm; seeds 1.8–2 mm; Great Plains, midwestern states . . . . .2. *Polanisia jamesii*
      4. Capsules 40–60 mm; seeds 0.7–0.9 mm; southeastern states. . . . . .4. *Polanisia tenuifolia*

1. **Polanisia dodecandra** (Linnaeus) de Candolle in A. P. de Candolle and A. L. P. P. de Candolle, Prodr. 1: 242. 1824 F W

Cleome dodecandra Linnaeus, Sp. Pl. 2: 672. 1753

Annuals, rarely perennials, (5–)10–60(–90+) cm. **Stems** branched; hairs stalked, glandular throughout. **Leaves:** petiole green to purple, angled, 1.5–4.5(–6) cm, glandular; leaflet blade oblanceolate to obovate or broadly elliptic, 1.9–4.5(–6.5) × 0.5–2(–3) cm, margins entire, surfaces sparsely glandular. **Racemes** (dense) 5–20 cm (10–30 cm in fruit); bracts unifoliate, lanceolate to orbiculate, 5–20 mm. **Pedicels** 10–25(–40) mm. **Flowers:** sepals purple, oblong, 3–6 × 1.5–2.5 mm, margins entire, apex acuminate, glandular; petals white, pink, rose, or purple, oblong-ovate, clawed, abaxial pair (3–)5–10 × 1.8–3(–5) mm, apex emarginate to lacerate, adaxial pair 8–14 × 3–5 mm, apex emarginate; nectary glands bright orange, not conspicuous (in fruit); stamens 10–20, exserted, purple, (6–)7–28(–40) mm; anthers purple, 1–1.3 mm; gynophore 0–2 mm in fruit; ovary 5–10 mm; style deciduous in fruit, 5–17 mm; stigma purple. **Capsules** (inflated) 40–70 × 5–9 mm, reticulate, glandular. **Seeds** 20–65, dark reddish brown, globose to oblong, 2–2.3 mm, roughened or tuberculate-rugose.

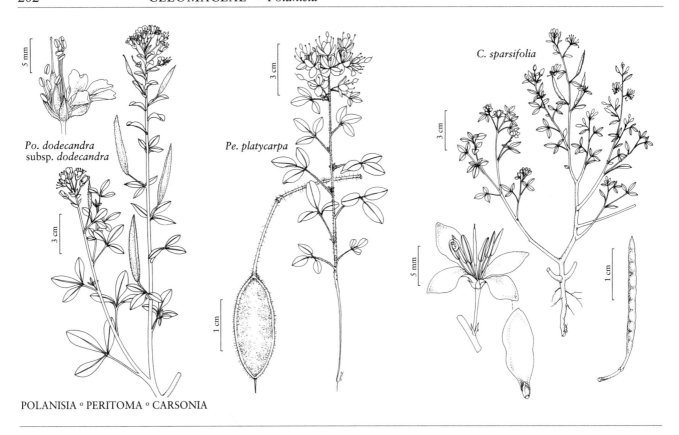

POLANISIA ○ PERITOMA ○ CARSONIA

Subspecies 3 (3 in the flora): North America, n Mexico.

SELECTED REFERENCE Iltis, H. H. 1966. Studies in the Capparidaceae. VIII. *Polanisia dodecandra* (L.) DC. Rhodora 68: 41–47.

1. Petals (3–)5–7(–8) mm; stamens ± equaling or slightly longer than petals ....
   .... 1a. *Polanisia dodecandra* subsp. *dodecandra*
1. Petals (6–)8–14(–17) mm; stamens ± equaling, or slightly longer to ca. 2 times or more as long as petals.
   2. Inflorescence bracts ovate to orbiculate; leaflet blade apex rounded; seeds tuberculate-rugose
      ..... 1b. *Polanisia dodecandra* subsp. *riograndensis*
   2. Inflorescence bracts lanceolate to ovate; leaflet blade apex acute to obtuse; seeds roughened
      ..... 1c. *Polanisia dodecandra* subsp. *trachysperma*

1a. Polanisia dodecandra (Linnaeus) de Candolle subsp. **dodecandra** • Redwhiskered clammyweed

E F

Polanisia graveolens Rafinesque

Leaflets: blade apex acute to obtuse. **Inflorescence bracts** lanceolate to ovate. **Petals** white (sometimes pinkish), (3–)5–7(–8) mm. **Stamens** (6–)7–10(–14) mm, ± equaling or slightly longer than petals. **Seeds** roughened. $2n = 20$.

Flowering summer–early fall. Roadsides, riverbanks, gravel bars, flood-scoured shorelines, grasslands, disturbed sites; 0–1100 m; Ont., Que.; Ala., Ariz., Ark., Calif., Colo., Conn., Ga., Idaho, Ill., Ind., Iowa, Kans., Ky., La., Maine, Md., Mass., Mich., Minn., Mo., Mont., Nebr., Nev., N.H., N.J., N.Mex., N.Y., N.Dak., Ohio, Okla., Oreg., Pa., S.Dak., Tex., Vt., Va., W.Va., Wis., Wyo.

Subspecies *dodecandra* is smaller-flowered than the other subspecies and is most often found in northeastern and midwestern states. 'White Spider' is a cultivar sometimes offered in seed catalogues. Occurrences in New England (except Vermont) probably represent adventives from farther west; this plant is common along weedy roadsides.

1b. Polanisia dodecandra (Linnaeus) de Candolle subsp. **riograndensis** H. H. Iltis, SouthW. Naturalist 14: 116, fig. 1[top]. 1969 • Rio Grande clammyweed

Leaflets: blade apex rounded. **Inflorescence bracts** ovate to orbiculate. **Petals** pink, rose, or purple, (6–)8–14(–17) mm. **Stamens** (6–)7–17 mm, equaling or slightly longer than petals. **Seeds** tuberculate-rugose. $2n = 20$.

Flowering spring–fall. River banks, coastal dunes, open woodlands, mesquite, and

semi-desert; 0–100 m; Tex.; Mexico (Nuevo León, Tamaulipas).

Subspecies *riograndensis* is found along the lower Rio Grande in Texas and adjacent Mexico.

1c. Polanisia dodecandra (Linnaeus) de Candolle subsp. **trachysperma** (Torrey & A. Gray) H. H. Iltis, Rhodora 68: 47. 1966 • Sandyseed or western clammyweed [W]

Polanisia trachysperma Torrey & A. Gray, Fl. N. Amer. 1: 669. 1840

Leaflets: blade apex acute to obtuse. **Inflorescence bracts** lanceolate to ovate. **Petals** white, (7–)8–13(–16) mm. **Stamens** (6–)12–28(–40) mm, longest ones ca. 2 or more times as long as petals. **Seeds** roughened. $2n = 20$.

Flowering summer–early fall. Gravelly or sandy, sunny places along streams, sandy places, open woodlands, grasslands, roadsides; 100–1500(–2100) m; Alta., Man., Ont., Sask.; Ala., Ariz., Ark., Calif., Colo., Conn., Idaho, Ill., Iowa, Kans., Ky., La., Md., Mass., Mich., Minn., Mo., Nebr., Nev., N.Mex., N.Dak., Ohio, Okla., Pa., S.Dak., Tex., Utah, Va., Wash., W.Va., Wyo.; Mexico.

Subspecies *trachysperma* is found mainly in the Great Plains and western states. Populations in the eastern United States are presumably adventive, as are those in eastern Ontario, where it has been found spreading in disturbed areas, such as quarries and roadsides (B. McBride 2006).

2. Polanisia jamesii (Torrey & A. Gray) H. H. Iltis, Brittonia 10: 54. 1958 • James's clammyweed [C][E]

Cristatella jamesii Torrey & A. Gray, Fl. N. Amer. 1: 124. 1838

Annuals, 15–40(–60) cm. **Stems** branched; hairs stalked, glandular throughout. **Leaves:** petiole green to purple, angled, (0.2–)0.5–0.8(–1) cm; leaflet blade linear to narrowly elliptic, 1.5–2.8 × 0.1–0.2 cm, margins entire, apex acute, mucronulate, surfaces densely glandular abaxially, sparsely glandular adaxially. **Racemes** 1–3 cm (6–8 cm in fruit); bracts trifoliate, ovate, 5–15 mm. **Pedicels** 5–15 mm. **Flowers:** sepals pale yellow, glandular, lanceolate, 1.6–2.6 × 0.8–1.2 mm, margins entire, apex acute to obtuse; petals white, oblong-ovate, clawed, abaxial pair 1.5–3 × 1–2 mm (6 or 7-lobed), apex emarginate to lacerate, adaxial pair 3–5 × 2–3 mm (4 or 5-lobed), apex emarginate to lacerate; nectary glands yellow (drying purple), 0.5–1 mm; stamens 6–9, slightly exserted, yellow, 3.2–5.2 mm; anthers yellow, 0.5–1 mm; gynophore 2–4 mm in fruit; ovary 2.5–5(–7) mm; style persistent in fruit, 2.5–4.5 mm; stigma red. **Capsules** 8–30 × 2–4 mm, reticulate, glandular. **Seeds** 18–36, bright reddish brown, globose to oblong, 1.8–2 mm, pebbled (without transverse ridges). $2n = 20$.

Flowering summer–early fall. Sandy soil, often in blowouts in prairies and stabilized dunes; of conservation concern; 200–1400 m; Colo., Ill., Ind., Iowa, Minn., Nebr., N.Mex., Ohio, Okla., S.Dak., Tex., Wis.

Populations of *Polanisia jamesii* in Indiana and Ohio are adventive.

3. Polanisia erosa (Nuttall) H. H. Iltis, Brittonia 10: 56. 1958 • Large clammyweed [E]

Cristatella erosa Nuttall, J. Acad. Nat. Sci. Philadelphia 7: 86, plate 9, fig. 1. 1834 (as Cristella)

Annuals, 10–60 cm. **Stems** branched or unbranched; hairs sessile or stalked, glandular throughout. **Leaves:** petiole purple, angled, 0.3–1.5 cm; leaflet blade (conduplicate) ,linear to narrowly elliptic, 0.9–3.5 × 0.1–0.2 cm, margins entire, apex acute, mucronulate, surfaces densely glandular abaxially, sparsely glandular adaxially, (fleshy). **Racemes** (glandular), 1–3 cm (6–8 cm in fruit); bracts uni- to trifoliate, ovate, 7–15 mm. **Pedicels** 10–23 mm. **Flowers:** sepals (± erect), green to purple, lanceolate to obovate, 2.5–3.6 × 0.8–1.2 mm, margins entire, apex acute, glabrous; petals white, oblong-ovate, clawed, abaxial pair 3–5 × 2.5–3 mm (9 or 10-lobed), apex lacerate, adaxial pair 6–11 × 3.5–5 mm (4–6-lobed), apex emarginate to lacerate; nectary glands yellow (drying purple), 1–5.5 mm; stamens 6–15, exserted, purple, 10–13 mm; anthers purple, 1–1.5 mm; gynophore 3–14 mm in fruit; ovary 5–8 mm; style persistent in fruit, 2.5–4.5 mm; stigma green. **Capsules** 20–60 × 2–2.5(–5) mm, reticulate, glabrous or sparsely glandular. **Seeds** 12–40, dark reddish brown, globose to oblong, 1.5–1.8 mm, finely pebbled or beaded.

Subspecies 2 (2 in the flora): sc United States.

1. Nectary glands (2.5–)3–5.5 mm; gynophore 7–14 mm in fruit; adaxial petals 7–11 mm; Arkansas, Louisiana, Oklahoma, eastern Texas. . . . . 3a. *Polanisia erosa* subsp. *erosa*
1. Nectary glands 1–2 mm; gynophore 3–6 mm in fruit; adaxial petals 6–9 mm; southernmost Texas . . . . 3b. *Polanisia erosa* subsp. *breviglandulosa*

3a. Polanisia erosa (Nuttall) H. H. Iltis subsp. erosa [E]

Nectary glands tubular, (2.5–)3–5.5 mm. **Adaxial petals** 7–11 mm. **Gynophores** 7–14 mm in fruit.

Flowering early spring–mid fall. Sandhills, prairies, woods, fields; 0–100 m; Ark., La., Okla., Tex.

Subspecies *erosa* is found primarily in Texas, barely extending into three adjacent states.

3b. Polanisia erosa (Nuttall) H. H. Iltis subsp. breviglandulosa H. H. Iltis, Brittonia 10: 58, fig. 3. 1958 [E]

Nectary glands cup-shaped, 1–2 mm. **Adaxial petals** 6–9 mm. **Gynophores** 3–6 mm in fruit.

Flowering early spring–mid fall. Sand plains; 0–100 m; Tex.

4. Polanisia tenuifolia Torrey & A. Gray, Fl. N. Amer. 1: 123. 1838 • Slender clammyweed, pineland catchfly [E]

Aldenella tenuifolia (Torrey & A. Gray) Greene; *Cleome aldenella* W. R. Ernst; *Jacksonia tenuifolia* (Torrey & A. Gray) Greene

Annuals, 20–45(–90) cm. **Stems** usually unbranched, sometimes branched (reddish purple); hairs stalked, glandular throughout. **Leaves:** petiole green to purple, subterete, (0.2–)1–3 cm; leaflet blade (conduplicate), linear, 1–5 × 0.05–0.2 cm, margins entire, apex obtuse, mucronulate, surfaces sparsely glandular abaxially, glabrous or sparsely glandular adaxially, (fleshy). **Racemes** 1–3 cm (6–8 cm in fruit); bracts trifoliate, elliptic, 5–15 mm. **Pedicels** 5–15 mm. **Flowers:** sepals (reflexed), pale yellow, lanceolate to deltate, 1.6–2.6 × 0.7–1.2 mm, clawed, margins entire, apex acute to obtuse, glandular; petals white, oblong-ovate, abaxial pair 1.5–3 × 1–2 mm (6 or 7-lobed), apex emarginate to lacerate, adaxial pair 3.5–5 × 2–3 mm (barely clawed, 4 or 5-lobed), apex emarginate to lacerate; nectary glands yellow (drying purple), 0.5 mm; stamens 8–13, slightly exserted, yellow, 3–6 mm; anthers yellow with maroon tip, 1–2 mm; gynophore 1.7–4 mm in fruit; ovary 2.5–5(–7) mm; style persistent in fruit, 2.5–4.5 mm; stigma red. **Capsules** 40–60 × 2–4 mm, reticulate, glandular or glabrous. **Seeds** 18–36, reddish brown, spheroidal, 0.7–0.9 mm, pebbled (without transverse ridges). $2n = 20$.

Flowering spring–late summer. Scrub, dry pinelands, oak-pine woods, sandhills, lakeshores; 0–100 m; Ala., Fla., Ga., Miss.

Polanisia tenuifolia is restricted to the Coastal Plain Province, from Florida to Mississippi, and Georgia. The common name catchfly refers to insects sticking to the viscid glandular secretions of the herbage. The same name is given also to some species of *Silene* (Caryophyllaceae) for the same reason.

5. Polanisia uniglandulosa (Cavanilles) de Candolle in A. P. de Candolle and A. L. P. P. de Candolle, Prodr. 1: 242. 1824 • Mexican clammyweed

Cleome uniglandulosa Cavanilles, Icon. 4: 3, plate 306. 1797; *Polanisia dodecandra* (Linnaeus) de Candolle subsp. *uniglandulosa* (Cavanilles) H. H. Iltis

Perennials, 40–80 cm. **Stems** sparsely or profusely branched; hairs stalked, glandular throughout. **Leaves:** petiole green to purple, angled, 1.5–4.5(–6) cm, (glandular); leaflet blade broadly elliptic to oblanceolate, 2–4 × 1–2 cm, margins entire, apex obtuse, mucronulate, surfaces sparsely glandular. **Racemes** 5–20 cm (10–30 cm in fruit); bracts unifoliate, ovate, 8–12 mm. **Pedicels** 10–25(–40) mm. **Flowers:** sepals purple, oblong, 3–6 × 1.5–2.5 mm, margins entire, apex acuminate, glandular; petals white, narrowly spatulate, clawed, abaxial pair 10–20 × 2–3(–5) mm, apex emarginate to lacerate, adaxial pair 15–30 × 3–5 mm, apex emarginate; nectary glands bright orange, not conspicuous (in fruit); stamens 20–27, exserted, purple, 20–50 mm; anthers purple, 1–1.3 mm; gynophore 0–2 mm in fruit; ovary 5–10 mm; style deciduous in fruit, 20–40 mm; stigma purple. **Capsules** (somewhat inflated), 60–100 × 7–10 mm, reticulate, glandular. **Seeds** 20–65, dark reddish brown, globose to oblong, 1.5–2 mm, smooth. $2n = 20$.

Flowering spring–fall. Pinyon, juniper, and oak woodlands, arroyos, riverbeds, roadsides, pastures; 300–800 m; N.Mex., Tex.; Mexico.

Polanisia uniglandulosa is the only perennial species of the genus, notable for its large, showy, white petals and silky-iridescent seed testa.

2. PERITOMA de Candolle in A. P. de Candolle and A. L. P. P. de Candolle, Prodr. 1: 237 1824 • Bee-plant [Greek *peri*, all around, and *tome*, cutting, perhaps alluding to dehiscence of fruit]

Staria S. Vanderpool

Hugh H. Iltis

Cleome Linnaeus [unranked] *Atalanta* Nuttall, Gen. N. Amer. Pl. 2: 73. 1818, not *Atalantia* Corrêa 1805; *Celome* Greene; *Isomeris* Nuttall

Herbs or shrubs, annual or (weak) perennial. **Stems** sparsely or profusely branched; glabrous, or glabrate, or glandular-pubescent. **Leaves:** stipules scalelike, bristlelike, or absent; petiole with pulvinus basally or distally; leaflets 3 or 5, (conduplicate and flat). **Inflorescences** terminal or axillary (from distal leaves), racemes (flat-topped or elongated); bracts usually present. **Flowers** zygomorphic; sepals persistent or deciduous, distinct or partly connate (⅓–½ of lengths), equal (each often subtending a nectary); petals equal; stamens 6; filaments inserted on cylindric androgynophore (usually expanded adaxially into a gibbous or flattened appendage), glabrous; anthers (linear), coiling as pollen is released; gynophore usually recurved in fruit (sometimes reflexed). **Fruits** capsules (erect to pendent), dehiscent, usually oblong (obovoid, subglobose, or fusiform in *P. arborea*). **Seeds** 5–38, globose, obovoid, triangular, or horseshoe-shaped, not arillate, (cleft fused between ends). $x = 10$.

Species 6 (6 in the flora): North America, Mexico.

Whether included in *Cleome* or treated as a separate genus, *Peritoma* comprises mostly distinct, western North American species, perhaps related to African genera. It is best treated as a taxon equivalent in rank to its three derivative genera, *Cleomella, Oxystylis,* and *Wislizenia* ($2n = 40$). These can be arranged in a much studied fruit and seed reduction series correlated with increasing aridity (H. H. Iltis 1955, 1956, 1957; K. Bremer and H. Wanntorp 1978; S. Keller 1979; S. S. Vanderpool et al. 1991). Some botanists may object to inclusion of the well-established *Isomeris arborea* in *Peritoma*. It is the only long-lived woody shrub species in the North American Cleomaceae. Except for the larger size, the flowers are basically identical (as are the fruits and seeds) to those of species such as *P. lutea*, a fact appreciated long ago by E. L. Greene when that notorious splitter lumped *Isomeris* with its relatives in *Cleome*.

With six sharply distinct species (H. H. Iltis 1957), *Peritoma* is an exceptionally robust and relatively ancient genus, usually characterized by the rather thick trifoliolate and glabrous leaves, yellow (except a rich purple in *P. multicaulis* and *P. serrulata*) petals, and, usually, well-developed nectary discs.

1. Shrubs; capsules usually inflated, 6–12 mm diam. . . . . 1. *Peritoma arborea*
1. Annuals; capsules not inflated, 1.5–12 mm diam.
    2. Leaflets 5 (proximally).
        3. Capsules 15–40 mm; gynophore 5–17 mm in fruit; petals light yellow, 5–8 mm; stamens 10–15 mm; plants (15–)25–30 cm . . . . 2. *Peritoma lutea*
        3. Capsules 40–60 mm; gynophore 15–25 mm in fruit; petals golden yellow, 10–13 mm; stamens 20–30 mm; plants 50–100(–200) cm . . . . 3. *Peritoma jonesii*
    2. Leaflets 3 (throughout).
        4. Petals golden yellow; capsules pendent, glandular-pubescent . . . . 6. *Peritoma platycarpa*
        4. Petals purple, pink, rose, or white; capsules erect or reflexed, glabrous.
            5. Leaflets 0.6–1.5 cm wide; petioles 1.5–3.5 cm; petals 7–12 mm; capsules 23–76 mm . . . . 4. *Peritoma serrulata*
            5. Leaflets 0.1 cm wide; petioles 0.5–1.5 cm; petals 4–5 mm; capsules 15–25 mm . . . . 5. *Peritoma multicaulis*

1. Peritoma arborea (Nuttall) H. H. Iltis, Novon 17: 449. 2007 • Bladderpod, bladderbush, burro-fat

Isomeris arborea Nuttall in J. Torrey and A. Gray, Fl. N. Amer. 1: 124. 1838; *Cleome isomeris* Greene

Shrubs (evergreen), 50–200 cm. **Stems** profusely branched; glabrate or puberulent (bark corky, twigs smooth). **Leaves:** petiole 1–3 cm; leaflets 3, blade oblong-elliptic, 1.5–4.5 × 0.4–1.3 cm, margins serrate, apex acuminate to obtuse, surfaces glaucous. **Racemes** 1–3 cm (6–40 cm in fruit); bracts unifoliate, obovate to spatulate, 2–15 mm. **Pedicels** 7–15 mm (thickened in fruit). **Flowers:** sepals persistent, connate ca. ½ of length, green, lanceolate, 4–7 × 2.2–4 mm, margins entire, glabrous; petals yellow, ovate-elliptic, 8–14 × 4.2–5 mm, (apex acute); stamens yellow, 15–25 mm; anthers 2–2.5 mm; gynophore (reflexed), 10–20 mm in fruit; ovary 3–6 mm (often aborting in bud); style 0.9–1.2 mm. **Capsules** (tardily dehiscent), usually inflated ,(valves sometimes 3), 20–30 × 6–12 mm, smooth. **Seeds** 5–25, dark brown, obovoid, 6–7 × 5–6 mm, smooth.

Varieties 3 (3 in the flora): sw United States, nw, w Mexico.

Peritoma arborea is woody and is variable in fruit size and shape. The large variability in fruit shape has led to the naming of varieties (L. D. Benson and R. A. Darrow 1945).

SELECTED REFERENCE Truesdale H. D. et al. 2004. Allozyme variability within and among varieties of *Isomeris arborea* (Capparaceae). Madroño 51: 364–371.

1. Capsules not inflated, narrowly fusiform . . . . 1b. *Peritoma arborea* var. *angustata*
1. Capsules inflated, obovoid or subglobose.
    2. Capsules inflated, obovoid . . . . 1a. *Peritoma arborea* var. *arborea*
    2. Capsules strongly inflated, subglobose . . . . 1c. *Peritoma arborea* var. *globosa*

1a. Peritoma arborea (Nuttall) H. H. Iltis var. **arborea**

Capsules inflated, obovoid. $2n = 40$.

Flowering summer. Coastal bluffs, hillsides, grasslands, desert washes and flats, roadsides; 0–1300 m; Calif.; Mexico (Baja California).

Variety *arborea*, which has capsules intermediate in shape between vars. *angustata* and *globosa*, is found throughout most of the species range; it is uncommon in Baja California and absent in the northern part of the range of the species.

1b. Peritoma arborea (Nuttall) H. H. Iltis var. **angustata** (Parish) H. H. Iltis, Novon 17: 449. 2007

Isomeris arborea Nuttall var. *angustata* Parish, Muhlenbergia 3: 128. 1907; *I. arborea* var. *insularis* Jepson

Capsules not inflated, narrowly fusiform (long-attenuate at both ends).

Flowering summer. Hillsides, grasslands, desert washes and flats, roadsides; 0–1300 m; Ariz., Calif.; Mexico (Baja California, Colima, Sonora).

Variety *angustata* is found in the Colorado and Mojave deserts and in western and northwestern Mexico, including offshore islands (e.g., Revillagigedo Islands). Plants with particularly narrow capsules from Cedros Islands (Baja California) have been distinguished as var. *insularis*.

1c. Peritoma arborea (Nuttall) H. H. Iltis var. **globosa** (Coville) H. H. Iltis, Novon 17: 449. 2007 [E]

Isomeris arborea var. *globosa* Coville, Proc. Biol. Soc. Wash. 7: 73. 1892

Capsules strongly inflated, subglobose.

Flowering summer. Coastal bluffs, hillsides, grasslands, desert washes and flats, roadsides; 0–1300 m; Calif.

Variety *globosa* is found in the northern half of the species range, south to San Diego County; it is absent from the deserts of eastern Kern County, although it is well-documented from deserts in the western part of the county.

2. Peritoma lutea (Hooker) Rafinesque, Sylva Tellur., 112. 1838 • Yellow bee-plant [W]

Cleome lutea Hooker, Fl. Bor.-Amer. 1: 70, plate 25. 1830; *Peritoma breviflora* Wooton & Standley

Annuals, (15–)25–30 cm. **Stems** sparsely branched distally; glabrous or glabrate. **Leaves:** petiole 1.5–4.5 cm; leaflets 5 (proximal ones sometimes early deciduous), blade linear to elliptic, 1.5–4(–6) × 0.4–1.3 cm, margins serrate, apex long-acuminate, surfaces glabrous. **Racemes** 1–3 cm (6–40 cm in fruit); bracts unifoliate, spatulate, 2–15 mm. **Pedicels** 7–15 mm. **Flowers:** sepals persistent, connate ca. ½ of length, yellow, lanceolate, 1.6–2.6 × 0.8–1.2 mm, margins denticulate, glabrous; petals light yellow, oblong to ovate, 5–8 × 2–4 mm; stamens yellow, 10–15 mm; anthers 1.9–2.6 mm; gynophore 5–17 mm in fruit; ovary 3–6 mm; style 0.5–0.8 mm. **Capsules** not inflated, 15–40 × 2–5 mm, striate. **Seeds** 10–20, gray to black, triangular (sharply angled), 3–4 × 2.5–3 mm, rugose. $2n = 34$.

Flowering spring–late summer. Dry sandy flats, desert scrub, roadsides; (100–)600–2400 m; Calif., Colo., Idaho, Mont., Nebr., Nev., N.Mex., Oreg., Utah, Wash., Wyo.; Mexico (Baja California).

3. **Peritoma jonesii** (J. F. Macbride) H. H. Iltis, Novon 17: 449. 2007 • Jones bee-plant

Cleome lutea Hooker var. *jonesii* J. F. Macbride, Contr. Gray Herb. 65: 39. 1922; *C. jonesii* (J. F. Macbride) Tidestrom

Annuals, 50–100(–200) cm. **Stems** sparsely branched distally; glabrous or glabrate. **Leaves:** petiole 1.5–4.5 cm; leaflets 5, (proximal ones sometimes early deciduous), blade linear to elliptic, 1.5–4(–6) × 0.4–1.3 cm, margins serrate, apex long-acuminate, surfaces glabrous. **Racemes** 1–3 cm (6–40 cm in fruit); bracts unifoliate, obovate to spatulate, 2–15 mm. **Pedicels** 7–15 mm. **Flowers:** sepals persistent, connate ca. ½ of length, yellow, lanceolate, 1.6–2.6 × 0.8–1.2 mm, margins denticulate, glabrous; petals golden yellow, oblong to ovate, 10–13 × 2–4 mm; stamens yellow, 20–30 mm; anthers 1.9–2.6 mm; gynophore 15–25 mm in fruit; ovary 3–6 mm; style 0.5–0.8 mm. **Capsules** not inflated, 40–60 × 2–5 mm, striate. **Seeds** 15–30, gray to black, triangular (sharply angled), 3–4 × 2.5–3 mm, rugose.

Flowering summer. Dry sandy flats, desert scrub, roadsides; 300–1200 m; Ariz., Calif.; Mexico (Baja California).

Often treated as a variety of *Peritoma lutea*, *P. jonesii* grows at lower elevations, has a more southerly (though overlapping) range (T. H. Kearney and R. H. Peebles 1960), and differs in morphological features. Most notable are its larger, showier flowers and longer capsules.

4. **Peritoma serrulata** (Pursh) de Candolle in A. P. de Candolle and A. L. P. P. de Candolle, Prodr. 1: 237. 1824 (as serrulatum) • Rocky Mountain bee-plant, guaco [E][W]

Cleome serrulata Pursh, Fl. Amer. Sept. 2: 441. 1813; *C. serrulata* subsp. *angusta* (M. E. Jones) Tidestrom; *Peritoma inornata* (Greene) Greene; *P. serrulata* var. *albiflora* Cockerell; *P. serrulata* var. *clavata* Lunell

Annuals, 30–80 cm. **Stems** sparsely branched; glabrous or glabrate. **Leaves:** (stipules bristlelike), petiole 1.5–3.5 cm; leaflets 3, blade elliptic, 2–6 × 0.6–1.5 cm, margins entire, weakly sinuate, or serrulate, apex acute, long-acuminate, or mucronate, surfaces glabrate (margins with sparse, relatively long hairs when young). **Racemes** 1–4 cm (4–30 cm in fruit); bracts unifoliate, obovate, 4–22 mm. **Pedicels** (green to purple), 8–20 mm. **Flowers:** sepals persistent, connate ½–⅔ of length, purple to green, lanceolate, 1.7–4 × 1–2 mm, margins denticulate, glabrous; petals purple (rarely white), oblong to ovate, 7–12 × 3–6 mm; stamens purple, 18–24 mm; anthers (green), 2–2.3 mm; gynophore 1–15 mm in fruit; ovary 5–7 mm; style 0.1–0.5 mm. **Capsules** (erect) not inflated, 23–76 × 3–6(–7) mm, striate, (glabrous). **Seeds** 12–38, black, globose or horseshoe-shaped, 2.8–4 × 2.5–3 mm, rugose. $2n = 34, 60$.

Flowering summer. Shortgrass and mixed grass prairies, pastures, pinyon pine and juniper woodland, desert scrub, roadsides, stabilized sand dunes; (100–)300–2500(–2900) m; Alta., B.C., Man., Ont., Sask.; Ariz., Calif., Colo., Idaho, Ill., Ind., Iowa, Kans., Mass., Mich., Minn., Mo., Mont., Nebr., Nev., N.Mex., N.Dak., Ohio, Okla., Oreg., S.Dak., Tex., Utah, Wash., Wyo.

Most collections of *Peritoma serrulata* from the northeastern and midwestern United States apparently represent non-persistent waifs or garden escapes. The species has been cultivated as a source of nectar for honeybees since ca. 1880 (L. H. Bailey 1900–1902). It shows considerable variation in fruit size, even within populations. The variation may reflect environmental influences, especially water availability, rather than genetics (H. H. Iltis 1952).

The seeds and leaves of *Peritoma serrulata* are consumed by the Navajo as food and provide a source of black dye. The leaves have been used as a remedy for insect bites, inflammation, and intestinal upsets (L. S. M. Curtin 1947).

5. **Peritoma multicaulis** (de Candolle) H. H. Iltis, Novon 17: 449. 2007 • Spiderflower [C]

Cleome multicaulis de Candolle in A. P. de Candolle and A. L. P. P. de Candolle, Prodr. 1: 240. 1824; *C. sonorae* A. Gray; *Peritoma sonorae* (A. Gray) Rydberg

Annuals, 20–60 cm. Stems unbranched or sparsely branched; glabrous. Leaves: petiole 0.5–1.5 cm; leaflets 3, blade linear to elliptic, 1–2 × 0.1 cm, margins entire, apex long-acuminate, surfaces glabrous. Racemes 1–3 cm (6–40 cm in fruit); bracts unifoliate, obovate to spatulate, 2–15 mm. Pedicels 7–15 mm. Flowers: sepals persistent, distinct or slightly connate basally, yellow, lanceolate, 1.6–2.6 × 0.8–1.2 mm, margins denticulate, glabrous; petals white, pink, or rose, oblong to ovate, 4–5 × 1–1.3 mm; stamens yellow, 4–5 mm; anthers 1.9–2.6 mm; gynophore 10–20 mm in fruit; ovary 3–6 mm; style 0.5–0.8 mm. Capsules (reflexed) not inflated, 15–25 × 1.5 mm, striate, (glabrous). Seeds 10–20, gray to black, triangular (sharply angled), 6–15 × 2–2.5 mm, (acute at both ends) rugose.

Flowering summer. Dry to moist open ground, often in saline or volcanic soils; of conservation concern; 700–2000(–2300) m; Ariz., Colo., N.Mex., Tex., Wyo.; Mexico (Coahuila, Distrito Federal, Jalisco, México, Sonora).

6. **Peritoma platycarpa** (Torrey) H. H. Iltis, Novon 17: 449. 2007 • Golden bee-plant [E][F]

Cleome platycarpa Torrey in C. Wilkes et al., U.S. Expl. Exped. 17: 235, plate 2. 1874; *Celome platycarpa* (Torrey) Greene

Annuals, 10–60 cm. Stems densely branched basally; purple-tinged; glandular-pubescent. Leaves: (stipules scalelike, minute, obscured by pubescence); petiole 1.5–4.5 cm; leaflets 3, blade flat, ovate to obovate, 1–3.5 × 0.5–1.3 cm, margins serrulate-denticulate or entire, apex obtuse, surfaces glandular-pubescent. Racemes 1–3.5 cm (5–40 cm in fruit); bracts unifoliate, obovate to spatulate, 3–25 mm. Pedicels 7–17 mm. Flowers: sepals deciduous, distinct, yellow, awl-shaped, (3–)4–5(–6) × 3–7 mm, margins entire, densely glandular-hairy; petals golden yellow, oblong, 6–12 × 0.3–0.4(–0.6) mm; stamens yellow, 10–17 mm; anthers 1.8–2 mm; gynophore 10–18 mm in fruit; ovary (compressed), 3.5–5 mm; style 1–3 mm. Capsules (pendent) not inflated, 12–25 × 8–12 mm, striate, (glandular-pubescent). Seeds 10–20, brownish black, spheroidal, 3–3.2 mm, (glossy) smooth. $2n = 40$.

Flowering summer. Alkaline clay soils, volcanic tuff, dry foothills, among junipers, in sagebrush scrub, fields, railroads; 800–1200 m; Calif., Idaho, Nev., Oreg.

Peritoma platycarpa resembles some species of *Cleomella* in the absence of a nectary-disc, shape of the replum, and indument.

3. CARSONIA Greene, Pittonia 4: 211. 1900 • Spiderflower, few-leaved beeplant [For Carson Desert of Nevada] [E]

Gordon C. Tucker

Herbs, annual. Stems profusely branched; glabrous. Leaves: stipules scalelike or absent; petiole with pulvinus basally or distally; leaflets 1 or 3. Inflorescences terminal or axillary (from distal leaves), racemes (flat-topped or elongated); bracts present. Flowers weakly zygomorphic; sepals deciduous, distinct, equal; petals equal (each with an epipetalous nectiferous scale); stamens 6; filaments inserted on 4-lobed androgynophore, glabrous; anthers (linear), coiling as pollen is released; gynophore erect in fruit. Fruits capsules, dehiscent, oblong. Seeds 10–13, oblong, not arillate, (cleft fused between ends). $x = 16$.

Species 1: sw United States.

A peculiar desert xerophyte with subsessile, erect capsules, and unique petal glands, *Carsonia* is segregated from *Peritoma* by its unique cytology ($2n = 32$). Whether it is autochthonously specialized or actually related to, and derived from, some Old World ancestors, such as the Central Asiatic-Near Eastern *Cleome* sect. *Thylacophora* Franch, the dozen or so species of which, probably through convergence, also bear epipetaloid glands, remains to be determined.

1. Carsonia sparsifolia (S. Watson) Greene, Pittonia 4: 212. 1900 [E] [F]

Cleome sparsifolia S. Watson, Botany (Fortieth Parallel), 32, plate 5. 1871

Plants 10–60(–90) cm. **Stems** glaucous. **Leaves:** petiole 4–15 (–30) cm; leaflets 3 proximally, 1 distally, blade obovate, 4–15 × 0.1–7 mm, margins serrate, apex acute to obtuse. **Racemes** 2–8 cm (3–10 cm in fruit); bracts (expanded) unifoliate or trifoliate, 4–15 mm. **Pedicels** 4–8 mm. **Flowers:** sepals brownish green, ovate, 1.4–2.5(–3) × 1–2 mm, base cuneate, margins serrulate, apex acuminate, glabrous; petals recurved, yellow with brown central streak, strap-shaped, 9–13 × 1.5–3 mm; stamens staggered, yellow, 9–15 mm; anthers 3–5 mm; gynophore 2–5 mm in fruit; ovary 4–7 mm; style 0.1–0.4 mm. **Capsules** 15–45 × 1–3 mm, smooth. **Seeds** silver-gray, compressed, 2–3 × 1–1.5 mm, smooth. $2n = 32$.

Flowering spring–summer. Sand dunes, beaches, desert valleys, typically with *Larrea divaricata* and *Artemisia* species; 900–2000 m; Calif., Nev.

4. CLEOMELLA de Candolle in A. P de Candolle and A. L. P. P. de Candolle, Prodr. 1: 237. 1824 • Stinkweed [Generic name *Cleome*, and Latin *-ella*, dimunitive]

Staria S. Vanderpool

Herbs, usually annual [perennial] (usually unpleasantly scented). **Stems** sparsely or profusely branched (usually erect); glabrous, pubescent, or scabrous. **Leaves:** stipules (sometimes deciduous), 3–8-palmatifid, threadlike, or setaceous; petiole without pulvinus (usually longer than leaflets); leaflets (1 or) 3. **Inflorescences** terminal, racemes, corymbs, or flowers solitary in distal leaf axils (sometimes remaining compact in fruit); bracts usually present (pedicels often anthocyanic). **Flowers** actinomorphic; sepals persistent, connate basally (ca. ¼ of length) [distinct], equal (each often subtending a nectary); petals equal; stamens 6 (usually exserted, distinct, equal); filaments inserted on a discoid or conical androgynophore, glabrous; anthers (ellipsoid), coiling as pollen is released; gynophore reflexed, spreading, or ascending in fruit. **Fruits** capsules, dehiscent (as long as or shorter than wide), rhomboidal, globose, deltoid, or ovoid (valves laterally expanded, falling away from round, persistent septum). **Seeds** 1–16, globose or reniform, not arillate, (cleft fused between ends). $x = 20$.

Species 10 (8 in the flora): United States, Mexico.

The two extraterritorial species of *Cleomella* are *C. mexicana* de Candolle and *C. perennis* H. H. Iltis, both known from central and northern Mexico.

SELECTED REFERENCE Payson, E. B. 1922b. A synoptical revision of the genus *Cleomella*. Univ. Wyoming Publ. Sci., Bot. 1: 29–46.

1. Gynophore 0.5–3 mm in fruit; petals 1.5–2.8 mm.
   2. Pedicels recurved in fruit (± equal to gynophore), 1–2 mm; inflorescences solitary flowers in axils of closely spaced leaves .... 2. *Cleomella brevipes*
   2. Pedicels spreading in fruit (much longer than gynophore), 12–20 mm (in fruit); inflorescences racemes, terminating stems and branches .... 7. *Cleomella parviflora*
1. Gynophore 2.5–17 mm in fruit; petals 3–9 mm.
   3. Leaf blade surfaces (and sometimes stems) spreading-hispidulous; styles 2–3(–5) mm; plants usually spreading or matlike, sometimes erect .... 5. *Cleomella obtusifolia*
   3. Leaf blade surfaces (and stems) glabrous; styles 0.3–2 mm; plants ± erect.
      4. Gynophore 2.5–6(–8) mm in fruit; bracts ± rudimentary; plants without dominant central stem .... 6. *Cleomella palmeriana*
      4. Gynophore 3.5–17 mm in fruit; bracts unifoliate or trifoliate (5–25 mm); plants with dominant central stem.

[5. Shifted to left margin.—Ed.]

5. Leaflet blades ovate-elliptic, oblanceolate, or oblong-lanceolate, 0.4–1.4 cm wide (usually 1.5–2.5 times longer than wide).
    6. Petioles 2–6(–8) cm; petals 1.8–2.2 mm wide; anthers 2.5–3 mm; California, Idaho, Nevada, Oregon, Utah . . . . 3. *Cleomella hillmanii*
    6. Petioles 0.8–2 cm; petals 2–4 mm wide; anthers 1.5–1.9 mm; Arizona, New Mexico, Texas. . . . . 4. *Cleomella longipes*
5. Leaflet blades linear to elliptic, 0.2–0.8 cm wide (usually 3–10 times longer than wide).
    7. Sepals deltate, 0.5–1 mm wide; styles 0.3–0.5 mm; gynophore reflexed in fruit; capsules 4–6 × 6–12 mm, rhomboidal; seeds 2–3 mm, rugose . . . . 1. *Cleomella angustifolia*
    7. Sepals lanceolate, 0.5–0.7 mm wide; styles 0.8–1.2 mm; gynophore ascending in fruit; capsules 2.5–5 × 2.5–6(–8) mm, ovoid to rhomboidal; seeds 2–2.2 mm, smooth . . . . 8. *Cleomella plocasperma*

1. **Cleomella angustifolia** Torrey, Hooker's J. Bot. Kew Gard. Misc. 2: 255. 1850 • Narrowleaf rhombopod [E] [F]

Plants (30–)40–150(–200) cm. **Stems** sparsely branched distally (central stem dominant); glaucous, glabrous. **Leaves:** stipules (± deciduous) 2–4-fid, 0.5–1 mm (scarious); petiole 0.7–2.8 cm; leaflet blade linear to elliptic, 2.3–5.5 × 0.2–0.8 cm, thin, margins entire, apex acute, surfaces glabrous. **Inflorescences** racemes, terminating stems and branches, 1–6 cm (10–40 cm in fruit); bracts trifoliate proximally, unifoliate distally, 7–15 mm. **Pedicels** reflexed in fruit, 5–10 mm (6–13 mm in fruit). **Flowers:** sepals yellow, deltate, 1–2.2 × 0.5–1 mm, glabrous; petals orange-yellow, ovate, (3–)4–5(–6) × 1–2 mm, glabrous; stamens yellow, 7–10 mm; anthers 1.5–2.2 mm; gynophore reflexed, 4–7 mm in fruit; ovary rhomboidal, 1–1.5 mm; style 0.3–0.5 mm. **Capsules** rhomboidal, 4–6 × 6–12 mm, glabrous. **Seeds** 2–4(–6), gray with black mottling, reniform, 2–3 mm, rugose.

Flowering mid spring–summer. Pond shores, stream bottoms, roadsides, grasslands, heavy alkaline clayey or sandy soils; 200–1200 m; Colo., Kans., Nebr., Okla., Tex.

2. **Cleomella brevipes** S. Watson, Proc. Amer. Acad. Arts 17: 365. 1882 • Shortstalk stinkweed [E]

Plants 3–15(–35) cm. **Stems** profusely branched (erect or spreading and matlike); glaucous, scabrous. **Leaves:** stipules finely divided into crinkled threads, 2 mm; petiole 0.1–0.3 cm; leaflet blade linear-obovate, 0.5–1.5 × 0.1–0.3 cm, fleshy, margins entire, apex obtuse to acute, mucronate, surfaces scabrous. **Inflorescences** solitary flowers in axils of closely spaced leaves throughout; bracts unifoliate, 8–10 mm. **Pedicels** reflexed in fruit, 1–2 mm. **Flowers:** sepals green, ovate, 0.8–1.2 × 0.3–0.7 mm, glabrous; petals pale yellow, oblong, 1.5–2 × 0.5–1 mm, glabrous; stamens yellow, 1.5–2.2 mm; anthers 0.3–0.5 mm; gynophore strongly reflexed, 0.5–3 mm in fruit; ovary globose, 1–1.5 mm; style 2–3(–5) mm. **Capsules** globose or deltoid, 2–3 × 2–3.2 mm, glabrous, (valves triangular). **Seeds** 1–4, yellowish, globose, 1.3–1.7 mm, smooth.

Flowering spring–fall. Alkaline marshes, wet saline soils around thermal springs; 400–1400 m; Calif., Nev.

3. **Cleomella hillmanii** A. Nelson, Proc. Biol. Soc. Wash. 18: 171. 1905 (as hillmani) • Hillman's or desert stinkweed [E]

Plants 10–55(–85) cm. **Stems** sparsely branched proximally (central stem dominant); glabrous. **Leaves:** stipules 3–5-fid, 0.5–1 mm (scarious); petiole 2–6(–8) cm; leaflet blade elliptic, ovate, or ovate-oblong, 0.8–2.4 × 0.4–1.4 cm, thin, margins entire, apex obtuse to retuse, mucronulate, surfaces glabrous. **Inflorescences** racemes, terminating stems, 3–10 cm (2–20 cm in fruit); bracts unifoliate, 5–9 mm. **Pedicels** divergent in fruit, 5–12 mm (4.5–17 mm in fruit). **Flowers:** sepals green, ovate-lanceolate, 1–2 × 0.5–0.7 mm, glabrous; petals yellow, oblong, (3.5–)4–8 × 1.8–2.2 mm, glabrous; stamens (well-exserted), yellow, 8–12 mm; anthers 2.5–3 mm; gynophore ascending, 3.5–15 mm in fruit; ovary ovoid, 1–1.3 mm; style (0.7–)1–2 mm. **Capsules** rhomboidal, 3.5–6 × 4–10.5 mm, glabrous. **Seeds** (1 or) 2–6, stramineous, globose, 2–2.7 mm, smooth.

Varieties 2 (2 in the flora): w United States.

1. Pedicels (6.5–)9–17 mm in fruit; gynophore (6–)7–15 mm in fruit; capsules 4–7(–8) mm diam. . . . . 3a. *Cleomella hillmanii* var. *hillmanii*
1. Pedicels 4.5–8.5(–10) mm in fruit; gynophore 3.5–6.5 mm in fruit; capsules (5–)6.5–10.5 mm diam. . . . . 3b. *Cleomella hillmanii* var. *goodrichii*

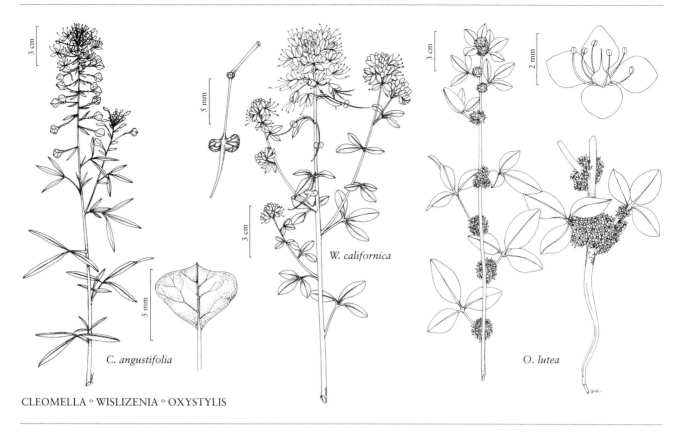

CLEOMELLA ∘ WISLIZENIA ∘ OXYSTYLIS

3a. Cleomella hillmanii A. Nelson var. **hillmanii** [E]

Cleomella longipes Torrey var. *grandiflora* S. Watson

Pedicels (6.5–)9–17 mm in fruit; gynophore (6–)7–15 mm in fruit (conspicuously longer than capsule). **Capsules** 4–7(–8) mm diam.

Flowering spring. Dry open usually alkaline meadows and flats; 800–1900 m; Calif., Idaho, Nev., Oreg.

3b. Cleomella hillmanii A. Nelson var. **goodrichii** (S. L. Welsh) P. K. Holmgren, Brittonia 56: 105. 2004 [E]

Cleomella palmeriana M. E. Jones var. *goodrichii* S. L. Welsh, Great Basin Naturalist 46: 263. 1986 (as palmerana); *C. macbrideana* Payson

Pedicels 4.5–8.5(–10) mm in fruit; gynophore 3.5–6.5 mm in fruit (about as long as capsule). **Capsules** (5–)6.5–10.5 mm diam. $2n = 34$.

Flowering spring. Dry open usually alkaline meadows and flats; 800–1900 m; Idaho, Utah.

Following P. K. Holmgren and A. Cronquist (2005), *Cleomella macbrideana* is treated as a synonym of var. *goodrichii*. That entity is disjunct from the main body of the species range; its closest morphological match is with this variety.

4. Cleomella longipes Torrey, Hooker's J. Bot. Kew Gard. Misc. 2: 255. 1850

Plants (rarely perennial) 30–80 cm. **Stems** sparsely branched (central stem dominant); glabrous. **Leaves:** stipules linear, 0.5–1 mm (scarious); petiole 0.8–2 cm; leaflet blade oblanceolate to oblong-lanceolate, 1.5–3 × 0.4–1 cm, thin, margins entire, apex acute to rounded or slightly emarginate, surfaces glabrous. **Inflorescences** racemes, terminating stems and branches, 2–12 cm (10–50 cm in fruit); bracts unifoliate, 5–12 mm (distalmost flowers often absent). **Pedicels** ascending in fruit, 4–6 mm (5–8 mm in fruit). **Flowers:** sepals green, lanceolate, 0.9–2.2 × 0.5–0.7 mm, glabrous; petals yellow, oblong, 6–9 × 2–4 mm, glabrous; stamens yellow, 8–12 mm; anthers 1.5–1.9 mm; gynophore ascending, 6–17 mm in fruit; ovary rhomboidal, 1–1.5 mm; style 1–2 mm. **Capsules** rhomboidal, 4–8 × 6–10 mm, glabrous. **Seeds** 6–16, dark brown, globose, 2–2.2 mm, smooth. $2n = 34$.

Flowering late spring–late summer. Saline or alkaline flats; 500–1000 m; Ariz., N.Mex., Tex.; Mexico.

5. **Cleomella obtusifolia** Torrey & Frémont in J. C. Frémont, Rep. Exped. Rocky Mts., 311. 1845 • Mojave stinkweed [E]

Cleomella obtusifolia var. *florifera* Crum ex Jepson; *C. obtusifolia* var. *jonesii* Crum ex Jepson; *C. obtusifolia* var. *pubescens* A. Nelson; *C. taurocranos* A. Nelson

Plants 10–40(–120) cm. **Stems** profusely branched at base (usually spreading or matlike, sometimes erect); moderately to densely hairy (sometimes spreading-hispidulous). **Leaves:** stipules finely divided into crinkled threads, 5–7 mm; petiole 0.7–0.8(–2) cm; leaflet blade elliptic to obovate, 0.5–1.5 × 0.2–0.8 cm, thin, margins entire, apex obtuse to rounded, mucronulate, surfaces spreading-hispidulous. **Inflorescences** racemes, terminating stems and branches (solitary flowers in leaf axils of distal stems), 0.5–1.2 cm (0.5–1 cm in fruit); bracts unifoliate, 2–7 mm. **Pedicels** ascending in fruit, 4–10 mm (3–12 mm in fruit). **Flowers:** sepals green, ovate-deltate, 1–1.5(–2.3) × 0.9–1.2 mm, sparsely hairy; petals yellow or orange, oblong, 3.5–6 × 1.2–2 mm, sparsely hairy abaxially; stamens yellow, 8–14 mm; anthers 1.5–2.2 mm; gynophore reflexed, 4–5 mm in fruit; ovary ± rhomboidal, 1–1.5 mm; style 2–3(–5) mm. **Capsules** rhomboidal, 3.5–4 × 7–10 mm (striate), strigose. **Seeds** (1–)2–6, light brown to gray with black mottling, globose, 1.5 mm, smooth.

Flowering late winter–fall. Sandy, often alkaline flats and desert playas; 300–1300(–2000) m; Calif., Nev., N.Mex.

Cleomella obtusifolia is variable in degree of pubescence, elaboration of stipules, and fruit shape. Four varieties have been recognized; additional taxonomic investigation is desirable.

6. **Cleomella palmeriana** M. E. Jones, Zoë 2: 236. 1891 (as palmerana) • Palmer's cleomella, Rocky Mountain stinkweed [E]

Cleomella cornuta Rydberg; *C. montrosae* Payson; *C. nana* Eastwood

Plants 10–38 cm. **Stems** usually sparsely branched proximally (without a dominant central stem); glabrous. **Leaves:** stipules 3–5-fid, (larger, more conspicuous in inflorescences), 0.5–1.5 mm (scarious); petiole 0.4–2(–2.7) cm; leaflet blade narrowly oblong-elliptic, 0.8–2.6 × 0.1–0.8 cm, thin, margins entire, apex obtuse to retuse, mucronulate, surfaces glabrous. **Inflorescences** corymbs or racemes, terminating stems, 1–2 cm (2–3.5 cm in fruit); bracts ± rudimentary. **Pedicels** slightly ascending in fruit, 6–7 mm (7–11 mm in fruit). **Flowers:** sepals green, deltate, 1–1.5 × 0.5–0.6 mm, glabrous; petals yellow, oblong, 3–5 × 1–1.3 mm, glabrous; stamens (well-exserted), yellow, 7–9 mm; anthers 1.4–1.6 mm; gynophore spreading, 2.5–6(–8) mm in fruit; ovary obovoid, 1.3–1.5 mm; style 0.7–1 mm. **Capsules** rhomboidal (widest distal to midpoint), 2–5 × 3–5(–6.5) mm, glabrous. **Seeds** (1 or) 2–4(–7), stramineous becoming black, globose, 2–2.5 mm, smooth.

Flowering spring–early fall. Dry open alkaline, gravelly or sandy flats; 1100–2000 m; Ariz., Colo., N.Mex., Utah.

7. **Cleomella parviflora** A. Gray, Proc. Amer. Acad. Arts 6: 520. 1865 • Slender stinkweed [E]

Plants 3–45 cm. **Stems** sometimes branched distally (erect or somewhat spreading); glabrous. **Leaves:** stipules finely divided into crinkled threads (scarious), 1–2 mm; petiole 0.2–0.6(–0.9) cm; leaflet blade linear-elliptic, planoconvex, 0.5–3.5 × 0.1–0.4 cm, slightly fleshy, margins entire, apex acute, surfaces glabrous. **Inflorescences** racemes, terminating stems and branches (from all but 2 or 3 proximalmost nodes), 0.5–20 cm (5–30 cm in fruit); bracts trifoliate or unifoliate, 1–3 cm. **Pedicels** spreading in fruit, 2–6 mm (12–20 mm in fruit). **Flowers:** sepals green, broadly lanceolate, 0.5–1 × 0.3–0.6 mm, glabrous; petals pale yellow, oblong, 1.8–2.8 × 0.8–1.2 mm, glabrous; stamens yellow, 1.5–3 mm; anthers 0.4–0.6 mm; gynophore spreading, 0.5–2 mm in fruit; ovary spherical, 0.8—1.3 mm; style 0.1–0.2 mm. **Capsules** rhomboidal, 2.5–5 × 2.5–6 mm, (smooth), glabrous, (valves scarcely conical). **Seeds** 3–12, light brown, globose, 1.2–2 mm, smooth.

Flowering spring–summer. Wet alkaline meadows around thermal springs in sagebrush desert; 1200–2000 m; Calif., Idaho, Nev.

Cleomella parviflora is often found growing with *C. brevipes* and *C. plocasperma*.

8. **Cleomella plocasperma** S. Watson, Botany (Fortieth Parallel), 33. 1871 [E]

Cleomella oöcarpa A. Gray;
C. plocasperma var. *mojavensis* (Payson) Crum ex Jepson;
C. plocasperma var. *stricta* Crum ex Jepson

Plants 10–55(–85) cm. Stems sparsely branched distally (central stem dominant); glabrous. Leaves: stipules (scarious) 0.5–1 mm; petiole 0.8–2 cm; leaflet blade linear-elliptic, 1.5–4.5 × 0.2–0.7 cm, thin, margins entire, apex acute, surfaces glabrous. Inflorescences racemes, terminating stems and branches, 1–15 cm (2–20 cm in fruit); bracts unifoliate, 5–25 mm. Pedicels ascending in fruit, 5–10 mm (6–12 mm in fruit). Flowers: sepals green, lanceolate, 0.9–2.2 × 0.5–0.7 mm, glabrous; petals yellow or orange, oblong, 3.5–5(–7) × 1.4–1.6(–2) mm, hairy abaxially; stamens yellow, 8–12 mm; anthers 1.5–1.9 mm; gynophore ascending, 6–10 mm in fruit; ovary rhomboidal, 1–1.5 mm; style 0.8–1.2 mm. Capsules ovoid to rhomboidal, 2.5–5 × 2.5–6(–8) mm, glabrous. Seeds 2–4, silver-gray with black mottling, globose, 2–2.2 mm, smooth.

Flowering mid spring–fall. Wet alkaline meadows, greasewood flats, around thermal springs; 800–1400 m; Calif., Idaho, Nev., Oreg., Utah.

5. **WISLIZENIA** Engelmann in F. A. Wislizenus, Mem. Tour N. Mexico, 99. 1848 • Jackass-clover, spectacle-fruit [For Friedrich (later Frederick) Adolph Wislizenus, 1810–1889, botanical collector in southwestern United States and adjacent Mexico]

Gordon C. Tucker

Herbs, shrubs, or subshrubs, annual or perennial. Stems sparsely branched laterally; puberulent, glabrous, or glabrate (scabrid when dry). Leaves: stipules minute tufts of filiform hairs; petiole without pulvinus (equal to or longer than leaflets); leaflets (1 or) 3, (conduplicate). Inflorescences terminal, racemes (dense); bracts usually absent; (pedicels 3–5 mm, glabrous). Flowers: zygomorphic; sepals usually persistent (eventually deciduous), distinct, equal; petals slightly unequal (in pairs); stamens 6 (well-exserted, ± equal); filaments inserted on gynophore, glabrous; anthers (oblong), coiling after dehiscence; gynophore reflexed in fruit; (pistil 2-lobed basally, style filiform, eventually spinelike). Fruits schizocarps, indehiscent, obovate, (reticulate, ± tuberculate, valves disarticulating from persistent, round septum). Seeds 2(–4), globose, not arillate. $x = 10$.

Species 3 (3 in the flora): sc, sw United States, nw Mexico.

Wislizenia has been the object of diverse taxonomic opinions; as many as eight or as few as one species have been recognized. S. Keller (1979) documented three geographically and morphologically distinct groups within the genus. Although she treated those as subspecies, recognition as species seems more appropriate (H. H. Iltis, pers. comm.). The differentiation among the taxa seems equal to or greater than that among species in other genera of the family.

SELECTED REFERENCES Greene, E. L. 1906b. Revision of the genus *Wislizenia*. Proc. Biol. Soc. Wash. 19: 127–132. Keller, S. 1979. A revision of the genus *Wislizenia* (Capparidaceae) based on population studies. Brittonia 31: 333–351.

1. Subshrubs or shrubs; leaflets 3 proximally, 1 distally, blade ovate to obovate, 3–8(–12) times as long as wide; sepals ovate . . . . 2. *Wislizenia palmeri*
1. Annuals; leaflets 3 throughout, blade ovate to obovate, 1.5–4(–5.5) times as long as wide; sepals lanceolate.
    2. Leaflets less than 2.8 times as long as wide; gynophore ca. 5 mm in fruit; anthers 1.3–2.5 mm; petals 2.5–5 mm . . . . 1. *Wislizenia californica*
    2. Leaflets 1.5–4(–5.5) times as long as wide; gynophore 2–12 mm in fruit; anthers 0.5–1.7 mm; petals 1.4–4.6 mm . . . . 3. *Wislizenia refracta*

1. **Wislizenia californica** Greene, Proc. Biol. Soc. Wash. 19: 130. 1906 E F

Wislizenia refracta Engelmann subsp. *californica* (Greene) S. Keller

Annuals, 40–240 cm. **Stems** puberulent or glabrous. **Leaves:** petiole 0.1–3 cm; leaflets 3, gray-green abaxially, yellowish green adaxially, blade obovate, 0.4–2(–3.5) × 0.4–2 cm, less than 2.8 times as long as wide, margins entire, apex rounded, surfaces puberulent. **Racemes** 1–1.5 cm (2–3 cm in fruit). **Flowers:** sepals green, lanceolate, 0.5–1.6 × 0.3–0.7 mm, more than 1.75 times as long as wide, margins entire, glabrous; petals yellow, oblong, 2.5–5 × 0.6–1.9 mm; stamens yellow, 2–9 mm; anthers 1.3–2.5 mm; gynophore ca. 5 mm in fruit, usually longer than pedicel. **Schizocarps** 1.4–3.4 × 1 mm. **Seeds** 0.5 × 0.3 mm. $2n = 40$.

Flowering late Jul–Oct. Sandy washes, banks of irrigation ditches, roadsides, croplands; 20–300 m; Calif.

Wislizenia californica is known only from the San Joaquin Valley.

2. **Wislizenia palmeri** A. Gray, Proc. Amer. Acad. Arts 8: 622. 1873

Wislizenia divaricata Greene; *W. fruticosa* Greene; *W. mamillata* Rose ex Greene; *W. refracta* Engelmann var. *mamillata* (Rose ex Greene) I. M. Johnston; *W. refracta* subsp. *palmeri* (A. Gray) S. Keller; *W. refracta* var. *palmeri* (A. Gray) I. M. Johnston

Subshrubs or shrubs, 50–200 cm. **Stems** glabrous or glabrate. **Leaves:** petiole 0.1–2.1 cm; leaflets 3 proximally, 1 distally, gray-green to green, blade ovate to obovate, 0.6–4 × 0.4–2 cm, 3–8(–12) times as long as wide, margins entire, apex emarginate to rounded. **Racemes** 1–1.5 cm (2–3 cm in fruit). **Flowers:** sepals green, ovate, 0.5–1.7 × 0.4–1.1 mm, less than 1.75 times as long as wide, margins entire, glabrous; petals yellow, oblong, 2.5–6.3 × 1–2.5 mm; stamens yellow, 4–10 mm; anthers (1–)1.5–2.5 mm; gynophore 5 mm in fruit, usually longer than pedicel. **Schizocarps** 1–5.6 × 1 mm. **Seeds** 0.5 × 0.3 mm. $2n = 40$.

Flowering and fruiting year-round. Sandy washes, coastal dunes, saline flats, desert scrub; 100–500 m; Ariz., Calif.; Mexico (Baja California, Baja California Sur, Sonora).

Essentially a Mexican species, *Wislizenia palmeri* occurs in the flora area in Riverside and San Diego counties in California, and in Organ Pipe National Monument in southern Arizona.

3. **Wislizenia refracta** Engelmann in F. A. Wislizenus, Mem. Tour N. Mexico, 99. 1848 W

Wislizenia costellata Rose ex Greene; *W. melilotoides* Greene; *W. refracta* var. *melilotoides* (Greene) I. M. Johnston; *W. scabrida* Eastwood

Annuals, 40–200 cm. **Stems** glabrous or glabrate (sometimes smooth when dry). **Leaves:** petiole 0.2–3.1 cm; leaflets 3, green, blade ovate or obovate, (0.4–)1–4.9 × 0.4–2 cm, 1.5–4(–5.5) times as long as wide, margins entire, apex acute to rounded. **Racemes** 1–1.5 cm (2–3 cm in fruit). **Flowers:** sepals green, lanceolate, 0.4–1.7 × 0.2–0.7 mm, more than 1.75 times as long as wide, margins entire, glabrous; petals yellow, oblong, 1.4–4.6 × 0.5–1.8 mm; stamens yellow, 2–6 mm; anthers 0.5–1.7 mm; gynophore (reflexed in fruit), 2–12 mm in fruit, usually shorter than pedicel. **Schizocarps** 1.2–3.3 × 1 mm. **Seeds** 0.5 × 0.3 mm. $2n = 40$.

Flowering spring–fall. Dry sandy flats, desert scrub, roadsides; 500–2400 m; Ariz., Calif., Nev., N.Mex., Tex., Utah; Mexico (Chihuahua, Sonora).

Wislizenia refracta is known from trans-Pecos Texas, and from the Mojave Desert (San Bernardino County and Little San Bernardino Mountains in Riverside County) in California.

6. **OXYSTYLIS** Torrey & Frémont in J. C. Frémont, Rep. Exped. Rocky Mts., 312. 1845
• Spiny-caper [Greek *oxys*, sharp, and *stylos*, pillar, alluding to style] E

Staria S. Vanderpool

Herbs, annual (unpleasantly scented). **Stems** unbranched, or sparsely branched laterally; glabrous or glabrate. **Leaves:** stipules 3–8-palmatifid, minute; petiole without pulvinus (slightly longer than leaflets); leaflets 3. **Inflorescences** axillary (from distal leaves), racemes (flat-topped); bracts present. **Flowers** actinomorphic; sepals persistent, distinct, equal (each often subtending a nectary); petals equal; stamens 6; filaments inserted on a spheroidal gynophore, glabrous; anthers (ellipsoid), coiling as pollen is released; gynophore sharply deflexed in fruit (stout, style spinelike in fruit). **Fruits** schizocarps (didymous, deflexed), indehiscent, obovoid. **Seeds** 2, reniform, not arillate. $x = 10$.

Species 1: sw United States.

1. **Oxystylis lutea** Torrey & Frémont in J. C. Frémont, Rep. Exped. Rocky Mts., 313. 1845 E F

Plants 10–60(–150) cm. **Leaves:** stipules crinkled; petiole 1.4–7 cm; petiolules 2–4 mm; leaflet blade elliptic to elliptic-obovate, 1.5–5 × 0.5–2.5 cm, margins entire, apex obtuse. **Racemes** dense, ca. 1 cm (2–3 cm in fruit); bracts unifoliate, 2–4 mm. **Pedicels** sparsely glandular, 1–2 mm. **Flowers:** sepals green, lanceolate, 1.4–1.7 × 0.2–0.3 mm, margins entire, glabrous; petals yellow, oblong to ovate, 2–4 × 1–1.6 mm; stamens yellow, 4–4.5 mm; anthers 1–1.2 mm; gynophore 0.1–0.2 mm in fruit; ovary 0.3 mm; style 0.8–1 mm. **Schizocarps** 1 × 0.6 mm, smooth. **Seeds** 0.5 × 0.3 mm. $2n = 20, 40$.

Flowering spring–summer. Dry, sandy, gravelly desert flats, scrub; 1200–1600 m; Calif., Nev.

7. **CLEOME** Linnaeus, Sp. Pl. 2: 671. 1753; Gen. Pl. ed. 5, 302. 1754 • Spiderflower [Origin obscure, perhaps from Greek *kleos*, glory, or after Kleo, Greek muse of history, first used by Priscian, fourteenth-century medical writer] I

Gordon C. Tucker

Herbs, annual or perennial. **Stems** unbranched or sparsely branched; glandular-pubescent, glabrous, glabrescent, or scabrous. **Leaves:** stipules absent or scalelike; petiole with pulvinus basally or distally, (petiolule basally adnate, forming pulvinar disc); leaflets 1 or 3[–11] (flat). **Inflorescences** terminal or axillary (from distal leaves), racemes (flat-topped or elongated); bracts present [absent]. **Flowers** zygomorphic; sepals persistent, basally connate (½ of length), equal (each often subtending a nectary); petals equal; stamens [4] 6; filaments inserted on a discoid or conical androgynophore, glabrous; anthers (oblong to linear), coiling as pollen is released; gynophore recurved in fruit [obsolete]. **Fruits** capsules, dehiscent, oblong. **Seeds** 4–25, reniform or ovoid-spheroidal, arillate or not, (cleft fused between ends). $x = 10$ (?).

Species ca. 20 (2 in the flora): introduced; Old World; warm temperate and tropical areas.

The center of diversity of *Cleome* is in southwestern Asia. There are only two true *Cleome* in North America. Other native and adventive species formerly included in *Cleome* are placed in *Arivela, Cleoserrata, Gynandropsis, Hemiscola, Peritoma,* and *Tarenaya*.

1. Capsules 25–35 mm; bracts mostly unifoliate; leaflet blade linear to elliptic . . . . . 1. *Cleome ornithopodioides*
1. Capsules 40–70 mm; bracts trifoliate; leaflet blade oblanceolate to rhombic-elliptic . . . . .
. . . . 2. *Cleome rutidosperma*

1. Cleome ornithopodioides Linnaeus, Sp. Pl. 2: 672. 1753 • Bird spiderflower, Levant mustard [F][I]

Cleome iberica de Candolle

Annuals, 15–60 cm. **Stems** unbranched or sparsely branched; glandular-pubescent. **Leaves**: stipules absent; petiole (0.6–)1.5–4.5 cm; leaflets 3 proximally, 1 distally, blade linear to elliptic, 0.5–3 × 0.1–0.7 cm, margins entire, apex acute to obtuse, surfaces stipitate-glandular throughout (sparingly hairy abaxially). **Racemes** 1–3 cm (6–15 cm in fruit); bracts mostly unifoliate, sometimes proximalmost trifoliate, 2–15 mm. **Pedicels** 7–15 mm. **Flowers**: sepals greenish with purplish margin, deltate, 0.6–2 × 0.8–1.2 mm, margins denticulate, glabrous; petals whitish or yellow, sometimes red-striped, oblong to ovate, 3–4 × 1.5–2 mm; stamens reddish, 5–6 mm; anthers 0.6–0.9 mm; gynophore (1–)3–5 mm in fruit (subtended by clavate, purple-black gland adaxially); ovary 3–6 mm; style 0.2–0.4 mm. **Capsules** 25–35 × 1.5–2 mm. **Seeds** (4–)10–20, reddish brown or mottled black and brown, ovoid-spheroidal, 1.2–1.5(–1.8), not arillate, (finely papillose).

Flowering summer. Railroad yards, ore piles (chrome, vanadium), roadsides; 100–200 m; introduced; Ky., Md., Ohio, Pa.; se Europe; sw Asia.

SELECTED REFERENCES Reed, C. F. 1965b. *Cleome ornithopodioides* L. on vanadium-slag at Canton, Baltimore, Maryland with notes on the biochemistry of vanadium. Phytologia 11: 423–431. Thieret, J. W. and R. L. Thompson. 1984. *Cleome ornithopodioides* (Capparaceae): Adventive and spreading in North America. Bartonia 50: 25–26.

2. Cleome rutidosperma de Candolle in A. P. de Candolle and A. L. P. P. de Candolle, Prodr. 1: 241. 1824 • Fringed spiderflower [I]

Cleome ciliata Schumacher & Thonning

Annuals or perennials, 30–100 cm. **Stems** sparsely branched, (often decumbent); glabrous or glabrescent to slightly scabrous (sometimes glandular-pubescent). **Leaves**: stipules 0–0.5 mm; petiole (winged proximally), 0.5–3.5 cm; leaflets 3, blade oblanceolate to rhombic-elliptic, 1–3.5 × 0.5–1.7 cm, margins entire or serrulate-ciliate, apex usually acute to obtuse, sometimes acuminate, surfaces with curved hairs on veins abaxially, glabrous adaxially. **Racemes** 2–4 cm (8–15 cm in fruit); bracts trifoliate, 10–35 mm. **Pedicels** 11–21 mm (18–30 mm in fruit). **Flowers**: sepals yellow, lanceolate, 2.5–4 × 0.2–0.3 mm, margins denticulate, ciliate, glabrous; petals white or purple-speckled (2 central ones with yellow transverse band abaxially), oblong to narrowly ovate, 7–10 × 1.5–2.3 mm; stamens yellow, 5–7 mm; anthers 1–2 mm; gynophore 4–12 mm in fruit; ovary 2–3 mm, glabrous; style 0.5–1.4 mm. **Capsules** 40–70 × 3–4 mm. **Seeds** 4–25, reddish brown to black, reniform, 1–1.5 mm, arillate.

Flowering ± year-round. Roadsides, vacant lots, canal banks, lawn edges in sun or shade; 0–200 m; introduced; Fla., S.C.; tropical Asia; Africa; introduced also in Mexico, West Indies, Central America, South America.

Cleome rutidosperma has sometimes been misidentified as *Hemiscola aculeata* (*Cleome aculeata*); it lacks the nodal spines of that species.

8. CLEOSERRATA H. H. Iltis, Novon 17: 447. 2007 • [Genus *Cleome* and *serrata*, serrate, alluding to leaflet margins] [I]

Gordon C. Tucker

Herbs, annual [perennial]. **Stems** unbranched or sparsely [profusely] branched; glabrous or glandular-pubescent. **Leaves**: stipules absent; petiole with pulvinus basally or distally, (petiolule base adnate, forming pulvinar disc; leaflets 3 or 5–9. **Inflorescences** terminal or axillary (from distal leaves), racemes (flat-topped or elongated); bracts present. **Flowers** (often appearing unisexual due to incomplete development), zygomorphic; sepals persistent or deciduous, distinct, equal (each often subtending a nectary); petals equal; stamens 6; filaments inserted on a discoid

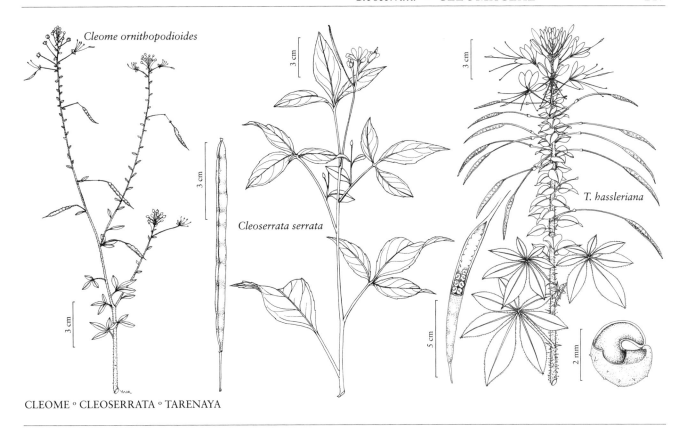

CLEOME ° CLEOSERRATA ° TARENAYA

or conical androgynophore, usually glabrous; anthers coiling as pollen is released; gynophore recurved in fruit. **Fruits** capsules, dehiscent, oblong. **Seeds** 10–30+, subglobose [reniform], not arillate, (cleft fused between ends). $x = 12$.

Segregation of *Cleoserrata* from *Tarenaya* is based on the absence of prickles on the minutely serrulate leaf, the $x = 12$ chromosomal complement, and molecular evidence suggesting a more basal clade than that of *Tarenaya* (L. A. Inda et al. 2008).

Species 5 (2 in the flora): introduced; Mexico, West Indies, Central America, South America.

1. Petals usually brilliant pink to purple, fading to pink or white, rarely initially white; bracts ovate-cordate, 3–18 mm; gynophore 30–85 mm in fruit . . . . 1. *Cleoserrata speciosa*
1. Petals white or whitish, with pinkish or red distally; bracts subulate, to 1 mm; gynophore 1(–2) mm in fruit. . . . . 2. *Cleoserrata serrata*

1. **Cleoserrata speciosa** (Rafinesque) H. H. Iltis, Novon 17: 448. 2007 • Showy or garden spiderflower, volantines-preciosos [I]

Cleome speciosa Rafinesque, Fl. Ludov., 86. 1817; *C. speciosissima* Deppe ex Lindley; *Gynandropsis speciosa* de Candolle

Plants 50–150 cm. **Stems** unbranched or sparsely branched; (fluted), glabrous or sparsely glandular-pubescent. **Leaves**: petiole 2–12 cm ,(glandular-pubescent); leaflets 5–9, blade narrowly lanceolate-elliptic, 6–15 × 1–5 cm, margins entire or serrulate, apex subobtuse, surfaces glabrate to glandular-pubescent. **Racemes** 15–50 cm (20–60 cm in fruit, glandular-pubescent); bracts unifoliate, ovate-cordate, 3–18 mm. **Pedicels** 10–50 mm, (glabrous). **Flowers**: sepals persistent, green, lanceolate, 4–7 × 0.8–1.2 mm, glandular-pubescent; petals usually brilliant pink to purple, fading to pink or white, rarely initially white, ovate, 15–42 × 8–11 mm, clawed; stamens green, 40–85 mm, (filaments adnate to gynophore $\frac{1}{3}$–$\frac{1}{2}$ of length); anthers 6–10 mm; gynophore 30–85 mm in fruit (filament scars visible ca. $\frac{1}{4}$ of length); ovary 6–10 mm; style 1–1.2 mm. **Capsules** (irregularly contracted between seeds) 60–150 × 3–5 mm. **Seeds** pale green to brown, 2.5–3.5 × 1–1.2 mm, tuberculate. $2n = 48$.

Flowering summer. Disturbed roadsides, vacant lots; 0–50 m; introduced; Fla.; Mexico; West Indies; Central America; South America.

Cleoserrata speciosa is widespread horticulturally (with white-flowered plants not uncommon). It strongly resembles *Tarenaya hassleriana* (*Cleome hassleriana*) to the untrained eye (W. R. Ernst 1963b); considering its unique floral morphology and cytology, it remains difficult to place.

2. **Cleoserrata serrata** (Jacquin) H. H. Iltis, Novon 17: 448. 2007 • Spiderflower [F] [I] [W]

Cleome serrata Jacquin, Enum. Syst. Pl., 26. 1760; *C. polygama* Linnaeus

Plants [perennial], 15–50 cm. **Stems** unbranched or slightly branched; glabrous. **Leaves:** petiole 2–5 cm; leaflets 3, blade lanceolate to ovate-elliptic, 5–12 × 0.4–1.4 cm, margins serrulate, apex acute to long-acuminate, surfaces glabrous. **Racemes** 6–25 cm; bracts unifoliate, subulate, to 1 mm. **Pedicels** 4–8 mm. **Flowers:** sepals deciduous, green, subulate, 2–2.5 × 0.4–0.5 mm, glabrous; petals white or whitish, with pinkish or red distally, obovate, 5–6 × 1.8–2 mm, shortly clawed; stamens yellow, 2–5 mm; anthers 2–3 mm; gynophore 1(–2) mm in fruit; ovary 2.5–4 mm; style 0.2–0.3 mm. **Capsules** 30–50 × 2.5–3 mm. **Seeds** pale green to brown, 1.4–1.5 mm, inconspicuously ridged. $2n = 24$.

Flowering summer. Disturbed areas; ca. 100 m; introduced; Ga.; Mexico; West Indies; Central America; South America.

9. **TARENAYA** Rafinesque, Sylva Tellur., 111. 1838 • Spiderflower [Origin obscure] [I]

Gordon C. Tucker

Hugh H. Iltis

Cleome Linnaeus sect. *Tarenaya* (Rafinesque) H. H. Iltis

Herbs [shrubs], annual. **Stems** sparsely to profusely branched; glabrous or glandular-pubescent (sometimes spiny). **Leaves:** stipular spines present [absent]; petiole with pulvinus basally or distally, (petiolule base adnate, forming pulvinar disc; spines present); leaflets [1 or](3) 5 or 7[–11] (with tiny prickles terminating teeth). **Inflorescences** terminal or axillary (from distal leaves), racemes (flat-topped or elongated); bracts present [absent]. **Flowers** (often appearing unisexual due to incomplete development), zygomorphic; sepals persistent, distinct, equal (each often subtending a nectary); petals equal; stamens 6; filaments inserted on a discoid or conical androgynophore, glabrous; anthers (linear), coiling as pollen is released; gynophore recurved in fruit. **Fruits** capsules, dehiscent, oblong. **Seeds** 10–30+, triangular to subglobose, not arillate, (cleft fused between ends). $x = 10$.

Species 33 (2 in the flora): introduced; South America; introduced worldwide in tropical and warm-temperate regions.

Traditionally included in a broad circumscription of *Cleome*, *Tarenaya* is distinguished by its stipular and petiolar spines, absence of arils, and seeds with a large cleft cavity. All species are native to tropical America, with the exception of *T. afrospina* (H. H. Iltis) H. H. Iltis, of western Africa. *Tarenaya hassleriana*, long known as *C. spinosa* (and commonly by the misapplied name *C. houtteana*) and more recently as *C. hassleriana*, is a popular garden ornamental and, probably, the most widely distributed member of the family.

1. Sepals, ovary, and capsules glabrous; petals usually pink or purple, sometimes white .... 1. *Tarenaya hassleriana*
1. Sepals, ovary, and capsules ± glandular-pubescent; petals white or greenish white. .... 2. *Tarenaya spinosa*

1. Tarenaya hassleriana (Chodat) H. H. Iltis, Novon 17: 450. 2007 • Pink-queen F I W

Cleome hassleriana Chodat, Bull. Herb. Boissier 6(app. 1): 12. 1898

Plants (50–)100–200 cm. **Stems** sparsely branched; glandular-pubescent. **Leaves:** stipular spines 1–3 mm; petiole 2.5–7.5 cm, glandular-pubescent, with scattered spines 1–3 mm; leaflets 5 or 7, blade elliptic to oblanceolate, 2–6(–12) × 1–3 cm, margins serrulate-denticulate, apex acute, surfaces glandular-pubescent abaxially, glandular adaxially. **Racemes** 5–30 cm (10–80 cm in fruit); bracts unifoliate, ovate, 10–25 mm. **Pedicels** 20–45 mm, glandular-pubescent. **Flowers:** sepals (reflexed after anthesis), green, linear-lanceolate, 5–7 × 0.8–1.3 mm, glabrous; petals usually pink or purple, sometimes white (or fading to white by second day), oblong to ovate, 20–30(–45) × 8–12 mm; stamens purple, 30–50 mm; anthers 9–10 mm; gynophore 45–80 mm in fruit; ovary 6–10 mm, glabrous; style 0.1 mm. **Capsules** (25–)40–80 × 2.5–4 mm, glabrous (in straight alignment with gynophore and pedicel). **Seeds** 10–20, 1.9–2.1 × 1.9–2.1 mm, tuberculate. $2n = 20$.

Flowering late spring–late summer. Disturbed roadsides, vacant lots, waste areas, gravel pits, lakeshores, streambeds; 0–200(–800) m; introduced; Que.; Ala., Ark., Conn., D.C., Fla., Ga., Ill., Ind., Iowa, Ky., La., Md., Mass., Mich., Miss., Mo., Nebr., N.J., N.Y., N.C., Ohio, Okla., Pa., R.I., S.C., Tenn., Tex., Va., W.Va., Wis.; South America; introduced also in Mexico, West Indies, Central America.

Tarenaya hassleriana is native to Argentina, Brazil, and Paraguay. It is often cultivated and has sometimes escaped and naturalized. In cultivation and various floras, it has long been treated under the name *Cleome spinosa*; that name properly applies to the next species.

2. Tarenaya spinosa (Jacquin) Rafinesque, Sylva Tellur., 111. 1838 • Espuela de caballero I W

Cleome spinosa Jacquin, Enum. Syst. Pl., 26. 1760; *C. pubescens* Sims; *C. pungens* Willdenow; *C. tonduzii* Briquet; *Neocleome spinosa* (Jacquin) Small

Plants [perennial], 50–200 cm. **Stems** sparsely to profusely branched; glandular-pubescent (spiny). **Leaves:** stipular spines 2–5 mm; petiole 2.5–10 cm, glandular-pubescent, with scattered, antrorse spines 1–3 mm; leaflets (3) 5 or 7, blade lanceolate to elliptic-oblanceolate, 4–10(–12) × 1–3 cm, margins serrulate-denticulate, apex acute to acuminate, surfaces glandular-pubescent abaxially, glandular adaxially. **Racemes** 1–20 cm (10–40 cm in fruit); bracts unifoliate, ovate to broadly elliptic, 8–25 mm. **Pedicels** 15–30 mm, glandular-pubescent. **Flowers:** sepals green, lanceolate, 6–8 × 0.8–1.3 mm, glandular-pubescent; petals white or greenish white (somewhat reddish abaxially), oblong to ovate, 11–21(–30) × 8–12 mm, (± glandular-pubescent especially apically); stamens white, 20–50 mm; anthers 9–10 mm; gynophore 45–70 mm in fruit; ovary 6–10 mm, ± glandular-pubescent; style 0.1 mm. **Capsules** 40–110(–150) × 3–5 mm, ± glandular-pubescent. **Seeds** 20–30, 1.8–2.2 × 1.9–2.1 mm, smooth. $2n = 20$.

Flowering summer. Disturbed roadsides, beaches, vacant lots, waste areas; 0–100 m; introduced; Ala., Fla., Pa., W.Va.; South America; introduced also in Mexico, West Indies, Central America.

10. HEMISCOLA Rafinesque, Sylva Tellur., 111. 1838 • Spiderflower [Greek, *hemi*-, half, and *skolios*, curved, alluding to seed shape]

Gordon C. Tucker

Hugh H. Iltis

Herbs, annual [perennial]. **Stems** (sometimes ± procumbent, weak), sparsely to moderately (or delicately) branched; glabrous or glandular-pubescent. **Leaves:** stipular spines present (recurved); petiole with pulvinus basally or distally; leaflets (1 or)3. **Inflorescences** terminal or axillary (from distal leaves), racemes (flat-topped or elongated); bracts present. **Flowers** (often appearing unisexual due to incomplete development), zygomorphic; sepals persistent or deciduous, distinct, equal (each often subtending a nectary); petals oblong to ovate, equal; stamens 6; filaments inserted on a discoid or conical gynophore, glabrous; anthers (ellipsoid to linear), coiling as pollen is released; gynophore reflexed in fruit. **Fruits** capsules, dehiscent, fusiform to linear-cylindric. **Seeds** 10–20, oblong or obovoid, prominently arillate, (cleft fused between ends). $x = 10$.

Species 6 (2 in the flora); introduced; Mexico, Central America, South America (south to Argentina).

1. Leaflet blade lanceolate-elliptic to ovate or rhombic; sepals lanceolate; anthers 2.5–4 mm; capsules (15–)25–40(–65) mm..1. *Hemiscola aculeata*
1. Leaflet blade obovate; sepals ovate; anthers 0.3–0.5 mm; capsules 15–20 mm..........2. *Hemiscola diffusa*

1. **Hemiscola aculeata** (Linnaeus) Rafinesque, Sylva Tellur., 111. 1838 • Prickly spiderflower [F][I]

Cleome aculeata Linnaeus, Syst. Nat. ed. 12, 3: 232. 1768

Varieties 3 (1 in the flora): introduced; Mexico, West Indies.

Variety *potosina* (B. L. Robinson) H. H. Iltis is endemic to Argentina and Mexico and is distinguished by relatively long gynophores; var. *affinis* (de Candolle) H. H. Iltis is endemic to southeastern Brazil and is spineless.

1a. **Hemiscola aculeata** (Linnaeus) Rafinesque var. **aculeata** [F][I]

Plants 30–50 cm. Stems sparsely to moderately branched; glabrate to glandular-pubescent. **Leaves:** stipular spines (recurved) 1–3 mm; petiole 1–6 cm (glandular-pubescent); leaflets 3, blade lanceolate-elliptic to ovate or rhombic, 2–4(–6) × 1.5–3 (–4) cm, margins serrulate-denticulate, apex acute or obtuse, surfaces glandular-pubescent abaxially, glandular adaxially. **Racemes** 1–20 cm (10–40 cm in fruit); bracts unifoliate, elliptic, 10–20 mm. **Pedicels** 10–25 mm, (glandular-pubescent). **Flowers:** sepals persistent, green, lanceolate, 2–6 × 0.8– 1.3 mm, margins entire, glandular-pubescent; petals white, oblong to ovate, 5–10 × 2–3 mm; stamens purple, 5–10(–25) mm; anthers 2.5–4 mm; gynophore 1–3 mm in fruit; ovary 4–6 mm; style 0.1 mm. **Capsules** (15–)25–40(–65) × 2.5–4 mm, glabrous. **Seeds** 10–20, yellowish (black), obovoid, 2.4–2.8 × 1.9–2.1 mm, transversely rugose. $2n = 18$.

Flowering late spring–late summer. Disturbed roadsides, vacant lots, waste areas, gravel pits, lakeshores, streambeds; 0–200 m; introduced; Ala., Tex.; Mexico; West Indies; Central America; South America.

2. **Hemiscola diffusa** (Banks ex de Candolle) H. H. Iltis, Novon 17: 448. 2007 • Spreading spiderflower [I]

Cleome diffusa Banks ex de Candolle in A. P. de Candolle and A. L. P. P. de Candolle, Prodr. 1: 241. 1824; *C. aculeata* Linnaeus var. *diffusa* (Banks ex de Candolle) Kuntze

Plants 10–30(–90) cm. **Stems** delicately branched; glandular-papillose. **Leaves:** stipular spines 0.5 mm; petiole 4–15 cm; leaflets 3 proximally, 1 distally, blade obovate, 1–2.5 × 0.5–1.2 cm, margins serrate, apex acute to obtuse or acuminate, surfaces glabrous or sparsely glandular-pubescent abaxially, glabrate

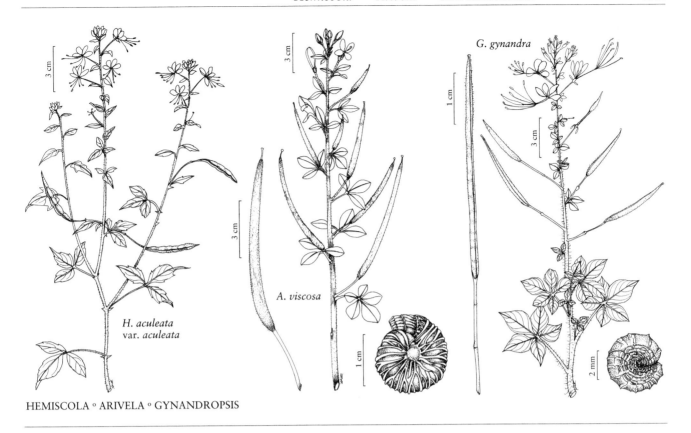

adaxially. **Racemes** 5–10 cm (8–12 cm in fruit); bracts unifoliate, elliptic, 4–15 mm. **Pedicels** 4–6 mm. **Flowers:** sepals deciduous, brownish green, ovate, 1–2.5(–3) × 1–2 mm, margins serrulate, glabrous; petals yellow or creamy white, 5–10 × 1.5–3 mm; stamens (staggered) ,yellow, 5–9 mm; anthers 0.3–0.5 mm; gynophore 2–4 mm in fruit; ovary 3 mm; style 0.1–0.2 mm. **Capsules** 15–20 × 3–6 mm, glabrous. **Seeds** 10–13, dull reddish brown to black, oblong or comma-shaped (compressed), 2–3 × 2 mm, tuberculate, (aril white).

Flowering summer. Disturbed areas; 0–10 m; introduced; Ala.; South America.

The only record of *Hemiscola diffusa* from the flora area is a collection from ballast in Mobile in 1893 (*Mohr s.n.*, NY, US).

11. ARIVELA Rafinesque, Sylva Tellur., 110. 1838 • Spiderflower [Origin obscure]

Gordon C. Tucker

Herbs, annual. **Stems** sparsely branched [unbranched]; glandular-hirsute [glabrous]. **Leaves:** stipules absent; petiole with pulvinus basally or distally (petiolule base adnate, forming pulvinar disc); leaflets 3 or 5. **Inflorescences** terminal or axillary (from distal leaves), racemes (flat-topped or elongated); bracts present [absent]. **Flowers:** zygomorphic; sepals persistent, distinct, equal; petals equal; stamens 14–25[–35]; filaments inserted on a discoid or conical androgynophore, glandular-hirsute; anthers (oblong), coiling as pollen is released; gynophore obsolete. **Fruits** capsules, partly dehiscent, oblong, (glandular). **Seeds** [10–]25–40[–100], spheroidal, not arillate, (cleft fused between ends). $x = 10$.

Species ca. 10 (1 in the flora): introduced; Asia, Africa.

1. Arivela viscosa (Linnaeus) Rafinesque, Sylva Tellur., 110. 1838 • Wild mustard, tickweed, yellow cleome F I W

Cleome viscosa Linnaeus, Sp. Pl. 2: 672. 1753; *Polanisia icosandra* (Linnaeus) Wight & Arnott; *P. microphylla* Eichler; *P. viscosa* (Linnaeus) de Candolle

Plants (10–)30–100(–160) cm. **Stems** viscid. **Leaves:** petiole 1.5–4.5(–8) cm, glandular-hirsute; leaflet blade ovate to oblanceolate-elliptic, (0.6–)2–6 × 0.5–3.5 cm, margins entire and glandular-ciliate, apex acute to obtuse, surfaces glandular-hirsute. **Racemes** 5–10 cm (10–15 cm in fruit); bracts (often deciduous), trifoliate, 10–25 mm, glandular-hirsute. **Pedicels** 6–30 mm, glandular-hirsute. **Flowers:** sepals green, lanceolate, 5–10 × 0.8–1.2 mm, glandular-hirsute; petals arranged in adaxial semicircle before anthesis, radially arranged at anthesis, bright yellow, sometimes purple basally, oblong to ovate, 7–14 × 3–4 mm; stamens dimorphic, 4–10 adaxial ones much shorter with swelling proximal to anthers, green, 5–9 mm; anthers 1.4–3 mm; ovary 6–10 mm, densely glandular; style 1–1.2 mm. **Capsules** dehiscing only partway from apex to base, 30–100 × 2–4 mm, glandular-hirsute. **Seeds** light brown, 1.2–1.8 × 1–1.2 mm, finely ridged transversely. $2n = 20$.

Flowering summer. Disturbed roadsides, vacant lots, citrus groves, railroads, ballasts; 0–50 m; introduced; Fla., Ga., La., N.J., Pa., S.C.; s, se Asia, Africa; introduced also in Mexico, West Indies, Central America, South America.

Arivela viscosa seeds have a high oil content (26%), are rich in linoleic acid, and are eaten in India. The leaves and seeds have been used medicinally in Asia (southern China, Guam, India) for the eyes and intestines, and as a poultice for headaches and rheumatism (R. N. Chopra et al. 1986; N. D. Mandahar 2002). Records from New Jersey and Pennsylvania are based on non-persistent ballast plants from the late 1800s.

12. **GYNANDROPSIS** de Candolle in A. P. de Candolle and A. L. P. P. de Candolle, Prodr. 1: 237. 1824, name conserved • Spider-wisp [Genus *Gynandra* (Orchidaceae), and Greek *opsis*, resemblance] I

Gordon C. Tucker

Herbs, annual [perennial]. **Stems** unbranched or sparsely branched; glabrate or glandular-pubescent. **Leaves:** stipules absent; petiole with pulvinus basally or distally, (petiolule base adnate, forming pulvinar disc); leaflets 3 or 5. **Inflorescences** terminal, racemes (elongated); bracts present. **Flowers** zygomorphic; sepals persistent, distinct, equal (each often subtending a nectary); petals equal; stamens 6; filaments adnate basally to gynophore (about as long as petals), glabrous; anthers coiling as pollen is released; gynophore recurved in fruit (filament scars visible ca. ⅓–½ its length). **Fruits** capsules, dehiscent, oblong. **Seeds** 10–20[–40], subglobose, not arillate, (cleft fused between ends). $x = 10, 17$.

Species 2 (1 in the flora): introduced; sw Asia; tropical and warm temperate climates.

Gynandropsis is allied to *Cleome*; it is distinguished by relatively long androgynophores. It has been included in *Cleome*; most regional accounts of Cleomaceae (including Capparaceae) in the Old World have given this taxon generic status, an approach followed here.

1. **Gynandropsis gynandra** (Linnaeus) Briquet, Annuaire Conserv. Jard. Bot. Genève 17: 382. 1914 • Cat's-whiskers [F][I][W]

Cleome gynandra Linnaeus, Sp. Pl. 2: 671. 1753; *C. heterotricha* Burchell; *C. pentaphylla* Linnaeus; *Gynandropsis heterotricha* (Burchell) de Candolle; *G. pentaphylla* (Linnaeus) de Candolle

Plants (50–)90–150 cm. **Leaves:** petiole 3.5–4.5(–8) cm, glandular-pubescent; leaflet blade oblanceolate to rhombic, 2.5–4.5 × 1.2–2.5 cm, margins serrulate-denticulate, apex acute, surfaces glabrate to glandular-pubescent. **Racemes** 5–20 cm (10–40 cm in fruit); bracts trifoliate, 10–25 mm. **Pedicels** purple, 8–15 mm, glabrous. **Flowers:** sepals green, lanceolate, 3.5–5 × 0.8–1.2 mm, glandular-pubescent; petals purple or white, oblong to ovate, 7–14 × 3–4 mm; stamens purple, 8–30 mm; anthers 1–2 mm; gynophore purple, 10–14 mm in fruit; ovary 6–10 mm; style 1–1.2 mm. **Capsules** 45–95 × 3–4 mm, glandular-pubescent. **Seeds** reddish brown to black, 1.4–1.6 × 1–1.2 mm, rugose to tuberculate. $2n = 34, 36$.

Flowering late spring–late summer. Disturbed roadsides, vacant lots, waste areas, railroads; 0–200 m; introduced; Ala., Fla., Ga., La., Miss., N.Y., N.C., Okla., Pa., S.C., Tex.; Old World tropics; introduced also in Mexico, West Indies, Central America, South America.

The C_4 photosynthetic pathway has been reported from *Gynandropsis gynandra* (S. K. Imbamba and L. T. Tieszen 1977). This species is sometimes grown as an ornamental. B. S. Barton (1836, p. 317) provided a detailed and accurate illustration of the flower, obviously drawn from life; this indicates that the species was cultivated (perhaps escaped) in Pennsylvania at that time. In some tropical countries, it is cultivated as a potherb (K. Waithaka and Chweya 1991; J. A. Chweya and N. A. Mnzava 1997). It is also used medicinally. The fresh plant has a peculiar odor that is sometimes described as similar to burning *Cannabis*.

BRASSICAEAE Burnett
• Mustard Family

Cruciferae Jussieu

Ihsan A. Al-Shehbaz

Herbs or subshrubs [shrubs or, rarely, lianas or trees], annual, biennial, or perennial; usually terrestrial, rarely submerged aquatics; with pungent watery juice; scapose or not; pubescent or glabrous, usually without papillae or tubercles (multicellular glandular papillae or tubercles present in *Bunias*, *Chorispora*, and *Parrya*); taprooted or rhizomatous (rarely stoloniferous), caudex simple or branched, sometimes woody, rhizomes slender or thick. **Trichomes** unicellular, simple, stalked, or sessile; forked, stellate, dendritic, malpighiaceous (medifixed, 2-fid, appressed), or peltate and scalelike, eglandular. **Stems** (absent in *Idahoa*, sometimes *Leavenworthia*) usually erect, sometimes ascending, descending, prostrate, decumbent, or procumbent; branched or unbranched. **Leaves** (sometimes persistent) cauline usually present, basal present or not (sometimes rhizomal present in *Cardamine*), rosulate or not, usually alternate (sometimes opposite or whorled in *Cardamine angustata*, *C. concatenata*, and *C. diphylla* and in *Lunaria annua*; sometimes subopposite in *C. dissecta* and *C. maxima* and in *Draba ogilviensis*), usually simple, rarely trifoliolate or pinnately, palmately, or bipinnately compound; stipules absent [with tiny, stipulelike glands at base of petioles and pedicels]; petiolate, sessile, or subsessile (sessile auriculate or not, sometimes amplexicaul); blade margins entire, dentate, crenate, sinuate, repand, or dissected. **Inflorescences** terminal, usually racemose (racemes often corymbose or paniculate) or flowers solitary on pedicels from axils of rosette leaves; bracts usually absent, sometimes present. **Pedicels** present (persistent or caducous [rarely geotropic]). **Flowers** bisexual [unisexual], usually actinomorphic (zygomorphic in *Iberis*, sometimes in *Pennellia*, *Streptanthus*, and *Teesdalia*); perianth and androecium hypogynous; sepals usually caducous, rarely persistent, 4, in 2 decussate pairs (1 pair lateral, 1 median), distinct [connate], not saccate or lateral (inner) pair (or, rarely, both pairs) saccate, forming tubular, campanulate, or urceolate calyx; petals 4, alternate with sepals, usually cruciform, rarely in abaxial and adaxial pairs, rarely rudimentary or absent, claw differentiated or not from blade, blade sometimes reduced and much smaller than well-developed claw, basally unappendaged, or, rarely, appendaged, margins entire or emarginate to 2-fid, rarely pinnatifid [fimbriate or filiform]; stamens (2 or 4) 6 [8–24], in 2 whorls, usually tetradynamous (lateral outer pair shorter than median inner 2 pairs), rarely equal in length or in 3 pairs of unequal length; filaments (slender, sometimes winged, appendaged, or toothed): median pairs usually distinct, rarely connate; anthers dithecal, dehiscing by longitudinal slits, pollen grains 3(–11)-colpate, trinucleate; nectar glands receptacular, variable in number, shape, size, and disposition around filament base, always present opposite bases of lateral filaments, median glands present or absent; disc absent; pistil 1, 2-carpellate; ovary 2-locular with false septum connecting 2 placentae, rarely 1-locular and eseptate, placentation usually parietal, rarely apical; gynophore usually absent; style 1, persistent [caducous], sometimes obsolete or absent; stigma capitate or conical, entire or 2-lobed, lobes spreading or connivent, sometimes decurrent, distinct or connate, rarely elongated into horns or spines; ovules 1–300 per ovary, anatropous or campylotropous, bitegmic, usually crassinucellate, rarely tenuinucellate. **Fruits** usually capsular, usually 2-valved ((3 or) 4(–6) in *Rorippa barbareifolia*, (2 or) 4 in *Tropidocarpum capparideum*), termed siliques if length 3+ times width, or silicles if length less than 3 times width, sometimes nutletlike, lomentaceous, samaroid, or schizocarpic and [with] without a carpophore carrying the 1-seeded mericarp, dehiscent or indehiscent, segmented or not, torulose or smooth, terete, angled, or flat, often latiseptate (flattened parallel to septum) or angustiseptate (flattened at right angle to septum); gynophore usually absent, sometimes distinct; valves each not or obscurely veined, or prominently 1–7-veined, usually dehiscing acropetally, rarely basipetally, sometimes spirally or circinately coiled, glabrous or pubescent [spiny or glochidiate]; replum (persistent placenta) rounded, flattened, or indistinct (obsolete in *Crambe*, often perforate in *Thysanocarpus*); septum complete, perforated, reduced to a rim, or absent (obsolete in *Crambe* and *Thysanocarpus*, not differentiated from replum in *Raphanus*), sometimes with a midvein or anastomosing veins. **Seeds** usually yellow or brown, rarely black or white, flattened or plump, winged or not, or narrowly margined, ovoid, oblong, globose, or ovate, usually uniseriate or biseriate, sometimes aseriate, per locule, mucilaginous or not when wetted; embryo usually strongly curved, rarely straight with tiny radicle; cotyledons entire, emarginate, 3[2]-fid to base, orientation to radicle: incumbent (embryo notorrhizal: radicle lying along back of 1 cotyledon), accumbent (embryo pleurorrhizal: radicle applied to margins of both cotyledons), conduplicate (embryo orthoplocal: cotyledons folded longitudinally around radicle), or spirally coiled (embryo spirolobal) [twice transversely folded (embryo diplecolobal)]; endosperm absent (germination epigeal).

Genera ca. 338, species ca. 3780 (97 genera, 744 species in the flora): nearly worldwide, especially temperate areas, with the highest diversity in the Irano-Turanian region, Mediterranean area, and western North America.

Of the 634 species of Brassicaceae (mustards or crucifers) native in the flora area, 616 (418 endemic) grow in the United States, 140 (12 endemic) in Canada, and 31 (1 endemic) in Greenland.

The latest comprehensive account of the Brassicaceae for North America (R. C. Rollins 1993) included Mexico and Central America and excluded Greenland. In that account, 667 native species were recognized for the continent; I place 37 of those in the synonymy of other species. Of the remaining 630 species, 111 are restricted to Mexico and Central America, and 519 are native to the flora area. This last number falls 114 species short of the 634 native species that I recognize in the flora area. Since Rollins's account, 50 species were added to the flora in the past 15 years. Of these, 35 species were described as new, ten were added as native but previously overlooked or misidentified, and five have since become naturalized. Additionally, 72 species recognized in this treatment were treated by Rollins as either synonyms or infraspecific taxa of other species. The generic placement of 158 species in this account differs drastically from that in Rollins, though most of the changes involve the transfer of most of his species of *Arabis* to *Boechera* (59 spp.) and of *Lesquerella* to *Physaria* (54 spp.). The generic circumscriptions adopted herein are fully compatible with the rapidly accumulating wealth of molecular data, and all genera recognized here are monophyletic. Some examples demonstrate the differences between the two treatments. *Arabis*, in the sense of Rollins, included 80 species and 64 varieties; in this account, those 144 taxa are assigned to six genera in five tribes: *Arabidopsis* (2 spp.; tribe Camelineae), *Arabis* (16 spp.; tribe Arabideae), *Boechera* (109 spp.; tribe Boechereae), *Pennellia* (2 spp.; tribe Halimolobeae), *Streptanthus* (1 sp.; tribe Thelypodieae), and *Turritis* (1 sp.; tribe Camelineae). A similar division involves *Thlaspi*, a genus recognized by Rollins to include nine species, of which two are retained here in *Thlaspi* (tribe Thlaspideae), one is placed in *Microthlaspi*, and three in *Noccaea* (both in tribe Noccaeeae), two are reduced to synonymy of the latter genus, and one species of *Noccaea* is endemic to Mexico. *Lepidium* in this treatment includes Rollins's *Cardaria*, *Coronopus*, and *Stroganowia*; *Hesperidanthus* includes his *Caulostramina*, *Glaucocarpum*, and *Schoenocrambe* (excluding its type).

The Brassicaceae include important crop plants that are grown as vegetables (e.g., *Brassica*, *Eruca*, *Lepidium*, *Nasturtium*, *Raphanus*) and condiments (*Armoracia*, *Brassica*, *Eutrema*, *Sinapis*). Vegetable oils of some species of *Brassica*, including *B. napus* (canola), probably rank first in terms of the world's tonnage production. The Eurasian weed *Arabidopsis thaliana* (thale or mouse-ear cress) has become the model organism in experimental and molecular biology. The family also includes ornamentals in the genera *Aethionema*, *Alyssum*, *Arabis*, *Aubrieta*, *Aurinia*, *Erysimum*, *Hesperis*, *Iberis*, *Lobularia*, *Lunaria*, *Malcolmia*, and *Matthiola*. Finally, the flora includes 106 species of weeds from southwest Asia and Europe (R. C. Rollins and I. A. Al-Shehbaz 1986), of which 11 species of *Lepidium* have become noxious weeds in western North America.

The Brassicaceae have been regarded as a natural group for over 250 years, beginning with their treatment by Linnaeus in 1753 as the "Klass" Tetradynamia. More recently and based on a limited sampling of genera, W. S. Judd et al. (1994) recommended that the Brassicaceae and Capparaceae (including Cleomaceae) be united into one family, Brassicaceae. Molecular studies (J. C. Hall et al. 2002) suggested that three closely related families be recognized, with Brassicaceae sister to Cleomaceae, and both sister to Capparaceae. All three families have consistently been placed in one order (e.g., Capparales or Brassicales) by A. Cronquist (1988), A. L. Takhtajan (1997), and J. E. Rodman et al. (1996, 1998), as well as by the Angiosperm Phylogeny Group (APG) (http://www.mobot.org/MOBOT/research/APweb/). Brassicales includes families uniquely containing glucosinolates (mustard-oil glucosides), myrosin cells, racemose inflorescences, superior ovaries, often-clawed petals, and a suite of other characteristics (see the APG website).

Tribal classification of Brassicaceae has been subject to controversy. O. E. Schulz's (1936) classification has been used for over 70 years, though many botanists (e.g., E. Janchen 1942; I. A. Al-Shehbaz 1984; M. Koch et al. 1999; O. Appel and Al-Shehbaz 2003; Koch et al. 2003; M. A. Beilstein et al. 2006; Al-Shehbaz et al. 2006) amply demonstrated the artificiality of that system. Schulz divided the family into 19 tribes and 30 subtribes based on characters (e.g., fruit length-to-width ratio, compression, dehiscence; cotyledonary position; sepal orientation) that exhibit tremendous convergence throughout the family. Of these, only the tribe Brassiceae was previously shown to be monophyletic.

Several molecular studies (e.g., R. A. Price et al. 1994; J. C. Hall et al. 2002; M. Koch 2003; T. Mitchell-Olds et al. 2005; C. D. Bailey et al. 2006; M. A. Beilstein et al. 2006, 2008) have demonstrated that the Brassicaceae are split into two major clades: the Mediterranean-Southwest

Asian *Aethionema* and its sister clade that includes the rest of the family. Although Beilstein et al. showed that the family, excluding *Aethionema*, is divided into three major clades, such subdivision was based on only ca. 30% of the total number of genera. These three major clades still hold when nearly all genera of the family are investigated (S. I. Warwick et al., unpubl.).

Tribal assignments in the flora area are based on critical evaluation of morphology in connection with all published molecular data. To date, about 230 of the 338 genera of the family are placed in 35 tribes, including all large genera, which account for over 70% of the total species. Most of the remaining 108 genera would likely be assigned to the 35 tribes, be placed in new, smaller tribes, or be reduced to synonymy of larger genera. The delimitation of tribes for the flora area follows I. A. Al-Shehbaz et al. (2006), Al-Shehbaz and S. I. Warwick (2007), and D. A. German and Al-Shehbaz (2008) and differs from that of O. E. Schulz (1936) and the subsequent adjustments proposed by E. Janchen (1942) and Al-Shehbaz (1984, 1985, 1985b, 1986, 1987, 1988, 1988b, 1988c). Some of the tribes (e.g., Brassiceae and Lepidieae) are easily distinguished by relatively few characters; others (e.g., Arabideae, Camelineae, and Thelypodieae) are more difficult to separate unless a larger suite of characters is used. Because of the incomplete molecular knowledge on all genera of the family, the tribes, their genera, and species are listed herein alphabetically. Both R. C. Rollins (1993) and O. Appel and Al-Shehbaz (2003) arranged the genera alphabetically throughout, and the only difference in this account is the placement together of closely related genera within well-established monophyletic tribes.

Morphological data alone are sometimes unreliable in establishing phylogenetic relationships within Brassicaceae. Convergence is common throughout the family, and almost all morphological characters, especially of the fruits and embryos, which are quite heavily utilized in the delimitation of the genera and tribes, evolved independently. For example, rare character states, such as the spirolobal cotyledons, are known in at least three genera of three tribes (*Bunias*, Buniadeae; *Erucaria*, Brassiceae; *Heliophila*, Heliophileae), and lianas evolved independently in the South American *Cremolobus* (Cremolobeae), the South African *Heliophila*, and the Australian *Lepidium* (Lepidieae). Similarly, the reduction of chromosome number in the family to $n = 4$ occurred independently in two species of the Australian *Stenopetalum* and in at least 11 species of the North American *Physaria*. Other character states (e.g., zygomorphy, apetaly, reduction of stamen number to four, connation of median filaments, etc.) also evolved independently. Reexamination of morphology in light of molecular data is essential in order to understand the role of homoplasy and the evolution of various character states.

The literature on chromosome numbers of Brassicaceae is rather extensive, and rarely is an individual work cited herein in that regard. Instead, the recently compiled cytological data for the entire family (S. I. Warwick and I. A. Al-Shehbaz 2006) are consulted for all species.

Because the size of ovules is relatively small, it is very difficult to determine the number of ovules per ovary. The number of ovules per ovary is based on the sum of mature seeds and aborted ovules in the fruit. The length of style and type of stigma are also taken from the fruits, and the length of fruiting pedicels is measured from several proximal pedicles of the infructescence. Elevation ranges are normally given for a taxon; unfortunately, the range is not known for some taxa.

Generic delimitation in Brassicaceae is often difficult because most genera are distinguished primarily by fruit characters. The following artificial keys emphasize either flowering or fruiting characters, and the most reliable identification of a given plant to a genus can be achieved when specimens have both flowers and fruits, and when both keys are successfully used to identify it to the same genus. The keys are based on species rather than generic descriptions so that all of the morphological manifestations in a given genus are covered and, therefore, a genus may appear multiple times within one of the first four key groups. For example, genera with highly diversified vegetative and floral morphology (e.g., *Cardamine*, *Caulanthus*, *Lepidium*, and *Streptanthus*) appear in the keys to groups multiple times. Because of such coverage, keys to flowering material incorporate characteristics of a species, or groups of species, rather than of genera. Leads marked (¤) in keys for groups 1–4 indicate that mature fruits and seeds are needed for the identification of genera in their subordinate couplet(s).

SELECTED REFERENCES Al-Shehbaz, I. A. 1977. Protogyny in the Cruciferae. Syst. Bot. 2: 327–333. Al-Shehbaz, I. A. 1984. The tribes of Cruciferae (Brassicaceae) in the southeastern United States. J. Arnold Arbor. 65: 343–373. Al-Shehbaz, I. A. 1985. The genera of Brassiceae (Cruciferae; Brassicaceae) in the southeastern United States. J. Arnold Arbor. 66: 279–351. Al-Shehbaz, I. A. 1985b. The genera of Thelypodieae (Cruciferae; Brassicaceae) in the southeastern United States. J. Arnold Arbor. 66: 95–111. Al-Shehbaz, I. A. 1986. The genera of Lepidieae (Cruciferae; Brassicaceae) in the southeastern United States. J. Arnold Arbor. 67: 265–311. Al-Shehbaz, I. A. 1987. The genera of Alysseae (Cruciferae; Brassicaceae) in the southeastern United States. J. Arnold Arbor. 68: 185–240. Al-Shehbaz, I. A. 1988. The genera of Arabideae (Cruciferae; Brassicaceae) in the southeastern United States. J. Arnold Arbor. 69: 85–166. Al-Shehbaz, I. A. 1988b. The genera of Anchonieae (Cruciferae; Brassicaceae) in the southeastern United States. J. Arnold Arbor. 69: 193–212. Al-Shehbaz, I. A. 1988c. The genera of Sisymbrieae (Cruciferae; Brassicaceae) in the southeastern United States. J. Arnold Arbor. 69: 213–237. Al-Shehbaz, I. A., M. A. Beilstein, and E. A. Kellogg. 2006. Systematics and phylogeny of the Brassicaceae (Cruciferae): An overview. Pl. Syst. Evol. 259: 89–120. Al-Shehbaz, I. A., S. L. O'Kane, and R. A. Price. 1999. Generic placement of species excluded from *Arabidopsis*. Novon 9: 296–307. Al-Shehbaz, I. A. and S. I. Warwick. 2007. Two new tribes (Dontostemoneae and Malcolmieae) in the Brassicaceae (Cruciferae). Harvard Pap. Bot. 12: 429–433. Appel, O. and I. A. Al-Shehbaz. 2003. Cruciferae. In: K. Kubitzki et al., eds. 1990+. The Families and Genera of Vascular Plants. 9+ vols. Berlin etc. Vol. 5, pp. 75–174. Bailey, C. D. et al. 2006. Toward a global phylogeny of the Brassicaceae. Molec. Biol. Evol. 23: 2142–2160. Bailey, C. D., R. A. Price, and J. J. Doyle. 2002. Systematics of the halimolobine Brassicaceae: Evidence from three loci and morphology. Syst. Bot. 27: 318–332. Bailey, C. D., I. A. Al-Shehbaz, and G. Rajanikanth. 2007. Generic limits in the tribe Halimolobeae and the description of the new genus *Exhalimolobos* (Brassicaceae). Syst. Bot. 32: 140–156. Beilstein, M. A., I. A. Al-Shehbaz, and E. A. Kellogg. 2006. Brassicaceae phylogeny and trichome evolution. Amer. J. Bot. 93: 607–619. Beilstein, M. A., I. A. Al-Shehbaz, S. Mathews, and E. A. Kellogg. 2008. Brassicaceae phylogeny inferred from phytochrome A and *ndh*F sequence data: Tribes and trichomes revisited. Amer. J. Bot. 95: 1307–1327. Bowman, J. L. 2006. Molecules and morphology: Comparative developmental genetics of the Brassicaceae. Pl. Syst. Evol. 259: 199–215. German, D. A. and I. A. Al-Shehbaz. 2008. Five additional tribes (Aphragmeae, Biscutelleae, Calepineae, Conringieae, and Erysimeae) in the Brassicaceae (Cruciferae). Harvard Pap. Bot. 13: 165–170. Hall, J. C., K. J. Sytsma, and H. H. Iltis. 2002. Phylogeny of Capparaceae and Brassicaceae based on chloroplast sequence data. Amer. J. Bot. 89: 1826–1842. Hauser, L. A. and T. J. Crovello. 1982. Numerical analysis of generic relationships in Thelypodieae (Brassicaceae). Syst. Bot. 7: 249–268. Janchen, E. 1942. Das System der Cruciferen. Oesterr. Bot. Z. 91: 1–18. Koch, M. 2003. Molecular phylogenetics, evolution and population biology in Brassicaceae. In: A. K. Sharma and A. Sharma, eds. 2003+. Plant Genome: Biodiversity and Evolution. 2+ vols. in parts. Enfield, N. H. Vol. 1, part A, pp. 1–35. Koch, M. et al. 1999b. Molecular systematics of *Arabidopsis* and *Arabis*. Pl. Biol. (Stuttgart) 1: 529–537. Koch, M. et al. 2003b. Molecular systematics, evolution, and population biology in the mustard family (Brassicaceae). Ann. Missouri Bot. Gard. 90: 151–171. Koch, M., B. Haubold, and T. Mitchell-Olds. 2000. Comparative analysis of chalcone synthase and alcohol dehydrogenase loci in *Arabidopsis*, *Arabis* and related genera (Brassicaceae). Molec. Biol. Evol. 17: 1483–1498. Koch, M., B. Haubold, and T. Mitchell-Olds. 2001. Molecular systematics of the Brassicaceae: Evidence from coding plastidic *mat*K and nuclear *Chs* sequences. Amer. J. Bot. 88: 534–544. Lysak, M. A. and C. Lexer. 2006. Towards the era of comparative evolutionary genomics in Brassicaceae. Pl. Syst. Evol. 259: 175–198. Mitchell-Olds, T., I. A. Al-Shehbaz, M. Koch, and T. F. Sharbel. 2005. Crucifer evolution in the post-genomic era. In: R. J. Henry, ed. 2005. Plant Diversity and Evolution: Genotypic and Phenotypic Variation in Higher Plants. Wallingford and Cambridge, Mass. Pp. 119–137. Payson, E. B. 1923. A monographic study of *Thelypodium* and its immediate allies. Ann. Missouri Bot. Gard. 9: 233–324. Rollins, R. C. 1993. The Cruciferae of Continental North America: Systematics of the Mustard Family from the Arctic to Panama. Stanford. Rollins R. C. and I. A. Al-Shehbaz. 1986. Weeds of south-west Asia in North America with special reference to the Cruciferae. Proc. Roy. Soc. Edinburgh, B 89: 289–299. Rollins, R. C. and U. C. Banerjee. 1976. Trichomes in studies of the Cruciferae. In: J. G. Vaughn et al., eds. 1976. The Biology and Chemistry of the Cruciferae. London and New York. Pp. 145–166. Rollins, R. C. and U. C. Banerjee. 1979. Pollen of the Cruciferae. Publ. Bussey Inst. Harvard Univ. 1979: 33–64. Sabourin, A. et al. 1991. Guide des Crucifères Sauvages de l'Est du Canada (Québec, Ontario et Maritimes). Montréal. Schulz, O. E. 1936. Cruciferae. In: H. G. A. Engler et al., eds. 1924+. Die natürlichen Pflanzenfamilien..., ed. 2. 26+ vols. Leipzig and Berlin. Vol. 17b, pp. 227–658. Warwick, S. I. et al. 2006. Phylogenetic position of *Arabis arenicola* and generic limits of *Eutrema* and *Aphragmus* (Brassicaceae) based on sequences of nuclear ribosomal DNA. Canad. J. Bot. 84: 269–281. Warwick, S. I. et al. 2006b. Brassicaceae: Species checklist and database on CD-ROM. Pl. Syst. Evol. 259: 249–258. Warwick, S. I. and L. D. Black. 1991. Molecular systematics of *Brassica* and allied genera (subtribe Brassicinae, Brassiceae)—Chloroplast genome and cytodeme congruence. Theor. Appl. Genet. 82: 81–92. Warwick, S. I. and L. D. Black. 1993. Molecular relationships in subtribe Brassicinae (Cruciferae, tribe Brassiceae). Canad. J. Bot. 71: 906–918. Warwick, S. I. and C. A. Sauder. 2005. Phylogeny of tribe Brassiceae (Brassicaceae) based on chloroplast restriction site polymorphisms and nuclear ribosomal internal transcribed spacer and chloroplast *trn*L intron sequences. Canad. J. Bot. 83: 467–483. Warwick, S. I., C. A. Sauder, and I. A. Al-Shehbaz. 2008. Phylogenetic relationships in the tribe Alysseae (Brassicaceae) based on nuclear ribosomal ITS DNA sequences. Canad. J. Bot. 86: 315–336. Warwick, S. I., C. A. Sauder, I. A. Al-Shehbaz, and F. Jacquemoud. 2007. Phylogenetic relationships in the tribes Anchonieae, Chorisporeae, Euclidieae, and Hesperideae (Brassicaceae) based on nuclear ribosomal ITS DNA sequences. Ann. Missouri Bot. Gard. 94: 56–78.

Key to Genera Based Primarily on Flowering Specimens

1. Trichomes simple or absent.
   2. Cauline leaves absent or, when present, petiole or blade bases sometimes auriculate or amplexicaul .... Group 1, p. 229
   2. Cauline leaves present, petiole or blade bases not auriculate .... Group 2, p. 231
1. Trichomes (at least some) branched.
   3. Ovaries and young fruits linear .... Group 3, p. 234
   3. Ovaries and young fruits not linear .... Group 4, p. 236

Key to Genera of Group 1

1. Plants scapose; cauline leaves absent.
   2. Flowers solitary (pedicellate from basal rosettes).
      3. Petals lavender, orange, yellow, or white, considerably longer than sepals, differentiated into blade and claw; se United States .... 42. *Leavenworthia*, p. 485
      3. Petals white, slightly longer than sepals, not differentiated into blade and claw; British Columbia, Pacific and some Mountain states .... 65. *Idahoa*, p. 566
   2. Flowers in racemes or corymbs.
      4. Petals yellow .... 22. *Diplotaxis* (in part), p. 432
      4. Petals white, lavender, or purple.
         5. Perennials; petals 4–23 mm.
            6. Petals often white, 0.8–1.5(–2) mm wide; plants eglandular .... 40. *Cardamine* (in part), p. 464
            6. Petals usually purple or lavender, rarely white, 3–8 mm wide; plants often with glandular papillae .... 49. *Parrya*, p. 511
         5. Annuals; petals 0.5–2.5 mm or absent.
            7. Leaf blades linear or subulate, margins entire; plants littoral or aquatic .... 47. *Subularia*, p. 509
            7. Leaf blades lanceolate to oblanceolate or obovate, margins sometimes lobed; plants terrestrial .... 64. *Teesdalia* (in part), p. 564
1. Plants not scapose; cauline leaves present, some blade bases auriculate to amplexicaul.
   8. Racemes bracteate throughout .... 91. *Streptanthus* (in part), p. 700
   8. Racemes ebracteate or proximalmost flowers bracteate.
      9. Cauline leaves petiolate.
         10. Cauline leaves compound.
            11. Plants terrestrial, not rooting from proximal nodes .... 40. *Cardamine* (in part), p. 464
            11. Plants aquatic, rooting from proximal nodes .... 43. *Nasturtium* (in part), p. 489
         10. Cauline leaves simple.
            12. Cauline leaf blade margins usually dentate or serrate, rarely subentire .... 41. *Iodanthus* (in part), p. 484
            12. Cauline leaf blade margins runcinate or pinnatifid.
               13. Bases of lateral sepals saccate; petals (12–)16–25(–32) mm .... 26. *Orychophragmus*, p. 437
               13. Bases of lateral sepals not saccate; petals 6–10 mm .... 84. *Coelophragmus*, p. 687
      9. Cauline leaves sessile.
         14. Stamens 2 .... 68. *Lepidium* (in part), p. 570
         14. Stamens 6, rarely 4.
            15. Stamens exserted; ovaries and young fruits on distinct gynophores.
               16. Racemes not or slightly elongated in fruit; filaments often puberulent basally; se United States .... 95. *Warea* (in part), p. 742
               16. Racemes elongated in fruit; filaments glabrous; Pacific and Mountain states.
                  17. Stigmas 2-lobed; stamens tetradynamous ...... 92. *Thelypodiopsis* (in part), p. 723
                  17. Stigmas entire; stamens subequal.

18. Sepals spreading to reflexed; petals usually yellow, rarely white, 12–25 mm .... 89. *Stanleya* (in part), p. 695
18. Sepals erect to ascending; petals purple or white, 6–12(–16) mm .... 93. *Thelypodium* (in part), p. 728

[15. Shifted to left margin.—Ed.]

15. Stamens (at least some) included; ovaries and young fruits sessile or on obsolete gynophores.
    19. Stamens often in 3 pairs of unequal length; petal margins crisped and/or channeled (◻).
        20. Fruits terete; seeds not winged .... 82. *Caulanthus* (in part), p. 677
        20. Fruits latiseptate; seeds often winged .... 91. *Streptanthus* (in part), p. 700
    19. Stamens tetradynamous; petal margins not or rarely crisped, not channeled.
        21. Petals yellow.
            22. Ovules 1 or 2 per ovary.
                23. Basal and proximal cauline leaf blade margins 2- or 3-pinnatifid or pinnatisect .... 68. *Lepidium* (in part), p. 570
                23. Basal and proximal cauline leaf blade margins usually entire, dentate, sinuate, or runcinate, rarely pinnatifid.
                    24. Fruiting pedicels reflexed .... 66. *Isatis*, p. 567
                    24. Fruiting pedicels erect (appressed) .... 67. *Myagrum*, p. 568
            22. Ovules (4–)10–94 per ovary.
                25. Calyces urceolate; petal margins crisped .... 92. *Thelypodiopsis* (in part), p. 723
                25. Calyces not urceolate; petal margins not crisped.
                    26. Fruiting pedicels reflexed; petal claws not attenuate .... 82. *Caulanthus* (in part), p. 677
                    26. Fruiting pedicels divaricate to erect; petal claws attenuate.
                        27. Petals (6–)8–30 mm, claws distinct .... 17. *Brassica* (in part), p. 419
                        27. Petals 0–9(–12) mm, claws absent.
                            28. Cauline leaf blade bases cordate-amplexicaul .... 51. *Conringia*, p. 517
                            28. Cauline leaf blade bases auriculate.
                                29. Ovaries and young fruits linear; stems angular distally .... 39. *Barbarea*, p. 460
                                29. Ovaries and young fruits usually globose, ovoid, or oblong (if linear, petals absent or 0.5–4 mm); stems seldom angular distally .... 45. *Rorippa* (in part), p. 493
        21. Petals white, lavender, pink, purple, or violet.
            30. Ovaries and young fruits ovate, orbicular, or cordate.
                31. Ovules usually 1 or 2 per ovary, if 4 (some *Lepidium*), style distinct.
                    32. Plants usually pubescent; ovules 2(–4) per ovary .... 68. *Lepidium* (in part), p. 570
                    32. Plants glabrous, rarely sparsely pubescent; ovules 1 per ovary.
                        33. Cauline leaf blade margins dentate or subentire; fruits terete, ovoid or ellipsoid; plants glabrous .... 31. *Calepina*, p. 446
                        33. Cauline leaf blade margins usually subentire, dentate, pinnatifid, or pinnatisect, rarely entire; ovaries latiseptate, orbicular to ovate, or elliptic; plants glabrous or pubescent .... 94. *Thysanocarpus* (in part), p. 739
                31. Ovules (4–)6–16 per ovary.
                    34. Perennials or, rarely, biennials .... 72. *Noccaea*, p. 600
                    34. Annuals (◻).
                        35. Seed coats smooth; cauline leaf blade margins entire; plants not fetid .... 71. *Microthlaspi*, p. 599
                        35. Seed coats striate or coarsely reticulate; cauline leaf blade margins often dentate; plants often fetid .... 97. *Thlaspi*, p. 745
            30. Ovaries and young fruits linear.
                36. Petals not differentiated into blade and claw, gradually narrowed basally.
                    37. Biennials or perennials; young fruits latiseptate .... 11. *Boechera* (in part), p. 348
                    37. Annuals; young fruits terete.

38. Cauline leaf blade bases sagittate-amplexicaul; stigmas entire .... 59. *Eutrema* (in part), p. 555
38. Cauline leaf blade bases auriculate; stigmas 2-lobed, rarely entire .... 92. *Thelypodiopsis* (in part), p. 723

[36. Shifted to left margin.—Ed.]

36. Petals differentiated into blade and claw.
   39. Petal claws not attenuate; fruits reflexed .... 82. *Caulanthus* (in part), p. 677
   39. Petal claws attenuate; fruits erect to divaricate.
     40. Petals magenta or purple with deeper purple center .... 91. *Streptanthus* (in part), p. 700
     40. Petals white, lavender, or purple, without deeper purple center.
       41. Petals 15–30 × 6–12 mm, claws 7–15 mm .... 17. *Brassica* (in part), p. 419
       41. Petals 2.5–15(–22) × 1–5(–9) mm, claws 1–6(–11) mm.
         42. Perennials; e, c United States .... 41. *Iodanthus* (in part), p. 484
         42. Annuals, biennials, or perennials; sw, w United States.
           43. Stigmas 2-lobed .... 92. *Thelypodiopsis* (in part), p. 723
           43. Stigmas entire.
             44. Annuals (biennials); petioles not ciliate; w Texas .... 85. *Dryopetalon* (in part), p. 688
             44. Biennials or, rarely, perennials; petioles often ciliate; Mountain and Pacific states .... 93. *Thelypodium* (in part), p. 728

Key to Genera of Group 2

1. Racemes bracteate throughout.
   2. Perennials; leaf blade margins entire .... 5. *Aphragmus* (in part), p. 256
   2. Annuals; leaf blade margins pinnately lobed, sinuate, or dentate.
     3. Plants pubescent (trichomes retrorse or recurved) .... 24. *Erucastrum* (in part), p. 435
     3. Plants glabrous.
       4. Petals pink or purple; cauline leaves compound (3–5-foliolate) .... 40. *Cardamine* (in part), p. 464
       4. Petals yellow; cauline leaves not compound.
         5. Basal leaf blade margins 1 or 2 (or 3)-pinnatisect; petals 4–20 mm ...... 46. *Selenia*, p. 506
         5. Basal leaf blade margins dentate, or sinuate- to runcinate-pinnatifid; petals 1.5–2 mm .... 80. *Sisymbrium* (in part), p. 667
1. Racemes ebracteate or proximalmost flowers bracteate.
   6. Petals absent.
     7. Sepals 0.7–1.2 mm; ovules 6–22 per ovary .... 40. *Cardamine* (in part), p. 464
     7. Sepals 1.2–3 mm; ovules 70–242 per ovary .... 45. *Rorippa* (in part), p. 493
   6. Petals present.
     8. Flowers zygomorphic.
       9. Ovaries and young fruits linear; ovules 10–120 per ovary; stamens in 3 pairs of unequal length .... 91. *Streptanthus* (in part), p. 700
       9. Ovaries and young fruits ovate, obcordate, or suborbicular; ovules 2 or 4 per ovary; stamens tetradynamous.
         10. Annuals or perennials (without basal rosettes); abaxial petals 5–16 mm ... 63. *Iberis*, p. 563
         10. Annuals (with basal rosettes); abaxial petals 0.5–1.5 mm .... 64. *Teesdalia* (in part), p. 564
     8. Flowers actinomorphic.
       11. Stamens 2 or 4.
         12. Stamens 2 .... 68. *Lepidium* (in part), p. 570
         12. Stamens 4.
           13. Ovaries linear .... 40. *Cardamine* (in part), p. 464
           13. Ovaries not linear.
             14. Ovules 10–24 per ovary .... 53. *Hornungia* (in part), p. 530
             14. Ovules 2 or 4 per ovary.
               15. Filaments appendaged .... 64. *Teesdalia* (in part), p. 564
               15. Filaments not appendaged .... 68. *Lepidium* (in part), p. 570

[11. Shifted to left margin.—Ed.]

11. Stamens 6.
    16. Ovaries ovate, ovoid, obovate, orbicular, globose, oblong, or elliptic (sometimes in 2 segments with such shapes).
        17. Petals (15–)17–25(–30) mm, usually purple or lavender, rarely white; styles and gynophores well-developed .... 69. *Lunaria*, p. 596
        17. Petals 0.5–12(–15) mm, usually yellow, cream, or white, rarely lavender; styles and gynophores often absent.
            18. Ovules 24–80 per ovary .... 45. *Rorippa* (in part), p. 493
            18. Ovules 1–12(–14) per ovary.
                19. Ovaries and young fruits 2-segmented.
                    20. Petals cream or pale yellow with dark purple or brown veins; ovules 4–6 per ovary .... 19. *Carrichtera*, p. 429
                    20. Petals white, lavender, or yellow, without dark veins; ovules 1 or 2(–4) per ovary.
                        21. Plants hispid proximally, not succulent; style distinct; petals yellow; disturbed habitats, waste places, roadsides .... 28. *Rapistrum*, p. 440
                        21. Plants glabrous throughout, succulent; style absent; petals white, lavender, or pale yellow; coastal sandy beaches.
                            22. Cauline leaves to 10 cm; filaments not toothed or winged .... 18. *Cakile* (in part), p. 424
                            22. Cauline leaves 10–40 cm; filaments toothed and winged ... 21. *Crambe*, p. 430
                19. Ovaries and young fruits unsegmented.
                    23. Ovules 1 or 2(–4) per ovary.
                        24. Ovules 2(–4) per ovary; young fruits angustiseptate .... 68. *Lepidium* (in part), p. 570
                        24. Ovules 1 per ovary; young fruits latiseptate .... 94. *Thysanocarpus* (in part), p. 739
                    23. Ovules 6–14 per ovary.
                        25. Perennials with fleshy roots; stems 5–12(–20) dm; basal leaf blades 10–60 cm; racemes corymbose-paniculate .... 38. *Armoracia*, p. 459
                        25. Annuals, biennials, or perennials without fleshy roots; stems 0.1–3(–5) dm; basal leaf blades 1–10 cm; racemes subumbellate or corymbose.
                            26. Racemes bracteate basally; petals with dark veins .... 5. *Aphragmus* (in part), p. 256
                            26. Racemes ebracteate; petals without darker veins .... 50. *Cochlearia*, p. 514
    16. Ovaries usually linear, rarely narrowly lanceolate.
        27. Cauline leaves compound.
            28. Plants terrestrial, not rooting from proximal stem nodes .... 40. *Cardamine* (in part), p. 464
            28. Plants aquatic, rooting from proximal stem nodes .... 43. *Nasturtium* (in part), p. 489
        27. Cauline leaves simple (sometimes divided, blade continuous between lobes).
            29. Petal blades pinnately lobed .... 85. *Dryopetalon* (in part), p. 688
            29. Petal blades entire.
                30. Plants with glandular papillae .... 48. *Chorispora*, p. 510
                30. Plants without glandular papillae.
                    31. Stamens equal or subequal in length (well-exserted).
                        32. Petals yellow, claws and/or filaments usually pubescent, rarely both glabrous .... 89. *Stanleya* (in part), p. 695
                        32. Petals white, lavender, pink, or purple, claws and filaments glabrous (except *Warea carteri*).
                            33. Inflorescences corymbose racemes; se United States .... 95. *Warea* (in part), p. 742
                            33. Inflorescences elongated racemes; Pacific and Mountain states, Oklahoma, Texas.

34. Petal claws dentate to sublaciniate; sepals ascending; stamens well-exserted . . . . 83. *Chlorocrambe*, p. 685
34. Petal claws entire; sepals spreading or erect; stamens (spreading) slightly exserted (◘).
 35. Fruits subsessile on equally broad gynophores. . . . . 82. *Caulanthus* (in part), p. 677
 35. Fruits stipitate on narrower gynophores . . . . 93. *Thelypodium* (in part), p. 728

[31. Shifted to left margin.—Ed.]
31. Stamens tetradynamous or in 3 pairs of unequal length.
 36. Stamens in 3 pairs of unequal length (◘).
  37. Fruits terete; seeds not winged . . . . 82. *Caulanthus* (in part), p. 677
  37. Fruits slightly to distinctly latiseptate; seeds winged all around or only distally.
   38. Cotyledons accumbent. . . . . 91. *Streptanthus* (in part), p. 700
   38. Cotyledons incumbent.
    39. Adaxial stamens with connate filaments and sterile anthers; fruit valves each with an obscure midvein . . . . 88. *Sibaropsis*, p. 694
    39. Adaxial stamens with distinct filaments and fertile anthers; fruit valves each with a prominent midvein . . . . 90. *Streptanthella*, p. 699
 36. Stamens tetradynamous.
  40. Petal blades with purple or brown veins darker than rest of blade.
   41. Perennials, glaucous, glabrous; leaf blade margins entire or dentate.
    42. Stigmas prominently 2-lobed; ovules 76–210 per ovary . . . . . 82. *Caulanthus* (in part), p. 677
    42. Stigmas obscurely 2-lobed; ovules 24–62 per ovary . . . 86. *Hesperidanthus* (in part), p. 689
   41. Annuals, not glaucous, often pubescent; leaf blade margins pinnately lobed (◘).
    43. Fruits indehiscent, corky, valvular segments rudimentary, seedless. . . . . 27. *Raphanus*, p. 438
    43. Fruits dehiscent, not corky, valvular segments well-developed, seeded.
     44. Fruit valves 3-veined; terminal segments usually 1–6-seeded, sometimes seedless. . . . . 20. *Coincya*, p. 429
     44. Fruit valves 1-veined; terminal segments seedless. . . . . 23. *Eruca*, p. 434
  40. Petals blades with veins same color intensity as rest of blade.
   45. Cauline leaf blade margins entire.
    46. Plants biennial . . . . 11. *Boechera* (in part), p. 348
    46. Plants perennial (with rhizomes or caudices).
     47. Plants with slender or thick rhizomes or tubers. . . . . 40. *Cardamine* (in part), p. 464
     47. Plants with caudices.
      48. Petals yellow, lavender, or purple . . . . 86. *Hesperidanthus* (in part), p. 689
      48. Petals white.
       49. Ovules 28–90 per ovary . . . . 11. *Boechera* (in part), p. 348
       49. Ovules 6–12(–18) per ovary.
        50. Basal leaf blades ciliate, leathery . . . . 13. *Nevada*, p. 414
        50. Basal leaf blades glabrous, not leathery (◘).
         51. Fruits latiseptate; cotyledons accumbent . . . . 40. *Cardamine* (in part), p. 464
         51. Fruits terete; cotyledons incumbent . . . . . . 59. *Eutrema* (in part), p. 555
   45. Cauline leaf blade margins pinnately divided, sinuate, or dentate.
    52. Petals yellow.
     53. Racemes bracteate basally. . . . . 24. *Erucastrum* (in part), p. 435
     53. Racemes usually ebracteate or, rarely, proximalmost 1 or 2 flowers bracteate.
      54. Ovaries and young fruits 2-segmented (◘).
       55. Fruit valves 1-veined; sepals usually erect or ascending, rarely spreading . . . . 17. *Brassica* (in part), p. 419
       55. Fruit valves 3–7-veined; sepals spreading to reflexed.

56. Seeds ovoid or oblong; stigmas entire . . . . 25. *Hirschfeldia*, p. 436
56. Seeds globose; stigmas 2-lobed. . . . . 29. *Sinapis*, p. 441
54. Ovaries and young fruits unsegmented.
57. Stigmas entire. . . . . 45. *Rorippa* (in part), p. 493
57. Stigmas 2-lobed (¤).
58. Seeds biseriate; fruit valves 1-veined; cotyledons conduplicate . . . . 22. *Diplotaxis* (in part), p. 432
58. Seeds uniseriate; fruit valves 3-veined; cotyledons incumbent . . . . 80. *Sisymbrium* (in part), p. 667

[52. Shifted to left margin.—Ed.]

52. Petals white, pink, lavender, or purple.
59. Ovaries and young fruits 2-segmented.
60. Ovules (1 or) 2(–4) per ovary . . . . .18. *Cakile* (in part), p. 424
60. Ovules 20–60 per ovary . . . . 22. *Diplotaxis* (in part), p. 432
59. Ovaries and young fruits unsegmented.
61. Cauline leaf blades reniform or cordate, margins crenate or dentate. . . . . .96. *Alliaria*, p. 744
61. Cauline leaf blades usually not reniform or cordate, at least some with margins pinnatifid, pinnatisect, runcinate, or pectinate.
62. Leaf blades pectinate or pinnatisect, lateral lobes filiform to linear; racemes: rachises flexuous . . . . .87. *Sibara* (in part), p. 692
62. Leaf blades dentate or variously divided, not pectinate, rarely pinnatisect, lobes (when present) usually ovate to oblong, rarely linear; racemes: rachises usually not flexuous (except some *Cardamine*).
63. Petals usually purple, if not, then margins crisped and/or channeled . . . . .82. *Caulanthus* (in part), p. 677
63. Petals white or lavender, margins neither crisped nor channeled.
64. Leaf blade margins pinnatisect or pinnatifid throughout (¤).
65. Seeds not winged; replum flattened; fruit valves dehiscing elastically . . . . 40. *Cardamine* (in part), p. 464
65. Seeds winged; replum rounded; fruit valves not dehiscing elastically . . . . .44. *Planodes*, p. 492
64. Leaf blade margins entire or dentate ± distally (¤).
66. Fruits terete; seeds not winged; cotyledons incumbent . . . . .82. *Caulanthus* (in part), p. 677
66. Fruits latiseptate; seeds winged; cotyledons accumbent. . . . . .91. *Streptanthus* (in part), p. 700

Key to Genera of Group 3

1. Racemes bracteate basally or throughout.
   2. Annuals; leaf blade margins pinnatifid; petals yellow . . . . 54. *Tropidocarpum* (in part), p. 531
   2. Perennials; leaf blade margins usually entire or dentate, rarely lobed; petals white . . . . 56. *Braya* (in part), p. 546
1. Racemes ebracteate or proximal flowers bracteate.
   3. Plants scapose; cauline leaves absent.
      4. Sepals 1.5–2.5 mm; petioles of previous years not persistent . . . . .9. *Draba* (in part), p. 269
      4. Sepals 4–5 mm; petioles of previous years persistent (stramineous). . . . 10. *Anelsonia* (in part), p. 347
   3. Plants not scapose; at least 1 cauline leaf present.
      5. Cauline leaves sessile, blade bases auriculate, sagittate, or amplexicaul.
         6. Trichomes finely dendritic . . . . .14. *Phoenicaulis* (in part), p. 415
         6. Trichomes stellate, forked, simple, and dendritic (never exclusively dendritic).
            7. Trichomes stalked, stellate, rarely few simple . . . . 6. *Arabis*, p. 257
            7. Trichomes of a different type or a mixture of more than one.
               8. Annuals; stigmas 2-lobed; petals with purple veins darker than blades . . . . .82. *Caulanthus* (in part), p. 677
               8. Biennials or perennials; stigmas entire or subentire; petals with veins not purple, color intensity same as blades.

9. Petals usually yellow or creamy white, rarely pink; young fruits appressed to rachises .... 37. *Turritis*, p. 458
9. Petals white, pink, lavender, or purple; young fruits not or rarely appressed to rachises (some *Boechera*) (◘).
   10. Fruits latiseptate; cotyledons accumbent .... 11. *Boechera* (in part), p. 348
   10. Fruits terete; cotyledons incumbent .... 36. *Transberingia*, p. 456

[5. Shifted to left margin.—Ed.]

5. Cauline leaves usually petiolate, if sessile, blade bases not auriculate, sagittate, or amplexicaul.
   11. Trichomes sessile, medifixed, malpighiaceous or stellate .... 55. *Erysimum*, p. 534
   11. Trichomes stalked or simple, not medifixed, malpighiaceous, stellate, forked, or dendritic.
      12. Stigmas 2-lobed (conical, connivent, decurrent).
         13. Sepals spreading to reflexed .... 76. *Nerisyrenia*, p. 609
         13. Sepals erect (sometimes connivent).
            14. Plants often with multiseriate glands; stigmas 2-horned; petal margins crisped... 4. *Matthiola* (in part), p. 253
            14. Plants without multiseriate glands; stigmas not horned; petal margins not crisped.
               15. Petals (6.5–)8–10(–12) mm .... 58. *Strigosella*, p. 553
               15. Petals (13–)15–30 mm.
                  16. Plants tomentose, trichomes dendritic; sepals 10–15 mm ......
                  .... 4. *Matthiola* (in part), p. 253
                  16. Plants not tomentose, trichomes simple and/or forked; sepals 5–8 mm... 62. *Hesperis*, p. 562
      12. Stigmas entire (capitate).
         17. Cauline leaf blade margins usually 1–3-pinnatisect, pectinate, or, rarely, pinnatifid.
            18. Perennials (cespitose), woody basally; leaf blades pectinate, rigid, persisting in subsequent years as pectinate spines .... 15. *Polyctenium* (in part), p. 415
            18. Annuals, biennials, or perennials, not woody basally; leaf blades 1–3-pinnatisect or pinnatifid, pliable, not persisting in subsequent years as spines.
               19. Petals yellow; trichomes dendritic .... 52. *Descurainia* (in part), p. 518
               19. Petals white; trichomes simple, 2-forked, dendritic, and, sometimes, cruciform.
                  20. Annuals; petals 2–3.5 mm; plants puberulent, some trichomes dendritic; California, Nevada .... 87. *Sibara* (in part), p. 692
                  20. Biennials or perennials; petals 4–10 mm; plants pubescent, trichomes forked or cruciform; n, e, c United States, Canada, Alaska (◘).
                     21. Fruits pubescent; cotyledons incumbent .... 16. *Sandbergia* (in part), p. 417
                     21. Fruits glabrous; cotyledons accumbent .... 32. *Arabidopsis* (in part), p. 447
         17. Cauline leaf blade margins entire, dentate, sinuate, or lyrate.
            22. Flowers cup-shaped; petals white with purple tips .... 61. *Pennellia*, p. 559
            22. Flowers not cup-shaped; petals white, pink, lavender, purple, yellow, or orange.
               23. Proximalmost flowers bracteate.
                  24. Basal and proximalmost cauline leaf blade margins entire .....
                  .... 9. *Draba* (in part), p. 269
                  24. Basal and proximalmost cauline leaf blade margins dentate ....
                  .... 56. *Braya* (in part), p. 546
               23. Proximalmost flowers ebracteate.

[25. Shifted to left margin.—Ed.]

25. Leaf blade margins sometimes lobed, sinuate, or lyrate.
    26. Trichomes forked; fruits terete or latiseptate, glabrous . . . . 32. *Arabidopsis* (in part), p. 447
    26. Trichomes subdendritic; fruits terete, pubescent . . . . .60. *Halimolobos* (in part), p. 557
25. Leaf blade margins entire or dentate.
    27. Petals yellow. . . . . .9. *Draba* (in part), p. 269
    27. Petals white, lavender, pink, or purple.
        28. Ovaries and young fruits pubescent (¤).
            29. Seeds winged; cotyledons accumbent; fruits strongly flattened . . . . 11. *Boechera* (in part), p. 348
            29. Seeds not winged; cotyledons incumbent; fruits slightly flattened. . . . .
            . . . . 16. *Sandbergia* (in part), p. 417
        28. Ovaries and young fruits glabrous.
            30. Petals (4–)5–18 mm; seeds winged . . . . 11. *Boechera* (in part), p. 348
            30. Petals 2–4(–5 mm); seeds not winged (¤).
                31. Fruits latiseptate; seeds biseriate; cotyledons accumbent . . . . .9. *Draba* (in part), p. 269
                31. Fruits usually terete, rarely slightly latiseptate; seeds uniseriate; cotyledons incumbent.
                    32. Basal leaves often petiolate; branched trichomes 2–4-rayed, stalked; ovules (30–)40–70 per ovary. . . . . 32. *Arabidopsis* (in part), p. 447
                    32. Basal leaves not petiolate; branched trichomes submalpighiaceous, subsessile; ovules 20–28 per ovary . . . . 56. *Braya* (in part), p. 546

Key to Genera of Group 4

1. Plants scapose; cauline leaves absent.
   2. Ovules 4 per ovary . . . . 12. *Cusickiella* (in part), p. 412
   2. Ovules 10 or more per ovary.
      3. Petals yellow or orange. . . . . .9. *Draba* (in part), p. 269
      3. Petals white, lavender, rose, or purple.
         4. Sepals 4–5 mm; petioles of previous years persistent (stramineous) . . . .
         . . . . 10. *Anelsonia* (in part), p. 347
         4. Sepals 1.5–3(–3.7) mm; petioles of previous years not persistent (¤).
            5. Cotyledons accumbent . . . . 9. *Draba* (in part), p. 269
            5. Cotyledons incumbent. . . . . 56. *Braya* (in part), p. 546
1. Plants not scapose; at least 1 cauline leaf present.
   6. Cauline leaves sessile, blade bases auriculate, sagittate, or amplexicaul.
      7. Plants perennial (caudex well-developed, woody); trichomes dendritic. . . . .
      . . . . .14. *Phoenicaulis* (in part), p. 415
      7. Plants annual, biennial or, rarely, perennial; trichomes often mixed simple, forked, stellate, or subdendritic.
         8. Petals white; at least some branched trichomes sessile, stellate . . . . 34. *Capsella*, p. 453
         8. Petals usually yellow, rarely creamy white or white with yellow claws; branched trichomes stalked, not stellate.
            9. Styles 0.5–0.9 mm; petals 2–2.5 mm; ovules 2–4 per ovary . . . . .35. *Neslia*, p. 455
            9. Styles 1–3 mm; petals (2.5–)4–12 mm; ovules 4–32(–40) per ovary.
               10. Stamens in 3 pairs of unequal length; nectar glands distinct (4); fruit valves extending into stylar area . . . . 33. *Camelina*, p. 451
               10. Stamens tetradynamous; nectar glands confluent; fruit valves not extending into stylar area . . . . .77. *Paysonia*, p. 611
   6. Cauline leaves usually petiolate, if sessile, blade bases neither auriculate nor amplexicaul.
      11. Racemes bracteate throughout . . . . 54. *Tropidocarpum* (in part), p. 531
      11. Racemes ebracteate or proximalmost flowers bracteate.
          12. Plants with multicellular glandular tubercles or papillae . . . . 30. *Bunias*, p. 444
          12. Plants without multicellular glandular tubercles or papillae.

[13. Shifted to left margin.—Ed.]

13. Petal apices strongly 2-fid .... 3. *Berteroa*, p. 252
13. Petal apices rounded, retuse, or emarginate.
    14. Cauline leaf blade trichomes exclusively malpighiaceous.
        15. Petals yellow; ovules 14–18 per ovary .... 9. *Draba* (in part), p. 269
        15. Petals white or purplish violet; ovules 2 per ovary .... 70. *Lobularia*, p. 597
    14. Cauline leaf blade trichomes not exclusively malpighiaceous.
        16. Some filaments winged, dentate, or appendaged.
            17. Plants annual or, rarely, perennial; ovules 1 or 2 per ovary .... 1. *Alyssum* (in part), p. 247
            17. Plants perennial; ovules 4–40 per ovary.
                18. Petals yellow, 3–6 mm .... 2. *Aurinia*, p. 251
                18. Petals usually purple to violet, rarely white, (10–)15–28 mm .... 8. *Aubrieta*, p. 268
        16. All filaments not winged, dentate, or appendaged.
            19. Ovules 1 or 2 per ovary.
                20. Ovaries and young fruits angustiseptate, suborbicular, didymous; petals (3.5–)4–15 mm.
                    21. Sepals spreading or reflexed; stigmas conical, 2-lobed; petals 3.5–10(–12) mm .... 73. *Dimorphocarpa*, p. 604
                    21. Sepals erect, connivent; stigmas capitate, entire; petals (10–)12–15 mm .... 74. *Dithyrea*, p. 607
                20. Ovaries and young fruits latiseptate (subterete or 4-angled), orbicular to obovate or ovoid, not didymous; petals 0.9–3(–4) mm.
                    22. Ovaries and young fruits ovoid, subterete or 4-angled; petals 0.9–1.3 mm .... 57. *Euclidium*, p. 552
                    22. Ovaries and young fruits orbicular to obovate, latiseptate; petals 1.5–3(–4) mm.
                        23. Sepals persistent; fruiting pedicels divaricate to ascending .... 1. *Alyssum* (in part), p. 247
                        23. Sepals caducous; fruiting pedicels reflexed .... 7. *Athysanus* (in part), p. 267
            19. Ovules 4–100 per ovary.
                24. Sepals erect, linear to oblong-linear, connivent; petals linear-oblanceolate to linear, twisted with age .... 75. *Lyrocarpa*, p. 608
                24. Sepals erect to spreading, oblong to ovate, not connivent; petals obovate, spatulate, oblong, or oblanceolate, not twisted with age.
                    25. Cauline leaf blade margins 1–3-pinnatisect, pectinate, pinnatifid, or palmately 3–7-lobed.
                        26. Leaf blades pectinate, rigid, persisting as pectinate spines .... 15. *Polyctenium* (in part), p. 415
                        26. Leaf blades 1–3-pinnatisect, pinnatifid, or palmately 3–7-lobed, not rigid, not persisting as spines.
                            27. Perennials (caudex well-developed) .... 81. *Smelowskia* (in part), p. 671
                            27. Annuals.
                                28. Petals yellow; most or all trichomes dendritic .... 52. *Descurainia* (in part), p. 518
                                28. Petals white; trichomes simple or forked ... 53. *Hornungia* (in part), p. 530
                    25. Cauline leaf blade margins usually entire or dentate, rarely sinuate.
                        29. Trichomes stellate or lepidote, webbed or not .... 78. *Physaria*, p. 616
                        29. Trichomes simple, forked, cruciform, pectinate, dendritic, or malpighiaceous, often mixed (never exclusively stellate or lepidote).
                            30. Ovaries and young fruits strongly angustiseptate.
                                31. Petals 1.5–2.5 × 0.7–1.2 mm, claws indistinct; styles 0.3–0.8(–1) mm; anthers oblong, 0.7–1 mm ...... 60. *Halimolobos* (in part), p. 557
                                31. Petals (7–)8–11(–13) × (5–)6–9 mm, claws distinct; styles 2–5 mm; anthers linear, 2.5–4 mm .... 79. *Synthlipsis*, p. 665
                            30. Ovaries and young fruits latiseptate (terete or 4-angled, except *Hornungia*).

[32. Shifted to left margin.—Ed.]

32. Plants annuals.
　　33. Racemes secund; fruiting pedicels reflexed ....7. *Athysanus* (in part), p. 267
　　33. Racemes not secund; fruiting pedicels erect or ascending to spreading or divaricate.
　　　　34. Petals usually yellow, (1.2–)1.5–8.5 mm (if white, late-season flowers often cleistogamous and apetalous) .... 9. *Draba* (in part), p. 269
　　　　34. Petals white, 0.6–1.2 mm .... 53. *Hornungia* (in part), p. 530
32. Plants perennials.
　　35. Basal leaf blade margins often apically 3–5-lobed, surfaces densely silvery villous, trichomes simple or densely grayish tomentose (mostly dendritic and simple).......
　　　　.. 81. *Smelowskia* (in part), p. 671
　　35. Basal leaf blade margins not apically 3–5-lobed, surfaces glabrate or sparsely to densely pubescent, trichomes seldom silvery or grayish (¤).
　　　　36. Cotyledons accumbent .... 9. *Draba* (in part), p. 269
　　　　36. Cotyledons incumbent.
　　　　　　37. Fruits 4-angled; ovules 4 per ovary...................... 12. *Cusickiella* (in part), p. 412
　　　　　　37. Fruits terete; ovules 10–30 per ovary .... 56. *Braya* (in part), p. 546

Key to Genera Based Primarily on Fruiting Specimens

1. Fruits silicles (lengths less than 3 times widths).
　　2. Trichomes simple or absent... Group 5, p. 238
　　2. Trichomes (at least some) branched... Group 6, p. 240
1. Fruits siliques (at least 3 times as long as wide).
　　3. Trichomes (at least some) branched... Group 7, p. 241
　　3. Trichomes simple or absent... Group 8, p. 243

Key to Genera of Group 5

1. Fruits 2-segmented.
　　2. Fruits corky; plants glabrous; strand and seashore beaches.
　　　　3. Fruits with well-developed proximal segments, usually seeded, terminal segments ovoid to lanceolate; cotyledons usually accumbent, sometimes incumbent ......
　　　　　　... 18. *Cakile* (in part), p. 424
　　　　3. Fruits with stalklike proximal segments, seedless, terminal segments globose; cotyledons conduplicate .... 21. *Crambe*, p. 430
　　2. Fruits not corky; plants pubescent; inland.
　　　　4. Cauline leaf blade margins 1- or 2-pinnatisect; fruiting pedicels strongly recurved; fruit terminal segments flattened, cochleariform, or short-lingulate, seedless ....
　　　　　　...19. *Carrichtera*, p. 429
　　　　4. Cauline leaf blade margins dentate; fruiting pedicels erect to ascending (straight); fruit terminal segments globose or ovoid, 1(–3)-seeded................... 28. *Rapistrum*, p. 440
1. Fruits unsegmented.
　　5. Fruits angustiseptate.
　　　　6. Cauline leaves usually petiolate, if sessile, blade bases not auriculate, sagittate, or amplexicaul.
　　　　　　7. Seeds 8–24 per fruit.
　　　　　　　　8. Perennials, 5–12(–20) dm, roots fleshy........................ 38. *Armoracia*, p. 459
　　　　　　　　8. Annuals, 0.2–3 dm, roots not fleshy.................... 53. *Hornungia* (in part), p. 530
　　　　　　7. Seeds 2(–4) per fruit.
　　　　　　　　9. Racemes not or slightly elongated in fruit; seeds winged .... 63. *Iberis*, p. 563
　　　　　　　　9. Racemes elongated in fruit; seeds not winged.
　　　　　　　　　　10. Seeds 4 per fruit; cotyledons accumbent .... 64. *Teesdalia*, p. 564
　　　　　　　　　　10. Seeds 2 per fruit; cotyledons incumbent, rarely accumbent ....
　　　　　　　　　　　　... 68. *Lepidium* (in part), p. 570
　　　　6. Cauline leaves sessile, blade bases auriculate, sagittate, or amplexicaul.

11. Fruits indehiscent, samaroid, 1-seeded . . . . .66. *Isatis* (in part), p. 567
11. Fruits dehiscent, not samaroid, 2–16-seeded.
    12. Seeds usually 2 per fruit . . . . 68. *Lepidium* (in part), p. 570
    12. Seeds 4–16 per fruit.
        13. Fruits not keeled, not winged, strongly reticulate; seeds biseriate . . .
        . . . . 50. *Cochlearia* (in part), p. 514
        13. Fruits keeled, winged, not reticulate; seeds uniseriate.
            14. Seeds blackish brown, concentrically striate or alveolate; plants
            fetid . . . . .97. *Thlaspi*, p. 745
            14. Seeds yellow-brown or brown, usually smooth, rarely minutely
            reticulate; plants not fetid.
                15. Annuals; styles 0–0.3 mm . . . . .71. *Microthlaspi*, p. 599
                15. Perennials or, rarely, biennials; styles (0.2–)0.4–4 mm . . . . . 72. *Noccaea*, p. 600

[5. Shifted to left margin.—Ed.]

5. Fruits terete or latiseptate.
   16. Cauline leaves absent.
      17. Fruits solitary on pedicles from basal rosette; seeds winged.
         18. Fruits oblong to globose; styles 0.7–7 mm; seeds uniseriate; se United States,
         s Missouri, e Texas . . . . 42. *Leavenworthia* (in part), p. 485
         18. Fruits orbicular or orbicular-ovate; styles 0.3–0.8 mm; seeds biseriate; British
         Columbia, Pacific and some Mountain states . . . . 65. *Idahoa*, p. 566
      17. Fruits racemes, corymbose; seeds not winged.
         19. Annuals, aquatic, without multicellular glandular papillae; basal leaf blades
         linear or subulate . . . . 47. *Subularia*, p. 509
         19. Perennials, terrestrial, with multicellular glandular papillae; basal leaf blades
         obovate to broadly spatulate . . . . 49. *Parrya* (in part), p. 511
   16. Cauline leaves present.
      20. Racemes bracteate throughout, if only basally bracteate, then septums absent.
         21. Perennials; basal leaf blade margins entire; septums absent; styles 0.2–1 mm;
         seeds not winged . . . . 5. *Aphragmus* (in part), p. 256
         21. Annuals; basal leaf blade margins 1 or 2 (or 3)-pinnatisect; septums present;
         styles 1–12 mm; seeds winged . . . . 46. *Selenia*, p. 506
      20. Racemes ebracteate.
         22. Fruits 30–45(–50) mm; gynophores 7–18 mm . . . . 69. *Lunaria*, p. 596
         22. Fruits 1–15 mm; gynophores 0–0.5 mm.
            23. Fruits dehiscent, seeds 8–90 per fruit.
                24. Fruit valves each with an obscure midvein; seeds 18–90 per fruit;
                petals yellow . . . . 45. *Rorippa* (in part), p. 493
                24. Fruit valves each with a distinct midvein; seeds 8–14 per fruit;
                petals white . . . . 50. *Cochlearia* (in part), p. 514
            23. Fruits indehiscent, seeds 1 or 2(–4) per fruit.
                25. Fruits strongly latiseptate, winged, with radiating rays . . . . . . 94. *Thysanocarpus*, p. 739
                25. Fruits terete, not winged, without radiating rays.
                    26. Perennials (rhizomatous), hirsute; fruits globose to obovoid . . . .
                    . . . . 68. *Lepidium* (in part), p. 570
                    26. Annuals, glabrous; fruits ovoid to ellipsoid or obpyriform to
                    clavate-obcordiform.
                        27. Fruits ovoid to ellipsoid; styles obsolete . . . . 31. *Calepina*, p. 446
                        27. Fruits obpyriform to clavate-obcordate; styles (0.8–)1.2–2 mm
                        . . . . 67. *Myagrum*, p. 568

Key to Genera of Group 6

1. Plants scapose; cauline leaves absent.
   2. Seeds 4 per fruit .... 12. *Cusickiella*, p. 412
   2. Seeds 10 or more per fruit.
      3. Seeds silvery papillate; fruit valves each with distinct (anastomosing) lateral veins
         .... 10. *Anelsonia* (in part), p. 347
      3. Seeds minutely reticulate; fruit valves each with obscure lateral veins.
         4. Seeds biseriate; cotyledons accumbent .... 9. *Draba* (in part), p. 269
         4. Seeds usually uniseriate, rarely biseriate; cotyledons incumbent ...... 56. *Braya* (in part), p. 546
1. Plants not scapose; at least 1 cauline leaf present.
   5. Cauline leaves sessile, blade bases auriculate, sagittate, or amplexicaul.
      6. Fruits indehiscent, 1-seeded, lenticular or compressed-globose; valves woody ..... 35. *Neslia*, p. 455
      6. Fruits dehiscent, 4–40-seeded, not lenticular or compressed-globose; valves papery or leathery.
         7. Fruits obdeltoid to obdeltoid-cordiform, strongly angustiseptate; styles 0.2–0.7 mm; stellate trichomes sessile .... 34. *Capsella*, p. 453
         7. Fruits globose, pyriform, obovoid, orbicular, or subovate, usually latiseptate, rarely angustiseptate (some *Paysonia*); styles 1–3.5 mm; stellate trichomes absent or short-stalked.
            8. Fruit valves with caudate apex extending 1–2.5 mm onto style; cotyledons incumbent .... 33. *Camelina*, p. 451
            8. Fruit valves not extending onto style; cotyledons accumbent .... 77. *Paysonia*, p. 611
   5. Cauline leaves usually petiolate, if sessile, blade bases not auriculate, sagittate, or amplexicaul.
      9. Fruits angustiseptate.
         10. Racemes bracteate throughout .... 54. *Tropidocarpum* (in part), p. 531
         10. Racemes usually ebracteate, rarely proximalmost flowers bracteate.
            11. Seeds 2 per fruit; fruits didymous, breaking into 1-seeded segments; valves indurate around margin.
               12. Stigmas conical, 2-lobed (decurrent); sepals spreading to reflexed...
                  .... 73. *Dimorphocarpa*, p. 604
               12. Stigmas capitate, entire; sepals erect .... 74. *Dithyrea*, p. 607
            11. Seeds 4–100 per fruit; fruits not didymous or, if didymous (some *Physaria*), inflated and not breaking into 1-seeded segments; valves not indurate around margin.
               13. Leaf blades pectinate, persisting as pectinate spines.... 15. *Polyctenium* (in part), p. 415
               13. Leaf blades entire or divided, neither pectinate nor persisting as spines.
                  14. Fruits obcordiform to panduriform; stigmas prominently 2-lobed
                     .... 75. *Lyrocarpa*, p. 608
                  14. Fruits not obcordiform or panduriform; stigmas entire or subentire.
                     15. Leaf blade margins palmately (3 or) 5 (or 7)-lobed; sepals persistent .... 81. *Smelowskia* (in part), p. 671
                     15. Leaf blade margins entire, dentate, sinuate, or pinnatisect; sepals caducous.
                        16. Annuals or biennials; styles 0–0.8(–1) mm; cotyledons incumbent.
                           17. Fruits 10–24-seeded; valves glabrous... 53. *Hornungia* (in part), p. 530
                           17. Fruits (60–)70–100-seeded; valves pubescent......
                              .... 60. *Halimolobos* (in part), p. 557
                        16. Perennials or, rarely, annuals; styles 1.8–7 mm; cotyledons accumbent.

18. Cauline leaf blade margins entire; trichomes stellate
.... 78. *Physaria* (in part), p. 616
18. Cauline leaf blade margins sinuate or dentate; trichomes dendritic mixed with simple and forked ones.................................. 79. *Synthlipsis*, p. 665

[9. Shifted to left margin.—Ed.]
9. Fruits latiseptate (terete or 4-angled).
   19. Racemes secund; fruits reflexed.. 7. *Athysanus*, p. 267
   19. Racemes not secund; fruits not reflexed.
      20. Plants with multicellular glandular papillae or tubercles; fruits woody, indehiscent; cotyledons spirolobal .... 30. *Bunias*, p. 444
      20. Plants without multicellular glandular papillae or tubercles; fruits not woody, mostly dehiscent (except *Euclidium*); cotyledons incumbent or accumbent.
         21. Cauline leaf blade margins 1- or 2-pinnatifid or pinnatisect.
            22. Annuals or biennials; petals yellow .... 52. *Descurainia* (in part), p. 518
            22. Perennials (cespitose); petals white, lavender, or purple.... 81. *Smelowskia* (in part), p. 671
         21. Cauline leaf blade margins entire, crenate, dentate, or sinuate.
            23. Seeds 1 or 2 per fruit.
               24. Fruits indehiscent, subterete or 4-angled .... 57. *Euclidium*, p. 552
               24. Fruits dehiscent, latiseptate.
                  25. Trichomes stellate (when branched) .... 1. *Alyssum*, p. 247
                  25. Trichomes malpighiaceous .... 70. *Lobularia*, p. 597
            23. Seeds 4 or more per fruit.
               26. Seeds winged or margined; filaments winged, toothed, or appendaged.
                  27. Perennials; cotyledons incumbent; petals emarginate .... 2. *Aurinia*, p. 251
                  27. Annuals or biennials; cotyledons accumbent; petals deeply 2-fid
                  .... 3. *Berteroa*, p. 252
               26. Seeds usually neither winged nor margined, if narrowly margined (some *Physaria*), trichomes exclusively stellate; filaments not winged, toothed, or appendaged.
                  28. Trichomes almost exclusively stellate (rays simple or divided, webbed or not), sometimes lepidote; septums often with apical midvein................................ 78. *Physaria* (in part), p. 616
                  28. Trichomes often mixed, not exclusively stellate, webbed, or lepidote; septums without apical midvein.
                     29. Cotyledons accumbent........................ 9. *Draba* (in part), p. 269
                     29. Cotyledons incumbent.
                        30. Seeds 10–30 per fruit, biseriate or uniseriate; plants not white-canescent .... 56. *Braya* (in part), p. 546
                        30. Seeds 4 per fruit, not in rows; plants white-canescent throughout .... 81. *Smelowskia* (in part), p. 671

Key to Genera of Group 7

1. Plants scapose; cauline leaves absent.
   2. Seeds minutely reticulate; fruit valves each with obscure lateral veins .... 9. *Draba* (in part), p. 269
   2. Seeds silvery papillate; fruit valves each with distinct (anastomosing) lateral veins . . .
   .... 10. *Anelsonia* (in part), p. 347
1. Plants not scapose; at least 1 cauline leaf present.
   3. Cauline leaves sessile, blade bases auriculate, sagittate, or amplexicaul.
      4. Fruits latiseptate; cotyledons accumbent.
         5. Plants tomentose, trichomes finely dendritic; seeds (6–)8–16(–18) per fruit
         .... 14. *Phoenicaulis*, p. 415
         5. Plants pubescent, trichomes usually stellate, forked, or 2–14-rayed; (if trichomes dendritic) seeds more than 24 per fruit.

6. Branched trichomes stellate, 3–5-rayed; fruits straight, erect (appressed to rachises) to divaricate, 0.6–2 mm wide .... 6. *Arabis*, p. 257
6. Branched trichomes forked, 2–14-rayed, or dendritic; fruits arcuate, curved, or straight, usually reflexed, pendent, or spreading, rarely erect, 1–6 mm wide.................................... 11. *Boechera* (in part), p. 348
4. Fruits terete, or 4-angled; cotyledons usually incumbent, rarely accumbent (*Turritis*).
7. Seeds uniseriate; stigmas 2-lobed; branched trichomes 2-rayed. . . 82. *Caulanthus* (in part), p. 677
7. Seeds biseriate; stigmas entire; branched trichomes dendritic or substellate.
8. Fruiting pedicels divaricate to ascending; fruits terete, not appressed to rachises .... 36. *Transberingia*, p. 456
8. Fruiting pedicels erect; fruits subterete or 4-angled, appressed to rachises .... 37. *Turritis*, p. 458

[3. Shifted to left margin.—Ed.]
3. Cauline leaves petiolate, if sessile, blade bases not auriculate, sagittate, or amplexicaul.
9. Racemes bracteate throughout.
10. Plants annual; fruits angustiseptate; petals yellow............ 54. *Tropidocarpum* (in part), p. 531
10. Plants perennial; fruits terete; petals white, pink, or purple .... 56. *Braya* (in part), p. 546
9. Racemes ebracteate or only proximalmost flowers bracteate.
11. Trichomes sessile, malpighiaceous, sometimes mixed with sessile 3–5-rayed, stellate ones (simple trichomes absent) .... 55. *Erysimum*, p. 534
11. Trichomes stalked, cruciform, dendritic, stellate, submalpighiaceous, or forked (simple trichomes sometimes present).
12. Stigmas conical, 2-lobed, lobes connivent, often decurrent.
13. Seeds winged; cotyledons accumbent; stigmas usually with 2 horns (if not, fruits (4–)6–12(–15) cm) .... 4. *Matthiola*, p. 253
13. Seeds not winged; cotyledons incumbent; stigmas without horns.
14. Fruits torulose, glabrous; trichomes simple and forked; glandular (glands unicellular, on uniseriate stalks)....................... 62. *Hesperis*, p. 562
14. Fruits not torulose, pubescent; some trichomes dendritic; not glandular.
15. Annuals; fruits 4-angled; sepals erect; styles obsolete .... 58. *Strigosella*, p. 553
15. Perennials or subshrubs; fruits angustiseptate or terete; sepals spreading; styles 0.9–4.3 mm .... 76. *Nerisyrenia*, p. 609
12. Stigmas capitate, usually entire, rarely slightly 2-lobed, lobes neither connivent nor decurrent.
16. Cauline leaf blade margins usually pectinate, pinnatifid, or pinnatisect, rarely lyrate.
17. Fruits latiseptate.
18. Fruits strongly torulose; cotyledons incumbent..... 16. *Sandbergia* (in part), p. 417
18. Fruits not torulose; cotyledons accumbent.
19. Biennials or perennials; branched trichomes forked and 3-rayed, stellate .... 32. *Arabidopsis* (in part), p. 447
19. Annuals; at least some branched trichomes dendritic ....87. *Sibara* (in part), p. 692
17. Fruits terete, 4-angled, or angustiseptate.
20. Leaf blades pectinate, rigid; fruits slightly angustiseptate .... 15. *Polyctenium* (in part), p. 415
20. Leaf blades pinnatifid or pinnatisect, not rigid; fruits terete or 4-angled.
21. Branched trichomes submalpighiaceous or 2-forked........ .... 56. *Braya* (in part), p. 546
21. Branched trichomes mostly dendritic.

22. Annuals, biennials, or, rarely, perennials (without caudex); seeds mucilaginous when wetted; petals yellow . . . . 52. *Descurainia* (in part), p. 518
22. Perennials (caudex well-developed); seeds not mucilaginous when wetted; petals white, lavender, or pink . . . . 81. *Smelowskia* (in part), p. 671

[16. Shifted to left margin.—Ed.]

16. Cauline leaf blade margins usually entire or dentate, rarely sinuate.
    23. Seeds winged.
        24. Fruiting racemes usually not secund; styles distinct; flowers not cup-shaped; fruiting pedicels reflexed, pendent, erect, or divaricate . . . . 11. *Boechera* (in part), p. 348
        24. Fruiting racemes secund; styles obsolete; flowers cup-shaped; fruiting pedicels divaricate-ascending . . . . 61. *Pennellia* (in part), p. 559
    23. Seeds not winged.
        25. Seeds biseriate.
            26. Fruiting pedicels pendent.
                27. Cotyledons accumbent; seeds 8–90 per fruit . . . . 11. *Boechera* (in part), p. 348
                27. Cotyledons incumbent; seeds 150–250 per fruit . . . . 61. *Pennellia* (in part), p. 559
            26. Fruiting pedicels erect, ascending, or spreading.
                28. Styles 4–6 mm; fruits pubescent, trichomes setiform mixed with smaller, stellate ones . . . . 8. *Aubrieta*, p. 268
                28. Styles to 2 mm; fruits glabrous or pubescent, trichomes 2- or 4-rayed, not setiform . . . . 9. *Draba* (in part), p. 269
        25. Seeds uniseriate.
            29. Cotyledons accumbent; fruits latiseptate.
                30. Basal leaf margins entire; seeds 14–22(–30) per fruit . . . . 11. *Boechera* (in part), p. 348
                30. Basal leaf margins coarsely dentate; seeds 24–48 per fruit . . . . 32. *Arabidopsis* (in part), p. 447
            29. Cotyledons incumbent; fruits terete or 4-angled.
                31. Fruits ellipsoid to oblong; seeds 4–8 per fruit . . . . 81. *Smelowskia* (in part), p. 671
                31. Fruits linear; seeds (16–)20–140 per fruit.
                    32. Fruits pubescent, trichomes subsessile, with coarser, stalked ones; cauline leaf blade margins dentate to sinuate . . . . 60. *Halimolobos* (in part), p. 557
                    32. Fruits glabrous or pubescent, trichomes dendritic, forked, cruciform, stellate, or submalpighiaceous (not mixed); cauline leaf blade margins usually entire, rarely dentate.
                        33. Seeds 90–140 per fruit; trichomes dendritic . . . . 61. *Pennellia* (in part), p. 559
                        33. Seeds 20–70 per fruit; trichomes forked, cruciform, stellate, or submalpighiaceous.
                            34. Trichomes submalpighiaceous . . . . 56. *Braya* (in part), p. 546
                            34. Trichomes forked, cruciform, or stellate.
                                35. Fruits densely pubescent; seedss 20–30 per fruit . . . . 16. *Sandbergia* (in part), p. 417
                                35. Fruits glabrous; seeds 30–70 per fruit . . . . 32. *Arabidopsis* (in part), p. 447

Key to Genera of Group 8

1. Plants scapose; cauline leaves absent.
   2. Fruits (some or all) on solitary pedicels (from basal rosettes); seeds winged or margined . . . . 42. *Leavenworthia* (in part), p. 485
   2. Fruits in corymbs or racemes; seeds not winged, not margined.
      3. Annuals or, rarely, perennials; cotyledons conduplicate; seeds 20–36 per fruit; petals yellow . . . . 22. *Diplotaxis* (in part), p. 432
      3. Perennials (cespitose); cotyledons accumbent; seeds 8–20 per fruit; petals white, lavender, or purple.

      4. Stigmas entire; plants eglandular . . . . 40. *Cardamine* (in part), p. 464
      4. Stigmas lobed (lobes prominent, decurrent); plants with multicellular glandular papillae or tubercles (except *Parrya arctica*) . . . . 49. *Parrya* (in part), p. 511
1. Plants not scapose; at least 1 cauline leaf present.
   5. Fruits indehiscent or breaking into 1-seeded segments.
      6. Plants with multicellular glandular papillae or tubercles. . . . . 48. *Chorispora*, p. 510
      6. Plants eglandular.
         7. Fruits not segmented, 1-seeded (samaroid) . . . . .66. *Isatis* (in part), p. 567
         7. Fruits segmented, usually more than 1-seeded.
            8. Styles obsolete; cotyledons accumbent; petal veins not darker than rest of blade; plants glabrous; beaches and strands. . . . . .18. *Cakile* (in part), p. 424
            8. Styles 1–5 mm; cotyledons conduplicate; petal veins darker than rest of blade; plants pubescent; inland . . . . .27. *Raphanus*, p. 438
   5. Fruits dehiscent, more than 1-seeded.
      9. Fruits usually segmented, if unsegmented (*Orychophragmus*), stigma lobes decurrent and petiole bases strongly auriculate; cotyledons conduplicate.
         10. Cauline leaf blade bases auriculate to amplexicaul.
            11. Fruits segmented; stigmas entire or lobes not decurrent; seeds globose . . . . 17. *Brassica* (in part), p. 419
            11. Fruits unsegmented; stigmas lobed, lobes decurrent; seeds oblong . . . . .26. *Orychophragmus*, p. 437
         10. Cauline leaf blade bases neither auriculate nor amplexicaul.
            12. Seeds biseriate.
               13. Terminal fruit segments beaklike, not ensiform; petals veins same color as blade . . . . 22. *Diplotaxis* (in part), p. 432
               13. Terminal fruit segments ensiform; petal veins darker than rest of blade. . . . . .23. *Eruca*, p. 434
            12. Seeds uniseriate.
               14. Fruit valves each prominently 3–7-veined.
                  15. Fruits (30–)40–150-seeded; sepals erect . . . . 20. *Coincya*, p. 429
                  15. Fruits 4–16(–24)-seeded; sepals spreading or reflexed. . . . . 29. *Sinapis*, p. 441
               14. Fruit valves each obscurely veined or midvein prominent.
                  16. Racemes bracteate throughout or at least basally . . . . 24. *Erucastrum*, p. 435
                  16. Racemes ebracteate.
                     17. Fruit valves each prominently 1-veined; fruits divaricate to erect, if appressed to rachises, terminal segments seedless . . . . 17. *Brassica* (in part), p. 419
                     17. Fruit valves obscurely 3–7-veined; fruits erect, appressed to rachises, terminal segments 1- or 2-seeded . . . . 25. *Hirschfeldia*, p. 436
      9. Fruits unsegmented; cotyledons incumbent or accumbent.
         18. Fruits latiseptate.
            19. Replums strongly flattened; fruit valves dehiscing elastically, becoming spirally or circinately coiled. . . . . 40. *Cardamine* (in part), p. 464
            19. Replums terete; fruit valves neither dehiscing elastically nor coiled.
               20. Seeds winged at least distally.
                  21. Cotyledons incumbent.
                     22. Fruits erect; cauline leaf blade margins entire. . . . . .88. *Sibaropsis*, p. 694
                     22. Fruits reflexed; cauline leaf blade margins dentate or pinnatifid . . . . .90. *Streptanthella*, p. 699
                  21. Cotyledons accumbent.
                     23. Cauline leaf blade margins pinnatifid to pinnatisect into dentate lobes . . . . .44. *Planodes*, p. 492
                     23. Cauline leaf blade margins sometimes entire or dentate.

24. Stigmas usually prominently 2-lobed, if entire, stamens in 3 pairs of unequal length and/or calyces urceolate . . . . 91. *Streptanthus* (in part), p. 700
24. Stigmas entire, stamens tetradynamous, calyces not urceolate.
    25. Cauline leaf blade bases not auriculate or sagittate . . . . 11. *Boechera* (in part), p. 348
    25. Cauline leaf blade bases auriculate or sagittate.
        26. Seeds 10–28 per fruit; fruits 6–11.7 cm. . . . . 11. *Boechera* (in part), p. 348
        26. Seeds 50–86 per fruit; fruits 1.5–4.7 cm; (Texas) . . . . 85. *Dryopetalon* (in part), p. 688
20. Seeds not winged.
    27. Cauline leaf blade margins pectinate (lobes filiform). . . . . 87. *Sibara* (in part), p. 692
    27. Cauline leaf blade margins not pectinate, entire, dentate, or pinnately lobed.
        28. Plants tomentose; gynophores 10–20 mm . . . . 89. *Stanleya* (in part), p. 695
        28. Plants glabrous or sparsely pubescent; gynophores (0–)0.1–11 (–16 sometimes in *Warea*) mm.
            29. Leaf blade margins dentate or pinnately lobed.
                30. Cauline leaves sessile, blade bases auriculate or amplexicaul; petals yellow. . . . . 39. *Barbarea* (in part), p. 460
                30. Cauline leaves petiolate, blade bases neither auriculate nor amplexicaul; petals white, lavender, or purple . . . . 93. *Thelypodium* (in part), p. 728
            29. Leaf blade margins entire in distalmost leaves.
                31. Perennials; septums absent; cotyledons incumbent . . . . 5. *Aphragmus* (in part), p. 256
                31. Annuals, biennials, or, rarely, perennials; septums complete; cotyledons accumbent.
                    32. Gynophores 0.1–3 mm; seeds smooth or minutely reticulate; stamens in 3 pairs of unequal length; California . . . . 91. *Streptanthus* (in part), p. 700
                    32. Gynophores 3–16 mm; seeds concentrically striate; stamens subequal; se United States . . . . . 95. *Warea*, p. 742
[18. Shifted to left margin.—Ed.]
18. Fruits terete or 4-angled.
    33. Cauline leaves pinnately compound; stems rooting from proximal nodes . . . . 43. *Nasturtium*, p. 489
    33. Cauline leaves simple, sometime pinnately lobed; stems not rooting from proximal nodes (except some *Rorippa*).
        34. Fruits with septums partitioned between seeds; cauline leaves with strongly auriculate petioles. . . . . 84. *Coelophragmus*, p. 687
        34. Fruits with septums not partitioned between seeds; cauline leaves either sessile or petioles not auriculate.
            35. Cauline leaf blade bases auriculate, sagittate, or amplexicaul.
                36. Stigmas prominently 2-lobed.
                    37. Fruits sessile or subsessile; petal margins often crisped . . . . 82. *Caulanthus* (in part), p. 677
                    37. Fruits usually on gynophores (1.5–)3–9.5 mm, if gynophores shorter, styles clavate; petal margins seldom crisped. . . . . 92. *Thelypodiopsis* (in part), p. 723
                36. Stigmas usually entire, rarely obscurely 2-lobed.
                    38. Gynophores (6–)10–25 mm. . . . . 89. *Stanleya* (in part), p. 695
                    38. Gynophores 0–5 mm, if longer, petals purple.
                        39. Cauline leaf blade margins dentate, serrate, or pinnately lobed.
                            40. Seeds biseriate, colliculate or foveolate. . . . . 45. *Rorippa* (in part), p. 493
                            40. Seeds uniseriate, reticulate or papillate.

41. Annuals; cotyledons incumbent; petal margins often crisped . . . . 82. *Caulanthus* (in part), p. 677
41. Biennials or perennials; cotyledons accumbent; petal margins not crisped.
    42. Stems angled; petals yellow; nectar glands separate (4) . . . . 39. *Barbarea* (in part), p. 460
    42. Stems not angled; petals purple, pink, or white; nectar glands confluent . . . . 41. *Iodanthus*, p. 484
39. Cauline leaf blade margins entire or repand.
    43. Fruits 4-angled; seeds papillose, mucilaginous when wetted . . . . 51. *Conringia*, p. 517
    43. Fruits terete; seeds reticulate, not mucilaginous when wetted.
        44. Gynophores (0.5–)1–6(–8.5) mm; anthers (1–)1.5–6 mm, often coiled after dehiscence . . . . 93. *Thelypodium* (in part), p. 728
        44. Gynophores 0–0.4 mm; anthers 0.2–4 mm, not coiled after dehiscence.
            45. Stigmas entire; cauline leaf blade bases sagittate-amplexicaul . . . . 59. *Eutrema* (in part), p. 555
            45. Stigmas 2-lobed; cauline leaf blade bases auriculate . . . . 92. *Thelypodiopsis* (in part), p. 723

[35. Shifted to left margin.—Ed.]

35. Cauline leaf blade bases not auriculate, sagittate, or amplexicaul.
  46. Stigmas 2-lobed.
    47. Fruit valves each distinctly 3-veined; petals yellow . . . . 80. *Sisymbrium*, p. 667
    47. Fruit valves each with only midvein distinct; petals purple or white with purple veins.
        48. Lateral sepals not saccate, median sepals not cucullate . . . . 82. *Caulanthus* (in part), p. 677
        48. Lateral sepals saccate, median sepals cucullate . . . . 86. *Hesperidanthus* (in part), p. 689
  46. Stigmas entire.
    49. Gynophores (0.5–)1.5–28 mm; anthers coiled after dehiscence.
        50. Petals yellow, claws and/or filaments pubescent or papillate; sepals spreading to reflexed . . . . 89. *Stanleya* (in part), p. 695
        50. Petals white, lavender, or purple, claws and filaments glabrous; sepals erect or ascending.
            51. Perennials; cauline leaf blades often hastate, margins entire; petal claws dentate or incised . . . . 83. *Chlorocrambe*, p. 685
            51. Biennials; cauline leaf blades linear, lanceolate, or oblong, margins entire, laciniate, or dentate; petal claws entire . . . . 93. *Thelypodium* (in part), p. 728
    49. Gynophores 0–1 mm; anthers not coiled after dehiscence.
      52. Perennials (with caudex).
        53. Seeds usually 16–110 per fruit (as few as 8 in *Hesperidanthus suffrutescens*); petals usually yellow, purple, or lavender, rarely white with purple veins . . . . 86. *Hesperidanthus* (in part), p. 689
        53. Seeds 6–12(–14) per fruit; petals white without purple veins.
            54. Leaf blades rigid, ciliate; fruit valves each with obscure midvein . . . . . 13. *Nevada*, p. 414
            54. Leaf blades not rigid, glabrous; fruit valves each with prominent midvein . . . . 59. *Eutrema* (in part), p. 555
      52. Annuals or, rarely, biennials.
        55. Seeds dark brown or black, longitudinally striate; fruiting pedicels as broad as fruit . . . . 96. *Alliaria*, p. 744
        55. Seeds yellow or reddish brown, sculptured, not striate; fruiting pedicels usually narrower than, rarely as broad as, fruit.
            56. Petals pinnatifid; filaments and petal claws papillate basally . . . . 85. *Dryopetalon* (in part), p. 688
            56. Petals entire; filaments and petal claws not papillate basally.
                57. Seeds biseriate; cotyledons accumbent . . . . 45. *Rorippa*, p. 493
                57. Seeds uniseriate; cotyledons incumbent . . . . 82. *Caulanthus* (in part), p. 677

a. BRASSICACEAE Burnett tribe ALYSSEAE de Candolle, Mém. Mus. Hist. Nat. 7: 231. 1821 (as Alyssineae)

Annuals, biennials, or perennials [shrubs]; eglandular. **Trichomes** short-stalked or sessile and stellate, or distinctly stalked and subdendritic or forked, sometimes mixed with simple ones. **Cauline leaves** petiolate or sessile; blade base not auriculate, margins usually entire. **Racemes** ebracteate [bracteate], often elongated in fruit. **Flowers** actinomorphic; sepals erect to spreading, lateral pair seldom saccate basally; petals white or yellow [orange, pink, or purple], claw present, usually distinct; filaments appendaged, winged (toothed), or unappendaged; pollen 3-colpate. **Fruits** silicles, dehiscent, unsegmented, latiseptate or terete; ovules 2–16(–20) per ovary; style distinct; stigma usually entire, rarely 2-lobed. **Seeds** biseriate or aseriate, rarely uniseriate; cotyledons accumbent or incumbent, rarely oblique.

Genera 13, species ca. 255 (3 genera, 9 species in the flora): North America, Eurasia, n Africa.

1. ALYSSUM Linnaeus, Sp. Pl. 2: 650. 1753; Gen. Pl. ed. 5, 293. 1754 • Madwort [Greek, *a-*, not or without, and *lyssa*, rabies or madness; name used for plants reputed in ancient times as remedy for hydrophobia, cure for madness, and calmative for anger]

Ihsan A. Al-Shehbaz

Annuals or perennials [biennials, subshrubs]; not scapose; trichomes sessile, stellate, with 2–6 minute basal branches (branches as many as 3–25), rays branched or not, sometimes trichomes simple [lepidote]. **Stems** erect, ascending, or decumbent, unbranched or branched. **Leaves** basal and cauline; petiolate or sessile; basal rosulate or not, petiolate or sessile, blade margins entire; cauline petiolate or sessile, blade (base cuneate or attenuate), margins entire. **Racemes** (few- to several-flowered, sometimes corymbose or paniculate). **Fruiting pedicels** ascending, divaricate, or reflexed, slender or stout. **Flowers:** sepals ovate or oblong, lateral pair not saccate; petals yellow or white [rarely pink], suborbicular, spatulate, oblanceolate, linear-oblanceolate, or, obovate (apex obtuse or emarginate); stamens tetradynamous; filaments not winged, uni- or bilaterally winged, appendaged, or toothed; anthers ovate or oblong; nectar glands (4), 1 on each side of lateral stamen, median glands absent; (placentation apical or parietal). **Fruits** sessile, ovate-oblong, obovate, or elliptic [obcordate, rarely globose], usually strongly flattened, latiseptate, rarely inflated; valves each not veined (smooth), pubescent or glabrous; replum (visible), rounded; septum complete, (membranous, translucent, veinless); ovules 1 or 2 [or 4–8] per ovary; stigma capitate. **Seeds** biseriate or aseriate, flattened, winged or not, orbicular or suborbicular to ovoid; seed coat (smooth or minutely reticulate), mucilaginous or not when wetted; cotyledons accumbent or incumbent.

Species ca. 170 (6 in the flora): North America, se Europe, Asia, n Africa.

Alyssum has five introduced and one native species in North America. It is taxonomically difficult and is centered in Turkey and adjacent countries. For the determination of most species, both flowers and mature fruits are needed. *Alyssum* has been split into nine or more segregates; the segregates are based on the presence of staminal appendages, petal color, and number of ovules per ovary. In the absence of thorough molecular studies on *Alyssum* and its immediate relatives, it is more practical to delimit the genus broadly, as done here.

SELECTED REFERENCES Dudley, T. R. 1964. Studies in *Alyssum*: Near Eastern representatives and their allies, I. J. Arnold Arbor. 45: 57–100. Dudley, T. R. 1964b. Synopsis of the genus *Alyssum*. J. Arnold Arbor. 45: 358–373. Dudley, T. R. 1965. Studies in *Alyssum*: Near Eastern representatives and their allies, II. Section *Meniocus* and section *Psilonema*. J. Arnold Arbor. 46: 181–217. Dudley, T. R. 1966. Ornamental madworts (*Alyssum*) and the correct name of the goldentuft alyssum. Arnoldia (Jamaica Plain) 26: 33–48. Dudley, T. R. 1968. *Alyssum* (Cruciferae) introduced in North America. Rhodora 70: 298–300.

1. Perennials; ovules 1 or 2 per ovary.
   2. Cauline leaf blades broadly oblanceolate, obovate-spatulate, or obovate; stems 0.7–1.5(–2) dm; ovules (1 or) 2 per ovary; seeds not winged or margined, 1–1.7 mm .... 5. *Alyssum obovatum*
   2. Cauline leaf blades narrowly oblanceolate to linear; stems (2.5–)3–6(–7) dm; ovules 1 per ovary; seeds broadly winged, 3–3.8 mm .... 6. *Alyssum murale*
1. Annuals; ovules 2 per ovary.
   3. Fruits usually glabrous, rarely sparsely pubescent (when young) .... 2. *Alyssum desertorum*
   3. Fruits pubescent throughout.
      4. Sepals persistent; filaments not appendaged, toothed, or winged (slender) ..... 1. *Alyssum alyssoides*
      4. Sepals caducous; filaments at least some appendaged, toothed, or winged (expanded basally).
         5. Fruits orbicular; fruiting pedicels divaricate .... 3. *Alyssum simplex*
         5. Fruits ovate-oblong; fruiting pedicels ascending to suberect .... 4. *Alyssum szowitsianum*

1. **Alyssum alyssoides** (Linnaeus) Linnaeus, Syst. Nat. ed. 10, 2: 1130. 1759 [I] [W]

Clypeola alyssoides Linnaeus, Sp. Pl. 2: 652. 1753; *Alyssum calycinum* Linnaeus

Annuals; canescent throughout, trichomes appressed, 6–10-rayed, mixed with simple and forked on pedicels and sepals. **Stems** simple or few to several from base, erect, ascending, or decumbent, (unbranched or branched distally), 0.5–3.5(–5) dm. **Cauline leaves** subsessile or (proximal) shortly petiolate; blade usually narrowly oblanceolate to linear, sometimes spatulate or obovate, 3–4(–4.5) cm × (0.5–)1–3.5(–5) mm, base attenuate or cuneate, apex obtuse or acute. **Fruiting pedicels** divaricate or ascending, straight, slender, 2–5(–6) mm, trichomes stellate, with fewer, simple and forked ones. **Flowers:** sepals (persistent) oblong, (1.5–)2–3 × 0.7–1.1 mm, pubescent as pedicels; petals (often persistent) white or pale yellow, usually linear to linear-oblanceolate, rarely obovate, 2–3(–4) × 0.3–0.7(–1) mm, apex emarginate, glabrous or sparsely stellate-pubescent abaxially; filaments (slender) not appendaged, toothed, or winged, 1–1.5 mm; anthers ovate, 0.15–0.2 mm. **Fruits** orbicular, (2–)3–4(–5) mm diam., apex emarginate or truncate; valves uniformly inflated at middle, strongly flattened at margins, sparsely stellate-pubescent; ovules 2 per ovary; style (slender), 0.3–0.6(–1) mm, basally stellate-pubescent or glabrous. **Seeds** oblong to ovoid, compressed, 1.1–2 × 0.7–1.1 mm, margins narrow, ca. 0.1 mm wide. $2n = 32$.

Flowering May–Jul. Roadsides, railways, waste grounds, disturbed sites, grassy areas, fields, sagebrush flats, limestone ledges or bluffs; 0–1800 m; introduced; Alta., B.C., Man., Nfld. and Labr. (Nfld.), Ont., Que., Sask.; Alaska, Ariz., Calif., Colo., Conn., Idaho, Ill., Ind., Iowa, Kans., Maine, Md., Mass., Mich., Mont., N.J., N.Mex., N.Y., Ohio, Oreg., Pa., R.I., S.Dak., Utah, Vt., Va., Wash., W.Va., Wis., Wyo.; Europe; sw Asia; n Africa.

2. **Alyssum desertorum** Stapf, Denkschr. Kaiserl. Akad. Wiss., Wien. Math.-Naturwiss. Kl. 51: 302. 1886 [I] [W]

Annuals; canescent throughout except fruit, trichomes appressed, 8–20-rayed. **Stems** simple or few to several from base, erect, ascending, or decumbent, (0.2–)0.5–1.8(–2.8) dm. **Cauline leaves** subsessile or (proximal) attenuate to petiole-like base (to 0.5 cm); blade linear to oblanceolate-linear, (0.3–)0.5–2.5(–3) cm × (0.5–)1–3(–4) mm (gradually smaller distally), base attenuate, apex acute. **Fruiting pedicels** ascending or subdivaricate, straight, stout, (1–)1.5–3.5(–4.5) mm, trichomes uniformly stellate. **Flowers:** sepals oblong, 1.4–1.8(–2) × 0.4–0.5 mm, stellate-pubescent; petals pale yellow, oblanceolate, 2–2.5 × 0.5–0.6 mm, base attenuate, apex obtuse or retuse, sparsely stellate-pubescent abaxially; filaments: median pairs not toothed, gradually expanded from apex to narrowly winged base, lateral pair with broadly winged appendage apically notched into 2 teeth, 1–1.8(–2) mm; anthers ovate, 0.1–0.2 mm. **Fruits** orbicular, 2.5–4(–4.5) mm diam., apex shallowly emarginate, usually glabrous; valves uniformly inflated at middle, broadly flattened at margins, often glaucous, rarely sparsely pubescent (when young);

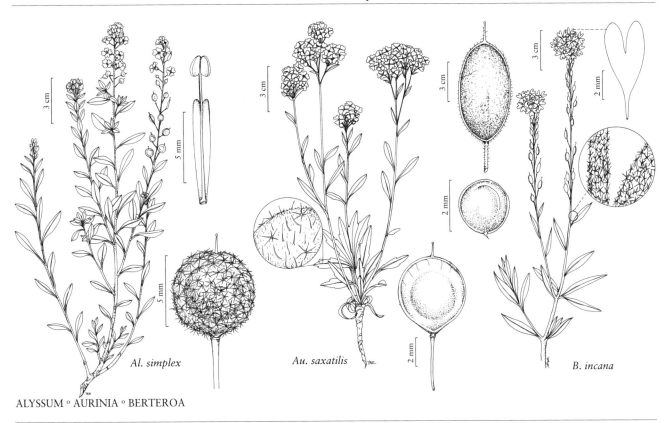

ALYSSUM · AURINIA · BERTEROA

ovules 2 per ovary; style 0.3–0.7 mm, glabrous. **Seeds** ovoid, slightly compressed, 1.2–1.5 × 0.9–1.1 mm, margined or not. $2n = 32$.

Flowering Apr–Jul. Deserts, rocky areas, disturbed sites, roadsides, fields, meadows, sagebrush flats; 800–2000 m; introduced; Alta., B.C., Sask.; Calif., Colo., Idaho, Mont., Nebr., Nev., N.Dak., Oreg., S.Dak., Utah, Wash., Wyo.; Europe; c Asia.

3. Alyssum simplex Rudolphi, J. Bot. (Schrader) 1799(2): 290. 1799 [F] [I]

Alyssum parviflorum M. Bieberstein var. *micranthum* (C. A. Meyer) Dorn

Annuals; canescent throughout, trichomes appressed, coarse, (3–)5–10-rayed, rays sometimes unequal. **Stems** often several from base, usually erect or ascending, rarely decumbent, (unbranched or branched distally), (0.3–)0.7–3(–4) dm. **Cauline leaves** subsessile or (proximal) attenuate to petiolelike base (to 0.4 cm); blade oblanceolate, obovate-spatulate, or elliptic-lanceolate, (0.5–)0.7–2.2 (–3) cm × (1–)2–6(–8) mm (gradually smaller distally), base attenuate, apex acute. **Fruiting pedicels** divaricate, straight, stout, (2–)3–5(–6) mm, trichomes stellate (some with unequal rays). **Flowers:** sepals oblong, 1.7–2.3 × 0.8–1.1 mm, trichomes stellate; petals pale yellow, oblanceolate, (1.8–)2–2.8(–3.2) × 0.6–0.8 mm, base attenuate, apex often emarginate, sparsely stellate-pubescent abaxially or glabrous; filaments: median pairs broadly winged, apically 1- or 2-toothed, lateral pair with broadly winged appendage, apically 2-cleft, 1.4–2 mm; anthers ovate, 0.3–0.4 mm. **Fruits** orbicular, (3.5–)4–6.5(–7) mm diam., apex truncate or shallowly emarginate; valves uniformly inflated at middle, broadly flattened at margins, densely stellate-pubescent (trichomes coarse, rays equal or unequal); ovules 2 per ovary; style 0.7–1.6 mm, basally stellate-pubescent or glabrous. **Seeds** ovoid, 1.6–2 × 1.2–1.5 mm, slightly compressed, margined or not. $2n = 16$.

Flowering Apr–Jun. Sagebrush areas, fields, roadsides, sandy slopes; 1400–2400 m; introduced; Ariz., Calif., Colo., Kans., Mont., Nebr., Nev., N.Mex., Oreg., Utah, Wyo.; Europe; c, sw Asia; n Africa.

Almost all European and North American authors, including T. R. Dudley (1964) and R. C. Rollins (1993), have used the name *Alyssum minus* (Linnaeus) Rothmaler for this species. Rothmaler made his new combination based on the nomen nudum *Clypeola minus* Linnaeus and cited *A. campestre* Linnaeus as a synonym. Therefore, that combination is illegitimate, and the species should be known as *A. simplex*, the earliest valid and legitimate name.

Most of the North American records of *Alyssum strigosum* are based on misidentified plants of *A. simplex*. The two species are similar in nearly all aspects except for trichome morphology of the fruit. In *A. simplex*, all fruit trichomes are stellate and occasionally some trichomes have rays coarser than the others. By contrast, the fruit trichomes in *A. strigosum* are markedly dimorphic and consist of stellate hairs mixed with tuberculate-based, much coarser, forked ones. I have not yet seen any material of *A. strigosum* from North America, and those annotated as such belong to *A. simplex*. T. R. Dudley (1968) and R. C. Rollins (1993) based their record of *A. strigosum* on a single collection, *Davis 108* (DS), made in 1920 from Santa Clara County, California. Although I have not examined that specimen to verify its identity, it appears that the species has not become established in the flora area, and it is not included in the present account.

4. **Alyssum szowitsianum** Fischer & C. A. Meyer, Index Seminum (St. Petersburg) 4: 31. 1837 [I]

Annuals; canescent throughout, trichomes appressed, coarse, 8–16-rayed, rays often subequal. **Stems** often several from base, erect, ascending, or decumbent, (unbranched or few-branched distally), (0.2–)0.4–1.6(–2) dm. **Cauline leaves** subsessile; blade oblanceolate or linear-oblanceolate, (0.5–)0.7–1.5(–2) cm × (1–)2–4(–6) mm (gradually smaller distally), base attenuate, apex subacute. **Fruiting pedicels** ascending to suberect, straight, stout, 1–3(–4) mm, trichomes uniformly stellate. **Flowers:** sepals oblong, 1.2–1.5 × 0.5–0.6 mm, stellate-pubescent (trichomes with longer rays and appearing simple near apex); petals pale yellow, spatulate, 1.7–2 × 0.4–0.5 mm, base abruptly narrowed to claw, apex deeply emarginate, glabrous; filaments: median pairs not toothed, slightly expanded basally, lateral pair with basal appendage to 0.5 × 0.1 mm, often apically 2-toothed, 1.4–1.7 mm; anthers ovate, 0.1–0.2 mm. **Fruits** ovate-oblong, (3–)3.5–5 × (2.5–)3–4 mm, apex obtuse; valves uniformly inflated at middle, broadly flattened at margins, densely stellate-pubescent (rays equal); ovules 2 per ovary; style (slender), 0.3–0.5 mm, basally stellate-pubescent. **Seeds** ovoid, compressed, 1.6–1.9 × 1.2–1.6 mm, margins winglike, 0.1–0.3 mm wide. $2n = 16$.

Flowering May. Rocky soil; 1500–2000 m; introduced; Utah; sw Asia.

Alyssum szowitsianum is known in the flora area only from Salt Lake County.

5. **Alyssum obovatum** (C. A. Meyer) Turczaninow, Bull. Soc. Imp. Naturalistes Moscou 10: 57. 1837

Odontarrhena obovata C. A. Meyer in C. F. von Ledebour, Fl. Altaica 3: 61. 1831; *Alyssum americanum* Greene; *A. biovulatum* N. Busch; *A. fallax* Nyárády

Perennials; (cespitose, caudex often woody); canescent throughout, trichomes 10–25-rayed. **Stems** often several from base (often with sterile shoots), usually erect or ascending, 0.7–1.5(–2) dm. **Cauline leaves** subsessile; blade broadly oblanceolate, obovate-spatulate, or obovate, 0.6–1.4(–1.7) cm × (1–)2–6 mm (gradually smaller distally), base attenuate, apex rounded, obtuse, or subacute. (**Racemes** corymbose, simple, or in panicles terminating each stem.) **Fruiting pedicels** divaricate, straight, slender, (2–)3–7(–9) mm, trichomes uniformly stellate. **Flowers:** sepals oblong, 1.5–2 × 0.7–1 mm, stellate-pubescent; petals yellow, broadly spatulate, 2.5–3.5(–4) × 1–1.5 mm, base attenuate, apex often obtuse or rounded, glabrous abaxially; filaments: median pairs unilaterally broadly winged, apically 1-toothed, lateral pair with lanceolate or narrowly oblong basal appendage, apically obtuse or subacute, 1.5–2 mm; anthers oblong, 0.3–0.4 mm. **Fruits** broadly elliptic or obovate, 3–4.5(–5) × 2.5–3.5 mm, apex obtuse to rounded; valves inflated at middle or on 1 side, flattened at margins, densely stellate-pubescent; ovules (1 or) 2 per ovary; style (slender), 1.5–2 mm, glabrous. **Seeds** ovoid, compressed, 1–1.7 × 0.8–1.2 mm, not winged or margined. $2n = 16, 32$.

Flowering May–Jul. Mountain slopes, cliffs, gravel, rocky places; 500–1500 m; Yukon; Alaska; c, e Asia.

6. **Alyssum murale** Waldstein & Kitaibel, Descr. Icon. Pl. Hung. 1: 5. 1799 [I]

Perennials; (cespitose, sometimes caudex woody); canescent or not, trichomes stellate, 5–20-rayed. **Stems** often several from caudex (sterile shoots absent or few), usually erect or ascending, (2.5–)3–6(–7) dm. **Cauline leaves** subsessile; blade narrowly oblanceolate to linear, (0.5–)0.7–2.5(–3) cm × 1–5(–8) mm (gradually smaller distally), base cuneate to attenuate, apex obtuse to subacute, (surfaces often more grayish abaxially than adaxially). (**Racemes** corymbose, in panicles terminating each stem.) **Fruiting pedicels** divaricate-ascending, straight or slightly curved distally, slender, (2–)2.5–6(–8) mm, trichomes uniformly stellate. **Flowers:** sepals oblong, 1.2–2 × 0.4–0.7 mm, stellate-pubescent; petals yellow, spatulate, 2.5–3.5 × 0.5–1 mm, base attenuate, apex often obtuse or rounded, glabrous abaxially; filaments: median pairs broadly winged, apically 1-toothed, lateral pair with lanceolate or narrowly oblong basal appendage, apically subacute, 1.5–2.5 mm; anthers oblong, 0.2–0.3 mm. **Fruits** broadly elliptic to orbicular, 3.5–5 × 2.5–5 mm, apex obtuse to rounded; valves not inflated, flattened and slightly undulate, stellate-pubescent; ovules 1 per ovary; style (slender) 1–2 mm, sparsely pubescent. **Seeds** suborbicular to broadly ovoid, compressed, 3–3.8 × 2–3.2 mm, margins winged, wing 0.5–0.9 mm wide. $2n = 16, 32$.

Flowering May–Jul. Waste areas, roadsides; 1500–1800 m; introduced; Alta., B.C., Ont., Que.; Colo., Oreg., Utah; Europe; sw Asia.

2. AURINIA Desvaux, J. Bot. Agric. 3: 162. 1815 • Goldentuft, rock-alyssum [Latin *aurum*, gold, and *-inia*, colored, alluding to flower]

Ihsan A. Al-Shehbaz

Perennials [biennials, subshrubs]; (caudex woody); not scapose; pubescent, trichomes minutely stalked to sessile, stellate, 6–10-rayed [lepidote]. **Stems** erect or ascending, often (paniculately) branched distally. **Leaves** basal and cauline; petiolate; basal rosulate, petiolate (petioles deeply grooved), blade margins repand, sinuate, dentate, or pinnatifid; cauline petiolate, blade (much smaller than basal), margins entire [dentate, sinuate]. **Racemes** (corymbose, several-flowered, buds globose), slightly or considerably elongated in fruit. **Fruiting pedicels** divaricate or ascending, slender. **Flowers:** sepals spreading, ovate, lateral pair not saccate basally; petals yellow [white], obovate to spatulate, claw slightly differentiated from blade, (apex emarginate, [2-fid or obtuse]); stamens tetradynamous; filaments dilated (winged or minutely appendaged) basally; anthers ovate; nectar glands lateral, 1 on each side of lateral stamen. **Fruits** sessile, ellipsoid to obovoid, or broadly obovate to orbicular [globose, elliptic], not torulose, terete or latiseptate; valves each not veined, glabrous [pubescent]; replum rounded; septum usually complete, rarely perforate; ovules 4–8(–16) per ovary; stigma capitate, usually 2-lobed, rarely subentire. **Seeds** uniseriate or biseriate, flattened, winged [not winged], suborbicular [orbicular or elliptic]; seed coat not mucilaginous when wetted; cotyledons incumbent or oblique. $x = 8$.

Species 10 (2 in the flora): introduced; c, se Europe (Caucasus), sw Asia.

SELECTED REFERENCE Dudley, T. R. 1966. Ornamental madworts (*Alyssum*) and the correct name of the goldentuft alyssum. Arnoldia (Jamaica Plain) 26: 33–48.

1. Fruits ellipsoid to obovoid, inflated; seeds 1.5–1.8 mm diam.; wing 0.1–0.3 mm wide .... 1. *Aurinia petraea*
1. Fruits broadly obovate to orbicular, flattened; seeds 2–3 mm diam.; wing 0.3–1.1 mm wide. .... 2. *Aurinia saxatilis*

1. Aurinia petraea (Arduino) Schur, Enum. Pl. Transsilv., 61. 1866

Alyssum petraeum Arduino, Animadv. Bot. Spec. Alt., 30. 1764

Plants slightly woody at base; finely stellate-pubescent. **Stems:** usually several, 1.5–6 dm. **Basal leaves:** petiole 0.3–2.5(–3.5) cm; blade obovate-oblong, 2.5–8 cm × 3–10 mm, base cuneate to attenuate, margins entire, sinuate, or pinnatifid, apex acute or obtuse, surfaces densely pubescent. **Cauline leaves:** blade oblanceolate to linear. **Racemes** considerably elongated in fruit. **Fruiting pedicels** 3–8 mm. **Flowers:** sepals yellowish, 1.5–2 × 0.5–1 mm, scarious at margins, finely pubescent; petals 3–4.5 × 1.2–2 mm, attenuate to clawlike base; filaments with basal wing or appendage to 0.5 mm, 1.2–2 mm; anthers 0.3–0.5 mm. **Fruits** divaricate-ascending, ellipsoid to obovoid, inflated at middle, slightly flattened at margin, 3–6 × 2–4 mm; style slender. **Seeds** winged throughout, 1.5–1.8 mm diam.; wing 0.1–0.3 mm wide. $2n = 16$.

Flowering May–Jul. Cliffs, gorges, rocky slopes; introduced; N.Y.; s Europe.

Aurinia petraea is cultivated as an ornamental and is known to be naturalized in the gorge, Oneida (near Trenton Falls) and Ulster counties.

2. Aurinia saxatilis (Linnaeus) Desvaux, J. Bot. Agric. 3: 162. 1815 • Basket-of-gold, gold-dust, rock-madwort F I

Alyssum saxatile Linnaeus, Sp. Pl. 2: 650. 1753; *A. arduini* Fritsch

Plants often woody at base; finely stellate-pubescent. **Stems:** usually several, 1–5 dm. **Basal leaves:** petiole 0.5–3(–4) cm, blade spatulate to obovate or lanceolate, (2.5–)4–8(–12) × 0.5–1.5(–2.5) cm, base cuneate to attenuate, margins entire, repand, or sinuate-dentate, apex acute or obtuse, surfaces densely pubescent. **Cauline leaves:** blade oblanceolate to linear. **Racemes** slightly elongated in fruit. **Fruiting pedicels** (3–)4.5–10 (–13) mm. **Flowers:** sepals yellowish, scarious at margins, finely pubescent, 1.5–2.3 × 0.5–1 mm; petals 3–6 × 1–2.5 mm, attenuate to clawlike base; filaments with basal wing or appendage to 0.5 mm, 1–2 mm; anthers 0.3–0.5 mm. **Fruits** divaricate-ascending, broadly obovate to orbicular, flattened, strongly latiseptate, 3.5–9 mm in diam.; style 0.5–1.5(–2.5) mm. **Seeds** winged throughout, 2–3 mm diam.; wing 0.3–1.1 mm wide. $2n = 16$.

Flowering May–Jul. Ledges, cliffs, rocky slopes; introduced; Que.; Calif., Mass., Mich., Miss., N.H., Ohio, Vt.; c, se Europe; Asia (Turkey).

Aurinia saxatilis is widely cultivated as an ornamental and likely naturalized in other states. It is highly variable in its native range (Turkey, southern and central Europe) and has been divided into three subspecies, of which subsp. *saxatilis* is naturalized in the flora area.

3. BERTEROA de Candolle, Mém. Mus. Hist. Nat. 7: 232: 290. 1821 • Hoary-alyssum [For Carlo Giuseppe Bertero, 1789–1831, Italian physician and botanist who settled in Chile] I

Ihsan A. Al-Shehbaz

Annuals or biennials [perennials]; not scapose; pubescent, trichomes stellate, mixed with simple ones. **Stems** erect [ascending], usually branched distally. **Leaves** basal and cauline; petiolate or sessile; basal not rosulate, petiolate, blade margins entire or repand [dentate, sinuate]; cauline (middle and distal) sessile. **Racemes** (corymbose, several-flowered, dense), considerably elongated in fruit. **Fruiting pedicels** erect or divaricate, slender. **Flowers:** sepals erect-ascending [suberect, spreading], oblong, lateral pair not saccate basally; petals usually white, rarely yellow, obcordate, apex deeply 2-fid; stamens tetradynamous; filaments: median pair flattened basally, unappendaged, [laterally 1-toothed], lateral pair with basal toothlike appendage; anthers oblong, (apex obtuse); nectar glands (4), lateral, 1 on each side of lateral stamen. **Fruits** sessile, oblong, or elliptic [ovate, obovate, or orbicular], smooth, slightly inflated [or not inflated], latiseptate; valves each not veined or with obscure midvein, stellate-hairy [glabrous]; replum rounded; septum complete, (membranous); ovules 4–16 per ovary; stigma capitate, obscurely 2-lobed. **Seeds** biseriate, flattened [plump], margined [winged or not], lenticular or ovoid-lenticular [suborbicular]; seed coat (minutely reticulate), not mucilaginous when wetted; cotyledons accumbent. $x = 8$.

Species 5 (1 in the flora): introduced; Europe, Asia.

Berteroa mutabilis (Ventenat) de Candolle, native to northeastern Italy, the Balkan Peninsula, and northern Turkey, is known in North America only from a handful of collections almost all made more than a century ago as garden escapes in Kansas and Massachusetts. Although it was included by R. C. Rollins (1993), who indicated that it had not been collected for 60 years, the species apparently did not become naturalized in North America and, therefore, is not included here. From *B. incana*, *B. mutabilis* is easily distinguished by having winged instead of margined seeds, and flat and glabrous versus inflated and pubescent fruits. As indicated by I. A. Al-Shehbaz (1987), the record of *B. obliqua* (Smith) de Candolle from the Catskill region, New York, was based on misidentified plants of *B. incana*.

1. **Berteroa incana** (Linnaeus) de Candolle, Syst. Nat. 2: 291. 1821 [F] [I] [W]

Alyssum incanum Linnaeus, Sp. Pl. 2: 650. 1753

Plants densely pubescent, trichomes appressed, stellate mixed with simple ones. **Stems** simple or few from base, (2–)3–8(–11) dm. **Basal leaves** (withered by flowering): blade oblanceolate, (2.5–)3.5–8(–10) cm, base cuneate or attenuate, apex obtuse. **Cauline leaves**: blade apex acute or obtuse. **Fruiting pedicels** appressed to rachis, (4–)5–9(–12) mm. **Flowers**: sepals 2–2.5 mm; petals narrowly obcordate, (4–)5–6.5(–8) mm (lobes oblong, (1–)1.5–3 × 0.5–1.5 mm); filaments white, median pair 2–3.3 mm, lateral pair 0.3–1 mm; anthers 0.5–1 mm. **Fruits** (4–)5–8.5(–10) × (2–)2.5–4 mm; valves obtuse at both ends, trichomes sometimes with unequal rays; style 1–4 mm, sparsely pubescent basally. **Seeds** slightly flattened, narrowly margined, 1–2.3 mm diam. $2n = 16$.

Flowering May–Sep. Flood plains, meadows, waste places, railroad embankments, woodlands, grasslands, roadsides, fields, stream banks, pastures, hillsides, forest floor; 0–2800 m; introduced; B.C., Man., N.B., Ont., Que., Sask.; Ariz., Calif., Colo., Conn., D.C., Idaho, Ill., Ind., Iowa, Kans., Maine, Mass., Mich., Minn., Mo., Mont., Nebr., Nev., N.H., N.J., N.Y., N.Dak., Ohio, Oreg., Pa., R.I., S.Dak., Utah, Vt., Va., Wash., W.Va., Wis., Wyo.; Europe; Asia.

b. BRASSICACEAE Burnett tribe ANCHONIEAE de Candolle, Mém. Mus. Hist. Nat. 7: 242. 1821 [I]

Annuals, biennials, or perennials [subshrubs]; usually glandular (glands multicellular on multiseriate stalks). **Trichomes** stalked [sessile], dendritic and forked, sometimes mixed with simple ones. **Cauline leaves** petiolate or sessile [subsessile]; blade base not auriculate, margins usually entire, dentate, or pinnatisect, rarely subentire. **Racemes** ebracteate, often elongated in fruit. **Flowers** actinomorphic; sepals erect, lateral pair often saccate basally; petals white, yellow, pink, purple, violet, or brown [yellowish green or orange], claw present, usually distinct; filaments unappendaged, not winged; pollen 3-colpate. **Fruits** siliques, dehiscent, unsegmented, latiseptate, terete, or 4-angled; ovules [4–]15–60 per ovary; style distinct or obsolete; stigma 2-lobed (lobes often connivent). **Seeds** uniseriate; cotyledons accumbent [incumbent].

Genera 6, species 68 (1 genus, 2 species in the flora): introduced; Eurasia, Africa, Atlantic Islands (Canary Islands).

4. MATTHIOLA W. T. Aiton in W. Aiton and W. T. Aiton, Hortus Kew. 4: 119. 1812 (as Mathiola), name and orthography conserved • Stock [For Pietro Andrea Matthioli, 1500–1577, Italian artist and botanist] [I]

Ihsan A. Al-Shehbaz

Lonchophora Durieu

Annuals, biennials, or perennials [subshrubs]; (base sometimes woody); not scapose; (glandular or not), pubescent, trichomes stalked, stellate or dendritic, or, rarely, forked or simple. **Stems** erect to ascending or decumbent, unbranched or branched. **Leaves** basal and cauline; petiolate or sessile; basal rosulate or not, petiolate, blade margins entire, dentate, or pinnatisect. **Racemes** (corymbose, several-flowered). **Fruiting pedicels** divaricate or ascending [erect], slender or stout. **Flowers:** sepals (connivent), oblong to lanceolate or linear, lateral pair strongly saccate basally; petals yellow, white, pink, purple, violet, or brown [yellowish green], broadly obovate, spatulate, oblong, or linear (much longer than sepals, flat or circinately rolled inwards, margins crisped or not), claw differentiated from blade, (apex obtuse, rounded, subacute, or emarginate); stamens strongly tetradynamous; filaments not dilated basally; anthers linear [oblong], (apex obtuse); nectar glands (4 and) lateral, 1 on each side of lateral stamen, or (2 and) semiannular, intrastaminal. **Fruits** sessile, linear, subtorulose [torulose], often straight, terete or latiseptate; valves each with prominent midvein, pubescent; replum rounded; septum complete, (often opaque, veinless); style obsolete or distinct; stigma conical, 2-lobed (lobes prominent, distinct or connate, decurrent, with or without 2 or 3 hornlike appendages). **Seeds** flattened, narrowly winged or not winged, oblong, ovate, suborbicular, or orbicular; seed coat (minutely reticulate), not mucilaginous when wetted; cotyledons accumbent. $x = 7$.

Species ca. 50 (2 in the flora); introduced; Europe, Asia, n, e Africa, Atlantic Islands (Canary Islands); introduced also in Australia.

1. Fruits compressed, 3–6 mm wide, without stigmatic horns; cauline leaf blade margins entire or repand, rarely sinuate; fruiting pedicels (6–)10–20(–25) mm, thinner than fruit; petals 7–15 mm wide. . . . . 1. *Matthiola incana*
1. Fruits terete, 1–2 mm wide, with 2 stigmatic horns; cauline leaf blade margins dentate to pinnatifid, rarely entire; fruiting pedicels (0.5–)1–2(–3) mm, nearly as thick as fruit; petals 2–5 mm wide. . . . . 2. *Matthiola longipetala*

1. **Matthiola incana** (Linnaeus) W. T. Aiton in W. Aiton and W. T. Aiton, Hortus Kew. 4: 119. 1812 (as Mathiola) [I] [W]

Cheiranthus incanus Linnaeus, Sp. Pl. 2: 662. 1753

Biennials or perennials, rarely annuals; usually densely tomentose. **Stems** erect, (1–)2.5–6(–9) dm, (unbranched or branched distally), often tomentose. **Basal leaves** often in vegetative rosettes. **Cauline leaves** shortly petiolate or sessile; blade linear-oblanceolate, narrowly oblong, or lanceolate, (2.5–)4–16(–22) cm × (5–)8–18(–25) mm (smaller distally), base attenuate to cuneate, margins usually entire or repand, rarely sinuate. **Fruiting pedicels** ascending, straight or slightly curved, (6–)10–20(–25) mm, thinner than fruit. **Flowers:** sepals linear-lanceolate to narrowly oblong, 10–15 × 2–3 mm; petals purple, violet, pink, or white, obovate to ovate, 20–30 × 7–15 mm, claw 10–17 mm (margin not crisped), apex rounded or emarginate; filaments 5–8 mm; anthers 3–4 mm. **Fruits** divaricate-ascending to suberect, latiseptate, (4–)6–12(–15) cm × (3–)4–6 mm; valves densely pubescent; style 1–5 mm; stigma without horns. **Seeds** orbicular or nearly so, 2.5–3.2 mm diam.; wing 0.2–0.5 mm. $2n = 14$.

Flowering Mar–Jun. Ocean cliffs and bluffs, sandy areas near beaches, roadsides, abandoned gardens; 0–300 m; introduced; Calif., Tex.; Europe; introduced also elsewhere in the New World, Australia.

Matthiola incana is widely cultivated worldwide for its attractive, highly scented flowers.

2. **Matthiola longipetala** (Ventenat) de Candolle, Syst. Nat. 2: 174. 1821 [F] [I]

Cheiranthus longipetalus Ventenat, Descr. Pl. Nouv., plate 93. 1802; *C. bicornis* Sibthorp & Smith; *Matthiola bicornis* (Sibthorp & Smith) de Candolle; *M. longipetala* subsp. *bicornis* (Sibthorp & Smith) P. W. Ball

Annuals; sparsely to moderately pubescent, (glandular papillae present or not). **Stems** erect or ascending to decumbent, (1–)1.5–5(–6) dm, pubescent, (glandular or not). **Basal leaves** not forming vegetative rosettes. **Cauline leaves:** petiole to 2 cm or (distal) sessile; blade linear-lanceolate to lanceolate or oblanceolate, (2–)3.5–8(–11) cm × 2–10(–20) mm (smaller distally), base attenuate to cuneate, margins usually pinnatisect to sinuate or dentate, rarely entire or subentire. **Fruiting pedicels** divaricate to divaricate-ascending, straight, (0.5–)

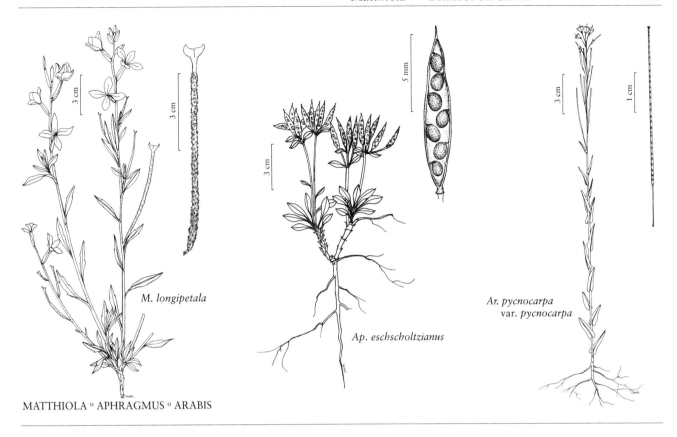

MATTHIOLA ∘ APHRAGMUS ∘ ARABIS

1–2(–3) mm, nearly as thick as fruit. **Flowers:** sepals narrowly oblong, (7–)8–11(–12.5) × 1–2 mm; petals usually purple, pink, yellow, or brown, rarely white, oblong to linear-lanceolate, (15–)18–23(–27) × 2–4(–5) mm, claw 7–13 mm, (margin crisped), apex subacute to obtuse; filaments 4–6 mm; anthers 2.5–3.5 mm. **Fruits** ascending to divaricate or, rarely, descending, straight, terete, (2.5–)4–8.5(–10) cm × 1–2 mm; valves pubescent, (often glandular); style obsolete to 3 mm; stigma horns 2, straight or curved upward, sometimes reflexed, 2–8 (–12) mm. **Seeds** suborbicular to oblong or ovate, 1–2 × 0.8–1.2 mm; wing 0.1–0.2 mm. $2n = 14$.

Flowering Mar–Jun. Roadsides, disturbed areas, waste grounds, fields; 700–1300 m; introduced; Alta., Ont., Sask.; Ariz., Calif., Tex.; Europe; c, w Asia; n Africa; introduced also in Australia.

Matthiola longipetala is sporadically naturalized in North America and Australia.

c. BRASSICACEAE Burnett tribe APHRAGMEAE D. A. German & Al-Shehbaz, Harvard Pap. Bot. 13: 168. 2008

Perennials [annuals]; eglandular. **Trichomes** simple or forked, (minute). **Cauline leaves** petiolate or sessile; blade base not auriculate, (cuneate or attenuate), margins entire. **Racemes** bracteate, sometimes elongated in fruit. **Flowers** actinomorphic; sepals erect or ascending, lateral pair seldom saccate basally; petals white or lavender [pink, blue, or purple], claw present [absent], distinct; filaments unappendaged, not winged; pollen 3-colpate. **Fruits** silicles or siliques, dehiscent, unsegmented, latiseptate [terete]; ovules 6–8[–32] per ovary; style obsolete or distinct; stigma entire. **Seeds** biseriate [uniseriate]; cotyledons incumbent.

Genus 1, species 11 (1 in the flora): nw North America, Asia (Himalaya, Russian Far East).

5. **APHRAGMUS** Andrzejowski ex de Candolle in A. P. de Candolle and A. L. P. P. de Candolle, Prodr. 1: 209. 1824 • [Greek *a-*, not or without, and *phragma*, septum, alluding to its lack in fruit of some species]

Ihsan A. Al-Shehbaz

Oreas Chamisso & Schlechtendal; *Orobium* Reichenbach; *Staintoniella* H. Hara

Plants not scapose; (sometimes rhizomatous, caudex thick, covered with persistent petiolar remains); glabrous or pubescent, trichomes simple or forked, (less than 0.1 mm). **Stems** erect or ascending, branched basally. **Leaves** basal and cauline; petiolate or sessile; basal rosulate, petiolate, blade margins entire; cauline petiolate or sessile, blade margins entire. **Racemes** (few- to several-flowered), bracteate throughout or basally. **Fruiting pedicels** [erect] ascending, divaricate, sometimes recurved, slender. **Flowers:** sepals (sometimes persistent), oblong; petals broadly obovate or spatulate, claw subequaling sepal, (apex rounded [obtuse]); stamens slightly tetradynamous; filaments dilated or not basally; anthers obtuse; nectar glands confluent, subtending bases of stamens. **Fruits** silicles or siliques, sessile or subsessile, oblong, elliptic, or lanceolate [ovate, linear]; valves each with obscure or distinct midvein, (smooth), glabrous; replum flattened basally; septum absent [complete and membranous]; style obsolete or distinct (to 2 mm); stigma capitate. **Seeds** plump, not winged, oblong or ovoid (on filiform funicles often longer than seeds); seed coat (minutely reticulate), not mucilaginous when wetted; cotyledons incumbent. $x = 7$.

Species 11 (1 in the flora): nw North America, Asia (Himalaya, Russian Far East).

SELECTED REFERENCES Al-Shehbaz, I. A. 2003. *Aphragmus bouffordii*, a new species from Tibet and a synopsis of *Aphragmus* (Brassicaceae). Harvard Pap. Bot. 8: 25–27. Ebel, A. L. 1998. Notes on genus *Aphragmus* Andrz. (Brassicaceae). Turczaninowia 1(4): 20–27.

1. **Aphragmus eschscholtzianus** Andrzejowski ex de Candolle in A. P. de Candolle and A. L. P. P. de Candolle, Prodr. 1: 210. 1824 [F]

Eutrema eschscholtzianum (Andrzejowski ex de Candolle) B. L. Robinson

Stems 2–6 cm; glabrous or sparsely puberulent, trichomes simple. **Basal leaves:** petiole 0.5–3 cm (base persistent, papery, broadly expanded, narrowly triangular, 0.2–0.4 cm wide); blade (subfleshy), spatulate, oblanceolate, or ovate, 0.5–1.2 cm × 2–4 mm, base cuneate to attenuate, apex obtuse, surfaces glabrous or puberulent. **Cauline leaves** (bracts), sessile; blade similar to basal, narrower, smaller distally. **Racemes** subumbellate. **Fruiting pedicels** 1.5–5 mm, glabrous or sparsely puberulent adaxially. **Flowers:** sepals often purplish, 1.5–2 × 0.7–1 mm, glabrous; petals purple with darker veins, 2–3 × 1–2 mm; filaments 1.5–2 mm; anthers 0.2–0.4 mm; gynophore 0–0.7 mm. **Fruits** compressed, 5–15 × 2–3 mm; style 0.2–1 mm. **Seeds** light brown, 1.2–1.7 × 0.8–1 mm; funicles 2–3 mm. $2n = 14$.

Flowering Jun–Jul. Limestone slopes, subalpine valleys, wet cliffs, ridges, fine gravel saturated by snowmelt; 0–2100 m; Yukon; Alaska; Asia (Russian Far East [Chukchi Peninsula]).

d. **BRASSICACEAE** Burnett tribe **ARABIDEAE** de Candolle, Mém. Mus. Hist. Nat. 7: 229. 1821

Annuals, biennials, or perennials [subshrubs]; eglandular. **Trichomes** stalked or sessile, usually stellate, dendritic, cruciform, or forked, sometimes mixed with simple ones, rarely malpighiaceous. **Cauline leaves** petiolate or sessile; blade base auriculate or not, margins entire or dentate. **Racemes** usually ebracteate, often elongated in fruit. **Flowers** actinomorphic;

sepals erect, ascending, or spreading, lateral pair seldom saccate basally; petals white, yellow, orange, pink, or purple, claw usually present, usually distinct; filaments unappendaged, not winged; pollen 3-colpate. **Fruits** silicles or siliques, usually dehiscent, very rarely indehiscent, unsegmented, latiseptate or terete; ovules 2–70(–88)[–110+] per ovary; style usually distinct, sometimes obsolete; stigma usually entire, rarely 2-lobed. **Seeds** biseriate or uniseriate [rarely aseriate]; cotyledons accumbent.

Genera 8, species ca. 460 (4 genera, 139 species in the flora): North America, Mexico, South America, Europe, Asia, n Africa.

6. ARABIS Linnaeus, Sp. Pl. 2: 664. 1753; Gen. Pl. ed. 5, 298. 1754 • Rockcress [Latin *Arabia*]

Ihsan A. Al-Shehbaz

Annuals, biennials, or perennials; (sometimes stoloniferous with vegetative rosettes, or caudex simple or branched); not scapose; often pubescent or hirsute, sometimes glabrous or glabrate, trichomes stalked, stellate, sometimes mixed with fewer, simple or forked ones. **Stems** erect, ascending, or decumbent, unbranched or branched distally. **Leaves** basal and cauline; petiolate or sessile; basal rosulate, petiolate or sessile, blade margins usually entire or dentate to denticulate, rarely lyrate-pinnatifid; cauline usually sessile, rarely shortly petiolate, blade (base often auriculate, sagittate, or amplexicaul), margins entire or dentate. **Racemes** (sometimes paniculate, usually simple, sometimes branched). **Fruiting pedicels** erect, ascending, or divaricate, (not reflexed or secund), slender. **Flowers:** sepals erect or ascending, ovate or oblong, lateral pair saccate or not basally, (margins membranous); petals white, pink, or purple, usually spatulate, oblong, or oblanceolate, rarely obovate, claw differentiated from blade, (shorter than sepals, apex obtuse or rounded); stamens tetradynamous; filaments usually not dilated basally; anthers ovate, oblong, or linear, (apex obtuse); nectar glands confluent, subtending bases of stamens, lateral glands semiannular or annular, median glands rarely absent, (sometimes toothlike and distinct). **Fruits** siliques, usually sessile, rarely shortly stipitate, linear, smooth or torulose, (usually straight, sometimes slightly curved), flattened, latiseptate; valves (papery), each with obscure or prominent midvein, glabrous; replum (visible), rounded; septum complete, (membranous, translucent, veinless); ovules 10–86[–110] per ovary; style obsolete or distinct; stigma capitate, (sometimes slightly 2-lobed). **Seeds** uniseriate, flattened, winged or margined, oblong or orbicular; seed coat (smooth or minutely reticulate), not mucilaginous when wetted; cotyledons accumbent. $x = 8$.

Species ca. 70 (15 in the flora): North America, Europe, Asia, n, alpine, c, e Africa.

As treated by most North American authors (e.g., M. Hopkins 1937; R. C. Rollins 1941, 1993; G. A. Mulligan 1996), *Arabis* was so broadly delimited that it included species presently assigned to six genera in five tribes (I. A. Al-Shehbaz et al. 2006). These are: *Arabidopsis* and *Turritis* (Camelineae), *Arabis* (Arabideae), *Boechera* (Boechereae), *Pennellia* (Halimolobeae), and *Streptanthus* (Thelypodieae) (see Al-Shehbaz 2003b; M. D. Windham and Al-Shehbaz 2006, 2007).

R. C. Rollins (1993) recognized 80 species of *Arabis* in North America, of which 24 were divided into 40 varieties. By contrast, G. A. Mulligan (1996) recognized 30 species and six varieties in Canada alone. The combination of characteristics that they used to circumscribe the genus (e.g., linear and latiseptate fruits, accumbent cotyledons, and often branched trichomes) evolved repeatedly in Brassicaceae and cannot be relied on in the delimitation of genera (O. Appel and I. A. Al-Shehbaz 2003). The vast majority of North American species have been transferred to *Boechera* ($x = 7$), a genus distinct morphologically, cytologically, and molecularly (Al-Shehbaz 2003b; M. D. Windham and Al-Shehbaz 2006, 2007, 2007b). As currently circumscribed, *Arabis* ($x = 8$) is a primarily Eurasian genus with only 15 species in North America.

SELECTED REFERENCES Hopkins, M. 1937. *Arabis* in eastern and central North America. Rhodora 39: 63–98, 106–148, 155–186. Mulligan, G. A. 1996. Synopsis of the genus *Arabis* (Brassicaceae) in Canada, Alaska and Greenland. Rhodora 97: 109–163. Rollins, R. C. 1936. The genus *Arabis* in the Pacific Northwest. Res. Stud. State Coll. Wash. 4: 1–52. 1936. Rollins, R. C. 1941. A monographic study of *Arabis* in western North America. Rhodora 43: 289–325, 348–411, 425–485.

1. Cauline leaves: blade bases auriculate, subcordate, amplexicaul or subamplexicaul.
   2. Perennials, stoloniferous; lateral sepals conspicuously saccate basally; fruits ascending to spreading.
      3. Petals 5–8(–9) × 2–3.5 mm; fruits 1.7–3.5(–4) cm; basal leaves: blade trichomes stellate with simple rays; sepals 2.5–4(–4.7) mm . . . . 1. *Arabis alpina*
      3. Petals 10–19 × 4–8 mm; fruits (3–)4–7 cm; basal leaves: blade trichomes stellate with some rays branched; sepals 4.5–8 mm . . . . 2. *Arabis caucasica*
   2. Biennials or perennials, not stoloniferous; lateral sepals not or slightly saccate basally, or not conspicuously so; fruits erect, erect-ascending, or nearly so (except *A. patens*).
      4. Fruits divaricate to erect-ascending (not appressed to rachises); ovules 16–28 per ovary . . . . 6. *Arabis patens*
      4. Fruits usually erect, rarely erect-ascending (often appressed to rachises); ovules 38–86 per ovary.
         5. Fruits 0.7–0.8 mm wide; petals 6–9 mm; basal leaf blades: abaxial surface with predominantly subsessile, stellate trichomes; Alabama, Georgia . . . . . . 5. *Arabis georgiana*
         5. Fruits 0.8–1.8(–2) mm wide; petals 3.5–5(–5.5) mm (longer in *A. eschscholtziana*); basal leaf blades: abaxial surface usually with mix of simple, forked, and stalked, stellate trichomes, rarely subglabrate; not Alabama, not Georgia.
            6. Petals 3.5–5(–5.5) mm; lateral sepals not saccate basally; fruits 0.8–1 (–1.2)mm wide; cauline leaves (7–)10–45(–61), overlapping or not . . . . 3. *Arabis pycnocarpa*
            6. Petals (5.5–)6.5–9(–10) mm; lateral sepals saccate basally; fruits 1.2–1.8(–2) mm wide; cauline leaves (2–)4–12(–18), not overlapping or rarely so. . . . . .4. *Arabis eschscholtziana*
1. Cauline leaves: blade bases not auriculate or amplexicaul, sometimes cuneate.
   7. Petals white, 3.5–11 mm.
      8. Trichomes of basal leaf blades short-stalked, stellate . . . . 7. *Arabis crucisetosa*
      8. Trichomes of basal leaf blades simple and forked, or only simple.
         9. Fruits and fruiting pedicels erect, (sub)appressed to rachises; petals 3.5–4 mm . . . . .9. *Arabis olympica*
         9. Fruits and fruiting pedicels ascending to divaricate or suberect, not appressed to rachises; petals (5–)6–11 mm.
            10. Fruits 1.7–2.2 mm wide; seeds winged, (1.5–)1.8–2.5(–3) mm; petals 7–11 . . . . 8. *Arabis furcata*
            10. Fruits 0.8–1.2(–1.5) mm wide; seeds not winged, 0.6–1.2(–1.5) mm; petals (5–)6–8 mm. . . . . .10. *Arabis nuttallii*
   7. Petals purple or pink, 8–18(–20) mm.
      11. Basal leaf blades: surfaces with short-stalked, stellate trichomes. . . . . 11. *Arabis modesta*
      11. Basal leaf blades: surfaces with simple, forked, and sometimes 2-, 3-, or 4-rayed trichomes, or, rarely, with fewer, stellate ones.
         12. Seeds orbicular or nearly so, 2–2.5 mm diam.; fruits 2–3 mm wide . . . 15. *Arabis blepharophylla*
         12. Seeds oblong, 1–1.3 mm wide; fruits 1.5–2 mm wide.
            13. Plants glabrous or with simple trichomes (terminating teeth of basal leaf blades); simple trichomes not bulbous-based; fruits 2–4 cm . . . . . . 14. *Arabis mcdonaldiana*
            13. Plants moderately pubescent with simple and forked trichomes; simple trichomes often bulbous-based; fruits (3–)3.5–6.5 cm.

[14. Shifted to left margin.—Ed.]

14. Plants with at least some 3-rayed, stellate trichomes; cauline leaves 1–3(–6) cm; fruits slightly torulose .... 12. *Arabis oregana*
14. Plants without 3-rayed, stellate trichomes; cauline leaves 0.4–1(–1.5) cm; fruits not torulose .... 13. *Arabis aculeolata*

1. **Arabis alpina** Linnaeus, Sp. Pl. 2: 664. 1753

Arabis alpina var. *glabrata* A. Blytt

Perennials; (stoloniferous, with vegetative rosettes, loosely cespitose to somewhat pulvinate); sparsely to moderately pubescent, trichomes stalked, cruciform, stellate, mixed with simple and forked-stalked ones. **Stems** usually simple from base, erect to ascending, often branched proximally, (0.6–)1–2(–2.5) dm. **Basal leaves**: petiole 0–1 cm; blade spatulate, oblanceolate, oblong, or obovate, (0.4–)1–4(–5) cm × (3–)6–15(–20) mm, margins dentate to denticulate, apex obtuse or acute, surfaces usually pubescent, rarely subglabrate, trichomes stellate with simple rays. **Cauline leaves** 3–5(–6); blade oblong or ovate, 1–3 cm × 5–15 mm, base subcordate or auriculate, margins usually dentate, rarely subentire, apex acute or obtuse. **Racemes** simple, (lax). **Fruiting pedicels** ascending to divaricate, 4–10(–12) mm. **Flowers**: sepals oblong, 2.5–4(–4.7) × 1–2 mm, lateral pair conspicuously saccate basally; petals white, spatulate to obovate, 5–8(–9) × 2–3.5 mm, apex obtuse; filaments 3–5 mm; anthers oblong, 0.7–1.2 mm. **Fruits** ascending to spreading, torulose, 1.7–3.5(–4) cm × 1–1.7 mm; valves each with midvein absent or obscure, along proximal ½; ovules 34–50 per ovary; style 0.3–0.6 mm. **Seeds** narrowly winged throughout, ovate, 1–1.4 × 0.9–1.1 mm; wing 0.1–0.2 mm wide. $2n = 16$.

Flowering Jun–Aug. Crevices of limestone rocks, along streams, calcareous alpine meadows, *Salix* scrub on slopes with scree; 0–2400 m; Greenland; Nfld. and Labr., Nunavut, Que.; N.Y., Wash.; Europe; sw Asia; n, tropical Africa.

Arabis alpina, the generic type, is variable. It is cultivated extensively for its attractive flowers, and it sometimes escapes from cultivation. There is disagreement as to whether one or more species should be recognized in this complex. European, some African, and most North American plants are recognized as *A. alpina*; most of the larger-flowered, southwestern Asian plants, which are most commonly cultivated, are recognized as *A. caucasica* or as *A. alpina* subsp. *caucasica* (Willdenow) Briquet. In my opinion, the morphological differences between the two (see key, couplet 3) support recognition of two species, as did R. C. Rollins (1993) and G. A. Mulligan (1996).

2. **Arabis caucasica** Willdenow, Enum. Pl., suppl., 45. 1814 • Garden rockcress [I]

Perennials; (stoloniferous, with vegetative rosettes, loosely cespitose to somewhat pulvinate); sparsely to moderately pubescent, trichomes stalked, cruciform, stellate, mixed with simple and forked-stalked ones. **Stems** several to many from base, erect, ascending to decumbent, branched (several), 1.5–3.5(–5) dm. **Basal leaves**: petiole 0–1 cm; blade spatulate, oblanceolate, oblong, or obovate, 2–5.5(–8) cm × 9–18(–25) mm, margins dentate to denticulate, apex obtuse or acute, surfaces usually pubescent, rarely subglabrate, trichomes stellate with some rays branched. **Cauline leaves** 4–8; blade oblong or ovate, 1–5 cm × 5–20 mm, base subcordate or auriculate, margins usually dentate, rarely subentire, apex acute or obtuse. **Racemes** simple, (lax). **Fruiting pedicels** ascending to divaricate, 7–17(–20) mm. **Flowers**: sepals oblong, 4.5–8 × 1–2 mm, lateral pair conspicuously saccate basally; petals white, spatulate to obovate, 10–19 × 4–8 mm, apex obtuse; filaments 4–7 mm; anthers oblong, 0.7–1.2 mm. **Fruits** ascending to spreading, torulose, (3–)4–7 cm × 1–2 mm; valves each with midvein absent or obscure, along proximal ½; ovules 40–60 per ovary; style 0.5–1 mm. **Seeds** narrowly winged throughout, ovate, 1–1.4 × 0.9–1.1 mm; wing 0.1–0.2 mm wide. $2n = 16$.

Flowering Jun–Aug. Rare garden escape; introduced; B.C., N.B., Ont., Que., Yukon; Mich., N.Y.; sw Asia.

3. **Arabis pycnocarpa** M. Hopkins, Rhodora 39: 112, plate 458, figs. 1–3. 1937 [F]

Arabis hirsuta (Linnaeus) Scopoli subsp. *pycnocarpa* (M. Hopkins) Hultén; *A. hirsuta* var. *pycnocarpa* (M. Hopkins) Rollins

Biennials or perennials; (caudex branched); usually densely hirsute (at least basally), rarely glabrescent, trichomes simple mixed with stalked or sessile, forked ones. **Stems** simple or several from base (rosette), erect, often branched distally, 1–8 dm, (pilose with trichomes appressed, malpighiaceous, or minutely stalked, forked, or hirsute basally with trichomes simple and minutely stalked, forked, sometimes almost exclusively pubescent with forked submalpighiaceous trichomes). **Basal leaves:** petiole 0.5–2 cm, (ciliate or not); blade spatulate, oblanceolate, or oblong, (0.8–)1.5–8 cm × (5–)10–25 mm, margins entire, repand, or dentate, apex obtuse or acute, surfaces sparsely to densely pubescent, trichomes sessile or stalked, simple or forked, and/or stellate. **Cauline leaves** (7–)10–45(–61), (overlapping or not); blade ovate to oblong or lanceolate, rarely linear, (1–)1.5–6(–8) cm × (1–)3–20(–25) mm, base subcordate or auriculate (auricles obtuse or subacute), margins dentate or entire, apex acute or obtuse, surfaces hirsute or adaxially glabrescent. **Racemes** often simple. **Fruiting pedicels** erect to erect-ascending, (2–)3–8(–12) mm (glabrous or sparsely pubescent). **Flowers:** sepals oblong, 2.5–4 × 0.5–1.5 mm, lateral pair not saccate basally; petals white, linear-oblanceolate or narrowly spatulate, 3.5–5(–5.5) × 1–2(–2.5) mm, apex obtuse; filaments 2.5–4 mm; anthers oblong, 0.7–1 mm. **Fruits** erect to erect-ascending, (often appressed to rachis), torulose, (3.5–)4–6(–6.5) cm × 0.8–1(–1.2) mm; valves each with obscure or somewhat prominent midvein extending to the middle; ovules (54–)60–86 per ovary; style (0.2–)0.5–1(–1.3) mm, (slender). **Seeds** narrowly winged throughout, oblong or suborbicular, (0.8–)1–1.5(–1.7) × 0.8–1.3 mm; wing to 0.2 mm wide distally.

Varieties 2 (2 in the flora): North America, Asia.

M. Hopkins (1937) synthesized earlier works on *Arabis pycnocarpa* and concluded that it is different from the European *A. hirsuta*. Both R. C. Rollins (1941, 1993) and G. A. Mulligan (1996) considered the North American and European plants different varieties of *A. hirsuta*; Mulligan recognized var. *hirsuta* in North America and Rollins did not. After examining thousands of specimens from Europe, Asia, and North America, I conclude that Hopkins was correct in treating the North American plants as a different species, *A. pycnocarpa*. The European *A. hirsuta* is a diploid ($2n = 16$) that has fruits 1.5–4 cm, stout styles 0.1–0.5 mm, prominent midvein extending the full length of the fruit valve, and 30–40(–44) ovules/seeds per ovary/fruit. By contrast, *A. pycnocarpa* is a tetraploid ($2n = 32$) that has fruits (3.5–)4–6(–6.5) cm, often slender styles (0.2–)0.5–1(–1.3) mm, obscure midvein hardly extending to the middle of the fruit valve, and ovules/seeds (54–)60–86 per ovary/fruit. Hopkins listed other differences in the number of cauline leaves and the development of seed wing, but these do not hold. Mulligan treated the perennial North American plants with minute styles as var. *hirsuta* and the biennial ones with longer styles as var. *pycnocarpa*. Habit and style length are not correlated at all, and one finds both biennials and perennials flowering in the same population.

1. Stems usually hirsute basally, rarely glabrescent, trichomes often simple, sometimes branched; basal leaf blades: surfaces with simple and stalked trichomes . . . . . . 3a. *Arabis pycnocarpa* var. *pycnocarpa*
1. Stems pilose, trichomes appressed, malpighiaceous, or minutely stalked, forked; basal leaf blades: surfaces with sessile, forked, and/or stellate trichomes . . . . . 3b. *Arabis pycnocarpa* var. *adpressipilis*

3a. Arabis pycnocarpa M. Hopkins var. pycnocarpa [F]

Arabis hirsuta (Linnaeus) Scopoli var. *minshallii* B. Boivin; *A. pycnocarpa* var. *reducta* M. Hopkins

Stems usually hirsute basally, rarely glabrescent, trichomes often simple, sometimes branched. **Basal leaves:** blade surfaces sparsely to densely pubescent, trichomes simple and stalked. **Fruits** (3.5–)4–5.8(–6) cm; style (0.2–)0.5–1 mm, (rarely stout). $2n = 32$.

Flowering Mar–Jul. Bluffs, cliffs, ledges, rocky hillsides, open woods, bottom lands, gravel bars, meadows, streamsides, upland prairies, grassy swales, hillsides, stream bottoms; 0–2500 m; Alta., B.C., Man., N.B., Nfld. and Labr. (Labr.), N.W.T., N.S., Ont., Que., Sask., Yukon; Alaska, Ariz., Calif., Colo., Conn., Idaho, Iowa, Kans., Maine, Mass., Mich., Minn., Mont., Nebr., Nev., N.H., N.J., N.Mex., N.Y., N.Dak., Ohio, Oreg., Pa., S.Dak., Utah, Vt., Va., W.Va., Wis., Wyo.; Asia (China, Japan, Russian Far East).

Some populations in northern Indiana and northern Illinois, [e.g., *Herman 8790* (Jo Davies County) and *Friesner 19072* (Elkhart County), both at GH] most likely represent hybrids between the two varieties of *Arabis pycnocarpa*. Trichomes borne proximally on stems of these plants are a mixture of forked and simple. It is expected that hybrids can be found where the ranges of the two taxa overlap.

3b. Arabis pycnocarpa M. Hopkins var. adpressipilis M. Hopkins, Rhodora 39: 117. 1937 [E]

Arabis hirsuta (Linnaeus) Scopoli var. *adpressipilis* (M. Hopkins) Rollins

Stems pilose, trichomes appressed, malpighiaceous, or minutely stalked, forked, plants rarely glabrescent. **Basal leaves:** blade surfaces pubescent, trichomes sessile, forked, and/or stellate. **Fruits** (4–)4.5–6(–6.5) cm; style 0.5–1.3 mm.

Flowering Mar–Jun. Ravines, pastures, cliffs, calcareous talus, dolomite glades, rich woods, bluffs, rocky ledges; 0–300 m; Ont.; Ill., Ind., Iowa, Kans., Mo., N.Y., Ohio, Pa., Tenn., Va.

R. C. Rollins (1941, 1993) recognized both *pycnocarpa* and *adpressipilis* as distinct varieties of *Arabis hirsuta*; G. A. Mulligan (1996) treated *adpressipilis* as a synonym of *A. hirsuta* var. *pycnocarpa*. In my opinion, the differences in trichome morphology of the stems and leaves are significant enough to justify recognition of infraspecific taxa of one species.

4. **Arabis eschscholtziana** Andrzejowski in C. F. von Ledebour, Fl. Altaica 3: 25. 1831 [E]

Arabis hirsuta (Linnaeus) Scopoli subsp. *eschscholtziana* (Andrzejowski) Hultén; *A. hirsuta* var. *eschscholtziana* (Andrzejowski) Rollins; *A. hirsuta* var. *glabrata* Torrey & A. Gray; *A. ovata* (Pursh) Poiret var. *glabrata* (Torrey & A. Gray) Farwell; *A. pycnocarpa* M. Hopkins var. *glabrata* (Torrey & A. Gray) M. Hopkins; *A. rupestris* Nuttall; *A. stelleri* de Candolle var. *eschscholtziana* (Andrzejowski) N. Busch; *Turritis spathulata* Nuttall

Biennials or perennials; (caudex branched); usually densely hirsute (at least basally), trichomes simple or stalked, forked, sometimes glabrescent with mostly simple ones. **Stems** simple or several from base (rosette), erect, often branched distally, 2–7(–10) dm, (hirsute basally, trichomes simple and minutely stalked, forked, rarely almost exclusively glabrous). **Basal leaves**: petiole 0.7–3 cm, (ciliate or not); blade broadly spatulate to narrowly oblanceolate, 1.2–10 cm × 7–30 mm, margins entire or dentate, apex obtuse or acute, surfaces usually pubescent, sometimes glabrous, trichomes simple, forked and stalked, stellate. **Cauline leaves** (2–)4–12(–18), (well-spaced or, rarely, overlapping); blade ovate to oblong or lanceolate, 1.5–6.5(–10) cm × 5–30 mm, base auriculate, margins dentate or entire, apex acute or obtuse, surfaces sparsely hirsute or glabrous. **Racemes** simple or branched. **Fruiting pedicels** erect to erect-ascending, 3.5–10(–15) mm, (glabrous or sparsely pubescent). **Flowers**: sepals oblong, 3.5–5 × 1.2–2 mm, lateral pair saccate basally; petals white, rarely pink, linear-oblanceolate or narrowly spatulate, (5.5–)6.5–9(–10) × 1.5–3 mm, apex obtuse; filaments 3.5–6 mm; anthers oblong, 0.7–1 mm. **Fruits** erect to erect-ascending, (often appressed to rachis), torulose, 3.5–6.5 cm × 1.2–1.8(–2) mm; valves each with prominent midvein extending full length or to middle; ovules 54–80 per ovary; style (0.1–)0.3–1 mm. **Seeds** usually narrowly winged throughout or only distally, rarely not winged, oblong or suborbicular, 1–1.8 × 0.9–1.3 mm; wing to 0.2 mm wide distally. $2n$ = 32, 64.

Flowering (Apr–)May–Jul. Rocky slopes, crevices, and ledges, meadows, moist banks and grounds, granitic soil, wooded slopes; 0–2800 m; Alta., B.C., Yukon; Alaska, Calif., Idaho, Mont., Nev., Oreg., Utah, Wash., Wyo.

G. A. Mulligan (1996) recognized *Arabis eschscholtziana* as a distinct species; R. C. Rollins (1941, 1993) treated it as a variety of *A. hirsuta*. As discussed under *A. pycnocarpa*, *A. hirsuta* does not occur in North America, and the characters separating all three species (see key to species), as well as the different ploidy levels, support Mulligan's conclusion.

Both M. Hopkins (1937) and R. C. Rollins (1941, 1993) recognized the glabrous or subglabrate forms native to North America as a distinct variety, var. *glabrata*; G. A. Mulligan (1996) did not accord such forms any taxonomic status. Glabrous and subglabrate forms occur in both *Arabis eschscholtziana* and *A. pycnocarpa* and sometimes even within a population that has moderately to densely pubescent forms. I support Mulligan's view in not recognizing the glabrous forms as an infraspecific taxon.

5. **Arabis georgiana** R. M. Harper, Torreya 3: 88. 1903 [C][E]

Biennials; sparsely to moderately hirsute (at least basally), trichomes simple, mixed with fewer, short-stalked, forked ones, subsessile cruciform or 3-rayed stellate trichomes commonly on abaxial blade surfaces, sometimes plants glabrous distally. **Stems** simple or few from base (rosette), erect, unbranched or branched (few) distally, 3–7 dm, (hirsute basally, glabrous distally). **Basal leaves**: petiole 0.5–2 cm, (ciliate or not); blade spatulate, oblanceolate, or obovate, 1.5–6 cm × 5–15 mm, margins dentate, apex obtuse or acute, abaxial surface moderately to sparsely pubescent, trichomes subsessile stellate, adaxial surface subglabrate or sparsely stellate. **Cauline leaves** 7–26; blade oblong, lanceolate, or linear-lanceolate, 1.5–7 cm × 3–18 mm, base auriculate to subamplexicaul, margins dentate or entire, apex acute or obtuse, pubescent as basal leaves except distalmost leaves often glabrous. **Racemes** often simple. **Fruiting pedicels** erect to erect-ascending, 7–16 mm, (glabrous). **Flowers**: sepals oblong, 2.5–4.5 × 1–1.5 mm, lateral pair subsaccate basally; petals white, narrowly spatulate or oblanceolate, 6–9 × 1–1.5 mm, apex obtuse; filaments 3–4.5 mm; anthers oblong, 0.8–1 mm. **Fruits** erect to erect-ascending, (often subappressed to rachis), smooth, 4–7 cm × 0.7–0.8 mm; valves each with midvein extending full length or to middle; ovules 38–44 per ovary; style 0.7–1.8 mm. **Seeds** narrowly winged throughout, oblong, 0.9–1.9 × 0.5–0.7 mm; wing to 0.1 mm wide distally.

Flowering Mar–Apr. Stream banks, roadsides; of conservation concern; 0–200 m; Ala., Ga.

Arabis georgiana is most closely related to *A. pycnocarpa*, from which it is easily distinguished by having narrower fruits, longer petals, and subsessile cruciform or 3-rayed trichomes on abaxial surfaces of basal leaves. It is known only in Alabama from Bibb and Elmore counties and in Georgia from Stewart County.

6. Arabis patens Sullivant, Amer. J. Sci. Arts 42: 49. 1842 [E]

Boechera patens (Sullivant) Al-Shehbaz

Biennials or, rarely, perennials; (short-lived, with caudex); usually densely hirsute (at least basally), trichomes simple (to 1 mm), sometimes mixed with stalked, forked ones. **Stems** simple or several from base (rosette), erect, often branched distally, (1.8–)2.5–4.5(–5.5) dm, (hirsute at least basally). **Basal leaves:** petiole (0.7–)2–5 cm; blade usually ovate to oblanceolate, rarely cordate, (0.8–)1.5–3 cm × 5–17 mm, margins dentate or serrate, surfaces sparsely to densely pubescent, trichomes simple or forked. **Cauline leaves** (5–)8–20 (–27), (overlapping or not); blade ovate to oblong or lanceolate, (2–)3–7.5(–9) cm × (10–)13–25(–35) mm, base amplexicaul, margins usually dentate or serrate, sometimes entire, surfaces hirsute. **Racemes** branched. **Fruiting pedicels** ascending to divaricate-ascending, (7–)10–20(–24) mm, (glabrous or sparsely pubescent). **Flowers:** sepals oblong, 2.5–4.5 × 1–1.5 mm, lateral pair slightly saccate basally; petals white, spatulate, (5–)6–9 × 2–3.5 mm, apex obtuse; filaments 3–5 mm; anthers oblong, 0.8–1.2 mm. **Fruits** divaricate to erect-ascending, (not appressed to rachis), torulose, 2.2–4.6 cm × 0.8–1.2 mm; valves each with prominent midvein extending to middle or full length; ovules 16–28 per ovary; style 0.5–2 mm, (slender). **Seeds** narrowly winged distally, oblong, 1–1.6 × 0.6–0.8 mm; wing 0.05–0.1 mm.

Flowering Apr–May. Rocky wooded slopes, shady stream banks, limestone ledges and bluffs; 0–500 m; D.C., Ind., Ky., Md., N.C., Ohio, Pa., Tenn., Va., W.Va.

7. Arabis crucisetosa Constance & Rollins, Proc. Biol. Soc. Wash. 49: 147. 1936 [E]

Perennials; (caudex usually simple, rarely branched, covered with persistent petiolar remains); sparsely to densely pubescent basally, trichomes short-stalked, cruciform or 3-rayed, stellate, (to 0.4 mm in diam.). **Stems** simple or few from caudex, erect, unbranched or rarely branched (few) distally, (1–)1.5–4(–5) dm, (glabrous throughout or sparsely pubescent proximally). **Basal leaves:** petiole 0.5–2.5(–3) cm; blade oblanceolate to obovate, (0.6–)1–3.5(–4.5) cm × (3–)5–15(–20) mm, margins entire or dentate, (not ciliate), apex obtuse, surfaces uniformly pubescent, trichomes short-stalked, stellate. **Cauline leaves** 3 or 4(–6); blade linear, oblong or oblanceolate, 1–3(–4) cm × (1.5–)3–7(–10) mm, base cuneate, not auriculate, margins usually entire, rarely few-toothed, apex obtuse, surfaces glabrous or sparsely pubescent at least along margin. **Racemes** simple. **Fruiting pedicels** ascending to divaricate, (5–)8–20 mm, (glabrous). **Flowers:** sepals (greenish), oblong, 2.5–4 × 1–1.5 mm, lateral pair saccate basally; petals white, oblanceolate, 7–9 × 2.5–3.5 mm, apex rounded; filaments 2.5–4 mm; anthers oblong, 0.8–1 mm; (gynophore to 0.5 mm). **Fruits** divaricate, slightly torulose, sometimes slightly curved, slightly flattened, (1.3–)1.7–3.5(–4) cm × 1–1.5 mm; valves each with obscure midvein extending to middle; ovules 14–28 per ovary; style 0.5–1.3 mm. **Seeds** not winged, oblong, 1–1.2 × 0.5–0.6 mm. $2n = 32$.

Flowering Apr–Jun. Open woods, rocky bluffs, steep banks, canyon margins; 400–1800 m; Idaho, Oreg., Wash.

Arabis crucisetosa is known from Idaho, Lewis, and Nez Perce counties in Idaho, from Wallowa County in Oregon, and from Asotin County in Washington.

8. Arabis furcata S. Watson, Proc. Amer. Acad. Arts 17: 362. 1882 [E]

Arabis suksdorfii Howell

Perennials; (caudex simple or branched, covered with persistent petiolar remains); glabrous or sparsely pubescent, trichomes simple or short-stalked, forked (0.3–1 mm), these rarely mixed with fewer, unequal 3-rayed ones. **Stems** simple or few from base (caudex), erect, usually unbranched, rarely branched (few) distally, (0.7–)1–3.8(–4.5) dm, (glabrous throughout or sparsely pubescent basally to distally). **Basal leaves:** petiole (0.4–)1–3(–4.5) cm, (glabrous or sparsely ciliate); blades oblanceolate, spatulate, or obovate, (0.7–)1.2–3 cm × 5–17(–22) mm, margins entire or dentate, (often ciliate), apex obtuse, surfaces glabrous or sparsely pubescent, trichomes short-stalked, forked and simple. **Cauline leaves** (2 or) 3–5 (or 6); blade linear, oblong, obovate, or oblanceolate, 0.7–3 (–4) cm × 3–8 mm, base cuneate, not auriculate, margins usually entire, rarely few-toothed, apex obtuse, surfaces usually glabrous, rarely margins ciliate. **Racemes** simple, (dense or lax). **Fruiting pedicels** suberect, ascending, or divaricate, (2–)4–17(–22) mm, (glabrous). **Flowers:** sepals (greenish), oblong, 3–4 × 1.5–2 mm, (usually glabrous, rarely with few trichomes subapically), lateral pair saccate basally; petals white, spatulate, 7–11 × 2.5–4 mm, apex rounded; filaments 3.5–5 mm; anthers oblong, 0.8–1 mm. **Fruits** ascending to suberect, (not appressed to rachis), slightly torulose, sometimes slightly curved, strongly flattened, (2–)2.5–4(–4.6) cm × 1.7–2.2 mm; valves each with prominent midvein extending full length; ovules 14–26 per ovary; style 0.5–1.2(–1.6) mm, (slender). **Seeds** winged distally, oblong to broadly ovate, (1.5–)1.8–2.5(–3) × 1–1.3 mm; wing (0.2–) 0.4–0.8(–1) mm wide.

Flowering May–Jul(–Aug). Open slopes, alpine meadows, cliffs, ridge crests; (50–)1000–2100 m; Oreg., Wash.

Arabis furcata is known in Washington from Chelan, Kittitas, Klickitat, Okanogan, Skamania, and Yakima counties, and in Oregon from Clackamas, Hood River, Multnomah, and Wasco counties.

9. **Arabis olympica** Piper, Contr. U.S. Natl. Herb. 16: 208. 1913 [E]

Arabis furcata S. Watson var. *olympica* (Piper) Rollins

Perennials; (caudex branched, covered with persistent petiolar remains); moderately pubescent, trichomes simple or short-stalked, forked, (to 0.6 mm). **Stems** simple or few from base (caudex), erect, unbranched, 0.7–1.1 dm, (pubescent throughout, trichomes short-stalked, forked). **Basal leaves:** petiole 0.3–0.6 cm, (sparsely ciliate); blades oblanceolate to obovate, 0.4–1 cm × 2–7 mm, margins entire or repand, (ciliate), apex obtuse, surfaces sparsely pubescent, trichomes simple and forked. **Cauline leaves** 4–6; blade oblong, 0.6–0.8 cm × 2–3 mm, base not auriculate, margins entire, (ciliate), apex obtuse. **Racemes** simple, (dense). **Fruiting pedicels** erect, (subappressed to rachis), 5–6.5 mm, (glabrous). **Flowers:** sepals (greenish), ovate, 1.5 × 0.5 mm, lateral pair not saccate basally, (glabrous); petals white, oblanceolate, 3.5–4 × 1 mm, apex rounded. **Fruits** erect, (appressed to rachis), slightly torulose, 2–2.3 cm × 1 mm; valves each with distinct midvein extending full length; ovules 28–34 per ovary; style 0.2–0.3 mm, (stout). **Seeds** winged distally, oblong, ca. 1 × 0.5 mm; wing ca. 0.1 mm wide.

Flowering Aug. Talus slopes near glacier; ca. 1400 m; Wash.

Arabis olympica, which is known from two collections, *J. B. Flett s.n.* (holotype, US; isotype, WS) and *N. Buckingham 1577* (WS) that were made from Jefferson and Clallam counties, respectively, was reduced by R. C. Rollins (1936, 1941, 1993) to a variety of *A. furcata*. An examination of the type collections of both species reveals that they are distinct. Although the fruits and seeds of *A. olympica* are not fully mature, they are clearly different in width and orientation from those of *A. furcata* at the same developmental stage. The striking differences in fruit width and orientation, stem indument, and seed and flower size support their maintenance as distinct species. Although both species grow in Washington, the range of *A. olympica* seems to be restricted to Clallam and Jefferson counties and is disjunct from Chelan, Kittitas, and Yakima counties, where *A. furcata* grows.

10. **Arabis nuttallii** (Kuntze) B. L. Robinson in A. Gray et al., Syn. Fl. N. Amer. 1(1,1): 160. 1895 [E]

Erysimum nuttallii Kuntze, Revis. Gen. Pl. 2: 933. 1891, based on *Arabis spathulata* Nuttall in J. Torrey and A. Gray, Fl. N. Amer. 1: 81. 1838, not de Candolle 1821; *A. bridgeri* M. E. Jones; *A. macella* Piper

Perennials; (caudex simple or branched, covered with persistent petiolar remains); glabrous or sparsely to densely pubescent, trichomes simple, sometimes subsetiform, (0.3–2 mm), these rarely mixed with fewer, short-stalked, forked ones. **Stems** simple or few to numerous from base (caudex), erect or ascending, unbranched, 0.5–2.5(–3.6) dm, (glabrous throughout or pubescent along proximal ½, trichomes simple). **Basal leaves:** petiole 0.3–2(–3.5) cm, (glabrous or ciliate); blade narrowly oblanceolate, spatulate, to obovate, 0.4–2(–3) cm × 3–10(–15) mm, margins entire, (ciliate or not), apex obtuse, surfaces glabrous or sparsely to densely pubescent, trichomes simple, rarely mixed with forked ones. **Cauline leaves** (1 or) 2–5(–7); blade linear, oblanceolate, oblong, or obovate, 0.5–2.2(–3) cm × 2–8 (–12) mm, base not auriculate, margins entire, (sometimes ciliate), apex obtuse, surfaces glabrous or, rarely, sparsely pubescent. **Racemes** simple, (dense or lax). **Fruiting pedicels** ascending to divaricate, (2–)5–20(–30) mm, (glabrous). **Flowers:** sepals (greenish) oblong, 2.5–3.5 × 1.2–1.5 mm, lateral pair saccate basally, (glabrous); petals white, oblanceolate to spatulate, (5–)6–8 × 2–4 mm, apex rounded; filaments 2.5–4 mm; anthers oblong, 0.6–1 mm. **Fruits** ascending to suberect, (not appressed to rachis), slightly torulose, sometimes slightly curved, (0.7–)1–2.5(–2.8) cm × 0.8–1.2(–1.5) mm; valves each with distinct midvein extending full length; ovules 10–24 per ovary; style 0.5–1.2(–2) mm, (slender). **Seeds** not winged, broadly ovate, 0.6–1.2(–1.5) × 0.5–1 mm. $2n = 32$.

Flowering Apr–Aug. Alpine and subalpine meadows and slopes, open woods, steep slopes and cliffs, mossy mats, dry or moist slopes and hillsides, ridge crests; 500–3200 m; Alta., B.C.; Idaho, Mont., Utah, Wash., Wyo.

Arabis nuttallii, the second most widespread native North American species of the genus, is variable in density of indumentum, types of trichomes, plant height, pedicel length, fruit width, leaf shape, flowering and fruiting time, and elevation. As indicated by R. C. Rollins (1941, 1993), the variation seems to be sporadic and does not correlate with clear-cut entities and, therefore, the species cannot be subdivided meaningfully.

11. Arabis modesta Rollins, Rhodora 43: 350. 1941 [E]

Perennials; (caudex usually simple, rarely branched, covered with persistent petiolar remains); sparsely to densely pubescent, trichomes short-stalked, cruciform, or 3-rayed stellate, (0.3–0.5 mm). **Stems** simple or few from base (caudex), erect, usually unbranched, rarely branched (few) distally, 2.2–5.5(–6.7) dm, (usually pubescent throughout, rarely subglabrate distally, trichomes stellate). **Basal leaves:** petiole 1–4(–5) cm, (rarely minutely ciliate near base); blade oblanceolate to obovate, (0.8–)2–4.5(–6) cm × (4–)10–20(–25) mm, margins entire, repand, or dentate, (not ciliate), apex obtuse, surfaces pubescent, trichomes short-stalked, stellate. **Cauline leaves** (2–)4–6(–9); blade oblong or ovate, 1–3.5(–4.5) cm × (3–)6–10(–20) mm, base not auriculate, margins entire, repand, or dentate, apex obtuse or acute, surfaces pubescent, trichomes short-stalked, stellate. **Racemes** simple, (dense). **Fruiting pedicels** ascending to divaricate, 7–18(–25) mm. **Flowers:** sepals (purple), oblong, (4–)5–6.5(–8) × 1.5–2 mm, lateral pair saccate basally; petals purple, spatulate, (10–)12–16(–20) × 4–6(–7) mm, apex rounded; filaments (4–)5–8 mm; anthers narrowly oblong, 1.3–2 mm. **Fruits** suberect to divaricate, sometimes slightly curved, slightly torulose, (2.8–)3.5–6 cm × 1.5–2 mm; valves each with prominent midvein extending to middle or full length; ovules 20–34 per ovary; style 0.5–1(–1.5) mm. **Seeds** winged distally, oblong, 1.7–2.2 × 0.9–1.2 mm; wing 0.2–0.5 mm wide. $2n = 32$.

Flowering Mar–Apr(–May). Moist shaded banks, slopes, rocky canyon walls, talus, basaltic bluffs; 150–500 m; Calif., Oreg.

Arabis modesta, known only from Napa, Siskiyou, and Yolo counties (California), and Jackson and Josephine counties (Oregon), and the following four species form a well-defined group easily separated from the first nine species above by their large, purple petals. *Arabis modesta* is readily distinguished from other purple-flowered *Arabis* by having very fine, short-stalked cruciform and 3-rayed, instead of simple and 2-rayed, forked trichomes. *Arabis blepharophylla* has broadly winged, orbicular seeds, and the other purple-flowered species (*A. aculeolata*, *A. mcdonaldiana*, *A. oregana*) have oblong seeds that are not winged or narrowly winged distally. The lines separating these three species are not as well-defined as those of *A. blepharophylla* and *A. modesta*, and further studies are needed to resolve them. I prefer to maintain all five species of purple-flowered *Arabis* as distinct until thorough molecular and hybridization studies are conducted. In the absence of such studies, it is unwise to make any nomenclatural changes.

12. Arabis oregana Rollins, Rhodora 43: 349. 1941 [E]

Arabis purpurascens Howell ex Greene, Pittonia 1: 161. 1888, not C. Presl 1825; *A. furcata* S. Watson var. *purpurascens* S. Watson

Perennials; (caudex simple or branched, with some persistent petiolar remains); sparsely to moderately pubescent, trichomes bulbous-based, simple and long-stalked, forked, (to 2 mm), these often mixed with fewer 3-rayed, stalked stellate ones. **Stems** simple or few from base (caudex), erect, unbranched distally, (0.6–)1.7–4.5(–5) dm, (usually pubescent throughout, rarely subglabrate, coarsely hirsute, trichomes simple, mixed with forked ones). **Basal leaves:** petiole 0.5–2(–3) cm, (ciliate); blade oblanceolate to obovate, (1–)2–5(–9) cm × (5–)8–18(–22) mm, margins entire or repand to dentate, (ciliate), apex obtuse, surfaces pubescent or glabrous, trichomes simple and forked, sometimes mixed with 3-rayed stellate ones. **Cauline leaves** 3–6 (or 7); blade oblong, 1–3(–6) cm × 2–7(–15) mm, base not auriculate, margins entire or dentate, (ciliate or not), apex obtuse, surfaces pubescent as basal leaves or glabrous. **Racemes** simple, (dense). **Fruiting pedicels** ascending to erect, 5–10 mm, (pubescent or glabrous). **Flowers:** sepals (purple), oblong, (5–)6–8 × 1.5–2 mm, lateral pair saccate basally; petals purple or pink, spatulate, (10–)12–15(–16) × (3–)4–5 mm, apex obtuse; filaments (4–)5–8 mm; anthers narrowly oblong, 1.2–1.8 mm. **Fruits** erect or nearly so, slightly torulose, (3–)4.5–6 cm × 1.5–2 mm; valves each with prominent midvein; ovules 24–30 per ovary; style 0.5–1 mm. **Seeds** narrowly winged distally or not winged, oblong, 1.8–2.2 × ca. 1 mm; wing ca. 0.1 mm wide. $2n = 32$.

Flowering Apr–May. Moist granitic soil, rocky hillsides, chaparrals, steep banks; 500–1000(–1400) m; Calif., Oreg.

Arabis oregana is known from Napa and Siskiyou counties (California) and Jackson County (Oregon).

13. Arabis aculeolata Greene, Leafl. Bot. Observ. Crit. 2: 69. 1910 [E]

Perennials; (caudex simple or branched, covered with persistent petiolar remains); sparsely to moderately pubescent, trichomes subsetiform, bulbous-based, simple, (to 1.5 mm), often mixed with fewer, forked, stalked ones. **Stems** simple or few from base (caudex), erect, unbranched, (0.6–)1.5–3.5(–4.5) dm, (sparsely to densely hirsute basally or throughout). **Basal leaves:** petiole 0.3–1.5 cm, (ciliate); blade oblanceolate to obovate, (0.5–)1–3(–4) cm × (2–)3–6(–10) mm, margins entire, repand, or obtusely dentate, (ciliate), apex obtuse, surfaces pubescent, trichomes usually simple, sometimes mixed with fewer forked ones, rarely subglabrate. **Cauline leaves** 3–6 (or 7); blade oblong, 0.4–1(–1.5) cm × (1–)2–4 mm, base not auriculate, margins entire or repand, apex obtuse. **Racemes** simple, (dense). **Fruiting pedicels** ascending to suberect, 8–12(–15) mm, (subglabrate or sparsely pubescent). **Flowers:** sepals (purple), oblong, 4–8 × 1.5–2.5 mm, lateral pair saccate basally; petals purple, spatulate, (9–)10–18(–20) × (2.5–)3.5–6(–8) mm, apex obtuse; filaments 4–9 mm; anthers oblong, 1.5–2 mm. **Fruits** ascending to suberect, not torulose, sometimes slightly curved, 3.5–6.5 cm × 1.5–2 mm; valves each with prominent midvein extending full length; ovules 24–36 per ovary; style (0.7–)1–2 mm. **Seeds** narrowly winged nearly throughout except wider distally, oblong, 1.5–2.3 × 1–1.3 mm; wing 0.1–0.3 mm wide. $2n = 32$.

Flowering Mar–May(–Jun). Serpentine slopes, rocky hillsides; 200–1800 m; Oreg.

Arabis aculeolata is known only from Curry, Douglas, and Josephine counties.

14. Arabis mcdonaldiana Eastwood, Bull. Torrey Bot. Club 30: 488, unnumb. fig. (p. 489). 1903 [C][E]

Arabis blepharophylla Hooker & Arnott var. *mcdonaldiana* (Eastwood) Jepson; *A. serpentinicola* Rollins

Perennials; (caudex simple or branched, covered with persistent petiolar remains); usually glabrous, rarely sparsely pubescent, trichomes simple, (to 0.5 mm), not bulbous-based. **Stems** simple or few from base (caudex), erect, unbranched, (0.6–)1.5–3(–4) dm, (glabrous). **Basal leaves:** petiole 0.3–1.5 cm; blade oblanceolate to obovate, (0.5–)1–3(–4) cm × (2–)3–6(–10) mm, margins entire, repand, or obtusely dentate, apex obtuse, surfaces sometimes with individual trichomes terminating some or all leaf teeth. **Cauline leaves** (2 or) 3–5 (or 6); blade oblong, 0.3–1(–1.2) cm × 1–3 mm, base not auriculate, margins entire or repand, apex obtuse, surfaces glabrous. **Racemes** simple, (dense). **Fruiting pedicels** ascending to suberect, 3–10(–13) mm. **Flowers:** sepals (purple), oblong, 4–8 × 1.5–2.5 mm, lateral pair saccate basally; petals purple, spatulate, 8–16 × 2–5 mm, apex obtuse; filaments 4–8 mm; anthers oblong, 1.5–2 mm. **Fruits** ascending to suberect, not torulose, sometimes slightly curved, 2–4 cm × 1.5–2 mm; valves each with prominent midvein extending full length; ovules 20–34 per ovary; style 0.3–1.5 mm. **Seeds** narrowly winged distally or, rarely, not winged, oblong, 1.5–2.2 × 1–1.3 mm, wing 0.1–0.2 mm wide.

Flowering May–Jun. Serpentine scrap and slopes, red serpentinized soil; of conservation concern; 200–1800 m; Calif., Oreg.

Arabis mcdonaldiana is known in California from Del Norte, Mendocino, and Siskiyou counties, and in Oregon from Curry and Jackson counties. It is in the Center for Plant Conservation's National Collection of Endangered Plants.

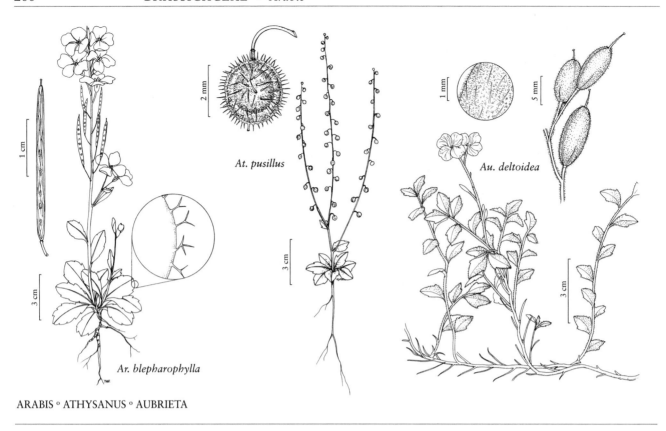

ARABIS ° ATHYSANUS ° AUBRIETA

15. **Arabis blepharophylla** Hooker & Arnott, Bot. Beechey Voy., 321. 1838 [E] [F]

Perennials; (caudex simple or branched, covered with persistent petiolar remains); sparsely to moderately pubescent, trichomes simple, forked-stalked, or rarely cruciform or 3-rayed, stalked, stellate. **Stems** simple or few from base (caudex), erect, unbranched or branched (few) distally, 0.6–2.5(–3) dm, (usually pubescent throughout, rarely subglabrate). **Basal leaves**: petiole 0.5–3(–6) cm, (ciliate); blade oblanceolate, spatulate, or obovate, (1.5–)2.1–3.5(–6) cm × (5–)8–20(–25) mm, margins entire or dentate, (ciliate), apex obtuse, surfaces pubescent or glabrous, trichomes simple and forked, sometimes mixed with 3- or 4-rayed stellate ones. **Cauline leaves** (2 or) 3–6 (or 7); blade oblong or ovate, 1–2(–4) cm × (2–)4–10 (–15) mm, base not auriculate, margins entire or dentate, (ciliate), apex obtuse, surfaces pubescent as basal leaves. **Racemes** simple, (dense). **Fruiting pedicels** ascending to erect, (3–)5–10(–15) mm. **Flowers**: sepals (purple), oblong, 5–7 × 1.5–2 mm, lateral pair saccate basally; petals purple, spatulate or broadly so, (12–)14–18 × 4–7 mm, apex obtuse or rounded; filaments 6–8 mm; anthers narrowly oblong, 1.2–1.5 mm. **Fruits** erect or nearly so, slightly torulose, 2–4 cm × 2–3 mm; valves each with prominent midvein extending full length or rarely to middle; ovules 20–28 per ovary; style 0.2–1(–1.5) mm. **Seeds** narrowly winged throughout, orbicular or suborbicular, 2–2.5 mm in diam.; wing 0.2–0.4 mm wide (wider distally). $2n = 16$.

Flowering Mar–Apr. Rocky hillsides and bluffs, grassy hillsides, slopes; 50–200 m; Calif.

Arabis blepharophylla is an attractive species on the gradual increase in cultivation as an ornamental. It is recorded from Contra Costa, Marin, San Francisco, San Mateo, Santa Cruz, and Sonoma counties; most of the records are based on older collections.

7. ATHYSANUS Greene, Bull. Calif. Acad. Sci. 1: 72. 1885 • [Greek *a*-, without, and *thysanos*, fringe, alluding to fruit margin]

Ihsan A. Al-Shehbaz

Heterodraba Greene

Annuals; not scapose; glabrous or pubescent, trichomes simple mixed with short-stalked, forked, 3-rayed, or cruciform ones. **Stems** (few from base), ascending, branched. **Leaves** basal and cauline; petiolate or sessile; basal not rosulate, shortly petiolate, blade margins entire or dentate; cauline shortly petiolate or sessile, blade (base not auriculate), margins entire or dentate. **Racemes** (corymbose, several-flowered, secund, lax), considerably elongated in fruit. **Fruiting pedicels** recurved or reflexed, slender or stout, (glabrous or pubescent). **Flowers** (cleistogamous and/or chasmogamous); sepals suberect, oblong, lateral pair not saccate basally, (glabrous or pubescent); petals white, (usually rudimentary, or well-developed and exceeding sepals, rarely absent), spatulate or oblong, claw indistinct; stamens subequal; filaments slightly dilated or not basally; anthers ovate to globose; nectar glands lateral, each side of lateral stamens. **Fruits** (pendulous), silicles, indehiscent or very tardily dehiscent, sessile, usually orbicular or obovate to elliptic, rarely oblong, twisted or flattened, latiseptate; valves each not veined or veins prominent, pubescent or glabrous; replum rounded; septum absent or complete; ovules 2–12 per ovary; style obsolete or distinct; stigma capitate. **Seeds** uniseriate, flattened, not winged, oblong; seed coat not mucilaginous when wetted; cotyledons accumbent.

Species 2 (2 in the flora): w North America, nw Mexico.

Athysanus and *Heterodraba* are strikingly similar in almost all aspects except for the differences in the key below. Although R. C. Rollins (1993) mentioned nothing about their relationship, he keyed them out in one couplet. Their species are more closely related to each other than to other Brassicaceae. In my opinion, the recognition of two independent monotypic genera obscures their sister phylogenetic relationship. Both genera were described simultaneously by Greene, and since W. L. Jepson (1901) transferred the type of *Heterodraba* to *Athysanus*, the latter name has nomenclatural priority.

1. Fruits 1-locular, not twisted; valves usually pubescent, rarely glabrous, some or all trichomes hooked; seeds 1 . . . . 1. *Athysanus pusillus*
1. Fruits 2-locular, flat or slightly twisted; valves glabrous or pubescent, trichomes not hooked; seeds often 2+ . . . . 2. *Athysanus unilateralis*

1. Athysanus pusillus (Hooker) Greene, Bull. Calif. Acad. Sci. 1: 72. 1885 F

Thysanocarpus pusillus Hooker, Icon. Pl. 1: plate 42. 1837; *Athysanus pusillus* var. *glabrior* S. Watson; *T. oblongifolius* Nuttall

Plants sparsely to densely pubescent proximally, sometimes glabrous distally. **Stems** (0.2–)0.5–3(–5) dm. **Basal leaves:** petiole 0.1–0.5 cm; blade oblanceolate, obovate, or oblong, 0.3–1.5(–2) cm × (1–)2–6(–8) mm, margins entire or dentate, apex acute to obtuse, surfaces uniformly pubescent, trichomes cruciform and 3-rayed, adaxially sometimes mixed with coarse, simple ones. **Cauline leaves** 1–6; blade similar and equaling or smaller than basal. **Fruiting pedicels** slender, (1.5–)2–4(–6) mm. **Flowers:** sepals 0.5–1 × 0.3–0.7 mm; petals oblong, 1.5–3 × 0.5–1 mm; filaments 0.4–0.6 mm; anthers (0.05–)0.1–0.2(–0.25) mm. **Fruits** usually orbicular to obovate, rarely broadly oblong, 1-locular, not twisted; valves thin, veins sometimes prominent, usually pubescent, rarely glabrous, trichomes simple, to 0.5 mm, hooked (sometimes restricted to margins, or mixed with smaller, sessile, cruciform or 3-rayed ones); septum absent; ovules 2(–4) per ovary; style 0.05–0.2 mm. **Seeds** 1, (0.8–)0.9–1.1(–1.2) × (0.5–)0.7–0.9(–1) mm. $2n = 26$.

Flowering Feb–Jun. Open grassy slopes, grassy glades in woodlands, chaparral, sandy and gravelly flats, flood plains, rock outcrops, cliffs and ledges, on limestone, serpentine, sandstone, granitic and basaltic substrates; 0–1800 m; B.C.; Ariz., Calif., Idaho, Mont., Nev., Oreg., Wash.; Mexico (Baja California).

Records of *Athysanus pusillus* for Montana appear to be solely from Ravalli County, and for Nevada from Storey and Washoe counties.

2. **Athysanus unilateralis** (M. E. Jones) Jepson, Fl. W. Calif., 224. 1901

Draba unilateralis M. E. Jones, Bull. Torrey Bot. Club 9: 124. 1882; *Heterodraba unilateralis* (M. E. Jones) Greene

Plants pubescent. **Stems** (0.3–)0.7–2.5(–3.5) dm. **Basal leaves** subsessile or petiole to 0.3 cm; blade oblanceolate, obovate, or oblong, (0.3–)0.7–2(–2.2) cm × (2–)4–7 mm, margins entire or with a tooth on each side, apex obtuse, surfaces uniformly pubescent, trichomes cruciform with fewer, 3-rayed ones. **Cauline leaves** 2–5; blade similar to but smaller than basal, (base cuneate, not auriculate). **Fruiting pedicels** stout, 1–2 (–3) mm. **Flowers:** sepals 0.6–1 × 0.4–0.6 mm; petals spatulate, 1.3–1.7 × 0.5–0.8 mm; filaments 0.5–0.9 mm; anthers 0.1–0.2 mm. **Fruits** elliptic, obovate, or orbicular, flattened or slightly twisted; valves thickened, veins often prominent, glabrous or uniformly pubescent, trichomes subsessile or short-stalked, cruciform and 3-rayed, sometimes mixed with stouter, simple or forked, subsetiform ones to 0.6 mm, not hooked; septum complete, (membranous); ovules 6–12 per ovary; style 0.1–0.2 mm. **Seeds** often 2 or more, 0.9–1.2 × 0.6–0.8 mm.

Flowering Feb–May. Open grassy slopes and flats, friable clay, flood plains, adobe and heavy gumbo clay, gypsum clay slopes; 100–900 m; Calif., Oreg.; Mexico (Baja California).

It appears that *Athysanus unilateralis* is known in Oregon only from Jackson County; it is more widespread in California.

8. **AUBRIETA** Adanson, Fam. Pl. 2: 420. 1763 • [For Claude Aubriet, 1663–1743, French artist] [I]

Ihsan A. Al-Shehbaz

Perennials; (often loosely pulvinate or cespitose, caudex many-branched); not scapose; pubescent, trichomes stellate, short-stalked or sessile, mixed with coarser, stalked, forked, and simple ones. **Stems** erect to decumbent, branched basally, (slender). **Leaves** basal and cauline; petiolate or sessile [subsessile]; basal rosulate, petiolate, blade margins entire or dentate; cauline petiolate or sessile [subsessile], blade (base not auriculate), margins entire or dentate. **Racemes** (few- to several-flowered), elongated in fruit. **Fruiting pedicels** erect to divaricate, slender. **Flowers:** sepals erect, oblong, lateral pair saccate basally, (glabrous or pubescent); petals usually purple to violet, rarely white [pink], obovate [spatulate], (apex obtuse); stamens tetradynamous; filaments narrowly winged, (lateral pair with toothed appendage); anthers oblong [ovate]; nectar glands lateral, semi-annular, extrastaminal. **Fruits** siliques or, rarely, silicles, sessile, ellipsoid, [linear, oblong, or elliptic], not torulose, terete or latiseptate; valves each with distinct midvein, pubescent or, rarely, glabrous; replum rounded; septum usually complete, sometimes perforate; ovules 10–40 per ovary; (style persistent); stigma capitate. **Seeds** biseriate, flattened, not winged, ovoid [elliptical]; seed coat not mucilaginous when wetted; cotyledons accumbent.

Species 15 (1 in the flora): introduced, California; Europe, sw Asia, nw Africa.

Aubrieta is taxonomically challenging and is centered primarily in Greece and Turkey. The delimitation of species is often difficult, possibly because species have resulted from hybridization.

SELECTED REFERENCES Mattfield, J. 1939. The species of the genus *Aubrieta* Adanson. Quart. Bull. Alpine Gard. Soc. 7: 157–181, 217–227. Rollins, R. C. 1982. Another alien in the California flora. Rhodora 84: 153–154.

1. Aubrieta deltoidea (Linnaeus) de Candolle, Syst. Nat. 2: 294. 1821 F I

Alyssum deltoideum Linnaeus, Sp. Pl. ed. 2, 2: 908. 1763

Plants forming mats or cushions; densely pubescent, trichomes stellate, mixed with fewer, setiform and forked ones. **Stems** several from base (caudex), ascending to procumbent, 0.7–3(–5) dm, pubescent. **Basal leaves:** petiole 0.1–1 cm; blade obovate, oblanceolate, or rhombic, (0.5–)1–3(–4.5) cm × (2–)4–13(–20) mm, base cuneate to attenuate, margins entire or 1–3 teeth on each side, surfaces densely pubescent. **Cauline leaves:** petiolate or (distalmost) sessile; blade similar to basal. **Racemes** 1–13-flowered, (lax). **Fruiting pedicels** erect to ascending, 5–12(–16) mm. **Flowers:** sepals 6–10 × 1–1.5 mm; petals (10–)15–28 × 4–7(–8) mm, (attenuate to claw, 5–12 mm); filaments 5–10 mm; anthers 1.2–1.6 mm. **Fruits** terete or slightly flattened, 0.7–2(–2.8) cm × 2–4(–4.8) mm; valves: trichomes long-setiform and forked, mixed with smaller, stellate ones; style 4–12 mm. **Seeds** 1.2–16 × 0.7–1 mm. $2n = 16$.

Flowering Apr–Jun. Rock crevices; ca. 1600 m; introduced; Calif.; s Europe (Mediterranean region); sw Asia; nw Africa.

Aubrieta deltoidea is known as an escape from the Mt. Hull area at the Mendocino-Lake counties boundary. It is highly variable in its native range, and several infraspecific taxa have been recognized.

9. DRABA Linnaeus, Sp. Pl. 2: 642. 1753; Gen. Pl. ed. 5, 291. 1754 • [Greek *drabe*, acrid, for taste of mustard plant]

Ihsan A. Al-Shehbaz

Michael D. Windham

Reidar Elven

Abdra Greene; *Erophila* de Candolle; *Nesodraba* Greene; *Tomostima* Rafinesque

Annuals, biennials, or perennials [subshrubs]; (caudex simple or branched); scapose or not; glabrous or pubescent, trichomes stalked or sessile, simple, forked, cruciform, stellate, malpighiaceous, or dendritic, often more than 1 kind present. **Stems** usually erect to ascending, sometimes decumbent or prostrate, unbranched or branched (usually distally). **Leaves** usually basal and cauline, sometimes cauline absent; petiolate or sessile; basal usually rosulate, usually petiolate, rarely sessile, blade margins usually entire or toothed, rarely pinnately lobed; cauline (when present), petiolate or sessile, blade (base cuneate [auriculate]), margins entire or dentate. **Racemes** (often corymbose, sometimes bracteate), elongated or not in fruit. **Fruiting pedicels** (proximalmost) erect or ascending to divaricate, slender. **Flowers:** sepals (rarely persistent), erect, ascending, or, rarely, spreading, ovate or oblong [elliptic], lateral pair not saccate or subsaccate basally; petals (erect or ascending to patent), yellow, white, pink, purple, or orange [red], obovate, spatulate, oblanceolate, or linear [orbicular, oblong], (longer than or, rarely, shorter than sepals), claw obscurely to well-differentiated from blade, (apex obtuse, rounded, notched, or, rarely, deeply 2-lobed); stamens slightly to strongly tetradynamous; filaments dilated or not basally, (glabrous); anthers ovate or oblong, (not apiculate); nectar glands (1, 2, or 4), distinct or confluent, subtending bases of stamens, median glands present or absent. **Fruits** silicles or siliques, sessile, ovate, lanceolate, elliptic, oblong, linear, suborbicular, ovoid, or subglobose, plane or spirally twisted, smooth, (not keeled, unappendaged), usually latiseptate, rarely terete; valves (papery), each with distinct or obscure midvein and lateral veins, glabrous or pubescent; replum rounded; septum complete; ovules 4–70(–88) per ovary; style obsolete or distinct; stigma capitate. **Seeds** biseriate, oblong, ovate, or orbicular, usually flattened (slightly flattened in

D. aleutica, *D. verna*), usually not winged (winged in *D. asterophora*, *D. brachycarpa*, *D. carnosula*, *D. pterosperma*); seed coat (minutely reticulate), not mucilaginous when wetted; cotyledons accumbent. $x = 6–12$.

Species ca. 380 (121 in the flora): North America, Mexico, South America (Andes, Colombia to Patagonia), Europe, Asia, nw Africa; alpine and boreal, rarely in temperate and low-elevation areas of North America and Eurasia.

Draba, the largest genus in Brassicaceae, is a well-defined, monophyletic, and complex group represented by native species on all continents except Australia and Antarctica. It is poorly developed in Africa. The four centers of highest diversity include western North America, subarctic regions, Himalaya, and high Andes.

The infrageneric classification of *Draba* is problematic, and preliminary molecular studies (M. Koch and I. A. Al-Shehbaz 2002) do not support the circumscriptions of most of the 17 sections recognized by O. E. Schulz (1927, 1936). A. I. Tolmatchew (1939, 1975) recognized representatives of 29 series occurring within the former Soviet Union, and his classification is far more practical that Schulz's, though it covers only a small portion of the genus. An overall comprehensive infrageneric taxonomy of *Draba* is needed.

Much of the complexity within *Draba* is attributed to hybridization, disploidy, polyploidy, and apomixis; the interested reader should consult E. Ekman (1932b), T. W. Böcher (1966), C. Brochmann (1992, 1992b, 1993), Brochmann and R. Elven (1992), Brochmann et al. (1992, 1992b, 1992c, 1992d, 1993), G. A. Mulligan (1970, 1971, 1971b, 1972, 1974, 1975, 1976), Mulligan and J. N. Findlay (1970), and M. D. Windham (2000, 2004). Except for the results of some studies (e.g., Mulligan and Findlay; Brochmann et al. 1992d; Brochmann 1993; H. H. Grundt et al. 2005), little is known about the reproductive biology of *Draba* species. The polyphyletic origins of some species (e.g., *D. cacuminum*, *D. corymbosa*, *D. lactea*) have been well-documented (Brochmann et al. 1992b, 1992c; Grundt et al. 2004).

Knowledge of chromosome numbers, combined with a critical study of morphology, is essential in the delimitation of *Draba* species and in understanding their origin. In addition to the works by T. W. Böcher, C. Brochmann, H. H. Grundt, G. A. Mulligan, and M. D. Windham cited above, those by O. Heilborn (1927) and G. Knaben (1966) also are pertinent. Although the base chromosome number in the genus ranges from six to twelve, the majority of the species (ca. 67%) are based on eight, as compiled from S. I. Warwick and I. A. Al-Shehbaz (2006). The highest chromosome number in the genus, $2n = 144$ (18-ploid), was reported by Mulligan (1974b) for *D. corymbosa*, and a complex, intraspecific disploid series (see below) was documented in *D. verna* (Ø. Winge 1940).

Although *Erophila* is recognized in recent floras as distinct from *Draba*, the sole characteristic on which it is based (petal apex deeply 2-fid versus emarginate to entire in *Draba*) is unreliable, and it is well-nested within the larger and highly diversified *Draba* (M. Koch and I. A. Al-Shehbaz 2002).

Because some superficially similar species of *Draba* grow sympatrically, extra care must be taken in recording flower color and in collecting voucher material for cytological or molecular studies. Cases in point are *Thompson & Thompson 578a* (DAO), a collection that consists of four species (see 2. *D. albertina*), and *Bruggeman 41* (DAO), one that includes *D. fladnizensis*, *D. lactea*, and *D. nivalis*.

The number of ovules per ovary is taxonomically important in *Draba*, but this is rather difficult to determine in the ovary. Counting total number of aborted ovules plus mature seeds per fruit should determine the overall ovule number. The use of seed number per fruit often is unreliable because tremendous variation can be found on the same plant as a result of the failure of some ovules to develop into mature seeds. The number of flowers is counted from the main raceme of a plant with branched stems, the length of style is measured from mature fruits, and the size of basal leaves is taken from the largest leaves of the rosette. In order to accurately determine a given collection to species, complete specimens with flowers and mature fruits are needed. Finally, although most authors of this account prefer to use the subspecific rather the varietal rank, no attempt was made here to enforce the choice of either; no new combinations have been made.

SELECTED REFERENCES Beilstein, M. A. and M. D. Windham. 2003. A phylogenetic analysis of western North American *Draba* (Brassicaceae) based on nuclear ribosomal DNA sequences from the ITS region. Syst. Bot. 28: 584–592. Ekman, E. 1929. Studies in the genus *Draba*. Svensk Bot. Tidskr. 23: 476–495. Ekman, E. 1930. Contribution to the *Draba* flora of Greenland. II. Svensk Bot. Tidskr. 24: 280–297. Ekman, E. 1931. Contribution to the *Draba* flora of Greenland. III. Some notes on the arctic, especially the Greenland drabas of the sections *Aizopsis* and *Chrysodraba* DC. Svensk Bot. Tidskr. 25: 465–494. Ekman, E. 1932. Contribution to the *Draba* flora of Greenland. IV. Svensk Bot. Tidskr. 26: 431–447. Fernald, M. L. 1934. *Draba* in temperate northeastern America. Rhodora 36: 241–261, 285–305, 314–344, 353–371, 392–404. Hitchcock, C. L. 1941. A Revision of the Drabas of Western North America. Seattle. [Univ. Wash. Publ. Biol. 11.] Koch, M. and I. A. Al-Shehbaz. 2002. Molecular data indicate complex intra- and intercontinental differentiation of American *Draba* (Brassicaceae). Ann. Missouri Bot. Gard. 89: 88–109. Mulligan, G. A. 1976. The genus *Draba* in Canada and Alaska: Key and summary. Canad. J. Bot. 54: 1386–1393. Payson, E. B. 1917. The perennial scapose drabas of North America. Amer. J. Bot. 4: 253–267. Schulz, O. E. 1927. Cruciferae—*Draba*, *Erophila*. In: H. G. A. Engler, ed. 1900–1953. Das Pflanzenreich.... 107 vols. Berlin. Vol. 89[IV,105], pp. 1–396.

Key to Species Groups of *Draba*

1. Annuals or biennials, without caudex or stolons, not pulvinate, not cespitose............. Group 1, p. 271
1. Perennials, with caudex or, rarely, stolons, sometimes pulvinate or cespitose.
    2. Cauline leaves of flowering stems 1+.
        3. Abaxial surface of leaf blades glabrous or with simple trichomes................ Group 2, p. 273
        3. Abaxial surface of leaf blades with only branched trichomes.
            4. Fruit valves glabrous .... Group 3, p. 275
            4. Fruit valves pubescent or puberulent (at least on margin).
                5. Abaxial surface of leaf blades with at least some 7–15-rayed trichomes ..... Group 4, p. 277
                5. Abaxial surface of leaf blades with 2–4(–6)-rayed trichomes .... Group 5, p. 278
    2. Cauline leaves of flowering stems 0.
        6. Rachises glabrous.
            7. Abaxial surface of leaf blades glabrous (sometimes trichomes only on margins and apices).. Group 6, p. 279
            7. Abaxial surface of leaf blades pubescent .... Group 7, p. 280
        6. Rachises sparsely to densely pubescent.
            8. Leaf blade margins not ciliate.................................. Group 8, p. 283
            8. Leaf blade margins ciliate.
                9. Abaxial surface of leaf blades usually with simple or simple and branched trichomes, rarely glabrous............................. Group 9, p. 284
                9. Abaxial surface of leaf blades with branched trichomes .... Group 10, p. 286

Group 1

1. Cauline leaves 0.
    2. Petals deeply 2-fid; rachises of fruiting racemes flexuous........................117. *Draba verna*
    2. Petals usually obtuse, rarely emarginate; rachises of fruiting racemes usually straight, if flexuous then petals yellow.
        3. Petals white; late-season flowers apetalous; fruit valves usually with antrorse, simple trichomes, rarely glabrous.
            4. Racemes 10–50(–70)-flowered, elongated in fruit, rachises densely pubescent; fruiting pedicels densely pubescent........................31. *Draba cuneifolia* (in part)
            4. Racemes (3–)5–12(–16)-flowered, subumbellate in fruit, rachises usually glabrous, rarely sparsely pubescent; fruiting pedicels glabrous or sparsely pubescent.......................... 92. *Draba reptans* (in part)
        3. Petals yellow; late-season flowers with petals; fruit valves usually glabrous, rarely puberulent.

5. Abaxial surface of leaf blades glabrous or sparsely pubescent, trichomes simple and 2-rayed; stems usually glabrous, rarely pubescent proximally, trichomes simple. . . . . 29. *Draba crassifolia* (in part)
5. Abaxial surface of leaf blades pubescent, trichomes 2–4-rayed; stems pubescent, trichomes 2–4-rayed (sometimes simple).
   6. Stems proximally with simple and 2-rayed trichomes; adaxial surface of leaf blades with simple and 2-rayed trichomes; anthers 0.15–0.25 mm; fruits (4–)6–12(–15) mm . . . . 2. *Draba albertina* (in part)
   6. Stems proximally with 2–4-rayed trichomes; adaxial surface of leaf blades with 2–4-rayed trichomes; anthers 0.25–0.4 mm; fruits (8–)10–17(–20) mm
. . . . .108. *Draba stenoloba* (in part)
1. Cauline leaves 1–54(–76).
   7. Styles (0.8–)1–3.5 mm.
      8. Adaxial surface of basal leaf blade pubescent with simple, subsetiform trichomes, rarely mixed with 2-rayed ones; fruit valves sparsely puberulent, trichomes simple
. . . . 15. *Draba bifurcata* (in part)
      8. Adaxial surface of basal leaf blade pubescent with at least some 2–4-rayed trichomes; fruit valves usually pubescent, rarely glabrous, trichomes 2–4-rayed.
         9. Petals linear, 0.2–0.5 mm wide; s California. . . . . 26. *Draba corrugata* (in part)
         9. Petals oblanceolate to spatulate, 1–3.5 mm wide; se Arizona, Colorado, New Mexico, Utah.
            10. Fruits plane, valves pubescent, trichomes 2–4-rayed; petals 3–5 × 1–1.5 mm, often fading whitish; Utah . . . . 94. *Draba santaquinensis* (in part)
            10. Fruits slightly to strongly twisted, valves glabrous, pubescent, or puberulent, trichomes simple and 2(–4)-rayed; petals 5–8.5 × 1.5–3.5 mm, not fading whitish; Arizona, Colorado, New Mexico.
               11. Cauline leaves (8–)12–31(–43); rachises pubescent; fruiting pedicels 4–10(–13) mm. . . . . .44. *Draba helleriana* (in part)
               11. Cauline leaves 1–3; rachises often glabrous or, rarely, sparsely pubescent; fruiting pedicels (6–)9–18(–22) mm . . . . .63. *Draba mogollonica* (in part)
   7. Styles 0.01–0.6 (–0.8) mm.
      12. Racemes bracteate basally; fruiting pedicels erect to ascending (subappressed to rachises); cauline leaves (15–)22–54(–76), densely overlapping proximally. . . . . 47. *Draba incana*
      12. Racemes ebracteate; fruiting pedicels horizontal to divaricate-ascending (not appressed to rachises); cauline leaves 1–15(–25), seldom overlapping.
         13. Rachises glabrous.
            14. Petals white or, rarely, absent; basal leaves not rosulate; racemes subumbellate in fruit . . . . 92. *Draba reptans* (in part)
            14. Petals yellow; basal leaves rosulate; racemes elongated in fruit.
               15. Cauline leaves 4–12(–15); fruiting pedicels 7–28(–35) mm, (1.5–)2–7 times longer than fruit. . . . . 67. *Draba nemorosa*
               15. Cauline leaves 1–3(–5); fruiting pedicels (1–)3–14(–19) mm, subequaling or shorter than fruit.
                  16. Abaxial surface of leaf blades glabrous or sparsely pubescent, trichomes simple and 2-rayed; stems usually glabrous, or, rarely, sparsely pubescent proximally, trichomes simple. . . . . 29. *Draba crassifolia* (in part)
                  16. Abaxial surface of leaf blades pubescent, trichomes 2–4-rayed; stems pubescent basally, trichomes 2–4-rayed (sometimes simple).
                     17. Stems proximally with simple and 2-rayed trichomes; adaxial surface of leaf blades with simple and 2-rayed trichomes; anthers 0.15–0.25 mm; fruits (4–)6–12(–15) mm . . . .
. . . . 2. *Draba albertina* (in part)
                     17. Stems proximally with 2–4-rayed trichomes; adaxial surface of leaf blades with 2–4-rayed trichomes; anthers 0.25–0.4 mm; fruits (8–)10–17(–20) mm . . . . .108. *Draba stenoloba* (in part)

[13. Shifted to left margin.—Ed.]
13. Rachises pubescent.
    18. Leaf blade surfaces with cruciform trichomes.
        19. Fruit valves pubescent; ovules 4–6(–8) per ovary .... 5. *Draba aprica*
        19. Fruit valves glabrous; ovules 8–16(–20) per ovary .... 17. *Draba brachycarpa*
    18. Leaf blade surfaces with simple and 2–7-rayed trichomes.
        20. Fruits obovate to obovate-oblong...85. *Draba platycarpa*
        20. Fruits elliptic, oblong, linear, lanceolate, or, rarely, ovate.
            21. Late-season flowers apetalous; basal leaves not rosulate.
                22. Racemes 10–50(–70)-flowered, elongated in fruit, rachises densely pubescent; fruiting pedicels densely pubescent...............31. *Draba cuneifolia* (in part)
                22. Racemes (3–)5–12(–16)-flowered, subumbellate in fruit, rachises glabrous or, rarely, sparsely pubescent; fruiting pedicels glabrous or sparsely pubescent... 92. *Draba reptans* (in part)
            21. Late-season flowers with petals; basal leaves rosulate.
                23. Abaxial surface of leaf blades pubescent, trichomes 3–6-rayed; petals white ....87. *Draba praealta* (in part)
                23. Abaxial surface of leaf blades pubescent, trichomes 2–4-rayed; petals yellow (sometimes fading white).
                    24. Styles 0.2–0.6(–0.8) mm; fruits 2–3.5 mm wide, valves pubescent, trichomes simple and 2-rayed .... 18. *Draba brachystylis* (in part)
                    24. Styles 0.01–0.15 mm; fruits 1.3–2.3 mm wide, valves glabrous or pubescent, trichomes simple.
                        25. Cauline leaves 3–10(–17); racemes (15–)20–51(–62)-flowered; fruits lanceolate, 6–9(–11) mm; Arizona, Colorado, New Mexico, Utah .... 91. *Draba rectifructa*
                        25. Cauline leaves 1 or 2(–4); racemes (2–)4–10(–15)-flowered; fruits usually linear, rarely linear-elliptic, (8–)10–17(–20) mm; Alaska, Alberta, British Columbia, Washington, Yukon ..... 108. *Draba stenoloba* (in part)

Group 2

1. Plants stoloniferous; basal leaves not rosulate (subopposite) .... 72. *Draba ogilviensis* (in part)
1. Plants not stoloniferous (usually cespitose); basal leaves rosulate.
    2. Abaxial surface of leaf blades glabrous.
        3. Fruit valves pubescent, trichomes simple and 2–4-rayed; styles (0.7–)1.6–3 mm; petals 5.5–8 mm; n California and s Oregon....................... 46. *Draba howellii* (in part)
        3. Fruit valves glabrous or puberulent, trichomes simple; styles 0.02–1.2(–1.8) mm; petals 1.5–5(–6) mm; not California nor Oregon.
            4. Racemes bracteate; basal leaf blades linear to linear-oblanceolate .... 40. *Draba graminea*
            4. Racemes ebracteate; basal leaf blades usually obovate to oblanceolate, rarely linear-lanceolate.
                5. Cauline leaves 1; fruits plane; styles 0.02–0.4(–0.6) mm.
                    6. Petals yellow (often fading white), oblanceolate, 1.5–2.5(–3) × 0.5–1 mm; stems glabrous or with simple trichomes; styles 0.02–0.1 mm .... 29. *Draba crassifolia* (in part)
                    6. Petals white, obovate, 3–5 × 1.8–3 mm; stems glabrous or with 2–8-rayed trichomes; styles 0.1–0.4(–0.6) mm................55. *Draba lactea* (in part)
                5. Cauline leaves (1 or) 2–8; fruits slightly twisted or plane; styles (0.4–)0.7–1.2(–1.8) mm.
                    7. Basal leaves undifferentiated into blade and petiole, 0.5–0.8 cm, persisting as a whole, imbricate, blade margins with setiform trichomes .... 43. *Draba heilii* (in part)
                    7. Basal leaves differentiated into blade and petiole, 1.2–8.5 cm, petiole persisting, not imbricate, blade margins without setiform trichomes.

8. Fruits lanceolate to ovate-lanceolate, (7–)8–14 × 3–5 mm; rachises pubescent (trichomes crisped); Colorado, Montana, Utah Wyoming .... 28. *Draba crassa*
8. Fruits linear-elliptic to elliptic, 5–10(–13) × 1.5–2.5 mm; rachises usually glabrous, rarely pubescent (trichomes non-crisped); se Arizona, s New Mexico, w Texas .... 107. *Draba standleyi* (in part)

[2. Shifted to left margin.—Ed.]

2. Abaxial surface of leaf blades pubescent.
   9. Fruit valves pubescent, trichomes 2–12-rayed.
      10. Styles 1.4–3.4 mm; cauline leaves (4–)6–10(–13); racemes (10–)18–55(–67)-flowered; petals yellow, linear .... 26. *Draba corrugata* (in part)
      10. Styles 0.1–0.8 mm; cauline leaves 1–3; racemes 2–20-flowered; petals white, spatulate to obovate.
         11. Fruits 2–5 × 1.5–2 mm; ovules 36–52 per ovary; petals 1.5–2 × 0.7–0.8 mm. .... 120. *Draba yukonensis*
         11. Fruits 4–9.5 × 2–4 mm; ovules 20–36 per ovary; petals 3–5 × 1.5–2.5 mm.
            12. Fruits with 2–4-rayed and, sometimes, fewer, simple trichomes; petals 3–4 mm; stems proximally with branched trichomes 2–6-rayed ..... 8. *Draba arctogena* (in part)
            12. Fruits with (2–)5–12-rayed trichomes; petals 3.5–5 mm; stems proximally with branched trichomes 4–10-rayed .... 71. *Draba oblongata* (in part)
   9. Fruit valves glabrous, pubescent, or puberulent, trichomes simple.
      13. Styles (0.7–)1–3 mm; fruits usually twisted, rarely plane; petals (4–)4.5–7.5 mm.
         14. Cauline leaf blade margins dentate; fruiting pedicels horizontal to divaricate; se Arizona .... 15. *Draba bifurcata* (in part)
         14. Cauline leaf blade margins entire; fruiting pedicels usually ascending, rarely divaricate-ascending; Arizona, Colorado, New Mexico, Texas, Wyoming.
            15. Fruits usually twisted (to 3 turns), rarely plane; stems proximally with trichomes 0.4–2.2 mm; ovules 20–34 per ovary; c Colorado, n New Mexico, se Wyoming .... 111. *Draba streptocarpa*
            15. Fruits usually twisted (to 1 turn), rarely plane; stems proximally with trichomes 0.1–0.7 mm; ovules 12–24 per ovary; se Arizona, s New Mexico, w Texas.
               16. Basal leaves undifferentiated into blade and petiole, 0.5–0.8 cm, persisting as a whole, imbricate, blade margins with setiform trichomes. .... 43. *Draba heilii* (in part)
               16. Basal leaves differentiated into blade and petiole, 1.2–8.5 cm, petiole persisting, not imbricate, blade margins without setiform trichomes .... 107. *Draba standleyi* (in part)
      13. Styles 0.01–1(–1.2) mm; fruits not or, rarely, slightly twisted; petals 1.5–4(–4.5) mm.
         17. Annuals, biennials, or perennials (not cespitose); fruiting pedicels (3–)5–14(–16) mm.
            18. Abaxial surface of leaf blades pubescent with 2–4-rayed trichomes; stems pubescent, trichomes simple and 2-rayed .... 2. *Draba albertina* (in part)
            18. Abaxial surface of leaf blades glabrous or with some simple and 2-rayed trichomes; stems usually glabrous, rarely pubescent basally, trichomes simple .... 29. *Draba crassifolia* (in part)
         17. Perennials (cespitose); fruiting pedicels 1–6 mm.
            19. Petals yellow, 1–2 mm wide; styles (0.2–)0.3–1.2 mm.
               20. Abaxial surface of leaf blades with sessile or subsessile, appressed, 2–4-rayed trichomes; fruits ovate, 4–8 mm; styles 0.2–0.3 mm ...... 119. *Draba weberi*
               20. Abaxial surface of leaf blades with stalked 2- or 3-rayed trichomes; fruits lanceolate to oblong-lanceolate, 4–13 mm; styles 0.3–1.2 mm.

21. Petals 2–2.5(–3) mm, erect, not clawed; anthers oblong, 0.6–0.9 mm; fruits plane; rachises usually glabrous, rarely sparsely pubescent .... 36. *Draba exunguiculata*
21. Petals 3–4.5 mm, not erect (flared, clawed); anthers ovate, 0.3–0.5 mm; fruits twisted or plane; rachises pubescent .... 42. *Draba grayana*

[19. Shifted to left margin.—Ed.]

19. Petals white, 0.4–1(–1.5) mm wide; styles 0.05–0.2(–0.3) mm.
   22. Fruit valves glabrous; stems glabrous or, rarely, sparsely pubescent (trichomes straight); petals 2–3.5 × 0.8–1.5 mm .... 37. *Draba fladnizensis* (in part)
   22. Fruit valves usually pubescent, rarely glabrous; stems pubescent (trichomes crisped); petals 1.5–2 × 0.4–0.8 mm.
       23. Racemes 2–7-flowered, rachises glabrous; sepals caducous; fruits lanceolate, flattened; Yukon .... 54. *Draba kluanei*
       23. Racemes (3–)6–13(–17)-flowered, rachises pubescent or, rarely, glabrous; sepals persistent; fruits ovoid to subellipsoid, slightly inflated basally; e California .... 64. *Draba monoensis* (in part)

Group 3

1. Basal leaf blade surfaces with malpighiaceous trichomes .... 61. *Draba malpighiacea*
1. Basal leaf blade surfaces with simple, 2-rayed, forked, stellate, or dendritic trichomes.
   2. Abaxial surface of leaf blades with some 7–15-rayed trichomes.
      3. Petals yellow; stem and basal leaf trichomes crisped .... 110. *Draba streptobrachia* (in part)
      3. Petals white or creamy white; stem and basal leaf trichomes usually non-crisped.
         4. Margins of basal and cauline leaf blades entire.
            5. Plants usually forming mats; rachises slightly flexuous.
               6. Fruits 6–15(–18) mm; adaxial surface of leaf blades with simple and long-stalked, 2-rayed trichomes .... 57. *Draba lonchocarpa* (in part)
               6. Fruits 3.5–9 mm; adaxial surface of leaf blades with short-stalked, 8–15-rayed trichomes .... 68. *Draba nivalis* (in part)
            5. Plants not forming mats; rachises not flexuous.
               7. Rachises pubescent; ovules 22–32 per ovary; sepals persistent (until fruiting); e California, w Nevada .... 21. *Draba californica* (in part)
               7. Rachises glabrous; ovules 12–16 per ovary; sepals caducous; Alaska, Alberta, w Montana, nw Wyoming.................... 86. *Draba porsildii* (in part)
         4. Margins of basal and, sometimes, cauline leaf blades usually dentate or denticulate, if entire, racemes not flexuous in fruit.
            8. Cauline leaves usually 0, rarely 1 (as a bract); basal leaf blades each with prominent midvein; sepals glabrous or, rarely, with simple trichomes subapically.. 55. *Draba lactea* (in part)
            8. Cauline leaves usually 2–25, rarely 1 (*D. chamissonis*); basal leaf blades each with obscure midvein; sepals with 2–5-rayed trichomes.
               9. Fruits 1.1–1.6(–2) mm wide; petals 2–3.5 mm; petiole margins not ciliate; proximalmost flowers bracteate...................... 24. *Draba chamissonis*
               9. Fruits 2–3.5 mm wide; petals 3.5–5.5 mm; petiole margins ciliate; proximalmost flowers usually ebracteate (sometimes bracteate in *D. glabella*).
                  10. Abaxial surface of leaf blades with sessile, stellate trichomes; styles (0.2–)0.4–1 mm; stems proximally with 8–12-rayed trichomes .... 6. *Draba arabisans* (in part)
                  10. Abaxial surface of leaf blades with short-stalked, stellate trichomes; styles 0.05–0.2(–0.5) mm; stems proximally with some simple trichomes .... 38. *Draba glabella* (in part)

[2. Shifted to left margin.—Ed.]
2. Abaxial surface of leaf blades with 2–5 (or 6)-rayed trichomes.
   11. Stem and leaf trichomes sessile, 2 rays parallel to long axis of stem or midvein, some malpighiaceous .... 104. *Draba spectabilis* (in part)
   11. Stem and leaf trichomes stalked, rays not parallel to long axis of stem or midvein, not malpighiaceous.
      12. Caudex fleshy, simple; petiole of basal leaves (1–)4–15 cm; fruits 4–7(–9) mm wide .... 41. *Draba grandis*
      12. Caudex not fleshy, usually branched, rarely simple; petiole of basal leaves obsolete or 1–2 cm; fruits 1–4(–6) mm wide.
         13. Abaxial surface of leaf blades with either cruciform or pectinate trichomes.
            14. Plants pulvinate, caudex densely branched; abaxial surface of leaf blades with pectinate trichomes; ovules 8–16(–20) per ovary .... 48. *Draba incerta* (in part)
            14. Plants not pulvinate, caudex loosely branched; abaxial surface of leaf blades with cruciform trichomes; ovules 16–30 per ovary ...... 52. *Draba juvenilis* (in part)
         13. Abaxial surface of leaf blades with cruciform and 2–6-rayed trichomes.
            15. Seeds 1.8–2.8 × 1.2–2 mm, wing 0.5–1 mm wide; fruits (3.5–)4–6 mm wide .... 12. *Draba asterophora* (in part)
            15. Seeds 0.7–1.4(–1.6) × 0.4–0.8(–1) mm, not winged; fruits 1–4(–5) mm wide.
               16. Styles (0.6–)0.8–3.6 mm; petals 4–8.5 mm; anthers usually oblong (ovate in *D. murrayi*).
                  17. Petals white; anthers ovate, 0.4–0.5 mm; fruits 1.5–2 mm wide; styles 0.6–1.5 mm .... 66. *Draba murrayi*
                  17. Petals yellow; anthers oblong, 0.6–1 mm; fruits 2–5 mm wide; styles (0.8–)1–3.6 mm.
                     18. Fruits plane; ovules 10–18 per ovary; basal leaf blades (2–)4–10(–14) mm wide; cauline leaves (3–)5–9(–11); petals 4–6 mm .... 1. *Draba abajoensis*
                     18. Fruits slightly twisted; ovules 24–36 per ovary; basal leaf blades 7–25(–33) mm wide; cauline leaves 1–3; petals (5–)6–8.5 mm .... 63. *Draba mogollonica* (in part)
               16. Styles 0.01–0.6 mm; petals 1.8–4 mm; anthers ovate.
                  19. Rachises usually glabrous, (rarely sparsely pubescent in *D. stenoloba*); fruits usually linear, elliptic, or linear-lanceolate, rarely lanceolate, 1–2(–2.5) mm wide; styles 0.01–0.15(–0.2) mm.
                     20. Sepals pubescent, trichomes simple and 2- or 3-rayed; abaxial surface of leaf blades with trichomes (2–)4(–6)-rayed .... 78. *Draba paucifructa* (in part)
                     20. Sepals glabrous or sparsely pubescent, trichomes simple; abaxial surface of leaf blades with trichomes 2–4-rayed.
                        21. Stems proximally with simple and 2-rayed trichomes; adaxial surface of leaf blades with simple and 2-rayed trichomes; anthers 0.15–0.25 mm; fruits (4–)6–12(–15) mm .... 2. *Draba albertina* (in part)
                        21. Stems proximally with 2–4-rayed trichomes; adaxial surface of leaf blades with 2–4-rayed trichomes; anthers 0.25–0.4 mm; fruits (8–)10–17(–20) mm .... 108. *Draba stenoloba* (in part)
                  19. Rachises pubescent; fruits lanceolate to oblong or ovoid, (1.8–)2–4 mm wide; styles 0.1–0.6 mm.
                     22. Petals yellow; fruits ovoid to ovoid-lanceolate, inflated (at least basally); ovules (4–)8–12 per ovary .... 103. *Draba sobolifera* (in part)
                     22. Petals white or creamy white; fruits oblong, lanceolate, or lanceolate-elliptic, flattened; ovules 18–32 per ovary.

[23. Shifted to left margin.—Ed.]

23. Fruiting pedicels pubescent, trichomes 4–8-rayed; sepals persistent; petals 0.8–1 mm wide; e California, w Nevada . . . . 21. *Draba californica* (in part)
23. Fruiting pedicels pubescent, trichomes simple and 2–4-rayed; sepals caducous; petals 1.7–2.5 mm wide; Canada, Greenland. . . . . .69. *Draba norvegica* (in part)

Group 4

1. Fruit trichomes 2–7-rayed (at least on replum); petals white.
   2. Racemes bracteate (at least basally).
      3. Racemes (10–)15–47(–63)-flowered, often considerably elongated in fruit; ovules 28–48 per ovary . . . . 22. *Draba cana*
      3. Racemes 2–5(–8)-flowered, slightly elongated in fruit; ovules 16–24 per ovary . . . . 50. *Draba inexpectata*
   2. Racemes ebracteate.
      4. Petals 2–3.5 × 0.7–1.4 mm; fruits 1.5–2.5 mm wide.
         5. Cauline leaves 1–3(–6); fruits usually twisted, rarely plane; rachises not flexuous in fruit; ovules 28–40 per ovary; California . . . . 19. *Draba breweri* (in part)
         5. Cauline leaf 1; fruits plane; rachises often slightly flexuous in fruit; ovules 12–24(–28) per ovary; Alaska, Canada, Greenland . . . . 68. *Draba nivalis* (in part)
      4. Petals 3.5–6 × 1.5–2.5 mm; fruits 2–3.5 mm wide.
         6. Basal leaf blade margins dentate or denticulate; fruits usually slightly twisted, rarely plane; styles (0.2–)0.4–1 mm . . . . 6. *Draba arabisans* (in part)
         6. Basal leaf blade margins usually entire, rarely 1-toothed on each side; fruits plane; styles 0.1–0.6 mm.
            7. Basal leaf blades with simple trichomes apically, abaxial surfaces with distinct midveins; seeds 0.8–1.1 × (0.6–)0.7–0.8 mm . . . . 7. *Draba arctica* (in part)
            7. Basal leaf blades without simple trichomes apically, abaxial surfaces with obscure midveins; seeds 0.6–0.8 × 0.4–0.6 mm . . . . 25. *Draba cinerea* (in part)
1. Fruit trichomes simple and 2–5-rayed; petals white or yellow.
   8. Petals yellow.
      9. Ovules 28–38(–44) per ovary; racemes basally bracteate, (10–)18–52(–72)-flowered; cauline leaves 5–20(–26); fruits (6–)9–14(–17) mm, often subappressed to rachis; stem trichomes non-crisped . . . . 13. *Draba aurea* (in part)
      9. Ovules 6–16(–18) per ovary; racemes ebracteate, 3–15(–18)-flowered; cauline leaves 1–4 (or 5); fruits 3–10 mm, not appressed to rachis; stem trichomes often crisped.
         10. Plants matted, caudex branches creeping; stem trichomes dendritic; ovules 6–12 per ovary; se Utah . . . . .90. *Draba ramulosa*
         10. Plants not matted, caudex branches not creeping; stem trichomes stellate; ovules 10–16(–18) per ovary; Colorado . . . . 110. *Draba streptobrachia* (in part)
   8. Petals white.
      11. Styles (0.7–)1–2(–2.3) mm; fruits twisted; sw Colorado . . . . .102. *Draba smithii*
      11. Styles 0.05–0.6(–0.8) mm; fruits plane or slightly twisted; not sw Colorado.
         12. Leaf blade margins dentate or denticulate; cauline leaves 2–25.
            13. Leaf blade surfaces pubescent, trichomes cruciform or (2–)4–6-rayed (rays usually simple, rarely branched); plants with rhizomatous caudex . . . . 16. *Draba borealis* (in part)
            13. Leaf blade surfaces pubescent, trichomes stellate-pectinate, 4–8(–12)-rayed; plants without rhizomatous caudex (sometimes cespitose) . . . . 38. *Draba glabella* (in part)
         12. Leaf blade margins entire; cauline leaves 1(–4).
            14. Rachises not flexuous; sepals persistent; fruits plane . . . . 21. *Draba californica* (in part)
            14. Rachises often slightly flexuous; sepals caducous; fruits slightly twisted or plane . . . . 57. *Draba lonchocarpa* (in part)

Group 5

1. Fruit trichomes 2-rayed and/or simple.
   2. Stem and leaf trichomes sessile, 2 rays parallel to long axis of stem or midvein, some trichomes malpighiaceous . . . . 104. *Draba spectabilis* (in part)
   2. Stem and leaf trichomes stalked, rays not parallel to long axis of stem or midvein, not malpighiaceous.
      3. Abaxial surface of basal leaf blades with cruciform trichomes.
         4. Fruits plane; styles 0.2–0.7 mm; Alaska, Canada . . . . 52. *Draba juvenilis* (in part)
         4. Fruits often strongly twisted; styles 0.8–1.8(–2.5) mm; Arizona . . . . 83. *Draba petrophila*
      3. Abaxial surface of basal leaf blades with 2–6-rayed trichomes.
         5. Fruits appressed or subappressed to rachises; proximalmost flowers bracteate . . . . 13. *Draba aurea* (in part)
         5. Fruits not appressed to rachises; proximalmost flowers ebracteate.
            6. Styles (1–)1.5–3.5 mm . . . . .44. *Draba helleriana* (in part)
            6. Styles usually 0.01–0.9 mm, if longer (*D. pennellii*), plants pulvinate or petals white.
               7. Plants not cespitose or pulvinate, caudex simple or poorly developed; fruits (7–)10–17(–20) mm; ovules 20–36(–42) per ovary.
                  8. Fruits narrowly elliptic to lanceolate, 2–3.5 mm wide; styles 0.2–0.6(–0.8) mm; racemes (5–)10–35(–47)-flowered; ne Nevada, Utah . . . . 18. *Draba brachystylis* (in part)
                  8. Fruits linear to linear-elliptic, 1.5–2 mm wide; styles 0.01–0.15 mm; racemes (2–)4–10(–15)-flowered; Alaska, Alberta, British Columbia, Washington, Yukon . . . . 108. *Draba stenoloba* (in part)
               7. Plants cespitose or pulvinate, caudex branched; fruits (3–)4–9 (–14) mm; ovules (4–)8–24 per ovary.
                  9. Seeds 1.8–2.8 × 1.2–2 mm, winged . . . . 12. *Draba asterophora* (in part)
                  9. Seeds 0.7–1.8 × 0.4–1 mm, not winged.
                     10. Petals white; cauline leaves 1–4; styles (0.7–)1–1.8(–2.1) mm; fruits slightly twisted or plane; e Nevada . . . . 82. *Draba pennellii* (in part)
                     10. Petals yellow (or fading white); cauline leaves 1; styles 0.2–0.9 mm; fruits plane; Nevada and elsewhere.
                        11. Fruits flattened; abaxial surface of leaf blades with pectinate and sometime 4–6-rayed trichomes; petals yellow (fading white), 4–6 mm. . . . . 48. *Draba incerta* (in part)
                        11. Fruits inflated (at least basally); abaxial surface of leaf blades with 2–4-rayed trichomes; petals yellow, 3–4 mm . . . . 103. *Draba sobolifera* (in part)
1. Fruit trichomes 4-rayed (some) (except *D. saxosa* 2- or 3-rayed).
   12. Caudices simple, plants not cespitose.
      13. Petals white; styles 0.03–0.1(–0.15) mm . . . . 87. *Draba praealta* (in part)
      13. Petals yellow, rarely fading whitish; styles (0.5–)0.9–3.5 mm.
         14. Cauline leaves 1–5; sepals with 2–4-rayed trichomes.
            15. Petals (5–)6–8.5 × 2.2–3.5 mm; fruits slightly twisted; ovules 24–36 per ovary; styles (0.8–)1–2.2 mm; se Arizona, w New Mexico. . . . . .63. *Draba mogollonica* (in part)
            15. Petals 3–5 × 1–1.5 mm; fruits plane; ovules 14–22 per ovary; styles 0.9–1.2 mm; nc Utah . . . . 94. *Draba santaquinensis* (in part)
         14. Cauline leaves 5–33(–43); sepals glabrous or with some simple trichomes.
            16. Fruits 3–5 mm wide, plane; stems 0.3–1.5 dm; petals 0.5–1(–1.2) mm wide. . . . . 14. *Draba aureola*
            16. Fruits 2–3.5 mm wide, usually twisted, rarely plane; stems (0.5–)1–4.3 (–5.2) dm; petals 1.5–2.5 mm wide.

17. Styles 0.5–1.2(–1.5) mm; petals 3.5–5 mm; ovules 28–38(–44) per ovary; fruits often subappressed to rachises.................13. *Draba aurea* (in part)
17. Styles (1–)1.5–3.5 mm; petals 5–7 mm; ovules 14–28 per ovary; fruits not appressed to rachises.........................44. *Draba helleriana* (in part)

[12. Shifted to left margin.—Ed.]
12. Caudices branched, plants cespitose (branches sometimes surculose or rhizomatous).
    18. Basal and often cauline leaf blade margins dentate; petals white.
        19. Styles (1–)1.5–3(–4) mm (sparsely pubescent); ovules 4–10(–12) per ovary; fruits often strongly twisted....................................... 89. *Draba ramosissima*
        19. Styles (0.01–)0.1–0.6(–0.8) mm (glabrous); ovules 16–28(–30) per ovary; fruits plane or slightly twisted.
            20. Petals 4–6 mm; fruits (5–)7–12 × 2.5–4.5 mm, slightly twisted or plane; Alaska, w Canada... 16. *Draba borealis* (in part)
            20. Petals 2.5–4 mm; fruits 4–8 × 2–3 mm, plane; e Canada, Greenland........69. *Draba norvegica*
    18. Basal and cauline leaf blade margins usually entire, sometimes denticulate (*D. viridis*); petals yellow (unknown in *D. trichocarpa*).
        21. Styles 0.3–1.2(–1.4) mm; ovules 8–12 per ovary; Idaho, Nevada.
            22. Basal leaf blades 6–19(–24) × 2–5(–8) mm; styles 0.5–1.2(–1.4) mm; petals 3.5–5 mm; fruiting pedicels (3–)4–8 mm; Nevada .... 10. *Draba arida*
            22. Basal leaf blades 2–4 × 0.5–1.5 mm; styles 0.3–0.7 mm; petals 2–4 mm; fruiting pedicels 1–4.5 mm; Idaho............................. 115. *Draba trichocarpa* (in part)
        21. Styles (0.7–)1.2–3.5 mm; ovules 8–24 per ovary; s Arizona, California, s Oregon.
            23. Cauline leaves 5–16; fruits 3.5–8 mm; leaf blade surfaces with cruciform trichomes; s Arizona.. 118. *Draba viridis*
            23. Cauline leaves 1–3 (or 4); fruits 6–11(–15) mm; leaf blade surfaces with 2- or 3-rayed trichomes; California, Oregon.
                24. Petals 5.5–8 mm; cauline leaves 1–3 (or 4); fruits lanceolate to broadly ovate, plane; n California, s Oregon .... 46. *Draba howellii* (in part)
                24. Petals 3–3.5 mm; cauline leaves 1; fruits oblong to linear-oblong, slightly twisted or plane; s California............................... 95. *Draba saxosa* (in part)

Group 6

1. Branched trichomes sessile, malpighiaceous; Greenland..................... 99. *Draba sibirica* (in part)
1. Branched trichomes stalked, not malpighiaceous, sometimes only simple cilia present; not Greenland (except *D. crassifolia* and *D. lactea*).
    2. Fruits 10–23 mm; seeds orbicular, 3–4.5 mm, broadly winged .... 23. *Draba carnosula*
    2. Fruits 2.5–11(–16) mm; seeds ovoid, oblong, or elliptic, 0.7–2(–2.6) mm, not winged.
        3. Plants stoloniferous; basal leaves not rosulate.................... 72. *Draba ogilviensis* (in part)
        3. Plants not stoloniferous (usually cespitose or pulvinate); basal leaves rosulate.
            4. Petals white.
                5. Petals 5–6.5 mm; styles 0.6–1.8(–2.5) mm; Nevada................. 97. *Draba serpentina*
                5. Petals 2.5–4.5(–5) mm; styles 0.1–0.9(–1.2) mm; not Nevada.
                    6. Fruits inflated (at least basally); racemes (6–)10–28(–40)-flowered .... 112. *Draba subalpina* (in part)
                    6. Fruits flattened; racemes 2–8(–12)-flowered.
                        7. Ovules (10–)14–22(–26) per ovary; petals 1.8–3 mm wide; Alaska, Canada, Greenland.
                            8. Petiole and midvein of basal leaves thickened; basal leaves (1–)2–6 mm wide; styles 0.1–0.4 mm............. 55. *Draba lactea* (in part)
                            8. Petiole and midvein of basal leaves not thickened; basal leaves 1–2 mm wide; styles 0.4–1 mm .... 65. *Draba mulliganii* (in part)

7. Ovules 8–16 per ovary; petals 1.2–2 mm wide; Idaho, Montana, Utah, Wyoming.
   9. Basal leaf blade apex acute; fruits ovate, plane, 2.5–4 mm wide; fruiting pedicels 2–6 mm; sepals persistent ... 39. *Draba globosa* (in part)
   9. Basal leaf blade apex obtuse; fruits ± oblong, slightly twisted, 1.5–3 mm wide; fruiting pedicels 4–13 mm; sepals caducous .... 74. *Draba oreibata*

[4. Shifted to left margin.—Ed.]

4. Petals yellow (rarely fading white).
   10. Fruits inflated (at least basally).
      11. Adaxial surface of leaf blades with simple and 2-rayed trichomes; styles 0.1–0.5 mm .... 59. *Draba macounii* (in part)
      11. Adaxial surface of leaf blades with 2–4-rayed trichomes; styles 0.3–1(–1.3) mm .... 106. *Draba sphaeroides* (in part)
   10. Fruits flattened.
      12. Styles (0.5–)0.7–3 mm; petals (4–)5–8 mm; sepals 2–4 mm.
         13. Leaf blade margins ciliate with subsetiform, simple trichomes .... 20. *Draba burkei*
         13. Leaf blade margins not ciliate, often with some branched trichomes.
            14. Fruits 6–11(–15) × 3–5 mm; styles (0.7–)1.6–3 mm; ovules 8–22 per ovary; California, Oregon .... 46. *Draba howellii* (in part)
            14. Fruits (3–)4–6.5(–8) × 2–3.2 mm; styles 0.6–1.7 mm; ovules 4–8 per ovary; Utah .... 60. *Draba maguirei* (in part)
      12. Styles 0.02–0.8(–1) mm; petals 1.5–4(–6) mm; sepals 1–3 mm.
         15. Plants without sterile rosettes or shoots, not pulvinate or matted, caudex simple or branched; petals 0.5–1 mm wide; ovules (8–)16–24(–30) per ovary; seeds 0.7–0.8 mm .... 29. *Draba crassifolia* (in part)
         15. Plants often with sterile rosettes or shoots, pulvinate or loosely matted, caudex branched; petals 1–3 mm wide; ovules 4–16 per ovary; seeds 1–2(–2.6) mm.
            16. Fruiting pedicels decurrent basally .... 49. *Draba incrassata*
            16. Fruiting pedicels not decurrent basally.
               17. Plants loosely matted; racemes considerably elongated in fruit .... 30. *Draba cruciata* (in part)
               17. Plants pulvinate; racemes not or slightly elongated in fruit.
                  18. Leaf blades with prominent midvein abaxially; fruit valves pubescent or puberulent .... 35. *Draba densifolia* (in part)
                  18. Leaf blades with obscure midvein; fruit valves glabrous.
                     19. Leaves fleshy, blade apex obtuse; fruiting pedicels 4–10 mm; fruit valves obscurely veined .... 34. *Draba daviesiae*
                     19. Leaves not fleshy, blade apex acute; fruiting pedicels 2–6 mm; fruit valves distinctly veined .... 39. *Draba globosa* (in part)

Group 7

1. Abaxial surface of leaf blades with some 7–16-rayed trichomes.
   2. Trichomes pectinate; leaf blades with midvein prominent abaxially; fruits inflated (at least basally) .... 73. *Draba oligosperma* (in part)
   2. Trichomes stellate or, rarely, subdendritic; leaf blades with midvein obscure abaxially (prominent in *D. lactea*); fruits flattened.
      3. Leaf blades with midvein strongly thickened abaxially .... 55. *Draba lactea* (in part)
      3. Leaf blades with midvein obscure abaxially.
         4. Petals white, 2–4 mm.
            5. Fruits linear to narrowly lanceolate or oblong, 6–15(–18) mm; ovules 16–24(–28) per ovary; rachises often slightly flexuous ...... 57. *Draba lonchocarpa* (in part)
            5. Fruits ovate to oblong, 4–8 mm; ovules 12–16 per ovary; rachises not flexuous.

6. Fruits not appressed to rachises; petals 2–3 mm wide; fruiting pedicels (4–)5–9 mm; style 0.4–1 mm.................... 65. *Draba mulliganii* (in part)

6. Fruits appressed to rachises; petals 0.7–1.2 mm wide; fruiting pedicels 0.5–2(–4) mm; style 0.05–0.1 mm.................. 86. *Draba porsildii* (in part)

4. Petals yellow (white fading yellowish in *D. palanderiana*), 4–6 mm.

7. Ovules 18–32 per ovary; seeds 0.9–1.1 × 0.5–0.7 mm; fruit valves glabrous .... 76. *Draba palanderiana* (in part)

7. Ovules 8–16 per ovary; seeds 1.2–2 × 0.8–1.3 mm; fruit valves often puberulent, sometimes glabrous.

8. Styles (0.7–)1–1.8(–2.4) mm; petiole bases and leaf blade margins not ciliate; sepals pubescent, trichomes 4–6-rayed; Idaho .... 9. *Draba argyrea* (in part)

8. Styles 0.5–0.9(–1.1) mm; petiole bases and leaf blade margins ciliate; sepals pubescent, trichomes simple and 2-rayed; Alaska, British Columbia, Washington, Yukon.............................. 93. *Draba ruaxes*

1. Abaxial surface of leaf blades with simple and/or 2–5 (or 6)-rayed or malpighiaceous trichomes.

9. Leaf blades with malpighiaceous trichomes........................ 99. *Draba sibirica* (in part)

9. Leaf blades with simple, forked, stellate, or subdendritic trichomes.

10. Abaxial surface of leaf blades with mixture of simple and 2(–4)-rayed trichomes.

11. Petals yellow (rarely fading white); fruits not inflated basally (flattened).

12. Perennials with poorly developed caudex, not stoloniferous; petals 1.5–3.2 × 0.5–1.2 mm; styles 0.01–0.12 mm; ovules (20–)24–38(–44) per ovary.

13. Abaxial surface of leaf blades pubescent with 2–4-rayed trichomes; stems pubescent, trichomes simple and 2–4-rayed......... 2. *Draba albertina* (in part)

13. Abaxial surface of leaf blades glabrous or with some simple and 2-rayed trichomes; stems usually glabrous, rarely sparsely pubescent basally, trichomes simple....................... 29. *Draba crassifolia* (in part)

12. Perennials with branched caudex or stoloniferous; petals 3.5–6 × 2–3.5 mm; styles 0.4–1 mm; ovules 8–20 per ovary.

14. Plants stoloniferous; basal leaves not rosulate, blades with obscure midvein; fruits oblong, 2–3 mm wide .... 72. *Draba ogilviensis* (in part)

14. Plants not stoloniferous; basal leaves rosulate, blades with prominent, persistent midvein; fruits elliptic to lanceolate, 3–4 mm wide .... 84. *Draba pilosa* (in part)

11. Petals white; fruits inflated basally (flattened in *D. fladnizensis*).

15. Basal leaf blades with a prominent midvein.

16. Fruits flattened, elliptic-lanceolate to oblong, 1.5–2 mm wide; rachises glabrous; petals 2–3.5 mm .... 37. *Draba fladnizensis* (in part)

16. Fruits slightly inflated, ovoid to oblong, 2–3 mm wide; rachises usually pubescent, rarely glabrous; petals 1.5–2(–2.5) mm .... 113. *Draba subcapitata* (in part)

15. Basal leaf blades with obscure midvein.

17. Sepals persistent; fruiting pedicels 1–2.5(–4) mm; petals 1.5–2 mm; fruits (2–)3–5 × (1.2–)1.5–2.5 mm; e California .... 64. *Draba monoensis* (in part)

17. Sepals caducous; fruiting pedicels (3–)5–10(–17) mm; petals 3–5 mm; fruits 4–8(–10) × 2.5–4 mm; s Utah.................112. *Draba subalpina* (in part)

10. Abaxial surface of leaf blades with 2–6-rayed trichomes.

18. Caudex simple or branched, without sterile rosettes; styles 0.01–0.15(–0.2) mm.

19. Sepals pubescent with simple and 2- or 3-rayed trichomes; abaxial surface of leaf blades with (2–)4(–6)-rayed trichomes; s Nevada .... 78. *Draba paucifructa* (in part)

19. Sepals glabrous or with simple trichomes; abaxial surface of leaf blades with 2–4-rayed trichomes; not s Nevada.

20. Stems proximally with simple and 2-rayed trichomes; adaxial surface of leaf blades with simple and 2-rayed trichomes; anthers 0.15–0.25 mm; fruits (4–)6–12(–15) mm .... 2. *Draba albertina* (in part)
20. Stems proximally with 2–4-rayed trichomes; adaxial surface of leaf blades with 2–4-rayed trichomes; anthers 0.25–0.4 mm; fruits (8–)10–17(–20) mm................. 108. *Draba stenoloba* (in part)

[18. Shifted to left margin.—Ed.]

18. Caudex often branched, with sterile rosettes; styles (0.1–)0.2–3 mm.
    21. Abaxial surface of leaf blades with some pectinate or submalpighiaceous trichomes.
        22. Abaxial surface of leaf blades with pectinate and some 4–6-rayed trichomes; ovules 8–16(–20) per ovary.................48. *Draba incerta* (in part)
        22. Abaxial surface of leaf blades with submalpighiaceous trichomes; ovules 20–28 per ovary ....53. *Draba kassii*
    21. Abaxial surface of leaf blades with cruciform or 2–5-rayed trichomes.
        23. Fruits inflated (at least basally).
            24. Styles 0.1–0.5 mm; fruiting pedicels not expanded basally; anthers 0.2–0.3 mm; Alaska, Alberta, British Columbia, Montana, Yukon........ 59. *Draba macounii* (in part)
            24. Styles 0.3–1(–1.3) mm; fruiting pedicels usually expanded basally; anthers 0.4–0.5 mm; Nevada, Utah.
                25. Plants densely pulvinate; petals 2.7–4 mm; fruiting pedicels divaricate-ascending......................... 106. *Draba sphaeroides* (in part)
                25. Plants loosely cespitose; petals 3.5–5.5 mm; fruiting pedicels divaricate, horizontal, or slightly descending.................81. *Draba pedicellata* (in part)
        23. Fruits flattened.
            26. Seeds 1.8–2.8 × 1.2–2 mm, winged ....12. *Draba asterophora* (in part)
            26. Seeds 0.7–2(–2.6) × 0.4–1(–1.4) mm, not winged.
                27. Fruits twisted, curved .... 98. *Draba sharsmithii*
                27. Fruits plane, not curved (straight).
                    28. Leaf blades with prominent midveins, margins ciliate (trichomes simple).................................... 35. *Draba densifolia* (in part)
                    28. Leaf blades with obscure midveins, margins not ciliate.
                        29. Fruits 2.5–4.5(–5.5) mm wide; rachises often flexuous; fruiting pedicels expanded basally or not................. 81. *Draba pedicellata* (in part)
                        29. Fruits 1.5–3 mm wide; rachises not flexuous; fruiting pedicels not expanded basally.
                            30. Ovules 4–10(–12) per ovary; seeds 0.8–1.3 mm wide.
                                31. Styles (0.1–)0.3–0.8 mm; petals 4–5(–6) × 1.2–2 mm; sepals 1.5–2 mm; ec California .... 30. *Draba cruciata* (in part)
                                31. Styles 0.6–1.7 mm; petals 5–7 × 2–3 mm; sepals 2.5–4 mm; n Utah .... 60. *Draba maguirei* (in part)
                            30. Ovules 12–30 per ovary; seeds 0.4–0.8 mm wide.
                                32. Abaxial surface of leaf blades pubescent, trichomes cruciform; racemes (2–)4–13(–18)-flowered; petals 3–5 mm; Alaska, Alberta, British Columbia, Northwest Territories, Yukon .... 52. *Draba juvenilis* (in part)
                                32. Abaxial surface of leaf blades pubescent, trichomes (2–)4 (or 5)-rayed; racemes 14–36-flowered; petals 5–6 mm; sw Utah .... 121. *Draba zionensis*

Group 8

1. Abaxial surface of leaf blades with pectinate, subpectinate, subdendritic, or stellate with 7–16-rayed trichomes.
    2. Leaves, pedicels, and sepals with sessile or subsessile, pectinate trichomes.
        3. Fruit valves pubescent, trichomes pectinate; ovules 4–8 per ovary .... 80. *Draba pectinipila*
        3. Fruit valves glabrous or pubescent, trichomes simple or 2-rayed; ovules (6–)8–16 (–20) per ovary.
            4. Fruits 5–9(–11) mm, flattened; petals 4–6 mm; sepals 2.5–3.5(–4) mm; leaf midvein usually obscure .... 48. *Draba incerta* (in part)
            4. Fruits 3–6(–7) mm, inflated; petals 2.5–4 mm; sepals 1.5–3 mm; leaf midvein prominent .... 73. *Draba oligosperma* (in part)
    2. Leaves, pedicels, and sepals with stalked, stellate to subdendritic trichomes.
        5. Petals white or creamy white.
            6. Sepals caducous; fruit valves glabrous; Alaska, Northwest Territories, Yukon.
                7. Stems pubescent throughout, trichomes 8–15-rayed; petals spatulate to oblanceolate, 2–3.5 × 0.8–1.4 mm; rachises and fruiting pedicels pubescent; styles 0.1–0.4 mm .... 68. *Draba nivalis* (in part)
                7. Stems pubescent proximally, trichomes 2–9-rayed; petals obovate, 4.4–5.5 × 2–3 mm; rachises and fruiting pedicels usually glabrous, rarely sparsely pubescent; styles 0.3–0.8 mm .... 76. *Draba palanderiana*
            6. Sepals persistent; fruit valves usually pubescent, rarely glabrous (*D. californica*); e California, w Nevada.
                8. Fruits usually twisted, rarely plane, with 2–5-rayed trichomes; fruiting pedicels 1.5–3(–4) mm........................... 19. *Draba breweri* (in part)
                8. Fruits plane, with simple and 2- or 3-rayed trichomes, rarely glabrous; fruiting pedicels (2–)4–9(–12) mm .... 21. *Draba californica* (in part)
        5. Petals yellow.
            9. Fruits 4–11 mm, flattened, twisted or plane.
                10. Styles (0.7–)1–1.8(–2.4) mm; fruits symmetric; fruiting pedicels (4–)5–12(–19) mm; c Idaho .... 9. *Draba argyrea* (in part)
                10. Styles (0.1–)0.3–0.7(–1) mm; fruits often asymmetric; fruiting pedicels 2–5(–7) mm; ec California................................ 100. *Draba sierrae* (in part)
            9. Fruits 2–5(–6) mm, inflated (at least basally), plane.
                11. Racemes (4–)6–18(–23)-flowered, elongated in fruit; petals 4–6 mm; fruiting pedicels 4–9(–13) mm; c Idaho .... 105. *Draba sphaerocarpa*
                11. Racemes 2–5(–10)-flowered (subumbellate), not or slightly elongated in fruit; petals 2.8–4 mm; fruiting pedicels 1.5–3(–6) mm; e California, w Nevada .... 114. *Draba subumbellata*
1. Abaxial surface of leaf blades with simple, cruciform, or 2–4(–6)-rayed trichomes.
    12. Plants stoloniferous; basal leaves not rosulate...................... 72. *Draba ogilviensis* (in part)
    12. Plants not stoloniferous (usually cespitose); basal leaves usually rosulate.
        13. Petals white.
            14. Sepals persistent; fruits slightly inflated basally; petals 1.5–2 × 0.5–0.6 mm .... 64. *Draba monoensis* (in part)
            14. Sepals caducous; fruits flattened; petals 2–6 × 0.8–2.8 mm.
                15. Styles 0.1–0.7 mm; fruit valves usually glabrous, rarely pubescent, trichomes simple (along margin); Alaska, Alberta, British Columbia, Northwest Territories, Yukon .... 52. *Draba juvenilis* (in part)
                15. Styles (0.7–)1–1.8(–2.1) mm; fruit valves with simple and 2-rayed trichomes; e Nevada .... 82. *Draba pennellii* (in part)
        13. Petals yellow.
            16. Abaxial surface of leaf blades with cruciform trichomes; ovules 16–30 per ovary .... 52. *Draba juvenilis* (in part)
            16. Abaxial surface of leaf blades with 2–6-rayed trichomes; ovules 4–18(–22) per ovary.

[17. Shifted to left margin.—Ed.]
17. Racemes (15–)30–75-flowered; Arizona .... 11. *Draba asprella*
17. Racemes 3–20(–27)-flowered; not Arizona.
    18. Styles (0.7–)1.6–3 mm; petals 5.5–8 mm; anthers 0.7–0.9 mm .... 46. *Draba howellii* (in part)
    18. Styles 0.3–1.4 mm; petals 3.5–5.5(–6) mm; anthers 0.4–0.6 mm.
        19. Fruit valves glabrous or pubescent, trichomes simple and 2-rayed; fruiting pedicels usually expanded basally.
            20. Petals 3.5–5.5 mm; fruiting pedicels horizontal to divaricate; ovules 8–16 per ovary .... 81. *Draba pedicellata* (in part)
            20. Petals 2.7–4 mm; fruiting pedicels divaricate-ascending; ovules 4–10 (–12) per ovary .... 106. *Draba sphaeroides* (in part)
        19. Fruit valves pubescent, trichomes 2–6-rayed; fruiting pedicels not expanded basally.
            21. Fruits flattened, 2–4.5 mm wide, 10–18-seeded; styles 0.4–0.8(–1) mm; racemes 10–20(–25)-flowered; surface of leaf blades pubescent with (2–) 4-rayed trichomes; se Oregon .... 32. *Draba cusickii*
            21. Fruits inflated basally, 3.5–5 mm wide, 8–12-seeded; styles (0.5–)0.7–1.4 mm; racemes 5–10(–16)-flowered; surface of leaf blades pubescent with 2–6-rayed trichomes; Alberta, British Columbia, Colorado, Utah, Wyoming, Yukon .... 116. *Draba ventosa*

Group 9

1. Abaxial surface of leaf blades usually with simple and/or 2-rayed trichomes, rarely glabrous.
    2. Sepals persistent.
        3. Petals yellow, 3.5–5 × 1.5–2 mm; styles 0.4–1 mm; fruiting pedicels 3–7 mm; fruits 2.5–5 mm wide, flattened .... 58. *Draba longisquamosa*
        3. Petals white, 1.5–2 × 0.5–0.6 mm; styles 0.1–0.2 mm; fruiting pedicels 1–2.5(–4) mm; fruits (1.2–)1.5–2.5 mm wide, slightly inflated basally .... 64. *Draba monoensis* (in part)
    2. Sepals caducous.
        4. Abaxial surface of leaf blades glabrous, midvein prominent; plants pulvinate.... .... 35. *Draba densifolia* (in part)
        4. Abaxial surface of leaf blades usually pubescent, sometimes glabrous (*D. cyclomorpha*), midvein usually obscure (except *D. subcapitata*); plants often not pulvinate.
            5. Petals 1.4–4 × 0.5–1 mm; racemes 1–8 (or 9)-flowered, not or slightly elongated in fruit; Alaska, Canada, Greenland.
                6. Petals yellow or yellowish green, linear-oblanceolate, 3–4 × 0.5–0.8 mm; leaf blades each with midvein usually obscured; ovules 4–8 per ovary; Alaska .... 3. *Draba aleutica*
                6. Petals white or creamy white, spatulate, 1.5–2(–2.5) × 0.7–1 mm; leaf blades each with midvein strongly thickened; ovules 14–24 per ovary; Alaska, Greenland, Northwest Territories, Nunavut, Quebec .... 113. *Draba subcapitata* (in part)
            5. Petals 4–6 × 1.5–2.5 mm; racemes 4–15(–21)-flowered, elongated in fruit; California, Oregon.
                7. Fruits plane, valves glabrous; fruiting pedicels softly pilose; ovules 8–12 per ovary; seeds 1.4–1.9 × 1–1.2 mm; ne Oregon .... 33. *Draba cyclomorpha*
                7. Fruits slightly twisted, valves pubescent; fruiting pedicels hirsute; ovules 10–16 per ovary; seeds 1–1.4 × 0.6–1 mm; ec California............. 56. *Draba lemmonii*
1. Abaxial surface of leaf blades with simple and 2–6-rayed trichomes, sometimes subdendritic and up to 12-rayed.

[8. Shifted to left margin.—Ed.]

8. Racemes not or slightly elongated (sometimes subumbellate) in fruit.
   9. Abaxial surface of leaf blades with 2–5-rayed trichomes; fruit valves glabrous; petals 1–2 mm wide .... 59. *Draba macounii* (in part)
   9. Abaxial surface of leaf blades with some simple trichomes; fruit valves pubescent or puberulent (sometimes glabrous in *D. stenopetala* and *D. subcapitata*); petals 0.3–1.5 mm wide.
      10. Basal leaves densely imbricate; fruits 3–6 mm, inflated (at least basally).
         11. Petals usually yellow, rarely purple, linear, 2.5–5 × 0.3–0.7 mm; ovules 4–10 per ovary .... 109. *Draba stenopetala* (in part)
         11. Petals white or creamy white, spatulate, 1.5–2(–2.5) × 0.7–1 mm; ovules 14–24 per ovary .... 113. *Draba subcapitata* (in part)
      10. Basal leaves not imbricate; fruits 5–10 mm, flattened.
         12. Fruits ovate-elliptic, 2–3.2 mm wide, valves often densely pubescent; ovules (16–)18–28 per ovary; leaf blades with apices obtuse to rounded, surfaces with subcruciform trichomes .... 62. *Draba micropetala* (in part)
         12. Fruits obovate, (3–)3.5–5 mm wide, valves often glabrate; ovules 8–16 (–20) per ovary; leaf blades with apices acute or subacute, surfaces with simple and/or 2-branched trichomes .... 77. *Draba pauciflora*
8. Racemes elongated in fruit.
   13. Plants stoloniferous; basal leaves not rosulate; fruit valves glabrous ...... 72. *Draba ogilviensis* (in part)
   13. Plants cespitose, not stoloniferous; basal leaves rosulate; fruit valves pubescent or glabrous.
      14. Petals linear, 0.3–0.7 mm wide; fruits inflated basally; ovules 4–10 per ovary. . . .
         .... 109. *Draba stenopetala* (in part)
      14. Petals obovate, spatulate, or narrowly oblanceolate, (0.7–)1–4.6 mm wide; fruits flattened; ovules 12–36 per ovary.
         15. Petals white; fruit valves pubescent, trichomes 2–5-rayed; stigmas distinctly wider than styles; abaxial surface of leaf blades with some 6–12-rayed trichomes.
            16. Fruits with 2–4-rayed and, sometimes, simple trichomes; petals 3–4 mm; stems proximally with branched trichomes 2–6-rayed .... 8. *Draba arctogena* (in part)
            16. Fruits with (2–)5–12-rayed trichomes; petals 3.5–5 mm; stems proximally with branched trichomes 4–10-rayed .... 71. *Draba oblongata* (in part)
         15. Petals pale or bright yellow (sometimes creamy white in *D. oxycarpa*); fruit valves usually pubescent, sometimes glabrous, trichomes simple, spurred, or 2-rayed; stigmas about as wide as styles; abaxial surface of leaf blades with 2–5-rayed trichomes (rays sometimes spurred in *D. oxycarpa*).
            17. Fruiting pedicels 1–3(–4) mm; petals 2–3 × (0.7–)1–1.5 mm; styles 0.05–0.3 mm; racemes slightly elongated in fruit .... 62. *Draba micropetala* (in part)
            17. Fruiting pedicels (2.5–)4–13(–30) mm; petals 3.5–6 × 1.7–4.6 mm; styles (0.1–)0.2–0.9 mm; racemes often considerably elongated in fruit.
               18. Basal leaf blades linear to linear-oblanceolate, 1–2.5(–4) mm wide, midveins prominent .... 84. *Draba pilosa* (in part)
               18. Basal leaf blades oblong, lanceolate, oblanceolate, or obovate, 2–9 mm wide, midveins obscure.
                  19. Petals 1.7–2.5 mm wide; sepals purplish tinged; fruits elliptic; abaxial surface of leaf blades with 2–4-rayed trichomes; seeds pale brown... 4. *Draba alpina*
                  19. Petals 2.5–4.6 mm wide; sepals gray-green; fruits ovate, ovate-oblong, or lanceolate; abaxial surface of leaf blades with 2–5-rayed and subdendritic trichomes; seeds dark brown or black.
                     20. Basal leaf blade margins usually denticulate; fruits ovate to ovate-oblong, 3–5 mm wide; styles 0.2–0.6 mm; Greenland
                        .... 75. *Draba oxycarpa* (in part)
                     20. Basal leaf blade margins entire; fruits lanceolate, 2.3–3.8 mm wide; styles 0.1–0.3 mm; Canada .... 101. *Draba simmonsii*

Group 10

1. Abaxial surface of leaf blades with pectinate, subdendritic, or some 7–12-rayed, stellate trichomes.
   2. Styles 1.5–3(–3.8) mm; seeds winged; anthers 0.7–0.9 mm . . . . 88. *Draba pterosperma*
   2. Styles (0.05–)0.1–1(–1.2) mm; seeds not winged; anthers 0.2–0.7 mm.
      3. Abaxial surface of leaf blades with some pectinate trichomes.
         4. Fruits 5–9(–11) mm, flattened; petals 4–6 mm; sepals 2.5–3.5(–4) mm; leaf blades usually with obscure midveins . . . . 48. *Draba incerta* (in part)
         4. Fruits 3–6(–7) mm, inflated; petals 2.5–4 mm; sepals 1.5–3 mm; leaf blades with prominent midveins . . . . 73. *Draba oligosperma* (in part)
      3. Abaxial surface of leaf blades with stellate or subdendritic trichomes.
         5. Fruit valves pubescent, trichomes 2–6-rayed.
            6. Plants not pulvinate; leaf blades 0.4–2.3(–3) cm × 1–6.5 mm; petals white; ovules 20–42 per ovary.
               7. Basal leaf blades apically with simple trichomes, midveins distinct abaxially; seeds 0.8–1.1 × (0.6–)0.7–0.8 mm. . . . . 7. *Draba arctica* (in part)
               7. Basal leaf blades apically without simple trichomes, midveins obscure abaxially; seeds 0.6–0.8 × 0.4–0.6 mm . . . . 25. *Draba cinerea* (in part)
            6. Plants pulvinate; leaf blades 0.2–0.8 cm × 0.5–1.5(–2) mm; petals yellow; ovules 4–12(–16) per ovary.
               8. Fruits inflated basally (symmetric), plane, (2.5–)3–4(–5) mm; Alberta, British Columbia, California, Montana, Utah, Wyoming . . . . 70. *Draba novolympica* (in part)
               8. Fruits flattened (asymmetric), twisted or plane, 4–8(–10) mm; ec California . . . . 100. *Draba sierrae* (in part)
         5. Fruit valves glabrous or pubescent to puberulent, trichomes primarily simple and 2-rayed.
            9. Seeds 1.5–2.2 × 1–1.4 mm; fruits 3–5 mm wide; styles 0.5–1.2 mm; petals yellow; ovules 8–16 per ovary.
               10. Leaf blades each with prominent midvein; basal blades linear to linear-oblanceolate, 0.8–1.5 mm wide; petals pale yellow; fruits slightly inflated basally; w Montana, nw Wyoming . . . . 79. *Draba paysonii* (in part)
               10. Leaf blades each with obscure midvein; basal blades oblanceolate to obovate or suborbicular, 2–4.5 mm wide; petals bright yellow; fruits flattened; Alaska, British Columbia, Washington, Yukon . . . . 93. *Draba ruaxes* (in part)
            9. Seeds 0.7–1.3 × 0.5–0.8 mm; fruits 1–2.5(–3) mm wide (–5 mm wide in *D. oxycarpa*); styles 0.05–0.6(–1 in *D. scotteri*) mm; petals white or creamy white (yellow in *D. scotteri* and, sometimes, *D. oxycarpa*); ovules (10–)12–28 per ovary.
               11. Rachises often slightly flexuous; petals 2–3.5 × 1–1.5 mm; fruits linear to narrowly lanceolate or narrowly oblong, slightly twisted or plane . . . . 57. *Draba lonchocarpa* (in part)
               11. Rachises not flexuous; petals 3–5.5 × 1.5–3.5 mm; fruits ovate, oblong, lanceolate, or broadly so, rarely narrowly lanceolate, plane.
                  12. Fruits 3–5 mm wide; fruiting pedicels often curved upwards; leaf blade margins usually denticulate . . . . 75. *Draba oxycarpa* (in part)
                  12. Fruits 1.5–3 mm wide; fruiting pedicels straight; leaf blade margins usually entire, rarely denticulate (*D. lactea*).
                     13. Petals white; fruits 4–8 mm, valves glabrous . . . . 55. *Draba lactea* (in part)
                     13. Petals yellow; fruits 5–11 mm, valves pubescent . . . . 96. *Draba scotteri*
1. Abaxial surface of leaf blades with 2–6-rayed trichomes.
   14. Ovules (16–)18–36 per ovary.
      15. Styles (0.8–)1.2–3.5 mm; racemes 12–43-flowered; s California . . . . 95. *Draba saxosa* (in part)
      15. Styles 0.01–0.7 mm; racemes 2–12(–23)-flowered; Alaska, Canada, Greenland.

16. Fruits linear to linear-elliptic, (8–)10–17(–20) × 1.5–2 mm; plants not cespitose, caudex simple....................................108. *Draba stenoloba* (in part)
16. Fruits oblong, ovate, or lanceolate, 4–12 × 2–5.5 mm; plants often cespitose, caudex branched.
 17. Leaf blade margins entire.
  18. Petals 4–6 × 3–5 mm; fruits 3.5–5.5 mm wide (stigmas distinctly wider than styles); racemes not elongated (corymbose) in fruit ....
   ....27. *Draba corymbosa* (in part)
  18. Petals 2–3 × (0.7–)1–1.5 mm; fruits 2–3.2 mm wide (stigmas narrower than styles); racemes mostly elongated in fruit.........62. *Draba micropetala* (in part)
 17. Leaf blade margins dentate or denticulate.
  19. Fruits oblong to lanceolate-elliptic, 4–8 × 2–3 mm; fruiting pedicels straight, suberect to ascending, often subappressed to rachises; petals white, 2.5–4 mm...........................69. *Draba norvegica* (in part)
  19. Fruits ovate to ovate-oblong, 5–10 × 3–5 mm; fruiting pedicels curved upwards, subhorizontal to divaricate-ascending, not appressed to rachises; petals yellow or creamy white, 3.5–5 mm.......75. *Draba oxycarpa* (in part)

[14. Shifted to left margin.—Ed.]
14. Ovules 4–14(–16) per ovary.
 20. Fruits inflated (at least basally), usually ovoid, subglobose, or ellipsoid, rarely lanceolate.
  21. Leaf blades (1–)2–4(–6) mm wide; fruit valves glabrous or pubescent, trichomes simple and 2(–4)-rayed.
   22. Racemes 3–10(–13)-flowered; fruiting pedicels horizontal to divaricate, curved, glabrous (at least adaxially); ovules 10–16 per ovary; Alaska, Alberta, British Columbia, Montana, Yukon ....59. *Draba macounii* (in part)
   22. Racemes (5–)9–18(–23)-flowered; fruiting pedicels divaricate-ascending, usually straight, rarely curved, pubescent; ovules (4–)8–12 per ovary; sc Utah
    ....103. *Draba sobolifera* (in part)
  21. Leaf blades 0.5–1.5 mm wide; fruit valves pubescent, trichomes 2–6-rayed.
   23. Adaxial surface of leaf blades with simple and 4–6-rayed trichomes; fruit valves with 4-rayed trichomes (some rays spurred)...........115. *Draba trichocarpa* (in part)
   23. Adaxial surface of leaf blades with simple and 2-rayed trichomes; fruit valves with 2–6-rayed and simple trichomes.
    24. Fruits (2.5–)3–4(–5) × 1.5–3.5 mm; petals bright yellow, 2–3.5(–4) mm; styles 0.2–0.6(–0.8) mm ....70. *Draba novolympica* (in part)
    24. Fruits (5–)6–9 × (3–)3.7–5 mm; petals pale yellow, (4–)5–6 mm; styles (0.6–)0.8–1.2 mm ....79. *Draba paysonii* (in part)
 20. Fruits flattened, ovate, lanceolate, oblong, or elliptic.
  25. Abaxial surface of leaf blades with only 2-rayed trichomes...........56. *Draba lemmonii* (in part)
  25. Abaxial surface of leaf blades glabrous or with 2–4-rayed and simple trichomes.
   26. Adaxial surface of leaf blades pubescent with simple and 2–4-rayed trichomes.
    27. Styles (0.8–)1.2–3.5 mm; petals linear-oblanceolate, 0.5–1 mm wide; racemes 12–43-flowered; s California ....95. *Draba saxosa* (in part)
    27. Styles 0.2–0.9(–1.1) mm; petals obovate, 2–3.5 mm wide; racemes 2–10(–14)-flowered; Alaska, Canada, Greenland, Washington.
     28. Racemes (corymbose) not or slightly elongated in fruit; fruits 6–12 mm; seeds 1–1.3 × 0.6–0.9 mm ....27. *Draba corymbosa* (in part)
     28. Racemes elongated in fruit; fruits 4–8(–10) mm; seeds 1.5–2 × 1–1.3 mm ....93. *Draba ruaxes* (in part)
   26. Adaxial surface of leaf blades glabrous (at least proximally), usually pubescent distally with simple trichomes.

[29. Shifted to left margin.—Ed.]

29. Styles 0.3–0.6(–1) mm; petals pale yellow, 2–5 mm; leaf blades each with prominent midvein . . . . .35. *Draba densifolia*
29. Styles (0.8–)1–4.5 mm; petals white, 4.5–6.5 mm; leaf blades each with obscure midvein.
    30. Styles (0.8–)1–1.7(–2) mm; leaf blades oblong-linear to narrowly oblanceolate, 1–2 mm wide; anthers ovate, 0.5–0.6 mm; c Idaho . . . . .45. *Draba hitchcockii*
    30. Styles (1.2–)2–4(–4.5) mm; leaf blades oblanceolate to obovate, 1.5–3.5 mm wide; anthers oblong, 0.8–1 mm; s Nevada . . . . .51. *Draba jaegeri*

1. **Draba abajoensis** Windham & Al-Shehbaz, Harvard Pap. Bot. 12: 416. 2007 [E]

Draba spectabilis Greene var. *glabrescens* O. E. Schulz

Perennials; caudex simple or branched, (sometimes with persistent leaf bases); not scapose. **Stems** unbranched, (0.6–)1–2.5(–3.5) dm, pubescent proximally, usually glabrous distally, rarely sparsely pubescent, trichomes simple, 0.1–0.5 mm, often with stalked, 2(–4)-rayed ones. **Basal leaves** rosulate; petiolate; petiole (0.2–)1(–2) cm), proximal margin ciliate, (trichomes simple, 0.2–0.7 mm); blade oblanceolate to spatulate, (0.6–)1–3.5(–4.3) cm × (2–)4–10(–14) mm, margins entire or dentate, surfaces pubescent with short-stalked, (2–)4-rayed trichomes, 0.05–0.35 mm, sometimes adaxially with simple ones. **Cauline leaves** (3–)5–9(–11); sessile; blade broadly ovate to lanceolate or oblong, margins entire or dentate, surfaces pubescent as basal or adaxially also with mostly simple trichomes. **Racemes** (11–)15–32(–47)-flowered, ebracteate, elongated in fruit; rachis not flexuous, glabrous or pubescent, trichomes simple and short-stalked, 2–4-rayed. **Fruiting pedicels** horizontal to divaricate-ascending, straight or curved, (5–)7–18(–23) mm, glabrous or sparsely pubescent abaxially, trichomes simple and 2-rayed. **Flowers:** sepals ovate-oblong, 2.2–3.5 mm, subapically sparsely pubescent, (trichomes simple, short-stalked, and 2-rayed); petals yellow, oblanceolate, 4–6 × 1.5–2 mm; anthers oblong, 0.6–1 mm. **Fruits** elliptic to elliptic-lanceolate, plane, flattened, 5–9(–12) × 2–3 mm; valves glabrous; ovules 10–18 per ovary; style (0.8–)1.4–2.7(–3.6) mm. **Seeds** ovoid, 1–1.2 × 0.7–0.8 mm. $2n = 20$.

Flowering May–Aug. Spruce, fir, or pine forests, subalpine meadows; 1900–3800 m; Ariz., N.Mex., Utah.

Plants of *Draba abajoensis* were treated by C. L. Hitchcock (1941), R. C. Rollins (1993), and N. H. Holmgren (2005b) as *D. spectabilis*. The differences between these taxa in chromosome number and trichome morphology strongly support their recognition as separate species (I. A. Al-Shehbaz and M. D. Windham 2007). *Draba abajoensis* has been collected from the Chuska and Lukachukai mountains in Apache County, Arizona, Chuska Mountains in San Juan County, New Mexico, and Abajo and La Sal mountains in San Juan and Grand counties, Utah.

2. **Draba albertina** Greene, Pittonia 4: 312. 1901 [E]

Draba crassifolia Graham var. *albertina* (Greene) O. E. Schulz; *D. crassifolia* var. *nevadensis* C. L. Hitchcock; *D. deflexa* Greene; *D. nitida* Greene; *D. nitida* var. *nana* O. E. Schulz; *D. stenoloba* Ledebour var. *nana* (O. E. Schulz) C. L. Hitchcock; *D. stenoloba* var. *ramosa* C. L. Hitchcock

Annuals, biennials, or perennials; caudex (or base) simple or branched (poorly developed); rarely scapose. **Stems** branched distally, (0.3–)0.5–3(–4.2) dm, pubescent proximally, often glabrous distally, trichomes simple, 0.1–1 mm, sometimes with fewer, stalked, 2-rayed ones. **Basal leaves** rosulate; shortly petiolate; petiole (0–0.5 cm), ciliate throughout; blade obovate to oblanceolate or linear-lanceolate, (0.3–)1–2.8(–3.5) cm × (1–)2–6(–9) mm, margins entire or denticulate, (ciliate, trichomes simple, 0.4–1 mm), surfaces usually pubescent, abaxially with stalked, 2–4-rayed trichomes, 0.05–0.4(–0.5) mm, (rarely with simple trichomes along midvein), adaxially with simple trichomes sometimes also with 2-rayed ones, 0.07–0.4 mm, rarely glabrous. **Cauline leaves** (0 or) 1–3(–5); sessile; blade lanceolate to elliptic or ovate, margins entire or denticulate, surfaces pubescent as basal. **Racemes** (2–)6–30(–50)-flowered, ebracteate, elongated in fruit; rachis straight or, rarely, flexuous, glabrous. **Fruiting pedicels** divaricate-ascending or horizontal, (not appressed to rachis), usually straight, rarely curved upward, (3–)5–14(–16) mm (subequaling or shorter than fruit), usually glabrous, rarely sparsely puberulent. **Flowers** (chasmogamous, petaliferous); sepals ovate, 1.4–2.1 mm, glabrous or sparsely pubescent, (trichomes simple); petals yellow, spatulate to oblanceolate, 2–3.2 × 0.7–1.2 mm; anthers ovate, 0.15–0.25 mm. **Fruits** lanceolate to narrowly elliptic or linear, plane, flattened, (4–)6–12(–15) × (1–)1.4–2.1 mm; valves glabrous; ovules (20–)24–38(–44) per ovary; style 0.01–0.12 mm. **Seeds** oblong, 0.7–1 × 0.4–0.5 mm. $2n = 24$.

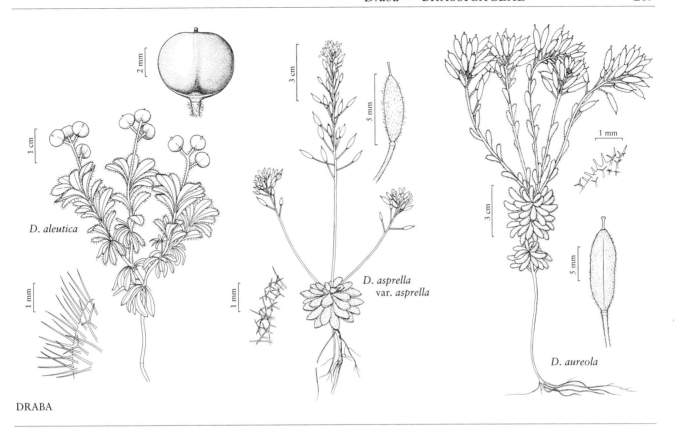

DRABA

Flowering Jun–Aug. Open woodlands, pine forests, meadows, rocky knolls, alpine slopes, stream banks, disturbed areas; 900–3700 m; Alta., B.C., N.W.T., Yukon; Alaska, Ariz., Calif., Colo., Idaho, Mont., Nev., N.Mex., Oreg., Utah, Wash., Wyo.

O. E. Schulz (1927) and C. L. Hitchcock (1941) confused the limits of *Draba albertina*, *D. crassifolia*, and *D. stenoloba*; the latter author treated the first two species as conspecific. G. A. Mulligan (1975) demonstrated that the three taxa are chromosomally and morphologically distinct (see *D. crassifolia* and *D. stenoloba* for differences from *D. albertina*). The ranges of *D. albertina* and *D. crassifolia* overlap extensively, and occasional sterile hybrids are encountered.

The Alaskan record of *Draba albertina* is based on *Minard 4* (ALA), collected on the northern coast of Afognak Island (58°22'N, 152°28'W). The record from New Mexico is based on *O'Kane & Hedin 3871* (ISTC, SJC), collected in Chuska Mountains (36°8'11"N, 108°54'19"W).

3. **Draba aleutica** Ekman, Svensk Bot. Tidskr. 30: 522, figs. 3c,d. 1936 C F

Draba behringii Tolmatchew

Perennials; (densely matted); caudex branched (with persistent leaf remains, some branches terminating in sterile rosettes); scapose. **Stems** unbranched, 0.04–0.4 dm, usually pubescent throughout, sometimes glabrous basally, trichomes simple, 0.1–1 mm, with 2-rayed ones. **Basal leaves** (densely imbricate); rosulate; petiolate; petiole base and margin ciliate, (trichomes simple, to 1.4 mm); blade oblanceolate to spatulate or obovate, 0.4–0.8(–1) cm × 2–3.5(–4.5) mm, margins entire, surfaces usually sparsely to densely pilose with simple trichomes, to 1.4 mm, with much fewer, stalked, 2-rayed ones, (sometimes surfaces glabrous, or adaxially only, midvein obscure abaxially). **Cauline leaves** 0. **Racemes** (1 or) 2–5(–9)-flowered, ebracteate, not or slightly elongated in fruit; rachis not flexuous, pubescent as stem. **Fruiting pedicels** divaricate, straight, 1.5–4(–6) mm, pubescent, trichomes simple with fewer, 2-rayed, stalked ones. **Flowers:** sepals oblong, 2–3 mm, glabrous or sparsely pubescent, (trichomes simple); petals yellowish green to pale yellow, linear-oblanceolate, 3–4 × 0.5–0.8 mm; anthers oblong, 0.4–0.5 mm. **Fruits** broadly obovoid to subglobose, plane, slightly flattened, 3–5 × 3–4.5 mm; valves glabrous or pubescent, trichomes simple, 0.1–0.3 mm; ovules 4–8 per ovary; style 0.1–0.4 mm. Seeds oblong, (slightly flattened), 1.4–1.8 × 0.9–1.1 mm.

Flowering May–Jul. Gravelly soil, retreating snow banks, fellfields; of conservation concern; 200–400 m; Alaska; e Asia (Russian Far East).

Draba aleutica is known in the flora area from the Aleutian Islands (e.g., Atka, Attu) and on the Pribilov Islands (St. Paul), as well as in the Alaskan Peninsula near Ugashik.

4. **Draba alpina** Linnaeus, Sp. Pl. 2: 642. 1753

Draba alpina var. *hydeana* B. Boivin; *D. alpina* var. *inflatisiliqua* Polunin

Perennials; (cespitose); caudex branched (covered with persistent leaves or leaf remains); scapose. **Stems** unbranched, (0.3–)0.5–1.7(–2.8) dm, pubescent throughout, trichomes simple and 2-rayed, 0.3–0.8 mm, with 3–5-rayed ones, 0.1–0.3 mm. **Basal leaves** rosulate; petiolate; petiole base (not thickened), ciliate, (trichomes simple, 0.3–1 mm); blade oblanceolate to obovate or lanceolate to oblong, 0.8–3(–4.5) cm × 2.5–6(–9) mm, margins entire, surfaces abaxially pubescent with stalked, 2–4-rayed trichomes, 0.1–0.5 mm, with simple ones (midvein obscure, not thickened), adaxially glabrous or sparsely pubescent with simple and stalked, 2–4-rayed trichomes. **Cauline leaves** 0. **Racemes** 6–18-flowered, ebracteate, considerably elongated in fruit; rachis not flexuous, pubescent as stem. **Fruiting pedicels** ascending to divaricate-ascending, straight or, sometimes, slightly curved upwards, 4–14(–30) mm, pubescent, trichomes simple and 2–4-rayed. **Flowers:** sepals (purplish tinged), narrowly ovate, 2.5–3 mm, pubescent, (trichomes simple and fewer, stalked, 2-rayed); petals bright yellow, narrowly obovate, 3.5–5 × 1.7–2.5 mm; anthers ovate, 0.3–0.4 mm. **Fruits** elliptic, plane, flattened, 6–10 × 2–3 mm; valves glabrous or glabrescent, trichomes simple, (not confined to replum); ovules 12–24 per ovary; style 0.2–0.3 mm (stigma about as wide as style). **Seeds** (pale brown), ovoid, 0.9–1.3 × 0.6–0.9 mm. $2n = 80$.

Flowering Jun–Aug. Moist tundra and ridges, sand and gravel flats or beaches; 0–1000 m; Greenland; Man., Nfld. and Labr. (Labr.), Nunavut, Ont., Que.; Europe (Finland, Norway, Russia, Sweden).

The synonymy above includes two North American names overlooked by C. L. Hitchcock (1941) and R. C. Rollins (1993). *Draba alpina* was broadly delimited by O. E. Schulz (1927) and included 17 varieties, some of which (e.g., *corymbosa, oxycarpa, pilosa*) are recognized herein as distinct species. The name *D. alpina* was so misapplied that it was used for any circumpolar or alpine, scapose, yellow-flowered, perennial *Draba*. Various chromosome numbers (e.g., $2n = 64, 80, 112, 120$; S. I. Warwick and I. A. Al-Shehbaz 2006) have been reported for the species. As circumscribed here, it has the narrow distribution outlined above and includes plants with $2n = 80$. Reports of the species from Alaska, Canadian Northwest Territories and Yukon, Siberia, eastern Asia, Russian Far East, and the Central Asian republics are either suspect or very unlikely. The entire *D. alpina* complex (including the above three species, *D. glacialis* Adams, *D. macounii*, etc.) is in need of critical molecular, cytological, and morphological study.

5. **Draba aprica** Beadle in J. K. Small, Fl. S.E. U.S. ed. 2, 1336, 1375. 1913 [E]

Draba brachycarpa Nuttall ex Torrey & A. Gray var. *fastigiata* Nuttall ex Torrey & A. Gray

Annuals; not scapose. **Stems** usually branched distally, (0.3–)0.7–3(–3.6) dm, pubescent throughout, trichomes short-stalked to subsessile, cruciform, 0.1–0.5 mm. **Basal leaves** not rosulate; petiolate; petiole (to 1 cm), ciliate throughout; blade obovate to spatulate or ovate to suborbicular, 0.6–2.5 cm × 3–10 mm, margins usually entire, rarely toothed, surfaces pubescent with stalked, cruciform trichomes, 0.1–0.5 mm. **Cauline leaves** (5–)8–20(–25); sessile; blade oblong to lanceolate, margins entire, (not ciliate), surfaces pubescent as basal. **Racemes** 7–46-flowered, ebracteate, elongated in fruit, (lateral racemes from axils of distalmost leaves subumbellate); rachis not flexuous, pubescent as stem. **Fruiting pedicels** horizontal to divaricate-ascending, straight, 2–5 mm, pubescent, trichomes subsessile, 4-rayed. **Flowers:** (late ones cleistogamous, apetalous); sepals oblong, 0.6–1.2 mm, pubescent; petals white, spatulate, 1.3–2.5 × 0.5–1 mm; anthers ovate, 0.1–0.2(–0.05 in cleistogamous flowers) mm. **Fruits** linear-ellipsoid to ellipsoid, plane, flattened, (3–)4–6 × 0.8–1.1 mm; valves pubescent, trichomes subsessile, (2–)4-rayed, 0.03–0.12 mm; ovules 4–6(–8) per ovary; style 0.07–0.13 mm. **Seeds** ovoid, 0.9–1.1 × 0.5–0.6 mm.

Flowering Mar–May. Open knolls, rocky roadsides, rock outcrops, igneous glades, woods, alluvial areas near streams; 100–400 m; Ark., Ga., Mo., Okla., S.C.

6. **Draba arabisans** Michaux, Fl. Bor.-Amer. 2: 28. 1803 E

Draba arabisans var. *superiorensis* Butters & Abbe; *D. incana* Linnaeus var. *arabisans* (Michaux) S. Watson; *D. incana* var. *glabriuscula* A. Gray

Perennials; (sometimes cespitose); caudex simple or branched; not scapose. **Stems** branched or unbranched, (0.5–)0.7–3(–4.5) dm, pubescent proximally, glabrous or pubescent distally, trichomes sessile (non-crisped), 4- or 5-rayed, stellate, or subpectinate, 0.07–0.4 mm. **Basal leaves** rosulate; petiolate; petiole base and margin sparsely ciliate, (trichomes simple and 2-rayed, 0.2–0.8 mm); blade oblanceolate to spatulate or linear-oblanceolate, 1–3.8(–5) cm × 2–5(–8) mm, margins dentate or denticulate, surfaces pubescent with sessile, non-crisped, 4- or 5-rayed, stellate trichomes, 0.07–0.5 mm, (some rays branched, appearing 2–12-rayed, midvein obscure abaxially). **Cauline leaves** (2 or) 3–10 (–13); sessile; blade ovate to oblong, margins dentate to subentire, surfaces pubescent as basal. **Racemes** 5–18(–25)-flowered, ebracteate, elongated in fruit; rachis not flexuous, glabrous or pubescent as stem. **Fruiting pedicels** divaricate-ascending to suberect, straight or slightly curved, (3–)4–10(–13) mm, glabrous or pubescent as stem. **Flowers:** sepals oblong, 1.7–2.5 mm, pubescent, trichomes short-stalked, 3- or 4-rayed; petals white, broadly obovate, 3.5–5 × 1.5–2.5 mm; anthers ovate, 0.4–0.5 mm. **Fruits** ovate to lanceolate or linear-lanceolate, usually slightly twisted, rarely plane, flattened, 5–11(–14) × 2–3.5 mm; valves usually glabrous, rarely pubescent, trichomes 3- or 4-rayed, 0.07–0.2 mm; ovules 20–28(–32) per ovary; style (0.2–)0.4–1 mm. **Seeds** oblong, 0.8–1.1 × 0.6–0.8 mm. $2n = 96$.

Flowering Jun–Aug. Rock outcrops and talus (usually limestone or dolomite), open woods, rocky shores; 0–300 m; Man., N.B., Nfld. and Labr. (Nfld.), N.S., Ont., Que.; Maine, Mich., Minn., N.Y., Vt., Wis.

Draba arabisans is often confused with *D. glabella* ($2n = 80$); it is easily distinguished by having sessile (versus short-stalked) stellate trichomes on abaxial leaf blade surfaces and longer styles (0.2–)0.4–1 [versus 0.1–0.2(–0.4)] mm. Both species have twisted and plane fruits; *D. arabisans* has mostly twisted fruits; *D. glabella* has predominantly plane fruits. The complete absence of simple trichomes proximally on stems of *D. arabisans* can be a useful feature separating it from *D. glabella*, which often has at least some simple trichomes on the proximal half of the stem.

Draba arabis Persoon is an illegitimate name, sometimes found in synonymy with *D. arabisans*.

7. **Draba arctica** J. Vahl in G. C. Oeder et al., Fl. Dan. 13(39): plate 5, 2294. 1840

Draba arctica subsp. *ostenfeldii* Böcher ex Kartesz & Gandhi; *D. arctica* var. *ostenfeldii* Böcher ex Kartesz & Gandhi; *D. cinerea* Adams var. *arctica* (J. Vahl) Pohle; *D. ostenfeldii* E. Ekman 1929, not *D. ×ostenfeldii* O. E. Schulz 1927; *D. ostenfeldii* var. *ovibovina* E. Ekman; *D. ovibovina* (E. Ekman) E. Ekman ex Gelting

Perennials; (cespitose, not pulvinate); caudex simple or branched; rarely scapose. **Stems** unbranched, 0.3–1.7 (–2.4) dm, pubescent, trichomes 4–10-rayed (rays sometimes branched), 0.1–0.3 mm, with fewer, simple ones, to 0.7 mm. **Basal leaves** rosulate; petiolate; petiole base ciliate, (trichomes simple, 0.2–0.8 mm); blade oblanceolate to narrowly obovate, 0.4–2.3(–3) cm × 1.5–6.5 mm, margins usually entire, rarely with 1 tooth on each side, surfaces densely pubescent with minutely stalked, 8–12-rayed, stellate trichomes, 0.2–0.4 mm, sometimes with coarser, simple or 2-rayed ones, (midvein distinct abaxially). **Cauline leaves** 0 or 1(–3); sessile; blade ovate or oblong to lanceolate, margins entire, surfaces pubescent as basal. **Racemes** 3–18(–25)-flowered, ebracteate, elongated in fruit; rachis not flexuous, pubescent as stem. **Fruiting pedicels** divaricate-ascending or ascending, straight, (1.5–)2–6 mm, pubescent, trichomes stalked, stellate, and, sometimes, simple. **Flowers:** sepals ovate, 2–2.5 mm, pubescent, (trichomes short-stalked, 2- or 3-rayed, and simple); petals white, spatulate to obovate, 3.5–6 × 1.8–2 mm; anthers ovate, 0.3–0.4 mm. **Fruits** oblong to lanceolate, plane, slightly flattened, (5–)6–11 × 2–3 mm; valves pubescent, trichomes short-stalked, 2–5-rayed (some rays branched), 0.05–0.3 mm; ovules 20–32 per ovary; style 0.1–1 mm (stigma distinctly wider than style). **Seeds** ovoid, 0.8–1.1 × (0.6–)0.7–0.8 mm. $2n = 80$.

Flowering Jun–Jul. Clay flats, gravel, beaches, limestone outcrops, talus; 0–500 m; Greenland; Nunavut; Europe (Norway [Svalbard]).

As circumscribed here, *Draba arctica* consists of two distinct elements often recognized as subspecies (*arctica* and *ostenfeldii*; T. W. Böcher 1966); perhaps they are better treated as distinct species (*D. arctica* and *D. ovibovina*). Lack of adequate material and inability to examine all of the types in this complex prevent us from doing so.

8. Draba arctogena (E. Ekman) E. Ekman, Svensk Bot. Tidskr. 25: 492. 1932 E

Draba groenlandica E. Ekman var. *arctogena* E. Ekman, Svensk Bot. Tidskr. 23: 486. 1930; *D. cinerea* Adams var. *arctogena* (E. Ekman) B. Boivin

Perennials; (cespitose); caudex branched (sometimes with persistent leaf remains); sometimes scapose. **Stems** unbranched, 0.3–1.4 dm, pubescent throughout, trichomes simple, 0.2–1 mm, and 2–6-rayed, 0.05–0.3 mm. **Basal leaves** rosulate; petiolate; petiole base and margin ciliate, (trichomes mostly simple, 0.2–1 mm); blade oblanceolate to narrowly obovate, 0.4–1.7 cm × 1.5–6 mm, margins entire or with 1–3 teeth on each side, surfaces pubescent with primarily simple trichomes, 0.3–1 mm, abaxially mixed with stalked, 4–10-rayed ones, 0.1–0.2 mm. **Cauline leaves** 0 or 1; sessile; blade ovate to oblong, margins entire or toothed, surfaces pubescent as basal. **Racemes** 2–14(–17)-flowered, ebracteate, elongated in fruit; rachis not flexuous, pubescent as stem. **Fruiting pedicels** suberect to ascending, straight, 1–5(–7) mm, pubescent, trichomes simple and 2–4-rayed. **Flowers**: sepals ovate, 1.8–2.5 mm, pubescent, (trichomes simple and short-stalked, 2–4-rayed); petals white, spatulate, 3–4 × 1.5–2 mm; anthers ovate, 0.3–0.4 mm. **Fruits** oblong, plane, flattened, 4–9 × 2–3.5 mm; valves pubescent, trichomes short-stalked, 2–4-rayed, 0.1–0.5 mm, sometimes with simple ones; ovules 20–32 per ovary; style 0.1–0.4 mm (stigma distinctly wider than style). **Seeds** ovoid, 0.7–1 × 0.5–0.6 mm. $2n = 48$.

Flowering Jun–Jul. Scree, sandy hills, gravel flats; 0–450 m; Greenland; N.W.T., Nunavut.

Draba arctogena is sometimes confused with *D. norvegica*, from which it differs by having less elongated fruiting racemes, abaxial leaf surfaces with 4–10-rayed (versus cruciform or, rarely, 2–6-rayed) trichomes, and fruits with 2–4-rayed trichomes, 0.1–0.5 mm, (versus glabrous or with simple and 2–4-rayed trichomes, 0.05–0.25 mm). The limits of *D. arctica*, *D. arctogena*, *D. cinerea*, and *D. oblongata* need to be critically evaluated in light of molecular and cytological data.

9. Draba argyrea Rydberg, Bull. Torrey Bot. Club 30: 251. 1903 E

Draba argyrea var. *glabrescens* O. E. Schulz

Perennials; caudex branched (branches loose, creeping, sometimes terminating in sterile shoots); scapose. **Stems** unbranched, (0.2–)0.3–0.8 (–1.1) dm, pubescent proximally, glabrous distally, trichomes short-stalked to subsessile, 5–8-rayed, 0.05–0.2 mm. **Basal leaves** rosulate; subsessile; petiole base and margin not ciliate, (midvein obscure abaxially); blade oblanceolate to obovate, 0.3–0.7(–1.2) cm × 1–2.5(–4) mm, margins entire, surfaces canescent with short-stalked, 8–12-rayed, stellate trichomes, 0.1–0.2 mm. **Cauline leaves** 0. **Racemes** 2–8(–12)-flowered, ebracteate, elongated in fruit; rachis slightly flexuous, usually glabrous, rarely sparsely pubescent as stem. **Fruiting pedicels** divaricate-ascending, straight or slightly curved upward, (4–)5–12(–19) mm, glabrous. **Flowers**: sepals ovate, 2–3 mm, pubescent, (trichomes 4–6-rayed); petals yellow, oblanceolate, 4–6 × 1.5–3 mm; anthers ovate, 0.5–0.6 mm. **Fruits** lanceolate to elliptic or ovate, plane or slightly twisted, flattened, 5–11 × 2–3.5(–4.5) mm, (symmetric); valves often puberulent, sometimes glabrous, trichomes simple and subsessile, 2–4-rayed, 0.03–0.2 mm; ovules 8–14 per ovary; style (0.7–)1–1.8(–2.4) mm. **Seeds** oblong, 1.2–1.8 × 0.8–1.1 mm. $2n = 36$.

Flowering Apr–Jul. Rock crevices, ledges, talus slopes; 1000–3100 m; Idaho.

M. A. Beilstein and M. D. Windham (2003) hypothesized that *Draba argyrea* is an allopolyploid derived through hybridization between *D. lonchocarpa* and *D. sphaerocarpa*. The species is restricted to Blaine, Boise, Custer, and Elmore counties in central Idaho.

10. Draba arida C. L. Hitchcock, Revis. Drabas W. N. Amer., 52, plate 3, fig. 23. 1941 C E

Perennials; (cespitose); caudex branched (with persistent leaf remains, branches relatively short); not scapose. **Stems** usually unbranched, (0.3–)0.4–1 dm, pubescent throughout, trichomes simple and 2-rayed, 0.4–1.4 mm, and shorter, 2–4-rayed ones, 0.05–0.3 mm. **Basal leaves** rosulate; subsessile; petiole not ciliate; blade obovate to spatulate or oblanceolate, 0.6–1.9(–2.4) cm ×

2–5(–8) mm, (base ciliate, trichomes simple and 2-rayed), margins entire, surfaces abaxially pubescent with stalked, (2–)4-rayed trichomes, 0.3–0.6 mm, adaxially with simple and stalked, 2-rayed trichomes, 0.3–0.6 mm, and smaller, 3- or 4-rayed ones. **Cauline leaves** (1 or) 2–5; sessile; blade oblong to oblanceolate or ovate, margins entire, surfaces pubescent as basal. **Racemes** 12–40-flowered, ebracteate, slightly elongated in fruit; rachis not flexuous, pubescent, trichomes simple and 2–4-rayed. **Fruiting pedicels** horizontal to divaricate-ascending, straight, (3–)4–8 mm, pubescent, trichomes simple and 2–4-rayed (0.1–0.5 mm). **Flowers:** sepals broadly ovate, 2–2.7 mm, pubescent, (trichomes simple and short-stalked, 2–4-rayed); petals yellow, spatulate to oblanceolate, 3.5–5 × 1.4–2 mm; anthers oblong, 0.4–0.6 mm. **Fruits** elliptic or ovate to lanceolate, plane, slightly inflated basally, 4–7 × 2.5–3.5 mm; valves pubescent, trichomes simple and stalked, 2–4-rayed, 0.05–0.4 mm; ovules 8–12 per ovary; style 0.5–1.2(–1.4) mm. **Seeds** ovoid, 1–1.4 × 0.6–0.8 mm. $2n = 24$.

Flowering May–Jul. Rock crevices and gravelly soil in conifer and subalpine shrub communities; of conservation concern; 2100–3400 m; Nev.

Draba arida is a sexually reproducing, diploid member of the *D. ventosa* complex that may have been involved in the origin of the more widespread apomictic triploids (M. A. Beilstein and M. D. Windham 2003). The species is known from the Monitor, Toiyabe, and Toquima ranges in Lander and Nye counties.

11. Draba asprella Greene, Bull. Torrey Bot. Club 10: 125. 1883 [E] [F]

Perennials; (cespitose); caudex branched (with some persistent leaf bases); scapose. **Stems** unbranched, (0.4–)0.6–1.5(–2.7) dm, pubescent throughout, trichomes simple, 0.5–1.5 mm, and 2-rayed ones, 0.05–0.3 mm, (sometimes 2–4-rayed ones distally). **Basal leaves** rosulate; subsessile or shortly petiolate; petiole ciliate throughout; blade oblanceolate to spatulate or obovate, (0.6–)0.8–4.5(–6) cm × 2–12(–16) mm, margins usually entire, rarely obscurely dentate, (not ciliate), surfaces abaxially pubescent with stalked, (2–)4-rayed trichomes, 0.2–1 mm, adaxially with stalked, 2–4-rayed trichomes, 0.3–0.8 mm, and simple ones, 0.5–1.9 mm. **Cauline leaves** 0. **Racemes** (15–)30–75-flowered, ebracteate, elongated in fruit; rachis not flexuous, pubescent, trichomes 2–4-rayed, sometimes with simple ones. **Fruiting pedicels** horizontal to divaricate-ascending, straight, 3–11(–16) mm, pubescent, trichomes 2–4-rayed, sometimes with simple ones. **Flowers:** sepals broadly ovate, 1.8–3.5 mm, pubescent, (trichomes simple and short-stalked, 2–4-rayed); petals pale yellow, oblanceolate, 4–7 × 1.5–2.5 mm; anthers ovate to oblong, 0.4–0.7 mm. **Fruits** ovoid-ellipsoid to oblong, plane, inflated basally, 3–8 × 1.8–3.5 mm; valves hirsute, trichomes simple and spurred, (0.2–)0.3–1 mm, or puberulent, trichomes short-stalked, 2–4-rayed, 0.1–0.2 mm; ovules 8–12(–18) per ovary; style 0.6–2(–2.5) mm. **Seeds** oblong, 0.9–1.2 × 0.6–0.8 mm. $2n = 30$.

Varieties 2 (2 in the flora): Arizona.

Both R. C. Rollins (1993) and N. H. Holmgren (2005b) indicated that the chromosome number of *Draba asprella* is $2n = 32$, but repeated counts by one of us (M. D. Windham 2000, unpubl.) show that the species consistently has $2n = 30$. Rollins divided *D. asprella* into four varieties encompassing tremendous variation in trichome morphology. One of these (var. *zionensis*) is more closely related to *D. sobolifera* (Windham and L. Allphin, unpubl.) and is treated herein at species rank. Within *D. asprella*, in the strict sense, we recognize two varieties separated primarily by the type of trichomes found on the fruits. The distinctions are not absolute and there appear to be forms connecting the two, especially in Coconino County.

1. Fruits hirsute, trichomes simple and unequally 2-rayed, (0.2–)0.3–1 mm. . . . . 11a. *Draba asprella* var. *asprella*
1. Fruits puberulent, trichomes 2–4-rayed, 0.1–0.2 mm . . . . 11b. *Draba asprella* var. *stelligera*

11a. Draba asprella Greene var. **asprella** [E] [F]

Fruit valves hirsute, trichomes simple and unequally 2-rayed, (0.2–)0.3–1 mm. $2n = 30$.

Flowering Mar–May. Rocky igneous soil in pine-oak woodlands; 1500–2500 m; Ariz.

Variety *asprella* is known only from Coconino and Yavapai counties in central Arizona.

11b. Draba asprella Greene var. **stelligera** O. E. Schulz in H. G. A. Engler, Pflanzenr. 89[IV,105]: 106. 1927 [C] [E]

Draba asprella var. *kaibabensis* C. L. Hitchcock

Fruit valves puberulent, trichomes 2–4-rayed, 0.1–0.2 mm.

Flowering Apr–Jul. On and around limestone and sandstone rock outcrops in pine-oak and pinyon-juniper communities; of conservation concern; 1500–2400 m; Ariz.

Variety *stelligera* is known from Coconino, Gila, and Yavapai counties in northern and central Arizona. The difference between vars. *kaibabensis* and *stelligera* rests

12. Draba asterophora Payson, Amer. J. Bot. 4: 263. 1917 [E]

Draba asterophora var. *macrocarpa* C. L. Hitchcock

Perennials; (loosely cespitose); caudex branched (somewhat surculose, with persistent leaf bases, branches sometimes terminating in sterile rosettes); scapose. **Stems** unbranched, 0.3–1.1 dm, glabrous throughout or sparsely pubescent proximally, trichomes (2–)4-rayed, 0.1–0.3 mm. **Basal leaves** rosulate; petiolate; petiole obsolete, margin rarely ciliate proximally; blade (somewhat fleshy), broadly obovate to suborbicular or spatulate, 0.4–1.4(–1.7) cm × (2–)3–6(–7) mm, margins entire, surfaces pubescent, trichomes stalked, cruciform, and 2-, 3-, or 5-rayed, 0.2–0.6 mm. **Cauline leaves** 0 (or 1, as a bract). **Racemes** (5–)8–20(–27)-flowered, usually ebracteate, rarely proximalmost flowers bracteate, elongated in fruit; rachis not flexuous, glabrous. **Fruiting pedicels** horizontal to divaricate-ascending, straight or curved upward, 3–9 mm, glabrous. **Flowers:** sepals oblong, 3–4 mm, glabrous or sparsely pubescent, (trichomes subapical, short-stalked, 2–4-rayed); petals bright yellow, oblanceolate, 5–7 × 1.5–2.5 mm; anthers ovate, 0.5–0.6 mm. **Fruits** lanceolate-ovate to broadly ovate or oblong, slightly twisted or plane, strongly flattened, 5–11(–14) × (3.5–)4–6 mm; valves glabrous or, rarely, puberulent, trichomes simple and short-stalked, 2- or 3-rayed, 0.05–0.3 mm; ovules 12–18 per ovary; style 0.2–1.6(–2) mm. **Seeds** (winged), ovate, 1.8–2.8 × 1.2–2 mm; (wing 0.5–1 mm wide). $2n = 40$.

Flowering Jun–Aug. Granitic rock outcrops, talus, gravelly soil; 2600–3300 m; Calif., Nev.

C. L. Hitchcock (1941) and R. C. Rollins (1993) divided *Draba asterophora* into two varieties based on minor differences in style and fruit lengths. These do not appear to define genetically discrete taxa. The species apparently is restricted to El Dorado County, California, and the Carson Range in Washoe County, Nevada.

13. Draba aurea Vahl ex Hornemann, Fors. Oecon. Plantel. ed. 2, 599. 1806 [E]

Draba aurea var. *aureiformis* (Rydberg) O. E. Schulz; *D. aurea* var. *leiocarpa* (Payson & H. St. John) C. L. Hitchcock; *D. aurea* var. *luteola* (Greene) O. E. Schulz; *D. aureiformis* Rydberg; *D. aureiformis* var. *leiocarpa* Payson & H. St. John; *D. bakeri* Greene; *D. decumbens* Rydberg; *D. henneana* Schlechtendal var. *maccallae* (Rydberg) O. E. Schulz; *D. luteola* Greene; *D. luteola* var. *minganensis* Victorin; *D. maccallae* Rydberg; *D. minganensis* (Victorin) Fernald; *D. surculifera* A. Nelson; *D. uber* A. Nelson

Perennials; (not cespitose); caudex simple or branched; not scapose. **Stems** sometimes branched distally, (0.5–)1–3.5(–5.2) dm, pubescent throughout, trichomes simple (non-crisped), 0.4–1.3 mm, and 3–6-rayed ones, 0.1–0.5 mm. **Basal leaves** rosulate; petiolate; petiole (distinct or obscure) ciliate, (trichomes simple, to 0.8 mm); blade oblanceolate to obovate, (0.4–)1–3.7(–5) cm × (1–)2–7(–10) mm, margins entire or denticulate, surfaces pubescent, trichomes stalked, (2–)4–7 (or 8)-rayed, 0.2–0.5(–0.6) mm. **Cauline leaves** 5–20(–26); sessile; blade oblong to lanceolate or ovate, margins entire or dentate, surfaces pubescent as basal, sometimes adaxially with simple trichomes. **Racemes** (10–)18–52(–72)-flowered, usually bracteate on proximalmost 1–12(–17) flowers, rarely ebracteate, elongated in fruit; rachis not flexuous, pubescent as stem. **Fruiting pedicels** divaricate-ascending to ascending or suberect, straight, 3–13(–20) mm, pubescent as stem. **Flowers:** sepals (green or yellowish), oblong, 2.2–3 mm, pubescent, (trichomes simple and branched); petals yellow, oblanceolate, 3.5–5 × 1.5–2.5 mm; anthers ovate, 0.4–0.5 mm. **Fruits** (often subappressed to rachis), lanceolate to linear-lanceolate or narrowly oblong, slightly twisted or plane, flattened, (6–)9–14(–17) × 2–3.5 mm; valves pubescent, trichomes simple and short-stalked, 2–4-rayed, 0.05–0.3 mm; ovules 28–38(–44) per ovary; style 0.5–1.2(–1.5) mm. **Seeds** oblong, 0.9–1.3 × 0.5–0.7 mm. $2n = 74$.

Flowering Jun–Aug. Rock outcrops, talus, damp gullies and meadows, subalpine conifer woodlands, alpine slopes and turf, tundra, road banks, river gravel; (0–)700–4200 m; Greenland; Alta., B.C., Man., Nfld. and Labr. (Nfld.), N.W.T., Nunavut, Ont., Que., Sask., Yukon; Alaska, Ariz., Colo., Idaho, Mont., Nev., N.Mex., S.Dak., Utah, Wash., Wyo.

Draba aurea is extremely variable in plant size, number of cauline leaves, number of bracteate flowers, style length, and fruit size, shape, orientation, twisting, and indumentum. Much of the variation in the number of bracts, style length, fruit twisting, and growth habit occurs in Greenland, where the type specimen was collected and where the species is found near sea level.

The highly deviant chromosome counts (e.g., $2n = 40 + 1, 64, 82$) listed by R. C. Rollins (1993) and S. I. Warwick and I. A. Al-Shehbaz (2006) are mostly unvouchered and have to be disregarded; counts of $2n$ = ca. 80 have been re-assigned to *Draba glabella*. Published (G. A. Mulligan 2002) and unpublished counts made by Mulligan and M. D. Windham from Alaska, British Columbia, Colorado, Quebec, Utah, and Yukon indicate that the most common chromosome number of *D. aurea* is $2n = 74$ (or 72). This suggests that the species is an allopolyploid (hexaploid or higher), incorporating genomes from both euploid and aneuploid lineages (M. A. Beilstein and Windham 2003). Detailed cytological and molecular studies are much needed to fully understand this widely distributed and highly variable species.

14. Draba aureola S. Watson in W. H. Brewer et al., Bot. California 2: 430. 1880 [E] [F]

Draba aureola var. *paniculata* L. F. Henderson

Perennials; (short-lived); caudex simple or branched (poorly developed, with persistent dry leaves); not scapose. **Stems** unbranched or branched, 0.3–1.5 dm, hirsute throughout, trichomes simple and long-stalked, 2–4-rayed, 0.2–1.5 mm. **Basal leaves** (forming dense clusters); rosulate; sessile; blade oblanceolate to linear, (0.7–)1–2.5(–3.2) cm × 2–3.5(–5) mm, margins entire, (ciliate, trichomes simple and 2-rayed, 0.5–1.5 mm), surfaces densely hirsute, abaxially with stalked, 3–5-rayed trichomes, 0.1–0.5 mm, adaxially with simple and long-stalked, 2-rayed trichomes, to 1 mm, with smaller, 3–5-rayed ones. **Cauline leaves** 5–33; sessile; blade oblong to linear, margins entire, surfaces pubescent as basal. **Racemes** 12–83-flowered, ebracteate or proximalmost 1–9 flowers bracteate, slightly or not elongated in fruit; rachis not flexuous, hirsute as stem. **Fruiting pedicels** horizontal to divaricate, usually straight, rarely curved upward, (3–)5–12(–19) mm, hirsute as stem. **Flowers:** sepals ovate, 2.5–4 mm, glabrous; petals bright yellow, linear to linear-oblanceolate, 4–6 × 0.5–1(–1.2) mm; anthers narrowly oblong, 0.8–1 mm. **Fruits** oblong or, rarely, oblong-ovate, plane, flattened, (6–)9–14(–16) × 3–5 mm; valves pubescent, trichomes stalked, 2–4-rayed, 0.07–0.5 mm; ovules 10–20 per ovary; style (0.6–)1–1.6(–2) mm. **Seeds** oblong, 1.4–1.9 × 0.8–1.1 mm. $2n = 20$.

Flowering Jul–Aug. Open conifer forests, alpine meadows and moraines, talus slopes; 2200–3200 m; Calif., Oreg., Wash.

Draba aureola is known from Lassen, Shasta, and Siskiyou counties in California, Deschutes County in Oregon, and Lewis and Pierce counties in Washington.

15. Draba bifurcata (C. L. Hitchcock) Al-Shehbaz & Windham, Harvard Pap. Bot. 12: 411. 2007 [E]

Draba helleriana Greene var. *bifurcata* C. L. Hitchcock, Revis. Drabas W. N. Amer., 39, plate 3, fig. 21e. 1941

Biennials or, rarely, perennials; caudex simple or branched; not scapose. **Stems** branched or unbranched, 1–3.9 dm, pubescent throughout or proximally, trichomes simple, 0.3–1.4 mm, often with smaller, 2 (or 3)-rayed ones. **Basal leaves** (soon withered); rosulate; petiolate; petiole ciliate, (trichomes simple, 0.4–1.1 mm); blade oblanceolate to spatulate, 1–7 cm × 2–12(–26) mm, margins entire or dentate (pubescent as petiole), surfaces pubescent, abaxially with simple trichomes, 0.6–1.1 mm, and stalked, 2- (or 3-)rayed ones, often smaller, adaxially with subsetiform, simple trichomes, 0.5–1.2 mm (rarely with fewer, stalked, 2-rayed ones). **Cauline leaves** (2–)6–15; sessile; blade lanceolate to ovate or oblong, margins dentate, surfaces pubescent as basal. **Racemes** 10–51-flowered, ebracteate, elongated in fruit; rachis glabrous or pubescent as stem, not flexuous. **Fruiting pedicels** horizontal to divaricate, straight, 4–15 mm, glabrous or pubescent, trichomes simple and 2-rayed. **Flowers:** sepals ovate, 2.5–3.2 mm, pubescent, (trichomes simple and short-stalked, 2-rayed); petals yellow, oblanceolate, 4.5–6.5 × 1.5–2 mm; anthers oblong, 0.8–1.1 mm. **Fruits** elliptic-lanceolate, twisted (1 turn or plane, flattened, 6–10 × 2–2.5 mm; valves glabrous or sparsely puberulent, trichomes simple, 0.05–0.15 mm; ovules 14–20 per ovary; style (1–)2–3 mm. **Seeds** oblong, 1.1–1.4 × 0.7–0.9 mm.

Flowering Jun–Jul. Rocky areas, damp shady ravines, aspen-spruce communities; 1800–3600 m; Ariz.

Draba bifurcata is distinguished from other North American species by a combination of yellow flowers with styles (1–)2–3 mm and almost exclusively simple trichomes on the abaxial surface of cauline leaf blades. It is known only from Chiricahua, Santa Catalina, and White mountains in Apache, Cochise, and Pima counties.

16. Draba borealis de Candolle, Syst. Nat. 2: 342. 1821

Draba borealis var. *maxima* (Hultén) S. L. Welsh; *D. maxima* Hultén; *D. unalaschkiana* de Candolle

Perennials; caudex branched (branches usually slender, elongated, rhizomatous, sometimes with persistent leaf bases); not scapose. **Stems** usually unbranched, rarely branched, (0.4–)1–3.6 (–5.5) dm, often hirsute proximally, trichomes usually simple and 2-rayed, 0.5–1.1 mm, with short-stalked, 3–8-rayed ones, 0.1–0.4 mm (rarely simple trichomes absent distally). **Basal leaves** rosulate; blade ovate or obovate to oblanceolate, (0.5–)1–4.2(–6) cm × 3–10(–25) mm, margins dentate or denticulate, surfaces pubescent, trichomes short-stalked, cruciform, or (2–)4–6-rayed, 0.2–0.6 mm (principal rays usually simple, rarely 1 or 2 with a lateral branch, sometimes appearing to 10-rayed). **Cauline leaves** (2 or) 3–7(–12); sessile; blade ovate, margins entire or dentate, surfaces pubescent as basal or adaxially with some simple trichomes. **Racemes** (6–)8–20(–35)-flowered, ebracteate, elongated in fruit; rachis not flexuous, pubescent as stem. **Fruiting pedicels** divaricate or ascending, straight, (2–)4–8(–13) mm, pubescent as stem. **Flowers:** sepals ovate, 2–3 mm, pubescent, (trichomes simple); petals white, obovate, 4–6 × 2–3 mm; anthers ovate, 0.3–0.5 mm. **Fruits** ovate to broadly oblong or lanceolate, slightly twisted or plane, flattened, (5–)7–12 × 2.5–4.5 mm; valves pubescent, trichomes simple and short-stalked, 2–4-rayed, 0.1–0.4 mm; ovules 16–28(–30) per ovary; style 0.2–0.6(–0.8) mm, glabrous. **Seeds** oblong, 1–1.5 × 0.7–1 mm. $2n = 64, 80$.

Flowering Jun–Jul. Rock outcrops and talus, gravelly terraces, meadows, forest edges and thickets, roadsides, grassy areas, alpine tundra; 0–2400 m; Alta., B.C., N.W.T., Nunavut, Yukon; Alaska; e Asia (Japan, Russian Far East).

Draba borealis is highly variable in leaf and stem indumentum, leaf shape and margin, number of cauline leaves, and fruit shape, size, and twisting. North American plants yielded decaploid chromosome counts; octoploid populations were reported from the Russian Far East. This suggests that more than one taxon is present, and the species is much in need of detailed molecular, cytogenetic, and morphological study.

In the absence of flowers, *Draba borealis* is occasionally confused with some forms of *D. aurea*. The latter usually has proximally bracteate (versus ebracteate) racemes, generally longer styles [0.5–1.5 (–1.7) versus 0.2–0.6(–0.8) mm], and more ovules [28– 38(–44) versus 16–28(–30)] per ovary. *Draba borealis* occasionally is confused with *D. glabella*, but the latter has pectinate-stellate trichomes on abaxial leaf blade surfaces. R. C. Rollins (1993) indicated that *D. borealis* occurs in Colorado, but we have not seen any material from the United States outside of Alaska.

17. Draba brachycarpa Nuttall ex Torrey & A. Gray, Fl. N. Amer. 1: 108. 1838 E

Abdra brachycarpa (Nuttall ex Torrey & A. Gray) Greene

Annuals; not scapose. **Stems** usually branched, rarely unbranched, (0.3–)0.4–1.9 (–2.2) dm, pubescent throughout, trichomes sessile, cruciform, 0.2–0.6(–0.8) mm (rays often equal, or those parallel to stem axis longer). **Basal leaves** not rosulate; petiolate; petiole (to 0.5 cm), not ciliate; blade oblanceolate to obovate, 0.5–2 cm × 2–8 mm, margins entire, surfaces pubescent, trichomes cruciform, 0.2–0.6(–0.8) mm. **Cauline leaves** (4–)6–11; sessile; blade lanceolate to oblong or linear, margins entire, surfaces pubescent as basal. **Racemes** (main branch) (20–)25–65(–74)-flowered, ebracteate, elongated in fruit; rachis not flexuous, pubescent as stem. **Fruiting pedicels** horizontal to divaricate-ascending, straight, (1–)1.5–4(–5) mm, pubescent, trichomes cruciform. **Flowers:** (late ones cleistogamous, apetalous); sepals (green or pink), oblong, 0.7–1.2(–1.5) mm, pubescent; petals white, spatulate, 2–3 × 0.8–1.1 mm; anthers ovate, 0.15–0.25 mm. **Fruits** usually elliptic to linear-elliptic, rarely ovate-elliptic, plane, flattened, (2–)2.5–5(–6) × 0.9–1.4(–1.9) mm; valves glabrous; ovules 8–16(–20) per ovary; style 0.05–0.1 mm. **Seeds** (winged), ovoid, 0.5–0.7 × 0.4–0.5 mm. $2n = 16$.

Flowering Feb–May. Open woods, cedar glades, pastures and lawns, roadsides, disturbed sites; 0–300 m; Ala., Ariz., Ark., Fla., Ga., Ill., Ind., Kans., Ky., La., Miss., Mo., N.C., Ohio, Okla., Oreg., S.C., Tenn., Tex., Va.

Draba brachycarpa is closely related to *D. aprica* and the two are sometimes confused. It is readily distinguished by having leaves with sessile (versus stalked) trichomes, glabrous (versus pubescent) fruits, and smaller (0.5–0.7 versus 0.9–1.1 mm) seeds.

The records from Arizona and Oregon are based on old collections, *Porter 802* (Devil's Canyon, 22 Feb 1926, US) and *Howell s.n.* (near Cobarg, Willamette Valley, 7 Apr 1887, US); it is not known if these records represent introductions or remnants of a previously wider distribution.

18. **Draba brachystylis** Rydberg, Bull. Torrey Bot. Club 29: 240. 1902 [C][E]

Annuals or perennials; (short-lived); caudex often simple (poorly developed); not scapose. Stems unbranched or branched, 0.6–3(–3.7) dm, pubescent throughout, trichomes simple, 0.4–0.9 mm, and 2–4-rayed, 0.05–0.4 mm. Basal leaves rosulate; subsessile; petiole base ciliate, (trichomes simple, 0.3–0.9 mm); blade oblanceolate, 1–3.5 cm × 4–7(–12) mm, margins entire or denticulate, surfaces pubescent, abaxially with stalked, (2–)4-rayed trichomes, 0.1–0.5 mm, adaxially with simple trichomes, 0.2–0.6 mm, with smaller, 2–4-rayed ones. Cauline leaves (1 or) 2–6(–8); sessile; blade lanceolate or oblanceolate to ovate, margins entire or denticulate, surfaces pubescent as basal. Racemes (5–)10–35(–47)-flowered, ebracteate, elongated in fruit; rachis not flexuous, pubescent as stem. Fruiting pedicels divaricate-ascending, straight or only slightly curved upward, (1.5–)3–8(–10) mm, pubescent throughout, trichomes simple and stalked, 2–4-rayed. Flowers: sepals ovate, 2–2.7 mm, pubescent, (trichomes simple and short-stalked, 2-rayed); petals pale yellow (fading white), oblanceolate, 2.5–3.7 × 0.9–1.2 mm; anthers ovate, 0.2–0.3 mm. Fruits narrowly elliptic to lanceolate, plane, flattened, (7–)10–16(–19) × 2–3.5 mm; valves pubescent, trichomes simple, spurred, and short-stalked, 2-rayed, 0.05–0.2(–0.3) mm; ovules 20–36(–42) per ovary; style 0.2–0.6(–0.8) mm. Seeds oblong, 0.9–1.3 × 0.6–0.7 mm. $2n = 44$.

Flowering Jun–Jul. Fir and aspen communities, moist areas on rocky slopes; of conservation concern; 1700–3000 m; Utah.

Based on morphological and chromosomal data (L. Allphin and M. D. Windham, unpubl.), *Draba brachystylis* is hypothesized to be an allopolyploid derived through hybridization between *D. albertina* and *D. santaquinensis*. It is known to us only from Cache, Duchesne, Juab, Salt Lake, and Utah counties. C. L. Hitchcock (1941) indicated that the species grows in the Charleston Mountains of Clark County, Nevada, but we have not seen unequivocal material of it from that area.

19. **Draba breweri** S. Watson, Proc. Amer. Acad. Arts 23: 260. 1888 [E]

Draba breweri var. *sublaxa* Jepson

Perennials; (cespitose, grayish pubescent); caudex branched (sometimes covered with persistent leaf bases, branches short, compact); sometimes scapose. Stems unbranched, (0.1–)0.2–1(–1.5) dm, pubescent throughout, trichomes stalked, 4–10-rayed, 0.1–0.3 mm. Basal leaves rosulate; petiolate; petiole base ciliate, margin not ciliate, (trichomes simple and 2-rayed, 0.3–0.8 mm); blade oblanceolate to obovate, (0.3–)0.4–1.5(–2.5) cm × 1.5–3(–5) mm, margins usually entire, rarely dentate, surfaces densely pubescent with stalked, 4–10-rayed, stellate trichomes, 0.1–0.2 mm (sometimes 1 or more rays spurred). Cauline leaves 0–3(–6); sessile; blade oblong or lanceolate to ovate, margins entire (sometimes ciliate at base), surfaces usually pubescent as basal. Racemes (5–)7–18(–24)-flowered, ebracteate, slightly to considerably elongated in fruit; rachis not flexuous, pubescent with stalked, 4–10-rayed trichomes, (0.1–0.3 mm). Fruiting pedicels ascending, (sometimes slightly appressed to rachis), straight, 1.5–3(–4) mm, pubescent as rachis. Flowers: sepals (persistent), ovate, 1.2–2 mm, pubescent, (trichomes short-stalked, 2–6-rayed); petals white, spatulate to oblanceolate, 2–3 × 0.7–1.1 mm; anthers ovate, 0.2–0.25 mm. Fruits lanceolate or oblong to linear, usually slightly to strongly twisted, rarely plane, flattened, 3.5–9(–11) × 1.5–2.5 mm; valves pubescent, trichomes short-stalked, 2–5-rayed, 0.05–0.25 mm; ovules 28–40 per ovary; style 0.1–0.3(–0.4) mm. Seeds ovoid, 0.5–0.7 × 0.3–0.5 mm. $2n = 32$.

Flowering Jul–Aug. Rock outcrops, talus, exposed ridges, alpine areas; 3100–4100 m; Calif.

The circumscription of *Draba breweri* was expanded by R. C. Rollins (1993) to include *D. cana*. Plants of *D. cana* differ from those of *D. breweri* by being noncespitose (versus cespitose) and taller [(4–)10–30(–38) versus (1–)2–9(–15) cm], and by having basally bracteate (versus ebracteate) racemes, and stems, pedicels, and sepals pubescent with a mixture of simple and branched (versus exclusively branched) trichomes. *Draba breweri* is known to us from Alpine, Fresno, Inyo, Mono, Plumas, Tulare, and Tuolumne counties.

20. Draba burkei (C. L. Hitchcock) Windham & Beilstein, Madroño 50: 221. 2004 [C][E]

Draba maguirei C. L. Hitchcock var. *burkei* C. L. Hitchcock, Revis. Drabas W. N. Amer., 72, plate 5, fig. 37c. 1941

Perennials; (cespitose, forming loose mats); caudex branched (with persistent leaves, branches sometimes terminating in sterile rosettes); scapose. **Stems** unbranched, 0.3–0.6(–0.9) dm, glabrous. **Basal leaves** subrosulate; subsessile; blade oblanceolate to obovate, 0.3–0.8(–1.3) cm × 1–2.5 mm, margins entire, (ciliate, trichomes simple, subsetiform, 0.25–0.8 mm), surfaces glabrous. **Cauline leaves** 0. **Racemes** 4–10-flowered, ebracteate, elongated in fruit; rachis not flexuous, glabrous. **Fruiting pedicels** divaricate-ascending to ascending, straight, 4–9(–15) mm, glabrous. **Flowers:** sepals broadly ovate, 2–3.5 mm, glabrous or sparsely pubescent, (trichomes simple, 0.07–0.35 mm); petals yellow, oblanceolate, 4–6 × 1.5–2 mm; anthers ovate, 0.4–0.5 mm. **Fruits** ovate, plane, flattened, 3–5.5 × 2–3.2 mm; valves glabrous or puberulent, trichomes simple, 0.02–0.08 mm; ovules 4–10 per ovary; style 0.5–1.7 mm. **Seeds** ovoid, 1–1.4 × 0.7–1 mm. $2n = 20$.

Flowering May–Jul. Rocky ridges, steep talus slopes, rock outcrops and crevices; of conservation concern; 1600–3000 m; Utah.

Draba burkei was treated by C. L. Hitchcock (1941) and R. C. Rollins (1993) as a variety of *D. maguirei*. Chromosome numbers ($2n = 20$ versus $2n = 32$), plant morphology, and molecular data support the recognition of these taxa as independent species (M. D. Windham 2004). *Draba burkei* is easily distinguished from *D. maguirei* by having exclusively simple trichomes confined to leaf blade margins (versus mostly branched trichomes on margins and surfaces) and smaller seeds (1–1.4 × 0.7–1 versus 1.6–2 × 1–1.3 mm). It is known from Box Elder, Cache, Morgan, and Weber counties, where it approaches but does not overlap the range of *D. maguirei*. *Draba burkei* (as *D. maguirei* var. *burkei*) is in the Center for Plant Conservation's National Collection of Endangered Plants and is listed in NatureServe as a plant of conservation concern.

21. Draba californica (Jepson) Rollins & R. A. Price, Aliso 12: 19. 1988 [E]

Draba cuneifolia Nuttall ex Torrey & A. Gray var. *californica* Jepson, Man. Fl. Pl. Calif., 443. 1925

Perennials; (short-lived); caudex simple or branched; sometimes scapose. **Stems** unbranched or branched (few), (0.2–)0.4–0.9(–1.2) dm, pubescent throughout, trichomes stellate, 4–8-rayed, 0.05–0.5 mm. **Basal leaves** rosulate; shortly petiolate; petiole ciliate, margin not ciliate, (trichomes simple and 2-rayed, 0.3–1 mm); blade oblanceolate, 0.6–2 cm × 1.5–4(–6) mm, margins entire, surfaces pubescent with short-stalked, 4–8-rayed trichomes, 0.1–0.4 mm. **Cauline leaves** 0–3; sessile; blade ovate to oblong, margins entire, surfaces pubescent as basal. **Racemes** 3–13-flowered, ebracteate, elongated in fruit; rachis not flexuous, pubescent as stem. **Fruiting pedicels** ascending to divaricate-ascending, straight, (2–)4–9(–12) mm, pubescent, trichomes short-stalked, 4–8-rayed, (0.1–0.4 mm). **Flowers:** sepals (persistent until fruit maturity), oblong to ovate, 1.7–2.5 mm, pubescent, (trichomes short-stalked, 2–5-rayed); petals white or creamy white, spatulate to oblanceolate, 2.2–3 × 0.8–1 mm; anthers ovate, 0.2–0.35 mm. **Fruits** lanceolate to oblong, plane, slightly flattened, (5–)6–9(–11) × 1.8–2.5 mm; valves usually pubescent, rarely glabrous, trichomes simple and short-stalked, 2- or 3-rayed, 0.05–0.25 mm; ovules 22–32 per ovary; style 0.1–0.3(–0.4) mm. **Seeds** oblong, 0.8–1 × 0.5–0.6 mm.

Flowering Jun–Aug. Grassy meadows, alpine areas, fellfields; 3200–4000 m; Calif., Nev.

Although originally treated as a variety of *Draba cuneifolia*, *D. californica* is but distantly related to that species. Instead, ongoing studies (L. Allphin and M. D. Windham, unpubl.) suggest that it may be an allopolyploid resulting from hybridization between *D. albertina* and *D. breweri*. The species is restricted to the White Mountains in Mono County, California, and adjacent Esmeralda County, Nevada.

22. Draba cana Rydberg, Bull. Torrey Bot. Club 29: 241. 1902 [E]

Draba breweri S. Watson var. *cana* (Rydberg) Rollins; *D. valida* Goodding

Perennials; caudex simple or branched (branches short); not scapose. **Stems** unbranched or branched distally, (0.6–)1–3(–3.8) dm, pubescent throughout, trichomes simple, 0.5–1 mm, with 4–10-rayed ones, 0.05–0.2 mm (mostly branched on basal parts). **Basal leaves** rosulate; petiolate;

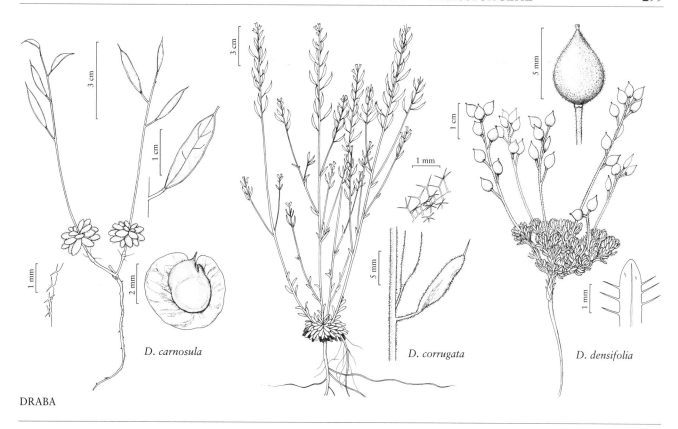

petiole ciliate throughout; blade linear to oblanceolate or oblong, (0.5–)0.8–2(–3.5) cm × 1.5–4(–11) mm, margins entire or dentate, (base and apex ciliate, trichomes simple, 0.3–0.8 mm), surfaces pubescent with short-stalked, 4–12-rayed trichomes, 0.1–0.3 mm. **Cauline leaves** 3–10(–17), (not overlapping); sessile; blade lanceolate to ovate or oblong, margins entire or dentate, surfaces pubescent as basal (adaxially with simple and forked trichomes near blade base). **Racemes** (10–)15–47 (–63)-flowered, basally bracteate, often considerably elongated in fruit; rachis not flexuous, densely pubescent, trichomes 4–10-rayed (0.05–0.2 mm), and fewer simple ones. **Fruiting pedicels** suberect or ascending, straight, 2–5(–10) mm, pubescent as rachis. **Flowers:** sepals (green or lavender), oblong to ovate, 1.5–2 mm, pubescent, (trichomes simple and few-rayed); petals white, oblanceolate to spatulate, 2.3–3.7(–4.5) × 0.7–1.7 mm; anthers ovate, 0.1–0.2 mm. **Fruits** (subappressed to rachis), linear-lanceolate to linear or, rarely, ovate-oblong, slightly twisted or plane, flattened, (5–)6–11 × 1.5–2(–2.5) mm; valves pubescent, trichomes short-stalked, 3–7-rayed, 0.05–0.3 mm; ovules 28–48 per ovary; style 0.1–0.6 mm. **Seeds** ovoid, 0.5–0.7(–0.9) × 0.3–0.5 mm. $2n = 32$.

Flowering (May–)Jun–Aug. Rock outcrops and talus, open prairie benchlands, roadsides, meadows, alpine tundra; 0–4100 m; Greenland; Alta., B.C., Man., N.B., Nfld. and Labr. (Nfld.), N.W.T., Ont., Que., Sask., Yukon; Alaska, Calif., Colo., Idaho, Maine, Mich., Mont., Nev., N.H., N.Mex., S.Dak., Utah, Vt., Wash., Wis., Wyo.

The limits of *Draba cana* have long been confused, and the species was treated as a synonym of the Himalayan *D. lanceolata* Royle (M. L. Fernald 1934; C. L. Hitchcock 1941) or as a variety of the western North American *D. breweri* (R. C. Rollins 1993). However, G. A. Mulligan (1971) clearly demonstrated that all three are distinct and should be maintained. Some Utah plants corresponding to the type of *D. valida* have shorter and wider oblong-ovate fruits. In all other aspects, they are indistinguishable from *D. cana*. Additional studies are needed to establish whether such plants should be formally recognized.

23. **Draba carnosula** O. E. Schulz in H. G. A. Engler, Pflanzenr. 89[IV,105]: 82. 1927 [C][E][F]

Draba howellii S. Watson var. *carnosula* (O. E. Schulz) C. L. Hitchcock

Perennials; (loosely cespitose); caudex branched (somewhat surculose, with persistent leaf bases, branches sometimes terminating in sterile rosettes); scapose. **Stems** unbranched, 0.3–1.2 dm, glabrous throughout. **Basal leaves** rosulate; petiolate; petiole (persistent, midvein prominent), ciliate, (trichomes sparse, short-stalked, 2–4-rayed, sometimes with simple ones, 0.2–0.5 mm); blade (somewhat fleshy), oblanceolate or spatulate to obovate, 0.3–1.5 cm × 1.5–5 mm, margins entire, (pubescent as petiole), surfaces glabrous. **Cauline leaves** 0. **Racemes** 2–6-flowered, ebracteate, elongated in fruit; rachis not flexuous, glabrous. **Fruiting pedicels** divaricate-ascending, straight, 3–8 mm, usually glabrous, rarely trichomes simple. **Flowers:** sepals ovate, 3–4 mm, glabrous; petals yellow, oblanceolate, 5–7 × 1.3–2 mm; anthers oblong, 0.7–0.9 mm. **Fruits** lanceolate to elliptic-lanceolate, plane, strongly flattened, 10–23 × 4–6 mm; valves (each with distinct midvein), glabrous; ovules 8–12 per ovary; style 2–3 mm. **Seeds** (broadly winged), orbicular, 3–4.5 in diam.; (wing 1–1.5 mm wide).

Flowering Jun–Jul. Open rocky hillsides; of conservation concern; 2800–3100 m; Calif.

Although *Draba carnosula* was reduced by Hitchcock to a variety of *D. howellii*, it differs significantly by having leafless flowering stems, 2–6-flowered racemes, and broadly winged seeds 3–4.5 mm in diam. *Draba howellii* almost always has 1–3-leaved stems, (5–)7–18(–25)-flowered racemes, and not winged, distally appendaged seeds 1–1.6 mm in diam. *Draba carnosula* is known only from a few collections from Mount Eddy in Trinity County and the north side of Mount Shasta in Siskiyou County.

24. **Draba chamissonis** G. Don, Gen. Hist. 1: 184. 1831 (as chamissoni)

Draba frigida Sauter var. *kamtschatica* Ledebour; *D. kamtschatica* (Ledebour) N. Busch; *D. lonchocarpa* Rydberg subsp. *kamtschatica* (Ledebour) Calder & Roy L. Taylor; *D. nivalis* Liljeblad var. *kamtschatica* (Ledebour) Pohle

Perennials; (cespitose, sometimes forming mats); caudex branched (covered with persistent leaf bases); not scapose. **Stems** unbranched, 0.4–1.2 dm, pubescent throughout, trichomes minutely stalked, stellate, 8–12-rayed, (non-crisped), 0.03–0.3 mm, sometimes with simple and 2-rayed ones, to 0.3 mm. **Basal leaves** rosulate; petiole (obsolete), margin usually not ciliate, (trichomes, when present, simple and 2-rayed, 0.2–0.35 mm); blade oblanceolate to obovate, 0.4–1 cm × 1–3 mm, margins denticulate or entire, surfaces pubescent with short-stalked, stellate, (non-crisped), 8–12-rayed trichomes, 0.05–0.2 mm, (midvein obscure abaxially). **Cauline leaves** (1 or) 2–7; sessile; blade broadly ovate, margins usually denticulate, rarely entire. **Racemes** 4–10-flowered, proximalmost 1–5 flowers bracteate, elongated; rachis not flexuous, pubescent as stem. **Fruiting pedicels** ascending, straight, (4–)6–13(–20) mm, pubescent as stem. **Flowers:** sepals ovate, 1.5–2.2 mm, pubescent, (trichomes simple and short-stalked, 2–5-rayed); petals white, obovate, 2–3.5 × 1.5–2 mm; anthers ovate, 0.2–0.3 mm. **Fruits** linear to narrowly lanceolate, slightly twisted or plane, flattened, 5–11 × 1.1–1.6(–2) mm; valves glabrous; ovules 16–20 per ovary; style 0.1–0.3 mm. **Seeds** ovoid, 0.6–0.9 × 0.3–0.5 mm. $2n = 16$.

Flowering Jun–Jul. Rock cliffs, bluffs, wind-swept tundra; 0–500 m; Alaska; e Asia (Russian Far East, n Siberia).

C. L. Hitchcock (1941) treated *Draba chamissonis* as a variety of *D. nivalis* and listed it from British Columbia; R. C. Rollins (1993) did not mention the taxon at any rank. The limited North American material of *D. chamissonis* that we examined is all from Alaska. The species strongly resembles *D. nivalis*, from which it is distinguished by having (1 or) 2–7 broadly ovate cauline leaves, proximalmost fruiting pedicels (4–)6–13(–20) mm, petals 1.5–2 mm wide, and fruits 1.1–1.6(–2) mm wide. By contrast, *D. nivalis* has one ovate or oblong cauline leaf, or none, proximalmost fruiting pedicels 1–4.5(–8) mm, petals 0.8–1.4 mm wide, and fruits 1.5–2.2 mm wide.

25. **Draba cinerea** Adams, Mém Soc. Imp. Naturalistes Moscou 5: 103. 1817

Draba magellanica Lamarck subsp. *cinerea* (Adams) E. Ekman

Perennials; (cespitose); caudex simple or branched; rarely scapose. **Stems** unbranched, (0.3–)0.5–1.6(–2.7) dm, pubescent throughout, trichomes 4–10-rayed, 0.1–0.3 mm, (often some rays branched, simple trichomes sparse, to 0.6 mm). **Basal leaves** rosulate; petiolate; petiole base and proximalmost margin ciliate, (trichomes simple, 0.2–0.8 mm); blade oblanceolate to narrowly obovate or linear-lanceolate, 0.4–1.5 cm × 1–5 mm, margins usually entire, rarely with 1 tooth on each side, surfaces densely pubescent with minutely stalked, stellate, 8–12-rayed, trichomes 0.1–0.25 mm, (midvein obscure abaxially, apex trichomes simple). **Cauline leaves** 0–3(–5); sessile; blade ovate or oblong to lanceolate, margins entire, surfaces pubescent as basal. **Racemes** (3–)5–17(–24)-flowered, ebracteate, elongated in fruit; rachis not flexuous, pubescent as stem. **Fruiting pedicels** divaricate-ascending or ascending, straight, (3–)4–7(–9) mm, pubescent as stem. **Flowers:** sepals ovate, 1.7–2.5 mm, pubescent, (trichomes simple and short-stalked, 2- or 3-rayed); petals white, spatulate to obovate, 3.5–4.5 × 1.5–2 mm; anthers ovate, 0.3–0.4 mm. **Fruits** oblong to elliptic, plane, slightly flattened, 5–8 × 2–3 mm; valves pubescent, trichomes short-stalked, 2–5-rayed,

0.05–0.3 mm, (some rays branched); ovules 20–36(–42) per ovary; style (0.1–)0.2–1 mm. **Seeds** ovoid, 0.6–0.8 × 0.4–0.6 mm. $2n = 48$.

Flowering Jun–Jul. Rock outcrops, ridges, meadows, gravel beaches, stream banks, alluvial fans; 0–500 m; Greenland; B.C., N.W.T., Nunavut, Ont., Que., Sask., Yukon; Alaska; Europe (Finland, Norway, n Russia); e Asia (Russian Far East, Siberia).

Draba cinerea is a polymorphic species in which tetraploid, hexaploid, heptaploid, and octoploid chromosome counts (all based on $x = 8$) have been reported (e.g., R. C. Rollins 1993). Although we have not seen the type collection, we tentatively consider the taxon to be hexaploid, which is the predominant ploidy level throughout the species range, including North America, and also from its type region in Siberia.

Hexaploid *Draba cinerea* is closely related to decaploid ($2n = 80$) *D. arctica*, from which it is distinguished with difficulty. In general, *D. cinerea* has 1–3(–5)-leaved stems, basal leaf blades usually not terminated with simple trichomes and with obscure midveins abaxially, petals 3.5–4.5 mm, and seeds 0.6–0.8 × 0.4–0.6 mm. By contrast, *D. arctica* has leafless or 1(–3)-leaved stems, basal leaf blades terminated with simple or 2-rayed trichomes and with midveins distinct abaxially, petals 3.5–6 mm, and seeds (0.8–)0.9–1.1 × (0.6–)0.7–0.8 mm.

26. Draba corrugata S. Watson in W. H. Brewer et al., Bot. California 2: 430. 1880 [C] [E] [F]

Draba vestita Davidson

Biennials or perennials; (short-lived); caudex simple (covered with persistent leaves); not scapose. **Stems** branched, (0.3–)0.4–1.7(–2.5) dm, pubescent throughout, trichomes simple and long-stalked, 2-rayed, 0.4–1.4 mm, with smaller, 2–4-rayed ones, 0.1–0.6 mm, (simple ones usually fewer distally). **Basal leaves** (densely imbricate); rosulate; shortly petiolate; petiole base and margin ciliate, (trichomes simple, 0.6–2 mm); blade oblong to narrowly oblanceolate, (0.5–)1–2.2(–4.5) cm × 2–5 mm, margins entire, surfaces densely pubescent, abaxially with stalked, 2–4-rayed trichomes, 0.4–1.2 mm, (simple trichomes often along midvein), adaxially with mostly simple and long-stalked, 2-rayed trichomes, 0.6–1.3 mm, sometimes with 3- or 4-rayed ones. **Cauline leaves** (4–)6–10(–13); sessile; blade oblong to ovate, margins entire, surfaces pubescent as basal. **Racemes** (10–)18–55(–67)-flowered, ebracteate or proximalmost flowers bracteate, slightly or considerably elongated in fruit; rachis not flexuous, pubescent as stem. **Fruiting pedicels** divaricate to ascending, straight, 2–6(–8) mm, pubescent as stem. **Flowers:** sepals oblong, 2–2.7 mm, pubescent, (trichomes short-stalked, 2–4-rayed); petals yellow, linear, 2–3.5 × 0.2–0.5 mm; anthers oblong, 0.6–0.8 mm (exserted). **Fruits** elliptic to oblong or linear- to oblong-elliptic, slightly twisted or plane, flattened, 5–13(–17) × 2–3(–4) mm; valves pubescent, trichomes short-stalked, cruciform, 0.1–0.5 mm, (sometimes with 2- or 3-rayed ones); ovules 16–28 per ovary; style 1.4–3.4 mm. **Seeds** oblong, 1–1.2 × 0.6–0.8 mm.

Flowering Jun–Jul. Alpine fellfields, talus, open pine woodlands; of conservation concern; 2000–3500 m; Calif.

Draba corrugata was broadly circumscribed by R. C. Rollins (1993) to include three varieties that we treat as distinct species. For a discussion of species limits and distinguishing features, see I. A. Al-Shehbaz and M. D. Windham (2007). *Draba corrugata*, in the strict sense, is known from the San Antonio, San Bernardino, and San Gabriel mountains in Los Angeles and San Bernardino counties.

27. Draba corymbosa R. Brown ex de Candolle, Syst. Nat. 2: 343. 1821

Draba alpina Linnaeus var. *bellii* (Holm) O. E. Schulz; *D. alpina* var. *corymbosa* Durand; *D. barbata* Pohle; *D. bellii* Holm; *D. kjellmanii* Lid ex E. Ekman; *D. macrocarpa* Adams; *D. vestita* Davidson

Perennials; (cespitose); caudex branched (with persistent leaves or leaf remains, branches sometimes terminating in sterile rosettes); scapose. **Stems** unbranched, (0.05–)0.2–0.8(–1.5) dm, pubescent throughout, trichomes simple and 2-rayed, 0.4–1 mm, with 3–5-rayed ones, 0.05–0.4 mm, (sometimes trichomes mostly simple). **Basal leaves** rosulate; petiolate; petiole base and margin ciliate, (trichomes often course, simple, 0.4–1.3 mm); blade oblanceolate to obovate, 0.6–1.8 cm × 1.5–5 mm, margins entire, surfaces pubescent, abaxially with stalked, 2–6-rayed trichomes, 0.2–0.6 mm, adaxially with primarily simple and stalked, 2-rayed trichomes, to 1.1 mm, with 3–5-rayed ones, 0.2–0.4 mm. **Cauline leaves** 0. **Racemes** 2–9(–12)-flowered, ebracteate, (corymbose), slightly elongated in fruit; rachis not flexuous, pubescent as stem. **Fruiting pedicels** divaricate-ascending, straight or slightly curved upward, 4–11(–16) mm, pubescent, trichomes simple and 2–4-rayed. **Flowers:** sepals (grayish green), broadly ovate, 2.2–3 mm, pubescent, (trichomes simple, to 1 mm, sometimes with stalked, smaller, 2–4-rayed ones); petals (broadly patent), yellow, obovate, 4–6 × 3–5 mm; anthers ovate, 0.3–0.4 mm. **Fruits** oblong or ovate, plane, flattened, 6–12 × 3.5–5.5 mm; valves pubescent or puberulent, trichomes simple, 0.1–0.4 mm, (sometimes with short-stalked,

2- or 3-rayed ones); ovules 12–24 per ovary; style 0.6–1 mm (stigma distinctly wider than style). **Seeds** (brown), ovoid, 1–1.3 × 0.6–0.9 mm. $2n = 128, 144$.

Flowering Jun–Aug. Moist tundra, among calcareous or dolomitic rocks, gravel beaches, silt and clay terraces; 0–1700 m; Greenland; B.C., N.W.T., Nunavut, Yukon; Alaska; Europe (Norway, n Russia); e Asia (Russian Far East, Siberia).

Draba corymbosa, which is 16-ploid or 18-ploid with $x = 8$, is an extremely variable species of polyphyletic, allopolyploid origin. Most individuals appear to have decaploid *D. alpina* ($2n = 80$) in their parentage, but the other genomes are provided by hexaploid ($2n = 48$) and octoploid ($2n = 64$) species (C. Brochmann et al. 1993). O. E. Schulz reduced *D. corymbosa* to a variety of *D. alpina*; R. C. Rollins (1993) treated it as a distinct species. *Draba corymbosa* is distinguished from *D. alpina* by having pubescent or puberulent (versus glabrous or glabrescent) fruits, corymbose (versus usually elongated) fruiting racemes, and abaxial leaf blade surfaces always lacking (versus usually with some) simple trichomes.

28. **Draba crassa** Rydberg, Mem. New York Bot. Gard. 1: 182. 1900 E

Draba chrysantha S. Watson var. *crassa* (Rydberg) O. E. Schulz

Perennials; (cespitose); caudex simple or branched (well-developed, with persistent leaf bases); not scapose. **Stems** (decumbent to ascending), unbranched, (0.4–)0.6–1.3(–1.5) dm, usually glabrous proximally, rarely sparsely pubescent, pubescent distally, trichomes simple and short-stalked, 2-rayed, (crisped), 0.1–0.6(–0.8) mm. **Basal leaves** rosulate; petiolate; petiole ciliate, (trichomes mostly simple, 0.3–0.8 mm); blade oblanceolate, 2–6(–7) cm × 2.5–8(–10) mm, margins entire, (pubescent as petiole), surfaces glabrous. **Cauline leaves** 2–4(–6); sessile; blade ovate to oblong, margins entire, (ciliate, trichomes simple and stalked, 2-rayed). **Racemes** (4–)8–20(–25)-flowered, ebracteate, elongated in fruit; rachis not flexuous, pubescent, trichomes simple and short-stalked, 2-rayed, (crisped), (0.1–0.8 mm). **Fruiting pedicels** divaricate-ascending, straight, 5–10(–15) mm, pubescent as rachis. **Flowers:** sepals (green tinged purplish), ovate to oblong, 2–3(–3.3) mm, pubescent, (trichomes simple and 2-rayed); petals yellow, spatulate to subobovate, 3.5–5(–6) × (2–)2.5–3.5(–4) mm; anthers ovate, 0.5–0.6 mm. **Fruits** lanceolate to ovate-lanceolate, slightly twisted, flattened, (7–)8–14 × 3–5 mm; valves glabrous; ovules 16–20 per ovary; style (0.4–)0.7–1.2(–1.5) mm. **Seeds** oblong, 1.2–1.7 × 0.8–1.1 mm. $2n = 24$.

Flowering Jul–Aug. Rock outcrops and talus, alpine tundra, rocky meadows; 2900–4300 m; Colo., Mont., N.Mex., Utah, Wyo.

Draba crassa is easily recognized by its combination of fleshy, distinctly petiolate leaves with glabrous surfaces and mostly decumbent flowering/fruiting stems. N. H. Holmgren (2005b) indicated that it has 52 seeds per fruit (26 per locule), but in the plants we examined, the seed count did not exceed 20 per fruit.

29. **Draba crassifolia** Graham, Edinburgh New Philos. J. 7: 182. 1829

Draba crassifolia var. *parryi* (Rydberg) O. E. Schulz; *D. parryi* Rydberg

Annuals or **perennials**; (short-lived); caudex branched (when present); usually scapose. **Stems** unbranched or, rarely, branched distally, (0.1–)0.3–1.1(–1.5) dm, usually glabrous throughout, rarely pubescent proximally, trichomes simple, 0.3–0.7 mm. **Basal leaves** rosulate; petiolate; petiole ciliate throughout; blade oblanceolate to obovate, (0.2–)0.5–2.5(–3) cm × (1–)2–4(–6) mm, margins usually entire, rarely denticulate, (sometimes ciliate), surfaces glabrous or sparsely pubescent, with simple and 2-rayed trichomes, 0.3–0.9 mm. **Cauline leaves** usually 0, rarely 1; sessile; blade oblong to ovate, margins entire, surfaces glabrous. **Racemes** (2–)4–15(–25)-flowered, ebracteate, elongated in fruit; rachis slightly flexuous or straight, glabrous. **Fruiting pedicels** horizontal to divaricate-ascending, usually straight, rarely curved upward, 3–8(–10) mm (subequaling or shorter than fruit), glabrous. **Flowers** (chasmogamous, petaliferous); sepals (green or purplish), ovate, 1–2 mm, glabrous; petals yellow (often fading white), oblanceolate, 1.5–2.5(–3) × 0.5–1 mm; anthers ovate, 0.15–0.25 mm. **Fruits** usually narrowly elliptic to lanceolate, rarely linear-lanceolate, plane, flattened, (3–)5–10 × 1.5–2.5 mm; valves glabrous; ovules (8–)16–24(–30) per ovary; style 0.02–0.1 mm. **Seeds** elliptic, 0.7–0.8 × 0.4–0.5 mm. $2n = 40$.

Flowering Jun–Sep. Rock outcrops and talus, subalpine meadows, alpine summits and tundra, bare snow-melt areas; (50–)1000–4300 m; Greenland; Alta., B.C., Nfld. and Labr. (Labr.), N.W.T., Nunavut, Que., Yukon; Alaska, Ariz., Colo., Idaho, Mont., Nev., N.Mex., Utah, Wash., Wyo.; n Europe (Norway, Sweden).

M. D. Windham (2004) presented morphological and chromosomal data suggesting that *Draba crassifolia* is an allopolyploid produced by hybridization between *D. albertina* and *D. fladnizensis*. Although the species is distinctive in large part, some individuals can be difficult to place and there is evidence of rare backcrossing (Windham, unpubl.). The attribution to Arizona is based on *Schaack 345* (US) and *Kearney & Peebles 12156* (US), both collected on the San Francisco Peaks in Coconino County. *Draba crassifolia* is found at elevations as low as 50 m in Greenland and the islands of Nunavut.

30. **Draba cruciata** Payson, Amer. J. Bot. 4: 265. 1917
C E

Perennials; (cespitose, loosely matted); caudex branched (with persistent leaf remains, branches sometimes terminating in sterile rosettes); scapose. **Stems** unbranched, (0.3–)0.5–1.4(–1.8) dm, often glabrous throughout, sometimes pubescent proximally, trichomes stalked, cruciform, 0.07–0.3 mm, with 3–5-rayed ones. **Basal leaves** rosulate; petiolate; petiole ciliate throughout; blade oblanceolate to narrowly obovate, (0.4–)0.6–1.1(–1.6) cm × 1.5–3 mm, margins entire or minutely denticulate, surfaces pubescent with stalked, cruciform trichomes, 0.07–0.4 mm, (rarely with fewer, 3–5-rayed ones, sometimes both surfaces glabrous and only margins pubescent, not ciliate, midvein obscure abaxially). **Cauline leaves** 0. **Racemes** (3–)5–18(–22)-flowered, ebracteate, considerably elongated in fruit; rachis not flexuous, glabrous. **Fruiting pedicels** divaricate-ascending, (not decurrent basally), straight, (4–)5–10(–13) mm, glabrous. **Flowers:** sepals ovate, 1.5–2 mm, glabrous or pubescent distally, (trichomes short-stalked, 4-rayed); petals yellow, spatulate to oblanceolate, 4–5(–6) × 1.2–2 mm; anthers ovate, 0.4–0.6 mm. **Fruits** narrowly lanceolate to elliptic-lanceolate, plane (not curved), flattened, (4–)6–12(–16) × 1.5–3 mm; valves glabrous or puberulent, trichomes simple and short stalked, 2–4-rayed, 0.03–0.15 mm; ovules 6–10(–12) per ovary; style (0.1–)0.3–0.8 mm. **Seeds** ovoid, 1–1.7 × 0.8–1.3 mm.

Flowering Jul–Aug. Subalpine areas, ridges in pine and fir forests; of conservation concern; 2500–3100 m; Calif.

Draba cruciata is known only from the Sierra Nevada of Tulare County.

31. **Draba cuneifolia** Nuttall ex Torrey & A. Gray, Fl. N. Amer. 1: 108. 1838

Annuals; scapose or subscapose. **Stems** (simple to many from or near base), unbranched, (0.2–)0.3–2.7(–3.7) dm, hirsute or pubescent throughout, trichomes 3–4-rayed, 0.05–0.4 mm, (sometimes mixed proximally with simple or spurred ones, 0.5–1.2 mm). **Basal leaves** not rosulate; petiole (obscure), not ciliate; blade oblanceolate to spatulate or broadly obovate, (0.4–)1–3.5(–5) cm × (2–)6–20(–28) mm, margins dentate (in distal ½), surfaces pubescent, abaxially with stalked, 2–4-rayed trichomes, 0.1–0.7 mm, adaxially similar or also with fewer, simple trichomes, 0.4–0.7 mm. **Cauline leaves** 0–6 (on proximal ⅓ of stem); blade similar to basal. **Racemes** 10–50(–70)-flowered (throughout or on distal ⅓ of scape), ebracteate, elongated in fruit; rachis not flexuous, densely pubescent, trichomes 2–4-rayed. **Fruiting pedicels** horizontal to divaricate-ascending, straight, (1–)2–7(–10) mm, pubescent as rachis. **Flowers:** (late ones cleistogamous, apetalous); sepals (green or pink), oblong, 1.5–2.5 mm, glabrous or pubescent, (trichomes simple); petals white, spatulate, (2–)2.5–4.5(–5) × 1–2 mm, (emarginate or obtuse); anthers ovate to oblong, (0.1–)0.25–0.4 mm. **Fruits** oblong to linear or lanceolate to broadly ovate, plane, flattened, (3–)6–12(–16) × 1.7–2.7(–3) mm; valves usually puberulent, rarely glabrous, trichomes simple, antrorse, 0.1–0.3 mm, (rarely with 2-rayed ones, or all trichomes short-stalked, 4-rayed, cruciform); ovules (12–)24–66(–72) per ovary; style 0.01–0.3(–0.4) mm. **Seeds** broadly ovoid, 0.5–0.7 × 0.4–0.5 mm. $2n = 30, 32$.

Varieties 3 (3 in the flora): United States, n Mexico.

Draba cuneifolia is highly variable in fruit shape and size, style length, ovule number per ovary, type of fruit indumentum, and length of the fruiting raceme in relation to the rest of the scape. Reported chromosome numbers also vary, but it is currently unclear whether this is the result of biologically relevant processes or erroneous counts. R. L. Hartman et al. (1975) divided the species into three varieties, a position followed by subsequent authors (e.g., R. C. Rollins 1993). Typical var. *sonorae* is rather distinct (see key below) and may deserve recognition as a separate species, as by O. E. Schulz (1927). Critical, population-based, molecular studies are needed to assess this. In the absence of such studies, we tentatively follow Hartman et al., recognizing three varieties that are distinguished primarily by the types of trichomes found on the fruits. Glabrous-fruited forms occasionally encountered in all three varieties can be difficult to identify.

1. Fruit valves: trichomes usually simple, sometimes with 2-rayed ones; stems hirsute with at least some simple trichomes .... 31a. *Draba cuneifolia* var. *cuneifolia*
1. Fruit valves: trichomes stalked and 4-rayed, sometimes with 2- or 3-rayed ones; stems pubescent with 3- or 4-rayed trichomes.
   2. Fruits (5–)7–12 mm; ovules 20–44 per ovary; racemes to ¾ of scape .... 31b. *Draba cuneifolia* var. *integrifolia*
   2. Fruits 3–6(–8) mm; ovules 12–24 per ovary; racemes on all or most of scape .... 31c. *Draba cuneifolia* var. *sonorae*

31a. Draba cuneifolia Nuttall ex Torrey & A. Gray var. **cuneifolia**

Draba cuneifolia var. *foliosa* Mohlenbrock & J. W. Voigt; *D. cuneifolia* var. *helleri* (Small) O. E. Schulz; *D. cuneifolia* var. *leiocarpa* O. E. Schulz; *D. helleri* Small

Stems hirsute with at least some simple trichomes. **Racemes** often on distal ⅓–½ of scape. **Fruits** oblong to linear or lanceolate, (5–)7–12(–16) mm; valves: trichomes simple, sometimes with 2-rayed ones; ovules (20–)32–66(–72) per ovary.

Flowering Mar–May. Rock outcrops and rocky slopes, glades and barrens, floodplains, meadows, thickets, sandy soil in pastures and prairies, roadsides, desert scrub, pinyon-juniper woodlands; 0–2500 m; Ala., Ariz., Ark., Calif., Colo., Fla., Ill., Kans., Ky., La., Miss., Mo., Nebr., Nev., N.Mex., N.C., Ohio, Okla., Pa., S.Dak., Tenn., Tex., Utah; Mexico (Baja California, Chihuahua, Coahuila, Zacatecas).

Draba ammophila A. Heller is an illegitimate name, sometimes found in synonymy with var. *cuneifolia*.

31b. Draba cuneifolia Nuttall ex Torrey & A. Gray var. **integrifolia** S. Watson, Proc. Amer. Acad. Arts 23: 256. 1888

Draba integrifolia (S. Watson) Greene; *D. sonorae* Greene var. *integrifolia* (S. Watson) O. E. Schulz

Stems pubescent with 3- or 4-rayed trichomes. **Racemes** to ¾ of scape. **Fruits** oblong to linear or lanceolate, (5–)7–12 mm; valves: trichomes stalked and 4-rayed, sometimes with 2- or 3-rayed ones; ovules 20–44 per ovary.

Flowering Jan–Mar. Rocky slopes and gravelly soil in desert scrub communities; 200–1800 m; Ariz., Calif., Nev., Tex., Utah; Mexico (Baja California, Nuevo León, Sonora).

31c. Draba cuneifolia Nuttall ex Torrey & A. Gray var. **sonorae** (Greene) Parish, Bull. S. Calif. Acad. Sci. 2: 81. 1903

Draba sonorae Greene, Bull. Calif. Acad. Sci. 2: 59. 1886; *D. cuneifolia* var. *brevifolia* S. Watson ex Munz

Stems pubescent with 3- or 4-rayed trichomes. **Racemes** on all or most of scape. **Fruits** oblong to broadly ovate, 3–6(–8) mm; valves: trichomes stalked and 4-rayed, sometimes with 2- or 3-rayed ones; ovules 12–24 per ovary.

Flowering Feb–Mar. Rocky slopes, gravelly soil in desert scrub communities, grassy slopes, arroyos, gullies, washes, ledges, under shrubs; 200–1800 m; Ariz., Calif.; Mexico (Baja California, Sonora).

32. Draba cusickii B. L. Robinson ex O. E. Schulz in H. G. A. Engler, Pflanzenr. 89[IV,105]: 105. 1927 E

Draba sphaeroides Payson var. *cusickii* (B. L. Robinson ex O. E. Schulz) C. L. Hitchcock

Perennials; (loosely cespitose); caudex branched (with few, persistent leaf remains); scapose. **Stems** unbranched, 0.3–1.2 dm, pubescent throughout, trichomes (2–)4-rayed, 0.05–0.5 (–0.6) mm. **Basal leaves** rosulate; shortly petiolate; petiole base ciliate (margin not ciliate, trichomes 2-rayed, 0.3–0.6 mm); blade oblanceolate to obovate, 0.4–1.2 cm × 2–5 mm, margins entire or, rarely, denticulate, surfaces pubescent with stalked, (2–)4-rayed trichomes, 0.15–0.4 mm. **Cauline leaves** 0. **Racemes** 10–20(–25)-flowered, ebracteate, considerably elongated in fruit; rachis not flexuous, pubescent as stems proximally. **Fruiting pedicels** divaricate-ascending, straight (not expanded basally), 4–9 mm, pubescent, trichomes (2–)4-rayed, (0.05–0.25 mm). **Flowers:** sepals broadly ovate, 2–2.5 mm, pubescent, (trichomes short-stalked, 3- or 4-rayed); petals yellow, obovate to oblanceolate, 3.7–6 × 1.5–2.5 mm; anthers ovate, 0.5–0.6 mm. **Fruits** lanceolate to broadly ovate, plane, flattened, 6–9 × 2–4.5 mm; valves pubescent, trichomes short-stalked, (2–)4-rayed, 0.05–0.25 mm; ovules 10–18 per ovary; style 0.4 0.8(–1) mm. **Seeds** oblong, 1.2–1.5 × 0.7–0.8 mm. $2n = 26$.

Flowering Jun–Jul. Basaltic outcrops, talus, rocky soil; 2000–2700 m; Oreg.

Draba cusickii was divided by R. C. Rollins (1993) into two varieties with significantly different morphologies and chromosome numbers. The Nevada endemic var. *pedicellata* ($2n = 20$) is here treated as a distinct species following M. D. Windham (2004). *Draba cusickii* has the same chromosome number as *D. sobolifera* but is readily distinguished from that species by having 4-rayed (versus simple or 2-rayed) fruit trichomes. *Draba cusickii* is known from Steens Mountain (Harney County).

33. Draba cyclomorpha Payson, Amer. J. Bot. 4: 263. 1917 E

Draba lemmonii S. Watson var. *cyclomorpha* (Payson) O. E. Schulz

Perennials; (cespitose, not pulvinate); caudex branched (with persistent petiole remains, branches sometimes terminating in sterile rosettes); scapose. **Stems** unbranched, 0.1–0.6(–1) dm, (thinly) pilose throughout, trichomes simple and stalked, 2- (or 3-)rayed, (often crisped), 0.1–0.8 mm. **Basal leaves** rosulate; shortly petiolate; petiole base and margin ciliate, (trichomes simple, 0.2–1 mm); blade (somewhat fleshy), oblanceolate to obovate, 0.4–1 cm × 2–5 mm, margins entire, surfaces glabrous or hirsute, abaxially usually with stalked, 2-rayed trichomes, 0.1–0.6 mm, rarely with fewer, simple ones, (midvein obscure), adaxially with almost exclusively simple trichomes, to 0.6 mm. **Cauline leaves** 0. **Racemes** 5–14(–20)-flowered, ebracteate, elongated in fruit; rachis not flexuous, pubescent as stem. **Fruiting pedicels** horizontal to divaricate-ascending, often curved upward or straight, 3–8(–10) mm, pilose as stem. **Flowers:** sepals ovate, 1.5–2.2 mm, pubescent, (trichomes simple and short-stalked, 2-rayed); petals yellow, spatulate, 4–5.5 × 1.5–2.5 mm; anthers oblong, 0.6–0.7 mm. **Fruits** oblong to oblong-ovate, plane, flattened, 3.5–9 × 2.5–4 mm; valves glabrous; ovules 8–12 per ovary; style 0.3–1 mm. **Seeds** ovoid, 1.4–1.9 × 1–1.2 mm.

Flowering Jul–Aug. Basaltic talus, stony ridges; 2500–3100 m; Oreg.

Draba cyclomorpha was treated by Schulz, C. L. Hitchcock (1941), and R. C. Rollins (1993) as a variety of *D. lemmonii*. The two taxa are quite distinct morphologically and separated by nearly 1600 kilometers, justifying their treatment as distinct species (I. A. Al-Shehbaz and M. D. Windham 2007). *Draba cyclomorpha* is known to us only from the Wallowa Mountains.

34. Draba daviesiae (C. L. Hitchcock) Rollins, Contr. Gray Herb. 214: 5. 1984 E

Draba apiculata C. L. Hitchcock var. *daviesiae* C. L. Hitchcock in C. L. Hitchcock et al., Vasc. Pl. Pacif. N.W. 2: 489, fig. p. 493 [upper right]. 1964; *D. densifolia* Nuttall var. *daviesiae* (C. L. Hitchcock) S. L. Welsh & Reveal

Perennials; (densely pulvinate); caudex branched (branches elongated, loose, with persistent leaf remains, terminating in flowering or sterile shoots); scapose. **Stems** unbranched, (0.05–)0.2–0.6 dm, glabrous. **Basal leaves** (densely imbricate); rosulate; petiolate; petiole ciliate throughout; blade (fleshy), oblong to obovate or oblanceolate, 0.3–0.7(–1) cm × 1–2(–2.5) mm, margins entire, (ciliate, trichomes simple, 0.1–0.5 mm, apex obtuse), surfaces glabrous (midvein obscure abaxially). **Cauline leaves** 0. **Racemes** 2–8(–10)-flowered, ebracteate, (subcorymbose), slightly elongated in fruit; rachis not flexuous, glabrous. **Fruiting pedicels** divaricate-ascending (not decurrent basally), straight, 4–10 mm, glabrous. **Flowers:** sepals oblong, 1.5–2.2 mm, glabrous; petals pale to bright yellow, spatulate, 3.5–4 × 1–2 mm; anthers ovate, 0.3–0.4 mm. **Fruits** ovate to oblong-elliptic, plane, flattened, 4–8 × 2–4 mm; valves (obscurely veined), glabrous; ovules 6–14 per ovary; style 0.1–0.5 mm. **Seeds** ovoid, 1.2–1.5 × 0.8–1 mm.

Flowering Jul–Aug. Talus slopes, rock crevices and cracks, rocky ridges and slides, alpine meadows; 2700–2900 m; Mont.

Although originally described as a variety of *Draba apiculata* (= *D. globosa*), *D. daviesiae* is distinct morphologically. It is easily distinguished from the former by its densely pulvinate habit, obtuse leaf blades, and obscurely veined fruit valves. By contrast, *D. globosa* exhibits a cespitose but non-pulvinate habit, acute leaf blades, and prominently veined fruit valves. *Draba daviesiae* is known from the Bitterroot Mountains in Ravalli County.

35. Draba densifolia Nuttall in J. Torrey and A. Gray, Fl. N. Amer. 1: 104. 1838 [E][F]

Draba caeruleomontana Payson & H. St. John; *D. caeruleomontana* var. *piperi* Payson & H. St. John; *D. glacialis* Adams var. *pectinata* S. Watson; *D. globosa* Payson var. *sphaerula* (J. F. Macbride & Payson) O. E. Schulz; *D. mulfordiae* Payson; *D. nelsonii* J. F. Macbride & Payson; *D. oligosperma* Hooker var. *pectinata* (S. Watson) Jepson; *D. pectinata* (S. Watson) Rydberg; *D. sphaerula* J. F. Macbride & Payson

Perennials; (cespitose, pulvinate); caudex branched (dense with persistent leaf remains, branches sometimes terminating in sterile rosettes); scapose. **Stems** unbranched, (0.05–)0.2–1(–1.7) dm, usually glabrous, rarely pubescent, trichomes usually simple, 0.3–0.8 mm, and 2–4 (or 5)-rayed, 0.1–0.6 mm, (rarely predominantly simple ones). **Basal leaves** rosulate; sessile; blade linear to oblong or oblanceolate-linear, 0.3–0.9(–1.4) cm × 0.5–1.5(–2) mm, margins entire, (ciliate, trichomes simple, 0.3–1.2 mm), surfaces glabrous or sparsely pubescent, abaxially with short-stalked, 2–4-rayed trichomes, 0.1–0.3 mm (midvein prominent), adaxially rarely with subapical, simple trichomes. **Cauline leaves** 0. **Racemes** 2–10(–22)-flowered, ebracteate, not or slightly elongated in fruit; rachis not or slightly flexuous, glabrous or pubescent as stem. **Fruiting pedicels** divaricate-ascending to ascending (not decurrent basally), straight, (0.7–)1.5–10(–25) mm, usually glabrous, rarely sparsely pubescent, trichomes predominantly simple (0.2–0.7 mm), sometimes 2–4 (or 5)-rayed, (0.1–0.5 mm). **Flowers:** sepals ovate, 2–3 mm, usually sparsely pubescent, rarely glabrous, (trichomes simple and short-stalked, 2–4-rayed); petals pale yellow (sometimes fading white), oblanceolate to obovate, 2–5 × 1–1.7(–2) mm; anthers ovate or oblong, 0.4–0.6 mm. **Fruits** ovoid or ovate-lanceolate, plane (not curved), flattened, (2.5–)3–6(–8) × 2–3 mm; valves pubescent or puberulent, trichomes simple and short-stalked, 2–5-rayed, 0.1–0.4 mm; ovules 4–12 per ovary; style 0.3–0.6(–1) mm. **Seeds** oblong to ovoid, 1.2–2 (–2.6) × 0.9–1.2(–1.4) mm. $2n = 36$.

Flowering Jun–Aug. Rock outcrops and talus, rocky knolls, alpine ridges; 800–3700 m; Alta., B.C.; Alaska, Calif., Idaho, Nev., Oreg., Utah, Wash., Wyo.

Draba densifolia is one of the most highly variable North American members of the genus, as evidenced by the extensive synonymy. The species is an apomict (G. A. Mulligan 1976) and it occupies tremendous geographic, edaphic, and altitudinal ranges. In forms corresponding to the type of *D. sphaerula*, the flowering stems are 2-flowered, and the stems, leaves, and fruiting pedicels are to 5, 1.5, and 0.6 mm, respectively. By contrast, forms comparable to the type of *D. caeruleomontana* var. *piperi* have stems, leaves, and fruiting pedicels to 17, 1.3, and 2.5 cm, respectively. Between these remarkable extremes fall all of the other populations of the species.

Draba densifolia is sometimes confused with *D. paysonii*, which it resembles, in being a cespitose, scapose, and densely pulvinate perennial having narrowly linear or oblong to linear-oblanceolate leaf blades with strongly ciliate margins. It is easily distinguished from the latter by having glabrous adaxial leaf blade surfaces, only sparsely pubescent abaxial surfaces, and often glabrous fruiting pedicels and rachises.

36. Draba exunguiculata (O. E. Schulz) C. L. Hitchcock, Revis. Drabas W. N. Amer., 46. 1941 [C][E]

Draba chrysantha S. Watson var. *exunguiculata* O. E. Schulz in H. G. A. Engler, Pflanzenr. 89[IV,105]: 194. 1927; *D. chrysantha* S. Watson 1882, not K. Koch 1847

Perennials; (densely cespitose); caudex branched (covered with persistent, somewhat thickened, dry petioles); not scapose. **Stems** unbranched, 0.15–0.7 dm, glabrous throughout or sparsely pubescent, trichomes simple, 0.3–0.9 mm. **Basal leaves** rosulate; petiolate; petiole ciliate, (trichomes simple, 0.3–1.1 mm); blade linear to linear-oblanceolate, (0.5–)0.8–2(–2.5) cm × 1–3 mm, margins entire, (pubescent as petiole), surfaces abaxially sparsely pubescent with simple and stalked, 2- or 3-rayed trichomes, 0.1–0.4 mm, adaxially glabrous or subapically sparsely pubescent with simple trichomes. **Cauline leaves** 1–4; sessile; blade linear to linear-oblanceolate, margins entire, surfaces pubescent as basal. **Racemes** 4–13(–20)-flowered, ebracteate, elongated in fruit; rachis not flexuous, usually glabrous, rarely sparsely pubescent, trichomes simple, to 1 mm. **Fruiting pedicels** divaricate-ascending, straight or slightly curved upward, 1–6 mm, usually glabrous, rarely with few, simple trichomes. **Flowers:** sepals ovate, 2–2.5 mm, pubescent, (trichomes simple); petals (erect), yellow, obovate, 2–2.5(–3) × 1.5–2 mm, (not clawed); anthers oblong, 0.6–0.9 mm. **Fruits** lanceolate to oblong-lanceolate, plane, flattened, 5–13 × 1.5–3 mm; valves glabrous; ovules 20–24 per ovary; style 0.3–1 mm. **Seeds** ovoid, 1–1.3 × 0.6–0.8 mm. $2n = 56 \pm 5$.

Flowering Jul–Aug. Open knolls, talus, gravelly alpine slopes, rocky alpine tundra; of conservation concern; 3600–4300 m; Colo.

Draba exunguiculata occupies alpine areas of Clear Creek, El Paso, Grand, and Summit counties in central Colorado. *Draba grayana* is found in the same general area; both species are apomicts that occasionally grow sympatrically. *Draba exunguiculata* is easily distinguished from *D. grayana* by having stems and fruiting pedicels glabrous or sparsely pubescent with simple trichomes (versus moderately to densely pubescent with simple and 2- or 3-rayed trichomes), and by having erect, non-clawed petals, 2–3 mm and slightly longer than sepals (versus flared, clawed petals 3–4.5 mm and nearly twice as long as sepals).

37. Draba fladnizensis Wulfen in N. J. Jacquin, Misc. Austriac. 1: 147, plate 17, fig. 1. 1778

Draba fladnizensis var. *pattersonii* (O. E. Schulz) Rollins; *D. pattersonii* O. E. Schulz; *D. pattersonii* var. *hirticaulis* O. E. Schulz; *D. wahlenbergii* Hartman

Perennials; (sometimes cespitose); caudex simple or branched (with persistent leaf bases); usually scapose. **Stems** unbranched, (0.2–)0.3–1(–1.3) dm, glabrous. **Basal leaves** rosulate; petiole (obscure), margin ciliate, (trichomes simple or 2-rayed, 0.25–0.6 mm); blade linear to oblanceolate or narrowly obovate, (0.3–)0.4–1.2(–1.6) cm × 1–3(–4) mm, margins usually entire, rarely toothed, surfaces abaxially pubescent or glabrous, trichomes simple, sometimes with fewer, short-stalked, 2-rayed ones, (midvein prominent), adaxially often glabrous. **Cauline leaves** 0–2; sessile; blade oblong to ovate, margins entire, (ciliate). **Racemes** (2 or) 3–11 (–14)-flowered, usually ebracteate, rarely proximalmost flowers bracteate, elongated in fruit; rachis not flexuous, glabrous. **Fruiting pedicels** divaricate-ascending, often straight, (1–)2–5(–6) mm, glabrous. **Flowers:** sepals (green or purplish), ovate, 1.2–2.2 mm, glabrous or pubescent, (trichomes simple); petals white, spatulate, 2–2.5 × 0.8–1.5 mm; anthers ovate, 0.2–0.25 mm. **Fruits** elliptic-lanceolate to oblong, plane, flattened, 3–8 (–9) × 1.5–2 mm; valves glabrous; ovules 12–24 per ovary; style 0.05–0.2(–0.3) mm. **Seeds** oblong to elliptic, 0.8–1 × 0.5–0.6 mm. $2n = 16$.

Flowering Jun–Aug. Rock outcrops and talus, alpine meadows, sandy gravel; 0–1400 m at higher latitudes, 3000–3800 m at lower latitudes; Greenland; B.C., N.W.T., Nunavut, Que., Yukon; Alaska, Colo., Utah, Wyo.; c, s Europe; Asia; circumpolar and high alpine areas.

Rollins reduced *Draba pattersonii* to a variety of *D. fladnizensis* and separated the two primarily on plant size and minor differences in fruit shape. Examination of *D. fladnizensis* specimens collected throughout Europe and North America reveals that the alleged differences between the two taxa are artificial. The type material of *D. pattersonii*, which was collected in Colorado, is a mixture of plants highly variable in their type of indumentum. The specimens have no flowers, but the habit, fruits, and leaves are nearly indistinguishable from those of *D. fladnizensis* from higher latitudes.

In the absence of flowers, the white-flowered *Draba fladnizensis* ($2n = 16$) is often confused with the yellow-flowered *D. crassifolia* ($2n = 40$). The latter is an annual or short-lived perennial that rarely forms a well-developed caudex, whereas *D. fladnizensis* almost always produces a distinct caudex. Although most individuals of both species are scapose, they occasionally produce one or two cauline leaves. The cauline leaves are usually glabrous in *D. crassifolia* and ciliate in *D. fladnizensis*; in the latter, the distalmost cauline leaf usually subtends the proximalmost flower. Finally, the seeds in *D. fladnizensis* are slightly larger (0.8–0.1 × 0.5–0.6 versus 0.7–0.8 × 0.4–0.5 mm) than those of *D. crassifolia*, though the reliability of this distinction needs to be examined in greater detail. N. H. Holmgren (2005b) reported *D. fladnizensis* from central Nevada, but we have not seen any material of the species from that state.

38. Draba glabella Pursh, Fl. Amer. Sept. 2: 434. 1813

Draba arabisans Michaux var. *canadensis* (Brunet) Fernald & Knowlton; *D. arabisans* var. *orthocarpa* Fernald & Knowlton; *D. canadensis* Brunet; *D. canadensis* var. *pycnosperma* (Fernald & Knowlton) O. E. Schulz; *D. daurica* de Candolle; *D. glabella* var. *megasperma* (Fernald & Knowlton) Fernald; *D. glabella* var. *orthocarpa* (Fernald & Knowlton) Fernald; *D. glabella* var. *pycnosperma* (Fernald & Knowlton) G. A. Mulligan; *D. henneana* Schlechtendal; *D. hirta* Linnaeus var. *laurentiana* (Fernald) B. Boivin; *D. hirta* var. *pycnosperma* (Fernald & Knowlton) B. Boivin; *D. laurentiana* Fernald; *D. megasperma* Fernald & Knowlton; *D. norvegica* Gunnerus var. *pleiophylla* Fernald; *D. pycnosperma* Fernald & Knowlton; *D. sornborgeri* Fernald

Perennials; (sometimes cespitose); caudex simple or branched; not scapose. **Stems** branched or unbranched, (0.4–)1–3.5(–4.7) dm, often pubescent throughout (sometimes sparsely so distally), sometimes glabrous, trichomes simple and 2-rayed, (non-crisped), 0.2–1 mm, or subsessile, stellate-pectinate, and 3–8-rayed, 0.1–0.3 mm. **Basal leaves** rosulate; petiolate; petiole ciliate, (trichomes simple, 0.1–0.8 mm); blade oblanceolate to spatulate or linear-oblanceolate, (0.6–)1–3.5(–5) cm × 2–8(–10) mm, margins dentate or denticulate, (pubescent as petiole), surfaces pubescent with non-crisped, minutely stalked, 4–8(–12)-rayed, stellate-pectinate trichomes, 0.15–0.6 mm, (midvein obscure abaxially), adaxially sometimes also with simple trichomes, or glabrous. **Cauline leaves** 2–17(–25); sessile; blade ovate to oblong, margins dentate to subentire, surfaces often pubescent as basal, or predominantly with simple trichomes adaxially. **Racemes** (5–)8–26(–34)-flowered, ebracteate or proximalmost 1 or 2 flowers bracteate, elongated in fruit; rachis not flexuous, usually glabrous, rarely pubescent as stem. **Fruiting pedicels** divaricate-ascending to suberect, straight, (1–)3–10(–16) mm, glabrous or pubescent as stem. **Flowers:** sepals oblong, 2–3.5 mm, pubescent, (trichomes simple and short-stalked, 2–4-rayed); petals white, broadly obovate, 4–5.5 × 1.5–3 mm; anthers ovate, 0.3–0.5 mm. **Fruits** oblong to ovate or ovoid to lanceolate or linear-lanceolate, usually plane, rarely slightly twisted, flattened or inflated, (3–)5–12(–16) × 2–3.5 mm; valves glabrous or pubescent, trichomes simple or 2–4-rayed, 0.05–0.2(–0.4) mm; ovules (20–)24–36 per ovary; style 0.05–0.2(–0.5) mm. **Seeds** oblong, 0.9–1.1 × 0.5–0.7 mm. $2n = 64, 80$.

Flowering May–Aug. Rock outcrops, talus, rocky ridges and knolls, meadows, tundra, gravelly beaches sandy river margins, disturbed soils; 0–1400 m; Greenland; B.C., Man., N.B., Nfld. and Labr., N.W.T., N.S., Nunavut, Ont., Que., Yukon; Alaska, Maine, Vt., Wis.; n Europe (n Russia); e Asia (Russian Far East, Siberia).

Draba glabella was reported by J. V. Freudenstein and J. K. Marr (1986) from Michigan, but that record likely was based on plants of *D. arabisans*, a highly variable species that occurs in that state.

Draba glabella is extremely variable in indumentum, number of cauline leaves, fruiting pedicel length, fruit shape and size, style length, and seed number. M. L. Fernald (1934) divided it into five species and three varieties; R. C. Rollins (1993) recognized three species. Of these, G. A. Mulligan (1970, 1976) reduced *D. laurentiana* to synonymy of *D. glabella* and treated *D. pycnosperma* as a variety. Of all the segregates of *D. glabella*, var. *pycnosperma* might merit recognition. It is restricted to northwestern Newfoundland and northeastern Quebec, where var. *glabella* also grows. The main difference between the two varieties is the presence in var. *pycnosperma* of plump (versus flattened), ovoid to oblong fruits. *Draba sornborgeri*, recognized by Rollins as a distinct species, is merely a glabrescent form of *D. glabella*. Because of the tremendous morphological variability, wide distribution, extensive synonymy, and different chromosome numbers, *D. glabella* will require extensive molecular, cytological, and morphological studies to properly delimit the species and any potential infraspecific taxa.

Some forms of *Draba glabella* approach both *D. borealis* and *D. praealta*, but these can be distinguished by examining the trichomes on the abaxial surfaces of basal leaf blades. In *D. glabella*, these trichomes are minutely stalked or subsessile and have branched rays. In the other two species, the trichomes have long stalks and the rays are always unbranched.

The Linnaean name *Draba hirta* was applied to this species previously, and still is in Russia. The name is not typified, and the material at LINN is in bad condition and probably belongs to two species, *D. glabella* and *D. norvegica*, as recognized here. A typification of *D. hirta* may necessitate its re-introduction for this species.

39. Draba globosa Payson, Amer. J. Bot. 4: 257. 1917 E

Draba apiculata C. L. Hitchcock; *D. densifolia* Nuttall var. *apiculata* (C. L. Hitchcock) S. L. Welsh; *D. densifolia* var. *decipiens* S. L. Welsh; *D. densifolia* var. *globosa* (Payson) S. L. Welsh

Perennials; (cespitose, pulvinate); caudex branched (with persistent leaves, branches sometimes terminating in sterile rosettes); scapose. **Stems** unbranched, 0.1–0.5 dm, glabrous. **Basal leaves** rosulate; sessile; blade (not fleshy), narrowly oblanceolate or lanceolate to linear, (0.2–)0.3–0.8 cm × 0.5–1.6(–2) mm, margins entire (ciliate, trichomes simple, 0.1–0.8 mm, apex acute, trichomes usually longer), surfaces glabrous, (midvein obscure abaxially). **Cauline leaves** 0. **Racemes** 2–5(–7)-flowered, ebracteate, slightly elongated in fruit; rachis not flexuous, glabrous. **Fruiting pedicels** divaricate-ascending, straight or slightly curved, 2–6 mm, glabrous. **Flowers:** sepals (persistent to near fruit maturity), ovate to broadly oblong, 2–3 mm, glabrous; petals white to pale yellow, obovate, 2.5–4 × 1.2–2 mm; anthers oblong, 0.4–0.6 mm. **Fruits** ovate, plane, flattened, 4.5–8 × 2.5–4 mm; valves (distinctly veined), glabrous; ovules 8–16 per ovary; style (0.1–)0.2–0.6 mm. **Seeds** oblong, 1.1–1.4 × 0.8–1 mm.

Flowering Jun–Aug. Ridges, talus, alpine tundra and meadows; 2700–3900 m; Idaho, Mont., Utah, Wyo.

Draba globosa is an apomictic species closely related to *D. burkei* (M. D. Windham, unpubl.). Though often treated as a variety of *D. densifolia*, it is morphologically and phyletically distinct from that species. Both R. C. Rollins (1993) and N. H. Holmgren (2005b) indicated that the species occurs in Colorado, but we have not seen material for that state.

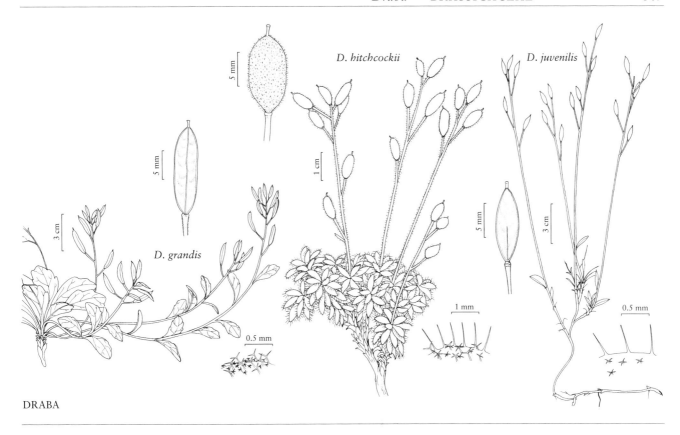

40. Draba graminea Greene, Pl. Baker. 3: 5. 1901 C E

Braya graminea (Greene) M. E. Jones; *Draba chrysantha* S. Watson var. *graminea* (Greene) O. E. Schulz; *D. chrysantha* var. *hirticaulis* O. E. Schulz

Perennials; (cespitose); caudex simple or branched (with persistent leaf bases); not scapose. **Stems** unbranched, 0.1–0.5(–0.8) dm, usually pubescent throughout, rarely glabrous, trichomes simple or subsessile, 2-rayed or spurred, (crisped), 0.1–0.5 mm. **Basal leaves** rosulate; petiolate; petiole ciliate, (trichomes usually straight and simple, rarely also 2-rayed, 0.2–0.6(–0.8) mm); blade linear to linear-oblanceolate, (0.5–)1–4 cm × 0.3–2(–3) mm, margins entire (pubescent as petiole), surfaces glabrous. **Cauline leaves** (1–)3–9(–12) (as bracts); sessile; blade linear to oblanceolate or lanceolate, margins entire, (ciliate proximally, similar to basal). **Racemes** 3–15-flowered, bracteate, elongated in fruit; rachis not flexuous, usually pubescent as stem, rarely glabrous. **Fruiting pedicels** divaricate-ascending, usually straight, rarely curved upward, 3–10(–15) mm, usually pubescent as stem, rarely glabrous. **Flowers:** sepals ovate, 1.5–2.5 mm, glabrous or pubescent, (trichomes simple and short-stalked, 2-rayed); petals yellow, spatulate to obovate, 3–5 × 1.5–3 mm; anthers ovate, 0.3–0.5 mm. **Fruits** ovate-elliptic to lanceolate, slightly twisted or plane, flattened, 5–11 × 2.5–5 mm; valves glabrous; ovules 8–16 per ovary; style 0.2–0.7 mm. **Seeds** ovoid, 1.2–1.5 × 0.7–1 mm. $2n = 18$.

Flowering Jul–Sep. Rocky areas and ridges, alpine tundra, gravel bars in streams; of conservation concern; 3000–4100 m; Colo.

Draba graminea is occasionally confused with *D. crassa*, which occupies similar habitats and elevations. It is easily distinguished from the latter by having narrower [0.3–2(–3) versus 2.5–8(–10) mm wide] basal leaves and bracteate (versus ebracteate) racemes. It is known to us only from Hinsdale, La Plata, Ouray, San Juan, and San Miguel counties.

41. Draba grandis Langsdorff ex de Candolle, Syst. Nat. 2: 355. 1821 F

Cochlearia siliquosa Schlechtendal ex de Candolle; *C. spathulata* Schlechtendal ex de Candolle; *Draba greenei* Pohle; *D. hatchiae* G. A. Mulligan; *D. hyperborea* (Linnaeus) Desvaux var. *spathulata* (Schlechtendal ex de Candolle) A. Gray ex S. Watson; *D. spathulata* (Schlechtendal ex de Candolle) Sprengel; *Nesodraba grandis* (Langsdorff ex de Candolle) Greene; *N. megalocarpa* Greene; *N. siliquosa* (Schlechtendal ex de Candolle) Greene

Perennials; caudex simple or branched (fleshy, with persistent leaf bases); not scapose. **Stems** (decumbent), unbranched, (0.2–)0.5–2.7(–3.7) dm, pubescent throughout, trichomes 2–4-rayed, 0.05–0.2 mm. **Basal leaves** rosulate; long-petiolate; petiole (winged, (1–)4–15 cm), often not ciliate (or ciliate to blade apex, trichomes simple, 0.3–0.8 mm); blade (somewhat fleshy), oblanceolate to spatulate or obovate, (1–)2–11(–17) cm × (5–)8–30(–45) mm, margins often coarsely dentate, (pubescent as petiole), surfaces usually pubescent, abaxially with stalked, 2–4-rayed trichomes, 0.05–0.25 mm, adaxially similar, or also with simple and long-stalked, 2-rayed trichomes, to 0.8 mm, rarely glabrescent, with mostly simple and 2-rayed trichomes. **Cauline leaves** 2–12(–16); sessile or petiolate; blade oblanceolate to obovate, margins dentate or entire, surfaces pubescent as basal. **Racemes** 5–26(–32)-flowered, ebracteate, elongated in fruit; rachis not flexuous, pubescent as stem. **Fruiting pedicels** horizontal to divaricate-ascending or ascending, usually straight, rarely curved upward, (5–)10–22(–27) mm, pubescent, trichomes 2–4-rayed (0.05–0.3 mm), sometimes with simple and spurred ones. **Flowers:** sepals broadly ovate, 3–4 mm, pubescent, (trichomes simple and short-stalked, 2-rayed); petals yellow, oblanceolate to spatulate, 4.5–7 × 1.8–3 mm; anthers oblong, 0.7–1 mm. **Fruits** oblong to lanceolate, or ovate to suborbicular, slightly twisted or plane, flattened, (6–)10–20(–25) × 4–7(–9) mm; valves glabrous; ovules 24–52 per ovary; style (0.2–)0.4–1.6(–2) mm. **Seeds** ovoid, 1.4–2 × 0.8–1.2 mm. $2n = 36$.

Flowering May–Jul. Rocky bluffs above salt-water beaches, loamy seaside banks, sea bird rookeries, coastal herbaceous tundra and sandy blowouts; 0–50(–200) m; B.C.; Alaska; e Asia (Kuril and Ratmanov islands, Russian Far East).

Almost all recent North American authors have used the name *Draba hyperborea* for this species. A. N. Berkutenko (1995) clearly showed that the type of that name belongs to an entirely different species that she placed in the genus *Schivereckia* Andrzejowski ex de Candolle. *Draba grandis* thus becomes the correct name for the North American taxon. Except for its fleshy leaves, *Nesodraba* is indistinguishable morphologically and molecularly from other species of *Draba*.

42. Draba grayana (Rydberg) C. L. Hitchcock, Revis. Drabas W. N. Amer., 29. 1941 [C] [E]

Draba streptocarpa A. Gray var. *grayana* Rydberg, Bull. Torrey Bot. Club 31: 555. 1904; *D. alpicola* Osterhout 1923, not Klotzsch 1862

Perennials; (densely cespitose); caudex branched (covered with persistent, somewhat thickened, dry petioles); not scapose. **Stems** unbranched, 0.08–0.6 dm, densely to moderately pubescent throughout, trichomes simple and stalked, 2- or 3-rayed, 0.2–1 mm. **Basal leaves** rosulate; petiolate; petiole ciliate, (trichomes simple, 0.2–1 mm); blade linear to linear-oblanceolate, (0.4–)0.6–1.5(–2) cm × 1–2 mm, margins entire, (pubescent as petiole), surfaces pubescent abaxially with simple and stalked, 2- or 3-rayed trichomes, 0.1–0.4 mm, adaxially glabrate or subapically sparsely pubescent with simple trichomes. **Cauline leaves** 1–4; sessile; blade linear to linear-oblanceolate, margins entire, surfaces pubescent as basal. **Racemes** 5–12(–16)-flowered, usually ebracteate, rarely proximalmost flowers bracteate, slightly elongated in fruit; rachis not flexuous, pubescent as stem. **Fruiting pedicels** divaricate-ascending, straight or slightly curved upward, 1.5–6 mm, pubescent as stem. **Flowers:** sepals ovate, 1.5–2 mm, pubescent, (trichomes simple and short-stalked, 2- or 3-rayed); petals bright yellow, spatulate, 3–4.5 × 1–2 mm, (flared, clawed); anthers ovate, 0.3–0.5 mm. **Fruits** lanceolate to oblong-lanceolate, slightly twisted or plane, flattened, 4–12 × 1.5–3 mm; valves glabrous or pubescent, trichomes simple, 0.07–0.2 mm; ovules 16–20 per ovary; style 0.4–1.2 mm. **Seeds** ovoid, 1–1.4 × 0.7–1 mm. $2n = 24 \pm 3$.

Flowering Jul–Aug. Alpine tundra, fellfields, gravelly wet meadows; of conservation concern; 3500–4000 m; Colo.

Draba grayana is an apomictic species clearly related to *D. exunguiculata*, with which it is occasionally sympatric (see discussion of 36. *D. exunguiculata* for distinguishing features). It is known from Chaffe, Clear Creek, Larimer, Park, and Summit counties.

43. Draba heilii Al-Shehbaz, Harvard Pap. Bot. 14: 83, fig. 1. 2009 E

Perennials; (densely cespitose); caudex few-branched (compact, with persistent, thickened leaves); not scapose. Stems unbranched or branched distally, 0.3–0.7 dm, glabrous throughout or, rarely, pilose, trichomes simple and stalked, 2-rayed, 0.2–0.5 mm. Basal leaves (persistent, imbricate); rosulate; undifferentiated into blade and petiole (becoming indurate at base); blade linear-lanceolate, 0.5–0.8 cm × 1–1.5(–2) mm, margins entire, surfaces glabrous or, rarely, pilose with simple and stalked, 2-rayed trichomes, (margins ciliate with simple and, rarely, 2-rayed setiform trichomes, 0.5–1.3 mm, midvein prominent). Cauline leaves 6–8; sessile; blade linear-oblong, 0.5–0.8 cm × 1–2 mm (widest about base), margins entire, (ciliate). Racemes 10–26-flowered, ebracteate, elongated in fruit; rachis not flexuous, usually glabrous, rarely pubescent, (trichomes noncrisped, simple and 2-rayed). Fruiting pedicels ascending, straight, 3–5(–7) mm, usually glabrous, rarely pilose, (trichomes simple and 2-rayed, stalked). Flowers: sepals ovate, 1.7–2.7 mm, usually glabrous, rarely pilose, (trichomes simple); petals yellow, oblong-oblanceolate, 4–6 × 1.5–2 mm; anthers oblong, 0.5–0.6 mm. Fruits narrowly lanceolate, twisted ½ turn or plane, flattened, 7–10 × 1.3–1.6 mm; valves glabrous; ovules 16–20 per ovary; style 0.7–1.3 mm. Seeds ovoid, 1 × 0.6 mm.

Flowering Jun–Jul. Alpine tundra; ca. 3700 m; N.Mex.

Draba heilii is restricted to Mora and Rio Arriba counties.

44. Draba helleriana Greene, Pittonia 4: 17. 1899

Draba aurea Vahl ex Hornemann var. *stylosa* A. Gray; *D. helleriana* var. *leiocarpa* O. E. Schulz; *D. helleriana* var. *neomexicana* (Greene) O. E. Schulz; *D. helleriana* var. *patens* (A. Heller) O. E. Schulz; *D. helleriana* var. *pinetorum* (Greene) O. E. Schulz; *D. neomexicana* Greene; *D. neomexicana* var. *robusta* A. Heller; *D. pallida* A. Heller; *D. patens* A. Heller; *D. pinetorum* Greene; *D. stylosa* (A. Gray) A. Heller 1897, not Turczaninow 1854

Biennials or perennials; (short-lived); caudex simple or branched; not scapose. Stems branched, (0.1–)1.5–4.3(–5.1) dm, hirsute throughout, trichomes simple, 0.4–1.3(–1.8) mm, with short-stalked to subsessile, 3–5-rayed ones, 0.2–0.4 mm, (some slightly coarser, 2-rayed). Basal leaves rosulate; petiolate; petiole base ciliate or not; blade oblanceolate to obovate, 0.9–4.1(–5.2) cm × 2–7(–10) mm, margins entire or dentate, surfaces pubescent, abaxially with stalked, cruciform, and fewer 3–5-rayed trichomes, 0.1–0.6 mm, adaxially with cruciform and simple trichomes, 0.4–1.3 mm, and smaller 2-rayed ones. Cauline leaves (8–)12–31(–43); sessile; blade ovate to lanceolate or oblong, margins usually dentate, rarely subentire, surfaces pubescent as basal. Racemes 10–52(–83)-flowered, usually ebracteate, rarely proximalmost 1 or 2 flowers bracteate, elongated in fruit; rachis not flexuous, pubescent as stem. Fruiting pedicels horizontal to divaricate-ascending, straight or slightly curved upward, 4–10(–13) mm, pubescent as rachis abaxially, glabrous adaxially. Flowers: sepals oblong, 2.5–4 mm, pubescent, (trichomes simple and stalked, 2-rayed); petals yellow, oblanceolate, 5–7 × 1.5–2.2 mm; anthers oblong, 0.5–0.8 mm. Fruits (not appressed to rachis), lanceolate to ovate or oblong-lanceolate, slightly to strongly twisted or plane, flattened, 5–15 × 2–3.5 mm; valves puberulent, trichomes simple and subsessile, 2(–4)-rayed, 0.03–0.25(–0.8) mm; ovules 14–28 per ovary; style (1–)1.5–3.5 mm. Seeds oblong, 1–1.3 × 0.6–0.8 mm. $2n = 18$.

Flowering Jun–Sep. Oak and pine-fir woodlands, aspen groves, rocky meadows; 2100–3600 m; Ariz., Colo., N.Mex.; Mexico (Nuevo León).

Draba helleriana is highly variable and was divided by C. L. Hitchcock (1941) into four varieties. For a discussion of those and the circumscription of the species, see I. A. Al-Shehbaz and M. D. Windham (2007).

45. Draba hitchcockii Rollins, J. Arnold Arbor. 64: 500. 1983 E F

Perennials; (densely cespitose); caudex branched (densely covered with persistent leaves, branches sometimes terminating in sterile rosettes); scapose. Stems unbranched, (0.1–)0.3–1(–1.3) dm, hirsute throughout, trichomes simple, 0.4–1 mm, and 2–4-rayed, 0.1–0.6 mm. Basal leaves (densely imbricate); rosulate; sessile; blade narrowly oblanceolate to oblong-linear, 0.3–1.2(–1.5) cm × 1–2 mm, margins entire, (ciliate, trichomes simple, 0.3–1.2 mm), surfaces pubescent abaxially with stalked, 2–4-rayed trichomes, 0.08–0.45 mm, (midvein obscure), adaxially glabrous proximally, sparsely pubescent distally with mostly simple trichomes. Cauline leaves 0. Racemes 4–15-flowered, ebracteate, elongated in fruit; rachis not flexuous, hirsute as stem. Fruiting pedicels ascending, straight, (2–)4–13(–18) mm, hirsute as stem. Flowers: sepals oblong, 2.5–3.5 mm, hirsute, (trichomes simple and stalked, 2–4-rayed); petals white, obovate, 5–6.5 × 2–3.5 mm; anthers ovate, 0.5–0.6 mm. Fruits ovate to broadly oblong or elliptic, plane, flattened, (3–)

4–7(–10) × (2.5–)3.5–5 mm; valves pubescent, trichomes simple and short-stalked, 2(–4)-rayed, 0.06–0.5 mm; ovules 8–12 per ovary; style (0.8–)1–1.7(–2) mm. Seeds oblong, 1.4–1.8 × 0.8–1 mm. $2n = 54$.

Flowering May–Jun. Limestone outcrops and gravelly soil; 1800–2200 m; Idaho.

Draba hitchcockii is known from the Lost River Range in Butte and Custer counties. Based on morphological and chromosomal evidence, M. D. Windham (2004) suggested that it may be an allopolyploid resulting from hybridization between *D. oreibata* and *D. paysonii*. *Draba hitchcockii* is superficially similar to *D. jaegeri*, a taxon known from the Charleston Mountains of Clark County, Nevada. Both are cespitose perennials with relatively large, white flowers and a chromosome number ($2n = 54$) otherwise unknown in *Draba* (Windham). Features distinguishing these two taxa are provided in the discussion of 51. *D. jaegeri*.

46. **Draba howellii** S. Watson, Proc. Amer. Acad. Arts 20: 354. 1885 [E]

Perennials; (loosely cespitose); caudex branched (somewhat surculose, with persistent leaf remains, branches sometimes terminating in sterile rosettes); often scapose. **Stems** unbranched, (0.2–)0.4–1.1(–1.5) dm, usually pubescent throughout, rarely glabrous, trichomes simple and 2–4-rayed, 0.1–0.6 mm. **Basal leaves** rosulate; blade (somewhat fleshy), oblanceolate or spatulate to obovate, 0.4–1.6(–2.5) cm × (1.5–)3–6(–10) mm, margins entire, surfaces pubescent with stalked, cruciform, and fewer 2- or 3-rayed trichomes, 0.07–0.5 mm, rarely both surfaces glabrous and trichomes on margins, (midvein obscure abaxially). **Cauline leaves** 0–3 (or 4); sessile; blade ovate to oblong, margins entire, pubescent as basal. **Racemes** (5–)7–18(–25)-flowered, usually ebracteate, sometimes proximalmost 1 or 2 flowers bracteate, elongated in fruit; rachis not flexuous, usually pubescent as stem, rarely glabrous. **Fruiting pedicels** divaricate-ascending, straight, (4–)7–10 mm, usually pubescent as stem, rarely glabrous. **Flowers:** sepals ovate, 2.5–3.2 mm, glabrous or pubescent, (trichomes 2–4-rayed); petals yellow, oblanceolate, 5.5–8 × 1–2 mm; anthers oblong, 0.7–0.9 mm. **Fruits** lanceolate to elliptic-lanceolate or broadly ovate, plane (not curved), strongly flattened, 6–11(–15) × 3–5 mm; valves usually pubescent, rarely glabrous, trichomes simple and 2–4-rayed, 0.05–0.3 mm; ovules 8–22 per ovary; style (0.7–)1.6–3 mm. **Seeds** oblong, 1–1.6 × 0.8–1 mm, (sometimes distally appendaged).

Flowering Jul–Aug. Rocky summits, cracks in granite walls, rock crevices; 1900–2700 m; Calif., Oreg.

Draba howellii is known from Siskiyou and Trinity counties, California, and Josephine County, Oregon. One collection, *Tracy 14623* (DS, GH, UC), is unusual in having glabrous stems, pedicels, and leaf blade surfaces. In this regard, it resembles *D. carnosula*, but in all other respects (bracts, seeds, inflorescences, etc.), it is indistinguishable from *D. howellii*. For characteristics distinguishing the two species, see 23. *D. carnosula*.

47. **Draba incana** Linnaeus, Sp. Pl. 2: 643. 1753

Draba confusa Ehrhart; *D. contorta* Ehrhart; *D. incana* subsp. *confusa* (Ehrhart) E. Ekman; *D. incana* var. *confusa* (Ehrhart) Liljeblad; *D. incana* var. *contorta* (Ehrhart) Liljeblad

Biennials; not scapose. **Stems** usually branched, rarely unbranched, (0.3–)0.6–3(–4) dm, densely pubescent proximally, trichomes simple, 0.4–1.3 mm, with fewer 2–4-rayed ones, to 0.3 mm, (distally with mostly branched trichomes, with fewer, simple ones). **Basal leaves** rosulate; petiolate; petiole base and margin ciliate, (trichomes simple, 0.2–0.7 mm); blade lanceolate or oblanceolate to spatulate, 0.5–2.9 cm × 1–5(–9) mm, margins dentate, surfaces pubescent, abaxially with stalked, cruciform, or 2- or 3-rayed trichomes, 0.1–0.4 mm, (rarely with a branched ray), adaxially with simple trichomes, 0.1–0.7 mm, and stalked, 2–4-rayed ones. **Cauline leaves** (15–)22–54(–76), (densely imbricate at least basally); sessile; blade ovate to oblong or lanceolate (base often densely bearded abaxially, trichomes simple, to 1.3 mm), margins dentate, surfaces pubescent as basal. **Racemes** (11–)17–36(–48)-flowered, bracteate basally, elongated or not in fruit; rachis not flexuous, pubescent as stems distally. **Fruiting pedicels** erect to ascending, (subappressed to rachis), straight, (1.5–)2.5–6(–13) mm, pubescent, trichomes 2–4-rayed and simple. **Flowers:** sepals ovate, 2–2.5 mm, pubescent, (trichomes simple and short-stalked, 2–4-rayed); petals white, spatulate, 3.5–4.5 × 1–2 mm; anthers ovate, 0.3–0.4 mm. **Fruits** oblong to linear-lanceolate or elliptic, slightly twisted or plane, flattened, 5–10(–12) × 1.5–2.5(–3) mm; valves glabrous or pubescent, trichomes simple and short-stalked, 2–4-rayed, 0.05–0.3 mm; ovules (24–)30–40 per ovary; style 0.01–0.25 mm. **Seeds** oblong, 0.7–1.1 × 0.5–0.6 mm. $2n = 32$.

Flowering Jul–Aug. Rock outcrops, talus, gravelly shores; 0–200 m; Greenland; Man., N.B., Nfld. and Labr., Ont., P.E.I., Que.; Mich., Minn.; n, c Europe (Russia).

48. Draba incerta Payson, Amer. J. Bot. 4: 261. 1917 E

Draba exalata E. Ekman; *D. incerta* var. *laevicapsula* (Payson) Payson & H. St. John; *D. incerta* var. *peasei* (Fernald) Rollins; *D. laevicapsula* Payson; *D. peasei* Fernald

Perennials; (cespitose, often pulvinate); caudex branched (dense with persistent leaf remains, branches sometimes terminating in sterile rosettes); scapose. **Stems** unbranched, (0.2–)0.4–1.4(–2.1) dm, often pubescent throughout, sometimes glabrous distally, trichomes often simple and 2–5-rayed, 0.1–0.5 mm, (sometimes with mostly subpectinate ones). **Basal leaves** rosulate; petiolate; petiole (0–1 cm), ciliate throughout; blade narrowly oblanceolate to linear, (0.4–)0.6–1.7(–2.5) cm × (1–)1.5–3.5(–5) mm, margins entire, (ciliate, trichomes usually simple, rarely 2-rayed, 0.2–1.1 mm), surfaces usually pubescent with short-stalked, pectinate trichomes, 0.15–0.5 mm, sometimes also with 4–6-rayed ones, (midvein usually obscure abaxially), sometimes glabrous adaxially. **Cauline leaves** usually 0 (or 1, as a bract); sessile; blade linear to oblong, margins entire, surfaces pubescent as basal. **Racemes** 3–14(–30)-flowered, usually ebracteate, rarely proximalmost flowers bracteate, elongated in fruit; rachis not flexuous, glabrous or pubescent as stem. **Fruiting pedicels** ascending, straight, (2.5–)4–11(–27) mm, glabrous or pubescent, trichomes 2–5-rayed or pectinate. **Flowers:** sepals broadly ovate, 2.5–3.5(–4) mm, pubescent, (trichomes simple and 2- or 3-rayed); petals yellow (fading white), oblanceolate to obovate, 4–6 × 1.5–2.5 mm; anthers ovate, 0.3–0.5 mm. **Fruits** broadly ovate to lanceolate, plane, flattened, 5–9(–11) × 2–4 mm; valves glabrous or puberulent, trichomes simple and 2-rayed, 0.05–0.3 mm; ovules 8–16(–20) per ovary; style 0.2–0.9 mm. **Seeds** oblong, 1.1–1.5 × 0.7–1 mm. $2n = 112$.

Flowering Jun–Aug. Rock outcrops, talus, gravelly areas, tundra; 0–3300 m; Alta., B.C., Que., Yukon; Alaska, Colo., Idaho, Mont., Nev., Utah, Wash., Wyo.

Draba incerta was shown by G. A. Mulligan (1972) to be sexually reproducing and 14-ploid with $x = 8$. It is often confused with the apomict *D. oligosperma* ($2n = 32, 64$). *Draba incerta* is readily separated from *D. oligosperma* by having well-formed (versus abortive) anthers and pollen, stalked (versus sessile) leaf trichomes, and ciliate (versus non-ciliate) basal leaves with obscure (versus prominent) midveins. Although both species have leafless scapes, one often finds a bract adnate to, or subtending, the proximalmost pedicel in *D. incerta*.

Draba incerta is found near sea level in Alaska.

49. Draba incrassata (Rollins) Rollins & R. A. Price, Harvard Pap. Bot. 1(3): 73. 1991 E

Draba lemmonii S. Watson var. *incrassata* Rollins, Rhodora 55: 323. 1953

Perennials; (cespitose, loosely matted); caudex branched (with persistent leaf bases, branches often creeping, sometimes terminating in sterile rosettes); scapose. **Stems** unbranched, 0.2–0.8 dm, glabrous throughout. **Basal leaves** rosulate; subsessile; petiole ciliate; blade (thick and somewhat fleshy), obovate to spatulate or oblanceolate, 0.3–1 cm × 1.5–4.5 mm, margins entire, (sometimes sparsely ciliate), surfaces glabrous or pubescent proximal to apex, with simple and fewer, short-stalked, 2-rayed trichomes, 0.2–0.7 mm. **Cauline leaves** 0. **Racemes** 8–22-flowered, ebracteate, elongated in fruit; rachis not flexuous, glabrous. **Fruiting pedicels** horizontal to divaricate-ascending (distinctly decurrent basally), curved upward, 3–7(–10) mm, glabrous. **Flowers:** sepals oblong, 1.7–3 mm, glabrous or pubescent, (trichomes subapical, simple or 2-rayed); petals bright yellow, oblanceolate to spatulate, 3–5 × 1.7–2.5 mm; anthers oblong, 0.4–0.6 mm. **Fruits** ovate to ovate-lanceolate or narrowly elliptic-lanceolate, plane, flattened, 3–7(–10) × 2.2–4.5 mm; valves usually glabrous, rarely sparsely puberulent, trichomes simple, 0.03–0.1 mm; ovules 8–12 per ovary; style 0.2–0.8 mm. **Seeds** oblong, 1.5–1.8 × 0.8–1.1 mm. $2n = 24$.

Flowering Jun–Aug. Rocky and gravelly slopes, alpine fellfields; 2500–3500 m; Calif.

Although originally treated as a variety of *Draba lemmonii*, *D. incrassata* is quite distinct both morphologically (R. A. Price and R. C. Rollins 1991) and chromosomally (M. D. Windham, unpubl.). The species is known from the Sweetwater Mountains in Mono County.

50. Draba inexpectata S. L. Welsh in S. L. Welsh et al., Utah Fl. ed. 3, 272. 2003 C E

Perennials; (cespitose); caudex simple or branched (with persistent leaf remains); not scapose. **Stems** unbranched, 0.1–0.4 dm, pubescent, trichomes 5–10-rayed, 0.05–0.2 mm, and fewer, simple and 2-rayed ones, 0.3–0.6 mm. **Basal leaves** rosulate; subsessile; petiole ciliate; blade obovate to oblanceolate, 0.2–0.4 cm × 0.7–1.5 mm, (base sparsely ciliate, trichomes simple and 2-rayed, 0.4–0.6 mm), margins entire, surfaces densely pubescent with subsessile, 6–12-rayed, stellate trichomes, 0.25–0.7 mm. **Cauline leaves** 1–3 (or 4); sessile; blade ovate to lanceolate, margins entire or denticulate, surfaces pubescent as basal. **Racemes** 2–5(–8)-flowered, proximalmost 1–3 flowers bracteate, slightly elongated in fruit; rachis not flexuous, pubescent, trichomes 5–10-rayed. **Fruiting pedicels** ascending, straight, 0.7–2(–3) mm, pubescent as rachis. **Flowers:** sepals (sometimes persistent), ovate, 1.2–1.7 mm, pubescent, (trichomes simple and branched); petals white, obovate, 2–2.5 × 1–1.2 mm; anthers ovate, 0.1–0.2 mm. **Fruits** elliptic to lanceolate, slightly twisted or plane, flattened, 3.5–5 × 1.5–2 mm; valves pubescent, trichomes short-stalked, (3 or) 4(–6)-rayed, 0.1–0.2 mm; ovules 16–24 per ovary; style 0.2–0.4 mm. **Seeds** oblong, 0.7–1 × 0.4–0.5 mm.

Flowering Jul–Aug. Rocky ridges and slopes, on accumulated soil among boulders, subalpine fir and juniper communities; of conservation concern; 3100–3700 m; Utah.

Draba inexpectata is known from the Uinta Mountains in Summit County. The measurements above are based on the holotype (*Franklin 6328*) and two of the three paratypes (*Franklin 6293, 6331*), all at BRY. One paratype (*Goodrich 26166*, BRY) is a robust, completely sterile plant of uncertain identity and was not used for the description above.

Although compared in the original description with *Draba lonchocarpa* (with which it grows sympatrically), *D. inexpectata* appears most closely related to *D. cana*. *Draba inexpectata* is distinguished from the latter by having stems 1–4 cm, basal leaves non-ciliate, racemes that are 2–5(–8)-flowered and scarcely elongated in fruit, fruits elliptic to lanceolate, 3.5–5 mm, and ovules 16–24 per ovary. By contrast, *D. cana* has stems (6–)10–30 cm, basal leaves ciliate at the apex, racemes that are (10–)15–47(–63)-flowered and often considerably elongated in fruit, fruits linear-lanceolate or very rarely ovate-oblong, (5–)6–11 mm, and ovules 28–48 per ovary.

51. **Draba jaegeri** Munz & I. M. Johnston, Bull. Torrey Bot. Club 56: 164. 1929 [C][E]

Perennials; (densely cespitose); caudex branched (densely covered with persistent leaves, branches sometimes terminating in sterile rosettes); scapose. **Stems** unbranched, (0.05–)0.15–0.5(–0.6) dm, hirsute throughout, trichomes simple, 0.1–0.8 mm, and 2–4-rayed, 0.05–0.4 mm. **Basal leaves** (densely imbricate); rosulate; sessile; blade oblanceolate to ovate, 0.4–1.5 cm × 1.5–3.5 mm, margins entire, (ciliate, trichomes simple, 0.3–1.1 mm), surfaces pubescent abaxially with stalked, (2–)4–6-rayed trichomes, 0.1–5 mm, (midvein obscure), adaxially glabrous proximally, sparsely pubescent distally with mostly simple trichomes. **Cauline leaves** 0. **Racemes** 3–12(–18)-flowered, ebracteate, slightly elongated in fruit; rachis not flexuous, hirsute as stem. **Fruiting pedicels** ascending, straight, 2–6(–8) mm, hirsute as stem. **Flowers:** sepals oblong, 2.5–3 mm, hirsute, (trichomes simple and stalked, 2–4-rayed); petals white, spatulate, 4.5–6 × 1.5–2 mm; anthers oblong, 0.8–1 mm. **Fruits** ovate to elliptic, plane, slightly flattened, 4–8(–11) × 2.5–4.5 mm; valves pubescent, trichomes stalked (2–)4-rayed, 0.05–0.5 mm; ovules 8–16 per ovary; style (1.2–)2–4(–4.5) mm. **Seeds** oblong, 1.4–2 × 0.9–1 mm. $2n = 54$.

Flowering Jun–Aug. Limestone outcrops and gravelly soil; of conservation concern; 2900–3600 m; Nev.

Draba jaegeri is superficially similar to *D. hitchcockii*, a narrow endemic of the Lost River Range in central Idaho. Both species are cespitose perennials with relatively large, white flowers, and a chromosome number ($2n = 54$) otherwise unknown in *Draba* (M. D. Windham 2004). *Draba jaegeri* is readily distinguished from *D. hitchcockii* by having fruits pubescent with (2–)4-rayed trichomes, styles (1.2–)2–4(–4.5) mm, fruiting pedicels 2–6(–8) mm, and spatulate petals 1.5–2 mm wide. By contrast, *D. hitchcockii* has fruits pubescent with mostly simple and 2-rayed trichomes (with 3- or 4-rayed ones), styles (0.8–)1–1.7(–2) mm, fruiting pedicels (2–)4–13(–18) mm, and obovate petals 2–3.5 mm wide. *Draba jaegeri* is known only from the Charleston Mountains in Clark County.

52. **Draba juvenilis** Komarov, Repert. Spec. Nov. Regni Veg. 13: 167. 1914 [F]

Draba hirta Linnaeus var. *tenella* Eastwood; *D. kananaskis* G. A. Mulligan; *D. longipes* Raup

Perennials; (not pulvinate); caudex branched (branches slightly loose, with persistent leaf remains, sometimes terminating in sterile rosettes); often scapose. **Stems** unbranched, (0.3–)0.8–2.3(–3) dm, pubescent proximally, glabrous or sparsely pubescent distally, trichomes simple and 2–4-rayed, 0.07–0.4 mm. **Basal leaves** (loosely) rosulate; petiolate; petiole base ciliate, margin not ciliate, (midvein obscure, trichomes simple and 2-rayed, 0.2–0.7 mm); blade oblanceolate to lanceolate, (0.3–)0.6–2.5(–3.5) cm × (1–)2–7(–10) mm, margins entire or denticulate, surfaces pubescent, abaxially with stalked, cruciform trichomes, 0.2–0.5 mm, adaxially with cruciform and/or simple and 2-rayed ones. **Cauline leaves** 0–2 (or 3); sessile; blade elliptic to ovate or lanceolate, margins usually entire, rarely denticulate, surfaces pubescent as basal adaxially or trichomes predominantly simple. **Racemes** (2–)4–13(–18)-flowered, ebracteate, elongated in fruit; rachis not flexuous, glabrous or sparsely pubescent as stem. **Fruiting pedicels** divaricate-ascending or ascending, straight or often curved upward (not expanded basally), (3–)5–17(–22) mm, usually glabrous, rarely pubescent, trichomes simple and stalked, 2–4-rayed. **Flowers:** sepals ovate, 2.2–3 mm, glabrous or pubescent subapically, (trichomes simple and short-stalked, 2-rayed); petals pale yellow to creamy white, spatulate, 3–5 × 1.5–2.5 mm; anthers ovate, 0.4–0.5 mm. **Fruits** elliptic to oblong or linear-lanceolate, plane (not curved), flattened, 5–11(14) × 2–3 mm; valves usually glabrous, rarely margin pubescent, trichomes simple; ovules 16–30 per ovary; style 0.2–0.7 mm. **Seeds** oblong, 0.7–1.2(–1.5) × 0.4–0.7(–0.8) mm. $2n = 64$.

Flowering Jun–Jul. Rock outcrops, talus, gravelly beaches and stream banks, meadows, tundra; 0–2700 m; Alta., B.C., N.W.T., Yukon; Alaska; e Asia (Russian Far East, n Siberia).

Although *Draba kananaskis* and *D. longipes* have often been treated as distinct species, we find no basis for maintaining them. Both exhibit variations in petal color (white to pale yellow) and leaf trichomes (short-stalked to sessile) characteristic of *D. juvenilis* from the Russian Far East. Perhaps most importantly, all three are octoploids ($2n = 64$) with $x = 8$. We conclude that *D. kananaskis* is nothing more than a minor variant of *D. juvenilis* and it is treated herein, for the first time, as a synonym of that species.

Draba juvenilis is occasionally confused with *D. borealis*, which also has stalked, cruciform trichomes with unbranched rays. Typical *D. juvenilis* is easily distinguished from that species by having narrower (2–3 mm) fruits that are glabrous (rarely pubescent) and untwisted, and 0–2 (or 3)-leaved stems. By contrast, *D. borealis* has wider (2.5–4.5 mm) fruits that are usually pubescent and/or twisted (rarely neither) and (2 or) 3–7(–12)-leaved stems.

53. **Draba kassii** S. L. Welsh, Great Basin Naturalist 46: 264. 1986 [C] [E]

Perennials; caudex branched (densely covered with persistent petiole remains, branches sometimes terminating in sterile rosettes); scapose. **Stems** unbranched, 0.5–1.8 dm, glabrous. **Basal leaves** rosulate; long-petiolate; petiole ciliate throughout; blade oblanceolate to linear-oblanceolate, 1.5–5.5 cm × (1.5–)2–5.5 mm, margins entire (ciliate, trichomes simple, 0.2–1 mm), surfaces abaxially sparsely pubescent with short-stalked, submalpighiaceous trichomes, 0.3–1.3 mm, sometimes with 3- or 4-rayed ones, adaxially usually glabrous, rarely sparsely pubescent. **Cauline leaves** 0. **Racemes** 3–8(–10)-flowered, ebracteate, considerably elongated in fruit; rachis not flexuous, glabrous. **Fruiting pedicels** ascending to divaricate-ascending, straight, 5–14(–17) mm, glabrous. **Flowers:** sepals broadly ovate, 2–3 mm, subapically sparsely pubescent, (trichomes simple); petals yellow, oblanceolate, 4.5–7 × 1.2–2 mm; anthers ovate, 0.4–0.5 mm. **Fruits** elliptic to oblong-elliptic, plane, flattened, 3–10(–14) × 2–3 mm; valves glabrous; ovules 20–28 per ovary; style 0.6–1 mm. **Seeds** oblong, 1.2–1.4 × 0.7–0.8 mm. $2n = 22$.

Flowering Apr–Jun. Quartzite and granitic outcrops; of conservation concern; 2100–2600 m; Utah.

Draba kassii is a distinctive species that is known from the Deep Creek Mountains in Juab and Tooele counties.

54. **Draba kluanei** G. A. Mulligan, Canad. J. Bot. 57: 1873. 1979 [C] [E]

Perennials; (cespitose); caudex branched (with persistent leaf remains); not scapose. **Stems** unbranched, 0.1–0.35 dm, pilose throughout, trichomes simple, (slightly crisped), 0.4–0.9 mm. **Basal leaves** rosulate; petiolate; petiole base and margin ciliate, (trichomes usually simple, rarely some 2-rayed, 0.2–0.7 mm); blade oblanceolate to linear-oblanceolate, 0.6–1.5 cm × 0.5–2 mm, margins entire or sparsely dentate, surfaces pubescent, abaxially with simple trichomes, 0.3–0.9 mm, with fewer, 2 (or 3)-rayed ones, 0.07–0.3 mm, adaxially with simple trichomes. **Cauline leaves** 1–3; sessile; blade lanceolate, margins entire, surfaces pubescent as basal. **Racemes** 2–7-flowered, ebracteate or proximalmost 1 or 2 flowers bracteate, slightly elongated in fruit; rachis not flexuous, glabrous. **Fruiting pedicels** divaricate-ascending, straight, 1–2 mm, glabrous. **Flowers:** sepals ovate, 0.7–1.3 mm, subapically sparsely pubescent, (trichomes simple); petals white, spatulate, 1.5–2 × 0.4–0.8 mm; anthers ovate, 0.2–0.3 mm. **Fruits** lanceolate, plane, flattened, 3–4 × 1–1.5 mm; valves pubescent, trichomes simple, 0.7–0.2 mm; ovules 14–16 per ovary; style 0.1–0.25 mm. **Seeds** oblong, ca. 0.6 × 0.4 mm.

Flowering Jul. Rocky alpine slopes; of conservation concern; ca. 2000 m; Yukon.

Draba kluanei is known only from the type collection made in Kluane National Park, southwestern Yukon Territory.

55. Draba lactea Adams, Mém. Soc. Imp. Naturalistes Moscou 5: 104. 1817

Draba allenii Fernald; *D. boecheri* Gjaerevoll & Ryvarden; *D. fernaldiana* Polunin

Perennials; (cespitose); caudex branched (with persistent petiole remains, branches sometimes terminating in sterile rosettes); scapose. **Stems** unbranched, 0.2–1.1(–1.5) dm, glabrous throughout or sparsely pubescent proximally, trichomes short-stalked, substellate, 2–8-rayed, (non-crisped), 0.5–0.3 mm. **Basal leaves** rosulate; petiolate; petiole (persistent, strongly thickened), margin usually ciliate, (trichomes usually simple and 2-rayed, 0.3–1 mm); blade oblanceolate to obovate, (0.3–)0.5–1.1(–1.7) cm × (1–)2–6 mm, margins usually entire, rarely denticulate, (sometimes ciliate), surfaces sometimes pubescent with stellate to subdendritic, 4–12-rayed, (non-crisped) trichomes, 0.1–0.4 mm, (midvein persistent, prominent, strongly thickened). **Cauline leaves** 0 (or, rarely, 1 as a bract). **Racemes** 2–8(–12)-flowered, ebracteate, elongated in fruit; rachis not flexuous, usually glabrous, rarely sparsely pubescent as stem basally. **Fruiting pedicels** divaricate-ascending, straight, (1–)2–5(–10) mm, glabrous. **Flowers:** sepals ovate, 1.8–3 mm, usually glabrous, rarely sparsely pubescent subapically, (trichomes simple); petals white, obovate, 3–5 × 1.8–3 mm; anthers ovate, 0.3–0.4 mm. **Fruits** oblong to elliptic-lanceolate or ovate to broadly so, plane, flattened, 4–8 × (1.5–)2–3 mm; valves glabrous; ovules (10–)14–22(–26) per ovary; style 0.1–0.4 mm. **Seeds** ovoid, 0.8–1.1 × 0.5–0.6 mm. $2n$ = 32, 48.

Flowering Jun–Aug. Rock outcrops, talus, rocky hillsides and ridges, open gravelly areas, seepage swales, meadows; 0–2000 m; Greenland; Nfld. and Labr., N.W.T., Nunavut, Que., Yukon; Alaska; Europe (n Finland, Norway, w Sweden); Asia (Russian Far East, c, n Siberia); Atlantic Islands (Iceland); circumpolar.

T. W. Böcher (1966) postulated that *Draba lactea* originated from hybridization between *D. fladnizensis* and *D. nivalis*, but A.-C. Scheen et al. (2002) showed that it is more closely allied to *D. subcapitata*. By contrast, H. H. Grundt et al. (2004) concluded that hexaploid *D. lactea* originated from tetraploids of the same species, which in turn originated from the diploid *D. palanderiana* lineage. They suggested that *D. lactea* probably originated multiple times in the Beringian area and migrated to reach its present circumpolar distribution. The hexaploids are distributed throughout the species range, whereas the tetraploids are known only from Alaska and the Russian Far East (Grundt et al. 2005b).

Draba fernaldiana, which was collected from Southampton Island (Nunavut), was not mentioned by R. C. Rollins (1993). The plants are completely glabrous except for leaf margins, which are ciliate with simple and sparse 2-rayed trichomes. The taxon resembles some forms of *D. lactea* and is tentatively herein included within that species. The only conflict in such placement is petal color, which was listed in the original description of *D. fernaldiana* as pale yellow instead of white.

Glabrous or glabrescent forms of *Draba lactea* are quite common in the Canadian Arctic Archipelago, whereas pubescent forms predominate in Alaska and the Russian Far East.

56. Draba lemmonii S. Watson in W. H. Brewer et al., Bot. California 2: 430. 1880 (as lemmoni) E

Perennials; (cespitose, not pulvinate); caudex branched (covered with persistent petiole remains, branches sometimes terminating in sterile rosettes); scapose. **Stems** unbranched, 0.3–1(–1.5) dm, hirsute throughout, trichomes simple, 0.2–0.7 mm, with short-stalked, 2-rayed ones, (smaller). **Basal leaves** rosulate; shortly petiolate; petiole base and margin ciliate, (trichomes simple, 0.5–1 mm, midvein obscure); blade (somewhat fleshy), oblanceolate to obovate, 0.4–1(–1.8) cm × 1.5–4(–6) mm, margins entire, surfaces hirsute, abaxially mostly with stalked, 2-rayed trichomes, 0.1–0.8 mm, rarely with fewer, simple ones, adaxially with mostly simple ones, 0.5–1 mm. **Cauline leaves** 0. **Racemes** 4–15(–21)-flowered, ebracteate, elongated in fruit; rachis not flexuous, hirsute as stem. **Fruiting pedicels** horizontal to divaricate-ascending (often somewhat decurrent basally), straight or curved upward, 4–10(–14) mm, hirsute as stem. **Flowers:** sepals ovate, 2–2.7 mm, pubescent, (trichomes simple, short-stalked, 2-rayed); petals yellow, spatulate, 4–6 × 1.5–2.5 mm; anthers oblong, 0.4–0.6 mm. **Fruits** ovate to ovate-lanceolate, slightly twisted, flattened, 4–9 × 3.5–5 mm; valves pubescent, trichomes usually simple, 0.1–0.45 mm, rarely with fewer, short-stalked, 2-rayed ones; ovules 10–14(–16) per ovary; style 0.1–0.6(–0.8) mm. **Seeds** ovoid, 1–1.4 × 0.6–1 mm. $2n$ = 50.

Flowering Jul–Aug. Granitic rock outcrops, boulder slopes, alpine fellfields; 3000–4000 m; Calif.

Draba lemmonii was so broadly circumscribed by C. L. Hitchcock (1941) and R. C. Rollins (1993) that it included plants here assigned to three different species. For a list of features distinguishing *D. lemmonii* from the recently-segregated *D. longisquamosa* and *D. cyclomorpha*, see I. A. Al-Shehbaz and M. D. Windham (2007). *Draba lemmonii* is apparently restricted to alpine areas of the Sierra Nevada in Fresno, Inyo, Madera, Mono, and Tuolumne counties.

57. Draba lonchocarpa Rydberg, Mem. New York Bot. Gard. 1: 181. 1900

Draba nivalis Liljeblad var. *elongata* S. Watson, Proc. Amer. Acad. Arts 23: 258. 1888, not *D. elongata* Host 1831; *D. lonchocarpa* var. *denudata* O. E. Schulz; *D. lonchocarpa* var. *exigua* O. E. Schulz; *D. lonchocarpa* var. *semitonsa* Payson & H. St. John; *D. lonchocarpa* var. *thompsonii* (C. L. Hitchcock) Rollins; *D. lonchocarpa* var. *vestita* O. E. Schulz; *D. nivalis* var. *denudata* (O. E. Schulz) C. L. Hitchcock; *D. nivalis* var. *exigua* (O. E. Schulz) C. L. Hitchcock; *D. nivalis* subsp. *lonchocarpa* (Rydberg) Hultén; *D. nivalis* var. *thompsonii* C. L. Hitchcock

Perennials; (cespitose, sometimes forming mats); caudex branched (covered with persistent leaf bases); often scapose. **Stems** unbranched, (0.1–)0.3–1.1 dm, glabrous throughout or pubescent proximally or throughout, trichomes minutely stalked, 8–12-rayed, (non-crisped), 0.08–0.3 mm. **Basal leaves** rosulate; shortly petiolate; petiole (obsolete), margin ciliate proximally, (trichomes soft, simple, 2-rayed, 0.2–0.6 mm); blade oblanceolate to obovate, (0.2–)0.3–1.5 cm × 1–3(–5) mm, margins entire, surfaces pubescent abaxially with short-stalked, stellate, 8–12-rayed, (non-crisped) trichomes, 0.15–0.4 mm, (midvein obscure), adaxially glabrous basally or with simple and long-stalked, branched trichomes. **Cauline leaves** 0 or 1(–4); sessile; blade ovate or oblong, margins entire, surfaces pubescent as basal. **Racemes** 3–9-flowered, usually ebracteate, rarely proximalmost 1(–4) flowers bracteate, elongated in fruit; rachis often slightly flexuous (in fruit), glabrous or pubescent as stem. **Fruiting pedicels** ascending, usually straight, rarely curved upward, 2–9(–15) mm, glabrous or pubescent as stem. **Flowers:** sepals ovate, 1.5–2 mm, pubescent, (trichomes simple and short-stalked, 2–5-rayed); petals white, oblanceolate, 2–3.5 × 1–1.5 mm; anthers ovate, 0.2–0.3 mm. **Fruits** linear to narrowly lanceolate or narrowly oblong, slightly twisted or plane, flattened, 6–15(–18) × 1–2(–3) mm; valves glabrous or sparsely puberulent, trichomes simple and minutely stalked, 2-rayed, 0.1–0.2 mm; ovules 16–24(–28) per ovary; style (0.05–)0.1–0.25(–0.35) mm. **Seeds** ovoid, 0.7–1 × 0.5–0.6 mm. $2n = 16$.

Flowering Jun–Jul. Rocky outcrops and ridges, loose talus, tundra; (300–)2800–4000 m; Alta., B.C., N.W.T., Yukon; Alaska, Calif., Colo., Idaho, Mont., Nev., Oreg., Utah, Wash., Wyo.; e Asia (Russian Far East, Siberia).

Draba lonchocarpa is a highly variable species within which O. E. Schulz (1927), G. A. Mulligan (1974), and R. Rollins (1993) recognized three to five varieties. By contrast, C. L. Hitchcock (1941) united it with *D. nivalis* and recognized six varieties (see 68. *D. nivalis* for differences). Some of the infraspecific taxa of *D. lonchocarpa* are based on trivial characteristics and are listed in the synonymy above without further comment. The most problematic are briefly discussed below.

Authors recognizing var. *vestita* claim that it differs from var. *lonchocarpa* by having pubescent (versus glabrous) stems and pedicels, 1- or 2-leaved (versus 0 or 1-leaved) scapes, and fruits appressed (versus not appressed) to the rachises. These characteristics do not appear to be strongly correlated. A case in point is the holotype sheet of var. *semitonsa*, which includes plants with puberulent or glabrous fruits, as well as with pubescent and glabrous stems that are 0–4-leaved. Leafless and densely pubescent scapes are found in *Trelease 3913* (MO), whereas completely glabrous, 0–2-leaved stems, and fully appressed fruits are found in *Calder 5617a* (DAO). Other exceptions can be cited, though the vast majority of the plants examined have leafless, glabrous scapes.

An examination of the type collections of var. *thompsonii*, *Thompson 9512* (holotype, UC; isotypes, DS, GH, MO, NY, RSA, US), clearly shows that the taxon usually has oblong to lanceolate fruits 2–3.2 mm wide, as opposed to linear fruits less than 2 mm wide in var. *lonchocarpa*. Indeed, a casual observation would immediately justify the recognition of var. *thompsonii*. Both fruit types can be found in plants of the same population (e.g., the RSA isotype) or even on the same plant (e.g., *Thompson 10816*, MO). Furthermore, fruits to 2.5 mm wide occur sporadically in various parts of the species range. For these reasons, and in the absence of a comprehensive study of the species, we choose to not recognize var. *thompsonii* at present.

Draba lonchocarpa is found at elevations of 300–1200 m in Alaska.

58. Draba longisquamosa O. E. Schulz in H. G. A. Engler, Pflanzenr. 89[IV,105]: 94. 1927 [E]

Perennials; (cespitose); caudex branched (covered with persistent petiole remains, branches sometimes terminating in sterile rosettes); scapose. **Stems** unbranched, 0.2–0.9 dm, hirsute throughout, trichomes simple, 0.2–0.8 mm. **Basal leaves** rosulate; shortly petiolate; petiole base and margin ciliate, (trichomes simple, 0.1–1 mm); blade oblanceolate to obovate, 0.5–2 cm × 2–5 mm, margins entire, surfaces pubescent with simple trichomes, 0.15–0.8 mm. **Cauline leaves** 0. **Racemes** 4–16-flowered, ebracteate, (subumbellate), slightly elongated in fruit; rachis not flexuous, hirsute as stem. **Fruiting pedicels** divaricate to ascending (not decurrent basally), straight, 3–7 mm, hirsute as stem. **Flowers:** sepals (persistent) ovate, 1.5–2.2 mm, pubescent, (trichomes simple); petals (persistent), yellow, spatulate, 3.5–5 × 1.5–2 mm; anthers oblong, 0.5–0.6 mm. **Fruits** ovate to suborbicular, plane, flattened, 3.5–7 × 2.5–5 mm; valves pubescent, trichomes simple, 0.1–0.3 mm; ovules 10–16 per ovary; style 0.4–1 mm. **Seeds** ovoid, 1–1.2 × 0.7–0.8 mm.

Flowering Jul. Gravelly areas; 3000–3900 m; Calif.

Draba longisquamosa was treated as a synonym of *D. lemmonii* by both C. L. Hitchcock (1941) and R. C. Rollins (1993). We feel that the two taxa show sufficient morphological divergence and merit recognition as separate species (see I. A. Al-Shehbaz and M. D. Windham 2007 for more detailed discussion). *Draba longisquamosa* is currently known only from the southern Sierra Nevada in Fresno, Inyo, and Tulare counties.

59. Draba macounii O. E. Schulz in H. G. A. Engler, Pflanzenr. 89[IV,105]: 97. 1927 [E]

Perennials; (cespitose); caudex simple or branched (with persistent leaf bases, branches sometimes terminating in sterile rosettes); scapose. **Stems** unbranched, (0.06–)0.1–0.4 (–0.6) dm, usually pubescent, rarely glabrous, trichomes simple and stalked, 2–4-rayed, 0.15–0.9 mm. **Basal leaves** rosulate; petiole (obscure), margin ciliate, (trichomes simple, 0.3–1 mm); blade oblanceolate to obovate, (0.4–)0.6–1(–1.5) cm × (1–)2–4 mm, margins entire, surfaces usually pubescent abaxially, rarely glabrous, with short-stalked, 2–5-rayed trichomes, 0.1–0.5 mm, adaxially glabrous or pubescent with simple and short-stalked, 2-rayed trichomes. **Cauline leaves** 0. **Racemes** 3–10(–13)-flowered, ebracteate, (subumbellate), slightly elongated in fruit; rachis not flexuous, usually pubescent as stem, rarely glabrous. **Fruiting pedicels** horizontal to divaricate (not expanded basally), curved upward, 1.5–4.5(–6) mm, abaxially pubescent as stem, adaxially usually glabrous, rarely throughout. **Flowers:** sepals ovate, 1.5–2.5 mm, glabrous or pubescent, (trichomes simple and short-stalked, 2- or 3-rayed); petals pale yellow, spatulate, 2.7–4 × 1–2 mm; anthers ovate, 0.2–0.3 mm. **Fruits** subglobose to ovoid or ellipsoid, plane, inflated at least basally, 4–8 × 2–4.5 mm; valves glabrous; ovules 10–14(–16) per ovary; style 0.1–0.5 mm. **Seeds** ovoid, 1–1.2 × 0.6–0.8 mm. **2n** = 64.

Flowering Jun–Aug. Rock outcrops, talus, tundra; 700–2800 m; Alta., B.C., Yukon; Alaska, Mont.

Draba macounii is often confused with the circumpolar decaploid (2n = 80) *D. alpina*. Plants of the former have scapes (0.6–)1–4(–6) cm, fruiting pedicels divaricate to horizontal, gently curved, and 1.5–4.5(–6) mm, and petals pale yellow, spatulate, and 2.7–4 × 1–2 mm. By contrast, *D. alpina* has scapes (3–)5–17(–28) cm, fruiting pedicels divaricate-ascending to ascending, often straight, and (3–)4–14(–30) mm, and petals bright yellow, obovate, and 3.5–5 × 1.7–2.5 mm. R. C. Rollins (1993) indicated that the species occurs in Colorado, but we have not seen any material from that state.

60. Draba maguirei C. L. Hitchcock, Revis. Drabas W. N. Amer., 70, plate 5, figs. 37a–c. 1941 [E]

Perennials; (cespitose, forming loose mats); caudex branched (with some persistent leaf bases, branches often creeping, sometimes terminating in sterile rosettes); scapose. **Stems** unbranched, (0.4–)0.7–1.7(–2.2) dm, usually glabrous, rarely proximalmost parts and sterile shoots pubescent, trichomes short-stalked, 2–4-rayed, 0.2–0.5 mm. **Basal leaves** rosulate; sessile; blade oblanceolate, (0.5–)0.7–1.4(–2) cm × 1.5–3.5(–5) mm, margins entire, surfaces pubescent with stalked, cruciform, and 2- or 3-rayed trichomes, 0.2–0.6 mm, (sometimes trichomes only on margins or apex, not ciliate, midvein obscure abaxially). **Cauline leaves** 0. **Racemes** 5–18(–23)-flowered, ebracteate, elongated in fruit; rachis not flexuous, glabrous. **Fruiting pedicels** divaricate-ascending to ascending (not expanded basally), straight, 5–13(–18) mm, glabrous. **Flowers:** sepals ovate, 2.5–4 mm, glabrous; petals yellow, oblanceolate, 5–7 × 2–3 mm; anthers ovate, 0.5–0.6 mm. **Fruits** broadly ovate to lanceolate, plane (not curved), flattened, (3–)4–6.5(–8) × 2–3 mm; valves glabrous or puberulent, trichomes simple, 0.05–0.2 mm; ovules 4–8 per ovary; style 0.6–1.7 mm. **Seeds** ovoid to oblong, 1.6–2 × 1–1.3 mm. **2n** = 16, 32.

Flowering May–Jul. Dolomite outcrops, talus, rocky slopes; 1600–2900 m; Utah.

Draba maguirei is known from the Bear River Range in Cache County. Despite this very narrow distribution, the species includes two ploidy levels (diploid and tetraploid) that are morphologically and ecologically distinct (M. D. Windham, unpubl.).

61. Draba malpighiacea Windham & Al-Shehbaz, Harvard Pap. Bot. 12: 417. 2007 [E][F]

Perennials; caudex simple or branched (sometimes with persistent leaf bases); not scapose. **Stems** unbranched, 0.5–1.5 dm, pubescent throughout, trichomes malpighiaceous, 0.2–0.5 mm, and sometimes simple. **Basal leaves** rosulate; petiolate; petiole not ciliate; blade oblanceolate to spatulate, 0.5–1.8 cm × 2–4 mm, margins entire or dentate, (not ciliate

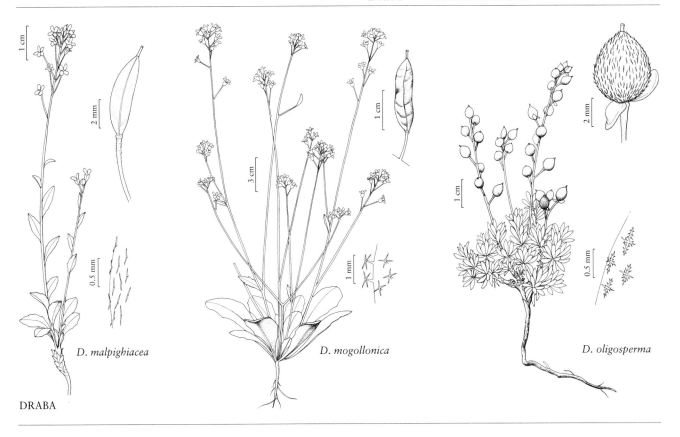

proximally), surfaces pubescent with malpighiaceous trichomes, 0.3–0.6 mm. **Cauline leaves** 4–13; sessile; blade broadly ovate to lanceolate, margins denticulate, surfaces pubescent as basal. **Racemes** 10–33-flowered, ebracteate, elongated in fruit; rachis not flexuous, glabrous or pubescent, trichomes simple and/or malpighiaceous. **Fruiting pedicels** divaricate-ascending, straight or curved upward, 6–13 mm, pubescent abaxially, trichomes simple and/or malpighiaceous. **Flowers:** sepals ovate, 2.5–3 mm, subapically pubescent, (trichomes simple and malpighiaceous); petals yellow, oblanceolate, 4–5 × 1.5–2 mm; anthers oblong, 0.7–1 mm. **Fruits** elliptic to elliptic-lanceolate, plane, flattened, 5–10 × 1.5–2.2 mm; valves glabrous; ovules 14–18 per ovary; style 0.8–1.2 mm. **Seeds** ovoid, 1–1.4 × 0.7–0.9 mm.

Flowering Jun–Aug. Rock outcrops and rocky slopes in subalpine conifer forests; 3000–3500 m; Colo.

Draba malpighiacea, which is restricted to higher elevations in Hinsdale, La Plata, and Montezuma counties, has been included within *D. spectabilis* by previous authors. Preliminary data suggest that it is one of the diploid parents of tetraploid *D. spectabilis*. It is readily distinguished from all species of *Draba* in Canada and the United States by having a leaf indumentum of exclusively malpighiaceous trichomes. The only other species in the flora area with such trichomes is *D. sibirica*, a stoloniferous, scapose perennial known only from Greenland.

62. Draba micropetala Hooker in W. E. Parry, App. Parry J. Sec. Voy., 385. 1825

Draba alpina Linnaeus var. *micropetala* Durand; *D. oblongata* R. Brown ex de Candolle subsp. *minuta* V. V. Petrovsky; *D. pauciflora* R. Brown var. *micropetala* (Hooker) O. E. Schulz

Perennials; caudex branched (with persistent leaf remains); scapose. **Stems** unbranched, (0.05–)0.15–0.7(–1.1) dm, pubescent throughout, sometimes sparsely so distally, trichomes stalked, 2–4-rayed, and simple, 0.05–0.6 mm. **Basal leaves** (not imbricate); rosulate; petiolate; petiole (base thickened), margin ciliate, (trichomes coarse, simple and 2-rayed, 0.3–1.3 mm); blade broadly oblanceolate to broadly obovate, 0.6–2 cm × 4–12 mm, margins entire, (apex obtuse to rounded, trichomes multibranched, subcruciform), surfaces pubescent abaxially with stalked, 2–4-rayed and, rarely, simple trichomes, 0.1–0.6 mm, adaxially glabrous or with mostly simple trichomes. **Cauline leaves** 0. **Racemes** 2–7(–10)-flowered, ebracteate, often elongated in fruit; rachis often flexuous, pubescent as stem. **Fruiting pedicels** divaricate to ascending, straight, 1–3(–4) mm, pubescent as stem. **Flowers:** sepals oblong, 1.8–2.5 mm, pubescent, (trichomes simple, with short-stalked, 2-rayed ones); petals pale yellow, narrowly spatulate or oblanceolate, 2–3 × (0.7–)1–1.5 mm; anthers ovate, 0.3–0.4 mm. **Fruits** elliptic-ovate, plane, slightly flattened, 5–10 × 2–3.2 mm; valves often densely pubescent, trichomes simple and spurred, 0.1–0.5 mm); ovules (16–)18–28 per ovary; style 0.05–0.3 mm (stigma about as wide as style). **Seeds** oblong, 1–1.1 × 0.5–0.6 mm. $2n = 48$.

Flowering Jun–Jul. Meadows, mesic tundra, swales, gravelly soils; 0–900 m; Greenland; N.W.T., Nunavut; Alaska; Europe (Norway [Spitsbergen]); e Asia (Russian Far East, Siberia).

Draba micropetala, which was not listed by C. L. Hitchcock (1941) or R. C. Rollins (1993), is frequently confused with the closely related tetraploid ($2n = 32$) *D. pauciflora*. It is distinguished from the latter by having elongated (versus congested) fruiting racemes, narrower [2–3.2 versus (3–)3.5–5 mm] and elliptic-ovate (versus obovate) fruits, more [(16–)18–28 versus 8–16(–20)] seeds per fruit, and obtuse to rounded (versus subacute to acute) leaf apices with multibranched, subcruciform (versus simple and/or 2-branched) trichomes.

63. **Draba mogollonica** Greene, Bot. Gaz. 6: 157. 1881
    [E] [F]

Annuals, biennials, or, rarely, perennials; (short-lived); not scapose. **Stems** usually branched, (0.6–)1.2–2.8(–3.6) dm, pubescent proximally, glabrous distally, trichomes simple, 0.3–1.3 mm, and 2–4-rayed, 0.1–0.6 mm. **Basal leaves** rosulate; petiole (0–2 cm), densely pubescent, (trichomes simple and 2–4-rayed, not ciliate); blade spatulate to obovate or oblanceolate, (1.5–)3–8(–10) cm × 7–25(–33) mm, margins dentate to denticulate, surfaces pubescent with long-stalked, cruciform, and fewer 2- or 3-rayed trichomes, 0.1–0.6 mm. **Cauline leaves** 1–3; sessile; blade oblong to ovate, margins entire, surfaces pubescent as basal. **Racemes** 12–40(–51)-flowered, ebracteate, elongated in fruit; rachis not flexuous, usually glabrous, rarely proximally with relatively few trichomes. **Fruiting pedicels** horizontal to divaricate, straight, (6–)9–18(–22) mm, usually glabrous, rarely sparsely pubescent as stem proximally. **Flowers:** sepals oblong, 2.5–3.5 mm, sparsely pubescent, (trichomes simple and 2–4-rayed); petals yellow, spatulate, (5–)6–8.5 × 2.2–3.5 mm; anthers oblong, 0.7–0.9 mm. **Fruits** linear-elliptic to elliptic-lanceolate, slightly twisted, flattened, 6–19 × 2.5–3.5 mm; valves glabrous or pubescent, trichomes simple and short-stalked, 2(–4)-rayed, 0.07–0.3 mm; ovules 24–36 per ovary; style (0.8–)1–2.2 mm. **Seeds** oblong, 1–1.4 × 0.7–0.8 mm. $2n = 22$.

Flowering Apr–May. Rock outcrops and rocky slopes in oak-pine woodlands and mixed conifer communities; 1500–2300 m; Ariz., N.Mex.

Draba mogollonica is distributed in New Mexico primarily in the Mogollon and neighboring mountains in Catron, Grant, Sierra, and Socorro counties. It is reported herein for the first time from Arizona from Greenlee County, based on *Barneby 2297* (NY), *Rollins & Rollins 93113* (CAS, GH), and *Windham 2918* (MO, UT). The only published chromosome count ($n = 16$, compiled in S. I. Warwick and I. A. Al-Shehbaz 2006) is erroneous; counts by M. D. Windham (unpubl.) from five populations representing all of the above-mentioned counties in Arizona and New Mexico consistently agree on $n = 11$.

64. **Draba monoensis** Rollins & R. A. Price, Aliso 12: 22, figs. 1f–j, 3. 1988 [C] [E]

Perennials; (usually cespitose); caudex simple or branched (poorly developed, with persistent leaf remains); sometimes scapose. **Stems** unbranched, (0.05–)0.1–0.4 dm, usually pubescent throughout, rarely sparsely pubescent or glabrous distally, trichomes simple and 2-rayed, (often crisped), 0.1–0.6 mm. **Basal leaves** rosulate; subsessile or shortly petiolate; petiole margin ciliate or not; blade narrowly oblanceolate, (0.3–)0.5–1.6(–2) cm × (1–)1.5–3(–4) mm, margins usually entire, rarely subapically denticulate, (ciliate or not), surfaces pubescent with simple and stalked, 2-rayed trichomes, 0.1–0.7 mm, (midvein obscure abaxially). **Cauline leaves** 0–2 (or 3); sessile; blade ovate to oblong, margins entire, surfaces pubescent as basal. **Racemes** (3–)6–13(–17)-flowered, ebracteate or proximalmost 1 (or 2) flowers bracteate, not or slightly elongated in fruit; rachis not flexuous, usually pubescent as stem, rarely glabrous. **Fruiting pedicels** divaricate-ascending, straight, 1–2.5(–4) mm, usually pubescent as stem, rarely glabrous. **Flowers:** sepals (persistent), oblong, 1–1.5 mm, subapically sparsely pubescent, (trichomes simple); petals white, spatulate, 1.5–2 × 0.5–0.6 mm; anthers ovate, 0.15–0.2 mm. **Fruits** ovoid to subellipsoid, plane, slightly inflated basally, (2–)3–5 × (1.2–)1.5–2.5 mm; valves usually puberulent, rarely glabrous, trichomes simple, 0.05–0.2 mm; ovules 12–20 per ovary; style 0.1–0.2 mm. **Seeds** ovoid, 0.6–0.8 × 0.4–0.6 mm.

Flowering Jul–Aug. Gravelly alpine meadows; of conservation concern; 3600–4000 m; Calif.

Draba monoensis appears to be polyploid (M. D. Windham, unpubl.), and its morphological similarity to *D. fladnizensis* (R. C. Rollins 1993) may indicate that the latter is one of its progenitors. The species is known from the White Mountains of Mono County.

65. Draba mulliganii Al-Shehbaz, Harvard Pap. Bot. 14: 85. 2009 [E]

Perennials; (cespitose); caudex branched (with persistent leaves); often scapose. Stems unbranched, 0.3–1.2 dm, glabrous throughout or sparsely pubescent proximally, trichomes short-stalked, 4–10-rayed, (non-crisped). Basal leaves rosulate; subsessile or shortly petiolate (petiolar base and proximal margin ciliate with stiff, subsetiform, simple trichomes, 0.3–0.6 mm); blade oblanceolate or lanceolate, 5–9 × 1–2 mm, margins entire, surfaces usually pubescent, abaxially with short-stalked, 8–12-rayed, stellate, (non-crisped) trichomes, 0.1–0.3 mm (often without spurred rays), rarely glabrescent, trichomes confined to margin, (midvein and petiole not thickened), adaxially with similar stellate and some simple trichomes. Cauline leaves 0. Racemes 3–11-flowered, ebracteate, elongated in fruit; rachis not flexuous, glabrous. Fruiting pedicels divaricate to ascending (not appressed to rachis), straight, (4–)5–9 mm, glabrous. Flowers: sepals (caducous), ovate, 1.7–2.2 mm, pubescent subapically, (trichomes simple and 2-rayed); petals white, broadly obovate, 3.2–4 × 2–3 mm, (apex obtuse); anthers ovate, 0.4 mm. Fruits ovate-lanceolate to oblong, not twisted, flattened, 5–8 × 2–2.5 mm; valves glabrous; ovules 12–16 per ovary; style 0.4 1 mm. Seeds ovoid, 0.7–1.1 × 0.5–0.6 mm.

Flowering Jun–Jul. Limestone gravel around outcrops, limestone crevices, dry serpentine knolls in heath vegetation with patches of exposed slide rock, shale rubble slopes; 300–1600 m; Alaska.

66. Draba murrayi G. A. Mulligan, Canad. J. Bot. 57: 1874. 1979 [C][E]

Perennials; (cespitose); caudex often branched (covered with persistent leaves or leaf remains, branches sometimes terminating in sterile rosettes); not scapose. Stems unbranched, (0.4–)0.9–2.7(–3.5) dm, pubescent proximally, glabrous distally, trichomes simple, 0.2–1 mm, and often smaller, fewer short-stalked, 2–4-rayed ones. Basal leaves rosulate; petiolate; petiole (0–0.7 cm), margin ciliate; blade oblanceolate to obovate, (0.3–)0.6–2.1(–2.8) cm × (1.5–)2.5–5(–9) mm, margins entire or sparsely denticulate, (ciliate, trichomes simple, 0.2–1.1 mm), surfaces pubescent with stalked, cruciform, and 2- or 3-rayed trichomes, (rays sometimes spurred), 0.1–0.6 mm, sometimes mostly simple trichomes adaxially. Cauline leaves 1–6; sessile; blade ovate or oblong to lanceolate, margins usually entire, rarely sparsely denticulate, surfaces pubescent as basal. Racemes 7–25-flowered, ebracteate, elongated in fruit; rachis not flexuous, glabrous. Fruiting pedicels divaricate-ascending, usually straight, rarely curved, (4–)7–18 mm, glabrous. Flowers: sepals ovate, 1.5–2.2 mm, pubescent, (trichomes simple and short-stalked, 2-rayed); petals white, obovate, 4–6 × 1.5–2.5 mm; anthers ovate, 0.4–0.5 mm. Fruits linear-lanceolate to narrowly elliptic, plane, flattened, 5–13 × 1.5–2 mm; valves glabrous; ovules 20–30 per ovary; style 0.6–1.5 mm. Seeds oblong, 0.9–1.2 × ca. 0.5 mm. $2n = 48$.

Flowering Jun–Jul. Rock outcrops and talus, open forests, dry slopes, alluvial gravels; of conservation concern; 150–900 m; Yukon; Alaska.

67. Draba nemorosa Linnaeus, Sp. Pl. 2: 643. 1753 [I][W]

Draba dictyota Greene; D. nemoralis Ehrhart; D. nemorosa var. brevisilicula Zapałowicz; D. nemorosa var. hebecarpa Lindblom; D. nemorosa var. leiocarpa Lindblom; Tomostima nemorosa (Linnaeus) Lunell

Annuals; not scapose. Stems unbranched or branched proximally, (0.3–)0.6–4.5(–6) dm, densely pubescent proximally, glabrous distally, trichomes simple, 0.5–1.3 mm, and stalked, 2–4-rayed, 0.1–0.5 mm. Basal leaves rosulate or not, petiolate; petiole ciliate throughout; blade oblong-obovate or oblanceolate, (0.4–)1–3.5(–5) cm × (2–)5–15(–20) mm, margins usually dentate to denticulate, rarely subentire, surfaces sparsely to densely pubescent with stalked, 2- or 3-rayed, and cruciform trichomes, 0.1–0.5 mm. Cauline leaves 4–12(–15); sessile; blade broadly ovate to oblong, margins dentate or denticulate, surfaces pubescent, abaxially as basal, adaxially with primarily simple, and 2- or 3-rayed trichomes. Racemes (10–)18–60(–92)-flowered, ebracteate, considerably elongated in fruit; rachis not flexuous (straight), glabrous. Fruiting pedicels divaricate, straight, 7–28(–35) mm [(1.5–)2–7 times longer than fruit], glabrous. Flowers: sepals ovate, (0.7–)0.9–1.6 mm, sparsely pubescent, (trichomes simple); petals yellow, spatulate to oblanceolate, (1.2–)1.7–2.2(–2.5) × (0.4–)0.6–1 mm, (deeply notched); anthers broadly ovate to subreniform, 0.1–0.2 mm. Fruits oblong to elliptic or, rarely, oblong-oblanceolate, plane, slightly flattened, (3–)5–8(–10) × 1.5–2.5 mm; valves glabrous or puberulent, trichomes simple, antrorse, 0.05–0.2 mm; ovules (30–)36–60(–72) per ovary; style 0.01–0.1 mm. Seeds ovoid, 0.5–0.7(–0.8) × 0.3–0.4(–0.5) mm. $2n = 16$.

Flowering Apr–Jul. Rock outcrops, open wooded slopes, meadows and prairies, stream banks, roadsides, disturbed gravelly grounds, waste places; 0–2700 m; introduced; Alta., B.C., Man., N.W.T., Nunavut, Ont., Que., Sask., Yukon; Alaska, Calif., Colo., Idaho, Iowa, Mich., Minn., Mont., Nebr., Nev., N.Dak., Oreg., S.Dak., Utah, Wash., Wis., Wyo.; Europe; Asia; introduced also in Australia.

68. Draba nivalis Liljeblad, Utkast Sv. Fl., 236. 1792

Draba caesia Adams

Perennials; (cespitose, usually forming mats); caudex branched (covered with persistent leaf bases); usually scapose. **Stems** unbranched, 0.2–0.8(–1.2) dm, pubescent throughout, trichomes minutely stalked, 8–15-rayed, stellate, (non-crisped), 0.03–0.15 mm. **Basal leaves** rosulate; petiole (obsolete), ciliate throughout; blade oblanceolate to obovate, 0.2–0.9(–1.5) cm × 1–3(–5) mm, margins entire, (not ciliate), surfaces usually pubescent, rarely glabrescent, with short-stalked, 8–15-rayed, stellate, (non-crisped) trichomes, 0.05–0.15 mm. **Cauline leaves** 0 or 1; sessile; blade ovate or oblong, margins entire, surfaces pubescent as basal. **Racemes** 3–9(–11)-flowered, ebracteate, elongated in fruit; rachis often slightly flexuous, pubescent as stem. **Fruiting pedicels** usually ascending, rarely divaricate, straight, 1–4.5(–8) mm, pubescent as stem. **Flowers:** sepals ovate, 1.5–2 mm, pubescent, (trichomes simple and short-stalked, 2–5-rayed); petals white, spatulate to oblanceolate, 2–3.5 × 0.8–1.4 mm; anthers ovate, 0.2–0.3 mm. **Fruits** elliptic to narrowly oblong-elliptic, twisted or not, flattened, 3.5–9 × 1.5–2.2 mm; valves usually glabrous, rarely with 3–6-rayed trichomes on replum; ovules 12–24(–28) per ovary; style 0.1–0.4 mm. **Seeds** ovoid, 0.6–1 × 0.5–0.6 mm. $2n = 16$.

Flowering Jun–Aug. Rock outcrops and fellfields, meadows, open tundra, stream banks, gravelly beaches, roadsides; 0–2500 m; Greenland; Alta., B.C., Man., Nfld. and Labr. (Nfld.), N.W.T., Nunavut, Ont., Que., Yukon; Alaska; Europe (Finland, Norway [including Svalbard], n, w Sweden); e Asia (Russian Far East, n Siberia); Atlantic Islands (Iceland); circumpolar.

Draba nivalis is most often confused with *D. lonchocarpa*, and C. L. Hitchcock (1941) reduced the latter to a variety (var. *elongata*) of *D. nivalis*. A critical comparison reveals that there are enough differences to warrant their recognition as distinct species. *Draba nivalis* is distinguished by having non-ciliate leaf bases, basal leaf blades pubescent with trichomes 0.05–0.15 mm, stems and pedicels that are always pubescent, and fruits that are elliptic to narrowly elliptic-oblong, plane, and 3.5–9 mm. By contrast, *D. lonchocarpa* has ciliate leaf bases, basal leaf blades pubescent with trichomes 0.15–0.5 mm, stems and pedicels that are glabrous or sparsely pubescent, and fruits that are linear to lanceolate or oblong, slightly twisted or plane, and 6–15(–18) mm. Except for Alaska, *D. nivalis* does not grow in the continental United States, whereas *D. lonchocarpa* grows in nearly all Mountain and Pacific states.

69. Draba norvegica Gunnerus, Fl. Norveg. 2: 106. 1772

Draba clivicola Fernald; *D. hirta* Linnaeus var. *norvegica* (Gunnerus) Liljeblad; *D. norvegica* var. *clivicola* (Fernald) B. Boivin; *D. norvegica* var. *hebecarpa* (Lindblom) O. E. Schulz; *D. rupestris* R. Brown; *D. scandinavica* Lindblom var. *hebecarpa* Lindblom; *D. trichella* Fries

Perennials; (cespitose); caudex branched (sometimes with persistent leaf remains); often scapose. **Stems** branched or unbranched, (0.2–)0.4–1.4(–2) dm, pubescent throughout, trichomes simple and 2–4-rayed, 0.2–0.5 mm. **Basal leaves** rosulate; petiolate; petiole (0–0.5 cm), ciliate throughout; blade oblanceolate to narrowly obovate or narrowly lanceolate, 0.4–2.1 cm × 1.5–7 mm, margins with 1–3 teeth on each side, (ciliate, trichomes simple and 2-rayed, 0.2–0.8 mm), surfaces pubescent abaxially with stalked, (2–)4(–6)-rayed trichomes, 0.1–0.4 mm, adaxially glabrous or pubescent, with simple and stalked, 2-rayed trichomes, 0.1–0.5 mm. **Cauline leaves** 0–3(–5); sessile; blade ovate to oblong, margins often dentate, (ciliate). **Racemes** 5–23-flowered, ebracteate or proximalmost 1(–3) flowers bracteate, considerably elongated in fruit; rachis slightly or not flexuous, pubescent as stem. **Fruiting pedicels** suberect to ascending (often appressed to rachis), straight, (1.5–)2.5–5(–9) mm, pubescent, trichomes simple and 2–4-rayed. **Flowers:** sepals ovate, 1.7–2.5 mm, pubescent, (trichomes simple and short-stalked, 2–4-rayed); petals white, spatulate, 2.5–4 × 1.7–2.5 mm; anthers ovate, 0.3–0.4 mm. **Fruits** (sometimes subappressed to rachis), oblong to lanceolate-elliptic, plane, flattened, 4–8 × 2–3 mm; valves glabrous or pubescent, trichomes simple and short-stalked, 2-rayed, 0.05–0.25 mm; ovules 18–26 per ovary; style (0.01–)0.1–0.4(–0.5) mm, (glabrous, stigma about as wide as style). **Seeds** oblong, 0.9–1.2 × 0.5–0.7 mm. $2n = 48$.

Flowering Jun–Aug. Rock outcrops and sea cliffs, talus, gravelly and sandy terraces, moist bluffs, turfy limestone shores; 0–500 m; Greenland; Nfld. and Labr., N.W.T., N.S., Que.; Europe (Finland, Norway, n Russia, Scotland, Sweden); Atlantic Islands (Iceland).

Draba norvegica is a highly variable hexaploid taxon and, as delimited herein, it probably represents two taxa with the same distribution and same chromosome number. It is related to hexaploid *D. arctogena*, from which it is distinguished by characteristics listed under 8. *D. arctogena*.

Draba norvegica was divided by O. E. Schulz (1927) into eight varieties, whereas R. C. Rollins (1993) recognized just two. In the latter treatment, var. *clivicola* is distinguished by having predominantly branched, appressed trichomes on stems proximally, lanceolate, glabrous fruits, and slender, erect-appressed pedicels. By contrast, var. *norvegica* is said to have predominantly simple, spreading trichomes on stems proximally, narrowly ovate to elliptical, glabrous or pubescent fruits, and stout, divaricately ascending pedicels. These differences are extremes of a continuum, and we agree with G. A. Mulligan (1970) in reducing *D. clivicola* to synonymy of *D. norvegica*. Rollins indicated that *D. norvegica* grows in Minnesota, but we have not seen any material from that state.

Draba norvegica is found in Europe in arctic and subarctic areas.

70. **Draba novolympica** Payson & H. St. John, Proc. Biol. Soc. Wash. 43: 113. 1930 E

Draba barbata Pohle var. *treleasei* O. E. Schulz; *D. paysonii* J. F. Macbride var. *treleasei* (O. E. Schulz) C. L. Hitchcock

Perennials; (cespitose, densely pulvinate); caudex branched (covered with persistent leaves, branches creeping, sometimes terminating in sterile rosettes); scapose. **Stems** unbranched, 0.05–0.4 dm, densely pubescent throughout, trichomes simple, 0.4–0.8 mm, and stalked, 2–5-rayed, 0.1–0.5 mm. **Basal leaves** (densely imbricate); rosulate; sessile; blade oblong to linear-oblanceolate, 0.2–0.8 cm × 0.5–1.5 mm, margins entire, (ciliate, trichomes simple and spurred, 0.3–1.2 mm), surfaces densely pubescent, abaxially with stalked, 2–12-rayed stellate trichomes, 0.1–0.6 mm, adaxially with simple and 2-rayed ones, 0.3–0.8 mm. **Cauline leaves** 0. **Racemes** 2–12-flowered, ebracteate, slightly elongated in fruit; rachis not flexuous, pubescent as stem. **Fruiting pedicels** divaricate-ascending, straight, 1–5 mm, pubescent, trichomes simple (0.3–0.9 mm) and stalked, 2–5-rayed (0.1–0.5 mm). **Flowers:** sepals oblong, 1.5–2.5 mm, pubescent, (trichomes simple and short-stalked, 2–4-rayed); petals bright yellow, oblanceolate to spatulate, 2–3.5(–4) × 1.5–2 mm; anthers oblong, 0.5–0.6 mm. **Fruits** often ovoid, plane, slightly inflated basally (symmetric), (2.5–)3–4(–5) × 1.5–3.5 mm; valves densely pubescent, trichomes 2–6-rayed, 0.05–0.4 mm, occasionally some simple; ovules 4–8(–12) per ovary; style 0.2–0.6(–0.8) mm. **Seeds** oblong, 1.2–1.8 × 0.8–1.1 mm. $2n = 42$.

Flowering Jun–Aug. Alpine crests, open knolls, fellfields, talus, weathered shale, calcareous shale scree, rocky grounds and cliffs, subalpine conifer forests; 1500–3700 m; Alta., B.C.; Calif., Idaho, Mont., Nev., Oreg., Utah, Wash., Wyo.

Draba novolympica is the same taxon that C. L. Hitchcock (1941) and R. C. Rollins (1993) called *D. paysonii* var. *treleasei*, and G. A. Mulligan (2002) called *D. paysonii*. The two are amply distinct and should be recognized as separate species. *Draba novolympica* is easily distinguished from *D. paysonii* by having fruit valves pubescent with 2–6-rayed (occasionally some simple) trichomes 0.05–0.4 mm, sepals 1.5–2.5 mm, petals 2–3.5(–4) × 1.5–2 mm, fruits (2.5–)3–4(–5) × 1.5–3.5 mm, styles 0.2–0.6(–0.8) mm, and ovules 1.2–1.8 × 0.8–1.1 mm. By contrast, *D. paysonii* has fruit valves pubescent with simple and 2-rayed (some 4- or 5-rayed) trichomes (0.2–)0.4–1 mm, sepals 2.8–3.5 mm, petals (4–)5–6 × (1.5–)2–3 mm, fruits (5–)6–9 × (3–)3.5–5 mm, styles (0.6–)0.8–1.2 mm, and ovules 1.7–2.2 × 1–1.4 mm.

Both R. C. Rollins (1993) and N. H. Holmgren (2005b) indicated that *Draba novolympica* (as *D. paysonii* var. *treleasei*) occurs in Alaska and Yukon, but we have not seen any material from there, and it is likely that their records were based on misidentified plants. Previous reports of *D. paysonii* from Canada (e.g., G. A. Mulligan 1971b) pertain instead to *D. novolympica*.

71. **Draba oblongata** R. Brown ex de Candolle, Syst. Nat. 2: 342. 1821

Draba arctica J. Vahl subsp. *groenlandica* (E. Ekman) Böcher; *D. cinerea* Adams subsp. *groenlandica* (E. Ekman) Böcher; *D. groenlandica* E. Ekman

Perennials; (cespitose); caudex branched (with persistent leaf remains); sometimes scapose. **Stems** unbranched, (0.2–)0.4–1.5(–1.9) dm, pubescent throughout, trichomes simple, 0.4–0.8 mm, and short-stalked, 4–10-rayed, 0.05–0.3 mm, (mostly branched distally). **Basal leaves** rosulate; petiolate; petiole base and margin ciliate, (trichomes simple, 0.3–1 mm); blade oblanceolate to lanceolate, 0.4–1.5 cm × 1.5–5 mm, margins entire, surfaces pubescent with simple and stalked, 2- or 3-rayed trichomes, 0.4–1 mm, with short-stalked, 8–12-rayed, stellate ones, 0.1–0.2 mm. **Cauline leaves** 0–2; sessile; blade oblong to broadly ovate, margins entire, surfaces pubescent as basal. **Racemes** 5–13(–18)-flowered, ebracteate, elongated in fruit; rachis not flexuous, pubescent as stem distally. **Fruiting pedicels** divaricate to ascending, straight, 2–6(–8) mm, pubescent as rachis. **Flowers:** sepals ovate, 2–3 mm, pubescent, (trichomes simple and short-stalked, mostly 2-rayed); petals white, spatulate to obovate, 3.5–5 × 1.5–2.5 mm; anthers ovate, 0.2–0.3 mm. **Fruits** elliptic to oblong, plane, slightly flattened, 4–8(–9.5) × 2–3(–4) mm; valves pubescent, trichomes short-stalked, (2–)5–12-rayed, 0.05–3 mm; ovules 24–36 per ovary; style 0.2–0.8 mm (stigma considerably wider than style, often appearing 2-lobed). **Seeds** ovate, 0.7–1 × 0.5–0.6 mm. $2n = 64$.

Flowering Jun–Jul. Ridges and hillsides, sand and gravel flood plains, swales, tundra; 0–400 m; Greenland; N.W.T., Nunavut; Alaska; e Asia (Russian Far East, Siberia).

Plants of *Draba oblongata* are occasionally misidentified as decaploid ($2n = 80$) *D. arctica*, but the former is readily separated by a preponderance of simple (versus stellate) trichomes on both leaf blade surfaces. *Draba oblongata* is sometimes confused with *D. micropetala*, but the latter has pale yellow (versus white) flowers, fruits pubescent with simple and spurred (versus (2–)5–12-rayed) trichomes, and stigmas as wide as (versus much wider than) the style.

72. Draba ogilviensis Hultén, Bot. Not. 119: 315, fig. 2. 1966 E

Perennials; (stoloniferous); caudex branched (sparsely covered with petiole remains, branches slender, creeping, loosely matted); often scapose. **Stems** unbranched, 0.4–1.5 dm, glabrous or pubescent throughout, trichomes simple and short-stalked, 2- or 3-rayed, 0.1–0.5 mm. **Basal leaves** not rosulate; (subopposite); petiolate; petiole base and margin ciliate or not, (trichomes simple, 0.2–0.8 mm); blade oblanceolate or lanceolate, 0.5–1.5 cm × 1.5–5 mm, margins entire, surfaces glabrous or sparsely pubescent with simple and stalked, 2–4-rayed trichomes, 0.1–0.5 mm, (midvein obscure abaxially). **Cauline leaves** (0 or) 1 or 2; (subopposite); sessile; blade ovate or oblong, margins entire, surfaces pubescent as basal. **Racemes** 5–13-flowered, ebracteate, elongated in fruit; rachis not or slightly flexuous, glabrous or pubescent as stem. **Fruiting pedicels** horizontal to divaricate, often curved upward, 6–13(–17) mm, glabrous or sparsely pubescent, trichomes simple and 2-rayed. **Flowers:** sepals ovate, 2–3 mm, glabrous or pubescent, (trichomes simple); petals golden yellow, obovate, 3.5–6 × 2–3 mm; anthers ovate, 0.3–0.5 mm. **Fruits** oblong, plane, flattened, 6–9 × 2–3 mm; valves glabrous; ovules 8–20 per ovary; style 0.4–1 mm. **Seeds** ovoid, 0.9–1 × ca. 0.6 mm. $2n = 16$.

Flowering Jun–Aug. Tundra, river flats and banks, exposed talus slopes, hummocks in wet sedge meadows; 900–2200 m; N.W.T., Yukon; Alaska.

The limits of *Draba ogilviensis* were confused by authors who reduced it to synonymy of *D. juvenilis* or *D. sibirica*. A thorough discussion of the three species and their distinguishing characteristics was provided by D. F. Murray and C. L. Parker (1999).

73. Draba oligosperma Hooker, Fl. Bor.-Amer. 1: 51. 1830 E F

Draba calcifuga Lesica; *D. oligosperma* var. *andina* Nuttall; *D. oligosperma* var. *leiocarpa* O. E. Schulz; *D. oligosperma* var. *microcarpa* Blankinship; *D. oligosperma* var. *saximontana* (A. Nelson) O. E. Schulz; *D. oligosperma* subsp. *subsessilis* (S. Watson) A. E. Murray; *D. oligosperma* var. *subsessilis* (S. Watson) O. E. Schulz; *D. saximontana* A. Nelson; *D. subsessilis* S. Watson

Perennials; (cespitose, densely pulvinate); caudex branched (with persistent leaf bases, branches congested or somewhat creeping, sometimes terminating in sterile rosettes); scapose. **Stems** unbranched, (0.1–)0.2–0.6 (–1) dm, glabrous throughout or pubescent, trichomes sessile, pectinate, 0.1–0.3 mm, (their length parallel to long axis of stem). **Basal leaves** rosulate; sessile; blade linear to linear-oblanceolate, (0.2–)0.4–1.1(–1.5) cm × 0.4–1.5(–1.8) mm, margins entire, (not or, rarely, ciliate, trichomes simple, 0.2–0.4 mm), surfaces pubescent with sessile, pectinate trichomes, 0.1–0.4 mm, (their long axis parallel to prominent abaxial midvein), sometimes glabrous adaxially. **Cauline leaves** 0. **Racemes** 4–12(–17)-flowered, ebracteate, elongated in fruit; rachis not flexuous, glabrous or pubescent as stem. **Fruiting pedicels** divaricate-ascending, straight, (2–)3–10 (–13) mm, glabrous or sparsely pubescent, trichomes pectinate. **Flowers:** sepals ovate, 1.5–3 mm, glabrous or sparsely pubescent, (trichomes pectinate, 2-rayed, or simple); petals usually yellow, rarely creamy white, obovate, 2.5–4 × 1.5–3 mm; anthers ovate, 0.4–0.5 mm, (not producing pollen). **Fruits** ovoid to lanceolate, plane, inflated at least basally, sometimes slightly flattened distally, 3–6(–7) × 2–3.5(–4) mm; valves usually puberulent, rarely glabrous, trichomes simple and sessile, often unequally 2-rayed, 0.07–0.35 mm; ovules 6–12 per ovary; style 0.1–0.8(–1.1) mm. **Seeds** ovoid, 1.1–1.5 × 0.7–1 mm. $2n = 32, 64$.

Flowering May–Jul. Rock outcrops, talus, gravel benches, tundra; 200–3900 m; Alta., B.C., N.W.T., Nunavut, Yukon; Alaska, Calif., Colo., Idaho, Mont., Nev., Oreg., Wyo.

Draba oligosperma is a highly variable and widespread species that has been shown to be apomictic (G. A. Mulligan and J. N. Findlay 1970; Mulligan 1972). It has been divided into species and infraspecific taxa by previous authors; the variation is continuous in every character; there are no clear geographical and morphological patterns that support its division. For characteristics separating *D. oligosperma* from the closely related *D. pectinipila*, see 80. *D. pectinipila*.

Draba andina (Nuttall) A. Nelson (1899), not Philippi (1858) is an illegitimate name, sometimes found in synonymy under *D. oligosperma*.

74. **Draba oreibata** J. F. Macbride & Payson, Amer. J. Bot. 4: 257. 1917 [E]

Perennials; (cespitose); caudex branched (with persistent leaves, branches sometimes terminating in sterile rosettes); scapose. **Stems** unbranched, 0.2–0.9 dm, glabrous throughout. **Basal leaves** rosulate; sessile; blade oblong to narrowly oblanceolate, 0.2–1 cm × 0.6–2(–4) mm, margins entire, (ciliate, trichomes simple, 0.1–0.5 mm, apex obtuse), surfaces glabrous. **Cauline leaves** 0. **Racemes** 3–8-flowered, ebracteate, considerably elongated in fruit; rachis slightly flexuous, glabrous. **Fruiting pedicels** divaricate-ascending to ascending, straight, 4–13 mm, glabrous. **Flowers:** sepals oblong, 1.7–2.5 mm, glabrous; petals white, oblanceolate to obovate, 3.5–4.5 × 1.2–2 mm; anthers ovate, 0.3–0.4 mm. **Fruits** oblong to narrowly so, slightly twisted, flattened, 5–9 × 1.5–3 mm; valves glabrous; ovules 8–16 per ovary; style 0.3–0.9(–1.2) mm. **Seeds** oblong, 0.9–1.3 × 0.6–0.8 mm. $2n = 32$.

Flowering May–Jul. Limestone cliffs, talus; 1800–2700 m; Idaho.

R. C. Rollins (1993) and N. H. Holmgren (2005b) divided *Draba oreibata* into two varieties: the Idaho endemic var. *oreibata* and the Nevada endemic var. *serpentina*. As indicated by I. A. Al-Shehbaz and M. D. Windham (2007), the two taxa are morphologically distinct, have different chromosome numbers, and are separated by over 480 kilometers. They are treated herein as separate species, and distinguished by characteristics discussed by Al-Shehbaz and Windham. *Draba oreibata*, in the strict sense, is known from Blaine, Butte, Clark, Custer, and Lemhi counties.

75. **Draba oxycarpa** Sommerfelt, Mag. Naturvidensk., n. s. 1: 240. 1833

Draba alpina Linnaeus var. *oxycarpa* (Sommerfelt) Th. Fries; *D. gredinii* E. Ekman

Perennials; (cespitose); caudex branched (covered with persistent petiole remains); scapose. **Stems** unbranched, 0.2–1.1 dm, pubescent proximally, trichomes short-stalked, 2–5-rayed, 0.1–0.5 mm, (some rays spurred). **Basal leaves** rosulate; petiolate; petiole base and margin ciliate, (trichomes simple, 0.3–1 mm); blade oblanceolate to lanceolate, 0.8–2.2 cm × 2–6 mm, margins often entire, surfaces pubescent abaxially with stalked, 2–5-rayed, stellate trichomes, 0.1–0.4 mm, often some rays spurred and trichomes appearing subdendritic, rarely glabrate, or with simple and 2-rayed ones, (midvein obscure abaxially, not thickened), adaxially glabrous or sparsely pubescent with simple trichomes, sometimes with long-stalked, 2- or 3-rayed ones. **Cauline leaves** 0. **Racemes** (2–)4–13-flowered, ebracteate, (subcorymbose), not elongated in fruit; rachis not flexuous, pubescent as stem. **Fruiting pedicels** subhorizontal to divaricate-ascending (not appressed to rachis), often slightly curved upward, 4–7(–10) mm, pubescent, trichomes simple (0.4–1 mm), and 2–4-rayed (0.1–0.4 mm). **Flowers:** sepals (grayish white), ovate to broadly so, 2.5–3 mm, pubescent, (trichomes simple and stalked, 2-rayed); petals creamy white to yellow, broadly obovate, 3.5–5 × 2.5–3.5 mm; anthers ovate, 0.4–0.5 mm. **Fruits** ovate to ovate-oblong, plane, flattened, 5–10 × 3–5 mm; valves puberulent along replum, trichomes simple, 0.1–0.3 mm; ovules (16–)18–28 per ovary; style 0.2–0.6 mm (stigma as wide as style). **Seeds** (black), ovoid, 1–1.3 × 0.6–0.8 mm. $2n = 64$.

Flowering Jun–Jul. Rocky soils; Greenland; Europe (Norway [including Spitsbergen], Sweden); Atlantic Islands (Iceland).

Draba oxycarpa is an octoploid that most closely resembles the decaploid *D. alpina*. It is distinguished from that species by having creamy white to pale yellow (versus bright yellow) petals, gray-green (versus purplish tinged) sepals, petals 2.5–3.5 (versus 1.7–2.5) mm wide, stems pubescent proximally with primarily branched (versus primarily simple) trichomes, nearly black (versus pale brown) seeds, and ovate to ovate-oblong (versus elliptic) fruits 3–5 (versus 2–3) mm wide.

76. Draba palanderiana Kjellman in A. E. Nordenskiöld, Vega Exp. Vetensk. Iakttag. 2: 45. 1883

Perennials; (cespitose); caudex branched (with persistent leaf remains, branches sometimes terminating in sterile rosettes); scapose. **Stems** unbranched, 0.2–1.1(–1.5) dm, pubescent proximally, trichomes 2–9-rayed, 0.05–0.2 mm, glabrous or sparsely pubescent distally. **Basal leaves** rosulate; petiolate; petiole (thickened), ciliate proximally, (margin not ciliate, trichomes simple and 2-rayed, 0.2–0.5 mm); blade oblanceolate to obovate, 0.3–0.9(–1.5) cm × 1–4 mm, margins usually entire, rarely denticulate, surfaces pubescent with minutely stalked, 8–16-rayed, stellate trichomes, 0.08–0.3 mm, (midvein obscure abaxially). **Cauline leaves** 0. **Racemes** 5–17-flowered, ebracteate, elongated in fruit; rachis not flexuous, usually glabrous, rarely sparsely pubescent as stem. **Fruiting pedicels** divaricate-ascending, straight, (slender), 4–10 mm, usually glabrous, rarely sparsely pubescent as stem. **Flowers:** sepals ovate, 2–2.5 mm, pubescent, (trichomes simple and short-stalked, 2–6-rayed); petals white or creamy white, obovate, 4.5–5.5 × 2–3 mm; anthers ovate, 0.4–0.6 mm. **Fruits** (often aborting, becoming lopsided), oblong to elliptic or ovate, plane, flattened, 4–8 × 2–3 mm; valves glabrous; ovules 18–32 per ovary; style 0.3 0.8 mm. **Seeds** ovoid, 0.9–1.1 × 0.5–0.7 mm. $2n$ = 16, 32, 64.

Flowering Jun–Jul. Rock outcrops, talus, *Dryas* fellfields, tundra; 0–1800 m; N.W.T., Yukon; Alaska; e Asia (Russian Far East, Siberia).

North American plants of *Draba palanderiana* are diploid and appear to be self-incompatible, an unusual situation among arctic members of the genus. By contrast, some plants from the Russian Far East are tetraploid ($2n$ = 32) or octoploid ($2n$ = 64) and appear to be self-compatible. Detailed studies are needed to establish whether one or two taxa are involved.

Many North American collections of *Draba palanderiana* have been misidentified as *D. nivalis*. The species is easily separated from *D. nivalis* by having ciliate (versus non-ciliate) bases of basal leaves, larger petals (4.5–5.5 × 2–3 versus 2–3.5 × 0.8–1.4 mm) that are pale yellow or cream (versus white), and longer fruiting pedicels 4–10 [versus 1–4.5(–8)] mm.

77. Draba pauciflora R. Brown, Chlor. Melvill., 266. 1823

Draba adamsii Ledebour

Perennials; caudex branched (with persistent, thickened leaf midveins); scapose. **Stems** unbranched, (0.05–)0.1–0.8 dm, pubescent throughout, sometimes sparsely so distally, trichomes simple, 0.3–0.7 mm, and 2–4-rayed, 0.05–0.2 mm. **Basal leaves** (not imbricate); rosulate; petiolate; petiole ciliate, (trichomes simple and 2-rayed, 0.4–1.3 mm); blade oblanceolate, 0.5–1.1 cm × 1.5–4 mm, margins entire, (pubescent as petiole, apex acute to subacute, trichomes simple and/or branched), surfaces pubescent, abaxially with simple trichomes, to 1 mm, and stalked, 2–4-rayed ones, 0.1–0.5 mm, adaxially with simple trichomes, 0.4–1 mm, (sometimes glabrous). **Cauline leaves** 0. **Racemes** 2–8-flowered (congested), ebracteate, slightly elongated in fruit; rachis not flexuous, pubescent as stem. **Fruiting pedicels** divaricate to ascending, straight or slightly curved upward, 1.5–4 mm, pubescent as stem or trichomes branched. **Flowers:** sepals ovate, 1.8–2.3 mm, pubescent, (trichomes simple with fewer, short-stalked ones); petals pale yellow, narrowly spatulate to oblanceolate, 2.5–3 × 0.8–1.5 mm; anthers ovate, 0.2–0.3 mm. **Fruits** often obovate, plane, slightly flattened, 5–10 × (3–)3.5–5 mm; valves glabrate or sparsely pubescent, 0.05–0.2 mm; ovules 8–16(–20) per ovary; style 0.05–0.15 mm. **Seeds** ovoid, 1.1–1.6 × 0.7–1 mm. $2n$ = 32.

Flowering Jun–Aug. Damp rocky slopes, tundra, swales, dry silt plains; 0–1000 m; Greenland; N.W.T., Nunavut; Alaska; Europe (Norway [Svalbard], Russia); e Asia (Russian Far East, n Siberia).

O. E. Schulz (1927) recognized five varieties within *Draba pauciflora*, of which four were listed from North America. Schulz's concept of *D. pauciflora* encompassed multiple taxa that we recognize as separate species, including *D. micropetala* and *D. subcapitata*. C. L. Hitchcock (1941) and R. C. Rollins (1993) did not mention *D. pauciflora*; the latter (and G. A. Mulligan 1974b) referred the North American material to *D. adamsii*.

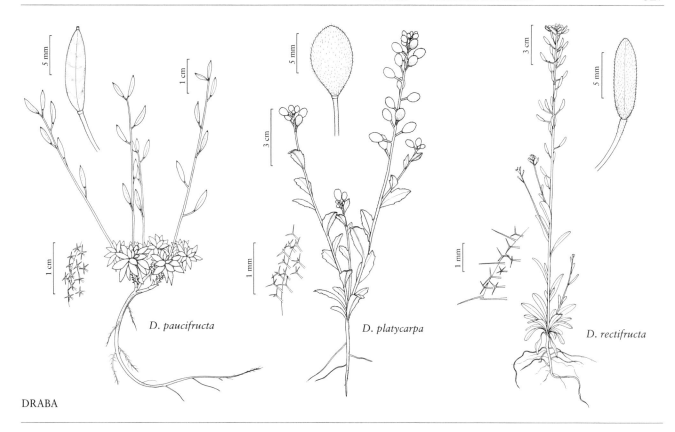

DRABA

78. **Draba paucifructa** Clokey & C. L. Hitchcock, Madroño 5: 127. 1939 (as paucifructus) [C][E][F]

Perennials; caudex simple or branched (not fleshy, usually with some persistent leaf remains); sometimes scapose. **Stems** unbranched, (0.1–)0.3–0.9 dm, glabrous throughout or sparsely pubescent proximally, trichomes simple and 2–6-rayed, 0.1–0.5 mm. **Basal leaves** rosulate; petiolate; petiole (0–0.5 cm), ciliate throughout; blade oblanceolate to obovate, 0.4–1.7(–2.3) cm × 1.6–4(–6) mm, margins usually entire, rarely denticulate, (sparsely ciliate, trichomes simple, 0.3–0.8 mm), surfaces pubescent, abaxially with stalked, (2–)4(–6)-rayed trichomes, (0.05–)0.1–0.6 mm, adaxially with simple and fewer, stalked, 2–4-rayed ones. **Cauline leaves** 0 or 1; sessile; blade ovate to lanceolate, margins entire, surfaces pubescent as basal. **Racemes** (2–)4–9(–12)-flowered, ebracteate, elongated in fruit; rachis slightly flexuous, glabrous. **Fruiting pedicels** divaricate-ascending, straight, 2–5(–8) mm, glabrous. **Flowers:** sepals ovate, 1.2–1.7 mm, pubescent, (trichomes simple and short-stalked, 2- or 3-rayed); petals yellowish (quickly fading white), spatulate, 1.8–2.2 × 0.6–1 mm; anthers ovate, 0.25–0.3 mm. **Fruits** elliptic to linear lanceolate, plane, flattened, 5–10(–12) × 1.7–2(–2.5) mm; valves glabrous; ovules 20–30 per ovary; style 0.08–0.2 mm. **Seeds** oblong, 0.9–1.2 × 0.5–0.8 mm. $2n = 40$.

Flowering Jun–Jul. Moist, shaded slopes among limestone and dolomite rocks; of conservation concern; 2600–3500 m; Nev.

Draba paucifructa is known from the Charleston Mountains in Clark County. M. D. Windham (2004) suggested that it is an allopolyploid resulting from hybridization between *D. albertina* and *D. lonchocarpa*, and he discussed the morphological features that distinguish this species from its putative parents.

79. **Draba paysonii** J. F. Macbride, Contr. Gray Herb. 56: 52. 1918 [E]

Draba vestita Payson, Amer. J. Bot. 4: 261. [May] 1917, not Davidson [Jan] 1917

Perennials; (cespitose, densely pulvinate); caudex branched (covered with persistent leaves, branches creeping, sometimes terminating in sterile rosettes); scapose. **Stems** unbranched, 0.05–0.3 dm, densely pubescent throughout, trichomes simple, 0.4–1.3 mm, and stalked, 2–5-rayed, stellate, 0.1–0.6 mm. **Basal leaves** (imbricate); rosulate; sessile; blade linear to linear-oblanceolate, 0.4–1.2 cm ×

0.8–1.5 mm, margins entire, (ciliate, trichomes simple and spurred, (0.4–)0.6–1.4 mm), surfaces pubescent, abaxially (dense) with stalked, 2–12-rayed trichomes, 0.1–0.8 mm, adaxially with simple and 2-rayed ones, 0.3–1 mm, (midvein prominent). **Cauline leaves** 0. **Racemes** 2–7-flowered, ebracteate, slightly elongated in fruit; rachis not flexuous, pubescent as stem. **Fruiting pedicels** divaricate-ascending, straight, 3–9 mm, pubescent, trichomes simple (0.4–1.4 mm), and stalked, 2–5-rayed (0.1–0.6 mm). **Flowers:** sepals oblong, 2.8–3.5 mm, pubescent, (trichomes simple and short-stalked, 2–4-rayed); petals pale yellow, oblanceolate to spatulate, (4–)5–6 × (1.5–)2–3 mm; anthers oblong, 0.5–0.7 mm. **Fruits** often broadly ovoid to lanceolate, plane, slightly inflated basally, (5–)6–9 × (3–)3.7–5 mm; valves densely pubescent, trichomes simple and 2-rayed, (0.2–)0.4–1 mm, (occasionally with fewer, smaller, 4- or 5-rayed ones); ovules 8–12 per ovary; style (0.6–)0.8–1.2 mm. **Seeds** oblong, 1.7–2.2 × 1–1.4 mm.

Flowering Jun–Jul. Limestone rock outcrops, talus, gravelly calcareous soil; 1800–3500 m; Mont., Wyo.

Draba paysonii is known from southwestern Montana (Gallatin, Glacier, Madison, and Meagher counties) and northwestern Wyoming (Fremont and Park counties). Reports of its occurrence in Alberta (e.g., G. A. Mulligan 1971b) are based on plants of *D. novolympica*, a species formerly treated as *D. paysonii* var. *treleasei*. Features distinguishing these two species are discussed under 70. *D. novolympica*.

80. **Draba pectinipila** Rollins, Rhodora 55: 231. 1953 [E]

Draba juniperina Dorn;
D. oligosperma Hooker var. *juniperina* (Dorn) S. L. Welsh;
D. oligosperma var. *pectinipila* (Rollins) C. L. Hitchcock

Perennials; (cespitose, sometimes forming mats); caudex branched (with persistent leaf bases, branches creeping, terminating in scapes or sterile rosettes); scapose. **Stems** unbranched, (0.3–)0.4–1.6(–1.9) dm, pubescent, trichomes sessile, pectinate, 0.1–0.4 mm, (parallel to long axis of stem, sometimes with irregularly 2–4-rayed ones, 0.2–0.6 mm). **Basal leaves** rosulate; subsessile; blade narrowly oblanceolate to linear, (0.4–)0.6–1.3(–2) cm × 0.9–2.2 mm, margins entire (not ciliate), surfaces pubescent with subsessile or sessile, pectinate trichomes, 0.2–0.5 mm. **Cauline leaves** 0. **Racemes** 5–22-flowered, ebracteate, considerably elongated in fruit; rachis not flexuous, pubescent as stem. **Fruiting pedicels** divaricate-ascending, straight, (5–)7–14 mm, sparsely pubescent, trichomes pectinate. **Flowers:** sepals broadly ovate, 2–3.2 mm, pubescent, (trichomes pectinate); petals yellow, obovate to spatulate, 4–6.5 × 1.5–2.5 mm; anthers ovate 0.4–0.5 mm. **Fruits** ovoid, plane, slightly inflated basally, 4–6(–7) × 2–3 mm; valves pubescent, trichomes usually sessile, pectinate, 0.2–0.5 mm, rarely with simple ones; ovules 4–8 per ovary; style 0.5–1.5 mm. **Seeds** oblong, 1.1–1.5 × 0.6–0.8 mm. $2n = 22$.

Flowering Apr–Jun. Rocky slopes in sagebrush scrub and pinyon-juniper woodlands; 1700–2400 m; Colo., Mont., Utah, Wyo.

Draba pectinipila was treated as a synonym of *D. oligosperma* by G. A. Mulligan (1972), R. C. Rollins (1993), and N. H. Holmgren (2005b). The latter recognized *D. juniperina* as a distinct species but overlooked the fact that it is indistinguishable from the type collections of the earlier-published *D. pectinipila*. The species (including *D. juniperina*) differs significantly from *D. oligosperma* in both chromosome number and morphology. *Draba pectinipila* is easily distinguished by having fruit valves pubescent with pectinate trichomes, fruiting pedicels (5–)7–14 mm, petals 4–6.5 mm, ovules 4–8 per ovary, and styles 0.5–1.5 mm. By contrast, *D. oligosperma* has fruit valves glabrous or pubescent with simple or 2-rayed trichomes, fruiting pedicels (2–)3–10(–13) mm, petals 2.5–4 mm, ovules 6–12 per ovary, and styles 0.1–0.8(–1.1) mm. *Draba pectinipila* was previously known only from the type locality in northwestern Wyoming (Park County). Its range is now expanded to include that of *D. juniperina* in northwestern Colorado (Moffat County), northeastern Utah (Daggett and Uintah counties), and southwestern Wyoming (Sweetwater County). The record from Uintah County is based on *Goodrich 22275* (NY).

81. **Draba pedicellata** (Rollins & R. A. Price) Windham, Madroño 50: 221. 2004 [E]

Draba cusickii B. L. Robinson ex O. E. Schulz var. *pedicellata* Rollins & R. A. Price, Harvard Pap. Bot. 1(3): 73. 1991;
D. pedicellata var. *wheelerensis* N. H. Holmgren

Perennials; (loosely cespitose); caudex branched (with persistent leaf bases, branches sometimes terminating in sterile rosettes); scapose. **Stems** unbranched, (0.15–)0.3–1.3(–1.6) dm, pubescent proximally, glabrous or nearly so distally, trichomes 2- or 3-rayed and simple, 0.1–1 mm. **Basal leaves** rosulate; subsessile; blade oblanceolate to obovate, 0.4–1.6(–2) cm × 2–4(–6) mm, margins entire, (not ciliate), surfaces pubescent with stalked, 2–4-rayed trichomes, 0.2–0.9 mm, rarely adaxially with simple trichomes, (midvein obscure abaxially). **Cauline leaves** 0. **Racemes** 3–20(–27)-flowered, ebracteate, slightly to considerably elongated in fruit; rachis straight or flexuous, usually glabrous, rarely pubescent, trichomes simple and 2–4-rayed. **Fruiting pedicels** divaricate to horizontal, straight or curved upward (expanded basally or not), 4–9(–13) mm, glabrous or pubescent, trichomes simple and 2–4-rayed. **Flowers:** sepals broadly ovate, 1.8–2.5 mm, glabrous or pubescent, (trichomes simple and branched); petals yellow, spatulate, 3.5–5.5 × 1.5–2.2 mm; anthers ovate, 0.4–0.5 mm. **Fruits** lanceolate to ellipsoid or ovoid, plane (not curved), inflated or slightly flattened, (4.2–)5.3–10(–13) × 2.5–4.5(–5.5) mm; valves glabrous or pubescent, trichomes simple and short-stalked, 2-rayed, 0.05–0.3 mm; ovules 8–16 per ovary; style 0.3–1(–1.2) mm. **Seeds** oblong, 1.2–1.6 × 0.7–0.9 mm. $2n = 20$.

Flowering May–Jul. Rock outcrops, talus, steep gravel slopes in pinyon-juniper, mixed conifer, and subalpine meadow communities; 2300–3800 m; Nev., Utah.

Draba pedicellata is known from Elko, Eureka, Nye, and White Pine counties in northeastern Nevada and to Tooele County in northwestern Utah. It was considered a variety of *D. cusickii* by R. C. Rollins (1993); chromosome number, morphology, and geographic distribution suggest a closer relationship to *D. sphaeroides*. For a detailed discussion and comparison of these species, consult M. D. Windham (2004). N. H. Holmgren (2005b) divided *D. pedicellata* into two varieties: var. *pedicellata*, which is found on limestone throughout the species range, and var. *wheelerensis*, which is found on non-calcareous substrates in White Pine County. Although var. *wheelerensis* has slightly more spreading pedicels on somewhat more flexuous rachises, it appears to be little more than a dwarf alpine form of the species.

82. **Draba pennellii** Rollins, J. Arnold Arbor. 64: 502. 1983 [C] [E]

Perennials; (cespitose, often forming mats); caudex branched (elongated, with persistent leaf bases, branches sometimes terminating in sterile rosettes); sometimes scapose. **Stems** unbranched, (0.15–)0.3–0.7(–1) dm, pubescent proximally, trichomes simple and spurred, 0.4–0.8 mm, and short-stalked, 2–4-rayed, 0.1–0.4 mm, (sometimes with simple trichomes distally). **Basal leaves** subrosulate; sessile; blade oblanceolate to oblong, (0.3–)0.5–0.8 cm × 1–3 mm, (base ciliate, trichomes simple, 0.2–0.8 mm), margins entire, (not ciliate), surfaces usually pubescent, with stalked, (2–)4 (or 5)-rayed trichomes, 0.2–0.5 mm, adaxially sometimes glabrate, or primarily with simple trichomes, 0.3–0.8 mm. **Cauline leaves** (0 or) 1–4; sessile; blade ovate to oblong, margins entire, surfaces pubescent as basal. **Racemes** (3–)7–15(–20)-flowered, ebracteate, elongated in fruit; rachis not flexuous, pubescent as stem. **Fruiting pedicels** divaricate-ascending to ascending, straight, 3–10 mm, pubescent, trichomes simple and short-stalked, 2–4-rayed, (0.1–0.4 mm). **Flowers:** sepals ovate to oblong, 2–3 mm, pubescent, (trichomes simple, with fewer, short-stalked, 2-rayed ones); petals white, obovate, 3.5–6 × 1.5–2.8 mm; anthers oblong, 0.5–0.6 mm. **Fruits** broadly or, rarely, narrowly lanceolate to elliptic or ovate, plane or slightly twisted, flattened, (4)5–8(–10) × 2–3.2 mm; valves (each with distinct midvein), pubescent, trichomes simple and short-stalked, 2-rayed, 0.08–0.2 mm; ovules 12–24 per ovary; style (0.7–)1–1.8(–2.1) mm. **Seeds** oblong, 0.9–1.3 × 0.7–1 mm. $2n = 32$.

Flowering May–Jul. Rock outcrops and talus slopes in pinyon-juniper, sagebrush, mountain shrub, and mixed conifer communities; of conservation concern; 1900–3500 m; Nev.

Draba pennellii is known from White Pine County in east-central Nevada. It is easily distinguished from other white-flowered species in the region by its elongated, many-branched caudices, long styles, usually few-leaved or, rarely, leafless flowering stems, and distinct midvein on fruit valves.

83. **Draba petrophila** Greene, Pittonia 4: 17. 1899 [E]

Draba helleriana Greene var. *blumeri* C. L. Hitchcock; *D. helleriana* var. *petrophila* (Greene) O. E. Schulz

Perennials; (cespitose, long-lived); caudex simple or branched (covered with persistent petioles), not scapose. **Stems** usually unbranched, rarely branched, 0.3–1.9(–2.8) dm, moderately to densely hirsute throughout, trichomes simple, 0.2–1.3 mm, with short-stalked to subsessile, cruciform, 0.02–0.4 mm, and 2-rayed ones, 0.3–0.9 mm. **Basal leaves** rosulate; petiolate; petiole ciliate, (trichomes simple, 0.4–1.5 mm); blade oblanceolate, 1–5(–6) cm × 2–5(–10) mm, margins usually entire, rarely denticulate, surfaces pubescent, abaxially with stalked, cruciform trichomes, 0.07–0.5 mm, adaxially often similar, sometimes with fewer, simple and 2-rayed trichomes, 0.4–1.3 mm. **Cauline leaves** 3–10; sessile; blade ovate to lanceolate or oblong, margins entire or denticulate, surfaces pubescent as basal. **Racemes** 10–37(–58)-flowered, ebracteate, elongated in fruit; rachis not flexuous, pubescent as stem. **Fruiting pedicels** divaricate-ascending, straight, 4–10 mm, pubescent throughout, trichomes simple (0.1–0.5 mm) and subsessile, 2–4-rayed, (0.03–0.2 mm). **Flowers:** sepals oblong, 2–3.5 mm, pubescent, (trichomes simple and 2-rayed); petals yellow, oblanceolate, 3.5–6 × 1.2–1.8 mm; anthers oblong, 0.8–1 mm. **Fruits** lanceolate to elliptic, often strongly twisted, flattened, 5–11 × 2–3 mm; valves puberulent at least along margin, trichomes simple, antrorse, 0.03–0.15 mm; ovules 14–24 per ovary; style 0.8–1.8(–2.5) mm. **Seeds** ovoid, 1–1.4 × 0.6–0.9 mm.

Flowering Jul–Sep. Crevices, ledges of cliffs; 1200–2800 m; Ariz.

Draba petrophila is often broadly circumscribed to include the taxon herein called *D. viridis*. For a discussion of the differences between these species and the closely related *D. helleriana*, see I. A. Al-Shehbaz and M. D. Windham (2007). *Draba petrophila*, in the strict sense, is known only from Cochise, Pima, and Santa Cruz counties in southeastern Arizona.

84. **Draba pilosa** Adams ex de Candolle in A. P. de Candolle, Syst. Nat. 2: 336. 1821

Draba alpina Linnaeus var. *pilosa* (Adams ex de Candolle) O. E. Schulz; *D. aspera* Adams ex de Candolle

Perennials; (not stoloniferous); caudex branched (covered with persistent, dry leaves and midveins); scapose. **Stems** unbranched, 0.4–1.7 dm, glabrous or pubescent, trichomes simple, 0.2–0.9 mm, and fewer, 2–4-rayed, 0.1–0.6 mm. **Basal leaves** (imbricate); rosulate; petiolate; petiole (base thickened), ciliate throughout; blade linear to linear-oblanceolate, 0.5–1.5(–2.2) cm × 1–2.5(–4) mm, margins entire (thickened, ciliate, trichomes simple and 2-rayed, to 1.1 mm), surfaces pubescent, abaxially with simple trichomes, 0.3–1 mm, and 2–4-rayed ones, 0.1–0.5 mm, adaxially similar, or with only simple trichomes, (midvein prominent, thickened). **Cauline leaves** 0. **Racemes** (2–)4–12-flowered, ebracteate, usually elongated in fruit; rachis not flexuous, glabrous or pubescent as stem. **Fruiting pedicels** divaricate-ascending, usually straight, rarely curved upward, 6–13 mm, glabrous or pubescent as stem. **Flowers:** sepals ovate, 2–3.5 mm, pubescent, (trichomes simple and short-stalked, 2-rayed); petals pale to bright yellow, obovate, 3.5–6 × 2–3.5 mm; anthers ovate, 0.4–0.5 mm. **Fruits** elliptic to lanceolate, 5–11 × 3–4 mm, plane, flattened; valves glabrous or puberulent with simple and short-stalked, 2-rayed trichomes, 0.07–0.3 mm; ovules 12–20 per ovary; style 0.4–0.9 mm (stigma about as wide as style). **Seeds** ovoid, 1.2–1.4 × 0.8–1 mm.

Flowering Jul–Aug. Dry gravelly slopes, sandy places, wet tundra; 0–1300 m; N.W.T., Nunavut, Yukon; Alaska; e Asia (ne Russian Far East, n Siberia).

O. E. Schulz (1927) reduced *Draba pilosa* to a variety of the decaploid *D. alpina* and cited North American collections. The former species was not listed by C. L. Hitchcock (1941) or R. C. Rollins (1993). From *D. alpina*, *D. pilosa* is easily distinguished by having strongly thickened and persistent (versus not thickened) midveins and margins, and usually narrower leaf blades [1–2.5(–5) versus 2.5–6(–9) mm wide].

As recognized herein, *Draba pilosa* is broadly circumscribed to include perhaps two or three closely related taxa. All are scapose plants with large, yellow flowers and prominent, persistent midveins and petioles. Some Alaskan forms (e.g., *Parker 7596*, ALA), which grow in moist heath habitats, have leaves 3–5 mm wide, whereas the majority have narrower leaf blades rarely reaching 2.5 mm in width. Most plants assigned to this species have leaves with exclusively simple trichomes and often glabrous scapes. Others (e.g., *Chesemore & Davies 13*, *Murray 3371*, *Parker & Batten 8954*, all at ALA) have leaf blade surfaces and scapes with 2–4-rayed trichomes and blade margins ciliate with simple trichomes. One collection (*Walker s.n.*, ALA) has a mixture of plants of both trichome types but no intermediates were found. Detailed molecular and cytological studies are needed on this complex to determine if more than one taxon is represented.

85. **Draba platycarpa** Torrey & A. Gray, Fl. N. Amer. 1: 108. 1838 [E] [F]

Draba cuneifolia Nuttall ex Torrey & A. Gray var. *platycarpa* (Torrey & A. Gray) S. Watson; *D. viperensis* H. St. John

Annuals; scapose or subscapose. **Stems** often unbranched, sometimes branched proximally, (0.3–)0.5–3.2(–4) dm, densely hirsute proximally, trichomes simple, 0.9–1.5 mm, and stalked, 2–4-rayed, 0.1–0.6 mm, pubescent distally, trichomes usually branched, rarely simple ones present. **Basal leaves** rosulate; petiole obscure; blade obovate to spatulate or oblanceolate, 1–4.5 cm × 5–25(–30) mm, margins dentate, surfaces pubescent, abaxially with stalked, (2 or) 3 or 4-rayed trichomes, 0.1–0.6 mm, adaxially with simple, 0.7–1.3 mm, and stalked, 2–4-rayed ones, 0.1–0.5 mm. **Cauline leaves** (3 or) 4–8(–14); sessile; blade similar to basal. **Racemes** 12–45(–60)-flowered, ebracteate, elongated in fruit; rachis not flexuous, pubescent, trichomes 2–4-rayed. **Fruiting pedicels** horizontal to divaricate, straight, 4–9(–11) mm, pubescent as stem. **Flowers:** (mid- and late-season ones cleistogamous, apetalous); sepals oblong, 1.4–2.2 mm, pubescent; petals white, oblanceolate, (1.7–)2.5–3.5(–4) × 1–2 mm; anthers ovate, 0.2–0.35 mm (shorter in cleistogamous flowers).

Fruits obovate-oblong to obovate, plane, flattened, 5–8.2(–9.5) × (2.4–)2.8–3.7(–4) mm; valves pubescent, trichomes simple, antrorse, 0.2–0.5 mm; ovules (28–)40–70(–82) per ovary; style 0.01–0.2 mm. **Seeds** ovoid, 0.5–0.7 × 0.3–0.5 mm. $2n = 16, 32$.

Flowering Feb–Mar. Rocky hillsides, gravelly and sandy areas, dry plains, roadsides; 0–500 m; Ariz., Ark., Idaho, La., Okla., Oreg., Tex., Wash.

Draba platycarpa is occasionally treated as a variety of *D. cuneifolia*, but is amply distinct from that species (R. L. Hartman et al. 1975).

86. Draba porsildii G. A. Mulligan, Canad. J. Bot. 52: 1795, fig. 8. 1974 E

Draba nivalis Liljeblad var. *brevicula* Rollins; *D. porsildii* var. *brevicula* (Rollins) Rollins

Perennials; (cespitose); caudex branched (with persistent leaf remains); often scapose. **Stems** unbranched, 0.2–1.2 dm, sparsely pubescent proximally, trichomes short-stalked, 3–5-rayed, (non-crisped), 0.2–0.4 mm, glabrous distally. **Basal leaves** rosulate; subsessile or shortly petiolate; petiole base and margin ciliate proximally, (trichomes simple, subsetiform, stiff, 0.3–0.9 mm); blade oblanceolate to obovate, 0.4–1.5 cm × 1–3 mm, margins entire, surfaces pubescent, abaxially with stalked, 6–12-rayed, (non-crisped) trichomes, 0.2–0.5 mm, (midvein obscure), adaxially similar, or with simple trichomes on proximal ½. **Cauline leaves** 0 or 1; sessile; blade oblong to lanceolate, margins entire, surfaces pubescent as basal. **Racemes** 3–13-flowered, ebracteate, elongated in fruit; rachis not flexuous, glabrous. **Fruiting pedicels** divaricate-ascending, straight, 1–4(–8) mm, glabrous. **Flowers:** sepals ovate, 1.5–2 mm, pubescent, (trichomes simple and 2-rayed); petals white, spatulate, 2–4 × 1–1.7 mm; anthers ovate, ca. 0.2 mm. **Fruits** ovate to oblong, plane, flattened, 4–7.5 × 1.7–3 mm; valves glabrous; ovules 12–16 per ovary; style 0.1–0.3(–0.5) mm. **Seeds** ovoid, 0.7–1 × 0.5–0.6 mm. $2n = 32$.

Flowering Jun–Jul. Rock outcrops, talus, meadows, gravel slopes; 600–3000 m; Alta., Yukon; Alaska, Mont., Wyo.

According to H. H. Grundt et al. (2004), tetraploid *Draba porsildii* probably originated from a diploid ancestor such as *D. lonchocarpa*. The Wyoming plants are a close match to the type of the species, though they have distinctly shorter pedicels. Some of the Alaskan plants are more problematic; they have finer and more branched trichomes, larger flowers, longer pedicels, and longer styles than the type. It is likely that they belong to another species; detailed molecular and cytological studies on this complex are needed before any meaningful conclusions are reached.

87. Draba praealta Greene, Pittonia 3: 306. 1898

Draba cascadensis Payson & H. St. John; *D. columbiana* Rydberg; *D. dolichopoda* O. E. Schulz; *D. lapilutea* A. Nelson; *D. lonchocarpa* Rydberg var. *dasycarpa* O. E. Schulz; *D. praealta* var. *yellowstonensis* (A. Nelson) O. E. Schulz; *D. yellowstonensis* A. Nelson

Annuals or **perennials**; (short-lived); caudex simple or branched; not scapose. **Stems** branched or unbranched, (0.5–)0.8–3.2(–3.8) dm, pubescent throughout, trichomes 3–5-rayed, 0.1–0.4 mm, with simple and 2-rayed ones, 0.4–1 mm. **Basal leaves** rosulate; subsessile; petiole base ciliate, (trichomes simple and 2-rayed, 0.2–0.8 mm); blade oblanceolate, (0.7–)1–3.5(–4.5) cm × 1.5–7(–9) mm, margins usually dentate, rarely entire, surfaces pubescent, abaxially with stalked, 3–6-rayed trichomes, 0.1–0.5 mm, adaxially with stalked, 3–5-rayed, 0.1–0.6 mm, and simple ones, to 0.8 mm. **Cauline leaves** (1 or) 2–5(–9); sessile; blade ovate to lanceolate, margins dentate or entire, surfaces pubescent as basal. **Racemes** (5–)8–30(–37)-flowered, ebracteate, elongated in fruit; rachis not flexuous, pubescent as stem. **Fruiting pedicels** divaricate-ascending to ascending or suberect (not appressed to rachis), straight, (3–)4–10(–12) mm, pubescent, trichomes 3–5-rayed (0.1–0.4 mm), and sometimes simple. **Flowers:** (late-season ones petaliferous); sepals ovate, 1.7–2.5 mm, pubescent, (trichomes simple and stalked, 2-rayed); petals white, oblanceolate to spatulate, 2.8–3.5(–4) × 0.8–1.2 mm; anthers ovate, 0.2–0.3 mm. **Fruits** (erect, not appressed to rachis) lanceolate to linear-lanceolate, plane, flattened, (5–)7–12(–15) × 1.5–2.5(–3) mm; valves puberulent, trichomes simple and short-stalked, 2–4-rayed, 0.05–0.25 mm; ovules (24–)30–52 per ovary; style 0.03–0.1(0.15) mm. **Seeds** oblong, 0.8–1.1 × 0.5–0.6 mm, (often apiculate). $2n = 56$.

Flowering Jun–Aug. Rocky pine slopes, subalpine areas, alpine meadows, grassy volcanic ridges, woodlands, damp rocky areas, limestone talus, shale cliffs, steep hillsides; 1200–2900 m; Alta., B.C., N.W.T., Yukon; Calif., Idaho, Mont., Nev., Oreg., Wash., Wyo.; e Asia (Russian Far East [Chukchi Peninsula]).

Morphological and chromosomal evidence (M. D. Windham, unpubl.) suggests that *Draba praealta* may be an allopolyploid originating through hybridization between *D. albertina* and *D. cana*. Both R. C. Rollins (1993) and N. H. Holmgren (2005b) placed *D. lonchocarpa* var. *dasycarpa* in the synonymy of *D. lonchocarpa*. G. A. Mulligan (1974) indicated that the type of var. *dasycarpa* does not belong to that species and he correctly annotated its isotype (*Macoun 64454*, CAN) as *D. praealta*. Mulligan (1976) and Rollins indicated that *D. praealta* occurs in Alaska, but we have not seen unequivocal material from the state.

We have not seen the material on which A. E. Katenin and V. V. Petrovsky (1995) based their record from the Russian Far East.

88. **Draba pterosperma** Payson, Amer. J. Bot. 4: 266. 1917 [E]

Perennials; (densely cespitose to pulvinate); caudex branched (with persistent leaves, branches sometimes terminating in sterile rosettes); scapose. **Stems** unbranched, 0.3–1.1 dm, pubescent throughout, trichomes subdendritic, 2–7-rayed, (often crisped), 0.2–0.7 mm. **Basal leaves** (densely imbricate); rosulate; petiolate; petiole base and margin ciliate, (trichomes simple or branched, often crisped, 0.3–1 mm); blade oblanceolate to obovate or spatulate, 0.2–0.7 cm × 1–2.8 mm, margins entire, surfaces densely pubescent (often grayish) with stellate-dendritic, stalked, 5–12-rayed, (often crisped) trichomes, 0.2–0.6 mm. **Cauline leaves** 0. **Racemes** 4–12-flowered, ebracteate, elongated in fruit; rachis not flexuous, pubescent as stem. **Fruiting pedicels** erect (and subappressed to rachis) or ascending, straight, (3–)4–9(–12) mm, pubescent as stem. **Flowers:** sepals oblong, 3–4 mm, pubescent, (trichomes stalked, branched, crisped); petals yellow, oblanceolate, 6–7 × 1.5–2.5 mm; anthers oblong, 0.7–0.9 mm. **Fruits** broadly ovate to lanceolate or subelliptic, plane, flattened, 5–10(–13) × (2.5–)3.5–5(–6) mm; valves pubescent, trichomes short stalked, 2–7-rayed, (straight or crisped), 0.15–0.6 mm; ovules 8–12 per ovary; style 1.5–3(–3.8) mm. **Seeds** (winged), ovate, 1.6–3 × 1.3–1.8 mm; (wing 0.3–0.9 mm wide).

Flowering Jun–Sep. Limestone and marble outcrops, talus, gravel slopes; 1500–2500 m; Calif.

Draba pterosperma is a distinctive species that is restricted to the Marble Mountains of northern California (Siskiyou County).

89. **Draba ramosissima** Desvaux, J. Bot. Agric. 3: 186. 1815 [E]

Alyssum dentatum Nuttall; *Draba dentata* (Nuttall) Hooker & Arnott; *D. ramosissima* var. *glabrifolia* E. L. Braun

Perennials; caudex branched (well-developed, with persistent leaf bases); not scapose. **Stems** unbranched proximally, paniculately branched distally, 0.7–3.8(–5) dm, pubescent proximally, trichomes short-stalked, 2–4-rayed, 0.1–0.6 mm, sometimes with simple ones, 0.4–0.8 mm. **Basal leaves** rosulate; petiolate; petiole (to 2 cm), ciliate (trichomes simple, 0.3–0.5 mm); blade oblanceolate, (0.7–)1–5.5 cm × 2–8 mm, margins dentate, surfaces pubescent with short-stalked to subsessile, 3- or 4-rayed trichomes, 0.1–0.5 mm, adaxially sometimes also simple, 0.4–1.3 mm. **Cauline leaves** (5–)7–15(–21); sessile; blade ovate to oblong or elliptic to lanceolate, margins usually coarsely dentate or serrate, rarely subpinnatifid, surfaces pubescent as basal. **Racemes** 9–28-flowered, ebracteate, paniculate; rachis not flexuous, pubescent as stem. **Fruiting pedicels** horizontal to divaricate-ascending, straight, (4–)6–12(–15) mm, pubescent, trichomes 2–4-rayed. **Flowers:** sepals oblong, 2.5–3 mm, pubescent, (trichomes simple and subsessile, 2–4-rayed); petals white, spatulate, 4–6 × 1.5–2.5 mm; anthers oblong, 0.5–0.6 mm. **Fruits** narrowly oblong to elliptic or oblong-lanceolate, often strongly twisted, flattened, 5–11 × 1.5–2.2 mm; valves pubescent, trichomes subsessile or sessile, 2–4-rayed, 0.07–0.3 mm; ovules 4–10(–12) per ovary; style (1)1.5–3(–4) mm, (sparsely pubescent). **Seeds** oblong, 1.2–1.8 × 0.6–0.9 mm. $2n = 16$.

Flowering Apr–Jun. Limestone cliffs and outcrops, shale barrens, rocky wooded areas; 0–1100 m; D.C., Ky., Md., N.C., Tenn., Va., W.Va.

90. **Draba ramulosa** Rollins, Contr. Gray Herb. 214: 6. 1984 [C][E]

Perennials; (loosely matted, grayish); caudex branched (with persistent leaf bases, branches creeping, sometimes terminating in sterile rosettes); not scapose. **Stems** unbranched, 0.4–0.6 dm, densely pubescent throughout, trichomes dendritic, 3–6-rayed, (often crisped), 0.1–0.4 mm. **Basal leaves** (imbricate); not rosulate; sessile; blade obovate to oblanceolate, 0.4–1.1 cm × 2–3.2 mm, margins entire, (base and margins not ciliate), surfaces pubescent with stalked, 4–8-rayed trichomes, 0.1–0.5 mm, (sometimes 1 or more rays spurred), adaxially sometimes trichomes simple. **Cauline leaves** 2 or 3 (sometimes basal leaves spaced, flowering stem appearing to 8-leaved); sessile; blade oblong to ovate, margins entire, surfaces pubescent as basal. **Racemes** 4–15-flowered, ebracteate or proximalmost 1 or 2 flowers bracteate, slightly elongated in fruit; rachis not flexuous, pubescent as stem. **Fruiting pedicels** divaricate-ascending, usually straight, rarely curved upward, 3–6(–10) mm, pubescent, trichomes 3–6-rayed, (crisped, 0.1–0.4 mm), and, sometimes, simple. **Flowers:** sepals broadly ovate, 1.7–2.4 mm, pubescent, (trichomes short-stalked, 2–5-rayed); petals yellow, obovate to oblanceolate, 3–4.5 × 1.5–2.5 mm; anthers ovate, 0.4–0.5 mm. **Fruits** (not appressed to rachis), ovate to elliptic, plane, flattened, 4–6.5 × 2.5–4 mm; valves densely pubescent, trichomes simple and short-stalked, 2–5-rayed, 0.08–0.35 mm; ovules 6–12 per ovary; style (0.1–)0.3–0.7 mm. **Seeds** oblong, 1.4–1.8 × 0.8–1.2 mm.

Flowering Jun–Aug. Rock outcrops, talus, gravelly soils; of conservation concern; 3300–3600 m; Utah.

Molecular and chromosomal data (M. D. Windham, unpubl.) strongly suggest that *Draba ramulosa* is an allopolyploid species. It is thought to have originated through hybridization between *D. sobolifera* and a member of the white-flowered, euploid lineage of M. A. Beilstein and M. D. Windham (2003). It is easily distinguished from *D. sobolifera* by having pale yellow to whitish (versus bright yellow) petals, grayish (versus green) foliage, non-ciliate (versus ciliate) basal leaves pubescent with 4–8-rayed (versus 2–4-rayed) trichomes, often proximally bracteate (versus ebracteate) racemes, and flattened (versus inflated basally) fruits. *Draba ramulosa* is known from the Tushar Mountains in south-central Utah (Beaver and Piute counties).

91. Draba rectifructa C. L. Hitchcock, Revis. Drabas W. N. Amer., 110. 1941 E F

Draba montana S. Watson, Smithsonian Misc. Collect. 258: 60. 1878, not Bergeret 1786

Annuals; not scapose. **Stems** (usually simple, rarely more, from base), unbranched or branched distally, (0.5–)0.7–3 (–3.7) dm, pubescent, proximally trichomes simple and stalked, 2-rayed, 0.4–1.3 mm, with 3- or 4-rayed ones, 0.05–0.2 mm, distally trichomes stalked and 2–4-rayed. **Basal leaves** rosulate; petiole (obscure), margin ciliate, (trichomes simple, to 1.5 mm); blade oblanceolate to obovate, 1–2(–3) cm × 2–7(–10) mm, margins entire or subapically denticulate, surfaces pubescent, abaxially with simple, 0.6–1.3 mm, and stalked, 2–4-rayed trichomes, 0.1–0.8 mm, adaxially with simple and stalked, 2-rayed ones, 0.4–1.2 mm. **Cauline leaves** 3–10(–17); sessile; blade lanceolate to ovate or oblong, margins entire, surfaces pubescent as basal. **Racemes** (15–)20–51(–62)-flowered, ebracteate, elongated in fruit; rachis not flexuous, pubescent as stem. **Fruiting pedicels** horizontal to divaricate-ascending, straight or curved upward, (2–)3–8(–10) mm, pubescent, trichomes simple and stalked, 2–4-rayed. **Flowers:** (late-season ones petaliferous); sepals oblong, 1.3–2 mm, pubescent, (trichomes simple); petals yellow (sometimes fading white), oblanceolate, 1.5–3 × 0.5–1 mm; anthers ovate, 0.2–0.3 mm. **Fruits** lanceolate, plane, flattened, 6–9(11) × 1.3–2.3 mm; valves pubescent, trichomes simple, antrorse, 0.1–0.2 mm; ovules 32–60 per ovary; style 0.01–0.1 mm. **Seeds** ovoid, 0.5–0.7 × 0.4–0.5 mm. $2n = 24$.

Flowering Apr–Aug. Open forests, meadows, gravelly knolls, disturbed sites; 1800–3200 m; Ariz., Colo., N.Mex., Utah.

Although O. E. Schulz (1927), C. L. Hitchcock (1941), and R. C. Rollins (1993) indicated that *Draba montana* was published in 1879; the name was first and validly published in 1878, with a reference to Rothrock's description of a misapplied Linnaean name (*D. nemorosa*), which cited specimens that served as Watson's type collection. Populations of *D. rectifructa* from northern Utah often grow with *D. albertina* and there is some evidence of hybridization. The only published chromosome count for *D. rectifructa* (M. D. Windham 2000) is identical to that of *D. albertina* and was derived from one of these mixed populations.

92. Draba reptans (Lamarck) Fernald, Rhodora 36: 368. 1934 E

Arabis reptans Lamarck in J. Lamarck et al., Encycl. 1: 222. 1783; *Draba caroliniana* Walter; *D. caroliniana* var. *dolichocarpa* O. E. Schulz; *D. caroliniana* var. *hunteri* Payson & H. St. John; *D. caroliniana* var. *micrantha* (Nuttall) A. Gray; *D. caroliniana* subsp. *stellifera* (O. E. Schulz) Payson & H. St. John; *D. caroliniana* var. *umbellata* Torrey & A. Gray; *D. coloradensis* Rydberg; *D. hispidula* Michaux; *D. micrantha* Nuttall; *D. reptans* var. *micrantha* (Nuttall) Fernald; *D. reptans* subsp. *stellifera* (O. E. Schulz) Abrams; *D. reptans* var. *stellifera* (O. E. Schulz) C. L. Hitchcock; *D. umbellata* Muhlenberg; *Tomostima caroliniana* (Walter) Rafinesque; *T. hispidula* (Michaux) Rafinesque; *T. micrantha* (Nuttall) Lunell

Annuals; scapose or subscapose. **Stems** (simple or few from, or distal to, base), unbranched, (0.1–)0.3–1.2 (–1.6) dm, sparsely to densely pubescent proximally, trichomes 2 (or 3)-rayed, 0.1–0.6 mm, sometimes with simple or spurred ones, to 0.9 mm, distally usually glabrous, rarely with few trichomes. **Basal leaves** not rosulate; petiole obscure; blade elliptic or spatulate to obovate or suborbicular, 0.5–2.3(–3) cm × 1.5–8(–13) mm, margins entire, surfaces pubescent abaxially with stalked, 2–4-rayed trichomes, 0.1–0.5 mm, adaxially with simple trichomes, 0.6–1 mm, with stalked, 2-rayed ones, to 0.7 mm. **Cauline leaves** usually 0, rarely 1–3; sessile; blade similar to basal. **Racemes** (3–)5–12(–16)-flowered, ebracteate, (subumbellate), not elongated in fruit; rachis not flexuous, usually glabrous, rarely sparsely pubescent. **Fruiting pedicels** horizontal to divaricate-ascending, straight, (1–)2–7(–9) mm, glabrous or glabrate. **Flowers:** (late-season ones cleistogamous, apetalous); sepals oblong, 1.5–2.3 mm, pubescent, (trichomes simple); petals (rarely absent), white, spatulate, 2–4.5 × 1–1.5 mm; anthers ovate, 0.4–0.5 mm. **Fruits** linear to linear-oblong, plane, flattened, (5–)7–16(–20) × 1.2–2.3 mm; valves glabrous or pubescent, trichomes usually simple and antrorse, 0.1–0.3 mm, rarely with fewer, spurred or 2-rayed ones; ovules 32–88 per ovary; style 0.02–0.1 mm. **Seeds** oblong to ovoid, 0.5–0.8 × 0.3–0.5 mm. $2n$ = 16, 30, 32.

Flowering Feb–Jun(–Aug). Rock outcrops, dry slopes and hillsides, prairies, glades, roadsides, disturbed sites; 0–3000 m; Alta., B.C., Man., Ont., Sask.; Ala., Ariz., Ark., Calif., Colo., Ga., Idaho, Ill., Iowa, Kans., Mass., Mich., Minn., Mo., Mont., Nebr., N.Mex., N.Y., N.C., N.Dak., Ohio, Okla., Oreg., Pa., S.C., S.Dak., Tenn., Tex., Utah, Wash., Wis., Wyo.

Draba reptans is often confused with *D. cuneifolia*, but the two are easily separated. The rachises and pedicels of *D. reptans* are usually glabrous (rarely with a few isolated trichomes); those of *D. cuneifolia* are always densely pubescent. Interestingly, both species show parallel variations in chromosome number; it is currently unclear whether this variation is real or the result of misidentified specimens and/or erroneous counts.

93. **Draba ruaxes** Payson & H. St. John, Proc. Biol. Soc. Wash. 43: 117. 1930 [E]

Draba ventosa A. Gray var. *ruaxes* (Payson & H. St. John) C. L. Hitchcock

Perennials; (cespitose, forming tufts); caudex branched (with persistent leaf bases, branches some terminating in sterile rosettes); scapose. **Stems** unbranched, 0.2–0.6(–0.8) dm, often pubescent throughout, sometimes glabrate distally, trichomes simple, 0.4–1 mm, and 2–4-rayed, 0.1–0.4 mm. **Basal leaves** rosulate; subsessile; petiole base and margin ciliate, (trichomes simple, 0.4–1.4 mm); blade oblanceolate to obovate or suborbicular, 0.3–1 cm × 2–4.5 mm, margins entire, surfaces pubescent, abaxially with stalked, 2–10-rayed, stellate trichomes, 0.2–0.8 mm, adaxially with simple trichomes, 0.4–1 mm, sometimes with smaller, 2–4-rayed ones, (midvein obscure). **Cauline leaves** 0. **Racemes** (2–)4–10(–14)-flowered, ebracteate, elongated in fruit; rachis not flexuous, pubescent as stem or glabrous. **Fruiting pedicels** ascending to divaricate-ascending, straight or slightly curved upward, 3–7(–9) mm, pubescent or glabrous, trichomes simple and 2–4-rayed. **Flowers:** sepals ovate, 2–3 mm, pubescent, (trichomes simple with fewer, 2-rayed ones); petals bright yellow, obovate, 4–6 × 2–3.5 mm; anthers oblong, 0.4–0.5 mm. **Fruits** elliptic to lanceolate or ovate to suborbicular, plane, flattened, 4–8(–10) × 3–4.5 mm; valves puberulent, trichomes simple with fewer 2-rayed ones, 0.1–0.35 mm; ovules 12–16 per ovary; style 0.5–0.9(–1.1) mm. **Seeds** oblong, 1.5–2 × 1–1.3 mm. $2n$ = 72.

Flowering Jun–Jul. Rock outcrops, talus slopes, ridges, alpine summits; 500–2400 m; Alta., B.C., Yukon; Alaska, Wash.

C. L. Hitchcock (1941) treated *Draba ruaxes* as a variety of *D. ventosa*; as demonstrated by G. A. Mulligan (1971b), the two are quite distinct. *Draba ruaxes* is an outcrossing hexaploid with well-formed anthers and pollen, and abundant, simple trichomes on leaves, stems, sepals, and fruits. By contrast, *D. ventosa* is an apomictic triploid with abortive anthers and/or pollen, and no simple trichomes anywhere on the plant.

94. **Draba santaquinensis** Windham & Allphin, Harvard Pap. Bot. 12: 410. 2007 [E]

Biennials or perennials; (short-lived); caudex simple (poorly developed, without persistent leaf bases); not scapose. **Stems** usually branched, 1.1–3.4 dm, hirsute proximally, trichomes mostly simple and 2-rayed, 0.2–1.5 mm, pubescent distally, trichomes mostly 2- or 3-rayed, 0.2–1 mm. **Basal leaves** rosulate; petiole (obscure), ciliate, (trichomes simple and 2-rayed); blade obovate to oblanceolate, 1.6–2.5 cm × 5–10 mm, margins usually entire, rarely denticulate, surfaces pubescent, abaxially with stalked, mostly cruciform trichomes, 0.5–1 mm, adaxially with simple and stalked, 2–4-rayed ones. **Cauline leaves** usually 1–5; sessile or subsessile; blade oblanceolate to oblong, margins often denticulate, (ciliate proximally, with 2- or 3-rayed trichomes). **Racemes** 9–25-flowered, ebracteate, elongated in fruit; rachis not flexuous, pubescent, trichomes 2- or 3-rayed. **Fruiting pedicels** horizontal to divaricate-ascending, usually straight, rarely curved upward, 7–10 mm, pubescent, trichomes 2- or 3-rayed. **Flowers:** sepals oblong-obovate, 2–3.5 mm, pubescent, (trichomes short- to long-stalked, 2–4-rayed, 0.2–0.8 mm); petals yellow (often fading whitish), oblanceolate, 3–5 × 1–1.5 mm; anthers ovate, 0.4–0.5 mm. **Fruits** narrowly elliptic to slightly falcate, plane, flattened, (7–)10–16 × (2–)2.5–3.7 mm; valves pubescent, trichomes short-stalked, 2–4-rayed, 0.2–0.5 mm; ovules 14–22 per ovary; style 0.9–1.2 mm. **Seeds** oblong, 1.1–1.5 × 0.7–0.9 mm. $2n$ = 20.

Flowering late Apr–early Jun. Limestone outcrops and rocky slopes in mixed conifer communities; 1800–2400 m; Utah.

Although *Draba santaquinensis* was included within *D. brachystylis* by previous authors, I. A. Al-Shehbaz and M. D. Windham (2007) have shown that it is distinct both morphologically and chromosomally. It is currently known only from Utah County (American Fork, Provo, and Santaquin canyons) in north-central Utah.

95. Draba saxosa Davidson, Bull. S. Calif. Acad. Sci. 19: 11. 1920 [C][E]

Draba corrugata S. Watson var. *saxosa* (Davidson) Munz & I. M. Johnston

Perennials; (cespitose); caudex branched (covered with persistent leaves); scapose. **Stems** unbranched, 0.5–1.5 dm, pubescent throughout, trichomes simple and long-stalked, 2-rayed, 0.3–1 mm, with smaller, 2–4-rayed ones, 0.1–0.5 mm. **Basal leaves** rosulate; petiolate; petiole base and margin ciliate, (trichomes simple, 0.5–1.2 mm); blade oblanceolate, 1–3.1 cm × 2–7 mm, margins entire, surfaces densely pubescent, abaxially with stalked, (2-)4-rayed trichomes, 0.2–0.8 mm, adaxially mostly with long-stalked, 2-rayed trichomes, 0.6–1.3 mm, sometimes with simple and 3- or 4-rayed ones. **Cauline leaves** 0 (or 1); sessile; blade oblong, margins entire, surfaces pubescent as basal. **Racemes** 12–43-flowered, ebracteate, elongated in fruit; rachis not flexuous, pubescent as stem. **Fruiting pedicels** divaricate to ascending, straight or curved, (6–)10–17 mm, pubescent as stem. **Flowers:** sepals oblong, 2–2.5 mm, pubescent, (trichomes short-stalked, 2–4-rayed); petals yellow, linear-oblanceolate, 3–3.5 × 0.5–1 mm; anthers oblong, 0.5–0.6 mm, (exserted). **Fruits** oblong to linear-oblong, slightly twisted or plane, flattened, 6–10 × 2.5–4 mm; valves pubescent, trichomes short-stalked, cruciform, 0.07–0.25 mm, sometimes with fewer, 2- or 3-rayed ones; ovules 8–24 per ovary; style (0.8–)1.2–3.5 mm. **Seeds** oblong, 1–1.2 × 0.6–0.8 mm.

Flowering Jun–Jul. Among rocks; of conservation concern; 2600–3300 m; Calif.

Draba saxosa is related to *D. corrugata* and is often considered a variety of that species. The two taxa are easily distinguished (I. A. Al-Shehbaz and M. D. Windham 2007) and the available data support their recognition as distinct species. *Draba saxosa* is restricted to the San Jacinto and Santa Rosa mountains in southern California (Riverside and San Diego counties).

96. Draba scotteri G. A. Mulligan, Canad. J. Bot. 57: 1874. 1979 [C][E]

Perennials; (cespitose); caudex simple or branched (covered with persistent leaves); scapose. **Stems** unbranched, 0.2–1.4 dm, pubescent throughout, trichomes 2–8-rayed, 0.07–0.4 mm, and, sometimes, simple ones, 0.2–0.8 mm. **Basal leaves** rosulate; petiolate; petiole base ciliate, (trichomes simple, 0.2–1 mm); blade oblanceolate, 0.4–1.5 cm × 1–3 mm, margins entire, (ciliate as petiole base), surfaces pubescent with short-stalked, stellate, 8–12-rayed trichomes, 0.1–0.4 mm. **Cauline leaves** 0. **Racemes** 1–9-flowered, usually ebracteate, rarely proximalmost flower subtended by a tiny bract, usually considerably elongated in fruit; rachis not flexuous, pubescent as stem. **Fruiting pedicels** divaricate-ascending, straight, 3–8(–12) mm, pubescent as stem. **Flowers:** sepals oblong, 2–3 mm, subapically pubescent, (trichomes simple); petals yellow, obovate to spatulate, 3.5–5.5 × 1.5–2.5 mm; anthers ovate, 0.4–0.5 mm. **Fruits** lanceolate to narrowly so, plane, flattened, 5–11 × 1.5–2.5 mm; valves pubescent, trichomes simple, 0.1–0.4 mm, occasionally with some 2-rayed ones; ovules 12–18 per ovary; style 0.3–1 mm. **Seeds** ovoid, 0.8–1.2 × 0.6–0.7 mm. $2n = 96$.

Flowering Jun–Jul. Talus and gravelly summits in alpine communities; of conservation concern; 1200–2000 m; Yukon.

The description of *Draba scotteri* is based on collections from Kluane National Park, southwestern Yukon Territory.

97. Draba serpentina (Tiehm & P. K. Holmgren) Al-Shehbaz & Windham, Harvard Pap. Bot. 12: 414. 2007 [C][E]

Draba oreibata J. F. Macbride & Payson var. *serpentina* Tiehm & P. K. Holmgren, Brittonia 43: 20, fig. 1A–D. 1991

Perennials; (cespitose); caudex branched (with persistent leaves, branches sometimes terminating in sterile rosettes); scapose. **Stems** unbranched, 0.3–1.3 dm, glabrous throughout. **Basal leaves** rosulate; sessile; blade obovate to obovate-oblanceolate, 0.4–1 cm × (2–)2.5–4(–5) mm, margins entire, (ciliate, trichomes simple,

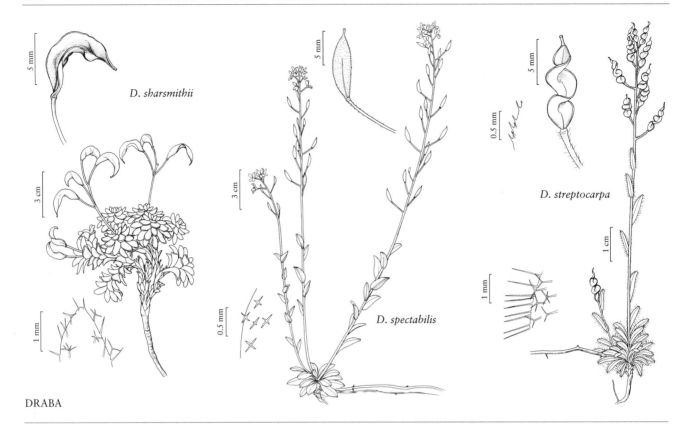

0.2–0.5 mm, apex acute), surfaces glabrous. **Cauline leaves** 0. **Racemes** 6–14-flowered, ebracteate, elongated in fruit; rachis not flexuous, glabrous. **Fruiting pedicels** divaricate-ascending, straight, 5–13 mm, glabrous. **Flowers:** sepals ovate, 2–3 mm, glabrous; petals white, spatulate to obovate, 5–6.5 × 2–3 mm; anthers ovate to oblong, 0.4–0.5 mm. **Fruits** lanceolate to narrowly oblong, slightly twisted or plane, flattened, 6–12 × 2–3.5 mm; valves glabrous; ovules 8–16 per ovary; style 0.6–1.8(–2.5) mm. **Seeds** oblong, 1.1–1.4 × 0.7–0.8 mm. $2n = 52$.

Flowering May–Jul. Rock outcrops, talus, and gravelly soil in mixed conifer and subalpine meadow communities; of conservation concern; 3200–3600 m; Nev.

Draba serpentina has been treated by other authors as a variety or synonym of *D. oreibata*. As indicated by I. A. Al-Shehbaz and M. D. Windham (2007), the two taxa are morphologically distinct, have different chromosome numbers, and are separated by over 480 kilometers. They are treated herein as separate species. *Draba serpentina* is known only from the Snake Range (White Pine County) and the Toiyabe Mountains (Lander County) in central Nevada.

98. Draba sharsmithii Rollins & R. A. Price, Harvard Pap. Bot. 1(3): 71. 1991 [C] [E] [F]

Draba cruciata Payson var. *integrifolia* C. L. Hitchcock & Sharsmith, Madroño 5: 151. 1940, not *D. integrifolia* (S. Watson) Greene 1900

Perennials; (densely cespitose); caudex branched (covered with persistent leaves or leaf remains, branches sometimes terminating in sterile rosettes); scapose. **Stems** unbranched, (0.15–)0.3–0.7(–1) dm, glabrous throughout or pubescent proximally, trichomes simple and 2-rayed, 0.1–0.4 mm. **Basal leaves** rosulate; petiolate; petiole base and margin ciliate, (trichomes simple and 2-rayed, 0.3–1 mm); blade oblanceolate to oblong-obovate or oblong, 0.4–0.7(–1) cm × 1–3(–5) mm, margins entire, surfaces pubescent abaxially with short-stalked, 2–5-rayed trichomes, 0.1–0.4 mm, adaxially glabrous or sparsely pubescent, with simple and short-stalked, 2-rayed trichomes. **Cauline leaves** 0. **Racemes** 2–10(–16)-flowered, ebracteate, elongated in fruit; rachis not flexuous, glabrous. **Fruiting pedicels** ascending, straight, (3–)5–12(–16) mm, glabrous. **Flowers:** sepals ovate, 2.8–3.5 mm, sparsely pubescent, (trichomes simple and short-stalked, 2–4-rayed); petals yellow, spatulate to oblanceolate, 4–6 × 2–3 mm; anthers oblong, 0.6–0.7 mm. **Fruits** oblong to lanceolate or narrowly ovate, twisted (curved), flattened, (5–)8–15(–20) × 2–4(–5) mm; valves usually glabrous, rarely pubescent, trichomes simple and short-stalked, 2-rayed, (marginal), 0.5–0.15 mm; ovules 16–20 per ovary; style (0.5–)0.9–1.8(–2) mm. **Seeds** oblong, 1–1.6 × 0.6–0.7 mm. $2n = 36$.

Flowering Jul–Aug. Granitic outcrops and rocky slopes in open subalpine conifer communities; of conservation concern; 3300–3800 m; Calif.

Draba sharsmithii is clearly related to *D. cruciata*; there is some evidence that it may be an allopolyploid incorporating a genome from that species. It is known only from the vicinity of Mount Whitney in the southern Sierra Nevada (Inyo County).

99. **Draba sibirica** (Pallas) Thellung, Neue Denkschr. Schweiz. Naturf. Ges. 41: 318. 1907

Lepidium sibiricum Pallas, Reise Russ. Reich. 3: 34. 1776; *Draba gmelinii* Adams; *D. repens* M. Bieberstein; *D. sibirica* subsp. *arctica* Böcher

Perennials; (stoloniferous); caudex branched (sparsely covered with persistent petiole remains, branches slender, creeping); scapose. **Stems** (subdecumbent when sterile), unbranched, 0.5–2.5 dm, sparsely to densely hispid, trichomes malpighiaceous, (flowering scapes sparsely pilose proximally with simple trichomes, often glabrous on distal ½). **Basal leaves** rosulate; petiolate; blade oblong-lanceolate or oblanceolate, 0.4–2.2 cm × 1–5(–10) mm, margins entire, surfaces often pubescent, sometimes glabrous (except margins), with appressed, malpighiaceous trichomes (these sometimes with 1 or 2 shorter, lateral branches, appearing 3-fid or cross-shaped). **Cauline leaves** 0. **Racemes** 7–20-flowered, ebracteate, considerably elongated in fruit; rachis not flexuous (straight), glabrous. **Fruiting pedicels** divaricate, straight or curved, (often filiform), 5–18 (–23) mm, glabrous. **Flowers:** sepals (erect), oblong or ovate, 2–2.7 mm (lateral pair subsaccate basally), glabrous or abaxially sparsely pilose, (trichomes simple); petals yellow, narrowly obovate, 4–6 × 2–3 mm, (apex emarginate); anthers ovate-subcordate, 0.4–0.5 mm. **Fruits** usually oblong to elliptic, rarely sublinear, plane, flattened, 4–8 × 1.5–2.2 mm; valves glabrous, (obscurely veined); ovules 24–30 per ovary; style 0.5–1 mm. **Seeds** (brown), ovoid, 0.9–1.2 × 0.5–0.6 mm. $2n = 16$.

Flowering May–Jul. Wet places on rocky slopes; 0–300 m; Greenland; Europe (Russia); Asia (Caucasus, Iran, Russian Far East, Siberia, Turkey, central republics).

Draba sibirica is one of three species of the genus in North America exhibiting malpighiaceous trichomes. It is easily distinguished from others (*D. malpighiacea* and some plants of *D. spectabilis*) by a complete absence of cauline leaves. It is found in the flora area in northeastern Greenland.

100. **Draba sierrae** Sharsmith, Madroño 5: 149. 1940
[C] [E]

Perennials; (cespitose, pulvinate, canescent); caudex branched (covered with persistent leaves, branches sometimes terminating in sterile rosettes); scapose. **Stems** unbranched, (0.05–)0.1–0.4 dm, pubescent throughout, trichomes pectinate or 2–4-rayed, 0.1–0.4 mm, and sometimes simple. **Basal leaves** (densely imbricate); rosulate; sessile; blade oblong, 0.2–0.6(–0.8) cm × 0.7–1.5(–2) mm, margins entire, (sometimes ciliate, trichomes simple, 0.5–1 mm), surfaces densely pubescent with stalked, subdendritic, 7–12-rayed trichomes, 0.2–0.5 mm. **Cauline leaves** 0. **Racemes** 2–6(–9)-flowered, ebracteate, elongated in fruit; rachis not flexuous, pubescent as stem. **Fruiting pedicels** divaricate to ascending, straight, 2–5(–7) mm, pubescent as stem. **Flowers:** sepals oblong, 2.5–3.5 mm, pubescent, (trichomes short-stalked, branched); petals yellow, spatulate, 4–5 × 1.2–1.8 mm; anthers oblong, 0.5–0.7 mm. **Fruits** lanceolate to elliptic or broadly oblong to ovate-orbicular, often twisted, occasionally plane, flattened, 4–8(–10) × (2–)3–4.5 mm (asymmetric); valves pubescent, trichomes short-stalked, subpectinate, or 2–4-rayed, 0.05–0.3 mm, sometimes with simple ones; ovules 8–14(–16) per ovary; style (0.1–)0.3–0.7(–1) mm. **Seeds** ovoid, 1.2–1.5 × 0.7–1 mm.

Flowering Jun–Aug. Granite outcrops, rocky slopes; of conservation concern; 3500–3900 m; Calif.

Draba sierrae is known only from the southern Sierra Nevada in east-central California (Inyo and Fresno counties).

101. **Draba simmonsii** Elven & Al-Shehbaz, Novon 18: 326. 2008 [E]

Draba alpina Linnaeus var. *gracilescens* Simmons, Vasc. Pl. Ellesmereland, 83, plate 6, figs. 1–3. 1906

Perennials; caudex branched (with persistent leaf bases and petioles); scapose. **Stems** unbranched, (0.15–)0.3–1.1 (–1.3) dm, often pubescent throughout, sometimes sparsely pubescent distally, trichomes stalked, 2–4-rayed, and fewer, simple ones, 0.1–0.3(–0.6) mm. **Basal leaves** (not imbricate); rosulate; petiolate; petiole (base somewhat thickened), margin ciliate, (trichomes simple, 0.5–1 mm); blade oblong to oblong-lanceolate, 0.5–1.6 cm × 2–5 mm, margins entire (apex subacute to obtuse), surfaces pubescent throughout or glabrous adaxially, with simple and stalked, 2–4(–6)-rayed, stellate to subdendritic trichomes. **Cauline leaves** 0. **Racemes** (2–)4–10(–14)-flowered, ebracteate, slightly elongated in fruit; rachis not flexuous (straight), pubescent as stem. **Fruiting pedicels** divaricate, straight, 2.5–10 mm, pubescent as stem. **Flowers:** sepals (purplish green), oblong, (2.5–)2.8–3.5(–3.8) mm, pubescent, (trichomes simple and sometimes short-stalked, 2-rayed); petals pale yellow, narrowly obovate, (sides non-parallel), (3.5–)3.8–5.5(–5.8) × (2.5–)2.8–4(–4.6) mm, (emarginate); anthers ovate, ca. 0.5 mm. **Fruits** lanceolate, plane, slightly flattened, 5.5–9(–11) × 2.3–3.8 mm; valves pubescent, trichomes simple, sometimes with fewer, 2-rayed ones, 0.1–0.3 mm; ovules 16–24(–28) per ovary; style 0.1–0.3 mm (stigma as wide as style). **Seeds** (dark brown), oblong-ovate, 1–1.3 × 0.7–0.8 mm.

Flowering late Jun–Jul. Dry open ground, open patches in meadows or heath vegetation, outwash plains; 0–100 m; N.W.T., Nunavut.

102. Draba smithii Gilg & O. E. Schulz in H. G. A. Engler, Pflanzenr. 89[IV,105]: 177. 1927 [C][E]

Perennials; (matted); caudex branched (with persistent leaf remains, branches slender, prostrate, sometimes terminating in sterile rosettes); not scapose. **Stems** usually unbranched, (0.5–)1–2.4(–3) dm, pubescent throughout, trichomes short-stalked, subdendritic, 4–12-rayed, 0.05–0.2 mm. **Basal leaves** rosulate; petiole (obscure), usually not ciliate, rarely pubescent proximally, (trichomes simple or spurred, to 0.25 mm); blade obovate to narrowly oblanceolate, 0.5–1.3(–2.5) cm × (1–)2–4(–7) mm, margins usually entire, rarely denticulate, surfaces pubescent with short-stalked, subdendritic, 5–12-rayed trichomes, 0.05–0.3 mm. **Cauline leaves** (2 or) 3–8; sessile; blade oblong to lanceolate, margins usually entire, rarely denticulate, surfaces pubescent as basal, or, rarely, adaxially with long-stalked, 2–5-rayed trichomes. **Racemes** 12–28-flowered, ebracteate, elongated in fruit; rachis not flexuous, pubescent as stem. **Fruiting pedicels** divaricate, straight, (3–)5–10(–13) mm, pubescent as stem. **Flowers:** sepals ovate, 2–2.5 mm, pubescent; petals white, spatulate, 4–6 × 2–3 mm; anthers ovate, 0.5–0.6 mm. **Fruits** ovate-lanceolate, twisted, flattened, 5–9 × 2–3 mm; valves pubescent, trichomes simple and short-stalked, 2–4-rayed, 0.05–0.2 mm; ovules 16–20 per ovary; style (0.7–)1–2(–2.3) mm. **Seeds** oblong, 0.8–1.2 × 0.5–0.7 mm. $2n = 32$.

Flowering May–Jul. Igneous outcrops and rocky slopes in mixed conifer, mountain shrub, and (rarely) sagebrush communities; of conservation concern; 2300–3700 m; Colo.

Draba smithii is a distinctive species known only from Archuleta, Las Animas, and Mineral counties in southern Colorado.

103. Draba sobolifera Rydberg, Bull. Torrey Bot. Club 30: 251. 1903 [C][E]

Draba sobolifera var. *uncinalis* (Rydberg) O. E. Schulz; *D. uncinalis* Rydberg

Perennials; (cespitose); caudex often branched (with persistent leaf bases, branches elongated, sometimes terminating in sterile rosettes); scapose. **Stems** unbranched, (0.15–)0.25–0.6(–0.7) dm, pubescent basally, trichomes simple and 2- or 3-rayed, 0.1–0.5 mm. **Basal leaves** rosulate; subsessile; petiole (0–0.5 cm) margin ciliate, (trichomes simple, 0.2–1 mm); blade oblanceolate to spatulate or obovate, (0.3–)0.4–1.4(–2) cm × (1.5–)2–4(–6) mm, margins entire, (ciliate as petiole), surfaces usually pubescent, sometimes glabrescent, abaxially with stalked, 2–4-rayed trichomes, 0.1–0.5 mm, adaxially often with simple and 2-rayed trichomes, to 0.9 mm. **Cauline leaves** 0 (or 1); sessile; blade similar to basal. **Racemes** (5–)9–18(–23)-flowered, ebracteate, slightly elongated in fruit; rachis not flexuous, pubescent as stems proximally (trichomes often crisped). **Fruiting pedicels** divaricate-ascending, usually straight, rarely curved upward, 3–6(–10) mm, pubescent, trichomes simple and 2- or 3-rayed, (crisped, 0.1–0.5 mm). **Flowers:** sepals ovate or broadly oblong, 1.7–2.5 mm, pubescent, (trichomes simple and stalked, 2-rayed); petals yellow, spatulate to oblanceolate, 3–4 × 1.5–2 mm; anthers ovate, 0.4–0.5 mm. **Fruits** ovoid or ovoid-lanceolate, plane, inflated at least basally, (3–)4–6(–7) × 2–4 mm; valves glabrous or pubescent, trichomes simple and 2-rayed, 0.07–0.3 mm; ovules (4–)8–12 per ovary; style (0.3–)0.4–0.6 mm. **Seeds** oblong, 1–1.4 × 0.6–0.8 mm. $2n = 26$.

Flowering Jun–Aug. Igneous rock outcrops, fellfields, rocky slopes in mixed conifer and alpine meadow communities; of conservation concern; 3100–3600 m; Utah.

Draba sobolifera is related to, and sympatric with, *D. ramulosa*, and occasional sterile hybrids are encountered. There is no evidence of introgression and the two species are easily distinguished (see discussion of 90. *D. ramulosa*). *Draba sobolifera* is known from the Tushar Mountains in south-central Utah (Beaver and Piute counties).

104. Draba spectabilis Greene, Pittonia 4: 19. 1899 E F

Draba oxyloba Greene; *D. spectabilis* var. *bella* O. E. Schulz; *D. spectabilis* var. *oxyloba* (Greene) Gilg & O. E. Schulz; *D. spectabilis* var. *purpusii* Gilg & O. E. Schulz

Perennials; caudex simple or branched (sometimes with persistent leaf bases); not scapose. **Stems** unbranched, (0.7–)1.1–3.7(–5.3) dm, pubescent throughout (sparsely so distally), trichomes simple, malpighiaceous, or sessile and 4-rayed, 0.15–0.5 mm, (2 rays parallel to stem axis longer). **Basal leaves** rosulate; petiolate; petiole sometimes ciliate, (trichomes simple); blade oblanceolate to spatulate or obovate, (1–)1.5–4.4(–6.3) cm × (3–)5–12(–15) mm, margins entire or dentate, surfaces pubescent with sessile, often 4-rayed trichomes, 0.1–0.6 mm (longest rays parallel to midvein, lateral rays sometimes reduced to tiny spurs), sometimes malpighiaceous trichomes present. **Cauline leaves** 4–12(–17); sessile; blade broadly ovate to lanceolate or oblong, margins entire or dentate, (ciliate or not), surfaces pubescent as basal or adaxially also with simple trichomes. **Racemes** (10–)16–49(–61)-flowered, ebracteate, elongated in fruit; rachis not flexuous, pubescent as stem. **Fruiting pedicels** horizontal to divaricate-ascending, straight or curved upward, (5–)7–20(–26) mm, glabrous or pubescent as rachis abaxially. **Flowers:** sepals ovate, (2.2–)2.5–4 mm, glabrous or pubescent, (trichomes simple and malpighiaceous); petals yellow, oblanceolate, 4–6.5 × 1–2.5 mm; anthers oblong, 0.5–0.7 mm. **Fruits** lanceolate to oblong or elliptic, plane, flattened, (6–)7–13 × 2–3.5 mm; valves glabrous or puberulent, trichomes simple and sessile, 2-rayed, 0.03–0.3 mm; ovules 12–24 per ovary; style (0.5–)1–2.7 mm. **Seeds** ovoid, 1–1.4 × 0.7–0.9 mm. $2n = 40$.

Flowering Jun–Aug. Talus, rocky hillsides, meadows in open conifer forests, aspen groves, and alpine communities; 2000–3900 m; Colo., N.Mex., Wyo.

Morphological and chromosomal data suggest that *Draba spectabilis* is an allopolyploid resulting from hybridization between *D. abajoensis* and *D. malpighiacea*. The characteristics distinguishing these three taxa were discussed in detail by I. A. Al-Shehbaz and M. D. Windham (2007). Chromosome counts attributed to *D. spectabilis* by Windham (2000) represent diploid *D. abajoensis*.

105. Draba sphaerocarpa J. F. Macbride & Payson, Amer. J. Bot. 4: 266. 1917 E

Draba argyrea Rydberg var. *sphaerocarpa* (J. F. Macbride & Payson) O. E. Schulz

Perennials; caudex branched (branches loose or compact, sometimes terminating in sterile rosettes); scapose. **Stems** unbranched, (0.15–)0.3–0.8 (–1.1) dm, usually densely pubescent throughout, trichomes stalked, 2–5-rayed, 0.1–0.4 mm, with subsessile, smaller, 4–8-rayed ones, rarely proximally with some simple trichomes. **Basal leaves** rosulate; subsessile; petiole margin not ciliate; blade oblanceolate to obovate or linear-oblanceolate, (0.2–)0.3–0.9(–1.5) cm × 1–2.5(–4) mm, margins entire, (not ciliate), surfaces canescent with short-stalked, 8–15-rayed, stellate trichomes, 0.1–0.25 mm. **Cauline leaves** 0. **Racemes** (4–)6–18(–23)-flowered, ebracteate, elongated in fruit; rachis not flexuous, pubescent as stem. **Fruiting pedicels** divaricate-ascending to ascending, straight, 4–9(–13) mm, pubescent as stem. **Flowers:** sepals oblong, 2–3.5 mm, pubescent, (trichomes 4–6-rayed); petals yellow or lemon, oblanceolate to spatulate, 4–6 × 1.5–3 mm; anthers ovate, 0.4–0.6 mm. **Fruits** ellipsoid to ovoid, plane, inflated at least basally, 2–5(–6) × 1.5–2.5(–3.5) mm; valves pubescent, trichomes 2–4-rayed, 0.03–0.2 mm (some or all rays spurred); ovules 4–10 (–12) per ovary; style 0.2–0.9(–1.2) mm. **Seeds** ovoid, 1–1.6 × 0.7–1.2 mm. $2n = 20$.

Flowering May–Jul. Granite outcrops and rocky slopes in open pine forests; 2100–2800 m; Idaho.

Draba sphaerocarpa is a distinctive species known only from Boise, Custer, Idaho, and Valley counties in central Idaho.

106. Draba sphaeroides Payson, Amer. J. Bot. 4: 265. 1917 C E

Draba oligosperma Hooker var. *sphaeroides* (Payson) O. E. Schulz

Perennials; (cespitose, densely pulvinate); caudex branched (branches often tangled with persistent leaf remnants, prostrate, slender, sometimes terminating in sterile rosettes); scapose. **Stems** unbranched, (0.2–)0.3–0.8 dm, sparsely pubescent proximally, glabrous distally, trichomes usually 2- or 3-rayed, rarely some simple, 0.1–0.25 mm. **Basal leaves** not rosulate; subsessile; petiole margins not ciliate; blade oblanceolate to oblong or elliptic, 0.3–0.8(–0.9) cm × (1–)1.5–3(–4) mm, margins entire, (not ciliate), surfaces usually pubescent, rarely both surfaces glabrous (except margin), with stalked, 2–4-rayed trichomes, 0.07–0.4 mm (or, rarely, trichomes mostly simple and 2-rayed adaxially). **Cauline leaves** 0. **Racemes** (3–)5–11(–14)-flowered, ebracteate, slightly elongated in fruit; rachis not flexuous, glabrous or sparsely pubescent as stem. **Fruiting pedicels** divaricate-ascending, straight or curved upward, (sometimes slightly expanded basally), 3–8 mm, usually glabrous, rarely sparsely pubescent, trichomes simple and 2- or 3-rayed. **Flowers:** sepals broadly ovate, 1.7–2.5 mm, glabrous or sparsely pubescent, (trichomes simple and short-stalked, 2- or 3-rayed); petals yellow, spatulate, 2.7–4 × 1.5–2.2 mm; anthers ovate, 0.4–0.5 mm. **Fruits** broadly lanceolate to ellipsoid or ovoid, plane, inflated at least basally, (2.5–)3.5–7(–8) × (1.5–)2–3.5 mm; valves glabrous or sparsely pubescent, trichomes simple and short-stalked, 2(–4)-rayed, 0.07–0.2 mm; ovules 4–10(–12) per ovary; style 0.3–1(–1.3) mm. **Seeds** oblong, 1–1.5 × 0.8–1 mm. $2n = 20$.

Flowering Jun–Aug. Rock outcrops, talus, soil pockets on rocky slopes; of conservation concern; 2500–3300 m; Nev.

Draba sphaeroides is closely related to *D. pedicellata*; the two are easily distinguished using the characteristics discussed by M. D. Windham (2004). The species is known from Elko County (East Humboldt, Ruby and Jarbidge mountains) in northeastern Nevada.

107. Draba standleyi J. F. Macbride & Payson, Ann. Missouri Bot. Gard. 5: 150. 1918 [C][E]

Draba gilgiana Wooton & Standley, Contr. U.S. Natl. Herb. 16: 124. 1913, not Muschler 1906; *D. chrysantha* S. Watson var. *gilgiana* O. E. Schulz

Perennials; (densely pulvinate); caudex branched (with persistent, thickened petioles, branches compact); not scapose. **Stems** unbranched, 0.3–1.3(–1.7) dm, usually glabrous throughout or sparsely pubescent proximally, rarely sparsely pubescent distally, trichomes simple and 2-rayed, 0.1–0.7 mm. **Basal leaves** (not imbricate); rosulate; petiolate; petiole ciliate, (trichomes usually simple, rarely 2-rayed, not setiform); blade narrowly oblanceolate to linear-lanceolate, (strongly differentiated into blade and petiole), (1.2–)1.8–6(–8.5) cm × 1–5(–7) mm, margins entire or sparsely denticulate, (ciliate as petiole; midvein not prominent), surfaces glabrous or pubescent, usually with simple trichomes 0.1–0.8 mm, rarely 2-rayed. **Cauline leaves** 1–8; sessile; blade lanceolate to narrowly oblong, margins usually entire, surfaces pubescent as basal. **Racemes** 5–17(–23)-flowered, ebracteate, elongated in fruit; rachis not flexuous, usually glabrous, rarely pubescent, trichomes simple and 2-rayed, (non-crisped). **Fruiting pedicels** divaricate-ascending or ascending, straight, 3–9(–13) mm, usually glabrous, rarely sparsely pubescent, trichomes simple. **Flowers:** sepals ovate, 2–2.5 mm, pubescent, (trichomes simple); petals yellow, oblanceolate, 4–6 × 1.5–2 mm; anthers oblong, 0.6–0.8 mm. **Fruits** linear-elliptic to elliptic, twisted or plane, flattened, 5–10(–13) × 1.5–2.5 mm; valves usually glabrous, rarely puberulent, trichomes simple, 0.05–0.1 mm; ovules 12–24 per ovary; style 0.7–1.4(–1.8) mm. **Seeds** ovoid, 1–1.2 × 0.7–0.8 mm.

Flowering Jun–Aug. Igneous rock outcrops, stabilized talus slopes; of conservation concern; 1800–3100 m; Ariz., N.Mex., Tex.

Draba standleyi is a distinctive species that is sporadically distributed in the mountains of southwestern United States. It is known from the Chiricahua Mountains (Cochise County, southeastern Arizona), the Organ Mountains and Black Range (Dona Ana and Sierra counties, south-central New Mexico), and the Davis Mountains (Jeff Davis County, western Texas). It has not been reported from Mexico, though it is very likely to occur there.

108. Draba stenoloba Ledebour, Fl. Ross. 1: 154. 1841 [E]

Draba acinacis H. St. John; *D. hirta* Linnaeus var. *siliquosa* Chamisso & Schlechtendal; *D. macouniana* Rydberg; *D. nemorosa* Linnaeus var. *stenoloba* (Ledebour) M. E. Jones; *D. nitida* Greene var. *praelonga* O. E. Schulz; *D. oligantha* Greene; *D. stenoloba* var. *oligantha* (Greene) O. E. Schulz

Biennials or perennials; (short-lived, not cespitose); caudex simple (not fleshy); scapose. **Stems** often unbranched, (0.2–)0.4–3(–3.4) dm, pubescent proximally, usually glabrous distally, rarely sparsely pubescent, trichomes short-stalked, 2–4-rayed, 0.1–0.4(–0.5) mm. **Basal leaves** rosulate; petiole (obscure), margin ciliate, (trichomes simple, 0.2–0.4 mm); blade oblanceolate to obovate, 0.4–2.6(–3.1) cm × 1.5–7(–10) mm, margins entire or denticulate, (ciliate as petiole), surfaces pubescent, abaxially with short-stalked, 3- or 4-rayed trichomes, 0.1–0.5 mm, adaxially with simple and stalked, 2–4-rayed trichomes, 0.1–0.5 mm. **Cauline leaves** (0 or) 1 or 2 (3 or 4); sessile; blade elliptic to lanceolate or ovate, margins toothed, surfaces pubescent, abaxially with short-stalked 3- or 4-rayed trichomes, adaxially with simple and short-stalked, 2–4-rayed trichomes. **Racemes** (2–)4–10(–15)-flowered, ebracteate, elongated in fruit; rachis not flexuous, usually glabrous, rarely sparsely pubescent. **Fruiting pedicels** horizontal to divaricate-ascending, straight,

3–14(–19) mm (subequaling or shorter than fruit), usually glabrous, rarely pubescent, trichomes 2–4-rayed. **Flowers** (chasmogamous, petaliferous); sepals (green or purplish), oblong, 1.5–2.5 mm, glabrous or pubescent, (trichomes simple); petals yellow, oblanceolate, 2.5–3.5 × 0.6–1 mm; anthers ovate, 0.25–0.4 mm. **Fruits** linear or, rarely, linear-elliptic, plane, flattened, (8–)10–17 (–20) × 1.5–2 mm; valves usually glabrous, rarely sparsely puberulent, trichomes simple; ovules 24–36 per ovary; style 0.01–0.15 mm. **Seeds** ovoid, 0.9–1 × 0.5–0.6 mm. $2n = 40$.

Flowering Jun–Jul. Grassy knolls, glacial moraines and creek banks, mesic meadows, alpine thickets; 600–2300 m; Alta., B.C., Yukon; Alaska, Oreg., Wash.

Draba stenoloba is occasionally confused with *D. albertina*, but is easily recognized by having exclusively 2–4-rayed (versus mostly simple) trichomes on stems proximally. It is rarely encountered and apparently confined to the Pacific Northwest. In contrast, *D. albertina* is common and widespread in the mountains of western North America.

109. Draba stenopetala Trautvetter, Trudy Imp. S.-Petersburgsk. Bot. Sada 6: 11. 1879

Draba stenopetala var. *purpurea* Hultén

Perennials; (pulvinate); caudex branched (covered with persistent leaves, branches sometimes terminating in sterile rosettes); scapose. **Stems** unbranched, (0.02–)0.07–0.2 (–0.3) dm, pubescent throughout, trichomes simple and short-stalked, 2–5-rayed, 0.1–0.5 mm, (sometimes simple ones very sparse). **Basal leaves** (densely imbricate); rosulate; petiolate; petiole base and margin ciliate, (trichomes simple and 2-rayed, 0.2–0.5 mm); blade obovate to oblong-spatulate, 0.2–0.6 (–0.8) cm × 1–2(–3) mm, margins entire, surfaces often pubescent with simple and stalked, 2–5-rayed trichomes, 0.2–0.9 mm, sometimes glabrous and trichomes on margins only. **Cauline leaves** 0. **Racemes** 2–5-flowered, ebracteate, not or slightly elongated in fruit; rachis slightly flexuous, pubescent as stem. **Fruiting pedicels** divaricate, straight, 2–5(–7) mm, pubescent as stem. **Flowers:** sepals (spreading or reflexed), oblong, 2–3.5 mm, pubescent, (trichomes simple and 2-rayed); petals yellow or purple, linear, 2.5–5 × 0.3–0.7 mm; anthers ovate, 0.3–0.4 mm. **Fruits** suborbicular, plane, inflated basally, flattened distally, 3–5 × 3–4 mm; valves glabrous or puberulent, trichomes simple, 0.02–0.3 mm; ovules 4–10 per ovary; style 0.2–0.6 mm. **Seeds** ovoid, 1–1.3 × 0.8–1 mm. $2n = 24, 64$.

Flowering May–Jul. Rock outcrops, talus, rocky ridges, alpine tundra; 100–1900 m; Yukon; Alaska; e Asia (Russian Far East).

Different chromosome counts for *Draba stenopetala* were reported as $2n = 24$ from North America and $2n = 64$ from the Russian Far East (S. I. Warwick and I. A. Al-Shehbaz 2006). It is unlikely that a single species is involved, and further work is needed to verify the counts from North America.

110. Draba streptobrachia R. A. Price, Brittonia 32: 168. 1980 [E]

Draba chrysantha S. Watson forma *dasycarpa* O. E. Schulz in H. G. A. Engler, Pflanzenr. 89[IV,105]: 195. 1927, not *D. dasycarpa* Bernhardi 1800; *D. spectabilis* Greene var. *dasycarpa* (O. E. Schulz) C. L. Hitchcock

Perennials; caudex branched (often with persistent leaf remains, branches not creeping); not scapose. **Stems** unbranched, (0.1–)0.2–1(–1.3) dm, pubescent throughout, trichomes subsessile (often crisped), 3–5-rayed, stellate, 0.03–0.25(–0.4) mm, (rays sometimes forked). **Basal leaves** rosulate; petiole (obscure), usually not ciliate, rarely sparsely pubescent, (trichomes simple, to 0.6 mm); blade oblanceolate to linear-oblanceolate, (0.4–)0.5–3(–4) cm × 1–5 mm, margins entire, surfaces pubescent with short-stalked (crisped), 3–8-rayed trichomes, 0.05–0.4 mm. **Cauline leaves** (1 or) 2–4 (or 5); sessile; blade oblong to ovate or linear, margins entire, surfaces pubescent as basal. **Racemes** 4–10 (–18)-flowered, ebracteate, elongated in fruit; rachis not flexuous, pubescent as stem. **Fruiting pedicels** ascending, usually straight, rarely curved upward, (2–)3–8(–12) mm, pubescent as stem. **Flowers:** sepals ovate, 2–3 mm, pubescent, (trichomes simple and short-stalked, 2–4-rayed); petals yellow, spatulate, 3–5 × 1.5–3 mm, (clawed); anthers ovate, 0.25–0.4 mm. **Fruits** (not appressed to rachis), ovate to elliptic or lanceolate, slightly twisted or plane, flattened, (3–)5–10 × 2–4 mm; valves often pubescent, occasionally glabrous, trichomes simple and minutely stalked, 2–4-rayed, 0.03–0.25 mm; ovules 10–16(–18) per ovary; style 0.3–0.8(–1.2) mm. **Seeds** oblong, 1–1.6 × 0.6–1 mm. $2n =$ ca. 64.

Flowering Jul–Aug. Alpine tundra, scree, ridges and alpine slopes, turf, fellfields, talus slopes, crevices in rock ledges, loose soils; 3200–4000 m; Colo.

As indicated by Price, *Draba streptobrachia* is an apomict, yielding abundant, well-developed seed despite producing only abortive pollen. Morphological studies (M. D. Windham, unpubl.) suggest that the species may be an allopolyploid containing a genome from *D. crassa*.

111. Draba streptocarpa A. Gray, Amer. J. Sci. Arts, ser. 2, 33: 242. 1862 [E] [F]

Draba streptocarpa var. *tonsa* (Wooton & Standley) O. E. Schulz; *D. tonsa* Wooton & Standley

Perennials; (tufted); caudex simple or branched (covered with persistent leaf remains); not scapose. Stems usually unbranched, (0.2–)0.4–2.5 dm, hirsute proximally, sparsely pubescent or glabrous distally or throughout, trichomes simple, 0.4–2.2 mm, often with stalked, 2(–4)-rayed ones, 0.2–1 mm. Basal leaves rosulate; blade oblong-oblanceolate to linear-oblanceolate, 0.5–3.8 cm × 1.5–6 mm, margins entire, (ciliate, trichomes setiform, simple, 0.6–2.8 mm), surfaces strigose to hirsute abaxially with long-stalked, 2(–4)-rayed trichomes, 0.4–1.4 mm, usually with simple ones, strigose adaxially with simple trichomes, 0.7–2.4 mm. Cauline leaves 2–15; sessile; blade oblong to lanceolate, margins entire, surfaces pubescent as basal. Racemes 9–62-flowered, ebracteate or proximalmost 1–3 flowers bracteate, elongated in fruit; rachis not flexuous, sparsely pubescent as stem or glabrous. Fruiting pedicels ascending, slightly curved or straight, 4–12 mm, glabrous or pubescent abaxially, trichomes simple (0.4–1.6 mm). Flowers: sepals oblong, 2.5–3.5 mm, usually pubescent, rarely glabrous, (trichomes simple, sometimes with stalked, 2-rayed ones); petals yellow, oblanceolate to spatulate, (4–)5–7.5 × 1–2(–2.5) mm; anthers ovate, 0.5–0.7 mm. Fruits ovate to linear-lanceolate, usually strongly twisted (to 3 turns), rarely plane, flattened, 5–16 × 2–3 mm; valves puberulent along margin, trichomes simple, 0.05–0.3 mm; ovules 20–34 per ovary; style (0.8–)1–2(–2.5) mm. Seeds ovoid, 0.9–1.2 × 0.6–0.7 mm. $2n = 20, 40$.

Flowering May–Jul. Rock outcrops, hillsides and meadows in open mixed conifer forests, aspen groves, and alpine tundra communities; 2400–4000 m; Colo., N.Mex., Wyo.

Draba streptocarpa is easily recognized by its strongly twisted fruits and broad-based, simple cilia on the basal leaves. It is found primarily in Colorado, but has also been collected in Larimer County, Wyoming, and Mora, San Miguel, and Taos counties, New Mexico.

112. Draba subalpina Goodman & C. L. Hitchcock, Ann. Missouri Bot. Gard. 19: 77. 1932 [E]

Perennials; (cespitose); caudex simple or branched (with some persistent leaf bases); scapose. Stems unbranched, 0.3–1(–1.3) dm, often glabrous throughout, sometimes sparsely pubescent proximally, trichomes simple and stalked, 2-rayed, 0.2–1 mm. Basal leaves rosulate; sessile; blade (fleshy), oblanceolate to linear, (0.4–)0.5–1.4(–2) cm × (1–)1.5–3(–4) mm, margins entire, (ciliate at least apically, trichomes simple and 2-rayed), surfaces usually glabrous, rarely sparsely pubescent with simple and stalked, 2- (or 3-)rayed trichomes, 0.2–1 mm, (midvein obscure). Cauline leaves 0. Racemes (6–)10–28(–40)-flowered, ebracteate, elongated in fruit; rachis not flexuous, glabrous. Fruiting pedicels divaricate-ascending, straight, (3–)5–10(–17) mm, glabrous. Flowers: sepals broadly ovate, 1.7–2.5 mm, usually glabrous, rarely pubescent, (trichomes simple and short-stalked, 2-rayed); petals white, spatulate to obovate, 3–5 × 2–3 mm; anthers oblong, 0.5–0.7 mm. Fruits ovoid to lanceolate, plane, inflated at least basally, 4–8(–10) × 2.5–4 mm; valves glabrous or sparsely puberulent, trichomes simple, 0.02–0.1 mm; ovules 6–12 per ovary; style 0.2–0.9 mm. Seeds ovoid, 1–1.5 × 0.7–1.1 mm. $2n = 26$.

Flowering May–Jun. Rocky knolls and marly limestone soil in pine-oak-juniper woodlands, edges of spruce-fir forests; 1800–3400 m; Utah.

Molecular studies (M. A. Beilstein and M. D. Windham 2003) and chromosomal data (Windham 2000, 2004) suggest that *Draba subalpina* is most closely related to *D. cusickii* and *D. sobolifera*. From those, it is easily distinguished by having white (versus yellow) petals, glabrous (versus pubescent) rachises and stems distally, and glabrous or, rarely, sparsely pubescent (versus always pubescent) abaxial leaf blade surfaces. *Draba subalpina* is known from Garfield, Iron, Kane, and Wayne counties in south-central Utah.

113. Draba subcapitata Simmons, Vasc. Pl. Fl. Ellesmereland, 87, fig. 3, plate 1, figs. 3–8. 1906

Perennials; (cespitose, pulvinate); caudex often branched (covered with persistent, thickened petioles, branches sometimes terminating in sterile rosettes); scapose. Stems unbranched, (0.04–)0.07–0.5(–0.6) dm, usually pubescent throughout, rarely glabrous distally, trichomes simple and 2-rayed, 0.2–0.6 mm, with smaller, dendritic ones. Basal leaves (densely imbricate); rosulate;

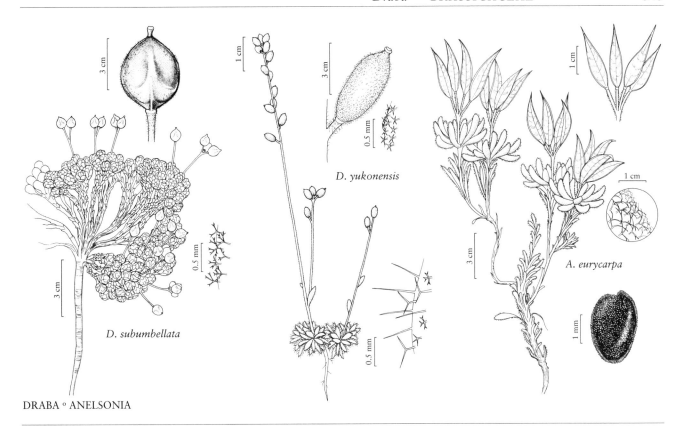

petiolate; petiole margin ciliate, (trichomes simple and 2-rayed, 0.3–0.8 mm); blade oblanceolate to obovate or nearly linear, 0.2–0.8(–1.1) cm × 1–3 mm, margins entire, (pubescent as abaxial surface), surfaces pubescent abaxially, glabrous adaxially, with simple and stalked, 2-rayed trichomes, 0.3–0.8 mm, sometimes with subdendritic ones, 0.1–0.3 mm, (midvein persistent, strongly thickened). **Cauline leaves** 0. **Racemes** 2–8-flowered, ebracteate, (corymbose), slightly elongated in fruit; rachis not flexuous, usually pubescent as stem, (trichomes often crisped), rarely glabrous. **Fruiting pedicels** horizontal to divaricate, straight, 1–5(–7) mm, usually pubescent as rachis, rarely glabrous. **Flowers:** sepals oblong, 1.2–1.8 mm, pubescent, (trichomes simple); petals white or creamy white, spatulate, 1.5–2(–2.5) × 0.7–1 mm; anthers ovate, 0.2–0.3 mm. **Fruits** ovoid to oblong, plane, slightly inflated, 3.5–6 × 2–3 mm; valves usually glabrous, rarely puberulent, trichomes simple, 0.05–0.15 mm; ovules 14–24 per ovary; style 0.1–0.3 mm. **Seeds** oblong, 0.9–1.1 × 0.5–0.6 mm. $2n = 16$.

Flowering Jul–Aug. Talus slopes, limestone barrens, peaty bluffs, silty slopes, seepage areas, disturbed gravel; 0–1000 m; Greenland; N.W.T., Nunavut, Que.; Alaska; Europe (Norway, n Sweden); e Asia (Russian Far East, Siberia).

114. Draba subumbellata Rollins & R. A. Price, Aliso 12: 25, figs. 1k–n, 4. 1988 E F

Perennials; (cespitose, densely pulvinate); caudex branched (densely covered with persistent leaves and remains, branches tightly grouped, sometimes terminating in sterile rosettes); scapose. **Stems** unbranched, (0.05–)0.1–0.25 dm, densely pubescent throughout, trichomes (grayish), stalked, dendritic, 5–12-rayed, 0.1–0.25 mm. **Basal leaves** (densely imbricate); rosulate; sessile; blade obovate to broadly oblong, 0.2–0.4 cm × 0.6–1.5 mm, margins entire, (not ciliate), surfaces densely pubescent, trichomes (grayish), stalked, dendritic, 5–12-rayed, 0.1–0.3 mm, (sometimes with long-stalked and spurred trichomes adaxially). **Cauline leaves** 0. **Racemes** 2–5(–10)-flowered, ebracteate, subumbellate or slightly elongated in fruit; rachis not flexuous, pubescent as stem. **Fruiting pedicels** divaricate-ascending, straight, 1.5–3(–6) mm, glabrous or pubescent as stem. **Flowers:** sepals broadly oblong, 1.8–2.8 mm, pubescent, (trichomes dendritic); petals yellow, spatulate, 2.8–4 × 1–1.5 mm; anthers ovate, 0.3–0.4 mm. **Fruits** ovoid to ovoid-lanceolate, plane, inflated and sometimes subgibbous basally, flattened distally, 2–5 × 2–3 mm; valves pubescent, trichomes short-stalked, dendritic,

4–12-rayed, (sometimes spurred), 0.05–0.2 mm; ovules 6–12 per ovary; style 0.2–0.6 mm. **Seeds** oblong, 1–1.2 × 0.5–0.7 mm.

Flowering Jul–Aug. Wind-eroded areas, alpine fellfields, rock crevices; 3300–4100 m; Calif., Nev.

Draba subumbellata is similar to, and sympatric with, *D. oligosperma*. It is distinguished from the latter by having stalked, stellate to dendritic leaf trichomes, subumbellate racemes, fruiting pedicels 1.5–3.0(–6.0) mm, fruits pubescent with mostly 4–12-rayed trichomes, and well-formed anthers and pollen. By contrast, *D. oligosperma* has sessile or subsessile, pectinately-branched leaf trichomes, elongated racemes, fruiting pedicels (2–)3–10(–13) mm, fruits glabrous or pubescent with simple and 2-rayed trichomes, and abortive anthers and/or pollen. *Draba subumbellata* is restricted to the White Mountains (Esmeralda County, Nevada, and Mono County, California) and to Coyote Ridge in the Sierra Nevada (Inyo County, California).

115. Draba trichocarpa Rollins, Contr. Gray Herb. 214: 4. 1984 [C][E]

Perennials; (cespitose, densely pulvinate); caudex branched (covered with persistent leaves, branches sometimes terminating in sterile rosettes); scapose. **Stems** unbranched, 0.07–0.35 dm, pubescent throughout, trichomes (soft), stalked, subdendritic, (somewhat crisped), 0.1–0.5 mm, (simple ones absent). **Basal leaves** (densely imbricate); rosulate; sessile; blade oblong to obovate, 0.2–0.4 cm × 0.5–1.5 mm, margins entire, (ciliate, trichomes simple and branched, subdendritic, or spurred, 0.3–0.8 mm), surfaces sparsely pubescent, abaxially with stalked, 4–6-rayed stellate trichomes, 0.1–0.4 mm, adaxially with simple and 4–6-rayed trichomes, mainly on distal ½. **Cauline leaves** 0 (or 1); sessile; blade similar to basal. **Racemes** 2–9-flowered, ebracteate, slightly elongated in fruit; rachis not flexuous, pubescent as stem. **Fruiting pedicels** divaricate-ascending to ascending, straight, 1–4.5 mm, pubescent as stem. **Flowers:** sepals ovate, 2–3 mm, pubescent, (trichomes short-stalked); petal color unknown, broadly obovate, 2–4 × 2–2.5 mm; anthers oblong, 0.5–0.6 mm. **Fruits** ovoid, plane, slightly inflated basally, flattened distally, 2–6 × 2–3.5 mm; valves densely pubescent, trichomes 4-rayed, 0.1–0.4 mm, (often some rays spurred or branched); ovules 4–10 per ovary; style 0.3–0.7 mm. **Seeds** oblong, 1.4–2 × 0.8–1.2 mm.

Flowering Jun–Jul. Gravelly metamorphic soil at ecotone between sagebrush steppe and open conifer forests; of conservation concern; ca. 2000 m; Idaho.

Draba trichocarpa is an apomictic polyploid that appears to be closely related to *D. novolympica*. It is readily distinguished from that species by the primarily dendritic trichomes (and absence of simple trichomes) on the stems, pedicels, and fruits. *Draba trichocarpa* is known from the Stanley Basin of central Idaho (Custer County).

116. Draba ventosa A. Gray, Amer. Naturalist 8: 212. 1874 [E]

Perennials; (cespitose); caudex branched (covered with persistent leaves, branches creeping, sometimes terminating in sterile rosettes); scapose. **Stems** unbranched, (0.1–)0.2–0.4(–0.6) dm, densely pubescent throughout, trichomes 2–6-rayed, 0.1–0.6 mm. **Basal leaves** (imbricate); rosulate; subsessile; petiole base and margin not ciliate; blade obovate to oblanceolate, (0.4–)0.5–1 cm × 1.5–4.5 mm, margins entire, surfaces densely pubescent with stalked, 2–6-rayed trichomes, (0.1–)0.2–0.9 mm. **Cauline leaves** 0. **Racemes** 5–10(–16)-flowered, ebracteate, slightly elongated in fruit; rachis not flexuous, densely pubescent as stem. **Fruiting pedicels** horizontal to divaricate-ascending, straight or slightly curved upward, 2–7(–9) mm, densely pubescent, trichomes 2–6-rayed, (0.1–0.6 mm). **Flowers:** sepals broadly ovate, 2–2.5 mm, pubescent, (trichomes short-stalked, 2–6-rayed); petals yellow, obovate, 3.5–5.5 × 1.5–3 mm; anthers oblong, 0.5–0.6 mm. **Fruits** suborbicular to broadly ovate, plane, inflated basally, flattened distally, 4–7.5(–9) × 3.5–5 mm; valves densely pubescent, trichomes short-stalked, 2–6-rayed, 0.15–0.5 mm; ovules 8–12 per ovary; style (0.5–)0.7–1.4 mm. **Seeds** oblong, 1.4–1.9 × 0.9–1.2 mm. $2n = 36$.

Flowering Jul–Aug. Talus slopes and alpine tundra; 2000–4000 m; Alta., B.C., Yukon; Colo., Utah, Wyo.

G. A. Mulligan (1971b) first reported apomixis in *Draba ventosa* based on studies of Canadian populations. This has now been confirmed in one of the southernmost populations (Duchesne County, Utah; M. D. Windham, unpubl.) as well. The species is easily overlooked and the large geographic gap between the Canadian and United States populations is likely to be narrowed or eliminated by additional collecting in western Montana. The limits of this species were expanded by C. L. Hitchcock (1941) to include *D. ruaxes*, but there are clear differences between them that support their recognition as distinct species.

117. Draba verna Linnaeus, Sp. Pl. 2: 642. 1753
• Whitlow grass, whitlow wort [I][W]

Draba boerhaavii (H. C. Hall) Dumortier; *D. praecox* Steven; *D. verna* var. *aestivalis* Lejeune; *D. verna* var. *boerhaavii* H. C. Hall; *Erophila boerhaavii* (H. C. Hall) Dumortier; *E. krockeri* Andrzejowski; *E. praecox* (Steven) de Candolle; *E. verna* (Linnaeus) L. Chevallier; *E. verna* subsp. *praecox* (Steven) Walters

Annuals; scapose. Stems (few to many from base), unbranched, (0.2–)0.5–2(–3) dm, pubescent proximally, glabrous distally, trichomes simple and 2(–4)-rayed, 0.1–0.4 mm. Basal leaves rosulate; petiolate; blade obovate, spatulate, oblanceolate, lanceolate, oblong, or, rarely, linear, 0.2–1.8(–3) cm × (0.5–)1–5(–10) mm, margins entire or 1–5-toothed on each side, surfaces pubescent with simple or stalked, 2–4-rayed trichomes, 0.1–0.5 mm. Cauline leaves 0. Racemes 4–20(–30)-flowered, ebracteate, usually considerably elongated in fruit; rachis usually flexuous, glabrous. Fruiting pedicels divaricate to ascending, straight or slightly curved upward, (2–)5–20(–35) mm, glabrous. Flowers: sepals (green or purplish), oblong, 1–2.5 mm, glabrescent or pubescent, (trichomes simple or 2-rayed); petals white, deeply 2-fid, (1.5–)2–4.5(–6) × 1–2 mm; anthers ovate, 0.2–0.4 mm. Fruits obovate, oblanceolate, lanceolate, elliptic, oblong, or linear, plane, flattened, (2.5–)4–9(–12) × 1.5–2.5(–3.5) mm; valves glabrous; ovules (20–)32–70(–84) per ovary; style 0.02–0.2 mm. Seeds ovoid (slightly flattened), 0.3–0.6(–0.8) × 0.2–0.4 mm. $2n$ = 14, 16, 20, 24, 28, 30, 32, 34, 36, 38, 40, 52, 54, 58, 60, 64.

Flowering Feb–May. Cedar glades, lawns, fields, pastures, waste places, grassy hillsides, disturbed sites, roadsides; 0–2500 m; introduced; Alta., B.C., N.B., Ont., Que.; Ala., Ark., Calif., Conn., Del., D.C., Ga., Idaho, Ill., Ind., Iowa, Ky., Maine, Md., Mass., Mich., Miss., Mo., Mont., Nev., N.H., N.J., N.Y., N.C., Ohio, Oreg., Pa., R.I., S.C., Tenn., Utah, Vt., Va., Wash., W.Va., Wyo.; Europe; Asia; nw Africa; introduced also in Central America, South America, Australia.

Draba verna represents a highly variable and taxonomically difficult complex within which species, subspecies, varieties, and forms have been named (O. E. Schulz 1927); only those synonyms pertaining to North America are listed above. Most of the taxonomic difficulties are the results of disploidy, autogamy, and hybridization. The morphological extremes are connected by intermediate forms in every conceivable character. Furthermore, there appears to be no correlation between morphology, cytology, geography, and ecology to support the division of this complex into meaningful taxa. A complex cytological picture was presented by Ø. Winge (1940), including the highest count of $2n$ = 94, which has not been confirmed by subsequent botanists.

Erophila vulgaris de Candolle is an illegitimate name for *Draba verna*.

118. Draba viridis A. Heller, Muhlenbergia 1: 27. 1901 [C]

Draba petrophila Greene var. *viridis* (A. Heller) C. L. Hitchcock

Perennials; (long-lived, cespitose); caudex simple or branched (densely covered with persistent petioles); not scapose. Stems unbranched or branched, 0.35–3 dm, pubescent throughout, trichomes simple, 0.2–0.8 mm, with short-stalked to subsessile, cruciform ones, 0.05–0.2 mm. Basal leaves rosulate; petiolate; petiole ciliate, (trichomes simple, 0.4–1 mm); blade oblanceolate, 1–6 cm × 2–6 mm, margins usually entire, rarely denticulate, surfaces pubescent with stalked, cruciform trichomes, 0.1–0.5 mm. Cauline leaves 5–16; sessile; blade ovate to lanceolate or oblong, margins entire or denticulate, surfaces pubescent as basal. Racemes 11–28-flowered, ebracteate, elongated in fruit; rachis not flexuous, pubescent as stem. Fruiting pedicels divaricate, straight, 4–7 mm, pubescent throughout, trichomes simple (0.1–0.4 mm) and subsessile, 2–4-rayed (0.03–0.25 mm). Flowers: sepals oblong, 2–3 mm, pubescent, (trichomes simple and 2-rayed); petals yellow, oblanceolate, 3.5–5 × 1–1.5 mm; anthers ovate, 0.5–0.6 mm. Fruits lanceolate to ovate, often plane, flattened, 3.5–8 × 2–3 mm; valves pubescent, trichomes (vertical), simple and 2-rayed, (0.1–)0.3–0.8 mm; ovules 12–20 per ovary; style (1.5–)2–3 mm. Seeds ovoid, 1.2–1.5 × 0.7–0.9 mm.

Flowering Aug–Sep. Open stands of pines; of conservation concern; ca. 2600 m; Ariz.; Mexico (Sonora).

Draba viridis has been reported from the Huachuca and Santa Catalina mountains in southern Arizona, and the adjacent San Jose Mountains in Sonora, Mexico. It was reduced to a variety of *D. petrophila* by Hitchcock and a form of *D. helleriana* by O. E. Schulz, but is easily separated from both (I. A. Al-Shehbaz and M. D. Windham 2007).

119. **Draba weberi** R. A. Price & Rollins, Harvard Pap. Bot. 1(3): 75, fig. 3. 1991 [C][E]

Perennials; (cespitose); caudex branched (covered with persistent, somewhat thickened, petioles); not scapose. Stems unbranched, 0.2–0.6(–1) dm, glabrous or sparsely pubescent throughout, trichomes simple, sessile or subsessile, 2- or 3-rayed, 0.2–0.6 mm. Basal leaves rosulate; petiolate; petiole margin ciliate, (trichomes simple, 0.2–0.6 mm); blade linear-oblanceolate, 0.4–1.5 cm × 0.8–1.7 mm, margins entire, (pubescent as petiole), surfaces pubescent abaxially with (appressed), simple, subsessile or sessile, 2–4-rayed trichomes, 0.1–0.4 mm, glabrous or subapically sparsely pubescent adaxially with simple trichomes. Cauline leaves 1–3; sessile; blade linear-oblong, margins entire, surfaces pubescent as basal. Racemes 5–15-flowered, ebracteate, elongated in fruit; rachis not flexuous, glabrous or pubescent as stem. Fruiting pedicels divaricate-ascending, straight, 2–6 mm, glabrous or sparsely pubescent as stem. Flowers: sepals ovate, 1.5–2 mm, pubescent, (trichomes simple and 2-rayed); petals yellow, spatulate, 3–4 × 1.2–1.8 mm, (flared, clawed); anthers ovate, 0.3–0.4 mm. Fruits ovate, plane, flattened, 4–8 × 2–3 mm; valves glabrous; ovules 16–18 per ovary; style 0.2–0.3 mm. Seeds ovoid, 1–1.2 × 0.6–0.7 mm.

Flowering Jun–Jul. Rock crevices along streamlets near timberline; of conservation concern; ca. 3500 m; Colo.

Draba weberi is an apomictic species allied to *D. exunguiculata*, *D. grayana*, and *D. streptobrachia*. From those, it is distinguished by having ovate fruits, clawed petals, and sessile, 2–4-rayed trichomes with untwisted rays often appressed to leaf and stem surfaces. *Draba weberi* is known from near North Star Peak in central Colorado (Summit County).

120. **Draba yukonensis** A. E. Porsild, Publ. Bot. (Ottawa) 4: 37, plate 7. 1975 [C][E][F]

Perennials; (short-lived); caudex simple or branched (with persistent leaf remains); not scapose. Stems branched, 0.4–2 dm, pubescent throughout, trichomes simple, 0.3–1.2 mm, and 2–6-rayed, 0.05–0.6 mm. Basal leaves rosulate; subsessile; blade oblanceolate to narrowly so, 0.3–1 cm × 0.5–2 mm, margins entire, (ciliate, trichomes simple, to 1.3 mm), surfaces pubescent, abaxially with (rigid), simple and 2-rayed trichomes, 0.4–1.2 mm, sometimes with short-stalked, 3–6-rayed ones, 0.1–0.3 mm, adaxially with simple trichomes. Cauline leaves 1–3; sessile; blade ovate to oblong, margins entire, surfaces pubescent as basal. Racemes 5–20-flowered, ebracteate, considerably elongated in fruit; rachis not flexuous, pubescent as stem. Fruiting pedicels erect to ascending, straight, 0.5–3 mm, pubescent as stem. Flowers: sepals ovate, 1.2–1.6 mm, pubescent, (trichomes simple and 2–4-rayed); petals white, spatulate, 1.5–2 × 0.7–0.8 mm; anthers ovate, ca. 0.2 mm. Fruits ovoid to oblong, plane, not flattened, 2–5 × 1.5–2 mm; valves pubescent, trichomes short-stalked, 4- or 5-rayed, 0.05–0.3 mm; ovules 36–52 per ovary; style 0.1–0.3 mm. Seeds oblong, 0.6–0.8 × 0.3–0.4 mm.

Flowering late May–Jun. Open stony ridges on ancient beach in grassy areas and aspen stands; of conservation concern; ca. 600 m; Yukon.

Draba yukonensis is known only from Kluane National Park, southwestern Yukon Territory.

121. **Draba zionensis** C. L. Hitchcock, Revis. Drabas W. N. Amer., 49, plate 2, fig. 16. 1941 [E]

Draba asprella Greene var. *zionensis* (C. L. Hitchcock) S. L. Welsh & Reveal

Perennials; (cespitose); caudex branched (with persistent leaf bases, branches sometimes terminating in sterile rosettes); scapose. Stems unbranched, 0.5–1.5(–1.8) dm, usually sparsely pubescent proximally, usually glabrous distally, rarely throughout, trichomes simple, 0.3–0.8 mm, often with smaller, 2–4-rayed ones. Basal leaves rosulate; shortly petiolate; petiole base sometimes ciliate (margin not ciliate, trichomes simple and 2-rayed, 0.2–0.5 mm); blade spatulate to obovate, 0.7–3.5 cm × 3–10(–12) mm, margins usually entire, rarely obscurely dentate (near apex), surfaces sparsely pubescent with stalked, (2–)4 (or 5)-rayed trichomes, 0.05–0.5 mm, (midvein obscure). Cauline leaves 0. Racemes 14–36-flowered, ebracteate, elongated in fruit; rachis not flexuous, glabrous. Fruiting pedicels ascending to divaricate-ascending, straight, (not expanded basally), (4–)5–12 (–15) mm, glabrous. Flowers: sepals ovate, 2–3 mm, glabrous or sparsely pubescent, (trichomes simple and short-stalked, 2–4-rayed); petals orange-yellow (fading pale yellow), broadly spatulate to subovate, 5–6 × 1.8–2.8 mm; anthers ovate, 0.4–0.5 mm. Fruits lanceolate to linear-lanceolate or ovate, plane (not curved), flattened, 6–12(–17) × 2–3 mm; valves glabrous; ovules 12–20 per ovary; style 0.4–1 mm. Seeds oblong, 1–1.3 × 0.6–0.8 mm. $2n = 26$.

Flowering Mar–May. Sandstone (rarely limestone) rock outcrops and sandy slopes in pinyon-juniper or pine communities; 1000–2500 m; Utah.

R. C. Rollins (1993) treated *Draba zionensis* as a variety of *D. asprella*, but its true relationships appear to lie with two other southern Utah endemics, *D. sobolifera* and *D. subalpina*. *Draba zionensis* is easily distinguished from *D. subalpina* by having orange-yellow (versus white) petals, and from *D. asprella* and *D. sobolifera* by its glabrous (versus pubescent) pedicels and stems distally. Nearly all populations of the species are found in and around Zion National Park in southwestern Utah (Iron, Kane, and Washington counties). A specimen supposedly from the Deep Creek Mountains (Juab County) may be mislabeled.

e. BRASSICACEAE Burnett tribe BOECHEREAE Al-Shehbaz, Beilstein & E. A. Kellogg, Pl. Syst Evol. 259: 111. 2006

Perennials or, rarely, biennials; eglandular. **Trichomes** often short-stalked, sessile, or subsessile, usually forked or dendritic, rarely malpighiaceous, sometimes simple or absent. **Cauline leaves** (sometimes absent); petiolate, subsessile, or sessile; blade base auriculate or not, margins usually entire or dentate, rarely pinnatifid. **Racemes** ebracteate, often elongated in fruit. **Flowers** actinomorphic; sepals erect, ascending, or spreading, lateral pair usually not saccate basally; petals white, yellowish, pink, lavender, or purple, claw present, often indistinct; filaments unappendaged, not winged; pollen 3-colpate. **Fruits** usually siliques, rarely silicles, dehiscent, unsegmented, usually latiseptate, rarely terete or slightly angustiseptate; ovules [2–]4–250 per ovary; style usually distinct, rarely obsolete; stigma usually entire, rarely 2-lobed. **Seeds** biseriate, sub-biseriate, uniseriate, or, rarely, aseriate; cotyledons accumbent or incumbent.

Genera 7, species 119 (7 genera, 117 species in the flora): North America, Asia (Russian Far East).

10. ANELSONIA J. F. Macbride & Payson, Bot. Gaz. 64: 81. 1917 • [For Aven Nelson, 1859–1952, American botanist who studied the flora of Wyoming and neighboring states] E

Ihsan A. Al-Shehbaz

Perennials; (cespitose, deep-rooted); scapose; pubescent, trichomes short-stalked, dendritic or irregularly forked, (soft). **Stems** erect, unbranched. **Leaves** (persistent) basal; rosulate; petiolate; blade margins entire. **Racemes** (corymbose, few- to several-flowered), not or slightly elongated in fruit. **Fruiting pedicels** ascending to suberect, slender. **Flowers:** sepals (early caducous, erect), oblong, (pubescent), base of lateral pair not saccate; petals white to purplish, oblanceolate, (slightly longer than sepals); stamens tetradynamous; filaments not dilated basally; anthers ovate; nectar glands confluent, subtending bases of stamens. **Fruits** (erect, siliques or silicles), sessile or short-stipitate, lanceolate, broadly oblong to narrowly ovate, not torulose, latiseptate; valves each with prominent midvein and somewhat anastomosing lateral veins, glabrous; replum rounded; septum complete; ovules 10–24 per ovary; stigma capitate. **Seeds** biseriate, somewhat flattened, not winged, oblong to ovoid; seed coat (silvery, papillate), not mucilaginous when wetted; cotyledons accumbent. $x = 7$.

Species 1: w United States.

Anelsonia is most closely related to *Boechera* and *Phoenicaulis*, from which it is readily distinguished by its scapose habit and papillate seeds.

1. **Anelsonia eurycarpa** (A. Gray) J. F. Macbride & Payson, Bot. Gaz. 64: 81. 1917 E F

Draba eurycarpa A. Gray, Proc. Amer. Acad. Arts 6: 520. 1865; *Parrya eurycarpa* (A. Gray) Jepson; *P. huddelliana* A. Nelson; *Phoenicaulis eurycarpa* (A. Gray) Abrams

Plants with caudex multistemmed, ultimate slender stems covered by persistent petiolar remains, terminating in rosettes; sparsely to densely pubescent. **Stems** 1–4 cm (leafless). **Basal leaves** tufted; petiole persistent becoming stramineous, 0.5–1 cm; blades narrowly oblanceolate to broadly linear, 0.5–1.8 cm × 1–2 mm, base attenuate, margins entire, apex obtuse. **Fruiting pedicels** 4–15 mm, pilose. **Flowers:** sepals purple, 4–5 × 1.5–2 mm; petals 4.5–6 × 1.7–2.3 mm; filaments 3.5–4.5 mm; anthers 0.5–0.7 mm; gynophore obsolete or to 1 mm. **Fruits** 1.5–3 cm × 5–9 mm; valves purplish, often glaucous, base obtuse, apex acute to acuminate; style 1–2 mm. **Seeds** brown, 2–3 × 1.2–1.5 mm. $2n = 14$.

Flowering Jun–Jul. Rock slides of metamorphics, whitish ash, subalpine and alpine ridges, rock and talus slides, disintegrated volcanic rock; 1600–4000 m; Calif., Idaho, Nev.

Anelsonia eurycarpa has been collected from multiple counties in California; in Nevada it is known only from Washoe County and in Idaho only from Blaine and Custer counties.

11. BOECHERA Á. Löve & D. Löve, Bot. Not. 128: 513. 1976 • [For Tyge Wittrock Böcher, 1909–1983, Danish cytogeneticist who worked on subarctic flowering plants]

Ihsan A. Al-Shehbaz

Michael D. Windham

Perennials or, rarely, biennials; (sexual or apomictic, caudex usually present, rarely absent); not scapose; usually glabrous or pubescent, rarely hirsute or hispid, trichomes simple or branched, 2–14-rayed, sometimes dendritic, not stellate. **Stems** erect, ascending, or decumbent, unbranched or branched distally. **Leaves** basal and cauline; petiolate or sessile; basal rosulate or not, petiolate, blade margins usually entire or dentate, rarely lyrate-pinnatifid; cauline usually sessile, rarely shortly petiolate, blade (base sometimes auriculate or sagittate), margins entire or dentate. **Racemes** (sometimes paniculate), often elongated in fruit. **Fruiting pedicels** erect, ascending, divaricate, or reflexed (secund or not), slender. **Flowers:** sepals ovate or oblong, (lateral pair slightly saccate or not basally, margins membranous); petals usually white, pink, lavender, or purple, rarely yellowish, red, or magenta, spatulate or oblanceolate, (claw shorter than sepals or undifferentiated from blade, apex obtuse); stamens tetradynamous; filaments not dilated basally; anthers ovate or oblong, (apex obtuse), [pollen ellipsoid (sexual plants) or spheroid (apomictic)]; nectar glands confluent, subtending bases of stamens, lateral glands semi-annular or annular. **Fruits** usually sessile, rarely shortly stipitate, usually linear, rarely oblong or lanceolate, straight or falcate, smooth or torulose; valves (papery), each with obscure or prominent midvein, usually glabrous, rarely pubescent; replum (visible), rounded; septum complete, (membranous, veinless); ovules 8–250 per ovary; (style sometimes obsolete); stigma capitate. **Seeds** usually uniseriate or sub-biseriate, rarely biseriate, flattened, winged, not winged, or margined, oblong or orbicular; seed coat (usually smooth or minutely reticulate, rarely papillate), not mucilaginous when wetted; cotyledons accumbent.

Species 111+ (109 in the flora): North America, n Mexico, e Asia (Russian Far East).

Boechera falcata (Turczaninow) Al-Shehbaz is known from eastern Asia (Russian Far East).

Boechera often is treated as a synonym of *Arabis* (e.g., R. C. Rollins 1993; S. L. Welsh et al. 2003) but it has become clear that morphological similarities between these groups are due to evolutionary convergence, not shared ancestry. Molecular analyses by M. Koch et al. (2001)

and T. Mitchell-Olds et al. (2005) revealed that *Arabis* and *Boechera* belong to distantly related clades of Brassicaceae that diverged some 19–25 million years ago. A new tribal classification of the family (I. A. Al-Shehbaz et al. 2006) places them in different tribes (Arabideae and Boechereae, respectively), reflecting their substantial molecular divergence.

The taxonomic complexity of *Arabis*, in the broad sense, is legendary (R. C. Rollins 1941, 1993; G. A. Mulligan 1996). When the genus is split, most of the problematic taxa come to reside in *Boechera*. A rare confluence of hybridization, apomixis, and polyploidy makes this one of the most difficult genera in the North American flora. The sexual diploid species are relatively distinct from one another, but they hybridize wherever they come into contact. Through apomixis and polyploidy, the hybrids become stable, self-propagating lineages. Most of the hybrid derivatives in *Boechera* are triploids, but apomictic diploids are known as well. Thus, for any pair of sexual diploid species (e.g., AA and BB), this process can yield different intermediates, including AB apomicts and both possible apomictic triploids (AAB and ABB). The situation becomes even more challenging when a third sexual diploid enters the picture. To date, we have identified three taxa (*B. divaricarpa*, *B. pinetorum*, *B. tularensis*) that appear to be trigenomic triploids. Under these circumstances, even the most distinctive sexual diploid progenitors can become lost in a seemingly continuous range of morphological variability.

In a genus characterized by the presence of polyploids and apomicts, it is essential to know which taxa represent the products of primary, divergent evolution (i.e., sexual diploids) and which are the result of secondary, reticulate evolution. Fortunately, a strong correlation between pollen morphology and ploidy level/reproductive mode facilitates the separation of sexual diploids from polyploids and apomicts in *Boechera*. Because of differences in meiosis, sexual diploids produce small (13–16 μm diam.), ellipsoid pollen grains with symmetrical colpi. In apomictic individuals, the pollen grains are significantly larger (20–30 μm diam.), and spheroid with asymmetrical colpi. The differences in pollen size and shape are so pronounced that the ploidy level and reproductive mode of most plants with flowers can be determined using a medium power (40×) dissecting microscope (see Fig. 1 in M. D. Windham and I. A. Al-Shehbaz 2006).

To facilitate the study of ploidy level and reproductive mode in *Boechera* and to allow direct comparison of taxa named by previous authors, we assembled at the Missouri Botanical Garden the holotypes and isotypes of all taxa originally described in *Arabis* and currently placed in *Boechera* (over 160 published basionyms). In addition to the types, another 12,000 specimens were examined to document morphological variability and geographic distribution. During this process, we identified additional morphological features (e.g., trichome branching patterns, number of ovules per ovary, pollen and seed morphology) overlooked or underused by previous authors. The result is a substantially revised taxonomy for the genus, the nomenclatural foundation for which was established in a series of papers (M. D. Windham and I. A. Al-Shehbaz 2006, 2007, 2007b).

In many cases, the species circumscriptions adopted here deviate significantly from those of previous authors. Where R. C. Rollins (1993) accepted 63 species with varieties, we recognize 109 species and two with two subspecies. Our treatment includes a total of 71 sexual species. These represent the morphological extremes of the complex and are often easily distinguished (when separated from the apomicts using pollen characters). Although it is likely that some diploid species remain to be discovered, we feel that this portion of the treatment is relatively complete.

Our coverage of the apomictic hybrids is much less comprehensive. There are literally hundreds of hybrids in *Boechera* with unique genomic combinations, each of which could be recognized at species level. Our treatment includes just 38 apomictic species, primarily taxa recognized at some level by other authors. Because some hybrid combinations are not formally recognized, it is inevitable that some names will be misapplied to superficially similar hybrids of different parentage. For example, plants of *B. goodrichii* (= *B. retrofracta* × *B. gracilipes*), *B. consanguinea* (= *B. retrofracta* × *B. fendleri*), and *B. pinetorum* (= *B. retrofracta* × *B. rectissima* × *B. sparsiflora*) are sufficiently similar that they might be considered a single taxon if their respective parentages and disjunct geographic ranges were not taken into account. Indeed, all three have been called *Arabis holboellii* var. *pinetorum* (e.g., R. C. Rollins 1993; S. L. Welsh et al. 2003), although the epithet *consanguinea* has priority.

The best way to avoid such misidentification is to pay close attention to the geographic distribution of apomictic taxa and their sexual progenitors. Apomictic hybrids in *Boechera* appear to be of relatively recent origin and generally have not migrated beyond regions where their parents are sympatric. Thus, users of this treatment should be wary of major range extensions for apomictic taxa; in most cases, these will turn out to be unique hybrid combinations not represented in the keys or descriptions. Because the use of hybrid binomials is potentially misleading, the best approach to identifying a hybrid is to provide a formula name based on the hypothesized parentage (e.g., *B. fendleri* × *B. stricta* or *B. fendleri* hybrid). This requires an accurate understanding of the sexual diploids occurring in the region of interest, which we hope the following keys and descriptions will provide.

Given the inherent taxonomic complexity of *Boechera*, it has been necessary to incorporate micromorphological characters such as pollen morphology and trichome branching patterns in the identification keys. Whenever possible, we have restricted such characters to later couplets, but microscopic observations are required to distinguish some species. Effective use of the keys also depends on having complete specimens bearing both flowers and fruits. In all cases, measurements of stem length are taken from fruiting plants, those of basal leaves from the largest in the basal rosette, for the fruiting pedicels from the longest in the infructescence, and for the stem trichomes from the largest near the base. Descriptions of the pedicels, flowers, and fruits are taken from the main inflorescence rather its lateral branches, the number of seed rows per locule is determined near the middle of the fruit, and the number of ovules is observed in mature fruits by counting the number of seeds plus the abortive ovules.

SELECTED REFERENCES Al-Shehbaz, I. A. 2003b. Transfer of most North American species of *Arabis* to *Boechera* (Brassicaceae). Novon 13: 381–391. Windham, M. D. and I. A. Al-Shehbaz. 2006. New and noteworthy species of *Boechera* (Brassicaceae) I: Sexual diploids. Harvard Pap. Bot. 11: 61–88. Windham, M. D. and I. A. Al-Shehbaz. 2007. New and noteworthy species of *Boechera* (Brassicaceae) II: Apomictic hybrids. Harvard Pap. Bot. 11: 257–274. Windham, M. D. and I. A. Al-Shehbaz. 2007b. New and noteworthy species of *Boechera* (Brassicaceae) III: Additional sexual diploids and apomictic hybrids. Harvard Pap. Bot. 12: 235–257.

Key to Species Groups of *Boechera*

1. Cauline leaf blades not auriculate .... Group 1, p. 351
1. Cauline leaf blades auriculate.
    2. Fruit valves pubescent .... Group 2, p. 353
    2. Fruit valves glabrous.
        3. Stems glabrous proximally .... Group 3, p. 354
        3. Stems pubescent proximally.
            4. Fruits reflexed, pendent, or descending .... Group 4, p. 356
            4. Fruits erect, ascending, or horizontal .... Group 5, p. 360

Group 1

1. Basal leaf blade surfaces glabrous or with simple trichomes only.
    2. Biennials, without caudices; stems (2–)3–10 dm; cauline leaves 18–80.
        3. Racemes usually unbranched; cauline leaves 18–28; ovules 64–80 per ovary; seeds 1.2–1.4 mm wide . . . . 6. *Boechera burkii*
        3. Racemes highly branched; cauline leaves 30–80; ovules 30–42 per ovary; seeds 0.7–1 mm wide. . . . . .93. *Boechera serotina*
    2. Perennials, with caudices; stems 0.3–2.5 dm; cauline leaves 1–12.
        4. Fruits pendent; petals white to pale lavender; (c Colorado). . . . . 62. *Boechera oxylobula* (in part)
        4. Fruits ascending to erect; petals usually purple to lavender (white in *B. davidsonii*).
            5. Leaf blade surfaces puberulent (trichomes simple); petals 4–5 mm; fruits 1–1.5 mm wide . . . . .95. *Boechera shevockii*
            5. Leaf blade surfaces glabrous; petals 6–10 mm; fruits 1.5–2.5 mm wide.
                6. Caudex branches with persistent, crowded leaf bases; fruits usually ascending, not appressed to rachises; seeds uniseriate . . . . 18. *Boechera davidsonii*
                6. Caudex branches without persistent, crowded leaf bases; fruits erect, appressed to rachises; seeds biseriate or sub-biseriate. . . . . .56. *Boechera lyallii* (in part)
1. Basal leaf blade surfaces with at least some branched trichomes.
    7. Styles 2.5–8 mm.
        8. Petals white, 6–8 mm; basal leaf blade surfaces glabrous; fruits pendent, 4–6(–7.5) cm; n California . . . . .14. *Boechera constancei*
        8. Petals lavender to purple, 8–13 mm; basal leaf blade surfaces pubescent; fruits ascending, 1.5–2.5 cm; s California . . . . 65. *Boechera parishii*
    7. Styles 0.05–2 mm.
        9. Plants usually glabrous throughout (except margins of basal leaf blades).
            10. Fruits 3–5(–7) mm wide; seeds 3–6 × 2–4 mm.
                11. Basal leaf blade margins with 2–4-rayed and simple trichomes; ovules 18–30 per ovary; plants apomictic, pollen spheroid . . . . .15. *Boechera covillei* (in part)
                11. Basal leaf blade margins with only 2-rayed and simple trichomes; ovules 10–20 per ovary; plants sexual, pollen ellipsoid. . . . . .46. *Boechera howellii* (in part)
            10. Fruits 1.4–2.5 mm wide; seeds 1.5–2.2 × 1–1.5 mm.
                12. Petals 6–8.5 mm; fruiting pedicels erect; ovules 34–64 per ovary; seeds biseriate to sub-biseriate, with continuous wing. . . . . .56. *Boechera lyallii* (in part)
                12. Petals 4–5 mm; fruiting pedicels ascending; ovules 14–22 per ovary; seeds uniseriate, not winged or with distal wing . . . . .103. *Boechera tiehmii* (in part)
        9. Plants usually sparsely to densely pubescent proximally (sometimes throughout).
            13. Fruit valves pubescent.
            14. Basal leaf blades 10–25(–50) mm wide; petals 3.5–6 mm . . . .85. *Boechera repanda* (in part)
            14. Basal leaf blades 1–5 mm wide; petals 6–12 mm.
                15. Fruiting pedicels usually ascending to widely pendent, rarely horizontal . . . . 55. *Boechera lincolnensis* (in part)
                15. Fruiting pedicels reflexed, pendent, or recurved.
                    16. Fruiting pedicels 1.5–2 mm; petals yellow proximally, brick-red distally . . . . 109. *Boechera yorkii*
                    16. Fruiting pedicels 4–20 mm; petals purple, lavender, or white.
                        17. Stems arising at ground surface, usually without woody bases; seeds uniseriate; ovules 26–64 per ovary.
                            18. Basal leaf blade margins entire; stems proximally with trichomes 0.05–0.15 mm; cauline leaves 4–10; racemes 5–12-flowered; ovules 26–36 per ovary . . . 52. *Boechera lasiocarpa* (in part)
                            18. Basal leaf blade margins usually dentate; stems proximally with trichomes 0.1–0.3(–0.5) mm; cauline leaves 7–45 (–65); racemes 10–40(–64)-flowered; ovules 38–64 per ovary . . . . 79. *Boechera puberula* (in part)

17. Stems elevated above ground surface on woody bases; seeds biseriate; ovules 68–106 per ovary.
    19. Fruiting pedicels not abruptly recurved at base; petals white to lavender; fruits 1.6–3 mm wide, usually not appressed to rachises; seeds 1.2–1.6 × 1–1.2 mm; Colorado Plateau .... 32. *Boechera formosa*
    19. Fruiting pedicels abruptly recurved at base; petals usually purple, rarely white; fruits 2.5–4 mm wide, usually appressed to rachises; seeds 1.7–2.8 × 1.5–2.2 mm; California, w Nevada.................. 80. *Boechera pulchra* (in part)

[13. Shifted to left margin.—Ed.]

13. Fruit valves glabrous.
    20. Fruits reflexed, pendent, or descending.
        21. Biennials; basal leaf blades 10–30 mm wide .... 9. *Boechera canadensis* (in part)
        21. Perennials; basal leaf blades 1–7 mm wide.
            22. Fruits 5–7(–8) mm wide; seeds 4–5 mm wide; primarily Mojave Desert..... .... 35. *Boechera glaucovalvula*
            22. Fruits 0.9–2.3 mm wide; seeds (0.1–)0.6–1.5 mm wide; not Mojave Desert.
                23. Fruiting pedicels abruptly recurved at bases; fruits reflexed, usually appressed to rachises; cauline leaves 15–40; racemes 15–80 (–140)-flowered .... 86. *Boechera retrofracta* (in part)
                23. Fruiting pedicels not abruptly recurved at bases; fruits divaricate-descending to pendent, not appressed to rachises; cauline leaves 2–17; racemes 2–13 (–17)-flowered.
                    24. Fruits divaricate-descending to slightly descending, strongly secund.
                        25. Basal leaf blade surfaces with 3–9-rayed trichomes; stems proximally with 2–6-rayed trichomes; seeds 0.9–1.5 mm wide, with continuous wing 0.1–0.5 mm wide; plants sexual, pollen ellipsoid .... 53. *Boechera lemmonii* (in part)
                        25. Basal leaf blade surfaces with 2- or 3-rayed trichomes; stems proximally with simple and 2-rayed trichomes; seeds 0.8–0.9 mm wide, not winged or with distal wing 0.05–0.1 mm wide; plants apomictic, pollen spheroid .... 81. *Boechera pusilla* (in part)
                    24. Fruits pendent, usually not secund, rarely weakly so.
                        26. Basal leaf blade surfaces with 4–8-rayed trichomes.
                            27. Petals 6–10 mm; stems proximally with 4–8-rayed trichomes 0.05–0.15 mm; ovules 26–36 per ovary; seeds uniseriate ... . 52. *Boechera lasiocarpa* (in part)
                            27. Petals 4–6 mm; stems proximally with simple and 2–4-rayed trichomes 0.1–0.4 mm; ovules 66–92 per ovary; seeds sub-biseriate .... 70. *Boechera pendulocarpa*
                        26. Basal leaf blade surfaces with simple and 2- or 3-rayed trichomes.
                            28. Basal leaf blade surfaces with 2- or 3-rayed trichomes 0.1–0.4 mm; ovules 28–44 per ovary; seeds uniseriate, with continuous wing; c Colorado .... 62. *Boechera oxylobula* (in part)
                            28. Basal leaf blade surfaces with simple and 2-rayed trichomes 0.3–0.8 mm; ovules 40–70(–100) per ovary; seeds biseriate, usually not winged; not c Colorado .... 69. *Boechera pendulina* (in part)

[20. Shifted to left margin.—Ed.]
20. Fruits erect, ascending, or horizontal.
    29. Fruits secund.
        30. Basal leaf blade surfaces with 3–9-rayed trichomes; stems proximally with 2–6-rayed trichomes; seeds 1–1.5 mm wide, with continuous wing 0.1–0.5 mm wide; plants sexual, pollen ellipsoid . . . . 53. *Boechera lemmonii* (in part)
        30. Basal leaf blade surfaces with 2- or 3-rayed trichomes or glabrous; stems proximally with simple and 2-rayed trichomes; seeds 0.8–0.9 mm wide, not winged or with distal wing 0.05–0.1 mm wide; plants apomictic, pollen spheroid . . . . . 81. *Boechera pusilla* (in part)
    29. Fruits not secund.
        31. Stems proximally with simple trichomes 0.5–1.5 mm; ovules 56–106 per ovary . . . . 17. *Boechera cusickii* (in part)
        31. Stems proximally with simple and/or branched trichomes usually less than 0.5 mm; ovules 8–52 per ovary.
            32. Fruits 1.3–1.7(–2) mm wide, erect, appressed to rachises . . . . . 67. *Boechera paupercula* (in part)
            32. Fruits 2–5.5 mm wide, divaricate-ascending to erect, ± appressed to rachises.
                33. Basal leaf blades 7–25(–50) mm wide, margins usually repand to dentate, rarely entire . . . . 85. *Boechera repanda* (in part)
                33. Basal leaf blades 0.8–7(–10) mm wide, margins entire.
                    34. Basal leaf blade surfaces with at least some 8–14-rayed trichomes.
                        35. Petals 3.5–6 mm; styles to 0.1 mm; ovules 44–52 per ovary; cauline leaves 1–5 . . . . 21. *Boechera dispar*
                        35. Petals 9–14 mm; styles (0.7–)1–2 mm; ovules 26–34 per ovary; cauline leaves 4–10 . . . . 48. *Boechera johnstonii*
                    34. Basal leaf blade surfaces with 2–6 (or 7)-rayed trichomes only.
                        36. Stems proximally with simple and branched trichomes; fruits (1.3–)2–3.3 cm; ovules 8–12 per ovary . . . . 82. *Boechera pygmaea*
                        36. Stems proximally with branched trichomes only; fruits (2.5–)3–8.5 cm; ovules 16–44 per ovary.
                            37. Fruits 3–5.5 mm wide; seeds 3–6.5(–8) × 2–4.5 mm, wing (0.8–)1.2–2.5 mm wide.
                                38. Basal leaf petioles with simple or spurred cilia to 1 mm, blade surfaces with short-stalked, 2–4 (or 5)-rayed trichomes 0.1–0.3 mm; fruits without parallel edges (undulate and constricted between seeds) . . . . 75. *Boechera platysperma*
                                38. Basal leaf petioles without cilia, blade surfaces with long-stalked, 3–6 (or 7)-rayed trichomes 0.2–0.5 mm; fruits with parallel edges (not undulate and constricted between seeds) . . . . 105. *Boechera ultra-alsa*
                            37. Fruits 2–3.2 mm wide; seeds 2.5–3.5 × 1.5–2.5 mm, wing 0.2–0.9 mm wide.
                                39. Sepals glabrous; basal leaf blades 2–6 mm wide, with sessile or subsessile trichomes; fruiting pedicels 4–8 mm; fruits erect-ascending, often appressed to rachises . . . . 25. *Boechera elkoensis* (in part)
                                39. Sepals sparsely pubescent; basal leaf blades 1–2.5(–3) mm wide, with stalked trichomes; fruiting pedicels 2–6 mm; fruits ascending, not appressed to rachises . . . 74. *Boechera pinzliae* (in part)

Group 2

1. Petals 2–4 mm; fruits 0.7–1 mm wide; seeds not winged; basal leaf blades (5–)10–45 mm wide.
    2. Basal leaf blade adaxial surfaces with simple trichomes; petals white to cream; fruiting pedicels 1–3(–6) mm . . . . 19. *Boechera dentata* (in part)
    2. Basal leaf blade adaxial surfaces with 3- or 4(–6)-rayed trichomes; petals purplish; fruiting pedicels (5–)7–15 mm . . . . 72. *Boechera perstellata*

1. Petals 4–16 mm; fruits 0.9–3(–4) mm wide; seeds usually winged; basal leaf blades 1–7(–11) mm wide.
    3. Fruiting pedicels reflexed, pendent, or descending.
        4. Petals white to lavender; fruits 0.9–2.2 mm wide; seeds 0.9–1.8 × 0.7–1.4 mm.
            5. Fruiting pedicels not abruptly recurved at bases; fruits 1.9–2.2 mm wide, pendent, rarely appressed to rachises; ovules 38–64 per ovary; seeds 1.4–1.8 mm, wing 0.1–0.3 mm wide. . . . . 79. *Boechera puberula* (in part)
            5. Fruiting pedicels abruptly recurved at bases; fruits 0.9–1.5 mm wide, reflexed, usually appressed to rachises; ovules 60–116 per ovary; seeds 0.9–1.4 mm, wing 0.05–0.1 mm wide.
                6. Fruit valves densely pubescent; petals sparsely pubescent abaxially . . . . 76. *Boechera polyantha*
                6. Fruit valves sparsely pubescent; petals glabrous abaxially . . . . .86. *Boechera retrofracta* (in part)
        4. Petals usually purple; fruits (1.6–)2–4 mm wide; seeds 1.7–3.5 × 1.3–2.2 mm.
            7. Stems from near ground surface, usually without woody bases; seeds uniseriate; ovules 20–42 per ovary; north of 40°N latitude.
                8. Basal leaf blade margins entire; petals 6–8 mm; styles 1.5–2 mm; fruits divaricate-descending, valves sparsely pubescent . . . . 94. *Boechera serpenticola* (in part)
                8. Basal leaf blade margins (at least some) prominently dentate or subpinnatifid; petals 9–14 mm; styles 0.5–1 mm; fruits pendent, valves moderately to densely pubescent . . . . 100. *Boechera subpinnatifida* (in part)
            7. Stems elevated above ground surface on woody bases; seeds biseriate to sub-biseriate; ovules 68–126 per ovary; south of 40°N latitude.
                9. Fruiting pedicels abruptly recurved at bases; fruits 2.5–4 mm wide, valves densely pubescent proximally; basal leaf blades 1–3 mm wide. . . . . 80. *Boechera pulchra* (in part)
                9. Fruiting pedicels not abruptly recurved at bases; fruits 2–2.5 mm wide, valves glabrous proximally; basal leaf blades 3–7 mm wide. . . . . . . .108. *Boechera xylopoda*
    3. Fruiting pedicels suberect, ascending, or horizontal.
        10. Stems proximally with simple and 2-rayed trichomes 0.4–1 mm; basal leaf blade surfaces with long-stalked, 2–4-rayed trichomes. . . . . 5. *Boechera breweri* (in part)
        10. Stems proximally with 2–12-rayed trichomes 0.1–0.5 mm; basal leaf blade surfaces with short-stalked, (3- or) 4–12-rayed trichomes.
            11. Fruits (4.5–)6–11 cm, usually curved; petals 0.8–1.2 mm wide; ovules 140–190 per ovary . . . . .96. *Boechera shockleyi* (in part)
            11. Fruits 3–5.8 cm, straight; petals 1.5–4.5 mm wide; ovules 34–120 per ovary.
                12. Styles 1–1.5 mm; seeds uniseriate; ovules 34–68 per ovary; stems proximally with 5–10-rayed trichomes; w Montana . . . . .29. *Boechera fecunda*
                12. Styles 0.1–0.3 mm; seeds biseriate or sub-biseriate; ovules 72–120 per ovary; stems proximally with 2–7-rayed trichomes; sw United States.
                    13. Fruit valves glabrous proximally; cauline leaves 3–8; basal leaves 2–5 mm wide; fruiting pedicels 7–14 mm. . . . . 24. *Boechera duchesnensis*
                    13. Fruit valves pubescent proximally; cauline leaves 10–25; basal leaves 1–2 mm wide; fruiting pedicels 10–20(–25) mm . . . . . 55. *Boechera lincolnensis* (in part)

Group 3

1. Fruits erect, usually appressed to rachises.
    2. Basal leaf blade surfaces with 4–8-rayed trichomes; petals 3–4 mm; seeds 0.7–0.9 × 0.5–0.6 mm . . . . 107. *Boechera williamsii*
    2. Basal leaf blade surfaces glabrous or with 2–4(–6)-rayed trichomes; petals 5–11 mm; seeds 1.3–2.2 × 1–1.8 mm.
        3. Basal leaf blade surfaces with malpighiaceous (usually along margins) and sessile, 3- or 4-rayed trichomes. . . . . .7. *Boechera calderi*
        3. Basal leaf blade surfaces either with malpighiaceous or with short-stalked 2–4(–6)-rayed trichomes (rarely completely glabrous).

4. Basal leaves usually with short-stalked, 2–4(–6)-rayed trichomes (at least on margins), malpighiaceous trichomes absent; petals lavender to purple; cauline leaves 1–5; ovules 34–64 per ovary . . . . 56. *Boechera lyallii* (in part)
4. Basal leaves with exclusively malpighiaceous trichomes (rarely with simple cilia on petioles); petals usually white, rarely lavender; cauline leaves 6–52; ovules 110–216 per ovary . . . . 99. *Boechera stricta* (in part)
1. Fruits reflexed, pendent, horizontal, or ascending, not appressed to rachises or, if so, reflexed.
   5. Fruits pendent to reflexed.
      6. Fruiting pedicels (at least some) abruptly recurved at bases.
         7. Stems elevated above ground surface on woody bases; fruits 3–5(–6) mm wide; seeds with wings 0.8–1.5 mm wide; racemes 6–12-flowered. . . . . 101. *Boechera suffrutescens* (in part)
         7. Stems from ground surface, usually without woody bases; fruits 1–2.5 mm wide; seeds with wings 0.05–0.25 mm wide; racemes 10–50-flowered.
            8. Petals lavender, 5–6 mm; basal leaf blade surfaces glabrous or with 3–6-rayed trichomes; fruits 1–1.5 mm wide; ne Oregon. . . . . 42. *Boechera hastatula*
            8. Petals white, 3–4 mm; basal leaf blade surfaces with simple and 2- or 3-rayed trichomes; fruits 1.8–2.5 mm wide; California to sc Oregon. . . . . 84. *Boechera rectissima* (in part)
      6. Fruiting pedicels not abruptly recurved at bases.
         9. Fruits 1.5–2.5 mm wide; stems arising from margins of rosettes or arising laterally to sterile shoots.
            10. Fruiting pedicels 10–27 mm; basal leaf blade surfaces with 2–6-rayed trichomes; cauline leaf blades with auricles 0.5–3.5(–5.5) mm; s New Mexico . . . . 77. *Boechera porphyrea* (in part)
            10. Fruiting pedicels 5–9 mm; basal leaf blade surfaces glabrous or with 2- or 3-rayed trichomes; cauline leaf blades with auricles 0.2–0.5 mm; c Nevada to w Utah. . . . . 92. *Boechera schistacea*
         9. Fruits 2.5–5(–6) mm wide; stems arising from centers of rosettes.
            11. Basal leaf blade surfaces with trichomes 0.4–0.6 mm; ovules 80–130 per ovary; seeds biseriate, wing 0.1–0.15 mm wide; New Mexico, Texas. . .102. *Boechera texana*
            11. Basal leaf blade surfaces glabrous or with trichomes 0.07–0.4 mm; ovules 14–30 per ovary; seeds uniseriate, wing 0.3–1.5 mm wide; California, Idaho, Nevada, Oregon, Washington.
               12. Petals 8–11 mm; fruits 2.5–3.5 mm wide; seeds with wing 0.3–0.6 mm wide; racemes 3–7-flowered. . . . . 88. *Boechera rollei*
               12. Petals 4.5–6 mm; fruits 3–5(–6) mm wide; seeds with wing 0.8–1.5 mm wide; racemes 6–12-flowered . . . . 101. *Boechera suffrutescens* (in part)
   5. Fruits ascending, horizontal, or slightly descending.
      13. Basal leaf blade surfaces glabrous or with simple trichomes only.
         14. Basal leaf blade margins entire; fruits 1.9–4 cm; seeds biseriate; Nevada . . . . .60. *Boechera nevadensis*
         14. Basal leaf blade margins dentate to lyrate-pinnatifid; fruits (4–)6–11.7 cm; seeds uniseriate; not Nevada.
            15. Basal leaf blade margins dentate or serrate; petals 3–5 mm . . . . 50. *Boechera laevigata*
            15. Basal leaf blade margins lyrate-pinnatifid; petals 5–10 mm . . . . .59. *Boechera missouriensis* (in part)
      13. Basal leaf blade surfaces with branched trichomes only.
         16. Basal leaf blade trichomes sessile or subsessile.
            17. Fruits secund, 2–3.5 mm wide; ovules 44–76(–104) per ovary; racemes 8–15-flowered . . . . 23. *Boechera drepanoloba* (in part)
            17. Fruits not secund, 1–2.5 mm wide; ovules 84–146 per ovary; racemes 12–88-flowered.

18. Petals usually purple, rarely lavender, 1.5–3 mm wide; fruits 1.7–2.5 mm wide; seeds 1–1.5 mm wide .... 22. *Boechera divaricarpa* (in part)
18. Petals usually white, rarely lavender, 1–2 mm wide; fruits 1–1.8 mm wide; seeds 0.8–1.2 mm wide .... 39. *Boechera grahamii* (in part)

[16. Shifted to left margin.—Ed.]

16. Basal leaf blade trichomes stalked.
    19. Fruits 6–10 cm; seeds biseriate; ovules 170–220 per ovary; cauline leaves 15–65; Channel Islands, California.................................... 43. *Boechera hoffmannii* (in part)
    19. Fruits 1.5–5.5(–6.5) cm; seeds uniseriate; ovules 10–86 per ovary; cauline leaves 2–12; not Channel Islands, California.
        20. Fruits 3–5(–7) mm wide; seeds 3–6 × 2–4 mm.
            21. Basal leaf blade margins with some 4-rayed trichomes; ovules 18–30 per ovary; plants apomictic, pollen spheroid.......................... 15. *Boechera covillei* (in part)
            21. Basal leaf blade margins without 4-rayed trichomes; ovules 10–20 per ovary; plants sexual, pollen ellipsoid............................. 46. *Boechera howellii* (in part)
        20. Fruits 1–3 mm wide; seeds 1–2.2 × 0.8–1.6 mm.
            22. Fruits secund.
                23. Basal leaf blade surfaces glabrous or with 2- or 3-rayed trichomes; seeds 0.8–0.9 mm wide, not winged or with distal wing 0.05–0.1 mm wide... .... 81. *Boechera pusilla* (in part)
                23. Basal leaf blade surfaces with 3–9-rayed trichomes; seeds 0.9–1.5 mm wide, with continuous wing 0.1–0.5 mm wide.
                    24. Cauline leaf blades with auricles to 0.5 mm; ovules 28–44 per ovary; basal leaves densely or sparsely pubescent, petioles ciliate.......... .... 53. *Boechera lemmonii* (in part)
                    24. Cauline leaf blades with auricles 1–2 mm; ovules 42–54 per ovary; basal leaves sparsely pubescent, petioles not ciliate .... 63. *Boechera paddoensis*
            22. Fruits not secund.
                25. Basal leaf blade surfaces glabrous, margins with simple or 2- or 3-rayed trichomes; ovules 14–22 per ovary; plants sexual, pollen ellipsoid...... .... 103. *Boechera tiehmii* (in part)
                25. Basal leaf blade surfaces and margins with 2–10-rayed trichomes; ovules 48–86 per ovary; plants apomictic, pollen spheroid.
                    26. Fruiting pedicels 8–12 mm; basal leaf blade surfaces with 2–5-rayed trichomes; stems from margins of rosettes; n Utah .... 41. *Boechera harrisonii* (in part)
                    26. Fruiting pedicels 2–6 mm; basal leaf blade surfaces with 4–7-rayed trichomes; stems from centers of rosettes; s California .... 68. *Boechera peirsonii* (in part)

Group 4

1. Fruiting pedicels usually abruptly recurved at bases.
    2. Fruits 3–6 mm wide; ovules 20–30 per ovary; racemes 6–12-flowered .... 101. *Boechera suffrutescens* (in part)
    2. Fruits 0.9–2.5 mm wide; ovules 46–126 per ovary; racemes 10–140-flowered.
        3. Stems proximally with branched trichomes only.
            4. Fruits (2–)2.2–2.5 mm wide, descending, not appressed to rachises, strongly secund; Greenland .... 44. *Boechera holboellii*
            4. Fruits 0.9–1.5 mm wide, reflexed, usually appressed to rachises, usually not secund, rarely so; not Greenland .... 86. *Boechera retrofracta* (in part)
        3. Stems proximally with simple and branched trichomes, or simple trichomes only.
            5. Stems proximally with simple trichomes only; petals 3–4 mm; ovules 46–80 per ovary .... 84. *Boechera rectissima* (in part)
            5. Stems proximally with simple and branched trichomes; petals 4–6 mm; ovules 70–126 per ovary.

6. Basal leaf blade surfaces with 5–8-rayed trichomes; fruits 0.9–1.5 mm wide; seeds 1–1.4 × 0.8–1 mm; plants sexual, pollen ellipsoid.............12. *Boechera collinsii*
6. Basal leaf blade surfaces with 2–4 (or 5)-rayed trichomes; fruits 1.5–2 mm wide; seeds 1.5–1.8 × 1.2–1.5 mm; plants apomictic, pollen spheroid ... .73. *Boechera pinetorum* (in part)

1. Fruiting pedicels usually not abruptly recurved at bases (except *B. subpinnatifida*).
   7. Fruiting pedicels sparsely to densely pubescent.
      8. Stems proximally with simple and branched trichomes, or simple trichomes only.
         9. Stems proximally with simple trichomes only; fruits 2.5–4 mm wide; seeds with wing 0.4–1 mm wide.............................9. *Boechera canadensis* (in part)
         9. Stems proximally with simple and branched trichomes; fruits 0.8–2.5 mm wide; seeds with wing 0.05–0.3 mm wide.
            10. Petals purple to magenta, 8–12 × 2.5–4 mm; stems elevated above ground surface on woody base; caudices with persistent (peg-like) leaf bases.... ....49. *Boechera koehleri* (in part)
            10. Petals lavender to white, 4–8 × 0.5–2 mm; stems from near ground surface, without woody base; caudices without persistent leaf bases.
               11. Fruits 2.5–4 cm; ovules 36–54 per ovary; stems usually 2–6 per caudex branch, arising from margins of rosettes; racemes 7–15-flowered; plants sexual, pollen ellipsoid ....40. *Boechera gunnisoniana* (in part)
               11. Fruits 4–10.5 cm; ovules 56–162 per ovary; stems usually 1 per caudex branch, arising from centers of rosettes; racemes 10–63-flowered; plants apomictic, pollen spheroid.
                  12. Fruits reflexed; fruiting pedicels with few simple and 2-rayed trichomes ....73. *Boechera pinetorum* (in part)
                  12. Fruits pendent, horizontal, divaricate, ascending, or descending; fruiting pedicels with many 2- or 3-rayed trichomes.
                     13. Petals 4–5 mm; fruits 0.8–1.2 mm wide; ovules 56–78 per ovary; cauline leaf blades with auricles 0.5–1.5 mm; basal leaf blade surfaces with 3–6-rayed trichomes 0.07–0.3 mm ....57. *Boechera macounii* (in part)
                     13. Petals 5–8 mm; fruits 1.5–2.2 mm wide; ovules 80–162 per ovary; cauline leaf blades with auricles (1–)3–10 mm; basal leaf blade surfaces with 2–5-rayed trichomes 0.3–0.6 mm ....66. *Boechera pauciflora* (in part)
      8. Stems proximally with branched trichomes only.
         14. Stems arising from margins of rosettes, elevated above ground surface on woody bases... 89. *Boechera rollinsiorum*
         14. Stems arising from centers of rosettes, produced near ground surface, usually without woody bases (except *B. californica*).
            15. Ovules 14–44 per ovary (–64 in *B. cobrensis*); plants sexual, pollen ellipsoid.
               16. Fruits 0.7–0.8 mm wide; seeds 0.7–1.1 × 0.5–0.6 mm, not winged; basal leaf blade surfaces with simple trichomes adaxially .... 19. *Boechera dentata* (in part)
               16. Fruits 1.6–3 mm wide; seeds 1.3–3.5 × 1–2.2 mm, winged; basal leaf blade surfaces with (2–)4–10-rayed trichomes adaxially.
                  17. Petals 9–14 mm; basal leaf blade margins (at least some) prominently dentate or subpinnatifid; cauline leaves (10–)20–60 ....100. *Boechera subpinnatifida* (in part)
                  17. Petals 3.5–8 mm; basal leaf blade margins entire; cauline leaves 2–12(–14).

18. Petals lavender to white, 3.5–6 mm; styles to 0.2 mm; ovules 34–64 per ovary; sandy habitats in California, Idaho, Nevada, Oregon, Wyoming . . . . 11. *Boechera cobrensis* (in part)
18. Petals purple, 6–8 mm; styles 1.5–2 mm; ovules 20–24 per ovary; serpentine outcrops in nw California . . . . .94. *Boechera serpenticola* (in part)
15. Ovules (56–)68–180 per ovary; plants apomictic, pollen spheroid.
19. Petals 9–14 mm; fruits (6–)8–12 cm; ovules 140–180 per ovary; s California . . . . .8. *Boechera californica* (in part)
19. Petals 5–8.5 mm; fruits 3–6.5 cm; ovules 56–134 per ovary; California, Colorado, Nevada, New Mexico, Utah.
20. Cauline leaves 5–9, not concealing stem proximally; styles 0.5–1 mm; ovules 62–80 per ovary; seeds uniseriate. . . . . 28. *Boechera falcifructa*
20. Cauline leaves (7–)12–72, often concealing stem proximally; styles 0.05–0.5 mm; ovules 74–136 per ovary; seeds usually biseriate or sub-biseriate (rarely uniseriate in *B. inyoensis*).
21. Stems proximally with 7–12-rayed trichomes. . . 47. *Boechera inyoensis* (in part)
21. Stems proximally with 2–8-rayed trichomes.
22. Fruiting pedicels 8–14 mm; stems proximally with 2–6-rayed trichomes; fruits 1–2 mm wide; Four Corners region . . . . .13. *Boechera consanguinea* (in part)
22. Fruiting pedicels (10–)15–28 mm; stems proximally with 4–8-rayed trichomes; fruits 1.5–2.9 mm wide; e Nevada, w Utah. . . . . 36. *Boechera goodrichii* (in part)

[7. Shifted to left margin.—Ed.]
7. Fruiting pedicels usually glabrous, rarely with few isolated trichomes present.
23. Seeds biseriate or sub-biseriate.
24. Stems proximally with subsessile, submalpighiaceous trichomes; seeds 2–2.5 × 1.1–1.5 mm; (s Sierra Nevada). . . . . 104. *Boechera tularensis*
24. Stems proximally with simple and/or stalked, branched trichomes; seeds 0.7–1.8 × 0.5–1.2 mm.
25. Stems proximally with 2–8-rayed trichomes; plants apomictic, pollen spheroid.
26. Fruiting pedicels divaricate-ascending to horizontal; stems proximally with simple and 2–4-rayed trichomes.
27. Sepals pubescent; fruiting pedicels divaricate-ascending; fruits 1.7–2 mm wide; ovules 60–96 per ovary; Colorado Plateau and vicinity. . . . . 37. *Boechera gracilenta* (in part)
27. Sepals glabrous; fruiting pedicels horizontal; fruits 1.8–2.5 mm wide; ovules 100–160 per ovary; s New Mexico, w Texas . . . 77. *Boechera porphyrea* (in part)
26. Fruiting pedicels reflexed or divaricate-descending; stems proximally with 2–8-rayed trichomes.
28. Fruiting pedicels 8–14 mm; stems proximally with 2–6-rayed trichomes; fruits 1–2 mm wide; Four Corners region. . . . . .13. *Boechera consanguinea* (in part)
28. Fruiting pedicels (10–)15–28 mm; stems proximally with 4–8-rayed trichomes; fruits (1.5–)2–3 mm wide; e Nevada, w Utah. . . . . 36. *Boechera goodrichii* (in part)
25. Stems proximally with simple, spurred, or 2- or 3-rayed trichomes; plants sexual, pollen ellipsoid (except *B. languida*).
29. Cauline leaves 30–65; fruiting pedicels (15–)20–47 mm; ovules 130–210 per ovary . . . . .38. *Boechera gracilipes*
29. Cauline leaves 3–25; fruiting pedicels 3–15(–23) mm; ovules 40–128 per ovary.
30. Cauline leaves usually concealing stem proximally; ovules 90–128 per ovary.

31. Petals 5–9 × 1–2 mm; fruiting pedicels 9–18(–23) mm; Colorado
     Plateau, s Nevada . . . . 30. *Boechera fendleri*
31. Petals 3–3.7(–4) × 0.5–0.8 mm; fruiting pedicels 6–10(–15) mm;
     mountains of c Colorado, nc North Mexico . . . . .98. *Boechera spatifolia*
  30. Cauline leaves not concealing stem proximally; ovules 40–70(–90) per
       ovary.
    32. Basal leaf blade surfaces densely pubescent, trichomes simple and
         2–4-rayed; seeds sub-biseriate; cauline leaf blades with auricles
         0.5–1.5 mm; racemes 10–20-flowered; plants apomictic, pollen
         spheroid; s Wyoming . . . . 51. *Boechera languida*
    32. Basal leaf blade surfaces sparsely pubescent, trichomes simple
         and 2-rayed; seeds biseriate; cauline leaf blades with auricles to
         0.7 mm; racemes 4–14-flowered; plants sexual, pollen ellipsoid;
         Arizona, California, Colorado, Nevada, Utah, Wyoming . . . .
         . . . . 69. *Boechera pendulina* (in part)

[23. Shifted to left margin.—Ed.]
23. Seeds usually uniseriate.
  33. Stems proximally with simple trichomes only . . . . .9. *Boechera canadensis* (in part)
  33. Stems proximally usually with simple and/or branched trichomes (rarely all simple in
       B. pusilla).
    34. Fruits 3–5(–6) mm wide; seeds winged, wings 0.8–1.5 mm wide; stems elevated above
         ground surface on woody bases . . . . 101. *Boechera suffrutescens* (in part)
    34. Fruits 0.7–3.5 mm wide; seeds not winged, or wings 0.05–1 mm wide; stems
         usually not from above ground surface on woody bases (except *B. perennans* and
         B. lignifera).
      35. Fruits reflexed to closely pendent . . . . 73. *Boechera pinetorum* (in part)
      35. Fruits descending to ascending or widely pendent.
        36. Fruits 0.7–0.8 mm wide; petals 2–3.5 mm; basal leaf blade surfaces with
             simple trichomes adaxially. . . . . 19. *Boechera dentata* (in part)
        36. Fruits 1–3.5 mm wide; petals 3.5–8(–9) mm; basal leaf blade surfaces with
             branched trichomes adaxially.
          37. Stems proximally with simple and branched trichomes.
            38. Stems proximally with sessile or subsessile, 2- or 3(–6)-rayed
                 trichomes . . . . 39. *Boechera grahamii* (in part)
            38. Stems proximally with stalked, 2–5-rayed trichomes.
              39. Basal leaf blade margins dentate, 3–15(–20) mm wide; fruiting
                   pedicels (6–)10–25 mm; ovules 60–96 per ovary; stems
                   (1.5–)2–7 dm . . . . 71. *Boechera perennans* (in part)
              39. Basal leaf blade margins entire, 1–3(–4) mm wide; fruiting
                   pedicels 2–7 mm; ovules 20–54 per ovary; stems 0.5–2(–2.5)
                   dm.
                40. Basal leaf blade surfaces densely pubescent, trichomes
                     3–6-rayed; fruits not secund, 1–1.5 mm wide; ovules
                     36–54 per ovary; seeds with continuous wing 0.1–0.15
                     mm wide; plants sexual, pollen ellipsoid; c Colorado. . .
                     . . . . .40. *Boechera gunnisoniana* (in part)
                40. Basal leaf blade surfaces sparsely pubescent, trichomes
                     2- or 3-rayed; fruits secund, 1.5–2 mm wide; ovules
                     20–32 per ovary; seeds not winged or with distal wing
                     0.05–0.1 mm wide; plants apomictic, pollen spheroid;
                     c Wyoming. . . . . .81. *Boechera pusilla* (in part)

[37. Shifted to left margin.—Ed.]
37. Stems proximally with branched trichomes only.
    41. Fruits secund.
        42. Petals white; Quebec.................................83. *Boechera quebecensis* (in part)
        42. Petals purple to lavender; not Quebec.
            43. Stems proximally with sessile and subsessile, 2- or 3-rayed trichomes; petals 6–8 mm; plants usually short-lived perennials, without woody caudices ....23. *Boechera drepanoloba* (in part)
            43. Stems proximally with stalked, 2–6-rayed trichomes; petals 3.5–6 mm; plants long-lived perennials, with woody caudices.
                44. Cauline leaf blades with auricles 0.5–1.5 mm; petals 1.5–2 mm wide; fruiting pedicels 4–11 mm; fruits 2–3 mm wide; ovules 40–54 per ovary; seeds 2–2.5 × 1.7–2 mm; sc Oregon..................45. *Boechera horizontalis* (in part)
                44. Cauline leaf blades with auricles to 0.5 mm; petals 1–1.5 mm wide; fruiting pedicels 2–6 mm; fruits 1.6–2.3 mm wide; ovules 28–44 per ovary; seeds 1.3–2 × 1–1.5 mm; Alaska, Alberta, British Columbia, California, Colorado, Idaho, Montana, Nevada, Oregon, Utah, Washington, Wyoming, Yukon................................53. *Boechera lemmonii* (in part)
    41. Fruits not secund.
        45. Basal leaf blade margins dentate; stems proximally with 2–5-rayed trichomes.
            46. Basal leaf blade surfaces with trichomes 0.05–0.2 mm; fruiting pedicels 5–10 mm; fruits 1.5–1.8 mm wide; seeds with distal wing; plants apomictic, pollen spheroid; nw Wyoming .... 33. *Boechera fructicosa*
            46. Basal leaf blade surfaces with trichomes 0.2–0.4 mm; fruiting pedicels (6–)10–25 mm; fruits 1.7–2.1 mm wide; seeds with continuous wing; plants sexual, pollen ellipsoid; Arizona, California, Nevada, New Mexico, Texas, Utah......... ....71. *Boechera perennans* (in part)
        45. Basal leaf blade margins entire; stems proximally with (3 or) 4–10-rayed trichomes.
            47. Stems usually 1 per caudex branch, arising from centers of rosettes; petals 0.7–1 mm wide; seeds with wings 0.25–0.5 mm wide; fruits 1.7–2.5 mm wide ....11. *Boechera cobrensis* (in part)
            47. Stems usually 2–5 per caudex branch, arising laterally proximal to sterile shoots; petals 1–1.5 mm wide; seeds with wings 0.1–0.15 mm wide; fruits 1.2–2 mm wide .... 54. *Boechera lignifera*

Group 5

1. Basal leaf blade surfaces with branched trichomes sessile, 2-rayed (malpighiaceous) ....99. *Boechera stricta* (in part)
1. Basal leaf blade surfaces with branched trichomes stalked or, if sessile, some more than 2-rayed.
    2. Stems proximally with branched trichomes only.
        3. Fruits secund.
            4. Petals white; Quebec ....83. *Boechera quebecensis* (in part)
            4. Petals purple to lavender; not Quebec.
                5. Stems proximally with sessile and subsessile, 2- or 3-rayed trichomes; petals 6–8 mm; seeds sub-biseriate; plants usually short-lived perennials, without woody caudices .... 23. *Boechera drepanoloba* (in part)
                5. Stems proximally with stalked, 2–6-rayed trichomes; petals 3.5–6 mm; seeds uniseriate; plants long-lived perennials, with woody caudices.

6. Cauline leaf blades with auricles 0.5–1.5 mm; petals 1.5–2 mm wide; fruiting pedicels 4–11 mm; fruits 2–3 mm wide; ovules 40–54 per ovary; seeds 2–2.5 × 1.7–2 mm; sc Oregon .... 45. *Boechera horizontalis* (in part)
6. Cauline leaf blades with auricles to 0.5 mm; petals 1–1.5 mm wide; fruiting pedicels 2–6 mm; fruits 1.6–2.3 mm wide; ovules 28–44 per ovary; seeds 1.3–2 × 1–1.5 mm; Alaska, Alberta, British Columbia, California, Colorado, Idaho, Montana, Nevada, Oregon, Utah, Washington, Wyoming, Yukon .... 53. *Boechera lemmonii* (in part)

[3. Shifted to left margin.—Ed.]

3. Fruits not secund.
    7. Biennials, without caudices; petals 2–3.5 mm; fruits 0.7–0.8 mm wide.... 19. *Boechera dentata* (in part)
    7. Perennials or, rarely, biennials, often with caudices (well-developed); petals 3.5–14 mm; fruits 0.9–3.2 mm wide.
        8. Stems proximally with 4–12-rayed trichomes only.
            9. Cauline leaves concealing stems proximally; stems proximally with 7–12-rayed trichomes; ovules 74–190 per ovary.
                10. Petals 1.2–2 mm wide; basal leaves ciliate proximally, surfaces with trichomes 0.2–0.7 mm; ovules 74–134 per ovary; seeds 1.7–2 × 1–1.5 mm, with continuous wing; plants apomictic, pollen spheroid.... 47. *Boechera inyoensis* (in part)
                10. Petals 0.8–1.2 mm wide; basal leaves not ciliate proximally, surfaces with trichomes 0.1–0.2 mm; ovules 140–190 per ovary; seeds 1–1.3 × 0.7–0.8 mm, usually with distal wing, rarely not winged; plants sexual, pollen ellipsoid.. 96. *Boechera shockleyi* (in part)
            9. Cauline leaves not concealing stems proximally; stems proximally with 4–8 (–10)-rayed trichomes; ovules 30–84 per ovary.
                11. Petals 3.5–5 mm; fruiting pedicels 1.5–5(–7) mm; plants apomictic, pollen spheroid; e California, w Nevada.
                    12. Stems 1.5–3.5 dm; fruits 4–6.2 cm; ovules 48–68 per ovary ...... 4. *Boechera bodiensis*
                    12. Stems 0.5–1.5(–2) dm; fruits 1.7–3.5(–4.5) cm; ovules 32–44 per ovary... 20. *Boechera depauperata*
                11. Petals 5–12 mm; fruiting pedicels 5–20 mm; plants sexual, pollen ellipsoid; Colorado, c Nevada, Utah.
                    13. Petals 5–7 mm; basal leaf blade surfaces with trichomes 0.1–0.2 mm; seeds 0.8–1 mm; c Colorado .... 16. *Boechera crandallii*
                    13. Petals 7–12 mm; basal leaf blade surfaces with trichomes 0.04–0.1 mm; seeds 1–1.2 mm; nw Colorado, Nevada, Utah... 31. *Boechera fernaldiana* (in part)
        8. Stems proximally with at least some 2- or 3-rayed trichomes, sometimes mixed with 4–6-rayed ones.
            14. Fruits erect to suberect, usually appressed to rachises.
                15. Racemes (7–)14–45-flowered; petals 6–10 mm; ovules 60–140 per ovary .... 78. *Boechera pratincola*
                15. Racemes 4–15-flowered; petals 4–7 mm; ovules 24–60 per ovary.
                    16. Basal leaf blade surfaces with sessile or subsessile trichomes; fruits 2–3 mm wide; stems 1–3 dm; plants apomictic, pollen spheroid .... 25. *Boechera elkoensis* (in part)
                    16. Basal leaf blade surfaces with stalked trichomes; fruits 1.2–1.8 mm wide; stems 0.4–1.4 dm; plants sexual, pollen ellipsoid.... 61. *Boechera ophira* (in part)
            14. Fruits horizontal to ascending, or pendent, not appressed to rachises.
                17. Racemes 30–120-flowered; fruits 6–12 cm; ovules 140–220 per ovary; s California.

18. Petals purple, 9–14 mm; fruits 1.5–2.5 mm wide; ovules 140–180 per ovary; fruiting pedicels 4–15(–20) mm; plants apomictic, pollen spheroid . . . . 8. *Boechera californica* (in part)

18. Petals white or lavender, 8–10 mm; fruits 2.5–3 mm wide; ovules 170–220 per ovary; fruiting pedicels 10–45 mm; plants sexual, pollen ellipsoid. . . . . 43. *Boechera hoffmannii* (in part)

17. Racemes 4–25(–36)-flowered; fruits 2–6.5(–7.5) cm; ovules 26–112 per ovary; not s California (except *B. peirsonii*).

19. Stems usually 2–5 per caudex branch, arising from margins of rosettes or laterally proximal to sterile shoots.

20. Fruits 2–3.2 mm wide; ovules 26–42 per ovary; seeds 2.5–3.5 × 1.5–2.5 mm; e California, Nevada.

21. Sepals glabrous; basal leaf blades 2–6 mm wide, surfaces with sessile or subsessile trichomes; fruiting pedicels 4–8 mm . . . . 25. *Boechera elkoensis* (in part)

21. Sepals sparsely pubescent; basal leaf blades 1–2.5(–3) mm wide, surfaces with stalked trichomes; fruiting pedicels 2–6 mm. . . . . 74. *Boechera pinzliae* (in part)

20. Fruits 1–2 mm wide; ovules 44–86 per ovary; seeds 1.1–1.3 × 0.8–1.1 mm; Colorado Plateau, nc Utah.

22. Basal leaf blades 2–4 mm wide, surfaces with 2–5-rayed trichomes; ovules 64–86 per ovary; plants apomictic, pollen spheroid; nc Utah . . . . 41. *Boechera harrisonii* (in part)

22. Basal leaf blades 5–11(–13) mm wide, surfaces with 4–8-rayed trichomes; ovules 44–62 per ovary; plants sexual, pollen ellipsoid; Colorado Plateau. . . . . 64. *Boechera pallidifolia*

19. Stems usually 1 per caudex branch, arising from centers of rosettes.

23. Petals 7–12 mm . . . . 31. *Boechera fernaldiana* (in part)

23. Petals 4–7 mm.

24. Basal leaf blade surfaces with 2- or 3-rayed trichomes only; plants sexual, pollen ellipsoid; Nevada . . . . 61. *Boechera ophira* (in part)

24. Basal leaf blade surfaces with 3–8-rayed trichomes; plants apomictic, pollen spheroid; not Nevada.

25. Plants cespitose, long-lived; fruits 2–2.8 mm wide; petals purple; s California . . . . 68. *Boechera peirsonii* (in part)

25. Plants not cespitose, short-lived; fruits 1.2–1.8 mm wide; petals white or lavender; n California, Idaho, Oregon, Wyoming.

26. Basal leaf blade surfaces with sessile trichomes 0.15–0.35 mm; petals 5–7 mm; seeds 1.8–2.2 × 1–1.4 mm; n California, Oregon . . . . 1. *Boechera acutina*

26. Basal leaf blade surfaces with short-stalked trichomes 0.04–0.15 mm; petals 4–5 mm; seeds 1.1–1.4 × 0.7–0.9 mm; Idaho, Wyoming . . . . 91. *Boechera saximontana*

[2. Shifted to left margin.—Ed.]

2. Stems proximally with simple and/or branched trichomes or glabrous.

27. Stems proximally with simple trichomes only.

28. Basal leaf blade margins lyrate-pinnatifid, surfaces glabrous or sparsely pubescent, with simple trichomes 0.1–0.5 mm; sepals and distalmost cauline leaf blade surfaces glabrous . . . . 59. *Boechera missouriensis* (in part)

28. Basal leaf blade margins entire or dentate, surfaces pubescent, with simple and/or branched trichomes 0.4–1.7 mm; sepals and distalmost cauline leaf blade surfaces pubescent.

29. Biennials, caudices not evident; petals 3–5 mm; basal leaf blades 10–30 mm wide; fruiting pedicels horizontal; seeds with wings 0.4–1 mm wide distally . . . . .9. *Boechera canadensis* (in part)
29. Perennials (long-lived), caudices well-developed; petals 7–12 mm; basal leaf blades 0.5–7(–11) mm wide; fruiting pedicels ascending to divaricate-ascending; seeds with wings 0.1–0.2 mm wide distally.
 30. Basal leaf blades 3–7(–11) mm wide, surfaces with 2–4-rayed trichomes; fruits 1.5–2.2 mm wide; seeds 1–1.2 mm wide. . . . . 5. *Boechera breweri* (in part)
 30. Basal leaf blades 0.5–2 mm wide, surfaces with simple and 2-rayed trichomes; fruits 2.3–3 mm wide; seeds 1.5–2 mm wide . . . . . 17. *Boechera cusickii* (in part)

[27. Shifted to left margin.—Ed.]

27. Stems proximally with simple and/or branched trichomes or glabrous (in *B. rigidissima*).
 31. Biennials or perennials (short-lived), without woody caudices.
  32. Stems usually 2–5 per caudex branch, arising from margins of rosettes or laterally proximal to sterile shoots; cauline leaves 3–9, not concealing stems proximally; petals 4–7 mm.
   33. Basal leaf blade surfaces with 4–6(–8)-rayed trichomes; petals 6–7 mm; cauline leaf blades with auricles 1–3 mm; fruiting pedicels (7–)10–18 mm; seeds subbiseriate; plants apomictic, pollen spheroid . . . . 37. *Boechera gracilenta* (in part)
   33. Basal leaf blade surfaces with simple and 2- or 3-rayed trichomes; petals 4–6 mm; cauline leaf blades with auricles usually less than 1 mm; fruiting pedicels 5–10 mm; seeds uniseriate; plants sexual, pollen ellipsoid.
    34. Distalmost cauline leaf blade surfaces glabrous; fruiting pedicels divaricate-ascending to horizontal, glabrous; nw Colorado, ne Utah . . . .34. *Boechera glareosa* (in part)
    34. Distalmost cauline leaf blade surfaces pubescent; fruiting pedicels ascending, glabrous or sparsely pubescent; nc New Mexico. . . . . 106. *Boechera villosa*
  32. Stems usually 1 per caudex branch, arising from centers of rosettes; cauline leaves (7–)15–52, usually concealing stems proximally; petals 6–13 mm.
   35. Basal leaf blade surfaces with sessile or subsessile branched trichomes.
    36. Fruits (1.7–)2–2.5 mm wide; petals purple to lavender; fruiting pedicels 5–10(–12) mm; basal leaf blade surfaces with some 5- or 6-rayed trichomes. . . . . 22. *Boechera divaricarpa* (in part)
    36. Fruits 1–1.5(–1.8) mm wide; petals usually white; fruiting pedicels (5–)8–30 mm; basal leaf blade surfaces with 2–4-rayed trichomes. . . . .
. . . . 39. *Boechera grahamii* (in part)
   35. Basal leaf blade surfaces with stalked branched trichomes.
    37. Fruiting pedicels with 2- or 3-rayed trichomes; fruits horizontal; petals 5–8 mm; plants apomictic, pollen spheroid. . . . . 66. *Boechera pauciflora* (in part)
    37. Fruiting pedicels glabrous or with predominantly simple trichomes; fruits usually ascending to divaricate-ascending, rarely horizontal; petals 7–13 mm; plants sexual, pollen ellipsoid.
     38. Petals dark reddish purple (drying indigo), 1.5–2 mm wide; stems proximally sparsely pubescent, trichomes to 0.15 mm; ovules 80–100 per ovary. . . . . 3. *Boechera atrorubens*
     38. Petals purple to lavender or white, 2–5 mm wide; stems proximally densely pubescent, trichomes to 1.5 mm; ovules 90–170 per ovary . . . . 97. *Boechera sparsiflora* (in part)
 31. Perennials (long-lived), usually with woody caudices (except *B. sparsiflora*).
  39. Stems usually 2–5(–7) per caudex branch, arising from margins of rosettes.
   40. Fruits secund; ovules 20–32 per ovary; basal leaf blade surfaces with 2- or 3-rayed trichomes. . . . . .81. *Boechera pusilla* (in part)
   40. Fruits not secund; ovules 36–84 per ovary; basal leaf blade surfaces with 2–8-rayed trichomes.
    41. Fruits 1.7–2 mm wide, usually strongly curved; seeds 1.5–2 × 1.1–1.5 mm . . . . 27. *Boechera falcatoria*
    41. Fruits 0.8–1.5 mm wide, straight or slightly curved; seeds 1–1.5 × 0.7–1 mm.

42. Petals 0.5–0.8 mm wide; cauline leaves (5–)9–16; racemes 10–33 -flowered; plants apomictic, pollen spheroid . . . . 57. *Boechera macounii* (in part)
42. Petals 1–2 mm wide; cauline leaves 2–9; racemes 7–15-flowered; plants sexual, pollen ellipsoid.
    43. Stems proximally with 2–4-rayed trichomes 0.08–0.3 mm; fruits horizontal; ovules 36–54 per ovary; (c Colorado) . . . . 40. *Boechera gunnisoniana* (in part)
    43. Stems proximally with simple and 2-rayed trichomes 0.3–0.6 mm; fruits usually ascending, rarely horizontal (*B. glareosa*); ovules 50–80 per ovary.
        44. Basal leaf blade surfaces with simple and 2- or 3-rayed trichomes 0.3–0.5 mm; cauline leaf blades with auricles 0.3–0.6 mm . . . . 34. *Boechera glareosa* (in part)
        44. Basal leaf blade surfaces with 4–8-rayed trichomes 0.05–0.1 mm; cauline leaf blades with auricles 0.5–1.5 mm . . . . 58. *Boechera microphylla* (in part)

[39. Shifted to left margin.—Ed.]

39. Stems usually 1 per caudex branch, arising from centers of rosettes.
    45. Fruits 2.5–3.5 mm wide; seeds 2.5–3.2 × 1.8–2.5 mm . . . . 87. *Boechera rigidissima*
    45. Fruits 0.8–2.2(–2.5) mm wide; seeds 1–2(–2.5) × 0.9–1.5 mm.
        46. Fruits erect, appressed to rachises; ovules 24–40 per ovary . . . . 67. *Boechera paupercula* (in part)
        46. Fruits ascending, divaricate, or horizontal, not appressed to rachises; ovules 46–250 per ovary.
            47. Petals 0.5–2 mm wide; usually e of Sierra Nevada-Cascade crest.
                48. Basal leaf blade surfaces with trichomes 0.3–0.6 mm; fruits 1.5–2.2 mm wide; ovules 80–162 per ovary; cauline leaves (8–)14–60, blades with auricles (1–)3–10 mm . . . . 66. *Boechera pauciflora* (in part)
                48. Basal leaf blade surfaces with trichomes 0.05–0.3 mm; fruits 0.8–1.5 mm wide; ovules 46–80 per ovary; cauline leaves 3–16, blades with auricles 0.5–1.5 mm.
                    49. Distal cauline leaf blades with surfaces pubescent; petals white; stems proximally with some 4- or 5-rayed trichomes; sc California . . . . . 26. *Boechera evadens*
                    49. Distal cauline leaf blades with surfaces glabrous; petals lavender or purplish; stems proximally with simple and 2- or 3-rayed trichomes only; not sc California.
                        50. Fruits 0.8–1.2 mm wide; petals 0.5–0.8 mm wide; cauline leaves (5–)9–16; racemes 10–33-flowered . . . . 57. *Boechera macounii* (in part)
                        50. Fruits 1.2–1.5 mm wide; petals 1–1.8 mm wide; cauline leaves 2–6; racemes 3–15-flowered.
                            51. Stems proximally with trichomes simple and 2- or 3-rayed 0.1–0.2 mm; basal leaf blade surfaces with (2- or) 3–6-rayed trichomes usually 0.05–0.2(–0.3) mm; seeds with continuous wing; styles 0.8–1.5 mm; plants apomictic, pollen spheroid . . . . 10. *Boechera cascadensis*
                            51. Stems proximally with trichomes simple and 2-rayed 0.3–0.6 mm; basal leaf blade surfaces with some 7- or 8-rayed trichomes 0.05–0.1 mm; seeds with distal wing; styles 0.05–0.3 mm; plants sexual, pollen ellipsoid . . . . 58. *Boechera microphylla* (in part)
            47. Petals 2–5 mm wide; w of Sierra Nevada-Cascade crest (except *B. sparsiflora*).
                52. Fruiting pedicels with subappressed, 2–4-rayed trichomes . . . . 2. *Boechera arcuata*
                52. Fruiting pedicels with spreading, simple and 2-rayed trichomes or glabrous.

[53. Shifted to left margin.—Ed.]

53. Basal leaf blades 1–3 mm wide; caudices woody, with persistent, crowded, peg-like leaf bases . . . . 49. *Boechera koehleri* (in part)
53. Basal leaf blades 3–12 mm wide; caudices woody or not, without persistent, crowded, peg-like leaf bases.
　　54. Distal cauline leaf blades with glabrous surfaces (rarely margins ciliate); cauline leaf blades with auricles 3–10 mm; stems 3–8 dm; usually e of Sierra Nevada-Cascade axis . . . . 97. *Boechera sparsiflora* (in part)
　　54. Distal cauline leaf blades with pubescent surfaces; cauline leaf blades with auricles 0.5–3(–5) mm; stems 0.6–3.5(–4.5) dm; w of Sierra Nevada-Cascade axis.
　　　　55. Fruiting pedicels ascending to divaricate-ascending; basal leaf blade surfaces with trichomes 0.4–0.8 mm; petals 7–12 mm; plants sexual, pollen ellipsoid . . . . 5. *Boechera breweri* (in part)
　　　　55. Fruiting pedicels horizontal; basal leaf blade surfaces with trichomes 0.1–0.5 mm; petals 6–8 mm; plants apomictic, pollen spheroid. 90. *Boechera rubicundula*

1. **Boechera acutina** (Greene) Windham & Al-Shehbaz, Harvard Pap. Bot. 12: 236. 2007 E

Arabis acutina Greene, Leafl. Bot. Observ. Crit. 2: 82. 1910; *A. divaricarpa* A. Nelson var. *interposita* (Greene) Rollins; *A. drummondii* A. Gray var. *interposita* (Greene) Rollins; *A. interposita* Greene

Biennials or perennials; short-lived; apomictic; caudex present or absent. **Stems** usually 1 per caudex branch, arising from center of rosette near ground surface, 1.5–6 dm, densely pubescent proximally, trichomes sessile, 2–4-rayed, 0.15–0.4 mm, usually sparsely pubescent distally, rarely glabrous. **Basal leaves:** blade oblanceolate, 1.5–6 mm wide, margins usually entire, rarely denticulate, ciliate proximally, trichomes (simple), to 0.5 mm, surfaces sparsely to moderately pubescent, trichomes sessile, 3–5 (or 6)-rayed, 0.15–0.35 mm. **Cauline leaves:** 2–20(–38), often concealing stem proximally; blade auricles 0.3–2 mm, surfaces of distalmost leaves sparsely pubescent or glabrous. **Racemes** 5–20(–36)-flowered, usually unbranched. **Fruiting pedicels** erect to ascending, usually straight, 3–10 mm, pubescent or glabrous, trichomes appressed, branched. **Flowers** ascending at anthesis; sepals pubescent; petals white or lavender, 5–7 × 0.9–1.3 mm, glabrous; pollen spheroid. **Fruits** ascending, not appressed to rachis, not secund, straight, edges parallel, 2.5–7.5 cm × 1.2–1.8 mm; valves glabrous; ovules 46–100 per ovary; style 0.3–1 mm. **Seeds** uniseriate, 1.8–2.2 × 1–1.4 mm; wing continuous, 0.1–0.4 mm wide.

Flowering Jun–Jul. Gravelly slopes in meadows, open forests; 1400–1900 m; Calif., Idaho, Oreg.

Boechera acutina is an apomictic hybrid that clearly contains a genome derived from *B. stricta*; the other parent remains uncertain. Plants of *B. acutina* are superficially similar to *B. pratincola* and are often misidentified as *Arabis* (*Boechera*) *divaricarpa*; the species is easily distinguished from both of those (see M. D. Windham and I. A. Al-Shehbaz 2007b for detailed comparison).

2. **Boechera arcuata** (Nuttall) Windham & Al-Shehbaz, Harvard Pap. Bot. 11: 64. 2006 E F

Streptanthus arcuatus Nuttall in J. Torrey and A. Gray, Fl. N. Amer. 1: 77. 1838; *Arabis holboellii* Hornemann var. *arcuata* (Nuttall) Jepson; *A. maxima* Greene; *A. sparsiflora* Nuttall var. *arcuata* (Nuttall) Rollins

Perennials; usually long-lived; sexual; caudex often woody (well-developed). **Stems** usually 1 per caudex branch, arising from center of rosette, elevated on woody base or from ground surface, (2–)3–8 dm, densely pubescent proximally, trichomes short-stalked, 2-rayed and simple, to 1 mm, pubescent distally. **Basal leaves:** blade linear to oblanceolate, 2–7(–12) mm wide, margins usually entire, rarely denticulate, ciliate along petiole, trichomes to 1.5 mm, surfaces densely pubescent, trichomes usually short-stalked, 2–5-rayed (rarely some simple), 0.4–0.8 mm. **Cauline leaves:** 10–30(–45), often concealing stem proximally; blade auricles 2–5(–6) mm, surfaces of distalmost leaves pubescent. **Racemes** 12–50(–70)-flowered, usually unbranched. **Fruiting pedicels** usually divaricate-ascending, rarely horizontal, gently recurved or straight, 8–22 mm, pubescent, trichomes subappressed, 2–4-rayed. **Flowers** ascending at anthesis; sepals pubescent; petals purple, 9–14 × 2–4 mm, glabrous or sparsely pubescent (trichomes abaxially); pollen ellipsoid. **Fruits** usually divaricate-ascending, rarely horizontal, not appressed to rachis,

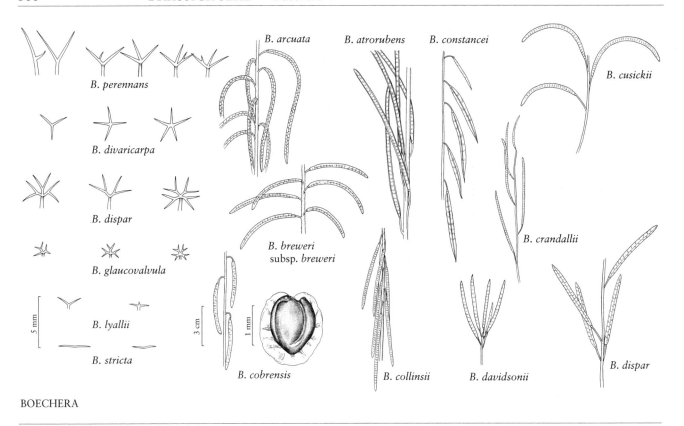

not secund, usually curved, edges parallel, (6–)8–13 cm × 1.5–2.2 mm; valves glabrous or trichomes relatively few, scattered; ovules 90–250 per ovary; style 0.01–0.5 mm. Seeds uniseriate or sub-biseriate, 1.5–1.7 × 1–1.2 mm; wing continuous, 0.1–0.2 mm wide.

Flowering Mar–Jun. Rocky hillsides and cliffs in pine forests and chaparral; 300–1800 m; Calif.

Although usually treated as a variety of *Arabis* (*Boechera*) *sparsiflora* (e.g., R. C. Rollins 1993), *B. arcuata* is easily distinguished from that species by having rachises and fruiting pedicels pubescent with subappressed, 2–4-rayed trichomes and a geographic range limited to southern and western California. By contrast, *B. sparsiflora* has rachises and fruiting pedicels with spreading, usually simple trichomes (sometimes glabrescent) and an allopatric distribution north and east of the Sierra Nevada.

3. Boechera atrorubens (Suksdorf ex Greene) Windham & Al-Shehbaz, Harvard Pap. Bot. 11: 64. 2006 [E] [F]

Arabis atrorubens Suksdorf ex Greene, Erythea 1: 223. 1893; *A. atriflora* Suksdorf; *A. sparsiflora* Nuttall var. *atrorubens* (Suksdorf ex Greene) Rollins

Perennials; usually short-lived; sexual; caudex not woody. Stems usually 1 per caudex branch, arising from center of rosette near ground surface, 0.8–6 dm, sparsely pubescent proximally, trichomes short-stalked and simple, (scattered), 2-rayed, 0.1–0.15 mm, glabrous distally. Basal leaves: blade oblanceolate, 4–10 mm wide, margins usually dentate, not ciliate, surfaces densely pubescent, trichomes short-stalked, 3- or 4-rayed, 0.1–0.2 mm. Cauline leaves: 7–25, concealing stem proximally; blade auricles 1–4 mm, surfaces of distalmost leaves glabrous. Racemes 6–20-flowered, usually unbranched. Fruiting pedicels ascending, usually straight, 5–10 mm, glabrous or pubescent, trichomes spreading, simple. Flowers ascending at anthesis; sepals pubescent; petals dark reddish purple (drying indigo), 7–9 × 1.5–2 mm, glabrous; pollen ellipsoid. Fruits ascending, not appressed to rachis, not secund, curved or straight, edges parallel, (4–)6–12 cm × 1.7–2.2 mm; valves glabrous; ovules 80–100 per ovary; style 0.2–0.5 mm. Seeds uniseriate, 1.2–1.7 × 0.9–1.2 mm; wing lateral and distal, 0.1–0.2 mm wide.

Flowering Apr–May. Rocky summits and sandy loam on sagebrush slopes; ca. 600 m; Oreg., Wash.

Boechera atrorubens is often treated as a variety of *B. sparsiflora* (e.g., R. C. Rollins 1993), it is readily separated from that species by having proximal stems sparsely (versus densely) pubescent with much smaller (0.15 versus 1.5 mm) trichomes. The two taxa rarely grow in proximity and, in areas where they are sympatric, *B. atrorubens* is further distinguished by its narrower (1.5–2 versus 2–5 mm) petals that are dark reddish purple to indigo (versus lavender or white).

4. **Boechera bodiensis** (Rollins) Al-Shehbaz, Novon 13: 384. 2003 [C] [E]

Arabis bodiensis Rollins, Contr. Gray Herb. 212: 113. 1982

Perennials; long-lived; (cespitose); apomictic; caudex somewhat woody. **Stems** usually 1 per caudex branch, arising from center of rosette near ground surface, 1.5–3.5 dm, densely pubescent proximally, trichomes short-stalked, 5–10-rayed, to 0.7 mm, sparsely pubescent or glabrous distally. **Basal leaves:** blade narrowly oblanceolate, 1–3 mm wide, margins entire, ciliate proximally, trichomes to 0.7 mm, surfaces densely pubescent, trichomes short-stalked, 5–10-rayed, 0.08–0.25 mm. **Cauline leaves:** 4–9, not concealing stem; blade auricles 0.5–2 mm, surfaces of distalmost leaves pubescent. **Racemes** 8–25-flowered, usually unbranched. **Fruiting pedicels** ascending to divaricate-ascending, straight, 3–7 mm, pubescent, trichomes appressed, branched. **Flowers** ascending at anthesis; sepals pubescent; petals lavender to purple, 4–5 × ca. 1 mm, glabrous; pollen spheroid. **Fruits** divaricate-ascending, not appressed to rachis, not secund, straight or curved, edges parallel, 4–6.2 cm × 1.2–1.8 mm; valves glabrous; ovules 48–68 per ovary; style 0.1–0.2 mm. **Seeds** uniseriate, 1–1.5 × 0.8–1.2 mm; wing continuous, 0.1–0.15 mm wide.

Flowering Jul–Aug. Loose soil, crevices of igneous rock; of conservation concern; 2400–2900 m; Calif., Nev.

In his original description, Rollins suggested that *Boechera bodiensis* is closely related to *B. cobrensis* and *B. falcifructa*. Subsequent studies (M. D. Windham and I. A. Al-Shehbaz, unpubl.) indicate that both *B. bodiensis* and *B. falcifructa* are apomictic hybrids containing one or more genomes derived from *B. cobrensis*. In the case of *B. falcifructa*, the second parent is clearly *B. fernaldiana*, but the second parent of *B. bodiensis* remains uncertain. *Boechera bodiensis* is most similar to *B. falcifructa*. The latter differs by having gently recurved fruiting pedicels 6–12 mm, 62–80 ovules per ovary, and pendulous fruits; it is known from Elko and Lander counties, Nevada.

By contrast, *B. bodiensis* has straight fruiting pedicels 3–7 mm long, divaricate-ascending fruits, 48–68 ovules per ovary, and is known from Mono County, California, and Lyon and Mineral counties, Nevada.

5. **Boechera breweri** (S. Watson) Al-Shehbaz, Novon 13: 384. 2003 [E] [F]

Arabis breweri S. Watson, Proc. Amer. Acad. Arts 11: 123. 1876; *A. breweri* var. *figularis* Jepson; *A. epilobioides* Greene

Perennials; long-lived; sexual; caudex woody. **Stems** usually 1 per caudex branch, arising from center of rosette, somewhat elevated on woody base or from ground surface, 0.6–3.5(–4.5) dm, densely pubescent proximally, trichomes long-stalked and simple, 2-rayed (sometimes all simple), 0.4–1 mm, sparsely pubescent or glabrous distally. **Basal leaves:** blade oblanceolate, 3–7(–11) mm wide, margins entire or dentate, ciliate proximally, trichomes to 1 mm, surfaces densely pubescent, trichomes long-stalked, 2–4-rayed, 0.4–0.8 mm. **Cauline leaves:** 5–15(–28), often concealing stem proximally; blade auricles 0.5–3(–5) mm, surfaces of distalmost leaves pubescent. **Racemes** 7–20(–30)-flowered, usually unbranched. **Fruiting pedicels** ascending to divaricate-ascending, straight, 3–25 mm, pubescent or glabrous, trichomes spreading, simple and 2-rayed. **Flowers** ascending at anthesis; sepals pubescent; petals usually purple (rarely lavender), 7–12 × 2–4 mm, glabrous; pollen ellipsoid. **Fruits** ascending to divaricate-ascending, not appressed to rachis, not secund, usually curved, rarely straight, edges parallel, 3.5–10 cm × 1.5–2.2 mm; valves usually glabrous, rarely sparsely pubescent; ovules 48–96 per ovary; style 0.05–0.3 mm. **Seeds** uniseriate, 1.2–1.7 × 1–1.2 mm; wing continuous, 0.1–0.2 mm wide distally.

Subspecies 2 (2 in the flora): w United States.

Relatively few other sexual diploid species of *Boechera* co-occur with *B. breweri*, and the only currently recognized hybrid involving this taxon is *B. rubicundula* (= *B. arcuata* × *B. breweri*; M. D. Windham and I. A. Al-Shehbaz 2007). *Boechera breweri* appears to be related to *B. sparsiflora*, and some specimens of subsp. *shastaensis* can be difficult to separate from that species. The two subspecies appear to intergrade on the western side of the Sacramento Valley from Tehama to Solano counties.

1. Stems 0.6–2 dm, pubescent distally; fruiting pedicels 3–8(–12) mm, pubescent; fruits 3.5–7.5 cm............5a. *Boechera breweri* subsp. *breweri*
1. Stems 1.8–3.5(–4.5) dm, glabrous distally; fruiting pedicels (7–)10–25 mm, usually glabrous; fruits (4.5–)7–10 cm ....
............5b. *Boechera breweri* subsp. *shastaensis*

5a. Boechera breweri (S. Watson) Al-Shehbaz subsp. breweri E F

Stems 0.6–2 dm, often densely pubescent proximally, moderately to sparsely pubescent distally. **Fruiting pedicels** 3–8 (–12) mm, moderately to sparsely pubescent. **Fruits** 3.5–7.5 cm.

Flowering Mar–Jul. Rocky outcrops, ledges, talus; 500–2300 m; Calif., Oreg.

5b. Boechera breweri (S. Watson) Al-Shehbaz subsp. shastaensis Windham & Al-Shehbaz, Harvard Pap. Bot. 11: 65. 2006 E

Arabis austiniae Greene; *A. breweri* var. *austiniae* (Greene) Rollins; *A. rostellata* Greene

Stems 1.8–3.5(–4.5) dm, moderately to sparsely pubescent proximally, usually glabrous distally. **Fruiting pedicels** (7–)10–25 mm, usually glabrous (or sometimes with a few scattered trichomes). **Fruits** (4.5–)7–10 cm. $2n = 14$.

Flowering Mar–Jun. Rocky areas in woodlands; 300–1200 m; Calif., Oreg.

6. Boechera burkii (Porter) Windham & Al-Shehbaz, Harvard Pap. Bot. 12: 237. 2007 E

Arabis laevigata (Muhlenberg ex Willdenow) Poiret var. *burkii* Porter, Bull. Torrey Bot. Club 17: 15. 1890; *A. burkii* (Porter) Small

Biennials; short-lived; sexual; caudex not evident. **Stems** usually 1 per plant, arising from center of rosette near ground surface, (2–)3–7 dm, glabrous throughout, often glaucous. **Basal leaves:** blade oblanceolate to obovate, 4–10 mm wide, margins dentate, not ciliate, surfaces glabrous or sparsely pubescent subapically, trichomes simple, 0.3–0.8 mm. **Cauline leaves:** 18–28, often concealing stem proximally; blade auricles absent, surfaces of distalmost leaves glabrous. **Racemes** 20–50-flowered, usually unbranched. **Fruiting pedicels** ascending to divaricate-ascending, usually straight, 4–12 mm, glabrous. **Flowers** ascending at anthesis; sepals glabrous; petals white, 3–5 × 0.5–0.7 mm, glabrous; pollen ellipsoid. **Fruits** divaricate-ascending, not appressed to rachis, not secund, usually curved, rarely straight, edges parallel, 5–10 cm × 1.5–1.8 mm; valves glabrous; ovules 64–80 per ovary; style 0.2–0.6 mm. **Seeds** uniseriate, 1.5–1.8 × 1.2–1.4 mm; wing continuous, 0.3–0.5 mm wide distally.

Flowering Apr–May. Rocky areas, wooded slopes, stream banks; Md., Pa., Tenn., Va., W.Va.

Boechera burkii usually has been treated as a variety of *Arabis laevigata* (e.g., R. C. 1993). The two taxa differ substantially in leaf morphology, and no intermediates have been found despite broad overlap in their geographic ranges. *Boechera burkii* typically has 18–28 cauline leaves per plant, with blades linear, non-auriculate, and margins entire, whereas *B. laevigata* has 7–15 cauline leaves per plant, blades lanceolate, auriculate, and margins often dentate. The two are also separable on petal width, with *B. burkii* having significantly narrower petals (0.5–0.7 versus 1–1.5 mm wide).

7. Boechera calderi (G. A. Mulligan) Windham & Al-Shehbaz, Harvard Pap. Bot. 11: 259. 2007 E

Arabis calderi G. A. Mulligan, Rhodora 97: 144, fig. 4. 1996

Perennials; usually short-lived; apomictic; caudex not woody. **Stems** usually 1 per caudex branch, arising from center of rosette near ground surface, 1–4.5 dm, glabrous throughout. **Basal leaves:** blade oblanceolate, 1.5–6 mm wide, margins entire, ciliate at petiole base, trichomes to 0.8 mm, surfaces sparsely pubescent, trichomes sessile, malpighiaceous (usually at margin) and 3- or 4-rayed, 0.15–0.4 mm. **Cauline leaves:** 5–17, often concealing stem proximally; blade auricles 1–3 mm, surfaces of distalmost leaves glabrous. **Racemes** 10–25-flowered, unbranched. **Fruiting pedicels** erect, straight, 5–10 mm, glabrous. **Flowers** erect at anthesis; sepals glabrous; petals purple, 6–9 × 1.5–2.5 mm, glabrous; pollen spheroid. **Fruits** erect, appressed to rachis, not secund, straight, edges parallel, 3.5–6.5 cm × 1.8–2.5 mm; valves glabrous; ovules 64–134 per ovary; style 0–0.2 mm. **Seeds** biseriate to sub-biseriate, 1.8–2.2 × 1–1.8 mm; wing continuous, 0.2–0.5 mm wide.

Flowering Jun–Aug. Exposed rocky ridges, meadows, open forests near timberline; 1500–3500 m; Alta., B.C., N.W.T., Yukon; Calif., Idaho, Mont., Oreg., Wash., Wyo.

Morphological evidence suggests that *Boechera calderi* is an apomictic species that arose through hybridization between *B. lyallii* and *B. stricta* (see M. D. Windham and I. A. Al-Shehbaz 2007 for detailed comparison).

8. Boechera californica (Rollins) Windham & Al-Shehbaz, Harvard Pap. Bot. 11: 260. 2007

Arabis sparsiflora Nuttall var. *californica* Rollins, Rhodora 43: 402, fig. 14. 1941

Perennials; usually long-lived; apomictic; caudex woody. **Stems** usually 1 per caudex branch, arising from center of rosette, elevated above ground surface on woody base, (2–)3.5–13 dm, densely pubescent proximally, trichomes short-stalked, 2–4-rayed, 0.3–0.9 mm, moderately to sparsely pubescent distally. **Basal leaves:** blade linear to oblanceolate, 3–10(–13) mm wide, margins entire or denticulate, ciliate along petiole, trichomes to 1.3 mm, surfaces densely pubescent, trichomes short-stalked, 4–8-rayed, 0.2–0.5 mm. **Cauline leaves:** 12–35(–55), often concealing stem proximally; blade auricles 1–5(–6) mm, surfaces of distalmost leaves pubescent. **Racemes** 30–120-flowered, usually unbranched. **Fruiting pedicels** descending to horizontal, usually recurved, rarely straight, 4–20 mm, pubescent, trichomes appressed, 3–7-rayed. **Flowers** divaricate-ascending at anthesis; sepals pubescent; petals usually purple, rarely pinkish, 9–14 × 1.5–3 mm, glabrous or sparsely pubescent (trichomes scattered abaxially); pollen spheroid. **Fruits** widely pendent, not appressed to rachis, not secund, usually curved, rarely straight, edges parallel, (6–)8–12 cm × 1.5–2.5 mm; valves glabrous or trichomes sometimes scattered; ovules 140–180 per ovary; style 0.05–0.3 mm. **Seeds** uniseriate or sub-biseriate, 1.7–2 × 1.2–1.7 mm; wing continuous, 0.2–0.4 mm wide.

Flowering Mar–Jun. Rocky slopes and gravelly soil in desert chaparral and oak woodlands; 300–2300 m; Calif.; Mexico (Baja California).

Morphological evidence suggests that *Boechera californica* is an apomictic species that arose through hybridization between *B. arcuata* and *B. pulchra* (see M. D. Windham and I. A. Al-Shehbaz 2007 for detailed comparison). In the flora area, *B. californica* is known only from Los Angeles, Riverside, San Bernardino, and San Diego counties.

9. Boechera canadensis (Linnaeus) Al-Shehbaz, Novon 13: 384. 2003 [E]

Arabis canadensis Linnaeus, Sp. Pl. 2: 665, 1200. 1753; *A. falcata* Michaux; *A. hirsuta* (Linnaeus) Scopoli var. *ovata* (Pursh) Torrey & A. Gray; *A. ovata* (Pursh) Poiret

Biennials; short-lived; sexual; caudex not evident. **Stems** 1 per plant, arising from center of rosette near ground surface, 2.5–10(–12.5) dm, sparsely to densely pubescent proximally, trichomes simple, to 1 mm, glabrous or sparsely pubescent distally. **Basal leaves:** blade obovate to oblanceolate, 10–30 mm wide, margins dentate, ciliate proximally, trichomes simple, to 1 mm, surfaces moderately pubescent, trichomes simple, 0.5–1.5 mm, these mixed with smaller, 2- or 3-rayed ones. **Cauline leaves:** 12–20, often concealing stem proximally; blade auricles absent or proximalmost blades with auricles 0.5–2(–5) mm, surfaces of distalmost leaves sparsely pubescent. **Racemes** 15–65(–82)-flowered, sometimes branched. **Fruiting pedicels** descending to horizontal, usually curved, rarely straight, 6–13(–17) mm, pubescent or glabrous, trichomes subappressed, usually 2-rayed. **Flowers** divaricate-ascending to horizontal at anthesis; sepals pubescent; petals white, 3–5 × 0.7–1.2 mm, glabrous; pollen ellipsoid. **Fruits** pendent to horizontal, not appressed to rachis, often secund, curved, edges parallel, 4–10 cm × 2.5–4 mm; valves glabrous; ovules 40–62 per ovary; style 0.3–0.8(–1.1) mm. **Seeds** uniseriate, 1.7–3.5 × 1.5–2.2 mm; wing continuous, 0.4–1 mm wide distally. $2n = 14$.

Flowering Apr–Jul. Bluffs, rocky slopes, ravines, open woods; 0–1200 m; Que.; Ala., Ark., Conn., D.C., Fla., Ga., Ill., Iowa, Kans., Ky., Maine, Md., Mass., Mich., Minn., Miss., Mo., Nebr., N.Y., N.C., Ohio, Okla., Pa., Tenn., Tex., Vt., Va., W.Va., Wis.

10. Boechera cascadensis Windham & Al-Shehbaz, Harvard Pap. Bot. 11: 260. 2007 [C] [E]

Arabis microphylla Nuttall var. *thompsonii* Rollins, Rhodora 43: 429. 1941, not *Boechera thompsonii* (S. L. Welsh) N. H. Holmgren 2005

Perennials; long-lived; (cespitose); apomictic; caudex somewhat woody. **Stems** usually 1 per caudex branch, arising from center of rosette near ground surface, 0.5–2.2 dm, sparsely pubescent proximally, trichomes simple and short-stalked, 2- or 3-rayed, 0.1–0.2 mm, glabrous distally. **Basal leaves:** blade linear-oblanceolate, 0.7–2 mm wide, margins entire, ciliate proximally, trichomes to 0.4 mm, surfaces densely pubescent, trichomes short-stalked, (2- or) 3–6-rayed, 0.05–0.2(–0.3) mm. **Cauline leaves:** 4–6, not concealing stem; blade auricles 0.5–1 mm, surfaces of distalmost leaves glabrous. **Racemes** 3–11-flowered, unbranched. **Fruiting pedicels** ascending to divaricate-ascending, straight, 3–8 mm, glabrous. **Flowers** ascending at anthesis; sepals glabrous or sparsely pubescent; petals lavender, 5–6 × 1–1.7 mm, glabrous; pollen spheroid. **Fruits** ascending to divaricate-ascending, not appressed to rachis, not secund, straight, edges parallel, 3.5–6.2 cm × 1.2–1.5 mm; valves glabrous; ovules 58–80 per ovary; style 0.8–1.5 mm. **Seeds** uniseriate, 1.1–1.3 × 0.9–1 mm; wing continuous, 0.05–0.1 mm wide.

Flowering Jun. Basaltic cliffs and rocky slopes in subalpine areas; of conservation concern; ca. 1900 m; Oreg., Wash.

Morphological evidence suggests that *Boechera cascadensis* is an apomictic species that arose through hybridization between *B. microphylla* and *B. paupercula* (see M. D. Windham and I. A. Al-Shehbaz 2007 for detailed comparison). It is known from two collections: the type specimens from Kittitas County, Washington, and a more recent collection from Baker County, Oregon.

11. Boechera cobrensis (M. E. Jones) Dorn, Vasc. Pl. Wyoming ed. 3, 375. 2001 [E] [F]

Arabis cobrensis M. E. Jones, Contr. W. Bot. 12: 1. 1908

Perennials; long-lived; (often cespitose); sexual; caudex somewhat woody (usually without persistent leaf bases). **Stems** usually 1 per caudex branch, arising from center of rosette near ground surface, (1.2–)2.5–6 dm, densely pubescent proximally, trichomes short-stalked, 4–10-rayed, 0.1–0.2 mm, sparsely pubescent or glabrous distally. **Basal leaves:** blade linear to narrowly oblanceolate, 1–4 mm wide, margins entire, ciliate near petiole base, trichomes (simple), to 0.6 mm, surfaces densely pubescent, trichomes short-stalked, 4–10-rayed, 0.1–0.2 mm. **Cauline leaves:** 5–10, not concealing stem; blade auricles 1–1.5 mm, surfaces of distalmost leaves usually sparsely pubescent. **Racemes** 10–25-flowered, often sparingly branched. **Fruiting pedicels** divaricate-ascending to horizontal, often straight proximally, usually strongly recurved or reflexed distally, 4–17 mm, usually pubescent, rarely glabrous, trichomes appressed, branched. **Flowers** divaricate-ascending at anthesis; sepals pubescent; petals white to lavender, 3.5–6 × 0.7–1 mm, glabrous; pollen ellipsoid. **Fruits** pendent, not appressed to rachis, not secund, straight or slightly curved, edges parallel, 2.5–5.5 cm × 1.7–2.5 mm; valves glabrous; ovules 34–64 per ovary; style 0.05–0.2 mm. **Seeds** uniseriate, 1.4–1.8 × 1–1.2 mm; wing continuous, 0.25–0.5 mm wide. $2n = 14$.

Flowering May–Jun. Sandy soil, usually under shelter of shrubs in semi-desert communities; 1200–2800 m; Calif., Idaho, Nev., Oreg., Wyo.

Boechera cobrensis is a sexual diploid commonly found on stabilized sand dunes. The related *B. lignifera* often is found in similar habitats, and the two species can be difficult to distinguish where their ranges overlap in Idaho and Wyoming. It is possible that *B. cobrensis* was involved in the origin of some apomictic populations currently assigned to *B. lignifera*.

12. Boechera collinsii (Fernald) Á. Löve & D. Löve, Taxon 31: 125. 1982 [E] [F]

Arabis collinsii Fernald, Rhodora 7: 32. 1905; *A. holboellii* Hornemann var. *collinsii* (Fernald) Rollins; *A. retrofracta* Graham var. *collinsii* (Fernald) B. Boivin

Perennials; usually short-lived; sexual; caudex not woody. **Stems** usually 1 per caudex branch, arising from center of rosette near ground surface, (1.5–)2.5–8 dm, densely pubescent proximally, trichomes simple and short-stalked, 2-rayed, to 1 mm, sparsely pubescent or glabrous distally. **Basal leaves:** blade oblanceolate, 1–7 mm wide, margins entire or dentate, ciliate along petiole, trichomes (simple), to 0.7 mm, surfaces moderately to densely pubescent, trichomes short-stalked, 5–8-rayed, 0.15–0.4 mm. **Cauline leaves:** (8–)15–40, concealing stem proximally; blade auricles 1–3 mm, surfaces of distalmost leaves usually sparsely pubescent. **Racemes** (15–)30–90-flowered, usually unbranched. **Fruiting pedicels** reflexed, abruptly recurved at base, otherwise straight, 4–13 mm, glabrous or sparsely pubescent, trichomes usually simple. **Flowers** descending to pendent at anthesis; sepals pubescent; petals usually white, rarely lavender, 4–6 × 0.8–1.5 mm, glabrous; pollen ellipsoid. **Fruits** strongly reflexed, usually appressed to rachis, rarely somewhat secund, straight, edges parallel, 3.5–6 cm × 0.9–1.5 mm; valves glabrous; ovules 70–126 per ovary; style 0.2–0.5 mm. **Seeds** uniseriate, 1–1.4 × 0.8–1 mm; wing continuous (rarely absent), to 0.1 mm wide. $2n = 14$.

Flowering May–Jun. Rocky and gravelly hillsides, prairies, open woods, floodplains; 600–1400 m; Alta., Man., Que., Sask.; Idaho, Minn., Mont., Nebr., N.Dak., S.Dak., Wyo.

Though often treated as a variety of *Arabis* (*Boechera*) *holboellii* (e.g., R. C. Rollins 1993; G. A. Mulligan 1996), *B. collinsii* is easily distinguished from that species by having simple and 2-rayed (versus 4–8-rayed)

13. Boechera consanguinea (Greene) Windham & Al-Shehbaz, Harvard Pap. Bot. 11: 261. 2007 [E]

Arabis consanguinea Greene, Pittonia 4: 190. 1900; *A. holboellii* Hornemann var. *consanguinea* (Greene) G. A. Mulligan

Biennials or perennials; short-lived; apomictic; caudex present or absent. **Stems** usually 1 per caudex branch, arising from center of rosette near ground surface, 1.5–5 dm, densely pubescent proximally, trichomes short-stalked, 2–6-rayed, 0.2–0.5 mm, glabrous or sparsely pubescent distally. **Basal leaves**: blade oblanceolate, 2–10 mm wide, margins usually dentate, sometimes ciliate near petiole base, trichomes (usually spurred), to 0.7 mm, surfaces moderately to densely pubescent, trichomes short-stalked, 4–8-rayed, 0.1–0.4 mm. **Cauline leaves**: 15–36, sometimes concealing stem proximally; blade auricles 1–3 mm, surfaces of distalmost leaves sparsely pubescent. **Racemes** 20–55-flowered, usually unbranched. **Fruiting pedicels** reflexed to divaricate-descending, usually curved downward, 8–14 mm, usually pubescent, rarely glabrous, trichomes subappressed, branched. **Flowers** divaricate-descending at anthesis; sepals pubescent; petals pale lavender, 5–8.5 × 1–2 mm, glabrous; pollen spheroid. **Fruits** reflexed to pendent, rarely appressed to rachis, not secund, straight or slightly curved, edges parallel, 4–6 cm × 1–2 mm; valves glabrous; ovules 100–128 per ovary; style 0.05–0.5 mm. **Seeds** biseriate to sub-biseriate, mature seeds not seen.

Flowering May. Rocky slopes and sandy soil in ponderosa pine, pinyon-juniper, and sagebrush communities; 1900–2500 m; Colo., N.Mex., Utah.

Morphological evidence suggests that *Boechera consanguinea* is an apomictic species that arose through hybridization between *B. fendleri* and *B. retrofracta*. It is most similar to *B. goodrichii*, another apomictic hybrid involving *B. retrofracta* (see M. D. Windham and I. A. Al-Shehbaz 2007 for detailed comparison).

14. Boechera constancei (Rollins) Al-Shehbaz, Novon 13: 384. 2003 [E] [F]

Arabis constancei Rollins, Contr. Gray Herb. 201: 5, plate 1. 1971; *A. suffrutescens* S. Watson var. *perstylosa* Rollins

Perennials; long-lived; sexual; caudex woody. **Stems** usually 1 per caudex branch, arising from center of rosette, elevated above ground surface on woody base, 1.2–3 dm, glabrous throughout. **Basal leaves**: blade narrowly oblanceolate, 1.5–4 mm wide, margins entire, ciliate, trichomes (simple, mixed with fewer stalked, 2-rayed), 0.3–0.8 mm, surfaces glabrous. **Cauline leaves**: 6–12, not concealing stem; blade auricles absent, surfaces of distalmost leaves glabrous. **Racemes** 5–15-flowered, unbranched. **Fruiting pedicels** arched, strongly recurved proximally, 4–12 mm, glabrous. **Flowers** divaricate-ascending at anthesis; sepals glabrous; petals creamy white, 6–8 × 1.5–2 mm, glabrous; pollen ellipsoid. **Fruits** pendent, not appressed to rachis, usually secund, straight or slightly curved, edges often somewhat undulate (not parallel), 4–7.5 cm × 3–3.5 mm; valves glabrous; ovules 18–28 per ovary; style 2.5–5 mm. **Seeds** uniseriate, 3–4 × 2.5–3 mm; wing continuous, 0.5–1 mm wide. $2n = 14$.

Flowering May. Steep ridges, serpentine slopes; 1200–1900 m; Calif.

Boechera constancei was originally treated as a variety of *Arabis* (*Boechera*) *suffrutescens*; it is easily separated from that species by its non-auriculate cauline leaf blades, longer (6–8 versus 4.5–6 mm) petals that are creamy white (versus purple or rose-tipped), much longer style (2.5–5 versus 0.4–1.2 mm), and exerted anthers. This distinctive sexual diploid is thus far known only from the west slope of the Sierra Nevada in Plumas County.

15. Boechera covillei (Greene) Windham & Al-Shehbaz, Harvard Pap. Bot. 12: 237. 2007 [E]

Arabis covillei Greene, Repert. Spec. Nov. Regni Veg. 5: 243. 1908; *A. leibergii* Greene

Perennials; long-lived; apomictic; caudex woody. **Stems** usually 1 per caudex branch, arising from center of rosette near ground surface, 0.5–2.5 dm, glabrous throughout. **Basal leaves**: blade oblanceolate, 1.5–5 mm wide, margins entire, ciliate proximally, trichomes (simple), to 0.4 mm, and distally, trichomes short-stalked, 2 or 3 (or 4)-rayed, 0.15–0.4 mm, surfaces glabrous. **Cauline leaves**: 2–7, not concealing stem; blade with auricles 0.2–1 mm, auricles rarely absent, surfaces of distalmost leaves glabrous. **Racemes** 2–9-flowered, unbranched. **Fruiting pedicels** ascending to suberect, straight, 6–20 mm, glabrous. **Flowers** ascending at anthesis; sepals glabrous; petals lavender to purplish, 5–6 × 1–2 mm, glabrous; pollen spheroid. **Fruits** ascending to suberect, rarely appressed to rachis, not secund, straight, edges slightly undulate (rarely parallel), 3.5–5 cm × 3–5 mm; valves glabrous; ovules 18–30 per ovary; style 0.2–1 mm. **Seeds** uniseriate, 4–5 × 3–4 mm; wing continuous, 1–1.5 mm wide.

Flowering Jul–Aug. Rocky slopes in alpine meadows and open coniferous forests; 2200–3500 m; Calif., Nev., Oreg.

Morphological evidence suggests that *Boechera covillei* is an apomictic species that arose through hybridization between *B. howellii* and *B. lyallii* (see M. D. Windham and I. A. Al-Shehbaz 2007b for detailed comparison).

16. Boechera crandallii (B. L. Robinson) W. A. Weber, Phytologia 51: 369. 1982 [C][E][F]

Arabis crandallii B. L. Robinson, Bot. Gaz. 28: 135. 1899; *A. stenoloba* Greene

Perennials; long-lived; (often cespitose); sexual; caudex somewhat woody. **Stems** usually 2–5 per caudex branch, arising from margin of rosette near ground surface, (1–)1.5–4 dm, densely pubescent proximally, trichomes short-stalked, 5–8-rayed, 0.1–0.2 mm, moderately to sparsely pubescent distally. **Basal leaves:** blade narrowly oblanceolate, 1.5–3(–5) mm wide, margins entire, ciliate along petiole base, trichomes (simple), to 0.6 mm, surfaces densely pubescent, trichomes short-stalked, 5–8-rayed 0.1–0.2 mm. **Cauline leaves:** 5–14, not concealing stem; blade auricles 0.1–0.5 mm, surfaces of distalmost leaves pubescent. **Racemes** 10–30-flowered, usually unbranched. **Fruiting pedicels** ascending to divaricate-ascending, straight, 5–10 mm, pubescent, trichomes branched. **Flowers** ascending at anthesis; sepals pubescent; petals usually white, 5–7 × 1–2 mm, glabrous; pollen ellipsoid. **Fruits** ascending to divaricate-ascending, not appressed to rachis, not secund, straight, edges parallel, 3–5.5 cm × 0.9–1.2 mm; valves glabrous; ovules 56–84 per ovary; style 0.1–0.5 mm. **Seeds** uniseriate, 0.8–1 × 0.6–0.9 mm; wing continuous (rarely absent), to 0.1 mm wide. $2n = 14$.

Flowering May–Jun. Rocky slopes and gravelly soil in sagebrush, mountain shrub, open conifer forests; of conservation concern; 2000–2700 m; Colo.

Boechera crandallii is a sexual diploid known only from the Gunnison Basin of west-central Colorado. Despite its narrow range, the species appears to hybridize frequently, most notably with *B. pallidifolia*.

17. Boechera cusickii (S. Watson) Al-Shehbaz, Novon 13: 384. 2003 [E][F]

Arabis cusickii S. Watson, Proc. Amer. Acad. Arts 17: 363. 1882

Perennials; long-lived; (often cespitose); sexual; caudex usually woody. **Stems** usually 1 per caudex branch, arising from center of rosette near ground surface, 1–2.5(–3.8) dm, densely pubescent proximally, trichomes simple, 0.5–1.5 mm, sparsely pubescent or glabrous distally. **Basal leaves:** blade linear to narrowly oblanceolate, 0.5–2 mm wide, margins entire, ciliate, trichomes (simple), 1–1.7 mm, surfaces moderately pubescent, trichomes simple and long-stalked, 2-rayed, 0.5–1.7 mm. **Cauline leaves:** 10–30, often concealing stem proximally; blade auricles 0–0.5 mm, surfaces of distalmost leaves sparsely pubescent. **Racemes** 5–14-flowered, usually unbranched. **Fruiting pedicels** ascending, straight, 4–15(–20) mm, glabrous or sparsely pubescent, trichomes simple. **Flowers** ascending at anthesis; sepals pubescent; petals white to lavender, 6–10 × 2–3 mm, glabrous; pollen ellipsoid. **Fruits** ascending to divaricate-ascending, not appressed to rachis, not secund, straight or curved, edges parallel, 4–8(–9.5) cm × 2.3–3 mm; valves glabrous; ovules 56–106 per ovary; style 0.1–0.5 mm. **Seeds** uniseriate, 1.5–2.5 × 1.5–2 mm; wing continuous, 0.1–0.2 mm wide distally.

Flowering Apr–mid May. Basaltic bluffs, rocky slopes, rock crevices, gravelly hillsides, sagebrush hills, outcrops of volcanic rock; 600–1800 m; Idaho, Nev., Oreg., Wash.

Peripheral populations of *Boechera cusickii* in south-central Idaho and northern Nevada have a higher proportion of branched hairs, possibly resulting from hybridization with *B. sparsiflora*.

18. Boechera davidsonii (Greene) N. H. Holmgren in A. Cronquist et al., Intermount. Fl. 2(B): 368. 2005 [E][F]

Arabis davidsonii Greene, Leafl. Bot. Observ. Crit. 2: 159. 1911; *A. bruceae* M. E. Jones; *A. cognata* Jepson; *A. davidsonii* var. *parva* Rollins; *A. lyallii* S. Watson var. *davidsonii* (Greene) Smiley

Perennials; long-lived; (cespitose); sexual; caudex woody (thickly covered by persistent, crowded leaf bases). **Stems** usually 1 per caudex branch, arising from center of rosette near ground surface, 0.6–2.3 dm, glabrous throughout. **Basal leaves:** blade oblanceolate, 3.5–14 mm wide, margins usually entire, rarely slightly dentate, not ciliate, surfaces glabrous.

Cauline leaves: 3–10, not concealing stem; blade auricles absent, glabrous. Racemes 4–24-flowered, usually unbranched. Fruiting pedicels ascending, straight, 3–18 mm, glabrous. Flowers ascending at anthesis; sepals glabrous; petals white to lavender, 6–10 × 2.5–4 mm, glabrous; pollen ellipsoid. Fruits ascending to divaricate-ascending, not appressed to rachis, not secund, straight to slightly curved, edges parallel, 2.5–7 cm × 1.5–2.5 mm; valves glabrous; ovules 28–50 per ovary; style 0.1–0.8 mm. Seeds uniseriate, 1.8–2.2 × 0.8–1.7 mm; wing distal or continuous, 0.1–0.5 mm wide.

Flowering Apr–Jul. Ledges and crevices of rock outcrops; 1200–3400 m; Calif., Nev., Oreg.

Plants of *Boechera davidsonii* sometimes are confused with completely glabrous individuals of *B. lyallii*, and G. A. Mulligan (1996) treated it as a synonym of the latter. It is easily distinguished from *B. lyallii* by the thick covering of persistent leaf bases on caudex branches, ascending (versus erect and appressed) fruits, and uniseriate (versus biseriate to sub-biseriate) seeds. *Boechera davidsonii* is nearly unique among western North American species of the genus in showing no evidence of hybridization with other species.

19. **Boechera dentata** (Rafinesque) Al-Shehbaz & Zarucchi, Harvard Pap. Bot. 13: 293. 2008 E

Shortia dentata Rafinesque, Autik. Bot., 17. 1840; *Arabis dentata* Torrey & A. Gray var. *phalacrocarpa* M. Hopkins; *A. perstellata* E. L. Braun var. *shortii* Fernald; *A. shortii* (Fernald) Gleason; *A. shortii* var. *phalacrocarpa* (M. Hopkins) Steyermark; *Boechera shortii* (Fernald) Al-Shehbaz; *Iodanthus dentatus* (Rafinesque) Greene

Biennials; short-lived; sexual; caudex present or absent. Stems usually 2–4 per plant or caudex branch, arising from margin of rosette near ground surface, or arising laterally proximal to sterile shoots, 2–6 dm, pubescent proximally, trichomes short-stalked, 2–4-rayed, 0.15–0.4 mm, sparsely pubescent or glabrous distally. Basal leaves: blade obovate to oblanceolate, (6–)10–45 mm wide, margins dentate, not ciliate, surfaces sparsely pubescent, abaxially trichomes short-stalked, 3- or 4-rayed, 0.15–0.4 mm, adaxially simple ones, 0.4–1 mm. Cauline leaves: 4–20, rarely concealing stem; blade auricles 1–8 mm, surfaces of distalmost leaves sparsely pubescent. Racemes 14–66-flowered, sparingly branched. Fruiting pedicels divaricate-ascending to slightly descending, straight, 1–6 mm, glabrous or pubescent, trichomes appressed, branched. Flowers divaricate at anthesis; sepals glabrous or sparsely pubescent; petals white or cream, 2–3.5 × 0.7–1 mm, glabrous; pollen ellipsoid. Fruits divaricate-ascending to slightly descending, not appressed to rachis, not secund, straight or slightly curved, edges parallel, 1.3–4.2 cm × 0.7–0.8 mm; valves usually sparsely pubescent throughout, rarely glabrous; ovules 14–44 per ovary; style 0.1–0.5 mm. Seeds uniseriate, 0.7–1.1 × 0.5–0.6 mm; not winged.

Flowering Apr–Jun. Bluffs and rocky ledges, wooded slopes, floodplains; 100–300 m; Ont.; Ala., Ark., D.C., Ill., Ind., Iowa, Kans., Ky., Md., Mich., Minn., Mo., Nebr., N.Y., Ohio, Okla., Pa., S.Dak., Tenn., Va., W.Va., Wis.

Sisymbrium dentatum Torrey is an illegitimate name that pertains to *Iodanthus dentatus*.

20. **Boechera depauperata** (A. Nelson & P. B. Kennedy) Windham & Al-Shehbaz, Harvard Pap. Bot. 11: 262. 2007 E

Arabis depauperata A. Nelson & P. B. Kennedy, Proc. Biol. Soc. Wash. 19: 36. 1906; *A. lemmonii* S. Watson var. *depauperata* (A. Nelson & P. B. Kennedy) Rollins

Perennials; usually long-lived; (cespitose); apomictic; caudex woody. Stems usually 1 per caudex branch, arising from center of rosette near ground surface, 0.5–1.5(–2) dm, densely pubescent proximally, trichomes short-stalked, 4–8-rayed, 0.07–0.4 mm, usually sparsely pubescent distally, rarely glabrous. Basal leaves: blade narrowly oblanceolate, 0.7–3(–5) mm wide, margins entire, sometimes ciliate proximally, trichomes to 0.4 mm, surfaces densely pubescent, trichomes short-stalked, 5–8-rayed, 0.05–0.25 mm. Cauline leaves: 3–7, not concealing stem; blade auricles 0.5–1.5 mm, surfaces of distalmost leaves usually pubescent, rarely glabrous. Racemes 9–23-flowered, unbranched. Fruiting pedicels divaricate-ascending, straight, 1.5–7 mm, usually pubescent, rarely glabrous, trichomes appressed, branched. Flowers ascending at anthesis; sepals pubescent; petals lavender to purplish, 3.5–5 × 0.8–1.2 mm, glabrous; pollen spheroid. Fruits divaricate-ascending, not appressed to rachis, not secund, straight, edges parallel, 1.7–3.5(–4.5) cm × 1–1.5 mm; valves glabrous; ovules 32–44 per ovary; style 0.05–0.2 mm. Seeds uniseriate, 1.2–1.5 × 1–1.2 mm; wing continuous, 0.1–0.15 mm wide.

Flowering Jun–Jul. Exposed ridges and talus slopes in subalpine and alpine habitats; 2700–3900 m; Calif., Nev.

Morphological evidence suggests that *Boechera depauperata* is an apomictic species that arose through hybridization between *B. lemmonii* and *B. paupercula* (see M. D. Windham and I. A. Al-Shehbaz 2007 for detailed comparison). It is known in California from Fresno, Mono, and Tulare counties and in Nevada from Washoe County.

21. Boechera dispar (M. E. Jones) Al-Shehbaz, Novon 13: 384. 2003 [E][F]

Arabis dispar M. E. Jones, Contr. W. Bot. 8: 41. 1898; *A. juniperina* M. E. Jones; *A. nardina* Greene; *A. salubris* M. E. Jones

Perennials; usually long-lived; sexual; caudex often woody. **Stems** usually 1 per caudex branch, arising from center of rosette near ground surface, 0.9–3 dm, densely pubescent proximally, trichomes short-stalked, 5–12-rayed, 0.1–0.3 mm, sparsely pubescent to glabrescent distally. **Basal leaves**: blade linear-oblanceolate, 2–5 mm wide, margins entire, not ciliate, surfaces densely pubescent, trichomes short-stalked, 5–12(–16)-rayed, 0.1–0.3 mm. **Cauline leaves**: 1–5, not concealing stem; blade without auricles, surfaces of distalmost leaves pubescent. **Racemes** 4–15(–20)-flowered, unbranched. **Fruiting pedicels** ascending, straight, 4–15(–25) mm, pubescent, trichomes appressed, branched. **Flowers** ascending at anthesis; sepals pubescent; petals purple to lavender, 3.5–6 × 1–1.5 mm, glabrous; pollen ellipsoid. **Fruits** divaricate-ascending, not appressed to rachis, not secund, straight, edges parallel, 4–7.3 cm × 2.7–4 mm; valves glabrous; ovules 44–52 per ovary; style 0.05–0.1 mm. **Seeds** uniseriate, 1.9–2.3 mm in diam.; wing continuous, 0.3–0.5 mm wide.

Flowering Apr–May. Rocky slopes and gravelly soil in desert scrub and pinyon-juniper communities; 1500–2300 m; Calif., Nev.

Boechera dispar is a distinctive sexual diploid that is known from Inyo, Mono, and San Bernardino counties, California, and Nye County, Nevada.

22. Boechera divaricarpa (A. Nelson) Á. Löve & D. Löve, Bot. Not. 128: 513. 1976 [E][F]

Arabis divaricarpa A. Nelson, Bot. Gaz. 30: 193. 1900; *A. stokesiae* Rydberg

Biennials or perennials; short-lived; apomictic; caudex present or absent. **Stems** usually 1 per caudex branch, arising from center of rosette near ground surface, (1.5–)3–9 dm, glabrous or pubescent proximally, trichomes sessile, 2–4-rayed (some simple), to 0.7 mm, glabrous distally. **Basal leaves**: blade oblanceolate, 2–10 mm wide, margins usually entire, rarely denticulate, ciliate along petiole, trichomes (simple), to 0.8 mm, surfaces sparsely to densely pubescent, trichomes sessile, 2–6-rayed, 0.1–0.4 mm. **Cauline leaves**: (10–)15–56, concealing stem proximally; blade auricles 1–5 mm, surfaces of distalmost leaves glabrous. **Racemes** 12–40(–65)-flowered, usually unbranched. **Fruiting pedicels** divaricate-ascending to horizontal, straight, 5–12 mm, glabrous. **Flowers** divaricate-ascending at anthesis; sepals glabrous or with scattered trichomes; petals usually purple, rarely lavender, 6–9 × 1.5–3 mm, glabrous; pollen spheroid. **Fruits** divaricate-ascending to horizontal, not appressed to rachis, not secund, straight, edges parallel, (4.5–)5.5–11 cm × 1.7–2.5 mm; valves glabrous; ovules 114–142 per ovary; style 0.05–0.2 mm. **Seeds** uniseriate to subbiseriate, 1.4–2 × 1–1.5 mm; wing continuous, 0.1–0.2 mm wide. $2n = 21$.

Flowering May–Jul. Rock outcrops, talus slopes and gravelly hillsides in sagebrush, mountain shrub, and open conifer forests; 900–2500 m; Calif., Idaho, Mont., Nev., Utah, Wash., Wyo.

The name *Arabis* (*Boechera*) *divaricarpa* has been applied to nearly every hybrid containing a genome derived from *B. stricta*. This presents a serious barrier to understanding the evolution of *Boechera* and also is contrary to the International Code of Botanical Nomenclature, because some names usually placed in synonymy (i.e., *B. grahamii* and *B. brachycarpa*) have priority at species level (M. D. Windham and I. A. Al-Shehbaz 2007b). To address this problem, we treat the following as distinct species: *B. acutina*, *B. grahamii* (= *B. brachycarpa* of R. D. Dorn 2001), and *B. pratincola* (all considered synonyms of *A. divaricarpa* by R. C. Rollins 1993), and *B. calderi*, *B. elkoensis*, and *B. quebecensis* (taxa described after 1993). Detailed comparison among these taxa are provided by Windham and Al-Shehbaz (2007, 2007b). The narrow concept of *B. divaricarpa* advocated here encompasses apomictic triploid populations containing three distinct genomes, one each derived from *B. retrofracta*, *B. sparsiflora*, and *B. stricta*. If the species is defined more broadly, the name *B. grahamii* has priority.

23. Boechera drepanoloba (Greene) Windham & Al-Shehbaz, Harvard Pap. Bot. 11: 263. 2007 [E]

Arabis drepanoloba Greene, Pittonia 3: 306. 1898; *A. drummondii* A. Gray var. *oreophila* (Rydberg) M. Hopkins; *A. lemmonii* S. Watson var. *drepanoloba* (Greene) Rollins; *A. oreophila* Rydberg

Perennials; short- to long-lived; apomictic; caudex usually not woody. **Stems** usually 1 per caudex branch, arising from center of rosette near ground surface, 1–4 dm, glabrous or sparsely pubescent proximally, trichomes sessile or subsessile, 2- or 3-rayed, 0.1–0.3 mm, glabrous distally. **Basal leaves**: blade oblanceolate, 2–6 mm wide, margins entire, ciliate along petiole, trichomes (submalpighiaceous or simple), to 0.6 mm, surfaces sparsely to densely pubescent, trichomes subsessile, 2–6-rayed, 0.1–0.3 mm. **Cauline leaves:** 3–15, often concealing stem proximally; blade auricles 0.7–2.5 mm, surfaces of distalmost leaves usually glabrous. **Racemes** 5–25-flowered, usually unbranched. **Fruiting pedicels** horizontal to ascending, straight or curved, 4–8 mm, glabrous. **Flowers** ascending at anthesis; sepals glabrous or pubescent with scattered trichomes; petals purple to lavender, 6–8 × 1.5–2.5 mm, glabrous; pollen spheroid. **Fruits** horizontal to ascending, secund, straight to slightly curved, edges parallel, 3–5.8 cm × 2–3.5 mm; valves glabrous; ovules 44–76(–104) per ovary; style 0.05–0.5 mm. **Seeds** uniseriate, 1.5–2.2 × 1.2–1.6 mm; wing continuous, 0.3–0.7 mm wide.

Flowering Jul–Aug. Rocky slopes and talus in alpine and subalpine habitats; 2000–3800 m; Alta.; Alaska, Colo., Idaho, Mont., Oreg., Utah, Wash., Wyo.

Morphological evidence suggests that *Boechera drepanoloba* is an apomictic species that arose through hybridization between *B. lemmonii* and *B. stricta* (see M. D. Windham and I. A. Al-Shehbaz 2007 for detailed comparison).

24. **Boechera duchesnensis** (Rollins) Windham, Al-Shehbaz & Allphin, Harvard Pap. Bot. 12: 239. 2007 [E]

Arabis pulchra M. E. Jones ex S. Watson var. *duchesnensis* Rollins, Syst. Bot. 6: 59. 1981; *Boechera pulchra* (M. E. Jones ex S. Watson) W. A. Weber var. *duchesnensis* (Rollins) Dorn

Perennials; usually long-lived; apomictic; caudex often woody. **Stems** usually 1 per caudex branch, arising from center of rosette, elevated on woody base or from near ground surface, 1.5–4.5 dm, densely pubescent proximally, trichomes short-stalked, 2–7-rayed, 0.2–0.5 mm (sometimes mixed with some larger trichomes proximally), sparsely pubescent distally. **Basal leaves:** blade linear-oblanceolate, 2–5 mm wide, margins entire or denticulate, ciliate at petiole base, trichomes (simple), to 1 mm, surfaces densely pubescent, trichomes short-stalked, 4–8-rayed, 0.15–0.4 mm. **Cauline leaves:** 3–8, not concealing stem; blade auricles 0.7–2 mm, surfaces of distalmost leaves sparsely pubescent. **Racemes** 11–22-flowered, usually unbranched. **Fruiting pedicels** horizontal, straight or slightly curved, 7–14 mm, pubescent, trichomes appressed, branched. **Flowers** divaricate-ascending at anthesis; sepals pubescent; petals whitish to pale lavender, 7–10 × 1.5–3 mm, glabrous; pollen spheroid. **Fruits** horizontal, not appressed to rachis, not secund, usually straight, edges parallel, 3.5–5 cm × 1.7–2 mm; valves glabrous proximally, pubescent distally; ovules 72–92 per ovary; style 0.1–0.3 mm. **Seeds** sub-biseriate, 1.1–1.3 × 0.7–1 mm; wing continuous, ca. 0.1 mm wide.

Flowering May. Sandy soil on rocky slopes, mostly in pinyon-juniper woodlands; 1300–1800 m; Colo., N.Mex., Utah.

Morphological and isozyme analyses indicate that *Boechera duchesnensis* is an apomictic species that arose through hybridization between *B. formosa* and *B. pallidifolia* (M. D. Windham and Allphin, unpubl.). It is easily distinguished from those species by having fruits that are glabrous proximally and sparsely pubescent distally. It shares this distinctive fruit pubescence with *B. xylopoda*, an apomictic hybrid between *B. perennans* and *B. pulchra* (see Windham and I. A. Al-Shehbaz 2007b for detailed comparison).

25. **Boechera elkoensis** Windham & Al-Shehbaz, Harvard Pap. Bot. 11: 263. 2007 [E]

Perennials; usually long-lived; apomictic; caudex often woody. **Stems** 1–5 per caudex branch, arising from center or margin of rosette near ground surface, 1–3 dm, sparsely pubescent proximally, trichomes subsessile, 2–5-rayed, 0.1–0.4 mm, glabrous or sparsely pubescent distally. **Basal leaves:** blade narrowly oblanceolate, 2–6 mm wide, margins entire, ciliate at petiole base, trichomes (simple), to 0.4 mm, surfaces moderately pubescent, trichomes sessile or subsessile, 2–5-rayed, 0.1–0.4 mm. **Cauline leaves:** 3–6, not concealing stem; blade auricles absent or 0.2–0.5 mm, surfaces of distalmost leaves glabrous, or sparsely pubescent along margins. **Racemes** 4–11-flowered, unbranched. **Fruiting pedicels** erect to ascending, straight, 4–8 mm, usually glabrous, rarely sparsely pubescent, trichomes appressed, branched. **Flowers** erect to ascending at anthesis; sepals glabrous; petals usually lavender, rarely whitish, 4–7 × 1.2–2 mm, glabrous; pollen spheroid. **Fruits** erect to ascending, often appressed to rachis, not secund, straight, edges often slightly undulate (not parallel), 3–7 cm × 2–3 mm; valves glabrous; ovules 26–42 per ovary; style 0.05–0.3 mm. **Seeds** uniseriate, 2.5–3.5 × 1.8–2.5 mm; wing continuous, 0.4–0.9 mm wide.

Flowering Jul. Gravelly soil among rocks in open forests and subalpine meadows; 2000–3200 m; Calif., Nev.

Morphological evidence suggests that *Boechera elkoensis* is an apomictic species that arose through hybridization between *B. platysperma* and *B. stricta* (see M. D. Windham and I. A. Al-Shehbaz 2007 for detailed comparison).

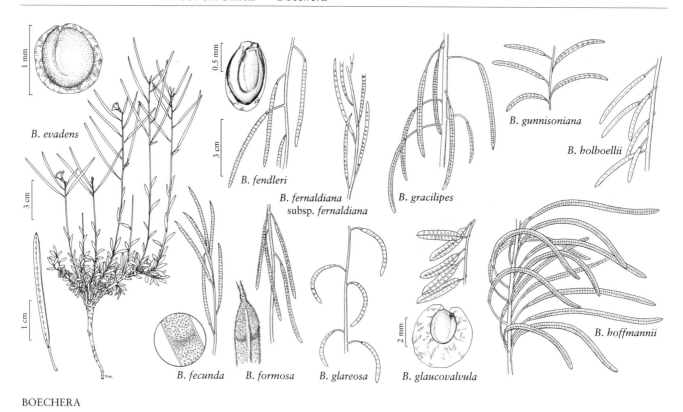

26. Boechera evadens Windham & Al-Shehbaz, Harvard Pap. Bot. 11: 66. 2006 [E][F]

Perennials; long-lived; (cespitose); sexual; caudex woody. Stems usually 1 per caudex branch, arising from center of rosette near ground surface, 1–2.5 dm, densely pubescent proximally, trichomes short-stalked, 2–5-rayed, 0.2–0.6 mm, and sparse, simple ones to 1 mm, glabrous distally. Basal leaves: blade narrowly oblanceolate, 1.5–3 mm wide, margins entire, ciliate along petiole, trichomes (simple), to 1 mm, surfaces densely pubescent, trichomes short-stalked, 4–8-rayed, 0.1–0.2 mm. Cauline leaves: 3–5, not concealing stem; blade auricles 0.5–1 mm, surfaces of distalmost leaves pubescent. Racemes 10–22-flowered, unbranched. Fruiting pedicels divaricate-ascending, straight, 4–8 mm, sparsely pubescent, trichomes appressed, branched. Flowers ascending at anthesis; sepals pubescent; petals white, 3–4 × 0.5–0.8 mm, glabrous; pollen ellipsoid. Fruits divaricate-ascending, not appressed to rachis, not secund, straight, edges parallel, 3–4 cm × 1.2–1.5 mm; valves glabrous; ovules 46–52 per ovary; style ca. 0.1 mm. Seeds uniseriate, 1–1.1 × 0.9–1 mm; wing continuous, ca. 0.1 mm wide.

Flowering Jun. Rock outcrops; ca. 2600 m; Calif.

Boechera evadens is a sexual taxon that was originally interpreted as a disjunct population of *Arabis* (*Boechera*) *fernaldiana*, but is easily distinguished from that species by having much shorter (3–4 versus 7–12 mm) petals, larger (0.2–0.6 versus 0.04–0.1 mm) trichomes proximally on stems, and shorter (3–4 versus 4–7.5 cm) fruits (see M. D. Windham and I. A. Al-Shehbaz 2006 for additional distinctions). It is known only from the holotype specimen collected near Sherman Pass in Tulare County.

27. Boechera falcatoria (Rollins) Dorn, Brittonia 55: 3. 2003 [C][E]

Arabis falcatoria Rollins, Contr. Gray Herb. 212: 106. 1982

Perennials; short- to long-lived; apomictic; caudex sometimes woody. Stems usually 2–7 per caudex branch, arising from margin of rosette near ground surface, 0.5–2(–3) dm, densely pubescent proximally, trichomes simple and stalked, 2-rayed, 0.3–0.8 mm, glabrous distally. Basal leaves: blade narrowly oblanceolate, 1–3 mm wide, margins entire, ciliate, trichomes to 1 mm, surfaces pubescent, trichomes stalked, 2–4-rayed, 0.3–0.8 mm. Cauline leaves: 3–7(–10), not concealing stem; blade auricles 0.2–1 mm, surfaces of distalmost leaves glabrous. **Racemes** 6–15-flowered, unbranched. **Fruiting pedicels** horizontal to divaricate-ascending, straight, 3–9 mm, glabrous. **Flowers** ascending at anthesis; sepals pubescent; petals white or lavender, 5–7 × 1.5–2.5 mm, glabrous; pollen spheroid. **Fruits** horizontal to divaricate-ascending, not appressed to rachis, not secund, usually strongly curved, edges parallel, 3.5–6.5 cm × 1.7–2 mm; valves glabrous; ovules 50–84 per ovary; style 0.05–0.5 mm. **Seeds** uniseriate, 1.5–2 × 1.1–1.5 mm; wing continuous, 0.1–0.2 mm wide.

Flowering May–Jun. Rock outcrops and gravelly soil in sagebrush and mountain shrub communities; of conservation concern; 2000–2200 m; Utah.

Although the exact parentage of *Boechera falcatoria*, an apomictic hybrid, remains in doubt, it is virtually certain that *B. cusickii* contributed at least one genome. *Boechera falcatoria* has been confused with both *B. pendulina* and *B. perennans*, but is amply distinct from both (see M. D. Windham and I. A. Al-Shehbaz 2007 for detailed comparison). Despite the proximity of the type locality to both Idaho and Nevada, the species is known only from northwestern Box Elder County.

28. **Boechera falcifructa** (Rollins) Al-Shehbaz, Novon 13: 385. 2003 [C][E]

Arabis falcifructa Rollins, Contr. Gray Herb. 212: 112. 1982

Perennials; usually long-lived; apomictic; caudex often woody. **Stems** usually 1 per caudex branch, arising from center of rosette near ground surface, 1.5–4.5 dm, densely pubescent proximally, trichomes short-stalked, 4–9-rayed, 0.05–0.2 mm, glabrous or sparsely pubescent distally. **Basal leaves**: blade linear to narrowly oblanceolate, 1–4 mm wide, margins entire, ciliate at petiole base, trichomes (simple), to 0.5 mm, surfaces densely pubescent, trichomes short-stalked, 4–9-rayed, 0.05–0.2 mm. **Cauline leaves**: 5–9, not concealing stem; blade auricles 0.5–1 mm, surfaces of distalmost leaves sparsely pubescent. **Racemes** 10–18-flowered, sparingly branched. **Fruiting pedicels** divaricate-descending, gently curved downward, 6–12 mm, pubescent, trichomes appressed, branched. **Flowers** divaricate-ascending at anthesis; sepals pubescent; petals lavender, 5–7 × 0.7–1.2 mm, glabrous; pollen spheroid. **Fruits** divaricate-descending, not appressed to rachis, not secund, gently curved, edges parallel, 4.5–6 cm × 1.2–1.5 mm; valves glabrous; ovules 62–80 per ovary; style 0.5–1 mm. **Seeds** uniseriate, 1.2–1.7 × 0.8–1.2 mm; wing mostly distal, 0.05–0.1 mm wide.

Flowering May–Jun. Rocky slopes and sandy soil in sagebrush or pinyon-juniper woodlands; of conservation concern; 1500–1800 m; Nev.

Morphological evidence suggests that *Boechera falcifructa* is an apomictic species that arose through hybridization between *B. cobrensis* and *B. fernaldiana*. It is more similar to the former, but is distinguished by its gently curved fruiting pedicels (versus strongly recurved or reflexed distally), narrower (1.2–1.5 versus 1.7–2.5 mm) fruits, more (62–80 versus 34–64) ovules per ovary, longer (0.5–1 versus 0.05–0.2 mm) styles, and narrower (0.05–0.1 versus 0.25–0.5 mm) seed wings. Although independently reproducing, *B. falcifructa* is known only from Elko and Lander counties, where the parental species grow in proximity.

29. **Boechera fecunda** (Rollins) Dorn, Brittonia 55: 3. 2003 [C][E][F]

Arabis fecunda Rollins, Contr. Gray Herb. 214: 1. 1984

Perennials; usually long-lived; sexual; caudex often woody. **Stems** 1 to several per caudex branch, arising from center and margin of rosette, slightly elevated on woody base or from near ground surface, 0.8–3 dm, densely pubescent throughout, trichomes short-stalked, 5–10-rayed, 0.1–0.2 mm. **Basal leaves**: blade linear-oblanceolate, 1–5 mm wide, margins entire or with few teeth subapically, not ciliate, surfaces densely pubescent, trichomes short-stalked, 5–10-rayed, 0.1–0.2 mm. **Cauline leaves**: 5–18, concealing stem proximally; blade auricles 0.5–2.5 mm, surfaces of distalmost leaves densely pubescent. **Racemes** 8–35-flowered, usually unbranched. **Fruiting pedicels** erect to ascending, straight, 5–15 mm, pubescent, trichomes appressed, branched. **Flowers** erect to ascending at anthesis; sepals pubescent; petals purplish, 8–12 × 2.5–4.5 mm; pollen ellipsoid. **Fruits** erect to ascending, often appressed to rachis, not secund, straight, edges parallel, 3–5.8 cm × 1.4–1.8 mm; valves densely pubescent throughout; ovules 34–68 per ovary; style 1–1.5 mm. **Seeds** uniseriate, 1–1.5 × 0.8–1.2 mm; wing continuous, 0.5–0.1 mm wide.

Flowering Apr–Jun. Rocky ground in sagebrush areas; of conservation concern; 1200–1800 m; Mont.

Boechera fecunda is a distinctive sexual species known from Ravali and Silver Bow counties.

Boechera fecunda is in the Center for Plant Conservation's National Collection of Endangered Plants.

30. Boechera fendleri (S. Watson) W. A. Weber, Phytologia 51: 370. 1982 [E] [F]

Arabis holboellii Hornemann var. *fendleri* S. Watson in A. Gray et al., Syn. Fl. N. Amer. 1(1,1): 164. 1895; *A. fendleri* (S. Watson) Greene

Perennials; short- to long-lived; sexual; caudex usually not woody (sometimes with persistent, crowded leaf bases). **Stems** 1–7 per caudex branch, arising from center of rosette near ground surface, or arising laterally proximal to sterile shoots, 1.5–5.5(–8) dm, densely pubescent proximally, trichomes simple and long-stalked, 2-rayed, 0.2–0.9 mm, glabrous distally. **Basal leaves**: blade broadly oblanceolate, 5–15(–20) mm wide, margins dentate, ciliate, trichomes (simple), to 1.2 mm, surfaces glabrous or pubescent, trichomes simple and long-stalked, usually 2-rayed, rarely some 3-rayed, 0.2–0.6 mm. **Cauline leaves**: 8–25, often concealing stem proximally; blade auricles 0.8–3 mm, surfaces of distalmost leaves often glabrous. **Racemes** 6–40(–74)-flowered, usually unbranched. **Fruiting pedicels** horizontal to divaricate-ascending, curved or angled downward, 9–18(–23) mm, usually glabrous, rarely sparsely pubescent, trichomes spreading, simple. **Flowers** ascending at anthesis; sepals sparsely hirsute; petals usually lavender, rarely white, 5–9 × 1–2 mm; pollen ellipsoid. **Fruits** widely pendent, not appressed to rachis, not secund, curved to nearly straight, edges parallel, 3–5.8 cm × 1.5–2 mm; valves glabrous; ovules 98–128 per ovary; style 0.2–0.5 mm. **Seeds** biseriate, 0.9–1.2 × 0.6–0.8 mm; wing distal (sometimes absent), 0.05–0.15 mm wide. $2n = 14$.

Flowering Apr–Jul. Rocky slopes in pine forests, pinyon-juniper woodlands, scrub oak; 1800–2800 m; Ariz., Colo., Nev., N.Mex., Utah.

As circumscribed by R. C. Rollins (1993), *Boechera fendleri* included distinctive elements, segregated here as *B. porphyrea*, *B. spatifolia*, and *B. texana* (see M. D. Windham and I. A. Al-Shehbaz 2006, 2007 for detailed comparison). *Boechera fendleri* in the strict sense is a sexual diploid extending from the Four Corners region through northern Arizona to southern Nevada.

31. Boechera fernaldiana (Rollins) W. A. Weber, Phytologia 51: 370. 1982 [E] [F]

Arabis fernaldiana Rollins, Rhodora 43: 430, fig. 3. 1941; *A. canescens* Nuttall var. *stylosa* S. Watson; *A. fernaldiana* var. *stylosa* (S. Watson) Rollins

Perennials; long-lived; (cespitose); sexual; caudex somewhat woody (sometimes with persistent, crowded leaf bases). **Stems** usually 1 per caudex branch, arising from center of rosette, near ground surface or slightly elevated on woody base, 1–3.8 dm, densely pubescent proximally, trichomes short-stalked, 4–8-rayed, 0.04–0.1 mm, sometimes mixed with stalked, 2-rayed ones, to 0.5 mm, usually sparsely pubescent distally. **Basal leaves**: blade narrowly oblanceolate, 1–4 mm wide, margins entire, ciliate along petiole, trichomes (simple), to 1 mm, surfaces densely pubescent, trichomes short-stalked, 4–8-rayed, 0.04–0.1 mm. **Cauline leaves**: 5–10, not concealing stem; blade auricles 0.3–2 mm, surfaces of distalmost leaves usually sparsely pubescent. **Racemes** 10–20-flowered, usually unbranched. **Fruiting pedicels** divaricate-ascending, straight, 5–20 mm, sparsely pubescent, trichomes appressed, branched. **Flowers** ascending at anthesis; sepals (purplish or greenish) glabrous or sparsely or moderately pubescent; petals white or purple to lavender, 7–12 × 1.5–4 mm, glabrous; pollen ellipsoid. **Fruits** divaricate-ascending, not appressed to rachis, not secund, straight or curved, edges parallel, 3.5–6.5(–7.5) cm × 1–1.6 mm; valves glabrous; ovules 30–72 per ovary; style 0.1–1 mm. **Seeds** uniseriate, 1–1.2 × 0.8–1 mm; wing continuous, ca. 0.1 mm wide.

Subspecies 2 (2 in the flora): w United States.

Taxa treated here as subspecies of *Boechera fernaldiana* share many morphological traits, and most recent authors (R. C. Rollins 1993; N. H. Holmgren 2005b) did not recognize them as distinct. They are consistently separated by the characters listed below, show some degree of molecular divergence (C. D. Bailey et al., unpubl.), and their ranges are separated by ca. 500 km. They clearly represent genetically isolated population systems that warrant taxonomic recognition.

1. Petals purple to lavender; sepals purplish, moderately to sparsely pubescent; Nevada ....31a. *Boechera fernaldiana* subsp. *fernaldiana*
1. Petals white; sepals greenish, usually glabrous, rarely sparsely pubescent; Colorado, Utah .... 31b. *Boechera fernaldiana* subsp. *vivariensis*

31a. Boechera fernaldiana (Rollins) W. A. Weber subsp. fernaldiana E F

Petals purple to lavender; sepals purplish, moderately to sparsely pubescent.

Flowering May–Jul. Igneous rock outcrops, talus, rocky slopes in sagebrush areas; 2200–3400 m; Nev.

Subspecies *fernaldiana* is known from Elko, Humboldt, Lander, and Nye counties.

31b. Boechera fernaldiana (Rollins) W. A. Weber subsp. vivariensis (S. L. Welsh) Windham & Al-Shehbaz, Harvard Pap. Bot. 11: 67. 2006 E

Arabis vivariensis S. L. Welsh, Great Basin Naturalist 46: 263. 1986; *Boechera vivariensis* (S. L. Welsh) W. A. Weber

Petals white; sepals greenish, usually glabrous, rarely sparsely pubescent.

Flowering Apr–Jun. Sandstone rock outcrops, rocky slopes, sandy soil, mostly pinyon-juniper woodlands; 1500–2400 m; Colo., Utah.

Subspecies *vivariensis* is known only from Moffat County, Colorado, and Uintah County, Utah.

32. Boechera formosa (Greene) Windham & Al-Shehbaz, Harvard Pap. Bot. 11: 68. 2006 E F

Arabis formosa Greene, Pittonia 4: 198. 1900; *A. pulchra* M. E. Jones ex S. Watson var. *pallens* M. E. Jones; *Boechera pulchra* (M. E. Jones ex S. Watson) W. A. Weber subsp. *pallens* (M. E. Jones) W. A. Weber; *B. pulchra* var. *pallens* (M. E. Jones) Dorn

Perennials; long-lived; sexual; caudex woody. Stems usually 1 per caudex branch, arising from center of leaf tuft, above ground surface on woody base, 2–5.5 dm, densely pubescent proximally, trichomes short-stalked, 4–7-rayed, 0.1–0.4 mm, similarly pubescent distally. Basal leaves: blade linear to linear-oblanceolate, 2–3(–4) mm wide, margins entire, sometimes ciliate near petiole base, trichomes (simple or 2-rayed), to 0.8 mm, surfaces densely pubescent, trichomes short-stalked, 4–8-rayed, 0.1–0.3 mm. Cauline leaves: 7–18, concealing stem proximally; blade auricles absent, surfaces of distalmost leaves densely pubescent. Racemes 6–26-flowered, usually unbranched. Fruiting pedicels horizontal to descending, usually recurved, 5–10(–20) mm, pubescent, trichomes appressed, branched. Flowers divaricate at anthesis; sepals pubescent; petals white to pale lavender, 8–18 × 2.5–4.4(–5.5) mm, glabrous; pollen ellipsoid. Fruits divaricate-descending to reflexed, usually not appressed to rachis, rarely somewhat secund, straight, edges parallel, 4.3–7 cm × 1.6–3 mm; valves pubescent throughout; ovules 72–98 per ovary; style ca. 0.1 mm. Seeds biseriate, 1.2–1.6 × 1–1.2 mm; wing continuous, 0.1–0.2 mm wide. $2n = 14$.

Flowering Apr–Jun. Rocky slopes and sandy soil in blackbrush, sagebrush, and pinyon-juniper communities; 1300–1900 m; Ariz., Colo., N.Mex., Utah, Wyo.

Although *Boechera formosa* is usually subsumed within *Arabis* (*Boechera*) *pulchra*, the two species are easily separated. *Boechera formosa* has white to lavender (versus mostly purple) petals, fruiting pedicels that are not abruptly recurved at the base, narrower (1.6–3 versus 2.5–4 mm wide), mostly non-appressed fruits, and narrower (1–1.2 versus 1.5–2.2 mm wide) seeds. The two taxa are separated by more than 500 km, with *B. formosa* found only on the Colorado Plateau and *B. pulchra*, in the strict sense, restricted to southern California, western Nevada, and northwestern Mexico.

33. Boechera fructicosa (A. Nelson) Al-Shehbaz, Novon 13: 385. 2003 (as fruticosa) E

Arabis fructicosa A. Nelson, Bot. Gaz. 30: 190. 1900

Perennials; short-lived; apomictic; caudex usually not woody. Stems 3 to many per caudex branch, arising from margin of rosette near ground surface, 3.5–6 dm, pubescent proximally, trichomes short-stalked, 2–4-rayed, 0.2–0.3 mm, glabrous distally. Basal leaves: blade oblanceolate, 3–7 mm wide, margins dentate, ciliate proximally, trichomes (simple), to 0.3 mm, surfaces pubescent, trichomes short-stalked, 4–8-rayed, 0.05–0.2 mm. Cauline leaves: 5–10, not concealing stem; blade auricles 0.5–1 mm, surfaces of distalmost leaves glabrous. Racemes 10–18-flowered, usually unbranched. Fruiting pedicels divaricate-descending, straight, 5–10 mm, glabrous. Flowers divaricate-ascending at anthesis; sepals pubescent; petals lavender, 4–6 × 1–1.5 mm, glabrous; pollen spheroid. Fruits divaricate-descending, not appressed to rachis, not secund, curved, edges parallel, 4–6 cm × 1.5–1.8 mm; valves glabrous; ovules 60–70 per ovary; style 0.1–0.4 mm. Seeds uniseriate, 1.5–2 × 1–1.3 mm; wing distal, ca. 0.1 mm wide.

Flowering Jun–Jul. Dry, disturbed soil; ca. 2000 m; Wyo.

Boechera fructicosa is known only from the type collection (*A. & E. Nelson 5681*) taken near Undine Falls in Park County. Multiple plants are represented (comprising the holotype and four isotypes), though one specimen with the same number at RM is *B. stricta*. Although the exact parentage of this apomictic hybrid remains in doubt, it is virtually certain that *B. microphylla* contributed at least one genome.

34. **Boechera glareosa** Dorn, Brittonia 55: 1, fig. 1. 2003 E F

Perennials; short- to long-lived; sexual; caudex usually not woody. **Stems** usually 2–6 per caudex branch, arising from margin of rosette near ground surface, 0.8–4 dm, densely pubescent proximally, trichomes simple and short-stalked, 2-rayed, 0.3–0.6 mm, glabrous distally. **Basal leaves:** blade narrowly oblanceolate, 1–5 mm wide, margins entire or few-toothed, ciliate proximally, trichomes (simple), to 0.9 mm, surfaces densely pubescent, trichomes simple and short-stalked, 2- or 3-rayed, 0.3–0.5 mm. **Cauline leaves:** 3–6, not concealing stem; blade auricles 0.3–0.6 mm, surfaces of distalmost leaves glabrous. **Racemes** 7–15-flowered, usually unbranched. **Fruiting pedicels** divaricate-ascending to horizontal, straight or gently recurved, 4–10 mm, glabrous or sparsely pubescent, trichomes spreading. **Flowers** ascending at anthesis; sepals pubescent; petals lavender, 3.5–6 × 1–1.5 mm, glabrous; pollen ellipsoid. **Fruits** divaricate-ascending to horizontal, not appressed to rachis, not secund, usually curved or straight, edges parallel, 2.5–4 cm × 1.2–1.5 mm; valves glabrous; ovules 50–80 per ovary; style 0.05–0.3 mm. **Seeds** uniseriate, 1–1.5 × 0.7–1 mm; wing usually distal, ca. 0.1 mm wide.

Flowering May–Jun. Conglomerate and limestone outcrops in pinyon-juniper and mountain shrub-conifer communities; 2000–2600 m; Colo., Utah.

Specimens of *Boechera glareosa* originally were identified as *Arabis microphylla* var. *macounii* (= *B. macounii*) by R. C. Rollins. The former (a sexual species) is distinguished from the latter (an apomictic hybrid) by having wider (1–2 versus 0.5–0.8 mm) petals, fewer (2–9 versus mostly 9–16) cauline leaves, and fewer (7–15 versus 10–33) flowers per inflorescence. *Boechera glareosa* is known only from Moffat County, Colorado, and Summit and Uintah counties, Utah.

35. **Boechera glaucovalvula** (M. E. Jones) Al-Shehbaz, Novon 13: 385. 2003 E F

Arabis glaucovalvula M. E. Jones, Contr. W. Bot. 8: 40. 1898

Perennials; short- to long-lived; sexual; caudex usually not woody. **Stems** usually 1 per caudex branch, arising from center of rosette near ground surface, (0.6–)1–4.5 dm, densely pubescent proximally, trichomes short-stalked, 4–8-rayed, 0.1–0.4 mm, similarly pubescent distally. **Basal leaves:** blade linear to linear-oblanceolate, 2–4(–6) mm wide, margins entire, usually ciliate at petiole base, trichomes (simple or spurred), to 1.5 mm, surfaces densely pubescent, trichomes short-stalked, 4–8-rayed, 0.1–0.4 mm. **Cauline leaves:** 6–10, not concealing stem; blade auricles absent, surfaces of distalmost leaves densely pubescent. **Racemes** (5–)10–25-flowered, usually unbranched. **Fruiting pedicels** reflexed, strongly curved at base, 2–10 mm, pubescent, trichomes appressed, branched. **Flowers** divaricate at anthesis; sepals pubescent; petals light purple to lavender, 6–9 × 1.5–2.5 mm, glabrous; pollen ellipsoid. **Fruits** strongly reflexed, sometimes appressed to rachis, often secund, straight, edges parallel, (1.8–)2.5–4.5 cm × 5–8 mm; valves glabrous; ovules 24–62 per ovary; style 0.2–0.6 mm. **Seeds** biseriate, 5–6 × 4–5 mm; wing continuous, 1.8–2.5 mm wide. $2n = 14$.

Flowering Mar–Apr. Rocky slopes and gravelly soil, usually under shelter of desert shrubs; 600–1600 m; Calif., Nev.

Boechera glaucovalvula, a distinctive sexual diploid that has the widest fruits in the genus, is known from Inyo, Mono, Riverside, and San Bernardino counties, California, and Clark and Nye counties, Nevada.

36. **Boechera goodrichii** (S. L. Welsh) N. H. Holmgren in A. Cronquist et al., Intermount. Fl. 2(B): 364. 2005 E

Arabis goodrichii S. L. Welsh in S. L. Welsh et al., Utah Fl. ed. 3, 255. 2003

Biennials or perennials; short-lived; apomictic; caudex present or absent. **Stems** usually 1 per caudex branch, arising from center of rosette near ground surface, 2.6–7.4(–10.2) dm, densely pubescent proximally, trichomes short-stalked, 4–8-rayed, 0.1–0.5 mm, sometimes mixed with 2- or 3-rayed ones, to 1 mm, usually sparsely pubescent distally, rarely glabrous. **Basal leaves:** blade oblanceolate, 3–8(–10) mm wide, margins entire or shallowly dentate, ciliate near petiole base, trichomes (simple), to 1 mm, surfaces moderately to densely pubescent, trichomes short-stalked, 4–8-rayed, 0.15–0.4 mm. **Cauline leaves:** 22–72, concealing stem proximally; blade auricles 1–4 mm, surfaces of distalmost leaves sparsely pubescent or glabrous. **Racemes** 25–112-flowered, usually unbranched. **Fruiting pedicels** divaricate-descending, straight or gently curved, (10–)15–28 mm, usually sparsely pubescent, rarely glabrous, trichomes appressed, branched. **Flowers** divaricate-ascending at anthesis; sepals pubescent; petals white to lavender, 6–8 × 1–1.5 mm, glabrous; pollen spheroid. **Fruits** pendent, not appressed to rachis, not secund, straight or gently curved, edges parallel, 4.5–6.5 cm × 1.5–3 mm; valves glabrous; ovules 100–136 per ovary; style to 0.2 mm. **Seeds** biseriate or sub-biseriate, 1–1.4 × 0.9–1 mm; wing continuous, 0.05–0.15 mm.

Flowering Apr–May. Rocky slopes in sagebrush, pinyon-juniper woodlands, oak thickets; 1300–2300 m; Nev., Utah.

Morphological evidence suggests that *Boechera goodrichii* is an apomictic species that arose through hybridization between *B. gracilipes* and *B. retrofracta*. Some specimens of it originally were identified as *Arabis holboellii* var. *pinetorum* (= *B. pinetorum*), but they are most similar to *B. consanguinea* (see M. D. Windham and I. A. Al-Shehbaz 2007, 2007b for detailed comparison).

37. **Boechera gracilenta** (Greene) Windham & Al-Shehbaz, Harvard Pap. Bot. 12: 240. 2007 [E]

Arabis gracilenta Greene, Pittonia 4: 194. 1900; *A. selbyi* Rydberg; *Boechera selbyi* (Rydberg) W. A. Weber

Perennials; short- or (rarely) long-lived; apomictic; caudex usually not woody. **Stems** 1–9 per caudex branch, arising from center or margin of rosette near ground surface, (1.5–)2.5–5.5 dm, densely pubescent proximally, trichomes short-stalked, 2–4-rayed, 0.1–0.5 mm, mixed with some simple ones, to 0.8 mm, glabrous distally. **Basal leaves:** blade oblanceolate, 2–9 mm wide, margins shallowly dentate or sometimes entire, ciliate proximally, trichomes (simple and spurred), to 1.2 mm, surfaces moderately to densely pubescent, trichomes short-stalked, 4–6(–8)-rayed, 0.1–0.5 mm. **Cauline leaves:** 4–9, not concealing stem; blade auricles 1–3 mm, surfaces of distalmost leaves usually glabrous. **Racemes** 7–20-flowered, usually unbranched. **Fruiting pedicels** divaricate-ascending, gently curved downward, (7–)10–18 mm, glabrous. **Flowers** ascending at anthesis; sepals pubescent; petals lavender, 6–7 × 1.5–2 mm, glabrous; pollen spheroid. **Fruits** usually widely pendent, rarely horizontal, not appressed to rachis, not secund, slightly curved, edges parallel, (3–)4.5–7 cm × 1.7–2 mm; valves glabrous; ovules 60–96 per ovary; style 0.1–0.3 mm. **Seeds** sub-biseriate, 1.2–1.8 × 0.9–1.2 mm; wing continuous, 0.1–0.25 mm wide.

Flowering Apr–May. Rocky slopes and sandy soil in pinyon-juniper woodlands and mountain shrub communities; 1900–2300 m; Ariz., Colo., N.Mex., Utah.

Morphological evidence suggests that *Boechera gracilenta* is an apomictic species that arose through hybridization between *B. fendleri* and *B. pallidifolia*; it is superficially similar to the sexual diploid *B. perennans* but is not closely related (see M. D. Windham and I. A. Al-Shehbaz 2007b for detailed comparison).

38. **Boechera gracilipes** (Greene) Dorn, Brittonia 55: 3. 2003 [E] [F]

Arabis gracilipes Greene, Pittonia 4: 193. 1900; *A. arcuata* (Nuttall) A. Gray var. *longipes* S. Watson; *A. perennans* S. Watson var. *longipes* (S. Watson) Jepson

Biennials or perennials; short-lived; sexual; caudex present or absent. **Stems** usually 1 per caudex branch, arising from center of rosette near ground surface, rarely arising laterally proximal to sterile shoots, 2.5–8.5 dm, densely pubescent proximally, trichomes simple or spurred, 0.9–1.5 mm, glabrous distally. **Basal leaves:** blade oblanceolate, 5–12 mm wide, margins shallowly dentate, sometimes ciliate near petiole base, surfaces sparsely to densely pubescent, trichomes short- to long-stalked, 2- or 3-rayed, usually 0.3–0.6 mm. **Cauline leaves:** 30–65, often concealing stem proximally; blade auricles 0.8–3 mm, surfaces of distalmost leaves usually glabrous. **Racemes** (12–)15–50-flowered, usually unbranched. **Fruiting pedicels** divaricate-ascending or horizontal, recurved, (15–)20–47 mm, glabrous. **Flowers** ascending at anthesis; sepals pubescent; petals white to pale lavender, 6–9 × 0.6–1 mm; valves glabrous; pollen ellipsoid. **Fruits** widely pendent, not appressed to rachis, not secund, curved, edges parallel, (2.5–)3–7.5 cm × 1.5–2.8 mm, glabrous; ovules 130–210 per ovary; style 0.2–0.5 mm. **Seeds** biseriate, 1.2–1.4 × 0.7–0.9 mm; wing continuous or distal, to 0.2 mm wide.

Flowering Apr–Jun. Basalt, limestone, and sandy soils in ponderosa pine forests and pinyon-juniper woodlands; 1700–2300 m; Ariz., Nev., Utah.

Fruiting pedicel length, considered diagnostic by R. C. Rollins (1993) and N. H. Holmgren (2005b), occasionally fails to separate *Boechera gracilipes* from the closely-related *B. fendleri*. The two are consistently distinguished by trichome characters. In *B. gracilipes*, basal leaves usually lack prominent cilia and surfaces are persistently pubescent with at least some 3-rayed trichomes. Also, stems are rather densely pilose proximally, with the largest trichomes more than 0.9 mm. By contrast, basal leaves of *B. fendleri* always have prominent cilia, surfaces are often glabrescent, and 3-rayed trichomes usually are rare or absent. Stems of the latter are proximally hirsute to hispid, with the largest trichomes less than 0.9 mm.

39. Boechera grahamii (Lehmann) Windham & Al-Shehbaz, Harvard Pap. Bot. 12: 241. 2007 E

Turritis grahamii Lehmann, Index Seminum (Hamburg) 1831: 7. 1831, based on *T. patula* Graham, Edinburgh New Philos. J. 7: 350. 1829, not Ehrhart 1792; *Arabis boivinii* G. A. Mulligan; *A. bourgovii* Rydberg; *A. confinis* S. Watson; *A. confinis* var. *brachycarpa* (Torrey & A. Gray) S. Watson & J. M. Coulter; *A. dacotica* Greene; *A. divaricarpa* A. Nelson var. *dacotica* (Greene) B. Boivin; *A. divaricarpa* var. *hemicylindrica* B. Boivin; *A. divaricarpa* var. *stenocarpa* M. Hopkins; *A. drummondii* A. Gray var. *brachycarpa* (Torrey & A. Gray) A. Gray; *A. holboellii* Hornemann var. *brachycarpa* (Torrey & A. Gray) S. L. Welsh; *Boechera brachycarpa* (Torrey & A. Gray) Dorn; *T. brachycarpa* Torrey & A. Gray

Biennials or perennials; short-lived; apomictic; caudex present or absent. **Stems** usually 1 per caudex branch, arising from center of rosette near ground surface, (1.2–)2–12 dm, glabrous or pubescent proximally, trichomes simple and sessile or subsessile, 2 or 3(–6)-rayed, to 1 mm, glabrous distally. **Basal leaves:** blade oblanceolate, 1.5–10(–20) mm wide, margins entire or dentate, ciliate proximally, trichomes (simple), to 1 mm, surfaces sparsely to densely pubescent, trichomes sessile or subsessile, 2–4(–7)-rayed, 0.2–0.6 mm. **Cauline leaves:** (10–)13–52, often concealing stem proximally; blade auricles 1–5 mm, surfaces of distalmost leaves glabrous. **Racemes** 16–88-flowered, usually unbranched. **Fruiting pedicels** divaricate-ascending to descending, usually gently curved downward, rarely straight, 5–22(–30) mm, glabrous. **Flowers** divaricate at anthesis; sepals glabrous or pubescent; petals usually white, rarely lavender, 5.5–8 × 1–2 mm, glabrous; pollen spheroid. **Fruits** divaricate-ascending to pendent, not appressed to rachis, not secund, straight or slightly curved, edges parallel, 3.5–9 cm × 1–1.8 mm; valves glabrous; ovules 84–146 per ovary; style 0.05–0.5 mm. **Seeds** uniseriate to sub-biseriate, 1–1.8 × 0.8–1.2 mm; wing continuous, 0.1–0.25 mm wide. $2n = 21$.

Flowering May–Aug. Rocky slopes and sandy soil in prairies and open forests; 100–3200 m; Alta., B.C., Man., N.W.T., Ont., Que., Sask., Yukon; Alaska, Colo., Mich., Minn., Mont., N.Y., N.Dak., S.Dak., Vt., Wash., Wis., Wyo.

Morphological evidence suggests that *Boechera grahamii* is an apomictic species that arose through hybridization between *B. collinsii* and *B. stricta*. Previous authors have assigned these specimens to *Arabis* (*Boechera*) *divaricarpa* (see M. D. Windham and I. A. Al-Shehbaz 2007b for detailed comparison); if these taxa are treated as conspecific, the name *B. grahamii* has priority.

40. Boechera gunnisoniana (Rollins) W. A. Weber, Phytologia 51: 370. 1982 E F

Arabis gunnisoniana Rollins, Rhodora 43: 434, fig. 3. 1941

Perennials; usually long-lived; (cespitose); sexual; caudex often woody. **Stems** usually 2–6 per caudex branch, arising from margin of rosette near ground surface, 0.8–2(–2.5) dm, densely pubescent proximally, trichomes short-stalked, 2–4-rayed, 0.08–0.3 mm, mixed with fewer simple ones, usually sparsely pubescent distally, rarely glabrous. **Basal leaves:** blade linear-oblanceolate, 1–3(–4) mm wide, margins entire, ciliate proximally, trichomes to 0.8 mm, surfaces densely pubescent, trichomes short-stalked, 3–6-rayed, 0.08–0.3 mm. **Cauline leaves:** 3–9, not concealing stem; blade auricles 0.2–1 mm, surfaces of distalmost leaves glabrous or sparsely pubescent. **Racemes** 7–15-flowered, usually unbranched. **Fruiting pedicels** divaricate, straight or slightly curved, 4–7 mm, glabrous or sparsely pubescent, trichomes appressed, branched. **Flowers** divaricate-ascending at anthesis; sepals pubescent; petals white or lavender, 4–6 × 1–2 mm, glabrous; pollen ellipsoid. **Fruits** horizontal to divaricate-descending, not appressed to rachis, not secund, straight to slightly curved, edges parallel, 2.5–4 cm × 1–1.5 mm; valves glabrous; ovules 36–54 per ovary; style 0.1–0.2 mm. **Seeds** uniseriate, 1–1.2 × 0.8–1 mm; wing continuous, 0.1–0.15 mm wide. $2n = 14$.

Flowering May–Jun. Rocky slopes and knolls with sagebrush; 2100–2700 m; Colo.

Boechera gunnisoniana is most similar to *B. glareosa*; it is distinguished from that species by having trichomes smaller (0.08–0.3 versus 0.3–0.6 mm) and more highly divided (mostly 2–4-rayed versus simple and 2-rayed) proximally on stems, and fewer (36–54 versus 50–80) ovules per ovary. The two species are allopatric, with *B. glareosa* known only from eastern Utah and extreme northwestern Colorado and *B. gunnisoniana* apparently endemic to Gunnison County in west-central Colorado.

41. Boechera harrisonii (S. L. Welsh) Windham & Al-Shehbaz, Harvard Pap. Bot. 11: 266. 2007 E

Arabis harrisonii S. L. Welsh in S. L. Welsh et al., Utah Fl. ed. 3, 256. 2003; *Boechera microphylla* (Nuttall) Dorn var. *harrisonii* (S. L. Welsh) N. H. Holmgren

Perennials; long-lived; (somewhat cespitose); apomictic; caudex often woody. **Stems** usually 3–7 per caudex branch, arising from margin of rosette near ground surface, 0.5–2.5 dm, usually sparsely pubescent proximally, rarely glabrous, trichomes short-stalked, 2- or 3-rayed, 0.06–0.2 mm, glabrous distally. **Basal leaves:** blade narrowly oblanceolate, 2–4 mm wide, margins entire or somewhat dentate, ciliate near petiole base, trichomes (simple), to 0.5 mm, surfaces sparsely pubescent, trichomes short-stalked, 2–5-rayed, 0.1–0.25 mm. **Cauline leaves:** 3–7, not concealing stem; blade auricles 0.5–1.5 mm, surfaces of distalmost leaves glabrous. **Racemes** 5–12-flowered, usually unbranched. **Fruiting pedicels** divaricate-ascending, straight, 8–12 mm, glabrous. **Flowers** ascending at anthesis; sepals pubescent; petals lavender, 5–7.5 × 0.8–1.5 mm, glabrous; pollen spheroid. **Fruits** usually divaricate-ascending, rarely nearly horizontal, not appressed to rachis, not secund, curved to straight, edges parallel, 3–4.7 cm × 1–1.5 mm; valves glabrous; ovules 64–86 per ovary; style 0.2–0.5 mm. **Seeds** uniseriate, 1.1–1.3 × 0.8–1.1 mm; wing continuous, 0.08–0.1 mm wide.

Flowering May–Jun. Limestone and quartzite cliffs; 1500–1600 m; Utah.

Morphological evidence suggests that *Boechera harrisonii* is an apomictic species that arose through hybridization between *B. microphylla* and *B. perennans* (see M. D. Windham and I. A. Al-Shehbaz 2007 for detailed comparison). *Boechera harrisonii* is known only from Utah and Wasatch counties in north-central Utah.

42. Boechera hastatula (Greene) Al-Shehbaz, Novon 13: 386. 2003 C E

Arabis hastatula Greene, Leafl. Bot. Observ. Crit. 2: 79. 1910

Perennials; short- to long-lived; apomictic or sexual; caudex often woody. **Stems** 1–4 per caudex branch, arising from margin of rosette near ground surface, 1–3 dm, glabrous throughout. **Basal leaves:** blade linear-oblanceolate to oblanceolate, 1–3(–5) mm wide, margins entire or minutely toothed, rarely ciliate near petiole base, trichomes (simple), to 0.8 mm, surfaces pubescent or glabrous, trichomes short-stalked, 3–6-rayed, 0.07–0.2 mm. **Cauline leaves:** 4–15, not concealing stem; blade auricles 0.5–2.5 mm, surfaces of distalmost leaves glabrous. **Racemes** 12–40-flowered, usually unbranched. **Fruiting pedicels** reflexed, abruptly recurved near base, 4–8 mm, glabrous. **Flowers** divaricate to pendent at anthesis; sepals pubescent; petals lavender, 5–6 × ca. 1 mm, glabrous; pollen ellipsoid or spheroid. **Fruits** reflexed to closely pendent, usually appressed to rachis, often secund, straight, edges parallel, 0.3–0.42 cm × 1–1.5 mm; valves glabrous; ovules 46–60 per ovary. **Seeds** uniseriate, 1–1.3 × 0.8–1 mm; wing continuous, 0.05–0.1 mm wide.

Flowering Jun–Jul. Igneous rock outcrops; of conservation concern; 1900–2100 m; Oreg.

Both sexual and apomictic collections of *Boechera hastatula* have been identified; further study is needed to determine whether they truly are conspecific. The species is known only from Baker and Wallowa counties.

43. Boechera hoffmannii (Munz) Al-Shehbaz, Novon 13: 386. 2003 (as hoffmanii) C E F

Arabis maxima Greene var. *hoffmannii* Munz, Bull. S. Calif. Acad. Sci. 31: 63. 1932; *A. hoffmannii* (Munz) Rollins

Perennials; long-lived; sexual; caudex woody (sometimes with persistent, crowded leaf bases). **Stems** usually 1 per caudex branch, arising from center of rosette, often elevated above ground surface on woody base, 5–7 dm, glabrous or sparsely pubescent proximally, trichomes short-stalked, 2- or 3-rayed, 0.15–0.4 mm, glabrous or sparsely pubescent distally. **Basal leaves:** blade narrowly oblanceolate, 3–8 mm wide, margins coarsely dentate, ciliate, trichomes (simple and 2-rayed), to 0.6 mm, surfaces sparsely to densely pubescent, trichomes short-stalked, 2–7-rayed, 0.15–0.4 mm. **Cauline leaves:** 15–65, concealing stem; blade auricles 1–4 mm, surfaces of distalmost leaves pubescent. **Racemes** 30–70-flowered, sometimes branched. **Fruiting pedicels** divaricate-ascending, straight, 10–45 mm, glabrous. **Flowers** ascending at anthesis; sepals pubescent; petals white or pale lavender, 8–10 × 1.5–2 mm, glabrous; pollen ellipsoid. **Fruits** divaricate-ascending to horizontal, not appressed to rachis, not secund, usually curved, rarely straight, edges parallel, 6–10 cm × 2.5–3 mm; valves glabrous; ovules 170–220 per ovary; style 0.05–0.5 mm. **Seeds** biseriate, 1.2–1.6 × 0.9–1.2 mm; wing continuous, 0.1–0.2 mm wide.

Flowering Feb–Mar. Sea cliff ledges and crevices of volcanic outcrop in chaparral communities; of conservation concern; 0–100 m; Calif.

Boechera hoffmannii is a distinctive sexual species that appears to be restricted to Santa Cruz Island in Santa Barbara County. It is in the Center for Plant Conservation's National Collection of Endangered Plants.

44. Boechera holboellii (Hornemann) Á. Löve & D. Löve, Bot. Not. 128: 513. 1976 E F

Arabis holboellii Hornemann in G. C. Oeder, Fl. Dan. 11(32): 5, plate 1879. 1827; *A. holboellii* var. *tenuis* Böcher; *Boechera tenuis* (Böcher) Á. Löve & D. Löve

Biennials or perennials; short-lived; apomictic or sexual; caudex present or absent. **Stems** usually 1 per caudex branch, arising from center of rosette near ground surface, 2–6.7 dm, densely pubescent proximally, trichomes short-stalked, 4–8-rayed, 0.1–0.3 mm, glabrous or sparsely pubescent distally. **Basal leaves:** blade oblanceolate, 2–5 mm wide, margins entire, sometimes ciliate near petiole base, trichomes (usually simple), to 0.7 mm, surfaces sparsely to moderately pubescent, trichomes short-stalked, 4–8-rayed, 0.1–0.3 mm. **Cauline leaves:** 7–40, often concealing stem proximally; blade auricles 1–4 mm, surfaces of distalmost leaves glabrous or sparsely pubescent. **Racemes** 10–60-flowered, usually unbranched. **Fruiting pedicels** descending to reflexed, most abruptly recurved near base, 5–10 mm, glabrous or sparsely pubescent, trichomes appressed, branched. **Flowers** divaricate to pendent at anthesis; sepals pubescent; petals white, 4–8 × 1–2 mm, glabrous; pollen ellipsoid or spheroid. **Fruits** descending to reflexed, not appressed to rachis, strongly secund, often curved, edges parallel, 0.35–0.65 cm × (2–)2.2–2.5 mm; valves glabrous; ovules 80–102 per ovary. **Seeds** uniseriate to sub-biseriate, 1.4–1.7 × 1–1.2 mm; wing continuous, to 0.2 mm wide. $2n = 14, 21$.

Flowering Jul–Aug. Rocky slopes and gravelly soil near coast; 100–300 m; Greenland.

Both sexual diploid and apomictic triploid collections of *Boechera holboellii* are known; available data suggest that the triploid is an autopolyploid derivative of the sexual diploid but further study is needed. In the narrow circumscription adopted here, it appears to be restricted to Greenland (see discussion of *B. quebecensis* in M. D. Windham and I. A. Al-Shehbaz 2007b).

45. Boechera horizontalis (Greene) Windham & Al-Shehbaz, Harvard Pap. Bot. 11: 266. 2007 E

Arabis horizontalis Greene, Leafl. Bot. Observ. Crit. 2: 74. 1910; *A. suffrutescens* S. Watson var. *horizontalis* (Greene) Rollins

Perennials; long-lived; (cespitose); apomictic; caudex woody. **Stems** usually 1 per caudex branch, arising from center of rosette, near ground surface or somewhat elevated on woody base, or 1–3.5 dm, sparsely pubescent proximally, trichomes short-stalked, 3–6-rayed, 0.1–0.3 mm, glabrous distally. **Basal leaves:** blade narrowly oblanceolate, 1–5 mm wide, margins entire, ciliate near petiole base, trichomes (simple), to 0.4 mm, surfaces densely pubescent, trichomes short-stalked, 3–6-rayed, 0.1–0.3 mm. **Cauline leaves:** 3–13, often concealing stem proximally; blade auricles 0.5–1.5 mm, surfaces of distalmost leaves glabrous. **Racemes** 5–32-flowered, usually unbranched. **Fruiting pedicels** horizontal to descending, straight or slightly curved downward, 4–11 mm, glabrous. **Flowers** divaricate-ascending at anthesis; sepals pubescent; petals lavender to purple, 5–6 × 1.5–2 mm, glabrous; pollen spheroid. **Fruits** horizontal or descending, not appressed to rachis, secund, straight, edges slightly undulate (not parallel), 2–4 cm × 2–3 mm; valves glabrous; ovules 40–54 per ovary; style 0.2–0.5 mm. **Seeds** uniseriate, 2–2.5 × 1.7–2 mm; wing continuous, 0.5–1 mm wide.

Flowering Jul–Aug. Dry pumice slopes; Oreg.

Morphological evidence suggests that *Boechera horizontalis* is an apomictic species that arose through hybridization between *B. lemmonii* and *B. suffrutescens* (M. D. Windham and I. A. Al-Shehbaz 2007). *Boechera horizontalis* is known only from the vicinity of Crater Lake in south-central Oregon.

46. Boechera howellii (S. Watson) Windham & Al-Shehbaz, Harvard Pap. Bot. 11: 69. 2006 E F

Arabis howellii S. Watson, Proc. Amer. Acad. Arts 25: 124. 1890; *A. conferta* Greene; *A. inamoena* Greene var. *acutata* Jepson; *A. platyloba* Greene; *A. platysperma* A. Gray var. *howellii* (S. Watson) Jepson; *A. platysperma* var. *imparata* Jepson

Perennials; long-lived; sexual; caudex usually woody. **Stems** usually 1 per caudex branch, arising from center of rosette near ground surface, 0.6–2(–3) dm, glabrous throughout. **Basal leaves:** blade narrowly oblanceolate, 1–7 mm wide, margins entire, ciliate proximally, trichomes (simple mixed with a few short-stalked,

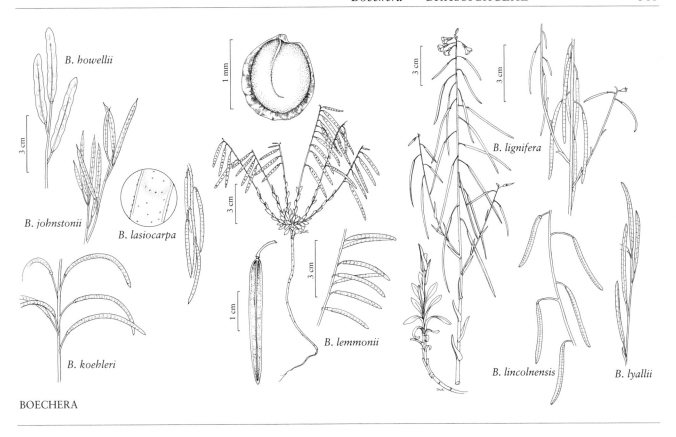

2-rayed ones) 0.1–0.4 mm, surfaces glabrous. **Cauline leaves:** 2–4, not concealing stem; blade auricles (0–)0.2–1 mm, surfaces of distalmost leaves usually glabrous, rarely ciliate proximally. **Racemes** 2–5-flowered, unbranched. **Fruiting pedicels** ascending, straight, 4–10 mm, glabrous. **Flowers** erect to ascending at anthesis; sepals glabrous; petals white to dark lavender, 4–8 × 1–2.5 mm, glabrous; pollen ellipsoid. **Fruits** suberect to ascending, usually not appressed to rachis, not secund, straight to curved, edges often somewhat undulate (not parallel), 2.5–6.5 cm × 3–7 mm; valves glabrous; ovules 10–20 per ovary; style 0.05–0.3 mm. **Seeds** uniseriate, 3–6 × 2–4 mm; wing continuous, 1.3–2.5 mm wide.

Flowering Jun–Aug. Rock outcrops, talus slopes and gravelly soil in alpine and subalpine habitats; 1800–3800 m; Calif., Nev., Oreg.

Boechera howellii usually is subsumed under *B. platysperma*, but the two taxa are distinct (see M. D. Windham and I. A. Al-Shehbaz 2006 for detailed comparison).

47. Boechera inyoensis (Rollins) Al-Shehbaz, Novon 13: 386. 2003 [E]

Arabis inyoensis Rollins, Rhodora 43: 457. 1941; *A. holboellii* Hornemann var. *derensis* S. L. Welsh; *Boechera selbyi* (Rydberg) W. A. Weber var. *inyoensis* (Rollins) N. H. Holmgren

Perennials; short-lived; apomictic; caudex usually not woody. **Stems** usually 1 per caudex, arising from center of rosette near ground surface, (1–)2.5–6.5 dm, densely pubescent proximally, trichomes short-stalked, 7–12-rayed, 0.1–0.2 mm, densely to sparsely pubescent distally. **Basal leaves:** blade oblanceolate, 1–4(–8) mm wide, margins entire, ciliate along petiole and proximally, trichomes (simple), to 1 mm, surfaces densely pubescent, trichomes short-stalked, 3–10-rayed, 0.2–0.7 mm. **Cauline leaves:** (7–)12–35, concealing stem proximally; blade auricles 0.5–2 mm, surfaces of distalmost leaves densely pubescent. **Racemes** 10–65-flowered, usually unbranched. **Fruiting pedicels** divaricate-ascending to horizontal, straight or slightly curved downward, 5–15 mm, sparsely pubescent, trichomes appressed, branched. **Flowers** ascending at anthesis; sepals pubescent; petals lavender to purplish, 5–8 × 1.2–2 mm, glabrous; pollen spheroid. **Fruits** divaricate-ascending to widely pendent, not appressed to rachis, not secund, usually curved, rarely straight, edges parallel, 3.7–6.5 cm × 1.5–2.2 mm; valves glabrous; ovules 74–134 per ovary; style 0.05–0.2 mm. **Seeds** usually sub-biseriate, rarely uniseriate, 1.7–2 × 1–1.5 mm; wing continuous, 0.1–0.2 mm wide. $2n = 21$.

Flowering Apr–Jun. Limestone and volcanic rock outcrops and clay soils in desert scrub and pinyon-juniper woodlands; 1400–2400 m; Calif., Nev., Utah.

Some plants here assigned to *Boechera inyoensis* were treated by R. C. Rollins (1993) and N. H. Holmgren (2005b) as *Arabis* (or *B.*) *beckwithii*. Closer examination of the type specimens of *A. beckwithii* reveals that they are conspecific with *B. puberula*, a sexual diploid widespread in the Great Basin of western North America. *Boechera inyoensis* is an apomictic triploid that clearly contains at least one genome derived from *B. shockleyi* (M. D. Windham and I. A. Al-Shehbaz 2007b); the other parent has not been determined.

48. **Boechera johnstonii** (Munz) Al-Shehbaz, Novon 13: 386. 2003 [C][E][F]

Arabis johnstonii Munz, Bull. S. Calif. Acad. Sci. 31: 63. 1932; *A. hirshbergiae* S. Boyd; *Boechera hirshbergiae* (S. Boyd) Al-Shehbaz

Perennials; long-lived; (cespitose); sexual; caudex woody. **Stems** usually 1 per caudex branch, arising from center of rosette near ground surface, 0.5–2 dm, densely pubescent proximally, trichomes short-stalked, 4–10-rayed, 0.07–0.15 mm, sparsely to densely pubescent distally. **Basal leaves:** blade narrowly oblanceolate, 1.5–4 mm wide, margins entire, not ciliate, surfaces densely pubescent, trichomes short-stalked, 6–14-rayed, 0.07–0.15 mm. **Cauline leaves:** 4–10, often concealing stem proximally; blade auricles absent, surfaces of distalmost leaves densely pubescent. **Racemes** 10–18-flowered, unbranched. **Fruiting pedicels** divaricate-ascending, straight, 5–14 mm, pubescent, trichomes appressed, branched. **Flowers** ascending at anthesis; sepals pubescent; petals purple, 9–14 × 2–4 mm, glabrous; pollen ellipsoid. **Fruits** divaricate-ascending, not appressed to rachis, not secund, straight, edges parallel, 4–6 cm × 2.5–4 mm; valves glabrous; ovules 26–34 per ovary; style (0.7–)1–2 mm. **Seeds** uniseriate, 1.9–2.7 × 1.5–2.2 mm; wing continuous, 0.3–0.7 mm wide.

Flowering Feb–Mar. Rocky areas and gravelly soil in chaparral and oak-pine savannas; of conservation concern; 1300–1700 m; Calif.

Boechera johnstonii and *B. hirshbergiae* are too similar to be considered distinct species. Even with the expanded circumscription, this distinctive sexual species is known only from the area of Cuyamaca Lake (San Diego County) and the San Jacinto Mountains (Riverside County).

49. **Boechera koehleri** (Howell) Al-Shehbaz, Novon 13: 386. 2003 [E][F]

Arabis koehleri Howell, Fl. N.W. Amer., 44. 1897; *A. arbuscula* Greene; *A. koehleri* var. *stipitata* Rollins

Perennials; long-lived; sexual; caudex woody (with persistent, peg-like leaf bases). **Stems** usually 1 per caudex branch, arising from center of rosette, elevated above ground surface on woody base, 0.8–4.5 dm, pubescent proximally, trichomes stalked, 2–4-rayed, to 0.5 mm, mixed with simple ones, to 1 mm, glabrous distally. **Basal leaves:** blade narrowly oblanceolate, 1–3 mm wide, margins entire, ciliate proximally, trichomes (simple), to 1 mm, surfaces moderately to sparsely pubescent, trichomes stalked, 2–4-rayed, 0.1–0.3(–0.5) mm. **Cauline leaves:** 3–17(–30), rarely concealing stem; blade auricles 0.5–2.5 mm, surfaces of distalmost leaves glabrous. **Racemes** 6–20(–35)-flowered, usually unbranched. **Fruiting pedicels** divaricate-ascending to horizontal, straight, 10–18 mm, glabrous or sparsely pubescent, trichomes spreading, simple and 2-rayed. **Flowers** ascending at anthesis; sepals pubescent; petals deep purple to magenta, 8–12 × 2.5–4 mm, glabrous; pollen ellipsoid. **Fruits** divaricate-ascending to widely pendent, not appressed to rachis, not secund, usually curved, rarely straight, edges parallel, 5–7.5 cm × 1.8–2.5 mm; valves glabrous; ovules 58–94 per ovary; style 0.05–0.2 mm. **Seeds** uniseriate, 1.3–1.8 × 1.2–1.5 mm; wing continuous, 0.1–0.2 mm wide.

Flowering Apr–May. Serpentine and limestone outcrops; 100–500 m; Calif., Oreg.

Boechera koehleri var. *koehlerei* is in the Center for Plant Conservation's National Collection of Endangered Plants.

50. **Boechera laevigata** (Muhlenberg ex Willdenow) Al-Shehbaz, Novon 13: 386. 2003 [E]

Turritis laevigata Muhlenberg ex Willdenow, Sp. Pl. 3: 543. 1801; *Arabis hastata* Eaton; *A. heterophylla* Nuttall; *A. laevigata* (Muhlenberg ex Willdenow) Poiret; *A. lyrifolia* de Candolle

Biennials; short-lived; sexual; caudex not evident. **Stems** 1 per plant, arising from center of rosette near ground surface, (1.5–)3–11 dm, glabrous throughout. **Basal leaves:** blade obovate to oblanceolate, (4–)10–40 mm wide, margins serrate or dentate, ciliate-mucronate on teeth, trichomes often minute, surfaces glabrous or sparsely pubescent, trichomes simple, 0.1–0.6 mm. **Cauline leaves:** 7–15, often concealing stem proximally; blade auricles 3–12 (–17) mm, surfaces of distalmost leaves glabrous. **Racemes** 16–45-flowered, sometimes branched. **Fruiting pedicels** suberect to divaricate-ascending, straight to slightly curved, 5–23 mm, glabrous. **Flowers** ascending at anthesis; sepals glabrous; petals white, 3–5 × 1–1.5 mm, glabrous; pollen ellipsoid. **Fruits** divaricate-ascending, not appressed to rachis, rarely somewhat secund, curved, edges parallel, (4–)6–11.7 cm × 1–2 mm; valves glabrous; ovules 50–80 per ovary; style 0.1–0.7(–1) mm. **Seeds** uniseriate, 1.2–2.2 × 0.8–1.4 mm; wing continuous, 0.1–0.3 mm wide distally. $2n$ = 14.

Flowering Mar–May. Rocky bluffs, cedar glades, wooded hillsides, floodplains; 100–500 m; Ont., Que.; Ala., Ark., Conn., Del., D.C., Ga., Ill., Ind., Iowa, Ky., Md., Mass., Mich., Mo., N.H., N.J., N.Y., N.C., Ohio, Okla., Pa., S.C., Tenn., Vt., Va., W.Va., Wis.

The taxon sometimes treated as *Arabis laevigata* var. *missouriensis* is here recognized as a separate species (*Boechera missouriensis*) based on its significantly longer petals (5–10 versus 3–5 mm) and distinctive, lyrate-pinnatifid basal leaves that persist well beyond anthesis.

51. **Boechera languida** (Rollins) Windham & Al-Shehbaz, Harvard Pap. Bot. 12: 244. 2007 [E]

Arabis demissa Greene var. *languida* Rollins, Rhodora 43: 388. 1941; *Boechera demissa* (Greene) W. A. Weber var. *languida* (Rollins) Dorn

Perennials; short- to long-lived; apomictic; caudex usually not, or, rarely, woody. **Stems** usually 2–5 per caudex branch, arising from margin of rosette near ground surface, or arising laterally proximal to sterile shoots, 1–3 dm, pubescent proximally, trichomes simple, to 1 mm, mixed with short-stalked, 2-rayed ones, 0.15–0.5 mm, glabrous or sparsely pubescent distally. **Basal leaves**: blade linear-oblanceolate to oblanceolate, 1–5(–9) mm wide, margins entire, ciliate, trichomes (simple), to 1 mm, surfaces densely pubescent, trichomes simple and 2–4-rayed, 0.15–0.5 mm. **Cauline leaves**: 3–7, not concealing stem; blade auricles 0.5–1.5 mm, surfaces of distalmost leaves sparsely pubescent or glabrous. **Racemes** 10–20-flowered, usually unbranched. **Fruiting pedicels** horizontal, slightly to strongly recurved near apex, 3–13 mm, glabrous or, rarely, with simple trichomes. **Flowers** divaricate-ascending at anthesis; sepals pubescent; petals white to pale lavender, 4–6 × 1.5–2 mm, glabrous; pollen spheroid. **Fruits** pendent, not appressed to rachis, not secund, straight to slightly curved, edges parallel, 3–4.5 cm × 1.8–2 mm; valves glabrous; ovules 42–70 per ovary; style 0.1–0.3 mm. **Seeds** sub-biseriate, 1.2–1.5 × 0.8–1 mm; wing absent or distal, 0.05–0.12 mm wide.

Flowering May–Jun. Rocky slopes in sagebrush and pinyon-juniper communities; 1800–2300 m; Wyo.

Boechera languida is an apomictic hybrid that is often subsumed under *Arabis demissa* (= *B. oxylobula*). Morphological evidence suggests a closer relationship to *B. spatifolia*, which probably is one of its parents (see M. D. Windham and I. A. Al-Shehbaz 2007b for detailed comparison).

52. **Boechera lasiocarpa** (Rollins) Dorn, Brittonia 55: 3. 2003 [E] [F]

Arabis lasiocarpa Rollins, Syst. Bot. 6: 58, fig. 1. 1981

Perennials; long-lived; sexual; caudex often woody. **Stems** 1 or few per caudex branch, arising from center or margin of rosette, somewhat elevated on woody base or near ground surface, (0.7–)2–3.5 dm, densely pubescent proximally, trichomes short-stalked, 4–8-rayed, 0.05–0.15 mm, moderately to sparsely pubescent distally. **Basal leaves**: blade narrowly oblanceolate, 2–5 mm wide, margins entire, not ciliate, surfaces densely pubescent, trichomes short-stalked, 4–8-rayed, 0.05–0.15 mm. **Cauline leaves**: 4–10, often concealing stem proximally; blade auricles absent, surfaces of distalmost leaves sparsely to moderately pubescent. **Racemes** 5–12-flowered, usually unbranched. **Fruiting pedicels** descending, recurved, 4–14 mm, pubescent, trichomes appressed, branched. **Flowers** divaricate at anthesis; sepals pubescent; petals purple to lavender, 6–10 × 1.5–2.5 mm, glabrous; pollen ellipsoid. **Fruits** pendent, usually not appressed to rachis, rarely secund, straight or curved, 2–5 cm × 1.7–2.2 mm; valves glabrous or pubescent throughout; ovules 26–36 per ovary; style 0.2–0.9 mm. **Seeds** uniseriate, 1.5–2 × 1–1.5 mm; wing distal or continuous, 0.1–0.2 mm wide.

Flowering May–Jun. Rocky ridges and slopes with dwarf sagebrush; 1800–2800 m; Utah.

Boechera lasiocarpa is a sexual species that appears to be restricted to the Bear River Mountains and northern Wasatch Range in north-central Utah. Reports of *Arabis* (*Boechera*) *lasiocarpa* from central Idaho (e.g., R. C. Rollins 1993; N. H. Holmgren 2005b) represent *B. rollinsiorum* (see M. D. Windham and I. A. Al-Shehbaz 2006 for detailed comparison).

53. Boechera lemmonii (S. Watson) W. A. Weber, Phytologia 51: 370. 1982 [E] [F]

Arabis lemmonii S. Watson, Proc. Amer. Acad. Arts 22: 467. 1887 (as lemmoni); *A. bracteolata* Greene; *A. canescens* Nuttall var. *latifolia* S. Watson; *A. codyi* G. A. Mulligan; *A. egglestonii* Rydberg; *A. kennedyi* Greene; *A. latifolia* (S. Watson) Piper; *A. oreocallis* Greene; *A. polyclada* Greene; *A. semisepulta* Greene

Perennials; long-lived; (somewhat cespitose); sexual or apomictic; caudex woody. **Stems** usually 1 per caudex branch, arising from center of rosette near ground surface, or arising laterally proximal to sterile shoots, 0.5–2 (–2.5) dm, glabrous or sparsely pubescent proximally, trichomes short-stalked, 2–6-rayed, 0.1–0.2 mm, glabrous distally. **Basal leaves**: blade oblanceolate to obovate, 1.5–5 mm wide, margins usually entire, rarely slightly dentate, ciliate along petiole, surfaces densely to sparsely pubescent, trichomes short-stalked, 3–9-rayed, 0.1–0.2 mm. **Cauline leaves**: 2–8(–12), not concealing stem; blade auricles absent or 0.1–0.5 mm, surfaces of distalmost leaves glabrous or sparsely pubescent. **Racemes** 3–12(–17)-flowered, usually unbranched. **Fruiting pedicels** divaricate-ascending to slightly descending, usually slightly recurved, rarely straight, 2–6 mm, glabrous or sparsely pubescent, trichomes appressed, branched. **Flowers** divaricate-ascending at anthesis; sepals glabrous or sparsely pubescent; petals purple to lavender, 3.5–6 × 1–1.5 mm; pollen ellipsoid or spheroid. **Fruits** divaricate-ascending to slightly descending, not appressed to rachis, secund, straight or curved, edges parallel, (1.6–)2–4.4 cm × 1.6–2.3 mm; valves glabrous; ovules 28 40(–44) per ovary; style 0.1–0.2 mm. **Seeds** uniseriate, 1.3–2 × 1–1.5 mm; wing continuous, 0.1–0.5 mm wide. $2n = 14$.

Flowering Jun–Aug. Cliffs, talus slopes, and gravelly soil in alpine and subalpine habitats; 2100–4400 m; Alta, B.C., Yukon; Alaska, Calif., Colo., Idaho, Mont., Nev., Oreg., Utah, Wash., Wyo.

Boechera lemmonii is easily recognized by its combination of secund fruits, mat-forming habit, purplish sepals, and obovate-oblanceolate cauline leaves. Both sexual and apomictic collections are known; further study is needed to determine whether they truly are conspecific. The taxa traditionally treated as *Arabis* (*Boechera*) *lemmonii* vars. *depauperata*, *drepanoloba*, and *paddoensis* are apomictic hybrids here recognized as distinct species (see M. D. Windham and I. A. Al-Shehbaz 2007 for detailed comparison).

54. Boechera lignifera (A. Nelson) W. A. Weber, Phytologia 51: 370. 1982 [E] [F]

Arabis lignifera A. Nelson, Bull. Torrey Bot. Club 26: 123. 1899

Perennials; short- to long-lived; apomictic or sexual; caudex usually woody. **Stems** usually 2–5 per caudex branch, arising laterally proximal to sterile shoots, often elevated on woody base, 1.2–4(–5) dm, densely pubescent proximally, trichomes short-stalked, 4–7-rayed, 0.1–0.3 mm, sparsely pubescent or glabrous distally. **Basal leaves**: blade narrowly oblanceolate, 2–5(–8) mm wide, margins entire, rarely ciliate near petiole base, trichomes (simple or spurred), to 1 mm, surfaces densely pubescent, trichomes short-stalked, 3–7-rayed, 0.05–0.3 mm. **Cauline leaves**: 4–12(–17), rarely concealing stem proximally; blade auricles 0.5–2 mm, surfaces of distalmost leaves sparsely pubescent. **Racemes** 6–15(–25)-flowered, usually unbranched. **Fruiting pedicels** slightly descending, gently recurved, 5–16 mm, glabrous or sparsely pubescent, trichomes appressed, branched. **Flowers** divaricate-ascending at anthesis; sepals pubescent; petals whitish (often aging pale lavender), 5–7 × 1–1.5 mm, glabrous; pollen ellipsoid or spheroid. **Fruits** widely pendent, not appressed to rachis, not secund, usually curved, edges parallel, 2.5–5.6 cm × 1.2–2 mm; valves glabrous; ovules 48–74 per ovary; style 0.1–0.2 mm. **Seeds** uniseriate, 1–1.3 × 0.8–1 mm; wing often continuous, 0.1–0.15 mm wide. $2n = 14$.

Flowering Apr–May. Rocky slopes and sandy soil in sagebrush and pinyon-juniper woodlands; 1700–2300 m; Ariz., Colo., Idaho, N.Mex., Utah, Wyo.

Boechera lignifera is commonly found on stabilized sand dunes. The related *B. cobrensis* often is found in similar habitats, and the two species can be difficult to distinguish where their ranges overlap in Idaho and Wyoming. It is possible that *B. cobrensis* was involved in the origin of some apomictic populations currently assigned to *B. lignifera*.

55. Boechera lincolnensis Windham & Al-Shehbaz, Harvard Pap. Bot. 11: 71. 2006 [E] [F]

Arabis pulchra M. E. Jones ex S. Watson var. *munciensis* M. E. Jones; *Boechera pulchra* (M. E. Jones ex S. Watson) W. A. Weber var. *munciensis* (M. E. Jones) Dorn

Perennials; long-lived; sexual; caudex woody. **Stems** usually 1 per caudex branch, arising from center of leaf tuft, elevated above ground surface on woody base, 2–4.2 dm, densely pubescent proximally, trichomes short-stalked,

3–6-rayed, 0.1–0.5 mm, similarly pubescent distally. **Basal leaves:** blade linear or linear-oblanceolate, 1–2 mm wide, margins entire, rarely ciliate near petiole base, trichomes (simple), to 1 mm, surfaces densely pubescent, trichomes short-stalked, 3–8-rayed, 0.1–0.4 mm. **Cauline leaves:** 10–25, concealing stem proximally; blade auricles absent or, rarely, to 1 mm, surfaces of distalmost leaves densely pubescent. **Racemes** 7–15-flowered, usually unbranched. **Fruiting pedicels** divaricate-ascending, straight to slightly curved proximally, recurved distally, 10–20(–25) mm, pubescent, trichomes appressed, branched. **Flowers** divaricate-ascending at anthesis; sepals pubescent; petals lavender to purple, 10–12 × 2–3 mm, glabrous or sparsely pubescent (occasionally some trichomes abaxially); pollen ellipsoid. **Fruits** usually widely pendent to ascending, rarely horizontal, not appressed to rachis, not secund, straight, edges parallel, (3.2–)4–5.5 cm × 2–2.5 mm; valves pubescent throughout; ovules 86–120 per ovary; style 0.1–0.3 mm. **Seeds** biseriate, 1–1.5 × 0.7–0.9 mm; wing continuous, 0.07–0.12 mm wide. $2n = 14$.

Flowering Apr–May. Rocky slopes and gravelly soil with sagebrush and other shrubs; 1400–1900 m; Calif., Nev., Utah.

Recent studies suggest that *Boechera lincolnensis* is a distinct species (see M. D. Windham and I. A. Al-Shehbaz 2006 for detailed comparison).

56. Boechera lyallii (S. Watson) Dorn, Vasc. Pl. Wyoming ed. 3, 376. 2001 [E][F]

Arabis lyallii S. Watson, Proc. Amer. Acad. Arts 11: 122. 1876; *A. armerifolia* Greene; *A. densa* Greene; *A. drummondii* A. Gray var. *alpina* S. Watson; *A. drummondii* var. *lyallii* (S. Watson) Jepson; *A. multiceps* Greene; *A. murrayi* G. A. Mulligan

Perennials; long-lived; (cespitose); sexual or apomictic; caudex woody. **Stems** usually 1 per caudex branch, arising from center of rosette near ground surface, 0.3–1.5(–2) dm, glabrous throughout. **Basal leaves:** blade linear-oblanceolate, 1–5(–8) mm wide, margins entire, usually ciliate at least proximally, trichomes (simple and/or short-stalked, 2- or 3-rayed), 0.1–0.3 mm, surfaces (sometimes the entire leaf) usually glabrous or (rarely, the youngest leaves of sterile shoots) pubescent, trichomes 4–6-rayed, 0.05–0.1 mm. **Cauline leaves:** 1–5, usually not concealing stem; blade auricles 0.5–1.5 mm or, rarely, absent, surfaces glabrous. **Racemes** 2–10(–15)-flowered, unbranched. **Fruiting pedicels** erect, straight, 3–8(–15) mm, glabrous. **Flowers** erect at anthesis; sepals glabrous; petals lavender to purplish, 6–8.5 × 1.5–3 mm, glabrous; pollen ellipsoid or spheroid. **Fruits** erect, appressed to rachis, not secund, straight, edges parallel, 3–5.6 cm × 1.5–2.5 mm; valves glabrous; ovules 34–64 per ovary; style 0.1–0.7 mm. **Seeds** biseriate or sub-biseriate, 1.5–2 × 1–1.5 mm; wing continuous, 0.3–0.5 mm wide.

Flowering Jun–Aug. Cliffs, talus slopes, gravelly soil in alpine and subalpine habitats; 1400–3700 m; B.C., Yukon; Calif., Idaho, Mont., Nev., Oreg., Utah, Wash.

The taxon traditionally treated as *Arabis* (*Boechera*) *lyallii* var. *nubigena* is here segregated as *B. paupercula* (see M. D. Windham and I. A. Al-Shehbaz 2006 for detailed comparison). Completely glabrous individuals of *B. lyallii* are sometimes confused with *B. davidsonii*, but they are easily distinguished by the absence of persistent leaf bases on caudex branches, erect and appressed (versus ascending) fruits, and biseriate to sub-biseriate (versus uniseriate) seeds. Both sexual and apomictic collections are known; further study is needed to determine whether they truly are conspecific.

57. Boechera macounii (S. Watson) Windham & Al-Shehbaz, Harvard Pap. Bot. 11: 267. 2007 [E]

Arabis macounii S. Watson, Proc. Amer. Acad. Arts 26: 124. 1891; *A. densicaulis* A. Nelson; *A. microphylla* Nuttall var. *macounii* (S. Watson) Rollins

Perennials; usually long-lived; apomictic; caudex often woody. **Stems** 1 or several per caudex branch, arising from center or margin of rosette near ground surface, 2–3.6 dm, densely pubescent proximally, trichomes simple and short-stalked, 2-rayed, 0.1–0.8 mm, glabrous distally. **Basal leaves:** blade oblanceolate to linear-oblanceolate, 1–3(–5) mm wide, margins entire or minutely toothed, ciliate proximally, trichomes (simple), to 0.7 mm, surfaces densely pubescent, trichomes short-stalked, 3–6-rayed, 0.07–0.3 mm. **Cauline leaves:** (5–)9–16, often concealing stem proximally; blade auricles 0.5–1.5 mm, surfaces of distalmost leaves glabrous. **Racemes** 10–33-flowered, usually unbranched. **Fruiting pedicels** divaricate-ascending to horizontal, straight, 5–13(–17) mm, usually densely pubescent, rarely glabrous, trichomes appressed, 2- or 3-rayed. **Flowers** divaricate-ascending at anthesis; sepals pubescent; petals lavender to purple, 4–5 × 0.5–0.8 mm, glabrous; pollen spheroid. **Fruits** divaricate-ascending to descending, not appressed to rachis, not secund, usually curved, edges parallel, 3.5–6.5 cm × 0.8–1.2 mm; valves glabrous; ovules 56–78 per ovary; style 0.1–0.4 mm. **Seeds** uniseriate, 1–1.2 × 0.7–1 mm; wing often continuous, 0.05–0.15 mm wide.

Flowering May–Jul. Rocky hillsides in open pinewoods; 2200–2900 m; B.C.; Idaho, Mont., Wyo.

Morphological evidence suggests that *Boechera macounii* is an apomictic species that arose through hybridization between *B. collinsii* and *B. microphylla* (see M. D. Windham and I. A. Al-Shehbaz 2007 for detailed comparison).

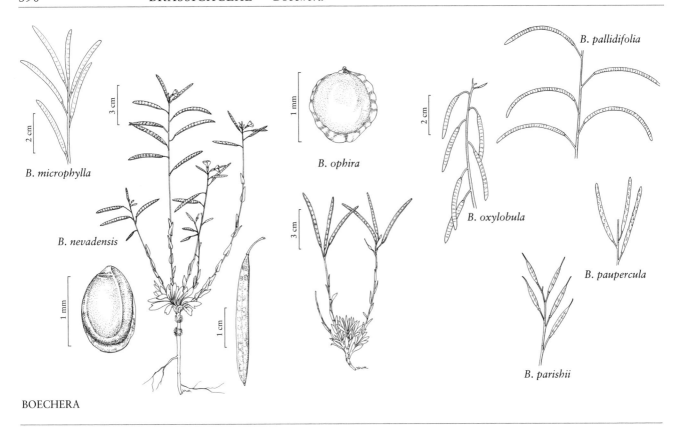

58. Boechera microphylla (Nuttall) Dorn, Vasc. Pl. Wyoming ed. 3, 376. 2001 [E] [F]

Arabis microphylla Nuttall in J. Torrey and A. Gray, Fl. N. Amer. 1: 82. 1838; *A. tenuicula* Greene

Perennials; long-lived; (cespitose); sexual or apomictic; caudex woody. **Stems** 1 or several per caudex branch, arising from center or margin of rosette near ground surface, 0.8–2.8(–3.5) dm, sparsely pubescent proximally, trichomes simple and subsessile, 2-rayed, 0.3–0.6 mm, glabrous distally. **Basal leaves**: blade oblanceolate to linear-oblanceolate, 1–4(–6) mm wide, margins entire or subapically toothed, ciliate proximally, trichomes (simple), to 0.6 mm, surfaces densely pubescent, trichomes short-stalked, 4–8-rayed, 0.05–0.1 mm. **Cauline leaves**: 2–6, not concealing stem; blade auricles 0.5–1.5 mm, surfaces of distalmost leaves glabrous. **Racemes** 5–17-flowered, usually unbranched. **Fruiting pedicels** ascending to divaricate-ascending, straight, 4–14 mm, glabrous. **Flowers** ascending at anthesis; sepals glabrous or pubescent; petals lavender to purple, 3.5–5.5 × 1–1.8 mm, glabrous; pollen ellipsoid or spheroid. **Fruits** ascending to divaricate-ascending, not appressed to rachis, not secund, straight to slightly curved, edges parallel, 3–7 cm × 1–1.5 mm; valves glabrous; ovules 52–72 per ovary; style 0.05–0.3 mm. **Seeds** uniseriate, 1–1.5 × 0.7–1 mm; wing often distal, ca. 0.1 mm wide. $2n = 14, 15$.

Flowering Apr–Jun. Cliffs and rocky slopes in sagebrush, mountain shrub, and open conifer forests; 400–2400 m; B.C.; Idaho, Mont., Nev., Oreg., Utah, Wash., Wyo.

Boechera microphylla is recognizable by its minute (0.05–0.1 mm), 4–8-rayed leaf trichomes, mat-forming habit, simple and 2-rayed trichomes on stems proximally, and ascending fruits. Both sexual and apomictic collections are known; further study is needed to determine whether they truly are conspecific. The taxa traditionally treated as *Arabis* (*Boechera*) *microphylla* vars. *macounii* and *thompsonii* are here recognized as *B. macounii* and *B. cascadensis*, respectively (see M. D. Windham and I. A. Al-Shehbaz 2007 for detailed comparison).

59. Boechera missouriensis (Greene) Al-Shehbaz, Novon 13: 387. 2003 [E]

Arabis missouriensis Greene, Repert. Spec. Nov. Regni Veg. 5: 244. 1908; *A. laevigata* (Muhlenberg ex Willdenow) Poiret var. *heterophylla* (Nuttall) Farwell; *A. laevigata* var. *missouriensis* (Greene) H. E. Ahles; *A. missouriensis* var. *deamii* (M. Hopkins) M. Hopkins; *A. viridis* Harger; *A. viridis* var. *deamii* M. Hopkins

Biennials; short-lived; sexual; caudex not evident. **Stems** usually 1 per plant, arising from center of rosette near ground surface, 2–7.5 dm, sparsely pubescent proximally, trichomes simple, to 0.5 mm, or glabrous throughout. **Basal leaves:** blade oblanceolate, 5–18 mm wide, margins lyrate-pinnatifid, often ciliate-mucronate on lobes, trichomes minute, surfaces glabrous or sparsely pubescent, trichomes simple, 0.1–0.5 mm. **Cauline leaves:** 10–45, often concealing stem proximally; blade auricles 1–5 mm, surfaces glabrous. **Racemes** 18–47-flowered, sometimes branched. **Fruiting pedicels** suberect to divaricate-ascending, straight or slightly curved, 5–13 mm, glabrous. **Flowers** ascending at anthesis; sepals glabrous; petals white, 5–10 × 1–1.5 mm, glabrous; pollen ellipsoid. **Fruits** ascending to divaricate-ascending, not appressed to rachis, not secund, curved, edges parallel, 6–11 cm × 1.5–2 mm; valves glabrous; ovules 60–86 per ovary; style 0.1–1 mm. **Seeds** uniseriate, 1.2–2 × 1–1.2 mm; wing continuous, 0.1–0.4 mm wide.

Flowering Mar–Jun. Rocky bluffs, wooded slopes, valley bottoms; 50–300 m; Ark., Conn., Ga., Ill., Ky., Mass., Mich., Minn., Mo., N.Y., N.C., Okla., S.C., Wis.

Boechera missouriensis occasionally has been treated as a variety of *B. laevigata*. With its significantly longer petals (5–10 versus 3–5 mm) and distinctive, lyrate-pinnatifid basal leaves that persist well beyond anthesis, it warrants recognition as a species.

60. **Boechera nevadensis** (Tidestrom) Windham & Al-Shehbaz, Harvard Pap. Bot. 11: 73. 2006 E F

Arabis nevadensis Tidestrom, Proc. Biol. Soc. Wash. 36: 182. 1923

Perennials; short- to long-lived; sexual; caudex usually not woody. **Stems** usually 2–4 per caudex branch, arising from center of rosette near ground surface, or arising laterally proximal to sterile shoots, 0.6–2 dm, glabrous throughout. **Basal leaves:** blade oblanceolate, 2–6 mm wide, margins entire, rarely ciliate along petiole, trichomes (simple), 0.5–0.7 mm, surfaces glabrous. **Cauline leaves:** 5–9, rarely concealing stem proximally; blade auricles 0.7–2.5 mm, surfaces of distalmost leaves glabrous. **Racemes** 4–12-flowered, usually unbranched. **Fruiting pedicels** divaricate-ascending to horizontal, straight, 2.5–8 mm, glabrous. **Flowers** ascending at anthesis; sepals glabrous; petals lavender, 4–5.5 × 1.5–2 mm, glabrous; pollen ellipsoid. **Fruits** usually divaricate-ascending, rarely slightly descending, usually secund, straight to slightly curved, 1.9–4 cm × 2–3 mm; valves glabrous; ovules 52–72 per ovary; style 0.05–0.2 mm. **Seeds** biseriate, 1.1–1.4 × 0.8–1 mm; wing continuous, 0.07–0.15 mm wide.

Flowering Jun. Ledges and talus of limestone cliffs; 3000–3400 m; Nev.

Boechera nevadensis usually is treated as a synonym of *Arabis pendulina* (R. C. Rollins 1993) or *B. demissa* (N. H. Holmgren 2005b), but is amply distinct from both (see M. D. Windham and I. A. Al-Shehbaz 2006 for detailed comparison). It is known only from Clark and Nye counties.

61. **Boechera ophira** (Rollins) Al-Shehbaz, Novon 13: 387. 2003 C E F

Arabis ophira Rollins, Syst. Bot. 6: 56. 1981

Perennials; long-lived; (cespitose); sexual; caudex often woody. **Stems** usually 1 per caudex branch, arising from center of rosette near ground surface, 0.4–1.4 dm, densely pubescent proximally, trichomes short-stalked, 2- or 3-rayed, 0.2–0.3 mm, sparsely to moderately pubescent distally. **Basal leaves:** blade narrowly oblanceolate to linear, 0.8–1.8 mm wide, margins entire, ciliate along petiole, trichomes (simple), to 0.8 mm, surfaces densely pubescent, trichomes short-stalked, 2- or 3-rayed, 0.2–0.3 mm. **Cauline leaves:** 6–10, somewhat concealing stem proximally; blade auricles to 0.5 mm, surfaces of distalmost leaves pubescent. **Racemes** 6–15-flowered, usually unbranched. **Fruiting pedicels** ascending, straight, 3–8 mm, pubescent, trichomes appressed, branched. **Flowers** ascending at anthesis; sepals pubescent; petals purplish, 4–5 × 1–1.8 mm, glabrous; pollen ellipsoid. **Fruits** suberect or ascending, sometimes appressed to rachis, not secund, straight, edges parallel, 2.5–4 cm × 1.2–1.8 mm; valves glabrous; ovules 50–60 per ovary; style 0.05–0.1 mm. **Seeds** uniseriate, 1.2–1.8 × 1–1.5 mm; wing continuous, 0.1–0.2 mm wide.

Flowering May–Jun. Rocky slopes, gravelly soil in subalpine meadows; of conservation concern; 3000–3200 m; Nev.

Boechera ophira is known only from the Toiyabe Range in west-central Nevada.

62. **Boechera oxylobula** (Greene) W. A. Weber, Phytologia 51: 370. 1982 E F

Arabis oxylobula Greene, Pittonia 4: 195. 1900; *A. aprica* Osterhout ex A. Nelson; *A. demissa* Greene; *A. rugocarpa* Osterhout; *Boechera demissa* (Greene) W. A. Weber

Perennials; short- to long-lived; (cespitose); sexual; caudex usually not woody. **Stems** usually 3–7 per caudex branch, arising from margin of rosette near ground surface, or arising laterally proximal to sterile shoots, 0.4–2.5

dm, glabrous or pubescent proximally, trichomes simple and short-stalked, 2-rayed, 0.1–0.4 mm, glabrous distally. **Basal leaves:** blade linear to linear-oblanceolate, 1–2.5 mm wide, margins usually entire, rarely denticulate, often ciliate, trichomes (simple), 0.3–0.7 mm, surfaces glabrous or sparsely pubescent, trichomes short-stalked, 2- or 3-rayed, 0.1–0.4 mm. **Cauline leaves:** 3–12, not concealing stem; blade auricles absent, surfaces of distalmost leaves usually glabrous, rarely sparsely pubescent. **Racemes** 2–12-flowered, unbranched. **Fruiting pedicels** divaricate-ascending to horizontal, slightly to strongly recurved, 3–8 mm, glabrous or sparsely pubescent, trichomes (isolated), simple. **Flowers** ascending-divaricate at anthesis; sepals glabrous or pubescent; petals white to pale lavender, 4–5 × 1.5–2 mm, glabrous; pollen ellipsoid. **Fruits** pendent, not appressed to rachis, not or, rarely, weakly secund, straight, edges parallel, 1.5–3.5 cm × 1.2–2 mm; valves glabrous; ovules 28–44 per ovary; style 0.1–0.4 mm. **Seeds** uniseriate, 0.9–1.2 × 0.6–1 mm; wing often continuous, 0.07–0.1 mm wide.

Flowering May–Jul. Cliffs, rocky slopes, gravelly soil in sagebrush and open conifer forests; 2100–3600 m; Colo.

Most of the collections assigned here have been called *Arabis* (*Boechera*) *demissa* by other authors (e.g., R. C. Rollins 1993; N. H. Holmgren 2005b). Because the holotype of *A. demissa* is identical to *B. oxylobula* in nearly every way (see M. D. Windham and I. A. Al-Shehbaz 2006 for detailed comparison), we treat them as conspecific. *Boechera oxylobula* is restricted to Garfield, Gunnison, Hinsdale, Lake, Mineral, Park, and Saguache counties in central Colorado. The taxon traditionally treated as *A.* (*Boechera*) *demissa* var. *languida* is here recognized as an apomictic species of hybrid origin (see Windham and Al-Shehbaz 2007b for detailed comparison).

63. **Boechera paddoensis** (Rollins) Windham & Al-Shehbaz, Harvard Pap. Bot. 11: 268. 2007 E

Arabis lemmonii S. Watson var. *paddoensis* Rollins, Rhodora 43: 384. 1941

Perennials; long-lived; (cespitose); apomictic; caudex usually woody. **Stems** 1–3 per caudex branch, arising from center of rosette near ground surface, or arising laterally proximal to sterile shoots, 1–2.5 dm, glabrous throughout. **Basal leaves:** blade narrowly oblanceolate, 1–3(–6) mm wide, margins usually entire, rarely apically few-toothed, not ciliate, surfaces sparsely pubescent, trichomes short-stalked, 4–8-rayed, 0.08–0.2 mm. **Cauline leaves:** 4–8, not concealing stem; blade auricles 1–2 mm, surfaces of distalmost leaves glabrous. **Racemes** 5–12(–18)-flowered, usually unbranched. **Fruiting pedicels** usually divaricate-ascending, rarely horizontal, straight, 2–5 (–10) mm, glabrous. **Flowers** ascending at anthesis; sepals glabrous; petals lavender to purple, 4–5.5 × 1–1.5 mm, glabrous; pollen spheroid. **Fruits** usually divaricate-ascending, rarely horizontal, not appressed to rachis, secund (often weakly so), straight, edges parallel, 3–5.5 cm × 1.5–2 mm; valves glabrous; ovules 42–54 per ovary; style 0.05–0.1 mm. **Seeds** uniseriate, 1.1–1.5 × 0.9–1.3 mm; wing continuous, 0.1–0.3 mm wide.

Flowering Jul–Aug. Rocky ridges; 1500–2100 m; Oreg., Wash.

Morphological evidence suggests that *Boechera paddoensis* is an apomictic species that arose through hybridization between *B. lemmonii* and *B. lyallii* (see M. D. Windham and I. A. Al-Shehbaz 2007 for detailed comparison). *Boechera paddoensis* is known only from the mountains of central Washington and northeastern Oregon.

64. **Boechera pallidifolia** (Rollins) W. A. Weber, Phytologia 79: 65. 1996 E F

Arabis pallidifolia Rollins, Cruciferae Continental N. Amer., 181. 1993; *A. thompsonii* S. L. Welsh; *Boechera thompsonii* (S. L. Welsh) N. H. Holmgren

Perennials; short-lived; sexual; caudex usually not woody. **Stems** usually 2–5 per caudex branch, arising from margin of rosette near ground surface, or arising laterally proximal to sterile shoots, (0.5–)1.5–4 dm, densely pubescent proximally, trichomes short-stalked, 2–6-rayed, 0.1–0.3 mm, glabrous or sparsely pubescent distally. **Basal leaves:** blade oblanceolate to obovate, 5–11(–13) mm wide, margins shallowly dentate or, sometimes, entire, often ciliate along petiole, trichomes (simple and 2-rayed), surfaces moderately pubescent, trichomes short-stalked, 4–8-rayed, 0.1–0.3 mm. **Cauline leaves:** 3–8(–11), not concealing stem; blade auricles (0.5–)1–2 mm, surfaces of distalmost leaves sparsely pubescent. **Racemes** 4–15(–20)-flowered, usually unbranched. **Fruiting pedicels** ascending to divaricate-ascending, straight or, sometimes, recurved, 7–15 mm, glabrous or sparsely pubescent. **Flowers** ascending at anthesis; sepals pubescent; petals usually lavender, rarely whitish, 5–9 × 1.2–2.5 mm, glabrous; pollen ellipsoid. **Fruits** usually ascending, rarely horizontal, not appressed to rachis, not secund, straight to curved, edges parallel, 2.5–5(–6.5) cm × 1–2 mm; valves glabrous; ovules 44–62 per ovary; style 0.05–0.4 mm. **Seeds** uniseriate, 1.1–1.3 × 1–1.1 mm; wing nearly continuous, 0.07–0.15 mm wide.

Flowering Apr–Jun. Rocky slopes and sandy soil in pinyon-juniper, sagebrush communities; 1600–2500 m; Ariz., Colo., N.Mex., Utah, Wyo.

Although *Boechera pallidifolia* was originally described as endemic to west-central Colorado, recent studies favor a broader circumscription that includes sexual populations previously assigned to *Arabis* (*Boechera*) *selbyi* (see M. D. Windham and I. A. Al-Shehbaz 2006 for detailed comparison). The species has hybridized with nearly every sexual diploid within its range (including *B. crandallii*, *B. fendleri*, *B. formosa*, and *B. pendulina*), producing a confusing array of apomictic triploids including the type specimen of *A. selbyi*.

65. Boechera parishii (S. Watson) Al-Shehbaz, Novon 13: 388. 2003 [C] [E] [F]

Arabis parishii S. Watson, Proc. Amer. Acad. Arts 22: 468. 1887

Perennials; short- to long-lived; sexual; caudex often woody. **Stems** usually 1 per caudex branch, arising from center of rosette near ground surface, 0.3–1.4 dm, densely pubescent proximally, trichomes short-stalked, 2–8-rayed, 0.2–0.4 mm, sparsely pubescent distally. **Basal leaves**: blade linear to linear-oblanceolate, 0.5–2 mm wide, margins entire, often ciliate along petiole, trichomes (branched), to 0.6 mm, surfaces densely pubescent, trichomes short-stalked, 6–12-rayed, 0.07–0.15 mm. **Cauline leaves**: 2–8, rarely concealing stem proximally; blade auricles absent, surfaces of distalmost leaves densely pubescent. **Racemes** 5–20-flowered, unbranched. **Fruiting pedicels** ascending, straight, 3–7 mm, pubescent, trichomes appressed, branched. **Flowers** ascending at anthesis; sepals pubescent; petals lavender to purple, 8–13 × 2.5–4 mm, glabrous; pollen ellipsoid. **Fruits** ascending, not appressed to rachis, not secund, straight, edges parallel, 1.5–2.5 cm × 1.8–2.5 mm; valves glabrous; ovules 12–20 per ovary; style 3–8 mm. **Seeds** uniseriate, 1.5–2 × 1–1.5 mm; wing distal or continuous, 0.05–0.2 mm wide. $2n = 14$.

Flowering Mar–May. Gravelly hillsides in sagebrush-juniper-pine areas; of conservation concern; 1900–2300 m; Calif.

Boechera parishii is a distinctive, long-styled, sexual diploid known only from San Bernardino Mountains in San Bernardino County.

66. Boechera pauciflora (Nuttall) Windham & Al-Shehbaz, Harvard Pap. Bot. 11: 268. 2007 [E]

Sisymbrium pauciflorum Nuttall in J. Torrey and A. Gray, Fl. N. Amer. 1: 93. 1838; *Arabis arcuata* (Nuttall) A. Gray var. *subvillosa* S. Watson; *A. columbiana* Macoun; *A. perelegans* A. Nelson; *A. sparsiflora* Nuttall var. *columbiana* (Macoun) Rollins; *A. sparsiflora* var. *subvillosa* (S. Watson) Rollins; *Boechera sparsiflora* (Nuttall) Dorn var. *subvillosa* (S. Watson) Dorn

Perennials; short- to long-lived; apomictic; caudex sometimes woody. **Stems** usually 1 per caudex branch, arising from center of rosette near ground surface, (1.4–)3–11.2 dm, densely pubescent proximally, trichomes simple, 0.6–1.5 mm, mixed with stalked, 2- (or 3-)rayed ones, 0.2–0.4 mm, sparsely pubescent distally. **Basal leaves**: blade oblanceolate, 3–10 mm wide, margins usually dentate, rarely entire, sometimes ciliate, trichomes (simple or branched), to 1 mm, surfaces densely pubescent, trichomes stalked, 2–5-rayed, 0.3–0.6 mm. **Cauline leaves**: (8–)14–60, often concealing stem proximally; blade auricles (1–)3–10 mm, surfaces of distalmost leaves glabrous or sparsely pubescent. **Racemes** 17–60-flowered, usually unbranched. **Fruiting pedicels** horizontal to divaricate-descending, usually straight, rarely slightly recurved, 4–13 mm, pubescent, trichomes spreading, 2- or 3-rayed. **Flowers** divaricate at anthesis; sepals pubescent; petals lavender to whitish, 5–8 × 1–2 mm, glabrous; pollen spheroid. **Fruits** horizontal, divaricate-descending or widely pendent, not appressed to rachis, not secund, curved, edges parallel, 5.5–10.5 cm × 1.5–2.2 mm; valves glabrous; ovules 80–162 per ovary; style 0.05–0.5 mm. **Seeds** uniseriate, 1.4–1.8 × 1–1.4 mm; wing continuous, 0.1–0.25 mm wide. $2n = 21$.

Flowering May–Jun. Rocky soil in sagebrush areas, mountain shrub communities, edges of conifer forests; 600–2500 m; B.C.; Calif., Idaho, Mont., Nev., Oreg., Utah, Wash., Wyo.

Morphological evidence suggests that *Boechera pauciflora* is an apomictic species that arose through hybridization between *B. retrofracta* and *B. sparsiflora*. Specimens of *B. pauciflora* are commonly identified as *Arabis holboellii* var. *pinetorum* (= *B. pinetorum*), a superficially similar species restricted to the northern Sierra Nevada and southern Cascade Range (see M. D. Windham and I. A. 2007 for detailed comparison).

Arabis elegans A. Nelson (1900), not Tineo & Lojacono (1886) is an illegitimate name, sometimes found in synonymy with *Boechera pauciflora*.

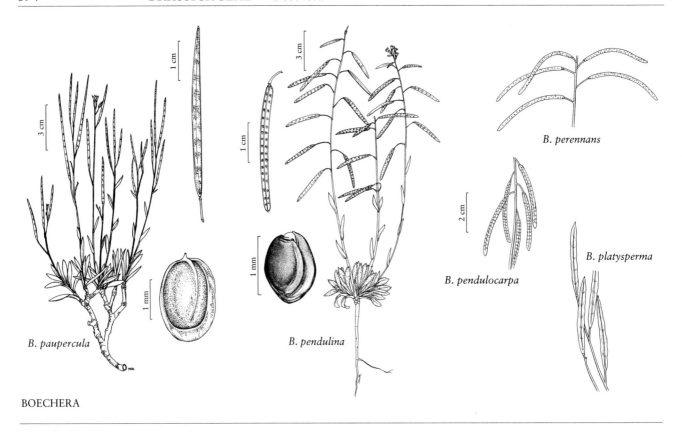

67. Boechera paupercula (Greene) Windham & Al-Shehbaz, Harvard Pap. Bot. 11: 75. 2006 E F

Arabis paupercula Greene, Leafl. Bot. Observ. Crit. 2: 77. 1910; A. lyallii S. Watson var. nubigena (J. F. Macbride & Payson) Rollins; A. microphylla Nuttall var. nubigena (J. F. Macbride & Payson) Rollins; A. nubigena J. F. Macbride & Payson

Perennials; long-lived; (cespitose); sexual; caudex often woody. **Stems** usually 1 per caudex branch, arising from center of rosette near ground surface, 0.3–1.5 dm, densely pubescent proximally, trichomes simple and short-stalked, 2–4-rayed, (0.07–)0.1–0.2 mm, glabrous or sparsely pubescent distally. **Basal leaves:** blade linear-oblanceolate, 1–3(–5) mm wide, margins entire, ciliate proximally, trichomes to 0.4 mm, surfaces densely pubescent, trichomes short-stalked, 2–6-rayed, (0.07–)0.1–0.2 mm. **Cauline leaves:** 2–6, rarely concealing stem; blade auricles 0 (or 0.2–1.5) mm, surfaces of distalmost leaves glabrous. **Racemes** 3–8-flowered, unbranched. **Fruiting pedicels** erect, straight, 3–9 mm, glabrous or sparsely pubescent. **Flowers** erect at anthesis; sepals glabrous or sparsely pubescent; petals lavender to purplish, 4–6(–7) × 1–2 mm, glabrous; pollen ellipsoid. **Fruits** erect, appressed to rachis, not secund, straight, edges parallel, 2.5–5.5 cm × 1.3–1.7(–2) mm; valves glabrous; ovules 24–40 per ovary; style 0.2–1 mm. **Seeds** usually uniseriate, rarely sub-biseriate, 1.5–2(–2.5) × 1–1.4 mm; wing continuous, 0.3–1 mm wide.

Flowering Jun–Aug. Rock outcrops, talus slopes, gravelly soil in alpine and subalpine habitats; 2400–3400 m; Calif., Idaho, Oreg., Wash., Wyo.

Boechera paupercula is usually subsumed under *Arabis* (*Boechera*) *lyallii* but is amply distinct (see M. D. Windham and I. A. Al-Shehbaz 2006 for detailed comparison).

68. Boechera peirsonii Windham & Al-Shehbaz, Harvard Pap. Bot. 11: 270. 2007 E

Arabis breweri S. Watson var. pecuniaria Rollins

Perennials; long-lived; (cespitose); apomictic; caudex woody. **Stems** usually 1 per caudex branch, arising from center of rosette near ground surface, 1–2.5 dm, glabrous or sparsely pubescent proximally, trichomes short-stalked, 2–5-rayed (rarely, some simple), 0.4–0.6 mm, glabrous distally. **Basal leaves:** blade oblanceolate, 2.5–6 mm wide, margins entire, ciliate proximally, trichomes (simple), surfaces densely pubescent, trichomes short-stalked, 4–7-rayed, 0.3–0.5 mm. **Cauline leaves:** 3–12, often concealing stem proximally; blade auricles

0.5–2 mm, surfaces of distalmost leaves usually glabrous, sometimes margins ciliate. **Racemes** 12–25-flowered, usually unbranched. **Fruiting pedicels** ascending to divaricate-descending, straight, 2–6 mm, glabrous. **Flowers** ascending at anthesis; sepals pubescent; petals purple, 5–6 × 1.5–2 mm, glabrous; pollen spheroid. **Fruits** ascending to divaricate-ascending, not appressed to rachis, not secund, straight, edges parallel, 2–3.7 cm × 2–2.8 mm; valves glabrous; ovules 56–80 per ovary; style 0.1–0.3 mm. **Seeds** uniseriate, 1–1.5 × 0.8–1 mm; wing distal or continuous, 0.05–0.1 mm wide.

Flowering Jun–Sep. Granitic ledges and talus slopes; 2700–3400 m; Calif.

Boechera peirsonii is an apomictic species of unknown origin; although previously treated as a variety of *Arabis* (*B.*) *breweri*, it is quite distinct from that taxon (see M. D. Windham and I. A. Al-Shehbaz 2007 for detailed comparison). All known collections come from the San Bernardino Mountains of southern California, ca. 200 km distant from the nearest populations of *B. breweri*.

69. Boechera pendulina (Greene) W. A. Weber, Phytologia 51: 370. 1982 [E] [F]

Arabis pendulina Greene, Leafl. Bot. Observ. Crit. 2: 81. 1910; *A. demissa* Greene var. *russeola* Rollins; *A. diehlii* M. E. Jones; *A. pendulina* var. *russeola* (Rollins) Rollins; *A. setulosa* Greene; *Boechera demissa* (Greene) W. A. Weber var. *pendulina* (Greene) N. H. Holmgren; *B. demissa* var. *russeola* (Rollins) N. H. Holmgren; *B. pendulina* (Greene) W. A. Weber var. *russeola* (Rollins) Dorn

Perennials; short- to long-lived; sexual or apomictic; caudex sometimes woody. **Stems** usually 2–6 per caudex branch, arising from margin of rosette near ground surface, or arising laterally proximal to sterile shoots, 0.6–3(–3.7) dm, sparsely to densely pubescent proximally, trichomes simple, 0.3–0.8 mm, glabrous distally. **Basal leaves**: blade oblanceolate to obovate, 1.5–6 mm wide, margins usually entire, rarely dentate, ciliate throughout, trichomes (usually simple), 0.4–1 mm, surfaces pubescent, trichomes simple and short- and long-stalked, 2-rayed, 0.3–0.8 mm. **Cauline leaves**: 2–10(–13), not concealing stem; blade auricles absent or, rarely, to 0.7 mm, surfaces of distalmost leaves glabrous or margins sparsely ciliate. **Racemes** 4–14-flowered, usually unbranched. **Fruiting pedicels** divaricate-ascending to horizontal, curved or angled downward, 3–7(–10) mm, usually glabrous, rarely with some simple trichomes. **Flowers** divaricate-ascending at anthesis; sepals usually sparsely pubescent, rarely glabrous; petals whitish to pale lavender, 4–6 × 1–1.5 mm, glabrous; pollen ellipsoid or spheroid. **Fruits** widely pendent, not appressed to rachis, not secund, curved to nearly straight, edges parallel, 2.2–4 cm × 1.2–2.1 mm; valves glabrous; ovules 40–70(–90) per ovary; style 0.1–0.3(–0.5) mm. **Seeds** biseriate, 0.9–1.2 × 0.6–0.9 mm, usually not winged. $2n = 14, 21$.

Flowering Apr–Jun. Rock outcrops, open gravelly flats and hillsides in sagebrush, pinyon-juniper, mountain mahogany, open conifer forests; 1600–3100 m; Ariz., Calif., Colo., Nev., Utah, Wyo.

Although included in *Arabis* (*Boechera*) *demissa* by some authors (e.g., S. L. Welsh et al. 2003; N. H. Holmgren 2005b), *B. pendulina* is readily distinguished from that species (see M. D. Windham and I. A. Al-Shehbaz 2006 for detailed comparison). Typical collections are sexual diploids, whereas the type of var. *russeola* is a triploid apomict; further study is needed to determine if the two are conspecific.

70. Boechera pendulocarpa (A. Nelson) Windham & Al-Shehbaz, Harvard Pap. Bot. 11: 77. 2006 [E] [F]

Arabis pendulocarpa A. Nelson, Bot. Gaz. 30: 192. 1900; *A. holboellii* Hornemann var. *pendulocarpa* (A. Nelson) Rollins; *Boechera holboellii* (Hornemann) Á. Löve & D. Löve var. *pendulocarpa* (A. Nelson) N. W. Snow

Perennials; short- to long-lived; sexual; caudex usually not woody. **Stems** usually 1 per caudex branch, arising from center of rosette near ground surface, 0.6–3 dm, densely pubescent proximally, trichomes simple and stalked, 2–4-rayed, 0.1–0.4 mm, glabrous or sparsely pubescent distally. **Basal leaves**: blade narrowly oblanceolate, 1.5–4(–5) mm wide, margins entire, often ciliate along petiole, trichomes (simple), to 0.6 mm, surfaces densely pubescent, trichomes short-stalked, 4–8-rayed, 0.08–0.2 mm. **Cauline leaves**: 6–17, sometimes concealing stem proximally; blade auricles absent, surfaces of distalmost leaves glabrous or sparsely pubescent. **Racemes** 4–11-flowered, unbranched. **Fruiting pedicels** arched, gently recurved, 3–8(–12) mm, glabrous or sparsely pubescent, trichomes appressed, branched. **Flowers** ascending at anthesis; sepals pubescent; petals white or lavender, 4–6 × 1–1.5 mm, glabrous; pollen ellipsoid. **Fruits** pendent, not appressed to rachis, not secund, usually straight, edges parallel, (2–)2.5–3.8 cm × 1.5–2.2 mm; valves glabrous; ovules 66–92 per ovary; style 0.3–0.5 mm. **Seeds** sub-biseriate, 1–1.2 × 0.7–0.9 mm; wing distal, 0.05–0.1 mm wide. $2n = 14$.

Flowering Apr–Jul. Rock outcrops and gravelly slopes in sagebrush, open conifer forests, subalpine meadows; 1000–3300 m; B.C., Yukon; Calif., Idaho, Mont., Nev., Oreg., Utah, Wyo.

Though often treated as a variety of *Arabis* (*Boechera*) *holboellii* (e.g., R. C. Rollins 1993), *B. pendulocarpa* is easily distinguished from that species by having simple and 2–4-rayed (versus 4–8-rayed) trichomes proximally on stems, cauline leaves without auricles, fruiting pedicels gently (versus sharply) recurved, and shorter (2–)2.5–3.8 (versus 3.5–6.5) cm, non-secund fruits. The two taxa have allopatric distributions, with *B. pendulocarpa* found in the mountains of western North America and *B. holboellii* apparently confined to Greenland. Recent use of the name *A.* (*Boechera*) *exilis* for this taxon (e.g., G. A. Mulligan 1996; R. D. Dorn 2001; N. H. Holmgren 2005b) is based on misinterpretation of the type (M. D. Windham and I. A. Al-Shehbaz 2006).

71. Boechera perennans (S. Watson) W. A. Weber, Phytologia 51: 370. 1982 [F]

Arabis perennans S. Watson, Proc. Amer. Acad. Arts 22: 467. 1887; *A. angulata* Greene ex Wooton & Standley; *A. arcuata* (Nuttall) A. Gray var. *perennans* (S. Watson) M. E. Jones; *A. eremophila* Greene; *A. recondita* Greene

Perennials; long-lived; sexual; caudex woody. **Stems** usually 2–5 per caudex branch, arising laterally proximal to sterile shoots or rosette, often elevated above ground surface on woody base, (1.5–)2–7 dm, densely pubescent proximally, trichomes short-stalked, Y-shaped, 0.2–0.4 mm, often mixed with 3–5-rayed or (rarely) simple ones, usually glabrous distally. **Basal leaves**: blade oblanceolate to obovate, 3–15(–20) mm wide, margins dentate, ciliate proximally, trichomes to 1.2 mm, surfaces moderately to densely pubescent, trichomes short-stalked, 3–6-rayed, 0.2–0.4 mm. **Cauline leaves**: 4–12(–17), not concealing stem; blade auricles 0.5–3.5 mm, surfaces of distalmost leaves glabrous. **Racemes** 16–35-flowered, usually unbranched. **Fruiting pedicels** usually horizontal, straight or slightly recurved, (6–)10–25 mm, usually glabrous. **Flowers** ascending at anthesis; sepals pubescent; petals white to purplish, 5–9 × 1–1.5 mm, glabrous or sparsely pubescent (occasionally some trichomes abaxially); pollen ellipsoid. **Fruits** widely pendent, not appressed to rachis, not secund, usually curved, edges parallel, (3–)4–7 cm × 1.7–2.1 mm; valves glabrous; ovules 60–96 per ovary; style 0.05–0.4 mm. **Seeds** uniseriate, 1.1–1.5 × 0.9–1.2 mm; wing continuous, 0.1–0.2 mm wide. $2n = 14$.

Flowering Feb–May. Rocky slopes and gravelly soil in warm desert, chaparral, low montane habitats; 200–1700 m; Ariz., Calif., Nev., N.Mex., Tex., Utah; Mexico (Baja California, Chihuahua, Sonora).

The circumscription of *Boechera perennans* followed here is much narrower than that adopted by R. C. Rollins (1993). The most significant difference is the removal of *Arabis gracilenta*. Recent studies (M. D. Windham and Allphin, in prep.) indicate that the latter represents an older name for the species usually called *A.* (*Boechera*) *selbyi*. Although superficially similar, *B. perennans* is distinguished from *B. gracilenta* by having stems arising from above ground surface on woody bases, proximal stems with mostly Y-shaped trichomes, strongly dentate margins on basal leaves, and uniseriate seeds. Typically, *B. perennans* is a sexual diploid largely restricted to the warm deserts of Arizona, California, Nevada, and southern New Mexico. Plants of *B. gracilenta* are scattered across the Colorado Plateau (northeast of the range of *B. perennans*) and are apomictic triploids, apparently produced by hybridization between *B. fendleri* and *B. pallidifolia*.

72. Boechera perstellata (E. L. Braun) Al-Shehbaz, Novon 13: 388. 2003 [C] [E]

Arabis perstellata E. L. Braun, Rhodora 42: 47. 1940; *A. perstellata* var. *ampla* Rollins

Perennials; usually short-lived; sexual; caudex not woody. **Stems** usually 2–5 per plant, arising from margin of rosette near ground surface, (1–)2–8 dm, densely pubescent proximally, trichomes short-stalked, 3- or 4-rayed, 0.1–0.5 mm, similarly pubescent distally. **Basal leaves**: blade obovate to oblanceolate, (5–)10–45 mm wide, margins usually coarsely dentate, rarely sinuate, ciliate on petiole base, surfaces moderately pubescent, trichomes short-stalked, 3- or 4(–6)-rayed, 0.1–0.5 mm. **Cauline leaves**: 6–25, concealing stem proximally; blade auricles 1–6 mm, surfaces of distalmost leaves pubescent. **Racemes** 10–40-flowered, usually unbranched. **Fruiting pedicels** usually horizontal, usually straight, rarely slightly curved, 5–15 mm, pubescent, trichomes spreading, simple. **Flowers** ascending at anthesis; sepals pubescent; petals purplish, 3–4 × 0.7–1 mm, glabrous; pollen ellipsoid. **Fruits** usually horizontal, not appressed to rachis, not secund, straight, edges parallel, (1.5–)2–3.3 cm × 0.8–1 mm; valves sparsely pubescent throughout; ovules 12–20 per ovary; style 0.4–0.8 mm. **Seeds** uniseriate, 0.8–1.1 × 0.6–0.8 mm, not winged. $2n = 14$.

Flowering Apr–May. Calcareous bluffs and wooded hillsides; of conservation concern; 200–400 m; Ky., Tenn.

Boechera perstellata is in the Center for Plant Conservation's National Collection of Endangered Plants.

73. Boechera pinetorum (Tidestrom) Windham & Al-Shehbaz, Harvard Pap. Bot. 11: 271. 2007 [E]

Arabis pinetorum Tidestrom, Proc. Biol. Soc. Wash. 36: 182. 1923; *A. divaricarpa* A. Nelson var. *pinetorum* (Tidestrom) B. Boivin; *A. holboellii* Hornemann var. *pinetorum* (Tidestrom) Rollins; *Boechera holboellii* (Hornemann) Á. Löve & D. Löve var. *pinetorum* (Tidestrom) Dorn

Perennials; short-lived; apomictic; caudex not woody. **Stems** usually 1 per caudex branch, arising from center of rosette near ground surface, 2–8(–9.6) dm, densely pubescent proximally, trichomes simple and short-stalked, 2–4-rayed, to 1 mm, sparsely pubescent or glabrous distally. **Basal leaves:** blade oblanceolate, 2–8(–11) mm wide, margins entire or denticulate, ciliate proximally, trichomes (simple), 0.5–1 mm, surfaces sparsely to densely pubescent, trichomes short-stalked, 2–4 (or 5)-rayed, 0.2–0.5 mm, these mixed with fewer simple ones. **Cauline leaves:** 7–33, often concealing stem proximally; blade auricles 1–3 mm, surfaces of distalmost leaves sparsely pubescent or glabrous. **Racemes** 15–63-flowered, usually unbranched. **Fruiting pedicels** reflexed, abruptly recurved near base, otherwise straight, 4–12 mm, sparsely pubescent or glabrous, trichomes spreading, simple and 2-rayed. **Flowers** divaricate to pendent at anthesis; sepals pubescent; petals white to lavender, 5–6 × 1–1.5 mm, glabrous; pollen spheroid. **Fruits** reflexed to closely pendent, not appressed to rachis, not secund, straight or curved, edges parallel, (4.5–)5.5–8.5 cm × 1.5–2 mm; valves glabrous; ovules 70–110 per ovary; style 0.05–0.4 mm. **Seeds** uniseriate, 1.5–1.8 × 1.2–1.5 mm; wing continuous, 0.15–0.3 mm wide.

Flowering May–Jul. Rock outcrops and gravelly soil in meadows and open conifer forests; 1100–3200 m; Calif., Nev., Oreg.

Most authors (e.g., R. C. Rollins 1993; R. D. Dorn 2001; S. L. Welsh et al. 2003; N. H. Holmgren 2005b) have treated *Boechera pinetorum* as a variety of *Arabis* (*Boechera*) *holboellii*. Under this guise, the name has been applied to a vast array of plants collected throughout western North America. This includes a diversity of sexual diploids plus nearly every hybrid containing a genome from *B. retrofracta*. Based on re-examination of the type collection, we have adopted a much narrower concept of the species. Morphological evidence suggests that *B. pinetorum* is an apomictic triploid hybrid containing three different genomes, derived from *B. rectissima*, *B. retrofracta*, and *B. sparsiflora*. Plants closely resembling the type of *B. pinetorum* are confined to the northern Sierra Nevada and adjacent southern Cascades. The majority of collections previously associated with the epithet *pinetorum* represent *B. pauciflora* (see M. D. Windham and I. A. Al-Shehbaz 2007 for detailed comparison).

74. Boechera pinzliae (Rollins) Al-Shehbaz, Novon 13: 388. 2003 (as pinzliae) [C][E]

Arabis pinzliae Rollins, Contr. Gray Herb. 212: 110. 1982 (as pinzliae)

Perennials; long-lived; apomictic; caudex usually woody. **Stems** 1–3 per caudex branch, arising from margin of rosette near ground surface, 0.4–1.6 dm, densely pubescent proximally, trichomes short-stalked, 2–5-rayed, 0.05–0.2 mm, glabrous or sparsely pubescent distally. **Basal leaves:** (petiole glabrous); blade linear-oblanceolate, 1–3 mm wide, margins entire, rarely ciliate proximally, trichomes to 0.5 mm, surfaces densely pubescent, trichomes short-stalked, 2–5-rayed, 0.05–0.2 mm. **Cauline leaves:** 3–6, not concealing stem; blade auricles 0(–0.3) mm, surfaces of distalmost leaves sparsely to moderately pubescent. **Racemes** 5–8-flowered, unbranched. **Fruiting pedicels** ascending to divaricate-ascending, straight, 2–6 mm, glabrous or sparsely pubescent, trichomes appressed, branched. **Flowers** ascending at anthesis; sepals pubescent; petals purple, 4–5 × 1–1.5 mm, glabrous; pollen spheroid. **Fruits** ascending to divaricate-ascending, not appressed to rachis, not secund, straight, edges parallel, 2.5–4.8 cm × 2.5–3.2 mm; valves glabrous; ovules 26–34 per ovary; style 0.1–0.3 mm. **Seeds** uniseriate, 2.5–3.5 × 1.5–2.5 mm; wing continuous, 0.2–0.9 mm wide.

Flowering Jul. Gravelly granitic soils in alpine and subalpine areas; of conservation concern; 3000–3400 m; Calif., Nev.

Morphological evidence suggests that *Boechera pinzliae* is an apomictic hybrid that contains at least one genome derived from *B. platysperma*; the other parent has not yet been determined. It is most similar to *B. elkoensis*, from which it differs in having sparsely pubescent (versus glabrous) sepals and basal leaves with stalked (versus sessile or subsessile) trichomes. *Boechera pinzliae* is known only from the White Mountains in extreme western Nevada (Esmeralda County) and east-central California (Mono County).

75. Boechera platysperma (A. Gray) Al-Shehbaz, Novon 13: 388. 2003 [E][F]

Arabis platysperma A. Gray, Proc. Amer. Acad. Arts 6: 519. 1865; *A. inamoena* Greene; *A. oligantha* Greene

Perennials; long-lived; sexual; caudex woody. **Stems** usually 1 per caudex branch, arising from center of rosette, elevated on woody base above or near ground surface, 0.6–2.8(–3.5) dm, densely pubescent proximally, trichomes short-stalked, 2–5-rayed,

0.1–0.3 mm, glabrous or sparsely pubescent distally. **Basal leaves:** blade narrowly oblanceolate, 3–7(–10) mm wide, margins entire, ciliate near petiole base, trichomes (simple or spurred), to 1 mm, surfaces densely pubescent, trichomes short-stalked, 2–4 (or 5)-rayed 0.1–0.3 mm. **Cauline leaves:** 3–7(–12), not concealing stem; blade auricles absent, surfaces of distalmost leaves glabrous or sparsely pubescent. **Racemes** 2–7-flowered, unbranched. **Fruiting pedicels** ascending, straight, 3–13 mm, glabrous or sparsely pubescent, trichomes appressed, branched. **Flowers** ascending at anthesis; sepals pubescent; petals creamy white to purplish, 4–6 × 1–2 mm, sparsely pubescent (often some trichomes abaxially); pollen ellipsoid. **Fruits** ascending, not appressed to rachis, not secund, straight to slightly curved, edges undulate (not parallel), (2.5–)4–8.5 cm × 3–5.5 mm; valves glabrous; ovules 16–44 per ovary; style 0.05–1 mm. **Seeds** uniseriate, 3–6(–8) × 2–3.5 mm; wing continuous, 0.8–2.5 mm wide. $2n = 14$.

Flowering Jun–Aug. Rock outcrops and gravelly soil in dry pine forests and lodgepole-chaparral woodlands; 1600–3000 m; Calif., Oreg.

Previous authors have adopted broader circumscriptions of *Boechera platysperma*, including plants herein called *B. covillei*, *B. elkoensis*, and *B. howellii*. In our opinion, the differences between these taxa are sufficient to justify their treatment as distinct species (see M. D. Windham and I. A. Al-Shehbaz 2006, 2007, 2007b for detailed comparison). Reports of *B. platysperma* from Nevada (e.g., R. C. Rollins 1993; N. H. Holmgren 2005b) are based on plants of *B. howellii* or apomictic hybrids with other species.

76. **Boechera polyantha** (Greene) Windham & Al-Shehbaz, Harvard Pap. Bot. 11: 78. 2006 [E] [F]

Arabis polyantha Greene, Leafl. Bot. Observ. Crit. 2: 80. 1910; *A. macdougalii* Rydberg

Perennials; usually short-lived; sexual; caudex not woody. **Stems** usually 1 per caudex branch, arising from center of rosette near ground surface, 2.5–9 dm, densely pubescent proximally, trichomes short-stalked, 2–6-rayed, 0.1–0.7 mm, similarly pubescent distally. **Basal leaves:** blade oblanceolate, 1.5–7 mm wide, margins entire or shallowly dentate, ciliate at petiole base, trichomes to 0.9 mm, surfaces densely pubescent, trichomes short-stalked, 5–10-rayed, 0.1–0.25 mm. **Cauline leaves:** 16–42, concealing stem proximally; blade auricles 0.5–1.5 mm, surfaces of distalmost leaves pubescent. **Racemes** 17–96-flowered, usually unbranched. **Fruiting pedicels** reflexed, abruptly recurved at base, otherwise straight, 7–13 mm, pubescent, trichomes appressed, branched. **Flowers** descending at anthesis; sepals pubescent; petals white, 6–9 × 1–2.5 mm, sparsely pubescent abaxially; pollen ellipsoid. **Fruits** strongly reflexed, usually appressed to rachis, sometimes somewhat secund, straight, edges parallel, 3.5–7.2 cm × 1.2–1.5 mm; valves densely pubescent; ovules 74–110 per ovary; styles 0.05–0.2 mm. **Seeds** uniseriate, 0.9–1.2 × 0.7–1 mm; wing continuous, 0.05–0.1 mm wide.

Flowering late Apr–early Jun. Rocky, gravelly, and sandy slopes in open areas; 900–2500 m; Idaho, Mont., Oreg., Wash.

Boechera polyantha is closely related to *B. retrofracta*, but is easily distinguished by densely pubescent fruits and scattered trichomes on abaxial petal surfaces.

77. **Boechera porphyrea** (Wooton & Standley) Windham, Al-Shehbaz & P. J. Alexander, Harvard Pap. Bot. 11: 272. 2007 [E]

Arabis porphyrea Wooton & Standley, Contr. U.S. Natl. Herb. 16: 123. 1913

Perennials; usually long-lived; apomictic; caudex often woody. **Stems** 1 to several per caudex branch, arising from center of rosette or arising laterally proximal to sterile shoots, somewhat elevated on woody base or from ground surface, 2–5(–9) dm, usually glabrous proximally, rarely sparsely pubescent, trichomes simple and short-stalked, 2- or 3-rayed, 0.1–0.5 mm, glabrous distally. **Basal leaves:** blade oblanceolate, (3–)5–15 mm wide, margins dentate, ciliate at least along petiole, trichomes to 1 mm, surfaces pubescent, trichomes short-stalked, 2–6-rayed, 0.1–0.5 mm. **Cauline leaves:** 5–17(–21), sometimes concealing stem proximally; blade auricles 0.5–3.5(–5.5) mm, surfaces of distalmost leaves glabrous. **Racemes** 10–30(–70)-flowered, usually unbranched. **Fruiting pedicels** horizontal, gently curved downward, 10–27 mm, glabrous. **Flowers** ascending at anthesis; sepals glabrous; petals lavender, 6.5–8.5 × 1–2.5 mm, glabrous; pollen spheroid. **Fruits** widely pendent, not appressed to rachis, not secund, usually curved, edges parallel, 3–7 cm × 1.8–2.5 mm; valves glabrous; ovules 100–160 per ovary; style 0.2–0.5 mm. **Seeds** sub-biseriate, 1.1–1.6 × 0.9–1.1 mm; wing continuous, 0.1–0.2 mm wide.

Flowering late Mar–early May. Rocky slopes in evergreen woodlands or desert scrub; 1200–2100 m; N.Mex., Tex.

Morphological evidence suggests that *Boechera porphyrea* is an apomictic species that arose through hybridization between *B. texana* and *B. perennans* (see M. D. Windham and I. A. Al-Shehbaz 2007 for detailed comparison).

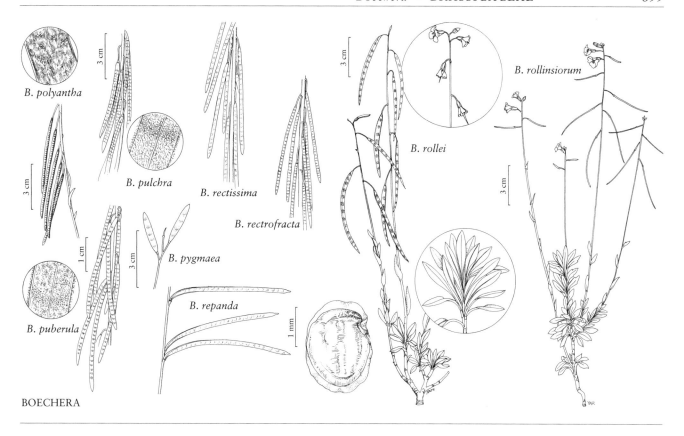

78. Boechera pratincola (Greene) Windham & Al-Shehbaz, Harvard Pap. Bot. 12: 245. 2007 E

Arabis pratincola Greene, Repert. Spec. Nov. Regni Veg. 5: 244. 1908; *A. drummondii* A. Gray var. *pratincola* (Greene) M. Hopkins; *A. nemophila* Greene

Perennials; usually short-lived; apomictic; caudex not woody. **Stems** usually 1 per caudex branch, arising from center of rosette near ground surface, 2–6 dm, pubescent proximally, trichomes sessile, 2- or 3-rayed, 0.15–0.6 mm, glabrous distally. **Basal leaves:** blade oblanceolate, 1.5–5(–7) mm wide, margins usually entire, rarely denticulate, ciliate at petiole base, trichomes to 0.4 mm, surfaces pubescent, trichomes sessile, 2–4(–6)-rayed, 0.1–0.45 mm. **Cauline leaves:** (7–)13–42, concealing stem proximally; blade auricles 1–4 mm, surfaces of distalmost leaves sparsely pubescent apically. **Racemes** (7–)14–45-flowered, usually unbranched. **Fruiting pedicels** erect, straight, 5–12 mm, glabrous. **Flowers** ascending at anthesis; sepals glabrous; petals lavender, 6–10 × 1.5–2 mm, glabrous; pollen spheroid. **Fruits** erect, appressed to rachis, not secund, straight, edges parallel, 4–6.5 cm × 1.5–2 mm; valves glabrous; ovules 60–140 per ovary; style 0.1–0.3 mm. **Seeds** sub-biseriate, 1.4–2 × 1–1.4 mm; wing continuous, 0.1–0.35 mm wide.

Flowering Jun–Aug. Rocky slopes and soil patches in open areas and along forest edges; 1900–3200 m; Calif., Idaho, Nev., Oreg.

Morphological evidence suggests that *Boechera pratincola* is an apomictic species that arose through hybridization between *B. paupercula* and *B. stricta* (see M. D. Windham and I. A. Al-Shehbaz 2007b for detailed comparison).

79. Boechera puberula (Nuttall) Dorn, Brittonia, 55: 3. 2003 E F

Arabis puberula Nuttall in J. Torrey and A. Gray, Fl. N. Amer. 1: 82. 1838; *A. arida* Greene; *A. beckwithii* S. Watson; *A. lignipes* A. Nelson var. *impar* A. Nelson; *A. sabulosa* M. E. Jones; *A. subpinnatifida* S. Watson var. *beckwithii* (S. Watson) Jepson; *A. subpinnatifida* var. *impar* (A. Nelson) Rollins; *Boechera beckwithii* (S. Watson) Dorn

Perennials; short-lived; sexual; caudex not woody. **Stems** usually 1 per caudex branch, arising from center of rosette near ground surface, (1–)2–6.3 dm, densely pubescent proximally, trichomes short-stalked, 3–8-rayed, 0.1–0.3(–0.5) mm, similarly pubescent distally. **Basal leaves:** blade oblanceolate to linear-oblanceolate, 1.5–5 mm wide, margins usually dentate, not ciliate, surfaces densely pubescent, trichomes short-stalked, 5–12-rayed, 0.05–0.2 mm. **Cauline leaves:** 7–45(–65), concealing stem proximally; blade auricles absent or 0.7–3 mm, surfaces of distalmost leaves pubescent. **Racemes** 10–40(–64)-flowered, usually unbranched. **Fruiting pedicels** pendent, recurved distal to horizontal to ascending base, 4–10 mm, pubescent, trichomes appressed, branched. **Flowers** ascending to descending at anthesis; sepals pubescent; petals white to lavender, 5–9 × 0.8–1.8 mm, glabrous; pollen ellipsoid. **Fruits** closely pendent, rarely appressed to rachis, sometimes somewhat secund, usually straight, edges parallel, 3–6.5 cm × 1.9–2.2 mm; valves pubescent throughout; ovules 38–64 per ovary; style 0.05–0.1 mm. **Seeds** uniseriate, 1.4–1.8 × 1–1.4 mm; wing continuous, 0.1–0.3 mm wide. $2n = 14$.

Flowering Apr–Jul. Ledges, rocky slopes, gravelly hillsides in sagebrush, pinyon-juniper, and mountain shrub communities; 1300–2900 m; Calif., Nev., Oreg., Utah.

Boechera puberula is a diploid species that appears to intergrade with both *B. retrofracta* and *B. subpinnatifida*. The glabrous-fruited specimens discussed by R. C. Rollins (1993) represent apomictic hybrids with other species, primarily *B. pendulocarpa*.

80. **Boechera pulchra** (M. E. Jones ex S. Watson) W. A. Weber, Phytologia 51: 370. 1982 [F]

Arabis pulchra M. E. Jones ex S. Watson, Proc. Amer. Acad. Arts 22: 468. 1887

Perennials; long-lived; sexual; caudex woody. **Stems** usually 1 per caudex branch, arising from center of leaf tufts, elevated above ground surface on woody base, (1.5–)3–7.5 dm, densely pubescent proximally, trichomes short-stalked, usually 4–7-rayed, 0.1–0.3 mm, similarly pubescent distally. **Basal leaves:** blade linear to linear-oblanceolate, 1–3 mm wide, margins entire, not ciliate, surfaces densely pubescent, trichomes short-stalked, 4–9-rayed, 0.1–0.3 mm. **Cauline leaves:** 10–30, often concealing stem proximally; blade auricles absent or, rarely, to 0.5 mm, surfaces of distalmost leaves pubescent. **Racemes** 8–25-flowered, usually unbranched. **Fruiting pedicels** reflexed, abruptly recurved at base, otherwise straight, 8–16 mm, pubescent, trichomes appressed, branched. **Flowers** descending at anthesis; sepals pubescent; petals usually purple, rarely white, 9–16 × 2–4(–5) mm, sparsely pubescent or glabrous (trichomes scattered abaxially); pollen ellipsoid. **Fruits** strongly reflexed, often appressed to rachis, sometimes somewhat secund, straight, edges parallel, 3.3–8 cm × 2.5–4 mm; valves pubescent throughout; ovules 68–106 per ovary; style 0.1–0.3 mm. **Seeds** biseriate, 1.7–2.8 × 1.5–2.2 mm; wing continuous, 0.25–0.65 mm wide.

Flowering Mar–Jun. Rocky, gravelly or sandy slopes in chaparral, sagebrush and desert scrub communities; 800–2800 m; Calif., Nev.; Mexico (Baja California).

Boechera pulchra is a distinctive diploid separated from related sexual species (*B. formosa* and *B. lincolnensis*) by abruptly recurved fruiting pedicels and purple flowers, and from apomictic hybrids (e.g., *B. xylopoda*) by having fruits densely pubescent throughout.

81. **Boechera pusilla** (Rollins) Dorn, Vasc. Pl. Wyoming ed. 3, 376. 2001 [C][E]

Arabis pusilla Rollins, Contr. Gray Herb. 212: 107. 1982

Perennials; long-lived; (cespitose); apomictic; caudex often woody. **Stems** usually 2–6 per caudex branch, arising from margin of rosette near ground surface, 0.5–2 dm, glabrous or sparsely pubescent proximally, trichomes simple and short-stalked, 2-rayed, to 0.2 mm, glabrous distally. **Basal leaves:** blade linear-oblanceolate, 1–2.5 mm wide, margins entire, ciliate along petiole, trichomes (simple), 0.4–0.7 mm, surfaces usually sparsely pubescent, rarely glabrous, trichomes short-stalked, 2- or 3-rayed, 0.1–0.4 mm. **Cauline leaves:** 3–5, not concealing stem; blade auricles 0–0.2 mm, surfaces of distalmost leaves usually glabrous or, rarely, margins sparsely ciliate. **Racemes** 6–13-flowered, unbranched. **Fruiting pedicels** horizontal to divaricate-descending, straight or slightly curved downward, 2–5 mm, glabrous. **Flowers** divaricate-ascending at anthesis; sepals glabrous or sparsely pubescent, trichomes spreading, 2-rayed; petals white to lavender, 4–5 × 1.5–1.8 mm, glabrous; pollen spheroid. **Fruits** horizontal or divaricate-descending, not appressed to rachis, secund, straight, edges parallel, 1.6–3.2 cm × 1.5–2 mm; valves glabrous; ovules 20–32 per ovary; style 0.1–0.4 mm. **Seeds** uniseriate, 1.2–1.5 × 0.8–0.9 mm; not winged or with distal wing 0.05–0.1 mm wide.

Flowering May–Jun. Cracks and crevices of granitic rock outcrops; of conservation concern; 2400–2500 m; Wyo.

Morphological evidence suggests that *Boechera pusilla* is an apomictic species that arose through hybridization between *B. lemmonii* and *B. pendulina*; it is known only from the type locality in southwestern Wyoming.

82. **Boechera pygmaea** (Rollins) Al-Shehbaz, Novon 13: 388. 2003 [C][E][F]

Arabis pygmaea Rollins, Rhodora 43: 476, fig. 1941

Perennials; usually long-lived; sexual; caudex woody, (often with persistent, crowded leaf bases). **Stems** usually 2–5 per caudex branch, arising from margin of rosette near ground surface, or arising laterally proximal to sterile shoots, 0.2–0.8 dm, pubescent proximally, trichomes simple and short-stalked, 2- or 3-rayed, to 0.4 mm, glabrescent distally. **Basal leaves**: blade linear, 0.8–1.5 mm wide, margins entire, ciliate, trichomes (simple and spurred), to 0.8 mm, surfaces moderately pubescent, trichomes short-stalked, 2–4-rayed, 0.05–0.4 mm. **Cauline leaves**: 2–4, not concealing stem; blade auricles absent, surfaces of distalmost leaves pubescent. **Racemes** 2–5-flowered, unbranched. **Fruiting pedicels** erect to ascending, straight, 2–7 mm, usually pubescent, rarely glabrous, trichomes subappressed, branched. **Flowers** erect at anthesis; sepals pubescent; petals white, 3.5–5 × 0.7–1 mm, glabrous; pollen ellipsoid. **Fruits** erect to ascending, often appressed to rachis, not secund, straight, edges parallel, (1.3–)2–3.3 cm × 4–5 mm; valves glabrous; ovules 8–12 per ovary; style 0.05–0.4 mm. **Seeds** uniseriate, 3–5 × 2.5–4.5 mm; wing continuous, 0.8–2 mm wide distally.

Flowering May–Jul. Barren flats of arkosic gravel; of conservation concern; 2400–3200 m; Calif.

Despite the distinctive morphology of *Boechera pygmaea*, there is evidence that it is capable of producing fertile hybrids with *B. stricta*. It is known only from Inyo and Tulare counties in the southern Sierra Nevada.

83. **Boechera quebecensis** Windham & Al-Shehbaz, Harvard Pap. Bot. 12: 246. 2007 [E]

Arabis divaricarpa A. Nelson var. *dechamplainii* B. Boivin

Biennials or perennials; short-lived; apomictic; caudex present or absent. **Stems** usually 1 per caudex branch, arising from center of rosette near ground surface, 1–4.5 dm, densely pubescent proximally, trichomes sessile, 2–4-rayed, 0.15–0.5 mm, glabrous distally. **Basal leaves**: blade oblanceolate, 5–15 mm wide, margins denticulate, ciliate proximally, trichomes (simple), to 1 mm, surfaces moderately pubescent, trichomes subsessile, (2- or) 3–7-rayed, 0.1–0.3 mm. **Cauline leaves**: 4–15, not concealing stem; blade auricles 1–3.5 mm, surfaces of distalmost leaves glabrous. **Racemes** 11–41-flowered, usually unbranched. **Fruiting pedicels** horizontal to slightly descending, curved to straight, 3–8(–14) mm, glabrous or with some subappressed, branched trichomes. **Flowers** divaricate at anthesis; sepals pubescent; petals white, 6–7 × 1–2 mm, glabrous; pollen spheroid. **Fruits** horizontal to slightly descending, not appressed to rachis, secund, straight, edges parallel, 3–6 cm × 1.5–2 mm; valves glabrous; ovules 56–94 per ovary. **Seeds** uniseriate, 1.2–1.5 × 1–1.3 mm; wing continuous, 0.1–0.15 mm wide.

Flowering Jun–Jul. Calcareous rock outcrops and talus slopes; 0–300 m; Que.

Morphological evidence suggests that *Boechera quebecensis* is an apomictic species that arose through hybridization between *B. holboellii* and *B. stricta*; it is most likely to be confused with *B. grahamii* (see M. D. Windham and I. A. Al-Shehbaz 2007b for detailed comparison).

84. **Boechera rectissima** (Greene) Al-Shehbaz, Novon 13: 388. 2003 [E][F]

Arabis rectissima Greene, Pittonia 4: 191. 1900; *A. setigera* Greene; *A. wyndii* L. F. Henderson

Biennials or perennials; short-lived; sexual; caudex present or absent. **Stems** usually 1 per caudex branch, arising from center of rosette near ground surface, 2–8(–10) dm, usually glabrous proximally, rarely sparsely pubescent, trichomes simple, to 0.7 mm, glabrous distally. **Basal leaves**: blade oblanceolate, 3–8(–13) mm wide, margins entire or denticulate, ciliate, trichomes (simple), 0.5–1.2 mm, surfaces sparsely to densely pubescent, trichomes simple, to 0.8 mm, often mixed with short-stalked, 2- or 3-rayed ones, 0.15–0.5 mm. **Cauline leaves**: 6–30(–45), often concealing stem proximally; blade auricles 1–1.5 mm, surfaces of distalmost leaves sparsely pubescent, sometimes margins ciliate. **Racemes** 10–30(–50)-flowered, usually unbranched. **Fruiting pedicels** reflexed, abruptly recurved at base, otherwise straight, 4–10 mm, glabrous. **Flowers** pendent at anthesis; sepals pubescent; petals white, 3–4 × 0.7–1.2 mm, glabrous; pollen ellipsoid. **Fruits** strongly reflexed, usually appressed to rachis, not secund, straight, edges parallel, 5–9 cm × 1.8–2.5 mm; valves glabrous; ovules 46–80 per ovary; style 0.2–0.8 mm. **Seeds** uniseriate, 1.8–2.1 × 1–1.5 mm; wing continuous, 0.15–0.25 mm wide.

Flowering May–Jul. Rocky slopes in open conifer forests; 1500–2500 m; Calif., Oreg.

Boechera rectissima is easily recognized by its combination of glabrous stems, basal leaves with large, predominantly simple trichomes, short (3–4 mm), white petals, and strongly reflexed, appressed fruits. Plants assigned to var. *simulans* appear to be hybrids between this species and *B. retrofracta*.

85. Boechera repanda (S. Watson) Al-Shehbaz, Novon 13: 388. 2003 E F

Arabis repanda S. Watson, Proc. Amer. Acad. Arts 11: 122. 1876; *A. repanda* var. *greenei* Jepson

Perennials; short- to long-lived; sexual; caudex usually not woody. **Stems** usually 1 per caudex branch, arising from center of rosette near ground surface, (1–)2–9 dm, densely pubescent proximally, trichomes short- to long-stalked, 2–6-rayed, 0.2–0.5 mm, rarely mixed with simple ones, to 1.5 mm, glabrous or sparsely pubescent distally. **Basal leaves:** blade broadly oblanceolate to obovate, 7–25(–50) mm wide, margins usually repand to coarsely dentate, rarely entire, not ciliate, surfaces moderately to densely pubescent, trichomes short- to long-stalked, 3–6-rayed, 0.2–0.5 mm. **Cauline leaves:** (3–)8–30, not concealing stem; blade auricles absent, surfaces of distalmost leaves sparsely pubescent. **Racemes** 7–25-flowered, sometimes branched. **Fruiting pedicels** erect to divaricate-ascending, straight, 3–10 mm, glabrous or pubescent, trichomes spreading, usually 2–5-rayed, rarely simple. **Flowers** erect to ascending at anthesis; sepals glabrous or sparsely pubescent; petals white, 3.5–6 × 0.8–1 mm, glabrous; pollen ellipsoid. **Fruits** divaricate-ascending, not appressed to rachis, not secund, usually curved, rarely straight, edges parallel, (3.5–)5–13.5 cm × 2.5–4 mm; valves glabrous or pubescent throughout; ovules 34–50 per ovary; style 0.5–1.5 mm. **Seeds** uniseriate, 2.5–5(–6) × 2–3(–3.5) mm; wing continuous, 0.7–1.8 mm wide. $2n = 14$.

Flowering Jun–Jul. Rock outcrops, talus, and gravelly soil in meadows and open pine forests; 1600–3500 m; Calif., Nev.

Boechera repanda is a sexual diploid that is distinguished from all other western North American species by its wide (10–50 mm), repand to coarsely dentate basal leaves. The form recognized as var. *greenei* appears to be a growth response to ecologically marginal habitats. *Boechera repanda* occurs sporadically in the Sierra Nevada and Transverse Ranges of California, and barely extends into the Carson Range in western Nevada.

Arabis inamoena Greene (1911), not Greene (1908) is an illegitimate name, sometimes found in synonymy with *Boechera repanda*.

86. Boechera retrofracta (Graham) Á. Löve & D. Löve, Taxon 31: 125. 1982 E F

Arabis retrofracta Graham, Edinburgh New Philos. J. 7: 344. 1829; *A. arcuata* (Nuttall) A. Gray var. *secunda* (Howell) B. L. Robinson ex S. Watson; *A. exilis* A. Nelson; *A. holboellii* Hornemann var. *retrofracta* (Graham) Rydberg; *A. holboellii* var. *secunda* (Howell) Jepson; *A. lignipes* A. Nelson; *A. retrofracta* var. *multicaulis* B. Boivin; *A. secunda* Howell; *A. sparsiflora* Nuttall subsp. *secunda* (Howell) Piper; *A. tenuis* Greene; *Boechera exilis* (A. Nelson) Dorn; *B. holboellii* (Hornemann) Á. Löve & D. Löve var. *secunda* (Howell) Dorn

Biennials or perennials; short-lived; sexual; caudex present or absent. **Stems** usually 1 per caudex branch, arising from center of rosette near ground surface, 1.5–7(–10.5) dm, densely pubescent proximally, trichomes short-stalked, 2–8-rayed, 0.1–0.2 mm, sparsely pubescent distally. **Basal leaves:** blade oblanceolate, 2–7 mm wide, margins entire or shallowly dentate, sometimes ciliate at petiole base, surfaces densely pubescent, trichomes short-stalked, 5–10-rayed, 0.05–0.2 mm. **Cauline leaves:** 15–40, concealing stem proximally; blade auricles 0.5–2.5 mm or, rarely, absent, surfaces of distalmost leaves pubescent. **Racemes** 15–80(–140)-flowered, usually unbranched. **Fruiting pedicels** reflexed, abruptly recurved at base, otherwise usually straight, 7–12(–18) mm, pubescent, trichomes appressed, branched. **Flowers** pendent at anthesis; sepals pubescent; petals white to lavender, 4–8 × 0.8–2.2 mm, glabrous; pollen ellipsoid. **Fruits** strongly reflexed, usually appressed to rachis, sometimes somewhat secund, straight, edges parallel, 3.5–9 cm × 0.9–1.8 mm; valves glabrous or sparsely pubescent throughout; ovules 60–116 per ovary; style 0.5–2 mm. **Seeds** uniseriate, 1–1.4 × 0.9–1.2 mm; wing continuous, to 0.1 mm wide. $2n = 14$.

Flowering Apr–Aug. Rock outcrops, open hillsides, gravel bars and sandy banks in grassland, sagebrush, oak woodlands, and open conifer forests; 300–3300 m; Alta., B.C., Man., N.W.T., Ont., Que., Sask.; Alaska, Calif., Idaho, Mich., Mont., Nev., Oreg., Utah, Wash., Wyo.

Though often treated as a variety of *Arabis* (*Boechera*) *holboellii* (e.g., R. C. Rollins 1993; G. A. Mulligan 1996; N. H. Holmgren 2005b), *B. retrofracta* is easily distinguished from that species by having narrower (0.9–1.8 versus 2–2.5 mm), mostly non-secund fruits that are almost always appressed to rachises. The two taxa have allopatric distributions, with *B. retrofracta* found on the North American continent (mostly west of the Great Plains) and *B. holboellii* apparently confined to Greenland. *Boechera retrofracta* has formed hybrids with at least 12 other species. Besides differing in macromorphological characters, all those hybrids are distinct from *B. retrofracta* in the strict sense in having wider (20–30 versus 13–16 µm), spheroid pollen grains with asymmetric colpi.

Arabis kochii Blankinship is an illegitimate name, sometimes found in synonymy with *Boechera retrofracta*.

87. Boechera rigidissima (Rollins) Al-Shehbaz, Novon 13: 389. 2003 C E

Arabis rigidissima Rollins, Rhodora 43: 380, fig. 2. 1941; *A. rigidissima* var. *demota* Rollins

Perennials; long-lived; apomictic; caudex woody. **Stems** usually 1 per caudex branch, arising from center of rosette, slightly elevated on woody base or from near ground surface, 2–6 dm, pubescent proximally, trichomes usually subsessile, 2- or 3-rayed, rarely simple, 0.25–0.4 mm, glabrous distally. **Basal leaves:** blade oblanceolate, 2–6 mm wide, margins entire, ciliate proximally, trichomes to 0.7 mm, surfaces sparsely pubescent, trichomes subsessile to short-stalked, 2–5-rayed, 0.1–0.4 mm. **Cauline leaves:** 5–10, rarely concealing stem proximally; blade auricles 0.5–2 mm, surfaces of distalmost leaves glabrous. **Racemes** 5–16-flowered, usually unbranched. **Fruiting pedicels** divaricate-ascending, straight, 4–10 mm, glabrous. **Flowers** ascending at anthesis; sepals glabrous; petals purple, 6–8 × 1.5–2 mm, glabrous; pollen spheroid. **Fruits** divaricate-ascending, not appressed to rachis, not secund, straight, edges somewhat undulate (not parallel), 4–6.5(–7.6) cm × 2.5–3.5 mm; valves glabrous; ovules 24–54 per ovary; style 0.2–0.8 mm. **Seeds** uniseriate, 2.5–3.2 × 1.8–2.5 mm; wing distal, 0.3–1 mm wide.

Flowering Jul–Aug. Rocky places in open conifer forests; of conservation concern; 1800–2000 m; Calif., Nev.

Plants assigned to *Boechera rigidissima* appear to represent a series of apomictic hybrids, possibly involving *B. stricta* and *B. suffrutescens*. The two varieties recognized by Rollins are well isolated geographically (in Trinity County, California, and Washoe County, Nevada, respectively) and show enough morphological divergence to suggest that they represent distinct origins, if not different combinations of parental genomes.

88. Boechera rollei (Rollins) Al-Shehbaz, Novon 13: 389. 2003 E F

Arabis rollei Rollins, Harvard Pap. Bot. 1(4): 43. 1993

Perennials; long-lived; sexual; caudex woody. **Stems** usually 1 per caudex branch, arising from center of rosette, elevated above ground surface on woody base, 1.5–2.5 dm, glabrous throughout. **Basal leaves:** blade oblanceolate, 3–8 mm wide, margins entire, ciliate proximally, trichomes (simple), 0.2–0.7 mm, surfaces glabrous or sparsely pubescent, trichomes short-stalked, 2–4-rayed, 0.2–0.4 mm. **Cauline leaves:** 6–12, sometimes concealing stem proximally; blade auricles 0.5–2.5 mm, surfaces of distalmost leaves glabrous. **Racemes** 3–7-flowered, unbranched. **Fruiting pedicels** strongly curved proximally, 4–6 mm, glabrous. **Flowers** divaricate-ascending at anthesis; sepals glabrous; petals creamy white, 8–11 × 2–2.5 mm, glabrous; pollen ellipsoid. **Fruits** pendent, not appressed to rachis, not secund, straight to slightly curved, edges undulate (not parallel), 3.5–6 cm × 2.5–3.5 mm; valves glabrous; ovules 14–22 per ovary; style 0.5–1 mm. **Seeds** uniseriate, 3–4 × 1.5–2 mm; wing distal and proximal, 0.3–0.6 mm wide.

Flowering Aug. Peridotite rocks on sparsely forested slopes; 1600–1800 m; Calif.

Boechera rollei is easily distinguished from the habitally similar *B. suffrutescens* by its longer (8–11 versus 4.5–6 mm) petals, narrower (2.5–3.5 versus 3–6 mm) fruits, and narrower (0.3–0.6 versus 0.8–1.5 mm) seed wings. This distinctive sexual species is known only from the vicinity of the type locality on the divide between the Applegate and Klamath Rivers in Siskiyou County.

89. Boechera rollinsiorum Windham & Al-Shehbaz, Harvard Pap. Bot. 11: 81. 2006 E F

Perennials; long-lived; sexual; caudex woody. **Stems** usually 3–5 per caudex branch, arising from margin of rosette, elevated above ground surface on woody base, 1–2.5 dm, densely pubescent proximally, trichomes short-stalked, 2–4 (or 5)-rayed, 0.2–0.4 mm, moderately to sparsely pubescent distally. **Basal leaves:** blade oblanceolate, 3–5 mm wide, margins entire, ciliate proximally, trichomes (simple), to 0.6 mm, surfaces densely pubescent, trichomes short-stalked, 2–4 (or 5)-rayed, 0.15–0.4 mm. **Cauline leaves:** 5–9, not concealing stem; blade auricles 0.5–1 mm, surfaces of distalmost leaves sparsely to moderately pubescent. **Racemes**

5–11-flowered, usually unbranched. **Fruiting pedicels** descending, straight to slightly recurved, 4–6 mm, pubescent, trichomes appressed, branched. **Flowers** divaricate-ascending at anthesis; sepals pubescent; petals pale purple, 6–8 × 1.5–2 mm, glabrous; pollen ellipsoid. **Fruits** (immature) pendent, not appressed to rachis, not secund, straight, edges parallel; valves glabrous. **Seeds** not seen.

Flowering Jun. Metamorphosed igneous gravel on steep slopes; ca. 2100 m; Idaho.

Boechera rollinsiorum shows superficial similarities to both *B. glareosa* and *B. lasiocarpa* (see M. D. Windham and I. A. Al-Shehbaz 2006 for detailed comparison). It is known thus far only from the type collection below Galena Summit in central Idaho.

90. Boechera rubicundula (Jepson) Windham & Al-Shehbaz, Harvard Pap. Bot. 11: 273. 2007 [E]

Arabis arcuata (Nuttall) A. Gray var. *rubicundula* Jepson, Fl. Calif. 2: 69. 1936

Perennials; long-lived; apomictic; caudex woody. **Stems** usually 1 per caudex branch, arising from center of rosette, often somewhat elevated above ground surface on woody base, 1–5 dm, densely pubescent proximally, trichomes simple, 0.5–1 mm, mixed with long-stalked, 2- or 3-rayed ones, 0.1–0.5 mm, moderately to sparsely pubescent distally. **Basal leaves**: blade oblanceolate, 4–8 mm wide, margins denticulate, ciliate along petiole, trichomes to 1 mm, surfaces densely pubescent, trichomes long-stalked, 2–5-rayed, 0.1–0.5 mm. **Cauline leaves**: 8–25, often concealing stem proximally; blade auricles 1–3 mm, surfaces of distalmost leaves moderately pubescent. **Racemes** 12–34-flowered, usually unbranched. **Fruiting pedicels** divaricate-ascending to horizontal, straight, 5–10 mm, pubescent, trichomes spreading, simple and 2-rayed. **Flowers** ascending at anthesis; sepals pubescent; petals purplish, 6–8 × 2–2.5 mm, glabrous; pollen spheroid. **Fruits** divaricate-ascending to horizontal, not appressed to rachis, not secund, straight or slightly curved, edges parallel, 4–8 cm × 1.7–2.2 mm; valves glabrous; ovules 70–102 per ovary; style 0.1–0.3 mm. **Seeds** uniseriate, 1.4–1.8 × 1–1.3 mm; wing continuous, 0.1–0.2 mm wide.

Flowering Apr–May. Mountain slopes; ca. 1200 m; Calif.

Morphological evidence suggests that *Boechera rubicundula* is an apomictic species that arose through hybridization between *B. arcuata* and *B. breweri* (see M. D. Windham and I. A. Al-Shehbaz 2007 for detailed comparison). It is known only from the type locality on Mt. Day in west-central California.

91. Boechera saxmontana (Rollins) Windham & Al-Shehbaz, Harvard Pap. Bot. 12: 248. 2007 [E]

Arabis microphylla Nuttall var. *saxmontana* Rollins, Rhodora 43: 429. 1941; *A. pendulocarpa* A. Nelson var. *saxmontana* (Rollins) Dorn; *A. williamsii* Rollins var. *saxmontana* (Rollins) Rollins; *Boechera williamsii* (Rollins) Dorn var. *saxmontana* (Rollins) Dorn

Perennials; short-lived; apomictic; caudex usually not woody. **Stems** usually 1 per caudex branch, arising from center of rosette near ground surface, 0.7–3 dm, densely pubescent proximally, trichomes short-stalked, 2–6-rayed, 0.06–0.25 mm, glabrous distally. **Basal leaves**: blade narrowly oblanceolate, 1–3 mm wide, margins entire, ciliate near petiole base, trichomes (simple and spurred), to 0.4 mm, surfaces sparsely to densely pubescent, trichomes short-stalked, 4–8-rayed, 0.04–0.15 mm. **Cauline leaves**: 5–9, rarely concealing stem proximally; blade auricles to 0.3–1 mm, surfaces of distalmost leaves glabrous. **Racemes** 5–13-flowered, usually unbranched. **Fruiting pedicels** divaricate-ascending, straight, 3–8 mm, glabrous. **Flowers** ascending at anthesis; sepals pubescent; petals lavender, 4–5 × 0.5–0.8 mm, glabrous; pollen spheroid. **Fruits** divaricate-ascending, not appressed to rachis, not secund, straight, edges parallel, (2.5–)3–4.7 cm × 1.2–1.7 mm; valves glabrous; ovules 78–112 per ovary; style 0.1–0.5 mm. **Seeds** uniseriate to sub-biseriate, 1.1–1.4 × 0.7–0.9 mm; wing continuous, 0.05–0.12 mm wide.

Flowering Jun–Jul. Rocky soil in sagebrush and open conifer forests; 2400–2900 m; Idaho, Wyo.

Boechera saxmontana is an apomictic taxon that has been treated as a variety of either *Arabis microphylla* (R. C. Rollins 1941) or *A.* (*Boechera*) *williamsii* (Rollins 1993; R. D. Dorn 2001). It is easily distinguished from typical collections of those species (see M. D. Windham and I. A. Al-Shehbaz 2007b for detailed comparison), but it is likely that one (or both) were involved in its hybrid origin. It is known from Blaine, Custer, and Lemhi counties in Idaho, and Big Horn and Fremont counties in Wyoming.

92. Boechera schistacea (Rollins) Dorn, Brittonia 55: 3. 2003 [E][F]

Arabis schistacea Rollins, Contr. Dudley Herb. 3: 370. 1946

Perennials; long-lived; sexual; caudex woody. **Stems** usually 2–5 per caudex branch, arising from margin of rosette, elevated above ground surface on woody base, 0.8–2.5(–4) dm, glabrous

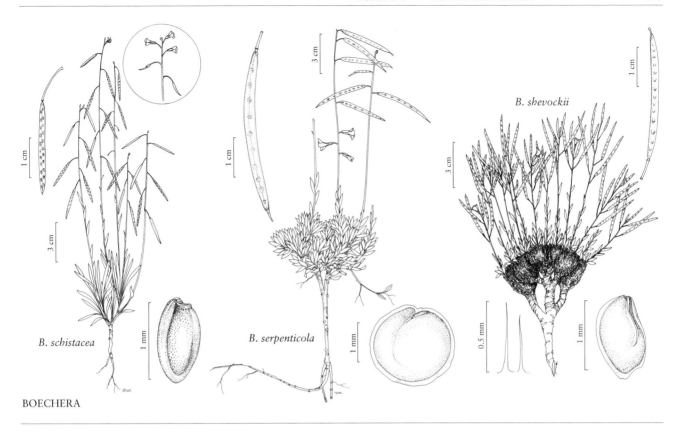

throughout. **Basal leaves:** blade linear to narrowly oblanceolate, 1.2–2.5 mm wide, margins entire, ciliate proximally, trichomes (simple), 0.3–0.7 mm, surfaces glabrous or sparsely pubescent, trichomes short-stalked, 2- or 3-rayed, 0.2–0.5 mm. **Cauline leaves:** 2–5(–9), not concealing stem; blade auricles 0.2–0.5 mm, surfaces of distalmost leaves glabrous. **Racemes** 5–12-flowered, usually unbranched. **Fruiting pedicels** reflexed, curved, 5–9 mm, glabrous. **Flowers** divaricate-ascending at anthesis; sepals glabrous; petals white to lavender, 4–6 × 1–1.5 mm, glabrous; pollen ellipsoid. **Fruits** pendent, not appressed to rachis, not secund, straight, edges parallel, 1.8–4.5 cm × 1.5–2 mm; valves glabrous; ovules 24–36 per ovary; style 0.2–0.5 mm. **Seeds** uniseriate, 0.9–1.2 × 0.7–0.9 mm; not winged. $2n = 14$.

Flowering May–Jul. Rocky slopes and gravelly soil in sagebrush, open conifer forests, and subalpine meadows; 2400–3200 m; Nev., Utah.

Boechera schistacea is a sexual diploid that appears to intergrade with *B. pendulina* where their geographic ranges overlap in southwestern Utah.

93. Boechera serotina (E. S. Steele) Windham & Al-Shehbaz, Harvard Pap. Bot. 12: 249. 2007 [C][E]

Arabis serotina E. S. Steele, Contr. U.S. Natl. Herb. 13: 365. 1911

Biennials; short-lived; sexual; caudex not evident. **Stems** usually 1 per plant, arising from center of rosette near ground surface, 4–10 dm, glabrous throughout. **Basal leaves:** blade obovate to oblanceolate, 5–20 mm wide, margins dentate, not ciliate, surfaces glabrous or subapically puberulent, trichomes simple, 0.1–0.2 mm. **Cauline leaves:** 30–80, not concealing stem; blade auricles absent, surfaces of distalmost leaves glabrous. **Racemes** 70–150-flowered, multi-branched. **Fruiting pedicels** divaricate-ascending to horizontal, straight or gently recurved, 6–15 mm, glabrous. **Flowers** divaricate-ascending at anthesis; sepals glabrous; petals white, 2.8–4 × 0.6–1 mm, glabrous; pollen ellipsoid. **Fruits** divaricate-ascending to widely pendent, not appressed to rachis, not secund, straight or slightly curved, edges parallel, 4.3–7.9 cm × 1.5–1.8 mm; valves glabrous; ovules 30–42 per ovary; style 0.1–0.2 mm. **Seeds** uniseriate, 1.2–1.7 × 0.7–1 mm; wing continuous or, sometimes, distal (rarely absent), 0.1–0.2 mm wide. $2n = 14$.

Flowering Jul–Sep. Shale barrens and wooded slopes of crumbling shale; of conservation concern; 100–500 m; Va., W.Va.

Boechera serotina is unique among members of the genus in its highly branched inflorescences and late (summer–fall) anthesis. It is in the Center for Plant Conservation's National Collection of Endangered Plants.

94. Boechera serpenticola Windham & Al-Shehbaz, Harvard Pap. Bot. 11: 82. 2006 E F

Perennials; long-lived; (cespitose); sexual; caudex woody. **Stems** usually 1 per caudex branch, arising from center of rosette near ground surface, 0.5–1.8 dm, densely pubescent proximally, trichomes short-stalked, 5–8-rayed, 0.1–0.2, sparsely to moderately pubescent distally. **Basal leaves**: blade narrowly oblanceolate, 1–2 mm wide, margins entire, ciliate along petiole, trichomes (branched), to 0.5 mm, surfaces densely pubescent, trichomes short-stalked, 5–8-rayed, 0.1–0.2 mm. **Cauline leaves**: 4–14, rarely concealing stem proximally; blade auricles 0.1–0.5 mm, surfaces of distalmost leaves densely pubescent. **Racemes** 10–20-flowered, unbranched. **Fruiting pedicels** divaricate-descending or horizontal, nearly straight, 3–7 mm, pubescent, trichomes appressed, branched. **Flowers** ascending at anthesis; sepals pubescent; petals purple, 6–8 × 1.5–2.5 mm, glabrous; pollen ellipsoid. **Fruits** usually divaricate-descending or horizontal, rarely widely pendent, not appressed to rachis, not secund, straight or slightly curved, edges parallel, 3–4.5 cm × 2.5–3 mm; valves sparsely pubescent to glabrate; ovules 20–24 per ovary; style 1.5–2 mm. **Seeds** uniseriate, 2–2.5 × 1.5–1.8 mm; wing continuous, to 0.2 mm wide.

Flowering Mar–Jun. Serpentine ridges and talus; ca. 1100 m; Calif.

Boechera serpenticola has been confused with *B. subpinnatifida* but is easily distinguished by its entire (versus subpinnatifid to dentate) leaves, shorter (6–8 versus 9–14 mm) petals, and longer (1.5–2 versus 0.5–1 mm) styles. It is known only from Shasta and Trinity counties in north-central California.

95. Boechera shevockii Windham & Al-Shehbaz, Harvard Pap. Bot. 11: 82. 2006 E F

Perennials; long-lived; (pulvinate); sexual; caudex woody (with persistent, crowded leaf bases). **Stems** 1 per caudex branch, arising from center of rosette near ground surface, 0.5–1 dm, puberulent proximally, trichomes (scattered), simple, 0.02–0.1 mm, similarly pubescent distally. **Basal leaves**: blade oblong-oblanceolate, 1–2 mm wide, margins entire, ciliate distally, trichomes (simple), 0.02–0.1 mm, surfaces glabrous. **Cauline leaves**: 3–7, not concealing stem; blade auricles absent, surfaces of distalmost leaves glabrous. **Racemes** 4–7-flowered, unbranched. **Fruiting pedicels** ascending, straight, 5–9 mm, glabrous. **Flowers** ascending at anthesis; sepals glabrous; petals lavender, 4–5 × 1–1.2 mm, glabrous; pollen ellipsoid. **Fruits** ascending, not appressed to rachis, not secund, straight, edges parallel, 2.5–3 cm × 1–1.5 mm; valves glabrous; ovules 30–34 per ovary; style ca. 0.5 mm. **Seeds** uniseriate, ca. 1.2 × 0.8 mm; wing distal or continuous, ca. 0.1 mm wide.

Flowering Jun. Ledges of rock outcrops; ca. 2500 m; Calif.

Boechera shevockii is a close relative of *B. davidsonii*, from which it is readily distinguished by having puberulent (versus glabrous) stems, shorter (4–5 versus 6–10 mm) petals, and much shorter (5–7 versus 30–80 mm) basal leaves. It is known only from the type specimen, from the southern Sierra Nevada in Tulare County.

96. Boechera shockleyi (Munz) Dorn, Brittonia 55: 3. 2003 E F

Arabis shockleyi Munz, Bull. S. Calif. Acad. Sci. 31: 62. 1932

Perennials; usually short-lived; sexual; caudex not woody, (rarely with persistent, crowded leaf bases). **Stems** usually 1 per caudex branch, arising from center of rosette near ground surface, (0.8–)2–5 dm, densely pubescent proximally, trichomes short-stalked, 7–12-rayed, 0.1–0.2 mm, densely to sparsely pubescent distally. **Basal leaves**: blade oblanceolate, 3–10 mm wide, margins entire, not ciliate, surfaces densely pubescent, trichomes short-stalked, 7–12-rayed, 0.1–0.2 mm. **Cauline leaves**: 14–60, concealing stem for most of length; blade auricles 0.5–4 mm, surfaces of distalmost leaves densely pubescent. **Racemes** 20–70-flowered, usually unbranched. **Fruiting pedicels** divaricate-ascending, straight, 7–28 mm, pubescent, trichomes appressed, branched. **Flowers** ascending at anthesis; sepals pubescent; petals lavender, 6–9 × 0.8–1.2 mm, glabrous; pollen ellipsoid. **Fruits** divaricate-ascending, not appressed to rachis, not secund, usually curved, rarely straight, edges parallel, 4.5–11 cm × 1.5–2 mm; valves glabrous or sparsely pubescent throughout; ovules 140–190 per ovary; style 0.05–0.6 mm. **Seeds** sub-biseriate, 1–1.3 × 0.7–0.8 mm; wing distal or, rarely, absent, 0.05–0.1 mm wide.

Flowering Apr–May. Rock outcrops (primarily dolomite) and gravelly soil in desert scrub, sagebrush, and pinyon-juniper woodlands; 1200–2200 m; Calif., Nev., Utah.

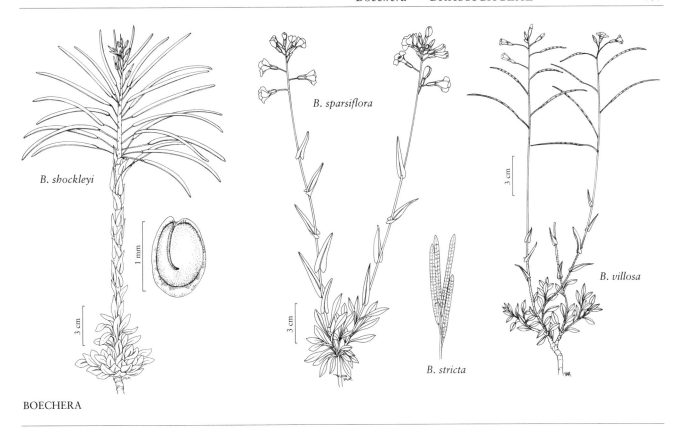

Boechera shockleyi is a distinctive sexual species recognizable by the combination of strongly overlapping cauline leaves, a dense covering of minute, 7–12-rayed trichomes, and relatively long, curved, divaricate-ascending fruits with sub-biseriate seeds. It is most similar to *B. inyoensis*, but differs from that species in its narrower (0.8–1.2 versus 1.2–2 mm) petals, eciliate basal leaves with smaller (0.1–0.2 versus 0.2–0.7 mm) trichomes, greater number (140–190 versus 74–134) of ovules per ovary, and shorter (1–1.3 versus 1.7–2 mm) seeds with distal (versus continuous) wings.

97. **Boechera sparsiflora** (Nuttall) Dorn, Vasc. Pl. Wyoming ed. 3, 376. 2001 [E] [F]

Arabis sparsiflora Nuttall in J. Torrey and A. Gray, Fl. N. Amer. 1: 81. 1838; *A. arcoidea* A. Nelson; *A. campyloloba* Greene; *A. peramoena* Greene; *A. polytricha* Greene; *A. sparsiflora* var. *peramoena* (Greene) Rollins

Biennials or perennials; short-lived; sexual; caudex present or absent. **Stems** usually 1 per caudex branch, arising from center of rosette near ground surface, 3–8 dm, densely pubescent proximally, trichomes simple mixed with fewer short-stalked, 2-rayed ones, 0.4–1.5 mm, glabrous or sparsely pubescent distally. **Basal leaves**: blade oblanceolate or linear-oblanceolate, 3–12 mm wide, margins usually entire, rarely dentate, ciliate proximally, trichomes (usually simple), to 1 mm, surfaces densely pubescent, trichomes short-stalked, 2–5-rayed, 0.3–0.8 mm. **Cauline leaves**: (9–)15–35, often concealing stem proximally; blade auricles 3–10 mm, surfaces of distalmost leaves glabrous, sometimes margins ciliate. **Racemes** 12–50-flowered, usually unbranched. **Fruiting pedicels** usually ascending, rarely almost horizontal, straight or slightly recurved, 3–10(–18) mm, usually pubescent, rarely glabrous, trichomes spreading, usually simple. **Flowers** ascending at anthesis; sepals sparsely pubescent; petals usually lavender to purple, rarely white, 7–13 × 2–5 mm, glabrous; pollen ellipsoid. **Fruits** usually ascending, rarely almost horizontal, not appressed to rachis, not secund, usually curved, edges parallel, 5–13 cm × 1.7–2 mm; valves glabrous; ovules 90–170 per ovary; style 0.05–0.3 mm. **Seeds** uniseriate, 1.5–2 × 1–1.5 mm; wing continuous, 0.1–0.2 mm wide. $2n = 14$.

Flowering Apr–Jun. Rocky slopes, clay hills, sandy soil in sagebrush and mountain shrub communities, meadows, and open conifer forests; 400–2800 m; B.C.; Calif., Idaho, Nev., Oreg., Utah, Wash.

As circumscribed by R. C. Rollins (1993), *Boechera sparsiflora* included six varieties encompassing three sexual diploids and a number of apomictic hybrids. The most distinctive of those elements are recognized here as the separate species *B. arcuata*, *B. atrorubens*, *B. californica*, and *B. pauciflora*. The narrow circumscription of *B. sparsiflora* adopted here includes only sexual diploids. It is distinguished from other taxa previously assigned to it by having proximal stems densely pubescent with predominantly simple (some 2-rayed) trichomes to 1.5 mm, usually glabrous distal stems, and ascending fruiting pedicels with spreading, usually simple trichomes (rarely glabrous).

98. Boechera spatifolia (Rydberg) Windham & Al-Shehbaz, Harvard Pap. Bot. 11: 84. 2006 [E]

Arabis spatifolia Rydberg, Fl. Rocky Mts., 361, 1062. 1917; *A. fendleri* (S. Watson) Greene var. *spatifolia* (Rydberg) Rollins; *Boechera fendleri* (S. Watson) W. A. Weber subsp. *spatifolia* (Rydberg) W. A. Weber; *B. fendleri* var. *spatifolia* (Rydberg) Dorn

Perennials; short- to long-lived; sexual; caudex usually not woody (rarely with persistent, crowded leaf bases). **Stems** simple or few to several per caudex branch, arising from center of rosette near ground surface, 1.5–3.5(–5) dm, densely pubescent proximally, trichomes simple mixed with short-stalked ones, 2-rayed, 0.3–0.7 mm. **Basal leaves:** blade narrowly oblanceolate, 1.5–3(–4) mm wide, margins entire, strongly ciliate at least along petiole, trichomes (simple), to 1 mm, surfaces glabrous or sparsely pubescent, trichomes simple and short-stalked, 2-rayed, 0.3–0.7 mm. **Cauline leaves:** 5–15(–20), often concealing stem proximally; blade auricles 0.5–1.5 mm, surfaces of distalmost leaves usually glabrous. **Racemes** 10–30-flowered, usually unbranched. **Fruiting pedicels** horizontal or slightly descending, curved or angled downward, 6–10(–15) mm, usually glabrous, rarely sparsely pubescent, trichomes spreading, simple. **Flowers** divaricate-ascending at anthesis; sepals pubescent; petals usually white, rarely pale lavender, 3–3.7(–4) × 0.5–0.8 mm, glabrous; pollen ellipsoid. **Fruits** pendent, not appressed to rachis, rarely slightly secund, straight or gently curved, edges parallel, 3.3–5.7 cm × 1.2–1.8 mm; valves glabrous; ovules 90–126 per ovary; style 0.1–0.4 mm. **Seeds** biseriate, 0.7–0.9 × 0.5–0.6 mm; wing distal or, sometimes, absent, 0.05–0.1 mm wide. $2n = 14$.

Flowering May–Aug. Rocky slopes and gravelly soil in sagebrush, pinyon-juniper woodlands, open conifer forests and subalpine meadows; 1800–2800 m; Colo., N.Mex.

Boechera spatifolia is a sexual diploid that usually has been treated as a variety of *Arabis* (*Boechera*) *fendleri* but appears to be sufficiently distinct to warrant recognition at species level (see M. D. Windham and I. A. Al-Shehbaz 2006 for detailed comparison). There is little geographic overlap between the two, with *B. spatifolia* confined to the mountains of central Colorado and north-central New Mexico and *B. fendleri* ranging from western New Mexico and the Four Corners region through northern Arizona to southern Nevada.

99. Boechera stricta (Graham) Al-Shehbaz, Novon 13: 389. 2003 [E] [F]

Turritis stricta Graham, Edinburgh New Philos. J. 7: 350. 1829; *Arabis connexa* Greene; *A. drummondii* A. Gray; *A. drummondii* var. *connexa* (Greene) Fernald; *A. drummondii* var. *oxyphylla* (Greene) M. Hopkins; *A. oxyphylla* Greene; *Boechera angustifolia* (Nuttall) Dorn; *B. drummondii* (A. Gray) Á. Löve & D. Löve; *Erysimum drummondii* (A. Gray) Kuntze; *Streptanthus angustifolius* Nuttall; *T. drummondii* (A. Gray) Lunell

Biennials or perennials; usually short-lived; sexual; caudex usually not woody. **Stems** usually 1–4 per caudex branch, usually arising from center of rosette near ground surface, rarely arising laterally proximal to sterile shoots, 1.5–8(–10.2) dm, glabrous or sparsely pubescent proximally, trichomes sessile, usually 2-rayed, rarely simple, 0.3–0.7 mm, glabrous distally. **Basal leaves:** blade oblanceolate, 1.5–8(–14) mm wide, margins usually entire, rarely dentate, ciliate, trichomes (sessile, usually 2-rayed, rarely simple ones at petiole base), surfaces usually pubescent, sometimes glabrous, trichomes sessile, 2-rayed (malpighiaceous), 0.3–0.7 mm. **Cauline leaves:** 6–52, concealing stem proximally; blade auricles 0.5–3 mm, surfaces of distalmost leaves glabrous. **Racemes** 8–35(–80)-flowered, usually unbranched. **Fruiting pedicels** erect, straight, 5–18 (–25) mm, glabrous. **Flowers** erect at anthesis; sepals glabrous; petals white (usually aging pale lavender), (5–)7–11 × 1.5–2.7 mm, glabrous; pollen ellipsoid. **Fruits** erect, appressed to rachis, not secund, straight, edges parallel, 4–9(–10.2) cm × 1.5–3.5 mm; valves glabrous; ovules 110–216 per ovary; style 0.05–0.3 mm. **Seeds** biseriate, 1.3–2.2 × 1–1.6 mm; wing lateral and distal, 0.3–0.8 mm wide. $2n = 14$.

Flowering May–Aug. Rocky slopes and gravelly soil in sagebrush and mountain shrub communities, open conifer and hardwood forests, alpine meadows; 700–3900 m; Alta., B.C., N.W.T., Que., Sask., Yukon; Alaska, Ariz., Calif., Colo., Idaho, Iowa, Maine, Mass., Mich., Minn., Mont., Nev., N.H., N.Mex., N.Dak., Ohio, Oreg., R.I., Utah, Vt., Wash., Wis., Wyo.

Arabis drummondii is the correct name for this species in that genus; the epithet *stricta* has priority in *Boechera*. This very distinctive species is easily recognized by having basal leaves with branched trichomes, all sessile and 2-rayed (malpighiaceous). It is also the most promiscuous, having formed apomictic hybrids with at least 15 other species of *Boechera*.

100. Boechera subpinnatifida (S. Watson) Al-Shehbaz, Novon 13: 389. 2003 [E]

Arabis subpinnatifida S. Watson, Proc. Amer. Acad. Arts 20: 353. 1885

Perennials; long-lived; sexual; caudex woody (often with persistent, crowded leaf bases). **Stems** usually 1 per caudex branch, arising from center of rosette near ground surface, 1–4(–5) dm, densely pubescent proximally, trichomes short-stalked, 2–6-rayed, 0.1–0.2 mm, sparsely pubescent distally. **Basal leaves:** blade narrowly oblanceolate, 1–4(–5) mm wide, margins prominently dentate to subpinnatifid (leaf margins of sterile shoots often entire), ciliate near petiole base, trichomes (simple or 2-rayed), 0.4–0.6 mm, surfaces densely pubescent, trichomes short-stalked, (2–)4–9-rayed, 0.05–0.2 mm. **Cauline leaves:** (10–)20–60, often concealing stem throughout; blade auricles 0.5–3 mm, surfaces of distalmost leaves moderately to sparsely pubescent. **Racemes** 8–30-flowered, usually unbranched. **Fruiting pedicels** reflexed, strongly recurved, 5–15 mm, pubescent, trichomes appressed, branched. **Flowers** divaricate-ascending to pendent at anthesis; sepals pubescent; petals usually purple, rarely lavender, 9–14 × 1.5–3 mm, glabrous; pollen ellipsoid. **Fruits** pendent, not appressed to rachis, not secund, straight to slightly curved, edges parallel, (3.5–)5–8 cm × (1.6–)2–3 mm; valves pubescent throughout; ovules 24–42 per ovary; style 0.5–1 mm. **Seeds** uniseriate, 2.5–3.5 × 1.5–2.2 mm; wing continuous or at both ends, 0.4–0.8 mm wide. $2n = 14$.

Flowering Mar–May. Rock outcrops, talus, gravelly soil, often in sagebrush-grassland communities; 800–2400 m; Calif., Idaho, Nev., Oreg., Utah.

Originally thought to be restricted to northern California and adjacent Oregon, *Boechera subpinnatifida* is a sexual species that recently has been found in central Idaho, northern Nevada, and northwestern Utah. It appears to intergrade with both *B. puberula* and *B. retrofracta*, and species boundaries within this complex need further study.

101. Boechera suffrutescens (S. Watson) Dorn, Brittonia 55: 3. 2003 [E]

Arabis suffrutescens S. Watson, Proc. Amer. Acad. Arts 17: 362. 1882; *A. dianthifolia* Greene; *A. duriuscula* Greene

Perennials; long-lived; sexual or apomictic; caudex woody. **Stems** usually 1 per caudex branch, arising from center of rosette, elevated above ground surface on woody base, 1–4(–5) dm, usually glabrous proximally, rarely pubescent, trichomes short-stalked, 2-rayed, 0.1–0.3 mm, glabrous distally. **Basal leaves:** blade narrowly oblanceolate, 1.5–6 mm wide, margins entire, not ciliate or, rarely, with simple trichomes to 0.7 mm, surfaces glabrous or sparsely to densely pubescent, trichomes short-stalked, 2–6-rayed, 0.07–0.4 mm. **Cauline leaves:** (4–)6–12, sometimes concealing stem proximally; blade auricles 0.5–2 mm, surfaces of distalmost leaves glabrous. **Racemes** 6–12-flowered, usually unbranched. **Fruiting pedicels** reflexed, usually abruptly recurved at base, otherwise straight, 4–18 mm, glabrous. **Flowers** pendent at anthesis; sepals glabrous; petals purple or whitish with rose tips, 4.5–6 × 2–2.5 mm, glabrous; pollen ellipsoid or spheroid. **Fruits** reflexed to closely pendent, sometimes appressed to rachis, often secund, straight to somewhat curved, edges undulate (not parallel), 3–7(–8) cm × 3–5(–6) mm; valves glabrous; ovules 20–30 per ovary; style 0.4–1.2 mm. **Seeds** uniseriate, 2.5–5.5 × 1.8–3.5 mm; wing continuous, 0.8–1.5 mm wide.

Flowering Jun–Aug. Rocky slopes and gravelly soil, often with sagebrush; 1800–3000 m; Calif., Idaho, Nev., Oreg., Wash.

Boechera suffrutescens is recognizable by its prominently suffrutescent habit and wide (greater than 3 mm), reflexed fruits. Both sexual and apomictic collections are known; further study is needed to determine whether they truly are conspecific. The taxon previously known as *Arabis suffrutescens* var. *horizontalis* appears to be of hybrid origin; it is treated here as a distinct species (see M. D. Windham and I. A. Al-Shehbaz 2007 for detailed comparison).

102. Boechera texana Windham & Al-Shehbaz, Harvard Pap. Bot. 11: 85. 2006 [E]

Perennials; long-lived; sexual; caudex woody. **Stems** usually 1 per caudex branch, arising from center of rosette, often slightly elevated above ground surface on woody base, 2–5 dm, glabrous throughout. **Basal leaves:** blade oblanceolate, 5–12 mm wide, margins dentate, ciliate (at least proximally), trichomes (usually simple), 1–2 mm, surfaces moderately pubescent, trichomes short-stalked, 2–4-rayed, 0.4–0.6 mm. **Cauline leaves:** 5–12, often concealing stem proximally; blade auricles 1–2 mm, surfaces distally glabrous. **Racemes** 10–35-flowered, usually unbranched. **Fruiting pedicels** horizontal to divaricate-ascending, curved or angled downward, 10–20 mm, glabrous. **Flowers** divaricate-ascending at anthesis; sepals glabrous; petals white to lavender, 5–8 × 1.5–2 mm, glabrous; pollen ellipsoid. **Fruits** widely pendent, not appressed to rachis, not secund, curved to nearly straight, edges parallel, 3.5–5 cm × 2.5–3 mm; valves glabrous; ovules 80–130 per ovary; style 0.2–0.7 mm. **Seeds** biseriate, 1.1–1.3 × 0.9–1 mm; wing distal or continuous, 0.1–0.15 mm wide.

Flowering Apr. Rock outcrops and gravelly soils in desert grassland and evergreen woodlands; 1200–1700 m; N.Mex., Tex.

Boechera texana is a sexual diploid included within *Arabis fendleri* var. *fendleri* by R. C. Rollins (1993), but it appears to be sufficiently distinct to warrant recognition at species level (see M. D. Windham and I. A. Al-Shehbaz 2006 for detailed comparison). There is little or no geographic overlap between the two, with *B. texana* confined to western Texas and southern New Mexico and *B. fendleri* ranging from western New Mexico and the Four Corners region through northern Arizona to southern Nevada.

103. **Boechera tiehmii** (Rollins) Al-Shehbaz, Novon 13: 389. 2003 [C] [E]

Arabis tiehmii Rollins, J. Arnold Arbor. 64: 496, fig. 2. 1983

Perennials; long-lived; sexual; caudex woody (covered by persistent leaf bases). **Stems** usually 2–5 per caudex branch, arising from margin of rosette near ground surface, 0.8–2 (–3) dm, glabrous throughout. **Basal leaves:** blade oblanceolate, (2–)3–7 mm wide, margins usually entire, rarely denticulate, ciliate, trichomes (simple and short-stalked) 2- or 3-rayed, 0.2–0.4 mm, surfaces glabrous. **Cauline leaves:** 3–5, not concealing stem; blade auricles absent or, rarely, to 0.5 mm, surfaces of distalmost leaves glabrous. **Racemes** 10–16-flowered, usually unbranched. **Fruiting pedicels** ascending, straight, 3–7 mm, glabrous. **Flowers** ascending at anthesis; sepals glabrous; petals white to lavender, 4–5 × 1.2–1.7 mm, glabrous; pollen ellipsoid. **Fruits** ascending, not appressed to rachis, not secund, straight, edges parallel, 1.5–3.7 cm × 1.4–1.7 mm; valves glabrous; ovules 14–22 per ovary; style 0.2–0.6 mm. **Seeds** uniseriate, 1.7–2.2 × 1–1.2 mm; wing absent or distal, 0–0.5 mm wide.

Flowering Jun–Aug. Rock outcrops and gravelly soils in alpine and subalpine habitats; of conservation concern; 3000–3500 m; Calif., Nev.

Boechera tiehmii is a sexual species considered by R. C. Rollins (1993) to be related to *Arabis* (*Boechera*) *davidsonii*. It is known only from Mono County, California, and Washoe County, Nevada.

104. **Boechera tularensis** Windham & Al-Shehbaz, Harvard Pap. Bot. 12: 249. 2007 [E]

Biennials or perennials; short-lived; apomictic; caudex present or absent. **Stems** usually 1 per caudex branch, arising from center of rosette near ground surface, 2–7 dm, sparsely to densely pubescent proximally, trichomes subsessile, submalpighiaceous, 0.3–0.6 mm, glabrous distally. **Basal leaves:** blade oblanceolate, 3–7 mm wide, margins entire, ciliate proximally, trichomes (simple), to 0.8 mm, surfaces sparsely to densely pubescent, trichomes subsessile, 2–5-rayed, 0.2–0.55 mm. **Cauline leaves:** 7–17, often concealing stem proximally; blade auricles 2–5 mm, surfaces of distalmost leaves sparsely pubescent or glabrous. **Racemes** 19–39-flowered, usually unbranched. **Fruiting pedicels** reflexed, recurved proximally, 5–13 mm, glabrous. **Flowers** divaricate to pendent at anthesis; sepals pubescent; petals white to pale lavender, 6–7 × 1.2–2 mm, glabrous; pollen spheroid. **Fruits** reflexed, rarely appressed to rachis, not secund, straight, edges parallel, 4–7(–8.5) cm × 2–2.3 mm; valves glabrous; ovules 88–104 per ovary; style 0.3–0.7 mm. **Seeds** sub-biseriate, 2–2.5 × 1.1–1.5 mm; wing continuous, 0.15–0.25 mm wide.

Flowering Jun–Jul. Rocky slopes in montane and subalpine habitats; 2400–3200 m; Calif.

Morphological evidence suggests that *Boechera tularensis* is an apomictic species that contains three different genomes, one each from *B. rectissima*, *B. retrofracta*, and *B. stricta*. It is most often confused with *B. pinetorum* and *B. retrofracta*, but is easily distinguishable from both (see M. D. Windham and I. A. Al-Shehbaz 2007b for detailed comparison). It is known from the southern Sierra Nevada (Fresno and Tulare counties).

105. **Boechera ultra-alsa** Windham & Al-Shehbaz, Harvard Pap. Bot. 11: 86. 2006 (as ultraalsa) [E]

Perennials; long-lived; reproductive mode unknown; caudex woody. **Stems** 1 per caudex branch, arising from center of rosette near ground surface, ca. 1 dm, densely pubescent proximally, trichomes stalked, (2 or) 3–6-rayed, 0.2–0.5 mm, glabrate distally. **Basal leaves:** blade oblanceolate, 4–6 mm wide, margins entire, not ciliate, surfaces moderately pubescent, trichomes long-stalked, 3–6 (or 7)-rayed, 0.2–0.5 mm. **Cauline leaves:** 2–5, concealing stem proximally; blade auricles absent, surfaces of distalmost leaves pubescent. **Racemes** 3- or 4-flowered, unbranched. **Fruiting pedicels** erect-ascending, straight, 4–5 mm, glabrous or sparsely pubescent, trichomes appressed, branched. **Flowers** not seen. **Fruits** erect-ascending, sometimes appressed to rachis, not secund, straight, edges parallel, 3–4 cm × ca. 5 mm; valves glabrous; ovules ca. 16 per ovary; style 0.5–0.7 mm. **Seeds** uniseriate, 5.5–6.5 × 4–4.5 mm; wing continuous, 1–2 mm wide.

Flowering Jun–Jul. Rocky soil; ca. 1800 m; Calif.

Boechera ultra-alsa is known only from the holotype, from Snow Mountain in Lake County. Originally identified as *Arabis* (*Boechera*) *platysperma*, this specimen differs from that taxon by having petioles without obvious cilia, basal leaves with long-stalked (0.1–0.2 versus less than 0.1 mm) trichomes, and fruits edges parallel (not undulate or constricted between the seeds).

106. **Boechera villosa** Windham & Al-Shehbaz, Harvard Pap. Bot. 11: 86. 2006 E F

Perennials; short-lived; sexual; caudex not woody. **Stems** 1–3 per caudex branch, arising laterally proximal to sterile shoots near ground surface, ca. 2.5 dm, densely pubescent proximally, trichomes simple and 2-rayed, 0.25–0.7 mm, glabrescent distally. **Basal leaves**: blade oblanceolate to obovate, 2.5–5 mm wide, margins entire, ciliate along petiole, trichomes (simple), to 1 mm, surfaces densely pubescent, trichomes simple and short-stalked, 2- or 3-rayed, 0.25–0.5 mm. **Cauline leaves**: 4–6, not concealing stem; blade auricles ca. 1 mm, surfaces of distalmost leaves sparsely pubescent. **Racemes** 6–10-flowered, unbranched. **Fruiting pedicels** ascending, straight, 6–10 mm, glabrous or sparsely pubescent, trichomes subappressed, branched. **Flowers** ascending at anthesis; sepals pubescent; petals lavender, 4–5 × 1–1.5 mm, glabrous; pollen ellipsoid. **Fruits** divaricate-ascending, not appressed to rachis, not secund, slightly curved, edges parallel, 4–5 cm × ca. 1 mm; valves glabrous; ovules ca. 64 per ovary; style ca. 0.2 mm. **Seeds** uniseriate, (none mature).

Flowering May. Basalt outcrop in pinyon-juniper woodlands; ca. 2100 m; N.Mex.

Boechera villosa is known only from the type collection from the Rio Grande Gorge in Taos County. The holotype was originally identified as *Arabis* (*Boechera*) *perennans* but clearly is more closely related to *B. pallidifolia* (see M. D. Windham and I. A. Al-Shehbaz 2006 for detailed comparison).

107. **Boechera williamsii** (Rollins) Dorn, Vasc. Pl. Wyoming ed. 3, 376. 2001 E

Arabis williamsii Rollins, Syst. Bot. 6: 62, fig. 2. 1981

Perennials; short-lived; sexual; caudex not woody. **Stems** usually 1 per caudex branch, arising from center of rosette near ground surface, 0.7–2.2 dm, glabrous throughout. **Basal leaves**: blade oblanceolate, 1–2 mm wide, margins entire, ciliate near petiole base, trichomes (simple), to 0.4 mm, surfaces sparsely to densely pubescent, trichomes short-stalked, 4–8-rayed, 0.03–0.1 mm. **Cauline leaves**: 6–11, sometimes concealing stem proximally; blade auricles to 0.5 mm, surfaces of distalmost leaves glabrous. **Racemes** 5–17-flowered, unbranched. **Fruiting pedicels** erect, straight, 2–4(–7) mm, glabrous. **Flowers** erect at anthesis; sepals glabrous; petals white, 3–4 × 0.5–0.8 mm, glabrous; pollen ellipsoid. **Fruits** erect, usually appressed to rachis, not secund, straight, edges parallel, 2–3.5 cm × 2.2–2.5 mm; valves glabrous; ovules 60–96 per ovary; style 0.05–0.2 mm. **Seeds** sub-biseriate, 0.7–0.9 × 0.5–0.6 mm, not winged or minutely winged distally.

Flowering May–Jun. Gravelly soil in sagebrush-grassland communities; 2300–2800 m; Wyo.

Boechera williamsii is distinguished from all other sexual species of the genus by its combination of relatively short (3–4 mm), white petals, minute, 4–8-rayed trichomes on basal leaves, erect, appressed fruits, and relatively small, wingless seeds. It is known only from Fremont and Sublette counties in west-central Wyoming. The taxon traditionally treated as *Arabis* (*Boechera*) *williamsii* var. *saximontana* is here recognized as an apomictic species of hybrid origin (see M. D. Windham and I. A. Al-Shehbaz 2007b for detailed comparison).

108. **Boechera xylopoda** Windham & Al-Shehbaz, Harvard Pap. Bot. 12: 250. 2007 E

Arabis pulchra M. E. Jones ex S. Watson var. *gracilis* M. E. Jones, Contr. W. Bot. 8: 41. 1898; *A. pulchra* var. *glabrescens* Wiggins; *A. pulchra* var. *viridis* Jepson; *Boechera pulchra* (M. E. Jones ex S. Watson) W. A. Weber var. *gracilis* (M. E. Jones) Dorn

Perennials; long-lived; apomictic; caudex woody. **Stems** usually 1 from per caudex branch, arising from center of leaf tuft, elevated above ground surface on woody base, (3–)5–7.5 dm, pubescent proximally, trichomes short-stalked, 2–5-rayed, 0.25–0.9 mm., similarly pubescent distally. **Basal leaves**: blade linear-oblanceolate to oblanceolate, 3–7 mm wide, margins entire or somewhat dentate, ciliate near petiole base, trichomes (simple), to 1.5 mm, surfaces densely pubescent, trichomes short-stalked, 4–8-rayed, 0.15–0.5 mm. **Cauline leaves:** 8–18, usually not concealing stem; blade auricles 1–2 mm, surfaces of distalmost leaves sparsely pubescent. **Racemes** 20–38-flowered, usually unbranched. **Fruiting pedicels** descending-divaricate, usually recurved, 9–20 mm, usually pubescent, rarely glabrous, trichomes subappressed, branched. **Flowers** divaricate to pendent at anthesis; sepals pubescent; petals purple, 9–12 × 2–3.5 mm, sparsely pubescent (often some trichomes abaxially); pollen spheroid. **Fruits** reflexed, not appressed to rachis, not secund, straight, edges parallel, 5–7 cm × 2–2.5 mm; valves glabrous proximally, sparsely pubescent distally; ovules 98–126 per ovary; style 0.1–0.5 mm. **Seeds** sub-biseriate, 2–2.5 × 1.3–1.6 mm; wing continuous, 0.2–0.25 mm wide.

Flowering Mar–May. Rock outcrops and gravelly slopes, often under shelter of shrubs in desert scrub, sagebrush, and chaparral communities; 900–2000 m; Calif., Nev.

Although usually treated as a variety of *Arabis* (*Boechera*) *pulchra*, *B. xylopoda* is an apomictic taxon that is morphologically intermediate between *B. perennans* and *B. pulchra* (see M. D. Windham and I. A. Al-Shehbaz 2007b for detailed comparison) and probably arose through hybridization between those two sexual diploids.

Arabis trichopoda Greene (1908), not Turczaninow (1840) is an illegitimate name, sometimes found in synonymy with *Boechera xylopoda*.

109. **Boechera yorkii** S. Boyd, Madroño 51: 387, figs. 1–3. 2004 [E]

Perennials; long-lived; sexual; caudex woody. **Stems** usually 1 per caudex branch, arising from center of rosette near ground surface, 1–3 dm, densely pubescent proximally, trichomes short-stalked, 4–7-rayed, 0.1–0.5 mm, mixed proximally with simple and short- to long-stalked, 2- or 3-rayed ones, to 1.5 mm, moderately pubescent distally. **Basal leaves:** blade linear-oblanceolate, 1.5–3 mm wide, margins entire, ciliate proximally, trichomes (simple and 2- or 3-rayed), to 1.5 mm, surfaces moderately pubescent, trichomes short-stalked, 4–7-rayed, 0.3–0.6 mm. **Cauline leaves:** 9–17, concealing stem proximally; blade auricles absent, surfaces of distalmost leaves moderately pubescent. **Racemes** 8–35-flowered, usually unbranched. **Fruiting pedicels** reflexed, straight, 1.5–2 mm, pubescent, trichomes appressed, branched. **Flowers** pendent at anthesis; sepals pubescent; petals yellowish proximally and brick-red distally or, rarely, one color throughout, 9–10 × 0.8–1 mm, glabrous; pollen ellipsoid. **Fruits** (immature) reflexed, often appressed to rachis, not secund, straight, edges parallel, ca. 4 cm; valves pubescent throughout; style ca. 0.3 mm. **Seeds** not seen.

Flowering May. Crevices and ledges of calcareous rock outcrops; 2200–2400 m; Calif.

Boechera yorkii is immediately recognizable by its yellowish to brick-red petals and extremely short (1.5–2 mm), reflexed fruiting pedicels. It is known only from Last Chance Mountains in Inyo County.

12. **CUSICKIELLA** Rollins, J. Jap. Bot. 63: 68. 1988 • [For William C. Cusick, 1842–1922, Oregon plant collector] [E]

Ihsan A. Al-Shehbaz

Perennials; (cespitose or pulvinate, taprooted, caudex much-branched, woody, with persistent leaf remains); scapose or subscapose; glabrous or pubescent, trichomes simple, often mixed with stalked, forked, or subdendritic ones. **Stems** erect, unbranched. **Leaves** basal and sometimes cauline; sessile; basal (terminating caudex branches, thickened at base, persistent, erect), rosulate, blade margins entire; cauline (absent or few as bracts), blade (base not auriculate), margins entire. **Racemes** (corymbose, proximal flowers sometimes bracteate), elongated in fruit. **Fruiting pedicels** divaricate or ascending, slender. **Flowers:** sepals (erect to ascending), oblong; petals white or yellowish, spatulate or oblanceolate, (slightly longer than sepals, base attenuate to short claw); stamens slightly tetradynamous; filaments dilated basally; anthers ovate; nectar glands: lateral annular, median glands confluent with lateral. **Fruits** sessile or minutely stipitate, ovoid or ellipsoid, not torulose, terete or 4-angled, (thick and leathery); valves each with prominent or obscure midvein, (sometimes keeled), puberulent or glabrous;

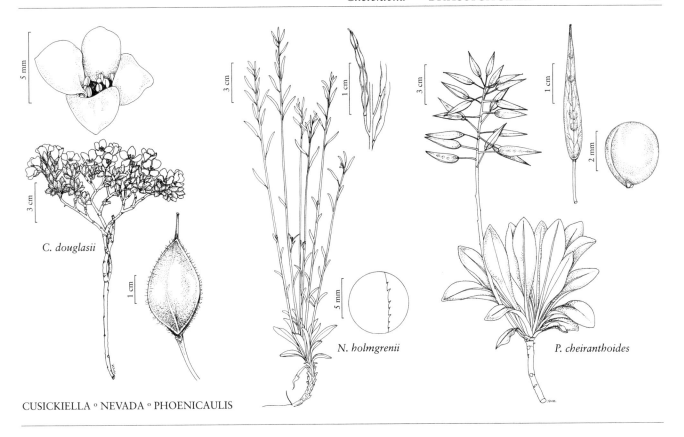

CUSICKIELLA ∘ NEVADA ∘ PHOENICAULIS

replum rounded; septum complete; ovules 4 per ovary; stigma capitate. **Seeds** aseriate, plump, not winged, ovoid to oblong; seed coat (smooth), not mucilaginous when wetted; cotyledons incumbent or obliquely so.

Species 2 (2 in the flora): w United States.

Species of *Cusickiella* superficially resemble those of *Draba*, but the two genera are clearly unrelated. *Cusickiella* is easily distinguished from *Draba* by having incumbent cotyledons and four seeds per fruit, whereas *Draba* has accumbent cotyledons and more seeds per fruit.

1. Racemes ebracteate; fruit valves rounded on back, each with obscure midvein; leaf blades (0.3–)0.5–1.2(–1.4) cm, trichomes on surfaces not setiform; petals white .... 1. *Cusickiella douglasii*
1. Racemes bracteate basally; fruit valves keeled on back, each with prominent midvein; leaf blades 0.2–0.5 cm, trichomes on surfaces setiform; petals pale yellow .... 2. *Cusickiella quadricostata*

1. **Cusickiella douglasii** (A. Gray) Rollins, J. Jap. Bot. 63: 69. 1988 [E] [F]

Draba douglasii A. Gray, Proc. Amer. Acad. Arts 7: 328. 1867; *Braya oregonensis* A. Gray; *Cusickia douglasii* (A. Gray) O. E. Schulz; *D. crockeri* Lemmon; *D. douglasii* var. *crockeri* (Lemmon) C. L. Hitchcock

Plants scapose; glabrous or puberulent, trichomes crisped, simple or branched. **Stems** 0.1–0.5 dm, glabrous or puberulent. **Basal leaves** thick, leathery; blade oblanceolate to oblong, (0.3–)0.5–1.2(–1.4) cm × (0.5–)1–2(–2.5) mm, margins ciliate, (trichomes simple and forked, to 0.3 mm), apex acute, surfaces glabrous or sparsely puberulent, trichomes simple and stalked, forked or subdendritic, crisped (not setiform); midvein prominent. **Cauline leaves** absent. **Racemes** 2–20-flowered. **Fruiting pedicels** 2–6(–8) mm, glabrous or puberulent. **Flowers:** sepals 2.5–4 × 1–1.5 mm; petals white, spatulate, 3.5–6 × 1.5–2 mm, attenuate to clawlike base to 1 mm; filaments 2–3 mm; anthers 0.5–0.6 mm; gynophore 0–0.5 mm. **Fruits** ovoid to ellipsoid, terete, (2–)3–6(–7) × 2.5–3 mm; valves each with obscure midvein, rounded on back, glabrous or puberulent; style (0.5–)1–2(–2.5) mm. **Seeds** brown, 2–3 × 1.5–2 mm.

Flowering Apr–Aug. Scree, serpentine ridges, red barren hillsides, rocky ridges, loose volcanic hillsides; 1500–2500 m; Calif., Idaho, Nev., Oreg., Utah, Wash.

2. Cusickiella quadricostata (Rollins) Rollins, J. Jap. Bot. 63: 69. 1988 [C][E]

Draba quadricostata Rollins, Contr. Dudley Herb. 3: 366. 1946

Plants subscapose; puberulent, trichomes crisped, branched, mixed with setiform ones. **Stems** 0.2–0.6 dm, pubescent. **Basal leaves** thin; blade linear to oblong, 0.2–0.5 cm × 0.5–1(–1.5) mm, margins ciliate, (trichomes setiform, to 0.8 mm), apex acute, often apiculate, surfaces densely puberulent, trichomes minute, crisped, subdendritic, mixed with fewer setiform and simple ones; midvein obscure. **Cauline leaves** 1–5 (as bracts). **Racemes** 5–17-flowered, (bracteate basally). **Fruiting pedicels** 2–5 mm, glabrous or puberulent. **Flowers:** sepals 2.5–4 × 1–1.5 mm; petals pale yellow, spatulate or oblanceolate, 3–4.5 × 1.5–2 mm, attenuate to clawlike base to 1 mm; filaments 2–3 mm; anthers 0.4–0.5 mm; gynophore obsolete. **Fruits** ovoid, 4-angled, 3–5 × 2–3 mm; valves each with prominent midvein, keeled on back, puberulent; style 0.5–1 mm. **Seeds** brown, 1.8–2.5 × 1–1.5 mm.

Flowering May–Jun. Rocky slopes and ridges, pinyon-juniper woods, loose granitic rocks, rocky crevices and flats, dwarf sagebrush slopes; of conservation concern; 2400–2800 m; Calif., Nev.

Cusickiella quadricostata is narrowly restricted in California to Mono County and in Nevada to Lyon and Mineral counties.

13. NEVADA N. H. Holmgren, Brittonia 56: 240, fig. 1. 2004 • [For the state of Nevada, where endemic] [C][E]

Ihsan A. Al-Shehbaz

Perennials; (cespitose, caudex simple or few-branched); not scapose; pubescent or glabrous, trichomes simple. **Stems** erect, unbranched or branched distally. **Leaves** basal and cauline; petiolate or sessile; basal rosulate, petiolate, blade margins entire; cauline sessile, blade (base not auriculate), margins entire. **Racemes** (corymbose, several-flowered), considerably elongated in fruit. **Fruiting pedicels** ascending to suberect, slender. **Flowers:** sepals (erect), oblong-ovate, (glabrous); petals white, obovate to spatulate, (longer than sepals, claw obscurely differentiated from blade, apex obtuse); stamens slightly tetradynamous; filaments not or slightly dilated basally; anthers ovate, (apex obtuse); nectar glands confluent, subtending bases of stamens. **Fruits** sessile or subsessile, linear to linear-lanceolate, torulose, terete; valves each with obscure midvein extending to middle, glabrous; replum rounded; septum complete; ovules 6–12 per ovary; stigma capitate. **Seeds** uniseriate, plump, not winged, narrowly oblong; seed coat not mucilaginous when wetted; cotyledons incumbent. $x = 7$.

Species 1: Nevada.

1. Nevada holmgrenii (Rollins) N. H. Holmgren, Brittonia 56: 241. 2004 [C][E][F]

Smelowskia holmgrenii Rollins, Contr. Gray Herb. 171: 50. 1950

Stems (0.8–)1–2.5(–3.2) dm, glabrous. **Basal leaves:** petiole to 2 cm (ciliate, trichomes to 0.4 mm); blade (rigid) narrowly to broadly oblanceolate, 1–5 cm × 3–6(–9) mm, base attenuate to cuneate, apex acute, surfaces glabrous or, rarely, margins ciliate to apex. **Cauline leaves** 3–5; blade linear to lanceolate, (0.6–)1–1.8(–2.5) cm × 1–3 mm, surfaces glabrous, sometimes margins ciliate, apex acute. **Racemes** somewhat lax in fruit. **Fruiting pedicels** sometimes appressed to rachis, straight, 3–9 mm, glabrous. **Flowers:** sepals 1.7–2.5 × 0.8–1.5 mm; petals 2.8–4 × 1.5–2 mm, attenuate to clawlike base; filaments 1.5–2.5 mm; anthers 0.4–0.6 mm. **Fruits** ascending to erect, (0.7–)1–1.8(–2.4) cm × 1–1.5 mm; style 0.1–0.4 mm. **Seeds** 2–3 × 0.7–1 mm. $2n = 14$.

Flowering Jun–Jul. Rock outcrops, crevices, metamorphic rocks, cracks, ledges, talus of schist; of conservation concern; 1900–3500 m; Nev.

Nevada holmgrenii is restricted to central Nevada in Humboldt, Lander, Nye, Pershing, and White Pine counties.

14. PHOENICAULIS Nuttall in J. Torrey and A. Gray, Fl. N. Amer. 1: 89. 1838 • [Greek *phoenix*, date palm, and *kaulos*, stem, alluding to petiolar remains] E

Ihsan A. Al-Shehbaz

Perennials; (caudex well-developed, woody, covered with persistent petiolar remains); not scapose; glabrous or pubescent, trichomes finely dendritic. **Stems** erect, usually unbranched, rarely branched distally. **Leaves** basal and cauline; petiolate or sessile; basal (persistent), rosulate, long-petiolate, blade margins entire; cauline sessile, blade (base auriculate), margins entire. **Racemes** (corymbose, several-flowered), considerably elongated in fruit. **Fruiting pedicels** divaricate, slender. **Flowers**: sepals (erect), oblong, (lateral pair saccate basally); petals purple or lavender, spatulate to oblanceolate, (longer than sepals, apex obtuse); stamens tetradynamous; filaments not dilated basally; anthers oblong, (apex obtuse); nectar glands confluent, lateral annular. **Fruits** sessile or stipitate, lanceolate to linear, not torulose, latiseptate; valves each with prominent midvein (lateral veins often conspicuous), glabrous; replum rounded; septum complete, (opaque); ovules (6–)8–16(–18) per ovary; stigma capitate. **Seeds** uniseriate, slightly flattened, not winged, oblong to broadly ovate; seed coat (smooth), not mucilaginous when wetted; cotyledons accumbent. $x = 7$.

Species 1: w United States.

1. **Phoenicaulis cheiranthoides** Nuttall in J. Torrey and A. Gray, Fl. N. Amer. 1: 89. 1838 E F

Arabis pedicellata A. Nelson; *Parrya cheiranthoides* (Nuttall) Jepson; *P. cheiranthoides* var. *lanuginosa* (S. Watson) M. Peck; *P. menziesii* (Hooker) Greene var. *glabra* Jepson; *P. menziesii* var. *lanuginosa* S. Watson; *P. pedicellata* (A. Nelson) Tidestrom; *Phoenicaulis cheiranthoides* var. *glabra* (Jepson) Abrams; *P. cheiranthoides* subsp. *lanuginosa* (S. Watson) Abrams; *P. cheiranthoides* var. *lanuginosa* (S. Watson) Rollins; *P. pedicellata* (A. Nelson) A. Heller

Stems simple or few from base (caudex), (0.8–)1.2–2.5(–3) dm, glabrous. **Basal leaves**: petiole (1–)2–5.5(–8) cm, thickened, glabrous at base; blade linear-oblanceolate to obovate, (2–)3–7(–10) cm × (5–)8–20(–26) mm, base attenuate, apex obtuse to acute, surfaces tomentose. **Cauline leaves** 5–12, not overlapping; blade ovate to oblong or linear lanceolate, 0.7–2 cm × 2–6 mm, surfaces usually densely to sparsely pubescent, rarely glabrate. **Racemes** glabrous. **Fruiting pedicels** divaricate, (6–)10–30(–35) mm. **Flowers**: sepals 4–6 × 1.5–2.2 mm, glabrous; petals 9–13(–15) × 2.5–4 mm, attenuate to base; filaments 4–5 mm; anthers 1–1.5 mm; gynophore obsolete or, rarely, to 2 mm. **Fruits** divaricate, (2–)3–6(–9) cm × (2–)3–5(–6) mm; style 0.5–2 mm. **Seeds** (2–)2.5–3.2 × (1.5–)1.8–2.5 mm. $2n = 28$.

Flowering mid Apr–mid Jun. Rocky scree, rocky basaltic slopes, volcanic rubble, barren clay slopes, rimrocks, scablands, ledges of metamorphic rock outcrops, rocky crevices and open knolls, sandy banks, gravelly meadows, grassy or gravelly hillsides, sagebrush scrub and slopes, alpine slopes, volcanic boulders; 700–2800 m; Calif., Idaho, Nev., Oreg., Utah, Wash.

15. POLYCTENIUM Greene, Leafl. Bot. Observ. Crit. 2: 219. 1912 • [Greek *polys*, many, and *ctenos*, comb, alluding to leaves] E

Ihsan A. Al-Shehbaz

Perennials; (cespitose); not scapose; pubescent at least basally, trichomes dendritic, mixed with fewer, 2-rayed and larger, simple ones. **Stems** erect to ascending, branched distally. **Leaves** cauline; petiolate or subsessile; blade (base not auriculate), not rosulate; margins pinnatifid. **Racemes** (corymbose, several-flowered), congested or considerably elongated in fruit. **Fruiting**

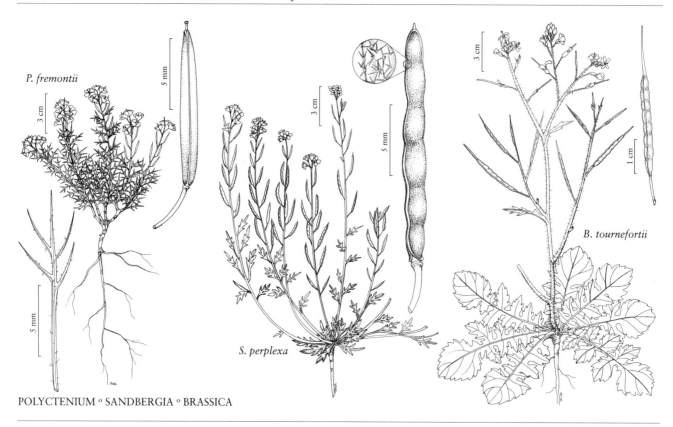

POLYCTENIUM ○ SANDBERGIA ○ BRASSICA

pedicels divaricate-ascending, slender. **Flowers:** sepals (erect or slightly ascending), oblong-ovate; petals white or purplish, obovate to oblanceolate, (longer than sepals, claw undifferentiated from blade, apex obtuse); stamens slightly tetradynamous; filaments not dilated basally; anthers ovate, (apex obtuse); nectar glands confluent, subtending bases of stamens. **Fruits** (siliques or silicles), sessile, usually linear to oblong, rarely ovate-oblong, not torulose, slightly to strongly angustiseptate; valves each not veined, glabrous or, rarely, sparsely puberulent; replum rounded; septum complete; ovules 30–46 per ovary; stigma capitate, entire. **Seeds** uniseriate, plump, not winged, oblong; seed coat not mucilaginous when wetted; cotyledons incumbent.

Species 1: nw United States.

SELECTED REFERENCE Rollins, R. C. 1938. *Smelowskia* and *Polyctenium*. Rhodora 40: 294–305.

1. Polyctenium fremontii (S. Watson) Greene, Leafl. Bot. Observ. Crit. 2: 219. 1912 [E] [F]

Smelowskia fremontii S. Watson, Proc. Amer. Acad. Arts 11: 123. 1876; *Braya pectinata* Greene; *Polyctenium bisulcatum* Greene; *P. fremontii* var. *bisulcatum* (Greene) Rollins; *P. fremontii* var. *confertum* Rollins; *P. glabellum* Greene; *P. williamsiae* Rollins; *S. fremontii* var. *bisulcata* (Greene) O. E. Schulz; *S. fremontii* var. *glabella* (Greene) O. E. Schulz

Plants with woody base, sparsely to densely pubescent. **Stems** few to numerous from base, (0.2–)0.5–1.6(–2) dm, usually sparsely to densely pubescent proximally, sometimes glaucous and glabrous distally. **Cauline leaves** often persistent in subsequent seasons as pectinate spines; petiolate proximally, subsessile distally; petiole winged, 0–3(–5) mm; blade linear (in outline), (0.5–)1–2.2(–2.8) cm, rigid, surfaces sparsely to densely pubescent, trichomes dendritic, margin and apex often with simple or forked ones; lobes (2–)3–4(–5) each side, with prominent midvein and narrow band of blade; terminal lobe (0.2–)0.4–0.7(–1) cm × (0.3–)0.5–1(–1.5) mm, as long as or slightly longer than adjacent lateral lobes. **Fruiting pedicels** (2–)4–7(–10) mm, sparsely pubescent or glabrous. **Flowers:** sepals 1.5–2 × 0.9–1.1 mm, hyaline-margined, glabrous or sparsely pubescent; petals 3–4 × 1.5–2 mm, attenuate to base; filaments 1.2–2 mm; anthers 0.4–0.5 mm. **Fruits** erect to ascending, (0.2–)0.4–1.3(–1.8) cm × 0.9–2 mm; style (0.2–)0.4–0.7(–1) mm. **Seeds** 0.7–0.9 × 0.3–0.5 mm.

Flowering late May–early Aug. Mud flats, dry meadows, sagebrush areas, edge of vernal pools, shallow soil on basalt, dry streambeds and swales, gravel bars, rocky wash; 1000–2700 m; Calif., Idaho, Nev., Oreg.

Polyctenium fremontii is highly variable in fruit size and the compactness of the fruiting raceme, but in habit, flower size and color, leaf morphology, indumentum, fruiting pedicel length and orientation, number of ovules per ovary, and basically every other aspect of the plants, it is quite constant. If one examines only the types of those two taxa and that of *P. fremontii*, it seems that perhaps two or three taxa might be recognized. Upon careful study of extensive material, one realizes that only one taxon, instead of three or more, is represented. The alleged differences between *P. fremontii* and *P. williamsiae* in characters other than fruit morphology do not hold. As for fruit size, it was said to be 2–4 × 2–2.5 mm in *P. williamsiae* and (4–)6–13(–20) × 1–2 mm in *P. fremontii*. Fruit lengths in material annotated by Rollins as *P. fremontii* are 2–7 mm in *Tiehm 8108* and 3.5–11 mm in *Ertter 5726*, both at GH. Furthermore, the compactness of the infructescence can be equally variable, and in the holotype of var. *confertum* there are 12–15 pedicels along 1 cm in the middle of the rachis, whereas in *Ertter 5726* (GH) there are 6–12. On one sheet, *Schoolcraft 1287* (GH), compact and lax racemes and relatively short (3 mm) and longer (7 mm) fruits are represented. The variation in fruit length and width depends largely upon the number of ovules maturing into seeds, and in plants with very short fruits, including the type collection of *P. williamsiae*, none of the ovules matured into seeds, whereas in those with longest and narrowest fruits almost all ovules matured into seeds. Regardless of how long the fruit is or how many ovules mature into seeds, the ovule number is fairly constant throughout the range of the species. In my opinion, except for the type species of *Polyctenium*, all of the other taxa recognized in this genus do not represent biologically distinct entities.

To my knowledge, *Polyctenium fremontii* is known from counties in California (Lassen, Modoc, Mono, Siskiyou), Idaho (Gooding), Nevada (Churchill, Douglas, Humboldt, Lyon, Mineral, Washoe), and Oregon (Crook, Deschutes, Harney, Klamath, Lake, Malheur).

16. SANDBERGIA Greene, Leafl. Bot. Observ. Crit. 2: 136. 1911 • [For John Herman Sandberg, 1848–1917, Swedish-born American botanist and physician who collected extensively in the Pacific Northwest] [E]

Ihsan A. Al-Shehbaz

Biennials or perennials; (caudex simple or branched, covered with persistent leaf remains); not scapose; pubescent throughout, trichomes shortly stalked or subsessile, cruciform, Y-shaped, or forked. **Stems** erect or decumbent, unbranched or branched distally. **Leaves** basal and cauline; petiolate or sessile; basal usually rosulate, petiolate, blade margins entire, dentate, or lyrate-pinnatifid, (apex obtuse to acute); cauline sessile, blade (base attenuate, not auriculate), margins entire, subentire, dentate, or pinnatifid. **Racemes** (corymbose, several-flowered), considerably elongated in fruit. **Fruiting pedicels** ascending to subdivaricate, (straight), slender, (terete). **Flowers:** sepals (erect), oblong; petals white, oblanceolate-spatulate, (longer than sepals, claw obscurely differentiated from blade, apex rounded); stamens slightly tetradynamous; filaments not dilated basally, (slender); anthers ovate or oblong, (apex obtuse); nectar glands confluent, subtending bases of stamens. **Fruits** subsessile or shortly stipitate (gynophore less than 1 mm), linear, slightly to strongly torulose, subterete to strongly latiseptate; valves each without midvein or with obscure one on proximal ½, sparsely to densely pubescent or glabrescent; replum rounded; septum complete; ovules 12–30 per ovary; (style obsolete or distinct); stigma capitate. **Seeds** uniseriate, plump, not winged, oblong; seed coat (minutely reticulate), not mucilaginous when wetted; cotyledons incumbent. $x = 7$.

Species 2 (2 in the flora): nw North America.

Although both species of *Sandbergia* were placed by R. C. Rollins (1993) in *Halimolobos*, the two genera are not closely related. *Sandbergia* is most closely related to *Boechera*, whereas *Halimolobos* is sister to *Mancoa* and *Sphaerocardamum* Schauer in the tribe Halimolobeae. For an account of the generic boundaries of *Sandbergia*, see under 60. *Halimolobos*.

SELECTED REFERENCE Al-Shehbaz, I. A. 2007b. The North American genus *Sandbergia* (Boechereae, Brassicaceae). Harvard Pap. Bot. 12: 425–427.

1. Cauline leaf blade margins usually entire, rarely denticulate; fruits slightly torulose . . . 1. *Sandbergia whitedii*
1. Cauline leaf blade margins coarsely dentate to pinnatifid; fruits strongly torulose . . . . . 2. *Sandbergia perplexa*

1. Sandbergia whitedii (Piper) Greene, Leafl. Bot. Observ. Crit. 2: 137. 1911 [E]

Arabis whitedii Piper, Bull. Torrey Bot. Club 28: 39. 1901; *Halimolobos whitedii* (Piper) Rollins

Perennials; densely and uniformly pubescent, trichomes shortly stalked, minute, Y-shaped or cruciform, mixed with simple or 1 or more ray-forked ones. **Stems** simple or few from caudex, erect, branched (several) distally, (1.5–)2.5–5.5(–6.5) dm, densely pubescent throughout. **Basal leaves** often rosulate; petiole (1–)2–5(–7) cm; blade spatulate to linear-oblanceolate, (1–)1.7–5 cm × (3–)4–9(–12) mm, margins entire or dentate, surfaces densely pubescent. **Cauline leaves** 3–7 (widely spaced); blade linear-oblanceolate, 2–7 cm × 1–5(–8) mm, margins usually entire, rarely minutely sparsely denticulate, apex obtuse, surfaces densely pubescent. **Fruiting pedicels** (5–)8–14(–17) mm, densely pubescent. **Flowers**: sepals 2.5–3 × 1–1.5 mm; petals (4–)5–6.5(–7.5) × (1.2–)1.5–2(–2.2) mm; filaments 1.5–3 mm; anthers oblong, 0.6–0.8 mm. **Fruits** suberect, (not appressed to rachis), straight, slightly torulose, subterete, (1–)1.3–1.8(–2) cm × 0.8–1 mm; valves each with obscure midvein on proximal ½, densely pubescent; ovules 20–30 per ovary; style 0.2–0.4 mm. **Seeds** 1–1.2 × 0.6–0.7 mm.

Flowering Apr–Jul. Dry sagebrush scabland, gravelly hillsides, basaltic talus, dry sandy slopes, alpine meadows, cliffs, ridge crests; 500–1200 m; B.C.; Wash.

In Washington state, *Sandbergia whitedii* appears to be restricted to Chelan, Douglas, Grant, Kittitas, Lincoln, and Okanogan counties.

2. Sandbergia perplexa (L. F. Henderson) Al-Shehbaz, Harvard Pap. Bot. 12: 426. 2007 [E] [F]

Sisymbrium perplexum L. F. Henderson, Bull. Torrey Bot. Club 27: 342. 1900; *Halimolobos perplexus* (L. F. Henderson) Rollins; *H. perplexus* var. *lemhiensis* C. L. Hitchcock; *Sophia perplexa* (L. F. Henderson) Rydberg

Biennials or perennials; densely to sparsely pubescent, trichomes subsessile, minute, Y-shaped or cruciform, mixed with simple or forked ones. **Stems** usually several from caudex, rarely simple, erect to decumbent, unbranched or branched distally, (1–)1.3–4(–5) dm, densely pubescent throughout or glabrescent distally. **Basal leaves** rosulate; petiole 0.5–2.5 cm; blade narrowly to broadly oblanceolate, 2–5 cm × 5–15 mm, margins lyrate-pinnatifid to coarsely dentate, apex obtuse to acute, surfaces densely pubescent. **Cauline leaves** 4–12; blade oblong to oblanceolate, 0.6–2 cm × 2–6 mm (base attenuate, not auriculate), margins coarsely dentate to pinnatifid, or (distal) subentire or dentate, apex obtuse to subacute, surfaces densely to sparsely pubescent. **Fruiting pedicels** 4–15 mm, densely pubescent to glabrescent. **Flowers**: sepals 1.5–2 × 0.8–1.2 mm; petals (3–)4–7 × 1.2–2 mm; filaments 2–3 mm; anthers ovate, 0.5–0.6 mm. **Fruits** ascending to suberect, straight to slightly tortuous, strongly torulose, strongly latiseptate, (1–)1.5–2(–2.5) cm × 0.8–1 mm; valves each with obscure midvein, densely pubescent to glabrescent; ovules 12–20 per ovary; style 0.2–1.3 mm. **Seeds** 1–1.3 × 0.5–0.6 mm. $2n = 14$.

Flowering Apr–Jul. Sagebrush flats, pine woods, basaltic gravel and outcrop, sandy banks, rocky hillsides, granitic talus; 300–1500 m; Idaho, Mont., Wash.

I have seen limited material of var. *lemhiensis*, and all the differences given by R. C. Rollins (1993) to separate it from var. *perplexa* (e.g., style and pedicel length, density of indumentum) are quantitative characters that show continuous, uncorrelated variation. *Sandbergia perplexa* is known from counties in Idaho (Adams, Butte, Custer, Idaho, Lemhi, Valley), Montana (Beaverhead), and Washington (Douglas).

f. BRASSICACEAE Burnett tribe BRASSICEAE de Candolle, Mém. Mus. Hist. Nat. 7: 242. 1821

Annuals, biennials, or perennials [shrubs]; eglandular. **Trichomes** absent or simple. **Cauline leaves** petiolate or sessile; blade base auriculate or not, margins entire, dentate, serrate, or pinnately lobed. **Racemes** usually ebracteate, often elongated in fruit. **Flowers** actinomorphic; sepals erect, ascending, or spreading, lateral pair saccate or not basally; petals white, cream, yellow, pink, lilac, lavender, or purple, claw present, often distinct; filaments unappendaged, not winged; pollen 3-colpate. **Fruits** silicles or siliques, dehiscent or indehiscent, usually segmented, usually latiseptate or terete (subterete or 4-angled in *Erucastrum*) [angustiseptate]; ovules (1–)2–276[–numerous] per ovary; style usually distinct (absent in *Cakile*, obscure in *Carrichtera*, obsolete in *Eruca*); stigma entire or strongly 2-lobed (sometimes slightly 2-lobed in *Cakile*). **Seeds** biseriate, uniseriate, or aseriate; cotyledons usually conduplicate, rarely accumbent or incumbent (in *Cakile*).

Genera 46, species ca. 245 (13 genera, 28 species in the flora): North America, Eurasia, n Africa; introduced widely.

The generic boundaries in Brassiceae are largely artificial, and the number of genera may be substantially reduced.

17. BRASSICA Linnaeus, Sp. Pl. 2: 666. 1753; Gen. Pl. ed. 5, 299. 1754 • Cabbage, cole, mustard, turnip [Latin name for cabbage]

Suzanne I. Warwick

Annuals, biennials, or, rarely, perennials; not scapose; glabrous, glabrescent, or pubescent. **Stems** erect, unbranched or branched distally. **Leaves** basal and cauline; petiolate or sessile; basal (persistent in *B. tournefortii*), rosulate or not, petiolate, blade margins entire, dentate, or lyrate-pinnatifid; cauline petiolate or sessile, blade (base sometimes auriculate or amplexicaul), margins entire, dentate, lobed, or sinuate-serrate. **Racemes** (corymbose), considerably elongated in fruit. **Fruiting pedicels** erect, spreading, ascending or divaricately-ascending, often slender. **Flowers:** sepals usually erect or ascending, rarely spreading, oblong [ovate], lateral pair usually saccate basally; petals yellow to orange-yellow [rarely white], obovate, ovate, elliptic, or oblanceolate, claw often differentiated from blade, (sometimes attenuate basally, apex rounded or emarginate); stamens tetradynamous; filaments slender; anthers oblong or ovate, (apex obtuse); nectar glands confluent or not, median glands present. **Fruits** siliques, dehiscent, sessile or stipitate, segments 2, linear, torulose or smooth, terete, 4-angled, or latiseptate; (terminal segment seedless or 1–3-seeded, usually filiform or conic, rarely cylindrical); valves each prominently 1-veined, glabrous; replum rounded; septum complete; ovules [4–]10–50 per ovary; stigma entire or 2-lobed. **Seeds** uniseriate, plump, not winged, globose; seed coat (reticulate or reticulate-alveolate), mucilaginous or not when wetted; cotyledons conduplicate. $x = 7, 8, 9, 10, 11$.

Species 35 (8 in the flora): introduced; sw Europe, sw Asia, e, nw Africa; introduced also in Mexico, West Indies, Central America, South America, Atlantic Islands, Pacific Islands (New Zealand), Australia.

Crops of *Brassica* are the most important economic plants of the family. Probably, the earliest known utilization of mustards dates from Sanskrit records in India to 3000 B.C., but there is archaeological evidence suggesting that cultivation of cabbage in coastal northern Europe was occurring nearly 8000 years ago. *Brassica* crops include oilseeds, food crops (e.g., *B. juncea*,

Asian vegetables; *B. oleracea*, cole crops; *B. rapa*, Chinese cabbages), fodder for animals, and condiments (*B. juncea* or *B. nigra*). The latter two species have also been used for medicinal purposes (I. A. Al-Shehbaz 1985). In addition to being noxious weeds, some species of *Brassica* are harmful or poisonous to humans and livestock (Al-Shehbaz).

Historically, native peoples of North America have used a number of "wild" *Brassica* species for both food and medicinal purposes (T. Arnason et al. 1981; H. A. Jacobson et al. 1988): *Brassica* species—young shoots cooked as greens by Iroquois and Malecite Indian tribes; *B. nigra*—seeds ground and used as snuff to cure head colds by the Meskwaki, and leaves used to relieve toothaches and headaches by the Mohegans; *B. napus*—bark used to treat colds, cough, grippe, and smallpox by the Micmac, and used for chilblains by the Rappahannock; *B. oleracea*—used for headaches by the Rappahannock; and *B. rapa*—used as medicine by the Bois Fort Chippewa.

SELECTED REFERENCES Diederichsen, A. 2001. *Brassica*. In: P. Hanelt, ed. 2001. Mansfeld's Encyclopedia of Agricultural and Horticultural Crops.... 6 vols. Berlin and New York. Vol. 3, pp. 1435–1465. Gómez-Campo, C. 1999. Taxonomy. In: C. Gómez-Campo, ed. 1999b. Biology of *Brassica* Coenospecies. Amsterdam. Pp. 3–32. Prakash, S. and K. Hinata. 1980. Taxonomy, cytogenetics and origin of crop brassicas, a review. Opera Bot. 55: 1–57. Snogerup, S., M. Gustafsson, and R. von Bothmer. 1990. *Brassica* sect. *Brassica* (Brassicaceae). 1. Taxonomy and variation. Willdenowia 19: 271–365.

1. Cauline leaves sessile, blade bases auriculate and/or amplexicaul.
   2. Biennials or perennials; petals (15–)18–25(–30) mm; terminal segments of fruits (3–)4–11 mm. . . . . 6. *Brassica oleracea*
   2. Annuals or biennials; petals 6–16 mm; terminal segments of fruits (5–)8–22 mm.
      3. Flowers usually not overtopping buds, rarely at same level, when open; petals pale yellow, 10–16 mm; terminal segments of fruits (5–)9–16 mm . . . . 4. *Brassica napus*
      3. Flowers overtopping or equaling buds when open; petals deep yellow, 6–11(–13) mm; terminal segments of fruits 8–22 mm . . . . 7. *Brassica rapa*
1. Cauline leaves petiolate or sessile, blade bases tapered, not auriculate or amplexicaul.
   4. Fruits and pedicels erect, ± appressed to rachises; fruits 10–25(–27) mm, not torulose; fruiting pedicels (2–)3–5(–6) mm . . . . 5. *Brassica nigra*
   4. Fruits and pedicels spreading to ascending, not appressed to rachises; fruits often 2 cm+, torulose; fruiting pedicels (6–)8–20 mm.
      5. Fruits stipitate, gynophores 1.5–4(–5) mm, terminal segments 0.5–2.5(–3) mm; basal leaf blade margins entire or dentate. . . . . 1. *Brassica elongata*
      5. Fruits sessile or stipitate, gynophores to 1 mm, terminal segments (4–)5–20 mm; basal leaf blade margins lyrate to pinnatisect, or pinnatifid to pinnately lobed.
         6. Basal leaves persistent, blades with 4–10 lobes each side, surfaces hirsute; petals 4–7 × 1.5–2(–2.5) mm . . . . 8. *Brassica tournefortii*
         6. Basal leaves deciduous, blades with 1–3 (or 4) lobes each side, surfaces glabrous or nearly so; petals (7–)9–13 × 3–7.5 mm.
            7. Fruits stipitate (gynophore 1–1.5 mm), 1.5–3 cm × 1.5–2 mm, terminal segment 3–6 mm . . . . 2. *Brassica fruticulosa*
            7. Fruits sessile, (2–)3–5(–6) cm × 2–5 mm, terminal segment (4–)5–10 (–15) mm. . . . . 3. *Brassica juncea*

1. **Brassica elongata** Ehrhart, Beitr. Naturk. 7: 159. 1792 [I]

Biennials or perennials; (short-lived, often woody basally); glabrous or hirsute. **Stems** (several from base), branched basally, 5–10 dm, (usually glabrous, rarely sparsely hirsute). **Basal leaves**: blade (usually bright green), obovate to elliptic (not lobed), (3–)5–20(–30) cm × (5–)10–35(–60) mm, (base cuneate), margins subentire to dentate, (surfaces glabrous or often with trichomes minute, tubercled-based, curved, coarse). **Cauline leaves** (distal) shortly petiolate; blade (oblong or lanceolate, to 10 cm) base not auriculate or amplexicaul. **Racemes** paniculately branched. **Fruiting pedicels** spreading to divaricately ascending, (6–)8–18 mm. **Flowers:** sepals 3–4(–4.5) × 1–1.5 mm; petals bright yellow to orange-yellow, obovate, (5–)7–10 × 2.5–3.5(–4) mm, claw 2.5–4 mm, apex rounded; filaments 3.5–4.5 mm; anthers 1–1.5 mm; gynophore 1.5–4(–5) mm in fruit. **Fruits**

(stipitate), spreading to ascending (not appressed to rachis), torulose, terete, (1.5–)2–4(–4.8) cm × (1–)1.5–2 mm; valvular segment with (2–)5–11(–13) seeds per locule, (1.2–)1.6–4(–4.5) cm, terminal segment seedless, 0.5–2.5(–3) mm. **Seeds** grey to brown, 1–1.6 mm diam.; seed coat reticulate, mucilaginous when wetted. $2n = 22$.

Flowering Jun–Jul. Roadsides, disturbed ground, adjacent open juniper and sagebrush desert areas; 0–2700 m; introduced; Nev., Oreg., Wash.; Europe; Asia; n Africa; introduced also in Australia.

The earliest North American collections of *Brassica elongata* were from ballast at Linnton, near Portland, Oregon, in 1911, and from a garden in Bingen, Klickitat County, Washington, in 1915. The species does not appear to have persisted at, or spread from, either location (R. C. Rollins and I. A. Al-Shehbaz 1986). It was next collected in 1968 from east-central Nevada, where it is now well-established in Eureka and White Pine counties, and just into Lander County, and spreading rapidly along both roadsides and adjacent high desert (Rollins 1980; Rollins and Al-Shehbaz; Rollins 1993). The semiarid region of North America appears to be a well-suited habitat for *B. elongata* and the species appears destined to become a permanent part of the flora of the Intermountain Basin (Rollins and Al-Shehbaz).

According to R. C. Rollins (1980), the Nevada plants belong to subsp. *integrifolia* (Boissier) Breistroffer, but the species is so variable that dividing it into infraspecific taxa is not practical.

2. Brassica fruticulosa Cirillo, Pl. Rar. Neapol. 2: 7. 1792 ⓘ Ⓦ

Annuals, biennials, or perennials; glabrous or nearly so. **Stems** branched distally, 3–9 dm. **Basal leaves** (early deciduous); petiole 1.5–6 cm; blade lyrate-pinnatifid, 3–10(–15) cm × 10–65 mm, lobes 1–3 (or 4) each side. **Cauline leaves** shortly petiolate; blade (often lanceolate, reduced in size distally), base tapered or cuneate, not auriculate, (apex acute). **Racemes** paniculately branched. **Fruiting pedicels** spreading to divaricately ascending, (slender), (5–)10–15(–25) mm. **Flowers:** sepals 3–8 × 1–1.7 mm; petals pale yellow, narrowly obovate, 7–15 × 3–4 mm, claw 2–3 mm, apex rounded; filaments 3–6 mm; anthers 1.5–2 mm; gynophore 1–1.5 mm in fruit. **Fruits** (stipitate), spreading to divaricately ascending, strongly torulose, linear, subcylindric, 1.5–3 cm × 1.5–2 mm; valvular segment with 5–13 seeds per locule, 1.2–2.5 cm, terminal segment seedless or 1-seeded, (conic), 3–6 mm. **Seeds** brown or yellow, 0.6–1.2 mm diam.; seed coat finely reticulate-alveolate, not mucilaginous when wetted. $2n = 16$.

Flowering Dec–Mar. Coastal plains and basins, deserts, valleys; 0–300 m; introduced; Calif.; s Europe; nw Africa.

Brassica fruticulosa is naturalized in Los Angeles, Riverside, San Bernardino, and San Mateo counties.

3. Brassica juncea (Linnaeus) Czernajew, Consp. Pl. Charcov., 8. 1859 • Chinese or brown or Indian or leaf mustard, mustard-greens ⓘ Ⓦ

Sinapis juncea Linnaeus, Sp. Pl. 2: 668. 1753; *Brassica japonica* (Thunberg) Siebold ex Miquel; *B. juncea* var. *crispifolia* L. H. Bailey; *B. juncea* var. *japonica* (Thunberg) L. H. Bailey

Annuals; (±glaucous), ±glabrous. **Stems** branched distally, 2–10 dm. **Basal leaves** (early deciduous); petiole (1–)2–8(–15) cm; blade pinnatifid to pinnately lobed, (4–)6–30(–80) cm × 15–150(–280) mm, lobes 1–3 each side. **Cauline leaves** usually shortly petiolate, rarely sessile; blade (oblong or lanceolate, reduced in size distally), base tapered or cuneate, not auriculate or amplexicaul, (margins dentate to lobed). **Racemes** not paniculately branched. **Fruiting pedicels** spreading to divaricately ascending, (slender), (5–)10–15(–20) mm. **Flowers:** sepals (3.5–)4–6(–7) × 1–1.7 mm; petals pale yellow, ovate to obovate, (7–)9–13 × 5–7.5 mm, claw 3–6 mm, apex rounded or emarginate; filaments 4–7 mm; anthers 1.5–2 mm. **Fruits** (sessile); spreading to divaricately ascending to nearly erect (not appressed to rachis), torulose, subcylindrical or somewhat flattened, (2–)3–5(–6) cm × 2–5 mm; valvular segment with 6–15(–20) seeds per locule, (1.5–)2–4.5 cm, terminal segment seedless (conic), (4–)5–10(–15) mm, (tapering to slender style). **Seeds** brown or yellow, 1.2–2 mm diam.; seed coat finely reticulate-alveolate, not mucilaginous when wetted. $2n = 36$.

Flowering May–Sep. Roadsides, disturbed areas, waste places, cultivated and abandoned fields, garden escape from cultivation; 0–3000 m; introduced; Alta., B.C., Man., N.B., Nfld. and Labr., N.W.T., N.S., Ont., P.E.I., Que., Sask.; Ala., Alaska, Ariz., Ark., Calif., Colo., Conn., Del., D.C., Fla., Ga., Idaho, Ill., Ind., Iowa, Kans., Ky., La., Maine, Md., Mass., Mich., Minn., Miss., Mo., Mont., Nebr., Nev., N.H., N.J., N.Mex., N.Y., N.C., N.Dak., Ohio, Okla., Oreg., Pa., R.I., S.C., S.Dak., Tenn., Tex., Utah, Vt., Va., Wash., W.Va., Wis., Wyo.; Europe; Asia; Africa; introduced also in Mexico, West Indies, Central America, South America, Australia.

Brassica juncea is cultivated in North America primarily as a vegetable and condiment, and is currently being developed as an oilseed crop in western Canada. Its greatest diversity of forms occurs in Asia, where the species is widely cultivated as a vegetable and as an oilseed crop (I. A. Al-Shehbaz 1985). Two main variants are distinguished on the basis of seed color: oriental mustard is yellow-seeded, and brown or Indian mustard is brown-seeded. The species is an allotetraploid derived from hybridization between *B. nigra* (*n* = 8) and *B. rapa* (*n* = 10). Its center of origin is uncertain but is most likely the Middle East, with possibly independent multiple origins within overlapping ranges of the putative parental taxa (S. I. Warwick and A. Francis 1994). Specimens from Delaware, District of Columbia, and Mississippi have not been observed, but are still listed here.

4. **Brassica napus** Linnaeus, Sp. Pl. 2: 666. 1753
   • Canola, oilseed rape, rape, rapeseed, rutabaga, swede, swede rape, Swedish turnip [I]

Brassica napobrassica (Linnaeus) Miller; *B. napus* var. *oleifera* de Candolle; *B. oleracea* Linnaeus var. *napobrassica* Linnaeus

Annuals or biennials; (taproot slender or swollen); (glaucous), glabrous, glabrescent, or pubescent, (trichomes coarse). **Stems** branched distally, 3–13 dm. **Basal leaves** (rosulate when biennial); petiole (often winged), to 15 cm; blade lyrate-pinnatifid, ± pinnately lobed, 5–25(–40) cm × 20–70(–100) mm, lobes 0–6 each side, (smaller than terminal), surfaces (glaucous), glabrous or sparsely hairy when immature, glabrescent, or, rarely, pubescent. **Cauline leaves** (middle and distal) sessile; blade base auriculate or amplexicaul, (margins entire). **Racemes** not paniculately branched, (buds overtopping or equal to open flowers). **Fruiting pedicels** spreading to ascending (slender), 1–3 cm. **Flowers:** sepals (5–)6–10 × 1.5–2.5 mm; petals golden or creamy to pale yellow, broadly obovate, 10–16 × (5–)6–9(–10) mm, claw 5–9 mm, apex rounded; filaments (5–)7–10 mm; anthers 1.5–2.5 mm. **Fruits** spreading to ascending, smooth or slightly torulose, terete, (3.5–)5–10(–11) cm × (2.5–)3.5–5 mm; valvular segment with 12–20(–30) seeds per locule, (3–)4–8.5(–9.5) cm, terminal segment usually seedless, rarely 1 or 2-seeded (attenuate-conic, thin), (5–)9–16 mm. **Seeds** dark brown to black, light brown, or reddish, 1.8–2.7(–3) mm diam.; seed coat finely reticulate-alveolate, not mucilaginous when wetted. $2n = 38$.

Flowering May–Sep. Roadsides, disturbed areas, waste places, cultivated and abandoned fields, escape from cultivation; 0–500 m; introduced; Alta., B.C., Man., N.B., Nfld. and Labr., N.W.T., N.S., Ont., P.E.I., Que., Sask.; Alaska, Ariz., Ark., Calif., Colo., Conn., Del., D.C., Fla., Ga., Idaho, Ill., Ind., Iowa, Ky., La., Maine, Md., Mass., Mich., Miss., Mo., N.H., N.J., N.Y., N.C., Ohio, Okla., Oreg., Pa., Tenn., Utah, Vt., Va., Wash., W.Va., Wis.; Europe; Asia; Africa; introduced also in Mexico, Central America, South America, Atlantic Islands, Australia.

Brassica napus is both a crop and a sporadically occurring naturalized weed in North America, grown in two forms recognized by some as subspecies. Subspecies *napus* (rape, rapeseed, or canola) is an annual with slender roots widely cultivated as an oil crop and is the most commonly naturalized. Subspecies *rapifera* Metzger [= subsp. *napobrassica* (Linnaeus) Hanelt] (rutabaga, swede, or Swedish turnip) is a biennial with fleshy roots that rarely escapes from cultivation.

Although *Brassica napus* has been reported as a weed from most southeastern states, it is very likely that most reports represent misidentifications of *B. rapa* (I. A. Al-Shehbaz 1985). It is difficult to distinguish between plants of *B. napus* and *B. rapa* that lack flowers and proximal leaves.

Brassica napus is an allotetraploid derived from hybridization between the *B. oleracea* complex (*n* = 9) and *B. rapa* (*n* = 10). Its center of origin is uncertain but likely Mediterranean Europe, with molecular data supporting evidence of multiple independent origins between the parental taxa *B. oleracea* and *B. rapa* and its related *n* = 9 species (Song K. et al. 1993). Specimens from West Virginia have not been observed.

5. **Brassica nigra** (Linnaeus) W. D. J. Koch in J. C. Röhling, Deutschl. Fl. ed. 3, 4: 713. 1833
   • Black mustard [I][W]

Sinapis nigra Linnaeus, Sp. Pl. 2: 668. 1753

Annuals; sparsely to densely hirsute-hispid (at least basally, proximally rarely subglabrate). **Stems** usually branched distally, (widely spreading), 3–20 dm. **Basal leaves:** petiole to 10 cm; blade lyrate-pinnatifid to sinuate-lobed, 6–30 cm × 10–100 mm, lobes 1–3 each side, (smaller than terminal, terminal lobe ovate, obtuse). **Cauline leaves** sessile or subsessile; blade (ovate-elliptic to lanceolate, similar to basal, reduced distally and less divided), base tapered, not auriculate or amplexicaul, (margins entire to sinuate-serrate). **Racemes** not paniculately branched. **Fruiting pedicels** erect (straight), (2–)3–5(–6) mm. **Flowers:** sepals 4–6(–7) × 1–1.5 mm; petals yellow, ovate, 7–11(–13) × (2.5–)3–4.5(–5.5) mm, claw 3–6 mm, apex rounded; filaments 3.5–5 mm; anthers 1–1.5 mm. **Fruits** erect-ascending (± appressed to rachis), smooth, ± 4-angled, 1–2.5(–2.7) cm × (1.5–)2–3(–4) mm; valvular segment 2–5(–8)-seeded per locule, (0.4–)0.8–2(–2.5) cm, terminal segment seedless (linear, narrow), (1–)2–5(–6) mm. **Seeds** brown to black, 1.2–1.5(–2) mm diam.; seed coat coarsely reticulate, minutely alveolate, not mucilaginous when wetted. $2n = 16$.

Flowering Apr–Sep. Roadsides, disturbed areas, waste places, fields, orchards; 0–1500 m; introduced; Alta., B.C., N.B., Nfld. and Labr., N.S., Ont., P.E.I., Que., Sask.; Ala., Alaska, Ariz., Ark., Calif., Colo., Conn., Del., D.C., Fla., Ga., Idaho, Ill., Ind., Iowa, Kans., Ky., La., Maine, Md., Mass., Mich., Minn., Miss., Mo., Mont., Nebr., Nev., N.H., N.J., N.Mex., N.Y., N.C., N.Dak., Ohio, Okla., Oreg., Pa., R.I., S.C., S.Dak., Tenn., Tex., Utah, Vt., Va., Wash., W.Va., Wis., Wyo.; Europe; Asia; Africa; introduced also in Mexico, Central America, South America, Atlantic Islands, Australia.

Brassica nigra is widely cultivated as a condiment mustard. It is also a cosmopolitan weed especially common in the valleys of California (R. C. Rollins 1993). It occurs only sporadically in southern Canada but most frequently in Ontario and along the St. Lawrence River. Specimens from Alberta, Arkansas, Delaware, and South Carolina have not been observed.

6. Brassica oleracea Linnaeus, Sp. Pl. 2: 667. 1753
• Cabbage [I]

Brassica alboglabra L. H. Bailey

Biennials or perennials; (with slender taproot or woody caudex, becoming suffrutescent and covered with conspicuous leaf scars); (glaucous), glabrous. **Stems** branched distally, 5–10 dm. **Basal leaves:** petiole to 30 cm; blade oblong or obovate, to 45 cm × 150 mm, (fleshy), blades pinnatifid or margins dentate. **Cauline leaves** (distal) sessile; blade (oblong to lanceolate), base auriculate and amplexicaul, (margins entire). **Racemes** not paniculately branched. **Fruiting pedicels** spreading to ascending, (8–)14–25(–40) mm. **Flowers:** sepals 8–15 × 1.5–2.7 mm; petals yellow, white, or lemon yellow, ovate or elliptic, (15–)18–25(–30) × (6–)8–12 mm, claw 7–15 mm, apex rounded; filaments 8–12 mm; anthers 2.5–4 mm. **Fruits** spreading to ascending, smooth, ± 4-angled or subterete, (2.5–)5–8(–10) cm × (2.5–)3–4(–5) mm; valvular segment with 10–20 seeds per locule, (2–)3–7.5(–9) cm, terminal segment usually seedless, rarely 1 or 2-seeded, (conic), (3–)4–10 mm. **Seeds** brown, 1.7–2.5 mm diam.; seed coat reticulate, not mucilaginous when wetted. $2n = 18$.

Flowering May–Aug. Maritime slopes or sea-facing cliffs, weedy escape, gardens, abandoned fields, waste places; 0–100 m; introduced; Nfld. and Labr., Ont., P.E.I., Que.; Calif., Conn., Ill., Iowa, Ky., Mass., N.Y., Ohio, Pa., R.I., Tex., Vt.; Europe; Asia; Africa; introduced also in Australia.

Brassica oleracea is widely cultivated worldwide as a vegetable crop, and its various forms are generally recognized as varieties instead of subspecies; these include var. *acephala* de Candolle (kale and collards), var. *botrytis* Linnaeus (cauliflower), var. *capitata* Linnaeus (cabbage), var. *gemmifera* Zenk (Brussels sprouts), var. *gongylodes* Linnaeus (kohlrabi), and var. *italica* Plenk (broccoli). It also occurs sporadically as a weedy escape from cultivation and seems unlikely to persist for long periods of time. It is reported to be naturalized on coastal cliffs (maritime slopes) in the northern Central Coastal Region and the central and southern North Coastal Region in California (Marin, San Francisco, San Mateo, Santa Barbara, and Ventura counties) (J. T. Howell et al. 1958; Howell 1970; H. G. Baker 1972; R. C. Rollins 1993b).

7. Brassica rapa Linnaeus, Sp. Pl. 2: 666. 1753
• Bird-rape, wild-rape, rapeseed, canola, turnip, wild-turnip, turnip-rape, field-mustard [I][W]

Brassica campestris Linnaeus; *B. campestris* var. *oleifera* de Candolle; *B. chinensis* Linnaeus; *B. pekinensis* (Loureiro) Ruprecht; *B. rapa* subsp. *chinensis* (Linnaeus) Hanelt; *B. rapa* subsp. *pekinensis* (Loureiro) Hanelt; *Sinapis pekinensis* Loureiro

Annuals or biennials; (roots fleshy or slender); (green to slightly glaucous), glabrous or sparsely hairy. **Stems** unbranched or branched distally, 3–10 dm. **Basal leaves:** petiole (winged), (1–)2–10(–17) cm; blade ± lyrate-pinnatifid to pinnate to pinnatisect, (5–)10–40(–60) cm × 30–100(–200) mm, (margins sinuate-dentate, sometimes ciliate), lobes 2–4(–6) each side, (terminal lobe oblong-obovate, obtuse, large, blade surfaces usually setose). **Cauline leaves** (middle and distal) sessile; base auriculate to amplexicaul, (margins subentire). **Racemes** not paniculately branched, (with open flowers overtopping or equal to buds). **Fruiting pedicels** ascending to spreading, (5–)10–25(–30) mm. **Flowers:** sepals (3–)4–6.5(–8) × 1.5–2 mm; petals deep yellow to yellow, obovate, 6–11(–13) × (2.5–)3–6(–7) mm, claw 3–7 mm, apex rounded; filaments 4–6(–7) mm; anthers 1.5–2 mm. **Fruits** ascending to somewhat spreading, torulose, terete, (2–)3–8(–11) cm × 2–4(–5) mm; valvular segment with 8–15 seeds per locule, (1.3–)2–5(–7.5) cm, terminal segment seedless, 8–22 mm. **Seeds** black, brown, or reddish, 1.1–2 mm diam.; seed coat very finely reticulate-lightly alveolate, not mucilaginous when wetted. $2n = 20$.

Flowering Apr–Sep. Roadsides, disturbed areas and waste places, cultivated fields, grain fields, orchards, gardens; 0–1500 m; introduced; Alta., B.C., Man., N.B., Nfld. and Labr., N.W.T., N.S., Ont., P.E.I., Que., Sask., Yukon; Ala., Alaska, Ariz., Ark., Calif., Colo., Conn., Del., D.C., Fla., Ga., Idaho, Ill., Ind., Iowa, Kans., Ky., La., Maine, Md., Mass., Mich., Minn., Miss., Mo., Mont., Nebr., Nev., N.H., N.J., N.Mex.,

N.Y., N.C., N.Dak., Ohio, Okla., Oreg., Pa., R.I., S.C., S.Dak., Tenn., Tex., Utah, Vt., Va., Wash., W.Va., Wis., Wyo.; Europe; Asia; Africa; introduced also in Mexico, West Indies, Central America, South America, Atlantic Islands, Australia.

Brassica rapa is widely cultivated as an oil crop and vegetable, and cultivars, especially in Asia, have been recognized as species, subspecies, and varieties. The most important crops include: rapeseed or canola, turnip (subsp. *rapa*), Chinese mustard or pakchoi [subsp. *chinensis* (Linnaeus) Hanelt], and Chinese cabbage or petsai [subsp. *pekinensis* (Loureiro) Hanelt]. The species is also a widespread naturalized weed [subsp. *sylvestris* (Linnaeus) Janchen] throughout temperate North America and elsewhere. It is self-incompatible. Hybridization in the field in Europe has been described between *B. napus* and *B. rapa* (R. B. Jørgensen and B. Andersen 1994).

8. **Brassica tournefortii** Gouan, Ill. Observ. Bot., 44, plate 20A. 1773 • Sahara mustard [F] [I] [W]

Annuals; densely hirsute proximally, glabrescent distally. **Stems** usually branched basally, (widely) branched distally, (1–)3–7(–10) dm. **Basal leaves:** (rosettes persistent); petiole (broad) 2–10 cm; blade lyrate to pinnatisect, 2–30 cm × 10–50 (–100) mm, (margins serrate-dentate), 4–10 lobes each side. **Cauline leaves** sessile; blade (reduced in size distally, distalmost bractlike), base tapered, not auriculate or amplexicaul. **Racemes** not paniculately branched. **Fruiting pedicels** widely spreading, 8–15 mm. **Flowers:** sepals 5–4.5 × 1–1.5 mm; petals pale yellow, fading or, sometimes, white, oblanceolate, 4–7 × 1.5–2(–2.5) mm, claw 1–3 mm, apex rounded; filaments 2.5–4 mm; anthers 1–1.3 mm; gynophore to 1 mm. **Fruits** (shortly stipitate); widely spreading to ascending (not appressed to rachis), torulose, cylindric, 3–7 cm × 2–4(–5) mm; valvular segment with 6–12(–15) seeds per locule, 2.2–5 cm, terminal segment 1(–3)-seeded, (cylindric, stout), 10–20 mm. **Seeds** light reddish brown or black, 1–1.2 mm diam.; seed coat prominently reticulate, mucilaginous when wetted. $2n = 20$.

Flowering Feb–Apr. Roadsides, waste places, old fields, washes, open desert areas intermixed with desert shrubs; 0–800 m; introduced; Ariz., Calif., Nev., Tex., Utah; Europe; Asia; Africa; introduced also in nw Mexico, Australia.

Brassica tournefortii was first reported from California (Imperial, Riverside, and western San Bernardino counties) by W. L. Jepson ([1923–1925]), with the first collections appearing from southern California in 1941 (R. C. Rollins and I. A. Al-Shehbaz 1986), Arizona in 1959 (T. H. Kearney and R. H. Peebles 1960), Nevada in 1977, and Texas in 1978 (D. E. Lemke and R. D. Worthington 1991).

18. CAKILE Miller, Gard. Dict. Abr. ed. 4, vol. 1. 1754 • Sea-rockets [Arabic name *qaqulleh*]

James E. Rodman

Annuals or, rarely, perennials; (succulent, taproot woody, with relatively long, horizontal roots); not scapose; glabrous or, sometimes, sparsely pubescent. **Stems** erect, ascending, prostrate, or divaricate, branched basally. **Leaves** cauline; usually petiolate, rarely sessile; blade (often fleshy), not rosulate, margins entire, crenate, dentate, sinuate, or pinnately lobed. **Racemes** considerably elongated in fruit. **Fruiting pedicels** (rachis) geniculate or not, slender or stout. **Flowers:** sepals erect, ovate or oblong, lateral pair saccate or not basally; petals (rarely aborted, reflexed), white to lavender, obovate to spatulate, claw differentiated from blade or not; stamens tetradynamous; filaments not dilated basally; anthers (introrse), ovate to oblong; nectar glands (4), distinct, median glands present. **Fruits** siliques or silicles, indehiscent, stipitate, segments

2, (fleshy and green becoming corky and dry), obovoid, oblong, fusiform, or lanceoloid, rarely hastate, (proximal segment) terete or laterally horned, (terminal segment) terete, 4-angled, or 8-ribbed; (segments each falsely 1-loculed, septum papery, appressed to one side, usually 1-seeded; proximal segment remaining attached to pedicel; terminal segment deciduous by transverse articulation, beaked); valves and replum not distinguishable; ovules (1 or) 2(–4) per ovary; (style absent); stigma entire or slightly 2-lobed. **Seeds** aseriate or uniseriate, plump or flattened, not winged, (brown), oblong; seed coat (smooth), not mucilaginous when wetted; cotyledons accumbent or incumbent, rarely contorted. *x* = 9.

Species 7 (5 in the flora): North America, Mexico, West Indies, Central America, Europe, Asia (Near East), n Africa; introduced in e Asia (Japan), Australia.

Cakile is common on sandy beaches of the North Atlantic Ocean, the Baltic, Black, Mediterranean, North, and White seas, the Caribbean and Gulf of Mexico, and the Great Lakes, and is naturalized in Australia, Japan, and on the Pacific Coast of North America; one species, *C. arabica* Velenovsky & Bornmüller, is found in deserts of the Middle East (s Iraq, Kuwait, Saudi Arabia).

SELECTED REFERENCES Barbour, M. G. and J. E. Rodman. 1970. Saga of the West Coast sea-rockets: *Cakile edentula* ssp. *californica* and *C. maritima*. Rhodora 72: 370–386. Rodman, J. E. 1974. Systematics and evolution of the genus *Cakile* (Cruciferae). Contr. Gray Herb. 205: 3–146. Rodman, J. E. 1980. Population variation and hybridization in sea-rockets (*Cakile*, Cruciferae): Seed glucosinolate characters. Amer. J. Bot. 67: 1145–1159.

1. Proximal fruit segments with 2 opposite, lateral horns distally; leaf blades broadly ovate to lanceolate, margins sinuately lobed or deeply pinnatifid; petals 3–6 mm wide, usually lavender, rarely white. . . . . 1. *Cakile maritima*
1. Proximal fruit segments without lateral horns distally; leaf blades ovate, ovate-lanceolate, or spatulate, margins entire, dentate, sinuate, crenate, or pinnatisect; petals 1.2–4.5 mm wide, white or lavender.
    2. Plants usually sprawling; leaf blades not especially fleshy; racemes often 3+ dm; petals usually white, rarely lavender, 3–4.5 mm wide . . . . 2. *Cakile lanceolata*
    2. Plants not or, rarely, sprawling (erect to prostrate); leaf blades usually fleshy; racemes 1–2 dm; petals lavender to white, less than 3 mm wide.
        3. Rachises geniculate in fruit; petals 1.2–1.9 mm wide . . . . 3. *Cakile geniculata*
        3. Rachises not geniculate in fruit; petals 1.3–3 mm wide.
            4. Fruits terete to 4-angled, 3–4 mm wide; beak conic, apex acute . . . . 4. *Cakile constricta*
            4. Fruits 8-ribbed, or 4-angled, 5–9 mm wide; beak flattened, apex usually retuse or blunt, rarely acute . . . . 5. *Cakile edentula*

1. Cakile maritima Scopoli, Fl. Carniol. ed. 2, 2: 35. 1772 ⟦I⟧ ⟦W⟧

Bunias cakile Linnaeus, Sp. Pl. 2: 670. 1753

Subspecies 3 (1 in the flora): introduced; Europe; introduced also in Asia, n Africa, Australia.

1a. Cakile maritima Scopoli subsp. **maritima** ⟦I⟧ ⟦W⟧

Annuals or, rarely, perennials. Stems erect to prostrate, to 8 dm. **Cauline leaves:** blade broadly ovate to lanceolate, (to 10 cm, smaller distally), margins sinuately lobed or deeply pinnatifid, (distal) often subentire; (lateral lobes 4–8, 5–15 mm). **Racemes** to 4 dm; rachis straight. **Fruiting pedicels** 1.5–7 mm. **Flowers:** sepals 4–5.5 mm, lateral pair not saccate basally; petals usually lavender, rarely white, 8–14 × 3–6 mm, claw distinct. **Fruits** (often reflexed), hastate, 12–27 × 4–8 mm; proximal segment conical (becoming pedicel-like when aborted, with 2 opposite, lateral horns distally); terminal segment usually 4-angled, mitriform or conic, (margins scarious basally), apex acute or blunt, (usually compressed). **Seeds:** cotyledons accumbent to, sometimes, incumbent or contorted. 2*n* = 18.

Flowering late spring–fall(–winter if mild). Sandy beaches; ca. 0 m; introduced; B.C.; Calif., Md., Oreg., Wash.; Europe; introduced also in nw Mexico.

Subspecies *maritima* is naturalized in Pacific North America (M. G. Barbour and J. E. Rodman 1970); it is also reported on the eastern shores of Chesapeake Bay, Maryland.

2. Cakile lanceolata (Willdenow) O. E. Schulz in I. Urban, Symb. Antill. 3: 504. 1903

Raphanus lanceolatus Willdenow, Sp. Pl. 3: 562. 1800

Annuals, (usually sprawling). **Stems** erect to prostrate, (much-branched), often 5+ dm. **Cauline leaves** (shortly petiolate or sessile); blade broadly ovate to ovate-lanceolate, or (distal) oblanceolate, (not especially fleshy, smaller distally), margins entire, dentate, or pinnatisect. **Racemes** often 3+ dm; rachis straight. **Fruiting pedicels** 1.5–4 mm. **Flowers:** sepals 3.5–5 mm, lateral pair saccate basally; petals usually white, rarely lavender, 4.9–9.4 × 3–4.5 mm, claw distinct. **Fruits** (weakly 4-angled to terete, striate or sulcate), fusiform or lanceoloid [turbinate], 15–31 × 3–4 mm; proximal segment terete, (5–10 mm); terminal segment slenderly conical, (9–18 mm), apex usually acute. **Seeds:** cotyledons accumbent or, occasionally, incumbent.

Subspecies 4 (3 in the flora): s United States, Mexico, West Indies, Central America.

Subspecies *alacranensis* is known from the Yucatán Peninsula.

1. Fruits lanceoloid, apex not tapering abruptly, terminal segment at least 2 times length of proximal segment; leaf blades: margins usually sinuately or crenately lobed, rarely pinnatisect . . . . 2a. *Cakile lanceolata* subsp. *lanceolata*
1. Fruits fusiform or lanceoloid, apex tapering abruptly, terminal segment less than 2 times length of proximal segment; leaf blades: margins pinnatifid or entire.
   2. Fruits fusiform, not constricted at articulation, 4- or 8-sulcate . . . . 2b. *Cakile lanceolata* subsp. *fusiformis*
   2. Fruits lanceoloid, usually constricted at articulation, weakly 4-angled or terete . . . . 2c. *Cakile lanceolata* subsp. *pseudoconstricta*

2a. Cakile lanceolata (Willdenow) O. E. Schulz subsp. **lanceolata**

Cakile aequalis L'Héritier ex de Candolle; *C. cubensis* Kunth; *C. maritima* Scopoli var. *cubensis* (de Candolle) Chapman

Plants sprawling. **Leaves:** blade (ovate to lanceolate), margins usually sinuately or crenately lobed, rarely pinnatisect. **Fruits** (slenderly) lanceoloid, 19–31 mm, usually not constricted at articulation; terminal segment at least 2 times length of proximal, apex acute to blunt, relatively long. $2n = 18$.

Flowering year-round. Sandy beaches; ca. 0 m; introduced; Fla.; West Indies.

2b. Cakile lanceolata (Willdenow) O. E. Schulz subsp. **fusiformis** (Greene) Rodman, Contr. Gray Herb. 205: 114. 1974

Cakile fusiformis Greene, Pittonia 3: 346. 1898; *C. chapmanii* Millspaugh

Plants sometimes sprawling. **Leaves:** blade (ovate), margins entire or pinnatifid. **Fruits** (oval in cross-section) fusiform, 16–25 mm, not constricted at articulation, (4- or 8-sulcate or striate); terminal segment less than 2 times length of proximal, apex tapering abruptly, acute. $2n = 18$.

Flowering year-round. Sandy shores; ca. 0 m; Fla.; Mexico; Central America.

2c. Cakile lanceolata (Willdenow) O. E. Schulz subsp. **pseudoconstricta** Rodman, Contr. Gray Herb. 205: 116, plate 1. 1974

Plants sprawling. **Leaves:** blade margins deeply pinnatifid. **Fruits** (slenderly) lanceoloid, 15–24 mm, constricted at articulation, (weakly 4-angled or terete); terminal segment less than 2 times length of proximal, apex acute.

Flowering year-round. Coastal strand; ca. 0 m; Fla., Tex.; Mexico (Tamaulipas).

Subspecies *lanceolata* in Texas and Mexico are suspected to be introduced from ballast and naturalized locally, sometimes occurring with *Cakile geniculata*.

3. Cakile geniculata (B. L. Robinson) Millspaugh, Publ. Field Columbian Mus., Bot. Ser. 2: 126. 1900

Cakile maritima var. *geniculata* B. L. Robinson in A. Gray et al., Syn. Fl. N. Amer. 1(1,1): 132. 1895; *C. lanceolata* (Willdenow) O. E. Schulz var. *geniculata* (B. L. Robinson) Shinners

Annuals. Stems erect, (sometimes suffrutescent, much-branched), to 10 dm. **Cauline leaves:** blade broadly ovate or spatulate, margins sinuate, dentate, or pinnately lobed. **Racemes:** 1–2 dm; rachis geniculate. **Fruiting pedicels** (rachis of equal width), 2–5 mm, (widely spaced). **Flowers:** sepals 3–4 mm, lateral pair not saccate basally; petals white to pale lavender, 4–6.1 × 1.2–1.9 mm, claw distinct. **Fruits** (8-ribbed), lanceoloid,

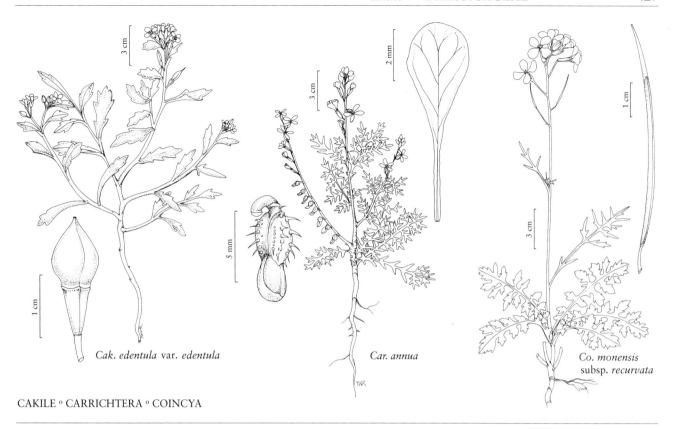

20–27 × 3–5 mm; proximal segment terete, (6–11 mm); terminal segment slenderly conical, (13–18 mm), apex acute to blunt, (often slightly curved). **Seeds:** cotyledons accumbent to obliquely incumbent. $2n = 18$.

Flowering year-round (peak Mar–Aug). Sandy beaches; ca. 0 m; La., Tex.; Mexico (Tamaulipas, Veracruz).

4. Cakile constricta Rodman, Contr. Gray Herb. 205: 131. 1974 [E]

Annuals. Stems erect, 1–3 dm. **Cauline leaves:** blade ovate to spatulate, margins crenate, dentate, or sinuate (not pinnatifid). **Racemes** usually 1–2 dm, rarely longer; rachis not geniculate. **Fruiting pedicels** 2.5–7 mm. **Flowers:** sepals ca. 4 mm, lateral pair not saccate basally; petals white or lavender, 5.3–8 × 1.3–2.6 mm, claw distinct. **Fruits** (terete or 4-angled), lanceoloid, 14–22 × 3–4 mm, (constricted at articulation of segments); proximal segment terete, long-cylindrical, (5–8.5 mm); terminal segment fusiform, (9–15 mm, beak conic), apex acute. **Seeds:** cotyledons accumbent. $2n = 18$.

Flowering spring–fall. Sandy shores; ca. 0 m; Fla., La., Tex.

Cakile constricta is disjunct in Florida (not known from the Keys or southwestern Florida, where immature plants of *C. lanceolata* subspp. *fusiformis* and *pseudoconstricta* may be confused with it).

5. Cakile edentula (Bigelow) Hooker, Fl. Bor.-Amer. 1: 59. 1830 [E] [F]

Bunias edentula Bigelow, Fl. Boston., 157. 1814

Annuals or, rarely, **perennials**, (compact, rarely sprawling). **Stems** erect, or, rarely, prostrate, (much-branched), to 8 dm. **Cauline leaves:** blade ovate to spatulate, or (distal) oblanceolate (less lobed), (smaller distally), margins crenate, dentate, or sinuate (not pinnatifid). **Racemes** 1–2 dm, (congested); rachis straight. **Fruiting pedicels** 1.8–8 mm. **Flowers:** sepals 3.5–5 mm, lateral pair not saccate basally; petals white to pale lavender, (well-developed or reduced to bristles), 4.6–9.7 × 1.5–3 mm, claw not distinct. **Fruits** (4-angled or 8-ribbed), cylindrical, 12–29 × 3–9 mm; proximal segment terete, (5–10 mm); terminal segment fusiform to turbinate, (7–15 mm), (beak flattened), apex acute, blunt, or retuse. **Seeds:** cotyledons accumbent to, sometimes, incumbent.

Subspecies 2 (2 in the flora): North America, introduced in e Asia (Japan), Australia.

Elements of *Cakile edentula* are native in North America; the typical variety is introduced on the Pacific Coast and probably on beaches of the Great Lakes Region.

1. Fruits 4-angled, terminal segment apex usually acute or blunt, rarely retuse .... 5a. *Cakile edentula* subsp. *edentula*
1. Fruits 8-ribbed, terminal segment apex blunt or retuse .... 5b. *Cakile edentula* subsp. *harperi*

5a. Cakile edentula (Bigelow) Hooker subsp. **edentula**
E F

Leaf blades ovate or spatulate, margins sinuate. **Flowers:** petals sometimes aborted or reduced to bristles. **Fruits** 4-angled; terminal segment turbinate or 4-angled, 12–26 × 3–9 mm, apex usually acute or blunt, rarely retuse.

Varieties 2 (2 in the flora): North America; introduced in e Asia (Japan), Australia.

1. Fruits: terminal segment turbinate, 5–9 mm wide, apex relatively short ....5a1. *Cakile edentula* var. *edentula*
1. Fruits: terminal segment 4-angled, 3–5 mm wide, apex relatively long ....5a2. *Cakile edentula* var. *lacustris*

5a1. Cakile edentula (Bigelow) Hooker var. **edentula**
E F

Cakile americana Nuttall; *C. californica* A. Heller; *C. edentula* subsp. *californica* (A. Heller) Hultén; *C. edentula* var. *californica* (A. Heller) Fernald; *C. maritima* Scopoli var. *americana* (Nuttall) Torrey & A. Gray

Leaf blades fleshy. **Petals** sometimes aborted or reduced to bristles early. **Fruits** 4-angled, 12–24 × 5–9 mm; terminal segment turbinate, apex usually acute or blunt, rarely retuse, relatively short. $2n = 18$.

Flowering Jun–Nov. Sandy beaches; ca. 0 m; B.C., N.B., Nfld. and Labr., N.S., P.E.I., Que.; Alaska, Calif., Conn., Del., Ill., Ind., Maine, Md., Mass., Mich., N.H., N.J., N.Y., N.C., Oreg., R.I., Va., Wash., Wis.; introduced in e Asia (Japan), Australia.

Variety *edentula* is introduced and naturalized on the Pacific Coast of North America, Japan (Y. Asai 1982, 1996) and Australia (M. G. Barbour and J. E. Rodman 1970; Rodman 1986), and sporadic on beaches of the Great Lakes, where it is likely introduced (Rodman 1974); it hybridizes with subsp. *harperi* on the Outer Banks of North Carolina (Rodman 1980).

5a2. Cakile edentula (Bigelow) Hooker var. **lacustris** Fernald, Rhodora 24: 23. 1922 E

Cakile lacustris (Fernald) Pobedimova

Leaf blades not especially fleshy. **Petals** rarely aborted. **Fruits** 4-angled, 16–26 × 3–5 mm; terminal segment 4-angled, apex usually blunt, (flattened), relatively long. $2n = 18$.

Flowering May–Oct. Great Lakes beaches; ca. 0 m; Ont.; Ill., Ind., Mich., N.Y., Ohio, Pa., Wis.

Variety *lacustris* is sometimes sympatric with var. *edentula*; hybridization has not been demonstrated.

5b. Cakile edentula (Bigelow) Hooker subsp. **harperi** (Small) Rodman, Contr. Gray Herb. 205: 125. 1974 E

Cakile harperi Small, Fl. S.E. U.S., 478, 1331. 1903

Leaf blades ovate or spatulate, margins sinuate, fleshy. **Flowers:** petals rarely aborted. **Fruits** (cylindrical), 8-ribbed, 20–29 × 5–9 mm; terminal segment cylindrical, (beak sometimes flattened, blunt), apex usually retuse. $2n = 18$.

Flowering Mar–Oct. Sandy beaches; ca. 0 m; Fla., Ga., N.C., S.C.

19. CARRICHTERA de Candolle, Mém. Mus. Hist. Nat. 7: 244, 250. 1821, name conserved • [For Bartholomaeus Carrichter, sixteenth-century herbalist, alchemist, and physician to Emperor Maximilian II]

Ihsan A. Al-Shehbaz

Annuals; not scapose; hirsute [glabrous], (trichomes often retrorse). **Stems** usually erect or ascending, rarely decumbent, usually branched, rarely simple. **Leaves** basal and cauline; petiolate or subsessile; basal not rosulate, petiolate, blade 1- or 2-pinnatisect, margins entire; cauline subsessile or shortly petiolate, blade 1- or 2-pinnatisect, (base not auriculate), margins entire. **Racemes** (lax), considerably elongated in fruit. **Fruiting pedicels** strongly recurved, often secund, not thickened. **Flowers:** sepals erect, oblong-linear, lateral pair slightly saccate basally; petals cream or pale yellow (with dark purple or brown veins, longer than sepals), spatulate to obovate, claw strongly differentiated from blade; stamens tetradynamous, (erect); filaments not dilated basally; anthers ovate to oblong, (apiculate); nectar glands: lateral intrastaminal, globose to semi-annular, median glands present (distinct). **Fruits** silicles, dehiscent, sessile, segments 2, ovoid to ellipsoid, terete; (proximal segment not torulose, 4–6-seeded; terminal segment indehiscent, flattened, cochleariform to short-lingulate, seedless); valves each with prominent midvein and distinct marginal veins, glabrous except setose or hirsute on veins; replum rounded; septum complete; ovules 4–6 per ovary; (style obscure); stigma capitate, 2-lobed. **Seeds** biseriate, plump, not winged, subglobose; seed coat (smooth), mucilaginous when wetted; cotyledons conduplicate. $x = 8$.

Species 1: introduced, California; s Europe, sw Asia, n Africa; introduced also in Australia.

1. Carrichtera annua (Linnaeus) de Candolle, Mém. Mus. Hist. Nat. 7: 250. 1821

Vella annua Linnaeus, Sp. Pl. 2: 641. 1753

Stems branched basally and distally, (0.5–)0.7–3.6(–4.5) dm. **Basal leaves:** petiole 1–4.5 cm; blade 3–6 lobes each side, 1.5–4 cm, terminal lobe linear to oblong, 0.3–1.5 cm × 0.5–1.5 mm, margins entire. **Cauline leaves** similar to basal. **Fruiting pedicels** 2.5–4 mm. **Flowers:** sepals 4–5 mm; petals 6.5–8 × 1–2 mm; median filaments 3.5–4 mm; anthers 0.5–0.8 mm. **Fruits** reflexed; proximal segment broadly ovoid to subglobose, 3–4 × 2.5–3.5 mm; terminal segment cochleariform or ovate, 3–5 × 3–4 mm (margin incurved). **Seeds** 1.1–1.6 mm diam. $2n = 16$.

Flowering Apr–May. Deserts; ca. 50 m; introduced; Calif.; sw Europe; sw Asia; n Africa; introduced also in Australia.

Carrichtera annua was found as a naturalized weed in southern California in 2007.

20. COINCYA Porta & Rigo ex Rouy, Naturaliste, sér. 2, 13: 248. 1891 • [For Auguste Henri Cornut de Coincy, 1837–1903, Spanish botanist, discoverer of first species described]

Suzanne I. Warwick

Brassicella Fourreau; *Hutera* Porta & Rigo; *Rhynchosinapis* Hayek

Annuals [biennials or perennials]; not scapose; glabrous or hispid. **Stems** usually erect, rarely prostrate, [unbranched] branched basally. **Leaves** basal and cauline; petiolate; basal rosulate, blade margins lobed, pinnatipartite to lyrate, or 1- or 2-pinnatisect, [entire or dentate]; cauline blade similar to basal, (base not auriculate). **Racemes** (corymbose), greatly elongated in fruit. **Fruiting pedicels** usually ascending to patent [reflexed], rarely erect, slender. **Flowers:** sepals erect, connivent, obtuse [narrowly oblong], lateral pair usually saccate basally, (setulose proximal to apex); petals yellow (usually with dark brown, maroon, or purple veins), obovate, claw differentiated from blade, (apex obtuse); stamens tetradynamous; filaments not dilated basally, slender; anthers linear, (base sagittate); nectar glands: lateral glands lobed, median glands usually absent. **Fruits** siliques, dehiscent, sessile or subsessile, segments 2, linear, torulose, terete; (terminal segment 1–6-seeded, indehiscent, ensiform to cylindrical, linear, smooth); valves each 3(–5)-veined, glabrous; replum rounded; septum complete; ovules 4–100 per ovary; stigma capitate, 2-lobed. **Seeds** uniseriate, plump, not winged, subglobose [globose, oblong]; seed coat (reticulate), slightly mucilaginous when wetted; cotyledons conduplicate. $x = 12$.

Species 6 (1 in the flora): introduced; w Europe, nw Africa.

SELECTED REFERENCE Leadlay, E. A. and V. H. Heywood. 1990. The biology and systematics of the genus *Coincya* Porta & Rigo ex Rouy (Cruciferae). Bot. J. Linn. Soc. 102: 313–398.

1. **Coincya monensis** (Linnaeus) Greuter & Burdet, Willdenowia 13: 87. 1983 F I

Sisymbrium monense Linnaeus, Sp. Pl. 2: 658. 1753

Subspecies 5 (1 in the flora): introduced; w Europe, Africa (Morocco).

Coincya monensis is a variable species, within which, in spite of lack of clear discontinuities in variation patterns, E. A. Leadlay and V. H. Heywood (1990) recognized five subspecies and divided subsp. *recurvata* into four varieties, not recognized here.

1a. **Coincya monensis** (Linnaeus) Greuter & Burdet subsp. **recurvata** (Allioni) Leadlay, Bot. J. Linn. Soc. 102: 370. 1990 • Wallflower-cabbage, tall wallflower-cabbage F I

Sinapis recurvata Allioni, Fl. Pedem. 1: 265. 1785; *Brassica cheiranthos* Villars; *Hutera cheiranthos* (Villars) Gómez-Campo; *Rhynchosinapis cheiranthos* (Villars) Dandy

Stems (0.8–)1–10 dm, usually sparsely to densely hispid basally, rarely glabrous, trichomes 0.5–3.6 mm. **Basal leaves** long-petiolate; blade (3–)5–20 cm × 25–100 mm, lobes 3–9(–10) each side, surfaces sparsely to densely hispid, rarely coriaceous or glaucous, trichomes patent, rarely appressed. **Cauline leaves:** blade similar to basal, (lobes fewer, narrower than basal). **Racemes** 2–8(–15)-flowered, open at one time. **Flowers:** sepals: median pair apex cucullate, (setulose below apex), lateral pair broader, saccate basally; petals 12.5–22(–26) × 2.5–7(–9) mm, claw nearly as long as sepal. **Fruits** usually straight, rarely curved, (1–)3–9 cm; proximal segment (15–)20–75(–90)-seeded, (8–)25–75 × 1.5–3 mm, apex obtuse; terminal segment (0 or) 1–5-seeded, (5–)7–23(–34) × 1.5–3 mm; style relatively short. **Seeds** black to brown, subglobose, 0.8–1.6 × 0.8–1.4 mm. $2n = 24, 48$.

Flowering May–Aug. Fields, roadsides, mountain road cuts, cliff ledges; introduced; Mich., N.C., Pa.; w Europe; nw Africa.

Subspecies *recurvata* was first recorded from North America in 1880 on ballast in New Jersey (R. C. Rollins 1961, 1981, 1993; I. A. Al-Shehbaz 1985); it was first reported from North Carolina (as *Brassica erucastrum*) in 1958 from Yancey County and in 1968 from Jackson County (H. E. Ahles and A. E. Radford 1964; Al-Shehbaz). From Pennsylvania, subsp. *recurvata* was reported from Luzerne County in 1964 and from Bradford County in 1983.

21. **CRAMBE** Linnaeus, Sp. Pl. 2: 671. 1753; Gen. Pl. ed. 5, 301. 1754 • Sea-cabbage [Greek *krambe*, cabbage] I

Suzanne I. Warwick

Perennials [annuals, subshrubs]; not scapose; (glaucous), usually glabrous. **Stems** erect, branched distally. **Leaves** basal and cauline; petiolate; basal rosulate, blade (base not auriculate), margins sinuate-dentate to shallowly lobed, or subpinnatifid and coarsely dentate, [lyrate-pinnatipartite,

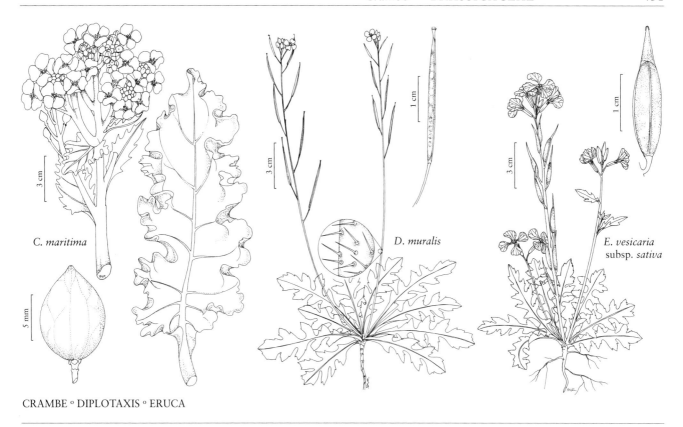

CRAMBE · DIPLOTAXIS · ERUCA

pinnatisect, less frequently undivided]; cauline blade margins sinuate-dentate to lobed, less divided than basal. **Racemes** (corymbose, several-flowered, often in panicles), considerably elongated in fruit. **Fruiting pedicels** suberect to divaricate-ascending, [erect, divaricate], stout [slender]. **Flowers:** sepals spreading to ascending, ovate [oblong], lateral pair not saccate basally; petals white, cream, or yellow, obovate or oblong, claw differentiated from blade, (much shorter than sepal); stamens tetradynamous; filaments not dilated basally, (lateral filiform, median (inner) winged and/or toothed, rarely filiform and not toothed); anthers ovate or oblong, (apex obtuse); gynophore absent; nectar glands (4), distinct, median glands present. **Fruits** silicles, nutlike, indehiscent, segments 2, smooth, terete or 4-angled; (proximal segment stalklike, seedless, much shorter than terminal; terminal segment caducous, 1 (or 2)-seeded, subglobose to ovoid [globose], corky, rugose, or reticulate); replum and septum obsolete; ovules 2 per ovary; (style 0(–1) mm); stigma capitate, entire. **Seeds** aseriate, plump, not winged, globose, ellipsoid, or ovoid [oblong], (pendulous on basal funicle); seed coat (smooth), not mucilaginous when wetted; cotyledons conduplicate. $x = 15$.

Species 35 (1 in the flora): introduced, Oregon; c Europe to w Asia (Mediterranean region), n Africa, Atlantic Islands (Canary Islands).

1. **Crambe maritima** Linnaeus, Sp. Pl. 2: 671. 1753

   · Sea-kale

Plants succulent; roots thick. Stems 2–5(–7.5) dm, stout. Basal leaves (cabbagelike); petiole 4–16 cm; blade oblong, or elliptic-ovate to ovate, 10–40 cm × 80–300 mm. Cauline leaves similar to basal, (proximal) blade margins irregularly pinnate or sinuate-dentate. Fruiting pedicels (10–)15–30(–37) mm, stout. **Flowers:** sepals 3–4 × 2–3.5 mm; petals (6–)8–12(–15) × (4–)5–7 mm; filaments 3–4 mm; anthers 1–1.5 mm. **Fruits:** proximal segment 0.1–0.4 cm, slightly thicker than pedicel; terminal segment 1-seeded, subglobose to ovoid, 0.7–1.2(–1.4) cm × 6–8 mm, thick. **Seeds** 4–5(–6) mm. $2n = 30, 60$.

Flowering May–Jul. Coastal beaches; ca. 0 m; introduced; Oreg.; Europe; n Africa.

In the flora area, *Crambe maritima* has been reported only from Yaquina Head in Lincoln County; it was not treated by R. C. Rollins (1993).

22. DIPLOTAXIS de Candolle, Mém. Mus. Hist. Nat. 7: 243. 1821 • Wall-rocket [Greek *diplo*-, double, and *taxis*, arrangement, alluding to number of seed rows in each locule of fruit] [I]

Juan B. Martínez-Laborde

Annuals, biennials, or perennials; (sometimes suffrutescent); scapose or not; glabrous, glabrescent, or pubescent. **Stems** erect or ascending, branched. **Leaves** basal and, sometimes, cauline; petiolate or sessile; basal rosulate or not, petiolate, blade margins dentate, sinuate, or pinnatisect; cauline petiolate or sessile, blade (base not auriculate), margins entire or dentate. **Racemes** (corymbose, sometimes shortly bracteate basally), considerably elongated in fruit. **Fruiting pedicels** ascending, divaricate, or reflexed, stout to slender. **Flowers:** sepals ascending to spreading, oblong, lateral pair not saccate basally; petals yellow or white [purple], obovate, (apex rounded or truncate); stamens tetradynamous; filaments not dilated basally; anthers oblong to ovate, (apex obtuse); nectar glands (4): lateral cushionlike, median cylindrical. **Fruits** siliques, dehiscent, sessile or (long-)stipitate, segments 1 or 2, linear to linear-oblong, torulose, latiseptate or terete; (proximal segment numerous-seeded, 1-veined; terminal segment 0–2-seeded); valves glabrous; replum rounded; septum complete; ovules [12–]20–36(–46)[–276] per ovary; (style obsolete or distinct); stigma capitate or somewhat decurrent, 2-lobed. **Seeds** usually biseriate, rarely uniseriate, not winged, ovoid or ellipsoid; seed coat (smooth or minutely reticulate), slightly mucilaginous or not when wetted; cotyledons conduplicate. x = 7, [8, 9, 10,] 11, [13,] 21.

Species 25–30 (3 in the flora): introduced; Europe, Asia, n Africa; introduced also in n Mexico, West Indies (Bahamas), Bermuda, South America, Pacific Islands (New Zealand), Australia.

SELECTED REFERENCES Eschmann-Grupe, G., H. Hurka, and B. Neuffer. 2003. Species relationships within *Diplotaxis* (Brassicaceae) and the phylogenetic origin of *D. muralis*. Pl. Syst. Evol. 243: 13–29. Martínez-Laborde, J. B. 1988. Estudio Sistemático del Género *Diplotaxis* DC. (Cruciferae, Brassiceae). Ph.D. dissertation. Universidad Politécnica de Madrid. Mummenhoff, K., G. Eschmann-Grupe, and K. Zunk. 1993. Subunit polypeptide composition of RUBISCO indicates *Diplotaxis viminea* as maternal parent species of amphidiploid *Diplotaxis muralis*. Phytochemistry 34: 429–431. Sánchez-Yélamo, M. D. and J. B. Martínez-Laborde. 1991. Chemotaxonomic approach to *Diplotaxis muralis* (Cruciferae, Brassiceae) and related species. Biochem. Syst. & Ecol. 19: 477–482.

1. Stems densely pubescent throughout; leaf blade surfaces shortly pubescent throughout; sepals pubescent, trichomes ± flexuous; petals white (turning purple); fruits: distal segment 1- or 2-seeded . . . . 3. *Diplotaxis erucoides*
1. Stems glabrescent to moderately pubescent basally or distally; leaf blade surfaces glabrescent, or margins and veins glabrescent to sparsely pubescent; sepals glabrous or pubescent, trichomes straight; petals yellow; fruits: distal segment seedless.
    2. Perennials, with adventitious buds on roots; stems frequently foliose, glabrescent or sparsely pubescent basally; fruits erect; gynophores 0.5–3 mm . . . . 1. *Diplotaxis tenuifolia*
    2. Annuals or perennials (short-lived), without buds on roots; stems frequently scapose, moderately pubescent; fruits erect-patent; gynophores obsolete or to 0.5 mm . . . . . . 2. *Diplotaxis muralis*

1. **Diplotaxis tenuifolia** (Linnaeus) de Candolle, Syst. Nat. 2: 632. 1821 • Perennial or slimleaf wall-rocket, flixweed [I][W]

Sisymbrium tenuifolium Linnaeus, Cent. Pl. I, 18. 1755

Perennials, (usually suffrutescent, roots with shoots from adventitious buds), strongly scented (with glucosinolates), (glaucescent). **Stems** erect, 2–7(–10) dm, glabrescent or sparsely pubescent basally. **Basal leaves**: blade elliptic to obovate, 2–15 cm × 10–60(–80) mm, margins sinuate to deeply pinnatifid, (2–5 lobes each side). **Cauline leaves** petiolate; similar to basal, (distal cauline shortly petiolate, blade similar, with narrower segments), surfaces usually glabrescent. **Fruiting pedicels** 8–35 mm. **Flowers:** sepals 4–6 mm, glabrous or pubescent, trichomes straight; petals yellow, 7–11(–13) × 5–8 mm, (apex rounded); filaments 4–8 mm; anthers 2.5–3 mm; gynophore 0.5–3 mm. **Fruits** usually erect, rarely ascending, (somewhat torulose, slightly compressed), 2–5 cm × 1.5–2.5 mm; terminal segment (stout), beaklike, 1.5–3 mm, seedless; (ovules 20–32(–46) per ovary). **Seeds** 1–1.3 × 0.6–0.9 mm. $2n = 22$.

Flowering spring–fall. Waste places, disturbed areas, wharf and railroad ballast, sandy beaches, muddy shores, wet woods, mountain slopes; 0–2100 m; introduced; N.S., Ont., Que.; Ariz., Calif., Conn., Fla., Ga., Ind., Maine, Mass., Mich., Mo., N.J., N.Mex., N.Y., Ohio, Oreg., Pa., Tex.; Eurasia; Africa; introduced also in South America (Argentina), Australia.

Diplotaxis tenuifolia was introduced from Europe as a ballast plant in the last century. It may have failed to persist in some of the recorded provinces and states.

2. **Diplotaxis muralis** (Linnaeus) de Candolle, Syst. Nat. 2: 634. 1821 • Annual or stinking wall-rocket, stinkweed, cross-weed, sand-rocket, wall-mustard [F][I][W]

Sisymbrium murale Linnaeus, Sp. Pl. 2: 658. 1753; *Sinapis muralis* (Linnaeus) W. T. Aiton

Annuals or perennials, (short-lived, frequently scapose or subscapose, taprooted), strongly scented (with glucosinolates). **Stems** ascending to suberect, (0.5–)2–5(–6) dm, moderately pubescent (trichomes predominantly patent basally, retrorse distally to near racemes). **Basal leaves** (rosulate); blade elliptic to obovate, 2–9 cm × 10–35 mm, margins sinuate to pinnatifid, lyrate, [2–4(–6) lobes each side], (margins and veins glabrescent to sparsely pubescent). **Cauline leaves** shortly petiolate to sessile; blade margins entire or dentate. **Fruiting pedicels** (3–)8–20(–37) mm. **Flowers:** sepals 3–5.5 mm, pubescent or glabrous, trichomes straight; petals yellow, 5–8(–10) × 3–5 mm; filaments 3.5–6 mm; anthers 1.5–2 mm; gynophore obsolete or to 0.5 mm. **Fruits** erect-patent, (1.5–)2–4 cm × 1.5–2.5 mm; terminal segment beaklike, (1–)1.5–3 mm, seedless; (ovules 20–36 per ovary). **Seeds** 0.9–1.3 × 0.6–0.9 mm. $2n = 42$.

Flowering spring–fall. Waste ground, disturbed or cultivated soil, ballast places, wharves, roadsides, railroads, around buildings, grazed grasslands; 80–2000 m; introduced; Alta., N.B., N.S., Ont., P.E.I., Que., Sask.; Ala., Ariz., Ark., Calif., Conn., Fla., Ill., Ind., Iowa, Kans., Mass., Mich., Minn., Mont., Nebr., N.J., N.Mex., N.Y., Ohio, Oreg., Pa., Tex., Utah, Wis.; Eurasia; Africa; introduced also in Mexico (Coahuila, Nuevo León), West Indies (Bahamas), Bermuda, South America, Pacific Islands (New Zealand), Australia.

Diplotaxis muralis was introduced from Europe as a ballast plant in the last century and may have failed to persist in some of the recorded provinces and states. It is an allopolyploid arisen from *D. tenuifolia* and the Eurasian *D. viminea* (Linnaeus) de Candolle with $2n = 20$ (M. D. Sánchez-Yélamo and J. B. Martínez-Laborde 1991; K. Mummenhoff et al. 1993; G. Eschmann-Grupe et al. 2003). There does not seem to be a sound basis for attributing *D. viminea* to the flora area, as done by V. I. Dorofeev (1998), because most specimens cited by him belong, in fact, to *D. muralis*.

3. **Diplotaxis erucoides** (Linnaeus) de Candolle, Syst. Nat. 2: 631. 1821 [I]

Sinapis erucoides Linnaeus, Cent. Pl. II, 24. 1756

Annuals or biennials, not scented. **Stems** erect, 1–4(–8) dm, densely pubescent throughout, (trichomes retrorse, appressed). **Basal leaves:** blades elliptic to obovate, 2.5–8 cm × 10–40 mm, margins sinuate to pinnatifid or lyrate, (2–5 lobes each side), (surfaces pubescent throughout, trichomes antrorse). **Cauline leaves** (distal) sessile; blade (base cuneate to broad, truncate), margins similar to basal, (distally reduced, subtending proximal flowers). **Fruiting pedicels** 3–10(–22) mm. **Flowers:** sepals 4–5.5 mm, pubescent, trichomes ± flexuous; petals white (turning purple when dried), 7–10 × 4–5 mm; filaments 4–6.5 mm; anthers 1.5–2 mm; gynophore obsolete or to 0.5 mm. **Fruits** erect-patent, 2–3.5(–4) cm × 1.5–2(–2.5) mm; terminal segment beaklike, 2–5 mm, 1- or 2-seeded. **Seeds** 1–1.2 × 0.5–0.8 mm. $2n = 14$.

Flowering summer. Ballast and waste places; introduced; Que.; Calif., Mass., N.J.; Eurasia; Africa; introduced also in South America (Argentina).

Diplotaxis erucoides was introduced from Europe as a ballast plant in the last century and may have failed to persist in some of the recorded provinces and states.

23. ERUCA Miller, Gard. Dict. Abr. ed. 4, vol. 1. 1754 • [Latin *uro*, burn, alluding to the burning taste of seeds]

Suzanne I. Warwick

Annuals; not scapose; glabrous, hirsute, or hispid, (trichomes often retrorse). **Stems** erect, branched [unbranched]. **Leaves** basal and cauline; petiolate or sessile; basal rosulate or not, petiolate, blade margins usually lyrate-pinnatifid or pinnatipartite, rarely bipinnatisect or undivided; cauline shortly petiolate or sessile, blade (base not auriculate), margins entire, dentate, or pinnatifid. **Racemes** (corymbose), greatly elongated in fruit. **Fruiting pedicels** erect to ascending, stout. **Flowers:** sepals (sometimes persistent), erect, oblong [linear], (connivent), lateral pair saccate basally; petals cream or yellow (with dark brown or purple veins), broadly obovate, claw differentiated from blade (± equal to sepals, apex obtuse or emarginate); stamens strongly tetradynamous; filaments not dilated basally; anthers oblong or linear, (base sagittate, apex obtuse); nectar glands (4), distinct, median pair present. **Fruits** siliques, dehiscent, sessile, segments 2, linear or oblong [elliptic], not torulose, terete or slightly 4-angled; (terminal segment indehiscent, flattened and ensiform, seedless); valves each with prominent midvein, (coriaceous), glabrous, hirsute, or hispid; replum rounded; septum complete, (membranous); ovules 10–50 per ovary; (style obsolete); stigma conical, 2-lobed (lobes connivent, decurrent). **Seeds** biseriate, plump, not winged, [sub]globose or ovoid; seed coat (minutely reticulate), mucilaginous when wetted; cotyledons conduplicate. $x = 11$.

Species 1: introduced; Eurasia, Africa; introduced also in Mexico, Central America, South America, Asia, Atlantic Islands, Australia.

1. **Eruca vesicaria** (Linnaeus) Cavanilles, Descr. Pl., 426. 1802

Brassica vesicaria Linnaeus, Sp. Pl. 2: 668. 1753

Subspecies 2 (1 in the flora): introduced; Europe (Spain), Asia, nw Africa; introduced also in South America, Australia.

1a. **Eruca vesicaria** (Linnaeus) Cavanilles subsp. **sativa** (Miller) Thellung in G. Hegi et al., Ill. Fl. Mitt.-Eur. 4: 201. 1918 • Garden- or salad-rocket, rocket-salad, rocket, edible-rocket, arugula, roquette

Eruca sativa Miller, Gard. Dict. ed. 8, Eruca no. 1. 1768; *Brassica eruca* Linnaeus

Stems usually branched basally, (1–)2–8(–10) dm, glabrous, hirsute, or hispid. **Basal leaves** (often withered by fruiting); petiole (1–)2–5(–7) cm; blade widely oblanceolate or pinnatisect, (2–)4–15(–20) cm × (10–)20–40(–60) mm, lobes 3–9 on each side, lobe margins entire or dentate. **Cauline leaves** (distal) subsessile; blade lobed or not, similar to basal. **Fruiting pedicels** (subappressed to rachis), 2–8(–10) mm. **Flowers:** sepals (6–)7–10 (–12) × 1.5–2.2 mm, outer pair cucullate or not, glabrous or with subapical tuft of trichomes; petals broadly obovate, (12–)15–20(–26) × (4–)5–7(–9) mm; median filaments (8–)10–13(–15) mm; anthers 2–3 mm. **Fruits** (1.1–)1.5–3.5(–4) cm × (2.5–)3–5 mm; valves (0.7–) 1–2.5(–3.2) mm; terminal segment 5-veined, (4–)5–10 (–11) mm, as long as or slightly shorter than valves. **Seeds** pale or grayish brown, 1.6–2.5 mm. $2n = 22$.

Flowering Mar–Sep. Roadsides, disturbed areas, waste places, cultivated fields, dry ditches, rocky outcrops, gravelly slopes, sandy plains, open rangelands; 0–1200 m; introduced; Alta., B.C., Man., Ont., Que., Sask.; Ariz., Calif., Colo., Conn., Ill., Iowa, Kans., Md., Mich., Mo., Mont., Nebr., Nev., N.J., N.Mex., N.Y., N.Dak., Oreg., Pa., S.Dak., Tex., Utah, Vt., Wash., W.Va.; Europe; Africa; introduced also in Mexico, Central America, South America, Asia, Atlantic Islands, Australia.

Subspecies *sativa*, widely naturalized and cultivated, was first introduced as a weed in North America in Flathead County, Montana, in 1898, with additional reports from 1900 to the 1920s as a seed contaminant of alfalfa fields in the United States.

Subspecies *vesicaria* and *pinnatifida* (Desfontaines) Emberger & Maire are endemic to Spain and North Africa and have escaped from cultivation in Europe; they seem not to have become adventive in North America (R. C. Rollins 1993). Recent molecular studies by S. I. Warwick and L. D. Black (1993) support the treatment of subsp. *vesicaria* and its presumed derivative subsp. *sativa* as a single species; subsp. *pinnatifida* is maintained as *Eruca pinnatifida* (Desfontaines) Pomel.

The earliest cultivation of subsp. *sativa* dates back to the ancient Romans and Greeks. It is currently grown in Europe and North America as a salad plant and in Asia for cooking oil and as food for animals. The oil is also used as an industrial lubricant and for cosmetic and medicinal purposes (I. A. Al-Shehbaz 1985). The seed cake and the entire plant are used as fodder for domestic animals. The oil is high in erucic acid and glucosinolates and is known to cause various skin allergies.

24. ERUCASTRUM C. Presl, Fl. Sicul., 92. 1826 • [Genus *Eruca* and Latin *-astrum*, resembling] I

Suzanne I. Warwick

Annuals or biennials [perennials, subshrubs]; not scapose; pubescent. **Stems** erect or ascending, unbranched or branched. **Leaves** basal and cauline; petiolate or sessile; basal rosulate or not, petiolate, blade margins dentate to lyrate-pinnatifid, [pinnatipartite]; cauline shortly petiolate or sessile, blade (base not auriculate), similar to basal. **Racemes** (corymbose, bracteate throughout or basally), greatly elongated in fruit. **Fruiting pedicels** ascending to divaricate [spreading], slender. **Flowers:** sepals erect, oblong [linear], lateral pair not saccate basally; petals yellow or white, obovate to oblanceolate or oblong, gradually attenuated to short claw, (apex rounded); stamens tetradynamous; filaments not dilated basally; anthers oblong or linear, (base sagittate, apex obtuse); nectar glands distinct, median glands present. **Fruits** siliques, dehiscent, subsessile, segments 2, linear, often torulose, subterete or 4-angled; (terminal segment stylelike, seedless); valves each with prominent midvein, glabrous; replum rounded; septum complete; ovules 10–60 per ovary; stigmas capitate, entire or 2-lobed. **Seeds** uniseriate, plump, not winged, elliptic [oblong, ovoid]; seed coat (usually reticulate), slightly mucilaginous when wetted; cotyledons conduplicate. $x = 7, 8, 9$.

Species 19 (1 in the flora): introduced; Europe, Asia, Africa.

SELECTED REFERENCE Luken, J. O., J. W. Thieret, and J. T. Kartesz. 1993. *Erucastrum gallicum* (Brassicaceae): Invasion and spread in North America. Sida 15: 569–582.

1. **Erucastrum gallicum** (Willdenow) O. E. Schulz, Bot. Jahrb. Syst. 54(Beibl. 119): 56. 1916 • Dog mustard, rocket-weed, bracted-rocket, French rocket F I W

Sisymbrium gallicum Willdenow, Enum. Pl., 678. 1809; *Erucastrum pollichii* Schimper & Spenner

Plants sparsely to densely pubescent, trichomes stiff, recurved (or retrorsely appressed). **Stems** erect or ascending, unbranched or branched (few to several), 0.9–6.5(–8) dm. **Basal leaves:** blade oblanceolate, 3–28 cm × 8–110 mm, margins dentate to deeply lobed or pinnatifid, lobes 3–10 each side, smaller than terminal, lobe margins crenate or dentate, surfaces sparsely pubescent. **Cauline leaves** similar to basal, distal shortly petiolate or sessile, blade smaller (distalmost 1–2 cm, passing into bracts, leaflike, linear, margins entire). **Fruiting pedicels** (3–)5–10(–20) mm. **Flowers:** sepals 3–5 × 1–2 mm, sparsely hispid apically; petals white to pale yellow, 4–8 × 1.5–3 mm; filaments 3.5–5.5 mm. **Fruits** slightly torulose, 1–4.5 cm × 1–2(–2.7) mm; terminal segment 1.5–4 mm; style 1–3 mm. **Seeds** reddish brown, 1.1–1.5 × 0.7–0.8 mm, alveolate. $2n = 30$.

Flowering Mar–Sep(–Dec in south, fruiting shortly after). Roadsides, waste places, disturbed sites, along railroads, fields, gardens, orchards, beaches of Great Lakes; 0–2000 m; introduced; St. Pierre and Miquelon; Alta., B.C., Man., N.B., Nfld. and Labr., N.W.T., N.S., Ont., P.E.I., Que., Sask.; Ala., Calif., Conn., Fla., Idaho, Ill., Ind., Iowa, Kans., Ky., Maine, Md., Mass., Mich., Minn., Mo., Mont., N.H., N.Y., N.Dak., Ohio, Oreg., Pa., S.Dak., Tex., Vt., Wash., W.Va., Wis., Wyo.; Europe.

A European native, *Erucastrum gallicum* was first recorded for North America from Massachusetts and Wisconsin (see J. O. Luken et al. 1993 for history of introduction and spread). It is naturalized in all the provinces of Canada and in parts of the United States, particularly the Midwest. It is an allopolyploid, with the $n = 7$ component from *Diplotaxis erucoides/ D. cossoniana* and $n = 8$ from the *E. nasturtiifolium* complex (S. I. Warwick and L. D. Black 1993). I have not seen specimens from Maryland.

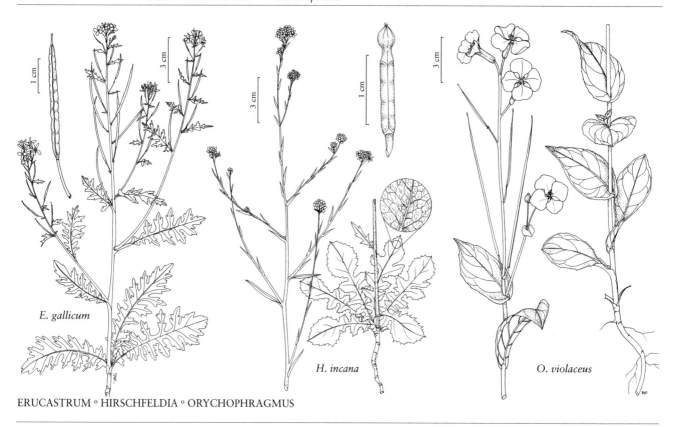

ERUCASTRUM · HIRSCHFELDIA · ORYCHOPHRAGMUS

25. HIRSCHFELDIA Moench, Methodus, 264. 1794 • [For Christian Cajus Lorenz Hirschfeldt, 1742–1792, Austrian botanist/horticulturist] [I]

Suzanne I. Warwick

Annual or biennials; not scapose; pubescent. **Stems** (simple or several from base), erect, branched basally and distally. **Leaves** basal and cauline; petiolate or subsessile; basal rosulate, petiolate, blade lyrate to pinnatifid, margins crenate-dentate; cauline subsessile or petiolate, blade (base not auriculate), margins dentate or pinnatifid. **Racemes** (corymbose, several-flowered), considerably elongated in fruit. **Fruiting pedicels** erect, stout. **Flowers:** sepals widely spreading or reflexed, oblong, lateral pair not saccate basally; petals yellow, obovate to spatulate, claw differentiated from blade, (apex obtuse); stamens tetradynamous; filaments not dilated basally; anthers oblong or ovate, (apex obtuse); nectar glands not confluent, median glands present. **Fruits** siliques, dehiscent, sessile, segments 2, linear, slightly torulose, terete or slightly 4-angled; (proximal segment not torulose, somewhat corky at maturity, 8–20-seeded; terminal segment indehiscent, 1- or 2-seeded, slightly swollen apically); valves 3(–7)-veined, usually glabrous, rarely sparsely pubescent; replum rounded; septum complete; ovules 10–22 per ovary; (style present); stigma capitate, entire. **Seeds** uniseriate, plump, not winged, globose; seed coat (smooth to finely reticulate), mucilaginous when wetted; cotyledons conduplicate. $x = 7$.

Species 1: introduced; Eurasia, nw Africa; introduced also in South America, s Africa, Atlantic Islands, Pacific Islands (Hawaii), Australia.

It is with some hesitation that I recognize this genus; it should perhaps be united with *Erucastrum*, as recently proposed for conservation by I. A. Al-Shehbaz (2005b). As clearly shown by S. I. Warwick and L. D. Black (1993), *Brassica*, *Diplotaxis*, and *Erucastrum* are artificially delimited genera, and a substantial revision of their boundaries is needed.

1. **Hirschfeldia incana** (Linnaeus) Lagrèze-Fossat, Fl. Tarn Garonne, 19. 1847 • Mediterranean mustard, summer- or hoary-mustard F I W

Sinapis incana Linnaeus, Cent. Pl. I, 19. 1755; *Brassica adpressa* (Moench) Boissier; *B. geniculata* (Desfontaines) Ball; *Hirschfeldia adpressa* Moench; *Sinapis geniculata* Desfontaines

Stems (2–)4–15(–20) dm, densely pubescent proximally, trichomes retrorse. **Basal leaves:** petiole 1–4(–10) cm; blade (3–)4–22(–35) cm × 15–60 (–80) mm, lobes 1–6(–9) each side, ovate or lanceolate, (smaller than terminal), terminal lobe broadly ovate, surfaces densely pubescent. **Cauline leaves** (distal) ± sessile; blade oblong to lanceolate, similar to basal, (smaller distally). **Fruiting pedicels** (appressed to rachis, almost as thick as fruit), 2–4(–5) mm. **Flowers:** sepals 3–5 × 1.2–2 mm; petals 5–10 × 2.5–4.5 mm; filaments 3–5 mm; anthers 1–1.5 mm. **Fruits** 0.7–1.5(–1.7) cm × 1–1.7 mm; valves 6–10 mm; terminal segment 3–6 mm. **Seeds** 0.9–1.5 mm diam. $2n = 14$.

Flowering Apr–Nov. Roadsides, waste places, disturbed areas, canyons, creek bottoms, dry fields, open desert; 100–1600 m; introduced; Calif., Nev., Oreg.; Eurasia; nw Africa; introduced also in South America, s Africa, Atlantic Islands, Pacific Islands (Hawaii), Australia.

Hirschfeldia incana was first collected in North America in 1895 in the San Bernardino region, and by 1936 it was described as "already a serious agricultural pest, spreading freely over dry, unbroken ground and flourishing chiefly during the arid summer season" (W. L. Jepson 1909–1943, vol. 2).

Hirschfeldia incana can be confused with *Brassica nigra* because both have fruits appressed to the rachis. The former is distinguished from the latter by its distinctly shorter fruit, seeded and often swollen beak, and smaller petals.

26. ORYCHOPHRAGMUS Bunge, Enum. Pl. China Bor., 7. 1833 • [Greek *oryche*, pit, and *phragmos*, partition, alluding to fruit septum] I

Ihsan A. Al-Shehbaz

Annuals or biennials [perennials]; (sometimes rhizomatous); not scapose; pubescent or glabrous. **Stems** erect or ascending, unbranched or branched distally. **Leaves** basal and cauline; petiolate or sessile; basal not rosulate, petiolate or sessile, blade (simple or pinnatisect with 1–6 leafletlike lobes on each side), margins crenate; cauline petiolate or sessile, blade (base sometimes auriculate or amplexicaul), margins entire, dentate, or, sometimes, with 1–4 lateral lobes [crenate or serrate]. **Racemes** (corymbose, several-flowered). **Fruiting pedicels** divaricate-ascending [divaricate, recurved], slender or stout. **Flowers:** sepals erect [ascending], linear [oblong], lateral pair strongly [slightly] saccate basally; petals purple, lavender, or white, broadly obovate [narrowly obcordate], claw obscurely to considerably differentiated from blade ([shorter than] as long as sepals, apex rounded [emarginate]); stamens tetradynamous; filaments dilated basally; anthers linear [oblong], (apiculate); nectar glands (2), lateral, semiannular or annular. **Fruits** siliques, dehiscent, sessile or shortly stipitate, linear, torulose, terete or somewhat 4-angled; valves (leathery), each with prominent midvein, usually glabrous, rarely densely hirsute; replum rounded; septum complete; ovules [20–]40–70 per ovary; stigma capitate, 2-lobed (lobes distinct, decurrent). **Seeds** uniseriate, plump, not winged, oblong; seed coat (alveolate-reticulate), not mucilaginous when wetted; cotyledons conduplicate. $x = 12$.

Species 2 (1 in the flora): introduced, Virginia; Asia (China, Japan, Korea).

Orychophragmus is fairly similar to, and closely related to, *Moricandia*, and they can easily be misidentified. The former has petiolate, pinnately divided, broadly auriculate proximalmost leaves, non-cucullate median sepals, apiculate anthers, and non-connivent stigma lobes. By contrast, *Moricandia* has sessile, entire, non-auriculate proximalmost leaves, cucullate median sepals, non-apiculate anthers, and connivent stigma lobes.

Orychophragmus limprichtianus (Pax) Al-Shehbaz & G. Yang is cultivated as an ornamental, rarely escaping.

SELECTED REFERENCE Al-Shehbaz, I. A. and Yang G. 2000. A revision of the Chinese endemic *Orychophragmus* (Brassicaceae). Novon 10: 349–353.

1. **Orychophragmus violaceus** (Linnaeus) O. E. Schulz, Bot. Jahrb. Syst. 54(Beibl. 119): 56. 1916 [F] [I]

Brassica violacea Linnaeus, Sp. Pl. 2: 667. 1753

Stems (0.6–)1.5–6(–9) dm, often branched distally, glabrous or sparsely to densely pilose. **Basal leaves**: petiole (1–)2–8(–11) cm; blade or terminal lobe cordate, reniform, broadly ovate, or suborbicular, (0.4–)1.5–10(–14) cm × (3–)10–40(–70) mm, base usually cordate, rarely obtuse, margins coarsely crenate with teeth ending in apiculae, apex acute or obtuse; lobes (0 or) 1–6 each side, sessile or petiolulate (to 3 × 2 cm), glabrous or pilose. **Cauline leaves** sessile or petiolate; (distal) blade (0.5–)2–9(–15) cm × (2–)10–60(–90) mm, base sometimes auriculate or amplexicaul, margins dentate or entire, apex acute or acuminate (auricles to 3 × 4 cm); lobes (0 or) 1–4 each side, sessile or petiolulate, glabrous or pilose. **Fruiting pedicels** (0.6–)0.8–2(–3) cm, narrower than fruit, glabrous or pilose. **Flowers**: sepals connivent, (6–)8–13(–16) × 1.5–2.5 mm; petals (12–)16–25(–32) × (4–)5–9(–11) mm; filaments 8–18 mm; anthers (3–)4–6(–8) mm. **Fruits** (3–)4.5–11(–13) cm × 1.5–3 mm; style (0.3–)0.7–3(–5.5) cm; stigma slightly to distinctly 2-lobed. **Seeds** 2–3(–3.5) × 1–2 mm. $2n = 24$.

Flowering Mar–Jun. Railroad tracks; introduced; Va.; Asia (China, Japan, Korea).

This is the first report of *Orychophragmus violaceus* as naturalized in North America, though I have seen the plant cultivated as an ornamental in multiple places. The record is based on *Wright 3145* (GH), a collection made on 12 May 1987 in the woods along Southern Railroad tracks in Richmond, Virginia; it is the very same collection on which R. C. Rollins (1993) based his record of *Moricandia arvensis* (Linnaeus) de Candolle for North America. The species appears to be spreading in other locations in the neighboring areas (Rollins).

27. **RAPHANUS** Linnaeus, Sp. Pl. 2: 669. 1753; Gen. Pl. ed. 5, 300. 1754 • Radish [Greek *raphanos*, radish] [I]

Suzanne I. Warwick

Quidproquo Greuter & Burdet

Annuals or **biennials**; (roots slender or fleshy, size, shape, and color variable in cultivated forms); not scapose; glabrous or pubescent. **Stems** erect, unbranched or branched. **Leaves** basal and cauline; petiolate or subsessile; basal not rosulate, petiolate, blade margins lyrately lobed or pinnatifid to pinnatisect; cauline shortly petiolate or subsessile, blade (base not auriculate), margins dentate or lobed, (smaller and fewer-lobed than basal). **Racemes** (corymbose, several-flowered), usually greatly elongated in fruit. **Fruiting pedicels** divaricate, ascending, or spreading [reflexed]. **Flowers**: sepals erect, narrowly oblong [linear], lateral pair slightly saccate basally; petals white, creamy white, yellow, pink, or purple [lilac] (usually with darker veins), broadly obovate [suborbicular], claw differentiated from blade, (± longer than sepals, apex obtuse or emarginate [rounded]); stamens strongly tetradynamous; filaments not dilated basally; anthers oblong or oblong-linear, (apex obtuse); nectar glands (4), median pair present. **Fruits** siliques or silicles, indehiscent, sessile, segments 2, (lomentaceous, often breaking into 1-seeded units), cylindrical, fusiform, lanceolate, or ovoid, [linear, oblong, ellipsoid], smooth or torulose to strongly moniliform, (constricted or not between seeds), terete or polygonal; (valvular segment seedless, rudimentary, or aborted, nearly as wide as pedicel; terminal segment several-seeded, corky); valves glabrous, antrorsely scabrous, or hispid; replum and septum not differentiated; ovules 2–22 per ovary; (style slender); stigma capitate, slightly 2-lobed. **Seeds** uniseriate, plump, not winged, oblong, ovoid, or globose [subglobose]; seed coat (nearly smooth to reticulate), not mucilaginous when wetted; cotyledons conduplicate. $x = 9$.

Species 3 (2 in the flora): introduced; Eurasia; introduced also nearly worldwide.

Natural hybridization between *Raphanus raphanistrum* and *R. sativus* has been known since 1788, and the hybrid has been named *R.* ×*micranthus* (Uechtritz) O. E. Schulz. The transfer of some of the weedy characters from *R. raphanistrum* to *R. sativus* through natural hybridization may have played a major role in converting the latter from a crop plant into a successful weed near the coastal areas of central California (C. A. Panetsos and H. G. Baker 1968). *Raphanus confusus* (Greuter & Burdet) Al-Shehbaz & Warwick is known from Asia (Israel, Lebanon).

1. Petals pale or creamy white; fruits (2.5–)3–8(–11) mm wide, strongly constricted between seeds and usually breaking, strongly ribbed, beak narrowly conical . . . . 1. *Raphanus raphanistrum*
1. Petals usually purple or pink, sometimes white; fruits (5–)7–13(–15) mm wide, rarely slightly constricted between seeds and usually not breaking, not ribbed, beak narrowly to broadly conical to linear . . . . 2. *Raphanus sativus*

1. Raphanus raphanistrum Linnaeus, Sp. Pl. 1: 669. 1753 • Wild-radish, jointed charlock or radish or wild-radish [F][I][W]

Annuals, roots not fleshy; sparsely to densely pubescent. **Stems** usually simple from base, (2–)3–8 dm, (retrorsely hispid). **Basal leaves:** petiole 1–6 cm; blade oblong, obovate, or oblanceolate in outline, lyrate or pinnatifid, sometimes undivided, 3–15(–22) cm × 10–50 mm, margins dentate, apex obtuse or acute; lobes 1–4 each side, oblong or ovate, to 4 cm × 20 mm (smaller than terminal). **Cauline leaves** (distal) subsessile; blade often undivided. **Fruiting pedicels** divaricate or ascending, 7–25 mm, (straight). **Flowers:** sepals 7–11 × 1–2 mm, sparsely pubescent; petals yellow or creamy white (veins dark brown or purple), 15–25 × 4–7 mm, claw to 15 mm; filaments (slender), 7–12 mm; anthers 2–2.5 mm. **Fruits** cylindrical or narrowly lanceolate; valvular segment 1–1.5 mm; terminal segment (1.5–)2–11(–14) cm × (2.5–)3–8(–11) mm, (base rounded), strongly constricted between seeds (usually breaking), strongly ribbed, beak narrowly conical; style 10–50 mm. **Seeds** (reddish brown or dark brown to black), oblong or ovoid, 2.5–3.5 × 1.8–2.5 mm. $2n = 18$.

Flowering May–Jul. Disturbed waste places, cultivated fields, roadsides, orchards, hill slopes; 0–800 m; introduced; Greenland; St. Pierre and Miquelon; Alta., B.C., Man., N.B., Nfld. and Labr., N.S., Ont., P.E.I., Que., Sask.; Ala., Alaska, Ariz., Ark., Calif., Colo., Conn., Del., D.C., Fla., Ga., Idaho, Ill., Ind., Iowa, Kans., Ky., La., Maine, Md., Mass., Mich., Minn., Miss., Mo., Mont., Nebr., Nev., N.H., N.J., N.Mex., N.Y., N.C., N.Dak., Ohio, Okla., Oreg., Pa., R.I., S.C., S.Dak., Tenn., Tex., Utah, Vt., Va., Wash., W.Va., Wis., Wyo.; Europe; Asia; introduced also in Mexico, Central America, South America, Africa, Atlantic Islands, Australia.

North American representatives of *Raphanus raphanistrum* are referable to subsp. *raphanistrum*. Four other subspecies are restricted to Europe.

2. Raphanus sativus Linnaeus, Sp. Pl. 2: 669. 1753 • Cultivated radish [I][W]

Annuals or biennials, roots often fleshy in cultivated forms; often sparsely scabrous or hispid, sometimes glabrous. **Stems** often simple from base, (1–)4–13 dm. **Basal leaves:** petiole 1–30 cm; blade oblong, obovate, oblanceolate, or spatulate in outline, lyrate or pinnatisect, sometimes undivided, 2–60 cm × 10–200 mm, margins dentate, apex obtuse or acute; lobes 1–12 each side, oblong or ovate, to 10 cm × 50 mm. **Cauline leaves** (distal) subsessile; blade often undivided. **Fruiting pedicels** spreading to ascending, 5–40 mm. **Flowers:** sepals 5.5–10 × 1–2 mm, glabrous or sparsely pubescent; petals usually purple or pink, sometimes white (veins

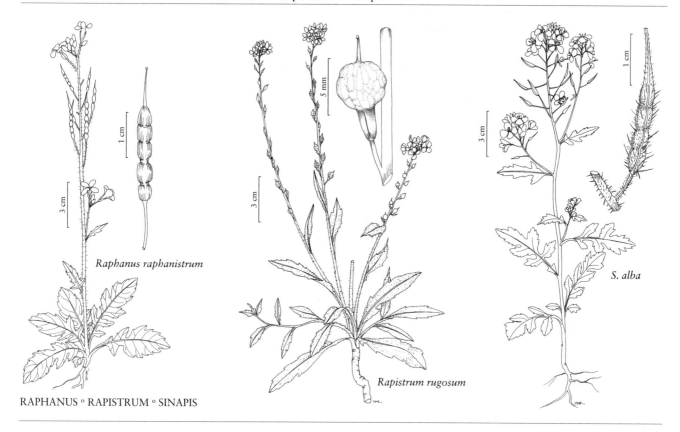

RAPHANUS ∘ RAPISTRUM ∘ SINAPIS

often darker), 15–25 × 3–8 mm, claw to 14 mm; filaments 5–12 mm; anthers 1.5–2 mm. **Fruits** usually fusiform or lanceolate, sometimes ovoid or cylindrical; valvular segment 1–3.5 mm; terminal segment (1–)3–15(–25) cm × (5–)7–13(–15) mm, smooth or, rarely, slightly constricted between seeds, not ribbed, beak narrowly to broadly conical to linear; style 10–40 mm. **Seeds** globose or ovoid, 2.5–4 mm diam. $2n = 18$.

Flowering May–Jul. Roadsides, disturbed areas, waste places, cultivated fields, gardens, orchards; 0–1000 m; introduced; B.C., Man., N.B., Nfld. and Labr., N.S., Ont., P.E.I., Que., Sask.; Ala., Alaska, Ariz., Ark., Calif., Colo., Conn., Del., D.C., Fla., Ga., Idaho, Ill., Ind., Iowa, Kans., Ky., La., Maine, Md., Mass., Mich., Minn., Miss., Mo., Mont., Nebr., Nev., N.H., N.J., N.Mex., N.Y., N.C., N.Dak., Ohio, Okla., Oreg., Pa., R.I., S.C., S.Dak., Tenn., Tex., Utah, Vt., Va., Wash., W.Va., Wis., Wyo.; Europe; Asia; introduced also in Mexico, Bermuda, South America, Africa, Atlantic Islands, Australia.

Raphanus sativus is an important crop plant that is cultivated and/or weedy in most temperate regions worldwide. It is unknown as a wild plant, but suggested to be derived from *R. raphanistrum* subsp. *landra*, which is endemic to the Mediterranean region (L. J. Lewis-Jones et al. 1982).

28. **RAPISTRUM** Crantz, Cl. Crucif. Emend., 105. 1769, name conserved • Wild-turnip, turnipweed [Latin *rapa*, turnip, and *-astrum*, incomplete resemblance] ⓘ

Suzanne I. Warwick

Annuals [perennials]; not scapose; glabrous or pubescent. **Stems** erect, unbranched or branched basally. **Leaves** basal and cauline; petiolate or subsessile; basal not rosulate, petiolate, blade margins usually pinnate to lyrately pinnatifid, rarely undivided, dentate; cauline subsessile or shortly petiolate, blade (base not auriculate), margins lobed, subentire, or dentate. **Racemes** (corymbose, several-flowered), greatly elongated in fruit. **Fruiting pedicels** erect [ascending], (usually appressed to rachis), slender to stout. **Flowers:** sepals ascending, oblong, lateral pair not saccate basally, (hispid [glabrous or with subapical tuft of hairs]); petals yellow, obovate, claw differentiated from blade, (apex ± truncate); stamens tetradynamous; filaments not dilated basally; anthers ovate to suboblong, (apex obtuse); nectar glands confluent, median glands present. **Fruits** silicles, indehiscent, sessile, segments 2, elliptic to oblong, torulose, (slightly to strongly constricted at transverse joint), terete or angular; (valvular segment persistent, dehiscent [indehiscent], 1(–3)-seeded, longitudinally striate or smooth, occasionally seedless and nearly as wide as pedicel; terminal segment indehiscent, caducous at maturity, usually 1-seeded, rarely seedless); valves glabrous or pubescent; septum complete; ovules 2–4; (style persistent, filiform); stigma capitate, (flattened), 2-lobed. **Seeds** uniseriate, slightly compressed, not winged, ovoid [oblong]; seed coat (smooth), not mucilaginous when wetted; cotyledons conduplicate. $x = 8$.

Species 2 (1 in the flora): introduced; s Europe (Mediterranean region); introduced also nearly worldwide.

Both species of *Rapistrum* have been introduced into North America; only *R. rugosum* has persisted with naturalized populations. *Rapistrum perenne* (Linnaeus) Allioni was first collected in 1922 from southeastern Saskatchewan but has not been seen or collected from there since 1932. It can be distinguished by being a perennial with a conical style shorter than the strongly 8-ribbed terminal segment, whereas *R. rugosum* is an annual with a slender, filiform style longer than the slightly ribbed terminal segment.

1. **Rapistrum rugosum** (Linnaeus) Allioni, Fl. Pedem. 1: 640. 1785 • Bastard-cabbage [F][I][W]

Myagrum rugosum Linnaeus, Sp. Pl. 2: 257. 1753; *Rapistrum hispanicum* (Linnaeus) Crantz

Plants hispid proximally, glabrous distally. **Stems** (1–)2–10(–15) dm. **Basal leaves:** petiole 1–5 cm; blade with 1–5 lobes each side, 2–25 cm, margins irregularly dentate, lateral lobe oblong or ovate, terminal lobe suborbicular or ovate, larger than lateral. **Cauline leaves:** blade simple or sinuately lobed, margins subentire or dentate. **Fruiting pedicels** erect, (appressed to rachis), 1.5–5 mm. **Flowers:** sepals 2.5–5 mm; petals pale yellow, 6–11 × 2.5–4 mm; filaments 4–7 mm; anthers 1.2–1.5 mm. **Fruits:** valvular segment ellipsoid, 0.7–3 × 0.5–1.5 mm; terminal segment globose or ovoid, 1.5–3.5 × 1–2.8 mm, usually rugose or ribbed, rarely smooth; style 1–3(–5) mm. **Seeds** 1.5–2.5 mm. $2n = 16$.

Flowering Apr–Jul (Nov–Jun in Texas). Roadsides, disturbed sites, waste places, fields, grassy banks, ballast; 0–2000 m; introduced; Ont., Que.; Calif., Ind., La., Mass., Nev., N.J., N.Y., Oreg., Pa., Tenn., Tex., Wash., Wis.; Europe; introduced also in Mexico, South America, Asia, Africa, Atlantic Islands, Australia.

Three subspecies of *Rapistrum rugosum* have been recognized, including: subsp. *rugosum*, subsp. *linnaeanum* Rouy & Foucaud, and subsp. *orientale* (Linnaeus) Arcangeli. They are distinguished mainly on the basis of silique shape, rugosity, prominence of ribs, and the length and thickness of fruiting pedicels. These features vary considerably and intergradation occurs in such a wide array of combinations that it seems rather meaningless to recognize them without further studies. All three subspecies and some of their intermediates have been found in the United States (I. A. Al-Shehbaz 1985; R. C. Rollins 1993). The species was first collected in the flora area in 1873 as a ballast plant from Boston, Massachusetts (Al-Shehbaz).

29. SINAPIS Linnaeus, Sp. Pl. 2: 668. 1753; Gen. Pl. ed. 5, 299. 1754 • Mustard [Greek *sinapi*, mustard] [I]

Suzanne I. Warwick

Annuals [perennials]; not scapose; glabrous or pubescent. **Stems** erect, unbranched or branched distally. **Leaves** basal and cauline; petiolate or sessile; basal usually not rosulate, petiolate, blade margins usually lyrate, pinnatifid, or 1- or 2-pinnatisect, rarely undivided, (lobes usually coarsely dentate); cauline shortly petiolate or subsessile [sessile], blade (base not auriculate), margins often dentate or shallowly lobed [entire], rarely subentire. **Racemes** (corymbose, several-flowered), considerably elongated in fruit. **Fruiting pedicels** ascending, divaricate, or suberect [erect, reflexed], stout [slender]. **Flowers:** sepals usually spreading, rarely reflexed, narrowly oblong [linear], lateral pair not saccate basally; petals (spreading), yellow, obovate, claw differentiated from blade, (claw subequaling sepal, apex obtuse or emarginate); stamens tetradynamous; filaments not dilated basally; anthers oblong, (apex obtuse); nectar glands (4), distinct, lateral pair usually prismatic, rarely lobed, median pair present, (ovoid). **Fruits** siliques, dehiscent, sessile, segments 2, linear or lanceolate [oblong], torulose, terete or slightly flattened [4-angled, latiseptate]; (valvular segment dehiscent, longer or shorter than terminal segment, 2–5(–12)-seeded; terminal segment indehiscent, seedless or 1- [2-]seeded, flattened and ensiform, or terete and conical or subulate, sometimes corky); valves each with 3–5(–7) prominent, longitudinal veins, (thin or thick), glabrous or pubescent; replum rounded; septum complete; ovules 4–20 per ovary; stigma capitate, 2-lobed. **Seeds** uniseriate, usually plump, rarely slightly flattened, not winged, globose; seed coat (finely reticulate [smooth or alveolate]), mucilaginous or not when wetted; cotyledons conduplicate. x = [7], 9, 12.

Species 5 (2 in the flora): introduced; Eurasia, n Africa; introduced also nearly worldwide.

Sinapis has often been merged with *Brassica* in North American taxonomic treatments, whereas taxonomists elsewhere maintain both genera and separate them by the number of veins of the fruit valves and the orientation of sepals. The two genera also differ in their mustard oils and seed proteins, thus supporting their maintenance (I. A. Al-Shehbaz 1985). Molecular studies (S. I. Warwick and L. D. Black 1991, 1993) have suggested that neither genus is monophyletic and, except for *S. aucheri* (Boissier) O. E. Schulz, the remaining taxa of *Sinapis* show a very close relationship to *B. nigra*, consistent with Linnaeus's original description of this species as *S. nigra*.

SELECTED REFERENCE Baillargeon, G. 1986. Eine taxonomische Revision der Gattung *Sinapis* (Cruciferae: Brassiceae). Doctoral thesis. Freie Universität Berlin.

1. Fruiting pedicels divaricate; fruits lanceolate, hispid, trichomes of 2 types; terminal segment flattened, equal to or longer than valves, seedless .... 1. *Sinapis alba*
1. Fruiting pedicels ascending to suberect; fruits linear, glabrous or pubescent, trichomes of 1 type; terminal segment terete, much shorter than valves, seedless or 1-seeded .... 2. *Sinapis arvensis*

1. **Sinapis alba** Linnaeus, Sp. Pl. 2: 668. 1753 • White mustard [F][I][W]

Brassica alba (Linnaeus) Rabenhorst; *B. hirta* Moench; *Rorippa coloradensis* Stuckey

Plants usually hispid, rarely glabrous. **Stems** often branched distally, (0.15–)0.25–1(–2.2) dm. **Basal leaves:** petiole 1–3 (–6) cm; blade oblong, ovate, or lanceolate (in outline), (3.5–)5–14(–20) cm × 20–60(–80) mm, margins lyrate, pinnatifid, pinnatisect; lobes 1–3 each side, (oblong, ovate, or lanceolate, 1.5–2.5 cm), margins usually dentate or repand, rarely pinnatifid. **Cauline leaves** (distal) shortly petiolate; blade (ovate or oblong-ovate, 2–4.5 cm), margins coarsely dentate, rarely subentire. **Fruiting pedicels** divaricate, (3–)6–12(–17) mm. **Flowers:** sepals (3.8–)4–7(–8) × 1–1.8 mm; petals pale yellow, (7–)8–12(–14) × (3–)4–6(–7) mm; filaments (3–)4–7(–8) mm; anthers 1.2–1.5 mm. **Fruits** lanceolate, (1.5–)2–4.2(–5) cm × (2–)3–5.5(–6.5) mm; valvular segment terete or slightly flattened, (0.5–)0.7–1.7(–2) cm, 2–5-seeded per locule; terminal segment ensiform, flattened, (1–)1.5–2.5(–3) cm, equal to or longer than valves, seedless; valves hispid, trichomes of 2 types (subsetiform mixed with shorter, slender ones). **Seeds** pale yellow to pale brown, (1.7–)2–3(–3.5) mm diam. $2n$ = 24.

Flowering Mar–Sep. Escape from cultivation, roadsides, waste places, disturbed areas, grain fields, cultivated areas, gardens, orchards; 0–1000 m; introduced; Greenland; Alta., B.C., Man., N.B., N.S., Ont., P.E.I., Que., Sask., Yukon; Ariz., Calif., Colo., Conn., Del., D.C., Ill., Ind., Iowa, Maine, Mass., Mich., Minn., Mo., Mont., Nebr., N.H., N.J., N.Mex., N.Y., N.C., N.Dak., Ohio, Okla., Oreg., Pa., R.I., S.C., S.Dak., Tenn., Tex., Utah, Vt., Wash., W.Va., Wis.;

Eurasia; introduced also in Mexico, Central America, South America (Argentina), Europe, sw Asia, n Africa, Atlantic Islands.

Of the three subspecies of *Sinapis alba* recognized in European and North African floras, only subsp. *alba* is naturalized in the New World. This taxon is an important crop plant, and is occasionally reported as a weedy escape from cultivation. Its seeds are used for the manufacture of condiment mustard (see also 17. *Brassica*), pickling spice, and production of oils for soap and mayonnaise, lubrication, and cooking (I. A. Al-Shehbaz 1985).

2. **Sinapis arvensis** Linnaeus, Sp. Pl. 2: 668. 1753 · Charlock, wild mustard [I][W]

Brassica arvensis (Linnaeus) Rabenhorst; *B. kaber* (de Candolle) L. C. Wheeler; *B. sinapistrum* Boissier; *Sinapis kaber* de Candolle

Plants hirsute, hispid, or glabrous. **Stems** unbranched or branched, (0.5–)2–10(–21) dm, (often hirsute or hispid, sometimes glabrous, trichomes retrorse or spreading). **Basal leaves**: (proximal) petiole 1–4(–7) cm; blade obovate, oblong, or lanceolate, (3–)4–18(–25) cm × 15–50(–70) mm, margins lyrate, pinnatifid, or, sometimes, undivided; lobes 1–4 each side, margins of terminal and smaller lateral lobes coarsely toothed, (surfaces sparsely pubescent). **Cauline leaves** usually shortly petiolate, rarely subsessile; blade margins often not divided, coarsely toothed. **Fruiting pedicels** ascending or suberect, (2–)3–7 mm. **Flowers**: sepals (yellow or green), (4.5–)5–6(–7) × 1–1.8 mm; petals bright yellow, (8–)9–12(–17) × (3–)4–6(–7.5) mm; filaments (3–)4–6 mm; anthers 1.2–1.5 mm. **Fruits** linear, (1.5–)2–4.5(–5.7) cm × (1.5–)2.5–3.5(–4) mm; valvular segment terete, (0.6–)1.2–3.5(–4.3) cm, (2–)4–8(–12)-seeded per locule; terminal segment (straight or upcurved), conical or subulate, terete, (0.7–)1–1.6 cm, shorter than valves, seedless or 1-seeded; valves glabrous or pubescent, trichomes of 1 type. **Seeds** reddish brown to black, (1–)1.5–2 mm diam. $2n = 18$.

Flowering Mar–Oct. Roadsides, waste places, disturbed areas, fields, grain fields, orchards; 0–1800 m; introduced; Greenland; St. Pierre and Miquelon; Alta., B.C., Man., N.B., Nfld. and Labr., N.W.T., N.S., Ont., P.E.I., Que., Sask., Yukon; Alaska, Ariz., Ark., Calif., Colo., Conn., Del., D.C., Fla., Ga., Idaho, Ill., Ind., Iowa, Kans., Ky., La., Maine, Md., Mass., Mich., Minn., Miss., Mo., Mont., Nebr., Nev., N.H., N.J., N.Mex., N.Y., N.C., N.Dak., Ohio, Okla., Oreg., Pa., R.I., S.C., S.Dak., Tenn., Tex., Utah, Vt., Va., Wash., W.Va., Wis., Wyo.; Eurasia; introduced also in Mexico, West Indies, Central America, South America (Argentina), nw, s Africa, Atlantic Islands, Australia.

Infraspecific taxa have been recognized in *Sinapis arvensis* on the basis of minor variation in fruit and basal leaf morphology, but the species is extremely variable, and none of the variants is recognized here.

Sinapis arvensis is one of the most widespread and abundant weeds of cultivated grain fields in North America, causing crop losses and acting as host for viruses and fungi that also attack some cruciferous vegetable crops (G. A. Mulligan and L. G. Bailey 1975; I. A. Al-Shehbaz 1985; R. C. Rollins and Al-Shehbaz 1986). It is generally considered a native of Eurasia and is thought to have been introduced into the New World by European settlers about 400 years ago. Recent archaeological and ethnobotanical studies (H. A. Jacobson et al. 1988) indicate that it (as *Brassica kaber*) grew in the northeastern United States as early as 8000 years ago and suggest that it originally had a semi-circumboreal distribution.

SELECTED REFERENCE Mulligan, G. A. and L. G. Bailey. 1975. The biology of Canadian weeds. 8. *Sinapis arvensis* L. Canad. J. Pl. Sci. 55: 171–183.

g. BRASSICACEAE Burnett tribe BUNIADEAE de Candolle, Mém. Mus. Hist. Nat. 7: 245. 1821 [I]

Annuals, biennials, or perennials; glandular (glands multicellular on multiseriate stalks). **Trichomes** stalked, forked, or simple. **Cauline leaves** sessile or subsessile [petiolate]; blade base not auriculate, margins dentate or entire. **Racemes** ebracteate, often elongated in fruit. **Flowers** actinomorphic; sepals spreading or ascending [erect], lateral pair not saccate basally; petals yellow, claw usually present [absent], often distinct; filaments unappendaged [winged]; pollen 3-colpate. **Fruits** silicles [siliques], indehiscent [dehiscent], unsegmented, terete or 4-angled [latiseptate]; ovules 2–4[–numerous] per ovary; style obsolete or distinct; stigma entire [2-lobed]. **Seeds** aseriate [biseriate]; cotyledons spirolobal.

Genus 1, species 2 (2 in the flora): introduced; Europe, Asia, Africa.

30. BUNIAS Linnaeus, Sp. Pl. 2: 669. 1753; Gen. Pl. ed. 5, 300. 1754 • Hill-mustard [Greek and Latin *bunias*, a kind of common mustard or turnip]

Ihsan A. Al-Shehbaz

Plants not scapose; glabrous or sparsely to densely pilose (multicellular glandular tubercles or papillae present throughout, except flowers). **Stems** erect, often branched (many) distally. **Leaves** basal and cauline; petiolate, sessile, or subsessile; basal not rosulate [rosulate], petiolate, blade margins entire, pinnatifid, or lyrate; cauline sessile (subsessile distally), blade (base cuneate, attenuate), margins dentate or entire. **Racemes** (corymbose or paniculate), considerably elongated in fruit. **Fruiting pedicels** divaricate, slender. **Flowers:** sepals (yellowish green), oblong, (margins membranous), (glabrous, [pubescent or glandular]); petals obovate, (longer than sepals), claw distinct or absent, (apex obtuse to emarginate); stamens strongly tetradynamous; filaments (yellowish), not dilated basally [dilated]; anthers oblong [ovate], (apex obtuse); nectar glands confluent, subtending bases of stamens, median glands present. **Fruits** nutletlike, sessile, (readily detached from pedicel), oblong, ovoid, or subglobose, smooth, terete, 4-angled, or with 4 cristate wings, (1–4-loculed), (woody); valves (not distinct) not veined, glabrous; replum not distinct; septum subwoody or absent; ovules 2–4 per ovary; style obsolete or distinct, (slender, filiform or subconical); stigma capitate. **Seeds** plump [flattened], not winged, subglobose to ovoid [oblong]; seed coat (smooth), not mucilaginous when wetted; cotyledons spirolobal. $x = 7$.

Species 2 (2 in the flora): introduced; Europe, Asia, n Africa.

Bunias erucago and *B. orientalis* have the same chromosome number; the former has only about 0.8-fold of the DNA amount of *B. orientalis* (J. Greilhuber and R. Obermayer 1999). Although both species are widespread weeds in Europe, they have not spread much in North America.

SELECTED REFERENCE Greilhuber, J. and R. Obermayer. 1999. Cryptopolyploidy in *Bunias* (Brassicaceae) revisited—A flowcytometric and densitometric study. Pl. Syst. Evol. 218: 1–4.

1. Fruits 4-winged, 4-loculed, (0.8–)0.9–1.3(–1.4) cm; seeds 3 or 4; petals (8–)10–13 mm; styles 3.5–5(–6) mm. . . . . 1. *Bunias erucago*
1. Fruits not winged, 1- or 2-loculed, (0.5–)0.6–0.7(–0.8) cm; seeds 1 or 2; petals (4–)4.5–7(–8) mm; styles obsolete or, rarely, to 1 mm . . . . 2. *Bunias orientalis*

1. **Bunias erucago** Linnaeus, Sp. Pl. 2: 670. 1753

Annuals; glabrous or pubescent. **Stems** unbranched or branched basally and distally, 1.5–7(–10) dm, (often white-hispid proximally and trichomes simple, glabrous distally). **Basal leaves:** petiole 1–4(–6) cm; blade lanceolate (in outline), lyrate to sinuate-pinnatifid, or runcinate-pinnatisect, (1–)3–15(–19) cm, lateral lobes oblong or lanceolate, (smaller than terminal lobe). **Cauline leaves:** (distalmost) blade lanceolate or sublinear. **Fruiting pedicels** divaricate, straight, 1.5–2.5(–3) cm. **Flowers:** sepals ascending to spreading, 3–4 × 1–1.5 mm; petals (8–)10–13 × 4.5–6 mm, claw minute or absent; filaments 3.5–6 mm; anthers 1–1.3 mm. **Fruits** oblong to subglobose, subquadrangular, with 4 irregularly dentate wings, 4-loculed, (0.8–)0.9–1.3(–1.4) cm × 5–6.5 mm; style 3.5–5(–6) mm. **Seeds** 3 or 4 per fruit, 2–3.5 mm. $2n = 14$.

Flowering Apr–Jun. Roadsides, fields, pastures, disturbed areas, waste places; introduced; Pa., Va.; s Europe; sw Asia; n Africa.

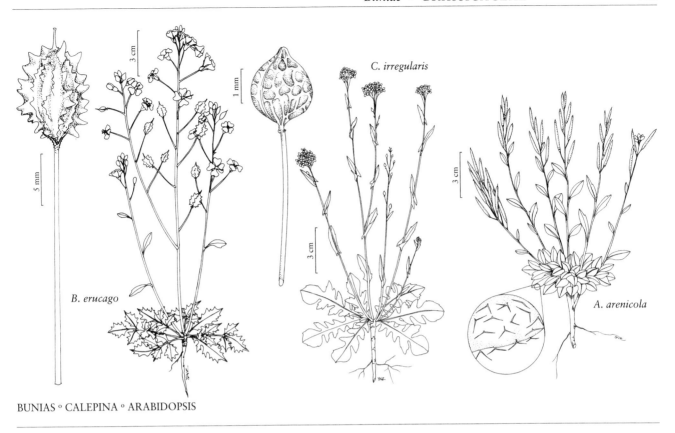

BUNIAS ∘ CALEPINA ∘ ARABIDOPSIS

2. **Bunias orientalis** Linnaeus, Sp. Pl. 2: 670. 1753
   • Turkish rocket

Biennials or perennials; sparsely to densely pilose. **Stems** usually branched distally, rarely basally, (2.5–)4–10(–15) dm. **Basal leaves**: petiole (1–)2–10(–15) cm; blade lanceolate, 10–45 cm, margins coarsely dentate or pinnatifid, lateral lobes oblong or lanceolate, (terminal lobe deltate or lanceolate, larger than lateral lobes). **Cauline leaves**: (distalmost) blade lanceolate or sublinear. **Fruiting pedicels** divaricate, straight, 1–2(–2.3) cm. **Flowers**: sepals spreading, 2.5–4 × 1–1.5 mm; petals (4–)4.5–7(–8) × (2–)3–5 mm, claw (slender), 1–2 mm; filaments 1.5–3.5 mm; anthers 0.8–1 mm. **Fruits** ovoid, or, rarely, suboblong and slightly constricted at middle, terete, not winged, 1- or 2-loculed, (0.5–)0.6–0.7(–0.8) cm × 3–4(–5) mm, (gradually tapering to apex); style obsolete or, rarely, to 1 mm. **Seeds** 1 or 2 per fruit, 2–3.5 mm. $2n = 14$.

Flowering Jun–Aug. Roadsides, fields, pastures, meadows, disturbed areas, waste places; introduced; B.C., N.B., N.S., Que.; Mass., Mich., N.Y, Ohio, Pa.; Europe; Asia.

h. BRASSICACEAE Burnett tribe CALEPINEAE Horaninov, Char. Ess. Fam., 169. 1847

Annuals or, rarely, **biennials**; eglandular. **Trichomes** absent [simple]. **Cauline leaves** sessile [petiolate]; blade base auriculate, sagittate, or amplexicaul, margins dentate or subentire. **Racemes** ebracteate [bracteate], elongated in fruit. **Flowers** actinomorphic; sepals ascending or spreading, lateral pair seldom saccate basally; petals white [pink, purple, or yellowish], claw often present, seldom distinct; filaments unappendaged, not winged; pollen 3-colpate. **Fruits** silicles [siliques], indehiscent, unsegmented, terete or slightly latiseptate, (woody); ovules 1[–3] per ovary; style obsolete (relatively short or beaklike); stigma entire. **Seeds** aseriate; cotyledons [incumbent] subconduplicate.

Genera 3, species 9 (1 in the flora): introduced; Europe, e, c, sw Asia, n Africa.

31. CALEPINA Adanson, Fam. Pl. 2: 423. 1763 • [Greek *chalepaino*, term used by Theophrastus probably in connection with weedy plants; some authors believe it derived from Arabic *Haleb* (erroneously rendered *Chaleb* by some), name for the Syrian city Aleppo, but highly unlikely since Adanson based it on Bauhin's *Myagrum monospermum minus*, collected in southern France]

Ihsan A. Al-Shehbaz

Plants not scapose. **Stems** erect to ascending, unbranched or branched. **Leaves** basal and cauline; petiolate or sessile; basal usually rosulate, petiolate, blade margins dentate to lyrate-pinnatifid; cauline blade (base sagittate or amplexicaul), margins dentate or subentire. **Racemes** (corymbose, several-flowered), considerably elongated in fruit. **Fruiting pedicels** divaricate or ascending, slender. **Flowers:** sepals ascending to spreading; petals oblanceolate, (apex obtuse or slightly emarginate); stamens slightly tetradynamous; filaments dilated basally; anthers ovate; nectar glands: lateral intrastaminal, median glands present (distinct). **Fruits** subsessile or shortly stipitate (gynophore relatively slender, fruit readily detached at maturity), nutlike, ovoid or ellipsoid, terete; valves reticulate and usually longitudinally 4-ribbed; replum rounded; septum absent; stigma capitate. **Seeds** (pendulous) plump, not winged, ovoid; seed coat (smooth), not mucilaginous when wetted; cotyledons involute in distal ½. $x = 7$.

Species 2 (1 in the flora): introduced; Europe, e, c, sw Asia, n Africa.

Calepina cochlearioides (Murray) Dumortier is distributed in Asia (China, Kazakhstan, Mongolia, and adjacent Russia).

SELECTED REFERENCES Blake, S. F. 1957. A new cruciferous weed, *Calepina irregularis*, in Virginia. Rhodora 59: 278–280. Hardin, J. W. 1958. *Calepina irregularis* in North Carolina. Castanea 23: 111.

1. **Calepina irregularis** (Asso) Thellung in H. Schinz and R. Keller, Fl. Schweiz ed. 2, 1: 218. 1905 F I

Myagrum irregulare Asso, Syn. Stirp. Aragon., 82. 1779; *Calepina corvini* (Allioni) Desvaux; *Crambe corvini* Allioni

Stems several from base, (1–)2–8 dm, unbranched or branched distally. **Basal leaves** early rosulate; petiole (0.2–)1–3.5(–6) cm; blade obovate, spatulate, or oblanceolate, (0.8–)2–5(–9) cm × (3–)10–30 mm, base attenuate. **Cauline leaves:** blade oblong to lanceolate, (1–)2–7(–8) cm × (4–)10–20(–30) mm, base sagittate or amplexicaul, apex obtuse to acute. **Racemes** with straight rachis. **Fruiting pedicels** straight or curved upward, (3–)5–10(–13) mm. **Flowers:** sepals oblong or ovate, 1.2–2 × 0.5–1 mm; petals (1.5–)2–3 × 0.5–1.5 mm, unequal, abaxial pair larger than adaxial (longer than sepals); filaments 1–1.5 mm; anthers ca. 0.3 mm; gynophore 0.1–0.2 mm. **Fruits** 2.5–3.5(–4) × 2–3 mm, abruptly tapering to a conical, beaklike apex, 0.5–0.8 mm. **Seeds** brown, 1.3–1.6 mm. $2n = 14, 28$.

Flowering Mar–May. Fields, waste places, roadsides; 0–200 m; introduced; Md., N.C., Va.; Europe; c, sw Asia; n Africa.

The fruits of *Calepina irregularis* and *Neslia apiculata* are remarkably similar; the two can easily be separated by the presence of white instead of yellow petals and by the absence of indumentum in *Calepina*.

Brassicaceae. Extensive molecular data and critical evaluation of morphology have shown that *Arabidopsis* is polyphyletic (S. L. O'Kane and Al-Shehbaz 2003). Nine of the ten species in the genus are native to Europe; only *A. arenicola* is endemic to North America.

SELECTED REFERENCES O'Kane, S. L. and I. A. Al-Shehbaz. 1997. A synopsis of *Arabidopsis* (Brassicaceae). Novon 7: 323–327. O'Kane, S. L. and I. A. Al-Shehbaz. 2003. Phylogenetic position and generic limits of *Arabidopsis* (Brassicaceae) based on sequences of nuclear ribosomal DNA. Ann. Missouri Bot. Gard. 90: 603–612. Price, R. A., J. D. Palmer, and I. A. Al-Shehbaz. 1994. Systematic relationship of *Arabidopsis*: A molecular and morphological approach. In: E. M. Meyerowitz and C. R. Somerville, eds. 1994. *Arabidopsis*. New York. Pp. 7–19.

1. Annuals; petals 2–3.5(–4) mm; fruits 0.5–0.8 mm wide; seeds 0.3–0.5 mm . . . . 4. *Arabidopsis thaliana*
1. Biennials or perennials; petals 4–10 mm; fruits 0.8–2.5 mm wide; seeds 0.8–1.4 mm.
   2. Fruits terete or slightly flattened, (0.8–)1–2(–2.8) cm × 1.5–2.2(–2.5) mm; cotyledons incumbent . . . . 1. *Arabidopsis arenicola*
   2. Fruits flattened, (1.5–)2–4(–4.5) cm × 0.8–1.8(–2) mm; cotyledons accumbent.
      3. Basal leaf blade margins pinnatisect or pinnatipartite; petals with 2 lateral teeth on claws . . . . 2. *Arabidopsis arenosa*
      3. Basal leaf blade margins entire, dentate, lyrate, or lyrate-pinnatifid; petals without teeth on claws . . . . 3. *Arabidopsis lyrata*

1. **Arabidopsis arenicola** (Richardson ex Hooker) Al-Shehbaz, Elvin, D. F. Murray & Warwick, Canad. J. Bot. 84: 279. 2006 E F

Eutrema arenicola Richardson ex Hooker, Fl. Bor.-Amer. 1: 67, plate 24. 1830; *Arabis arenicola* (Richardson ex Hooker) Gelert; *A. arenicola* var. *pubescens* (S. Watson) Gelert; *A. humifusa* S. Watson; *A. humifusa* var. *pubescens* S. Watson

Perennials; (caudex usually branched, rarely simple, with persistent petiolar remains); glabrous throughout or sparsely to moderately pubescent, trichomes simple and short-stalked, forked, (to 0.5 mm). **Stems** often several from base (caudex), erect, ascending, or decumbent, unbranched or branched (few) distally, 0.5–2(–3) dm, glabrous or sparsely pubescent distally. **Basal leaves**: petiole 0.5–2 cm, (ciliate or glabrous); blade (fleshy), oblanceolate to spatulate, 0.5–2 cm × 2–6 mm, margins entire or dentate, (not ciliate), apex obtuse or acute, surfaces pubescent or glabrous. **Cauline leaves** [(1–)2–4(–6)] sessile; blade oblong to oblanceolate, 0.7–1.8(–2.5) cm × 2–5 mm, (base not auriculate), margins usually entire, rarely dentate, (apex obtuse or acute, surfaces glabrous or sparsely pubescent). **Fruiting pedicels** ascending to divaricate, 5–10(–12) mm, (glabrous). **Flowers:** sepals 1.7–3 mm, lateral pair not saccate basally; petals white, usually oblanceolate, rarely spatulate, 4–5 × 1–2 mm; filaments 1.7–3 mm. **Fruits** (suberect to divaricate), straight, smooth, terete or slightly flattened, (0.8–)1–2(–2.8) cm × 1.5–2.2(–2.5) mm; valves each with obscure midvein; ovules 30–54 per ovary; style 0.4–0.8 mm. **Seeds** (sometimes subbiseriate), reddish brown, narrowly oblong, 0.9–1.1 mm, pointed distally; cotyledons incumbent. $2n = 32$.

Flowering Jun–Aug. Sandy beaches, gravel, flats of streambeds; 0–1500 m; Greenland; Man., Nfld. and Labr. (Labr.), Nunavut, Ont., Que., Sask.

2. **Arabidopsis arenosa** (Linnaeus) Lawalrée, Bull. Soc. Roy. Bot. Belgique 42: 242. 1969

Sisymbrium arenosum Linnaeus, Sp. Pl. 2: 658. 1753

Subspecies 2 (1 in the flora): North America, Europe.

Subspecies *arenosa* is widespread in Europe; subsp. *borbasii* (Zapałowicz) Pawłowski is restricted to central Europe.

2a. **Arabidopsis arenosa** (Linnaeus) Lawalrée subsp. **arenosa**

Biennials or perennials; (caudex short); glabrous or pubescent, trichomes simple with or stellate or forked ones. **Stems** often several from base, erect or decumbent, usually branched distally, (0.5–)1.5–6(–8) dm, pubescent basally with simple, 1-forked, or 3-rayed stellate trichomes, sparsely pubescent or glabrous apically. **Basal leaves**: petiole (0.6–)1.5–3.5(–5) cm; blade oblanceolate, (0.6–)2–6(–9) cm × (3–)10–20(–30) mm, margins pinnatisect or pinnatipartite, with 3–10 lobes on each side, apex obtuse to acute, surfaces sparsely to densely pubescent. **Cauline leaves** shortly petiolate or sessile; blade oblanceolate, (0.4–)1.5–5(–6.5) cm × (2–)8–20(–30) mm (gradually smaller distally), margins usually pinnatifid to coarsely toothed, rarely subentire. **Fruiting pedicels** divaricate or ascending, (4–)5–12(–15) mm. **Flowers:** sepals (1.7–)2–3 mm, lateral pair

i. BRASSICACEAE Burnett tribe CAMELINEAE de Candolle, Mém. Mus. Hist. Nat. 7: 239. 1821

Annuals, biennials, or perennials; eglandular. **Trichomes** stalked or sessile, stellate, dendritic, or forked, sometimes mixed with simple ones. **Cauline leaves** usually sessile, rarely petiolate or subsessile; blade base auriculate or not, margins usually entire, sometimes dentate or, rarely, lyrate. **Racemes** ebracteate [bracteate], often elongated in fruit. **Flowers** actinomorphic; sepals erect, spreading, or ascending, lateral pair usually not saccate basally; petals [absent] white, yellow, pink, lavender, or purple [orange], claw present, often distinct; filaments unappendaged, not winged; pollen 3-colpate. **Fruits** silicles or siliques, dehiscent or indehiscent, unsegmented, terete, latiseptate, or angustiseptate; ovules 2–200[–numerous] per ovary; style often distinct; stigma usually entire [2-lobed (subentire in *Turritis*)]. **Seeds** biseriate or uniseriate [aseriate]; cotyledons accumbent or incumbent.

Genera 7, species 37 (6 genera, 12 species in the flora): North America, Europe, Asia, Australia.

Camelineae appeared as a monophyletic lineage in M. A. Beilstein et al. (2006); the sampling included seven genera and, with study of further genera, the boundaries of tribe may well need to be redefined.

32. ARABIDOPSIS (de Candolle) Heynhold in F. Holl and G. Heynhold, Fl. Sachsen 1: 538. 1842, name conserved • Mouse-ear or thale cress [Genus *Arabis* and Greek *opsis*, resembling]

Ihsan A. Al-Shehbaz

Annuals, biennials, or perennials; (stoloniferous or with woody caudex); not scapose; glabrous or pubescent, trichomes simple, mixed with stalked, 1–3-forked ones. **Stems** erect, ascending, or decumbent, unbranched or branched distally, (usually glabrous distally). **Leaves** basal and cauline; petiolate, sessile, or subsessile; basal rosulate, petiolate, blade margins entire, toothed, or pinnately lobed; cauline blade margins usually entire or dentate, rarely lyrate. **Racemes** (few- to several-flowered), not elongated in fruit. **Fruiting pedicels** ascending, divaricate, or slightly reflexed, slender. **Flowers:** sepals erect or ascending, usually oblong (or ovate), (lateral pair sometimes saccate or subsaccate basally); petals white, lavender, or purplish [pink, purple], obovate, spatulate, or oblanceolate, claw differentiated from blade or not, (apex obtuse or emarginate); stamens slightly tetradynamous; filaments not dilated basally; anthers oblong or ovate, (apex obtuse); nectar glands confluent, subtending bases of stamens. **Fruits** siliques, dehiscent, shortly stipitate or subsessile, linear, smooth or somewhat torulose, terete or latiseptate; valves (papery), each not veined or midvein prominent or obscure, glabrous; replum rounded; septum complete; ovules 15–80 per ovary; (style obsolete or distinct, to 1 mm); stigma capitate. **Seeds** usually uniseriate, plump or flattened, not winged or margined, oblong, ovoid, or ellipsoid; seed coat (minutely reticulate), mucilaginous or not when wetted; cotyledons usually accumbent, rarely incumbent. $x = 5, 8$.

Species 10 (4 in the flora): North America, Europe, n, e Asia.

The limits of *Arabidopsis* have been the subject of long controversy, and more than 50 species were placed in the genus (I. A. Al-Shehbaz et al. 1999). The delimitation of the genus was based primarily on the presence of branched trichomes, linear fruits, and accumbent or incumbent cotyledons. That combination of characteristics evolved independently multiple times in

subsaccate basally, (glabrous or pubescent); petals white or lavender, spatulate or obovate, (5–)6–8(–9) × (2.5–)3–5 mm, (claw to 2 mm, with a tooth on each side); filaments 2.5–4.5 mm. **Fruits** torulose, flattened, (1.5–)2–4(–4.8) cm × 0.8–1.2 mm; valves each with distinct midvein; style (0.2–)0.5–0.9(–1.3) mm. **Seeds** light brown, (flattened), ovoid, 0.8–1 mm; cotyledons accumbent. $2n$ = 16, 32.

Flowering Jul. Above fjords; 0 m; Greenland; Europe.

This is the first record of subsp. *arenosa* from North America. It is based on *M. P. Porsild s.n.* (MO) collected from western Greenland, 28 July 1935. It is not known if the occurrence in Greenland represents an introduction or the westernmost extent of its native range.

3. **Arabidopsis lyrata** (Linnaeus) O'Kane & Al-Shehbaz, Novon 7: 325. 1997

Arabis lyrata Linnaeus, Sp. Pl. 2: 665. 1753; *Cardaminopsis lyrata* (Linnaeus) Hiitonen

Biennials or perennials; (caudex branched or not, sometimes root crown present); glabrous or pubescent, trichomes simple, with stellate, forked, and rayed ones. **Stems** simple or few to several from base, erect or decumbent, usually branched distally, 0.5–5 dm, pubescent basally, trichomes mixed simple, 1-forked, and (fewer) stellate ones, rarely 3-rayed, glabrous or pubescent apically. **Basal leaves:** petiole 0.5–6 cm; blade oblanceolate or ovate, 0.5–8.5 cm × 2–18 mm, margins entire, dentate, or lyrate-pinnatifid (when lobed, terminal lobes larger than lateral), apex obtuse, surfaces glabrous or sparsely to densely pubescent. **Cauline leaves** shortly petiolate or sessile; blade oblanceolate, 0.4–4.2 cm × 1–8(–10) mm (smaller distally), margins usually entire, repand, or obscurely toothed, rarely lobed. **Fruiting pedicels** divaricate or ascending, 2–15 mm. **Flowers:** sepals 2–4.5 mm, lateral pair saccate basally, (glabrous or densely pubescent); petals white or purplish, spatulate or obovate, 4–10 × 1.5–4 mm, (claw to 2 mm, toothless); filaments 2–4 mm. **Fruits** torulose, flattened, (1.5–)2–4.5 cm × 0.8–1.8 mm; valves each with distinct midvein; ovules 24–46 per ovary; style 0–1 mm. **Seeds** light brown, (flattened), oblong, 0.8–1.4 mm; cotyledons accumbent.

Subspecies 3 (3 in the flora): North America, n Europe, n, e Asia.

Arabidopsis lyrata and its infraspecific taxa recognized below were treated in *Arabis* by R. C. Rollins (1993) and G. A. Mulligan (1996). As shown by S. L. O'Kane and I. A. Al-Shehbaz (1997, 2003), the two genera are relatively distantly related and *A. lyrata* should be placed in *Arabidopsis* with the rest of its relatives.

Arabidopsis lyrata is highly variable. It has been divided (see synonymy below) into independent species. The three subspecies recognized here are rather difficult to separate, especially where their ranges overlap. Cytological data are not helpful; both diploid and tetraploid populations have been reported for each of the three subspecies.

1. Basal leaf blade margins entire or toothed; caudices well-developed, often branched, somewhat thickened . . . .
 . . . . 3b. *Arabidopsis lyrata* subsp. *petraea*
1. Basal leaf blade margins lyrate or lyrate-pinnatifid; caudices often slender, unbranched, or with a root crown.
 2. Basal leaf blade surfaces usually pubescent; petioles often hirsute; petals 6–8 mm; fruits 0.8–1 mm wide; styles 0.5–1 mm . . . .
 . . . . 3a. *Arabidopsis lyrata* subsp. *lyrata*
 2. Basal leaf blade surfaces glabrous or sparsely pubescent; petioles often glabrous; petals (4–)5–5.5(–6) mm; fruits 1.2–1.5(–1.8) mm wide; styles 0–0.5 mm . . . .
 . . . . 3c. *Arabidopsis lyrata* subsp. *kamchatica*

3a. **Arabidopsis lyrata** (Linnaeus) O'Kane & Al-Shehbaz subsp. **lyrata** E

Biennials or perennials; caudex unbranched or with a root crown, often slender. **Basal leaves:** (petiole often hirsute); blade margins lyrate or lyrate-pinnatifid, terminal lobes distinct, (surfaces pubescent). **Petals** 6–8 mm. **Fruits** 0.8–1 mm wide; style 0.5–1 mm. $2n$ = 16, 32.

Flowering Apr–Jul. Cliffs, ledges, thickets, stream and river banks, woods, limestone crevices and bluffs, sandstone hills and outcrops, serpentine rocks and barrens, shale, talus, sand dunes; 0–1000 m; Alta., B.C., Man., N.W.T., Ont., Sask.; Alaska, Conn., Del., D.C., Ill., Ind., Iowa, Md., Mass., Mich., Minn., Miss., Mo., N.J., N.Y., N.C., Ohio, Pa., Tenn., Vt., Va., W.Va., Wis.

3b. Arabidopsis lyrata (Linnaeus) O'Kane & Al-Shehbaz subsp. **petraea** (Linnaeus) O'Kane & Al-Shehbaz, Novon 7: 326. 1997

Cardamine petraea Linnaeus, Sp. Pl. 2: 654. 1753; *Arabis media* N. Busch

Perennials; caudex well-developed, often branched, somewhat thickened. **Basal leaves**: blade margins entire or toothed, without distinct terminal lobes. Petals 4–6 mm. Fruits 1.2–1.8 mm wide; style 0–0.5 mm. $2n = 16, 32$.

Flowering Jun–Jul. Sand dunes, sand bars, tussocks, gravel, tundra, volcanic ash, stream terraces; 0 m; Yukon; Alaska; n Europe; n, e Asia.

The occurrence of subsp. *petraea* in North America was pointed out by S. L. O'Kane and I. A. Al-Shehbaz (1997) and Al-Shehbaz and O'Kane (www.aspb.org/publications/arabidopsis, 2002); no specimens were cited. G. A. Mulligan (1996), who treated the taxon as *Arabis media*, did not cite specimens either. Representative Alaskan collections were examined, including *Parker 7995* (MO) and *Hardy 134* (CAS, MO).

3c. Arabidopsis lyrata (Linnaeus) O'Kane & Al-Shehbaz subsp. **kamchatica** (Fischer ex de Candolle) O'Kane & Al-Shehbaz, Novon 7: 326. 1997

Arabis lyrata Linnaeus var. *kamchatica* Fischer ex de Candolle, Syst. Nat. 2: 231. 1821; *A. ambigua* de Candolle var. *glabra* de Candolle; *A. ambigua* var. *intermedia* de Candolle; *A. kamchatica* (Fischer ex de Candolle) Ledebour; *A. lyrata* var. *glabra* (de Candolle) M. Hopkins; *A. lyrata* var. *intermedia* (de Candolle) Farwell; *A. lyrata* subsp. *kamchatica* (Fischer ex de Candolle) Hultén; *A. lyrata* var. *occidentalis* S. Watson; *A. occidentalis* (S. Watson) A. Nelson; *Cardaminopsis kamchatica* (Fischer ex de Candolle) O. E. Schulz

Biennials or perennials; caudex unbranched or with a root crown, often slender. **Basal leaves**: (petiole often glabrous); blade margins lyrate or lyrate-pinnatifid, terminal lobes distinct, (surfaces glabrous or sparsely pubescent). Petals (4–)5–5.5(–6) mm. Fruits 1.2–1.5 (–1.8) mm wide; style 0–0.5 mm. $2n = 16, 32$.

Flowering Jun–Jul. Gravelly slopes, forests, gravel bars, high-elevation scree and talus, roadsides; 50–2000 m; Alta., B.C., N.W.T., Sask., Yukon; Alaska, Wash.; ne Asia.

G. A. Mulligan (1996) treated subsp. *kamchatica* as a species of *Arabis*; R. C. Rollins (1993) treated it as a variety of *A. lyrata*.

4. Arabidopsis thaliana (Linnaeus) Heynhold in F. Holl and G. Heynhold, Fl. Sachsen 1: 538. 1842 • Thale cress ⊞ Ⓦ

Arabis thaliana Linnaeus, Sp. Pl. 2: 665. 1753; *Sisymbrium thalianum* (Linnaeus) J. Gay & Monnard

Annuals; glabrous or pubescent, trichomes usually simple, sometimes mixed with stalked, forked ones. Stems simple or few from base, erect, unbranched or branched distally, (0.2–)0.5–3(–5) dm, pubescent basally, trichomes predominantly simple, glabrous apically. **Basal leaves** shortly petiolate; blades obovate, spatulate, ovate, or elliptic, 0.8–3.5(–4.5) cm × (1–)2–10(–15) mm, margins entire, repand, or dentate, apex obtuse, adaxial surface with predominantly simple and stalked, 1-forked trichomes. **Cauline leaves** subsessile; blade lanceolate, linear, oblong, or elliptic, (0.4–)0.6–1.8(–2.5) cm × 1–6(–10) mm, margins usually entire, rarely toothed. **Fruiting pedicels** divaricate, 3–10(–15) mm. **Flowers**: sepals 1–2(–2.5) mm, lateral pair not saccate basally, (glabrous or sparsely pubescent distally, trichomes simple); petals white, spatulate, 2–3.5(–4) × 0.5–1.5 mm, (base attenuate to claw); filaments 1.5–2 mm. **Fruits** cylindric or linear, smooth, terete, (0.8–)1–1.5(–1.8) cm × 0.5–0.8 mm; valves each with distinct midvein; ovules 40–70 per ovary; style to 0.5 mm. **Seeds** light brown, (plump), ellipsoid 0.3–0.5 mm; cotyledons incumbent. $2n = 10$.

Flowering Feb–May. Sandy areas along roadsides, stream banks, railroad tracks and embankments, open pastures, grassy flats, fields, prairies, floodplains, woods, lawns, limestone ledges and crevices, bluffs, shale and serpentine barrens, gravel, sandstone; 0–1000 m; introduced; B.C., Ont., Que.; Ala., Ark., Calif., Conn., Del., D.C., Ga., Idaho, Ill., Ind., Iowa, Ky., La., Maine, Md., Mass., Mich., Miss., Mo., Mont., N.H., N.J., N.Y., NC., Ohio, Okla., Oreg., Pa., R.I., S.C., Tenn., Tex., Utah, Vt., Va., Wash., W.Va., Wis.; Europe; sw, c Asia; introduced also nearly worldwide.

Arabidopsis thaliana is the most widely used model organism in plant biology. Its small genome size, fully sequenced in the year 2000, chromosome number, fast growth cycle (from seed germination to set in four to six weeks), small size (hundreds can be grown in a pot and thousands in a growth chamber), autogamous breeding system (induced mutations are expressed in two generations), and ability to grow on various synthetic media, all make the species an ideal system in experimental biology.

33. CAMELINA Crantz, Stirp. Austr. Fasc. 1: 17. 1762 • Flaxweed, false-flax, gold-of-pleasure [Greek *chamai*, dwarf or on the ground, and *linon*, flax, alluding to suppressing influence on growth of flax] 🄸

Ihsan A. Al-Shehbaz

Mark A. Beilstein

Annuals or biennials; not scapose; pubescent, glabrescent, or glabrous, trichomes simple or short-stalked, with forked to substellate or subdendritic (smaller) ones. **Stems** erect, unbranched basally, branched distally, (basally hirsute with simple trichomes or sparsely pubescent with branched ones). **Leaves** basal and cauline; petiolate or subsessile; basal (often withered by flowering), rosulate or not, petiolate, blade margins entire or toothed or, rarely, lobed; cauline blade (base auriculate or sagittate), margins entire, dentate to lobed, or denticulate. **Racemes** (corymbose, several-flowered), considerably elongated in fruit, (rachis straight, rarely strongly flexuous). **Fruiting pedicels** ascending to divaricate, slender. **Flowers:** sepals erect to ascending, oblong or ovate; petals usually yellow, rarely white, oblanceolate [spatulate], (longer than sepals), claw and blade somewhat differentiated, (apex obtuse); stamens in 3 pairs of unequal length; filaments not dilated basally; anthers ovate or oblong, (apex obtuse); nectar glands (4), lateral, 1 on each side of lateral stamen. **Fruits** silicles or, rarely, siliques, dehiscent, shortly stipitate, pyriform, obovoid, or depressed globose [linear], keeled or not, slightly latiseptate; valves each with prominent or obscure midvein, (leathery, smooth, margins of each flattened and connate, apex abruptly caudate and extending 1–2.5 mm onto, and appearing as part of, style), pubescent; replum concealed by connate margins of valves; septum complete; ovules 8–25 per ovary; stigma capitate. **Seeds** biseriate or, rarely, uniseriate, plump or slightly flattened, not winged or narrowly margined, oblong; seed coat (minutely reticulate), copiously mucilaginous when wetted; cotyledons incumbent or, rarely, accumbent. $x = 6, 7, 10, 13$.

Species 8 (4 in the flora): introduced; Europe, Asia, n Africa; introduced also in South America, Australia.

Some authors have studied allelopathic and other effects of *Camelina* on the growth and production of flax, and the interested reader should consult I. A. Al-Shehbaz (1987) for leads. Some species, especially *C. sativa*, were cultivated for their fibers and seed oil by the Romans as early as 600 B.C., and remain in cultivation in some parts of eastern Europe and Russia.

Camelina alyssum and *C. sativa* may no longer be established in the United States.

SELECTED REFERENCES McGregor, R. L. 1985. Current status of the genus *Camelina* (Brassicaceae) in the prairies and plants of central North America. Contr. Univ. Kansas Herb. 15: 1–13. Tedin, O. 1925. Vererbung, Variation und Systematik der Gattung *Camelina*. Hereditas (Lund) 6: 275–386.

1. Fruits 3.5–7 mm, valves obscurely veined; seeds 0.8–1.5 mm; stems basally with simple trichomes (to 2.5–3.5 mm), these often mixed with branched ones.
    2. Petals pale yellow, (2.5–)3–4(–6) mm; basal leaves withered by anthesis . . . . 1. *Camelina microcarpa*
    2. Petals white or creamy white, (5–)6–8(–9) mm; basal leaves persistent after anthesis
       . . . . 2. *Camelina rumelica*
1. Fruits 7–13 mm, valves prominently veined; seeds (1.5–)1.8–3 mm; stems basally glabrous or trichomes almost exclusively minute and branched, rarely trichomes simple.
    3. Fruits pyriform to broadly obovoid, 4–5(–6) mm wide, distinctly longer than wide; cauline leaf blade margins entire or denticulate . . . . 3. *Camelina sativa*
    3. Fruits depressed globose, (5.5–)6.5–8(–9) mm wide, nearly as long as wide or slightly longer; cauline leaf blade margins coarsely dentate to lobed . . . . 4. *Camelina alyssum*

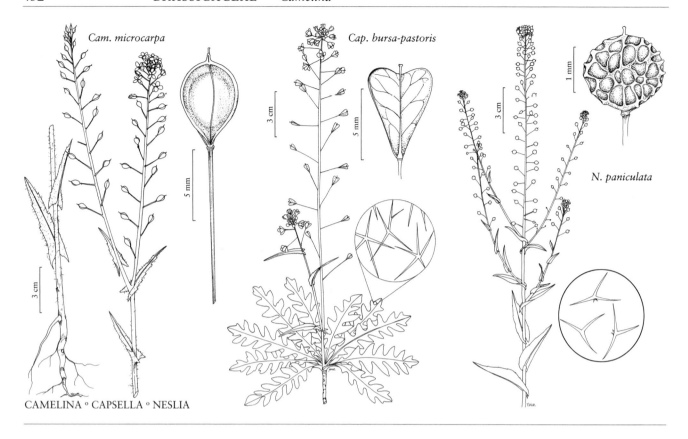

CAMELINA · CAPSELLA · NESLIA

1. **Camelina microcarpa** de Candolle, Syst. Nat. 2: 517. 1821 F I W

Camelina sativa (Linnaeus) Crantz subsp. *microcarpa* (de Candolle) E. Schmid

Annuals. Stems unbranched or branched distally, (0.8–)2–8 (–10) dm, densely to moderately hirsute basally, trichomes simple, to 2.5 mm, often mixed branched ones, (glabrescent distally). **Basal leaves** withered by anthesis. **Cauline leaves:** blade lanceolate, narrowly oblong, or linear-lanceolate, (0.8–)1.5–5.5(–7) cm × 1–10(–20) mm, base sagittate or minutely auriculate, margins entire or, rarely, remotely denticulate, (often subciliate), apex acute, surfaces pubescent, trichomes primarily simple. **Fruiting pedicels** ascending, 4–14(–17) mm. **Flowers:** sepals 2–3.5 × 0.5–1 mm; petals pale yellow, (2.5–)3–4(–6) × 1–2 mm; filaments 1.5–3 mm; anthers ca. 0.5 mm. **Fruits** pyriform to narrowly so, 3.5–5(–7) × 2–4(–5) mm, apex acute; valves each often with obscure midvein, margin narrowly winged; style 1–3.5 mm. **Seeds** reddish brown or brown, 0.8–1.4(–1.5) × 0.5–0.6 mm. $2n = 40$.

Flowering May–Jul. Farms, fields, meadows, prairies, roadsides, forest margins, open woods; 0–2500 m; introduced; Alta., B.C., Man., N.B., Nfld. and Labr., N.W.T., N.S., Nunavut, Ont., P.E.I., Que., Sask., Yukon; Ala., Ariz., Calif., Colo., Conn., Del., D.C., Idaho, Ill., Ind., Iowa, Kans., Ky., La., Maine, Md., Mass., Mich., Minn., Mo., Mont., Nebr., Nev., N.H., N.J., N.Mex., N.Y., N.C., N.Dak., Ohio, Okla., Oreg., Pa., R.I., S.Dak., Tenn., Tex., Utah, Vt., Va., Wash., W.Va., Wis., Wyo.; Europe; Asia; n Africa; introduced also in South America.

2. **Camelina rumelica** Velenovsky, Sitzungsber. Königl. Böhm. Ges. Wiss., Math.-Naturwiss. Cl. 1886: 448, fig. 13a. 1887 I

Annuals. Stems unbranched or branched distally, 1.5–4(–6) dm, densely to moderately hirsute-hispidulous basally, trichomes simple, to 3.5 mm, mixed with fewer, branched ones, (glabrescent distally). **Basal leaves** persistent after anthesis (into fruiting). **Cauline leaves:** blade lanceolate to oblong, (1–)2–6(–9) cm × 2–10(–20) mm, base sagittate or minutely auriculate, margins entire or irregularly denticulate, (often subciliate), apex acute, surfaces pubescent, trichomes primarily simple. **Fruiting pedicels** ascending to divaricate, 7–10(–14) mm. **Flowers:** sepals (2.7–)3–4(–4.5) × 0.5–1 mm; petals white or creamy white, (5–)6–8(–9) × 1.5–2 mm; filaments 2–3.5 mm; anthers ca. 0.5 mm. **Fruits** pyriform to obovoid, 5–7 × 3.5–5 mm, apex acute; valves each obscurely veined, margin narrowly winged; style 2–3 mm. **Seeds** brown, 1.2–1.5 × 0.5–0.6 mm. $2n = 12, 26$.

Flowering May–Jun. Fields, roadsides, waste places; 100–1700 m; introduced; Colo., Kans., Nev., Okla., Oreg., Tex.; Europe; sw Asia.

R. L. McGregor (1984, 1985) and R. C. Rollins (1993) stated that *Camelina rumelica* is naturalized also in Texas; we have not seen material that supports those reports.

3. Camelina sativa (Linnaeus) Crantz, Stirp. Austr. Fasc. 1: 17. 1762 [I][W]

Myagrum sativum Linnaeus, Sp. Pl. 2: 641. 1753

Annuals or biennials. Stems unbranched or branched distally, (1.2–)3–10(–13) dm, glabrous or sparsely pubescent basally, trichomes branched, rarely mixed with fewer, simple ones. **Basal leaves** often withered by anthesis. **Cauline leaves:** blade lanceolate, narrowly oblong, or linear-lanceolate, (1–)2–7(–9) cm × 2–10(–15) mm, base sagittate or strongly auriculate, margins entire or, rarely, remotely denticulate, apex acute, surfaces glabrescent or sparsely pubescent, trichomes usually forked. **Fruiting pedicels** ascending, (8–)12–20(–27) mm. **Flowers:** sepals 2–3(–4) × 0.5–1 mm; petals yellow, (3.5–)4–6 × 1–1.5 mm; filaments 1.5–3 mm; anthers ca. 0.5 mm. **Fruits** pyriform or broadly obovoid, 7–9 (–13) × 4–5(–6) mm (distinctly longer than wide), apex often subtruncate; valves each with prominent midvein, margin narrowly winged; style 1.2–2.5 mm. **Seeds** dark brown, (1.5–)1.8–2.5 × 0.7–1 mm. $2n = 40$.

Flowering May–Jun. Farms, grassy areas, fields, waste places; 0–1500 m; introduced; Alta., B.C., Man., N.W.T., Nunavut, Ont., Que., Yukon; Colo., Conn., D.C., Ill., Ind., Iowa, Kans., Maine, Mass., Mich., Minn., Mo., Mont., Nebr., N.H., N.J., N.Mex., N.Y., N.Dak., Ohio, Oreg., Pa., S.C., S.Dak., Vt., Va., W.Va., Wyo.; Europe; Asia; introduced also in South America, Australia.

R. L. McGregor (1985) indicated that *Camelina sativa* is no longer established in North America; we tend to agree because we have not seen any collections made within the past 40 years.

4. Camelina alyssum (Miller) Thellung, Index Seminum (Zürich) 1906: 10. 1906 [I]

Myagrum alyssum Miller, Gard. Dict. ed. 8, Myagrum no. 2. 1768

Annuals or biennials. Stems unbranched or branched distally, 2–7(–10) dm, glabrous or pubescent basally, trichomes branched, minute. **Basal leaves** often withered by anthesis. **Cauline leaves:** blade lanceolate, narrowly oblong, or linear-lanceolate, pinnatifid or sinuate-dentate, (1.5–)2.5–7(–10) cm × 2–10(–20) mm, base sagittate or strongly auriculate, margins usually coarsely dentate to lobed, rarely entire, apex acute, surfaces glabrescent or sparsely pubescent, trichomes primarily forked. **Fruiting pedicels** ascending to divaricate, 15–30(–40) mm. **Flowers:** sepals 2–4.2 × 0.7–1.5 mm; petals pale yellow, (3.5–)4–6.5 × 1.5–2 mm; filaments 2–3.5 mm; anthers ca. 0.5 mm. **Fruits** depressed globose, 7–11 × (5.5–)6.5–8(–9) mm (almost as long as wide, or longer), apex subtruncate or, rarely, rounded; valves each with prominent midvein, margin obscurely or narrowly winged; style 1.2–2.5 mm. **Seeds** reddish or yellowish brown, (1.8–)2–3 × 0.7–1 mm. $2n = 40$.

Flowering May–June. Fields, roadsides, prairies; introduced; Alta., Man., Sask.; Calif., Colo., Mont., N.Dak., S.Dak., Wyo.; Europe; introduced also in Australia.

As indicated by R. L. McGregor (1985), *Camelina alyssum* is known from the above states only from old collections, and, apparently, the species has not been collected again during the past five decades. Although we have not examined all of his cited specimens, those on which the Missouri and Nebraska records are based belong to *C. sativa*.

34. **CAPSELLA** Medikus, Pfl.-Gatt., 85. 1792, name conserved • Shepherd's-purse [Latin *capsa*, box or case, alluding to fruit resembling medieval wallet or purse] [I]

Ihsan A. Al-Shehbaz

Bursa Boehmer; *Rodschiedia* P. Gaertner, B. Meyer & Scherbius; *Solmsiella* Borbás

Annuals or biennials; not scapose; mostly pubescent, trichomes sessile and stellate, sometimes mixed with simple or forked ones. **Stems** erect or ascending, unbranched or branched. **Leaves** basal and cauline; petiolate or sessile; basal rosulate, petiolate, blade margins pinnately lobed, lyrate, runcinate, or, rarely, entire, dentate, or repand [sinuate]; cauline blade (base sagittate, amplexicaul, or auriculate), margins entire, dentate, or sinuate. **Racemes** (corymbose, several-flowered), considerably elongated in fruit. **Fruiting pedicels** divaricate, slender. **Flowers:** sepals erect or ascending, [ovate-]oblong, (glabrous or pubescent); petals [sometimes absent] usually white or pink [reddish], obovate [spatulate], (much longer or shorter than sepals), claw differentiated from blade, (apex obtuse); stamens tetradynamous; filaments not dilated basally; anthers ovate [oblong], (apex obtuse); nectar glands (4), lateral, 1 on each side of lateral stamen. **Fruits** silicles, dehiscent, sessile, obdeltoid to obdeltoid-obcordiform, strongly flattened, strongly keeled, angustiseptate; valves (papery), each prominently veined, glabrous; replum rounded; septum complete; ovules (12–)20–40 per ovary; (style included or exserted from apical notch); stigma capitate. **Seeds** uniseriate, plump, not winged, oblong; seed coat (reticulate), mucilaginous when wetted; cotyledons incumbent. $x = 8$.

Species 4 (1 in the flora): introduced; Europe, Asia, n Africa; introduced also in Mexico, West Indies, Central America, South America, Atlantic Islands, Pacific Islands, Australia.

The number of species to be recognized in *Capsella* is controversial, and some authors (e.g., S. Svensson 1983; R. C. Rollins 1993; O. Appel and I. A. Al-Shehbaz 2003) treated the genus as monotypic, whereas others (e.g., H. Hurka and B. Neuffer 1997; P. Nutt et al. 2003) recognized three or four species.

SELECTED REFERENCES Hurka, H. and B. Neuffer. 1997. Evolutionary processes in the genus *Capsella* (Brassicaceae). Pl. Syst. Evol. 206: 295–316. Neuffer, B. and H. Hurka. 1999. Colonizing history and introduction dynamics of *Capsella bursa-pastoris* in North America: Isozymes and quantitative traits. Molec. Ecol. 8: 1667–1681. Nutt, P. et al. 2006. *Capsella* as a model system to study the evolutionary relevance of floral homeotic mutants. Pl. Sys. Evol. 259: 217–235. Slotte, T. et al. 2006. Intrageneric phylogeny of *Capsella* (Brassicaceae) and the origin of the tetraploid *C. bursa-pastoris* based on chloroplast and nuclear DNA sequences. Amer. J. Bot. 93: 1714–1724.

1. Capsella bursa-pastoris (Linnaeus) Medikus, Pfl.-Gatt., 85. 1792 [F] [I] [W]

Thlaspi bursa-pastoris Linnaeus, Sp. Pl. 2: 647. 1753

Plants mostly sparsely to densely pubescent, trichomes sessile, 3–5-rayed, stellate (base of plant often mixed with much longer, simple ones). **Stems** (0.2–)1–5(–7) dm. **Basal leaves:** petiole 0.5–4(–6) cm; blade oblong or oblanceolate, (0.5–)1.5–10(–15) cm × 2–25(–50) mm, base cuneate or attenuate, apex acute or acuminate. **Cauline leaves:** blade narrowly oblong, lanceolate, or linear, 1–5.5(–8) cm × 1–15(–20) mm, base sagittate, amplexicaul, or, rarely, auriculate. **Fruiting pedicels** usually straight, (0.3–)0.5–1.5(–2) cm, glabrous. **Flowers:** sepals green or reddish, 1.5–2 × 0.7–1 mm (margins membranous); petals (1.5–)2–4(–5) × 1–1.5 mm; filaments 1–2 mm; anthers to 0.5 mm. **Fruits** (0.3–)0.4–0.9(–1) cm × (2–)3–7(–9) mm, base cuneate, apex emarginate or truncate; valves each with subparallel lateral veins; style 0.2–0.7 mm. **Seeds** brown, 0.9–1.1 × 0.4–0.6 mm. $2n = 32$.

Flowering and fruiting Jan–Oct. Roadsides, gardens, fields, barren gravel, pastures, plantations, lawns, orchards, cultivated ground, waste areas, vineyards, mountain slopes; 0–2800 m; introduced; Greenland; Alta., B.C., Man., N.B., Nfld. and Labr., N.W.T., N.S., Nunavut, Ont., Que., Sask., Yukon; Ala., Alaska, Ariz., Ark., Calif., Colo., Conn., Del., D.C., Fla., Ga., Idaho, Ill., Ind., Iowa, Kans., Ky., La., Maine, Md., Mass., Mich., Minn., Miss., Mo., Mont., Nebr., Nev., N.H., N.J., N.Mex., N.Y., N.C., N.Dak., Ohio, Okla., Oreg., Pa., R.I., S.C., S.Dak., Tenn., Tex., Utah, Vt., Va., Wash., W.Va., Wis., Wyo.; Europe; Asia; n Africa; introduced also in Mexico, West Indies, Central America, South America, Atlantic Islands, Pacific Islands, Australia.

According to M. Coquillat (1951), *Capsella bursa-pastoris* is the second most common weed on earth, after *Polygonum aviculare*.

35. NESLIA Desvaux, J. Bot. Agric. 3: 162. 1815, name conserved • [For J. A. N. de Nesle, eighteenth-century French gardener at Poitiers]

Ihsan A. Al-Shehbaz

Annuals; not scapose; mostly pubescent, trichomes short-stalked, forked or substellate, mixed (on stem) with simple ones. **Stems** erect, unbranched basally, branched distally. **Leaves** basal and cauline; petiolate or sessile; basal not rosulate, shortly petiolate, blade margins entire, dentate, or denticulate; cauline blade (base sagittate or strongly auriculate), margins usually entire, rarely denticulate. **Racemes** (corymbose, several-flowered, forming panicles), considerably elongated in fruit. **Fruiting pedicels** divaricate-ascending, slender. **Flowers:** sepals erect, oblong-ovate (pubescent); petals yellow, spatulate, (longer than sepals), claw not differentiated from blade, (apex obtuse); stamens slightly tetradynamous; filaments not dilated basally; anthers ovate, (apex obtuse); nectar glands lateral, 1 on each side of lateral stamen. **Fruits** silicles, nutletlike, indehiscent, subsessile, woody, compressed globose or sublenticular (readily detached from pedicel at maturity, apex truncate [umbonate]); valves (1-seeded), prominently reticulate, glabrous; replum rounded (obscured by valve margin); septum complete; ovules 2–4 per ovary; (style distinct, cylindrical, readily caducous at fruit maturity, leaving umbo or apicula); stigma capitate. **Seeds** uniseriate, plump, not winged, ovoid; seed coat (minutely reticulate), not mucilaginous when wetted; cotyledons incumbent. $x = 7$.

Species 1: introduced; Europe, Asia, n Africa; introduced also in South America (Argentina), Australia.

Some authors recognize two species in *Neslia*, while others recognize only *N. paniculata* with two subspecies somewhat separated geographically, though intermediates are common in areas of overlap (P. W. Ball 1961). The sole difference between them is whether the fruit apex is truncate (subsp. *paniculata*) or apiculate [subsp. *thracica* (Velenovský) Bornmüller or *N. apiculata* Fischer, C. A. Meyer & Avé-Lallemant].

SELECTED REFERENCE Ball, P. W. 1961. The taxonomic status of *Neslia paniculata* (L.) Desv. and *N. apiculata* Fisch., Mey. & Avé-Lall. Feddes Repert. Spec. Nov. Regni Veg. 64: 11–13.

1. Neslia paniculata (Linnaeus) Desvaux, J. Bot. Agric. 3: 162. 1815 • Ball-mustard F I W

Myagrum paniculatum Linnaeus, Sp. Pl. 2: 641. 1753

Plants sparsely to moderately pubescent, fruits glabrous. **Stems** (1.4–)2.5–7.5(–9) dm. **Basal leaves** shortly petiolate; blade oblanceolate to oblong, 2–7.5 cm × 5–20 mm. **Cauline leaves:** blade lanceolate, narrowly oblong, or linear-lanceolate, (1.5–)2.5–7(–9) cm × (2–)3–15(–25) mm, apex acute or acuminate, surfaces sparsely pubescent, trichomes mostly forked. **Fruiting pedicels** straight or slightly curved upwards, (4–)6–10(–14) mm. **Flowers:** sepals 1.5–1.7 × 0.5–0.7(–1) mm; petals 2–2.5 × 0.5–0.7(–1) mm; filaments 1.5–2 mm; anthers 0.2–0.4 mm. **Fruits** (1.7–)2–2.2 × (2–)2.2–2.5 mm; style slender, 0.5–0.9 mm. **Seeds** 1.2–1.4 × 1–1.1 mm. $2n = 14$.

Flowering May–Sep. Fields, grassy mountain slopes, plains, roadsides, cultivated fields; 0–1000 m; introduced; Alta., B.C., Man., N.B., Nfld. and Labr. (Nfld.), N.W.T., N.S., Ont., Que., Yukon; Alaska, Conn., Ind., Maine, Mass., Mich., Minn., Mont., N.H., Ohio, Pa., Vt.; Europe; Asia; n Africa; introduced also in South America (Argentina), Australia.

36. TRANSBERINGIA Al-Shehbaz & O'Kane, Novon 13: 396. 2003 • [Latin *trans-*, across, and Bering Sea, alluding to distribution]

Ihsan A. Al-Shehbaz

Beringia R. A. Price, Al-Shehbaz & O'Kane, Novon 11: 333. 2001, not Perestenko 1975

Perennials; not scapose; glabrous or pubescent, trichomes simple, forked, stalked, or dendritic. **Stems** erect or ascending, unbranched or branched basally and distally. **Leaves** basal and cauline; petiolate or sessile; basal rosulate, petiolate, blade margins often entire, sometimes dentate, rarely sinuate; cauline blade (base not auriculate, sagittate), margins usually entire, rarely dentate. **Racemes** (corymbose, several-flowered), elongated in fruit. **Fruiting pedicels** ascending to subdivaricate, slender (terete). **Flowers:** sepals erect, oblong, (pubescent); petals white, oblanceolate, (longer than sepals), claw obscurely differentiated from blade, (apex obtuse); stamens slightly tetradynamous; filaments not dilated basally; anthers oblong, (apiculate); nectar glands confluent, subtending bases of stamens. **Fruits** siliques, dehiscent, sessile or subsessile, linear, smooth, terete; valves each with prominent midvein and marginal veins, secondary veins anastomosing and distinct or obscure, glabrous; replum rounded; septum complete; ovules 70–150 per ovary; (style obsolete or distinct); stigma capitate. **Seeds** biseriate, plump, not winged, narrowly oblong; seed coat (minutely reticulate), not mucilaginous when wetted; cotyledons incumbent. $x = 8$.

Species 1: n, w North America, e Asia (Russian Far East).

For discussion of the generic limits of *Transberingia*, see the original description and also under 60. *Halimolobos*.

1. **Transberingia bursifolia** (de Candolle) Al-Shehbaz & O'Kane, Novon 13: 396. 2003 F

Nasturtium bursifolium de Candolle, Syst. Nat. 2: 194. 1821; *Arabidopsis bursifolia* (de Candolle) Botschantzev; *Beringia bursifolia* (de Candolle) R. A. Price, Al-Shehbaz & O'Kane

Plants densely hirsute basally, trichomes primarily simple, to 1 mm, these often mixed with smaller, forked, and/or dendritic ones, glabrescent or sparsely pubescent distally. **Stems** simple or few to several from caudex, (0.5–)1–4.5(–6.5) dm. **Basal leaves:** petiole 0.3–3 cm, flattened at base, ciliate; blade oblanceolate to obovate, or linear-lanceolate, (0.5–)1.2–6.5(–10) cm × (2–)3–8(–17) mm, base cuneate to attenuate, apex obtuse to subacute, surfaces pubescent, trichomes usually branched. **Cauline leaves:** blade lanceolate to linear-lanceolate or oblong, (0.5–)1–3(–4.2) cm × (1–)2.5–7(–10) mm, base strongly sagittate. **Fruiting pedicels** not appressed to rachis, straight, (3–)5–14(–18) mm, glabrous. **Flowers:** sepals 1.2–2.5 × 0.6–1 mm, sparsely to densely pubescent; petals (2.5–)3–4(–4.5) × (0.8–)1–1.5 mm; filaments 2–2.5 mm; anthers 0.4–0.5 mm. **Fruits** (1.5–)2–3.3(–4) cm × (0.7–)1–1.5(–2) mm; style slender, 0.2–0.8(–1) mm. **Seeds** 0.8–1 × 0.4–0.5 mm.

Subspecies 2 (2 in the flora): n, w North America, e Asia (Russian Far East).

R. C. Rollins (1993) treated subspp. *bursifolia* and *virgata* as distinct species of *Halimolobos*, but the alleged differences (stems branched basally versus distally, plants cespitose versus non-cespitose, fruits slightly compressed versus terete) do not hold, and the only reliable difference between the two taxa is in the type of trichomes. Therefore, a subspecific, rather than specific, rank is adopted here.

1. Plants glabrous or sparsely pubescent on stems distally and on rachises, trichomes simple, spreading ....
   .... 1a. *Transberingia bursifolia* subsp. *bursifolia*
1. Plants usually pubescent, rarely glabrescent, on stems distally and on rachises, trichomes branched, subappressed, these sometimes mixed with simple ones ....
   .... 1b. *Transberingia bursifolia* subsp. *virgata*

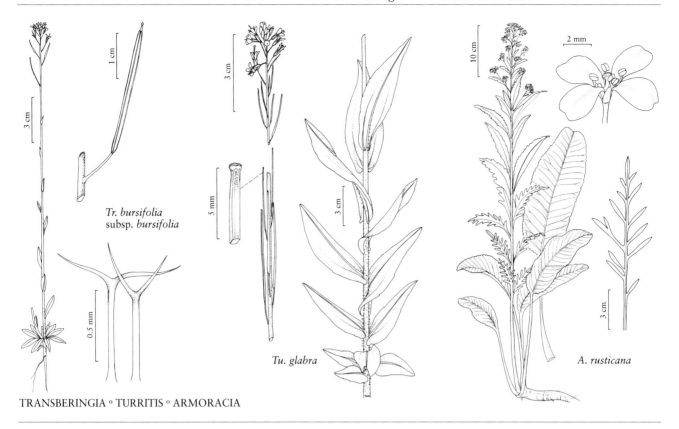

TRANSBERINGIA ○ TURRITIS ○ ARMORACIA

1a. Transberingia bursifolia (de Candolle) Al-Shehbaz & O'Kane subsp. **bursifolia** [F]

Arabidopsis bursifolia (de Candolle) Botschantzev var. *beringensis* Jurtzev; *A. mollis* (Hooker) O. E. Schulz; *A. trichopoda* (Turczaninow) Botschantzev; *A. tschuktschorum* (Jurtzev) Jurtzev; *Arabis hookeri* Lange; *A. hookeri* var. *breviramosa* Abromeit; *A. hookeri* var. *multicaulis* Simmons; *A. trichopoda* Turczaninow; *A. tschuktschorum* Jurtzev; *Halimolobos mollis* (Hooker) Rollins; *Turritis mollis* Hooker

Plants glabrous or sparsely pubescent on stems distally and raceme rachis, trichomes simple, spreading. $2n = 16$.

Flowering May–Jul. Rocky knolls, limestone hills, grassy slopes, sandy flats, shaley cliffs, stream banks, roadsides, open woodlands; 0–1200 m; Greenland; N.W.T., Nunavut, Yukon; Alaska; Asia (Russian Far East).

1b. Transberingia bursifolia (de Candolle) Al-Shehbaz & O'Kane subsp. **virgata** (Nuttall) Al-Shehbaz & O'Kane, Novon 13: 396. 2003 [E]

Sisymbrium virgatum Nuttall in J. Torrey and A. Gray, Fl. N. Amer. 1: 93. 1838; *Arabidopsis stenocarpa* (Rydberg) Rydberg; *A. virgata* (Nuttall) Rydberg; *Arabis brebneriana* A. Nelson; *Beringia bursifolia* (de Candolle) R. A. Price, Al-Shehbaz & O'Kane subsp. *virgata* (Nuttall) R. A. Price, Al-Shehbaz & O'Kane; *Halimolobos virgata* (Nuttall) O. E. Schulz; *Hesperis virgata* (Nuttall) Kuntze; *Pilosella stenocarpa* Rydberg; *P. virgata* (Nuttall) Rydberg; *Stenophragma virgatum* (Nuttall) Greene; *Transberingia virgata* (Nuttall) N. H. Holmgren

Plants usually pubescent, rarely glabrescent, on stems distally and raceme rachis, trichomes branched, subappressed, these sometimes mixed with simple ones.

Flowering May–Jul. Alkaline flats, plains, meadows, brushy hillsides, alpine areas, foothills, mountain slopes, ridges, dry knolls, grassy areas; 1000–3600 m; Alta., Sask.; Calif., Colo., Idaho, Mont., Nev., Utah, Wyo.

Apart from its distribution in Wyoming, subsp. *virgata* appears to be limited to counties in California (Inyo, Mono), Colorado (Eagle, Gunnison, Park), Idaho (Butte, Clark), Montana (Beaverhead, Phillips), Nevada (Esmeralda, Nye), and Utah (Dagget, Wasatch).

37. TURRITIS Linnaeus, Sp. Pl. 2: 666. 1753; Gen. Pl. ed. 5, 298. 1754 • Tower-mustard, towercress [Latin *turris*, tower, alluding to pyramidal shape of plants due to overlap of leaves and fruits]

Ihsan A. Al-Shehbaz

Biennials or, rarely, perennials (short-lived); not scapose; (glaucous distally), glabrous or pubescent (mostly proximally), trichomes simple and/or stalked, forked, or substellate. **Stems** erect, unbranched basally, often branched distally. **Leaves** basal and cauline; petiolate or sessile; basal rosulate, petiolate, margins usually dentate, sinuate, repand, or pinnatifid, rarely entire; cauline blade (base auriculate or sagittate [amplexicaul]), margins dentate or entire. **Racemes** (corymbose), considerably elongated in fruit. **Fruiting pedicels** erect [divaricate], (appressed to rachis), slender. **Flowers:** sepals erect, oblong or linear, (margins membranous); petals usually yellowish or creamy white, rarely pink [purplish], usually narrowly spatulate or linear-oblanceolate, rarely linear, claw not differentiated from blade, (apex obtuse); stamens (erect), tetradynamous; filaments not dilated basally; anthers narrowly oblong [linear], (apex obtuse); nectar glands confluent, subtending bases of stamens, lateral glands annular, median glands present. **Fruits** siliques, dehiscent, sessile, linear, smooth, often subterete-quadrangular, (leathery); valves each with prominent midvein, glabrous; replum rounded; septum complete, (membranous, veinless); ovules 130–200 per ovary; (style distinct, short, stout); stigma capitate, (subentire). **Seeds** biseriate, flattened, not winged or, rarely, narrowly winged, oblong or subglobose [elliptic, orbicular]; seed coat not mucilaginous when wetted; cotyledons accumbent. $x = 6, 8$.

Species 2 (1 in the flora): North America, Eurasia, n Africa.

Turritis laxa (Smith) Hayek is distributed in eastern Europe and the Middle East.

SELECTED REFERENCE Dvořák, F. 1967. A note on the genus *Turritis* L. Oesterr. Bot. Z. 114: 84–87.

1. Turritis glabra Linnaeus, Sp. Pl. 2: 666. 1753 F W

Arabis glabra (Linnaeus) Bernhardi; *A. glabra* var. *furcatipilis* M. Hopkins; *A. macrocarpa* (Nuttall) Torrey; *A. perfoliata* Lamarck; *A. pseudoturritis* Boissier & Heldreich; *Turritis glabra* var. *lilacina* O. E. Schulz; *T. macrocarpa* Nuttall; *T. pseudoturritis* (Boissier & Heldreich) Velenovsky

Plants (3–)4–12(–15) dm, sparsely to densely pilose basally, glabrous distally, trichomes simple and short-stalked, forked. **Basal leaves:** blades spatulate, oblanceolate, or oblong, (4–)5–12(–15) cm × 10–30 mm, apex obtuse, surfaces usually pubescent, rarely glabrous. **Cauline leaves:** blade lanceolate, oblong-elliptic, or ovate, 2–9(–12) cm × (5–)10–25(–40) mm, apex acute. **Fruiting pedicels** appressed to rachis, (6–)7–16(–20) mm, glabrous. **Flowers:** sepals (2.5–)3–5 × 0.5–1.2 mm, glabrous; petals 5–8.5 × 1.3–1.7 mm; filaments slender, median pairs 3.5–6.5 mm, lateral pair 2.5–4.5 mm; anthers 0.7–1.5 mm. **Fruits** (3–)4–10(–12.5) cm × 0.7–1.5 mm; style 0.5–0.8(–1) mm. **Seeds** 0.6–1.2 × 0.5–0.9 mm. $2n$ = 12, 16, 32.

Flowering Apr–Jul. Forest margins, fields, roadsides, stream banks, disturbed sites, mountain slopes, woods, meadows; 0–2800 m; Alta., B.C., Man., N.B., N.W.T., Ont., Que., Sask., Yukon; Alaska, Ariz., Calif., Colo., Conn., Idaho, Ill., Ind., Iowa, Maine, Mass., Mich., Minn., Mo., Mont., Nebr., Nev., N.H., N.Mex., N.Y., N.C., Ohio, Oreg., Pa., S.Dak., Tenn., Utah, Vt., Va., Wash., W.Va., Wis., Wyo.; Europe; sw Asia; n Africa; introduced in Australia.

j. BRASSICACEAE Burnett tribe CARDAMINEAE Dumortier, Fl. Belg., 124. 1827

Annuals, biennials, or perennials; eglandular. **Trichomes** absent or simple. **Cauline leaves** (rarely absent), usually petiolate, sometimes sessile; blade (simple or compound), base auriculate or not, margins entire or dentate to pinnately lobed. **Racemes** usually ebracteate (*Selenia* bracteate throughout), often elongated in fruit. **Flowers** actinomorphic; sepals erect to spreading or ascending, lateral pair seldom saccate basally; petals white, yellow, pink, lilac, or purple, claw usually present, rarely absent, often distinct; filaments unappendaged, not winged; pollen 3-colpate. **Fruits** silicles or siliques, dehiscent, unsegmented, terete, 4-angled, or latiseptate, rarely angustiseptate; ovules 4–300[–numerous] per ovary; style distinct or obsolete (absent in *Subularia*); stigma usually entire. **Seeds** biseriate or uniseriate; cotyledons usually accumbent, sometimes incumbent.

Genera 12, species ca. 335 (10 genera, 85 species in the flora): nearly worldwide.

The assignment of *Subularia* to Cardamineae is provisional and based solely on morphology.

38. ARMORACIA P. Gaertner, B. Meyer & Scherbius, Oekon. Fl. Wetterau 2: 426. 1800, name conserved • Horseradish [Ancient Greek name for horseradish, or perhaps Celtic *ar*, near, *mor*, sea, and *rich*, against, alluding to habitat]

Ihsan A. Al-Shehbaz

Perennials; (aquatic or of mesic habitats, with rootstocks); not scapose; glabrous. **Stems** erect, branched distally. **Leaves** basal and cauline; petiolate and sessile; basal rosulate, long-petiolate, blade margins crenate or pinnatifid [entire]; cauline petiolate or sessile distally, blade margins crenate, serrate, pinnatifid, pinnatisect [laciniate]. **Racemes** ([often corymbose-paniculate], several-flowered), considerably elongated in fruit. **Fruiting pedicels** ascending, divaricate, or slightly reflexed, slender. **Flowers:** sepals spreading or ascending, ovate or oblong, lateral pair not saccate basally, (glabrous); petals white, obovate, or oblanceolate [spatulate, oblong], claw somewhat differentiated from blade (relatively short, apex obtuse); stamens slightly tetradynamous; filaments slightly dilated basally; anthers ovate [oblong or linear], (apex obtuse); nectar glands confluent, subtending bases of stamens, median glands present. **Fruits:** silicles, sessile, oblong, ovate, elliptic, or orbicular, angustiseptate; valves each not veined; replum rounded; septum perforated or reduced to a rim; ovules 8–12[–20] per ovary; style obsolete or distinct; stigma capitate, (sometimes 2-lobed). **Seeds** biseriate, plump, not winged, ovate [oblong]; seed coat (punctate) not mucilaginous when wetted; cotyledons accumbent.

Species 3 (1 in the flora): introduced; c, s Europe, Asia (Russian Far East, Siberia).

SELECTED REFERENCES Fosberg, F. R. 1965. Nomenclature of the horseradish (Cruciferae). Baileya 13: 1–4. Lawrence, G. H. M. 1971. The horseradish *Armoracia rusticana*. Herbarist 37: 17–19.

1. Armoracia rusticana P. Gaertner, B. Meyer & Scherbius, Oekon. Fl. Wetterau 2: 426. 1800 [F][I][W]

Cochlearia armoracia Linnaeus; *Rorippa armoracia* (Linnaeus) A. S. Hitchcock

Roots fusiform or cylindrical, fleshy or woody. **Stems** 5–12(–20) dm. **Basal leaves:** petiole to 60 cm (broadly expanded basally); blade broadly oblong, oblong-lanceolate, or ovate, (10–)20–45(–60) cm × (30–)50–120(–170) mm, margins usually coarsely crenate, rarely pinnatifid. **Cauline leaves:** proximal shortly petiolate, blade oblong to linear-oblong (lobed), smaller than basal, margins pinnatifid or pinnatisect; distal sessile or shortly petiolate, blade linear to linear-lanceolate, base cuneate or attenuate, margins usually serrate or crenate, rarely entire. **Racemes** to 40 cm. **Fruiting pedicels** ascending, 8–20 mm. **Flowers:** sepals 2–4 mm; petals obovate or oblanceolate, 5–7(–8) mm, claw to 1.5 mm; filaments 1–2.5 mm; anthers 0.5–0.8 mm. **Fruits** (rarely produced), 4–6 mm; style obsolete or to 0.5 mm; stigma well-developed. **Seeds** compressed (often not produced, rarely to 4 per locule). $2n = 32$.

Flowering Mar–Jul. Fields, moist stream banks, roadsides, ditches, disturbed sites, open woods, along railroads, shallow ponds, marshes, waste places; 0–1100 m; introduced; B.C., Man., N.B., N.S., Ont., P.E.I., Que., Sask.; Calif., Colo., Conn., Idaho, Ill., Ind., Iowa, Maine, Mass., Mich., Minn., Mo., Mont., Nebr., N.H., N.J., N.Mex., N.Y., N.C., N.Dak., Ohio, Oreg., Pa., R.I., S.Dak., Tenn., Utah, Vt., Va., Wash., W.Va., Wis., Wyo.; Europe; Asia; introduced also nearly worldwide.

Armoracia rusticana has been widely cultivated for about 2000 years for its fleshy roots that are grated to produce the pungent horseradish sauce. The plant is also a noxious weed that is very difficult to eradicate, even tiny root fragments are capable of regenerating new plants.

39. BARBAREA W. T. Aiton in W. Aiton and W. T. Aiton, Hortus Kew. 4: 109. 1812, name conserved • Wintercress, scurvygrass, rocket, uplandcress, corn-mustard [For Saint Barbara, fourth-century, or perhaps alluding to being the only plants available for food on Saint Barbara's Day (4 December)]

Ihsan A. Al-Shehbaz

Campe Dulac

Biennials or perennials [annuals]; (rhizomatous or with woody caudex); not scapose; glabrous or sparsely pubescent. **Stems** erect [prostrate], branched distally, (angular [not angular]). **Leaves** basal and cauline; petiolate and sessile; basal rosulate or not, (and proximal cauline) petiolate, blade margins usually entire, crenate or lobed, rarely dentate or repand; cauline sessile, blade (base auriculate or amplexicaul) margins entire, dentate, or lobed. **Racemes** (corymbose, several-flowered), considerably [slightly] elongated in fruit, (rachis striate). **Fruiting pedicels** (sometimes absent), erect to divaricate, slender or stout. **Flowers:** sepals (sometimes persistent), erect [spreading], oblong [ovate, linear], lateral pair saccate or not basally, (apex often cucullate); petals yellow or pale yellow [creamy white], spatulate or oblanceolate, (longer than sepals), claw obscurely differentiated from blade, (apex obtuse or rounded); stamens tetradynamous; filaments (yellow), not dilated basally; anthers oblong, (apex obtuse); nectar glands (4): lateral annular, median toothlike. **Fruits** siliques, sessile or shortly stipitate, usually linear, rarely elliptic-linear, smooth or torulose, terete, 4-angled, or latiseptate; valves each with prominent midvein and distinct marginal veins, usually glabrous, rarely pubescent; replum rounded; septum complete; ovules 16–52 per ovary; style obsolete or distinct; stigma capitate, (sometimes slightly 2-lobed). **Seeds** uniseriate [sub-biseriate], plump or slightly flattened, not winged [winged or margined], oblong, ovoid, or orbicular; seed coat (reticulate or, rarely, tuberculate), not mucilaginous when wetted; cotyledons accumbent. $x = 8$.

Species 22 (4 in the flora): North America, Europe, Asia, n Africa, Australia

Barbarea is a difficult genus much in need of systematic and phylogenetic studies throughout its range. Although some of its species are easily recognizable, some complexes, especially the widespread or weedy taxa, remain problematic. It is likely that hybridization is involved, but no studies have confirmed that.

The determination of flowering material is not always possible, and most workers have relied heavily on the distalmost leaves to separate species. Both *Barbarea orthoceras* and *B. verna* are said to have pinnatisect to pinnatifid distalmost leaves, whereas *B. stricta* and *B. vulgaris* are said to have undivided, entire, or dentate leaves. This separation can be misleading because *B. orthoceras* sometimes has entire or dentate distal leaves, whereas in some *B. vulgaris* the distal leaves are deeply divided. In *B. vulgaris*, the style length and its thickness in relation to the fruits are useful in separating it from the remaining species, though, on rare occasions, both *B. orthoceras* and *B. stricta* have slender styles to 2 mm. In these cases, both *B. orthoceras* and *B. stricta* can be separated from *B. vulgaris* by the ciliate auricles of cauline leaves and subapically pubescent sepals. Both sepals and auricles of distalmost leaves are always glabrous in *B. vulgaris*. Although all the Eurasian specimens of *B. stricta* that I examined have pubescent auricles and sepal apices, the number of trichomes can be quite variable and ranges from one to many. Most of the naturalized North American plants of *B. stricta* have glabrescent or glabrous sepals. However, that species can be further distinguished by having petals 2.5–4.5 × 0.5–1 (–1.2) mm, the smallest and narrowest among the four species growing in North America. In the absence of mature fruits, one needs to be aware of the variation in the other characters.

SELECTED REFERENCE   Fernald, M. L. 1909. The North American species of *Barbarea*. Rhodora. 11: 134–141.

1. Fruits (4.5–)5.3–7(–8) cm; ovules (34–)38–48(–52) per ovary; fruiting pedicels as broad as fruit; cauline leaf blades pinnatifid to pinnatisect. . . . . 4. *Barbarea verna*
1. Fruits (0.7–)1.5–4(–4.5) cm; ovules 16–36 per ovary; fruiting pedicels narrower than fruit; cauline leaf blades undivided or lyrate-pinnatifid.
    2. Styles (1–)1.5–3(–3.5) mm, slender; auricles of cauline leaves glabrous; fruits usually not appressed to rachises. . . . . 1. *Barbarea vulgaris*
    2. Styles 0.2–1.5(–2) mm, sometimes stout; auricles of cauline leaves usually sparsely pubescent, rarely glabrous, sometimes margins ciliate; fruits (erect to erect-ascending) sometimes appressed to rachises.
        3. Fruits (1.2–)1.8–2.8(–3) cm; ovules (16–)20–28 per ovary; petals 2.5–4.5 × 0.5–1 (–1.2) mm; cauline leaf blade margins dentate . . . . 2. *Barbarea stricta*
        3. Fruits (2.5–)3.1–4(–4.5) cm; ovules (24–)26–36 per ovary; petals 5–7(–8) × (1.5–)2–3 mm; cauline leaf blade margins incised or pinnatifid . . . . 3. *Barbarea orthoceras*

1. Barbarea vulgaris W. T. Aiton in W. Aiton and W. T. Aiton, Hortus Kew. 4: 109. 1812 • Cress, yellow-rocket, cressy-greens [I] [W]

Erysimum barbarea Linnaeus, Sp. Pl. 2: 660. 1753; *Barbarea arcuata* (Opiz ex C. Presl) Reichenbach; *B. vulgaris* var. *arcuata* (Opiz ex C. Presl) Fries; *E. arcuatum* Opiz ex C. Presl

Biennials or, rarely, perennials; glabrous throughout or margins ciliate. **Stems** (1.5–)2–9(–12) dm. **Basal leaves:** petiole (0.5–)2–10(–17) cm; blade lyrate-pinnatifid, (1–)2–8(–10) cm, lobes 1–3(–5) on each side (rarely early ones undivided), lateral lobes oblong or ovate, 0.3–2(–4) cm × 1–8(–15) mm, sometimes slightly fleshy, margins entire, repand, crenate, or dentate, terminal lobe (ovate or suborbicular), (0.7–)1.5–4.5(–7) cm × (4–)10–30(–50) mm, (surfaces glabrous or margins ciliate). **Cauline leaves:** blade ovate or suborbicular (undivided), margins usually coarsely dentate, rarely subentire; conspicuously auriculate, auricles ovate or narrowly oblong (to 10 × 5 mm), glabrous. **Fruiting pedicels** divaricate to ascending or erect, 3–7 mm, terete or subquadrangular, thickened (narrower than fruit). **Flowers:** sepals 3–4.5(–5) × 1–1.5 mm, lateral pair slightly saccate basally, margins scarious; petals yellow, spatulate or oblanceolate, (5–)6–9(–10) × 1.5–2.5(–3.5) mm, base attenuate, apex rounded; filaments 3–4.5 mm; anthers 0.7–1.2 mm; ovules 18–24(–28) per ovary; gynophore to 0.5 mm. **Fruits** erect to erect-ascending, rarely appressed to rachis, torulose, terete, somewhat compressed, or 4-angled, (0.7–)1.5–3 cm × 1.2–2 mm; style slender, (1–)1.5–3(–3.5) mm. **Seeds** dark brown, plump, broadly ovoid to oblong or subglobose, 1.2–1.5 × 1–1.2 mm. $2n = 16$.

Flowering Apr–Jul. Waste places, ditches, riverbanks, damp grasslands, roadsides, fields, disturbed sites; 0–3000 m; introduced; Alta., B.C., Man., N.B., Nfld. and Labr., N.S., Ont., P.E.I., Que., Sask.; Ala., Alaska, Ark., Calif., Colo., Conn., Ga., Idaho, Ill., Ind., Iowa, Kans., Ky., Maine, Md., Mass., Mich., Minn., Mo., Mont., Nebr., N.H., N.J., N.Mex., N.Y., N.C., Ohio, Okla., Oreg., Pa., R.I., S.Dak., Tenn., Utah, Vt., Va., Wash., W.Va., Wis., Wyo.; Europe; Asia; n Africa.

Barbarea vulgaris, which is sometimes grown as a potherb, is highly variable in length and orientation of fruit and fruiting pedicel, style length, and the division of cauline leaves. Several varieties have been recognized, and they represent some of the many points along one continuum. In my opinion, it is better not to recognize any infraspecific taxa in North America.

2. Barbarea stricta Andrzejowski in W. S. J. G. von Besser, Enum. Pl., 72. 1821

Biennials or perennials; mostly glabrous, except blade auricles ciliate or sparsely pubescent. **Stems** 2–7.5(–10) dm. **Basal leaves:** petiole (1–)2–6(–9) cm, (sometimes ciliate); blade lyrate-pinnatifid, 1.5–5(–7) cm, lobes (0 or) 1 or 2(–4) on each side, lateral lobes oblong or ovate, 0.05–1.5 cm × 1–5 mm, not fleshy, margins entire, terminal lobe 2–7 cm × 10–55 mm. **Cauline leaves:** blade sometimes lyrate-pinnatifid, lateral lobes 1–3, margins (distalmost) dentate; conspicuously auriculate, auricles ovate or narrowly oblong, (margins entire). **Fruiting pedicels** erect, 2–4(–5) mm, terete or subquadrangular, slender (narrower than fruit, glabrous). **Flowers:** sepals 2–3 × 0.5–1 mm, lateral pair not saccate basally, margins scarious, (glabrous, apex sparsely pubescent to, rarely, glabrescent, subapically sparsely pubescent or, rarely, glabrous); petals yellow, narrowly oblanceolate, 2.5–4.5 × 0.5–1(–1.2) mm, base attenuate, apex rounded; filaments 2–3.5 mm; anthers 0.5 mm; ovules (16–)20–28 per ovary; gynophore to 0.5 mm. **Fruits** erect, appressed to rachis, torulose, terete to subquadrangular, (1.2–)1.8–2.8(–3) cm × 1–1.5 mm; style stout or slender, 0.2–1.5(–2) mm. **Seeds** brown, somewhat plump, ovoid or oblong, 0.8–1.5 × 0.5–1 mm.

Flowering Apr–Jul. Waste places, ditches, riverbanks, damp grasslands, roadsides, fields, disturbed sites; 0–1000 m; introduced; Ont., Que.; Colo., Conn., Maine, Mass., Mich., N.H., N.Y., R.I., Vt., Wis.; Europe; Asia.

M. L. Fernald (1909) treated some plants of *Barbarea stricta* as *B. vulgaris* var. *longisiliqua* Carion. Although G. A. Mulligan (1978) was the first to record *B. stricta* for North America, neither his record nor the species was accounted for by R. C. Rollins (1993). Forms of *B. vulgaris* with fruits appressed to rachis can easily be confused with *B. stricta* and some forms of *B. orthoceras*. The latter is separated from both *B. stricta* and *B. vulgaris* by having distalmost cauline leaves either pinnatifid or with at least a pair of well-developed lateral lobes. The distalmost leaves in both *B. stricta* and *B. vulgaris* have margins entire or obscurely to coarsely dentate. *Barbarea stricta* has shorter petals (2.5–4.5 mm) and styles (0.2–1.5(–2) mm), strictly erect and straight fruiting pedicels, and sepals and auricles of distalmost leaves often with a few apical hairs. By contrast, *B. vulgaris* has longer petals [(5–)6–9(–10) mm] and styles [(1–)1.5–3(–3.5) mm], erect or widely spreading and often curved fruiting pedicels, and glabrous sepals and leaf auricles. The boundaries between the two species can be somewhat blurred in parts of New England, and it is not known if hybridization is involved. In that region, plants of *B. stricta* tend to have slightly longer styles (to 2 mm), and the number of hairs at the auricle apices can be reduced to one and those of the sepals can be absent.

3. Barbarea orthoceras Ledebour, Index Seminum (Dorpat) 2. 1824

Barbarea americana Rydberg; *B. orthoceras* var. *dolichocarpa* Fernald; *Campe orthoceras* (Ledebour) A. Heller

Biennials or perennials; usually glabrous, except blade auricles often ciliate. **Stems** (1–)2–6(–10) dm. **Basal leaves:** petiole (0.5–)1–7(–15) cm, (rarely ciliate basally); blade usually lyrate-pinnatifid, (1–)1.5–6(–11) cm, lobes (0, 1, or) 2–4 (or 5) on each side, lateral lobes oblong or ovate, 0.2–1 cm × 1–5 mm, not fleshy, margins entire, terminal lobe 1.5–5 cm × 10–25 mm. **Cauline leaves:** blade usually lyrate-pinnatifid (sometimes undivided and margins coarsely toothed or subentire), lateral lobes 1–4, (usually oblong or ovate, rarely lanceolate, to 2 × 1 cm), margins entire, (terminal lobe to 5 × 3 cm, margins usually entire or repand, rarely dentate); conspicuously auriculate, auricles ovate or narrowly oblong, (to 8 × 5 mm, margins entire). **Fruiting pedicels** erect or ascending, (2–)3–6(–7) mm, terete or subquadrangular, stout, (narrower than fruit). **Flowers:** sepals 2.5–3.5 × 1–1.5 mm, lateral pair slightly saccate basally, margins scarious, (subapically sparsely pubescent or glabrous); petals yellow or pale yellow, oblanceolate, 5–7(–8) × (1.5–)2–3 mm, base attenuate, apex rounded; filaments 3–4.5 mm; anthers 1 mm; ovules (24–)26–36 per ovary; gynophore to 0.5 mm. **Fruits** erect to erect-ascending, rarely appressed to rachis, torulose, terete to subquadrangular, (2.5–)3.1–4(–4.5) cm × 1.5–2 mm; style stout, 0.5–1.2(–2) mm. **Seeds** brown, somewhat plump, ovoid or oblong, 1.2–1.5 × 0.9–1 mm. $2n = 16$.

Flowering Mar–Jul. Open grassland, sandbars, scree, alpine meadows, ledges, rocky cliffs, forests, streamsides, railroad embankment, boggy ground, gravel pits, moist grassy slopes; 0–3400 m; Alta., B.C., Man., N.B., Nfld. and Labr., N.W.T., Ont., Que., Sask., Yukon; Alaska, Ariz., Calif., Colo., Idaho, Maine, Mich., Minn., Mont., Nev., N.H., N.Mex., Oreg., S.Dak., Utah, Wash., Wyo.; e, c Asia.

M. L. Fernald (1909) reported the southwestern Asian *Barbarea plantaginea* de Candolle from Alaska, but it is certain that his records were based on plants of *B. orthoceras*, which is rather rare in New England, where it appears to be restricted to portions of northern Maine and New Hampshire.

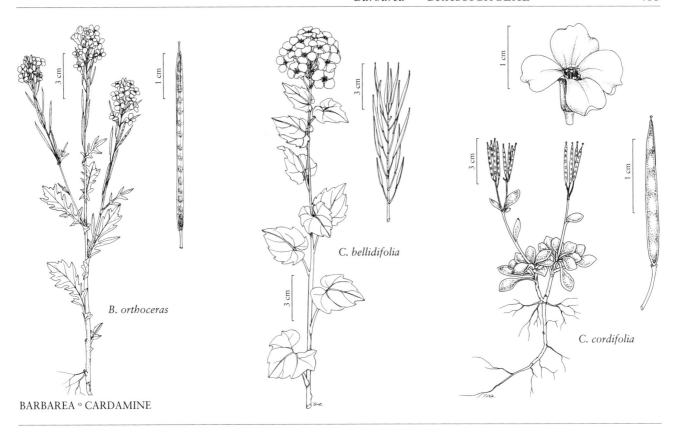

4. Barbarea verna (Miller) Ascherson, Fl. Brandenb. 1: 36. 1860 • Early winter or Belle Isle or American or land cress, scurvygrass [I]

Erysimum vernum Miller, Gard. Dict. ed. 8, Erysimum no. 3. 1768; Barbarea praecox (Smith) R. Brown; Campe verna (Miller) A. Heller; E. praecox Smith

Biennials or, rarely, perennials; mostly glabrous, except blade auricles ciliate. **Stems** (1–)2.5–8 dm. **Basal leaves**: petiole (0.5–)1–6(–8) cm; blade pinnatifid to pinnatisect, 1.5–11 cm, lobes (3–)6–10 on each side, lateral lobes oblong or ovate, 0.4–3 cm × 1–10 mm, not fleshy, margins entire, terminal lobe considerably larger than lateral ones, 1–5 cm × 10–350 mm. **Cauline leaves**: blade pinnatisect, lateral lobes 1–4, (oblong to lanceolate), margins often entire, rarely coarsely toothed; conspicuously auriculate, auricles ovate or narrowly oblong, (to 10 × 4 mm, margins entire). **Fruiting pedicels** divaricate to ascending, (2–)3–6(–7) mm, terete, stout (almost as broad as fruit). **Flowers**: sepals 3–5 × 0.7–1.5 mm, lateral pair slightly saccate basally, margins scarious; petals yellow or pale yellow, oblanceolate to spatulate, (5–)6–7(–8.5) × 1.5–3 mm, base cuneate, apex truncate or emarginate; filaments 3–5 mm; anthers 0.8–1.2 mm; ovules (34–)38–48(–52) per ovary; gynophore to 0.3 mm. **Fruits** erect to ascending, not appressed to rachis, torulose, subterete to slightly latiseptate, (4.5–)5.3–7(–8) cm × 1.5–2 mm; style stout, 0.2–1(–2) mm. **Seeds** dark brown, somewhat plump, oblong or quadrate, 1.8–2.5 × 1.4–1.6 mm. $2n = 16$.

Flowering Mar–Jul. Waste grounds, fields, limestone glades, rocky outcrops, fields, railroad embankments, disturbed sites, roadsides; 0–1600 m; introduced; B.C., Nfld. and Labr. (Nfld.); Ala., Alaska, Ark., Calif., Colo., Conn., Del., D.C., Ga., Ill., Ind., Ky., Maine, Md., Mass., Mich., Miss., Mo., N.J., N.Y., N.C., Ohio, Oreg., Pa., R.I., S.C., Tenn., Va., Wash., W.Va.; Europe; Asia; n, s Africa; introduced also in Mexico, South America (Argentina, Brazil, Chile), Australia.

40. CARDAMINE Linnaeus, Sp. Pl. 2: 654. 1753; Gen. Pl. ed. 5, 295. 1754 • Bittercress

[Greek *kardamon*, name for a cress]

Ihsan A. Al-Shehbaz

Karol Marhold

Judita Lihová

Dentaria Linnaeus; *Dracamine* Nieuwland; *Loxostemon* Hooker f. & Thomson

Annuals, biennials, or perennials; (perennials usually rhizomatous, sometimes tuberiform, stolons present in *C. flagellifera*, caudex present in *C. bellidifolia*); not scapose (subscapose in *C. bellidifolia*); glabrous or pubescent. **Stems** erect, ascending, decumbent, or prostrate, unbranched or branched. **Leaves**: cauline, rhizomal, or basal; rhizomal and basal rosulate or not, petiolate, blade margins entire, toothed, or 1–3-pinnatisect, or palmately lobed, sometimes trifoliolate, pinnately, palmately, or bipinnately compound (leaflets petiolulate, subsessile, or sessile); cauline (usually alternate, rarely opposite or whorled) petiolate or sessile, blade (base cuneate, attenuate, or auriculate to sagittate), margins entire, dentate, or variously lobed, (leaflets petiolulate or sessile). **Racemes** (corymbose or paniculate, bracteate in *C. pattersonii*), elongated in fruit. **Fruiting pedicels** erect, ascending, divaricate, or reflexed, slender or stout. **Flowers**: sepals (caducous), usually erect, rarely spreading or ascending, ovate or oblong, lateral pair saccate or not basally, (usually glabrous, rarely pubescent); petals (rarely absent), white, pink, purple, or lilac, obovate, spatulate, or oblanceolate, claw absent or strongly differentiated from blade, (apex obtuse, rounded, emarginate, or subemarginate); stamens (6, rarely 4), equal in length; filaments not dilated basally; anthers ovate, oblong, or linear, (apex obtuse), glabrous [rarely pubescent]; nectar glands confluent, lateral glands annular or semiannular, subtending bases of stamens, median glands present (2, rarely 4) or absent. **Fruits** siliques, sessile, usually linear, rarely narrowly oblong or narrowly lanceolate, smooth or torulose, latiseptate; valves (papery, elastically dehiscent, becoming spirally or circinately coiled) each not veined, glabrous or, rarely, pubescent; replum strongly flattened; septum complete, (membranous); ovules 4–80 per ovary; style usually distinct, rarely obsolete; stigma capitate. **Seeds** uniseriate, flattened, usually not winged, rarely margined or winged, oblong, ovoid, or globose; seed coat (smooth, minutely reticulate, colliculate, or rugose) mucilaginous or not when wetted; cotyledons accumbent, rarely incumbent. $x = 7, 8$.

Species ca. 200 (39 in the flora): nearly worldwide.

R. C. Rollins (1993) recognized 50 species and 18 varieties of *Cardamine* in North America minus Greenland. Of these, 14 species are restricted to Central America and one, *C.* ×*incisa* (Eames) K. Schumann, is not recognized here. The oversplitting of some species into poorly delimited varieties is avoided in this account, and four species (*C. blaisdellii*, *C. holmgrenii*, *C. nymanii*, *C. umbellata*) are added to the flora. *Cardamine corymbosa* Hooker f., a native of New Zealand, apparently is becoming naturalized in greenhouses and in some gardens in British Columbia, as evidenced from *Lomer 4368* (MO), a specimen collected in Sechelt Peninsula on 6 June 2002.

SELECTED REFERENCES Detling, L. E. 1936. The genus *Dentaria* in the Pacific states. Amer. J. Bot. 23: 570–576. Detling, L. E. 1937. The Pacific coast species of *Cardamine*. Amer. J. Bot. 24: 70–76. Franzke, A. and H. Hurka. 2000. Molecular systematics and biogeography of the *Cardamine pratensis* complex (Brassicaceae). Pl. Syst. Evol. 224: 213–234. Franzke, A. et al. 1998. Molecular systematics of *Cardamine* and allied genera (Brassicaceae): ITS and non-coding chloroplast DNA. Folia Geobot. 33: 225–240. Harriman, N. A. 1965. The Genus *Dentaria* L. (Cruciferae) in Eastern North America. Ph.D. dissertation. Vanderbilt University. Hart, T. W. and W. H. Eshbaugh. 1976. The biosystematics of *Cardamine bulbosa* (Muhl.) B.S.P. and *C. douglasii* Britt. Rhodora 78: 329–419. Lihová, J. and K. Marhold. 2006. Phylogenetic and diversity patterns in *Cardamine*

(Brassicaceae)—A genus with conspicuous polyploid and reticulate evolution. In: A. K. Sharma and A. Sharma, eds. 2003+. Plant Genome: Biodiversity and Evolution. 2+ vols. in parts. Enfield, N.H. Vol. 1, part C, pp. 149–186. Lihová, J., K. Marhold, H. Kudoh, and M. Koch. 2006. Worldwide phylogeny and biogeography of *Cardamine flexuosa* (Brassicaceae) and its relatives. Amer. J. Bot. 93: 1206–1221. Lövkvist, B. 1956. The *Cardamine pratensis* complex—Outlines of its cytogenetics and taxonomy. Symb. Bot. Upsal. 14(2): 1–131. Montgomery, F. H. 1955. Preliminary studies in the genus *Dentaria* in eastern North America. Rhodora 57: 161–173. Schulz, O. E. 1903. Monographie der Gattung *Cardamine*. Bot. Jahrb. Syst. 32: 280–623. Sweeney, P. W. and R. A. Price. 2000. Polyphyly of the genus *Dentaria* (Brassicaceae): Evidence from *trn*L intron and *ndh*F sequence data. Syst. Bot. 25: 468–478.

1. Racemes bracteate . . . . 32. *Cardamine pattersonii*
1. Racemes ebracteate.
    2. Cauline leaves (at least some) simple (rarely absent).
        3. Petioles of cauline leaves auriculate basally . . . . 8. *Cardamine clematitis* (in part)
        3. Petioles of cauline leaves not auriculate basally (rarely these leaves absent).
            4. Plants with distinct caudices; rhizomes absent; cauline leaves 0 or 1 (or 2). . .
                . . . . 3. *Cardamine bellidifolia*
            4. Plants without caudices; rhizomes present; cauline leaves 2–23 (sometimes 1 in
               C. purpurea).
                5. Petals 0–0.7 mm; fruiting pedicels 0.5–2(–4) mm. . . . . 21. *Cardamine longii*
                5. Petals 3.5–28 mm; fruiting pedicels (5–)7–50 mm.
                    6. Ovules 40–80 per ovary . . . . 37. *Cardamine rotundifolia*
                    6. Ovules 10–24 per ovary.
                        7. Fruits 0.8–1.6 cm × 0.8–1 mm; anthers ovate, ca. 0.2 mm; petals
                           3.5–5 × 1.2–1.8 mm . . . . 24. *Cardamine micranthera*
                        7. Fruits (1.5–)2–6 cm × 1.2–4 mm; anthers oblong to linear, 0.6–3
                           mm; petals (5–)7–28 × 3–7 mm.
                            8. Rhizomes (not fleshy), 1–3 mm diam.; rhizomal leaves usually
                               absent.
                                9. Petals 15–28 mm; sepals 6–8 mm . . . . 10. *Cardamine constancei*
                                9. Petals 5–12 mm; sepals 2–4.5 mm.
                                    10. Cauline leaves (3–)5–17(–23); petals white, 7–12
                                        mm; fruits (2–)2.5–3.7(–4) cm; stems 2–7(–10.2) dm
                                        . . . . 11. *Cardamine cordifolia*
                                    10. Cauline leaves 1–3; petals usually purple to pink,
                                        rarely white, 5–7(–9) mm; fruits 1.5–2.5 cm; stems to
                                        1.5 dm . . . . 36. *Cardamine purpurea* (in part)
                            8. Rhizomes (fleshy), (3–)4–18 mm diam.; rhizomal leaves
                               present.
                                11. Petals 3–5 mm wide; fruits 1.4–2 mm wide; e, c North
                                    America.
                                    12. Petals usually white, rarely pale pink; stem distally
                                        glabrous or trichomes 0.02–0.1 mm . . . . 6. *Cardamine bulbosa*
                                    12. Petals usually purple or pink, rarely white; stem
                                        distally with trichomes (0.2–)0.3–0.6(–0.8) mm . . .
                                        . . . . 15. *Cardamine douglassii*
                                11. Petals 4–8 mm wide; fruits 2–4 mm; California.
                                    13. Petals white or pale rose, rarely purple, 8–13(–17)
                                        mm; ovules 12–22 per ovary; stems (2–)2.7–6(–7) dm;
                                        rhizomal leaves usually 3(–7)-foliolate, rarely simple
                                        . . . . 7. *Cardamine californica* (in part)
                                    13. Petals usually purple or pink, rarely white, 14–18
                                        mm; ovules 10–14 per ovary; stems 1–3 dm; rhizomal
                                        leaves simple . . . . 30. *Cardamine pachystigma*
    2. Cauline leaves pinnately or palmately compound, 3–25-foliolate, sometimes pinnatisect
       and appearing compound.
        14. Petiole of cauline leaves basally auriculate.
            15. Cauline leaves 3-foliolate; perennials (rhizomatous) . . . . 8. *Cardamine clematitis* (in part)
            15. Cauline leaves (9–)13–25-foliolate; annuals or biennials . . . . 20. *Cardamine impatiens*
        14. Petiole of cauline leaves not basally auriculate.

[16. Shifted to left margin.—Ed.]

16. Annuals or biennials, not rhizomatous.
    17. Rachises strongly flexuous to geniculate; petals 6–8 mm; fruits (2.5–)3–4.6 × 1.7–2.1 mm . . . . 22. *Cardamine macrocarpa*
    17. Rachises not flexuous (slightly so in *C. flexuosa*); petals 1.5–4(–5) mm; fruits (0.5–)1–2.5(–3.2) × 0.6–1.5 mm.
        18. Basal leaves rosulate, persistent to anthesis.
            19. Stamens usually 4, rarely 5 or 6; fruiting pedicels erect to ascending; fruits often appressed to rachis, valves glabrous; seeds narrowly margined . . . . . 18. *Cardamine hirsuta*
            19. Stamens 6; fruiting pedicels divaricate-ascending; fruits often not appressed to rachis, valves sparsely pubescent or glabrous; seeds often not margined (wingless) . . . . 29. *Cardamine oligosperma*
        18. Basal leaves not rosulate, often withered by anthesis.
            20. Lateral leaflets of cauline leaves narrowly oblong, linear or filiform, 1–3 mm wide; fruits 0.6–0.9 mm wide. . . . . 31. *Cardamine parviflora*
            20. Lateral leaflets of cauline leaves orbicular, ovate, elliptic, or oblong, 4–25 mm wide; fruits 0.8–1.5 mm wide.
                21. Rachises slightly flexuous; ovules 18–40 per ovary; plants of waste grounds, disturbed sites . . . . 17. *Cardamine flexuosa*
                21. Rachises straight; ovules 40–80 per ovary; plants of margins of wet habitats . . . . 34. *Cardamine pensylvanica*
16. Perennials, rhizomatous.
    22. Rhizomes tuberiform, usually segmented, fragile (segments fusiform, ovoid, oblong, or globose).
        23. Cauline leaves biternate, distal lobes linear. . . . . 14. *Cardamine dissecta*
        23. Cauline leaves 3–11-foliolate, distal lobes not linear.
            24. Cauline leaves 5–11-foliolate . . . . 33. *Cardamine penduliflora*
            24. Cauline leaves usually 3-, sometimes 5–7-foliolate, rarely simple.
                25. Rhizomes moniliform; e, c United States.
                    26. Cauline leaves different from rhizomal leaves, usually alternate, rarely opposite . . . . 2. *Cardamine angustata*
                    26. Cauline leaves similar to rhizomal leaves, usually whorled or opposite, rarely alternate . . . . 9. *Cardamine concatenata*
                25. Rhizomes not moniliform; Pacific North America.
                    27. Middle cauline leaves 5- or 7-foliolate; sepals 1.7–2 mm; petals 4–6 mm . . . . 28. *Cardamine occidentalis*
                    27. Middle cauline leaves 3(–5)-foliolate; sepals 3.5–5.5 mm; petals 8–15 mm.
                        28. Cauline leaves 2–5; stems (2–)2.7–6(–7) dm; ovules 12–22 per ovary; petals white or pale rose; California. . . . . . 7. *Cardamine californica* (in part)
                        28. Cauline leaves 1–3; stems 0.5–2(–3) dm; ovules 8–16 per ovary; petals usually purple to pale pink, rarely white; British Columbia, California, Oregon, Washington. . . . . 26. *Cardamine nuttallii*
    22. Rhizomes cylindrical, segments not fragile.
        29. Cauline leaves pinnatisect or (7–)9–21-foliolate.
            30. Leaves thick, veins impressed. . . . . 27. *Cardamine nymanii*
            30. Leaves thin, veins raised. . . . . 35. *Cardamine pratensis*
        29. Cauline leaves 3–5(–7)-foliolate.
            31. Rhizomes compact at stem base, stolons present . . . . 16. *Cardamine flagellifera*
            31. Rhizomes slender, stolons absent.
                32. Rhizomes fleshy, with dentate leaf scars; plants of North America east of the 110th meridian.
                    33. Rhizomes somewhat uniform in diameter; cauline leaves (sub)opposite . . . . 13. *Cardamine diphylla*
                    33. Rhizomes not uniform in diameter (distinctly constricted at intervals); cauline leaves usually alternate, rarely subopposite . . . . 23. *Cardamine maxima*

[32. Shifted to left margin.—Ed.]

32. Rhizomes not fleshy, without dentate leaf scars; plants of North America west of the 110th meridian.

    34. Plants cespitose, hirsute; petals purple or pink, rarely white . . . . 36. *Cardamine purpurea* (in part)

    34. Plants not cespitose, glabrous or sparsely pubescent or pilose; petals usually white, rarely pink.

        35. Basal leaves rosulate; fruiting racemes subumbellate. . . . . 39. *Cardamine umbellata*

        35. Basal leaves absent; fruiting racemes not subumbellate.

            36. Sepals 1.2–1.5 mm; petals 2–2.5 × 0.8–1 mm; anthers 0.2–0.3 mm . . . 19. *Cardamine holmgrenii*

            36. Sepals 2–5 mm; petals 3.5–15 × 1.5–8 mm; anthers 0.7–1.5 mm.

                37. Rhizomal leaves absent; ovules 14–28 per ovary . . . . 5. *Cardamine breweri*

                37. Rhizomal leaves present; ovules 6–16(–24) per ovary.

                    38. Stems hirsute at base; cauline leaves (3 or) 4–8 . . . . 1. *Cardamine angulata*

                    38. Stems usually glabrous at base, rarely sparsely pilose; cauline leaves 1–3 (or 4).

                        39. Rhizomal leaves palmately or subpalmately compound, fleshy; Montana . . . . 38. *Cardamine rupicola*

                        39. Rhizomal leaves pinnately compound, not fleshy; Alaska, Northwest Territories, Yukon.

                            40. Rhizomal leaves with terminal leaflet blade broadly obovate to suborbicular, margins often 3–5-toothed . . . . 4. *Cardamine blaisdellii*

                            40. Rhizomal leaves with terminal leaflet blade linear, narrowly oblong, narrowly oblanceolate, ovate, or elliptic, margins entire.

                                41. Cauline leaves with terminal leaflet narrowly lanceolate to linear, 1–3.2(–4) cm. . . . . 12. *Cardamine digitata*

                                41. Cauline leaves with terminal leaflet ovate to elliptic, 0.6–1.3 cm. . . . . 25. *Cardamine microphylla*

1. Cardamine angulata Hooker, Fl. Bor.-Amer. 1: 44. 1829 E

Cardamine angulata var. *alba* Torrey & A. Gray; *C. angulata* var. *hirsuta* O. E. Schulz; *C. angulata* var. *pentaphylla* O. E. Schulz; *Dentaria grandiflora* Rafinesque

Perennials; glabrous or sparsely pubescent. **Rhizomes** cylindrical, slender, to 2 mm diam. **Stems** erect, unbranched, (1.5–)2.5–8.5(–10) dm, sparsely to densely hirsute at base. **Rhizomal leaves** 3 (or 5)-foliolate, (4–)7–20(–22) cm, leaflets petiolulate or subsessile; petiole (2–)4–12(–14) cm; lateral leaflets subsessile, blade similar to terminal, larger or smaller in size; terminal leaflet petiolulate [(0.3–)0.5–1.5 cm], blade ovate to broadly lanceolate, 1.5–7(–9) cm, base usually cuneate, rarely subreniform or obtuse, margins 3–5(–7)-lobed or -toothed, surfaces puberulent. **Cauline leaves** (3 or) 4–8, 3 (or 5)-foliolate, petiolate, leaflets petiolulate or sessile; petiole 1–4 cm, base not auriculate; lateral leaflets sessile, blade similar to terminal, smaller, margins usually dentate, rarely entire; terminal leaflet sessile or petiolulate, blade broadly ovate to narrowly lanceolate, 2–7 cm × 6–40 mm, margins minutely pubescent. **Racemes** ebracteate.

Fruiting pedicels ascending to divaricate, (9–)12–25 mm. **Flowers:** sepals oblong, 2.5–4 × 1.3–2 mm, lateral pair saccate basally; petals usually white, rarely pinkish, obovate, 8–15 × 4–8 mm (clawed, apex rounded or emarginate); filaments: median pairs 3.5–6 mm, lateral pair 2–3.5 mm; anthers oblong, 0.8–1.2 mm. **Fruits** linear, 1.5–3.2 cm × 1.4–2 mm; ovules 10–16 per ovary; style (0.5–)1–4 mm. **Seeds** dark brown, oblong, 1.8–2.3 × 1–1.2 mm. $2n = 40$.

Flowering Apr–Jun. Moist ground, stream banks, swampy or damp woods, thickets, wet meadows; 0–900 m; B.C.; Alaska, Calif., Oreg., Wash.

2. Cardamine angustata O. E. Schulz, Bot. Jahrb. Syst. 32: 349. 1903 E

Cardamine angustata var. *ouachitana* E. B. Smith; *Dentaria heterophylla* Nuttall

Perennials; usually glabrous, rarely sparsely pubescent. **Rhizomes** (tuberiform, fragile) moniliform, segments fusiform, 3–6 mm diam. (fleshy). **Stems** erect, unbranched, 1.2–3(–4) dm, glabrous or pubescent. **Rhizomal leaves** 3-foliolate, to 24 cm, leaflets petiolulate or subsessile; petiole (3–)

5–12(–16) cm; lateral leaflets subsessile or petiolulate (0.2–1 cm), blade similar to terminal leaflet or smaller; terminal leaflet (petiolule (0.2–)0.5–1.5(–2) cm, blade broadly ovate to rhombic-obovate, 1.5–6(–8) cm, base usually cuneate, rarely subtruncate, margins coarsely dentate to crenate or 3-lobed, surfaces puberulent or not. **Cauline leaves** 2 (or 3), 3-foliolate (usually alternate, rarely opposite, different in morphology from rhizomal), petiolate, leaflets petiolulate or sessile; petiole 0.5–2 cm, base not auriculate; lateral leaflets sessile, blade similar to terminal, smaller, margins usually dentate, rarely entire; terminal leaflet sessile or petiolulate, blade narrowly lanceolate to narrowly oblong or oblong-lanceolate, 2–7 × 0.3–0.6 cm, margins minutely puberulent. **Racemes** ebracteate. **Fruiting pedicels** ascending to divaricate, 15–40 mm. **Flowers:** sepals oblong, 5–7.5 × 1–2 mm, lateral pair slightly saccate basally; petals purple to pale pink, oblanceolate, 9–18 × 2–5 mm (clawed, apex rounded); filaments: median pairs 5–10 mm, lateral pair 3.5–8 mm; anthers linear, 1.5–3 mm. **Fruits** linear, 2.5–4 cm × 1.5–2.5 mm; ovules 8–12 per ovary; style (5–)7–11 mm. **Seeds** dark brown, oblong, 2–2.5 × 1–1.5 mm. $2n$ = ca. 128.

Flowering Mar–May. Moist woods, wooded ridges and bottomlands, floodplains, shady ravines, streambeds; 300–1300 m; Ala., Ark., Del., D.C., Ga., Ind., Ky., Md., Miss., N.J., N.C., Ohio, Pa., S.C., Tenn., Va., W.Va.

Cardamine heterophylla (Nuttall) Alph. Wood (1870), not Host (1797) is an illegitimate name, sometimes found in synonymy with *C. angustata*.

3. **Cardamine bellidifolia** Linnaeus, Sp. Pl. 2: 654. 1753 [F]

Cardamine bellidifolia var. *beringensis* A. E. Porsild; *C. bellidifolia* var. *laxa* Lange; *C. bellidifolia* var. *pachyphylla* Coville & Leiberg; *C. bellidifolia* var. *pinnatifida* Hultén

Perennials (somewhat cespitose, subscapose); glabrous throughout. **Rhizomes** absent (caudex usually with slender, rarely stout, rhizomelike branches). **Stems** erect to ascending, branched (several), 0.1–0.8 (–1.4) dm, glabrous. **Basal leaves** rosulate, simple, (0.6–)1.2–5(–7) cm; petiole (3–)1–3.5(–5.5) cm; blade usually ovate to oblong, rarely oblanceolate to obovate, (0.4–)0.8–1.7(–2.5) cm × (2–)5–10(–16) mm, base cuneate to obtuse, margins usually entire, rarely repand or obtusely small-toothed (apex obtuse). **Cauline leaves** 0 or 1 (or 2), simple, shortly petiolate; petiole base not auriculate; blade similar to basal leaves, much smaller. **Racemes** ebracteate. **Fruiting pedicels** ascending to erect, 3–6(–8) mm. **Flowers:** sepals oblong, 2–3(–4) × 0.8–1.5(–2) mm, lateral pair not saccate basally; petals white, oblanceolate, 4–5.5(–7) × 1.3–2(–2.5) mm (not clawed, apex rounded or emarginate); filaments: median pairs 2.5–4 mm, lateral pair 2–3 mm; anthers oblong, 0.5–0.9 mm. **Fruits** linear, (0.8–)1.3–2.8(–3.7) cm × 1.3–2 mm; ovules 8–18 per ovary; style 0.5–3 mm. **Seeds** brown, oblong, 1.5–2 × 0.9–1.2 mm. $2n$ = 16.

Flowering Jun–Sep. Mossy areas, tundra, marshes at stream headwaters, cliffs, talus slopes, barren chert slopes, moist rock crevices, rocky slopes, alpine streams, sandy beaches, moist rocky streambeds; 0–2300 m; Greenland; Alta., B.C., Nfld. and Labr. (Labr.), N.W.T., Nunavut, Que., Yukon; Alaska, Calif., Maine, N.H., Oreg., Wash.; n Europe; Asia (Russian Far East, Siberia).

4. **Cardamine blaisdellii** Eastwood, Bot. Gaz. 33: 146. 1902

Cardamine microphylla Adams subsp. *blaisdellii* (Eastwood) D. F. Murray & S. Kelso; *C. microphylla* var. *blaisdellii* (Eastwood) Khatri

Perennials; usually glabrous throughout, rarely pilose. **Rhizomes** cylindrical, slender, 0.7–1.5 mm diam. **Stems** erect or ascending, unbranched, 0.5–2(–2.5) dm. **Rhizomal leaves** pinnately 5- or 7-foliolate, 2.5–12 cm (not fleshy), leaflets petiolulate; petiole 1.5–4(–9) cm; lateral leaflets similar to terminal, or smaller and margins usually toothed, rarely entire; terminal leaflet (petiolule 0.15–0.3 cm) blade suborbicular to broadly obovate, 0.4–1.5 cm × 2.5–14 mm, base obtuse to subcordate, margins 3–5-toothed, (apiculate). **Cauline leaves** 1–3, (3–)5-foliolate (alternate), petiolate, leaflets petiolulate or subsessile; petiole 0.2–2(–6.5) cm, base not auriculate; lateral leaflets similar to terminal; terminal leaflet subsessile or petiolulate (to 0.2 cm), blade obovate to oblanceolate, 0.5–1.5 cm × 3–10 mm, base cuneate, margins 3-toothed or entire. **Racemes** ebracteate. **Fruiting pedicels** erect to ascending, 0.7–22 mm. **Flowers:** sepals oblong, 2–3.5 × 2–2.5 mm, lateral pair slightly saccate basally; petals white, broadly obovate, 7–10 × 3–6 mm (clawed, apex rounded); filaments: median pairs 3–4 mm, lateral pair 2–3 mm; anthers oblong, 1–1.5 mm. **Fruits** linear, 1.6–4 cm × 1–1.3 mm; ovules 14–24 per ovary; style 0.7–3 mm. **Seeds** brown, oblong, ca. 1.5 × 1 mm. $2n$ = 28, 42.

Flowering Jul–Aug. Moist streamsides, meadows, river gravel, mesic grounds, wet tundra, moist humus, scree slopes, calcareous fellfields; 50–1000 m; Alaska; n, e Europe (Russian Far East, Siberia).

Cardamine blaisdellii was treated by R. C. Rollins (1993) as a synonym of *C. microphylla*, but the morphological differences (see key) and molecular data (R. B. Jørgensen et al., 2008) clearly demonstrate that they are distinct.

5. **Cardamine breweri** S. Watson, Proc. Amer. Acad. Arts 10: 339. 1875 E

Cardamine breweri var. *leibergii* (Holzinger) C. L. Hitchcock; *C. breweri* var. *orbicularis* (Greene) Detling; *C. breweri* var. *oregana* (Piper) Detling; *C. callosicrenata* Piper; *C. foliacea* Greene; *C. hederifolia* Greene; *C. leibergii* Holzinger; *C. modocensis* Greene; *C. orbicularis* Greene; *C. oregana* Piper; *C. vallicola* Greene; *C. vallicola* subsp. *leibergii* (Holzinger) O. E. Schulz

Perennials; usually glabrous, rarely sparsely pubescent basally. **Rhizomes** cylindrical, slender (rarely slightly thickened at stem base), 1–3(–4) mm diam. **Stems** erect or decumbent basally, unbranched or branched, (0.6–)1.5–6(–7) dm. **Rhizomal leaves** absent. **Cauline leaves** 3–8(–11), 3 or 5-foliolate (rarely only terminal leaflet present), petiolate, leaflets petiolulate or subsessile; petiole (0.7–)1–4(–6) cm, base not auriculate; lateral leaflets (when present) subsessile or petiolulate (to 0.4 cm), blade similar to terminal, often much smaller and narrower than terminal; terminal leaflet (petiolule 0.4–1.6 cm), blade usually ovate to orbicular, rarely subcordate, 1.5–4(–5) cm × 15–35(–50) mm, base truncate, rounded, or cordate, margins crenate, dentate, sinuate, or to 11-lobed. **Racemes** ebracteate. **Fruiting pedicels** ascending to divaricate-ascending, (7–)10–20 mm. **Flowers:** sepals oblong, 2–3(–3.8) × 1–1.5 mm, lateral pair not saccate basally; petals white, oblanceolate, 3.5–6(–7) × 1.5–2.5(–3) mm (not clawed, apex rounded or ± emarginate); filaments: median pairs 2.5–3.5 mm, lateral pair 2–2.5 mm; anthers oblong, 0.7–1 mm. **Fruits** linear, 1.5–3.5 cm × 1–1.5 mm; ovules 14–28 per ovary; style 0.2–1.5(–2.5) mm. **Seeds** brown, oblong, 1–1.6 × 0.9–1.1 mm.

Flowering Jun–Jul. Stream banks, seepage, lakeshores, creeks, wet meadows, swamps, ponds; 1200–3000 m; B.C.; Calif., Colo., Idaho, Mont., Nev., Oreg., Utah, Wash., Wyo.

6. **Cardamine bulbosa** (Schreber ex Muhlenberg) Britton, Sterns & Poggenburg, Prelim. Cat., 4. 1888 E

Arabis bulbosa Schreber ex Muhlenberg, Trans. Amer. Philos. Soc. 3: 174. 1793; *A. rhomboidea* Persoon; *Cardamine rhomboidea* (Persoon) de Candolle; *C. rhomboidea* var. *hirsuta* O. E. Schulz; *C. rhomboidea* var. *parviflora* O. E. Schulz; *C. rhomboidea* var. *pilosa* O. E. Schulz; *Dentaria rhomboidea* (Persoon) Greene; *Dracamine bulbosa* (Schreber ex Muhlenberg) Nieuwland

Perennials; glabrous or sparsely pubescent distally. **Rhizomes** (tuberous at stem base, sometimes also at intervals) subglobose, lobed, 4–15 mm diam. (fleshy). **Stems** erect, unbranched, (1–)2–6 dm, glabrous or sparsely pubescent on distal ½ (trichomes 0.02–0.1 mm). **Rhizomal leaves** simple, (2–)4–13(–16) cm; petiole (1.5–)2.5–10(–13) cm; blade usually reniform to cordate or ovate, rarely oblong, (1–)2–4(–6) cm, base obtuse to cordate, margins usually repand or entire, rarely shallowly dentate. **Cauline leaves** (2–)4–10(–14), simple, petiolate or sessile; (middle ones) shortly petiolate or (distally) sessile, base not auriculate; blade ovate to oblong, or oblong-linear to lanceolate, 3–6(–9) cm × 10–30(–45) mm, margins entire, repand, or dentate (margins minutely pubescent). **Racemes** ebracteate. **Fruiting pedicels** ascending to divaricate, (10–)15–22(–30) mm. **Flowers:** sepals oblong, 2.5–4.5 × 1.5–2 mm, lateral pair not saccate basally, (glabrous); petals usually white, rarely pale pink, obovate, (6–)7–12(–16) × 3–5 mm, (short-clawed, apex rounded); filaments: median pairs 4.5–7 mm, lateral pair 2–3.5 mm; anthers oblong, 1–1.5 mm. **Fruits** linear, 2–3.5(–4) cm × 1.4–1.7 mm; ovules 14–24 per ovary; style 2–4(–5) mm. **Seeds** dark orange to greenish yellow, oblong or globose, 1.7–2.1 × 1–1.4 mm. $2n = 16, 56, 64, 80, 96, 112$.

Flowering Mar–Jun. Wet grounds, low woodland, moss hummocks, alluvial woods, grassy floodplains, wet pastures, meadows, pinelands, creek bottoms, stream banks, sandy bottoms, ditches, mesic or wet forests, swamps, marshes, seepy bluffs; 0–900 m; Man., Ont., Que.; Ala., Ark., Conn., Del., D.C., Fla., Ga., Ill., Ind., Iowa, Kans., Ky., La., Maine, Md., Mass., Mich., Minn., Miss., Mo., Nebr., N.H., N.J., N.Y., N.C., Ohio, Okla., Pa., S.C., S.Dak., Tenn., Tex., Vt., Va., W.Va., Wis.

7. **Cardamine californica** (Nuttall) Greene, Fl. Francisc., 266. 1891

Dentaria californica Nuttall in J. Torrey and A. Gray, Fl. N. Amer. 1: 88. 1838; *Cardamine californica* var. *brevistyla* O. E. Schulz; *C. californica* var. *cardiophylla* (Greene) Rollins; *C. californica* subsp. *cuneata* (Greene) O. E. Schulz; *C. californica* var. *cuneata* (Greene) Rollins; *C. californica* var. *fecunda* O. E. Schulz; *C. californica* var. *integrifolia* (Nuttall) Rollins; *C. californica* var. *pubescens* O. E. Schulz; *C. californica* var. *robinsoniana* O. E. Schulz; *C. californica* var. *sinuata* (Greene) O. E. Schulz; *C. cardiophylla* Greene; *C. cuneata* Greene; *C. integrifolia* (Nuttall) Greene; *C. integrifolia* var. *sinuata* (Greene) C. L. Hitchcock; *C. pachystigma* (S. Watson) Rollins var. *dissectifolia* (Detling) Rollins; *C. paucisecta* Bentham; *C. sinuata* Greene; *D. californica* var. *cardiophylla* (Greene) Detling; *D. californica* var. *cuneata* (Greene) Detling; *D. californica*

var. *integrifolia* (Nuttall) Detling; *D. californica* var. *sinuata* (Greene) Detling; *D. cardiophylla* (Greene) B. L. Robinson; *D. cuneata* (Greene) Greene; *D. integrifolia* Nuttall; *D. integrifolia* var. *californica* (Nuttall) Jepson; *D. integrifolia* var. *cardiophylla* (Greene) Jepson; *D. integrifolia* var. *traceyi* Jepson; *D. pachystigma* (S. Watson) S. Watson var. *dissectifolia* Detling; *D. sinuata* (Greene) Greene

Perennials; usually glabrous, rarely minutely pubescent. **Rhizomes** (tuberiform, fragile), globose to ovoid or suboblong, (3–)4–10 mm diam., (fleshy, deeply underground). **Stems** erect, unbranched, (2–)2.7–6(–7) dm, usually glabrous, rarely pubescent. **Rhizomal leaves** 3 (or 5–7)-foliolate, sometimes simple, 8–25(–38) cm, leaflets petiolulate or subsessile; petiole (5–)8–25(–32) cm; lateral leaflets (when present) petiolulate to subsessile, blade similar to terminal, sometimes smaller; terminal leaflet (petiolule (0.7–)2–5(–11) cm), blade (of simple leaf) ovate to orbicular to broadly cordate or reniform, (1.5–)2.5–7.5(–10) cm × (12–)20–90(–130) mm, base obtuse to cordate, margins entire or dentate to shallowly sinuate, (often with apiculae at veins ending at margin, sometimes minutely pubescent on veins). **Cauline leaves** 2–5, usually 3 (or 5)-foliolate, rarely simple, petiolate, leaflets petiolulate or sessile; petiole 1–5(–9) cm, base not auriculate; lateral leaflets sessile, blade similar to terminal, smaller, margins usually dentate, rarely entire; terminal leaflet sessile or petiolulate, blade usually broadly ovate to suborbicular or lanceolate, rarely narrowly oblong, 1–7 cm × (5–)10–47(–65) mm. **Racemes** ebracteate. **Fruiting pedicels** ascending to divaricate, 10–33(–41) mm. **Flowers:** sepals (erect to ascending), oblong, 3.5–4.5(–5.5) × 1.5–2(–2.5) mm, lateral pair saccate basally; petals white to pale rose, often broadly obovate, 8–13(–15) × 4–8 mm (clawed, apex rounded); filaments: median pairs 4–6.5 mm, lateral pair 3–4 mm; anthers oblong, 1.2–1.7 mm. **Fruits** linear, 2.2–5.4(–6) cm × 2–3 mm; ovules 12–22 per ovary; style 2–5(–6) mm. **Seeds** dark brown, oblong to broadly ovoid, 1.7–2.8 × 1.2–1.8 mm. $2n = 32$.

Flowering Jan–May. Wooded ravines, forest floors, shady slopes, open woods, shady rock crevices, stream banks and bottoms, canyons, moist hillsides, cliffs; 0–1400 m; Calif.; Mexico (Baja California).

The synonymy above suggests that Schulz, Detling, and Rollins did not agree on the characters used or the number and rank of taxa recognized. The taxonomy of this complex is based solely on differences in the number of leaflets, their division, size, and shape, all other aspects of these plants (e.g., rhizomes, flowers, fruits, and seeds) being fairly uniform. Indeed, the infraspecific taxa recognized represent only part of the overall variation in the species, and one is faced with either recognizing poorly defined infraspecific taxa or treating the entire complex as a single polymorphic species. In the absence of thorough biosystematic and molecular studies on this group, we prefer not to recognize any infraspecific taxa.

8. Cardamine clematitis Shuttleworth ex A. Gray, Proc. Amer. Acad. Arts 15: 45. 1880 [E]

Perennials; glabrous throughout. **Rhizomes** cylindrical, slender, 1–3 mm diam. **Stems** erect, unbranched or rarely branched distally, (0.8–)1–2.5(–3.5) dm. **Rhizomal leaves** simple or 3-foliolate, (1.5–)3–8 cm, leaflets petiolulate or subsessile; petiole (1–)2–6 cm; lateral leaflets subsessile, blade similar to terminal, much smaller; terminal leaflet petiolulate (to 1 cm), blade reniform to cordate, (0.5–)1–2 cm, base cordate, margins entire or 3-lobed. **Cauline leaves** 3–7, 3-foliolate (or distalmost simple), petiolate, leaflets petiolulate or sessile; petiole 0.7–3.5 cm, base auriculate (auricle 0.7–5 mm); lateral leaflets sessile or petiolulate (to 0.5 cm), blade oblong to ovate or oblong; terminal leaflet (petiolule 0.5–1 cm), blade broadly ovate to suborbicular or reniform, 1.5–4 cm × 15–35 mm, margins often 3 or 5-lobed (lobe apex minutely apiculate, glabrous on margin). **Racemes** ebracteate. **Fruiting pedicels** divaricate-ascending, (7–)10–17 mm. **Flowers:** sepals oblong, 2.5–3 × 1–1.5 mm, lateral pair not saccate basally; petals white, oblanceolate, 6–8 × 2–3 mm (not clawed, apex obtuse to submarginate); filaments: median pairs 3.5–4 mm, lateral pair 2–2.5 mm; anthers ovate, ca. 0.7 mm. **Fruits** linear, (1.5–)2–3.5(–4) cm × 1.3–1.7 mm; ovules 10–16 per ovary; style 2–4 mm. **Seeds** brown, oblong, 1.7–2 × 1–1.2 mm.

Flowering May–Jun. Wet areas, springs, moist slopes; 1300–1800 m; Ala., Fla., Ga., N.C., Tenn., Va.

9. Cardamine concatenata (Michaux) O. Schwarz, Repert. Spec. Nov. Regni Veg. 46: 188. 1939 [E]

Dentaria concatenata Michaux, Fl. Bor.-Amer. 2: 30. 1803; *Cardamine laciniata* (Muhlenberg ex Willdenow) Alph. Wood var. *integra* O. E. Schulz; *C. laciniata* var. *lasiocarpa* O. E. Schulz; *D. laciniata* Muhlenberg ex Willdenow; *D. laciniata* var. *alterna* Farwell; *D. laciniata* var. *coalescens* Fernald; *D. laciniata* var. *integra* (O. E. Schulz) Fernald; *D. laciniata* var. *latifolia* Farwell; *D. laciniata* var. *opposita* Farwell

Perennials; usually sparsely pubescent, sometimes glabrous. **Rhizomes** (tuberiform, fragile), moniliform, segments fusiform, 2–10(–20) mm diam. (fleshy). **Stems** erect, unbranched, (1–)2–4(–5.5) dm, glabrous or pubescent distally. **Rhizomal leaves** 3-foliolate, (7–)10–20(–30) cm, leaflets sessile; petiole (4–)7–18(–25) cm; lateral leaflets similar to terminal, blade sometimes smaller; terminal leaflet blade oblong, lanceolate, oblanceolate, or linear, 2.5–6 cm, base cuneate, margins coarsely dentate to incised, laciniate, or 3-lobed (lobes usually toothed to incised, rarely entire, surfaces puberulent or not, trichomes 0.2–0.3 mm). **Cauline leaves** (2 or) 3, 3-foliolate (usually whorled or opposite, rarely alternate, similar in morphology to rhizomal leaves), petiolate, leaflets petiolulate or subsessile; petiole (1–)1.5–6(–7.5) cm, base not auriculate; lateral leaflets sessile, blade similar to terminal, sometimes smaller; terminal leaflet subsessile or petiolulate (to 3 cm), blade lanceolate, linear, or oblanceolate, (3–)4–10(–12) cm × (3–)5–20 (–25) mm, margins usually coarsely dentate to incised, rarely subentire (margins minutely puberulent). **Racemes** ebracteate. **Fruiting pedicels** ascending to divaricate, (6–) 10–27(–33) mm. **Flowers:** sepals oblong, (4–)5–8 × 2–4 mm, lateral pair slightly saccate basally; petals white to pale pink, oblanceolate, (8–)10–20 × (3–)4–7(–9) mm, (short-clawed, apex rounded); filaments: median pairs 8–12 mm, lateral pair 6–8 mm; anthers oblong-linear, 1.5–2.5 mm. **Fruits** linear-lanceolate, (2–)2.5–3.8(–4.8) cm × 1.5–3 mm; (valves glabrous or sparsely pubescent); ovules 10–14 per ovary; style (2–)5–9(–12) mm. **Seeds** brown, oblong, 1.6–3 × 1.8–2 mm. $2n$ = 128–256.

Flowering Feb–May. Wooded bottoms and bluffs, rich woods, limestone cliffs and outcrops, rocky banks, mesic forests, moist areas with leaf litter, floodplain woods; 0–1000 m; Ont., Que.; Ala., Ark., Conn., Del., D.C., Fla., Ga., Ill., Ind., Iowa, Kans., Ky., La., Maine, Md., Mass., Mich., Minn., Miss., Mo., Nebr., N.H., N.J., N.Y., N.C., Ohio, Okla., Pa., S.C., Tenn., Tex., Vt., Va., W.Va., Wis.

10. **Cardamine constancei** Detling, Madroño 3: 176, plate 9. 1935 E

Perennials; glabrous or sparsely hirsute. **Rhizomes** cylindrical, slender, to 2 mm diam. **Stems** erect, unbranched, 1.5–5 dm, glabrous or sparsely hirsute basally. **Rhizomal leaves** absent. **Cauline leaves** 4–7, simple, (crowded distally), petiolate; petiole (0.5–)1–4(–5) cm, base not auriculate; blade usually broadly ovate or ovate-lanceolate to ovate-elliptic, rarely obovate, 5–13 cm × 20–65 mm, base cuneate, margins coarsely serrate (with apiculae), or repand to undulate (margins minutely pubescent, apex acute). **Racemes** ebracteate. **Fruiting pedicels** divaricate-ascending to suberect, 10–22 mm. **Flowers:** sepals oblong, 6–8 × 2–2.5 mm, lateral pair slightly saccate basally; petals pink, oblanceolate, 15–28 × 5–8 mm, (claw to 10 mm, apex rounded); filaments: median pairs 8–10 mm, lateral pair 4–6 mm; anthers linear, 2–3 mm. **Fruits** linear, 2.5–3.5(–5) cm × 1.9–2.1 mm; ovules 12–16 per ovary; style 2–3.5 mm. **Seeds** brown, oblong, 2–2.5 × 1.2–1.5 mm.

Flowering May–Jun. Moist cliffs, wooded creek bottoms, shaded draws, hillsides, moist woods, mixed coniferous forests, granitic soils; 400–600 m; Idaho.

Cardamine constancei is known from Clearwater, Idaho, Kootenai, Nez Perce, and Shoshone counties.

11. **Cardamine cordifolia** A. Gray, Mem. Amer. Acad. Arts, n. s., 4: 8. 1849 E F

Cardamine cordifolia var. *cardiophylla* O. E. Schulz; *C. cordifolia* var. *diversifolia* O. E. Schulz; *C. cordifolia* var. *incana* A. Gray ex M. E. Jones; *C. cordifolia* subsp. *lyallii* (S. Watson) O. E. Schulz; *C. cordifolia* var. *lyallii* (S. Watson) A. Nelson & J. F. Macbride; *C. cordifolia* var. *pubescens* A. Gray ex O. E. Schulz; *C. incana* (A. Gray ex M. E. Jones) A. Nelson; *C. infausta* Greene; *C. lyallii* S. Watson; *C. uintahensis* F. J. Hermann

Perennials; glabrous or densely puberulent. **Rhizomes** cylindrical, slender or stout, 1.5–3 mm diam., (not fleshy). **Stems** erect, unbranched or branched distally, 2–7(–10.2) dm, glabrous or sparsely to densely puberulent basally or throughout. **Rhizomal leaves** usually absent, rarely present, simple, 5–15 cm; petiole 2.5–12 cm; blade reniform to cordate, 1.5–4.5(–6) cm. **Cauline leaves** (3–)5–17(–23), simple or, rarely, proximalmost 3-foliolate, petiolate; petiole 1–6.5(–8.5) cm, base not auriculate; blade reniform, deltate-cordate, or ovate-cordate, (1–)2–7.2(–9.7) cm × 10–55(–85) mm, (somewhat fleshy), base cordate or truncate, margins crenate to slightly sinuate, (veins ending in apiculae, surfaces glabrous or sparsely to densely pubescent). **Racemes** ebracteate. **Fruiting pedicels** divaricate to ascending, (7–)10–20 mm. **Flowers:** sepals oblong, 2.5–4.5 × 1.5–2 mm, lateral pair slightly saccate basally; petals white, broadly obovate, 7–12 × 4–6 mm (claw to 6 mm, apex rounded to subtruncate or emarginate); filaments: median pairs 3.5–5 mm, lateral pair 2–3.5 mm; anthers oblong, 1–1.5 mm. **Fruits** linear, (2–)2.5–3.7(–4) cm × 1.2–2 mm; (valves glabrous or sparsely puberulent); ovules 14–24 per ovary; style 0.5–3(–6) mm. **Seeds** brown, oblong, 1.6–2 × 1–1.3 mm. $2n$ = 24.

Flowering May–Aug. Stream banks, springs, shady gullies, creek bottoms, lakeshores, ponds, cold springs, meadows, moist hillsides, mossy areas, alpine streams, mixed coniferous forests; 600–3600 m; B.C.; Ariz., Calif., Colo., Idaho, Nev., N.Mex., Oreg., Utah, Wash., Wyo.

Cardamine cordifolia is highly variable in leaf morphology, especially in leaf width, depth of the cordate base, and indumentum. This variation occurs throughout the species range and is rather weakly or not at all correlated with geography. In the absence of a detailed biosystematic study over the entire species range, we follow N. H. Holmgren (2005b) in not recognizing any infraspecific taxa, instead of accepting the three rather poorly defined varieties recognized by R. C. Rollins (1993).

12. Cardamine digitata Richardson in J. Franklin, Narr. Journey Polar Sea, 743. 1823

Cardamine digitata var. *oxyphylla* Trautvetter; *C. hyperborea* O. E. Schulz; *C. hyperborea* var. *oxyphylla* (Trautvetter) O. E. Schulz; *C. richardsonii* Hultén

Perennials; glabrous throughout. **Rhizomes** cylindrical, slender, 0.5–1.5 mm diam. **Stems** erect, unbranched, (0.6–)1–2(–3) dm. **Rhizomal leaves** pinnately 5 or 7-foliolate, (2–)4–10 cm, (not fleshy), leaflets sessile; petiole (1.5–)3–7(–9) cm; lateral leaflets similar to terminal, blade sometimes smaller; terminal leaflet blade linear to narrowly oblong or narrowly lanceolate, (0.5–)1–2.5(–3.3) cm × 0.5–4(–8) mm, base cuneate to attenuate, margins entire. **Cauline leaves** (1 or) 2 or 3, 5 or 7-foliolate, petiolate, leaflets sessile; petiole (0.2–)0.5–1.2(–1.5) cm, base not auriculate; lateral leaflets similar to terminal, blade smaller; terminal leaflet blade narrowly lanceolate to linear, 1–3.2(–4) cm × 0.8–2.5(–3.5) mm, base cuneate to attenuate, margins entire (apex acute to acuminate). **Racemes** ebracteate. **Fruiting pedicels** suberect to divaricate-ascending, (7–)10–25 mm. **Flowers:** sepals ovate, 2.5–3.5 × 1.5–2.5 mm, lateral pair not saccate basally; petals white, obovate, 5–9 × 2.5–5 mm (clawed, apex rounded); filaments: median pairs 3–4.5 mm, lateral pair 1.5–2.5 mm; anthers oblong, 1–1.5 mm. **Fruits** linear, (1.5–)2–4 cm × 1.5–2 mm; ovules 6–12 per ovary; style (1–)1.5–2.5(–4) mm. **Seeds** brown, oblong, 1.8–2.2 × 0.9–1.1 mm. $2n$ = 28, 42.

Flowering Jun–Aug. Damp flats, stream banks, tundra, meadows, bluffs, hummocks, sandy beaches, slopes, mossy mats, sedge swales; 0–1400 m; N.W.T., Yukon; Alaska; ne Asia (Russian Far East, Siberia).

13. Cardamine diphylla (Michaux) Alph. Wood, Amer. Bot. Fl., 37. 1870 E

Dentaria diphylla Michaux, Fl. Bor.-Amer. 2: 30. 1803; *D. bifolia* Stokes; *D. incisa* Small

Perennials; usually glabrous, rarely sparsely pubescent. **Rhizomes** (unsegmented), cylindrical, 2–10 mm diam., (somewhat uniform, fleshy, not fragile, with dentate leaf scars). **Stems** erect, unbranched, (1.2–)1.5–3.5(–4) dm, rarely sparsely pubescent distally. **Rhizomal leaves** 3-foliolate, (5.5–)8–22(–26) cm, leaflets petiolulate or subsessile; petiole (3–)4.5–16(–20) cm; lateral leaflets subsessile or petiolulate, blade often similar to terminal, base sometimes oblique; terminal leaflet (petiolule 0.5–1.2 cm), blade ovate-elliptic to broadly ovate, (2–)3.5–8(–10) cm × (5–)20–65(–80) mm, base cuneate to obtuse, margins coarsely crenate or dentate, (surfaces puberulent, trichomes to 0.1 mm). **Cauline leaves** 2 (or 3) [(sub)opposite], 3-foliolate, (similar to rhizomal leaves), petiolate, leaflets petiolulate or subsessile; petiole (1–)2–4.5 cm, base not auriculate; lateral leaflets similar to terminal; terminal leaflet subsessile or petiolulate (0.2–1 cm), blade broadly elliptic to ovate, (2–)4–8(–10) cm × 10–50 mm, margins coarsely dentate or crenate, (margins minutely puberulent). **Racemes** ebracteate. **Fruiting pedicels** ascending to divaricate, (10–)15–30(–36) mm. **Flowers:** sepals oblong, (4–)5–8 × 2–3 mm, lateral pair slightly saccate basally; petals white or pink to purple, obovate to oblanceolate, (7–)9–15(–17) × (3–)4–7 mm, (short-clawed, apex rounded); filaments: median pairs 5–8 mm, lateral pair 3.5–6 mm; anthers linear to oblong, 2.5–3 mm. **Fruits** linear-lanceolate, 1.5–4 cm × 1.5–2.5 mm; ovules 10–14 per ovary; style 4–8(–10) mm. **Seeds** (rarely produced) brown, oblong, 2–2.2 × 1.2–1.5 mm (cotyledons incumbent). $2n$ = 96.

Flowering Apr–Jun. Wooded bottoms and ravines, cliffs, bluffs, ledges, shaded slopes, meadows, moist fields, alluvial banks, rich woods; 50–1300 m; N.B., N.S., Ont., Que.; Ala., Ark., Conn., Ga., Ill., Ind., Ky., Maine, Mass., Mich., Minn., N.H., N.J., N.Y., N.C., Ohio, Pa., S.C., Tenn., Vt., Va., W.Va., Wis.

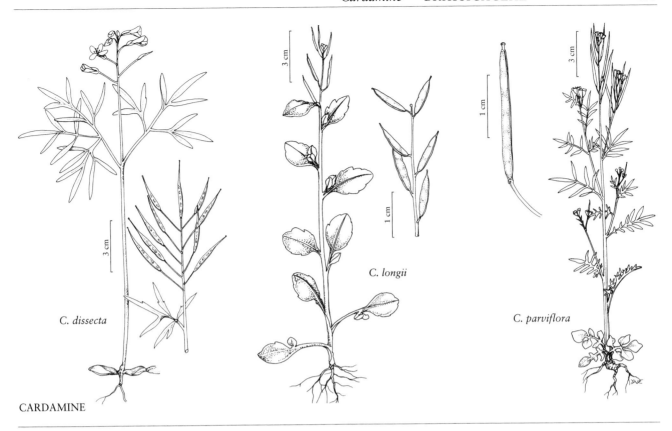

14. Cardamine dissecta (Leavenworth) Al-Shehbaz, J. Arnold Arbor. 69: 82. 1988 [E] [F]

Dentaria dissecta Leavenworth, Amer. J. Sci. Arts 7: 62. 1824; Cardamine angustata O. E. Schulz var. multifida (Muhlenberg ex Elliott) H. E. Ahles; C. laciniata (Muhlenberg ex Willdenow) Alph. Wood subsp. multifida (Muhlenberg ex Elliott) O. E. Schulz; D. furcata Small; D. heterophylla Nuttall var. multifida (Muhlenberg ex Elliott) H. E. Ahles; D. laciniata Muhlenberg ex Willdenow var. multifida (Muhlenberg ex Elliott) S. Watson & J. M. Coulter; D. multifida Muhlenberg ex Elliott

Perennials; glabrous throughout. **Rhizomes** (tuberiform, fragile), moniliform, segments oblong, 3–5 mm diam., (fleshy). **Stems** erect, unbranched, 1–3.5 dm. **Rhizomal leaves** palmately compound (biternate), 7–20 cm, leaflets petiolulate; petiole 4–13 cm; lateral leaflets similar to terminal, blade sometimes smaller; terminal leaflet (petiolule 1–3(–4) cm), blade ternately or pinnately lobed, (distal segment) linear, 0.7–3.5 cm × 0.6–3 mm, base attenuate to cuneate, margins entire (apiculate). **Cauline leaves** (2 or) 3, palmately compound (biternate, similar to rhizomal leaves, alternate to subopposite), petiolate, leaflets petiolulate; petiole (0.5–)1–4(–5.5) cm, base not auriculate; lateral leaflets similar to terminal; terminal leaflet (petiolule (0.3–)0.7–1.5(–2.5) cm), blade (ternate, distal segment) linear, (0.4–)1.5–5(–6.5) cm × 0.7–4(–6) mm, base cuneate to attenuate, margins entire. **Racemes** ebracteate. **Fruiting pedicels** ascending to divaricate, 10–25(–35) mm. **Flowers:** sepals oblong, 4–6 × 1.5–2 mm, lateral pair not saccate basally; petals white to pale pink, oblanceolate, 9–15 × 2–4 mm, (not clawed, apex rounded); filaments: median pairs 6–7.5 mm, lateral pair 3–5 mm; anthers linear, 1.5–2.5 mm. **Fruits** linear-lanceolate, 2–3.5 cm × 1.5–2 mm; ovules 10–14 per ovary; style 4–7(–10) mm. **Seeds** brown, oblong, 1.5–2.5 × 1–1.5 mm. $2n = 64, 112$.

Flowering Mar–May. Oak-hickory woods, moist loamy areas, floodplain woods, bluffs, rocky calcareous woods, limestone slopes, along streams; 0–300 m; Ala., Ga., Ind., Ky., N.C., Ohio, Tenn.

15. Cardamine douglassii Britton, Trans. New York Acad. Sci. 9: 8. 1889 [E]

Arabis rhomboidea Persoon var. purpurea Torrey, Amer. J. Sci. Arts 4: 66. 1822; Dentaria douglassii (Britton) Greene; Dracamine purpurea (Torrey) Nieuwland; Thlaspi tuberosum Nuttall

Perennials; hirsute throughout or glabrous proximally. **Rhizomes** (tuberous at stem base), subglobose, (lobed or not), (3–)4–10 mm diam., (fleshy). **Stems** erect, unbranched, (0.7–)1–2.5(–3) dm, sparsely to densely hirsute, or glabrous basally, (trichomes (0.2–)0.3–0.6(–0.8) mm). **Rhizomal leaves** simple, (3–)5–15(–18) cm; petiole (2–)4–12(–16) cm; blade often orbicular to cordate, sometimes reniform or ovate, (1–)2–6 cm × (7–)17–50 mm, base obtuse to cordate, margins repand or entire. **Cauline leaves** 3–6(–8), simple, petiolate or sessile; (middle) shortly petiolate or (distal) sessile, base not auriculate; blade oblong to ovate or lanceolate, 2–5 cm × 5–25 mm, margins entire, repand, or coarsely dentate. **Racemes** ebracteate. **Fruiting pedicels** ascending to divaricate, (10–)15–35(–50) mm, sparsely pubescent or glabrous. **Flowers:** sepals oblong, 2.5–4(–6) × 1.5–2.5 mm, lateral pair not saccate basally, (surfaces often hirsute); petals usually rose-purple to pink, rarely white, obovate, (7–)8–13(–15) × 3–5 mm, (short-clawed, apex rounded); filaments: median pairs 4–7 mm, lateral pair 2–4 mm; anthers oblong, 1.3–1.7 mm. **Fruits** linear, (1.5–)2–4 cm × 1.5–2 mm; ovules 10–16 per ovary; style 2–5 mm. **Seeds** brown, oblong to ovoid, 1.7–2.5 × 1–5 mm. $2n = 56, 64, 96, 112, 144$.

Flowering Mar–May. Rich woods, bluffs, mesic bottomland forests, rocky hillsides, floodplains, seepage of bogs, springy areas; 50–400 m; Ont.; Ala., Conn., D.C., Ill., Ind., Iowa, Ky., Md., Mass., Mich., Minn., Mo., N.J., N.Y., N.C., Ohio, Pa., S.C., Tenn., Va., W.Va., Wis.

16. Cardamine flagellifera O. E. Schulz, Bot. Jahrb. Syst. 32: 405. 1903 [E]

Cardamine flagellifera var. *hugeri* (Small) Rollins; *C. hugeri* Small

Perennials; hirsute or pilose proximally, sparsely pubescent or glabrous distally. **Rhizomes** swollen, compact at stem base, (fleshy, stolons many, slender, ca. 1 mm diam., pubescent or glabrous). **Stems** erect, usually unbranched, rarely branched distally, 1–2.5 dm, sparsely to densely hirsute or pilose proximally, sparsely so or glabrous distally. **Rhizomal leaves** usually simple, rarely 3-foliolate, (3–)6–16 cm, leaflets petiolulate or subsessile; petiole (1.5–)4–13 cm; lateral leaflets subsessile or petiolulate (to 0.5 cm), blade similar to terminal, much smaller; terminal leaflet (petiolule 0.5–2 cm), blade orbicular to reniform or broadly ovate, (1–)1.5–4.5 cm, base cordate, margins repand to coarsely crenate, (apiculate at vein endings, surfaces sparsely pubescent or glabrous). **Cauline leaves** 2–5, 3 or 5-foliolate, petiolate, leaflets petiolulate or subsessile; petiole 0.8–5 cm, base not auriculate; lateral leaflets sessile or petiolulate (to 0.5 cm), blade similar to terminal, smaller; terminal leaflet petiolulate (0.5–1.5 cm), blade broadly ovate to suborbicular, 1.5–4(–5) cm × 12–35 mm, margins repand to coarsely crenate or slightly lobed, (apiculate at vein endings, margins glabrous). **Racemes** ebracteate. **Fruiting pedicels** ascending to divaricate-ascending, 7–20 mm. **Flowers:** sepals (ascending) oblong, 3–3.5 × 1–1.5 mm, lateral pair not saccate basally; petals white, oblanceolate, 5–6.5(–8) × 1.5–2.5 mm, (not clawed, apex obtuse to submarginate); filaments: median pairs 3.5–5 mm, lateral pair 2.5–3 mm; anthers oblong, 1.2–1.6 mm. **Fruits** linear, 1.5–2.5 cm × 1–1.2 mm; ovules 10–14 per ovary; style 1.2–2.5 mm. **Seeds** brown, oblong, 1.2–1.7 × 0.8–1.1 mm.

Flowering Mar–Jun. Moist wooded slopes, gorges, wooded ravines, seepage places; 300–1000 m; Ga., N.C., S.C., Tenn., W.Va.

17. Cardamine flexuosa Withering, Arr. Brit. Pl. ed. 3, 3: 578. 1796 [I]

Cardamine flexuosa subsp. *debilis* O. E. Schulz; *C. flexuosa* var. *debilis* (O. E. Schulz) T. Y. Cheo & R. C. Fang; *C. hirsuta* Linnaeus subsp. *flexuosa* (Withering) F. B. Forbes & Hemsley; *C. scutata* Thunberg subsp. *flexuosa* (Withering) H. Hara

Annuals or biennials; sparsely to densely hirsute basally or throughout, or glabrous. **Rhizomes** absent. **Stems** erect, ascending, or decumbent, branched or unbranched, (0.6–)1–5 dm, (slightly flexuous). **Basal leaves** (often withered by anthesis), not rosulate, 5–15-foliolate, (2.7–)4–14 (–19) cm, leaflets petiolulate; petiole 0.7–5 cm, (ciliate or not); lateral leaflet blade oblong, ovate, or elliptic, smaller than terminal, margins entire, repand, crenate, or 3 (or 5)-lobed; terminal leaflet (petiolule 0.3–1.7 cm), blade reniform, broadly ovate, or suborbicular, 0.5–2.5 cm × 4–30 mm, margins repand, crenate, or 3 or 5-lobed. **Cauline leaves** 3–15, 5–15-foliolate [leaves (2–)3.5–5.5(–7) cm, including petiole], petiolate, leaflets petiolulate; petiole base not auriculate; lateral leaflets similar to basal, (0.4–2.5 mm wide). **Racemes** ebracteate. **Fruiting pedicels** divaricate or ascending, (5–)6–14(–17) mm. **Flowers:** sepals oblong, 1.5–2.5 × 0.7–1 mm, lateral pair not saccate basally; petals white, spatulate, 2.5–4(–5) × 1–1.7 mm; (stamens rarely 4, lateral pair absent); filaments 2–3 mm; anthers ovate, 0.3–0.5 mm. **Fruits** linear, (torulose), (0.8–)1.2–2.8 cm × 1–1.5 mm; ovules 18–40 per ovary; style 0.3–1(–1.5) mm. **Seeds** brown, oblong or subquadrate, 0.9–1.5 × 0.6–1 mm, (narrowly margined or not). $2n = 32$.

Flowering Apr–Jul. Disturbed areas, fields, nurseries, plantations, gardens, flower beds, lawns, roadsides; 0–1100 m; introduced; B.C., Nfld. and Labr. (Nfld.), Ont.; Ala., Calif., Fla., Ga., Ill., Ind., La., Md., Mich., N.Y., N.C., Ohio, Oreg., R.I., Tex., Va., Wash.; Europe; e Asia; introduced also in Mexico, Central America, South America, Australia.

According to J. Lihová et al. (2006), the populations referred to *Cardamine flexuosa* in North America comprise two taxa of different polyploid origins and evolutionary histories: tetraploid *C. flexuosa* ($2n$ = 32), native to Europe, and the octoploid taxon informally called "Asian *C. flexuosa*" ($2n$ = 64), native to eastern Asia. For the latter, the name *C. flexuosa* subsp. *debilis* can be used. Nevertheless, these two taxa should be recognized at species level and the correct name for the Asian species should be sought. Based on available data, both taxa occupy the same habitats in North America, but the Asian taxon is much more widespread. The occurrence of European *C. flexuosa* was, until now, confirmed only for Washington, where both taxa have been recorded. More detailed studies of the North American distributions of both these weeds are needed.

18. **Cardamine hirsuta** Linnaeus, Sp. Pl. 2: 655. 1753
    I W

Annuals; sparsely hirsute basally (at least on petiole of basal leaves), often glabrous distally. **Rhizomes** absent. **Stems** erect, ascending, or decumbent, unbranched or branched basally and/or distally, (0.3–)1–3.5 (–4.5) dm, (not flexuous). **Basal leaves** (persistent to anthesis), rosulate, (5–)8–15(–22)-foliolate, (2–)3.5–15(–17) cm, leaflets petiolulate; petiole 0.5–5 cm, (ciliate); lateral leaflet blade oblong, ovate, obovate, or orbicular, smaller than terminal, margins entire, repand, crenate, or 3-lobed; terminal leaflet (petiolule 0.2–1 cm), blade reniform or orbicular, 0.4–2 cm × 6–30 mm, margins entire, repand, dentate, or 3 or 5-lobed. **Cauline leaves** 1–4(–6), compound as basal, petiolate, [(0.5–)1.2–5.5 (–7) cm, including petiole], leaflets petiolulate; blade base not auriculate; leaflets similar to basal. **Racemes** ebracteate. **Fruiting pedicels** erect to ascending, (2–)3–10(–14) mm. **Flowers:** sepals oblong, 1.5–2.5 × 0.3–0.7 mm, lateral pair not saccate; petals (sometimes absent) white, spatulate, 2.5–4.5(–5) × 0.5–1.1 mm; (stamens usually 4, lateral pair often absent, rarely 5 or 6); filaments 1.8–3 mm; anthers ovate, 0.3–0.5 mm. **Fruits** linear, (torulose), (0.9–)1.5–2.5(–2.8) cm × (0.8–)1–1.4 mm, (often appressed to rachis); ovules 14–40 per ovary; style 0.1–0.6(–1) mm. **Seeds** light brown, oblong or subquadrate, 0.9–1.3(–1.5) × 0.6–0.9 (–1.1) mm, (narrowly margined). $2n$ = 16.

Flowering Feb–Jul. Roadsides, clearings, disturbed sites, slopes, cedar glades, mixed woods, meadows, fields, waste grounds, damp places, grassy areas; 0–700 m; introduced; B.C., Ont.; Ala., Ark., Calif., Conn., Del., D.C., Fla., Ga., Ill., Ky., La., Md., Mass., Mich., Miss., Mo., N.J., N.Y., N.C., Ohio, Okla., Oreg., Pa., S.C., Tenn., Tex., Utah, Va., Wash., W.Va.; w Eurasia; introduced also in Central America, South America, e Asia (Japan), South Africa, Australia.

Herbarium specimens of *Cardamine hirsuta* have been misidentified as *C. oligosperma*.

19. **Cardamine holmgrenii** Al-Shehbaz, Harvard Pap. Bot. 11: 275. 2007 E

Perennials; glabrous throughout. **Rhizomes** slender, 0.5–1 mm diam. **Stems** erect to ascending, branched, 1.5–2 dm. **Rhizomal and basal leaves** absent. **Cauline leaves** 3 or 4, 3- or 5-foliolate, petiolate, leaflets petiolulate; petiole 0.5–1.7 cm, base not auriculate; lateral leaflet blade linear to linear-lanceolate, 0.9–1.5 cm × 1–3 mm, margins entire; terminal leaflet (petiolule 0.3–0.5 cm), blade obovate to lanceolate (linear distally), 1.3–2.3 cm × 2–8 mm, base cuneate, margins entire or subapically 1- or 2-toothed. **Racemes** ebracteate. **Fruiting pedicels** ascending to suberect, 5–14 mm. **Flowers:** sepals oblong, 1.2–1.5 × 0.6–0.8 mm, lateral pair not saccate basally; petals white, oblanceolate, 2–2.5 × 0.8–1 mm, (not clawed, apex rounded); filaments: median pairs 1.5–1.8 mm, lateral pair 1–1.2 mm; anthers ovate, 0.2–0.3 mm. **Fruits** linear, 1–2 cm × ca. 1 mm; ovules 16–24 per ovary; style 0.5–0.7 mm. **Seeds** brown, oblong, 1.2–1.5 × 0.8–1 mm.

Flowering Jul–Aug. Boggy slopes; Oreg.

Cardamine holmgrenii is known only from the type collection, from the Blue Mountains in Baker County.

20. Cardamine impatiens Linnaeus, Sp. Pl. 2: 655. 1753 [I]

Cardamine impatiens var. *angustifolia* O. E. Schulz

Biennials or, rarely, **annuals**; usually glabrous, rarely sparsely pubescent basally. **Rhizomes** absent. **Stems** erect, (angled, sometimes flexuous), unbranched basally, usually branched distally, (1.2–)2–6.5(–9) dm. **Basal leaves** (often withered by flowering), rosulate, similar to cauline, with fewer distal leaflets. **Cauline leaves** 9–24, (9–)13–25-foliolate, petiolate, leaflets petiolulate; petiole 2–6 cm, base auriculate (auricles to 10 × 2.2 mm); lateral leaflets similar to terminal, blade often smaller; terminal leaflet (petiolule to 0.5 cm), blade orbicular, obovate, ovate, or lanceolate, 1–4(–5) cm × 5–17 mm, margins entire or 3–5(–9)-toothed or -lobed. **Racemes** ebracteate. **Fruiting pedicels** divaricate or ascending, 3.5–12(–15) mm. **Flowers:** sepals oblong, 1.2–2(–2.5) × 0.7–1(–1.2) mm, lateral pair not saccate basally; petals (rarely absent), white, oblanceolate, 1.5–4(–5) × 0.6–1.2 mm; filaments 2–3(–4) mm; anthers ovate, 0.3–0.5 mm. **Fruits** linear, (torulose), (1–)1.6–3(–3.5) cm × 0.9–1.5 mm; (valves glabrous or, rarely, pilose); ovules 10–30 per ovary; style 0.6–1.6(–2) mm. **Seeds** brown, oblong, 1.1–1.5 × 0.8–1 mm, (compressed, sometimes narrowly winged apically). $2n = 16$.

Flowering May–Jul. Streamsides, slopes, roadsides, fields, disturbed areas; 0–200 m; introduced; Ont.; Conn., Ky., Mich., Minn., N.H., Ohio, Pa., Va., W.Va.; Eurasia; introduced also in South Africa.

21. Cardamine longii Fernald, Rhodora 19: 91. 1917 [E][F]

Perennials; glabrous throughout. **Rhizomes** cylindrical, slender, 0.8–1.5 mm diam. **Stems** prostrate to decumbent or erect, unbranched or branched, 0.5–4(–6) dm. **Rhizomal leaves** absent. **Cauline leaves** 3–10, usually simple, rarely 3 or 5-foliolate, petiolate; petiolulate or sessile; petiole 0.4–2.5 cm, base not auriculate; lateral leaflets (when present) sessile or petiolulate (to 0.2 cm), blade similar to terminal, considerably smaller; terminal leaflet blade orbicular to reniform or ovate to oblong, 0.4–3 cm × 3–20 mm, (somewhat fleshy), base subcordate to rounded or subtruncate, margins entire or repand, rarely undulate, (apex rounded). **Racemes** ebracteate. **Fruiting pedicels** divaricate-ascending to spreading, 0.5–2(–4) mm. **Flowers:** sepals ovate, 0.7–1.2 × 0.5–0.7 mm, lateral pair not saccate basally; petals absent or rudimentary, to 0.7 mm; filaments 0.5–0.8 mm; anthers ovate, 0.1–0.2 mm. **Fruits** narrowly oblong to linear, (3–)5–10(–15) × 0.8–1.2 mm; ovules 6–22 per ovary; style 0.2–0.5(–1) mm. **Seeds** light or yellowish brown, oblong or ovoid, 1–1.4 × 0.7–1 mm.

Flowering Jun–Sep. Tidal marshes, mud flats, tidal shores of rivers, shallow water, swampy areas, shady rocky crevices covered at high tide; 0–10 m; Fla., Maine, Md., Mass., N.J., N.Y., N.C., S.C., Va.

22. Cardamine macrocarpa Brandegee, Zoë 5: 233. 1906

Cardamine macrocarpa var. *texana* Rollins

Annuals; glabrous or sparsely puberulent. **Rhizomes** absent. **Stems** (simple or several from base), erect or decumbent, (flexuous or straight, narrowly winged-angled), unbranched or branched distally, (1.4–)2–4.5(–5.3) dm. **Basal leaves** (soon withered), not rosulate. **Cauline leaves** 3–9, middle ones 5–9-foliolate, petiolate, leaflets petiolulate or subsessile; petiole 1–3 cm, base not auriculate; lateral leaflets similar to terminal, blade often smaller, with oblique base, distal leaflets subsessile, blade smaller and narrower distally; terminal leaflet (petiolule 0.2–0.8 cm), blade usually broadly ovate to narrowly lanceolate, rarely oblong, 0.7–2 cm × 2–10 mm, base cuneate to rounded, margins repand, crenate, or 3-lobed. **Racemes** ebracteate, (rachis slightly to strongly flexuous or geniculate). **Fruiting pedicels** horizontal to divaricate or ascending, (3–)4–9(–12) mm. **Flowers:** sepals oblong, 2–3.5 × 0.7–1.2 mm, lateral pair not saccate basally; petals white, linear, 6–8 × 0.7–1 mm; filaments: median pairs 4–5 mm, lateral pair 3.5–4 mm; anthers oblong, 0.7–1 mm. **Fruits** linear, (2.5–)3–4.6 cm × 1.7–2.1 mm; ovules 14–22 per ovary; style 1–3 mm. **Seeds** dark brown, oblong, 2–2.5 × 0.9–1.2 mm.

Flowering Mar–Sep. Rock crevices and ledges, gravel bars of mountain streams, moist rocky stream banks, shaded loamy forest floors; Tex.; Mexico (Coahuila, Nuevo León).

Cardamine macrocarpa is known from the Chisos Mountains, Brewster County.

The characters by which var. *texana* is said to differ from var. *macrocarpa* are artificially drawn, and the style length, presence or absence of indumentum on the pedicels, and degree of flexuosity of the raceme rachises do not correlate and can vary within a given area. For these reasons, we do not recognize infraspecific taxa in *Cardamine macrocarpa*.

23. Cardamine maxima (Nuttall) Alph. Wood, Amer. Bot. Fl., 38. 1870 [E]

Dentaria maxima Nuttall, Gen. N. Amer. Pl. 2: 66. 1818; *Cardamine anomala* (Eames) K. Schumann; *D. anomala* Eames

Perennials; glabrous (except leaflet margins and, sometimes, rachis). Rhizomes cylindrical, 3–6 mm diam., (distinctly constricted at intervals, non-uniform diam., fleshy, slightly fragile, with dentate leaf scars). Stems erect, unbranched, 0.9–3(–4) dm. Rhizomal leaves 3-foliolate, 7–20 cm, leaflets petiolulate or subsessile; petiole 4–15 cm; lateral leaflets subsessile or petiolulate, blade often similar to terminal, base often oblique; terminal leaflet (petiolule 0.2–1(–1.7) cm), blade broadly ovate to oblong, 2–7.5 cm × 12–37 mm, base cuneate to obtuse, margins coarsely dentate to sharply incised, or deeply cleft into 2 or 3 lobes (lobes dentate or incised, margins puberulent). Cauline leaves 2 or 3, 3-foliolate, (rarely subopposite), petiolate, leaflets petiolulate or subsessile; petiole (0.5–)1–4(–6.5) cm, base not auriculate; lateral and terminal leaflets similar to rhizomal, distalmost sometimes much smaller. Racemes ebracteate. Fruiting pedicels: (flowering ones) horizontal to divaricate or deflexed, 7–20 mm. Flowers: sepals (erect to ascending), oblong, 5–7 × 2–3 mm, lateral pair slightly saccate basally; petals white or pink, oblanceolate, 10–17 × 3–6 mm, (not clawed, apex rounded); filaments: median pairs 4–8 mm, lateral pair 3–6.5 mm; anthers linear, 1.7–2.7 mm. Fruits (undeveloped), linear-lanceolate, to 3 cm × 2 mm; ovules 10–14 per ovary; style 3.5–7 mm. Seeds not known. $2n$ = 120, 124, 132, 138, 156, 161, ca. 208.

Flowering Apr–Jun. Rich woods, shady ravines, ledges, moist alluvial bottoms, steep forested slopes, stream banks; N.B., Ont., Que.; Conn., Maine, Mass., Mich., N.J., N.Y., Ohio, Pa., Vt.

Cardamine maxima has not been found with mature fruits and seeds and it has long been suspected to be a hybrid between *C. concatenata* and *C. diphylla*. Molecular studies (P. W. Sweeney and R. A. Price 2000) indicate that *C. maxima* is distinct from both those species. Although we hesitate to maintain it as a species, its wide distribution and morphological distinctness warrant its recognition.

24. Cardamine micranthera Rollins, Castanea 5: 87. 1940 [C][E]

Perennials; glabrous throughout. Rhizomes (relatively short), 2–3 mm diam., (covered with extensive root system). Stems erect to ascending, unbranched or branched distally, 0.9–4 dm. Rhizomal leaves usually 3-foliolate, rarely simple, 1–8 cm, leaflets petiolulate or subsessile; petiole 0.5–5 cm; lateral leaflets subsessile, often minute; terminal leaflet (petiolule 0.4–2 cm), blade orbicular to broadly ovate, 0.5–3 cm × 5–25 mm, base rounded, margins entire, repand, or dentate. Cauline leaves 5–10, petiolate; middle leaves often simple, petiole 0.3–1.5 cm, blade rhombic to suborbicular or ovate, 1–3.5 cm × 6–22 mm, base obtuse to cuneate, margins entire, repand, or dentate; distal ones with shorter petiole, blade smaller. Racemes ebracteate. Fruiting pedicels divaricate, 9–17 mm. Flowers: sepals (ascending), oblong, 1.5–2.2 × 0.7–1 mm, lateral pair not saccate basally; petals (somewhat spreading), white, oblanceolate, 3.5–5 × 1.2–1.8 mm, (not clawed, apex rounded); filaments: median pairs 2.5–3 mm, lateral pair 2–2.5 mm; anthers ovate, ca. 0.2 mm. Fruits linear, 0.8–1.6 cm × 0.8–1 mm; ovules 16–22 per ovary; style 1.2–1.8 mm. Seeds brown, oblong to ovoid, 0.9–1.2 × 0.6–0.8 mm.

Flowering Apr–May. Wet grounds along streams, seepage, gravelly sandbars, moist crevices; of conservation concern; N.C.

Cardamine micranthera is known only from Stokes County. It is in the Center for Plant Conservation's National Collection of Endangered Plants.

25. Cardamine microphylla Adams, Mém. Soc. Imp. Naturalistes Moscou 5: 111. 1817

Cardamine minuta Willdenow ex O. E. Schulz

Perennials; usually glabrous, rarely sparsely hirtellous. Rhizomes cylindrical, slender, 0.7–1.5 mm diam. Stems erect or ascending, unbranched, 0.3–1.5(–2) dm. Rhizomal leaves pinnately 5 or 7-foliolate, (1.3–)2.5–6.5 cm, (not fleshy), leaflets petiolulate; petiole (0.5–)1.5–5.5 cm; lateral leaflets similar to terminal, blade sometimes smaller; terminal leaflet (petiolule 0.15–0.7 cm), blade ovate to elliptic, (0.2–)0.4–1.2(–1.5) cm × (1–)2.5–6 mm, base cuneate to obtuse, margins entire, (apiculate). Cauline leaves 1–3, 5 or 7-foliolate, petiolate, leaflets petiolulate or subsessile; petiole 0.2–1.5 cm, base not auriculate; lateral leaflets similar to terminal; terminal leaflet (subsessile or petiolule to 0.2 cm), blade ovate to elliptic, 0.6–1.3 × 0.1–0.7 cm, base cuneate, margins entire. **Racemes** ebracteate. **Fruiting pedicels** erect to ascending, 7–15(–25) mm. **Flowers:** sepals ovate to oblong, 3–4 × 1.8–2.5 mm, lateral pair slightly saccate basally; petals usually white, rarely lavender, broadly obovate, 7–10 × 3–6 mm, (clawed, apex rounded); filaments: median pairs 3–4.5 mm, lateral pair 2–3 mm; anthers oblong, 1–1.5 mm. **Fruits** linear, 2–3 cm × 1.2–1.7 mm; ovules 12–16 per ovary; style 1–2 mm. **Seeds** brown, oblong, ca. 1.5 × 1 mm. $2n = 28, 42, 64$.

Flowering Jul–Aug. Streamsides, sand and cobbles on gravel bars, shale banks, floodplains, alluvial sand between cobbles, moist moss, turf, seepage areas, meadows, wet grounds; 0–1600 m; N.W.T., Yukon; Alaska; e Asia (Russian Far East, Siberia).

26. Cardamine nuttallii Greene, Bull. Calif. Acad. Sci. 2: 389. 1887 [E]

Dentaria tenella Pursh, Fl. Amer. Sept. 2: 439. 1813, not *Cardamine tenella* E. D. Clarke 1812; *C. californica* (Nuttall) Greene var. *gemmata* (Greene) O. E. Schulz; *C. gemmata* Greene; *C. nuttallii* var. *covilleana* (O. E. Schulz) Rollins; *C. nuttallii* var. *dissecta* (O. E. Schulz) Rollins; *C. nuttallii* var. *gemmata* (Greene) Rollins; *C. nuttallii* var. *pulcherrima* (Greene) Roy L. Taylor & Macbryde; *C. pulcherrima* Greene; *C. pulcherrima* var. *tenella* (Pursh) C. L. Hitchcock; *C. quercetorum* Howell; *C. tenella* var. *covilleana* O. E. Schulz; *C. tenella* var. *dissecta* O. E. Schulz; *C. tenella* var. *quercetorum* (Howell) O. E. Schulz; *D. gemmata* (Greene) Howell; *D. macrocarpa* Nuttall; *D. macrocarpa* var. *pulcherrima* (Greene) B. L. Robinson ex S. Watson; *D. quercetorum* (Howell) Greene; *D. tenella* var. *palmata* Detling; *D. tenella* var. *pulcherrima* (Greene) Detling; *D. tenella* var. *quercetorum* (Howell) Detling

Perennials; glabrous or sparsely pubescent. **Rhizomes** (tuberiform, fragile), with ovoid to oblong or cylindrical nodal swellings, slender, 2–5 mm diam., (fleshy). **Stems** erect, unbranched, 0.5–2(–3) dm, glabrous or sparsely pubescent distally. **Rhizomal leaves** simple or 3 (or 5)-foliolate, (3–)4–20(–25) cm, leaflets petiolulate or subsessile; petiole (2–)3–18(–21) cm; lateral leaflets (when present) petiolulate to subsessile, blade similar to terminal, sometimes smaller; terminal leaflet (subsessile or petiolule 0.2–3 cm), blade (simple leaf or terminal leaflet) reniform to suborbicular or ovate to oblong, (0.9–)1.3–4(–5.2) cm × (8–)12–50(–70) mm, base cordate to obtuse, margins crenate, dentate, or 5–7-lobed, (apiculae terminating teeth or lobes, surfaces glabrous). **Cauline leaves** 1–3, 3 (or 5)-foliolate, (appearing palmate), petiolate, leaflets petiolulate or sessile; petiole (0.2–)0.5–2(–3) cm, base not auriculate; lateral leaflets sessile, blade similar to terminal, smaller; terminal leaflet petiolulate or sessile, blade broadly ovate to oblong or linear, (0.5–)1–3.5(–6) cm, margins usually entire or dentate, rarely lobed. **Racemes** ebracteate. **Fruiting pedicels** ascending to divaricate, 10–35 mm. **Flowers:** sepals oblong, 3.5–5 × 1.5–2 mm, lateral pair saccate basally; petals usually purple to pale pink, rarely white, obovate, 10–15 × 4–7.5 mm, (not clawed, apex rounded); filaments: median pairs 5–8 mm, lateral pair 3.5–5 mm; anthers oblong, 1.5–2 mm. **Fruits** linear, 2.5–5.6 cm × 2–2.3 mm; ovules 8–16 per ovary; style 4–8 mm. **Seeds** dark brown, oblong, 2–2.5 × 1.4–1.6 mm.

Flowering Mar–May. Open pine forests, damp woods, shaded bottomlands, mossy slopes, streamsides, shaded and moist hillsides; 150–1000 m; B.C.; Calif., Oreg., Wash.

The infraspecific taxonomy of *Cardamine nuttallii* has been based almost entirely on the division and margin of rhizomal leaves. The treatments by O. E. Schulz (1903), L. E. Detling (1937), and R. C. Rollins (1993), though utilizing the same characters, varied considerably, especially in the application of names to varieties. The absence of rhizomal leaves on most specimens makes varietal determination an almost impossible task. Furthermore, leaf morphology is so highly variable that it is not useful for formally recognizing some of the other variants in the species. We therefore prefer to not subdivide the species.

27. Cardamine nymanii Gandoger, Bull. Soc. Bot. France 72: 1043. 1925 (as nymani)

Cardamine pratensis Linnaeus var. *angustifolia* Hooker

Perennials; (cespitose); glabrous. **Rhizomes** absent. **Stems** erect, unbranched or, rarely, branched, 0.5–1.6(–3.5) dm. **Basal leaves** (7 or) 9–21-foliolate, (thick, veins impressed); leaflets petiolulate or sessile; lateral lobes or leaflets similar to terminal; terminal lobe or leaflet blade orbicular, broadly ovate to lanceolate, base rounded to cuneate, margins usually entire. **Cauline leaves** 2–4(–7), pinnatisect or pinnately compound, (7 or) 9–21-foliolate, (thick, veins impressed), petiolate, leaflets petiolulate or sessile; petiole base not auriculate; lobes or leaflets (of proximal leaves) (4–)7–10 each side of rachis, fewer distally, distal leaves with 4 or 5 lobes or leaflets each side of rachis; terminal leaflet petiolulate or sessile, blade (or lobe) narrowly lanceolate to lanceolate, base cuneate, margins entire. **Racemes** ebracteate. **Fruiting pedicels** erect-ascending, 5–15 mm. **Flowers:** sepals oblong or ovate, 3.6–4.4 mm, lateral pair saccate basally, (green with hyaline margins); petals white-lilac, 9–12.3 × 4.8–6.8 mm, (clawed, apex rounded or emarginate); filaments: median pairs 3.5–4.5 mm, lateral pair 2–3 mm; anthers narrowly oblong, 0.9–1.4 mm. **Fruits** linear, 1–1.8 cm × ca. 1.5 mm; ovules ca. 16 per ovary; style ca. 1 mm, (stout). **Seeds** brown, oblong, ca. 1.5 mm. $2n$ = 56, 60, 64, 80–100.

Flowering Jun–Aug. Wet meadows, marshes, margins of ponds, along streams, seacoasts, swamps; Greenland; Man., Nfld. and Labr., N.W.T., Nunavut, Que., Yukon; Alaska; n Eurasia.

28. **Cardamine occidentalis** (S. Watson) Howell, Fl. N.W. Amer., 50. 1897 E

Cardamine pratensis Linnaeus var. *occidentalis* S. Watson in A. Gray et al., Syn. Fl. N. Amer. 1(1,1): 158. 1895; *C. neglecta* Greene

Perennials; glabrous or hirsute. **Rhizomes** (tuberiform, fragile), ovoid or globose at base of stem, 3–10 mm diam., (fleshy). **Stems** (simple from base), erect to ascending, (not flexuous), unbranched or branched distally, 1–5 dm, glabrous or pubescent proximally. **Basal leaves** not rosulate, pinnately compound, (3 or) 5 (or 7)-foliolate, 2–10 cm, leaflets petiolulate or subsessile; petiole 0.5–6.5 cm; lateral leaflets petiolulate or subsessile, blade similar to terminal, ovate, smaller, margins entire; terminal leaflet (petiolule 0.03–0.18 cm), blade orbicular to broadly ovate or subcordate, 0.5–2 cm × 7–25 mm, base cordate to rounded, margins entire or repand, (surfaces glabrous). **Cauline leaves** 3–7, (3 or) 5 or 7-foliolate (middle ones 5 or 7-foliolate, smaller distally, becoming 3-foliolate), petiolate; petiole 0.5–3 cm, base not auriculate; lateral leaflets similar to terminal, smaller; terminal leaflet blade obovate to oblanceolate, 0.5–2.6 cm × 3–13 mm, margins shallowly toothed, entire, or repand. **Racemes** ebracteate. **Fruiting pedicels** divaricate-ascending, 7–18 mm. **Flowers:** sepals oblong, 1.7–2 × 1–1.2 mm, lateral pair not saccate basally; petals white, oblanceolate, 4–6 × 1.5–2 mm, (not clawed); filaments: median pairs 2–2.5 mm, lateral pair 1–1.5 mm; anthers ovate, 0.3–0.5 mm. **Fruits** linear, (torulose), 1.5–3.3 cm × 1.7–2.2 mm; (valves glabrous or sparsely pubescent); ovules 18–40 per ovary; style 0.5–1.5 mm. **Seeds** brown, ovoid, 1–1.6 × 1–1.2 mm. $2n$ = 64.

Flowering Apr–Jul. Muddy grounds, lake margins, shallow streams, meadows; 150–1500 m; B.C.; Alaska, Calif., Oreg., Wash.

29. **Cardamine oligosperma** Nuttall in J. Torrey and A. Gray, Fl. N. Amer. 1: 85. 1838 W

Cardamine acuminata (Nuttall) Rydberg; *C. hirsuta* Linnaeus var. *acuminata* Nuttall; *C. hirsuta* var. *bracteata* O. E. Schulz; *C. hirsuta* subsp. *oligosperma* (Nuttall) O. E. Schulz; *C. hirsuta* var. *parviflora* Nuttall; *C. oligosperma* var. *bracteata* (O. E. Schulz) G. S. Torrey; *C. oligosperma* var. *lucens* G. S. Torrey; *C. oligosperma* var. *unijuga* (Rydberg) G. S. Torrey; *C. unijuga* Rydberg

Annuals or biennials; usually sparsely hirsute (at least proximally), rarely glabrous. **Rhizomes** absent. **Stems** (simple or few from base), erect to ascending, (not flexuous), unbranched or branched distally, (0.5–)0.8–3.2(–4.1) dm, usually pubescent throughout or proximally, rarely glabrous. **Basal leaves** (persistent to anthesis), rosulate, pinnately compound, 5–9(–13)-foliolate, 2–8.5(–11) cm, leaflets petiolulate or subsessile; petiole 1–6 cm; lateral leaflets petiolulate or subsessile, blade obovate to oblanceolate, smaller than terminal, margins entire or crenate; terminal leaflet (petiolule 0.1–0.7 cm), blade usually orbicular to ovate, rarely oblong, 0.4–1.5(–2.3) cm × 3–10(–13) mm, base cordate to rounded, margins entire, or crenate-dentate to obscurely 3 or 5-lobed, (surfaces often hirsute, sometimes glabrous). **Cauline leaves** 3–8, pinnately compound, similar to basal, smaller and fewer distally, petiolate, leaflets petiolulate or subsessile; petiole 0.5–2 cm, base not auriculate. **Racemes** ebracteate. **Fruiting pedicels** divaricate-ascending, (2–)3–9(–12) mm. **Flowers:** sepals oblong, 1.3–1.8(–2) × 0.5–1 mm, lateral pair not saccate basally; petals white, narrowly spatulate to oblanceolate, 2.5–3.5 × 0.9–1.5 mm (not clawed); filaments: median pairs 1.7–2.5 mm, lateral pair 1.2–2 mm; anthers ovate, 0.3–0.5 mm. **Fruits** linear, (torulose), (1.3–)1.6–2.8 cm × 1–1.7 mm; (valves glabrous or sparsely pubescent); ovules 16–36(–42) per ovary; style 0.4–1(–1.5) mm. **Seeds** brown, oblong, 1–1.6 × 0.8–1.2 mm. $2n$ = 16.

Flowering Mar–Jul. Stream banks, shady banks, creek bottoms, lakeshores, meadows, moist areas, wooded slopes; 50–3300 m; B.C.; Calif., Colo., Idaho, Mont., Nev., Oreg., Utah, Wash., Wyo.; Mexico (Baja California).

30. Cardamine pachystigma (S. Watson) Rollins, Harvard Pap. Bot. 1(4): 45. 1993 [E]

Dentaria californica Nuttall var. *pachystigma* S. Watson, Proc. Amer. Acad. Arts 14: 289. 1879; *Cardamine californica* (Nuttall) Greene var. *pachystigma* (S. Watson) O. E. Schulz; *D. corymbosa* var. *grata* Jepson; *D. pachystigma* (S. Watson) S. Watson

Perennials; glabrous. **Rhizomes** ovoid to oblong, (3–)4–18 mm diam., (fleshy, deeply underground). **Stems** erect, unbranched, 1–3 dm. **Rhizomal leaves** simple, 1.2–2.3 cm; petiole 6.7–19.5 cm; blade orbicular to reniform, or cordate to broadly ovate, 2.7–5.5 cm × 30–68 mm, base cordate to subtruncate, margins with obtuse teeth terminating in apiculae. **Cauline leaves** 2–5, (crowded distally), simple; petiole 0.3–2.6 cm, base not auriculate; blade similar to basal, 1.8–5.5 cm × 12–47 mm, base cuneate or obtuse to subcordate, marginal teeth apiculate. **Racemes** ebracteate. **Fruiting pedicels** divaricate to ascending, 7–24 mm. **Flowers:** sepals oblong, 4–7 × 2–2.5 mm, lateral pair saccate basally; petals usually pink or purple, rarely white, obovate to oblanceolate, 14–18 × 5–7 mm, (claw to 9 mm, apex rounded); filaments: median pairs 7–8 mm, lateral pair 5.5–6.5 mm; anthers oblong, ca. 1.5 mm. **Fruits** linear, 3.2–5.4 cm × 2.2–4 mm; ovules 10–14 per ovary; style 4–7 mm. **Seeds** brown, broadly ovoid, 2–2.8 × 1.8–2.5 mm.

Flowering Mar–May. Forests, lava slides, talus, cliffs; 700–2900 m; Calif.

Cardamine pachystigma is known from the disjunct counties of Plumas, Tehama, and Tulare.

31. Cardamine parviflora Linnaeus, Syst. Nat. ed. 10, 2: 1131. 1759 [F][W]

Cardamine arenicola Britton; *C. flexuosa* Withering var. *gracilis* O. E. Schulz; *C. parviflora* var. *arenicola* (Britton) O. E. Schulz

Annuals; (slender); glabrous or sparsely to densely pubescent throughout. **Rhizomes** absent. **Stems** (simple or few to several from base), erect, (somewhat flexuous), often branched distally, (0.5–)1–3(–4) dm. **Basal leaves** (often withered by anthesis), usually not rosulate, pinnately (5 or) 7–13(–17)-foliolate, (2–)4–10 cm, leaflets sessile or petiolulate; petiole 0.5–2.5(–4.5) cm; lateral leaflets similar to terminal, sometimes smaller; terminal leaflet (sessile or petiolule to 0.5 cm), blade linear to oblong, oblanceolate to obovate, or suborbicular, (0.1–)0.3–1 cm × 1–7 mm, base cuneate, margins entire or 3(–5)-toothed or -lobed. **Cauline leaves** 5–10(–14), (5–)9–15(–17)-foliolate, petiolate, leaflets sessile; petiole 0.3–1 cm, base not auriculate; lateral leaflets similar to terminal, sometimes smaller; terminal leaflet blade filiform, linear, or narrowly oblong, 0.3–1(–1.6) cm × 0.3–3 mm, margins usually entire, rarely 1–3-toothed. **Racemes** ebracteate. **Fruiting pedicels** divaricate or ascending, 4–10 mm. **Flowers:** sepals oblong, 1–1.5(–2) × 0.3–0.5 mm, lateral pair not saccate basally, (margins membranous); petals white, oblanceolate, (1.5–)1.8–2.5(–3) × 0.4–0.8(–1) mm; filaments 1.4–2.5 mm; anthers ovate, 0.2–0.4 mm. **Fruits** linear, (torulose), (0.5–)1–2(–2.5) cm × 0.6–0.9 mm; ovules 20–50 per ovary; style 0.3–0.7(–1) mm. **Seeds** pale brown, oblong-ovoid, 0.6–0.9 × 0.4–0.6 mm, (narrowly margined or not). $2n = 16$.

Flowering Mar–Jun. Roadsides, stream banks, rocky crests and outcrops, crevices of granitic bedrock, dry woods, glades, fallow fields, disturbed ground, limestone barrens, marsh and swamp margins, floodplains, waste ground, slopes, ledges, cliffs, meadows; 0–1500 m; Alta., B.C., Man., N.B., Nfld. and Labr. (Nfld.), N.W.T., Ont., Que., Sask.; Ala., Ark., Conn., Del., D.C., Fla., Ga., Idaho, Ill., Ind., Iowa, Kans., Ky., La., Maine, Md., Mass., Mich., Minn., Mo., N.H., N.J., N.Y., N.C., Ohio, Okla., Oreg., Pa., R.I., S.C., Tenn., Tex., Vt., Va., W.Va., Wis.; Eurasia.

Within *Cardamine parviflora*, in the broad sense, two species or varieties have been recognized: the Eurasian *C. parviflora* (or *C. parviflora* var. *parviflora*) versus the North American *C. arenicola* (or *C. parviflora* var. *arenicola*). Nuclear DNA data suggested a sister relationship of the North American and Eurasian entities, while cpDNA data showed them intermingled (J. Lihová et al. 2006). There are no apparent morphological differences between these entities, although detailed morphological studies are still lacking. We currently prefer to treat them as a single taxon.

32. Cardamine pattersonii L. F. Henderson, Rhodora 32: 25. 1930 (as pattersoni) [C][E]

Annuals or perennials; (short-lived); glabrous throughout. **Rhizomes** cylindrical, slender, 0.5–1.5 mm diam. **Stems** (simple from base), erect, (not flexuous), unbranched or branched basally, 0.6–3 dm. **Basal leaves** rosulate or not, 3 or 5-foliolate, 1–6 cm, leaflets petiolulate; petiole 0.4–2.5 cm; lateral leaflets similar to terminal, considerably smaller; terminal leaflet (petiolule 0.1–0.4 cm), blade obovate to orbicular or subcordate, 0.3–1.5(–2.0) cm × 2.5–16(–18) mm, base obtuse to rounded or cordate, margins entire or dentate to slightly sinuately lobed.

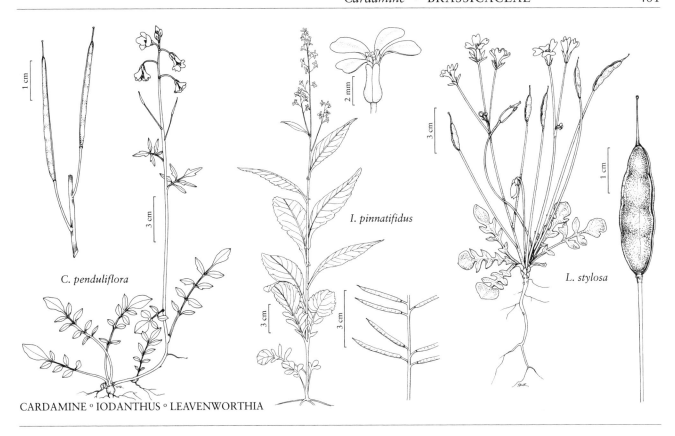

CARDAMINE ∘ IODANTHUS ∘ LEAVENWORTHIA

Cauline leaves 2–4, 3 or 5-foliolate, blade similar to basal, gradually reduced in size as bract, distalmost ones subtending pedicels of flowers (usually simple); leaflet blade or bract linear to linear-oblanceolate, 0.2–0.8 cm. **Racemes** bracteate throughout. **Fruiting pedicels** divaricate-ascending, (10–)15–30(–45) mm. **Flowers:** sepals oblong, (2–)3 × 0.7(–1) mm, lateral pair not saccate basally; petals purple or pink, obovate, 6–9 × 3–4 mm, (not clawed); filaments: median pairs 3–3.5 mm, lateral pair 2–2.5 mm; anthers oblong, 0.7–1 mm. **Fruits** linear, (torulose), 2–3 cm × 1–1.5 mm; ovules 14–20 per ovary; style 2–4 mm. **Seeds** brown, oblong to ovoid, 1.7–2.2 × 1–1.5 mm, (winged distally).

Flowering May–Jun. Moist mossy cliffs, rocky slopes, mossy banks; of conservation concern; 800–900 m; Oreg.

Cardamine pattersonii is known from Saddle Mountain and Onion Peak in Clastop County.

33. Cardamine penduliflora O. E. Schulz, Bot. Jahrb. Syst. 32: 538. 1903 [E] [F]

Cardamine rariflora M. Peck

Perennials; glabrous throughout. **Rhizomes** (tuberiform, fragile), 4–9(–11) mm diam., (fleshy). **Stems** erect or decumbent at base, unbranched, 2–6(–7.5) dm. **Rhizomal leaves** 5–13-foliolate, (4–)10–18(–25) cm, leaflets petiolulate or subsessile; petiole (3–)5–12(–17) cm; lateral leaflets subsessile, blade similar to terminal, sometimes smaller; terminal leaflet (petiolule 0.4–0.7 cm), blade oblong to elliptic or ovate, (0.4–)0.7–1.7(–2) cm, base cuneate or obtuse, margins entire or obscurely 3-lobed. **Cauline leaves** 2–6, 5–11-foliolate, petiolate, leaflets petiolulate or sessile; petiole 3–10 cm, base not auriculate; lateral leaflets sessile, blade similar to terminal, smaller; terminal leaflet (petiolule 0.5–1.5 cm), blade narrowly ovate or oblong to oblanceolate, 1.5–3.5 cm × 2–15 mm, margins entire or toothed to 3-lobed. **Racemes** ebracteate. **Fruiting pedicels** ascending to divaricate, (10–)20–40(–60) mm. **Flowers:** sepals oblong to ovate, 3.5–5 × 1.8–2.5 mm, lateral pair saccate basally; petals white, obovate, 12–16 × 6–8 mm, (not clawed, apex rounded or subemarginate); filaments: median pairs 6–7 mm, lateral pair 4–5 mm;

anthers oblong, 1.5–1.8 mm. **Fruits** linear, 2.5–4.5 cm × 1.4–2 mm; ovules 12–24 per ovary; style 4–6 mm. **Seeds** brown, oblong, 1.8–2 × 1–1.5 mm.

Flowering Mar–May. Shallow pools, wet grounds, marshes, meadows, creeks, channels, swampy woods; 50–150 m; Oreg.

Cardamine penduliflora is known from Douglas County north into Benton, Lane, Marion, Polk, and Yamhill counties.

34. Cardamine pensylvanica Muhlenberg ex Willdenow, Sp. Pl. 3: 486. 1801 [E] [W]

Cardamine breweri S. Watson var. *oregana* (Piper) Detling; *C. flexuosa* Withering subsp. *pensylvanica* (Muhlenberg ex Willdenow) O. E. Schulz; *C. hirsuta* Linnaeus var. *pensylvanica* (Muhlenberg ex Willdenow) P. W. Graff; *C. multifolia* Rydberg; *C. oregana* Piper; *C. pensylvanica* var. *brittoniana* Farwell; *C. rotundifolia* Michaux var. *diversifolia* O. E. Schulz; *Dracamine pensylvanica* (Muhlenberg ex Willdenow) Nieuwland

Annuals or biennials; sparsely hirsute basally, glabrous distally. **Rhizomes** absent. **Stems** (simple from base), erect, (not flexuous), unbranched or branched distally, (0.5–)1.5–5.5(–7) dm. **Basal leaves** (soon withered), not rosulate, similar to proximalmost cauline leaves, 4–15 cm. **Cauline leaves** (3–)5–20(–35), pinnately (5 or) 7–13(–19)-foliolate, sometimes appearing pinnatisect, lobe number similar to leaflets (middle and proximal ones 2–11 cm), petiolate, leaflets petiolulate, subsessile, or sessile; petiole (0.4–)1–3.5(–4.5) cm, base not auriculate, (often sparsely hirsute); lateral leaflets shortly petiolulate or sessile, (decurrent on rachis smaller than terminal, distalmost blades narrower, with fewer lobes or leaflets), margins entire or crenate; terminal leaflet (subsessile or petiolule to 1 cm), blade suborbicular, obovate to oblanceolate, or elliptic, 1.3–3(–4) cm × 6–25 mm, base often cuneate, margins entire, repand, or obscurely 3 or 5-lobed. **Racemes** ebracteate. **Fruiting pedicels** divaricate-ascending, (3–)4–10(–13) mm. **Flowers:** sepals oblong, (1–)1.3–2.3 × 0.5–1 mm, lateral pair not saccate basally; petals white, narrowly spatulate to oblanceolate, 2–3.5(–4) × 0.8–1.5 mm, (not clawed); filaments: median pairs 1.5–2.5 mm, lateral pair 1–2 mm; anthers ovate, 0.2–0.3 mm. **Fruits** linear, (torulose), (1.4–)1.7–2.7(–3.2) cm × 0.8–1.1 mm; ovules 40–80 per ovary; style 0.5–1 mm. **Seeds** brown, oblong to ovoid, 0.7–1.1 × 0.5–0.8 mm. $2n = 32, 64$.

Flowering Apr–Jul. Marshes, streams, swamps, ditches, seepage, springs, lake margins, mesic bottomland and upland forests, wet areas, ledges of sheltered bluffs, banks and shallow water of streams and spring branches, margins of crop fields, waste ground; 0–2800 m; Alta., B.C., Man., N.B., Nfld. and Labr., N.W.T., N.S., Ont., P.E.I., Que., Sask., Yukon; Ala., Alaska, Ark., Calif., Colo., Conn., Del., D.C., Fla., Ga., Idaho, Ill., Ind., Iowa, Kans., Ky., La., Maine, Md., Mass., Mich., Minn., Miss., Mo., Mont., Nev., N.H., N.J., N.Mex., N.Y., N.C., N.Dak., Ohio, Okla., Oreg., Pa., R.I., S.C., Tenn., Tex., Utah, Vt., Va., Wash., W.Va., Wis., Wyo.

We have not seen material of *Cardamine pensylvanica* from Nebraska or South Dakota; it is very likely that the species grows in these states as well.

35. Cardamine pratensis Linnaeus, Sp. Pl. 2: 656. 1753 [W]

Dracamine pratensis (Linnaeus) Nieuwland

Perennials; usually glabrous, rarely sparsely pilose basally. **Rhizomes** cylindrical, (not fragile), relatively short, (not fleshy). **Stems** erect, unbranched, (0.8–)1.5–5.5(–8) dm. **Rhizomal leaves** simple or 5–9(–31)-foliolate, to 30 cm, (thin, veins raised), leaflets petiolulate or sessile; petiole (1–)2–7(–10) cm; lateral leaflets petiolulate or sessile, blade similar to terminal or smaller, orbicular, ovate, or obovate, margins crenate or repand; terminal leaflet (petiolule to 1.5 cm), blade orbicular or broadly obovate, 0.3–2 cm diam., base usually rounded, rarely subreniform or cuneate, margins repand, (apex rounded). **Cauline leaves** 2–12(–18), pinnatisect, petiolate, leaflets petiolulate or sessile, (2–17 cm including petiole, thin, veins raised); petiole base not auriculate; lobes or leaflets (4–7(–13) each side of rachis), petiolulate or sessile and decurrent, blade similar to terminal lobe or leaflet, margins usually entire, rarely dentate; terminal lobe or leaflet (petiolulate or sessile), blade linear, oblong, ovate, or lanceolate, 1–2.5(–3.5) cm × 5–8(–10) mm, (surfaces glabrous). **Racemes** ebracteate. **Fruiting pedicels** erect-ascending or subdivaricate, (5–)12–25(–30) mm. **Flowers:** sepals (erect or spreading), oblong or ovate, (2.5–)3–5(–6) × 1–2 mm, lateral pair saccate basally; petals usually purple or lilac, rarely white, obovate, (6–)8–15(–18) × 3–7.5(–10) mm, (clawed, apex rounded or emarginate); filaments: median pairs 5–10 mm, lateral pair 3–6 mm; anthers narrowly oblong, (0.8–)1.2–2 mm. **Fruits** linear, (1.6–)2.5–4.5(–5) cm × (1.2–)1.5–2.3 mm; ovules 20–30 per ovary; style (0.5–)1–2.2(–2.7) mm, (stout). **Seeds** light brown, oblong, 1.2–1.8(–2) × 1–1.4 mm. $2n = 16$.

Flowering May–Aug. Moist grounds, stream sides, limestone shores, sedge and grass meadows, marshy pond margins, mossy areas, wet hollows, boggy areas, turfy shores, damp creek banks, swamps, brooks and ditches, moist ravines, springy swales; 0–1000 m; B.C.,

N.B., Nfld. and Labr. (Nfld.), N.S., Ont., Que.; Conn., Ind., Maine, Mass., Mich., N.H., N.J., N.Y., Ohio, Pa., Vt.; Eurasia.

The taxonomy of *Cardamine pratensis* in North America requires further detailed study. Most, if not all, populations of this species were introduced from Europe. Some specimens resemble the European *C. dentata* Schultes (high polyploid, characterized by all leaves, including distalmost, pinnate with petiolate and sometimes deciduous leaflets) and these populations might be native.

36. Cardamine purpurea Chamisso & Schlechtendal, Linnaea 1: 20. 1826

Cardamine purpurea var. *albiflora* Hultén; *C. purpurea* var. *albiflos* Hultén; *C. purpurea* var. *lactiflora* O. E. Schulz ex Steffen

Perennials; (often cespitose); hirsute. **Rhizomes** vertical, 1–3 mm diam., (not fleshy). **Stems** (simple or few to several from base), usually erect, unbranched, (0.3–)0.5–1.2(–1.5) dm. **Basal leaves** sometimes rosulate, usually 3 or 5-foliolate, rarely simple, (1.5–)2.5–7 cm, leaflets petiolulate or sessile; petiole (1.2–)2–5 cm; lateral lobes or leaflets sessile, blade similar to terminal, distinctly smaller; terminal lobe or leaflet petiolulate (0.05–0.6 cm), blade reniform or suborbicular to broadly ovate, 0.3–1 cm × 4–15 mm, base subcordate to rounded, margins entire, repand, or obscurely 2-toothed. **Cauline leaves** 1–3, usually compound, rarely simple, petiolate, leaflets petiolulate or sessile; petiole (0.2–)0.5–2 cm, base not auriculate; similar to basal, smaller. **Racemes** ebracteate. **Fruiting pedicels** suberect, ascending, or divaricate, 5–12 mm, pubescent. **Flowers:** sepals oblong, 2–3(–4) × 1.4–1.7 mm, lateral pair not saccate basally; petals usually purple to pink, rarely white, obovate, 5–7(–9) × 3–4(–5) mm, (clawed, apex rounded); filaments: median pairs 2.5–3.5 mm, lateral pair 1.5–3.5 mm; anthers oblong, 0.6–0.9 mm. **Fruits** linear, 1.5–2.5 cm × 1.5–1.8 mm; ovules 10–14 per ovary; style 1–2.5 mm. **Seeds** brown, oblong to broadly ovoid, 1.7–2.1 × 1.4–1.7 mm. $2n = 96$.

Flowering Jun–Aug. Moist tundra, damp woods and ravines, alpine turf, river flats, peaty subarctic meadows, streamsides, moist slopes; 600–1800 m; Yukon; Alaska; e Asia (Russian Far East).

37. Cardamine rotundifolia Michaux, Fl. Bor.-Amer. 2: 30. 1803 [E]

Dentaria rotundifolia (Michaux) Greene

Perennials; glabrous throughout. **Rhizomes** slender, to 2 mm diam. **Stems** usually procumbent, sometimes erect, (not flexuous), unbranched or branched, (1–)1.5–3(–5) dm. **Rhizomal leaves** absent. **Basal leaves** (soon withered), not rosulate. **Cauline leaves** simple, petiolate; petiole (0.3–)0.5–2.5(–4) cm, base not auriculate; blade oblong, ovate, suborbicular, or cordate, (0.5–)1–4.5(–5.5) cm × (5–)10–40(–54) mm, base cordate, rounded, or truncate, margins entire, repand, or sinuate, (distally with shorter petiole, blade smaller). **Racemes** ebracteate. **Fruiting pedicels** divaricate to ascending, (6–)10–15(–20) mm. **Flowers:** sepals oblong, 2–3 × 1–1.7 mm, lateral pair not saccate basally; petals white, broadly oblanceolate, spreading, 5–7(–8) × 2–3 mm, (not clawed, apex rounded); filaments: median pairs 3.5–4.5 mm, lateral pair 2.7–3.5 mm; anthers oblong, 0.8–1.2 mm. **Fruits** linear, (torulose), 1–1.5(–2) cm × 0.8–1.1 mm; ovules 40–80 per ovary; style 1.5–2.5 mm. **Seeds** brown, oblong to ovoid, 0.9–1.1 × 0.5–0.6 mm.

Flowering Apr–Jun. Stream banks, swamps, low woodland, wet rocky areas, seepage areas; 150–400 m; Del., Ga., Ky., Md., N.J., N.Y., N.C., Ohio, Pa., Tenn., Va., W.Va.

38. Cardamine rupicola (O. E. Schulz) C. L. Hitchcock in C. L. Hitchcock et al., Vasc. Pl. Pacif. N.W. 2: 474. 1964 [E]

Cardamine californica (Nuttall) Greene var. *rupicola* O. E. Schulz, Bot. Jahrb. Syst. 32: 388. 1903; *Dentaria rupicola* (O. E. Schulz) Rydberg

Perennials; glabrous throughout. **Rhizomes** cylindrical, slender, 1–2 mm diam. **Stems** erect or decumbent at base, unbranched, 0.6–2 dm. **Rhizomal leaves** palmately or subpalmately compound, 3 or 5 (or 7)-foliolate, 5.5–17(–22) cm, (fleshy), petiolate, leaflets petiolulate or subsessile; petiole 4–14(–17) cm; lateral leaflets subsessile, blade similar to terminal, sometimes smaller; terminal leaflet (petiolule 0.1–0.5 cm), blade ovate to lanceolate or elliptic-oblong, 1–3 cm × 6–17 mm, base cuneate or obtuse, margins entire, (apiculate). **Cauline leaves** 2 or 3, 3 or 5-foliolate, petiolate, leaflets petiolulate or sessile; petiole 0.7–4(–8) cm, base not auriculate; lateral leaflets sessile, blade similar to terminal, smaller; terminal leaflet petiolulate (0.1–0.5 cm), blade elliptic to oblong, or ovate, 1.2–3.5 cm × 4–25 mm, margins entire. **Racemes** ebracteate. **Fruiting pedicels** ascending to divaricate, 6–18 mm. **Flowers:** sepals oblong, 3–5 × 1.5–2 mm, lateral pair saccate basally; petals white, obovate, 8–14 × 4–7 mm, (short-clawed, apex rounded or subemarginate); filaments: median pairs 4–5 mm, lateral pair 2.5–3.5 mm; anthers oblong, 1–1.2 mm. **Fruits** linear, 2–4 cm × 1.5–2.2 mm; ovules 10–14 per ovary; style 1–5 mm. **Seeds** brown, oblong, 1.8–2.2 × 1.2–1.5 mm.

Flowering Jul–Aug. Limestone talus slopes, loose limey shale, moist banks; 2200–2700 m; Mont.

Cardamine rupicola is known from Flathead, Lewis and Clarke, and Missoula counties.

39. Cardamine umbellata Greene, Pittonia 3: 154. 1897

Cardamine hirsuta Linnaeus subsp. *kamtschatica* (Regel) O. E. Schulz; *C. kamtschatica* (Regel) Piper; *C. oligosperma* Nuttall var. *kamtschatica* (Regel) Detling; *C. sylvatica* Link var. *kamtschatica* Regel

Perennials; usually glabrous. **Rhizomes** often elongated, usually slender, rarely thickened, 1–2(–5) mm diam., (not fleshy). **Stems** (simple or few to several from base), erect to ascending, (not flexuous), unbranched basally, sometimes branched distally, (0.3–)0.8–2.5(–3) dm. **Basal leaves** (sometimes withered by anthesis), rosulate, pinnately compound, (3 or) 5 or 7 (or 9)-foliolate, 2–5(–9) cm, leaflets petiolulate or subsessile; lateral leaflets shortly petiolulate or subsessile, blade usually broadly ovate, rarely broadly obovate or orbicular, smaller than terminal, margins usually entire, rarely slightly 3 (or 5)-lobed or crenate; terminal leaflet subsessile, blade reniform or orbicular, 0.4–0.8(–1.2) cm × 5–9(–16) mm, margins entire or 3 (or 5)-lobed or crenate. **Cauline leaves** 3–5(–7), 3–7 (or 9)-foliolate, petiolate, leaflets subsessile or sessile; base not auriculate; lateral leaflets: blade narrowly obovate, oblanceolate to linear, margins similar to terminal; terminal leaflet blade narrowly obovate, ovate, oblanceolate, lanceolate, oblong, margins usually entire, sometimes 3-lobed or crenate. **Racemes** ebracteate, (subumbellate, 2–8(–14)-flowered, rachis usually 3–20 mm). **Fruiting pedicels** suberect to ascending, 3–8(–10) mm. **Flowers:** sepals (greenish or purplish), oblong, 1–2 × 0.5–1 mm, lateral pair not saccate basally; petals white, narrowly obovate, 2.5–5 × 1–3 mm. **Fruits** linear, (torulose), (1.3–)1.8–2.5(–3) cm × 0.8–1.5(–2) mm; (valves glabrous or sparsely pubescent); style 0.5–2 mm. **Seeds** brown, oblong, 1–1.5 × 0.8–1 mm. $2n$ = 32, 36, 48.

Flowering Jun–Sep. Stream banks, tundra, alpine slopes, wetlands, damp, swampy and mossy areas, beach gravel and sand, alpine stream margins; 0–1800 m; Alta., B.C., N.W.T., Yukon; Alaska, Wash.; e Asia (Russian Far East).

Recent molecular data (J. Lihová et al. 2006) indicate that *Cardamine umbellata*, often treated as a variety of *C. oligosperma*, represents a distinct lineage more closely related to taxa from New Zealand; this does not exclude *C. oligosperma* as one of the possible parents of this polyploid.

41. IODANTHUS (Torrey & A. Gray) Steudel, Nomencl. Bot. ed. 2, 1: 812. 1840

• [Greek *iodes*, violet-colored, and *anthos*, flower] E

Ihsan A. Al-Shehbaz

Cheiranthus Linnaeus [unranked] *Iodanthus* Torrey & A. Gray, Fl. N. Amer. 1: 72. 1838

Perennials; not scapose; glabrous or sparsely pubescent. **Stems** erect, branched distally. **Leaves** basal and cauline; petiolate or sessile; basal (withered by flowering), not rosulate, petiolate, blade margins lobed or not; cauline petiolate (petioles winged) or sessile, blade (base auriculate), margins dentate, entire, or lyrately lobed. **Racemes** (lax), considerably elongated in fruit. **Fruiting pedicels** divaricate to ascending, slender or stout. **Flowers:** sepals erect to ascending, oblong, lateral pair not or slightly saccate basally; petals purple, pink, or white, spatulate, (longer than sepals), claw differentiated from blade; stamens tetradynamous; filaments not dilated basally; anthers linear to narrowly oblong, (apiculate); nectar glands: lateral annular, median glands confluent with lateral. **Fruits** siliques, sessile or shortly stipitate, linear, smooth, terete; valves each with distinct midvein, glabrous; replum rounded; septum complete; ovules 22–36 per ovary; style distinct; stigma capitate. **Seeds** uniseriate, plump, not winged, oblong; seed coat (minutely reticulate), not mucilaginous when wetted; cotyledons accumbent.

Species 1: c, e United States.

R. C. Rollins (1993) recognized four species in *Iodanthus*, of which three are endemic to Mexico. As summarized by R. A. Price and I. A. Al-Shehbaz (2001), molecular data clearly support treatment of the genus as monospecific and closely allied to *Cardamine*, as well as recognition of the three Mexican species as members of *Chaunanthus*, a genus unrelated to *Iodanthus* and members of the Thelypodieae.

SELECTED REFERENCE Rollins, R. C. 1942. A systematic study of *Iodanthus*. Contr. Dudley Herb. 3: 209–215.

1. Iodanthus pinnatifidus (Michaux) Steudel, Nomencl. Bot., ed. 2, 1: 812. 1840 • Purple- or violet-rocket [E] [F]

Hesperis pinnatifida Michaux, Fl. Bor.-Amer. 2: 31. 1803; *Arabis hesperidoides* (Torrey & A. Gray) A. Gray; *Cheiranthus hesperidoides* Torrey & A. Gray; *Iodanthus hesperidoides* (Torrey & A. Gray) Torrey & A. Gray; *Thelypodium pinnatifidum* (Michaux) S. Watson

Stems 3–8(–10) dm, striate. **Cauline leaves:** (proximal) petiole (0.5–)1–4 cm, (narrowly to broadly winged), distal sessile; blade lanceolate, ovate, elliptic, or oblong, (3–)4–12(–15) cm × (12–)20–55(–70) mm, base cuneate to attenuate, or (distalmost) minutely to coarsely auriculate, margins usually minutely to coarsely, regularly or irregularly, dentate or serrate, rarely subentire, apex acute to acuminate, surfaces glabrous or pubescent. **Fruiting pedicels** straight or slightly curved upward, (2–)3–8 (–9) mm (nearly as thick as fruit). **Flowers:** sepals 3–6 × 1–1.5 mm, glabrous or subapically pilose; petals 7–12 (–14) × 1.5–2.5(–3) mm, attenuate to claw (claw 4–7 mm); filaments 3–6 mm; anthers 2–2.5 mm; gynophore obsolete or to 1 mm. **Fruits** usually divaricate to ascending, rarely erect, (1.5–)2–3.5(–4) cm × 1–1.5 mm; style (1–)2–4 mm. **Seeds** 1.2–1.5 × 0.7–1 mm.

Flowering late Apr–early Jul. Shaded banks, thickets, wooded ravines, limestone or sandstone bluffs, bottomland woods, swamps, flood plains, creeks, streamsides; 50–300 m; Ala., Ark., Ill., Ind., Iowa, Kans., Ky., Md., Minn., Mo., Ohio, Okla., Pa., Tenn., Tex., W.Va.

42. **LEAVENWORTHIA** Torrey, Ann. Lyceum Nat. Hist. New York 4: 87. 1837 • [For Melines Conkling Leavenworth, 1796–1862, American physician and botanist who collected in the southeastern United States] [E]

Ihsan A. Al-Shehbaz

James B. Beck

Annuals (winter); scapose; glabrous. **Stems** (sometimes absent), decumbent, branched basally and distally. **Leaves** basal and, sometimes, cauline; petiolate; basal rosulate, blade margins entire or lyrate-pinnatifid; cauline blade (base not auriculate), margins entire or lyrate-pinnatifid, (similar to basal). **Inflorescences** usually solitary flowers (on long peduncles from basal rosettes), sometimes racemes present in robust plants (corymbose, several-flowered), elongated in fruit. **Flowers:** sepals spreading or suberect, oblong or oblong-linear, lateral pair not saccate basally; petals white, lavender, orange, or yellow, obovate to spatulate, obcordate, or oblanceolate, (much longer than sepals), claw differentiated from blade, (much shorter than blade, apex shallowly to deeply emarginate or, rarely, truncate or obtuse); stamens strongly tetradynamous, (erect); filaments not dilated basally; anthers oblong, (apex obtuse); nectar glands: lateral annular, median glands present (distinct). **Fruits** siliques or silicles, sessile or shortly stipitate, linear to oblong or subglobose, torulose or smooth (or submoniliform), latiseptate, subterete, or terete; valves each obscurely veined; replum rounded; septum complete; ovules 5–25 per ovary; style distinct, (slender or stout); stigma capitate, (sometimes slightly 2-lobed). **Seeds** uniseriate, flattened, broadly winged or margined, suborbicular; seed coat (prominently reticulate), not mucilaginous when wetted; cotyledons obscurely accumbent, (radicle much shorter than cotyledon, straight or slightly bent). x = 11, 12, 15.

Species 8 (8 in the flora): s, se United States.

Leavenworthia has been subjected to extensive studies covering the taxonomy, breeding systems, evolution, and ecology of its species (I. A. Al-Shehbaz 1988). Monophyly of the genus and its sister relationship to *Selenia* are fairly well-established and need no further elaboration here. Although all eight species are fairly well-defined, it is often difficult to determine them based on material without mature fruits. In our opinion, it is far more difficult, and indeed impractical, to determine the five additional varieties recognized by R. C. Rollins (1963, 1993), because they are based solely on minor differences in the petal color, style length, and petal size, all of which are characters that show considerable variation of a continuous nature. We prefer not to recognize any infraspecific taxa at this stage. The varieties tend to have some geographical basis and might eventually be recognized as such with additional phylogeographic study.

SELECTED REFERENCES Beck, J. B., I. A. Al-Shehbaz, and B. A. Schaal. 2006. *Leavenworthia* (Brassicaceae) revisited: Testing classic systematic and mating system hypotheses. Syst. Bot. 31: 151–159. Charlesworth, D. and Z. Yang. 1998. Allozyme diversity in *Leavenworthia* populations with different inbreeding levels. Heredity 81: 453–461. Lloyd, D. G. 1965. Evolution of self-compatibility and racial differentiation in *Leavenworthia* (Cruciferae). Contr. Gray Herb. 195: 1–134. Lloyd, D. G. 1969. Petal color polymorphism in *Leavenworthia* (Cruciferae). Contr. Gray Herb. 198: 9–40. Rollins, R. C. 1963. The evolution and systematics of *Leavenworthia* (Cruciferae). Contr. Gray Herb. 192: 3–98. Solbrig, O. T. 1972. Breeding systems and genetic variation in *Leavenworthia* (Cruciferae). Evolution 26: 155–160. Solbrig, O. T. and R. C. Rollins. 1977. The evolution of autogamy in species of the mustard genus *Leavenworthia*. Evolution 31: 265–281.

1. Petals oblanceolate, 3.7–6.4 mm, apex obtuse or truncate; leaf blade lobe margins coarsely dentate, terminal lobes slightly larger than lateral lobes; ovules 18–26 per ovary . . . . 8. *Leavenworthia uniflora*
1. Petals obovate, broadly spatulate, to obcordate, (5–)6–14(–15) mm, apex emarginate; leaf blade lobe margins entire or shallowly dentate, terminal lobes considerably larger than lateral lobes; ovules 4–16(–18) per ovary.
    2. Fruits markedly torulose or submoniliform; seed wings obsolete or to 0.1 mm wide . . . . 7. *Leavenworthia torulosa*
    2. Fruits smooth, or, rarely, obscurely torulose; seed wings 0.2–0.5 mm wide.
        3. Styles 0.7–3 mm; petals 5–9 mm, apex shallowly emarginate, apical notch 0.1–0.4 (–0.6) mm deep.
            4. Fruits not margined; styles 0.7–1.7(–2.2) mm; Alabama, Georgia, Kentucky, Tennessee . . . . 4. *Leavenworthia exigua*
            4. Fruits margined; styles (1.5–)2–3 mm; Oklahoma, Texas.
                5. Leaf blade terminal lobes shorter than wide, margins slightly lobed or shallowly dentate; petals bright yellow, narrowly obovate . . . . 3. *Leavenworthia texana*
                5. Leaf blade terminal lobes equal to or slightly longer than wide, margins usually shallowly dentate; petals pale yellow, obcordate to broadly obovate . . . . 5. *Leavenworthia aurea*
        3. Styles (1.5–)3–7 mm; petals 9–16 mm, apex deeply emarginate, apical notch 0.5–1.3 mm deep.
            6. Fruits latiseptate, valves thin; styles 1.5–4.5 mm; petals white to lavender . . . . 1. *Leavenworthia alabamica*
            6. Fruits subterete, valves thick; styles 2.2–7 mm; petals usually yellow, sometimes lavender or white.
                7. Fruits (0.6–)0.8–1.2(–1.4) cm × (3.5–)4–5(–6) mm; ovules 4–6(–8) per ovary . . . . 2. *Leavenworthia crassa*
                7. Fruits (1–)1.5–3.4 cm × (2.5–)3–4(–4.5) mm; ovules (6–)8–12 per ovary . . . . 6. *Leavenworthia stylosa*

1. **Leavenworthia alabamica** Rollins, Contr. Gray Herb. 192: 68. 1963 [C][E]

Leavenworthia alabamica var. *brachystyla* Rollins

Stems (when present), 1–2.5 dm. Basal leaves: petiole (0.5–)1–3 cm; blade (2–)3.5–8.5(–10) cm, lobes 1–9 on each side, margins entire or shallowly dentate, terminal lobe orbicular to broadly ovate, 0.5–1.3 cm × 5–13 mm, considerably larger than lateral lobes, margins entire or shallowly dentate. Fruiting pedicels: solitary flowers 40–100 mm; racemes 20–60 mm. Flowers: sepals widely spreading, oblong-linear, (3.5–)4–6(–7) × 1–2 mm; petals spreading, white to lavender, obovate to broadly spatulate, 9–14 × (2.5–)3–6.5(–7.5) mm, claw yellow-orange, (2–)2.5–3.5 mm, apex deeply emarginate, apical notch 0.5–1 mm deep; filaments: median 3.5–6 mm, lateral 2–3.5 mm; anthers 1–1.5 mm. Fruits narrowly oblong, 1.5–3 cm × 3–4.5 mm, smooth, latiseptate; valves thin; ovules 6–12 per ovary; style 1.5–4.5 mm. Seeds 2.2–4 mm diam.; wing 0.2–0.5 mm wide; embryo straight. $2n = 22$.

Flowering Mar–Apr. Limestone outcrops and cedar glades, abandoned fields, pastures, rocky knolls, roadsides; of conservation concern; 150–300 m; Ala.

Leavenworthia alabamica is known only from Colbert, Franklin, Lawrence, and Morgan counties.

2. **Leavenworthia crassa** Rollins, Contr. Gray Herb. 192: 62. 1963 [C][E]

Leavenworthia crassa var. *elongata* Rollins

Stems (when present) 1–4 dm. Basal leaves: petiole (0.9–)1.5–3 cm; blade 2.8–6.5(–8) cm, lobes 1–8 on each side, margins entire or shallowly dentate, terminal lobe orbicular to broadly ovate, 0.5–1.2(–2) cm × 5–11(–18) mm, considerably larger than lateral lobes, margins entire or shallowly dentate. Fruiting pedicels: solitary flowers 40–80 mm; racemes 35–70 mm. Flowers: sepals widely spreading, oblong-linear, 4–5.7 × 1–2.1 mm; petals spreading, usually yellow, sometimes white, broadly spatulate to obovate, 9.5–14 × (2.5–)3–6.7(–8) mm, claw yellow to orange, 2.5–4 mm, apex deeply emarginate, apical notch 0.5–1 mm deep; filaments: median 4–6 mm, lateral 1.6–3 mm; anthers 0.9–1.3 mm. Fruits subglobose to oblong, (0.6–)0.8–1.2(–1.4) cm × (3.5–)4–5(–6) mm, smooth, subterete; valves thick; ovules 4–6(–8) per ovary; style 2.2–5(–6) mm. Seeds 1.7–3.4 mm diam.; wing 0.2–0.4 mm wide; embryo straight or nearly so. $2n = 22$.

Flowering Mar–Apr. Limestone cedar glades, pastures, fields, roadsides, near limestone sinks; of conservation concern; 150–300 m; Ala.

Leavenworthia crassa is known only from Lawrence and Morgan counties.

SELECTED REFERENCES Lloyd, D. G. 1967. The genetics of self-incompatibility in *Leavenworthia crassa* Rollins (Cruciferae). Genetica 38: 227–242. Lyons, E. E. and J. Antonovics. 1991. Breeding system evolution in *Leavenworthia*: Breeding system variation and reproductive success in natural populations of *Leavenworthia crassa* (Cruciferae). Amer. J. Bot. 78: 270–287.

3. **Leavenworthia texana** Mahler, Sida 12: 239, fig. 1. 1987 [C][E]

Leavenworthia aurea Torrey var. *texana* (Mahler) Rollins

Stems (when present) 1–2 dm. Basal leaves: petiole 1.5–4 cm; blade 3.5–5 cm, lobes 0–4 on each side, margins shallowly dentate, terminal lobe transversely broadly oblong, 0.6–0.7 cm × 8–11 mm, (distinctly shorter than wide), considerably larger than lateral lobes, margins slightly lobed or shallowly dentate. Fruiting pedicels: solitary flowers 30–70 mm; racemes 20–40 mm. Flowers: sepals widely spreading, oblong-linear, 3.9–4.5 × 1.4–1.8 mm; petals spreading, bright yellow, narrowly obovate, 7.3–9 × 2–3 mm, claw dark yellow, 2.6–3.3 mm, apex shallowly emarginate, apical notch 0.1–0.3 mm deep; filaments: median 4–4.5 mm, lateral 2.2–2.5 mm; anthers 0.8–1.3 mm. Fruits oblong, 1.5–2.3 cm × 4–5 mm, smooth, latiseptate, (margined); valves thin; ovules 4–14 per ovary; style 1.7–3 mm. Seeds 3–3.5 mm diam.; wing 0.2–0.4 mm wide; embryo straight. $2n = 22$.

Flowering Mar–Apr. Pastures, seepage areas of rock outcrops; of conservation concern; 70–150 m; Tex.

Leavenworthia texana is known only from San Augustine County. It is in the Center for Plant Conservation's National Collection of Endangered Plants.

Rollins reduced *Leavenworthia texana* to a variety of *L. aurea*. The differences in petal color, shape and margin of the terminal lobe, and chromosome number clearly support recognition as distinct species.

4. Leavenworthia exigua Rollins, Rhodora 58: 75, plate 1223, fig. 2. 1956 E

Leavenworthia exigua var. *laciniata* Rollins; *L. exigua* var. *lutea* Rollins

Stems (when present) 1–1.5 dm. **Basal leaves:** petiole 1.2–4 cm; blade 2.5–4.5(–6) cm, lobes 1–4 on each side, margins entire or shallowly dentate, terminal lobe orbicular to broadly ovate, 0.5–1 cm × 4–10 mm, considerably larger than lateral lobes, margins entire or shallowly dentate. **Fruiting pedicels:** solitary flowers 30–60 mm; racemes 10–25 mm. **Flowers:** sepals widely spreading, oblong-linear, 3–4.5(–5) × 0.9–1.5 mm; petals spreading, white, lavender, or yellow, spatulate, (5–)6–8(–9) × 1.7–3(–4) mm, claw yellow, 1.8–2.5 mm, apex shallowly emarginate, apical notch 0.1–0.4(–0.6) mm deep; filaments: median 2.2–4.2 mm, lateral 1–1.9 mm; anthers 0.5–0.9 mm. **Fruits** oblong, (1–)1.5–2.2 cm × (3.5–)4–5(–5.5) mm, smooth, latiseptate, (not margined); valves thin; ovules (4–)6–14 per ovary; style 0.7–1.7(–2.2) mm. **Seeds** 2.5–3.5 mm diam.; wing 0.2–0.5 mm wide; embryo straight. $2n = 22$.

Flowering Mar–May. Limestone glades, pastures, near limestone sinks, roadsides, old fields, thin soil over limestone beds; 70–300 m; Ala., Ga., Ky., Tenn.

5. Leavenworthia aurea Torrey, Ann. Lyceum Nat. Hist. New York 4: 88, plate 5. 1837 C E

Stems (when present) 1–3 dm. **Basal leaves:** petiole 1.8–3.9 cm; blade 3.5–6.5(–8) cm, lobes 1–3 on each side, margins shallowly dentate, terminal lobe suborbicular, 6–12 × 5–11 mm (equal to or slightly longer than wide), considerably larger than lateral lobes, margins usually shallowly dentate, sometimes slightly lobed. **Fruiting pedicels:** solitary flowers 30–90 mm; racemes 30–60 mm. **Flowers:** sepals widely spreading, oblong-linear, 2.9–4.6 × 1–1.9 mm; petals spreading, pale yellow, obcordate to broadly obovate, (5–)6–9 × 1.5–4.5 mm, claw bright yellow, 1.6–2.8 mm, apex shallowly emarginate, apical notch 0.1–0.4(–0.6) mm deep; filaments: median 3–4.5 mm, lateral 1.3–2.4 mm; anthers 0.6–1.1 mm. **Fruits** oblong, 2–3.1 cm × 3–5 mm, smooth, latiseptate, (margined); valves thin; ovules 6–14 per ovary; style (1.5–)2–3 mm. **Seeds** 2–4.2 mm diam.; wing 0.2–0.3 mm wide; embryo straight. $2n = 48$.

Flowering Mar–Apr. Pastures, roadsides, shallow limestone soils, seeps, creek margins; of conservation concern; 100–200 m; Okla.

Leavenworthia aurea is known only from Choctaw and McCurtain counties.

6. Leavenworthia stylosa A. Gray, Bot. Gaz. 5: 26. 1880 E F

Stems (when present) 1–2.5 dm. **Basal leaves:** petiole 1–2.5 cm; blade (2–)3–7.5(–10) cm, lobes 1–5 on each side, margins entire or shallowly dentate, terminal lobe orbicular to broadly ovate, 0.5–1.2(–1.8) cm × 5–11(–17) mm, considerably larger than lateral lobes, margins entire or shallowly dentate. **Fruiting pedicels:** solitary flowers 30–100 mm; racemes 15–70 mm. **Flowers:** sepals widely spreading, oblong-linear, (3.3–)3.7–5.8 × 1–1.6 mm; petals spreading, usually yellow, sometimes white or lavender, broadly spatulate to obovate, 9–13(–15) × 3.5–6(–9) mm, claw yellow to orange, 2–3.5 mm, apex deeply emarginate, apical notch 0.5–1.3 mm deep; filaments: median 4–6.3 mm, lateral 1.7–2.6 mm; anthers 0.9–1.5 mm. **Fruits** oblong to linear, (1–)1.5–3.4 cm × (2.5–)3–4(–4.5) mm, smooth, subterete; valves thick; ovules (6–)8–12 per ovary; style (3–)3.5–7 mm. **Seeds** 2.5–3.4 mm diam.; wing 0.2–0.3 mm wide; embryo nearly straight to slightly accumbent. $2n = 30$.

Flowering Mar–Apr. Pastures, roadsides, thin soil on limestone beds, limestone glades, streamsides, seeps, old fields; 150–300 m; Tenn.

Leavenworthia stylosa is known only from Bedford, Davidson, Rutherford, Smith, and Wilson counties.

7. Leavenworthia torulosa A. Gray, Bot. Gaz. 5: 26. 1880 E

Stems (when present) 1–2.5 dm. **Basal leaves:** petiole (1–)1.5–3.5 cm; blade (2–)2.7–7.5(–9) cm, lobes (0 or) 1–6 on each side, margins usually entire or shallowly dentate, rarely lobed, terminal lobe orbicular to broadly ovate, 0.5–1.1 cm × 5–12 mm, considerably larger than lateral lobes, margins entire or shallowly dentate. **Fruiting pedicels:** solitary flowers 30–80(–100) mm; racemes to 60 mm. **Flowers:** sepals suberect to spreading, narrowly oblong, 3–5.2 × 0.8–1.7 mm; petals spreading, white to lavender or, rarely, yellow, broadly spatulate, (5–)7–10 × (1.5–)2.5–4.5(–6) mm, claw yellow, (1.2–)2–3 mm, apex emarginate, apical notch 0.3–1.1 mm deep; filaments: median 3.5–4.7 mm, lateral 1.2–2.5 mm; anthers 0.7–1.4 mm. **Fruits** linear, (1.5–)2–3 cm × 2–4 mm, markedly torulose to submoniliform, latiseptate; valves somewhat thick; ovules 8–16(–18) per ovary; style 2.5–5 mm. **Seeds** 1.9–2.8 mm diam.; wing obsolete or to 0.1 mm wide; embryo slightly bent. $2n = 30$.

Flowering Mar–Apr. Cedar glades, pastures, seepy areas, thin soil over limestone beds, roadsides, old fields; 150–300 m; Ala., Ky., Tenn.

Leavenworthia torulosa is known in Alabama only from Madison County and in Kentucky only from Warren County.

8. **Leavenworthia uniflora** (Michaux) Britton, Mem. Torrey Bot. Club 5: 171. 1894 [E]

Cardamine uniflora Michaux, Fl. Bor.-Amer. 2: 29. 1803

Stems (when present) 1–2 dm. **Basal leaves:** petiole 0.3–2.6 cm; blade 2.3–8.5(–10) cm, lobes 3–10 on each side, margins coarsely dentate, terminal lobe suborbicular, 0.3–0.6 cm × 5–7 mm (slightly shorter than wide), slightly larger than lateral lobes, margins coarsely dentate. **Fruiting pedicels:** solitary flowers 30–120 mm; racemes to 40 mm. **Flowers:** sepals widely spreading, oblong, 2.5–4.5 × 0.9–1.5 mm; petals often erect, white, oblanceolate, 3.7–6.4 × 1–2(–2.5) mm, claw white or pale yellow, 1.3–2.5 mm, apex obtuse or truncate; filaments: median 2.5–4.2 mm, lateral 1.6–2.6 mm; anthers 0.5–0.8(–1) mm. **Fruits** narrowly oblong, (1.5–)2–3 cm × 3–4(–5) mm, smooth, latiseptate; valves thin; ovules 18–26 per ovary; style 0.9–2(–2.5) mm. **Seeds** 2.3–3.5 mm diam.; wing 0.2–0.4 mm wide; embryo slightly accumbent. $2n = 30$.

Flowering Mar–Apr. Rocky ledges, cedar glades, pastures, roadsides, old fields, thin soil on limestone beds, seeps on limestone rubble; 100–500 m; Ala., Ark., Ga., Ind., Ky., Mo., Ohio, Tenn.

Leavenworthia michauxii Torrey, which pertains here, is an illegitimate name based on *Cardamine uniflora* Michaux.

43. **NASTURTIUM** W. T. Aiton in W. Aiton and W. T. Aiton, Hortus Kew. 4: 110. 1812
    • Watercress [Latin *nasus*, nose, and *tortus*, distortion, alluding to pungency of plants]

Ihsan A. Al-Shehbaz

Perennials; (aquatic, rhizomatous, rooting at proximal nodes); not scapose; often glabrous, sometimes pubescent. **Stems** prostrate or decumbent, or erect in emergent plants, unbranched. **Leaves** cauline; not rosulate; petiolate; petiole base sometimes auriculate, blade (pinnately compound in emerged plants, or simple in deeply submerged plants, lateral leaflets petiolulate or sessile, 1–6(–12) pairs), margins entire, repand, or, rarely, dentate. **Racemes** (corymbose, several-flowered), elongated in fruit. **Fruiting pedicels** divaricate or descending, slender, (glabrous or adaxially puberulent). **Flowers:** sepals erect or ascending, oblong [ovate], lateral pair subsaccate or not saccate basally, (glabrous); petals usually white, rarely pink, obovate or narrowly spatulate, (longer than sepals), claw undifferentiated from blade, (attenuate to clawlike base, apex obtuse, acute, or rounded); stamens tetradynamous; filaments (white), not dilated basally; anthers oblong, (apex obtuse); nectar glands (2), lateral, annular or semiannular. **Fruits** siliques, sessile, usually linear, rarely narrowly oblong, smooth or slightly torulose, straight or slightly curved, terete; valves each obscurely veined, glabrous; replum rounded; septum complete; ovules 25–50 per ovary; style obsolete or distinct; stigma capitate. **Seeds** plump, not winged, oblong or ovoid; seed coat (minutely to coarsely reticulate), not mucilaginous when wetted; cotyledons accumbent. $x = 8$.

Species 5 (4 in the flora): North America, n Mexico, Central America, Europe, Asia, n Africa; introduced also in South America, tropical and s Africa, Australia, nearly worldwide.

There has been considerable disagreement as to whether *Nasturtium* should be maintained as a distinct genus or be united with *Rorippa*. Molecular data and a critical evaluation of morphology (I. A. Al-Shehbaz and R. A. Price 1998) clearly show that *Nasturtium* is much more closely related to *Cardamine* than it is to *Rorippa*, and that the two genera should not be united.

Plants of *Nasturtium floridanum*, *N. microphyllum*, and *N. officinale* typically produce compound leaves when submerged in shallow waters or when their branches are emergent. When submerged in deep waters, all three produce simple leaves, and, in that case, it is impossible to distinguish them. The hybrid between *N. officinale* and *N. microphyllum*, *N.* ×*sterilis* Airy Shaw, is uncommon in North America, having been reported from Connecticut, Idaho, Michigan, and New Hampshire (P. S. Green 1962), and is far more common in Europe, where it has recently been studied thoroughly (W. Bleeker et al. 1999 and references therein). *Nasturtium africanum* Braun-Blanquet is known from Morocco in northwestern Africa.

SELECTED REFERENCES Al-Shehbaz, I. A. and R. A. Price. 1998. Delimitation of the genus *Nasturtium* (Brassicaceae). Novon 8: 124–126. Al-Shehbaz, I. A. and R. C. Rollins. 1988. A reconsideration of *Cardamine curvisiliqua* and *C. gambellii* as species of *Rorippa* (Cruciferae). J. Arnold Arbor. 69: 65–71. Green, P. S. 1962. Watercress in the New World. Rhodora 64: 32–43. Rollins, R. C. 1978. Watercress in Florida. Rhodora 80: 147–153.

1. Seeds biseriate, coarsely reticulate, with 25–50(–60) areolae on each side; fruits (1.8–)2–2.5(–3) mm wide .... 1. *Nasturtium officinale*
1. Seeds uniseriate, minutely to moderately reticulate, with (75–)100–500 areolae on each side; fruits 0.8–1.2(–1.8) mm wide.
    2. Emergent leaves: petiolar bases non-auriculate; blades 3(–5)-foliolate; seeds light or yellowish brown; Florida .... 4. *Nasturtium floridanum*
    2. Emergent leaves: petiolar bases often minutely auriculate; blades (3–)5–17-foliolate; seeds reddish brown; not Florida.
        3. Leaflet margins entire or repand; seeds with (75–)100–150(–175) areolae on each side; filaments 2.5–3.5 mm; petals 4.5–6 mm; petioles and rachises not winged .... 2. *Nasturtium microphyllum*
        3. Leaflet margins coarsely dentate or, rarely, sinuate-repand; seeds with 300–450 areolae on each side; filaments 5–7 mm; petals 6–8 mm; petioles and rachises narrowly winged .... 3. *Nasturtium gambelii*

1. **Nasturtium officinale** W. T. Aiton in W. Aiton and W. T. Aiton, Hortus Kew. 4: 110. 1812 [I] [W]

Sisymbrium nasturtium-aquaticum Linnaeus, Sp. Pl. 2: 657. 1753; *Rorippa nasturtium-aquaticum* (Linnaeus) Hayek

Plants glabrous throughout or sparsely pubescent. **Stems** 1–11(–20) dm. **Cauline leaves:** petiole not winged, base auriculate; blade 3–9(–13)-foliolate, (1–)2–15(–22) cm; lateral leaflets sessile or petiolulate, rachis not winged, blade smaller than terminal; terminal leaflet (or simple blade) suborbicular to ovate, or oblong to lanceolate, (0.4–)1–4(–5) cm × (3–)7–25(–40) mm, base obtuse, cuneate, or subcordate, margins entire or repand, apex obtuse. **Fruiting pedicels** divaricate or descending, straight or recurved, 5–17(–24) mm. **Flowers:** sepals 2–3.5 × 0.9–1.6 mm; petals white or pink, spatulate or obovate, 2.8–4.5(–6) × 1.5–2.5 mm, (base to 1 mm), apex rounded; filaments 2–3.5 mm; anthers 0.6–1 mm. **Fruits** (0.6–)1–1.8(–2.5) cm × (1.8–)2–2.5(–3) mm; ovules (28–)36–60 per ovary; style 0.5–1(–1.5) mm. **Seeds** biseriate, reddish brown, ovoid, (0.8–)0.9–1.1(–1.3) × (0.6–)0.7–0.9(–1) mm, coarsely reticulate with 25–50(–60) areolae on each side. $2n = 32$.

Flowering Feb–Sep. Flowing streams, ditches, lake margins, swamps, marshes, seeps; 0–3000 m; introduced; Alta., B.C., Man., N.B., Nfld. and Labr. (Nfld.), N.S., Ont., P.E.I., Que.; Ala., Alaska, Ariz., Ark., Calif., Colo., Conn., Del., D.C., Fla., Ga., Idaho, Ill., Ind., Iowa, Kans., Ky., Maine, Md., Mass., Mich., Minn., Miss., Mo., Mont., Nebr., Nev., N.H., N.J., N.Mex., N.Y., N.C., N.Dak., Ohio, Okla., Oreg., Pa., R.I., S.C., S.Dak., Tenn., Tex., Utah, Vt., Va., Wash., W.Va., Wis., Wyo.; Europe; Asia; n Africa; introduced also elsewhere in the New World, tropical and s Africa, Australia.

2. **Nasturtium microphyllum** Boenninghausen ex Reichenbach, Fl. Germ. Excurs., 683. 1832 [I]

Nasturtium uniseriatum H. W. Howard & Manton; *Rorippa microphylla* (Boenninghausen ex Reichenbach) Hylander ex Á. Löve & D. Löve

Plants glabrous throughout or sparsely pubescent. **Stems** (1–)2–10(–18) dm. **Cauline leaves:** petiole not winged, base auriculate; blade (3–)5–9(–11)-foliolate, (1.5–)2.5–10(–15) cm; lateral leaflets sessile or petiolulate, rachis not winged, blade smaller than terminal; terminal

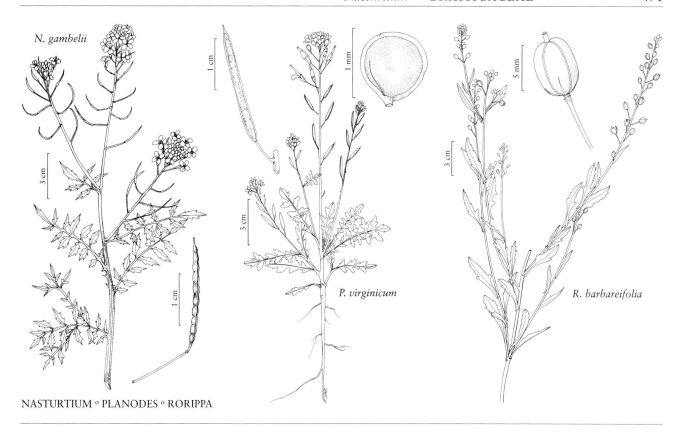

leaflet (or simple blade) suborbicular or oblong, (0.6–)1.5–3.5(–4.5) cm × (4–)8–20(–25) mm, base obtuse, cuneate, or subcordate, margins entire or repand, apex obtuse. **Fruiting pedicels** divaricate or descending, straight or, occasionally, recurved, (7–)9–21(–25) mm. **Flowers:** sepals 2.5–4 × 1–1.5 mm; petals white or pink, spatulate or obovate, 4.5–6 × 1.5–3 mm, (base to 1 mm), apex rounded; filaments 2.5–3.5 mm; anthers 0.7–1 mm. **Fruits** 1.4–2.2(–2.7) cm × 1–1.5(–1.8) mm; ovules 30–38(–40) per ovary; style 0.5–1.5(–2) mm. **Seeds** uniseriate, reddish brown, ovoid, (0.8–)1–1.2 × (0.6–)0.7–0.8(–0.9) mm, moderately reticulate with (75–)100–150(–175) areolae on each side. $2n = 64$.

Flowering Jan–Aug. Lake margins, streams, ponds, springs, river shores, seeps, swales, wet meadows; 0–1500 m; introduced; B.C., Man., N.B., Nfld. and Labr. (Nfld.), Ont., P.E.I., Que.; Conn., Idaho, Ind., Ky., Maine, Mass., Mich., N.H., N.Mex., N.Y., Oreg., Pa., R.I., Vt., Wyo.; Europe; introduced also elsewhere in the New World, Australia.

3. Nasturtium gambelii (S. Watson) O. E. Schulz, Bot. Jahrb. Syst. 66: 98. 1933 [C][F]

Cardamine gambelii S. Watson, Proc. Amer. Acad. Arts 11: 147. 1876 (as gambellii); *Rorippa gambelii* (S. Watson) Rollins & Al-Shehbaz

Plants sparsely to moderately pubescent. **Stems** 5–12 dm. **Cauline leaves:** petiole narrowly winged, base auriculate or not; blade 9–17-foliolate, 3–10 cm; lateral leaflets sessile, rachis narrowly winged, blade about same size as terminal; terminal leaflet blade linear to narrowly oblong or lanceolate, 0.5–2.5 cm × 2–8(–10) mm, base decurrent with adjacent pair of leaflets, margins often dentate, rarely sinuate-repand, apex acute. **Fruiting pedicels** divaricate or descending, straight, 9–25 mm. **Flowers:** sepals 3–4 × 1–1.5 mm; petals white, spatulate or oblanceolate, 6–8 × 2–2.5 mm, (base to 2 mm), apex obtuse to acute; filaments 5–7 mm; anthers 1–1.2 mm. **Fruits** 2–3 cm × 1–1.2 mm; ovules 24–40 per ovary; style 1–2.5 mm. **Seeds** uniseriate, reddish brown, ovate, 1–1.2 × 0.6–0.8 mm, minutely reticulate with 300–450 areolae on each side.

Flowering May–Aug. Lake margins, streams, swamps, marshes, ponds; of conservation concern; 5–30 m; Calif.; Mexico (Chiapas, Federal District); Central America (Guatemala).

Nasturtium gambelii is known from southern California in Los Angeles, San Bernardino, San Luis Obispo, Santa Barbara, and Ventura counties. It is likely to be found also in undisturbed, freshwater habitats elsewhere in southern California and in Mexico and Central America. Intermediates between *N. gambelii* and *N. officinale* have been found in southern California, suggesting hybridization between them. Experimental work is needed to confirm this hypothesis.

4. **Nasturtium floridanum** (Al-Shehbaz & Rollins) Al-Shehbaz & R. A. Price, Novon 8: 125. 1998 [E]

Rorippa floridana Al-Shehbaz & Rollins, J. Arnold Arbor. 69: 68. 1988, based on *Cardamine curvisiliqua* Shuttleworth ex Chapman Fl. South. U.S. ed. 2, 605. 1883, not *R. curvisiliqua* (Hooker) Bessey ex Britton 1894, not *Nasturtium curvisiliqua* (Hooker) Nuttall 1838; *N. stylosum* Shuttleworth ex O. E. Schulz

Plants usually glabrous throughout, rarely sparsely pubescent. **Stems** 1.5–9 dm. **Cauline leaves:** petiole not winged, base not auriculate; blade 3–5(–7)-foliolate, (1.5–)3–7(–10) cm; lateral leaflets petiolulate, rachis not winged, blade often much smaller than terminal; terminal leaflets (or simple blade) subreniform, orbicular, ovate to oblong, or obovate, 0.5–3.5 × 0.5–3 cm, base obtuse, subcordate, rounded, or cuneate, margins repand, entire, or, rarely, obtusely dentate, apex obtuse or rounded. **Fruiting pedicels** divaricate, straight or slightly recurved, 5–15 mm. **Flowers:** sepals 2–3 × 1–1.5 mm; petals white, spatulate, 4–5 × 1.5–2 mm, apex rounded; filaments 2–3 mm; anthers 0.6–0.9 mm. **Fruits** 1.5–3 cm × 0.9–1.2 mm; ovules 34–50 per ovary; style 0.5–2.5 mm. **Seeds** uniseriate, light or yellowish brown, ovoid, 0.6–0.8(–0.9) × 0.4–0.7 mm, minutely reticulate with 400–500 areolae on each side. $2n = 32$.

Flowering Feb–Oct. Flowing streams, hummock margins, springs, swamps, stream banks; 0–50 m; Fla.

Nasturtium floridanum is known from Brevard, Citrus, Clay, Collier, Columbia, Dade, Duval, Gilchrist, Hillsborough, Lake, Levy, Manatee, Marion, Seminole, Sumter, Taylor, Volusia, and Wakulla counties.

Nasturtium stylosum Shuttleworth ex O. E. Schulz (1936), not (de Candolle) O. E. Schulz ex Cheeseman (1911) is an illegitimate name, sometimes found in synonymy with *N. floridanum*.

44. **PLANODES** Greene, Leafl. Bot. Observ. Crit. 2: 220. 1912 • [Greek *planis*, wanderer, and *–odes*, resemblance, alluding to original assignment to another genus]

Ihsan A. Al-Shehbaz

Annuals (sometimes winter); not scapose; pubescent or glabrous. **Stems** erect or ascending to decumbent, often branched distally. **Leaves** basal and cauline; petiolate; basal (loosely) rosulate, blade margins pinnatifid to pinnatisect, (lobes) dentate or entire; cauline similar to basal. **Racemes** (corymbose, several-flowered), considerably elongated in fruit. **Fruiting pedicels** divaricate-ascending, slender. **Flowers:** sepals erect, oblong, lateral pair not saccate basally; petals white, oblanceolate, (longer than sepals), claw undifferentiated from blade, (apex obtuse); stamens slightly tetradynamous; filaments not dilated basally; anthers ovate, (apex obtuse); nectar glands lateral (minute), 1 on each side of lateral stamen. **Fruits** siliques, sessile or subsessile, linear, torulose, straight, latiseptate; valves each with obscure midvein, glabrous; replum rounded; septum complete; ovules (20–)28–44 per ovary; style distinct (conical); stigma capitate. **Seeds** uniseriate, flattened, narrowly winged, orbicular or suborbicular; seed coat not mucilaginous when wetted; cotyledons accumbent. $x = 8$.

Species 1: c, s United States, nw Mexico.

Planodes virginicum has floated among *Arabis*, *Cardamine*, and *Sibara*. However, it is distinct from all three both morphologically and phylogenetically, and it appears to be most closely allied to *Cardamine*. It is known from Baja California; in the southeastern and some of the central United States, it has become weedy, especially in cultivated or abandoned fields.

SELECTED REFERENCE Rollins, R. C. 1947. Generic revisions in the Cruciferae: *Sibara*. Contr. Gray Herb. 165: 133–143.

1. **Planodes virginicum** (Linnaeus) Greene, Leafl. Bot. Observ. Crit. 2: 221. 1912 F W

Cardamine virginica Linnaeus, Sp. Pl. 2: 656. 1753; *Arabis ludoviciana* (Hooker) C. A. Meyer; *A. virginica* (Linnaeus) Poiret; *C. ludoviciana* Hooker; *Sibara virginica* (Linnaeus) Rollins

Stems simple to many from base, (0.5–)1–3.5(–5.5) dm, hirsute or puberulent basally, often glabrous distally. **Basal leaves:** petiole (0.3–)0.8–1.5(–2) cm; blade oblong to oblanceolate in outline, (1–)1.5–7(–10) cm × 4–20(–30) mm, margins pinnatifid to pinnatisect; lobes (4–)6–12(–15) per side, ovate or oblong to linear, margins often coarsely dentate proximally, entire or minutely dentate distally, larger distally; terminal lobe equal to or larger than lateral, margins entire or laterally 1-toothed, surfaces usually pubescent, rarely glabrate. **Cauline leaves** shortly petiolate; blade smaller, narrower, (lobe) margins entire. **Racemes** lax in fruit. **Fruiting pedicels** straight, (1.5–)2.5–6(–8) mm, glabrous. **Flowers:** sepals 1–2 × 0.5–0.8 mm, glabrous or subapically with few hairs; petals 2–3 × 0.5–1 mm; filaments 1.5–2 mm; anthers 0.3–0.4 mm. **Fruits** divaricate-ascending, (1–)1.5–2.5(–3.2) cm × 1–1.5 mm, slightly torulose; style 0.2–0.7 mm. **Seeds** orbicular or suborbicular, 1–1.2 × 0.9–1 mm; wing 0.1–0.15 mm. $2n = 16$.

Flowering Feb–Apr. Fields, floodplains, waste places, lawns, railroad tracks, embankments, roadsides, cultivated ground, streamsides, open woods; 0–500 m; Ala., Ark., Calif., Fla., Ga., Ill., Ind., Iowa, Kans., Ky., La., Miss., Mo., N.C., Ohio, Okla., S.C., Tenn., Tex., Va., W.Va.; Mexico (Baja California).

Distribution of *Planodes virginicum* is primarily in the southeastern and central United States (Virginia west into Kansas, south into Texas, and east into Florida). Its disjunction in southern California and northern Baja California suggests that it was introduced there. Considering the weedy nature of the species, I expect that its range extends beyond that given above.

45. **RORIPPA** Scopoli, Fl. Carniol., 520. 1760 • Yellow-cress [Saxon *rorippen*, name cited by Euricius Cordus, 1515–1544]

Ihsan A. Al-Shehbaz

Brachiolobos Allioni; *Kardamoglyphos* Schlechtendal; *Neobeckia* Greene; *Radicula* Moench; *Tetrapoma* Turczaninow ex Fischer & C. A. Meyer

Annuals, biennials, or perennials; (usually aquatic or of mesic habitats, rhizomatous, sometimes with caudex); not scapose; glabrous or pubescent. **Stems** erect or prostrate, unbranched or branched. **Leaves** basal and/or cauline; petiolate or sessile; basal (usually withered early), rosulate or not, petiolate, blade margins entire, dentate, sinuate, lyrate, pectinate, or 1–3-pinnatisect; cauline petiolate or sessile, blade (base cuneate, attenuate, auriculate, or sagittate), margins entire, dentate, pinnatifid, or pinnatisect. **Racemes** slightly to considerably elongated in fruit. **Fruiting pedicels** erect, suberect, ascending, horizontal, reflexed, or divaricate, usually slender. **Flowers:** sepals (rarely persistent), erect or spreading, ovate or oblong, lateral pair not or, rarely, saccate basally, (margins often membranous); petals (rarely vestigial or absent), often yellow, sometimes white or pink, usually obovate, spatulate, oblong, or oblanceolate, rarely linear, claw undifferentiated or not from blade, (often shorter than sepals, apex obtuse or emarginate); stamens usually tetradynamous, rarely 4 and equal; anthers usually ovate or oblong, rarely linear, (apex usually obtuse, rarely apiculate); nectar glands confluent, often subtending bases of stamens, median present. **Fruits** siliques or silicles, usually sessile, rarely shortly stipitate, linear, oblong, ovoid, ellipsoid, pyriform, subglobose, or globose, smooth or torulose, terete or slightly latiseptate; valves (3–6 in *R. barbareifolia*) papery or leathery, each obscurely veined, glabrous or pubescent; replum (visible), rounded; septum usually complete, rarely perforate; ovules [10–]18–242[–300] per ovary; style obsolete or distinct; stigma capitate, (rarely slightly

2-lobed). **Seeds** usually biseriate, rarely uniseriate, plump, not or, rarely, winged, oblong, ovoid, ovate, orbicular, cordiform, subglobose, or globose; seed coat (reticulate, colliculate, rugose, tuberculate, or foveolate), mucilaginous or not when wetted; cotyledons accumbent.

Species ca. 85 (22 in the flora): nearly worldwide.

Rorippa coloradensis Stuckey was based on material collected in 1875 by T. S. Brandegee from Colorado. Although R. C. Rollins (1993) did not examine the holotype (*Brandegee 1180*, PH), he recognized the taxon as a distinct species. An examination of the holotype immediately reveals that it is the introduced European weed *Sinapis alba* Linnaeus. Although the holotype has no fruits, the presence of retrorse trichomes along the stems and on the pistil, and the spreading sepals and flower size match *S. alba*.

The infraspecific variation in some species of *Rorippa*, especially in *R. curvipes*, *R. curvisiliqua*, and *R. palustris*, has led some authors (e.g., F. K. Butters and E. C. Abbe 1940; R. L. Stuckey 1972; R. C. Rollins 1993) to recognize varieties or subspecies so poorly defined that they cannot be distinguished with some (or any) degree of confidence. I take the position of recognizing infraspecific taxa only if it leads to a practical taxonomy and if the infraspecific taxa have some delimited morphological and geographical attributes.

Rorippa cantoniensis (Loureiro) Ohwi, cited by R. L. Stuckey (1972) and R. C. Rollins (1993) as *R. microsperma* (de Candolle) L. H. Bailey, is known from a single collection made by J. Macoun from Vancouver Island in 1893 (NY). *Rorippa globosa* (Turczaninow ex Fischer & C. A. Meyer) Hayek is known from collections made over 120 years ago by Walter Dean from Cambridge, Massachusetts (GH). Finally, *R. islandica* (Oeder ex Murray) Borbás was collected from western Greenland by A. E. and M. Porsild in 1925 (GH). To my knowledge, no additional collections of any of these species have been made in the past 80 or more years, and they are not treated here.

Hybridization has been documented between species of *Rorippa* in both Europe and North America (B. Jonsell 1968; G. A. Mulligan and A. E. Porsild 1968; R. L. Stuckey 1972; W. Bleeker and H. Hurka 2001). At least in North America, the occurrence of these interspecific hybrids is rather rare.

Rorippa prostrata (Berteret) Schinz & Thellung apparently is an interspecific hybrid involving *R. amphibia* and *R. sylvestris* (R. L. Stuckey 1972); it is known from sporadic collections made in Connecticut, Massachusetts, New Jersey, New York, and Pennsylvania. Most of the specimens that I have examined were collected 60–70 years ago, but occasional ones have been collected recently from Massachusetts; it is not known if these hybrids were introduced or originated in the United States. *Rorippa prostrata* has not become established as a weedy member of our aquatic flora and therefore is not treated here.

The number of ovules per ovary is an important taxonomic character in *Rorippa*, and some closely related species can easily be separated using this feature. It is misleading to give the number of mature seeds per fruit because not all ovules develop into mature seeds, and the difference between ovule and seed number can be substantial. For example, R. L. Stuckey (1972) reported in *R. columbiae* 20–30 seeds per fruit, but the ovule number in this species is 48–64.

SELECTED REFERENCES Bleeker, W. and H. Hurka. 2001. Introgressive hybridization in *Rorippa* (Brassicaceae): Gene flow and its consequences in natural and anthropogenic habitats. Molec. Ecol. 10: 2013–2022. Bleeker, W., C. Weber-Sparenberg, and H. Hurka. 2002. Chloroplast DNA variation and biogeography in the genus *Rorippa* Scop. (Brassicaceae). Pl. Biol. (Stuttgart) 4: 104–111. Rollins, R. C. 1961b. Notes on American *Rorippa* (Cruciferae). Rhodora 63: 1–10. Stuckey, R. L. 1972. Taxonomy and distribution of the genus *Rorippa* (Cruciferae) in North America. Sida 4: 279–430.

1. Fruits silicles, less than 3 times longer than wide.
   2. Plants aquatic (submerged), rooting from proximal nodes; leaves: submerged pectinate, emergent usually not divided, rarely lobed.
      3. Petals yellow; anthers ovate, 0.7–1 mm; styles (0.5–)1–2.2(–2.5) mm; septums complete . . . . 2. *Rorippa amphibia*
      3. Petals white; anthers linear, 1.7–2.2 mm; styles 2–4 mm; septums reduced to rims . . . . 3. *Rorippa aquatica*
   2. Plants terrestrial or of wet habitats, not rooting from nodes; leaves divided or not.
      4. Fruit valves (3 or) 4(–6); septums fenestrate . . . . 5. *Rorippa barbareifolia*
      4. Fruit valves 2; septums complete.
         5. Fruit valves papillate, strigose, or pilose.
            6. Annuals, glabrous; cauline leaves not auriculate; sepals deciduous, 0.7–1.3 mm; petals 0.5–0.8 mm; fruit valves papillate . . . . 21. *Rorippa tenerrima* (in part)
            6. Perennials, hirsute or pilose; cauline leaves sometimes auriculate; sepals persistent, 2–3.5 mm; petals 2.2–4.2 mm; fruit valves strigose or pilose.
               7. Fruit valves strigose; styles glabrous; fruiting pedicels 2–5(–6) mm, glabrous or sparsely hirsute . . . . 6. *Rorippa calycina*
               7. Fruit valves pilose; styles pilose; fruiting pedicels (3–)4–10(–12) mm, densely pilose . . . . 7. *Rorippa columbiae*
         5. Fruit valves glabrous.
            8. Racemes not or slightly elongated (subumbellate) in fruit; sepals persistent; anthers narrowly oblong, 0.8–1.2 mm . . . . 19. *Rorippa subumbellata*
            8. Racemes elongated in fruit; sepals deciduous; anthers ovate, 0.2–0.6 mm.
               9. Fruits globose or subglobose, 1.2–3.2 mm.
                  10. Perennials, with thickened rhizomes; fruiting pedicels 4–15 mm; petals 3–5 mm; styles 1–1.5(–2) mm . . . . 4. *Rorippa austriaca*
                  10. Annuals or biennials, without thickened rhizomes; fruiting pedicels 1.5–3.7(–4.3) mm; petals 0.6–1.2 mm; styles 0.1–0.7(–1) mm . . . . 18. *Rorippa sphaerocarpa*
               9. Fruits lanceolate, oblong, ovoid, ellipsoid, or pyriform, 2–10(–14) mm.
                  11. Perennials (caudex well-developed); cauline leaf blades: bases not auriculate, not amplexicaul . . . . 1. *Rorippa alpina* (in part)
                  11. Annuals or, rarely, short-lived perennials (without caudex); cauline leaf blades: bases sometimes auriculate or amplexicaul.
                     12. Stems ascending, decumbent, or prostrate, few to several from base; sepals 0.8–1.8 mm; petals 0.5–1.8 mm . . . . . 9. *Rorippa curvipes* (in part)
                     12. Stems erect, often simple from base; sepals 1.5–2.4(–2.6) mm; petals (1.5–)1.8–2.5(–3) mm . . . . 14. *Rorippa palustris* (in part)
1. Fruits siliques, greater than 3 times longer than wide.
   13. Sepals 4–5 mm; petals 6–8 mm . . . . 8. *Rorippa crystallina*
   13. Sepals 0.7–3(–4.5) mm; petals (0–)0.5–5.5(–6) mm.
      14. Petals absent.
         15. Fruits linear, (15–)25–40 × 0.7–0.9(–1) mm; ovules 70–90 per ovary; fruiting pedicels (2–)3–8(–10) mm . . . . 11. *Rorippa dubia* (in part)
         15. Fruits oblong to oblong-linear, (4–)6–9(–12) × (1.4–)1.8–2.8(–3.5) mm; ovules 158–242 per ovary; fruiting pedicels 0.5–2(–4) mm . . . . 16. *Rorippa sessiliflora*
      14. Petals present (sometimes less than 1 mm).
         16. Fruits oblong to lanceolate, (1.5–)9–10 mm.
            17. Cauline leaf blades: bases not auriculate.
               18. Perennials (with caudex); fruit valves glabrous; petals (1.3–)1.5–2 mm . . . . 1. *Rorippa alpina* (in part)
               18. Annuals; fruit valves papillate; petals 0.5–0.8 mm . . . . . 21. *Rorippa tenerrima* (in part)

17. Cauline (distal) leaf blades: bases auriculate.
  19. Plants perennials, glabrous or pubescent, trichomes hemispherical, vesicular; petals 2.5–5.3(–6) mm; styles (0.8–)1–2.5(–3.5) mm.
    20. Fruiting pedicels straight or, rarely, curved-ascending, 3–5(–6.5) mm; petals 2.5–3.5 × 0.8–1.3 mm .... 15. *Rorippa ramosa*
    20. Fruiting pedicels sigmoid or recurved, 4–12(–14.5) mm; petals (2.7–)3.2–5.3(–6) × 1.5–2.5 mm .... 17. *Rorippa sinuata* (in part)
  19. Plants annuals or, rarely, perennials, glabrous or hirsute, trichomes cylindrical; petals 0.5–2.5(–3) mm; styles 0.1–1(–1.2) mm.
    21. Stems erect, (often simple from base); petals (1.5–)1.8–2.5(–3) mm .... 14. *Rorippa palustris* (in part)
    21. Stems ascending or decumbent to prostrate, (usually few- or several-branched from base); petals 0.5–1.8(–2) mm.
      22. Petals erect; fruits ovoid to pyriform .... 9. *Rorippa curvipes* (in part)
      22. Petals spreading; fruits oblong to linear ...... 10. *Rorippa curvisiliqua* (in part)
[16. Shifted to left margin.—Ed.]
16. Fruits linear or, oblong-linear, (8–)10–40 mm.
  23. Plants pubescent (trichomes vesicular, hemispherical or clavate).
    24. Perennials (rhizomatous); fruiting pedicels sigmoid or recurved, 4–12(–14.5) mm; petals (2.7–)3.2–5.3(–6) mm; ovules (30–)50–82(–98) per ovary ...... 17. *Rorippa sinuata* (in part)
    24. Annuals or biennials (not rhizomatous); fruiting pedicels straight or curved-ascending, (1.5–)2.3–4.7(–5.3) mm; petals 1–2 mm; ovules (100–)150–210 per ovary .... 22. *Rorippa teres*
  23. Plants usually glabrous, rarely pubescent (trichomes not vesicular).
    25. Perennials; cauline (distal) leaf blades pinnatisect; petals 1.5–2.5 mm wide; seeds rarely produced .... 20. *Rorippa sylvestris*
    25. Annuals or biennials; cauline leaf blades not pinnatisect; petals 0.2–1.5 mm wide; seeds usually produced.
      26. Seeds uniseriate; fruits straight, 0.7–0.9(–1) mm wide .... 11. *Rorippa dubia* (in part)
      26. Seeds biseriate, or often nearly so; fruits often curved, 1–2 mm wide.
        27. Petals 0.6–1.8(–2) mm, oblong to oblanceolate, shorter than sepals; median filaments 1–1.7 mm .... 10. *Rorippa curvisiliqua* (in part)
        27. Petals 2.5–4(–4.5) mm, obovate or spatulate, longer than sepals; median filaments 1.5–3 mm.
          28. Cauline (distal) leaves sessile, margins entire or denticulate; ovules (60–)70–110 per ovary; styles (0.5–)1–1.5(–2) mm .... 12. *Rorippa indica*
          28. Cauline (distal) leaves petiolate, margins pinnately lobed; ovules 42–68 per ovary; styles 0.5–1(–1.2) mm .... 13. *Rorippa microtitis*

1. **Rorippa alpina** (S. Watson) Rydberg, Mem. New York Bot. Gard. 1: 176. 1900 (as Roripa) [E]

Nasturtium obtusum Nuttall var. *alpinum* S. Watson, Botany (Fortieth Parallel), 15. 1871; *Radicula alpina* (S. Watson) Greene; *Rorippa curvipes* Greene var. *alpina* (S. Watson) Stuckey; *R. obtusa* (Nuttall) Britton var. *alpina* (S. Watson) Britton

Perennials; (terrestrial or of wet habitat, not submerged, caudex branched or simple, well-developed); usually glabrous. **Stems** usually decumbent to prostrate, rarely erect, much-branched basally and distally, (0.3–)0.4–1.9(–2.6) dm (rarely pilose basally). **Basal leaves** not rosulate; [petiole (0.3–)0.9–2(–2.5) cm]; blade [(0.6–)1–3.2(–4) cm × (2–)3–8(–15) mm], margins usually dentate to pinnatifid, rarely repand. **Cauline leaves** petiolate or subsessile; blade obovate to oblanceolate, smaller distally, base cuneate, attenuate, not auriculate, margins entire, crenate, or repand. **Racemes** elongated. **Fruiting pedicels** ascending to horizontal or slightly reflexed, straight or curved, (2–)3–6(–8) mm. **Flowers:** sepals ascending, oblong, 1–1.7(–2) × 0.6–0.8 mm; petals yellow, spatulate, (1.3–)1.5–2 × 0.5–0.8 mm; median filaments 0.8–1.2 mm; anthers ovate, 0.3–0.4 mm. **Fruits** often siliques, sometimes silicles, straight or slightly curved, oblong to lanceolate or ovoid, 3–7(–8) × 1.4–2.1(–2.7) mm; valves glabrous; ovules 26–42 per ovary; style (0.3–)0.5–1(–1.2) mm. **Seeds** biseriate, yellow-brown, ovoid to subglobose, 0.5–0.7 mm (0.4–0.5 mm diam.), reticulate. $2n = 16$.

Flowering Jun–Sep. Lakeshores, pond margins, streamsides, dried snow ponds, meadows, seep areas; 1400–3800 m; Colo., Idaho, Mont, Nev., Utah, Wyo.

2. **Rorippa amphibia** (Linnaeus) Besser, Enum. Pl., 27. 1821 (as Roripa)

Sisymbrium amphibium Linnaeus, Sp. Pl. 2: 657. 1753; *Armoracia amphibia* (Linnaeus) Petermann; *Brachiolobos amphibius* (Linnaeus) Allioni; *Cochlearia amphibia* (Linnaeus) Ledebour; *Myagrum amphibium* (Linnaeus) Loiseleur; *Nasturtium amphibium* (Linnaeus) W. T. Aiton; *Radicula amphibia* (Linnaeus) Druce

Perennials; (submerged aquatic, rooting from proximal nodes, forming clumps); glabrous or pubescent proximally on young leaves. **Stems** prostrate basally (hollow), branched, (2–)4–12(–13) dm. **Basal leaves** not rosulate; blade margins pinnatifid. **Cauline leaves** (proximal) petiolate, (distal) sessile; blade pectinate or pinnatisect to pinnatifid (especially when submerged), oblanceolate to elliptic, 4–12 cm × 5–30 mm, base auriculate or not, margins entire or denticulate to serrate (emergent ones usually undivided, rarely lobed). **Racemes** considerably elongated. **Fruiting pedicels** horizontal to reflexed, straight or recurved, (5–)7–15(–18) mm. **Flowers**: sepals erect, oblong, 2.5–3.5 × 1–1.7 mm; petals yellow, obovate, (3.3–)3.8–5.5(–6.2) × 1.5–2.5 mm; median filaments 3–4 mm; anthers ovate, 0.7–1 mm. **Fruits** silicles, straight, ovoid-oblong, (2.5–)3–5.5(–7) × 1.7–2.7(–3) mm; valves glabrous; ovules 36–62 per ovary; style (0.5–)1–2.2(–2.5) mm. **Seeds** biseriate, reddish brown, ovoid, 0.7–1 mm (0.5–0.6 mm diam.), colliculate. $2n = 16, 32$.

Flowering Jun–Aug. Lakeshores, marshes, ponds, streams; introduced; Que.; Maine, N.Y.; Europe; Asia.

3. **Rorippa aquatica** (Eaton) E. J. Palmer & Steyermark, Ann. Missouri Bot. Gard. 22: 550. 1935 (as Roripa) E

Cochlearia armoracia Linnaeus var. *aquatica* Eaton, Man. Bot. ed. 3, 243. 1822; *Armoracia lacustris* (A. Gray) Al-Shehbaz & V. M. Bates; *C. aquatica* (Eaton) Eaton; *Nasturtium lacustre* A. Gray; *N. natans* de Candolle var. *americanum* A. Gray; *Neobeckia aquatica* (Eaton) Greene; *Radicula aquatica* (Eaton) B. L. Robinson; *Rorippa americana* (A. Gray) Britton

Perennials; (submerged aquatic with emergent flowering branches, rhizomatous, rooting from proximal nodes); glabrous throughout. **Stems** erect, unbranched or often branched distally, 3–8.5(–11) dm. **Basal leaves** absent. **Cauline leaves**: submerged shortly petiolate, emergent sessile or petiolate (to 2 mm); blade: submerged pectinate, finely 1–4-pinnatisect into filiform or capillary segments, emergent lanceolate to oblong, usually undivided, rarely lobed, (1.5–)2–5.5(–6.7) cm × (5–)7–15(–20) mm, margins entire or dentate. **Racemes** elongated. **Fruiting pedicels** divaricate to horizontal or slightly reflexed, straight or curved, (5–)7–15 mm. **Flowers**: sepals (deciduous after anthesis), ascending, oblong, 2–4 × 1.4–1.8 mm; petals white, spatulate to obovate, 4–8 × 2–3.5 mm; median filaments 3–4 mm; anthers linear, 1.7–2.2 mm. **Fruits** silicles, straight, oblong to ellipsoid, 4–7 × 2.5–3 mm; (septum reduced to a rim); ovules 48–80 per ovary; style 2–4 mm; (stigma slightly 2-lobed). **Seeds** biseriate, brown to reddish, ovoid, 0.7–0.8 mm (0.5–0.6 mm diam.), reticulate. $2n = 24$.

Flowering Apr–Aug. Springs, lakes, ditches, streams, open sloughs, swamps; 0–200 m; Ont., Que.; Ala., Ark., D.C., Fla., Ga., Ill., Ind., Iowa, Ky., La., Mass., Mich., Minn., Miss., Mo., N.J., N.Y., Ohio, Okla., Tenn., Tex., Vt., Va., Wis.

Submerged leaves of *Rorippa aquatica* often detach readily and produce adventitious buds from which new plantlets develop. The infrequent production of seeds in this species may indicate that it is self-incompatible and that most or all plants within a given population might be the result of asexual reproduction. It appears to prefer slow, unpolluted, running water. The apparent wide distribution does not reflect how uncommon the species is in any given area. Based on the relatively small number of recent collections in major herbaria, it is likely that most of the county records given by I. A. Al-Shehbaz and V. M. Bates (1987) reflect populations that have disappeared.

4. **Rorippa austriaca** (Crantz) Besser, Enum. Pl., 103. 1821 (as Roripa)

Nasturtium austriacum Crantz, Stirp. Aust. Fasc. 1: 15. 1762; *Camelina austriaca* (Crantz) Persoon; *Cochlearia austriaca* (Crantz) Ledebour; *Myagrum austriacum* (Crantz) Jacquin

Perennials; (terrestrial or of wet habitat, not submerged, rhizomes thickened, short); usually glabrous, rarely pubescent proximally. **Stems** (simple from base), erect, much-branched distally, 4–11(–18) dm. **Basal leaves** not rosulate; blade margins pinnatifid. **Cauline leaves** sessile; blade lanceolate, (2.5–)4–12(–15) cm × 5–20(–25) mm, base auriculate to amplexicaul, margins entire or serrate. **Racemes** elongated. **Fruiting pedicels** divaricate-ascending to horizontal, straight, 4–15 mm. **Flowers**: sepals ascending, oblong, 2–3

× 1–1.3 mm; petals yellow, obovate, 3–5 × 1.7–2.5 mm; median filaments 2.3–3 mm; anthers ovate, 0.4–0.6 mm. **Fruits** silicles (rarely produced), straight, globose or subglobose, 2.5–3.2 × 1.5–2.7 mm; ovules 18–40 per ovary; style 1–1.5(–2) mm. **Seeds** biseriate, reddish brown, ovoid, 0.7–0.9 mm, finely colliculate. $2n = 16$.

Flowering May–Jul. Mud flats, floodplains, fields, roadsides, lakeshores, marshes, ditches, stream banks, wet grasslands, waste grounds; 100–1900 m; introduced; Alta., Man., Sask.; Calif., Conn., Idaho, Ill., Iowa, Nebr., Nev., N.J., N.Mex., N.Y., N.Dak., Pa., Utah, Wash., Wis.; Europe.

5. **Rorippa barbareifolia** (de Candolle) Kitagawa, J. Jap. Bot. 13: 137. 1937 F

Camelina barbareifolia de Candolle, Syst. Nat. 2: 517. 1821; *Radicula barbareifolia* (de Candolle) W. Wight ex P. S. Smith; *Rorippa hispida* (Desvaux) Britton var. *barbareifolia* (de Candolle) Hultén; *R. islandica* (Oeder ex Murray) Borbás var. *barbareifolia* (de Candolle) S. L. Welsh; *Tetrapoma barbareifolium* (de Candolle) Turczaninow ex Fischer & C. A. Meyer; *T. pyriforme* Seemann

Annuals; (terrestrial or of wet habitat, not submerged); densely villous or sparsely hirsute at least basally, sometimes glabrate distally. **Stems** (simple from base), erect, branched distally, (2–)3–9.5(–11) dm. **Basal leaves** rosulate; (petiole 1–7 cm); blade margins lyrate-pinnatifid or subruncinate. **Cauline leaves** sessile; blade lanceolate to oblanceolate or oblong, 2.5–10(–15) cm × 4–25(–45) mm, base auriculate or amplexicaul, margins: proximal lyrate-pinnatifid, (lobes 2–7 on each side), laciniate, irregularly serrate, repand, or entire, distal undivided and entire or obscurely denticulate, (apex acute). **Racemes** considerably elongated. **Fruiting pedicels** ascending, straight, (2–)4–12(–14) mm, (glabrous or hirsute). **Flowers:** sepals spreading, oblong, 1.6–2.8 × 0.6–1.2 mm; petals yellow, obovate or spatulate, (1.5–)1.8–3(–3.5) × 0.7–1.8(–2) mm; median filaments 1.5–2.5 mm; anthers oblong, 0.5–0.6 mm, (gynophore 0.3–0.8(–1) mm). **Fruits** silicles, straight, globose or subglobose, (2.5–)3.5–6(–6.5) × (2.3–)2.8–4(–4.3) mm; valves [(3 or) 4(–6), leathery, not veined], glabrous; (septum fenestrate at middle); ovules 60–85 per ovary; style (stout), 0.5–1(–1.4) mm. **Seeds** dark reddish brown, oblong-ovate, 0.5–0.7 mm (0.3–0.4 mm diam.), reticulate. $2n = 16$.

Flowering Jun–Aug. Forest borders, roadsides, waste grounds, moist areas, stream banks, gravel pits; 100–700 m; Sask., Yukon; Alaska; e Asia.

Rorippa barbareifolia is readily distinguished from other species of the genus by having fruits consistently with more than two valves. Other species (e.g., *R. calycina*, *R. palustris*) occasionally show three-valved fruits, but these always appear with more, normal, two-valved fruits on the same plant.

6. **Rorippa calycina** (Engelmann) Rydberg, Mem. New York Bot. Gard. 1: 175. 1900 (as Roripa) E

Nasturtium calycinum Engelmann ex Hayden in G. K. Warren, Prelim. Rep. Expl. Nebraska Dakota, 156. 1859; *N. sinuatum* Nuttall var. *calycinum* (Engelmann) S. Watson; *N. sinuatum* var. *pubescens* S. Watson; *Radicula calycina* (Engelmann) Greene; *Rorippa sinuata* (Nuttall) Hitchcock var. *pubescens* (S. Watson) Howell

Perennials; (terrestrial or of wet habitat, not submerged); densely hirsute throughout, (trichomes pointed, expanded basally). **Stems** (simple or few from base), erect to prostrate, branched distally, 1–4 dm. **Basal leaves** not rosulate; blade margins sinuate. **Cauline leaves** sessile; blade oblong to oblanceolate (lateral lobes oblong to ovate), 2.5–5.5(–7) cm × 5–13 mm, base auriculate to amplexicaul, margins sinuate. **Racemes** elongated. **Fruiting pedicels** ascending to suberect, (somewhat appressed to rachis), straight or curved, 2–5(–6) mm (glabrous or sparsely hirsute). **Flowers:** sepals (persistent after anthesis), ascending, oblong-ovate, 2–3.3 × 0.7–1.3 mm; petals yellow, oblanceolate, (2.2–)2.5–3.7 × 0.7–1.4 mm; median filaments 1.8–2.2 mm; anthers oblong, 0.5–0.6 mm. **Fruits** silicles, straight, globose or broadly oblong, 2–4 × 1.5–2.5 mm; valves short-strigose; ovules 30–44 per ovary; style 1–2.5 mm. **Seeds** biseriate, yellowish, ovoid, 0.6–0.7 mm (0.4–0.6 mm diam.), colliculate. $2n = 16$.

Flowering May–Aug. Stream edges, sandy riverbanks, reservoir, pond, and lake margins; 1000–2000 m; N.W.T.; Mont., N.Dak., Wyo.

G. A. Mulligan and A. E. Porsild (1966) reported *Rorippa calycina* from Northwest Territories and suggested that it may have been introduced to Canada from the main species range in the United States about 4000 km south.

7. **Rorippa columbiae** (S. Watson) Howell, Fl. N.W. Amer., 40. 1897 (as Roripa) [E]

Nasturtium sinuatum Nuttall var. *columbiae* S. Watson in A. Gray et al., Syn. Fl. N. Amer. 1(1,1): 147. 1895; *N. columbiae* (S. Watson) Suksdorf; *N. sinuatum* var. *pubescens* S. Watson; *Radicula columbiae* (S. Watson) Greene; *Rorippa calycina* (Engelmann) Rydberg var. *columbiae* (S. Watson) Rollins; *R. sinuata* (Nuttall) Hitchcock var. *pubescens* (S. Watson) Howell

Perennials; (terrestrial or of wet habitat, not submerged, with creeping roots and adventitious stems); pilose or hirsute. **Stems** suberect or decumbent to prostrate, branched distally, 1–3.2(–4.1) dm, (pilose to hirsute). **Basal leaves** absent. **Cauline leaves** shortly petiolate or sessile; blade oblanceolate to oblong, (lateral lobes oblong to ovate, often reaching midrib), 2.4–5.2 cm × 5–12 mm, base auriculate or not, margins sinuate to pinnatifid or (lateral lobes) entire or dentate, (surfaces pilose). **Racemes** elongated. **Fruiting pedicels** ascending, (subappressed to rachis), straight, (3–)4–10(–12) mm, (densely pilose). **Flowers:** sepals (persistent), ascending, oblong, 2–3.5 × 0.8–1.5 mm, (pilose); petals yellow, oblanceolate to spatulate, 2.7–4.2 × 0.7–1.7 mm; median filaments 2–3.5 mm; anthers ovate, ca. 0.8 mm. **Fruits** silicles, straight, subglobose to oblong ellipsoid, (1.5–)2.5–5.5(–7) × (1–)1.7–2.8(–3.5) mm; valves densely pilose; ovules 24–40 per ovary; style 0.7–3.2 mm, (pilose). **Seeds** biseriate, yellowish brown, ovoid-globose, 0.7–0.9 mm, prominently colliculate.

Flowering Jun–Aug. Stream banks, ditches, margins of lakes and ponds, meadows, roadsides, gravel bars, wet fields; 100–1600 m; Calif., Oreg., Wash.

8. **Rorippa crystallina** Rollins, Rhodora 64: 326, plate 1271. 1962 [E]

Nasturtium crystallinum (Rollins) G. A. Mulligan

Perennials; (rhizomatous); glabrous throughout. **Stems** (simple from base), erect, branched distally, 1–4 dm. **Basal leaves** rosulate; [petiole (1.5–)3–7.7(–12) cm]; blade [(2.5–)4–14(–22) cm × (10–)20–40(–70) mm], margins dentate to crenate. **Cauline leaves** petiolate or subsessile; blade ovate to lanceolate, 2.5–7 cm × 4–13 mm (smaller distally), base cuneate, not auriculate, margins dentate. **Racemes** elongated. **Fruiting pedicels** divaricate to horizontal, straight, 12–20(–26) mm. **Flowers:** sepals (deciduous after anthesis), erect, oblong, 4–5 × 1–1.5 mm; petals whitish, oblanceolate, 6–8 × 2.5–3 mm; median filaments 3–4.2 mm; anthers oblong, 0.8–1.2 mm. **Fruits** siliques, straight or slightly curved, linear, 14–26 × 2–2.8 mm; ovules 28–40 per ovary; style 0.3–1 mm. **Seeds** uniseriate, brown, ovoid, 1.6–2 mm (1.2–1.5 mm diam.), colliculate. $2n = 32$.

Flowering Jun–Jul. Meadows, marshes, peaty soils, ditches; N.W.T.

In his original description of *Rorippa crystallina*, Rollins considered the presence of calcium oxalate crystals in the plant to be a unique feature, but this was later found to be an artifact of treating the plants with formaldehyde (R. L. Stuckey 1972; R. C. Rollins 1993). Characterization of the species as native to Canada was questioned by G. A. Mulligan and W. L. Cody (1995), who believed that it was probably introduced from China. There is no species of mustard from elsewhere in the world that closely resembles *R. crystallina*, and it should be considered a Canadian endemic.

9. **Rorippa curvipes** Greene, Pittonia 3: 97. 1896 (as Roripa)

Cardamine palustris (Linnaeus) Kuntze var. *jonesii* Kuntze; *Radicula curvipes* (Greene) Greene; *R. integra* (Rydberg) A. Heller; *R. sinuata* (Nuttall) Greene var. *integra* Jepson; *R. sinuata* var. *truncata* Jepson; *R. underwoodii* (Rydberg) A. Heller; *Rorippa curvipes* var. *integra* (Rydberg) Stuckey; *R. curvipes* var. *truncata* (Jepson) Rollins; *R. integra* Rydberg; *R. obtusa* (Nuttall) Britton var. *integra* (Rydberg) Victorin; *R. truncata* (Jepson) Stuckey; *R. underwoodii* Rydberg

Annuals or, rarely, perennials; (short-lived, terrestrial or of wet habitat, not submerged); glabrous or hirsute, (trichomes cylindrical). **Stems** (few to several from base), usually ascending, decumbent, or prostrate, rarely erect, branched distally, 1–4.2(–5) dm, (hirsute proximally). **Basal leaves** not rosulate; blade margins pinnatifid. **Cauline leaves** shortly petiolate or sessile; blade oblong or oblanceolate to obovate, (terminal lobe oblong), (2–)3.5–10(–12) cm × (5–)10–30(–37) mm, (lateral lobe smaller than terminal), base usually auriculate, rarely amplexicaul, margins: proximal pinnatifid or sinuate, distal dentate or entire, (surfaces sparsely pubescent). **Racemes** elongated. **Fruiting pedicels** divaricate-ascending to horizontal, straight or recurved, (1.2–)1.7–5(–8) mm. **Flowers:** sepals erect, oblong, 0.8–1.8 × 0.5–1 mm; petals (erect), yellow, oblanceolate to spatulate, 0.5–1.8 × 0.2–1 mm; median filaments 0.9–1.3 mm; anthers ovate, 0.3–0.5 mm. **Fruits** silicles or siliques, curved, ovoid to pyriform, 2–8(–8.8) × (0.5–)1–2.5 mm; valves glabrous; ovules (20–)30–80 per ovary; style 0.3–1 mm. **Seeds** biseriate, brown, cordiform, 0.5–0.7 mm, colliculate. $2n = 16$.

Flowering May–Sep. Muddy shores of lakes and ponds, stream beds and banks, edges of cultivated fields, wet roadside, meadows, seepage areas, ditches, creeks, gravel bars; 100–3500 m; Alta., B.C., Sask.; Ariz., Ark., Calif., Colo., Idaho, Ill., Kans., Mich., Minn., Mo., Mont., Nebr., Nev., N.Mex., N.Dak., Oreg., S.Dak., Tex., Utah, Wash., Wyo.; Mexico (Coahuila).

10. **Rorippa curvisiliqua** (Hooker) Bessey ex Britton, Mem. Torrey Bot. Club 5: 169. 1894 E W

Sisymbrium curvisiliqua Hooker, Fl. Bor.-Amer. 1: 61. 1830; *Nasturtium curvisiliqua* (Hooker) Nuttall; *N. curvisiliqua* var. *lyratum* (Nuttall) S. Watson; *N. curvisiliqua* var. *nuttallii* S. Watson; *N. lyratum* Nuttall; *N. occidentale* Greene; *N. polymorphum* Nuttall; *Radicula curvisiliqua* (Hooker) Greene; *R. lyrata* (Nuttall) Greene; *R. multicaulis* (Greene) Greene; *R. nuttallii* (S. Watson) Greene; *R. occidentalis* (Greene) Greene; *R. pectinata* (A. Nelson) A. Heller; *R. polymorpha* (Nuttall) Greene; *Rorippa curvisiliqua* var. *lyrata* (Nuttall) C. L. Hitchcock; *R. curvisiliqua* var. *nuttallii* (S. Watson) Stuckey; *R. curvisiliqua* var. *occidentalis* (Greene) Stuckey; *R. curvisiliqua* var. *orientalis* Stuckey; *R. curvisiliqua* var. *procumbens* Stuckey; *R. curvisiliqua* var. *spatulata* Stuckey; *R. lyrata* (Nuttall) Greene; *R. multicaulis* Greene; *R. nuttallii* (S. Watson) Rydberg; *R. occidentalis* (Greene) Greene; *R. pectinata* A. Nelson; *R. polymorpha* (Nuttall) Howell

Annuals; usually glabrous, sometimes sparsely hirsute, (trichomes cylindrical). **Stems** (usually few-branched from base, rarely simple), ascending or decumbent to prostrate, (0.5–)1–4(–6) dm, branched distally, (glabrous or hirsute proximally). **Basal leaves** not rosulate; blade margins pinnatifid. **Cauline leaves** petiolate or sessile; blade oblong, oblanceolate to spatulate, or obovate, (lateral lobes linear to oblong or ovate), (2–)3–9(–13) cm × 8–20(–35) mm, (lateral lobe smaller than terminal), base auriculate, margins usually pinnatifid to pinnatisect, rarely pectinate, or (terminal lobe) entire or dentate. **Racemes** considerably elongated. **Fruiting pedicels** divaricate-ascending to horizontal, straight, 1–4.5(–9) mm, (glabrous or sparsely pubescent). **Flowers:** sepals (rarely persistent), ascending, oblong, 0.8–2(–2.5) × 0.6–1.4 mm; petals (spreading), yellow, oblong to oblanceolate, 0.6–1.8(–2) × 0.3–1.3 mm; median filaments 1–1.7 mm; anthers ovate, 0.4–0.5 mm. **Fruits** siliques, curved-ascending, oblong to linear, 4–13(–18) × 1–2 mm; valves glabrous; ovules (30–)42–106 per ovary; style 0.1–0.8 mm. **Seeds** biseriate, brown, cordiform, 0.5–0.7 mm, colliculate. $2n = 16$.

Flowering May–Oct. Shores of lakes, ponds, and reservoirs, edges of pools and sloughs, swales, marshy grounds, mud flats, streamsides, sandy banks, wet roadsides, meadows, seepage areas; 50–3100 m; B.C.; Alaska, Calif., Idaho, Mont., Nev., Oreg., Wash., Wyo.

Rorippa curvisiliqua is a highly variable species divided artificially by R. L. Stuckey (1972) into seven varieties. They were only reluctantly recognized by R. C. Rollins (1993) and N. H. Holmgren (2005b), though these authors felt, and I concur, that it is impossible to determine any of them reliably. A collection from New Brunswick, *Blaney s.n.* (DAO, MO, NBM, UNB), Northumberland County, 2 Sep 2004, was most likely introduced by migratory birds.

11. **Rorippa dubia** (Persoon) H. Hara, J. Jap. Bot. 30: 196. 1955 I

Sisymbrium dubium Persoon, Syn. Pl. 2: 199. 1806, based on *S. apetalum* Desfontaines, Tabl. École Bot., 130. 1804, not Loureiro 1790; *Nasturtium heterophyllum* Blume; *Rorippa heterophylla* (Blume) R. O. Williams; *R. indica* (Linnaeus) Hiern var. *apetala* (de Candolle) Hochreutiner

Annuals; usually glabrous, rarely sparsely pubescent. **Stems** (simple or branched from base), erect or ascending, branched proximally and distally, (0.4–)1.5–3.3(–4.5) dm, (glabrous or sparsely pubescent proximally). **Basal leaves** not rosulate; blade margins pinnatifid. **Cauline leaves** petiolate (to 4 cm) or (distal) sessile; blade lyrate-pinnatipartite or undivided, obovate, oblong, or lanceolate, (2–)3–11(–15) cm × (5–)10–30(–50) mm, (lateral lobes smaller than terminal), base auriculate or not, margins entire, irregularly crenate, or serrate (entire or serrulate distally). **Racemes** elongated. **Fruiting pedicels** divaricate, straight, (2–)3–8(–10) mm. **Flowers:** sepals erect, oblong-linear, (2–)2.5–3 × 0.5–0.7 mm; petals (usually absent), yellow, linear or narrowly oblanceolate, 1–1.5 × 0.2–0.7 mm, (shorter than sepals); median filaments 1.5–2.8 mm; anthers oblong, 0.5–0.8 mm. **Fruits** siliques, straight, linear, (15–)25–40 × 0.7–0.9(–1) mm; ovules 70–90 per ovary; style 0.2–1(–1.5) mm, (stout, as thick as fruit). **Seeds** uniseriate, reddish brown, ovate-orbicular, 0.5–0.8 mm (0.4–0.6 mm diam.), foveolate. $2n = 32, 48$.

Flowering often year-round. Waste areas, slopes, roadsides, wet grounds, grassy places, field margins; 0–500[–3700] m; introduced; La., Miss., Oreg.; se Asia; introduced also in Central America, South America.

Rorippa dubia was treated by R. L. Stuckey (1972) as *R. heterophylla* and by R. C. Rollins (1993) as *R. indica* var. *apetala*. Both authors overlooked H. Hara's (1955b) comprehensive account of its typification and that of *R. indica*.

12. Rorippa indica (Linnaeus) Hiern, Cat. Afr. Pl. 1: 24. 1896 [I]

Sisymbrium indicum Linnaeus, Sp. Pl. ed. 2, 2: 917. 1763; *Radicula indica* (Linnaeus) J. M. Macoun

Annuals; usually glabrous, rarely sparsely pubescent. Stems erect, unbranched or branched proximally or distally, (0.6–)2–6(–7.5) dm. Basal leaves not rosulate; blade margins pinnatifid. Cauline leaves petiolate (to 4 cm) or (distal) sessile; blade lyrate-pinnatipartite or undivided, obovate, oblong, or lanceolate, (lobes 0 or 1–5 (or 6) on each side), (2.5–)3.5–12(–16) cm × (8–)15–40(–50) mm, base auriculate or not, margins entire, irregularly crenate, or serrate, (entire or denticulate distally). Racemes elongated. Fruiting pedicels usually ascending or divaricate, rarely slightly reflexed, straight, (2–)3–10(–15) mm. Flowers: sepals ascending, oblong-ovate, 2–3 × 0.8–1.5 mm; petals yellow, obovate or spatulate, (2.5–)3–4(–4.5) × 1–1.5 mm; median filaments 1.5–3 mm; anthers oblong, 0.5–0.8 mm. Fruits siliques, often curved-ascending, linear, (7–)10–24(–30) × 1–1.5(–2) mm; ovules (60–)70–110 per ovary; style (0.5–)1–1.5(–2) mm, (slender, narrower than fruit). Seeds biseriate or nearly so, reddish brown, ovate or ovate-orbicular, 0.5–0.9 mm (0.4–0.6 mm diam.), foveolate. $2n = 16, 24, 32, 48$.

Flowering most of the year. Roadsides, wet places, field margins, gardens, streamsides, ditches, flood plains, waste grounds; 0–200[–3200] m; introduced; B.C.; La., Miss., N.Y., Oreg.; Asia; introduced also in Central America, South America.

Nasturtium indicum (Linnaeus) de Candolle 1821, not Garsault 1764 is a combination and a later homonym of *Rorippa indica*.

13. Rorippa microtitis (B. L. Robinson) Rollins, Rhodora 59: 71. 1957

Sisymbrium microtitis B. L. Robinson, Bot. Gaz. 30: 59. 1900; *Nasturtium microtitis* (B. L. Robinson) O. E. Schulz

Annuals or, rarely, biennials; usually glabrous throughout, rarely sparsely pubescent proximally. Stems (simple or few from base), erect, branched distally, 1.5–6 dm. Basal leaves often not rosulate; blade margins pinnatifid. Cauline leaves petiolate; blade pinnatifid to pinnatisect or pectinate, (lobes 4–10 on each side, linear or oblong to ovate), 4–13 cm × 9–35 mm; base auriculate, margins entire or dentate, or (distally) pinnately lobed. Racemes (subumbellate), considerably elongated. Fruiting pedicels divaricate-ascending to horizontal, straight, 3–11 mm. Flowers: sepals ascending to spreading, oblong, 2–3 × 0.7–1 mm; petals (spreading), yellow, spatulate, 2.5–4 × 0.7–1.2 mm; median filaments 1.8–3 mm; anthers oblong, 0.5–0.8 mm. Fruits siliques, usually curved inward, rarely straight, linear, (8–)10–17(–20) × 1–2 mm; ovules 42–68 per ovary; style 0.5–1(–1.2) mm. Seeds biseriate, reddish brown, cordiform, 0.6–0.9 mm (0.5–0.6 mm diam.), reticulate-foveolate.

Flowering Jun–Aug. Wet pastures, moist fields, ponds, ditches, meadows; 1700–2300 m; Ariz., N.Mex.; Mexico (Chihuahua).

14. Rorippa palustris (Linnaeus) Besser, Enum. Pl., 27. 1821 (as Roripa) [W]

Sisymbrium amphibium Linnaeus var. *palustre* Linnaeus, Sp. Pl. 2: 657. 1753; *Brachiolobos palustris* (Linnaeus) Clairville; *Cardamine palustris* (Linnaeus) Kuntze; *Caroli-Gmelina palustris* (Linnaeus) P. Gaertner, B. Meyer & Scherbius; *Myagrum palustre* (Linnaeus) Lamarck; *Nasturtium palustre* (Linnaeus) de Candolle; *Radicula palustris* (Linnaeus) Moench

Annuals or, rarely, perennials; (short-lived, terrestrial or of wet habitat, not submerged); usually glabrous, rarely hirsute, (trichomes cylindrical). Stems (often simple from base), erect, branched distally, (0.5–)1–10 (–14) dm, (sometimes hirsute proximally). Basal leaves rosulate; blade [(4–)6–20(–30) cm × 10–50(–80) mm] margins lyrate-pinnatisect, (abaxial surface sometimes hirsute). Cauline leaves petiolate or subsessile; blade lyrate-pinnatisect, (lateral lobes oblong or ovate when present), (1.5–)2.5–10(–18) cm × (5–)8–25(–30) mm, (lateral lobes smaller than terminal), base auriculate or amplexicaul, margins subentire or irregularly dentate, sinuate, serrate, or crenate, (abaxial surface sometimes hirsute). Racemes often considerably elongated. Fruiting pedicels divaricate or slightly to strongly reflexed, straight or curved, (2.5–)3–10(–14) mm. Flowers: sepals erect, oblong, 1.5–2.4(–2.6) × 0.5–1 mm; petals yellow or pale yellow, spatulate, (1.5–)1.8–2.5(–3) × 0.5–1.5(–2) mm; median filaments 1–2.5 mm; anthers ovate, 0.3–0.5 mm. Fruits usually silicles, rarely siliques, often slightly curved, oblong, ellipsoid, or oblong-ovoid, (2.5–)4–10 × (1.5–)1.7–3(–3.5) mm; ovules 20–90 per ovary; style 0.2–1(–1.2) mm. Seeds biseriate, brown to yellowish brown, ovoid or subglobose, 0.5–0.7(–0.9) mm (0.4–0.6(–0.7) mm diam.), colliculate.

Subspecies 2 (2 in the flora): North America, Europe, Asia; introduced in n Mexico, South America, Australia.

Rorippa palustris is a highly variable species with controversial infraspecific taxonomy. B. Jonsell (1968) recognized four subspecies, of which one (subsp. *palustris*) is cosmopolitan and three are North American. R. L. Stuckey (1972) followed Jonsell but further divided the North American plants into eleven, poorly defined varieties. Division of the species into subspecies based solely on stem height and fruit length is artificial. The variation is continuous in every character, and the recognized infraspecific taxa represent only some of the extremes. Some collections cannot be adequately assigned to a given subspecies or variety, and of all the infraspecific taxa recognized, only two can be consistently separated from each other; they are recognized here as subspecies.

1. Stems and abaxial leaf blade surfaces usually glabrous, rarely sparsely pubescent proximally . . . . 14a. *Rorippa palustris* subsp. *palustris*
1. Stems and abaxial leaf blades surfaces often densely hirsute . . . 14b. *Rorippa palustris* subsp. *hispida*

14a. Rorippa palustris (Linnaeus) Besser subsp. palustris

Nasturtium terrestre (Withering) W. T. Aiton var. *occidentale* S. Watson; *Radicula pacifica* (Howell) Greene; *Rorippa hispida* (Desvaux) Britton var. *glabrata* Lunell; *R. islandica* (Oeder ex Murray) Borbás subsp. *fernaldiana* (Butters & Abbe) Hultén; *R. islandica* var. *fernaldiana* Butters & Abbe; *R. islandica* var. *glabra* (O. E. Schulz) S. L. Welsh & Reveal; *R. islandica* var. *glabrata* (Lunell) Butters & Abbe; *R. islandica* var. *occidentalis* (S. Watson) Butters & Abbe; *R. pacifica* Howell; *R. palustris* subsp. *fernaldiana* (Butters & Abbe) Jonsell; *R. palustris* var. *fernaldiana* (Butters & Abbe) Stuckey; *R. palustris* subsp. *glabra* (O. E. Schulz) Stuckey; *R. palustris* var. *glabra* (O. E. Schulz) Roy L. Taylor & MacBryde; *R. palustris* var. *glabrata* (Lunell) Victorin; *R. palustris* subsp. *occidentalis* (S. Watson) Abrams; *R. palustris* var. *occidentalis* (S. Watson) Rollins; *R. palustris* var. *pacifica* (Howell) G. N. Jones; *R. terrerstris* (Withering) Fuss var. *globosa* A. Nelson

Stems usually glabrous, rarely sparsely pubescent proximally. **Leaf blades:** abaxial surface usually glabrous, rarely sparsely pubescent proximally. $2n = 32$.

Flowering Mar–Sep. Marshlands, pastures, prairies, meadows, swales, flats, sand bars, wet grounds, stream banks, moist depressions, ditches, estuaries, waste grounds, roadsides, sloughs, shores of lakes and ponds, bogs, thickets, grasslands; 0–3200[–4000] m; Alta., B.C., Man., Nfld. and Labr., N.W.T., N.S., Ont., P.E.I., Que., Sask., Yukon; Ala., Alaska, Ariz., Ark., Calif., Colo., Conn., Del., D.C., Fla., Ga., Idaho, Ill., Ind., Iowa, Kans., Ky., La., Maine, Md., Mass., Mich., Minn., Miss., Mo., Mont., Nebr., Nev., N.H., N.J., N.Mex., N.Y., N.C., N.Dak., Ohio, Okla., Oreg., Pa., R.I., S.C., S.Dak., Tenn., Tex., Utah, Vt., Va., Wash., W.Va., Wis., Wyo.; Europe; Asia; introduced in n Mexico, South America, Australia.

14b. Rorippa palustris (Linnaeus) Besser subsp. hispida (Desvaux) Jonsell, Symb. Bot. Upsal. 19(2): 159. 1968 E

Brachiolobos hispidus Desvaux, J. Bot. Agric. 3: 183. 1815 (as Brachilobus); *Cardamine palustris* (Linnaeus) Kuntze var. *hispida* (Desvaux) Fischer & C. A. Meyer ex Kuntze; *Nasturtium hispidum* (Desvaux) de Candolle; *N. palustre* (Linnaeus) de Candolle subsp. *hispidum* (Desvaux) Fischer & C. A. Meyer ex O. E. Schulz; *N. palustre* var. *hispidum* (Desvaux) A. Gray; *N. terrestre* (Withering) W. T. Aiton var. *hispidum* (Desvaux) S. Watson; *Radicula hispida* (Desvaux) Britton; *R. palustris* (Linnaeus) Moench var. *hispida* (Desvaux) B. L. Robinson; *Rorippa hispida* (Desvaux) Britton; *R. islandica* (Oeder ex Murray) Borbás var. *hispida* (Desvaux) Butters & Abbe; *R. palustris* var. *hispida* (Desvaux) Rydberg; *R. terrestris* (Withering) Fuss var. *hispida* (Desvaux) A. Nelson; *Sisymbrium hispidum* (Desvaux) Poiret

Stems often densely hirsute. **Leaf blades:** abaxial surface often densely hirsute.

Flowering Jun–Aug. Margins of ponds and streams, lake shores, gravelly beaches, roadside ditches, mud flats, wet meadows, stream banks, springy ledges, swamps; Alta., B.C., Man., N.B., Nfld. and Labr., N.W.T., Ont., Que., Sask., Yukon; Alaska, Calif., Colo., Conn., Del., Idaho, Ind., Iowa, Maine, Mass., Mich., Minn., Mont., Nebr., Nev., N.H., N.J., N.Mex., N.Y., N.Dak., Ohio, Oreg., Pa., R.I., Utah, Vt., Wash., Wis., Wyo.

15. Rorippa ramosa Rollins, Rhodora 63: 4, figs. A–C. 1961

Perennials; glabrous or sparsely pubescent, (trichomes hemispherical, vesicular). **Stems** (simple or several from base), prostrate, branched distally, (1–)2–5(–6) dm, (pubescent). **Basal leaves** not rosulate; blade margins pinnatifid. **Cauline leaves** sessile; blade oblong to oblanceolate or broadly lanceolate (lateral lobes oblong to ovate), (2–)3–5 cm × 5–15 mm, base auriculate, margins pinnatifid to deeply sinuate or entire, (surfaces sparsely pubescent abaxially with vesicular trichomes along veins, glabrous adaxially). **Racemes** slightly elongated. **Fruiting pedicels** ascending to horizontal, usually straight, rarely curved-ascending, 3–5(–6.5) mm,

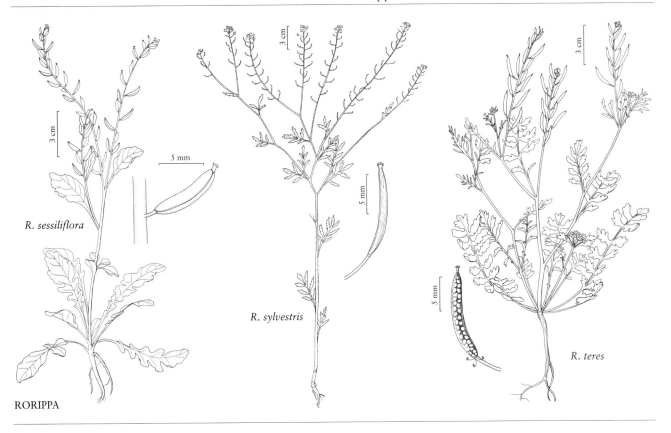

(sparsely pubescent). **Flowers:** sepals erect, oblong, 2–2.5 × 0.8–1.3 mm; petals pale yellow, spatulate, 2.5–3.5 × 0.8–1.3 mm; median filaments 2.7–3 mm; anthers narrowly oblong, 0.7–0.9 mm. **Fruits** siliques, curved, oblong to lanceolate, 6–10 × 2–3 mm; valves glabrous or puberulent; ovules 60–80 per ovary; style 1–2(–2.5) mm. **Seeds** biseriate, light brown, cordiform, 0.7–1.5 mm, colliculate.

Flowering Mar–May. Floodplains, sandy arroyo bottoms, lake shores; Tex.; Mexico (Coahuila, Durango).

Rorippa ramosa is known from the Big Bend area in Brewster County and in adjacent Mexico.

16. Rorippa sessiliflora (Nuttall) Hitchcock, Key Spring Fl. Manhattan, 18. 1894 (as Roripa) E F

Nasturtium sessiliflorum Nuttall in J. Torrey and A. Gray, Fl. N. Amer. 1: 73. 1838; *N. limosum* Nuttall; *Radicula limosa* (Nuttall) Greene; *R. sessiliflora* (Nuttall) Greene

Annuals; glabrous throughout. **Stems** (simple or several-branched from base) erect, branched distally, (0.9–)2–4.5(–6.3) dm. **Basal leaves** not rosulate; blade margins dentate. **Cauline leaves** shortly petiolate or sessile; blade oblong or oblanceolate to obovate, 1.5–7(–13.3) cm × (5–)10–28(–50) mm, (lateral lobes much smaller than terminal), base auriculate or not, margins dentate or less frequently sinuate, or (lateral lobes) denticulate or entire. **Racemes** considerably elongated. **Fruiting pedicels** divaricate-ascending to horizontal, straight, 0.5–2(–4) mm, (slender or stout). **Flowers:** sepals ascending to spreading, ovate, 1.2–2 × 0.5–0.8 mm; petals absent; median filaments 1–1.6 mm; anthers ovate, 0.2–0.3 mm. **Fruits** siliques, straight or curved, oblong to oblong-linear, (4–)6–9(–12) × (1.4–)1.8–2.8(–3.5) mm; ovules 158–242 per ovary; style 0.1–0.5 mm. **Seeds** biseriate, yellow-brown, cordiform, 0.4–0.5 mm, foveolate. $2n = 16$.

Flowering Mar–Oct. Mud flats, ditches, wet old fields, roadsides, sloughs, fallow fields, floodplains, stream banks, edges of pools, waste grounds, gardens; 0–300 m; Ala., Ark., D.C., Fla., Ga., Ill., Ind., Iowa, Kans., Ky., La., Mass., Miss., Mo., Nebr., N.C., Ohio, Okla., S.C., S.Dak., Tenn., Tex., Va., Wis.

Both R. L. Stuckey (1972) and R. C. Rollins (1993) indicated that *Rorippa sessiliflora* has strongly saccate sepals, but all specimens that I examined failed to show any sign of this character. The species is very distinctive and can easily be distinguished by an absence of petals and by having fruiting pedicels 0.5–2(–4) mm.

17. Rorippa sinuata (Nuttall) Hitchcock, Key Spring Fl. Manhattan, 18. 1894 (as Roripa) [E] [W]

Nasturtium sinuatum Nuttall in J. Torrey and A. Gray, Fl. N. Amer. 1: 73. 1838; *N. trachycarpum* A. Gray; *Radicula sinuata* (Nuttall) Greene; *R. trachycarpa* (A. Gray) Rydberg; *Rorippa trachycarpa* (A. Gray) Greene

Perennials; (rhizomatous or with creeping roots); sparsely to moderately pubescent, (trichomes hemispherical, vesicular). **Stems** (many to few from base), decumbent or prostrate, branched distally, 1–4.2(–5) dm, (pubescent proximally). **Basal leaves** not rosulate; blade margins pinnatifid. **Cauline leaves** sessile; blade oblong to oblanceolate or lanceolate (lateral lobes oblong or ovate), (1.5–)2.5–6.5(–9) cm × 5–18(–25) mm, base usually auriculate, rarely proximalmost not auriculate, margins pinnatifid to deeply sinuate, or (lateral lobes) dentate to sinuate or entire, (surfaces pubescent abaxially with vesicular trichomes on veins, glabrous adaxially). **Racemes** considerably elongated. **Fruiting pedicels** divaricate-ascending to horizontal, sigmoid or recurved, 4–12(–14.5) mm. **Flowers:** sepals ascending, oblong, 2.2–3.7(–4.5) × 1–1.8 mm; petals yellow, oblanceolate to spatulate, (2.7–)3.2–5.3(–6) × 1.5–2.5 mm; median filaments 3–5.5 mm; anthers narrowly oblong, 1.2–1.7 mm. **Fruits** siliques, curved, oblong to lanceolate or linear, (4–)5–12(–16) × (1–)1.5–2.5 mm; valves glabrous or pubescent; ovules (30–)50–82(–98) per ovary; style (0.8–)1–2.5(–3.5) mm. **Seeds** biseriate, yellow-brown, angled, cordiform, 0.7–1 mm (0.5–0.6 mm diam.), colliculate. $2n = 16$.

Flowering Apr–Aug. Shores of ponds and lakes, sandy banks, stream banks, fields, wet depressions, gravel banks, ditches, moist grounds; 0–2600 m; Alta., B.C., Ont., Sask.; Ariz., Ark., Calif., Colo., Idaho, Ill., Iowa, Kans., Ky., La., Minn., Mo., Mont., Nebr., Nev., N.Mex., N.Dak., Okla., Oreg., S.Dak., Tenn., Tex., Utah, Wash., Wis., Wyo.

18. Rorippa sphaerocarpa (A. Gray) Britton, Mem. Torrey Bot. Club 5: 170. 1894

Nasturtium sphaerocarpum A. Gray, Mem. Amer. Acad. Arts, n. s. 4: 6. 1849; *Cardamine globosa* (Turczaninow ex Fischer & C. A. Meyer) Kuntze var. *sphaerocarpa* (A. Gray) Kuntze; *Caroli-Gmelina sylvestris* (Linnaeus) P. Gaertner, B. Meyer & Scherbius; *N. obtusum* Nuttall var. *sphaerocarpum* (A. Gray) S. Watson; *Radicula obtusa* (Nuttall) Greene var. *sphaerocarpa* (A. Gray) B. L. Robinson; *R. sphaerocarpa* (A. Gray) Greene; *Rorippa obtusa* (Nuttall) Britton var. *sphaerocarpa* (A. Gray) Cory

Annuals or, rarely, **biennials**; (terrestrial or of wet habitat, not submerged); glabrous or hirsute. **Stems** (simple or few- to many-branched from base), decumbent or erect, branched distally, 1–4(–5.5) dm, (hirsute proximally). **Basal leaves** rosulate; blade margins pinnatifid. **Cauline leaves** shortly petiolate or sessile; blade oblong to oblanceolate, (lateral lobes oblong to ovate), (3.5–)4.5–9(–12) cm × 10–25(–33) mm, (lateral lobes smaller than terminal), base auriculate or not, margins pinnatifid to pinnatisect, or (lateral lobes) crenate to subentire. **Racemes** elongated. **Fruiting pedicels** divaricate to slightly reflexed, straight or recurved, 1.5–3.7(–4.3) mm. **Flowers:** sepals ascending, oblong to ovate, 0.7–1.3 × 0.4–0.7 mm; petals yellow, oblanceolate to spatulate, 0.6–1.2 × 0.2–0.5 mm; median filaments 0.9–1.2 mm; anthers ovate, 0.2–0.3 mm. **Fruits** silicles, straight, globose or subglobose, 1.2–2.5(–3) mm diam.; valves glabrous; ovules 20–42 per ovary; style 0.1–0.7(–1) mm. **Seeds** biseriate, brown, cordiform, 0.5–0.7 mm, finely colliculate.

Flowering May–Aug. Shores of ponds and lakes, mud flats, stream edges, moist grounds; 1200–3300 m; Ariz., Calif., Colo., Idaho, N.Mex., Tex., Utah, Wyo.; Mexico (Chihuahua).

19. Rorippa subumbellata Rollins, Contr. Dudley Herb. 3: 177, plate 46, fig. 2. 1941 [C] [E]

Perennials; (terrestrial or of wet habitat, not submerged, with underground rhizomes); glabrous or pilose, (trichomes crisped). **Stems** decumbent, much-branched distally, 0.5–2.5(–3) dm, (glabrous or pilose proximally). **Basal leaves** not rosulate; blade margins subpinnatifid to sinuate. **Cauline leaves** sessile or shortly petiolate; blade broadly oblanceolate to oblong, (lateral lobes oblong to ovate), 1–3.2 cm × 3–12 mm, base not or minutely auriculate, margins subpinnatifid to sinuate, or (lateral lobes) usually entire, (surfaces pilose or adaxially glabrous). **Racemes** (subumbellate), not or slightly elongated. **Fruiting pedicels** erect to divaricate-ascending, straight, 3–7(–9) mm, (pilose). **Flowers:** sepals (persistent), erect, oblong or ovate, 2–3 × 1–1.5 mm; petals yellow, spatulate to oblanceolate, 2.5–3.5 × 1–1.7 mm; median filaments 2–2.5 mm; anthers narrowly oblong, 0.8–1.2 mm. **Fruits** silicles, straight, subglobose to broadly oblong, 3–6 × 2–3.5 mm; valves glabrous; ovules 30–44 per ovary; style 0.8–1.5 mm. **Seeds** biseriate, yellowish brown, angled, cordiform, 0.8–1.1 mm, strongly colliculate.

Flowering Jun–Sep. Shores of lakes, beaches; of conservation concern; 1800–2000 m; Calif., Nev.

Rorippa subumbellata appears to be restricted to Tallac and Truckee lakes and Lake Tahoe in eastern California (El Dorado and Placer counties) and western Nevada (Douglas County).

The record by Rollins of $n = 5$ for the species (see S. I. Warwick and I. A. Al-Shehbaz 2006) is highly unlikely; no such count is known in the entire tribe Cardamineae, to which *Rorippa* belongs.

Rorippa subumbellata is in the Center for Plant Conservation's National Collection of Endangered Plants.

20. **Rorippa sylvestris** (Linnaeus) Besser, Enum. Pl., 27. 1821 (as Roripa) [F] [I] [W]

Sisymbrium sylvestre Linnaeus, Sp. Pl. 2: 657. 1753; *Brachiolobos sylvestris* (Linnaeus) Allioni; *Nasturtium sylvestre* (Linnaeus) W. T. Aiton; *Radicula sylvestris* (Linnaeus) Druce

Perennials; glabrous or sparsely pubescent. **Stems** prostrate, decumbent, ascending, or suberect, branched mainly basally, (0.5–)1.5–8(–10) dm. **Basal leaves** not rosulate; similar to cauline. **Cauline leaves** petiolate, or (distal) often subsessile; blade deeply pinnatisect, (lobes 3–6 on each side, sublinear, lanceolate, oblong, elliptic, or ovate), (2–)3.5–15(–20) cm × (7–)10–45(–60) mm, base usually not auriculate, rarely minutely auriculate, margins dentate, serrate, subentire, or (distally) pinnatisect, (lobes 1–3 on each side). **Racemes** elongated. **Fruiting pedicels** divaricate, straight, (3–)4–10(–12) mm. **Flowers:** sepals ascending or spreading, oblong, 1.8–3(–3.5) × 0.7–1.5 mm; petals yellow, spatulate or obovate, (2.2–)2.8–5.5(–6) × 1.5–2.5 mm; median filaments (1.5–)1.8–3.5(–4) mm; anthers oblong, 0.7–1 mm. **Fruits** siliques, straight, usually linear, rarely oblong-linear, 10–20(–25) × (0.7–)1–1.3(–1.6) mm; valves glabrous; ovules 24–80 per ovary; style 0.5–1(–1.5) mm. **Seeds** (rarely produced), usually uniseriate, rarely sub-biseriate, reddish brown, ovoid, 0.5–0.9 mm (0.4–0.5 mm diam.), colliculate. $2n = 32, 40, 48$.

Flowering May–Aug. Along ditches, damp areas, shores of ponds and lakes, sandy beaches, waste grounds, ditches, wet roadsides, meadows, washes, fields, gardens; 0–2500 m; introduced; Alta., B.C., N.B., Nfld. and Labr. (Nfld.), Ont., P.E.I., Que.; Ala., Ark., Colo., Conn., Del., D.C., Idaho, Ill., Ind., Iowa, Kans., Ky., La., Maine, Md., Mass., Mich., Minn., Miss., Mo., Mont., N.H., N.J., N.Y., N.C., N.Dak., Ohio, Oreg., Pa., R.I., Tenn., Utah, Vt., Va., Wash., W.Va., Wis.; Europe; sw Asia; introduced also in South America.

21. **Rorippa tenerrima** Greene, Erythea 3: 46. 1895 (as Roripa)

Radicula tenerrima (Greene) Greene

Annuals; (terrestrial or of wet habitat, not submerged); glabrous throughout. **Stems** (several-branched from base), prostrate to decumbent, branched distally, 0.7–3.5 (–4) dm. **Basal leaves** not rosulate; blade margins pinnatifid. **Cauline leaves** shortly petiolate, or (distal) sessile; blade oblong to oblanceolate or lanceolate, (lateral lobes linear, oblong to ovate, or obovate), 2–9(–11) cm × 7–20(–30) mm, (lateral lobes smaller than terminal), base not auriculate, margins pinnatifid, or (lateral lobes) entire, dentate, or sinuate. **Racemes** elongated. **Fruiting pedicels** ascending to divaricate, (1–)1.5–3.2(–4.2) mm, straight. **Flowers:** sepals ascending, oblong, 0.7–1.3 × 0.4–0.7 mm; petals yellow, oblong to oblanceolate or spatulate, 0.5–0.8 × 0.1–0.3 mm; median filaments 0.8–1.2 mm; anthers ovate, 0.1–0.2 mm. **Fruits** siliques or silicles, curved-ascending, lanceolate to narrowly ovoid or oblong-lanceolate, 3–7(–9) × (0.8–)1–1.7(–2) mm (often slightly constricted at middle); valves papillate; ovules 20–80 per ovary; style (0.2–)0.5–1 mm. **Seeds** biseriate, reddish brown, cordiform, 0.5–0.7 mm, colliculate.

Flowering Jun–Oct. Shores of lakes and ponds, mud flats, marshes, sand bars, moist grounds, streamsides; 1300–3000 m; Alta., B.C., Sask.; Calif., Colo., Idaho, Mo., Mont., Nebr., Nev., N.Mex., N.Dak., Oreg., S.Dak., Tex., Utah, Wash., Wyo.; Mexico (Baja California, Chihuahua).

22. **Rorippa teres** (Michaux) Stuckey, Sida 2: 409. 1966 [F] [W]

Cardamine teres Michaux, Fl. Bor.-Amer. 2: 29. 1803; *Erysimum walteri* (Elliott) Eaton; *Nasturtium micropetalum* Fischer & C. A. Meyer; *N. obtusum* Nuttall; *N. palustre* (Linnaeus) de Candolle var. *tanacetifolium* de Candolle; *N. tanacetifolium* (de Candolle) Hooker & Arnott; *N. walteri* (Elliott) Alph. Wood; *Rorippa obtusa* (Nuttall) Britton; *R. teres* var. *rollinsii* Stuckey; *R. walteri* (Elliott) C. Mohr; *Sisymbrium teres* (Michaux) Torrey & A. Gray; *S. walteri* Elliott

Annuals or, rarely, biennials; puberulent, at least proximally, or glabrous, (some trichomes clavate to hemispherical, vesicular). **Stems** (simple or several from base), usually prostrate or decumbent, rarely erect, branched distally, 1–4 dm, (glabrous or pubescent, trichomes vesicular). **Basal leaves** rosulate; blade margins pinnatifid. **Cauline leaves** shortly petiolate; blade oblong, oblanceolate to obovate, or lyrate-pinnatisect, (lateral lobes oblong to ovate), (2–)3.5–10(–13.5) cm × 10–40(–53) mm, base auriculate or not, margins usually pinnatifid to pinnatisect, rarely 2-pinnatifid, (lateral lobes) dentate to crenate or sinuate, (surfaces glabrous or adaxially pubescent, trichomes vesicular). **Racemes** elongated. **Fruiting pedicels** ascending to horizontal, straight or curved-ascending, (1.5–)2.3–4.7(–5.3) mm. **Flowers:** sepals erect, oblong, 1.5–2.5 × 0.5–1 mm; petals yellow, spatulate, 1–2 × 0.4–0.7 mm; median filaments 1.2–1.7 mm; anthers ovate, 0.2–0.3 mm. **Fruits** siliques, straight or curved, linear to oblong-linear, 8–14(–21) × 1–2.5 mm; valves glabrous or pubescent; ovules (100–)150–210 per ovary; style (0.2–)0.5–1.1 mm. **Seeds** biseriate, reddish brown, cordiform, 0.4–0.5 mm, foveolate.

Flowering Dec–May. Wet areas, muddy grounds, edges of canals and ditches, sandy fields, margins of ponds, streamsides, peat; 0–600 m; Ala., Ark., Fla., Ga., La., Miss., N.C., Okla., S.C., Tex.; Mexico; Central America (Honduras, Nicaragua).

46. SELENIA Nuttall, J. Acad. Nat. Sci. Philadelphia 5: 132, plate 6. 1825 • [Greek *selene*, moon, alluding to resemblance to *Lunaria*]

Ihsan A. Al-Shehbaz

Annuals (sometimes winter); not scapose; glabrous. **Stems** (rarely absent, base of plant forming inflated crown), erect, ascending, subdecumbent, or decumbent, unbranched or branched distally. **Leaves** basal and cauline; petiolate or subsessile; basal rosulate or not, petiolate, blade margins 1- or 2- (or 3-)pinnatisect, (terminal lobe margin entire or dentate); cauline sessile [subsessile], blade (base not auriculate), margins similar to basal. **Racemes** (sometimes pedicels originating between basal leaves, bracteate throughout, rachis straight), considerably elongated in fruit. **Fruiting pedicels** divaricate to ascending, slender. **Flowers:** sepals (sometimes persistent), erect, spreading, or ascending, (yellowish), oblong, oblong-linear or -lanceolate, lateral pair slightly saccate basally; petals yellow, spatulate to broadly obovate, claw differentiated from blade, (apex rounded or emarginate [obtuse]); stamens tetradynamous; filaments not dilated basally; anthers linear, ovate, or oblong [sagittate]; nectar glands confluent (extending into spreading teeth). **Fruits** silicles, sessile or shortly stipitate, usually oblong, elliptical, suborbicular, or globose, rarely obovoid, smooth, latiseptate, terete, or slightly inflated; valves (vesicular in *S. grandis*, thin-papery to leathery), each without midvein, obscurely to prominently reticulate-veined; replum rounded or flattened; septum complete or perforated; ovules 8–44 per ovary; style distinct, (1–12 mm, sometimes flattened basally); stigma capitate. **Seeds** biseriate, flattened, winged throughout, orbicular; seed coat (coarsely reticulate), not mucilaginous when wetted; cotyledons accumbent. $x = 7, 12, 13$.

Species 5 (4 in the flora): c, sw United States, ne Mexico.

Species of *Selenia* might be difficult to separate when the plants are only in flower. *Selenia aurea* is easily distinguished by its 1-pinnatisect leaves and absence of flowers from the base of the plant. The remaining three species have 2- or 3-pinnatisect leaves and at least some pedicels from the basal rosette. In *S. jonesii*, the sepal appendage is absent or less than 1 mm; in both *S. dissecta* and *S. grandis*, it is 1–4 mm. The last two can be distinguished in flower by the presence in *S. grandis* of terete styles and ovary margins (replum) and in *S. dissecta* by flattened style bases and winged ovary margins. *Selenia mexicana* Standley is endemic to Nuevo León, Mexico.

SELECTED REFERENCES Beck, J. B. 2009. The phylogeny of *Selenia* (Brassicaceae) inferred from chloroplast and nuclear sequence data. J. Bot. Res. Inst. Texas 3: 169–176. Martin, R. F. 1940. A review of the cruciferous genus *Selenia*. Amer. Midl. Naturalist 23: 455–462.

1. Leaf blade margins 1-pinnatisect; fruiting pedicels from racemes; sepals with unappendaged apex .... 1. *Selenia aurea*
1. Leaf blade margins 2- or 3-pinnatisect; fruiting pedicels (at least some) from basal leaf axils or rosettes; sepals with appendaged apex.
    2. Fruit valves vesicular, leathery; sepals persistent; fruits 5.5–7.5 mm wide .... 3. *Selenia grandis*
    2. Fruit valves not vesicular, papery; sepals caducous or tardily so; fruits 8–17 mm wide.
        3. Fruits oblong to elliptical, latiseptate; replums and styles flattened; petals (12–)15–20 × (5–)6–9 mm; ovules 28–40 per ovary; median filament pairs 6–10 mm; sepal apex appendage (1–)1.5–3 mm .... 2. *Selenia dissecta*
        3. Fruits usually globose, rarely obovoid, terete; replums and styles not flattened; petals 4–7 × 2.5–4 mm; ovules 8–14 per ovary; median filament pairs 3–4 mm; sepal apex appendage 0.5–0.8 mm .... 4. *Selenia jonesii*

1. Selenia aurea Nuttall, J. Acad. Nat. Sci. Philadelphia 5: 132, plate 6. 1825 [E] [F]

Selenia aptera (S. Watson) Small; *S. aurea* var. *aptera* S. Watson

Plants not winter annuals. **Stems** (simple or few to many from base), usually erect to ascending, rarely subdecumbent, (slender), (0.5–)0.8–2.7(–3.5) dm. **Basal leaves** not rosulate; petiole 0.5–1 cm; blade margins 1-pinnatisect, (1–)2.5–7(–10) cm; lobes (3–)6–12(–18) on each side, (smaller than terminal), linear to oblong or ovate, (1–)2–10 × 0.5–1(–2.5) mm, margins entire or coarsely dentate. **Cauline leaves** (and bracts) similar to basal, smaller distally. **Fruiting pedicels** from racemes, (6–)10–22(–30) mm, (slender). **Flowers:** sepals spreading to ascending, oblong-linear, (3.5–)5–7 × 1–1.5 mm, apex appendage not developed; petals spatulate, (8–)10–13 × 3–4.5 mm, apex rounded; median filament pairs 5–7(–8) mm, not dilated basally; anthers oblong, 1–1.5 mm; gynophore (0.5–)1–2(–5) mm, or, rarely, obsolete. **Fruits** usually oblong to elliptical, rarely suborbicular, usually latiseptate, rarely inflated, (0.5–)1–2.5(–3) cm × (3.5–)5–8(–11) mm, (not fleshy, thin-papery), base and apex acute; valves faintly reticulate-veined; replum flattened; septum complete or perforated; ovules (8–)10–20 per ovary; style 3–9(–12) mm, slender or flattened basally. **Seeds** 3–4 mm diam.; wing 0.5–1 mm. $2n = 46, 138$.

Flowering Mar–May. Sandy open grounds, barren rocky sandstone or clay, granite soil, rocky grounds, shale barren, open areas in mixed juniper and oak, chert barrens, pastures, sandstone glades, fields, rocky prairies; 100–400 m; Ark., Kans., Mo., Okla.

Although R. C. Rollins (1993) and R. F. Martin (1940) suggested that *Selenia aurea* probably occurs in northeastern Texas, I have not seen any material from that state.

2. Selenia dissecta Torrey & A. Gray in War Department [U.S.], Pacif. Railr. Rep. 2(2): 160. 1855

Plants winter annuals, (often nearly acaulescent). **Stems** (often inflated into 2.3 cm thick crown), usually ascending, rarely decumbent, 0.8–2.2 dm (when formed). **Basal leaves** rosulate; petiole 0.5–3(–5) cm; blade margins usually 2-, rarely 3-pinnatisect, (2–)3–10(–15) cm; lobes 5–10(–15) on each side, (smaller than terminal); apical segment linear to oblong or ovate, 1–8(–12) × 0.5–1(–2.5) mm, margins entire. **Cauline leaves** (and bracts, when present) similar to basal, smaller distally. **Fruiting pedicels** usually from basal leaf axil, (20–)30–80(–100) mm. **Flowers:** sepals (caducous or tardily so), spreading, oblong, (6–)7–12(–14) × 2–3.5 mm, apex appendage well-developed, (1–)1.5–3 mm; petals broadly spatulate to obovate, (12–)15–20 × (5–)6–9 mm, apex rounded; median filament pairs 6–10 mm, not dilated basally; anthers linear, 2–3 mm; gynophore (1–)1.5–3(–4) mm. **Fruits** oblong to elliptical, latiseptate, 1.4–3.5(–4) cm × (8–)10–17 mm, (slightly fleshy when green, thick, papery), base and apex acute; valves prominently reticulate-veined; replum strongly flattened; septum complete; ovules 28–40 per ovary; style (2–)3.5–6(–7) mm, strongly flattened basally. **Seeds** 5–7 mm diam.; wing 1–2 mm. $2n = 14$.

Flowering Feb–Apr. Grassy banks, pastures, salt draws, gypseous llano, roadsides, sandy alluvium, limestone or sandy areas, creosote bush scrubland, open flats; 600–1900 m; N.Mex., Tex.; Mexico (Chihuahua, Coahuila, Nuevo León).

Selenia dissecta rarely produces racemes, and most flowers originate from the axils of basal leaves that cover an inflated stem reduced to a crown.

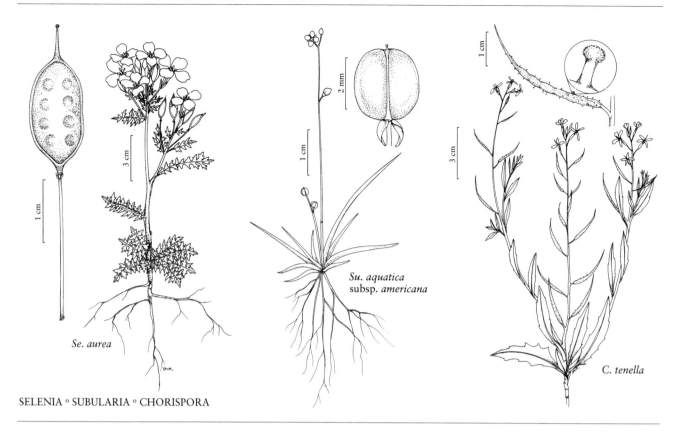

SELENIA ° SUBULARIA ° CHORISPORA

3. **Selenia grandis** R. F. Martin, Rhodora 40: 183. 1938 E

Selenia oinosepala Steyermark

Plants winter annuals. **Stems** erect to ascending, (slender or stout), 1.5–6.5 dm. **Basal leaves** (soon withered), early rosulate; petiole 1–4 cm; blade margins 2- or 3-pinnatisect, 4–15 cm; lobes 8–16 on each side; apical segment oblong to ovate, 1–10(–12) × 0.5–2(–3.5) mm, margins entire. **Cauline leaves** (and bracts) similar to basal, smaller distally (lobes fewer). **Fruiting pedicels:** some from basal rosette (straight or slightly recurved), (30–)50–100 (–180) mm. **Flowers:** sepals (persistent), erect, oblong-lanceolate, 9–12(–15) × 2–3.5 mm, apex appendage well-developed, (1–)2–4 mm; petals broadly obovate, 12–15(–20) × (5–)7–11 mm, apex rounded; median filament pairs 4–6 mm, slightly dilated basally; anthers linear, 3–4 mm; gynophore usually obsolete, rarely to 2 mm. **Fruits** oblong, somewhat inflated, (0.8–)1–2(–2.5) cm × 5.5–7.5 mm, (fleshy green, thick, leathery), base and apex obtuse to subacute; valves (covered with well-developed vesicles), not veined; replum rounded; septum complete; ovules 16–44 per ovary; style 2–5(–7) mm, not flattened basally. **Seeds** 4–5 mm diam.; wing 1–1.5 mm. $2n = 24$.

Flowering Dec–Mar. Open grounds, fields, flood plains, roadsides, slightly saline alluvial silt, ditch banks; 0–100 m; Tex.

Selenia grandis, which is restricted to the lower valley of the Rio Grande, is easily distinguished from other species of the genus by the presence of vesicles on fruits and by the persistent sepals.

4. **Selenia jonesii** Cory, Rhodora 33: 142. 1931 E

Selenia jonesii var. *obovata* Rollins

Plants winter annuals, (not or rarely subacaulescent). **Stems** (not inflated into crown), ascending or subdecumbent, (0.5–)1–3(–4) dm. **Basal leaves** rosulate; petiole 1–2.5 cm; blade margins 2-pinnatisect, 4–8 cm; lobes 4–11 on each side, (smaller than terminal); apical segment linear to oblong or ovate, 1–6 × 0.5–2.5 mm, margins entire. **Cauline leaves** (and bracts) similar to basal, smaller distally. **Fruiting pedicels:** some from basal leaf axils, (10–)20–40(–50) mm. **Flowers:** sepals spreading, oblong, 4–6(–7) × 1.5–2.5 mm, apex appendage developed, 0.5–0.8 mm; petals spatulate, 4–7 × 2.5–4 mm, apex rounded or emarginate; median filament pairs 3–4 mm, not dilated basally; anthers ovate, 1–1.5 mm; gynophore 1–3 mm. **Fruits** globose or, rarely, obovoid, terete, 0.8–1.5 cm

× 8–14 mm, (not fleshy, papery), base usually obtuse, rarely cuneate, apex obtuse; valves obscurely reticulate-veined; replum not flattened; septum complete; ovules 8–14 per ovary; style 1–2(–3) mm, not flattened basally. Seeds 4–5 mm diam.; wing 1–1.5 mm. $2n = 24$.

Flowering Mar–Apr. Dry lake beds, draws, moist swales, prairie plateaus, playa lakes, buffalo wallows; 0–1100 m; Tex.

Selenia jonesii is endemic to six counties in the western Edwards Plateau. Variety *obovata* is reduced herein to synonymy because the alleged difference in fruit shape does not hold; it appears to be an artifact of pressing inflated fruit.

47. **SUBULARIA** Linnaeus, Sp. Pl. 2: 642. 1753; Gen. Pl. ed. 5, 290. 1754 • Awlwort

[Latin *subula*, awl, alluding to leaf shape of type species]

Ihsan A. Al-Shehbaz

Annuals; (littoral or aquatic); scapose; glabrous throughout. **Stems** erect, unbranched. **Leaves** (persistent); basal; rosulate; sessile; blade margins entire; cauline absent. **Racemes** (lax or somewhat congested), slightly or considerably elongated in fruit. **Fruiting pedicels** usually ascending, rarely divaricate, divaricate-ascending, or suberect, slender or stout. **Flowers:** sepals (sometimes persistent), ascending to erect, ovate-oblong, lateral pair not saccate basally; petals (rarely absent), white, narrowly oblanceolate to lingulate, (slightly exceeding sepals), claw undifferentiated from blade, apex obtuse; stamens subequal; filaments not dilated basally; anthers ovate; nectar glands confluent, subtending bases of stamens. **Fruits** silicles, shortly stipitate, obovoid to ellipsoid [oblong], smooth, terete or slightly inflated; valves each not veined; replum rounded; septum complete; ovules 4–18 per ovary; style absent; stigma capitate. **Seeds** biseriate, slightly compressed, not winged, oblong; seed coat not mucilaginous when wetted; cotyledons incumbent. $x = 14, 15$.

Species 2 (1 in the flora): n North America, Europe (n, Russia), Africa.

The second species of the genus, *Subularia monticola* A. Brown ex Schweinfurth, is endemic to high elevations of tropical Africa. For a discussion and distinguishing characteristics of all taxa of the genus, as well as a map of the North American distribution, see G. A. Mulligan and J. A. Calder (1964).

SELECTED REFERENCE Mulligan, G. A. and J. A. Calder. 1964. The genus *Subularia* (Cruciferae). Rhodora 66: 127–135.

1. **Subularia aquatica** Linnaeus, Sp. Pl. 2: 642. 1753 F

Subspecies 2 (1 in the flora): North America; Europe (n, Russia).

1a. **Subularia aquatica** Linnaeus subsp. **americana** G. A. Mulligan & Calder, Rhodora 66: 132, plate 1295, fig. 1. 1964 • Awlwort E F

Subularia aquatica var. *americana* (G. A. Mulligan & Calder) B. Boivin

Stems (0.05–)0.15–1.5(–2.5) dm. **Basal leaves** erect or ascending, blade subulate, (0.5–)1–7(–10) cm, margins entire; cauline absent. **Racemes** (1 or) 2–12(–18)-flowered. **Fruiting** pedicels usually forming 30–50° angle with rachis, slender or slightly stout, 1–7(–10) mm, (sometimes terminal flower on longer pedicel to 18 mm). **Flowers** chasmogamous or cleistogamous; sepals usually persistent, rarely caducous, (0.5–)0.7–1(–1.3) × 0.2–0.5 (–0.7) mm; petals 1.2–1.5 × 0.2–0.5 mm; filaments 0.7–1 mm; anthers 0.1–0.2 mm; gynophore 0.2–0.7(–1) mm. **Fruits** (0.15–)0.2–0.35(–0.55) cm × 1.2–2(–2.5) mm. **Seeds** light brown, 0.8–1 × 0.5–0.8 mm. $2n = 30$.

Flowering Jul–Oct. Muddy pool margins, rocky gravelly bottoms, shallow stream pools, wet sedge meadows, shallow sandy water flats, muddy tidal flats, salt marshes, gravelly lake beaches, stream shorelines, pools and lakes to 1.5 m; 0–3200 m; Greenland; B.C., Man., Nfld. and Labr., N.W.T., N.S., Ont., Que., Sask., Yukon; Alaska, Calif., Colo., Idaho, Maine, Mich., Minn., N.H., N.Y., Utah, Vt., Wash., Wyo.

Subspecies *americana* appears to be the sole representative of *Subularia* in North America. It differs from subsp. *aquatica*, which is restricted to northern Europe and Russia, by having persistent (versus caducous) sepals, fruiting pedicels ascending at 30–50° (versus 50–90°) angles, and broadly ellipsoid to broadly obovoid (versus ellipsoid) fruits. G. A. Mulligan and J. A. Calder (1964) indicated that plants slightly intermediate between the two subspecies grow sporadically in North America, and it is not known whether they represent hybrids.

k. BRASSICACEAE Burnett tribe CHORISPOREAE Ledebour, C. A. Meyer & Bunge in C. F. von Ledebour, Fl. Altaica 3: 104. 1831

Annuals or perennials; usually glandular, rarely eglandular (glands multicellular on multiseriate stalks). **Trichomes** simple or absent. **Cauline leaves** (sometimes absent); petiolate; blade base not auriculate, margins entire or dentate to pinnately lobed. **Racemes** usually ebracteate, usually elongated in fruit. **Flowers** actinomorphic; sepals erect [ascending], lateral pair saccate [not saccate] basally; petals white, pink, lavender, or purple [yellow], claw present, often distinct; filaments unappendaged, not winged; pollen 3-colpate. **Fruits** usually siliques [silicles], dehiscent (or breaking into 1-seeded segments), rarely indehiscent, segmented or not, terete or latiseptate; ovules [2–]5–50[–numerous] per ovary; style distinct; stigma strongly 2-lobed [entire]. **Seeds** uniseriate [aseriate]; cotyledons accumbent.

Genera 3, species 47 (2 genera, 5 species in the flora): North America, Europe, Asia, n Africa.

48. CHORISPORA R. Brown ex de Candolle, Mém. Mus. Hist. Nat. 7: 237. 1821, name conserved • [Greek *choris*, asunder or separate, and *spora*, seed, alluding to fruit breaking at constrictions into one-seeded segments] □

Ihsan A. Al-Shehbaz

Annuals [perennials]; not scapose; usually glandular, rarely eglandular, glabrous or pubescent. **Stems** erect or decumbent, branched basally [and distally] (leafy or not). **Leaves** basal and sometimes cauline; petiolate; basal not rosulate [rosulate], blade margins sinuate-dentate, [pinnatifid, or pinnatisect, rarely entire]; cauline absent or shortly petiolate, blade (base not auriculate) margins often entire. **Racemes** (corymbose [or, rarely, flowers solitary on long pedicels from axils of rosettes], several-flowered), slightly or considerably elongated in fruit. **Fruiting pedicels** divaricate, stout [slender] (nearly as thick as fruit). **Flowers:** sepals linear [ovate, or oblong]; petals usually purple or lavender [yellow], rarely white, (much longer than sepals), oblanceolate [broadly obovate or obcordate], claw strongly differentiated from blade, (apex obtuse [emarginate]); stamens strongly tetradynamous; filaments not dilated basally; anthers narrowly oblong [linear], (apex obtuse); nectar glands (2 or 4), lateral, intrastaminal or each side of lateral stamen. **Fruits** sessile, segments 2, linear, slightly [strongly] torulose or submoniliform, terete; (segments breaking into 1-seeded units, lomentaceous with thick, corky, or woody wall); valves usually glandular, rarely eglandular; replum flattened, (persistent after segments fall off); septum becoming corky, splitting at middle; ovules 5–30 per ovary; (style beaklike); stigma conical, 2-lobed (lobes decurrent, strongly connivent). **Seeds** flattened, not winged, oblong; seed coat not mucilaginous when wetted; cotyledons accumbent. $x = 7$.

Species 11 (1 in the flora): introduced; Europe, Asia, n Africa.

1. **Chorispora tenella** (Pallas) de Candolle, Syst. Nat. 2: 435. 1821 [F][I][W]

Raphanus tenellus Pallas, Reise Russ. Reich. 3: 741. 1776; *Chorispermum tenellum* (Pallas) R. Brown

Plants papillose, sometimes pubescent, papillae sometimes mixed with simple trichomes. **Stems** (0.5–)1–4(–5.6) dm. **Basal leaves** (often withered by flowering); petiole (0.5–)1–2(–4) cm; blade oblanceolate or oblong, (1.5–)2.5–8(–13) cm × (4–)8–20(–30) mm, base cuneate or attenuate, apex acute, surfaces glandular. **Cauline leaves** similar to basal, distalmost subsessile, blades smaller distally. **Fruiting pedicels** (2–)3–5 mm, glandular. **Flowers:** sepals purplish, (3–)4–5(–6) × 0.5–0.7 mm; petals 8–10(–12) × 1–2 mm, claw 6–7 mm; filaments 4–6(–7) mm; anthers ca. 1.5 mm. **Fruits** slightly curved-ascending, (1.4–)1.8–2.5(–3) cm × 1.5–2 mm, with 8–12 constrictions on each side; style (6–)10–18(–22) mm. **Seeds** brown, 1–1.4 × 0.8–1 mm. $2n = 14$.

Flowering Apr–Jul. Waste places, pastures, fields, roadsides, railroad embankments, grassy slopes; 0–2300 m; introduced; Alta., B.C., Sask.; Ariz., Calif., Colo., Idaho, Ill., Ind., Iowa, Kans., La., Mass., Mich., Mo., Mont., Nebr., Nev., N.Mex., N.Y., N.Dak., Ohio, Okla., Oreg., S.Dak., Tenn., Tex., Utah, Wash., Wyo.; Europe; Asia; n Africa; introduced also in South America.

Chorispora tenella appears to be most widely distributed in Colorado, Nevada, and Wyoming, of all the provinces and states listed above.

49. **PARRYA** R. Brown, Chlor. Melvill., 10, plate B. 1823 • [For William E. Parry, 1790–1855, arctic explorer during whose first expedition to the North American Arctic (1819–1820) specimens of the genus were first collected]

Ihsan A. Al-Shehbaz

Achoriphragma Soják; *Neuroloma* Andrzejowski ex de Candolle

Perennials [subshrubs]; (caudex well-developed, often covered with persistent petiolar remains or leaves); scapose [not scapose]; glandular or eglandular, glabrous [pubescent]. **Stems** erect, unbranched. **Leaves** basal [sometimes cauline]; rosulate; petiolate; blade margins entire, subentire, or dentate [pinnately lobed]. **Racemes** (corymbose, 3–20-flowered, rarely proximalmost flowers bracteate), considerably elongated in fruit. **Fruiting pedicels** ascending or divaricate-ascending [erect]. **Flowers:** sepals ovate or oblong [linear], (unequal, glandular or eglandular); petals purple, lavender, or white [pink], obovate, claw differentiated from blade, (subequaling or longer than sepals, apex rounded or emarginate); stamens tetradynamous; filaments dilated or not basally; anthers oblong [linear], (apex obtuse); nectar glands lateral, annular or semi-annular. **Fruits** sessile or shortly stipitate (gynophore persistently attached to pedicel), not segmented, linear, oblong, or lanceolate, smooth or torulose, strongly latiseptate or, rarely, subterete or 4-angled; valves (leathery), each with prominent midvein and with obscure to distinct lateral and marginal veins, eglandular or glandular; replum almost always flattened (visible); septum complete; ovules 6–20[–50] per ovary; stigma conical or cylindric, 2-lobed (lobes prominent, connate, decurrent). **Seeds** often broadly winged, suborbicular to broadly ovate [oblong], strongly flattened; seed coat (smooth), not mucilaginous when wetted; cotyledons accumbent. $x = 7$.

Species 25–30 (4 in the flora): North America, Asia (w China, Himalayas, Russian Far East, Siberia).

Parrya is a distinctive genus reduced by V. P. Botschantzev (1972) to being monospecific, including only *P. arctica*, with almost all of the other species transferred to *Neuroloma*. Except for the four North American species, the genus is centered in central Asia, adjacent western China, and the Himalayas.

SELECTED REFERENCE Botschantzev, V. P. 1972. On *Parrya* R. Br., *Neuroloma* Andrz. and some other genera (Cruciferae). Bot. Zhurn. (Moscow & Leningrad) 57: 664–673.

1. Sepals 2.5–3.5 mm; petals 6–7 mm, apex rounded; ovules 6–8 per ovary; fruits obovate to oblong, 0.8–1.4(–1.7) cm . . . . 1. *Parrya nauruaq*
1. Sepals (3–)4–9 mm; petals (8–)10–23 mm, apex emarginate; ovules 10–20 per ovary; fruits narrowly oblong to linear-lanceolate, (1–)1.5–5 cm.
   2. Plants eglandular; leaf blades linear-oblanceolate to narrowly oblanceolate, 2–5(–7) mm wide; sepals (3–)4–5 mm; petals (8–)10–13 × 3–5 mm, claws 3.5–4.5 mm; ovules 14–20 per ovary . . . . 2. *Parrya arctica*
   2. Plants glandular or eglandular; leaf blades obovate, spatulate, broadly oblanceolate, lanceolate, or oblong, (6–)10–28 mm wide; sepals 5–9 mm; petals (14–)16–23 × 7–12 mm, claws 6–12 mm; ovules 10–16 per ovary.
      3. Plants glandular or eglandular, not cespitose; leaf blade margins entire, minutely to coarsely dentate, or, rarely, incised; fruiting pedicels (10–)15–40(–60) mm; filaments 6–10 mm; Alaska, British Columbia, Northwest Territories, Yukon . . . . 3. *Parrya nudicaulis*
      3. Plants usually densely glandular, rarely eglandular, densely cespitose; leaf blade margins incised to coarsely dentate; fruiting pedicels 4–15(–20) mm; filaments 4–6 mm; Utah, Wyoming . . . . 4. *Parrya rydbergii*

1. **Parrya nauruaq** Al-Shehbaz, J. R. Grant, R. Lipkin, D. F. Murray & C. L. Parker, Novon 17: 277. 2007 [E]

Plants densely cespitose, caudex branched; leaves and scapes densely glandular. **Stems** 0.4–0.8 dm. **Leaves:** petiole 0.2–1 cm, 2–4 mm wide at base; blade obovate to broadly spatulate, 0.6–1.5(–2) cm × 4–8(–10) mm, base cuneate, margins coarsely dentate to subentire, apex acute. **Racemes** 6–16-flowered. **Fruiting pedicels** (proximalmost) 5–9(–12) mm. **Flowers:** sepals ovate, 2.5–3.5 × 1.2–1.7 mm, eglandular; petals lavender to purple, 6–7 × 3–5 mm, claw 2–3.5 mm, apex rounded; median filaments 2.5–3 mm; anthers 0.9–1 mm. **Fruits** obovate to oblong, 0.8–1.4(–1.7) cm × 5–7 mm; valves eglandular or sparsely glandular; ovules 6–8 per ovary; style 0.2–1 mm. **Seeds** 4–6 × 3.5–5 mm; wing 0.8–1.5 mm wide.

Flowering Jun. Marbleized impure carbonate landscapes, shallow decomposed granite slopes, outcrops and hogback ridges; 0–100 m; Alaska.

Parrya nauruaq is known from the Moon Mountains, Seward Peninsula.

2. **Parrya arctica** R. Brown, Chlor. Melvill., 11, plate B. 1823 [E]

Plants cespitose, caudex simple or branched; eglandular. **Stems** (0.3–)0.5–1.7(–2.5) dm. **Leaves:** petiole (0.5–)1–3(–4) cm, to 4 mm wide at base; blade linear-oblanceolate to narrowly oblanceolate, (0.7–)1.2–4(–5.2) cm × 2–5(–7) mm, base attenuate, margins entire or, rarely, obscurely dentate, apex acute or obtuse. **Racemes** 3–12-flowered. **Fruiting pedicels** (proximalmost) 6–20(–30) mm. **Flowers:** sepals broadly oblong, (3–)4–5 × 1.5–2 mm; petals white to lavender or purple, (8–)10–13 × 3–5 mm, claw 3.5–4.5 mm, apex rounded to shallowly emarginate; median filaments 3.5–5 mm; anthers 0.8–1.5 mm. **Fruits** narrowly oblong, 1–2.5(–3.5) cm × 3–5 mm; valves eglandular; ovules 14–20 per ovary; style 0.2–0.5(–0.7) mm. **Seeds** 3.5–4.5 × 3–3.5 mm; wing 0.7–1 mm wide. $2n = 14$.

Flowering late Jun–early Aug. Rock and cliff crevices, arctic tundra, rocky grounds, gravelly slopes, wet meadows, mounds in deltas, stream banks, sandy grassy areas between rocks, dry calcareous gravel; 0–100 m; N.W.T., Nunavut, Yukon.

Parrya arctica is distributed in the Canadian arctic and subarctic islands and rarely has been collected inland, such as from the Great Bear Lake area (Northwest Territories). The major center of distribution appears to fall between 93–125°W and 67–76°N.

3. **Parrya nudicaulis** (Linnaeus) Regel, Bull. Soc. Imp. Naturalistes Moscou 34: 176. 1861

Cardamine nudicaulis Linnaeus, Sp. Pl. 2: 654. 1753; *Achoriphragma nudicaule* (Linnaeus) Soják; *Arabis nudicaulis* (Linnaeus) de Candolle; *C. articulata* Pursh; *Matthiola nudicaulis* (Linnaeus) Trautvetter; *Neuroloma nudicaule* (Linnaeus) Andrzejowski ex de Candolle; *Parrya macrocarpa* R. Brown; *P. nudicaulis* var. *grandiflora* Hultén; *P. nudicaulis* subsp. *interior* Hultén; *P. nudicaulis* var. *interior* (Hultén) B. Boivin; *P. nudicaulis* subsp. *septentrionalis* Hultén

Plants often not cespitose, caudex branched; glandular throughout or eglandular. **Stems** (0.4–)0.7–2.7(–3.5) dm. **Leaves:** petiole (0.7–)1.5–7(–10) cm, to 5 mm wide at base (glandular or not); blade narrowly spatulate

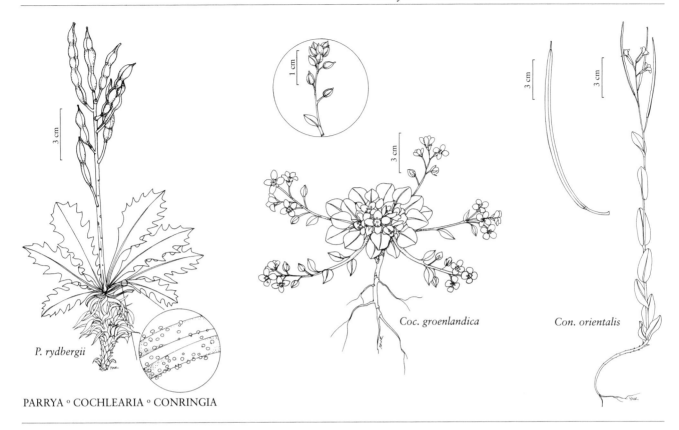

PARRYA · COCHLEARIA · CONRINGIA

or oblanceolate to lanceolate or oblong, (1–)1.7–7 cm × (6–)10–23(–28) mm, base cuneate or attenuate, margins entire or minutely to coarsely dentate, or sometimes incised, apex acute. **Racemes** 3–20-flowered. **Fruiting pedicels** (proximalmost) (10–)15–40(–60) mm (glandular or not). **Flowers:** sepals oblong, 5–8 × 1.5–3 mm (glandular or not); petals lavender to white, purple, (14–)16–20(–22) × 7–10(–12) mm, claw 6–10 mm, apex emarginate; median filaments 6–10 mm; anthers 1.5–2.5 mm. **Fruits** narrowly oblong to linear-lanceolate, (2–)3–4(–4.7) cm × (3.5–)5–7 mm; valves glandular or eglandular; ovules 12–16 per ovary; style (0.5–)1–2.5(–3.5) mm. **Seeds** 3.5–6 × 3–5 mm; wing 0.7–1.5 mm wide. $2n = 14, 28$.

Flowering early Jun–early Aug. Tundra, alpine stream valleys, flats and flood banks, limestone or schist mountain slopes and tops, grassy summits, disturbed gravel, moist open areas, meadows, sandy shores, mossy carpets, hillsides, alpine stony slopes, stable sand ridges, turfy snow flushes; 0–1800 m; B.C., N.W.T., Yukon; Alaska; e Asia (Russian Far East).

Parrya nudicaulis is the most variable species in the genus, especially in leaf shape, size, and margin, as well as in the presence versus absence of the extrafloral glands. Much of the confusion about its limits resulted from different emphases on various characters. For example, E. Hultén (1971), who recognized six subspecies, expanded the range of *P. nudicaulis* to extend from the Canadian arctic and Alaska into the Russian Far East, Siberia, Central Asia, China, and the Himalayas. Three of his four North American subspecies, sometimes growing together, are rather poorly defined morphologically and appear to have been based primarily on the degree of development of leaf teeth. By contrast, R. C. Rollins (1993) recognized a single polymorphic species that included the Utah-Wyoming endemic *P. rydbergii*. Within a given population of *P. nudicaulis*, one finds both glandular and eglandular plants with leaf margins entire or variously dentate. In my opinion, these variables alone are unreliable, and flower size, in combination with other characters, can give a better indication of taxon identity.

4. Parrya rydbergii Botschantzev, Bot. Mater. Gerb. Bot. Inst. Komarova Akad. Nauk S.S.S.R. 17: 178. 1955
E F

Parrya platycarpa Rydberg, Bull. Torrey Bot. Club 39: 326. 1912, not Hooker f. & Thomson 1861; *Neuroloma rydbergii* (Botschantzev) Botschantzev; *P. nudicaulis* (Linnaeus) Regel subsp. *rydbergii* (Botschantzev) Hultén; *P. nudicaulis* var. *rydbergii* (Botschantzev) N. H. Holmgren

Plants densely cespitose, caudex branched; densely glandular or, rarely, eglandular. **Stems** (0.3–)0.5–1.5

(–2.6) dm. **Leaves:** petiole (0.5–)1.5–5(–7) cm, to 5 mm wide at base; blade obovate to spatulate or broadly oblanceolate to lanceolate, (1–)2–5(–8) cm × (7–)13–26 mm, base cuneate or attenuate, margins incised to coarsely dentate, apex acute. **Racemes** 3–15-flowered. **Fruiting pedicels** (proximalmost) 4–15(–20) mm. **Flowers:** sepals oblong, 5.5–9 × 1.5–2.5 mm; petals lavender to purple, (15–)16–20(–23) × 7–9 mm, claw 8–12 mm, apex emarginate; median filaments 4–6 mm; anthers 1.5–2 mm. **Fruits** narrowly oblong to linear-lanceolate, (2–)3–5 cm × 5–7 mm; valves densely glandular; ovules 10–16 per ovary; style 0.5–1.5 mm. **Seeds** 4–5 × 3.5–4.5 mm; wing 1–2 mm wide.

Flowering late Jun–Jul. Barren rock slides, alpine tundra, finely divided limestone rubble, calcareous talus, shale slopes, alpine rocky slopes, crevices among boulders; 2900–3700 m; Utah, Wyo.

Parrya rydbergii, which is restricted to Uinta and Emmons mountains in Utah (Daggett, Duchesne, Summit, and Uintah counties) and Beartooth and Big Sheep mountains in Wyoming (Park and Sublette counties), was treated by E. Hultén (1971) as a subspecies of *P. nudicaulis* and reduced by R. C. Rollins (1993) to synonymy of that species. It is geographically disjunct from *P. nudicaulis* and can be easily distinguished from that nearest relative.

l. BRASSICACEAE Burnett tribe COCHLEARIEAE Buchenau, Fl. Nordwestdeut. Tiefebene, 245. 1894 (as Cochleariinae)

Annuals, biennials, or perennials; eglandular. **Trichomes** absent. **Cauline leaves** petiolate or sessile; blade base auriculate or not, margins usually entire or dentate. **Racemes** ebracteate, elongated [not elongated] in fruit. **Flowers** actinomorphic; sepals ascending to spreading, lateral pair not saccate basally; petals white, claw present, indistinct; filaments unappendaged, not winged; pollen 3-colpate. **Fruits** silicles, dehiscent, unsegmented, subterete or angustiseptate; ovules [4–]8–14[–numerous] per ovary; style distinct; stigma entire. **Seeds** biseriate; cotyledons usually accumbent, rarely incumbent.

Genus 1, species 21 (3 in the flora): North America, Europe, Asia, nw Africa.

50. COCHLEARIA Linnaeus, Sp. Pl. 2: 647. 1753; Gen. Pl. ed. 5, 292. 1754
    • Scurvy-grass [Latin *cochlear*, spoon, alluding to leaf shape of some species]

Ihsan A. Al-Shehbaz

Marcus Koch

Cochleariopsis Á. Löve & D. Löve; *Glaucocochlearia* (O. E. Schulz) Pobedemova

Annuals, biennials, or, rarely, perennials; not scapose. **Stems** erect or decumbent, unbranched or branched. **Leaves** basal and cauline; petiolate or sessile; basal rosulate or not, petiolate, blade margins usually entire, repand, or dentate, rarely sinuate-dentate; cauline petiolate or sessile, blade (base auriculate or not), margins entire, repand, or dentate. **Racemes** (corymbose), slightly to greatly elongated in fruit. **Fruiting pedicels** erect, divaricate, or ascending, slender. **Flowers:** sepals ovate or oblong; petals oblanceolate or spatulate, [oblong, lingulate, elliptic], claw not differentiated from blade; stamens subequal or slightly tetradynamous; filaments not dilated basally; anthers ovate; nectar glands lateral, 1 on each side lateral stamens. **Fruits** sessile, ovoid, ellipsoid, obovoid, orbicular, ovoid-orbicular, or elliptic to oblong, [sub]terete or angustiseptate; valves each with distinct midvein (sometimes inflated); replum rounded; septum complete, fenestrate, or absent; ovules [5–]8–14[–32] per ovary; stigma capitate. **Seeds** plump, not winged, ovoid-oblong or ovoid to subglobose [ellipsoid]; seed coat (smooth or papillose), not mucilaginous when wetted; cotyledons accumbent, rarely incumbent. $x = 6, 7$.

Species 21 (3 in the flora): North America, Europe, Asia, nw Africa.

Molecular data (M. Koch et al. 1999) provide some evidence that *Ionopsidium* Reichenbach could be integrated into *Cochlearia* as a section, consisting of at least five species in Eurasia.

Molecular studies (M. Koch 2002; Koch et al. 1996, 1998, 1999, 2003) as well as cytological ones (D. M. Pegtel 1999, and references therein) have demonstrated that in *Cochlearia* hybridization and polyploidization have created in Europe extensive complexes based on $x = 6$; the circumpolar and subarctic taxa form different complexes based on $x = 7$. It appears that all North American taxa belong to the latter group. The North American taxa have not been studied as comprehensively as the European ones.

SELECTED REFERENCES Koch, M. et al. 1998. Isozymes, speciation and evolution in the polyploid complex *Cochlearia* L. (Brassicaceae). Bot. Acta 111: 411–425. Koch, M. et al. 1999. Molecular phylogenetics of *Cochlearia* (Brassicaceae) and allied genera based on nuclear ribosomal ITS DNA sequence analysis contradict traditional concepts of their evolutionary relationship. Pl. Syst. Evol. 216: 207–230. Koch, M., H. Hurka, and K. Mummenhoff. 1996. Chloroplast DNA restriction site variation and RAPD-analyses in *Cochlearia* (Brassicaceae): Biosystematics and speciation. Nordic J. Bot. 16: 585–603. Pobedimova, E. 1969. Revisio generis *Cochlearia* L., 1. Novosti Sist. Vyssh. Rast. 6: 67–106. Pobedimova, E. 1970. Revisio generis *Cochlearia* L., 2. Novosti Sist. Vyssh. Rast. 7: 167–195.

1. Annuals; fruits oblong or elliptic, 8–12 mm; stems simple from base; cauline leaves sessile, blade margins entire, rarely repand . . . . 2. *Cochlearia sessilifolia*
1. Biennials or perennials; fruits ovoid, obovoid, orbicular, or ellipsoid, rarely to 9 mm; stems usually few to several from base; (proximal) cauline leaves petiolate, blade margins dentate, repand, or entire.
   2. Fruits obovoid, ovoid, or ellipsoid, terete or slightly angustiseptate, valves usually not or obscurely reticulate, rarely distinctly so; proximal and median cauline blades: margins entire, repand, or slightly dentate. . . . . 1. *Cochlearia groenlandica*
   2. Fruits orbicular or ovoid-orbicular, distinctly angustiseptate, valves prominently reticulate; proximal and median cauline blades: margins dentate . . . . 3. *Cochlearia tridactylites*

1. **Cochlearia groenlandica** Linnaeus, Sp. Pl. 2: 647. 1753 E F

Cochlearia arctica Schlechtendal ex de Candolle; *C. arctica* subsp. *oblongifolia* (de Candolle) V. V. Petrovsky; *C. fenestrata* R. Brown; *C. oblongifolia* de Candolle; *C. officinalis* Linnaeus subsp. *arctica* (Schlechtendal ex de Candolle) Hultén; *C. officinalis* subsp. *oblongifolia* (de Candolle) Hultén; *C. officinalis* var. *oblongifolia* (de Candolle) Gelert; *Cochleariopsis groenlandica* (Linnaeus) Á. Löve & D. Löve; *C. groenlandica* subsp. *oblongifolia* (de Candolle) Á. Löve & D. Löve

Biennials or perennials. Stems usually few to several from base, rarely simple, erect to decumbent, branched distally, (0.1–)0.5–3(–4) dm. **Basal leaves** rosulate; petiole (0.2–)1–7(–10) cm; blade deltate to ovate, (0.3–)0.7–2(–2.5) cm × (2–)5–20 mm, base cuneate or truncate, margins entire, repand, or obscurely dentate, apex obtuse. **Cauline leaves** petiolate (shortly petiolate or sessile distally); blade 0.4–2 cm × 1–15 mm, base cuneate (not auriculate), margins entire, repand, or slightly dentate, apex acute. **Racemes** many-flowered. **Fruiting pedicels** divaricate to ascending, rarely erect, (2–)5–15 (–20) mm. **Flowers:** sepals ovate or oblong, 1–2(–3) × 0.5–1.5 mm; petals oblanceolate to spatulate, 2–4(–5) × (0.5–)0.8–2(–3) mm; filaments 1–2(–2.5) mm; anthers 0.2–0.3 mm. **Fruits** ovoid to ellipsoid or obovoid, 3–6 (–7) × 2–4(–5) mm, terete or slightly angustiseptate; valves usually not or obscurely reticulate, rarely distinctly so; septum complete or fenestrate; ovules 8–14 per ovary; style 0.1–0.4 mm. **Seeds** brown, ovoid to subglobose, 1–1.5 × 0.8–1.3 mm, papillate. $2n = 14$.

Flowering Jun–Aug. Tidal flats, beaches, dunes, lagoons, stream banks, seepage, peat hammocks, meadows, herb mats, tundra, maritime rocky beaches and slopes, bare areas of polygons, mud flat bird nesting sites; 0–100 m; Greenland; B.C., Nfld. and Labr., N.W.T., Nunavut, Que., Yukon; Alaska, Calif., Oreg., Wash.

R. C. Rollins (1993) treated the North American plants with $2n = 14$ as members of *Cochlearia officinalis*. That species is a strictly European tetraploid with $2n = 24$. In our opinion, plants of the arctic and subarctic *C. groenlandica* complex represent an evolutionary lineage with $x = 7$, which is entirely distinct from that including the European *C. officinalis* and its relatives with $x = 6$. The systematic relationships of the $x = 7$ group to the $2n = 14$ Icelandic plants of the *C. pyrenaica* complex are still unresolved.

The North American plants are extremely variable in flower size, petal shape, and fruit shape and size. They are much in need of detailed cytological, morphological, and molecular studies.

Cochlearia groenlandica is known in California from nesting areas on off-shore rocks in Del Norte County; in Oregon it occurs on ocean bluffs in Coos and Curry counties (A. Liston, pers. comm.). It appears to be naturally occurring in both states.

2. **Cochlearia sessilifolia** Rollins, Contr. Dudley Herb. 3: 182, plate 46, fig. 1. 1941 [C][E]

Cochlearia officinalis var. *sessilifolia* (Rollins) Hultén

Annuals. Stems simple from base, erect, branched (few to several) distally, 0.2–0.8(–1.5) dm. **Basal leaves** not rosulate (representing cotyledons); petiole 0.2–0.6 cm; blade spatulate to oblong, 0.8–1.6 cm × 3–5 mm, base cuneate, margins entire, apex obtuse. **Cauline leaves** sessile; blade 0.5–1.7(–2) cm × 3–5(–7) mm, base obtuse (not auriculate), margins usually entire, rarely repand, apex obtuse. **Racemes** 3–7(–15)-flowered, (slightly elongated in fruit). **Fruiting pedicels** ascending to erect, 3.5–8(–12) mm. **Flowers:** sepals oblong, 1.8–2.2 × 0.7–1 mm; petals spatulate, (3–)4–6 × 1.5–2.5 mm; filaments 1.5–2 mm; anthers 0.3–0.4 mm. **Fruits** (crowded) elliptic to oblong, 8–12 × 3.5–4.5 mm, angustiseptate, (obtuse at both ends); valves not reticulate; septum complete; ovules 10–14 per ovary; style 0.2–0.5 mm. **Seeds** light brown, ovoid-oblong, 1.1–1.3 × 0.9–1 mm, coarsely papillate.

Flowering Jul–Aug. Gravel bars, gravel spit in lagoon outlets submerged at high tide, seashores; of conservation concern; 0 m; Alaska.

Although *Cochlearia sessilifolia* was reduced to a variety of the European *C. officinalis*, it appears to be sufficiently distinct in habit, leaves, and fruit shape and size from the other North American species. In the absence of extensive field and experimental work, it is better to maintain this taxon as a distinct species. It belongs to the *C. groenlandica* rather than the *C. officinalis* lineage. Its annual life cycle has to be regarded as a derived character.

3. **Cochlearia tridactylites** Banks ex de Candolle, Syst. Nat. 2: 367. 1821 [E]

Cochlearia cyclocarpa S. F. Blake

Biennials or perennials. Stems few to several from base, decumbent, branched distally, 0.6–3 dm. **Basal leaves** rosulate; petiole 0.5–5(–9) cm; blade cordate to broadly ovate, 0.5–1.5(–2) cm × 3–15(–20) mm, base cordate or truncate, margins entire or dentate, rarely sinuate-dentate, apex obtuse. **Cauline leaves** petiolate (shortly petiolate or sessile distally); blade 4–20 × 3–17 mm, base cuneate to subtruncate (distalmost auriculate), margins coarsely dentate, apex acute. **Racemes** many-flowered, (more elongated proximally). **Fruiting pedicels** divaricate or ascending, (3–)4–11(–15) mm. **Flowers:** sepals ovate, 1–2(–2.5) × 0.7–1.2 mm; petals oblanceolate, 1.8–3(–4) × 1–2 mm; filaments 1–2 mm; anthers 0.3–0.5 mm. **Fruits** orbicular or ovoid-orbicular, 3.5–7(–9) × 4–7(–9) mm, distinctly angustiseptate; valves prominently reticulate; septum complete; ovules 8–12 per ovary; style 0.3–0.9 mm. **Seeds** brown, ovoid-oblong, 1.4–1.6 × 1.1–1.3 mm, papillate.

Flowering Jul–Aug. Coastal areas, calcareous sandstone, sea cliffs, rocky seashores, limestone gravel beds, serpentine barrens, talus or gravel, escarpments; 0–100 m; Nfld. and Labr., N.S., Que.

m. BRASSICACEAE Burnett tribe CONRINGIEAE D. A. German & Al-Shehbaz, Harvard Pap. Bot. 13: 169. 2008 [I]

Annuals [biennials]; eglandular. **Trichomes** absent [simple]. **Cauline leaves** sessile; blade base cordate or auriculate [sagittate], margins usually entire, rarely crenulate. **Racemes** ebracteate, often elongated in fruit. **Flowers** actinomorphic; sepals erect [ascending], lateral pair saccate basally; petals white or yellow, claw present, distinct or not; filaments unappendaged, not winged; pollen 3-colpate. **Fruits** siliques, dehiscent, unsegmented, terete, angled, or latiseptate; ovules 10–50[–numerous] per ovary; style distinct; stigma (often decurrent), capitate [conical], entire [2-lobed]. **Seeds** uniseriate; cotyledons usually incumbent, rarely subconduplicate.

Genera 2, species 9 (1 in the flora): introduced; Europe, Asia.

51. CONRINGIA Heister ex Fabricius, Enum., 160. 1759 • Hare's-ear mustard

[For Hermann Conring, 1606–1681, German professor of medicine and philosophy at Helmstedt] [I]

Suzanne I. Warwick

Plants not scapose; (usually glaucous). **Stems** erect, unbranched or branched proximally. **Leaves** basal and cauline; subsessile or sessile; basal not rosulate, subsessile, blade margins usually entire; cauline blade (base cordate-amplexicaul or, rarely, auriculate), margins usually entire, rarely crenulate. **Racemes** (corymbose, several-flowered). **Fruiting pedicels** ascending, stout (almost as thick as fruit, or, rarely, much narrower). **Flowers:** sepals oblong; petals usually narrowly obovate, rarely oblanceolate, claw differentiated from blade [undifferentiated], (apex obtuse); stamens slightly tetradynamous; filaments not dilated, slender; anthers oblong (base slightly sagittate); nectar glands lateral, median glands often absent. **Fruits** sessile, linear, torulose, 4-angled or terete; valves each with prominent midvein; replum rounded; septum complete; stigmas capitate-flattened, entire. **Seeds** not winged, oblong [ellipsoid]; seed coat (papillose), copiously mucilaginous (granular) when wetted; cotyledons incumbent. $x = 7$ [9].

Species 6 (1 in the flora): introduced; c Europe, e Mediterranean region, Asia (Afghanistan, Pakistan); introduced also in Mexico, nw Africa, Australia.

1. **Conringia orientalis** (Linnaeus) Dumortier, Fl. Belg., 123. 1827 • Rabbit's-ear, rabbit-ears, treacle mustard, slinkweed [F] [I] [W]

Brassica orientalis Linnaeus, Sp. Pl. 2: 666. 1753; *Erysimum orientale* (Linnaeus) Crantz 1769, not Miller 1768

Plants sometimes winter annuals. **Stems** mostly simple, (1–)3–7 dm. **Basal leaves:** blade (slightly fleshy), pale green, oblanceolate to obovate, 5–9 cm, margins ± entire. **Cauline leaves:** blade oblong to elliptic or lanceolate, (1–)3–10(–15) cm × (5–)20–25 (–50) mm, base deeply cordate-amplexicaul, apex rounded. **Fruiting pedicels** ascending, straight or curved-ascending, (8–)10–15(–20) mm. **Flowers:** sepals 6–8 × 1–1.5 mm, median pair narrower than lateral, apex acute; petals 7–12 × 2–3 mm, base attenuate, claw usually as long as sepal; filaments 5–7 mm; anthers 1.5–2 mm. **Fruits** ± torulose, strongly 4-angled to ± cylindrical, 1-nerved, keeled, (5–)8–14 cm × 2–2.5 mm; style cylindrical, 0.5–4 mm. **Seeds** brown, 2–2.9 × 1.2–1.5 mm. $2n = 14$.

Flowering (Mar in Texas) May–Aug. Cultivated lands, grain fields, disturbed areas, waste places, roadsides, gardens; 0–3500 m; introduced; Alta., B.C., Man., N.B., Nfld. and Labr., N.S., Ont., P.E.I., Que., Sask.; Ala., Alaska, Ariz., Ark., Calif., Colo., Conn., Del., D.C., Fla., Ga., Idaho, Ill., Ind., Iowa, Kans., Ky., La., Maine, Md., Mass., Mich., Minn., Miss., Mo., Mont., Nebr., Nev., N.H., N.J., N.Mex., N.Y., N.C., N.Dak., Ohio, Okla., Oreg., Pa., R.I., S.C., S.Dak., Tenn., Tex., Utah, Vt., Va., Wash., W.Va., Wis., Wyo.; Europe; Asia; introduced also in Mexico, nw Africa, Australia.

Conringia orientalis was collected on ballast in New York as early as 1879. It is most abundant in the plains and prairies of both the United States and Canada (I. A. Al-Shehbaz 1985; R. C. Rollins and Al-Shehbaz 1986). In disturbed places, it has penetrated into the native vegetation over a wide area.

n. BRASSICACEAE Burnett tribe DESCURAINIEAE Al-Shehbaz, Beilstein & E. A. Kellogg, Pl. Syst. Evol. 259: 111. 2006

Annuals or perennials [shrubs]; glandular or eglandular (glands unicellular papillae). **Trichomes** stalked, dendritic or forked, sometimes simple, rarely absent. **Cauline leaves** usually petiolate, sometimes sessile; blade base not auriculate, margins usually pinnatisect or dentate, sometimes entire. **Racemes** ebracteate or bracteate, often elongated in fruit. **Flowers** actinomorphic;

sepals erect, ascending, spreading, or reflexed, lateral pair not saccate basally; petals usually yellow, sometimes white [pink or purple], claw usually present, sometimes absent, often obscure, obsolete, or distinct; filaments unappendaged, not winged; pollen 3-colpate. **Fruits** silicles or siliques, dehiscent, unsegmented, terete or angustiseptate; ovules 4–100[–numerous] per ovary; style usually distinct, sometimes obsolete or absent; stigma entire. **Seeds** usually biseriate or uniseriate (rarely 4-seriate in *Tropidocarpum*); cotyledons usually incumbent, rarely accumbent.

Genera 6, species ca. 60 (3 genera, 18 species in the flora): North America, Mexico, South America, Europe, Asia, n Africa, Atlantic Islands (Canary Islands).

52. DESCURAINIA Webb & Berthelot, Hist. Nat. Îsles Canaries 3(2,3): 72. 1836, name conserved • Tansy mustard [For François Descurain, 1658–1740, French botanist and apothecary]

Barbara E. Goodson

Ihsan A. Al-Shehbaz

Huguenenia Reichenbach, name rejected; *Sophia* Adanson, name rejected

Annuals, biennials, or perennials [shrubs, subshrubs]; not scapose; glabrous, glabrate, or sparsely to densely pubescent, trichomes usually short-stalked, dendritic, rarely also simple, sometimes mixed with unicellular, glandular, clavate papillae. **Stems** erect or prostrate, unbranched or branched basally and/or distally. **Leaves** basal and cauline; petiolate or sessile; basal (often withered by flowering), rosulate, petiolate, blade (1–3-pinnate), margins entire or toothed; cauline petiolate or sessile, blade often similar to basal. **Racemes** (proximally sometimes bracteate), elongated or not in fruit. **Fruiting pedicels** divaricate or erect, slender. **Flowers**: sepals erect to spreading, ovate to oblong or linear; petals usually obovate or oblanceolate, rarely oblong, (shorter to longer than sepals), claw obsolete or distinct, (apex obtuse); stamens tetradynamous; filaments not dilated basally; anthers oblong to ovate, (apex obtuse); nectar glands confluent, subtending bases of stamens, median glands present. **Fruits** siliques or silicles, sessile, usually linear, oblong, clavate, or fusiform, rarely ellipsoid or obovoid, smooth or torulose, terete; valves each often with distinct midvein, usually glabrous, rarely pubescent; replum rounded; septum complete or perforated (membranous, not veined or with 1–3 longitudinal veins); ovules 5–100 per ovary; style usually absent, rarely distinct; stigma capitate. **Seeds** uniseriate or biseriate, plump, wingless, oblong or ellipsoid; seed coat (minutely reticulate), usually mucilaginous when wetted; cotyledons incumbent. $x = 7$.

Species 45–47 (14 in the flora): North America, Mexico, South America, Eurasia, n Africa, Atlantic Islands (Canary Islands).

Descurainia species are distributed in three major centers: North America (17 species), South America (ca. 20 species), and the Canary Islands (7 species). Excluding *D. sophioides*, which extends into Siberia from arctic North America, three additional species are found outside those regions: *D. kochii* (Petri) O. E. Schulz (Caucasus, Turkey), *D. sophia* (a cosmopolitan weed of Eurasian origin), and *D. tanacetifolia* (Linnaeus) Prantl (Alps and the Iberian Peninsula).

Descurainia is taxonomically difficult throughout most of the New World, especially the United States. Extensive morphological variation exists within numerous species and many wide-ranging and widely-overlapping taxa appear to intergrade endlessly. Inter- and infraspecific hybridization is probably extensive given the lack of reproductive barriers between the recently evolved New World species, the ready dispersibility of the mucilaginous seeds, and the weedy tendencies of the majority of taxa that readily occupy disturbed habitats. The frequent occurrence of hybrid populations, suggested by the intergrading morphology and widespread polyploidy, is supported by recent molecular evidence (B. E. Goodson 2007). A second factor contributing to the complexity of North American *Descurainia* is that the number of taxonomically reliable characters is somewhat limited, with a general lack of agreement among various authors on the characters to be emphasized for circumscription of taxa. A taxonomic nightmare has resulted from the recognition of numerous poorly defined species, subspecies, and varieties. Although an extensive molecular systematic study of the genus was recently conducted (Goodson), some relationships and species boundaries remain unclear. Because detailed population-level, morphological, cytological, and molecular studies of North American *Descurainia* are still needed, this treatment represents an early "pit stop" on the road to full understanding of the complexity in the genus.

The keys and circumscriptions in both L. E. Detling (1939) and R. C. Rollins (1993) are unreliable for the identification of a given collection to an infraspecific taxon. In a genus where interspecific hybridization is so extensive, it is sometimes impossible to identify reliably every single specimen, especially if mature fruits are lacking. Unfortunately, most North American material in major herbaria consists of either specimens without mature fruits or unrecognized hybrids between various taxa. As a consequence, up to 60% of the holdings of *Descurainia* species in the major North American herbaria are misidentified. Because mature fruits are critical to proper identification, the following key relies primarily on fruiting material.

Almost all taxonomists have used the presence versus absence of glandular papillae as an important character to subdivide a given complex into species, subspecies, varieties, or forms. Both glandular and eglandular forms sometimes occur within a given population of *Descurainia incana*, *D. incisa*, *D. paradisa*, *D. pinnata*, or *D. sophioides*. Some authors were inconsistent in according formal recognition to the glandular versus eglandular plants of a given species. For example, R. C. Rollins (1993) treated the glandular and eglandular forms of *D. obtusa* and *D. paradisa*, but not of *D. incana* and *D. sophioides*, as distinct subspecies; N. H. Holmgren (2005b) recognized such forms as varieties in *D. paradisa* but not *D. incana*. Although the presence versus absence of glands appears to be consistent within certain taxa, reliance on the use of glands as a diagnostic character should be applied with extreme caution.

SELECTED REFERENCES Bramwell, D. 1977. A revision of *Descurainia* Webb & Berth. section *Sisymbriodendron* (Christ) O. E. Schulz in the Canary Islands. Bot. Macar. 4: 31–53. Detling, L. E. 1939. A revision of the North American species of *Descurainia*. Amer. Midl. Naturalist 22: 481–520. Goodson, B. E. 2007. Molecular Systematics and Biogeography of *Descurainia* Webb & Berthel. (Brassicaceae). Ph.D. dissertation. University of Texas.

1. Fruits sparsely to densely pubescent at least when young.
    2. Perennials; fruiting pedicels 1.5–3 mm; fruits subappressed to rachises; styles sparsely pubescent; ovules 10–18 per ovary; Wyoming. . . . . 14. *Descurainia torulosa*
    2. Biennials; fruiting pedicels 6–31 mm; fruits not appressed to rachises; styles glabrous; ovules 16–64 per ovary; Arizona, California, Nevada, New Mexico.
        3. Seeds biseriate; ovules 48–64 per ovary; fruits 1–1.3 mm wide; fruiting pedicels 13–31 mm . . . . 1. *Descurainia adenophora*
        3. Seeds uniseriate; ovules 16–40 per ovary; fruits 0.7–1 mm wide; fruiting pedicels 6–15 mm . . . . 9. *Descurainia obtusa*
1. Fruits glabrous.

[4. Shifted to left margin.—Ed.]

4. Fruits usually fusiform, obovate, clavate, or broadly ellipsoid, rarely broadly linear (wider distally).
   5. Fruits usually clavate, rarely broadly linear (wider distally); seeds biseriate; ovules 16–40 per ovary; valves each with distinct midvein.................... 11. *Descurainia pinnata*
   5. Fruits fusiform, obovate, or broadly ellipsoid (sometimes clavate in *D. paradisa*); seeds uniseriate (sometimes biseriate in *D. paradisa*); ovules 4–12 per ovary; valves each with obscure midvein.
      6. Fruits long-acute basally and apically; styles (0.2–)0.3–0.6(–0.8) mm; stems (1.3–)2–10.5(–13.5) dm, unbranched basally, branched distally.............. 3. *Descurainia californica*
      6. Fruits acute basally, obtuse apically; styles 0.05–0.3 mm; stems (1–)1.5–3.2(–4.1) dm, branched basally and distally..................................... 10. *Descurainia paradisa*
4. Fruits linear (sometimes oblong in *D. brevisiliqua*).
   7. Perennials (short-lived); stems 0.1–0.15 dm; racemes not elongated in fruit; ovules 4–8 per ovary .... 6. *Descurainia kenheilii*
   7. Annuals or biennials; stems (0.5–)0.9–12(–18) dm; racemes elongated considerably in fruit (sometimes slightly elongated in *D. sophioides*); ovules 10–62 per ovary (6–12 in *D. nelsonii*).
      8. Fruits often strictly appressed to rachises; fruiting pedicels erect to erect-ascending .... 4. *Descurainia incana*
      8. Fruits not appressed to rachises; fruiting pedicels horizontal, divaricate, or ascending.
         9. Leaf blades 2- or 3-pinnate; fruit septums appearing 2- or 3-veined...... 12. *Descurainia sophia*
         9. Leaf blades usually 1-pinnate (rarely 2-pinnate in *D. sophioides*); fruit septums not veined.
            10. Plants eglandular; fruits 3–8(–10) mm; petals 0.7–1.2 mm; seeds 0.5–0.8 mm.
               11. Biennials; stems 6–11 dm, unbranched; ovules 10–28 per ovary; seeds biseriate ....2. *Descurainia brevisiliqua*
               11. Annuals; stems (0.7–)0.9–3.2(–4.5) dm, usually branched at or near base, rarely unbranched; ovules 6–12 per ovary; seeds uniseriate ... .8. *Descurainia nelsonii*
            10. Plants glandular or eglandular; fruits 8–30(–34) mm; petals 1.7–2.8 mm; seeds 0.9–1.5 mm.
               12. Flowers overtopped by developing fruits; ovules 30–62 per ovary... ....13. *Descurainia sophioides*
               12. Flowers not overtopped by developing fruits; ovules 14–32 per ovary.
                  13. Plants canescent or not; cauline leaf blades: distal segments oblong to lanceolate, or linear, margins dentate, denticulate or entire; fruits straight or strongly curved inward .... 5. *Descurainia incisa*
                  13. Plants not canescent; cauline leaf blades: distal segments linear or oblong, margins entire; fruits straight or slightly curved inward .... 7. *Descurainia longepedicellata*

1. **Descurainia adenophora** (Wooton & Standley) O. E. Schulz in H. G. A. Engler, Pflanzenr. 86[IV,105]: 321. 1924

Sophia adenophora Wooton & Standley, Contr. U.S. Natl. Herb. 16: 127. 1913; *Descurainia obtusa* (Greene) O. E. Schulz subsp. *adenophora* (Wooton & Standley) Detling

Biennials; glandular (at least distally); finely pubescent, often canescent, trichomes dendritic, sometimes mixed with simple ones. **Stems** erect, unbranched basally, branched distally, 4.5–13 dm. **Basal leaves:** petiole 1–3 cm; blade pinnate, oblanceolate to obovate or ovate in outline, 2–10 cm, lateral lobes (2–5 pairs), oblanceolate to lanceolate, (4–12 × 1–5 mm), margins entire or serrate to crenate, (apex obtuse). **Cauline leaves** sessile or shortly petiolate; blade smaller distally, distal lobes often narrower, surfaces densely pubescent. **Racemes** considerably elongated in fruit. **Fruiting pedicels** divaricate, straight, 13–31 mm. **Flowers:** sepals ascending, greenish to yellowish, oblong, 2–2.9 mm, pubescent, (trichomes dendritic, mixed with glandular papillae); petals oblanceolate, 1.8–2.6 × 0.5–0.7

mm; median filaments 1.8–2.4 mm; anthers 0.3–0.5 mm. **Fruits** divaricate to erect, linear, slightly torulose, 8–16(–20) × 1–1.3 mm, (abruptly acute at both ends); valves each with distinct midvein, (sparsely pubescent or glabrescent); septum not veined; ovules 48–64 per ovary; style 0.1–0.2 mm, glabrous. **Seeds** biseriate, light brown, ellipsoid, 0.9–1.1 × 0.5–0.6 mm. 2*n* = 42.

Flowering Jun–Aug. Open forests, sandy grounds, gravelly flats, disturbed areas; 1100–2000 m; Ariz., Calif., Nev., N.Mex., Tex.; Mexico (Baja California, Chihuahua, Coahuila).

Both L. E. Detling (1939) and R. C. Rollins (1993) treated *Descurainia adenophora* as a subspecies of *D. obtusa*, but the differences are so substantial that they should be recognized as distinct species. From the latter, *D. adenophora* is distinguished by being hexaploid (versus diploid) with densely glandular (versus eglandular) distal parts, longer sepals (2–2.9 versus 1–2 mm) and petals (1.8–2.6 versus 1.2–2 mm), longer fruiting pedicels (13–31 versus 6–15 mm), biseriate (versus uniseriate) seeds, and more ovules (42–64 versus 16–40) per ovary.

2. **Descurainia brevisiliqua** (Detling) Al-Shehbaz & Goodson, Harvard Pap. Bot. 12: 421. 2007 [E]

Descurainia obtusa (Greene) O. E. Schulz subsp. *brevisiliqua* Detling, Amer. Midl. Naturalist 22: 499. 1939

Biennials; eglandular; finely pubescent, often canescent, trichomes dendritic, rarely also mixed with simple ones. **Stems** erect, unbranched basally, branched distally, 6–11 dm. **Basal leaves:** petiole 0.5–1.5 cm; blade pinnate, oblanceolate to obovate or ovate in outline, 1–5 cm, lateral lobes (2–5 pairs), linear, margins entire or serrate to incised (apex obtuse). **Cauline leaves** sessile or shortly petiolate; blade smaller distally, distal lobes often narrower, surfaces densely pubescent. **Racemes** (paniculate, often with short branches), considerably elongated in fruit. **Fruiting pedicels** divaricate-ascending, straight, 4–7(–11) mm. **Flowers:** sepals spreading, yellowish, oblong, 1.4–2 mm, pubescent, (trichomes dendritic, sometimes mixed with simple ones); petals oblanceolate, 0.7–1 × 0.1–0.2 mm; median filaments 1.5–2 mm; anthers 0.2–0.3 mm. **Fruits** divaricate to erect, linear to oblong, not torulose, 3–8(–10) × 1–1.2 mm; valves each with distinct midvein; septum not veined; ovules 10–28 per ovary; style obsolete to 0.1 mm, glabrous. **Seeds** biseriate, light brown, ellipsoid, 0.5–0.7 × 0.3–0.4 mm. 2*n* = 42.

Flowering (late Jun–)Jul–Aug(–early Oct). Pine and juniper communities, rocky washes, roadsides, grasslands, gravelly mesa; 1900–2500 m; Ariz., N.Mex., Tex.

R. C. Rollins (1993) reduced *Descurainia brevisiliqua* to synonymy of the diploid *D. obtusa*. This hexaploid species differs from the latter in having biseriate (versus uniseriate) seeds and glabrous (versus densely to moderately pubescent) fruits. Molecular data (B. E. Goodson 2007) show that it is consistently separate from, and unrelated to, *D. obtusa*, instead being affiliated with *D. incana* and *D. incisa*. *Descurainia brevisiliqua* differs from the latter two species by having biseriate (versus uniseriate) seeds and oblong, non-torulose fruits that, unlike those of *D. incana*, are not appressed to rachises. Although the fruits of *D. brevisiliqua* resemble those of some *D. pinnata* subspecies, it can be readily distinguished from the latter by its tall, strict growth habit with numerous short branches in the distal part of the plant, more linear leaf segments, stems that are usually purple, and late-summer flowering period.

3. **Descurainia californica** (A. Gray) O. E. Schulz in H. G. A. Engler, Pflanzenr. 86[IV,105]: 330. 1924 [E] [F]

Smelowskia californica A. Gray, Proc. Amer. Acad. Arts 6: 520. 1865; *Sisymbrium californicum* (A. Gray) S. Watson; *Sophia leptostylis* Rydberg

Annuals or **biennials**; eglandular; usually pubescent, trichomes dendritic, sometimes glabrous distally. **Stems** erect, unbranched basally, branched distally, (1.3–)2–10.5 (–13.5) dm. **Basal leaves:** petiole 0.4–4.2 cm; blade pinnate, oblanceolate to obovate in outline, 1.5–6 cm, lateral lobes [2–4 (or 5) pairs], lanceolate, (5–22 × 1–5 mm), margins usually entire or crenate to incised, rarely lobed. **Cauline leaves** sessile or shortly petiolate; blade smaller distally, distal lobes often narrower, surfaces sparsely pubescent. **Racemes** considerably elongated in fruit. **Fruiting pedicels** divaricate to ascending or suberect, often straight, 3–9(–11) mm. **Flowers:** sepals spreading, yellowish, oblong, 0.9–1.5 mm, glabrous; petals oblanceolate, 1.1–1.8 × 0.4–0.6 mm; median filaments 0.8–1.4 mm; anthers 0.3–0.4 mm. **Fruits** divaricate to erect, fusiform, not torulose, (2–)3–5(–6) × (0.8–)1–1.3 mm, (long-acute at both ends); valves each with obscure midvein; septum not veined; ovules 4–12 per ovary; style (0.2–)0.3–0.6(–0.8) mm, glabrous. **Seeds** uniseriate, light brown, ellipsoid, 1–1.4 × 0.6–0.8 mm. 2*n* = 14.

Flowering Jun–Aug. Disturbed areas in pinyon-juniper, dry hillsides, decomposed granite slopes, sagebrush, moist roadsides, open woods, fir-spruce or aspen communities, gravel and talus slopes; 1700–3400 m; Ariz., Calif., Colo., Nev., N.Mex., Oreg., Utah, Wyo.

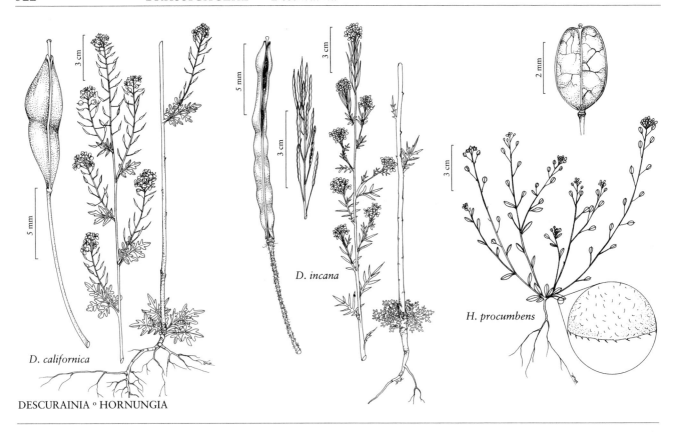

DESCURAINIA · HORNUNGIA

4. **Descurainia incana** (Bernhardi ex Fischer & C. A. Meyer) Dorn, Vasc. Pl. Wyoming, 296. 1988 [E][F]

Sisymbrium incanum Bernhardi ex Fischer & C. A. Meyer, Index Seminum (St. Petersburg) 1: 38. 1835; *Descurainia incana* var. *brevipes* (Nuttall) S. L. Welsh; *D. incana* var. *macrosperma* (O. E. Schulz) Dorn; *D. incana* var. *major* (Hooker) Dorn; *D. incana* subsp. *procera* (Greene) Kartesz & Gandhi; *D. richardsonii* var. *alpestris* (Cockerell) O. E. Schulz; *D. richardsonii* var. *brevipes* (Nuttall) S. L. Welsh & Reveal; *D. richardsonii* var. *macrosperma* O. E. Schulz; *D. richardsonii* subsp. *procera* (Greene) Detling; *D. richardsonii* var. *procera* (Greene) Breitung; *S. canescens* Nuttall var. *alpestre* Cockerell; *S. canescens* var. *brevipes* Nuttall; *S. canescens* var. *major* Hooker; *S. procerum* (Greene) K. Schumann; *Sophia brevipes* (Nuttall) Rydberg; *S. procera* Greene

Biennials; usually eglandular, rarely glandular; finely pubescent, sometimes canescent, trichomes dendritic. **Stems** erect, unbranched basally, often many-branched distally, (1.5–)2.5–12 dm. **Basal leaves:** petiole 1–3.5(–5.5) cm; blade pinnatifid, broadly lanceolate to oblanceolate or obovate in outline, 1.5–10(–13) cm, lateral lobes linear to oblong or narrowly lanceolate, [3–10(–15) × 1–3(–5) mm], margins entire. **Cauline leaves** sessile or shortly petiolate; blade smaller distally, distal lobes often narrower. **Racemes** considerably elongated in fruit. **Fruiting pedicels** erect to erect-ascending, straight, 2–8(–11) mm. **Flowers:** sepals erect, yellowish, oblong, 1–1.8 mm, sparsely pubescent; petals oblanceolate, 1.2–2 × 0.3–0.6 mm; median filaments 1.4–2 mm; anthers 0.3–0.4 mm. **Fruits** erect, (often strictly appressed to rachis), linear, slightly torulose, (4–)5–10(–15) × 0.7–1.2(–1.5) mm, (acute at both ends); valves each with distinct midvein; septum often with distinct midvein; ovules 14–22 per ovary; style 0.1–0.4 mm, glabrous. **Seeds** uniseriate, reddish brown, ellipsoid to narrowly oblong, 0.8–1.2 × 0.4–0.5 mm. $2n = 14, 28$.

Flowering May–Sep. Alpine and subalpine areas, gravel and sand bars, scree, grassy slopes, prairies, steep rocky slopes, roadsides, disturbed sites, waste grounds, meadows, spruce-fir, pine, aspen, or sagebrush communities; 100–3500 m; Alta., B.C., Man., N.W.T., Nunavut, Ont., Que., Sask., Yukon; Alaska, Calif., Colo., Idaho, Maine, Minn., Mont., Nev., N.Mex., N.Dak., S.Dak., Utah, Wyo.

Descurainia incana is a distinctive species readily separated from the other North American taxa of the genus by having fruits and fruiting bases strictly appressed to rachises, and septums with a distinct midvein. Collections identified as such, but with fruits and pedicels not or only weakly appressed to the rachis, most likely represent hybrids between this species and others.

5. Descurainia incisa (Engelmann ex A. Gray) Britton, Mem. Torrey Bot. Club 5: 173. 1894

Sisymbrium incisum Engelmann ex A. Gray, Mem. Amer. Acad. Arts, n. s. 4: 8. 1849

Annuals; glandular or eglandular; densely to sparsely pubescent, glabrous or pubescent distally, sometimes canescent, trichomes dendritic. **Stems** erect, usually unbranched basally, branched distally or sometimes throughout, (1.3–)2–8.2(–10.7) dm. **Basal leaves**: petiole 0.5–4.7 cm; blade pinnate, obovate to oblanceolate in outline, 1.5–10.3 cm, lateral lobes (2–9 pairs), ovate or oblong to lanceolate or linear, margins usually dentate to incised or entire, rarely pinnatifid or crenate. **Cauline leaves** sessile or shortly petiolate; blade smaller distally, distal lobes oblong, lanceolate, linear, (margins dentate to denticulate or entire), surfaces pubescent or glabrous. **Racemes** considerably elongated in fruit, (glandular or eglandular). **Fruiting pedicels** ascending to divaricate or horizontal, straight, (3–)5–25(–30) mm. **Flowers**: sepals erect to ascending, yellowish, oblong to ovate, 1.6–2.4 mm, glabrous or pubescent; petals narrowly oblanceolate, 1.7–2.8 × 0.5–0.9 mm; median filaments 1.6–2.4 mm; anthers 0.3–0.4 mm. **Fruits** erect to ascending, linear, slightly torulose, 8–20 × 0.9–1.3 mm, (straight or slightly to strongly curved inward); valves each not veined or with distinct midvein; septum not veined; ovules 14–26 per ovary; style 0.1–0.3 mm, glabrous. **Seeds** uniseriate, reddish brown, oblong, 0.9–1.3 × 0.5–0.6 mm.

Subspecies 2 (2 in the flora): North America, n Mexico.

As delimited by various authors and as represented in all major herbaria consulted, *Descurainia incisa* is highly variable in almost all features. The variation is most likely the result of hybridization with all species of the genus that have overlapping ranges. Forms with few-seeded, short fruits tapering at both ends most likely represent hybrids with *D. californica*; those with sub-biseriate seeds most likely resulted from crossing with *D. pinnata*, and the origin of forms with somewhat subappressed fruits almost certainly involved *D. incana*. The recognition of glandular versus eglandular forms as distinct varieties or subspecies is completely artificial. The lectotype (*Fendler 29*, GH) and isolectotype (MO) are eglandular; a syntype (*Fendler 31*, MO) is densely glandular.

1. Fruiting pedicels ascending to divaricate, (3–)5–10(–12) mm; lateral lobes of basal and proximal cauline blades (3–)5–9 pairs, margins usually coarsely dentate to incised, rarely crenate or pinnatifid; lobes of distal cauline blades oblong to lanceolate, margins dentate to denticulate; fruits straight or curved inward. . . . . 5a. *Descurainia incisa* subsp. *incisa*
1. Fruiting pedicels horizontal to divaricate, (10–)13–25(–30) mm; lateral lobes of basal and proximal cauline blades 2 or 3 (or 4) pairs, margins usually entire; lobes of distal cauline blades linear, margins entire; fruits curved inward . . . . 5b. *Descurainia incisa* subsp. *paysonii*

5a. Descurainia incisa (Engelmann ex A. Gray) Britton subsp. **incisa**

Descurainia incana (Bernhardi ex Fischer & C. A. Meyer) Dorn subsp. *incisa* (Engelmann ex A. Gray) Kartesz & Gandhi; *D. incana* subsp. *viscosa* (Rydberg) Kartesz & Gandhi; *D. incana* var. *viscosa* (Rydberg) Dorn; *D. incisa* var. *leptophylla* (Rydberg) O. E. Schulz; *D. incisa* subsp. *viscosa* (Rydberg) Rollins; *D. incisa* var. *viscosa* (Rydberg) G. A. Mulligan; *D. richardsonii* O. E. Schulz subsp. *incisa* (Engelmann ex A. Gray) Detling; *D. richardsonii* var. *sonnei* (B. L. Robinson) C. L. Hitchcock; *D. richardsonii* subsp. *viscosa* (Rydberg) Detling; *D. richardsonii* var. *viscosa* (Rydberg) M. Peck; *D. rydbergii* O. E. Schulz; *D. serrata* (Greene) O. E. Schulz; *Hesperis incisa* (Engelmann ex A. Gray) Kuntze; *Sisymbrium incisum* var. *sonnei* B. L. Robinson; *S. leptophyllum* (Rydberg) A. Nelson & J. F. Macbride; *S. serratum* (Greene) Tidestrom; *S. viscosum* (Rydberg) Blankinship; *Sophia glandulifera* Rydberg; *S. incisa* (Engelmann ex A. Gray) Greene; *S. leptophylla* Rydberg; *S. purpurascens* Rydberg; *S. serrata* Greene; *S. sonnei* (B. L. Robinson) Greene; *S. viscosa* Rydberg

Plants glandular or eglandular, not canescent. **Basal leaves**: lateral lobes (3–)5–9 pairs, margins usually coarsely dentate to incised, rarely crenate or pinnatifid. **Cauline leaves**: lobes oblong to lanceolate, margins dentate to denticulate. **Racemes** glandular or eglandular. **Fruiting pedicels** ascending to divaricate, (3–)5–10(–12) mm. **Fruits** straight or curved inward. $2n = 14$.

Flowering late Jun–Aug. Alpine, stream banks, disturbed sites, roadsides, meadows, sagebrush and juniper communities, open woods, rocky cliffs, sandy areas, talus slopes; 900–2900 m; Alta., B.C., Yukon; Ariz., Calif., Colo., Idaho, Mont., Nev., N.Mex., Oreg., Utah, Wash., Wyo.; Mexico (Chihuahua, Coahuila).

5b. Descurainia incisa (Engelmann) Britton subsp. **paysonii** (Detling) Rollins, Harvard Pap. Bot. 1(4): 46. 1993 [E]

Descurainia pinnata (Walter) Britton subsp. *paysonii* Detling, Amer. Midl. Naturalist 22: 515. 1939; *D. incisa* var. *paysonii* (Detling) N. H. Holmgren; *D. pinnata* var. *paysonii* (Detling) S. L. Welsh & Reveal

Plants eglandular, often canescent. **Basal leaves:** lateral lobes 2 or 3 (or 4) pairs, margins usually entire. **Cauline leaves:** lobes linear, margins entire. **Racemes** eglandular. **Fruiting pedicels** horizontal to divaricate, (10–)13–25 (–30) mm. **Fruits** curved inward.

Flowering May–Jun. Disturbed grounds, rocky hillsides, pinyon-juniper woodlands, grassy benches, flats; 1500–2300 m; Colo., N.Mex., Utah.

Subspecies *paysonii* and *incisa* are reliably separated by their pedicel length, orientation, and shape and margin of lateral leaf lobes.

6. Descurainia kenheilii Al-Shehbaz, Harvard Pap. Bot. 12: 395, fig. 1. 2007 [E]

Perennials; (short-lived, dwarf); eglandular; sparsely pubescent throughout, trichomes dendritic. **Stems** (simple from base), erect, usually unbranched, rarely branched distally, 0.1–0.15 dm. **Basal leaves:** petiole 0.3–0.6 cm; blade pinnate, oblanceolate in outline, lateral lobes (2–4 pairs), oblanceolate to obovate, (1–2.5 × 0.5–1 mm), margins entire. **Cauline leaves** subsessile; blade smaller distally, distal lobes narrower, surfaces sparsely pubescent. **Racemes** not elongated in fruit. **Fruiting pedicels** erect to ascending, straight, 1–1.5 mm. **Flowers:** sepals ascending, yellowish, ovate, 1–1.4 mm, pubescent; petals narrowly oblanceolate, 1–1.5 × 0.3–0.4 mm; median filaments 0.6–1 mm; anthers (broadly ovate), ca. 0.1 mm. **Fruits** erect, linear, torulose, 6–10 × 1–1.3 mm, (straight); valves each with distinct midvein; septum with distinct midvein; ovules 4–8 per ovary; style obsolete, 0.1–0.2 mm, glabrous. **Seeds** uniseriate, reddish brown, oblong, 1–1.2 × 0.5–0.6 mm.

Flowering Aug–Sep. Alpine tundra, talus slopes; 3600–3800 m; Colo.

Descurainia kenheilii is known from the high alpine areas in Archuleta and San Juan counties.

7. Descurainia longepedicellata (E. Fournier) O. E. Schulz in H. G. A. Engler, Pflanzenr. 86[IV,105]: 324. 1924 (as longipedicellata) [E]

Sisymbrium longepedicellatum E. Fournier, Recherches Anat. Taxon. Fam. Crucifèr., 59. 1865; *Descurainia brachycarpa* (Richardson) O. E. Schulz var. *eglandulosa* (Thellung) O. E. Schulz; *D. incisa* (Engelmann ex A. Gray) Britton subsp. *filipes* (A. Gray) Rollins; *D. incisa* var. *filipes* (A. Gray) N. H. Holmgren; *D. longepedicellata* var. *glandulosa* O. E. Schulz; *D. pinnata* (Walter) Britton subsp. *filipes* (A. Gray) Detling; *D. pinnata* var. *filipes* (A. Gray) M. Peck; *D. rydbergii* O. E. Schulz var. *eglandulosa* O. E. Schulz; *Hesperis longepedicellata* (E. Fournier) Kuntze; *S. brachycarpum* Richardson var. *filipes* (A. Gray) J. F. Macbride; *S. incisum* Engelmann ex A. Gray var. *filipes* A. Gray; *S. incisum* var. *xerophilum* E. Fournier; *S. longepedicellatum* var. *glandulosum* (O. E. Schulz) H. St. John; *Sophia filipes* (A. Gray) A. Heller; *S. glandifera* Osterhout; *S. gracilis* Rydberg; *S. longepedicellata* (E. Fournier) Howell

Annuals; usually eglandular, rarely glandular; moderately to sparsely pubescent, often glabrous distally, not canescent, trichomes dendritic. **Stems** erect, unbranched or branched proximally, often branched distally, (1.5–)3–6.2(–8.5) dm. **Basal leaves:** petiole 0.4–3.5 cm; blade pinnate, ovate to oblanceolate in outline, 1.5–7 cm, lateral lobes linear or oblong, margins entire or dentate to incised. **Cauline leaves** sessile or shortly petiolate; blade smaller distally, lobes linear to filiform, margins entire, surfaces usually glabrous, rarely pubescent. **Racemes** considerably elongated in fruit. **Fruiting pedicels** horizontal to divaricate, straight, (8–)10–15(–20) mm. **Flowers:** sepals ascending, yellow, oblong, 1.5–2 mm, glabrous; petals narrowly oblanceolate, 1.7–2.6 cm × 0.5–1 mm; median filaments 1.5–2 mm; anthers 0.3–0.4 mm. **Fruits** erect, linear, not torulose, (9–)12–17 × 0.8–1.1 mm, (straight or slightly curved inward); valves each with obscure midvein; septum not veined; ovules 18–32 per ovary; style 0.1–0.2 mm, glabrous. **Seeds** uniseriate, reddish brown, oblong, 1–1.3 × 0.6–0.7 mm. $2n = 14$.

Flowering Apr–Jul. Sandy plains and banks, dry washes, open hillsides, sagebrush and juniper or pine communities, grasslands; 200–2100 m; B.C.; Ariz., Colo., Idaho, Mont., Nev., Oreg., Utah, Wash., Wyo.

L. E. Detling (1939) treated *Descurainia longepedicellata* as subsp. *filipes* of *D. pinnata*, whereas R. C. Rollins (1993) and N. H. Holmgren (2005b) treated it as a subspecies and variety, respectively, of *D. incisa*. Molecular data, both nuclear and plastidic (B. E. Goodson 2007), place the three taxa in different, well-supported clades.

R. C. Rollins (1993) and N. H. Holmgren (2005b) reported $2n = 28$ and 42 for *Descurainia longepedicellata* (as *D. pinnata* var. *filipes*), but these counts are not vouchered. Rollins indicated that the taxon range extends into California and New Mexico; we have not seen material from those states.

Descurainia longepedicellata resembles *D. incisa* subsp. *paysonii* in having long fruiting pedicels and linear leaf lobes with entire margins. The latter is easily distinguished by being canescent (versus not canescent) and having fruits strongly curved inward (versus straight). Because the two taxa are not closely related (B. E. Goodson 2007), the similarities in fruiting pedicels and distalmost leaf segments represent convergence.

8. **Descurainia nelsonii** (Rydberg) Al-Shehbaz & Goodson, Harvard Pap. Bot. 12: 422. 2007 [E]

Sophia nelsonii Rydberg, Bull. Torrey Bot. Club 34: 436. 1907; *Descurainia brachycarpa* (Richardson) O. E. Schulz var. *nelsonii* (Rydberg) O. E. Schulz; *D. pinnata* (Walter) Britton subsp. *nelsonii* (Rydberg) Detling; *D. pinnata* var. *nelsonii* (Rydberg) M. Peck

Annuals; eglandular; sparsely to moderately pubescent, sometimes glabrous distally, not canescent, trichomes dendritic. **Stems** erect, usually branched basally or slightly distally, rarely unbranched, (0.7–)0.9–3.2(–4.5) dm. **Basal leaves**: petiole 0.5–1.5 cm; blade pinnate, ovate or oblong in outline, 0.8–2.5 cm, lateral lobe (2–5 pairs), margins dentate or entire. **Cauline leaves** sessile or shortly petiolate; blade smaller distally, distal lobes often narrower, sparsely to moderately pubescent. **Racemes** considerably elongated in fruit. **Fruiting pedicels** divaricate-ascending (often at 20–45° angle), straight, (1.5–)2.5–7(–10) mm. **Flowers**: sepals ascending, yellowish, oblong, 0.7–1.2 mm, pubescent; petals narrowly oblanceolate, 0.8–1.2 × 0.2–0.4 mm; median filaments 1–1.5 mm; anthers 0.1–0.2 mm. **Fruits** erect or ascending, linear, not or slightly torulose, (0.4–)5–8(–10) × 0.7–1 mm; valves each with distinct midvein; septum not veined; ovules 6–12 per ovary; style 0.1–0.2 mm, glabrous. **Seeds** uniseriate, brown, oblong, 0.6–0.8 × 0.4–0.5 mm. $2n = 14$.

Flowering late May–mid Jul. Roadsides, sagebrush, wash bottoms, silty flats, gravelly grounds; 800–3000 m; B.C.; Calif., Idaho, Mont., Nev., Oreg., Wash., Wyo.

Descurainia nelsonii was treated by L. E. Detling (1939) and R. C. Rollins (1993) as a subspecies of *D. pinnata*, but the latter in the sense of these authors is not monophyletic, comprising instead either four or two unrelated species, respectively. ITS molecular data (B. E. Goodson 2007) suggest that *D. nelsonii* is most closely related to *D. longepedicellata* and *D. paradisa*. It can be distinguished from the latter species by its linear fruits with cuneate tips; *D. paradisa* has obovoid fruits with rounded tips. *Descurainia nelsonii* resembles *D. pinnata* subsp. *brachycarpa* in the orientation of fruiting pedicels and in having short styles (to 0.3 mm) and small seeds (to 1 × 0.5 mm). It differs in being branched (versus simple) at base and in having smaller flowers (petals 0.7–1 versus 1.5–2.6 mm), fewer ovules (6–12 versus 16–40) per ovary, linear (versus subclavate) fruits, and uniseriate (versus biseriate) seeds.

9. **Descurainia obtusa** (Greene) O. E. Schulz in H. G. A. Engler, Pflanzenr. 86[IV,105]: 321. 1924

Sophia obtusa Greene, Leafl. Bot. Observ. Crit. 1: 96. 1904; *Sisymbrium obtusum* (Greene) A. Nelson & J. F. Macbride

Biennials; glandular or eglandular; finely pubescent, often canescent, trichomes dendritic, sometimes mixed with simple ones. **Stems** erect, unbranched basally or branched proximally and/or distally, 4–12(–15) dm. **Basal leaves**: petiole 0.5–3.7 cm; blade pinnate, oblanceolate to obovate or ovate in outline, 1–6 cm, lateral lobes (2–5 pairs), oblanceolate to linear or narrowly lanceolate, (7–25 × 2–10 mm), margins usually entire or serrate, rarely incised, (apex obtuse). **Cauline leaves** sessile or shortly petiolate; blade smaller distally, distal lobes often narrower, surfaces densely pubescent. **Racemes** considerably elongated in fruit. **Fruiting pedicels** ascending to divaricate, straight, 6–15 mm. **Flowers**: sepals spreading or sometimes ascending, greenish to yellowish, oblong, 1–2 mm, densely pubescent, (trichomes dendritic, sometimes mixed with glandular papillae); petals oblanceolate, 1–2 × 0.5–0.7 mm (equaling or shorter than sepals); median filaments 1.4–2 mm; anthers 0.2–0.3 mm. **Fruits** divaricate to suberect, linear, slightly torulose, 10–20(–23) × 0.7–1 mm, (acute at both ends); valves each with distinct midvein, (sparsely to densely pubescent); septum not veined; ovules 16–40 per ovary; style 0.1–0.2 mm, glabrous. **Seeds** uniseriate or biseriate, light brown, oblong, 0.7–1.1 × 0.5–0.6 mm. $2n = 14$.

Flowering May–Sep(–Oct). Gravelly grounds, sandy areas, disturbed sites, open forests, plateaus, abandoned mine areas, dry streams and washes; 1500–2600 m; Ariz., Nev., N.Mex.; Mexico (Baja California)

As circumscribed here, *Descurainia obtusa* is a relatively uniform, diploid species. It probably was involved as a parent of *D. adenophora*, which is a hexaploid readily distinguished by characters discussed thereunder.

10. **Descurainia paradisa** (A. Nelson & P. B. Kennedy) O. E. Schulz in H. G. A. Engler, Pflanzenr. 86[IV,105]: 331. 1924 [E]

Sophia paradisa A. Nelson & P. B. Kennedy, Proc. Biol. Soc. Wash. 19: 155. 1906; *Descurainia paradisa* subsp. *nevadensis* Rollins; *D. paradisa* var. *nevadensis* (Rollins) N. H. Holmgren; *D. pinnata* (Walter) Britton subsp. *paradisa* (A. Nelson & P. B. Kennedy) Detling; *D. pinnata* var. *paradisa* (A. Nelson & P. B. Kennedy) M. Peck; *Sisymbrium paradisum* (A. Nelson & P. B. Kennedy) A. Nelson & J. F. Macbride

Annuals; glandular or eglandular; sparsely to densely pubescent, trichomes dendritic. **Stems** erect, branched basally and distally, (often purplish), (1–)1.5–3.2(–4.1) dm. **Basal leaves** (soon withered); petiole 0.3–1.3 cm; blade pinnate, oblanceolate to obovate in outline, 1.5–3 cm, lateral lobes oblong to linear or lanceolate, (1–5 × 0.3–1 mm), margins entire or dentate. **Cauline leaves** sessile; blade smaller distally, distal lobes often narrower, surfaces moderately to densely pubescent. **Racemes** considerably elongated in fruit. **Fruiting pedicels** divaricate to ascending, straight, 2.5–7(–9) mm. **Flowers**: sepals spreading to ascending, pale yellow, oblong, 0.8–1.2 mm, pubescent; petals oblanceolate, 0.9–1.3 × 0.2–0.5 mm; median filaments 0.8–1.2 mm; anthers 0.1–0.2 mm. **Fruits** divaricate to erect, usually obovoid to clavate, rarely broadly ellipsoid, not torulose, 2–5 × 1–2 mm, (acute basally, obtuse apically); valves each with obscure midvein; septum not veined; ovules 4–10 per ovary; style 0.05–0.3 mm, glabrous. **Seeds** uniseriate or biseriate, brown, oblong, 0.8–1.2 × 0.5–0.6 mm.

Flowering Apr–Jun. Shrub communities, sandy washes and dunes, roadsides; 1000–2300 m; Calif., Nev., Oreg.

Although L. E. Detling (1939) reduced *Descurainia paradisa* to a subspecies of *D. pinnata*, molecular data (B. E. Goodson 2007) clearly show that it should not be included in that species. The boundaries of *D. paradisa* in its northern and southern ranges tend to be blurred relative to *D. nelsonii* and *D. pinnata*, respectively.

R. C. Rollins (1993) and N. H. Holmgren (2005b) recognized some of the Nevada plants of *Descurainia paradisa* that have eglandular racemes as a subspecies and variety (*nevadensis*), respectively. This poorly delimited division of the species is artificial, and both glandular and eglandular forms are sometimes found within individual populations of most species. They also indicated that *nevadensis* has styles 0.2–0.3 mm (versus 0.05–0.15 mm in *D. paradisa*), but this distinction is equally unreliable. Indeed, the style length and the presence versus absence of glands are not inherited together. Some of the eglandular plants have styles to 0.1 mm (e.g., *Williams & Tiehm 86-51-1*, GH). It is likely that some of the plants identified as *nevadensis* are of hybrid origin involving other species, especially *D. nelsonii*.

11. **Descurainia pinnata** (Walter) Britton, Mem. Torrey Bot. Club 5: 173. 1894

Erysimum pinnatum Walter, Fl. Carol. 174. 1788; *Sophia pinnata* (Walter) Howell

Annuals; glandular or eglandular; sparsely to densely pubescent, sometimes glabrous distally, canescent or not, trichomes dendritic. **Stems** erect, unbranched or branched basally and/or distally, (0.8–)1.3–5.7(–9.2) dm. **Basal leaves**: petiole 0.5–3.6 cm; blade 1- or 2-pinnate, ovate or oblong to oblanceolate in outline, 1–15 cm, lateral lobes (4–9 pairs), linear or oblanceolate to ovate, margins entire or dentate. **Cauline leaves** sessile or shortly petiolate; blade smaller distally, distal lobes often narrower, surfaces densely pubescent. **Racemes** considerably elongated in fruit. **Fruiting pedicels** usually ascending to divaricate or horizontal, rarely descending (at 20–110° angle), straight or slightly recurved, 4–18 (–23) mm. **Flowers**: sepals spreading to ascending, yellow, purple, or rose, oblong, 0.8–2.6 mm, pubescent; petals (whitish or yellow), narrowly oblanceolate, 1–3 × 0.3–1 mm; median filaments 1–2.8 mm; anthers 0.2–0.4 mm. **Fruits** erect to ascending, usually clavate, rarely broadly linear (wider distally), not torulose, 4–13(–17) × 1.2–2.2 mm; valves each with distinct midvein; septum not veined; ovules 16–40 per ovary; style obsolete, 0.02–0.2 mm, glabrous. **Seeds** biseriate, reddish brown, oblong, 0.6–0.9 × 0.4–0.5 mm.

Subspecies 4 (4 in the flora): North America, n Mexico.

Plants assigned to *Descurainia pinnata* by various authors represent one of the most complex assemblages of any North American Brassicaceae. L. E. Detling (1939) divided it into eleven subspecies, of which R. C. Rollins (1993) accepted eight. The species includes series of populations that show tremendous variability in every conceivable character, and the variation is continuous between extremes. Some of the taxa recognized by these authors were based solely on trivial variations in the degree of pubescence on the distal parts, shape of the distalmost leaf segments, and presence or absence of glands. The latter character can vary within a given population and, therefore, its utility in distinguishing taxa is questionable at best. Cases in point are *Cully s.n.* (spring 1984) and *Worthington 26353*, both at UNM, collected from Bernalillo and

Luna counties (New Mexico), respectively; each sheet has both glandular and eglandular forms.

As delimited by L. E. Detling (1939) and R. C. Rollins (1993), subspp. *halictorum* and *menziesii* represent a heterogeneous assemblage of intermediates within *Descurainia pinnata* and between it and other species. In the absence of thorough studies on these subspecies, we prefer to avoid placing them in the synonymy of a putative parental species or recognizing them as subspecies of a parent. Readers interested in the synonymies of subspp. *halictorum* and *menziesii* may consult Detling and Rollins.

Molecular studies by B. E. Goodson (2007) indicated that some taxa previously included in *Descurainia pinnata* (i.e., subspp. *filipes*, *nelsonii*, *paradisa*, and *paysonii*) should be placed elsewhere. While the molecular data were not able to resolve the remaining subspecies, a critical evaluation of morphology shows that the entire *D. pinnata* complex falls into at least four relatively distinct groups recognized here as subspecies.

1. Rachises glabrous, eglandular..............
   .... 11d. *Descurainia pinnata* subsp. *glabra*
1. Rachises sparsely to densely pubescent, glandular or eglandular.
   2. Plants canescent, often eglandular; stems branching basally or distal to base; sepals purple or rose ....
      .... 11c. *Descurainia pinnata* subsp. *ochroleuca*
   2. Plants not canescent, often glandular; stems unbranched basally; sepals yellow or rose.
      3. Fruiting pedicels 4–14(–17) mm, forming (60–)70–90(–110)° angles; sepals rose (at least apically), 0.8–2 mm; petals 1–1.8 mm
         ....11a. *Descurainia pinnata* subsp. *pinnata*
      3. Fruiting pedicels (7–)10–18(–23) mm, forming 20–60(–80)° angles; sepals yellow, 1.5–2.6 mm; petals (1.7–)2–3 mm ......
         ....11b. *Descurainia pinnata* subsp. *brachycarpa*

11a. Descurainia pinnata (Walter) Britton subsp. **pinnata** [E]

Cardamine multifida Pursh; *Descurainia canescens* (Nuttall) Prantl; *D. multifida* (Pursh) O. E. Schulz; *D. multifoliata* Cory; *Sisymbrium canescens* Nuttall; *S. canescens* var. *californicum* Torrey & A. Gray; *S. incisum* Engelmann ex A. Gray var. *californicum* (Torrey & A. Gray) Blankinship; *S. multifidum* (Pursh) MacMillan; *S. multifidum* subsp. *canescens* (Nuttall) Thellung; *Sophia californica* (Torrey & A. Gray) Rydberg; *S. millefolia* Rydberg; *S. myriophylla* Rydberg

Plants usually glandular, rarely eglandular, usually not canescent. **Stems** unbranched basally, branched distally. **Racemes:** rachis sparsely to densely pubescent, often glandular. **Fruiting pedicels** divaricate to horizontal or descending, forming (60–)70–90(–110)° angle, 4–14 (–17) mm. **Flowers:** sepals rose (at least apically), 0.8–2 mm; petals 1–1.8 × 0.3–0.7 mm. $2n = 14, 28$.

Flowering Feb–Apr. Roadsides, waste grounds, disturbed sites, railroad tracks and embankments, grassy areas, sandy knolls, stream banks, abandoned fields; 0–600 m; Ala., Fla., Ga., La., Miss., N.C., S.C., Tex.

R. C. Rollins (1993) indicated that subsp. *pinnata* grows in Oklahoma and Virginia. We have not seen material of it from those states, and it is likely that those records are based on plants of subsp. *brachycarpa*.

11b. Descurainia pinnata (Walter) Britton subsp. **brachycarpa** (Richardson) Detling, Amer. Midl. Naturalist 22: 509. 1939 [E]

Sisymbrium brachycarpum Richardson in J. Franklin et al., Narr. Journey Polar Sea, 744. 1823 (as brachycarpon); *Descurainia intermedia* (Rydberg) Daniels; *D. magna* (Rydberg) F. C. Gates; *D. pinnata* Britton var. *brachycarpa* (Richardson) Fernald; *D. pinnata* subsp. *intermedia* (Rydberg) Detling; *D. pinnata* var. *intermedia* (Rydberg) C. L. Hitchcock; *D. ramosissima* Rollins; *Hesperis brachycarpa* (Richardson) Kuntze; *S. brachycarpum* var. *intermedium* (Rydberg) J. F. Macbride; *S. canescens* Nuttall var. *brachycarpum* (Richardson) S. Watson; *S. intermedium* (Rydberg) Garrett; *S. multifidum* (Pursh) MacMillan subsp. *brachycarpum* (Richardson) Thellung; *S. pinnatum* (Walter) Greene var. *brachycarpum* (Richardson) Jepson; *Sophia brachycarpa* (Richardson) Rydberg; *S. intermedia* Rydberg; *S. magna* Rydberg; *S. pinnata* (Walter) Howell var. *brachycarpa* (Richardson) Farwell

Plants usually glandular, rarely eglandular, usually not canescent. **Stems** unbranched basally, branched distally. **Racemes:** rachis sparsely to densely pubescent, often glandular. **Fruiting pedicels** divaricate to ascending, forming 20–60(–80)° angle, (7–)10–18(–23) mm. **Flowers:** sepals yellow, 1.5–2.6 mm; petals (1.7–)2–3 × 0.6–1 mm. $2n = 14, 28$.

Flowering Mar–Jul. Roadsides, sagebrush and pinyon-juniper communities, disturbed sites, sandy fields, dry washes, limestone ledges, foothills, canyon margins, gravel washes, dry slopes, cliffs, streamsides, railroad tracks and embankments, prairies, sandy grounds; 100–2100 m; Alta., B.C., Man., N.W.T., Nunavut, Ont., Que., Sask.; Ark., Calif., Colo., Idaho, Ill., Ind., Iowa, Kans., Ky., Mich., Minn., Mo., Mont., Nebr., Nev., N.H., N.Y., N.Dak., Ohio, Okla., Oreg., S.Dak., Tenn., Tex., Utah, Vt., Va., Wash., Wis., Wyo.

Except for having numerous instead of one or few stems from the base, *Descurainia ramosissima* is indistinguishable from *D. pinnata* subsp. *brachycarpa* in every aspect of indumentum, leaf morphology, flower size, seed size and arrangement, and fruit size and orientation. Plants of the type collections (*Rollins et al. 8349*, holotype GH, isotypes GH, MO) are infected with white rust, a fungus disease rarely encountered elsewhere in North American *Descurainia*. It is not known if this unusual branching is inherited or disease-related, but it should not be used as the main reason for recognizing a species separate from *D. brachycarpa*.

Although subsp. *brachycarpa* is almost always glandular in the inflorescence, both glandular and eglandular forms are found in some populations. Cases in point are *J. & H. Massey 1931* and *Dunn 2* (both at TEX) that were collected from Cleveland and Caddo counties, Oklahoma, respectively.

11c. Descurainia pinnata (Walter) Britton subsp. **ochroleuca** (Wooton) Detling, Amer. Midl. Naturalist 22: 504. 1939

Sophia ochroleuca Wooton, Bull. Torrey Bot. Club 25: 455. 1898; *Descurainia andrenarum* (Cockerell) Cory; *D. halictorum* (Cockerell) O. E. Schulz var. *andrenarum* (Cockerell) O. E. Schulz; *D. menziesii* (de Candolle) O. E. Schulz var. *ochroleuca* (Wooton) O. E. Schulz; *D. pinnata* var. *ochroleuca* (Wooton) Shinners; *Sisymbrium ochroleucum* (Wooton) K. Schumann; *Sophia andrenarum* Cockerell

Plants usually eglandular, rarely glandular, distinctly canescent. **Stems** branched basally or just distal to base, branched distally. **Racemes:** rachis often densely pubescent, eglandular. **Fruiting pedicels** divaricate to ascending, forming 30–60° angle, (4–)6–12 mm. **Flowers:** sepals purple or rose, 1–2 mm; petals 1.5–2 × 0.3–0.5 mm.

Flowering mid Mar–Apr. Gravelly and stony hills, desert grasslands, roadsides; 1400–2000 m; Ariz., N.Mex., Tex.; Mexico (Chihuahua).

11d. Descurainia pinnata (Walter) Britton subsp. **glabra** (Wooton & Standley) Detling, Amer. Midl. Naturalist 22: 507. 1939

Sophia glabra Wooton & Standley, Contr. U.S. Natl. Herb. 16: 127. 1913; *Descurainia pinnata* var. *glabra* (Wooton & Standley) Shinners

Plants eglandular, not canescent. **Stems** unbranched basally, branched distally. **Racemes:** rachis glabrous, eglandular. **Fruiting pedicels** divaricate to horizontal or descending, forming 70–90(–100)° angle, 4–10(–15) mm. **Flowers:** sepals rose (at least apically), 0.8–1.5 mm; petals 1–1.8 × 0.3–0.7 mm.

Flowering Feb–Apr(–Jun). Sandy washes and beds, scrub and bush communities, granitic sand, under junipers, oak and pine woodlands, stream beds, sagebrush, floodplains, limestone outcrops; 300–2400 m; Ariz., Calif., Nev., N.Mex., Utah; Mexico (Baja California, Chihuahua, Coahuila, Hidalgo, Sonora).

12. Descurainia sophia (Linnaeus) Webb ex Prantl in H. G. A. Engler and K. Prantl, Nat. Pflanzenfam. 55(III,2): 192. 1891 • Flixweed, tansy mustard [I][W]

Sisymbrium sophia Linnaeus, Sp. Pl. 2: 659. 1753; *Hesperis sophia* (Linnaeus) Kuntze; *S. parviflorum* Lamarck; *Sophia parviflora* (Lamarck) Standley

Annuals; eglandular; sparsely to densely pubescent, sometimes glabrous distally, trichomes dendritic. **Stems** erect, unbranched or branched distally, (1–)2–7(–10) dm. **Basal leaves:** petiole 0.1–2(–3) cm; blade 2- or 3-pinnate, ovate or oblong to obovate in outline, to 15 cm, lateral lobes linear or oblong, (to 10 × 2 mm), margins entire. **Cauline leaves** sessile or shortly petiolate; blade smaller distally, distal lobes often narrower, surfaces often glabrous. **Racemes** considerably elongated in fruit. **Fruiting pedicels** divaricate to ascending, straight, (5–)8–15(–20) mm. **Flowers:** sepals erect to ascending, yellowish, oblong, 1.8–2.8 mm, glabrate to sparsely pubescent; petals narrowly oblanceolate, 2–3 × 0.4–0.6 mm; median filaments 2–3 mm; anthers 0.3–0.4 mm. **Fruits** divaricate-ascending to erect, narrowly linear, torulose, (12–)15–27(–30) × 0.5–0.8(–1) mm, (straight or curved upward); valves each with distinct midvein; septum with a broad central longitudinal band appearing as 2 or 3 veins; ovules 20–48 per ovary; style obsolete, 0.05–0.2 mm, glabrous. **Seeds** uniseriate, reddish brown, oblong, 0.7–1.3 × 0.3–0.6 mm. $2n = 28$.

Flowering Mar–Jul. Roadsides, waste places, disturbed sites, railroad embankments, hillsides, mountain slopes, canyon bottoms, stream banks, fields, lawns, pastures, deserts, sagebrush and pinyon-juniper communities; 0–3000 m; introduced; Alta., B.C., Man., N.B., Nfld. and Labr. (Nfld.), N.W.T., N.S., Ont., P.E.I., Que., Sask., Yukon; Alaska, Ariz., Ark., Calif., Colo., Conn., Del., D.C., Ga., Idaho, Ill., Ind., Iowa, Kans., Ky., La., Maine, Mass., Mich., Minn., Mo., Mont., Nebr., Nev., N.H., N.Mex., N.Y., N.C., N.Dak., Okla., Oreg., Pa., R.I., S.Dak., Tenn., Tex., Utah, Vt., Va., Wash., W.Va., Wis., Wyo.; Eurasia; introduced also in Mexico, Central America, South America, South Africa, Australia.

Deviant chromosome counts (e.g., $2n = 12, 14, 20, 38$; see R. C. Rollins 1993, N. H. Holmgren 2005b, S. I. Warwick and I. A. Al-Shehbaz 2006) are most certainly erroneous, and the species appears to be exclusively tetraploid based on $x = 7$.

SELECTED REFERENCE Best, K. F. 1977. The biology of Canadian weeds. 22. *Descurainia sophia* (L.) Webb. Canad. J. Pl. Sci. 57: 499–507.

13. **Descurainia sophioides** (Fischer ex Hooker) O. E. Schulz in H. G. A. Engler, Pflanzenr. 86[IV,105]: 316. 1924

Sisymbrium sophioides Fischer ex Hooker, Fl. Bor.-Amer. 1: 61, plate 20. 1830; *Hesperis arctica* (E. Fournier) Kuntze; *S. arcticum* E. Fournier; *Sophia sophioides* (Fischer) A. Heller

Annuals or biennials; eglandular or glandular distally; glabrate to moderately pubescent, trichomes dendritic, sometimes mixed with simple ones. **Stems** erect, unbranched or sometimes branched distally, (0.5–)1.5–11(–18) dm. **Basal leaves:** petiole 0.5–5 cm; blade pinnate or, sometimes, 2-pinnate, broadly oblanceolate to ovate in outline, 2.5–11.4(–15.2) cm, lateral lobes lanceolate, (to 10 × 4 mm), margins incised. **Cauline leaves** sessile or shortly petiolate; blade smaller distally, distal lobes often narrower, surfaces often glabrous or sparsely pubescent. **Racemes** elongated or not in fruit, (flowers overtopped by developing fruits). **Fruiting pedicels** divaricate to ascending, (often recurved in age), slender, (3–)4–9(–13) mm. **Flowers:** sepals erect, yellowish, oblong, 1.6–2.7 mm, glabrous; petals narrowly oblanceolate, 2–2.5 × 0.3–0.6 mm; median filaments 2.5–3.5 mm; anthers 0.3–0.4 mm. **Fruits** erect to widely spreading, narrowly linear, slightly torulose, (9–)14–30(–34) × 0.6–1.1 mm, (usually terete, rarely slightly flattened, often curved inward); valves each with obscure midvein; septum not veined; ovules 30–62 per ovary; style obsolete, 0.07–0.3 mm, glabrous. **Seeds** uniseriate, light brown, narrowly oblong, 1–1.5 × 0.3–0.5 mm. $2n = 14$.

Flowering Jun–Sep. Open meadows, eroded peat, roadsides, disturbed and waste sites, rocky outcrops, mining dumps, gravelly grounds, stream banks, gullies; 0–1000 m; B.C., Man., N.W.T., Nunavut, Yukon; Alaska; Asia (Russian Far East, Siberia).

14. **Descurainia torulosa** Rollins, J. Arnold Arbor. 64: 499. 1983 [C][E]

Perennials; (short-lived); eglandular; moderately to densely pubescent throughout, trichomes dendritic. **Stems** (several to numerous from base), decumbent, usually unbranched, rarely branched distally, 0.4–1.5 dm. **Basal leaves:** petiole 0.3–1.4 cm; blade pinnate, oblanceolate in outline, 0.9–3.5 cm, lateral lobes (3–5 pairs), oblanceolate to oblong, (2–5 × 0.5–1.5 mm), margins entire. **Cauline leaves** sessile or shortly petiolate; blade smaller distally, distal lobes often narrower, surfaces moderately pubescent. **Racemes** considerably elongated in fruit. **Fruiting pedicels** erect to erect-ascending, straight, 1.5–3 mm. **Flowers:** sepals ascending, yellowish, oblong, 0.9–1.2 mm, pubescent; petals narrowly oblanceolate, 1.4–2 × 0.4–0.6 mm; median filaments 1.6–2 mm; anthers 0.2–0.3 mm. **Fruits** erect, (subappressed to rachis basally), narrowly linear, strongly torulose, 6–15 × 0.6–0.8 mm, (curved outward distally); valves each with obscure midvein, (pubescent); septum not veined; ovules 10–18 per ovary; style obsolete, 0.1–0.3 mm, sparsely pubescent. **Seeds** uniseriate, reddish brown, oblong, 1–1.3 × 0.5–0.6 mm.

Flowering Jun–Jul. Rocky slopes at bases of cliffs; of conservation concern; ca. 3100 m; Wyo.

Descurainia torulosa is known from Fremont, Park, Sweetwater, and Teton counties. It was studied by J. S. Bricker et al. (2000), who concluded that its status and relationship to *D. incana* remain unresolved as they were unable to find a set of morphological characters that consistently separated the two taxa. We believe that *D. torulosa* should be maintained, and that it is easily distinguished from *D. incana* by having fruits strongly torulose (versus weakly or not torulose), pubescent (versus glabrous), and distinctly curved (versus straight) distally, sparsely pubescent (versus glabrous) styles, fruit valves each with obscure (versus distinct) midvein, and stems decumbent, unbranched, and often several (versus stems erect, branched, and simple) from the base.

Descurainia torulosa is in the Center for Plant Conservation's National Database of Endangered Plants.

SELECTED REFERENCE Bricker, J. S., G. K. Brown, and T. L. Patts Lewis. 2000. Status of *Descurainia torulosa* (Brassicaceae). W. N. Amer. Naturalist 60: 426–432.

53. HORNUNGIA Reichenbach in H. G. L. Reichenbach et al., Deutschl. Fl. 1: 33. 1837 • [For Ernst Gottfried Hornung, 1795–1862, German pharmacist in Schwarzburg]

Ihsan A. Al-Shehbaz

Hutchinsia W. T. Aiton; *Hutchinsiella* O. E. Schulz; *Hymenolobus* Nuttall; *Microcardamum* O. E. Schulz; *Pritzelago* Kuntze

Annuals [perennials]; [caudex branched]; not scapose; glabrous or puberulent, trichomes minutely branched and subsessile, mixed with simple ones. **Stems** erect, ascending, or decumbent [procumbent], branched or, rarely, unbranched. **Leaves** basal and cauline [or cauline absent]; petiolate or subsessile; basal rosulate or not, petiolate, blade margins entire, dentate, or pinnatisect; cauline petiolate or subsessile, blade margins pinnatisect, pinnatifid, dentate, or entire. **Racemes** (corymbose), elongated or not in fruit. **Fruiting pedicels** divaricate, slender. **Flowers:** sepals spreading or reflexed, ovate [or oblong], (glabrous or puberulent); petals white, spatulate, [obovate, oblong, or oblanceolate], (longer or shorter than sepals), claw absent, (apex obtuse or rounded); stamens (rarely 4), subtetradynamous; filaments often dilated basally; anthers ovate, (apex obtuse); nectar glands lateral, 1 on each side of lateral stamen, median glands present or absent. **Fruits** silicles, sessile, oblong, elliptic, obovoid [ovoid, suborbicular, lanceoloid], keeled, angustiseptate; valves each with prominent midvein, glabrous; replum rounded; septum complete; ovules [4–]10–24 per ovary; style usually obsolete (rarely to 0.5 mm); stigma capitate. **Seeds** biseriate or aseriate, plump, not winged, oblong; seed coat (obscurely reticulate) mucilaginous or not when wetted; cotyledons incumbent, rarely accumbent. $x = 6$.

Species 3 (1 in the flora): introduced; Europe, Asia, n Africa; introduced also in Mexico, South America, s Africa, Australia.

R. C. Rollins (1993) recognized *Hutchinsia* to include the more commonly used *Hymenolobus*. As shown by F. K. Meyer (1982), *Hutchinsia* is an illegitimate name because, when described, it included the type of the earlier published *Iberis* Linnaeus, and he suggested that the name be replaced by *Pritzelago*. O. Appel and I. A. Al-Shehbaz (1998) demonstrated that the differences separating *Hymenolobus* and *Pritzelago* from *Hornungia* are trivial, and they adopted the latter for the combined genus. Molecular data (K. Mummenhoff et al. 2001) clearly support the placement of these taxa in one genus.

SELECTED REFERENCE Appel, O. and I. A. Al-Shehbaz. 1998. Generic limits and taxonomy of *Hornungia*, *Pritzelago*, and *Hymenolobus* (Brassicaceae). Novon 7: 338–340.

1. Hornungia procumbens (Linnaeus) Hayek, Repert. Spec. Nov. Regni Veg. Beih. 30: 480. 1925

Lepidium procumbens Linnaeus, Sp. Pl. 2: 643. 1753; *Capsella elliptica* C. A. Meyer; *C. procumbens* (Linnaeus) Fries; *Hutchinsia procumbens* (Linnaeus) Desvaux; *Hymenolobus divaricatus* Nuttall; *H. erectus* Nuttall; *H. procumbens* (Linnaeus) Nuttall; *Noccaea procumbens* (Linnaeus) Reichenbach; *Thlaspi procumbens* (Linnaeus) Wallroth

Plants glabrous throughout or puberulent, trichomes usually minutely forked, some simple. **Stems** (0.2–)0.5–2.2(–3) dm. **Basal leaves:** petiole (0.2–)0.5–1.2(–2) cm; blade obovate, oblanceolate, or oblong, (0.2–)1–2.5(–4) cm × (1–)5–11(–16) mm, base cuneate or attenuate, apex acute or obtuse. **Cauline leaves** (distal) subsessile; blade similar to basal (smaller distally). **Racemes** few- to several-flowered (rachis straight or slightly flexuous in fruit). **Fruiting pedicels** 3–8(–12) mm. **Flowers:** sepals 0.6–1.1 × 0.4–0.6 mm; petals 0.6–1.2 × 0.3–0.6 mm; filaments (white), 0.5–1 mm; anthers ca. 0.1 mm. **Fruits** (0.2–)0.3–0.4(–0.5) cm × (1–)1.4–2.2 mm, apex subtruncate to slightly emarginate. **Seeds** 0.5–0.6 × 0.3–0.4 mm. $2n = 12, 24$.

Flowering Feb–Jul. Deserts, meadows, shade of bushes, disturbed habitats, waste places, saline banks, salt marshes, sagebrush plains, alkaline flats, slopes; 0–2600 m; introduced; B.C., Man., Nfld. and Labr.

(Nfld.), Sask.; Calif., Colo., Idaho, Mont., Nev., Oreg., Utah, Wash., Wyo.; Europe; Asia; n Africa; introduced also in Mexico, South America, s Africa, Australia.

Hornungia procumbens is highly variable, especially in fruit size and shape, number of seeds per fruit, indumentum, plant size, and shape and number of leaf divisions. Many of its morphological extremes were recognized at specific and infraspecific ranks, and more than 40 synonyms exist.

54. TROPIDOCARPUM Hooker, Icon. Pl. 1: plate 43. 1836 • [Greek *tropis*, keel, and *karpos*, fruit, alluding to fruit shape]

Ihsan A. Al-Shehbaz

Agallis Philippi; *Twisselmannia* Al-Shehbaz

Annuals; not scapose; pubescent or glabrous, trichomes simple, these mixed with smaller, short-stalked, Y-shaped, forked ones, rarely rays branched. **Stems** (usually several from base, rarely simple), usually ascending, decumbent, or prostrate, rarely erect, unbranched or branched [basally] distally, (slender). **Leaves** basal and cauline; petiolate or sessile; basal (withered by fruiting), not rosulate, petiolate, blade margins pinnatifid or pinnatisect; cauline sessile or subsessile, blade margins pinnatifid. **Racemes** (several-flowered, bracteate throughout), considerably elongated in fruit. **Fruiting pedicels** erect, ascending, or divaricate, slender. **Flowers:** sepals erect to ascending, oblong, (glabrous or pubescent); petals yellow, sometimes with purple tinge, oblanceolate, obovate, or spatulate, (slightly longer than sepals), claw undifferentiated from blade; stamens slightly tetradynamous; filaments sometimes slightly dilated basally; anthers ovate, (apex obtuse); nectar glands confluent, subtending bases of stamens. **Fruits** siliques or silicles, sessile, linear, oblong, or obdeltoid, smooth, strongly angustiseptate; valves [(2 or) 4 in *T. capparideum*, keeled], not veined, (thin-leathery throughout, or distally subwoody or thick-leathery and tuberculate-rugose), pubescent; replum rounded; septum complete or absent; ovules 4–70 per ovary; style distinct, (slender); stigma capitate. **Seeds** uniseriate or (in 4-valved fruits) 4-seriate, not winged, oblong; seed coat (reticulate), mucilaginous when wetted; cotyledons incumbent. $x = 8$.

Species 4 (3 in the flora): California, c Mexico, South America (c Chile).

Tropidocarpum lanatum (Barnéoud) Al-Shehbaz & R. A. Price is known from central Chile.

SELECTED REFERENCES Al-Shehbaz, I. A. 2003c. A synopsis of *Tropidocarpum* (Brassicaceae). Novon 13: 392–395. Al-Shehbaz, I. A. and R. A. Price. 2001. The Chilean *Agallis* and Californian *Tropidocarpum* (Brassicaceae) are congeneric. Novon 11: 292–293. Robinson, B. L. 1896. The fruit of *Tropidocarpum*. Erythea 4: 108–119.

1. Fruits obdeltoid, valves thick-leathery and tuberculate-rugose distally, puberulent, trichomes antrorse; ovules 4–8 per ovary...3. *Tropidocarpum californicum*
1. Fruits linear or oblong, valves thin-leathery and smooth distally, usually pubescent or puberulent, rarely glabrescent, trichomes retrorse; ovules 25–70 per ovary.
    2. Fruits narrowly linear, (25–)30–60(–70) × 1.5–2(–3) mm, length 13–46 times width; valves 2...1. *Tropidocarpum gracile*
    2. Fruits oblong, (5–)9–20 × (3–)4–5 mm, length (1.6–)2.8–5 times width; valves (2 or) 4..2. *Tropidocarpum capparideum*

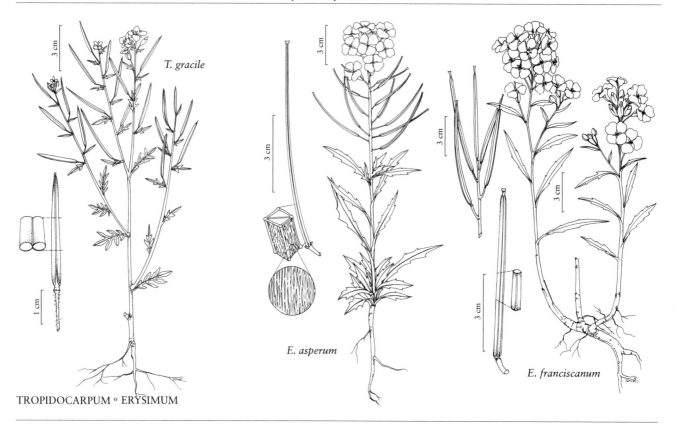

TROPIDOCARPUM ○ ERYSIMUM

1. **Tropidocarpum gracile** Hooker, Icon. Pl. 1: plate 43. 1836 [F] [W]

Tropidocarpum dubium Davidson; *T. gracile* var. *dubium* (Davidson) Jepson; *T. gracile* var. *scabriusculum* (Hooker) Greene; *T. macrocarpum* Hooker & Harvey ex Greene; *T. scabriusculum* Hooker

Plants sparsely to densely hirsute basally, trichomes simple, to 1.5 mm, these mixed with fewer, forked, stalked ones, sparsely pubescent distally. **Stems** usually ascending, prostrate, or decumbent, rarely erect, (0.4–)1–4.5(–6) dm. **Basal leaves:** petiole (0.3–)1–2.5(–3.5) cm; blade (1.5–)2.5–10(–15) cm, margins pinnatifid to pinnatisect; lobes 3–8(–12) on each side, ovate, oblong, or linear, (0.1–)0.6–2(–3) cm × (0.5–)1–3(–6) mm, shorter than terminal, margins entire, dentate, or pinnatifid. **Cauline leaves:** (proximal) petiolate or (distal and bracts) sessile; blade similar to basal, smaller and less divided distally. **Fruiting pedicels** erect to ascending, straight, (4–)6–17(–35) mm, pubescent. **Flowers:** sepals 2.5–4 × 0.7–1.5 mm, glabrous or pubescent; petals obovate to spatulate, 3–6 × 1.5–4 mm, not clawed; filaments 2.5–4 mm; anthers 0.5–0.6 mm. **Fruits** narrowly linear, (25–)30–60(–70) × 1.5–2(–3) mm, length 13–46 times width; valves 2, thin-leathery, smooth, usually pubescent or puberulent, rarely glabrescent, trichomes simple, retrorse; septum incomplete or absent; ovules 30–70 per ovary; style (0.3–)0.8–2.5(–4) mm. **Seeds** dark brown, 1.2–1.6 × 0.7–1 mm. $2n = 16$.

Flowering Mar–May. Roadsides, fields, hillsides, grassy pastures, sandy washes, sage scrub, chaparral, ravines, beaches, gravel flats; 0–1200 m; Calif.; Mexico (Baja California).

Tropidocarpum gracile is known from central and coastal to southernmost California.

2. **Tropidocarpum capparideum** Greene, Pittonia 1: 217. 1888 [C] [E]

Plants sparsely to densely pilose basally, trichomes (soft), simple, to 1.5 mm, these mixed with fewer, forked, stalked ones, sparsely pubescent distally. **Stems** usually ascending, prostrate, or decumbent, rarely erect, 1.5–7 dm. **Basal leaves:** petiole 1–3 cm; blade 1.2–7 cm, margins pinnatifid; lobes 3–6 on each side, oblong to linear, 0.3–1.5 cm × 1–3 mm, shorter than terminal, margins entire or dentate. **Cauline leaves:** (proximal) petiolate or (distal and bracts) sessile; blade similar to basal, smaller and less divided distally. **Fruiting pedicels** divaricate to ascending, straight or slightly recurved, 5–17(–25) mm, pubescent. **Flowers:** sepals 2.5–3.5 × 1–1.5 mm, sparsely pubescent; petals spatulate, 3–5 ×

1.5–2 mm, not clawed; filaments 2–2.5 mm; anthers ca. 0.5 mm. **Fruits** oblong, (5–)9–20 × (3–)4–5 mm, length (1.6–)2.8–5 times width; valves (2 or) 4, thin-leathery, smooth, puberulent, trichomes simple, retrorse; septum absent; ovules 25–40 per ovary; style 1–2 mm. **Seeds** dark brown, 1.2–1.6 × 0.7–1 mm.

Flowering and fruiting Mar–Apr. Flats, grassland, moderately alkaline areas, hillsides; of conservation concern; 300–400 m; Calif.

Tropidocarpum capparideum was believed to be restricted to Mt. Diablo (Contra Costa County) and to have become extinct, but new collections have been made from Fort Hunter Leggett in Monterey County.

Tropidocarpum capparideum is in the Center for Plant Conservation's National Collection of Endangered Plants.

3. **Tropidocarpum californicum** (Al-Shehbaz) Al-Shehbaz, Novon 13: 393. 2003 [C] [E]

Twisselmannia californica Al-Shehbaz, Novon 9: 133, fig, 1. 1999

Plants pilose basally, trichomes (soft), simple, to 1.5 mm, these rarely mixed with fewer, forked, stalked ones, sparsely pubescent distally. **Stems** ascending or decumbent, 0.3–2.5 dm. **Basal leaves** not seen. **Cauline leaves:** (proximal) petiole to 1 cm or (distal and bracts) subsessile; blade 2.5–4.5 cm (smaller distally), margins pinnatifid to pinnatisect (less divided distally); lobes 2–4 on each side, oblong to linear, 0.3–1.5 cm × 1–3 mm, shorter than terminal, margins entire. **Fruiting pedicels** divaricate-ascending, straight, 3–10(–28) mm, pilose. **Flowers:** sepals 1.2–2 × 0.7–0.9 mm, glabrous; petals (not with purple tinge), oblanceolate to spatulate, 1.6–2.5 × 0.7–0.9 mm, cuneate into short claw; filaments 1.2–1.7 mm; anthers 0.3–0.4 mm. **Fruits** obdeltoid, 4–5 × 4–5 mm, length equal to width; valves 2, thin-leathery proximally, thick-leathery to subwoody and tuberculate-rugose on outside distally, puberulent, trichomes simple, antrorse, these sometimes mixed with minute, 1-forked ones; septum present; ovules 4–8 per ovary; style 0.3–0.9 mm. **Seeds** brown, 1.2–1.5 × 0.6–0.9 mm.

Flowering Mar. Subalkaline clay in scrub; of conservation concern (as *Twisselmannia californica*); ca. 70 m; Calif.

Tropidocarpum californicum is known from adjacent parts of Kern and King counties in San Joaquin Valley.

o. **BRASSICACEAE** Burnett tribe **ERYSIMEAE** Dumortier, Fl. Belg., 123. 1827

Annuals, biennials, perennials, or subshrubs [shrubs]; eglandular. **Trichomes** sessile, stellate or malpighiaceous. **Cauline leaves** petiolate or sessile; blade base not auriculate, margins entire, dentate, denticulate, dentate-sinuate, or repand. **Racemes** usually ebracteate [bracteate], elongated [not elongated] in fruit. **Flowers** actinomorphic; sepals erect, lateral pair saccate or not basally; petals usually yellow or orange, rarely pink or purple [white], claw present, distinct; filaments unappendaged, not winged; pollen 3-colpate. **Fruits** siliques [silicles], dehiscent, unsegmented, 4-angled, terete, or latiseptate, rarely angustiseptate; ovules [10–]15–100[–numerous] per ovary; style obsolete or distinct; stigma entire or 2-lobed. **Seeds** biseriate or uniseriate; cotyledons usually accumbent, rarely incumbent.

Genus 1, species ca. 150 (19 in the flora): North America, Mexico, Central America, Europe, Asia, n Africa, Atlantic Islands (Macaronesia).

55. ERYSIMUM Linnaeus, Sp. Pl. 2: 660. 1753; Gen. Pl. ed. 5, 296. 1754 • Wallflower
[Greek *eryso*, to ward off or to cure, alluding to the supposed medicinal properties of some species]

Ihsan A. Al-Shehbaz

Cheiranthus Linnaeus; *Cheirinia* Link; *Cuspidaria* (de Candolle) Besser; *Syrenia* Andrzejowski ex Besser

Plants not scapose; pubescent, trichomes sessile, medifixed, appressed, 2-rayed (malpighiaceous) or 3–5(–8)-rayed (stellate), rays (when 2) parallel to long axis of stems, leaves, sepals, and fruits. **Stems** erect or ascending [decumbent], unbranched or branched basally and/or distally. **Leaves** basal and cauline; petiolate or sessile; basal rosulate or not, petiolate, blade margins usually entire, dentate, sinuate-dentate, or denticulate, rarely pinnatifid or pinnatisect; cauline petiolate or sessile, blade (base cuneate or attenuate [auriculate]), margins entire, dentate, denticulate, dentate-sinuate, or repand. **Racemes** (densely flowered, *E. pallasii* bracteate basally). **Fruiting pedicels** erect, ascending, divaricate, reflexed, horizontal, or spreading, slender or stout (nearly as wide as fruit). **Flowers:** sepals oblong or linear, lateral pair saccate or not basally (pubescent); petals suborbicular, obovate, or spatulate, claw differentiated from blade (subequaling or longer than sepals, apex rounded [emarginate]); stamens (erect), tetradynamous; filaments not dilated basally; anthers oblong or linear; nectar glands (1, 2, or 4), distinct or confluent, subtending bases of stamens, median glands present or absent. **Fruits** usually sessile, rarely shortly stipitate (gynophore to 4 mm), usually linear or narrowly so [oblong], smooth or torulose, (keeled or not); valves each with obscure to prominent midvein, pubescent outside, usually glabrous inside; replum rounded; septum complete, (not veined); ovules [15–]20–120 per ovary; (style relatively short, rarely ½ as long as or subequaling fruit, often pubescent); stigma capitate. **Seeds** plump or flattened, winged, margined, or not winged, oblong, ovoid, obovate, or suborbicular; seed coat (minutely reticulate), mucilaginous when wetted; cotyledons incumbent, rarely accumbent. $x = (6)$ 7, 8 (9–17).

Species ca. 150 (19 in the flora): North America, n Mexico, Central America, Europe, Asia, n Africa, Atlantic Islands (Macaronesia); introduced in South America, Australia.

Erysimum is found in the northern hemisphere, primarily Asia and Europe, with eight species in northern Africa and Macaronesia, and one each endemic to Baja California (*E. moranii* Rollins) and Costa Rica and Guatemala (*E. ghiesbreghtii* J. D. Smith). Of the 21 species found in North America, four are naturalized. Most of the native species have $x = 9$ and are believed to represent a monophyletic group (R. A. Price 1987).

Erysimum is a taxonomically difficult genus much in need of comprehensive phylogenetic and systematic studies covering its entire range. The principal sources of difficulty are the inflation in the number of species described, the heavy reliance on vegetative morphological characters in the delimitation of species, and the inadequacy of most herbarium specimens. In order to reliably identify a given sample, one often needs a complete specimen that has basal leaves, flowers, mature fruits, and seeds. Unfortunately, plants of most species shed their basal leaves or have no flowers when at full fruit maturity. Another complicating factor in North America is that almost all of the native species readily hybridize in areas of overlap to produce wide arrays of intermediates that backcross with the parents and blur species boundaries.

SELECTED REFERENCES Price, R. A. 1987. Systematics of the *Erysimum capitatum* Alliance (Brassicaceae) in North America. Ph.D. dissertation. University of California, Berkeley. Rossbach, G. B. 1940. *Erysimum* in North America. Ph.D. dissertation. Stanford University. Rossbach, G. B. 1958. The genus *Erysimum* (Cruciferae) in North America north of Mexico—A key to the species and varieties. Madroño 14: 261–267.

1. Petals 3–10(–15) × 1.5–3 mm; median filaments 2–7(–10) mm.
   2. Fruit valves densely pubescent inside; sepals 1.8–3.2 mm; petal claws 1.5–3.5 mm... .... 5. *Erysimum cheiranthoides*
   2. Fruit valves usually glabrous inside, rarely sparsely pubescent; sepals 4–7(–8) mm; petal claws 3–8 mm.
      3. Annuals; fruiting pedicels as wide as fruit .... 17. *Erysimum repandum*
      3. Biennials or perennials (short-lived); fruiting pedicels narrower than fruit.
         4. Stigmas entire; fruits appressed to rachises; leaf blade surfaces and fruit valves with 3- and 4-rayed trichomes .... 10. *Erysimum hieracifolium*
         4. Stigmas strongly 2-lobed; fruits not appressed or subappressed to rachises; leaf blade surfaces and fruit valves with 2- or 3-rayed trichomes.
            5. Petals narrowly obovate to spatulate, (8–)10–15 × (2–)2.5–4 mm; fruits 1.8–2.5 mm wide; seeds 1.5–2 mm .... 7. *Erysimum coarctatum*
            5. Petals oblanceolate, 6–9(–11) × 1–2 mm; fruits 1.2–1.7 mm wide; seeds 1.2–1.7 × 0.8–1 mm .... 11. *Erysimum inconspicuum*
1. Petals (10–)13–30(–35) × 3–10(–15) mm; median filaments (6–)7–15 mm.
   6. Subshrubs (stems woody at base).
      7. Fruit valve trichomes 2-rayed; stigmas strongly 2-lobed, lobes much longer than wide; petals orange, yellow, brown, red, purple, violet, or white .... 6. *Erysimum cheiri*
      7. Fruit valve trichomes 2–4-rayed; stigmas not strongly 2-lobed, lobes as long as wide; petals yellow or cream.
         8. Fruits angustiseptate; leaf trichomes 2- or 3-rayed .... 12. *Erysimum insulare*
         8. Fruits latispetate or 4-angled; leaf trichomes 2–5-rayed.
            9. Distal cauline leaves petiolate; fruits latiseptate, not 4-angled; fruiting pedicels stout, 5–17(–22) mm .... 9. *Erysimum franciscanum*
            9. Distal cauline leaves sessile; fruits slightly latiseptate or 4-angled; fruiting pedicels slender, (3–)5–10 mm .... 18. *Erysimum suffrutescens*
   6. Biennials or perennials (stems not woody at base).
      10. Petals purple or lilac; leaf trichomes 2-rayed; sepals 5–9 mm; anthers 1–1.5 mm .... 15. *Erysimum pallasii*
      10. Petals usually yellow or orange, rarely lavender or purplish; leaf trichomes (at least some) 3–7-rayed; sepals 7–14 mm; anthers 2–4 mm.
         11. Basal leaf blades filiform to narrowly linear, (somewhat revolute, appearing terete) .... 19. *Erysimum teretifolium*
         11. Basal leaf blades not filiform or narrowly linear.
            12. Fruits 4-angled, longitudinally 4-striped; valves densely pubescent between midvein and replum with 2-rayed trichomes; ovules 72–120 per ovary .... 3. *Erysimum asperum*
            12. Fruits latiseptate, rarely 4-angled, not longitudinally striped; valves pubescent with 2–6-rayed trichomes; ovules 24–86 per ovary.
               13. Leaf blades: surfaces with 2 (or 3)-rayed trichomes.
                  14. Fruits usually 4-angled, rarely latiseptate; ovules (40–)54–82 per ovary; seeds not winged, 1.5–2(–2.4) mm .... 4. *Erysimum capitatum* (in part)
                  14. Fruits strongly latiseptate; ovules 24–46 per ovary; seeds usually broadly winged all around or apically, rarely not winged, 2–3.5 mm.
                     15. Perennials; seeds not winged or winged distally; fruits 1.5–2.7 mm wide .... 2. *Erysimum arenicola*
                     15. Biennials; seeds winged all around; fruits (2–)2.4–3.7 mm wide .... 14. *Erysimum occidentale*
               13. Leaf blades: surfaces with 2–5(–7)-rayed trichomes.
                  16. Fruits torulose; ovules 26–44 per ovary; petals 3.5–6 mm wide .... 16. *Erysimum perenne*
                  16. Fruits not torulose; ovules (32–)42–86 per ovary; petals (5–)6–16 mm wide.

[17. Shifted to left margin.—Ed.]

17. Fruiting pedicels 2–4(–6) mm; seeds broadly obovate to suborbicular, 1.5–3 mm wide . . . . 8. *Erysimum concinnum*
17. Fruiting pedicels 4–17(–25) mm; seeds oblong, 1–2 mm wide.
  18. Fruits divaricate, ascending, or erect; petals orange or orange-yellow to yellow; seeds winged distally; Midwestern, Mountain, Pacific states . . . . 4. *Erysimum capitatum* (in part)
  18. Fruits spreading; petals yellow; seeds winged all around; California (Humboldt, Mendocino, Monterrey, Santa Cruz counties).
    19. Basal leaf blades linear-oblanceolate, 2–9 mm wide; stems 0.4–9(–13) dm; fruit valves each with prominent midvein; ovules 50–86 per ovary . . . . 1. *Erysimum ammophilum*
    19. Basal leaf blades spatulate, 5–15 mm wide; stems 0.2–2.5(–3.5) dm; fruit valves each with obscure midvein; ovules 32–74 per ovary . . . . 13. *Erysimum menziesii*

1. Erysimum ammophilum A. Heller, Muhlenbergia 1: 51. 1904 [C] [E]

Cheiranthus ammophilus A. Heller; *Cheirinia ammophila* (A. Heller) A. Heller

Biennials or perennials; (short-lived). **Trichomes** of leaves 2–4(–7)-rayed. **Stems** erect, unbranched or branched distally, 0.4–9(–13) dm. **Basal leaves:** blade (somewhat fleshy), linear-oblanceolate, 3.5–15.5 cm × 2–9 mm, base attenuate, margins entire or obscurely dentate, apex acute. **Cauline leaves:** blade (oblanceolate), margins entire. **Racemes** considerably elongated in fruit. **Fruiting pedicels** divaricate to divaricate-ascending, stout, slightly narrower than fruit, 4–10(–13) mm. **Flowers:** sepals oblong to linear-oblong, 7.5–11(–13) mm, lateral pair slightly saccate basally; petals bright yellow, broadly obovate to suborbicular, 14–24 × 6–11(–14) mm, claw 8–14 mm, apex rounded; median filaments 8.5–12 mm; anthers linear, 3–4 mm. **Fruits** spreading, narrowly linear, straight, not torulose, (2–)3.5–12 cm × 1.5–3.5 mm, latiseptate, not striped; valves with prominent midvein, pubescent outside, trichomes 2–4-rayed, glabrous inside; ovules 50–86 per ovary; style cylindrical, stout, 0.3–1.5(–2) mm, sparsely pubescent; stigma slightly 2-lobed, lobes as long as wide. **Seeds** oblong, 1.5–3 × 1–1.8 mm; winged distally (narrowly so around). $2n = 36$.

Flowering Mar–Apr. Sand dunes; of conservation concern; 0–50 m; Calif.

Erysimum ammophilum is restricted to coastal and nearby sand dunes of Monterey and Santa Cruz counties.

2. Erysimum arenicola S. Watson, Proc. Amer. Acad. Arts 26: 124. 1891 [E]

Cheiranthus arenicola (S. Watson) Greene; *Erysimum arenicola* var. *torulosum* (Piper) C. L. Hitchcock; *E. torulosum* Piper

Perennials; (caudex simple or many-branched). **Trichomes** of leaves 2- or 3-rayed. **Stems** erect, unbranched or branched (several) basally, 0.4–3 dm. **Basal leaves:** blade usually oblanceolate, rarely spatulate, 2–8 cm × 3–10 mm, base attenuate, margins entire or dentate, apex acute or obtuse. **Cauline leaves** (distal) sessile; blade margins entire or dentate-sinuate. **Racemes** considerably elongated in fruit. **Fruiting pedicels** divaricate-ascending, slender, narrower than fruit, 5–8 mm. **Flowers:** sepals oblong to linear-oblong, 7–10 mm, lateral pair saccate basally; petals yellow, obovate, 14–27 × 3.5–6 mm, claw 9–13 mm, apex rounded; median filaments 9–12 mm; anthers linear, 2–3.5 mm. **Fruits** erect to ascending, narrowly linear, often straight, sometimes twisted, strongly torulose, 3–10(–12) cm × 1.5–2.7 mm, latiseptate, not striped; valves with somewhat prominent midvein, pubescent outside, trichomes 2–4-rayed, glabrous inside; ovules 24–42 per ovary; style cylindrical, slender, 1.5–5 mm, sparsely pubescent; stigma slightly 2-lobed, lobes as long as wide. **Seeds** oblong, 2–3.5 × 0.8–1.5 mm; not winged or winged distally. $2n = 36$.

Flowering Jun–Aug. Rock crevices, talus slopes, alpine areas, open ridges, gravelly ground; 900–2200 m; B.C.; Oreg., Wash.

Erysimum arenicola is distributed at the higher elevations of northern Oregon northward into the Olympic and Cascade mountains in Washington and Vancouver Island.

Both G. B. Rossbach (1958) and R. C. Rollins (1993) recognized *Erysimum arenicola* as a distinct species. It is closely related to *E. perenne* and both can be easily distinguished from *E. capitatum*, with which they hybridize where their ranges meet, by the strongly torulose (versus not torulose) fruits and the longer styles 1.5–5.5 versus 0.2–2.5(–3) mm.

3. **Erysimum asperum** (Nuttall) de Candolle, Syst. Nat. 2: 505. 1821 [E] [F] [W]

Cheiranthus asper Nuttall, Gen. N. Amer. Pl. 2: 69. 1818; *Cheirinia aspera* (Nuttall) Rydberg; *Erysimum asperum* var. *dolichocarpum* O. E. Schulz

Biennials. Trichomes of leaves 2- or 3-rayed. **Stems** erect, unbranched or branched distally, (0.6–)1.2–6.5(–8) dm. **Basal leaves** (often withered by fruiting); blade oblanceolate, 2–10 cm × (0.2–)0.5–1.5(–2.4) mm, base attenuate, margins dentate, apex acute. **Cauline leaves** (distal) sessile; blade margins entire or denticulate. **Racemes** considerably elongated in fruit. **Fruiting pedicels** horizontal to divaricate, slender, narrower than fruit, 5–16(–25) mm. **Flowers:** sepals oblong to linear-oblong, 8–12 mm, lateral pair slightly saccate basally; petals yellow, obovate to suborbicular, 13–22 × 4–9 mm, claw 8–15 mm, apex rounded; median filaments 8–14 mm; anthers linear, 2.5–4 mm. **Fruits** widely spreading or divaricate, narrowly linear, usually straight, rarely curved upward, not torulose, (3–)5–12(–14) cm × 1.2–2.7 mm, 4-angled, strongly (longitudinally) 4-striped; valves with prominent midvein and replum, densely pubescent outside, trichomes 2-rayed between midvein and replum, glabrous inside; ovules 72–120 per ovary; style cylindrical, slender, 1–4 mm, sparsely pubescent; stigma slightly 2-lobed, lobes as long as wide. **Seeds** ovoid, (1–)1.3–2.3 × 0.7–1.2 mm; usually not winged, rarely winged distally. $2n = 36$.

Flowering Apr–Jun(–Aug). Prairies, sand dunes, roadsides, bluffs, sandhills along stream banks, knolls, open plains; 200–2000 m; Alta., Man., Sask.; Ark., Colo., Ill., Kans., Minn., Mont., Nebr., N.Mex., N.Dak., Okla., S.Dak., Tex., Wyo.

4. **Erysimum capitatum** (Douglas ex Hooker) Greene, Fl. Francisc., 269. 1891

Cheiranthus capitatus Douglas ex Hooker, Fl. Bor.-Amer. 1: 38. 1829, based on *C. asper* Chamisso & Schlechtendal, Linnaea 1: 14. 1826, not Nuttall 1818

Biennials or perennials; (short-lived). **Trichomes** of leaves 2–4(–7)-rayed. **Stems** erect, often branched distally, sometimes proximally, (0.5–)1.2–10(–12) dm. **Basal leaves** (often withered by flowering); blade spatulate to narrowly oblanceolate or linear, 2–18(–27) cm × 3–15(–30) mm, base attenuate, margins entire or dentate to denticulate, apex acute. **Cauline leaves** (distal) sessile; blade margins entire or denticulate. **Racemes** considerably elongated in fruit. **Fruiting pedicels** divaricate to ascending, stout or slender, narrower than fruit, 4–17(–25) mm. **Flowers:** sepals narrowly oblong, 7–14 mm, lateral pair saccate basally; petals usually orange to yellow, rarely lavender or purplish, suborbicular to obovate, 12–25(–30) × (5–)6–10(–13) mm, claw 8–16 mm, apex rounded; median filaments 9–18 mm; anthers linear, 3–4 mm. **Fruits** divaricate or ascending to erect, narrowly linear, straight or curved upward, not torulose, 3.5–11(–15) cm × 1.3–3.3 mm, 4-angled to latiseptate, not striped; valves with prominent midvein, pubescent outside, trichomes 2–5-rayed, glabrous inside; ovules (40–)54–82 per ovary; style cylindrical, usually stout, rarely slender, 0.2–2.5(–3) mm, sparsely pubescent; stigma 2-lobed, lobes as long as wide. **Seeds** oblong, 1.5–4 × 1–2 mm; winged apically or not winged. $2n = 36$.

Varieties 2 (2 in the flora): North America, n Mexico.

Erysimum capitatum is extremely widespread, ecologically diverse, morphologically highly variable, and nomenclaturally complex. It has been divided into infraspecific taxa; G. B. Rossbach (1958) recognized three varieties, R. A. Price (1987) eight subspecies, R. C. Rollins (1993) five varieties, Price (1993) four subspecies, and N. H. Holmgren (2005b) three varieties. The majority of these taxa were based on highly variable characters with considerable overlap, including width, margin, shape, and apex of basal leaves, fruit orientation, and development of elongated versus short caudices. Furthermore, there is a substantial degree of disagreement among these authors as to the limits of a given infraspecific taxon, its distribution, and synonymies involved. Except for two distinctive species that were previously treated as infraspecific taxa (*arenicola*, *perenne*) the vast majority of the remaining variation in *E. capitatum* is divided herein into two varieties.

1. Seeds winged (at least distally, rarely wing rudimentary), 2–4 × (0.8–)1–2 mm; fruits latiseptate, rarely 4-angled; petals usually orange, sometimes orange-yellow or yellow; adaxial surfaces of basal and proximalmost cauline leaf blades with mostly 3(–7)-rayed trichomes . . . . 4a. *Erysimum capitatum* var. *capitatum*
1. Seeds not winged, 1.5–2(–2.4) × 0.7–1.2 mm; fruits 4-angled, rarely latiseptate; petals yellow, rarely lavender or purplish; adaxial surfaces of basal and proximalmost cauline leaf blades with mostly 2- or 3-rayed trichomes . . . . 4b. *Erysimum capitatum* var. *purshii*

4a. Erysimum capitatum (Douglas ex Hooker) Greene var. capitatum C

Cheiranthus angustatus Greene;
C. arkansanus (Nuttall) Greene;
C. californicus (Greene) Greene;
C. elatus (Nuttall) Greene;
C. pacificus E. Sheldon; *C. wheeleri* (Rothrock) Greene; *Cheirinia arkansana* (Nuttall) Moldenke;
C. elata (Nuttall) Rydberg;
C. wheeleri (Rothrock) Rydberg;
Erysimum arkansanum Nuttall; *E. asperum* (Nuttall) de Candolle var. *arkansanum* (Nuttall) A. Gray; *E. asperum* var. *bealianum* Jepson; *E. asperum* var. *capitatum* (Douglas ex Hooker) B. Boivin; *E. asperum* var. *elatum* (Nuttall) Torrey; *E. asperum* var. *stellatum* J. T. Howell; *E. californicum* Greene; *E. capitatum* var. *angustatum* (Greene) Rossbach; *E. capitatum* var. *bealianum* (Jepson) Rossbach; *E. capitatum* var. *stellatum* (J. T. Howell) Twisselmann; *E. capitatum* var. *washoense* Rossbach; *E. elatum* Nuttall; *E. insulare* Greene var. *angustatum* (Greene) Jepson; *E. moniliforme* Eastwood; *E. suffrutescens* (Abrams) Rossbach var. *lompocense* Rossbach; *E. tilimii* J. Gay; *E. wheeleri* Rothrock

Proximal leaves with mostly 3(–7)-rayed trichomes adaxially. **Flowers:** petals usually orange, sometimes orange-yellow or yellow. **Fruits** usually latiseptate, rarely 4-angled. **Seeds** 2–4 × (0.8–)1–2 mm; winged at least distally or, rarely, wing rudimentary. $2n = 36$.

Flowering (Jan–)Mar–Sep. Hillsides, open slopes, valley bottoms, alpine areas, deserts, woodlands, sandy mesas, chaparral clearings; of conservation concern; 0–1700 m; Ariz., Ark., Calif., Idaho, Ill., Ind., Iowa, Mo., Nev., Ohio, Okla., Oreg., Tenn., Tex., Wash.; Mexico (Durango, San Luis Potosí).

Although its overall distribution is extensive, var. *capitatum* has been collected only sporadically outside the main range in western Idaho, western Nevada, and the Pacific states. There is some local differentiation in California that has been recognized formally. For example, some populations in the Mohave desert in Kern, Los Angeles, and San Bernardino counties, as well as disjunct ones in eastern San Luis Obispo County, differ from typical var. *capitatum* by having yellow petals, fruits to 3.3 mm wide, and seeds to 4 × 2 mm; these were recognized by G. B. Rossbach (1958) and R. C. Rollins (1993) as var. *bealianum*. Variety *angustatum*, which is highly localized in Contra Costa County and was recognized by both Rossbach and Rollins, differs from typical var. *capitatum* by having elongated (versus not elongated) woody caudices, 4-angled (versus latiseptate) fruits, and much-branched (versus moderately-branched or simple) fruiting racemes.

4b. Erysimum capitatum (Douglas ex Hooker) Greene var. purshii (Durand) Rollins, Cruciferae Continental N. Amer., 482. 1993

Erysimum asperum (Nuttall) de Candolle var. *purshii* Durand, Trans. Amer. Philos. Soc., n. s. 11: 159. 1859; *Cheiranthus alpestris* (Cockerell) A. Heller; *C. argillosus* Greene; *C. aridus* A. Nelson; *C. asperrimus* Greene; *C. bakeri* Greene; *C. nivalis* Greene; *C. nivalis* var. *amoenus* Greene; *C. oblanceolatus* (Rydberg) A. Heller; *C. radicatus* (Rydberg) A. Heller; *Cheirinia amoena* (Greene) Rydberg; *C. argillosa* (Greene) Rydberg; *C. arida* (A. Nelson) Rydberg; *C. asperrima* (Greene) Rydberg; *C. bakeri* (Greene) Rydberg; *C. brachycarpa* Rydberg; *C. cockerelliana* (Daniels) Cockerell; *C. desertorum* Wooton & Standley; *C. nivalis* (Greene) Rydberg; *C. nivalis* var. *radicata* (Rydberg) Cockerell; *C. oblanceolata* (Rydberg) Rydberg; *C. radicata* (Rydberg) Rydberg; *E. amoenum* (Greene) Rydberg; *E. angustatum* Rydberg; *E. argillosum* (Greene) Rydberg; *E. aridum* (A. Nelson) A. Nelson; *E. asperrimum* (Greene) Rydberg; *E. asperum* var. *amoenum* (Greene) Reveal; *E. asperum* var. *angustatum* (Rydberg) B. Boivin; *E. asperum* var. *pumilum* S. Watson; *E. bakeri* (Greene) Rydberg; *E. capitatum* var. *amoenum* (Greene) R. J. Davis; *E. capitatum* var. *argillosum* (Greene) R. J. Davis; *E. capitatum* var. *nivale* (Greene) N. H. Holmgren; *E. cockerellianum* Daniels; *E. desertorum* (Wooton & Standley) Rossbach; *E. nivale* (Greene) Rydberg; *E. oblanceolatum* Rydberg; *E. radicatum* Rydberg

Proximal leaves with mostly 2- or 3-rayed trichomes adaxially. **Flowers:** petals usually yellow, rarely lavender or purplish. **Fruits** usually 4-angled, rarely latiseptate. **Seeds** 1.5–2(–2.4) × 0.7–1.2 mm; not winged. $2n = 36$.

Flowering Apr–Sep. Meadows, dry slopes, hillsides; 1000–3800 m; Yukon; Alaska, Ariz., Calif., Colo., Idaho, Mont., Nev., N.Mex., Tex., Utah, Wyo.; Mexico (Chihuahua, Michoacán, Sonora, Zacatecas).

Plants of var. *purshii* were placed by G. B. Rossbach (1958) in three species (*Erysimum angustatum*, *E. argillosum*, *E. desertorum*) and by R. C. Rollins (1993) in two species (*E. angustatum*, *E. capitatum*). The considerable overlap in morphology and the frequent hybridization between vars. *capitatum* and *purshii* in areas where their geographical ranges meet argue against such delimitation. Forms with linear leaves to 3 mm wide, which are restricted to eastern Alaska and western Yukon, were recognized by Rossbach and Rollins as *E. angustatum*, but these hardly differ from typical var. *purshii* in all other characters.

5. Erysimum cheiranthoides Linnaeus, Sp. Pl. 2: 661. 1753 • Wormseed mustard [I][W]

Cheiranthus cheiranthoides (Linnaeus) A. Heller; *Cheirinia cheiranthoides* (Linnaeus) Link

Annuals. Trichomes of leaves primarily 3- or 4-rayed, sometimes mixed with 5-rayed ones, 2-rayed trichomes primarily on stem and pedicels. **Stems** erect, often branched distally, (ribbed), (0.7–)1.5–10(–15) dm. **Basal leaves** (withered by fruiting), similar to cauline. **Cauline leaves:** (median and distal) shortly petiolate or sessile; blade (lanceolate, linear, or elliptic-oblong, (1–)2–7(–11) cm × (2–)5–12 (–23) mm, base cuneate) margins usually subentire or denticulate, rarely sinuate-dentate, (apex acute or obtuse). **Racemes** considerably elongated in fruit. **Fruiting pedicels** divaricate or ascending, slender, much narrower than fruit, 5–13(–16) mm. **Flowers:** sepals oblong, 1.8–3.2 mm, lateral pair not saccate basally; petals yellow, narrowly spatulate, 3–5.5 × 1.5–2 mm, claw 1.5–3.5 mm, apex rounded; median filaments 2–3.5 mm; anthers oblong, 0.5–0.7 mm. **Fruits** suberect or divaricate-ascending, linear, straight, somewhat torulose, (1–)1.5–2.5(–4) cm × 1–1.3 mm, 4-angled, not striped; valves with prominent midvein, pubescent outside, trichomes 3–5-rayed, densely pubescent inside; ovules (20–)30–55 per ovary; style cylindrical, slender, 0.5–1.5 mm; stigma entire or slightly 2-lobed, lobes as long as wide. **Seeds** oblong, 1–1.5 × 0.4–0.6 mm; not winged. $2n = 16$.

Flowering Jun–Aug. Dry stream beds, moist areas, waste ground, fields, pastures, disturbed ground; 0–3000 m; introduced; Alta., B.C., Man., N.B., Nfld. and Labr. (Nfld.), N.W.T., N.S., Nunavut, Ont., P.E.I., Que., Sask., Yukon; Alaska, Ark., Calif., Colo., Conn., D.C., Fla., Idaho, Ill., Ind., Iowa, Kans., Maine, Md., Mass., Mich., Minn., Mo., Mont., Nebr., Nev., N.H., N.J., N.Mex., N.Y., N.C., N.Dak., Ohio, Oreg., Pa., R.I., S.Dak., Tenn., Utah, Vt., Va., Wash., W.Va., Wis., Wyo.; Europe; Asia; n Africa.

6. Erysimum cheiri (Linnaeus) Crantz, Cl. Crucif. Emend., 116. 1769 • Wallflower [I]

Cheiranthus cheiri Linnaeus, Sp. Pl. 2: 661. 1753

Biennials or subshrubs. Trichomes of leaves 2-rayed, rarely mixed with fewer 3-rayed ones apically. **Stems** erect, unbranched or branched distally, (woody at base when subshrubs), 1.5–8 dm. **Basal leaves** (rosulate when biennial, often withered by fruiting), similar to cauline. **Cauline leaves** petiolate; blade (obovate to oblanceolate, 4–22 cm × 3–12 mm, base cuneate to attenuate), margins entire to repand. **Racemes** considerably elongated in fruit. **Fruiting pedicels** divaricate-ascending to ascending, slender, narrower than fruit, 7–13 mm. **Flowers:** sepals oblong, 6–10 mm, lateral pair not or slightly saccate basally; petals orange, yellow, brown, red, purple, violet, or white, broadly obovate to suborbicular, 20–35 × 5–10 mm, claw 7–12 mm, apex rounded; median filaments 7–9 mm; anthers linear, 2.5–3.5 mm. **Fruits** ascending, narrowly linear, straight, not torulose, 3–10 cm × 2–7 mm, latiseptate to terete, not striped; valves with prominent midvein, pubescent outside, trichomes 2-rayed, glabrous inside; ovules 32–44 per ovary; style cylindrical or subconical, slender, 0.5–4 mm, pubescent; stigma strongly 2-lobed, lobes much longer than wide. **Seeds** ovate, 2–4 × 1.5–3 mm; wing continuous or distal. $2n = 12$.

Flowering Apr–Jul. Disturbed sites, lawns, abandoned gardens; 0–1500 m; introduced; B.C., Que., Yukon; Calif.; Europe.

Erysimum cheiri is a widely cultivated ornamental of European origin.

7. Erysimum coarctatum Fernald, Rhodora 29: 141. 1927 [E]

Erysimum inconspicuum (S. Watson) MacMillan var. *coarctatum* (Fernald) Rossbach

Biennials or perennials; (caudex thickened, usually simple, rarely branched). **Trichomes** of leaves 2- (or 3-) rayed. **Stems** erect, unbranched or branched basally and/or distally, (0.8–) 1.5–7(–9.4) dm. **Basal leaves:** blade oblanceolate to linear-oblanceolate, 2–7 cm × 2–7 mm, base attenuate, margins entire or denticulate, apex acute. **Cauline leaves** sessile; blade margins entire or denticulate. **Racemes** (densely flowered), elongated or not in fruit. **Fruiting pedicels** usually ascending, rarely divaricate-ascending, stout, slightly narrower than fruit, (often crowded), 4–7(–10) mm. **Flowers:** sepals linear-oblong, 5–8 mm, lateral pair saccate basally; petals yellow, narrowly obovate to spatulate, (8–)10–15 × (2–)2.5–4 mm, claw 6–9 mm, apex rounded; median filaments 6–10 mm; anthers narrowly linear, 1.7–2 mm. **Fruits** suberect to ascending, (subappressed to rachis or not), narrowly linear, straight, not torulose, 2.5–5.8(–6.4) cm × 1.8–2.5 mm, slightly 4-angled, not striped; valves with somewhat prominent midvein, pubescent outside, trichomes 2–4-rayed, glabrous inside; ovules 40–56 per ovary; style stout, 0.5–1.5 mm, moderately to sparsely pubescent; stigma strongly 2-lobed, lobes as long as wide. **Seeds** ovoid to oblong, 1.5–2 × 1–1.2 mm; not winged. $2n = 54$.

Flowering Jul–Aug. Calcareous talus and cliffs, meadows, bluffs, spruce forests, roadsides, crevices, ledges; 200–3000 m; Alta., Nfld. and Labr. (Nfld.), N.W.T., Que., Yukon; Alaska.

Both G. B. Rossbach (1958, 1958b) and R. C. Rollins (1993) treated *Erysimum coarctatum* as a variety of *E. inconspicuum*. However, the morphological and cytological (G. A. Mulligan 2002) differences clearly support its maintenance as a distinct species. The vast majority of the collections from Yukon belong to *E. coarctatum*, and the species is reported herein from Alberta [Banff, *Clark s.n.*, July-August (GH)] for the first time. The records from Alaska are based on *Scamman 5062* (GH) from Eklutna Lake and *Wiggins 12886* (GH) from Umiat on Colville River.

8. Erysimum concinnum Eastwood, Zoë 5: 103. 1901 E

Biennials or perennials; (short-lived). **Trichomes** of leaves 2- or 3(–7)-rayed. **Stems** erect, unbranched or branched distally, 0.4–5(–7) dm. **Basal leaves**: blade (slightly fleshy), spatulate to oblanceolate, 2–11 cm × 4–20 mm, base attenuate, margins sinuate-dentate to coarsely dentate, apex rounded to subacute. **Cauline leaves** (distal) sessile; blade margins entire or denticulate. **Racemes** considerably elongated in fruit. **Fruiting pedicels** ascending, stout, narrower than fruit, 2–4(–6) mm. **Flowers**: sepals oblong, 8–19 mm, lateral pair saccate basally; petals yellow to cream, suborbicular to broadly obovate, 15–32 × 6–16 mm, claw 8–12 mm, apex rounded; median filaments 8–11 mm; anthers linear, 3–4 mm. **Fruits** usually ascending to suberect, rarely divaricate-ascending, narrowly linear, straight or curved inwards, not torulose, (3–)5–13 cm × 2.2–5 mm, terete when immature, becoming strongly latiseptate, not striped; valves with obscure midvein, pubescent outside, trichomes 2–5-rayed, glabrous inside; ovules 42–68 per ovary; style cylindrical or flattened, stout, 0.5–2.5 mm, sparsely pubescent; stigma 2-lobed, lobes as long as wide. **Seeds** broadly ovate to suborbicular, (1.5–)2–4 × 1.5–3 mm; wing continuous. $2n = 36$.

Flowering Mar–Jun. Coastal bluffs, dunes, prairies; 0–400 m; Calif., Oreg.

Erysimum concinnum is a coastal species known from Curry County in Oregon, and from Del Norte, Humboldt, Marin, Mendocino, and Sonoma counties in California. Both G. B. Rossbach (1958) and R. C. Rollins (1993) treated it as a distinct species, but R. A. Price (1993) reduced it (invalidly) to a subspecies of *E. menziesii*.

9. Erysimum franciscanum Rossbach, Aliso 4: 118. 1958 E F

Erysimum franciscanum var. *crassifolium* Rossbach

Perennials or subshrubs. **Trichomes** of leaves 2-rayed mixed with 3(–5)-rayed ones. **Stems** erect, often branched distally, (woody at base), 0.6–5(–6) dm. **Basal leaves** (often withered in suffrutescent plants); blade oblanceolate to oblanceolate-linear, 2.5–17 cm × (2–)3–16(–20) mm, base attenuate, margins sinuate-dentate or dentate, apex acute. **Cauline leaves** (distal) petiolate; blade margins usually dentate, rarely denticulate. **Racemes** considerably elongated in fruit. **Fruiting pedicels** divaricate to ascending, stout, narrower than fruit, 5–17(–22) mm. **Flowers**: sepals oblong to linear-oblong, 8–12(–15) mm, lateral pair saccate basally; petals yellow to cream, obovate to suborbicular, 14–29 × 5–12(–15) mm, claw 9–17 mm, apex rounded; median filaments 9–15 mm; anthers linear, 2.5–4 mm. **Fruits** usually ascending, rarely spreading, narrowly linear, straight or curved upward, not or, rarely, slightly torulose, (3.8–)4–11(–14) cm × 2–4 mm, latiseptate, not striped; valves with somewhat prominent midvein, pubescent outside, trichomes (2 or) 3 (or 4)-rayed, glabrous inside; ovules 32–64 per ovary; style cylindrical, slender, 0.5–3.5 mm, sparsely pubescent; stigma 2-lobed, lobes as long as wide. **Seeds** oblong, 2–3.5(–4) × 1.2–2.2(–2.5) mm; wing distal, present on 1 or both margins. $2n = 36$.

Flowering Jan–Apr. Serpentine outcrops, coastal scrub or sand dunes, granitic hillsides; 0–500 m; Calif.

Erysimum franciscanum is known from Marin, San Francisco, San Mateo, and Santa Cruz counties. Historical records indicate that it grew previously in Sonoma County.

10. Erysimum hieracifolium Linnaeus, Cent. Pl. I, 18. 1755 • Wormseed mustard I W

Biennials. **Trichomes** of leaves mostly 3- or 4-rayed, 2-rayed ones usually on stem and pedicel. **Stems** erect, often branched distally, (slightly ribbed), (1–)3–9(–11) dm. **Basal leaves** (withered by fruiting; petiole 1–4 cm); blade elliptic-oblong or oblanceolate, (1–)2–6(–8) cm × (0.3–)5–10 mm, base cuneate, margins entire or obscurely denticulate, apex acute. **Cauline leaves** sessile or subsessile; blade (linear or elliptic-linear), margins

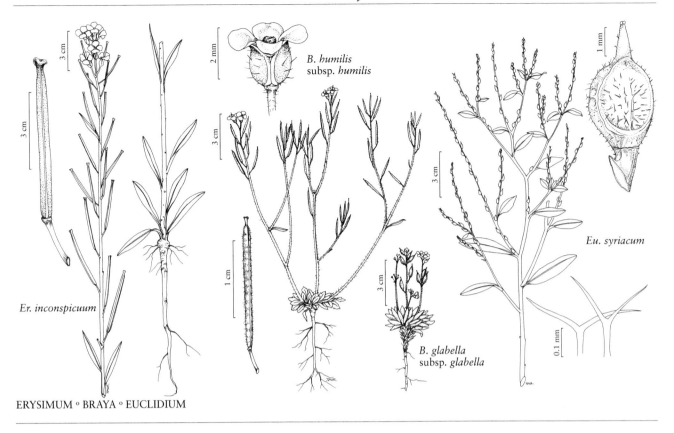

entire or obscurely denticulate. **Racemes** considerably elongated in fruit. **Fruiting pedicels** suberect or ascending, slender, narrower than fruit, (3–)4–8(–10) mm. **Flowers:** sepals oblong, 4–7 mm, lateral pair not saccate basally; petals yellow, obovate, (6–)7–9(–10) × 2–3 mm, claw 5–8 mm, apex rounded; median filaments 4–6 mm; anthers oblong, 1.5–2 mm. **Fruits** erect, (often appressed to rachis), linear, straight, somewhat torulose, (1–)1.5–2.5(–4) cm × 1–1.3 mm, 4-angled, not striped; valves with prominent midvein, pubescent outside, trichomes 3- or 4-rayed, fewer 2-rayed ones, usually glabrous inside, rarely sparsely pubescent; ovules 40–60 per ovary; style cylindrical, slender, 0.5–2 mm; stigma capitate, entire. **Seeds** oblong, 1–1.5 × 0.4–0.6 mm; not winged.

Flowering Jun–Sep. Moist areas, disturbed sites, grasslands, roadsides, waste ground; 50–300 m; introduced; Man., N.B., N.S., Ont., Que., Sask.; Mass., Vt., Wis.; Eurasia.

11. **Erysimum inconspicuum** (S. Watson) MacMillan, Metasp. Minnesota Valley, 268. 1892 [E] [F] [W]

Erysimum asperum (Nuttall) de Candolle var. *inconspicuum* S. Watson, Botany (Fortieth Parallel), 24. 1871; *Cheiranthus inconspicuus* (S. Watson) Greene; *C. syrticola* (E. Sheldon) Greene; *Cheirinia inconspicua* (S. Watson) Rydberg; *C. parviflora* Farwell; *C. syrticola* (E. Sheldon) Rydberg; *E. asperum* var. *parviflorum* M. E. Jones; *E. inconspicuum* var. *syrticola* (E. Sheldon) Farwell; *E. syrticola* E. Sheldon

Biennials or perennials; (short-lived, caudex thickened, usually simple, rarely branched). **Trichomes** of leaves 2- or 3-rayed. **Stems** erect, unbranched or branched distally, 1.5–7 dm. **Basal leaves:** blade linear to linear-oblanceolate, 1.5–6(–8) cm × 2–8 mm, base attenuate, margins entire or dentate, apex acute. **Cauline leaves** (distal) sessile; blade margins entire or denticulate. **Racemes** elongated or not in fruit. **Fruiting pedicels** divaricate-ascending to ascending, stout, slightly narrower than fruit, 4–9(–15) mm. **Flowers:** sepals oblong to linear-oblong, 4–6 mm, lateral pair slightly saccate basally; petals yellow, oblanceolate, 6–9(–11) × 1–2 mm, claw 5–7 mm, apex rounded; median filaments 5–7 mm; anthers narrowly oblong to linear, 1–2 mm. **Fruits** ascending to divaricate-ascending, (not appressed to rachis), narrowly linear, straight, not torulose, 3–5.8(–7) cm × 1–1.5(–1.8) mm, slightly 4-angled or terete, not striped; valves with somewhat prominent midvein, pubescent outside, trichomes 2–4-rayed, glabrous inside; ovules 36–66 per ovary; style cylindrical, stout, 0.7–3 mm, moderately to sparsely pubescent; stigma strongly 2-lobed, lobes as long as wide. **Seeds** ovoid, 1.2–1.7 × 0.8–1 mm; not winged or wing apiculate. $2n = 81$.

Flowering Apr–Aug. Calcareous talus and cliffs, roadsides, railroad embankments, abandoned fields, hillsides, alkaline ground, bluffs, crevices and ledges, gravel, prairies, rocky pastures, among brush, waste sites; 100–2700 m; Alta., B.C., Man., N.S., Ont., Que., Sask., Yukon; Alaska, Ark., Colo., Idaho, Ill., Ind., Kans., Maine, Mich., Minn., Mo., Mont., Nev., N.H., N.Y., N.Dak., Ohio, Okla., Oreg., S.Dak., Utah, Wis., Wyo.

Erysimum inconspicuum is known in Yukon from *Raup & Correll 11255* (A), which was collected 7 miles east of Little Atlin Lake, and *Malte s.n.* (GH), which was collected from Dawson on 10 August 1916. Most of the other collections from Yukon identified as *E. inconspicuum* belong to *E. coarctatum*. The record from Quebec is based on *Cayouette J80-8* (GH), from Cté de Charlevoix, and *Victorin, Germain, & Meilleur 43128* (GH), from Du Lac-Saint-Jean.

12. **Erysimum insulare** Greene, Bull. Torrey Bot. Club 13: 218. 1886 [E]

Cheiranthus insulare (Greene) Greene

Perennials or subshrubs. Trichomes of leaves primarily 2-rayed, sometimes mixed with 3-rayed ones. **Stems** ascending to sprawling, proximal branches terminating in sterile rosettes, (woody at base), 0.5–3 dm. **Basal leaves:** blade linear to linear-oblanceolate, (2–)3.7–9 cm × 2–5(–10) mm, base attenuate, margins usually entire, rarely sparsely denticulate, apex acute. **Cauline leaves** (distal) sessile; similar to basal. **Racemes** considerably elongated in fruit. **Fruiting pedicels** widely spreading, slender, narrower than fruit, (8–)10–22 mm. **Flowers:** sepals oblong to linear-oblong, 6–10 mm, lateral pair slightly saccate basally; petals yellow, broadly obovate to suborbicular, (11–)14–20(–22) × (3–)4–11.5 mm, claw 7–10 mm, apex rounded; median filaments 7–10 mm; anthers linear, 2–3 mm. **Fruits** ascending to spreading, narrowly linear, straight or only slightly curved inward, not torulose, 2–5(–7) cm × 2–3 mm, angustiseptate, not striped; valves with prominent midvein, pubescent outside, trichomes 2–4-rayed, glabrous inside; ovules 36–50 per ovary; style cylindrical or slightly flattened, stout, 1–4 mm, sparsely pubescent; stigma slightly 2-lobed, lobes as long as wide. **Seeds** oblong, 1–2 × 1.2–2 mm; not winged. $2n = 36$.

Flowering Mar–May. Sandy areas along coastal bluffs, coastal scrub; 0–300 m; Calif.

Erysimum insulare is restricted to the northern Channel Islands (San Miguel and Santa Rosa islands) of western Ventura County.

13. **Erysimum menziesii** (Hooker) Wettstein, Oesterr. Bot. Z. 39: 283. 1889 [C][E]

Hesperis menziesii Hooker, Fl. Bor.-Amer. 1: 60. 1830; *Cheiranthus grandiflorus* A. Heller

Biennials or perennials; (short-lived). Trichomes of leaves 2–5(–7)-rayed on adaxial leaf surface. **Stems** erect, unbranched or branched distally, 0.2–2.5(–3.5) dm. **Basal leaves:** blade (fleshy or not), spatulate, 2–10 cm × 5–15 mm, base cuneate, margins dentate, entire, or lobed, apex obtuse. **Cauline leaves** (distal) sessile or short-petiolate; blade margins entire or dentate. **Racemes** elongated slightly in fruit. **Fruiting pedicels** horizontal to divaricate, slender, narrower than fruit, 4–15 mm. **Flowers:** sepals oblong, 7–14 mm, lateral pair saccate basally; petals yellow, usually suborbicular, rarely obovate, 15–30 × 6–14 mm, claw 10–15 mm, apex rounded; median filaments 10–13 mm; anthers linear, 3–4 mm. **Fruits** widely spreading or divaricate, linear, straight, not torulose, 3–14 cm × 2–4 mm, terete when green, becoming latiseptate when dry, not striped; valves with obscure midvein, pubescent outside, trichomes (2 or) 3 or 4 (or 6)-rayed, glabrous inside; ovules 32–74 per ovary; style cylindrical, slender, 0.3–2 mm, sparsely pubescent; stigma 2-lobed, lobes as long as wide. **Seeds** oblong (compressed), 1.8–2.8(–3.5) × 1–2 mm; winged distally (wing narrow all around). $2n = 36$.

Flowering Jan–Aug. Stabilized coastal sand dunes; of conservation concern; 0–300 m; Calif.

Erysimum menziesii is restricted to the coasts of Humboldt, Mendocino, and Monterey counties. R. A. Price (1993) divided the species into four subspecies, three of which were invalidly published.

14. Erysimum occidentale (S. Watson) B. L. Robinson in A. Gray et al., Syn. Fl. N. Amer. 1(1,1): 144. 1895 E

Cheiranthus occidentalis S. Watson, Proc. Amer. Acad. Arts 23: 261. 1888; *Cheirinia occidentalis* (S. Watson) Tidestrom

Biennials. Trichomes of leaves 2- or 3-rayed. **Stems** erect, often branched distally, 0.5–3(–6.5) dm. **Basal leaves** (often withered by fruiting); blade linear-oblanceolate, 2–11(–15) cm × (1.5–)2–6 mm, margins usually entire, rarely denticulate. **Cauline leaves** (distal) sessile; blade margins entire. **Racemes** (simple or branched), considerably elongated in fruit. **Fruiting pedicels** divaricate-ascending, slender, narrower than fruit, 4–11(–15) mm. **Flowers:** sepals linear-oblong, 8–13 mm, lateral pair saccate basally; petals bright or pale yellow, obovate to broadly so, 14–22(–25) × 3–7.5 mm, claw 11–16 mm, apex rounded; median filaments 10–14 mm; anthers linear, 3–4 mm. **Fruits** ascending, narrowly linear, straight, not torulose, 3–12 cm × (2–)2.4–3.7 mm, strongly latiseptate, not striped; valves with prominent midvein, pubescent outside, trichomes 2- and 3-rayed, glabrous inside; ovules 34–46 per ovary; style cylindrical, slender, (2–)2.5–5 mm, sparsely pubescent; stigma entire. **Seeds** ovoid, (2–)2.5–3.5 × 1.3–2.5 mm; wing continuous distally, (0.3–0.8 mm wide). $2n = 36$.

Flowering Apr–Jun. Sand deposits; 50–700 m; Oreg., Wash.

Erysimum occidentale is restricted to sand deposits along or near the Columbia River and its tributaries. It is distributed in Gilliam, Hood River, Morrow, Sherman, and Umatilla counties in Oregon, and in Franklin, Grant, Kittitas, Klickitat, Lincoln, Walla Walla, and Yakima counties in Washington.

15. Erysimum pallasii (Pursh) Fernald, Rhodora 27: 171. 1925

Cheiranthus pallasii Pursh, Fl. Amer. Sept. 2: 436. 1813; *C. pygmaeus* Adams; *Erysimum pallasii* var. *bracteosum* Rossbach; *E. pygmaeum* (Adams) J. Gay; *Hesperis hookeri* Ledebour

Perennials; (caudex simple or branched, covered with persistent leaf bases). **Trichomes** of leaves 2-rayed. **Stems** erect, unbranched, (0.1–)0.4–4 dm. **Basal leaves:** blade linear to narrowly linear-oblanceolate, 3–5 cm × 1–2 mm, base attenuate, margins entire or denticulate. **Cauline leaves:** blade reduced distally, often as bracts. **Racemes** (sometimes bracteate basally), elongated in fruit. **Fruiting pedicels** divaricate to ascending, slender, narrower than fruit, (4–)7–24(–35) mm. **Flowers:** sepals oblong or narrowly so, 5–9 mm, lateral pair slightly saccate basally; petals usually purple, rarely lilac, oblanceolate to broadly obovate, 10–20 × 3–5 mm, claw 5–9 mm, apex rounded; median filaments 6–9 mm; anthers narrowly oblong, 1–1.5 mm. **Fruits** erect to divaricate-ascending, narrowly linear, straight or curved inward, not torulose, (3–)5–11(–13) cm × 2–4 mm, latiseptate, not striped; valves with prominent midvein, pubescent outside, trichomes 2-rayed, glabrous inside; ovules (28–)36–74 per ovary; style cylindrical, stout, 0.5–3 mm, sparsely pubescent; stigma strongly 2-lobed, lobes as long as wide. **Seeds** oblong, 1.7–2.2 × 0.9–0.1.4 mm; not winged. $2n = 24, 36$.

Flowering Jun–Aug. Cliffs, shaley slopes, talus, sandy ground, alpine areas, clay banks, pebbles, gravel; 0–1300 m; Greenland; N.W.T., Nunavut, Yukon; Alaska; e Asia (Russian Far East).

R. C. Rollins (1993) and G. A. Mulligan (2002) listed haploid chromosome numbers of $n = 12, 14, 18$, and 21 for *Erysimum pallasii*, but it is doubtful that those deviating from $x = 6$ are correct. Tetraploid counts ($2n = 24$) are known from both Canada and Russian Far East, whereas hexaploid counts are known only from the latter region. G. B. Rossbach (1958) and Rollins recognized plants with racemes bracteate to the middle as var. *bracteosum*, but this distinction is rather artificial, and the species shows considerable variation in the number of bracteate flowers.

16. Erysimum perenne (S. Watson ex Coville) Abrams in L. Abrams and R. S. Ferris, Ill. Fl. Pacific States 2: 318. 1944 E

Erysimum asperum (Nuttall) de Candolle var. *perenne* S. Watson ex Coville, Proc. Biol. Soc. Wash. 7: 70. 1892; *Cheiranthus perennis* (S. Watson ex Coville) Greene; *Cheirinia nevadensis* (A. Heller) A. Heller; *E. capitatum* (Douglas ex Hooker) Greene var. *perenne* (S. Watson ex Coville) R. J. Davis; *E. nevadense* A. Heller

Perennials or, rarely, biennials; (caudex slender). **Trichomes** of leaves 2–5-rayed. **Stems** erect, unbranched or branched (few to several) basally, 0.4–6.5 dm. **Basal leaves:** blade spatulate to broadly oblanceolate, 2.5–7 cm × 3–10 mm, base attenuate, margins dentate or subentire, apex often obtuse, (surfaces pubescent adaxially, trichomes 2 or 3–5-rayed). **Cauline leaves** (distal) sessile; blade margins often entire. **Racemes** considerably elongated in fruit. **Fruiting pedicels** divaricate-ascending, slender, narrower than fruit, 4–12 mm. **Flowers:** sepals linear-oblong to oblong, 8–12 mm, lateral pair saccate basally; petals yellow, broadly obovate to suborbicular, 15–22 × 3.5–6 mm, claw 8–14 mm, apex rounded; median filaments 7–14 mm; anthers linear, 3–4 mm. **Fruits** erect to ascending, narrowly linear, straight, torulose, 3.8–14 cm × 1.2–3 mm, latiseptate, not striped; valves with prominent midvein, pubescent outside, trichomes 2 or 3 (or 4)-rayed, glabrous inside; ovules 26–44 per ovary; style cylindrical, slender, (1.5–)2–5.5 mm, sparsely pubescent; stigma subentire to slightly 2-lobed, lobes as long as wide. **Seeds** ovoid, 2–3.4 × 1–2 mm; not winged or, rarely, winged distally. $2n = 36$.

Flowering Jun–Sep. Alpine fellfields, decomposing marble, gravelly ground and knolls, rocky slopes, talus, granitic sand; 2000–4000 m; Calif., Nev., Oreg.

Erysimum perenne is a high alpine species of the western sierras in California from Fresno, Inyo, and Madera counties northward into Plumas, Siskiyou, and Trinity counties. Its range in Nevada appears to be restricted to Douglas and Washoe counties.

The limits of *Erysimum perenne* have been controversial, and it is with some hesitation that I recognize it as a species. G. B. Rossbach (1958) accepted it as a distinct species, R. A. Price (1993) transferred it (invalidly) to a subspecies of *E. capitatum*, R. C. Rollins (1993) treated it as a variety of *E. capitatum*, and N. H. Holmgren (2005b) treated the name as a synonym of *E. capitatum*. It is readily distinguished from *E. capitatum* by having torulose (versus not torulose) and flattened (versus 4-angled or flattened) fruits, slender (versus stout or, rarely, slender) and longer styles (1.5–)2–5.5 mm (versus 0.2–2.5(–3) mm), and yellow (versus orange to, rarely, yellow) petals. Where the two species are allopatric, they remain consistently distinct, but at lower elevations, where their ranges overlap, the distinction becomes blurred. In such areas of overlap, one finds fruit variation ranging from distinctly torulose to non-torulose, as well as continuity in the other characters above.

17. **Erysimum repandum** Linnaeus, Demonstr. Pl., 17. 1753 [I][W]

Cheirinia repanda (Linnaeus) Link

Annuals. **Trichomes** of leaves 2-rayed, mixed with fewer 3-rayed ones. **Stems** erect, unbranched or branched basally, (0.4–)1.5–4.5(–7) dm. **Basal leaves** (often withered by fruiting), similar to cauline. **Cauline leaves:** (proximal and median) petiolate and (distal) sessile, (petiole (0.3–)0.5–2(–3) cm); blade [linear, narrowly oblanceolate, elliptic, or oblong, (1–)2–8(–11) cm × (2–)5–12(–17) mm, base attenuate], margins sinuate or coarsely dentate to denticulate or repand, (distal) entire or denticulate, (apex acute). **Racemes** considerably elongated in fruit. **Fruiting pedicels** divaricate, stout, as wide as fruit, 2–4(–6) mm. **Flowers:** sepals linear-oblong, 4–6 mm, lateral pair not saccate basally; petals yellow, narrowly oblanceolate to spatulate, 6–8 × 1.5–2 mm, claw 3–6 mm, apex rounded; median filaments 4–6 mm; anthers narrowly oblong, 0.8–1.3 mm. **Fruits** widely spreading to divaricate-ascending, narrowly linear, straight or curved upward, somewhat torulose, (2–)3–8(–10) cm × 1.5–2 mm, 4-angled, not striped; valves with prominent midvein, pubescent outside, trichomes 2-rayed and, fewer, 3-rayed, often glabrous, sometimes pubescent inside; ovules (40–)50–80(–90) per ovary; style cylindrical or subclavate, stout, 1–4 mm, sparsely pubescent; stigma slightly 2-lobed, lobes as long as wide. **Seeds** oblong, 1.1–1.5 × 0.6–0.7 mm; not winged or, rarely, winged distally. $2n = 16$.

Flowering Apr–Jun. Disturbed sites, roadsides, fields, waste places, barren hillsides, brush communities, pastures; 0–2100 m; introduced; B.C., Ont., Que.; Ala., Ariz., Ark., Calif., Colo., Idaho, Ill., Ind., Iowa, Kans., Ky., La., Md., Mass., Mich., Miss., Mo., Mont., Nebr., Nev., N.J., N.Mex., N.Y., N.C., Okla., Oreg., Pa., Tenn., Tex., Utah, Va., Wash., W.Va., Wyo.; Eurasia; n Africa; introduced also in South America, Australia.

18. **Erysimum suffrutescens** (Abrams) Rossbach, Aliso 4: 121. 1958 [E]

Cheiranthus suffrutescens Abrams, Bull. S. Calif. Acad. Sci. 2: 41. 1903; *Erysimum concinnum* Eastwood subsp. *suffrutescens* (Abrams) Abrams; *E. suffrutescens* var. *grandifolium* Rossbach

Perennials or subshrubs. **Trichomes** of leaves 2-rayed, sometimes mixed with 3- or 4-rayed ones. **Stems** ascending, proximal branches terminating in sterile rosettes, (woody at base), 1.5–8.1 dm. **Basal leaves:** blade linear to linear-oblanceolate, 3.7–15 cm × 1.5–6(–7) mm, base attenuate, margins usually entire, rarely sparsely denticulate, apex acute. **Cauline leaves** (distal) sessile; blade similar to basal. **Racemes** considerably elongated in fruit. **Fruiting pedicels** ascending, slender, narrower than fruit, (3–)5–10 mm. **Flowers:** sepals oblong to linear-oblong, 6–11 mm, lateral pair saccate basally; petals yellow, obovate to suborbicular, (11–)14–20(–22) × (3–)4–11.5 mm, claw 8–13 mm, apex rounded; median filaments 7–11 mm; anthers linear, 2.5–4 mm. **Fruits** ascending to spreading, narrowly linear, straight or only slightly curved inward, not torulose, (2–)3–8.4(–11) cm × 1.5–2.4(–3.5) mm, 4-angled to only slightly latiseptate, not striped; valves with prominent midvein, pubescent outside, trichomes 2–4-rayed, glabrous inside; ovules 48–82 per ovary; style cylindrical, stout, 0.5–4 mm, sparsely pubescent; stigma slightly 2-lobed, lobes as long as wide. **Seeds** oblong, 1.5–2.5(–3) × (0.7–)1–1.2(–1.4) mm; not winged or winged apically. $2n = 36$.

Flowering Dec–Aug. Stabilized coastal sand dunes, coastal scrub vegetation; 0–150 m; Calif.

Erysimum suffrutescens is restricted to the coastal regions of Los Angeles County northward into San Luis Obispo, Santa Barbara, and Ventura counties. Both G. B. Rossbach (1958, 1958b) and R. C. Rollins (1993) treated it as a distinct species, but R. A. Price (1993) transferred it (invalidly) to a subspecies of *E. insulare*. The latter species has angustiseptate (versus 4-angled to slightly latiseptate) fruits, and the two are sufficiently distinct to be recognized as independent species.

19. Erysimum teretifolium Eastwood, Leafl. W. Bot. 2: 144. 1938 [C][E]

Erysimum filifolium Eastwood, Leafl. W. Bot. 2: 73. 1838, not F. Mueller 1852

Biennials or perennials; (short-lived, caudex often woody). **Trichomes** of leaves 2-rayed, mixed with 3-rayed ones. **Stems** erect, unbranched, (1.4–)2.5–8(–10) dm. **Basal leaves:** blade filiform to narrowly linear, 5–17 cm × 0.4–3 mm, (somewhat revolute, appearing terete), base attenuate, margins denticulate, apex subacuminate. **Cauline leaves** sessile; blade margins entire or remotely denticulate. **Racemes** considerably elongated in fruit. **Fruiting pedicels** divaricate to ascending, slender, narrower than fruit, 5–14 mm. **Flowers:** sepals oblong to linear-oblong, 7–11 mm, lateral pair saccate basally; petals orange-yellow to yellow, broadly obovate to suborbicular, 15–20(–25) × 5–10 mm, claw 6–13 mm, apex rounded; median filaments 7–14 mm; anthers linear, 2.3–3.5 mm. **Fruits** widely spreading to ascending, narrowly linear, curved or slightly twisted, somewhat torulose, (4–)7–12(–15) cm × 1.2–2.2(–2.5) mm, slightly latiseptate, not striped; valves with prominent midvein, pubescent outside, trichomes 2 and 3 (or 4)-rayed, glabrous inside; ovules 40–72 per ovary; style cylindrical, slender, 0.5–2(–2.5) mm, sparsely pubescent; stigma subentire to slightly 2-lobed, lobes as long as wide. **Seeds** ovoid, 1.5–2.3(–2.7) × 0.9–1.5 mm; wing appendage-like, distal. $2n = 36$.

Flowering Feb–May. Sandy areas bordering sage scrub or chaparral, sand deposits derived from sandstone; of conservation concern; 100–400 m; Calif.

Erysimum teretifolium is a highly endangered species known only from Santa Cruz County.

p. BRASSICACEAE Burnett tribe EUCLIDIEAE de Candolle, Mém. Mus. Hist. Nat. 7: 236. 1821

Annuals or perennials [shrubs or subshrubs]; eglandular. **Trichomes** short-stalked or sessile, stellate, forked, dendritic, malpighiaceous, or simple, rarely absent. **Cauline leaves** (sometimes absent), usually petiolate or sessile, sometimes subsessile; blade base not auriculate, margins usually entire or dentate, rarely lobed. **Racemes** usually ebracteate, usually elongated in fruit. **Flowers** actinomorphic; sepals erect [ascending to spreading], lateral pair seldom saccate basally; petals white, pink, or purple [yellow], claw present, distinct or obscure; filaments unappendaged, not winged; pollen 3-colpate. **Fruits** silicles or siliques, usually dehiscent, unsegmented, terete or latiseptate [angustiseptate]; ovules 2–80[–numerous] per ovary; style distinct or obsolete; stigma entire or strongly 2-lobed. **Seeds** usually biseriate or uniseriate (aseriate in *Euclidium*); cotyledons accumbent or incumbent.

Genera 13, species 115 (3 genera, 9 species in the flora): North America, Europe, Asia, n Africa.

56. BRAYA Sternberg & Hoppe, Denkschr. Königl.-Baier. Bot. Ges. Regensburg 1(1): 65. 1815 • [For Franz Gabriel de Bray, 1765–1832, French ambassador to Bavaria, head of Regensberg Botanical Society]

James G. Harris

Platypetalum R. Brown

Perennials; (sometimes pulvinate, caudex simple or many-branched); scapose or not; usually pubescent or pilose, sometimes glabrous, trichomes short-stalked, forked, subdendritic, or submalpighiaceous, mixed with simple ones (rarely exclusively). **Stems** erect to decumbent or ascending, unbranched or branched. **Leaves** basal and, sometimes, cauline; petiolate or sessile; basal rosulate, petiolate, blade margins entire, sinuate, dentate, or, rarely, pinnately lobed; cauline usually absent, rarely few present, (sub)sessile, blade margins usually entire, rarely dentate or pinnately lobed. **Racemes** (corymbose, sometimes bracteate basally or throughout), elongated or not in fruit. **Fruiting pedicels** erect, divaricate, or ascending, slender (much narrower than fruit). **Flowers:** sepals [sometimes persistent], oblong [ovate], lateral pair not saccate basally (sometimes slightly so in *B. humilis* and *B. linearis*); petals white, pink, or purple [rarely pale yellow], obovate, oblanceolate, or spatulate, (slightly to much longer than sepals), claw distinct or not, (shorter than sepal, apex obtuse or rounded); stamens tetradynamous; filaments dilated or not basally; anthers ovate or oblong, (apex usually obtuse, sometimes apiculate); nectar glands (4), lateral, 1 on each side of lateral stamen. **Fruits** siliques or silicles, sessile, linear, oblong, cylindrical, oval-elliptic, ovoid, lanceoloid, lanceoloid-subulate, or globose, smooth or torulose, terete or slightly latiseptate; valves each often with prominent midvein, glabrous or pubescent; replum rounded; septum complete, (membranous, translucent); ovules (5–)14–44 per ovary; stigma capitate, entire or slightly 2-lobed. **Seeds** plump, not winged, oblong or ovoid; seed coat (minutely reticulate), not mucilaginous when wetted; cotyledons incumbent. $x = 7$.

Species 17 (7 in the flora): North America, n Europe, Asia.

SELECTED REFERENCES Abbe, E. C. 1948. *Braya* in boreal eastern America. Rhodora 50: 1–15. Harris, J. G. 1985. A Revision of the Genus *Braya* (Cruciferae) in North America. Ph.D. thesis. University of Alberta. Parsons, K. A. 2002. Reproductive Biology and Floral Variation in the Endangered *Braya longii* and Threatened *B. fernaldii* (Brassicaceae): Implications for Conservation Management of Rare Plants. M.S. thesis. Memorial University of Newfoundland. Warwick, S. I. et al. 2004. Phylogeny of *Braya* and *Neotorularia* (Brassicaceae) based on nuclear ribosomal internal transcribed spacer and chloroplast *trn*L intron sequences. Canad. J. Bot. 82: 376–392.

1. Plants not scapose; cauline leaves (1 or) 2–4; fruits linear; seeds usually uniseriate, rarely weakly biseriate.
    2. Cauline leaves 3 or more; basal leaves (0.3–)0.5–2(–3.5) cm × 1–8(–10) mm; racemes elongated in fruit; fruits (0.9–)1.2–2.5(–3.2) cm . . . . 3. *Braya humilis*
    2. Cauline leaves 1–4; basal leaves 0.5–3 cm × 0.5–2(–3) mm; racemes not elongated in fruit; fruits (0.5–)0.9–1.2(–1.4) cm . . . . 4. *Braya linearis*
1. Plants scapose; cauline leaves 0 or 1 (or with leaflike bract subtending proximalmost pedicel); fruits ovoid, globose, oval-elliptic, oblong, oblong-elliptic, oblong-lanceoloid, or lanceoloid-subulate, not linear; seeds usually biseriate (uniseriate in *B. fernaldii* and, sometimes, *B. longii*).
    3. Fruits ovoid or globose.
        4. Petals 4.7–6.6 × 3–5.1 mm; styles 1.2–2(–2.5) mm; stems erect to ascending . . . . 6. *Braya pilosa*
        4. Petals 2–3.7 × 1–1.5 mm; styles obsolete to 0.7(–1) mm; stems usually decumbent to prostrate, sometimes ascending . . . . 7. *Braya thorild-wulffii*
    3. Fruits oval-elliptic, oblong, oblong-elliptic, oblong-lanceoloid, or lanceoloid-subulate.

[5. Shifted to left margin.—Ed.]

5. Fruits oval-elliptic, oblong-cylindrical, or lanceoloid; septum margins not expanded basally (not forming sacklike pouch around proximalmost seeds); seeds biseriate . . . . 2. *Braya glabella*
5. Fruits lanceoloid-subulate; septum margins broadly expanded basally (forming sacklike pouch around proximalmost seeds); seeds somewhat to nearly uniseriate.
    6. Fruit valves pubescent; petals 2.4–3.8(–4) × (0.8–)1–1.3(–2) mm, (claws often not well-differentiated from blades) . . . . 1. *Braya fernaldii*
    6. Fruit valves glabrous or sparsely pubescent; petals (3–)3.3–4.8(–5) × (1.2–)1.4–2.5(–3) mm, (claws usually well-differentiated from blades) . . . . 5. *Braya longii*

1. **Braya fernaldii** Abbe, Rhodora 50: 12, plate 1090, fig. 2. 1948 [C][E]

Braya purpurascens (R. Brown) Bunge ex Ledebour var. *fernaldii* (Abbe) B. Boivin

Plants scapose; moderately to densely pubescent, trichomes 2-forked and simple. **Stems** simple or few to several from base, erect, 0.2–0.7(–10) dm. **Basal leaves**: blade narrowly spatulate to oblanceolate, (0.5–)1–3(–4) cm × 1–3 mm, base (membranous), broadly expanded near point of attachment, margins entire, (sparsely ciliate proximally), apex obtuse. **Cauline leaves**: 0 or 1 (or a leaflike bract subtending proximalmost pedicel). **Fruiting pedicels** erect or ascending, 1–2.6 mm. **Flowers**: sepals 2–2.6 × 1–1.5 mm; petals white or rose-purple, 2.4–3.8(–4) × (0.8–)1–1.3(–2) mm, (claw and blade often gradually and not well-differentiated, apex rounded); filaments 1.7–2 mm; anthers oblong, 0.3–0.4 mm. **Fruits** lanceoloid-subulate, not torulose, (straight), (0.3–)0.4–0.7 cm × (0.7–)1–1.5(–1.7) mm; valves densely to moderately pubescent, trichomes 2-forked and simple; septum margin broadly expanded basally (forming sacklike pouch around proximalmost seed in each locule); ovules 10–16 per ovary; style (0.5–)0.6–1(–1.2) mm; stigma 2-lobed or entire. **Seeds** uniseriate, oblong, (0.9–)1–1.3 × (0.4–)0.5–0.6(–0.7) mm. $2n = 56$.

Flowering Jun–Jul. Limestone barrens; of conservation concern; 0–60 m; Nfld. and Labr. (Nfld.).

Braya fernaldii and *B. longii* are closely related northwestern Newfoundland endemics that differ from the remaining species of *Braya* by their lanceoloid-subulate fruits and septum margins that broadly expand basally to produce a sacklike pouch around bases of proximalmost seeds in each locule.

2. **Braya glabella** Richardson in J. Franklin et al., Narr. Journey Polar Sea, 743. 1823 [F]

Braya alpina Sternberg & Hoppe var. *glabella* (Richardson) S. Watson

Plants scapose; sparsely to densely pubescent, trichomes simple, 2- or 3-forked. **Stems** simple or few to several from base, ascending or erect, rarely decumbent to prostrate, (0.1–)0.3–1.7(–2.3) dm. **Basal leaves**: blade (often somewhat fleshy), linear-oblanceolate to broadly spatulate, (0.4–)0.8–6(–7.9) cm × (0.3–)0.6–4(–6) mm, base (membranous), broadly expanded near point of attachment, margins usually entire, sometimes weakly dentate with 1 or 2 teeth per side, apex obtuse, (often with a tuft of long, simple hairs, surfaces sparsely to moderately pubescent). **Cauline leaves** 0 or 1 (or a leaflike bract subtending proximalmost pedicel). (Racemes elongated or not in fruit.) **Fruiting pedicels** ascending-erect to divaricate, (0.9–)1.9–7.5(–8.6) mm. **Flowers**: sepals (1.6–)1.9–3.7 × (0.7–)1–2 mm; petals white or purplish (broadly obovate or spatulate), (2.1–)2.4–4.5(–4.7) × (0.7–)1–3(–3.2) mm, (apex rounded); filaments 1.7–2.2 mm; anthers oblong, 0.3–0.5 mm. **Fruits** oval-elliptic, oblong-elliptic, oblong, or oblong-lanceoloid, sometimes slightly torulose, (straight or somewhat curved), (0.3–)0.5–1.2(–1.5) cm × (0.8–)1.1–3(–3.6) mm; valves sparsely to densely pubescent or glabrous, trichomes simple or 2 (or 3)-forked; septum margin not expanded or not basally; ovules (5–)16–20 per ovary; style (0.3–)0.5–1.8(–2) mm; stigma entire or strongly 2-lobed, (narrow or broad). **Seeds** biseriate, oblong, (0.7–)0.9–1.6 × 0.4–0.8(–0.9) mm.

Subspecies 3 (3 in the flora): n North America; Asia.

Braya glabella is extremely variable, sometimes even within a population, and the species has often been split into taxa based primarily on fruit attributes. When the species is examined from its entire range, the perceived morphological gaps blur into a bewildering array of overlapping forms. On the basis of morphological and molecular evidence (S. I. Warwick et al. 2004) it is difficult to justify the recognition of more than one species in this very plastic group, but it does seem useful to divide the species into three fairly distinctive subspecies. Some populations will be readily separable into one of these subspecies, but others will likely defy unequivocal placement, particularly those from areas where the ranges of the subspecies meet.

1. Fruits oblong to narrowly oblong-lanceoloid, 3.5–8.3 times as long as wide; racemes often loosely elongated in fruit . . . .
. . . . 2a. *Braya glabella* subsp. *glabella*
1. Fruits oval-elliptic to oblong-elliptic, rarely broadly oblong-lanceoloid, 2.5–3.7 times as long as wide; racemes not elongated in fruit, often compact.
  2. Stems decumbent to prostrate, sometimes weakly ascending; leaf blades not fleshy, to 6 mm wide; fruits 0.8–1.2 cm; styles 0.8–1.8 mm . . . . 2b. *Braya glabella* subsp. *prostrata*
  2. Stems ascending to erect; leaf blades often fleshy, to 4 mm wide; fruits 0.5–1 cm; styles 0.5–1.2 mm . . . . . 2c. *Braya glabella* subsp. *purpurascens*

2a. Braya glabella Richardson subsp. **glabella** [F]

Braya alpina Sternberg & Hope var. *americana* Hooker; *B. americana* (Hooker) Fernald; *B. bartlettiana* Jordal; *B. bartlettiana* var. *vestita* Hultén; *B. henryae* Raup; *B. humilis* (C. A. Meyer) B. L. Robinson var. *americana* (Hooker) B. Boivin

Stems ascending to erect, 0.5–1.7(–2.3) dm. **Leaves:** blade often somewhat fleshy, to 4 mm wide, margins often entire, sometimes 1 or 2 weak teeth per side. **Racemes** often loosely elongated in fruit. **Fruits** oblong to narrowly oblong-lanceoloid, often curved, (0.3–)0.5–1.2(–1.5) cm × (0.8–)1.1–3(–3.6) mm, 3.5–8.3 times as long as wide; style slender or stout, (0.3–)0.5–1.6(–2) mm. $2n = 56$.

Flowering Jun–Jul. Barren, usually calcareous soils and gravel, gravel bars, disturbed sites, lake and sea shores, scree slopes, solifluction lobes; 0–3700 m; Alta., B.C., N.W.T., Nunavut, Que., Yukon; Alaska, Colo., Mont., Wyo.; Asia.

2b. Braya glabella Richardson subsp. **prostrata** J. G. Harris, Novon 16: 350. 2006 [E]

Stems decumbent to prostrate or weakly ascending, 0.3–1.5 dm. **Leaves:** blade not fleshy, to 6 mm wide, margins usually entire. **Racemes** not elongated in fruit (often compact). **Fruits** oblong-elliptic to broadly oblong-lanceoloid, straight or curved, (0.7–)0.8–1.2 cm × 2.5–3.6 mm, 2.5–3.7 times as long as wide; style stout, 0.8–1.8 mm. $2n = 56$.

Flowering Jun–Jul. Dry or damp sand and silt on barren slopes and plains; 0–200 m; Nunavut.

Subspecies *prostrata* is known only from Ellesmere Island.

2c. Braya glabella Richardson subsp. **purpurascens** (R. Brown) W. J. Cody, Canad. Field-Naturalist 108: 93. 1964

Platypetalum purpurascens R. Brown, Chlor. Melvill., 9, 50. 1823; *Braya arctica* Hooker; *B. purpurascens* (R. Brown) Bunge ex Ledebour; *B. purpurascens* var. *dubia* (R. Brown) O. E. Schulz; *P. dubium* R. Brown

Stems ascending to erect, 0.3–1(–1.3) dm. **Leaves:** blade often fleshy, to 4 mm wide, margins usually entire, rarely 1 or 2 weak teeth per side. **Racemes** not elongated in fruit (often compact). **Fruits** oval-elliptic to oblong-elliptic, usually straight, 0.5–1 cm × 1.5–3 mm, 2.5–3.7 times as long as wide; style slender or stout, (0.3–)0.5–1.2 mm. $2n = 56$.

Flowering Jun–Jul. Barren, usually calcareous soils and gravel on solifluction lobes, scree, gravel bars, disturbed sites, rocky slopes, seashores; 0–1800 m; B.C., N.W.T., Nunavut; Alaska; n Eurasia.

3. Braya humilis (C. A. Meyer) B. L. Robinson in A. Gray et al., Syn. Fl. N. Amer. 1(1,1): 141. 1895 [F]

Sisymbrium humile C. A. Meyer in C. F. von Ledebour, Icon. Pl. 2: 16. 1830; *Neotorularia humilis* (C. A. Meyer) Hedge & J. Léonard; *Torularia humilis* (C. A. Meyer) O. E. Schulz

Plants not scapose; sparsely to densely pubescent throughout, or, rarely, glabrescent, trichomes short-stalked or subsessile, submalpighiaceous or, rarely, 2-forked, often mixed along petioles and stem base with simple ones. **Stems** usually few to several from base, rarely simple, ascending or erect, rarely subdecumbent, (0.4–)0.8–2.5(–3.5) dm. **Basal leaves:** blade obovate, spatulate, oblanceolate, oblong, or sublinear, (0.3–)0.5–2 (–3.5) cm × 1–8(–10) mm, base attenuate or cuneate, margins entire, sinuate-dentate, or pinnatifid, apex acute or obtuse, (surfaces sparsely to densely pubescent or, rarely, glabrous). **Cauline leaves:** 3 or more; blade similar to basal, smaller distally, distalmost sessile or subsessile. (**Racemes** bracteate proximally, very rarely throughout, elongated in fruit.) **Fruiting pedicels** erect, ascending, or divaricate, (2.5–)3–8(–12) mm. **Flowers:** sepals 2–3 × 0.8–1.2 mm, (sometimes slightly saccate basally); petals white, pink, or purple, (broadly obovate or spatulate), 3–5(–8) × (1–)1.5–2.5(–4) mm, (apex rounded); filaments 2–3(–4) mm; anthers oblong, 0.4–0.7 mm, (apex apiculate). **Fruits** linear, torulose or not, (mostly straight), (0.9–)1.2–2.5(–3.2) cm × 0.6–1.8(–2) mm (uniform in width); valves pubescent or, rarely, glabrescent, trichomes submalpighiaceous, rarely mixed with fewer, simple ones; septum fenestrate or not; ovules 20–44 per ovary; style 0.3–0.8(–1) mm; stigma entire or strongly 2-lobed. **Seeds** uniseriate, oblong, 0.6–0.9 × 0.4–0.5 mm.

Subspecies 4 (4 in the flora): North America, e, c Asia.

Braya humilis was recognized in Asian floristic accounts as a member of *Neotorularia* Hedge & J. Léonard (= *Torularia* O. E. Schulz), but molecular studies (S. I. Warwick et al. 2004) clearly support its assignment to *Braya*, as done by all North American authors (e.g., M. L. Fernald 1918; E. C. Abbe 1948; T. W. Böcher 1956, 1973; J. G. Harris 1985; R. C. Rollins 1993). The species is highly variable in leaf shape and margin, flower size and color, pubescence, fruit length and orientation, chromosome number, and length of the bracteate portion of the raceme. Occurrence of "races" with various ploidy levels is one of the reasons for variability that led to recognition of infraspecific taxa. The synonymy below pertains only to North America, with nearly as many names given to the Asian variants. Numerous morphological extremes were described in North America, Russia, and China, but most of those represent only part of an otherwise continuous variation. For example, fully bracteate racemes, though rare, appear sporadically in populations that otherwise have racemes only basally bracteate. Three morphological forms are more sharply distinct from the general subsp. *humilis* amalgam and seem to have some biological significance. All of them are restricted to areas in or near regions believed to have served as glacial refugia during the Pleistocene. They are recognized here as additional subspecies.

1. Fruits (1–)1.2–1.8(–2) mm wide, not or weakly torulose; stems unbranched, ascending (prostrate in fruit) . . . . 3b. *Braya humilis* subsp. *ellesmerensis*
1. Fruits 0.6–1.2(–1.3) mm wide, usually somewhat torulose; stems unbranched or branched, ascending to erect.
    2. Petals 2.5–6.9(–7.5) mm; fruits usually fertile and fully developed; leaf blade margins often sinuate-dentate, pinnatifid, or entire . . . . 3a. *Braya humilis* subsp. *humilis*
    2. Petals (4.4–)4.9–6.9(–7.2) mm; fruits often abortive; leaf blade margins entire or sinuate-dentate, not pinnatifid.
        3. Leaves and stems glabrescent or moderately pubescent . . . . 3c. *Braya humilis* subsp. *maccallae*
        3. Leaves and stems densely pubescent . . . . 3d. *Braya humilis* subsp. *porsildii*

3a. Braya humilis (C. A. Meyer) B. L. Robinson subsp. **humilis** [F]

Arabidopsis novae-angliae (Rydberg) Britton; *Braya humilis* var. *abbei* (Böcher) B. Boivin; *B. humilis* subsp. *arctica* (Böcher) Rollins; *B. humilis* var. *arctica* (Böcher) B. Boivin; *B. humilis* var. *interior* (Böcher) B. Boivin; *B. humilis* var. *laurentiana* (Böcher) B. Boivin; *B. humilis* var. *leiocarpa* (Trautvetter) Fernald; *B. humilis* var. *novae-angliae* (Rydberg) Fernald; *B. humilis* subsp. *richardsonii* (Rydberg) Hultén; *B. humilis* subsp. *ventosa* Rollins; *B. humilis* var. *ventosa* (Rollins) B. Boivin; *B. intermedia* T. J. Sørenson; *B. novae-angliae* (Rydberg) T. J. Sørenson; *B. novae-angliae* subsp. *abbei* Böcher; *B. novae-angliae* var. *interior* Böcher; *B. novae-angliae* var. *laurentiana* Böcher; *B. novae-angliae* subsp. *ventosa* (Rollins) Böcher; *B. richardsonii* (Rydberg) Fernald; *Pilosella novae-angliae* Rydberg; *P. richardsonii* Rydberg; *Torularia humilis* subsp. *arctica* Böcher

Stems ascending to erect, branched or unbranched, 0.4–3.3 dm, sparsely to moderately pubescent. **Leaves**: blade margins sinuate-dentate, shallowly pinnatifid, or entire, surfaces moderately pubescent. **Flowers**: petals white, pink, or purple, 2.5–6.9(–7.5) × (0.7–)0.9–4(–4.2) mm. **Fruits** usually fertile and fully developed, somewhat torulose, 0.6–1.2(–1.3) mm wide; septum not fenestrate or split longitudinally. $2n$ = 28, 42, 56, 70.

Flowering May–Jul. Sandy, gravelly soil along streams, lakeshores, roadsides, moraines, open stony slopes, dolomite cliffs and slopes, limestone ledges, solifluction soils; 0–4000 m; Greenland; Alta., B.C., Man., Nfld. and Labr. (Nfld.), N.W.T., Nunavut, Ont., Que., Yukon; Alaska, Colo., Mich., Mont., Vt., Wyo.; e, c Asia.

Subspecies *humilis* is extremely variable morphologically. In a general way, morphological form correlates with ploidy level, e.g., tetraploids, octoploids, and decaploids tend to be short in stature with small leaves. Hexaploids are less predictable. They range from short plants with small leaves to large, robust, multi-branched plants with large, pinnatifid leaves. Attempting to segregate most morphological forms of *Braya humilis* into logical infraspecific taxa is an exercise in futility. Populations that appear distinctive in the field almost always blur imperceptibly into the larger subsp. *humilis* continuum when compared with other populations from across the range of distribution. Subspecies *humilis* is broadly distributed on calcareous substrates in arctic, subarctic, alpine, and boreal regions of North America and Asia.

3b. Braya humilis (C. A. Meyer) B. L. Robinson subsp. ellesmerensis J. G. Harris, Novon 16: 345. 2006 [E]

Stems usually ascending (prostrate in fruit), unbranched, 0.3–1.6 dm, moderately pubescent. Leaves: blade margins pinnatifid or entire, surfaces moderately pubescent. Flowers: petals white or purple-tinged, (3–)4–5.6 × (1.3–)2–3.3(–3.8) mm. Fruits usually fertile and fully developed, not or weakly torulose, (1–)1.2–1.8(–2) mm wide; septum often fenestrate (with circular perforations at regular intervals longitudinally or with a narrow, elliptical, longitudinal split at base or both). $2n = 42$.

Flowering Jun–Jul, fruiting Jul–Aug. Sand, clay, and gravel slopes and plains; 0–200 m; Nunavut.

Prostrate fruiting stems, exceptionally broad, non-torulose fruits, and fenestrate silique septae distinguish subsp. *ellesmerensis* from other subspecies of *Braya humilis*. It is known only from northern Ellesmere Island.

3c. Braya humilis (C. A. Meyer) B. L. Robinson subsp. maccallae J. G. Harris, Novon 16: 346. 2006 [E]

Stems ascending to erect, often unbranched, 0.4–2.3 dm, moderately pubescent. Leaves: blade margins entire or, sometimes, sinuate-dentate, surfaces glabrescent to moderately pubescent. Flowers: petals white, (4.4–)4.9–6.7(–7.2) × (2.1–)2.3–4(–4.2) mm. Fruits often abortive, somewhat torulose, 0.7–1.2 mm wide when fully developed; septum not fenestrate or split longitudinally. $2n = 28$.

Flowering May–Jul. Sandy gravelly riverbanks and floodplains, sometimes on slopes and glacial moraines; 1500–3000 m; Alta., B.C.

Subspecies *maccallae* differs from subspp. *elesmerensis* and *humilis* in having leaf margins mostly entire, flowers exceptionally large, and a high percentage of abortive fruit. Cauline leaves are much reduced and arise from the base of the stem, giving an almost acaulescent appearance to the plant. Breeding studies (J. G. Harris 1985) indicate that subspp. *maccallae* and *porsildii* are self-incompatible, while most subspecies of *Braya humilis* are strongly autogamous.

3d. Braya humilis (C. A. Meyer) B. L. Robinson subsp. porsildii J. G. Harris, Novon 16: 348. 2006 [E]

Stems ascending to erect or, rarely, decumbent, usually unbranched, 0.3–1.7(–2.5) dm, densely pubescent. Leaves: blade margins usually entire, sometimes weakly sinuate-dentate, surfaces densely pubescent. Flowers: petals white, 4.4–6.9 × (1.9–)2.3–4.2 mm. Fruits often abortive, somewhat torulose, 0.8–1.2 mm wide when fully developed; septum not fenestrate or split longitudinally. $2n = 28$.

Flowering May–Jul. Dry alpine scree slopes, glacial moraines, and gravel bars, often on limestone gravels and soils; 500–3000 m; Alta., B.C., N.W.T.

Subspecies *maccallae* and *porsildii* share the large flowers, high percentage of abortive fruits, and leaf margins mostly entire, but they differ in indumentum and habitats.

4. Braya linearis Rouy, Ill. Pl. Eur. 11: 84. 1899

Plants not scapose; sparsely to moderately pubescent throughout, trichomes subsessile or sessile, submalpighiaceous, 2-forked, and simple. Stems simple or few to several from base, erect, (usually unbranched), (0.4–)0.7–1.4(–1.8) dm. Basal leaves (without distinct petiole); blade linear to narrowly spatulate, 0.5–3 cm × 0.5–2(–3) mm, base attenuate or cuneate, margins dentate (with 1 or 2 teeth per side) or entire, apex obtuse, (surfaces glabrous or sparsely pubescent). Cauline leaves: 1–4 (each stem); (subsessile); blade similar to basal, smaller distally. (Racemes not elongated in fruit.) Fruiting pedicels erect to divaricate, 1–4.5 mm. Flowers: sepals 1.8–2.8 × 1–1.4 mm (sometimes slightly saccate basally); petals white or purplish (especially claw), (broadly obovate or spatulate), 2.5–3.5(–4) × 1.3–2 mm; filaments 1.5–2 mm; anthers oblong, 0.4–0.6 mm. Fruits linear, more or less torulose, (straight or somewhat curved), (0.5–)0.9–1.2(–1.4) cm × 0.9–1.3 mm (uniform in width); valves glabrous or sparsely pubescent, trichomes simple and submalpighiaceous; septum margin not expanded, or not basally; ovules 20–28 per ovary; style 0.3–0.5 mm; stigma weakly 2-lobed to entire. Seeds uniseriate to weakly biseriate, oblong, 0.8–1 × 0.3–0.6 mm. $2n = 42$.

Flowering Jun–Jul. Dry or moist calcareous soils and alkaline clays and sands at the margins of evaporation pools on stream bank terraces and moraines; 0–200 m; Greenland; Europe (Scandinavia).

Braya linearis appears to be related to the European and Asian species of the genus. Hybridization studies with *B. alpina* from the European Alps (T. W. Böcher 1973) produced fertile F_1 hybrids. DNA sequence evidence (S. I. Warwick et al. 2004) confirmed the close relationship between *B. linearis* and *B. alpina*, but suggested that *B. linearis* is even more closely related to *B. humilis* and the Asian *B. siliquosa* Bunge.

5. **Braya longii** Fernald, Rhodora 28: 202. 1926 [C] [E]

Braya purpurascens (R. Brown) Bunge ex Ledebour var. *longii* (Fernald) B. Boivin

Plants scapose; sparsely pubescent, trichomes simple and 2-forked. **Stems** simple or few to several from base, erect, (unbranched), 0.3–1.1(–1.5) dm. **Basal leaves:** blade narrowly spatulate to oblanceolate, (0.5–)1–3(–5) cm × 1–3.5 mm, base (membranous), broadly expanded near point of attachment, margins entire, (sparsely ciliate proximally), apex obtuse. **Cauline leaves:** 0 or 1 (or a leaflike bract subtending proximalmost pedicel). **Fruiting pedicels** erect or ascending, (1.2–)1.5–4.5(–4.9) mm. **Flowers:** sepals 2–3(–3.5) × (1.1–)1.2–1.7 mm; petals white (or claw purplish), (3–)3.3–4.8(–5) × (1.2–)1.4–2.5(–3) mm, (claw and blade usually well-differentiated, apex rounded); filaments 1.7–2 mm; anthers oblong, 0.3–0.4 mm. **Fruits** lanceoloid-subulate, not torulose, (straight), (0.3–)0.4–0.8(–1) cm × (0.7–)1–1.5(–1.7) mm; valves glabrous or, sometimes, sparsely pubescent; septum margin broadly expanded basally (forming sacklike pouch around proximalmost seed in each locule); ovules 10–16 per ovary; style 0.5–1(–1.2) mm; stigma 2-lobed or entire. **Seeds** somewhat uniseriate, oblong, 1–1.4 × 0.5–0.9(–1) mm. $2n = 56$.

Flowering Jun–Jul. Limestone barrens; of conservation concern; 0–60 m; Nfld. and Labr. (Nfld.).

6. **Braya pilosa** Hooker, Fl. Bor.-Amer. 1: 65, plate 17A. 1830 [C] [E]

Braya purpurascens (R. Brown) Bunge ex Ledebour subsp. *pilosa* (Hooker) Hultén

Plants scapose; moderately to densely lanate-pilose, trichomes simple and 2-forked. **Stems** simple or few to several from base, erect to ascending, (branched or unbranched), 0.4–1.1(–1.3) dm. **Basal leaves:** blade linear-spatulate, 0.7–2(–3) cm × 0.7–2.5(–3.5) mm, base (membranous), broadly expanded near point of attachment, margins entire, (ciliate, trichomes long, simple), apex obtuse, often with tuft of hairs. **Cauline leaves:** 0 or 1 (or a leaflike bract subtending proximalmost pedicel). **Fruiting pedicels** divaricate to erect, (1–)2.5–6 mm. **Flowers:** sepals 2.8–3.5 × (1.3–)1.6–2.5 mm; petals white, (obovate), 4.7–6.6 × 3–5.1 mm, (claw short, apex rounded); filaments 2–3 mm; anthers ovoid, 0.4–0.6 mm. **Fruits** ovoid or globose, not torulose, (0.4–)0.5–0.6 cm × (2.5–)3–4 mm; valves densely to moderately pubescent, trichomes relatively short, simple and 2-forked; septum margin not expanded, or not basally; ovules 14–20 per ovary; style 1.2–2(–2.5) mm; stigma (broadly expanded), strongly or weakly 2-lobed. **Seeds** biseriate, oblong, 0.7–0.9 × 0.4–0.6 mm.

Flowering Jul–Aug. Sandy, calcareous seashores; of conservation concern; 0–10 m; N.W.T.

Braya pilosa is distinguished from other members of the genus by a combination of large flowers, broad fruits, and exceptionally long styles. This striking *Braya* was known from only three pre-1850 collections until the type locality near Cape Bathurst in the Northwest Territories was rediscovered in 2004. The species is apparently restricted to the type locality, where it is found on bare patches of soil disturbed by caribou hooves. Preliminary DNA sequence analyses indicate that it is most closely related to *B. thorild-wulffii*.

SELECTED REFERENCE Harris, J. G. 2006. Pilose braya, *Braya pilosa* Hooker (Cruciferae; Brassicaceae), an enigmatic endemic of arctic Canada. Canad. Field-Naturalist 118: 550–557.

7. **Braya thorild-wulffii** Ostenfeld, Meddel. Grønland 64: 176. 1923 [E]

Braya pilosa Hooker subsp. *thorild-wulffii* (Ostenfeld) V. V. Petrovsky; *B. purpurascens* (R. Brown) Bunge ex Ledebour subsp. *thorild-wulffii* (Ostenfeld) Hultén; *B. purpurascens* var. *thorild-wulffii* (Ostenfeld) B. Boivin

Plants scapose; densely pubescent throughout or glabrous, trichomes simple and 2-forked. **Stems** simple or few to several from base, decumbent to prostrate or, sometimes, ascending, (unbranched), (0.3–)0.5–0.9(–1.4) dm. **Basal leaves:** blade spatulate to linear-spatulate, (0.5–)1–3(–4) cm × 1–4 mm, base (membranous), broadly expanded near point of attachment, margins entire, (ciliate, trichomes long, mostly simple), apex obtuse, often with tuft of hairs. **Cauline leaves:** 0 or 1 (or a leaflike bract subtending proximalmost pedicel). **Fruiting pedicels** divaricate to erect, 1.5–4 mm. **Flowers:** sepals 2–3.5 × 1–2 mm; petals white to purplish, (narrowly spatulate), 2–3.7 × 1–1.5 mm, (claw and blade not differentiated), apex rounded; filaments 2–2.5(–3) mm; anthers ovoid, 0.4–0.6 mm. **Fruits** ovoid to globose, not torulose,

(0.4–)0.5–0.8(–1) cm × (2.5–)3–5 mm; valves densely pubescent or glabrous, trichomes 2-forked and simple; septum margin not expanded, or not basally; ovules 16–30 per ovary; style obsolete to 0.7(–1) mm; stigma strongly 2-lobed or entire. **Seeds** biseriate, oblong, (1.1–)1.2–1.4(–1.5) × (0.5–)0.7–0.8(–1) mm.

Subspecies 2 (2 in the flora): n North America.

1. Plants densely pubescent. . . . .
   . . . . 7a. *Braya thorild-wulffii* subsp. *thorild-wulffii*
1. Plants glabrous or glabrescent. . . . .
   . . . . 7b. *Braya thorild-wulffii* subsp. *glabrata*

7a. Braya thorild-wulffii Ostenfeld subsp. **thorild-wulffii** E

Plants densely pubescent. $2n = 28$.

Flowering Jul. Dry, often calcareous gravel, sand, and clay barrens, often on south-facing slopes; 0–200 m; Greenland; N.W.T., Nunavut.

Subspecies *thorild-wulffii* is known only from Axel Heiberg, Ellesmere, Melville, and Prince Patrick islands.

7b. Braya thorild-wulffii Ostenfeld subsp. **glabrata** J. G. Harris, Novon 16: 352. 2006 E

Plants glabrous or glabrescent. $2n = 28$.

Flowering Jul. Dry, often calcareous soils; 0–200 m; N.W.T., Nunavut.

Subspecies *glabrata* is known only from Banks and Victoria islands.

57. **EUCLIDIUM** W. T. Aiton in W. Aiton and W. T. Aiton, Hortus Kew. 4: 74. 1812, name conserved • [Greek *eu-*, well, and *kleio*, shut, alluding to indehiscent fruits] I

Ihsan A. Al-Shehbaz

Annuals; not scapose; scabrous, trichomes stalked, 2-forked, submalpighiaceous, mixed with fewer, simple, and, rarely, 3-forked ones, usually different sizes. **Stems** erect or ascending, unbranched or branched. **Leaves** basal and cauline; petiolate, sessile, or subsessile; basal (often withered by flowering), not rosulate, petiolate, blade margins entire, dentate, or repand, rarely pinnatifid; cauline sessile or subsessile, blade similar to basal, margins entire, dentate, or repand. **Racemes** (corymbose), elongated in fruit. **Fruiting pedicels** erect, stout. **Flowers**: sepals ovate to oblong; petals white, narrowly spatulate, (slightly longer than sepals), claw slightly differentiated from blade (shorter than sepals, apex emarginate); stamens slightly tetradynamous; filaments not dilated basally; anthers ovate, (apiculate); nectar glands (4), lateral, 1 on each side of lateral stamen. **Fruits** silicles (nutletlike), indehiscent, sessile, ovoid, subterete to slightly 4-angled; valves not veined, (thickened, woody), scabrous; replum strongly expanded laterally; septum complete, (thickened); ovules 2 per ovary, (subapical); style (persistent), obsolete or distinct; stigma capitate, 2-lobed (lobes not decurrent). **Seeds** aseriate, plump, not winged, oblong; seed coat (smooth), not mucilaginous when wetted; cotyledons accumbent or obliquely so.

Species 1: introduced; e Europe, c, w Asia; introduced also in Australia.

1. **Euclidium syriacum** (Linnaeus) W. T. Aiton in W. Aiton and W. T. Aiton, Hortus Kew. 4: 74. 1812 F I W

Anastatica syriaca Linnaeus, Sp. Pl. ed. 2, 2: 895. 1763; *Bunias syriaca* (Linnaeus) M. Bieberstein

Plants scabrous throughout, trichomes to 1 mm. **Stems** often ascending, (rigid), usually branched basally and near middle, (0.4–)1–4(–4.5) dm. **Basal leaves** similar to cauline. **Cauline leaves** petiolate [(0.2–)0.5–2(–2.5) cm] or (distal) sessile or subsessile; blade oblong, oblong-lanceolate, or elliptic, (1–)1.5–7(–9) cm × (3–)7–20(–30) mm (smaller distally), base cuneate, margins usually entire, dentate, or repand, rarely pinnatifid, apex acute or obtuse. **Fruiting pedicels** appressed to rachis, 0.5–1(–1.2) mm (ca. ½ as wide as fruit). **Flowers:** sepals 0.6–0.9 × 0.2–0.4 mm, sparsely pubescent; petals 0.9–1.3 × 0.1–0.2 mm, claw 0.4–0.6 mm; filaments 0.5–0.8 mm; anthers 0.1–0.2 mm. **Fruits** erect, 0.2–0.3 cm × 1.5–2 mm, 2-seeded; replum expanded, to 1.5 mm wide basally, narrowed to apex; style curved away from rachis, subconical, 1–1.8 mm, sparsely pubescent. **Seeds** 1.3–1.7 × 0.8–1.2 mm. $2n = 14$.

Flowering May–Jun. Waste places, roadsides, flats; 0–2500 m; introduced; Calif., Colo., Idaho, Mass., Oreg., Utah, Wash., Wyo.; Europe; Asia; introduced also in Australia.

Euclidium syriacum is known in the flora area from relatively few collections. The Massachusetts record, *Knowlton s.n.* (GH), was collected nearly a century ago; it is not known if the species has persisted in that state.

58. **STRIGOSELLA** Boissier, Diagn. Pl. Orient. 3(1): 22. 1854 • [Latin *strigosus*, covered with short, bristly trichomes, and *–ella*, diminutive] I

Ihsan A. Al-Shehbaz

Annuals; not scapose; pubescent [glabrous], trichomes simple and stalked, forked or dendritic. **Stems** erect or ascending, unbranched or branched. **Leaves** basal and cauline; petiolate or subsessile; basal not rosulate, petiolate, blade margins entire, dentate, or pinnatifid; cauline petiolate or subsessile, blade margins usually entire or dentate, rarely sinuate [lobed]. **Racemes** (few- to several-flowered), considerably elongated in fruit. **Fruiting pedicels** ascending or divaricate, stout [slender] (about equal to fruits). **Flowers:** sepals narrowly oblong [ovate], pubescent; petals pink or purple [rarely white], oblanceolate [spatulate, oblong], (longer than sepals), claw undifferentiated from blade, (apex obtuse or rounded); stamens tetradynamous, (erect); filaments not dilated basally [sometimes median 4 connate in 2 pairs]; anthers oblong [ovate], (apiculate or not); nectar glands (4), lateral, 1 on each side of lateral stamen, median glands absent. **Fruits** siliques, subsessile, linear, smooth [torulose], 4-angled [terete]; valves each with obscure [prominent] midvein, pubescent or glabrous; replum rounded; septum complete; ovules 40–80 per ovary; style obsolete; stigma conical, 2-lobed (lobes connivent or connate, opposite replum). **Seeds** uniseriate, plump or slightly flattened, not winged, oblong [ovate]; seed coat (reticulate), not mucilaginous when wetted; cotyledons incumbent. $x = 7$.

Species 20 (1 in the flora): introduced; Europe, c, w Asia, n Africa.

SELECTED REFERENCE Botschantzev, V. P. 1972b. The genus *Strigosella* Boiss. and its relation to the genus *Malcolmia* R. Br. (Cruciferae). Bot. Zhurn. (Moscow & Leningrad) 57: 1033–1046.

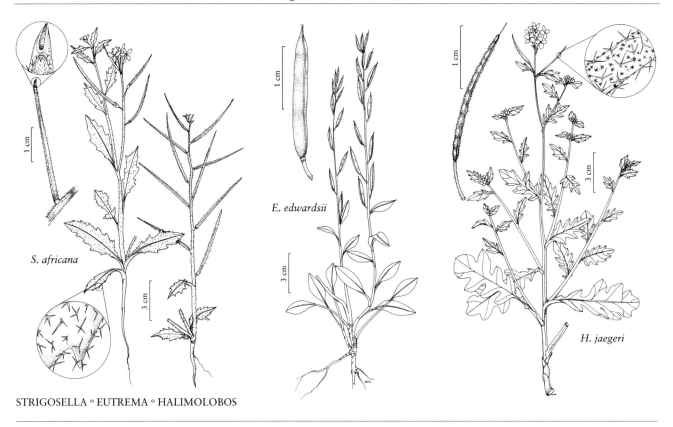

STRIGOSELLA ○ EUTREMA ○ HALIMOLOBOS

1. **Strigosella africana** (Linnaeus) Botschantzev, Bot. Zhurn. (Moscow & Leningrad) 57: 1038. 1972

F I W

Hesperis africana Linnaeus, Sp. Pl. 2: 663. 1753; *Malcolmia africana* (Linnaeus) W. T. Aiton

Plants usually sparsely to densely pubescent, rarely glabrescent, trichomes short-stalked, forked or subdendritic, these, sometimes, with simple, subsetiform ones. **Stems** unbranched or branched proximally, (0.4–)1.5–3(–5) dm, pubescent. **Basal leaves** soon withered. **Cauline leaves** petiolate or (distal) subsessile; petiole (0.1–)0.6–2(–3) cm; blade elliptic, oblanceolate, or oblong, (0.5–)1.5–6(–10) cm × (3–)10–25(–35) mm (smaller distally), base cuneate, apex acute. **Racemes:** rachis straight or slightly flexuous. **Fruiting pedicels** 0.5–2(–4) mm. **Flowers:** sepals sometimes persistent, (3.5–)4–5 × 0.5–0.7 mm; petals narrowly oblanceolate, (6.5–)8–10(–12) × 1–2 mm; filaments distinct, 2.5–5 mm; anthers narrowly oblong, 0.9–1.1 mm. **Fruits** divaricate-ascending, straight, (2.5–)3.5–5.5(–7) cm × 1–1.3 mm; valves usually pubescent, rarely glabrous, trichomes coarse and forked, these mixed with smaller, forked, subdendritic, or simple, subsetiform ones; stigma to 1 mm. **Seeds** 1–1.2 × 0.5–0.6 mm. $2n = 14, 28$.

Flowering May–Aug. Fields, disturbed areas, roadsides, deserts, sandy flats, vacant lots, sagebrush and greasewood areas, grasslands, railroad tracks, shale outcrops, alkaline flats, juniper woodlands, plains; 600–2400 m; introduced; Ariz., Calif., Colo., Idaho, Mont., Nev., Oreg., Utah, Wyo.; Europe; Asia; n Africa; introduced also in South America (Argentina).

q. BRASSICACEAE Burnett tribe EUTREMEAE Al-Shehbaz, Beilstein & E. A. Kellogg, Pl. Syst. Evol. 259: 112. 2006

Annuals or perennials; eglandular. **Trichomes** absent [simple]. **Cauline leaves** petiolate or sessile; blade base not auriculate [auriculate], margins entire or repand [dentate, pinnately or palmately lobed]. **Racemes** ebracteate or bracteate, elongated [not elongated] in fruit. **Flowers** actinomorphic; sepals erect [ascending to spreading], lateral pair not saccate basally; petals usually white, rarely pink, claw present, obscure [distinct]; filaments unappendaged, not winged; pollen 3-colpate. **Fruits** siliques or silicles, dehiscent, unsegmented, terete, 4-angled, latiseptate, or angustiseptate; ovules 2–96[–numerous] per ovary; style obsolete or distinct; stigma usually entire [rarely 2-lobed]. **Seeds** usually uniseriate, rarely biseriate; cotyledons usually incumbent, rarely accumbent.

Genus 1, species 26 (2 in the flora): North America, Europe, Asia.

59. EUTREMA R. Brown, Chlor. Melvill., 9, plate A. 1823 • [Greek *eu-*, well, and *trema*, hole, alluding to perforation in fruit septum]

Ihsan A. Al-Shehbaz

Neomartinella Pilger; *Platycraspedum* O. E. Schulz; *Thellungiella* O. E. Schulz; *Wasabia* Matsumura

Annuals or perennials; (caudex simple or branched), [rhizomatous]; not scapose. **Stems** erect or ascending [decumbent], unbranched or branched distally. **Leaves** basal and cauline; petiolate or sessile; basal rosulate or not, petiolate, blade margins usually entire or palmately lobed, rarely dentate or pinnatifid, (pinnately or palmately veined); cauline petiolate or sessile, blade (base cuneate or sagittate-amplexicaul), margins entire or repand [dentate, crenate], (pinnately or palmately veined, ultimate veins sometimes ending with apiculate callosities). **Racemes** (corymbose). **Fruiting pedicels** divaricate to ascending, slender or stout. **Flowers:** sepals ovate or oblong; petals usually spatulate, rarely obovate, claw not differentiated from blade, (apex obtuse to emarginate); stamens slightly tetradynamous; filaments slightly dilated basally; anthers ovate or oblong, (apex obtuse); nectar glands confluent, often subtending bases of stamens, median glands present or absent. **Fruits** sessile or shortly stipitate, linear or oblong [ovoid, lanceoloid], smooth or torulose, terete, slightly 4-angled, latiseptate, or angustiseptate; valves each with obscure or prominent midvein; replum rounded; septum complete or perforated; ovules [2–](6–)8–96 per ovary; stigma capitate. **Seeds** usually uniseriate, rarely biseriate, plump, not winged, oblong or ovoid; seed coat (usually obscurely reticulate, rarely foveolate or papillate), often not mucilaginous when wetted; cotyledons incumbent [rarely accumbent].

Species 26 (2 in the flora): North America, c, e Asia (Himalayas).

The fleshy rhizomes of the eastern Asian *Eutrema japonicum* (Siebold) Maximowicz are pungent and are the main source of true wasabi (W. H. Hodge 1974). Paste of horseradish (*Armoracia rusticana*) is often used as an inexpensive substitute.

Molecular studies, combined with a critical evaluation of morphology, have shown that *Thellungiella* is nested within, and hardly distinct from, the earlier-published *Eutrema*, and therefore the two genera have been combined (I. A. Al-Shehbaz and S. I. Warwick 2005; Warwick et al. 2006).

SELECTED REFERENCE Al-Shehbaz, I. A. and S. I. Warwick. 2005. A synopsis of *Eutrema* (Brassicaceae). Harvard Pap. Bot. 10: 129–135.

1. Perennials; cauline leaf blades: bases cuneate, not auriculate or sagittate; ovules (6–)8–12 (–14) per ovary .... 1. *Eutrema edwardsii*
1. Annuals; cauline leaf blades: bases usually sagittate-amplexicaul, rarely auriculate; ovules 55–96 per ovary .... 2. *Eutrema salsugineum*

1. **Eutrema edwardsii** R. Brown, Chlor. Melvill., 9, plate A. 1823 F

Eutrema edwardsii subsp. *penlandii* (Rollins) W. A. Weber; *E. labradoricum* Turczaninow; *E. penlandii* Rollins

Perennials; (roots fleshy). **Stems** simple or few from caudex, erect, unbranched, (0.3–)0.8–3(–4.5) dm. **Basal leaves** (soon withered); rosulate; petiole 1–4.5(–6) cm; blade (somewhat fleshy), ovate, lanceolate, or oblong, (0.3–)0.8–2(–2.5) cm × (1–)4–14(–18) mm, (base truncate, obtuse, or cuneate, sometimes oblique), margins entire, apex obtuse. **Cauline leaves** 3–7(–10); petiolate or (distal) sessile; blade ovate, lanceolate, oblong, or linear, (0.7–)1–3(–4) cm × (1–)3–10(–14) mm, base cuneate, margins entire, apex subacute. **Racemes** considerably elongated in fruit. **Fruiting pedicels** divaricate to ascending, (curved-ascending or straight), (1.5–)3–10(–15) mm. **Flowers**: sepals ovate, 1.5–3 × 1–1.5 mm, (margins membranous); petals spatulate, 3–5 × 1.5–3 mm; filaments 2–2.5 mm; anthers ovate, 0.2–0.4 mm; gynophore usually 0.2–1 mm, rarely obsolete. **Fruits** stipitate, not torulose, linear to narrowly oblong, (0.7–)1–2(–2.5) cm × 2–3 mm slightly 4-angled; valves (cuneate basally and distally), each with prominent midvein; septum mostly perforate; ovules (6–)8–12(–14) per ovary; style 0.2–1 mm. **Seeds** usually oblong, rarely ovoid, (1.5–)2–3 × (0.7–)1–1.5 mm. $2n = 18, 28, 42, 56$.

Flowering early Jun–early Aug, fruiting Jul–early Sep. Tundra, talus slopes, glaciated hills, grassy margins of streams, wet areas of peat ridges; 0–3900 m; Greenland; B.C., N.W.T., Nunavut, Que., Yukon; Alaska, Colo.; Asia (Mongolia, Russia).

Plants of *Eutrema edwardsii* from Alaska, northern Canada, and Greenland grow at elevations of 0–1900 m; those of Colorado, treated by R. C. Rollins (1993) as *E. penlandii*, grow at 3700–3900 m.

An examination of collections of *Eutrema edwardsii* from its entire range amply demonstrates that it is a highly variable species that can easily accommodate *E. penlandii*. Although the latter is known from the high mountains of Park County (Colorado), its plants are indistinguishable from those of *E. edwardsii* from Greenland, and the higher latitudes in Alaska, British Columbia, Northwest Territories, Nunavut, and Yukon. In general, plants of *E. edwardsii* from lower latitudes tend to be far more robust than those of higher latitudes. R. C. Rollins (1993) indicated that *E. penlandii* is distinguished from *E. edwardsii* by having stems less than 10 (versus 10–45) cm, lingulate (versus obovate) petals, fruiting pedicels less that 4 (versus 5–10) mm, infructescence less than 5 (versus to 20) cm, and cordate (versus truncate or cuneate) basal leaves. In fact, the petals of *E. penlandii* are spatulate, just as those of most plants of *E. edwardsii*, and the basal leaves in *E. penlandii* (e.g., *Penland 3909* and *Weber 13315*, both at GH) are cuneate. Furthermore, some plants from Greenland (e.g., *Raup et al. 465*, GH) are almost identical to those of *E. penlandii*, and they are only 3–8 cm with narrowly obovate petals, fruiting pedicels 2–4 mm, and infructescence 2–3 cm. Other examples of *E. edwardsii* that show some or all of the above aspects of *E. penlandii* could be cited. Weber treated *E. penlandii* as a subspecies of *E. edwardsii*, but the only difference I see between the two is geographic. Such distinction is unacceptable because the widely disjunct populations of *E. edwardsii* within Russia would have to be treated that way too.

As *Eutrema penlandii*, *E. edwardsii* is in the Center for Plant Conservation's National Collection of Endangered Plants.

2. **Eutrema salsugineum** (Pallas) Al-Shehbaz & Warwick, Harvard Pap. Bot. 10: 134. 2005

Sisymbrium salsugineum Pallas, Reise Russ. Reich. 2: 466. 1773; *Arabidopsis salsuginea* (Pallas) N. Busch; *Hesperis salsuginea* (Pallas) Kuntze; *Pilosella glauca* (Nuttall) Rydberg; *S. glaucum* Nuttall; *Stenophragma salsugineum* (Pallas) Prantl; *Thellungiella salsuginea* (Pallas) O. E. Schulz; *Thelypodium salsugineum* (Pallas) B. L. Robinson

Annuals; (often glaucous). **Stems** simple or few to several from caudex, erect or ascending, branched basally, (0.6–)1–3(–4) dm. **Basal leaves** rosulate or not; petiole 0.5–1 cm; blade (not fleshy), obovate, spatulate, or oblong, 0.5–1.5(–2.5) cm × 2–5 mm, margins usually entire, rarely dentate or pinnatifid, apex obtuse. **Cauline leaves** sessile; blade cordate, ovate, or oblong, 0.4–1.7(–2.5) cm × 1–7(–10) mm, base deeply sagittate-amplexicaul, margins entire or repand, apex acute or obtuse. **Racemes** elongated in fruit, (rachis straight). **Fruiting pedicels** divaricate to divaricate-ascending, (slender), 3–10 mm. **Flowers**: sepals oblong, 1–1.5 × 0.5–0.6 mm; petals obovate, 2–3 × 1–1.7 mm; filaments 1–1.5 mm; anthers oblong, 0.2–0.4 mm, (apiculate). **Fruits** sessile, linear, distinctly torulose, 0.7–1.6(–2) cm × (0.7–)0.8–1 mm, terete; valves each obscurely veined; septum complete; ovules 55–96 per ovary; style 0.1–0.3 mm. **Seeds** (biseriate) oblong, 0.4–0.5 × 0.2–0.3 mm. $2n = 14$.

Flowering May–Jun. Sandy alkaline ground, saline flats and fields, alkaline sloughs, stream banks, salt meadows and plains, steppes; 600–2500 m; B.C., N.W.T., Sask., Yukon; Colo., Mont.; c, e Asia.

r. BRASSICACEAE Burnett tribe HALIMOLOBEAE Al-Shehbaz, Beilstein & E. A. Kellogg, Pl. Syst. Evol. 259: 111. 2006

Annuals, biennials, or perennials [subshrubs]; eglandular. **Trichomes** stalked to subsessile, forked, stellate, or dendritic [simple]. **Cauline leaves** petiolate or sessile; blade base not auriculate [auriculate], margins entire, dentate, or lobed. **Racemes** ebracteate, elongated [not elongated] in fruit. **Flowers** usually actinomorphic, rarely zygomorphic; sepals spreading or erect [ascending], lateral pair seldom saccate basally; petals white, lavender, or purple [pink], claw present, obscure [distinct]; filaments unappendaged, not winged; pollen 3-colpate. **Fruits** silicles or siliques, dehiscent, unsegmented, terete, latiseptate, or angustiseptate; ovules [4–]16–250[–numerous] per ovary; style distinct or obscure; stigma entire. **Seeds** biseriate or uniseriate; cotyledons accumbent or incumbent.

Genera 5, species 39 (2 genera, 6 species in the flora): North America, Mexico, Central America, South America.

60. HALIMOLOBOS Tausch, Flora 19: 410. 1836 • [Greek *halimos*, of salt, and *lobos*, rounded protuberance, alluding to superficial resemblance of fruit indumentum to salt]

Ihsan A. Al-Shehbaz

C. Donovan Bailey

Poliophyton O. E. Schulz

Annuals, biennials, or perennials; (base often woody); not scapose; pubescent throughout, trichomes sessile, subsessile, or short-stalked, dendritic and subdendritic, 3–6-rayed. **Stems** erect or ascending, unbranched or branched distally. **Leaves** cauline and sometimes basal; petiolate or sessile; basal (often withered by anthesis or fruiting), not rosulate, petiolate, blade margins coarsely dentate or pinnatifid [sinuate]; cauline petiolate or sessile, blade margins dentate or sinuate-dentate. **Racemes** (corymbose), slightly to considerably elongated in fruit. **Fruiting pedicels** divaricate or descending, slender, (densely pubescent). **Flowers:** sepals slightly or widely spreading, oblong; petals white, oblanceolate or spatulate, (± equal to sepals), claw obscurely differentiated from blade, (apex obtuse); stamens slightly tetradynamous, (slightly spreading, equal to or longer than petals); filaments not dilated basally; anthers oblong [ovate], (apex obtuse); nectar glands confluent, subtending bases of stamens. **Fruits** siliques or silicles, sessile, not or obscurely torulose, linear or ovoid to oblong, terete or angustiseptate; valves each obscurely veined, densely pubescent, (trichomes appressed, [sub]sessile, subdendritic, mixed with coarser, short-stalked, larger ones); replum rounded; septum complete; ovules 16–110 per ovary; style distinct; stigma capitate. **Seeds** uniseriate in siliques, biseriate in silicles, plump, not winged, oblong; seed coat (minutely reticulate), mucilaginous when wetted; cotyledons incumbent. $x = 8$.

Species 6 (3 in the flora): sw United States, n Mexico.

As delimited by R. C. Rollins (1943, 1976, 1993), *Halimolobos* consists of 18 species distributed in North America and South America. Detailed molecular studies (C. D. Bailey et al. 2002) and critical evaluation of morphology (Bailey et al. 2007) indicate that Rollins's concept of the genus corresponds to at least four well-defined genera, *Halimolobos* in the narrow sense, *Sandbergia*, *Transberingia*, and *Exhalimolobos*, of which only the last is not represented in the flora area. Both *Exhalimolobos* and *Halimolobos* are members of the tribe Halimolobeae, *Sandbergia* belongs to the tribe Boechereae; *Transberingia* is sister to the Asian *Crucihimalaya* Al-Shehbaz, O'Kane & R. A. Price of the tribe Camelineae; see Bailey et al. (2002, 2007).

SELECTED REFERENCE Rollins, R. C. 1943. Generic revisions in the Cruciferae: *Halimolobos*. Contr. Dudley Herb. 3: 241–265.

1. Annuals or biennials (bases not woody); fruits oblong to ovoid, 3–4(–5) mm wide, slightly angustiseptate; ovules (60–)70–110 per ovary; seeds biseriate . . . . 3. *Halimolobos pubens*
1. Perennials (bases often woody); fruits linear, 0.5–0.9 mm wide, terete; ovules 16–38 per ovary; seeds uniseriate.
    2. Fruits 0.6–1.4(–1.7) cm; ovules 16–24 per ovary; sepals 1.2–2 mm; petals 1.8–2.5 mm; se Arizona, New Mexico, sw Texas . . . . 1. *Halimolobos diffusa*
    2. Fruits (1–)1.5–2.6 cm; ovules 28–38 per ovary; sepals 2–4 mm; petals (3.5–)4.5–6 mm; e California, w Nevada . . . . 2. *Halimolobos jaegeri*

1. **Halimolobos diffusa** (A. Gray) O. E. Schulz in H. G. A. Engler, Pflanzenr. 86[IV,105]: 228. 1924 (as Halimolobus diffusus)

Sisymbrium diffusum A. Gray, Smithsonian Contr. Knowl. 3(5): 8. 1852; *Hesperis diffusa* (A. Gray) Kuntze; *Sophia diffusa* (A. Gray) Tidestrom

Perennials. Stems erect to ascending, often paniculately branched distally, (1.2–)3–7.5(–12) dm, trichomes sessile or subsessile. **Basal leaves** absent on older plants. **Cauline leaves** petiolate or (distal) sessile; petiole 0.5–1 cm; blade oblanceolate, lanceolate to oblong, or elliptic, (1–)2–5.5(–7) cm × (4–)7–20(–30) mm (smaller distally), base cuneate, margins sinuately lobed or dentate, surfaces with minutely stalked to subsessile trichomes. **Racemes** slightly to considerably elongated in fruit. **Fruiting pedicels** usually divaricate, rarely slightly descending, (2–)3–6(–9) mm. **Flowers:** sepals slightly spreading, 1.2–2 × 0.4–0.7 mm; petals (slightly spreading), spatulate, (slender), 1.8–2.5 × 0.7–1 mm, claw distinctly differentiated from blade, 0.5–1 mm; filaments spreading, 2.2–3 mm, longer than petals; anthers 0.3–0.5 mm. **Fruits** divaricate or slightly descending, straight, subtorulose, linear, terete, 0.6–1.4(–1.7) cm × 0.5–0.8 mm; ovules 16–24 per ovary; style 0.7–1(–1.5) mm. **Seeds** uniseriate, 0.8–1 × 0.4–0.6 mm.

Flowering mid Jul–mid Nov. Shaded talus, ravines, granite outcrops, rock crevices, bluffs, steep canyons, limestone slopes, oak-juniper communities, igneous slopes; 1100–2300 m; Ariz., N.Mex., Tex.; Mexico (Chihuahua, Coahuila).

2. **Halimolobos jaegeri** (Munz) Rollins, Harvard Pap. Bot. 1: 46. 1993 [E] [F]

Sisymbrium diffusum A. Gray var. *jaegeri* Munz, Bull. S. Calif. Acad. Sci. 31: 61. 1932; *Halimolobos diffusa* (A. Gray) O. E. Schulz var. *jaegeri* (Munz) Rollins

Perennials. Stems erect to ascending, often paniculately branched distally, 1.5–7.5 dm, trichomes sessile or subsessile. **Basal leaves** absent on older plants. **Cauline leaves** petiolate or (distal) sessile; petiole 0.3–2 cm; blade oblanceolate or lanceolate to obovate, (1.5–)3–8(–11.5) cm × (7–)15–35(–46) mm (smaller distally), base cuneate, margins coarsely dentate to shallowly sinuate, surfaces with stalked to subsessile trichomes. **Racemes** slightly to considerably elongated in fruit. **Fruiting pedicels** divaricate or slightly ascending, (3–)4–9(–12) mm. **Flowers:** sepals slightly spreading, 2–4 × 0.8–1.5 mm; petals (slightly spreading), spatulate or oblanceolate, (3.5–)4.5–6 × 1–1.5 mm, claw obscurely differentiated from blade, to 1.5 mm; filaments spreading, 3.5–6 mm, equal to or longer than petals; anthers 0.7–1 mm. **Fruits** divaricate or slightly ascending, curved or straight, torulose, linear, terete, (1–)1.5–2.6 cm × 0.6–0.9 mm; ovules 28–38 per ovary; style (1–)1.5–2.2 mm. **Seeds** uniseriate, 0.9–1.2 × 0.5–0.6 mm.

Flowering early Jun–late Aug. Limestone cliffs, canyon walls, granitic slopes, rocky outcrops, crevices of volcanic cliffs; 1300–2400 m; Calif., Nev.

Halimolobos jaegeri is known in California from Inyo and San Bernardino counties and in Nevada from Esmeralda and Nye counties.

3. Halimolobos pubens (A. Gray) Al-Shehbaz & C. D. Bailey, Syst. Bot. 32: 146. 2007

Hymenolobus pubens A. Gray, Smithsonian Contr. Knowl. 3(5): 9. 1852; *Capsella pubens* (A. Gray) S. Watson; *Mancoa pubens* (A. Gray) Rollins; *Poliophyton pubens* (A. Gray) O. E. Schulz

Annuals or biennials; (base not woody). Stems erect, unbranched or branched distally, (2–)3–7.5(–11.5) dm, trichomes sessile. Basal leaves: petiole 0.4–2 cm; blade oblanceolate to oblong, 2–8(–11) cm × 05–20(–30) mm, margins coarsely dentate or somewhat pinnatifid, surfaces with sessile trichomes. Cauline leaves sessile; blade oblong to oblanceolate, 1.5–5(–6) cm × 3–12(–15) mm (smaller distally), base truncate or broadly cuneate, margins coarsely dentate, pubescent as basal. Racemes considerably elongated in fruit. Fruiting pedicels divaricate, (slightly curved upward or straight), (6–)8–15(–18) mm, (trichomes appressed). Flowers: sepals widely spreading, 1.5–2.5 × 1–1.5 mm; petals spatulate, 1.5–2.5 × 0.7–1.2 mm, claw obscurely differentiated from blade, to 1 mm; filaments spreading, 1.5–2.5 mm, equal to petals; anthers 0.7–1 mm. Fruits ascending, straight, not torulose, oblong to ovoid, slightly angustiseptate to subterete, 0.5–0.8(–0.9) cm × 3–4(–5) mm; ovules (60–)70–110 per ovary; style 0.3–0.8(–1) mm. Seeds biseriate, 0.6–0.8 × 0.4–0.5 mm.

Flowering late Apr–early Aug. Spring-fed areas, canyons, draws, streamsides, shaded moist grounds, grassy plains, creek beds, tobosa flats; 1200–1500 m; Ariz., Tex.; Mexico (Coahuila, Sonora).

Halimolobos pubens appears to be restricted in Texas to Brewster, Jeff Davis, Presidio, and Reeves counties. It has recently been reported from Cochise County, Arizona. For a detailed discussion of its generic disposition, see C. D. Bailey et al. (2007).

61. PENNELLIA Nieuwland, Amer. Midl. Naturalist 5: 224. 1918 • [For Francis Whittier Pennell, 1886–1952, American botanist]

Sara Fuentes-Soriano

Ihsan A. Al-Shehbaz

Heterothrix (B. L. Robinson) Rydberg, Bull. Torrey Bot. Club 34: 435. 1905, based on *Thelypodium* Endlicher subgen. *Heterothrix* B. L. Robinson in A. Gray et al., Syn. Fl. N. Amer. 1(1,1): 178. 1895, not Müller Argoviensis 1860; *Lamprophragma* O. E. Schulz

Perennials; not scapose; pubescent or glabrous, trichomes simple, with stalked, forked or dendritic ones. Stems erect [ascending], often branched distally. Leaves basal and cauline; petiolate or sessile; basal (withered with age), rosulate, petiolate, blade margins entire, dentate to sinuate, or runcinate; cauline subsessile or sessile [shortly petiolate], blade margins entire [dentate to sinuate]. Racemes (several-flowered, sometimes secund, lax), considerably elongated in fruit. Fruiting pedicels erect, descending, ascending, or divaricate-ascending, slender, (glabrous). Flowers (usually actinomorphic, rarely zygomorphic, cup-shaped); sepals erect, oblong to ovate, lateral pair slightly to strongly saccate basally, (usually pubescent, rarely glabrous); petals white (lavender or purple apically), spatulate to oblanceolate, (± longer than sepals), claw undifferentiated from blade, (apex obtuse); stamens slightly tetradynamous; filaments dilated basally; anthers ovate to oblong, (apex obtuse); nectar glands confluent, subtending bases of stamens. Fruits siliques, sessile or shortly stipitate, linear, not torulose, straight or curved, terete or latiseptate; valves each with prominent or obscure midvein, glabrous or pubescent; replum rounded; septum complete; ovules 40–250 per ovary; style distinct or obsolete; stigma capitate. Seeds uniseriate or biseriate, plump or flattened, winged or not, oblong to ovate; seed coat (reticulate-areolate), mucilaginous or not when wetted; cotyledons accumbent or incumbent. $x = 8$.

Species 8 (3 in the flora): sw United States, Mexico, Central America, South America.

SELECTED REFERENCES Fuentes-Soriano, S. 2004. A taxonomic revision of *Pennellia* (Brassicaceae). Harvard Pap. Bot. 8: 173–202. Rollins, R. C. 1980b. The genus *Pennellia* (Cruciferae) in North America. Contr. Gray Herb. 210: 5–21.

1. Fruits strongly latiseptate, 2–2.5 mm wide; seeds 2.4–2.8 mm, winged (broadly); septums opaque; ovules 40–50 per ovary; cotyledons accumbent . . . . 3. *Pennellia tricornuta*
1. Fruits terete or slightly latiseptate, 0.6–1.8 mm wide; seeds 0.7–1.3 mm, not winged; septums transparent; ovules 90–250 per ovary; cotyledons incumbent.
    2. Racemes not secund; fruiting pedicels straight, erect or ascending; fruits (1.5–)2.2–5.8 cm, terete; ovules 90–140 per ovary; sepals 2.7–4 mm; petals 1.5–4(–5) mm . . . . . 1. *Pennellia micrantha*
    2. Racemes secund; fruiting pedicels arcuate, reflexed; fruits 5–10 cm, latiseptate; ovules 150–250 per ovary; sepals 4–9 mm; petals (4–)5–7(–12) mm . . . . 2. *Pennellia longifolia*

1. **Pennellia micrantha** (A. Gray) Nieuwland, Amer. Midl. Naturalist 5: 224. 1918

Streptanthus micranthus A. Gray, Mem. Amer. Acad. Arts, n.s. 4: 7. 1849; *Heterothrix micrantha* (A. Gray) Rydberg; *Pennellia robinsonii* Rollins; *Thelypodium longifolium* (Bentham) S. Watson var. *catalinense* M. E. Jones; *T. micranthum* (A. Gray) S. Watson

Stems simple or several from base, 2–12.3 dm, pubescent proximally, trichomes predominantly dendritic, mixed with simple and long-stalked, 2-rayed ones, glabrous or sparsely pubescent distally, trichomes dendritic. **Basal leaves** caducous; petiole 0.5–4 cm; blade oblanceolate, (1–)3–5 cm × 6–34 mm, margins sinuate, lobed, dentate, or, rarely, subentire, surfaces pubescent, trichomes dendritic. **Cauline leaves** (distal) subsessile; blade oblanceolate, 3–10 cm × 1–20 mm, surfaces glabrous or glabrate. **Racemes** not secund, 1.5–6.7 dm in fruit. **Fruiting pedicels** slightly erect to divaricate-ascending, straight, (3–)3.5–8.5(–11) mm. **Flowers:** sepals purple or green, broadly oblong, 2.7–4 × 1.2–2 mm; petals purplish apically, spatulate to oblanceolate, 1.5–4(–5) × 0.5–2.2 mm; filaments 0.9–2.6 mm; anthers 1–1.2 mm. **Fruits** erect to ascending, straight, (1.5–)2.2–5.8 cm × 0.6–1.8 mm, terete; valves glabrous or sparsely pubescent; septum transparent; ovules 90–140 per ovary; style obsolete to 0.5 mm. **Seeds** biseriate or uniseriate, plump, not winged, oblong, 0.7–0.8 × 0.4–0.5 mm; seed coat mucilaginous when wetted; cotyledons incumbent.

Flowering Jun–Sep. Pine-oak forests, open grasslands, rolling andesitic loam soils, among volcanic rocks, igneous cliffs; 1100–3600 m; Ariz., N.Mex., Tex.; Mexico (Chihuahua, Coahuila, Distrito Federal, Durango, state of México, Michoacán, Querétaro, San Luis Potosí, Sonora).

2. **Pennellia longifolia** (Bentham) Rollins, Rhodora 62: 16. 1960

Streptanthus longifolius Bentham, Pl. Hartw., 10. 1839; *Heterothrix longifolia* (Bentham) Rydberg; *Lamprophragma longifolium* (Bentham) O. E. Schulz; *Pennellia hunnewellii* Rollins; *P. mcvaughii* Rollins; *Thelypodium anisopetalum* Greene; *T. longifolium* (Bentham) S. Watson

Stems simple or few from base, 4–15 dm, pubescent proximally, trichomes simple and long-stalked, 2-rayed, mixed with smaller, dendritic ones. **Basal leaves** caducous; petiole to 3 cm; blade spatulate to oblanceolate, 1.8–3 cm × 2–4 mm, margins dentate, sinuate, incised, or runcinate, surfaces pubescent, trichomes stalked, forked and dendritic, mixed with fewer, simple ones. **Cauline leaves** (distal) sessile; blade linear, 3–9.2 cm × 0.8–5 mm, surfaces glabrous or glabrate. **Racemes** secund, 1.6–7.9 dm in fruit. **Fruiting pedicels** slightly descending, recurved, 5–13 mm. **Flowers** (slightly zygomorphic, receptacle asymmetric); sepals purple or green, oblong, 4–9 × 1.5–4 mm, (adaxial largest); petals lavender to purple apically, spatulate, (4–)5–7(–12) × 1–3 mm, (adaxial largest); filaments 4–9 mm; anthers 1–1.2(–2) mm. **Fruits** pendent, straight or nearly so, 5–10 cm × 1–1.2 mm, slightly latiseptate; valves glabrous; septum transparent; ovules 150–250 per ovary; style 0.5–1.2 mm. **Seeds** biseriate, plump, not winged, oblong, 0.7–1.3 × 0.4–0.5 mm; seed coat mucilaginous when wetted; cotyledons incumbent. $2n = 16$.

Flowering Jul–Oct. Mixed pine-oak forest, limestone hills, rocky outcrops and meadows, stream banks, shaded ravines; 1400–3400 m; Ariz., N.Mex., Tex.; Mexico; Central America (Costa Rica, Guatemala).

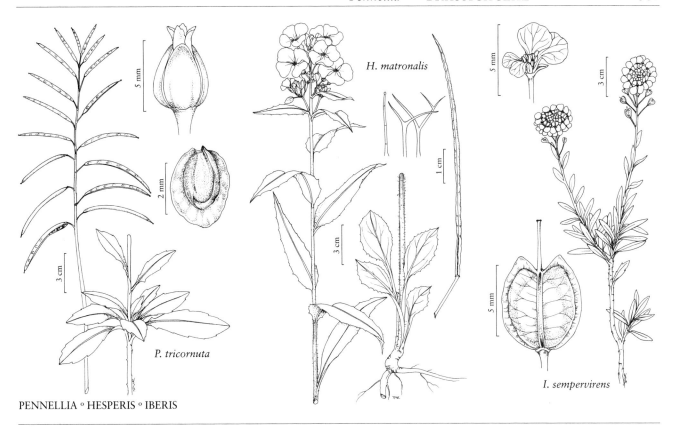

PENNELLIA · HESPERIS · IBERIS

3. **Pennellia tricornuta** (Rollins) R. A. Price, C. D. Bailey & Al-Shehbaz, Novon 11: 339. 2001 [C][E][F]

Arabis tricornuta Rollins, J. Wash. Acad. Sci. 29: 478. 1939

Stems simple or few from base, 1.5–9 dm, pubescent proximally, trichomes simple and long-stalked, 2–3-rayed. **Basal leaves** usually persistent; petiole 0.5–3 cm; blade oblanceolate, 3–9 cm × 3–23 mm, margins dentate, surfaces pubescent, trichomes stalked, 2–4-rayed. **Cauline leaves** (distal) sessile; blade linear, 2.7–7.5 cm × 1–9 mm, surfaces glabrous or sparsely pubescent. **Racemes** secund, 2–6 dm in fruit. **Fruiting pedicels** ascending, arcuate, 7–11 mm. **Flowers:** sepals purple or green, oblong, 2.3–4 × 0.8–2.1 mm, (subequal); petals lavender apically, spatulate, (2.7–)3.3–4.5 × (0.5–)1–1.5(–1.7) mm, (equal); filaments (1.7–)2–2.8(–3.3) mm; anthers 0.9–1(–1.3) mm. **Fruits** divaricate-ascending, straight or curved, 3.3–6.5 cm × 2–2.5 mm, latiseptate; valves glabrous; septum opaque; ovules 40–50 per ovary; style obsolete to 0.2 mm. **Seeds** uniseriate, flattened, winged, oblong, 2.4–2.8 × 1.2–1.9 mm; seed coat not mucilaginous when wetted; cotyledons accumbent.

Flowering Jul–Sep. Steep rocky slopes, under pines, road and sand banks; of conservation concern; 1300–2800 m; Ariz.

Pennellia tricornuta is known from Cochise, Pima, and Santa Cruz counties. It was recognized by R. C. Rollins (1941, 1993) under *Arabis*, but as shown by Price, Bailey & Al-Shehbaz and by S. Fuentes-Soriano (2004), it clearly belongs to *Pennellia*.

s. **BRASSICACEAE** Burnett tribe **HESPERIDEAE** Prantl in H. G. A. Engler and K. Prantl, Nat. Pflanzenfam. 55[III,2]: 154. 1891 [I]

Biennials [annuals, perennials]; glandular (glands unicellular on a few-celled, uniseriate stalk). **Trichomes** stalked, forked, and simple. **Cauline leaves** petiolate [sessile]; blade base not auriculate [auriculate], margins usually dentate. **Racemes** ebracteate or bracteate, elongated [not elongated] in fruit. **Flowers** actinomorphic; sepals erect, lateral pair often saccate basally; petals white, lavender, or purple [yellow, orange, green, or brown], claw present, distinct; filaments unappendaged, not winged; pollen 3-colpate. **Fruits** siliques, dehiscent, unsegmented, terete [4-angled, latiseptate]; ovules 4–40[–numerous] per ovary; style obsolete or distinct; stigma strongly 2-lobed. **Seeds** uniseriate; cotyledons incumbent.

Genus 1, species 25 (1 in the flora): introduced; Europe, Asia, n Africa.

62. HESPERIS Linnaeus, Sp. Pl. 2: 663. 1753; Gen. Pl. ed. 5, 297. 1754 • [Greek *hesperos*, evening, alluding to time when flowers of some species are most fragrant] I

Ihsan A. Al-Shehbaz

Plants with caudex; not scapose; pubescent or glabrous, trichomes simple and/or forked, often mixed with unicellular glands on uniseriate stalks. **Stems** erect, unbranched or branched. **Leaves** basal and cauline; petiolate [sessile]; basal rosulate [not rosulate], blade margins entire, dentate, or pinnatifid; cauline similar to basal. **Racemes** (corymbose), considerably elongated in fruit. **Fruiting pedicels** divaricate or ascending [reflexed], slender or stout. **Flowers:** sepals oblong [linear], (sometimes connivent), lateral pair strongly saccate basally, (pubescent or glabrous); petals obovate [oblong], (much longer than sepals), claw distinctly differentiated from blade, (apex rounded [obtuse]); stamens strongly tetradynamous; filaments (erect), slender or dilated basally; anthers linear [oblong], (apex obtuse); nectar glands (2), lateral, annular or lunar. **Fruits** tardily dehiscent, sessile, linear, torulose; valves each with prominent midvein, glabrous; replum rounded; septum complete; ovules 4–40 per ovary; style obsolete or distinct (relatively short); stigma conical (lobes prominent, connivent or distinct, decurrent). **Seeds** plump, not winged, oblong; seed coat (reticulate), not mucilaginous when wetted; cotyledons incumbent.

Species 25 (1 in the flora): introduced; se Europe, c, sw Asia, n Africa; introduced also in South America (Argentina, Chile).

SELECTED REFERENCES Dvořák, F. 1966. A contribution to the study of the evolution of *Hesperis* series *Matronales* Cvěl. emend. Dvořák. Feddes Repert. 73: 94–99. Dvořák, F. 1973. Infrageneric classification of *Hesperis*. Feddes Repert. 84: 259–272.

1. **Hesperis matronalis** Linnaeus, Sp. Pl. 2: 663. 1753
   • Dame's-rocket, rocket, dame's-violet F I W

Stems unbranched basally, often branched distally, 4–8(–11) dm, often eglandular, glabrous distally. **Basal leaves** withered by flowering, long-petiolate. **Cauline leaves** short-petiolate; blade narrowly oblong, lanceolate, or broadly ovate, (2–)4–15(–20) cm × (4–)8–40 (–60) mm, base cuneate, margins denticulate or entire, apex acute or acuminate, surfaces pubescent. **Fruiting pedicels** (5–)7–17(–25) mm. **Flowers:** sepals 5–8 × 1.5–2 mm; petals (13–)15–20(–22) × 3.5–9 mm, claw 6–12 mm; filaments 2.5–6 mm; anthers 2.5–4 mm. **Fruits** (4–)6–10(–14) cm × 2–2.5 mm. **Seeds** (2.5–)3–4 × 1–1.5 mm. $2n = 24$.

Flowering Apr–Jul. Gardens, roadsides, oak glades, waste areas, bluffs, floodplains, abandoned fields, railroad embankments, thickets, woodland; 0–2200 m; introduced; Alta., B.C., Man., Nfld. and Labr. (Nfld.), N.S., Ont., P.E.I., Que., Sask.; Ark., Calif., Colo., Conn., Del., D.C., Idaho, Ill., Ind., Iowa, Kans., Ky., Maine, Md., Mass., Mich., Minn., Mo., Mont., Nebr., Nev., N.H., N.J., N.Y., N.C., N.Dak., Ohio, Oreg., Pa., R.I., S.Dak., Tenn., Utah, Vt., Va., Wash., W.Va., Wis., Wyo.; se Europe; c, sw Asia; n Africa; introduced also in South America (Argentina, Chile).

t. BRASSICACEAE Burnett tribe IBERIDEAE Webb & Berthelot, Hist. Nat. Îles Canaries 3(2,1): 92. 1837 ⓘ

Annuals, perennials, or subshrubs [biennials]; eglandular. **Trichomes** simple or absent. **Cauline leaves** (sometimes absent), petiolate or sessile; blade base not auriculate, margins usually entire, sometimes dentate or lobed. **Racemes** ebracteate, usually elongated in fruit. **Flowers** usually zygomorphic, rarely actinomorphic; sepals erect, ascending, or spreading, lateral pair not saccate basally; petals white, pink, or purple, claw present, often distinct; filaments not or rarely appendaged; pollen 3-colpate. **Fruits** silicles, dehiscent, unsegmented, angustiseptate; ovules 2 or 4 per ovary; style usually distinct (sometimes absent in *Teesdalia*); stigma entire or 2-lobed. **Seeds** usually aseriate, rarely uniseriate; cotyledons accumbent.

Genera 2, species 30 (2 genera, 5 species in the flora): introduced; Europe, Asia, Africa.

The position of *Teesdalia* in Iberideae is not entirely resolved; further studies are needed to finalize its tribal assignment.

63. IBERIS Linnaeus, Sp. Pl. 2: 648. 1753; Gen. Pl. ed. 5, 292. 1754 • Candytuft [Name used by Dioscorides for an Iberian plant] ⓘ

Ihsan A. Al-Shehbaz

Annuals, perennials, or subshrubs [biennials]; not scapose; glabrous or pubescent. **Stems** erect or decumbent, often branched distally. **Leaves** cauline and sometimes basal; petiolate or sessile; basal rosulate or not, sessile [petiolate], blade (somewhat fleshy), margins entire [dentate to pinnatifid]; cauline petiolate or sessile, blade margins entire, dentate, or pinnatifid. **Racemes** (corymbose), elongated or not in fruit. **Fruiting pedicels** spreading, divaricate, descending, or ascending, slender. **Flowers**: sepals ascending [erect], ovate or oblong; petals (zygomorphic, outer [abaxial] pair larger than inner [adaxial] pair), white or pink to purple, obovate [oblanceolate], claw often distinct; stamens tetradynamous; filaments not dilated basally; anthers ovate or oblong, (apex obtuse); nectar glands (4), lateral, 1 on each side of lateral stamen. **Fruits** sessile, (winged), suborbicular or ovate [obcordate], not torulose, keeled, strongly angustiseptate; valves not veined, (winged), glabrous; replum rounded; septum complete; ovules 2 per ovary; stigma capitate, entire or 2-lobed. **Seeds** flattened, often winged, ovate [orbicular to reniform]; seed coat mucilaginous or not when wetted; cotyledons accumbent. $x = 7, 8, 9, 11$.

Species 27 (3 in the flora): introduced; Europe, sw Asia.

Iberis is a well-defined genus readily distinguished by having often corymbose infructescences, zygomorphic flowers, angustiseptate and often distally winged fruits, and exclusively accumbent cotyledons. N. H. Holmgren (2005b) erroneously indicated that the cotyledons in *Iberis* are incumbent.

All three species treated here are ornamentals that sometimes escape from cultivation. Their distributions are based on verified records, but it is very likely that the species are more widely naturalized than the records show.

1. Perennials or subshrubs (with sterile shoots) . . . . 2. *Iberis sempervirens*
1. Annuals (without sterile shoots).
    2. Racemes considerably elongated in fruit; abaxial petals 5–8 mm; styles 0.8–2 mm; cauline leaf blades: margins pinnatifid or dentate . . . . 1. *Iberis amara*
    2. Racemes not elongated in fruit; abaxial petals 10–16 mm; styles 2–4.5 mm; cauline leaf blades: margins entire . . . . 3. *Iberis umbellata*

1. **Iberis amara** Linnaeus, Sp. Pl. 2: 649. 1753
   * Rocket candytuft [I]

Annuals; glabrous or pubescent distal to base. **Stems** erect, branched distally, 1–4 dm. **Basal leaves** absent. **Cauline leaves** petiolate or (distal) sessile; blade spatulate or oblanceolate to oblong, 2–6 cm × 5–15 mm, margins pinnatifid or dentate, or (distalmost) entire or dentate. **Racemes** considerably elongated in fruit. **Fruiting pedicels** spreading to slightly descending, 3–10 mm. **Flowers:** sepals oblong, 1.5–3 mm; petals white or pink to purple, abaxial pair 5–8 × 2–5 mm, adaxial pair 2–5 × 1.5–3 mm. **Fruits** suborbicular to broadly ovate, 4–6(–7) × 3–6 mm, apically notched; valves extending into acute to obtuse wing; style 0.8–2 mm, exserted or included in apical notch. **Seeds** narrowly winged, 2–3 mm. $2n = 14$.

Flowering Jun–Aug. Escape from cultivation, waste places, old fields, lawns; introduced; N.S., Ont.; Mich.; w Europe; introduced also in South America.

2. **Iberis sempervirens** Linnaeus, Sp. Pl. 2: 648. 1753
   * Evergreen candytuft [F][I]

Perennials or **subshrubs;** (evergreen, with sterile shoots); glabrous. **Stems** decumbent, branched basally and distally, (0.5–)1–2.5 dm. **Basal leaves** (of sterile shoots) sessile; blade linear-oblanceolate, 1–3(–5) cm × 2–5 mm. **Cauline leaves** (of flowering shoots) similar to basal, blade smaller, margins entire. **Racemes** slightly elongated in fruit. **Fruiting pedicels** divaricate-ascending, 4–11 mm. **Flowers:** sepals oblong or ovate, 2–4 mm; petals white or sometimes pink, abaxial pair 5–13 × 2–7 mm, adaxial pair 3–6 × 1–3 mm. **Fruits** suborbicular to broadly ovate, 6–8 × 5–6 mm, apically notched; valves extending into subacute wing; style 1–3 mm, exserted beyond apical notch. **Seeds** narrowly winged, 2–3 mm. $2n = 22$.

Flowering Apr–Jul. Escape from cultivation, abandoned gardens; introduced; Calif., Mich., N.Y.; s Europe; introduced also in South America.

3. **Iberis umbellata** Linnaeus, Sp. Pl. 2: 649. 1753
   * Globe candytuft [I]

Annuals; glabrous. **Stems** erect, branched distally, 1–5.5(–7) dm. **Basal leaves** absent. **Cauline leaves** petiolate or (distal) sessile; blade linear-oblanceolate, 2–5 (–6.5) cm, margins entire or subentire. **Racemes** not elongated in fruit. **Fruiting pedicels** divaricate to ascending, 4–10(–12) mm. **Flowers:** sepals oblong, 3–5.5 mm; petals white or pink to purple, abaxial pair 10–16 × 5–7 mm, adaxial pair 6–10 × 2–5 mm. **Fruits** ovate, (7–)8–10.5 × 5–8 mm, apically notched; valves extending into subacuminate wing; style 2–4.5 mm, included or exserted beyond apical notch. **Seeds** narrowly winged, 2–3 mm.

Flowering Jun–Sep. Garden escape, stream banks, abandoned gardens and lawns, urban areas, docks; introduced; N.S., Ont., P.E.I., Que.; Calif., Mich., N.Y., Utah; se Europe.

64. TEESDALIA W. T. Aiton in W. Aiton and W. T. Aiton, Hortus Kew. 4: 83. 1812

* [For Robert Teesdale, 1740–1804, British botanist and gardener at Yorkshire] [I]

Ihsan A. Al-Shehbaz

Annuals [perennials]; scapose; glabrous or sparsely pubescent. **Stems** (simple or few to several from base), erect or ascending, unbranched. **Leaves** basal and cauline; petiolate or sessile; basal (persistent), rosulate, petiolate, blade margins usually lyrate-pinnatifid or pinnatisect, rarely entire or dentate; cauline (0–4), sessile, blade margins entire or dentate. **Racemes** (corymbose, several-flowered). **Fruiting pedicels** divaricate, slender. **Flowers** (actinomorphic or zygomorphic);

sepals ascending to spreading, ovate, glabrous; petals white, oblong or obovate, (equal to or longer than sepals, or lateral pair much larger), claw undifferentiated from blade, (apex obtuse); stamens (6) tetradynamous, or (4) equal; filaments dilated basally (appendaged); anthers ovate; nectar glands lateral, 1 on each side of lateral stamen, median glands absent. **Fruits** (often divaricate), sessile, broadly obcordate to suborbicular, slightly keeled, angustiseptate (strongly compressed), (apex notched); valves prominently veined, (apex narrowly winged), glabrous; replum rounded; septum complete; ovules 4 per ovary; style absent or not; stigma capitate, entire. **Seeds** uniseriate, slightly compressed, not winged, broadly ovate; seed coat (rugulose), copiously mucilaginous when wetted; cotyledons accumbent. $x = 9$.

Species 3 (2 in the flora): introduced; Europe, sw Asia (Middle East), Africa; introduced also in South America (Chile), n, c Europe, nw Africa, Australia.

Teesdalia conferta (Lagasca) O. Appel is known from Europe (Portugal, Spain). For a summary of the nomenclature of *Teesdalia* and references on its species introduced in North America, see I. A. Al-Shehbaz (1986).

SELECTED REFERENCE Appel, O. 1998. The status of *Teesdaliopsis* and *Teesdalia* (Brassicaceae). Novon 8: 218–219.

1. Flowers zygomorphic; stamens 6, tetradynamous; basal leaf blades: lateral lobes obtuse; styles ca. 0.1 mm .... 1. *Teesdalia nudicaulis*
1. Flowers actinomorphic; stamens 4, equal; basal leaf blades: lateral lobes acute; styles 0 mm .... 2. *Teesdalia coronopifolia*

1. **Teesdalia nudicaulis** (Linnaeus) W. T. Aiton in W. Aiton and W. T. Aiton, Hortus Kew. 4: 83. 1812

   • Shepherd's-cress

   Iberis nudicaulis Linnaeus, Sp. Pl. 2: 650. 1753

   Stems 0.5–1.5(–2) dm. **Basal leaves:** petiole 0.5–2.5(–3.5) cm; blade oblanceolate to obovate in outline, 0.3–1.5(–2) cm × 2–4(–7) mm, margins usually lyrate-pinnatifid, rarely dentate or entire; lateral lobes obtuse. **Cauline leaves** 0–4; blade linear to oblong. **Fruiting pedicels** 3–7 mm. **Flowers** zygomorphic; sepals 0.5–0.8 (–1) × 0.4–0.6(–0.8) mm; petals oblong to obovate, abaxial pair 1.5–2.5 × 0.6–1 mm, adaxial pair 0.5–1 × 0.3–0.4 mm; stamens 6, tetradynamous; filaments 0.8–1 mm, (basal appendage white, ca. ½ its length); anthers 0.1–0.2 mm. **Fruits** 3–4.5 × 3–4.5 mm; style ca. 0.1 mm. **Seeds** 1–1.2 × 0.9–1 mm. $2n = 36$.

   Flowering Apr–Jun. Sandy areas along trails, near beaches, pine clearings, waste grounds, fields, roadsides, gravelly slopes; 0–300 m; introduced; B.C.; Calif., Conn., Md., Mass., Mich., N.J., N.Y., N.C., Oreg., Pa., S.C., Va., Wash.; Europe; Asia (Middle East); introduced also in South America (Chile), nw Africa, Australia.

2. **Teesdalia coronopifolia** (Bergeret) Thellung, Repert. Sp. Nov. Regni Veg. 10: 289. 1912

   Thlaspi coronopifolium Bergeret, Phytonom. Univ. 3: 29. 1786

   Stems (0.6–)1–2(–2.8) dm. **Basal leaves:** petiole 0.2–2 cm; blade lanceolate in outline, 0.5–3(–4) cm × 0.5–2.5 mm, margins usually lyrate-pinnatifid to pinnatisect, rarely dentate or entire; lateral lobes acute. **Cauline leaves** 0–3; blade linear. **Fruiting pedicels** 3–6 mm. **Flowers** actinomorphic; sepals 0.2–0.6 × 0.2–0.4 mm; petals oblong, 0.5–1.5 × 0.3–0.6 mm; stamens 4, equal; filaments 0.5–1 mm (basal appendage white, ca. ½ its length); anthers 0.1–0.2 mm. **Fruits** 3–3.5 × 3–5 mm; style 0 m. **Seeds** 0.8–1.2 × 0.7–1 mm. $2n = 36$.

   Flowering Feb–Apr. Roadsides, abandoned fields; 50–500 m; introduced; Calif.; s Europe; n Africa; introduced also in n, c Europe.

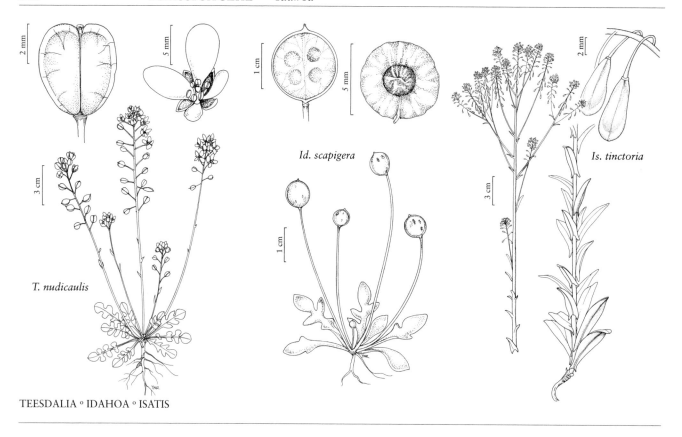

TEESDALIA ∘ IDAHOA ∘ ISATIS

u. BRASSICACEAE Burnett IDAHOA Group E

Annuals; eglandular. **Trichomes** absent. **Cauline leaves** absent. **Inflorescences** (solitary flowers), ebracteate, not elongated in fruit. **Flowers** actinomorphic; sepals ascending to spreading, lateral pair not saccate basally; petals white, claw absent; filaments unappendaged, not winged; pollen 3-colpate. **Fruits** silicles, dehiscent, unsegmented, latiseptate; ovules 6–10(–12) per ovary; style obsolete or distinct; stigma entire. **Seeds** biseriate; cotyledons accumbent.

Genus 1, species 1: w North America

Despite repeated sequencing attempts and cloning of the entire *ndh*F gene of *Idahoa scapigera* (M. A. Beilstein et al. 2006, 2008), the genus showed no affiliation to any of the other genera studied, and its systematic position remains unresolved.

65. IDAHOA A. Nelson & J. F. Macbride, Bot. Gaz. 56: 474. 1913 • [For the state Idaho] E

Ihsan A. Al-Shehbaz

Platyspermum Hooker, Fl. Bor.-Amer. 1: 68, plate 18, fig. B. 1830, not Hoffmann 1814

Plants scapose. **Stems** absent. **Leaves** basal, rosulate, petiolate, blade margins entire or lyrately lobed. **Inflorescences** (few to many) on peduncle from basal rosette. **Fruiting pedicels** ascending or erect, slender. **Flowers:** sepals ovate or oblong; petals oblanceolate, (slightly longer than sepals, apex obtuse); stamens (erect), slightly tetradynamous; filaments not dilated basally; anthers ovate or broadly so, (apex obtuse); nectar glands lateral, semiannular, intrastaminal, median glands present (smaller, distinct). **Fruits** sessile or, rarely, stipitate, orbicular or orbicular-ovate, smooth, strongly latiseptate; valves not veined or obscurely veined; replum slightly flattened; septum complete, (hyaline); stigma capitate. **Seeds** flattened, broadly winged, orbicular; seed coat (coarsely reticulate at center), not mucilaginous when wetted; cotyledons accumbent. $x = 8$.

Species 1: w North America.

1. Idahoa scapigera (Hooker) A. Nelson & J. F. Macbride, Bot. Gaz. 56: 474. 1913 • Flatpod [E] [F]

Platyspermum scapigerum Hooker, Fl. Bor.-Amer. 1: 68, plate 18, fig. B. 1830

Basal leaves: petiole (0.2–)0.5–2(–3) cm; blade ovate, obovate, or spatulate, (0.3–)0.7–2(–2.5) cm × 2–8(–12) mm. **Fruiting pedicels** (2–)4–15(–22), straight, (15–)25–110(–130) mm. **Flowers:** sepals purplish, 1.5–1.8 × 0.5–0.8 mm, margins not hyaline; petals 2–2.5 × 0.5–0.8 mm, attenuate to base; filaments 1.4–1.6 mm; anthers 0.15–0.2 mm, gynophore obsolete or, rarely, to 1 mm. **Fruits** (0.5–)0.6–0.9(–1) cm (slightly longer than wide); valves rounded at each end; style subconical, 0.3–0.8 mm. **Seeds** strongly compressed, 3.5–5 mm diam., coarsely reticulate (except for wing); wing 0.8–1.2 mm wide. $2n = 16$.

Flowering Mar–May. Sagebrush slopes and scabland, dry ground, rocky volcanic flats and outcrops, open moist slopes and banks, gravelly hills, meadows, rocky hillsides, seepy sunny slopes; 300–1700 m; B.C.; Calif., Idaho, Mont., Nev., Oreg., Wash.

v. **BRASSICACEAE** Burnett tribe **ISATIDEAE** de Candolle, Mém. Mus. Hist. Nat. 7: 241. 1821 [I]

Annuals or biennials [perennials]; eglandular. **Trichomes** simple or absent. **Cauline leaves** sessile [petiolate]; blade base auriculate [not auriculate], margins usually entire [dentate]. **Racemes** ebracteate, elongated in fruit. **Flowers** actinomorphic; sepals erect or ascending [spreading], lateral pair not saccate or subsaccate basally; petals yellow [white], claw usually present, rarely absent, obscure; filaments unappendaged, not winged; pollen 3-colpate. **Fruits** siliques or silicles (samaroid), indehiscent, unsegmented, angustiseptate, (woody); ovules 1 or 2 per ovary; style usually absent, rarely distinct; stigma entire. **Seeds** aseriate; cotyledons accumbent or incumbent.

Genera 2–4, species 90–95 (2 genera, 2 species in the flora): introduced; Europe, Asia, n Africa.

66. **ISATIS** Linnaeus, Sp. Pl. 2: 670. 1753; Gen. Pl. ed. 5, 301. 1754 • Woad [Greek *isatis*, name used for a dye plant, most likely woad] [I]

Ihsan A. Al-Shehbaz

Biennials [annuals, perennials]; not scapose; (often glaucous), glabrous or pubescent. **Stems** erect, often unbranched basally, paniculately branched distally. **Leaves** basal and cauline; petiolate or sessile; basal rosulate [or not rosulate], petiolate [rarely sessile], blade margins entire, repand, or dentate [rarely pinnately lobed]; cauline blade (base auriculate, sagittate, [or amplexicaul, rarely attenuate]), margins entire [dentate]. **Racemes** (corymbose, in panicles, several-flowered), considerably elongated in fruit. **Fruiting pedicels** reflexed, slender, (filiform, often thickened and clavate apically). **Flowers:** sepals erect or ascending, oblong [ovate]; petals oblanceolate [obovate, spatulate, or oblong], (equal to or longer than sepals), claw absent, (apex obtuse [subemarginate]); stamens slightly tetradynamous; filaments not dilated basally; anthers oblong [ovate], (apex obtuse or apiculate); nectar glands (6) confluent, or (4) lateral and median. **Fruits** siliques or silicles (samaroid), sessile, oblong, oblanceolate, elliptic, or obovate [ovate, cordate, spatulate, orbicular], 1- (or 2-)seeded, smooth, strongly angustiseptate, (prominently winged all around or distally; seed-bearing locule papery or corky, distinctly or obscurely 1–3-veined, sometimes keeled or shortly winged), glabrous or pubescent; valves and replum united; septum absent; ovules 1 (or 2) per ovary, (subapical); stigma capitate. **Seeds** plump, not winged, narrowly oblong; seed coat (smooth), not mucilaginous when wetted; cotyledons incumbent or accumbent.

Species 50 (1 in the flora): introduced; Europe, c, sw Asia, n Africa; introduced also in South America (Chile, Peru).

1. **Isatis tinctoria** Linnaeus, Sp. Pl. 2: 670. 1753
   • Dyers woad F I W

Plants glaucous, usually glabrous, sometimes pubescent proximally. **Stems** (3–)4–10(–15) dm. **Basal leaves:** petiole 0.5–5.5 cm; blade oblong or oblanceolate, (2.5–)5–15(–20) cm × (5–)15–35(–50) mm, base attenuate, margins entire, repand, or dentate, apex obtuse. **Cauline leaves:** blade usually oblong or lanceolate, rarely linear-oblong, base sagittate or auriculate, apex acute. **Fruiting pedicels** 5–10 mm. **Flowers:** sepals 1.5–2.8 × 1–1.5 mm, glabrous; petals 2.5–4 × 0.9–1.5 mm, base attenuate; filaments 1–2.5 mm; anthers 0.5–0.7 mm. **Fruits** black or dark brown, often broader distal to middle, (0.9–)1.1–2(–2.7) cm × 3–6(–10) mm, base cuneate, margins sometimes slightly constricted, apex usually subacute or rounded, rarely subemarginate; locule with distinct midvein, lateral veins inconspicuous, 3–6(–10) mm; apical wing 3.5–5(–7) mm wide. **Seeds** light brown, 2.3–3.5(–4.5) × 0.8–1 mm. $2n = 14, 28$.

Flowering Apr–Jun. Roadsides, fields, pastures, sagebrush hillsides, prairies, railroad embankments, waste places; 300–2200 m; introduced; B.C., Nfld. and Labr. (Nfld.), Ont., Que.; Calif., Idaho, Ill., Mo., Mont., Nev., N.Mex., N.Y., Oreg., Utah, Va., Wash., W.Va., Wyo.; Europe; c, sw Asia; n Africa; introduced also in South America (Chile, Peru).

Isatis tinctoria has been cultivated since ancient times as a source of a blue dye (woad) obtained by fermenting the ground leaves and proximal portions of the plant.

67. **MYAGRUM** Linnaeus, Sp. Pl. 2: 640. 1753; Gen. Pl. ed. 5, 289. 1754 • [Greek *muagron*, name used by Dioscorides and Pliny for a species of mustard] I

Ihsan A. Al-Shehbaz

Annuals; not scapose; (glaucous), glabrous. **Stems** erect, branched distally. **Leaves** basal and cauline; petiolate or sessile; basal (soon withered), not rosulate, petiolate, blade margins pinnatifid, runcinate, sinuate, or dentate; cauline blade (base auriculate, sagittate, or amplexicaul), margins entire or denticulate. **Racemes** (corymbose, several-flowered), considerably elongated in fruit. **Fruiting pedicels** appressed to rachis, stout. **Flowers:** sepals oblong to ovate, lateral pair subsaccate basally; petals oblong to oblanceolate, (slightly longer than sepals), claw undifferentiated from blade, (apex obtuse); stamens tetradynamous; filaments not dilated basally; anthers oblong, (apex obtuse); nectar glands (4) lateral, and (2) median (sometimes obscurely confluent with lateral). **Fruits** silicles, sessile, obpyriform to clavate-obcordiform, (woody, broadest distal to middle), slightly angustiseptate; (seed-bearing locule basal, 1 (or 2)-seeded, distal 2 locules

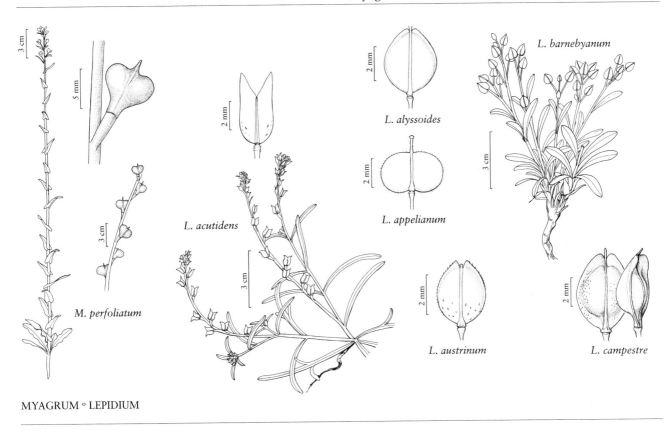

seedless); replum rounded; ovules 2 per ovary; style distinct, (flattened, triangular); stigma capitate. **Seeds** plump, not winged, oblong to ovoid; seed coat not mucilaginous when wetted; cotyledons incumbent. $x = 7$.

Species 1: introduced; s Europe, sw Asia, introduced also in Africa, Australia.

1. **Myagrum perfoliatum** Linnaeus, Sp. Pl. 2: 640. 1753

Stems branched at or distal to middle, rigid, 2–5(–6) dm. **Basal leaves** (withered by anthesis); shortly petiolate; blade oblanceolate, 2–7 cm × 5–15 mm. **Cauline leaves** sessile; blade oblong to lanceolate, (1–)1.5–6(–8)cm × (3–)5–20(–25) mm (smaller distally), apex acute or obtuse. **Fruiting pedicels** 3–5 mm, (much wider distally, hollow). **Flowers:** sepals 2.5–3.5 × 0.6–1.4 mm; petals 3–5 × 0.7–1.4 mm, base attenuate; filaments 2.5–3.7 mm; anthers 0.5–0.7 mm. **Fruits** 4.5–7.5(–8.5) × 2.5–4.5(–5) mm; style (0.8–)1.2–2 mm. **Seeds** 2–3 × 1–2 mm. $2n = 14$.

Flowering Apr–Jun. Roadsides, waste grounds, disturbed sites; introduced; Tex., Va.; s Europe; sw Asia; introduced also in nw, s Africa, Australia.

w. **Brassicaceae** Burnett tribe **Lepidieae** de Candolle, Mém. Mus. Hist. Nat. 7: 240. 1821 (as Lepidineae)

Annuals, biennials, or perennials [lianas, shrubs]; eglandular. **Trichomes** simple or absent. **Cauline leaves** petiolate or sessile; blade base auriculate or not, margins usually entire, dentate, or pinnately divided. **Racemes** ebracteate, elongated or not in fruit. **Flowers** actinomorphic;

sepals ascending to spreading, lateral pair not saccate basally; petals white, yellow, pink, or purple [orange], claw usually present, sometimes absent, distinct; filaments unappendaged, not winged; pollen 3-colpate. **Fruits** silicles, usually dehiscent, unsegmented, usually angustiseptate, rarely terete; ovules 2(–4) per ovary; style distinct, obsolete, or absent; stigma entire or, rarely, 2-lobed. **Seeds** aseriate; cotyledons accumbent or incumbent.

Genera 4, species ca. 235 (1 genus, 42 species in the flora): nearly worldwide.

68. LEPIDIUM Linnaeus, Sp. Pl. 2: 643. 1753; Gen. Pl. ed. 5, 291. 1754 • Peppergrass, pepperwort, cress, peppercress [Greek *lepidion* or *lepidos*, scale, alluding to appearance of fruit]

Ihsan A. Al-Shehbaz

John F. Gaskin

Carara Medikus; *Cardaria* Desvaux; *Coronopus* Zinn; *Neolepia* W. A. Weber; *Physolepidion* Schrenk; *Senebiera* de Candolle; *Sprengeria* Greene; *Stroganowia* Karelin & Kirilow

Plants not scapose; glabrous, pubescent, hirsute, or pilose. **Stems** usually erect or ascending, sometimes procumbent, decumbent, or prostrate, unbranched or branched. **Leaves** usually basal and cauline (basal absent in *L. fremontii*); petiolate or sessile; basal rosulate or not, petiolate (or petiole undifferentiated from blade), blade margins entire, dentate, denticulate, serrate, crenate, or lobed; cauline petiolate or sessile, blade (base auriculate or not), margins entire, dentate, or pinnately divided. **Racemes** (usually corymbose), elongated or not in fruit. **Fruiting pedicels** erect to divaricate, slender or stout. **Flowers:** sepals (usually deciduous, sometimes persistent), usually ovate or oblong, rarely suborbicular; petals (erect or spreading, sometimes rudimentary or absent), obovate, spatulate, oblong, oblanceolate, orbicular, linear, or filiform, claw absent or differentiated from blade, (apex obtuse, rounded, or emarginate); stamens 2 or 4 and equal in length, lateral or median, or 6 and tetradynamous; filaments not dilated basally; anthers ovate or oblong; (ovary placentation apical); nectar glands (4 or 6), distinct, median glands often present. **Fruits** schizocarps or silicles, (rarely indehiscent), sessile, didymous, oblong, ovate, obovate, cordate, obcordate, elliptic, orbicular, ovoid, obovoid, or globose, strongly angustiseptate or inflated and terete; valves each with prominent veins or not veined, (keeled or rounded, apex winged or not, thin or strongly thickened and ornamented, enclosing or readily releasing seed), glabrous or pubescent; replum rounded, (visible); septum complete or perforated; style absent, obsolete, or distinct, (included or exserted from apical notch); stigma capitate, usually entire, rarely 2-lobed. **Seeds** oblong or ovate [obovate], plump or flattened, winged, margined, or not winged; seed coat (smooth, minutely reticulate, or papillate), usually copiously mucilaginous when wetted, rarely not; cotyledons usually incumbent (accumbent in *L. virginicum*) [diplecolobal].

Species ca. 220 (42 in the flora): North America, Mexico, Central America, South America, Europe, Asia, n, s Africa, Australia.

The limits of *Lepidium* were expanded by I. A. Al-Shehbaz et al. (2002) to include *Cardaria*, *Coronopus*, and *Stroganowia*. Molecular data (J. L. Bowman et al. 1999; K. Mummenhoff et al. 2001) provide overwhelming support for this circumscription. A. Thellung (1906) was the first to take such a position, but subsequent authors largely ignored his work. Indeed, three Linnaean species, originally described as *L. chalepense*, *L. didymum*, and *L. draba*, were the core of the genera *Cardaria* and *Coronopus*. Both are nested within *Lepidium*, and *Coronopus* was clearly shown to be polyphyletic (C. D. Bailey et al. 2007). All three taxa differ from *Lepidium* by trivial fruit characters; the interested reader should consult Al-Shehbaz et al. for discussions.

Lepidium species are reasonably well-defined worldwide, and taxonomists often find no problems in telling them apart. The infraspecific taxonomy, especially some of the North American species, is in a state of disarray. The main reason is dependence on characters of questionable value to establish most varieties. For example, glabrous and pubescent forms were almost always recognized as distinct varieties despite the fact that they occur within the same population. The painstaking work of R. C. Rollins (1958) on *Dithyrea* clearly demonstrated that the presence versus absence of trichomes can result from a minor genetic difference. Rollins (1993) and others (e.g., N. H. Holmgren 2005b) were influenced by the excellent work of C. L. Hitchcock (1936, 1945, 1950), who then had no idea about the genetic basis of such differences. A similar situation exists in *Draba*, where tens of varieties were established based on glabrous versus pubescent fruits. In this account, we do not recognize such varieties, and the interested reader is referred to the discussion under 35. *L. ramosissimum*.

Due to their release of copious mucilage upon wetting, the seeds of most species of *Lepidium* can be transported easily by animals, especially migratory birds, for long distances. This most likely accounts for the transcontinental dispersals in the genus discussed by K. Mummenhoff et al. (2001).

Lepidium africanum (Burman f.) de Candolle, *L. bonariense* Linnaeus, *L. graminifolium* Linnaeus, and *L. schinzii* Thellung have been recorded from North America as ballast introductions (I. A. Al-Shehbaz 1986b; R. C. Rollins 1993). Apparently, they never became naturalized in the flora area and, therefore, are not included in this account. For a recent report of *L. bonariense*, see the discussion under 28. *L. oblongum*.

SELECTED REFERENCES Al-Shehbaz, I. A., K. Mummenhoff, and O. Appel. 2002. *Cardaria*, *Coronopus*, and *Stroganowia* are united with *Lepidium* (Brassicaceae). Novon 12: 5–11. Bowman, J. L., H. Brüggemann, J.-Y. Lee, and K. Mummenhoff. 1999. Evolutionary changes in floral structure within *Lepidium* L. (Brassicaceae). Int. J. Pl. Sci. 160: 917–929. Hitchcock, C. L. 1936. The genus *Lepidium* in the United States. Madroño 3: 265–300. Mulligan, G. A. 1961. The genus *Lepidium* in Canada. Madroño 16: 77–89. Mulligan, G. A. and C. Frankton. 1962. Taxonomy of the genus *Cardaria* with particular reference to the species introduced into North America. Canad. J. Bot. 40: 1411–1425. Mummenhoff, K., H. Brüggemann, and J. L. Bowman. 2001. Chloroplast DNA phylogeny and biogeography of *Lepidium* (Brassicaceae). Amer. J. Bot. 88: 2051–2063. Muschler, R. 1908. Die Gattung *Coronopus* (L.) Gaertn. Bot. Jahrb. Syst. 41: 111–147. Thellung, A. 1906. Die Gattung *Lepidium* (L.) R. Br. Mitt. Bot. Mus. Univ. Zürich 28: 1–340.

1. Cauline leaves (at least some) sessile, blade bases auriculate, sagittate, or amplexicaul.
　2. Plants glabrous or sparsely pubescent proximally; uppermost cauline leaf blades: bases cordate-amplexicaul; basal leaf blades: margins 2- or 3-pinnatifid or pinnatisect; petals yellow . . . . 33. *Lepidium perfoliatum*
　2. Plants usually puberulent or hirsute proximally, rarely glabrate; uppermost cauline leaf blades: bases auriculate or sagittate; basal leaf blades: margins usually entire, subentire, dentate, denticulate, lyrate, or sinuate, rarely 1- or 2-pinnatifid; petals (when present) white.
　　3. Stamens 2; petals absent or rudimentary; basal leaf blades: margins 1- or 2-pinnatifid; styles usually obsolete (rarely to 0.1 mm) . . . . 28. *Lepidium oblongum* (in part)
　　3. Stamens 6; petals present (distinct); basal leaf blades: margins usually entire, subentire, dentate, denticulate, lyrate, or sinuate, rarely pinnatifid; styles distinct (0.2 mm or longer).
　　　4. Plants not rhizomatous; fruits dehiscent, apically broadly winged, notch present; racemes much-elongated in fruit.
　　　　5. Annuals; fruit valves papillate; styles 0.2–0.5(–0.7) mm, included or only slightly exserted beyond apical notch of fruits . . . . 6. *Lepidium campestre*
　　　　5. Perennials (with caudex); fruit valves often not papillate; styles (0.6–)1–1.5 mm, well-exserted beyond apical notch of fruits . . . . 18. *Lepidium heterophyllum*

4. Plants rhizomatous; fruits indehiscent, apically not winged, notch absent; racemes not much-elongated (corymbose panicles) in fruit.
    6. Fruits flattened, cordate to subreniform, valves reticulate-veined. . . . . 14. *Lepidium draba*
    6. Fruits inflated, globose, subglobose, obovoid, or obcompressed globose, valves not reticulate-veined.
        7. Fruits usually globose, rarely subglobose, (2–)3–4.4(–5) mm wide, valves puberulent; styles 0.5–1.5 mm . . . . 3. *Lepidium appelianum*
        7. Fruits obovoid to subglobose, or obcompressed globose, (3.5–) 4–6.2(–7) mm wide, valves glabrous; styles (0.8–)1.2–2(–2.3) mm . . . . 7. *Lepidium chalepense*
1. Cauline leaves usually petiolate, blade bases (when sessile) not auriculate, sagittate, or amplexicaul (absent in *L. nanum*).
    8. Valves rugose or rugose-verrucose.
        9. Fruits reniform to ovate-cordate, 2.3–3.4 mm, apical notches absent, ridged; stamens 6. . . . . 8. *Lepidium coronopus*
        9. Fruits didymous, 1.3–1.7 mm, apical notches present, not ridged; stamens 2 . . . . 13. *Lepidium didymum*
    8. Valves usually smooth (rarely minutely papillate in *L. jaredii*).
        10. Subshrubs or perennials (caudex woody, sometimes with persistent petiolar remains).
            11. Styles included in apical notch of fruits; stamens 2; petals absent or rudimentary. . . . . 32. *Lepidium paysonii*
            11. Styles usually exserted beyond apical notch of fruits, rarely subequaling notch, or notch absent; stamens 6 (or 2 or 4 in *L. integrifolium*); petals well-developed.
                12. Basal and proximalmost cauline leaf blade margins pinnately lobed.
                    13. Plants glabrous throughout; basal leaves absent; fruits 4.2–7(–8) mm wide . . . . 17. *Lepidium fremontii*
                    13. Plants puberulent (at least on fruiting pedicels or rachises); basal leaves present; fruits (1.5–)1.8–3.6(–4) mm wide.
                        14. Fruits broadly obovate; fruiting pedicels puberulent throughout . . . . .19. *Lepidium huberi*
                        14. Fruits ovate to suborbicular or oblong; fruiting pedicels puberulent adaxially, rarely glabrous.
                            15. Fruits usually ovate to suborbicular, rarely oblong; basal leaf blades: margins 1- or 2-pinnatifid to -pinnatisect; cauline leaf blades: margins often pinnately lobed. . . . . . . 25. *Lepidium montanum* (in part)
                            15. Fruits broadly ovate; basal leaf blades: margins pinnately lobed; cauline leaf blades: margins usually entire, rarely dentate.
                                16. Perennials or subshrubs; stems (0.7–)1–4.8(–6.1) dm; middle cauline leaf blades (0.7–)1–2(–3) mm wide. . . 2. *Lepidium alyssoides*
                                16. Perennials; stems (3.5–)4.5–16(–18) dm; middle cauline leaf blades (2.5–)4–10 mm wide. . . . . . . 15. *Lepidium eastwoodiae* (in part)
                12. Basal and proximalmost cauline leaf blade margins usually entire, crenate, or dentate, rarely palmately 3–5-lobed (at apex).
                    17. Plants pulvinate or cespitose; basal leaf blades: margins usually subapically or apically 3–5-toothed or -lobed, rarely entire (*L. barnebyanum*); fruiting pedicels puberulent throughout.
                        18. Basal leaves rosulate; cauline leaves absent; anthers 1.4–2 mm . . . . 26. *Lepidium nanum*
                        18. Basal leaves not rosulate; cauline leaves present; anthers 0.3–0.8 mm.
                            19. Basal leaf blades: margins entire; petals 3.2–4.6 mm; fruits 4–5.5(–6.2) mm . . . . .5. *Lepidium barnebyanum*
                            19. Basal leaf blades: margins subapically 3–5-lobed or toothed, rarely entire (*L. davisii*); petals 2–3.5 (–4) mm; fruits 2.5–4(–5) mm.

20. Plants cespitose; fruits suborbicular to broadly ovate; styles 0.5–1 mm; seeds (1.8–)2–2.3 mm . . . . 10. *Lepidium davisii*

20. Plants pulvinate; fruits ovate; styles 0.3–0.6 mm; seeds 1.2–1.5 mm . . . . 29. *Lepidium ostleri*

17. Plants not pulvinate, not cespitose; basal leaf blades: margins entire or uniformly dentate; fruiting pedicels usually puberulent adaxially, rarely glabrous.

21. Fruits obovate to somewhat rhomboid, 7–11 mm; fruiting pedicels glabrous; seeds 3.7–4.5 mm . . . . 41. *Lepidium tiehmii*

21. Fruits oblong-elliptic, ovate, oblong, or suborbicular, 1.6–4(–4.4) mm; fruiting pedicels usually pubescent or puberulent, sometimes glabrous (less than 6 mm); seeds 0.8–2 mm.

22. Plants rhizomatous; fruits (1.6–)1.8–2.4(–2.7) mm, apically not winged, notch usually absent; styles 0.05–0.15 mm . . . . 23. *Lepidium latifolium*

22. Plants not rhizomatous; fruits (2.5–)3–4(–4.4) mm, apically winged, notch present; styles 0.2–1 mm.

23. Perennials or subshrubs (woody base aboveground); stamens 6; racemes slightly elongated (subcorymbose panicles) in fruit . . . . 9. *Lepidium crenatum*

23. Perennials (woody base not aboveground); stamens usually 4, rarely 2 or 6; racemes elongated (not corymbose) in fruit . . . . 20. *Lepidium integrifolium*

[10. Shifted to left margin.—Ed.]

10. Annuals or biennials (without woody caudex, without persistent petiolar remains).

24. Stamens 2.

25. Fruit valves distinctly reticulate-veined; sepals persistent . . . . 39. *Lepidium strictum*

25. Fruit valves usually not veined (rarely weakly veined); sepals deciduous or, rarely, tardily so.

26. Fruits 1.7–2.1 × 1.2–1.6 mm; plants puberulent with clavate trichomes . . . . 38. *Lepidium sordidum*

26. Fruits 1.8–7 × 1.5–5(–5.6) mm; plants puberulent, hirsute, or hispid with cylindrical trichomes.

27. Fruiting pedicels often strongly flattened, 0.2–0.7 mm wide; fruit valves hirsute to hispid (sometimes only on margins) . . . . 22. *Lepidium lasiocarpum*

27. Fruiting pedicels terete or only slightly flattened, 0.1–0.3(–4) mm wide; fruit valves glabrous or puberulent (at least on margins).

28. Fruits elliptic.

29. Basal leaf blades pinnatifid; racemes slightly elongated in fruit, rachises with curved trichomes; fruiting pedicels usually puberulent adaxially, rarely throughout . . . . 35. *Lepidium ramosissimum*

29. Basal leaf blades (1- or) 2- or 3-pinnatisect; racemes considerably elongated in fruit, rachises with straight trichomes; fruiting pedicels puberulent throughout . . . . 36. *Lepidium ruderale*

28. Fruits obovate, suborbicular, or orbicular.

30. Plants hirsute; basal leaf blade margins pinnatifid.

31. Stems often simple from base; rachises pubescent, trichomes curved with fewer, longer, straight ones . . . . 4. *Lepidium austrinum*

31. Stems often several from base; rachises hirsute, trichomes mostly straight . . . . 28. *Lepidium oblongum* (in part)

30. Plants puberulent or glabrous; basal leaf blade margins dentate, serrate, lyrate, or pinnatifid.

32. Fruits obovate to obovate-suborbicular, widest beyond middle; rachises with straight, slender to subclavate trichomes; petals absent or rudimentary, 0.3–0.9 mm; cotyledons incumbent. . . . . 11. *Lepidium densiflorum*

32. Fruits orbicular, widest at middle; rachises usually with curved, cylindrical trichomes, rarely glabrous; petals usually present, rarely rudimentary, 1–2(–2.5) mm; cotyledons usually accumbent, rarely incumbent . . . . 42. *Lepidium virginicum*

[24. Shifted to left margin.—Ed.]

24. Stamens 4 or 6.

33. Stamens 4; styles usually obsolete, rarely to 0.1 mm, usually included in, or equaling, apical notch of fruits.

34. Fruiting pedicels terete or slightly flattened, 0.1–0.3 mm wide.

35. Fruits 2.4–3.6 mm, apically winged, apical notches 0.3–0.8 mm deep; stamens all median; petals absent . . . . .30. *Lepidium oxycarpum*

35. Fruits 1.8–2 mm, apically not winged, apical notches 0.05–0.1 mm deep; stamens 2 median, 2 lateral; petals present (rudimentary, 0.4–0.6 mm) . . . . 34. *Lepidium pinnatifidum*

34. Fruiting pedicels strongly flattened, 0.4–1.4 mm wide.

36. Sepals somewhat persistent; fruit valves hirsute (trichomes spreading, mixed with smaller ones); petals usually pubescent on outside, rarely glabrescent, with fringed margin; racemes compact, subcapitate to cylindrical in fruit . . . . 24. *Lepidium latipes*

36. Sepals deciduous; fruit valves glabrous or (sparsely, uniformly) puberulent, pubescent, or hirsute (sometimes only on margin); petals usually absent, when present, glabrous, without fringed margin; racemes lax or somewhat dense, not compact in fruit.

37. Petals usually present, rarely rudimentary or absent; fruit valves slightly veined or not; fruiting pedicels not appressed to rachises. . . . . 27. *Lepidium nitidum*

37. Petals absent; fruit valves strongly reticulate-veined; fruiting pedicels appressed to rachises or distally recurved.

38. Fruits (3–)4–6 × 2.5–3.5 mm, apical notches (0.8–)1–2 mm deep, V-shaped; fruiting pedicels (2–)3–4.2 mm . . . . 1. *Lepidium acutidens*

38. Fruits 2.5–3.5 × 2–2.8 mm, apical notches 0.5–0.7(–0.8) mm deep (closed), often U-shaped; fruiting pedicels (1.6–)1.9–2.5(–3) mm . . . . .12. *Lepidium dictyotum*

33. Stamens 6; styles (0.1–)0.2–1.6 mm, usually exserted beyond, rarely subequaling or included in, apical notch of fruits.

39. Fruiting pedicels suberect to ascending, subappressed to rachises; fruits (4–)5–6.4(–7) mm; styles usually included in, rarely subequaling, apical notch of fruits; cotyledons 3-lobed . . . . 37. *Lepidium sativum*

39. Fruiting pedicels divaricate-ascending to horizontal or descending, not appressed to rachises; fruits 2–3.8(–4.2) mm; styles exserted from apical notch of fruits; cotyledons entire.

40. Petals usually yellow (rarely creamy white at early anthesis, or fading whitish).

41. Fruiting pedicels (2.7–)3–4.4(–5) mm, glabrous adaxially; racemes dense, slightly elongated (corymbose to subcapitate) in fruit, rachises glabrous adaxially; fruits divergently winged apically, apical notches 0.2–0.6 mm deep. . . . . 16. *Lepidium flavum*

41. Fruiting pedicels 5–15 mm, pilose adaxially; racemes lax, much-elongated in fruit, rachises usually pilose, rarely glabrous; fruits not winged apically, apical notches 0(–0.1) mm deep. . . . . .21. *Lepidium jaredii*

40. Petals white.

42. Filaments puberulent . . . . 31. *Lepidium papilliferum*

42. Filaments glabrous.

[43. Shifted to left margin.—Ed.]

43. Annuals; rachises pilose, trichomes straight.................................... 40. *Lepidium thurberi*
43. Annuals or biennials; rachises puberulent, trichomes straight or curved.
    44. Petals suborbicular, 1.5–2.5 mm wide; cauline leaf blades lanceolate or oblanceolate to linear; stems (3.5–)4.5–16(–18) dm, often simple from base........ 15. *Lepidium eastwoodiae* (in part)
    44. Petals spatulate to oblanceolate, 1.3–1.8 mm wide; cauline leaf blades often pinnatifid to pinnatisect, sometimes linear; stems 0.4–5(–7) dm, simple or few to several from base
    .... 25. *Lepidium montanum* (in part)

1. **Lepidium acutidens** (A. Gray) Howell, Fl. N.W. Amer., 64. 1897 [F]

Lepidium dictyotum A. Gray var. *acutidens* A. Gray, Proc. Amer. Acad. Arts 12: 54. 1877; *L. oxycarpum* Torrey & A. Gray var. *acutidens* (A. Gray) Jepson

Annuals; hirsute or puberulent. **Stems** few to several from base, erect to ascending, unbranched, (0.5–)0.8–3 dm. **Basal leaves** (soon withered); not rosulate; petiole 1–4 cm; blade linear or pinnatisect, (2–)2.5–6.1(–7.2) cm × (0.5–)1–2(–3) mm, (lobes linear to narrowly oblong), margins entire. **Cauline leaves** sessile; blade linear, 1.2–5.8 cm × 0.5–2(–3) mm, base attenuate, not auriculate, margins entire. **Racemes** elongated, (dense or lax) in fruit; rachis puberulent or hirsute, trichomes straight, cylindrical. **Fruiting pedicels** erect to slightly ascending, straight and appressed to rachis or distally slightly recurved, (strongly flattened), (2–)3–4.2 × 0.4–0.6 mm (width proximal to apex), puberulent throughout. **Flowers:** sepals oblong to ovate, 0.7–1.1 × 0.2–0.6 mm; petals absent; stamens 4, median; filaments 0.5–1 mm; anthers ca. 0.1 mm. **Fruits** ovate to ovate-oblong, (3–)4–6 × 2.5–3.5 mm, apically winged, apical notch (V-shaped), (0.8–)1–2 mm deep; valves thin, smooth, strongly reticulate-veined, usually glabrous, rarely pubescent; style absent or obsolete, included in apical notch. **Seeds** ovate, 1.3–2 × 0.8–1.2 mm.

Flowering Feb–Apr. Alkaline flats, gullies, or fields, saline vernal flats, grassy fields; 0–400 m; Calif., Oreg.; Mexico (Baja California).

Lepidium acutidens was treated by C. L. Hitchcock (1936) and R. C. Rollins (1993) as a variety of *L. dictyotum*. The differences in the fruits as well as the absence of intermediates between them, despite the overlap of their ranges in California, justify their recognition as independent species.

2. **Lepidium alyssoides** A. Gray, Mem. Amer. Acad. Arts, n. s. 4: 10. 1849 [F]

Lepidium alyssoides var. *angustifolium* (C. L. Hitchcock) Rollins; *L. alyssoides* var. *junceum* Rollins; *L. alyssoides* var. *minus* Thellung; *L. alyssoides* var. *polycarpum* Thellung; *L. alyssoides* var. *streptocarpum* Thellung; *L. montanum* Nuttall subsp. *alyssoides* (A. Gray)
C. L. Hitchcock; *L. montanum* var. *alyssoides* (A. Gray) M. E. Jones; *L. montanum* subsp. *angustifolium* (C. L. Hitchcock) C. L. Hitchcock; *L. montanum* var. *angustifolium* C. L. Hitchcock; *L. tortum* L. O. Williams

Perennials or subshrubs; (woody base often aboveground); glabrous or minutely puberulent. **Stems** few to several from base, erect to ascending, branched throughout, (0.7–)1–4.8(–6.1) dm. **Basal leaves** often not rosulate; petiole 1–6 cm; blade pinnately lobed, (1–)1.5–8(–11) cm × (5–)10–35 mm, margins (of lobes) entire or denticulate. **Cauline leaves** sessile; blade linear, (0.8–)1.3–7(–9.5) cm × (0.7–)1–2(–3) mm, base attenuate, not auriculate, margins entire. **Racemes** elongated in fruit; rachis puberulent or glabrous. **Fruiting pedicels** divaricate to horizontal, straight or recurved to somewhat sigmoid, (terete), 3.5–8(–11) × 0.2 mm, glabrous or puberulent adaxially. **Flowers:** sepals ovate to oblong, 1–2 × 0.8–1 mm; petals white, suborbicular, 2–3 × 1–2 mm, claw 0.5–1.5 mm; stamens 6; filaments 1.5–2 mm, (glabrous); anthers 0.2–0.4 mm. **Fruits** broadly ovate, 2–3.7(–4.3) × (1.5–)1.8–2.9(–3.4) mm, apically winged, apical notch 0.1–0.3(–0.4) mm deep; valves thin, smooth, not veined, glabrous; style 0.2–0.6 mm, exserted beyond apical notch. **Seeds** ovate, 1.5–1.8(–2) × 0.9–1.2(–1.5) mm. $2n = 32$.

Flowering May–Jul. Pinyon-juniper or sagebrush communities, prairies, grasslands, sandstone outcrops, gypsum flats, sand dunes, dry flats and river bottoms, gravelly roadsides; 1200–2800 m; Ariz., Colo., Nev., N.Mex., Tex., Utah, Wyo.; Mexico (Chihuahua, Coahuila, Nuevo León, San Luis Potosí).

Of the five varieties of *Lepidium alyssoides* recognized by R. C. Rollins (1993), one (var. *mexicanum* Rollins) is a short-tufted form of the species restricted to Mexico that does not seem to merit recognition, another (var. *junceum*) is a glabrescent form of the type variety, a third (var. *eastwoodiae*) is treated below as a distinct species, and the fourth (var. *angustifolium*) is included here within *L. alyssoides*.

3. **Lepidium appelianum** Al-Shehbaz, Novon 12: 7. 2002 F I W

Hymenophysa pubescens C. A. Meyer in C. F. Ledebour, Icon. Pl. 2: 20. 1830, not *Lepidium pubescens* Desvaux 1814; *Cardaria pubescens* (C. A. Meyer) Jarmolenko; *C. pubescens* var. *elongata* Rollins

Perennials; (rhizomatous); often densely hirsute. **Stems** simple or several from base, erect or ascending, branched distally, (1–)1.5–3.5(–5) dm. **Basal leaves** (often withered by anthesis); not rosulate; petiole 0.5–1.5 cm; blade obovate to oblanceolate, (1–)2–6(–7) cm × 3–20 mm, margins dentate to sinuate. **Cauline leaves** sessile; blade oblong or lanceolate, 1–5(–8) cm × (3–)5–15(–30) mm, base sagittate, margins dentate or subentire, (surfaces pubescent). **Racemes** (usually corymbose, rarely paniculate), rarely elongated in fruit; rachis pubescent, trichomes often curved. **Fruiting pedicels** divaricate to ascending, straight or slightly curved, (terete), 3–9(–12) × 0.2–0.3 mm, pubescent. **Flowers:** sepals oblong, 1.4–2 × 0.7–1 mm; petals white, broadly obovate, (2.2–)2.8–4 × 1–3 mm, claw 1–1.4 mm; stamens 6; filaments 2–2.5 mm, (glabrous); anthers 0.4–0.5 mm. **Fruits** (indehiscent), globose or, rarely, subglobose, (2–)3–4.4(–5) mm diam., (inflated), apically not winged, apical notch absent; valves thin, smooth, not veined, densely puberulent; style 0.5–1.5 mm. **Seeds** (brown or dark brown), ovoid, 1.5–2 × 1–1.5 mm. $2n$ = 16.

Flowering May–Sep. Roadsides, sagebrush communities, alkaline meadows, waste grounds, ditch and stream sides, fields, pastures; 400–2400 m; introduced; Alta., B.C., Man., Sask.; Calif., Colo., Idaho, Mich., Mo., Mont., Nebr., Nev., N.Mex., N.Dak., Oreg., Pa., Utah, Wash., Wis., Wyo.; c Asia; introduced also in South America, other parts of Asia.

Lepidium appelianum has become a noxious weed in most of its range in North America.

4. **Lepidium austrinum** Small, Fl. S.E. U.S., 468, 1331. 1903 F W

Lepidium austrinum var. *orbiculare* Thellung; *L. lasiocarpum* var. *orbiculare* (Thellung) C. L. Hitchcock

Annuals or biennials; often densely hirsute, (trichomes cylindrical). **Stems** often simple from base, erect, branched distally, (1.5–)2–6.7(–9.4) dm. **Basal leaves** (later withered); rosulate; petiole (0.7–)1.5–4.5 cm; blade pinnatifid, 2–8.3 cm × 9–26 mm, margins (of lobes) entire or dentate. **Cauline leaves** shortly petiolate; blade oblanceolate to nearly linear, 1–4.5(–6.2) cm × 3–10(–17) mm, base attenuate to cuneate, not auriculate, margins entire or dentate. **Racemes** much-elongated in fruit; rachis pubescent, trichomes curved, with fewer and longer, straight ones. **Fruiting pedicels** usually divaricate, rarely horizontal, straight or slightly recurved, (terete), (2.5–)3–4.1(–4.7) × 0.2 mm, puberulent adaxially. **Flowers:** sepals oblong, 0.8–1 × 0.2–0.4 mm; petals (sometimes absent), white, oblanceolate, 0.4–1.6 × 0.1–0.8 mm, claw absent; stamens 2, median; filaments 0.8–1 mm; anthers 0.1–0.2 mm. **Fruits** elliptic-obovate to obovate-orbicular, 2.4–3.2 × 1.8–2.5 mm, apically winged, apical notch 0.2–0.5 mm deep; valves thin, smooth, not veined, sparsely puberulent, (trichomes often antrorsely appressed, sometimes restricted to margin); style 0.05–0.1 mm, included in apical notch. **Seeds** ovate, 1.4–1.6 × 0.7–0.9 mm.

Flowering Mar–Jun. Disturbed grounds, railroad tracks and embankments, fields, knolls, stream banks, waste areas, open banks, roadsides, sandy terraces; Kans., La., Miss., N.Mex., Okla., Tex.; Mexico (Coahuila, Nuevo León, San Luis Potosí).

5. **Lepidium barnebyanum** Reveal, Great Basin Naturalist 27: 178. 1967 C E F

Lepidium montanum Nuttall subsp. *demissum* C. L. Hitchcock, Madroño 10: 157. 1950, not *L. demissum* C. L. Hitchcock 1945

Perennials; (cespitose, caudex thick, woody, numerous-branched, with persistent petiolar remains); puberulent. **Stems** simple or few from base (caudex), erect to ascending, unbranched or branched distally, (0.3–)0.5–1.5(–1.7) dm. **Basal leaves** not rosulate; petiole 0.3–0.8 cm, (papery); blade linear, (0.5–)1–7(–8) cm × 10–36 mm, margins entire. **Cauline leaves** sessile; blade linear, base attenuate, not auriculate, margins entire, (similar to basal, smaller distally). **Racemes** slightly elongated in fruit; rachis puberulent, trichomes straight. **Fruiting pedicels** divaricate-ascending, straight, (terete), 3–8 × 0.3–0.4 mm, puberulent throughout. **Flowers:** sepals (sometimes somewhat persistent), oblong-obovate, 1.8–2.8 × 1.3–1.8 mm; petals white to pale yellow, suborbicular to broadly obovate, 3.2–4.6 × 2.5–3.2 mm, claw 0.5–1.5 mm; stamens 6; filaments 1.8–2.6 mm, (glabrous); anthers 0.5–0.8 mm. **Fruits** ovate, 4–5.5(–6.2) × 3–3.8 mm, apically winged, apical notch 0–0.2 mm deep; valves thin, smooth, not veined, glabrous; style 0.5–1.2 mm, exserted beyond apical notch. **Seeds** ovate, 2.2–2.8 × 1.4–1.6 mm.

Flowering May–Jun. Pinyon-juniper and sagebrush communities, white sandy shale; of conservation concern; 1800–2000 m; Utah.

Lepidium barnebyanum is known from the Green River and Uinta Shale Formations in Duchesne County. It is in the Center for Plant Conservation's National Collection of Endangered Plants.

6. **Lepidium campestre** (Linnaeus) W. T. Aiton in W. Aiton and W. T. Aiton, Hortus Kew. 4: 88. 1812

F I W

Thlaspi campestre Linnaeus, Sp. Pl. 2: 646. 1753; *Neolepia campestris* (Linnaeus) W. A. Weber

Annuals; densely hirsute. **Stems** simple from base, erect, unbranched or branched distally, (0.8–)1.2–5(–6.3) dm. **Basal leaves** rosulate; petiole (0.5–)1.5–6 cm; blade oblanceolate or oblong, (1–)2–6(–8) cm × 5–15 mm, margins entire, lyrate, or pinnatifid. **Cauline leaves** sessile; oblong, lanceolate, or narrowly deltate-lanceolate, (0.7–)1–4(–6.5) cm × (2–)5–10(–15) mm, base sagittate or auriculate, margins dentate or subentire. **Racemes** much-elongated in fruit; rachis hirsute, trichomes spreading, straight. **Fruiting pedicels** horizontal, straight or slightly recurved, (terete), (3–)4–8(–10) × 0.3–0.4 mm, hirsute. **Flowers:** sepals oblong, (1–)1.3–1.8 × 0.6–0.8 mm; petals white, spatulate, (1.5–)1.8–2.5(–3) × (0.2–)0.5–0.7 mm, claw 0.6–1 mm; stamens 6; filaments (1.2–)1.5–1.8(–2) mm, (glabrous); anthers 0.3–0.5 mm. **Fruits** broadly oblong to ovate, (4–)5–6(–6.5) × (3–)4–5 mm, (curved adaxially), apically broadly winged, apical notch (0.2–)0.4–0.6 mm deep; valves thin, papillate except for wing, not veined; style 0.2–0.5(–0.7) mm, slightly exserted beyond, or included in, apical notch. **Seeds** (dark brown), ovoid, 2–2.3(–2.8) × 1–1.4 mm. $2n = 16$.

Flowering May–Jun. Roadside, pastures, gardens, open flats, pine woodlands, rocky slopes, forests, waste grounds, disturbed areas, meadows, fields; 0–2600 m; introduced; B.C., Nfld. and Labr. (Nfld.), N.S., Ont., P.E.I., Que., Sask.; Ala., Ark., Calif., Colo., Conn., Del., D.C., Fla., Ga., Idaho, Ill., Ind., Iowa, Kans., Ky., La., Maine, Md., Mass., Mich., Miss., Mo., Mont., Nebr., Nev., N.H., N.J., N.Mex., N.Y., N.C., Ohio, Oreg., Pa., R.I., S.C., Tenn., Utah, Vt., Va., Wash., W.Va., Wis., Wyo.; Europe; Asia; introduced also in South America, South Africa.

7. **Lepidium chalepense** Linnaeus, Cent. Pl. II, 23. 1756

F I

Cardaria chalepensis (Linnaeus) Handel-Mazzetti; *C. draba* (Linnaeus) Desvaux subsp. *chalepensis* (Linnaeus) O. E. Schulz; *C. draba* var. *repens* (Schrenk) O. E. Schulz; *C. repens* (Schrenk) Jarmolenko; *Cochlearia draba* (Linnaeus) Linnaeus; *Lepidium draba* Linnaeus subsp. *chalepense* (Linnaeus) Thellung; *L. draba* var. *repens* (Schrenk) Thellung; *L. repens* (Schrenk) Boissier; *Physolepidion repens* Schrenk

Perennials; (rhizomatous); densely hirsute to glabrate or glabrous. **Stems** several from base, erect or decumbent basally, branched (several) distally, (0.8–)2.1–6.6(–9.2) dm. **Basal leaves** (early withered); not rosulate; petiole 0.9–4.4 cm; blade obovate, spatulate, or ovate, (1.8–)2.5–8.6(–14) cm × 10–37 mm, margins subentire or dentate. **Cauline leaves** sessile; blade obovate to oblong or lanceolate to oblanceolate, (1.5–)2.6–9.3(–13.2) cm × (7–)12–31(–45) mm, base sagittate-amplexicaul or auriculate, margins dentate or entire, (surfaces pubescent or glabrous). **Racemes** (corymbose panicles), elongated in fruit; rachis glabrous or puberulent, trichomes cylindrical, straight or curved. **Fruiting pedicels** ascending to horizontal, straight, (terete), 5–16(–19) × 0.2–0.3 mm, glabrous or sparsely puberulent adaxially. **Flowers:** sepals oblong to ovate, 1.7–3 × 1–1.6 mm; petals white, obovate, 3–5 × 1.2–2.4 mm, claw 1.2–2 mm; stamens 6; filaments 2–3.3 mm, (glabrous); anthers 0.5–0.6 mm. **Fruits** (indehiscent), obovoid to subglobose or obcompressed globose, 3.5–5.8(–7) × (3.5–)4–6.2 (–7) mm, apically not winged, apical notch absent; valves thin, smooth, often not veined, glabrous; style (0.8–)1.2–2(–2.3) mm. **Seeds** (dark reddish brown), ovate, 1.5–2.3 × 1–1.3 mm. $2n = 48, 80, 128$.

Flowering May–Jun. Mountain slopes, roadsides, fields, agricultural lands, stream banks, pastures, waste areas; 300–4200 m; introduced; Alta., B.C., Man., Ont., Sask.; Ariz., Calif., Colo., Idaho, Ill., Iowa, Kans., Mont., Nebr., Nev., N.Mex., N.Dak., Oreg., S.Dak., Tex., Utah, Wash., Wyo.; Asia; introduced also in South America (Argentina), Europe.

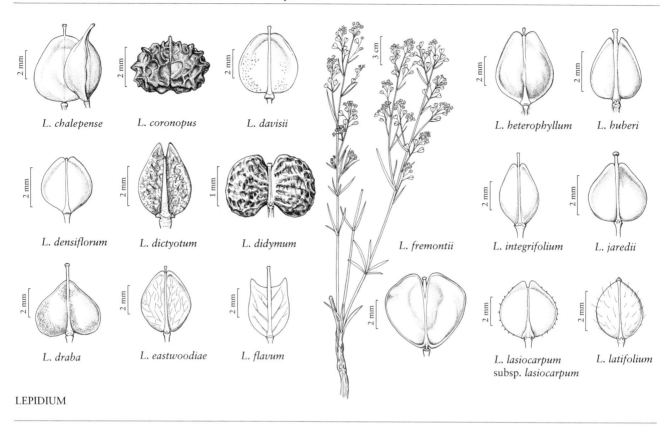

LEPIDIUM

From the synonymy above, it is evident that the disposition of *Lepidium chalepense* has varied: more than one species (e.g., R. C. Rollins 1940; G. A. Mulligan and C. Frankton 1962), one species (e.g., Rollins 1993), a variety of *Lepidium (Cardaria) draba* (N. H. Holmgren 2005b), or a synonym of the latter species (C. L. Hitchcock 1936). In our opinion, the differences in fruit morphology and chromosome number justify its recognition as a distinct species.

8. **Lepidium coronopus** (Linnaeus) Al-Shehbaz, Novon 14: 156. 2004 F I

Cochlearia coronopus Linnaeus, Sp. Pl. 2: 648. 1753; *Carara coronopus* (Linnaeus) Medikus; *Coronopus procumbens* Gilibert; *C. ruellii* Allioni; *C. squamatus* (Forsskål) Ascherson; *C. verrucarius* (Garsault) Muschler & Thellung; *Lepidium squamatum* Forsskål; *Senebiera coronopus* (Linnaeus) Poiret

Annuals; glabrous or puberulent. **Stems** often several from base, usually procumbent to decumbent, rarely ascending, branched distally, (0.3–)0.6–2.5(–3.5) dm. **Basal leaves** rosulate; petiole (1–)2–5(–5.8) cm; blade 1- or 2-pinnatisect, (3–)4–10(–15) cm, margins (of lobes) entire or dentate. **Cauline leaves** shortly petiolate; blade pinnatisect, base cuneate, not auriculate, margins (of lobes) entire or dentate, (similar to basal, smaller and less divided distally). **Racemes** (leaf-opposed), ± slightly elongated in fruit; rachis glabrous. **Fruiting pedicels** ascending, straight, (stout, terete), (0.7–)1–2(–2.4) × 0.4–0.5 mm, glabrous. **Flowers:** sepals (persistent), oblong, 1–1.5 × 0.5–0.6 mm; petals white, obovate to obovate-oblong, 1–2 × 0.4–0.6 mm, claw absent; stamens 6; filaments (median pairs) 0.7–1 mm, (glabrous); anthers 0.15–0.25 mm. **Fruits** (indehiscent), reniform to ovate-cordate, 2.3–3.4 × 3–4.4 mm, apically not winged, apical notch absent; valves thick, rugose-verrucose, with distinct ridges, prominently veined, glabrous; style 0.2–0.7 mm. **Seeds** ovate-oblong, (curved, not winged), 1.2–1.6 × 0.7–1 mm. $2n = 32$.

Flowering May–Aug. Waste grounds, abandoned fields, pastures, roadsides, disturbed sites; introduced; N.B., N.S., Ont., Que.; Ala., Calif., La., Mo., N.J., Tenn.; Europe; sw Asia; n Africa; introduced also in South America (Chile), s Africa, Australia.

9. **Lepidium crenatum** (Greene) Rydberg, Bull. Torrey Bot. Club 33: 141. 1906 [C][E]

Thelypodium crenatum Greene, Pittonia 4: 20. 1899; *Lepidium montanum* Nuttall subsp. *spatulatum* (B. L. Robinson) C. L. Hitchcock; *L. montanum* var. *spatulatum* (B. L. Robinson) C. L. Hitchcock; *L. scopulorum* M. E. Jones var. *spatulatum* B. L. Robinson; *L. vaseyanum* Thellung

Perennials or subshrubs; (woody base aboveground); puberulent. Stems simple from base, erect, branched distally, (2–)3–8(–11) dm. Basal leaves rosulate (on sterile shoots); petiole (1.5–)2.5–8(–10) cm; blade oblanceolate to spatulate, (2–)3–7(–9) cm × 5–23(–32) mm, margins crenate to serrate-crenate. Cauline leaves shortly petiolate or sessile; blade oblong to oblanceolate, 1–3.5 cm × 4–15 mm, base cuneate, not auriculate, margins entire. Racemes (subcorymbose panicles), slightly elongated in fruit; rachis puberulent, trichomes straight. Fruiting pedicels divaricate-ascending to horizontal, straight, (terete), 3–6(–8) × 0.2–0.25 mm, puberulent adaxially. Flowers: sepals oblong to ovate, (1–)1.3–1.8 (–2) × 0.9–1.2 mm; petals white, suborbicular to broadly obovate, (1.8–)2–3 × 1.3–2 mm, claw 0.5–1 mm; stamens 6; filaments 1.4–1.6 mm, (glabrous); anthers 0.5–0.7 mm. Fruits broadly ovate, (2.5–)3–3.5(–4) × (2–)2.3–2.8 mm, apically winged, apical notch 0.1–0.2 mm deep; valves thin, smooth, not veined, glabrous; style 0.2–0.6 mm, exserted beyond apical notch. Seeds ovate, 1.5–2 × 1–1.3 mm.

Flowering Jun–Aug. Pinyon-juniper and brush communities, clay bluff of sandstone mesa, arroyo banks; of conservation concern; 1800–2000 m; Colo., N.Mex.

Lepidium crenatum is known from Delta, Moffatt, Montezuma, and Montrose counties in Colorado. It was reported from Utah by C. L. Hitchcock (1936) and R. C. Rollins (1993), but we have not seen any material to confirm its occurrence there.

10. **Lepidium davisii** Rollins, Madroño 9: 164. 1948 [E][F]

Lepidium montanum Nuttall subsp. *davisii* (Rollins) C. L. Hitchcock

Perennials; (cespitose, caudex woody, many-branched, with persistent petiolar remains); puberulent. Stems simple from base, erect, unbranched or branched (few) distally, (0.2–)0.5–1(–1.4) dm. Basal leaves (often deciduous); not rosulate; blade spatulate to oblanceolate, 1–2.5(–3.2) cm × 2–6(–9) mm, margins entire or apically 3(–5)-toothed or -lobed. Cauline leaves sessile; blade usually oblanceolate or oblong, rarely obovate, (0.8–)1.3–2.5 cm × (2–)4–7 mm, base obtuse or cuneate, not auriculate, margins entire or apically 3(–5)-toothed. Racemes slightly elongated in fruit; rachis puberulent, trichomes straight or curved. Fruiting pedicels divaricate-ascending, straight, (slender or slightly stout, terete), (2.5–)3–4.2(–5) × 0.4–0.5 mm, usually puberulent throughout, rarely glabrate. Flowers: sepals suborbicular to oblong-ovate, 1.2–2 × 1–1.5 mm; petals white, obovate, 2–3.2(–4) × 1.5–2 mm, claw 0.5–1 mm; stamens 6; filaments 1.7–2.3 mm, (glabrous); anthers 0.4–0.7 mm. Fruits suborbicular to broadly ovate, (2.5–)3–4.3(–5) × (2–)2.3–4 mm, apically winged, apical notch 0.1–0.4 mm deep; valves thin, smooth, not veined, glabrous or sparsely puberulent; style 0.5–1 mm, exserted beyond apical notch. Seeds oblong-ovate, (1.8–)2–2.3 × 1–1.2 mm. $2n = 32$.

Flowering May–Jun. Playas of sagebrush plains and mesa, vernal ponds; 800–1600 m; Idaho, Nev., Oreg.

Lepidium davisii is restricted to six counties in Idaho (Ada, Elmore, Owyhee, Twin Falls), Nevada (Elko), and Oregon (Malheur).

11. **Lepidium densiflorum** Schrader, Index Seminum (Göttingen) 1832: 4. 1832 [F][W]

Lepidium densiflorum var. *elongatum* (Rydberg) Thellung; *L. densiflorum* var. *macrocarpum* G. A. Mulligan; *L. densiflorum* var. *pubicarpum* (A. Nelson) Thellung; *L. densiflorum* var. *ramosum* (A. Nelson) Thellung; *L. elongatum* Rydberg; *L. neglectum* Thellung; *L. pubicarpum* A. Nelson; *L. ramosum* A. Nelson

Annuals or biennials; puberulent or glabrous, (trichomes cylindrical). Stems simple from base, erect, branched distally, (1–)2.5–5(–6.5) dm. Basal leaves (early withered); rosulate; petiole 0.5–1.5(–2) cm; blade oblanceolate, spatulate, or oblong, (1.5–)2.5–8(–11) cm × 5–10(–20) mm, margins coarsely serrate or pinnatifid. Cauline leaves shortly petiolate; blade narrowly oblanceolate or linear, (0.7–)1.3–6.2(–8) cm × (0.5–)1.5–10 (–18) mm, base attenuate to cuneate, not auriculate, margins usually entire or irregularly serrate to dentate, rarely pinnatifid. Racemes considerably elongated in fruit; rachis puberulent, trichomes straight, slender to subclavate. Fruiting pedicels divaricate-ascending to horizontal, straight or slightly recurved, (terete), (1.5–)2–3.5(–4) × 0.15–0.25 mm, puberulent adaxially. Flowers: sepals oblong, 0.5–0.8(–1) × 0.3–0.5 mm; petals (absent or rudimentary) white, filiform, 0.3–0.9 mm, claw absent; stamens 2, median; filaments 0.6–1(–1.8) mm; anthers 0.1–0.2 mm. Fruits obovate to obovate-suborbicular, (2–)2.5–3(–3.5) × 1.5–2.5(–3) mm (widest beyond middle), apically winged, apical notch 0.2–0.4 mm deep; valves thin, smooth, not veined, glabrous or sparsely puberulent (at least on margin); style 0.1–0.2 mm, included in apical notch. **Seeds** ovate, 1.1–1.3 (–1.5) × 0.8–0.9 mm. $2n$ = 32.

Flowering May–Jul. Waste places, disturbed sites, prairies, fields, pastures, grasslands, chaparral, meadows, sagebrush flats, flood plains, gravelly hillsides, rock crevices, seashores, sandy places, shaley barrens, open mesa, roadsides; 0–3500 m; Alta., B.C., Man., N.B., Nfld. and Labr., N.W.T., N.S., Ont., P.E.I., Que., Sask., Yukon; Ala., Alaska, Ariz., Ark., Calif., Colo., Conn., Del., D.C., Fla., Ga., Idaho, Ill., Ind., Iowa, Kans., Ky., La., Maine, Md., Mass., Mich., Minn., Miss., Mo., Mont., Nebr., Nev., N.H., N.J., N.Mex., N.Y., N.C., N.Dak., Ohio, Okla., Oreg., Pa., R.I., S.C., S.Dak., Tenn., Tex., Utah, Vt., Va., Wash., W.Va., Wis., Wyo.; Mexico (Chihuahua, Nuevo León); introduced in Europe, Asia.

North American records of *Lepidium apetalum* Willdenow mostly represent misidentifications of *L. densiflorum*. The latter has obovate fruits widest beyond the middle, whereas *L. apetalum* has elliptic fruits widest at the middle.

The number and limits of the varieties recognized in *Lepidium densiflorum*, as well as the characters used to delimit them, vary among authors (A. Thellung 1906; C. L. Hitchcock 1936; G. A. Mulligan 1961; R. C. Rollins 1993; N. H. Holmgren 2005b). The variation almost always does not correlate with geography, and the recognition of varieties in this species is neither practical nor very useful. All of those authors admitted that these varieties are "very weak at best" (Rollins, p. 554). Of them, perhaps var. *pubicarpum* (including var. *elongatum*) might merit recognition. It is distributed in almost all of the Mountain and Pacific states and is distinguished from the other varieties solely by the presence of trichomes or minute papillae on the fruit valves. The density of these trichomes ranges from moderate and covering the entire valve surface to very sparse and represented by individual papillate trichomes restricted to the valve margin. Furthermore, the length of these trichomes may vary from ca. 0.01 to 0.3 mm. In some species (e.g., *L. dictyotum*) both glabrous- and pubescent-fruited forms occur, yet none of the above authors gave formal recognition to both forms. It is not known if both glabrous and puberulent fruits occur within the same population in *L. densiflorum*. The species is autogamous, but nothing is known about the rates of gene flow between and within populations.

12. Lepidium dictyotum A. Gray, Proc. Amer. Acad. Arts 7: 329. 1868 [F]

Lepidium acutidens (A. Gray) Howell var. *microcarpum* Thellung; *L. dictyotum* var. *macrocarpum* Thellung; *Nasturtium dictyotum* (A. Gray) Kuntze

Annuals; hirsute. **Stems** few to several from base, erect to ascending, or (outer ones) decumbent, unbranched, (0.15–) 0.3–1.3(–2.1) dm. **Basal leaves** (soon withered); not rosulate; petiole 0.5–2 cm; blade pinnatifid to pinnatisect (lobes linear to narrowly oblong), (1.5–)2.2–5.7(–7) cm × (0.5–)1–2(–3) mm, margins entire. **Cauline leaves** sessile; blade usually linear, rarely with linear lobes, 1–5 cm × 0.5–2 mm, base attenuate, not auriculate, margins entire. **Racemes** elongated, (dense) in fruit; rachis hirsute, trichomes straight, cylindrical. **Fruiting pedicels** erect to slightly ascending, straight and appressed to rachis or distally slightly recurved, (strongly flattened), (1.6–) 1.9–2.5(–3) × 0.4–0.8 mm (width proximal to apex), usually hirsute or, rarely, only adaxially. **Flowers:** sepals oblong to ovate, 0.7–1.1 × 0.2–0.6 mm; petals absent; stamens 4, median; filaments 0.5–1 mm; anthers ca. 0.1 mm. **Fruits** ovate, 2.5–3.5 × 2–2.8 mm, apically winged, apical notch (closed, often U-shaped), 0.5–0.7(–0.8) mm deep; valves thin, smooth, strongly reticulate-veined, usually hirsute, rarely glabrous; style absent or obsolete, included in apical notch. **Seeds** ovate, 1.2–1.8 × 0.8–1.2 mm.

Flowering Mar–Jun. Margins of playas, saline areas, meadows, gypsum hills, dried pools, alkaline and clay flats and sinks, near hot springs, roadsides, borders of springs and ponds, sandy flats; 0–1600 m; Calif., Idaho, Nev., Oreg., Utah, Wash.; Mexico (Baja California).

13. Lepidium didymum Linnaeus, Syst. Nat. ed. 12, 2: 433. 1767; Mant Pl. 1: 92. 1767 [F] [I]

Carara didyma (Linnaeus) Britton; *Coronopus didymus* (Linnaeus) Smith; *Senebiera didyma* (Linnaeus) Persoon; *S. incisa* Willdenow; *S. pinnatifida* de Candolle

Annuals; (fetid); glabrous or pilose. **Stems** few to several from base, erect to ascending or decumbent, branched distally, 1–4.5(–7) dm. **Basal leaves** (soon withered); not rosulate; petiole 0.5–4(–6) cm; blade 1- or 2-pinnatisect, 1–6(–8) cm, margins (of lobes) entire or dentate (sometimes deeply lobed). **Cauline leaves** shortly petiolate to subsessile; blade similar to basal, smaller and less divided distally, lobes lanceolate to oblong or elliptic, 1.5–3.5(–4.5) cm ×

5–12 mm, base not auriculate, margins (of lobes) entire, serrate, or incised. **Racemes** elongated in fruit; rachis glabrous or pubescent, trichomes straight, cylindrical. **Fruiting pedicels** divaricate to horizontal, straight slightly recurved, (terete), 1.4–2.5(–4) × 0.15–2 mm, glabrous or sparsely pubescent adaxially. **Flowers:** sepals (tardily deciduous), ovate, 0.5–0.7(–0.9) mm; petals white, elliptic to linear, 0.4–0.5 × ca. 0.1 mm, claw absent; stamens 2, median; filaments 0.3–0.6 mm; anthers 0.1–0.2 mm. **Fruits** schizocarpic, didymous, 1.3–1.7 × 2–2.5 mm, apically not winged, apical notch 0.2–0.4 mm deep; valves thick, rugose, strongly veined, glabrous; style absent or obsolete, included in apical notch. **Seeds** ovate, 1–1.2 × 0.7–0.8 mm. $2n = 32$.

Flowering Mar–Jul. Roadsides, waste areas, lawns, pastures, fields, gardens, disturbed areas; 0–1000 m; introduced; B.C., N.B., Nfld. and Labr. (Nfld.), N.S., Que.; Ala., Ariz., Ark., Calif., Conn., Fla., Ga., La., Maine, Md., Mass., Miss., Mo., N.J., N.Y., N.C., Ohio, Oreg., Pa., S.C., Tenn., Tex., Va., W.Va., Wis.; South America; introduced also in Mexico (Sinaloa), Central America (Honduras), Europe, Asia, s Africa, Australia.

14. **Lepidium draba** Linnaeus, Sp. Pl. 2: 645. 1753
    F I W

Cardaria draba (Linnaeus) Desvaux; *Cochlearia draba* (Linnaeus) Linnaeus; *Nasturtium draba* (Linnaeus) Crantz

Perennials; (rhizomatous); hirsute or glabrate. **Stems** often simple from base, erect or decumbent basally, branched (several) distally, (0.8–)2–6.5 (–9) dm. **Basal leaves** (early withered); not rosulate; petiole 1–4 cm; blade obovate, spatulate, or ovate, (1.5–)3–10(–15) cm × 10–40 mm, margins sinuate to dentate or entire. **Cauline leaves** sessile; blade ovate, elliptic, oblong, or lanceolate, oblanceolate, or obovate, (1–)3–9(–15) cm × (5–)10–20(–50) mm, base sagittate-amplexicaul or auriculate, margins dentate or entire, (surfaces pubescent or glabrous). **Racemes** (corymbose panicles), slightly or considerably elongated in fruit; rachis glabrous or puberulent, trichomes straight or curved, cylindrical. **Fruiting pedicels** ascending to horizontal, straight, (terete), 5–10(–15) × 0.2–0.3 mm, glabrous or sparsely puberulent adaxially. **Flowers:** sepals oblong to ovate, 1.5–2.5 × 0.7–1.2 mm; petals white, obovate, (2.5–)3–4(–4.5) × (1–)1.3–2(–2.2) mm, claw 1–1.7 mm; stamens 6; filaments 2–3 mm, (glabrous); anthers 0.4–0.5 mm. **Fruits** (indehiscent), cordate to subreniform, (2–)2.5–3.7(–4.3) × (3.2–)3.7–5(–5.6) mm, apically (obtuse to subacute), not winged, apical notch absent; valves thin, smooth, reticulate-veined, glabrous; style (0.6–)1–1.8(–2) mm. **Seeds** ovate, 1.5–2.3 × 1–1.3 mm. $2n = 32, 64$.

Flowering Apr–Aug. Mountain slopes, roadsides, fields, agricultural lands, stream sides, disturbed grounds, pastures, waste areas; 0–3300 m; introduced; Alta., B.C., Man., N.S., Ont., Sask.; Ariz., Calif., Colo., Conn., D.C., Idaho, Ill., Ind., Iowa, Kans., Md., Mass., Mich., Minn., Mo., Mont., Nebr., Nev., N.J., N.Mex., N.Y., N.Dak., Ohio, Okla., Oreg., Pa., R.I., S.Dak., Utah, Wash., Wyo.; s Europe; sw Asia; introduced also in Mexico (Distrito Federal), South America, s Africa, Australia.

Although *Lepidium draba* is poorly established and known from old collections in the eastern part of the United States, it has become a noxious weed in several western states.

Lepidium draba and its nearest relatives, *L. appelianum* and *L. chalepense*, form a monophyletic clade most closely related to *L. campestre* (K. Mummenhoff et al. 2001). A. Thellung (1906) and C. L. Hitchcock (1936) correctly placed *L. draba* in *Lepidium*, as did Linnaeus. The recognition of the first three species in *Cardaria* and the maintenance of their nearest relative, *L. campestre*, in *Lepidium* do not make any sense on both phylogenetic and taxonomic grounds.

15. **Lepidium eastwoodiae** Wooton, Bull. Torrey Bot. Club 25: 258. 1898   E F

Lepidium alyssoides A. Gray var. *eastwoodiae* (Wooton) Rollins; *L. moabense* S. L. Welsh; *L. montanum* Nuttall var. *eastwoodiae* (Wooton) C. L. Hitchcock

Annuals, biennials, or perennials; (base woody); glabrous or pubescent. **Stems** simple from base, erect, branched distally, (3.5–)4.5–16(–18) dm. **Basal leaves** (soon deciduous), not rosulate; petiole (1–)2–5.5(–7.5) cm; blade pinnatifid, (2–)3–6.8(–9) cm × 10–30 mm, margins (of lobes) dentate to serrate. **Cauline leaves** shortly petiolate or sessile; blade narrowly lanceolate or oblanceolate to linear, 3–7 cm × (2.5–)4–10 mm (smaller distally), base attenuate to cuneate, not auriculate, margins usually entire, rarely dentate. **Racemes** elongated in fruit; rachis puberulent, trichomes straight or curved. **Fruiting pedicels** divaricate-ascending to horizontal, usually slightly recurved or somewhat sigmoid, rarely straight, (not winged), (3–)3.5–7.5(–8) × 0.2–0.3 mm, puberulent adaxially. **Flowers:** sepals suborbicular to oblong, 0.8–1.5 × 0.7–1.2 mm; petals white, suborbicular, 2.2–3.5(4) × 1.5–2.5 mm, claw 0.7–1.5 mm; stamens 6; filaments 1.5–2.5 mm, (glabrous); anthers 0.3–0.4 mm. **Fruits** broadly ovate, 2–3.5(–4) × 1.8–2.6(–3) mm, apically winged, apical notch 0.1–0.2 mm deep; valves thin, smooth, not veined, glabrous; style (0.2–)0.3–0.6(–0.7) mm, exserted beyond apical notch. **Seeds** (dark brown), ovate, 1.4–1.8 × 0.8–1.1 mm.

Flowering Jul–Sep. Pinyon-juniper, sagebrush, or mixed desert shrub communities; 900–2200 m; Ariz., Colo., N.Mex., Utah.

C. L. Hitchcock (1936) and R. C. Rollins (1993) reduced *Lepidium eastwoodiae* to a variety of *L. montanum* and *L. alyssoides*, respectively. However, the differences in morphology and flowering periods support its recognition as an independent species.

We have not examined the holotype of *Lepidium moabense* and follow N. H. Holmgren (2005b) in reducing it to synonymy of *L. eastwoodiae*.

16. **Lepidium flavum** Torrey in War Department [U.S.], Pacif. Railr. Rep. 4(5): 67. 1857 F

Lepidium flavum var. *apterum* Henrard; *L. flavum* var. *felipense* C. L. Hitchcock; *Nasturtium flavum* (Torrey) Kuntze; *Sprengeria flava* (Torrey) Greene; *S. minuscula* Greene; *S. watsoniana* Greene

Annuals; glabrous. **Stems** few to several from base, prostrate or decumbent, branched, (0.2–)0.4–3(–4.6) dm. **Basal leaves** rosulate; petiole (0.3–)0.6–2.5(–3.2) cm; blade spatulate to oblanceolate or linear, pinnatifid or lobed (lobes 3–9 pairs, ovate to oblong), (0.7–)1.3–5.2(–6) cm × 3–8 mm, margins (of lobes) entire. **Cauline leaves**: petioles 0.1–0.5 cm; blade obovate to spatulate or oblanceolate, (0.6–)1–1.8(–2.3) cm × 2–8(–14) mm, base attenuate-cuneate, not auriculate, margins dentate-sinuate to crenate or entire. **Racemes** (corymbose to subcapitate), slightly elongated in fruit; rachis glabrous. **Fruiting pedicels** divaricate to horizontal, straight or slightly sigmoid, (terete), (2.7–)3–4.4(–5) × 0.2–0.3 mm, glabrous. **Flowers**: sepals obovate-oblong, 1–2 × 0.6–1 mm; petals yellow (rarely creamy white at early anthesis), spatulate, (1.8–)2–2.8(–3) × 0.6–1 mm, claw 0.5–1 mm; stamens 6; filaments (median pairs) 1.2–2 mm, (glabrous); anthers 0.4–0.5 mm. **Fruits** usually narrowly to broadly ovate, rarely suborbicular, (2.2–)2.5–3.8(–4.2) × (1.6–)2.2–3.2(–3.5) mm, apically divergently winged, apical notch 0.2–0.6 mm deep; valves thin, smooth, moderately reticulate-veined, usually glabrous, rarely pubescent; style 0.7–1.6 mm, exserted beyond apical notch. **Seeds** ovate, 1–1.6 × 0.6–0.9 mm. $2n = 32$.

Flowering Mar–Jun. Sagebrush communities, open mesas, sandy flats and deserts, alluvial fans, dry valley floors, floodplains, washes, alkaline flats, roadsides, playa margins; 600–1600 m; Calif., Nev.; Mexico (Baja California).

The slightly smaller-fruited form of *Lepidium flavum*, recognized by C. L. Hitchcock (1936) and R. C. Rollins (1993) as var. *felipense*, does not merit recognition.

17. **Lepidium fremontii** S. Watson, Botany (Fortieth Parallel), 30, plate 4, figs. 3, 4. 1871 E F

Lepidium fremontii var. *stipitatum* Rollins; *Nasturtium fremontii* (S. Watson) Kuntze

Perennials or subshrubs; (woody base aboveground); (glaucous), glabrous throughout. **Stems** several from base, erect or ascending, branched (several) distally, 2–5.5(–10) dm. **Basal leaves** absent. **Cauline leaves** sessile; blade linear and undivided or pinnately lobed (lobes 3–7(–9), linear), (1.5–)2.2–8.4(–10.2) cm × (0.7–)1–2.8(–4.2) mm, base attenuate, not auriculate, margins entire, (similar, smaller distally). **Racemes** (panicles), elongated in fruit. **Fruiting pedicels** divaricate-ascending to horizontal, usually straight, rarely slightly curved, (terete), (3.5–)4.3–7.6(–8.5) × 0.15–0.2 mm. **Flowers**: sepals obovate, 1.5–2.5(–3) × 1–1.6 mm; petals white, spatulate, 2.5–4.2 × 1.5–2.2 mm, claw 1–2 mm; stamens 6; filaments 1.8–3 mm; anthers 0.4–0.7 mm. **Fruits** obovate to orbicular, (4–)4.5–7(–8) × 4.2–7(–8) mm, apically winged, apical notch (0.1–)0.2–0.5 mm deep; valves thin, smooth, not veined; style 0.2–0.8(–1) mm, exserted beyond apical notch. **Seeds** ovate, 1.6–2.1 × 0.9–1.1 mm. $2n = 32$.

Flowering Mar–Jun. Desert shrub communities, pinyon-juniper woodlands, sandy washes, gravelly deserts, barren knolls, bluffs, roadsides, steep limestone outcrops, rocky ledges and slopes; 400–2100 m; Ariz., Calif., Nev., Utah.

Rollins established var. *stipitatum* on the basis of fruits obovate with gynophores 0.5–0.75 mm versus (var. *fremontii*) fruits orbicular to ovate and sessile. These two characters often do not covary and stipitate fruits occur sporadically throughout the species range, including Arizona (*Lemmon s.n.*, GH), California (*Morefield 3427*, GH), and Nevada (*Comanor 18*, GH; *Hitchcock 3078*, GH).

18. **Lepidium heterophyllum** Bentham, Cat. Pl. Pyrénées, 95. 1826 F I

Perennials; (caudex branched); hirsute. **Stems** branched from base, erect to ascending, often decumbent basally, unbranched or branched (few) distally, 1–5 dm. **Basal leaves** rosulate; petiole 1–6.2 cm; blade oblanceolate or oblong-elliptic, 1–4.5 cm × 4–14 mm, margins entire, repand, or denticulate. **Cauline leaves** sessile; blade oblong to deltate-lanceolate, 1–3.5 cm × 3–8 mm, base sagittate or auriculate, margins dentate to denticulate. **Racemes** much-elongated in fruit; rachis hirsute, trichomes spreading, straight. **Fruiting pedicels** horizontal, straight or slightly recurved, (terete), 2.8–5 × 0.3–0.4 mm, hirsute. **Flowers:** sepals oblong, 1.6–2.2 × 0.6–1.1 mm; petals white, spatulate, 1.8–2.8 × 0.8–1.4 mm, claw 1–2 mm; stamens 6; filaments 1.8–2.6 mm, (glabrous); anthers 0.4–0.5 mm. **Fruits** broadly oblong to ovate, 4–5.5 × 3.5–4 mm, (curved adaxially), apically broadly winged, apical notch 0.2–0.3 mm deep; valves thin, often not papillate, not veined; style (0.6–)1–1.5 mm, well-exserted beyond apical notch. **Seeds** (dark brown), ovoid, 1.8–2.2 × 1–1.2 mm. $2n = 48$.

Flowering May–Jun. Gravel mounds, roadsides, abandoned fields, waste grounds, disturbed sites, gardens, hillsides; 0–300 m; introduced; B.C., Nfld. and Labr. (Nfld.); Calif., Colo., Maine, Mass., N.Y., Oreg., Pa., Wash; Europe.

The records from Maine and Massachusetts are based on old collections, and it is not known if *Lepidium heterophyllum* has become established as part of the weedy flora of those states.

19. Lepidium huberi S. L. Welsh & Goodrich, Great Basin Naturalist 55: 359, fig. 1. 1995 [C][E][F]

Perennials or subshrubs; (caudex to 2 cm diam., woody base aboveground); puberulent. **Stems** several from base, erect to ascending, branched distally, (1.8–)3–6(–7) dm. **Basal leaves** not rosulate; petiole 1–3 cm; blade ovate to lanceolate, pinnatifid, 1–4.5 cm × 8–25 mm, margins (of lobes) entire. **Cauline leaves** shortly petiolate or sessile; blade lanceolate to oblanceolate, 1.8–3.5 cm × 3–10 mm (smaller distally), base cuneate, not auriculate, margins entire or coarsely serrate. **Racemes** (subcorymbose panicles), individual racemes slightly elongated; rachis puberulent, trichomes straight. **Fruiting pedicels** divaricate to horizontal, straight or slightly curved, (not winged), 4–7 × 0.2 mm, puberulent throughout. **Flowers:** sepals suborbicular to broadly ovate, 1.2–2 × 1.2–1.5 mm; petals white, suborbicular to obovate, 2–3.2 × 1.5–2 mm, claw 0.5–1 mm; stamens 6; filaments 1.4–1.7 mm, (glabrous); anthers 0.4–0.5 mm. **Fruits** often broadly obovate, 2–3.3 × 1.8–2.5 mm, apically winged, apical notch 0.1–0.2 mm deep; valves thin, smooth, not veined, glabrous; style 0.2–0.8 mm, exserted beyond apical notch. **Seeds** ovate, 1.4–1.6 × 0.8–1 mm.

Flowering Jul–Aug. Pine and sagebrush communities; of conservation concern; 1500–3000 m; Colo., Utah.

Lepidium huberi is known in Colorado from Rio Blanco County and in Utah from Uintah County.

20. Lepidium integrifolium Nuttall in J. Torrey and A. Gray, Fl. N. Amer. 1: 116. 1838 [C][E][F]

Lepidium montanum Nuttall subsp. *integrifolium* (Nuttall) C. L. Hitchcock; *L. montanum* var. *integrifolium* (Nuttall) C. L. Hitchcock; *L. utahense* M. E. Jones; *L. zionis* A. Nelson; *Nasturtium integrifolium* (Nuttall) Kuntze

Perennials; (caudex often thick, not aboveground, covered with persistent petiolar remains); puberulent. **Stems** several from base (caudex), ascending, branched distally, (1–)1.5–3.5(–4) dm. **Basal leaves** rosulate; petiole (0.5–)1.5–6(–7.5) cm; blade oblanceolate to obovate, (1.5–)2.5–7(–9) cm × (10–)15–25 mm, margins usually entire, rarely denticulate subapically. **Cauline leaves** shortly petiolate or sessile; blade narrowly lanceolate to broadly oblanceolate, 1–5 cm × 2–9(–12) mm, base cuneate, not auriculate, margins usually entire, rarely denticulate subapically. **Racemes** elongated in fruit; rachis puberulent, trichomes straight, sometimes clavate. **Fruiting pedicels** divaricate-ascending to horizontal, straight, (not winged), (4–)5–10 × 0.3–0.5 mm, puberulent adaxially. **Flowers:** sepals oblong-obovate, (1.5–)1.8–2.5 × 0.8–1.3 mm; petals white, obovate, (2.3)2.5–3.6(–4) × 1.5–2.2 mm, claw 0.5–1 mm; stamens (2 or) 4 (or 6), median and lateral when 4, (erect); filaments 1.7–2.5 mm, (glabrous); anthers 0.5–0.8 mm. **Fruits** ovate, (3–)3.2–4(–4.4) × 2–3.5 mm, apically winged, apical notch 0.1–0.3 mm deep; valves thin, smooth, not veined, glabrous; style 0.5–0.8(–1) mm, exserted beyond apical notch. **Seeds** ovate, 1.8–2 × 0.9–1.1 mm.

Flowering Jun–Jul. Alkaline and saline meadows; of conservation concern; 1300–2000 m; Ariz., Nev., Utah, Wyo.

The circumscription of *Lepidium integrifolium* is somewhat controversial. C. L. Hitchcock (1936) treated it as two varieties or (Hitchcock 1950) two subspecies of *L. montanum*, whereas R. C. Rollins (1993) treated it as a distinct species with two varieties. Rollins indicated that the species has two stamens, but such occurrence is rather rare. Most commonly, it has four stamens and is readily distinguished from related species by having four nectar glands and sepals sparsely pubescent subapically with crisped trichomes. Nothing is known about the populational variation of stamen number in the species and whether one or more taxa are involved.

21. Lepidium jaredii Brandegee, Zoë 4: 398. 1894 (as jaredi) [C][E][F]

Lepidium jaredii subsp. *album* Hoover

Annuals; sparsely pilose (at least distally). **Stems** simple or several from base, erect to ascending, branched distally, 1–6(–7) dm. **Basal leaves** (withered by anthesis); not rosulate. **Cauline leaves** sessile; blade lanceolate to linear, 2–7.5(–10) cm × 2–10 mm, base cuneate to attenuate, not auriculate, margins entire or sparsely dentate subapically. **Racemes** considerably elongated in fruit, (lax); rachis usually pilose, rarely glabrous, trichomes usually curved, cylindrical. **Fruiting pedicels** divaricate to horizontal, straight or slightly sigmoid, (terete), 5–15 × 0.15–0.2 mm, pilose adaxially. **Flowers**: sepals oblong, 1.8–2.5 × 0.9–1.5 mm; petals lemon yellow (fading whitish), spatulate, 2.8–4 × 1.2–1.8 mm, claw 1–1.4 mm; stamens 6; filaments 1.8–2.5 mm, (glabrous); anthers 0.5–0.6 mm. **Fruits** broadly ovate, 3–3.8(–4) × 2.8–3.2(–3.5) mm, apically not winged, apical notch 0(–0.1) mm deep; valves thin, obscurely veined, smooth or minutely papillate, glabrous; style 0.3–0.8(–1) mm, exserted beyond apical notch. **Seeds** (reddish brown), oblong, 1.8–2.2 × 1–1.4 mm. $2n = 16$.

Flowering Mar–Apr. Arroyos, washes, alkaline bottoms and meadows, dry hillsides; of conservation concern; 500–700 m; Calif.

Lepidium jaredii is known from Fresno, San Benito, and San Luis Obispo counties.

22. Lepidium lasiocarpum Nuttall in J. Torrey and A. Gray, Fl. N. Amer. 1: 115. 1838 [F][W]

Lepidium ruderale Linnaeus var. *lasiocarpum* (Nuttall) Engelmann ex A. Gray; *Nasturtium lasiocarpum* (Nuttall) Kuntze

Annuals; hirsute or hispid, (trichomes cylindrical). **Stems** usually few to several, rarely simple from base, erect to ascending or (outer ones) decumbent, branched distally, (0.15–)0.6–3(–3.8) dm. **Basal leaves** (later withered); not rosulate; petiole (0.4–)1–3.5(–5) cm; blade spatulate to oblanceolate, lyrate-pinnatifid, pinnatisect, or 2-pinnatifid, (0.7–)1.5–4.5(–7.5) cm × (9–)12–20(–30) mm, margins rarely dentate, (lobes) entire or dentate. **Cauline leaves** subsessile or petioles 0.8–2.2 cm, blade lanceolate to oblanceolate, (0.7–)1.2–3.3(–5) cm × (2–)4–12 mm, base cuneate, not auriculate, margins subentire to dentate. **Racemes** often considerably elongated in fruit; rachis hirsute or hispid, trichomes straight, cylindrical. **Fruiting pedicels** divaricate-ascending to horizontal, straight or slightly curved, (often strongly flattened), (1.8–)2–4(–4.6) × 0.2–0.7 mm (to 0.3 mm thick), hirsute to hispid throughout or adaxially. **Flowers**: sepals oblong, 1–1.3(–1.5) × 0.5–0.8 mm; petals (sometimes absent), white, oblanceolate to linear, (0.3–)0.6–1.5(–2) × (0.1–)0.2–0.5 mm, claw absent; stamens 2, median; filaments 1–1.4 mm; anthers 0.2–0.3 mm. **Fruits** ovate to ovate-orbicular, 2.8–4(–4.6) × 2.4–3.6(–4) mm, (base broadly cuneate to rounded), apically winged, apical notch (0.2–)0.3–0.6(–0.7) mm deep; valves thin, smooth, not veined, hirsute to hispid (on surface or margin); style obsolete or to 0.1 mm, included in apical notch. **Seeds** ovate, 1.4–2.2 × 0.9–1.4 mm.

Subspecies 3 (2 in the flora): w United States, n Mexico.

R. C. Rollins (1993) and C. L. Hitchcock (1945) divided *Lepidium lasiocarpum* into four and seven varieties, respectively. Hitchcock admitted that the species is highly variable in every aspect and that his varieties cover some of the "more conspicuous" variations. In our opinion, perhaps two or three groups are somewhat sufficiently defined to be recognized formally, though the lines separating them blur in some parts of the species range. We are reluctantly treating them as subspecies; subsp. *palmeri* (S. Watson) Thellung is restricted to Mexico (Baja California). The other infraspecific taxa recognized by those authors are based on trivial differences in leaf and indumentum, and we do not believe that they merit recognition.

The reports by R. C. Rollins (1993) and N. H. Holmgren (2005b) of four and six stamens in *Lepidium lasiocarpum* were most likely repeated from C. L. Hitchcock (1936, 1945b) without further verification. We have examined hundreds of specimens that definitely belong to this species, and in not a single case did we find a flower with more than two stamens. It is quite likely that the reports of more than two stamens were based on misidentified plants.

1. Fruit valves hirsute (or fringed on margin), trichomes not pustular-based; fruiting pedicels 0.2–0.4(–0.6) mm wide, usually less than 3 times as wide as thick; nectar glands toothlike, to 0.2 mm . . . . 22a. *Lepidium lasiocarpum* subsp. *lasiocarpum*
1. Fruit valves hispid, trichomes pustular-based; fruiting pedicels (0.4–)0.5–0.7 mm wide, often more than 3 times as wide as thick; nectar glands not toothlike (subulate), 0.3–0.5 mm . . . . 22b. *Lepidium lasiocarpum* subsp. *wrightii*

22a. Lepidium lasiocarpum Nuttall subsp. lasiocarpum [F]

Lepidium georginum Rydberg; *L. lasiocarpum* subsp. *georginum* (Rydberg) Thellung; *L. lasiocarpum* var. *georginum* (Rydberg) C. L. Hitchcock; *L. lasiocarpum* var. *rosulatum* C. L. Hitchcock

Fruiting pedicels 0.2–0.4(–0.6) mm wide, usually less than 3 times as wide as thick. **Nectar glands** toothlike, 0.1–0.2 mm. **Fruit valves** hirsute or minutely fringed on margin, trichomes not pustular-based. $2n = 32$.

Flowering Mar–Jun. Pinyon-juniper woodlands, sagebrush and other shrub communities, open deserts, dry washes and flats, waste places, streambeds, roadsides, sandy areas, rock slides, stony slopes; -50–2700 m; Ariz., Calif., Colo., Nev., N.Mex., Tex., Utah; Mexico (Baja California, Coahuila, Nuevo León, San Luis Potosí, Sinaloa, Sonora).

In subsp. *lasiocarpum* there is considerable variation in the density and location of trichomes on the fruit valve; in some populations (e.g., *Rollins & Munz 6741*, GH), the trichomes are present over the entire valve; in others they are restricted as a fringe along part of the fruit margin.

22b. Lepidium lasiocarpum Nuttall subsp. wrightii (A. Gray) Thellung, Mitt. Bot. Mus. Univ. Zürich 28: 266. 1906

Lepidium wrightii A. Gray, Smithsonian Contr. Knowl. 5(6): 15. 1853; *L. lasiocarpum* var. *rotundum* C. L. Hitchcock; *L. lasiocarpum* var. *wrightii* (A. Gray) C. L. Hitchcock; *L. nelsonii* L. O. Williams

Fruiting pedicels (0.4–)0.5–0.7 mm wide, often more than 3 times as wide as thick. **Nectar glands** subulate, 0.3–0.5 mm. **Fruit valves** hispid, trichomes pustular-based. $2n = 32$.

Flowering Feb–May. Roadsides, rocky draws, dry washes, loose sand, stony or gravelly areas, clay flats; 200–1100 m; Ariz., N.Mex., Tex.; Mexico (Coahuila, Tamaulipas).

I. A. Al-Shehbaz (1986b) reported subsp. *wrightii* from Louisiana, but we have seen no additional material from that state, where it was likely introduced.

23. Lepidium latifolium Linnaeus, Sp. Pl. 2: 644. 1753 [F] [I] [W]

Cardaria latifolia (Linnaeus) Spach

Perennials; (rhizomes thick, caudex woody); glabrous or pubescent. **Stems** simple from base, erect, branched distally, (2–)3.5–12(–15) dm. **Basal leaves** not rosulate; petiole 1–9(–14) cm; blade elliptic-ovate to oblong, (2–)3.5–15(–25) cm × (5–)15–50(–80) mm, (leathery), margins entire or serrate. **Cauline leaves** sessile or shortly petiolate; blade oblong to elliptic-ovate or lanceolate, (1–)2–9(–12) cm × 3–45 mm, base cuneate, not auriculate, margins serrate or entire. **Racemes** (subcorymbose panicles), slightly elongated or not in fruit; rachis glabrous or sparsely puberulent, trichomes straight, cylindrical. **Fruiting pedicels** ascending to divaricate, straight or slightly curved, (terete), 2–5(–6) × 0.1–0.2 mm, glabrous or puberulent adaxially. **Flowers**: sepals suborbicular to ovate, 1–1.4 × 0.8–0.9 mm; petals white, obovate, 1.8–2.5 × (0.8–)1–1.3 mm, claw 0.7–1 mm; stamens 6; filaments 0.9–1.4 mm, (glabrous); anthers 0.4–0.5 mm. **Fruits** oblong-elliptic to broadly ovate or suborbicular, (1.6–)1.8–2.4(–2.7) × 1.3–1.8 mm, apically not winged, apical notch 0(–0.1) mm deep; valves thin, smooth, not veined, glabrous or sparsely pilose; style 0.05–0.15 mm, exserted beyond apical notch (when present). **Seeds** oblong, (0.8–)1–1.2 × 0.6–0.9 mm. $2n = 24$.

Flowering Jun–Sep. Pastures, grasslands, disturbed places, fields, roadsides, slopes, saline meadows, stream banks, waste grounds, dry flats, sagebrush and pinyon-juniper communities, barren hillsides, ditch banks, edge of marshes; 0–2500 m; introduced; Alta., B.C., Que.; Ariz., Calif., Colo., Conn., Idaho, Ill., Kans., Mass., Mo., Mont., Nebr., Nev., N.Mex., N.Y., Oreg., Tex., Utah, Wash., Wyo.; s Europe; Asia; n Africa; introduced also in Mexico (Durango, San Luis Potosí), Australia.

24. Lepidium latipes Hooker, Icon. Pl. 1: plate 41. 1836 [F]

Lepidium latipes var. *heckardii* Rollins; *Nasturtium latipes* (Hooker) Kuntze

Annuals; puberulent or hirsute. **Stems** simple or several from base, erect to ascending or (outer ones) decumbent, unbranched or branched, 0.2–1.5(–3.8) dm. **Basal leaves** (soon withered); not rosulate; petiole often undifferentiated (to 3 cm); blade linear, 2–10 cm × 1–4 mm, margins entire, dentate, or pinnatisect (lobes 2–10 pairs, margins entire or dentate). **Cauline leaves** similar to basal, smaller,

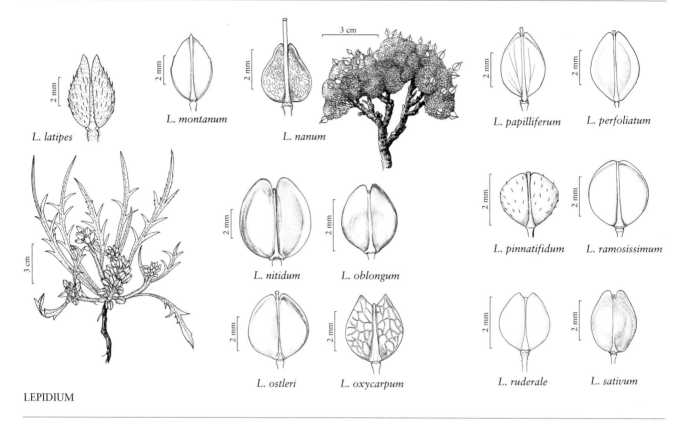

blade base attenuate, not auriculate, margins entire. **Racemes** (subcapitate to cylindrical), elongated or not in fruit, (compact); rachis puberulent, trichomes straight, cylindrical. **Fruiting pedicels** erect to slightly ascending, straight and appressed to rachis or distally slightly recurved, (strongly flattened), 2.5–5 × 0.9–1.4 mm, usually puberulent throughout, rarely only adaxially. **Flowers:** sepals (somewhat persistent), ovate, 1.1–1.4 × 0.6–0.8 mm; petals greenish, obovate-oblong, 1.9–3 × 0.8–1.3 mm, claw absent, (usually pubescent outside, with fringed margin, rarely glabrescent); stamens 4, median; filaments 0.8–1.1 mm; anthers 0.15–0.2 mm. **Fruits** oblong-ovate, 5–7 × 2.8–4 mm, apically winged, apical notch 1.4–2.8 mm deep; valves thick, smooth, strongly reticulate-veined, hirsute and puberulent, (trichomes spreading, mixed with smaller ones); style obsolete, included in apical notch. **Seeds** oblong, 2–2.4 × 1.1–1.3 mm.

Flowering Mar–Jun. Margins of vernal pools, edges of salt marshes, alkaline flats and adobe, pastures, mud-wet fields; 0–700 m; Calif.; Mexico (Baja California).

Variety *heckardii*, which is said to differ from var. *latipes* mainly by having elongated stems simple at base (instead of short and branched basally), grows mixed with var. *latipes* in single populations. It appears that the difference is trivial and may well be controlled by a few-gene difference. In our opinion, formal distinction is unwarranted; similar conditions exist in other species (e.g., 27. *Lepidium nitidum*).

25. **Lepidium montanum** Nuttall in J. Torrey and A. Gray, Fl. N. Amer. 1: 116. 1838 [E] [F]

Lepidium albiflorum A. Nelson & P. B. Kennedy; *L. alyssoides* A. Gray var. *jonesii* (Rydberg) Thellung; *L. alyssoides* var. *stenocarpum* Thellung; *L. brachybotryum* Rydberg; *L. corymbosum* Hooker & Arnott; *L. crandallii* Rydberg; *L. integrifolium* Nuttall var. *heterophyllum* S. Watson; *L. jonesii* Rydberg; *L. montanum* subsp. *alpinum* (S. Watson) C. L. Hitchcock; *L. montanum* var. *alpinum* S. Watson; *L. montanum* subsp. *canescens* (Thellung) C. L. Hitchcock; *L. montanum* var. *canescens* (Thellung) C. L. Hitchcock; *L. montanum* subsp. *cinereum* (C. L. Hitchcock) C. L. Hitchcock; *L. montanum* var. *cinereum* (C. L. Hitchcock) Rollins; *L. montanum* var. *claronense* S. L. Welsh; *L. montanum* subsp. *glabrum* (C. L. Hitchcock) C. L. Hitchcock; *L. montanum* var. *glabrum* C. L. Hitchcock; *L. montanum* subsp. *heterophyllum* (S. Watson) C. L. Hitchcock; *L. montanum* var. *heterophyllum* (S. Watson) C. L. Hitchcock; *L. montanum* subsp. *jonesii* (Rydberg) C. L. Hitchcock; *L. montanum* var. *jonesii* (Rydberg) C. L. Hitchcock; *L. montanum* var. *neeseae* S. L. Welsh & Reveal; *L. montanum* var. *nevadense* Rollins; *L. montanum* var. *saliarborense* S. L. Welsh; *L. montanum* var. *stellae*

S. L. Welsh & Reveal; *L. montanum* var. *stenocarpum* Thellung; *L. montanum* var. *todiltoense* N. D. Atwood & S. L. Welsh; *L. montanum* var. *wyomingense* (C. L. Hitchcock) C. L. Hitchcock; *L. philonitrum* A. Nelson & J. F. Macbride; *L. scopulorum* M. E. Jones; *L. utahense* Rydberg

Annuals, biennials, or perennials; (cespitose or not, woody or not at base); glabrous or pubescent. **Stems** simple to several from base, erect to ascending, often branched (several) distally, 0.4–5(–7) dm. **Basal leaves** rosulate or not; petiole (0.5–)1.2–5.3(–7.6) cm; blade 1- or 2-pinnatifid to pinnatisect, or, rarely, undivided, (0.8–)1.5–4(–6) cm, margins (of lobes) entire or dentate. **Cauline leaves** shortly petiolate; blade similar to basal, or undivided and linear, smaller distally, base cuneate to attenuate, not auriculate. **Racemes** often much-elongated in fruit; rachis usually puberulent, rarely glabrous, trichomes straight or curved, cylindrical. **Fruiting pedicels** divaricate to horizontal, slightly recurved or somewhat sigmoid, (terete), (2.7–)3.3–8.5(–10) × 0.2–0.3 mm, sparsely to densely puberulent adaxially. **Flowers:** sepals oblong to broadly ovate, 1.2–1.8(–2.1) × 0.8–1.2 mm; petals white, spatulate to oblanceolate, 2.2–3.7(–4.3) × 1.3–1.8 mm, claw 1–1.4 mm; stamens 6; filaments 1.4–2.2 mm, (glabrous); anthers 0.4–0.5 mm. **Fruits** usually ovate to suborbicular, rarely oblong, 2–4.3(–5) × (1.5–)1.8–3.6(–4) mm, apically winged, apical notch 0.1–0.3 mm deep; valves thin, smooth, not veined, usually glabrous, rarely puberulent; style obsolete 0.2–0.7(–0.9) mm, usually exserted beyond, rarely subequaling, apical notch. **Seeds** oblong, 1.2–1.8 × 0.7–1 mm. $2n = 32$.

Flowering Apr–Aug. Pinyon-juniper woodlands, sagebrush and other shrub communities, rocky hillsides and crevices, bajadas, spring seepages, washes, gypseous grounds, sandstone cliffs, limestone gravel, playas, knolls, gumbo hills, sandy areas, alkaline flats and lowlands, roadsides; 1200–2700 m; Ariz., Calif., Colo., Idaho, Mich., Mont., Nev., N.Mex., Oreg., Utah, Wyo.

There is little agreement among North American authors as to the characters emphasized, the number of infraspecific taxa, their ranks, and their synonymies in treatments of *Lepidium montanum*. C. L. Hitchcock (1936) divided the species into 13 varieties and two forms, and later (Hitchcock 1950) further expanded its limits to include 15 subspecies and two varieties. His circumscription was so broad that it involved at least eight taxa accepted as different species by R. C. Rollins (1993), N. H. Holmgren (2005b), and us. These include *L. alyssoides, L. barnebyanum, L. crenatum, L. davisii, L. eastwoodiae, L. integrifolium, L. montanum,* and *L. papilliferum.*

Although the recognition of seven of Hitchcock's infraspecific taxa as distinct species helped somewhat in reducing heterogeneity in *Lepidium montanum,* this species complex remains far from being adequately treated in recent monographic (R. C. Rollins 1993)

or floristic (N. H. Holmgren 2005b) works. In all, 12 varieties remain accepted in the species, and these only partially cover its overall complexity. The characters that various authors emphasized in the delimitation of infraspecific taxa include duration, habit, plant height, indumentum density, division of basal and cauline leaves, leaf shape, and fruit size. Indeed, the species is so highly variable in all of these aspects that the overall number of forms is mind boggling. Holmgren's assessment that there are intermediates between the various varieties, even in areas where they do not overlap, is correct. The species is badly in need of thorough biosystematic and molecular studies to determine the number and range of infraspecific taxa, to discern the patterns of variation, and to determine how distinct the species is from the seven segregates mentioned in the previous paragraph.

Some of the varieties (e.g., *alpinum, coloradense, montanum, neeseae,* and *stellae*), all of which were accepted by both R. C. Rollins (1993) and N. H. Holmgren (2005b), are distinct enough, and merit recognition at some rank. As indicated by Rollins, var. *neeseae* is quite distinct from the rest and should perhaps be treated as a separate species, but we have seen only two isotypes, and neither had fully developed fruits. Without a painstaking study of this entire species complex, we prefer not to accept formally only a small fraction of its enormous variation. This does not mean that leaving the species as such is the ideal solution, and the listing of synonyms above is meant solely to guide the reader into the complexity involved. A workable and satisfactory key to all of the varieties is not possible at the current stage of knowledge, and attempts to make one (e.g., Rollins, Holmgren) have been unsuccessful. The Michigan record is based on an introduction more than 70 years ago.

SELECTED REFERENCE Hitchcock, C. L. 1950. On the subspecies of *Lepidium montanum*. Madroño 10: 155–158.

26. Lepidium nanum S. Watson, Botany (Fortieth Parallel), 30, plate 4, figs. 5–7. 1871 E F

Perennials; (forming pincushion-like, pulvinate mounds, caudex woody, to 1.5 cm diam., buried, much-branched, covered with persistent leaves); puberulent. **Stems** simple from base (caudex branches), erect to ascending, unbranched distally, 0.05–0.2 dm. **Basal leaves** rosulate; petiole undifferentiated; blade obovate, 2.5–5 cm × 15–25(–35) mm, margins entire, (ciliolate), apex deeply 3-lobed (lobes ovate to suborbicular, margins entire). **Cauline leaves** absent. **Racemes** slightly elongated in fruit, (2–7-fruited); rachis puberulent, trichomes straight, cylindrical. **Fruiting pedicels** suberect to ascending, often straight, (terete), 2–4.5 × 0.2–0.3 mm, puberulent throughout. **Flowers:** sepals (tardily deciduous), obovate, 1.3–4 × 0.8–1.1 mm; petals pale yellow or creamy white, spatulate, 1.8–2.9 × 0.8–1.2 mm, claw 0.8–1.1 mm; stamens 6; filaments 1.4–2 mm, (glabrous); anthers 1.4–2 mm. **Fruits** ovate, 2–4.2 × 1.5–3 mm, often apically winged, apical notch 0.1–0.2 mm deep; valves thin, smooth, not veined, glabrous; style (0.4–)0.6–1(–1.2) mm, exserted beyond apical notch. **Seeds** oblong, 1–2 × 0.8–1 mm.

Flowering May–Jun. Gypsum knolls, tufa mounds around hotsprings, quartzite gravel, barren areas with shale and chalky soil, gravelly hillsides, white calcareous soils; 1500–2200 m; Nev., Utah.

Lepidium nanum is most common in Nevada and is known in Utah from collections in Tooele County.

27. Lepidium nitidum Nuttall in J. Torrey and A. Gray, Fl. N. Amer. 1: 116. 1838 [F] [W]

Lepidium leiocarpum Hooker & Arnott; *L. nitidum* var. *howellii* C. L. Hitchcock; *L. nitidum* var. *insigne* Greene; *L. nitidum* var. *oreganum* (Howell ex Greene) C. L. Hitchcock; *L. oreganum* Howell ex Greene; *L. strictum* (S. Watson) Rattan var. *oreganum* (Howell ex Greene) B. L. Robinson; *Nasturtium nitidum* (Nuttall) Kuntze

Annuals; puberulent. **Stems** simple to numerous from base, erect to ascending or decumbent, sometimes branched distally, (0.4–)1–3.5(–4.2) dm. **Basal leaves** (soon withered); not rosulate; petiole (0.4–)1–4.2(–5) cm; blade pinnatisect (lobes usually linear to filiform, rarely oblong to lanceolate), (0.8–)1.5–7.3(–8) cm, margins (of lobes) usually entire, rarely dentate. **Cauline leaves** petiolate or subsessile; similar to basal, blade smaller, sometimes undivided and linear, base attenuate, not auriculate. **Racemes** much-elongated, (lax) in fruit; rachis puberulent to hirsutulous, trichomes straight, cylindrical. **Fruiting pedicels** usually divaricate, rarely ascending or suberect, usually slightly or weakly recurved, rarely straight, (strongly flattened), 2.5–5(–6.5) × (0.4–)0.6–0.9(–1.1) mm, puberulent adaxially and sometimes on proximal ⅓ abaxially. **Flowers:** sepals oblong-ovate, (0.7–)0.9–1.3(–1.5) × 0.5–0.8 mm; petals (rarely rudimentary or absent), white, oblanceolate, (0.8–)1.2–2(–2.8) × 0.2–1(–1.6) mm, claw absent, (glabrous); stamens usually 4, median, rarely 6; filaments (0.5–)0.7–1(–1.3) mm; anthers 0.1–0.2 mm. **Fruits** orbicular to broadly ovate, (2.5–)3–5.5(–6.5) × (2–)2.6–5(–5.4) mm, apically winged, apical notch 0.3–0.7(–1) mm deep; valves thin, smooth, slightly veined or not, usually glabrous, rarely minutely and sparsely puberulent (along margin); style usually obsolete, rarely to 0.1 mm, included in apical notch. **Seeds** ovate-oblong, 1.6–2.6 × 0.9–1.2 mm.

Flowering Feb–Mar. Alkaline flats and sinks, meadows, pastures, dry vernal pools, fields, sandy beaches, grassy area, gravelly slopes, creosote bush desert; 0–1000 m; Calif., Oreg., Wash.; Mexico (Baja California).

C. L. Hitchcock (1936) and R. C. Rollins (1993) recognized three varieties within *Lepidium nitidum*, of which var. *howellii* was said to differ from var. *nitidum* by having puberulent (versus glabrous) fruit margins and stems densely (versus glabrous or sparsely to densely) pubescent. These characters do not covary, and some glabrate plants have puberulent fruit margins, whereas some very densely pubescent plants have glabrous fruits. As for var. *oreganum*, it is based on plants with divergent fruit wings, but this feature shows every degree of transition from divergent to non-divergent, and both those authors accepted in var. *nitidum* forms with slightly divergent fruit wings. The species is highly variable in leaf division, fruit size and shape, indumentum density, and flower morphology, especially in the number of stamens and presence or absence of petals. We prefer to maintain it without any further splitting.

C. L. Hitchcock (1945b) indicated that *Lepidium nitidum* occurs also in Chile but did not indicate whether the plant is introduced or native there. The material that he annotated and cited as *L. nitidum* from Chile clearly represents misidentified *L. chilense* Kunze ex Walpers. The latter always has two stamens, whereas *L. nitidum* usually has four, rarely six, and Hitchcock's report of two stamens in the latter species was likely based on misidentified plants of another species.

28. Lepidium oblongum Small, Fl. S.E. U.S., 468, 1331. 1903 [F] [W]

Lepidium greenei Thellung; *L. oblongum* var. *insulare* C. L. Hitchcock

Annuals; hirsute (trichomes cylindrical). **Stems** often several from base, erect to ascending or decumbent, branched distally, (5–)1–2.4(–3.2) dm. **Basal leaves** not rosulate; petiole (0.5–)1–3 cm; blade 1- or 2-pinnatifid, 0.7–3.5 cm, margins (of lobes) entire or dentate. **Cauline leaves** usually sessile, rarely shortly petiolate; blade obovate to oblanceolate (in outline), 0.8–2 cm × 2–9 mm, base cuneate, auriculate or not, margins dentate to laciniate or pinnatifid. **Racemes** elongated in fruit; rachis hirsute, trichomes mostly straight, cylindrical. **Fruiting pedicels** divaricate to horizontal, usually slightly recurved, rarely straight, (terete), 2–3.5(–5) × 0.2–0.3 mm, puberulent adaxially or, rarely, throughout. **Flowers:** sepals (tardily deciduous to somewhat persistent), ovate to broadly oblong, 0.7–1 × 0.4–0.6 mm; petals (absent or rudimentary), white, linear-oblanceolate, 0.1–0.7 × 0.05–0.15 mm, claw absent; stamens 2, median; filaments 0.7–1 mm; anthers 0.15–0.2 mm. **Fruits** orbicular to broadly obovate or elliptic, 2.2–3.5 × 2–3 mm, apically winged, apical notch 0.2–0.3 mm deep; valves thin, smooth, not veined, glabrous or sparsely puberulent (along margin); style to 0.1 mm, included in apical notch. **Seeds** ovate, 1.2–1.6 × 0.7–1 mm.

Flowering Mar–Aug. Prairies, pastures, floodplains, waste grounds, llanos, disturbed areas, roadsides, flats, calcareous sand, alluvial terraces; 0–1200 m; Ariz., Ark., Calif., Kans., La., Miss., Mo., N.Mex., Okla., Tex.; Mexico (Baja California, Coahuila, Hidalgo, Nuevo León, Puebla, San Luis Potosí, Tamascaltepec, Veracruz); Central America (El Salvador, Guatemala).

Hitchcock distinguished var. *insulare* from var. *oblongum* on the basis of having fruits smaller (2–2.5 versus 2.5–3 mm) and rotund to obovate (versus elliptic or obovate-elliptic) and fruiting pedicels puberulent (versus glabrous) abaxially. The shape and size of fruits almost never covary, and some of the insular plants (e.g., *Trask 28*, GH) have the largest fruits; some populations from Arizona, Oklahoma, and Texas have smaller and perfectly orbicular fruits. As for the pubescence of fruiting pedicels, some of the inland plants cited by Hitchcock as *L. oblongum* (e.g., *Brewer 27*, GH) have pedicels pubescent abaxially.

A recent report of *Lepidium bonariense* naturalized in Skagit County, Washington, was based on misidentified *L. oblongum*; all the vouchers for it belong to the latter species. *Lepidium oblongum* often has several stems from base, usually auriculate distalmost leaves, and fruits 2–3 mm wide. By contrast, *L. bonariense* has single stems from base, non-auriculate distalmost leaves, and fruits 2.7–3.5 mm wide.

29. **Lepidium ostleri** S. L. Welsh & Goodrich, Great Basin Naturalist 40: 80, fig. 3. 1980 [C] [E] [F]

Perennials; (pulvinate, caudex woody, many-branched, covered with persistent petiolar remains); densely (grayish) puberulent. **Stems** simple from base (caudex branch), erect to ascending, unbranched distally, (0.1–)0.3–0.8 dm. **Basal leaves** not rosulate; petiole often undifferentiated, (expanded base to 2 mm); blade (somewhat fleshy), linear to linear-oblanceolate (when margins entire), or spatulate in outline (when apically 3–5-lobed, lobes obovate to oblong), 3–12(–15) cm × 5–15 mm, margins entire. **Cauline leaves** sessile; similar to basal, smaller, base not auriculate. **Racemes** slightly elongated in fruit; rachis puberulent, trichomes straight or curved, cylindrical. **Fruiting pedicels** divaricate-ascending to subhorizontal, straight or slightly curved, (not winged), (2.5–)3–5(–6.5) × 04–0.5 mm, densely puberulent throughout. **Flowers:** sepals orbicular to broadly obovate, 1.5–2(–2.5) × 1–1.5(–2) mm; petals white to pale purple, suborbicular to obovate, 2–3.5 × 1.5–2.5(–2.8) mm, claw to 0.5 mm; stamens 6; filaments 1.5–1.8 mm, (glabrous); anthers 0.3–0.5 mm. **Fruits** ovate, 2.5–3.5(–4) × 2–2.5(–3) mm, apically winged, apical notch 0.05–0.2 mm deep; valves thin, smooth, not veined, glabrous; style 0.3–0.6 mm, exserted beyond apical notch. **Seeds** ovate, 1.2–1.5 × 0.9–1.1 mm.

Flowering May–Jun. White limestone outcrops and gravel, pinyon-juniper, sagebrush, or pine communities; of conservation concern; 1700–2100 m; Utah.

Lepidium ostleri is known only from the San Francisco Mountains in Beaver County.

30. **Lepidium oxycarpum** Torrey & A. Gray, Fl. N. Amer. 1: 116. 1838 [E] [F]

Nasturtium oxycarpum (Torrey & A. Gray) Kuntze

Annuals; glabrous or puberulent. **Stems** several from base, usually erect to ascending, rarely decumbent, branched, 0.4–1.5(–2) dm. **Basal leaves** (soon withered); not rosulate; petiole 0.5–1.5(–2) cm; blade margins entire or pinnatifid (lobes 2–5 pairs, linear to filiform), 1.5–5 cm × 0.5–2 mm. **Cauline leaves** petiolate; blade linear, 0.1–0.3 cm × 0.5–2 mm, base attenuate, not auriculate, margins entire. **Racemes** considerably elongated, (lax) in fruit; rachis glabrous or puberulent, trichomes straight, cylindrical. **Fruiting pedicels** divaricate to horizontal or descending, usually recurved, rarely straight, (terete or slightly flattened), 2–4(–6) × 0.2–0.3 mm, glabrous or puberulent adaxially. **Flowers:** sepals oblong, 0.6–0.8 × 0.4–0.5 mm; petals absent; stamens 4, median; filaments 0.6–0.7; anthers ca. 0.1 mm. **Fruit** ovate, 2.4–3.6 × 1.8–2.5 mm, apically winged, apical notch (V-shaped), 0.3–0.8 mm deep; valves thin, smooth, strongly reticulate-veined, glabrous; style obsolete or to 0.1 mm, included in apical notch. **Seeds** oblong, 1.4–1.8 × 0.8–0.9 mm.

Flowering Mar–May. Borders of vernal pools, grassy fields, roadsides ditches, alkaline flats, margins of salt marshes; 0–400 m; Calif.

Lepidium oxycarpum apparently did not persist in British Columbia following its introduction there over 110 years ago (G. A. Mulligan 2002b). That record is based on *Macoun s.n.* (GH, MO, NY, US), which was collected on 31 May 1893 from the vicinity of Victoria, Vancouver Island.

31. Lepidium papilliferum (L. F. Henderson) A. Nelson & J. F. Macbride, Bot. Gaz. 56: 474. 1913 [C][E][F]

Lepidium montanum Nuttall var. *papilliferum* L. F. Henderson, Bull. Torrey Bot. Club 27: 342. 1900; *L. montanum* subsp. *papilliferum* (L. F. Henderson) C. L. Hitchcock

Annuals or biennials; puberulent, (trichomes clavate). **Stems** several from base, ascending, branched (several) distally, (0.3–)0.5–2.3(–4) dm. **Basal leaves** (soon deciduous), not rosulate; blade often bipinnately divided. **Cauline leaves** attenuate to petiolelike base, to 1 cm; blade pinnatifid to pinnatisect, obovate to oblanceolate in outline (lobes oblong to linear), 0.5–4 cm × 3–15 mm (smaller distally), base not auriculate, margins (of lobes) entire or dentate. **Racemes** slightly elongated, (dense) in fruit; rachis puberulent, trichomes straight or curved, clavate. **Fruiting pedicels** divaricate to horizontal, straight or recurved, (not winged), (2.5–)3–6(–7.5) × 0.2–0.3 mm, puberulent adaxially or throughout. **Flowers:** sepals ovate to broadly oblong, 1.5–2.2 × 0.8–1.1 mm; petals white, orbicular to spatulate, 2.5–3.7(–4) × 1.5–2.5 mm, claw 0.8–1.5 mm; stamens 6; filaments 1.5–2 mm, (trichomes clavate); anthers 0.4–0.6 mm. **Fruits** orbicular to broadly ovate, 2.2–3.6 × 2–3.4 mm, apically winged, apical notch 0.1–0.3 mm deep; valves thin, smooth, often not veined, glabrous; style 0.2–0.8 mm, exserted beyond apical notch. **Seeds** ovate, 1.2–1.8 × 0.9–1.1 mm.

Flowering May–Jun. Sagebrush plains, desert flats, edge of playa; of conservation concern; 600–1700 m; Idaho.

Lepidium papilliferum is distributed in Ada, Canyon, Elmore, Gem, Owyhee, and Payette counties.

32. Lepidium paysonii Rollins, Cruciferae Continental N. Amer., 577. 1993 [E]

Perennials; (caudex woody, to 6 mm diam.); densely puberulent. **Stems** several from base (caudex), erect or ascending to decumbent, branched distally, 0.5–2.3 dm. **Basal leaves** not rosulate; petiole 1–2.5 cm; blade oblanceolate (rarely with 1 or 2 lateral lobes), 1–3.5 cm × 3–7 mm, margins serrate-dentate. **Cauline leaves** shortly petiolate or sessile; blade narrowly oblanceolate to linear, 0.7–3.5 cm × 1–4 mm, base attenuate, not auriculate, margins entire or distally serrulate. **Racemes** elongated in fruit; rachis puberulent, trichomes curved, cylindrical. **Fruiting pedicels** divaricate to horizontal, slightly recurved, (terete), 2–4(–5.5) × 0.15–0.2 mm, densely puberulent throughout. **Flowers:** sepals oblong, 0.6–0.9 × 0.3–0.4 mm; petals (absent or rudimentary), white, oblanceolate, 0.3–0.6 × 0.1–0.2 mm, claw absent; stamens 2, median; filaments 0.6–0.7 mm; anthers 0.1–0.2 mm. **Fruits** elliptic, 2.4–2.8 × 1.6–2 mm, apically winged, apical notch 0.2–0.3 mm deep; valves thin, smooth, not veined, puberulent (at least along margin); style obsolete or to 0.1 mm, included in apical notch. **Seeds** oblong, 1.3–1.4 × 0.7–0.8 mm.

Flowering Jun–Jul. Dry open woods, dry grounds; Colo., Idaho, Wyo.

Lepidium paysonii is known in Idaho from Bear Lake County, and in Wyoming from Park and Sublette counties.

Placing *Lepidium paysonii* in the synonymy of *L. densiflorum* var. *pubicarpum*, N. H. Holmgren (2005b) depended on the presence in both taxa of minute papillae at the fruit valve margin. However, *L. paysonii* is a perennial with elliptic fruits widest at the middle, curved rachis trichomes, and fruiting pedicels puberulent throughout. By contrast, *L. densiflorum* is a biennial or annual with obovate fruits widest beyond the middle, straight rachis trichomes, and fruiting pedicels usually puberulent adaxially, rarely glabrate; in our opinion, the two species are not closely related.

33. Lepidium perfoliatum Linnaeus, Sp. Pl. 2: 643. 1753 [F][I][W]

Nasturtium perfoliatum (Linnaeus) Besser

Annuals or biennials; (glaucous), glabrous or sparsely pubescent proximally. **Stems** simple from base, erect, branched distally, (0.7–)1.5–4.3(–5.6) dm. **Basal leaves** rosulate; petiole (0.5–)1–2(–4) cm; blade 2- or 3-pinnatifid or pinnatisect (lobes linear to oblong), (1–)3–8(–15) cm, margins entire. **Cauline leaves** sessile; blade ovate to cordate or suborbicular, (0.5–)1–3(–4) cm × (5–)10–25(–35) mm, base deeply cordate-amplexicaul, not auriculate, margins entire. **Racemes** considerably elongated in fruit; rachis glabrous. **Fruiting pedicels** divaricate-ascending to horizontal, straight, (terete), 3–6(–7) × 0.2–0.3 mm, glabrous. **Flowers:** sepals oblong, 0.8–1(–1.3) × 0.5–0.8 mm; petals pale yellow, narrowly spatulate, 1–1.5(–1.9) × 0.2–0.5 mm, claw 0.5–1 mm; stamens 6; filaments 0.6–0.9 mm, (glabrous); anthers 0.1–0.2 mm. **Fruits** orbicular to rhombic or broadly obovate, 3–4.5(–5) × 3–4.1 mm, apically winged, apical notch 0.1–0.3 mm deep; valves thin, smooth, not or obscurely veined, glabrous; style 0.1–0.4 mm, subequaling or slightly exserted beyond apical notch. **Seeds** (dark brown), ovate, 1.6–2(–2.3) × 1.2–1.4 mm. $2n = 16$.

Flowering Mar–Jun. Waste places, dry sandy slopes, pinyon-juniper woodlands, sagebrush flats, open deserts, roadsides, pastures, meadows, open grasslands, alkaline flats and sinks, fields, disturbed sites; 0–2500 m; introduced; Alta., B.C., Man., Ont., Sask.; Ariz., Ark., Calif., Colo., Conn., Ga., Idaho, Ind., Iowa, Kans., Maine, Mass., Mich., Miss., Mo., Mont., Nebr., Nev., N.Mex., N.Y., N.C., Ohio, Okla., Oreg., Pa., S.Dak., Tenn., Utah, Wash., Wis., Wyo.; Europe; Asia; n Africa; introduced also in Mexico (Baja California), South America (Argentina), Australia.

34. **Lepidium pinnatifidum** Ledebour, Fl. Ross. 1: 206. 1841 F I

Annuals; puberulent. Stems simple from base, erect, (paniculately) branched beyond base or distally, 2–6 dm. Basal leaves (soon withered, often before anthesis); not rosulate; blade dentate to pinnatifid. Cauline leaves shortly petiolate to subsessile; blade narrowly oblanceolate to linear, 1–3.3 cm × 1–4 mm, base attenuate, not auriculate, margins entire. Racemes (often paniculate), considerably elongated in fruit; rachis glabrous or puberulent, trichomes straight, cylindrical. Fruiting pedicels divaricate-ascending to horizontal, straight, (terete), 2–3.5 × 0.1–0.15 mm, puberulent adaxially. Flowers: sepals oblong, 0.7–0.8 × 0.3–0.4 mm; petals (rudimentary), white, linear, 0.4–0.6 × 0.05–0.1 mm, claw absent; stamens 4, median and lateral; filaments 0.6–0.8 mm; anthers ca. 0.2 mm. Fruits orbicular to broadly elliptic, 1.8–2 × 1.7–1.8 mm, apically not winged, apical notch 0.05–0.1 mm deep; valves thin, smooth, not veined, sparsely pilose; style ca. 0.1 mm, equaling apical notch. Seeds oblong, 1–1.2 × 0.7–0.8 mm.

Flowering May–Jun. Waste places, disturbed sites; 0–600 m; introduced, Calif.; e Europe; w Asia.

Lepidium pinnatifidum apparently has not become a serious weed of the Californian flora.

35. **Lepidium ramosissimum** A. Nelson, Bull. Torrey Bot. Club 26: 124. 1899 F

Lepidium bourgeauanum Thellung;
L. densiflorum Schrader var. *bourgeauanum* (Thellung) C. L. Hitchcock; *L. divergens* Osterhout;
L. fletcheri Rydberg;
L. ramosissimum var. *bourgeauanum* (Thellung) Rollins;
L. ramosissimum var. *divergens* (Osterhout) Rollins;
L. ramosissimum var. *robustum* Thellung

Perennials; puberulent, (trichomes cylindrical). Stems simple from base, erect, branched (several) distally, (0.6–)1–5.3(–7.7) dm. Basal leaves (soon withered); not rosulate; petiole 1–4 cm; blade oblanceolate or pinnatifid, 2–5 cm × 8–15 mm, margins (of lobes) entire serrate or dentate. Cauline leaves shortly petiolate or sessile; blade oblanceolate or (distal) linear, (0.6–)1.2–4.8(–6) cm × 1–8(–10) mm, base attenuate to cuneate, not auriculate, margins dentate, (distal) entire, or, rarely, lobed. Racemes slightly elongated in fruit; rachis puberulent, trichomes curved, cylindrical to subclavate. Fruiting pedicels divaricate-ascending to horizontal, straight or recurved, (terete), (1.6–)2–3.8(–5) × 0.2–0.3 mm, usually puberulent adaxially, rarely throughout. Flowers: sepals oblong, 0.6–0.9(–1.1) × 0.3–0.4 mm; petals (absent or rudimentary), white, linear, 0.2–0.8(–1) × 0.05–0.1 mm, claw absent; stamens 2, median; filaments 0.6–0.9 mm; anthers 0.15–0.2 mm. Fruits elliptic, 2.2–3.2 × 1.7–2.1 mm, apically winged, apical notch 0.1–0.3(–0.4) mm deep; valves thin, smooth, not veined, glabrous or puberulent at least along margin; style usually obsolete, rarely to 0.1 mm, included in apical notch. Seeds oblong, 1.2–1.6 × 0.8–0.9 mm. $2n = 32, 64$.

Flowering Jun–Aug. Sagebrush communities, pine woodlands, waste grounds, roadsides, railroad embankments, alkaline flats, abandoned fields; 0–2900 m; Alta., B.C., Man., N.B., Nfld. and Labr. (Nfld.), N.W.T., Ont., Que., Sask., Yukon; Alaska, Calif., Colo., Idaho, Maine, Minn., Mont., Nebr., Nev., N.Mex., N.Dak., S.Dak., Tex., Utah, Wyo.; Mexico (Chihuahua).

As noted by R. C. Rollins (1993, p. 581), the varieties of *Lepidium ramosissimum* are "weak at best." They are based largely on the branching habit and, most importantly, on the presence versus absence of trichomes on the fruit valve. In some collections (e.g., *Scoggan 4233*, GH; *Boivin et al., 13221*, GH), both puberulent- and glabrous-fruited forms occur. It is almost certain that the same situation exists not only in other populations of this species, but in other North American *Lepidium*. It is also clear that some populations might consist entirely of one of the two forms, but it is highly unlikely that this variation has any geographical basis. Therefore, we believe that the separation of varieties solely on the basis of presence or absence of the fruit trichomes is taxonomically meaningless.

We are reluctantly including *Lepidium divergens* in the synonymy of *L. ramosissimum* because we have not seen its type; the topotypes that we studied have broadly obovate to suborbicular fruits that appear more at home in the *L. densiflorum* or *L. virginicum* complexes.

36. Lepidium ruderale Linnaeus, Sp. Pl. 2: 645. 1753
F I W

Lepidium texanum Buckley;
L. virginicum Linnaeus subsp. *texanum* (Buckley) Thellung

Annuals or biennials; (fetid); puberulent (trichomes cylindrical). **Stems** simple from base, erect ascending, branched (several) distally, (0.5–)1–3.5 (–5.5) dm. **Basal leaves** rosulate; petiole 1–3.2(–5.3) cm; blade (1- or) 2- or 3-pinnatisect (lobes oblong), (1.5–)3–5(–7.2) cm, margins (of lobes) usually entire, rarely dentate. **Cauline leaves** sessile; blade linear, (0.4–)1–2(–3) cm × 0.5–2.5(–3.5) mm, base cuneate, not auriculate, margins entire. **Racemes** considerably elongated in fruit; rachis puberulent, trichomes straight, cylindrical. **Fruiting pedicels** divaricate to horizontal, straight, (terete), (1.5–)2–4(–5) × 0.1–0.15 mm, puberulent. **Flowers:** sepals oblong, 0.5–0.9(–1) × 0.2–0.4 mm; petals (absent or rudimentary), white, linear, 0.2–0.5 × 0.1 mm, claw absent; stamens 2, median; filaments 0.7–0.8 mm; anthers 0.1–0.2 mm. **Fruits** elliptic, (1.5–)1.8–2.5(–3) × 1.5–2(–2.3) mm, apically winged, apical notch 0.1–0.2 mm deep; valves thin, smooth, not veined, glabrous; style obsolete or to 0.1 mm, included in apical notch. **Seeds** oblong to ovate-oblong, 1–1.5 × 0.6–0.8 mm. $2n$ = 16, 32.

Flowering Apr–Jul. Fields, pastures, waste places, roadsides, gardens; 0–300 m; introduced; Alta., B.C., Man., N.B., Nfld. and Labr. (Nfld.), N.S., Ont., P.E.I., Que., Sask.; Ala., Ark., Calif., Del., Fla., Ind., La., Maine, Md., Mass., Mich., N.H., N.J., N.Mex., N.Y., N.C., Ohio, Oreg., Pa., R.I., Tenn., Tex.; Eurasia; introduced also in South America, Australia.

37. Lepidium sativum Linnaeus, Sp. Pl. 2: 644. 1753
F I

Annuals; (often glaucous), usually glabrous, rarely sparsely pilose. **Stems** simple from base, erect, branched distally, (1–)2–8 (–10) dm. **Basal leaves** (withered by anthesis); not rosulate; petiole 1–4 cm; blade 1- or 2-pinnatifid or pinnatisect (lobes ovate to oblong), 2–8(–10) cm, margins (of lobes) entire or dentate. **Cauline leaves** petiolate; blade similar to basal, usually less divided, rarely undivided, (distal) often linear, bases not auriculate, margins entire. **Racemes** considerably elongated in fruit; rachis glabrous. **Fruiting pedicels** suberect to ascending, appressed to rachis, straight, (terete or slightly flattened), 1.5–4(–6) × 0.4–0.6 mm, glabrous. **Flowers:** sepals oblong-obovate, 1–1.8 × 0.5–0.8 mm; petals white or lavender, spatulate to obovate, 2–3.5(–4) × 0.7–1.4 mm, claw 1–1.4 mm; stamens 6; filaments (median pairs) 1.5–2 mm, (glabrous); anthers 0.4–0.5 mm. **Fruits** broadly ovate or ovate-oblong, (4–)5–6.4(–7) × 3–4.5 (–5.6) mm, apically broadly winged, apical notch 0.2–0.8 mm deep; valves thin, smooth, not veined, glabrous; style 0.1–0.5(–0.8) mm, usually included in, rarely subequaling, apical notch. **Seeds** (reddish brown), ovate-oblong, 2–2.7(–3) × 1–1.5 mm, (3-lobed). $2n$ = 16, 32.

Flowering Apr–Aug. Gardens, old fields, vacant lots, disturbed areas, railroad embankments, waste grounds, roadsides, cultivated areas; introduced; Alta., B.C., Man., Nfld. and Labr. (Nfld.), N.W.T., N.S., Ont., P.E.I., Que., Sask.; Conn., Idaho, Iowa, Maine, Md., Mass., Mich., N.H., N.Y., Ohio, Oreg., Pa., R.I., Tenn., Wash., Wyo.; Europe; sw Asia; perhaps ne Africa; introduced also in South America (Argentina), Australia.

Lepidium sativum is cultivated as a salad green and is sporadically naturalized, though never as an aggressive weed. It is seldom collected; the above range may be incomplete.

38. Lepidium sordidum A. Gray, Smithsonian Contr. Knowl. 3(5): 10. 1852 F

Lepidium granulare Rose

Annuals; puberulent, (trichomes clavate). **Stems** simple or several from base, erect to decumbent, branched (several) distally, 0.5–2.4(–3.2) dm. **Basal leaves** (soon withered); not rosulate; petiole 0.8–2.5 cm; blade 1- or 2-pinnatifid, 3–5.6 cm, margins (of lobes) entire or dentate. **Cauline leaves** petiolate; blade pinnatifid (similar to basal), 0.7–2 cm × 2–10 mm, base attenuate to cuneate, not auriculate, margins dentate to incised or pinnately lobed. **Racemes** (paniculate), elongated, (often dense) in fruit; rachis puberulent, trichomes straight or curved, clavate. **Fruiting pedicels** ascending to divaricate, straight or slightly recurved, (terete), (1.4–)1.6–2.3(–2.9) × 0.1–0.15 mm, puberulent adaxially. **Flowers:** sepals (tardily deciduous), oblong, 0.5–0.7 × 0.3–0.4 mm; petals (absent or rudimentary), white, linear, 0.2–0.4 × 0.05 mm, claw absent; stamens 2, median; filaments 0.5–0.8 mm; anthers 0.1–0.15 mm. **Fruits** ovate-elliptic, 1.7–2.1 × 1.2–1.6 mm, apically winged, apical notch 0.1–0.2 mm deep; valves thin, smooth, not or weakly veined, glabrous; style 0.1–0.15 mm, included in or equaling apical notch. **Seeds** oblong, 0.8–1 × 0.5–0.6 mm.

Flowering May–Jul. Alluvial fans, sandy flats, rocky hillsides, grassy valleys, canyons; 1500–1900 m; Tex.; Mexico (Chihuahua, Durango, Federal District, Hidalgo, Sinaloa, Zacatecas).

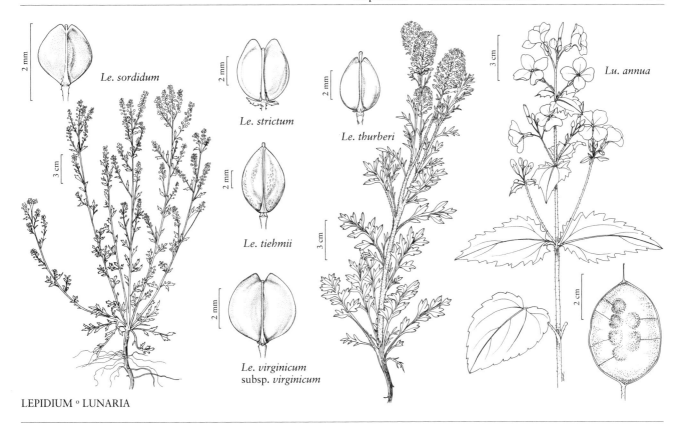

LEPIDIUM ○ LUNARIA

39. **Lepidium strictum** (S. Watson) Rattan ex B. L. Robinson in A. Gray et al., Syn. Fl. N. Amer. 1(1,1): 129. 1895 [F] [I]

Lepidium oxycarpum Torrey & A. Gray var. *strictum* S. Watson in W. H. Brewer et al., Bot. California 1: 46. 1876; *L. reticulatum* Howell

Annuals; hirsute. **Stems** often several from base, usually ascending or decumbent to prostrate, rarely erect, branched distally, (0.4–)0.7–1.7(–2) dm. **Basal leaves** not rosulate; petiole 0.5–3 cm; blade 2-pinnatifid (lobes lanceolate to oblong), 1.5–5.6 cm, margins (of lobes) entire. **Cauline leaves** shortly petiolate; blade pinnatifid, 0.8–3 cm × 0.3–8 mm, base cuneate to attenuate, not auriculate, margins (of lobes) entire. **Racemes** elongated, (dense) in fruit; rachis puberulent, trichomes straight, cylindrical. **Fruiting pedicels** suberect and subappressed at base, recurved and becoming divaricate distally, strongly curved, (often flattened and narrowly winged), (1–)1.4–2.5(–3) × 0.2–0.4 mm, puberulent adaxially. **Flowers:** sepals (persistent), oblong, 0.7–1(–1.2) × 0.3–0.4 mm; petals (rudimentary), white, linear, 0.2–0.5 × 0.05 mm, claw absent; stamens 2, median; filaments 0.5–0.8 mm; anthers 0.1–0.15 mm. **Fruits** ovate-orbicular to ovate, 2.5–3.3 × 2–3 mm, apically winged, apical notch 0.3–0.6 mm deep; valves (enclosing seeds), thin, smooth, reticulate-veined, glabrous or puberulent on margin; style obsolete or to 0.1 mm, included in apical notch. **Seeds** oblong, 1.2–1.6 × 0.7–0.8 mm. $2n = 32$.

Flowering Apr–Jun. Waste grounds, woodlands, hillsides; introduced; Calif., Oreg.; South America (Chile).

Lepidium strictum was reported from Utah (C. L. Hitchcock 1936) and Colorado (W. A. Weber 1989), but we have been unable to verify those records. The species is easily distinguished by a combination of reticulate-veined fruits, persistent sepals, flattened and narrowly winged fruiting pedicels, and filiform nectaries.

40. **Lepidium thurberi** Wooton, Bull. Torrey Bot. Club 25: 259. 1898 [F]

Annuals; pubescent, (trichomes cylindrical, to 1 mm, and much shorter, clavate ones). **Stems** often simple from base, erect, branched (several) distally, (0.8)1.2–4.9(–6) dm. **Basal leaves** (often withered at anthesis); rosulate; petiole 1–3 (–4.5) cm; blade pinnatifid (lobes oblong to ovate or lanceolate), (1.4–)2.2–7(–10) cm, margins (of lobes) dentate-sinuate. **Cauline leaves** shortly petiolate; 1.5–6 cm × 6–25 mm, base not auriculate, margins (of lobes) entire or dentate. **Racemes** considerably elongated in fruit; rachis pilose, trichomes straight, cylindrical (to 1 mm) with much smaller, clavate ones, sometimes one type present. **Fruiting pedicels** divaricate-ascending to horizontal, straight or slightly recurved, (terete), 4–8(–10) × 0.2–0.3 mm, puberulent or pilose adaxially. **Flowers:** sepals suborbicular to broadly ovate, 1–1.6 × 0.7–1 mm; petals white, broadly obovate to suborbicular, 3–4 × 1.2–2.2 mm, claw 0.7–1.3 mm; stamens 6; filaments (median pairs) 1–1.6 mm, (glabrous); anthers 0.4–0.5 mm. **Fruits** broadly ovate to orbicular, 2–2.9 × 2–2.8 mm, apically winged, apical notch 0.1–0.2 mm deep; valves thin, smooth, not veined, glabrous; style 0.3–0.8 mm, exserted beyond apical notch. **Seeds** ovate-oblong, 1.3–1.6 × 0.8–1.1 mm.

Flowering Apr–Aug. Salt flats, mesquite and creosote bush communities, playas, stream banks, sandy deserts, washes, clay bottoms, bluffs, gravelly granitic sand, grasslands, alluvial fans, roadsides, silty terraces, washes, gravelly flats; 600–1800 m; Ariz., Calif., N.Mex.; Mexico (Chihuahua, Coahuila, Sonora).

41. Lepidium tiehmii (Rollins) Al-Shehbaz, Novon 12: 9. 2002 C E F

Stroganowia tiehmii Rollins, Syst. Bot. 7: 215, fig. 1. 1982

Perennials; (caudex woody, to 1 cm diam., covered with persistent petiolar remains); glabrous throughout. **Stems** simple from base (caudex branch), erect, branched distally, 1–7.5 dm. **Basal leaves** rosulate; petiole (1.5–)2.5–10(–13) cm; blade oblong to lanceolate, (2.5–)4.5–9(–14) cm × 15–40 mm, margins entire. **Cauline leaves** shortly petiolate; blade oblanceolate, (much smaller than basal), base attenuate-cuneate, not auriculate, margins entire. **Racemes** (paniculate), considerably elongated in fruit. **Fruiting pedicels** divaricate-ascending, straight, (terete), 8–15 × 0.4–0.5 mm. **Flowers:** sepals oblong, 2.5–3(–4) × 1.2–1.8 mm; petals creamy white to pale yellow, obovate to oblanceolate, 4–5.5(–6.5) × 2–3.3 mm, claw 1.5–2.2 mm; stamens 6; filaments (median pairs) 3.5–4.5 mm; anthers 1–1.2 mm. **Fruits** obovate to somewhat rhomboid, 7–11 × 5–6.5 mm, apically not winged, apical notch absent; valves thin, smooth, obscurely veined; style 0.2–0.6 mm. **Seeds** oblong, 3.7–4.5 × 1.6–2 mm.

Flowering Apr–Jun. Rocky crevices and slopes in sagebrush communities; of conservation concern; 1400–1800 m; Nev.

Lepidium tiehmii is known from mountain ranges in Douglas and Lyon counties. It was described and has been maintained (N. H. Holmgren 2005b) in *Stroganowia*, a genus now united with *Lepidium* (I. A. Al-Shehbaz et al. 2002) that otherwise is disjunct and restricted to the central Asian states of the Former Soviet Union and adjacent western China. In our opinion, the similarity of this species to those Asian ones formerly placed in *Stroganowia* is superficial and is the result of convergence rather than descent.

The cotyledonary type was erroneously reported as conduplicate (R. C. Rollins 1993; N. H. Holmgren 2005b). In the several seeds that we dissected it was always incumbent.

42. Lepidium virginicum Linnaeus, Sp. Pl. 2: 645. 1753 F W

Annuals; puberulent, (trichomes cylindrical). **Stems** simple from base, erect, branched distally, (0.6–)1.5–5.5(–7) dm. **Basal leaves** (withered by anthesis); not rosulate; petiole 0.5–3.5 cm; blade obovate, spatulate, or oblanceolate, (1–)2.5–10(–15) cm × 5–30(–50) mm, margins pinnatifid to lyrate or dentate. **Cauline leaves** shortly petiolate; blade oblanceolate or linear, 1–6 cm × (1–)3–10 mm, base attenuate to subcuneate, not auriculate, margins serrate or entire. **Racemes** considerably elongated in fruit; rachis usually puberulent, rarely glabrous, trichomes curved, cylindrical. **Fruiting pedicels** divaricate-ascending to nearly horizontal, straight or slightly recurved, (slender, terete or flattened), 2.5–4(–6) × 0.15–0.4 mm, puberulent adaxially or, rarely, throughout or glabrous. **Flowers:** sepals oblong to ovate, (0.5–)0.7–1(–1.1) × 0.4–0.7 mm; petals (rarely rudimentary), white, spatulate to oblanceolate, 1–2(–2.5) × 0.3–0.8(–1) mm, claw undifferentiated or to 0.8 mm; stamens 2, median; filaments 0.6–1.2 mm; anthers 0.1–0.2 mm. **Fruits** orbicular or nearly so, 2.5–3.5(–4) mm diam. (widest at middle), apically winged, apical notch 0.2–0.5 mm deep; valves thin, smooth, not veined, glabrous; style 0.1–0.2 mm, included in apical notch. **Seeds** ovate, 1.3–1.9(–2.1) × 0.7–1(–1.2) mm; (cotyledons accumbent or incumbent).

Subspecies 2 (2 in the flora): North America, Mexico; introduced in South America, Europe, Asia, s Africa, Australia.

Lepidium virginicum is perhaps the only species in the genus with accumbent cotyledons; the other species have incumbent cotyledons, except for Australian ones with diplecolobal cotyledons. This contradicts N. H. Holmgren's (2005b) assertion that the genus characteristically has accumbent cotyledons.

Examination of thousands of specimens provides good evidence (e.g., Florida: *Lakela et al. 27038*, GH; Missouri: *Raven & Raven 27501*, GH, MO) that *Lepidium virginicum* hybridizes with *L. densiflorum*.

The ranges of both species overlap in much of the flora area, and this intergradation is, perhaps, the reason behind the recognition of some infraspecific taxa in both species. Molecular studies, along with a critical evaluation of morphology, are needed. Typical *L. virginicum* is easily distinguished by having well-developed or, rarely, rudimentary petals, accumbent cotyledons, orbicular fruits, and raceme rachises with curved cylindrical trichomes. By contrast, *L. densiflorum* has rudimentary or, often, no petals, incumbent cotyledons, obovate fruits, and raceme rachises with straight, often subclavate trichomes.

Of the seven varieties recognized by C. L. Hitchcock (1945) and R. C. Rollins (1993) in *Lepidium virginicum*, three do not occur in the flora area, and they most likely belong to other species. The four present in our area clearly fall into two groups. The first, which corresponds to the species lectotype, has accumbent cotyledons and terete fruiting pedicels. The second group, which includes the holotypes of vars. *medium*, *pubescens*, and *robinsonii*, has incumbent cotyledons and flattened fruiting pedicels. The type of var. *medium* has completely glabrous raceme rachises and fruiting pedicels, but most authors (e.g., Hitchcock 1936, 1945; Rollins) assigned this varietal name to glabrous plants regardless of whether they have accumbent or incumbent cotyledons. Indeed, the cotyledonary position and indumentum absence (or presence) do not always covary, and some glabrous plants have accumbent cotyledons (e.g., *Demaree 47912* (GH), from Arkansas) or incumbent cotyledons (e.g., *Demaree 43698* (GH) from Arizona). Therefore, var. *medium* does not merit recognition. As for var. *robinsonii*, it was based solely on being shorter plants with or without divided leaves. We believe that delimitation is artificial; such plants occur sporadically in the ranges of the two groups noted above. With the elimination of vars. *medium* and *robinsonii*, *L. virginicum* consists of two infraspecific taxa recognized herein as subspecies.

1. Cotyledons accumbent; fruit valves glabrous; fruiting pedicels terete, 0.15–0.2 mm wide . . . . 42a. *Lepidium virginicum* subsp. *virginicum*
1. Cotyledons incumbent or obliquely so; fruit valves glabrous or puberulent; fruiting pedicels flattened (at least proximal to apex), (0.2–)0.3–0.4 mm wide . . . . 42b. *Lepidium virginicum* subsp. *menziesii*

42a. Lepidium virginicum Linnaeus subsp. virginicum F

Lepidium virginicum var. *linearifolium* Farwell

Fruiting pedicels terete, 0.15–0.2 mm wide. **Fruit valves** glabrous. **Cotyledons** accumbent. $2n = 32$.

Flowering Mar–Sep. Fields, roadsides, waste places, disturbed sites, fields, grassy areas; 0–1000 m; B.C., Nfld. and Labr. (Nfld.), N.S., Ont., P.E.I., Que.; Ala., Ariz., Ark., Calif., Colo., Conn., Del., D.C., Fla., Ga., Ill., Ind., Iowa, Kans., Ky., La., Maine, Md., Mass., Mich., Minn., Miss., Mo., Mont., Nebr., N.H., N.J., N.Mex., N.Y., N.C., Ohio, Okla., Pa., R.I., S.C., Tenn., Tex., Vt., Va., Wash., W.Va., Wis.; Mexico; introduced in South America, Europe, Asia, s Africa, Australia.

42b. Lepidium virginicum Linnaeus subsp. menziesii (de Candolle) Thellung, Mitt. Bot. Mus. Univ. Zürich 28: 230. 1906

Lepidium menziesii de Candolle, Syst. Nat. 2: 539. 1821; *L. bernardinum* Abrams; *L. californicum* Nuttall; *L. glaucum* Greene; *L. hirsutum* Rydberg; *L. idahoense* A. Heller; *L. intermedium* var. *pubescens* Greene; *L. medium* Greene; *L. medium* var. *pubescens* (Greene) B. L. Robinson; *L. occidentale* Howell; *L. robinsonii* Thellung; *L. simile* A. Heller; *L. virginicum* var. *californicum* Jepson; *L. virginicum* var. *medium* (Greene) C. L. Hitchcock; *L. virginicum* var. *menziesii* (de Candolle) C. L. Hitchcock; *L. virginicum* var. *pubescens* (Greene) Thellung; *L. virginicum* var. *robinsonii* (Thellung) C. L. Hitchcock

Fruiting pedicels flattened (at least proximal to apex), (0.2–)0.3–0.4 mm wide. **Fruit valves** glabrous or puberulent. **Cotyledons** incumbent or obliquely so. $2n = 32$.

Flowering Mar–Jun. Roadsides, bottomlands, gravelly and sandy shores, waste grounds, stream banks, grassy meadows, dry flats and stream beds, abandoned fields, woods, cliffs, plains, pastures, sagebrush and other desert shrub communities, dry mountain slopes; 700–2600 m; B.C.; Ariz., Calif., Colo., Idaho, Mont., Nev., N.Mex., Oreg., Tex., Utah, Wash., Wyo.; Mexico.

x. BRASSICACEAE Burnett tribe LUNARIEAE Dumortier, Fl. Belg., 119. 1827 ⊡

Biennials or, rarely, annuals [perennials]; eglandular. **Trichomes** simple. **Cauline leaves** petiolate or sessile; blade base not auriculate, margins dentate. **Racemes** usually ebracteate, elongated in fruit. **Flowers** actinomorphic; sepals erect, lateral pair strongly saccate basally; petals usually purple, rarely lavender or white [violet], claw present, distinct; filaments unappendaged, not winged; pollen 3-colpate. **Fruits** silicles, dehiscent, unsegmented, latiseptate; ovules 4–8 per ovary; style distinct; stigma 2-lobed [entire]. **Seeds** biseriate, (funicle adnate to septum); cotyledons accumbent.

Genus 1, species 3 (1 in the flora): introduced; Europe.

69. LUNARIA Linnaeus, Sp. Pl. 2: 653. 1753; Gen. Pl. ed. 5, 294. 1754 • Honesty, money-plant, satin-flower, moonwort [Latin *luna*, moon, alluding to persistent, silvery, large fruit septum] ⊡

Ihsan A. Al-Shehbaz

Plants not scapose; pubescent, glabrate, or glabrous. **Stems** erect, unbranched or branched distally. **Leaves** basal and cauline; petiolate or sessile; basal (soon withered, opposite), rosulate, long-petiolate, blade margins coarsely dentate; cauline (opposite or alternate), petiolate or (distal) sessile, blade margins coarsely dentate. **Racemes** (corymbose, several-flowered, rarely proximalmost flowers bracteate), considerably elongated in fruit. **Fruiting pedicels** divaricate or ascending, slender. **Flowers:** sepals cucullate, (median pair) linear or (lateral pair) broadly oblong-elliptic; petals obovate, (much longer than sepals), claw strongly differentiated from blade, (nearly as long as sepal, apex obtuse); stamens tetradynamous; filaments usually not dilated basally (or slender); anthers oblong or linear, (apex obtuse); nectar glands lateral, annular or semi-annular. **Fruits** (often pendulous), long-stipitate [rarely subsessile], oblong to suborbicular [orbicular, lanceolate-elliptic], not torulose, strongly latiseptate; valves each not veined, glabrous; replum rounded; septum (persistent), complete, (broad, shiny); style slender; stigma capitate, 2-lobed (lobes opposite replum, connivent or not). **Seeds** strongly flattened, broadly winged [not winged], reniform [orbicular]; seed coat not mucilaginous when wetted; cotyledons accumbent. $x = 15$.

Species 3 (1 in the flora): introduced; Europe; introduced also in s South America.

Two species of *Lunaria* are ornamentals cultivated worldwide; *L. annua* often escapes from cultivation.

Although R. C. Rollins (1993) listed *Lunaria rediviva* Linnaeus as "rarely found in long persisting populations outside of cultivation," I have not seen any material that unequivocally proves that the species is naturalized in North America. It is easily distinguished from *L. annua* by being perennial with petiolate distalmost leaves and elliptic-lanceolate fruits acute at base and apex. By contrast, *L. annua* is biennial or, rarely, annual and has sessile distalmost leaves and suborbicular to broadly oblong fruits obtuse at base and apex.

1. Lunaria annua Linnaeus, Sp. Pl. 2: 653. 1753

* Bolbonac, silver-dollar, penny-flower [F] [I]

Lunaria biennis Moench; *L. inodora* Lamarck

Plants sparsely to densely hispid, glabrous, or glabrate. **Stems** (3–)4–10(–12) dm, pubescent or, rarely, glabrate. **Basal leaves**: petiole (1.5–)3–10(–17) cm; blade broadly cordate to narrowly cordate-ovate, (1.5–)3–12(–18) × (1–)2–8(–12) cm, base cordate, often pubescent. **Cauline leaves** similar to basal, petiole shorter (distal sessile); blade (proximal opposite, distal alternate), smaller distally. **Fruiting pedicels** (7–)10–15 mm, glabrous or pubescent. **Flowers:** sepals (5–)6–9 (–10) × 1–2 mm; petals (15–)17–25(–30) × 5–10 mm, claw 5–10 mm; filaments 5–8 mm; anthers oblong, 2–3 mm; gynophore relatively slender, 7–18 mm. **Fruits** 3–4.5(–5) × 2–3(–3.5) cm, strongly latiseptate; valves each rounded basally and apically; replum glabrous or sparsely ciliate; style 4–10 mm. **Seeds** grayish brown, (6–)7–10(–12) × 5–9 mm. $2n = 30$.

Flowering Apr–Jun. Roadsides, waste grounds, railroad embankments, thickets, woods, pasture margins; 0–1000 m; introduced; B.C., Man., N.S., Ont., Que.; Calif., Conn., Del., Idaho, Ind., Ky., Md., Mass., Mich., N.Y., Ohio, Oreg., Pa., R.I., Utah, Vt., Wash.; Europe; introduced also in South America (Argentina).

Lunaria annua is cultivated for its attractive flowers but especially for the infructescences, which are used in dry bouquets after removal of the fruit valves and seeds.

y. **BRASSICACEAE** Burnett tribe **MALCOLMIEAE** Al-Shehbaz & Warwick, Harvard Pap. Bot. 12: 432. 2007 [I]

Annuals or subshrubs [perennials, shrubs, or trees]; eglandular. **Trichomes** sessile, exclusively malpighiaceous [stellate]. **Cauline leaves** sessile [shortly petiolate]; blade base not auriculate, margins usually entire. **Racemes** usually ebracteate, often elongated in fruit. **Flowers** actinomorphic; sepals ascending to spreading [erect], lateral pair not saccate [strongly saccate] basally; petals white, cream, purple, or violet [pink, yellow, orange, brown], claw present, often distinct; filaments unappendaged, not winged; pollen 3-colpate. **Fruits** silicles [siliques], dehiscent, unsegmented, latiseptate [terete, 4-angled]; ovules 2–10[–numerous] per ovary; style distinct; stigma entire [strongly 2-lobed]. **Seeds** usually uniseriate; cotyledons accumbent.

Genera 10, species 68 (1 in the flora): introduced; Europe, Asia, Africa.

70. **LOBULARIA** Desvaux, J. Bot. Agric. 3: 162. 1815, name conserved • [Latin *lobulus*, small lobe, alluding to small silicles] [I]

Liv Borgen

Koniga R. Brown

Plants not scapose; pubescent, trichomes appressed, unicellular, medifixed. **Stems** erect, procumbent, or decumbent [ascending], branched basally [and distally]. **Leaves** cauline; not rosulate; blade (base not auriculate), margins entire. **Racemes** (several-flowered, sometimes bracteate at base). **Fruiting pedicels** ascending [divaricate], slender. **Flowers:** sepals oblong [ovate]; petals obovate [spatulate or orbicular], claw differentiated from blade, (margins entire, apex often rounded); stamens slightly tetradynamous; filaments dilated basally; anthers ovate, (apex obtuse); nectar glands (8): 4 lateral (rudimentary), 4 median (cylindrical). **Fruits** sessile or shortly stipitate, elliptic-suborbicular [orbicular, obovate, elliptic], smooth, convex, latiseptate;

valves (papery), each 1-veined, pubescent; replum rounded; septum complete; ovules 2–10 per ovary; stigma capitate. **Seeds** uniseriate or biseriate, strongly flattened, winged or not, lenticular or ovate [orbicular]; seed coat (reticulate), mucilaginous when wetted; cotyledons accumbent. *x* = 11, 12, 23.

Species 4 (1 in the flora): introduced; s Europe, sw Asia, n Africa (Mediterranean region); introduced also nearly worldwide.

SELECTED REFERENCE Borgen, L. 1987. *Lobularia* (Cruciferae). A biosystematic study with special reference to the Macaronesian region. Opera Bot. 91: 1–96.

1. **Lobularia maritima** (Linnaeus) Desvaux, J. Bot. Agric. 3: 162. 1815 • Sweet-alyssum [F] [I] [W]

Clypeola maritima Linnaeus, Sp. Pl. 2: 652. 1753; *Alyssum maritimum* (Linnaeus) Lamarck; *Koniga maritima* (Linnaeus) R. Brown

Plants suffruticose (when subshrubs). **Stems** 0.5–2.5(–4) dm. **Leaves:** blade linear or lanceolate-oblanceolate, (1–)1.6–2.5(–4.2) cm × (1–)2–3(–7) mm, base attenuate, apex acute. **Racemes** elongated in fruit, (1–)4–8(–16) cm, (dense). **Fruiting pedicels** (3–)4.5–6(–9.5) mm. **Flowers:** sepals often tinged purplish, (1.4–)1.5–1.7(–2.4) mm; petals broadly obovate, (1.9–)2.3–2.8(–3.1) × (1.2–)1.6–2(–2.6) mm, abruptly contracted into claw; filaments 1.2–2 mm; anthers 0.3–0.5 mm. **Fruits** (1.9–)2.3–2.7(–4.2) × (1.2–)1.6–2(–2.9) mm; valve margins thin, sparsely pubescent; style 0.4–0.6 mm. **Seeds** light to reddish brown, not winged, (1–)1.2–1.4(–2) × (0.7–)1–1.1(–1.6) mm. $2n = 24$.

Flowering year-round (peak spring–summer). Roadsides, waste places, vacant lots, cultivated fields, walls, coastal fir zone, mainly along Atlantic, Gulf, and Pacific coasts, also ephemeral inland; 0–800 m; introduced; B.C., N.S., Ont., Que.; Ariz., Calif., Conn., Del., Fla., Ga., Ill., Ind., Iowa, La., Maine, Md., Mass., Mich., Miss., N.H., N.J., N.Y., N.C., Ohio, Oreg., Pa., R.I., S.C., Tex., Utah, Vt., Wash.; Europe; Asia; Africa; introduced also in Mexico, West Indies, Central America, South America, Pacific Islands, Australia.

Lobularia maritima is widely cultivated as an ornamental; many cultivars are on the market. It was introduced to North America because of its drought tolerance and attractive, scented white flowers (R. Ornduff 1974). It has been reported as cultivated in the northern United States back to 1856 (A. Gray 1856). The cultivars naturalize very easily and have been known as locally established garden escapes in North America back to the end of the nineteenth century (N. L. Britton and A. Brown 1896–1898, vol. 2).

z. **BRASSICACEAE** Burnett tribe **NOCCAEEAE** Al-Shehbaz, Beilstein & E. A. Kellogg, Pl. Syst. Evol. 259: 112. 2006

Annuals, biennials, or perennials; eglandular. **Trichomes** [simple or] absent. **Cauline leaves** sessile; blade base usually auriculate, sometimes amplexicaul, margins usually entire. **Racemes** ebracteate, elongated or not in fruit. **Flowers** actinomorphic; sepals erect or ascending [spreading], lateral pair not saccate basally; petals white, pink, or purple, claw present, often obscure; filaments unappendaged, not winged; pollen 3-colpate. **Fruits** silicles, dehiscent, unsegmented, angustiseptate; ovules 4–24 per ovary; style distinct; stigma entire. **Seeds** uniseriate; cotyledons accumbent.

Genera 3, species ca. 90 (2 genera, 4 species in the flora): North America, Mexico, South America (Patagonia), Europe, Asia, n Africa.

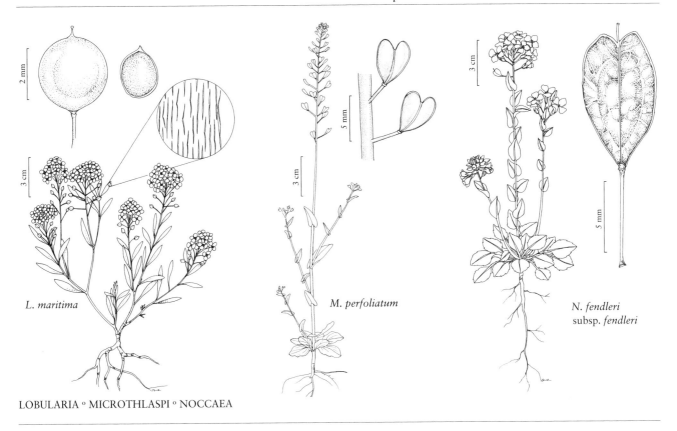

LOBULARIA ○ MICROTHLASPI ○ NOCCAEA

71. MICROTHLASPI F. K. Meyer, Feddes Repert. 84: 452. 1973 • [Greek *micro-*, small, and genus *Thlaspi*] I

Ihsan A. Al-Shehbaz

Annuals; not scapose; (glaucous). **Stems** erect or subdecumbent, unbranched or branched. **Leaves** basal and cauline; petiolate, sessile, or subsessile; basal (withered by fruiting), loosely rosulate or not; petiolate [subsessile], blade margins entire or obscurely dentate; cauline blade (base cordate to amplexicaul), margins entire or repand. **Racemes** (several-flowered), considerably elongated in fruit. **Fruiting pedicels** divaricate, straight or recurved, slender. **Flowers:** sepals ovate or oblong, (margins membranous); petals white, oblong or spatulate, (sometimes slightly unequal), claw (short or) undifferentiated from blade, (apex rounded [obtuse]); stamens slightly tetradynamous; filaments not dilated basally; anthers ovate, (apex obtuse); nectar glands (2 or 4), lateral, 1 on each side of lateral stamen (when 4). **Fruits** sessile, usually obcordate, rarely obovate, (apex emarginate), strongly angustiseptate; valves keeled, (winged around, much wider apically); replum rounded; septum complete, (not veined); ovules (4–)6–8(–10) per ovary; style obsolete or distinct, (included in apical notch of fruit); stigma capitate. **Seeds** plump, not winged, ovoid; seed coat (smooth), mucilaginous when wetted; cotyledons accumbent. $x = 7$.

Species 4 (1 in the flora): introduced; Europe, Asia, n Africa.

SELECTED REFERENCES Koch, M. et al. 1998b. Molecular biogeography and evolution of the *Microthlaspi perfoliatum* s.l. polyploid complex (Brassicaceae): Chloroplast DNA and nuclear ribosomal DNA restriction site variation. Canad. J. Bot. 76: 382–396. Meyer, F. K. 2003. Kritische Revision der "*Thlaspi*"-Arten Europas, Afrikas und Vorderasiens. Spezieller Teil. III. *Microthlaspi* F. K. Mey. Haussknechtia 9: 3–59.

1. **Microthlaspi perfoliatum** (Linnaeus) F. K. Meyer, Feddes Repert. 84: 453. 1973 F I

Thlaspi perfoliatum Linnaeus, Sp. Pl. 2: 646. 1753

Stems unbranched or branched basally or distally, (0.3–)0.5–2.8(–4) dm. **Basal leaves** (not rosulate or loosely so); petiole 0.5–1.3 cm; blade ovate or broadly elliptic, 0.3–2(–2.7) cm × 2–10 mm, base cuneate or attenuate, margins entire or, rarely, obtusely and sparsely dentate, apex rounded. **Cauline leaves:** blade ovate-cordate or suboblong, (1–)2–4(–5.5) cm × (2–)5–15(–20) mm, base cordate-amplexicaul, margins entire or repand, apex obtuse. **Fruiting pedicels** (2.5–)4–6(–8) mm. **Flowers:** sepals (0.8–)1–1.5(–2) × 0.5–1 mm, (margins white); petals 1.5–3.5(–4.7) × 0.5–1.3 mm; filaments 1–1.5(–2) mm; anthers 0.1–0.2 mm. **Fruits** 3–6.5(–8) × (2.5–)3–6(–7) mm, base obtuse, apical notch 1–1.5 mm deep; wings 1–2 mm wide apically, much narrower basally, style obsolete or 0.1–0.3 mm. **Seeds** yellow-brown, 0.9–1.3(–1.5) × 0.7–1(–1.2) mm. $2n$ = 14, 28, 42.

Flowering Mar–May. Waste grounds, roadsides, fields, plains, thickets; 0–500 m; introduced; Ill., Ind., Kans., Ky., Md., Mo., N.J., N.Y., N.C., Ohio, Pa., Tenn., Va., Wash., W.Va.; Europe; Asia; n Africa.

72. **NOCCAEA** Moench, Suppl. Meth., 89. 1802 • [For Domenico Nocca, 1758–1841, Italian clergyman, botanist, director of botanic garden at Pavia]

Ihsan A. Al-Shehbaz

Biennials or perennials; (stoloniferous or simple or several from caudex); not scapose; (often glaucous). **Stems** erect or decumbent, unbranched or branched distally. **Leaves** basal and cauline; petiolate or sessile; basal rosulate, petiolate, margins entire, denticulate, or dentate; cauline blade (base auriculate or subamplexicaul [sagittate]), margins entire or dentate. **Racemes** (corymbose, several-flowered), considerably elongated or congested in fruit. **Fruiting pedicels** horizontal or, rarely, ascending, slender. **Flowers:** sepals erect, oblong, or ovate [obovate]; petals white, pink, or purple, spatulate [obovate, oblanceolate, oblong, or, rarely, broadly linear], (longer than sepals), claw obscurely differentiated from blade, (apex obtuse or rounded); stamens slightly tetradynamous; filaments not dilated basally; anthers ovate [oblong], (apex obtuse); nectar glands lateral, 2 and subtending stamens, or 4 and 1 on each side of stamen. **Fruits** sessile, obcordate, obovate, obdeltate, elliptical, or oblong, smooth, strongly angustiseptate, (winged or not winged apically); valves each obscurely to prominently veined, strongly keeled; replum rounded; septum complete; ovules 4–14[–24] per ovary; style included in, or much exceeding, apical notch; stigma capitate. **Seeds** plump or slightly compressed, not winged, ovoid [oblong]; seed coat (longitudinally, minutely reticulate to nearly smooth), not mucilaginous when wetted; cotyledons accumbent. $x = 7$.

Species ca. 80 (3 in the flora): North America, Mexico, South America (Patagonia), Europe, Asia, n Africa.

Previous North American accounts (e.g., E. B. Payson 1926; P. K. Holmgren 1971; R. C. Rollins 1993) treated species of *Noccaea* as members of *Thlaspi*. As discussed under the latter genus (see references there) and as shown by M. Koch and I. A. Al-Shehbaz (2004), *Noccaea* is definitely distinct from *Thlaspi*. Excellent comments on the biology, variability, and distribution of the North American taxa were given by Holmgren and are not repeated here.

SELECTED REFERENCES Holmgren, P. K. 1971. A biosystematic study of North American *Thlaspi montanum* and its allies. Mem. New York Bot. Gard. 21(2): 1–106. Koch, M. and I. A. Al-Shehbaz. 2004. Taxonomic and phylogenetic evaluation of the American "*Thlaspi*" species: Identity and relationship to the Eurasian genus *Noccaea* (Brassicaceae). Syst. Bot. 29: 375–384.

1. Petals not flaring between blade and claw, 2–2.8(–3.6) mm; styles (0.2–)0.3–0.5(–0.6) mm; sepals (1–)1.3–1.8(–2.2) mm . . . . 3. *Noccaea parviflora*
1. Petals often flaring between blade and claw, 3–13 mm; styles (0.3–)0.6–4.2 mm; sepals 1.5–5.3 mm.
   2. Styles (0.3–)0.6–1 mm; fruits not winged apically; petals 3–5 mm; arctic Alaska, Yukon . . . . 1. *Noccaea arctica*
   2. Styles (0.4–)1.1–4.2 mm; fruits usually winged, rarely not winged apically; petals (3.4–)4.2–13 mm; Pacific and Mountain states . . . . 2. *Noccaea fendleri*

1. **Noccaea arctica** (A. E. Porsild) Holub, Preslia 70: 107. 1998 E

Thlaspi arcticum A. E. Porsild, Sargentia 4: 40. 1943

Perennials; (with persistent petiolar remains). **Stems** simple or several from caudex, erect, unbranched, 0.5–1.8 dm. **Basal leaves**: petiole 0.4–1.3 cm; blade oblanceolate to spatulate, 0.3–1.8 cm × 2–4 mm, base cuneate or attenuate, margins entire, apex obtuse. **Cauline leaves** 3–7; blade ovate or suboblong, 0.5–1 cm × 1–5 mm, base minutely auriculate, margins entire, apex obtuse to subacute. **Racemes** 0.5–6 cm. **Fruiting pedicels** ascending to horizontal, straight, 3–8 mm. **Flowers**: sepals 1.5–2.5 × 0.5–1 mm, (margins white); petals white, 3–5 × 2–2.5 mm, often flaring between blade and claw, apex rounded; filaments 2–3 mm; anthers ca. 0.5 mm. **Fruits** not winged, narrowly obovate, 6–10 × 2–3.5 mm, base cuneate, apex obtuse; ovules 8–14 per ovary; style (0.3–)0.6–1 mm. **Seeds** yellow-brown, 1.5–2 mm, nearly smooth. $2n = 14$.

Flowering Jun–Aug. Dry tundra ridges, arctic coast; 0–500 m; Yukon; Alaska.

Noccaea arctica is rare, known from a limited number of collections. P. K. Holmgren (1971) suggested that additional collections might indicate that it is a variety of *N. fendleri* (as *Thlaspi montanum*).

2. **Noccaea fendleri** (A. Gray) Holub, Preslia 70: 108. 1998 F

Thlaspi fendleri A. Gray, Smithsonian Contr. Knowl. 5(6): 14. 1853; *T. montanum* Linnaeus var. *fendleri* (A. Gray) P. K. Holmgren

Perennials; (cespitose or not). **Stems** simple or several from caudex, erect or ascending, unbranched or branched distally, (0.1–)0.4–3.2(–4.5) dm. **Basal leaves**: petiole 0.4–7.3 cm; blade linear, oblong, oblanceolate, ovate, obovate, or spatulate, 0.4–3 cm × 2–20 mm, base cuneate to attenuate, margins entire, denticulate, or dentate, apex obtuse to acute. **Cauline leaves** 2–21; blade ovate or suboblong, 0.4–2.8 × 0.1–1.7 mm, base auriculate to subamplexicaul, margins entire or dentate, apex obtuse to subacute. **Racemes** 0.5–25 cm, (congested or considerably elongated). **Fruiting pedicels** horizontal to ascending, straight, 2.5–15 mm. **Flowers**: sepals 1.6–5.3 × 0.5–1.5 mm, (margins membranous); petals white to pinkish purple, (3.4–)4.2–13 × 1–4.9 mm, often flaring between blade and claw, apex obtuse; filaments 2–7.5 mm; anthers 0.5–1 mm. **Fruits** obovate, obcordate, obdeltate, or elliptic, 2.5–16 × 1.5–9 mm, usually winged, rarely not winged apically, base cuneate, apex obtuse, truncate, or emarginate; ovules 4–10 per ovary; style (0.4–)1.1–4.2 mm. **Seeds** brown, 1.1–2.1 mm, minutely reticulate.

Subspecies 5 (5 in the flora): w North America, n Mexico.

Some North American authors (e.g., P. K. Holmgren 1971; R. C. Rollins 1993) treated *Noccaea fendleri* as conspecific with the European *Thlaspi montanum*, whereas others (e.g., E. B. Payson 1926) recognized the North American taxa as endemic to the continent. Molecular studies (M. Koch and I. A. Al-Shehbaz 2004) supported Payson's view of the independent status of the North American representatives. Although Payson and Rollins recognized more than one species here (e.g., *T. californicum*, *T. idahoense*, *T. glaucum*, *T. fendleri*), I agree with Holmgren in treating them as infraspecific taxa, but at the subspecific instead of varietal rank. The distinctions among them are not very sharp, and hybridization appears to have played some role in blurring their boundaries, as evidenced from Holmgren's report of morphologically intermediate forms between subsp. *glauca* (as *T. montanum* var. *montanum*) and each of subspp. *fendleri*, *idahoensis*, and *siskiyouensis*.

1. Fruits elliptic, 2.2–3.2 times as long as wide, apices acute; fruiting pedicels strongly ascending to divaricate-ascending, forming 15–70° angle with rachises . . . . 2b. *Noccaea fendleri* subsp. *californica*
1. Fruits obovate, obcordate, obdeltate, or, rarely, elliptic, 1–2 times as long as wide, apices obtuse, truncate, or emarginate; fruiting pedicels subhorizontal, horizontal, or descending, forming 70–130° angle with rachises.

[2. Shifted to left margin.—Ed.]

2. Basal leaf blades linear to narrowly oblanceolate, 1.1–4(–5) mm wide, bases attenuate; central Idaho . . . . 2d. *Noccaea fendleri* subsp. *idahoensis*
2. Basal leaf blades oblong, ovate, or spatulate, (3.5–)4–20 mm wide, bases cuneate; widespread in Pacific and Mountain states.
    3. Seeds 2–4 per fruit; petioles of basal leaves (1.7–)2–3.6(–4) times longer than blades; cauline leaves 2–6; Curry, Douglas, and Josephine counties, Oregon . . . . 2e. *Noccaea fendleri* subsp. *siskiyouensis*
    3. Seeds 4–6 per fruit; petioles of basal leaves 0.6–2.2(–3.3) times longer than blades; cauline leaves 4–16(–21); Arizona, California, Colorado, Idaho, Montana, Nevada, New Mexico, Oregon, Texas, Utah, Washington, Wyoming.
        4. Petals pinkish purple or, occasionally, white, (6–)6.5–11(–13) mm; styles (1.8–)2.5–3.5(–4.2) mm; fruits 7–12(–16) mm; racemes often compact . . . . 2a. *Noccaea fendleri* subsp. *fendleri*
        4. Petals white or, occasionally, pinkish purple, (3.4–)4–7(–8.5) mm; styles (0.4–)1–2.2(–3) mm; fruits (2.5–)5–8(–12) mm; racemes often lax . . . . 2c. *Noccaea fendleri* subsp. *glauca*

2a. Noccaea fendleri (A. Gray) Holub subsp. **fendleri**
E F

Thlaspi prolixum A. Nelson; *T. stipitatum* A. Nelson

Stems (0.6–)0.8–2.2(–2.8) dm. **Basal leaves:** petiole differentiated from blade, (0.6–)0.9–2.2(–3.3) times longer than blade; blade ovate to oblong, (3.5–)6–16(–20) mm wide, base cuneate. **Cauline leaves** (4–)7–16(–21). **Racemes** often compact. **Fruiting pedicels** horizontal or slightly descending, forming an angle 80° or more with rachis. **Flowers:** petals pinkish purple or, occasionally, white, (6–)6.5–11(–13) × (1.5–)2–4(–5) mm. **Fruits** often winged, obovate to obcordate, 7–12(–16) × (4–)4.5–6(–9) mm, 1–2 times as long as wide, apex truncate to emarginate; style (1.8–)2.5–3.5(–4.2) mm. **Seeds** 4–6 per fruit. $2n = 14$.

Flowering Mar–Jun. Shady areas on slopes, in canyons and gulches, igneous rocky ravines, steep banks, woods; 1300–2100 m; Ariz., N.Mex., Tex.

R. C. Rollins (1966, 1993) recorded a tetraploid count of $2n = 28$ from Hinsdale County, Colorado, but subsp. *fendleri* does not occur in that state, and his count is based on plants of subsp. *glauca*.

2b. Noccaea fendleri (A. Gray) Holub subsp. **californica** (S. Watson) Al-Shehbaz & M. Koch, Syst. Bot. 29: 382. 2004 C E

Thlaspi californicum S. Watson, Proc. Amer. Acad. Arts 17: 365. 1882; *T. alpestre* Linnaeus var. *californicum* (S. Watson) Jepson; *T. glaucum* (A. Nelson) A. Nelson subsp. *californicum* (S. Watson) Munz; *T. montanum* Linnaeus var. *californicum* (S. Watson) P. K. Holmgren

Stems (0.5–)1–1.2(–0.2) dm. **Basal leaves:** petiole differentiated from blade, 0.8–1.8 times longer than blade; blade spatulate to obovate, 5–7(–8) mm wide, base cuneate. **Cauline leaves** 2–5. **Racemes** lax. **Fruiting pedicels** strongly ascending to divaricate-ascending, forming an angle 15–70° with rachis. **Flowers:** petals white, 6–8 × 1.6–2.5 mm. **Fruits** not winged, elliptic, 0.7–10.5 × 2.7–4 mm, 2.2–3.2 times as long as wide, apex acute; style (1.3–)1.5–2(2.4–) mm. **Seeds** 2–6 per fruit.

Flowering May–Jun. Serpentine rock outcrops; of conservation concern; ca. 800 m; Calif.

Subspecies *californica* is known from one population at Kneeland Prairie in Humboldt County. It was recognized by Holmgren as a variety of *Thlaspi montanum*, whereas R. C. Rollins (1993) treated it as *T. californicum*. In my opinion, the differences in fruit apex and size between this taxon and *Noccaea fendleri* are rather small and do not seem to justify its recognition as an independent species.

2c. Noccaea fendleri (A. Gray) Holub subsp. **glauca** (A. Nelson) Al-Shehbaz & M. Koch, Syst. Bot. 29: 382. 2004

Thlaspi alpestre Linnaeus var. *glaucum* A. Nelson, Wyoming Agric. Exp. Sta. Bull. 28: 84. 1896; *Noccaea coloradensis* (Rydberg) Holub; *N. glauca* (A. Nelson) Holub; *T. alpestre* var. *purpurascens* (Rydberg) Ostenfeld; *T. australe* A. Nelson; *T. coloradense* Rydberg; *T. fendleri* A. Gray var. *coloradense* (Rydberg) Maguire; *T. fendleri* var. *glaucum* (A. Nelson) C. L. Hitchcock; *T. fendleri* var. *hesperium* (Payson) C. L. Hitchcock; *T. fendleri* var. *tenuipes* Maguire; *T. glaucum* (A. Nelson) A. Nelson; *T. glaucum* var. *hesperium* Payson; *T. glaucum* var. *pedunculatum* Payson; *T. hesperium* (Payson) G. N. Jones; *T. purpurascens* Rydberg

Stems (0.1–)0.5–3.2(–4.5) dm. **Basal leaves:** petiole differentiated from blade, 0.8–1(–2) times longer than blade; blade ovate to oblong, 4–9(–15) mm wide, base cuneate. **Cauline leaves** (4–)7–16(–21). **Racemes** often lax. **Fruiting pedicels** horizontal to strongly descending, forming an angle to 130° with rachis. **Flowers:** petals white or, occasionally, pinkish purple, (3.4–)4–7(–8.5) × (1–)1.5–2.7(–4.2) mm. **Fruits** winged or, occasionally, not winged, obovate to obcordate, (2.5–)5–8(–12) × (1.5–)2.6–4.5(–6.6) mm, 1–2 times as long as wide, apex obtuse, truncate, to emarginate; style (0.4–)1–2.2(–3) mm. **Seeds** 4–6 per fruit. $2n = 14, 28$.

Flowering Apr–Aug. Dry or moist, open alluvial flats or fans, rocky or talus slopes, scree, limestone cliffs, alpine or subalpine meadows, near snowbanks, streamsides, forest clearings; 300–4400 m; Ariz., Calif., Colo., Idaho, Mont., Nev., N.Mex., Oreg., Tex., Utah, Wash., Wyo.; Mexico (Chihuahua).

Subspecies *glauca*, which is the most morphologically variable and most widespread North American taxon in *Noccaea*, corresponds to *Thlaspi montanum* var. *montanum* in the sense of P. K. Holmgren (1971) and R. C. Rollins (1993). As indicated above, that variety is a strictly European taxon. Subspecies *glauca* exhibits a wide elevational range, and one collection, *Clements 427* (US), was made at 4350 m at the summit of Pike's Peak, Colorado.

2d. Noccaea fendleri (A. Gray) Holub subsp. **idahoensis** (Payson) Al-Shehbaz & M. Koch, Syst. Bot. 29: 383. 2004 (as idahoense) [E]

Thlaspi idahoense Payson, Univ. Wyoming Publ. Sci., Bot. 1: 159. 1926; *T. aileeniae* Rollins; *T. fendleri* A. Gray var. *idahoense* (Payson) C. L. Hitchcock; *T. idahoense* var. *aileeniae* (Rollins) Rollins; *T. montanum* Linnaeus var. *idahoense* (Payson) P. K. Holmgren

Stems (0.3–)0.4–1.1(–1.2) dm. **Basal leaves:** petiole not differentiated from blade, about as long as blade; blade linear to narrowly oblanceolate, 1.1–4(–5) mm wide, base attenuate. **Cauline leaves** (2 or)3–7(–9). **Racemes** lax or slightly compact. **Fruiting pedicels** horizontal or subhorizontal, forming an angle more than 70° with rachis. **Flowers:** petals white, (3.4–)3.6–6 × 1–2 mm. **Fruits** only slightly winged, obovate or, rarely, elliptic, (4–)5–8(–9.5) × (2.5–)2.7–4(–4.5) mm, 1–2 times as long as wide, apex obtuse, truncate, or emarginate; style (0.5–)0.8–2(–2.5) mm. **Seeds** 2–8 per fruit. $2n = 14, 28$.

Flowering Jun–Aug. Alpine or subalpine slopes, meadows, rocky crevices, rocky and gravelly soil in valleys, steep slopes, knolls; 2100–3700 m; Idaho.

Subspecies *idahoensis* is restricted to Blaine, Boise, Custer, Elmore, and Valley counties of central Idaho. Holmgren treated it as a variety of *Thlaspi montanum*, whereas R. C. Rollins (1993) recognized it as a species, *T. idahoense*, with two varieties. In my opinion, his varieties are based on continuous characters such as the loosely versus closely branched caudex, oblanceolate versus linear-oblanceolate basal leaves, and cuneate versus minutely auriculate distalmost cauline leaves.

2e. Noccaea fendleri (A. Gray) Holub subsp. **siskiyouensis** (P. K. Holmgren) Al-Shehbaz & M. Koch, Syst. Bot. 29: 383. 2004 (as siskiyouense) [E]

Thlaspi montanum Linnaeus var. *siskiyoense* P. K. Holmgren, Mem. New York Bot. Gard. 21(2): 84. 1971

Stems (0.6–)0.7–1.6(–2.4) dm. **Basal leaves:** petiole differentiated from blade, (1.7–)2–3.6(–4) times longer than blade; blade ovate to oblong, 4–8(–9) mm wide, base cuneate. **Cauline leaves** 2–6. **Racemes** lax to compact. **Fruiting pedicels** horizontal or slightly descending, forming an angle 80° or more with rachis. **Flowers:** petals white, 3.8–6 × 1.2–2.2(–2.9) mm. **Fruits** often strongly winged, obovate to obdeltate, (4–)4.8–7(–8) × (3–)3.5–5.5(–6) mm, 1–2 times as long as wide, apex truncate to strongly emarginate; style (0.8–)1–2 mm. **Seeds** 2–4 per fruit. $2n = 14$.

Flowering Apr–Jun. Moist, open rocky serpentine slopes; 300–500 m; Oreg.

Subspecies *siskiyouensis* is known from Curry, Douglas, and Josephine counties.

3. Noccaea parviflora (A. Nelson) Holub, Preslia 70: 108. 1998 [E]

Thlaspi parviflorum A. Nelson, Bull. Torrey Bot. Club 27: 265. 1900

Biennials or perennials; (short-lived, rarely cespitose, glaucous). **Stems** simple or 2 from base (rarely several), erect or ascending, (0.7–)1–2.5(–3.2) dm, unbranched or branched distally. **Basal leaves** (soon withered; subrosulate); petiole (0.2–)0.4–1.4(–2) cm; leaf blades oblanceolate to spatulate, 0.3–1.5(–1.9) cm × 2–9 mm, base cuneate, margins entire or repand, apex obtuse. **Cauline leaves** 4–10(–12); blade ovate or suboblong, 0.5–2(–2.3) cm × 2–7(–8) mm, base minutely auriculate, margins entire, apex obtuse to subacute. **Racemes** (1–)2–13(–17) cm. **Fruiting pedicels** horizontal, straight, (2–)2.5–5.5(–8) mm. **Flowers:** sepals (1–)1.3–1.8(–2.2) × 0.4–0.7 mm, (margins white); petals white, 2–2.8(–3.6) × 0.6–1.2 mm, (erect), not flaring between blade and claw, apex obtuse; filaments 1–2.2 mm; anthers ca. 0.3 mm.

Fruits not winged or, rarely, narrowly winged apically, obovate, (3.5–)5–7 × 2–3.5 mm, base cuneate, apex obtuse, truncate, or retuse; ovules 4–12 per ovary; style (0.2–)0.3–0.5(–0.6) mm. **Seeds** golden brown, 0.8–1.2 mm, smooth. $2n = 14$.

Flowering May–Jul. Meadows, sagebrush, limestone cliffs and outcrops, dry grassy slopes and ridges; 1200–2800 m; Idaho, Mont., Wyo.

Noccaea parviflora is readily distinguished from the other North American members of the genus by the smaller size of its flowers, fruits, and seeds. It is restricted to central Idaho, adjacent Montana, and northwestern Wyoming.

aa. BRASSICACEAE Burnett tribe PHYSARIEAE B. L. Robinson in A. Gray et al., Syn. Fl. N. Amer. 1(1,1): 100. 1895

Annuals, biennials, perennials, or subshrubs; eglandular. **Trichomes** usually short-stalked, subsessile, or sessile, sometimes long-stalked, stellate, scalelike, subdendritic, or forked, sometimes mixed with simple ones. **Cauline leaves** petiolate, sessile, or subsessile; blade base usually not auriculate (except *Paysonia*), margins entire, dentate, or sinuate. **Racemes** ebracteate, often elongated in fruit. **Flowers** actinomorphic; sepals erect, spreading, ascending, or reflexed, lateral pair seldom saccate basally; petals white, yellow, lavender, purple, violet, orange, or brown [pink], claw present, often distinct; filaments unappendaged, not winged; pollen (3 or) 4–11-colpate. **Fruits** silicles or siliques, dehiscent, unsegmented, terete, latiseptate, or angustiseptate; ovules 2–100 per ovary; style usually distinct; stigma entire or strongly 2-lobed. **Seeds** biseriate, uniseriate, or aseriate; cotyledons accumbent or incumbent.

Genera 7, species ca. 130 (7 genera, 105 species in the flora): North America, Mexico, South America, Asia (ne Russia).

73. DIMORPHOCARPA Rollins, Publ. Bussey Inst. Harvard Univ. 1979: 20, plates 3, 4. 1979 • Spectacle-pod [Latin *dimorphus*, having two forms, and *carpus*, fruit, alluding to production of two fruit types in some species]

Ihsan A. Al-Shehbaz

Annuals, biennials, or, rarely, perennials; not scapose; densely pubescent throughout, trichomes usually subsessile and stellate, mixed with minutely stalked, dendritic ones, rarely unbranched. **Stems** erect [ascending], unbranched or branched proximally, branched distally. **Leaves** basal and cauline; petiolate or sessile; basal not rosulate, petiolate, blade margins dentate, pinnatifid, or lobed; cauline sessile, subsessile, or shortly petiolate, blade (base not auriculate), margins entire or lobed. **Racemes** (corymbose), considerably elongated in fruit. **Fruiting pedicels** divaricate, sometimes ascending or slightly reflexed, slender. **Flowers:** sepals widely spreading to reflexed, oblong, (equal), lateral pair not saccate basally; petals (spreading), white or lavender, obovate (longer than sepals), claw well-differentiated from blade, (apex rounded); stamens (somewhat spreading), tetradynamous; filaments not dilated basally; anthers oblong [sagittate]; nectar glands distinct, lateral and median present; (gynophore to 0.5 mm). **Fruits** silicles, subsessile, breaking into two 1-seeded units at maturity, didymous, suborbicular, or broader than long, winged, strongly angustiseptate; valves keeled, enclosing seeds when falling off, not winged, narrowly winged, or margined, indurated around margin, glabrous or pubescent; replum concealed by valve margin; septum complete (ca. 0.1 mm wide); ovules 2 per ovary; style relatively short or obsolete, (terete or flattened); stigma conical, decurrently 2-lobed (appearing entire). Seeds aseriate, strongly flattened, not winged, broadly oblong to suborbicular; seed coat (smooth), not mucilaginous when wetted; cotyledons accumbent to obliquely incumbent. $x = 9$.

Species 4 (3 in the flora): sw United States, n Mexico.

Dimorphocarpa membranacea (Payson) Rollins is known from northern Mexico (San Luis Potosí, Tamaulipas).

SELECTED REFERENCES Rollins, R. C. 1958. The genetic evaluation of a taxonomic character in *Dithyrea* (Cruciferae). Rhodora 60: 145–152. Rollins, R. C. 1979. *Dithyrea* and a related genus (Cruciferae). Publ. Bussey Inst. Harvard Univ. 1979: 3–32.

1. Fruit valves (7–)8–10 mm; petals (7–)8–10(–12) mm, claws not expanded basally; distal cauline leaves sessile, blades usually ovate to narrowly oblong, rarely lanceolate, bases obtuse to truncate .... 1. *Dimorphocarpa candicans*
1. Fruit valves 4–5.5(–6.5) mm; petals (3.5–)4–7(–8) mm, claws expanded basally; distal cauline leaves shortly petiolate or subsessile, blades linear to lanceolate, bases cuneate.
   2. Distal cauline blades: margins pinnately lobed; fruit valves nearly as long as wide, rounded apically .... 2. *Dimorphocarpa pinnatifida*
   2. Distal cauline blades: margins usually entire, rarely dentate; fruit valves often longer than wide, truncate apically .... 3. *Dimorphocarpa wislizeni*

1. **Dimorphocarpa candicans** (Rafinesque) Rollins, Cruciferae Continental N. Amer., 361. 1993 [E]

Iberis candicans Rafinesque, Atlantic J. 1: 146. 1832; *Dimorphocarpa palmeri* (Payson) Rollins; *Dithyrea wislizeni* Engelmann var. *palmeri* Payson

Annuals or biennials. Stems unbranched proximally, branched distally, (3–)4–8(–10) dm. **Basal leaves:** petiole 1–4 (–6) cm; blade lanceolate to oblong or ovate, (2–)4–8 (–10) cm × (10–)15–25(–40) mm, base cuneate to obtuse, margins dentate. **Cauline leaves** (distal) sessile; blade usually ovate to narrowly oblong, rarely lanceolate, base obtuse to truncate, margins entire, sometimes repand. **Fruiting pedicels** divaricate, (8–)10–16(–20) mm. **Flowers:** sepals 3.5–5 × 1–1.5 mm, pubescent abaxially; petals white or lavender, (7–)8–10(–12) × 4–6(–7) mm, attenuate to claw, claw 2–3 mm, not expanded basally; filaments lavender or white, 3–4 mm; anthers 1–1.5 mm. **Fruits:** each valve suborbicular or orbicular, (7–)8–10 × (6–)7–10 mm, base rounded, apex truncate, with or without narrow margin beyond indurated part surrounding seeds, glabrous or pubescent; style (0.3–)0.6–1(–1.2) mm. **Seeds** suborbicular-ovoid, 3–4 × 2.5–3 mm. $2n = 18$.

Flowering Apr–Sep. Sandy hills and plains, prairies, sand dunes; 100–800 m; Kans., N.Mex., Okla., Tex.

2. **Dimorphocarpa pinnatifida** Rollins, Publ. Bussey Inst. Harvard Univ. 1979: 28. 1979

Annuals. Stems branched basally and distally, 2–9 dm. **Basal leaves:** petiole 0.5–2 cm; blade linear to linear-lanceolate, (2–)3–6(–8) cm × (2–)3–5(–8) mm, base attenuate, margins pinnatifid. **Cauline leaves** (distal) shortly petiolate or subsessile; blade linear to narrowly lanceolate, base cuneate, margins pinnately lobed. **Fruiting pedicels** divaricate to slightly reflexed, (8–)10–17(–22) mm, (straight). **Flowers:** sepals 2–3 × 1–1.5 mm, pubescent abaxially; petals white, (3.5–)4–6.5 × 2–3.5 mm, attenuate to claw, claw 1–1.5 mm, expanded basally; filaments white, 2–3 mm; anthers 0.8–1.1 mm. **Fruits:** each valve orbicular, 5–6.5 × 6–7 mm (nearly as long as wide), base and apex rounded, not margined beyond indurated part surrounding seeds, pubescent; style 0.5–1 mm. **Seeds** suborbicular, 2.5–3.5 × 2–3 mm. $2n = 18$.

Flowering Mar–Apr. Sandy hills and flats; 100–800 m; Ariz.; Mexico (Sonora).

Dimorphocarpa pinnatifida is known from Pima and Yuma counties.

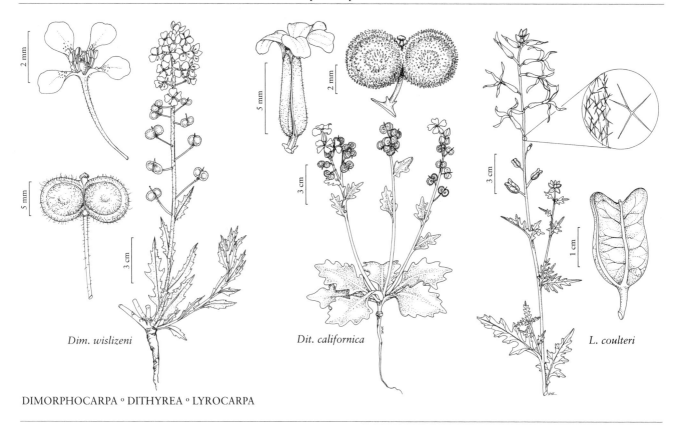

DIMORPHOCARPA ° DITHYREA ° LYROCARPA

3. **Dimorphocarpa wislizeni** (Engelmann) Rollins, Publ. Bussey Inst. Harvard Univ. 1979: 24. 1979 (as wislizenii)

F

Dithyrea wislizeni Engelmann in F. A. Wislizenius, Mem. Tour N. Mexico, 95. 1848; *Biscutella wislizeni* (Engelmann) Bentham & J. D. Hooker ex W. H. Brewer & S. Watson; *D. griffithsii* Wooton & Standley; *D. wislizeni* var. *griffithsii* (Wooton & Standley) Payson

Annuals. Stems unbranched or branched basally, branched distally, (1–)2–6(–8) dm. **Basal leaves:** petiole 1–4(–5) cm; blade lanceolate to linear-lanceolate, (2–)3–7(–10) cm × 4–15(–20) mm, base cuneate to attenuate, margins pinnately lobed to coarsely dentate. **Cauline leaves:** (proximalmost) petiole 1–4(–5) cm, (distalmost) shortly petiolate or subsessile; blade linear to narrowly lanceolate, base cuneate, margins usually entire, rarely dentate, or repand. **Fruiting pedicels** divaricate to slightly reflexed, (6–)8–14 (–22) mm. **Flowers:** sepals (2.5–)3–4 × 1–1.5 mm, pubescent abaxially; petals white or lavender, 4–7(–8) × (2.5–)3–4(–5) mm, attenuate to claw, claw 1–1.5 mm, expanded basally; filaments white, 2–3.5 mm; anthers 0.7–1 mm. **Fruits:** each valve usually transversely ovoid-oblong, rarely suborbicular, 4–5.5(–6) × (4–)5–7(–7.5) mm (often longer than wide), base slightly rounded, apex truncate, with or without narrow wing beyond indurated part surrounding seeds, glabrous or pubescent; style 0.5–1(–1.2) mm. **Seeds** suborbicular-ovoid, 2–3 × 1.5–2 mm. $2n = 18$.

Flowering Feb–Oct. Sandy roadsides, sandstone knolls, sand hills and dunes, sandy streambeds and dry washes, desert flats; 1000–2200 m; Ariz., Colo., Nev., N.Mex., Tex., Utah; Mexico (Chihuahua, Coahuila).

74. DITHYREA Harvey, London J. Bot. 4: 77, plate 5. 1845 • Spectacle-pod [Greek *di-*, two or double, and *thyreos*, shield, alluding to spectacle-shaped fruits]

Ihsan A. Al-Shehbaz

Annuals or perennials; (rhizomatous); not scapose; densely pubescent, trichomes stalked, stellate to subdendritic, sometimes as simple clavate papillae on fruits. **Stems** erect or decumbent, branched basally and distally. **Leaves** basal and cauline; petiolate or subsessile; basal rosulate, petiolate, blade margins dentate, sometimes repand, rarely pinnatifid; cauline petiolate, sessile, or subsessile, blade margins entire, dentate, or repand. **Racemes** (densely corymbose), considerably elongated or not elongated in fruit. **Fruiting pedicels** divaricate, stout, (straight, with 2 well-developed glands basally). **Flowers:** sepals erect, linear or oblong-linear, (forming tube, connivent), lateral pair saccate basally; petals (erect), white, lavender, or purplish, lingulate, (longer than sepals), claw undifferentiated from blade; stamens (erect), tetradynamous; filaments not dilated basally, (filiform); anthers narrowly linear or sagittate; nectar glands: lateral annular, median confluent with lateral; (gynophore to 0.5 mm). **Fruits** silicles, sessile or shortly stipitate, didymous, not winged, strongly angustiseptate; valves 1-seeded, ovate, suborbicular, or slightly broader than long, keeled, enclosing seeds when falling off, indurated around margin, (rim raised), pubescent (with branched or clavate, simple papillae); replum concealed by valve margin; septum complete; ovules 2 per ovary; style distinct, (pubescent); stigma capitate, entire. **Seeds** aseriate, flattened, not winged, oblong; seed coat (smooth), not mucilaginous when wetted; cotyledons accumbent. $x = 10$.

Species 2 (2 in the flora): sw United States, nw Mexico.

SELECTED REFERENCE Rollins, R. C. 1979. *Dithyrea* and a related genus (Cruciferae). Publ. Bussey Inst. Harvard Univ. 1979: 3–32.

1. Annuals; racemes elongated in fruit; fruit valves 3.5–5(–6) × 4–6(–7) mm, trichomes as simple, clavate papillae (stellate hairs absent or, rarely, restricted to fruit margins) .... 1. *Dithyrea californica*
1. Perennials; racemes not elongated in fruit; fruit valves 8–11 × 8–10 mm, trichomes sessile, stellate, sometimes mixed with forked or, rarely, simple ones (clavate papillae absent) ...
.... 2. *Dithyrea maritima*

1. Dithyrea californica Harvey, London J. Bot. 4: 77, plate 5. 1845 [F]

Biscutella californica (Harvey) Bentham & Hooker f. ex W. H. Brewer & S. Watson; *Dithyrea californica* var. *clinata* (J. F. Macbride & Payson) Wiggins; *D. clinata* J. F. Macbride & Payson

Annuals. Stems few to several from base, erect, branched distally, (0.7–)1–6(–7) dm. **Basal leaves:** petiole (0.5–)1–4(–5) cm; blade lanceolate, oblanceolate, or ovate, (1.2–)2–5(–6) cm × (5–)10–20 (–30) mm, base cuneate, margins usually dentate, sinuate, rarely pinnatifid. **Cauline leaves** (distal) shortly petiolate or sessile; blade ovate, lanceolate, or oblong, 1–4(–5) cm × (3–)10–20(–30) mm, base cuneate, margins entire, dentate, or repand. **Racemes** considerably elongated in fruit. **Fruiting pedicels** 1.5–2.5 mm. **Flowers:** sepals purple to lavender, linear, (6–)7–9(–10) × 0.8–1.2 mm; petals white to pale lavender, (10–)12–15 × (1.5–)2–3 mm; filaments 6–7 mm; anthers 1.5–2(–2.5) mm. **Fruits:** valves transversely ovate to suborbicular, 3.5–5 (–6) × 4–6(–7) mm, base and apex rounded, pubescent, trichomes clavate, simple, papillate (branched trichomes rarely present on fruit margin); style 0.1–0.5(–1) mm. **Seeds** 3–4 × 1.5–2.5 mm. $2n = 20$.

Flowering (depending on elevation) Mar–Oct. Sandy deserts, sand dunes; 50–1400 m; Ariz., Calif., Nev.; Mexico (Baja California, Sonora).

Dithyrea californica is known in Arizona from Maricopa, Mohave, and Yuma counties and in Nevada from Clark and Lincoln counties. In material with immature fruits and lacking the proximal portion of stems, the species is readily distinguished from *D. maritima* by the presence of clavate trichomes, quite visible even on the ovaries of developing fruits. By contrast, *D. maritima* has branched instead of clavate trichomes.

2. Dithyrea maritima (Davidson) Davidson in A. Davidson and G. L. Moxley, Fl. S. Calif., 151. 1923 (as Dithyraea) [C]

Biscutella californica (Harvey) Bentham & Hooker f. ex Brewer & S. Watson var. *maritima* Davidson, Erythea 2: 179. 1894; *Dithyrea californica* Harvey var. *maritima* Davidson ex B. L. Robinson

Perennials; (rhizomatous). **Stems** several from base, decumbent, unbranched or branched (few) distally, 1–2.1 dm. **Basal leaves**: petiole 1–6(–8) cm; blade broadly ovate to suborbicular or oblanceolate, 1–5(–6) cm × 7–40 mm, base cuneate, margins dentate, sometimes repand. **Cauline leaves** (distal) shortly petiolate to subsessile; blade ovate or oblong, 1–3 cm × 8–20 mm, base cuneate, margins entire, sometimes repand. **Racemes** not elongated in fruit. **Fruiting pedicels** 1.5–2.5 mm. **Flowers**: sepals lavender, oblong-linear, (6–)7–10 × 0.8–1.2 mm; petals white to purplish, 12–15 × 2.5–3.5 mm; filaments 5.5–7.5 mm; anthers 2–3 mm. **Fruits**: valves suborbicular, 8–11 × 8–10 mm, base and apex rounded, pubescent, trichomes stellate, sessile, sometimes mixed with basally forked, or, rarely, simple ones (clavate papillae absent); style 0.3–1 mm. **Seeds** 3–4 × 2–2.5 mm. $2n = 60$.

Flowering Mar–Aug. Seashores, coastal sand dunes; of conservation concern; 0–50 m; Calif.; Mexico (Baja California).

75. **LYROCARPA** Hooker & Harvey, London J. Bot. 4: 76, plate 4. 1845 • [Greek *lyra*, lyre, and *karpos*, fruit, alluding to fruit shape]

Ihsan A. Al-Shehbaz

Subshrubs [annuals]; (caudex woody); not scapose; pubescent [glabrate], trichomes minutely stalked or sessile, dendritic [rarely few-rayed]. **Stems** erect to decumbent, usually branched distally. **Leaves** basal and cauline; petiolate; basal not rosulate, caducous; cauline blade (base not auriculate), margins usually pinnatisect, pinnatifid, or dentate, rarely entire, (canescent to sparsely pubescent). **Racemes** (corymbose, several-flowered, lax), considerably elongated in fruit. **Fruiting pedicels** divaricate, ascending, or reflexed, slender. **Flowers**: sepals erect, (connivent), linear to oblong-linear, lateral pair saccate basally, (often densely pubescent); petals yellowish or yellow-green to brown, brownish purple, or salmon, linear or linear-oblanceolate [spatulate], (sometimes twisted with age, considerably longer than sepals), claw well-differentiated from blade, (equal to or longer than sepal, apex acute or acuminate [obtuse]); stamens (erect), strongly tetradynamous; filaments not dilated basally; anthers linear, (apiculate); nectar glands lateral, 1 on each side of lateral stamen. **Fruits** silicles or siliques, sessile, not torulose, obcordiform to panduriform, strongly angustiseptate; valves strongly keeled, pubescent or glabrate; replum rounded; septum complete; ovules 6–16[–20] per ovary; style obsolete or distinct, (short); stigma capitate, prominently 2-lobed (lobes spreading). **Seeds** uniseriate, flattened, not winged, orbicular [broadly ovate]; seed coat mucilaginous when wetted; cotyledons accumbent. $x = 10$.

Species 3 (1 in the flora): sw United States, nw Mexico.

Lyrocarpa xanti Brandegee (southern Baja California) is an annual with purple, spatulate petals, deltate and sparsely pubescent leaves, often reflexed fruiting pedicels, and panduriform fruits. By contrast, *L. linearifolia* Rollins (Isla Angel de la Guarda and San Esteban Island, Baja California) is a subshrub with brown, linear petals, linear or pinnatifid and canescent leaves, spreading fruiting pedicels, and broadly obcordate fruits.

SELECTED REFERENCE Rollins, R. C. 1941b. A revision of *Lyrocarpa*. Contr. Dudley Herb. 3: 169–173.

1. **Lyrocarpa coulteri** Hooker & Harvey, London J. Bot. 4: 76, plate 4. 1845 F

Lyrocarpa coulteri var. *apiculata* Rollins; *L. coulteri* var. *palmeri* (S. Watson) Rollins; *L. palmeri* S. Watson

Plants suffrutescent; moderately to densely pubescent throughout. **Stems** straw-colored, woody (at least proximally), many-branched, (2–)3–9(–11) dm. **Cauline leaves:** petiole 0.5–2.5(–4) cm (shorter distally); blades lanceolate, oblanceolate, or ovate in outline, (1–)2–9(–12) cm × (5–)10–50(–70) mm (smaller, less divided distally), base attenuate to cuneate, or truncate to hastate, margins usually pinnatisect to pinnatifid, or runcinate to dentate, rarely repand, teeth and lobes apiculate or not, sparsely to densely pubescent, nearly canescent. **Fruiting pedicels** divaricate to ascending or, rarely, reflexed, 2–7(–11) mm. **Flowers:** sepals (6–)8–10(–11) × 1–1.5 mm; petals (twisted with age), (13–)16–25 × 1–3 mm, claw 7–12 mm; filaments 4–6 mm; anthers 2.7–3.5 mm. **Fruits** (8–)10–25 × (8–)10–16 mm, base obtuse, apex deeply emarginate, (sinus to 8 mm deep), retuse, or truncate; style obsolete or to 0.5 mm. **Seeds** 2–3 mm diam. $2n$ = 36, 42.

Flowering late Sep–mid Apr. Gravelly or rocky slopes, desert washes, dry streambeds, sandy plains, shady ravines, granitic hills, mesa slopes, thorn scrub, stony ridges, gravelly arroyo beds, in open sun or under shrubs; 0–1300 m; Ariz., Calif.; nw Mexico (Baja California, Sonora).

Lyrocarpa coulteri, which is known from Pima and Yuma counties in Arizona and Imperial and San Diego counties in California, is extremely variable in nearly every aspect. The leaves range from relatively large, thin, and sparsely pubescent to smaller, thicker, and almost canescent; their shapes and margins are highly variable also. The flowers are strongly scented at night and appear to be moth-pollinated; their color varies a great deal. Fruit shape and size depend on the original number of ovules and how many of them mature into seeds, and one sometimes finds on the same plant fruits ranging from broadly obcordate to panduriform. Finally, the plants occupy a wide array of habitats; they can be tall, slender, and broad- and thin-leaved when grown in the shade of larger shrubs, whereas those growing in direct sunlight can be shorter, stout, and small- and thick-leaved.

With the above in mind, it is easy to understand why the two varieties recognized by R. C. Rollins (1941b, 1993) do not merit recognition. Variety *apiculata* was based on a fragmentary specimen with narrow petals and broad leaves apiculate at lobe apices and teeth. However, these two characters are uncorrelated. Similarly, var. *palmeri* was distinguished from var. *coulteri* by having broadly obcordate versus pandurifom fruits and 6–10 ovules, instead of 8–16 ovules, per ovary. As indicated above, fruit shape largely depends not on the number of ovules but on how many of them mature into seeds.

76. NERISYRENIA Greene, Pittonia 4: 225. 1900 • [Greek *neros*, flowing, and genus *Syrenia*, presumably alluding to resemblance]

Ihsan A. Al-Shehbaz

Greggia A. Gray, Smithsonian Contr. Knowl. 3(5): 8, plate 1. 1852, not Solander ex J. Gaertner 1788

Perennials [subshrubs]; (caudex woody); not scapose; usually pubescent, sometimes glabrous or glabrate, trichomes long-stalked to subsessile, dendritic. **Stems** (few to several from base), erect to ascending, branched distally. **Leaves** cauline; not rosulate; petiolate or sessile; blade (often fleshy), margins entire, dentate, or repand, sometimes revolute. **Racemes** (corymbose, several-flowered, lax). **Fruiting pedicels** divaricate to ascending, or, rarely, recurved, slender. **Flowers:** sepals spreading to reflexed, oblong or lanceolate to ovate [linear]; petals usually white (often fading lavender), rarely lavender, obovate to spatulate [broadly elliptic], (longer than sepals), claw well-differentiated from blade, (dilated and denticulate basally); stamens slightly tetradynamous, (somewhat spreading); anthers linear, (base sagittate, apex obtuse); nectar glands confluent, subtending bases of stamens. **Fruits** siliques or silicles, sessile, linear to oblong [obovoid], smooth [torulose], straight or curved, angustiseptate, terete, [or latiseptate]; valves each with distinct midvein, pubescent [glabrous]; replum rounded; septum complete; ovules 30–100 per ovary; style distinct; stigma conical, 2-lobed (lobes connivent, decurrent). **Seeds** uniseriate or biseriate, plump, not winged, oblong [elliptic]; seed coat mucilaginous when wetted; cotyledons incumbent. $x = 9, 10$

Species 7–9 (2 in the flora): s United States, n Mexico.

Parrasia Greene (1895), not Rafinesque (1837) is an illegitimate name, sometimes found in synonymy with *Nerisyrenia*.

SELECTED REFERENCE Bacon, J. D. 1978. Taxonomy of *Nerisyrenia* (Cruciferae). Rhodora 80: 159–227.

1. Leaf blades usually obovate to spatulate or oblanceolate, rarely elliptic, (4–)7–20(–30) mm wide, not fleshy .... 1. *Nerisyrenia camporum*
1. Leaf blades linear to linear-oblanceolate, 1–4.5 mm wide, fleshy .... 2. *Nerisyrenia linearifolia*

1. Nerisyrenia camporum (A. Gray) Greene, Pittonia 4: 225. 1900

Greggia camporum A. Gray, Smithsonian Contr. Knowl. 3(5): 9, plate 1. 1852; *Parrasia camporum* (A. Gray) Greene

Plants with woody caudex; moderately to densely pubescent or glabrate. **Stems** 1–6 dm, woody proximally, pubescent or glabrate with age. **Cauline leaves**: petiole 0.3–1.5 cm, often not sharply differentiated from blade, or (distal) sessile; blade usually obovate to spatulate or oblanceolate, rarely elliptic, 1–4(–4.5) cm × (4–)7–20(–30) mm (not fleshy), base attenuate, margins dentate, repand, or entire, apex acute to obtuse. **Racemes** 3.5 dm. **Fruiting pedicels** (rarely recurved), (7–)10–20 mm, densely pubescent. **Flowers**: sepals (spreading) oblong to oblong-lanceolate, 5–9 × 1–2 mm; petals obovate, (8–)10–14(–15) × 5–9 mm, claw (often flattened, margin dentate) to 2 mm; filaments 4.5–7.5 mm; anthers 2–3.5 mm. **Fruits** angustiseptate, 1.5–3(–4) cm × 1.5–3(–4) mm; ovules 40–100 per ovary; style (1.5–)2–3.8(–4.3) mm. **Seeds** 0.8–1.1 × 0.5–0.7 mm. $2n = 18, 19, 21, 22, 27, 32, 34, 36, 40, 41, 58$.

Flowering year-round, mostly Feb–Sep. Clay flats, gypseous clay, gravelly knolls, hillsides, sandy washes; 200–1300 m; N.Mex., Tex.; Mexico (Chihuahua, Coahuila, Durango, Nuevo León, Tamaulipas, Zacatecas).

2. Nerisyrenia linearifolia (S. Watson) Greene, Pittonia 4: 225. 1900 [F]

Greggia linearifolia S. Watson, Proc. Amer. Acad. Arts 18: 191. 1883; *G. camporum* A. Gray var. *angustifolia* J. M. Coulter; *G. camporum* var. *linearifolia* (S. Watson) M. E. Jones; *Parrasia linearifolia* (S. Watson) Greene

Plants with woody caudex; moderately to densely pubescent. **Stems** (0.5–)1–4 dm, woody proximally, densely to moderately pubescent. **Cauline leaves** sessile or nearly so; blade linear to linear-oblanceolate, 1.5–7 cm × 1–4.5 mm, (fleshy), base attenuate, margins usually entire, rarely dentate, apex acute to obtuse. **Racemes** to 3.5 dm in fruit. **Fruiting pedicels** 0.6–1.4 cm, densely pubescent. **Flowers**: sepals broadly lanceolate to ovate, 4.8–7.5 × 1–2 mm; petals obovate to spatulate, 8–12(–13) × 5–8.5 mm, (often flattened), claw to 2 mm, (margin dentate); filaments 4–6 mm; anthers 2.5–3.5 mm. **Fruits** terete to slightly angustiseptate, 0.9–3 cm × 1–2.2 mm; ovules 30–80 per ovary; style 0.9–4 mm. **Seeds** 0.5–0.7 × 0.4–0.5 mm. $2n = 18, 19, 20, 34, 36$.

Flowering Apr–Dec. Gypsum soils in knolls, bluffs, open flats; 1000–1200 m; N.Mex., Tex.; Mexico (Coahuila, Nuevo León, San Luis Potosí, Tamaulipas).

Both J. D. Bacon (1978) and R. C. Rollins (1993) recognized two weakly defined varieties of *Nerisyrenia linearifolia* distinguished primarily on the position of the widest portion of the fruit. Of those, var. *mexicana* Bacon (Coahuila, Mexico) has fruits widest near the base instead of the middle.

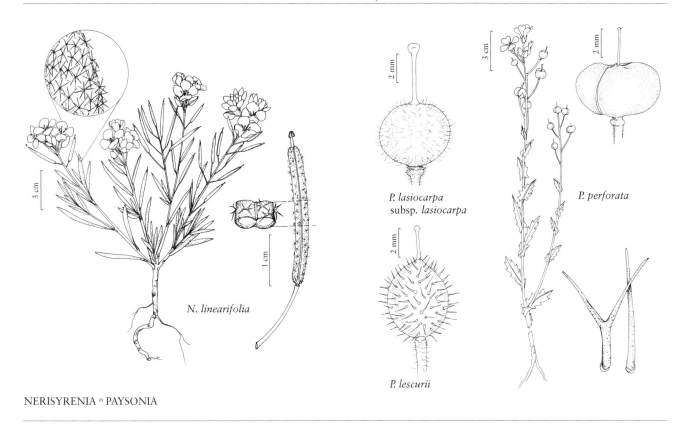

77. PAYSONIA O'Kane & Al-Shehbaz, Novon 12: 380. 2002 • [For Edwin Blake Payson, 1893–1927, American botanist and first monographer of *Lesquerella*]

Steve L. O'Kane Jr.

Annuals or, rarely, biennials or perennials; not scapose; pubescent or glabrous, trichomes short-stalked, forked or subdendritic, often mixed with coarse, simple ones. **Stems** erect, ascending, decumbent, or procumbent, unbranched or branched. **Leaves** basal and cauline; petiolate or sessile; basal rosulate, petiolate, blade margins entire, dentate, or pinnately lobed; cauline sessile, blade (base usually auriculate or sagittate, sometimes amplexicaul), margins dentate to lobed. **Racemes** (several-flowered), often strongly elongated in fruit, (rachis straight). **Fruiting pedicels** ascending to divaricate or recurved, (straight or slightly curved), slender. **Flowers:** sepals (deciduous), suberect, ascending, or spreading, oblong, (equal), lateral pair not saccate basally, (margins membranous); petals yellow or white, broadly obovate, claw slightly differentiated from blade, (apex rounded, truncate, retuse, or emarginate); stamens (erect), tetradynamous; filaments dilated or not basally; anthers oblong or sagittate, (not apiculate); nectar glands confluent, subtending bases of median stamens, surrounding lateral stamens. **Fruits** silicles, subsessile or sessile, globose, subglobose, subpyriform, suborbicular, orbicular, elliptic, cordate, or obovoid, terete, flattened, latiseptate, or angustiseptate; valves (papery or leathery), not veined, glabrous or pubescent, (trichomes branched and/or simple); replum not flattened; septum complete, perforated, or reduced to a rim, (membranous); ovules 4–40 per ovary; style distinct, (persistent, slender); stigma capitate, entire. **Seeds** sub-biseriate, flattened, margined, orbicular, suborbicular, or oval; cotyledons accumbent. $x = 7, 8, 9$.

Species 8 (8 in the flora): sc, se United States, n Mexico.

Paysonia was segregated from *Lesquerella* (now *Physaria*) based on its different base chromosome number, presence of auriculate stem leaves, typically annual duration, trichomes that are neither stellate nor tuberculate, and phylogenetic analysis of DNA sequences. Some species of the genus have very limited distributions and are of conservation concern. Five (*P. densipila*, *P. lescurii*, *P. lyrata*, *P. perforata*, and *P. stonensis*) are known to be interfertile under experimental conditions, and some combinations form fertile hybrids in the field (R. C. Rollins 1988).

SELECTED REFERENCES Rollins, R. C. 1955. The auriculate-leaved species of *Lesquerella* (Cruciferae). Rhodora 57: 241–264. Rollins, R. C. and E. A. Shaw. 1973. The Genus *Lesquerella* (Cruciferae) in North America. Cambridge, Mass.

1. Fruits orbicular, elliptic, or cordate.
   2. Fruits angustiseptate; fruiting pedicels recurved; cauline leaf blades: base cuneate to auriculate, not amplexicaul; filaments not dilated basally . . . . 4. *Paysonia lasiocarpa*
   2. Fruits latiseptate; fruiting pedicels divaricate or nearly straight; cauline leaf blades: base auriculate to sagittate, sometimes amplexicaul; filaments dilated basally . . . . 5. *Paysonia lescurii*
1. Fruits globose, subglobose, subpyriform, or obovoid.
   3. Petals white or pale lavender; septums perforate or nearly absent.
      4. Fruits subpyriform, valves sparsely hirsute or glabrate; styles glabrous . . . . 7. *Paysonia perforata*
      4. Fruits subglobose, valves densely hirsute; styles hirsute, at least proximally . . . . 8. *Paysonia stonensis*
   3. Petals yellow; septums complete.
      5. Fruit valves pubescent; styles glabrous or pubescent proximally . . . . 2. *Paysonia densipila*
      5. Fruit valves and styles glabrous.
         6. Fruits 3–4 mm; ovules 4–8 per ovary; septums opaque . . . . 6. *Paysonia lyrata*
         6. Fruits 4–6(–8) mm; ovules (8–)12–28(–40) per ovary; septums translucent.
            7. Stems hirsute proximally, trichomes simple; petals 4–5 mm wide . . . . . 1. *Paysonia auriculata*
            7. Stems pubescent proximally, trichomes branched; petals 6–9 mm wide
               . . . . 3. *Paysonia grandiflora*

1. **Paysonia auriculata** (Engelmann & A. Gray) O'Kane & Al-Shehbaz, Novon 12: 380. 2002 • Earleaf bladderpod E

Vesicaria auriculata Engelmann & A. Gray, Boston J. Nat. Hist. 5: 240. 1845; *Alyssum auriculatum* (Engelmann & A. Gray) Kuntze; *Lesquerella auriculata* (Engelmann & A. Gray) S. Watson

Annuals; hirsute or glabrous, trichomes relatively long, simple, with smaller, branched ones. **Stems** erect or decumbent at base, (stout), 0.5–2 dm, (hirsute proximally, trichomes simple). **Basal leaves**: blade 2–5 cm × 8–15 mm, margins lyrate to sinuate-dentate or entire, (apex obtuse to subacute), surfaces often pubescent (trichomes branched, smaller), or hirsute on much of adaxial surface, margins, and midrib (trichomes simple). **Cauline leaves** (overlapping); blade oblong to sagittate, 1–4 cm × 3–10 mm, base auriculate, margins entire or dentate. **Fruiting pedicels** divaricate-ascending to divergent (ca. 45°), slightly curved, 7–15 mm, hirsute (trichomes spreading, simple and branched). **Flowers**: sepals 4–6 × 1.5–2 mm, hirsute (trichomes spreading, simple and branched); petals yellow, 7–10 × 4–5 mm, (apex rounded or slightly emarginate); filaments abruptly dilated basally. **Fruits** subsessile, globose or subglobose, (longer than broad), 4–6(–8) × 4–6 mm; valves glabrous; replum not flattened; septum complete, (translucent); ovules (8–)12–16(–20) per ovary; style 1.5–2 mm, glabrous; stigma expanded. **Seeds** suborbicular, ca. 2 mm. $2n = 16$.

Flowering Mar–May. Bluffs, prairies, pastures, limestone outcrops, disturbed soils of banks and roadsides; Kans., Okla., Tex.

2. **Paysonia densipila** (Rollins) O'Kane & Al-Shehbaz, Novon 12: 380. 2002 • Duck River bladderpod E

Lesquerella densipila Rollins, Rhodora 54: 186, fig. 1. 1952

Annuals; pubescent or glabrous, trichomes simple, with branched, smaller ones. **Stems** erect, or outer ones decumbent at base, (purplish proximally), 1–4 dm. **Basal leaves**: blade 4–8 cm × 10–18 mm, margins lyrately pinnatifid to pinnately lobed, (apex obtuse, lateral lobes decurrent on rachis, terminal lobes relatively large), surfaces hirsute (abaxial with simple and branched trichomes, adaxial with mostly simple ones). **Cauline leaves:** blade broadly ovate to oblong, 1–3 cm × 5–15 mm (smaller distally), base auriculate, margins dentate to nearly lobed, (proximal with apex broadly obtuse, distal more acute, surfaces hirsute, trichomes mostly simple, spreading). **Fruiting pedicels** divaricate-ascending, straight, (expanded apically), 10–20 mm, pubescent. **Flowers:** sepals (yellowish), 2.5–4 × 1.5–2 mm, sparsely to densely pubescent (trichomes appressed, branched, often with some spreading, single ones); petals yellow, 6–8 × 4–5 mm, (not markedly differentiated into blade and claw), apex rounded; filaments strongly dilated basally, (anthers oblong, 1.5 mm, glandular tissue thin, continuous beneath stamens, forming projections between single and paired stamens and abbreviated ring around bases of single stamens). **Fruits** subsessile, subglobose, (slightly broader than long), 3–4 mm diam., (uncompressed); valves densely pubescent, trichomes spreading, simple or branched; replum ± orbicular; septum complete; ovules 4–8 per ovary; style 2–3 mm, glabrous or pubescent proximally; stigma slightly expanded. **Seeds** orbicular or suborbicular (or slightly longer than broad), 2–2.5 mm. $2n = 16$.

Flowering Mar–May. Cedar glades with thin soil over limestone, open alluvial sites, stream bottoms, fallow fields; Ala., Tenn.

3. **Paysonia grandiflora** (Hooker) O'Kane & Al-Shehbaz, Novon 12: 381. 2002 • Bigflower bladderpod [E]

Vesicaria grandiflora Hooker, Bot. Mag. 63: plate 3464. 1836; *Alyssum grandiflorum* (Hooker) Kuntze; *Lesquerella grandiflora* (Hooker) S. Watson; *V. brevistyla* Torrey & A. Gray; *V. grandiflora* var. *pinnatifida* A. Gray

Annuals; mostly densely pubescent, sometimes glabrous, trichomes usually 5-parted. **Stems** erect to decumbent at base, (unbranched or branched), 2–7 dm, (pubescent proximally, trichomes branched). **Basal leaves:** blade (oblanceolate), 5–15 cm, margins irregularly dentate to 2-pinnatifid, (apex acute to obtuse), surfaces densely pubescent (trichomes erect, 5-parted). **Cauline leaves** (usually overlapping); blade oblong to lanceolate, 1–4 cm, base (proximal) often narrowed and cuneate, (distal) auriculate, margins dentate to toothed, (apex acute to obtuse, densely pubescent on midrib, trichomes branched). **Fruiting pedicels** ascending to divaricate, straight to slightly curved upward, 10–20 mm, densely pubescent. **Flowers:** sepals 5–7 × 2–3 mm, densely pubescent; petals yellow, 8–12 × 6–9 mm, (abruptly narrowed to a short claw), apex rounded to slightly emarginate; filaments dilated basally, (3–4 mm, anthers sagittate, 2.5–3 mm, glandular tissue projecting, subtending paired stamens, nearly surrounding bases of single stamens). **Fruits** sessile or subsessile, globose or subglobose (slightly longer than broad), 4–6 × 4–5 mm; valves glabrous; replum not flattened; septum complete, (translucent); ovules (8–)16–28(–40) per ovary, (attached in proximal ½); style 1–1.5 mm, glabrous; stigmas expanded. **Seeds** orbicular, 2.5–3 mm. $2n = 18$.

Flowering Mar–May. Dry areas, roadsides, meadows, rocky slopes, scrubland, with sandy, loose, well-drained soils; Tex.

Paysonia grandiflora probably grows also in northeastern Mexico.

4. **Paysonia lasiocarpa** (Hooker ex A. Gray) O'Kane & Al-Shehbaz, Novon 12: 381. 2002 • Roughpod bladderpod [F]

Vesicaria lasiocarpa Hooker ex A. Gray, Smithsonian Contr. Knowl. 5(6): 13. 1853; *Lesquerella lasiocarpa* (Hooker ex A. Gray) S. Watson

Annuals, biennials, or perennials; pubescent, trichomes branched, or mixed with simple, larger ones. **Stems** decumbent to procumbent, (slender), 1–5 dm. **Basal leaves:** blade (oblanceolate), 3–10 cm × 10–30 mm, margins sinuate-dentate to somewhat lobed or incised, surfaces densely pubescent. **Cauline leaves:** blade obovate-elliptic to oblong, 1–4 cm × 5–20 mm, base cuneate to auriculate (narrowed toward base), margins sinuate-dentate to incised, (surfaces densely pubescent). **Fruiting pedicels** recurved, slightly curved, 1–3 cm, (racemes lax in fruit), densely pubescent. **Flowers:** sepals (green), 4–5.5 × 1.5–2 mm, (nearly acuminate), pubescent or hirsute (trichomes branched); petals light yellow (often drying purplish), 5–9 × 4–6 mm, (claw relatively short), apex rounded; filaments not dilated basally, (2.5–3.5 mm, anthers sagittate, 2–3 mm, glands with horn-like projections at base of filaments). **Fruits** sessile, orbicular, elliptic, or cordate, 5–9 × 4–9 mm, (angustiseptate or slightly so); valves (rounded or strongly keeled), densely pubescent, trichomes branched, sometimes with simple, larger ones; replum oblong to nearly elliptical; septum complete, (transparent); ovules 14–32 per ovary, (attached in proximal ½); style 1–1.5 mm; stigma expanded. **Seeds** suborbicular (slightly longer than broad), 1–2.2 mm.

Subspecies 2 (2 in the flora): s United States, n Mexico.

1. Annuals; trichomes simple and branched; fruits inflated, valves rounded ....
.... 4a. *Paysonia lasiocarpa* subsp. *lasiocarpa*
1. Annuals, biennials, or perennials; trichomes usually simple, sometimes mixed with smaller, branched ones; fruits not inflated, valves somewhat to strongly keeled...............
.... 4b. *Paysonia lasiocarpa* subsp. *berlandieri*

4a. Paysonia lasiocarpa (Hooker ex A. Gray) O'Kane & Al-Shehbaz subsp. lasiocarpa F

Annuals; trichomes simple and branched. **Cauline leaves:** blade base auriculate. **Fruits** inflated (often subglobose); valves rounded (not keeled). $2n = 14$.

Flowering Jan–Apr. Inland areas on sandy, gravelly, and silty clay soils, roadsides, brushland, open hills; 0–200 m; Tex.; Mexico (Coahuila, Nuevo León, Tamaulipas).

4b. Paysonia lasiocarpa (Hooker ex A. Gray) O'Kane & Al-Shehbaz subsp. berlandieri (A. Gray) O'Kane & Al-Shehbaz, Novon 12: 381. 2002 • Berlandier bladderpod

Synthlipsis berlandieri A. Gray in W. H. Emory, Rep. U.S. Mex. Bound. 2(1): 34. 1859; *Lesquerella lasiocarpa* (Hooker ex A. Gray) S. Watson subsp. *berlandieri* (A. Gray) Rollins & E. A. Shaw; *L. lasiocarpa* var. *ampla* Rollins; *L. lasiocarpa* var. *berlandieri* (A. Gray) Payson; *L. lasiocarpa* var. *hispida* (S. Watson) Rollins & E. A. Shaw; *S. berlandieri* var. *hispida* S. Watson

Annuals, biennials, or perennials; trichomes usually simple, sometimes mixed with branched, smaller ones. **Cauline leaves:** blade base cuneate to auriculate. **Fruits** not inflated; valves somewhat to strongly keeled. $2n = 14$.

Flowering Feb–May. Coastal plains, near beaches, open chaparral, sandy areas and flats with black, clayey soils; 0–200 m; Tex.; Mexico (Nuevo León, Tamaulipas, Veracruz).

5. Paysonia lescurii (A. Gray) O'Kane & Al-Shehbaz, Novon 12: 381. 2002 • Lescur's bladderpod E F

Vesicaria lescurii A. Gray, Manual ed. 2, 38. 1856; *Alyssum lescurii* (A. Gray) A. Gray; *Lesquerella lescurii* (A. Gray) S. Watson

Annuals; hirsute to coarsely pubescent or glabrous, trichomes simple, with branched, smaller ones. **Stems** erect or decumbent, 1–3 dm. **Basal leaves:** blade 3–7 cm × 5–20 mm, margins lyrate, deeply lobed (lateral lobes remote), abaxial surface hirsute (trichomes mostly simple), adaxial coarsely pubescent (trichomes simple, with branched, smaller ones). **Cauline leaves:** blade broadly oblong to ovate, 0.5–2 cm × 3–10 mm, base auriculate to sagittate, sometimes amplexicaul, margins dentate, (surfaces coarsely pubescent, trichomes simple, with branched, smaller ones, often simple ones absent distally). **Fruiting pedicels** divaricate, nearly straight (not expanded apically), 8–15 mm, densely pubescent (trichomes coarse, branched). **Flowers:** sepals (yellowish at anthesis), 3–4 × 1.5–2 mm, (outer pair slightly saccate, inner pair flat basally), apex rounded to slightly emarginate, coarsely pubescent; petals yellow, 5–7 × 3.5–4.5 mm, (gradually narrowed to base); filaments dilated basally, (2.5–3.5 mm, glandular tissue continuous, subtending filament base, surrounding attachment point of single stamens). **Fruits** sessile, orbicular or suborbicular (slightly longer than broad), 4–6 × 3–4 mm, (strongly latiseptate); valves coarsely, densely pubescent outside, trichomes simple, bulbous-based, mixed with branched ones, sparsely pubescent inside, trichomes branched; replum not flattened, (thick, glabrous); septum complete, (dense); ovules 4–8 per ovary; style 1.5–2 mm, glabrous; stigma expanded in dried material, same diameter as style when fresh. **Seeds** suborbicular (slightly longer than broad), 2–2.8 × 1.5–2 mm. $2n = 16$.

Flowering Mar–May. Hillsides, cedar glades, flood plains, fields, pastures; Ala., Ky., Tenn.

6. Paysonia lyrata (Rollins) O'Kane & Al-Shehbaz, Novon 12: 381. 2002 • Lyrate or lyreleaf bladderpod C E

Lesquerella lyrata Rollins, Rhodora 57: 252, plate 1209. 1955

Annuals; hirsute or glabrous, trichomes simple proximally, simple, with branched, smaller ones distally. **Stems** erect, outer decumbent at base, 1–3 dm. **Basal leaves:** blade 2–7 cm × 6–15 mm, margins lyrate, (terminal lobes relatively large, orbicular to elliptic), surfaces hirsute (trichomes simple, or mixed with branched, smaller ones). **Cauline leaves:** blade ovate to broadly oblong, 0.5–2 cm × 4–10 mm, base auriculate, clasping, margins nearly entire or coarsely dentate, (apex obtuse, surfaces densely to sparsely hirsute, trichomes simple, or simple and branched). **Fruiting pedicels** divaricate-ascending, straight, 10–15 mm, densely pubescent. **Flowers:** sepals (spreading, yellowish at anthesis), 3–4 × 1.2–1.5 mm, (outer pair slightly saccate basally), pubescent; petals yellow, 5–7 × 3.5–4 mm, (claw relatively short), apex slightly rounded, truncate, or shallowly retuse; filaments dilated basally, (glandular tissue subtending paired stamens, surrounding single stamens). **Fruits** sessile, subglobose, (often slightly depressed at base of style and slightly didymous), 3–4 × 3–4 mm; valves glabrous; replum orbicular or suborbicular, slightly broader than high; septum complete, (opaque); ovules 4–8 per ovary; style 1–1.5 mm, glabrous; stigma not expanded, nearly same diameter as style. **Seeds** oval to suborbicular, 1.5–2.5 mm. $2n = 16$.

Flowering Mar–Apr. Cedar glades, limestone hills, red soils, open pastures, disturbed soils of roadsides, bottom lands, old fields; of conservation concern; Ala.

7. **Paysonia perforata** (Rollins) O'Kane & Al-Shehbaz, Novon 12: 381. 2002 • Spring Creek bladderpod
C E F

Lesquerella perforata Rollins, Rhodora 54: 190. 1952

Annuals; pubescent, hirsute, or glabrate, trichomes simple, with branched, smaller ones. **Stems** erect, outer ones usually decumbent at base, 1–1.5 dm. **Basal leaves:** blade 2–5 cm × 5–15 mm, margins lyrately lobed, (lobes) entire or dentate, (lateral lobes broadly oblong, margins becoming remote basally, terminal lobes orbicular or ovate, apex obtuse to more pointed), surfaces hirsute (trichomes mostly simple, marginal ones branched, smaller). **Cauline leaves:** blade broadly oblong to nearly ovate, 0.8–2 cm × 4–8 mm, base auriculate, sagittate, margins dentate, (abaxial surface with simple and branched trichomes, adaxial surface hirsute, trichomes simple). **Fruiting pedicels** divaricate-ascending, straight, (scarcely swollen apically), 6–12 mm, uniformly pubescent (trichomes branched, or simple and branched). **Flowers:** sepals 3.5–5 × 1.5–2 mm, pubescent (trichomes relatively large and small, branched); petals white to pale lavender (with pale yellow claw, sometimes tinged light purple when dry), 7–9 × 5–6 mm, apex emarginate to rounded; filaments dilated basally, (anthers oblong, 1.3–1.5 mm, glandular tissue subtending all filaments, nearly surrounding those of single stamens, with projections between single and paired stamens). **Fruits** subsessile, broadly obovoid to subpyriform, (inflated, papery), 4–6 × 4 mm, (widest distal to middle); valves sparsely hirsute or glabrate outside, trichomes simple or branched, densely pubescent inside, trichomes dendritic; replum not flattened; septum perforated to nearly absent, consisting of a narrow rim; ovules 4–12 per ovary; style 1.5–2.5 mm, glabrous; stigma not or slightly expanded. **Seeds** suborbicular or orbicular, 1.2–2.5 × 1.2–2 mm. $2n = 16$.

Flowering Mar–May. Open fields, pastures, flood plains, roadsides; of conservation concern; 100–200 m; Tenn.

Paysonia perforata has a very limited distribution in Rutherford and Wilson counties. It is in the Center for Plant Conservation's National Collection of Endangered Plants.

8. **Paysonia stonensis** (Rollins) O'Kane & Al-Shehbaz, Novon 12: 381. 2002 • Stones River bladderpod
C E

Lesquerella stonensis Rollins, Rhodora 57: 255, plate 1210. 1955

Annuals; densely hirsute, trichomes simple proximally, densely pubescent distally, trichomes simple, or mixed simple, forked, and slightly branched. **Stems** erect, outer ones usually decumbent at base, 2–4 dm. **Basal leaves:** blade 3–6 cm × 8–15 mm, margins lyrately lobed to pinnatifid, (lobes) entire or sinuate-dentate (lateral lobes decurrent on leaf rachis, triangular to broadly oblong, terminal lobes ovate to nearly orbicular, relatively large, apex obtuse), surfaces densely hirsute (abaxial with simple, relatively long trichomes, mixed with forked or branched, shorter ones, adaxial with simple trichomes). **Cauline leaves:** blade broadly oblong to ovate, 1–5 cm × 5–15 mm, base auriculate, clasping, margins dentate, (proximal surfaces densely hirsute, abaxial with simple, forked, and branched trichomes, adaxial with predominantly simple ones, distal surfaces densely pubescent, trichomes predominantly forked or branched). **Fruiting pedicels** divaricate-ascending, straight, 10–25 mm, densely pubescent. **Flowers:** sepals 4.5–5.5 × 1.5–2 mm, (outer pair slightly saccate, inner pair narrower, not saccate), pubescent; petals white, (claw pale yellow, short), 7–9 × 5–6 mm, apex rounded to emarginate; filaments dilated basally, (glandular tissue continuous, surrounding insertion point of single stamens, subtending that of paired stamens). **Fruits** subsessile, subglobose, slightly didymous, 3–5 × 4–5 mm; valves densely hirsute, trichomes simple; replum orbicular or suborbicular (slightly wider than long, rounded at apex); septum usually perforate, sometimes complete; ovules 8–12 per ovary; style ca. 2 mm, hirsute at least proximally; stigma expanded. **Seeds** oval, 1.8–2 × 1.5 mm. $2n = 16$.

Flowering Mar–May. Flood plains, knoll tops, pastures, fields, roadsides, stream banks; of conservation concern; Tenn.

Paysonia stonensis has a very limited distribution in the watershed of the East Fork of the Stones River, where it is known to form hybrids with *P. densipila*. The hybrid has been named *Lesquerella* ×*maxima* Rollins.

Paysonia stonensis is in the Center for Plant Conservation's National Collection of Endangered Plants.

78. **PHYSARIA** (Nuttall ex Torrey & A. Gray) A. Gray, Gen. Amer. Bor. 1: 162. 1848
    • Bladderpod [Greek *physa*, bladder, alluding to inflated fruits of some species]

Steve L. O'Kane Jr.

Vesicaria Adanson sect. *Physaria* Nuttall ex Torrey & A. Gray, Fl. N. Amer. 1: 102. 1838; *Coulterina* Kuntze; *Lesquerella* S. Watson

Annuals, biennials, or perennials; (caudex often present, enlarged, usually branched); not scapose; usually pubescent, trichomes usually sessile, sometimes subsessile or shortly stalked, usually stellate, sometimes stellate-scalelike, rarely simple. **Stems** erect, spreading, decumbent, or prostrate, unbranched or branched distally. **Leaves** basal and cauline; petiolate or sessile; basal usually rosulate, petiolate, blade margins usually entire, sometimes repand to pinnatifid; cauline petiolate or sessile, blade margins usually entire, sometimes repand to dentate. **Racemes** (few- to several-flowered, proximalmost flowers rarely bracteate), elongated or not in fruit. **Fruiting pedicels** erect, horizontal, divaricate, spreading, ascending, or sigmoid, usually slender, sometimes stout. **Flowers:** sepals erect or spreading, linear, lanceolate, elliptic, oblong, ovate, or deltate, lateral pair usually saccate basally, sometimes subsaccate or not saccate; petals usually yellow, sometimes orange (occasionally drying purplish or maroon), rarely white or purple, spatulate, obovate, ovate, oblanceolate, or obdeltate, (longer than sepals), claw differentiated or not from blade, (apex usually rounded, rarely slightly emarginate); stamens tetradynamous; filaments usually not dilated basally; anthers ovate to narrowly oblong, (apex usually obtuse); nectar glands confluent, subtending bases of stamens, median glands present or absent. **Fruits** silicles, sessile, subsessile or, rarely, shortly stipitate, globose, subglobose, orbicular, suborbicular, ellipsoid, elliptic, lanceolate, obcordate, obdeltate, oblong, obpyriform, ovate, ovoid, or obovoid, not torulose, inflated or not, terete, latiseptate, or angustiseptate; valves each often with obscure midvein, (usually not retaining seeds after dehiscence), usually pubescent, sometimes glabrous, or, rarely, pubescent inside; replum rounded to narrowly oblong; septum usually complete, sometimes perforate, or, rarely, reduced to a rim (often with apical midvein extending to center); ovules (2–)4–32(–40[–80]) per ovary; style distinct; stigma entire. **Seeds** biseriate, often flattened, sometimes plump, rarely lenticular, usually not winged, rarely narrowly winged or margined, often suborbicular; seed coat (smooth), mucilaginous or not when wetted; cotyledons accumbent. $x = 4, 5, 6, 7, 8, 9, 10, 12, 15$.

Species 106 (88 in the flora): North America, n Mexico, s South America (Argentina, s Bolivia), Asia (ne Russia).

Seeds of *Physaria* contain hydroxy fatty acids, and some species, notably *P. fendleri*, are being intensively studied as a source of specialized, high-quality lubricants. The genus is notable for its relatively large number of local, often endangered, and edaphically-determined (usually calciphilic) endemics. Most of the genus *Lesquerella* (except for the auriculate-leaved species placed in *Paysonia* O'Kane & Al-Shehbaz) was recently united with *Physaria*, which is now much larger but monophyletic and morphologically coherent (I. A. Al-Shehbaz and S. L. O'Kane 2002). Where details of the trichomes are sparse, this is due to an absence of electron microscopy for these taxa (although the number of primary rays and some details are often visible at 10–30× with glancing light). The number of rays of the ubiquitous unicellular, stellate trichomes refers to the primary divisions immediately proximal to the center of the trichome. These primary rays are then typically furcate or bifurcate, often imperfectly so (2 + 1 branches, rather than 2 + 2). Trichome rays are usually appressed or parallel to surfaces on a short stalk; when the rays flare from surfaces, this is noted in the descriptions. Umbonate trichomes have a distinctive raised mound at the center; unless otherwise stated, trichomes are not umbonate. Tubercles are bumps or granules scattered along the rays and often over the center of the trichome. Flowering is likely to occur earlier than indicated in the descriptions, because specimens are typically (and optimally) collected when the fruits are nearly or fully mature, rather than when plants are only in flower. Raceme descriptions refer to mature infructescences unless otherwise noted. In *Physaria*, didymous is used as a term for fruit shape. A didymous fruit is inflated and appears as two balloons pressed together. The overall fruit shape is otherwise difficult to define and authors have traditionally referred to it this way. Replum shape and placement of the valve orifice are reported for only those taxa that were traditionally placed in *Physaria*, in the strict sense, where replum shape is sometimes helpful in separating species. The valves of didymous-fruited *Physaria* typically do not release their seeds; the valves and seeds disperse as a unit; species previously placed in *Lesquerella* freely release their seeds. The number of ovules reported, easily ascertained by counting funiculi, is most often greater than the number of seeds that occupy the fruit, either because of abortion or lack of fertilization. Species with mucilaginous seeds typically occupy steep and/or unstable habitats; when wetted, mucilaginous seeds can be "glued" onto optimal, local habitats.

SELECTED REFERENCES Al-Shehbaz, I. A. and S. L. O'Kane. 2002. *Lesquerella* is united with *Physaria*. Novon 12: 319–329. Mulligan, G. A. 1968. *Physaria didymocarpa*, *P. brassicoides*, and *P. floribunda* (Cruciferae) and their close relatives. Canad. J. Bot. 46: 735–740. Payson, E. B. 1921. A monograph of the genus *Lesquerella*. Ann. Missouri Bot. Gard. 8: 103–236. Rollins, R. C. 1939. The cruciferous genus *Physaria*. Rhodora 41: 392–415. Rollins, R. C. 1939b. Studies in the genus *Lesquerella*. Amer. J. Bot. 26: 419–421. Rollins, R. C. 1981b. Studies in the genus *Physaria* (Cruciferae). Brittonia 33: 332–341. Rollins, R. C. and U. C. Banerjee. 1975. Atlas of the trichomes of *Lesquerella* (Cruciferae). Publ. Bussey Inst. Harvard Univ. 1975: 1–48. Rollins, R. C. and U. C. Banerjee. 1979b. Trichome patterns in *Physaria*. Publ. Bussey Inst. Harvard Univ. 1979: 65–77. Rollins, R. C. and E. A Shaw. 1973. The Genus *Lesquerella* (Cruciferae) in North America. Cambridge, Mass.

1. Fruits papery (coriaceous in *P. bellii*, *P. rollinsii*), inflated (often double, didymous); valves retaining seeds after dehiscence, basal sinus usually present, sometimes absent; replum narrower than fruit (traditional *Physaria* in the strict sense).
    2. Fruits somewhat inflated, not didymous, angustiseptate . . . . 33. *Physaria geyeri*
    2. Fruits inflated, strongly didymous at least apically, not angustiseptate (except *P. alpestris*).
        3. Petals white . . . . 25. *Physaria eburniflora*
        3. Petals yellow (sometimes drying purplish) or yellow and purple-tinged.
            4. Trichomes stellate-scalelike [rays fused (webbed)].
                5. Fruits strongly didymous, highly inflated, basal sinus shallow; sc Utah
                . . . . 47. *Physaria lepidota*
                5. Fruits slightly didymous, slightly inflated, basal sinus absent; Piceance Basin, Colorado . . . . 58. *Physaria obcordata*
            4. Trichomes stellate.
                6. Fruiting pedicels recurved; fruits nearly pendent; basal leaf blades: margins dentate or pinnatifid (rarely subentire) . . . . 30. *Physaria floribunda*
                6. Fruiting pedicels not recurved; fruits erect, spreading, ascending, not pendent; basal leaf blades: margins entire, few-toothed, dentate, repand, or lyrate-lobed.

[7. Shifted to left margin.—Ed.]
7. Fruit valves keeled (2-keeled on side away from replum).
    8. Valves with sides flat or slightly convex, keels rounded, apical sinus V-shaped or convex; styles (4–)6–8 mm... 15. *Physaria chambersii*
    8. Valves with sides concave, keels sharp-angled, apical sinus concave; styles 2–9 mm
        .... 57. *Physaria newberryi*
7. Fruit valves rounded or irregular, not keeled.
    9. Styles less than 3 mm; fruits: basal sinus absent........................... 61. *Physaria oregona*
    9. Styles greater than 3 mm; fruits: basal sinus present or absent.
        10. Fruits strongly angustiseptate (at least toward replum); replums: apex acute to acuminate.. 2. *Physaria alpestris*
        10. Fruits not angustiseptate; replums: apex usually obtuse.
            11. Petals 10–12(–15) mm; fruits irregularly angled and roughened, not strongly inflated; plants of alpine Colorado.................................... 3. *Physaria alpina*
            11. Petals 6–14 mm; fruits irregular in shape, or suborbicular, subglobose, cordate, or angular, usually inflated; plants usually not alpine (if so, fruits rounded, strongly inflated).
                12. Fruits: basal sinus obscure or absent, apical sinus deeper.
                    13. Fruit valves: trichomes with spreading rays (ovaries and immature fruits fuzzy).
                        14. Blades of younger basal leaves tapered to petioles; apical sinus of fruits usually broad (often giving fruits a flared appearance) ...
                            .... 11. *Physaria brassicoides*
                        14. Blades of all basal leaves abruptly narrowed to petioles; apical sinus of fruits narrow........................... 78. *Physaria saximontana*
                    13. Fruit valves: trichomes with appressed or spreading rays (ovaries and immature fruits silvery, not fuzzy).
                        15. Plants compact; stems decumbent; fruiting pedicels straight ... 77. *Physaria rollinsii*
                        15. Plants not usually compact; stems decumbent to ascending; fruiting pedicels sigmoid.. 88. *Physaria vitulifera*
                12. Fruits: basal and apical sinuses well-developed, usually ± equal.
                    16. Ovules mostly 4 per ovary.
                        17. Fruit valves: trichomes with ascending rays (appearing fuzzy).
                            18. Replum obovate to broadly oblong; Idaho, Montana, Washington, Wyoming................ 22. *Physaria didymocarpa* (in part)
                            18. Replum oblong to oblanceolate; c, ec Utah............. 38. *Physaria grahamii*
                        17. Fruit valves: trichomes with appressed rays (not appearing fuzzy).
                            19. Basal leaf blades: base abruptly narrowed to petioles, margins usually entire, rarely with scattered teeth, apex rounded or obtuse (sometimes with apical mucro)................ 1. *Physaria acutifolia*
                            19. Basal leaf blades: base gradually tapering to petioles, margins shallowly dentate, apex obtuse...................... 10. *Physaria bellii*
                    16. Ovules (4–)8(–12) per ovary.
                        20. Plants compact (from condensed rosette); racemes barely exceeding leaves.
                            21. Basal leaves mostly horizontal, blades 0.5–1.5 cm × 40–80 mm; replums 3–4 mm......................................17. *Physaria condensata*
                            21. Basal leaves ascending or erect, blades (1.5–)5–7 cm × 12–20 mm; replums 1–1.8 mm............................. 23. *Physaria dornii*
                        20. Plants loose (cespitose); racemes greatly exceeding leaves.
                            22. Leaf blade margins usually dentate or repand, bases ± abruptly narrowing to petioles................ 22. *Physaria didymocarpa* (in part)
                            22. Leaf blade margins entire, bases mostly abruptly tapering to petioles..42. *Physaria integrifolia*

1. Fruits firm, not or slightly inflated (not didymous); valves not retaining seeds after dehiscence, basal sinus absent; replum usually as wide as or wider than fruit (traditional *Lesquerella* in the strict sense).
   23. Fruit valves glabrous outside.
       24. Annuals or biennials (usually with a fine taproot).
           25. Petals white . . . . 64. *Physaria pallida*
           25. Petals yellow (sometimes drying purplish), yellow and purple-tinged, or orange.
               26. Basal and cauline leaves similar in shape.
                   27. Fruits sessile or shortly stipitate. . . . . 4. *Physaria angustifolia*
                   27. Fruits stipitate (gynophores 0.5–2 mm).
                       28. Gynophores 0.5–1 mm; fruiting pedicels sigmoid . . . . 36. *Physaria gordonii* (in part)
                       28. Gynophores 1–2 mm; fruiting pedicels straight or slightly curved
                           . . . . 37. *Physaria gracilis*
               26. Basal and cauline leaves different in shape.
                   29. Fruiting pedicels recurved . . . . 75. *Physaria recurvata*
                   29. Fruiting pedicels usually erect, straight, ascending, or sigmoid.
                       30. Ovules 4 per ovary . . . . 29. *Physaria filiformis*
                       30. Ovules 8–20 per ovary.
                           31. Fruit valves papillose outside, densely pubescent inside . . . . 80. *Physaria sessilis*
                           31. Fruit valves smooth outside, glabrous inside.
                               32. Pedicels straight or slightly curved; trichomes with a
                                   U-shaped notch on one side . . . . 21. *Physaria densiflora*
                               32. Pedicels sigmoid, horizontal, or recurved; trichomes
                                   without a U-shaped notch . . . . 49. *Physaria lindheimeri*
       24. Perennials or, rarely, biennials (with caudex, except *P. gordonii*).
           33. Arctic and subarctic (Alaska, Alberta, British Columbia, Greenland, Manitoba, Newfoundland and Labrador, Northwest Territories, Nunavut, Quebec, Yukon).
               34. Petals spatulate, blade gradually narrowed to claw, 5–6(–7) mm; fruits mostly uncompressed. . . . . 5. *Physaria arctica* (in part)
               34. Petals obovate, blade abruptly narrowed to claw, (6–)7–10 mm; fruits usually compressed (angustiseptate). . . . . 13. *Physaria calderi*
           33. Dry temperate or subtropical areas (not Alaska, Alberta, British Columbia, Greenland, Manitoba, Newfoundland and Labrador, Northwest Territories, Nunavut, Quebec, Yukon).
               35. Plants pulvinate-cespitose, or forming mounds, mats, or tufts.
                   36. Petioles and leaf blades differentiated (or slightly so); plants forming soft mats or tufts. . . . . 40. *Physaria hitchcockii*
                   36. Petioles and leaf blades undifferentiated; plants forming hard mats.
                       37. Seeds strongly mucilaginous when wetted; petals deep yellow and slightly orange in center, blade and claw joined at right angle; sepals elliptic; San Juan and McKinley counties, New Mexico, Apache County, Arizona. . . . . 55. *Physaria navajoensis*
                       37. Seeds not mucilaginous when wetted; petals yellow, blade and claw joined in an arch; sepals elliptic or oblong; Kane County, Utah . . . . 85. *Physaria tumulosa*
               35. Plants not cespitose, not forming mounds, mats, or tufts.
                   38. Trichome rays fused (webbed) ½ or most of their length.
                       39. Basal and cauline leaf blade shapes similar . . . . 28. *Physaria fendleri*
                       39. Basal and cauline leaf blade shapes different. . . . . 52. *Physaria mcvaughiana*
                   38. Trichome rays not fused or only basally.
                       40. Petals white, sometimes purple-veined (fading purplish).
                           41. Basal leaf blades suborbicular to elliptic, or ovate or deltate, 0.5–2(–6.5) cm; petals 8–15 mm . . . . 62. *Physaria ovalifolia* (in part)
                           41. Basal leaf blades elliptic or obovate to oblong, 4–15 cm; petals 4.5–10(–12) mm. . . . . 72. *Physaria purpurea*

40. Petals yellow or yellow and orange (sometimes fading purplish).
  42. Fruits obcordate or obdeltate, compressed (angustiseptate) .... 39. *Physaria hemiphysaria* (in part)
  42. Fruits globose, subglobose, ellipsoid, ovoid, or obovoid, slightly compressed or terete.
    43. Trichomes asymmetrical with a deep notch (on one side) .... 26. *Physaria engelmannii*
    43. Trichomes symmetrical without a notch.
      44. Fruiting pedicels recurved; fruits ± pendent.
        45. Biennials or short-lived perennials, without a woody caudex.................... 9. *Physaria aurea* (in part)
        45. Perennials, with a woody caudex ...... 84. *Physaria thamnophila*
      44. Fruiting pedicels not recurved (sigmoid, straight, or curved); fruits not pendent.
        46. Ovules 16–32 per ovary; cauline leaves densely overlapping, erect and often appressed .... 7. *Physaria argyraea*
        46. Ovules 4–24(–26) per ovary; cauline leaves not overlapping, not erect and not appressed.
          47. Basal leaf blades and petioles differentiated, blades suborbicular, elliptic, ovate, deltate, or obovate to rhombic.
            48. Cauline leaf blades elliptic or obovate; racemes not elongated (subumbellate to densely corymbiform) .... 62. *Physaria ovalifolia* (in part)
            48. Cauline leaf blades obovate to rhombic; racemes elongated .... 70. *Physaria pruinosa*
          47. Basal leaf blades and petioles not differentiated (tapering), blades obovate, oblong, rhombic to elliptic, or spatulate to oblanceolate.
            49. Stems prostrate; cauline leaves densely overlapping .... 36. *Physaria gordonii* (in part)
            49. Stems ascending to erect; cauline leaves (relatively few), not or loosely overlapping...................... 68. *Physaria pinetorum*

[23. Shifted to left margin.—Ed.]
23. Fruit valves pubescent outside.
  50. Petals white or cream-white.
    51. Basal leaf blades not abruptly narrowed to petioles; Kaibab Plateau, n Arizona .... 44. *Physaria kingii* (in part)
    51. Basal leaf blades abruptly narrowed to petioles; Montrose and Ouray counties, Colorado .... 87. *Physaria vicina*
  50. Petals yellow or orange (sometimes drying purplish).
    52. Basal and cauline leaf blades similar in shape (usually narrow), blades less than 5 mm wide, not differentiated from (tapering to) petioles.
      53. Plants forming dense, hard mats, caudices highly branched (sw Colorado) .... 71. *Physaria pulvinata*
      53. Plants not forming dense, hard mats, caudices usually simple or few-branched.
        54. Fruiting pedicels usually recurved, sometimes divaricate-spreading or nearly horizontal.
          55. Annuals or perennials (short-lived); basal leaf blades flat, margins entire or dentate; racemes secund .... 6. *Physaria arenosa*
          55. Perennials; basal leaf blades involute, margins usually entire (rarely shallowly dentate); racemes not secund .... 50. *Physaria ludoviciana*
        54. Fruiting pedicels ascending, sigmoid, erect, spreading, recurved.

56. Fruit valves: trichomes with erect or spreading rays (± fuzzy in appearance).
   57. Basal leaf blades obovate to orbicular, margins folded; petioles differentiated from blades; caudices thickened (± 1 cm diam.); Sheep Mountain, Pioneer Range, Montana.............. 27. *Physaria eriocarpa*
   57. Basal leaf blades linear to oblanceolate, margins not folded; petioles not differentiated from blades; caudices not thickened; Arizona, Idaho, Montana, Wyoming.
      58. Styles 2.5–4 mm (as long as or longer than mature fruits); fruits 2.5–3(–4) mm............................. 56. *Physaria nelsonii*
      58. Styles (0.5–)1–3(–4) mm (shorter than mature fruits); fruits 3.5–7 mm.
         59. Stems simple or few to several from branched caudices (each with a sub-basal tuft of leaves); n Arizona, s Utah .... 8. *Physaria arizonica*
         59. Stems few to several from few-branched or unbranched caudices (each laterally from a basal tuft of rosette leaves); Idaho, se Montana........................ 73. *Physaria pycnantha*
56. Fruit valves: trichomes with appressed rays.
   60. Basal blades usually involute, sometimes flattened, linear to linear-oblanceolate, or narrowly spatulate.
      61. Sepals 4.5–7.5(–9) mm; styles (2–)3–4.5(–5.5) mm; pedicels straight or curved-ascending; n Arizona, n New Mexico, c, s Utah..................................... 43. *Physaria intermedia*
      61. Sepals 3.5–7 mm; styles 2–4 mm; pedicels sigmoid to curved-ascending; nw Colorado, ne Utah, sw Wyoming......... 66. *Physaria parvula*
   60. Basal blades flattened, not involute, linear, linear-oblanceolate, spatulate to nearly rhombic, oblanceolate, or elliptic.
      62. Racemes not or barely exceeding basal leaves; fruits compressed (latiseptate) on margins and at apices.
         63. Plants strongly condensed; stems lateral from a tight hemispherical tuft of leaves; Piceance Basin of Colorado .... 18. *Physaria congesta*
         63. Plants not condensed (loosely cespitose); stems lateral and also from within a basal tuft of leaves; nw Colorado, sw Nebraska, se Wyoming.................... 76. *Physaria reediana*
      62. Racemes exceeding basal leaves; fruits not compressed (or barely so on distal margins and apices) or strongly compressed (latiseptate) throughout.
         64. Fruits compressed throughout (strongly latiseptate, more so at apices); Montana, Wyoming (Big Horn Mountains) .... 20. *Physaria curvipes* (in part)
         64. Fruits not compressed at apices or distal margins; Arizona, s Colorado, n New Mexico, s Utah.
            65. Basal leaf blades linear; cauline leaves not secund; plants compact .... 12. *Physaria calcicola*
            65. Basal leaf blades narrowly oblanceolate to broadly elliptic; cauline leaves usually secund; plants not compact (loose, spreading) .... 74. *Physaria rectipes*

[52. Shifted to left margin.—Ed.]

52. Basal and cauline leaves dissimilar in shape, blades sometimes greater than 5(–6.5) mm wide, differentiated from petioles.
   66. Silicles strongly latiseptate (the valves strongly compressed parallel to the septum).
      67. Annuals or biennials; stems branched or unbranched; e Arizona, w New Mexico .... 35. *Physaria gooddingii*
      67. Perennials; stems usually unbranched; California, Idaho, Nevada, Oregon, n Utah, Washington... 60. *Physaria occidentalis*

[66. Shifted to left margin.—Ed.]

66. Silicles not compressed or slightly to strongly angustiseptate (the valves compressed perpendicular to the septum) or very slightly latiseptate.
  68. Fruits usually strongly compressed (angustiseptate, if slightly so, fruits obdeltate or obcordate).
    69. Fruits elliptic to orbicular; Idaho, Montana, Wyoming.................... 14. *Physaria carinata*
    69. Fruits cordate to obdeltate, obcordate, or obovate; California, Idaho, Nevada, Utah.
      70. Fruiting pedicels recurved; fruits obdeltate, usually pendent; septums nearly obsolete... 59. *Physaria obdeltata*
      70. Fruiting pedicels spreading, sigmoid, or recurved; fruits obcordate to obdeltate, not pendent; septums complete.
        71. Fruiting pedicels sigmoid; fruits strongly flattened; racemes sometimes elongated.. 19. *Physaria cordiformis*
        71. Fruiting pedicels spreading, or, occasionally, loosely sigmoid or recurved; fruits slightly flattened; racemes not elongated (dense, congested) .... 39. *Physaria hemiphysaria* (in part)
  68. Fruits usually not compressed (or slightly so at apices, if slightly compressed, fruits not obcordate or obdeltate).
    72. Annuals or, rarely, biennials... 83. *Physaria tenella*
    72. Perennials.
      73. Fruit valves sparsely pubescent or glabrous; plants of arctic and boreal regions of Canada and Greenland .... 5. *Physaria arctica* (in part)
      73. Fruit valves ± densely pubescent; plants usually of the continental United States (*P. spatulata* occurring to sw Canada).
        74. Fruiting pedicels recurved (*P. lesicii* sometimes with many pedicels arching); fruits pendent.
          75. Inner stems erect; fruits: septum complete.
            76. Cauline leaves 1–3.5 cm wide, not spreading (somewhat appressed to stem); fruit valves sparsely pubescent; ovules 4–6 per ovary .... 9. *Physaria aurea* (in part)
            76. Cauline leaves less than 6(–8) mm wide, spreading; fruit valves densely pubescent; ovules 8–12 per ovary................. 67. *Physaria pendula*
          75. Inner stems usually prostrate or decumbent; fruits: septum complete, fenestrate, strongly perforate, or obsolete (except *P. parviflora*).
            77. Fruits: septum complete; ovules 4 per ovary; nw Colorado...65. *Physaria parviflora*
            77. Fruits: septum fenestrate, perforate, or obsolete; ovules 4–12(–16) per ovary; Arizona, California, Idaho, Nevada, Oregon, Utah, Wyoming.
              78. Fruits papery, strongly inflated; basal leaf blades orbicular to broadly obovate, abruptly narrowing to petioles ..... 51. *Physaria macrocarpa*
              78. Fruits rigid, not or slightly inflated; basal leaf blades elliptic, rhombic, ovate, suborbicular, or broadly oblanceolate to broadly elliptic, or rhombic, gradually tapering to petioles.
                79. Fruits sometimes slightly obcompressed; styles 1–2 mm; racemes secund .... 31. *Physaria fremontii*
                79. Fruits slightly compressed (angustiseptate); styles 2.4–9 mm; racemes usually not secund .... 44. *Physaria kingii* (in part)
        74. Fruiting pedicels usually not recurved (sigmoid, horizontal, divaricate-ascending, ascending, or spreading); fruits horizontal to erect.
          80. Fruit valves: trichome rays strongly ascending, spreading (appearing fuzzy) .... 45. *Physaria klausii*
          80. Fruit valves: trichome rays appressed or somewhat spreading.

[81. Shifted to left margin.—Ed.]

81. Fruits globose, (1–)2–2.8 mm; styles 2–3.5(–4) mm............................. 34. *Physaria globosa*
81. Fruits globose, subglobose, orbicular, suborbicular, ovoid, obovoid, lanceolate, ellipsoid, or obpyriform, (2.5–)3–12 mm; styles 1.5–9 mm.
    82. Basal leaf blades broadly elliptic to suborbicular, distinctly rhombic, or deltate, petioles well-differentiated from blades.
        83. Stems prostrate or decumbent.
            84. Fruits wider than long, sparsely pubescent inside; Bitterroot Mountains, Montana ....41. *Physaria humilis*
            84. Fruits longer than wide, glabrous or sparsely pubescent inside; Idaho, Utah, sw Wyoming .... 69. *Physaria prostrata*
        83. Stems (inner) ascending to erect.
            85. Racemes lax, elongated; fruits not compressed apically.
                86. Stems to 4.5 dm; basal leaf blades: margins entire, sinuate, coarsely dentate, or lyrate-pinnatifid.. 24. *Physaria douglasii*
                86. Stems 1–1.5 dm; basal leaf blades: margins entire .... 48. *Physaria lesicii*
            85. Racemes dense, (subcorymbose to subumbellate, few-flowered); fruits compressed apically.
                87. Basal leaf blades rhombic; sw Colorado, ne Utah, sw Wyoming ....
                ....82. *Physaria subumbellata*
                87. Basal leaf blades oblanceolate to orbicular (usually noticeably thickened).
                    88. Fruits with slight apical constriction; Pryor Mountains, Montana...
                    ....63. *Physaria pachyphylla*
                    88. Fruits compressed at margins, rounded apically; Arizona ....16. *Physaria cinerea*
    82. Basal leaf blades ± spatulate, narrowly oblanceolate, or elliptic (if wider, blades gradually tapering to petioles), petioles usually weakly differentiated from blades.
        89. Fruits often somewhat compressed, elliptic to lanceolate, apices ± acute.
            90. Sepals 5–8.5 mm, median pair cucullate at apex; petals narrowly spatulate to obovate; ovules (8–)12–20(–24) per ovary .... 53. *Physaria montana*
            90. Sepals 3.5–5 mm, median pair not cucullate at apex; petals lingulate to narrowly oblanceolate; ovules 4–8 per ovary.
                91. Fruiting pedicels 4–7 mm; styles shorter than fruits......... 20. *Physaria curvipes* (in part)
                91. Fruiting pedicels 10–20 mm; styles ± equal to fruits............... 81. *Physaria spatulata*
        89. Fruits compressed or not, subglobose to ovoid (rarely obdeltate), apices rounded, truncate, or obtuse.
            92. Fruit valves: trichomes spreading rays (appearing fuzzy) .... 32. *Physaria garrettii*
            92. Fruit valves: trichomes appressed rays.
                93. Fruits inflated, slightly didymous, becoming purplish; sw Colorado ....
                ....79. *Physaria scrotiformis*
                93. Fruits not inflated, not didymous, remaining greenish; Arizona, California, Idaho, Nevada, New Mexico, Oregon, Texas, Utah, Wyoming.
                    94. Caudices thickened; fruit apices somewhat compressed............86. *Physaria valida*
                    94. Caudices not thickened; fruit apices not compressed (or fruits slightly compressed throughout).
                        95. Basal leaves with relatively long, slender petioles; sw New Mexico .... 46. *Physaria lata*
                        95. Basal leaves with relatively short, stout petioles; Arizona, California, Idaho, Nevada, Oregon, Utah, Wyoming.
                            96. Caudices simple; ovules 4–16 per ovary .... 44. *Physaria kingii* (in part)
                            96. Caudices branched (underground); ovules usually 4 per ovary .... 54. *Physaria multiceps*

1. Physaria acutifolia Rydberg, Bull. Torrey Bot. Club 28: 279. 1901 • Rydberg's or sharpleaf twinpod E F

Physaria acutifolia var. *stylosa* (Rollins) S. L. Welsh; *P. australis* (Payson) Rollins; *P. didymocarpa* (Hooker) A. Gray var. *australis* Payson; *P. stylosa* Rollins

Perennials; caudex branched, (sometimes forming a thick crown, cespitose); (silvery) pubescent throughout, trichomes several-rayed, rays furcate, (moderately tuberculate, rays weakly so). **Stems** several from base, usually somewhat decumbent, (unbranched), (0.4–)0.5–2 dm. **Basal leaves:** (petiole slender, often narrowly winged); blade obovate to orbicular or rhombic-orbicular, 2–9 cm, (base abruptly narrowed to petiole), margins usually entire, rarely with few scattered teeth, (apex rounded or obtuse, sometimes with apical mucro). **Cauline leaves:** blade spatulate to oblanceolate, 1–3 cm, margins entire, (apex usually obtuse). **Racemes** loose, (elongated in fruit). **Fruiting pedicels** (divaricate, slightly sigmoid or nearly straight), 6–12 mm. **Flowers:** sepals linear-oblong, 4–7.5 mm; petals spatulate, 6–11 mm. **Fruits** (erect), didymous, suborbicular, inflated, (4–)6–15 × (4–)8–20 mm, (papery, basal and apical sinuses similar, basal rarely shallower, apical deep, narrow and closed or nearly so); valves retaining seeds after dehiscence, pubescent, trichomes appressed; replum oblong, constricted, 2–3.5 mm, narrower than fruit, apex obtuse; ovules (2 or) 4 per ovary; style 4–6(–9) mm. **Seeds** (dark brown), flattened, (2–3 mm). $2n$ = 10, 16, 24.

Flowering May–Jun(–Jul). Hillsides, roadcuts, sagebrush, pinyon-juniper, Gambel oak, ponderosa pine communities; 1500–3500 m; Ariz., Colo., Idaho, Mont., Nev., N.Mex., S.Dak., Utah, Wyo.

Physaria acutifolia tends to be somewhat dwarfed, with a branched caudex and especially long styles (var. *stylosa*), where it grows at high elevations, especially at the western end of the Uinta Mountains in Utah. Intermediates form an uninterrupted cline and no infraspecific taxa are here recognized. In R. C. Rollins (1939), the discussion of *P. acutifolia* actually pertains to *P. rollinsii*. The discussion of *P. australis* pertains to what is now known as *P. acutifolia*. The plants are usually found in open soil patches, rarely into the subalpine or alpine tundra.

2. Physaria alpestris Suksdorf, W. Amer. Sci. 15: 58. 1906 • Washington or alpine twinpod E

Lesquerella alpestris (Suksdorf) G. A. Mulligan

Perennials; caudex usually simple, rarely branched, (cespitose); (silvery) pubescent throughout, trichomes several-rayed, rays (1- or) 2-bifurcate, (low-umbonate, tubercles relatively few, small). **Stems** several from base, decumbent to ascending, (unbranched), 0.5–1.5 dm. **Basal leaves:** (petiole slender); blade obovate, 3–5 cm (width 10–20 mm, base tapering abruptly to petiole), margins entire, (apex rarely slightly acute). **Cauline leaves:** blade oblanceolate, 0.5–1.5 cm (width 3–5 mm), margins entire. **Racemes** subcorymbose. **Fruiting pedicels** (divaricate, straight), 5–10 mm. **Flowers:** sepals oblong, 8–10 mm; petals spatulate, 12–14 mm. **Fruits** didymous, mostly highly inflated (strongly flattened at least in ½ toward replum), 14–18 × 14–18 mm, (papery, basal sinus slightly notched, apical open, shallow); valves (retaining seeds after dehiscence), evenly pubescent; replum lanceolate, 7–10 mm, width 1.5–2.5 mm, as wide as or wider than fruit, apex acute to acuminate; ovules 8–10 per ovary; style 5–7 mm. **Seeds** flattened, (2–3 mm). $2n$ = 48–52, 52, 64, 67–70.

Flowering May–Jun. Alpine scree, rocky ridges, talus slopes, volcanic sands and gravel, serpentine gravel, granitic slopes, mountain shrub, subalpine fir, and whitebark pine communities; (700–)1300–2400 m; Wash.

3. Physaria alpina Rollins, Brittonia 33: 339. 1981 • Avery Peak or alpine twinpod C E

Perennials; (with a long taproot), caudex usually buried, simple, (enlarged, covered with marcescent leaf bases, crown rosulate and horizontal to somewhat ascending, forming a dense crown at apex of caudex); (silvery) pubescent throughout, trichomes (sessile or stipitate), 5–8-rayed, rays furcate or bifurcate, (rounded to umbonate, strongly tuberculate, less so or smooth over center). **Stems** few from base, decumbent, (arising laterally proximal to current season's leaves), 0.3–0.8 dm. **Basal leaves:** (petiole slender); blade broadly obovate, or deltate to ovate or narrower, 1.5–3.5 cm, (base abruptly to gradually narrowed to petiole), margins entire or obscurely few-toothed, (apex usually obtuse, nearly acute in narrower leaves). **Cauline leaves:** (2–5 per stem); blade oblanceolate to spatulate, similar to

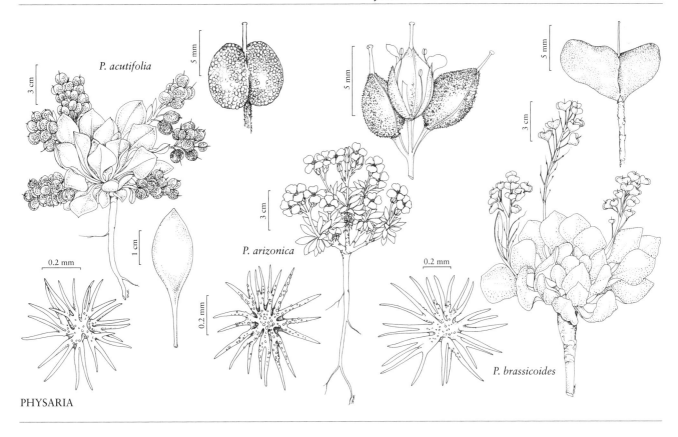

PHYSARIA

basal, margins entire, (apex acute). **Racemes** loose, (3–6-flowered). **Fruiting pedicels** (widely spreading to ascending, slightly curved or straight), 7–11 mm. **Flowers:** sepals narrowly oblong to linear, 7–9 mm; petals (erect), spatulate, 10–12(–15) mm. **Fruits** (usually purplish in age), didymous, irregular and somewhat angular, not highly inflated, 4–11 × 10–13 mm, (coriaceous, papery, shallowly grooved distally and on sides, tapered and narrowed toward replum, base obtuse to truncate, apex with broad sinus to nearly truncate); valves (retaining seeds after dehiscence), densely pubescent, not silvery; replum elliptic to obovate, as wide as or wider than fruit, base rounded, margins sparsely pubescent or glabrous, apex rounded with funicles); ovules 4 per ovary; style 5–7 mm, (glabrous). **Seeds** flattened.

Flowering Jun–Jul. Whitish or red substrates from limestone or dolomite, ridge crests, rocky alpine tundra and open areas; of conservation concern; 3500–4000 m; Colo.

4. **Physaria angustifolia** (Nuttall ex Torrey & A. Gray) O'Kane & Al-Shehbaz, Novon 12: 321. 2002 • Threadleaf bladderpod [E]

Vesicaria angustifolia Nuttall ex Torrey & A. Gray, Fl. N. Amer 1: 101. 1838; *Lesquerella angustifolia* (Nuttall ex Torrey & A. Gray) S. Watson; *L. longifolia* Cory

Annuals; with a fine taproot; ± densely pubescent, trichomes several-rayed, rays distinct or fused at base, bifurcate, (prominently tuberculate throughout). **Stems** simple or few to several from base, erect, (sometimes branched), to 4 dm. **Basal leaves:** blade elliptic to rhombic, 3–8 cm, (base narrowing gradually to petiole), margins entire, repand, coarsely toothed, or pinnatifid. **Cauline leaves:** (proximal often shortly petiolate, distal sessile); blade linear or narrowly obovate, 1.5–6(–10) cm, margins entire, repand, or shallowly toothed. **Racemes** usually loose. **Fruiting pedicels** (usually divaricate, sometimes horizontal, straight or slightly curved), 8–20 mm. **Flowers:** sepals elliptic or ovate, 4–6 mm, (lateral pair usually subsaccate); petals obovate to obdeltate, 6–10 mm, (apex often emarginate). **Fruits** not didymous, ± globose, slightly inflated, 4–6 mm; valves (not retaining seeds after dehiscence), glabrous throughout; replum as wide as or wider than fruit; ovules 4 per ovary; style 2–3.5 mm; (stigma expanded). **Seeds** flattened, (margined). $2n = 10$.

Flowering Apr(–May). Shallow limestone-derived soils, sometimes spreading to disturbed sites; 90–300 m; Okla., Tex.

5. Physaria arctica (Wormskjöld ex Hornemann) O'Kane & Al-Shehbaz, Novon 12: 321. 2002 • Arctic bladderpod

Alyssum arcticum Wormskjöld ex Hornemann in G. C. Oeder et al., Fl. Dan. 9(26): 3, plate 1520. 1816; *Lesquerella arctica* (Wormskjöld ex Hornemann) S. Watson; *L. arctica* subsp. *purshii* (S.Watson) A. E. Porsild; *Vesicaria arctica* (Wormskjöld ex Hornemann) Richardson; *V. leiocarpa* (Trautvetter) N. Busch

Perennials; caudex simple or branched, (woody, cespitose); ± densely pubescent, trichomes (sessile or short-stalked), several-rayed, rays distinct, furcate or bifurcate, (somewhat umbonate, finely tuberculate to ± smooth). **Stems** simple or few to several from base, erect to spreading or prostrate, 0.5–2(–3) dm. **Basal leaves** (usually ± rosulate); blade obovate to oblanceolate, (1–)2–6(–15) cm, margins entire. **Cauline leaves** (sessile or shortly petiolate); blade oblanceolate or lingulate, 0.5–1.5(–3) cm, margins entire. **Racemes** loose. **Fruiting pedicels** (erect to divaricate or ascending), (5–)10–20(–40) mm, (stout). **Flowers:** sepals ovate to elliptic, (3–)4–5(–6) mm, (median pair often thickened apically, cucullate); petals spatulate, 5–6(–7) mm, (blade gradually narrowed to claw). **Fruits** subglobose to ellipsoid, uncompressed, 4–6(–9) mm; valves (not retaining seeds after dehiscence), glabrous or sparsely pubescent outside, trichomes sessile; replum as wide as or wider than fruit; ovules (8–)10–14(–16) per ovary; style 1–2.5(–4) mm. **Seeds** plump. $2n = 60$.

Flowering May–Aug. Typically on sand and gravel from calcareous bedrock, river bars and terraces, cliff ledges, scree and talus slopes, often growing after disturbance; 0–1800 m; Greenland; Alta., B.C., Man., Nfld. and Labr., N.W.T., Nunavut, Que., Yukon; Alaska; circumarctic (except n Europe, ne Russia).

6. Physaria arenosa (Richardson) O'Kane & Al-Shehbaz, Novon 12: 321. 2002 • Great Plains bladderpod E

Vesicaria arenosa Richardson in J. Franklin, Narr. Journey Polar Sea, 743. 1823; *Lesquerella arenosa* (Richardson) Rydberg; *L. argentea* var. *arenosa* (Richardson) Rydberg; *L. ludoviciana* var. *arenosa* (Richardson) S. Watson

Perennials or, rarely, annuals; caudex simple or branched; ± densely pubescent, trichomes (sessile or short-stalked), few-rayed, rays (usually spreading), distinct or slightly fused at base, furcate or bifurcate, (tuberculate). **Stems** simple or few from base, prostrate or straggling to erect, (sometimes purplish, usually unbranched), (0.5–)1–2(–3) dm. **Basal leaves:** blade oblanceolate, 1.5–5(–7) cm, margins entire or shallowly dentate, (flat). **Cauline leaves:** blade elliptic to linear, (0.5–)1–2.5(–3) cm, margins usually entire. **Racemes** (secund), dense, (elongated in fruit). **Fruiting pedicels** (usually sharply recurved, sometimes divaricate-spreading or nearly horizontal), 5–15(–20) mm, (stout). **Flowers:** sepals elliptic or oblong, 4–6(–7) mm, (lateral pair subsaccate, median pair thickened apically, cucullate); petals (often red or lavender when dried), obovate, 6–8.5(–9.5) mm, (narrowing to broad claw). **Fruits** subglobose, obovoid, or broadly ellipsoid, slightly inflated, (3.5–)4–5.5(–6.5) mm; valves densely pubescent outside, trichomes spreading or closely appressed, rarely sparsely pubescent inside; ovules (4–)8(–10) per ovary; style (slender), 3–5.5(–6.5) mm. **Seeds** slightly flattened.

Subspecies 2 (2 in the flora): wc North America.

1. Perennials or, rarely, annuals, short-lived; fruit valves: trichomes spreading.................
.... 6a. *Physaria arenosa* subsp. *arenosa*
1. Perennials, long-lived; fruit valves: trichomes closely appressed............................
.... 6b. *Physaria arenosa* subsp. *argillosa*

6a. Physaria arenosa (Richardson) O'Kane & Al-Shehbaz subsp. **arenosa** E

Lesquerella lunellii A. Nelson; *L. macounii* Greene; *L. rosea* Greene

Perennials or, rarely, annuals; short-lived. **Fruit valves:** trichomes spreading. $2n = 10, 18$.

Flowering May–Jun. Open, dry, sandy or gravelly, limey soils, grasslands or sagebrush dominated sites, sandhills, dunes; 600–1600 m; Alta., Man., Sask.; Colo., Mont., Nebr., N.Dak., S.Dak., Wyo.

6b. Physaria arenosa (Richardson) O'Kane & Al-Shehbaz subsp. **argillosa** (Rollins & E. A. Shaw) O'Kane & Al-Shehbaz, Novon 12: 321. 2002

• Secund or sidesaddle bladderpod [E]

Lesquerella arenosa (Richardson) Rydberg var. *argillosa* Rollins & E. A. Shaw, Gen. Lesquerella, 178. 1973; *Physaria arenosa* var. *argillosa* (Rollins & E. A. Shaw) B. L. Turner

Perennials; long-lived. **Fruit valves**: trichomes closely appressed. $2n = 26, 30$.

Flowering May–Jun. Open, barren sites of grasslands and badlands with clayey or sandy-clay soils derived from limestone; 700–1600 m; Colo., Nebr., S.Dak., Wyo.

7. Physaria argyraea (A. Gray) O'Kane & Al-Shehbaz, Novon 12: 321. 2002 • Silver bladderpod

Vesicaria argyraea A. Gray, Boston J. Nat. Hist. 6: 146. 1850; *Alyssum argyraeum* (A. Gray) Kuntze; *Lesquerella argyraea* (A. Gray) S. Watson

Subspecies 2 (1 in the flora): sw United States, n Mexico.

7a. Physaria argyraea (A. Gray) O'Kane & Al-Shehbaz subsp. **argyraea**

Perennials; caudex simple or branched, (woody); ± densely pubescent, trichomes (sessile or short-stalked), few- to several-rayed, rays fused at base, (smooth or finely tuberculate). **Stems** simple or several from base, prostrate and spreading to decumbent or erect, 0.5–4(–7) dm. **Basal leaves** (usually shed early); blade oblanceolate, 2–6(–8) cm, margins entire or pinnatifid. **Cauline leaves** (densely overlapping, often appressed, erect; petiolate); blade elliptic or obovate to rhombic, 0.5–4.5 cm, margins entire or sinuate or remotely dentate, (apex usually acute). **Racemes** loose. **Fruiting pedicels** (spreading to erect or occasionally slightly recurved, sigmoid, straight, or slightly curved), 15–40 mm. **Flowers**: sepals linear to oblong or elliptic, (3–)4–8 mm, (median pair thickened apically, cucullate); petals (orange or yellow, rarely fading purplish, claw often orange), spatulate or obovate to obdeltate, (narrowing to relatively short claw), 4.5–8.5(–11) mm. **Fruits** subglobose to broadly ellipsoid, slightly compressed (or terete), 4–8 mm; valves (not retaining seeds after dehiscence), glabrous throughout; replum as wide as or wider than fruit; ovules 16–32 per ovary; style 1.5–5.5 mm. **Seeds** flattened, (sometimes narrowly margined).

Flowering Feb–May(–Jun). Sandy granite- or, usually, limestone-derived soils; 0–800 m; Tex.; n Mexico (Coahuila, Nuevo León).

Subspecies *argyraea* differs from subsp. *diffusa* (Rollins) O'Kane & Al-Shehbaz of the Sierra Madre Oriental of Mexico in being herbaceous (rather than at least somewhat woody throughout) and having decumbent to erect (rather than prostrate) stems.

8. Physaria arizonica (S. Watson) O'Kane & Al-Shehbaz, Novon 12: 321. 2002 • Arizona bladderpod [E] [F]

Lesquerella arizonica S. Watson, Proc. Amer. Acad. Arts 23: 254. 1888; *L. arizonica* var. *nudicaulis* Payson; *Physaria arizonica* var. *andrusensis* N. D. Atwood, S. L. Welsh & L.C. Higgins

Perennials; caudex branched, (cespitose); densely (silvery gray) pubescent, trichomes (sessile or subsessile), (4-), 6-, or 8-rayed, rays fused at base, furcate or bifurcate, (slightly umbonate, tuberculate throughout). **Stems** simple or few to several from base, erect, (unbranched, slender), 0.2–1(–1.5) dm. **Basal leaves** (densely tufted, not rosulate, reflexed in age); blade obovate to oblanceolate, 0.7–2(–3) cm, margins usually entire, sometimes repand or shallowly dentate, (apex acute). **Cauline leaves** similar to basal, becoming narrower distally, somewhat reflexed, (distal) blade linear or narrowly oblanceolate, 0.5–2.5(–5.5) cm. **Racemes** dense, often subcorymbiform. **Fruiting pedicels** (erect or divaricate-spreading, straight or slightly curved), (3–)5–10(–15) mm. **Flowers**: sepals (green or greenish yellow), ovate or broadly ovate, 3.5–6.5 mm, (lateral pair subsaccate, cucullate, median pair thickened, slightly cucullate apically); petals (spreading), oblanceolate to obovate, 5.5–8(–10) mm, (claw erect). **Fruits** (sessile or substipitate), suborbicular to ovoid or ellipsoid, slightly inflated, 4–7 mm; valves pubescent outside, trichomes substipitate, spreading, sometimes sparsely pubescent inside, trichomes sessile, smooth; ovules 4–10(–16) per ovary; style (0.5–)1–2(–4) mm (shorter than fruit). **Seeds** flattened. $2n = 10$.

Flowering Apr–Jun. Sandy and gravelly soils, limey knolls or limestone chip, often in open stands of sagebrush-pinyon, pinyon-juniper, Gambel oak and sometimes ponderosa pine; 1000–2200 m; Ariz., Utah.

The circumscription of *Physaria arizonica* here is quite broad and includes plants that have densely tufted basal leaves and relatively few or no cauline leaves; plants that are loosely tufted and have several cauline leaves; and plants that have a strongly branched caudex, leafy stems, and sterile shoots (var. *andrusensis*). Additional study is needed to understand the pattern of variation in this complex species; all of the characters given above vary considerably.

9. **Physaria aurea** (Wooton) O'Kane & Al-Shehbaz, Novon 12: 322. 2002 • Golden bladderpod C E

Lesquerella aurea Wooton, Bull. Torrey Bot. Club 25: 260. 1898

Biennials or perennials; (short-lived); caudex branched; densely pubescent, trichomes (sessile or short-stalked, simple or not), 5–9-rayed, rays furcate, (fine, smooth or finely tuberculate). **Stems** several from base, erect or outer ones decumbent or procumbent, (sometimes much-branched distally), to 6 dm. **Basal leaves:** blade obovate or rhombic, to ca. 2.5 cm, margins usually shallowly dentate, sometimes lyrate-pinnatifid. **Cauline leaves:** (proximal shortly petiolate, distal sessile); blade obovate to rhombic or oblanceolate, 2–4(–6) cm, margins entire or shallowly and remotely dentate. **Racemes** usually dense, (several-flowered). **Fruiting pedicels** (strongly recurved), to 20 mm. **Flowers:** sepals ovate or oblong (tapering at base), 3.6–4.8(–5.3) mm, (lateral pair subsaccate, median pair thickened apically, cucullate); petals obovate to spatulate, 4.5–7.5 mm, (blade narrowed to broad claw, margins sinuate). **Fruits** (± pendent), ovoid, obcompressed, or globose, compressed, 4–6(–8) mm; valves (not retaining seeds after dehiscence), sparsely pubescent or glabrous, sparsely pubescent inside; replum as wide as or wider than fruit; ovules usually 4, rarely 6, per ovary; style 2.5–3.6 mm. Seeds flattened. $2n = 14$.

Flowering Jun–Aug. Open sites and bare areas in rocky limestone soil in mountains, roadbanks, open woods; of conservation concern; 2000–2800 m; N.Mex.

Physaria aurea (known from the Jicarilla and Sacramento mountains) is similar to 35. *P. gooddingii*, which is found farther west in the mountains of Catron and Sierra counties, New Mexico, and Greenlee County, Arizona.

10. **Physaria bellii** G. A. Mulligan, Canad. J. Bot. 44: 1662, fig. 1, plate 1, fig. 3. 1966 • Bell's or Front Range twinpod C E

Perennials; caudex simple, (relatively large); densely (silvery) pubescent, trichomes (sessile, appressed), rays furcate, fused at base. **Stems** simple from base, decumbent to nearly prostrate, 0.5–1.3 dm. **Basal leaves** (strongly rosulate; shortly petiolate); blade broadly obovate, 1.5–7.5 (width 7.5–26 mm, base gradually tapering to petiole), margins shallowly dentate, (apex obtuse). **Cauline leaves:** blade oblanceolate to broadly obovate, 1–2.5 cm, margins entire. **Racemes** dense. **Fruiting pedicels** (divaricate-ascending to widely spreading, slightly sigmoid to curved), 7–12 mm. **Flowers:** sepals (pale yellow or yellow-green), narrowly lanceolate to narrowly deltate, 4–8 mm; petals yellow, broadly spatulate to obovate, 9–13 mm, (not clawed). **Fruits** didymous, slightly flattened (contrary to replum) to uncompressed, 4–9 × 2–8 mm, (strongly coriaceous, apical and basal sinuses narrow, deep); valves (retaining seeds after dehiscence), pubescent, trichomes appressed; replum narrowly oblanceolate to narrowly linear-oblong, as wide as or wider than fruit, apex obtuse; ovules 4 per ovary; style more than 3 mm. Seeds compressed. $2n = 8$.

Flowering Mar–Jun(–Jul). Dark shale, road cuts, ridge crests, washes; of conservation concern; 1500–1800 m; Colo.

Physaria bellii is often found in shale and limestone soils of the Fountain/Ingleside, Lykins, Niobrara, and Pierre formations. It is in the Center for Plant Conservation's National Collection of Endangered Plants.

11. **Physaria brassicoides** Rydberg, Bull. Torrey Bot. Club 29: 237. 1902 • Double bladderpod E F

Perennials; (somewhat compact); caudex branched, (relatively large); (silvery) pubescent throughout, trichomes (sessile), several-rayed, rays furcate, (slightly umbonate, tuberculate throughout). **Stems** several from base, decumbent to ascending (arising laterally, unbranched, stout), (0.2–)0.5–1.7 dm. **Basal leaves:** (petiole somewhat winged); blades orbicular to obovate, 2–6 cm (width 1–2.5 cm, thick), margins usually repand, rarely entire, (adaxial surface scurfy). **Cauline leaves:** blade oblanceolate to broadly spatulate, 1–2 cm (width 3–5 mm), margins entire, (apex obtuse to subacute). **Racemes** moderately dense (or elongated). **Fruiting pedicels** (divergent, straight to somewhat curved or sigmoid), 5–12 mm. **Flowers:** sepals linear-oblong, 6–8 mm; petals spatulate, 9–12 mm. **Fruits** (erect), didymous, cordate, moderately inflated, (6–)10–20 × 10–23 mm, (papery, base obtuse or with obscure sinus, apical sinus deep, broad); valves (retaining seeds after dehiscence), densely and loosely pubescent, trichomes spreading; replum linear-oblong, constricted, as wide as or wider than fruit; ovules 4 per ovary; style 4–5(–9) mm. Seeds plump, (broad). $2n = 8, 16$.

Flowering May–Jun. Bare hillsides, dry gravel and clay soil, badlands, clay knolls, banks; 900–1400 m; Colo., Mont., Nebr., N.Dak., S.Dak., Wyo.

12. Physaria calcicola (Rollins) O'Kane & Al-Shehbaz, Novon 12: 322. 2002 • Rocky Mountain bladderpod [C] [E]

Lesquerella calcicola Rollins, Amer. J. Bot. 26: 419, fig. 1A,B. 1939

Perennials; (compact); caudex branched; densely (silvery) pubescent, trichomes (sessile or short-stalked), 5–8-rayed, rays distinct, furcate or bifurcate, (umbonate, tuberculate and the center less so). Stems several from base, erect or outer ones decumbent, (unbranched, stout, usually sparsely leaved), 1–3 dm. Basal leaves: blade linear, 2–7(–10) cm, margins entire, repand, or shallowly dentate. Cauline leaves (sessile); blade (erect), spatulate to linear, (1–)2–3(–4.5) cm, margins entire, sometimes involute, (apex acute or subacute). Racemes dense, (exceeding basal leaves). Fruiting pedicels (spreading, sharply sigmoid), 8–15 mm. Flowers: sepals ovate or oblong, (4.5–)5–6(–7) mm, (lateral pair subsaccate, cucullate, median pair thickened, cucullate apically); petals spatulate, 7–9(–11) mm (widened at base, slightly retuse). Fruits (sessile or substipitate), ovate to oblong, not compressed at distal margins or apex, 5–9 mm; valves sparsely pubescent, trichomes appressed; ovules 4–8 per ovary; style 3–5 mm. Seeds flattened. $2n = 16$, ca. 20.

Flowering May–Jun. Shale bluffs, limestone hillsides, gypseous knolls and ravines, calcareous substrates, grasslands and pinyon-juniper communities; of conservation concern; 1400–2100 m; Colo., N.Mex.

13. Physaria calderi (G. A. Mulligan & A. E. Porsild) O'Kane & Al-Shehbaz, Novon 12: 322. 2002 • Calder's bladderpod [E]

Lesquerella calderi G. A. Mulligan & A. E. Porsild, Canad. J. Bot. 47: 215, plate 1. 1969; *L. arctica* (Wormskjold ex Hornemann) S. Watson subsp. *calderi* (G. A. Mulligan & A. E. Porsild) Hultén

Perennials; caudex simple or branched; densely pubescent throughout, trichomes (sessile or subsessile), rays distinct or slightly fused at base, furcate or bifurcate, (strongly umbonate, tuberculate, tubercles often relatively larger, fewer over center). Stems simple or few to several from base, usually erect to spreading, sometimes prostrate, 0.5–2 dm. Basal leaves: blade oblanceolate, 2–3 cm, margins entire. Cauline leaves (sessile or proximal shortly petiolate); blade narrowly oblanceolate, 0.5–1.5 cm, margins entire. Racemes loose. Fruiting pedicels (erect to divaricate or ascending, sometimes curved), (5–)10–20(–40) mm, (stout). Flowers: sepals ovate to elliptic, (3–)4–5(–6) mm, (median pair often thickened apically, cucullate); petals obovate, (6–)7–10 mm (nearly as wide, abruptly narrowed to claw, ca. 1 mm wide). Fruits subglobose to ellipsoid, compressed (usually angustiseptate), to 8 mm; (valves not retaining seeds after dehiscence; replum as wide as or wider than fruit; ovules 10–14 per ovary; style 1–2 mm. Seeds plump. $2n = 20$.

Flowering Jun–Aug. Dry rocky summits, limestone flats and slopes, alpine knolls; 600–1500 m; N.W.T., Yukon; Alaska.

Physaria calderi is known from the Ogilvie and Richardson mountains.

14. Physaria carinata (Rollins) O'Kane & Al-Shehbaz, Novon 12: 322. 2002 • Idaho bladderpod [E] [F]

Lesquerella carinata Rollins, Contr. Gray Herb. 171: 42, fig. 1E–H. 1950

Perennials; caudex simple, (often enlarged by persistent leaf bases); densely pubescent, trichomes (sessile or short-stalked), rays furcate or bifurcate, (nearly smooth to finely tuberculate). Stems simple from base, decumbent, (occasionally few-branched), 0.5–1.5(–2) dm. Basal leaves: blade elliptic to broadly obovate, triangular, rhombic, or round, 1.5–3(–4) cm, margins often sinuate or shallowly lobed. Cauline leaves (sessile or shortly petiolate); blade elliptic to oblanceolate to obovate, 0.5–1.5 cm, (base narrowed to petiole), margins entire. Racemes compact to loose. Fruiting pedicels (ascending or divaricate-spreading, straight to loosely sigmoid or curved), 4–10 mm. Flowers: sepals oblong to broadly elliptic, 4–7.5 mm, (lateral pair saccate or not); petals spatulate, 7.5–10 mm. Fruits (sessile or substipitate), elliptic, suborbicular, or oblong, strongly compressed (angustiseptate), 5–9 mm, (rounded to sharply keeled on 1 side, edges ± keeled); valves: (margins covering replum edges or not), usually pubescent throughout or, rarely, glabrous inside; ovules (4–)8–14(–16) per ovary; style 2–4.5(–5) mm. Seeds slightly flattened.

Subspecies 3 (3 in the flora): wc United States.

Differences in fruit morphology become blurred and the three subspecies are often indistinguishable where their ranges meet near the intersection of Idaho, Montana, and Wyoming.

1. Fruits elliptic, not keeled, valve margins (thin, rounded), not covering replum edges . . . . 14b. *Physaria carinata* subsp. *paysonii*
1. Fruits elliptic, suborbicular, or broadly oblong, keeled on one side, valve margins covering replum edges.

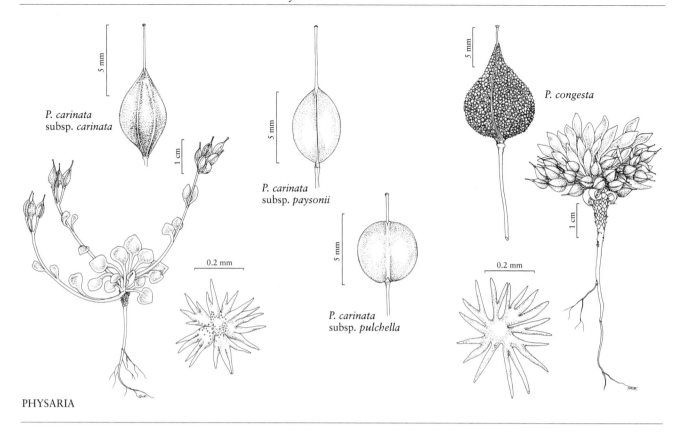

[2. Shifted to left margin.—Ed.]

2. Fruits elliptic, bases narrow-rounded to acute, apices ± acute ... 14a. *Physaria carinata* subsp. *carinata*
2. Fruits suborbicular to elliptic or broadly oblong, bases and apices rounded ....14c. *Physaria carinata* subsp. *pulchella*

14a. Physaria carinata (Rollins) O'Kane & Al-Shehbaz subsp. carinata E F

Lesquerella carinata var. *languida* Rollins

Fruits elliptic, sharply keeled on one side; valves: margins covering replum edges, (base narrow-rounded to acute, apex ± acute), pubescent inside.

Flowering Jun–early Aug. Grassy slopes, *Juniperus*, *Cercocarpus*, and mountain shrub communities, roadcuts, usually on limestone substrates, occasionally on decomposed basalt; 1600–2400 m; Idaho, Mont., Wyo.

14b. Physaria carinata (Rollins) O'Kane & Al-Shehbaz subsp. paysonii (Rollins) O'Kane, Novon 17: 379. 2007 • Payson's bladderpod E F

Lesquerella paysonii Rollins, Contr. Gray Herb. 171: 44. 1950; *Physaria paysonii* (Rollins) O'Kane & Al-Shehbaz

Fruits elliptic, not keeled; valves: margins (thin, rounded), not covering replum edges, pubescent inside.

Flowering Jun–Aug. Limestone substrates of rocky summits, talus, steep slopes, rocky ridges, fine soils, red sandstone, gravelly and gypseous soils, herbaceous communities on slopes, open Douglas-fir and limber pine communities; 1800–3200 m; Idaho, Wyo.

14c. **Physaria carinata** (Rollins) O'Kane & Al-Shehbaz subsp. **pulchella** (Rollins) O'Kane, Novon 17: 379. 2007 • Beautiful bladderpod [C][E][F]

Lesquerella pulchella Rollins, Novon 5: 72, figs. 1–4. 1995; *Physaria pulchella* (Rollins) O'Kane & Al-Shehbaz

Fruits suborbicular to elliptic or broadly oblong, keeled on one side; valves: margins covering replum edges, (base and apex rounded), glabrous inside.

Flowering late Jul–early Aug. Grass-forb communities on slopes, sagebrush, mountain shrub, limber pine communities, limestone gravel or cobbles; of conservation concern; 2100–2300 m; Idaho, Mont.

15. **Physaria chambersii** Rollins, Rhodora 41: 403, plate 556, figs. 15–18. 1939 • Chambers's twinpod [E]

Physaria chambersii var. *canaani* S. L. Welsh; *P. chambersii* var. *sobolifera* S. L. Welsh

Perennials; caudex usually simple, sometimes branched, (thick, cespitose); (silvery) pubescent throughout, trichomes few-rayed, rays furcate, sometimes slightly fused at base, (umbonate, lightly tuberculate to nearly smooth). **Stems** several from base, erect or decumbent (arising laterally, unbranched), 0.5–1.5 mm. **Basal leaves:** (petiole slender); blade obovate to orbicular, 3–6 cm (width 10–20 mm), margins entire or dentate. **Cauline leaves:** blade spatulate, 1–2 cm (width 3–6 mm), margins entire, (apex often acute). **Racemes** congested. **Fruiting pedicels** (divaricate, slightly sigmoid), 8–15 mm. **Flowers:** sepals narrowly lanceolate, 5–8(–9) mm; petals narrowly oblanceolate, 9–12 mm, (claw undifferentiated from blade). **Fruits** (often purplish in age), didymous, subreniform, strongly inflated, 9–18 × 11–21(–30) mm, (papery, base obtuse to slightly cordate, apical sinus V-shaped or convex, open crests rounded); valves (2-keeled on side away from replum, each 3-sided, keels rounded, sides flat or slightly convex, retaining seeds after dehiscence), evenly and densely pubescent; replum oblong, as wide as or wider than fruits, apex obtuse; ovules 4–12 per ovary; style (4–)6–8 mm (exceeding sinus). **Seeds** flattened. $2n$ = 8, 10, 16, 24.

Flowering Apr–Jul. Clay hillsides, limestone gravel, dolomite ridges, roadbanks, loose gravel, reddish clay, sagebrush and pinyon-juniper areas; 1500–3200 m; Ariz., Calif., Idaho, Nev., Oreg., Utah.

Physaria chambersii has been divided into three varieties based on whether the fruit is stipitate (var. *canaani*) or not, and whether the caudex elongates (var. *sobolifera*) or not (var. *chambersii*). In this species and in some others, e.g., *P. newberryi*, the latter character often depends on substrate and microclimate. Shifting substrates, such as moving sand and talus, often cause caudices to elongate. The species can be confused with 57. *P. newberryi*.

16. **Physaria cinerea** (S. Watson) O'Kane & Al-Shehbaz, Novon 12: 322. 2002 • Basin bladderpod [E]

Lesquerella cinerea S. Watson, Proc. Amer. Acad. Arts 23: 255. 1888

Perennials; caudex branched, (woody); densely pubescent, trichomes (appressed, except spreading on pedicels and fruits, sessile or short-stalked), several-rayed, rays furcate or bifurcate, (strongly tuberculate throughout). **Stems** few from base, erect, (stout), to 1.5 dm (sometimes greatly reduced). **Basal leaves:** blade suborbicular to rhombic or broadly elliptic, 1–4.5 cm, margins entire, (apex rounded to subacute). **Cauline leaves:** (proximal petiolate, distal subsessile); blade elliptic, 1–4 cm, (distal narrower), margins entire or remotely dentate. **Racemes** condensed, (subcorymbose to subumbellate, few-flowered). **Fruiting pedicels** (horizontal or divaricate-ascending, straight or slightly curved), 5–15 mm, (stout and rigid). **Flowers:** sepals (persistent), broadly ovate or oblong to narrowly elliptic, 5.5–8(–9.5) mm, (lateral pair slightly cucullate, median pair thickened apically, cucullate, usually keeled); petals (orange to yellow), oblong to obovate, 8–11.5 (–14.5) mm, (slightly narrowed to broad claw, margins sinuate, often retuse). **Fruits** (sessile or substipitate), globose, ovoid, or suborbicular, compressed (with marginal and apical constriction), 4–7 mm; valves pubescent, trichomes contiguous or overlapping, often spreading; ovules 16–24 per ovary; style 2–4 mm. **Seeds** lenticular, ovate in outline. $2n$ = 10.

Flowering Mar–May. Red soil, chiprock, gypsum or chalky knolls, limestone rubble; 900–2200 m; Ariz.

17. **Physaria condensata** Rollins, Rhodora 41: 407, plate 556, figs. 1, 9, 10. 1939 • Tufted twinpod [C][E]

Perennials; caudex usually simple, rarely branched, (enlarged with persistent leaf bases, cespitose); (silvery) pubescent throughout, trichomes several-rayed, rays typically furcate, (fused at base, arms slender, tuberculate throughout). **Stems** several from base, decumbent to ascending, (arising laterally beneath a dense rosette), less than 0.1 dm. **Basal leaves:** (petiole slender); blade (horizontal on the ground), obovate, 0.5–1.5 cm (width 4–8 mm, base tapering abruptly to petiole), margins entire, (apex usually acute, surfaces silvery from a dense incrustation of appressed, stellate trichomes). **Cauline leaves:** blade oblanceolate, 0.5–1 cm (width 2–3 mm), margins entire, (surfaces densely stellate pubescent). **Racemes** congested, (subumbellate, often almost sessile, barely exceeding basal leaves). **Fruiting pedicels** (divaricate, straight), 5–10 mm. **Flowers:** sepals (yellowish green), narrowly lanceolate, 4–5 mm; petals (erect), oblanceolate, 6–7 mm, (claw weakly differentiated from blade). **Fruits** didymous, ovate, inflated, 4.8–6 × 6–10 mm, (papery, basal and apical sinuses deep); valves (retaining seeds after dehiscence), pubescent, trichomes loosely spreading; replum obovate, 3–4 mm, as wide as or wider than fruit; ovules 8 per ovary (2–4 abortive); style 4–6 mm. **Seeds** flattened.

Flowering May–Jun. Calcareous knolls and ridges, clay banks, limey slopes, shaley hills, clay patches; of conservation concern; 1800–2400 m; Wyo.

18. **Physaria congesta** (Rollins) O'Kane & Al-Shehbaz, Novon 12: 322. 2002 • Dudley Bluffs bladderpod [C][E][F]

Lesquerella congesta Rollins, Contr. Gray Herb. 214: 8. 1984

Perennials; (relatively diminutive, strongly condensed); caudex (buried), simple or branched, (stout, thatched, thickened with persistent leaf bases); densely pubescent, trichomes (appressed, stiff), 4- or 5-rayed, rays fused at center, (mostly bifurcate). **Stems** simple or few from base, decumbent to ascending, (arising laterally from a tight hemispherical tuft of leaves), to 0.15 dm. **Basal leaves** similar to cauline, (erect, surfaces silvery). **Cauline leaves** (ascending, subsessile); blade linear-oblanceolate, (0.6–)0.8–1.3(–1.5) cm, margins entire, (apex acute to narrowly obtuse). **Racemes** strongly congested, (often sessile or nearly so, lateral to leaves). **Fruiting pedicels** (erect or ascending, straight to slightly curved), 3–6 mm. **Flowers:** sepals (loosely erect), narrowly oblong, 3–4 mm; petals spatulate, 5–6 mm. **Fruits** ovate, compressed (latiseptate) on margins and apically, 4–5 mm; valves pubescent, trichomes densely appressed; ovules 4 per ovary; style 1–1.5 mm. **Seeds** plump.

Flowering Apr–May. Barren knolls with pinyon-juniper; of conservation concern; 1800–2100 m; Colo.

Physaria congesta is found on white, decomposed shale of the Thirteenmile Creek Tongue of the Green River Formation.

19. **Physaria cordiformis** Rollins, Contr. Gray Herb. 171: 47. 1950 • Wassuk Range bladderpod [E]

Lesquerella cordiformis (Rollins) Rollins & E. A. Shaw; *L. kingii* S. Watson var. *cordiformis* (Rollins) Maguire & A. H. Holmgren; *L. kingii* var. *nevadensis* Maguire & A. H. Holmgren

Perennials; caudex simple or branched; densely pubescent, trichomes (short-stalked), several-rayed, rays furcate or bifurcate, (sometimes slightly umbonate, prominently tuberculate). **Stems** simple or few from base, prostrate to decumbent (arising laterally from a tuft of leaves, unbranched), 0.5–1.5 dm. **Basal leaves:** blade suborbicular, deltate to rhombic, or elliptic, margins entire or sparsely dentate, 2–4(–6) cm. **Cauline leaves** (shortly petiolate); blade oblanceolate to linear, 1–2(–3) cm, margins entire. **Racemes** loose, (sometimes elongated). **Fruiting pedicels** (sigmoid), 5–10 mm. **Flowers:** sepals lanceolate, 3.5–6(–8) mm; petals obovate to oblanceolate, (5–)7–8.5(–10) mm. **Fruits** obcordate to truncate or obcompressed, slightly compressed (angustiseptate, inflated at lobe tips), 3–6 mm (wider than long); valves densely pubescent, trichomes appressed or slightly spreading; (septum usually fenestrate); ovules 4–8 per ovary; style (slender), 3–6.5 mm, (often pubescent). **Seeds** flattened. $2n = 10$.

Flowering May–Aug. Dry sandy or gravelly soils, sagebrush, pinyon-juniper, and juniper communities, steep hillsides, rocky ridges, talus, whitish clay hills; 1500–2700 m; Calif., Idaho, Nev., Utah.

20. **Physaria curvipes** (A. Nelson) Grady & O'Kane, Novon 17: 183. 2007 • Curved bladderpod [E]

Lesquerella curvipes A. Nelson, Bull. Torrey Bot. Club 25: 205. 1898

Perennials; caudex simple; densely pubescent, trichomes (often wavy, closely appressed to blade surfaces), 4–5-rayed, rays furcate or bifurcate, slightly fused near base, (tuberculate throughout). **Stems** simple from base, loosely spreading, usually decumbent, (well exserted from basal leaves, often reddish purple), 0.8–2.4 dm. **Basal leaves:** blade (erect), spatulate to nearly rhombic, 2.5–5(–9) cm, (base gradually narrowed to petiole), margins entire, (flat). **Cauline leaves:** blade spatulate, similar to basal, margins entire. **Racemes** loose, (elongated, exceeding basal leaves). **Fruiting pedicels** (ascending, curved or sigmoid), 4–7 mm. **Flowers:** sepals (pale yellow), lingulate to spatulate, 3.5–4 mm; petals narrowly oblanceolate, 4–6 mm. **Fruits** ellipsoid, not inflated (strongly latiseptate, more so at apex), (3–)5–9 mm; valves pubescent, trichomes closely appressed to surface; ovules 4–8 per ovary; style 2.5–4.5 mm (never more than ½ fruit length). **Seeds** plump.

Flowering Jun–Jul. Limestone outcrops; 1600–2800 m; Mont., Wyo.

Physaria curvipes is known from the Big Horn Mountains.

21. **Physaria densiflora** (A. Gray) O'Kane & Al-Shehbaz, Novon 12: 322. 2002 • Denseflower or low bladderpod [E]

Vesicaria densiflora A. Gray, Boston J. Nat. Hist. 6: 145. 1850; *Lesquerella densiflora* (A. Gray) S. Watson

Annuals or biennials; caudex simple or branched, (relatively small, cespitose); densely pubescent, trichomes (spreading, sessile or short-stalked), 5–7-rayed, rays distinct and simple, (tuberculate, finely tubercled with a U-shaped notch on one side). **Stems** simple or few to several from base, erect or decumbent, (rarely branched, usually leafy), to 4 dm. **Basal leaves:** blade lyrate-pinnatifid, 1–7 cm, margins entire or shallowly dentate. **Cauline leaves** (sessile or shortly petiolate); blade narrowly obovate to elliptic, 0.5–6 cm, margins entire, repand, or shallowly dentate. **Racemes** dense, (elongated in fruit, often subtended by distal cauline leaves). **Fruiting pedicels** (usually divaricate-spreading, straight or slightly curved, delicate, sometimes drooping, especially on herbarium specimens), 7–10 mm, (somewhat rigid). **Flowers:** sepals elliptic, 3.7–7.2 mm, (lateral pair somewhat cucullate, median pair thickened apically); petals (yellow to orange-yellow), obovate to obdeltate, (4.5–)7–10(–11) mm, (tapering to short claw, apex often emarginate). **Fruits** (sessile or substipitate), globose or broadly obovate, not inflated, 4–6 mm, (smooth); valves (not retaining seeds after dehiscence), glabrous; replum as wide as or wider than fruit; ovules 8–16 per ovary; style 2–5 mm. **Seeds** flattened. $2n = 14$.

Flowering Mar–May. Sandy, granitic, or calcareous soils, sandy ledges, limestone outcrops, rocky prairies, uplands; 30–400 m; Tex.

Alyssum densiflorum (A. Gray) Kuntze (1891), not Desfontaines (1808) is an illegitimate name, sometimes found in synonymy with *Physaria densiflora*.

22. **Physaria didymocarpa** (Hooker) A. Gray, Gen. Amer. Bor. 1: 162. 1848 • Common twinpod [E]

Vesicaria didymocarpa Hooker, Fl. Bor.-Amer. 1: 49, plate 16. 1830; *Coulterina didymocarpa* (Hooker) Kuntze

Perennials; caudex branched, (cespitose); densely pubescent, trichomes (often stalked, appressed to wavy and spreading), several-rayed, rays furcate or simple, (slightly to strongly umbonate, nearly smooth to strongly tuberculate). **Stems** several from base, decumbent, (unbranched, leafy for the genus), ca. 1 dm. **Basal leaves** (forming a strong rosette; long-petiolate); blade obovate, 1.5–4(–8) cm, (base ± abruptly narrowing to petiole), margins usually repand or dentate, rarely entire, (apex usually angular, surfaces silvery). **Cauline leaves:** blade oblanceolate, 1–2 cm (width 4–8 mm), margins entire or with occasional tooth, (apex acute). **Racemes** congested, (elongated in fruit, greatly exceeding leaves). **Fruiting pedicels** (spreading, straight or slightly curved), 8–12 mm. **Flowers:** sepals lanceolate to oblong, 6–8 mm, (often keeled); petals spatulate, 10–12 mm. **Fruits** (erect), didymous, inflated, 10–20 × 10–20 mm, (papery or firm, basal sinus shallow to deep, sometimes barely notched, apical sinus deep, narrow, usually closed); valves (retaining seeds after dehiscence), loosely pubescent, trichomes spreading (appearing fuzzy; replum obovate to broadly oblong, not constricted, 3–4 mm, as wide as or wider than fruit, apex obtuse; ovules (4–)8 per ovary; style 7–9 mm. **Seeds** flattened.

Subspecies 3 (3 in the flora): w North America.

The characters used to differentiate *Physaria didymocarpa* from *P. saximontana* (especially subsp. *dentata*) appear to be weak at best: whether there are 4 or 8 ovules per ovary and whether the silicle lacks a basal sinus or one is present. There appears to be intergradation in each of those characters. A traditional circumscription of these species is followed here. Further work is needed at both the species and subspecies level in these taxa.

1. Trichomes spreading throughout; basal leaf blade margins dentate..22b. *Physaria didymocarpa* subsp. *lanata*
1. Trichomes appressed (except fruits in subsp. *didymocarpa*); basal leaf blade margins repand to dentate, ± lyrate or, rarely, entire.
    2. Fruits moderately inflated, trichomes spreading..22a. *Physaria didymocarpa* subsp. *didymocarpa*
    2. Fruits highly inflated (papery), trichomes appressed..22c. *Physaria didymocarpa* subsp. *lyrata*

22a. Physaria didymocarpa (Hooker) A. Gray subsp. didymocarpa [E]

Physaria macrantha Blankinship

Trichomes somewhat spreading (on herbage), loose and spreading (on fruits). **Basal leaves:** blade margins repand to dentate. **Fruits** moderately inflated, (firm); ovules 8 per ovary. $2n = 16, 24, 56?$

Flowering May–Aug. Gravel bars, steep shale outcrops, rocky flats, gravelly prairies, talus slopes, dry hillsides, road cuts, mountains and foothills; 500–3100 m; Alta., B.C.; Idaho, Mont., Wash., Wyo.

22b. Physaria didymocarpa (Hooker) A. Gray subsp. lanata (A. Nelson) O'Kane, Novon 17: 379. 2007 • Lanate common twinpod [E]

Physaria didymocarpa var. *lanata* A. Nelson, Bull. Torrey Bot. Club 31: 241. 1904; *P. lanata* (A. Nelson) Rydberg

Trichomes spreading throughout (tangled). **Basal leaves:** blade margins dentate. **Fruits** inflated; ovules (4–)8 per ovary. $2n = 8, 16$.

Flowering May–Jun. Steep limey banks, rock and sand road cuts; (2100–)3300–4100 m; Mont., Wyo.

22c. Physaria didymocarpa (Hooker) A. Gray subsp. lyrata (C. L. Hitchcock) O'Kane, Novon 17: 379. 2007 • Idaho or salmon twinpod [C][E]

Physaria didymocarpa var. *lyrata* C. L. Hitchcock in C. L. Hitchcock et al., Vasc. Pl. Pacif. N.W. 2: 530, fig. p. 534 [upper left]. 1964

Trichomes appressed throughout. **Basal leaves:** blade margins ± lyrate or, rarely, entire. **Fruits** highly inflated, (papery); ovules 8 per ovary. $2n = 24$.

Flowering May–Jun. Gravel, alluvial fans, steep banks; of conservation concern; 1300–2100 m; Idaho.

Subspecies *lyrata* is found only in the Salmon River drainage.

23. Physaria dornii Lichvar, Brittonia 35: 150, figs. 1–3. 1983 • Dorn's twinpod [C][E]

Perennials; (compact); caudex simple, (stout); densely (silvery) pubescent throughout (except style), trichomes several-rayed, rays furcate, fused at base, (umbonate, tuberculate throughout). **Stems** simple from base, erect, (arising from a condensed rosette), to 1 dm. **Basal leaves** (ascending or erect); blade elliptic to oblanceolate to obovate, (usually curled from middle to apex), (1.5–)5–7 cm (width 12–20 mm), margins entire. **Cauline leaves** (1–5); blade oblanceolate, 1–2.5 cm, margins entire. **Racemes** compact (or elongated in fruit, to 1 dm, barely exceeding leaves). **Fruiting pedicels** (divaricate-ascending, slightly curved), 7–18 mm. **Flowers:** sepals (erect), oblong to linear or spatulate, 5.5–7 mm; petals spatulate, 10–14 mm, (claw undifferentiated from blade). **Fruits** didymous, irregular, highly inflated, 8–11(–18) × 10–15 mm, (papery, basal sinus shallower than the deep apical sinus; valves retaining seeds after dehiscence); replum obovate, not constricted, 1–1.8 mm, apex obtuse, as wide as or wider than fruit; ovules (4–)8(–12) per ovary; style 4–6 mm, (glabrous). **Seeds** flattened, (oblong to elliptic, thin-margined or not).

Flowering May–Jun. Calcareous shale, slopes, ridges; of conservation concern; 1900–2200 m; Wyo.

24. Physaria douglasii (S. Watson) O'Kane & Al-Shehbaz, Novon 12: 322. 2002 • Douglas's bladderpod [E]

Lesquerella douglasii S. Watson, Proc. Amer. Acad. Arts 23: 255. 1888

Perennials; caudex simple; densely pubescent, trichomes (sessile or nearly so), 4–6(–10)-rayed, rays usually furcate near base, rarely bifurcate, (umbonate, tuberculate throughout). **Stems** simple from base, erect, (usually unbranched), to 4.5 dm. **Basal leaves:** blade suborbicular to elliptic, 2–9.5(–11.5) cm, margins entire, sinuate, coarsely dentate, or almost lyrate-pinnatifid. **Cauline leaves** similar to basal, blade narrowly linear or, sometimes, orbicular. **Racemes** loose (lax). **Fruiting pedicels** (recurved, straight, curved, or sigmoid), 6–20 mm. **Flowers:** sepals elliptic or ovate, (2–)3.5–7.5 mm, (cucullate); petals 6–11 mm. **Fruits** obovoid to subglobose, not inflated (not angustiseptate), 3–6 mm; valves sparsely pubescent, sometimes glabrous inside, trichomes sessile or stalked; ovules 4(–8) per ovary; style (1.6–)3–6 mm. **Seeds** flattened. $2n = 10, 30$.

Subspecies 2 (2 in the flora): w North America.

1. Fruit valves: trichomes sessile; cauline leaves loosely arranged, blades narrowly linear; British Columbia, Idaho, Montana, Oregon, Washington . . . . 24a. *Physaria douglasii* subsp. *douglasii*
1. Fruit valves: trichomes stalked; cauline leaves imbricate, blades sometimes orbicular; White Bluffs adjacent to Columbia River of Washington . . . . 24b. *Physaria douglasii* subsp. *tuplashensis*

24a. Physaria douglasii (S. Watson) O'Kane & Al-Shehbaz subsp. douglasii E

Cauline leaves loosely arranged, blade narrowly linear. Fruit valves: trichomes sessile.

Flowering Apr–Jun. Sandy and gravelly soil, stream banks, base of granitic cliffs, sagebrush and grassy slopes, pine woods; 100–1400 m; B.C.; Idaho, Mont., Oreg., Wash.

Subspecies *douglasii* appears to be restricted mainly to river valleys.

24b. Physaria douglasii (S. Watson) O'Kane & Al-Shehbaz subsp. tuplashensis (Rollins, K. A. Beck & Caplow) O'Kane & Al-Shehbaz, Novon 12: 322. 2002 • White Bluffs bladderpod C E

Lesquerella tuplashensis Rollins, K. A. Beck & Caplow, Rhodora 97: 203, fig. 107. 1995

Cauline leaves imbricate, blade sometimes orbicular. Fruit valves: trichomes stalked.

Flowering May–Jun(–Aug). Caliche soil with sagebrush; of conservation concern; 300 m; Wash.

It is possible that subsp. *tuplashensis* is simply an ecotype, or that its phenotype is in response to its severe habitat on the White Bluffs of the Columbia River.

25. Physaria eburniflora Rollins, Brittonia 33: 333. 1981 • Devils Gate twinpod E

Perennials; caudex usually simple; densely pubescent, trichomes (sessile), rays often furcate, fused toward base, (nearly smooth). Stems simple from base, prostrate, (arising lateral to rosette), 0.1–0.5 dm. Basal leaves: blade suborbicular, (1–)2.5(–3) cm, (base abruptly narrowed to petiole), margins entire, (flat), (surfaces densely silvery pubescent, trichomes in multiple layers, appressed). Cauline leaves (2–4); blade oblanceolate, ca. 1 cm, (base cuneate), margins entire, (apex acute). Racemes condensed. Fruiting pedicels (divaricate-ascending, nearly straight), 6–10 mm. Flowers: sepals (erect, purplish to greenish), linear-oblong or boat-shaped, 5.5–6.5 mm, (lateral pair more saccate than median); petals (white), spatulate, 9–12 mm, (claw undifferentiated from blade). Fruits strongly didymous, irregular in shape and size, (base slightly cordate, apex with a deep closed sinus), strongly to somewhat inflated, 6–8 × 6–8 mm (± bladderlike, papery); valves (retaining seeds after dehiscence), pubescent; replum elliptic to obovate, not constricted, as wide as or wider than fruit, apex obtuse; ovules 4–8 per ovary; style 4–5 mm, (sparsely pubescent or glabrous). Seeds plump.

Flowering May–Jun. Limestone hills, red soil, rocky calcareous slopes, clay depressions, granite and marble detritus; 1800–3000 m; Wyo.

26. Physaria engelmannii (A. Gray) O'Kane & Al-Shehbaz, Novon 12: 322. 2002 • Engelmann's bladderpod E

Vesicaria engelmannii A. Gray, Gen. Amer. Bor. 1: 162, plate 70. 1848; *Alyssum engelmannii* (A. Gray) Kuntze; *Lesquerella engelmannii* (A. Gray) S. Watson; *V. engelmannii* var. *elatior* A. Gray; *V. pulchella* Kunth & Bouché

Perennials; caudex simple or branched, (woody, aerial); densely pubescent, trichomes (sessile or short-stalked), several-rayed, rays simple or furcate, distinct or fused at base, (asymmetrical with deep notch on one side, often with a U-shaped gap between 2 of the rays, umbonate, strongly tuberculate). Stems few to several from base, erect, (usually unbranched), (1.5–)2.5–4(–6) dm. Basal leaves: blade elliptic to obovate, 2–6.5 cm, margins entire, sinuate, or remotely toothed, (surfaces occasionally sparsely pubescent). Cauline leaves: (proximal often petiolate, distal sessile or subsessile); blade oblanceolate to linear, 1–4 cm, margins entire. Racemes dense, (subumbellate). Fruiting pedicels (ascending), relatively short. Flowers: sepals ovate or elliptic, 5.5–10 mm, (median pair thickened apically, cucullate); petals (bright yellow), obovate to elliptic, 8–14 mm, (sometimes with distinct claw, often retuse). Fruits (shortly stipitate), ± globose or ellipsoid, not or slightly inflated, 5–8 mm; valves (not retaining seeds after dehiscence), glabrous throughout; replum as wide as or wider than fruit; ovules (8–)12–20 per ovary; style 3.5–5 mm. Seeds flattened. $2n$ = 12, 24, 36.

Flowering Apr–May. Limestone prairies, rocky ridges, pebbly shores, thin caliche soils, limestone outcrops; 150–400 m; Okla., Tex.

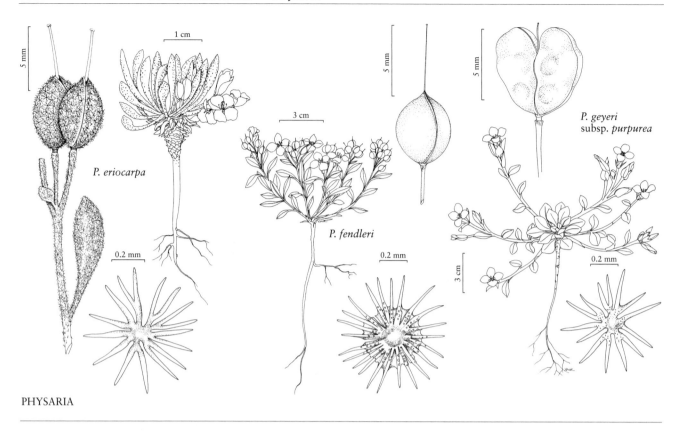

27. Physaria eriocarpa Grady & O'Kane, Novon 17: 184, fig. 3. 2007 • Sheep Mountain bladderpod C E F

Perennials; caudex branched, (thickened, cespitose); densely pubescent, trichomes 5- or 6-rayed, rays slightly fused at base, furcate or bifurcate, (tuberculate throughout). Stems few from base, well-exserted from basal leaves, 0.4–1(–1.2) dm. Basal leaves: blade (erect), obovate to orbicular, 1.5–2.5 cm, (base evidently distinct from petiole), margins entire, (folded). Cauline leaves: blade spatulate, margins entire. Racemes compact, subumbellate. Fruiting pedicels (ascending, curved to slightly sigmoid), 5–8 mm. Flowers: sepals (pale yellow), 4–5 mm; petals lingulate, 6–7 mm. Fruits ovoid to ellipsoid, slightly inflated, (apex not compressed), 3–4 mm; valves pubescent, trichomes erect on mature fruits, (± appearing fuzzy); ovules 8 per ovary; style 4–5 mm, (glabrous). Seeds plump.

Flowering Jun–Jul. Windswept ridge lines and mountain peaks in limestone rubble and cobbles; of conservation concern; 2600–3000 m; Mont.

Physaria eriocarpa is known from Sheep Mountain.

28. Physaria fendleri (A. Gray) O'Kane & Al-Shehbaz, Novon 12: 323. 2002 • Fendler's bladderpod F

Vesicaria fendleri A. Gray, Mem. Amer. Acad. Arts, n. s. 4: 9. 1849; *Alyssum fendleri* (A. Gray) Kuntze; *A. stenophyllum* (A. Gray) Kuntze; *Lesquerella fendleri* (A. Gray) S. Watson; *L. foliacea* Greene; *L. praecox* Wooton & Standley; *L. stenophylla* (A. Gray) Rydberg; *V. stenophylla* A. Gray; *V. stenophylla* var. *diffusa* A. Gray; *V. stenophylla* var. *humilis* A. Gray; *V. stenophylla* var. *procera* A. Gray

Perennials; caudex branched, (sometimes woody at base); densely (silvery) pubescent, trichomes (sessile or short-stalked), several-rayed, rays not furcate, fused (webbed) ca. ½ their length, (tuberculate throughout or tubercles scarce or absent over center). Stems several from base, erect or laterally decumbent, (usually unbranched), (0.3–)0.5–2.5(–4) dm. Basal leaves: blade linear to somewhat elliptic, 1–4(–8) cm, margins entire or coarsely dentate. Cauline leaves (shortly petiolate); blade usually linear to narrowly oblanceolate, rarely elliptic to rhombic, 0.5–2.5 cm, (base narrowing to petiole), margins entire or remotely dentate (sometimes involute). **Racemes** loose to somewhat dense. **Fruiting pedicels** (divaricate-spreading to erect, usually straight or slightly curved, occasionally sigmoid), 8–20(–40) mm. **Flowers:** sepals elliptic to oblong, 5–8 mm, (lateral pair not saccate, median pair often thickened apically, ± cucullate); petals (usually orange or orange-yellow at junction of blade and claw, sometimes also with orange guidelines), obdeltate to obovate, 8–12 mm, (claw relatively short). **Fruits** globose, broadly ellipsoid, or ovoid, not or slightly inflated, 5–8 mm, (firm, apex usually acute); valves (not retaining seeds after dehiscence, often reddish in age), glabrous throughout; replum as wide as or wider than fruit; ovules (12–)20–32(–40) per ovary; style (2–)3–6 mm. **Seeds** flattened. $2n = 12, 14, 24$.

Flowering Mar–May. Limestone outcrops, gypseous hills, gravels, sandy washes, rocky slopes, bluffs, shallow drainage areas, plains and desert shrub areas; 100–2000 m; Ariz., Colo., Kans., N.Mex., Okla., Tex., Utah; Mexico (Chihuahua, Coahuila, Nuevo León).

In dry areas, *Physaria fendleri* may flower following suitable rains at any time of the year.

29. Physaria filiformis (Rollins) O'Kane & Al-Shehbaz, Novon 12: 323. 2002 • Limestone glade or Missouri bladderpod C E

Lesquerella filiformis Rollins, Rhodora 58: 201. 1956

Annuals; with a fine taproot; densely pubescent, trichomes (sessile), usually 4-rayed, rays forked, rarely simple or tripartite, (finely tuberculate). **Stems** several from base, (slender), erect or outer decumbent, (usually branched, branches filiform, bud clusters of growing plants drooping), to 2.5 dm. **Basal leaves:** blade orbicular to broadly spatulate, 1–2.4 cm, margins entire or sinuate. **Cauline leaves:** (proximal often petiolate, distal sessile); similar to basal, blade spatulate to oblanceolate or (distal) linear, (base cuneate), margins entire or sinuate. **Racemes** loose. **Fruiting pedicels** (usually divaricate-ascending, straight or slightly curved), 7–11 mm. **Flowers:** sepals oblong or elliptic, 2.5–4.6 mm, (median pair slightly thickened apically); petals (pale yellow), spatulate to obovate, 5–9 mm, (apex ± emarginate). **Fruits** (sessile or shortly stipitate), globose, not inflated, 3–4 mm; valves (not retaining seeds after dehiscence), glabrous throughout; replum as wide as or wider than fruit; ovules 4 per ovary; style 3–5 mm. **Seeds** flattened. $2n = 14$.

Flowering Apr–May. Limestone, dolomite, and shale, sparsely vegetated or barren areas, cedar glades, old pastures, along roadsides; of conservation concern; 200–300 m; Ala., Ark., Mo.

30. Physaria floribunda Rydberg, Bull. Torrey Bot. Club 28: 279. 1901 • Point-tip twinpod E

Perennials; caudex branched, (cespitose); (silvery) pubescent throughout, trichomes several-rayed, rays mostly furcate, (arms of unequal lengths, finely tuberculate). **Stems** several from base, erect or lateral decumbent, (unbranched), 1–2 dm. **Basal leaves:** (petiole usually winged); blade broadly oblanceolate, 3–8 cm, margins usually dentate or pinnatifid, rarely subentire, (terminal lobe acute or obtuse, not rounded). **Cauline leaves:** blade spatulate to linear-oblanceolate, 1–3 cm, margins usually entire, rarely toothed, (apex acute). **Racemes** loose (and greatly elongated in fruit) to congested. **Fruiting pedicels** (recurved), 6–15 mm. **Flowers:** sepals linear-oblong, 5–7 mm; petals spatulate, 9–11 mm. **Fruits** (usually pendent on arching pedicels, less frequently widely divergent), irregular in shape, (base obtuse or slightly cordate, apex deeply and broadly notched), not strongly inflated, 8–11 × 8–12 mm, (papery; valves retaining seeds after dehiscence; replum linear-oblong, constricted, 2.5–4 mm, as wide as or wider than fruit, apex obtuse; ovules 4 per ovary; style 5–8 mm. **Seeds** flattened.

Subspecies 2 (2 in the flora): wc United States.

1. Fruit valves inflated, membranous; trichomes appressed . . . 30a. *Physaria floribunda* subsp. *floribunda*
1. Fruit valves usually not inflated, coriaceous; trichomes slightly spreading . . . . 30b. *Physaria floribunda* subsp. *osterhoutii*

30a. Physaria floribunda Rydberg subsp. **floribunda** E

Racemes usually loose, rarely somewhat congested. **Fruiting pedicels** 6–12 mm. **Fruit valves** slightly inflated, membranous; trichomes appressed. $2n = 8, 10, 16, 24$.

Flowering May–Jun. Steep hillsides, decomposed granite, rocky banks, shale hills, rocky ravines, sagebrush and pinyon-juniper areas; 1700–2900 m; Colo., N.Mex.

30b. Physaria floribunda Rydberg subsp. **osterhoutii** (Payson) O'Kane, Novon 17: 379. 2007 • Osterhout's twinpod [C][E]

Physaria osterhoutii Payson, Ann. Missouri Bot. Gard. 5: 146. 1918; *P. floribunda* var. *osterhoutii* (Payson) Rollins

Racemes congested (4–8 cm). **Fruiting pedicels** 10–15 mm. **Fruit valves** usually not inflated, coriaceous; trichomes slightly spreading.

Flowering May–Jun. Limey ridges, chiprock and clay knolls, shale hillsides, gravelly clay bluffs; of conservation concern; 2200–2500 m; Colo.

31. Physaria fremontii (Rollins & E. A. Shaw) O'Kane & Al-Shehbaz, Novon 12: 323. 2002 • Fremont's bladderpod [C][E]

Lesquerella fremontii Rollins & E. A. Shaw, Gen. Lesquerella, 228. 1973

Perennials; caudex simple; densely pubescent, trichomes (subsessile), 5–7-rayed, rays distinct or slightly fused at base, usually furcate, (roughly tuberculate). **Stems** few to several from base, prostrate, (arising proximal to a terminal cluster of erect leaves, usually unbranched, slender), 0.5–1(–1.5) dm. **Basal leaves:** (petiole slender); blade elliptic to rhombic, 1.5–4(–5) cm, (base gradually tapering to petiole), margins usually entire, rarely dentate. **Cauline leaves:** blade narrowly obovate to oblanceolate, 5–15 mm, margins entire. **Racemes** (relatively short), rather loose. **Fruiting pedicels** (usually secund, recurved), 5–8 mm. **Flowers:** sepals ± elliptic, 4–6 mm; petals ovate, 6–8 mm, (claw slightly expanded at base). **Fruits** (pendent), globose, subglobose, or slightly obcompressed, slightly or not inflated, 3–6 mm, (rigid, apex usually beaked); valves pubescent throughout; ovules 8–12 per ovary; style 1–2 mm, (pubescent). **Seeds** slightly flattened, (suborbicular).

Flowering May–Jun. Calcareous gravel, loose whitish rubble, limestone pavement, rocky calcareous ridges; of conservation concern; 2100–2800 m; Wyo.

Physaria fremontii is known from the area of the Wind River Mountains.

32. Physaria garrettii (Payson) O'Kane & Al-Shehbaz, Novon 12: 323. 2002 • Garrett's bladderpod [C][E]

Lesquerella garrettii Payson, Ann. Missouri Bot. Gard. 8: 213. 1922

Perennials; caudex simple or branched; densely pubescent, trichomes (sessile), 4–7-rayed, rays furcate or bifurcate, (smooth or, rarely, finely tuberculate). **Stems** simple or several from base, spreading, (unbranched, sparsely pubescent), to 1.5 dm. **Basal leaves:** blade narrowly elliptic or obovate, 1–3(–4) cm, margins entire or nearly so. **Cauline leaves** (sessile or shortly petiolate); blade narrowly obovate or oblanceolate, 0.4–1.2 cm, margins entire. **Racemes** loose, (few-flowered). **Fruiting pedicels** (spreading, straight or slightly curved), 4–7 mm. **Flowers:** sepals linear, lanceolate, or elliptic, 3.5–6.5 mm, (median pair thickened apically, cucullate); petals oblanceolate, 5.5–9(–10) mm. **Fruits** globose or subglobose, not or slightly compressed, 3.5–4.3 mm; valves densely pubescent, trichomes spreading, 3–6-rayed, (appearing shaggy); ovules 4–8 per ovary; style 4.5–7 mm. **Seeds** slightly flattened, (suborbicular).

Flowering Jun–Aug. Rock crevices, rocky slopes, ridges; of conservation concern; 3000–3700 m; Utah.

Physaria garrettii is known from the area of the Wasatch Mountains.

33. Physaria geyeri (Hooker) A. Gray, Gen. Amer. Bor. 1: 162. 1848 • Geyer's bladderpod [E][F]

Vesicaria geyeri Hooker, London J. Bot. 6: 70, plate 5. 1847; *Coulterina geyeri* (Hooker) Kuntze; *Lesquerella geyeri* (Hooker) G. A. Mulligan

Perennials; caudex usually simple, (cespitose); (silvery) pubescent throughout, trichomes (sessile), 6–8-rayed, rays mostly furcate, (tuberculate to nearly smooth). **Stems** several from base, decumbent, (arising laterally, unbranched), 1–3 dm. **Basal leaves** (numerous); (petiole slender, rarely with a few broad teeth); blade obovate, 3–7 cm, margins entire. **Cauline leaves:** blade oblanceolate, 1.5–3 cm, margins entire. **Racemes** loose. **Fruiting pedicels** (ascending or spreading, slightly curved or sigmoid), 1–2 cm. **Flowers:** sepals oblong, 5–7 mm; petals (yellow to purplish), spatulate, 8–12 mm. **Fruits** obcordate, angustiseptate, somewhat inflated, (not bladdery), 5–7 × 6–9 mm, (papery, basal sinus absent, apical sinus broad and open); valves (retaining seeds after dehiscence), loosely pubescent, trichomes spreading; replum ovate, 5–7 mm, as wide as or wider than fruit, apex acute or obtuse; ovules 4–6 per ovary; style 5–7 mm.

Subspecies 2 (2 in the flora): nw United States.

1. Replum apices acute; ovules 4 per ovary; petals yellow, drying yellow ....33a. *Physaria geyeri* subsp. *geyeri*
1. Replum apices obtuse; ovules usually 6 per ovary; petals purplish or light yellow, drying purplish .... 33b. *Physaria geyeri* subsp. *purpurea*

33a. Physaria geyeri (Hooker) A. Gray subsp. geyeri E

Trichomes tuberculate. **Petals** yellow, drying yellow. **Fruits:** replum apex acute; ovules 4 per ovary. $2n = 8$.

Flowering Apr–Jun. Gravel slides and banks, sandy bluffs, washes, stream and lake shores; 300–2000 m; Mont., Oreg., Wash.

33b. Physaria geyeri (Hooker) A. Gray subsp. purpurea (Rollins) O'Kane, Novon 17: 380. 2007 • Purple twinpod E F

Physaria geyeri var. *purpurea* Rollins, Rhodora 41: 401. 1939

Trichomes nearly smooth. **Petals** purplish or light yellow, drying purplish. **Fruits:** replum apex obtuse; ovules usually 6 per ovary. $2n = 8$.

Flowering May–Jul. Steep marly hillsides, steep slides, fine talus, scree, sandy road cuts; 1200–2200 m; Idaho.

34. Physaria globosa (Desvaux) O'Kane & Al-Shehbaz, Novon 12: 323. 2002 • Globe bladderpod C E

Vesicaria globosa Desvaux, J. Bot. Agric. 3: 184. 1815; *Alyssum globosum* (Desvaux) Kuntze; *A. shortii* (Torrey & A. Gray) Kuntze; *Lesquerella globosa* (Desvaux) S. Watson

Biennials or perennials; caudex branched, (± woody); densely pubescent, trichomes (sessile), 3–6-rayed, rays distinct and simple or furcate, (in 2 layers, lower layer umbonate, smooth to finely tuberculate, some often with a U-shaped notch). **Stems** several from base, erect, (arising among leaves of an elongated main axis), to 5 dm (± equal). **Basal leaves** (shortly petiolate); blade obovate to oblanceolate, (1.5–)2.5–5(–6) cm, margins entire, sinuate to shallowly toothed, or pinnatifid. **Cauline leaves** (sessile or shortly petiolate); blade oblanceolate to oblong, 1.3–3(–4) cm, (base cuneate), margins entire or repand to dentate. **Racemes** dense. **Fruiting pedicels** (usually spreading horizontally, straight), 7–14(–21) mm. **Flowers:** sepals elliptic or obovate, 2.6–4.1 mm, (median pair thickened apically, cucullate); petals (bright yellow), obovate, 3.5–6.5(–7.5) mm, (margins sinuate). **Fruits** (sessile or substipitate); globose, often slightly compressed apically, (1–)2–3 mm; valves sparsely pubescent, sometimes pubescent inside, trichomes spreading, 3–5-rayed; ovules 4 per ovary; style 2–3.5(–4) mm. **Seeds** flattened or plump, (often outer surface hemispherical, inner surface flattened, or both surfaces rounded). $2n = 14$.

Flowering Mar–May. Open rocky areas, shale at cliff bases, open talus, ledges, open cedar glades; of conservation concern; 100–300 m; Ind., Ky., Tenn.

Physaria globosa is possibly introduced in Indiana. A report for Ohio was based on a collection by "Jones," but that specimen cannot be located.

35. Physaria gooddingii (Rollins & E. A. Shaw) O'Kane & Al-Shehbaz, Novon 12: 323. 2002 • Goodding's bladderpod E

Lesquerella gooddingii Rollins & E. A. Shaw, Gen. Lesquerella, 164. 1973

Annuals or biennials; without caudex, (cespitose); densely pubescent, trichomes (sessile or short-stalked), few-rayed, rays (ascending or erect), simple or infrequently furcate near base, (long and slender, sometimes with U-shaped notch on one side, smooth or finely tuberculate). **Stems** several from base, erect (and stout) or outer ones decumbent, (unbranched or branched, stiff and densely foliate, sterile leaf-bearing branches sometimes present), to 4 dm. **Basal leaves:** blade obovate or elliptic, to ca. 3 cm, margins sinuate or shallowly dentate. **Cauline leaves:** (proximal usually shortly petiolate, distal sessile); blade obovate to broadly elliptic, 1–3 cm, margins sinuate or shallowly toothed. **Racemes** dense, compact, (elongated in fruit). **Fruiting pedicels** (recurved, curved or sigmoid), somewhat expanded apically. **Flowers:** sepals elliptic or narrowly elliptic or oblong, (3.8–)4.5–5.5 mm, (lateral pair cucullate, very convex, median pair tapering to base, thickened apically, cucullate, often slightly keeled); petals cuneate, 6.5–8 mm, (slightly expanded at base, margins sinuate, apex retuse or entire). **Fruits** (sessile or substipitate), oblong or broadly elliptic, compressed (latiseptate), 5–8 mm; valves pubescent, trichomes spreading, sparsely pubescent inside; ovules 4–6 per ovary; style 3–5 mm. **Seeds** flattened.

Flowering Jun–Sep. Mountainous areas, open areas in pinyon-juniper and ponderosa pine forests; 1800–2300 m; Ariz., N.Mex.

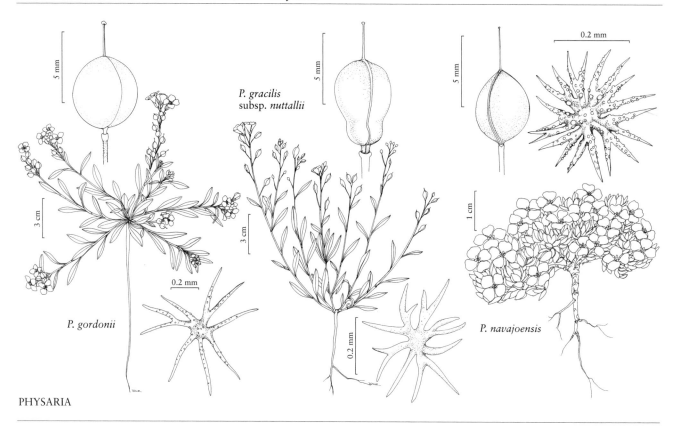

Physaria gooddingii (found in the mountains of Catron, Sierra, and western Socorro counties, New Mexico, and in Greenlee County, Arizona) is similar to 9. *P. aurea* (found farther east), but differs in having trichomes with ascending or erect rays (rather than appressed) and fruits that are strongly latiseptate (rather than not, or very little, compressed), a state that is infrequent in the genus.

36. **Physaria gordonii** (A. Gray) O'Kane & Al-Shehbaz, Novon 12: 323. 2002 • Gordon's bladderpod F W

Vesicaria gordonii A. Gray, Boston J. Nat. Hist. 6: 149. 1850 (as gordoni); *Alyssum gordonii* (A. Gray) Kuntze; *Lesquerella gordonii* (A. Gray) S. Watson; *L. gordonii* var. *densifolia* Rollins; *Physaria gordonii* subsp. *densifolia* (Rollins) O'Kane & Al-Shehbaz; *P. gordonii* var. *densifolia* (Rollins) B. L. Turner

Annuals, biennials, or perennials; (short-lived); with a fine taproot; usually densely pubescent, trichomes (sessile or short-stalked), 4–7-rayed, rays distinct and furcate or bifurcate, (nearly smooth to finely tuberculate). **Stems** several from base, erect to decumbent or prostrate, (unbranched or branched, sometimes densely leaved), 1–3.5(–4.5) dm. **Basal leaves:** blade obovate to broadly oblong, 1.5–5(–8) cm, margins lyrate-pinnatifid, dentate, or entire. **Cauline leaves:** (proximal sometimes petiolate, distal sessile); blade linear to oblanceolate, often falcate, 1–4(–7) cm, (proximal with base sometimes cuneate), margins entire, repand, or shallowly dentate. **Racemes** dense. **Fruiting pedicels** (divaricate-ascending, sigmoid or, sometimes, nearly straight), 5–15(–25) mm. **Flowers:** sepals elliptic or oblong, 3–6.5 mm, (lateral pair subsaccate, median pair thickened apically, cucullate); petals (widely spreading at anthesis, yellow to orange, claw sometimes whitish), cuneate, obdeltate, or obovate, (tapering to claw), 5–8(–10) mm, (claw often widened at base). **Fruits** (shortly stipitate), subglobose, not or slightly compressed, (3–)4–8 mm; valves (not retaining seeds after dehiscence), glabrous throughout; replum as wide as or wider than fruit; ovules (8–)12–20(–26) per ovary; style (1.5–)2–4(–5) mm. **Seeds** flattened. $2n = 12, 32$.

Flowering Feb–Jul. Sandy or light soils, rocky plains, caprock ledges, gravelly brushland, sandy desert washes, stream bottoms, pastures, roadsides, abandoned fields; 150–1700 m; Ariz., Kans., N.Mex., Okla., Tex., Va.; Mexico (Chihuahua, Sonora).

Physaria gordonii was reported from Virginia in 1987 by Robert Wright from a Hampton Shale roadcut along the Blue Ridge Parkway, where it was probably a short-lived waif.

Subspecies *densifolia*, of Lincoln County, New Mexico, of which there is now more material than Rollins had available in 1993, appears to represent a suite of environmentally determined, variable, and intergrading characteristics that does not merit taxonomic recognition.

37. **Physaria gracilis** (Hooker) O'Kane & Al-Shehbaz, Novon 12: 323. 2002 • Spreading bladderpod E F

Vesicaria gracilis Hooker, Bot. Mag. 63: plate 3533. 1836; *Alyssum gracile* (Hooker) Kuntze; *Lesquerella gracilis* (Hooker) S. Watson

Annuals or biennials; (delicate, wiry); with a fine taproot; pubescent, trichomes (sessile or subsessile), 4–7-rayed, rays distinct, usually furcate, occasionally bifurcate, (smooth to somewhat tuberculate). **Stems** simple to several from base, erect, often outer decumbent, (unbranched or branched distally), 1–7 dm. **Basal leaves**: blade oblanceolate to elliptic, 1.5–8(–11.5) cm, margins lyrate-pinnatifid to dentate or repand, (abaxial surface densely pubescent, adaxial sparsely pubescent). **Cauline leaves**: (proximal petiolate, distal sessile); blade oblanceolate to oblong, 1–7 cm, margins dentate to repand. **Racemes** loose, (elongated). **Fruiting pedicels** (usually divaricate-spreading, sometimes horizontal or shallowly recurved, straight or slightly curved), (7–)10–20(–25) mm, (slender or stout). **Flowers**: sepals elliptic or broadly ovate, 3–6.5(–8) mm, (median pair slightly thickened apically, cucullate); petals (yellow to orange), broadly obovate, 6–11 mm, (narrowing gradually to short claw). **Fruits** (stipitate or subsessile, gynophore 1–2 mm), globose, subglobose, obpyriform, or obovoid, not or slightly inflated, 3–9 mm; valves (not retaining seeds after dehiscence), glabrous throughout or sparsely pubescent inside; replum as wide as or wider than fruit; ovules 8–20(–28) per ovary; style 2–4.5 mm. **Seeds** slightly flattened.

Subspecies 2 (2 in the flora): c, se United States.

1. Stems to 7 dm; cauline leaves: blade margins usually deeply dentate, rarely repand; fruits ± sessile, globose or subglobose, 3–6 mm, bases rounded . . . . 37a. *Physaria gracilis* subsp. *gracilis*
1. Stems usually less than 3 dm; cauline leaves: blade margins frequently repand, occasionally dentate; fruits stipitate (gynophore slender), obpyriform to narrowly obovoid, (5–)6–9 mm, bases truncate. . . . . 37b. *Physaria gracilis* subsp. *nuttallii*

37a. **Physaria gracilis** (Hooker) O'Kane & Al-Shehbaz subsp. **gracilis** E

Lesquerella polyantha (Schlechtendal) Small; *Vesicaria polyantha* Schlechtendal

Stems to 7 dm; bud clusters erect to drooping. **Cauline leaves**: blade margins usually deeply dentate, rarely repand. **Fruiting pedicels** slender to stout. **Fruits** ± sessile, globose or subglobose, 3–6 mm, base rounded. $2n = 12$.

Flowering Mar–May. Prairies, pastures, grassy banks, rocky slopes, sandy loam, roadsides, old fields; 50–300 m; Ala., Ill., Iowa, La., Miss., Mo., Okla., Tenn., Tex.

Subspecies *gracilis* is adventive in Illinois, Iowa, Missouri, and western Tennessee.

The record from Iowa was based on *Gilly & MacDonald 827*, reported to be in the Parsons College herbarium, now at ISTC. The specimen has not been relocated.

37b. **Physaria gracilis** (Hooker) O'Kane & Al-Shehbaz subsp. **nuttallii** (Torrey & A. Gray) O'Kane & Al-Shehbaz, Novon 12: 323. 2002 • Nuttall's bladderpod E F

Vesicaria nuttallii Torrey & A. Gray, Fl. N. Amer. 1: 101. 1838; *Alyssum nuttallii* (Torrey & A. Gray) Kuntze; *A. repandum* (Nuttall ex Torrey & A. Gray) Kuntze; *Lesquerella gracilis* (Hooker) S. Watson subsp. *nuttallii* (Torrey & A. Gray) Rollins & E. A. Shaw; *L. gracilis* var. *repanda* (Nuttall ex Torrey & A. Gray) Payson; *L. nuttallii* (Torrey & A. Gray) S. Watson; *L. repanda* (Nuttall ex Torrey & A. Gray) S. Watson; *Physaria gracilis* var. *repanda* (Nuttall ex Torrey & A. Gray) B. L. Turner; *V. repanda* Nuttall ex Torrey & A. Gray

Stems usually less than 3 dm; bud clusters erect. **Cauline leaves**: blade margins frequently repand, occasionally dentate. **Fruiting pedicels** slender. **Fruits** stipitate (gynophore slender), obpyriform to narrowly obovoid, (5–)6–9 mm, base truncate. $2n = 12$.

Flowering Apr–May. Prairies, pastures, hillsides, rocky limestone, shallow soil of glades, roadsides, open fields; 150–500 m; Kans., Okla., Tex.

38. Physaria grahamii C. V. Morton, Ann. Carnegie Mus. 26: 220. 1937 • Graham's twinpod E

Physaria acutifolia Rydberg var. *purpurea* S. L. Welsh & Reveal; *P. acutifolia* var. *repanda* (Rollins) S. L. Welsh; *P. repanda* Rollins

Perennials; caudex branched, (thick, cespitose); densely pubescent, trichomes rays (appressed on leaves, ascending on pedicels and fruits), distinct, furcate or bifurcate. **Stems** several from base, decumbent to erect or ascending (unbranched), 1–2.5 dm. **Basal leaves:** (outer ones spreading, inner erect or ascending); blade ovate, often broadly so, 4–7 cm, margins repand to lyrate-lobed. **Cauline leaves** similar to basal, blade oblanceolate or narrowly oblong, reduced in size, (base gibbous). **Racemes** loose, (elongated). **Fruiting pedicels** (ascending to divaricate-ascending, sigmoid to nearly straight), 10–17 mm. **Flowers:** sepals lanceolate or narrowly oblong, 5.8–7.2 mm; petals (erect, sometime purplish or drying purple), narrowly oblong to oblanceolate, 7–10 mm, (not or weakly clawed). **Fruits** didymous, globose or subglobose, inflated, 10–13 mm, (papery, basal and apical sinuses deep); valves (retaining seeds after dehiscence), pubescent, trichomes ascending, appearing fuzzy; replum oblong to oblanceolate, as wide as or wider than fruit, apex obtuse; ovules 4 per ovary; style (4–)5–7 mm. **Seeds** plump, (suborbicular).

Flowering May–Jun. Sagebrush, pinyon-juniper, ponderosa pine, Douglas-fir, limber pine communities on clay, or a mixture of shale fragments and clay; 2100–2900 m; Utah.

Physaria grahamii is difficult to evaluate due to the paucity of collections. The tentative recognition by N. H. Holmgren (2005b) is followed here.

39. Physaria hemiphysaria (Maguire) O'Kane & Al-Shehbaz, Novon 12: 323. 2002 • Intermountain or skyline bladderpod E

Lesquerella hemiphysaria Maguire, Amer. Midl. Naturalist 27: 466. 1942

Perennials; caudex simple or branched (tightly); sparsely to densely pubescent, trichomes (sessile or short-stalked), 4–6-rayed, rays furcate or bifurcate, distinct or slightly fused at base, (sometimes umbonate, tuberculate throughout). **Stems** few to several from base, decumbent, 0.5–1(–2) dm, (rather stout, sparsely pubescent). **Basal leaves:** blade elliptic to suborbicular, 1.5–3.5(–5.5) cm, margins entire or shallowly dentate (at base, surfaces densely pubescent, silvery). **Cauline leaves** (petiolate or distal nearly sessile); blade elliptic to obovate, 0.5–1.5 cm, margins entire. **Racemes** dense, congested, (few-flowered). **Fruiting pedicels** (spreading or recurved, sometimes loosely sigmoid), 2–6.5 mm. **Flowers:** sepals lanceolate, oblanceolate, or narrowly elliptic, 3.8–5 mm, (median pair thickened apically, cucullate); petals narrowly lanceolate to linear, 6–10(–13) mm. **Fruits** (sessile or substipitate), broadly obcordate, obdeltate, or obcompressed, slightly compressed (angustiseptate), 3–5(–7) mm; valves (not retaining seeds after dehiscence), sparsely pubescent or glabrous, trichomes appressed; replum as wide as or wider than fruit; ovules 8–16 per ovary; style (1.8–)3–6(–7) mm. **Seeds** slightly flattened, (ellipsoid to suborbicular).

Subspecies 2 (2 in the flora): wc United States

1. Styles 3–5 mm; fruiting pedicels 4–6.5 mm; fruit valves pubescent, often sparsely so .... 39a. *Physaria hemiphysaria* subsp. *hemiphysaria*
1. Styles 1.8–2.5(–3) mm; fruiting pedicels 2–4(–5.5) mm; fruit valves glabrous or sparsely pubescent .... 39b. *Physaria hemiphysaria* subsp. *lucens*

39a. Physaria hemiphysaria (Maguire) O'Kane & Al-Shehbaz subsp. hemiphysaria E

Fruiting pedicels 4–6.5 mm. **Fruits:** valves pubescent, often sparsely so, trichomes uniform; style 3–5 mm.

Flowering mid May–early Aug. Low ridges of sand and sandstone, bluish shaley clay outcrops, stony soil, open limey knolls, ridges; 2100–3400 m; Utah.

Subspecies *hemiphysaria* is found only in stony meadows of the Wasatch Plateau.

39b. Physaria hemiphysaria (Maguire) O'Kane & Al-Shehbaz subsp. lucens (S. L. Welsh & Reveal) O'Kane & Al-Shehbaz, Novon 12: 323. 2002 • Tavaputs bladderpod E

Lesquerella hemiphysaria Maguire var. *lucens* S. L. Welsh & Reveal, Great Basin Naturalist 37: 338. 1978; *Physaria hemiphysaria* var. *lucens* (S. L. Welsh & Reveal) N. H. Holmgren

Fruiting pedicels 2–4(–5.5) mm. **Fruits:** valves glabrous or sparsely pubescent, trichomes not uniform; style 1.8–2.5(–3) mm.

Flowering May–Jun. Shale outcrops and sandy soils of sagebrush-woodland areas; 2100–3400 m; Utah.

Subspecies *lucens* is known from the West Tavaputs Plateau.

40. **Physaria hitchcockii** (Munz) O'Kane & Al-Shehbaz, Novon 12: 324. 2002 • Hitchcock's bladderpod [E]

Lesquerella hitchcockii Munz, Bull. Torrey Bot. Club 56: 163. 1929

Perennials; (forming loose mats); caudex (buried), branched; densely pubescent, trichomes (short-stalked), 4–6-rayed, rays distinct, bifurcate, (rough-tuberculate). **Stems** few to several from base, prostrate to erect or spreading, 0.05–0.5(–1.2) dm. **Basal leaves:** (petiole and blade differentiated or not); blade spatulate to elliptic or linear or narrowly oblanceolate, 0.5–1.5 (–2.5) cm, margins entire. **Cauline leaves** similar to basal, smaller. **Racemes** dense. **Fruiting pedicels** (ascending, straight or slightly curved), 2–6 mm. **Flowers:** sepals narrowly lanceolate to lanceolate, 2.8–6 mm; petals (pale to deep yellow), narrowly lanceolate to oblanceolate, 5–9 mm, (claw undifferentiated from blade). **Fruits** (sessile or substipitate), globose or subglobose to obovoid, not or slightly inflated, 3–6 mm, (firm, apex acute); valves (reddish in age, not retaining seeds after dehiscence), glabrous throughout; replum as wide as or wider than fruit; ovules 4–8 per ovary; style 1.7–6 mm. **Seeds** flattened.

Subspecies 3 (3 in the flora): w United States.

The taxonomic treatment of *Physaria hitchcockii* has varied widely over the years. Molecular study (pers. obs.) has shown no direct relationship to *P. tumulosa*; morphologically, though, *P. navajoensis* and *P. tumulosa* appear closely related. Infraspecific taxonomy is based on the presence of a discernable petiole and whether or not the caudex is elastically elongated. The subspecies recognized here are usually geographically coherent, except that collections from the Table Cliff Plateau are more similar to subsp. *hitchcockii*, disjunct in Nevada, than they are to the very nearly sympatric subsp. *rubicundula*.

1. Caudices elongated and elastic; basal leaves: petiole not differentiated from blade, blades linear-oblanceolate; Aquarius, Markagunt, and Paunsaugunt plateaus, Utah (limited to the pink member of the limestone Wasatch (Claron) Formation) . . . . 40c. *Physaria hitchcockii* subsp. *rubicundula*
1. Caudices elongated or not, elastic or not; basal leaves: petiole differentiated (sometimes weakly) from blade, blades oblanceolate to obovate; Nevada, Utah.

[2. Shifted to left margin.—Ed.]
2. Plants forming tufts; caudices not elongated, not elastic; fruits 2.6–3.8 mm wide; Table Cliff Plateau, Utah (limited to the white member of the limestone Wasatch (Claron) Formation) and limestones of the Sheep Range and Spring Mountains, Nevada . . . . 40a. *Physaria hitchcockii* subsp. *hitchcockii*
2. Plants forming soft mats; caudices elongated, elastic (creeping); fruits 1.7–3 mm wide; Grant, Quinn Canyon, and Schell Creek ranges, Nevada . . . . 40b. *Physaria hitchcockii* subsp. *confluens*

40a. **Physaria hitchcockii** (Munz) O'Kane & Al-Shehbaz subsp. **hitchcockii** [E]

Plants forming tufts; caudex not elongated, not elastic. **Basal leaves:** petiole differentiated from blade (sometimes weakly); blade oblanceolate to obovate. **Anthers** 1.4–1.8(–2) mm. **Fruits** 2.6–3.8 mm wide.

Flowering Jun–Jul. Gravelly or rocky limestone at or above timberline; 2300–3500 m; Nev., Utah.

It is possible that populations of subsp. *hitchcockii* on the Table Cliff Plateau, Utah, are consubspecific with the nearby subsp. *rubicundula*. The leaf blades are indistinguishable from the material from Nevada and the plants do not form elongated, elastic caudices.

Subspecies *hitchcockii* is found in the Sheep Range and Spring Mountains (Charleston Mountain), Nevada, and on the Table Cliff Plateau, Utah, where it is limited to the white member of the limestone Wasatch (Claron) Formation.

40b. **Physaria hitchcockii** (Munz) O'Kane & Al-Shehbaz subsp. **confluens** (Maguire & A. H. Holmgren) O'Kane, Novon 17: 380. 2007 [E]

Lesquerella hitchcockii Munz subsp. *confluens* Maguire & A. H. Holmgren, Madroño 11: 174. 1951; *L. confluens* (Maguire & A. H. Holmgren) Reveal

Plants forming soft mats; caudex elongated, elastic (creeping). **Basal leaves:** petiole differentiated from blade (sometimes weakly); blade oblanceolate to obovate. **Anthers** 1–1.5 mm. **Fruits** 1.7–3 mm wide.

Flowering Jun–Jul. Gravelly or rocky limestone; 2300–3500 m; Nev.

Subspecies *confluens* is found at or above timberline in the Grant, Quinn Canyon, and Schell Creek ranges.

40c. Physaria hitchcockii (Munz) O'Kane & Al-Shehbaz subsp. **rubicundula** (Rollins) O'Kane & Al-Shehbaz, Novon 12: 324. 2002 [C][E]

Lesquerella rubicundula Rollins, Contr. Dudley Herb. 3: 178. 1941; *L. hitchcockii* Munz subsp. *rubicundula* (Rollins) Maguire & A. H. Holmgren; *Physaria hitchcockii* var. *rubicundula* (Rollins) B. L. Turner; *P. rubicundula* (Rollins) S. L. Welsh

Plants forming a loose mat; caudex elongated, elastic. **Basal leaves:** petiole not differentiated from blade; blade linear-oblanceolate. **Anthers** 1.4–1.6 mm. **Fruits** 2.4–3.4 mm wide.

Flowering May–Jun. Pinyon-juniper communities on barren slopes; of conservation concern; 2100–3400 m; Utah.

Subspecies *rubicundula* is found in talus of the pink and white members of the Wasatch (Claron) Formation of the Aquarius, Markagunt, and Paunsaugunt plateaus.

41. Physaria humilis (Rollins) O'Kane & Al-Shehbaz, Novon 12: 324. 2002 • St. Mary's Peak or Bitterroot bladderpod [C][E]

Lesquerella humilis Rollins, Contr. Gray Herb. 214: 9. 1984

Perennials; caudex simple, (thickened with persistent leaf bases); densely (silvery) pubescent, trichomes 5- or 6-rayed, rays furcate or 3-branched. **Stems** simple or few to several from base, prostrate, (from below a terminal rosette of leaves, unbranched), 0.2–0.5 dm. **Basal leaves:** blade elliptic to broadly ovate or obovate, (1–)1.5–2.5(–3) cm, margins entire, (apex obtuse). **Cauline leaves:** blade spatulate, 3–7 mm, (base cuneate), margins entire. **Racemes** not loose, (scarcely elongated in fruit, 3–5-flowered). **Fruiting pedicels** (straight or slightly curved). **Flowers:** sepals (yellow-green), narrowly elliptic to narrowly long-triangular, 3.7–5 mm; petals oblanceolate to nearly obovate, 7–8.5 mm, (abruptly tapering to narrow claw). **Fruits** wider than long, apex truncate to shallowly notched, compressed (angustiseptate), 3–4 mm; valves densely pubescent, trichomes ascending to erect, sparsely pubescent inside; ovules 4 per ovary; style 2–3 mm. **Seeds** plump, (slightly compressed).

Flowering Jun–early Aug. Steep slopes, dry summits, rocky fellfields, dry ledges; of conservation concern; 2700–2900 m; Mont.

Physaria humilis is found in metamorphosed rock and detritus on the peaks of the Bitterroot Mountains.

42. Physaria integrifolia (Rollins) Lichvar, Madroño 31: 203. 1984 • Snake River or creeping twinpod [E]

Physaria didymocarpa (Hooker) A. Gray var. *integrifolia* Rollins, Rhodora 41: 407. 1939; *P. integrifolia* var. *monticola* Lichvar

Perennials; caudex usually branched, (rhizomelike, cespitose); densely (silvery) pubescent, trichomes (often stalked, appressed), several-rayed, rays furcate or bifurcate, (umbonate, strongly tuberculate throughout). **Stems** several from base, ± erect, exceeding basal rosette by ± 0.5 dm. **Basal leaves** (forming a strong rosette; long-petiolate); blade oblanceolate to ovate or orbicular, (1.5–)2–4(–8) cm, (base usually abruptly tapering to petiole), margins entire. **Cauline leaves:** blade oblanceolate, 1–2 cm, margin entire, (apex acute). **Racemes** congested, (greatly exceeding leaves). **Fruiting pedicels** (spreading, straight or slightly curved), 7–11 mm. **Flowers:** sepals often keeled, 6–8 mm; petals spatulate, 8–10 mm. **Fruits** didymous, highly inflated, 8–22 × 10–25 mm, (papery, basal and apical sinuses deep); valves (retaining seeds after dehiscence), densely pubescent, trichomes appressed; replum linear to oblong, as wide as or wider than fruit; ovules 8 per ovary; style 7–9 mm. **Seeds** flattened. $2n = 16$.

Flowering (May–)Jun–Jul(–Aug). Calcareous hills and slopes, shale-limestone cliffs, bare steep slopes, red clay banks, shale; 1900–2700 m; Idaho, Mont., Wyo.

Physaria integrifolia has traditionally been recognized as a variety of *P. didymocarpa*, but it is morphologically and ecologically quite distinctive. Variety *monticola* (no combination has been made at subspecific rank) is not recognized here; it is considered another example, in the genus, of caudices elongating in response to shifting substrates.

43. Physaria intermedia (S. Watson) O'Kane & Al-Shehbaz, Novon 12: 324. 2002 • Mid-bladderpod [E]

Lesquerella alpina (Nuttall) S. Watson var. *intermedia* S. Watson, Proc. Amer. Acad. Arts 23: 251. 1888; *L. intermedia* (S. Watson) A. Heller

Perennials; caudex (buried), branched, (thickened with persistent leaf bases, cespitose); densely pubescent (usually grayish-green), trichomes (sessile or short stalked, spreading), several-rayed, rays furcate or bifurcate, slightly fused at base, (tuberculate or finely tuberculate). Stems several from base, erect to decumbent, (unbranched, stout, densely leafy sterile shoots sometimes present), (0.5–)4–2.5 dm. Basal leaves (clustered at stem base); blade linear to linear-oblanceolate, 2–5 cm, margins entire, usually involute, sometimes flattened, (apex obtuse to subacute). Cauline leaves: blade linear-oblanceolate to linear, 1–3.5(–4.5) cm, margins entire, usually involute. Racemes compact, (often nearly subumbellate). Fruiting pedicels (often expanded distally, ascending or recurved, usually straight or slightly curved, rarely nearly sigmoid), 4–15 mm, (stout). Flowers: sepals (yellowish or greenish yellow), ovate or oblong, 4.5–7.5(–9) mm, (lateral pair sometimes cucullate, median pair tapering at both ends, thickened apically, cucullate); petals spatulate or oblong, 6.5–10.5(–15) mm, (base sometimes widened, apex rounded or retuse). Fruits (sessile or substipitate), subglobose to slightly ovoid, usually inflated, rarely compressed or obcompressed, 4–6(–10) mm, (apex acute, slightly flattened); valves sparsely pubescent, trichomes appressed; ovules (8–)12–16(–20) per ovary; style (2–)3–4.5(–5.5) mm. Seeds flattened. 2n = 18, 20, 36.

Flowering Apr–Aug. Dry sandy, gravelly, or rocky soil, claylike hillsides, open chiprock, dry stream beds, gravel bars, open knolls, open pinyon-juniper woods, open stands of sagebrush, Gambel oak or ponderosa pine communities, calcareous substrates; 1600–2400 m; Ariz., N.Mex., Utah.

N. H. Holmgren (2005b) pointed out that the lectotype and other material from New Mexico, where *Physaria intermedia* is very infrequent, is quite similar to *P. parvula* from northern Colorado and northeastern Utah; it is also quite similar to, but less robust than, *P. pulvinata* from southwestern Colorado. The material from Arizona, northwestern New Mexico, and Utah may represent an unnamed taxon; further study is needed.

44. **Physaria kingii** (S. Watson) O'Kane & Al-Shehbaz, Novon 12: 324. 2002 • King bladderpod

Vesicaria kingii S. Watson, Proc. Amer. Acad. Arts 20: 353. 1885; *Lesquerella kingii* (S. Watson) S. Watson

Perennials; caudex usually simple, sometimes branched, (not thickened); usually densely pubescent, trichomes (sessile or short-stalked), 3–7-rayed, rays distinct or slightly fused at base, typically furcate near base, bifurcate or 3-partite, (not to slightly umbonate, smooth or moderately to strongly tuberculate). **Stems** few to several from base, prostrate to decumbent or erect, 0.5–2(–4) dm. **Basal leaves:** blade suborbicular to narrowly or broadly oblanceolate to broadly elliptic or rhombic, (1.2–)2–6(–8) cm, (base usually abruptly narrowed to petiole), margins entire, sinuate, or lobed. **Cauline leaves:** (proximal petiolate, distal sessile); blade obovate or elliptic to spatulate, 0.5–2 cm, margins entire. **Racemes** (usually not secund), dense, (sometimes elongated in fruit). **Fruiting pedicels** (erect to divaricate-ascending or recurved, erect in distal 1/3, usually sigmoid, sometimes straight or slightly curved), 4.5–10(–15) mm. **Flowers:** sepals lanceolate, 4–6(–7) mm; petals (yellow, cream-yellow, cream-white, or white), obovate to oblanceolate, 6–13 mm, (claw weakly differentiated from blade). **Fruits** (sessile or substipitate), subglobose, obovoid, or ellipsoid, compressed (sometimes slightly angustiseptate), 3–9 mm, (rigid, apex truncate, retuse, or rounded-acute); valves sparsely or densely pubescent, sometimes sparsely pubescent inside; (septum sometimes fenestrate, perforate, or obsolete); ovules 4–16 per ovary; style 1–9 mm. **Seeds** flattened (sometimes slightly).

Subspecies 7 (7 in the flora): w United States, nw Mexico.

The *Physaria kingii* complex is in need of further study. It is widespread in the western United States, mostly in montane environments. This treatment recognizes a highly variable species with generally well-marked, geographically coherent subspecies. Hybridization may be involved in some of the subspecies, especially in subsp. *kaibabensis*, where molecular data indicate intra-individual genetic variation (pers. obs.).

1. Fruits slightly wider than long, apices truncate or retuse, valves pubescent inside.
   2. Basal leaf blades: margins ± entire (sometimes slightly lobed or widened at base); California, Idaho, Nevada, Oregon . . . . 44a. *Physaria kingii* subsp. *kingii*
   2. Basal leaf blades: margins sinuate or lobed, or, sometimes, lyrate; Wallowa and Elkhorn mountains, Oregon . . . . 44d. *Physaria kingii* subsp. *diversifolia*
1. Fruits as wide as or longer than wide, apices rounded-acute, valves glabrous inside.
   3. Fruiting pedicels recurved . . . . 44c. *Physaria kingii* subsp. *cobrensis*
   3. Fruiting pedicels not recurved (divaricate-ascending or ± erect, straight or sigmoid).
      4. Petals cream-white or white; styles 1–2 mm; Kaibab Plateau, n Arizona . . . . 44e. *Physaria kingii* subsp. *kaibabensis*
      4. Petals yellow (occasionally cream-yellow or cream-white on Kaibab Plateau, Arizona); styles (4–)4.5–9 mm; n Arizona (including Kaibab Plateau), California, Nevada, Utah.
         5. Plants erect; styles 6–9 mm; ovules 4–8 per ovary; se California . . . . 44b. *Physaria kingii* subsp. *bernardina*
         5. Plants ascending, erect, decumbent, or prostrate; styles (4–)4.5–7 mm; ovules (6–)8–16 per ovary; n Arizona, e California, s Nevada, Utah.

[6. Shifted to left margin.—Ed.]

6. Plants prostrate, decumbent, or erect; racemes not or somewhat secund in fruit; ovules usually 8–16 per ovary; n Arizona, e California, s Nevada, Utah .... 44f. *Physaria kingii* subsp. *latifolia*
6. Plants ascending; racemes secund in fruit; ovules (6–)8–12 per ovary; n Utah.................. .... 44g. *Physaria kingii* subsp. *utahensis*

44a. Physaria kingii (S. Watson) O'Kane & Al-Shehbaz subsp. **kingii** E

Plants prostrate and straggling to erect; trichomes tuberculate throughout, center low-mounded. **Basal leaves:** blade margins ± entire, (sometimes slightly lobed or widened at base). **Racemes** not secund, loose and elongated in fruit, or dense in alpine forms. **Fruiting pedicels** usually sigmoid. **Petals** yellow. **Fruits** slightly wider than long, apex truncate or retuse; valves pubescent inside; septum ± perforate; ovules 4(–8) per ovary; style to 7(–9) mm.

Flowering May–Jun. Granitic ridges, quartz and limestone chip, stream gravels, dry slopes, calcareous soils, sagebrush hillsides, pinyon-juniper woodlands; 1700–2200 m; Calif., Idaho, Nev., Oreg.

44b. Physaria kingii (S. Watson) O'Kane & Al-Shehbaz subsp. **bernardina** (Munz) O'Kane & Al-Shehbaz, Novon 12: 324. 2002 C E

Lesquerella bernardina Munz, Bull. S. Calif. Acad. Sci. 31: 62. 1932; *L. kingii* (S. Watson) S. Watson subsp. *bernardina* (Munz) Munz; *Physaria kingii* var. *bernardina* (Munz) B. L. Turner

Plants erect; trichomes tuberculate throughout, less densely so over the center, (rays often fused at base). **Basal leaves:** blade margins entire. **Racemes** not secund, fairly dense in fruit. **Fruiting pedicels** straight (nearly erect), or sigmoid. **Petals** yellow. **Fruits** as wide as or longer than wide, apex rounded-acute; valves glabrous inside; septum complete or $1/3–1/2$ perforate; ovules 4–8 per ovary; style 6–9 mm.

Flowering May–Jun. Pine woods; of conservation concern; 1800–2200 m; Calif.

Subspecies *bernardina* is found near Bear Lake in the San Bernardino Mountains.

44c. Physaria kingii (S. Watson) O'Kane & Al-Shehbaz subsp. **cobrensis** (Rollins & E. A. Shaw) O'Kane & Al-Shehbaz, Novon 12: 324. 2002 E

Lesquerella kingii (S. Watson) S. Watson var. *cobrensis* Rollins & E. A. Shaw, Gen. Lesquerella, 255. 1973; *Physaria cobrensis* (Rollins & E. A. Shaw) N. H. Holmgren; *P. kingii* var. *cobrensis* (Rollins & E. A. Shaw) B. L. Turner

Plants ± erect or ascending; trichomes tuberculate throughout, center low-mounded. **Basal leaves:** blade margins entire. **Racemes** secund, usually elongated and loose in fruit. **Fruiting pedicels** curved (recurved) or slightly sigmoid. **Petals** yellow. **Fruits** as wide as or longer than wide, apex rounded-acute; valves glabrous inside; septum ± perforate; ovules 4(–8) per ovary; style 3–6 mm.

Flowering May–Jun. Light-colored silt, limestone gravel, rocky areas with low sagebrush; 500–1000 m; Idaho, Nev., Oreg.

44d. Physaria kingii (S. Watson) O'Kane & Al-Shehbaz subsp. **diversifolia** (Greene) O'Kane & Al-Shehbaz, Novon 12: 325. 2002 E

Lesquerella diversifolia Greene, Pittonia 4: 309. 1901; *L. kingii* (S. Watson) S. Watson subsp. *diversifolia* (Greene) Rollins & E. A. Shaw; *L. kingii* var. *sherwoodii* (M. Peck) C. L. Hitchcock; *L. occidentalis* S. Watson subsp. *diversifolia* (Greene) Maguire & A. H. Holmgren; *L. occidentalis* var. *diversifolia* (Greene) C. L. Hitchcock; *L. sherwoodii* M. Peck; *Physaria kingii* var. *diversifolia* (Greene) B. L. Turner

Plants usually prostrate and straggling; trichomes (lower layer) smoother, (upper layer) moderately tuberculate, much less so over flat or mounded center. **Basal leaves:** blade margins sinuate or lobed, or, sometimes, lyrate. **Racemes** not secund, elongated and loose in fruit. **Fruiting pedicels** usually sigmoid. **Petals** yellow. **Fruits** slightly wider than long, apex truncate or retuse; valves pubescent inside; septum fenestrate; ovules 4–6 per ovary; style to 9 mm. $2n = 10$.

Flowering May–Aug. Talus slopes, gravelly flood banks, steep limestone cliffs, rock crevices, marble chiprock, sandy and gravelly soils; 1200–3000 m; Oreg.

Subspecies *diversifolia* is found in the Elkhorn and Wallowa mountains.

44e. Physaria kingii (S. Watson) O'Kane & Al-Shehbaz subsp. **kaibabensis** (Rollins) O'Kane, Novon 17: 380. 2007 • Kaibab bladderpod [C][E]

Lesquerella kaibabensis Rollins, Contr. Gray Herb. 211: 110. 1982; *Physaria kaibabensis* (Rollins) N. H. Holmgren

Plants decumbent to ascending; trichomes (no information available). Basal leaves: blade margins entire. Racemes not usually secund, hardly elongated in fruit. Fruiting pedicels (divaricate-ascending), straight or slightly sigmoid. Petals cream-white or white. Fruits as wide as or longer than wide, apex rounded-acute; valves glabrous inside; septum complete; ovules 10–14 per ovary; style 1–2 mm.

Flowering Jun. Limestone-clay knolls, in open, parklike meadows; of conservation concern; 2500–2700 m; Ariz.

Subspecies *kaibabensis* is found on the Kaibab Plateau.

44f. Physaria kingii (S. Watson) O'Kane & Al-Shehbaz subsp. **latifolia** (A. Nelson) O'Kane & Al-Shehbaz, Novon 12: 325. 2002

Lesquerella latifolia A. Nelson, Bot. Gaz. 42: 49. 1906; *L. barnebyi* Maguire; *L. kingii* (S. Watson) S. Watson subsp. *latifolia* (A. Nelson) Rollins & E. A. Shaw; *L. kingii* var. *parvifolia* (Maguire & A. H. Holmgren) S. L. Welsh & Reveal; *L. occidentalis* S. Watson var. *parvifolia* Maguire & A. H. Holmgren; *L. wardii* S. Watson; *Physaria wardii* (S. Watson) O'Kane & Al-Shehbaz

Plants prostrate, decumbent, or erect; trichomes with large tubercles throughout, ± flat across middle. Basal leaves: blade margins entire. Racemes not or somewhat secund, usually dense and compact, sometimes slightly elongated in fruit. Fruiting pedicels usually sigmoid. Petals usually yellow, sometimes cream-yellow, cream-white, or white. Fruits as wide as or longer than wide, apex rounded-acute; valves glabrous inside; septum complete; ovules 8–16 per ovary; style to 7 mm.

Flowering May–Jun. Gravelly loam soils, rocky basaltic slopes, limestone outcrops, ridges and flats, canyon bottoms, open pinyon-juniper woodlands; 2000–2500 m; Ariz., Calif., Nev., Utah; Mexico (Baja California).

Populations of subsp. *latifolia* with flowers cream-yellow, cream-white, or white are frequently encountered on the Kaibab Plateau of northern Arizona.

44g. Physaria kingii (S. Watson) O'Kane & Al-Shehbaz subsp. **utahensis** (Rydberg) O'Kane, Novon 17: 380. 2007 • Utah bladderpod [E]

Lesquerella utahensis Rydberg, Bull. Torrey Bot. Club 30: 252. 1903; *Physaria utahensis* (Rydberg) O'Kane & Al-Shehbaz

Plants ascending; trichomes tuberculate throughout, center low-mounded. Basal leaves: blade margins entire. Racemes secund, ± dense and compact in fruit. Fruiting pedicels sigmoid. Petals yellow. Fruits as wide as or longer than wide, apex rounded-acute; valves glabrous inside; septum complete; ovules (6–)8–12 per ovary; style (4–)4.5–6.5 mm.

Flowering Jun–Jul. Rocky ridges, gravel, sagebrush hillsides, exposed limestone, granitic rock areas, sandy soils; 2400–3400 m; Utah.

Subspecies *utahensis* is found in the Uinta and Wasatch mountains.

45. Physaria klausii (Rollins) O'Kane & Al-Shehbaz, Novon 12: 325. 2002 • Rogers Pass or Klaus's or Divide bladderpod [C][E]

Lesquerella klausii Rollins, Contr. Gray Herb. 214: 10. 1984

Perennials; caudex simple; densely pubescent, trichomes (loosely spreading), 3–5-rayed, rays distinct, furcate (with exceptionally long branches). Stems simple from base, erect to decumbent, (slender), 0.6–1.5 dm. Basal leaves: blades obovate to deltate, 1.5–3(–4) cm, margins entire or outer one with 1 or 2 broad teeth. Cauline leaves: blade oblanceolate to spatulate, 0.6–1.5 cm, margins entire. Racemes loose. Fruiting pedicels (sigmoid), 5–9 mm. Flowers: sepals (green-yellow, often tinged with purple), elliptic, 3–4.6 mm; petals oblanceolate, 6–8 mm (claw expanded). Fruits (depressed), broadly obovate, compressed (angustiseptate), 2–4 mm, (apex slightly bilobed to nearly truncate); valves densely pubescent, trichomes strongly ascending, spreading, long, (appearing fuzzy), pubescent inside; ovules 4 per ovary; style 3–4 mm, (pubescent or glabrous). Seeds flattened.

Flowering Jul. Open gravel slides, solifluction cross-stripes of shale rubble, barren shale-derived soil; of conservation concern; 1200–1900 m; Mont.

46. Physaria lata (Wooton & Standley) O'Kane & Al-Shehbaz, Novon 12: 325. 2002 • Lincoln County bladderpod [C] [E]

Lesquerella lata Wooton & Standley, Contr. U.S. Natl. Herb. 16: 126. 1913

Perennials; caudex simple, (not thickened); densely pubescent, trichomes (short-stalked), several-rayed, rays distinct, furcate or bifurcate, (tuberculate, much less so over center, often nearly smooth on lower layer). **Stems** simple from base, spreading or erect, (unbranched), ca. 1 dm. **Basal leaves:** (petiole long, slender); blade elliptic to obovate, 3–4 cm, (base narrowing to petiole), margins entire. **Cauline leaves** (shortly petiolate); blade elliptic to obovate, 1–2 cm, margins entire. **Racemes** dense. **Fruiting pedicels** (sigmoid), 5–8 mm. **Flowers:** sepals narrowly elliptic or oblong, ca. 4.5 mm, (median pair thickened apically, cucullate); petals narrowly spatulate, 7–8 mm. **Fruits** (erect, substipitate), globose, ellipsoid, or obovoid, not or slightly compressed, 3–4 mm; valves sparsely pubescent, sometimes few trichomes inside; ovules 10–12 per ovary; style 3–5 mm. **Seeds** flattened.

Flowering Apr–Jul. Limestone soils and rocky places, pinyon-juniper-oak woodland and montane coniferous forest; of conservation concern; 2100–2900 m; N.Mex.

Additional research is needed to determine whether *Physaria lata* is a variant of *P. pinetorum*, with which it sometimes grows.

47. Physaria lepidota Rollins, Brittonia 33: 335, figs. 1, 2. 1981 • Kane County twinpod [E]

Perennials; caudex simple, (with deep roots, thickened); densely (silvery) pubescent throughout (densely covering leaves with several appressed layers), less dense on stems, trichomes (stellate-scalelike), rays fused (webbed) in proximal ½ or to tips, (umbonate, nearly smooth to moderately tuberculate). **Stems** simple from base, erect or outer ones slightly decumbent toward base, (from below or in basal leaves, unbranched), (0.5–)0.8–1.6(–2) dm. **Basal leaves** (erect, petiole long, slender); blade spatulate to broadly oblanceolate, (3–)5–7(–12) cm, (base gradually tapering to petiole), margins entire, (apex rounded or obtuse). **Cauline leaves:** blade oblanceolate, similar to basal, (base cuneate), margins entire. **Racemes** dense. **Fruiting pedicels** (divaricate-ascending, straight or slightly curved), 10–15 mm. **Flowers:** sepals (erect), linear to linear-oblong, somewhat boat-shaped, 7–10 mm; petals (erect at anthesis), lingulate, 11–15 mm, (claw undifferentiated from blade). **Fruits** (purplish in age), strongly didymous, semiorbicular, highly inflated, 10–18 × 14–19 mm, (papery, basal sinus usually shallow, rarely absent, apical sinus deep, narrowly V-shaped; valves retaining seeds after dehiscence, sides flat, back rounded, margins keeled, base and apex obtuse; replum narrowly oblong to linear, as wide as or wider than fruit, base slightly narrowed, apex obtusely rounded; ovules 4 per ovary; style 3–5 mm, (slender). **Seeds** slightly flattened.

Subspecies 2 (2 in the flora): wc United States.

1. Trichomes: rays fused nearly to tips; fruits with deep sinuses, or shallow basally, deep apically .... 47a. *Physaria lepidota* subsp. *lepidota*
1. Trichomes: rays fused in proximal ½; fruits with sinuses absent or shallow basally, deep apically .... 47b. *Physaria lepidota* subsp. *membranacea*

47a. Physaria lepidota Rollins subsp. **lepidota** [E]

Trichomes: rays fused nearly to tips. **Basal leaves:** blades elliptic, apex usually obtuse. **Fruits:** sinuses deep, or shallow basally, deep apically. $2n = 16$.

Flowering Apr–Jun. Clay soil of road cuts, clayey knolls, steep chiprock slides, shalelike rocky outcrops, pinyon-juniper areas; 1500–2700 m; Utah.

47b. Physaria lepidota Rollins subsp. **membranacea** (Rollins) O'Kane, Novon 17: 531. 2007 [E]

Physaria chambersii Rollins var. *membranacea* Rollins, Rhodora 41: 405, plate 556, figs. 17, 18. 1939; *P. lepidota* var. *membranacea* (Rollins) Rollins

Trichomes: rays fused in proximal ½. **Basal leaves:** blades linear-oblong, apex usually acute. **Fruits:** sinuses absent or shallow basally, deep apically. $2n = 8$.

Flowering Jun–Jul. Steep slopes, red chiprock, whitish clay and shale, steep talus slopes; 1500–2500 m; Utah.

48. **Physaria lesicii** (Rollins) O'Kane & Al-Shehbaz, Novon 12: 325. 2002 • Pryor Mountains bladderpod C E

Lesquerella lesicii Rollins, Novon 5: 71. 1995

Perennials; (delicate, short-lived); caudex simple, (sometimes elongated, covered with persistent leaf bases); usually sparsely pubescent, trichomes 7–12-rayed, rays furcate near base. **Stems** simple from base, erect to decumbent, (unbranched, mostly filiform, slender), 1–1.5 dm. **Basal leaves** (erect, petiole slender); blades broadly ovate to elliptic, 0.5–1 cm, (base abruptly narrowing to petiole), margins entire. **Cauline leaves** (remote, distally shortly petiolate); blade ± spatulate, (base cuneate), margins entire. **Racemes** lax, (elongated, few-flowered). **Fruiting pedicels** (recurved to widely spreading, filiform, slender), 5–10 mm. **Flowers:** sepals (erect), oblong, 3.5–4 mm, (lateral pair not saccate); petals (often fading to light purple apically), spatulate to nearly lingulate, 6–7 mm. **Fruits** (pendent), globose or subglobose, compressed, 3–4 mm; valves ± densely pubescent; ovules 6–10 per ovary; style ca. 1.5 mm. **Seeds** not seen.

Flowering Jun(–early Jul). Pryor Mountains, on limestone soils in woodlands of Rocky Mountain juniper and/or mountain mahogany, and widely scattered Douglas-fir, fellfields dominated by bluebunch wheatgrass and cushion plants; of conservation concern; 1600–2000 m; Mont.

49. **Physaria lindheimeri** (A. Gray) O'Kane & Al-Shehbaz, Novon 12: 325. 2002 • Lindheimer's bladderpod

Vesicaria lindheimeri A. Gray, Boston J. Nat. Hist. 6: 145. 1850; *Alyssum lindheimeri* (A. Gray) Kuntze; *Lesquerella gracilis* (Hooker) S. Watson var. *pilosa* Lundell; *L. lindheimeri* (A. Gray) S. Watson

Annuals or biennials; with a fine taproot; densely pubescent, trichomes (sessile or short-stalked), 4–7-rayed, rays usually furcate at base, sometimes bifurcate, (rough-tuberculate throughout). **Stems** several from base, erect or outer decumbent, (often several-branched, branches slender and flexuous), to 8 dm. **Basal leaves:** blade pinnatisect to repand, 3–9(–14) cm, margins entire. **Cauline leaves** (sometimes secund, proximal usually petiolate, distal sessile); blade elliptic, 1–6 cm, (distal with cuneate base), margins entire or deeply dentate. **Racemes** dense. **Fruiting pedicels** (horizontal or recurved and ascending at tip, sometimes loosely sigmoid), (5–)10–20 mm. **Flowers:** sepals elliptic to oblong, 3–5.5 mm, (median pair slightly thickened apically, cucullate); petals (sometimes drying slightly purplish), suborbicular or broadly ovate, 4.5–7(–9) mm, (narrowing gradually to short claw). **Fruits** globose or broadly ellipsoid, not or slightly inflated, (4–)5–8 mm, (smooth); valves (not retaining seeds after dehiscence), glabrous; replum as wide as or wider than fruit; ovules (8–)12–16(–20) per ovary; style (1.5–)2–3(–4) mm. **Seeds** flattened. $2n = 12$.

Flowering Dec–Apr. Heavy, black, claylike soils, or lighter, sandy soils, thickets, field-margins, roadsides, coastal prairies; 20–800 m; Tex.; Mexico (Tamaulipas).

50. **Physaria ludoviciana** (Nuttall) O'Kane & Al-Shehbaz, Novon 12: 325. 2002 • Foothill bladderpod E

Alyssum ludovicianum Nuttall, Gen. N. Amer. Pl. 2: 63. 1818, based on *Myagrum argenteum* Pursh, Fl. Amer. Sept. 2: 434. 1813, not *A. argenteum* Vitman 1790; *Lesquerella ludoviciana* (Nuttall) S. Watson; *Vesicaria ludoviciana* (Nuttall) de Candolle

Perennials; caudex simple or branched; densely pubescent, trichomes (sessile or short-stalked), 4–7-rayed, rays usually furcate, sometimes bifurcate, (rough-tuberculate throughout). **Stems** few from base, erect with outer usually decumbent, 1–3.5(–5) dm. **Basal leaves** (erect); blade narrowly lanceolate to linear or (outer) oblanceolate, (1–)2–6(–9) cm, margins usually entire, rarely shallowly dentate, (inner involute, outer usually flat, base usually with some simple trichomes). **Cauline leaves:** blade narrowly oblanceolate to linear, (1–)2–4(–8) cm, margins flat or involute. **Racemes** compact, (elongated and loose in fruit, densely-flowered). **Fruiting pedicels** (usually recurved), (5–)10–20(–25) mm. **Flowers:** sepals oblong to broadly elliptic, 4–7(–8) mm, (lateral pair subsaccate, median pair cucullate); petals oblanceolate or obovate, (5–)6.5–9.5(–11) mm, (claw undifferentiated from blade, or blade gradually narrowed to claw). **Fruits** subglobose or obovoid, usually inflated, sometimes slightly compressed, (3–)4–6 mm; valves densely pubescent, trichomes spreading, usually pubescent inside, trichomes appressed, sessile; ovules (4–)8–12(–16) per ovary; style 3–4.5(–6.5) mm. **Seeds** slightly flattened. $2n = 10, 20, 30$.

Flowering Apr–Jul(–Aug). Sandy or gravelly soils, steep hillsides, prairie pastures, clay slopes, limestone outcrops, sand dunes, open plains, sandy bluffs, rocky flats, white tuff sands; 0–1900 m; Ariz., Calif., Colo., Ill., Iowa, Kans., Minn., Mont., Nebr., Nev., N.Mex., N.Dak., Okla., S.Dak., Utah, Wis., Wyo.

Material previously reported as *Physaria ludoviciana* from Canada (Alberta, Manitoba, Saskatchewan) is here included in 6a. *P. arenosa* subsp. *arenosa*. *Lesquerella argentea* (Pursh) MacMillan is a later homonym that has been used for *P. ludoviciana*.

51. Physaria macrocarpa (A. Nelson) O'Kane & Al-Shehbaz, Novon 12: 325. 2002 • Largefruit bladderpod [C][E]

Lesquerella macrocarpa A. Nelson, Bot. Gaz. 34: 366. 1902

Perennials; caudex branched; densely pubescent, trichomes (sessile or short-stalked), 4–6-rayed, rays distinct, usually furcate, rarely bifurcate, (finely tuberculate throughout). **Stems** few or several from base, prostrate to decumbent, (unbranched or branched), 0.5–1.5 dm. **Basal leaves**: blades orbicular to broadly obovate, 1.5–3 cm, margins usually entire, rarely remotely dentate. **Cauline leaves** (sessile or shortly petiolate); blade elliptic to oblanceolate, 1–1.5(–2.5) cm, margins entire, (apex obtuse). **Racemes** dense, (elongated in fruit). **Fruiting pedicels** (sharply recurved), 5–10 mm, (stout). **Flowers**: sepals ovate or oblong-elliptic, 5–5.5 mm, (lateral pair not saccate); petals cuneate or broadly obovate, ca. 7 mm, (sometimes slightly narrowed to a broad claw, apex sometimes retuse). **Fruits** subglobose to broadly obovoid, strongly inflated (often slightly angustiseptate), 5–7 mm, (papery); valves sparsely pubescent; (septum fenestrate, perforate, or obsolete); ovules 4–8 per ovary; style 2–3 mm. **Seeds** somewhat flattened.

Flowering May–Jun. Gypsum-clay hills and benches, naked clay flats and barren hills; of conservation concern; 2000–2400 m; Wyo.

Physaria macrocarpa is found in the Great Divide and Green River basins.

52. Physaria mcvaughiana (Rollins) O'Kane & Al-Shehbaz, Novon 12: 325. 2002 • McVaugh's bladderpod

Lesquerella mcvaughiana Rollins, Contr. Gray Herb. 171: 44. 1950

Perennials; caudex simple or branched, (sometimes enlarged); densely pubescent, trichomes (sessile), several-rayed, rays fused (webbed) most of their length, (umbonate, peltate, tuberculate throughout). **Stems** few to several from base, erect or outer ones decumbent, 0.5–4 dm. **Basal leaves** (long-petiolate); blade elliptic to obovate or rhombic, 2–6(–9) cm, margins entire. **Cauline leaves** (sessile or shortly petiolate); blade oblanceolate to spatulate, 1–3 cm, (proximal broader), margins entire. **Racemes** dense, (relatively short). **Fruiting pedicels** (erect to spreading, ascending, or (proximal) horizontal, straight to slightly curved, sometimes loosely sigmoid), 6–12(–20) mm. **Flowers**: sepals elliptic or narrowly oblong, 4–5.4 mm, (tapered to apex); petals (white, base and claw yellow, conspicuously purple-veined), usually broadly obovate or rhombic, 6–10 mm, (± equal to blade, tapering to slender claw). **Fruits** (sessile or substipitate, often reddish magenta), usually ovoid to subglobose, inflated, 4–6(–7) mm; valves (not retaining seeds after dehiscence), glabrous; replum as wide as or wider than fruit; septum perforate; ovules 8–12 per ovary; style 1.5–4 mm. **Seeds** somewhat flattened. $2n = 12$.

Flowering mid Mar–Apr(–Aug). Stream bed gravels, rocky limestone slopes and hills, canyon bottoms and slopes, limestone rubble; 1200–1600 m; Tex.; Mexico (Coahuila).

53. Physaria montana (A. Gray) Greene, Fl. Francisc., 249. 1891 • Mountain bladderpod [E]

Vesicaria montana A. Gray, Proc. Acad. Nat. Sci. Philadelphia 15: 58. 1864; *Alyssum grayanum* Kuntze; *Lesquerella montana* (A. Gray) S. Watson; *L. montana* var. *suffruticosa* Payson; *L. rosulata* A. Nelson

Perennials; caudex simple or branched, (often enlarged); densely pubescent, trichomes (sessile or short-stalked), 4–7-rayed, rays furcate or bifurcate, (tuberculate throughout). **Stems** simple or several from base, prostrate to erect, 0.5–2(–3.5) dm. **Basal leaves**: blade suborbicular or obovate to elliptic, (1–)2–5(–7) cm, margins entire, sinuate, or shallowly dentate. **Cauline leaves** (often secund, proximal shortly petiolate, distal sessile); blade linear to obovate or rhombic, 1–2.5(–4) cm, margins entire or shallowly dentate. **Racemes** dense, compact, (usually elongated in fruit). **Fruiting pedicels** (usually sharply sigmoid, rarely nearly divaricate-spreading and straight), 5–15(–20) mm, (stout). **Flowers**: sepals elliptic, 5–8.5 mm, (lateral pair boat-shaped, saccate, median pair thickened apically, cucullate); petals (yellow to orange, sometimes fading purplish), narrowly spatulate or obovate, (6–)7.5–12 mm, (claw undifferentiated from blade, or gradually narrowed to claw, slightly expanded basally). **Fruits** (erect), ellipsoid or ovoid, not or slightly obcompressed, (apex not compressed), (6–)7–12 mm; valves densely pubescent, sometimes sparsely pubescent inside; ovules (8–)12–20(–24) per ovary; style 3–7 mm, (sometimes pubescent). **Seeds** flattened. $2n = 10$.

Flowering Apr–Jun(–Aug). Banks, rock outcrops, stony slopes and benchlands, from plains into mountains, in sagebrush, open scrub oak, pinyon-juniper woodland, ponderosa pine, Douglas fir on granitic, often gravelly, non-calcareous soils, rarely on calcareous soils; 1000–3300 m; Ariz., Colo., Nebr., N.Mex., S.Dak., Wyo.

Physaria montana is a rather variable species that in southwestern Colorado morphologically approaches *P. rectipes* and in eastern Wyoming approaches *P. curvipes*; it is unusual in the genus for its frequent presence on igneous, non-calcareous soils.

54. **Physaria multiceps** (Maguire) O'Kane & Al-Shehbaz, Novon 12: 325. 2002 • Manyhead bladderpod [E]

Lesquerella multiceps Maguire, Amer. Midl. Naturalist 27: 465. 1942

Perennials; caudex (buried), branched, (not thickened); densely pubescent, trichomes (sessile or short-stalked), 5–8-rayed, rays furcate or bifurcate, (umbonate, rough to finely tuberculate throughout). **Stems** several from base, prostrate, (slender, sparsely pubescent), 0.5–2 dm. **Basal leaves:** blade obovate to narrowly elliptic, 1.5–6 cm, margins usually entire, rarely shallowly dentate, (surfaces densely pubescent, often silvery). **Cauline leaves:** blade oblanceolate to spatulate, 0.5–1 cm, margins entire, (surfaces often sparsely pubescent). **Racemes** (narrow), loose, (elongated in fruit). **Fruiting pedicels** (ascending to somewhat spreading, straight to slightly curved), 4–8(–12) mm. **Flowers:** sepals (greenish brown, sometimes magenta), linear or elliptic, 4.3–6(–7.5) mm, (median pair thickened apically, cucullate); petals (frequently pink or magenta in distal ⅓–½), spatulate to oblanceolate, 6–10(–12) mm, (claw undifferentiated from blade). **Fruits** broadly ovoid to suborbicular, inflated, (terete or, often, slightly angustiseptate), 3–6 mm; valves sparsely pubescent; ovules usually 4, rarely 6–8 per ovary; style 3–6.5 mm. **Seeds** plump.

Flowering May–Jul. Douglas-fir or spruce woodlands, limestone ridges, damp open slopes, soil pockets among rocks, crevices of rocks, decomposed calcareous rocks; 2400–2900 m; Idaho, Utah, Wyo.

55. **Physaria navajoensis** (O'Kane) O'Kane & Al-Shehbaz, Novon 12: 325. 2002 • Navajo bladderpod [C][E][F]

Lesquerella navajoensis O'Kane, Madroño 46: 88, figs. 1, 2. 1999

Perennials; caudex branched, (woody, pulvinate-cespitose, forming hard, hemispherical mats, basal parts covered with persistent leaf bases); densely (silvery gray) pubescent, trichomes mostly 5-rayed, rays bifurcate, slightly fused at base, (umbonate, strongly tuberculate except nearly smooth over umbo). **Stems** several from base (crowded), erect, not exceeding leaves. **Basal leaves:** usually absent. **Cauline leaves:** (petiole not differentiated from blade); blade linear-oblanceolate, 3–8(–13) mm, margins entire. **Racemes** (secund), dense, corymbose, (few-flowered, not or barely exceeding leaves). **Fruiting pedicels** (ascending to divaricate-ascending, straight), 3.5–6 mm. **Flowers:** sepals (yellow-green), linear to narrowly triangular, 3.7–4.8 mm, (lateral pair subsaccate); petals (deep yellow, slightly orange in center), spatulate, 5.2–6.5 mm, (claw joined at right angle). **Fruits** (becoming reddish or copper-colored in age), ovate, often slightly compressed (at margins apically), 3–5 mm, (apex acute); valves (not retaining seeds after dehiscence), glabrous; (septum perforate or not); ovules 4–8 per ovary; style 1.8–3 mm. **Seeds** plump or slightly flattened, (strongly mucilaginous).

Flowering May–early Jun. Pinyon-juniper communities on nearly barren outcrops of Todilto Limestone; of conservation concern; 2200–2400 m; Ariz., N.Mex.

Physaria navajoensis is morphologically similar to 85. *P. tumulosa* of southern Utah, differing subtly. *Physaria navajoensis* has petals slightly orange at the junction of blade and claw, a sharp bend at that junction giving the flower a flat-topped appearance, and strongly mucilaginous seeds. *Physaria tumulosa* has pure yellow petals that gently flex at the junction of blade and claw, and seeds that are not mucilaginous. Molecular data (pers. obs.) show that these two species are not directly related. A population of plants on Deer Spring Point, Kane County, Utah, appears to be this species, but molecular data indicate that it is probably a hybrid between *P. tumulosa* and, most likely, *P. intermedia*.

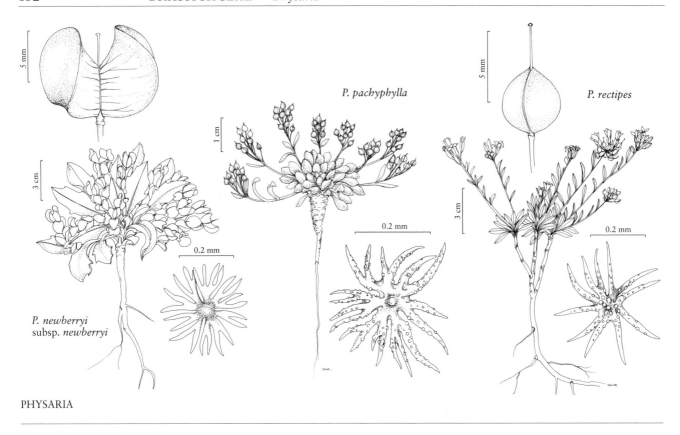

56. Physaria nelsonii O'Kane & Al-Shehbaz, Novon 12: 326. 2002 • Nelson's bladderpod E

Lesquerella condensata A. Nelson, Bull. Torrey Bot. Club 26: 238. 1899, not *Physaria condensata* Rollins 1939; *L. alpina* (Nuttall) S. Watson subsp. *condensata* (A. Nelson) Rollins & E. A. Shaw; *L. alpina* var. *condensata* (A. Nelson) C. L. Hitchcock

Perennials; (diminutive); caudex branched, (densely cespitose, mound-forming); densely pubescent, trichomes (flaring, giving a shaggy appearance), usually 4–5-rayed, rays furcate or bifurcate, (strongly tuberculate throughout). **Stems** few to several from base, erect, (from basal tuft), 0.1–0.2(–0.3) dm, (not or just barely exceeding leaves). **Basal leaves**: blade spatulate to oblanceolate, 0.5–1.5 cm, (base gradually narrowed to petiole), margins entire. **Cauline leaves** (absent or few); similar to basal, blade linear. **Racemes** dense, (few-flowered). **Fruiting pedicels** (loosely sigmoid), 3–5 mm. **Flowers:** sepals (pale yellow), oblong to elliptic, 4–5 mm, (median pair usually thickened apically, cucullate); petals (bright yellow), lingulate, 6–8 mm. **Fruits** lanceolate, compressed apically, 2.5–3(–4) mm; valves pubescent, trichomes spreading, (appearing fuzzy), rarely with trichomes inside; ovules 4–8 per ovary; style 2.5–4 mm. **Seeds** plump, (oblong).

Flowering May–mid Jun. Limestone, windswept knolls and cliffs, nearly barren areas with other cushion-forming plants; 1600–2300 m; Utah, Wyo.

Physaria nelsonii is morphologically similar to 73. *P. pycnantha*, which traditionally was included in a broader *P. nelsonii*. These allopatric species are distinguished by styles equal to or exceeding the length of fruit and fruiting stems overtopped by basal leaves (*P. nelsonii*) versus styles shorter than fruits and fruiting stems usually exserted beyond basal leaves (*P. pycnantha*).

57. Physaria newberryi A. Gray in J. C. Ives, Rep. Colorado R. 4: 6. 1861 • Newberry twinpod E F

Physaria didymocarpa (Hooker) A. Gray var. *newberryi* (A. Gray) M. E. Jones

Perennials; caudex simple or branched, (branches often covered with persistent leaf bases, cespitose); densely (silvery) pubescent, trichomes rays fused at least ½ their length. **Stems** several from base, ascending to erect (arising laterally, unbranched), 0.5–1(–2.5) dm. **Basal leaves** (ascending to erect, petiole slender); blade oblanceolate to obovate, 3–8 cm, (base tapering to petiole), margins incised or dentate with broad teeth, (apex acute to obtuse). **Cauline leaves:** blade linear-oblanceolate to oblanceolate, 1–2 cm, margins entire. **Racemes** dense (elongated or not in fruit, 2.5–8.5(–10) cm). **Fruiting pedicels** (divaricate, straight), 5–11(–15) mm, (rigid, fruits not pendent on arching pedicels). **Flowers:** sepals (greenish yellow), lanceolate, 6–8.5 mm, (saccate and cucullate); petals spatulate to narrowly oblanceolate, 7–10(–12) mm. **Fruits** didymous, sides curved and angular, highly inflated, 6–16 × 8–12 mm, (papery, apical sinus broad and concave); valves (retaining seeds after dehiscence, distinctly 2-keeled on side away from replum), pubescent, trichomes appressed; replum linear to linear-lanceolate, as wide as or wider than fruit, apex acute; ovules 4–8 per ovary; style 2–9 mm, (usually not exceeding sinus). **Seeds** slightly flattened, (ovate).

Subspecies 2 (2 in the flora): sw United States.

Physaria newberryi, with its unusual fruits, can be confused with 15. *P. chambersii*. In *P. chambersii*, the sides of the fruit are flat, the style always exceeds the top, or shoulders, of the fruit, and shoulders form an angle that does not curve in toward the style. In *P. newberryi*, the sides of the fruit are concave, the styles are shorter than shoulders of the silicle (except in subsp. *yesicola*), and shoulders of the silicle form a curved, inward arching crown on the fruit.

1. Styles less than 4 mm, shorter than fruit sinuses .... 57a. *Physaria newberryi* subsp. *newberryi*
1. Styles 5–9 mm, longer than fruit sinuses .... 57b. *Physaria newberryi* subsp. *yesicola*

57a. Physaria newberryi A. Gray subsp. newberryi E F

Physaria newberryi var. *racemosa* Rollins

Plants usually not mound-forming. **Caudex** simple or few-branched (some populations on shifting substrates with elongated branches). **Styles** less than 4 mm, shorter than fruit sinus. $2n = 8, 16$.

Flowering Mar–May. Low elevation blackbrush, pinyon-juniper, Gambel oak, and ponderosa pine communities; 800–2400 m; Ariz., Nev., N.Mex., Utah.

Supposed differences between subspp. *newberryi* and *racemosa* do not hold up in light of recent collections that show a continuum in replum and raceme lengths.

57b. Physaria newberryi A. Gray subsp. yesicola (Sivinski) O'Kane, Novon 17: 381. 2007 C E

Physaria newberryi var. *yesicola* Sivinski, Sida 18: 673, fig. 1. 1999

Plants mound-forming. **Caudex** diffusely branched. **Styles** 5–9 mm, longer than fruit sinus.

Flowering late Apr–May. Gypsiferous soils of the Yeso Formation in the Sierra Lucero Range and Mesa Lucero, sandy and silty soil, associated with singleseed juniper; of conservation concern; 1700–2100 m; N.Mex.

58. Physaria obcordata Rollins, J. Arnold Arbor. 64: 495, fig. 1. 1983 • Piceance twinpod C E

Perennials; caudex branched, (thick, covered with persistent, overlapping leaf bases); densely (silvery) pubescent, trichomes (stellate-scalelike), several-rayed, rays fused (webbed) to tips. **Stems** several from base, erect to decumbent, (unbranched, fertile stems from among basal leaves), 1.2–1.8 dm. **Basal leaves** (erect, not rosulate); blade broadly oblanceolate, 4–8 cm, margins entire or shallowly sinuate-dentate, (apex acute). **Cauline leaves:** (proximal petiolate); blade narrowly lanceolate, similar to basal in size, (distal with cuneate base), margins entire. **Racemes** loose, (elongated in fruit). **Fruiting pedicels** (widely spreading to recurved), 1–1.5 cm. **Flowers:** sepals (greenish yellow), elliptic, often broadly so, 4.8–7.1 mm; petals oblanceolate, 6.8–9.8 mm. **Fruits** (usually pendent), slightly didymous apically, obcordate, slightly inflated, 4–7 × 3–6 mm, (papery, basal sinus absent, apical sinus evident to nearly absent); valves (retaining seeds after dehiscence), pubescent; replum broadly obovate to suborbicular, 4–5 mm, as wide as or wider than fruit, margins entire; ovules usually 4 (rarely 6–8) per ovary; style (2.5–)3–4(–5) mm. **Seeds** plump.

Flowering May–Jun. Steep slopes, fine chiprock, shaley hillsides; of conservation concern; 1800–2300 m; Colo.

Physaria obcordata is known from the Thirteenmile Creek Tongue and the Parachute Creek Member of the Green River Formation. It is in the Center for Plant Conservation's National Collection of Endangered Plants.

59. Physaria obdeltata (Rollins) O'Kane & Al-Shehbaz, Novon 12: 326. 2002 • Middle Butte bladderpod E

Lesquerella obdeltata Rollins, Cruciferae Continental N. Amer., 640. 1993

Perennials; caudex simple; densely pubescent, trichomes (appressed in layers), 5–7-rayed, rays usually bifurcate, sometimes furcate, (thickened toward center). Stems simple from base, prostrate to decumbent, (unbranched, from within and below leaf clusters, slender), 0.2–0.8 dm. Basal leaves (tufted, erect or ascending, silvery); blade linear to oblanceolate or (outer) oblanceolate to obovate or rhombic, 1.5–3.3 cm, (base sometimes subhastate), margins entire or dentate, (often involute). Cauline leaves: blade nearly linear, to 1.5 cm, margins entire. Racemes dense, (subcorymbose). Fruiting pedicels (recurved), 5–8 mm. Flowers: sepals (yellow-green), lanceolate, (2.5–)3.5–4.5 mm; petals spatulate to oblanceolate, 4–6.5 mm. Fruits (usually pendent), obdeltate, compressed (angustiseptate), 2–4 mm, (wider than long, base tapered to acute angle, apex truncate with inflated shoulders); valves densely pubescent; (septum obsolete or with a narrow fringe inside of replum, funicles attached close to replum apex); ovules 4 per ovary; style 2–4 mm, (slender). Seeds plump, (mucilaginous when wetted).

Flowering Jun(–Jul). Clayey, silty, or gravelly soils, overlaying basalt lava flows, silty playas, sagebrush, barren areas; 1300–1700 m; Idaho.

Physaria obdeltata is known from the eastern Snake River Plain.

60. Physaria occidentalis (S. Watson) O'Kane & Al-Shehbaz, Novon 12: 326. 2002 • Western bladderpod E

Vesicaria occidentalis S. Watson, Proc. Amer. Acad. Arts 20: 353. 1885; *Lesquerella occidentalis* (S. Watson) S. Watson

Perennials; caudex simple or branched, (thickened, sometimes subterranean); densely pubescent, trichomes (sessile or short-stalked), 5–7-rayed, rays distinct, bifurcate, (moderately to prominently tuberculate). Stems few to several from base, prostrate to decumbent or erect, (usually unbranched), 0.3–1.5(–3) dm. Basal leaves: (petiole slender); blade suborbicular to obovate or elliptic, 1–8 cm, (base narrowing gradually or abruptly to petiole), margins sinuate-dentate or entire. Cauline leaves: (proximal shortly petiolate, distal sessile); blade oblanceolate, 0.5–1.5(–2.5) cm, margins entire or remotely dentate. Racemes dense or loose. Fruiting pedicels (sigmoid or curved), 5–10(–15) mm. Flowers: sepals elliptic or oblong-elliptic, 4.5–7 mm, (median pair cucullate); petals spatulate, 7–9(–14) mm, (claw undifferentiated from blade). Fruits (erect), ellipsoid to obovoid, compressed at apex and sometimes margins (strongly latiseptate), (5–)6–9 mm, (apex acute, often beaked); valves densely pubescent, trichomes appressed or spreading, sometimes sparsely pubescent inside, trichomes 4- or 5-rayed; ovules 4–12 per ovary; style (2–)3–6.5 mm, (often sparsely pubescent). Seeds: inner surface flattened, outer convex.

Subspecies 2 (2 in the flora): w United States.

1. Stems erect or decumbent; fruit margins and apices compressed, apices often beaked; valves pubescent inside; ovules 4–8 per ovary; styles frequently curved; n California, c Idaho, nw Nevada, Oregon, Washington................
   ....60a. *Physaria occidentalis* subsp. *occidentalis*
1. Stems prostrate to decumbent; fruit, margins not conspicuously compressed, apices somewhat compressed, usually not conspicuously beaked; valves usually glabrous, sometimes sparsely pubescent inside; ovules 8–12 per ovary; styles usually straight; Nevada, w Utah............
   ....60b. *Physaria occidentalis* subsp. *cinerascens*

60a. Physaria occidentalis (S. Watson) O'Kane & Al-Shehbaz subsp. occidentalis E

Lesquerella cusickii M. E. Jones; *L. occidentalis* subsp. *cusickii* (M. E. Jones) Maguire & A. H. Holmgren; *L. occidentalis* var. *cusickii* (M. E. Jones) C. L. Hitchcock

Caudex a thickened crown, branched. Stems several from base, exserted to nearly covered with basal leaves (densely clustered), erect or decumbent. Basal leaves: blade margins sinuate-dentate or entire. Fruits (strongly latiseptate), compressed at margins and apex, apex (acute), often beaked; valves pubescent inside; ovules 4 per ovary; style frequently curved. $2n = 10$.

Flowering (Apr–)Jun–Aug. Loose gravelly soils, talus slopes, volcanic rocks, rocky ridges, metamorphic rubble, marbleized rocks, decomposed limestone and schist, barren hillsides, whitish ash deposits; (600–)1800–3400 m; Calif., Idaho, Nev., Oreg., Wash.

60b. Physaria occidentalis (S. Watson) O'Kane & Al-Shehbaz subsp. **cinerascens** (Maguire & A. H. Holmgren) O'Kane & Al-Shehbaz, Novon 12: 326. 2002 E

Lesquerella occidentalis S. Watson var. *cinerascens* Maguire & A. H. Holmgren, Madroño 11: 178. 1951; *L. goodrichii* Rollins; *L. occidentalis* subsp. *cinerascens* (Maguire & A. H. Holmgren) Rollins & E. A. Shaw; *Physaria goodrichii* (Rollins) O'Kane & Al-Shehbaz; *P. occidentalis* var. *cinerascens* (Maguire & A. H. Holmgren) S. L. Welsh; *P. occidentalis* var. *goodrichii* (Rollins) S. L. Welsh

Caudex thickened, usually simple, rarely branched, (subterranean). **Stems** usually simple, rarely few from base, well-exserted, prostrate to decumbent. **Basal leaves:** blade margins entire. **Fruits** (slightly longer than broad), not conspicuously compressed at margins, apex somewhat compressed and usually not conspicuously beaked; valves usually glabrous, sometimes sparsely pubescent inside; ovules 8–12 per ovary; style usually straight.

Flowering May–Jul. Gravelly slopes, grassy ridges, outwash fans, steep limestone outcrops, limey gravel; 1500–3700 m; Nev., Utah.

61. Physaria oregona S. Watson, Proc. Amer. Acad. Arts 17: 363. 1882 • Oregon twinpod E

Coulterina oregona (S. Watson) Kuntze; *Lesquerella oregona* (S. Watson) G. A. Mulligan

Perennials; caudex simple, (cespitose); (silvery) pubescent throughout, trichomes (stalked), few-rayed, rays furcate or imperfectly so, (tuberculate throughout). **Stems** several from base, erect or somewhat decumbent, (unbranched), 1–3.5 dm. **Basal leaves:** (petiole slender, usually incised or with broad teeth along petiole); blade obovate, 4–6 cm, margins entire. **Cauline leaves:** blade oblanceolate or broader, 1.5–2.5 cm, margins entire or sparsely dentate, (apex acute). **Racemes** somewhat loose, (5–15 cm). **Fruiting pedicels** (spreading or ascending, curved, fruits not pendent), 10–20 mm. **Flowers:** sepals oblong, 5–7 mm; petals (lemon yellow), spatulate, 9–12 mm. **Fruits** didymous, obreniform, moderately inflated, angustiseptate, (8–)10–12(–15) × 10–14(–16) mm, (papery, not keeled, basal sinus absent, apical sinus broad and open); valves (retaining seeds after dehiscence, rounded or irregular), loosely pubescent, trichomes spreading; replum broadly lanceolate, as wide as or wider than fruit, apex acute; ovules 8 per ovary; style 1–2 mm. **Seeds** flattened. $2n = 8$.

Flowering Apr–Jun. Gravelly banks, stream shores, rocky slopes, dry hillsides, serpentine soils; 900–1900 m; Idaho, Oreg., Wash.

62. Physaria ovalifolia (Rydberg) O'Kane & Al-Shehbaz, Novon 12: 326. 2002 • Roundleaf bladderpod E

Lesquerella ovalifolia Rydberg in N. L. Britton and A. Brown, Ill. Fl. N. U. S. 2: 137, fig. 1749. 1897; *L. engelmannii* (A. Gray) S. Watson subsp. *ovalifolia* (Rydberg ex Britton) C. Clark

Perennials; caudex simple or branched, (thickened by persistent leaf bases); densely pubescent (foliage usually scabrous), trichomes (sessile or short-stalked), several-rayed, rays furcate near base, (usually strongly umbonate, roughly tuberculate, less so over umbo). **Stems** few to several from base, erect or outer decumbent, 0.5–2.5 dm. **Basal leaves:** blade suborbicular to elliptic or ovate or deltate, 0.5–2(–6.5) cm, margins entire or shallowly dentate. **Cauline leaves:** (proximal shortly petiolate, distal usually sessile); blade narrowly elliptic or obovate, (0.5–)1–2.5(–4) cm, margins entire. **Racemes** compact, (± subumbellate to densely corymbiform, elongated or not). **Fruiting pedicels** (usually spreading at right angles, sometimes nearly erect, ± straight), 5–15(–20) mm, (stout). **Flowers:** sepals ± elliptic, 4.5–7(–8.5) mm, (median pair thickened apically); petals (sometimes white), suborbicular to obovate or obdeltate, 6.5–15 mm, (base narrowing to broad claw, apex sometimes emarginated). **Fruits** (sessile or shortly stipitate, less than 1 mm), subglobose to broadly ellipsoid, inflated or slightly compressed (terete or subterete), (4–)5–8(–9) mm; valves (not retaining seeds after dehiscence), glabrous; replum as wide as or wider than fruit; ovules 8–16 per ovary; style 4–8(–9) mm. **Seeds** flattened.

Subspecies 2 (2 in the flora): wc United States.

1. Caudices branched (well-developed); petals usually yellow, rarely white, 6.5–12(–14) mm, usually 1.5 times or less as long as sepals; racemes usually not elongated (subumbellate) . . . . 62a. *Physaria ovalifolia* subsp. *ovalifolia*
1. Caudices usually simple; petals white, (9–)11–15 mm, often 2 times as long as sepals; racemes usually elongated . . . . 62b. *Physaria ovalifolia* subsp. *alba*

62a. Physaria ovalifolia (Rydberg) O'Kane & Al-Shehbaz subsp. **ovalifolia** E

Lesquerella ovata Greene

Caudex branched (well-developed). **Stems** several from base, stiffly erect, (unbranched, slender), 1–2 dm. **Basal leaves:** blade (outer) suborbicular or ovate to elliptic, base narrowing abruptly to petiole, margins entire. **Racemes** usually not elongated (subumbellate). **Petals** usually yellow, rarely white, 6.5–12(–14) mm, usually 1.5 times or less as long as sepals. $2n$ = 12, 24, 36, 48, 50, 72.

Flowering Mar–May. Bare limestone flats, rocky knolls and slopes, limestone chip, gypseous outcrops, rock crevices, exposed caprock; 500–1700 m; Colo., Kans., Nebr., N.Mex., Okla., Tex.

62b. Physaria ovalifolia (Rydberg) O'Kane & Al-Shehbaz subsp. **alba** (Goodman) O'Kane & Al-Shehbaz, Novon 12: 326. 2002 E

Lesquerella ovalifolia Rydberg var. *alba* Goodman, Rhodora 38: 239. 1936; *L. engelmannii* (A. Gray) S. Watson subsp. *alba* (Goodman) C. Clark; *L. ovalifolia* subsp. *alba* (Goodman) Rollins & E. A. Shaw; *Physaria ovalifolia* var. *alba* (Goodman) B. L. Turner

Caudex usually simple. **Stems** usually few from base, erect or outer decumbent, 1.5–2.5 dm. **Basal leaves:** blade (outer) often broadly elliptic, base narrowing gradually to petiole, margins sinuate to dentate (or blade deltate, less than 1 cm, base abruptly narrowed to petiole). **Racemes** usually elongated. **Petals** white, (9–)11–15 mm, often 2 times as long as sepals. $2n$ = 12.

Flowering Mar–May, usually in Apr. Limey and gravelly knolls, grassland hills, limestone hillsides and breaks, gypsum, shale, rocky calcareous soils, stony areas, prairie pastures, limestone roadcuts; 150–700 m; Kans., Okla.

63. Physaria pachyphylla O'Kane & Grady, Novon 17: 187, fig. 4. 2007 • Thick-leaf bladderpod C E F

Perennials; caudex usually simple, rarely branched, (loosely mounded, rosette-like growth); densely (silvery or gray) pubescent, trichomes (sessile), 5-rayed, rays bifurcate, slightly fused near base of main rays, (tuberculate throughout, less over umbo). **Stems** several from base, decumbent to prostrate, (well-exserted beyond basal leaves), 0.2–0.5 dm. **Basal leaves:** (petiole differentiated from blade); blade (slightly cupped, leathery, nearly 1 mm thick), oblanceolate to orbicular, 1.2–2 cm, margins entire, (apex acute). **Cauline leaves:** blade spatulate, similar to basal. **Racemes** dense, (subumbellate). **Fruiting pedicels** (ascending, curved), 5–7 mm. **Flowers:** sepals (pale yellow), elliptic to oblong, 3.5–4.0 mm, (median pair somewhat thickened apically, cucullate); petals lingulate, 5–6 mm. **Fruits** globose or ellipsoid, slightly inflated (with slight apical constriction), 3–6 mm; valves pubescent, trichomes closely appressed; ovules 8 per ovary; style 1–3 mm (shorter than mature fruit). **Seeds** plump, (oblong).

Flowering Jun–Jul. Barren areas of mixed white, pink, or reddish limestone and diatomaceous earth; of conservation concern; 1300–1600 m; Mont.

Physaria pachyphylla is known from the Pryor Mountain Desert near the Wyoming state line.

64. Physaria pallida (Torrey & A. Gray) O'Kane & Al-Shehbaz, Novon 12: 326. 2002 • White bladderpod C E

Vesicaria grandiflora Hooker var. *pallida* Torrey & A. Gray, Fl. N. Amer. 1: 101. 1838; *Alyssum pallidum* (Torrey & A. Gray) Kuntze; *Lesquerella pallida* (Torrey & A. Gray) S. Watson; *V. pallida* (Torrey & A. Gray) Nuttall ex Torrey & A. Gray

Annuals (winter); with a fine taproot; sparsely pubescent, trichomes (minute), 3- or 4-rayed, rays furcate or, sometimes, trifurcate. **Stems** few to several from base, erect, (from basal leaf cluster, branched distally, flowering branches slender, subtended by bracts), 3–6 dm. **Basal leaves:** blade oblanceolate or broadly obovate, to 10 cm, margins usually sinuate-dentate or entire, sometimes lobed. **Cauline leaves:** (proximal shortly petiolate, distal sessile); blade oblanceolate to narrowly oblong, similar to basal, (distal with base slightly cuneate). **Racemes** paniculate, (rachises and pedicels more densely pubescent than proximal leaves). **Fruiting pedicels** (widely divaricate-ascending and straight, or slightly recurved), 10–15 mm, (slender, pubescent). **Flowers:** sepals elliptic, 3–7 mm, (median pair slightly thickened apically, cucullate); petals (white), broadly ovate, to 12 mm, (narrowing gradually to short claw). **Fruits** (widely spreading to nearly pendent in age, shortly stipitate), globose or subglobose, not or slightly inflated, 3–10 mm; valves (not retaining seeds after dehiscence), glabrous; replum as wide as or wider than fruit; ovules 8–12 per ovary; style ca. 2 mm, (slender, fragile). **Seeds** flattened. $2n = 12$.

Flowering Apr–May. Grassy openings of small glade prairies, outcrops; of conservation concern; 90 m; Tex.

Physaria pallida is known from the Weches Formation in San Augustine County.

65. Physaria parviflora (Rollins) O'Kane & Al-Shehbaz, Novon 12: 326. 2002 • Picenace bladderpod C E

Lesquerella parviflora Rollins, J. Arnold Arbor. 64: 506. 1983

Perennials; caudex simple or branched; densely (silvery) pubescent, trichomes (irregularly radiate), 6–8-rayed, rays furcate or bifurcate, fused at base. **Stems** several from base, prostrate to decumbent, (usually unbranched, rarely branched distally), 1–3 dm. **Basal leaves** (tufted); blade broadly obovate, 1–2 cm, margins entire or with 1 or 2 broad teeth, (apex rounded to obtuse). **Cauline leaves:** blade oblanceolate to nearly oblong, similar to basal, (base cuneate), margins entire. **Racemes** (secund), loose, (elongated in fruit). **Fruiting pedicels** (recurved), 6–8(–12) mm. **Flowers:** sepals (yellowish), elliptic to lanceolate, (2–)3–4 mm; petals spatulate, (3.9–)5–7 mm. **Fruits** (usually pendent), elliptic to subglobose, usually slightly compressed (latiseptate), 3–4 mm; valves densely pubescent, sometimes with scattered trichomes inside; ovules 4 per ovary; style ca. 3 mm. **Seeds** somewhat flattened.

Flowering Jun–Jul. Shale of steep slopes, rock crevices, ledges, canyon sides, shale-marlstone; of conservation concern; 2100–2700 m; Colo.

Physaria parviflora is known from the Parachute Creek Member of the Green River Formation, Rio Blanco County.

66. Physaria parvula (Greene) O'Kane & Al-Shehbaz, Novon 12: 326. 2002 • Pygmy bladderpod E

Lesquerella parvula Greene, Pittonia 4: 308. 1901; *L. alpina* (Nuttall) S. Watson subsp. *parvula* (Greene) Rollins & E. A. Shaw; *L. alpina* var. *parvula* (Greene) S. L. Welsh & Reveal

Perennials; caudex (buried), usually branched, sometimes simple, (cespitose); densely pubescent, trichomes (appressed), 4–7-rayed, rays distinct, furcate or bifurcate near base. **Stems** few to several from base, erect, (unbranched, slender), 0.3–1.5(–3) dm. **Basal leaves** (tufted, erect); blade linear to very narrowly spatulate, 1–3(–4) cm, margins entire (involute). **Cauline leaves** similar to basal. **Racemes** relatively dense. **Fruiting pedicels** (ascending, curved or sigmoid), 2–10 mm. **Flowers:** sepals (greenish yellow), elliptic, 3.5–7 mm; petals spatulate, 5–6 mm, (not clawed). **Fruits** (erect), ovoid (or longer than broad), usually inflated, 4–5 mm, (apex acute, slightly flattened); valves pubescent, trichomes appressed; ovules 4–8 per ovary; style 2–4 mm. **Seeds** flattened, (mucilaginous). $2n = 10, 20$.

Flowering May–Jul. Exposed windblown ridges, gravelly hills, open rocky knolls, gravelly hilltops, clay hillsides, granitic sand, reddish soil, sagebrush, mountain scrub, and pinyon-juniper areas; 1800–2800 m; Colo., Utah, Wyo.

67. Physaria pendula (Rollins) O'Kane & Al-Shehbaz, Novon 12: 326. 2002 • Snake Range bladderpod C E

Lesquerella pendula Rollins, Cruciferae Continental N. Amer., 647. 1993

Perennials; caudex simple; densely pubescent, trichomes 5–7-rayed, rays bifurcate or trifurcate. **Stems** several from base, erect or outer usually decumbent toward base, (from below a terminal tuft of mostly erect leaves, unbranched), 1–2 dm. **Basal leaves:** blade ovate to elliptic, 2–4 cm, (base gradually narrowed to petiole), margins entire, (surfaces densely pubescent with trichome layers). **Cauline leaves** (remote, proximal shortly petiolate); blade spatulate to oblanceolate, similar to basal, (base often cuneate), margins entire. **Racemes** elongated. **Fruiting pedicels** (recurved), 7–10 mm. **Flowers:** sepals linear-oblong, 5–7 mm, (lateral pair subsaccate); petals (erect), lingulate, 8–10 mm, (claw barely differentiated from blade). **Fruits** (usually pendent, sessile), subglobose, slightly flattened (angustiseptate), 4–5 mm; valves densely pubescent, trichomes somewhat spreading; ovules 8–12 per ovary; style 4–5 mm. **Seeds** plump, (mucilaginous when wetted).

Flowering May–Jun. Limestone gravel and cobbles, typically with junipers; of conservation concern; 1700–2100 m; Nev.

68. **Physaria pinetorum** (Wooton & Standley) O'Kane & Al-Shehbaz, Novon 12: 327. 2002 • White Mountain bladderpod E

Lesquerella pinetorum Wooton & Standley, Contr. U.S. Natl. Herb. 16: 126. 1913

Perennials; caudex simple or branched; densely pubescent, trichomes (sessile or short-stalked), 6–8-rayed, rays furcate or bifurcate, (tuberculate, less so on outer layers). **Stems** simple or few from base, ascending to erect, (0.5–)1–2(–3.5) dm. **Basal leaves:** (petiole tapering to blade); blade rhombic to elliptic and irregularly angular, sometimes spatulate to oblanceolate, 1.5–7.5(–10) cm, margins entire. **Cauline leaves:** (not or loosely overlapping, petiolate or distal sessile); blade spatulate to oblanceolate, 1–4 cm, margins entire. **Racemes** crowded, elongated. **Fruiting pedicels** (ascending, curved or sigmoid), 6–12(–20) mm. **Flowers:** sepals ovate, oblong, or elliptic 4–7.5 mm, (median pair thickened apically, cucullate); petals spatulate or broadly cuneate, 6–13 mm, (claw slightly expanded at base). **Fruits** (substipitate), globose or obovoid to ellipsoid, sometimes slightly obcompressed, 4–9 mm; valves (not retaining seeds after dehiscence), glabrous throughout; replum as wide as or wider than fruit; ovules 4–24 per ovary; style (2–)4–7 mm. **Seeds** flattened. $2n = 10$.

Flowering Apr–Jul. Scrub oak, pinyon-juniper woodland, open ponderosa pine forests, these sometimes mixed with Douglas fir, white pine, white fir, Engelmann spruce, or Gambel oak, on limestone-derived or otherwise basic soils, often in rock crevices; 1400–2900 (–3400) m; Ariz., N.Mex.

Physaria pinetorum with reduced forms are found at high elevations; in disturbed, moist soils plants can become quite large, as in the Manzano Mountains. Densely cespitose plants with crowded racemes not exceeding the basal leaves are found at the crest (3200–3400 m) of the Sandia Mountains, New Mexico. These probably represent an undescribed taxon.

69. **Physaria prostrata** (A. Nelson) O'Kane & Al-Shehbaz, Novon 12: 327. 2002 • Low bladderpod C E

Lesquerella prostrata A. Nelson, Bull. Torrey Bot. Club 26: 124. 1899

Perennials; caudex branched; densely pubescent, trichomes (usually sessile, rarely short-stalked), several-rayed, rays furcate or bifurcate, (umbonate, tuberculate throughout). **Stems** several from base, usually prostrate, rarely decumbent, (unbranched, often purplish, sparsely pubescent), to 1.5 dm. **Basal leaves:** blade deltate, hastate, or, less often, rhombic to elliptic, 1–5 cm, margins entire (often partially involute). **Cauline leaves:** (proximal shortly petiolate); blade linear to oblanceolate, 0.5–1.5 cm, margins entire. **Racemes** loose, elongated. **Fruiting pedicels** (ascending, slightly sigmoid to straight), 5–10 mm. **Flowers:** sepals (often purplish), oblong, lanceolate, or ovate, 4–6 mm, (median pair thickened apically); petals spatulate or cuneate, 5–8(–9) mm, (margins undulate). **Fruits** ovoid or ellipsoid, slightly compressed, 3–5(–6) mm, (base often gibbous); valves pubescent, trichomes loose, furcate near their bases and spreading, sometimes sparsely pubescent inside; ovules 4(–8) per ovary; style 1.5–4 mm. **Seeds** flattened.

Flowering May–Jun. Whitish sand and small rocks on steep slopes, dry hillsides, windswept knolls, shaley slopes; of conservation concern; 1800–2500 m; Idaho, Utah, Wyo.

Physaria prostrata is sometimes found on igneous substrates, which is unusual for the genus.

70. **Physaria pruinosa** (Greene) O'Kane & Al-Shehbaz, Novon 12: 327. 2002 • Frosty or Pagosa bladderpod C E

Lesquerella pruinosa Greene, Pittonia 4: 307. 1901

Perennials; caudex simple or branched, (covered with persistent leaf bases); densely pubescent, trichomes (sessile or subsessile), 4–7-rayed, rays furcate or bifurcate, (tuberculate throughout). **Stems** simple or several from base, decumbent or erect, (unbranched), to 2 dm. **Basal leaves:** (petiole sharply differentiated from blade, slender); blade suborbicular or obovate to rhombic, 4–8 cm, (base abruptly narrowed to petiole), margins entire, sinuate, or shallowly dentate, (abaxial surface densely pubescent, adaxial lightly pubescent). **Cauline leaves:** (proximal petiolate, distal sessile); blade obovate to rhombic, 0.8–2.3 cm, margins entire or shallowly toothed. **Racemes** dense, (somewhat elongated in fruit). **Fruiting pedicels** (horizontal to ascending, sigmoid or slightly curved), 8–11 mm, (stout). **Flowers:** sepals elliptic or oblong, ca. 6 mm, (lateral pair not saccate or subsaccate, cucullate, median pair thickened apically, cucullate); petals spatulate, ca. 9 mm, (claw expanded at base). **Fruits** (sessile or substipitate, often becoming copper-red in age), subglobose or ellipsoid, inflated, 6–9 mm, (firm, glossy); valves (not retaining seeds after dehiscence), glabrous throughout; replum as wide as or wider than fruit; ovules 4–8(–12) per ovary; style 3.5–7 mm. **Seeds** somewhat flattened.

Flowering May–Jun(–Aug). Mancos slate or shale, meadows, gentle slopes, edges of ponderosa pine stands; of conservation concern; 2100–2600 m; Colo., N.Mex.

The one New Mexico population is near the border with Colorado, in Rio Arriba County.

71. **Physaria pulvinata** O'Kane & Reveal, Brittonia 58: 74, fig. 1. 2006 • Cushion bladderpod [C][E]

Perennials; caudex (buried), branched, (dense, forming hard mats); densely pubescent, trichomes (subsessile), 8–13-rayed, rays usually furcate, distinct, (umbonate, usually tuberculate, less so over umbo). **Stems** several (to several hundred) from base, erect, (each terminating in a tufted cluster of leaves), to 7 dm. **Basal leaves:** (petiole not differentiated from blade); blade narrowly elliptic to narrowly linear-oblanceolate, (0.8–)1–1.5 cm, (base cuneate), margins entire. **Cauline leaves** similar to basal, blade sometimes linear, (apex acute). **Racemes** dense, (often ± subumbellate, somewhat elongated in fruit). **Fruiting pedicels** (strongly sigmoid), 5–10 mm. **Flowers:** sepals narrowly elliptic, 2.5–3.5(–4) mm, (not keeled); petals narrowly spatulate, 4–7 mm. **Fruits** ellipsoid, compressed, 4–6 mm; valves densely pubescent, trichomes appressed; ovules 2 per ovary; style 2–3.5 mm. **Seeds** flattened, (oval).

Flowering late May–Jun. Gray, argillaceous shale outcrops with sagebrush and junipers; of conservation concern; 2300–2600 m; Colo.

Physaria pulvinata is known from an area surrounded by a pygmy forest of Utah juniper in Dolores and San Miguel Counties.

72. **Physaria purpurea** (A. Gray) O'Kane & Al-Shehbaz, Novon 12: 327. 2002 • Rose bladderpod

Vesicaria purpurea A. Gray, Smithsonian Contr. Knowl. 5(6): 14. 1853; *Lesquerella purpurea* (A. Gray) S. Watson; *L. purpurea* subsp. *foliosa* (Rollins) Rollins & E. A. Shaw; *L. purpurea* var. *foliosa* Rollins; *Physaria purpurea* var. *foliosa* (Rollins) B. L. Turner

Perennials; caudex simple, (usually woody); densely pubescent, trichomes (sessile or short-stalked), several-rayed, rays simple or furcate, (smooth or tuberculate). **Stems** simple from base, erect, (unbranched, sparsely leaved), to 7 dm. **Basal leaves:** blade elliptic or obovate to oblong, 4–15 cm, margins entire, dentate, or lyrate-pinnatifid. **Cauline leaves:** (proximal often narrowed to short petiole, distal sessile); blade broadly elliptic to obovate or rhombic, 0.5–3(–5) cm, margins entire. **Racemes** dense or slightly elongated. **Fruiting pedicels** (spreading or recurved, loosely sigmoid), 5–25 mm. **Flowers:** sepals elliptic to ovate, 3.5–6(–7) mm, (median pair usually thickened apically, cucullate); petals (white, often purple-veined, fading purplish), suborbicular to obovate, obdeltate, or cuneate, 4.5–10(–12) mm, (often narrowed to broad claw, apex emarginate, less frequently claw undifferentiated from blade). **Fruits** (pendent or horizontal, sessile or substipitate), subglobose to broadly ellipsoid, not or slightly inflated, (4–)5–8 mm; valves (not retaining seeds after dehiscence), glabrous throughout; replum as wide as or wider than fruit; ovules 4–8(–12) per ovary; style 1–3(–4) mm. **Seeds** flattened. $2n = 18, 36$.

Flowering Mar–Oct. Rocky draws, canyons, stony hills, ridges, rock crevices on limestone ledges, lava cliffs, sand and gravel of dry stream beds, rocky slopes, talus, shade of bushes or cactus clumps; 400–2400 m; Ariz., N.Mex., Tex.; Mexico (Chihuahua, Coahuila, Sonora).

73. **Physaria pycnantha** Grady & O'Kane, Novon 17: 188, fig. 5. 2007 • Mountain-view bladderpod [E]

Perennials; caudex branched, (densely cespitose and forming hemispheric mounds); densely pubescent, trichomes 5-rayed, rays bifurcate near base, fused at base, (strongly tuberculate throughout). **Stems** few to several from base, erect, (usually exceeding basal leaves), 0.3–0.7 dm. **Basal leaves:** blade linear-spatulate, 1.5–4 cm, (base narrowed gradually to petiole), margins entire. **Cauline leaves:** blade spatulate, similar to basal. **Racemes** crowded in distal ⅓, (4–10-flowered). **Fruiting pedicels** (loosely to strongly sigmoid), 6–10 mm. **Flowers:** sepals (pale yellow), oblong to elliptic, 3–4 mm, (median pair usually thickened apically, cucullate); petals (sometimes with slight tinge of orange basally), lingulate, 4–6 mm. **Fruits** ellipsoid, slightly inflated (somewhat latiseptate), 4–5 mm, (apex acute); valves pubescent, trichomes erect, appearing slightly shaggy; ovules 4–8 per ovary; styles 2.5–3 mm, (shorter than mature fruits). **Seeds** ± flattened, convex on outer side.

Flowering late May–Jun(–Jul). Dry, windswept knolls of limestone gravel, with other cushion-forming plants; 1600–2300 m; Idaho, Mont.

Physaria pycnantha is morphologically similar to 56. *P. nelsonii*.

74. **Physaria rectipes** (Wooton & Standley) O'Kane & Al-Shehbaz, Novon 12: 327. 2002 • Straight bladderpod E F

Lesquerella rectipes Wooton & Standley, Contr. U.S. Natl. Herb. 16: 127. 1913

Perennials; (loose, spreading); caudex simple or branched; densely pubescent, trichomes (subsessile), 4–6-rayed, rays furcate or bifurcate, (moderately tuberculate over arms, less so or smooth over center). **Stems** few to several from base, ascending or prostrate, (arising laterally, also from within basal leaves, usually unbranched, rarely branched), 0.5–3(–6) dm. **Basal leaves**: blade narrowly oblanceolate or broadly elliptic, 1–7(–12) cm, margins entire or shallowly toothed, sometimes repand, (inner blades usually flattened in age, surfaces often gray-green, scabrous). **Cauline leaves** (usually secund); blade spatulate or obovate, 1–2.5(–4.5) cm, margins entire or shallowly toothed, (flat or involute). **Racemes** somewhat crowded (to moderately elongated in fruit, exceeding basal leaves). **Fruiting pedicels** (often divaricate-spreading and straight, or horizontal and loosely sigmoid, sometimes slightly recurved), 5–15 mm. **Flowers:** sepals broadly elliptic or oblong, 4–7.5(–9) mm, (median pair thickened apically, cucullate); petals cuneate or obovate, 7–10(–16) mm, (tapering gradually to broad claw). **Fruits** subglobose to ovoid or ellipsoid, sometimes compressed, (4–)5–7(–9) mm; valves sparsely pubescent, trichomes appressed or erect, sometimes sparsely pubescent inside; ovules (8–)12–16(–20) per ovary; style 2–7 mm. **Seeds** somewhat flattened. $2n = 10 + 2, 18, 20, \pm 40$.

Flowering May–Jul. Sandy soils, limey knolls, rocky hills, clay hillsides, dry ridges, weathered rocks, gravelly outwashes, stony slopes, pinyon-juniper woodlands; 1500–2600 m; Ariz., Colo., N.Mex., Utah.

As here circumscribed, *Physaria rectipes* remains heterogeneous and may represent more than one taxon, even after the recent removal of *P. pulvinata*.

75. **Physaria recurvata** (Engelmann ex A. Gray) O'Kane & Al-Shehbaz, Novon 12: 327. 2002 • Gaslight bladderpod E

Vesicaria recurvata Engelmann ex A. Gray, Boston J. Nat. Hist. 6: 147. 1850; *Lesquerella recurvata* (Engelmann ex A. Gray) S. Watson

Annuals or, sometimes, biennials; with a fine taproot; sparsely pubescent, trichomes (sessile), 3–6-rayed, rays furcate, (tuberculate throughout). **Stems** several from base, erect or decumbent and straggling, (branched distally, branches usually filiform), to 5 dm. **Basal leaves:** blade obovate or rhombic to broadly elliptic, 1–4.5(–6.5) cm, margins entire or lyrate-pinnatifid. **Cauline leaves:** (proximal petiolate, distal sessile); blade rhombic or obovate to elliptic, 0.5–2(–3) cm, margins entire or sinuate to remotely toothed. **Racemes** loose. **Fruiting pedicels** (recurved in age), 5–10(–15) mm, (slender). **Flowers:** sepals elliptic or ovate, 2.5–5.5 mm, (median pair thickened apically, cucullate); petals (yellow to orange-yellow), obovate to cuneate, 4–7 mm, (apex sometimes retuse). **Fruits** globose or subglobose, not or slightly inflated, (2–)3–5(–7) mm; valves (not retaining seeds after dehiscence), glabrous throughout; replum as wide as or wider than fruit; ovules (4–)8–16(–20) per ovary; style (1–5–)2–4.5 mm. **Seeds** flattened. $2n = 10$.

Flowering Mar–Apr. Light dry soils, limestone chip, open rocky areas, among boulders, roadsides, pastures, stony open sandy prairies, dry streamside meadows, calcareous soils, limestone outcroppings, scrub-oak grassland flats; 150–700 m; Tex.

Physaria recurvata is known from the Edwards Plateau.

76. **Physaria reediana** O'Kane & Al-Shehbaz, Novon 12: 327. 2002 • Alpine or Rollins's bladderpod E

Vesicaria alpina Nuttall in J. Torrey and A. Gray, Fl. N. Amer. 1: 102. 1838, not *Physaria alpina* Rollins 1981; *Alyssum alpinum* (Nuttall) Kuntze; *Lesquerella alpina* (Nuttall) S. Watson; *L. alpina* var. *laevis* (Payson) C. L. Hitchcock; *L. condensata* A. Nelson var. *laevis* Payson

Perennials; caudex simple or branched, (covered with persistent leaf bases, loosely cespitose); densely pubescent, trichomes (appressed to ascending, plant appearing shaggy, always appressed on fruits), 4- or 5-rayed, rays furcate or bifurcate, (tuberculate throughout). **Stems** few to several from base, ± erect, (arising laterally, also from within basal leaves), 0.2–0.4 dm. **Basal leaves**

(erect); blade linear-oblanceolate, 1.2–2.8 cm, (base gradually narrowed to petiole), margins entire. **Cauline leaves:** blade linear, similar to basal. **Racemes** dense, (often subumbellate, not or barely exceeding basal leaves). **Fruiting pedicels** (ascending, curved), 3–5.5 mm. **Flowers:** sepals (pale green-yellow), oblong to elliptic, 4–5 mm, (median pair usually thickened apically, cucullate); petals lingulate, 6–9 mm. **Fruits** lanceolate in outline, compressed (latiseptate) on margins and at apex, 4–5 mm; valves pubescent, trichomes closely appressed; ovules 8–12 per ovary; style 3.5–4.5 mm (equaling or exceeding length of fruit, curved proximal to stigma). **Seeds** plump, (oblong).

Flowering May–Jul. Open areas of grasslands on calcareous soils; 1200–1900 m; Colo., Nebr., Wyo.

77. **Physaria rollinsii** G. A. Mulligan, Canad. J. Bot. 44: 1663, fig. 2, plate 1, fig. 4. 1966 • Rollins's twinpod [E]

Perennials; (compact); caudex usually simple, (cespitose); (silvery) pubescent throughout, trichomes 6–8-rayed, rays furcate near base, fused at base, (umbonate, strongly tuberculate throughout). **Stems** several from base, decumbent, (unbranched, slender), 0.5–1 dm. **Basal leaves** (strongly rosulate); blade usually oblanceolate or broader, sometimes triangular, 2–3.5 cm (width 5–10 mm), margins entire or with 1 or 2 broad teeth, (apex acute). **Cauline leaves:** blade oblanceolate, 1–1.5 cm (width 2–4 mm), margins entire, (apex acute). **Racemes** congested, (elongated moderately in fruit). **Fruiting pedicels** (spreading, straight or somewhat sigmoid), 5–8 mm. **Flowers:** sepals linear, 5–7 mm; petals spatulate, 8–10 mm, (apex often somewhat truncate). **Fruits** (erect), didymous, suborbicular, inflated, 2–5(–8) × 4–8(–10)mm, (coriaceous, base slightly cordate or nearly obtuse, sinus obsolete or absent, apical sinus broad and deep); valves (retaining seeds after dehiscence), pubescent, trichomes appressed, (silvery on ovaries and immature fruit; replum obovate to oblong, as wide as or wider than fruit, rarely somewhat constricted basally, apex obtuse; ovules 4 per ovary; style 5–7 mm. **Seeds** slightly flattened. $2n = 8$.

Flowering May–Jun. Sagebrush, granitic talus, open knolls, limestone chiprock, steep slopes, clay banks, near granite boulders; 2300–2500(–3900) m; Colo.

78. **Physaria saximontana** Rollins, Contr. Gray Herb. 214: 13. 1984 • Rocky Mountain or Fremont County twinpod [E]

Perennials; caudex usually simple; (silvery) pubescent throughout, trichome rays furcate. **Stems** several from base, prostrate to decumbent, 0.3–1 dm. **Basal leaves** (rosulate; petiole winged); blade orbicular to broadly obovate, 1.5–3 cm, margins entire or with broad, obscure toothlike angles each side at apex, (apex obtuse, surfaces densely pubescent, trichomes appressed). **Cauline leaves:** blade broadly spatulate to linear-oblanceolate, 1–1.5 cm, margins entire. **Racemes** condensed, (subumbellate to slightly more elongated, few-flowered). **Fruiting pedicels** (divaricate-ascending, straight to slightly curved), 6–10 mm. **Flowers:** sepals (yellowish, often with some purple), narrowly lanceolate, 5–6 mm; petals spatulate, 7.3–9.2 mm, (not clawed). **Fruits** didymous, irregular, suborbicular, deeply bilobed, inflated in age, 10–12 × 12–15 mm, (papery, basal sinus absent or obsolete, apical sinus deep); valves (retaining seeds after dehiscence), densely pubescent, trichomes spreading, (ovaries and immature fruit downy; replum narrowly ovate to broadly oblong, not narrowed at middle, as wide as or wider than fruit, apex acute to obtuse; ovules 4 per ovary; style 3–7 mm. **Seeds** flattened.

Subspecies 2 (2 in the flora): wc United States.

Physaria saximontana (especially subsp. *dentata*) is morphologically similar to 22. *P. didymocarpa*.

1. Basal leaf blades: margins entire, apices rounded to angled; styles 3–5 mm . . . .
 . . . . 78a. *Physaria saximontana* subsp. *saximontana*
1. Basal leaf blades: margins dentate (teeth broad), apices angled; styles 4–7 mm . . . .
 . . . . 78b. *Physaria saximontana* subsp. *dentata*

78a. **Physaria saximontana** Rollins subsp. **saximontana** • Fremont County bladderpod [E]

Basal leaves: blade margins entire, apex rounded to angled. **Styles** 3–5 mm. **Ovules** 4 per ovary.

Flowering May–Jun. Red shaley banks and ledges, heavy clay, hillsides; 1700–2600 m; Wyo.

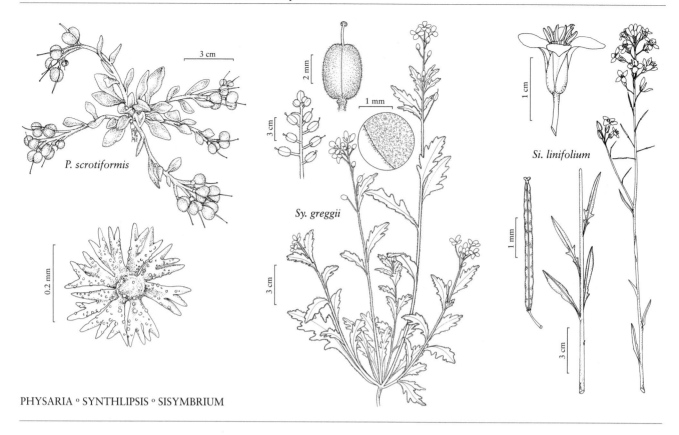

PHYSARIA ° SYNTHLIPSIS ° SISYMBRIUM

78b. Physaria saximontana Rollins subsp. dentata (Rollins) O'Kane, Novon 17: 381. 2007 [E]

Physaria saximontana var. *dentata* Rollins, Contr. Gray Herb. 214: 14. 1984

Basal leaves: blade margins dentate, (teeth broad), apex angled. **Styles** 4–7 mm. **Ovules** 4–8 per ovary.

Flowering Jun–Jul. Open gravelly slopes, scree, mountain fellfields on limestone; 1500–2700 m; Mont.

79. Physaria scrotiformis O'Kane, Novon 17: 376, fig. 1. 2007 • West silver bladderpod [E] [F]

Perennials; (diminutive); caudex simple or branched, (buried, with thatch of persistent leaf bases distally); (appearing silvery gray-green to silvery purple), densely pubescent, trichomes usually 5 or 6 (rarely 7)-rayed, rays bifurcate or incompletely so, (relatively short, stout, umbonate, moderately tuberculate to nearly smooth, lower layer smoother). **Stems** 1–5 from base, prostrate to slightly decumbent, (arising laterally, also erect or ascending from tuft of basal leaves, unbranched, purple-green), 0.08–0.3 dm. **Basal leaves:** (petiole slightly winged); blade oblanceolate, elliptic, or rhombic, (mostly flat, sometimes somewhat folded), 0.6–2.7 cm, (base tapering to petiole), margins entire, (apex rounded to rounded-acute). **Cauline leaves:** (3–7, shortly petiolate or sessile); blade elliptic to oblanceolate, 0.3–0.5 cm, margins entire. **Racemes** crowded, (ca. 3–7 fruits). **Fruiting pedicels** (ascending, straight), 1.8–3.4 mm. **Flowers:** sepals (greenish yellow), linear-triangular, 3.7–5 mm, (lateral pair subsaccate); petals oblanceolate to narrowly obovate, 4.5–9 mm. **Fruits** (shortly stipitate, purple or greenish purple in age), slightly didymous, ovoid to obpyriform, 3–5 mm (wider than long, base rounded-obtuse, apex rounded, flattened, or slightly emarginate); valves (inflated, slightly wider than replum), pubescent, trichomes scattered; replum obovate to orbicular-obdeltate, apex rounded, obtuse, or truncate; septum complete or medially small-perforate; ovules 4–6(–8) per ovary; style 2–3.6 mm. **Seeds** relatively plump, (ovate to suborbicular, usually rounded on one side, ± flat or concave on the other, not mucilaginous when wetted).

Flowering Jun–early Jul. Tundra areas with islands of Engelmann spruce on Leadville limestone, amidst limestone cobbles and gravel; 3500–3700 m; Colo.

Physaria scrotiformis is known only from La Plata County.

80. **Physaria sessilis** (S. Watson) O'Kane & Al-Shehbaz, Novon 12: 328. 2002 • Sessile bladderpod E

Lesquerella gracilis (Hooker) S. Watson var. *sessilis* S. Watson, Proc. Amer. Acad. Arts 23: 253. 1888; *L. sessilis* (S. Watson) Small

Annuals or biennials; with a fine taproot; densely pubescent, trichomes (sessile or short-stalked), 4–6-rayed, rays usually furcate, rarely bifurcate, (moderately tuberculate or nearly smooth). **Stems** simple or few to several from base, erect, (often distal ½ branched), to 6 dm. **Basal leaves**: blade oblanceolate, to 9 cm, margins entire, dentate, or sinuate to lyrate-pinnatifid. **Cauline leaves**: (proximal shortly petiolate, distal sessile); blade narrowly elliptic to linear, 2–4(–6) cm, margins entire or repand to shallowly dentate. **Racemes** loose. **Fruiting pedicels** (divaricate-ascending to widely spreading, straight), 8–20 mm. **Flowers**: sepals elliptic or elongate-ovate, 3.4–5.2(–6.5) mm, (lateral pair subsaccate, median pair thickened apically, cucullate); petals obovate or deltate, 5–10 mm, (sometimes with short claw, margins undulate). **Fruits** globose or subglobose, sometimes slightly compressed, 3–6 mm; valves (not retaining seeds after dehiscence), densely papillose, densely pubescent inside, trichomes raised; replum as wide as or wider than fruit; ovules 8–18 per ovary; style 1.5–3.5 mm. **Seeds** flattened. $2n = 12$.

Flowering and fruiting Apr–Jun. Limestone chip, black soils, grassy roadsides, fields, limestone, oak woodlands, mesquite brush lands, pastures, open dry hills; 30–700 m; Tex.

81. **Physaria spatulata** (Rydberg) Grady & O'Kane, Novon 17: 190. 2007 • Spatula-leaf bladderpod E

Lesquerella spatulata Rydberg, Contr. U.S. Natl. Herb. 3: 486. 1896; *L. alpina* (Nuttall) S. Watson var. *spatulata* (Rydberg) Payson; *L. nodosa* Greene; *Physaria reediana* O'Kane & Al-Shehbaz subsp. *spatulata* (Rydberg) O'Kane & Al-Shehbaz; *P. reediana* var. *spatulata* (Rydberg) B. L. Turner

Perennials; caudex simple, (relatively small); sparsely to moderately pubescent, trichomes 4- or 5-rayed, rays furcate or bifurcate, not fused, (tuberculate). **Stems** simple from base, erect to decumbent, (well-exserted beyond basal leaves, loosely spreading), 0.3–1.2 dm. **Basal leaves** (erect to prostrate, petiole distinct from blade); blade (inner) spatulate to oblanceolate, or (outer) oblanceolate or orbicular, 1.5–4 cm, margins entire (rarely folded). **Cauline leaves**: blade spatulate, distinctly different from basal. **Racemes** moderately dense, (6–20-flowered). **Fruiting pedicels** (strongly sigmoid), 10–20 mm (2 times longer than fruits). **Flowers**: sepals (pale yellow), elliptic, 3.5–5 mm; petals lingulate, 6–9 mm. **Fruits** lanceolate or orbicular, slightly inflated, (2.5–)3–6 mm, (apex usually strongly narrowed); valves pubescent, trichomes sparse and closely appressed to surface; ovules 4–8 per ovary; style 2.5–6 mm (usually ± equal in length to mature fruit). **Seeds** plump.

Flowering May–early Jul. Grasslands, subalpine meadows, sagebrush, scattered pines, fellfields, calcareous (sometimes alkaline) substrates; 900–2900 m; Alta., Sask.; Mont., Nebr., N.Dak., S.Dak., Wyo.

82. **Physaria subumbellata** (Rollins) O'Kane & Al-Shehbaz, Novon 12: 328. 2002 • Parasol bladderpod E

Lesquerella subumbellata Rollins, Amer. J. Bot. 26: 420, fig. 1C,D. 1939

Perennials; caudex simple or branched, (usually covered with persistent leaf bases, cespitose); densely pubescent, trichomes (closely appressed), rays distinct, usually bifurcate. **Stems** several from base, erect, (unbranched, slender), 0.1–0.6 dm. **Basal leaves**: blade rhombic to obovate, 2–4 cm, margins entire. **Cauline leaves**: blade linear-oblanceolate, similar to basal. **Racemes** dense (distally, subumbellate). **Fruiting pedicels** (divaricate-ascending), 3–5 mm, (densely pubescent). **Flowers**: sepals (yellowish), oblong to elliptic, 3.5–7 mm, (median pair usually thickened apically, cucullate); petals lingulate to spatulate, 4–7 mm. **Fruits** (erect), ovate to suborbicular, compressed apically (latiseptate), 3–4 mm; valves pubescent; replum ovate to obovate; ovules 4–6 per ovary; style 2–3 mm. **Seeds** plump. $2n = 10$.

Flowering Apr–Jun. Rocky high ridges, gravel and stony areas, juniper covered knolls, rock crevices, clay hillsides, pinyon-juniper areas, calcareous substrates; 1600–2700 m; Colo., Utah, Wyo.

83. **Physaria tenella** (A. Nelson) O'Kane & Al-Shehbaz, Novon 12: 328. 2002 • Moapa bladderpod

Lesquerella tenella A. Nelson, Bot. Gaz. 47: 426. 1909; *L. gordonii* (A. Gray) S. Watson var. *sessilis* S. Watson

Annuals or, rarely, biennials; with a taproot; densely pubescent, trichomes (simple or stellate, sessile or short-stalked), 4–7-rayed, rays usually furcate, rarely bifurcate, (nearly smooth to finely tuberculate). **Stems** several from base, decumbent to erect, (several-branched, frequently stout), 1.5–6 dm. **Basal leaves**: blade elliptic, (1.5–)3–6.5 cm, margins entire, repand, or shallowly dentate. **Cauline leaves:** (proximal shortly petiolate, distal sessile); blade linear to elliptic or obovate, (0.5–)1–3.5(–4.5) cm, margins entire or repand. **Racemes** loose. **Fruiting pedicels** (recurved, sigmoid), 5–15 mm. **Flowers:** sepals oblong, lanceolate, or elliptic, (3–)3.5–6(–7.5) mm, (lateral pair subsaccate, median pair thickened apically, cucullate); petals (yellow to orange), suborbicular or obovate, (5–)6.5–8(–11) mm, (narrowing gradually to broad claw, usually widened at base). **Fruits** (sessile or shortly stipitate), orbicular or obovoid, often slightly compressed, (3.5–)4–6 mm; valves sparsely pubescent, trichomes sessile and stellate, densely pubescent inside, trichomes simple or branched; ovules 4–12 per ovary; style 2–4.5 mm. **Seeds** flattened. $2n = 10, 20$.

Flowering Feb–May. Sandy soils, gravel, clayey loam, loose rocky slopes, washes, desert slopes and plains, lava hills, frequently in or near bushes; (0–)600–1900 m; Ariz., Calif., Nev., Utah; Mexico (Sonora).

84. **Physaria thamnophila** (Rollins & E. A. Shaw) O'Kane & Al-Shehbaz, Novon 12: 328. 2002 • Zapata bladderpod [C][E]

Lesquerella thamnophila Rollins & E. A. Shaw, Gen. Lesquerella, 86, plates 15, 16. 1973

Perennials; caudex simple or branched, (woody); densely (silvery) pubescent, trichomes (short-stalked), 4–8-rayed, rays furcate or bifurcate, (finely tuberculate throughout). **Stems** simple or few from base, decumbent, (straggling and flexuous, usually branched distally), 4–8 dm. **Basal leaves:** blade narrowly elliptic to oblanceolate, 4–12 cm, margins entire, sinuate, or shallowly dentate, (apex acute). **Cauline leaves** (sessile or proximal shortly petiolate); blade linear to narrowly elliptic, 3–4 cm, margins entire, sinuate, or remotely dentate, (apex acute). **Racemes** loose, (sometimes greatly elongated). **Fruiting pedicels** (recurved), 15–20(–25) mm. **Flowers:** sepals elliptic, 3.5–4 mm, (lateral pair subsaccate, median pair thickened apically, cucullate); petals broadly obovate, 4–5 mm, (sometimes with short, broad claw). **Fruits** (pendent), subglobose or broadly ovoid, slightly compressed, 5–7 mm; valves (not retaining seeds after dehiscence), glabrous throughout; ovules per ovary unknown; style 1.5–2 mm. **Seeds** flattened. $2n = 16$.

Flowering Apr. Sandy soils, entangled in shrubs, cactus clumps; of conservation concern; 1700–1800 m; Tex.

Physaria thamnophila is found in sandy areas with shrubs and cactus in sparse shrubland communities of Starr and Zapata counties. It sometimes flowers through September with sufficient moisture.

85. **Physaria tumulosa** (Barneby) O'Kane & Al-Shehbaz, Novon 12: 328. 2002 • Kodachrome bladderpod [C][E]

Lesquerella hitchcockii Munz subsp. *tumulosa* Barneby, Leafl. W. Bot. 10: 313. 1966; *L. tumulosa* (Barneby) Reveal; *Physaria rubicundula* Rollins var. *tumulosa* (Barneby) S. L. Welsh

Perennials; caudex (buried), branched, (forming hard mats); densely pubescent, trichomes several-rayed, rays furcate or bifurcate, (tuberculate). **Stems** several from base, erect, (unbranched), 0.2–0.3 dm. **Basal leaves** (few), similar to cauline. **Cauline leaves:** (petiole not differentiated from blade); blade (somewhat succulent), linear to narrowly oblanceolate, 5–12 mm, margins entire. **Racemes** dense, (few-flowered). **Fruiting pedicels** (ascending to divaricate-ascending, ± straight), 3.5–6 mm. **Flowers:** sepals (yellowish), elliptic, 3–4.5 mm; petals (erect or, more commonly, arching), spatulate to oblanceolate, 5.8–7 mm, (claw not or weakly differentiated from blade). **Fruits** (coppery or reddish brown in age), broadly ovoid, slightly inflated, 3–4 mm; valves (not retaining seeds after dehiscence), glabrous throughout; replum as wide as or wider than fruit; ovules 4–8 per ovary; style 1.8–3 mm. **Seeds** flattened.

Flowering May–Jun. Barren white knolls surrounded by sagebrush, pinyon pine, and Utah juniper; of conservation concern; 1600–1800 m; Utah.

Physaria tumulosa is morphologically similar to 55. *P. navajoensis* of northeastern Arizona and northwestern New Mexico, and differing very subtly. It has been long treated as an infraspecific taxon of *P. hitchcockii*; unpublished molecular data do not support that disposition. It is found on knolls of the Winsor Member of the Carmel Formation.

86. **Physaria valida** (Greene) O'Kane & Al-Shehbaz, Novon 12: 328. 2002 • Strong bladderpod [E]

Lesquerella valida Greene, Pittonia 4: 68. 1899; *L. lepidota* Cory

Perennials; caudex branched, (thickened); densely pubescent, trichomes (sessile), several-rayed, rays furcate or bifurcate, usually fused toward base, (strongly tuberculate). **Stems** several from base, erect or outer decumbent, (unbranched), to 2 dm. **Basal leaves:** blade elliptic to lanceolate or obovate, 3–8 cm, margins entire. **Cauline leaves:** (proximal shortly petiolate or sessile, distal sessile); blade elliptic or obovate, to 2 cm, margins entire. **Racemes** dense. **Fruiting pedicels** (divaricate-ascending to horizontal, straight to loosely curved), to 15 mm. **Flowers:** sepals narrowly elliptic or oblong, 4.5–5.3 mm, (tapering to the somewhat thickened, cucullate apex, lateral pair subsaccate); petals (bright yellow), lingulate or broadly obovate, 7.5–8.5 mm, (narrowing to broad claw, joining in an arch, margins lacerate). **Fruits** (sessile or substipitate), suborbicular to broadly ovate or ellipsoid, slightly compressed, 6–8 mm; valves pubescent; ovules 12–22 per ovary; style 2–3 mm. **Seeds** flattened. $2n = 10$.

Flowering Apr–May. Limestone soils, steep slopes, roadcuts, open woods; 1900–2200 m; N.Mex., Tex.

Physaria valida is known from the Sacramento and White mountains of south central New Mexico, and southward through the Guadalupe Mountains to Hudspeth County, Texas.

87. **Physaria vicina** (J. L. Anderson, Reveal & Rollins) O'Kane & Al-Shehbaz, Novon 12: 328. 2002 • Uncompaghre or good-neighbor bladderpod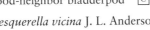

Lesquerella vicina J. L. Anderson, Reveal & Rollins, Novon 7: 9, fig. 1. 1997

Perennials; (flowering in the first year); caudex branched, (well-developed); densely pubescent, trichomes (subsessile, appressed to spreading), 3–6-rayed, rays furcate or bifurcate, usually slightly fused at base, less frequently distinct. **Stems** few to several from base, ascending (in flower) to nearly prostrate (in fruit), (unbranched), 1–2.5 dm. **Basal leaves:** (petiole 1–3.5 cm); blade ovate to rhombic or rotund, 2–7 cm, (base narrowed abruptly to petiole), margins usually entire, occasionally shallowly repand, (flat). **Cauline leaves:** blade elliptic or narrowly so, (0.7–)1–2.5 cm, (base narrowed gradually to petiole), margins entire. **Racemes** dense, (elongated in fruit). **Fruiting pedicels** (ascending, curved to slightly sigmoid), (4–)6–12 mm, (stout). **Flowers:** sepals (lavender under grayish trichomes), elliptic, 4–6 mm; petals (white, pale yellow basally, often tinged lavender abaxially), narrowly spatulate, 6–10 mm, (claw undifferentiated from blade). **Fruits** subglobose to ovoid, slightly compressed, 5–7 mm; valves densely pubescent, trichomes spreading; ovules 8–12 per ovary; style 4–6 mm. **Seeds** flattened, (faintly margined).

Flowering May–Jun. Nearly barren sites, soils derived from Mancos Shale or, less frequently, Jurassic sandstone, pinyon-juniper, sagebrush, Gambel oak; of conservation concern; 1800–2200 m; Colo.

88. **Physaria vitulifera** Rydberg, Bull. Torrey Bot. Club 28: 278. 1901 • Roundtip twinpod E

Perennials; caudex simple or branched, (cespitose); (silvery) pubescent throughout, trichomes several-rayed, rays furcate or bifurcate, (relatively massive, smooth to few-tubercled). **Stems** several from base, usually decumbent to ascending, (arising laterally, unbranched, coarse), 1–2 dm. **Basal leaves:** blade pandurate or obovate, 3–6 cm, margins usually deeply and broadly incised, rarely subentire, (apex obtuse). **Cauline leaves:** blade oblanceolate to spatulate, similar to basal, (3–6 mm wide), margins entire, (apex often somewhat acute). **Racemes** congested, (elongated in fruit). **Fruiting pedicels** (usually curving upward, sigmoid), 6–10 mm. **Flowers:** sepals oblong, 6–8 mm; petals spatulate, to 10 mm. **Fruits** didymous, irregular in shape, somewhat angular, inflated, 5–7 × 6–8 mm, (papery, often rigid, base obtuse or truncate, apical sinus broad, open and deep); valves (retaining seeds after dehiscence), pubescent, trichomes spreading, loose; replum oblong, often constricted, as wide as or wider than fruit, apex obtuse; ovules 4 per ovary; style 5–7 mm. **Seeds** flattened. $2n = 8, 16$.

Flowering Apr–Jun. Rocky hillsides, dry banks, gravel and sand, granitic slopes, soil scree, red shale; 1600–3000 m; Colo.

79. **SYNTHLIPSIS** A. Gray, Mem. Amer. Acad. Arts, n. s. 4: 116. 1849 • [Greek *synthlipsis*, compression, alluding to flattened fruits]

Ihsan A. Al-Shehbaz

Annuals, biennials, or perennials; (short-lived, sometimes cespitose); not scapose; pubescent throughout, trichomes dendritic, base of plant sometimes mixed with fewer, simple or forked, stalked ones. **Stems** ascending to decumbent, unbranched or branched distally. **Leaves** basal and cauline; petiolate or sessile; basal rosulate, petiolate, blade margins usually sinuately lobed to dentate, rarely repand; cauline petiolate or sessile, blade similar to basal. **Racemes** (corymbose, several-flowered), slightly or considerably elongated in fruit. **Fruiting pedicels** divaricate or divaricate-ascending, slender. **Flowers:** sepals spreading, narrowly oblong, (equal), lateral pair not saccate basally; petals white to violet [purple], broadly obovate, (much longer than sepals), claw differentiated from blade, (much shorter, apex rounded); stamens tetradynamous; filaments not dilated basally, (somewhat spreading); anthers linear [narrowly oblong]; nectar glands confluent, subtending bases of stamens. **Fruits** silicles, sessile, broadly oblong [broadly elliptic], smooth, angustiseptate; valves carinate; replum rounded; septum complete; ovules 10–50 per ovary; style distinct; stigma broadly capitate, entire. **Seeds** ± biseriate, flattened, not winged, narrowly margined, broadly ovate; seed coat (nearly smooth), copiously mucilaginous when wetted; cotyledons accumbent. $x = 10$.

Species 2 (1 in the flora): Texas, n Mexico.

Synthlipsis densiflora Rollins is known from Mexico (Coahuila).

SELECTED REFERENCE Rollins, R. C. 1959. The genus *Synthlipsis* (Cruciferae). Rhodora 61: 253–264.

1. **Synthlipsis greggii** A. Gray, Mem. Amer. Acad. Arts, n. s. 4: 116. 1849 [F]

Synthlipsis greggii var. *hispidula* Rollins

Stems few to several, (1–)2–6 (–7.5) dm, densely pubescent. **Basal leaves:** petiole (0.7–)1.5–5 (–7) cm; blade obovate to spatulate or broadly oblong, (1–)2–8(–10) cm × (5–)10–40 (–60) mm, base cuneate to truncate, apex acute to subobtuse, surfaces densely pubescent. **Cauline leaves** shortly petiolate or (distalmost) sessile; blade similar to basal, 1–5.5(–6.5) cm × 5–25(–40) mm (smaller distally). **Racemes** 0.5–4 cm. **Fruiting pedicels** usually straight or upcurved, rarely slightly sigmoid, (4–)5–15(–20) mm. **Flowers:** sepals (4–)5–8 × 1.5–2 mm; petals (7–)8–11(–13) × (5–)6–9 mm, claw to 2 mm; filaments 4–6 mm; anthers 2.5–4 mm, (base strongly sagittate). **Fruits** (0.7–)0.8–1.2(–1.5) cm × 5–8(–9) mm, apex deeply notched; valves densely pubescent; style 2–5 mm. **Seeds** 1.5–2 × 1.2–1.7 mm. $2n = 20$.

Flowering Oct–May. Creosote bush-mesquite scrubland, grasslands, gravelly hillsides, rocky sandstone or limestone slopes, rocky flats, limestone rubble, gypsum of llano, calcareous gravel, sandy or loamy areas; 50–2400 m; Tex.; Mexico (Chihuahua, Coahuila, Durango, Hidalgo, Nuevo León, San Luis Potosí, Tamaulipas, Zacatecas).

Synthlipsis greggii is known from Brewster, Hidalgo, and Zapata counties.

bb. BRASSICACEAE Burnett tribe SISYMBRIEAE de Candolle, Mém. Mus. Hist. Nat. 7: 237. 1821 (as Sisymbreae)

Annuals or **perennials** [biennials, subshrubs]; eglandular. **Trichomes** simple or absent, [rarely branched]. **Cauline leaves** petiolate or sessile; blade base not auriculate [auriculate], margins usually pinnately lobed or dentate, sometimes entire. **Inflorescences** (usually racemose, sometimes fasciculate, or flowers solitary in *Sisymbrium polyceratium*), usually ebracteate (bracteate in *S. polyceratium*), often elongated in fruit. **Flowers** actinomorphic; sepals erect to spreading, lateral pair seldom saccate basally; petals yellow [pink, white], claw present, often distinct; filaments unappendaged, not winged; pollen 3-colpate. **Fruits** siliques, dehiscent, unsegmented, usually terete [slightly latiseptate]; ovules 6–120(–140)[–160] per ovary; style usually distinct, rarely obsolete; stigma 2-lobed. **Seeds** uniseriate; cotyledons incumbent.

Genus 1, species 41 (8 in the flora): North America, Europe, Asia, Africa.

80. SISYMBRIUM Linnaeus, Sp. Pl. 2: 657. 1753; Gen. Pl. ed. 5, 296. 1754 • Hedge-mustard, rocket [Latinized ancient Greek name used by Dioscorides and Pliny for various species of mustards]

Ihsan A. Al-Shehbaz

Chamaeplium Wallroth; *Norta* Adanson; *Pachypodium* Webb & Berthelot; *Schoenocrambe* Greene; *Velarum* Reichenbach

Plants not scapose; pubescent or glabrous. **Stems** often erect, sometimes ascending, rarely subprostrate or decumbent, often branched distally, sometimes unbranched. **Leaves** basal and cauline; petiolate or sessile; basal rosulate or not, petiolate, blade margins dentate, sinuate, lyrate, runcinate, or pinnately lobed [entire]; cauline similar to basal, (blade smaller distally). **Racemes** (several-flowered), often considerably elongated in fruit. **Fruiting pedicels** ascending, divaricate, or erect, slender or stout (sometimes as wide as fruit). **Flowers:** sepals ovate or oblong, (glabrous or pubescent); petals yellow, obovate, spatulate, oblong, or suborbicular, (longer than sepals), claw differentiated from blade, (subequaling or longer than sepals, apex obtuse or emarginate); stamens tetradynamous; filaments not dilated basally; anthers oblong, (apex obtuse); nectar glands confluent, subtending bases of stamens, median glands present. **Fruits** usually sessile, rarely shortly stipitate (gynophore to 1 mm), usually linear, rarely lanceolate or subulate, smooth or torulose; valves each with prominent midvein and 2 conspicuous marginal veins, usually glabrous, rarely pubescent; replum rounded; septum complete; style subclavate [clavate, conical, cylindrical]; stigma capitate (lobes not decurrent). **Seeds** plump, not winged, oblong [ovoid]; seed coat (reticulate or papillate), not mucilaginous when wetted; cotyledons incumbent. $x = 7$.

Species 41 (8 in the flora): North America, Europe, Asia, Africa; introduced in Central America, South America, Australia.

All except one of the eight species of *Sisymbrium* in North America are introduced Eurasian weeds. *Sisymbrium linifolium* (Nuttall) Nuttall was retained in *Sisymbrium* by both E. B. Payson (1922) and O. E. Schulz (1924). Greene took that species as the type of his genus *Schoenocrambe*. R. C. Rollins (1982b, 1993) maintained *Schoenocrambe* and (1993) recognized *Sisymbrium auriculatum* as the only native North American species of the genus. Molecular studies (S. I. Warwick et al. 2002, 2005) clearly demonstrated that *Schoenocrambe* should be united with *Sisymbrium*, that *S. linifolium* is most closely related to the Eurasian *S. polymorphum* (Murray) Roth (as was suggested by both Payson and Schulz), that 40 of the 41 species of *Sisymbrium* are native to the Old World, and that *S. auriculatum* is a member of the New World Thelypodieae and is unrelated to *Sisymbrium*. See also Warwick and I. A. Al-Shehbaz (2003) and Al-Shehbaz (2005).

SELECTED REFERENCES Payson, E. B. 1922. Species of *Sisymbrium* native to America north of Mexico. Univ. Wyoming Publ. Sci., Bot. 1(1): 1–27. Rollins, R. C. 1982b. *Thelypodiopsis* and *Schoenocrambe* (Cruciferae). Contr. Gray Herb. 212: 71–102. Warwick, S. I., I. A. Al-Shehbaz, R. A. Price, and C. A. Sauder. 2002. Phylogeny of *Sisymbrium* (Brassicaceae) based on ITS sequences of nuclear ribosomal DNA. Canad. J. Bot. 80: 1002–1017. Warwick, S. I., C. A. Sauder, and I. A. Al-Shehbaz. 2005. Molecular phylogeny and cytological diversity of *Sisymbrium* (Brassicaceae). In: A. K. Sharma and A. Sharma, eds. 2003+. Plant Genome: Biodiversity and Evolution. 2+ vols. in parts. Enfield, N.H. Vol. 1, part C, pp. 219–250.

1. Racemes bracteate, flowers fasciculate..4. *Sisymbrium polyceratium*
1. Racemes ebracteate, flowers not fasciculate.
   2. Fruits subulate-linear, (0.7–)1–1.4(–1.8) cm; fruiting pedicels appressed to rachises
      ....8. *Sisymbrium officinale*
   2. Fruits narrowly linear, 2–10(–13) cm; fruiting pedicels not appressed to rachises.

[3. Shifted to left margin.—Ed.]

3. Fruiting pedicels nearly as wide as fruit.
    4. Distal cauline leaf blades divided into linear or filiform lobes; sepals cucullate; fruiting pedicels (4–)6–10(–13) mm . . . . 1. *Sisymbrium altissimum*
    4. Distal cauline leaf blades not divided into linear or filiform lobes; sepals not cucullate; fruiting pedicels 1–6 mm.
        5. Fruiting pedicels 1–2(–3) mm; petals 1.4–2(–2.5) mm; seeds 30–46(–54) per fruit . . . . 5. *Sisymbrium erysimoides*
        5. Fruiting pedicels 3–6 mm; petals (6–)7–9(–10) mm; seeds (60–)80–100(–140) per fruit . . . . 7. *Sisymbrium orientale*
3. Fruiting pedicels narrower than fruit.
    6. Perennials; distalmost cauline leaf blades usually filiform to linear, rarely oblanceolate, 1–3.5(–5) mm wide . . . . 2. *Sisymbrium linifolium*
    6. Annuals; distalmost cauline leaf blades oblanceolate or oblong (in outline), not filiform or linear, 10–30 mm wide.
        7. Plants usually densely hispid (at least proximally); petals 6–8 mm; young fruits not overtopping flowers . . . . 3. *Sisymbrium loeselii*
        7. Plants glabrous or sparsely pubescent; petals 2.5–3.5(–4) mm; young fruits overtopping flowers . . . . 6. *Sisymbrium irio*

1. **Sisymbrium altissimum** Linnaeus, Sp. Pl. 2: 659. 1753 • Tumbleweed [I][W]

Annuals; glabrous or pubescent. Stems erect, branched distally, (2–)4–12(–16) dm, sparsely to densely hirsute basally, glabrous or glabrate distally. Basal leaves rosulate; petiole 1–10(–15) cm; blade broadly oblanceolate, oblong, or lanceolate (in outline), (2–)5–20(–35) cm × (10–)20–80(–100) mm, margins pinnatisect, pinnatifid, or runcinate; lobes (3–)4–6(–8) on each side, oblong or lanceolate, smaller than terminal lobe, margins entire, dentate, or lobed. Cauline leaves similar to basal; distalmost blade with linear to filiform lobes. Fruiting pedicels usually divaricate, rarely ascending, stout, nearly as wide as fruit, (4–)6–10(–13) mm. Flowers: sepals ascending or spreading, oblong, (cucullate), 4–6 × 1–2 mm; petals spatulate, (5–)6–8(–10) × 2.5–4 mm, claw 3.5–6 mm; filaments 2–6 mm; anthers oblong, 1.5–2.2 mm. Fruits narrowly linear, usually straight, smooth, stout, (4.5–)6–9(–12) cm × 1–2 mm; valves glabrous; ovules 90–120 per ovary; style 0.5–2 mm; stigma prominently 2-lobed. Seeds 0.8–1 × 0.5–0.6 mm. $2n = 14$.

Flowering Apr–Sep. Roadsides, fields, pastures, waste grounds, disturbed sites, grasslands; 0–2700 m; introduced; Greenland; Alta., B.C., Man., N.B., Nfld. and Labr. (Nfld.), N.W.T., N.S., Ont., P.E.I., Que., Sask., Yukon; Alaska, Ariz., Calif., Colo., Conn., D.C., Fla., Idaho, Ill., Ind., Iowa, Kans., Maine, Md., Mass., Mich., Minn., Mo., Mont., Nebr., Nev., N.H., N.J., N.Mex., N.Y., N.C., N.Dak., Ohio, Okla., Oreg., Pa., R.I., S.Dak., Tenn., Tex., Utah, Vt., Va., Wash., W.Va., Wis., Wyo.; Europe; w Asia; nw Africa; introduced also in South America (Argentina, Chile).

2. **Sisymbrium linifolium** (Nuttall) Nuttall in J. Torrey and A. Gray, Fl. N. Amer. 1: 91. 1838 [E][F]

Nasturtium linifolium Nuttall, J. Acad. Nat. Sci. Philadelphia 7: 12. 1834; *Erysimum glaberrimum* Hooker & Arnott; *E. linifolium* (Nuttall) M. E. Jones; *N. pumilum* Nuttall; *Schoenocrambe decumbens* Rydberg; *S. linifolia* (Nuttall) Greene; *S. pinnata* Greene; *S. pygmaea* (Nuttall) Greene; *Sisymbrium decumbens* (Rydberg) Blankinship; *S. linifolium* var. *decumbens* (Rydberg) O. E. Schulz; *S. linifolium* var. *pinnatum* (Greene) O. E. Schulz; *S. pygmaeum* Nuttall

Perennials; (rhizomatous); usually glabrous basally, rarely sparsely pubescent, glabrous (and sometimes glaucous) distally. Stems erect or ascending, unbranched or branched (few) distally, (1.5–)3–7(–11) dm, glabrous distally. Basal leaves (soon withered); not rosulate; petiole (proximally) (0.5–)1–3 cm; blade similar to proximal cauline, 1.5–6 cm, margins entire, pinnatifid or pinnatisect; lateral lobes oblong to linear, 0.5–2.5 cm × 0.3–1(–2) mm, margins entire. Cauline leaves sessile or shortly petiolate; blade usually filiform to linear, rarely oblanceolate, 1–2.5–9(–12) cm × 1–3.5(–5) mm (smaller distally, base attenuate or cuneate), margins usually entire, rarely dentate or pinnately lobed. Fruiting pedicels divaricate to ascending, slender, narrower than fruit, (3–)5–9(–11) mm. Flowers: sepals ascending, oblong to oblong-linear, (3–)4–7 × (0.8–)12 mm; petals spatulate, (6–)8–12 × (1.5–)2–4 mm, claw 2–5 mm; filaments (yellowish), (4–)5–7 mm; anthers linear, 1.5–2.5 mm. Fruits (divaricate to erect), narrowly linear, smooth, slender, (2.5–)3.5–6.5 cm × 0.9–1.2 mm; valves glabrous; ovules 60–94 per ovary; style 0.5–1 mm;

stigma prominently 2-lobed. **Seeds** 1–2 × 0.4–0.6 mm. $2n = 14$.

Flowering Apr–Aug. Rocky or gravelly hillsides, sagebrush communities, pinyon-juniper areas, shady rock cliffs, abandoned fields, sandy prairies, steep banks; 700–2800 m; B.C.; Ariz., Colo., Idaho, Mont., Nev., N.Mex., Oreg., Utah, Wash., Wyo.

N. H. Holmgren (2005b) recognized *Sisymbrium linifolium* and others (see 86. *Hesperidanthus*) in *Schoenocrambe* even though the molecular evidence (S. I. Warwick et al. 2002) overwhelmingly shows that the latter is nested within *Sisymbrium*, whereas the species of *Hesperidanthus* are not closely related. Indeed, I. A. Al-Shehbaz et al. (2006) placed *Hesperidanthus* and *Sisymbrium* in different tribes. This is an example where the superficial resemblances in fruit morphology are the result of convergence and can easily mislead to erroneous taxonomy.

3. **Sisymbrium loeselii** Linnaeus, Cent. Pl. I, 18. 1755 [I][W]

Annuals; densely hispid at least proximally. **Stems** erect, branched distally, (2–)3.5–12 (–17.5) dm, often densely hispid proximally, (trichomes retrorse), usually glabrous distally. **Basal leaves** rosulate; petiole 1–4(–5) cm; blade broadly oblanceolate (in outline), (1.5–)2.5–8(–12) cm × (10–)20–50(–70) mm, margins runcinate to lyrate-pinnatifid; lobes 2–4 on each side, much smaller than terminal lobe, margins entire or dentate, (terminal lobe triangular, often hastate). **Cauline leaves** similar to basal; blade (much smaller than basal, to 1.5 cm wide), margins entire or toothed. **Fruiting pedicels** divaricate or ascending, slender, narrower than fruit, 5–12(–15) mm. **Flowers:** sepals ascending, oblong, 3–4 × 1–1.5 mm; petals spatulate, 6–8 × 2–3 mm, claw 2.5–3.5 mm; filaments 3–4.5 mm; anthers ovate, 0.6–1 mm. **Fruits** (ascending to suberect, young fruits not overtopping flowers), narrowly linear, curved or straight, subtorulose, slender, 2–3.5(–5) cm × 0.9–1.1 mm; valves often glabrous; ovules 40–60 per ovary; style stout, 0.3–0.7 mm; stigma prominently 2-lobed. **Seeds** 0.7–1 × 0.5–0.6 mm. $2n = 14$.

Flowering late May–early Nov. Valleys, stream banks, fields, roadsides, pastures, waste grounds, vacant lots, prairies, disturbed sites, railroad tracks; 0–2400 m; introduced; Alta., B.C., Man., N.B., Que., Sask.; Colo., Conn., Idaho, Ill., Ind., Iowa, Mass., Mich., Mont., Nebr., Nev., N.Mex., N.Dak., Ohio, Oreg., Pa., S.Dak., Utah, Vt., Wash., Wyo.; e Europe; w, c Asia.

4. **Sisymbrium polyceratium** Linnaeus, Sp. Pl. 2: 658. 1753 [I]

Chamaeplium polyceratium (Linnaeus) Wallroth

Annuals; (densely leafy throughout); often glabrous. **Stems** (simple or few to several from base), erect or ascending to subprostrate, branched distally, 1–5(–7) dm, glabrous. **Basal leaves** (soon withered); initially rosulate; petiole (1–)2–5(–6.5) cm; blade oblanceolate to lanceolate (in outline), 2–7(–10) cm × 10–30(–45) mm, margins sinuate- or runcinate-pinnatifid to coarsely dentate; lobes 3–6 on each side, often triangular, slightly smaller than terminal lobe, margins dentate or subentire. **Cauline leaves** (sometimes bracts, several), similar to basal, (shortly petiolate); blade margins dentate or subentire, (surfaces glabrous or sparsely puberulent at and near margin). (**Inflorescences** 2–4(–6)-fasciculate, or flowers solitary and axillary, bracteate throughout.) **Fruiting pedicels** ascending to erect, stout, narrower than fruit base, 0.5–1(–2) mm. **Flowers:** sepals erect, oblong, 1–1.5 × 0.3–0.5 mm; petals spatulate, 1.5–2 × 0.4–0.6 mm, claw 0.6–1 mm; filaments (erect, yellowish), 1.2–1.7 mm; anthers ovate, 0.3–0.5 mm. **Fruits** (erect to ascending), subulate-linear, recurved, straight, slightly torulose, stout (widest at base), 1–2(–2.3) cm × 0.9–1.7 mm; valves usually glabrous, rarely puberulent; ovules 20–44(–54) per ovary; style 0.5–1 mm; stigma slightly 2-lobed. **Seeds** 0.7–1 × 0.4–0.5 mm. $2n = 28$.

Flowering Mar–Jun. Waste grounds, disturbed sites, roadsides; 0–200 m; introduced; Pa., Tex.; Europe; w, c Asia; nw Africa.

5. **Sisymbrium erysimoides** Desfontaines, Fl. Atlant. 2: 84. 1798 [I][W]

Annuals; glabrous or pubescent. **Stems** erect, unbranched or branched distally, (1–)2–6(–8) dm, glabrous or sparsely pilose at least basally, usually glabrous distally. **Basal leaves** (soon withered); not rosulate; petiole (0.5–)1–4(–6) cm; blade broadly ovate to obovate or broadly oblanceolate, 2–8(–10) cm × 10–30(–50) mm, margins lyrate-pinnatipartite; lobes 2–4 on each side, oblong or lanceolate, smaller than terminal lobe, margins subentire or dentate, (terminal lobe ovate, margins dentate). **Cauline leaves** similar to basal, (distalmost with shorter petiole); blade smaller, margins dentate, lobes 1–3 on each side. **Fruiting pedicels** divaricate, stout, nearly as wide as fruit, 1–2(–3) mm, (pilose adaxially). **Flowers:** sepals ascending, oblong, 1.5–2(–2.5) × 0.8–1.5 mm; petals pale spatulate, 1.4–2(–2.5) × 0.2–0.5 mm, claw ca. 0.5 mm; filaments 1.7–2.5(–3) mm; anthers ovate, 0.1–0.2 mm. **Fruits** narrowly linear, straight, obscurely torulose, stout, (1.7–)2–4.5(–5.2) cm × 0.9–1.2 mm; valves glabrous; ovules 30–46(–54) per ovary; style (slender) 0.5–1(–2) mm; stigma prominently 2-lobed. **Seeds** 0.9–1.3 × 0.4–0.6 mm. $2n = 14$.

Flowering Jan–Oct. Waste grounds, abandoned fields, disturbed sites; 300–1900 m; introduced, Calif.; Europe; introduced also in South America (Argentina), Australia.

Sisymbrium erysimoides apparently is naturalized in the flora area in Riverside County.

6. **Sisymbrium irio** Linnaeus, Sp. Pl. 2: 659. 1753

Annuals; glabrous or sparsely pubescent. **Stems** erect, branched proximally and distally, (1–)2–6(–7.5) dm, glabrous or sparsely pubescent at least basally. **Basal leaves** not rosulate; petiole (0.5–)1–4.5(–6) cm; blade oblanceolate or oblong (in outline), (1.5–)3–12 (–15) cm × (5–)10–60(–90) mm, margins runcinate to pinnatisect; lobes (1–)2–6(–8) on each side, oblong or lanceolate, smaller than terminal lobe, margins entire, dentate, or lobed. **Cauline leaves** similar to basal; (distalmost) blade (smaller, to 2 cm wide), margins entire or 1–3-lobed. **Fruiting pedicels** divaricate or ascending, slender, much narrower than fruit, (5–)7–12(–20) mm. **Flowers:** sepals erect, oblong, 2–2.5 × 1–1.5 mm; petals oblong-oblanceolate, 2.5–3.5(–4) × 1–1.5 mm, claw 1–1.5 mm; filaments 2.5–4 mm; anthers ovate, 0.5–0.9 mm. **Fruits** (divaricate to ascending, young fruits overtopping flowers), narrowly linear, straight or slightly curved inward, slightly torulose, slender, (2.5–)3–4(–5) cm × 0.9–1.1 mm; valves glabrous; ovules 40–90 per ovary; style 0.2–0.5 mm; stigma prominently 2-lobed. **Seeds** 0.8–1 × 0.5–0.6 mm. $2n = 14$.

Flowering Dec–May. Rocky slopes, orchards, roadsides, fields, pastures, waste grounds, prairies, disturbed sites; 0–1700 m; introduced; Ariz., Calif., Colo., Conn., Fla., Idaho, Mass., Mich., Nev., N.Mex., Ohio, Pa., Tex., Utah, Wyo.; Europe; w, c Asia; n Africa; introduced also in South America, Australia.

7. **Sisymbrium orientale** Linnaeus, Cent. Pl. II, 24. 1756

Annuals; glabrous or pubescent. **Stems** erect, branched distally, (1–)2–7(–8.5) dm, sparsely to densely (soft) pubescent at least basally, usually glabrous distally. **Basal leaves** rosulate; petiole (1–)2–5(–9) cm; blade broadly oblanceolate to oblong-oblanceolate (in outline), 3–8 (–10) cm × (10–)20–40(–60) mm, margins runcinate-pinnatipartite; lobes 2–5 on each side, oblong or lanceolate, much smaller than terminal lobe, margins subentire or dentate, (terminal lobe lanceolate, deltate, or often hastate). **Cauline leaves** similar to basal; (distalmost) blade with 1 or 2 lobes on each side, much smaller than terminal lobe, (terminal lobe narrowly lanceolate, linear, or hastate). **Fruiting pedicels** ascending to subdivaricate, stout, nearly as wide as fruit, 3–6 mm. **Flowers:** sepals ascending, oblong, 3.5–5.5 × 1–2 mm; petals spatulate, (6–)7–9(–10) × 2.5–4 mm, claw 3–5.5 mm; filaments (4–)5–8 mm; anthers oblong, 1–1.8 mm. **Fruits** narrowly linear, straight, smooth, stout, (5–)6–10 (–13) cm × 1–1.5 mm; valves glabrous or pubescent; ovules (60–)80–100(–140) per ovary; style (subclavate), 1–3(–4) mm; stigma prominently 2-lobed. **Seeds** 1–1.5 × 0.7–0.9 mm. $2n = 14$.

Flowering Mar–early Jun. Waste grounds, roadsides, disturbed sites; 0–1300 m; introduced; B.C.; Ariz., Calif., Mass., Nev., Oreg., Tex., Wash.; Europe; sw Asia; n Africa; introduced also in Central America, South America, Australia.

8. **Sisymbrium officinale** (Linnaeus) Scopoli, Fl. Carniol. ed. 2, 2: 26. 1772

Erysimum officinale Linnaeus, Sp. Pl. 2: 660. 1753; *Sisymbrium officinale* var. *leiocarpum* de Candolle

Annuals; glabrous or pubescent. **Stems** erect, branched distally, 2.5–7.5(–11) dm, usually sparsely to densely hirsute, (trichomes retrorse), rarely glabrate distally. **Basal leaves** usually rosulate; petiole (1–)2–7(–10) cm; blade broadly oblanceolate or oblong-obovate (in outline), (2–)3–10(–15) cm × (10–)20–50 (–80) mm, margins lyrate-pinnatifid, pinnatisect, or runcinate; lobes (2)3 or 4(5) on each side, oblong or lanceolate, smaller than terminal lobe, margins entire, dentate, or lobed, (terminal lobe suborbicular or deltate, margins dentate). **Cauline leaves** similar to basal; blade with lobe margins dentate or subentire. **Fruiting pedicels**

erect, (appressed to rachis), stout, narrower than fruit, 1.5–3(–4) mm. **Flowers:** sepals erect, oblong-ovate, 2–2.5 × ca. 1 mm; petals spatulate, 2.5–4 × 1–2 mm; claw 1–2 mm; filaments (erect, yellowish), 2–3 mm; anthers ovate, 0.3–0.5 mm. **Fruits** (erect), subulate-linear, straight, slightly torulose or smooth, stout, (0.7–)1–1.4(–1.8) cm × 1–1.5 mm; valves glabrous or pubescent; ovules 10–20 per ovary; style (0.8–)1–1.5 (–2) mm; stigma slightly 2-lobed. **Seeds** 1–1.3 × 0.5–0.6 mm. $2n = 14$.

Flowering Apr–late Sep. Roadsides, fields, pastures, waste grounds, deserts; 0–2200 m; introduced; Alta., B.C., Man., N.B., Nfld. and Labr. (Nfld.), N.W.T., N.S., Ont., P.E.I., Que., Yukon; Ala., Alaska, Ark., Calif., Conn., Del., Fla., Ga., Idaho, Ill., Ind., Iowa, Kans., Ky., La., Maine, Md., Mass., Mich., Minn., Miss., Mo., Mont., N.H., N.J., N.Mex., N.Y., N.C., Ohio, Okla., Oreg., Pa., R.I., S.C., S.Dak., Tenn., Tex., Utah, Vt., Va., Wash., W.Va., Wis.; Europe; Asia; n Africa; introduced also in Central America, South America, Australia.

cc. BRASSICACEAE Burnett tribe SMELOWSKIEAE Al-Shehbaz, Beilstein & E. A. Kellogg, Pl. Syst. Evol. 259: 111. 2006

Perennials [annuals]; eglandular. **Trichomes** stalked, usually dendritic, stellate, or forked, sometimes mostly simple. **Cauline leaves** [rarely absent], petiolate, subsessile, or sessile; blade base not auriculate, margins usually pinnately or palmately lobed, sometimes entire [dentate]. **Racemes** usually ebracteate, often elongated in fruit. **Flowers** actinomorphic; sepals suberect to spreading [erect], lateral pair not saccate basally; petals white, pink, or purple [yellow], claw present, distinct; filaments unappendaged, not winged; pollen 3-colpate. **Fruits** silicles or siliques, dehiscent, unsegmented, 4-angled, angustiseptate, terete, or subterete [latiseptate]; ovules 4–30[–numerous] per ovary; style distinct [obsolete]; stigma entire [rarely slightly 2-lobed]. **Seeds** uniseriate [biseriate]; cotyledons accumbent or incumbent.

Genus 1, species 25 (7 in the flora): North America, Europe, Asia.

81. SMELOWSKIA C. A. Meyer in C. F. von Ledebour, Icon. Pl. 2: 17, plate 151. 1830, name conserved • [For Timotheus Smielowsky, 1769–1815, Russian botanist and pharmacist from St. Petersburg]

Ihsan A. Al-Shehbaz

Acroschizocarpus Gombócz; *Ermania* Chamisso ex Botschantzev; *Melanidion* Greene

Plants cespitose, caudex well-developed, thick, often-branched, covered with persistent petiolar remains; not scapose; usually pubescent. **Stems** erect to decumbent, unbranched or branched distally, densely pubescent. **Leaves** basal and cauline; petiolate or sessile; basal rosulate, petiolate, blade margins usually 1- or 2-pinnatisect, rarely entire. **Racemes** (corymbose, several-flowered), often considerably elongated in fruit. **Fruiting pedicels** ascending, spreading, suberect, or divaricate [recurved], slender. **Flowers:** sepals (sometimes persistent), oblong [ovate]; petals spatulate to obovate or suborbicular, (longer than sepals), claw differentiated from blade, (apex rounded); stamens slightly tetradynamous; filaments often dilated basally; anthers ovate or oblong, (apex obtuse); nectar glands usually confluent, subtending bases of stamens, median glands present or not. **Fruits** siliques or silicles, usually sessile, rarely shortly stipitate, linear, oblong, obovoid, ellipsoid, spatulate, oblanceolate, suboblong, or pyriform [fusiform, ovoid, suborbicular], smooth, 4-angled, angustiseptate, terete, or subterete [latiseptate]; valves each with prominent or obscure midvein, usually glabrous; replum rounded; septum complete or perforated; ovules 4–18 per ovary; stigma capitate. **Seeds** plump, not winged, usually oblong, rarely oblong-lanceolate; seed coat (minutely reticulate), not mucilaginous when wetted; cotyledons incumbent or accumbent.

Species 25 (7 in the flora): North America, e, c Asia.

SELECTED REFERENCES Al-Shehbaz, I. A. and S. I. Warwick. 2006. A synopsis of *Smelowskia* (Brassicaceae). Harvard Pap. Bot. 11: 91–99. Drury, W. H. Jr. and R. C. Rollins. 1952. The North American representatives of *Smelowskia* (Cruciferae). Rhodora 54: 85–119. Velichkin, E. M. 1979. *Smelowskia* C. A. Mey. (Cruciferae). Critical review and relation to close genera. Bot. Zhurn. (Moscow & Leningrad) 64: 153–171. Warwick, S. I. et al. 2004b. Phylogeny of *Smelowskia* and related genera (Brassicaceae) based on nuclear ITS DNA and chloroplast *trn*L intron DNA sequences. Ann. Missouri Bot. Gard. 91: 99–123.

1. Basal leaves: blade margins entire, apically 3 or 5-toothed or -lobed, or palmately lobed.
   2. Basal leaves: blade margins deeply palmately (3 or) 5 (or 7)-lobed; fruiting pedicels secund; sepals usually persistent; fruits angustiseptate, (8–)12–28 × 4–8 mm; ovules 10–18 per ovary . . . . .2. *Smelowskia borealis*
   2. Basal leaves: blade margins entire or apically 3 or 5-toothed or -lobed; fruiting pedicels not secund; sepals caducous; fruits subterete or slightly 4-angled, 5–10 × 1.5–3 mm; ovules 4–8 per ovary.
      3. Basal leaf blade surfaces densely silvery villous, trichomes mostly simple, 1–1.8 mm; petals lavender to purplish; fruiting pedicels 11–27 mm; fruit valves each with obscure midvein; ovules 4 per ovary; seeds 2.2–2.7 mm . . . . .3. *Smelowskia johnsonii*
      3. Basal leaf blade surfaces densely grayish tomentose, trichomes mostly dendritic with fewer simple ones, to 1 mm; petals white or creamy white; fruiting pedicels 4–12 mm; fruit valves each with prominent midvein; ovules 4–8 per ovary; seeds 1.5–2 mm . . . . 6. *Smelowskia porsildii*
1. Basal leaves: blade margins 1- or 2-pinnatisect or pinnatifid.
   4. Sepals persistent; fruits 2–6 mm, bases obtuse, valves each with obscure midvein . . . 5. *Smelowskia ovalis*
   4. Sepals caducous; fruits 5–13 mm, bases cuneate, valves each with prominent midvein.
      5. Fruits pyriform, subterete, apices rounded; ovules 4 per ovary; seeds 2.5–3.2 × 1–1.5 mm . . . . .7. *Smelowskia pyriformis*
      5. Fruits fusiform, ellipsoid, oblong, or linear, 4-angled, apices cuneate; ovules 8–12 (–14) per ovary; seeds 1.1–2.2 × 0.6–1.1 mm.
         6. Fruiting pedicels suberect to ascending, subappressed to rachises, forming less than 40° angle; seeds 1.1–1.9 × 0.6–0.9 mm; Alberta, British Columbia, Mountain and Pacific states . . . . 1. *Smelowskia americana*
         6. Fruiting pedicels usually spreading to divaricate, rarely divaricate-ascending, not appressed to rachises, often forming greater than 40° angle; seeds 1.7–2.2 × 0.9–1.1 mm; Alaska, Northwest Territories, Yukon . . . . 4. *Smelowskia media*

1. Smelowskia americana Rydberg, Bull. Torrey Bot. Club 29: 239. 1902 [E]

Hutchinsia calycina (Stephan) Desvaux var. *americana* Regel & Herder; *Smelowskia calycina* (Stephan) C. A. Meyer var. *americana* (Regel & Herder) W. H. Drury & Rollins; *S. lineariloba* Rydberg; *S. lobata* Rydberg

Plants sometimes canescent basally; caudex branched. **Stems:** several from base, often unbranched, (0.4–)0.6–2 (–2.7) dm, trichomes simple, 0.5–1.3 mm, mixed with smaller, dendritic ones. **Basal leaves:** petiole 1–5(–8) cm, ciliate, trichomes simple; blade oblanceolate to obovate, or ovate to oblong in outline, (terminal segments linear, oblong, or ovate), (0.5–)1–3.5(–5.2) cm × 4–17 mm, (terminal segments 0.2–1.4 cm × 0.5–4 mm), margins 1- or 2-pinnatifid or -pinnatisect, apex obtuse or subacute. **Cauline leaves** shortly petiolate or sessile; blade similar to basal, smaller distally. **Racemes** elongated in fruit. **Fruiting pedicels** suberect to ascending, (subappressed to rachis, often forming less than 40° angle), proximalmost bracteate, 4–10(–14) mm, pubescent, trichomes simple (to 1.5 mm), mixed with smaller, dendritic ones. **Flowers:** sepals 2–3.5 mm; petals usually white, rarely pinkish or lavender, suborbicular to obovate, 3.5–6.5 × 1.5–3.5 mm, narrowed to claw, 1.5–3 mm, apex rounded; anthers oblong, 0.5–0.7 mm. **Fruits** usually suberect, rarely ascending, ellipsoid or oblong to linear, 4-angled, 5–13 × 1.5–2 mm, base and apex cuneate; valves each with prominent midvein; ovules 8–12(–14) per ovary; style 0.2–0.8 mm. **Seeds** 1.1–1.9 × 0.6–0.9 mm. $2n = 12, 22$.

Flowering Jun–Aug. Talus and scree slopes, rock crevices, tundra, alpine meadows, fellfields; 2900–4000 m; Alta., B.C.; Colo., Idaho, Mont., Nev., Oreg., Utah, Wash., Wyo.

Both R. C. Rollins (1993) and N. H. Holmgren (2005b) listed $2n = 44$ for *Smelowskia americana* (as *S. calycina* var. *americana*), but no such number is known for any species of the genus (S. I. Warwick and I. A. Al-Shehbaz 2006). It is most likely that the first two authors erred in reporting $2n = 22$ for the species. The latter count is likely to represent a dysploid reduction of tetraploid populations based on $x = 6$.

Previous North American authors (e.g., W. H. Drury Jr. and R. C. Rollins 1952; Rollins 1993; N. H. Holmgren 2005b) believed that the central Asian *Smelowskia calycina* and the North American plants also attributed to it are conspecific. S. I. Warwick et al. (2004b) clearly demonstrated that they are different species. The North American plants, *S. americana*, are easily distinguished from *S. calycina* by having readily caducous instead of persistent calyces. As recognized by Rollins (1993), the North American *S. calycina* represented three distinct taxa (*S. americana*, *S. media*, *S. porsildii*) none of which belongs to that species.

2. **Smelowskia borealis** (Greene) W. H. Drury & Rollins, Rhodora 54: 111. 1952 E

Melanidion boreale Greene, Ottawa Naturalist 25: 146. 1912; *Acroschizocarpus kolianus* Gombócz; *Ermania borealis* (Greene) Hultén; *Smelowskia borealis* var. *jordalii* W. H. Drury & Rollins; *S. borealis* var. *koliana* (Gombócz) W. H. Drury & Rollins; *S. borealis* var. *villosa* W. H. Drury & Rollins

Plants sometimes canescent basally; caudex usually simple, rarely branched. **Stems:** several from base, often unbranched or branched proximally, 0.6–3(–3.5) dm, trichomes simple, to 1 mm, mixed with smaller, dendritic ones. **Basal leaves:** petiole 0.5–3.2 cm, ciliate or not; blade obovate to ovate in outline, (terminal segments linear, oblong, or ovate), 0.7–2.5 cm × 4–14 mm, (terminal segments 0.2–1.1 cm × 1–5 mm), margins palmately (3 or) 5 (or 7)-lobed, (surfaces densely pubescent, trichomes dendritic, often mixed with larger, simple ones, to 1 mm), apex obtuse or subacute. **Cauline leaves** shortly petiolate or sessile; blade often similar to basal, smaller distally, margins sometimes pinnatifid. **Racemes** considerably elongated in fruit. **Fruiting pedicels** spreading to divaricate-ascending, (secund, often curved), proximalmost bracteate, 6–15(–20) mm, pubescent, trichomes simple mixed with smaller, dendritic ones. **Flowers:** sepals (usually persistent), 2.5–3.2 mm; petals purple to lavender, suborbicular to obovate, 4–5 × 1.5–2 mm, narrowed to claw, 1.5–2.5 mm, apex rounded; anthers oblong, 0.4–0.5 mm. **Fruits** spreading to ascending, usually oblanceolate to spatulate or linear-oblanceolate, rarely oblong-obovate, angustiseptate, (8–)12–28 × 4–8 mm, base cuneate, apex obtuse; valves each with prominent midvein and distinct lateral veins, (rarely sparsely pubescent); ovules 10–18 per ovary; style 0.1–1.1 mm. **Seeds** 1.5–2.2 × 1–1.3 mm. $2n = 12$.

Flowering Jun–Aug. Loose talus, metamorphic slide rock, rocky slopes, scree, shale splinters, limestone rubble, alpine ridges, barren rocks, unstable talus in alpine glacial bowl; 600–1700 m; N.W.T., Yukon; Alaska.

W. H. Drury Jr. and R. C. Rollins (1952) and Rollins (1993) divided *Smelowskia borealis* into four varieties based on the duration of sepals (persistent versus caducous), style length (less than versus greater than 0.5 mm), and density of indumentum. The various states of these three characters are present in all possible combinations, and they are not inherited in the combinations presented by these authors. Some forms of the species are densely villous and resemble *S. johnsonii* in their leaf indumentum and perhaps merit recognition as an infraspecific taxon (var. *villosa*). As delimited here, *S. borealis* is quite variable; more studies are needed to determine whether or not any of the four varieties merit recognition at some rank. Although numerous collections of the species were examined, none of the types was available for preparing this account. The species is easily distinguished from the remaining North American *Smelowskia* by having secund infructescences and distinctly angustiseptate fruits.

3. **Smelowskia johnsonii** G. A. Mulligan, Canad. Field-Naturalist 115: 341. 2001 C E

Plants densely white-canescent throughout; caudex simple. **Stems:** several from base, unbranched or branched proximally, 0.4–1.6 dm, trichomes simple, 1–1.5 mm, mixed with smaller, dendritic ones. **Basal leaves:** petiole 0.8–1.3 cm, ciliate, trichomes simple; blade oblanceolate to spatulate, 0.8–1.8 cm × 3–7 mm, (terminal segments 0.2–0.7 cm × 1–4 mm), margins usually entire or apically 3-toothed or -lobed, rarely palmately 3-lobed, (terminal segments linear to ovate), apex obtuse (surfaces densely villous, silvery, trichomes primarily simple, 1–1.8 mm). **Cauline leaves** subsessile; blade similar to basal, smaller distally, margins entire. **Racemes** elongated in fruit. **Fruiting pedicels** ascending, (often forming less than 40° angle, straight), proximalmost bracteate, 11–27 mm, pubescent, trichomes primarily simple (to 1.5 mm). **Flowers:** sepals 3–3.5 mm; petals lavender to purplish, suborbicular to obovate, 4–5 × 3–4 mm, narrowed to claw, ca. 2 mm, apex rounded; anthers oblong, 0.5–0.7 mm. **Fruits** ascending, ellipsoid to obovoid-ellipsoid, subterete, 5–6 × 2–3 mm, base and sometimes apex cuneate; valves each with obscure midvein; ovules 4 per ovary; style 0.2–0.3 mm. **Seeds** 2.2–2.7 × ca. 1 mm.

Flowering Jun–Aug. Steep talus slopes, loose rocks, limestone rubble, talus; of conservation concern; 0–600 m; Alaska.

Smelowskia johnsonii is known only from the Bering Strait District, Alaska. It is a distinctive species resembling only superficially the villous forms of *S. borealis* with simple trichomes. From the latter, *S. johnsonii* is easily distinguished by having subterete (versus angustiseptate) fruits 2–3 (versus 4–8) mm wide, non-secund (versus secund) and straight (versus often curved) fruiting pedicels, 4 (versus 10–18) ovules per ovary, and simple or apically 3 (or 5)-toothed or -lobed (versus palmately (3 or) 5 (or 7)-lobed) basal leaf blades. It is readily distinguished from all species of the genus by having leaves densely silvery pubescent, with primarily simple trichomes 1–1.8 mm.

4. Smelowskia media (W. H. Drury & Rollins) Velichkin, Bot. Zhurn. (Moscow & Leningrad) 64: 167. 1979 E

Smelowskia calycina (Stephan) C. A. Meyer var. *media* W. H. Drury & Rollins, Rhodora 54: 100. 1952

Plants sometimes canescent basally; caudex branched. **Stems:** several from base, unbranched, (0.2–)0.5–1.4 (–1.8) dm, trichomes simple, 0.5–1.2 mm, mixed with smaller, dendritic ones. **Basal leaves:** petiole 0.7–3.5 cm, ciliate, trichomes simple; blade oblanceolate to obovate, or ovate to oblong in outline, (terminal segments linear, oblong, or ovate), 0.5–3 cm × 4–12 mm, (terminal segments 0.2–1.4 cm × 0.5–4 mm), margins usually pinnatifid, rarely apically 3 or 5-lobed, apex obtuse or subacute. **Cauline leaves** shortly petiolate or sessile; blade similar to basal, smaller distally. **Racemes** elongated in fruit. **Fruiting pedicels** usually spreading to divaricate, rarely divaricate-ascending, (often forming greater than 40° angle, straight or upcurved), proximalmost bracteate, 4–12 mm, pubescent, trichomes simple mixed with smaller, dendritic ones. **Flowers:** sepals 2–3 mm; petals white, suborbicular to obovate, 4–5 × 2–3 mm, narrowed to claw, 1.5–2 mm, apex rounded; anthers oblong, 0.5–0.6 mm. **Fruits** divaricate-ascending, ellipsoid to oblong, 4-angled, 5–11 × 2–3 mm, base and apex cuneate; valves each with prominent midvein; ovules 8–12 per ovary; style 0.1–0.5 mm. **Seeds** 1.7–2.2 × 0.9–1.1 mm. $2n = 12$.

Flowering Jun–Jul. Talus scree slopes, dry calcareous gravel flats, limestone scree, tundra turf, seepages, bluffs, stony slopes, grassy quartzite slopes, gravelly lake shores; 400–1500 m; N.W.T., Yukon; Alaska.

Smelowskia media resembles *S. americana* in aspects of habit, flowers, and fruits. It differs by having non-appressed (versus appressed) fruits, fruiting pedicels held at wider angles (greater than 40° versus less), and somewhat larger seeds (1.7–2.3 × 0.9–1.2 mm versus 1.1–1.9 × 0.6–0.9 mm).

5. Smelowskia ovalis M. E. Jones, Proc. Calif. Acad. Sci. ser. 2, 5: 624. 1895 E

Smelowskia ovalis var. *congesta* Rollins

Plants sometimes canescent basally; caudex branched. **Stems:** several from base, unbranched or branched distally, 0.3–1.8 dm, trichomes simple, 0.3–0.6 mm, mixed with smaller, dendritic ones. **Basal leaves:** petiole 1–6 cm, ciliate, trichomes simple; blade obovate, ovate, suborbicular, or oblong in outline, (terminal segments obovate or oblong), 0.5–2.5 cm × 5–15 mm, (terminal segments 0.2–1 cm × 1.5–5 mm), margins pinnatisect or pinnatifid, apex obtuse or rounded. **Cauline leaves** shortly petiolate or sessile; blade similar to basal, smaller distally. **Racemes** elongated and dense in fruit. **Fruiting pedicels** suberect to ascending, (subappressed to rachis, often forming less than 40° angle), proximalmost sometimes bracteate, 3–10 mm, pubescent, trichomes simple (to 1 mm), mixed with smaller, dendritic ones. **Flowers:** sepals (persistent), 2–2.5 mm; petals usually white, rarely pinkish, spatulate to obovate, 3.5–4.5 × 1.5–2.5 mm, narrowed to claw, 0.5–1.5 mm, apex rounded; anthers oblong, 0.5–0.6 mm. **Fruits** suberect to ascending, ovoid to suboblong, terete or slightly flattened, 2–6 × 2–3 mm, base and apex obtuse; valves each with obscure midvein; ovules 4–8 per ovary; style 0.2–1 mm. **Seeds** 1–1.5 × 0.6–0.7 mm.

Flowering Jul–Aug. Loose talus, mica shist, alpine rock slides, rocky moraines, rock crevices; 1500–3400 m; B.C.; Calif., Oreg., Wash.

Smelowskia ovalis appears to be rare in Oregon, common at Mt. Lassen (Shasta County, California), and widespread at high elevations in Washington.

6. **Smelowskia porsildii** (W. H. Drury & Rollins) Jurtsev, Novosti Sist. Vyssh. Rast. 6: 309. 1970

Smelowskia calycina (Stephan) C. A. Meyer var. *porsildii* W. H. Drury & Rollins, Rhodora 54: 105, fig. 3A. 1952; *Hutchinsia calycina* (Stephan) Desvaux var. *integrifolia* Seemann; *S. calycina* subsp. *integrifolia* (Seeman) Hultén; *S. calycina* var. *integrifolia* (Seeman) Rollins; *S. jurtzevii* Velichkin; *S. spathulatifolia* Velichkin

Plants sometimes canescent basally; caudex branched. **Stems:** several from base, unbranched, (0.3–)0.5–1.4 (–2) dm, trichomes simple, 0.3–1 mm, mixed with smaller, dendritic ones, (sometimes simple ones absent). **Basal leaves:** petiole 0.7–3(–4) cm, often ciliate, trichomes simple; blade usually linear to oblanceolate or spatulate, rarely obovate, (0.5–)0.8–2.2(–3.5) cm × 0.7–5 (–7) mm, margins usually entire, rarely apically 3 or 5-toothed or -lobed, apex obtuse to rounded, (surfaces densely tomentose, grayish, trichomes mostly dendritic, mixed with fewer, simple ones, to 1 mm). **Cauline leaves** shortly petiolate; blade similar to basal, smaller distally, margins usually entire, rarely lobed. **Racemes** elongated in fruit. **Fruiting pedicels** divaricate to ascending, (forming 30–70° angle), proximalmost bracteate, 4–12 mm, pubescent, trichomes simple mixed with smaller, dendritic ones. **Flowers:** sepals 2.5–3.5 mm; petals white or creamy white, suborbicular to obovate, 4–6 × 2–3(–4) mm, narrowed to claw, 1–2(–3) mm, apex rounded; anthers oblong, 0.5–0.7 mm. **Fruits** divaricate to ascending, ellipsoid to oblong, subterete or slightly 4-angled, 6–10 × 1.5–2.5 mm, base and apex cuneate; valves each with prominent midvein; ovules 4–8 per ovary; style 0.1–0.5 mm. **Seeds** 1.5–2 × 0.9–1.2 mm. $2n = 12, 22, 24$.

Flowering Jun–Jul. Scree, dry gravelly slopes, sandstone shale-scree, fellfields, gravel beaches and benches, slides, alpine ridges, rock crevices and outcrops, sedge tundra, *Dryas* heath meadows, dry slopes, tundra slopes, conglomerate outcrops, marshes, sedge meadows, windswept sandstone ridges; 0–1700 m; Alaska; e Asia (Russian Far East).

As delimited here, *Smelowskia porsildii* is taken in the broad sense to include plants recognized by W. H. Drury Jr. and R. C. Rollins (1952) and Rollins (1993) as *S. calycina* var. *integrifolia*. The latter was raised by Velichkin to species rank, named as *S. spathulatifolia* because of the earlier-published *S. integrifolia* Bunge. Leaf shape and pedicel orientation, used by those authors to distinguish the two taxa, vary continuously. The species is also highly variable in the relative occurrence of simple versus dendritic trichomes and leaf margins. Although I. A. Al-Shehbaz and S. I. Warwick (2006) reduced *S. spathulatifolia* to synonymy of *S. porsildii*, the latter most likely represents a complex in which *S. media* might be involved. Conflicting chromosome numbers (e.g., $2n = 12, 18, 22, 24, 32$) have been reported for this complex (Warwick and Al-Shehbaz 2006), and detailed cytological and molecular studies are needed to resolve the taxa involved. The complex involves diploid and tetraploid populations based on $x = 6$; the counts of $2n = 18$ and 32 most likely reflect misidentifications.

7. **Smelowskia pyriformis** W. H. Drury & Rollins, Rhodora 54: 108, fig. 3B. 1952 [C][E][F]

Plants sometimes canescent throughout; caudex simple. **Stems:** usually several, rarely simple from base, usually unbranched, rarely branched, 0.5–2.4(–3) dm, trichomes simple, to 1.3 mm, mixed with smaller, dendritic ones. **Basal leaves:** petiole 1–4 cm, often ciliate, trichomes simple; blade broadly ovate to oblong in outline, (terminal segments oblong to or ovate), 1–2.2 cm × 8–15 mm, (terminal segments 0.4–1 cm × 1.5–3 mm), margins pinnatifid, apex obtuse. **Cauline leaves** shortly petiolate or sessile; blade similar to basal, smaller distally. **Racemes** considerably elongated in fruit. **Fruiting pedicels** ascending to divaricate-ascending, (often forming less than 40° angle), proximalmost bracteate, 5–15(–20) mm, pubescent, trichomes simple (to 1.4 mm), mixed with smaller, dendritic ones. **Flowers:** sepals 1.2–2.2 mm; petals purple, lavender, or white, suborbicular to obovate, 2.5–4.5 × 1–2.5 mm, narrowed to claw, 1–2 mm, apex rounded; anthers ovate, 0.3–0.5 mm. **Fruits** ascending, pyriform, subterete, 5–9 × 2.5–4 mm, base cuneate, apex rounded; valves each with prominent midvein; ovules 4 per ovary; style 0.4–1.2 mm. **Seeds** (oblong-lanceolate) 2.5–3.2 × 1–1.5 mm. $2n = 12$.

Flowering Jun–Aug. Loose talus, scree slopes, shale ridgetops, limestone volcanic rubble, mixed sandstone-siltstone-carbonate scree; of conservation concern; 600–1700 m; Alaska.

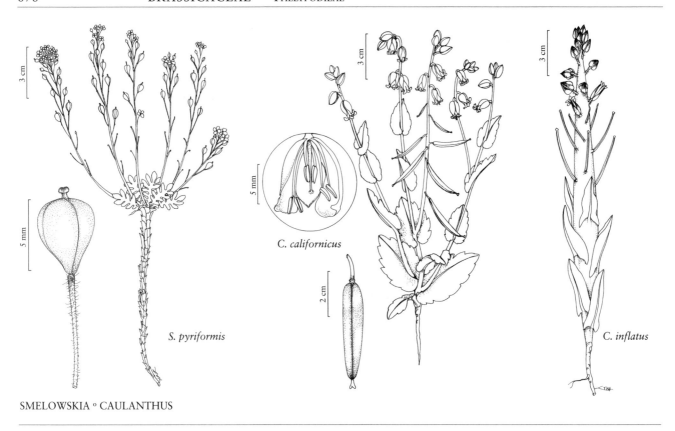

SMELOWSKIA ∘ CAULANTHUS

dd. BRASSICACEAE Burnett tribe THELYPODIEAE Prantl in H. G. A. Engler and K. Prantl, Nat. Pflanzenfam. 55[III,2]: 155. 1891

Annuals, biennials, perennials, shrubs, or subshrubs; eglandular. **Trichomes** usually simple, rarely forked or dendritic [subdendritic], sometimes absent. **Cauline leaves** petiolate or sessile; blade base auriculate or not, margins entire, dentate, or pinnately lobed. **Racemes** usually ebracteate, often elongated in fruit. **Flowers** usually actinomorphic, rarely zygomorphic; sepals erect, ascending, spreading, or reflexed, lateral pair saccate or not basally; petals white, yellow, orange, pink, lilac, lavender, purple, green, brown, or nearly black, claw present, often distinct; filaments unappendaged, not winged; pollen 3-colpate. **Fruits** usually siliques, rarely silicles, usually dehiscent, unsegmented, usually terete, 4-angled, or latiseptate; ovules 1–210[–numerous] per ovary; style obsolete, distinct, or absent; stigma usually entire or 2-lobed (subentire in *Sibaropsis*, *Streptanthella*). **Seeds** usually biseriate or uniseriate, rarely aseriate; cotyledons accumbent or incumbent.

Genera 27, species ca. 215 (14 genera, 105 species in the flora): North America, Mexico, Central America, South America.

82. CAULANTHUS S. Watson, Botany (Fortieth Parallel), 27, plate 3. 1871

• Wild-cabbage [Greek *kaulos*, stem, and *anthos*, flower, alluding to insertion of flowers along stem]

Ihsan A. Al-Shehbaz

Guillenia Greene; *Microsisymbrium* O. E. Schulz; *Stanfordia* S. Watson

Annuals, biennials, or perennials; (caudex woody); not scapose; (usually glaucous), glabrous or pubescent, trichomes usually simple, rarely mixed with fewer, stalked, 2-rayed ones. **Stems** usually erect or ascending, rarely subdecumbent, unbranched or branched distally, (sometimes inflated). **Leaves** basal and cauline; petiolate, subsessile, or sessile; basal rosulate or not, petiolate, blade margins entire, dentate, sinuate, lobed, or pinnatifid; cauline petiolate, subsessile, or sessile, blade (base sometimes sagittate or amplexicaul, rarely subauriculate), margins entire, dentate, or lobed. **Racemes** (corymbose, sometimes with a terminal cluster of sterile flowers from grouping of dark purple flower buds, rarely proximalmost flowers bracteate), elongated in fruit. **Fruiting pedicels** erect, ascending to divaricate, or reflexed, slender or stout. **Flowers:** sepals usually erect or ascending, rarely spreading, oblong or ovate to lanceolate, (usually forming urceolate calyx, sometimes keeled), lateral pair saccate or not basally; petals yellow, brown, purple, lavender, white, yellowish green, or, rarely, pinkish, linear to oblanceolate, (often channeled and crisped, rarely neither, equal or abaxial (lower) pair slightly shorter than adaxial (upper) pair, longer than sepals), claw often well-differentiated from blade (sometimes longer and wider than blade, entire); stamens tetradynamous or 3 unequal pairs (rarely subequal); filaments not dilated basally, (usually distinct, sometimes adaxial or both median pairs connate); anthers oblong, (equal or adaxial pair smaller, apex obtuse or apiculate); nectar glands usually confluent, subtending bases of stamens. **Fruits** sessile, linear, usually smooth, rarely torulose, usually terete, latiseptate, or angustiseptate, rarely 4-angled; valves each with distinct or obscure midvein, glabrous or sparsely pubescent; replum rounded; septum complete, (veinless); ovules 14–60[–210] per ovary; style obsolete or distinct; stigma capitate or not, entire, subentire, or 2-lobed (lobes opposite valves or replum). **Seeds** uniseriate, usually flattened (plump in *C. californicus*), usually not winged, rarely margined, oblong, ovoid, or, rarely, subglobose; seed coat not mucilaginous when wetted; cotyledons incumbent or obliquely so, (entire or, rarely, 3-fid). $x = 14$.

Species 17 (17 in the flora): w United States, nw Mexico.

The limits of *Caulanthus* have not changed for about 60 years following the excellent account by E. B. Payson (1923). R. C. Rollins (1993), however, resurrected *Stanfordia* mainly based on angustiseptate instead of terete fruits and on the presence of 3-fid cotyledons. The latter character is also present in *C. coulteri*, a species that Rollins maintained in *Caulanthus*. As emphasized by Payson and by I. A. Al-Shehbaz (1973), *Stanfordia* does not merit recognition, a position taken also by O. Appel and Al-Shehbaz (2003). R. E. Buck (1993, 1995) excluded three species of *Caulanthus* (*C. anceps*, *C. flavescens*, *C. lasiophyllus*) and placed them in the long-abandoned *Guillenia*. His primary reasoning was that they do not have the urceolate calyx or the crisped and channeled petals. However, *C. flavescens* has crisped and channeled petals, and sometimes has somewhat urceolate calyces, a feature not found in all species of *Caulanthus*. Extensive discussion on the maintenance of these three species in *Caulanthus* was given by Al-Shehbaz and needs not be repeated here. In my opinion, the recognition of *Guillenia* as distinct from *Caulanthus* is based on over-emphasis of a few petal characters and, without thorough molecular studies on the entire streptanthoid genera, it is more practical to maintain a more inclusive *Caulanthus*.

SELECTED REFERENCE Buck, R. E. 1995. Systematics of *Caulanthus* (Brassicaceae). Ph.D. dissertation. University California, Berkeley.

1. Median and/or distal cauline leaves sessile, blade bases sagittate or amplexicaul.
    2. Stems inflated (to 4 cm diam. at widest point)......................... 12. *Caulanthus inflatus*
    2. Stems not inflated.
        3. Fruits angustiseptate; seeds subglobose, plump, cotyledons 3-fid .... 4. *Caulanthus californicus*
        3. Fruits terete, latiseptate, or 4-angled; seeds ovoid to oblong, flattened, cotyledons usually entire (3-fid in C. coulteri).
            4. Racemes with a terminal cluster of sterile flowers; filaments in 3 unequal pairs, median pairs often connate.
                5. Fruits usually reflexed, rarely divaricate; basal leaf blade surfaces with simple and forked trichomes; cotyledons 3-fid; stigma lobes 0.5–1.5 mm .... 6. *Caulanthus coulteri*
                5. Fruits erect or ascending; basal leaf blade surfaces with simple trichomes; cotyledons entire; stigma lobes 1–4 mm...................... 14. *Caulanthus lemmonii*
            4. Racemes without a terminal cluster of sterile flowers; filaments tetradynamous (except in C. amplexicaulis), median pairs distinct.
                6. Plants glabrous throughout; filaments in 3 unequal pairs; fruits divaricate to ascending..1. *Caulanthus amplexicaulis*
                6. Plants hispid or puberulent at least basally; filaments tetradynamous; fruits usually reflexed, rarely divaricate.
                    7. Plants puberulent, trichomes appressed, 2-rayed; stems erect to ascending (often flexuous, weak)........................ 5. *Caulanthus cooperi*
                    7. Plants hispid, trichomes spreading, simple; stems erect.
                        8. Fruits latiseptate or 4-angled; cauline leaf blades linear-lanceolate ....11. *Caulanthus heterophyllus*
                        8. Fruits terete; cauline leaf blades ovate to oblong.......... 17. *Caulanthus simulans*
1. Median and/or distal cauline leaves usually petiolate or subsessile (sessile in C. hallii), blade bases not auriculate or sagittate.
    9. Perennials, with woody caudices; plants glabrous or sparsely pubescent; anthers 3–6.5 mm.
        10. Stems fusiform, strongly inflated (to 3 cm diam. at widest point) .... 7. *Caulanthus crassicaulis*
        10. Stems cylindrical, not or slightly inflated.
            11. Basal leaves rosulate; fruiting pedicels 1–6 mm; stigma lobes opposite valves ....15. *Caulanthus major*
            11. Basal leaves not rosulate; fruiting pedicels 5–35 mm; stigma lobes opposite replum.
                12. Petals prominently dark-veined; stigma lobes to 0.4 mm in fruit .... 3. *Caulanthus barnebyi*
                12. Petals not or faintly veined; stigma lobes 0.9–1.5 mm in fruit .... 9. *Caulanthus glaucus*
    9. Annuals, biennials, or, rarely, perennials (short-lived), without caudices; plants moderately to densely hispid, hirsute, or pilose (at least basally); anthers 0.4–3 mm.
        13. Sepals widely spreading; filaments subequal .... 2. *Caulanthus anceps*
        13. Sepals usually erect, rarely ascending to spreading; filaments tetradynamous or in 3 unequal pairs.
            14. Petals 2.5–5(–6.5) mm, not crisped or channeled, claw undifferentiated from blade; calyces not urceolate .... 13. *Caulanthus lasiophyllus*
            14. Petals 6–13 mm, crisped and channeled (except C. hallii), claw well-differentiated from blade; calyces urceolate (except C. flavescens).
                15. Petals purple; ovules 152–198 per ovary ....16. *Caulanthus pilosus*
                15. Petals usually creamy white, rarely pink; ovules 44–96 per ovary.
                    16. Filaments tetradynamous; fruits 3.8–7.7 cm; ovules 44–58 per ovary ....8. *Caulanthus flavescens*
                    16. Filaments in 3 unequal pairs; fruits 6.5–12.5 cm; ovules 78–96 per ovary... 10. *Caulanthus hallii*

1. **Caulanthus amplexicaulis** S. Watson, Proc. Amer. Acad. Arts 17: 364. 1882 [E]

Caulanthus amplexicaulis var. *barbarae* (J. T. Howell) Munz; *Euklisia amplexicaulis* (S. Watson) Greene; *Pleiocardia magna* Greene; *Streptanthus amplexicaulis* (S. Watson) Jepson; *S. amplexicaulis* var. *barbarae* J. T. Howell

Annuals; glabrous throughout. **Stems** erect, unbranched or branched distally, 0.4–11 dm. **Basal leaves** rosulate; petiole 0.5–3 cm; blade obovate to broadly oblanceolate, 1–7 cm × 5–37 mm, margins coarsely dentate. **Cauline leaves** (median) sessile; blade oblong to broadly ovate or obovate, 2–8 cm × 10–40 mm (smaller distally, base amplexicaul), margins coarsely dentate or entire. **Racemes** (lax), without a terminal cluster of sterile flowers. **Fruiting pedicels** divaricate to ascending, 2.5–19 mm. **Flowers:** sepals erect, (dark maroon-purple or ochroleucous to yellowish), ovate to ovate-lanceolate, 4–9.2 × 2–4 mm (equal); petals (abaxial pair) usually stramineous to yellowish green, rarely pale purple, (adaxial pair) purple, (slightly longer), 10–18 mm, blade 4–10 × 0.5–1.5 mm, crisped, claw narrowly oblong-oblanceolate, 6–10 × 1–3 mm; filaments in 3 unequal pairs, abaxial pair 3–7 mm, lateral pair 1–5 mm, adaxial pair 4.5–9 mm; anthers oblong, unequal, 1–3.5 mm (adaxial pair smaller). **Fruits** divaricate to ascending (often distinctly curved), terete or slightly latiseptate, 4.5–14(–16.7) cm × 1–1.5 mm; valves each with obscure midvein; ovules 40–92 per ovary; style 0–0.4 mm; stigma subentire. **Seeds** 1.4–2.2 × 0.9–1.2 mm. $2n = 28$.

Flowering Apr–Jul(–Aug). Chaparral, montane forests, serpentine areas, granitic and shale scree; 800–2900 m; Calif.

Caulanthus amplexicaulis is known from Los Angeles, San Bernardino, Santa Barbara, and Ventura counties. The serpentine populations of Santa Barbara County have sepals that are ochroleucous to yellowish instead of the dark maroon ones found elsewhere in the species range. They were recognized by R. C. Rollins (1993) and R. E. Buck (1995) as var. *barbarae*. Both vars. *barbarae* and *amplexicaulis* form a distinct monophyletic clade but they are without reproductive barriers (A. E. Pepper and L. E. Norwood 2001). Some other species of *Caulanthus* show comparable or even greater variation in sepal color, and it might be more practical to treat var. *barbarae* as a color form.

SELECTED REFERENCE Pepper, A. E. and L. E. Norwood. 2001. Evolution of *Caulanthus amplexicaulis* var. *barbarae* (Brassicaceae), a rare serpentine endemic plant: A molecular phylogenetic perspective. Amer. J. Bot. 88: 1479–1489.

2. **Caulanthus anceps** Payson, Ann. Missouri Bot. Gard. 9: 303. 1923 [E]

Thelypodium lemmonii Greene, W. Amer. Sci. 3: 156. 1887 (as lemmoni), not *Caulanthus lemmonii* S. Watson 1888; *Streptanthus anceps* (Payson) Hoover

Annuals; sparsely to densely hirsute. **Stems** erect, unbranched or branched distally (or, rarely, basally), 3.5–15 dm, at least sparsely hirsute basally. **Basal leaves** soon withered. **Cauline leaves** petiolate (median 0.4–3 cm); blade lanceolate to oblong, 1.5–9.5 cm × 3–30 mm (smaller distally), margins denticulate to subentire (proximal blade margins dentate). **Racemes** without a terminal cluster of sterile flowers, (considerably elongated in fruit). **Fruiting pedicels** ascending to strongly reflexed (slender or thickened), 3–10 mm. **Flowers:** sepals spreading, oblong, 3.5–5.5 × 1–1.7 mm; petals (spreading), white to lavender, 4–8 × 2–4 mm, not channeled or crisped, claw undifferentiated from blade; filaments (spreading), subequal, 3.5–5 mm; anthers narrowly oblong, equal, 1.5–2 mm, (coiled after dehiscence). **Fruits** erect or reflexed, (straight), terete, 3–6.7 cm × 1.2–2 mm; valves each with prominent midvein, (usually glabrous, rarely sparsely pubescent); ovules 40–54 per ovary; style (subconical or cylindrical), 1–4 mm; stigma subentire. **Seeds** (brown), 1.4–1.8 × 1–1.3 mm. $2n = 28$.

Flowering Mar–May. Grassy slopes, open flats, roadsides, fields, hillsides; 300–1700 m; Calif.

Caulanthus anceps is distributed in Kern, Monterey, San Benito, San Luis Obispo, Santa Barbara, and Ventura counties.

3. **Caulanthus barnebyi** Rollins & P. K. Holmgren, Brittonia 32: 148, fig. 1. 1980 [C][E]

Perennials; glabrous throughout. **Stems** erect or ascending, unbranched or branched distally, 5–11 dm. **Basal leaves** not rosulate; petiole 1.5–3.8 cm; blade broadly obovate to broadly ovate or oblong, 3.5–15.5 cm × 22–100 mm, margins entire or coarsely dentate-sinuate. **Cauline leaves** (distalmost) shortly petiolate to subsessile; blade linear to lanceolate or oblanceolate, margins entire. **Racemes** (densely flowered), without a terminal cluster of sterile flowers. **Fruiting pedicels** ascending, 5–17 mm. **Flowers:** sepals erect (purplish or sometimes green), ovate-oblong to lanceolate, 7.5–11 × 2.5–3 mm (equal); petals creamy white (with prominent purple or brown veins), 12.5–18.5 mm, blade 6–8 × 2–2.5 mm, not or slightly crisped, claw oblanceolate, 5–9 × 1.2–2 mm; filaments tetradynamous, median pairs 3–5 mm, lateral pair 2.5–4.5 mm; anthers narrowly oblong, equal, 3–4.5 mm. **Fruits** divaricate to ascending (often distinctly curved), terete, 5.7–12.5 cm × 1.7–2.3 mm; valves each with obscure midvein; ovules 76–90 per ovary; style usually obsolete, rarely to 2.5 mm; stigma strongly 2-lobed (lobes to 0.4 mm, connivent, opposite replum). **Seeds** 1.4–2.1 × 0.8–1.2 mm. $2n = 20$.

Flowering May–Jun. Dry, steep slopes, rocky outcrops, on slate, metamorphic, or igneous substrates; of conservation concern; 1300–1500 m; Nev.

Caulanthus barnebyi is known from Humboldt and Pershing counties.

4. **Caulanthus californicus** (S. Watson) Payson, Ann. Missouri Bot. Gard. 9: 299. 1923 E F

Stanfordia californica S. Watson in W. H. Brewer et al., Bot. California 2: 479. 1880; *Streptanthus californicus* (S. Watson) Greene

Annuals; (sometimes glaucous), hispid proximally, glabrous distally. **Stems** erect to subdecumbent, usually branched distally, 0.9–5.5 dm, sparsely hispid basally. **Basal leaves** rosulate; petiole 0.3–3.5 cm; blade oblanceolate, 0.7–8 cm × 3–25 mm, margins often coarsely dentate, sometimes somewhat pinnatifid. **Cauline leaves** (median) sessile; blade oblong or suborbicular to obovate, 0.6–7.5 cm × 3–55 mm (smaller distally, base amplexicaul), margins coarsely dentate or entire. **Racemes** (densely flowered), with a terminal cluster of sterile flowers. **Fruiting pedicels** ascending to reflexed, 2–11 mm, usually pubescent, rarely glabrous. **Flowers:** sepals erect to ascending (dark purple in bud, purplish green after anthesis), ovate-lanceolate, 4–9(–11) × 2.5–3.5 mm (unequal, adaxial one longest, keeled); petals white (with purple veins), 5.5–12 mm, blade 2–5 × 1–2 mm, crisped, claw narrowly oblong to lanceolate, 5–8 × 2.5–4 mm; filaments in 3 unequal pairs, abaxial pair 3–8 mm, lateral pair 2–7 mm, adaxial pair (connate), 5–9 mm; anthers oblong, equal, 1–3.5 mm. **Fruits** erect or reflexed (often straight), angustiseptate, 1.7–5 cm × 3.5–6 mm; valves each with prominent midvein; ovules 46–100 per ovary; style 0.2–2.7 mm; stigma strongly 2-lobed (lobes to 2 mm, opposite valves). **Seeds** (subglobose, plump), 1–1.6 mm diam., (cotyledons deeply 3-fid). $2n = 28$.

Flowering Feb–Apr. Grasslands, juniper woodlands; 100–1000 m; Calif.

According to R. E. Buck (1995), the range of *Caulanthus californicus* was highly reduced from conversion of habitats into agricultural land, and it is now restricted to portions of Fresno, Santa Barbara, and San Luis Obispo counties, whereas its previous range included also Kern, Kings, Monterey, Tulare, and Ventura counties.

5. **Caulanthus cooperi** (S. Watson) Payson, Ann. Missouri Bot. Gard. 9: 293. 1923

Thelypodium cooperi S. Watson, Proc. Amer. Acad. Arts 12: 246. 1877; *Guillenia cooperi* (S. Watson) Greene

Annuals; puberulent or glabrous (trichomes simple and subappressed, and 2-rayed). **Stems** erect to ascending (often flexuous, weak, often tangled with desert shrubs), usually branched distally, 1–8 dm, glabrous or puberulent. **Basal leaves** rosulate; petiole 0.3–2.5 cm; blade oblanceolate to spatulate, 0.7–6 cm × 2–27 mm, margins usually coarsely dentate or somewhat pinnatifid, rarely entire, (surfaces glabrous). **Cauline leaves** (median) sessile; blade lanceolate or oblong, 1.5–7.5 cm × 5–20 mm (smaller distally, base amplexicaul to sagittate), margins dentate or entire, (surfaces glabrous). **Racemes** (lax), without a terminal cluster of sterile flowers. **Fruiting pedicels** reflexed, 1–4.5 mm, usually glabrous, rarely puberulent. **Flowers:** sepals erect, (purplish or yellow-green), narrowly lanceolate, 3–6.5 × 0.8–1.5 mm (equal); petals yellow-green to purplish (often with purple veins), 4.5–9 mm, blade 2–3 × 0.7–1.5 mm, not crisped, claw narrowly oblong-oblanceolate, 2.5–7 × 1–1.5 mm; filaments slightly tetradynamous, median pairs 2–4.5 mm, lateral pair 1.5–3.5 mm; anthers oblong, equal, 1.5–2 mm. **Fruits** usually reflexed, rarely divaricate (often subfalcate), terete, 2–6 cm × 1.5–2.5 mm; valves each with prominent midvein, (glabrous or puberulent); ovules 24–48 per ovary; style 0.2–2.7 mm; stigma slightly 2-lobed. **Seeds** 1–2 × 1–1.2 mm. $2n = 28$.

Flowering (Jan–)Feb–Mar. Desert shrubs, woodlands; 600–2300 m; Ariz., Calif., Nev., Utah; Mexico (Baja California).

Caulanthus cooperi is distributed in the Colorado and Mojave deserts in western Arizona, central and southern California, southern Nevada, and southern Utah.

6. **Caulanthus coulteri** S. Watson, Botany (Fortieth Parallel), 27. 1871 E

Streptanthus coulteri (S. Watson) Greene

Annuals; usually hispid, rarely glabrous distally, (trichomes simple, stalked, or 2-rayed forked). **Stems** erect, usually branched distally, 1–16(–25) dm, sparsely hispid basally. **Basal leaves** rosulate; petiole 0.3–5 cm; blade narrowly oblong to oblanceolate, 0.7–12 cm × 4–25 mm, margins pinnatifid to coarsely dentate-sinuate, (surfaces pubescent, trichomes simple and 2-rayed forked). **Cauline leaves** (median) sessile;

blade lanceolate to linear-lanceolate, 1–18(–26) cm × 3–35(–45) mm (smaller distally, base amplexicaul), margins dentate or entire, (surfaces glabrous). **Racemes** (densely flowered), with a terminal cluster of sterile flowers. **Fruiting pedicels** reflexed or, rarely, spreading, 3–16 mm, usually pubescent, rarely glabrous. **Flowers:** sepals erect to ascending, (dark purple in bud becoming yellowish green with purplish or brown margins), ovate to narrowly lanceolate, 5–15(–19) × 2–4.5 mm (subequal, keeled, glabrous or pubescent, trichomes simple and forked); petals white or purplish (with dark purple veins), 8–25(–30) mm, blade 4–10 × 1.5–4 mm, crisped, claw oblanceolate, 4–12 × 2–2.5 mm; filaments in 3 unequal pairs, abaxial pair 3.5–8.5 mm, lateral pair 1.5–7 mm, adaxial pair (connate), 5.5–11 mm; anthers oblong to linear-oblong, unequal, 1–5 mm, (adaxial pair smaller). **Fruits** usually reflexed, rarely divaricate (often straight), terete or slightly latiseptate, 3.5–15 cm × 2.2–3.5 mm; valves each with prominent midvein basally; ovules 70–96 per ovary; style 0–4 mm; stigma strongly 2-lobed (lobes 0.5–1.5 mm, opposite valves). **Seeds** 1.5–3.5 × 1.5–2.2 mm, (cotyledons deeply 3-fid). $2n = 28$.

Flowering Mar–Jun(–Jul). Grasslands, scrub, woodlands, chaparral; 100–2100 m; Calif.

Caulanthus coulteri is highly variable in almost every morphological character. For a discussion of its limits, see 14. *Caulanthus lemmonii*. It is known from central and southern California.

7. Caulanthus crassicaulis (Torrey) S. Watson, Botany (Fortieth Parallel), 27. 1871 [E]

Streptanthus crassicaulis Torrey in H. Stansbury, Exped. Great Salt Lake, 383, plate 1. 1852; *Caulanthus crassicaulis* var. *glaber* M. E. Jones; *C. glaber* (M. E. Jones) Rydberg; *C. senilis* A. Heller

Perennials; glabrous or sparsely pubescent. **Stems** erect or ascending, usually unbranched, rarely branched, (hollow, strongly inflated, fusiform, to 3 cm diam. at widest point), 2–10 dm. **Basal leaves** rosulate; petiole 0.5–9 cm; blade obovate to oblanceolate (in outline), 1–12 cm × 3–45 mm, margins entire, dentate-sinuate, lyrate, runcinate, or pinnatifid. **Cauline leaves** (distalmost) shortly petiolate; blade linear to narrowly oblanceolate, margins entire. **Racemes** (densely flowered), with a terminal cluster of sterile flowers. **Fruiting pedicels** ascending, 1–5 mm, glabrous or pubescent, (trichomes flattened). **Flowers:** sepals erect (creamy white, purplish, or greenish), ovate to lanceolate, 7.5–14 × 2.5–4 mm (equal); petals brown or purple, 10–15 mm, blade 3–5 × 1.2–2.5 mm, not or hardly crisped, claw oblanceolate, 7–10 × 2–2.5 mm; filaments tetradynamous, median pairs 3–8 mm, lateral pair 2–7 mm; anthers narrowly oblong, equal, 3–6.5 mm. **Fruits** erect to ascending, terete or slightly latiseptate, 4.5–14 cm × 2–2.5 mm; valves each with obscure midvein; ovules 98–126 per ovary; style obsolete or, rarely, to 0.6 mm; stigma strongly 2-lobed (lobes to 1 mm, connivent, opposite valves). **Seeds** 1.5–4 × 1.5–2 mm. $2n = 28$.

Flowering Apr–Jul. Sagebrush scrub, pinyon-juniper woodland; 1200–2900 m; Ariz., Calif., Colo., Idaho, Nev., Utah, Wyo.

E. B. Payson (1923), R. C. Rollins (1993), and R. E. Buck (1995) recognized two varieties of *Caulanthus crassicaulis* and separated them solely on the basis of var. *crassicaulis* having pubescent sepals and glabrous or pubescent leaves, and var. *glaber* having glabrous or sparsely pubescent sepals and glabrous leaves. In my opinion, this distinction is impractical; as in some other species of *Caulanthus*, both glabrous and pubescent forms occur sometimes within the same population.

8. Caulanthus flavescens (Hooker) Payson, Ann. Missouri Bot. Gard. 9: 301. 1923 [C][E]

Streptanthus flavescens Hooker, Icon. Pl. 1: plate 44. 1837; *Caulanthus procerus* (Brewer ex A. Gray) S. Watson; *Guillenia flavescens* (Hooker) Greene; *G. hookeri* (Greene) Greene; *S. dudleyi* Eastwood; *S. lilacinus* Hoover; *S. procerus* Brewer ex A. Gray; *Thelypodium flavescens* S. Watson; *T. greenei* Jepson; *T. hookeri* Greene

Annuals; (not glaucous), glabrous or sparsely to moderately hirsute. **Stems** erect, unbranched or branched distally, 1.7–12 dm, usually hirsute basally. **Basal leaves** soon withered; blade lyrate, margins sinuate-pinnatifid. **Cauline leaves** petiolate (median 0.3–2 cm) or (distal) sessile; blade lanceolate to oblong or oblanceolate, 2–13.5 cm × 3–35 mm (smaller distally, base sometimes subauriculate), margins dentate. **Racemes** without a terminal cluster of sterile flowers, (considerably elongated in fruit). **Fruiting pedicels** ascending to strongly reflexed, (slender or thickened), 4.5–8 mm, glabrous or sparsely pubescent. **Flowers:** sepals usually erect, rarely ascending to spreading, oblong, 6–9 × 1.5–2 mm, (not saccate or urceolate); petals (erect), usually white to creamy white, rarely pinkish, (narrowly lanceolate), 7–13 mm, blade 3–5 × 1–1.5 mm, channeled, crisped, claw linear-oblanceolate, 4–8 × 1–1.7 mm; filaments tetradynamous, (erect), median pairs 5–7 mm, lateral pair 3–5 mm; anthers narrowly oblong, equal, 1.5–2.5 mm. **Fruits** erect, (straight), terete, 3.8–7.7 cm × 1.4–2 mm; valves each with obscure midvein, (glabrous or sparsely pubescent); ovules 44–58 per ovary; style 1–3.5 mm; stigma subentire. **Seeds** (brown), 1.3–1.8 × 0.9–1.2 mm. $2n = 28$.

Flowering Mar–May. Serpentine grounds, roadsides, gumbo clay slopes, open hillsides, grassy fields; of conservation concern; 100–800 m; Calif.

Caulanthus flavescens is distributed primarily from Monterey and San Benito counties north to Napa and San Joaquin counties.

9. **Caulanthus glaucus** S. Watson, Proc. Amer. Acad. Arts 17: 364. 1882 [C][E]

Streptanthus glaucus (S. Watson) Jepson

Perennials; glabrous throughout. **Stems** erect or ascending, unbranched or branched distally, 3–12 dm. **Basal leaves** not rosulate; petiole 1–8 cm; blade broadly obovate or oblong to elliptic, 2–10 cm × 20–120 mm, margins usually entire, rarely coarsely dentate-sinuate. **Cauline leaves** (distalmost) shortly petiolate; blade lanceolate to linear, 1–3.5 cm × 3–10 mm, margins entire. **Racemes** (densely flowered), without a terminal cluster of sterile flowers. **Fruiting pedicels** ascending, 5–35 mm. **Flowers:** sepals erect, (purplish or green to yellowish green), lanceolate, 7–12 × 2–3 mm (equal); petals purple or yellowish green, (not or faintly veined),11–17.5 mm, blade 3–6 × 1.8–2.7 mm, not crisped, claw narrowly oblong to narrowly oblanceolate, 8–12 × 3–4 mm; filaments tetradynamous, median pairs 3.5–8 mm, lateral pair 2–7 mm; anthers narrowly oblong, equal, 3.5–4.5 mm. **Fruits** divaricate to ascending (often distinctly curved), terete, 4.5–15.5 cm × 1–1.6 mm; valves obscurely veined; ovules 180–210 per ovary; style 0.2–1 mm; stigma strongly 2-lobed (lobes 0.9–1.5 mm, connivent, opposite replum). **Seeds** 1–2 × 0.8–1.2 mm. $2n = 20$.

Flowering Apr–Jun. Sagebrush scrub, dry slopes and rocky outcrops; of conservation concern; 1500–2400 m; Calif., Nev.

Caulanthus glaucus is known in California from Inyo and Mono counties and in Nevada from Esmeralda, Mineral, and Nye counties.

10. **Caulanthus hallii** Payson, Ann. Missouri Bot. Gard. 9: 290. 1923 [C][E]

Streptanthus hallii (Payson) Jepson

Annuals; sparsely to densely hispid or subglabrate. **Stems** erect or ascending, unbranched or branched distally, (hollow, sometimes slightly inflated), 2–12 dm. **Basal leaves** rosulate; petiole 0.5–3.5 cm; blade oblanceolate to oblong (in outline), 1.5–11.5 cm × 5–55 mm, margins pinnately lobed (lobes dentate). **Cauline leaves** (distalmost) sessile; blade lanceolate-linear, margins entire, (surfaces sparsely hispid). **Racemes** (somewhat lax), without a terminal cluster of sterile flowers. **Fruiting pedicels** ascending, 9–25 mm, hispid or subglabrate. **Flowers:** sepals erect, (creamy white), lanceolate to ovate, 3–6.5 × 1.8–2.5 mm (equal); petals creamy white, 6–10.5 mm, blade 3–4 × 1.5–2 mm, not crisped, claw narrowly oblanceolate or oblong, 3–6 × 2–3 mm; filaments in 3 unequal pairs, abaxial pair 2.5–6 mm, lateral pair 1.5–4.5, adaxial pair 4.5–8 mm; anthers narrowly oblong, unequal, 2–3 mm (adaxial pair slightly smaller). **Fruits** divaricate to ascending (sometimes curved), terete, 6.5–12.5 cm × 1.8–2.2 mm; valves each with prominent midvein; ovules 78–96 per ovary; style to 2 mm; stigma slightly 2-lobed. **Seeds** 1–1.6 × 0.7–1 mm.

Flowering Mar–May. Rocky areas, chaparral, scrub; of conservation concern; 150–1800 m; Calif.

Caulanthus hallii is known from Riverside and San Diego counties.

11. **Caulanthus heterophyllus** (Nuttall) Payson, Ann. Missouri Bot. Gard. 9: 298. 1923

Streptanthus heterophyllus Nuttall in J. Torrey and A. Gray, Fl. N. Amer. 1: 77. 1838; *Caulanthus stenocarpus* Payson; *Guillenia heterophylla* (Nuttall) O. E. Schulz; *S. repandus* Nuttall

Annuals; hispid basally, glabrous or subglabrate distally. **Stems** erect, usually branched distally, 2.5–12 dm, hispid. **Basal leaves** weakly rosulate; petiole 0.3–3 cm; blade linear-oblanceolate to linear-oblong, 0.7–7 cm × 2–18 mm, margins coarsely dentate or pinnately lobed. **Cauline leaves** (median) sessile; blade linear-lanceolate, 5–16 cm × 5–40 mm, (smaller distally, base amplexicaul to sagittate), margins dentate or (distalmost) entire. **Racemes** (densely flowered), without a terminal cluster of sterile flowers. **Fruiting pedicels** reflexed, 2–8 mm, glabrous or hispid. **Flowers:** sepals erect (purple or yellow to creamy white), lanceolate, 3–8 × 1–1.8 mm (equal); petals purple or yellowish (often with darker purple veins), 5–15 mm, blade 2–6 × 1–1.5 mm, not crisped, claw narrowly oblanceolate or oblong, 3–9 × 1–1.5 mm; filaments tetradynamous, median pairs 3–6 mm, lateral pair 2–5 mm; anthers oblong, equal, 1–3 mm. **Fruits** reflexed (often straight, rarely curved), latiseptate or 4-angled, 4.5–10 cm × 1–1.5 mm; valves each with prominent midvein; ovules 56–82 per ovary; style 0.5–3.5 mm; stigma slightly 2-lobed. **Seeds** 1.2–2 × 0.9–1.4 mm. $2n = 28$.

Flowering Mar–May. Coastal scrub, chaparral, rocky areas; 0–1400 m; Calif.; Mexico (Baja California).

Caulanthus heterophyllus is a common species distributed from Santa Barbara County southward into northwestern Baja California, Mexico.

R. E. Buck (1993) divided the species into two varieties, including one invalidly published, based on flower color, but these are treated here as mere color variants.

12. Caulanthus inflatus S. Watson, Proc. Amer. Acad. Arts 17: 364. 1882 [E] [F]

Streptanthus inflatus (S. Watson) Greene

Annuals; glabrous or sparsely (or densely) pilose basally. **Stems** erect, unbranched or branched distally, (hollow, strongly inflated, fusiform, to 4 cm diam. at widest point), 1.5–9.7 dm. **Basal leaves** rosulate; petiole 0.5–5 cm; blade obovate to oblanceolate, 1–13 cm × 5–60 mm, margins entire or dentate-sinuate. **Cauline leaves** (distalmost) sessile; blade ovate to lanceolate, 1–20 cm × 5–70 mm (base amplexicaul), margins entire. **Racemes** (densely flowered), without a terminal cluster of sterile flowers. **Fruiting pedicels** ascending to divaricate, 3–12 mm, sparsely to densely pilose, or glabrous. **Flowers:** sepals erect, (dark purple at least apically, whitish basally), narrowly ovate to lanceolate or oblong-lanceolate, 5.5–14 × 2–3 mm (unequal, adaxial one slightly smaller or narrower, keeled); petals purple, 9–14 mm, blade 3–4 × 1.5–2 mm, crisped, claw oblanceolate to oblong, 6–10 × 1.5–3 mm; filaments in 3 unequal pairs, (median pairs connate), abaxial pair 4–8 mm, lateral pair 3–5 mm, adaxial pair 7–10 mm; anthers narrowly oblong, equal, 2–4 mm. **Fruits** erect to ascending, terete, 3.5–12.7 cm × 2–3 mm; valves each with prominent midvein; ovules 58–82 per ovary; style obsolete or, rarely, to 0.4 mm; stigma strongly 2-lobed (lobes opposite valves). **Seeds** 1.5–3 × 0.7–1.7 mm. $2n = 28$.

Flowering Feb–Jun. Desert plains, dry hillsides; 150–1500 m; Calif.

Caulanthus inflatus is found in central and southern California.

13. Caulanthus lasiophyllus (Hooker & Arnott) Payson, Ann. Missouri Bot. Gard. 9: 303. 1923

Turritis lasiophylla Hooker & Arnott, Bot. Beechey Voy., 321. 1838; *Caulanthus lasiophyllus* var. *inalienus* (B. L. Robinson) Payson; *C. lasiophyllus* var. *rigidus* (Greene) Payson; *C. lasiophyllus* var. *utahensis* (Rydberg) Payson; *Erysimum retrofractum* Torrey; *Guillenia inaliena* (B. L. Robinson) Greene; *G. lasiophylla* (Hooker & Arnott) Greene; *G. rigida* (Greene) Greene; *Microsisymbrium lasiophyllum* (Hooker & Arnott) O. E. Schulz; *M. lasiophyllum* var. *dasycarpum* O. E. Schulz; *M. lasiophyllum* var. *dissectum* O. E. Schulz; *M. lasiophyllum* var. *inalienum* (B. L. Robinson) O. E. Schulz; *M. lasiophyllum* var. *integrifolium* O. E. Schulz; *M. lasiophyllum* var. *rigidum* (Greene) O. E. Schulz; *Sisymbrium acuticarpum* M. E. Jones; *S. deflexum* Harvey ex Torrey; *S. deflexum* var. *xerophilum* E. Fournier; *S. lasiophyllum* (Hooker & Arnott) K. Brandegee; *S. reflexum* Nuttall; *Streptanthus lasiophyllus* (Hooker & Arnott) Hoover; *S. lasiophyllus* var. *inalienus* (B. L. Robinson) Hoover; *S. lasiophyllus* var. *utahensis* (Rydberg) Hoover; *S. rigidus* (Greene) Hoover; *Thelypodium lasiophyllum* (Hooker & Arnott) Greene; *T. lasiophyllum* var. *inalienum* B. L. Robinson; *T. lasiophyllum* var. *rigidum* (Greene) B. L. Robinson; *T. lasiophyllum* var. *utahense* (Rydberg) Jepson; *T. rigidum* Greene; *T. utahense* Rydberg

Annuals; (not glaucous), usually sparsely to densely hispid or hirsute, rarely subglabrate. **Stems** erect, unbranched or branched distally, (0.8–)2–10(–16) dm, at least sparsely hirsute basally. **Basal leaves** soon withered. **Cauline leaves** petiolate (proximal and median 0.5–3 cm); blade lanceolate to oblong or oblanceolate, pinnatifid, 2–12 cm × 5–50 mm, (smaller with fewer lobes distally), margins of lateral lobes dentate or entire. **Racemes** without a terminal cluster of sterile flowers, (considerably elongated in fruit). **Fruiting pedicels** strongly reflexed or spreading, (slender or thickened), (0.7–)1–2.2(–3) mm. **Flowers:** sepals erect, oblong, 2–4 × 0.7–1 mm, (not saccate or urceolate); petals usually white to creamy white, rarely pinkish, (narrowly oblanceolate), 2.5–5(–6.5) × 0.9–1.1 mm, not channeled or crisped, claw undifferentiated from blade; filaments slightly tetradynamous, median pairs 3–4.5 mm, lateral pair 1.5–2.5 mm; anthers ovate, equal, 0.4–0.7 mm. **Fruits** ascending or descending (straight to slightly curved, sometimes subtorulose), terete, 2–4.8(–5.7) cm × 0.7–1.2 mm; valves each with obscure midvein, (glabrous or sparsely pubescent); ovules 14–60 per ovary; style 0.5–2 mm; stigma subentire. **Seeds** (brown), 0.9–1.5 × 0.6–1 mm. $2n = 28$.

Flowering Mar–May. Desert flats, gravelly areas, limestone rocks, talus slopes, shrub communities, hillsides, sandy banks, disturbed sites, grassy fields, ravines; 0–1400 m; Ariz., Calif., Nev., N.Mex., Oreg., Utah, Wash.; Mexico (Baja California, Sonora).

Caulanthus lasiophyllus is highly variable in flower size, leaf morphology, fruit morphology (length, width, curvature, presence or absence of indumentum) and orientation, number of ovules per ovary, and plant height. This species is badly in need of thorough studies at both populational and molecular levels, and it is very likely that some varieties recognized by E. B. Payson (1923), such as var. *rigidus*, may well represent distinct species or subspecies.

14. Caulanthus lemmonii S. Watson, Proc. Amer. Acad. Arts 23: 261. 1888 (as lemmoni) [C] [E]

Caulanthus coulteri S. Watson var. *lemmonii* (S. Watson) Munz; *Streptanthus coulteri* (S. Watson) Greene var. *lemmonii* (S. Watson) Jepson; *S. parryi* Greene

Annuals; hispid basally, subglabrate or glabrous distally. **Stems** erect or ascending, usually branched distally, 1–8 dm, sparsely hispid basally. **Basal leaves** rosulate; petiole 0.3–3 cm; blade oblanceolate, 0.7–9 cm × 4–25 mm, margins coarsely dentate-sinuate. **Cauline leaves** (median) sessile; blade lanceolate to narrowly ovate, 0.5–11 cm × 2–45 mm, (smaller distally, base amplexicaul), margins entire or denticulate. **Racemes** (densely flowered), with a terminal cluster of sterile flowers. **Fruiting pedicels** ascending to divaricate, 3–18(–27) mm, pubescent or glabrous. **Flowers:** sepals erect to ascending, (dark purple in bud, becoming greenish or creamy white and purplish or brown distally), ovate to narrowly lanceolate, 6–17 × 2.5–3.5 mm (subequal, keeled, usually glabrous, rarely pubescent, trichomes simple); petals white (with dark purple veins), 8–20 mm, blade 4–8 × 1.5–2 mm, crisped, claw oblanceolate, 4–11 × 2–3 mm; filaments in 3 unequal pairs, (median pairs often connate), abaxial pair 3.5–11 mm, lateral pair 2–7 mm, adaxial pair 5–12 mm; anthers oblong to linear-oblong, unequal, 1.5–4 mm, (adaxial pair smaller). **Fruits** erect or ascending (often straight), terete or slightly latiseptate, 5–12 cm × 2.5–3.5 mm; valves each with prominent midvein basally; ovules 52–72 per ovary; style 0–4 mm; stigma strongly 2-lobed (lobes 1–4 mm, opposite valves). **Seeds** 2–3.5 × 1.7–2.2 mm. $2n = 28$.

Flowering (Feb–)Mar–May. Grassland, chaparral, scrub; of conservation concern; 100–1100 m; Calif.

Both R. C. Rollins (1993) and R. E. Buck (1995) treated *Caulanthus lemmonii* as a variety of *C. coulteri*, whereas E. B. Payson (1923) treated the two as independent species. The differences between them clearly justify their separate recognition. In fact, those differences are far greater than those that distinguish the minor color form "*barbarae*" that both Rollins and Buck recognized as a distinct variety of *C. amplexicaulis*. The types of both *C. coulteri* and *C. lemmonii* are quite distinct. The slight intergradation between the two taxa, especially in occurrence of branched trichomes and lobing of cotyledons, most likely resulted from hybridization, but that needs to be verified experimentally and molecularly.

15. Caulanthus major (M. E. Jones) Payson, Ann. Missouri Bot. Gard. 9: 291. 1923 [E]

Caulanthus crassicaulis (Torrey) S. Watson var. *major* M. E. Jones, Proc. Calif. Acad. Sci., ser. 2, 5: 623. 1895; *C. major* var. *nevadensis* Rollins; *Streptanthus major* (M. E. Jones) Jepson

Perennials; glabrous (or petioles and sepals pubescent). **Stems** erect or ascending, unbranched or branched distally, (hollow, sometimes slightly inflated), 2–10 dm. **Basal leaves** rosulate; petiole 0.5–9 cm; blade obovate to oblanceolate or elliptic (in outline), 1–14 cm × 5–25 mm, margins entire, dentate-sinuate, lyrate, or pinnatifid-runcinate. **Cauline leaves** (distalmost) shortly petiolate; blade linear to narrowly oblanceolate, margins entire. **Racemes** (densely flowered), without a terminal cluster of sterile flowers. **Fruiting pedicels** ascending, 1–6 mm. **Flowers:** sepals erect, (creamy white or purple), ovate to lanceolate, 6.5–9.5 × 2.7–4 mm, (equal, pubescent); petals purple, 11–17 mm, blade 4–7 × 1–1.7 mm, not or hardly crisped, claw oblanceolate, 7–13 × 2.5–3.5 mm; filaments tetradynamous, median pairs 5–7 mm, lateral pair 4–6 mm; anthers narrowly oblong, equal, 4–6 mm. **Fruits** erect to ascending, terete or slightly latiseptate, 4.5–12 cm × 2.2–2.8 mm; valves each with obscure midvein; ovules 46–58 per ovary; style obsolete or, rarely, to 0.4 mm; stigma slightly 2-lobed (lobes opposite valves). **Seeds** 2–3.5 × 1.3–1.8 mm. $2n = 28$.

Flowering May–Aug. Margin of montane forests, sagebrush, pinyon-juniper woodland; 1500–3200 m; Calif., Nev., Oreg., Utah.

Caulanthus major is found in eastern and southern California, Nevada, southeastern Oregon, and Utah.

16. Caulanthus pilosus S. Watson, Botany (Fortieth Parallel), 27. 1871 [E]

Streptanthus pilosus (S. Watson) Jepson; *Thelypodium stamineum* Eastwood

Biennials; moderately to densely pilose. **Stems** erect or ascending, unbranched or branched distally, 2–12 dm. **Basal leaves** rosulate; petiole 1–8 cm; blade oblanceolate or oblong (in outline), 2–24 cm × 5–90 mm, margins usually pinnatifid to pinnatisect, rarely dentate-sinuate (lobes dentate). **Cauline leaves** (distalmost) shortly petiolate; blade linear to narrowly oblanceolate, margins entire or dentate. **Racemes** (densely flowered), without a terminal cluster of sterile flowers, (sometimes proximalmost flowers bracteate). **Fruiting pedicels** ascending, 4–18 mm, glabrous or pilose. **Flowers:** sepals erect, (dark purple in bud becoming paler or greenish), narrowly ovate to lanceolate, 4.5–9.5 × 1.5–2 mm, (equal); petals purple, 7–12 mm, blade 3–4 × 1–1.5 mm, crisped, claw oblanceolate to spatulate, 4–9 × 1–2 mm; filaments tetradynamous, median pairs 4.5–10 mm, lateral pair 3–8.5 mm; anthers narrowly oblong, equal, 2–3.5 mm. **Fruits** ascending to divaricate, terete, 2–18 cm × 1–1.5 mm; valves each with obscure midvein; ovules 152–198 per ovary; style obsolete or, rarely, to 1 mm; stigma slightly 2-lobed. **Seeds** 1–2 × 0.7–1 mm.

Flowering late Mar–early Jul. Flats, rocky slopes, scrub and sagebrush communities, pinyon-juniper woodland; 600–2800 m; Calif., Idaho, Nev., Oreg., Utah.

Caulanthus pilosus is found in northeastern California, southwestern Idaho, Nevada, eastern and southern Oregon, and western Utah.

17. Caulanthus simulans Payson, Ann. Missouri Bot. Gard. 9: 295. 1923 [E]

Streptanthus simulans (Payson) Jepson

Annuals; hispid. **Stems** erect, usually branched distally, 1–7 dm. **Basal leaves** rosulate; petiole 0.2–1 cm; blade oblanceolate, 1–7 cm × 4–18 mm, margins coarsely dentate or pinnately lobed. **Cauline leaves** (median) sessile; blade ovate to oblong, 2–8 cm × 5–20 mm, (smaller distally, base amplexicaul to sagittate), margins coarsely dentate or entire. **Racemes** (densely flowered), without a terminal cluster of sterile flowers. **Fruiting pedicels** reflexed, 2–5 mm. **Flowers:** sepals erect, (yellow), lanceolate, 3–6.5 × 1.5–2 mm (equal, keeled); petals creamy white or pale yellow (sometimes with purple midvein), 10–14 mm, blade 4–5 × 1.5–2 mm, not crisped, claw oblanceolate or oblong, 5–9 × 1–1.7 mm; filaments tetradynamous, median pairs 3–5 mm, lateral pair 2–4 mm; anthers oblong, equal, 1–3 mm. **Fruits** usually reflexed, rarely divaricate (often straight), terete, 3–7.5 cm × 1.2–1.5 mm; valves each with obscure midvein; ovules 48–62 per ovary; style 0–3 mm; stigma 2-lobed. **Seeds** 1–2 × 0.9–1.1 mm. $2n = 28$.

Flowering Mar–Jun. Chaparral, scrub, pinyon-juniper woodlands; 400–2100 m; Calif.

Caulanthus simulans is restricted to Riverside and San Diego counties.

83. CHLOROCRAMBE Rydberg, Bull. Torrey Bot. Club 34: 435. 1907 • [Greek *chlor-*, green, and *Crambe*, a genus of Brassicaceae] [E]

Ihsan A. Al-Shehbaz

Perennials; (caudex simple or few-branched, woody, without persistent leaf remains); not scapose; glabrous. **Stems** erect, unbranched or branched (few) distally. **Leaves** basal and cauline; petiolate; basal not rosulate, long-petiolate, (soon withered); cauline petiolate, blade (base hastate, not auriculate), margins entire or proximalmost lobed. **Racemes** (reflexed at anthesis, several-flowered, lax, proximal flowers sometimes bracteate), considerably elongated in fruit. **Fruiting pedicels** straight or curved upward, stout. **Flowers:** sepals ascending, narrowly lanceolate, lateral pair not saccate basally; petals (ascending), white, linear, (only slightly longer than sepals, slightly crisped), claw obovate or oblanceolate, (distinctly wider than blade); stamens subequal (well-exserted beyond petals); filaments slightly dilated basally; anthers linear, (apiculate, coiled after dehiscence); nectar glands confluent, lateral annular or semi-annular, median glands present. **Fruits** long-stipitate, linear, subtorulose, subterete or slightly compressed; valves each with prominent midvein extending full-length; replum rounded; septum complete; ovules 40–60 per ovary; style obsolete or distinct; stigma capitate, entire. **Seeds** uniseriate, plump, winged distally, oblong; seed coat not mucilaginous when wetted; cotyledons obliquely accumbent.

Species 1: w United States.

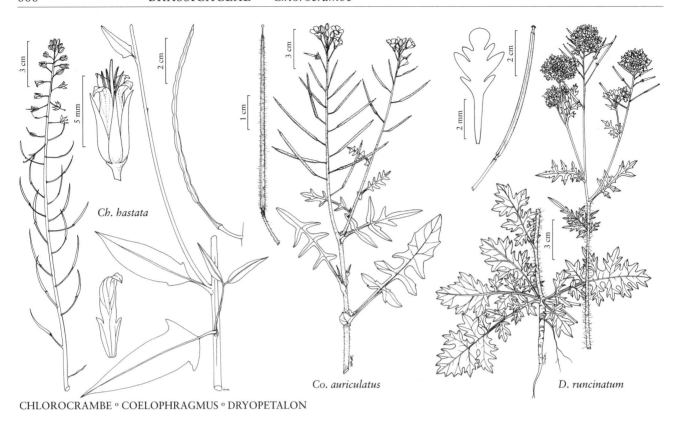

CHLOROCRAMBE ∘ COELOPHRAGMUS ∘ DRYOPETALON

1. **Chlorocrambe hastata** (S. Watson) Rydberg, Bull. Torrey Bot. Club 34: 436. 1907 E F

Caulanthus hastatus S. Watson, Botany (Fortieth Parallel), 28, plate 3. 1871; *Streptanthus hastatus* (S. Watson) M. Peck

Stems 5–15 dm, (stout). **Cauline leaves:** petiole 6–15 cm (shorter distally); blade deltate to lanceolate, lyrate, or sinuately lobed, or (distally) narrowly lanceolate, 5–20 cm × 30–70 mm (smaller distally), margins entire or subapically denticulate, apex acute. **Fruiting pedicels** 5–12 mm, (thicker than gynophore). **Flowers:** sepals yellowish green, 4–6 × 1–1.5 mm; petals white, 5–7 mm, blade 2.5–4 × 0.5–1 mm, claw 2–3.5 × 1–1.7 mm, margins dentate or incised (sublaciniate); filaments 5–9 mm; anthers 2.5–3.5 mm; gynophore (1–)2–8(–10) mm. **Fruits** divaricate, 4–13 cm × 1.5–2.5 mm; style obsolete or, rarely, to 0.5 mm. **Seeds** brown, 3–5 × 1.7–2.2 mm; wing to 0.7 mm wide.

Flowering Jun–Aug. Mountain slopes, canyons, alder thickets, shady damp areas, stony and brushy hillsides; 1500–2800 m; Idaho, Oreg., Utah.

Chlorocrambe hastata is rare and known from only a few counties in Idaho (Washington County), Oregon (Baker and Wallowa counties), and Utah (Cache, Salt Lake, Utah, and Wasatch counties). It is easily distinguished from the other North American species of Brassicaceae by having yellowish green, reflexed flowers in lax racemes, hastate and petiolate cauline leaves, and fruits borne on distinct gynophores.

84. COELOPHRAGMUS O. E. Schulz in H. G. A. Engler, Pflanzenr. 86[IV,105]: 157, fig. 25. 1924 • [Greek *kilos* (Latin *coelus*), hollow, and *phragmos*, partition, alluding to deep pits on sides of fruit septum where seeds are located]

Ihsan A. Al-Shehbaz

Annuals; not scapose; glabrous or pilose. **Stems** erect, unbranched or branched distally. **Leaves** basal and cauline; petiolate; basal not rosulate, blade margins pinnatifid or runcinate, lobes dentate or entire; cauline blade (base auriculate or amplexicaul), margins similar to basal. **Racemes** (corymbose), considerably elongated in fruit. **Fruiting pedicels** horizontal, stout. **Flowers:** sepals erect to ascending, oblong, lateral pair not saccate basally; petals white to lavender, (longer than sepals), spatulate, claw and blade obscurely differentiated; stamens tetradynamous; filaments dilated basally; anthers oblong; nectar glands: lateral annular, median glands confluent with lateral. **Fruits** sessile or shortly stipitate, linear, straight, terete; valves each with distinct midvein; replum rounded; septum complete, (deeply partitioned between seeds); ovules 78–120 per ovary; style distinct; stigma capitate, deeply 2-lobed. **Seeds** uniseriate, plump, not winged, ovoid; seed coat not mucilaginous when wetted; cotyledons incumbent.

Species 1: Texas, n Mexico.

Coelophragmus was described as consisting of two species. The type of one, *C. umbrosus* O. E. Schulz (1924), belongs in *Dryopetalon* and is conspecific with the type of *D. runcinatum* A. Gray (1853). Because removal of that species does not constitute effective lectotypification of the genus name, *C. auriculatus* O. E. Schulz is designated here as type of *Coelophragmus*.

Although *Coelophragmus auriculatus* was originally described as *Sisymbrium auriculatum* and recognized as such by every North American botanist, including E. B. Payson (1922) and R. C. Rollins (1993), the species clearly is not a *Sisymbrium* (S. I. Warwick et al. 2002, 2006b), a genus almost exclusively restricted to the Old World. *Coelophragmus* is recognized herein as a monotypic genus, and its affinities are nearest *Dryopetalon* (I. A. Al-Shehbaz et al. 2006). It is unique among all North American genera of the family by having fruit septums deeply partitioned between seeds.

1. **Coelophragmus auriculatus** (A. Gray) O. E. Schulz in H. G. A. Engler, Pflanzenr. 86[IV,105]: 157. 1924 [F]

Sisymbrium auriculatum A. Gray, Smithsonian Contr. Knowl. 3(5): 8. 1852; *Thelypodium lobatum* Brandegee

Plants glabrous or sparsely to densely pilose, at least proximally. **Stems** (2–)3–12(–15) dm. **Basal leaves** soon withered. **Cauline leaves:** (proximal and median) petiole 1–6 cm, shorter distally; blade oblanceolate to obovate, (2–)3–12(–17) cm × (15–)25–60(–85) mm, (smaller distally), base auriculate to amplexicaul, margins pinnatifid to runcinate, lobes: lateral ovate to lanceolate; terminal larger than lateral. **Racemes** densely flowered. **Fruiting pedicels** straight or slightly curved-ascending, 5–10(–14) mm, glabrous or pubescent. **Flowers:** sepals 4–5 × 1–1.5 mm, sparsely pilose; petals 6–10 × 1.5–2.5 mm, gradually narrowed to clawlike base, margins entire, glabrous; median filament pairs 2.5–4 mm, glabrous; anthers 1–1.5 mm. **Fruits** 2–4.2(–5.3) cm × 0.5–0.7 mm; valves glabrous or pilose; style 0.5–2.5 mm. **Seeds** yellowish, 0.4–0.5 × 0.3–0.4 mm.

Flowering Mar–Nov. Hillsides, shallow washes, calcareous grounds, thickets, scrubby pine and oak woodland, gravelly flats; 1200–2400 m; Tex.; Mexico (Chihuahua, Coahuila, Durango, Nuevo León, San Luis Potosí, Zacatecas).

Collections of *Coelophragmus auriculatus* from Texas are known primarily from Brewster and Hudspeth counties.

85. DRYOPETALON A. Gray, Smithsonian Contr. Knowl. 5(6): 11, plate 11. 1853

• [Greek *drys*, oak, and *petalon*, leaf, alluding to resemblance of petal shape to leaves of some oaks]

Ihsan A. Al-Shehbaz

Rollinsia Al-Shehbaz

Annuals or biennials; not scapose; usually pubescent or hirsute, rarely glabrous. **Stems** erect or ascending, unbranched or branched basally or distally. **Leaves** basal and cauline; petiolate or sessile; basal rosulate or not, petiolate, blade margins entire, dentate to runcinate, or pinnatifid; cauline petiolate or [sub]sessile, blade (base not auriculate, or auriculate to amplexicaul), margins entire or dentate to pinnatifid. **Racemes** (corymbose, initially congested), considerably elongated in fruit. **Fruiting pedicels** ascending, divaricate, or horizontal, slender. **Flowers:** sepals erect to ascending, oblong [ovate], lateral pair slightly saccate or not basally, (glabrous or pubescent); petals white or purplish, spatulate or obovate (longer than sepals), claw gradually narrowed from blade to base, (margins sometimes pinnatifid, or deeply 2-lobed); stamens tetradynamous; filaments not dilated basally, (glabrous or papillate basally); anthers oblong; nectar glands: lateral annular, median glands confluent with lateral. **Fruits** sessile or shortly stipitate, usually linear, rarely linear-oblong, torulose or smooth, terete or flattened (latiseptate); valves each with a distinct midvein, glabrous; replum rounded; septum complete, not veined; ovules 10–110 per ovary; style distinct; stigma capitate, usually entire, rarely slightly 2-lobed. **Seeds** uniseriate, plump, usually not winged, rarely narrowly so, ovate [oblong]; seed coat not mucilaginous when wetted; cotyledons accumbent [incumbent]. x = [10] 12, 14.

Species 8 (2 in the flora): sw United States, n Mexico.

SELECTED REFERENCES Al-Shehbaz, I. A. 2007. Generic limits of *Dryopetalon*, *Rollinsia*, *Sibara*, and *Thelypodiopsis* (Brassicaceae) and a synopsis of *Dryopetalon*. Novon 17: 397–402. Rollins, R. C. 1941c. The cruciferous genus *Dryopetalon*. Contr. Dudley Herb. 3: 199–207.

1. Cauline leaves shortly petiolate or sessile, blade bases not auriculate; fruits 0.5–1.2 mm wide; ovules 60–110 per ovary; seeds not winged; petal margins pinnatifid . . . . . . 1. *Dryopetalon runcinatum*
1. Cauline leaves sessile, blade bases auriculate; fruits 2–3 mm wide; ovules 10–28 per ovary; seeds winged (narrowly); petal margins usually entire, sometimes repand . . . . 2. *Dryopetalon viereckii*

1. Dryopetalon runcinatum A. Gray, Smithsonian Contr. Knowl. 5(6): 12, plate 11. 1853 [F]

Coelophragmus umbrosus (B. L. Robinson) O. E. Schulz; *Dryopetalon runcinatum* var. *laxiflorum* Rollins; *Sisymbrium umbrosum* B. L. Robinson

Annuals or biennials; usually sparsely to densely hirsute proximally, rarely glabrous throughout. **Stems** erect, unbranched or branched basally, branched distally, 2–8 dm. **Basal leaves:** petiole (1–)2–6(–8) cm; blade spatulate to obovate in outline, lyrate, pinnatifid, or bipinnately lobed, (2–)4–20(–25) cm × 30–80 mm, surfaces glabrous or hirsute. **Cauline leaves** shortly petiolate or sessile; blade base not auriculate, similar to basal, smaller distally. **Racemes** lax or dense. **Fruiting pedicels** divaricate-ascending, 5–15(–20) mm, glabrous or pubescent. **Flowers:** sepals (2.5–)3.5–4.5(–5) × 1.5–2 mm; petals white to purplish, spatulate, 6–9(–11) × 2.5–3.5(–4.5) mm, claw 2.5–4(–5) mm, margins pinnatifid, 5–7(–11)-lobed, papillate basally; filaments papillate basally, median pairs 3–5 mm, lateral pair 2.5–4 mm; anthers oblong, 0.7–1 mm. **Fruits** linear, (straight or arcuate), 2–6 cm × 0.5–1.2 mm; ovules 60–110 per ovary; style 0.1–0.7(–1) mm. **Seeds** 0.6–0.8 × 0.4–0.6 mm, not winged. $2n$ = 24.

Flowering Jan–May. Ledges, shade of boulders and cliffs, foothills, canyons, scrub woodlands, streambeds, rocky basalt, crevices; 800–1900 m; Ariz., N.Mex., Tex.; Mexico (Chihuahua, Sinaloa, Sonora).

Dryopetalon runcinatum is easily distinguished from other mustards in the flora area by having divided petals.

In the United States, it is restricted to Cochise, Pima, Pinal, and Santa Cruz counties in Arizona, Catron, Dona Ana, Hidalgo, and Luna counties in New Mexico, and El Paso County in Texas.

2. **Dryopetalon viereckii** (O. E. Schulz) Al-Shehbaz, Novon 17: 401. 2007

Arabis viereckii O. E. Schulz, Notizbl. Bot. Gart. Berlin-Dahlem 11: 389. 1932; *A. endlichii* O. E. Schulz; *Sibara runcinata* Rollins; *S. runcinata* var. *brachycarpa* Rollins; *S. viereckii* (O. E. Schulz) Rollins; *S. viereckii* var. *endlichii* (O. E. Schulz) Rollins

Annuals; sparsely to densely hirsute proximally, sparsely pubescent distally. **Stems** erect or ascending, usually branched at or slightly distal to base, branched distally, rarely unbranched, (0.8–)2–5.5(–7.5) dm. **Basal leaves**: petiole (0.5–)1–5 cm; blade spatulate to obovate in outline, pinnatifid-runcinate, (1.5–)2.5–10(–15) cm × (10–)15–35(–52) mm, surfaces hirsute. **Cauline leaves** (median) sessile; blade base auriculate to amplexicaul, margins less divided than basal, dentate or entire distally. **Racemes** dense. **Fruiting pedicels** divaricate-ascending to horizontal, (3–)5–10(–14) mm (straight or slightly curved distally or proximally), sparsely pubescent. **Flowers**: sepals (1.7–)2–3 × 0.7–1 mm; petals white to lavender (sometimes with purplish veins), spatulate, (3–)4–5.5(–7) × 1.5–2.5(–3) mm, claw 1–1.5 mm, margins usually entire, sometimes repand, usually glabrous, rarely minutely papillate basally; filaments: median pairs 1.4–4 mm, lateral pair 1–2.5 mm, (rarely minutely papillate basally); anthers ovate, 0.5–0.7 mm. **Fruits** linear to linear-oblong, (1.5–)2–4(–4.7) cm × 2–3 mm; ovules 10–28 per ovary; style (0.5–)1.5–4 mm. **Seeds** 2–3 × 1.2–2 mm, narrowly winged. $2n = 28$.

Flowering Aug–Apr. Roadsides, thickets, limestone ridges, steep slopes, shady cliffs, arroyos, gravelly hillsides, canyon walls, open woods, desert shrubs; 200–2200 m; Tex.; Mexico (Coahuila, Hidalgo, Nuevo León, Puebla, San Luis Potosí, Zacatecas).

In the flora area, *Dryopetalon viereckii* is known from Cameron, Hidalgo, Jim Hogg, Kinney, La Salle, McMullen, Starr, Webb, and Zapata counties. It is highly variable in its leaf, fruit, style, pedicel, and petal lengths, density of indumentum, development of auricles, and division of basal leaves. This variation, which is continuous, has been the basis for dividing the species into varieties.

86. **HESPERIDANTHUS** (B. L. Robinson) Rydberg, Bull. Torrey Bot. Club 34: 433. 1907 • [Genus *Hesperis* and Greek *anthos*, flower, alluding to resemblance of flowers]

Ihsan A. Al-Shehbaz

Thelypodium Endlicher subg. *Hesperidanthus* B. L. Robinson in A. Gray et al., Syn. Fl. N. Amer. 1(1,1): 174. 1895; *Caulostramina* Rollins; *Glaucocarpum* Rollins

Perennials or subshrubs; (caudex well-developed, woody); not scapose; (glaucous), glabrous. **Stems** (simple or few to several from base), erect or ascending, usually branched distally. **Leaves** cauline; not rosulate, petiolate, subsessile, or sessile, blade (somewhat fleshy, becoming leathery when dry, base not auriculate), margins entire, denticulate, or coarsely dentate. **Racemes** (corymbose, few- to several-flowered, sometimes bracteate proximally), elongated in fruit. **Fruiting pedicels** divaricate to ascending or suberect, slender. **Flowers**: sepals ascending to erect, oblong, lateral pair usually not saccate basally (slightly saccate in *H. linearifolius*, apex of median pair cucullate or not); petals white, lilac, lavender, purple, or yellow (sometimes with darker veins), spatulate, claw differentiated or not from blade; stamens tetradynamous or subequal; filaments not dilated basally; anthers oblong to linear, (apex obtuse or apiculate); nectar glands: lateral annular or lunar, median glands often confluent with lateral. **Fruits** sessile or stipitate, linear, not torulose, terete or slightly latiseptate; valves each with prominent midvein; replum rounded; septum complete; ovules 8–110 per ovary; style distinct; stigma capitate or conical, entire or 2-lobed (lobes prominent, connivent). **Seeds** uniseriate, plump, not winged, oblong; seed coat not or slightly mucilaginous when wetted; cotyledons incumbent or obliquely so. $x = 11$.

Species 5 (5 in the flora): w United States, n Mexico.

For a detailed account on the generic limits and affinities of *Hesperidanthus*, see I. A. Al-Shehbaz (2005). As delimited here, the genus includes what R. C. Rollins (1993) assigned to *Schoenocrambe* Greene (minus the type), *Caulostramina*, and *Glaucocarpum*. One species is endemic to Inyo County, California, three are highly localized endemics in Utah, and the fifth is widespread in the southwestern United States and northern Mexico.

SELECTED REFERENCES Al-Shehbaz, I. A. 2005. *Hesperidanthus* (Brassicaceae) revisited. Harvard Pap. Bot. 10: 47–51. Rollins, R. C. 1938. *Glaucocarpum*, a new genus in the Cruciferae. Madroño 4: 232–235. Rollins, R. C. 1982b. *Thelypodiopsis* and *Schoenocrambe* (Cruciferae). Contr. Gray Herb. 212: 71–102.

1. Petals yellow; fruits 1–2 cm; ovules 8–16 per ovary.................... 5. *Hesperidanthus suffrutescens*
1. Petals purple, lilac, lavender, or white; fruits (1.8–)2.5–11 cm; ovules 26–110 per ovary.
   2. Median sepals cucullate; stigmas conical, lobes connivent; petal veins not darker than blade; ovules (76–)80–110 per ovary; Arizona, Colorado, New Mexico, Texas ....1. *Hesperidanthus linearifolius*
   2. Median sepals not cucullate; stigmas flat, lobes entire or obscure; petal veins darker than blade; ovules 26–62 per ovary; California, Utah.
      3. Leaves petiolate, blades ovate to broadly so, 10–35 mm wide, margins coarsely and irregularly dentate; California.................................4. *Hesperidanthus jaegeri*
      3. Leaves sessile, subsessile, or petiolate, blades linear, linear-lanceolate, oblong, elliptic, or oblanceolate, (0.8–)1–24 mm wide, margins entire or obscurely denticulate; Utah.
         4. Leaves sessile or subsessile, blades linear or linear-lanceolate; sepals 4.2–6.5 mm; anthers oblong, 1–1.5 mm............................2. *Hesperidanthus argillaceus*
         4. Leaves petiolate, blades oblong, elliptic, or oblanceolate; sepals 5–8 mm; anthers linear, 2.5–3 mm..................................3. *Hesperidanthus barnebyi*

1. **Hesperidanthus linearifolius** (A. Gray) Rydberg, Bull. Torrey Bot. Club 34: 434. 1907 F

Streptanthus linearifolius A. Gray, Mem. Amer. Acad. Arts, n. s. 4: 7. 1849; *Pachypodium linearifolium* (A. Gray) A. Gray; *Schoenocrambe linearifolia* (A. Gray) Rollins; *Sisymbrium linearifolium* (A. Gray) Payson; *S. stenophyllus* Rollins; *Thelypodiopsis linearifolia* (A. Gray) Al-Shehbaz; *Thelypodium linearifolium* (A. Gray) S. Watson

Perennials. Stems simple or few from caudex, erect, (often much-branched distally), (2.5–)3.5–10(–15) dm. **Leaves** shortly petiolate; blade usually linear to linear-lanceolate, rarely lanceolate, (2.5–)3.5–12(–15) cm × (1–)2–6(–10) mm (smaller distally), base cuneate to attenuate (or petiolelike), margins entire or, very rarely, sparsely denticulate, apex acuminate to acute. **Racemes** several-flowered, (proximalmost flowers sometimes bracteate). **Fruiting pedicels** divaricate or ascending, straight, (5–)8–20(–25) mm. **Flowers:** sepals purplish, (4–)4.5–7 × 1–2 mm, (median pair apical cuculla to 0.7 mm); petals purple, (10–)12–16(–18) × (2–)3.5–5.5 mm, claw distinctly differentiated from blade, [5–7(–9) mm]; filaments 2–4 mm; anthers narrowly oblong to linear, 2–4 mm; gynophore 0.4–1(–2) mm. **Fruits** usually straight, rarely curved, terete, (3.5–)4–9(–11) cm × 1–1.5 mm; ovules (76–)80–110 per ovary; style subclavate to cylindrical, 0.5–1.8(–2.5) mm; stigma conical, 2-lobed (lobes prominent, connivent). **Seeds** 0.9–1.6 × 0.5–0.9 mm. $2n = 22, 44$.

Flowering Jul–Nov. Open woods, dry hillsides, oak woodland, mixed conifer forests, arroyos, canyons, rocky ridges, limestone ledges, sandstone crevices, roadsides; 700–3100 m; Ariz., Colo., N.Mex., Tex.; Mexico (Chihuahua, Coahuila, Durango, Nuevo León, San Luis Potosí, Sonora, Zacatecas).

2. **Hesperidanthus argillaceus** (S. L. Welsh & N. D. Atwood) Al-Shehbaz, Harvard Pap. Bot. 10: 50. 2005 C E

Thelypodiopsis argillacea S. L. Welsh & N. D. Atwood, Great Basin Naturalist 37: 95, fig. 1. 1977; *Schoenocrambe argillacea* (S. L. Welsh & N. D. Atwood) Rollins

Perennials. Stems simple or few from caudex, erect, (few-branched distally), 1.3–3 dm. **Leaves** subsessile; blade linear or linear-lanceolate, 0.9–4 cm × (0.8–)1–3(–6) mm, base cuneate to attenuate, margins entire, apex acute. **Racemes** 5+-flowered.

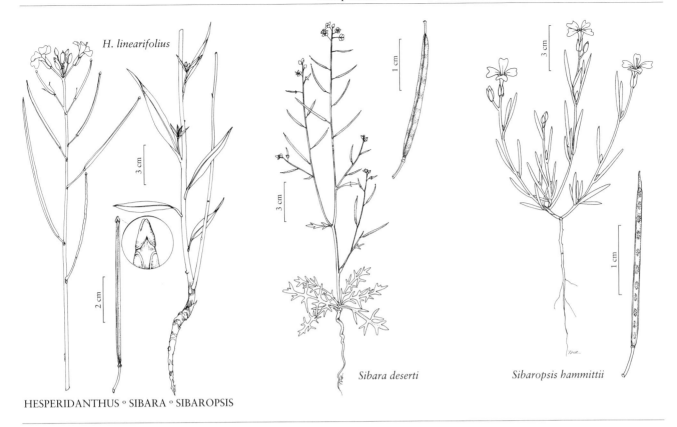

Sibara deserti

Sibaropsis hammittii

HESPERIDANTHUS ° SIBARA ° SIBAROPSIS

Fruiting pedicels suberect, ascending, or divaricate, straight or curved upward, 7–18 mm. Flowers: sepals purple, 4.2–6.5 × ca. 1.5 mm; petals white or lilac (with darker purple veins), 8–11 × 2.5–4 mm, claw undifferentiated from blade; filaments 1.5–2.5 mm; anthers oblong, 1–1.5 mm; gynophore 0–0.3 mm. Fruits curved, terete, 1.8–5.5 cm × 1–1.2 mm; ovules 54–62 per ovary; style subclavate, 0.5–2 mm; stigma flat, obscurely 2-lobed. Seeds 1.5–1.8 × 0.7–0.9 mm.

Flowering Apr–Jun. Desert shrub or sagebrush communities, rocky slopes, shale barrens; of conservation concern; 1400–1800 m; Utah.

Hesperidanthus argillaceus is restricted to the Uinta and upper Green River shale formation in Uintah County.

3. **Hesperidanthus barnebyi** (S. L. Welsh & N. D. Atwood) Al-Shehbaz, Harvard Pap. Bot. 10: 50. 2005 [C][E]

Thelypodiopsis barnebyi S. L. Welsh & N. D. Atwood, Brittonia 33: 300, fig. 6. 1981; *Schoenocrambe barnebyi* (S. L. Welsh & N. D. Atwood) Rollins

Perennials. Stems few to several from caudex, erect, 1–3.5 (–3.8) dm. Leaves subsessile or petiolate, (0.4–1 cm); blade oblong or elliptic to oblanceolate, 1.5–5 cm × 4–24 mm, base cuneate, margins entire or obscurely denticulate, apex obtuse to rounded. Racemes usually 2–8-flowered (rarely more). Fruiting pedicels ascending to divaricate, straight, 10–27 mm. Flowers: sepals green to purple, 5–8 × 2–3 mm; petals white or lilac (with darker purple veins), 9.5–12 × 2.5–3.5 mm, claw distinctly differentiated from blade, (narrower); filaments 2.5–3 mm; anthers linear, 2.5–3 mm; gynophore 0–1.5 mm. Fruits straight or curved, terete, 3.5–5.5 cm × 1–2 mm; ovules 24–42 per ovary; style subclavate, 1–2 mm; stigma flat, obscurely 2-lobed. Seeds 1.8–2.2 × 0.9–1.2 mm.

Flowering Apr–Jun. Mixed desert communities; of conservation concern; 1700–2000 m; Utah.

The species is restricted to the Chinle Formation in Emery and Wayne counties.

Hesperidanthus barnebyi, as *Schoenocrambe barnebyi*, is in the Center for Plant Conservation's National Collection of Endangered Plants.

4. **Hesperidanthus jaegeri** (Rollins) Al-Shehbaz, Harvard Pap. Bot. 10: 50. 2005 C E

Thelypodium jaegeri Rollins, Contr. Dudley Herb. 3: 174, plate 46, fig. 3. 1941; *Caulostramina jaegeri* (Rollins) Rollins

Perennials or subshrubs. Stems several from caudex, erect or ascending, (stiff, usually gyrose), 1–3 dm. **Leaves** (3–7 per stem); petiolate (1–2 cm); blade ovate to broadly so, (1–)1.5–4 cm × 10–35 mm, base obtuse, subtruncate, or cuneate, margins usually coarsely and irregularly dentate, rarely (distalmost) repand, apex obtuse or subacute. **Racemes** few-flowered, (lax in fruit). **Fruiting pedicels** divaricate or ascending, straight or slightly curved, 6–14 mm. **Flowers:** sepals purplish, 5–7 × 2–3 mm; petals white to lavender or purplish (with darker veins), 9–14 × 2.5–4 mm, attenuate to clawlike base; filaments 2–3 mm; anthers linear, 2–2.5 mm; gynophore obsolete to 0.5 mm. **Fruits** straight to strongly recurved, terete, (2–)3–5 cm × 1–1.2 mm; ovules 26–42 per ovary; style 0.7–1.5 mm; stigma flat, entire. **Seeds** 1.2–1.5 × 0.6–0.9 mm.

Flowering Apr–Jun. Rocky crevices, cliffs, limestone clefts; of conservation concern; 1500–2800 m; Calif.

Hesperidanthus jaegeri has been collected from Marble and Teufel canyons and Cerro Gordo Peak of the Inyo Mountains, Inyo County.

SELECTED REFERENCE Rollins, R. C. 1973. A reconsideration of *Thelypodium jaegeri*. Contr. Gray Herb. 204: 155–157.

5. **Hesperidanthus suffrutescens** (Rollins) Al-Shehbaz, Harvard Pap. Bot. 10: 50. 2005 C E

Thelypodium suffrutescens Rollins, Ann. Carnegie Mus. 26: 224. 1937; *Glaucocarpum suffrutescens* (Rollins) Rollins; *Schoenocrambe suffrutescens* (Rollins) S. L. Welsh & Chatterley

Perennials; (suffrutescent, woody proximally). **Stems** several from caudex, erect to ascending, (few-branched distally), 1–3.5 dm. **Cauline leaves** sessile or petiolate, (to 11 mm); blade oblanceolate to lanceolate or elliptic, (0.7–)1–2.5 cm × 3–10 mm, base cuneate to attenuate, margins entire or, rarely, obscurely denticulate, apex acute or obtuse. **Racemes** 5+-flowered, (sometimes 1 or 2 flowers bracteate). **Fruiting pedicels** ascending, straight or curved upward, 3–10(–12) mm. **Flowers:** sepals yellowish green, 4–6 × 1–1.5 mm; petals yellow, 7–11 × 1.5–2.5 mm, claw undifferentiated from blade; filaments 2–3 mm; anthers oblong, 0.7–0.9 mm; gynophore 0.4–0.8 mm. **Fruits** straight, latiseptate, 1–2 cm × 1.2–2.5 mm; ovules 8–16 per ovary; style cylindrical, (0.5–)1–2 mm; stigma flat, entire. **Seeds** 1.5–1.9 × 0.9–1.3 mm.

Flowering May–Jul. Mixed desert shrubs, shale barrens, decomposed chip-rock, limy shale; of conservation concern; 1500–2100 m; Utah.

Hesperidanthus suffrutescens is restricted to the Green River Formation in Duchesne and Uintah counties.

87. SIBARA Greene, Pittonia 3: 10. 1896 • [Anagram of generic name *Arabis*]

Ihsan A. Al-Shehbaz

Annuals; not scapose; (usually glaucous), pubescent or glabrous, trichomes simple, 2-forked, dendritic. **Stems** erect, unbranched or branched distally. **Leaves** basal and cauline; petiolate; basal rosulate or not, blade margins pectinate or pinnatisect; cauline blade (base not auriculate), margins pectinate or pinnatisect, (terminal lobes filiform and semiterete, or linear and flat). **Racemes** (several-flowered), considerably elongated in fruit. **Fruiting pedicels** divaricate-ascending, divaricate, horizontal, or reflexed, slender [stout]. **Flowers:** sepals erect, oblong [ovate], lateral pair not saccate basally (usually glabrous, rarely pubescent); petals white, purple, or lavender, spatulate [oblanceolate] (longer than sepals), claw differentiated from blade, (often oblanceolate, equal to or longer than sepals, apex obtuse to emarginate); stamens slightly tetradynamous; filaments not dilated basally; anthers ovate or oblong; nectar glands lateral, semi-annular or annular, median glands absent. **Fruits** sessile or shortly stipitate, linear, smooth or torulose, latiseptate [terete]; valves each with prominent or obscure midvein, pubescent or glabrous; replum rounded; septum complete; ovules [14–]16–40[–94] per ovary; style distinct; stigma capitate, entire. **Seeds** uniseriate or biseriate, (yellowish brown), flattened or plump, not winged, oblong; seed coat not mucilaginous when wetted; cotyledons accumbent [incumbent]. $x = 13, 14$.

Species 6 (2 in the flora): sw United States, n Mexico.

As recognized by R. C. Rollins (1947, 1993), the limits of *Sibara* are highly artificial. One of the ten species that he recognized, *S. virginica*, is here placed in the monotypic genus *Planodes* of the tribe Cardamineae. Another, *S. viereckii*, is assigned to *Dryopetalon*, and *S. grisea* is reduced to synonymy under *Thelypodium texanum*. The four Mexican species recognized in the genus are *S. angelorum* (S. Watson) Greene (Baja California, Sonora), *S. brandegeeana* (Rose) Greene (Baja California), *S. laxa* (S. Watson) Greene (Baja California), and *S. mexicana* (S. Watson) Rollins (Guanajuato).

SELECTED REFERENCE Rollins, R. C. 1947. Generic revisions in the Cruciferae: *Sibara*. Contr. Gray Herb. 165: 133–143.

1. Plants pubescent; petals 2–3.5 × 0.5–1 mm; fruits often curved, (0.8–)1.2–2.5(–3.2) cm × 1.2–1.5 mm; ovules 16–24 per ovary; deserts of e California, s Nevada . . . . 1. *Sibara deserti*
1. Plants glabrous; petals 3.5–6 × 2–3 mm; fruits straight, 2.5–4.1 cm × 0.7–0.9 mm; ovules 32–40 per ovary; Santa Cruz Island, California . . . . 2. *Sibara filifolia*

1. Sibara deserti (M. E. Jones) Rollins, Contr. Gray Herb. 165: 140. 1947 E F

Thelypodium deserti M. E. Jones, Contr. W. Bot. 12: 1. 1908; *Arabis deserti* (M. E. Jones) Abrams; *Sibara rosulata* Rollins

Plants sparsely to moderately puberulent, trichomes 0.05–0.3 mm. **Stems** often branched distally, 1–3.5(–4.5) dm, puberulent or glabrous. **Basal leaves** sometimes withered by flowering; rosulate or not. **Cauline leaves**: petiole 5–1.7 cm; blade (distal linear), margins pinnatisect: lobes linear to linear-lanceolate, 0.2–2 cm × 0.3–3 mm, margins entire, (surfaces glabrous or puberulent). **Racemes** (corymbose), rachis straight. **Fruiting pedicels** divaricate or horizontal to reflexed, straight, recurved, or curved upward, (2–)3–5.5(–10) mm, glabrous or puberulent. **Flowers**: sepals 1.5–2 × 0.5–0.7 mm, puberulent; petals white, 2–3.5 × 0.5–1 mm, claw ca. 1.5 mm; median filaments 1.2–1.5 mm; anthers ovate, 0.3–0.4 mm. **Fruits** divaricate-ascending to spreading or reflexed, often curved, smooth, (0.8–)1.2–2.5(–3.2) cm × 1.2–1.5 mm; valves puberulent or glabrous; ovules 16–24 per ovary; style 1–2.5(–3) mm. **Seeds** 0.9–1.4 × 0.6–0.9 mm. $2n = 26, 28$.

Flowering Mar–Apr. Steep talus or rocky slopes, dry creosote bush flats, deserts, rocky bluffs, rock detritus, calcareous debris, exposed crevices, steep banks; 150–1200 m; Calif., Nev.

California collections examined are from Inyo County, and the only Nevada material I have studied is the type collection, from Amargosa Desert in southern Nye County.

R. C. Rollins (1947, 1993) separated *Sibara deserti* from *S. rosulata* primarily by having pendent to reflexed fruits less than 1–1.5 cm (versus divaricate-ascending fruits 1.5–3 cm), slender (versus club-shaped) styles, and presence (versus absence) of basal rosettes. These characters show continuous variation from one extreme to the other and are found in various combinations. The length and orientation of fruits can vary substantially in a single population or even on the same plant. For example, in *Rollins & Munz 6743* and *Rollins & Munz 6744* (both at GH and collected from the same general area) the fruits are ascending, divaricate, or reflexed and are 1.5–2.5 cm. Styles can be slender to stout, cylindrical to slightly club-shaped, and 1–3 mm; their considerable variation is independent of fruit length and orientation. *Sibara deserti* and *S. rosulata* are identical in leaf and trichome morphology, flower color and size, seed size, the overall range of variation in fruit size and length, and the orientation of fruiting pedicels. The complex represents a single species easily distinguished from the remaining North American members of *Sibara* by having minute, dendritic to forked (instead of simple) or no trichomes.

2. Sibara filifolia (Greene) Greene, Pittonia 3: 11. 1896 C E

Cardamine filifolia Greene, Pittonia 1: 30. 1887; *Arabis filifolia* (Greene) Greene

Plants (glaucous), glabrous throughout. **Stems** unbranched or branched (few) distally, 1.5–3 dm. **Basal leaves** (not seen), withered by flowering. **Cauline leaves:** petiole 0.4–1.6 cm; blade margins pectinate or pinnatisect, lobes filiform to linear, 0.5–1.5 cm × 0.2–0.8 mm, margins entire. **Racemes** (lax); rachis flexuous. **Fruiting pedicels** divaricate-ascending, straight, (2–)3–10(–15) mm. **Flowers:** sepals 2.2–3 × 0.7–1 mm; petals purple or lavender, 3.5–6 × 2–3 mm, claw ca. 1.5 mm; median filaments 2–2.7 mm; anthers oblong, 0.6–0.8 mm. **Fruits** divaricate-ascending, straight, subtorulose, 2.5–4.1 cm × 0.7–0.9 mm; ovules 32–40 per ovary; style 0.5–0.8 mm. **Seeds** 1–1.3 × 0.6–0.9 mm.

Flowering Apr. Dry ridges; of conservation concern; ca. 0 m; Calif.

Sibara filifolia is known from Santa Cruz Island. It is in the Center for Plant Conservation's National Collection of Endangered Plants.

88. SIBAROPSIS S. Boyd & T. S. Ross, Madroño 44: 30, figs. 2–4. 1997 • [Genus *Sibara* and Greek *opsis*, appearance] C E

Steve Boyd

Annuals; not scapose; (often glaucous), usually glabrous, sometimes glabrate, trichomes minute, (proximalmost leaves with evanescent cilia). **Stems** erect, often branched basally. **Leaves** basal and cauline; sessile; blade (base not auriculate), not rosulate, margins entire. **Racemes** (corymbose, several-flowered, lax), considerably elongated in fruit. **Fruiting pedicels** usually ascending, rarely straight, slender. **Flowers:** sepals erect, lanceolate- to ovate-oblong, (subequal), lateral pair obscurely subsaccate basally; petals light purple- or pink-lavender (with darker purplish veins, adaxial pair slightly larger), spatulate, claw well-differentiated from blade, (apex slightly emarginate to obcordate); stamens in 3 unequal pairs, (adaxial pair sterile); filaments not dilated basally, (adaxial pair ± connate); anthers ovate-oblong, (not apiculate); nectar glands lateral, (minute), median glands absent. **Fruits** tardily dehiscent, sessile, linear, smooth, slightly latiseptate; valves each with obscure midvein, glabrate, (margins minutely scabrous); replum rounded; septum complete; ovules 24–44 per ovary; style distinct; stigma subentire. **Seeds** uniseriate, flattened, obscurely winged distally, oblong; seed coat not mucilaginous when wetted; cotyledons incumbent, (linear).

Species 1: California.

Sibaropsis is unusual in Brassicaceae in that the inflorescence axis disarticulates distal to each pedicel and subtending axis internode, thus fruits are dispersed as individual units, except that the proximalmost fruits remain persistent.

1. Sibaropsis hammittii S. Boyd & T. S. Ross, Madroño 44: 30, figs. 2–4. 1997 C E F

Stems 0.5–2 dm, unbranched or branched (several, ascending) near base. **Basal leaves** soon withered. **Cauline leaves:** blade somewhat fleshy, narrowly linear, (1–)1.5–3(–4.5) cm × 0.5–1 mm. **Fruiting pedicels** 2.5–4 mm. **Flowers:** sepals 2.8–3.2 × 0.5–1 mm; petals 8.5–10 × 2–2.5 mm, margins not crisped, claw attenuate to base, 5–6 mm, longer than blade; filaments: abaxial and lateral pairs distinct, shorter, adaxial pair ± connate, 4.5–5 mm; anthers ca. 0.8 mm. **Fruits** erect, (1.5–)2–2.5 cm × 0.7–0.9 mm; style (1.5–)3–4.5 mm. **Seeds** reddish to dark olive-brown, 1–1.3 × 0.5–0.6 mm. $2n = 28$.

Flowering Mar–Apr. Patches of open, relatively moist, heavy clay soil dominated by native grasses, geophytes, and annuals; of conservation concern; 700–1100 m; Calif.

Sibaropsis hammittii is known from two areas, separated by about 120 km, in the Peninsular Ranges of southern California: the Santa Ana Mountains in Riverside County, and the Viejas and Poser mountains in San Diego County.

89. **STANLEYA** Nuttall, Gen. N. Amer. Pl. 2: 71. 1818 • Prince's plume [For Edward Smith Stanley, 1775–1851, British statesman and ornithologist] [E]

Ihsan A. Al-Shehbaz

Annuals, perennials, shrubs, or subshrubs; (base usually woody); not scapose; glabrous or pubescent. **Stems** usually erect, rarely ascending, unbranched or branched. **Leaves** cauline and, sometimes, basal; petiolate or sessile; basal rosulate, petiolate, blade margins entire, lyrately lobed or 1- or 2-pinnatifid; cauline blade (base sometimes auriculate or amplexicaul), margins entire or dentate to pinnatifid. **Racemes** considerably elongated in fruit. **Fruiting pedicels** horizontal, divaricate, or divaricate-ascending, slender. **Flowers**: sepals spreading to reflexed, oblong-linear or linear, lateral pair not saccate basally; petals usually yellow or whitish, rarely white or yellow-orange, obovate, orbicular, oblong, linear, filiform, or oblanceolate, claw distinctly differentiated from blade (claw glabrous or papillose); stamens (exserted), equal; filaments not dilated basally, (often papillose basally); anthers linear, (strongly spirally coiled after dehiscence); lateral nectar glands annular, median present or absent, confluent with lateral ones. **Fruits** long-stipitate, linear, often torulose, terete or latiseptate; valves each with prominent midvein, glabrous; replum rounded; septum complete; ovules 22–70 per ovary; style obsolete or distinct (to 1.7 mm); stigma capitate, entire. **Seeds** uniseriate, plump, not winged, usually oblong, rarely ovoid; seed coat (obscurely reticulate), slightly mucilaginous when wetted; cotyledons accumbent to incumbent. $x = 14$.

Species 7 (7 in the flora): w, c United States.

Both R. C. Rollins (1993) and N. H. Holmgren (2005b) reported $n = 12$ and $2n = 24$ for various species of *Stanleya*. However, those counts, all reported previously by Rollins (1939c), are erroneous; no species of the genus has numbers deviating from $n = 14$ or 28.

All species of *Stanleya* are well-defined, and interspecific hybridization has not yet been reported. One species, *S. pinnata*, is a hyperaccumulator of selenium and is a good indicator for the presence of this element in soils. Poisoning of livestock results from their feeding on large quantities of plants of this species.

SELECTED REFERENCE Rollins, R. C. 1939c. The cruciferous genus *Stanleya*. Lloydia 2: 109–127.

1. Cauline leaves sessile, blade bases auriculate to sagittate.
    2. Annuals or biennials (without caudex); racemes dense; sepals 6–12 mm; petals linear to filiform, 0.5–1.5 mm wide, margins crisped; fruiting pedicels 10–20(–26) mm . . . 3. *Stanleya confertiflora*
    2. Perennials (with caudex); racemes loose; sepals 12–18 mm; petals narrowly oblanceolate, 1–3 mm wide, margins usually erose, rarely subentire and crisped; fruiting pedicels 4–9(–12) mm. . . . . 7. *Stanleya viridiflora*
1. Cauline leaves petiolate, blade bases not auriculate or sagittate.
    3. Basal leaf blades: surfaces densely tomentose; fruiting pedicels 11–22 mm; petals with glabrous claws; fruits flattened . . . . 6. *Stanleya tomentosa*
    3. Basal leaf blades: surfaces usually glabrous, rarely sparsely pubescent; fruiting pedicels 3–11(–15) mm; petals with pubescent claws (except *S. elata*); fruits terete or subterete.
        4. Cauline leaf blades: margins usually entire, rarely dentate proximally.
            5. Petals linear, 0.3–1 mm wide, claws glabrous; ovules 46–70 per ovary; filaments 5–13 mm; Arizona, California, s, w Nevada . . . . 4. *Stanleya elata*
            5. Petals oblanceolate to oblong, 2–3 mm wide, claws pubescent; ovules 10–38 per ovary; filaments 11–28 mm; Colorado, Kansas, Nevada, sw Texas, Utah, Wyoming . . . . 5. *Stanleya pinnata* (in part)
        4. Cauline leaf blades: margins often pinnatisect, pinnatifid, 2-pinnatifid, lyrate-pinnatifid, or runcinate.

[6. Shifted to left margin.—Ed.]

6. Biennials; petals orbicular to broadly obovate, (2.5–)3–6 mm wide; fruits suberect to ascending, slightly curved inward . . . . 1. *Stanleya albescens*
6. Perennials; petals oblanceolate or oblong, 0.8–3 mm wide; fruits usually spreading or divaricate, rarely ascending, sometimes curved downward.
    7. Cauline leaf blades: margins sometimes 2-pinnatifid; sepals 6.5–10 mm; petals 5–12 mm; filaments glabrous basally; gynophores 4–11 mm; fruits torulose, tortuous. . . . 2. *Stanleya bipinnata*
    7. Cauline leaf blades: margins not 2-pinnatifid; sepals 9–16 mm; petals 10–20 mm; filaments pilose basally; gynophores 7–28 mm; fruits smooth, not tortuous . . . .
    . . . . 5. *Stanleya pinnata* (in part)

1. **Stanleya albescens** M. E. Jones, Zoë 2: 17. 1891 [E]

Biennials; (glaucous), mostly glabrous. **Stems** unbranched or branched (few) proximally and distally, 2–8(–10) dm. **Basal leaves**: petiole 1–6.5 cm; blade (fleshy), broadly lanceolate to oblanceolate or ovate in outline, (1.6–)3–17(–20) cm × (10–)20–80 mm, margins lyrate-pinnatifid or runcinate. **Cauline leaves** petiolate; blade similar to basal, smaller distally, margins sometimes entire. **Racemes** somewhat dense. **Fruiting pedicels** horizontal to divaricate-ascending, (straight), (4–)5–11(–15) mm. **Flowers**: sepals oblong-linear, (9–)11–18 mm; petals pale yellow to whitish, orbicular to broadly obovate, 12–18 × (2.5–)3–6 mm, claw 5–8 mm, wider at base, pubescent inside apically; filaments 1–2.2 mm, base pilose; anthers 3–5.5 mm; gynophore 10–22(–26) mm, hirsute at base. **Fruits** suberect to ascending, slightly curved inward, subterete, 2.3–6 cm × 1–2 mm; ovules 24–44 per ovary; style 0.2–0.5 mm. **Seeds** ovoid to oblong, 1.4–2 × 0.8–0.9 mm.

Flowering May–Jun. Barren clay slopes and flats, open clay knolls, gumbo clay bluffs, Mancos shale; 1300–2100 m; Ariz., Colo., N.Mex.

2. **Stanleya bipinnata** Greene, Erythea 4: 173. 1896 [E]

Stanleya pinnata (Pursh) Britton var. *bipinnata* (Greene) Rollins; *S. pinnata* var. *gibberosa* Rollins

Perennials; (base sometimes woody); (glaucous), pubescent or glabrous. **Stems** erect to ascending, usually unbranched, rarely branched (few) proximally, 1.5–4.5 dm, (sparsely pubescent). **Basal leaves** (withered by flowering), similar to cauline. **Cauline leaves** petiole 1.5–3 cm; blade (fleshy), lanceolate to ovate in outline, 4–7.5(–9.5) cm (smaller distally), margins (proximalmost) often 2-pinnatifid, or (distal) pinnatifid or pinnatisect, (surfaces sparsely pubescent, trichomes crisped). **Racemes** dense. **Fruiting pedicels** horizontal to divaricate-ascending, 5–10 mm, (sparsely pubescent). **Flowers**: sepals linear, 6.5–10 mm, glabrous; petals yellow-orange, oblong or narrowly so, 5–12 × 0.8–2 mm, claw (nearly linear), 5–7 mm, distinctly wider at base, pubescent inside; filaments 10–15 mm, glabrous; anthers 3–4 mm; gynophore 4–11 mm, sparsely to densely pubescent. **Fruits** divaricate, tortuous, (torulose), terete, 2.5–4.6 cm × 1.5–2 mm; ovules 24–34 per ovary; style 0.02–0.4 mm. **Seeds** oblong, 2.2–2.6 × 0.9–1.2 mm. $2n = 28$.

Flowering Jun–Jul. Loose shale, clay hills, open plains, gumbo swales, dry draws; 1800–2400 m; Colo., Utah, Wyo.

Stanleya bipinnata is known from Larimer County in Colorado, Uinta County in Utah, and Albany, Carbon, and Uinta counties in Wyoming. R. C. Rollins (1993) and R. W. Lichvar (1983) treated it as a variety of *S. pinnata*; the morphological differences between these taxa strongly support the recognition of two species.

3. **Stanleya confertiflora** (B. L. Robinson) Howell, Fl. N.W. Amer., 59. 1897 [C][E]

Stanleya viridiflora Nuttall var. *confertiflora* B. L. Robinson in A. Gray et al., Syn. Fl. N. Amer. 1(1,1): 178. 1895; *S. annua* M. E. Jones; *S. rara* A. Nelson

Annuals or biennials; (glaucous), glabrous throughout. **Stems** erect, unbranched, (slightly ribbed), 2–6(–8) dm. **Basal leaves** (often withered by flowering); blade obovate to ovate, margins entire. **Cauline leaves** sessile; blade (fleshy), lanceolate, (2.2–)4–13(–16) cm × (5–)10–30(–40) mm, (smaller distally, base auriculate to sagittate), margins entire. **Racemes** dense, (slightly or not elongated in fruit). **Fruiting pedicels** horizontal to divaricate-ascending, 10–20(–26) mm. **Flowers**: sepals oblong-linear, 6–12 mm; petals yellow becoming whitish, narrowly linear to filiform, (12–)14–25 × 0.5–1.5 mm, (margins crisp), claw (nearly linear), 5–11 mm, slightly wider at base; filaments 12–17 mm; anthers 3–4.5 mm; gynophore (6–)10–18 mm. **Fruits** suberect to divaricate-ascending, slightly curved inward, (torulose), subterete, 2–5.2(–6) cm × 1.5–2.2 mm; ovules 40–62 per ovary; style 0.6–1.7 mm. **Seeds** oblong, 1.5–2.5 × 0.8–1 mm.

Flowering Apr–Jun. Barren clay slopes in sagebrush communities, heavy clay flats, loose soil mounds, dry sandy grounds, alkaline meadows and flats; of conservation concern; 600–1500 m; Idaho, Oreg.

Stanleya confertiflora is distributed in Gooding, Owyhee, and Washington counties in Idaho, and in Baker, Harney, and Malheur counties in Oregon.

4. **Stanleya elata** M. E. Jones, Zoë 2: 16. 1891 [E]

Perennials; (short-lived); (glaucous), mostly glabrous. **Stems** erect, unbranched or branched (few) proximally and distally, 6–15(–18) dm, (weakly striate). **Basal leaves** (withered by flowering); similar to cauline. **Cauline leaves:** petiole 3–9(–12) cm; blade broadly lanceolate or oblong to ovate, (5.5–)8–21(–26) cm × 20–80(–130) mm, (leathery, smaller distally), margins entire or, rarely, with small lobes just proximal to blade, (surfaces rarely sparsely pilose abaxially). **Racemes** dense. **Fruiting pedicels** slightly reflexed to horizontal or divaricate, (5–)7–11(–15) mm. **Flowers:** sepals linear, 7–11 mm, (sometimes sparsely pilose); petals whitish to lemon yellow, linear, 8–13 × 0.3–1 mm, claw (thickened), 4–7 mm, slightly wider at base; filaments 5–13 mm, papillate basally; anthers 2.5–4 mm; gynophore 7–20 mm. **Fruits** spreading to curved downward, slightly curved, subterete, 4–9(–10.5) cm × 1.5–2 mm; ovules 46–70 per ovary; style 0.2–1.5 mm. **Seeds** oblong, 1.5–2.6 × 1–1.3 mm. $2n = 28$.

Flowering May–Jul. Sandy or gravelly soil in sagebrush and mixed shrub communities, desert scrub, decomposing granite; 1200–2000 m; Ariz., Calif., Nev.

Stanleya elata is distributed from Coconino County in Arizona, to Inyo and Mono counties in California, and Churchill, Clark, Esmeralda, Lincoln, Lyon, Mineral, and Nye counties in Nevada.

5. **Stanleya pinnata** (Pursh) Britton, Trans. New York Acad. Sci. 8: 62. 1889 [E]

Cleome pinnata Pursh, Fl. Amer. Sept. 2: 739. 1813

Perennials, subshrubs, or shrubs; (sometimes suffrutescent); (glaucous or not), mostly glabrous. **Stems** erect, unbranched or branched (few) proximally and distally, (1.2–)3–12(–15.3) dm. **Basal leaves** (withered by flowering); similar to cauline. **Cauline leaves:** petiole 0.7–6.2 cm; blade (fleshy), oblanceolate to broadly lanceolate or ovate in outline, or (distally) lobed or linear to narrowly lanceolate, 3–15 cm (smaller distally), margins lyrate-pinnatifid or runcinate, or (distal) pinnately lobed or entire. **Racemes** somewhat dense. **Fruiting pedicels** horizontal to divaricate, 3–11 mm. **Flowers:** sepals oblong-linear, 8–16 mm; petals yellow, oblanceolate to oblong, 8–20 × 2–3 mm, claw 4–10 mm, wider at base, densely pubescent inside; filaments 11–28 mm, pilose at base; anthers 3–5 mm; gynophore 7–28 mm. **Fruits** usually spreading to divaricate, rarely ascending, often strongly curved downward, not tortuous, (smooth), terete, 3–9 cm × 1.5–3 mm; ovules 10–38 per ovary; style 0.2–0.6 mm. **Seeds** (sometimes black), oblong, 2.5–4.5 × 1.2–2 mm.

Varieties 3 (3 in the flora): w, c United States.

The three varieties of *Stanleya pinnata* are reasonably well-defined and can be easily separated from each other. Their maintenance here at varietal rank is tentative; a genus-wide phylogenetic study is much needed to establish whether or not they merit recognition as distinct species.

SELECTED REFERENCE Lichvar, R. W. 1983. Evaluation of varieties in *Stanleya pinnata* (Cruciferae). Great Basin Naturalist 43: 684–686.

1. Cauline leaf blades: margins (at least proximal ones) lyrate-pinnatifid to runcinate . . . . 5a. *Stanleya pinnata* var. *pinnata*
1. Cauline leaf blades: margins usually entire, rarely proximalmost dentate.
    2. Cauline leaf blades ovate to lanceolate; petals 11–16 mm, claws 6–9 mm; ovules 26–34 per ovary; gynophores pubescent basally; Colorado, Kansas, Nevada, Utah, Wyoming . . . . 5b. *Stanleya pinnata* var. *integrifolia*
    2. Cauline leaf blades linear-lanceolate; petals 8–9(–10) mm, claws 4–5 mm; ovules 10–20 per ovary; gynophores glabrous or nearly so; Texas . . . . 5c. *Stanleya pinnata* var. *texana*

5a. **Stanleya pinnata** (Pursh) Britton var. **pinnata** [E]

Stanleya arcuata Rydberg; *S. canescens* Rydberg; *S. fruticosa* Nuttall; *S. glauca* Rydberg; *S. heterophylla* Nuttall; *S. pinnata* subsp. *inyoensis* Munz & J. C. Roos; *S. pinnata* var. *inyoensis* (Munz & J. C. Roos) Reveal; *S. pinnatifida* Nuttall

Cauline leaves: blade oblanceolate to broadly lanceolate, margins (at least proximal ones) lyrate-pinnatifid or runcinate. **Flowers:** sepals 9–16 mm; petals 10–20 mm, claw 5–10 mm; gynophore 10–28 mm, pubescent basally. **Fruits** 3–9 cm; ovules 22–38 per ovary. **Seeds** 2.5–4 mm. $2n = 28, 56$.

Flowering Apr–Sep. Prairie hills, chalk bluffs, shale slopes, talus, sagebrush and mixed shrub communities, pinyon-juniper areas, selenium-rich areas, barren hillsides, sandstone breaks and cliffs, dry sandy or gravelly washes, sand dunes, canyon bottoms, soils derived from limestone, sandstone, basaltic, and volcanic rocks; 200–2500 m; Ariz., Calif., Colo., Idaho, Kans., Mont., Nev., N.Mex., N.Dak., Oreg., S.Dak., Utah, Wyo.

Subshrubby or shrubby forms of var. *pinnata* were recognized as a distinct var. or subsp. *inyoensis*. Although such growth forms occur primarily in southern California, western Nevada, and southwestern Utah, they occur sporadically throughout the species range, especially in desert habitats with sand dunes.

5b. Stanleya pinnata (Pursh) Britton var. **integrifolia** (E. James) Rollins, Lloydia 2: 118. 1939 [E]

Stanleya integrifolia E. James, Account Exped. Pittsburgh 2: 17. 1823 (as Stanleyea); *S. glauca* Rydberg var. *latifolia* Cockerell; *S. pinnatifida* Nuttall var. *integrifolia* (E. James) B. L. Robinson

Cauline leaves: blade broadly ovate to lanceolate, margins usually entire, rarely proximalmost dentate. **Flowers**: sepals 11–16 mm; petals 11–16 mm, claw 6–9 mm; gynophore 12–20 mm, pubescent basally. **Fruits** 3–8 cm; ovules 26–34 per ovary. **Seeds** 2.5–4.5 mm. $2n = 56$.

Flowering Apr–Jul. Barren rocky hillsides, gravelly knolls, badlands, dry prairies, arroyos, dry washes, shaley sandstone, sagebrush and pinyon-juniper communities, shale slopes and cliffs; 300–2700 m; Colo., Kans., Nev., Utah, Wyo.

5c. Stanleya pinnata (Pursh) Britton var. **texana** B. L. Turner, Lundellia 7: 39, figs. 2, 3. 2004 [C][E]

Cauline leaves: blade linear-lanceolate, margins entire. **Flowers**: sepals 8–10 mm; petals 8–9(–10) mm, claw 4–5 mm; gynophore 10–20 mm, glabrous or nearly so. **Fruits** 5–8 cm; ovules 10–20 per ovary. **Seeds** 2–3 mm.

Flowering Apr–Jun. Silty flats, desert washes, bare calcareous-gypseous outcrops, betonite clay hills; of conservation concern; Tex.

R. C. Rollins (1939c, 1993) treated plants of var. *texana* under both var. *integrifolia* and var. *pinnata*. As shown by Turner, var. *texana* is somewhat distinct morphologically and is disjunct by ca. 800 km from the nearest populations of var. *pinnata*. Also, it grows on gypseous instead of seleniferous soils. It is found only in Brewster County.

6. Stanleya tomentosa Parry, Amer. Naturalist 8: 212. 1874 (as Stanley atomentosa) [E]

Stanleya runcinata Rydberg; *S. tomentosa* var. *runcinata* (Rydberg) Rollins

Perennials; (caudex simple, covered with persistent petiolar remains); pubescent or glabrous. **Stems** erect, unbranched or branched distally, 5–15 dm, (tomentose throughout or glabrate distally). **Basal leaves**: petiole 0.3–1.2 cm; blade lanceolate to oblanceolate, 6.7–19(–23) cm × 20–50(–60) mm, margins runcinate, (surfaces densely tomentose). **Cauline leaves** petiolate; (proximalmost similar to basal), blade lanceolate, 2–4 cm × 5–10 mm (smaller distally), margins entire to hastate. **Racemes** dense, (considerably elongated in fruit). **Fruiting pedicels** horizontal to divaricate-ascending, 11–22 mm, (pilose or glabrous). **Flowers**: sepals linear, 10–16 mm, pubescent; petals pale lemon yellow, linear, 12–21 × 1–1.8 mm, claw 8–15 mm, (nearly linear), slightly wider at base, glabrous; filaments 12–22 mm, glabrous; anthers 4–5.5 mm; gynophore 10–20 mm, glabrous. **Fruits** suberect to ascending, straight, flattened, 4–7 cm × 1.2–2.2 mm; ovules 30–46 per ovary; style 0.02–0.3 mm. **Seeds** oblong, 1.8–2.6 × 1–1.5 mm.

Flowering Jun–Aug. Rocky limestone hillsides, knolls, steep grassy banks, sagebrush communities, stony clay slopes; 1300–2300 m; Idaho, Wyo.

Stanleya tomentosa is known in Wyoming from Big Horn, Hot Springs, and Park counties. R. C. Rollins (1993) recognized two varieties and distinguished them primarily on the basis of having sparsely pilose (var. *tomentosa*) versus glabrous (var. *runcinata*) distalmost stems. The distinction is artificial and unwarranted.

7. Stanleya viridiflora Nuttall in J. Torrey and A. Gray, Fl. N. Amer. 1: 98. 1838 [E][F]

Perennials; (caudex simple, covered with persistent petiolar remains); (glaucous), glabrous throughout. **Stems** erect, unbranched or branched distally, (2.5–)4–12(–14) dm. **Basal leaves**: petiole 2–10(–16) cm; blade lanceolate to oblanceolate or ovate, (2.2–)5–18(–22) cm × 10–40(–60) mm, margins often entire, sometimes dentate, rarely lyrate-pinnatifid. **Cauline leaves** sessile; blade lanceolate, (2–)3.5–8.5(–11) cm × (2–)5–19(–28) mm (smaller distally, base auriculate to sagittate), margins entire. **Racemes** loose. **Fruiting pedicels** horizontal to divaricate-ascending, 4–9(–12) mm.

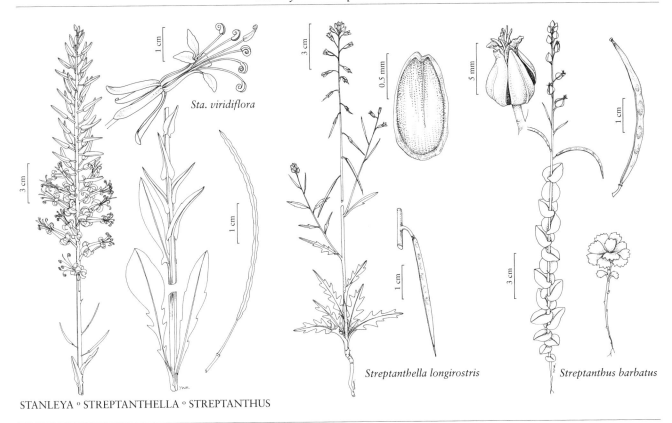

Flowers: sepals oblong-linear, 12–18 mm; petals whitish to lemon yellow, narrowly oblanceolate, 13–20 × 1–3 mm, claw 7–11 mm, (nearly linear-lanceolate), slightly wider at base, (margins usually erose, rarely subentire and crisped); filaments 11–20 mm; anthers 3.5–6 mm; gynophore (6–)11–22(–25) mm. Fruits divaricate or descending, curved inward, (torulose), subterete, 3–6(–7) cm × 1.2–2 mm; ovules 28–50 per ovary; style 0.04–0.3 mm. Seeds oblong, 2–3 × 1–1.2 mm. $2n = 28$.

Flowering May–Jul. Sagebrush and pinyon-juniper communities, limestone shale and rocks, red sandstone slopes, volcanic rocky slopes, clay knolls, steep bluffs; 1300–2700 m; Calif., Colo., Idaho, Mont., Nev., Oreg., Utah, Wyo.

Stanleya collina M. E. Jones is an illegitimate name that pertains to *S. viridiflora*.

90. STREPTANTHELLA Rydberg, Fl. Rocky Mts., 364, 1062. 1917 • [Genus *Streptanthus* and Latin *-ella*, diminutive]

Ihsan A. Al-Shehbaz

Annuals; not scapose; (often glaucous), glabrous throughout. Stems erect, usually branched distally, rarely unbranched. Leaves basal and cauline; petiolate or sessile; basal not rosulate, petiolate, blade similar to cauline; cauline petiolate or sessile, blade (base not auriculate), margins entire, dentate, sinuate, or pinnatifid. Racemes (corymbose, several-flowered, lax), considerably elongated in fruit. Fruiting pedicels usually reflexed, rarely divaricate, slender or stout. Flowers: sepals erect, oblong, (unequal), lateral pair subsaccate basally; petals white or yellow (often veins purplish), spatulate, (slightly exceeding sepals, margins crisp), claw well-differentiated from blade, (apex obtuse); stamens in 3 unequal pairs; filaments not dilated basally; anthers ovate to oblong, (apiculate); nectar glands lateral, semi-annular, subtending lateral stamens, median glands absent. **Fruits** sessile, linear, smooth, latiseptate; valves each with prominent midvein; replum rounded or slightly flattened; septum complete; ovules (12–)16–28(–34); style distinct, (well-developed, adnate with valves and replum apex); stigma subentire. **Seeds** uniseriate, flattened, winged, oblong; seed coat not mucilaginous when wetted; cotyledons incumbent. $x = 7$.

Species 1: w United States, nw Mexico.

1. Streptanthella longirostris (S. Watson) Rydberg, Fl. Rocky Mts., 364. 1917 [F]

Arabis longirostris S. Watson, Botany (Fortieth Parallel), 17. 1871; *Euklisia longirostris* (S. Watson) Rydberg; *Guillenia rostrata* Greene; *Streptanthella longirostris* var. *derelicta* J. T. Howell; *Streptanthus longirostris* (S. Watson) S. Watson; *Thelypodium longirostris* (S. Watson) Jepson

Stems often several, ascending distally, (1.2–)2–6(–7.5) dm. **Basal leaves** soon withered. **Cauline leaves** shortly petiolate or (distal) sessile; blade lanceolate to oblanceolate, or (distal) linear, 2–5.5(–6.5) cm × 3–10(–15) mm, (attenuate to petiolelike base), margins entire, dentate to sinuate, or pinnatifid, (distal cauline (1–)1.7–5.8(–7) cm × (0.7–)1.5–4(–7) mm, base cuneate to attenuate, margins entire, apex obtuse to acute). **Fruiting pedicels** usually curved, rarely straight, (1–)2–5(–7) mm. **Flowers:** sepals (2–)2.5–4(–5) × 0.7–1.5 mm; petals (3.5–)4–6(–7) × 0.7–1.1 mm, claw oblanceolate, 3–5 mm (longer than blade); filaments with at least dorsal pair exserted, (longest pair) 3–6 mm; anthers 0.5–1(–1.3) mm. **Fruits** slightly to strongly reflexed, sometimes secund, (2.5–)3.5–6(–7) cm × 1.5–2(–2.2) mm; valve apex adnate with style and replum, forming indehiscent tip, often 1-seeded; style (2–)3.5–6(–8) mm. **Seeds** light brown, 2–3 × 1–1.7(–2) mm; wing 0.3–0.7 mm wide. $2n = 28$.

Flowering early Jan–late Jun. Roadsides, rocky areas, sandy ridges, sagebrush and creosote bush deserts, dry slopes, dry washes, decomposed granitic soil, pinyon-juniper areas, alluvial fans, chaparral, sandstone gravel and outcrops, sandstone hills and cliffs; 60–2200 m; Ariz., Calif., Colo., Idaho, Mont., Nev., N.Mex., Oreg., Utah, Wash., Wyo.; Mexico (Baja California, Sonora).

Streptanthella longirostris is most widely distributed in southern California, Nevada, and southern and central Utah, and appears to be restricted elsewhere: Colorado (Mesa, Montezuma, Montrose, San Miguel), Idaho (Butte), Montana (Carbon), New Mexico (San Juan), Washington (Franklin, Grant), and Wyoming (Fremont, Natrona, Sweetwater, Uinta).

91. **STREPTANTHUS** Nuttall, J. Acad. Nat. Sci. Philadelphia 5: 134, plate 7. 1825 • Jewel-flower [Greek *streptos*, twisted, and *anthos*, flower, alluding to crisped petal margin]

Ihsan A. Al-Shehbaz

Agianthus Greene; *Cartiera* Greene; *Disaccanthus* Greene; *Euklisia* (Nuttall ex Torrey & A. Gray) Rydberg; *Icianthus* Greene; *Mesoreanthus* Greene; *Microsemia* Greene; *Mitophyllum* Greene; *Pleiocardia* Greene

Annuals, biennials, or perennials; (short-lived, caudex poorly developed, usually not woody); not scapose; glabrous or pubescent. **Stems** usually erect, rarely ascending, unbranched or branched, (often glaucous, glabrous distally). **Leaves** basal and cauline; petiolate or sessile; basal rosulate or not, petiolate or sessile, blade margins entire or dentate to lyrate-pinnatifid; cauline usually sessile, rarely petiolate, blade (base usually auriculate or amplexicaul), margins entire or dentate, pinnatifid, or pinnatisect. **Racemes** (rarely with a terminal cluster of sterile flowers, usually ebracteate, sometimes bracteate), usually elongated in fruit. **Fruiting pedicels** usually divaricate-ascending, rarely erect or suberect, slender or stout. **Flowers** (sometimes zygomorphic); sepals erect, (calyx often urceolate or campanulate), both pairs often saccate basally; petals white, yellow, pink, purple, or brownish, oblong to ovate, (narrow and margins crisped or channeled, or broad and margins neither crisped nor channeled), claw poorly differentiated from blade or distinct, (apex rounded); stamens usually in 3 unequal pairs, rarely tetradynamous; filaments not dilated basally, (adaxial pair longest, frequently connate and anthers sterile or partially sterile, sometimes all filaments distinct and all anthers fertile); anthers linear to oblong; nectar glands confluent. **Fruits** subsessile or shortly stipitate, linear, torulose or smooth, usually latiseptate, rarely subterete (flattened); valves each with prominent or obscure midvein, usually glabrous, rarely pubescent; replum rounded; septum complete; ovules (10–)12–120 per ovary; style distinct or obsolete; stigma entire or 2-lobed. **Seeds** uniseriate, usually flattened, rarely plump, usually winged, rarely not winged, oblong, ovoid, orbicular, or suborbicular; seed coat (smooth or minutely reticulate), not mucilaginous when wetted; cotyledons usually accumbent, rarely obliquely so.

Species ca. 35 (35 in the flora): c, w United States, n Mexico.

The infrageneric taxonomy of *Streptanthus* is somewhat controversial. W. L. Jepson (1909–1943, vol. 2) united it with *Caulanthus* and divided the combined genus into four subgenera, whereas J. E. Rodman et al. (1981) and A. R. Kruckeberg and J. L. Morrison (1983) recognized subgenera and sections. In the absence of comprehensive phylogenetic studies on the entire complex, I prefer not to recognize any infrageneric taxa.

Streptanthus includes species endemic to serpentine soils. Extensive experimental and molecular studies were conducted on the *S. glandulosus* complex (e.g., A. R. Kruckeberg 1957; M. S. Mayer and P. S. Soltis 1994, 1999; Mayer et al. 1994) and on the nickel hyperaccumulators (e.g., R. D. Reeves et al. 1981; Kruckeberg and Reeves 1995); the taxonomy of that complex has not yet been satisfactorily resolved.

In the following descriptions, characteristics described for basal leaves also apply to those of juvenile perennial plants. Measurements of floral parts are made from flowers in the proximal portion of racemes; flower size decreases distally, and distal and late-season flowers are often smaller than those of the proximal or early-season flowers. Measurements of fruiting pedicels and fruits are taken from proximalmost ones. Seed width measurements include the seed wing.

1. Cauline leaves usually petiolate, rarely sessile, blades with bases not auriculate, sagittate, or amplexicaul.
    2. Middle cauline leaf blades: margins runcinate- or lyrate-pinnatifid; fruits 3.5–5 mm wide.
        3. Stems glabrous proximally; petals unequal (abaxial pair 5–6 mm, adaxial pair 15–26 mm); stamens in 3 unequal pairs; styles 0.2–0.5 mm . . . . 12. *Streptanthus cutleri*
        3. Stems pilose proximally; petals equal (4–8 mm); stamens tetradynamous; styles 0.5–2.7 mm . . . . 29. *Streptanthus petiolaris*
    2. Middle cauline leaf blades: margins entire or repand; fruits 1.2–3.2(–3.5) mm wide.
        4. Perennials; middle cauline leaves spatulate-obovate to suborbicular; filament pairs distinct; fruits 2.5–3.2(–3.5) mm wide . . . . 21. *Streptanthus howellii*
        4. Annuals; middle cauline leaves usually linear to linear-lanceolate, rarely oblanceolate; filament pairs connate; fruits 1.2–2 mm wide.
            5. Calyces urceolate; petals 6–9 mm; ovules 22–38 per ovary; California. . . 2. *Streptanthus barbiger*
            5. Calyces campanulate; petals 13–20 mm; ovules 90–120 per ovary; Arkansas, Kansas, Louisiana, Oklahoma, Texas. . . . . .22. *Streptanthus hyacinthoides*
1. Cauline leaves sessile, blades with bases auriculate, sagittate, or amplexicaul.

[6. Shifted to left margin.—Ed.]
6. Racemes bracteate throughout or partly.
    7. Racemes bracteate throughout or proximalmost flowers bracteate; Texas.
        8. Petals 14–19 mm; stamens tetradynamous; fruits 8–14.5 cm × 2.5–4 mm; ovules 48–80 per ovary; styles 1–3.5 mm .... 6. *Streptanthus bracteatus*
        8. Petals 16–27 mm; stamens in 3 unequal pairs; fruits 4–9.5 cm × 4.5–6 mm; ovules 26–42 per ovary; styles 0.5–2 mm .... 30. *Streptanthus platycarpus* (in part)
    7. Racemes bracteate below or between proximalmost 1 or 2 flowers; California, Nevada, Oregon.
        9. Basal leaf blades pectinate, 1- or 2-pinnatifid, or pinnatisect; fruits smooth or slightly torulose.
            10. Middle cauline leaf blades pectinate or pinnatisect into filiform segments, bases not auriculate; fruits geniculately reflexed .... 13. *Streptanthus diversifolius*
            10. Middle cauline leaf blades usually not lobed, rarely pinnatifid, not forming filiform segments, bases auriculate; fruits usually ascending or divaricate, rarely reflexed.
                11. Petals white (veins violet); fruits 6–12 cm × 1.7–2.5 mm, valves each with prominent midvein; ovules 60–100 per ovary; seeds 2–2.5 × 1.5–2 mm
                .... 15. *Streptanthus farnsworthianus*
                11. Petals rose-purple; fruits 2–5 cm × 1.2–1.7 mm, valves each with obscure midvein; ovules 22–38 per ovary; seeds 1–1.5 × 0.7–0.9 mm..........
                .... 16. *Streptanthus fenestratus*
        9. Basal leaf blades not pectinate, pinnatifid, or pinnatisect (not lobed); fruits torulose.
            12. Annuals; sepals 4–5 mm; fruiting pedicels not expanded at receptacles .....
            ....18. *Streptanthus gracilis*
            12. Biennials or perennials (short-lived); sepals 6–10(–13) mm; fruiting pedicels expanded at receptacles.
                13. Cauline leaf blades linear-lanceolate; filaments: adaxial pair connate, 13–16 mm, lateral pair 7–9 mm; Trinity County, California........27. *Streptanthus oblanceolatus*
                13. Cauline leaf blades oblong to obovate or suborbicular; filaments: adaxial pair distinct, (5–)7–11 mm, lateral pair (1.5–)3–5 mm; California, Nevada, Oregon .... 33. *Streptanthus tortuosus*
6. Racemes ebracteate throughout.
    14. Racemes with terminal cluster of sterile flowers; fruit valves sparsely to densely hirsute or hispid.
        15. Petals purple (with dark purple veins), claws narrower than blades; fruits 1.3–2.5 cm × 2.5–3.5 mm; seeds not winged; Santa Clara County, California .... 8. *Streptanthus callistus*
        15. Petals purple (with white margins), purplish white, lemon-yellow, or yellowish white, claws as wide as or wider than blades; fruits 3.5–11.4 cm × 1.5–2.5 mm; seeds winged (wings continuous); Contra Costa, Fresno, Merced, Monterey, San Benito counties, California.
            16. Middle cauline leaf blade bases not auriculate; filaments: abaxial pair connate; Mt. Diablo, Contra Costa County, California .... 20. *Streptanthus hispidus*
            16. Middle cauline leaf blade bases auriculate; filaments: abaxial pair distinct; Fresno, Merced, Monterey, San Benito counties, California .... 23. *Streptanthus insignis*
    14. Racemes without terminal cluster of sterile flowers; fruit valves glabrous (sometimes hirsute in *S. glandulosus*).
        17. Perennials, with woody caudex (except *S. campestris*); calyces ± campanulate; filaments: adaxial pair distinct; anthers 2.5–5 mm, adaxial pair fertile.
            18. Cauline leaves overlapping, not smaller distally; petals with darker veins, margins slightly crisped .... 1. *Streptanthus barbatus*
            18. Cauline leaves not overlapping, smaller distally; petals without darker veins, margins not crisped.

19. Fruits descending, arcuate; petals purple or brownish with yellow-green claws .... 24. *Streptanthus longisiliquus*
19. Fruits spreading, divaricate-ascending, or ascending, straight or slightly curved; petals uniformly white, maroon-purple, or brownish, rarely claws yellow (*S. campestris*).
    20. Petals: claws white or pale yellow; stamens tetradynamous or nearly so; fruit valves each with obscure midvein.
        21. Plants with woody caudex (elevated); stems 2.5–8.6 dm; racemes lax in bud; sepals pale yellow to white; stigmas subentire .... 4. *Streptanthus bernardinus*
        21. Plants without woody caudex; stems (2.5–)6–15(–18) dm; racemes dense in bud; sepals purple; stigmas 2-lobed .... 9. *Streptanthus campestris*
    20. Petals: claws purplish; stamens in 3 unequal pairs; fruit valves each with prominent midvein.
        22. Basal leaf blades: margins dentate, petioles broadly winged; fruits (2.5–)3–6(–7) mm wide; stigmas slightly to strongly 2-lobed; ovules 20–38(–46) per ovary; seeds 2.5–5 × 2.2–5 mm ....11. *Streptanthus cordatus*
        22. Basal leaf blades: margins entire, petioles usually not winged, rarely narrowly winged; fruits 2–2.7(–3) mm wide; stigmas entire; ovules (42–)48–60 per ovary; seeds 2–2.5(–2.7) × 1.5–2 mm .... 28. *Streptanthus oliganthus*

[17. Shifted to left margin.—Ed.]

17. Annuals or biennials, without woody caudex; calyces often urceolate, sometimes campanulate; filaments: adaxial pair often connate, sometimes distinct; anthers 1–6 mm, adaxial pair often sterile, sometimes fertile.
    23. Calyces usually campanulate, rarely urceolate; filaments: adaxial pair distinct; anthers: adaxial pair fertile; stigmas strongly 2-lobed.
        24. Stamens in 3 unequal pairs; fruits 4.5–6 mm wide; ovules 26–54 per ovary; seeds 3–5 mm wide.
            25. Petals creamy white (with purple veins) or purple with white margins, 14–18 mm, margins crisped, 1–1.5 mm wide; fruit valves each with obscure midvein; Arizona, New Mexico, Texas .... 10. *Streptanthus carinatus*
            25. Petals lavender or purplish lavender, 16–27 mm, margins not crisped, 6–9 mm wide; fruit valves each with prominent midvein; Texas .... 30. *Streptanthus platycarpus* (in part)
        24. Stamens tetradynamous; fruits 2–3 mm wide; ovules 54–92 per ovary; seeds 1–1.3 mm wide.
            26. Sepals glabrous, lateral pairs usually without, rarely with, subapical callus, 0.1–0.3 mm .... 25. *Streptanthus maculatus*
            26. Sepals hirsute, lateral (and median) pairs with subapical callus, 1–2 mm .... 32. *Streptanthus squamiformis*
    23. Calyces urceolate; filaments: adaxial pair connate; anthers: adaxial pair sterile; stigmas entire or subentire.
        27. Replums constricted between seeds.
            28. Annuals; basal leaves not rosulate, blade not mottled; sepals green; petals without darker veins .... 34. *Streptanthus vernalis*
            28. Biennials; basal leaves rosulate, blade mottled; sepals rose-purple, yellow, or violet; petals with darker veins.
                29. Stems 1.4–5(–6) dm, usually branched basally, rarely unbranched; petals whitish, abaxial pairs with purple spot, adaxial pairs with faint purple veins; fruit valves each with obscure midvein .... 5. *Streptanthus brachiatus*
                29. Stems (2.5–)6.5–12(–15) dm, unbranched or paniculately branched distally; petals creamy white, abaxial and adaxial pairs with brownish purple veins; fruit valves each with prominent midvein .... 26. *Streptanthus morrisonii*
        27. Replums not constricted between seeds (straight).

[30. Shifted to left margin.—Ed.]

30. Sepals: abaxial ovate-cordate, lateral ovate-lanceolate, adaxial pair suborbicular to broadly ovate-cordate (forming a bannerlike hood), adaxial pair distinctly larger than abaxial and lateral sepals; fruits pendent .... 31. *Streptanthus polygaloides*

30. Sepals: all lanceolate to broadly ovate (not forming a bannerlike hood), subequal; fruits usually erect, ascending, divaricate, or spreading (rarely reflexed in *S. glandulosus*).

    31. Middle cauline leaf blades orbicular or orbicular-ovate, strongly overlapping; fruits strongly falcate .... 14. *Streptanthus drepanoides*

    31. Middle cauline leaf blades ovate, oblanceolate, lanceolate, or linear, not overlapping; fruits straight, curved, or arcuate.

        32. Plants usually hirsute basally, rarely glabrous; petals with crisped margins; fruit valves each with prominent midvein, glabrous or sparsely to moderately hirsute .... 17. *Streptanthus glandulosus*

        32. Plants glabrous basally; petals without crisped margins; fruit valves each with obscure midvein, glabrous.

            33. Basal leaf blades: margins entire; middle cauline leaf blades linear to narrowly linear-lanceolate; petals 2–3 mm wide .... 35. *Streptanthus vimineus*

            33. Basal leaf blades: margins usually dentate (*S. breweri* sometimes entire); middle cauline leaf blades ovate or lanceolate; petals 1–2 mm wide.

                34. Petals: abaxial pair purplish; filaments: adaxial pair 5–6 mm; fruits 1.3–3 cm; ovules 12–22 per ovary .... 3. *Streptanthus batrachopus*

                34. Petals: abaxial pair white (with purple veins); filaments: adaxial pair 6–10 mm; fruits 3–9 cm; ovules 24–54 per ovary.

                    35. Foliage green; rachises straight; petals 8–12 mm; fruits strongly arcuate (recurved) .... 7. *Streptanthus breweri*

                    35. Foliage mostly yellowish; rachises flexuous; petals 6–8 mm; fruits usually nearly straight, rarely arcuate .... 19. *Streptanthus hesperidis*

1. **Streptanthus barbatus** S. Watson, Proc. Amer. Acad. Arts 25: 125. 1890 [E] [F]

Cartiera barbata (S. Watson) Greene

Perennials; (caudex branched); glabrous, (sepals and leaf blade apices pubescent). **Stems** (sometimes ascending), unbranched or branched (few), 1.5–8(–9.2) dm. **Basal leaves** subrosulate (in juvenile plants); shortly petiolate; blade obovate, 1–2.7 cm (5–14 mm wide), margins dentate apically (teeth with relatively short, bristly trichomes). **Cauline leaves:** blade (fleshy), broadly ovate to orbicular, 0.7–3.5 cm × 5–24 mm (not smaller distally), base auriculate to amplexicaul, (proximally overlapping, distally not overlapping), margins usually entire, (apex obtuse). **Racemes** ebracteate, (lax in fruit). **Fruiting pedicels** erect to ascending, (straight), 3–8 mm. **Flowers:** calyx campanulate; sepals (ascending), yellowish green (in bud) or purple (in flower), (ovate), 4–7 mm, not keeled, (apex with short, bristly, flattened trichomes); petals purplish (with darker veins), 5–9 mm, blade 0.5–3 × 0.3–0.7 mm, margins slightly crisped, claw 4–7 mm, wider than blade; stamens in 3 unequal pairs; filaments (distinct): abaxial pair 4–6 mm, lateral pair 3–5 mm, adaxial pair 5–7 mm; anthers (all) fertile, 2.5–3.5 mm; gynophore 1–3 mm. **Fruits** erect to ascending, smooth, often strongly recurved distally, distinctly flattened, 2–7(–8.2) cm × 2–3 mm; valves each with obscure midvein (sometimes not basally, margins somewhat thickened); replum straight; ovules 16–30 per ovary; style 0.4–1 mm; stigma entire. **Seeds** oblong to broadly ovoid, 3–4 × 1.7–2 mm; wing (0–)0.1–0.25 mm wide at apex, (narrower at margin). $2n = 28, 56$.

Flowering Jun–Aug. Dry, open Jeffrey pine woods, gravelly serpentine grounds in forest openings; 800–2200 m; Calif.

Streptanthus barbatus is known from Siskiyou, Tehama, and Trinity counties.

2. **Streptanthus barbiger** Greene, Pittonia 1: 217. 1888 [E]

Mesoreanthus barbiger (Greene) Greene

Annuals; glabrous, (sometimes sepals pubescent). **Stems** branched basally, (0.7–)1–6.7 (–8) dm. **Basal leaves** (soon withered); not rosulate; petiolate; blade oblanceolate or oblong to lanceolate, 2–5(–7) cm (5–15 mm wide), margins remotely dentate. **Cauline leaves:** blade linear to linear-lanceolate, (1.5–)3–9(–10) cm × 0.5–2 mm (smaller distally), base not auriculate, margins entire. **Racemes** ebracteate, (lax, often secund). **Fruiting pedicels** ascending to divaricate, 1–2.5(–4) mm. **Flowers:** calyx urceolate; sepals (erect), green to purplish, (ovate), 4–6 mm, slightly keeled, (apex recurved, glabrous or hirsute, trichomes retrorse); petals white (abaxial pair with purplish veins), 6–9 mm, blade 2–4 × 1.5–2.5 mm, margins not crisped, claw 4–6 mm, narrower than blade; stamens in 3 unequal pairs; filaments: abaxial pair (connate to middle), 4.5–5.5 mm, lateral pair 2.5–3.5 mm, adaxial pair (connate, strongly recurved, purplish), 7–9 mm; anthers: abaxial and lateral pairs fertile, 1.5–2.2 mm, adaxial pair sterile, 0.4–1 mm; gynophore 0.3–1 mm. **Fruits** divaricate-ascending to reflexed, slightly torulose, curved or, rarely, straight, slightly flattened, 2–6(–7) cm × 1.2–1.5 mm; valves each with obscure midvein; replum straight; ovules 22–38 per ovary; style 0.1–0.7 mm; stigma entire. **Seeds** narrowly oblong, 1.3–1.8 × 0.6–0.8 mm; wing (0–)0.1–0.25 mm wide distally, (narrower at margin or absent). $2n = 28$.

Flowering May–Aug. Serpentine ridges and barrens, openings in chaparral, cypress or pine-oak woodlands; 200–1500 m; Calif.

Streptanthus barbiger is distributed in Lake, Mendocino, Sonoma, and Tehama counties. Its holotype has pubescent sepals; most of the collections examined have glabrous ones, though plants with pubescent sepals occur throughout the species range. R. E. Buck et al. (1993) suggested that the two forms apparently represent distinct taxa. This is highly unlikely because both forms occur sometimes in the same population, as evidenced by *Breedlove 5145* (GH).

3. **Streptanthus batrachopus** J. L. Morrison, Madroño 4: 204, plate 31, figs. 20–28. 1938 [C] [E]

Annuals; (glaucous), glabrous throughout. **Stems** unbranched or branched basally, 0.3–1.7(–2.8) dm. **Basal leaves** (soon withered); not rosulate; shortly petiolate; blade (succulent, mottled), obovate to oblanceolate, 0.5–2 cm, margins coarsely dentate. **Cauline leaves:** blade lanceolate, 0.5–2.5 cm × 1–7 mm (smaller distally), base auriculate, margins entire or dentate (entire distally). **Racemes** ebracteate (lax, sometimes secund). **Fruiting pedicels** divaricate-ascending, 1–2.5(–4) mm. **Flowers:** calyx urceolate; sepals (erect) reddish purple, (ovate), 3–5 mm, keeled, (apex spreading); petals whitish (abaxial pair with purplish veins, adaxial pair purple), 5–8 mm, blade 2–3 × 1–1.5 mm, margins not crisped, claw 3–4 mm; stamens in 3 unequal pairs; filaments: abaxial pair (connate to middle), 3–4 mm, lateral pair 1.5–2.5 mm, adaxial pair (completely connate, not recurved), 5–6 mm; anthers: abaxial and lateral pairs fertile, 1.4–1.7 mm, adaxial pairs sterile, 0.3–0.5 mm; gynophore 0.3–0.7 mm. **Fruits** divaricate-ascending, slightly torulose, straight to slightly curved, slightly flattened, 1.3–3 cm × 1–1.5 mm; valves each with obscure midvein; replum straight; ovules 12–22 per ovary; style 0.2–0.8 mm; stigma entire. **Seeds** oblong, 1.3–2 × 0.8–1.2 mm; wing 0.1–0.3 mm wide distally, (narrower at margin, rarely absent). $2n = 28$.

Flowering May–Jun. Serpentine barrens and outcrops in chaparral; of conservation concern; 100–600 m; Calif.

Streptanthus batrachopus is known from Mt. Tamalpais, Marin County.

4. **Streptanthus bernardinus** (Greene) Parish, Pl. World 20: 216. 1917

Agianthus bernardinus Greene, Leafl. Bot. Observ. Crit. 1: 228. 1906; *A. jacobaeus* Greene; *Streptanthus campestris* S. Watson var. *bernardinus* (Greene) I. M. Johnston; *S. campestris* var. *jacobaeus* (Greene) Jepson

Perennials; (caudex woody, often elevated); (glaucous), glabrous, (petioles and sepal apices pubescent). **Stems** unbranched or branched (few), 2.5–8.6 dm. **Basal leaves** not rosulate; petiolate (petioles ciliate); blade oblanceolate to spatulate, 2–8 cm (7–25 mm wide), margins remotely dentate apically. **Cauline leaves:** blade ovate to lanceolate, 2–10 cm × 5–28 mm (smaller distally), base amplexicaul, margins entire, (apex acute to acuminate). **Racemes** ebracteate, (lax in bud). **Fruiting pedicels** divaricate-ascending, (straight), 4–7.5 mm. **Flowers:** calyx campanulate; sepals (ascending), pale yellow to white, (broadly oblong to ovate) 5–9 mm, not keeled; petals white, 7–11 mm, blade 2–4 × 1–1.3 mm, margins not crisped, claw 6–9 mm, wider than blade; stamens nearly tetradynamous; filaments: median pairs (distinct), 6.5–8.5 mm, lateral pair 4.5–5.5 mm; anthers (all) fertile, 3.5–4.5 mm; gynophore 0.5–1.5 mm. **Fruits** ascending to spreading, smooth, straight or curved, flattened, 5–12.7 cm × 1.5–3 mm; valves each with obscure midvein (at least distally); replum straight; ovules 36–56 per ovary; style 0.8–2.5 mm; stigma subentire. **Seeds** oblong, 2–3 × 1.4–2 mm; wing 0.1–0.2 mm wide at apex. $2n = 28$.

Flowering Jun–Aug. Dry open pine or cypress woods, chaparral; 1200–2500 m; Calif.; Mexico (Baja California).

Streptanthus bernardinus is known from San Bernardino and San Diego counties; in Baja California it is restricted to Sierra Juárez and Sierra San Pedro Mártir.

The chromosome number $2n = 14$, reported by R. C. Rollins (1993) and by R. E. Buck et al. (1993), must be an error for $n = 14$. No other species of *Streptanthus* or its immediate generic relatives has such a low number.

5. Streptanthus brachiatus F. W. Hoffman, Madroño 11: 230, fig. 5. 1952 [C][E]

Streptanthus brachiatus subsp. *hoffmanii* R. W. Dolan & LaPré

Biennials; (somewhat glaucous), glabrous throughout. **Stems** usually branched basally, rarely unbranched, 1.4–5(–6) dm. **Basal leaves** (soon withered); rosulate; shortly petiolate; blade (fleshy, mottled), broadly obovate to suborbicular, 1.5–4 cm, margins coarsely dentate. **Cauline leaves:** blade ovate to cordate, 0.7–3.7 cm × 3–15 mm (smaller distally), base auriculate to amplexicaul, margins serrate-dentate or entire. **Racemes** ebracteate, (lax, sometimes secund). **Fruiting pedicels** divaricate-ascending, (straight), 1–2 mm. **Flowers:** calyx urceolate; sepals (erect), rose-purple (to yellowish at base), (ovate), 5–7 mm, keeled, (apex recurved); petals whitish, (abaxial pair with purple spot, adaxial pair faintly purple-veined), 7–10 mm, blade 1.5–3 × 1–1.5 mm, margins not crisped, claw 5–7 mm, as wide as or wider than blade; stamens in 3 unequal pairs; filaments: abaxial pair (connate to middle), 4–7 mm, lateral pair 2–4 mm, adaxial pair (completely connate, usually recurved), 8–10 mm; anthers: abaxial and lateral pairs fertile, 2–2.5 mm, adaxial pair sterile, 0.7–1.2 mm; gynophore 0.3–0.7 mm. **Fruits** divaricate-ascending, torulose, nearly straight, flattened, 4–6 cm × 1–1.3 mm; valves each with obscure midvein; replum constricted between seeds; ovules 22–30 per ovary; style 0.1–0.4 mm; stigma subentire. **Seeds** oblong, 1.8–2.5 × 0.7–0.9 mm; wing 0–0.1 mm wide distally. $2n = 28$.

Flowering Jun–Jul. Serpentine barrens in chaparral, pine-oak or cypress woodland openings; of conservation concern; 600–1000 m; Calif.

Streptanthus brachiatus is known from Lake and Sonoma counties. At the time that R. W. Dolan and L. F. LaPré (1989) studied it, infraspecific taxa based on differences in sepal color and pubescence seemed distinct (R. E. Buck et al. 1993). Populations discovered since then do not accord with the putative differences.

6. Streptanthus bracteatus A. Gray, Gen. Amer. Bor., 146, plate 60. 1848 [C][E]

Erysimum bracteatum (A. Gray) Kuntze

Annuals or biennials; (glaucous); usually glabrous, (sometimes pedicels pubescent). **Stems** often branched distally, (2.3–)4.5–12 dm. **Basal leaves** not rosulate; long-petiolate; blade oblanceolate to spatulate, 5–25 cm, margins lyrately lobed to irregularly dentate. **Cauline leaves:** blade oblong to ovate, 3–15 cm × 15–80 mm (smaller distally as bracts), base auriculate to amplexicaul, margins entire or shallowly dentate. **Racemes** bracteate throughout, (proximalmost bracts leaflike, distalmost much reduced). **Fruiting pedicels** divaricate-ascending, 7–19 mm, (glabrous or puberulent). **Flowers:** calyx campanulate; sepals (ascending to suberect), 8–12 mm, (not saccate basally), not keeled, (inner pair apiculate); petals purplish, 14–19 mm, blade 7–12 × 5–7 mm, margins not crisped, claw 6–8 mm, (slender), much narrower than blade; stamens tetradynamous; filaments: median pairs (distinct), 6–8 mm, lateral pair 4–6 mm; anthers (all) fertile, 4–6 mm; gynophore 1–2 mm. **Fruits** divaricate ascending, smooth, straight, flattened, 8–14.5 cm × 2.5–4 mm; valves each with prominent midvein; replum straight; ovules 48–80 per ovary; style 1–3.5 mm; stigma 2-lobed. **Seeds** oblong, 3–4 × 2–3 mm; wing 0.5–0.7 mm wide at apex, continuous.

Flowering Apr–Jun. Openings in oak-juniper woodlands, shallow, well-drained, gravelly clay-loam areas derived from limestone, dry rock hills, bluffs; of conservation concern; Tex.

Streptanthus bracteatus is restricted to Austin, Bandera, Blanco, Medina, Real, and Travis counties. It is in the Center for Plant Conservation's National Collection of Endangered Plants.

7. Streptanthus breweri A. Gray, Proc. Calif. Acad. Sci. 3: 101. 1864 [E]

Pleiocardia breweri (A. Gray) Greene

Annuals; (glaucous), usually glabrous, (sometimes sepals pubescent). **Stems** unbranched or branched basally, (0.5–)1.5–6.5(–8) dm. **Basal leaves** (soon withered); not rosulate; shortly petiolate; blade broadly ovate to obovate, 1.5–4 cm, margins entire or coarsely dentate. **Cauline leaves:** blade ovate or (distally) narrowly lanceolate, 1.5–10 cm × 3–45 mm, (much smaller distally), base amplexicaul, margins entire. **Racemes** ebracteate, (sometimes secund, rachis straight). **Fruiting pedicels** divaricate-ascending to suberect, 1.5–5 (–6) mm. **Flowers:** calyx urceolate; sepals (erect), purple or white, (ovate), 5–7 mm, keeled, (ribbed, apex recurved); petals (recurved), white (with purple veins or adaxial pair white), 8–12 mm, blade 2–5 × 1–2 mm, margins not crisped, claw 5–7 mm, about as wide as blade; stamens in 3 unequal pairs; filaments: abaxial pair (connate proximally), 5–6 mm, lateral pair 3–4 mm, adaxial pair (connate, recurved or not), 7–10 mm; anthers: abaxial and lateral pairs fertile, 2–3 mm, adaxial pair sterile, 0.6–1.3 mm; gynophore 0.2–0.5 mm. **Fruits** erect to ascending or recurved, usually strongly arcuate, rarely straight, slightly flattened, 3–9 cm × 1–1.5 mm; valves each with obscure midvein; replum straight; ovules 24–54 per ovary; style 0.2–0.5 mm; stigma entire. **Seeds** broadly oblong, 1–1.5 × 0.7–1 mm; wing (0–)0.1 mm wide distally. $2n = 28$.

Flowering May–Jul. Serpentine barrens in chaparral, oak and pine woodlands; 300–2100 m; Calif.

Streptanthus breweri is known from Alameda, Colusa, Fresno, Glenn, Lake, Mendocino, Napa, San Benito, Santa Clara, Stanislaus, Tehama, and Trinity counties in the North Coast and South Coast ranges. It is variable in fruit length and curvature, orientation of fruiting pedicels, and presence versus absence of sepal trichomes and seed wing. The type collection has straight fruits, glabrous sepals, and wingless seeds. Without further detailed studies, it is impractical to recognize infraspecific taxa.

8. **Streptanthus callistus** J. L. Morrison, Madroño 4: 205, plate 31, figs. 1–10. 1938 [C] [E]

Annuals; (somewhat glaucous), hirsute. **Stems** unbranched or branched, (often bristly proximally), 0.2–0.9 dm. **Basal leaves** (soon withered); not rosulate; shortly petiolate; blade oblong-orbicular to obovate, 0.7–1.5 cm, margins dentate. **Cauline leaves:** blade broadly ovate to obovate, 0.8–1.7 cm × 4–13 mm (smaller distally), base amplexicaul, margins dentate. **Racemes** ebracteate, (with a terminal cluster of sterile flowers). **Fruiting pedicels** divaricate, 2–3 mm. **Flowers:** calyx narrowly campanulate; sepals green to purplish or (sterile flowers) lilac-purple, (narrowly ovate, sterile flowers linear-lanceolate), 3–5 mm, (8–13 mm in sterile flowers), keeled basally (not keeled in sterile flowers; sparsely hirsute in fertile flowers, glabrous in sterile flowers); petals purple (with darker purple veins), 8–11 mm, blade 4–6 × 2.5–3.5 mm, margins not crisped, (flaring), claw 4–5 mm, narrower than blade; stamens in 3 unequal pairs, (purple); filaments: abaxial pair (connate), 4–5 mm, lateral pair 2–3 mm, adaxial pair (connate nearly to apex), 5–6.5 mm; anthers: abaxial and lateral pairs fertile, 1.4–1.8 mm, adaxial pair sterile, 0.4–0.8 mm; gynophore 0.1–0.3 mm. **Fruits** divaricate, smooth, curved upward, slightly flattened, 1.3–2.5 cm × 2.5–3.5 mm; valves each with prominent midvein, (hispid, trichomes setiform, 0.5–0.8 mm); replum straight; ovules 46–60 per ovary; style 0.2–0.5 mm; stigma slightly 2-lobed. **Seeds** ovoid (plump), 1.2–1.5 × 0.6–0.8 mm; wing absent. $2n = 28$.

Flowering Apr–May. Gravelly sedimentary scree and lag-barrens in chaparral-oak woodlands; of conservation concern; 500–900 m; Calif.

Streptanthus callistus is known from the Mount Hamilton Range of Santa Clara County and is considered to be the most endangered species in the genus.

9. **Streptanthus campestris** S. Watson, Proc. Amer. Acad. Arts 25: 125. 1890 [C]

Perennials; (short-lived, caudex not woody); usually glabrous, (basal leaf blade margins pubescent, sometimes sepals). **Stems** unbranched or branched, (few, glaucous), 6–15(–18) dm. **Basal leaves** often rosulate; petiolate; blade (fleshy), oblanceolate to obovate, 3.5–21 cm, margins dentate, (bristly ciliate throughout or only teeth and petiole ciliate). **Cauline leaves:** blade lanceolate to narrowly ovate, 3.5–11(–15) cm × 6–14 mm (smaller distally), base auriculate to amplexicaul, margins usually entire or undulate, rarely dentate. **Racemes** ebracteate, (with densely clustered buds, later lax). **Fruiting pedicels** divaricate-ascending, (straight), 5–18 mm. **Flowers:** calyx campanulate; sepals (suberect), purple, (broadly ovate or oblong), 7–10 mm, not keeled, (apically bristly or not); petals light purple (with pale yellow claw), 9–12 mm, blade 2–3.5 × 0.5–1 mm, margins not crisped, claw 6–9 mm, wider than blade; stamens tetradynamous; filaments: median pairs (distinct), 6–8 mm, lateral pair 4–6 mm; anthers (all) fertile, 3–4 mm; gynophore 0.5–1.5 mm. **Fruits** spreading to ascending, smooth, slightly curved to straight, flattened, 6–14 cm × 2–3.5 mm; valves each with obscure midvein; replum straight; ovules 50–102 per ovary; style 1–3 mm; stigma 2-lobed. **Seeds** oblong, 2–3 × 1.4–2 mm; wing 0.1–0.2 mm wide at apex.

Flowering May–Jun. Rocky openings in chaparral, open conifer forests, openings and after fires in chaparral-oak woodlands; of conservation concern; 900–2300 m; Calif.; Mexico (Baja California).

Streptanthus campestris is distributed in California in Riverside, San Bernardino, San Diego, Santa Barbara, and Ventura counties, and in Baja California in Sierra San Pedro Mártir and Sierra Juárez.

10. Streptanthus carinatus C. Wright ex A. Gray, Smithsonian Contr. Knowl. 5(6): 11. 1853

Disaccanthus carinatus (C. Wright ex A. Gray) Greene

Annuals or biennials; (glaucous), glabrous throughout. **Stems** branched basally and/or distally, (2.1–)3–6.5(–7.5) dm. **Basal leaves** rosulate; petiolate; blade pinnatifid or oblanceolate, 3.5–15(–30) cm, margins runcinate-pinnatifid, dentate, or entire. **Cauline leaves**: blade ovate to lanceolate, 2.5–14 cm × 4–55 mm, (smaller distally), base auriculate to amplexicaul, margins runcinate-pinnatifid, dentate, or entire (usually entire distally). **Racemes** ebracteate, (lax). **Fruiting pedicels** divaricate-ascending, (straight or curved upward), 7–22(–35) mm. **Flowers**: calyx urceolate or campanulate; sepals purple, or ochroleucous to yellowish, 8–11 mm, keeled; petals white (with purplish veins) or purple (with white margins, recurved), 14–18 mm, blade 5–8 × 1–1.5 mm, margins crisped, claw 9–12 mm, about as wide as blade; stamens in 3 unequal pairs; filaments (distinct): abaxial pair 6–10 mm, lateral pair 4–7 mm, adaxial pair (exserted), 9–12 mm; anthers (all) fertile, 4–5 mm; gynophore 0.5–2 mm. **Fruits** ascending, smooth, straight, strongly flattened, 3–8 cm × 4.5–6 mm; valves each with obscure midvein; replum straight; ovules 26–54 per ovary; style 0.5–2 mm; stigma strongly 2-lobed. **Seeds** orbicular, 3–5 mm diam.; wing 0.6–1.1 mm wide, continuous. $2n = 28$.

Subspecies 2 (2 in the flora): sw United States, n Mexico.

A. R. Kruckeberg et al. (1982) presented evidence for hybridization between subsp. *carinatus* and subsp. *arizonicus*, and showed that the lighter flower color in the latter correlates with degree of leaf division. Within the range of *Streptanthus carinatus*, four color forms are known: purple, white, ochroleucous, and yellow. Variability in petal color does not correlate with geography or habitat, and flowers are often poorly preserved on herbarium specimens, making it unclear whether the color forms are allopatric.

1. Basal and proximal cauline leaf blades: margins runcinate-pinnatifid; sepals purple; petals purple with white margins . . . .
 . . . . 10a. *Streptanthus carinatus* subsp. *carinatus*
1. Basal and proximal cauline leaf blades: margins entire, dentate, or runcinate-pinnatifid; sepals ochroleucous to yellowish; petals white with purplish veins . . . .
 . . . . 10b. *Streptanthus carinatus* subsp. *arizonicus*

10a. Streptanthus carinatus C. Wright ex A. Gray subsp. carinatus

Disaccanthus validus Greene; *Streptanthus validus* (Greene) Cory

Basal and proximal cauline leaves: blade margins runcinate-pinnatifid. **Flowers**: sepals purple; petals purple with white margins. $2n = 28$.

Flowering Mar–Apr. Cliff bases, talus, limestone, gravelly slopes, brushy areas, canyons and isolated in washes; 700–1500 m; Ariz., N.Mex., Tex.; Mexico (Chihuahua, Coahuila).

10b. Streptanthus carinatus C. Wright ex A. Gray subsp. arizonicus (S. Watson) Kruckeberg, Rodman & Worthington, Syst. Bot. 7: 298. 1982

Streptanthus arizonicus S. Watson, Proc. Amer. Acad. Arts 25: 125. 1890; *Disaccanthus arizonicus* (S. Watson) Greene; *D. luteus* Greene; *D. mogollonicus* Greene; *S. arizonicus* var. *luteus* Kearney & Peebles

Basal and proximal cauline leaves: blade margins entire, dentate, or runcinate-pinnatifid. **Flowers**: sepals ochroleucous to yellowish; petals white with purplish veins. $2n = 28$.

Flowering Feb–Apr. Well-drained sandy or rocky soils, gravelly bajada slopes, canyons, in scrub or open woodland, gravelly roadsides, desert scrub, sandy grasslands; 600–1700 m; Ariz., N.Mex., Tex.; Mexico (Baja California, Chihuahua).

The Texas record, which is disjunct from the rest of the subspecies range, is based on *Jones 29260* (MO) collected on 23 Mar 1932 from Eagle Pass in Maverick County.

11. Streptanthus cordatus Nuttall in J. Torrey and A. Gray, Fl. N. Amer. 1: 77. 1838 E

Cartiera cordata (Nuttall) Greene; *Erysimum cordatum* (Nuttall) Kuntze; *Euklisia cordata* (Nuttall) Rydberg

Perennials; (caudex simple or few-branched); (glaucous), usually glabrous, (petioles of basal leaves pubescent, sometimes sepals). **Stems** un-branched or branched, (1–)3–9(–11) dm. **Basal leaves** rosulate; petiolate (petioles broadly winged, setose-ciliate); blade usually spatulate to obovate, rarely narrowly oblanceolate, 1.5–8(–11) cm, margins dentate (at least distally). **Cauline leaves:** blade broadly oblong to ovate, suborbicular, or lanceolate, 2–9 cm × 7–45(–60) mm, (smaller distally), base auriculate to amplexicaul, margins entire or toothed, (apex rounded, obtuse, or acuminate to acute). **Racemes** ebracteate, (lax). **Fruiting pedicels** divaricate-ascending, (straight), 3–11(–14) mm. **Flowers:** calyx campanulate; sepals greenish brown to purple, (broadly oblong), 5–12 mm, (lateral pair subsaccate basally), not keeled, (glabrous or subapically bristly); petals purple to brownish, (claw purplish), 9–15 mm, blade 2–6 × 0.7–1 mm, (recurved or not), margins not crisped, claw 7–10 mm, wider than blade; stamens in 3 unequal pairs; filaments (distinct): abaxial pair 5–7 mm, lateral pair 4–5 mm, adaxial pair 7.5–10 mm; anthers (all) fertile, 2.5–5 mm; gynophore 0.5–1.5 mm. **Fruits** ascending to divaricate-ascending, smooth, straight, flattened, 5–10.5(–14.5) cm × (2.5–)3–6(–7) mm; valves each with prominent midvein; replum straight; ovules 20–38(–46) per ovary; style 0.2–3 mm; stigma subentire to strongly 2-lobed. **Seeds** broadly oblong to suborbicular, 2.5–5 × 2.2–5 mm; wing 0.1–0.9 mm wide, continuous.

Varieties 2 (2 in the flora): w United States.

Streptanthus cordatus is the most widespread species in the genus. It is highly variable in the shape and size of basal and cauline leaves, flower size, pedicel length, fruit length and width, style length, stigma lobing, seed shape and size, and width of the seed wing. The variation is continuous in almost every character, especially fruit width. R. E. Buck et al. (1993) indicated that in var. *cordatus* the fruits are more than 5 mm wide and in the Californian var. *duranii* they are less than 3 mm wide. One sporadically finds throughout the range of var. *cordatus* plants with mature fruits 2.5–3 mm wide. Such narrow-fruited forms are found in central Nevada (*Beatley 6037*, DS), Oregon (*Howell 28719*, CAS), Utah (*Goodman 1877*, MO), and Wyoming (*Rollins & Muñoz 2876*, GH).

1. Basal leaf blades obovate to spatulate, apices obtuse or rounded; cauline leaf blades broadly oblong to ovate or suborbicular, apices rounded to obtuse...... 11a. *Streptanthus cordatus* var. *cordatus*
1. Basal leaf blades narrowly oblanceolate, apices acuminate; cauline leaf blades lanHceolate, apices acuminate or acute................. .... 11b. *Streptanthus cordatus* var. *piutensis*

11a. Streptanthus cordatus Nuttall var. cordatus [E]

Cartiera arguta (Greene) Greene; *C. crassifolia* (Greene) Greene; *C. cuneata* Rydberg; *C. leptopetala* Greene; *C. multiceps* Greene; *Euklisia crassifolia* (Greene) Rydberg; *Streptanthus argutus* Greene; *S. crassifolius* Greene; *S. coloradensis* A. Nelson; *S. cordatus* var. *duranii* Jepson

Basal leaves: blade obovate to spatulate, apex obtuse or rounded. **Cauline leaves:** blade broadly oblong to ovate or suborbicular, apex rounded to obtuse. **Fruits** 2.5–6(–7) mm wide; stigma subentire to strongly 2-lobed. $2n = 28, 56$.

Flowering Apr–Jul. Dry stony slopes, sandy draws, limestone ledges, gravelly knolls, pinyon-juniper woodlands, sagebrush, open pine forests; 500–3100 m; Ariz., Calif., Colo., Idaho, Nev., N.Mex., Oreg., Utah, Wyo.

Some authors (e.g., R. E. Buck et al. 1993; R. C. Rollins 1993; N. H. Holmgren 2005b) reported var. *cordatus* (as *Streptanthus cordatus*) to have $2n = 24$. That count must be an error, one that has been perpetuated in the literature since it was first reported in 1939 (S. I. Warwick and I. A. Al-Shehbaz 2006).

11b. Streptanthus cordatus Nuttall var. piutensis J. T. Howell, Leafl. W. Bot. 10: 31. 1963 [E]

Basal leaves: blade narrowly oblanceolate, apex acuminate. **Cauline leaves:** blade lanceolate, apex acuminate or acute. **Fruits** 3–4.5 mm wide; stigma subentire.

Flowering Jun–Jul. Open chaparral, cypress woodland; 1200–1700 m; Calif.

Variety *piutensis* is known from the Piute Mountains, Kern County.

12. Streptanthus cutleri Cory, Rhodora 45: 259. 1943 [C] [F]

Annuals; glabrous throughout. **Stems** (simple from base), usually branched distally, 2–7 dm. **Basal leaves** subrosulate; petiolate; blade oblanceolate, 2.5–20 cm, margins runcinate-pinnatifid. **Cauline leaves** (petiolate); blade oblanceolate, 4–20 cm × 10–70 mm, (smaller distally), base not auriculate, margins runcinate-pinnatifid (undivided, usually entire distally). **Racemes** ebracteate, (lax). **Fruiting pedicels** divaricate-ascending,

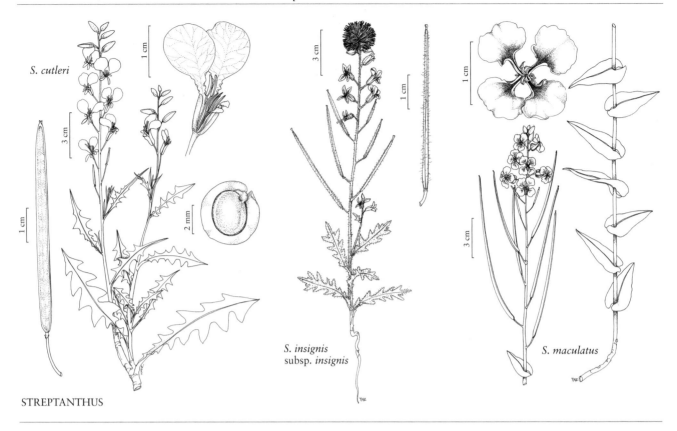

STREPTANTHUS

(5–)10–27 mm. **Flowers** (strongly zygomorphic); calyx campanulate; sepals (ascending), dark purple, 8–12 mm, not keeled; petals lavender to purple-lavender, (in 2 unequal pairs), abaxial pair undifferentiated into blade and claw, 5–6 mm, margins crisped, adaxial pair 15–26 mm, (not crisped), claw 7–11 mm, crisped, narrower than blade; stamens in 3 unequal pairs; filaments distinct: abaxial pair 6–7 mm, lateral pair 3–5 mm, adaxial pair 8–12 mm; anthers: abaxial and lateral pairs fertile, 4–5 mm, adaxial pair fertile, 3–3.5 mm; gynophore 0.2–0.5 mm. **Fruits** ascending, smooth, straight, strongly flattened, 3.2–7.5 cm × 4–5 mm; valves each with prominent midvein; replum straight; ovules 30–46 per ovary; style 0.2–0.5 mm; stigma 2-lobed. **Seeds** orbicular, 3–4 mm diam.; wing 0.5–1 mm wide, continuous.

Flowering Feb–Apr. Talus slopes, rocky hillsides, gravelly and dry stream beds, sand flats, limestone slopes, open scrub or woodland; of conservation concern; 400–700 m; Tex.; Mexico (Coahuila).

In Texas, *Streptanthus cutleri* is restricted to the Big Bend area in Brewster County.

13. **Streptanthus diversifolius** S. Watson, Proc. Amer. Acad. Arts 17: 363. 1882 [E]

Mitophyllum diversifolium (S. Watson) Greene; *Streptanthus linearis* Greene

Annuals; (somewhat glaucous), glabrous throughout. **Stems** branched distally, (1–)2–9 (–10.5) dm. **Basal leaves** not rosulate; petiolate; blade pinnatisect to pectinate into filiform segments, 2–8 cm. **Cauline leaves** (petiolate proximally, sessile distally); blade pectinate or pinnatisect into filiform segments, or (distally) linear-lanceolate to ovate or cordate, 2–12 cm, (segments 0.4–2.7 cm × 0.5–1.5 mm), base not auriculate proximally, amplexicaul distally, margins entire distally, becoming bracts. **Racemes** bracteate below or between proximalmost 1 or 2 flowers. **Fruiting pedicels** divaricate to divaricate-ascending, (straight), 2–10 mm. **Flowers:** calyx urceolate; sepals yellow or purplish, 5–7 mm, keeled, (apex recurved); petals pale yellow to whitish (with purple veins, recurved), 8–16 mm, blade 3–6 × 1.5–3.5 mm, margins not crisped, (undulate), claw 5–10 mm, narrower than blade; stamens in 3 unequal pairs; filaments (distinct): abaxial pair 2.5–6 mm, lateral pair 1.5–4 mm, adaxial pair 4–8 mm; anthers: abaxial and lateral pairs fertile, 3.5–4 mm, adaxial pair partially fertile, 1.5–2.3 mm;

gynophore 0.5–1.5 mm. **Fruits** geniculately reflexed, smooth, straight, flattened, 3–9 cm × 1–1.5(–2.2) mm; valves each with obscure midvein; replum straight; ovules (22–)38–80 per ovary; style 0.2–1 mm; stigma obscurely 2-lobed. **Seeds** oblong, 1–1.5(–2.2) × 0.6–1 (–1.5) mm; wing (0–)0.05–0.1(–0.4) mm wide, distal. $2n = 28$.

Flowering Apr–Jul. Dry rock outcrops, granitic domes, adjacent road cuts, disturbed areas, open oak woodland; 200–1900 m; Calif.

Streptanthus diversifolius is distributed in Butte, El Dorado, Fresno, Madera, Mariposa, Tuolumne, and Tulare counties.

14. **Streptanthus drepanoides** Kruckeberg & J. L. Morrison, Madroño 30: 230, figs. 1, 2b. 1983 E

Annuals; (glaucous), usually glabrous, (sometimes sepals pubescent). **Stems** unbranched or divaricately branched basally, 0.4–3.5(–4.5) dm. **Basal leaves** (soon withered); not rosulate; subsessile or shortly petiolate; blade orbicular, similar to cauline, margins with blunt teeth. **Cauline leaves:** blade (succulent), orbicular or orbicular-ovate, 1.3–9 cm × 10–75 mm (smaller distally), base auriculate-cordate, strongly overlapping, margins entire or shallowly dentate. **Racemes** ebracteate, (dense). **Fruiting pedicels** divaricate-ascending, (straight), 1.5–4 mm. **Flowers:** calyx urceolate; sepals greenish yellow, (ovate), 5–7 mm, keeled, (apex reflexed, glabrous or pilose); petals ochroleucous (with brownish purple veins), 7–10 mm, blade 2–3 × 1–1.5 mm, margins crisped, claw 5–7 mm, wider than blade, (apex recurved); stamens in 3 unequal pairs; filaments: abaxial pair (connate to middle), 5–6 mm, lateral pair 3–4 mm, adaxial pair (connate to apex), 8–10 mm; anthers: abaxial and lateral pairs fertile, 1.8–2.5 mm, adaxial pair sterile, 0.7–1 mm; gynophore 0.5–1 mm. **Fruits** divaricate-ascending, usually smooth, rarely slightly torulose, strongly falcate, slightly flattened, 3–9 cm × 1–1.2 mm; valves each with obscure midvein; replum straight; ovules 30–50 per ovary; style 0.5–1 mm; stigma entire. **Seeds** oblong, 1–1.5 × 0.7–1 mm; wing 0–0.1 mm wide, distal. $2n = 28$.

Flowering May–Jul. Steep mobile substrate on sedimentary or volcanic slopes, usually on serpentine outcrops, openings in chaparral, pine woodland; 200–1800 m; Calif.

Streptanthus drepanoides is distributed in Butte, Colusa, Glenn, Shasta, Tehama, and Trinity counties.

15. **Streptanthus farnsworthianus** J. T. Howell, Leafl. W. Bot. 10: 182. 1965 E

Annuals; (glaucous, usually with purplish cast), glabrous. **Stems** unbranched or branched distally, 2–10 dm. **Basal leaves** (soon withered); rosulate; petiolate; blade 2-pinnatifid or 2-pinnatisect, 3.5–15 cm. **Cauline leaves:** blade lanceolate, 3–16 cm × 10–40 mm, (smaller distally), base amplexicaul, margins dentate or entire (entire distally). **Racemes** bracteate below or between proximalmost 1 or 2 flowers. **Fruiting pedicels** divaricate-ascending, (straight), 4–15 mm. **Flowers:** calyx urceolate; sepals violet-purple, 6–10 mm, slightly keeled or not, (apex recurved); petals white (with violet veins), 8–11 mm, blade 3–4 × 1.5–2 mm, margins not crisped, (undulate), claw 6–8 mm, about as wide as blade, (apex reflexed); stamens in 3 unequal pairs; filaments (distinct): abaxial pair 4–6 mm, lateral pair 3–4 mm, adaxial pair 7–9 mm; anthers: abaxial and lateral pairs fertile, 3.5–5 mm, adaxial pair partially fertile, 2.5–3 mm; gynophore 1–2 mm. **Fruits** ascending, smooth, straight or slightly curved, flattened, 6–12 cm × 1.7–2.5 mm; valves each with prominent midvein; replum straight; ovules 60–100 per ovary; style 0.1–1 mm; stigma slightly 2-lobed. **Seeds** broadly ovoid to suborbicular, 2–2.5 × 1.5–2 mm; wing 0.1–0.3 mm wide, distal ½. $2n = 28$.

Flowering May–Jun. Sparsely rocky sites with vegetation (fractured slate, metamorphic gravels, gruss deposits), oak-pine woodlands; 400–1400 m; Calif.

Streptanthus farnsworthianus is known from Fresno, Kern, Madera, and Tulare counties.

16. **Streptanthus fenestratus** (Greene) J. T. Howell, Leafl. W. Bot. 9: 184. 1961 C E

Pleiocardia fenestrata Greene, Leafl. Bot. Observ. Crit. 1: 86. 1904

Annuals; (glaucous), glabrous throughout. **Stems** unbranched or branched proximally, (0.5–)1–2(–3.5) dm. **Basal leaves** rosulate; petiolate (petioles slender); blade 1- or 2-pinnatisect, 1.5–6 cm. **Cauline leaves:** blade ovate to lanceolate, similar to basal, base auriculate proximally, amplexicaul distally, margins usually coarsely dentate or entire, rarely pinnatifid (not divided into filiform segments, entire distally). **Racemes** bracteate below or between proximalmost 1 or 2 flowers. **Fruiting pedicels** divaricate-ascending, (straight), 2–7 mm. **Flowers:** calyx tubular-urceolate; sepals purple, 5–7 mm, slightly keeled or not, (apex recurved); petals rose-purple, 9–15 mm, blade 3–6 × 2.5–3.5 mm, margins not crisped, claw 6–9 mm, narrower than blade; stamens in 3 unequal pairs; filaments (distinct): abaxial pair 3–4 mm, lateral pair 1.5–2.5 mm, adaxial pair 5–7 mm, (recurved); anthers (all) fertile, 2–2.5 mm; gynophore 0.2–2.5 mm. **Fruits** usually ascending to divaricate, rarely reflexed, smooth or slightly torulose, slightly flattened, 2–5 cm × 1.2–1.7 mm; valves each with obscure midvein; replum straight; ovules 22–38 per ovary; style 0.1–1 mm; stigma entire. **Seeds** oblong, 1–1.5 × 0.7–0.9 mm; wing (0–)0.05–0.1 mm wide, distal. $2n = 28$.

Flowering May–Jun. Granitic ledges and sand, gruss deposits in open pine forest; of conservation concern; 1100–1800 m; Calif.

Streptanthus fenestratus is known from Kings Canyon, Fresno County.

17. **Streptanthus glandulosus** Hooker, Icon. Pl. 1: plate 40. 1836 [E]

Erysimum glandulosum (Hooker) Kuntze; *Euklisia glandulosa* (Hooker) Greene

Annuals; (usually glaucous distally), often sparsely to densely hirsute proximally, sometimes glabrous throughout. **Stems** unbranched or branched throughout, 1.5–12 dm. **Basal leaves** (soon withered); not rosulate; petiolate (petiole winged); blade lanceolate to oblanceolate, 1–15 cm, base narrowed to petiole, margins coarsely dentate to ± lobed. **Cauline leaves**: blade linear-lanceolate to lanceolate or oblanceolate, 1–12 cm × 1–11 mm, (smaller distally), blade auriculate to amplexicaul, margins entire or coarsely dentate, (often entire distally). **Racemes** ebracteate, (lax, secund or not, rachis usually straight, rarely flexuous). **Fruiting pedicels** ascending to divaricate or spreading, (straight or recurved), 2–32 mm, (glabrous or pubescent). **Flowers** (zygomorphic); calyx urceolate; sepals white, greenish white, cream, yellow, rose, purple, lilac, maroon, reddish purple, or purple-black, (lanceolate to broadly ovate), (3–)5–10(–13) mm, not keeled, (glabrous or sparsely hirsute); petals lavender, purple, or white (sometimes with brown or purple veins), 7–17 mm, blade 2–7 × 1–3 mm, (subequal or adaxial pair distinctly longer, more recurved), margins crisped, claw 5–13 mm, wider than blade; stamens in 3 unequal pairs; filaments: abaxial pair (distinct), 4–9 mm, lateral pair 2.5–7 mm, adaxial pair (exserted, connate ⅔ their length), 5–13 mm; anthers: abaxial and lateral pairs fertile, 1–2.5 mm, adaxial pair sterile, 0.3–1 mm; gynophore 0.2–1.5 mm. **Fruits** ascending to divaricate or spreading, smooth, straight, curved upward, or arcuate, flattened, 3–11 cm × 1.5–2.5 mm; valves each with prominent midvein, (glabrous or sparsely to moderately hirsute); replum straight; ovules 22–70 per ovary; style 0.1–2.5 mm; stigma subentire. **Seeds** ovoid to oblong, 1.5–2.1 × 1–1.5 mm; wing 0.1–0.5 mm wide, continuous.

Subspecies 8 (8 in the flora): w United States.

Except for the widespread subsp. *glandulosus*, the subspecies are highly restricted and six are of conservation concern.

1. Stems proximally and cauline leaf blade surfaces glabrous; sepals purple-black; rachises flexuous; fruiting pedicels 10–32 mm. . . . .
. . . . 17e. *Streptanthus glandulosus* subsp. *niger*
1. Stems proximally and cauline leaf blade surfaces usually sparsely to densely hirsute (cauline leaves glabrous in subsp. *albidus*); sepals white, greenish white, cream, yellow, lilac, lavender, rose, purple, maroon, or reddish purple; rachises straight; fruiting pedicels 2–15 mm.
  2. Racemes not secund.
    3. Sepals reddish purple, dark maroon, or lilac-lavender; San Luis Obispo County north into Lake, Mendocino, Santa Clara counties . . . . 17a. *Streptanthus glandulosus* subsp. *glandulosus*
    3. Sepals white, greenish white, or pale yellow; Santa Clara County . . . .
    . . . . 17b. *Streptanthus glandulosus* subsp. *albidus*
  2. Racemes secund.
    4. Sepals lavender to rose or purple.
      5. Fruiting pedicels 5–15 mm; fruits spreading to reflexed, arcuate; filaments: adaxial pair 7–10 mm; Sonoma County . . . . 17c. *Streptanthus glandulosus* subsp. *hoffmanii*
      5. Fruiting pedicels 2–5 mm; fruits divaricate to ascending, curved upward; filaments: adaxial pair 6–8 mm; Marin County . . . . 17f. *Streptanthus glandulosus* subsp. *pulchellus*
    4. Sepals greenish white, white, cream, or pale yellow.
      6. Petals 7–8 mm; filaments: adaxial pair 5–6 mm; fruits divaricate-ascending, straight or curved inwards; Josephine County, Oregon. . . . .
      . . . . 17d. *Streptanthus glandulosus* subsp. *josephinensis*
      6. Petals 10–17 mm; filaments: adaxial pair 7–10 mm; fruits spreading to reflexed, arcuate; Marin, Napa, Sonoma counties, California.
        7. Sepals greenish white with lavender or purplish base; petals with purple veins; Marin, Napa counties. . . . . . .
        . . . . .17g. *Streptanthus glandulosus* subsp. *secundus*
        7. Sepals cream or pale yellow, without purplish or lavender base; petals without purple veins; Sonoma County. . . . 17h. *Streptanthus glandulosus* subsp. *sonomensis*

17a. Streptanthus glandulosus Hooker subsp. glandulosus [E]

Euklisia aspera (Greene) Greene; *E. bakeri* Greene; *E. biolettii* (Greene) Greene; *E. eliator* Greene; *E. violacea* Greene; *Streptanthus albidus* Greene subsp. *peramoenus* (Greene) Kruckeberg; *S. asper* Greene; *S. biolettii* Greene; *S. mildrediae* Greene; *S. peramoenus* Greene; *S. versicolor* Greene

Stems 1–10 dm, densely to moderately hirsute proximally, moderately to sparsely so distally. Cauline leaves: blade conduplicate or flat distally, margins entire or denticulate, surfaces densely to sparsely hirsute. Racemes not secund; rachis straight. Fruiting pedicels 3–15 mm, sparsely pubescent or glabrous. Flowers: sepals reddish purple, dark maroon, or lilac-lavender, 5–13 mm, sparsely pubescent or glabrous; petals lavender to purple (with or without darker purple veins), 8–17 mm; adaxial filaments 8–13 mm. Fruits ascending to divaricate or reflexed, straight or arcuate; valves glabrous or sparsely pubescent. $2n = 28$.

Flowering Apr–Jul. Rocky, often barren, ultramafic or metamorphic (Franciscan formation) slopes, chaparral openings, steep slopes in woodland; 0–1400 m; Calif.

As delimited here, subsp. *glandulosus* is heterogeneous, with varying flower color, fruit orientation, and indumentum density. It grows in Alameda, Colusa, Contra Costa, Lake, Mendocino, Monterey, Napa, San Benito, San Luis Obispo, Santa Barbara, Santa Clara, Solano, Sonoma, Stanislaus, and Tehama counties.

17b. Streptanthus glandulosus Hooker subsp. albidus (Greene) Al-Shehbaz, M. S. Mayer & D. W. Taylor, Novon 18: 280. 2008 [C][E]

Streptanthus albidus Greene, Pittonia 1: 62. 1887; *Euklisia albida* (Greene) Greene; *S. glandulosus* var. *albidus* (Greene) Jepson

Stems 3–12 dm, sparsely hirsute proximally, glabrous distally. Cauline leaves: blade conduplicate distally, margins entire, surfaces glabrous. Racemes not secund; rachis straight. Fruiting pedicels 4–8 mm, glabrous. Flowers: sepals greenish white, white, or pale yellow, 7–11 mm, glabrous; petals white (with light brown or purplish veins), 11–17 mm; adaxial filaments 8–10 mm. Fruits divaricate to ascending, straight; valves glabrous. $2n = 28$.

Flowering Apr–Jul. Rocky, often barren ultramafic grassland openings; of conservation concern; 100–800 m; Calif.

Subspecies *albidus* is restricted to about ten populations at the western base of the Mount Hamilton Range, Santa Clara County. It intergrades locally with subsp. *glandulosus* in flower color.

17c. Streptanthus glandulosus Hooker subsp. hoffmanii (Kruckeberg) M. S. Mayer & D. W. Taylor, Novon 18: 280. 2008 [C][E]

Streptanthus glandulosus var. *hoffmanii* Kruckeberg, Madroño 14: 223. 1958

Stems 1.5–4.5(–7) dm, densely hirsute proximally, sparsely to moderately so distally. Cauline leaves: blade flat distally, margins entire or denticulate, surfaces moderately to sparsely hirsute. Racemes secund; rachis straight. Fruiting pedicels 5–15 mm, sparsely pubescent or glabrous. Flowers: sepals lavender or rose to purple, 6–7 mm, sparsely pubescent or glabrous; petals lavender to purplish, 10–12 mm; adaxial filaments 7–10 mm. Fruits spreading to reflexed, arcuate; valves sparsely pubescent or glabrous. $2n = 28$.

Flowering May–Jul. Serpentine or Franciscan formation outcrops; of conservation concern; Calif.

Subspecies *hoffmanii* is known from Sonoma County.

17d. Streptanthus glandulosus Hooker subsp. josephinensis Al-Shehbaz & M. S. Mayer, Novon 18: 280. 2008 [C][E]

Stems moderately hirsute proximally, glabrous distally, 1.7–3.8 dm. Cauline leaves: blade flat distally, margins dentate, surfaces sparsely hirsute. Racemes secund; rachis straight. Fruiting pedicels 5–10 mm, glabrous or sparsely pubescent. Flowers: sepals white or cream, 3–5 mm, glabrous or sparsely pubescent; petals white (with purple veins), 7–8 mm; adaxial filaments 5–6 mm. Fruits divaricate-ascending, straight or curved inwards; valves glabrous or sparsely pubescent.

Flowering Jun–Jul. Serpentine areas; of conservation concern; Oreg.

Subspecies *josephinensis*, known from Josephine County, is disjunct from the white-flowered Californian subspecies. R. C. Rollins (1993) and R. E. Buck et al. (1993) attributed the Oregonian plants to subspp. *glandulosus* and *secundus*, respectively.

17e. Streptanthus glandulosus Hooker subsp. **niger** (Greene) Al-Shehbaz, M. S. Mayer & D. W. Taylor, Novon 18: 280. 2008 [C][E]

Streptanthus niger Greene, Bull. Torrey Bot. Club 13: 141. 1886; *Euklisia nigra* (Greene) Greene; *S. glandulosus* var. *niger* (Greene) Munz

Stems 2–11 dm, glabrous throughout. **Cauline leaves:** blade conduplicate distally, margins entire, surfaces glabrous. **Racemes** not secund; rachis flexuous. **Fruiting pedicels** 10–32 mm, glabrous. **Flowers:** sepals dark purple-black, 7–8 mm, glabrous; petals dark purple, 10–12 mm; adaxial filaments 8–10 mm. **Fruits** divaricate-ascending, straight; valves glabrous. $2n = 28$.

Flowering May–Jul. Serpentine grasslands; of conservation concern; Calif.

Subspecies *niger* is known from the Tiburon Peninsula, Marin County.

17f. Streptanthus glandulosus Hooker subsp. **pulchellus** (Greene) Kruckeberg, Madroño 14: 222. 1958 [C][E]

Streptanthus pulchellus Greene, Pittonia 2: 225. 1892; *Euklisia pulchella* (Greene) Greene; *S. glandulosus* var. *pulchellus* (Greene) Jepson

Stems 1–3(–3.7) dm, densely hirsute proximally, glabrous distally. **Cauline leaves:** blade flat distally, margins dentate, surfaces hirsute. **Racemes** secund; rachis straight. **Fruiting pedicels** 2–5 mm, sparsely pubescent or glabrous. **Flowers:** sepals rose to lavender or purple, 5–7 mm, glabrous or sparsely pubescent; petals purple, 8–11 mm; adaxial filaments 6–8 mm. **Fruits** divaricate to ascending, curved upward; valves glabrous or sparsely pubescent. $2n = 28$.

Flowering May–Jun. Serpentine outcrops; of conservation concern; Calif.

Subspecies *pulchellus* is known from Mt. Tamalpais, Marin County.

17g. Streptanthus glandulosus Hooker subsp. **secundus** (Greene) Kruckeberg, Madroño 14: 223. 1958 [E]

Streptanthus secundus Greene, Fl. Francisc., 261. 1891; *Euklisia secunda* (Greene) Greene

Stems 1.5–9.3 dm, densely hirsute proximally, sparsely pubescent distally. **Cauline leaves:** blade flat distally, margins dentate, surfaces densely to moderately hirsute. **Racemes** secund; rachis straight. **Fruiting pedicels** 7–10 mm, sparsely pubescent or glabrous. **Flowers:** sepals greenish white (base lavender or purplish), 6–7 mm, glabrous; petals white (with purple veins), 12–16 mm; adaxial filaments 7–9 mm. **Fruits** spreading to reflexed, arcuate; valves glabrous. $2n = 28$.

Flowering Apr–Jun. Rocky, often barren, open slopes, talus, forest openings; 100–1200 m; Calif.

17h. Streptanthus glandulosus Hooker subsp. **sonomensis** (Kruckeberg) M. S. Mayer & D. W. Taylor, Novon 18: 280. 2008 [C][E]

Streptanthus glandulosus var. *sonomensis* Kruckeberg, Madroño 14: 223. 1958

Stems 2.3–7(–9) dm, moderately to densely hirsute proximally, glabrous distally. **Cauline leaves:** blade conduplicate distally, margins entire or denticulate, surfaces hirsute. **Racemes** secund; rachis straight. **Fruiting pedicels** 3–10 mm, sparsely pubescent or glabrous. **Flowers:** sepals cream or pale yellow, 5–6 mm, glabrous or sparsely pubescent; petals white (without purple veins), 10–17 mm; adaxial filaments 7–10 mm. **Fruits** spreading to slightly reflexed, arcuate; valves glabrous or sparsely pubescent. $2n = 28$.

Flowering May–Jul. Serpentine or Franciscan formation outcrops; of conservation concern; Calif.

Subspecies *sonomensis* is known from Sonoma County.

18. Streptanthus gracilis Eastwood, Proc. Calif. Acad. Sci., ser. 3, 2: 285. 1902 E

Pleiocardia gracilis (Eastwood) Greene

Annuals; (slender); (glaucous), glabrous throughout. **Stems** often branched basally, (0.6–)1–3.5 dm. **Basal leaves** rosulate; petiolate; blade oblanceolate to spatulate, 1–7 cm, margins usually sinuate to dentate, rarely lobed. **Cauline leaves** (shortly petiolate or sessile); blade oblong to ovate, 0.5–3 cm × 1–5 mm, (smaller distally), base auriculate or (distally) amplexicaul, margins entire or dentate apically, (entire distally). **Racemes** bracteate below or between proximalmost 1 or 2 flowers. **Fruiting pedicels** divaricate to ascending, (straight or recurved), 3–6 mm, (not expanded at receptacle). **Flowers:** calyx slightly urceolate; sepals rose-purple, 4–5 mm, not keeled; petals (spreading), pinkish, 7–10 mm, blade 3–5 × 1.5–2.5 mm, margins not crisped, claw 5–6 mm, narrower than blade; stamens in 3 unequal pairs; filaments (distinct): abaxial pair 3–4 mm, lateral pair 1.5–2.5 mm, adaxial pair 5–6 mm; anthers (all) fertile, 1.5–2 mm; gynophore 0.5–3 mm. **Fruits** ascending, torulose, straight, slightly flattened, 3–7 cm × 1–1.5 mm; valves each with obscure midvein; replum straight; ovules 24–52 per ovary; style 0.1–0.5 mm; stigma entire. **Seeds** oblong, 1–1.5 × 0.6–0.9 mm; wing (0–)0.05–0.25 mm, distal.

Flowering Jun–Sep. Rocky open subalpine or alpine vegetation, pockets of weathered granite sand and gruss; 2600–3600 m; Calif.

Streptanthus gracilis is known from the Kings-Kern Divide in the southern Sierra Nevada of Fresno, Inyo, and Tulare counties.

19. Streptanthus hesperidis Jepson, Erythea 1: 14. 1893 C E

Pleiocardia hesperidis (Jepson) Greene; *Streptanthus breweri* A. Gray var. *hesperidis* (Jepson) Jepson

Annuals; glabrous. **Stems** unbranched or branched basally, 1–3 dm. **Basal leaves** (soon withered); not rosulate; shortly petiolate; blade broadly obovate, 2–3 cm, margins coarsely and bluntly dentate distally. **Cauline leaves:** blade (mostly yellowish), ovate to lanceolate, 1–4 cm × 5–15 mm, (smaller distally), base amplexicaul, margins often entire. **Racemes** ebracteate, (secund, rachis flexuous). **Fruiting pedicels** divaricate-ascending, (straight), 1–2 mm. **Flowers:** calyx urceolate; sepals yellow-green, (ovate-lanceolate), 5–8 mm, keeled, (apex recurved or flaring); petals whitish (with purplish veins), 6–8 mm, blade 2–3 × 1–1.3 mm, margins not crisped, claw 4–5 mm, about as wide as blade; stamens in 3 unequal pairs; filaments: abaxial pair (connate ½ their length), 3–5 mm, lateral pair 2–3.5 mm, adaxial pair (connate entire length, recurved), 6–9 mm; anthers: abaxial and lateral pairs fertile, 1.7–2.5 mm, adaxial pair sterile, 0.7–1.2 mm; gynophore 0.2–0.5 mm. **Fruits** divaricate-ascending, somewhat torulose, nearly straight to, rarely, arcuate, flattened, 3–6 cm × 0.9–1.1 mm; valves each with obscure midvein; replum straight; ovules 26–38 per ovary; style 0.1–0.3 mm; stigma entire. **Seeds** oblong, 1.2–1.5 × 0.6–0.8 mm; wing (0–)0.05–0.1 mm wide, distal. $2n = 28$.

Flowering May–Jul. Serpentine barrens and associated openings in chaparral-oak woodland and cypress woodland; of conservation concern; 200–600 m; Calif.

Streptanthus hesperidis is known from Lake and Napa counties. R. E. Buck et al. (1993) reduced it to a variety of *S. breweri*, but the two are sufficiently distinct to be recognized as independent species. It differs from *S. breweri* by having flexuous (versus straight) raceme rachises, yellowish (versus glaucous green) foliage, smaller petals (6–8 versus 8–12 mm), and often straight (versus often arcuate) fruits.

20. Streptanthus hispidus A. Gray, Proc. Calif. Acad. Sci. 3: 101. 1864 C E

Euklisia hispida (A. Gray) Greene

Annuals; densely hirsute-hispid throughout, (trichomes to 1.4 mm). **Stems** unbranched or branched basally, 0.3–3 dm. **Basal leaves** not rosulate; shortly petiolate; blade obovate, 1–5 cm, margins coarsely to shallowly dentate. **Cauline leaves:** blade obovate to oblong, 0.7–6 cm × 2–25 mm, base cuneate or truncate, not auriculate or (distally) minutely ariculate, margins coarsely dentate. **Racemes** ebracteate, (not secund, with a terminal cluster of sterile flowers). **Fruiting pedicels** divaricate-ascending, (straight), 2–5 mm. **Flowers:** calyx subcampanulate; sepals pale green to purplish, ovate, 4–6 mm, not keeled; petals light purple (with white margins), 6–9 mm, blade 2–3 × 1–1.5 mm, margins crisped, claw 5–6 mm, wider than blade; stamens in 3 unequal pairs; filaments: abaxial pair (connate ca. ½ their length), 4–5 mm, lateral pair 3–4 mm, adaxial pair (exserted, connate to near apex), 5–6 mm; anthers: abaxial and adaxial pairs fertile, 1.5–1.8 mm, adaxial pair sterile, 0.3–0.5 mm; gynophore 0.1–0.3 mm. **Fruits** divaricate-ascending to suberect, straight, flattened, 4–8.5 cm × 2–2.5 mm; valves each with prominent midvein; replum straight; ovules 34–66 per ovary; style 0.4–1 mm; stigma slightly 2-lobed. **Seeds** ovoid to suborbicular, 1.6–2 × 1.2–1.8 mm; wing 0.2–0.35 mm wide, continuous. $2n = 28$.

Flowering Mar–Jun. Talus or rocky outcrops (Franciscan formation, largely on chert) and sparsely vegetated openings in grassland or chaparral; of conservation concern; 600–1200 m; Calif.

Streptanthus hispidus is known from Mt. Diablo in Contra Costa County.

21. Streptanthus howellii S. Watson, Proc. Amer. Acad. Arts 20: 353. 1885 C E

Cartiera howellii (S. Watson) Greene

Perennials; (caudex simple or branched); (glaucous), glabrous. **Stems** usually unbranched, rarely branched distally, 3–8 dm. **Basal leaves** (soon withered); not rosulate; petiolate; blade (somewhat fleshy), similar to cauline. **Cauline leaves** (petiolate); blade broadly spatulate-obovate to suborbicular, or (distally) narrowly oblong-oblanceolate to linear, 1.5–10 cm × 5–45 mm, base (distally) cuneate to attenuate, not auriculate, margins entire, repand, or obtusely dentate, (entire distally). **Racemes** ebracteate, (lax). **Fruiting pedicels** divaricate to ascending, (straight), 7–17 mm. **Flowers:** calyx subcampanulate; sepals purplish, oblong, 5–8 mm, not keeled; petals maroon to purple, 8–12 mm, blade 3–5 × 0.5–1 mm, margins not crisped, claw 5–7 mm, wider than blade; stamens in 3 unequal pairs; filaments (distinct): abaxial pair 5–6 mm, lateral pair 4–5 mm, adaxial pair 6–7 mm; anthers (all) fertile, 3–3.5 mm; gynophore 0.5–5 mm. **Fruits** divaricate-ascending, smooth, straight to slightly curved inwards, flattened, 5.5–12 cm × 2.5–3.2(–3.5) mm; valves each with prominent midvein; replum straight; ovules 24–44 per ovary; style 0.3–3 mm; stigma subentire. **Seeds** broadly oblong to suborbicular, 3–4 × 2–3 mm; wing 0.5–1.1 mm wide, continuous. $2n = 28$.

Flowering Jun–Jul. Dry serpentine slopes and ridges in open conifer-hardwood forests; of conservation concern; 600–800 m; Calif., Oreg.

Streptanthus howellii is known from the Klamath Mountains in California (Del Norte and Siskiyou counties) and Oregon (Curry and Josephine counties).

22. Streptanthus hyacinthoides Hooker, Bot. Mag. 63: plate 3516. 1836 E

Erysimum hyacinthoides (Hooker) Kuntze; *Euklisia hyacinthoides* (Hooker) Small; *Icianthus atratus* Greene; *I. glabrifolius* (Buckley) Greene; *I. hyacinthoides* (Hooker) Greene; *Streptanthus glabrifolius* Buckley

Annuals; usually glabrous throughout (sometimes stem bases pubescent). **Stems** unbranched or branched distally, 2.8–13 dm. **Basal leaves** not seen. **Cauline leaves** (shortly petiolate or sessile); blade usually linear to linear-lanceolate, rarely oblanceolate, 2.5–16 cm × 2–7(–20) mm, (smaller distally), base not auriculate, margins entire. **Racemes** ebracteate. **Fruiting pedicels** divaricate-ascending, (straight), 3–10 mm. **Flowers** (slightly zygomorphic, horizontal to pendent); calyx campanulate; sepals purple, lanceolate, 7–11 mm, not keeled; petals deep purple to magenta or whitish (with purple veins), 13–20 mm, blade 7–12 × 1–4 mm, margins not crisped, claw 5–12 mm, (strongly recurved); stamens in 3 unequal pairs; filaments: abaxial pair (connate less than ½ their length), 8–14 mm, lateral pair 4–7 mm, adaxial pair (connate ca. ¾ their length), 7–10 mm; anthers: abaxial and lateral pairs fertile, 2–3 mm, adaxial pairs sterile, 1–1.5 mm; gynophore 0.3–3 mm. **Fruits** divaricate-ascending, smooth, straight or slightly curved inwards, flattened, 3.7–11.2 cm × 1.5–2 mm; valves each with prominent midvein; replum straight; ovules 90–120 per ovary; style 0.3–1.5 mm; stigma subentire. **Seeds** broadly ovoid, 1.5–1.8 × 1–1.3 mm; wing 0.1–0.25 mm, continuous.

Flowering May–Jun. Sandy soils of prairies, open glades in woods, grassy roadsides; 100–800 m; Ark., Kans., La., Okla., Tex.

Streptanthus hyacinthoides is widespread in eastern Texas and central and northwestern Oklahoma; in Arkansas it is restricted to Nevada and Ouachita counties; in Kansas to Barber and Comanche counties; and in Louisiana to Bienville, Caddo, Natchitoches, and Winn parishes.

23. **Streptanthus insignis** Jepson, Man. Fl. Pl. Calif., 420. 1925 E F

Annuals; hirsute-hispid throughout. **Stems** unbranched or branched distally, 1.2–6 dm. **Basal leaves** not rosulate; shortly petiolate; blade lanceolate to oblanceolate, 2–8 cm, margins coarsely dentate to pinnatifid. **Cauline leaves:** blade lanceolate, 1.3–9 cm × 2–20 mm, (smaller distally), base auriculate, margins dentate. **Racemes** ebracteate, (lax, with a terminal cluster of dark purple or greenish yellow sterile flowers). **Fruiting pedicels** divaricate to ascending, (straight, curved upward, or recurved), 3–8 mm. **Flowers:** calyx campanulate; sepals (of fertile flowers) dark purple, purplish, or greenish yellow, (oblong to lanceolate), 4–6 mm, (8–14 mm in sterile flowers), not or slightly keeled, (hirsute-hispid in fertile flowers, glabrous in sterile flowers); petals: (abaxial pair spreading and reflexed, adaxial pair parallel and erect), purplish white, lemon-yellow, or yellowish white (with darker midvein), 7–12 mm, blade 2–4 × 1–2 mm, margins crisped, claw 5–8 mm, about as wide as blade; stamens in 3 unequal pairs; filaments: abaxial pair (distinct), 4–6 mm, lateral pair 3–4 mm, adaxial pair (connate their entire length), 6–9 mm; anthers: abaxial and lateral pairs fertile, 1.3–2 mm, adaxial pairs sterile, 0.2–0.7 mm; gynophore 0.4–1 mm. **Fruits** ascending or reflexed, smooth, straight or slightly curved, flattened, 3.5–11.4 cm × 1.5–2 mm; valves each with prominent midvein, (glabrate or sparsely to densely hispid, trichomes to 1.7 mm); replum straight; ovules 32–94 per ovary; style 1–2 mm; stigma entire. **Seeds** broadly oblong, 1.4–2 × 1–1.3 mm; wing 0.15–0.3 mm wide, continuous.

Subspecies 2 (2 in the flora): California.

1. Racemes with terminal cluster of sterile flowers dark purple; sepals of fertile flowers dark purplish; petals purplish white; fruit valves usually glabrate or sparsely hispid, rarely densely so . . . . 23a. *Streptanthus insignis* subsp. *insignis*
1. Racemes with terminal cluster of sterile flowers greenish yellow or purple; sepals of fertile flowers greenish yellow or purplish; petals lemon-yellow or yellowish white; fruit valves moderately to densely hispid . . . . .23b. *Streptanthus insignis* subsp. *lyonii*

23a. **Streptanthus insignis** Jepson subsp. **insignis** E F

Racemes: terminal cluster of sterile flowers dark purple. **Flowers:** sepals of fertile ones purple or purplish; petals purplish white. **Fruit valves** glabrate or sparsely to, rarely, densely hispid. $2n = 28$.

Flowering Mar–May. Steep scree or talus slopes of serpentine or Franciscan formation, openings in chaparral, oak woodland; 300–1100 m; Calif.

Subspecies *insignis* is known from Fresno, Merced, Monterey, and San Benito counties.

23b. **Streptanthus insignis** Jepson subsp. **lyonii** Kruckeberg & J. L. Morrison, Madroño 30: 234, fig. 3. 1983 C E

Racemes: terminal cluster of sterile flowers pale yellow or purplish. **Flowers:** sepals of fertile ones greenish yellow or purplish; petals lemon yellow or yellowish white. **Fruit valves** moderately to densely hispid. $2n = 28$.

Flowering Apr–May. Serpentine outcrops in arid grassland, oak woodland; of conservation concern; 200–900 m; Calif.

Subspecies *lyonii* is known from Merced County.

Kruckeberg and Morrison described subsp. *lyonii* based on plants with an exclusively greenish yellow terminal cluster of sterile flowers, and discussed a sympatric form with greenish yellow sepals and a purple terminal cluster as a 'bicolor' form. Subsequent collections show that plants with only a greenish yellow terminal sterile cluster are absent in some years, and the 'bicolor' form is here included in subsp. *lyonii*.

24. **Streptanthus longisiliquus** G. L. Clifton & R. E. Buck, Madroño 54: 94, fig. 1. 2007 (as longisiliqus) E

Perennials; (short-lived, caudex simple or few-branched); (glaucous), usually glabrous throughout, (except sepals pubescent, sometimes also petioles). **Stems** branched, 2.2–12(–15) dm. **Basal leaves** rosulate (in juvenile plants); petiolate (petioles usually glabrous, rarely ciliate); blade obovate to spatulate, 3.5–10 cm, margins entire. **Cauline leaves:** blade broadly oblong to ovate or suborbicular, 2.5–10 cm × 10–35 mm, (smaller distally), base amplexicaul, margins entire.

Racemes ebracteate, (lax). **Fruiting pedicels** divaricate-ascending, (straight), 5–10 mm. **Flowers:** calyx subcampanulate; sepals yellow-greenish proximally, purple distally, oblong, 6–8 mm, not keeled, (with subapical tuft of hairs); petals purple or brownish (claw yellow-green), 8–12 mm, blade 1–3 × 0.5–0.8 mm, margins not crisped, claw 6–10 mm, wider than blade; stamens in 3 unequal pairs; filaments (distinct): abaxial pair 6–8 mm, lateral pair 4–6 mm, adaxial pair 7–10 mm; anthers (all) fertile, 3.5–5 mm; gynophore 0.3–1 mm. **Fruits** descending, smooth or slightly torulose, arcuate, flattened, 5–13(–15) cm × 2–2.5 mm; valves each with prominent midvein; replum straight; ovules 50–82 per ovary; style 1.5–3.5 mm; stigma entire. **Seeds** oblong, 2.2–3 × 1.4–1.8 mm; wing 0.1–0.4 mm wide, continuous.

Flowering May–Jul. Openings in pine forests, oak woodland; 400–1700 m; Calif.

Streptanthus longisiliquus is known from Butte, Shasta, and Tehama counties.

25. Streptanthus maculatus Nuttall, J. Acad. Nat. Sci. Philadelphia 5: 134, plate 7. 1825 [E] [F]

Streptanthus maculatus subsp. *obtusifolius* (Hooker) Rollins; *S. obtusifolius* Hooker

Annuals; (glaucous), usually glabrous throughout (sometimes stem bases pubescent). **Stems** unbranched or branched distally, 1.7–10.6 dm, (trichomes 0.05–0.5 mm). **Basal leaves** (soon withered); rosulate; petiolate; blade similar to cauline. **Cauline leaves:** blade broadly ovate to broadly oblong or lanceolate, 2–15 cm × 10–80 mm, (smaller distally), base amplexicaul, margins entire, (apex acute to acuminate or rounded). **Racemes** ebracteate, (dense or lax). **Fruiting pedicels** divaricate-ascending, (straight), 4–7 mm. **Flowers:** calyx campanulate; sepals (erect), purplish, 4–9 mm, (lateral pair rarely with subapical callus, 0.1–0.3 mm), not keeled; petals (widely spreading to somewhat reflexed), magenta (with deep purple center), 11–21 mm, blade 6–12 × 5–11 mm, margins not crisped, claw 5–10 mm, narrower than blade; stamens tetradynamous; filaments: median pairs (distinct), 5–7 mm, lateral pair 3–5 mm; anthers (all) fertile, 3–4 mm; gynophore 0.7–1.5 mm. **Fruits** ascending, smooth, straight, flattened, 6–11.4 cm × 2–3 mm; valves each with prominent midvein; replum straight; ovules 56–92 per ovary; style 1–2.5 mm; stigma strongly 2-lobed. **Seeds** broadly oblong, 2–2.5 × 1–1.3 mm; wing 0.2–0.3 mm wide, continuous.

Flowering Apr–May. Near creeks, roadside banks, moist bottoms, open oak woods, moist land in woods, rocky bluffs, ledges, dry talus slopes, rock crevices, pine-hickory forests; 200–500 m; Ark., Okla., Tex.

Streptanthus maculatus is distributed in Arkansas in Garland, Hot Springs, Montgomery, Pike, Polk, Pulaski, and Saline counties; in Oklahoma in Latimer, Le Flore, McCurtain, and Pushmataha counties; and in Texas in Anderson, Cherokee, Gregg, and Smith counties.

R. C. Rollins (1993) divided *Streptanthus maculatus* into two subspecies based on the presence versus absence of the subapical callus on sepals and on leaf shape. Subspecies *maculatus* was said to differ from subsp. *obtusifolius* by the absence (versus presence) of the sepal callus and by having lanceolate (versus broadly ovate) cauline leaves. These highly variable characters do not covary.

26. Streptanthus morrisonii F. W. Hoffman, Madroño 11: 225, figs. 2–4. 1952 [C] [E]

Streptanthus morrisonii subsp. *elatus* F. W. Hoffman; *S. morrisonii* subsp. *hirtiflorus* F. W. Hoffman; *S. morrisonii* subsp. *kruckebergii* R. W. Dolan & LaPré

Biennials; (glaucous), usually glabrous throughout (sometimes sepals pubescent). **Stems** (simple from base), often paniculately branched distally, (2.5–)6.5–12(–15) dm. **Basal leaves** rosulate; petiolate; blade (fleshy, purplish abaxially, gray-green adaxially, often mottled distally), oblanceolate to broadly obovate or orbicular, 1–7 cm, margins dentate or repand. **Cauline leaves:** blade broadly ovate to narrowly lanceolate, 0.7–5 cm × 2–40 mm, (much smaller distally), base auriculate to amplexicaul, margins entire. **Racemes** ebracteate, (lax, secund). **Fruiting pedicels** divaricate to ascending, (straight), 1–3(–4) mm. **Flowers:** calyx suburceolate; sepals pale yellow to violet, (ovate-lanceolate), 5–8 mm, keeled, (apex usually recurved, glabrous or sparsely to densely hirsute, trichomes to 1.3 mm); petals (barely exserted), creamy white (with brownish purple veins), 7–10 mm, blade 2–3 × 1–2 mm, margins not crisped (channeled), claw 5–7 mm, about as wide as blade; stamens in 3 unequal pairs; filaments: abaxial pair (connate ca. ½ their length), 4–6 mm, lateral pair 2–4 mm, adaxial pair (connate to apex), 7–9 mm; anthers: abaxial and lateral pairs fertile, 2–3 mm, adaxial pair sterile, 0.6–1 mm; gynophore 0.1–1 mm. **Fruits** suberect to divaricate, strongly torulose, usually straight, sometimes slightly curved, flattened, 2.5–8 cm × 1.5–2 mm; valves each with prominent midvein; replum constricted between seeds; ovules 24–38 per ovary; style 0.1–0.2 mm; stigma entire. **Seeds** oblong, 1.5–2 × 0.9–1.3 mm; wing (0–)0.05–0.1 mm wide, distal. $2n = 28$.

Flowering May–Sep. Ridges, serpentine barrens, openings in serpentine chaparral, cypress-knob cone pine woodlands; of conservation concern; 150–1100 m; Calif.

Streptanthus morrisonii is known from Colusa, Lake, Monterey, Napa, and Sonoma counties.

R. W. Dolan and L. F. LaPré (1989) studied *Streptanthus morrisonii* in the North Coast Ranges and found fixed local population differentiation in sepal pubescence and color. Subsequent discovery of plants 200 km southward in the Santa Lucia Mountains (Monterey County) that span all of the characters alleged to distinguish the subspecies makes recognition of infraspecific taxa pointless.

27. Streptanthus oblanceolatus T. W. Nelson & J. P. Nelson, Madroño 56: 127, fig. 1. 2009 [C][E]

Biennials; (glaucous), glabrous throughout. **Stems**: (base woody), virgately branched distally, 5–10 dm. **Basal leaves** rosulate; petiolate; blade oblanceolate, 3.5–7 cm, margins coarsely dentate distally. **Cauline leaves**: blade linear-lanceolate, 2–10 cm, (much smaller distally), base auriculate, margins entire. **Racemes** bracteate below or between proximalmost 1 or 2 flowers, (lax). **Fruiting pedicels** ascending, 3–6 mm, (strongly expanded at receptacle). **Flowers**: calyx shape unknown; sepals yellow, 8–9 mm, slightly keeled, (apex acuminate, recurved); petals yellowish, 12–16 mm, blade 4–5 × 0.5–1 mm, (adaxial pair slightly longer), margins not crisped, claw 7–9 mm, wider than blade, (apex recurved); stamens in 3 unequal pairs; filaments: abaxial pair (distinct), 9–11 mm, lateral pair 7–9 mm, adaxial pair (connate distal to middle of their length), 13–16 mm; anthers: abaxial and lateral pairs fertile, 3–3.5 mm, adaxial pair sterile, ca. 1 mm. **Fruits** divaricate ascending, torulose, flattened, 4–8 cm × 1.5–2 mm; replum straight; ovules per ovary unknown; style to 1 mm; stigma entire. **Seeds** oblong, ca. 2 × 1 mm; wing distal.

Flowering Jun–Jul. Cliffs and canyon walls in conifer forests; of conservation concern; ca. 400 m; Calif.

Streptanthus oblanceolatus is known from Box Canyon, Trinity River, Trinity County.

28. Streptanthus oliganthus Rollins, Contr. Dudley Herb. 3: 372. 1946 [E]

Streptanthus cordatus Nuttall var. *exiguus* Jepson

Perennials; (caudex simple or few-branched); (glaucous), usually glabrous, (petioles of basal leaves pubescent). **Stems** usually unbranched, rarely branched, 1.5–4(–5) dm. **Basal leaves** rosulate; petiolate (petioles usually not or rarely narrowly winged, setose-ciliate); blade narrowly oblanceolate to lanceolate, 4–10 cm, margins entire. **Cauline leaves**: blade oblong-lanceolate, 2.5–8 cm × 5–16(–25) mm, (smaller distally), base auriculate, margins entire, (apex obtuse to acute). **Racemes** ebracteate, (lax). **Fruiting pedicels** divaricate-ascending, (straight), 3–10 mm. **Flowers**: calyx campanulate; sepals purple, (broadly oblong), 5–8 mm, not keeled; petals maroon-purple (claw purplish), 9–12 mm, blade 2–4 × 0.7–1 mm, margins not crisped, claw 6–8 mm, wider than blade; stamens in 3 unequal pairs; filaments (distinct): abaxial pair 5–7 mm, lateral pair 3–5 mm, adaxial pair 7–9 mm; anthers (all) fertile, 2.5–4 mm; gynophore 0.5–1 mm. **Fruits** ascending to divaricate-ascending, smooth, straight, flattened, 4.5–9.7(–10.5) cm × 2–2.7(–3) mm; valves each with prominent midvein; replum straight; ovules (42–)48–60 per ovary; style 0.1–1 mm; stigma entire. **Seeds** oblong, 2–2.5(–2.7) × 1.5–2 mm; wing 0.2–0.4 mm wide, continuous.

Flowering Jun–Aug. Dry open pinyon woodland, pine forest, rocky subalpine forests, sagebrush; 2000–3100 m; Calif., Nev.

Streptanthus oliganthus is known in California from Mono County and in Nevada from Esmeralda, Lyon, and Mineral counties. N. H. Holmgren (2005b) reported it as disjunct in Nye County (Nevada).

Streptanthus oliganthus is related to *S. cordatus*, from which it is readily distinguished by having basal leaves with margins entire versus dentate, petioles usually not winged, rarely narrowly so, versus broadly winged, stigmas entire versus slightly to strongly 2-lobed, ovules (41–)48–60 versus 20–38(–44) per ovary, fruits 2–2.7(–3) versus (2.5–)3–6(–7) mm wide, and seeds narrower (2–2.5(–2.7) × 1.5–2 versus 2.5–5 × 2.2–5 mm).

29. Streptanthus petiolaris A. Gray, Mem. Amer. Acad. Arts, n. s. 4: 7. 1849 [E]

Arabis petiolaris (A. Gray) A. Gray; *Erysimum petiolare* (A. Gray) Kuntze; *Streptanthus brazoensis* Buckley

Annuals; pilose proximally, glabrous distally. **Stems** unbranched or branched distally, 2–12 dm. **Basal leaves** subrosulate; long-petiolate; blade lyrate-pinnatifid, oblanceolate in outline, 3–21 cm, margins (of lobes) dentate, (surfaces pilose). **Cauline leaves** (petiolate); blade lyrate-pinnatifid, hastate, or (distally) linear lanceolate, 2–14 cm × 5–70 mm, margins (distally) entire. **Racemes** ebracteate, (lax). **Fruiting pedicels** divaricate to ascending, (straight or curved upward), 7–13 mm. **Flowers**: calyx subcampanulate; sepals lavender, (oblong), 3–4.5 mm, not keeled; petals white to lavender, 4–8 mm, blade 1.5–4 × 1.5–2 mm, margins not crisped, claw 2.5–4 mm, narrower than blade; stamens tetradynamous; filaments: median pairs (distinct), 2.5–4 mm, lateral pair 1.5–3 mm; anthers (all) fertile, 0.5–1 mm; gynophore 0.3–0.7 mm. **Fruits** suberect to divaricate-ascending, slightly torulose, straight, strongly flattened, 4–9 cm × 3.5–5 mm; valves each with obscure midvein; replum straight; ovules 28–50 per ovary; style 0.5–2.7 mm; stigma slightly 2-lobed. **Seeds** orbicular, 3.5–5 mm diam.; wing 0.8–1.4 mm wide, continuous. $2n = 28$.

Flowering Mar–May. Thickets, roadsides, canyons, grassy grounds, bluffs, knolls, rocky limestone prairies, juniper-oak woods; Tex.

Streptanthus petiolaris was treated by other North American authors under *Arabis*, but the species does not belong to that genus (I. A. Al-Shehbaz 2003b) or to *Boechera*; it has $2n = 28$, a chromosome number universal in *Streptanthus* and not found in any species of *Arabis* or *Boechera*, and it lacks the branched trichomes characteristic of those genera.

30. **Streptanthus platycarpus** A. Gray, Smithsonian Contr. Knowl. 5(6): 10. 1853

Erysimum platycarpum (A. Gray) Kuntze; *Streptanthus sparsiflorus* Rollins

Annuals; (glaucous), usually glabrous throughout (sometimes distalmost stems and pedicels pubescent). **Stems** (simple from base), usually branched, 1.2–10 dm, (sometimes sparsely pilose distally). **Basal leaves** (soon withered); subrosulate; petiolate; blade oblanceolate in outline, 4–20 cm, margins sinuate to lyrate or pinnatifid. **Cauline leaves:** blade ovate, 1.5–15 cm × 5–60 mm, (smaller distally), base auriculate to amplexicaul, margins entire or sinuate-dentate (entire distally). **Racemes** usually ebracteate or proximalmost flowers bracteate, rarely bracteate throughout, (lax). **Fruiting pedicels** divaricate to ascending, (straight), 7–35 mm, (glabrous or pilose). **Flowers:** calyx campanulate; sepals purple to deep maroon, (lanceolate), 8–12 mm, not keeled; petals (spreading to reflexed), lavender or purplish lavender, 16–27 mm, blade 10–18 × 6–9 mm, margins not crisped, claw 6–10 mm (crisped); stamens in 3 unequal pairs; filaments (distinct): abaxial pair 4–6 mm, lateral pair 3–5 mm, adaxial pair 5–8 mm; anthers (all) fertile, 4.5–6 mm; gynophore 0.3–1 mm. **Fruits** erect to divaricate-ascending, smooth, straight, strongly flattened, 4–9.5 cm × 4.5–6 mm; valves each with prominent midvein; replum straight; ovules 26–42 per ovary; style 0.5–2 mm; stigma strongly 2-lobed. **Seeds** orbicular to oblong-orbicular, 4.5–5 × 4–5 mm; wing 0.7–1.2 mm wide, continuous. $2n = 28$.

Flowering Mar–Apr. Cliff ledges, often on limestone, rocky sites, roadsides, rocky creek beds; 400–1200 m; Tex.; Mexico (Coahuila).

Streptanthus platycarpus is known in Texas from Brewster, Crockett, Culberson, Pecos, Presidio, Terrell, and Val Verde counties.

Although the racemes in *Streptanthus platycarpus* are usually ebracteate or have only the proximalmost 1–3 flowers bracteate, an exception is one collection (*Hinckley 3780*, GH) in which all flowers are bracteate.

31. **Streptanthus polygaloides** A. Gray, Proc. Amer. Acad. Arts 6: 519. 1865 [E] [F]

Microsemia polygaloides (A. Gray) Greene

Annuals; (sometimes glaucous), glabrous throughout. **Stems** unbranched or branched distally, (0.8–)2–8(–10) dm. **Basal leaves** (soon withered); rosulate; petiolate; blade 1- or 2-pinnatifid (with broadly linear to filiform lobes), 2–20 cm, margins sinuate-dentate. **Cauline leaves:** blade linear, 1–10 cm × 1–3 mm, (smaller distally), base auriculate, margins entire. **Racemes** ebracteate, (lax). **Fruiting pedicels** strongly recurved, 2–5 mm. **Flowers** (markedly zygomorphic); calyx urceolate; sepals greenish yellow or purplish, (abaxial broadly ovate, not keeled, 4–6 × 3–4 mm, lateral ovate-lanceolate, keeled, 4–6 × 1.5–2 mm, adaxial suborbicular to broadly ovate-cordate, forming a bannerlike hood, keeled, 4–6 × 6–8 mm); petals white (with brownish veins), 5–8 mm, blade 1–3 × 0.7–1.2 mm, margins crisped (channeled), claw 4–5 mm, wider than blade; stamens in 3 unequal pairs; filaments: abaxial pair (distinct), 4–5 mm, lateral pair 3–4 mm, adaxial pair (connate), 5–6 mm; anthers: abaxial and lateral pairs fertile, 1.5–2 mm, adaxial pairs sterile, 0.3–0.7 mm; gynophore 0.1–0.4 mm. **Fruits** pendent, smooth, straight, flattened, 2.4–5.6 cm × 1.2–1.7 mm; valves each with obscure or somewhat prominent midvein; replum straight; ovules (10–)18–50 per ovary; style 0.8–2 mm; stigma entire. **Seeds** oblong, 1.7–2 × 0.9–1.1 mm; wing 0.2–0.3 mm wide, distal. $2n = 28$.

Flowering May–Jul. Serpentine substrates in grasslands, openings chaparral, oak and pine woodlands; 200–1900 m; Calif.

Streptanthus polygaloides is highly variable in plant height, flower color, fruit size, and dissection of basal leaves. Further studies may lead to recognition of infraspecific taxa. The species is distributed in Butte, Calaveras, El Dorado, Fresno, Mariposa, Nevada, Sierra, Tuolumne, and Yuba counties.

Streptanthus polygaloides is one of the few nickel hyperaccumulators in the flora area, and it averages 2,430–18,600 µg/g dry weight (R. D. Reeves et al. 1981; A. R. Kruckeberg and Reeves 1995).

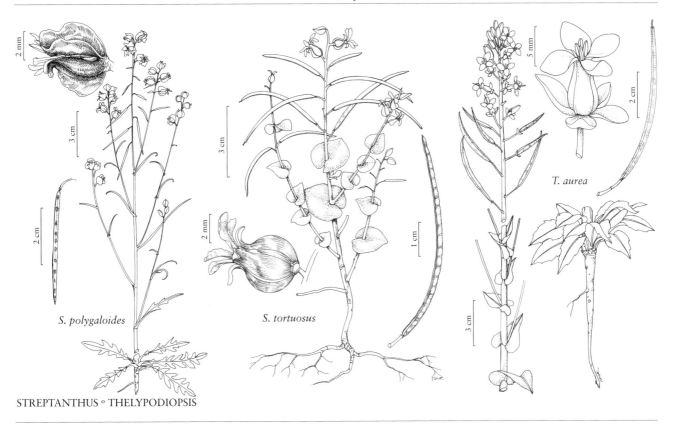

STREPTANTHUS ○ THELYPODIOPSIS

32. **Streptanthus squamiformis** Goodman, Rhodora 58: 354. 1957 [C][E]

Annuals; sparsely to densely pubescent or glabrous. Stems (simple from base), unbranched or branched distally, 1.5–8.5 (–11) dm, (sparsely to densely hirsute basally, usually glabrous distally, rarely throughout). Basal leaves (withered by flowering); not rosulate; petiolate; blade similar to cauline. Cauline leaves: blade ovate to oblong or lanceolate, 1–9 cm × 5–30 mm, (smaller distally), base amplexicaul, margins usually entire, rarely shallowly dentate, (midvein abaxially glabrous or pubescent, surfaces glabrous distally). Racemes ebracteate, (dense to lax). Fruiting pedicels divaricate-ascending, (straight), 5–16 mm, (glabrous or pubescent). Flowers: calyx campanulate; sepals purplish, 4–7 mm, not keeled, (hirsute, trichomes 0.7–2 mm, both pairs with subapical callosities, 1–2.5 mm); petals purple (with dark purple center), 9–17(–20) mm, blade 5–10 × 5–7 mm, margins not crisped, claw 4–7 mm, narrower than blade; stamens tetradynamous; filaments: median pairs (distinct), 5–6 mm, lateral pair 3–4 mm; anthers (all) fertile, 1.5–2.8 mm; gynophore 1–1.5 mm. Fruits ascending to suberect, smooth, straight, flattened, 0.6–1.4 cm × 2–3 mm; valves each with prominent midvein; replum straight; ovules 54–86 per ovary; style 1–1.5 mm; stigma strongly 2-lobed. Seeds broadly oblong, 2–2.7 × 1–1.2 mm; wing 0.1–0.25 mm wide, continuous.

Flowering Apr–May. Glades, steep slopes, sandstone and soft shale in ravines or rocky openings in oak-pine-hickory forests; of conservation concern; Ark., Okla.

Streptanthus squamiformis is known from the Ouachita Mountains in Arkansas (Howard, Polk, and Sevier counties) and Oklahoma (McCurtain County).

The similarities between *Streptanthus squamiformis* and *S. maculatus* in most aspects are truly remarkable, especially in foliage, fruits, seeds, and flower color and size. The former differs by having the sepals densely pubescent (versus glabrous) with trichomes to 2 mm, both sepal pairs with subapical callosities 1–2 mm (versus callosities absent or, rarely, present on lateral sepals and 0.1–0.3 mm), smaller anthers (1.5–2.7 versus 3–4 mm), and pubescent or, sometimes, glabrous (versus always glabrous) fruiting pedicels.

33. Streptanthus tortuosus Kellogg, Proc. Calif. Acad. Sci. 2: 152, plate 46. 1863 [E] [F]

Disaccanthus tortuosus (Kellogg) O. E. Schulz; *Erysimum tortuosum* (Kellogg) Kuntze; *Pleiocardia foliosa* (Greene) Greene; *P. orbiculata* (Greene) Greene; *P. suffrutescens* (Greene) Greene; *P. tortuosa* (Kellogg) Greene; *Streptanthus foliosus* Greene; *S. orbiculatus* Greene; *S. sanhedrensis* Eastwood; *S. suffrutescens* Greene; *S. tortuosus* var. *flavescens* Jepson; *S. tortuosus* var. *oblongus* Jepson; *S. tortuosus* var. *optatus* Jepson; *S. tortuosus* var. *orbiculatus* (Greene) H. M. Hall; *S. tortuosus* var. *pallidus* Jepson; *S. tortuosus* var. *suffrutescens* (Greene) Jepson

Biennials or perennials; (short-lived); (glaucous), glabrous throughout. **Stems** unbranched or branched (several) basally and distally, (0.5–)1.5–12(–15) dm. **Basal leaves** (soon withered); subrosulate; petiolate; blade broadly ovate, obovate or oblong, 1.5–6 cm, margins entire or repand, or dentate apically. **Cauline leaves:** blade oblong to obovate or suborbicular, or (distally) orbicular or oblong-ovate, (0.7–)1.5–6(–9) cm × (4–)10–45 mm, base auriculate to amplexicaul (amplexicaul distally), margins entire or repand. **Racemes** bracteate below or between proximalmost 1 or 2 flowers, (lax to compact, secund or not). **Fruiting pedicels** divaricate to ascending, (2–)3–12(–17) mm, (expanded at receptacle). **Flowers:** calyx urceolate; sepals purplish, gray-green, or yellowish, 6–10(–13) mm, keeled or not, (apex recurved); petals purplish or yellowish white (usually with purple veins), (6–)8–14 mm, blade 2–5 × 1–2.5 mm, margins not crisped, claw 4–10 mm, as wide as or wider than blade, (apex reflexed); stamens in 3 unequal pairs; filaments (distinct): abaxial pair (2.5–)4–7 mm, lateral pair (1.5–)3–5 mm, adaxial pair (5–)7–11 mm; anthers (all) fertile or adaxial pair with reduced fertility and shorter, (1.5–)2.5–4.5(–6) mm; gynophore 0.2–1 mm. **Fruits** arcuate-spreading to pendent, torulose or smooth, straight, flattened, (3–)4–13 (–16) cm × 1.5–2.5(–3) mm; valves each with obscure or somewhat prominent midvein; replum straight; ovules 26–76(–110) per ovary; style 0.4–0.7 mm; stigma entire. **Seeds** broadly oblong to ovoid or orbicular, 1.5–2.5 × 0.8–1.7 mm; wing 0.1–0.5 mm wide, continuous or on distal ½. $2n = 28$.

Flowering Apr–Sep. Rocky open slopes, open woodlands, montane forests, alpine areas; 200–4100 m; Calif., Nev., Oreg.

Streptanthus tortuosus is highly polymorphic, forming distinct local races. Variation in duration, stature, branching, development of the sepal keel, petal color, and bract size are not correlated with habitat or geography. Some authors (e.g., R. C. Rollins 1993; R. E. Buck et al. 1993) divided it into five varieties based on variations in those characters, but the variation is continuous and the purported differences do not hold. For example, those authors and N. H. Holmgren (2005b), recognized the highly branched, shorter plants of the alpine Sierra Nevada as var. *orbiculatus*, but intergradation downslope is completely clinal. The most distinctive variant (*S. foliosus*) is restricted to the central Sierra Nevada, where plants with large, brittle, subachlorophyllus bracts predominate. However, that condition reappears sporadically in the Klamath Ranges and elsewhere. Annual plants have never been documented in *S. tortuosus*, contrary to the claim by some authors (e.g., Rollins; Buck et al.; Holmgren); the alleged "annuals" produce rosettes that invariably overwinter before flowering. Without thorough molecular, experimental, and cytological study of this complex, it is impractical to recognize ill-defined varieties that represent only a minor fraction of the overall variation in the species.

R. C. Rollins (1993) placed *Streptanthus foliosus* in the synonymy of *S. diversifolius*, but its type clearly belongs to *S. tortuosus*.

34. Streptanthus vernalis R. O'Donnell & R. W. Dolan, Madroño 52: 202, fig. 1. 2005 [E]

Annuals; glabrous throughout. **Stems** unbranched or branched distally, 0.2–2 dm. **Basal leaves** (soon withered); not rosulate; shortly petiolate; blade (fleshy), broadly obovate to suborbicular, 1.5–4 cm, margins coarsely dentate. **Cauline leaves:** blade ovate to oblong, 0.5–1.8 cm × 1–7 mm, (much smaller distally), base auriculate to amplexicaul, margins entire. **Racemes** ebracteate, (lax, sometimes secund). **Fruiting pedicels** divaricate-ascending, (straight), 1–2 mm. **Flowers:** calyx urceolate; sepals (erect), green, (ovate-lanceolate), 5–7 mm, keeled, (apex recurved); petals white, 6.5–8 mm, blade 1.5–2 × ca. 1.5 mm, margins not crisped, claw 5–6 mm, about as wide as blade; stamens in 3 unequal pairs; filaments: abaxial pair (connate ½ their length), 3–4 mm, lateral pair 1.5–2.5 mm, adaxial pair (connate their whole length, usually recurved), 6–8 mm; anthers: abaxial and lateral pairs fertile, 1.5–2 mm, adaxial pairs sterile, ca. 0.7 mm; gynophore 0.2–0.5 mm. **Fruits** divaricate-ascending, torulose, straight, flattened, 3–5 cm × 1.5–2 mm; valves each with obscure midvein; replum constricted between seeds; ovules 16–20 per ovary; style 0.1–0.2 mm; stigma entire. **Seeds** oblong, 1.6–2 × 1–1.2 mm; wing 0.1–0.2 mm wide, distal.

Flowering Mar–May. Serpentine talus and gravel; ca. 600 m; Calif.

Streptanthus vernalis, which is known from the Three Peaks in Lake County, is most closely related to *S. brachiatus* and *S. morrisonii*. All three species have the replum constricted between the seeds, a feature not known elsewhere in *Streptanthus*. They also have auriculate-amplexicaul middle and distal cauline leaves, torulose fruits, urceolate calyces with keeled sepals, fruiting pedicels rarely to 4 mm, and connate median filaments. *Streptanthus vernalis* is distinguished from the other two by being annual (versus biennial), without (versus with) basal rosettes, and having non-mottled (versus mottled) basal leaves, green (versus rose-purple, violet, or yellowish) sepals, petals without (versus with) darkly veined blades, and fewer (16–20 versus 22–38) ovules per ovary.

35. **Streptanthus vimineus** (Greene) Al-Shehbaz & D. W. Taylor, Novon 18: 281. 2008 [C][E]

Mesoreanthus vimineus Greene, Leafl. Bot. Observ. Crit. 1: 90. 1904; *M. fallax* Greene

Annuals; (glaucous), usually glabrous throughout (sometimes sepals pubescent). **Stems** branched distal to base, 1–7.5 dm. **Basal leaves** (soon withered); not rosulate; petiolate; blade narrowly ovate to oblong, 2.5–5 cm, margins entire. **Cauline leaves**: blade linear to narrowly linear-lanceolate, 2–12 cm × 1–3(–4) mm, (smaller distally), base auriculate, margins entire. **Racemes** ebracteate, (lax, usually secund). **Fruiting pedicels** divaricate, (straight), 1–3(–6) mm. **Flowers**: calyx urceolate; sepals greenish to purplish, ovate-lanceolate, 6–8 mm, keeled, (apex recurved, usually glabrous, rarely sparsely pubescent); petals white (abaxial pair with purple veins), 8–12 mm, blade 3–5 × 2–3 mm, (unequal, abaxial longer than adaxial), margins not crisped, claw 5–7 mm, narrower than blade; stamens in 3 unequal pairs; filaments: abaxial pair (connate ½ their length), 4–7 mm, lateral pair 2–4 mm, adaxial pair (connate their whole length, recurved), 7–11 mm; anthers: abaxial and lateral pairs fertile, 2–2.7 mm, adaxial pairs sterile, 0.5–1.2 mm; gynophore 0.2–0.5 mm. **Fruits** divaricate, slightly torulose, recurved, flattened, 3.5–6.5 cm × 1–1.2 mm; valves each with obscure midvein; replum straight; ovules 28–40 per ovary; style 0.1–0.3 mm; stigma entire. **Seeds** narrowly oblong, 1–1.5 × 0.6–0.8 mm; wing 0–0.1 mm wide, distal.

Flowering May–Jul. Serpentine grasslands, ridges and barrens, openings in chaparral; of conservation concern; 200–800 m; Calif.

Streptanthus vimineus is restricted to Colusa, Lake, and Napa counties. The taxon has long been unrecognized by various authors, including recent ones (e.g., R. C. Rollins 1993; R. E. Buck et al. 1993). It remained under *Mesoreanthus* and was often treated as a synonym of *S. barbiger*. The latter always has non-auriculate (versus auriculate) cauline leaves.

92. THELYPODIOPSIS Rydberg, Bull. Torrey Bot. Club 34: 432. 1907 • [Genus *Thelypodium* and Greek *opsis*, appearance]

Ihsan A. Al-Shehbaz

Annuals, biennials, or, rarely, perennials; not scapose; glabrous or pubescent. **Stems** (simple or few to several from base), erect [ascending, decumbent], branched basally and/or distally, (glabrous or pubescent). **Leaves** basal and cauline; petiolate or sessile; basal (soon caducous), rosulate or not, petiolate, blade margins entire or dentate; cauline sessile, blade (base auriculate, sagittate, or amplexicaul), margins entire or dentate. **Racemes** (corymbose, dense or lax), considerably [slightly] elongated in fruit. **Fruiting pedicels** erect to ascending or divaricate, slender. **Flowers**: sepals (equal), usually erect, sometimes ascending or spreading, rarely reflexed, oblong, lateral pair not saccate basally; petals (erect basally), usually white, lavender, or purple, rarely yellow, oblanceolate, oblong, spatulate, or obovate, (margins rarely crisped), claws differentiated or not from blade, (apex rounded); stamens tetradynamous, (exserted); filaments distinct; anthers usually linear, sometimes ovate or oblong (apically coiled); nectar glands confluent, subtending bases of stamens. **Fruits** stipitate, linear, torulose or smooth, terete; valves each with prominent midvein, usually glabrous, rarely pilose; replum rounded; septum complete; ovules 20–114 per ovary; style distinct, (cylindrical or subclavate to clavate); stigma capitate, usually strongly

2-lobed, rarely entire, (lobes opposite replum). **Seeds** uniseriate, plump, not winged, oblong; seed coat (minutely reticulate), not mucilaginous when wetted; cotyledons incumbent or obliquely so. $x = 10, 11$.

Species 10 (9 in the flora): w United States, n Mexico.

As recognized by R. C. Rollins (1982b, 1993), *Thelypodiopsis* was artificially delimited, and no single characteristic or combination of characteristics reliably distinguished it from related genera. Rollins's circumscription was so broad that it included species of *Dryopetalon* and *Romanschulzia* O. E. Schulz.

Of the nine species of *Thelypodiopsis* that occur in the flora area, three (*T. purpusii, T. shinnersii,* and *T. vaseyi*) are quite anomalous and, together with the Mexican *T. versicolor* (Brandegee) Rollins (Coahuila, San Luis Potosí), eventually may be excluded from the genus. Unlike the other six species, these three have entire or obscurely (versus prominently) 2-lobed stigmas, oblong or ovate anthers 0.5–1 mm (versus linear and 2.5–4 mm), and petal claws obscurely (versus strongly) differentiated from blades. They are here retained in the genus only tentatively and currently are being subjected to detailed molecular and morphological studies, along with Mexican, Central American, and South American taxa.

SELECTED REFERENCE Rollins, R. C. 1982b. *Thelypodiopsis* and *Schoenocrambe* (Cruciferae). Contr. Gray Herb. 212: 71–102.

1. Anthers ovate or ovate-oblong, 0.5–1 mm; stigmas entire or obscurely 2-lobed; petal claws 0–3 mm.
    2. Cauline leaf blades: proximalmost with pinnatifid to sinuate-dentate margins, distal ones with subamplexicaul or auriculate bases . . . . 6. *Thelypodiopsis purpusii*
    2. Cauline leaf blades: proximalmost usually with entire or repand margins, rarely denticulate, distal ones with auriculate bases.
        3. Proximalmost cauline leaf blades pandurate to broadly obovate; sepals erect; fruits 3.2–7.5 cm; ovules 60–102 per ovary . . . . 7. *Thelypodiopsis shinnersii*
        3. Proximalmost cauline leaf blades oblong to lanceolate or oblanceolate; sepals spreading; fruits 1.5–2.5 cm; ovules 20–32 per ovary . . . . 8. *Thelypodiopsis vaseyi*
1. Anthers linear, 2.5–4 mm; stigmas 2-lobed; petal claws 3.5–8 mm.
    4. Gynophores slender, 2–6(–9.5) mm.
        5. Sepals yellow, spreading to reflexed; petals yellow . . . . 2. *Thelypodiopsis aurea*
        5. Sepals purple, erect; petals usually purple or lavender, rarely white.
            6. Stems glabrous basally; petals 10–14 mm; fruiting pedicels 6–9 mm; styles stout, 0.3–1 mm . . . . 1. *Thelypodiopsis ambigua*
            6. Stems pilose basally; petals 14–17 mm; fruiting pedicels 7–20 mm; styles slender, 2–3 mm . . . . 5. *Thelypodiopsis juniperorum*
    4. Gynophores stout, 0.2–1.5 mm.
        7. Sepals yellowish; petals yellow, 1–1.7 mm wide, margins crisped . . . . 3. *Thelypodiopsis divaricata*
        7. Sepals purplish, lavender, greenish, or white; petals purple, lavender, or white, 3–5 mm wide, margins not crisped.
            8. Proximal portions of stems and abaxial leaf surfaces often pilose; fruits not tortuous; styles often clavate . . . . 4. *Thelypodiopsis elegans*
            8. Proximal portions of stems and abaxial leaf surfaces usually glabrous, rarely sparsely pubescent; fruits tortuous; styles cylindrical . . . . 9. *Thelypodiopsis vermicularis*

1. **Thelypodiopsis ambigua** (S. Watson) Al-Shehbaz, Contr. Gray Herb. 204: 138. 1973 [C][E]

Thelypodium ambiguum S. Watson, Proc. Amer. Acad. Arts 14: 290. 1879; *Sisymbrium ambiguum* (S. Watson) Payson; *Thelypodiopsis ambigua* var. *erecta* Rollins

Annuals or biennials; (glaucous), glabrous throughout. **Stems** branched distally, (3–)5–10 dm. **Basal leaves** rosulate; petiole 0.3–3 cm; blade oblanceolate, (1.5–)2.5–14.5(–20) cm × 10–30(–42) mm, margins dentate to subpinnatifid. **Cauline leaves** sessile; blade lanceolate to oblong, base auriculate, margins usually entire, rarely dentate. **Racemes** lax or dense. **Fruiting pedicels** horizontal to divaricate-ascending, often upcurved, 6–9 mm. **Flowers:** sepals erect, purple, 5–7.5 × 1.7–2.5 mm; petals purple to lavender or white, spatulate, 10–14 × 1.2–1.7 mm, claw 4–6 mm; median filament pairs 4–6 mm; anthers linear, 3–4 mm; gynophore (slender), 3–6(–9.5) mm. **Fruits** ascending to divaricate or descending, straight or slightly recurved, torulose, 4.5–9 cm × 1.1–1.4 mm; ovules 80–112 per ovary; style cylindrical, (stout), 0.3–1 mm; stigma 2-lobed. **Seeds** 1.2–1.5 × 0.8–1.2 mm. $2n = 22$.

Flowering Mar–Jun. Pinyon-juniper desert shrub communities, dry hillsides; of conservation concern; 800–1600 m; Ariz., Utah.

Thelypodiopsis ambigua is known from Kane County in Utah and from Coconino, Mohave, and Yavapai counties in Arizona.

2. **Thelypodiopsis aurea** (Eastwood) Rydberg, Bull. Torrey Bot. Club 34: 432. 1907 [E][F]

Thelypodium aureum Eastwood, Zoë 2: 227. 1891; *Sisymbrium aureum* (Eastwood) Payson

Annuals or perennials; (short-lived); (glaucous), glabrous or sparsely pubescent basally. **Stems** branched basally and distally, (1.4–)2–5(–6) dm, (glabrous or sparsely pubescent basally). **Basal leaves** rosulate; petiole 0.5–4 cm; blade oblanceolate, 2–7 cm × 5–22 mm, margins irregularly dentate. **Cauline leaves** sessile; blade lanceolate to oblong, (smaller distally), base auriculate, margins entire, (surfaces glabrous). **Racemes** dense. **Fruiting pedicels** divaricate-ascending, straight, (5–)6.6–13(–15) mm. **Flowers:** sepals spreading to reflexed, yellow, 5–7.5(–8.5) × 1.7–2.5 mm; petals yellow, spatulate to broadly oblong, 7–11(–13) × 2–3.5 mm, claw 5–7 mm (to 2 mm wide); median filament pairs 5–6.5 mm; anthers linear, 3–4 mm; gynophore (slender), 2–6(–8) mm. **Fruits** erect to divaricate-ascending, straight or slightly curved, torulose, 5–7.5(–9) cm × 1.2–1.7 mm; ovules 72–98 per ovary; style subclavate, 0.5–2 mm; stigma 2-lobed. **Seeds** 1.2–1.5 × 0.6–0.8 mm. $2n = 22$.

Flowering Apr–May. Shrub communities on clay or, rarely, sandy soil; 1200–2200 m; Colo., N.Mex., Utah.

Thelypodiopsis aurea is restricted to the Four Corners area and is known only from Montezuma County in Colorado, Sandoval and San Juan counties in New Mexico, and San Juan County in Utah.

3. **Thelypodiopsis divaricata** (Rollins) S. L. Welsh & Reveal, Great Basin Naturalist 37: 355. 1978 [E]

Caulanthus divaricatus Rollins, Contr. Gray Herb. 201: 8, plate 2. 1971

Annuals or biennials; pubescent proximally, glabrate distally. **Stems** usually branched distally, rarely unbranched, 2–8.5(–10) dm, (often densely pubescent proximally, trichomes flat, crisped). **Basal leaves** (soon withered); not rosulate; blade oblong, 2–10 cm × 10–30 mm, margins entire or dentate, (surfaces glabrous or sparsely pubescent). **Cauline leaves** sessile; blade (proximalmost) oblong or (distal) oblong to ovate, base auriculate to amplexicaul, margins usually entire or dentate (rarely dentate distally). **Racemes** dense. **Fruiting pedicels** divaricate-ascending to spreading, straight, (slender), 6.5–14(–21) mm, (glabrous or pubescent). **Flowers:** sepals (calyx often urceolate), erect, yellowish, 3.5–5.5 × 1.5–2 mm, (sparsely pubescent); petals yellow, oblong, 6.5–9(–10) × 1–1.7 mm, (margins crisped), claw 3.5–5 mm (to 2 mm wide); median filament pairs 4–5 mm; anthers linear, 2.5–3.5 mm; gynophore (stout), 0.5–1.5 mm. **Fruits** suberect to divaricate-ascending or spreading, straight, slightly torulose, 4–8(–9.5) cm × 1.2–1.7 mm; (valves sparsely pubescent at least when immature); ovules 70–94 per ovary; style subclavate, 0.7–2 mm; stigma 2-lobed. **Seeds** 1–1.7 × 0.7–1 mm. $2n = 22$.

Flowering Apr–Jun. Rocky knolls, sandy or clay grounds in juniper and shrub communities; 1200–2100 m; Utah.

Thelypodiopsis divaricata is known from Carbon, Emery, Garfield, Grand, Kane, San Juan, and Wayne counties.

4. Thelypodiopsis elegans (M. E. Jones) Rydberg, Bull. Torrey Bot. Club 34: 432. 1907 [E]

Thelypodium elegans M. E. Jones, Zoë 4: 265. 1893; *Sisymbrium elegans* (M. E. Jones) Payson; *Streptanthus wyomingensis* A. Nelson; *Thelypodiopsis bakeri* (Greene) Rydberg; *T. wyomingensis* (A. Nelson) Rydberg; *Thelypodium bakeri* Greene

Annuals or biennials; pilose throughout or at least basally. Stems branched basally and distally, 1.5–8.5(–12) dm, (sparsely to densely pilose basally or throughout, trichomes crisped). Basal leaves (soon withered); rosulate; petiole 0.5–2.5(–5) cm; blade oblanceolate, 1–5.8 cm × 5–22 mm, margins entire or dentate, (surfaces glabrous or pilose abaxially). Cauline leaves sessile; blade ovate to oblong, base auriculate, margins usually entire, rarely dentate, (surfaces glabrous or pilose abaxially). Racemes dense. Fruiting pedicels divaricate, straight or slightly upcurved, (3.5–)5–16 mm, (glabrous or sparsely pilose). Flowers: sepals usually erect to ascending, rarely spreading, purplish to lavender or whitish, 4–7 × 1.2–2 mm; petals pale purple to white, spatulate, 7.5–14 × 3–5 mm, (margins not crisped), claw 4–8 mm; median filament pairs 4–8 mm; anthers linear, 2.5–4 mm; gynophore (stout), 0.3–1.5 mm. Fruits divaricate-ascending to spreading, straight or curved, torulose, 4–8(–9) cm × 1.2–1.5 mm; (valves glabrous or pilose); ovules 46–94 per ovary; style often clavate, 1–3 mm; stigma strongly 2-lobed. Seeds 1.2–1.5 × 0.8–1 mm.

Flowering Apr–Jun. Shale grounds, loose gypsum, barren areas, clay banks of rocky hillsides, shrub communities; 1400–2400 m; Ariz., Colo., Utah, Wyo.

5. Thelypodiopsis juniperorum (Payson) Rydberg, Fl. Rocky Mts. ed. 2, 1123. 1923 [C][E]

Sisymbrium juniperorum Payson, Univ. Wyoming Publ. Sci., Bot. 1: 12. 1922; *S. elegans* (M. E. Jones) Payson var. *juniperorum* (Payson) H. D. Harrington

Annuals; (glaucous), pilose throughout or at least basally, or glabrous distally. Stems (simple or few to several from base), branched distally, 1.5–10 dm, (pilose basally). Basal leaves (soon withered); rosulate; petiole 0.5–2.5 cm; blade oblanceolate, 5–15 cm × 10–20 mm, margins entire or dentate. Cauline leaves sessile; blade oblong, base auriculate to amplexicaul, margins entire. Racemes slightly dense. Fruiting pedicels horizontal to divaricate-ascending, often straight, 7–20 mm, (glabrous or sparsely pilose). Flowers: sepals erect, purple, 5–7 × 2–3 mm; petals purple, suborbicular to broadly obovate, 14–17 × 5–9 mm, claw 4–7 mm; median filament pairs 4–7 mm; anthers linear, 3–4 mm; gynophore (slender), 3–6 mm. Fruits erect to ascending, straight, torulose, 5–9 cm × 1–1.2 mm; style cylindrical, (slender), 2–3 mm; stigma 2-lobed. Seeds ca. 1.5 × 0.9 mm.

Flowering May–Jun. Pinyon-juniper woodlands, sagebrush communities; of conservation concern; Colo.

Thelypodiopsis juniperorum is known only from Gunnison and Montrose counties.

6. Thelypodiopsis purpusii (Brandegee) Rollins, Contr. Gray Herb. 206: 14. 1976

Thelypodium purpusii Brandegee, Zoë 5: 232. 1906 (as purpusi); *Sisymbrium kearneyi* Rollins; *S. purpusii* (Brandegee) O. E. Schulz; *S. vernale* (Wooton & Standley) O. E. Schulz; *T. vernale* Wooton & Standley

Annuals; (often glaucous), glabrous throughout. Stems unbranched or branched distally, 1.5–5(–7) dm. Basal leaves (soon withered); not rosulate; petiole 0.5–3 cm; blade oblanceolate, 1.5–9 cm × 5–30 mm, margins pinnatifid to dentate-sinuate. Cauline leaves (proximalmost) petiolate or (distal) sessile; blade (proximalmost) oblanceolate, (distal) ovate to oblong, base subamplexicaul or auriculate, margins (proximalmost) pinnatifid to dentate-sinuate, or (distal) entire. Racemes lax. Fruiting pedicels ascending to divaricate, straight, 3–7(–10) mm. Flowers: sepals erect, green or purplish, 3–4.5 × 0.7–1.2 mm; petals white, oblanceolate, 4–5.5 × 1–1.5 mm, claw 2–3 mm; median filament pairs 3–4 mm; anthers ovate, 0.5–0.8 mm; gynophore 0.2–0.4 mm. Fruits divaricate to ascending or spreading, straight or curved, torulose, 3–6.5 cm × 1–1.2 mm; ovules 46–114 per ovary; style subclavate, 0.7–1.5 mm; stigma entire or obscurely 2-lobed. Seeds 0.9–1.2 × 0.5–0.7 mm.

Flowering Feb–May. Juniper woodlands, rocky slopes, shale grounds, loose gypsum, barren areas, clay banks of rocky hillsides, shrub communities; 1400–2100 m; Ariz., N.Mex., Tex.; Mexico (Coahuila).

7. **Thelypodiopsis shinnersii** (M. C. Johnston) Rollins, Contr. Gray Herb. 206: 13. 1976

Sisymbrium shinnersii M. C. Johnston, SouthW. Naturalist 2: 129. 1957, based on *Thelypodium vaseyi* J. M. Coulter, Contr. U.S. Natl. Herb. 1: 30. 1890, not *S. vaseyi* S. Watson ex B. L. Robinson 1895

Annuals; (often glaucous), glabrous throughout. **Stems** unbranched or branched distally, 3.5–8 dm. **Basal leaves** (soon withered); not rosulate; petiole (winged), 0.5–3.5 cm; blade pandurate to broadly obovate, 2–10 cm × 10–50 mm, margins entire or repand. **Cauline leaves** (proximalmost) petiolate or (distal) sessile; blade (proximalmost) pandurate to broadly obovate or (distal) obovate, base auriculate, margins entire or repand. **Racemes** lax. **Fruiting pedicels** often divaricate, sometimes ascending, straight, 4–13 mm. **Flowers**: sepals erect, purplish, 2.5–3.5 × 0.7–1 mm; petals white, narrowly oblanceolate, 3.5–4.5 × 1–1.2 mm, claw not developed; median filament pairs 1.5–2.5 mm; anthers ovate-oblong, 0.7–1 mm; gynophore 0.2–0.4 mm. **Fruits** divaricate to ascending, straight or curved, obscurely torulose, 3.2–7.5 cm × 0.9–1.2 mm; ovules 60–102 per ovary; style subclavate, 0.4–2.5 mm; stigma entire. **Seeds** 0.9–1.1 × 0.5–0.7 mm.

Flowering Apr–Oct. Canyon sides, rocky arroyo floors, chaparral thickets, scrubs, dry banks; Tex.; Mexico (Tamaulipas).

Thelypodiopsis shinnersii is known from Cameron County.

8. **Thelypodiopsis vaseyi** (S. Watson ex B. L. Robinson) Rollins, Contr. Gray Herb. 206: 12. 1976 [E]

Sisymbrium vaseyi S. Watson ex B. L. Robinson in A. Gray et al., Syn. Fl. N. Amer. 1(1,1): 138. 1895

Annuals; (often glaucous), glabrous throughout. **Stems** unbranched or branched distally, 3–10 dm. **Basal leaves** (soon withered); not rosulate; blade oblong to lanceolate or oblanceolate, 3–6 cm × 5–15 mm, margins usually entire, rarely denticulate. **Cauline leaves** sessile; blade (proximalmost) oblong to lanceolate or oblanceolate, or (distal) linear to narrowly oblong or lanceolate, base auriculate, margins usually entire, rarely denticulate. **Racemes** lax. **Fruiting pedicels** horizontal to divaricate, straight or slightly recurved, (slender), 7–15 mm. **Flowers**: sepals spreading, whitish or purplish, 1.8–2.5 × 0.7–1 mm; petals white, obovate to spatulate, 2.5–4.5 × 1.7–2.5 mm, claw 1–1.5 mm; median filament pairs 1.7–2.5 mm; anthers ovate, 0.5–0.8 mm; gynophore 0.2–0.4 mm. **Fruits** erect to ascending, straight or curved, strongly torulose, 1.5–2.5 cm × 1–1.2 mm; ovules 20–32 per ovary; style cylindrical, 0.5–0.8 mm; stigma obscurely 2-lobed. **Seeds** 1–1.5 × 0.7–0.9 mm. $2n = 20$.

Flowering Jul–Aug. Open wooded slopes, mixed coniferous forests, canyons; 1900–2500 m; N.Mex., Tex.

Thelypodiopsis vaseyi is known from Lincoln, Otero, and San Miguel counties in New Mexico and Culberson County in Texas. R. C. Rollins (1982b, 1993) did not record it from Texas; the first report from there is based on *Johnston 3148* (MO), collected in Guadalupe Mountains. *Sisymbrium watsonii* Payson is an illegitimate name that pertains to *S. vaseyi*.

9. **Thelypodiopsis vermicularis** (S. L. Welsh & Reveal) Rollins, Contr. Gray Herb. 212: 81. 1982 [E]

Thelypodium sagittatum (Nuttall) Endlicher var. *vermicularis* S. L. Welsh & Reveal, Great Basin Naturalist 37: 358. 1978

Annuals or biennials; (glaucous), glabrous throughout or sparsely pubescent basally. **Stems** (simple or, often, several from base), branched distally, 1.5–5(–6) dm. **Basal leaves** rosulate; petiole 0.5–2.3 cm; blade broadly oblanceolate, (1.8–)2–3.5(–4) cm × 10–15 mm, margins entire or remotely denticulate. **Cauline leaves** sessile; blade broadly ovate to oblong, (slightly smaller distally), base strongly auriculate to amplexicaul, margins entire. **Racemes** dense. **Fruiting pedicels** horizontal to divaricate-ascending, straight or upcurved, (slender), 4–8.5(–10) mm. **Flowers**: sepals ascending to spreading, purplish to greenish or white, 4.5–5.5 × 1.2–1.5 mm; petals white, spatulate, 9–11 × 3–4 mm, (margins not crisped), claw 4–6 mm; median filament pairs 4–6 mm; anthers linear, 3–4 mm; gynophore (stout), 0.2–1.5 mm. **Fruits** erect to ascending, straight or slightly recurved, (distinctly tortuous), torulose, 2–4 cm × 1.2–1.5 mm; ovules 30–42 per ovary; style cylindrical, 1–3 mm; stigma slightly 2-lobed. **Seeds** 1.2–1.7 × 0.6–0.9 mm.

Flowering Apr–Jun. Brush communities, shale formations, clay or silty flat, juniper woodlands; 1300–2200 m; Nev., Utah.

Thelypodiopsis vermicularis is known in eastern Nevada from Elko and White counties.

93. THELYPODIUM Endlicher, Gen. Pl. 11: 876. 1839 • Thelypody [Greek *thelys*, female, and *podion*, little foot, alluding to gynophore carrying pistil]

Ihsan A. Al-Shehbaz

Pachypodium Nuttall in J. Torrey and A. Gray, Fl. N. Amer. 1: 96. 1838, not Lindley 1830; *Pleurophragma* Rydberg; *Stanleyella* Rydberg

Biennials, perennials, or, rarely, annuals; not scapose; glabrous or pubescent. **Stems** (simple or few to several from base), usually erect, rarely decumbent, branched basally and/or distally, (glabrous or pubescent). **Leaves** basal and cauline; petiolate or sessile; basal rosulate or not, petiolate, blade margins usually entire, dentate, lyrate or pinnately lobed, rarely laciniate; cauline petiolate or sessile, blade (base cuneate, attenuate, auriculate, sagittate, or amplexicaul), margins often entire, sometimes dentate or pinnately lobed. **Racemes** (corymbose, dense or lax), usually slightly to considerably elongated in fruit (sometimes not elongated in *T. integrifolium*). **Fruiting pedicels** usually horizontal, erect to ascending, or divaricate, rarely reflexed, slender or stout, (flattened or not basally, glabrous). **Flowers:** sepals usually erect or ascending, rarely spreading to reflexed, ovate to oblong, linear, lanceolate, or oblanceolate, lateral pair slightly saccate or not basally; petals (erect or spreading), white, lavender, or purple, spatulate to obovate, or oblanceolate to linear, (margins crisped or not), claw differentiated or not from blade, (apex rounded); stamens subequal or tetradynamous, (exserted or included); filaments (erect or spreading, usually distinct, very rarely median ones united), not dilated basally; anthers usually linear to linear-oblong, rarely oblong or ovate, (sometimes apiculate, often circinately coiled after dehiscence); nectar glands confluent and subtending bases of stamens, or 2 or 4 and lateral. **Fruits** stipitate, linear, torulose or smooth, terete, slightly 4-angled, or flattened; valves each with prominent midvein, glabrous; replum rounded; septum complete; ovules 12–128 per ovary; style distinct, (often cylindrical, rarely subclavate or subconical); stigma capitate, entire. **Seeds** uniseriate, plump or flattened, not winged, usually oblong, rarely ovate; seed coat (minutely reticulate), not mucilaginous when wetted; cotyledons oblique, rarely incumbent or accumbent. $x = 13$.

Species 16 (16 in the flora): w, wc North America, n Mexico.

As recognized herein and by recent authors (e.g., I. A. Al-Shehbaz 1973; R. C. Rollins 1993), *Thelypodium* is somewhat heterogeneous and the segregate *Stanleyella* might merit recognition, as by E. B. Payson (1923). *Thelypodium* has erect sepals, petals, and stamens, terete fruits, prominently veined fruit septa, and, often, cylindrical styles. In contrast, species of *Stanleylla* have spreading sepals, petals, and stamens, flattened fruits, veinless fruit septa, and clavate, subclavate, or subconical styles.

SELECTED REFERENCES Al-Shehbaz, I. A. 1973. The biosystematics of the genus *Thelypodium* (Cruciferae). Contr. Gray Herb. 204: 3–148. Payson, E. B. 1923. A monographic study of *Thelypodium* and its immediate allies. Ann. Missouri Bot. Gard. 9: 233–324.

1. Cauline leaves petiolate, (proximal blades: margins usually pinnately lobed, rarely dentate); petal claws widest at base.
    2. Racemes dense; sepals erect; styles usually cylindrical, rarely subclavate in fruit.
        3. Stems solid; basal leaves: petioles glabrous; fruiting pedicels straight, horizontal; petals linear .... 7. *Thelypodium laciniatum*
        3. Stems hollow; basal leaves: petioles ciliate; fruiting pedicels strongly curved upward; petals spatulate to oblanceolate .... 9. *Thelypodium milleflorum*
    2. Racemes often lax; sepals spreading or ascending; styles usually clavate, subclavate, or subconical, rarely cylindrical in fruit.

4. Stems hirsute or glabrous basally; fruits terete, submoniliform to strongly torulose (replum constricted between seeds). . . . . 8. *Thelypodium laxiflorum*
4. Stems glabrous basally; fruits flattened, torulose (replum not constricted between seeds).
   5. Basal leaf blades usually ovate or obovate, rarely orbicular or spatulate, margins usually sinuate and repand, or dentate; petals lavender or purple, (2.5–)3–4 mm. . . . . .11. *Thelypodium repandum*
   5. Basal leaf blades lanceolate, oblanceolate, or spatulate, margins mostly pinnately lobed; petals usually white, rarely lavender, (3.5–)4–7.5(–9) mm.
      6. Annuals; styles conical or subconical in fruit; petals spatulate; distalmost cauline leaf blade margins pinnately lobed (pectinate) . . . . 15. *Thelypodium texanum*
      6. Biennials; styles clavate to subclavate in fruit; petals oblong to linear; distalmost cauline leaf blade margins usually entire or dentate, rarely lobed . . . . .16. *Thelypodium wrightii*
1. Cauline leaves sessile (blade bases often auriculate, sagittate, or amplexicaul, proximal blades: margins often entire); petal claws narrowest at base.
   7. Perennials, with (woody) caudex, covered with persistent petiolar remains; stems flexuous. . . . . .4. *Thelypodium flexuosum*
   7. Biennials or perennials (short-lived), without caudex, not covered with persistent petiolar remains; stems straight.
      8. Cauline leaf blade bases cuneate to attenuate, not auriculate or sagittate . . . . .6. *Thelypodium integrifolium*
      8. Cauline leaf blade bases usually sagittate or amplexicaul, rarely auriculate.
         9. Terminal racemes often corymbose, not elongated in fruit.
            10. Basal leaves: petioles not ciliate, blade margins usually dentate to repand, rarely entire; cauline leaves erect (partly to completely appressed to stems); anthers exserted; fruits usually strongly incurved, overtopping buds . . . . .12. *Thelypodium rollinsii*
            10. Basal leaves: petioles ciliate, blade margins entire; cauline leaves ascending; anthers usually included; fruits often straight, not overtopping buds.
                11. Petals 2.5–5(–6) mm wide; fruits (0.8–)1.5–2.3 mm wide; seeds plump . . . . 10. *Thelypodium paniculatum*
                11. Petals (0.5–)1–3(–4) mm wide; fruits (0.5–)0.8–1(–1.2) mm wide; seeds flattened . . . . .13. *Thelypodium sagittatum* (in part)
         9. Terminal racemes not corymbose, considerably elongated in fruit.
            12. Racemes densely flowered; petal margins crisped throughout.
                13. Sepals lanceolate to linear-lanceolate or ovate; fruiting pedicels horizontal to divaricate, stout, base flattened, 1–2(–2.5) mm; seeds plump . . . . 1. *Thelypodium brachycarpum*
                13. Sepals oblong to linear-oblong; fruiting pedicels erect or erect-ascending (appressed to rachis at least basally), slender, base not flattened, (1.5–)2–5(–10) mm; seeds flattened . . . . 2. *Thelypodium crispum*
            12. Racemes lax; petal margins not crisped or crisped between blade and claw.
                14. Petals linear, 0.3–0.5(–0.8) mm wide; s California. . . . . . 14. *Thelypodium stenopetalum*
                14. Petals spatulate to oblanceolate, (0.5–)1–3(–4) mm wide; not s California.
                    15. Flower buds oblong-linear; anthers exserted; gynophores (1–)2.5–6 (–7.5) mm in fruit. . . . . 3. *Thelypodium eucosmum*
                    15. Flower buds ovate to lanceolate; anthers partially to wholly included; gynophores (0.2–)0.5–1(–3.5) mm in fruit.
                        16. Fruiting pedicels ascending, straight; fruits forming straight line with pedicels; Oregon . . . . 5. *Thelypodium howellii*
                        16. Fruiting pedicels horizontal to divaricate, often curved upward; fruits forming arc with pedicels; not Oregon . . . . .13. *Thelypodium sagittatum* (in part)

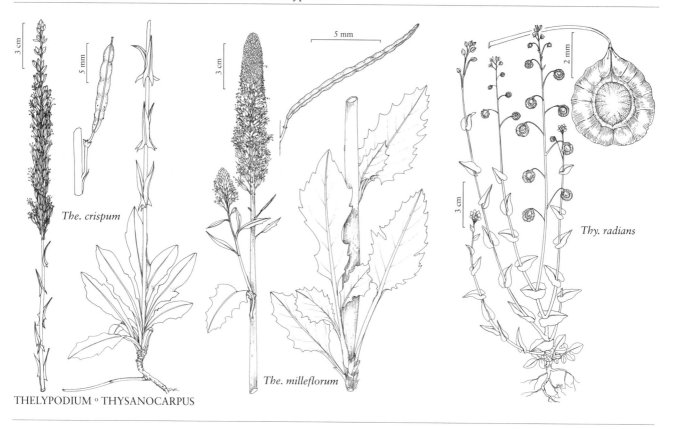

THELYPODIUM ○ THYSANOCARPUS

1. **Thelypodium brachycarpum** Torrey in C. Wilkes et al., U.S. Expl. Exped. 17(2): 231, plate 1. 1874 [E]

Thelypodiopsis brachycarpa (Torrey) O. E. Schulz; *T. brachypoda* O. E. Schulz

Biennials; often glaucous, glabrous or pubescent. **Stems** branched, (1.3–)3.4–8.3(–12) dm, (somewhat glaucous, glabrous throughout or pubescent proximally). **Basal leaves:** petiole (1.5–)2–5.2(–9) cm, glabrous or pubescent; blade usually oblanceolate to spatulate, rarely obovate or lanceolate, 3.3–14(–20) cm × (05–)10–32(–47) mm, margins often pinnately lobed to lyrate, sometimes dentate or entire. **Cauline leaves** sessile; blade linear to lanceolate, smaller distally, (base sagittate, appressed to stem), margins usually entire, rarely dentate. **Racemes** dense, slightly elongated in fruit, (flower buds oblong). **Fruiting pedicels** horizontal to divaricate, straight or slightly curved, stout, 1–2(–2.5) mm, flattened at base. **Flowers:** sepals erect, lanceolate to linear-lanceolate or ovate, (3–)3.5–5(–5.5) × 1–1.5(–2) mm; petals white, linear to narrowly oblanceolate, 8–12.5(–16) × 0.3–0.5(–1) mm, margins strongly crisped, claw differentiated from blade, (slender, 2.5–4 mm, narrowest at base); nectar glands surrounding bases of lateral stamens, median glands absent; filaments subequal or slightly tetradynamous, 2.5–6.5(–10) mm; anthers exserted, linear, 1.5.–2(–2.5) mm, slightly circinately coiled, (apiculate); gynophore (0.5–)1–2(–5) mm. **Fruits** divaricate to ascending, torulose, often straight, sometimes slightly curved, terete, (0.8–)1.2–2.7(–3) cm × 1–1.5(–2) mm; ovules 12–26 per ovary; style cylindrical, (0.2–)0.5–1(–2.5) mm. **Seeds** (plump), (1.3–)1.5–2 × 0.8–1(–1.3) mm.

Flowering Apr–Aug. Strongly alkaline meadows and desert flats; 600–2300 m; Calif., Oreg.

Thelypodium brachycarpum is known in California from Napa, Shasta, and Siskiyou counties and in Oregon from Klamath County.

2. **Thelypodium crispum** Greene ex Payson, Ann. Missouri Bot. Gard. 9: 264. 1923 [E] [F]

Thelypodiopsis crispa (Greene ex Payson) O. E. Schulz; *Thelypodium brachycarpum* Torrey var. *crispum* (Greene ex Payson) Jepson

Biennials or perennials; (short-lived); often glaucous, glabrous or pubescent. **Stems** branched, (0.9–)1.3–7(–12) dm, (somewhat glaucous, glabrous throughout or pubescent basally). **Basal leaves:** petiole (0.7–)2–8.5 cm, ciliate; blade usually oblanceolate to spatulate, rarely obovate or lanceolate, (1.5–)2.2–15(–25) cm × (6–)10–35(–50) mm, margins often pinnately lobed to lyrate, sometimes dentate or entire. **Cauline leaves** sessile; blade linear to lanceolate, smaller distally, (base sagittate or auriculate, appressed to stem), margins usually entire, rarely dentate. **Racemes** dense, elongated in fruit, (flower buds narrowly oblong). **Fruiting pedicels** erect to erect-ascending, (appressed to rachis at least basally), straight or slightly curved, slender, (1.5–)2–5 (–10) mm, not flattened at base. **Flowers:** sepals erect to ascending, oblong to linear-oblong, (3–)3.5–6(–8) × (0.8–)1–1.8(–2) mm; petals white to lavender, linear to narrowly oblanceolate, (6–)6.5–11(–14.5) × 0.5–0.7 (–1) mm, margins strongly crisped, claw differentiated from blade, (slender, 2–4 mm, narrowest at base); nectar glands surrounding bases of lateral stamens, median glands absent; filaments subequal, (2.5–)4.5–8.5 (–10.5) mm; anthers exserted, linear to narrowly oblong, 2–3.5(–4.5) mm, circinately coiled, (apiculate); gynophore 0.5–1.5(–3.5) mm. **Fruits** divaricate to ascending, torulose, straight or slightly curved, terete, (0.6–)1–2.5(–4.2) cm × 0.7–1(–1.8) mm; ovules 22–50 per ovary; style cylindrical, 0.5–1.5(–2.5) mm. **Seeds** (flattened), 1–1.5(–1.7) × 0.5–0.7(–1) mm. $2n = 26$.

Flowering Jun–Aug. Alkaline meadows and desert flats, mineralized ground near hot springs, desert shrub communities; 1200–3000(–3200) m; Calif., Nev.

3. **Thelypodium eucosmum** B. L. Robinson in A. Gray et al., Syn. Fl. N. Amer. 1(1,1): 175. 1895 [C][E]

Biennials or perennials; (short-lived); glaucous, glabrous (except petioles). **Stems** branched distally, 2–10 dm. **Basal leaves:** petiole (0.9–)1.4–3(–4.5) cm, ciliate; blade usually oblanceolate to oblong or lanceolate, rarely ovate or elliptic, (2.8–)3.5–8.8(–11) cm × (7–)10–25(–35) mm, margins often entire or repand, sometimes sinuate. **Cauline leaves** (ascending); sessile; blade lanceolate to oblong, 1.8–4.5(–6) cm × 5–16(–24) mm, (base amplexicaul to strongly auriculate), margins entire. **Racemes** lax, slightly elongated in fruit, (flower buds oblong-linear). **Fruiting pedicels** horizontal to divaricate, usually straight, rarely slightly incurved, slender, (2.5–)3–5.5(–6.5) mm, slightly flattened at base. **Flowers:** sepals erect, linear-oblong, 5–7(–8) × (0.8–)1–1.5(–1.8) mm; petals dark purple, spatulate to oblanceolate, (6.6–)7.5–10(–11.5) × 1–1.8(–2) mm, margins not crisped, claw differentiated from blade, [slender, (3–)3.5–5(–5.5) mm, narrowest at base]; nectar glands confluent, subtending bases of stamens; filaments subequal, (4.5–)6–9(–10) mm; anthers exserted, linear to narrowly oblong, 2.5–4(–4.5) mm, circinately coiled; gynophore (1–)2.5–6(–7.5) mm. **Fruits** divaricate to ascending, torulose, straight or slightly incurved, terete, (2–)2.4–5(–6.5) cm × 0.7–1(–1.3) mm; ovules 44–58 per ovary; style usually cylindrical, rarely subclavate, 0.5–1.5(–2) mm. **Seeds** 0.7–1.5 × 0.5–0.8 mm.

Flowering May–Jul. Shady slopes and canyons, pinyon-juniper and oak woodland communities, stream beds, streamsides; of conservation concern; 700–1000 m; Oreg.

Thelypodium eucosmum is known only from Baker, Grant, and Wheeler counties. It is in the Center for Plant Conservation's National Collection of Endangered Plants.

4. **Thelypodium flexuosum** B. L. Robinson in A. Gray et al., Syn. Fl. N. Amer. 1(1,1): 175. 1895 [E]

Perennials; (caudex woody, covered with persistent petiolar remains); somewhat glaucous, glabrous throughout. **Stems** (often subdecumbent), branched basally, (flexuous), 1.5–5.6(–8.5) dm. **Basal leaves:** petiole 1–6.5 (–13) cm; blade often lanceolate, sometimes oblong or oblanceolate, (2–)3.5–16.5(–20.5) cm × (5–)10–25(–45) mm, margins entire. **Cauline leaves** (ascending); sessile; blade lanceolate to linear, 1–7(–11) cm × 2–7(–14) mm, (base sagittate to somewhat amplexicaul), margins entire. **Racemes** (few-flowered, corymbose), elongated in fruit, (flower buds oblong). **Fruiting pedicels** usually horizontal to divaricate, rarely divaricate-ascending, straight or slightly curved upward, slender, (2.5–)4–9 (–16) mm, slightly flattened at base. **Flowers:** sepals erect, oblong, 3–4(–4.5) × 1–1.5(–1.7) mm; petals lavender or white, often spatulate, sometimes oblanceolate or obovate, 6–9(–10) × (1.5–)2–3(–3.5) mm, margins not crisped, claw strongly differentiated from blade, (slender, 2–3.5(–4) mm, narrowest at base); nectar glands lateral, median glands absent; filaments tetradynamous, median pairs 3–4(–5) mm, lateral pair 2–3.5(–4) mm; anthers included, oblong, 1–2(–2.5) mm, not circinately coiled; gynophore (stout), 0.5–1 mm. **Fruits** erect to ascending, torulose, slightly incurved or straight, terete, 1–2.5(–4.2) cm × 0.8–1(–1.5) mm; ovules 12–30 per ovary; style cylindrical, (0.3–)1–2(–3) mm. **Seeds** (1–)1.3–1.5 × 0.5–8(–1) mm. $2n = 26$.

Flowering Apr–Jun. Often tangled among woody shrubs in moderately to strongly alkaline sandy loam or clay, open deserts; 1000–2400 m; Calif., Idaho, Nev., Oreg.

5. Thelypodium howellii S. Watson, Proc. Amer. Acad. Arts 21: 445. 1886 C E

Cartiera howellii (S. Watson) Greene; *Thelypodiopsis howellii* (S. Watson) O. E. Schulz

Biennials or, rarely, **perennials**; (short-lived); glaucous, glabrous (except petioles). **Stems** branched distally, 1–8(–9.3) dm. **Basal leaves**: petiole 0.5–2 cm, ciliate; blade often oblanceolate to spatulate, sometimes oblong or lanceolate, 2–10(–13.5) cm × 10–23(–40) mm, margins often lyrate to dentate, sometimes entire or repand. **Cauline leaves** (ascending); sessile; blade often lanceolate to linear, sometimes oblong, (0.8–)1–9.7(–14.5) cm × 2–10(–30) mm, (base usually amplexicaul to strongly sagittate, rarely auriculate), margins entire. **Racemes** lax, elongated in fruit, (flower buds ovate to lanceolate). **Fruiting pedicels** usually ascending, rarely divaricate-ascending, usually straight, rarely slightly incurved, stout, (2.5–)3.5–8(–14.5) mm, not flattened at base. **Flowers**: sepals erect, linear or lanceolate, (3.5–)5–8.5(–11.5) × (0.7–)1–2(–2.5) mm; petals lavender to purple, linear or oblanceolate to spatulate, (6–)9–16.5(–22) × 0.5–2.7(–3) mm, margins partially crisped, claw differentiated from blade, [slender, (2.5–)4–7(–11) mm, narrowest at base]; nectar glands confluent, subtending bases of stamens; filaments slightly tetradynamous, median pairs (3–)5.5–9(–12) mm, lateral pair (2.5–)4.5–8(–10) mm; anthers slightly exserted to included, linear, (1.5–)2–4(–5.5) mm, not circinately coiled; gynophore (stout), 0.5–1(–3.5) mm. **Fruits** often ascending, sometimes erect or divaricate, slightly torulose, straight or slightly incurved (forming straight line with pedicel), terete, 1.5–6.5(–8.2) cm × 1–1.3(–2) mm; ovules 22–40 per ovary; style cylindrical or subclavate, (0.5–)1–2.7(–4.3) mm. **Seeds** (1–)1.2–1.7(–2) × 0.5–0.7 mm.

Subspecies 2 (2 in the flora): w United States.

Streptanthus howellii (S. Watson) M. E. Jones is a later homonym of *S. howellii* S. Watson 1885.

1. Petals usually linear to narrowly oblanceolate, rarely spatulate, 0.5–1.2(–3) mm wide; median filaments connate most of their lengths; ne California, c, s Oregon, s Washington . . . .
. . . . 5a. *Thelypodium howellii* subsp. *howellii*
1. Petals usually spatulate, rarely oblanceolate, (1.2–)1.5–2.7(–3) mm wide; median filaments distinct; e Oregon . . . 5b. *Thelypodium howellii* subsp. *spectabilis*

5a. Thelypodium howellii S. Watson subsp. **howellii** C E

Streptanthus coombsiae Eastwood; *Thelypodium simplex* Greene

Basal leaves: blade oblanceolate, margins usually lyrate, rarely dentate or entire. **Flowers**: petals usually linear to narrowly oblanceolate, rarely spatulate, 0.5–1.2(–3) mm wide; median filaments connate most of their lengths. **Fruits** 1.5–4.5(–7) cm. $2n = 26$.

Flowering May–Jul. Desert shrub communities, alkaline meadows and flats; of conservation concern; 100–1600 m; Calif., Oreg., Wash.

5b. Thelypodium howellii S. Watson subsp. **spectabilis** (M. Peck) Al-Shehbaz, Contr. Gray Herb. 204: 116. 1973 C E

Thelypodium howellii var. *spectabilis* M. Peck, Torreya 32: 150. 1932

Basal leaves: blade oblong-lanceolate to oblong, margins entire or repand. **Flowers**: petals usually spatulate, rarely oblanceolate, (1.2–)1.5–2.7(–3) mm wide; median filaments distinct. **Fruits** (1.8–)3–6.5(–8.2) cm. $2n = 26$.

Flowering Jun–Jul. Desert shrub communities, alkaline ground; of conservation concern; Oreg.

Subspecies *spectabilis* is known only from Baker, Harney, Malheur, and Union counties. It is in the Center for Plant Conservation's National Collection of Endangered Plants.

6. Thelypodium integrifolium (Nuttall) Endlicher in W. G. Walpers, Repert. Bot. Syst. 1: 172. 1842 E

Pachypodium integrifolium Nuttall in J. Torrey and A. Gray, Fl. N. Amer. 1: 96. 1838; *Pleurophragma integrifolium* (Nuttall) Rydberg

Biennials; somewhat glaucous throughout, glabrous. **Stems** paniculately branched distally, (2–)4.5–17(–28) dm. **Basal leaves**: petiole (0.5–)1.2–10(–15) cm; blade usually oblong to lanceolate, or oblanceolate to spatulate, rarely ovate or obovate, (3.7–)5–31(–54) cm × (12–)16–78(–140) mm, margins usually entire, rarely dentate or repand. **Cauline leaves** (ascending, not appressed to stem); sessile; blade linear to linear-lanceolate, (1–)2.2–8.3(–19.5) cm × (2–)3–11(–25) mm, (base cuneate to attenuate, not auriculate), margins entire. **Racemes** dense, slightly or considerably elongated in fruit, (flower buds narrowly oblong). **Fruiting pedicels** usually horizontal to divaricate, rarely ascending or reflexed, usually straight, rarely incurved, slender or stout, (2–)3–9(–13) mm, often strongly flattened, rarely not flattened at base. **Flowers:** sepals erect, linear to linear-oblong, (3–)3.5–5.5(–7.5) × (0.5–)0.8–1(–1.5) mm; petals white or lavender to purple, spatulate to oblanceolate, (4.5–)6–9(–13) × (0.5–)0.8–1.5(–2.3) mm, margins not crisped, claw differentiated from blade, [slender, (2–)3–4.5(–7) mm, narrowest at base]; nectar glands lateral, flat or toothlike, median glands absent; filaments subequal to slightly tetradynamous, (3–)4.5–7.5(–12.5) mm; anthers exserted, linear to narrowly oblong, (1–)1.5–2.5(–4.5) mm, not circinately coiled; gynophore 0.5–2(–5.5) mm. **Fruits** usually horizontal to divaricate-ascending, or ascending, rarely reflexed, usually torulose, rarely submoniliform, usually strongly incurved, sometimes straight, rarely arcuate, terete, (0.8–)1.4–6.5(–8) cm × 1–1.3(–2) mm; ovules 14–40 per ovary; style usually cylindrical, rarely subclavate, 0.5–1.3(–3.2) mm. **Seeds** (1–)1.2–2(–2.3) × 0.7–1(–1.3) mm.

Subspecies 5 (5 in the flora): w United States.

1. Petals white; fruiting pedicels whitish, stout, (5–)6–11(–13) mm.
    2. Petals (5–)6–9(–10) mm; fruiting pedicels horizontal to divaricate, (5–)6–9(–13) mm; fruits (1.4–)1.9–3.7(–4.5) cm; gynophores (0.5–)1–3 mm; se California, s Nevada, sw Utah. . . . 6b. *Thelypodium integrifolium* subsp. *affine*
    2. Petals (7.5–)8–11(–13) mm; fruiting pedicels usually horizontal to reflexed, rarely divaricate, (6–)7–11(–13) mm; fruits (2.2–)3.5–6.5(–8) cm; gynophores (1–)1.5–4(–5.5) mm; Grand Canyon, Colorado River, Arizona . . . . 6e. *Thelypodium integrifolium* subsp. *longicarpum*
1. Petals usually lavender to purple, rarely white; fruiting pedicels not whitish, slender and 2–10(–13) mm, or stout and 2–5(–6) mm.
    3. Fruiting pedicels often slender, sometimes stout, not or slightly flattened at base, (4–)6–10(–13) mm . . . . 6a. *Thelypodium integrifolium* subsp. *integrifolium*
    3. Fruiting pedicels stout, strongly flattened at base, 2–5(–6) mm.
        4. Racemes often strongly congested, central rachises (0.6–)1.2–7(–9) cm; gynophores stout, 0.5–1(–3) mm; e California, Nevada, s Oregon, nw Utah . . . . 6c. *Thelypodium integrifolium* subsp. *complanatum*
        4. Racemes somewhat congested, central rachises (3–)8–20(–28) cm; gynophores slender, (0.5–)0.8–3(–4) mm; ne Arizona, w Colorado, nw New Mexico, e Utah . . . . 6d. *Thelypodium integrifolium* subsp. *gracilipes*

6a. Thelypodium integrifolium (Nuttall) Endlicher subsp. **integrifolium** [E]

Pleurophragma lilacinum (Greene) Rydberg; *Thelypodium lilacinum* Greene; *T. lilacinum* var. *subumbellatum* Payson

Racemes elongated or subumbellate in fruit; central rachis (0.5–)1.2–10(–25) cm. **Fruiting pedicels** usually divaricate to divaricate-ascending, rarely horizontal, straight or slightly incurved, not whitish, often slender, sometimes stout, (4–)6–10(–13) mm, not or slightly flattened at base. **Flowers:** petals usually lavender to purple, rarely white, (4.5–)5.5–8(–10.5) mm; gynophore stout, 0.5–1.2(–2.5). **Fruits** divaricate-ascending to ascending, straight or incurved, (1–)1.5–3(–4) cm. $2n = 26$.

Flowering Jun–Sep. Alkaline grounds and flats, desert shrub communities, barren hillsides, mineralized grounds near hot springs; 600–2500 m; Colo., Idaho, Mont., Nebr., N.Dak., Oreg., S.Dak., Utah, Wash., Wyo.

6b. Thelypodium integrifolium (Nuttall) Endlicher subsp. **affine** (Greene) Al-Shehbaz, Contr. Gray Herb. 204: 110. 1973 [E]

Thelypodium affine Greene, Pittonia 4: 314. 1901; *Pleurophragma rhomboideum* (Greene) O. E. Schulz; *T. integrifolium* var. *affine* (Greene) S. L. Welsh & Reveal; *T. rhomboideum* Greene

Racemes (congested), elongated in fruit; central rachis (3.7–)4.5–12(–16.5) cm. **Fruiting pedicels** horizontal to divaricate, straight, whitish, stout, (5–)6–9(–13) mm, moderately flattened at base. **Flowers:** petals white, (5–)6–9(–10) mm; gynophore stout, (0.5–)1–3 mm. **Fruits** usually horizontal or divaricate, rarely divaricate-ascending, incurved, (1.4–)1.9–3.7(–4.5) cm. $2n = 26$.

Flowering Jun–Oct. Desert flats, shrub communities, alkaline grounds; 600–1800 m; Calif., Nev., Utah.

6c. Thelypodium integrifolium (Nuttall) Endlicher subsp. **complanatum** Al-Shehbaz, Contr. Gray Herb. 204. 105, plate 20, fig. 1. 1973 E

Thelypodium integrifolium var. *complanatum* (Al-Shehbaz) S. L. Welsh & Reveal

Racemes (often strongly congested), not elongated in fruit; central rachis (0.6–)1.2–7(–9) cm. **Fruiting pedicels** horizontal, straight, not whitish, stout, 3–5(–6) mm, strongly flattened at base. **Flowers:** petals usually lavender, rarely white, (5–)6–8.5(–9.5) mm; gynophore stout, 0.5–1(–3) mm. **Fruits** horizontal or divaricate-ascending to ascending, straight or incurved, (0.9–)1.4–2.7(–3.4) cm.

Flowering Jun–Aug. Alkaline areas, desert shrub communities, canyons; 800–2400 m; Calif., Nev., Oreg., Utah.

6d. Thelypodium integrifolium (Nuttall) Endlicher subsp. **gracilipes** (B. L. Robinson) Al-Shehbaz, Contr. Gray Herb. 204: 108. 1973 E

Thelypodium integrifolium var. *gracilipes* B. L. Robinson in A. Gray et al., Syn. Fl. N. Amer. 1(1,1): 176. 1895; *Pleurophragma gracilipes* (B. L. Robinson) Rydberg; *P. platypodum* Rydberg; *T. gracilipes* (B. L. Robinson) Rydberg; *T. rhomboideum* Greene var. *gracilipes* (B. L. Robinson) Payson

Racemes (somewhat congested), considerably elongated in fruit; central rachis (3–)8–20(–28) cm. **Fruiting pedicels** often horizontal, sometimes divaricate or slightly reflexed, straight, not whitish, stout, (2–)2.5–4.5(–6) mm, strongly flattened at base. **Flowers:** petals usually white or lavender, rarely purple, (5–)6.5–9(–10) mm; gynophore slender, (0.5–)0.8–3(–4) mm. **Fruits** horizontal to divaricate or slightly reflexed, often incurved, sometimes straight, (0.8–)1.5–2.5(–3.3) cm. $2n = 26$.

Flowering Jul–Aug. Canyons, desert shrub communities, alkaline flats; 1000–2500 m; Ariz., Colo., N.Mex., Utah.

6e. Thelypodium integrifolium (Nuttall) Endlicher subsp. **longicarpum** Al-Shehbaz, Contr. Gray Herb. 204: 111. 1973 C E

Thelypodium integrifolium var. *longicarpum* (Al-Shehbaz) N. H. Holmgren

Racemes (moderately to strongly congested), not elongated in fruit; central rachis (4.3–)8–20 cm. **Fruiting pedicels** usually horizontal to reflexed, rarely divaricate, straight, whitish, stout, (6–)7–11(–13) mm, strongly flattened at base. **Flowers:** petals white, (7.5–)8–11(–13) mm; gynophore stout, (1–)1.5–4(–5.5) mm. **Fruits** usually horizontal to reflexed, rarely divaricate, straight or incurved to arcuate, (2.2–)3.5–6.5(–8) cm.

Flowering Aug–Oct. Canyons, shrub communities; of conservation concern; 500–1300 m; Ariz.

Subspecies *longicarpum* is known from the Grand Canyon National Park and adjacent canyons of the Colorado River.

7. Thelypodium laciniatum (Hooker) Endlicher in W. G. Walpers, Repert. Bot. Syst. 1: 172. 1842 E

Macropodium laciniatum Hooker, Fl. Bor.-Amer. 1: 43. 1829; *Pachypodium laciniatum* (Hooker) Nuttall; *Thelypodium laciniatum* var. *streptanthoides* (Leiberg ex Piper) Payson; *T. leptosepalum* Rydberg; *T. streptanthoides* Leiberg ex Piper

Biennials; somewhat glaucous, glabrous throughout. **Stems** branched proximally and/or distally, (1.3–)2.6–10(–14.5) dm. **Basal leaves** (and proximal cauline): petiole (1–)1.5–10(–15) cm; blade lanceolate to deltate-lanceolate, or oblong to ovate (lateral lobes oblong to ovate), (4.3–)6.8–24(–45) cm × (10–)18–85(–140) mm, margins pinnately lobed or laciniate (lobes sinuate, laciniate, or dentate). **Cauline leaves** petiolate; blade lanceolate to linear-lanceolate, similar to basal, smaller distally, margins entire or sinuate to laciniate. **Racemes** dense, slightly elongated in fruit (to 9.5 dm, flower buds narrowly oblong). **Fruiting pedicels** horizontal to, rarely, divaricate, straight, often stout, (2–)3–6.5(–15) mm, often flattened at base. **Flowers:** sepals erect, oblong to linear-oblong, (3.5–)4–7.5(–9.5) × (0.8–)1–2(–2.5) mm; petals white or purple, often linear, sometimes linear-lanceolate, (6–)8.5–18(–20) × 0.3–0.8(–1.5) mm, margins slightly crisped, claw strongly differentiated from blade, [(2.5–)3–6(–7) mm, widest at base]; nectar glands confluent, lateral and median; filaments subequal, (4.5–)6.5–12(–15) mm; anthers exserted, linear, 2–4(–5.5) mm, circinately coiled, (apiculate); gynophore (0.5–)1–5(–8.5) mm. **Fruits** divaricate-ascending to slightly reflexed, torulose, somewhat tortuous, terete, (2.5–)3.5–10(–12) cm × (0.7–)1–1.5(–2) mm; ovules 56–108 per ovary; style usually cylindrical, rarely subclavate, (0.3–)0.7–2.5 (–4) mm. **Seeds** (0.7–)1–1.5(–1.8) × 0.5–0.7(–1) mm. $2n = 26$.

Flowering Apr–Aug. Rock crevices, cliffs, rocky outcrops, among boulders, serpentine rock, talus, canyon walls, limestone ledges; 0–2400 m; B.C.; Calif., Idaho, Nev., Oreg., Wash.

8. **Thelypodium laxiflorum** Al-Shehbaz, Contr. Gray Herb. 204: 129, plate 22A. 1973 E

Stanleyella wrightii (A. Gray) Rydberg var. *tenella* (M. E. Jones) Payson; *Thelypodium wrightii* A. Gray var. *tenellum* M. E. Jones

Biennials; somewhat glaucous, glabrous throughout or pubescent basally. **Stems** branched distally, (1.5–)3–14 (–23.5) dm, (glabrous throughout or sparsely to densely hirsute proximally). **Basal leaves** (and proximal cauline): petiole (1–)1.7–8 (–11) cm, pubescent or glabrous; blade oblanceolate in outline (lateral lobes usually oblong to linear, rarely deltate), (4–)7.2–20.5(–30) cm × (6–)10–40(–100) mm, margins pinnately lobed or lyrate (lobes entire or dentate). **Cauline leaves** petiolate; blade usually lanceolate to linear, sometimes oblanceolate to oblong, 1.5–7(–10) cm × 3–12(–20) mm, margins usually entire or repand, rarely lobed. **Racemes** corymbose, somewhat lax, considerably elongated in fruit. **Fruiting pedicels** often divaricate, sometimes reflexed, rarely horizontal or divaricate-ascending, straight, slender, (4–)5–13 (–19) mm, somewhat flattened at base. **Flowers:** sepals (equal), ascending, oblong, (2.5–)3–5(–5.5) × 1–1.7(–3) mm, (lateral pair not saccate); petals usually white, rarely lavender, usually spatulate, rarely obovate, 5–7.5(–9.5) × 1.5–2.5(–3.5) mm, margins not crisped, claw obscurely differentiated from blade (widest at base); nectar glands continuous, flat, subtending bases of stamens; filaments tetradynamous, median pairs 3–4.5(–5.5) mm, lateral pair 2.5–3.5(–4.5) mm; anthers included, oblong, (1–)1.5–2.2(–2.5) mm, not circinately coiled; gynophore 0.5–0.8(–1.5) mm. **Fruits** divaricate to reflexed, submoniliform to strongly torulose, straight or curved, terete, (2–)3–6.6(–7.4) cm × 0.7–1(–1.5) mm, (replum constricted between seeds); ovules 26–62 per ovary; style usually clavate to subclavate, rarely cylindrical, (0.5–)0.8–2(–5) mm. **Seeds** (oblong), (1–)1.3–1.8(–2) × 0.5–1 mm.

Flowering May–Sep. Talus, rocky slopes, cliffs, pinyon-juniper-brush communities; 1500–3100 m; Colo., Nev., Utah.

9. **Thelypodium milleflorum** A. Nelson, Bot. Gaz. 52: 263. 1911 E F

Thelypodium laciniatum (Hooker) Endlicher var. *milleflorum* (A. Nelson) Payson

Biennials; somewhat glaucous, glabrous (except petioles). **Stems** (hollow), branched distally, (1.8–) 4.5–13(–21) dm. **Basal leaves** (and proximal cauline): petiole (1–)1.8–9.5(–13) cm, ciliate; blade often oblong, sometimes ovate or lanceolate, (3.8–) 6–23(–28) cm × (9–)20–65(–100) mm, margins usually dentate, sometimes sinuate, rarely repand. **Cauline leaves** petiolate; blade lanceolate, similar to basal, smaller distally, margins dentate or entire. **Racemes** dense, slightly elongated in fruit (to 10 dm, flower buds narrowly oblong). **Fruiting pedicels** erect distally, strongly curved upward, stout, (1.5–)2.5–5(–7) mm, slightly flattened at base. **Flowers:** sepals erect, oblong to linear-oblong, (4–)4.5–8(–9) × 1–2 mm; petals white, spatulate to oblanceolate, (7–)9–15(–16) × 1–2 mm, claw strongly differentiated from blade [(3–)3.5–5.5(–6) mm, widest at base]; nectar glands confluent, median and lateral; filaments subequal, (6–)7–14.5(–15.5) mm; anthers exserted, linear to narrowly oblong, 2.5–4.5(–6) mm, circinately coiled, (apiculate); gynophore (0.5–) 1–4(–6) mm. **Fruits** erect, (subappressed to rachis), torulose, somewhat tortuous, terete, (1.8–)3.3–8(–10) cm × (0.7–)1–1.5(–1.8) mm; ovules 50–78 per ovary; style usually cylindrical, rarely subclavate, (0.5–)0.7–1.5 (–3) mm. **Seeds** (0.7–)1–1.5(–2) × 0.5–0.7(–1) mm. $2n = 26$.

Flowering Apr–Aug. Sand dunes, sagebrush and desert shrub communities; 150–2200 m; B.C.; Calif., Idaho, Nev., Oreg., Utah, Wash.

10. **Thelypodium paniculatum** A. Nelson, Bull. Torrey Bot. Club 26: 126. 1899 (as Thelepodium) C E

Thelypodiopsis paniculata (A. Nelson) O. E. Schulz; *Thelypodium sagittatum* (Nuttall) Endlicher var. *crassicarpum* Payson

Biennials or perennials; (short-lived); glaucous, glabrous or sparsely pubescent. **Stems** branched often proximally and distally, (1.4–)2–6.5(–7.5) dm, (glabrous or pubescent). **Basal leaves:** petiole (0.8–)2–4 (–6.5) cm, ciliate; blade usually oblanceolate, rarely spatulate or oblong, (3–)6–15(–22) cm × (6–)10–25 (–40) mm, margins entire, (surfaces glabrous). **Cauline leaves** (ascending); sessile; blade lanceolate to linear-lanceolate, (1.7–)2.2–6(–9) cm × (3–)5–15(–30) mm, (base sagittate to somewhat amplexicaul), margins entire, (surfaces glabrous). **Racemes** corymbose, dense, terminal not elongated in fruit, (flower buds ovate). **Fruiting pedicels** usually horizontal to divaricate, rarely divaricate-ascending, often curved upward, slender, 6–11.5(–17) mm, slightly flattened at base. **Flowers:** sepals erect, ovate to oblong-ovate, (3–)3.5–5(–6) × (1–)1.5–2 mm; petals lavender to purple, spatulate to broadly obovate, (6.5–)8–11.5(–14) × 2.5–5(–6) mm, margins not crisped, claw strongly differentiated from blade, [slender, (2–)2.5–4(–5) mm, narrowest at base]; nectar glands confluent, subtending bases of stamens; filaments tetradynamous, median pairs (3–)3.5–5(–6) mm, lateral pair (2–)2.5–4(–5) mm; anthers included, ovate to ovate-oblong, 1–2 mm, not circinately coiled; gynophore (stout), 0.5–0.8(–1.2) mm. **Fruits** erect, slightly torulose, usually slightly incurved, rarely straight, terete, (1–)1.3–3.2(–5) cm × (0.8–)1.5–2.3 mm; ovules 20–30 per ovary; style cylindrical, (0.5–)1.5–1.7(–3) mm. **Seeds** (plump), (1.3–)1.5–2 × 0.8–1 mm.

Flowering Jun–Jul. Wet sedge meadows, streamsides; of conservation concern; 1800–2800 m; Colo., Idaho, Mont., Wyo.

11. **Thelypodium repandum** Rollins, Contr. Dudley Herb. 3: 371. 1946 E

Biennials or perennials; (short-lived); distinctly glaucous throughout, glabrous. **Stems** usually branched distally, 1–4.4(–6) dm. **Basal leaves** (and proximal cauline): petiole 1–4(–6) cm; blade (fleshy), usually ovate or obovate, rarely orbicular or spatulate, (2.2–)4–10.5(–14) cm × (10–)15–43(–55) mm, margins usually lyrate, sinuate and repand, or dentate, rarely entire. **Cauline leaves** shortly petiolate; blade lanceolate or elliptic, much smaller than basal, margins entire or repand. **Racemes** somewhat lax, slightly elongated in fruit. **Fruiting pedicels** divaricate to divaricate-ascending, straight or incurved, slender, 4–12(–15) mm, slightly flattened at base. **Flowers:** sepals spreading to reflexed, oblong, (2.5–)3–4 × (0.8–)1–1.5(–1.8) mm; petals purple or lavender, usually oblanceolate, rarely spatulate, (2.5–)3–4.5 × 0.5–1 mm, margins not crisped, claw differentiated from blade; nectar glands confluent; filaments subequal, 2.5–3.8 mm; anthers oblong, 1–1.8 mm; gynophore 0.5–0.8(–1) mm. **Fruits** usually erect to ascending, rarely divaricate, torulose, straight or, sometimes, incurved, flattened, (2–)4–7.4 cm × 1–1.5(–1.8) mm, (replum not constricted between seeds); ovules 32–44 per ovary; style subclavate, 0.5–1.2(–1.5) mm. **Seeds** (0.7–)1–1.5 × 0.5–0.8 mm.

Flowering Jun. Decomposing shale banks; ca. 1700 m; Idaho.

Thelypodium repandum is known from the shale banks of the Salmon River and its tributaries in Custer County.

12. **Thelypodium rollinsii** Al-Shehbaz, Contr. Gray Herb. 204: 97, plates, figs. 1973 E

Biennials; glaucous, glabrous throughout. **Stems** branched distally, (4–)6–16(–20) dm. **Basal leaves:** petiole 0.5–1.3(–1.5) cm, not ciliate; blade spatulate or oblanceolate to obovate, 1.3–4(–7) cm × (4–)6–13(–18) mm, margins usually dentate to repand, rarely entire. **Cauline leaves** (erect, appressed to stem at least partly); sessile; blade linear to linear-lanceolate, (1.8–)1.3–4.5(–6) cm × (1–)1.7–7(–8) mm, (base auriculate or sagittate), margins entire. **Racemes** dense, terminal not elongated in fruit, (flower buds narrowly oblong). **Fruiting pedicels** usually horizontal to divaricate, rarely divaricate-ascending, straight, slender, (6–)7–14(–18) mm, often not flattened at base. **Flowers:** sepals erect, usually linear to linear-oblong, rarely oblong, 4–6(–7) × (0.8–)1–1.5 mm; petals lavender to purple, spatulate to obovate or oblong, (6–)7–9.5(–12) × (1.2–)1.5–2(–3) mm, margins not crisped, claw differentiated from blade, [slender, 3–4.5(–5.5) mm, narrowest at base]; nectar glands confluent, subtending bases of stamens; filaments tetradynamous, median pairs (5–)5.5–7.5(–8.5) mm, lateral pair (4–)5–6.5(7.5) mm; anthers exserted, linear to linear-oblong, 2–3(–4) mm, not circinately coiled; gynophore 0.5–2.5(–6) mm. **Fruits** erect distally, slightly torulose, usually strongly incurved, (overtopping buds), rarely straight and divaricate-ascending, terete, (1.8–)2.8–5(–6.3) cm × 0.7–1 mm; ovules 52–70 per ovary; style cylindrical, (0.5–)0.8–1.5(–2) mm. **Seeds** 1–1.3(–1.5) × 0.5–0.6(–0.8) mm. $2n = 26$.

Flowering Jun–Aug. Alkaline flats; 1500–2000 m; Utah.

Thelypodium rollinsii is restricted mostly to Beaver, Carbon, Juab, Millard, Piute, Sanpete, and Sevier counties.

13. **Thelypodium sagittatum** (Nuttall) Endlicher in W. G. Walpers, Repert. Bot. Syst. 1: 172. 1842 E

Streptanthus sagittatus Nuttall, J. Acad. Nat. Sci. Philadelphia 7: 12. 1834; *Pachypodium sagittatum* (Nuttall) Nuttall; *Thelypodiopsis nuttallii* (S. Watson) O. E. Schulz; *T. sagittata* (Nuttall) O. E. Schulz; *T. torulosa* (A. Heller) O. E. Schulz; *Thelypodium nuttallii* S. Watson; *T. torulosum* A. Heller

Biennials or perennials; (short-lived); sometimes glaucous, glabrous throughout or sparsely to densely pubescent. **Stems** branched often proximally and distally, (2–)3–8 (–12.5) dm, (glabrous or pubescent). **Basal leaves:** petiole (1–)2–8(–14) cm, ciliate; blade usually oblanceolate to ovate, rarely oblong, (2–)6.5–20(–29) cm × 10–42(–50) mm, margins entire, (surfaces glabrous or pubescent). **Cauline leaves** (ascending); sessile; blade lanceolate to oblong-lanceolate, (0.8–)1.6–8.5(–14) cm × 2–13(–28) mm, (base amplexicaul to sagittate), margins entire, (surfaces glabrous or pubescent). **Racemes** sometimes corymbose, lax or dense, elongated or not in fruit, (flower buds ovate to lanceolate). **Fruiting pedicels** horizontal to divaricate, often curved upward, stout or slender, (2.5–)3.5–11(–20) mm, not flattened at base. **Flowers:** sepals erect, ovate to oblong, (2.5–)3–6(–10) × (1–)1.5–2(–2.5) mm; petals white or lavender to purple, oblanceolate to spatulate or linear, (5–)7–14(–19) × (0.5–)1–3(–4) mm, margins slightly crisped between blade and claw, claw differentiated from blade, [slender, (2–)3–6.5(–10) mm, narrowest at base]; nectar glands confluent and subtending bases of stamens, sometimes only lateral; filaments tetradynamous, median pairs 3–5.5(–8) mm, lateral pair (2–)2.5–4.5(–7) mm; anthers included or partially exserted, oblong, (1.5–)2–3.2(–5) mm, not circinately coiled; gynophore (stout), 0.5–1(–2) mm. **Fruits** erect, torulose, straight or slightly incurved (forming arc with pedicel), usually terete, rarely 4-angled, 1–5.3(–6.9) cm × (0.5–)0.8–1(–1.2) mm; ovules 20–62 per ovary; style cylindrical, 0.5–1.5(–3.3) mm. **Seeds** (flattened), (0.7–)1–1.3(–1.5) × 0.5–0.7(–1) mm.

Subspecies 2 (2 in the flora): w United States.

1. Racemes often dense; fruiting pedicels 5–11(–20) mm; petals spatulate to oblanceolate, 1–3(–4) mm wide; median nectar glands often absent . . . . 13a. *Thelypodium sagittatum* subsp. *sagittatum*
1. Racemes lax; fruiting pedicels (2.5–)3.5–7(–8) mm; petals linear to linear-oblanceolate, 0.5–1 (–1.5) mm wide; median nectar glands present . . . . . . 13b. *Thelypodium sagittatum* subsp. *ovalifolium*

13a. Thelypodium sagittatum (Nuttall) Endlicher subsp. sagittatum [E]

Thelypodium amplifolium Greene; *T. macropetalum* Rydberg

Racemes (corymbose), often dense. **Fruiting pedicels** 5–11 (–20) mm. **Flowers:** petals spatulate to oblanceolate, (5–) 7–14(–19) × 1–3(–4) mm, claw 3–6.5(–10) mm; median nectar glands often absent. **Fruits** (1.3–) 1.8–5.3(–6.9) cm. 2*n* = 26.

Flowering May–Jul. Alkaline grounds and flats, meadows; 1300–2800 m; Colo., Idaho, Mont., Nev., Utah, Wash., Wyo.

13b. Thelypodium sagittatum (Nuttall) Endlicher subsp. ovalifolium (Rydberg) Al-Shehbaz, Contr. Gray Herb. 204: 121. 1973 [C][E]

Thelypodium ovalifolium Rydberg, Bull. Torrey Bot. Club 30: 253. 1903; *Thelypodiopsis sagittata* (Nuttall) O. E. Schulz var. *ovalifolia* (Rydberg) S. L. Welsh; *Thelypodium palmeri* Rydberg; *T. sagittatum* var. *ovalifolium* (Rydberg) S. L. Welsh & Reveal

Racemes lax. **Fruiting pedicels** (2.5–)3.5–7(–8) mm. **Flowers:** petals linear to linear-oblanceolate, 5–7.5(–8.5) × 0.5–1(–1.5) mm, claw 2–3.5(–4) mm; median nectar glands present. **Fruits** 1–3(–3.8) cm. 2*n* = 26.

Flowering May–Aug. Alkaline meadows and flats; of conservation concern; 1800–2600 m; Nev., Utah.

Subspecies *ovalifolium* is known from White Pine County, Nevada, and adjacent Garfield and Iron counties, Utah.

14. Thelypodium stenopetalum S. Watson, Proc. Amer. Acad. Arts 22: 468. 1887 [C][E]

Thelypodiopsis stenopetala (S. Watson) O. E. Schulz

Biennials; glaucous, glabrous (except petioles). **Stems** (often decumbent), branched basally, sometimes also distally, (2.6–)3–8 (–9) dm. **Basal leaves:** petiole 1–4.5(–6) cm, ciliate; blade often oblong or lanceolate, sometimes ovate to spatulate, 3.8–15(–18) cm × 15–35 (–42) mm, margins usually entire, rarely repand. **Cauline leaves** (ascending); sessile; blade usually lanceolate to oblong, rarely linear to linear-lanceolate, (1.3–)1.6–4.8(–6) cm × (3–)5–9(–15) mm, (base usually amplexicaul to sagittate, rarely auriculate), margins entire. **Racemes** lax, elongated in fruit, (flower buds oblong-linear). **Fruiting pedicels** usually ascending to divaricate-ascending, rarely horizontal or divaricate, usually straight, rarely slightly incurved, stout, (3.5–) 4–8 mm, not flattened at base. **Flowers:** sepals erect, linear to linear-oblong, (6–)6.5–9(–10) × 1–1.5(–1.8) mm; petals usually lavender, rarely white, linear, (8–) 9.5–15(–16.5) × 0.3–0.5(–0.8) mm, margins crisped between blade and claw, claw differentiated from blade, [slender, 4–6.5(–8.5) mm, narrowest at base]; nectar glands confluent, subtending bases of stamens; filaments slightly tetradynamous, median pairs (7–)8–12.5(–14) mm, lateral pair (5.5–)7–11(–12) mm; anthers exserted, linear to narrowly oblong, 3.5–5(–6) mm, circinately coiled; gynophore stout, 0.5–3.5(–5) mm. **Fruits** usually divaricate-ascending to ascending, rarely divaricate, slightly torulose, straight or slightly incurved, terete or slightly 4-angled, (2.2–)2.8–5(–6.3) cm × 1–1.5(–1.8) mm; ovules 50–82 per ovary; style cylindrical, 1–2(–2.5) mm. **Seeds** 1–1.3(–1.5) × 0.7–0.8 mm.

Flowering May–Aug. Alkaline meadows and flats; of conservation concern; 1900–2100 m; Calif.

Thelypodium stenopetalum is known from Bear Valley, San Bernardino County.

15. Thelypodium texanum (Cory) Rollins, Contr. Dudley Herb. 3: 371. 1946 [E]

Sisymbrium texanum Cory, Rhodora 39: 418. 1937; *Sibara grisea* Rollins; *Stanleyella texana* (Cory) Rollins; *Thelypodium tenue* Rollins

Annuals; slightly glaucous, glabrous. **Stems** usually branched distally, 1.3–4.8(–6.1) dm. **Basal leaves** (and proximal cauline): petiole 1–2.5(–4) cm; blade oblanceolate or spatulate in outline (lateral lobes often oblong, sometimes ovate), (3.4–)5–15(–24) cm × (11–)15–35(–55) mm, margins pinnately lobed (lobes dentate, entire, or repand). **Cauline leaves** petiolate; blade often pectinate (lobes linear), similar to basal, much smaller, margins pinnately lobed. **Racemes** somewhat lax, considerably elongated in fruit. **Fruiting pedicels** horizontal to divaricate, sometimes reflexed, straight or incurved, slender or stout, (4–)6–2(–3.5) mm, slightly flattened at base. **Flowers:** sepals spreading or ascending, oblong to linear-oblong, (2.5–)3.5–5 × 1–1.5 mm; petals white, spatulate to oblanceolate, (3.5–)4–6.5(–7) × (0.5–)1–2 mm, margins not crisped, claw differentiated from blade (widest at base); nectar glands confluent; filaments equal, (4–)4.5–6.5(–7) mm; anthers oblong, (1–)1.5–2 mm, circinately coiled; gynophore 0.5–1(–2) mm. **Fruits** usually divaricate, rarely ascending, torulose, straight or slightly recurved, flattened, (1.9–)2.8–6(–7) cm × (1–)1.3–1.5(–1.7) mm, (replum not constricted between seeds); ovules 28–48 per ovary; style usually conical to subconical, rarely subcylindrical, 0.8–1.5(–3) mm. **Seeds** (1–)1.2–1.5 × 0.8–1(–1.3) mm. $2n = 26$.

Flowering Feb–Apr. Barren hillsides, creek beds, stream banks; N.Mex., Tex.

The type of *Thelypodium tenue* is an immature, somewhat abnormal specimen with rather slender pedicels longer than those typical of *T. texanum*. In all other aspects, including habitat and geography, it belongs in *T. texanum*. Repeated attempts to re-collect *T. tenue* were unsuccessful. Plants of *T. texanum* disjunct in New Mexico were misidentified by R. C. Rollins (1993) as *Sibara grisea*.

16. Thelypodium wrightii A. Gray, Smithsonian Contr. Knowl. 3(5): 7. 1852

Stanleyella wrightii (A. Gray) Rydberg; *Thelypodium wrightii* subsp. *oklahomensis* Al-Shehbaz

Biennials; slightly glaucous, glabrous. **Stems** branched distally, (6.5–)9.5–22.5(–28) dm. **Basal leaves** (and proximal cauline): petiole (2–)4.5–8.5 (–13.7) cm; blade often lanceolate, sometimes oblanceolate in outline (lateral lobes oblong to linear or deltate), (6.5–)9.5–22.5(–28) cm × 28–55(–75) mm, margins pinnately lobed or lyrate (lobes entire or denticulate). **Cauline leaves** petiolate; blade lanceolate to linear-lanceolate, 3.5–8.5(–13.5) cm × 7–14(–23) mm, margins usually entire or dentate, rarely lobed. **Racemes** somewhat lax, considerably elongated in fruit. **Fruiting pedicels** horizontal to divaricate, sometimes reflexed, straight or curved, sometimes secund, slender, (5–)7–13(–17) mm, slightly flattened at base. **Flowers:** sepals spreading or ascending oblong to linear, (3–)4–6(–7) × (0.8–)1–1.3(–1.5) mm; petals usually white, rarely lavender, linear or oblong, 4–7.5(–9) × 1–1.8(–2) mm, margins not crisped, claw not differentiated from blade (base sometimes clawlike, to 2 mm, widest at base); nectar glands confluent; filaments equal, (3–)3.5–6.5(–8.5) mm; anthers (1–)1.5–2.5(–3) mm, circinately coiled; gynophore 0.2–2(–5) mm. **Fruits** horizontal or reflexed, torulose, straight to somewhat curved, flattened, (2.5–)3.8–7.4(–9) cm × 1–1.2(–1.5) mm, (replum not constricted between seeds); ovules 76–128 per ovary; style usually clavate to subclavate, rarely cylindrical, (0.5–)0.8–2(–3) mm. **Seeds** 0.7–1.3 (–1.5) × 0.5–0.8 mm.

Flowering (Mar–)Jun–Oct. Rock crevices, shady slopes, rocky hillsides, canyons, stream banks, creek beds, pinyon-juniper communities, oak woodlands; 1200–2300 m; Ariz., Colo., N.Mex., Tex., Utah; Mexico (Baja California, Chihuahua, Coahuila, Hidalgo).

Study of extensive material that has accumulated during the past three decades reveals that the differences between subsp. *oklahomensis* and subsp. *wrightii*, which are based solely on variations in gynophore and fruit lengths, do not hold as clearly as previously thought (I. A. Al-Shehbaz 1973; R. C. Rollins 1993).

94. THYSANOCARPUS Hooker, Fl. Bor.-Amer. 1: 69, plate 18, fig. A. 1830

• Fringepod, lacepod [Greek *thysanos*, fringe, and *karpos*, fruit, alluding to fruit margin]

Patrick J. Alexander

Michael D. Windham

Annuals; not scapose; (glaucous), glabrous or pubescent. **Stems** erect, unbranched or branched distally. **Leaves** basal and cauline; petiolate or sessile; basal (often withered by anthesis or fruiting), not or, rarely, rosulate, shortly petiolate, blade margins subentire, dentate, pinnatifid, or pinnatisect [rarely entire]; cauline sessile, blade (base usually auriculate), margins entire, dentate, or pinnatifid to pinnatisect. **Racemes** (corymbose, several-flowered), considerably elongated in fruit. **Fruiting pedicels** often recurved, sometimes divaricate-ascending, slender. **Flowers**: sepals ascending, oblong to ovate, lateral pair not saccate basally; petals white to purplish, spatulate to oblong, (subequaling or longer than sepals), claw not differentiated from blade; stamens slightly tetradynamous; filaments slightly dilated basally; anthers ovate; nectary glands each side of lateral stamen or semi-annular, median glands absent. **Fruits** (pendulous), sessile, cymbiform, orbicular, obovate [ovate, elliptic], smooth, strongly latiseptate; valves each with prominent midvein, glabrous or pubescent; replum entire or crenate, often perforated, (winged, wing flattened, with radiating rays); septum obsolete; ovule 1 per ovary; style distinct (relatively short) or obsolete; stigma entire. **Seeds** aseriate, flattened, not winged, elliptical to orbicular; seed coat not mucilaginous when wetted; cotyledons accumbent.

Species 5 (4 in the flora): w North America, nw Mexico.

Thysanocarpus erectus S. Watson occurs in northwestern Mexico on Cedros and Guadalupe islands and on the mainland of Baja California.

1. Fruits cymbiform, wings strongly incurved (toward flat side of fruit); raceme internodes ca. 0.7–1.5(–2) mm in fruit . . . . 1. *Thysanocarpus conchuliferus*
1. Fruits flat or plano-convex, wings not strongly incurved; raceme internodes (1.5–)2–18 mm in fruit.
   2. Fruiting pedicels weakly ascending, straight or nearly so (geniculately reflexed apically), 7–18 mm; fruits 7–10 mm wide, wings with distinct rays ± 0.1 mm wide . . . . . . 4. *Thysanocarpus radians*
   2. Fruiting pedicels smoothly recurved or straight and stiffly spreading, 2–7(–12) mm; fruits 2.5–6(–9) mm wide, wings with indistinct rays or (0–)0.2–0.5 mm wide.
      3. Cauline leaf blades lanceolate, widest at base, bases auriculate-clasping, auricles extending around stems (at least some leaves); basal leaf blade margins subentire to sinuate-dentate, never pinnatifid, surfaces often hirsute, sometimes glabrous . . . . 2. *Thysanocarpus curvipes*
      3. Cauline leaf blades linear to narrowly elliptic, widest near middle or equally wide throughout, bases not auriculate or inconspicuous auricles not extending around stems; basal leaf blade margins pinnatifid, sinuate-dentate, or subentire, surfaces usually glabrous, rarely sparsely hirsute . . . . 3. *Thysanocarpus laciniatus*

1. **Thysanocarpus conchuliferus** Greene, Bull. Torrey Bot. Club 13: 218. 1886 [C][E]

Thysanocarpus conchuliferus var. *planiusculus* B. L. Robinson; *T. laciniatus* Nuttall var. *conchuliferus* (Greene) Jepson

Stems 0.5–1.5 dm. **Basal leaves:** blade oblanceolate to elliptic, 1–2.5(–3.5) cm, margins often pinnatifid, sometimes sinuate-dentate, surfaces glabrous. **Cauline leaves:** blade lanceolate to narrowly elliptic, or nearly linear, widest near base or middle, base auriculate-clasping, auricles extending around stem (at least some leaves). **Racemes:** internodes 0.7–1.5(–2) mm in fruit. **Fruiting pedicels** stiffly divaricate-ascending to slightly recurved, (proximal) 3.5–6.5 mm. **Fruits** cymbiform, (wings strongly incurved toward flat side of fruit); valves glabrous; wing with spatulate lobes (these sometimes joined distally, 0.25–0.4 mm wide at narrowest), rays absent or indistinct.

Flowering Mar–Apr. Rocky ridges, slopes, cliffs, canyons; of conservation concern; 50–500 m; Calif.

Thysanocarpus conchuliferus is known from Santa Cruz Island, where *T. laciniatus* and *T. curvipes* also occur. Rarely, specimens of those two species will have slightly incurved wings, or fruits folded in pressing. In such cases, the very dense, short inflorescences of *T. conchuliferus* provide a useful distinguishing feature.

Thysanocarpus conchuliferus is in the Center for Plant Conservation's National Collection of Endangered Plants.

2. **Thysanocarpus curvipes** Hooker, Fl. Bor.-Amer. 1: 69, plate 18, fig. A. 1830

Thysanocarpus amplectens Greene; *T. curvipes* var. *cognatus* Jepson; *T. curvipes* var. *elegans* (Fischer & C. A. Meyer) B. L. Robinson; *T. curvipes* var. *emarginatus* (Greene) Jepson; *T. curvipes* var. *eradiatus* Jepson; *T. curvipes* var. *involutus* Greene; *T. curvipes* var. *longistylus* Jepson; *T. curvipes* subsp. *madocarpus* Piper; *T. curvipes* var. *pulchellus* (Fischer & C. A. Meyer) Greene; *T. emarginatus* Greene; *T. filipes* Greene; *T. foliosus* A. Heller; *T. hirtellus* Greene; *T. laciniatus* Nuttall var. *emarginatus* (Greene) Jepson; *T. pulchellus* Fischer & C. A. Meyer; *T. trichocarpus* Rydberg

Stems 1–6(–8) dm. **Basal leaves:** blade oblanceolate to obovate, 1–6(–13) cm, margins subentire to sinuate-dentate, surfaces often hirsute, sometimes glabrous, (trichomes white, 0.3–0.6 mm). **Cauline leaves:** blade lanceolate, widest at base, base auriculate-clasping, auricles extending around stem (at least some leaves). **Racemes:** internodes 3–6(–9) mm in fruit. **Fruiting pedicels** smoothly recurved, (proximal) 3–7(–12) mm. **Fruits** flat or plano-convex, obovate to nearly orbicular, [3–6(–9) mm wide]; valves pubescent or glabrous, trichomes clavate and 0.2–0.4 mm, or pointed and ± 0.2 mm; wing entire, perforate, or incised, rays absent or distinct, (0–)0.2–0.5 mm wide.

Flowering Feb–Jun. Rocky slopes, washes, oak woodlands, streamsides, meadows, sometimes serpentine soils; 150–2000 m; B.C.; Ariz., Calif., Colo., Idaho, Nev., N.Mex., Oreg., Utah, Wash.; Mexico (Baja California, Sonora).

Thysanocarpus curvipes is the most widespread and variable species in the genus. Variants have been named as varieties or species, but they grade into each other imperceptibly. Notable among these are var. *elegans*, a form with incised or perforate fruit wings, and var. *eradiatus*, a form with rayless, entire wings. Some of these may be the result of hybridization with other taxa. For instance, var. *elegans* has large fruits and occurs in the vicinity of *T. radians*, the largest-fruited member of the genus. Furthermore, fruits of var. *elegans* often have pointed hairs like those usually found on fruits of *T. radians*; such hairs are not found on fruits of any other members of the genus. *Thysanocarpus curvipes* includes both diploid and tetraploid populations (M. D. Windham, unpubl.), but these do not appear to segregate into recognizable groups. Although the variation in *T. curvipes* is considerable, its great complexity prevents recognition of infraspecific taxa at this time.

3. **Thysanocarpus laciniatus** Nuttall in J. Torrey and A. Gray, Fl. N. Amer. 1: 118. 1838

Stems 1–6 dm. **Basal leaves:** blade oblanceolate to elliptic, 1–6 cm, margins often pinnatifid with narrow lobes (lobes 0.5–1.5 mm), sometimes sinuate-dentate or subentire, surfaces usually glabrous, rarely sparsely hirsute, trichomes whitish, 0.3–0.4 mm. **Cauline leaves:** blade linear to narrowly elliptic, widest near middle or equally wide throughout, base not auriculate or with small, inconspicuous auricles (not extending around stem). **Racemes:** internodes (1.5–)2–4.5 mm in fruit. **Fruiting pedicels** smoothly recurved or straight and stiffly spreading, (proximal) 3–6(–10) mm. **Fruits** flat or plano-convex, obovate to nearly orbicular, (2.5–5 mm wide); valves often glabrous, sometimes pubescent, trichomes clavate, 0.05–0.4 mm; wing entire or deeply crenate, rays absent or indistinct.

Varieties 3 (3 in the flora): sw United States, nw Mexico.

Thysanocarpus laciniatus presents some of the same problems as does *T. curvipes*. Variety *laciniatus* contains both diploids and tetraploids (M. D. Windham, unpubl.) and varies in fruit characters, pubescence, and basal leaf shape. Specimens with sinuate-dentate basal leaf margins and small auricles on cauline leaves can be difficult to distinguish from *T. curvipes*. Preliminary molecular phylogenetic analyses support the distinction between *T. curvipes* and *T. laciniatus* var. *laciniatus*, but suggest that tetraploid populations of the latter may have arisen through hybridization between *T. curvipes* and a diploid member of the *T. laciniatus* clade (P. Alexander, unpubl.). Varieties *hitchcockii* and *rigidus* are distinctive diploids (Windham, unpubl.) with restricted ranges and may deserve specific rank. Variety *rigidus* (known to us from only four collections) can be difficult to distinguish from the more purplish specimens of var. *laciniatus*, but the latter have at least some recurved pedicels and often have pinnatifid leaves.

1. Fruit valves pubescent, trichomes 0.05–0.1 mm .... 3b. *Thysanocarpus laciniatus* var. *hitchcockii*
1. Fruit valves usually glabrous, or trichomes 0.2–0.4 mm.
   2. Foliage usually greenish throughout, sometimes purplish basally; basal leaf blade margins pinnatifid or sinuate-dentate; fruiting pedicels smoothly recurved .... 3a. *Thysanocarpus laciniatus* var. *laciniatus*
   2. Foliage purplish throughout; basal leaf blade margins subentire to sinuate-dentate; fruiting pedicels straight or nearly so .... 3c. *Thysanocarpus laciniatus* var. *rigidus*

3a. Thysanocarpus laciniatus Nuttall var. laciniatus

Thysanocarpus affinis Greene; *T. laciniatus* var. *affinis* (Greene) Munz; *T. laciniatus* subsp. *desertorum* (A. Heller) Abrams; *T. laciniatus* var. *eremicola* Jepson; *T. laciniatus* var. *ramosus* (Greene) Munz; *T. ramosus* Greene

Foliage usually greenish, sometimes purplish basally. **Basal leaves:** blade margins often pinnatifid, sometimes sinuate-dentate. **Fruiting pedicels** usually smoothly recurved, rarely nearly straight. **Fruit valves** usually glabrous, sometimes pubescent, trichomes clavate, 0.2–0.4 mm.

Flowering Mar–May. Chaparral, rocky slopes, canyons, oak woodlands, washes; 200–1800 m; Ariz., Calif.; Mexico (Baja California).

Variety *laciniatus* is found in Arizona on the extreme western side of the state.

3b. Thysanocarpus laciniatus Nuttall var. hitchcockii
Munz, Bull. S. Calif. Acad. Sci. 31: 62. 1932 [E]

Foliage greenish or purplish throughout. **Basal leaves:** blade margins subentire to pinnatifid. **Fruiting pedicels** smoothly recurved to nearly straight. **Fruit valves** pubescent, trichomes clavate, 0.05–0.1 mm.

Flowering Mar–May. Dry rocky slopes, sandy washes; 600–1900 m; Calif.

Variety *hitchcockii* is known from Inyo, Los Angeles, Riverside, and San Bernardino counties.

3c. Thysanocarpus laciniatus Nuttall var. rigidus
Munz, Bull. S. Calif. Acad. Sci. 31: 62. 1932

Foliage purplish throughout. **Basal leaves:** blade margins subentire to sinuate-dentate. **Fruiting pedicels** straight or nearly so. **Fruit valves** glabrous.

Flowering Feb–May. Dry rocky slopes, ridges, pine and oak woodlands; 600–2200 m; Calif.; Mexico (Baja California).

Variety *rigidus* is known from Riverside, San Bernardino, and San Diego counties.

4. Thysanocarpus radians Bentham, Pl. Hartw., 297. 1849 [E] [F]

Thysanocarpus radians var. *montanus* Jepson

Stems 1.5–6 dm. **Basal leaves:** blade oblanceolate, 1.5–4 cm, margins sinuate-dentate to runcinate-pinnatifid, surfaces usually glabrous, rarely sparsely hirsute, trichomes whitish, 0.3–0.4 mm. **Cauline leaves:** blade lance-ovate to lanceolate, widest at base, base auriculate-clasping, auricles extending around stem (at least some leaves). **Racemes:** internodes 9–18 mm in fruit. **Fruiting pedicels** weakly ascending, straight or nearly so, geniculately-reflexed apically, (proximal) 7–18 mm. **Fruits** flat, orbicular, (7–10 mm wide); valves pubescent or glabrous, trichomes pointed, ± 0.2 mm; wing entire or with undulate margins, rays distinct, ± 0.1 mm wide.

Flowering Mar–Apr. Meadows in oak woodlands, fields, swales; 20–400 m; Calif., Oreg.

The large (7–10 mm wide), strongly rayed fruits and geniculately reflexed apices of fruiting pedicels make *Thysanocarpus radians* a very distinctive species. Occasional plants in northern California appear to be hybrids between *T. curvipes* and *T. radians*.

95. WAREA Nuttall, J. Acad. Nat. Sci. Philadelphia 7: 83, plate 10. 1834 • [For Nathaniel A. Ware, 1789–1853, teacher in South Carolina and plant collector, especially in Florida] E

Ihsan A. Al-Shehbaz

Annuals; not scapose; (usually somewhat glaucous), glabrous throughout (rarely petal claws pubescent). **Stems** erect, often branched distally, rarely unbranched, (usually slender, rarely stout). **Leaves** cauline (basal not seen, soon withered, not rosulate); petiolate or sessile; blade (base cuneate, auriculate, or amplexicaul), margins entire. **Racemes** (corymbose, several-flowered, floral buds clavate), not or slightly elongated in fruit. **Fruiting pedicels** (often deciduous at maturity, leaving elevated discoid scars on rachis), divaricate, slender, (sometimes filiform, straight, with 2 lateral glands basally). **Flowers:** sepals spreading or reflexed, linear-oblanceolate, lateral pair not saccate basally; petals (spreading), white or pink to deep purple, obovate, orbicular, or spatulate, (margins entire or crisped), claw strongly differentiated from blade (slender, often dilated basally, usually minutely to coarsely papillate or pubescent, rarely nearly smooth, apex rounded); stamens (strongly exserted, spreading), subequal; filaments not dilated basally; anthers linear, (coiled after dehiscence); nectar glands confluent, subtending bases of stamens, (often 6 teeth alternating with filaments), median glands present. **Fruits** stipitate, narrowly linear, smooth, (recurved), latiseptate; valves each with prominent midvein throughout; replum rounded; septum complete; ovules 20–60 per ovary; style usually obsolete, rarely distinct; stigma entire. **Seeds** uniseriate, not winged, oblong; seed coat (concentrically striate), not mucilaginous when wetted; cotyledons accumbent. $x = 12$.

Species 4 (4 in the flora): se United States.

SELECTED REFERENCE Channell, R. B. and C. W. James. 1964. Nomenclatural and taxonomic corrections in *Warea* (Cruciferae). Rhodora 66: 18–26.

1. Cauline leaves petiolate or obsolete, blades linear-oblanceolate, oblancolate, or narrowly oblong.
   2. Petal claws nearly smooth or obscurely papillate, margins entire; gynophores (5–)7–11 mm . . . . .1. *Warea cuneifolia*
   2. Petal claws coarsely papillate to pubescent, margins crisped; gynophores 3–6(–7) mm . . . . 2. *Warea carteri*
1. Cauline leaves sessile, blades oblong, ovate, or lanceolate.
   3. Leaf blade base not clasping stem, obtuse to minutely auriculate . . . . 3. *Warea sessilifolia*
   3. Leaf blade base clasping stem, amplexicaul to strongly auriculate. . . . . 4. *Warea amplexifolia*

1. **Warea cuneifolia** (Muhlenberg ex Nuttall) Nuttall, J. Acad. Nat. Sci. Philadelphia 7: 84. 1834 E

Cleome cuneifolia Muhlenberg ex Nuttall, Gen. N. Amer. Pl. 2: 73. 1818; *Stanleya gracilis* de Candolle

Stems (2–)3–6.5(–8) dm. **Cauline leaves** petiolate (petiole (0.05–)0.1–0.2(–0.3) cm proximally, obsolete distally); blade usually linear-oblanceolate to oblanceolate, rarely linear, (0.7–)1–3(–4) cm × 1.5–6(–8) mm, base cuneate, apex rounded to retuse. **Racemes** 0.3–2(–3) cm in fruit. **Fruiting pedicels** (4–)5–9(–11) mm. **Flowers:** sepals white or purplish, spreading or reflexed, 3–5(–7) × 0.2–0.3 mm; petals white or pink, broadly obovate to spatulate, 4–9 mm, blade 2–5 × 1.5–3 mm, claw 2–4 mm, nearly smooth or obscurely papillate, margins entire; filaments 6–8(–10) mm; anthers 1–1.5 mm; gynophore slender, (5–)7–11 mm. **Fruits** 2–4(–5) cm × 0.7–1 mm; ovules 32–54 per ovary; style rarely to 0.1 mm. **Seeds** 0.6–0.8 × 0.4–0.5 mm.

Flowering Jul–Sep. Sandy areas, scrublands, sand hills, fields, open banks, oak-pinyon woods, roadside embankments; 0–150 m; Ala., Fla., Ga., N.C., S.C.

Although *Warea cuneifolia* is fairly widespread in Georgia and South Carolina, it is known in Alabama only from Pike County, in Florida from Gadsden and Liberty counties, and in North Carolina from Harnett and Hoke counties.

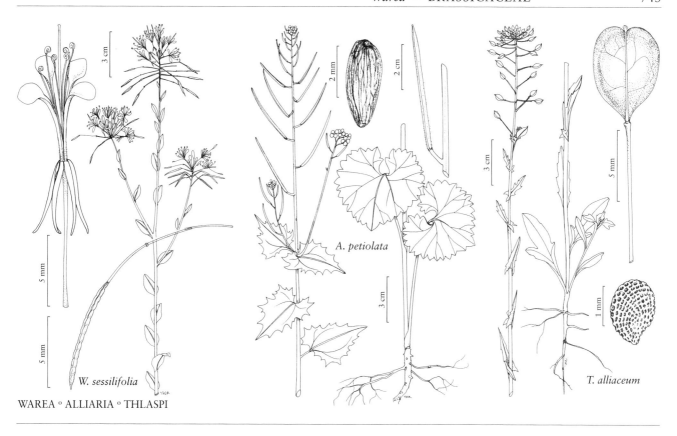

WAREA · ALLIARIA · THLASPI

2. **Warea carteri** Small, Bull. Torrey Bot. Club 36: 159. 1909 E

Stems (4–)5–14 dm. **Cauline leaves** petiolate (petiole 0.1–0.8 cm proximally, obsolete distally); blade usually linear-oblanceolate to oblanceolate or narrowly oblong, rarely linear, 1–4.5 cm × 1–6(–10) mm, base cuneate to attenuate, apex obtuse to subapiculate. **Racemes** 0.4–3(–4) cm in fruit. **Fruiting pedicels** 4–10 mm. **Flowers:** sepals white, spreading or reflexed, 3–5 × 0.3–0.5 mm; petals white, broadly obovate to suborbicular, 4–6 mm, blade 2–3 × 2–3 mm, claw 2–3 mm, coarsely papillate to pubescent, margins crisped; filaments 5–7 mm; anthers 1–1.5 mm; gynophore slender, 3–6(–7) mm. **Fruits** 3–5(–6) cm × 1–1.5 mm; ovules 22–34 per ovary; style rarely to 0.5 mm. **Seeds** 1.2–1.8 × 0.8–1 mm. $2n = 24$.

Flowering late Sep–Jan; fruiting Oct–late Jan. Sandy areas in open scrub oak, sand scrub; 0–50 m; Fla.

Warea carteri is known from Brevard, Glades, Highlands, Miami-Dade, and Polk counties. It is in the Center for Plant Conservation's National Collection of Endangered Plants.

3. **Warea sessilifolia** Nash, Bull. Torrey Bot. Club 23: 101. 1896 C E F

Stems (sometimes stout), (1.5–)2.5–6.5(–8) dm. **Cauline leaves** sessile; blade ovate to lanceolate, (0.8–)1–2.5(–4) cm × 3–15(–30) mm, base not clasping stem, obtuse or, rarely, minutely auriculate (auricles to 2 × 2 mm, those proximally on robust plants rarely larger), apex acute to obtuse. **Racemes** 1–3 cm in fruit. **Fruiting pedicels** 9–12 mm. **Flowers:** sepals white or purplish, strongly reflexed, 6–7 × 0.4–0.7 mm; petals purple or pink, broadly obovate to suborbicular, 7–11 mm, blade 2–5 × 2–5 mm, claw 4–6 mm, minutely papillate, margins entire; filaments 9–15 mm; anthers 1–1.5 mm; gynophore slender, 10–16 mm. **Fruits** 2.5–4.5 cm × 1–1.5 mm; ovules 22–40 per ovary; style rarely to 0.5 mm. **Seeds** 1.2–1.5 × 0.9–1 mm.

Flowering Aug–Sep. Sand hills, pine barrens, sandy pinewoods, scrub oak and pine, turkey oak hills; of conservation concern; 0–50 m; Ala., Fla.

Warea sessilifolia is widespread in the Florida panhandle; in Alabama it is known only from Pike County.

4. **Warea amplexifolia** (Nuttall) Nuttall, J. Acad. Nat. Sci. Philadelphia 7: 83. 1834 C E

Stanleya amplexifolia Nuttall, Amer. J. Sci. Arts 5: 297. 1822; *Warea auriculata* Shinners

Stems 3.5–7(–8) dm. **Cauline leaves** sessile; leaf blades ovate, oblong, to lanceolate, (1–)1.5–4 cm × 4–14(–16) mm, base always clasping stem, amplexicaul to strongly auriculate (auricles ovate, 3–9 × 2–6 mm), apex acute to obtuse. **Racemes** 1–5(–8) cm in fruit. **Fruiting pedicels** 8–15 mm. **Flowers:** sepals white or pinkish, spreading or strongly reflexed, 5–8 × 0.4–0.6 mm; petals white, pink, or purple, broadly obovate to orbicular, 7–10 mm, blade 3–5 × 2–4 mm, claw 4–5 mm, minutely papillate, margins entire; filaments 12–15 mm; anthers 1–1.5 mm; gynophore slender, (8–)10–15 mm. **Fruits** (3–)4–7 cm × 1.3–1.5 mm; ovules 24–32 per ovary. **Seeds** 1–1.5 × 0.6–0.8 mm.

Flowering Aug–Sep. Sandy areas, woods, oak scrub, pine barrens; of conservation concern; 0–50 m; Fla.

Warea amplexifolia is known from Lake, Orange, Osceola, and Polk counties. It is in the Center for Plant Conservation's National Collection of Endangered Plants.

ee. BRASSICACEAE Burnett tribe THLASPIDEAE de Candolle, Mém. Mus. Hist. Nat. 7: 234. 1821 I

Annuals or biennials [perennials]; eglandular. **Trichomes** absent or simple. **Cauline leaves** petiolate or sessile; blade base auriculate or not, margins entire or dentate [pinnately lobed]. **Racemes** usually ebracteate, often elongated in fruit. **Flowers** actinomorphic; sepals erect or ascending [spreading], lateral pair not saccate basally; petals white [pink, purple], claw present, distinct or indistinct (often short); filaments unappendaged, not winged; pollen 3-colpate. **Fruits** silicles or siliques, dehiscent [indehiscent], unsegmented, terete, 4-angled, or angustiseptate [latiseptate]; ovules [4–]6–22[–numerous]; style distinct or obsolete; stigma entire. **Seeds** uniseriate (striate or coarsely reticulate); cotyledons accumbent or incumbent.

Genera 8, species 31 (2 genera, 3 species in the flora): introduced; Eurasia, n Africa.

96. ALLIARIA Heister ex Fabricius, Enum., 161. 1759 • Garlic mustard [Genus *Allium*, garlic or onion, and Latin -*aria*, connection, alluding to odor of crushed plant] I

Ihsan A. Al-Shehbaz

Plants with garlic smell when crushed; not scapose; glabrous or pubescent, trichomes simple. **Stems** erect [decumbent], often branched distally. **Leaves** basal and cauline; petiolate; basal (often withered by anthesis or fruiting), rosulate, long-petiolate, blade margins crenate, dentate, or sinuate; cauline shortly petiolate, blade margins dentate. **Racemes** elongated in fruit. **Fruiting pedicels** divaricate or ascending, stout (almost as broad as fruit [slender, narrower than fruit]). **Flowers:** sepals erect, oblong, lateral pair not saccate basally, (glabrous); petals oblanceolate, (longer than sepals), claw obscurely differentiated from blade, (apex obtuse); stamens slightly tetradynamous; filaments not dilated basally; anthers ovate or oblong, (apex obtuse); nectar glands confluent, subtending bases of stamens. **Fruits** siliques, sessile, linear [oblong], torulose or subtorulose, terete, subterete, or 4-angled; valves each with prominent midvein and distinct marginal veins, glabrous [scabrous]; replum rounded; septum complete; ovules [4–]6–22 per ovary; style obsolete or distinct (to 6 mm); stigma capitate, entire. **Seeds** plump, not winged, oblong; seed coat (longitudinally striate), not mucilaginous when wetted; cotyledons incumbent.

Species 2 (1 in the flora): introduced; Eurasia, n Africa.

Alliaria brachycarpa M. Bieberstein is endemic to Caucasus.

SELECTED REFERENCE Cavers, P. B., M. I. Heagy, and R. F. Kokron. 1979. The biology of Canadian weeds. 35. *Alliaria petiolata* (M. Bieb.) Cavara and Grande. Canad. J. Pl. Sci. 59: 217–229.

1. **Alliaria petiolata** (M. Bieberstein) Cavara & Grande, Bull. Orto Bot. Regia Univ. Napoli 3: 418. 1913 F I W

Arabis petiolata M. Bieberstein, Fl. Taur.-Cauc. 2: 126. 1808; *Alliaria officinalis* Andrzejowski ex M. Bieberstein; *Erysimum alliaria* Linnaeus

Stems simple or branched distally, (1.5–)3–9(–13) dm; glabrous or pilose basally, trichomes to 1.5 mm. **Basal leaves**: petiole 3–16(–22) cm; blade reniform or cordate, (6–)15–88(–118) mm wide (shorter in length), surfaces glabrous or pilose. **Cauline leaves**: petiole shorter than basal; blade ovate, cordate, or deltate, to 15 × 15 cm, base cordate or truncate, margins acutely to obtusely toothed, apex acute. **Racemes** several-flowered. **Fruiting pedicels** terete, (2–)3–10(–15) mm. **Flowers**: sepals (2–)2.5–3.5(–4.5) × 0.7–1.5 mm; petals (2.5–)4–8(–9) × (1.5–)2–3(–3.5) mm, base attenuate to clawlike; filaments 2–3.5(–4.5) mm; anthers oblong, 0.7–1 mm. **Fruits** divaricate-ascending, subtorulose, quadrangular or subterete, (2–)3–7(–8) cm × 1.2–2.5 mm; style (0.2–)1–2(–3) mm. **Seeds** dark brown or black, narrowly oblong, 2–4.5 × 0.7–2 mm. $2n = 42$.

Flowering Apr–May. Roadsides, trails, railroad tracks, stream banks, waste places, fields, shaded woodlands, bluffs, thickets, steep slopes, disturbed fields, floodplains, woods, shaded forest floor; 0–1200 m; introduced; B.C., Ont.; Conn., Del., D.C., Ill., Ind., Iowa, Kans., Ky., Maine, Md., Mass., Mich., Mo., Nebr., N.H., N.J., N.Y., N.C., Ohio, Okla., Oreg., Pa., R.I., Tenn., Utah, Vt., Va., Wash., W.Va., Wis.; South America (Argentina); Eurasia; n Africa.

97. **THLASPI** Linnaeus, Sp. Pl. 2: 645. 1753; Gen. Pl. ed. 5, 292. 1754 • Pennycress [Greek *thalo*, *thals-* to compress, alluding to flattened fruits] I

Ihsan A. Al-Shehbaz

Plants often fetid when crushed; not scapose; glabrous or pubescent. **Stems** erect, unbranched or branched distally. **Leaves** basal and cauline; petiolate or sessile [subsessile]; basal (often withered in fruit), rosulate or not, petiolate [subsessile], margins entire, repand, dentate, or sinuate-dentate; cauline sessile, blade (base auriculate or sagittate), margins dentate, repand, or entire. **Racemes** (several-flowered), considerably elongated in fruit. **Fruiting pedicels** divaricate, (straight or slightly curved), slender. **Flowers**: sepals erect or ascending, ovate or oblong, (margins membranous); petals spatulate [oblong], claw differentiated or not from blade, (apex obtuse or emarginate); stamens slightly tetradynamous; filaments not dilated basally; anthers ovate, (apex obtuse); nectar glands (2 or 4), lateral, often 1 on each side of lateral stamen, median glands absent. **Fruits** silicles, sessile, oblong, obovate, obcordate, or suborbicular, (apex often notched), keeled, strongly angustiseptate; valves winged throughout or apically, glabrous; replum rounded; septum complete, (not veined); ovules 6–16 per ovary; style obsolete or not, (included in apical notch); stigma capitate. **Seeds** plump, not winged, ovoid; seed coat (coarsely reticulate, alveolate or concentrically striate), not mucilaginous when wetted; cotyledons accumbent. $x = 7$.

Species 6 (2 in the flora): introduced; Eurasia, n Africa; introduced also nearly worldwide.

Thlaspi was divided by F. K. Meyer (1973) into 12 genera largely based on seed-coat sculpture and anatomy. Molecular data (K. Mummenhoff et al. 1997; M. Koch and Mummenhoff 2001) showed that *Thlaspi* is polyphyletic and strongly supported some of Meyer's segregates, including *Microthlaspi* and *Noccaea*, both of which are recognized in this flora. As here delimited, *Thlaspi* consists of six species (Meyer), instead of over 75 recognized by various authors.

SELECTED REFERENCES Koch, M. and K. Mummenhoff. 2001. *Thlaspi* s.str. (Brassicaceae) versus *Thlaspi* s.l.: Morphological and anatomical characters in the light of ITS nDNA sequence data. Pl. Syst. Evol. 227: 209–225. Meyer, F. K. 1973. Conspectus der "*Thlaspi*"-Arten Europas, Afrikas und Vorderasiens. Feddes Repert. 84: 449–470. Meyer, F. K. 1979. Kritische Revision der "*Thlaspi*"-Arten Europas, Afrikas und Vorderasiens: I. Geschichte, Morphologie und Chorologie. Feddes Repert. 90: 129–154. Meyer, F. K. 2001. Kritische Revision der "*Thlaspi*"-Arten Europas, Afrikas und Vorderasiens. Spezieller Teil. I. *Thlaspi* L. Häussknechtia 8: 3–42. Mummenhoff, K., A. Franzke, and M. Koch. 1997. Molecular phylogenetics of *Thlaspi* s.l. (Brassicaceae) based on chloroplast DNA restriction site variation and sequences of the internal transcribed spacers of nuclear ribosomal DNA. Canad. J. Bot. 75: 469–482.

1. Plants glabrous throughout; fruits broadly winged throughout, wing 3.5–5 mm wide apically; seeds concentrically striate. . . . . 1. *Thlaspi arvense*
1. Plants sparsely to densely pubescent basally along stems; fruits obscurely winged basally, wing to 1 mm wide apically; seeds alveolate . . . . 2. *Thlaspi alliaceum*

1. Thlaspi arvense Linnaeus, Sp. Pl. 2: 646. 1753

• Fanweed, field pennycress, stinkweed [I] [W]

Teruncius arvensis (Linnaeus) Lunell

Plants (sometimes glaucous), glabrous throughout. **Stems** (0.9–)1.5–5.5(–8) dm. **Basal leaves** not rosulate; petiole 0.5–3 cm; blade oblanceolate, spatulate, or obovate, 1–5 cm × 4–23 mm, base attenuate or cuneate, margins entire, repand, or coarsely dentate, apex rounded. **Cauline leaves:** blade oblong, (0.5–)1.5–4(–8) cm × (2–)5–15(–25) mm, apex rounded, obtuse, or subacute. **Fruiting pedicels** straight or slightly upcurved, (5–)9–13(–15) mm. **Flowers:** sepals (1.5–)2–3(–3.3) × 1–1.5 mm; petals (2.4–)3–4.5(–5) × (0.8–)1.1–1.7 mm, narrowed to clawlike base (ca. 1 mm), apex obtuse or emarginate; filaments (1–)1.5–2.2 mm; anthers 0.3–0.5 mm. **Fruits** obovate or suborbicular, (0.6–)0.9–2 cm × (5–)7–20 mm, base obtuse or rounded, apex deeply emarginate, notch ca. 5 mm deep; wings broad throughout, 1–1.5 mm wide basally, 3.5–5 mm wide apically; ovules 6–16 per ovary; style 0.1–0.3 mm. **Seeds** (1.2–)1.6–2(–2.3) × 1.1–1.3 mm, concentrically striate. $2n = 14$.

Flowering Mar–Aug. Roadsides, fields, waste places, lawns, gardens, railroad tracks, stream banks, bluffs, thickets, slopes, floodplains, woods; 0–2000 m; introduced; Alta., B.C., Man., N.B., Nfld. and Labr., N.W.T., N.S., Ont., P.E.I., Que., Sask., Yukon; Ala., Alaska, Ariz., Ark., Calif., Colo., Conn., Del., D.C., Fla., Ga., Idaho, Ill., Ind., Iowa, Kans., Ky., La., Maine, Md., Mass., Mich., Minn., Miss., Mo., Mont., Nebr., Nev., N.H., N.J., N.Mex., N.Y., N.C., N.Dak., Ohio, Okla., Oreg., Pa., R.I., S.C., S.Dak., Tenn., Tex., Utah, Vt., Va., Wash., W.Va., Wis., Wyo.; Europe; Asia.

Thlaspi arvense is a cosmopolitan weed of Eurasian origin. Cattle feeding on it develop tainted milk.

SELECTED REFERENCE Best, K. F. and G. K. McIntyre. 1975. The biology of Canadian weeds. 9. *Thlaspi arvense* L. Canad. J. Pl. Sci. 55: 279–292.

2. Thlaspi alliaceum Linnaeus, Sp. Pl. 2: 646. 1753

[F] [I]

Plants mostly glabrous, sparsely to densely pubescent basally along stem, trichomes to 2 mm. **Stems** (1.2–)2.5–6.5(–7.5) dm. **Basal leaves** (withered in fruit) laxly rosulate; petiole (0.5–)1–4(–6) cm; blade spatulate, obovate, or oblong-oblanceolate, 0.5–3.5 cm × 6–20 mm, bases attenuate or cuneate, margins entire, repand, or sinuate-dentate, apex rounded. **Cauline leaves:** blade oblong, 1–4.5 cm × 2–10 mm, apex obtuse or subacute. **Fruiting pedicels** straight or slightly recurved, (6–)9–12(–16) mm. **Flowers:** sepals 1.2–2.6 × 0.7–1.2 mm; petals 2.5–4 × 1–1.7 mm, narrowed to clawlike base, apex obtuse; filaments 1.2–2 mm; anthers 0.2–0.4 mm. **Fruits** obovate or narrowly obcordate, 5–7.5 × 3.7–5.5 mm, base cuneate, apex shallowly emarginate, notch to 0.5 mm deep; wings obscure basally, to 1 mm wide apically; ovules 6–10 per ovary; style 0.1–0.3 mm. **Seeds** 1.2–1.4 × 0.9–1 mm, alveolate. $2n = 14$.

Flowering Apr–May. Roadsides, grassy shoulders, fields, waste ground, pastures; 0–600 m; introduced; Del., Ind., Ky., La., Md., N.C., Ohio, Pa., Tenn., Va., W.Va.; Europe.

Thlaspi alliaceum has a garlic-like smell when fresh.

Literature Cited

Robert W. Kiger, Editor

This is a consolidated list of all works cited in volume 7, whether as selected references, in text, or in nomenclatural contexts. In citations of articles, both here and in the taxonomic treatments, and also in nomenclatural citations, the titles of serials are rendered in the forms recommended in G. D. R. Bridson and E. R. Smith (1991). When those forms are abbreviated, as most are, cross references to the corresponding full serial titles are interpolated here alphabetically by abbreviated form. In nomenclatural citations (only), book titles are rendered in the abbreviated forms recommended in F. A. Stafleu and R. S. Cowan (1976–1988) and F. A. Stafleu et al. (1992–2009). Here, those abbreviated forms are indicated parenthetically following the full citations of the corresponding works, and cross references to the full citations are interpolated in the list alphabetically by abbreviated form. Two or more works published in the same year by the same author or group of coauthors will be distinguished uniquely and consistently throughout all volumes of *Flora of North America* by lower-case letters (b, c, d, ...) suffixed to the date for the second and subsequent works in the set. The suffixes are assigned in order of editorial encounter and do not reflect chronological sequence of publication. The first work by any particular author or group from any given year carries the implicit date suffix "a"; thus, the sequence of explicit suffixes begins with "b". There may be citations in this list that have dates suffixed "b" but that are not preceded by citations of "[a]" works for the same year, or that have dates suffixed "c" but that are not preceded by citations of "[a]" and/or "b" works for that year. In such cases, the missing "[a]" and/or "b" works are ones cited (and encountered first from) elsewhere in the *Flora* that are not pertinent in this volume.

Abbe, E. C. 1948. *Braya* in boreal eastern America. Rhodora 50: 1–15.

Abbott, R. J. and C. Brochmann. 2003. History and evolution of the arctic flora: In the footsteps of Eric Hultén. Molec. Ecol. 32: 299–313.

Abdallah, M. S. and H. C. D. de Wit. 1978. The Resedaceae: A taxonomical revision of the family (final installment). Meded. Landbouwhoogeschool 78.

Abrams, L. and R. S. Ferris. 1923–1960. Illustrated Flora of the Pacific States: Washington, Oregon, and California. 4 vols. Stanford. (Ill. Fl. Pacific States)

Account Exped. Pittsburgh—See: E. James 1823

Acta Bot. Neerl. = Acta Botanica Neerlandica.

Adanson, M. 1763[–1764]. Familles des Plantes. 2 vols. Paris. [Vol. 1, 1764; vol. 2, 1763.] (Fam. Pl.)

Ahles, H. E. and A. E. Radford. 1959. Species new to the flora of North Carolina. J. Elisha Mitchell Sci. Soc. 75: 140–147.

Airy Shaw, H. K. 1964. Diagnoses of new families, new names, etc., for the seventh edition of 'Willis's Dictionary.' Kew Bull. 18: 249–273.

Aiton, W. 1789. Hortus Kewensis; or, a Catalogue of the Plants Cultivated in the Royal Botanic Garden at Kew. 3 vols. London. (Hort. Kew.)

Aiton, W. and W. T. Aiton. 1810–1813. Hortus Kewensis; or a Catalogue of the Plants Cultivated in the Royal Botanic Garden at Kew. 5 vols. London. (Hortus Kew.)

Allertonia = Allertonia; a Series of Occasional Papers.

Allioni, C. 1785. Flora Pedemontana sive Enumeratio Methodica Stirpium Indigenarum Pedemontii. 3 vols. Turin. (Fl. Pedem.)

Al-Shehbaz, I. A. 1973. The biosystematics of the genus *Thelypodium* (Cruciferae). Contr. Gray Herb. 204: 3–148.

Al-Shehbaz, I. A. 1977. Protogyny in the Cruciferae. Syst. Bot. 2: 327–333.

Al-Shehbaz, I. A. 1984. The tribes of Cruciferae (Brassicaceae) in the southeastern United States. J. Arnold Arbor. 65: 343–373.

Al-Shehbaz, I. A. 1985. The genera of Brassiceae (Cruciferae; Brassicaceae) in the southeastern United States. J. Arnold Arbor. 66: 279–351.

Al-Shehbaz, I. A. 1985b. The genera of Thelypodieae (Cruciferae; Brassicaceae) in the southeastern United States. J. Arnold Arbor. 66: 95–111.

Al-Shehbaz, I. A. 1986. The genera of Lepidieae (Cruciferae; Brassicaceae) in the southeastern United States. J. Arnold Arbor. 57: 265–311.

Al-Shehbaz, I. A. 1986b. New wool-alien Cruciferae (Brassicaceae) in eastern North America: *Lepidium* and *Sisymbrium*. Rhodora 88: 347–355.

Al-Shehbaz, I. A. 1987. The genera of Alysseae (Cruciferae; Brassicaceae) in the southeastern United States. J. Arnold Arbor. 68: 185–240.

Al-Shehbaz, I. A. 1988. The genera of Arabideae (Cruciferae: Brassicaceae) in the southeastern United States. J. Arnold Arbor. 69: 85–166.

Al-Shehbaz, I. A. 1988b. The genera of Anchonieae (Cruciferae; rassicaceae) in the southeastern United States. J. Arnold Arbor. 69: 193–212.

Al-Shehbaz, I. A. 1988c. The genera of Sisymbrieae (Cruciferae; Brassicaceae) in the southeastern United States. J. Arnold Arbor. 69: 213–237.

Al-Shehbaz, I. A. 2003. *Aphragmus bouffordii*, a new species from Tibet and a synopsis of *Aphragmus* Andrz. (Brassicaceae). Harvard Pap. Bot. 8: 25–27.

Al-Shehbaz, I. A. 2003b. Transfer of most North American species of *Arabis* to *Boechera* (Brassicaceae). Novon 13: 381–391.

Al-Shehbaz, I. A. 2003c. A synopsis of *Tropidocarpum* (Brassicaceae). Novon 13: 392–395.

Al-Shehbaz, I. A. 2005. *Hesperidanthus* (Brassicaceae) revisited. Harvard Pap. Bot. 10: 47–51.

Al-Shehbaz, I. A. 2005b. Proposal to conserve the name *Erucastrum* against *Kibera* and *Hirschfeldia* (Brassicaceae). Taxon 54: 204–205.

Al-Shehbaz, I. A. 2007. Generic limits of *Dryopetalon*, *Rollinsia*, *Sibara*, and *Thelypodiopsis* (Brassicaceae) and a synopsis of *Dryopetalon*. Novon 17: 394–402.

Al-Shehbaz, I. A. 2007b. The North American genus *Sandbergia* (Boechereae, Brassicaceae). Harvard Pap. Bot. 12: 425–427.

Al-Shehbaz, I. A. and V. M. Bates. 1987. *Armoracia lacustris* (Brassicaceae), the correct name of the North American lake cress. J. Arnold Arbor. 68: 357–359.

Al-Shehbaz, I. A., M. A. Beilstein, and E. A. Kellogg. 2006. Systematics and phylogeny of the Brassicaceae (Cruciferae): An overview. Pl. Syst. Evol. 259: 89–120.

Al-Shehbaz, I. A., K. Mummenhoff, and O. Appel. 2002. *Cardaria*, *Coronopus*, and *Stroganowia* are united with *Lepidium* (Brassicaceae). Novon 12: 5–11.

Al-Shehbaz, I. A. and S. L. O'Kane. 2002. *Lesquerella* is united with *Physaria*. Novon 12: 319–329.

Al-Shehbaz, I. A., S. L. O'Kane, and R. A. Price. 1999. Generic placement of species excluded from *Arabidopsis*. Novon 9: 296–307.

Al-Shehbaz, I. A. and R. A. Price. 1998. Delimitation of the genus *Nasturtium* (Brassicaceae). Novon 8: 124–126.

Al-Shehbaz, I. A. and R. A. Price. 2001. The Chilean *Anagallis* and Californian *Tropidocarpum* are congeneric. Novon 11: 292–293.

Al-Shehbaz, I. A. and R. C. Rollins. 1988. A reconsideration of *Cardamine curvisiliqua* and *C. gambellii* as species of *Rorippa* (Cruciferae). J. Arnold Arbor. 69: 65–71.

Al-Shehbaz, I. A. and S. I. Warwick. 2005. A synopsis of *Eutrema* (Brassicaceae). Harvard Pap. Bot. 10: 129–135.

Al-Shehbaz, I. A. and S. I. Warwick. 2006. A synopsis of *Smelowskia* (Brassicaceae). Harvard Pap. Bot. 11: 91–99.

Al-Shehbaz, I. A. and S. I. Warwick. 2007. Two new tribes (Dontostemoneae and Malcolmieae) in the Brassicaceae (Cruciferae). Harvard Pap. Bot. 12: 429–433.

Al-Shehbaz, I. A. and M. D. Windham. 2007. New or noteworthy North American *Draba* (Brassicaceae). Harvard Pap. Bot. 12: 409–419.

Al-Shehbaz, I. A. and Yang G. 2000. A revision of the Chinese endemic *Orychophragmus* (Brassicaceae). Novon 10: 349–353.

Amer. Bot. Fl.—See: A. Wood 1870

Amer. J. Bot. = American Journal of Botany.

Amer. J. Sci. = American Journal of Science.

Amer. J. Sci. Arts = American Journal of Science, and Arts.

Amer. Midl. Naturalist = American Midland Naturalist; Devoted to Natural History, Primarily That of the Prairie States.

Amer. Naturalist = American Naturalist....

Amoen. Acad.—See: C. Linnaeus 1749[–1769]

Andersson, L. and S. Andersson. 2000. A molecular phylogeny of Tropaeolaceae and its systematic implications. Taxon 49: 721–736.

Andersson, N. J. 1867. Monographia Salicum.... Pars I. 1 fasc. only. Stockholm. (Monogr. Salicum)

Angiosperm Phylogeny Group. 1998. An ordinal classification for the families of flowering plants. Ann. Missouri Bot. Gard. 85: 531–553.

Angiosperm Phylogeny Group. 2003. An update of the Angiosperm Phylogeny Group classification for the orders and families of flowering plants: APG II. Bot. J. Linn. Soc. 141: 399–436.

Animadv. Bot. Spec. Alt.—See: P. Arduino 1764
Ann. Bot. Fenn. = Annales Botanici Fennici.
Ann. Bot. (Oxford) = Annals of Botany. (Oxford.)
Ann. Carnegie Mus. = Annals of the Carnegie Museum.
Ann. Lyceum Nat. Hist. New York = Annals of the Lyceum of Natural History of New York.
Ann. Missouri Bot. Gard. = Annals of the Missouri Botanical Garden.
Ann. Naturhist. Mus. Wien = Annalen des Naturhistorischen Museums in Wien.
Annuaire Conserv. Jard. Bot. Genève = Annuaire du Conservatoire et Jardin Botaniques de Genève.
App. Parry J. Sec. Voy.—See: W. E. Parry 1825
Appel, O. 1998. The status of *Teesdaliopsis* and *Teesdalia* (Brassicaceae). Novon 8: 218–219.
Appel, O. and I. A. Al-Shehbaz. 1998. Generic limits and taxonomy of *Hornungia*, *Pritzelago*, and *Hymenolobus* (Brassicaceae). Novon 7: 338–340.
Appel, O. and I. A. Al-Shehbaz. 2003. Cruciferae. In: K. Kubitzki et al., eds. 1990+. The Families and Genera of Vascular Plants. 9+ vols. Berlin etc. Vol. 5, pp. 75–174.
Arber, A. 1942. Studies in flower structure. VII. On the gynaecium of *Reseda*, with a consideration of paracary. Ann. Bot. (Oxford), n. s. 6(21): 43–48.
Arbor. Frutic. Brit.—See: J. C. Loudon [1835–]1838
Arbust. Amer.—See: H. Marshall 1785
Archer, W. A. 1965. *Salix* of Nevada. Beltsville. [Contr. Fl. Nevada 50.]
Arctic = Arctic; Journal of the Arctic Institute of North America.
Arctic Alpine Res. = Arctic and Alpine Research.
Arduino, P. 1764. Animadversionum Botanicarum Specimen Alterum. Venice. (Animadv. Bot. Spec. Alt.)
Argus, G. W. 1965. The taxonomy of the *Salix glauca* L. complex in North America. Contr. Gray Herb. 196: 1–142.
Argus, G. W. 1969. New combinations in the *Salix* of Alaska and Yukon. Canad. J. Bot. 57: 795–801.
Argus, G. W. 1973. The Genus *Salix* in Alaska and the Yukon. Ottawa. [Natl. Mus. Nat. Sci. Publ. Bot. 2.] (Salix Alaska Yukon)
Argus, G. W. 1974. An experimental study of hybridization and pollination in *Salix* (willows). Canad. J. Bot. 52: 1613–1619.
Argus, G. W. 1985. Computerized Catalogue of Herbarium Specimens of *Salix* in the Southeastern United States. Ottawa.
Argus, G. W. 1986. The genus *Salix* in the southeastern United States. Syst. Bot. Monogr. 9.
Argus, G. W. 1986b. Studies in the *Salix lucida* Muhl. and *S. reticulata* L. complexes in North America. Canad. J. Bot. 64: 541–551.
Argus, G. W. 1991. Salicaceae. In: G. W. Douglas et al., eds. 1989–1994. The Vascular Plants of British Columbia. 4 vols. Victoria. Vol. 3, pp. 55–67.
Argus, G. W. 1993. *Salix*. In: J. C. Hickman, ed. 1993. The Jepson Manual. Higher Plants of California. Berkeley, Los Angeles, and London. Pp. 990–999.
Argus, G. W. 1995. Arizona Salicaceae: Willow family. Part two: *Salix*. J. Arizona-Nevada Acad. Sci. 29: 39–62.
Argus, G. W. 1997. Infrageneric classification of *Salix* L. in the New World. Syst. Bot. Monogr. 52.
Argus, G. W. 2003. The identity of *Salix waghornei* (Salicaceae). Harvard Pap. Bot. 8: 111–114.
Argus, G. W. 2007. *Salix* L. (Salicaceae): Distribution maps and a synopsis of their classification in North America, north of Mexico. Harvard Pap. Bot. 12: 335–368.
Argus, G. W. and M. B. Davis. 1962. Macrofossils from a late-glacial deposit at Cambridge, Massachusetts. Amer. Midl. Naturalist 67: 106–117.
Argus, G. W., R. Elven, and A. K. Skvortsov. 1999. Salicaceae—A 'PAF' checklist example. Skr. Norske Vidensk.-Akad. Oslo, Mat.-Naturvidensk. Kl., n. s. 38: 387–418.
Argus, G. W. and C. L. McJannet. 1992. A taxonomic reconsideration of *Salix taxifolia* sensu lato (Salicaceae). Brittonia 44: 461–474.
Argus, G. W. and J. W. Steele. 1979. A reevaluation of the taxonomy of *Salix tyrrellii*, a sand dune endemic. Syst. Bot. 4: 163–177.
Ark. Bot. = Arkiv för Botanik Utgivet av K. Svenska Vetenskapsakademien.
Arnason, T., R. J. Hebda, and T. Johns. 1981. Use of plants for food and medicine by native peoples of eastern Canada. Canad. J. Bot. 59: 2189–2325.
Arnoldia (Jamaica Plain) = Arnoldia; a Continuation of the Bulletin of Popular Information.
Arr. Brit. Pl. ed. 3—See: W. Withering 1796
Arroyo, M. T. K. 1973. A taxometric study of infraspecific variation in autogamous *Limnanthes floccosa* (Limnanthaceae). Brittonia 25: 177–191.
Asai, Y. 1982. *Cakile edentula* (Bigel.) Hook. naturalizing in Japan. J. Jap. Bot. 57: 187–191.
Asai, Y. 1996. Additional notes on the distribution of *Cakile edentula* (Bigel.) Hook. in Japan. J. Jap. Bot. 71: 50–52.
Ascherson, P. F. A. 1859–1864. Flora der Provinz Brandenburg.... 3 vols. Berlin. (Fl. Brandenburg)
Asso y del Rio, I. J. de. 1779. Synopsis Stirpium Indigenarum Aragoniae.... Marseille. (Syn. Stirp. Aragon.)
Atlantic J. = Atlantic Journal, and Friend of Knowledge.
Autik. Bot.—See: C. S. Rafinesque 1840
Avetisian, E. M. and A. K. Mekhakian. 1980. Palynology of the genus *Reseda* L. Biol. Zhurn. Armenii 23: 472–479.
Azuma, T., T. Kajita, J. Yokoyama, and H. Ohashi. 2000. Phylogenetic relationships of *Salix* (Salicaceae) based on *rbc*L sequence data. Amer. J. Bot. 87: 67–75.
Bacon, J. D. 1978. Taxonomy of *Nerisyrenia* (Cruciferae). Rhodora 80: 159–227.
Badillo, V. M. 2000. *Carica* L. vs. *Vasconcella* [sic] St. Hil. (Caricaceae) con la rehabilitación de este último. Ernstia 10: 74–79.
Bailey, C. D. et al. 2006. Toward a global phylogeny of the Brassicaceae. Molec. Biol. Evol. 23: 2142–2160.
Bailey, C. D., I. A. Al-Shehbaz, and G. Rajanikanth. 2007. Generic limits in the tribe Halimolobeae and the description of the new genus *Exhalimolobos* (Brassicaceae). Syst. Bot. 32: 140–156.
Bailey, C. D., R. A. Price, and J. J. Doyle. 2002. Systematics of the halimolobine Brassicaceae: Evidence from three loci and morphology. Syst. Bot. 27: 318–332.
Bailey, L. H., ed. 1900–1902. Cyclopedia of American Horticulture.... 4 vols. New York and London.

Baileya = Baileya; a Quarterly Journal of Horticultural Taxonomy.

Baillargeon, G. 1986. Eine taxonomische Revision der Gattung *Sinapis* (Cruciferae: Brassiceae). Doctoral thesis. Freie Universität Berlin.

Baker, H. G. 1972. Migration of weeds. In: D. H. Valentine, ed. 1972. Taxonomy, Phytogeography and Evolution. London and New York. Pp. 327–347.

Baker, R. G., J. A. Mason, and L. J. Maher. 1999. Petaloid organs preserved in an arctic plant macrofossil assemblage from full-glacial sediments in southeastern Minnesota. Quatern. Res. 52: 388–392.

Ball, C. R. 1950. New combinations in southwestern *Salix*. J. Wash. Acad. Sci. 40: 324–335.

Ball, C. R. 1961. *Salix*. In: C. L. Lundell. 1942–1969. Flora of Texas. 3 vols. in parts. Dallas and Renner, Tex. Vol. 3, pp. 369–392.

Ball, P. W. 1961. The taxonomic status of *Neslia paniculata* (L.) Desv. and *N. apiculata* Fisch., Mey. & Avé-Lall. Feddes Repert. Spec. Nov. Regni Veg. 64: 11–13.

Barbour, M. G. and J. E. Rodman. 1970. Saga of the West Coast sea-rockets: *Cakile edentula* ssp. *californica* and *C. maritima*. Rhodora 72: 370–386.

Barnes, B. V. 1961. Hybrid aspens in the Lower Peninsula of Michigan. Rhodora 63: 311–324.

Barnes, B. V. 1966. The clonal growth habit of American aspens. Ecology 47: 439–447.

Barnes, B. V. 1969. Natural variation and delineation of clones of *Populus tremuloides* and *P. grandidentata* in northern Lower Michigan. Silvae Genet. 18: 130–142.

Barnes, B. V. 1975. Phenotypic variation of trembling aspen in western North America. Forest Sci. 21: 319–328.

Barratt, J. 1840. Salices Americanae.... Middletown, Conn. [Unpaginated, 8 unnumbered leaves.]

Barton, B. S. 1836. Elements of Botany..., ed. 4. Philadelphia.

Bartonia = Bartonia; a Botanical Annual.

Bay, C. 1992. A phytogeographical study of the vascular plants of northern Greenland, north of 74° northern latitude. Meddel. Grønland, Biosci. 36: 1–102.

Bayer, C. and O. Appel. 2003. Bataceae. In: K. Kubitzki et al., eds. 1990+. The Families and Genera of Vascular Plants. 9+ vols. Berlin etc. Vol. 5, pp. 30–32.

Bean, W. J. 1970–1988. Trees and Shrubs Hardy in the British Isles, ed. 8 rev. 5 vols. London.

Bebb, M. S. 1879. Willows of California. Cambridge, Mass. [Preprinted from W. H. Brewer et al. 1876–1880. Geological Survey of California.... Botany.... 2 vols. Cambridge, Mass. Vol. 2, pp. 82–92.] (Willows Calif.)

Bebb, M. S. 1890. White Mountain willows. IV. Bull. Torrey Bot. Club 17: 149–151.

Bebb, M. S. 1895. Notes on some arborescent willows of North America III. Gard. & Forest 8: 423–425.

Beck, J. B. 2009. The phylogeny of *Selenia* (Brassicaceae) inferred from chloroplast and nuclear sequence data. J. Bot. Res. Inst. Texas 3: 169–176.

Beck, J. B., I. A. Al-Shehbaz, and B. A. Schaal. 2006. *Leavenworthia* revisited: Testing classic systematic and mating system hypotheses. Syst. Bot. 31: 151–159.

Beerling, D. J. 1998. Biological flora of the British Isles. *Salix herbacea* L. J. Ecol. 86: 872–895.

Behnke, H.-D. and B. L. Turner. 1971. On specific sieve-tube plastids in Caryophyllales. Further investigations with special reference to the Bataceae. Taxon 20: 731–737.

Beilstein, M. A., I. A. Al-Shehbaz, and E. A. Kellogg. 2006. Brassicaceae phylogeny and trichome evolution. Amer. J. Bot. 93: 607–619.

Beilstein, M. A., I. A. Al-Shehbaz, S. Mathews, and E. A. Kellogg. 2008. Brassicaceae phylogeny inferred from phytochrome A and *ndh*F sequence data: Tribes and trichomes revisited. Amer. J. Bot. 95: 1307–1327.

Beilstein, M. A. and M. D. Windham. 2003. A phylogenetic analysis of western North American *Draba* (Brassicaceae) based on nuclear ribosomal DNA sequences from the ITS region. Syst. Bot. 28: 584–592.

Beismann, H., J. H. A. Barker, A. Karp, and T. Speck. 1997. AFLP analysis sheds light on distribution of two *Salix* species and their hybrids along a natural gradient. Molec. Ecol. 6: 989–993.

Beitr. Naturk.—See: J. F. Ehrhart 1787–1792

Belyaeva, I. V. 2009. Nomenclature of *Salix fragilis* L. and a new species, *S. euxina* (Salicaceae). Taxon 58: 1344–1348.

Bennion, G. C., R. K. Vickery, and W. P. Cottam. 1961. Hybridizationn of *Populus fremontii* and *Populus angustifolia* in Perry Canyon, Box Elder County, Utah. Proc. Utah Acad. Sci. 38: 31–35.

Benson, L. D. and R. A. Darrow. 1945. A Manual of Southwestern Desert Trees and Shrubs.... Tucson.

Bentham, G. 1826. Catalogue des Plantes Indigènes des Pyrénées et du Bas-Languedoc.... Paris. (Cat. Pl. Pyrénées)

Bentham, G. 1839[–1857]. Plantas Hartwegianas Imprimis Mexicanas.... London. [Issued by gatherings with consecutive signatures and pagination.] (Pl. Hartw.)

Berchtold, F. and J. S. Presl. 1823–1835. O Přirozenosti Rostlin aneb Rostlinář.... 3 vols. Prague. [Vols. 2 and 3 by Presl only.] (Prir. Rostlin)

Bergeret, J. 1783–1784[–1786]. Phytonomatotechnie Universelle.... 3 vols. Paris. (Phytonom. Univ.)

Berkutanko, A. N. 1995. Detective story about one Linnaean species of Cruciferae. Linzer Biol. Beitr. 27: 1115–1122.

Bernard, F. G. 1968. *Populus* ×*balsamifera* hybrida nova (Salicaceae). Naturaliste Canad. 95: 797–799.

Besser, W. S. J. G. von. 1821. Enumeratio Plantarum.... Vilna. (Enum. Pl.)

Bessey, C. E. 1915. The phylogenetic taxonomy of flowering plants. Ann. Missouri Bot. Gard. 2: 109–164.

Best, K. F. 1977. The biology of Canadian weeds. 22. *Descurainia sophia* (L.) Webb. Canad. J. Pl. Sci. 57: 499–507.

Best, K. F. and G. K. McIntyre. 1975. The biology of Canadian weeds. 9. *Thlaspi arvense* L. Canad. J. Pl. Sci. 55: 279–292.

Bigelow, J. 1814. Florula Bostoniensis. A Collection of Plants of Boston and Its Environs.... Boston. (Fl. Boston.)

Bijdr. Natuurk. Wetensch. = Bijdragen tot de Natuurkundige Wetenschappen.

Biochem. Syst. & Ecol. = Biochemical Systematics and Ecology.

Bioenergy Res. = BioEnergy Research.

Biol. Cent.-Amer., Bot.—See: W. B. Hemsley 1879–1888

Biol. Conservation = Biological Conservation; International Quarterly Journal Devoted to Scientific Protection of Plant and Animal Wildlife and All Nature throughout the World....

Biol. J. Linn. Soc. = Biological Journal of the Linnean Society.

Biol. Skr. = Biologiske Skrifter.

Biol. Zhurn. Armenii = Biologicheskii Zhurnal Armenii.

Blake, S. F. 1957. A new cruciferous weed, *Calepina irregularis*, in Virginia. Rhodora 59: 278–280.

Bleeker, W. and H. Hurka. 2001. Introgressive hybridization in *Rorippa* (Brassicaceae): Gene flow and its consequences in natural and anthropogenic habitats. Molec. Ecol. 10: 2103–2022.

Bleeker, W., M. Huthmann, and H. Hurka. 1999. Evolution of hybrid taxa in *Nasturtium* R. Br. (Brassicaceae). Folia Geobot. 34: 421–433.

Bleeker, W., C. Weber-Sparenberg, and H. Hurka. 2002. Chloroplast DNA variation and biogeography in the genus *Rorippa* Scop. (Brassicaceae). Pl. Biol. (Stuttgart) 4: 104–111.

Böcher, T. W. 1952. Contributions to the flora and plant geography of West Greenland III. Vascular plants collected or observed during the botanical expedition to West Greenland 1946. Meddel. Grønland 147(9): 1–85.

Böcher, T. W. 1956. Further studies in *Braya humilis* and allied species. Meddel. Grønland 124: 1–29.

Böcher, T. W. 1966. Experimental and cytological studies on plant species. IX. Some arctic and montane crucifers. Biol. Skr. 14(7): 2–74.

Böcher, T. W. 1973. Interspecific hybridization in *Braya* (Cruciferae). Ann. Bot. Fenn. 10: 57–65.

Böcher, T. W., K. Holmen, and K. Jakobsen. 1968. The Flora of Greenland, ed. 2. Copenhagen.

Boissier, P. E. 1842–1859. Diagnoses Plantarum Orientalium Novarum. 3 vols. in 19 parts. Geneva etc. [Parts paged independently.] (Diagn. Pl. Orient.)

Boivin, B. 1966b. Énumération des plantes du Canada, 2. Lignidées (suite). Naturaliste Canad. 93: 371–437.

Bolle, F. 1936. Resedaceae. In: H. G. A. Engler et al., eds. 1924+. Die natürlichen Pflanzenfamilien..., ed. 2. 26+ vols. Leipzig and Berlin. Vol. 17b, pp. 659–693.

Boom, B. K. 1957. *Populus canadensis* Moench versus *P. euramericana* Guinier. Acta Bot. Neerl. 6: 54–59.

Borgen, L. 1987. *Lobularia* (Cruciferae). A biosystematic study with special reference to the Macaronesian region. Opera Bot. 91: 1–96.

Boston J. Nat. Hist. = Boston Journal of Natural History.

Bot. Acta = Botanica Acta; Berichte der Deutschen botanischen Gesellschaft.

Bot. Beechey Voy.—See: W. J. Hooker and G. A. W. Arnott [1830–]1841

Bot. California—See: W. H. Brewer et al. 1876–1880

Bot. Gaz. = Botanical Gazette; Paper of Botanical Notes.

Bot. Helv. = Botanica Helvetica.

Bot. J. Linn. Soc. = Botanical Journal of the Linnean Society.

Bot. Jahrb. Syst. = Botanische Jahrbücher für Systematik, Pflanzengeschichte und Pflanzengeographie.

Bot. Mag. = Botanical Magazine; or, Flower-garden Displayed.... [Edited by Wm. Curtis.] [With vol. 15, 1801, title became Curtis's Botanical Magazine; or....]

Bot. Mag. (Tokyo) = Botanical Magazine. [Shokubutsu-gaku Zasshi.] (Tokyo.)

Bot. Mater. Gerb. Bot. Inst. Komarova Akad. Nauk S.S.S.R. = Botanicheskie Materialy Gerbariya Botanicheskogo Instituti Imeni V. L. Komarova, Akademii Nauk S S S R.

Bot. Misc. = Botanical Miscellany.

Bot. Not. = Botaniska Notiser.

Bot. Rev. (Lancaster) = Botanical Review, Interpreting Botanical Progress.

Bot. Zhurn. (Moscow & Leningrad) = Botanicheskii Zhurnal. (Moscow and Leningrad.)

Botany (Fortieth Parallel)—See: S. Watson 1871

Bothalia = Bothalia; a Record of Contributions from the National Herbarium, Union of South Africa.

Botschantzev, V. P. 1972. On *Parrya* R. Br., *Neuroloma* Andrz. and some other genera (Cruciferae). Bot. Zhurn. (Moscow & Leningrad) 57: 664–673.

Botschantzev, V. P. 1972b. The genus *Strigosella* Boiss. and its relation to the genus *Malcolmia* R. Br. (Cruciferae). Bot. Zhurn. (Moscow & Leningrad) 57: 1033–1046.

Boucher, L. D., S. R. Manchester, and W. S. Judd. 2003. An extinct genus of Salicaceae based on twigs with attached flowers, fruits and foliage from the Eocene Green River Formation of Utah and Colorado, U.S.A. Amer. J. Bot. 90: 1389–1399.

Bowman, J. L. 2006. Molecules and morphology: Comparative developmental genetics of the Brassicaceae. Pl. Syst. Evol. 259: 199–215.

Bowman, J. L., H. Brüggemann, J. Y. Lee, and K. Mummenhoff. 1999. Evolutionary changes in floral structure within *Lepidium* (Brassicaceae). Int. J. Pl. Sci. 160: 917–929.

Brayshaw, T. C. 1965. The status of the black cottonwood (*Populus trichocarpa* Torrey and Gray). Canad. Field-Naturalist 79: 91–95.

Brayshaw, T. C. 1965b. Native poplars of southern Alberta and their hybrids. Publ. Dept. Forest. Rural Developm. Canada 1109: 7–40.

Bremer, K. and H.-E. Wanntorp. 1978. Phylogenetic systematics in botany. Taxon 27: 317–329.

Brewer, W. H. et al. 1876–1880. Geological Survey of California.... Botany.... 2 vols. Cambridge, Mass. (Bot. California)

Bricker, J. S., G. K. Brown, and T. L. Patts Lewis. 2000. Status of *Descurainia torulosa* (Brassicaceae). W. N. Amer. Naturalist 60: 426–432.

Bridson, G. D. R. and E. R. Smith. 1991. B-P-H/S. Botanico-Periodicum-Huntianum/Supplementum. Pittsburgh.

Brinker, A. M., J. E. Ebinger, and D. S. Seigler. 1987. Purasin from *Salix interior*. Biochem. Syst. & Ecol. 15: 69–70.

Brit. Fl.—See: W. J. Hooker 1830

Britton, N. L. 1886. Leaf-forms of *Populus grandidentata*. Bull. Torrey Bot. Club 13: 89–91.

Britton, N. L. and A. Brown. 1896–1898. An Illustrated Flora of the Northern United States, Canada and the British Possessions from Newfoundland to the Parallel of the Southern Boundary of Virginia, and from the Atlantic Ocean Westward to the 102d Meridian.... 3 vols. New York. (Ill. Fl. N. U.S.)

Britton, N. L., E. E. Sterns, J. F. Poggenburg, et al. 1888. Preliminary Catalogue of Anthophyta and Pteridophyta Reported As Growing Spontaneously within One Hundred Miles of New York City. New York. [Authorship often attributed as B.S.P. in nomenclatural contexts.] (Prelim. Cat.)

Brittonia = Brittonia; a Journal of Systematic Botany....
Brochmann, C. 1992. Pollen and seed morphology of Nordic *Draba* (Brassicaceae): Phylogenetic and ecological implications. Nordic J. Bot. 12: 657–673.
Brochmann, C. 1992b. Polyploid evolution in arctic-alpine *Draba* (Brassicaceae). Sommerfeltia Suppl. 4: 1–37.
Brochmann, C. 1993. Reproductive strategies of diploid and polyploid populations of arctic *Draba* (Brassicaceae). Pl. Syst. Evol. 185: 55–83.
Brochmann, C. et al. 1992. Electrophoretic relationships and phylogeny of Nordic polyploids in *Draba* (Brassicaceae). Pl. Syst. Evol. 182: 35–70.
Brochmann, C. et al. 1992b. Recurrent formation and polyphyly of Nordic polyploids in *Draba* (Brassicaceae). Amer. J. Bot. 79: 673–688.
Brochmann, C. et al. 1992c. Multiple origins of the octoploid Scandinavian endemic *Draba cacuminum*: Electrophoretic and morphological evidence. Nordic J. Bot. 12: 257–272.
Brochmann, C. et al. 1992d. Gene flow across ploidal levels in *Draba* (Brassicaceae). Evol. Trends Pl. 6: 125–134.
Brochmann, C., L. Borgen, and B. Stedje. 1993. Crossing relationships and chromosome numbers of Nordic populations of *Draba* (Brassicaceae), with emphasis on the *D. alpina* complex. Nordic J. Bot. 13: 121–147.
Brochmann, C. and R. Elven. 1992. Ecological and genetic consequences of polyploidy in arctic *Draba* (Brassicaceae). Evol. Trends Pl. 6: 111–124.
Brotero, F. 1804[–1805]. Flora Lusitanica.... 2 vols. Lisbon. (Fl. Lusit.)
Brown, R. 1823. Chloris Melvilliana. A List of Plants Collected in Melville Island...in the Year 1820.... London. [Preprint with independent pagination from W. E. Parry. 1824. A Supplement to the Appendix of Captain Parry's Voyage.... London.] (Chlor. Melvill.)
Browne, P. 1756. The Civil and Natural History of Jamaica.... London. (Civ. Nat. Hist. Jamaica)
Brummitt, R. K. and C. E. Powell, eds. 1992. Authors of Plant Names. A List of Authors of Scientific Names of Plants, with Recommended Standard Forms of Their Names, Including Abbreviations. Kew.
Brunsfeld, S. J. and F. D. Johnson. 1985. Field Guide to the Willows of East-central Idaho. Moscow, Idaho.
Brunsfeld, S. J., D. E. Soltis, and P. S. Soltis. 1991. Patterns of genetic variation in *Salix* section *Longifoliae* (Salicaceae). Amer. J. Bot. 78: 855–869.
Brunsfeld, S. J., D. E. Soltis, and P. S. Soltis. 1992. Evolutionary patterns and processes in *Salix* section *Longifoliae*: Evidence from chloroplast DNA. Syst. Bot. 17: 239–256.
Buchenau, F. 1894. Flora der nordwestdeutschen Tiefebene. Leipzig. (Fl. Nordwestdeut. Tiefebene)
Büchler, W. 1996. Phyllotaxis and morphology of proximal leaves on vegetative shoots of *Salix* and their systematic implications. Bot. Helv. 106: 31–44.
Buck, R. E. 1993. *Caulanthus*. In: J. C. Hickman, ed. 1993. The Jepson Manual. Higher Plants of California. Berkeley, Los Angeles, and London. Pp. 410–412.
Buck, R. E. 1995. Systematics of *Caulanthus* (Brassicaceae). Ph.D. dissertation. University of California, Berkeley.
Buck, R. E., D. W. Taylor, and A. R. Kruckeberg. 1993. *Streptanthus*. In: J. C. Hickman, ed. 1993. The Jepson Manual. Higher Plants of California. Berkeley, Los Angeles, and London. Pp. 439–444.
Bull. Bot. Res., Harbin = Bulletin of Botanical Research. [Zhiwi Yanjiu.]
Bull. Bot. Surv. India = Bulletin of the Botanical Survey of India.
Bull. Calif. Acad. Sci. = Bulletin of the California Academy of Sciences.
Bull. Geol. Nat. Hist. Surv. Minnesota = Bulletin of the Geological and Natural History Survey, Minnesota.
Bull. Herb. Boissier = Bulletin de l'Herbier Boissier.
Bull. Mens. Soc. Linn. Lyon = Bulletin Mensuel de la Société Linnéenne de Lyon.
Bull. Natl. Mus. Canada = Bulletin of the National Museum of Canada.
Bull. New York Bot. Gard. = Bulletin of the New York Botanical Garden.
Bull. Orto Bot. Regia Univ. Napoli = Bulletino dell'Orto Botanico della Regia Università di Napoli.
Bull. S. Calif. Acad. Sci. = Bulletin of the Southern California Academy of Sciences.
Bull. Soc. Bot. France = Bulletin de la Société Botanique de France.
Bull. Soc. Dendrol. France = Bulletin de la Société Dendrologique de France.
Bull. Soc. Imp. Naturalistes Moscou = Bulletin de la Société Impériale des Naturalistes de Moscou.
Bull. Soc. Roy. Bot. Belgique = Bulletin de la Société Royale de Botanique de Belgique.
Bull. Torrey Bot. Club = Bulletin of the Torrey Botanical Club.
Bunge, A. A. [1833.] Enumeratio Plantarum, Quas in China Boreali Collegit.... St. Petersburg. [Preprinted from Mém. Acad. Imp. Sci. St.-Pétersbourg Divers Savans 2: 75–148. 1835.] (Enum. Pl. China Bor.)
Burman, N. L. 1768. Flora Indica ... Nec Non Prodromus Florae Capensis. Leiden and Amsterdam. (Fl. Indica)
Butters, F. K. and E. C. Abbe. 1940. The American varieties of *Rorippa islandica*. Rhodora 42: 25–32.
Buxton, E. G. 1998. Noteworthy collections, California. Madroño 45: 184.
Callihan, R. H., S. L. Carson, and R. T. Dobbins. 1995. NAWEEDS, Computer-aided Weed Identification for North America. Illustrated User's Guide plus Computer Floppy Disk. Moscow, Idaho.
Canad. Field-Naturalist = Canadian Field-Naturalist.
Canad. J. Bot. = Canadian Journal of Botany.
Canad. J. Forest Res. = Canadian Journal of Forest Research.
Canad. J. Pl. Sci. = Canadian Journal of Plant Science.
Candolle, A. P. de. [1817]1818–1821. Regni Vegetabilis Systema Naturale.... 2 vols. Paris, Strasbourg, and London. (Syst. Nat.)
Candolle, A. P. de and A. L. P. P. de Candolle, eds. 1823–1873. Prodromus Systematis Naturalis Regni Vegetabilis.... 17 vols. Paris etc. [Vols. 1–7 edited by A. P. de Candolle, vols. 8–17 by A. L. P. P. de Candolle.] (Prodr.)
Carlquist, S. 1978. Wood anatomy and relationships of Bataceae, Gyrostemonaceae, and Stylobasiaceae. Allertonia 1: 297–330.

Carlquist, S. 1998b. Wood anatomy of Resedaceae. Aliso 16: 127–135.

Carter, R. J. 1993. Rampion mignonette and its co-ordinated control. In: J. T. Swarbrick et al., eds. 1993. Proceedings: 10th Australian Weeds Conference and 14th Asian Pacific Weed Science Society Conference.... 2 vols. Brisbane. Vol. 1, pp. 505–509.

Caryologia = Caryologia; Giornale di Citologia, Citosistematica e Citogenetica.

Castanea = Castanea; Journal of the Southern Appalachian Botanical Club.

Cat. Afr. Pl.—See: W. P. Hiern 1896–1901

Cat. N. Amer. Pl. ed. 3—See: A. A. Heller 1909–1914

Cat. Pl. Pyrénées—See: G. Bentham 1826

Cavanilles, A. J. 1791–1801. Icones et Descriptiones Plantarum, Quae aut Sponte in Hispania Crescunt, aut in Hortis Hospitantur. 6 vols. Madrid. (Icon.)

Cavanilles, A. J. [1801–]1802. Descripción de las Plantas.... Madrid. (Descr. Pl.)

Cavers, P. B., M. I. Heagy, and R. F. Kokron. 1979. The biology of Canadian weeds. 35. *Alliaria petiolata* (M. Bieb.) Cavara and Grande. Canad. J. Pl. Sci. 59: 217–229.

Cent. Pl. I—See: C. Linnaeus 1755

Cent. Pl. II—See: C. Linnaeus [1756]

Ceulemans, R. et al. 1990. Crown architecture of *Populus* clones as determined by branch orientation and branch characteristics. Tree Physiol. 7: 157–167.

Channell, R. B. and C. W. James. 1964. Nomenclatural and taxonomic corrections in *Warea* (Cruciferae). Rhodora 66: 18–26.

Chapman, A. W. 1860. Flora of the Southern United States.... New York. (Fl. South. U.S.)

Chapman, A. W. 1883. Flora of the Southern United States..., ed. 2. New York. (Fl. South. U.S. ed. 2)

Char. Ess. Fam.—See: P. F. Horaninov 1847

Charlesworth, D. and Z. Yang. 1998. Allozyme diversity in *Leavenworthia* populations with different inbreeding levels. Heredity 81: 453–461.

Chase, M. W. et al. 2002. When in doubt, put it in Flacourtiaceae: A molecular phylogenetic analysis based on plastid *rbc*L DNA sequences. Kew Bull. 57: 141–181.

Chlor. Melvill.—See: R. Brown 1823

Chmelař, J. 1978. Taxonomic significance of bud scales in the genus *Salix* L. Folia Dendrol. 4: 5–19.

Chmelař, J. 1983. Weeping willows. Int. Dendrol. Soc. Year Book 1983: 107–110.

Chong, D. K. X., L. Zsuffa, and F. A. Aravanopoulos. 1995. Taxonomic relationships between *S. exigua* and other North American willows: Evidence from allozyme variation. Biochem. Syst. & Ecol. 23: 767–771.

Chopra, R. N., S. L. Nayar, and I. C. Chopra. 1986. Glossary of Indian Medicinal Plants (Including the Supplement). New Delhi.

Chweya, J. A. and N. A. Mnzava. 1997. Cat's Whiskers: *Cleome gynandra* L. Rome.

Cirillo, D. M. L. 1788–1792. Plantarum Rariorum Regni Neapolitani. 2 fascs. Naples. (Pl. Rar. Neapol.)

Civ. Nat. Hist. Jamaica—See: P. Browne 1756

Cl. Crucif. Emend.—See: H. J. N. von Crantz 1769b

Committee on the Status of Endangered Wildlife in Canada. 2004. COSEWIC Assessment and Update Status Report on the Macoun's Meadowfoam *Limnanthes macounii* in Canada. Ottawa.

Compt.-Rend. Trav. Carlsberg Lab., Sér. Physiol. = Comptes-Rendus des Travaux du Carlsberg Laboratoriet; Série Physiologique.

Conservation Biol. = Conservation Biology; Journal of the Society for Conservation Biology.

Consp. Pl. Charcov.—See: V. M. Czernajew 1859

Contr. Dudley Herb. = Contributions from the Dudley Herbarium of Stanford University.

Contr. Gray Herb. = Contributions from the Gray Herbarium of Harvard University. [Some numbers reprinted from (or in?) other periodicals, e.g. Rhodora.]

Contr. Lab. Bot. Univ. Montréal = Contributions du Laboratoire de Botanique de l'Université de Montréal.

Contr. U.S. Natl. Herb. = Contributions from the United States National Herbarium.

Contr. Univ. Kansas Herb. = Contributions from the University of Kansas Herbarium.

Contr. W. Bot. = Contributions to Western Botany.

Coquillat, M. 1951. Sur les plantes les plus communes à la surface du globe. Bull. Mens. Soc. Linn. Lyon, n. s. 20: 165–170.

Correll, D. S. 1960. A new cottonwood from Texas. Wrightia 2: 45–47.

Coulter, J. M. [1885.] Manual of the Botany (Phaenogamia and Pteridophyta) of the Rocky Mountain Region.... New York and Chicago. (Man. Bot. Rocky Mt.)

Crantz, H. J. N. von. 1762–1767. Stirpium Austriarum Fasciculus I [–III]. 3 fascs. Vienna and Leipzig. (Stirp. Austr. Fasc.)

Crantz, H. J. N. von. 1769b. Classis Cruciformium Emendata.... Leipzig. (Cl. Crucif. Emend.)

Crawford, D. J. 1974. A morphological and chemical study of *Populus acuminata*. Brittonia 26: 74–89.

Critchfield, W. B. 1960. Leaf dimorphism in *Populus trichocarpa*. Amer. J. Bot. 47: 699–711.

Cronquist, A. 1968. The Evolution and Classification of Flowering Plants. Boston.

Cronquist, A. 1981. An Integrated System of Classification of Flowering Plants. New York.

Cronquist, A. 1988. The Evolution and Classification of Flowering Plants, ed. 2. Bronx.

Cronquist, A. et al. 1972+. Intermountain Flora. Vascular Plants of the Intermountain West, U.S.A. 6+ vols. in 7+. New York and London. (Intermount. Fl.)

Cronquist, A. and R. D. Dorn. 2005. Salicaceae. In: A. Cronquist et al. 1972+. Intermountain Flora. Vascular Plants of the Intermountain West, U.S.A. 6+ vols. in 7+. New York and London. Vol. 2, part B, pp. 118–160.

Crovello, T. J. 1968. A numerical taxonomic study of the genus *Salix*, section *Sitchenses*. Univ. Calif. Publ. Bot. 44: 1–61.

Cruciferae Continental N. Amer.—See: R. C. Rollins 1993

Curtin, L. S. M. 1947. Healing Herbs of the Upper Río Grande. Santa Fe. [Reprinted 1965, Los Angeles.]

Curtis, J. D. and N. R. Lersten. 1978. Heterophylly in *Populus grandidentata* (Salicaceae) with emphasis on resin glands and extrafloral nectaries. Amer. J. Bot. 65: 1003–1010.

Czernajew, V. M. 1859. Conspectus Plantatum Circa Charcoviam et in Ucrania Sponte Crescentium et Vulgo Cultarum. Kharkov. (Consp. Pl. Charcov.)

Davidson, A. and G. L. Moxley. 1923. Flora of Southern California.... Los Angeles. (Fl. S. Calif.)

Dawson, T. E. 1987. Comparative Ecophysiological Adaptations in Arctic and Alpine Populations of a Dioecious Shrub, *Salix arctica* Pall. Ph.D. thesis. University of Washington.

De Cock, K. et al. 2003. Diversity of the willow complex *Salix alba–S. ×rubens–S. fragilis*. Silvae Genet. 52: 148–153.

Del Tredici, P. 2001. Sprouting in temperate trees: A morphological and ecological review. Bot. Rev. (Lancaster) 67: 121–140.

Demonstr. Pl.—See: C. Linnaeus [1753]

Denkschr. Kaiserl. Akad. Wiss., Wien. Math.-Naturwiss. Kl. = Denkschriften der Kaiserlichen Akademie der Wissenschaften, Wien. Mathematisch-naturwissenschaftliche Klasse.

Denkschr. Königl.-Baier. Bot. Ges. Regensburg = Denkschriften der Königlich-baierischen botanischen Gesellschaft in Regensburg.

Densmore, R. A., B. J. Neiland, J. C. Zasada, and M. A. Masters. 1987. Planting willow for moose habitat restoration on the North Slope of Alaska, U.S.A. Arctic Alpine Res. 19: 537–543.

Densmore, R. A. and J. C. Zasada. 1978. Rooting potential of Alaskan willow cuttings. Canad. J. Forest Res. 8: 477–479.

Densmore, R. A. and J. C. Zasada. 1983. Seed dispersal and dormancy patterns in northern willows: Ecological and evolutionary significance. Canad. J. Bot. 61: 3207–3216.

Descr. Icon. Pl. Hung.—See: F. Waldstein and P. Kitaibel [1799–]1802–1812

Descr. Pl.—See: A. J. Cavanilles [1801–]1802

Descr. Pl. Nouv.—See: É. P. Ventenat [1800–1803]

Desfontaines, R. L. [1798–1799.] Flora Atlantica sive Historia Plantarum, Quae in Atlante, Agro Tunetano et Algeriensi Crescunt. 2 vols. in 9 parts. Paris. (Fl. Atlant.)

Desfontaines, R. L. 1804. Tableau de l'École Botanique du Muséum d'Histoire Naturelle. Paris. (Tabl. École Bot.)

Detling, L. E. 1936. The genus *Dentaria* in the Pacific states. Amer. J. Bot. 23: 570–576.

Detling, L. E. 1937. The Pacific coast species of *Cardamine*. Amer. J. Bot. 24: 70–76.

Detling, L. E. 1939. A revision of the North American species of *Descurainia*. Amer. Midl. Naturalist 22: 481–520.

Deut. Dendrol.—See: B. A. E. Koehne 1893

Deutschl. Fl.—See: H. G. L. Reichenbach et al. 1837–1870

Deutschl. Fl. ed. 3—See: J. C. Röhling et al. 1823–1839

Diagn. Pl. Orient.—See: P. E. Boissier 1842–1859

Diederichsen, A. 2001. *Brassica*. In: P. Hanelt, ed. 2001. Mansfeld's Encyclopedia of Agricultural and Horticultural Crops.... 6 vols. Berlin and New York. Vol. 3, pp. 1435–1465.

Dolan, R. W. and L. F. LaPré. 1989. Taxonomy of *Streptanthus* sect. *Biennes*, the *Streptanthus morrisonii* complex (Brassicaceae). Madroño 36: 33–40.

Dole, J. A. and M. Sun. 1992. Field and genetic survey of endangered Butte County meadowfoam—*Limnanthes floccosa* subsp. *californica* (Limnanthaceae. Conservation Biol. 6: 549–558.

Don, G. 1831–1838. A General History of the Dichlamydeous Plants.... 4 vols. London. (Gen. Hist.)

Dorn, R. D. 1975. A systematic study of *Salix* section *Cordatae* in North America. Canad. J. Bot. 53: 1491–1522.

Dorn, R. D. 1975b. Cytological and taxonomic notes on North American *Salix*. Madroño 23: 99.

Dorn, R. D. 1976. A synopsis of American *Salix*. Canad. J. Bot. 54: 2769–2789.

Dorn, R. D. 1977b. Willows of the Rocky Mountain states. Rhodora 79: 390–429.

Dorn, R. D. 1988. Vascular Plants of Wyoming. Cheyenne. (Vasc. Pl. Wyoming)

Dorn, R. D. 1994. North American *Salix* (Salicaceae): Typifications and notes. Phytologia 77: 89–95.

Dorn, R. D. 1995. A taxonomic study of *Salix* section *Cordatae* subsection *Luteae* (Salicaceae). Brittonia 47: 160–174.

Dorn, R. D. 1997. Rocky Mountain Region Willow Identification Field Guide. Denver.

Dorn, R. D. 1998. A taxonomic study of *Salix* section *Longifoliae* (Salicaceae). Brittonia 50: 193–210.

Dorn, R. D. 2000. A taxonomic study of *Salix* sections *Mexicanae* and *Viminella* subsection *Sitchenses* (Salicaceae) in North America. Brittonia 52: 1–19.

Dorn, R. D. 2001. Vascular Plants of Wyoming, ed. 3. Cheyenne. (Vasc. Pl. Wyoming ed. 3)

Dorn, R. D. 2003. Environmental influence on leaf glaucescence in willows (*Salix*). Madroño 50: 141–144.

Dorofeev, V. I. 1998. The four new species of Brassicaceae in North America. Bot. Zhurn. (Moscow & Leningrad) 83(9): 133–135.

Douglas, D. A. 1989. Clonal growth of *Salix setchelliana* on glacial river gravel bars in Alaska. J. Ecol. 77: 112–126.

Douglas, G. W., G. B. Straley, and D. V. Meidinger. 1989–1994. The Vascular Plants of British Columbia. 4 vols. Victoria. [B.C. Minist. Forests, Special Rep. 1–4.]

Drury, W. H. Jr. and R. C. Rollins. 1952. The North American representatives of *Smelowskia* (Cruciferae). Rhodora 54: 85–119.

Dudley, T. R. 1964. Studies in *Alyssum*: Near Eastern representatives and their allies, I. J. Arnold Arbor. 45: 57–100.

Dudley, T. R. 1964b. Synopsis of the genus *Alyssum*. J. Arnold Arbor. 45: 358–373.

Dudley, T. R. 1965. Studies in *Alyssum*: Near Eastern representatives and their allies, II. Section *Meniocus* and section *Psilonema*. J. Arnold Arbor. 46: 181–217.

Dudley, T. R. 1966. Ornamental madworts (*Alyssum*) and the correct name of the goldentuft alyssum. Arnoldia (Jamaica Plain) 26: 33–48.

Dudley, T. R. 1968. *Alyssum* (Cruciferae) introduced in North America. Rhodora 70: 298–300.

Dumortier, B. C. J. 1827. Florula Belgica, Operis Majoris Prodromus.... Tournay. (Fl. Belg.)

Dvořák, F. 1966. A contribution to the study of the evolution of *Hesperis* series *Matronales* Cvĕl. emend. Dvořák. Feddes Repert. 73: 94–99.

Dvořák, F. 1967. A note on the genus *Turritis* L. Oesterr. Bot. Z. 114: 84–87.

Dvořák, F. 1973. Infrageneric classification of *Hesperis*. Feddes Repert. 84: 259–272.

Eaton, A. 1822. A Manual of Botany for the Northern States, ed. 3. Albany. (Man. Bot. ed. 3)

Ebel, A. L. 1998. Notes on genus *Aphragmus* Andrz. (Brassicaceae). Turczaninowia 1(4): 20–27.

Eckenwalder, J. E. 1977. North American cottonwoods (*Populus*, Salicaceae) of sections *Abaso* and *Aigeiros*. J. Arnold Arbor. 58: 193–208.

Eckenwalder, J. E. 1980c. Foliar heteromorphism in *Populus* (Salicaceae), a source of confusion in the taxonomy of Tertiary leaf remains. Syst. Bot. 5: 366–383.

Eckenwalder, J. E. 1982. *Populus* ×*inopina* hybr. nov. (Salicaceae), a natural hybrid between the native North American *P. fremontii* S. Watson and the introduced Eurasian *P. nigra* L. Madroño 29: 67–78.

Eckenwalder, J. E. 1984. Natural intersectional hybridization between North American species of *Populus* (Salicaceae) in sections *Aigeiros* and *Tacamahaca*. II. Taxonomy. Canad. J. Bot. 62: 325–335.

Eckenwalder, J. E. 1984b. Natural intersectional hybridization between North American species of *Populus* (Salicaceae) in sections *Aigeiros* and *Tacamahaca*. I. Population studies of *P.* ×*parryi*. Canad. J. Bot. 62: 317–324.

Eckenwalder, J. E. 1984c. Natural intersectional hybridization between North American species of *Populus* (Salicaceae) in sections *Aigeiros* and *Tacamahaca*. III. Paleobotany and evolution. Canad. J. Bot. 62: 336–342.

Eckenwalder, J. E. 1992. Vascular plants of Arizona: Salicaceae, willow family. Part one: *Populus*. J. Arizona-Nevada Acad. Sci. 26: 29–33.

Eckenwalder, J. E. 1996. Systematics and evolution of Populus. In: R. F. Stettler et al., eds. 1996. Biology of *Populus* and Its Implications for Management and Conservation. Ottawa. Pp. 7–32.

Eckenwalder, J. E. 1996b. Taxonomic signal and noise in multivariate interpopulational relationships in *Populus mexicana* (Salicaceae). Syst. Bot. 21: 261–271.

Ecology = Ecology, a Quarterly Journal Devoted to All Phases of Ecological Biology.

Econ. Bot. = Economic Botany; Devoted to Applied Botany and Plant Utilization.

Edinburgh New Philos. J. = Edinburgh New Philosophical Journal.

Edlund, S. A. and P. A. Egginton. 1984. Morphology and description of an outlier population of tree-sized willows on western Victoria Island, District of Franklin. In: Geological Survey of Canada. 1984. Current Research: Part A. Ottawa. Pp. 279–285.

Edmonson, J. R. 1993. Resedaceae. In: T. G. Tutin et al., eds. 1993+. Flora Europaea, ed. 2. 1+ vol. Cambridge and New York. Vol. 1, pp. 417–420.

Ehrhart, J. F. 1787–1792. Beiträge zur Naturkunde.... 7 vols. Hannover and Osnabrück. (Beitr. Naturk.)

Ekman, E. 1929. Studies in the genus *Draba*. Svensk Bot. Tidskr. 23: 476–495.

Ekman, E. 1930. Contribution to the *Draba* flora of Greenland. II. Svensk Bot. Tidskr. 24: 280–297.

Ekman, E. 1931. Contribution to the *Draba* flora of Greenland. III. Some notes on the arctic, especially the Greenland drabas of the sections *Aizopsis* and *Chrysodraba* DC. Svensk Bot. Tidskr. 25: 465–494.

Ekman, E. 1932. Contribution to the *Draba* flora of Greenland. IV. Svensk Bot. Tidskr. 26: 431–447.

Ekman, E. 1932b. Some notes on the hybridization in the genus *Draba*. Svensk Bot. Tidskr. 26: 198–200.

El Naggar, S. M. 2002. Taxonomic significance of pollen morphology in some taxa of Resedaceae. Feddes Repert. 113: 518–527.

Emory, W. H. 1857–1859. Report on the United States and Mexican Boundary Survey, Made under the Direction of the Secretary of the Interior. 2 vols. in parts. Washington. (Rep. U.S. Mex. Bound.)

Encycl.—See: J. Lamarck et al. 1783–1817

Endlicher, S. L. 1836–1840[–1850] Genera Plantarum Secundum Ordines Naturales Disposita. 18 parts + 5 suppls. in 6 parts. Vienna. [Paged consecutively through suppl. 1(2); suppls. 2–5 paged independently.] (Gen. Pl.)

Engler, H. G. A., ed. 1900–1953. Das Pflanzenreich.... 107 vols. Berlin. [Sequence of vol. (Heft) numbers (order of publication) is independent of the sequence of series and family (Roman and Arabic) numbers (taxonomic order).] (Pflanzenr.)

Engler, H. G. A. et al., eds. 1924+. Die natürlichen Pflanzenfamilien..., ed. 2. 26+ vols. Leipzig and Berlin.

Engler, H. G. A. and K. Prantl, eds. 1887–1915. Die natürlichen Pflanzenfamilien.... 254 fascs. Leipzig. [Sequence of fasc. (Lieferung) numbers (order of publication) is independent of the sequence of division (Teil) and subdivision (Abteilung) numbers (taxonomic order).] (Nat. Pflanzenfam.)

Enum.—See: P. C. Fabricius 1759

Enum. Pl.—See: W. S. J. G. von Besser 1821; C. L. Willdenow 1809–1813[–1814]

Enum. Pl. China Bor.—See: A. A. Bunge [1833]

Enum. Pl. Transsilv.—See: P. J. F. Schur 1866

Enum. Syst. Pl.—See: N. J. Jacquin 1760

Ernst, W. R. 1963b. The genera of Capparidaceae and Moringaceae in the southeastern United States. J. Arnold Arbor. 44: 81–95.

Erythea = Erythea; a Journal of Botany, West American and General.

Eschmann-Grupe, G., H. Hurka, and B. Neuffer. 2003. Species relationships within *Diplotaxis* (Brassicaceae) and the phylogenetic origin of *D. muralis*. Pl. Syst. Evol. 243: 13–29.

Evol. Trends Pl. = Evolutionary Trends in Plants.

Evolution = Evolution, International Journal of Organic Evolution.

Exempl. Rév. Salix—See: N. C. Seringe 1824

Exped. Great Salt Lake—See: H. Stansbury 1852

Fabricius, P. C. 1759. Enumeratio Methodica Plantarum Horti Medici Helmstadiensis.... Helmstedt. (Enum.)

Fairweather, M. L. 1993. Biological Evaluation of *Melampsora* Rust Disease of Arizona Willow (*Salix arizonica*) on the Apache-Sitgreaves National Forest. Albuquerque.

Fam. Pl.—See: M. Adanson 1763[–1764]
Fang, Z. F., Zhao S. D., and A. K. Skvortsov. 1999. Salicaceae. In: Wu Z. and P. H. Raven, eds. 1994+. Flora of China. 16+ vols. Beijing and St. Louis. Vol. 4, pp. 139–274.
Feddes Repert. = Feddes Repertorium.
Feddes Repert. Spec. Nov. Regni Veg. = Feddes Repertorium Specierum Novarum Regni Vegetabilis.
Felger, R. S. and M. B. Moser. 1985. People of the Desert and Sea: Ethnobotany of the Seri Indians. Tucson.
Fernald, M. L. 1909. The North American species of *Barbarea*. Rhodora 11: 134–141.
Fernald, M. L. 1918b. Some North American representatives of *Braya humilis*. Rhodora 20: 201–203.
Fernald, M. L. 1934. *Draba* in temperate northeastern America. Rhodora 36: 241–261, 285–305, 314–344, 353–371, 392–404.
Fernald, M. L. 1950. Gray's Manual of Botany, ed. 8. New York.
Fish and Wildlife Service [U.S.D.I.]. 1954. Gulf of Mexico: Its Origin, Waters, and Marine Life. Washington. [U.S.D.I. Fish Wildlife Serv., Fish. Bull. 55.]
Fisher, M. J. 1928. The morphology and anatomy of the flowers of the Salicaceae. Amer. J. Bot. 15: 307–326, 372–395.
Fl. Altaica—See: C. F. von Ledebour 1829–1833
Fl. Amer. Sept.—See: F. Pursh [1813]1814
Fl. Atlant.—See: R. L. Desfontaines [1798–1799]
Fl. Belg.—See: B. C. J. Dumortier 1827
Fl. Bor.-Amer.—See: W. J. Hooker [1829–]1833–1840; A. Michaux 1803
Fl. Boston.—See: J. Bigelow 1814
Fl. Brandenburg—See: P. F. A. Ascherson 1859–1864
Fl. Calif.—See: W. L. Jepson 1909–1943
Fl. Carniol.—See: J. A. Scopoli 1760
Fl. Carniol. ed. 2—See: J. A. Scopoli 1771–1772
Fl. Carol.—See: T. Walter 1788
Fl. Dan.—See: G. C. Oeder et al. [1761–]1764–1883
Fl. France—See: G. Rouy et al. 1893–1913
Fl. Francisc.—See: E. L. Greene 1891–1897
Fl. Germ. Excurs.—See: H. G. L. Reichenbach 1830[–1832]
Fl. Indica—See: N. L. Burman 1768
Fl. Ins. Austr.—See: G. Forster 1786
Fl. Lapp.—See: G. Wahlenberg 1812
Fl. Ludov.—See: C. S. Rafinesque 1817
Fl. Lusit.—See: F. Brotero 1804[–1805]
Fl. N. Amer.—See: J. Torrey and A. Gray 1838–1843
Fl. N.W. Amer.—See: T. J. Howell 1897–1903
Fl. Nordwestdeut. Tiefebene—See: F. Buchenau 1894
Fl. Norveg.—See: J. E. Gunnerus 1766–1772
Fl. Pedem.—See: C. Allioni 1785
Fl. Rocky Mts.—See: P. A. Rydberg 1917
Fl. Rocky Mts. ed. 2—See: P. A. Rydberg 1923
Fl. Ross.—See: C. F. von Ledebour [1841]1842–1853; P. S. Pallas 1784–1788[–1831]
Fl. S. Calif.—See: A. Davidson and G. L. Moxley 1923
Fl. S.E. U.S.—See: J. K. Small 1903
Fl. S.E. U.S. ed. 2—See: J. K. Small 1913
Fl. Sachsen—See: F. Holl and G. Heynhold 1842
Fl. Schweiz ed. 2—See: H. Schinz and R. Keller 1905
Fl. Sicul.—See: C. B. Presl 1826
Fl. South. U.S.—See: A. W. Chapman 1860
Fl. South. U.S. ed. 2—See: A. W. Chapman 1883

Fl. Tarn Garonne—See: A. R. A. Lagrèze-Fossat 1847
Fl. Taur.-Caucas.—See: F. A. Marschall von Bieberstein 1808–1819
Fl. URSS—See: V. L. Komarov et al. 1934–1964
Fl. W. Calif.—See: W. L. Jepson 1901
Flexner, S. B. and L. C. Hauck, eds. 1987. The Random House Dictionary of the English Language, ed. 2 unabridged. New York.
Floderus, B. G. O. 1923. Om Grönlands salices. (On the Salicaceae of Greenland.) Meddel. Grønland 63: 61–204.
Floderus, B. G. O. 1939. Two Linnaean species of *Salix* and their allies. Ark. Bot. 29A(2): 1–54.
Flora = Flora; oder (allgemeine) botanische Zeitung. [Vols. 1–16, 1818–1833, include "Beilage" and "Ergänzungsblätter"; vols. 17–25, 1834–1842, include "Beiblatt" and "Intelligenzblatt."]
Folia Dendrol. = Folia Dendrologica.
Folia Geobot. = Folia Geobotanica; a Journal of Plant Ecology and Systematics.
Forest Sci. = Forest Science.
Fors. Oecon. Plantel. ed. 2—See: J. W. Hornemann 1806
Forster, G. 1786. Florulae Insularum Australium Prodromus. Göttingen. (Fl. Ins. Austr.)
Fosberg, F. R. 1965. Nomenclature of the horseradish. Baileya 13: 1–4.
Fournier, E. 1865. Recherches Anatomiques et Taxonomiques sur la Famille des Crucifères et sur le Genre *Sisymbrium* en Particulier.... Paris. (Recherches Anat. Taxon. Fam. Crucifèr.)
Franklin, J. et al. 1823. Narrative of a Journey to the Shores of the Polar Sea, in the Years 1819, 20, 21 and 22. London. [Richardson: Appendix VII. Botanical appendix, pp. [729]–768, incl. bryophytes by Schwägrichen, algae and lichens by Hooker.] (Narr. Journey Polar Sea)
Franzke, A. et al. 1998. Molecular systematics of *Cardamine* and allied genera (Brassicaceae): ITS and non-coding chloroplast DNA. Folia Geobot. 33: 225–240.
Franzke, A. and H. Hurka. 2000. Molecular systematics and biogeography of the *Cardamine pratensis* complex (Brassicaceae). Pl. Syst. Evol. 224: 213–234.
Frémont, J. C. 1843–1845. Report of the Exploring Expedition to the Rocky Mountains in the Year 1842, and to Oregon and North California in the Year 1843–44. 2 parts. Washington. [Parts paged consecutively.] (Rep. Exped. Rocky Mts.)
Freudenstein, J. V. and J. K. Marr. 1986. The discovery of *Draba glabella* in Michigan with notes on *Draba* from Isle Royale National Park. Michigan Bot. 25: 109–111.
Fuentes-Soriano, S. 2004. A taxonomic revision of *Pennellia* (Brassicaceae). Harvard Pap. Bot. 8: 173–202.
Gaertner, P. G., B. Meyer, and J. Scherbius. 1799–1802. Oekonomisch-technische Flora der Wetterau. 3 vols. Frankfurt am Main. [Vols. 1 and 2 paged independently, vol. 3 in 2 parts paged independently.] (Oekon. Fl. Wetterau)
Gard. & Forest = Garden and Forest; a Journal of Horticulture, Landscape Art and Forestry.
Gard. Dict. ed. 8—See: P. Miller 1768
Gard. Dict. Abr. ed. 4—See: P. Miller 1754
Gen. Amer. Bor.—See: A. Gray 1848–1849

Gen. Hist.—See: G. Don 1831–1838

Gen. Lesquerella—See: R. C. Rollins and E. A. Shaw 1973

Gen. N. Amer. Pl.—See: T. Nuttall 1818

Gen. Pl.—See: S. L. Endlicher 1836–1840[–1850]

Gen. Pl. ed. 5—See: C. Linnaeus 1754

Genetics = Genetics; a Periodical Record of Investigations Bearing on Heredity and Variation.

Geological Survey of Canada. 1984. Current Research: Part A. Ottawa.

German, D. A. and I. A. Al-Shehbaz. 2008. Five additional tribes (Aphragmeae, Biscutelleae, Calepineae, Conringieae, and Erysimeae) in the Brassicaceae (Cruciferae). Harvard Pap. Bot. 13: 165–170.

Ges. Naturf. Freunde Berlin Neue Schriften = Der Gesellschaft naturforschender Freunde zu Berlin, neue Schriften.

Gilg, E. 1915. Zur Frage der Verwandtschaft der Salicaceae mit den Flacourtiaceae. Bot. Jahrb. Syst. 50(suppl.): 424–434.

Glatfelter, N. M. 1894. A study of the relations of *Salix nigra* and *Salix amygdaloides*, together with the hybrids arising from them as these species exhibit themselves in the vicinity of St. Louis. Trans. Acad. Sci. St. Louis 6: 427–431.

Glatfelter, N. M. 1898. Notes on *Salix* longipes, Shuttlw. and its relations to *S. nigra*, Marsh. Rep. (Annual) Missouri Bot. Gard. 9: 43–51.

Gleason, H. A. and A. Cronquist. 1991. Manual of Vascular Plants of Northeastern United States and Adjacent Canada, ed. 2. Bronx.

Goldblatt, P. 1976. Chromosome number and its significance in *Batis maritima* (Bataceae). J. Arnold Arbor. 57: 526–530.

Gómez-Campo, C., ed. 1999. Biology of *Brassica* Coenospecies. Amsterdam.

Gómez-Campo, C. 1999b. Taxonomy. In: C. Gómez-Campo, ed. 1999. Biology of *Brassica* Coenospecies. Amsterdam. Pp. 3–32.

González Aguilera, J. J. and A. M. Fernández Peralta. 1983. The nature of polyploidy in *Reseda* sect. *Leucoreseda* (Resedaceae). Pl. Syst. Evol. 142: 223–237.

González Aguilera, J. J. and A. M. Fernández Peralta. 1984. Phylogenetic relationships in the family Resedaceae. Genetica 64: 185–197.

Goodson, B. E. 2007. Molecular Systematics and Biogeography of *Descurainia* Webb & Berthel. (Brassicaceae). Ph.D. dissertation. University of Texas.

Gori, C. 1957. Sull'embriologia e citologia di alcune specie del genere *Reseda*. Caryologia 10: 391–401.

Gouan, A. 1773. Illustrationes et Observationes Botanicae.... Zürich. (Ill. Observ. Bot.)

Grant, V. 1981. Plant Speciation, ed. 2. New York.

Gray, A. 1848–1849. Genera Florae Americae Boreali-orientalis Illustrata. The Genera of the Plants of the United States.... 2 vols. Boston, New York, and London. (Gen. Amer. Bor.)

Gray, A. 1856. A Manual of the Botany of the Northern United States..., ed. 2. New York. (Manual ed. 2)

Gray, A., S. Watson, B. L. Robinson, et al. 1878–1897. Synoptical Flora of North America. 2 vols. in parts and fascs. New York etc. [Vol. 1(1,1), 1895; vol. 1(1,2), 1897; vol. 1(2), 1884; vol. 2(1), 1878.] (Syn. Fl. N. Amer.)

Green, P. S. 1962. Watercress in the New World. Rhodora 64: 32–43.

Greene, E. L. 1891–1897. Flora Franciscana. An Attempt to Classify and Describe the Vascular Plants of Middle California. 4 parts. San Francisco. [Parts paged consecutively.] (Fl. Francisc.)

Greene, E. L. [1901.] Plantae Bakerianae. 3 vols. [Washington.] (Pl. Baker.)

Greene, E. L. 1906b. Revision of the genus *Wislizenia*. Proc. Biol. Soc. Wash. 19: 127–132.

Greilhuber, J. and R. Obermayer. 1999. Cryptopolyploidy in *Bunias* (Brassicaceae) revisited—A flowcytometric and densitometric study. Pl. Syst. Evol. 218: 1–4.

Grundt, H. H. et al. 2004. Polyploid origins in a circumpolar complex in *Draba* (Brassicaceae) inferred from cloned nuclear DNA sequences and fingerprints. Molec. Phylogen. Evol. 32: 695–710.

Grundt, H. H. et al. 2005. A rare case of self-incompatibility in arctic plants: *Draba palanderiana* (Brassicaceae). Flora 200: 321–325.

Grundt, H. H. et al. 2005b. Ploidal levels in the arctic-alpine polyploid *Draba lactea* (Brassicaceae) and its low-ploid relatives. Bot. J. Linn. Soc. 147: 333–347.

Gunnerus, J. E. 1766–1772. Flora Norvegica.... 2 vols. Trondhiem and Copenhagen. (Fl. Norveg.)

Guppy, H. B. 1903–1906. Observations of a Naturalist in the Pacific between 1896 and 1899. 2 vols. London and New York.

Haissig, B. E. 1974. Origins of adventitious roots. New Zealand J. Forest. Sci. 4: 299–310.

Hall, J. C., H. H. Iltis, and K. J. Sytsma. 2004. Molecular phylogenetics of core Brassicales, placement of orphan genera *Emblingia*, *Forchhammeria*, *Tirania*, and character evolution. Syst. Bot. 29: 654–669.

Hall, J. C., K. J. Sytsma, and H. H. Iltis. 2002. Phylogeny of Capparaceae and Brassicaceae based on chloroplast sequence data. Amer. J. Bot. 89: 1826–1842.

Hallier, H. 1905. Phylogenetic studies of flowering plants. New Phytol. 5: 151–162.

Hanelt, P., ed. 2001. Mansfeld's Encyclopedia of Agricultural and Horticultural Crops.... 6 vols. Berlin and New York.

Hara, H. 1955b. Critical notes on some type specimens of east-Asiatic plants in foreign herbaria (4). J. Jap. Bot. 30: 193–198.

Hardig, T. M. et al. 2000. Morphological and molecular evidence for hybridization and introgression in a willow (*Salix*) hybrid zone. Molec. Ecol. 9: 9–24.

Hardin, J. W. 1958. *Calepina irregularis* in North Carolina. Castanea 23: 111.

Harriman, N. A. 1965. The Genus *Dentaria* L. (Cruciferae) in Eastern North America. Ph.D. dissertation. Vanderbilt University.

Harris, J. D., E. S. Davis, and D. M. Wichman. 1995. Yellow mignonette (*Reseda lutea*) in the United States. Weed Technol. 19: 196–198.

Harris, J. G. 1985. A Revision of the Genus *Braya* (Cruciferae) in North America. Ph.D. thesis. University of Alberta.

Harris, J. G. 2006. Pilose braya, *Braya pilosa* Hooker (Cruciferae; Brassicaceae), an enigmatic endemic of arctic Canada. Canad. Field-Naturalist 118: 550–557.

Hart, T. W. and W. H. Eshbaugh. 1976. The biosystematics of *Cardamine bulbosa* (Muhl.) B.S.P. and *C. douglasii* Britt. Rhodora 78: 329–419.

Hartman, R. L., J. D. Bacon, and C. F. Bohnstedt. 1975. Biosystematics of *Draba cuneifolia* and *D. platycarpa* (Cruciferae) with emphasis on volatile and flavonoid constituents. Brittonia 27: 317–327.

Harvard Pap. Bot. = Harvard Papers in Botany.

Haston, E. et al. 2007. A linear sequence of Angiosperm Phylogeny Group II families. Taxon 56: 7–12.

Hauptli, H. H., B. D. Webster, and S. K. Jain. 1978. Variation in nutlet morphology in *Limnanthes*. Amer. J. Bot. 65: 615–624.

Hauser, L. A. and T. J. Crovello. 1982. Numerical analysis of generic relationships in Thelypodieae (Brassicaceae). Syst. Bot. 7: 249–268.

Häussknechtia = Häussknechtia; Mitteilungen der Thüringischen botanischen Gesellschaft.

Hegi, G. et al. [1906–1931.] Illustrierte Flora von Mittel-Europa. 7 vols. in 13. Munich. (Ill. Fl. Mitt.-Eur.)

Heilborn, O. 1927. Chromosome numbers in *Draba*. Hereditas (Lund) 9: 59–68.

Heller, A. A. 1909–1914. Catalogue of North American Plants North of Mexico, Exclusive of the Lower Cryptogams, ed. 3. [Lancaster, Pa.] (Cat. N. Amer. Pl. ed. 3)

Hemsley, W. B. 1879–1888. Biologia Centrali-Americana.... Botany.... 5 vols. London. (Biol. Cent.-Amer., Bot.)

Hennig, L. 1929. Beiträge zur Kenntnis der Resedaceen—Blüte und Frucht. Planta 9: 507–563.

Henry, J. K. 1915. Flora of Southern British Columbia and Vancouver Island. Toronto.

Henry, R. J., ed. 2005. Plant Diversity and Evolution: Genotypic and Phenotypic Variation in Higher Plants. Wallingford and Cambridge, Mass.

Herbarist = The Herbarist; a Publication of the Herb Society of America.

Heredity = Heredity; an International Journal of Genetics.

Hickey, L. J. and J. A. Wolfe. 1975. The bases of angiosperm phylogeny: Vegetative morphology. Ann. Missouri Bot. Gard. 62: 538–589.

Hickman, J. C., ed. 1993. The Jepson Manual. Higher Plants of California. Berkeley, Los Angeles, and London.

Hiern, W. P. 1896–1901. Catalogue of the African Plants Collected by Dr. Friedrich Welwitsch in 1853–61.... 2 vols in parts. London. [Parts paged consecutively within volumes.] (Cat. Afr. Pl.)

Hillebrand, W. 1888. Flora of the Hawaiian Islands.... Annotated and Published after the Author's Death by W. F. Hillebrand. London, New York, and Heidelberg.

Hilu, K. W. 2003. Angiosperm phylogeny based on *mat*K sequence information. Amer. J. Bot. 90: 1758–1776.

Hist. Nat. Îles Canaries—See: P. B. Webb and S. Berthelot [1835–]1836–1850

Hitchcock, A. S. 1894. A Key to the Spring Flora of Manhattan.... Manhattan, Kans. (Key Spring Fl. Manhattan)

Hitchcock, C. L. 1936. The genus *Lepidium* in the United States. Madroño 3: 265–300.

Hitchcock, C. L. 1941. A Revision of the Drabas of Western North America. Seattle. [Univ. Wash. Publ. Biol. 11.] (Revis. Drabas W. N. Amer.)

Hitchcock, C. L. 1945. The Mexican, Central American, and West Indian lepidia. Madroño 8: 118–143.

Hitchcock, C. L. 1945b. The South American species of *Lepidium*. Lilloa 11: 75–134.

Hitchcock, C. L. 1950. On the subspecies of *Lepidium montanum*. Madroño 10: 155–158.

Hitchcock, C. L., A. Cronquist, M. Ownbey, and J. W. Thompson. 1955–1969. Vascular Plants of the Pacific Northwest. 5 vols. Seattle. [Univ. Wash. Publ. Biol. 17.] (Vasc. Pl. Pacif. N.W.)

Hodge, W. H. 1974. Wasabi—Native condiment plant of Japan. Econ. Bot. 28: 118–129.

Holarc. Ecol. = Holarctic Ecology.

Holl, F. and G. Heynhold. 1842. Flora von Sachsen. 1 vol. only. Dresden. (Fl. Sachsen)

Holmes, W. C., K. L. Yip, and A. E. Rushing. 2008. Taxonomy of *Koeberlinia* (Koeberliniaceae). Brittonia 60: 171–184.

Holmgren, N. H. 2005b. Brassicaceae. In: A. Cronquist et al. 1972+. Intermountain Flora. Vascular Plants of the Intermountain West, U.S.A. 6+ vols. in 7+. New York and London. Vol. 2, part B, pp. 174–418.

Holmgren, P. K. 1971. A biosystematic study of North American *Thlaspi montanum* and its allies. Mem. New York Bot. Gard. 21(2): 1–106.

Holmgren, P. K. and A. Cronquist. 2005. Cleomaceae. In: A. Cronquist et al. 1972+. Intermountain Flora. Vascular Plants of the Intermountain West, U.S.A. 6+ vols. in 7+. New York and London. Vol. 2, part B, pp. 160–174.

Hooker, W. J. [1829–]1833–1840. Flora Boreali-Americana; or, the Botany of the Northern Parts of British America.... 2 vols. in 12 parts. London, Paris, and Strasbourg. (Fl. Bor.-Amer.)

Hooker, W. J. 1830. The British Flora.... London. (Brit. Fl.)

Hooker, W. J. and G. A. W. Arnott. [1830–]1841. The Botany of Captain Beechey's Voyage; Comprising an Account of the Plants Collected by Messrs Lay and Collie, and Other Officers of the Expedition, during the Voyage to the Pacific and Bering's Strait, Performed in His Majesty's Ship Blossom, under the Command of Captain F. W. Beechey...in the Years 1825, 26, 27, and 28. 10 parts. London. [Parts paged and plates numbered consecutively.] (Bot. Beechey Voy.)

Hooker's J. Bot. Kew Gard. Misc. = Hooker's Journal of Botany and Kew Garden Miscellany.

Hopkins, M. 1937. *Arabis* in eastern and central North America. Rhodora 39: 63–98, 106–148, 155–186.

Horaninov, P. F. 1847. Characteres Essentiales Familiarum ac Tribuum Regni Vegetabilis.... St. Petersburg. (Char. Ess. Fam.)

Hornemann, J. W. 1806. Forsøg til en Dansk Oeconomisk Plantelaere, ed. 3. Copenhagen. (Fors. Oecon. Plantel. ed. 2)

Hornemann, J. W. 1813–1815. Hortus Regius Botanicus Hafniensis.... 2 vols. Copenhagen. (Hort. Bot. Hafn.)

Hornschuch, C. F. [1822–]1824–1828. Sylloge Plantarum Novarum.... 2 vols. Regensburg. (Syll. Pl. Nov.)

Hort. Bot. Hafn.—See: J. W. Hornemann 1813–1815

Hort. Kew.—See: W. Aiton 1789

Hortus Kew.—See: W. Aiton and W. T. Aiton 1810–1813

Houle, G., M. F. McKenna, and L. Lapointe. 2001. Spatiotemporal dynamics of *Floerkia proserpinacoides* (Limnanthaceae), an annual plant of the deciduous forest of eastern North America. Amer. J. Bot. 88: 594–607.

Howell, J. T. 1970. Marin Flora: Manual of the Flowering Plants and Ferns of Marin County, California, ed. 2 with suppl. Berkeley.

Howell, J. T., P. H. Raven, and P. Rubtzoff. 1958. A flora of San Francisco, California. Wasmann J. Biol. 16: 1–157.

Howell, T. J. 1897–1903. A Flora of Northwest America. 1 vol. in 8 fascs. Portland. [Fascs. 1–7 (text) paged consecutively, fasc. 8 (index) independently.] (Fl. N.W. Amer.)

Hultén, E. 1967. Comments on the flora of Alaska and Yukon. Ark. Bot., n. s. 7: 1–147.

Hultén, E. 1968. Flora of Alaska and Neighboring Territories: A Manual of the Vascular Plants. Stanford.

Hultén, E. 1971. The circumpolar plants. II. Dicotyledons. Kongl. Svenska Vetensk. Acad. Handl., ser. 4, 13: 1–463.

Humboldt, A. von, A. J. Bonpland, and C. S. Kunth. 1815[1816]–1825. Nova Genera et Species Plantarum Quas in Peregrinatione Orbis Novi Collegerunt, Descripserunt.... 7 vols. in 36 parts. Paris. (Nov. Gen. Sp.)

Hurka, H. and B. Neuffer. 1997. Evolutionary processes in the genus *Capsella* (Brassicaceae). Pl. Syst. Evol. 206: 295–316.

Huynh, K. L. 1982. Le pollen du *Limnanthes douglasii* (Limnanthaceae) en microscopie électronique. Pollen & Spores 24: 211–234.

Icon.—See: A. J. Cavanilles 1791–1801

Icon. Pl. = Icones Plantarum....

Icon. Pl.—See: C. F. von Ledebour 1829–1834

Ill. Fl. Mitt.-Eur.—See: G. Hegi et al. [1906–1931]

Ill. Fl. N. U.S.—See: N. L. Britton and A. Brown 1896–1898

Ill. Fl. Pacific States—See: L. Abrams and R. S. Ferris 1923–1960

Ill. Handb. Laubholzk.—See: C. K. Schneider [1904–]1906–1912

Ill. Observ. Bot.—See: A. Gouan 1773

Ill. Pl. Eur.—See: G. Rouy 1895–1905

Iltis, H. H. 1952. A Revision of the Genus *Cleome* in the New World. Ph.D. dissertation. Washington University.

Iltis, H. H. 1955. Capparidaceae of Nevada. Washington. [Contr. Fl. Nevada 35.]

Iltis, H. H. 1956. Studies in the Capparidaceae. II. The Mexican species of *Cleomella*: Taxonomy and evolution. Madroño 13: 177–189.

Iltis, H. H. 1957. Studies in the Capparidaceae. III. Evolution and phylogeny of the western North American Cleomoideae. Ann. Missouri Bot. Gard. 44: 77–119.

Iltis, H. H. 1958. Studies in the Capparidaceae. IV. *Polanisia* Raf. Brittonia 10: 33–58.

Iltis, H. H. 1960. Studies in the Capparidaceae. VII. Old World cleomes adventive in the New World. Brittonia 12: 279–294.

Iltis, H. H. 1966. Studies in the Capparidaceae. VIII. *Polanisia dodecandra* (L.) DC. Rhodora 68: 41–47.

Imbamba, S. K. and L. T. Tieszen. 1977. Influence of light and temperature on photosynthesis and transpiration in some C_3 and C_4 vegetable plants from Kenya. Physiol. Pl. (Copenhagen) 39: 311–316.

Inda, L. A., P. Torrecilla, P. Catalán, and T. Ruíz Z. 2008. Phylogeny of *Cleome* L. and its close relatives *Podandrogyne* Ducke and *Polanisia* Raf. (Cleomoideae, Cleomaceae) based on analysis of nuclear ITS sequences and morphology. Pl. Syst. Evol. 274: 111–126.

Index Seminum (Dorpat) = Index Seminum Horti Academici Dorpatensis.

Index Seminum (Göttingen) = Index Seminum Horti Academici Goettingensis Anno...Collecta.

Index Seminum (Hamburg) = Semina in Horto Botanico Hamburgensi...Collecta Quae pro Mutua Commutatione Offeruntur. [1823–1840 for years 1822–1840. Title after 1829: Delectus Seminum Quae in Horto Hamburgensium Botanico e Collectioni Anni...Mutuae Commutatione Offeruntur.]

Index Seminum (St. Petersburg) = Index Seminum, Quae Hortus Botanicus Imperialis Petropolitanus pro Mutua Commutatione Offert.

Index Seminum (Zürich) = Verzeichniss im Tausch abgebbarer Samaereien und Fruchte des Botanischen Gartens der Universität Zürich.

Int. Dendrol. Soc. Year Book = International Dendrology Society Year Book.

Int. J. Pl. Sci. = International Journal of Plant Sciences.

Intermount. Fl.—See: A. Cronquist et al. 1972+

Interpr. Herb. Amboin.—See: E. D. Merrill 1917

Ives, J. C. 1861. Report upon the Colorado River of the West, Explored in 1857 and 1858 by Lieutenant Joseph C. Ives.... 5 parts, appendices. Washington. (Rep. Colorado R.)

J. Acad. Nat. Sci. Philadelphia = Journal of the Academy of Natural Sciences of Philadelphia.

J. Arboric. = Journal of Arboriculture.

J. Arizona-Nevada Acad. Sci. = Journal of the Arizona-Nevada Academy of Science.

J. Arnold Arbor. = Journal of the Arnold Arboretum.

J. Beijing Forest. Univ. = Journal of Beijing Forestry University.

J. Biogeogr. = Journal of Biogeography.

J. Bot. Agric. = Journal de Botanique, Appliquée à l'Agriculture, à la Pharmacie, à la Médecine et aux Arts.

J. Bot. Res. Inst. Texas = Journal of the Botanical Research Institute of Texas.

J. Bot. (Schrader) = Journal für die Botanik. [Edited by H. A. Schrader.] [Volumation indicated by nominal year date and vol. number for that year (1 or 2); e.g. 1800(2).]

J. Chem. Ecol. = Journal of Chemical Ecology.

J. Ecol. = Journal of Ecology.

J. Elisha Mitchell Sci. Soc. = Journal of the Elisha Mitchell Scientific Society.

J. Heredity = Journal of Heredity.

J. Jap. Bot. = Journal of Japanese Botany.

J. Linn. Soc., Bot. = Journal of the Linnean Society. Botany.

J. Wash. Acad. Sci. = Journal of the Washington Academy of Sciences.

Jacobson, H. A., J. B. Petersen, and D. E. Putnam. 1988. Evidence of pre-Columbian *Brassica* in the northeastern U.S. Rhodora 90: 355–362.

Jacquemont, V. [1835–]1841–1844. Voyage dans l'Inde.... 4 vols. Paris. (Voy. Inde)

Jacquin, N. J. 1760. Enumeratio Systematica Plantarum, Quas in Insulis Caribaeis Vicinaque Americes Continente Detexit Novas.... Leiden. (Enum. Syst. Pl.)

Jacquin, N. J. 1778–1781. Miscellanea Austriaca.... 2 vols. Vienna. (Misc. Austriac.)

James, E. 1823. Account of an Expedition from Pittsburgh to the Rocky Mountains, Performed in the Years 1819 and '20...under the Command of Major Stephen H. Long. 2 vols. + atlas. Philadelphia. (Account Exped. Pittsburgh)

Janchen, E. 1942. Das System der Cruciferen. Oesterr. Bot. Z. 91: 1–18.

Jepson, W. L. 1901. A Flora of Western Middle California.... Berkeley. (Fl. W. Calif.)

Jepson, W. L. 1909–1943. A Flora of California.... 3 vols. in 12 parts. San Francisco etc. [Pagination consecutive within each vol.; vol. 1 page sequence independent of part number sequence (chronological); part 8 of vol. 1 (pp. 1–32, 579–index) never published.] (Fl. Calif.)

Jepson, W. L. [1923–1925.] A Manual of the Flowering Plants of California.... Berkeley. (Man. Fl. Pl. Calif.)

Jones, A. G. and D. S. Seigler. 1975. Flavonoid data and populational observations in support of hybrid status for *Populus acuminata*. Biochem. Syst. & Ecol. 2: 201–206.

Jonsell, B. 1968. Studies in the north-west European species of *Rorippa* s.str. Symb. Bot. Upsal. 19(2): 1–222.

Jonsell, B. and T. Karlsson, eds. 2000+. Flora Nordica. 3+ vols. Stockholm.

Jørgensen, M. H., T. Carlsen, I. Skrede, and R. Elven. 2008. Microsatellites resolve the taxonomy of the polyploid *Cardamine digitata* aggregate (Brassicaceae). Taxon 57: 882–892.

Jørgensen, R. B. and B. Andersen. 1994. Spontaneous hybridization between oilseed rape (*Brassica napus*) and weedy *B. campestris* (Brassicaceae): A risk of growing genetically modified oilseed rape. Amer. J. Bot. 81: 1620–1626.

Judd, W. S. 1997b. The Flacourtiaceae in the southeastern United States. Harvard Pap. Bot. 10: 65–79.

Judd, W. S., R. W. Sanders, and M. J. Donoghue. 1994. Angiosperm family pairs: Preliminary phylogenetic analyses. Harvard Pap. Bot. 5: 1–51.

Jurtzev, B. A. and P. G. Zhukova. 1982. Chromosome numbers of some plants of northeastern Yakutia (the drainage of the Indigirka Rover and its middle reaches). Bot. Zhurn. (Moscow & Leningrad) 67: 778–787.

Kalmia = Kalmia; Botanic Journal.

Katenin, A. E. and V. V. Petrovsky. 1995. On the finding in the Chukchi Peninsula of 2 species of Brassicaceae, new for Eurasia. Bot. Zhurn. (Moscow & Leningrad) 80(10): 94–99.

Kaye, T. N. et al. 1997. Conservation of Native Plants and Fungi. Corvallis.

Kearney, T. H. and R. H. Peebles. 1960. Arizona Flora, ed. 2. Berkeley.

Keim, P., K. N. Paige, T. G. Whitham, and K. G. Lark. 1989. Genetic analysis of an interspecific hybrid swarm of *Populus*: Occurrence of unidirectional introgression. Genetics 123: 557–565.

Keller, S. 1979. A revision of the genus *Wislizenia* (Capparidaceae) based on population studies. Brittonia 31: 333–351.

Kers, L. E. 2003. Capparaceae. In: K. Kubitzki et al., eds. 1990+. The Families and Genera of Vascular Plants. 9+ vols. Berlin etc. Vol. 5, pp. 36-56.

Kesseli, R. V. and S. K. Jain. 1984. An ecological genetic study of gynodioecy in *Limnanthes douglasii* (Limnanthaceae). Amer. J. Bot. 71: 775–786.

Kesseli, R. V. and S. K. Jain. 1984b. New variation and biosystematic patterns detected by allozyme and morphological comparison in *Limnanthes* sect. *Reflexae* (Limnanthaceae). Pl. Syst. Evol. 147: 133–164.

Kesseli, R. V. and S. K. Jain. 1987. Origin of gynodioecy in *Limnanthes*. Theor. Appl. Genet. 74: 379–386.

Kew Bull. = Kew Bulletin.

Key Spring Fl. Manhattan—See: A. S. Hitchcock 1894

Kiger, R. W. 2001. *Xylosma*. In: R. McVaugh and W. R. Anderson, eds. 1974+. Flora Novo-Galiciana: A Descriptive Account of the Vascular Plants of Western Mexico. 8+ vols. Ann Arbor. Vol. 3, pp. 328–334.

Kiger, R. W. and D. M. Porter. 2001. Categorical Glossary for the Flora of North America Project. Pittsburgh.

Kimura, A. 1928. Über *Toisusu*—eine neue Salicaceen-Gattung und die systematische Stellung derselben. Bot. Mag. (Tokyo) 45: 28.

Kishore, V. K. et al. 2004. Conserved simple sequence repeats for the Limnanthaceae (Brassicales). Theor. Appl. Genet. 108: 450–457.

Klein, D. R. and C. Bay. 1991. Diet selection by vertebrate herbivores in the high arctic of Greenland. Holarc. Ecol. 14: 152–155.

Knaben, G. 1966. Cytological studies in some *Draba* species. Bot. Not. 119: 427–444.

Knight, O. W. 1907. Three plants from Maine. Rhodora 9: 202–204.

Koch, M. 2002. Genetic differentiation and speciation in prealpine *Cochlearia*: Allohexaploid *Cochlearia bavarica* (Brassicaceae) compared to its diploid ancestor *Cochlearia pyrenaica* in Germany and Austria. Pl. Syst. Evol. 232: 35–49.

Koch, M. 2003. Molecular phylogenetics, evolution and population biology in Brassicaceae. In: A. K. Sharma and A. Sharma, eds. 2003+. Plant Genome: Biodiversity and Evolution. 2+ vols. in parts. Enfield, N.H. Vol. 1, part A, pp. 1–35.

Koch, M. et al. 1998. Isozymes, speciation and evolution in the polyploid complex *Cochlearia* L. (Brassicaceae). Bot. Acta 111: 411–425.

Koch, M. et al. 1998b. Molecular biogeography and evolution of the *Microthlaspi perfoliatum* s.l. polyploid complex (Brassicaceae): Chloroplast DNA and nuclear ribosomal DNA restriction site variation. Canad. J. Bot. 76: 382–396.

Koch, M. et al. 1999. Molecular phylogenetics of *Cochlearia* (Brassicaceae) and allied genera based on nuclear ribosomal ITS DNA sequence analysis contradict traditional concepts of their evolutionary relationship. Pl. Syst. Evol. 216: 207–230.

Koch, M. et al. 1999b. Molecular systematics of *Arabidopsis* and *Arabis*. Pl. Biol. (Stuttgart) 1: 529–537.

Koch, M. et al. 2003. *Cochlearia macrorrhiza*: A bridging species between *Cochlearia* taxa from the eastern Alps and the Carpathians. Pl. Syst. Evol. 242: 137–147.

Koch, M. et al. 2003b. Molecular systematics, evolution, and population biology in the mustard family (Brassicaceae). Ann. Missouri Bot. Gard. 90: 151–171.

Koch, M. and I. A. Al-Shehbaz. 2002. Molecular data indicate complex intra- and intercontinental differentiation of American *Draba* (Brassicaceae). Ann. Missouri Bot. Gard. 89: 88–109.

Koch, M. and I. A. Al-Shehbaz. 2004. Taxonomic and phylogenetic evaluation of the American "*Thlaspi*" species: Identity and relationship to the Eurasian genus *Noccaea* (Brassicaceae). Syst. Bot. 29: 375–384.

Koch, M., B. Haubold, and T. Mitchell-Olds. 2000. Comparative analysis of chalcone synthase and alcohol dehydrogenase loci in *Arabidopsis*, *Arabis* and related genera (Brassicaceae). Molec. Biol. Evol. 17: 1483–1498.

Koch, M., B. Haubold, and T. Mitchell-Olds. 2001. Molecular systematics of the Brassicaceae: Evidence from coding plastidic *mat*K and nuclear *Chs* sequences. Amer. J. Bot. 88: 534–544.

Koch, M., H. Hurka, and K. Mummenhoff. 1996. Chloroplast DNA restriction site variation and RAPD-analyses in *Cochlearia* (Brassicaceae): Biosystematics and speciation. Nordic J. Bot. 16: 585–603.

Koch, M. and K. Mummenhoff. 2001. *Thlaspi* s.str. (Brassicaceae) versus *Thlaspi* s.l.: Morphological and anatomical characters in the light of ITS nDNA sequence data. Pl. Syst. Evol. 227: 209–225.

Koehne, B. A. E. 1893. Deutsche Dendrologie. Stuttgart. (Deut. Dendrol.)

Komarov, V. L., B. K. Schischkin, B. K., and E. Bobrov, eds. 1934–1964. Flora URSS.... 30 vols. Leningrad. (Fl. URSS)

Kongl. Svenska Vetensk. Acad. Handl. = Kongl[iga]. Svenska Vetenskaps Academiens Handlingar.

Krebs, S. and S. K. Jain. 1985. Variation in morphological and physiological traits associated with yield in *Limnanthes* spp. New Phytol. 101: 717–719.

Kruckeberg, A. R. 1957. Variation in fertility of hybrids between isolated populations of the serpentine species, *Streptanthus glandulosus* Hook. Evolution 11: 185–211.

Kruckeberg, A. R. and J. L. Morrison. 1983. New *Streptanthus* taxa (Cruciferae) from California. Madroño 30: 230–244.

Kruckeberg, A. R. and R. D. Reeves. 1995. Nickel accumulation by serpentine species of *Streptanthus* (Brassicaceae): Field and greenhouse studies. Madroño 42: 458–469.

Kruckeberg, A. R., J. E. Rodman, and R. D. Worthington. 1982. Natural hybridization between *Streptanthus arizonicus* and *S. carinatus* (Cruciferae). Syst. Bot. 7: 291–299.

Kubitzki, K. 2003. Resedaceae. In: K. Kubitzki et al., eds. 1990+. The Families and Genera of Vascular Plants. 9+ vols. Berlin etc. Vol. 5, pp. 334–338.

Kubitzki, K. et al., eds. 1990+. The Families and Genera of Vascular Plants. 9+ vols. Berlin etc.

Kuntze, O. 1891–1898. Revisio Generum Plantarum Vascularium Omnium atque Cellularium Multarum.... 3 vols. Leipzig etc. [Vol. 3 in 3 parts paged independently; parts 1 and 3 unnumbered.] (Revis. Gen. Pl.)

Kus, B. E. and J. L. Beyers, coords. 2005. Planning for Biodiversity: Bringing Research and Management Together. Albany, Calif. [U.S.D.A. Forest Serv., Gen. Techn. Rep. PSW-GTR-195.]

Lagrèze-Fossat, A. R. A. 1847. Flore de Tarn et Garonne.... Montauban and Moissac. (Fl. Tarn Garonne)

Lamarck, J. et al. 1783–1817. Encyclopédie Méthodique. Botanique.... 13 vols. Paris and Liège. [Vols. 1–8, suppls. 1–5.] (Encycl.)

Larsson, G. 1995. Nomenclatural remarks on the *Salix viminalis* group in Norden. Nordic J. Bot. 15: 343–345.

Lawrence, G. H. M. 1971. The horseradish *Armoracia rusticana*. Herbarist 37: 17–19.

Leadlay, E. A. and V. H. Heywood. 1990. The biology and systematics of the genus *Coincya* Porta & Rigo ex Rouy (Cruciferae). Bot. J. Linn. Soc. 102: 313–398.

Leafl. Bot. Observ. Crit. = Leaflets of Botanical Observation and Criticism.

Leafl. W. Bot. = Leaflets of Western Botany.

Ledebour, C. F. von. 1829–1833. Flora Altaica. 4 vols. Berlin. (Fl. Altaica)

Ledebour, C. F. von. 1829–1834. Icones Plantarum...Floram Rossicam.... 5 vols. Riga etc. [Vols. paged independently, plates numbered consecutively.] (Icon. Pl.)

Ledebour, C. F. von. [1841]1842–1853. Flora Rossica sive Enumeratio Plantarum in Totius Imperii Rossici Provinciis Europaeis, Asiaticis, et Americanis Hucusque Observatarum.... 4 vols. Stuttgart. (Fl. Ross.)

Lemke, D. E. 1988. A synopsis of Flacourtiaceae. Aliso 12: 29–43.

Lemke, D. E. and R. D. Worthington. 1991. *Brassica* and *Rapistrum* (Brassicaceae) in Texas. SouthW. Naturalist 36: 194–197.

Leong, J. M. and R. W. Thorp. 2005. Bee diversity associated with *Limnanthes* floral patches in California vernal pool habitats. In: B. E. Kus and J. L. Beyers, coords. 2005. Planning for Biodiversity: Bringing Research and Management Together. Albany, Calif. Pp. 267–268.

Lepage, E. 1961. Études sur quelques plantes Américaines—X. Naturaliste Canad. 88: 44–51.

Lepage, E. 1962. Nouveautés dans la flore du bassin de la Baie d'Ungava, Québec. Naturaliste Canad. 89: 113–119.

Lepage, E. 1964. Hybrides nouveautés dans les genres *Carex* et *Salix*. Naturaliste Canad. 99: 165–174.

Leskinen, E. and C. Alström-Rapaport. 1999. Molecular phylogeny of Salicaceae and closely related Flacourtiaceae: Evidence from 5.8 S, ITS 1 and ITS 2 of the rDNA. Pl. Syst. Evol. 215: 209–227.

Lewis-Jones, L. J., J. P. Thorpe, and G. P. Wallis. 1982. Genetic divergence in four species of the genus *Raphanus*: Implications for the ancestry of the domestic radish *R. sativus*. Biol. J. Linn. Soc. 18: 35–48.

L'Héritier de Brutelle, C.-L. 1784[1785–1805]. Stirpes Novae aut Minus Cognitae.... 2 vols. in 9 fascs. Paris. [Fascicles paged and plates numbered consecutively.] (Stirp. Nov.)

Lichvar, R. W. 1983. Evaluation of varieties in *Stanleya pinnata* (Cruciferae). Great Basin Naturalist 43: 684–686.

Lihová, J. and K. Marhold. 2006. Phylogenetic and diversity patterns in *Cardamine* (Brassicaceae)—A genus with conspicuous polyploid and reticulate evolution. In: A. K. Sharma and A. Sharma, eds. 2003+. Plant Genome: Biodiversity and Evolution. 2+ vols. in parts. Enfield, N.H. Vol. 1, part C, pp. 149–186.

Lihová, J., K. Marhold, H. Kudoh, and M. Koch. 2006. Worldwide phylogeny and biogeography of *Cardamine flexuosa* (Brassicaceae) and its relatives. Amer. J. Bot. 93: 1206–1221.

Liljeblad, S. 1792. Utkast til en Svensk Flora.... Uppsala. (Utkast Sv. Fl.)

Lilloa = Lilloa; Revista de Botánica.

Lindman, C. A. M. 1926. Svensk Fanerogamflora..., ed. 2. Stockholm. (Sv. Fanerogamfl. ed. 2)

Link, D. A. 1992. The floral nectaries of Limnanthaceae. Pl. Syst. Evol. 179: 235–243.

Linnaea = Linnaea. Ein Journal für die Botanik in ihrem ganzen Umfange.

Linnaeus, C. 1737[1738]. Hortus Cliffortianus.... Amsterdam.

Linnaeus, C. 1749[–1769]. Amoenitates Academicae seu Dissertationes Variae Physicae, Medicae Botanicae.... 7 vols. Stockholm and Leipzig. (Amoen. Acad.)

Linnaeus, C. [1753.] Demonstrationes Plantarum in Horto Upsaliensi.... Uppsala. (Demonstr. Pl.)

Linnaeus, C. 1753. Species Plantarum.... 2 vols. Stockholm. (Sp. Pl.)

Linnaeus, C. 1754. Genera Plantarum, ed. 5. Stockholm. (Gen. Pl. ed. 5)

Linnaeus, C. 1755. Centuria I. Plantarum.... Uppsala. (Cent. Pl. I)

Linnaeus, C. [1756.] Centuria II. Plantarum.... Uppsala. (Cent. Pl. II)

Linnaeus, C. 1758[–1759]. Systema Naturae per Regna Tria Naturae..., ed. 10. 2 vols. Stockholm. (Syst. Nat. ed. 10)

Linnaeus, C. 1759. Plantarum Jamaicensium Pugillus.... Uppsala. (Pl. Jamaic. Pug.)

Linnaeus, C. 1762–1763. Species Plantarum..., ed. 2. 2 vols. Stockholm. (Sp. Pl. ed. 2)

Linnaeus, C. 1766–1768. Systema Naturae per Regna Tria Naturae..., ed. 12. 3 vols. Stockholm. (Syst. Nat. ed. 12)

Linnaeus, C. 1767[–1771]. Mantissa Plantarum. 2 parts. Stockholm. [Mantissa [1] and Mantissa [2] Altera paged consecutively.] (Mant. Pl.)

Linzer Biol. Beitr. = Linzer Biologische Beiträge.

Little, E. L. Jr. 1971. Atlas of United States Trees. I. Conifers and Important Hardwoods. Washington. [U.S.D.A., Misc. Publ. 1146.]

Little, E. L. Jr., K. A. Brinkman, and A. L. McComb. 1957. Two natural Iowa hybrid poplars. Forest Sci. 3: 253–262.

Lloyd, D. G. 1965. Evolution of self-compatibility and racial differentiation in *Leavenworthia* (Cruciferae). Contr. Gray Herb. 195: 1–134.

Lloyd, D. G. 1967. The genetics of self-incompatibility in *Leavenworthia crassa* Rollins (Cruciferae). Genetica 38: 227–242.

Lloyd, D. G. 1969. Petal color polymorphism in *Leavenworthia* (Cruciferae). Contr. Gray Herb. 198: 9–40.

Lloydia = Lloydia; a Quarterly Journal of Biological Science.

Lonard, R. I., J. H. Everitt, and F. W. Judd. 1991. Woody Plants of the Lower Rio Grande Valley, Texas. Austin.

London Edinburgh Philos. Mag. & J. Sci. = The London and Edinburgh Philosophical Magazine and Journal of Science.

London J. Bot. = London Journal of Botany.

Loudon, J. C. [1835–]1838. Arboretum et Fruticetum Britannicum; or, the Trees and Shrubs of Britain, Native and Foreign.... 8 vols. London. (Arbor. Frutic. Brit.)

Lövqvist, B. 1956. The *Cardamine pratensis* complex—Outlines of its cytogenetics and taxonomy. Symb. Bot. Upsal. 14(2): 1–131.

Luken, J. O., J. W. Thieret, and J. T. Kartesz. 1993. *Erucastrum gallicum* (Brassicaceae): Invasion and spread in North America. Sida 15: 569–582.

Lundell, C. L. 1942–1969. Flora of Texas. 3 vols. in parts. Dallas and Renner, Tex.

Lundellia = Lundellia; Journal of the Plant Resources Center of the University of Texas at Austin.

Lyons, E. E. and J. Antonovics. 1991. Breeding system evolution in *Leavenworthia*: Breeding system variation and reproductive success in natural populations of *Leavenworthia crassa* (Cruciferae). Amer. J. Bot. 78: 270–287.

Lysak, M. A. and C. Lexer. 2006. Towards the era of comparative evolutionary genomics in Brassicaceae. Pl. Syst. Evol. 259: 175–198.

Mabberley, D. J. 1997. The Plant-book: A Portable Dictionary of the Vascular Plants..., ed. 2. Cambridge.

Mabry, T. J. 1976. Pigment dichotomy and DNA-RNA hybridization data for centrospermous families. Pl. Syst. Evol. 126: 79–94.

Mabry, T. J. and B. L. Turner. 1964. Chemical investigations of the Batidaceae. Betaxanthins and their systematic implications. Taxon 13: 197–200.

MacMillan, C. 1892. The Metaspermae of the Minnesota Valley. List of the Higher Seed-producing Plants Indigenous to the Drainage-basin of the Minnesota River.... Minneapolis. (Metasp. Minnesota Valley)

Madroño = Madroño; Journal of the California Botanical Society [from vol. 3: a West American Journal of Botany].

Mag. Naturvidensk. = Magazin for Naturvidenskaberne.

Man. Bot. ed. 3—See: A. Eaton 1822

Man. Bot. Rocky Mt.—See: J. M. Coulter [1885]

Man. Fl. Pl. Calif.—See: W. L. Jepson [1923–1925]

Manandhar, N. P. 2002. Plants and People of Nepal. Portland.

Manchester, S. R., W. S. Judd, and B. Handley. 2006. Foliage and fruits of early poplars (Salicaceae: *Populus*) from the Eocene of Utah, Colorado, and Wyoming. Int. J. Pl. Sci. 167: 897–908.

Mant. Pl.—See: C. Linnaeus 1767[–1771]

Manual ed. 2—See: A. Gray 1856

Marcet, E. 1962. Über die geographische Variabilität blattmorphologischer Merkmale bei *Populus deltoides* Bartr. Silvae Genet. 10: 161–172.

Marie-Victorin, Frère 1930. Les variations Laurentiennes du *Populus tremuloides* et du *P. grandidentata*. Contr. Lab. Bot. Univ. Montréal 16.

Marschall von Bieberstein, F. A. 1808–1819. Flora Taurico-Caucasica.... 3 vols. Charkow. (Fl. Taur.-Caucas.)

Marshall, H. 1785. Arbustrum Americanum: The American Grove.... Philadelphia. (Arbust. Amer.)

Martin, R. F. 1940. A review of the cruciferous genus *Selenia*. Amer. Midl. Naturalist 23: 455–462.

Martín-Bravo, S. et al. 2007. Molecular systematics and biogeography of Resedaceae based on ITS and *trn*L-F sequences. Molec. Phylogen. Evol. 44: 1105–1120.

Martín-Bravo, S. et al. 2009. Is *Oligomeris* (Resedaceae) indigenous to North America: Molecular evidence for a natural colonization from the Old World. Amer. J. Bot. 96: 507–518.

Martínez-Laborde, J. B. 1988. Estudio Sistemático del Género *Diplotaxis* DC. (Cruciferae, Brassiceae). Ph.D. dissertation. Universidad Politécnica de Madrid.

Martini, F. and P. Paiero. 1988. I Salici d'Italia..., ed. 2. Trieste.

Maschinski, J. 2001. Impacts of ungulate herbivores on a rare willow at the southern edge of its range. Biol. Conservation 101: 119–130.

Mason, C. T. 1952. A systematic study of the genus *Limnanthes* R. Br. Univ. Calif. Publ. Bot. 25: 455–512.

Mattfield, J. 1939. The species of the genus *Aubrieta* Adanson. Quart. Bull. Alpine Gard. Soc. Gr. Brit. 7: 157–181, 217–227.

Maycock, P. F. and B. Matthews. 1966. An arctic forest in the tundra of northern Ungava, Quebec. Arctic 19: 114–144.

Mayer, M. S. 1994. The evolution of serpentine endemics: A chloroplast DNA phylogeny of the *Streptanthus glandulosus* complex (Cruciferae). Syst. Bot. 19: 557–574.

Mayer, M. S. and P. S. Soltis. 1999. Intraspecific phylogeny analysis using ITS sequences: Insights from studies of the *Streptanthus glandulosus* complex (Cruciferae). Syst. Bot. 24: 47–61.

Mayer, M. S., P. S. Soltis, and D. E. Soltis. 1994. The evolution of the *Streptanthus glandulosus* complex (Cruciferae): Genetic divergence and gene flow in serpentine endemics. Amer. J. Bot. 81: 1288–1299.

McBride, B. 2006. Clammyweed "at large" in Ottawa. Trail & Landscape 40: 25–31.

McGregor, R. L. 1984. *Camelina rumelica*, another weedy mustard established in North America. Phytologia 55: 227–228.

McGregor, R. L. 1985. Current status of the genus *Camelina* (Brassicaceae) in the prairies and plains of central North America. Contr. Univ. Kansas Herb. 15: 1–13.

McNeill, C. I. and S. K. Jain. 1983. Genetic differentiation studies and phylogenetic inference in the plant genus *Limnanthes* (section *Inflexae*). Theor. Appl. Genet. 66: 257–269.

McVaugh, R. and W. R. Anderson, eds. 1974+. Flora Novo-Galiciana: A Descriptive Account of the Vascular Plants of Western Mexico. 8+ vols. Ann Arbor.

Meddel. Grønland = Meddelelser om Grønland, af Kommissionen for Ledelsen af de Geologiske og Geografiske Undersølgeser i Grønland.

Meddel. Grønland, Biosci. = Meddelelser om Grønland. Bioscience.

Meded. Landbouwhoogeschool = Mededeelingen van de Landbouwhoogeschool te Wageningen.

Medikus, F. K. 1792. Pflanzen-Gattungen.... Mannheim. (Pfl.-Gatt.)

Meeuse, A. D. J. 1975. Taxonomic relationships of Salicaceae and Flacourtiaceae: Their bearing on interpretative floral morphology and dilleniid phylogeny. Acta Bot. Neerl. 24: 437–457.

Meikle, R. D. 1984. Willows and Poplars of Great Britain and Ireland. London. [B.S.B.I. Handb. 4.]

Mém. Acad. Imp. Sci. St.-Pétersbourg, Sér. 6, Sci. Math. = Mémoires de l'Académie Impériale des Sciences de St.-Pétersbourg. Sixième Série. Sciences Mathématiques, Physiques et Naturelles.

Mem. Amer. Acad. Arts = Memoirs of the American Academy of Arts and Science.

Mem. Boston Soc. Nat. Hist. = Memoirs Read before the Boston Society of Natural History; Being a New Series of the Boston Journal of Natural History.

Mém. Mus. Hist. Nat. = Mémoires du Muséum d'Histoire Naturelle.

Mem. New York Bot. Gard. = Memoirs of the New York Botanical Garden.

Mém. Soc. Imp. Naturalistes Moscou = Mémoires de la Société Impériale des Naturalistes de Moscou.

Mem. Torrey Bot. Club = Memoirs of the Torrey Botanical Club.

Mem. Tour N. Mexico—See: F. A. Wislizenus 1848

Merriam-Webster. 1988. Webster's New Geographical Dictionary. Springfield, Mass.

Merrill, E. D. 1917. An Interpretation of Rumphius's Herbarium Amboinense.... Manila. (Interpr. Herb. Amboin.)

Metasp. Minnesota Valley—See: C. MacMillan 1892

Methodus—See: C. Moench 1794

Meyer, F. K. 1973. Conspectus der "*Thlaspi*"-Arten Europas, Afrikas und Vorderasiens. Feddes Repert. 84: 449–470.

Meyer, F. K. 1979. Kritische Revision der "*Thlaspi*"-Arten Europas, Afrikas und Vorderasiens: I. Geschichte, Morphologie und Chorologie. Feddes Repert. 90: 129–154.

Meyer, F. K. 1982. Was ist *Hutchinsia* R. Br. in Ait.? Wiss. Z. Friedrich-Schiller-Univ. Jena, Math.-Naturwiss. Reihe 31: 267–276.

Meyer, F. K. 2001. Kritische Revision der "*Thlaspi*"-Arten Europas, Afrikas und Vorderasiens. Spezieller Teil. I. *Thlaspi* L. Häussknechtia 8: 3–42.

Meyer, F. K. 2003. Kritische Revision der "*Thlaspi*"-Arten Europas, Afrikas und Vorderasiens. Spezieller Teil. III. *Microthlaspi* F. K. Mey. Häussknechtia 9: 3–59.

Meyerowitz, E. M. and C. R. Somerville, eds. 1994. *Arabidopsis*. Plainview, N.Y.

Michaux, A. 1803. Flora Boreali-Americana.... 2 vols. Paris and Strasbourg. (Fl. Bor.-Amer.)

Michigan Bot. = Michigan Botanist.

Miller, P. 1754. The Gardeners Dictionary.... Abridged..., ed. 4. 3 vols. London. (Gard. Dict. Abr. ed. 4)

Miller, P. 1768. The Gardeners Dictionary..., ed. 8. London. (Gard. Dict. ed. 8)

Minnesota Forest. Notes = Minnesota Forestry Notes.

Misc. Austriac.—See: N. J. Jacquin 1778–1781

Mitchell-Olds, T., I. A. Al-Shehbaz, M. Koch, and T. F. Sharbel. 2005. Crucifer evolution in the post-genomic era. In: R. J. Henry, ed. 2005. Plant Diversity and Evolution: Genotypic and Phenotypic Variation in Higher Plants. Wallingford and Cambridge, Mass. Pp. 119–137.

Mitra, K. and S. N. Mitra. 1979. Pollen morphology in relation to taxonomy and geography of Resedaceae. Bull. Bot. Surv. India 18: 194–202.

Mitt. Bot. Mus. Univ. Zürich = Mitteilungen aus dem Botanischen Museum der Universität Zürich.

Moench, C. 1794. Methodus Plantas Horti Botanici et Agri Marburgensis.... Marburg. (Methodus)

Moench, C. 1802. Supplementum ad Methodum Plantas.... Marburg. (Suppl. Meth.)

Mohlenbrock, R. H. 1980. Flowering Plants, Willows to Mustards. Carbondale.

Molec. Biol. Evol. = Molecular Biology and Evolution.

Molec. Ecol. = Molecular Ecology.

Molec. Phylogen. Evol. = Molecular Phylogenetics and Evolution.

Monogr. Salicum—See: N. J. Andersson 1867

Montgomery, F. H. 1955. Preliminary studies in the genus *Dentaria* in eastern North America. Rhodora 57: 161–173.

Moore, E. 1909. The study of winter buds with reference to their growth and leaf contents. Bull. Torrey Bot. Club 36: 117–145.

Mosseler, A. 1989. Interspecific pollen-pistil incongruity in *Salix*. Canad. J. Forest Res. 19: 1161–1168.

Mosseler, A. 1990. Hybrid performance and species crossability relationships in willows (*Salix*). Canad. J. Bot. 68: 2329–2338.

Mosseler, A. and C. S. Papadopol. 1989. Seasonal isolation as a reproductive barrier among sympatric *Salix* species. Canad. J. Bot. 67: 2563–2570.

Muhlenbergia = Muhlenbergia; a Journal of Botany.

Müller Argoviensis, J. 1857. Monographie de la Famille des Résédacées.... Zürich.

Müller Argoviensis, J. 1868. Resedaceae. In: A. P. de Candolle and A. L. P. P. de Candolle, eds. 1823–1873. Prodromus Systematis Naturalis Regni Vegetabilis.... 17 vols. Paris etc. Vol. 16(2), pp. 548–589.

Mulligan, G. A. 1961. The genus *Lepidium* in Canada. Madroño 16: 77–89.

Mulligan, G. A. 1968. *Physaria didymocarpa*, *P. brassicoides*, and *P. floribunda* (Cruciferae) and their close relatives. Canad. J. Bot. 46: 735–740.

Mulligan, G. A. 1970. Cytotaxonomic studies of *Draba glabella* and its close allies in Canada and Alaska. Canad. J. Bot. 48: 1431–1437.

Mulligan, G. A. 1971. Cytotaxonomic studies of the closely allied *Draba cana*, *D. cinerea*, and *D. groenlandica* in Canada and Alaska. Canad. J. Bot. 49: 89–93.

Mulligan, G. A. 1971b. Cytotaxonomic studies of *Draba* species of Canada and Alaska: *D. ventosa*, *D. ruaxes*, and *D. paysonii*. Canad. J. Bot. 49: 1455–1460.

Mulligan, G. A. 1972. Cytotaxonomic studies of *Draba* species in Canada and Alaska: *D. oligosperma* and *D. incerta*. Canad. J. Bot. 50: 1763–1766.

Mulligan, G. A. 1974. Cytotaxonomic studies of *Draba nivalis* and its close allies in Canada and Alaska. Canad. J. Bot. 52: 1793–1801.

Mulligan, G. A. 1974b. Confusion in the names of three *Draba* species of the Arctic: *D. adamsii*, *D. oblongata*, and *D. corymbosa*. Canad. J. Bot. 52: 791–793.

Mulligan, G. A. 1975. *Draba crassifolia*, *D. albertina*, *D. nemorosa*, and *D. stenoloba* in Canada and Alaska. Canad. J. Bot. 53: 745–751.

Mulligan, G. A. 1976. The genus *Draba* in Canada and Alaska: Key and summary. Canad. J. Bot. 54: 1386–1393.

Mulligan, G. A. 1978. *Barbarea stricta* Andrz., a new introduction to Quebec. Naturaliste Canad. 105: 297–298.

Mulligan, G. A. 1996. Synopsis of the genus *Arabis* (Brassicaceae) in Canada, Alaska and Greenland. Rhodora 97: 109–163.

Mulligan, G. A. 2002. Chromosome numbers determined from Canadian and Alaskan material of native and naturalized mustards, Brassicaceae (Cruciferae). Canad. Field-Naturalist 116: 611–622.

Mulligan, G. A. 2002b. Weedy introduced mustards (Brassicaceae) of Canada. Canad. Field-Naturalist 116: 623–631.

Mulligan, G. A. and L. G. Bailey. 1975. The biology of Canadian weeds. 8. *Sinapis arvensis* L. Canad. J. Pl. Sci. 55: 171–183.

Mulligan, G. A. and J. A. Calder. 1964. The genus *Subularia* (Cruciferae). Rhodora 66: 127–135.

Mulligan, G. A. and W. J. Cody. 1995. New information on the problem of Asiatic cress, *Rorippa crystallina* Rollins (Brassicaceae). Canad. Field-Naturalist 109: 111–112.

Mulligan, G. A. and J. N. Findlay. 1970. Sexual reproduction and agamospermy in the genus *Draba*. Canad. J. Bot. 48: 269–270.

Mulligan, G. A. and C. Frankton. 1962. Taxonomy of the genus *Cardaria* with particular reference to the species introduced into North America. Canad. J. Bot. 40: 1411–1425.

Mulligan, G. A. and A. E. Porsild. 1966. *Rorippa calycina* in the Northwest Territories. Canad. J. Bot. 44: 1105–1106.

Mulligan, G. A. and A. E. Porsild. 1968. A natural first-generation hybrid between *Rorippa barbareifolia* and *R. islandica*. Canad. J. Bot. 46: 1079–1081.

Mummenhoff, K., H. Brüggemann, and J. L. Bowman. 2001. Chloroplast DNA phylogeny and biogeography of *Lepidium* (Brassicaceae). Amer. J. Bot. 88: 2051–2063.

Mummenhoff, K., G. Eschmann-Grupe, and K. Zunk. 1993. Subunit polypeptide composition of RUBISCO indicates *Diplotaxis viminea* as maternal parent species of amphidiploid *Diplotaxis muralis*. Phytochemistry 34: 429–431.

Mummenhoff, K., A. Franzke, and M. Koch. 1997. Molecular phylogenetics of *Thlaspi* s.l. (Brassicaceae) based on chloroplast DNA restriction site variation and sequences of the internal transcribed spacers of nuclear ribosomal DNA. Canad. J. Bot. 75: 469–482.

Murray, D. F. and C. L. Parker. 1999. On the taxonomic disposition of the whitlow-grass *Draba ogilviensis* Hultén. Canad. Field-Naturalist 113: 659–662.

Muschler, R. 1908. Die Gattung *Coronopus* (L.) Gaertn. Bot. Jahrb. Syst. 41: 111–147.

N. Amer. Sylv.—See: T. Nuttall 1842–1849

Nakai, T. 1928. Une nouvelle systématique des Salicacées de Corée. Bull. Soc. Dendrol. France 66: 37–51.

Nandi, O. I., M. W. Chase, and P. K. Endress. 1998. A combined cladistic analysis of angiosperms using *rbc*L and non-molecular data sets. Ann. Missouri Bot. Gard. 85: 137–212.

Narr. Journey Polar Sea—See: J. Franklin et al. 1823

Nat. Pflanzenfam.—See: H. G. A. Engler and K. Prantl 1887–1915

Naturaliste = Le Naturaliste; Journal des Échanges et des Nouvelles. [Sér. 2: Revue Illustrée des Sciences Naturelles.]

Naturaliste Canad. = Naturaliste Canadien. Bulletin de Recherches, Observations et Découvertes se Rapportant à l'Histoire Naturelle du Canada.

Nauchnye Dokl. Vysshei Shkoly Biol. Nauki = Nauchnye Doklady Vysshei Shkoly. Biologicheskie Nauki.

Neue Denkschr. Schweiz. Naturf. Ges. = Neue Denkschriften der Schweizerischen Naturforschenden Gesellschaft.

Neuffer, B. and H. Hurka. 1999. Colonizing history and introduction dynamics of *Capsella bursa-pastoris* in North America: Isozymes and quantitative traits. Molec. Ecol. 8: 1667–1681.

New Phytol. = New Phytologist; a British Botanical Journal.

New Zealand J. Forest. Sci. = New Zealand Journal of Forestry Science.

Nomencl. Bot. ed. 2—See: E. G. Steudel 1840–1841

Nordenskiöld, A. E. 1882–1887. Vega-expeditionens Vetenskapliga Iakttagelser Bearbetade af Deltagare i Resan och Andra Forskare.... 5 vols. Stockholm. (Vega Exp. Vetensk. Iakttag.)

Nordic J. Bot. = Nordic Journal of Botany.

Notizbl. Bot. Gart. Berlin-Dahlem = Notizblatt des Botanischen Gartens und Museums zu Berlin-Dahlem.

Nouv. Mém. Soc. Imp. Naturalistes Moscou = Nouveaux Mémoires de la Société Impériale des Naturalistes de Moscou.

Nov. Gen. Sp.—See: A. von Humboldt et al. 1815[1816]–1825

Novon = Novon; a Journal for Botanical Nomenclature.

Novosti Sist. Vyssh. Rast. = Novosti Sistematiki Vysshikh Rastenii.

Nutt, P. et al. 2006. *Capsella* as a model system to study the evolutionary relevance of floral homeotic mutants. Pl. Syst. Evol. 259: 217–235.

Nuttall, T. 1818. The Genera of North American Plants, and Catalogue of the Species, to the Year 1817.... 2 vols. Philadelphia. (Gen. N. Amer. Pl.)

Nuttall, T. 1842–1849. The North American Sylva.... 3 vols. Philadelphia. (N. Amer. Sylv.)

Obedzinski, R. A., C. G. Shaw, and D. G. Neary. 2001. Declining woody vegetation in riparian ecosystems of the western United States. W. J. Appl. Forest. 16: 169–181.

Oeder, G. C. et al., eds. [1761–]1764–1883. Icones Plantarum... Florae Danicae Nomine Inscriptum. 17 vols. in 51 fascs. Copenhagen. [Fascs. paged independently and numbered consecutively throughout vols.] (Fl. Dan.)

Oekon. Fl. Wetterau—See: P. G. Gaertner et al. 1799–1802

Oesterr. Bot. Z. = Oesterreichische botanische Zeitschrift. Gemeinütziges Organ für Botanik.

Öfvers. Kongl. Vetensk.-Akad. Förh. = Öfversigt af Kongl. Vetenskaps-Akademiens Förhandlingar.

Ohashi, H. 2001. Salicaceae of Japan. Sci. Rep. Tohoku Imp. Univ., Ser. 4, Biol. 40: 269–396.

O'Kane, S. L. and I. A. Al-Shehbaz. 1997. A synopsis of *Arabidopsis* (Brassicaceae). Novon 7: 323–327.

O'Kane, S. L. and I. A. Al-Shehbaz. 2003. Phylogenetic position and generic limits of *Arabidopsis* (Brassicaceae) based on sequences of nuclear ribosomal DNA. Ann. Missouri Bot. Gard. 90: 603–612.

Olson, M. E. 2002. Intergeneric relationships within the Caricaceae-Moringaceae clade (Brassicales), and potential morphological synapomorphies of the clade and its families. Int. J. Pl. Sci. 163: 51–65.

Olson, M. E. 2002b. Combining data from DNA sequences and morphology for a phylogeny of Moringaceae. Syst. Bot. 27: 55–73.

Olson, M. E. 2003. Developmental origins of floral bilateral symmetry in Moringaceae. Amer. J. Bot. 90: 49–71.

Opera Bot. = Opera Botanica a Societate Botanice Lundensi.

Organization for Flora Neotropica. 1968+. Flora Neotropica. 98+ nos. New York.

Ornduff, R. 1971. Systematic studies of Limnanthaceae. Madroño 21: 103–111.

Ornduff, R. 1974. An Introduction to California Plant Life. Berkeley.

Ornduff, R. and T. J. Crovello. 1968. Numerical taxonomy of Limnanthaceae. Amer. J. Bot. 55: 173–182.

Ottawa Naturalist = Ottawa Naturalist; Transactions of the Ottawa Field-Naturalists' Club.

Pacif. Railr. Rep.—See: War Department 1855–1860

Pallas, P. S. 1771–1776. Reise durch verschiedene Provinzen des russischen Reichs.... 3 vols. St. Petersburg. (Reise Russ. Reich.)

Pallas, P. S. 1784–1788[–1831]. Flora Rossica seu Stirpium Imperii Rossici par Europam et Asiam Indigenarum.... 2 vols. in 3 parts. St. Petersburg. [Parts 1 and 2 of vol. 1 paged independently.] (Fl. Ross.)

Palo, R. T. 1984. Distribution of birch (*Betula* spp.), willow (*Salix* spp.), and poplar (*Populus* spp.) secondary metabolites and their potential role as chemical defense against herbivores. J. Chem. Ecol. 10: 499–520.

Panasahatham, S., G. C. Fisher, J. T. DeFrancesco, and D. T. Ehrensing. 1999. Seasonal development of meadowfoam fly, *Scaptomyza apicalis* Hardy (Diptera: Drosophilidae) in the Willamette Valley. In: W. C. Young, ed. 1999. 1998 Seed Production Research at Oregon State University, USDA-ARS Cooperating. Corvallis. Pp. unnumb.

Panetsos, C. A. and H. G. Baker. 1967. The origin of variation in wild *Raphanus sativus* (Cruciferae) in California. Genetica 38: 243–274.

Pankhurst, R. J. 1991. Practical Taxonomic Computing. Cambridge.

Parker, W. H. 1981. Contrasting patterns of U.V. absorbance/reflectance in *Limnanthes* flowers: A novel mechanism of elaboration and evolutionary significance. [Abstract.] In: G. G. E. Scudder and J. L. Reveal, eds. 1981. Evolution Today: Proceedings of the Second International Congress of Systematic and Evolutionary Biology, University of British Columbia, Vancouver, Canada, 17–24 July, 1980. Pittsburgh. P. 304.

Parker, W. H. and B. A. Bohm. 1979. Flavonoids and taxonomy of Limnanthaceae. Amer. J. Bot. 66: 191–197.

Parolin, P. et al. 2002. Pioneer trees in Amazonia floodplains: Three key species form monospecific stands in different habitats. Folia Geobot. 37: 225–238.

Parry, W. E. 1825. Appendix to Captain Parry's Journal of a Second Voyage for the Discovery of a North-west Passage...1821-22-23. London. (App. Parry J. Sec. Voy.)

Parsons, K. A. 2002. Reproductive Biology and Floral Variation in the Endangered *Braya longii* and Threatened *B. fernaldii* (Brassicaceae): Implications for Conservation Management of Rare Plants. M.S. thesis. Memorial University of Newfoundland.

Patterson, D. T. et al. 1989. Composite List of Weeds. Champaign.

Pauley, S. S. 1957. Natural hybridization of the aspens. Minnesota Forest. Notes 47: 1–2.

Pax, F. A. and K. Hoffmann. 1936. Capparidaceae. In: H. G. A. Engler et al., eds. 1924+. Die natürlichen Pflanzenfamilien..., ed. 2. 26+ vols. Leipzig and Berlin. Vol. 17b, pp. 146–233.

Payson, E. B. 1917. The perennial scapose drabas of North America. Amer. J. Bot. 4: 253–267.

Payson, E. B. 1921. A monograph of the genus *Lesquerella*. Ann. Missouri Bot. Gard. 8: 103–236.

Payson, E. B. 1922. Species of *Sisymbrium* native to America north of Mexico. Univ. Wyoming Publ. Sci., Bot. 1(1): 1–27.

Payson, E. B. 1922b. A synoptical revision of the genus *Cleomella*. Univ. Wyoming Publ. Sci., Bot. 1: 29–46.

Payson, E. B. 1923. A monographic study of *Thelypodium* and its immediate allies. Ann. Missouri Bot. Gard. 9: 233–324.

Payson, E. B. 1926. The genus *Thlaspi* in North America. Univ. Wyoming Publ. Sci., Bot. 1(6): 145–163.

Pegtel, D. M. 1999. Effect of ploidy level on fruit morphology, seed germination and juvenile growth in scurvy grass (*Cochlearia officinalis* L. s.l., (Brassicaceae). Pl. Spec. Biol. 14: 201–215.

Pepper, A. E. and L. E. Norwood. 2001. Evolution of *Caulanthus amplexicaulis* var. *barbarae* (Brassicaceae), a rare serpentine endemic plant: A molecular phylogenetic perspective. Amer. J. Bot. 88: 1479–1489.

Perkins, K. D. and W. W. Payne. 1978. Guide to the Poisonous and Irritant Plants of Florida. Gainesville.

Persoon, C. H. 1805–1807. Synopsis Plantarum.... 2 vols. Paris and Tubingen. (Syn. Pl.)

Pfl.-Gatt.—See: F. K. Medikus 1792

Pflanzenr.—See: H. G. A. Engler 1900–1953

Physiogr. Sällsk. Årsberätt. = Physiographiska Sällskapets Årsberättelse.

Physiol. Pl. (Copenhagen) = Physiologia Plantarum.

Phytologia = Phytologia; Designed to Expedite Botanical Publication.

Phytonom. Univ.—See: J. Bergeret 1783–1784[–1786]

Pl. Baker.—See: E. L. Greene 1901

Pl. Biol. (Stuttgart) = Plant Biology. [Stuttgart.]

Pl. Ecol. = Plant Ecology.

Pl. Hartw.—See: G. Bentham 1839[–1857]

Pl. Jamaic. Pug.—See: C. Linnaeus 1759

Pl. Rar. Neapol.—See: D. M. L. Cirillo 1788–1792

Pl. Spec. Biol. = Plant Species Biology; an International Journal.

Pl. Syst. Evol. = Plant Systematics and Evolution.

Pl. Wilson.—See: C. S. Sargent 1913–1917

Pl. World = Plant World.

Planta = Planta. Archiv für wissenschaftliche Botanik. (Zeitschrift für wissenschaftliche Biologie. Abt. E.)

Plotkin, M. S. 1998. Phylogeny and Biogeography of Limnanthaceae. M.S. thesis. University of California, Davis.

Pobedimova, E. 1969. Revisio generis *Cochlearia* L., 1. Novosti Sist. Vyssh. Rast. 6: 67–106.

Pobedimova, E. 1970. Revisio generis *Cochlearia* L., 2. Novosti Sist. Vyssh. Rast. 7: 167–195.

Pollen & Spores = Pollen et Spores.

Polunin, N. 1940. Botany of the Canadian eastern Arctic. Part 1. Pteridophyta and Spermatophyta. Bull. Natl. Mus. Canada 92: 1–408.

Polunin, N. 1940b. The flora of Devon Island in arctic Canada. Canad. Field-Naturalist 54: 31–37.

Polunin, N. 1943. Contributions to the flora and phytogeography of south-western Greenland: An enumeration of the vascular plants, with critical notes. J. Linn. Soc., Bot. 52: 349–406.

Prakash, S. and K. Hinata. 1980. Taxonomy, cytogenetics and origin of crop brassicas, a review. Opera Bot. 55: 1–57.

Prelim. Cat.—See: N. L. Britton et al. 1888

Prelim. Rep. Expl. Nebraska Dakota—See: G. K. Warren 1859

Presl, C. B. 1826. Flora Sicula, Exhibens Plantas Vasculosas in Sicilia.... Prague. (Fl. Sicul.)

Preslia = Preslia. Věstník (Časopis) Československé Botanické Společnosti.

Price, R. A. 1987. Systematics of the *Erysimum capitatum* Alliance (Brassicaceae) in North America. Ph.D. dissertation. University of California, Berkeley.

Price, R. A. 1993. *Erysimum*. In: J. C. Hickman, ed. 1993. The Jepson Manual. Higher Plants of California. Berkeley, Los Angeles, and London. Pp. 421–422.

Price, R. A. and I. A. Al-Shehbaz. 2001. A reconsideration of *Chaunanthus* (Brassicaceae). Novon 11: 329–331.

Price, R. A., J. D. Palmer, and I. A. Al-Shehbaz. 1994. Systematic relationship of *Arabidopsis*: A molecular and morphological approach. In: E. M. Meyerowitz and C. R. Somerville, eds. 1994. *Arabidopsis*. Plainview, N.Y. Pp. 7–19.

Price, R. A. and R. C. Rollins. 1991. New taxa of *Draba* (Cruciferae) from California, Nevada, and Colorado. Harvard Pap. Bot. 1(3): 71–77.

Prijanto, B. 1970. Batidaceae. World Pollen Fl. 3: 1–15.

Prir. Rostlin—See: F. Berchtold and J. S. Presl 1823–1835

Proc. Acad. Nat. Sci. Philadelphia = Proceedings of the Academy of Natural Sciences of Philadelphia.

Proc. Amer. Acad. Arts = Proceedings of the American Academy of Arts and Sciences.

Proc. Biol. Soc. Wash. = Proceedings of the Biological Society of Washington.

Proc. Calif. Acad. Sci. = Proceedings of the California Academy of Sciences.

Proc. Montana Acad. Sci. = Proceedings of the Montana Academy of Sciences.

Proc. Natl. Acad. Sci. U.S.A. = Proceedings of the National Academy of Sciences of the United States of America.

Proc. Ohio Acad. Sci. = Proceedings of the Ohio Academy of Science.

Proc. Roy. Soc. Edinburgh, B = Proceedings of the Royal Society of Edinburgh. Series B, Biology [later: Biological Sciences].

Proc. Utah Acad. Sci. = Proceedings of the Utah Academy of Sciences.

Proc. Wash. Acad. Sci. = Proceedings of the Washington Academy of Sciences.

Prodr.—See: A. P. de Candolle and A. L. P. P. de Candolle 1823–1873

Prodr. Stirp. Chap. Allerton—See: R. A. Salisbury 1796

Prosp. Hist. Pl. Dauphiné—See: D. Villars 1779

Publ. Bot. (Ottawa) = Publications in Botany, National Museum of Natural Sciences, Canada.

Publ. Bussey Inst. Harvard Univ. = Publicaitons of the Bussey Institution, Harvard University.

Publ. Dept. Forest. Rural Developm. Canada = Publications, Department of Forestry and Rural Development, Canada.

Publ. Field Columbian Mus., Bot. Ser. = Publications of the Field Columbian Museum. Botanical Series.

Purdy, B. G. and R. J. Bayer. 1995. Allozyme variation in the Athabasca sand dune endemic, *Salix silicicola*, and the closely related, widespread species, *S. alaxensis*. Syst. Bot. 20: 179–190.

Pursh, F. [1813]1814. Flora Americae Septentrionalis; or, a Systematic Arrangement and Description of the Plants of North America. 2 vols. London. (Fl. Amer. Sept.)

Quart. Bull. Alpine Gard. Soc. Gr. Brit. = Quarterly Bulletin of the Alpine Garden Society of Great Britain.

Quatern. Res. = Quaternary Research; Interdisciplinary Journal.

Rafinesque, C. S. 1817. Florula Ludoviciana; or, a Flora of the State of Louisiana. Translated, Revised, and Improved, from the French of C. C. Robin.... New York. (Fl. Ludov.)

Rafinesque, C. S. 1838b. Sylva Telluriana. Mantis. Synopt. ...Being a Supplement to the Flora Telluriana. Philadelphia. (Sylva Tellur.)

Rafinesque, C. S. 1840. Autikon Botanikon. 3 parts. Philadelphia. [Parts paged consecutively.] (Autik. Bot.)

Raup, H. M. 1943. The willows of the Hudson Bay Region and the Labrador Peninsula. Sargentia 4: 81–135.

Raup, H. M. 1959. The willows of boreal western America. Contr. Gray Herb. 185: 1–95.

Raup, H. M. 1965. The flowering plants and ferns of the Mesters Vig District, northeast Greenland. Meddel. Grønland 166(2): 1–119.

Recherches Anat. Taxon. Fam. Crucifèr.—See: E. Fournier 1865

Rechinger, K. H. 1964b. *Salix*. In: T. G. Tutin et al., eds. 1964–1980. Flora Europaea. 5 vols. Cambridge. Vol. 1, pp. 43–55.

Rechinger, K. H. 1993. *Salix* (rev. J. R. Akeroyd). In: T. G. Tutin et al., eds. 1993+. Flora Europaea, ed. 2. 1+ vol. Cambridge and New York. Vol. 1, pp. 53–64.

Reed, C. F. 1965b. *Cleome ornithopodioides* L. on vanadium-slag at Canton, Baltimore, Maryland with notes on the biochemistry of vanadium. Phytologia 11: 423–431.

Reeves, R. D., R. R. Brooks, and R. M. Macfarlane. 1981. Nickel uptake by Californian *Streptanthus* and *Caulanthus* with particular reference to the hyperaccumulator *S. polygaloides* Gray (Brassicaceae). Amer. J. Bot. 68: 708–712.

Reichenbach, H. G. L. 1830[–1832]. Flora Germanica Excursoria.... 2 parts. Leipzig. [Parts paged consecutively.] (Fl. Germ. Excurs.)

Reichenbach, H. G. L. et al. 1837–1870. Deutschlands Flora.... 24 vols. Leipzig. [German-language issue of Icon. Fl. Germ. Helv. vols. 2–25; some corresp. vols. publ. simultaneously, others not.] (Deutschl. Fl.)

Reise Russ. Reich.—See: P. S. Pallas 1771–1776

Rep. (Annual) Missouri Bot. Gard. = Report (Annual) of the Missouri Botanical Garden.

Rep. Colorado R.—See: J. C. Ives 1861

Rep. Exped. Rocky Mts.—See: J. C. Frémont 1843–1845

Rep. U.S. Geogr. Surv., Wheeler—See: J. T. Rothrock 1878[1879]

Rep. U.S. Mex. Bound.—See: W. H. Emory 1857–1859

Repert. Bot. Syst.—See: W. G. Walpers 1842–1847

Repert. Spec. Nov. Regni Veg. = Repertorium Specierum Novarum Regni Vegetabilis.

Repert. Spec. Nov. Regni Veg. Beih. = Repertorium Specierum Novarum Regni Vegetabilis. Beihefte.

Res. Stud. State Coll. Wash. = Research Studies of the State College of Washington.

Revis. Drabas W. N. Amer.—See: C. L. Hitchcock 1941

Revis. Gen. Pl.—See: O. Kuntze 1891–1898

Rhodora = Rhodora; Journal of the New England Botanical Club.

Ritland, K. and S. K. Jain. 1984. A comparative study of floral and electrophoretic variation with life history variation in *Limnanthes alba* (Limnanthaceae). Oecologia 63: 243–251.

Robinson, B. L. 1896. The fruit of *Tropidocarpum*. Erythea 4: 108–119.

Rodman, J. E. 1974. Systematics and evolution of the genus *Cakile* (Cruciferae). Contr. Gray Herb. 205: 3–146.

Rodman, J. E. 1980. Population variation and hybridization in sea-rockets (*Cakile*, Cruciferae): Seed glucosinolate characters. Amer. J. Bot. 67: 1145–1159.

Rodman, J. E. 1986. Introduction, establishment and replacement of sea-rockets (*Cakile*, Cruciferae) in Australia. J. Biogeogr. 13: 159–171.

Rodman, J. E. et al. 1993. Nucleotide sequences of the *rbc*L gene indicate monophyly of mustard oil plants. Ann. Missouri Bot. Gard. 80: 686–699.

Rodman, J. E. et al. 1998. Parallel evolution of glucosinolate biosynthesis inferred from congruent nuclear and plastid gene phylogenies. Amer. J. Bot. 85: 997–1006.

Rodman, J. E., K. G. Karol, R. A. Price, and K. J. Sytsma. 1996. Molecules, morphology, and Dahlgren's expanded order Capparales. Syst. Bot. 21: 289–307.

Rogers, G. K. 1982c. The Bataceae in the southeastern United States. J. Arnold Arbor. 63: 375–386.

Röhling, J. C., F. C. Mertens, and W. D. J. Koch. 1823–1839. Deutschlands Flora, ed. 3. 5 vols. Frankfurt am Main. (Deutschl. Fl. ed. 3)

Rollins, R. C. 1936. The genus *Arabis* in the Pacific Northwest. Res. Stud. State Coll. Wash. 4: 1–52.

Rollins, R. C. 1938. *Smelowskia* and *Polyctenium*. Rhodora 40: 294–305.

Rollins, R. C. 1939. The cruciferous genus *Physaria*. Rhodora 41: 392–415.

Rollins, R. C. 1939b. Studies in the genus *Lesquerella*. Amer. J. Bot. 26: 419–421.

Rollins, R. C. 1939c. The cruciferous genus *Stanleya*. Lloydia 2: 109–127.

Rollins, R. C. 1940. On two weedy Crucifers. Rhodora 42: 302–306.

Rollins, R. C. 1941. A monographic study of *Arabis* in western North America. Rhodora 43: 289–325, 348–411, 425–485.

Rollins, R. C. 1941b. A revision of *Lyrocarpa*. Contr. Dudley Herb. 3: 169–173.

Rollins, R. C. 1941c. The cruciferous genus *Dryopetalon*. Contr. Dudley Herb. 3: 199–207.

Rollins, R. C. 1942. A systematic study of *Iodanthus*. Contr. Dudley Herb. 3: 209–215.

Rollins, R. C. 1943. Generic revisions in the Cruciferae: *Halimolobos*. Contr. Dudley Herb. 3: 241–265.

Rollins, R. C. 1947. Generic revisions in the Cruciferae: *Sibara*. Contr. Gray Herb. 165: 133–143.

Rollins, R. C. 1955. The auriculate-leaved species of *Lesquerella* (Cruciferae). Rhodora 57: 241–264.

Rollins, R. C. 1958. The genetic evaluation of a taxonomic character in *Dithyrea* (Cruciferae). Rhodora 60: 145–152.

Rollins, R. C. 1959. The genus *Synthlipsis* (Cruciferae). Rhodora 61: 253–264.

Rollins, R. C. 1961. A weedy crucifer again reaches North America. Rhodora 63: 345–346.

Rollins, R. C. 1961b. Notes on American *Rorippa* (Cruciferae). Rhodora 63: 1–10.

Rollins, R. C. 1963. The evolution and systematics of *Leavenworthia* (Cruciferae). Contr. Gray Herb. 192: 3–98.

Rollins, R. C. 1966. Chromosome numbers of Cruciferae. Contr. Gray Herb. 197: 43–65.

Rollins, R. C. 1973. A reconsideration of *Thelypodium jaegeri*. Contr. Gray Herb. 204: 155–157.

Rollins, R. C. 1976. Studies on Mexican Cruciferae. Contr. Gray Herb. 206: 3–18.

Rollins, R. C. 1978. Watercress in Florida. Rhodora 80: 147–153.

Rollins, R. C. 1979. *Dithyrea* and a related genus (Cruciferae). Publ. Bussey Inst. Harvard Univ. 1979: 3–32.

Rollins, R. C. 1980. Another cruciferous weed establishes itself in North America. Contr. Gray Herb. 210: 1–3.

Rollins, R. C. 1980b. The genus *Pennellia* (Cruciferae) in North America. Contr. Gray Herb. 210: 5–21.

Rollins, R. C. 1981. Weeds of the Cruciferae (Brassicaceae) in North America. J. Arnold Arbor. 62: 517–561.

Rollins, R. C. 1981b. Studies in the genus *Physaria* (Cruciferae). Brittonia 33: 332–341.

Rollins, R. C. 1982. Another alien in the California flora. Rhodora 84: 153–154.

Rollins, R. C. 1982b. *Thelypodiopsis* and *Schoenocrambe* (Cruciferae). Contr. Gray Herb. 212: 71–102.

Rollins, R. C. 1988. A population of interspecific hybrids of *Lesquerella* (Cruciferae). Syst. Bot. 13: 60–63.

Rollins, R. C. 1993. The Cruciferae of Continental North America: Systematics of the Mustard Family from the Arctic to Panama. Stanford. (Cruciferae Continental N. Amer.)

Rollins, R. C. 1993b. *Brassica*. In: J. C. Hickman, ed. 1993. The Jepson Manual. Higher Plants of California. Berkeley, Los Angeles, and London. P. 406.

Rollins, R. C. and I. A. Al-Shehbaz. 1986. Weeds of southwest Asia in North America with special reference to the Cruciferae. Proc. Roy. Soc. Edinburgh, B 89: 289–299.

Rollins, R. C. and U. C. Banerjee. 1975. Atlas of the trichomes of *Lesquerella* (Cruciferae). Publ. Bussey Inst. Harvard Univ. 1975: 1–48.

Rollins, R. C. and U. C. Banerjee. 1976. Trichomes in studies of the Cruciferae. In: J. G. Vaughan et al., eds. 1976. The Biology and Chemistry of the Cruciferae. London and New York. Pp. 145–166.

Rollins, R. C. and U. C. Banerjee. 1979. Pollen of the Cruciferae. Publ. Bussey Inst. Harvard Univ. 1979: 33–64.

Rollins, R. C. and U. C. Banerjee. 1979b. Trichome patterns in *Physaria*. Publ. Bussey Inst. Harvard Univ. 1979: 65–77.

Rollins, R. C. and E. A. Shaw. 1973. The Genus *Lesquerella* (Cruciferae) in North America. Cambridge, Mass. (Gen. Lesquerella)

Ronald, W. G., L. M. Lenz, and W. A. Cumming. 1973. Biosystematics of the genus *Populus* L. I. Distribution and morphology of native Manitoba species and variants. Canad. J. Bot. 51: 2431–2442.

Ronald, W. G. and J. W. Steele. 1974. Biosystematics of the genus *Populus* L. III. Naturally occurring Manitoba hybrids of introduced P. ×petrowskyana with native P. deltoides var. occidentalis and P. balsamifera. Canad. J. Bot. 52: 1883–1887.

Rood, S. B., J. S. Campbell, and T. Despins. 1985. Natural poplar hybrids from southern Alberta. I. Continuous variation for foliar characteristics. Canad. J. Bot. 64: 1382–1388.

Rood, S. B., C. Hillman, T. Sanche, and J. M. Mahoney. 1994. Clonal reproduction of riparian cottonwoods in southern Alberta. Canad. J. Bot. 72: 1766–1774.

Rossbach, G. B. 1940. *Erysimum* in North America. Ph.D. dissertation. Stanford University.

Rossbach, G. B. 1958. The genus *Erysimum* (Cruciferae) in North America north of Mexico—A key to the species and varieties. Madroño 14: 261–267.

Rossbach, G. B. 1958b. New taxa and new combinations in the genus *Erysimum* in North America. Aliso 4: 115–1246.

Rothrock, J. T. 1878[1879]. Report upon United States Geographical Surveys West of the One Hundredth Meridian, in Charge of First Lieut. Geo. M. Wheeler.... Vol. 6—Botany. Washington. (Rep. U.S. Geogr. Surv., Wheeler)

Rouleau, E. 1944. Notes taxonomiques sur la flore phanérogamique du Québec—I. Naturaliste Canad. 71: 265–272.

Rouleau, E. 1948. Two new names in *Populus*. Rhodora 50: 233–236.

Rouy, G. 1895–1905. Illustrationes Plantarum Europae Rariorum. 20 fasc. Paris. [Fascs. paged and plates numbered consecutively.] (Ill. Pl. Eur.)

Rouy, G., J. Foucaud, and E. G. Camus. 1893–1913. Flore de France.... 14 vols. Asnières. (Fl. France)

Rowlee, W. W. 1900. North American willows. I. *Longifoliae*. Bull. Torrey Bot. Club 27: 247–257.

Rowlee, W. W. and K. M. Wiegand. 1896. *Salix candida* Willd. and its hybrids. Bull. Torrey Bot. Club 23: 194–201.

Rydberg, P. A. 1917. Flora of the Rocky Mountains and Adjacent Plains. New York. (Fl. Rocky Mts.)

Rydberg, P. A. 1923. Flora of the Rocky Mountains and Adjacent Plains, ed. 2. New York. (Fl. Rocky Mts. ed. 2)

Sabourin, A. et al. 1991. Guide des Crucifères Sauvages de l'Est du Canada (Québec, Ontario et Maritimes). Montréal.

Sacchi, C. F. and P. W. Price. 1992. The relative roles of abiotic and biotic factors in seedling demography of arroyo willow (*Salix lasiolepis*: Salicaceae). Amer. J. Bot. 79: 395–405.

Salick, J. and E. Pfeffer. 1999. The interplay of hybridization and clonal reproduction in the evolution of willows. Pl. Ecol. 141: 163–178.

Salisbury, R. A. 1796. Prodromus Stirpium in Horto ad Chapel Allerton Vigentium.... London. (Prodr. Stirp. Chap. Allerton)

Salix Alaska Yukon—See: G. W. Argus 1973

Sánchez-Acebo, L. 2005. A phylogenetic study of the New World *Cleome* (Brassicaceae, Cleomoideae). Ann. Missouri Bot. Gard. 92: 179–201.

Sánchez-Yélamo, M. D. and J. B. Martínez-Laborde. 1991. Chemotaxonomic approach to *Diplotaxis muralis* (Cruciferae, Brassiceae) and related species. Biochem. Syst. & Ecol. 19: 477–482.

Santamour, F. S. Jr. and A. J. McArdle. 1988. Cultivars of *Salix babylonica* and other weeping willows. J. Arboric. 14: 180–184.

Sargent, C. S., ed. 1913–1917. Plantae Wilsonianae: An Enumeration of the Woody Plants Collected in Western China for the Arnold Arboretum...During the Years 1907, 1908, and 1910 by E. H. Wilson. 3 vols. in parts. Cambridge, Mass. [Parts paged consecutively within vols.] (Pl. Wilson.)

Sargentia = Sargentia; Continuation of the Contributions from the Arnold Arboretum of Harvard University.

Savile, D. B. O. 1964. General ecology and vascular plants of the Hazen Camp area. Arctic 17: 237–258.

Savile, D. B. O. 1979b. Ring counts in *Salix arctica* from northern Ellesmere Island. Canad. Field-Naturalist 93: 81–82.

Savolainen, V. et al. 2000. Phylogenetics of flowering plants based on combined analysis of plastid *atp*B and *rbc*L gene sequences. Syst. Biol. 49: 306–362.

Scheen, A.-C., R. Elven, and C. Brochmann. 2002. A molecular-morphological approach solves taxonomic controversy in arctic *Draba* (Brassicaceae). Canad. J. Bot. 80: 59–71.

Schinz, H. and R. Keller. 1905. Flora der Schweiz, ed. 2. 2 vols. Zürich. (Fl. Schweiz ed. 2)

Schneider, C. K. [1904–]1906–1912. Illustriertes Handbuch der Laubholzkunde.... 2 vols. in 12 fascs. Jena. (Ill. Handb. Laubholzk.)

Schneider, C. K. 1918. A conspectus of Mexican, West Indian, Central and South American varieties of *Salix*. Bot. Gaz. 65: 1–41.

Schneider, C. K. 1919. Notes on American willows. V. The species of the *Pleoandrae* group. J. Arnold Arbor. 1: 1–31.

Schneider, C. K. 1921. Notes on American willows. XII. J. Arnold Arbor. 3: 61–125.

Schulz, O. E. 1903. Monographie der Gattung *Cardamine*. Bot. Jahrb. Syst. 32: 280–623.

Schulz, O. E. 1924. Cruciferae–Sisymbrieae. In: H. G. A. Engler, ed. 1900–1953. Das Pflanzenreich.... 107 vols. Berlin. Vol. 86[IV,105], pp. 1–388.

Schulz, O. E. 1927. Cruciferae—*Draba*, *Erophila*. In: H. G. A. Engler, ed. 1900–1953. Das Pflanzenreich.... 107 vols. Berlin. Vol. 89[IV,105], pp. 1–396.

Schulz, O. E. 1936. Cruciferae. In: H. G. A. Engler et al., eds. 1924+. Die natürlichen Pflanzenfamilien..., ed. 2. 26+ vols. Leipzig and Berlin. Vol. 17b, pp. 227–658.

Schur, P. J. F. 1866. Enumeratio Plantarum Transsilvaniae.... Vienna. (Enum. Pl. Transsilv.)

Sci. Rep. Tohoku Imp. Univ., Ser. 4, Biol. = Science Reports of the Tohoku Imperial University. Ser. 4, Biology. [Tohoku Teikoku-daigaku Rikwa Hokoku.]

Science = Science; an Illustrated Journal [later: a Weekly Journal Devoted to the Advancement of Science]. [American Association for the Advancement of Science.]

Scopoli, J. A. 1760. Flora Carniolica.... Vienna. (Fl. Carniol.)

Scopoli, J. A. 1771–1772. Flora Carniolica..., ed. 2. 2 vols. Vienna. (Fl. Carniol. ed. 2)

Scudder, G. G. E. and J. L. Reveal, eds. 1981. Evolution Today: Proceedings of the Second International Congress of Systematic and Evolutionary Biology, University of British Columbia, Vancouver, Canada, 17–24 July, 1980. Pittsburgh.

Seringe, N. C. 1824. Exemplaires Desséchés de la Révision Inédite du Genre *Salix*. Geneva. [Exsiccata with 18 sheets of printed text.] (Exempl. Rév. Salix)

Shafroth, P. B., M. L. Scott, and J. M. Friedman. 1994. Establishment, sex structure and breeding system of an exotic riparian willow, *Salix* ×*rubens*. Amer. Midl. Naturalist 132: 159–172.

Sharma, A. K. and A. Sharma, eds. 2003+. Plant Genome: Biodiversity and Evolution. 2+ vols. in parts. Enfield, N.H.

Sida = Sida; Contributions to Botany.

Silvae Genet. = Silvae Genetica.

Simmons, H. G. 1906. The Vascular Plants in the Flora of Ellesmereland.... Oslo. (Vasc. Pl. Ellesmereland)

Sitzungsber. Königl. Böhm. Ges. Wiss., Math.-Naturwiss. Cl. = Sitzungsberichte der Königlichen böhmischen Gesellschaft der Wissenschaften. Mathematisch-naturwissenschaftliche Classe.

Skr. Norske Vidensk.-Akad. Oslo, Mat.-Naturvidensk. Kl. = Skrifter Utgitt av det Norske Videnskaps-Akademi i Oslo. Matematisk-naturvidenskapelig Klasse.

Skvortsov, A. K. 1971. Zur Taxonomie der Wieden von Grönland, Island und den Färöern. Ann. Naturhist. Mus. Wien 75: 223–233.

Skvortsov, A. K. 1999. Willows of Russia and Adjacent Countries: Taxonomical and Geographical Revision, transl. I. N. Kadis, ed. A. G. Zinovjev et al. Joensuu.

Skvortsov, A. K. and M. D. Golysheva. 1966. On some structural peculiarities of leaves that are important for systematics and phylogeny of genus *Salix* L. Nauchnye Dokl. Vysshei Shkoly Biol. Nauki 5: 91–97.

Sleumer, H. 1980b. *Xylosma*. In: Organization for Flora Neotropica. 1968+. Flora Neotropica. 98+ nos. New York. No. 22, pp. 128–182.

Slotte, T. et al. 2006. Intrageneric phylogeny of *Capsella* (Brassicaceae) and the origin of the tetraploid *C. bursa-pastoris* based on chloroplast and nuclear DNA sequences. Amer. J. Bot. 93: 1714–1724.

Small, J. K. 1903. Flora of the Southeastern United States.... New York. (Fl. S.E. U.S.)

Small, J. K. 1913. Flora of the Southeastern United States..., ed. 2. New York. (Fl. S.E. U.S. ed. 2)

Smith, R. L. and K. J. Sytsma. 1990. Evolution of *Populus nigra* (sect. *Aigeros*): Introgressive hybridization and the chloroplast contribution of *Populus alba* (sect. *Populus*). Amer. J. Bot. 77: 1176–1187.

Smithsonian Contr. Knowl. = Smithsonian Contributions to Knowledge.

Smithsonian Misc. Collect. = Smithsonian Miscellaneous Collections.

Snogerup, S., M. Gustafsson, and R. von Bothmer. 1990. *Brassica* sect. *Brassica* (Brassicaceae). 1. Taxonomy and variation. Willdenowia 19: 271–365.

Sobick, U. 1983. Blutenentwicklungsgeschichtliche Untersuchungen an Resedaceen unter besonderer Berucksichtigung von Androeceum und Gynoeceum. Bot. Jahrb. Syst. 104: 203–248.

Solbrig, O. T. 1972. Breeding systems and genetic variation in *Leavenworthia* (Cruciferae). Evolution 26: 155–160.

Solbrig, O. T. and R. C. Rollins. 1977. The evolution of autogamy in species of the mustard genus *Leavenworthia*. Evolution 31: 265–281.

Soltis, D. E. and P. S. Soltis. 1990. Isozyme evidence of ancient polyploidy in primitive angiosperms. Syst. Bot. 15: 328–337.

Sommerfeltia Suppl. = Sommerfeltia Supplement.

Song, K., Tang K., and T. C. Osborn. 1993. Development of synthetic *Brassica* amphidiploids by reciprocal hybridization and comparison to natural amphidiploids. Theor. Appl. Genet. 86: 811–821.

Soper, J. H. and J. M. Powell. 1985. Botanical Studies in the Lake Hazen Region, Northern Ellesmere Island, Northwest Territories, Canada. Ottawa. [Natl. Mus. Canada, Publ. Nat. Sci. 5.]

SouthW. Naturalist = Southwestern Naturalist.

Southworth, D. and J. Seevers. 1997. Taxonomic status of *Limnanthes floccosa* subsp. *bellingeriana* (Limnanthaceae). In: T. N. Kaye et al., eds. 1997. Conservation of Native Plants and Fungi. Corvallis. Pp. 147–152.

Sp. Pl.—See: C. Linnaeus 1753; C. L. Willdenow et al. 1797–1830

Sp. Pl. ed. 2—See: C. Linnaeus 1762–1763

Spies, T. A. and B. V. Barnes. 1982. Natural hybridization between *Populus alba* L. and the native aspens in southeastern Michigan. Canad. J. Forest Res. 14: 789–793.

Spisok Rast. Gerb. Fl. S.S.S.R. Bot. Inst. Vsesoyuzn. Akad. Nauk = Spisok Rastenii Gerbariya Flory S S S R Izdavaemogo Botanicheskim Institutom Vsesoyuznogo Akademii Nauk.

Stafleu, F. A. and R. S. Cowan. 1976–1988. Taxonomic Literature: A Selective Guide to Botanical Publications and Collections with Dates, Commentaries and Types, ed. 2. 7 vols. Utrecht, Antwerp, The Hague, and Boston.

Stafleu, F. A., E. A. Mennega, L. J. Dorr, and D. H. Nicolson. 1992–2009. Taxonomic Literature: A Selective Guide to Botanical Publications and Collections with Dates, Commentaries and Types. Supplement. 8 vols. Königstein.

Stansbury, H. 1852. An Expedition to the Valley of the Great Salt Lake of Utah.... Philadelphia. [Botanical appendix by J. Torrey, pp. 381–397, plates 1–9.] (Exped. Great Salt Lake)

Stettler, R. F., H. D. Bradshaw, P. E. Heilman, and T. M. Hinckley. 1996. Biology of *Populus* and Its Implications for Management and Conservation. Ottawa.

Stettler, R. F., R. C. Fenn, P. E. Heilman, and B. J. Stanton. 1988. *Populus trichocarpa* × *Populus deltoides* hybrids for short rotation culture: Variation patterns and four-year field performance. Canad. J. Forest Res. 18: 745–753.

Steudel, E. G. 1840–1841. Nomenclator Botanicus Enumerans Ordine Alphabetico Nomina atque Synonyma tum Generica tum Specifica..., ed. 2. 2 vols. Stuttgart and Tübingen. (Nomencl. Bot. ed. 2)

Steyn, E. M., G. F. Smith, and A. E. van Wyk. 2004. Functional and taxonomic significance of seed structure in *Salix mucronata* (Salicaceae). Bothalia 34: 53–59.

Stirp. Austr. Fasc.—See: H. J. N. von Crantz 1762–1767

Stirp. Nov.—See: C.-L. L'Héritier de Brutelle 1784[1785–1805]

Stout, A. B. and E. J. Schreiner. 1933. Results of a project in hybridizing poplars. J. Heredity 24: 216–229.

Stuckey, R. L. 1972. Taxonomy and distribution of the genus *Rorippa* (Cruciferae) in North America. Sida 4: 279–430.

Suda, Y. and G. W. Argus. 1968. Chromosome numbers of some North American *Salix*. Brittonia 20: 191–197.

Sudworth, G. B. 1934. Poplars, Principal Tree Willows and Walnuts of the Rocky Mountain Region. Washington. [U.S.D.A., Techn. Bull. 420.]

Sugaya, S. 1960. Bearing of the cataphyllotaxy on the interpretation of the nectary structures in the flowers of the Salicaceae. Sci. Rep. Tohoku Imp. Univ., Ser. 4, Biol. 26: 9–24.

Suppl. Meth.—See: C. Moench 1802

Sv. Fanerogamfl. ed. 2—See: C. A. M. Lindman 1926

Svensk Bot. Tidskr. = Svensk Botanisk Tidskrift Utgifven af Svenska Botaniska Föreningen.

Svensson, S. 1983. Chromosome numbers and morphology in the *Capsella bursa-pastoris* complex (Brassicaceae) in Greece. Willdenowia 13: 267–276.

Swarbrick, J. T. et al., eds. 1993. Proceedings: 10th Australian Weeds Conference and 14th Asian Pacific Weed Science Society Conference.... 2 vols. Brisbane.

Sweeney, P. W. and R. A. Price. 2000. Polyphyly of the genus *Dentaria* (Brassicaceae): Evidence from *trn*L intron and *ndh*F sequence data. Syst. Bot. 25: 468–478.

Sydnor, T. D. and W. F. Cowen. 2000. Ohio Trees. Columbus.
Syll. Pl. Nov.—See: C. F. Hornschuch [1822–]1824–1828
Sylva Tellur.—See: C. S. Rafinesque 1838b
Symb. Antill.—See: I. Urban 1898–1928
Symb. Bot. Upsal. = Symbolae Botanicae Upsalienses; Arbeten från Botaniska Institutionen i Uppsala.
Syn. Fl. N. Amer.—See: A. Gray et al. 1878–1897
Syn. Pl.—See: C. H. Persoon 1805–1807
Syn. Stirp. Aragon.—See: I. J. de Asso y del Rio 1779
Syst. Biol. = Systematic Biology.
Syst. Bot. = Systematic Botany; Quarterly Journal of the American Society of Plant Taxonomists.
Syst. Bot. Monogr. = Systematic Botany Monographs; Monographic Series of the American Society of Plant Taxonomists.
Syst. Nat.—See: A. P. de Candolle [1817]1818–1821
Syst. Nat. ed. 10—See: C. Linnaeus 1758[–1759]
Syst. Nat. ed. 12—See: C. Linnaeus 1766[–1768]
Tabl. École Bot.—See: R. L. Desfontaines 1804
Takhtajan, A. L. 1997. Diversity and Classification of Flowering Plants. New York.
Taxon = Taxon; Journal of the International Association for Plant Taxonomy.
Taylor, R. L. and G. A. Mulligan. 1968. Flora of the Queen Charlotte Islands. Part 2. Cytological Aspects of the Vascular Plants. Ottawa.
Taylor, R. L. and S. Taylor. 1977. Chromosome numbers of vascular plants of British Columbia. Syesis 10: 125–138.
Tedin, O. 1925. Vererbung, Variation und Systematik der Gattung *Camelina*. Hereditas (Lund) 6: 275–386.
Thellung, A. 1906. Die Gattung *Lepidium* (L.) R. Br. Mitt. Bot. Mus. Univ. Zürich 28: 1–340.
Theor. Appl. Genet. = Theoretical and Applied Genetics; International Journal of Breeding Research and Cell Genetics.
Thieret, J. W. and R. L. Thompson. 1984. *Cleome ornithopodioides* (Capparaceae): Adventive and spreading in North America. Bartonia 50: 25–26.
Thompson, J. T., R. Van Buren, and K. T. Harper. 2003. Genetic analysis of the rare species *Salix arizonica* (Salicaceae) and associated tree willows in Arizona and Utah. W. N. Amer. Naturalist 63: 273–282.
Thorne, R. F. 1954. Flowering plants of the waters and shores of the Gulf of Mexico. In: Fish and Wildlife Service [U.S.D.I.]. 1954. Gulf of Mexico: Its Origin, Waters, and Marine Life. Washington. Pp. 193–202.
Thorne, R. F. 1992b. Classification and geography of the flowering plants. Bot. Rev. (Lancaster) 58: 225–348.
Tolmatchew, A. I. 1939. *Draba*. In: V. L. Komarov et al., eds. 1934–1964. Flora URSS.... 30 vols. Leningrad. Vol. 8, pp. 371–454, 649–650.
Tolmatchew, A. I., ed. 1960–1987. Flora Arctica URSS. 10 vols. Moscow and Leningrad.
Tolmatchew, A. I. 1975. *Draba*. In: A. I. Tolmatchew, ed. 1960–1987. Flora Arctica URSS. 10 vols. Moscow and Leningrad. Vol. 7, pp. 106–155.
Torrey, J. and A. Gray. 1838–1843. A Flora of North America.... 2 vols. in 7 parts. New York, London, and Paris. (Fl. N. Amer.)
Torreya = Torreya; a Monthly Journal of Botanical Notes and News.
Trail & Landscape = Trail and Landscape; Publication Concerned with Natural History and Conservation.
Trans. Acad. Sci. St. Louis = Transactions of the Academy of Science of St. Louis.
Trans. Amer. Philos. Soc. = Transactions of the American Philosophical Society Held at Philadelphia for Promoting Useful Knowledge.
Trans. Linn. Soc. = Transactions of the Linnean Society.
Trans. New York Acad. Sci. = Transactions of the New York Academy of Sciences.
Tree Physiol. = Tree Physiology.
Triest, L. 2001. Hybridization in staminate and pistillate *Salix alba* and *S. fragilis* (Salicaceae): Morphology versus RAPDs. Pl. Syst. Evol. 226: 143–154.
Trudy Imp. S.-Peterburgsk. Bot. Sada = Trudy Imperatorskago S.-Peterburgskago Botanicheskago Sada.
Truesdale, H. D. et al. 2004. Allozyme variability within and among varieties of *Isomeris arborea* (Capparaceae). Madroño 51: 364–371.
Trybush, S., S. Jahodová, W. Macalpine, and A. Karp. 2008. A genetic study of a *Salix* germplasm resource reveals new insights into relationships among subgenera, sections, and species. Bioenergy Res. 1: 67–79.
Turner, B. L., H. Nichols, G. Denny, and O. Doron. 2003. Atlas of the Vascular Plants of Texas. 2 vols. Fort Worth. [Sida Bot. Misc. 24.]
Tuskan, G. A. et al. 2006. The genome of black cottonwood, *Populus trichocarpa* (Torr. & Gray). Science, n. s. 313: 1596–1604.
Tutin, T. G. et al., eds. 1964–1980. Flora Europaea. 5 vols. Cambridge.
Tutin, T. G. et al., eds. 1993+. Flora Europaea, ed. 2. 1+ vol. Cambridge and New York.
U.S. Expl. Exped.—See: C. Wilkes et al. 1854–1876
Univ. Calif. Publ. Bot. = University of California Publications in Botany.
Univ. Wyoming Publ. Sci., Bot. = University of Wyoming Publications in Science. Botany.
University of Chicago Press. 1993. The Chicago Manual of Style, ed. 14. Chicago.
Urban, I., ed. 1898–1928. Symbolae Antillanae seu Fundamenta Florae Indiae Occidentalis.... 9 vols. Berlin etc. (Symb. Antill.)
Utah Fl. ed. 3—See: S. L. Welsh et al. 2003
Utkast Sv. Fl.—See: S. Liljeblad 1792
Valentine, D. H., ed. 1972. Taxonomy, Phytogeography and Evolution. London and New York.
Van Splunder, I. H., L. A. Coops, C. J. Voesenek, and C. W. P. M. Blom. 1995. Establishment of alluvial forest species in floodplains: The role of dispersal timing, germination characteristics and water level fluctuation. Acta Bot. Neerl. 44: 269–278.
Vanderpool, S. S., W. J. Elisens, and J. R. Estes. 1991. Pattern, tempo, and mode of evolutionary and biogeographic divergence in *Oxystylis* and *Wislizenia* (Capparaceae). Amer. J. Bot. 78: 925–937.
Vasc. Pl. Ellesmereland—See: H. G. Simmons 1906
Vasc. Pl. Pacif. N.W.—See: C. L. Hitchcock et al. 1955–1969
Vasc. Pl. Wyoming—See: R. D. Dorn 1988
Vasc. Pl. Wyoming ed. 3—See: R. D. Dorn 2001
Vaughan, J. G., A. J. MacLeod, and B. M. G. Jones, eds. 1976. The Biology and Chemistry of the Cruciferae. London and New York.

Vega Exp. Vetensk. Iakttag.—See: A. E. Nordenskiöld 1882–1887

Velichkin, E. M. 1979. *Smelowskia* C. A. Mey. (Cruciferae). Critical review and relation to close genera. Bot. Zhurn. (Moscow & Leningrad) 64: 153–171.

Ventenat, É. P. [1800–1803.] Description des Plantes Nouvelles et Peu Connues Cultivés dans le Jardin de J. M. Cels.... 10 parts. Paris. [Plates numbered consecutively.] (Descr. Pl. Nouv.)

Verdcourt, B. 1985. A synopsis of the Moringaceae. Kew Bull. 40: 1–23.

Verh. K. K. Zool.-Bot. Ges. Wien = Verhandlungen der Kaiserlich-königlichen zoologisch-botanischen Gesellschaft in Wien.

Viereck, L. A. and J. M. Foote. 1970. The status of *Populus balsamifera* and *P. trichocarpa* in Alaska. Canad. Field-Naturalist 84: 169–173.

Viereck, L. A. and E. L. Little Jr. 1972. Alaska Trees and Shrubs. Washington. [Agric. Handb. 410.]

Villars, D. 1779. Prospectus de l'Histoire des Plantes de Dauphiné.... Grenoble. (Prosp. Hist. Pl. Dauphiné)

Voss, E. G. 1972–1996. Michigan Flora.... 3 vols. Bloomfield Hills and Ann Arbor.

Voy. Inde—See: V. Jacquemont [1835–]1841–1844

W. Amer. Sci. = West American Scientist.

W. J. Appl. Forest. = Western Journal of Applied Forestry.

W. N. Amer. Naturalist = Western North American Naturalist.

Wagner, W. H. Jr. 1970. The Barnes hybrid aspen, *Populus* ×*barnesii*, hybr. nov.—a nomenclatural case in point. Michigan Bot. 9: 53–54.

Wahlenberg, G. 1812. Flora Lapponica Exhibens Plantas Geographice et Botanice Consideratas.... Berlin. (Fl. Lapp.)

Waithaka, K. 1991. *Gynandropsis gynandra* (L.) Briq., a Tropical Leafy Vegetable: Its Cultivation and Utilization. Rome.

Waldstein, F. and P. Kitaibel. [1799–]1802–1812. Descriptiones et Icones Plantarum Rariorum Hungariae. 3 vols. in parts. Vienna. (Descr. Icon. Pl. Hung.)

Walker, L. R., J. C. Zasada, and F. S. Chapin. 1986. The role of life history processes in the primary succession on an Alaskan floodplain. Ecology 67: 1243–1253.

Walpers, W. G. 1842–1847. Repertorium Botanices Systematicae.... 6 vols. Leipzig. (Repert. Bot. Syst.)

Walter, T. 1788. Flora Caroliniana, Secundum Systema Vegetabilium Perillustris Linnaei Digesta.... London. (Fl. Carol.)

Wang, R. and Wang J. 1991. A study on the chromosome numbers and their evolution in Salicaceae. J. Beijing Forest. Univ. 13: 32–38.

War Department [U.S.]. 1855–1860. Reports of Explorations and Surveys, to Ascertain the Most Practicable and Economical Route for a Railroad from the Mississippi River to the Pacific Ocean. Made under the Direction of the Secretary of War, in 1853[–1856].... 12 vols. in 13. Washington. (Pacif. Railr. Rep.)

Warren, G. K. 1859. Preliminary Report of Explorations in Nebraska and Dakota, in the Years 1855– '56– '57.... Washington. (Prelim. Rep. Expl. Nebraska Dakota)

Warwick, S. I. et al. 2004. Phylogeny of *Braya* and *Neotorularia* (Brassicaceae) based on nuclear ribosomal internal transcribed spacer and chloroplast *trn*L intron sequences. Canad. J. Bot. 82: 376–392.

Warwick, S. I. et al. 2004b. Phylogeny of *Smelowskia* and related genera (Brassicaceae) based on nuclear ITS DNA and chloroplast *trn*L intron DNA sequences. Ann. Missouri Bot. Gard. 91: 99–123.

Warwick, S. I. et al. 2006. Phylogenetic position of *Arabis arenicola* and generic limits of *Eutrema* and *Aphragmus* (Brassicaceae) based on sequences of nuclear ribosomal DNA. Canad. J. Bot. 84: 269–281.

Warwick, S. I. et al. 2006b. Brassicaceae: Species checklist and database on CD-ROM. Pl. Syst. Evol. 259: 249–258.

Warwick, S. I. and I. A. Al-Shehbaz. 2003. Nomenclatural notes on *Sisymbrium* (Brassicaceae). Novon 13: 265–267.

Warwick, S. I. and I. A. Al-Shehbaz. 2006. Brassicaceae: Chromosome number index and database on CD-ROM. Pl. Syst. Evol. 259: 237–248.

Warwick, S. I., I. A. Al-Shehbaz, R. A. Price, and C. A. Sauder. 2002. Phylogeny of *Sisymbrium* (Brassicaceae) based on ITS sequences of nuclear ribosomal DNA. Canad. J. Bot. 80: 1002–1017.

Warwick, S. I. and L. D. Black. 1991. Molecular systematics of *Brassica* and allied genera (subtribe Brassicinae, Brassicaceae)—Chloroplast genome and cytodeme congruence. Theor. Appl. Genet. 82: 81–92.

Warwick, S. I. and L. D. Black. 1993. Molecular relationships in subtribe Brassicinae (Cruciferae, tribe Brassiceae). Canad. J. Bot. 71: 906–918.

Warwick, S. I. and A. Francis. 1994. Guide to the Wild Germplasm of *Brassica* and Allied Crops. Part V. Life History and Geographical Data for Wild Species in the Tribe Brassiceae (Cruciferae). Ottawa. [Agric. Canad. Res. Branch, Techn. Bull. 1993-14E.]

Warwick, S. I. and C. A. Sauder. 2005. Phylogeny of tribe Brassiceae (Brassicaceae) based on chloroplast restriction site polymorphisms and nuclear ribosomal internal transcribed spacer and chloroplast *trn*L intron sequences. Canad. J. Bot. 83: 467–483.

Warwick, S. I., C. A. Sauder, and I. A. Al-Shehbaz. 2005. Molecular phylogeny and cytological diversity of *Sisymbrium* (Brassicaceae). In: A. K. Sharma and A. Sharma, eds. 2003+. Plant Genome: Biodiversity and Evolution. 2+ vols. in parts. Enfield, N.H. Vol. 1, part C, pp. 219–250.

Warwick, S. I., C. A. Sauder, and I. A. Al-Shehbaz. 2008. Phylogenetic relationships in the tribe Alysseae (Brassicaceae) based on nuclear ribosomal ITS DNA sequences. Canad. J. Bot. 86: 315–336.

Warwick, S. I., C. A. Sauder, I. A. Al-Shehbaz, and F. Jacquemoud. 2007. Phylogenetic relationships in the tribes Anchonieae, Chorisporeae, Euclidieae, and Hesperideae (Brassicaceae) based on nuclear ribosomal ITS DNA sequences. Ann. Missouri Bot. Gard. 94: 56–78.

Wasmann J. Biol. = Wasmann Journal of Biology.

Watson, S. 1871. United States Geological Expolration [sic] of the Fortieth Parallel. Clarence King, Geologist-in-charge. [Vol. 5] Botany. By Sereno Watson.... Washington. [Botanical portion of larger work by C. King.] [Botany (Fortieth Parallel)]

Webb, P. B. and S. Berthelot. [1835–]1836–1850. Histoire Naturelle des Îles Canaries.... 3 vols. in 9. Paris. [Tome troisième, Botanique: Première partie, 1 vol; deuxième partie, 4 vols.] (Hist. Nat. Îles Canaries)

Weber, W. A. 1989. Additions to the Flora of Colorado: 12. Phytologia 67: 429–437.

Weberling, F. 1968. Über die Rudimentärstipeln der Resedaceae. Acta Bot. Neerl. 17: 360–372.

Weed Technol. = Weed Technology; a Journal of the Weed Science Society of America.

Welsh, S. L., N. D. Atwood, S. Goodrich, and L. C. Higgins, eds. 2003. A Utah Flora, ed. 3. Provo. (Utah Fl. ed. 3)

West, C. J., comp. 1994. Wild Willows in New Zealand: Proceedings of a Willow Control Workshop.... Wellington.

White, W. W. 1951. Native cottonwoods of Montana. Proc. Montana Acad. Sci. 9: 33–39.

Wilkes, C. et al. 1854–1876. United States Exploring Expedition. During the years 1838, 1839, 1840, 1841, 1842. Under the Command of Charles Wilkes, U.S.N..... 18 vols. (1–17, 19). Philadelphia. [Vol. 15: Botany, Phanerogamia (A. Gray), 1854; Atlas, 1856. Vol. 16: Botany, Cryptogamia, Filices (W. D. Brackenridge), 1854; Atlas, 1855. Vol. 17: incl. Phanerogamia of Pacific North America (J. Torrey), 1874. Vol. 19 (2 parts): Geographical Distribution of Animals and Plants (C. Pickering), 1854. Vol. 18: Botany, Phanerogamia, part 2 (A. Gray) not published.] (U.S. Expl. Exped.)

Wilkinson, H. P. 2007. Leaf teeth in certain Salicaceae and 'Flacourtiaceae.' Bot. J. Linn. Soc. 155: 241–256.

Willdenow, C. L. 1809–1813[–1814]. Enumeratio Plantarum Horti Regii Botanici Berolinensis.... 2 parts + suppl. Berlin. [Parts paged consecutively.] (Enum. Pl.)

Willdenow, C. L., C. F. Schwägrichen, and J. H. F. Link. 1797–1830. Caroli a Linné Species Plantarum.... Editio Quarta.... 6 vols. Berlin. [Vols. 1–5(1), 1797–1810, by Willdenow; vol. 5(2), 1830, by Schwägrichen; vol. 6, 1824–1825, by Link.] (Sp. Pl.)

Willows Calif.—See: M. S. Bebb 1879

Windham, M. D. 2000. Chromosome counts and taxonomic notes on *Draba* (Brassicaceae) of the Intermountain West. 1: Utah and vicinity. Madroño 47: 21–28.

Windham, M. D. 2004. Chromosome counts and taxonomic notes on *Draba* (Brassicaceae) of the Intermountain West. 2: Idaho, Nevada and vicinity. Madroño 50: 221–231.

Windham, M. D. and I. A. Al-Shehbaz. 2006. New and noteworthy species of *Boechera* (Brassicaceae) I: Sexual diploids. Harvard Pap. Bot. 11: 61–88.

Windham, M. D. and I. A. Al-Shehbaz. 2007. New and noteworthy species of *Boechera* (Brassicaceae) II: Apomictic hybrids. Harvard Pap. Bot. 11: 257–274.

Windham, M. D. and I. A. Al-Shehbaz. 2007b. New and noteworthy species of *Boechera* (Brassicaceae) III: Additional sexual diploids and apomictic hybrids. Harvard Pap. Bot. 12: 235–257.

Winge, Ø. 1940. Taxonomic and evolutionary studies in *Erophila* based on cytogenetic investigations. Compt.-Rend. Trav. Carlsberg Lab., Sér. Physiol. 23: 41–74.

Wislizenus, F. A. 1848. Memoir of a Tour to Northern Mexico, Connected with Col. Doniphan's Expedition, in 1846 and 1847.... Washington. (Mem. Tour N. Mexico)

Wiss. Z. Friedrich-Schiller-Univ. Jena, Math.-Naturwiss. Reihe = Wissenschaftliche Zeitschrift der Friedrich-Schiller-Universität Jena/Thüringen. Mathematisch-naturwissenschaftliche Reihe.

Withering, W. 1796. An Arrangement of British Plants..., ed. 3. 4 vols. London. (Arr. Brit. Pl. ed 3)

Wood, A. 1870. The American Botanist and Florist; Including Lessons in the Structure, Life, and Growth of Plants; Together with a Simple Analytical Flora, Descriptive of the Native and Cultivated Plants Growing in the Atlantic Division of the American Union.... New York and Chicago. (Amer. Bot. Fl.)

Woodson, R. E. Jr. 1948. *Gynandropsis*, *Cleome*, and *Podandrogyne*. Ann. Missouri Bot. Gard. 35: 139–148.

World Pollen Fl. = World Pollen Flora.

Wrightia = Wrightia; a Botanical Journal.

Wu, Z. and P. H. Raven, eds. 1994+. Flora of China. 16+ vols. Beijing and St. Louis.

Wyoming Agric. Exp. Sta. Bull. = Wyoming Agricultural Experiment Station Bulletin.

Young, W. C., ed. 1999. 1998 Seed Production Research at Oregon State University, USDA-ARS Cooperating. Corvallis.

Zhang, M. L. 1994. A preliminary cladistic study on the multistaminal willows (*Salix*) in China. Bull. Bot. Res., Harbin 14: 299–305.

Zoë = Zoë; a Biological Journal.

Index

Names in *italics* are synonyms, casually mentioned hybrids, or plants not established in the flora. Page numbers in **boldface** indicate the primary entry for a taxon. Page numbers in *italics* indicate an illustration. Roman type is used for all other entries, including author names, vernacular names, and accepted scientific names for plants treated as established members of the flora.

Abbe, E. C., 494, 549
Abbott, R. J., 28
Abdallah, M. S., 190, 193
Abdra, 269
   brachycarpa, 296
Achoriphragma, 511
   nudicaule, 512
Acroschizocarpus, 671
   kolianus, 673
Aethionema, 226, 227
Agallis, 531
Agianthus, 700
   bernardinus, 705
   jacobaeus, 705
Ahles, H. E., 430
Airy Shaw, H. K., 195
Akaniaceae, 165
Álamillo, 19
Álamo cimarron, 20
Álamo temblón, 22
Alaska bog willow, 72
Aldenella
   tenuifolia, 204
Alexander, P. J., 739
Alliaria, 234, 246, *743*, **744**
   brachycarpa, 745
   officinalis, 745
   petiolata, *743*, **745**
Allthorn, 184
Allthorn Family, 184
Almond leaf willow, 50
Alnus, 64
Alpine bladderpod, 660
Alpine twinpod, 624
Al-Shehbaz, I. A., 224, 226, 227, 247, 251, 252, 253, 256, 257, 258, 267, 268, 269, 270, 288, 290, 295, 301, 305, 311, 316, 318, 320, 325, 330, 335, 336, 339, 341, 345, 347, 348, 349, 365, 367, 368, 369, 370, 371, 372, 373, 374, 375, 376, 377, 378, 381, 382, 383, 384, 385, 386, 387, 388, 389, 390, 391, 392, 393, 394, 395, 396, 397, 398, 399, 401, 404, 408, 409, 410, 411, 412, 414, 415, 417, 420, 421, 422, 424, 429, 430, 435, 436, 437, 441, 442, 443, 444, 446, 447, 448, 449, 450, 451, 453, 454, 455, 456, 458, 459, 460, 464, 484, 485, 486, 489, 492, 493, 497, 505, 506, 509, 510, 511, 514, 517, 518, 529, 530, 531, 534, 552, 553, 555, 557, 559, 561, 562, 563, 564, 565, 566, 567, 568, 570, 571, 585, 594, 596, 599, 600, 601, 604, 607, 608, 609, 617, 665, 667, 669, 671, 673, 675, 677, 685, 687, 688, 689, 690, 692, 695, 699, 700, 709, 720, 723, 728, 738, 742, 744, 745
Alström-Rapaport, C., 4, 23, 30, 51, 61
Alysseae, **247**
Alyssum, 226, 237, 241, **247**
   alpinum, 660
   alyssoides, **248**
   americanum, 250
   arcticum, 626
   arduini, 252

   argenteum, 649
   argyraeum, 627
   auriculatum, 612
   biovulatum, 250
   calycinum, 248
   campestre, 249
   deltoideum, 269
   densiflorum, 633
   dentatum, 332
   desertorum, **248**
   engelmannii, 635
   fallax, 250
   fendleri, 636
   globosum, 639
   gordonii, 640
   gracile, 641
   grandiflorum, 613
   grayanum, 650
   incanum, 253
   lescurii, 614
   lindheimeri, 649
   ludovicianum, 649
   maritimum, 598
   minus, 249
   murale, 248, **250**
   nuttallii, 641
   obovatum, 248, **250**
   pallidum, 656
   parviflorum
      micranthum, 249
   petraeum, 251
   repandum, 641
   saxatile, 252
   shortii, 639
   simplex, 248, **249**, 250
   stenophyllum, 636
   strigosum, 250
   szowitsianum, 248, **250**
American cress, 463

Anastatica
   syriaca, 553
Anchonieae, **253**
Andersen, B., 424
Andersson, L., 165
Andersson, S., 165
Andrena
   limnanthis, 174
   pulverea, 174
Anelsonia, 234, 236, 240, 241, **347**
   eurycarpa, *343*, **348**
Annual wall-rocket, 433
Añu, 166
Aphragmeae, **255**
Aphragmus, 231, 232, 239, 245, **256**
   eschscholtzianus, *255*, **256**
Appel, O., 226, 227, 258, 454, 530, 677
Arabideae, 226, 227, **256**, 257, 349
Arabidopsis, 226, 235, 236, 242, 243, 257, **447**, 448, 449
   arenicola, *445*, 448
   arenosa, **448**
      arenosa, **448**, 449
      borbasii, 448
      bursifolia, 456
      beringensis, 457
   lyrata, 448, **449**, 450
      kamchatica, 449, **450**
      lyrata, **449**
      petraea, 449, **450**
   mollis, 457
   novae-angliae, 549
   salsuginea, 556
   stenocarpa, 457

Arabidopsis (*continued*)
 thaliana, 226, 448, **450**
 trichopoda, 457
 tschuktschorum, 457
 virgata, 457
Arabis, 226, 234, 242, **257**, 258, 264, 348, 349, 449, 450, 492, 561, 720
 aculeolata, 259, 264, **265**
 acutina, 365
 alpina, 258, **259**
  caucasica, 259
  glabrata, 259
 ambigua
  glabra, 450
  intermedia, 450
 angulata, 396
 aprica, 391
 arbuscula, 386
 arcoidea, 407
 arcuata
  longipes, 381
  perennans, 396
  rubicundula, 404
  secunda, 402
  subvillosa, 393
 arenicola, 448
  pubescens, 448
 arida, 399
 armerifolia, 389
 atriflora, 366
 atrorubens, 366
 austiniae, 368
 beckwithii, 386, 399
 blepharophylla, 258, 264, **266**
  mcdonaldiana, 265
 bodiensis, 367
 boivinii, 382
 bourgovii, 382
 bracteolata, 388
 brebneriana, 457
 breweri, 367, 395
  austiniae, 368
  figularis, 367
  pecuniaria, 394
 bridgeri, 263
 bruceae, 372
 bulbosa, 469
 burkii, 368
 calderi, 368
 campyloloba, 407
 canadensis, 369
 canescens
  latifolia, 388
  stylosa, 378
 caucasica, 258, **259**
 cobrensis, 370
 codyi, 388
 cognata, 372
 collinsii, 370
 columbiana, 393
 conferta, 384
 confinis, 382
  brachycarpa, 382
 connexa, 408
 consanguinea, 371
 constancei, 371
 covillei, 371
 crandallii, 372

crucisetosa, 258, **262**
cusickii, 372
dacotica, 382
davidsonii, 372, 410
 parva, 372
demissa, 387, 391, 392, 395
 languida, 387, 392
 russeola, 395
densa, 389
densicaulis, 389
dentata
 phalacrocarpa, 373
depauperata, 373
deserti, 693
dianthifolia, 409
diehlii, 395
dispar, 374
divaricarpa, 365, 374, 382
 dacotica, 382
 dechamplainii, 401
 hemicylindrica, 382
 interposita, 365
 pinetorum, 397
 stenocarpa, 382
drepanoloba, 374
drummondii, 408
 alpina, 389
 brachycarpa, 382
 connexa, 408
 interposita, 365
 lyallii, 389
 oreophila, 374
 oxyphylla, 408
 pratincola, 399
duriuscula, 409
egglestonii, 388
elegans, 393
endlichii, 689
epilobioides, 367
eremophila, 396
eschscholtziana, 258, **261**
exilis, 396, 402
falcata, 369
falcatoria, 376
falcifructa, 377
fecunda, 377
fendleri, 378, 408
 fendleri, 410
 spatifolia, 408
fernaldiana, 376, 378
 stylosa, 378
filifolia, 694
formosa, 379
fructicosa, 379
furcata, 258, **262**, 263
 olympica, 263
 purpurascens, 264
georgiana, 258, **261**, 262
glabra, 458
 furcatipilis, 458
glaucovalvula, 380
goodrichii, 380
gracilenta, 381, 396
gracilipes, 381
gunnisoniana, 382
harrisonii, 383
hastata, 386
hastatula, 383
hesperidoides, 485
heterophylla, 386

hirshbergiae, 386
hirsuta, 260, 261
 adpressipilis, 260
 eschscholtziana, 261
 glabrata, 261
 hirsuta, 260
 minshallii, 260
 ovata, 369
 pycnocarpa, 259, 261
hoffmannii, 383
holboellii, 370, 384, 396, 397, 402
 arcuata, 365
 brachycarpa, 382
 collinsii, 370
 consanguinea, 371
 derensis, 385
 fendleri, 378
 pendulocarpa, 395
 pinetorum, 350, 381, 393, 397
 retrofracta, 402
 secunda, 402
 tenuis, 384
hookeri, 457
 breviramosa, 457
 multicaulis, 457
horizontalis, 384
howellii, 384
humifusa, 448
 pubescens, 448
inamoena, 397, 402
 acutata, 384
 interposita, 365
inyoensis, 385
johnstonii, 386
juniperina, 374
kamchatica, 450
kennedyi, 388
koehleri, 386
 stipitata, 386
laevigata, 368, 386
 burkii, 368
 heterophylla, 390
 missouriensis, 387, 390
lasiocarpa, 387
latifolia, 388
leibergii, 371
lemmonii, 388
 depauperata, 373, 388
 drepanoloba, 374, 388
 paddoensis, 388, 392
lignifera, 388
lignipes, 402
 impar, 399
longirostris, 700
ludoviciana, 493
lyallii, 389, 394
 davidsonii, 372
 nubigena, 389, 394
lyrata, 449, 450
 glabra, 450
 intermedia, 450
 kamchatica, 450
 occidentalis, 450
lyrifolia, 386
macdougalii, 398
macella, 263
macounii, 389
macrocarpa, 458

maxima, 365
 hoffmannii, 383
mcdonaldiana, 258, 264, **265**
media, 450
microphylla, 390, 404
 macounii, 380, 389, 390
 nubigena, 394
 saximontana, 404
 thompsonii, 369, 390
missouriensis, 390
 deamii, 390
modesta, 258, **264**
multiceps, 389
murrayi, 389
nardina, 374
nemophila, 399
nevadensis, 391
nubigena, 394
nudicaulis, 512
nuttallii, 258, **263**, 264
occidentalis, 450
oligantha, 397
olympica, 258, **263**
ophira, 391
oregana, 259, **264**, 265
oreocallis, 388
oreophila, 374
ovata, 369
 glabrata, 261
oxylobula, 391
oxyphylla, 408
pallidifolia, 392
parishii, 393
patens, 258, **262**
paupercula, 394
pedicellata, 415
pendulina, 391, 395
 russeola, 395
pendulocarpa, 395
 saximontana, 404
peramoena, 407
perelegans, 393
perennans, 396, 411
 longipes, 381
perfoliata, 458
perstellata, 396
 ampla, 396
 shortii, 373
petiolaris, 719
petiolata, 745
pinetorum, 397
pinzliae, 397
platyloba, 384
platysperma, 397, 411
 howellii, 384
 imparata, 384
polyantha, 398
polyclada, 388
polytricha, 407
porphyrea, 398
pratincola, 399
pseudoturritis, 458
puberula, 399
pulchra, 379, 400, 412
 duchesnensis, 375
 glabrescens, 411
 gracilis, 411
 munciensis, 388
 pallens, 379
 viridis, 411

purpurascens, 264
pusilla, 400
pycnocarpa, 255, 258, **259**, 260, 261, 262
    adpressipilis, **260**, 261
    glabrata, 261
    pycnocarpa, 255, **260**, 261
    reducta, 260
pygmaea, 401
recondita, 396
rectissima, 401
repanda, 401
    greenei, 401
reptans, 333
retrofracta, 402
    collinsii, 370
    multicaulis, 402
rhomboidea, 469
    purpurea, 473
rigidissima, 403
    demota, 403
rollei, 403
rostellata, 368
rugocarpa, 391
rupestris, 261
sabulosa, 399
salubris, 374
schistacea, 404
secunda, 402
selbyi, 381, 393, 396
semisepulta, 388
serotina, 405
serpenticola, 265
setigera, 401
setulosa, 395
shockleyi, 406
shortii, 373
    phalacrocarpa, 373
sparsiflora, 366, 407
    arcuata, 365
    atrorubens, 366
    californica, 369
    columbiana, 393
    peramoena, 407
    secunda, 402
    subvillosa, 393
spathulata, 263
spatifolia, 408
stelleri
    eschscholtziana, 261
stenoloba, 372
stokesiae, 374
subpinnatifida, 409
    beckwithii, 399
    impar, 399
suffrutescens, 371, 409
    horizontalis, 384, 409
    perstylosa, 371
suksdorfii, 262
tenuicula, 390
tenuis, 402
thaliana, 450
thompsonii, 392
tiehmii, 410
trichopoda, 412, 457
tricornuta, 561
tschuktschorum, 457
viereckii, 689
virginica, 493
viridis, 390
    deamii, 390
    vivariensis, 379
    whitedii, 418
    williamsii, 404, 411
        saximontana, 404, 411
    wyndii, 401
Archer, W. A., 37
Arctic bladderpod, 626
Arctic hare, 80
Arctic seashore willow, 75
Arctic willow, 78
Argus, G. W., 3, 23, 24, 26, 28, 29, 30, 32, 35, 40, 43, 46, 49, 51, 54, 61, 66, 68, 69, 76, 79, 80, 81, 86, 87, 88, 89, 90, 91, 92, 109, 110, 111, 115, 117, 121, 127, 128, 130, 131, 133, 135, 136, 139, 140, 149, 153, 155, 156, 157, 158, 161
Arivela, 200, 215, **221**
    viscosa, 200, *221*, **222**
Arizona bladderpod, 627
Arizona willow, 112
Armoracia, 226, 232, 238, **459**
    amphibia, 497
    lacustris, 497
    rusticana, 457, **459**, 460, 555
Arnason, T., 420
Arroyo willow, 156
Arroyo, M. T. K., 183
Arugula, 434
Asai, Y., 428
Aspen, 5, 7, 22
Astrocarpeae, 190
Atamisquea, 194, **195**
    emarginata, *196*
Athabasca willow, 84
Athysanus, 237, 238, 241, **267**
    pusillus, 266, **267**, 268
        glabrior, 267
    unilateralis, 267, **268**
Aubrieta, 226, 237, 243, **268**
    deltoidea, 266, **269**
Aurinia, 226, 237, 241, **251**
    petraea, **251**, 252
    saxatilis, *251*, **252**
        saxatilis, 252
Autumn willow, 45
Avery Peak twinpod, 624
Awlwort, 509
Azuma, T., 4, 24, 30, 51, 61

Bacon, J. D., 610
Badillo, V. M., 171
Bailey, C. D., 226, 557, 558, 559, 561, 571
Bailey, L. G., 443
Bailey, L. H., 207
Baker, H. G., 423, 439
Baker, R. G., 68
Baker's meadowfoam, 175
Ball, C. R., 37, 50, 58
Ball, P. W., 455
Ball's willow, 111
Ball-mustard, 455
Balm of Gilead, 9, 15
Balsam poplar, 5, 14, 15, 16, 17
Balsam willow, 115
Bam tree, 14
Barbarea, 230, 245, 246, **460**
    americana, 462
    arcuata, 461
    orthoceras, 460, 461, **462**, 463
        dolichocarpa, 462
    plantaginea, 462
    praecox, 463
    stricta, 460, 461, **462**
    verna, 460, 461, **463**
    vulgaris, 460, **461**, 462
        arcuata, 461
        longisiliqua, 462
Barbour, M. G., 425, 428
Barclay's willow, 113
Barnes, B. V., 7, 21, 22
Barratt's willow, 148
Barren-ground willow, 88
Barrens willow, 78
Barton, B. S., 223
Basin bladderpod, 631
Basket willow, 50, 149, 162
Basket-of-gold, 252
Bastard-cabbage, 441
Batacaeae, 184, **186**
Bates, V. M., 497
Batis, **187**
    argillicola, 187
    californica, 187
    maritima, *187*, 188
Baumier, 14
Baumier de l'ouest, 13
Bay willow, 44
Bay, C., 80
Bayer, R. J., 148
Bay-leaf willow, 44
Bay-leaved caper-tree, 197
Bean, W. J., 40
Bearberry willow, 73
Beautiful bladderpod, 631
Bebb, M. S., 42, 43, 152
Bebb's willow, 134
Beck, J. B., 485
Bee-plant, 205
Beerling, D. J., 68
Behnke, H.-D., 186
Beilstein, M. A., 226, 292, 293, 295, 333, 342, 447, 451
Beissmann, H., 43
Bell's twinpod, 628
Belle Isle cress, 463
Belyaeva, I. V., 43
Ben tree, 168
Bennion, G. C., 16
Benson, L. D., 206
Berg, T., 43
Beringia, 456
    bursifolia, 456
    virgata, 457
Berkutenko, A. N., 310
Berlandier bladderpod, 614
Bernard, F. G., 15, 22
Berteroa, 237, 241, **252**
    incana, 249, 252, **253**
    mutabilis, 252
    obliqua, 253
Bessey, C. E., 3
Bigflower bladderpod, 613
Bignoniaceae, 168
Bigtooth aspen, 20, 21
Bird spiderflower, 216
Bird-rape, 423
Biscutella
    californica, 607
        maritima, 608
    wislizeni, 606
Bittercress, 464
Bitterroot bladderpod, 644
Black cottonwood, 13
Black mustard, 422
Black willow, 36, 198
Black, L. D., 434, 435, 436, 442
Bladderbush, 206
Bladderpod, 206, 616
Blanket-leaf willow, 147
Bleeker, W., 490, 494
Blue-leaf willow, 119
Blue-stem willow, 155
Böcher, T. W., 73, 92, 270, 291, 316, 549, 551
Boechera, 226, 230, 233, 235, 236, 242, 243, 245, 257, 258, 347, 348, 349, 350, 367, 374, 408, 418, 720
    acutina, 362, **365**, 374
    angustifolia, 408
    arcuata, 364, **365**, 366, 367, 369, 404, 407
    atrorubens, 363, **366**, 367, 407
    beckwithii, 386, 399
    bodiensis, 361, **367**
    brachycarpa, 374, 382
    breweri, 354, 363, 365, 366, **367**, 395, 404
        breweri, *366*, 367, **368**
        shastaensis, 367, **368**
    burkii, 351, **368**
    calderi, 354, **368**, 374
    californica, 357, 358, 362, **369**, 407
    canadensis, 352, 357, 359, 363, **369**
    cascadensis, 364, **369**, 370, 390
    cobrensis, 357, 358, 360, 366, 367, **370**, 377, 388
    collinsii, 357, 366, **370**, 371, 382, 389
    consanguinea, 350, 358, **371**, 381
    constancei, 351, 366, **371**
    covillei, 351, 356, **371**, 372, 398
    crandallii, 361, 366, **372**, 393
    cusickii, 353, 363, 366, **372**, 377
    davidsonii, 351, 366, **372**, 373, 389, 406, 410
    demissa, 391, 392, 395
        languida, 387
        pendulina, 395
        russeola, 395
    dentata, 353, 357, 359, 361, **373**
    depauperata, 361, **373**
    dispar, 353, 366, **374**

778   INDEX

Boechera (continued)
    divaricarpa, 349, 356, 363, 365, *366*, **374**, 382
    drepanoloba, 355, 360, **374**, 375
    drummondii, 408
    duchesnensis, 354, **375**
    elkoensis, 353, 361, 362, 374, **375**, 397, 398
    evadens, 364, **376**
    exilis, 396, 402
    falcata, 348
    falcatoria, 363, **376**, 377
    falcifructa, 358, 367, **377**
    fecunda, 354, **376**, 377
    fendleri, 350, 359, 371, *376*, **378**, 381, 382, 393, 396, 408, 410
        spatifolia, 408
    fernaldiana, 361, 362, 367, *376*, 377, **378**
        fernaldiana, 376, 378, **379**
        vivariensis, 378, **379**
    formosa, 352, 376, *375*, **379**, 393, 400
    fructicosa, 360, **379**, 380
    glareosa, 363, 364, *376*, **380**, 382, 404
    glaucovalvula, 352, *366*, 376, **380**
    goodrichii, 350, 358, 371, **380**, 381
    gracilenta, 358, 363, **381**, 396
    gracilipes, 350, 358, *376*, **381**
    grahamii, 356, 359, 363, 374, **382**, 401
    gunnisoniana, 357, 359, 364, *376*, **382**
    harrisonii, 356, 362, **383**
    hastatula, 355, **383**
    hirshbergiae, 386
    hoffmannii, 356, 362, *376*, **383**, 384
    holboellii, 356, 370, 371, *376*, **384**, 396, 397, 401, 402
        pendulocarpa, 395
        pinetorum, 397
        secunda, 402
    horizontalis, 360, 361, **384**
    howellii, 351, 356, 372, **384**, *385*, 398
    inyoensis, 358, 361, **385**, 386, 407
    johnstonii, 353, *385*, **386**
    koehleri, 357, 365, *385*, **386**
        koehlerei, 386
    laevigata, 355, 368, **386**, 391
    languida, 358, 359, **387**
    lasiocarpa, 351, 352, *385*, **387**, 404
    lemmonii, 352, 353, 356, 360, 361, 373, 375, 384, *385*, **388**, 392, 400
    lignifera, 359, 360, 370, *385*, **388**
    lincolnensis, 351, 354, *385*, **388**, 389, 400
    lyallii, 351, 355, *366*, 368, 372, 373, *385*, **389**, 392, 394
    macounii, 357, 364, 380, **389**, 390
    microphylla, 364, 370, 380, 383, 389, *390*
        harrisonii, 383
    missouriensis, 355, 362, 387, **390**, 391
    nevadensis, 355, *390*, **391**
    ophira, 361, 362, *390*, **391**
    oxylobula, 351, 352, 387, *390*, **391**, 392
    paddoensis, 356, **392**
    pallidifolia, 362, 372, 375, 381, *390*, **392**, 393, 396, 411
    parishii, 351, *390*, **393**
    patens, 262
    pauciflora, 357, 363, 364, **393**, 407
    paupercula, 353, 364, 370, 373, 389, *390*, **394**, 399
    peirsonii, 356, 362, **394**, 395
    pendulina, 352, 359, 377, *393*, *394*, **395**, 400, 405
        russeola, 395
    pendulocarpa, 352, *394*, **395**, 396, 400
    perennans, 359, 360, 366, 375, 377, 381, 383, *394*, **396**, 398, 411, 412
    perstellata, 353, **396**
    pinetorum, 349, 350, 357, 359, 381, 393, **397**, 410
    pinzliae, 353, 362, **397**
    platysperma, 353, 375, 385, *394*, **397**, 398, 411
    polyantha, 354, **398**, 399
    porphyrea, 355, 358, 378, **398**
    pratincola, 361, 365, 374, **399**
    puberula, 351, 354, 386, **399**, 400, 409
    pulchra, 352, 354, 369, 375, 379, *399*, **400**, 412
        duchesnensis, 375
        gracilis, 411
        munciensis, 388
        pallens, 379
    pusilla, 352, 353, 356, 359, 363, **400**
    pygmaea, 353, 399, **401**
    quebecensis, 360, 374, 384, **401**
    rectissima, 350, 355, 356, 397, 399, **401**, 410
        simulans, 401
    repanda, 351, 353, 399, **402**
        greenei, 402
    retrofracta, 350, 352, 354, 356, 371, 374, 381, 393, 397, 398, 399, 400, 401, **402**, 403, 409, 410
    rigidissima, 363, 364, **403**
    rollei, 355, 399, **403**
    rollinsiorum, 357, 387, 399, **403**, 404
    rubicundula, 365, 367, **404**
    saximontana, 362, **404**
    schistacea, 355, **404**, *405*
    selbyi, 381, 393, 396
        inyoensis, 385
    serotina, 351, **405**, 406
    serpenticola, 354, 358, *405*, **406**
    shevockii, 351, *405*, **406**
    shockleyi, 354, 361, 386, **406**, *407*
    shortii, 373
    sparsiflora, 350, 363, 364, 365, 366, 367, 372, 374, 393, 397, **407**
        subvillosa, 393
    spatifolia, 359, 378, 387, **408**
    stricta, 350, 355, 360, 365, 366, 368, 374, 375, 380, 382, 399, 401, 403, **407**, **408**, 410
    subpinnatifida, 354, 357, 400, 406, **409**
    suffrutescens, 355, 356, 359, 371, 384, 403, **409**
    tenuis, 384
    texana, 355, 378, 398, **409**, 410
    thompsonii, 369, 392
    tiehmii, 351, 356, **410**
    tularensis, 349, 358, **410**
    ultra-alsa, 353, **410**, 411
    villosa, 363, *407*, **411**
        vivariensis, 379
    williamsii, 354, 404, **411**
        saximontana, 404, 411
    xylopoda, 354, 375, 400, **411**, 412
    yorkii, 351, **412**
Boechereae, 226, 257, **347**, 349, 558
Bog willow, 83
Bohm, B. A., 173, 175
Boivin, B., 16, 17, 22, 148
Bolbonac, 597
Bolleana poplar, 21
Bonpland's willow, 33
Boom, B. K., 18
Booth's willow, 113
Borgen, L., 597
Botschantzev, V. P., 511
Boucher, L. D., 4, 8
Bowman, J. L., 570
Boyd, S., 694
Brachiolobos, 493
    amphibius, 497
    hispidus, 502
    palustris, 501
    sylvestris, 505
Bracted-rocket, 435
Brassica, 226, 230, 231, 233, 244, **419**, 420, 436, 442, 443
    adpressa, 437
    alba, 442
    alboglabra, 423
    arvensis, 443
    campestris, 423
        oleifera, 423
    cheiranthos, 430
    chinensis, 423
    elongata, **420**, 421
        integrifolia, 421
    eruca, 434
    erucastrum, 430
    fruticulosa, 420, **421**
    geniculata, 437
    hirta, 420, 442
    japonica, 421
    juncea, 419, 420, **421**
        crispifolia, 421
        japonica, 421
    kaber, 443
    napobrassica, 422
    napus, 226, 420, **422**, 424
        napobrassica, 422
        napus, 422
        oleifera, 422
        rapifera, 422
    nigra, 420, **422**, 423, 437, 442
    oleracea, 420, 422, **423**
        acephala, 423
        botrytis, 423
        capitata, 423
        gemmifera, 423
        gongylodes, 423
        italica, 423
        napobrassica, 422
    orientalis, 517
    pekinensis, 423
    rapa, 420, 422, **423**, 424
        chinensis, 423, 424
        pekinensis, 423, 424
        rapa, 424
        sylvestris, 424
    sinapistrum, 443
    tournefortii, *416*, 419, 420, **424**
    vesicaria, 434
    violacea, 438
Brassicaceae, 190, 193, 194, 195, **224**, 225, 226, 227, 257, 270, 349, 448, 526, 267, 686, 694
    [group] Idahoa, 566
    tribe Alysseae, **247**
    tribe Anchonieae, **253**
    tribe Aphragmeae, **255**
    tribe Arabideae, 226, 227, **256**, 257, 349
    tribe Boechereae, 226, 257, **347**, 349
    tribe Brassiceae, 226, 227, **419**
    tribe Buniadeae, 227, **443**
    tribe Calepineae, **445**
    tribe Camelineae, 226, 227, 257, **447**
    tribe Cardamineae, **458**, 459, 505, 693
    tribe Chorisporeae, **510**
    tribe Cochlearieae, **514**
    tribe Conringieae, **516**
    tribe *Cremolobeae*, 227
    tribe Descurainieae, **517**
    tribe Erysimeae, **533**
    tribe Euclidieae, **545**
    tribe Eutremeae, **554**
    tribe Halimolobeae, 226, 257, **557**

tribe *Heliophileae*, 227
tribe Hesperideae, **561**
tribe Iberideae, **563**
tribe Isatideae, **567**
tribe Lepidieae, 227, **569**
tribe Lunarieae, **596**
tribe Malcolmieae, **597**
tribe Noccaeeae, 226, **598**
tribe Physarieae, **604**
tribe Sisymbrieae, **666**
tribe Smelowskieae, **671**
tribe Thelypodieae, 226, 227, 257, 485, 667, **676**
tribe Thlaspideae, 226, **744**
Brassicales, 165, 172, 186, 189, 226
Brassiceae, 226, 227, **419**
Brassicella, 429
Braya, 234, 235, 236, 238, 240, 241, 242, 243, **546**, 547, 549, 551
 alpina, 551
  americana, 548
  glabella, 547
 americana, 548
  arctica, 548
  bartlettiana, 548
   vestita, 548
 fernaldii, 546, **547**
 glabella, *541*, **547**
  glabella, *541*, 548
  prostrata, 548
  purpurascens, **548**
 graminea, 309
 henryae, 548
 humilis, *541*, 546, **548**, 549, 550, 551
  abbei, 549
  americana, 548
  arctica, 549
  ellesmerensis, 549, 550
  humilis, *541*, 549, 550
  interior, 549
  laurentiana, 549
  leiocarpa, 549
  maccallae, 549, 550
  novae-angliae, 549
  porsildii, 549, 550
  richardsonii, 549
  ventosa, 549
 intermedia, 549
 linearis, 546, **550**, 551
 longii, 546, 547, **551**
 novae-angliae, 549
  abbei, 549
  interior, 549
  laurentiana, 549
  ventosa, 549
 oregonensis, 413
 pectinata, 416
 pilosa, 546, **551**
  thorild-wulffii, 551
 purpurascens, 548
  dubia, 548
  fernaldii, 547
  longii, 551
  pilosa, 551
  thorild-wulffii, 551
 richardsonii, 549
 siliquosa, 551

thorild-wulffii, 546, **551**
 glabrata, **552**
 thorild-wulffii, **552**
Brayshaw, T. C., 14, 15, 16, 17, 22
Bremer, K., 205
Brewer's willow, 160
Bricker, J. S., 529
Brinker, A. M., 51
Brittle willow, 43
Britton, N. L., 7, 598
Broccoli, 423
Brochmann, C., 28, 270
Brown mustard, 421, 422
Brown, A., 598
Brunsfeld, S. J., 51, 52, 56, 107, 139
Brush-holly, 164
Brussels sprouts, 423
Büchler, W., 51
Buck, R. E., 677, 679, 680, 681, 683, 684, 705, 706, 709, 714, 715, 722, 723
Buniadeae, 227, **443**
Bunias, 224, 227, 236, 241, **444**
 cakile, 425
 edentula, 427
 erucago, **444**, *445*
 orientalis, 444, **445**
 syriaca, 553
Burro-fat, 206
Bursa, 453
Butters, F. K., 494
Buxton, E. G., 176

Cabbage, 419, 423
Cabomba
 pinnata, 173
Cakile, 232, 234, 238, 244, 419, **424**, 425
 aequalis, 426
 americana, 428
 arabica, 425
 californica, 428
 chapmanii, 426
 constricta, 425, **427**
 cubensis, 426
 edentula, 425, **427**, 428
  californica, 428
  edentula, 427, **428**
  harperi, **428**
  lacustris, **428**
 fusiformis, 426
 geniculata, 425, **426**
 harperi, 428
 lacustris, 428
 lanceolata, 425, **426**
  alacranensis, 426
  fusiformis, **426**, 427
  geniculata, 426
  lanceolata, 426
  pseudoconstricta, **426**, 427
 maritima, **425**
  americana, 428
  cubensis, 426
  geniculata, 426
  maritima, 425
Calder, J. A., 509, 510
Calder's bladderpod, 629

Calepina, 230, 239, **446**
 cochlearioides, 446
 corvini, 446
 irregularis, *445*, **446**
Calepineae, **445**
Cambess, 191
Camelina, 236, 240, **451**
 alyssum, 451, **453**
 austriaca, 497
 barbareifolia, 498
 microcarpa, 451, *452*
 rumelica, 451, *452*, 453
 sativa, 451, **453**
  microcarpa, 452
Camelineae, 226, 227, 257, **447**
Campe, 460
 orthoceras, 462
 verna, 463
Candytuft, 563
Cannabis, 223
Canola, 226, 422, 423, 424
Caper, 198
Caper Family, 194
Caper-tree, 196
Capparaceae, 184, 190, **194**, 195, 200, 222, 226
 subfam. Capparoideae, 194
 subfam. Cleomoideae, 194
Capparales, 186, 189, 226
Capparis
 sect. *Cynophalla*, 196
 sect. *Quadrella*, 197
 subg. *Cynophalla*, 196
 subg. *Quadrella*, 197
 atamisquea, 196
 brevisiliqua, 197
 emarginata, 198
 flexuosa, 197
 incana, 198
 jamaicensis, 198
 karwinskiana, 198
 longifolia, 198
 pauciflora, 198
Capparoideae, 194
Capsella, 236, 240, **453**, 454
 bursa-pastoris, *452*, **454**
 elliptica, 530
 procumbens, 530
 pubens, 559
Carara, 570
 coronopus, 578
 didyma, 580
Cardamine, 224, 228, 229, 231, 232, 233, 234, 244, **464**, 485, 489, 492
 acuminata, 479
 angulata, 467
  alba, 467
  hirsuta, 467
  pentaphylla, 467
 angustata, 224, 466, **467**, 468
  multifida, 473
  ouachitana, 467
 anomala, 477
 arenicola, 480
 articulata, 512
 bellidifolia, *463*, 464, 465, **468**
  beringensis, 468

 laxa, 468
 pachyphylla, 468
 pinnatifida, 468
 blaisdellii, 464, 467, **468**
 breweri, 467, **469**
  leibergii, 469
  orbicularis, 469
  oregana, 469, 482
 bulbosa, 465, **469**
 californica, 465, 466, **469**
  brevistyla, 469
  cardiophylla, 469
  cuneata, 469
  fecunda, 469
  gemmata, 478
  integrifolia, 469
  pachystigma, 480
  pubescens, 469
  robinsoniana, 469
  rupicola, 483
  sinuata, 469
 callosicrenata, 469
 cardiophylla, 469
 clematitis, 465, **470**
 concatenata, 224, 466, **470**, 477
 constancei, 465, **471**
 cordifolia, *463*, 465, **471**, 472
  cardiophylla, 471
  diversifolia, 471
  incana, 471
  lyallii, 471
  pubescens, 471
 corymbosa, 464
 cuneata, 469
 curvisiliqua, 492
 dentata, 483
 digitata, 467, **472**
  oxyphylla, 472
 diphylla, 224, 466, **472**, 477
 dissecta, 224, 466, **473**
 douglassii, 465, **473**
 filifolia, 694
 flagellifera, 464, 466, **474**
  hugeri, 474
 flexuosa, 466, **474**, 475
  debilis, 474, 475
  gracilis, 480
  pensylvanica, 482
 foliacea, 469
 gambelii, 491
 gemmata, 478
 globosa
  sphaerocarpa, 504
 hederifolia, 469
 heterophylla, 468
 hirsuta, 466, **475**
  acuminata, 479
  bracteata, 479
  flexuosa, 474
  kamtschatica, 484
  oligosperma, 479
  parviflora, 479
  pensylvanica, 482
 holmgrenii, 464, 467, **475**
 hugeri, 474
 hyperborea, 472
  oxyphylla, 472
 impatiens, 465, **476**

Cardamine impatiens (continued)
    angustifolia, 476
  incana, 471
  ×*incisa*, 464
  infausta, 471
  integrifolia, 469
    sinuata, 469
  kamtschatica, 484
  laciniata
    integra, 470
    lasiocarpa, 470
    multifida, 473
  leibergii, 469
  longii, 465, 473, **476**
  ludoviciana, 493
  lyallii, 471
  macrocarpa, 466, **476**
    macrocarpa, 476
    texana, 476
  maxima, 224, 466, **477**
  micranthera, 465, **477**
  microphylla, 467, 468, **477**
    blaisdellii, 468
  minuta, 477
  modocensis, 469
  multifida, 527
  multifolia, 482
  neglecta, 479
  nudicaulis, 512
  nuttallii, 466, **478**
    covilleana, 478
    dissecta, 478
    gemmata, 478
    pulcherrima, 478
  nymanii, 464, 466, **478**
  occidentalis, 466, **479**
  oligosperma, 466, 475, **479**, 484
    bracteata, 479
    kamtschatica, 484
    lucens, 479
    unijuga, 479
  orbicularis, 469
  oregana, 469, 482
  pachystigma, 465, **480**
    dissectifolia, 469
  palustris, 501
    hispida, 502
    jonesii, 499
  parviflora, 466, 473, **480**
    arenicola, 480
    parviflora, 480
  pattersonii, 464, 465, **480**, 481
  paucisecta, 469
  penduliflora, 466, *481*, 482
  pensylvanica, 466, **482**
    brittoniana, 482
  petraea, 450
  pratensis, 466, **482**, 483
    angustifolia, 478
    occidentalis, 479
    pulcherrima, 478
    tenella, 478
  purpurea, 465, 467, **483**
    albiflora, 483
    albiflos, 483
    lactiflora, 483
    quercetorum, 478
  rariflora, 481
  rhomboidea, 469
    hirsuta, 469
    parviflora, 469
    pilosa, 469
  richardsonii, 472
  rotundifolia, 465, **483**
    diversifolia, 482
  rupicola, 467, **483**, 484
  scutata
    flexuosa, 474
    sinuata, 469
  sylvatica
    kamtschatica, 484
  tenella, 478
    covilleana, 478
    dissecta, 478
    quercetorum, 478
  teres, 505
  uintahensis, 471
  umbellata, 464, 467, **484**
  uniflora, 489
  unijuga, 479
  vallicola, 469
    leibergii, 469
  virginica, 493
Cardamineae, **458**, 459, 505, 693
Cardaminopsis
  kamchatica, 450
  lyrata, 449
Cardaria, 226, 570, 571, 581
  chalepensis, 577
  draba, 578, 581
    chalepensis, 577
    repens, 577
    latifolia, 585
  pubescens, 576
    elongata, 576
  repens, 577
Carica, **171**
  papaya, 169, 170, **171**
Caricaceae, 167, **170**
Carlquist, S., 186
Caroli-Gmelina
  palustris, 501
  sylvestris, 504
Carolina poplar, 18
Carolina willow, 34
Carrichtera, 232, 238, 419, **429**
  annua, 427, **429**
Carsonia, 200, **208**
  sparsifolia, 202, **209**
Cartiera, 700
  arguta, 709
  barbata, 704
  cordata, 708
  crassifolia, 709
  cuneata, 709
  howellii, 716, 732
  leptopetala, 709
  multiceps, 709
Caryophyllaceae, 204
Caryophyllales, 3
Cascade willow, 82
Cat's-whiskers, 223
Caulanthus, 228, 230, 231, 233, 234, 242, 245, 246, 677, 679, 681, 701
  amplexicaulis, 678, **679**, 684
    amplexicaulis, 679
    barbarae, 679, 684
  anceps, 677, 678, **679**
  barnebyi, 678, **679**, 680
  californicus, 676, 677, 678, 680
  cooperi, 678, **680**
  coulteri, 677, 678, **680**, 681, 684
    lemmonii, 684
  crassicaulis, 678, **681**
    crassicaulis, 681
    glaber, 681
    major, 684
  divaricatus, 725
  flavescens, 677, 678, **681**, 682
  glaber, 681
  glaucus, 678, **682**
  hallii, 678, **682**
  hastatus, 686
  heterophyllus, 678, **682**
  inflatus, 676, 678, **683**
  lasiophyllus, 677, 678, **683**
    inalienus, 683
    rigidus, 683
    utahensis, 683
  lemmonii, 678, 679, 681, **684**
  major, 678, **684**
    nevadensis, 684
  pilosus, 678, **684**, 685
  procerus, 681
  senilis, 681
  simulans, 678, **685**
  stenocarpus, 682
Cauliflower, 423
Caulostramina, 226, 689, 690
  jaegeri, 692
Cayluseae, 190
Celome, 205
  platycarpa, 208
Centrospermae, 186
Ceulemans, R., 52
Chamaeplium, 667
  polyceratium, 669
Chambers's twinpod, 631
Chamisso willow, 71
Charlock, 443
Chase, M. W., 4, 8, 9
Chaunanthus, 485
Cheiranthus, 534
  [unranked] *Iodanthus*, 484
  alpestris, 538
  ammophilus, 536
  angustatus, 538
  arenicola, 536
  argillosus, 538
  aridus, 538
  arkansanus, 538
  asper, 537
  asperrimus, 538
  bakeri, 538
  bicornis, 254
  californicus, 538
  capitatus, 537
  cheiranthoides, 539
  cheiri, 539
  elatus, 538
  grandiflorus, 542
  hesperidoides, 485
  incanus, 254
  inconspicuus, 541
  insulare, 542
  longipetalus, 254
  nivalis, 538
    amoenus, 538
    oblanceolatus, 538
    occidentalis, 543
    pacificus, 538
    pallasii, 543
    perennis, 543
    pygmaeus, 543
    radicatus, 538
    suffrutescens, 544
    syrticola, 541
    wheeleri, 538
Cheirinia, 534
  ammophila, 536
  amoena, 538
  argillosa, 538
  arida, 538
  arkansana, 538
  aspera, 537
  asperrima, 538
  bakeri, 538
  brachycarpa, 538
  cheiranthoides, 539
  cockerelliana, 538
  desertorum, 538
  elata, 538
  inconspicua, 541
  nevadensis, 543
  nivalis, 538
    radicata, 538
  oblanceolata, 538
  occidentalis, 543
  parviflora, 541
  radicata, 538
  repanda, 544
  syrticola, 541
  wheeleri, 538
Chenopodiaceae, 186
Chenopodiales, 186
Chinese cabbage, 420, 424
Chinese mustard, 421, 424
Chlorocrambe, 233, 246, **685**
  hastata, **686**
Chmelař, J., 41, 42
Chong, D. K. X., 51
Chopra, R. N., 222
Chorispermum
  tenellum, 511
Chorispora, 224, 232, 244, **510**
  tenella, 508, **511**
Chorisporeae, **510**
Chosenia, 4, 23, 24
Chweya, J. A., 223
Clammyweed, 201
Cleomaceae, 195, **199**, 200, 205, 222, 226
Cleome, 201, 205, **215**, 218, 222
  [unranked] *Atalanta*, 205
  sect. *Tarenaya*, 218
  sect. *Thylacophora*, 208
  aculeata, 216, 220
    diffusa, 220
  aldenella, 204
  ciliata, 216

cuneifolia, 742
diffusa, 220
dodecandra, 201
gynandra, 223
hassleriana, 218, 219
heterotricha, 223
houtteana, 218
iberica, 216
isomeris, 206
jonesii, 207
lutea, 206
  jonesii, 207
multicaulis, 208
ornithopodioides, **216**, 217
pentaphylla, 223
pinnata, 697
platycarpa, 208
polygama, 218
pubescens, 219
pungens, 219
rutidosperma, **216**
serrata, 218
serrulata, 207
  angusta, 207
sonorae, 208
sparsifolia, 209
speciosa, 217
speciosissima, 217
spinosa, 218, 219
tonduzii, 219
uniglandulosa, 204
viscosa, 222
Cleomella, 200, 205, 208, **209**
  angustifolia, **210**, *211*
  brevipes, 209, **210**, 212
  cornuta, 212
  hillmanii, **210**
    goodrichii, 210, **211**
    hillmanii, 210, **211**
  longipes, 210, **211**
    grandiflora, 211
  macbrideana, 211
  mexicana, 209
  montrosae, 212
  nana, 212
  obtusifolia, 209, **212**
    florifera, 212
    jonesii, 212
    pubescens, 212
  oöcarpa, **213**
  palmeriana, 209, **212**
    goodrichii, 211
  parviflora, 209, **212**
  perennis, 209
  plocasperma, 210, 212, **213**
    mojavensis, 213
    stricta, 213
  taurocranos, 212
Cleomoideae, 194
Cleoserrata, 200, 215, **216**, 217
  serrata, **217**, **218**
  speciosa, **217**, 218
Clypeola
  alyssoides, 248
  maritima, 598
  minus, 249
Coastal plain willow, 34
Coastal willow, 127
Cochlearia, 232, 239, **514**, 515
  amphibia, 497

aquatica, 497
arctica, 515
  oblongifolia, 515
armoracia, 459
  aquatica, 497
austriaca, 497
coronopus, 578
cyclocarpa, 516
draba, 577, 581
fenestrata, 515
groenlandica, *513*, **515**, 516
  oblongifolia, 515
officinalis, **515**, 516
  arctica, 515
  oblongifolia, 515
  sessilifolia, 516
pyrenaica, 515
sessilifolia, 515, **516**
siliquosa, 309
spathulata, 309
tridactylites, 515, **516**
Cochlearieae, **514**
Cochleariopsis, 514
  groenlandica, 515
    oblongifolia, 515
Cody, W. L., 499
Coelophragmus, 229, 245, **687**
  auriculatus, *686*, **687**
  umbrosus, 687, **688**
Coincya, 233, 244, **429**
  monensis, *427*, **430**
    recurvata, *427*, **430**
Coin-leaf willow, 68
Cole, 419, 420
Collards, 423
Collared lemming, 80
Columbia River willow, 59
Common nasturtium, 166
Common twinpod, 633
Conringia, 230, 246, **517**
  orientalis, *513*, **517**
Conringieae, **516**
Coquillat, M., 454
Corn-mustard, 460
Corona-de-Cristo, 184
Coronilla, 164
Coronillo, 164
Coronopus, 226, 570, **571**
  didymus, 580
  procumbens, 578
  ruellii, 578
  squamatus, 578
  verrucarius, 578
Correll, D. S., 20
Cotonier, 17
Cottonwood, 5, 7, 15, 16, 19, 20
Coulterina, 616
  didymocarpa, 633
  geyeri, 638
  oregona, 655
Cowen, W. F., 50
Coyote willow, 55
Crack willow, 43
Crambe, 225, 232, 238, **430**
  corvini, 446
  maritima, *431*
Crawford, D. J., 16
Creeping twinpod, 644
Creeping willow, 75

Cremolobeae, 227
Cremolobus, 227
Cress, 461, 570
Cressy-greens, 461
Cristatella, 201
  erosa, 203
  jamesii, 203
Critchfield, W. B., 7, 25
Cronquist, A., 4, 8, 28, 37, 165, 172, 184, 194, 211, 226
Cross-weed, 433
Crovello, T. J., 157
Crown-of-thorns, 184
Crucifer, 225
Cruciferae, 224
Crucihimalaya, 558
Cucurbitaceae, 170
Cultivated radish, 439
Curtin, L. S. M., 207
Curtis, J. D., 7
Curved bladderpod, 632
Cushion bladderpod, 659
Cusickia
  douglasii, **413**
Cusickiella, 236, 238, 240, **412**, 413
  douglasii, *413*
  quadricostata, 413, **414**
Cuspidaria, 534
Cutleaf mignonette, 192
Cynophalla, 195, **196**, 197
  flexuosa, *196*, **197**
    flexuosa, 197
    polyantha, 197

Dallwitz, M. J., 28, 165
Dame's-rocket, 562
Dame's-violet, 562
Daniel, T. F., 189
Daphne willow, 161
Dark-leaved willow, 125
Darrow, R. A., 206
Davis, M. B., 68
Dawson, T. E., 80
De Cock, K., 43
de Wit, H. C. D., 190, 193
Del Norte willow, 161
Del Tredici, P., 25
Denseflower bladderpod, 633
Densmore, R. A., 24, 25, 146
Dentaria, 464
  anomala, 477
  bifolia, 472
  californica, 469
    cardiophylla, 469
    cuneata, 469
    integrifolia, 470
    pachystigma, 480
    sinuata, 470
  cardiophylla, 470
  concatenata, 470
  corymbosa
    grata, 480
  cuneata, 470
  diphylla, 472
  dissecta, 473
  douglassii, 473
  furcata, 473
  gemmata, 478

  grandiflora, 467
  heterophylla, 467
  multifida, 473
  incisa, 472
  integrifolia, 470
    californica, 470
    cardiophylla, 470
    traceyi, 470
  laciniata, 470
    alterna, 470
    coalescens, 470
    integra, 470
    latifolia, 470
    multifida, 473
    opposita, 470
  macrocarpa, 478
    pulcherrima, 478
  maxima, 477
  multifida, 473
  pachystigma, 480
    dissectifolia, 470
  quercetorum, 478
  rhomboidea, 469
  rotundifolia, 483
  rupicola, 483
  sinuata, 470
  tenella, 478
    palmata, 478
    pulcherrima, 478
    quercetorum, 478
Descurainia, 224, 235, 237, 241, 243, **518**, 519, 528
  adenophora, 519, **520**, **521**, 525
  andrenarum, 528
  brachycarpa, 527
    eglandulosa, 524
    nelsonii, 525
  brevisiliqua, 520, **521**
  californica, 520, **521**, *522*, 523
  canescens, 527
  halictorum
    andrenarum, 528
  incana, 519, 520, 521, **522**, 523, 529
    brevipes, 522
    incisa, 523
    macrosperma, 522
    major, 522
    procera, 522
    viscosa, 523
  incisa, 519, 520, 521, **523**
    filipes, 524
    incisa, **523**, 524
    leptophylla, 523
    paysonii, 523, **524**, 525
    viscosa, 523
  intermedia, 527
  kenheilii, 520, **524**
  kochii, 518
  longepedicellata, 520, **524**, 525
    glandulosa, 524
  magna, 527
  menziesii
    ochroleuca, 528
  multifida, 527
  multifoliata, 527
  nelsonii, 520, **525**, 526

Descurainia (continued)
obtusa, 519, 521, **525**
adenophora, 520
brevisiliqua, 521
paradisa, 519, 520, 525, **526**
nevadensis, 526
pinnata, 519, 520, 521, 523, 525, **526**, 527
brachycarpa, 525, **527**, 528
filipes, 524, 525, 527
glabra, 527, **528**
halictorum, 527
intermedia, 527
menziesii, 527
nelsonii, 525, 527
ochroleuca, 527, **528**
paradisa, 526, 527
paysonii, 524, 527
pinnata, **527**
ramosissima, 527
richardsonii
alpestris, 522
brevipes, 522
incisa, 523
macrosperma, 522
procera, 522
sonnei, 523
viscosa, 523
rydbergii, 523
eglandulosa, 524
serrata, 523
sophia, 518, 520, **528**
sophioides, 518, 519, 520, **529**
tanacetifolia, 518
torulosa, 519, **529**
Descurainieae, **517**
Desert stinkweed, 210
Desert-spike, 191
Detling, L. E., 478, 519, 521, 524, 525, 526, 527
Devils Gate twinpod, 635
Dewy-stemmed willow, 155
Diamond willow, 120
Diamond-leaf willow, 138
Dicotyledoneae, 3
Dimorphocarpa, 237, 240, **604**
candicans, **605**
membranacea, 605
palmeri, 605
pinnatifida, **605**
wislizeni, 605, **606**
Diplotaxis, 229, 234, 243, 244, **432**, 436
cossoniana, 435
erucoides, 432, **433**, 435
muralis, 431, 432, **433**
tenuifolia, 432, **433**
viminea, 433
Disaccanthus, 700
arizonicus, 708
carinatus, 708
luteus, 708
mogollonicus, 708
tortuosus, 722
validus, 708
Dithyrea, 237, 240, 571, **607**
californica, 606, **607**
clinata, 607

maritima, 608
clinata, 607
griffithsii, 606
maritima, 607, **608**
wislizeni, 606
griffithsii, 606
palmeri, 605
Divide bladderpod, 647
Dodge's willow, 71
Dog mustard, 435
Dolan, R. W., 706, 719
Dole, J. A., 183
Dorn, R. D., 23, 26, 28, 30, 33, 54, 56, 57, 58, 59, 65, 104, 110, 111, 115, 117, 120, 121, 122, 127, 128, 131, 136, 157, 159, 160, 161, 374, 396, 397, 404
Dorn's twinpod, 634
Dorofeev, V. I., 433
Double bladderpod, 628
Douglas, D. A., 67
Douglas's bladderpod, 634
Downy poplar, 12
Draba, 234, 235, 236, 237, 238, 240, 241, 243, **269**, 270, 271, 290, 310, 312, 314, 319, 413, 571
abajoensis, 276, **288**, 339
acinacis, 340
adamsii, 326
albertina, 270, 272, 274, 276, 281, 282, **288**, 289, 297, 298, 303, 327, 331, 333, 341
aleutica, 270, 284, **289**, 290
allenii, 316
alpicola, 310
alpina, 285, **290**, 302, 318, 325, 330
bellii, 301
corymbosa, 301
gracilescens, 337
hydeana, 290
inflatisiliqua, 290
micropetala, 319
oxycarpa, 325
pilosa, 330
ammophila, 304
andina, 325
apiculata, 305, 308
daviesiae, 305
aprica, 273, **290**, 296
arabis, 291
arabisans, 275, 277, **291**, 308
canadensis, 307
orthocarpa, 307
superiorensis, 291
arctica, 277, 286, **291**, 292, 301, 324
arctica, 291
groenlandica, 323
ostenfeldii, 291
arctogena, 274, 285, **292**, 323
argyrea, 281, 283, **292**
glabrescens, 292
sphaerocarpa, 339
arida, 279, **292**, 293

aspera, 330
asprella, 284, **289**, 293, 347
asprella, **289**, **293**
kaibabensis, 293
stelligera, **293**
zionensis, 293, 346
asterophora, 270, 276, 278, 282, **294**
macrocarpa, 294
aurea, 277, 278, 279, **294**, 295, 296
aureiformis, 294
leiocarpa, 294
luteola, 294
stylosa, 311
aureiformis, 294
leiocarpa, 294
aureola, 278, **289**, **295**
paniculata, 295
bakeri, 294
barbata, 301
treleasei, 323
behringii, 289
bellii, 301
bifurcata, 272, 274, **295**
boecheri, 316
boerhaavii, 345
borealis, 277, 279, **296**, 308, 315
maxima, 296
brachycarpa, 270, 273, **296**
fastigiata, 290
brachystylis, 273, 278, **297**, 335
breweri, 277, 283, **297**, 298, 299
cana, 298
sublaxa, 297
burkei, 280, **298**, 308
cacuminum, 270
caeruleomontana, 306
piperi, 306
caesia, 322
calcifuga, 324
californica, 275, 277, 283, **298**
cana, 277, 297, **298**, 299, 314, 331
canadensis, 307
pycnosperma, 307
carnosula, 270, 279, **299**, 300, 312
caroliniana, 333
dolichocarpa, 333
hunteri, 333
micrantha, 333
stellifera, 333
umbellata, 333
cascadensis, 331
chamissonis, 275, **300**
chrysantha, 306
crassa, 302
dasycarpa, 341
exunguiculata, 306
gilgiana, 340
graminea, 309
hirticaulis, 309
cinerea, 277, 286, 292, **300**, 301
arctica, 291

arctogena, 292
groenlandica, 323
clivicola, 322, 323
coloradensis, 333
columbiana, 331
confusa, 312
contorta, 312
corrugata, 272, 274, 299, **301**, 335
saxosa, 335
corymbosa, 270, 287, 290, **301**, 302
crassa, 274, **302**, 309, 341
crassifolia, 272, 273, 274, 279, 280, 281, 289, **302**, 303, 307
albertina, 288
nevadensis, 288
parryi, 302
crockeri, 413
cruciata, 280, 282, **303**, 337
integrifolia, 336
cuneifolia, 271, 273, 298, **303**, 331, 334
brevifolia, 304
californica, 298
cuneifolia, **304**
foliosa, 304
helleri, 304
integrifolia, **304**
leiocarpa, 304
platycarpa, 330
sonorae, 303, **304**
cusickii, 284, **304**, 305, 329, 342
pedicellata, 304, 328
cyclomorpha, 284, **305**, 316
dasycarpa, 341
daurica, 307
daviesiae, 280, **305**
decumbens, 294
deflexa, 288
densifolia, 280, 282, 284, 288, 299, **306**, 308
apiculata, 308
daviesiae, 305
decipiens, 308
globosa, 308
dentata, 332
dictyota, 321
dolichopoda, 331
douglasii, 413
crockeri, 413
elongata, 317
eurycarpa, 348
exalata, 313
exunguiculata, 275, **306**, 307, 310, 346
fernaldiana, 316
fladnizensis, 270, 275, 281, 303, **307**, 316, 320
pattersonii, 307
frigida
kamtschatica, 300
gilgiana, 340
glabella, 275, 277, 291, 295, 296, **307**, 308
glabella, 307
megasperma, 307
orthocarpa, 307

pycnosperma, 307, 308
glacialis, 290
    pectinata, 306
globosa, 280, 305, **308**
    sphaerula, 306
gmelinii, 337
graminea, 273, **309**
grandis, 276, **309**, 310
grayana, 275, 306, 307, **310**, 346
gredinii, 325
greenei, 309
groenlandica, 323
    arctogena, 292
hatchiae, 309
heilii, 273, 274, **311**
helleri, 304
helleriana, 272, 278, 279, **311**, 330, 345
    bifurcata, 295
    blumeri, 329
    leiocarpa, 311
    neomexicana, 311
    patens, 311
    petrophila, 329
    pinetorum, 311
henneana, 307
    maccallae, 294
hirta, 308
    laurentiana, 307
    norvegica, 322
    pycnosperma, 307
    siliquosa, 340
    tenella, 314
hispidula, 333
hitchcockii, 288, **309**, **311**, 312, 314
howellii, 273, 279, 280, 284, 300, **312**
    carnosula, 299
hyperborea, 310
    spathulata, 309
incana, 272, **312**
    arabisans, 291
    confusa, 312
    contorta, 312
    glabriuscula, 291
incerta, 276, 278, 282, 283, 286, **313**
    laevicapsula, 313
    peasei, 313
incrassata, 280, **313**
inexpectata, 277, **313**, 314
integrifolia, 304, 336
jaegeri, 288, 312, **314**
juniperina, 328
juvenilis, 276, 278, 282, 283, *309*, **314**, 315, 324
kamtschatica, 300
kananaskis, 314, 315
kassii, 282, **315**
kjellmanii, 301
kluanei, 275, **315**
lactea, 270, 273, 275, 279, 280, 286, **316**
laevicapsula, 313
lanceolata, 299
lapilutea, 331
laurentiana, 307, 308
lemmonii, 284, 287, 305, 313, **316**, 318
cyclomorpha, 305
incrassata, 313
lonchocarpa, 275, 277, 280, 286, 292, 314, **317**, 322, 327, 331
    dasycarpa, 331
    denudata, 317
    exigua, 317
    kamtschatica, 300
    lonchocarpa, 317
    semitonsa, 317
    thompsonii, 317
    vestita, 317
longipes, 314, 315
longisquamosa, 284, 316, **317**, 318
luteola, 294
    minganensis, 294
maccallae, 294
macouniana, 340
macounii, 280, 282, 285, 287, 290, **318**
macrocarpa, 301
magellanica
    cinerea, 300
maguirei, 280, 282, 298, **318**
    burkei, 298
malpighiacea, 275, **318**, *319*, 337, 339
maxima, 296
megasperma, 307
micrantha, 333
micropetala, 285, 287, **319**, 320, 324, 326
minganensis, 294
mogollonica, 272, 276, 278, *319*, **320**
monoensis, 275, 281, 283, 284, **320**
montana, 333
mulfordiae, 306
mulliganii, 279, 281, **321**
murrayi, 276, **321**
nelsonii, 306
nemoralis, 321
nemorosa, 272, **321**, 333
    brevisilicula, 321
    hebecarpa, 321
    leiocarpa, 321
    stenoloba, 340
neomexicana, 311
    robusta, 311
nitida, 288
    nana, 288
    praelonga, 340
nivalis, 270, 275, 277, 283, 300, 316, 317, **322**, 326
    brevicula, 331
    denudata, 317
    elongata, 317, 322
    exigua, 317
    kamtschatica, 300
    lonchocarpa, 317
    thompsonii, 317
norvegica, 277, 279, 287, 292, 308, **322**, 323
    clivicola, 322, 323
    hebecarpa, 322
    norvegica, 323
pleiophylla, 307
novolympica, 286, 287, **323**, 328, 344
oblongata, 274, 285, 292, **323**, 324
    minuta, 319
ogilviensis, 224, 273, 279, 281, 283, 285, **324**
oligantha, 340
oligosperma, 280, 283, 286, 313, *319*, **324**, 325, 328, 344
    andina, 324
    juniperina, 328
    leiocarpa, 324
    microcarpa, 324
    pectinata, 306
    pectinipila, 328
    saximontana, 324
    sphaeroides, 339
    subsessilis, 324
oreibata, 280, 312, **325**, 336
    oreibata, 325
    serpentina, 325, 335
×*ostenfeldii*, 291
ostenfeldii, 291
    ovibovina, 291
    ovibovina, 291
oxycarpa, 285, 286, 287, 290, **325**
oxyloba, 339
palanderiana, 281, 283, 316, **326**
pallida, 311
parryi, 302
patens, 311
pattersonii, 307
    hirticaulis, 307
pauciflora, 285, 320, **326**
    micropetala, 319
paucifructa, 276, 281, **327**
paysonii, 286, 287, 306, 312, 323, **327**, 328
    treleasei, 323, 328
peasei, 313
pectinata, 306
pectinipila, 283, 325, **328**
pedicellata, 282, 284, **328**, 329, 340
    pedicellata, 329
    wheelerensis, 328, 329
pennellii, 278, 283, **329**
petrophila, 278, **329**, 330, 345
    viridis, 345
pilosa, 281, 285, 290, **330**
pinetorum, 311
platycarpa, 273, *327*, **330**, 331
porsildii, 275, 281, **331**
    brevicula, 331
praealta, 273, 278, 308, **331**, 332
    yellowstonensis, 331
praecox, 345
pterosperma, 270, 286, **332**
pycnosperma, 307, 308
quadricostata, 414
ramosissima, 279, **332**
    glabrifolia, 332
ramulosa, 277, **332**, 333, 338
rectifructa, 273, *327*, **333**
repens, 337
reptans, 271, 272, 273, **333**, 334
    micrantha, 333
    stellifera, 333
ruaxes, 281, 286, 287, **334**, 344
rupestris, 322
santaquinensis, 272, 278, 297, **334**, 335
saximontana, 324
saxosa, 278, 279, 286, 287, **335**
scandinavica
    hebecarpa, 322
scotteri, 286, **335**
serpentina, 279, **335**, 336
sharsmithii, 282, *336*, 337
sibirica, 279, 281, 319, 324, **337**
    arctica, 337
sierrae, 283, 286, **337**
simmonsii, 285, **337**
smithii, 277, **338**
sobolifera, 276, 278, 287, 293, 305, 333, **338**, 342, 347
    uncinalis, 338
sonorae, 304
    integrifolia, 304
sornborgeri, 307, 308
spathulata, 309
spectabilis, 276, 278, 288, 319, *336*, 337, **339**
    bella, 339
    dasycarpa, 341
    glabrescens, 288
    oxyloba, 339
    purpusii, 339
sphaerocarpa, 283, 292, **339**
sphaeroides, 280, 282, 284, 329, **339**, 340
    cusickii, 304
sphaerula, 306
standleyi, 274, **340**
stenoloba, 272, 273, 276, 278, 282, 287, 289, **340**, 341
    nana, 288
    oligantha, 340
    ramosa, 288
stenopetala, 285, **341**
    purpurea, 341
streptobrachia, 275, 277, **341**, 346
streptocarpa, 274, *336*, **342**
    grayana, 310
    tonsa, 342
stylosa, 311
subalpina, 279, 281, **342**, 347
subcapitata, 281, 284, 285, 316, 326, **342**
subsessilis, 324
subumbellata, 283, **343**, 344
surculifera, 294
tonsa, 342
trichella, 322

Draba (continued)
  trichocarpa, 279, 287, **344**
  uber, 294
  umbellata, 333
  unalaschkiana, 296
  uncinalis, 338
  unilateralis, 268
  valida, 298, 299
  ventosa, 284, 293, 334, **344**
    ruaxes, 334
  verna, 270, 271, **345**
    aestivalis, 345
    boerhaavii, 345
  vestita, 301, 327
  viperensis, 330
  viridis, 279, 330, **345**
  wahlenbergii, 307
  weberi, 274, **346**
  yellowstonensis, 331
  yukonensis, 274, *343*, **346**
  zionensis, 282, **346**, 347
Dracamine, 464
  bulbosa, 469
  pensylvanica, 482
  pratensis, 482
  purpurea, 473
Drummond's willow, 135
Drumstick tree, 168
Drumstick Tree Family, 167
Drury, W. H. Jr., 673, 675
Dryopetalon, 231, 232, 245,
    246, 687, **688**, 693, 724
  runcinatum, *686*, 687, **688**
    laxiflorum, 688
  viereckii, 688, **689**
Duck River bladderpod, 612
Dudley Bluffs bladderpod, 632
Dudley, T. R., 249, 250
Dusky willow, 58
Dwarf meadowfoam, 183
Dwarf prairie willow, 130
Dwarf snow willow, 66
Dyer's-rocket, 193
Dyers woad, 568

Eared willow, 133
Earleaf bladderpod, 612
Early winter cress, 463
Eastern cottonwood, 17
Eckenwalder, J. E., 3, 5, 7,
    8, 14, 15, 16, 17, 19,
    20
Edible-rocket, 434
Edlund, S. A., 147
Egginton, P. A., 147
Ekman, E., 270
Elven, R., 269, 270
Engelmann's bladderpod, 635
Engler, H. G. A., 3
Ermania, 671
  borealis, 673
Ernst, W. R., 218
Erophila, 269, 270
  boerhaavii, 345
  krockeri, 345
  praecox, 345
  verna, 345
    praecox, 345
  vulgaris, 345
Eruca, 226, 233, 244, 419, **434**

  pinnatifida, 434
  sativa, 434
  vesicaria, *431*, **434**
    pinnatiafida, 434
    sativa, *431*, **434**, 435
    vesicaria, 434
Erucaria, 227
Erucastrum, 231, 233, 244,
    419, **435**, 436
  gallicum, **435**, *436*
  nasturtiifolium, 435
  pollichii, 435
Erysimeae, 533
Erysimum, 226, 235, 242, **534**
  alliaria, 745
  ammophilum, **536**
  amoenum, 538
  angustatum, 538
  arcuatum, 461
  arenicola, 535, **536**, 537
    torulosum, 536
  argillosum, 538
  aridum, 538
  arkansanum, 538
  asperrimum, 538
  asperum, *532*, 535, **537**
    amoenum, 538
    angustatum, 538
    arkansanum, 538
    bealianum, 538
    capitatum, 538
    dolichocarpum, 537
    elatum, 538
    inconspicuum, 541
    parviflorum, 541
    perenne, 543
    pumilum, 538
    purshii, 538
    stellatum, 538
  bakeri, 538
  barbarea, 461
  bracteatum, 706
  californicum, 538
  capitatum, 535, 536, **537**,
    538, 544
    amoenum, 538
    angustatum, 538
    argillosum, 538
    bealianum, 538
    capitatum, 537, **538**
    nivale, 538
    perenne, 543
    purshii, 537, **538**
    stellatum, 538
    washoense, 538
  cheiranthoides, 535, **539**
  cheiri, 535, **539**
  coarctatum, 535, **539**, 540,
    542
  cockerellianum, 538
  concinnum, 536, **540**
    suffrutescens, 544
  cordatum, 708
  desertorum, 538
  drummondii, 408
  elatum, 538
  filifolium, 545
  franciscanum, *532*, 535, **540**
    crassifolium, 540
  ghiesbreghtii, 534

  glaberrimum, 668
  glandulosum, 712
  hieracifolium, 535, **540**
  hyacinthoides, 716
  inconspicuum, 535, 540,
    541, 542
    coarctatum, 539
    syrticola, 541
  insulare, 535, **542**, 545
    angustatum, 538
  linifolium, 668
  menziesii, 536, 540, **542**
  moniliforme, 538
  moranii, 534
  nevadense, 543
  nivale, 538
  nuttallii, 263
  oblanceolatum, 538
  occidentale, 535, **543**
  officinale, 670
  orientale, 517
  pallasii, 534, 535, **543**
    bracteosum, 543
  perenne, 535, 536, 537, **543**,
    544
  petiolare, 719
  pinnatum, 526
  platycarpum, 720
  praecox, 463
  pygmaeum, 543
  radicatum, 538
  repandum, 535, **544**
  retrofractum, 683
  suffrutescens, 535, **544**, 545
    grandifolium, 544
    lompocense, 538
  syrticola, 541
  teretifolium, 535, **545**
  tilimii, 538
  tortuosum, 722
  torulosum, 536
  vernum, 463
  walteri, 505
  wheeleri, 538
Eschmann-Grupe, G., 433
Espuela de caballero, 219
Euclidieae, **545**
Euclidium, 237, 241, 545, 552
  syriacum, *541*, 553
Eugenei, 9, 18
Euklisia, 700
  albida, 713
  amplexicaulis, 679
  aspera, 713
  bakeri, 713
  biolettii, 713
  cordata, 708
  crassifolia, 709
  eliator, 713
  glandulosa, 712
  hispida, 715
  hyacinthoides, 716
  longirostris, 700
  nigra, 714
  pulchella, 714
  secunda, 714
  violacea, 713
Eutrema, 226, 231, 233, 246,
    555
  arenicola, 448

  edwardsii, *554*, 555, **556**
    penlandii, 556
  eschscholtzianum, 256
  japonicum, 555
  labradoricum, 556
    penlandii, 556
  salsugineum, 555, **556**
Eutremeae, **554**
Evergreen candytuft, 564
Exhalimolobos, 557, 558

Fairweather, M. L., 112
False mermaid, 173
False mountain willow, 117
False-flax, 451
Fang, Z. F., 40, 46
Fanweed, 746
Farr's willow, 116
Felger, R. S., 185
Felt-leaf willow, 146
Fendler's bladderpod, 636
Fernald, M. L., 37, 42, 46,
    47, 84, 116, 119, 121,
    127, 132, 135, 142, 153,
    158, 299, 308, 462, 549
Few-leaved beeplant, 208
Field pennycress, 746
Field-mustard, 423
Findlay, J. N., 270, 325
Fisher, M. J., 26
Flacourtia, 5, 8, *163*, *164*
  celastrina, 164
  flexuosa, 164
  indica, *163*, *164*
  ramontchi, 163
Flacourtiaceae, 3, 4, 8, 9, 23
  tribe Flacourtieae, 8, 9
Flacourtieae, 8, 9
Flatpod, 567
Flax, 451
Flaxleaf-whitepuff, 191
Flaxweed, 451
Flixweed, 433, 528
Floderus, B. G. O., 27, 127,
    139
Floerkea, 172, **173**
  proserpinacoides, *169*, **173**,
    175
  versicolor, 181
Floerkée fausse, 173
Florida willow, 32
Foote, J. M., 14
Foothill bladderpod, 649
Forchhammeria, 190
Fragrant mignonette, 193
Francis, A., 422
Frankton, C., 578
Fremont cottonwood, 19
Fremont County bladderpod,
  661
Fremont County twinpod, 661
Fremont's bladderpod, 638
French rocket, 435
Freudenstein, J. V., 308
Fringed spiderflower, 216
Fringepod, 739
Front Range twinpod, 628
Frosty bladderpod, 658
Fuentes-Soriano, S., 559, 561
Garden mignonette, 193

Garden nasturtium, 166
Garden rockcress, 259
Garden spiderflower, 217
Garden-rocket, 434
Garlic mustard, 744
Garrett's bladderpod, 638
Gaskin, J. F., 570
Gaslight bladderpod, 660
Geraniaceae, 165, 172
Geraniales, 172
German, D. A., 227
Geyer's bladderpod, 638
Geyer's willow, 153
Gilg, E., 3
Glatfelter, N. M., 35, 37, 38
Glaucae, 86
Glaucocarpum, 226, 689, 690
   suffrutescens, 692
Glaucocochlearia, 514
Gleason, H. A., 37
Globe bladderpod, 639
Globe candytuft, 564
Gmelina
   indica, 163
Goat willow, 131
Goldblatt, P., 186
Gold-dust, 252
Golden bee-plant, 208
Golden bladderpod, 628
Goldentuft, 251
Gold-of-pleasure, 451
Golysheva, M. D., 51
Goodding's black willow, 36
Goodding's bladderpod, 639
Good-neighbor bladderpod, 665
Goodson, B. E., 518, 519, 521, 524, 525, 526, 527
Gordon's bladderpod, 640
Governor's plum, 163
Graham's twinpod, 642
Grand tremble, 20
Grant, V., 51
Gray poplar, 21
Gray willow, 89, 129, 132, 134
Gray, A., 598
Gray-leaf Sierra willow, 108
Gray-leaf willow, 89
Great Plains bladderpod, 626
Green, P. S., 490
Green-bract willow, 85
Greggia, 609
   camporum, 610
      angustifolia, 610
      linearifolia, 610
   linearifolia, 610
Greilhuber, J., 444
Grundt, H. H., 270, 316, 331
Guaco, 207
Guilandina
   moringa, 168
Guillenia, 677
   cooperi, 680
   flavescens, 681
   heterophylla, 682
   hookeri, 681
   inaliena, 683
   lasiophylla, 683
   rigida, 683
   rostrata, 700

Guppy, H. B., 186
Gynandropsis, 199, 200, 215, **222**
   gynandra, *221*, **223**
   heterotricha, 223
   pentaphylla, 223
   speciosa, 217
Gyrostemonaceae, 186, 190

Hackmatack, 14
Haissig, B. E., 25
Halberd willow, 116
Halimolobeae, 226, 257, **557**
Halimolobos, 236, 237, 240, 243, 418, 456, **557**, 558
   diffusa, **558**
      jaegeri, 558
   jaegeri, *554*, **558**
   mollis, 457
   perplexus, 418
      lemhiensis, 418
   pubens, 558, **559**
   virgata, 457
   whitedii, 418
Hall, J. C., 190, 194, 195, 226
Hallier, H., 3
Hara, H., 500
Hardig, T. M., 27, 121
Hare's-ear mustard, 517
Harris, J. G., 546, 549, 550
Hartman, R. L., 303, 331
Hastatae, 104
Hauptli, H. H., 174, 176
Heart-leaf willow, 106, 120
Hedge-mustard, 667
Heilborn, O., 270
Heliophila, 227
Heliophileae, 227
Hemiscola, 195, 199, 200, 215, **220**
   aculeata, 216, **220**, *221*
      aculeata, **220**, *221*
      affinis, 220
      potosina, 220
   diffusa, **220**, 221
Henry, J. K., 129, 153, 160
Hesperidanthus, 226, 233, 246, 669, **689**, 690
   argillaceus, **690**, 691
   barnebyi, 690, **691**, 692
   jaegeri, 690, **692**
   linearifolius, 689, **690**, *691*
   suffrutescens, 246, 690, **692**
Hesperideae, **561**
Hesperis, 226, 235, 242, **562**
   africana, 554
   arctica, 529
   brachycarpa, 527
   diffusa, 558
   hookeri, 543
   incisa, 523
   longepedicellata, 524
   matronalis, *561*, **562**
   menziesii, 542
   pinnatifida, 485
   salsuginea, 556
   sophia, 528
   virgata, 457
Heterodraba, 267
   unilateralis, 268

Heterothrix, 559
   longifolia, 560
   micrantha, 560
Heywood, V. H., 430
Hickey, L. J., 4
Hillebrand, W., 188
Hillman's stinkweed, 210
Hill-mustard, 444
Hilu, K. W., 4
Hind's willow, 56
Hirschfeldia, 234, 244, **436**
   adpressa, 437
   incana, *436*, **437**
Hitchcock, C. L., 288, 289, 290, 294, 297, 298, 299, 300, 305, 311, 316, 317, 318, 320, 322, 323, 326, 330, 333, 334, 344, 345, 571, 575, 578, 579, 580, 581, 582, 583, 584, 587, 588, 593, 595
Hitchcock's bladderpod, 643
Hoary willow, 131, 150
Hoary-alyssum, 252
Hoary-mustard, 437
Hodge, W. H., 555
Hoffmann, K., 194
Holmes, W. C., 170, 185
Holmgren, N. H., 288, 293, 302, 307, 308, 323, 325, 328, 329, 331, 378, 381, 386, 387, 391, 392, 395, 396, 397, 398, 402, 472, 500, 519, 524, 525, 526, 529, 537, 544, 563, 571, 578, 580, 582, 584, 587, 590, 594, 642, 645, 669, 673, 695, 709, 719, 722
Holmgren, P. K., 211, 600, 601, 603
Honesty, 596
Hooker's willow, 127
Hopkins, M., 257, 260, 261
Hornungia, 231, 237, 238, 240, **530**
   procumbens, *522*, **530**, 531
Horseradish, 459, 555
Horseradish tree, 168
Howell, J. T., 423
Huguenenia, 518
Hultén, E., 76, 79, 81, 90, 111, 116, 513, 514
Humboldt willow, 34
Hungry willow, 121
Hurka, H., 454, 494
Hutchinsia, 530
   calycina
      americana, 672
      integrifolia, 675
      procumbens, 530
Hutchinsiella, 530
Hutera, 429
   cheiranthos, 430
Huynh, K. L., 174
Hymenolobus, 530
   divaricatus, 530
   erectus, 530
   procumbens, 530
   pubens, 559

Hymenophysa
   pubescens, 576
Hyperanthera
   moringa, 168

Iberideae, **563**
Iberis, 224, 226, 231, 238, 530, **563**
   amara, **563**, **564**
   candicans, 605
   nudicaulis, 565
   sempervirens, *561*, 563, **564**
   umbellata, 563, **564**
Icianthus, 700
   atratus, 716
   glabrifolius, 716
   hyacinthoides, 716
Idaho bladderpod, 629
Idaho twinpod, 634
Idaho willow, 107
Idahoa, 224, 229, 239, **566**
   scapigeram, **566**, *567*
Idesia, 4, 8
Iltis, H. H., 197, 205, 207, 218, 220
Imbamba, S. K., 223
Indian cress, 166
Indian mustard, 421, 422
Indian plum, 163
Intermountain bladderpod, 642
Iodanthus, 229, 231, 246, **484**, 485
   dentatus, 373
   hesperidoides, 485
   pinnatifidus, *481*, **485**
Ionopsidium, 515
Isatideae, **567**
Isatis, 230, 239, 244, **567**
   tinctoria, 193, *566*, **568**
Isomeris, 205
   arborea, 205, 206
   angustata, 206
   globosa, 206
   insularis, 206
Italica, 17
Itoa, 8

Jacaratia, 170
Jackass-clover, 213
Jacksonia
   tenuifolia, 204
Jacobson, H. A., 420, 443
Jahns, T. R., 174
Jain, S. K., 174, 175, 176, 178, 180, 183
Jamaican caper tree, 198
James's clammyweed, 203
Janchen, E., 226, 227
Jarilla, 170
Jepson, W. L., 267, 424, 437, 701
Jepson's willow, 160
Jewel-flower, 700
Johnson, F. D., 107, 139
Jointed charlock, 439
Jointed radish, 439
Jointed wild-radish, 439
Jones bee-plant, 207
Jones, A. G., 16
Jonsell, B., 43, 83, 90, 116, 132, 494, 502

Jørgensen, R. B., 424
Judd, W. S., 4, 8, 163, 194, 226
Junco, 184
Jurtzev, B. A., 70

Kaibab bladderpod, 647
Kale, 423
Kane County twinpod, 648
Kardamoglyphos, 493
Karlsson, T., 43, 84, 90, 116, 132
Katenin, A. E., 332
Kearney, T. H., 207, 424
Keim, P., 16
Keller, S., 205, 213
Kesseli, R. V., 174, 175, 176, 178
Kiger, R. W., 3, 163
Kimura, A., 30
King bladderpod, 645
Kishore, V. K., 174
Klaus's bladderpod, 647
Klein, D. R., 80
Knaben, G., 270
Knight, O. W., 121
Koch, M., 226, 270, 348, 514, 515, 600, 601, 746
Kodachrome bladderpod, 664
Koeberlinia, **184**, 185
  holacantha, 185
  spinosa, *177*, **185**
    spinosa, *177*, **185**
    tenuispina, **185**
    verniflora, 185
    wivaggii, **185**
Koeberliniaceae, **184**
Kohlrabi, 423
Koniga, 597
  maritima, 598
Krebs, S., 174
Kruckeberg, A. R., 701, 708, 720

Labrador willow, 151
Lacepod, 739
Lamprophragma, 559
  longifolium, 560
Lanate common twinpod, 634
Land cress, 463
LaPré, L. F., 706, 719
Large clammyweed, 203
Large gray willow, 132
Large pussy willow, 126
Large-flowered dwarf meadowfoam, 183
Largefruit bladderpod, 650
Larsson, G., 132
Laurel willow, 44
Leadlay, E. A., 430
Leaf mustard, 421
Least willow, 70
Leavenworthia, 224, 229, 239, 243, **485**, 486
  alabamica, 486, **487**
    brachystyla, 487
  aurea, 486, 487, **488**
    texana, 487
  crassa, 486, **487**
    elongata, 487
  exigua, 486, **488**
    laciniata, 488
    lutea, 488
  michauxii, 489
  stylosa, *481*, 486, **488**
    texana, 486, **487**
  torulosa, 486, **488**, 489
  uniflora, 486, **489**
Lechosa, 171
Lemke, D. E., 8, 424
Lemmon's willow, 154
Leong, J. M,, 174
Lepage, E., 15, 22, 84, 111
Lepidieae, 227, **569**
Lepidium, 226, 227, 228, 229, 230, 231, 232, 238, 239, **570**, 571, 581, 591, 594
  acutidens, 569, 574, **575**
    microcarpum, 580
  africanum, 571
  albiflorum, 586
  alyssoides, 569, 572, **575**, 576, 582, 587
    angustifolium, 575, 576
    eastwoodiae, 576, 581
    jonesii, 586
    junceum, 575, 576
    mexicanum, 575
    minus, 575
    polycarpum, 575
    stenocarpum, 586
    streptocarpum, 575
  apetalum, 580
  appelianum, 569, 572, **576**, 581
  austrinum, 569, 573, **576**
    orbiculare, 576
  barnebyanum, 569, 572, **576**, 577, 587
  bernardinum, 595
  bonariense, 571, 589
  bourgeauanum, 591
  brachybotryum, 586
  californicum, 595
  campestre, 569, 571, **577**, 581
  chalepense, 570, 572, **577**, *578*, 581
  chilense, 588
  coronopus, 572, **578**
  corymbosum, 586
  crandallii, 586
  crenatum, 573, **579**, 587
  davisii, 572, 573, *578*, **579**, 587
  demissum, 576
  densiflorum, 574, *578*, **579**, 580, 590, 591, 594, 595
    bourgeauanum, 591
    elongatum, 579, 580
    macrocarpum, 579
    pubicarpum, 579, 580, 590
    ramosum, 579
  dictyotum, 574, 575, *578*, **580**, 581
    acutidens, 575
    macrocarpum, 580
  didymum, 570, 572, *578*, **580**
  divergens, 591
  draba, 570, 572, *578*, **581**
    chalepense, 577
    repens, 577
  eastwoodiae, 572, 575, *578*, 581, 582, 587
  elongatum, 579
  flavum, 574, *578*, **582**
    apterum, 582
    felipense, 582
  fletcheri, 591
  fremontii, 570, 572, *578*, **582**
    fremontii, 582
    stipitatum, 582
  georginum, 585
  glaucum, 595
  graminifolium, 571
  granulare, 592
  greenei, 588
  heterophyllum, 571, *578*, **582**, 583
  hirsutum, 595
  huberi, 572, *578*, **583**
  idahoense, 595
  integrifolium, 572, 573, *578*, **583**, 587
    heterophyllum, 586
  intermedium
    pubescens, 595
  jaredii, 572, 574, *578*, **584**
    album, 584
  jonesii, 586
  lasiocarpum, 573, *578*, **584**
    georginum, 585
    lasiocarpum, 584, **585**
    orbiculare, 576
    palmeri, 584
    rosulatum, 585
    rotundum, 585
    wrightii, 584, **585**
  latifolium, 573, *578*, **585**
  latipes, 574, **585**, 586
    heckardii, 585, 586
    latipes, 586
  leiocarpum, 588
  medium, 595
    pubescens, 595
  menziesii, 595
  moabense, 581, 582
  montanum, 572, 575, 582, 583, **586**, 587
    alpinum, 586, 587
    alyssoides, 575
    angustifolium, 575
    canescens, 586
    cinereum, 586
    claronense, 586
    coloradense, 587
    davisii, 579
    demissum, 576
    eastwoodiae, 581
    glabrum, 586
    heterophyllum, 586
    integrifolium, 583
    jonesii, 586
    montanum, 587
    neeseae, 586, 587
    nevadense, 586
    papilliferum, 590
    saliarborense, 586
    spatulatum, 579
    stellae, 586, 587
    stenocarpum, 587
    todiltoense, 587
    wyomingense, 587
  nanum, 572, 586, **587**, 588
  neglectum, 579
  nelsonii, 585
  nitidum, 574, *586*, **588**
    howellii, 588
    insigne, 588
    nitidum, 588
    oreganum, 588
  oblongum, 571, 573, *586*, **588**, 589
    insulare, 588, 589
    oblongum, 589
  occidentale, 595
  oreganum, 588
  ostleri, 573, *586*, **589**
  oxycarpum, 574, *586*, **589**
    acutidens, 575
    strictum, 593
  papilliferum, 574, *586*, 587, **590**
  paysonii, 572, **590**
  perfoliatum, 571, *586*, **590**
  philonitrum, 587
  pinnatifidum, 574, *586*, **591**
  procumbens, 530
  pubescens, 576
  pubicarpum, 579
  ramosissimum, 571, 573, *586*, **591**
    bourgeauanum, 591
    divergens, 591
    robustum, 591
    ramosum, 579
    repens, 577
  reticulatum, 593
  robinsonii, 595
  ruderale, 573, *586*, **592**
    lasiocarpum, 584
  sativum, 574, *586*, **592**
  schinzii, 571
  scopulorum, 587
    spatulatum, 579
  sibiricum, 337
  simile, 595
  sordidum, 573, *592*, **593**
  squamatum, 578
  strictum, 573, **593**
    oreganum, 588
    texanum, 592
  thurberi, 575, **593**
  tiehmii, 573, **593**, 594
  tortum, 575
  utahense, 583, 587
  vaseyanum, 579
  virginicum, 570, 574, 591, *593*, **594**, 595
    californicum, 595
    linearifolium, 595
    medium, 595
    menziesii, **595**
    pubescens, 595
    robinsonii, 595
    texanum, 592
    virginicum, 593, **595**
  wrightii, 585
  zionis, 583
Lersten, N. R., 7

Lescur's bladderpod, 614
Leskinen, E., 4, 23, 30, 51, 61
Lesquerella, 226, 612, 616, 617, 619
  alpestris, 624
  alpina, 660
    condensata, 652
    intermedia, 644
    laevis, 660
    parvula, 657
    spatulata, 663
  angustifolia, 625
  arctica, 626
    calderi, 629
    purshii, 626
  arenosa, 626
    argillosa, 627
  argentea, 650
    arenosa, 626
  argyraea, 627
  arizonica, 627
    nudicaulis, 627
  aurea, 627
  auriculata, 612
  barnebyi, 647
  bernardina, 646
  calcicola, 629
  calderi, 629
  carinata, 629
    languida, 630
  cinerea, 631
  condensata, 652
    laevis, 660
  confluens, 643
  congesta, 632
  cordiformis, 632
  curvipes, 632
  cusickii, 654
  densiflora, 633
  densipila, 612
  diversifolia, 646
  douglasii, 634
  engelmannii, 635
    alba, 656
    ovalifolia, 655
  fendleri, 636
  filiformis, 637
  foliacea, 636
  fremontii, 638
  garrettii, 638
  geyeri, 638
  globosa, 639
  gooddingii, 639
  goodrichii, 655
  gordonii, 640
    densifolia, 640
    sessilis, 663
  gracilis, 641
    nuttallii, 641
    pilosa, 649
    repanda, 641
    sessilis, 663
  grandiflora, 613
  hemiphysaria, 642
    lucens, 642
  hitchcockii, 643
    confluens, 643
    rubicundula, 644
    tumulosa, 664
  humilis, 644

  intermedia, 644
  kaibabensis, 647
  kingii, 645
    bernardina, 646
    cobrensis, 646
    cordiformis, 632
    diversifolia, 646
    latifolia, 647
    nevadensis, 632
    parvifolia, 647
    sherwoodii, 646
  klausii, 647
  lasiocarpa, 613
    ampla, 614
    berlandieri, 614
    hispida, 614
  lata, 648
  latifolia, 647
  lepidota, 664
  lescurii, 614
  lesicii, 649
  lindheimeri, 649
  longifolia, 625
  ludoviciana, 649
    arenosa, 626
  lunellii, 626
  lyrata, 614
  macounii, 626
  macrocarpa, 650
  ×*maxima*, 616
  mcvaughiana, 650
  montana, 650
    suffruticosa, 650
  multiceps, 651
  navajoensis, 651
  nodosa, 663
  nuttallii, 641
  obdeltata, 654
  occidentalis, 654
    cinerascens, 655
    cusickii, 654
    diversifolia, 646
    parvifolia, 647
  oregona, 655
  ovalifolia, 655
    alba, 656
  ovata, 656
  pallida, 656
  parviflora, 657
  parvula, 657
  paysonii, 630
  pendula, 657
  perforata, 615
  pinetorum, 658
  polyantha, 641
  praecox, 636
  prostrata, 658
  pruinosa, 658
  pulchella, 631
  purpurea, 659
    foliosa, 659
  rectipes, 660
  recurvata, 660
  repanda, 641
  rosea, 626
  rosulata, 650
  rubicundula, 644
  sessilis, 663
  sherwoodii, 646
  spatulata, 663

  stenophylla, 636
  stonensis, 615
  subumbellata, 663
  tenella, 663
  thamnophila, 664
  tumulosa, 664
  tuplashensis, 635
  utahensis, 647
  valida, 664
  vicina, 665
  wardii, 647
Levant mustard, 216
Lewis-Jones, L. J., 440
Liard amer, 16
Lichvar, R. W., 696
Lihová, J., 464, 475, 480, 484
Limber-caper, 197
Limestone glade bladderpod, 637
Limestone willow, 143
Limnanthaceae, **172**
Limnanthes, 172, **173**, 174, 175
  sect. Inflexae, 173, 175, **179**
  sect. Limnanthes, 173, **175**
  sect. *Reflexae*, 173, 175
  alba, 174, 179, **180**
    alba, 179, 180, **181**, 183
    gracilis, 175, 179, 180, **181**
    parishii, 174, 175, 179, 180, **181**
    versicolor, 179, 180, **181**
  bakeri, **175**, 178
  bellingeriana, 182
  douglasii, 175, **176**, *177*, 178, 179
    douglasii, 174, 176, **177**, 178
    nivea, 174, 176, 177, **178**
    rosea, 174, 176, 177, **179**
    striata, 174, 176, 177, **178**, 179
    sulphurea, 175, 176, 177, **178**
  floccosa, 174, *177*, 179, 180, **182**, 183
    bellingeriana, 174, 175, 179, **182**, 183
    californica, 174, 175, *177*, **182**, **183**
    floccosa, 174, *177*, 179, **182**, 183
    grandiflora, 174, 175, 182, **183**
    pumila, 175, 182, **183**
  gracilis, 180, 181
    gracilis, 180
    parishii, 180, 181
  macounii, 172, 174, 175, **176**
  montana, 174, 179, **180**
  pumila, 183
  rosea, 179
  striata, 178
  versicolor
    parishii, 181
  vinculans, 174, 175, **176**, 178
Lincoln County bladderpod, 648

Lindheimer's bladderpod, 649
Lineleaf whitepuff, 191
Link, D. A., 174
Little, E. L. Jr., 21, 37, 110, 111
Little-tree willow, 141
Lobularia, 226, 237, 241, **597**
  maritima, **598**, *599*
Lombardy poplar, 17, 18
Lonard, R. I., 34
Lonchophora, 253
Long-beaked willow, 134
Low bladderpod, 633, 658
Low blueberry willow, 109
Loxostemon, 464
Luken, J. O., 435
Lunaria, 226, 232, 239, **596**
  annua, 224, *593*, *596*, **597**
  biennis, 597
  inodora, 597
  rediviva, 596
Lunarieae, *596*
Lyrate bladderpod, 614
Lyreleaf bladderpod, 614
Lyrocarpa, 237, 240, **608**
  coulteri, *606*, **609**
    apiculata, 609
    coulteri, 609
    palmeri, 609
  linearifolia, 608
  palmeri, 609
  xanti, 608

Mabberly, D. J., 8
Mabry, T. J., 186
MacCalla's willow, 49
Mackenzie's willow, 125
Macoun's meadowfoam, 176
Macropodium
  laciniatum, 734
Madagascar plum, 163
Madwort, 247
Magallana, 165
Malcolmia, 226
  africana, 554
Malcolmieae, 597
Malpighiales, 4
Manchester, S. R., 8
Mancoa, 418
  pubens, 559
Mandahar, N. D., 222
Manyhead bladderpod, 651
Marcet, E., 18
Marhold, K., 464
Marie-Victorin, Frère, 21
Marr, J. K., 308
Martin, R. F., 507
Martín-Bravo, S., 189, 190, 193
Martínez-Laborde, J. B., 432, 433
Martini, F., 50, 133
Maschinski, J., 112
Mason, C. T., 178, 179, 180
Matthews, J. B., 146
Matthiola, 226, 235, 242, **253**
  bicornis, 254
  incana, **254**
  longipetala, **254**, *255*
    bicornis, 254
    nudicaulis, 512

Maycock, P. F., 146
Mayer, M. S., 701
McArdle, A. J., 41
McBride, B., 203
McGregor, R. L., 453
McJannet, C. L., 54
McNeill, C. I., 180, 183
McVaugh's bladderpod, 650
Meadow willow, 152
Meadowfoam, 173
Meadowfoam Family, 172
Meadowfoam fly, 174
Mediterranean mustard, 437
Meeuse, A. D. J., 3, 8
Meikle, R. D., 40, 41, 42
Melampsora, 112
Melanidion, 671
 boreale, 673
Mesoreanthus, 700, 723
 barbiger, 704
 fallax, 723
 vimineus, 723
Mexican clammyweed, 204
Meyer, F. K., 530, 746
Microcardamum, 530
Microsemia, 700
 polygaloides, 720
Microsisymbrium, 677
 lasiophyllum, 683
  dasycarpum, 683
  dissectum, 683
  inalienum, 683
  integrifolium, 683
  rigidum, 683
Microthlaspi, 226, 230, 239, 599, 746
 perfoliatum, 599, **600**
Mid-bladderpod, 644
Middle Butte bladderpod, 654
Mignonette, 191
Mignonette Family, 189
Mignonette-blanche, 192
Mignonette-jaunâtre, 193
Mignonette-odorante, 193
Mimbre del monte, 197
Missouri bladderpod, 637
Missouri willow, 120
Mitchell-Olds, T., 226, 349
Mitophyllum, 700
 diversifolium, 710
Mnzava, N. A., 223
Moapa bladderpod, 663
Mohlenbrock, R. H., 38
Mojave stinkweed, 212
Money-plant, 596
Moonwort, 596
Moore, E., 25
Moricandia, 438
 arvensis, 438
Morin, N. R., 173
Moringa, **168**
 concanensis, 168
 longituba, 168
 oleifera, **168**, *169*
 pterygosperma, 169
 stenopetala, 168
Moringaceae, **167**
Morisonia
 flexuosa, 197
Morrison, J. L., 701

Moser, M. B., 185
Mosseler, A., 27, 37, 47, 51, 57, 121, 127, 135, 153
Mountain bladderpod, 650
Mountain cottonwood, 16
Mountain willow, 118, 131
Mountain-view bladderpod, 659
Mouse-ear cress, 226, 447
Müller Argoviensis, J., 190
Mulligan, G. A., 128, 257, 259, 260, 261, 270, 289, 295, 299, 306, 308, 313, 317, 323, 325, 326, 328, 331, 334, 344, 349, 370, 373, 396, 402, 443, 449, 450, 462, 494, 498, 499, 509, 510, 540, 543, 578, 580, 589
Mummenhoff, K., 433, 530, 570, 571, 581, 746
Murray, D. F., 324
Musk ox, 80
Mustard, 225, 419, 423, 441
Mustard Family, 224
Mustard-greens, 421
Myagrum, 230, 239, **568**
 alyssum, 453
 amphibium, 497
 argenteum, 649
 austriacum, 497
 irregulare, 446
 palustre, 501
 paniculatum, 455
 perfoliatum, *569*
 rugosum, 441
 sativum, 453

Nakai, T., 30
Nandi, O. I., 4
Narrowleaf cottonwood, 16
Narrowleaf rhombopod, 210
Narrow-leaf willow, 55
Nasturtium, 166, 226, 229, 232, 245, **489**
 africanum, 490
 amphibium, 497
 austriacum, 497
 bursifolium, 456
 calycinum, 498
 columbiae, 499
 crystallinum, 499
 curvisiliqua, 492, 500
  lyratum, 500
  nuttallii, 500
 dictyotum, 580
 draba, 581
 flavum, 582
 floridanum, 490, **492**
 fremontii, 582
 gambelii, 490, ***491***, 492
 heterophyllum, 500
 hispidum, 502
 indicum, 501
 integrifolium, 583
 lacustre, 497
 lasiocarpum, 584
 latipes, 585
 limosum, 503
 linifolium, 668

 lyratum, 500
 micropetalum, 505
 microphyllum, **490**
 microtitis, 501
 natans
  americanum, 497
 nitidum, 588
 obtusum, 505
  alpinum, 496
  sphaerocarpum, 504
 occidentale, 500
 officinale, **490**, 492
 oxycarpum, 589
 palustre, 501
  hispidum, 502
  tanacetifolium, 505
 perfoliatum, 590
 polymorphum, 500
 pumilum, 668
 sessiliflorum, 503
 sinuatum, 504
  calycinum, 498
  columbiae, 499
  pubescens, 498, 499
  sphaerocarpum, 504
 ×*sterilis*, 490
 stylosum, 492
 sylvestre, 505
 tanacetifolium, 505
 terrestre
  hispidum, 502
  occidentale, 502
 trachycarpum, 504
 uniseriatum, 490
 walteri, 505
Nasturtium Family, 165
Navajo bladderpod, 651
Nectris
 pinnata, 173
Nelson, A., 393
Nelson's bladderpod, 652
Neobeckia, 493
 aquatica, 497
Neocleome
 spinosa, 219
Neolepia, 570
 campestris, 577
Neomartinella, 555
Neotorularia, 549
 humilis, 548
Nerisyrenia, 235, 242, **609**, 610
 camporum, **610**
 linearifolia, **610**, *611*
 mexicana, 610
Neslia, 236, 240, **455**
 apiculata, 446, 455
 paniculata, *452*, **455**
  paniculata, 455
  thracica, 455
Nesodraba, 269, 310
 grandis, 309
 megalocarpa, 309
 siliquosa, 309
Net-leaf willow, 65
Neuffer, B., 454
Neuroloma, 511
 nudicaule, 512
 rydbergii, 513
Nevada, 233, 246, **414**
 holmgrenii, *413*, **414**

Newberry twinpod, 652
Noccaea, 226, 230, 239, **600**, 603, 746
 arctica, **601**
 coloradensis, 602
 fendleri, *599*, **601**, 602
  californica, 601, **602**
  fendleri, *599*, 601, **602**
  glauca, 601, **602**, 603
  idahoensis, 601, 602, **603**
  siskiyouensis, 601, 602, **603**
 glauca, 602
 parviflora, 601, **603**, 604
 procumbens, 530
Noccaeeae, 226, **598**
Norta, 667
Northern willow, 73
Northwest sandbar willow, 59
Norwood, L. E., 679
Nutt, P., 454
Nuttall's bladderpod, 641

O'Kane, S. L. Jr., 448, 449, 450, 611, 616, 617
Obedzinski, R.A., 112
Obermayer, R., 444
Octanema
 incana, 198
Odontarrhena
 obovata, 250
Ohashi, H., 4
Oilseed, 419, 421
Oilseed rape, 422
Oligomeris, **190**
 linifolia, *187*, 190, **191**
Olson, M. E., 167
One-color willow, 124
Oregon twinpod, 655
Oriental mustard, 422
Ornduff, R., 173, 178, 180, 183, 598
Orobium, 256
Orychophragmus, 229, 244, **437**, 438
 limprichtianus, 438
 violaceus, **436**, **438**
Osier, 149
Osterhout's twinpod, 638
Oval-leaf willow, 75
Oxystylidaceae, 200
Oxystylis, 199, 200, 205, **215**
 lutea, *211*, **215**

Pachypodium, 667, 728
 integrifolium, 732
 laciniatum, 734
 linearifolium, 690
 sagittatum, 736
Pacific willow, 47
Pagosa bladderpod, 658
Paiero, P., 50, 133
Pakchoi, 424
Palmer's cleomella, 212
Palo, R. T., 4
Palo-zorillo, 196
Panasahatham, S., 174
Panetsos, C. A., 439
Pankhurst, R. J., 28
Papadopol, C. S., 27, 57

INDEX

Papaw, 171
Papaya, 170, 171
Papaya Family, 170
Parasol bladderpod, 663
Park willow, 118
Parker, C. L., 324
Parker, W. H., 173, 174, 175
Parolin, P., 24
Parrasia, 610
   camporum, 610
   linearifolia, 610
Parrya, 224, 229, 239, 244, **511**
   arctica, 244, 511, **512**
   cheiranthoides, 415
      lanuginosa, 415
   eurycarpa, 348
   huddelliana, 348
   macrocarpa, 512
   menziesii
      glabra, 415
      lanuginosa, 415
   nauruaq, **512**
   nudicaulis, **512**, 513, 514
      grandiflora, 512
      interior, 512
      rydbergii, 513
      septentrionalis, 512
   pedicellata, 415
   platycarpa, 513
   rydbergii, 512, *513*, 514
Passifloraceae, 170
Pauley, S. S., 21
Pawpaw, 171
Pax, F., 194
Payne, W. W., 188
Payson, E. B., 600, 601, 667, 677, 681, 683, 684, 687, 728
Payson's bladderpod, 630
Paysonia, 224, 236, 240, 604, **611**, 612, 617
   auriculata, **612**
   densipila, **612**, 616
   grandiflora, 612, **613**
   lasiocarpa, *611*, 612, **613**
      berlandieri, **614**
      lasiocarpa, *611*, **614**
   lescurii, *611*, 612, **614**
   lyrata, 612, **614**
   perforata, *611*, 612, **615**
   stonensis, 612, **615**, 616
Peach leaf willow, 37
Peebles, R. H., 207, 424
Pegtel, D. M., 515
Pennellia, 224, 226, 235, 243, 257, **559**, 561
   hunnewellii, 560
   longifolia, **560**
   mcvaughii, 560
   micrantha, **560**
   robinsonii, 560
   tricornuta, 560, *561*
Pennycress, 745
Penny-flower, 597
Pepper, A. E., 679
Peppercress, 570
Peppergrass, 570
Pepperwort, 570
Perennial wall-rocket, 433

Peritoma, 195, 200, 201, **205**, 208, 215
   arborea, 199, 205, **206**
      angustata, **206**
      arborea, **206**
      globosa, **206**
      insularis, 206
   breviflora, 206
   inornata, 207
   jonesii, 205, **207**
   lutea, 205, **206**, 207
   multicaulis, 205, **208**
   platycarpa, 202, 205, **208**
   serrulata, 205, **207**
      albiflora, 207
      clavata, 207
   sonorae, 208
Perkins, K. D., 188
Petrovsky, V. V., 332
Petsai, 424
Pfeffer, E., 27, 57
Phoenicaulis, 234, 236, 241, 347, **415**
   cheiranthoides, *413*, **415**
      glabra, 415
      lanuginosa, 415
      eurycarpa, 348
      pedicellata, 415
Phylicifoliae, 135
Physaria, 226, 227, 237, 240, 241, 612, **616**, 617
   acutifolia, 618, **624**, 625
      purpurea, 642
      repanda, 642
      stylosa, 624
   alpestris, 617, 618, **624**
   alpina, 618, **624**, 660
   angustifolia, 619, **625**
   arctica, 619, 622, **625**
   arenosa, 620, **626**
      arenosa, **626**, 650
      argillosa, 626, **627**
   argyraea, 620, **627**
      argyraea, **627**
      diffusa, 627
   arizonica, 621, 625, **627**
      andrusensis, 627
   aurea, 620, 622, **628**, 640
      australis, 624
   bellii, 617, 618, **628**
   brassicoides, 618, 625, **628**
   calcicola, 621, **629**
   calderi, 619, **629**
   carinata, 622, **629**, *630*
      carinata, **630**
      paysonii, 629, *630*
      pulchella, 630, **631**
   chambersii, 618, **631**, 653
      canaani, 631
      chambersii, 631
      membranacea, 648
      sobolifera, 631
   cinerea, 623, **631**
   cobrensis, 646
   condensata, 618, **631**, 652
   congesta, 621, *630*, **632**
   cordiformis, 622, **632**
   curvipes, 621, 623, **632**, 633, 651
   densiflora, 619, **633**

   didymocarpa, 618, **633**, 644, 661
      australis, 624
      didymocarpa, 633, **634**
      integrifolia, 644
      lanata, 633, **634**
      lyrata, 633, **634**
      newberryi, 652
   dornii, 618, **634**
   douglasii, 623, **634**
      douglasii, **635**
      tuplashensis, **635**
   eburniflora, 617, **635**
   engelmannii, 620, **635**
   eriocarpa, 621, **636**
   fendleri, 616, 619, *636*, 637
   filiformis, 619, **637**
   floribunda, 617, **637**
      floribunda, **637**
      osterhoutii, 637, **638**
   fremontii, 622, **638**
   garrettii, 623, **638**
   geyeri, 617, *636*, **638**
      geyeri, **639**
      purpurea, *636*, **639**
   globosa, 623, **639**
   gooddingii, 621, 628, **639**, 640
   goodrichii, 655
   gordonii, 619, 620, *640*
      densifolia, 640, 641
   gracilis, 619, *640*, **641**
      gracilis, **641**
      nuttallii, *640*, **641**
      repanda, **641**
   grahamii, 618, **642**
   hemiphysaria, 620, 622, **642**
      hemiphysaria, **642**
      lucens, **642**
   hitchcockii, 619, **643**, 664
      confluens, **643**
      hitchcockii, **643**
      rubicundula, 643, **644**
   humilis, 623, **644**
   integrifolia, 618, **644**
      monticola, **644**
   intermedia, 621, 644, 645, 651
      kaibabensis, 647
   kingii, 620, 622, 623, **645**
      bernardina, 645, **646**
      cobrensis, 645, **646**
      diversifolia, 645, **646**
      kaibabensis, 645, **647**
      kingii, 645, **646**
      latifolia, 646, **647**
      utahensis, 646, **647**
   klausii, 622, **647**
   lanata, 634
   lata, 623, **648**
   lepidota, 617, **648**
      lepidota, **648**
      membranacea, **648**
   lesicii, 622, 623, **649**
   lindheimeri, 619, **649**
   ludoviciana, 620, **649**, 650
   macrantha, 634
   macrocarpa, 622, **650**
   mcvaughiana, 619, **650**
   montana, 623, **650**, 651

   multiceps, 623, **651**
   navajoensis, 619, *640*, 643, **651**, 664
   nelsonii, 621, **652**, 660
   newberryi, 618, 631, **652**, 653
      newberryi, **652**, 653
      racemosa, **653**
      yesicola, **653**
   obcordata, 617, **653**
   obdeltata, 622, **654**
   occidentalis, 621, **654**
      cinerascens, 654, **655**
      goodrichii, 655
      occidentalis, **654**
   oregona, 618, **655**
   osterhoutii, 638
   ovalifolia, 619, 620, **655**
      alba, 655, **656**
      ovalifolia, 655, **656**
   pachyphylla, 623, *652*, **656**
   pallida, 619, **656**, 657
   parviflora, 622, **657**
   parvula, 621, 645, **657**
   paysonii, 630
   pendula, 622, **657**
   pinetorum, 620, 648, **658**
   prostrata, 623, **658**
   pruinosa, 620, **658**
   pulchella, 631
   pulvinata, 620, 645, **659**, 660
   purpurea, 619, **659**
      foliosa, 659
   pycnantha, 621, 652, **659**, 660
   rectipes, 621, *652*, 651, **660**
   recurvata, 619, **660**
   reediana, 621, **660**
      spatulata, 663
   repanda, 642
   rollinsii, 617, 618, 624, **661**
   rubicundula, 644
      tumulosa, 664
   saximontana, 618, 633, **661**
      dentata, 633, 661, **662**
      saximontana, **661**
   scrotiformis, 623, **662**
   sessilis, 619, **663**
   spatulata, 622, **663**
   stylosa, 624
   subumbellata, 622, 623, **663**
   tenella, 622, **663**
   thamnophila, 620, **664**
   tumulosa, 619, 643, 651, **664**
   utahensis, 647
   valida, 623, **664**, 665
   vicina, 620, **665**
   vitulifera, 618, **665**
   wardii, 647
Physarieae, **604**
Physolepidion, 570
   repens, 577
Piceance twinpod, 653
Picenace bladderpod, 657
Pilosella
   glauca, 556
   novae-angliae, 549
   richardsonii, 549
   stenocarpa, 457

Pilosella (continued)
  virgata, 457
Pineland catchfly, 204
Pink-queen, 219
Plains cottonwood, 18
Planodes, 234, 244, **492**, 693
  virginicum, *491*, **492**, **493**
Platanus
  ×*acerifolia,* 50
Plateau cottonwood, 20
Platycraspedum, 555
Platypetalum, 546
  dubium, 548
  purpurascens, 548
Platyspermum, 566
  scapigerum, 567
Pleiarina
  lucida, 46
Pleiocardia, 700
  breweri, 706
  fenestrata, 711
  foliosa, 722
  gracilis, 715
  hesperidis, 715
  magna, 679
  orbiculata, 722
  suffrutescens, 722
  tortuosa, 722
Pleurophragma, 728
  gracilipes, 734
  integrifolium, 732
  lilacinum, 733
  platypodum, 734
  rhomboideum, 733
Plotkin, M. S., 173, 175, 176, 180, 183
Point Reyes meadowfoam, 178
Point-tip twinpod, 637
Polanisia, 199, 200, **201**
  dodecandra, **201**, *202*
    dodecandra, ***202***
    riograndensis, **202**, 203
    trachysperma, 202, **203**
    uniglandulosa, 204
  erosa, 201, **203**
    breviglandulosa, 203, **204**
    erosa, 203, **204**
  graveolens, 202
  icosandra, 222
  jamesii, 201, **203**
  microphylla, 222
  tenuifolia, 201, **204**
  trachysperma, 203
  uniglandulosa, 201, **204**
  viscosa, 222
Polar willow, 69
Poliophyton, 557
  pubens, 559
Poliothyrsis, 8
Polished willow, 34
Polunin, N., 73, 80
Polyctenium, 235, 237, 240, 242, **415**, 417
  bisulcatum, 416
  fremontii, **416**, 417
    bisulcatum, 416
    confertum, 416, 417
    glabellum, 416
    williamsiae, 416, 417
Polygonum

  aviculare, 454
Poplar, 7, 8, 9, 14, 18
Populus, 4, 5, 8, 9, 10, 11, 21, 25, 26, 51, 52, 67
  sect. *Abaso,* 8
  sect. *Aigeiros,* 5, 14, 15, 16, 17
  sect. *Leucoides,* 5, 17
  sect. *Populus,* 5, 17
  sect. *Tacamahaca,* 5, 14, 15, 16, 17, 19
  sect. *Turanga,* 8
  subg. *Turanga,* 51
  ×*acuminata,* 16, 17
    rehderi, 19
  adenopoda, 21
  alba, 7, 9, 10, 11, 21, 22
    ×*andrewsii,* 15
  angustifolia, 5, 8, 10, 11, 12, *13*, 14, 15, **16**, 17, 19, 22
  arizonica, 20
  aurea, 22
  balsamifera, 5, 10, 11, 12, *13*, **14**, 15, 16, 17, 22
    subcordata, 14
    trichocarpa, 13
  ×*barnesii,* 21
  ×*bernardii,* 15, 17
  ×*berolinensis,* 16
  besseyana, 18
  ×*brayshawii,* 15, 16
  ×*canadensis,* 7, 9, 11, 12, 15, 17, 18
  candicans, 14
  ×*canescens,* 9, 10, 11, 21
  deltoides, 5, 7, 10, 11, 12, *13*, 14, 15, 16, **17**, 18, 22
    deltoides, *13*, 14, 15, **18**, 19
    mesetae, 20
    missouriensis, 18
    monilifera, 14, 15, **18**, 19
    occidentalis, 18
    wislizeni, 17, 18, **19**
  ×*dutillyi,* 15
  euphratica, 8
  ×*euramericana,* 18
  fremontii, 5, 7, 10, 11, 12, *13*, 14, 16, 17, **19**, 20
    arizonica, 20
    fremontii, *13*, 16, **20**
    macdougalii, 20
    mesetae, 16, **20**
    pubescens, 20
    thornberi, 20
    toumeyi, 20
    wislizeni, 19
  ×*generosa,* 7, 14, 17, 18
  ×*gileadensis,* 15
  grandidentata, 5, 7, 10, 11, 12, *13*, **20**, 21, 22
  ×*hastata,* 14, 15
  ×*heimburgeri,* 21, 22
  heterophylla, 5, 7, 10, 11, **12**, *13*, 17, 21
  ×*hinckleyana,* 16, 19
  ilicifolia, 8
  ×*inopina,* 17, 20

  ×*interamericana,* 14
  ×*intercurrens,* 16, 19
  ×*jackii,* 9, 10, 11, 12, 14, 15, 17
  laurifolia, 16
  macdougalii, 20
  maximowiczii, 14
  mexicana, 8
  monilifera, 18
  nigra, 9, 11, 12, 15, 16, 17, 18, 20
  ×*parri,* 14, 19, 20
  ×*polygonifolia,* 22
  pruinosa, 8
  ×*rollandii,* 15
  ×*rouleauiana,* 21
  sargentii, 18
    texana, 18, 19
  ×*senni,* 16
  simonii, 9, 10, 12, 15, 17
  ×*smithii,* 21, 22
  tacamahacca, 14
  ×*tomentosa,* 21
  tremula, 21, 22
    grandidentata, 20
    tremuloides, 22
  tremuloides, 5, 7, 10, 11, 12, *13*, 15, 16, 17, 21, **22**
    aurea, 22
    magnifica, 22
    vancouveriana, 22
  trichocarpa, 5, 7, 8, 10, 11, 12, *13*, 14, 15, 16, 17, 18, 19, 20
    ingrata, 13
    tweedyi, 16
  ×*wettsteinii,* 22
  wislizeni, 19
Porsild, A. E., 494, 498
Powell, J. M., 79
Prairie willow, 129
Prantl, K., 3
Price, P. W., 24
Price, R. A., 226, 313, 477, 485, 489, 534, 537, 540, 542, 544, 545, 561
Prickly spiderflower, 220
Prijanto, B., 186
Prince's plume, 695
Pritzelago, 530
Pryor Mountains bladderpod, 649
Pseudosalix, 4, 8
Purdy, B. G., 148
Purple osier, 162
Purple twinpod, 639
Purple willow, 162
Purple-rocket, 485
Pussy willow, 126
Pygmy bladderpod, 657

Quadrella, 195, **197**, 198
  cynophallophora, 198
  incana, 194, *196*, **198**
  jamaicensis, **198**
Quaking aspen, 22
Quidproquo, 438
Rabbit's-ear, 517
Rabbit-ears, 517
Radford, A. E., 430

Radicula, 493
  alpina, 496
  amphibia, 497
  aquatica, 497
  barbareifolia, 498
  calycina, 498
  columbiae, 499
  curvipes, 499
  curvisiliqua, 500
  hispida, 502
  indica, 501
  integra, 499
  limosa, 503
  lyrata, 500
  multicaulis, 500
  nuttallii, 500
  obtusa
    sphaerocarpa, 504
  occidentalis, 500
  pacifica, 502
  palustris, 501
    hispida, 502
  pectinata, 500
  polymorpha, 500
  sessiliflora, 503
  sinuata, 504
    integra, 499
    truncata, 499
  sphaerocarpa, 504
  sylvestris, 505
  tenerrima, 505
  trachycarpa, 504
  underwoodii, 499
Radish, 438
Rape, 422
Rapeseed, 422, 423, 424
Raphanus, 225, 226, 233, 244, **438**
  confusus, 439
  lanceolatus, 426
  ×*micranthus,* 439
  raphanistrum, **439**, *440*
    landra, 440
    raphanistrum, 439
  sativus, **439**, 440
  tenellus, 511
Rapistrum, 232, 238, **440**, 441
  hispanicum, 441
  perenne, 441
  rugosum, **440**, **441**
    linnaeanum, 441
    orientale, 441
    rugosum, 441
Raup willow, 85
Raup, H. M., 27, 49, 80
Rechinger, K. H., 42, 50, 90, 133
Red willow, 34
Redwhiskered clammyweed, 202
Reeves, R. D., 701, 720
Reseda, 190, **191**, 192
  sect. *Glaucoreseda,* 190
  alba, *187*, **191**, **192**
  linifolia, 191
  lutea, 191, **192**, 193
  luteola, 192, **193**
  odorata, 192, **193**
  phyteuma, 192
Réséda, 191

Réséda blanc, 192
Resedaceae, **189**, 190
  tribe Astrocarpeae, 190
  tribe Cayluseae, 190
  tribe Resedeae, 190
Resedeae, 190
Rhynchosinapis, 429
  cheiranthos, 430
Richardson's willow, 144
Rio Grande clammyweed, 202
Rio Grande cottonwood, 19
Rock willow, 64
Rock-alyssum, 251
Rockcress, 257
Rocket, 434, 460, 562, 667
Rocket candytuft, 564
Rocket-salad, 434
Rocket-weed, 435
Rock-madwort, 252
Rocky Mountain bee-plant, 207
Rocky Mountain bladderpod, 629
Rocky Mountain stinkweed, 212
Rocky Mountain twinpod, 661
Rocky Mountain willow, 81
Rodman, J. E., 186, 194, 226, 424, 425, 428, 701
Rodschiedia, 453
Rogers Pass bladderpod, 647
Rogers, G. K., 188
Rollins, R. C., 225, 226, 227, 249, 250, 252, 257, 259, 260, 261, 263, 264, 267, 288, 290, 293, 294, 295, 296, 297, 298, 299, 300, 301, 302, 303, 304, 305, 307, 308, 313, 316, 317, 318, 320, 323, 325, 326, 328, 329, 330, 331, 333, 347, 348, 349, 350, 366, 367, 368, 370, 374, 378, 380, 381, 386, 387, 391, 392, 396, 397, 398, 400, 402, 404, 407, 410, 418, 421, 423, 424, 430, 431, 434, 438, 441, 443, 449, 450, 453, 454, 456, 462, 464, 468, 472, 478, 485, 486, 494, 499, 500, 503, 507, 513, 514, 515, 517, 519, 521, 524, 525, 526, 527, 529, 530, 536, 537, 538, 540, 543, 544, 545, 549, 556, 557, 561, 571, 575, 578, 579, 580, 582, 583, 584, 587, 588, 591, 594, 595, 596, 600, 601, 602, 603, 609, 610, 612, 624, 667, 673, 675, 677, 679, 681, 684, 687, 690, 693, 695, 696, 698, 706, 709, 714, 718, 722, 723, 724, 727, 728, 738
Rollins's bladderpod, 660
Rollins's twinpod, 661
Rollinsia, 688
Romanschulzia, 724
Ronald, W. G., 15, 16
Rood, S. B., 7, 16

Roquette, 434
Rorippa, 230, 231, 232, 234, 239, 245, 246, 489, **493**, 494, 505
  alpina, 495, **496**
  americana, 497
  amphibia, 494, 495, **497**
  aquatica, 495, **497**
  armoracia, 459
  austriaca, 495, **497**
  barbareifolia, 225, *491*, 493, 495, **498**
  calycina, 495, **498**
    columbiae, 499
  cantoniensis, 494
  coloradensis, 442, 494
  columbiae, 494, 495, **499**
  crystallina, 495, **499**
  curvipes, 494, 495, 496, **499**
    alpina, 496
    integra, 499
    truncata, 499
  curvisiliqua, 492, 494, 496, **500**
    lyrata, 500
    nuttallii, 500
    occidentalis, 500
    orientalis, 500
    procumbens, 500
    spatulata, 500
  dubia, 495, 496, **500**
  floridana, 492
  gambelii, 491
  globosa, 494
  heterophylla, 500
  hispida, 502
    barbareifolia, 498
    glabrata, 502
  indica, 496, 500, **501**
    apetala, 500
  integra, 499
  islandica, 494
    barbareifolia, 498
    fernaldiana, 502
    glabra, 502
    glabrata, 502
    hispida, 502
    occidentalis, 502
  lyrata, 500
  microphylla, 490
  microsperma, 494
  microtitis, 496, **501**
  multicaulis, 500
  nasturtium-aquaticum, 490
  nuttallii, 500
  obtusa, 505
    alpina, 496
    integra, 499
    sphaerocarpa, 504
  occidentalis, 500
  pacifica, 502
  palustris, 494, 495, 496, 498, 501, 502
    fernaldiana, 502
    glabra, 502
    glabrata, 502
    hispida, **502**
    occidentalis, 502
    pacifica, 502
    palustris, **502**

pectinata, 500
polymorpha, 500
prostrata, 494
ramosa, 496, **502**, 503
sessiliflora, 495, *503*
sinuata, 496, **504**
  pubescens, 498, 499
sphaerocarpa, 495, **504**
subumbellata, 495, **504**, 505
sylvestris, 494, 496, *503*, **505**
tenerrima, 495, **505**
teres, 496, *503*, **505**
  rollinsii, 505
terrestris
  globosa, 502
  hispida, 502
trachycarpa, 504
truncata, 499
underwoodii, 499
walteri, 505
Rose bladderpod, 659
Rossbach, G. B., 536, 537, 538, 540, 543, 544, 545
Roughpod bladderpod, 613
Rouleau, E., 15
Roundleaf bladderpod, 655
Round-leaf willow, 70
Roundtip twinpod, 665
Rowlee, W. W., 58, 142
Rusty willow, 133
Rutabaga, 422
Rydberg's twinpod, 624

Sacchi, C. F., 24
Sage willow, 141
Sage-leaf willow, 141
Sahara mustard, 424
Salad-rocket, 434
Salicaceae, 3, 4, 8, 9, 23, 29, 51
  tribe Flacourtieae, 8, 9
Salicales, 8
Saliceae, 8, 9
Salick, J., 27, 57
Salix, 4, 5, 8, 9, **23**, 24, 25, 26, 27, 28, 29, 30, 31, 49, 51, 157
  [unranked] *Griseae*, 157
  [unranked] *Lanatae*, 143
  [unranked] *Longifoliae*, 50
  [unranked] *Myrtilloides*, 83
  [unranked] *Ovalifoliae*, 75
  [unranked] *Sitchenses*, 159
  [unranked] *Villosae*, 146
  sect. *Amygdalinae*, 49
  sect. Arbuscella, **140**
  sect. Argyrocarpae, **150**
  sect. *Breweriana*, 159
  sect. Canae, **150**
  sect. Candidae, **141**
  sect. Chamaetia, 29, 60, **64**
  sect. Cinerella, 94, 97, **126**
  sect. Cordatae, 101, 104, **120**, 122, 157
  sect. Daphnella, **161**
  sect. Diplodictyae, 29, 78
  sect. Floridanae, 30, 31, **32**
  sect. Fulvae, **134**
  sect. Geyerianae, **152**
  sect. Glaucae, 49, 61, **86**
  sect. Griseae, 157

sect. Hastatae, 61, 101, **104**
sect. Helix, 51, **162**
sect. Herbella, 29, **67**, 69
sect. Humboldtianae, 30, **32**, 51
sect. Lanatae, 95, **143**, 149
sect. *Longifoliae*, 23
sect. Maccallianae, 39, **48**
sect. Mexicanae, 95, **155**, 157
sect. Myrtilloides, 29, 61, **83**, 85, 86
sect. Myrtosalix, 29, 69, **71**
sect. Nigricantes, **125**
sect. Ovalifoliae, 29, **75**
sect. *Pentandrae*, 44
sect. Phylicifoliae, 96, 97, 98, **135**
sect. *Retusae*, 67
sect. Salicaster, 24, 29, 30, 31, 38, 39, **44**, 49
sect. Salix, 41
sect. Setchellianae, 61, **66**
sect. Sitchenses, 93, 96, 100, **159**
sect. Subalbae, 38, **40**
sect. *Tetraspermae*, 32
sect. Triandrae, 29, 39, **49**, 51
sect. *Vetrix*, 93
sect. Villosae, **146**, 149
sect. Viminella, **149**
subg. Chamaetia, 23, 24, 29, 49, 51, **60**, 61, 64, 66, 67, 71, 75, 78, 83, 86
subg. *Chosenia*, 4, 51
subg. Longifoliae, 23, 24, 26, 29, 30, 47, **50**, 51, 52, 67
subg. *Pleuradinea*, 4, 24
subg. Protitea, 23, 24, 25, 29, 30, 31, 32, 38, 47, 51, 57
subg. Salix, 23, 24, 29, 30, 31, 38, 39, 40, 41, 44, 48, 49, 51, 57, 61
subg. Vetrix, 23, 24, 29, 47, 49, 51, 57, 60, 61, 86, **93**, 104, 107, 120, 125, 126, 133, 134, 135, 140, 141, 143, 146, 149, 150, 151, 152, 155, 157, 159, 161, 162
subsect. Arbusculae, 140
subsect. Bicolores, 135
subsect. Cordatae, 104
subsect. Hastatae, 104
subsect. Luteae, 104, 120
subsect. Myrtilloides, 71
subsect. Sitchenses, 159
subsect. Substriatae, 134
tribe *Glaucae*, 86
tribe *Hastatae*, 104
tribe *Phylicifoliae*, 135
tribe Saliceae, 8, 9
adenophylla, 106
alaxensis, 25, 98, **146**, 148, 159
  alaxensis, 136, 137, 139, 144, *145*, **146**, 147, 148

Salix alaxensis (*continued*)
longistylis, 146, **147**, 160
silicicola, 147
alba, 23, 29, 30, 36, 39, 40, **42**, 43, 47, 153
caerulea, 42
chermesina, 42
sericea, 42
vitellina, 41, 42
×*amoena*, 92
amplifolia, 127, 128
amygdalina, 50
discolor, 50
amygdaloides, 23, 24, 30, 31, **35**, 36, **37**, 38, 47, 121
ancorifera, 126
angustata, 120
angustifolia, 160
arbusculoides, 46, 99, *137*, **141**
arctica, 24, 28, 63, 68, 69, 70, 72, 73, 74, 75, 76, **78**, 79, 80, 81, 82, 84, 90, 91, 114
antiplasta, 78
araioclada, 78
arctica, 79
brownei, 78
crassijulis, 78, 79
kophophylla, 78
pallasii, 78
petraea, 81
petrophila, 81
torulosa, 78, 79
arctolitoralis, 77
arctophila, 63, 72, **73**, 80, 90, 91, 92, 152
lejocarpa, 73
argophylla, 55
×*argusii*, 87
argyrocarpa, 68, 84, 93, 99, 107, 139, *151*, 152
arizonica, 28, 87, 102, 107, *109*, 110, **112**, 113, 123
athabascensis, 29, 60, 61, **84**, 85, 86, 128
atrocinerea, 94, 127, 130, 132, **133**
aurita, 94, **133**
babylonica, 38, **40**, 41, 42
bakeri, 156
ballii, 90, 92, 101, 110, **111**
balsamifera, 115
barclayi, 75, 80, 83, 87, 90, 103, 105, **113**, **114**, 115, 116, 117, 128, 145, 149
padophylla, 118
pseudomonticola, 117
barrattiana, 5, 23, 98, 105, 111, 114, 115, *145*, **148**, 149
marcescens, 148
tweedyi, 145
bebbiana, 25, 57, 99, 119, 121, 127, *130*, 133, **134**, 135, 142, 153
capreifolia, 134
depilis, 134
luxurians, 134
perrostrata, 134

projecta, 134
bigelovii, 156
boganidensis, 46
bolanderiana, 58
bonplandiana, 29, 30, 31, *33*, 37
laevigata, 34
toumeyi, 33
boothii, 87, 90, 92, 101, 102, 107, 108, *109*, 110, 112, **113**
brachycarpa, 25, 62, 84, 85, **86**, 90, 92, 107, 112, 113, 114
antimima, 87
brachycarpa, 84, 85, **87**, 112, 113, 139, 142
fullertonensis, 88, 89
glabellicarpa, 87
mexiae, 88
niphoclada, 88, 89
psammophila, **87**, 88, 116, 122, 148
sansonii, 87
brachypoda, 111
×*brachypurpurea*, 88
breweri, 93, 96, 98, 107, 156, 157, *158*, **160**, 161
delnortensis, 161
calcicola, *95*, 142, **143**, 144, 145, 147
calcicola, *142*, **143**, 144
glandulosior, 143, **144**
nicholsiana, 143, 144
californica, 106
callicarpaea, 92
cana, 152
candida, 25, 84, 87, 93, 99, 100, 110, 121, 122, 127, 135, 139, **141**, *142*, 144, 153
denudata, 141
capitata, 40
caprea, 94, **131**, 132, 133, 149
caroliniana, 27, 31, **34**, *35*, 36, 37
cascadensis, 63, 79, **82**, 83, 114, 115
thompsonii, 82
chaenomeloides, 24
chamissonis, 63, *68*, **71**, 72
chilensis, 34
chlorolepis, 61, 84, **85**, 87
chlorophylla
[unranked] *pellita*, 136
chosenia, 51
cinerea, 28, 94, 127, **132**, 133
atrocinerea, 133
oleifolia, 133
×*clarkei*, 142
coactilis, 157
columbiana, 52, 53, 56, 58, **59**, 60
commutata, 101, 102, *105*, 106, 107, 114, 115, 149
denudata, 105
mixta, 105
puberula, 105

rubicunda, 108
sericea, 105
cordata, 25, 101, **106**, 120, 142
abrasa, 120
mackenzieana, 125
monticola, 118
watsonii, 123
cordifolia, 92
callicarpaea, 92
eucycla, 92
intonsa, 92
macounii, 92
tonsa, 92
coulteri, 159
crassijulis, 78
×*cryptodonta*, 135
cyclophylla, 77
daphnoides, 94, **161**, 162
delnortensis, 96, 98, 157, 159, **161**
depressa
rostrata, 134
desertorum, 91
discolor, 47, 57, 97, 98, 116, 119, 121, *124*, **126**, 127, 130, 133, 135, 137, 139, 142, 153
overi, 126
prinoides, 126
divaricata
pulchra, 138
dodgeana, 71
drummondiana, 97, **135**, 136, 137, 139, 147, 155, 156, 159
bella, 135
subcoerulea, 135
×*dutillyi*, 152
eastwoodiae, 48, 90, 100, *105*, **106**, 107, 109, 112, 113
×*ehrhartiana*, 43
elaeagnos, 94, **150**
eriocephala, 27, 37, 47, 57, 84, 95, 96, 101, *118*, 119, **120**, 121, 122, 127, 130, 135, 142, 153, 158
famelica, 121
ligulifolia, 122
mackenzieana, 125
monochroma, 124
watsonii, 123
euxina, 23, 29, 39, 40, 41, 42, 43, 44
exigua, 23, 51, 52, 53, **54**, *55*, 57
angustissima, 58
columbiana, 59
exigua, *55*, 56, 57, 59
exterior, 56
gracilipes, 58
hindsiana, 52, 55, **56**
interior, 56
luteosericea, 55
melanopsis, 58
nevadensis, 55
parishiana, 56
pedicellata, 56
sericans, 56

sessilifolia, 59
stenophylla, 55
tenerrima, 58
virens, 55
exilifolia, 54
fallax, 84
famelica, 102, 103, 104, 118, **121**, 122, 142, 153
farriae, 103, 114, 115, **116**, 117
microserrulata, 116
walpolei, 116
flagellaris, 77
floridana, 27, 29, 30, 31, **32**, *33*
fluviatilis, 52, 58
argophylla, 55
exigua, 54
sericans, 56
sessilifolia, 59
tenerrima, 58
×fragilis, 28, 39, 42, 43
fragilis, 43, 44
sphaerica, 43
franciscana, 156
fullertonensis, 88
fuscescens, 25, 61, 68, **72**, 73, 74, 76, 80
hebecarpa, 81
reducta, 72
×*gaspeensis*, 87
geyeriana, 100, 123, 135, 136, *151*, **153**, 154, 155, 156, 160
argentea, 153
meleina, 153
glacialis, 77
×*glatfelteri*, 38
glauca, 25, 28, 49, 61, 80, 85, 86, **88**, **89**, 90, 91, 92, 107, 109, 151
acutifolia, 86, *88*, 89, 90, **91**
alicea, 91
callicarpaea, 92
cordifolia, 73, 80, 84, 87, 88, 90, **92**, 93, 111, 119, 139
desertorum, 91
glabrescens, 91
glauca, 90, 92
intonsa, 92
niphoclada, 88
orestera, 108
perstipula, 91
poliophylla, 91
sericea, 91
stipulata, *88*, **90**, 91
tonsa, 92
villosa, 86, 87, *88*, 90, **91**, 92, 109, 113
glaucophylla
albovestita, 119
glaucophylloides, 119
albovestita, 119
glaucophylla, 119
gooddingii, 31, 34, **36**, 37, 48
vallicola, 36
variabilis, 36

gracilis, 152
　textoris, 152
×*grayi*, 152
groenlandica
　lejocarpa, 73
hastata, 103, *114*, 115, **116**, 117
　farriae, 116, 117
　subintegrifolia, 116
×*hebecarp*, 81
herbacea, 62, **67**, *68*, 72, 73, 74, 80, 152
hindsiana, 52, 56
　leucodendroides, 56
　parishiana, 56
　tenuifolia, 55
hookeriana, 95, 96, 97, 98, 100, 103, 114, 115, **127**, 128, 129, 131, 156
　laurifolia, 127, 128
　tomentosa, 127, 128
×*hudsonensis*, 80
humaensis, 46
humboldtiana, 31, 34
humilis, 25, 97, 98, 121, 127, **129**, *130*, 131, 133, 135, 136, 139, 158
　humilis, **129**, *130*
　hyporhysa, 129
　keweenawensis, 129
　microphylla, 130
　rigidiuscula, 129
　tristis, 129, *130*
idahoensis, 108
interior, 24, 37, 47, 51, 52, 53, *55*, **56**, 57, 58, 121, 127, 135
　angustissima, 58
　exterior, 56, 57
　pedicellata, 56
　wheeleri, 56, 57
irrorata, 95, 96, 98, 136, 153, *154*, **155**, 156
ivigtutiana, 73
×*jamesensis*, 84
jejuna, 63, 76, **78**
jepsonii, 100, **160**
×*jesupii*, 39, 42, 43
krylovii, 141
labradorica, 92
laevigata, 31, *33*, **34**, 37
　angustifolia, 34
　araquipa, 34
　congesta, 34
lanata, 90
　calcicola, 143
　richardsonii, 144
lasiandra, 25, 37, 39, 46, **47**, 107, 121
　caudata, 46, 47, **48**
　lasiandra, 46, **47**, 48
　lyallii, 47
　macrophylla, 47
　recomponens, 47
lasiolepis, 96, 128, *154*, **155**, **156**, 157, 161
　bigelovii, 156
　bracelinae, 156
　lasiolepis, 156
　sandbergii, 156

×*laurentiana*, 127
leiocarpa, 70
leiolepis, 64
lemmonii, 100, 107, 109, 136, 153, *154*, 155
ligulifolia, 104, **122**, 123, 153, 154
linearifolia, 51, 56
longifolia
　angustissima, 58
　argophylla, 55
　exigua, 54
　interior, 57
　opaca, 55
　pedicellata, 57
　sericans, 57
　sessilifolia, 59
　tenerrima, 58
　wheeleri, 57
longipes
　venulosa, 34
　wardii, 34
longistylis, 147
lucida, 36, 37, 39, 42, 43, *45*, **46**, 47, 48, 49
　angustifolia, 46
　caudata, 48
　intonsa, 46
　lasiandra, 47
　serissima, 45
lutea, 102, 103, 104, 112, 113, **123**, *124*
　famelica, 121, 122
　ligulifolia, 122
　nivaria, 156
　turnorii, 122
　watsonii, 123
luteosericea, 55
maccalliana, 27, 39, *45*, **49**
mackenzieana, 125
　macrogemma, 125
macrostachya, 59
　cusickii, 59
　leucodendroides, 56
matsudana, 40
melanopsis, 52, *55*, 56, **58**, 59, 60, 160
　bolanderiana, 58
　gracilipes, 58
　kronkheitii, 58
　tenerrima, 58
microphylla, 51, 53, 54, 60
missouriensis, 120
monica, 138
monochroma, 102, **124**
monticola, 95, 104, **118**
muriei, 88
myricoides, 90, 92, 93, 95, 96, 98, 101, *118*, **119**, 120, 121, 127, 135
　albovestita, 119
myrsinifolia, 94, **125**, 126
myrsinites
　parvifolia, 73
myrtillifolia, 101, 102, **109**, 110, 111, 142
　brachypoda, 111
　cordata, 110, 111
　pseudomyrsinites, 110, 111
myrtilloides, 83

hypoglauca, 83
nevadensis, 55
nigra, 24, 31, 34, *35*, **36**, 37, 38, 42, 43, 47
　falcata, 36
　lindheimeri, 36
　vallicola, 36
nigricans, 125
niphoclada, 62, 67, **88**, 89, 90, 91
　fullertonensis, 88
　mexiae, 88
　muriei, 88
nivalis, 62, *65*, **66**
　saximontana, 66
novae-angliae, 111
　aequalis, 113
　cordata, 110
nummularia, 27, 62, **68**, 69
　tundricola, 68
orestera, 92, 100, 107, **108**, 109, 151
ovalifolia, frontispiece, 27, 62, 63, 72, 73, **75**, 76, 78, 80, 81
　arctolitoralis, 76, 77
　cyclophylla, 76, 77
　glacialis, 76, 77, 81
　ovalifolia, 76, 77
padophylla, 118
paraleuca, 127
parishiana, 56
parksiana, 58
×*peasei*, 68
pedicellaris, 25, 61, 80, 81, **83**, *84*, 85, 90, 92, 93, 121, 128, 137, 152, 153, 154
　athabascensis, 84
　hypoglauca, 83
　tenuescens, 83
×*pedunculata*, 93, 127
×*pellicolor*, 127
pellita, 80, 84, 97, 127, 132, **136**, *137*, 139, 147, 149, 152, 153
　angustifolia, 160
×*pendulina*, 38, 40, 41
　blanda, 41
　elegantissima, 41
pentandra, 23, 27, 39, 43, **44**, 46
　caudata, 48
　intermedia, 46
petiolaris, 26, 37, 42, 47, 57, 99, 121, 122, 130, 135, 137, 142, *151*, **152**, 153, 158
　gracilis, 152
petraea, 81
petrophila, 63, 79, **81**, 82
　caespitosa, 81
phlebophylla, 27, 63, 70, 72, 73, **74**, 80, 81
phylicifolia
　monica, 138
　planifolia, 138
　pulchra, 138
　subglauca, 138
piperi, 127, 128

planifolia, 27, 80, 87, 90, 92, 93, 97, 98, 119, 126, 127, 130, 131, 136, 137, **138**, 139, 140, 142, 147, 152
　monica, 87, 138, 139
　pulchra, 138
　tyrrellii, 140
　yukonensis, 138
polaris, 62, 63, **68**, **69**, 70, 74, 80, 81
　leiocarpa, 70
　pseudopolaris, 69
　selwynensis, 69
prolixa, 104, *124*, **125**
pseudocordata
　aequalis, 113
pseudolapponum, 91
pseudolasiogyne, 40
pseudomonticola, 95, 96, **117**, *118*, 122
　padophylla, 118
pseudomyrsinites, 25, 101, 102, **110**, 111, 149
pseudopentandra, 46
pulchra, 98, 131, *137*, **138**, 139
　yukonensis, 138
purpurea, 3, 94, **162**
pyrifolia, 88, 101, 102, *114*, **115**, 116, 118, 127
　lanceolata, 115
raupii, 61, **85**, 86
reticulata, 24, 62, **65**, 66
　gigantifolia, 65
　glabellicarpa, 65, 66
　nivalis, 66
　saximontana, 66
　semicalva, 65
richardsonii, 95, 114, 115, 142, **144**, 145
　macouniana, 143
　mckeandii, 144
rigida, 120
　angustata, 120
　mackenzieana, 125
　macrogemma, 125
　vestita, 120
　watsonii, 123
rostrata, 134
rotundifolia, 27, 62, **69**, 70, 74, 80, 81
　dodgeana, 70, **71**, 83
　rotundifolia, 70
×*rubella*, 142
×*rubens*, 43
rubra, 56
safsaf, 24
×*salamonii*, 40
×*schneideri*, 47
　nothovar. *chrysocoma*, 40, 41
scouleriana, 25, 97, 98, 115, 128, 129, *130*, **131**, 136, 138, 139, 159
　poikila, 131
×*sepulcralis*, 38, 40, 41, 42
×*sericans*, 132
sericea, 27, 99, 121, 142, 153, **157**, *158*

INDEX

Salix (continued)
  serissima, 24, 26, 27, 39, **45**, 46, 47
  sessilifolia, 52, 53, **59**, 60
    hindsiana, 56
    leucodendroides, 56
    vancouverensis, 58
    villosa, 59
  setchelliana, 29, 51, 61, *65*, **66**, 67, 89
  silicicola, 25, 88, 98, **147**, 148
  sitchensis, 24, 59, 100, 136, 147, 153, *158*, **159**, 160, 161
    angustifolia, 160
    parvifolia, 159
    ralphiana, 160
  ×smithiana, 94, 132, 133, 137, 149
  speciosa
    alaxensis, 146
  sphenophylla, 63, **82**
    pseudotorulosa, 82
  stenophylla, 55
  stipulifera, 90
  stolonifera, 24, 27, 62, 75, 80, 81, 114, 115
  subcoerulea, 135
  subfragilis, 24
  ×subsericea, 152, 153
  syrticola, 106
  taxifolia, 52, 53, **54**, 60
    lejocarpa, 54
    limitanea, 54
    sericocarpa, 54
  tenera, 82
  tenerrima, 58
  tetrasperma, 24
  thurberi, 51, 52, 53, 54, **58**
  tracyi, 95, 98, 102, 103, **156**, 157
  triandra, 24, 39, 49, **50**
    discolor, 50
  tristis, 130
  turnorii, 88, 103, 120, **122**
  tweedyi, 95, *145*
  tyrrellii, 27, 97, **140**
  ×ungavensis, 111
  uva-ursi, 60, 62, 68, 72, 73, 74
  vestita, 46, 55, 61, **64**
    leiolepis, 64
    psilophylla, 64
  villosa, 91
    acutifolia, 91
  viminalis, 94, 132, 137, **149**
  ×waghornei, 80
  walpolei, 116
  wheeleri, 57
  ×wiegandii, 144
  wolfii, 93, 100, 101, *105*, **107**, 113, 151
    idahoensis, *105*, 107, **108**
    pseudolapponum, 91
    wolfii, *105*, 107, **108**
  ×wrighti, 37
  ×wyomingensi, 87
Salmon twinpod, 634
Saltwort, 187

Saltwort Family, 186
Salvadoraceae, 184, 186
Sánchez-Yélamo, M. D., 433
Sand dune willow, 106
Sandbar willow, 56
Sandbergia, 235, 236, 242, 243, **417**, 418, 557, 558
  perplexa, **416**, **418**
    lemhiensis, 418
    perplexa, 418
  whitedii, **418**
Sand-rocket, 433
Sandyseed clammyweed, 203
Santamour, F. S. Jr., 41
Satin-flower, 596
Satiny willow, 136
Saule, 23
Savile, D. B. O., 79, 80
Savolainen, V., 4
Scaptomyza
  apicalis, 174
Scheen, A.-C., 316
Schivereckia, 310
Schizopetaleae, 227, 257
Schneider, C. K., 23, 34, 37, 48, 92, 116, 127, 132, 135, 139, 142, 153, 158
Schoenocrambe, 226, 667, 669, 690
  argillacea, 690
  barnebyi, 691, 692
  decumbens, 668
  linearifolia, 690
  linifolia, 668
  pinnata, 668
  pygmaea, 668
  suffrutescens, 692
Schreiner, E. J., 20
Schulz, O. E., 226, 227, 270, 289, 290, 303, 305, 317, 323, 326, 330, 333, 345, 478, 667
Scouler's willow, 131
Scurvygrass, 460, 463, 514
Sea-cabbage, 430
Sea-kale, 431
Sea-rockets, 424
Sebastopol meadowfoam, 176
Secund bladderpod, 627
Seevers, J., 182
Seigler, D. S., 16
Selenia, 231, 239, 458, 486, **506**
  aptera, 507
  aurea, 506, **507**, *508*
    aptera, 507
  dissecta, 506, **507**
  grandis, 506, 507, **508**
  jonesii, 506, 507, **508**, 509
    obovata, 508, 509
  mexicana, 506
  oinosepala, 508
Senebiera, 570
  coronopus, 578
  didyma, 580
  incisa, 580
  pinnatifida, 580
Serviceberry willow, 118
Sessile bladderpod, 663
Setchell's willow, 66

Shafroth, P. B., 43
Sharpleaf twinpod, 624
Sheep Mountain bladderpod, 636
Shepherd's-cress, 565
Shepherd's-purse, 453
Shining willow, 46
Shippee meadowfoam, 183
Shortia
  dentata, 373
Shortstalk stinkweed, 210
Showy spiderflower, 217
Shrubby dog caper, 198
Sibara, 234, 235, 242, 245, 492, **692**, 693
  angelorum, 693
  brandegeeana, 693
  deserti, 691, **693**
  filifolia, 693, **694**
  grisea, 693, 738
  laxa, 693
  mexicana, 693
  rosulata, 693
  runcinata, 689
    brachycarpa, 689
  viereckii, 689, 693
    endlichii, 689
  virginica, 493, 693
Sibaropsis, 233, 244, 676, **694**
  hammittii, 691, **694**
Sidesaddle bladderpod, 627
Sierra willow, 106
Silene, 204
Silky osier, 149
Silky willow, 157
Silver bladderpod, 627
Silver-dollar, 597
Sinapis, 226, 234, 244, **441**, 442
  alba, 440, **442**, 443, 494
    alba, 443
  arvensis, 442, **443**
  aucheri, 442
  erucoides, 433
  geniculata, 437
  incana, 437
  juncea, 421
  kaber, 443
  muralis, 433
  nigra, 422, 442
  pekinensis, 423
  recurvata, 430
Sisymbrieae, **666**
Sisymbrium, 231, 234, 246, **667**, 669, 687
  acuticarpum, 683
  altissimum, **668**
  ambiguum, 725
  amphibium, 497
    palustre, 501
  apetalum, 500
  arcticum, 529
  arenosum, 448
  aureum, 725
  auriculatum, 667, 687
  brachycarpum, 527
    filipes, 524
    intermedium, 527
  californicum, 521
  canescens, 527

  alpestre, 522
  brachycarpum, 527
  brevipes, 522
  californicum, 527
    major, 522
  curvisiliqua, 500
  decumbens, 668
  deflexum, 683
    xerophilum, 683
  dentatum, 373
  diffusum, 558
    jaegeri, 558
  dubium, 500
  elegans, 726
    juniperorum, 726
  erysimoides, 668, **669**, 670
  gallicum, 435
  glaucum, 556
  hispidum, 502
  humile, 548
  incanum, 522
  incisum, 523
    californicum, 527
    filipes, 524
    sonnei, 523
    xerophilum, 524
  indicum, 501
  intermedium, 527
  irio, 668, **670**
  juniperorum, 726
  kearneyi, 726
  lasiophyllum, 683
  leptophyllum, 523
  linearifolium, 690
  linifolium, 662, 667, **668**, 669
    decumbens, 668
    pinnatum, 668
  loeselii, 668, **669**
  longepedicellatum, 524
    glandulosum, 524
  microtitis, 501
  monense, 430
  multifidum, 527
    brachycarpum, 527
    canescens, 527
  murale, 433
  nasturtium-aquaticum, 490
  obtusum, 525
  ochroleucum, 528
  officinale, 667, **670**
    leiocarpum, 670
  orientale, 668, **670**
  paradisum, 526
  parviflorum, 528
  pauciflorum, 393
  perplexum, 418
  pinnatum
    brachycarpum, 527
  polyceratium, 666, 667, **669**
  polymorphum, 667
  procerum, 522
  purpusii, 726
  pygmaeum, 668
  reflexum, 683
  salsugineum, 556
  serratum, 523
  shinnersii, 727
  sophia, 528
  sophioides, 529

stenophyllus, 690
sylvestre, 505
tenuifolium, 433
teres, 505
texanum, 738
thalianum, 450
umbrosum, 688
vaseyi, 727
vernale, 726
virgatum, 457
viscosum, 523
walteri, 505
watsonii, 727
Sitka willow, 159
Skeleton-leaf willow, 74, 152
Skvortsov, A. K., 4, 25, 27, 28, 30, 40, 46, 50, 51, 60, 73, 80, 90, 92, 116, 133, 157
Skyline bladderpod, 642
Slender clammyweed, 204
Slender stinkweed, 212
Slimleaf wall-rocket, 433
Slinkweed, 517
Small pussy willow, 129
Small-fruit willow, 86
Small-fruit sand dune willow, 87
Smelowskia, 237, 238, 240, 241, 243, **671**, 673
  americana, **672**, 673, 674
  borealis, 672, **673**, 674
    jordalii, 673
    koliana, 673
    villosa, 673
  californica, 521
  calycina, 673
    americana, 672, 673
    integrifolia, 675
    media, 674
    porsildii, 675
  fremontii, 416
    bisulcata, 416
    glabella, 416
  holmgrenii, 414
  integrifolia, 675
  johnsonii, 672, **673**, 674
  jurtzevii, 675
  lineariloba, 672
  lobata, 672
  media, 672, 673, **674**, 675
  ovalis, 672, **674**
    congesta, 674
  porsildii, 672, 673, **675**
  pyriformis, 672, **675**, 676
  spathulatifolia, 675
Smelowskieae, **671**
Smith, R. L., 8
Snake Range bladderpod, 657
Snake River twinpod, 644
Snow willow, 88
Snowbed willow, 67
Solitary bee, 174
Solmsiella, 453
Soltis, D. E., 4
Soltis, P. S., 4, 701
Song, K., 422
Soper, J. H., 79
Sophia, 518
  adenophora, 520
  andrenarum, 528
  brachycarpa, 527
  brevipes, 522
  californica, 527
  diffusa, 558
  filipes, 524
  glabra, 528
  glandifera, 524
  glandulifera, 523
  gracilis, 524
  incisa, 523
  intermedia, 527
  leptophylla, 523
  leptostylis, 521
  longepedicellata, 524
  magna, 527
  millefolia, 527
  myriophylla, 527
  nelsonii, 525
  obtusa, 525
  ochroleuca, 528
  paradisa, 526
  parviflora, 528
  perplexa, 418
  pinnata, 526
    brachycarpa, 527
  procera, 522
  purpurascens, 523
  serrata, 523
  sonnei, 523
  sophioides, 529
  viscosa, 523
Southern cottonwood, 18
Southworth, D., 182
Spatula-leaf bladderpod, 663
Spectacle-fruit, 213
Spectacle-pod, 604, 607
Sphaerocardamum, 418
Spiderflower, 208, 215, 218, 220, 221
Spiderflower Family, 199
Spider-wisp, 222
Spies, T. A., 21, 22
Spiny-caper, 215
Spreading bladderpod, 641
Spreading spiderflower, 220
Sprengeria, 570
  flava, 582
  minuscula, 582
  watsoniana, 582
Spring Creek bladderpod, 615
St. Mary's Peak bladderpod, 644
Staintoniella, 256
Stanfordia, 677
  californica, 680
Stanleya, 230, 232, 245, 246, 695
  albescens, **696**
  amplexifolia, 744
  annua, 696
  arcuata, 697
  bipinnata, **696**
  canescens, 697
  collina, 699
  confertiflora, 695, **696**, 697
  elata, 695, **697**
  fruticosa, 697
  glauca, 697
  latifolia, 698
  gracilis, 742
  heterophylla, 697
  integrifolia, 698
  pinnata, 695, 696, **697**
    bipinnata, 696
    gibberosa, 696
    integrifolia, 697, **698**
    inyoensis, 697, 698
    pinnata, **697**, 698
    texana, 697, **698**
  pinnatifida, 697
    integrifolia, 698
  rara, 696
  runcinata, 698
  tomentosa, 695, **698**
    runcinata, 698
    tomentosa, 698
  viridiflora, 695, **698**, 699
    confertiflora, 696
Stanleyella, 728
  texana, 738
  wrightii, 738
  tenella, 735
Steele, J. W., 16, 140
Stenopetalum, 227
Stenophragma
  salsugineum, 556
  virgatum, 457
Stettler, R. F., 14
Stevens, P. F., 184
Steyn, E. M. A., 24
Stinking wall-rocket, 433
Stinkweed, 209, 443, 746
Stock, 253
Stones River bladderpod, 615
Stout, A. B., 20
Straight bladderpod, 660
Strap-leaf willow, 122
Streptanthella, 233, 244, 676, **699**
  longirostris, **699**, **700**
    derelicta, 700
Streptanthus, 224, 226, 228, 229, 230, 231, 233, 234, 245, 257, **700**, 701, 706, 720, 723
  albidus, 713
    peramoenus, 713
  amplexicaulis, 679
    barbarae, 679
  anceps, 679
  angustifolius, 408
  arcuatus, 365
  argutus, 709
  arizonicus, 708
    luteus, 708
  asper, 713
  barbatus, 699, 702, **704**
  barbiger, 701, **704**, 705, 723
  batrachopus, 704, **705**
  bernardinus, 703, **705**
  biolettii, 713
  brachiatus, 703, **706**, 723
    hoffmanii, 706
  bracteatus, 702, **706**
  brazoensis, 719
  breweri, 704, **706**, 707, 715
    hesperidis, 715
  californicus, 680
  callistus, 702, **707**
  campestris, 702, 703, **707**
    bernardinus, 705
    jacobaeus, 705
  carinatus, 703, **708**
    arizonicus, **708**
    carinatus, **708**
  coloradensis, 709
  coombsiae, 732
  cordatus, 703, **708**, 709, 719
    cordatus, **709**
    duranii, 709
    exiguus, 719
    piutensis, **709**
  coulteri, 680
    lemmonii, 684
  crassicaulis, 681
  crassifolius, 709
  cutleri, 701, **709**, 710
  diversifolius, 702, **710**, 711, 722
  drepanoides, 704, **711**
  dudleyi, 681
  farnsworthianus, 702, **711**
  fenestratus, 702, **711**, 712
  flavescens, 681
  foliosus, 722
  glabrifolius, 716
  glandulosus, 701, 702, 704, **712**
    albidus, 712, **713**
    glandulosus, 712, **713**, 714
    hoffmanii, 712, **713**
    josephinensis, 712, **713**, 714
    niger, 712, **714**
    pulchellus, 712, **714**
    secundus, 712, **714**
    sonomensis, 712, **714**
  glaucus, 682
  gracilis, 702, **715**
  hallii, 682
  hastatus, 686
  hesperidis, 704, **715**
  heterophyllus, 682
  hispidus, 702, **715**, 716
  howellii, 701, **716**, 732
  hyacinthoides, 701, **716**
  inflatus, 683
  insignis, 702, 710, **717**
    insignis, 710, **717**
    lyonii, **717**
  lasiophyllus, 683
    inalienus, 683
    utahensis, 683
  lilacinus, 681
  linearifolius, 690
  linearis, 710
  longifolius, 560
  longirostris, 700
  longisiliquus, 703, **717**, 718
  maculatus, 703, 710, **718**, 721
    maculatus, 718
    obtusifolius, 718
  major, 684
  micranthus, 560
  mildrediae, 713
  morrisonii, 703, **718**, 719, 723
    elatus, 718

Streptanthus morrisonii (cont.)
  hirtiflorus, 718
  kruckebergii, 718
 niger, 714
 oblanceolatus, 702, **719**
 obtusifolius, 718
 oliganthus, 703, **719**
 orbiculatus, 722
 parryi, 684
 peramoenus, 713
 petiolaris, 701, **719**, 720
 pilosus, 684
 platycarpus, 702, 703, **720**
 polygaloides, 704, **720**, 721
 procerus, 681
 pulchellus, 714
 repandus, 682
 rigidus, 683
 sagittatus, 736
 sanhedrensis, 722
 secundus, 714
 simulans, 685
 sparsiflorus, 720
 squamiformis, 703, **721**
 suffrutescens, 722
 tortuosus, 702, 721, **722**
   flavescens, 722
   oblongus, 722
   optatus, 722
   orbiculatus, 722
   pallidus, 722
   suffrutescens, 722
 validus, 708
 vernalis, 703, **722**, 723
 versicolor, 713
 vimineus, 704, **723**
 wyomingensis, 726
Strigosella, 235, 242, **553**
 africana, **554**
Stroganowia, 226, 570, 594
 tiehmii, 594
Strong bladderpod, 664
Stuckey, R. L., 494, 499, 500, 502, 503
Subularia, 229, 239, 459, **509**, 510
 aquatica, 508, **509**
   americana, 508, **509**, 510
     aquatica, 510
   monticola, 509
Suda, Y., 131, 136
Sugaya, S., 26
Summer-mustard, 437
Sun, M., 183
Svensson, S., 454
Swamp poplar, 5, 12, 13
Swede, 422
Swede rape, 422
Swedish turnip, 422
Sweeney, P. W., 477
Sweet mignonette, 193
Sweet-alyssum, 598
Sydnor, T. D., 50
Synthlipsis, 237, 241, **665**
 berlandieri, 614
   hispida, 614
 densiflora, 666
 greggii, 662, **666**
   hispidula, 666
Syrenia, 534
Sytsma, K. J., 8

Tacamahaca, 14
Tail-leaf willow, 48
Takhtajan, A. L., 226
Tall blueberry willow, 110
Tall wallflower-cabbage, 430
Tansy mustard, 518, 528
Tarenaya, 199, 200, 215, 217, **218**
 afrospina, 218
 hassleriana, 217, 218, **219**
 spinosa, **219**
Tavaputs bladderpod, 642
Taylor, R. L., 128
Taylor, S., 128
Tea-leaf willow, 138
Teesdalia, 224, 229, 231, 238, 563, **564**, 565
 conferta, 565
 coronopifolia, **565**
 nudicaulis, 566, **565**
Teruncius
 arvensis, 746
Tetracentraceae, 4
Tetradynamia, 226
Tetrapoma, 493
 barbareifolium, 498
 pyriforme, 498
Thale, 226
Thale cress, 447, 450
Thellung, A., 570, 580, 581
Thellungiella, 555
 salsuginea, 556
Thelypodieae, 226, 227, 257, 485, 667, **676**
Thelypodiopsis, 229, 230, 231, 245, 246, **723**, 724
 ambigua, 724, **725**
   erecta, 725
 argillacea, 690
 aurea, 721, 724, **725**
 bakeri, 726
 barnebyi, 691
 brachycarpa, 730
 brachypoda, 730
 crispa, 730
 divaricata, 724, **725**
 elegans, 724, **726**
 howellii, 732
 juniperorum, 724, **726**
 linearifolia, 690
 nuttallii, 736
 paniculata, 735
 purpusii, 724, **726**
 sagittata, 736
   ovalifolia, 737
 shinnersii, 724, **727**
 stenopetala, 737
 torulosa, 736
 vaseyi, 724, **727**
 vermicularis, 724, **727**
 versicolor, 724
 wyomingensis, 726
Thelypodium, 230, 231, 233, 245, 246, **728**
 subg. Hesperidanthus, 689
 subg. Heterothrix, 559
 affine, 733
 ambiguum, 725
 amplifolium, 737
 anisopetalum, 560
 aureum, 725
 bakeri, 726
 brachycarpum, 729, **730**
   crispum, 730
 cooperi, 680
 crenatum, 579
 crispum, 729, **730**
 deserti, 693
 elegans, 726
 eucosmum, 729, **731**
 flavescens, 681
 flexuosum, 729, **731**
 gracilipes, 734
 greenei, 681
 hookeri, 681
 howellii, 729, **732**
   howellii, **732**
   spectabilis, **732**
 integrifolium, 728, 729, **732**
   affine, **733**
   complanatum, 733, **734**
   gracilipes, 733, **734**
   integrifolium, **733**
   longicarpum, 733, **734**
 jaegeri, 692
 laciniatum, 728, **734**
   milleflorum, 735
   streptanthoides, 734
 lasiophyllum, 683
   inalienum, 683
   rigidum, 683
   utahense, 683
 laxiflorum, 729, **735**
 lemmonii, 679
 leptosepalum, 734
 lilacinum, 733
   subumbellatum, 733
 linearifolium, 690
 lobatum, 687
 longifolium, 560
   catalinense, 560
 longirostris, 700
 macropetalum, 737
 micranthum, 560
 milleflorum, 728, 730, **735**
 nuttallii, 736
 ovalifolium, 737
 palmeri, 737
 paniculatum, 729, **735**
 pinnatifidum, 485
 purpusii, 726
 repandum, 729, **736**
 rhomboideum, 733
   gracilipes, 734
 rigidum, 683
 rollinsii, 729, **736**
 sagittatum, 729, **736**
   crassicarpum, 735
   ovalifolium, **737**
   sagittatum, **737**
   vermicularis, 727
 salsugineum, 556
 simplex, 732
 stamineum, 684
 stenopetalum, 729, **737**, 738
 streptanthoides, 734
 suffrutescens, 692
 tenue, 738
 texanum, 693, 729, **738**
 torulosum, 736

 utahense, 683
 vaseyi, 727
 vernale, 726
 wrightii, 729, **738**
   oklahomensis, 738
   tenellum, 735
   wrightii, 738
Thelypody, 728
Thick-leaf bladderpod, 656
Thlaspi, 226, 230, 239, 600, **745**, 746
 aileeniae, 603
 alliaceum, **743**, **746**
 alpestre
   californicum, 602
   glaucum, 602
   purpurascens, 602
 arcticum, 601
 arvense, **746**
 australe, 602
 bursa-pastoris, 454
 californicum, 601, 602
 campestre, 577
 coloradense, 602
 coronopifolium, 565
 fendleri, 601
 coloradense, 602
   glaucum, 602
   hesperium, 602
   idahoense, 603
   tenuipes, 602
 glaucum, 601, 602
   californicum, 602
   hesperium, 602
   pedunculatum, 602
 hesperium, 602
 idahoense, 601, 603
   aileeniae, 603
 montanum, 601, 602, 603
   californicum, 602
   fendleri, 601
   idahoense, 603
   montanum, 601, 603
   siskiyoense, 603
 parviflorum, 603
 perfoliatum, 600
 procumbens, 530
 prolixum, 602
 purpurascens, 602
 stipitatum, 602
 tuberosum, 473
Thlaspideae, 226, **744**
Thompson, J. T., 112, 113
Thorne, R. F., 4, 186, 188
Thorp, R. W., 174
Threadleaf bladderpod, 625
Thurber's willow, 58
Thysanocarpus, 225, 230, 232, 239, **739**
 affinis, 741
 amplectens, 740
 conchuliferus, 739, **740**
   planiusculus, 740
 curvipes, 739, **740**, 741
   cognatus, 740
   elegans, 740
   emarginatus, 740
   eradiatus, 740
   involutus, 740
   longistylus, 740
   madocarpus, 740

pulchellus, 740
emarginatus, 740
erectus, 739
filipes, 740
foliosus, 740
hirtellus, 740
laciniatus, 739, **740**, 741
　affinis, 741
　conchuliferus, 740
　desertorum, 741
　emarginatus, 740
　eremicola, 741
　hitchcockii, **741**
　laciniatus, **741**
　ramosus, 741
　rigidus, **741**
oblongifolius, 267
pulchellus, 740
pusillus, 267
radians, 730, 739, 740, **741**
　montanus, 741
　ramosus, 741
　trichocarpus, 740
Tickweed, 222
Tieszen, L. T., 223
Tirania, 190
Toisusu, 4, 24
Tolmatchew, A. I., 270
Tomostima, 269
　caroliniana, 333
　hispidula, 333
　micrantha, 333
　nemorosa, 321
Torularia, 549
　humilis, 548
　arctica, 549
Towercress, 458
Tower-mustard, 458
Tracy's willow, 156
Transberingia, 235, 242, **456**, 557, 558
　bursifolia, **456**, *457*
　　bursifolia, 456, **457**
　　virgata, 456, **457**
　virgata, 457
Treacle mustard, 517
Tremble, 22
Trembling aspen, 22
Triest, L. L., 43
Tropaeolaceae, **165**
Tropaeolum, 165, **166**
　majus, *164*, **166**
　minus, 166
　peltophorum, 166
　peregrinum, 166
　tuberosum, 166
Trophaeastrum, 165
Tropidocarpum, 234, 236, 240, 242, **531**
　californicum, 531, **533**
　capparideum, 225, 531, **532**, *533*
　dubium, 532
　gracile, 531, ***532***
　　dubium, 532
　　scabriusculm, 532
　lanatum, 531
　macrocarpum, 532
　scabriusculm, 532
Trybush, S., 49

Tucker, G. C., 165, 172, 173, 184, 189, 194, 195, 199, 201, 208, 213, 215, 216, 218, 220, 221, 222
Tufted twinpod, 631
Tumbleweed, 668
Turkish rocket, 445
Turner, B. L., 54, 186
Turnip, 419, 423, 424
Turnip-rape, 423
Turnipweed, 440
Turnor's willow, 122
Turritis, 226, 235, 242, 257, **458**
　brachycarpa, 382
　drummondii, 408
　glabra, *457*, **458**
　　lilacina, 458
　grahamii, 382
　laevigata, 386
　lasiophylla, 683
　laxa, 458
　macrocarpa, 458
　mollis, 457
　patula, 382
　pseudoturritis, 458
　spathulata, 261
　stricta, 408
Tuskan, G. A., 8
Tweedy's willow, 145
Twisselmannia, 531
　californica, 533
Tyrrell's willow, 140

Uncompahgre bladderpod, 665
Under-green willow, 105
Upland willow, 129
Uplandcress, 460
Utah bladderpod, 647
Van Splunder, I., 24
Vanderpool, S. S., 199, 205, 209, 215
Vasconcellea, 171
Velarum, 667
Vella
　annua, 429
Vesicaria
　sect. *Physaria*, 616
　alpina, 660
　angustifolia, 625
　arctica, 626
　arenosa, 626
　argyraea, 627
　auriculata, 612
　brevistyla, 613
　densiflora, 633
　didymocarpa, 633
　engelmannii, 635
　　elatior, 635
　fendleri, 636
　geyeri, 638
　globosa, 639
　gordonii, 640
　gracilis, 641
　grandiflora, 613
　　pallida, 656
　　pinnatifida, 613
　kingii, 645
　lasiocarpa, 613
　leiocarpa, 626

　lescurii, 614
　lindheimeri, 649
　ludoviciana, 649
　montana, 650
　nuttallii, 641
　occidentalis, 654
　pallida, 656
　polyantha, 641
　pulchella, 635
　purpurea, 659
　recurvata, 660
　repanda, 641
　stenophylla, 636
　　diffusa, 636
　　humilis, 636
　　procera, 636
Vidrillos, 187
Viereck, L. A., 14, 110, 111
Violales, 4, 8
Violet willow, 161
Violet-rocket, 485
Volantines-preciosos, 217
Voss, E. G., 153

Wagner, W. H. Jr., 21
Waithaka, K., 223
Walker, L. R., 24
Wallflower, 534, 539
Wallflower-cabbage, 430
Wall-mustard, 433
Wall-rocket, 432
Wang, J., 4
Wang, R., 4
Wanntorp, H., 205
Warea, 229, 232, 245, **742**
　amplexifolia, 742, **744**
　auriculata, 744
　carteri, 232, 742, **743**
　cuneifolia, **742**
　sessilifolia, 742, **743**
Warwick, S. I., 227, 270, 290, 295, 320, 341, 419, 422, 429, 430, 434, 435, 436, 438, 440, 441, 442, 505, 517, 529, 547, 549, 551, 555, 667, 669, 673, 675, 687, 709
Wasabi, 555
Wasabia, 555
Washington twinpod, 624
Wassuk Range bladderpod, 632
Watercress, 489
Watson, L., 165
Weber, W. A., 593
Weberling, F., 190
Wedge-leaf willow, 82
Weeping willow, 40, 41
Welsh, S. L., 348, 350, 395, 397
West silver bladderpod, 662
West, C. J., 28
Western bladderpod, 654
Western clammyweed, 203
White bladderpod, 656
White Bluffs bladderpod, 635
White mignonette, 192
White Mountain bladderpod, 658
White mustard, 442
White poplar, 21
White Spider, 202

White willow, 42
White, W. W., 16
Whitepuff, 190
Whitlow grass, 345
Whitlow wort, 345
Wiegand, K. M., 142
Wild mignonette, 192
Wild mustard, 222, 443
Wild-cabbage, 677
Wild-radish, 439
Wild-rape, 423
Wild-turnip, 423, 440
Wilkinson, H. P., 4
Willow, 8, 23, 24, 25, 27, 28, 42, 146, 147, 157
Willow Family, **3**
Windham, M. D., 257, 258, 269, 270, 288, 292, 293, 295, 298, 301, 303, 305, 311, 312, 314, 316, 318, 325, 327, 329, 330, 333, 335, 336, 339, 340, 342, 345, 348, 349, 365, 367, 368, 369, 370, 371, 372, 373, 374, 375, 376, 377, 378, 381, 382, 383, 384, 385, 386, 387, 388, 389, 390, 391, 392, 393, 394, 395, 396, 397, 398, 399, 401, 404, 408, 409, 410, 411, 412, 739
Winge, Ø., 270, 345
Wintercress, 460
Wislizenia, 199, 200, 205, **213**
　californica, *211*, 213, **214**
　costellata, 214
　divaricata, 214
　fruticosa, 214
　mamillata, 214
　melilotoides, 214
　palmeri, 213, **214**
　refracta, 213, **214**
　　californica, 214
　　mamillata, 214
　　melilotoides, 214
　　palmeri, 214
　scabrida, 214
Woad, 193, 567, 568
Wolfe, J. A., 4
Woolly meadowfoam, 182
Wormseed mustard, 539, 540
Worthington, R. D., 424
Xylosma, 5, 8, **163**, 164
　blepharodes, 164
　celastrina, 164
　flexuosa, *164*
　pringlei, 164

Yellow bee-plant, 206
Yellow cleome, 222
Yellow mignonette, 192
Yellow willow, 123
Yellow-cress, 493
Yellow-rocket, 461
Yew-leaf willow, 54
Ysaño, 166
Zapata bladderpod, 664
Zasada, J. C., 24, 25
Zebrawood, 198
Zhang, M. L., 30, 195
Zhukova, P. G., 70

Political Map of North America North of Mexico

Canadian Provinces

Alta.	Alberta	N.S.	Nova Scotia
B.C.	British Columbia	Nunavut	
Man.	Manitoba	Ont.	Ontario
N.B.	New Brunswick	P.E.I.	Prince Edward Island
Nfld. and Labr.	Newfoundland and Labrador	Que.	Quebec
		Sask.	Saskatchewan
N.W.T.	Northwest Territories	Yukon	

United States

Ala.	Alabama	Mont.	Montana
Alaska		Nebr.	Nebraska
Ariz.	Arizona	Nev.	Nevada
Ark.	Arkansas	N.H.	New Hampshire
Calif.	California	N.J.	New Jersey
Colo.	Colorado	N.Mex.	New Mexico
Conn.	Connecticut	N.Y.	New York
Del.	Delaware	N.C.	North Carolina
D.C.	District of Columbia	N.Dak.	North Dakota
Fla.	Florida	Ohio	
Ga.	Georgia	Okla.	Oklahoma
Idaho		Oreg.	Oregon
Ill.	Illinois	Pa.	Pennsylvania
Ind.	Indiana	R.I.	Rhode Island
Iowa		S.C.	South Carolina
Kans.	Kansas	S.Dak.	South Dakota
Ky.	Kentucky	Tenn.	Tennessee
La.	Louisiana	Tex.	Texas
Maine		Utah	
Md.	Maryland	Vt.	Vermont
Mass.	Massachusetts	Va.	Virginia
Mich.	Michigan	Wash.	Washington
Minn.	Minnesota	W.Va.	West Virginia
Miss.	Mississippi	Wis.	Wisconsin
Mo.	Missouri	Wyo.	Wyoming